AP* EDITION

COLLEGE PHYSICS

A STRATEGIC APPROACH

SECOND EDITION

RANDALL D. KNIGHT

California Polytechnic State University, San Luis Obispo

BRIAN JONES

Colorado State University

STUART FIELD

Colorado State University

PEARSON

Boston Columbus Indianapolis New York San Francisco Upper Saddle River
Amsterdam Cape Town Dubai London Madrid Milan Munich Paris Montreal Toronto
Delhi Mexico City Sao Paulo Sydney Hong Kong Seoul Singapore Taipei Tokyo

Publisher:	Jim Smith
Sr. Development Editor:	Alice Houston, Ph.D.
Editorial Manager:	Laura Kenney
Sr. Project Editor:	Martha Steele
Editorial Assistant:	Dyan Menezes
Media Producer:	David Huth
Director of Marketing:	Christy Lawrence
Executive Marketing Manager:	Kerry Chapman
Managing Editor:	Corinne Benson
Production Supervisors:	Beth Collins, Nancy Tabor, and Camille Herrera
Production Management:	Rose Kernan of RPK Editorial Services, Inc. and Nesbitt Graphics, Inc.
Compositor and Interior Designer:	Nesbitt Graphics, Inc.
Cover Designer:	Riezebos Holzbaur Group
Illustrators:	Rolin Graphics
Photo Researcher:	Eric Schrader
Manufacturing Buyer:	Jeffrey Sargent
Printer and Binder:	Courier Kendallville
Cover Photo Credit:	Stephen Dalton/Minden Pictures

Credits and acknowledgments borrowed from other sources and reproduced, with permission, in this textbook appear on p. C-1

Library of Congress Cataloging-in-Publication Data

Knight, Randall Dewey.
College physics : a strategic approach / Randall D. Knight, Brian Jones,
Stuart Field. — 2nd ed.
p. cm.
Includes bibliographical references and index.
ISBN 978-0-321-59549-2
1. Physics—Textbooks. I. Jones, Brian. II. Field, Stuart. III. Title.
QC23.2.K649 2010
530—dc22

1 2 3 4 5 6 7 8 9 10—CRK—13 12 11 10 09

www.PearsonSchool.com/Advanced

ISBN 10: 0-132-83211-9 (High School Binding)
ISBN 13: 978-0-13-283211-3 (High School Binding)

About the Authors

Randy Knight has taught introductory physics for 28 years at Ohio State University and California Polytechnic University, where he is currently Professor of Physics and Director of the Minor in Environmental Studies. Randy received a Ph.D. in physics from the University of California, Berkeley and was a post-doctoral fellow at the Harvard-Smithsonian Center for Astrophysics before joining the faculty at Ohio State University. It was at Ohio that he began to learn about the research in physics education that, many years later, led to *Five Easy Lessons: Strategies for Successful Physics Teaching, Physics for Scientists and Engineers: A Strategic Approach,* and now to this book. Randy's research interests are in the field of lasers and spectroscopy. He also directs the environmental studies program at Cal Poly. When he's not in the classroom or in front of a computer, you can find Randy hiking, sea kayaking, playing the piano, or spending time with his wife Sally and their six cats.

Brian Jones has won several teaching awards at Colorado State University during his 20 years teaching in the Department of Physics. His teaching focus in recent years has been the College Physics class, including writing problems for the MCAT exam and helping students review for this test. Brian is also Director of the *Little Shop of Physics,* the Department's engaging and effective hands-on outreach program, which has merited coverage in publications ranging from the *APS News* to *People* magazine. Brian has been invited to give workshops on techniques of science instruction throughout the United States and internationally, including Belize, Chile, Ethiopia, Azerbaijan, Mexico and Slovenia. Previously, he taught at Waterford Kamhlaba United World College in Mbabane, Swaziland, and Kenyon College in Gambier, Ohio. Brian and his wife Carol have dozens of fruit trees and bushes in their yard, including an apple tree that was propagated from a tree in Isaac Newton's garden, and they have traveled and camped in most of the United States.

Stuart Field has been interested in science and technology his whole life. While in school he built telescopes, electronic circuits, and computers. After attending Stanford University, he earned a Ph.D. at the University of Chicago, where he studied the properties of materials at ultralow temperatures. After completing a postdoctoral position at the Massachusetts Institute of Technology, he held a faculty position at the University of Michigan. Currently at Colorado State University, Stuart teaches a variety of physics courses, including algebra-based introductory physics, and was an early and enthusiastic adopter of Knight's *Physics for Scientists and Engineers.* Stuart maintains an active research program in the area of superconductivity. His hobbies include woodworking; enjoying Colorado's great outdoors; and ice hockey, where he plays goalie for a local team.

BUILDS PROBLEM-SOLVING SKILLS AND CONFIDENCE

Clear, consistent instruction

Build confidence and success through a consistent three-step approach: **Prepare** the problem, try to **Solve** it, and **Assess** the answer.

Topic-specific ***Problem-Solving Strategies*** follow the same three-step framework and provide more detailed guidance.

1. **PREPARE** reinforces the value of gathering information, drawing figures, making assumptions, and planning—key steps that research shows are often skipped.
2. **SOLVE** carefully works through the mathematical steps of the solution, explaining algebraic manipulations and use of key information.
3. **ASSESS** verifies whether the answer makes sense—numerically and in context.

PROBLEM-SOLVING STRATEGY 24.1 Magnetic field problems

PREPARE Because current-carrying wires do not lie in the same plane as the fields they produce, you'll need to prepare an especially careful drawing. Generally, you should choose the plane of your drawing so that the magnetic field vectors lie either in the plane of the paper or perpendicular to it.

- Straight wires are usually easiest to draw as seen from their ends. Then the field vectors will lie in the plane of the paper.
- Us...
- Sol...
- If t... tha... can...

SOLVE ... noid) ... field ... noid. ...

ASSESS ...

PROBLEM-SOLVING STRATEGY 9.1 Conservation of momentum problems

PREPARE Clearly define *the system.*

- If possible, choose a system that is isolated ($\vec{F}_{net} = \vec{0}$) or within which the interactions are sufficiently short and intense that you can ignore external forces for the duration of the interaction (the impulse approximation). Momentum is then conserved.
- If it's not possible to choose an isolated system, try to divide the problem into parts such that momentum is conserved during one segment of the motion. Other segments of the motion can be analyzed using Newton's laws or, as you'll learn in Chapter 10, conservation of energy.

Following Tactics Box 9.1, draw a before-and-after visual overview. Define symbols that will be used in the problem, list known values, and identify what you're trying to find.

SOLVE The mathematical representation is based on the law of conservation of momentum: $\vec{P}_f = \vec{P}_i$. In component form, this is

$$(p_{1x})_f + (p_{2x})_f + (p_{3x})_f + \cdots = (p_{1x})_i + (p_{2x})_i + (p_{3x})_i + \cdots$$
$$(p_{1y})_f + (p_{2y})_f + (p_{3y})_f + \cdots = (p_{1y})_i + (p_{2y})_i + (p_{3y})_i + \cdots$$

ASSESS Check that your result has the correct units, is reasonable, and answers the question.

Exercise 20

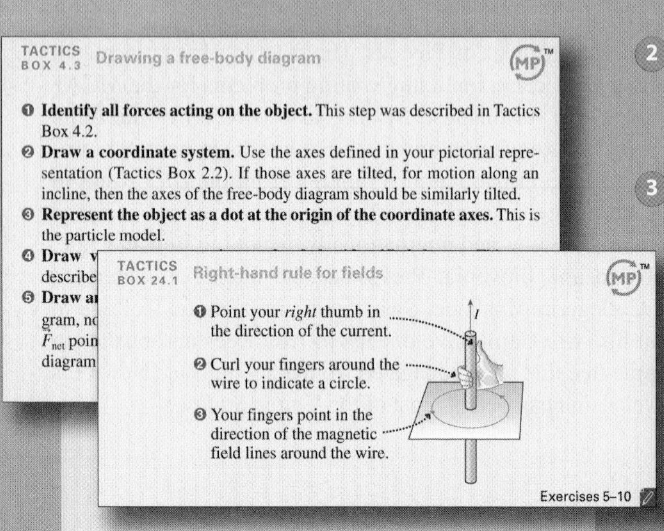

TACTICS BOX 4.3 Drawing a free-body diagram
❶ Identify all forces acting on the object. This step was described in Tactics Box 4.2.
❷ Draw a coordinate system. Use the axes defined in your pictorial representation (Tactics Box 2.2). If those axes are tilted, for motion along an incline, then the axes of the free-body diagram should be similarly tilted.
❸ Represent the object as a dot at the origin of the coordinate axes. This is the particle model.
TACTICS BOX 24.1 Right-hand rule for fields
❶ Point your right thumb in the direction of the current.
❷ Curl your fingers around the wire to indicate a circle.
❸ Your fingers point in the direction of the magnetic field lines around the wire.
I
Exercises 5–10

Tactics Boxes provide step-by-step procedures that build key skills that will be used over and over—such as drawing free-body diagrams and using ray tracing.

Inverse-square relationships

Two quantities have an **inverse-square relationship** if y is inversely proportional to the *square* of x. We write the mathematical relationship as:

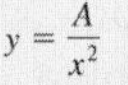

$$y = \frac{A}{x^2}$$

y is inversely proportional to x^2

When x is 1, y is A.

When x is 2, y is $A/4$. That is, doubling x reduces y by a factor of four (two squared). We don't need to know A to know this.

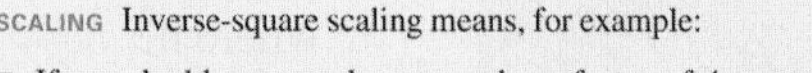
SCALING Inverse-square scaling means, for example:

- If you double x, you decrease y by a factor of 4, as you can see in the graph.
- If you increase x by a factor of 3, you decrease y by a factor of 9.
- If you decrease x by a factor of 3, you increase y by a factor of 9.

Generally: **An *increase* in x by a factor C results in a *decrease* of y by a factor C^2.**

LIMITS As x becomes large, y becomes very small; as x becomes small, y becomes very large.

Exercise 23

Math Relationship Boxes ensure confidence with the key mathematical relationships most common in this course. Each relationship is consolidated in words, math, and graphics, along with tips on reasoning with limiting cases and scaling. Icons in the text refer back to these boxes to reinforce connections.

Explicit, guided practice

Worked Examples implement the Strategies and follow the same **PREPARE/SOLVE/ASSESS** framework as part of developing good problem-solving habits. They carefully walk through the underlying reasoning and pitfalls to avoid.

Integrated Examples at the end of each chapter demonstrate problem-solving in the context of a capstone, multi-concept real-world scenario.

INTEGRATED EXAMPLE 24.15 Making music with magnetism

A loudspeaker makes sound by pushing air back and forth with a paper cone that is driven by a magnetic force on a wire coil at the base of the cone. FIGURE 24.61 shows the details. The bottom of the cone is wrapped with several turns of fine wire. This coil of wire sits in the gap between the poles of a circular magnet, the black disk in the photo. The magnetic field exerts a force on a current in the wire, pushing the cone and thus pushing the air.

FIGURE 24.61 The arrangement of the coil and magnet poles in a loudspeaker.

(a)

(b)

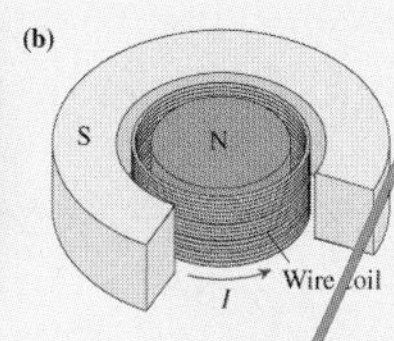

There is a 0.18 T field in the gap between the poles. The coil of wire that sits in this gap has a diameter of 5.0 cm, contains 20 turns of wire, and has a resistance of 8.0 Ω. The speaker is connected to an amplifier whose instantaneous output voltage of 6.0 V creates a clockwise current in the coil as seen from above. What is the magnetic force on the coil at this instant?

1 **PREPARE** The current in the coil experiences a force due to the magnetic field between the poles. Let's start with a sketch of the field to determine the direction of this force. Magnetic field lines go from the north pole to the south pole of a magnet, so the field lines for the loudspeaker magnet appear as in FIGURE 24.62. The field is at all points perpendicular to the current, and the right-hand rule shows us that, for a clockwise current, the force at each point of the wire is out of the page.

FIGURE 24.62 The magnetic field in the gap and the current in the coil.

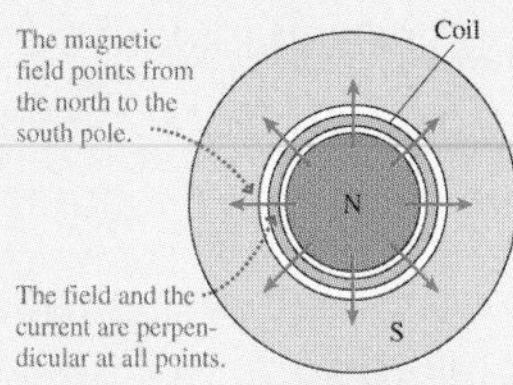

2 **SOLVE** The current in the wire is produced by the amplifier. The current is related to the potential difference and the resistance of the wire by Ohm's law:

$$I = \frac{\Delta V}{R} = \frac{6.0\text{ V}}{8.0\ \Omega} = 0.75\text{ A}$$

Because the current is perpendicular to the field, we can use Equation 24.10 to determine the force on this current. We know the field and the current, but we need to know the length of the wire in the field region. The coil has diameter 5.0 cm and thus circumference $\pi(0.050\text{ m})$. The coil has 20 turns, so the total length of the wire in the field is

$$l = 20\pi(0.050\text{ m}) = 3.1\text{ m}$$

The magnitude of the force is then given by Equation 24.10 as

$$F = IlB = (0.75\text{ A})(3.1\text{ m})(0.18\text{ N/A}\cdot\text{m}) = 0.42\text{ N}$$

out of the page, as already noted.

3 **ASSESS** The force is small, but this is reasonable. A loudspeaker cone is quite light, so only a small force is needed for a large acceleration. The force for a clockwise current is out of the page, but when the current switches direction, to counterclockwise, the force will be directed in. A current that alternates direction will cause the cone to oscillate in and out—just what is needed for making music.

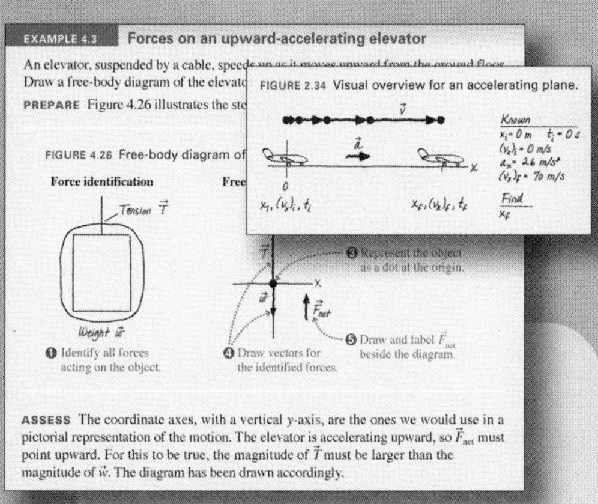

EXAMPLE 4.3 Forces on an upward-accelerating elevator

An elevator, suspended by a cable, speeds up as it moves upward from the ground floor. Draw a free-body diagram of the elevator.

PREPARE Figure 4.26 illustrates the ste

FIGURE 4.26 Free-body diagram of

FIGURE 2.34 Visual overview for an accelerating plane.

ASSESS The coordinate axes, with a vertical *y*-axis, are the ones we would use in a pictorial representation of the motion. The elevator is accelerating upward, so $\vec{F}_{net}$ must point upward. For this to be true, the magnitude of $\vec{T}$ must be larger than the magnitude of $\vec{w}$. The diagram has been drawn accordingly.

Pencil Sketches provide an explicit and accessible example of what to draw in solving a problem—steps often outlined in the Tactic Boxes.

Conceptual Examples target qualitative reasoning skills. Since no math is involved, they follow a **REASON** and **ASSESS** approach.

CONCEPTUAL EXAMPLE 6.2 Who has the larger acceleration?

FIGURE 6.8 Top view of a merry-go-round.

Jacob Emma

Two children are riding in circles on a merry-go-round, as shown in Figure 6.8. Which child experiences the larger acceleration?

1 **REASON** All points on the merry-go-round move at the same angular speed. The second expression for the acceleration in Equation 6.7 tells us that $a = \omega^2 r$. As the two children are moving with the same angular speed, Emma, with a larger value of r, experiences a larger acceleration.

3 **ASSESS** In the previous example, we saw that points farther from the center move at a higher speed. This would imply a higher acceleration as well, so our answer makes sense.

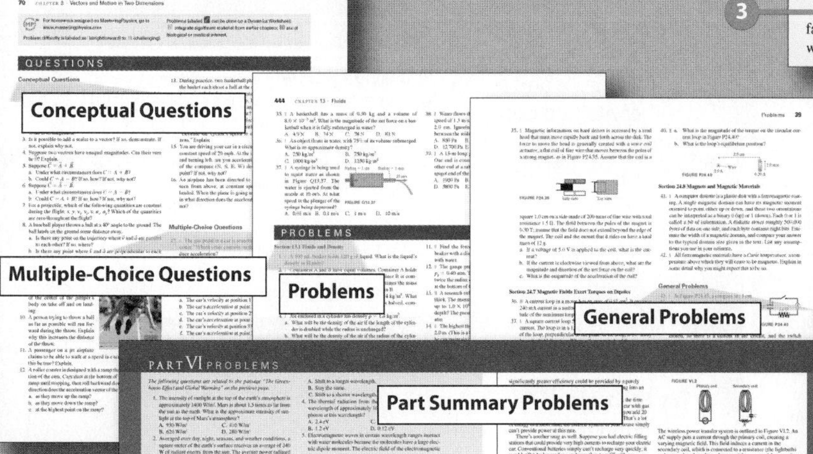

Conceptual Questions require thoughtful reasoning and can be used for group discussions or individual work. ***Multiple-Choice Questions*** use carefully chosen distractors to elicit common misconceptions. ***Problems***, are keyed to sections and draw on real-world applications to provide motivational examples. More advanced ***General Problems*** require the simplification and modeling of more complex real-world situations. ***Part Summary Problems*** close each of the book's seven parts and take problem-solving one step further by covering topics that span several chapters.

INTEGRATES INTERESTING AND RELEVANT TOPICS

An active, inductive approach

Drawn from various fields of study and the world around us, relevant examples and interesting topics are carefully woven into the text. These provide motivation, a means to consolidate understanding, and a clear context for understanding physics.

New concepts are introduced through observations about the real world, an inductive approach shown to improve learning (see the magnetism introduction in the sample chapter that follows).

Worked Examples incorporate scenarios from everyday life and the world around us.

EXAMPLE 10.13 Protecting your head

A bike helmet is basically a shell of hard, crushable foam 3.0 cm thick. In testing, the helmet is strapped onto a 5.0 kg headform that is dropped from a height of 2.0 m onto a hard anvil. What force is encountered by the head in such a fall?

PREPARE A visual overview of the test is shown in Figure 10.31. We can use the law of conservation of en-ergy, ... ike helmet ... upon ... e head- ... V on the ... n above ... hen the

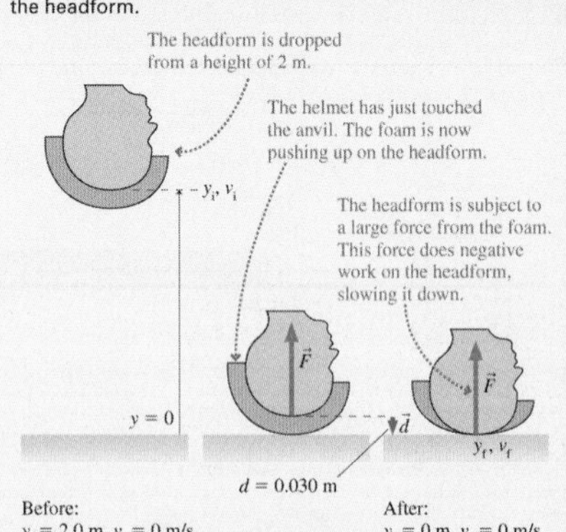

The headform is dropped from a height of 2 m.
The helmet has just touched the anvil. The foam is now pushing up on the headform.
The headform is subject to a large force from the foam. This force does negative work on the headform, slowing it down.
y_i, v_i
$\vec{F}$
$\vec{F}$
$y = 0$
$\vec{d}$
y_f, v_f
$d = 0.030$ m
Before:
$y_i = 2.0$ m, $v_i = 0$ m/s
After:
$y_f = 0$ m, $v_f = 0$ m/s
Find: F

FIGURE 10.31 The foam in the helmet does negative work on the headform.

EXAMPLE 3.8 Speed of a roller coaster

A classic wooden coaster has cars that go down a big first hill, gaining speed. The cars then ascend a second hill with a slope of 30°. If the cars are going 25 m/s at the bottom and it takes them 2.0 s to climb this hill, how fast are they going at the top?

PREPARE We start with the visual overview in Figure 3.24, which includes a motion diagram, a pictorial representation, and a list of values. Notice how the motion diagram of Figure 3.24 differs from that of the previous example: The velocity decreases as the car moves up the hill, so the acceleration vector is opposite the direction of the velocity vector. The motion is along the x-axis, as before, but the acceleration vector points in the negative-x direction, so the component a_x is negative.

FIGURE 3.24 The coaster's speed decreases as it goes up the hill.

Known: $x_i = 0$ m, $(v_x)_i = 25$ m/s, $t_i = 0$ s, $t_f = 2.0$ s, $\theta = 30°$
Find: $(v_x)_f$

SOLVE To determine the final speed, we need to know the acceleration. We will assume that there is no friction or air resistance.

Taking a picture in a flash When you take a flash picture, the flash is fired using electric potential energy stored in a capacitor. Batteries are unable to deliver the required energy rapidly enough, but capacitors can discharge all their energy in only microseconds. A battery is used to slowly charge up the capacitor, which then rapidly discharges through the flashlamp. This slow recharging process is why you must wait some time between taking flash pictures.

Free-standing Applications, found in the margin with photographs and a self-contained caption, connect the physical principles with the real world.

TRY IT YOURSELF

HEINZ
TOMATO KETCHUP
57

Getting the ketchup out The ketchup stuck at the bottom of the bottle is initially at rest. If you hit the bottom of the bottle, the bottle suddenly moves down, taking the ketchup on the bottom of the bottle with it, so that the ketchup just stays stuck to the bottom. But if instead you hit *up* on the bottle, as shown, you force the bottle rapidly upward. By the first law, the ketchup that was stuck to the bottom stays at rest, so it separates from the upward-moving bottle: the ketchup has moved forward with respect to the bottle!

TRY IT YOURSELF

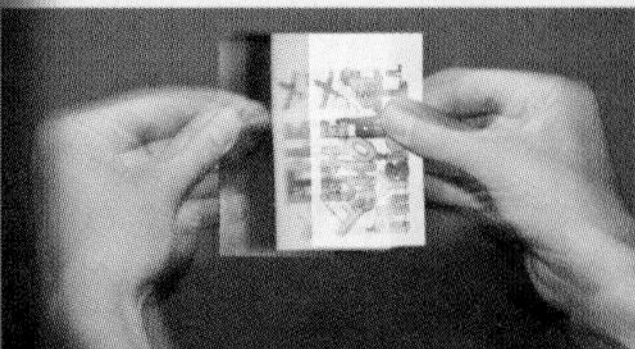

Buzzing magnets You can use two *identical* flexible refrigerator magnets for a nice demonstration of their alternating pole structure. Place the two magnets together, back to back, then quickly pull them across each other, noting the alternating attraction and repulsion from the alternating poles. If you pull them quickly enough, you will hear a buzz as the magnets are rapidly pushed apart and then pulled together.

Try It Yourself Activities throughout the text provide simple real-world experiments designed to quickly reinforce a key idea through direct experience.

Engaging treatment

Optional sections provide **in-depth coverage of key topics**—such as electrical conduction in the nervous system, the workings of an EKG, and how to correctly measure blood pressure.

FIGURE 20.35 The beating heart generates a dipole electric field.

(a) The electric dipole of the heart

This line separates cells that have depolarized and those that have not.

(b) The field of the heart in the body

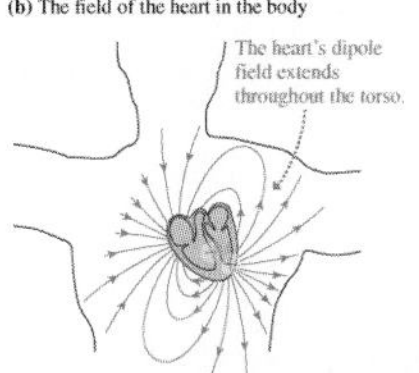

- The electric field is created by charges. Field lines start on a positive charge and end on a negative charge.

You can use the above information as the basis of a technique for sketching a field line picture for an arrangement of charges. Draw field lines starting on positive charges and moving toward negative charges. Draw the lines tangent to the field vector at each point. Make the lines close together where the field is strong, far apart where the field is weak. For example, Figure 20.34 pictures the electric field of a dipole using electric field lines. You should compare this to Figure 20.29b, which illustrated the field with field vectors.

The Electric Field of the Heart

Nerve and muscle cells have a prominent electrical nature. As we will see in detail in Chapter 23, a cell membrane is an insulator that encloses a conducting fluid and is surrounded by conducting fluid. While resting, the membrane is *polarized* with positive charges on the outside of the cell, negative charges on the inside. When a nerve or a muscle cell is stimulated, the polarity of the membrane switches; we say that the cell *depolarizes*. Later, when the charge balance is restored, we say that the cell *repolarizes*.

All nerve and muscle cells generate an electrical signal when depolarization occurs, but the largest electrical signal in the body comes from the heart. The rhythmic beating of the heart is produced by a highly coordinated wave of depolarization that sweeps across the tissue of the heart. As Figure 20.35a shows, the surface of the heart is positive on one side of the boundary between tissue that is depolarized and tissue that is not yet depolarized, negative on the other. In other words, the heart is a large electric dipole. The orientation and strength of the dipole change during each beat of the heart as the depolarization wave sweeps across it.

The electric dipole of the heart generates a dipole electric field that extends throughout the torso, as shown in Figure 20.35b. As we will see in Chapter 21, an *electrocardiogram* measures the changing electric field of the heart as it beats. Measurement of the heart's electric field can be used to diagnose the operation of the heart.

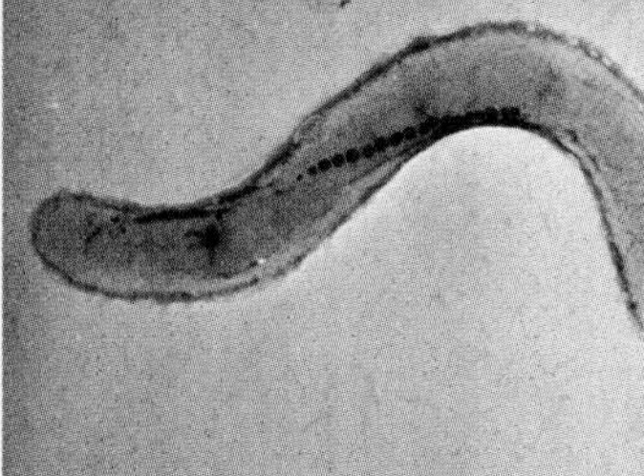

Magnetotactic bacteria BIO Several organisms use the earth's magnetic field navigate. The clearest example of this is *magnetotactic bacteria*. The dark dots in image are small pieces of iron; each piec single domain and hence a very strong ma Such a bacterium possesses a very strong magnetic moment: The bacterium itself like a bar magnet, and lines up with the earth's magnetic field. In the temperate regions where such bacteria live, the earth field has a large vertical component. The bacteria use their alignment with this ver field component to navigate up and dow

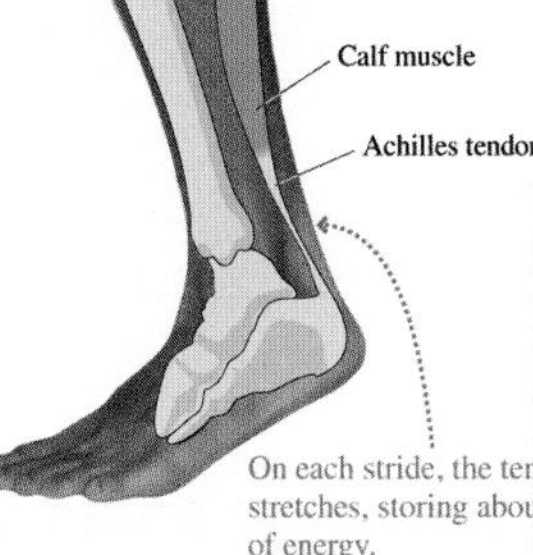

Calf muscle
Achilles tendon
On each stride, the ten stretches, storing abou of energy.

Spring in your step BIO As you run, yo lose some of your mechanical energy eac time your foot strikes the ground; this en is transformed into unrecoverable therma energy. Luckily, about 35% of the decrea your mechanical energy when your foot is stored as elastic potential energy in the stretchable Achilles tendon of the lower On each plant of the foot the tendon is stretched, storing some energy. The tend springs back as you push off the ground again, helping to propel you forward. Th recovered energy reduces the amount of internal chemical energy you use, increasing your efficiency.

Dinner at a distance BIO A chameleon's tongue is a powerful tool for catching prey. Certain species can extend the tongue to a distance of over 1 ft in less than 0.1 s! A study of the kinematics of the motion of the chameleon tongue, using techniques like those in this chapter, reveals that the tongue has a period of rapid acceleration followed by a period of constant velocity. This knowledge is a very valuable clue in the analysis of the evolutionary relationships between chameleons and other animals.

Fascinating, self-contained ***Life-science Applications*** throughout the text illustrate how physics relates to the real world.

Passage Problems

Kangaroo Locomotion BIO

Kangaroos have very stout tendons in their legs that can be used to store energy. When a kangaroo lands on its feet, the tendons stretch, transforming kinetic energy of motion to elastic potential energy. Much of this energy can be transformed back into kinetic energy as the kangaroo takes another hop. The kangaroo's peculiar hopping gait is not very efficient at low speeds but is quite efficient at high speeds.

Figure P11.61 shows the energy cost of human and kangaroo locomotion. The graph shows oxygen uptake (in mL/s) per kg of body mass, allowing a direct comparison between the two species.

Oxygen uptake (mL/kg·s)
0, 0.2, 0.4, 0.6, 0.8, 1, 1.2, 1.4, 1.6, 1.8, 2
Human running
Red kangaroo, hopping
Speed (m/s)
0, 2, 4, 6, 8, 10, 12

FIGURE P11.61 Oxygen uptake (a measure of energy use per second) for a running human and a hopping kangaroo.

For humans, the energy used per second (i.e., power) is proportional to the speed. That is, the human curve nearly passes through the origin, so running twice as fast takes approximately twice as much power. For a hopping kangaroo, the graph of energy use has only a very small slope. In other words, the energy used per second changes very little with speed. Going faster requires very little additional power. Treadmill tests on kangaroos and observations in the wild have shown that they do not become winded at any speed at which they are able to hop. No matter how fast they hop, the necessary power is approximately the same.

61. | A person runs 1 km. How does his speed affect the total energy needed to cover this distance?
    A. A faster speed requires less total energy.
    B. A faster speed requires more total energy.
    C. The total energy is about the same for a fast speed and a slow speed.
62. | A kangaroo hops 1 km. How does its speed affect the total energy needed to cover this distance?
    A. A faster speed requires less total energy.
    B. A faster speed requires more total energy.
    C. The total energy is about the same for a fast speed and a slow speed.
63. | At a speed of 4 m/s,
    A. A running human is more efficient than an equal-mass hopping kangaroo.
    B. A running human is less efficient than an equal-mass hopping kangaroo.
    C. A running human and an equal-mass hopping kangaroo have about the same efficiency.
64. | At approximately what speed would a human use half the power of an equal-mass kangaroo moving at the same speed?
    A. 3 m/s B. 4 m/s C. 5 m/s D. 6 m/s
65. | At approximately what speed would a human use twice the power of a kangaroo of half the mass moving at the same speed?
    A. 3 m/s B. 5 m/s C. 7 m/s D. 9 m/s

Multiple-choice and Passage Problems carefully test understanding by targeting common misconceptions and providing context-rich situations.

PROMOTES DEEPER UNDERSTANDING

Structured learning path

This text incorporates many subtle but powerful techniques that improve learning and retention, including a self-evident and structured learning path and a unique visual pedagogy.

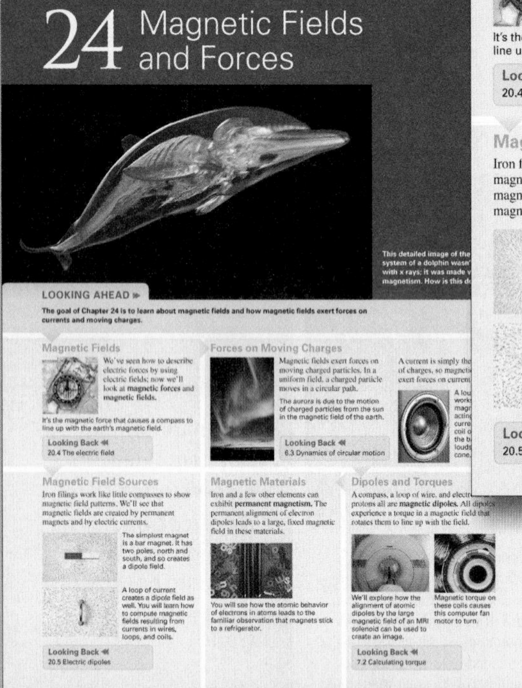

LOOKING AHEAD ▸▸

The goal of Chapter 24 is to learn about magnetic fields and how magnetic fields exert forces on currents and moving charges.

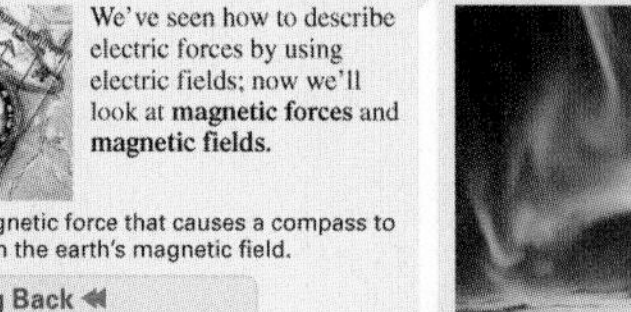

Magnetic Fields

We've seen how to describe electric forces by using electric fields; now we'll look at **magnetic forces** and **magnetic fields.**

It's the magnetic force that causes a compass to line up with the earth's magnetic field.

Looking Back ◂◂
20.4 The electric field

Forces on Moving Charges

Magnetic fields exert forces on moving charged particles. In a uniform field, a charged particle moves in a circular path.

The aurora is due to the motion of charged particles from the sun in the magnetic field of the earth.

Looking Back ◂◂
6.3 Dynamics of circular motion

A current is simply the motion of charges, so magnetic fields exert forces on currents.

A loudspeaker works by the magnetic force acting on a current in a coil of wire at the base of the loudspeaker's cone.

Magnetic Field Sources

Iron filings work like little compasses to show magnetic field patterns. We'll see that magnetic fields are created by permanent magnets and by electric currents.

The simplest magnet is a bar magnet. It has two poles, north and south, and so creates a dipole field.

A loop of current creates a dipole field as well. You will learn how to compute magnetic fields resulting from currents in wires, loops, and coils.

Looking Back ◂◂
20.5 Electric dipoles

Magnetic Materials

Iron and a few other elements can exhibit **permanent magnetism.** The permanent alignment of electron dipoles leads to a large, fixed magnetic field in these materials.

You will see how the atomic behavior of electrons in atoms leads to the familiar observation that magnets stick to a refrigerator.

Dipoles and Torques

A compass, a loop of wire, and electrons and protons all are **magnetic dipoles.** All dipoles experience a torque in a magnetic field that rotates them to line up with the field.

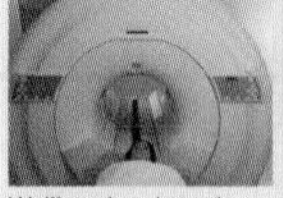

We'll explore how the alignment of atomic dipoles by the large magnetic field of an MRI solenoid can be used to create an image.

Magnetic torque on these coils causes this computer fan motor to turn.

Looking Back ◂◂
7.2 Calculating torque

Chapter Previews are based on the educational psychology concept of an "advance organizer." Each chapter begins with an illustrated preview of the upcoming ideas, setting them in context, explaining their utility, and tying them to existing knowledge (through ***Looking Back*** references).

Stop to Think Questions at the end of a section allow for a quick comprehension check before moving on. Using powerful ranking-task and graphical techniques, they efficiently probe key misconceptions and encourage active reading. (Answers are provided at the end of the chapter.)

STOP TO THINK 10.4 Rank in order, from largest to smallest, the gravitational potential energies of identical balls 1 to 4.

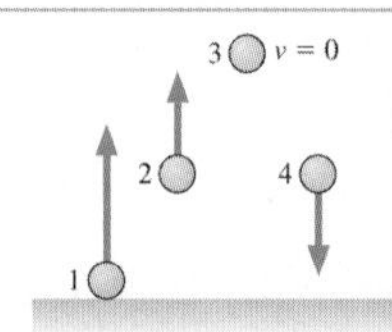

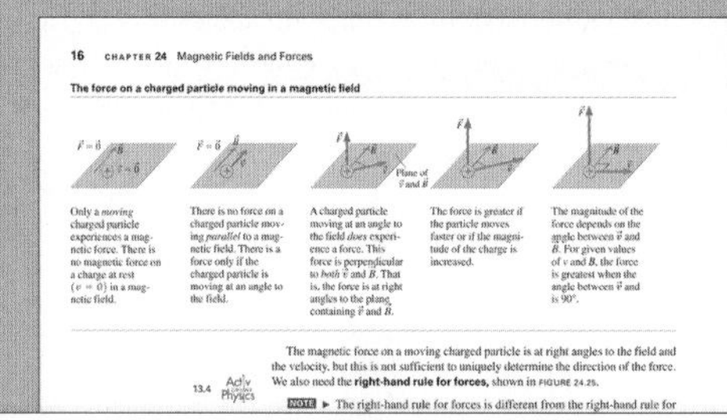

NOTE ▸ The right-hand rule for forces gives the direction of the force on a *positive* charge. For a negative charge, the force is in the opposite direction. ◂

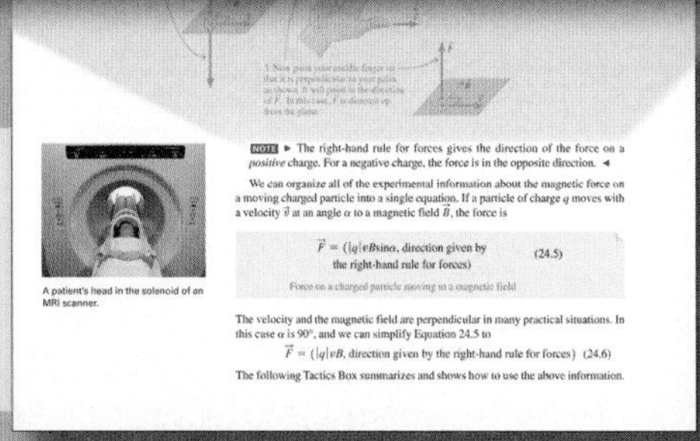

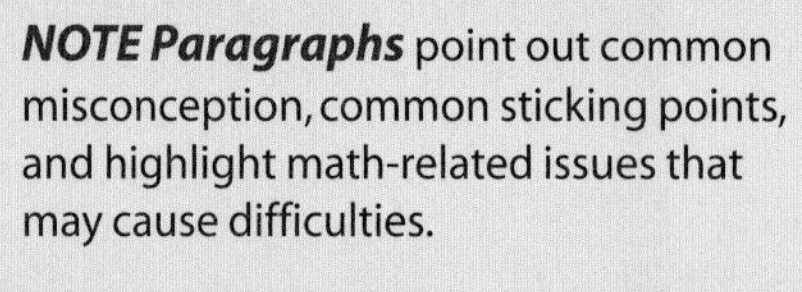

NOTE Paragraphs point out common misconception, common sticking points, and highlight math-related issues that may cause difficulties.

Proven visual pedagogy

Figures are carefully streamlined in detail and color. In addition, chalkboard-like dialogue boxes (versus lengthy captions in other textbooks) are effectively used in:

- interpreting a graph, equation, or figure
- understanding a process
- translating between text, math, graphs, and figures
- grasping a difficult concept through a visual analogy

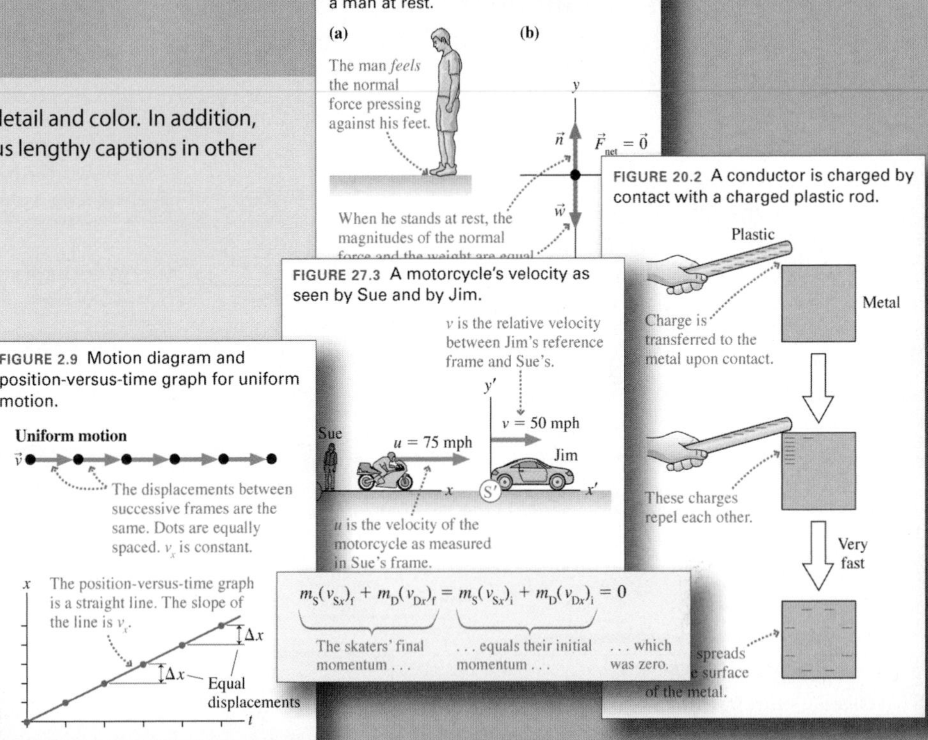

Critically acclaimed ***Visual Chapter Summaries*** consolidate understanding by providing each concept in words, math, and figures and organizing these into a coherent hierarchy—from General Principles to Applications.

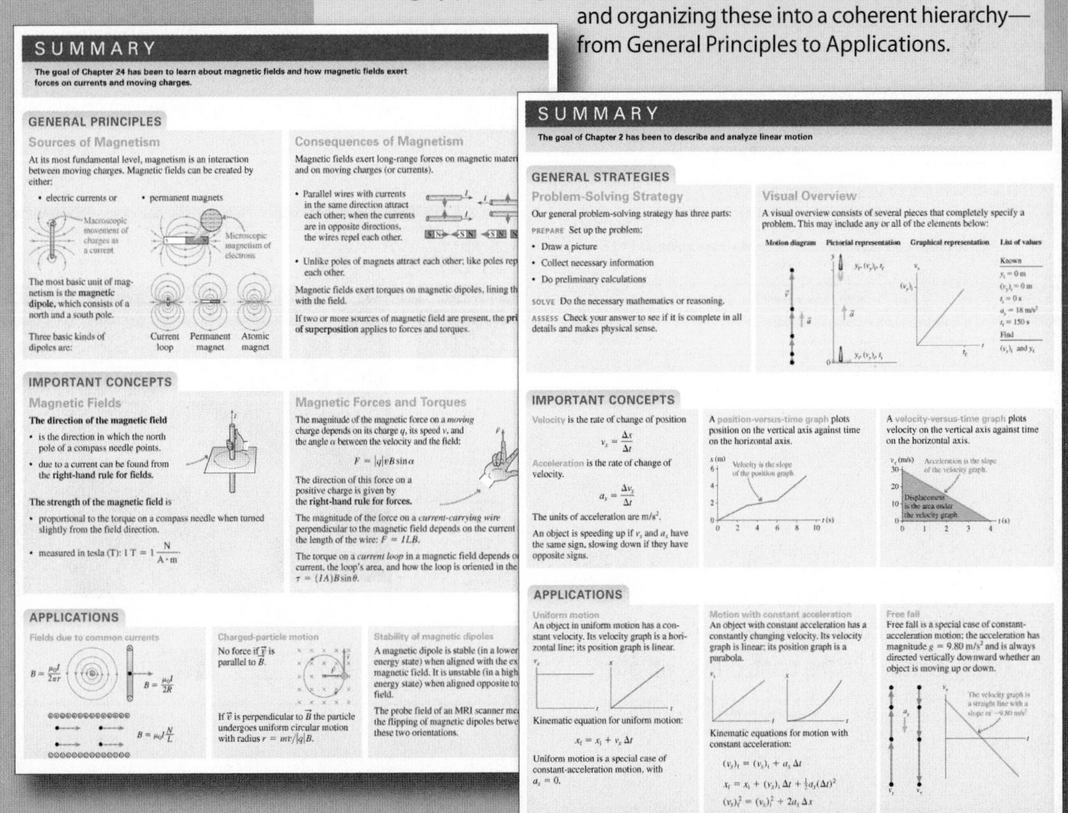

Make a Difference with **MasteringPhysics®**

MasteringPhysics is the most effective and widely used online science tutorial, homework, and assessment system available.

www.masteringphysics.com

Self-Paced, Individualized Coaching

For every chapter of the book, MasteringPhysics® provides assignable, in-depth tutorials designed to coach students with hints and feedback specific to their misconceptions.

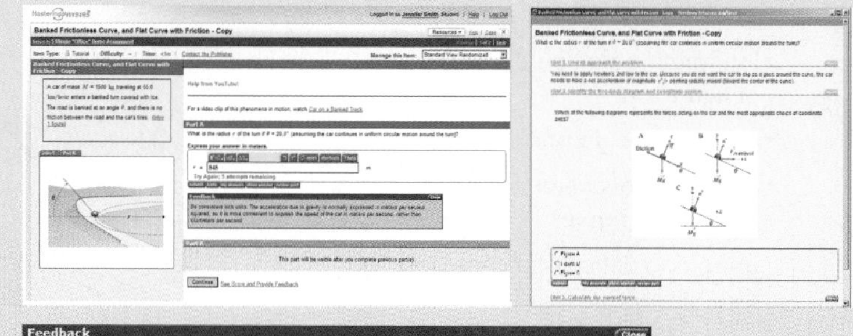

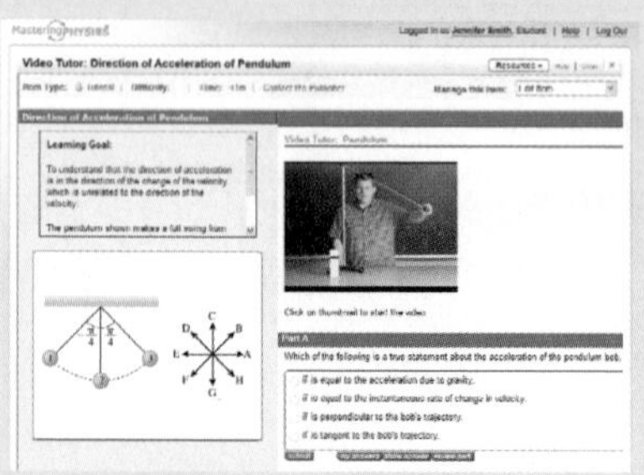

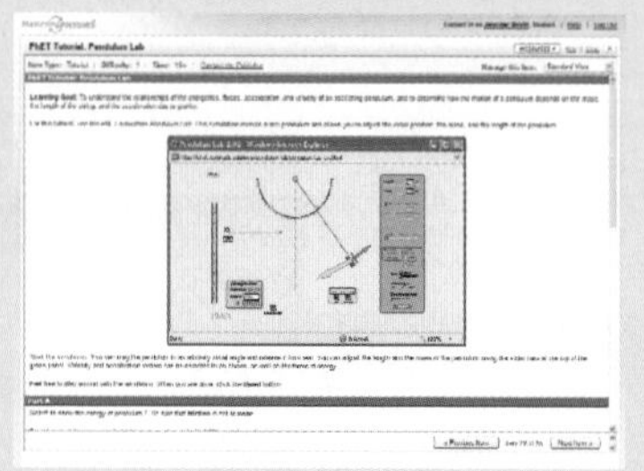

Bring Physics to Life with Dynamic Visuals

Video tutor demonstration and PhET simulations help students connect physics concepts to physics in the real world.

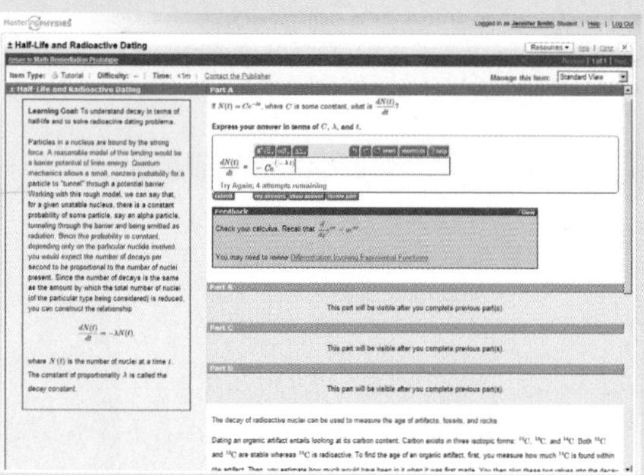

Get Up to Speed on Math

Math remediation found within selected tutorials provide just-in-time math help, allowing students to brush up on the most important concepts—while making connections between math and physics.

Access When You Need It

Access an interactive eText of the Student Edition 24/7. eText pages look exactly like the printed text, yet offer additional functionality.

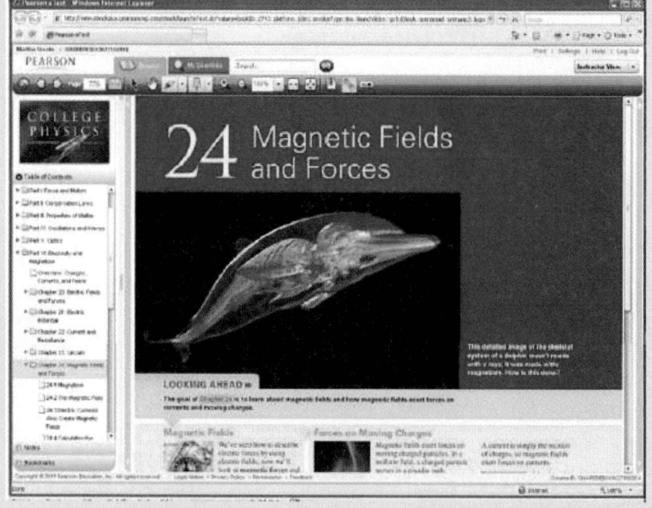

NEW! Pre-Built Assignments

For every chapter in the book, MasteringPhysics now provides pre-built assignments that cover the material with a tested mix of tutorials and end-of-chapter problems of graded difficulty. Instructors may use these assignments as-is or take them as a starting point for modification.

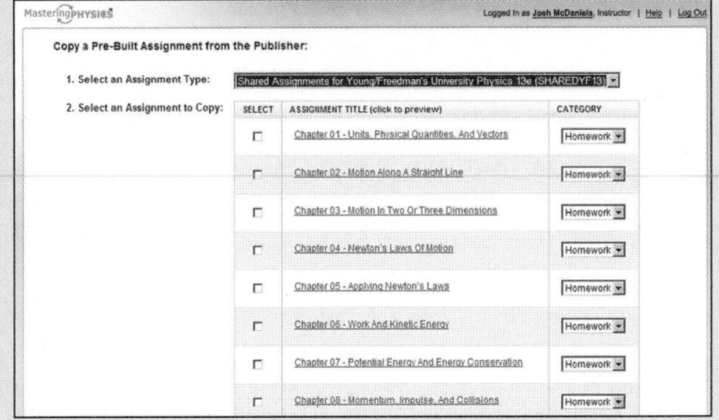

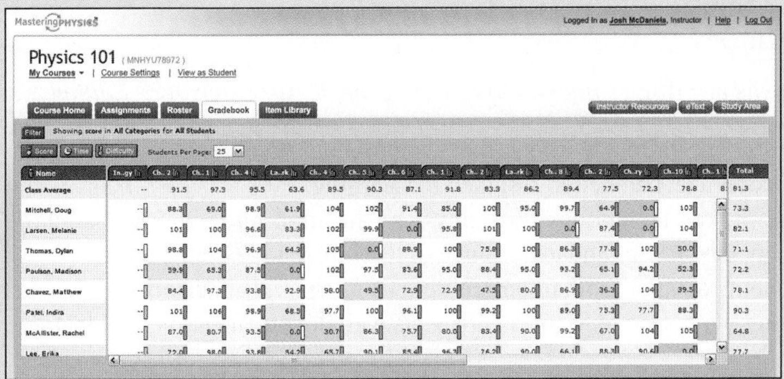

Gradebook

- Every assignment is graded automatically.
- Shades of red highlight vulnerable students and challenging assignments.

Class Performance on Assignment

Click on a problem to see which step your students struggled with most, and even their most common wrong answers. Compare results at every stage with the national average or with your previous class.

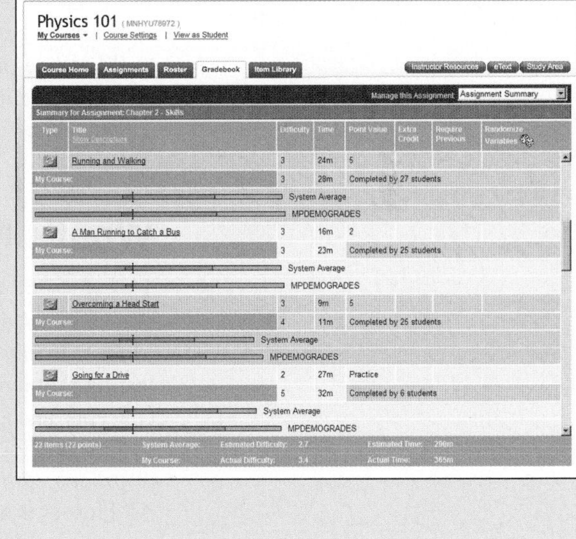

Gradebook Diagnostics

This screen provides your favorite weekly diagnostics. With a single click, charts summarize the most difficult problems, vulnerable students, grade distribution, and even improvement in scores over the course.

Preface to the AP* Edition

In 2006, we published *College Physics: A Strategic Approach,* a new algebra-based physics textbook for college students majoring in the biological and life sciences, architecture, natural resources, and other disciplines. Our goal from the beginning has been a textbook combining the best results from physics education research with inspiring photographs and examples connecting physics to the real world. Our commitment to this goal is undiminished, and with the publication of the AP* Edition of *College Physics: A Strategic Approach, 2nd Edition,* we are pleased to provide both the motivation and the tools required for AP Physics B students to succeed in the classroom, on the AP UC Exam, and in college-level courses.

Objectives

Our primary goals in writing *College Physics: A Strategic Approach, AP* Edition* are:

- To provide students with a textbook that's a more manageable size, less encyclopedic in its coverage, and better designed for learning.
- To integrate proven techniques from physics education research into the classroom in a way that accommodates a range of teaching and learning styles.
- To help students develop both quantitative reasoning skills and solid conceptual understanding, with special focus on concepts well documented to cause learning difficulties.
- To help students develop problem-solving skills and confidence in a systematic manner using explicit and consistent tactics and strategies.
- To motivate students by integrating real-world examples relevant to their everyday experiences.
- To utilize proven techniques of visual instruction and design from educational research and cognitive psychology that improve student learning and retention and address a range of learner styles.

What's New to This Edition

This AP* Edition leverages the hallmarks of the First Edition—effective conceptual explanations and problem-solving instruction—with new pedagogical features. More than any other book, *College Physics* leads students to proficient and long-lasting problem-solving skills, a deeper and better-connected understanding of the concepts, and a broader picture of the relevance of physics to the world around them.

First and foremost, the content of the AP* Edition of *College Physics: A Strategic Approach, 2nd Edition* is the same as the college edition—as all parties agree that AP Physics is a college-level course, and as such, its content and rigor should match the college offering. So, you ask, how is the AP* Edition different from the college edition?

- The front matter of the AP* Edition includes a detailed topic guide that correlates the Second Edition content to the current College Board AP Physics B curriculum guidelines (pp. xix–xxi). It also includes a comprehensive listing of all print and media supplements for AP students and teachers (pp. xvi–xvii).
- The AP* Edition of *College Physics: A Strategic Approach, 2nd Edition* includes multi-year access to **MasteringPhysics® with Pearson eText**—a next-generation, one-source learning and assessment system.
- The AP* Edition has a reinforced binding, which meets NASTA requirements, to withstand multiple years of high school use.
- New illustrated **Chapter Previews** at the start of each chapter provide visual, hierarchical, and non-technical previews proven to help students organize their thinking and improve their understanding of the upcoming material.
- New **Integrated Examples** at the end of each chapter give students additional help in solving general problems not tied to particular sections. Many integrate material from other chapters.
- New **Part Summary Problems** at the end of each of the seven parts of the book test students' abilities to draw on concepts and techniques from multiple chapters.
- **More streamlined presentations** throughout the text. Based on extensive feedback, we've pared some topics, reconfigured others, and provided a more readable, student-friendly text.
- **Improved and more varied end-of-chapter problems.** Using data from MasteringPhysics, we have reworked the problem sets to enhance clarity, topic coverage, and variety—adding, in particular, more problems based on real-world situations and more problems using ratio reasoning.

The more significant content changes include:

- The treatment of Newton's third law in Chapters 4 and 5 has been better focused on the types of problems that students will be asked to solve.
- Angular position and angular velocity are now developed together in Chapter 6, rather then being divided between Chapters 3 and 6. More emphasis has been given to angular position and angular velocity graphs, emphasizing the analogy with the linear position and velocity graphs of Chapter 2.
- The Chapter 10 presentation of work and energy has been streamlined and clarified. The problem-solving strategy for conservation of energy problems now plays a more prominent role.
- Chapter 11, Using Energy, is now more focused on concrete applications of energy use. All discussions of thermal properties have matter have been moved to Chapter 12, which has been reorganized to emphasize the single theme, "What happens to matter when you heat or cool it?"
- The ordering of topics within Chapters 18 and 19 has been revised. Ray tracing and the thin-lens equation are now paired together in Chapter 18; the pinhole camera and color/dispersion have moved to Chapter 19.
- Chapter 21 has been significantly rewritten to make the difficult idea of electric potential more concrete and usable.
- The section on household electricity has been moved from Chapter 23 to Chapter 26. Chapter 23 is now better focused on resistors and capacitors while Chapter 26, AC Circuits, has become a more practical chapter with sections on household electricity and electrical safety.
- Chapters 28–30 on quantum, atomic, and nuclear physics have been significantly streamlined in the hope that more instructors will be able to teach these important topics.

Textbook Organization

College Physics: A Strategic Approach is a 30-chapter text intended for use in a full-year AP Physics B course. The textbook is divided into seven parts: Part I: *Force and Motion*, Part II: *Conservation Laws*, Part III: *Properties of Matter*, Part IV: *Oscillations and Waves*, Part V: *Optics*, Part VI: *Electricity and Magnetism*, and Part VII: *Modern Physics*.

Part I covers Newton's laws and their applications. The coverage of two fundamental conserved quantities, momentum and energy, is in Part II, for two reasons. First, the way that problems are solved using conservation laws—comparing an *after* situation to a *before* situation—differs fundamentally from the problem-solving strategies used in Newtonian dynamics. Second, the concept of energy has a significance far beyond mechanical (kinetic and potential) energies. In particular, the key idea in thermodynamics is energy, and moving from the study of energy in Part II into thermal physics in Part III allows the uninterrupted development of this important idea.

Optics (Part V) is covered directly after oscillations and waves (Part IV), but *before* electricity and magnetism (Part VI). Further, we treat wave optics before ray optics. Our motivations for this organization are twofold. First, wave optics is largely just an extension of the general ideas of waves; in a more traditional organization, students will have forgotten much of what they learned about waves by the time they get to wave optics. Second, optics as it is presented in introductory physics makes no use of the properties of electromagnetic fields. The documented difficulties that students have with optics are difficulties with waves, not difficulties with electricity and magnetism. There's little reason other than historical tradition to delay optics. However, the optics chapters are easily deferred until after Part VI for instructors who prefer that ordering of topics.

AP Teacher Supplements

NOTE ►For convenience, all of the following teacher supplements (except for the Instructor Resource DVD) can be downloaded from the "Instructor Resources" area within MasteringPhysics (www.masteringphysics.com). In addition, many of the teacher supplements and resources for this text are available electronically to qualified adopters on the Instructor Resource Center (IRC). Upon adoption or to preview, please go to www.PearsonSchool.com/Access_Request and select "Instructor Resource Center." You will be required to complete a brief one-time registration. Upon verification of educator status, access information and instructions will be sent to you via email. Once logged into the IRC, enter your text ISBN in the "Search Our Catalog" box to locate your resources. ◄

- The **Instructor Guide** provides chapter-by-chapter creative ideas and teaching tips. In addition, it contains an extensive review of what has been learned from physics education research, and provides guidelines for using active-learning techniques.
- The **Instructor Solutions Manual,** provides *complete* solutions to all the end-of-chapter questions and problems. All solutions follow the Prepare/Solve/Assess problem-solving strategy used in the textbook for quantitative problems, and Reason/ Assess strategy for qualitative ones.

MP **MasteringPhysics®** (www.masteringphysics.com) is a homework, tutorial, and assessment system designed to assign, assess, and track each student's progress using a wide diversity of tutorials and extensively pre-tested problems. In addition to the textbook's end-of-chapter and new end-of-part problems, MasteringPhysics for *College Physics, Second Edition,* also includes author-selected prebuilt assignments, specific tutorials for all the textbook's Problem-Solving Strategies, Tactics Boxes, and Math Relationship boxes, as well as Reading Quizzes and Test Bank questions for each chapter.

MasteringPhysics provides instructors with a fast and effective way to assign uncompromising, wide-ranging online homework assignments of just the right difficulty and duration. The tutorials coach 90% of students to the correct answer with specific wrong-answer feedback. The powerful post-assignment diagnostics allow instructors to assess the progress of their class as a whole or to quickly identify individual student's areas of difficulty.

- Upon textbook purchase, students and teachers are granted access to MasteringPhysics®. High school teachers can obtain preview or adoption access for MasteringPhysics in one of the following ways:

Preview Access

- Teachers can request preview access online by visiting PearsonSchool.com/Access_Request (choose option 4). Preview Access information will be sent to the teacher via email.

Adoption Access

- A Pearson Adoption Access Card, with codes and complete instructions, will be delivered with your textbook purchase. (ISBN: 0-13-034391-9)
  OR
- Visit PearsonSchool.com/Access_Request (choose option 2/3). Adoption access information will be sent to the teacher via email.

- The cross-platform **Instructor Resource DVD** (ISBN 978-0-321-59628-4) provides invaluable and easy-to-use resources for your class, organized by textbook chapter. The contents include a comprehensive library of more than 220 applets from **ActivPhysics OnLine™,** as well as all figures, photos, tables, and summaries from the textbook in JPEG format. In addition, all the Problem-Solving Strategies, Math Relationships Boxes, Tactics Boxes, and Key Equations are provided in editable Word and JPEG formats. The **Instructor Guide** is also included as editable Word files, along with pdfs of answers to the **Student Workbook** exercises, and **Lecture Outlines (with Classroom Response System "Clicker" Questions)** in PowerPoint.
- The **Test Bank**, contains more than 2,000 high-quality problems, with a range of multiple-choice, true/false, short-answer, and regular homework-type questions. Test files are provided in both TestGen (an easy-to-use, fully networkable program for creating and editing quizzes and exams) and Word format, and can also be downloaded from the Instructor Resource Center.
- ActivPhysics OnLine **ActivPhysics OnLine™** (accessed through the Self Study area within www.masteringphysics.com) provides a comprehensive library of more than 420 tried and tested *ActivPhysics* applets updated for web delivery using the latest online technologies. In addition, it provides a suite of highly regarded applet-based tutorials developed by education pioneers Professors Alan Van Heuvelen and Paul D'Alessandris. The *ActivPhysics* margin icon directs students to specific exercises that complement the textbook discussion.

  The online exercises are designed to encourage students to confront misconceptions, reason qualitatively about physical processes, experiment quantitatively, and learn to think critically. They cover all topics from mechanics to electricity and magnetism and from optics to modern physics. More than 220 applets from the *ActivPhysics OnLine* library are also available on the *Instructor Resource DVD.*

Student Supplements

(available for purchase)

- The **Student Workbooks** (Volume 1, Chapters 1–16, ISBN: 978-0-321-59632-1 and Volume 2, Chapters 17–30, ISBN: 978-0-321-59633-8) bridge the gap between textbook and homework problems. The workbook exercises, which are keyed to each section of the textbook, focus on developing specific skills ranging from identifying forces and drawing free-body diagrams to interpreting field diagrams.
- The **Student Solutions Manuals Chapters 1–16** (ISBN 978-0-321-59629-1) and **Chapters 17–30** (ISBN 978-0-321-59630-7), provide *detailed* solutions to more than half of the odd-numbered end-of-chapter problems. Following the problem-solving strategy presented in the text, thorough solutions are provided to carefully illustrate both the qualitative (Reason/Assess) and quantitative (Prepare/Solve/Assess) steps in the problem-solving process.
- **MasteringPhysics®** (www.masteringphysics.com) is a homework, tutorial, and assessment system based on years of research into how students work physics problems and precisely where they need help. Studies show that students who use MasteringPhysics significantly increase their final scores compared to hand-written homework. MasteringPhysics achieves this improvement by providing students with instantaneous feedback specific to their wrong answers, simpler sub-problems upon request when they get stuck, and partial credit for their method(s) used. This individualized, 24/7 Socratic tutoring is recommended by nine out of ten students to their peers as the most effective and time-efficient way to study.
- **Pearson eText** is available through MasteringPhysics. Allowing students access to the text wherever they have access to the Internet, Pearson eText comprises the full student text, including figures that can be enlarged for better viewing. Students are also able to pop up definitions and terms to help with vocabulary and the reading of the material, as well as take notes using the annotation feature at the top of each page.
- ActivPhysics OnLine **ActivPhysics OnLine™** (accessed via www.masteringphysics.com), provides students with a suite of highly regarded applet-based tutorials. The *ActivPhysics* margin icons throughout the book direct students to specific exercises that complement the textbook discussion.

How to Succeed in AP Physics

The most incomprehensible thing about the universe is that it is comprehensible.
—Albert Einstein

What can you expect to learn in this course? Let's start by talking about what physics is. Physics is a way of thinking about the physical aspects of nature. Physics is not about "facts." It's far more focused on discovering *relationships* between facts and the *patterns* that exist in nature than on learning facts for their own sake. Our emphasis will be on thinking and reasoning. We are going to look for patterns and relationships in nature, develop the logic that relates different ideas, and search for the reasons *why* things happen as they do. Once we've figured out a pattern, a set of relationships, we'll look at applications to see where this understanding takes us.

Like any subject, physics is best learned by doing. "Doing physics" in this course means solving problems, applying what you have learned to answer questions at the end of the chapter. When you are given a homework assignment, you may find yourself tempted to simply solve the problems by thumbing through the text looking for a formula that seems like it will work. This isn't how to do physics—you want to learn to **reason**, not to "plug and chug."

How do you learn to reason in this way? There's no single strategy for studying physics that will work for all students, but we can make some suggestions that will certainly help:

- **Read each chapter *before* it is discussed in class.** Class attendance is much less effective if you have not prepared. When you first read a chapter, focus on learning new vocabulary, definitions, and notation. You won't understand what's being discussed or how the ideas are being used if you don't know what the terms and symbols mean.
- **Participate actively in class.** Take notes, ask and answer questions, take part in discussion groups. There is ample scientific evidence that *active participation* is far more effective for learning science than is passive listening.
- **After class, go back for a careful rereading of the chapter.** In your second reading, pay close attention to the details and the worked examples. Look for the *logic* behind each example, not just at what formula is being used. We have a three-step process by which we solve all of the worked examples in the text. Most chapters have detailed Problem-Solving Strategies to help you see how to apply this procedure to particular topics, and Tactics Boxes that explain specific steps in your analysis.
- **Apply what you have learned to the homework problems at the end of each chapter.** By following the techniques of the worked examples, applying the tactics and problem-solving strategies, you'll learn how to apply the knowledge you are gaining. In short, you'll learn to reason like a physicist.
- **Form a study group with two or three classmates.** There's good evidence that students who study regularly with a group do better than the rugged individualists who try to go it alone.

And we have one final suggestion. As you read the book, take part in class, and work through problems, step back every now and then to appreciate the big picture. You are going to study topics that range from motions in the solar system to the electrical signals in the nervous system that let you order your hand to turn the pages of this book. You will learn quantitative methods to calculate things such as how far a car will move as it brakes to a stop and how to build a solenoid for an MRI machine. It's a remarkable breadth of topics and techniques that is based on a very compact set of organizing principles. It's quite remarkable, really, well worthy of your study.

Now, let's get down to work.

AP Topic Correlation

This chart correlates the Advanced Placement Physics B topics as outlined by the College Board with the corresponding page numbers in Knight/Jones/Field AP* Edition of *College Physics: A Strategic Approach,* 2nd edition
(SE = Student Edition)

I. NEWTONIAN MECHANICS

A. Kinematics (including vectors, vector algebra, components of vectors, coordinate systems, displacement, velocity, and acceleration)

1. Motion in one dimension	SE: 32–38, 40, 42–46, 48–52, 58, 59–66
2. Motion in two dimensions, including projectile motion	SE: 75–78, 82–83, 85–88, 93–94, 97–100

B. Newton's Laws of Motion

1. Static equilibrium (first law)	SE: 103–104, 124, 125–126, 129, 131–135, 159, 161, 163
2. Dynamics of a single particle (second law)	SE: 115–117, 118–120, 124, 128–130, 135–138, 142–150, 158, 160–164
3. Systems of two or more objects (third law)	SE: 120–123, 124, 125–126, 128–130, 150–160, 162–165

C. Work, Energy, Power

1. Work and the work-energy theorem	SE: 294–297, 315, 316–317, 319–320
2. Forces and potential energy	SE: 301–304, 317–318
3. Conservation of energy	SE: 239–241, 294–296, 306–310, 317–318, 320, 447–448
4. Power	SE: 312–313, 315, 318–319

D. Systems of Particles, Linear Momentum

1. Center of mass	SE: 209–213, 224, 227, 229
2. Impulse and momentum	SE: 261–266, 284–286
3. Conservation of linear momentum, collisions	SE: 268–276, 284, 888, 917–918

E. Circular Motion and Rotation

1. Uniform circular motion	SE: 167–177, 179–182, 193, 196–197
2. Torque and rotational statics	SE: 204–209, 213–217, 224, 226–229, 233–237, 247, 249–250
3. Rotational kinematics and dynamics	SE: 168–170, 202–204, 217–222, 224, 226, 228, 300–301, 315
4. Angular momentum and its conservation	SE: 204–208, 214–220, 226, 276–279, 285

F. Oscillations and Gravitation

1. Simple harmonic motion (dynamics and energy relationships)	SE: 449–460, 466–467, 469, 472–474
2. Mass on a spring	SE: 457–458, 472–474
3. Pendulum and other oscillations	SE: 460–463, 473, 475
4. Newton's law of gravity	SE: 185–189, 197
5. Orbits of planets and satellites	SE: 189–191, 277–279

II. FLUID MECHANICS AND THERMAL PHYSICS

A. Fluid Mechanics	
1. Hydrostatic pressure	SE: 407–409, 411–413, 431, 434, 436–437
2. Buoyancy	SE: 415–418, 431, 435
3. Fluid flow continuity	SE: 420–422, 435
4. Bernoulli's equation	SE: 425–427, 435
B. Temperature and Heat	
1. Mechanical equivalent of heat	SE: 336–337, 349, 352
2. Heat transfer and thermal expansion	SE: 378–381, 390–394, 396, 400–401
C. Kinetic Theory and Thermodynamics	
1. Ideal gases	SE: 365–369, 372–378, 399, 400–402
2. Laws of thermodynamics	SE: 336–337, 339–340, 344–346, 350, 352

III. ELECTRICITY AND MAGNETISM

A. Electrostatics	
1. Charge and Coulomb's Law	SE: 643–645, 648, 651–655, 667–668, 670–671
2. Electric field and electric potential (including point charges)	SE: 655–661, 667, 670–671, 677–678, 682, 684–693, 703, 705–707, 710
3. Gauss's law	TECH: www.masteringphysics.com
4. Fields and potentials of other charge distributions	SE: 655–661, 684–690
B. Conductors, Capacitors, Dielectrics	
1. Electrostatics with conductors	SE: 648, 662–663, 667, 671, 819–820
2. Capacitors	SE: 695–701, 703, 708
3. Dielectrics	SE: 698–699, 703, 708
C. Electric Circuits	
1. Current, resistance, power	SE: 714–716, 720–723, 732, 735–736
2. Steady-state direct current circuits with batteries and resistors only	SE: 718–719, 724, 735–736, 741–749, 766, 770
3. Capacitors in circuits	SE: 753–757, 766, 771–772
D. Magnetic Fields	
1. Forces on moving charges in magnetic fields	SE: 789–793, 806, 808, 810–811
2. Forces on current-carrying wires in magnetic fields	SE: 796, 799–800, 806, 811–812
3. Fields of long current-carrying wires	SE: 796–798, 806, 811
4. Biot-Savart law and Ampere's law	SE: 784–788
E. Electromagnetism	
1. Electromagnetic induction (including Faraday's law and Lenz's law)	SE: 818–828, 843, 847–848
2. Inductance (including LR and LC circuits)	SE: 741–748, 818–821
3. Maxwell's equations	SE: 830–831, 843

IV. WAVES AND OPTICS

A. Wave Motion (including Sound)

1. Traveling waves	SE: 479–484, 486–487, 491–492, 495–496, 500, 503–505, 512
2. Wave propagation	SE: 479, 500, 503–504
3. Standing waves	SE: 513–520, 530, 533–534
4. Superposition	SE: 508–510, 530, 532–533

B. Physical Optics

1. Interference and diffraction	SE: 523–527, 530, 534, 549–555, 557, 558–564, 567, 569–570
2. Dispersion of light and the electromagnetic spectrum	SE: 489,547–548, 836–841, 843

C. Geometric Optics

1. Reflection and refraction	SE: 547–548, 578–584, 602, 605
2. Mirrors	SE: 579–580, 583, 594–600, 602, 604–605
3. Lenses	SE: 587–593, 597–600, 602, 605–606

V. ATOMIC AND NUCLEAR PHYSICS

A. Atomic Physics and Quantum Effects

1. Photons, the photoelectric effect, Compton scattering, x-rays	SE: 840, 849, 925–930, 932, 947–950, 952–953,
2. Atomic energy levels	SE: 939–940, 950, 962, 975–977, 984, 987–988
3. Wave-particle duality	SE: 933–936, 946, 947, 950–952

B. Nuclear Physics

1. Nuclear reactions (including conservation of mass number and charge)	SE: 913, 992, 996–997, 999–1002, 1019–1020
2. Mass-energy equivalence	SE: 912–913

Upon publication, this text was correlated to the College Board's Physics B Course Description dated Fall 2011. We continually monitor the College Board's AP Course Description for updates to exam topics.

For the most current AP Exam Topic correlation for this textbook, visit PearsonSchool.com/AdvancedCorrelations.

Detailed Contents

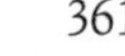

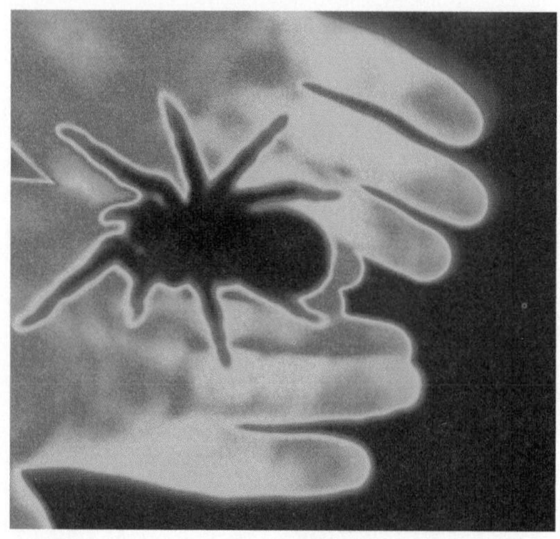

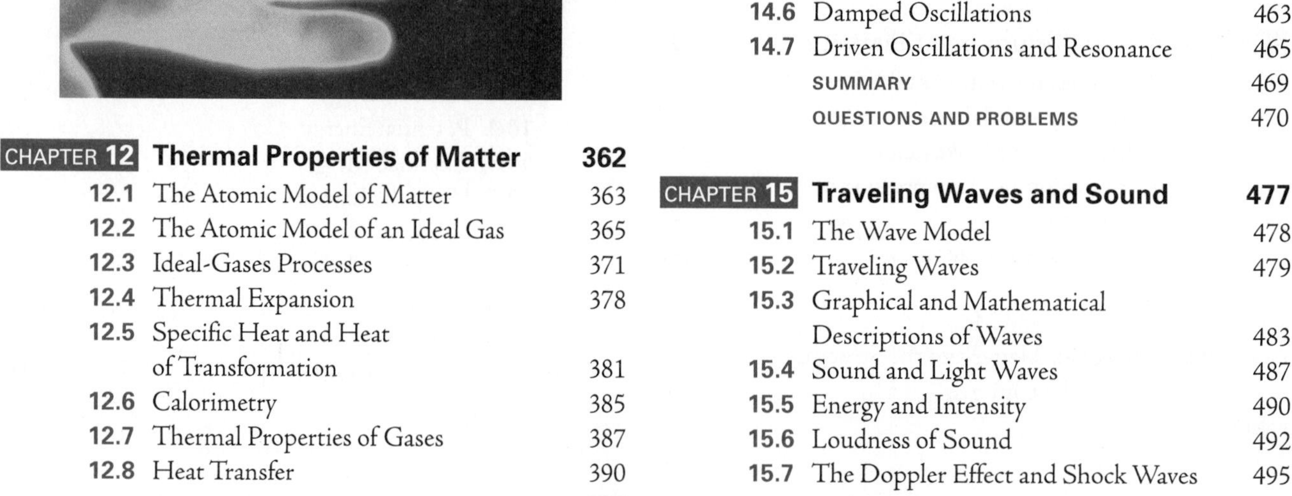

PART VII Modern Physics

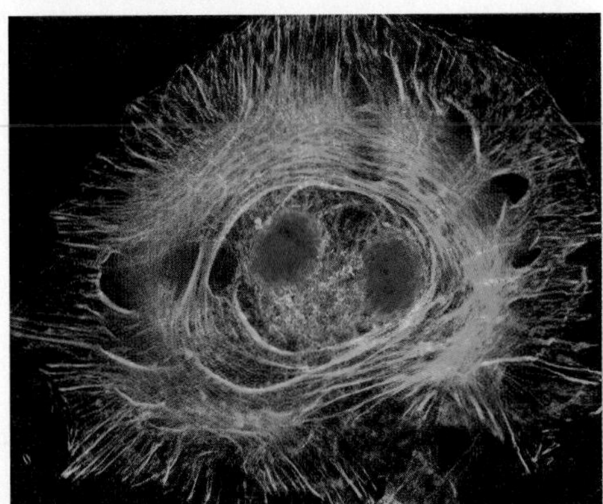

PART I

Force and Motion

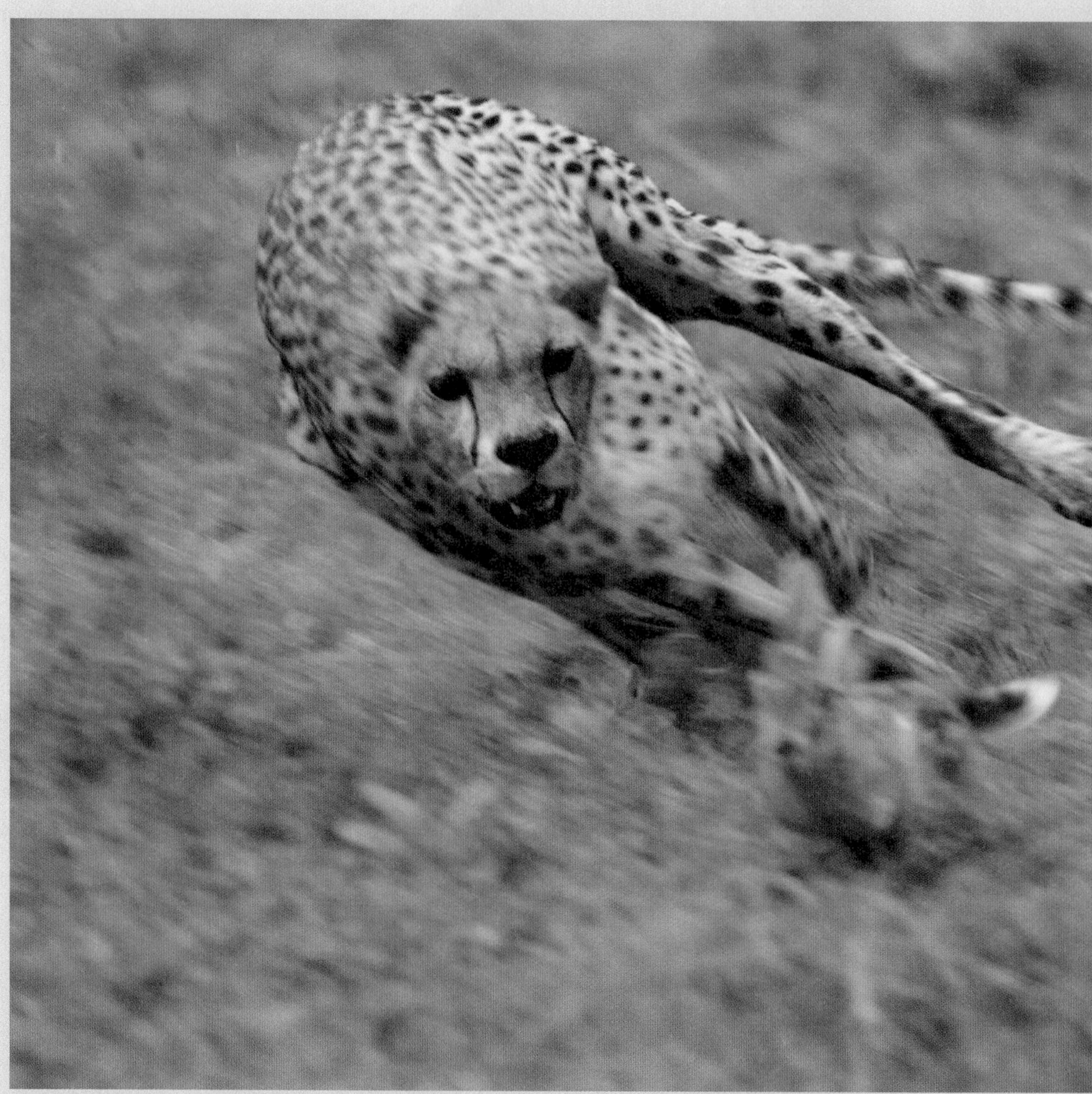

The cheetah is the fastest land animal, able to run at speeds exceeding 60 miles per hour. Nonetheless, the rabbit has an advantage in this chase. It can *change* its motion more quickly and will likely escape. How can you tell, by looking at the picture, that the cheetah is changing its motion?

OVERVIEW

Why Things Change

Each of the seven parts of this book opens with an overview that gives you a look ahead, a glimpse of where your journey will take you in the next few chapters. It's easy to lose sight of the big picture while you're busy negotiating the terrain of each chapter. In Part I, the big picture is, in a word, *change*.

Simple observations of the world around you show that most things change. Some changes, such as aging, are biological. Others, such as sugar dissolving in your coffee, are chemical. We will look at changes that involve *motion* of one form or another—running and jumping, throwing balls, lifting weights.

There are two big questions we must tackle to study how things change by moving:

- **How do we describe motion?** How should we measure or characterize the motion if we want to analyze it mathematically?
- **How do we explain motion?** Why do objects have the particular motion they do? Why, when you toss a ball upward, does it go up and then come back down rather than keep going up? What are the "laws of nature" that allow us to predict an object's motion?

Two key concepts that will help answer these questions are *force* (the "cause") and *acceleration* (the "effect"). Our basic tools will be three laws of motion elucidated by Isaac Newton. Newton's laws relate force to acceleration, and we will use them to explain and explore a wide range of problems. As we learn to solve problems dealing with motion, we will learn basic techniques that we can apply in all the parts of this book.

Simplifying Models

Reality is extremely complicated. We would never be able to develop a science if we had to keep track of every detail of every situation. Suppose we analyze the tossing of a ball. Is it necessary to analyze the way the atoms in the ball are connected? Do we need to analyze what you ate for breakfast and the biochemistry of how that was translated into muscle power? These are interesting questions, of course. But if our task is to understand the motion of the ball, we need to simplify!

We can do a perfectly fine analysis if we treat the ball as a round solid and your hand as another solid that exerts a force on the ball. This is a *model* of the situation. A model is a simplified description of reality—much as a model airplane is a simplified version of a real airplane—that is used to reduce the complexity of a problem to the point where it can be analyzed and understood.

Model building is a major part of the strategy that we will develop for solving problems in all parts of the book. We will introduce different models in different parts. We will pay close attention to where simplifying assumptions are being made, and why. Learning *how* to simplify a situation is the essence of successful modeling—and successful problem solving.

1 Representing Motion

As this skier moves in a graceful arc through the air, the direction of his motion, and the distance between each of his positions and the next, are constantly changing. What language should we use to describe this motion?

LOOKING AHEAD ➧

The goals of Chapter 1 are to introduce the fundamental concepts of motion and to review the related basic mathematical principles.

The Chapter Preview

Each chapter will start with an overview of the material to come. You should read these chapter previews carefully to get a sense of the content and structure of the chapter.

Arrows will show the connections and flow between different topics in the preview.

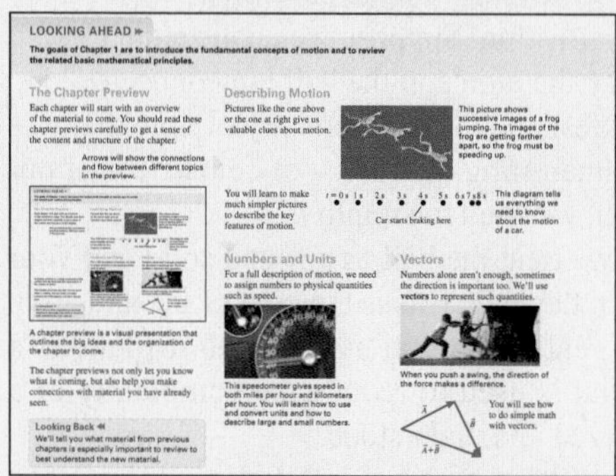

A chapter preview is a visual presentation that outlines the big ideas and the organization of the chapter to come.

The chapter previews not only let you know what is coming, but also help you make connections with material you have already seen.

Looking Back ◀◀
We'll tell you what material from previous chapters is especially important to review to best understand the new material.

Describing Motion

Pictures like the one above or the one at right give us valuable clues about motion.

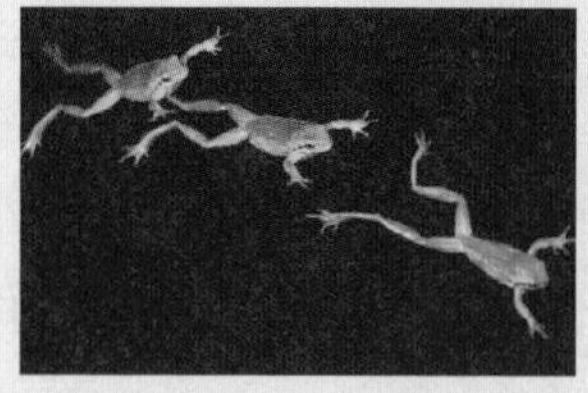

This picture shows successive images of a frog jumping. The images of the frog are getting farther apart, so the frog must be speeding up.

You will learn to make much simpler pictures to describe the key features of motion.

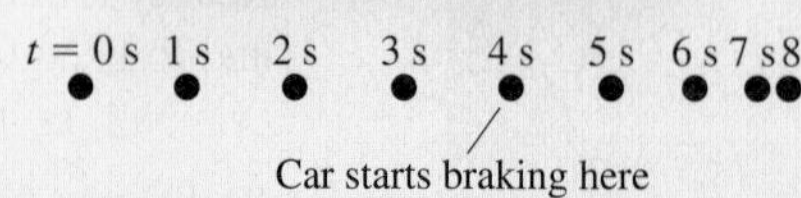

This diagram tells us everything we need to know about the motion of a car.

Numbers and Units

For a full description of motion, we need to assign numbers to physical quantities such as speed.

This speedometer gives speed in both miles per hour and kilometers per hour. You will learn how to use and convert units and how to describe large and small numbers.

Vectors

Numbers alone aren't enough, sometimes the direction is important too. We'll use **vectors** to represent such quantities.

When you push a swing, the direction of the force makes a difference.

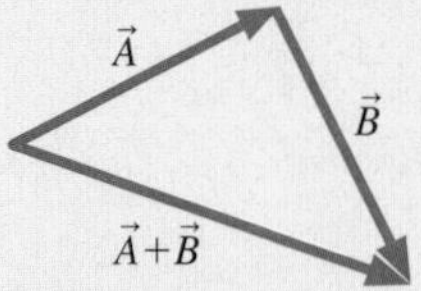

You will see how to do simple math with vectors.

1.1 Motion: A First Look

The concept of motion is a theme that will appear in one form or another throughout this entire book. You have a well-developed intuition about motion, based on your experiences, but you'll see that some of the most important aspects of motion can be rather subtle. We need to develop some tools to help us explain and understand motion, so rather than jumping immediately into a lot of mathematics and calculations, this first chapter focuses on visualizing motion and becoming familiar with the concepts needed to describe a moving object.

One key difference between physics and other sciences is how we set up and solve problems. We'll often use a two-step process to solve motion problems. The first step is to develop a simplified *representation* of the motion so that key elements stand out. For example, the photo of the skier at the start of the chapter allows us to observe his position at many successive times. It is precisely by considering this sort of picture of motion that we will begin our study of this topic. The second step is to analyze the motion with the language of mathematics. The process of putting numbers on nature is often the most challenging aspect of the problems you will solve. In this chapter, we will explore the steps in this process as we introduce the basic concepts of motion.

Types of Motion

As a starting point, let's define **motion** as the change of an object's position or orientation with time. Examples of motion are easy to list. Bicycles, baseballs, cars, airplanes, and rockets are all objects that move. The path along which an object moves, which might be a straight line or might be curved, is called the object's **trajectory.**

FIGURE 1.1 shows four basic types of motion that we will study in this book. In this chapter, we will start with the first type of motion in the figure, motion along a straight line. In later chapters, we will learn about circular motion, which is the motion of an object along a circular path; projectile motion, the motion of an object through the air; and rotational motion, the spinning of an object about an axis.

FIGURE 1.1 Four basic types of motion.

Straight-line motion

Circular motion

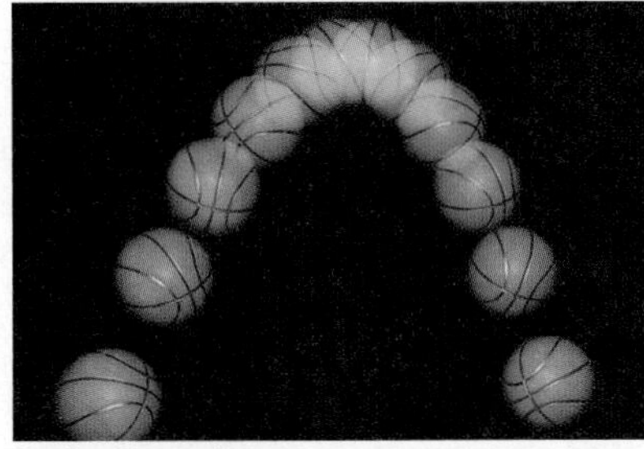

Projectile motion

Rotational motion

Making a Motion Diagram

An easy way to study motion is to record a video of a moving object with a stationary camera. A video camera takes images at a fixed rate, typically 30 images every second. Each separate image is called a *frame*. As an example, FIGURE 1.2 shows several frames from a video of a car going past. Not surprisingly, the car is in a different position in each frame.

NOTE ► It's important to keep the camera in a *fixed position* as the object moves by. Don't "pan" it to track the moving object. ◄

FIGURE 1.2 Several frames from the video of a car.

Suppose we now edit the video by layering the frames on top of each other and then look at the final result. We end up with the picture in FIGURE 1.3. This composite image, showing an object's positions at several *equally spaced instants of time,* is called a **motion diagram.** As simple as motion diagrams seem, they will turn out to be powerful tools for analyzing motion.

The same amount of time elapses between each image and the next.

FIGURE 1.3 A motion diagram of the car shows all the frames simultaneously.

Now let's take our camera out into the world and make some motion diagrams. The following table illustrates how a motion diagram shows important features of different kinds of motion.

Examples of motion diagrams

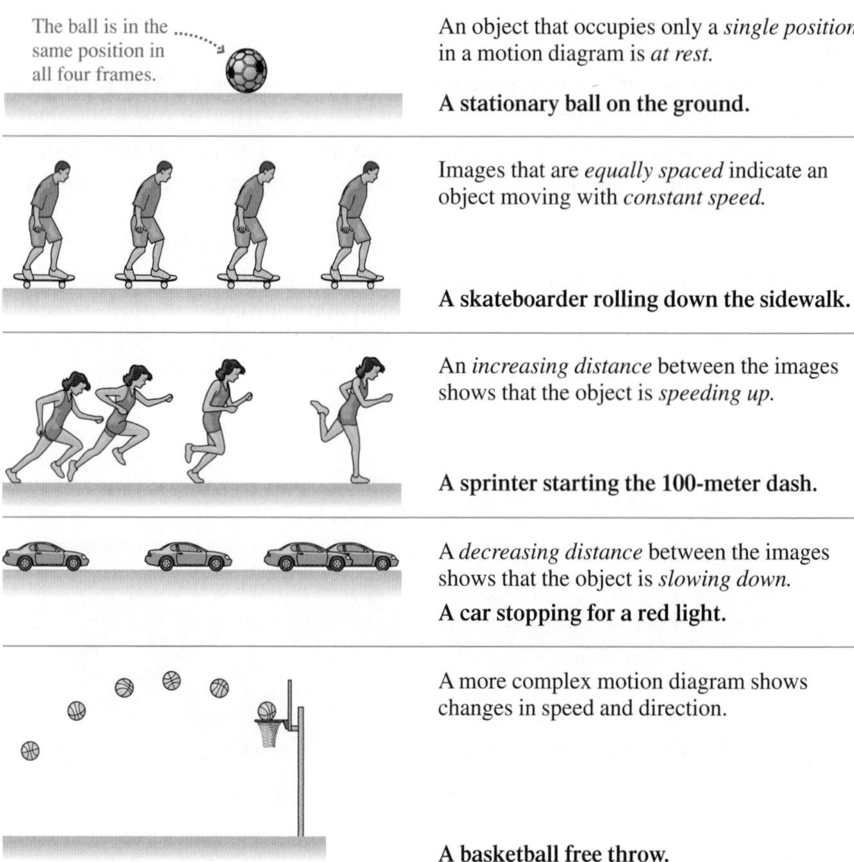

The ball is in the same position in all four frames.

A stationary ball on the ground.

A skateboarder rolling down the sidewalk.

A sprinter starting the 100-meter dash.

A basketball free throw.

We have defined several concepts (at rest, constant speed, speeding up, and slowing down) in terms of how the moving object appears in a motion diagram. These are called **operational definitions,** meaning that the concepts are defined in terms of a particular procedure or operation performed by the investigator. For example, we could answer the question Is the airplane speeding up? by checking whether or not the images in the plane's motion diagram are getting farther apart. Many of the concepts in physics will be introduced as operational definitions. This reminds us that physics is an experimental science.

STOP TO THINK 1.1 Which car is going faster, A or B? Assume there are equal intervals of time between the frames of both videos.

NOTE ► Each chapter in this textbook has several *Stop to Think* questions. These questions are designed to see if you've understood the basic ideas that have just been presented. The answers are given at the end of the chapter, but you should make a serious effort to think about these questions before turning to the answers. If you answer correctly and are sure of your answer rather than just guessing, you can proceed to the next section with confidence. But if you answer incorrectly, it would be wise to reread the preceding sections carefully before proceeding onward. ◄

FIGURE 1.4 Simplifying a motion diagram using the particle model.

(a) Motion diagram of a car stopping

(b) Same motion diagram using the particle model

The same amount of time elapses between each frame and the next.
0
1
2
3
Numbers show the order in which the frames were taken.
A single dot is used to represent the object.

The Particle Model

For many objects, the motion of the object *as a whole* is not influenced by the details of the object's size and shape. To describe the object's motion, all we really need to keep track of is the motion of a single point: You could imagine looking at the motion of a dot painted on the side of the object.

In fact, for the purposes of analyzing the motion, we can often consider the object *as if* it were just a single point, without size or shape. We can also treat the object *as if* all of its mass were concentrated into this single point. An object that can be represented as a mass at a single point in space is called a **particle.**

If we treat an object as a particle, we can represent the object in each frame of a motion diagram as a simple dot. FIGURE 1.4 shows how much simpler motion diagrams appear when the object is represented as a particle. Note that the dots have been numbered 0, 1, 2, . . . to tell the sequence in which the frames were exposed. These diagrams still convey a complete understanding of the object's motion.

Treating an object as a particle is, of course, a simplification of reality. Such a simplification is called a **model.** Models allow us to focus on the important aspects of a phenomenon by excluding those aspects that play only a minor role. The **particle model** of motion is a simplification in which we treat a moving object as if all of its mass were concentrated at a single point. Using the particle model may allow us to see connections that are very important but that are obscured or lost by examining all the parts of an extended, real object. Consider the motion of the two objects shown in FIGURE 1.5. These two very different objects have exactly the same motion diagram. As we will see, all objects falling under the influence of gravity move in exactly the same manner if no other forces act. The simplification of the particle model has revealed something about the physics that underlies both of these situations.

Not all motions can be reduced to the motion of a single point, as we'll see. But for now, the particle model will be a useful tool in understanding motion.

FIGURE 1.5 The particle model for two falling objects.

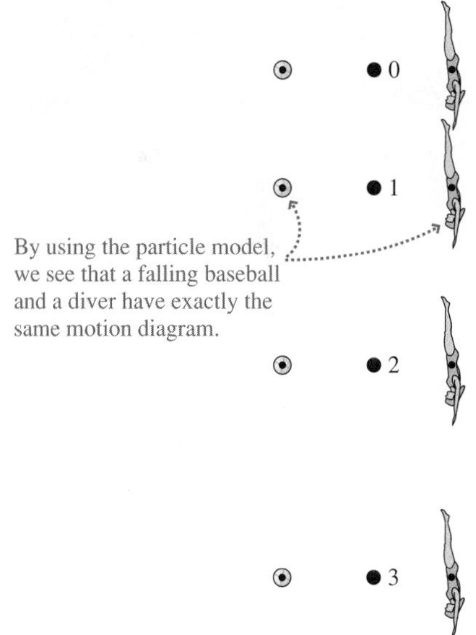

STOP TO THINK 1.2 Three motion diagrams are shown. Which is a dust particle settling to the floor at constant speed, which is a ball dropped from the roof of a building, and which is a descending rocket slowing to make a soft landing on Mars?

A. 0● 1● 2● 3● 4● 5●

B. 0● 1● 2● 3● 4● 5●

C. 0● 1● 2● 3● 4● 5●

1.2 Position and Time: Putting Numbers on Nature

To develop our understanding of motion further, we need to be able to make quantitative measurements: We need to use numbers. As we analyze a motion diagram, it is useful to know where the object is (its *position*) and when the object was at that position (the *time*). We'll start by considering the motion of an object that can move only along a straight line. Examples of this **one-dimensional** or "1-D" motion are a bicyclist moving along the road, a train moving on a long straight track, and an elevator moving up and down a shaft.

Position and Coordinate Systems

FIGURE 1.6 Describing your position.

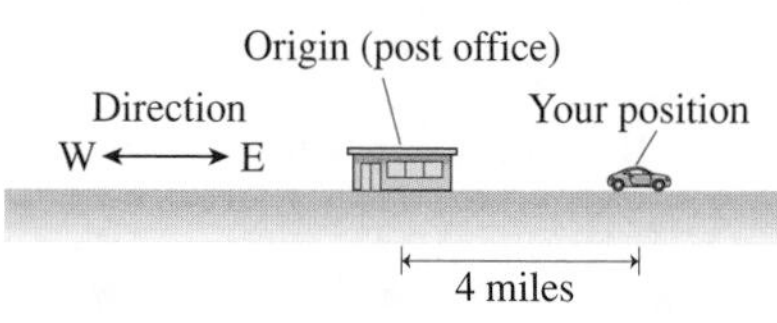

Suppose you are driving along a long, straight country road, as in FIGURE 1.6, and your friend calls and asks where you are. You might reply that you are 4 miles east of the post office, and your friend would then know just where you were. Your location at a particular instant in time (when your friend phoned) is called your **position.** Notice that to know your position along the road, your friend needed three pieces of information. First, you had to give her a reference point (the post office) from which all distances are to be measured. We call this fixed reference point the **origin.** Second, she needed to know how far you were from that reference point or origin—in this case, 4 miles. Finally, she needed to know which side of the origin you were on: You could be 4 miles to the west of it or 4 miles to the east.

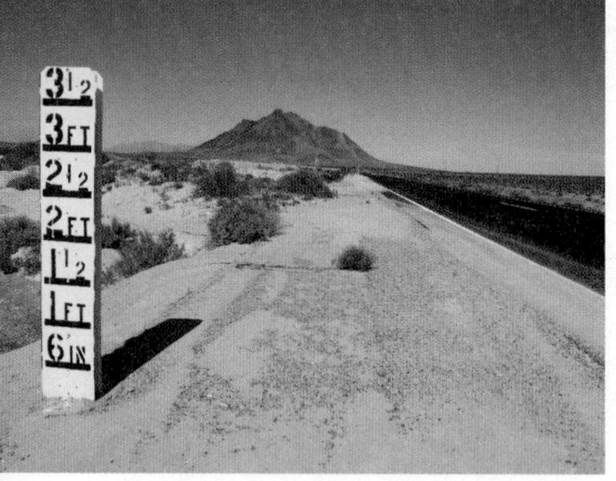

Sometimes measurements have a natural origin. This snow depth gauge has its origin set at road level.

We will need these same three pieces of information in order to specify any object's position along a line. We first choose our origin, from which we measure the position of the object. The position of the origin is arbitrary, and we are free to place it where we like. Usually, however, there are certain points (such as the well-known post office) that are more convenient choices than others.

In order to specify how far our object is from the origin, we lay down an imaginary axis along the line of the object's motion. Like a ruler, this axis is marked off in equally spaced divisions of distance, perhaps in inches, meters, or miles, depending on the problem at hand. We place the zero mark of this ruler at the origin, allowing us to locate the position of our object by reading the ruler mark where the object is.

Finally, we need to be able to specify which side of the origin our object is on. To do this, we imagine the axis extending from one side of the origin with increasing positive markings; on the other side, the axis is marked with increasing *negative* numbers. By reporting the position as either a positive or a negative number, we know on what side of the origin the object is.

These elements—an origin and an axis marked in both the positive and negative directions—can be used to unambiguously locate the position of an object. We call this a **coordinate system.** We will use coordinate systems throughout this book, and we will soon develop coordinate systems that can be used to describe the positions of objects moving in more complex ways than just along a line. FIGURE 1.7 shows a coordinate system that can be used to locate various objects along the country road discussed earlier.

FIGURE 1.7 The coordinate system used to describe objects along a country road.

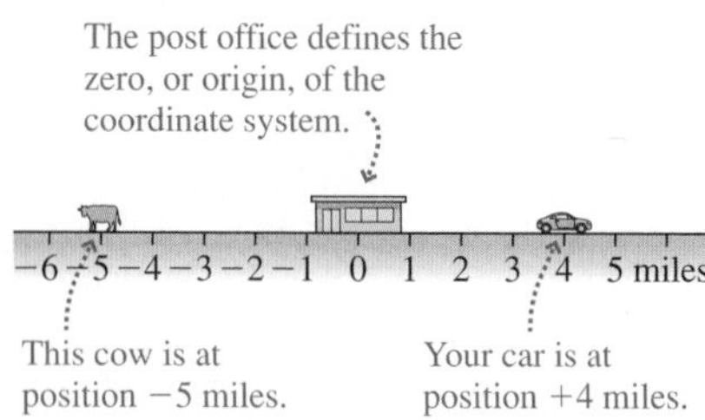

Although our coordinate system works well for describing the positions of objects located along the axis, our notation is somewhat cumbersome. We need to keep saying things like "the car is at position +4 miles." A better notation, and one that will become particularly important when we study motion in two dimensions, is to use a symbol such as x or y to represent the position along the axis. Then we can say "the cow is at $x = -5$ miles." The symbol that represents a position along an axis is called a **coordinate.** The introduction of symbols to represent positions (and, later, velocities and accelerations) also allows us to work with these quantities mathematically.

FIGURE 1.8 below shows how we would set up a coordinate system for a sprinter running a 50-meter race (we use the standard symbol "m" for meters). For horizontal motion like this we usually use the coordinate x to represent the position.

FIGURE 1.8 A coordinate system for a 50-meter race.

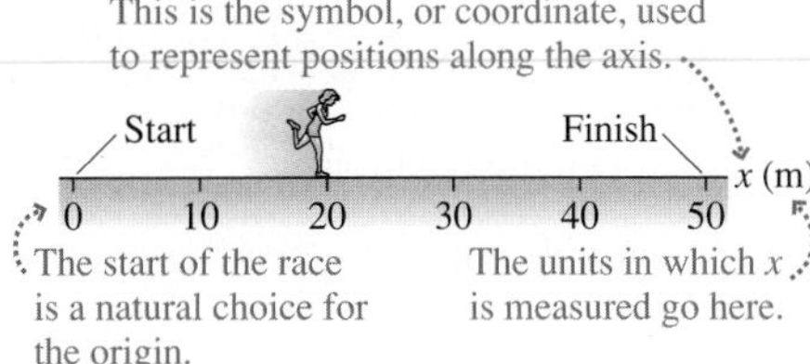
This is the symbol, or coordinate, used to represent positions along the axis.
Start
Finish
x (m)
0 10 20 30 40 50
The start of the race is a natural choice for the origin.
The units in which x is measured go here.

FIGURE 1.9 Examples of one-dimensional motion.

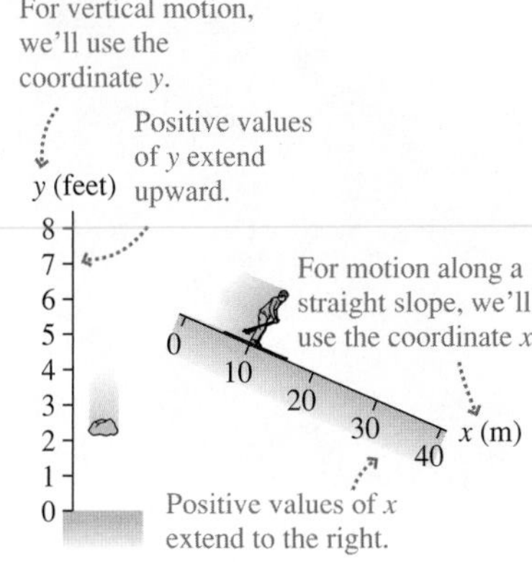
For vertical motion, we'll use the coordinate y.
Positive values of y extend upward.
y (feet)
8
7
6
5
4
3
2
1
0
For motion along a straight slope, we'll use the coordinate x.
0
10
20
30
40
x (m)
Positive values of x extend to the right.

Motion along a straight line need not be horizontal. As shown in **FIGURE 1.9**, a rock falling vertically downward and a skier skiing down a straight slope are also examples of straight-line or one-dimensional motion.

Time

The pictures in Figure 1.9 show the position of an object at just one instant of time. But a full motion diagram represents how an object moves as time progresses. So far, we have labeled the dots in a motion diagram by the numbers 0, 1, 2, . . . to indicate the order in which the frames were exposed. But to fully describe the motion, we need to indicate the *time,* as read off a clock or a stopwatch, at which each frame of a video was made. This is important, as we can see from the motion diagram of a stopping car in **FIGURE 1.10**. If the frames were taken 1 second apart, this motion diagram shows a leisurely stop; if 1/10 of a second apart, it represents a screeching halt.

FIGURE 1.10 Is this a leisurely stop or a screeching halt?

0
1
2
3
4 5

For a complete motion diagram, we thus need to label each frame with its corresponding time (symbol t) as read off a clock. But when should we start the clock? Which frame should be labeled $t = 0$? This choice is much like choosing the origin $x = 0$ of a coordinate system: You can pick any arbitrary point in the motion and label it "$t = 0$ seconds." This is simply the instant you decide to start your clock or stopwatch, so it is the origin of your time coordinate. A video frame labeled "$t = 4$ seconds" means it was taken 4 seconds after you started your clock. We typically choose $t = 0$ to represent the "beginning" of a problem, but the object may have been moving before then.

To illustrate, **FIGURE 1.11** shows the motion diagram for a car moving at a constant speed and then braking to a halt. Two possible choices for the frame labeled $t = 0$ seconds are shown; our choice depends on what part of the motion we're interested in. Each successive position of the car is then labeled with the clock reading in seconds (abbreviated by the symbol "s").

FIGURE 1.11 The motion diagram of a car that travels at constant speed and then brakes to a halt.

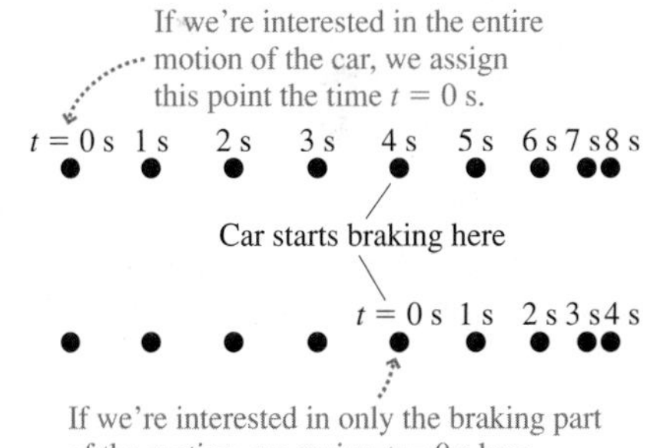
If we're interested in the entire motion of the car, we assign this point the time t = 0 s.
t = 0 s 1 s 2 s 3 s 4 s 5 s 6 s 7 s 8 s
Car starts braking here
t = 0 s 1 s 2 s 3 s 4 s
If we're interested in only the braking part of the motion, we assign t = 0 s here.

Changes in Position and Displacement

Now that we've seen how to measure position and time, let's return to the problem of motion. To describe motion we'll need to measure the *changes* in position that occur with time. Consider the following:

> Sam is standing 50 feet (ft) east of the corner of 12th Street and Vine. He then walks to a second point 150 ft east of Vine. What is Sam's change of position?

FIGURE 1.12 Sam undergoes a displacement Δx from position x_i to position x_f.

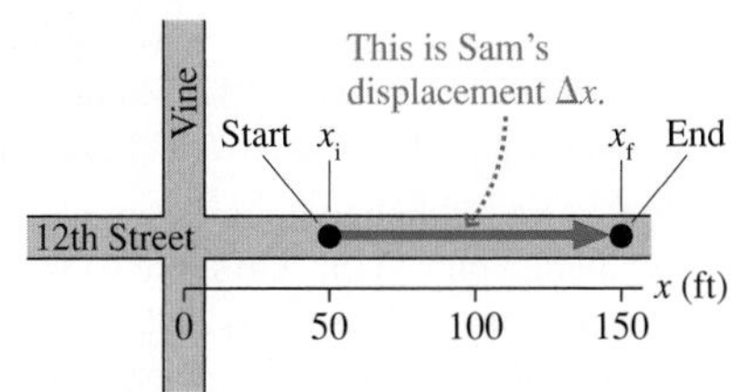
This is Sam's displacement Δx.
Vine
Start x_i
x_f End
12th Street
x (ft)
0
50
100
150

FIGURE 1.12 shows Sam's motion on a map. We've placed a coordinate system on the map, using the coordinate x. We are free to place the origin of our coordinate system wherever we wish, so we have placed it at the intersection. Sam's initial position is then at $x_i = 50$ ft. The positive value for x_i tells us that Sam is east of the origin.

The size and the direction of the displacement both matter. Roy Riegels (pursued above by teammate Benny Lom) found this out in dramatic fashion in the 1928 Rose Bowl when he recovered a fumble and ran 69 yards—toward his own team's end zone. An impressive distance, but in the wrong direction!

NOTE ► We will label special values of x or y with subscripts. The value at the start of a problem is usually labeled with a subscript "i," for *initial,* and the value at the end is labeled with a subscript "f," for *final.* For cases having several special values, we will usually use subscripts "1," "2," and so on. ◄

Sam's final position is $x_f = 150$ ft, indicating that he is 150 ft east of the origin. You can see that Sam has changed position, and a *change* of position is called a **displacement.** His displacement is the distance labeled Δx in Figure 1.12. The Greek letter delta (Δ) is used in math and science to indicate the *change* in a quantity. Thus Δx indicates a change in the position x.

NOTE ► Δx is a *single* symbol. You cannot cancel out or remove the Δ in algebraic operations. ◄

To get from the 50 ft mark to the 150 ft mark, Sam clearly had to walk 100 ft, so the change in his position—his displacement—is 100 ft. We can think about displacement in a more general way, however. **Displacement is the *difference* between a final position x_f and an initial position x_i.** Thus we can write

$$\Delta x = x_f - x_i = 150 \text{ ft} - 50 \text{ ft} = 100 \text{ ft}$$

NOTE ► A general principle, used throughout this book, is that the change in any quantity is the final value of the quantity minus its initial value. ◄

FIGURE 1.13 A displacement is a signed quantity. Here Δx is a negative number.

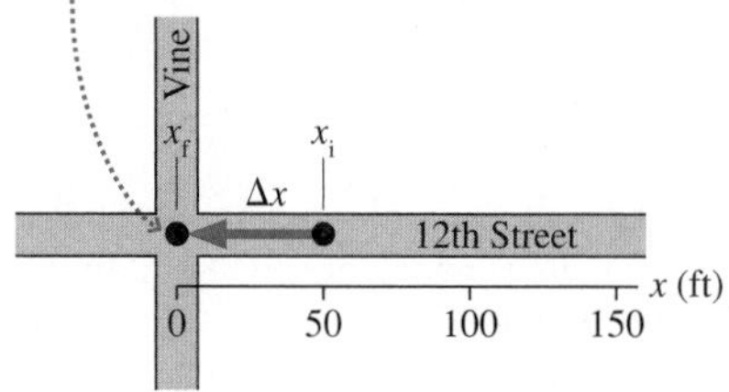

Displacement is a *signed quantity;* that is, it can be either positive or negative. If, as shown in FIGURE 1.13, Sam's final position x_f had been at the origin instead of the 150 ft mark, his displacement would have been

$$\Delta x = x_f - x_i = 0 \text{ ft} - 50 \text{ ft} = -50 \text{ ft}$$

The negative sign tells us that he moved to the *left* along the x-axis, or 50 ft *west.*

Change in Time

A displacement is a change in position. In order to quantify motion, we'll need to also consider changes in *time,* which we call **time intervals.** We've seen how we can label each frame of a motion diagram with a specific time, as determined by our stopwatch. FIGURE 1.14 shows the motion diagram of a bicycle moving at a constant speed, with the times of the measured points indicated.

The displacement between the initial position x_i and the final position x_f is

$$\Delta x = x_f - x_i = 120 \text{ ft} - 0 \text{ ft} = 120 \text{ ft}$$

FIGURE 1.14 The motion diagram of a bicycle moving to the right at a constant speed.

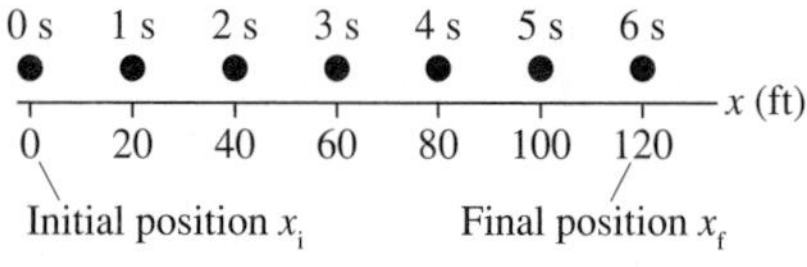

Similarly, we define the time interval between these two points to be

$$\Delta t = t_f - t_i = 6 \text{ s} - 0 \text{ s} = 6 \text{ s}$$

A time interval Δt measures the elapsed time as an object moves from an initial position x_i at time t_i to a final position x_f at time t_f. Note that, unlike Δx, Δt is always positive because t_f is always greater than t_i.

EXAMPLE 1.1 How long a ride?

Carol is enjoying a bicycle ride on a country road that runs east-west past a water tower. Define a coordinate system so that increasing x means moving east. At noon, Carol is 3 miles (mi) east of the water tower. A half-hour later, she is 2 mi west of the water tower. What is her displacement during that half-hour?

FIGURE 1.15 A drawing of Carol's motion.

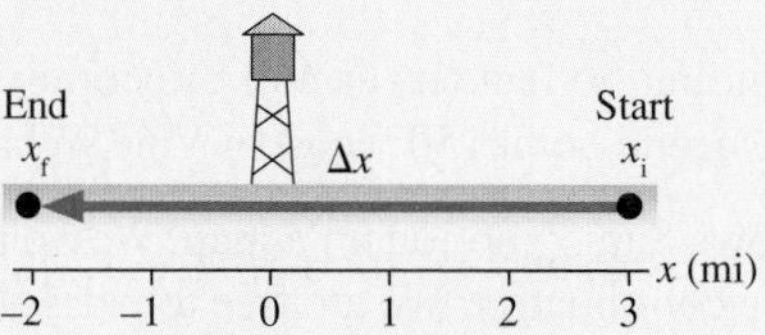

PREPARE Although it may seem like overkill for such a simple problem, you should start by making a drawing, like the one in FIGURE 1.15, with the x-axis along the road. Distances are measured with respect to the water tower, so it is a natural origin for the coordinate system. Once the coordinate system is established, we can show Carol's initial and final positions and her displacement between the two.

SOLVE We've specified values for Carol's initial and final positions in our drawing. We can thus compute her displacement:

$$\Delta x = x_f - x_i = (-2 \text{ mi}) - (3 \text{ mi}) = -5 \text{ mi}$$

ASSESS Once we've completed the solution to the problem, we need to go back to see if it makes sense. Carol is moving to the west, so we expect her displacement to be negative—and it is. We can see from our drawing in Figure 1.15 that she has moved 5 miles from her starting position, so our answer seems reasonable.

NOTE ► All of the numerical examples in the book are worked out with the same three-step process: Prepare, Solve, Assess. It's tempting to cut corners, especially for the simple problems in these early chapters, but you should take the time to do all of these steps now, to practice your problem-solving technique. We'll have more to say about our general problem-solving strategy in Chapter 2. ◄

STOP TO THINK 1.3 Sarah starts at a positive position along the x-axis. She then undergoes a negative displacement. Her final position

A. Is positive. B. Is negative. C. Could be either positive or negative.

1.3 Velocity

We all have an intuitive sense of whether something is moving very fast or just cruising slowly along. To make this intuitive idea more precise, let's start by examining the motion diagrams of some objects moving along a straight line at a *constant* speed, objects that are neither speeding up nor slowing down. This motion at a constant speed is called **uniform motion.** As we saw for the skateboarder in Section 1.1, for an object in uniform motion, successive frames of the motion diagram are *equally spaced.* We know now that this means that the object's displacement Δx is the same between successive frames.

To see how an object's displacement between successive frames is related to its speed, consider the motion diagrams of a bicycle and a car, traveling along the same street as shown in FIGURE 1.16. Clearly the car is moving faster than the bicycle: In any 1-second time interval, the car undergoes a displacement $\Delta x = 40$ ft, while the bicycle's displacement is only 20 ft.

FIGURE 1.16 Motion diagrams for a car and a bicycle.

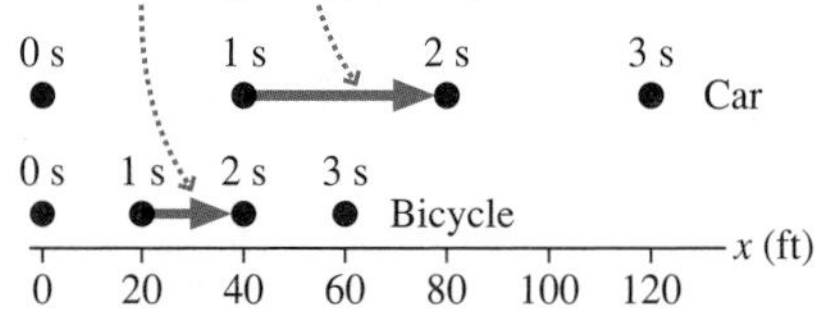

The distances traveled in 1 second by the bicycle and the car are a measure of their speeds. The greater the distance traveled by an object in a given time interval, the greater its speed. This idea leads us to define the speed of an object as

$$\text{speed} = \frac{\text{distance traveled in a given time interval}}{\text{time interval}} \tag{1.1}$$

Speed of a moving object

For the bicycle, this equation gives

$$\text{speed} = \frac{20 \text{ ft}}{1 \text{ s}} = 20 \frac{\text{ft}}{\text{s}}$$

while for the car we have

$$\text{speed} = \frac{40 \text{ ft}}{1 \text{ s}} = 40 \frac{\text{ft}}{\text{s}}$$

The speed of the car is twice that of the bicycle, which seems reasonable.

NOTE ► The division gives units that are a fraction: ft/s. This is read as "feet per second," just like the more familiar "miles per hour." ◄

FIGURE 1.17 Two bicycles traveling at the same speed, but with different velocities.

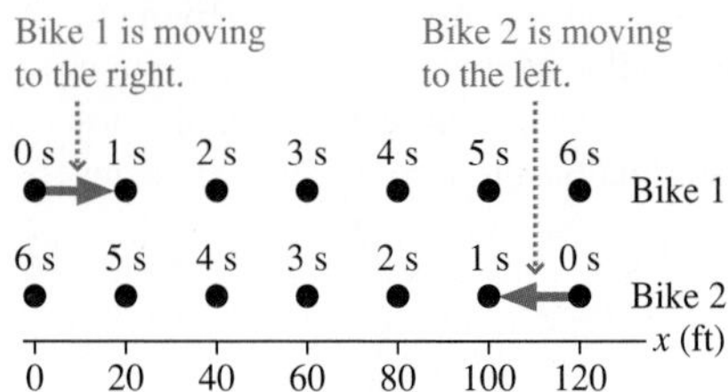

To fully characterize the motion of an object, it is important to specify not only the object's speed but also the *direction* in which it is moving. For example, FIGURE 1.17 shows the motion diagrams of two bicycles traveling at the same speed of 20 ft/s. The two bicycles have the same speed, but something about their motion is different—the *direction* of their motion.

The problem is that the "distance traveled" in Equation 1.1 doesn't capture any information about the direction of travel. But we've seen that the *displacement* of an object does contain this information. We can then introduce a new quantity, the **velocity,** as

$$\text{velocity} = \frac{\text{displacement}}{\text{time interval}} = \frac{\Delta x}{\Delta t} \tag{1.2}$$

Velocity of a moving object

The velocity of bicycle 1 in Figure 1.17, computed using the 1 second time interval between the $t = 2$ s and $t = 3$ s positions, is

$$v = \frac{\Delta x}{\Delta t} = \frac{x_3 - x_2}{3\text{ s} - 2\text{ s}} = \frac{60\text{ ft} - 40\text{ ft}}{1\text{ s}} = +20\,\frac{\text{ft}}{\text{s}}$$

while the velocity for bicycle 2, during the same time interval, is

$$v = \frac{\Delta x}{\Delta t} = \frac{x_3 - x_2}{3\text{ s} - 2\text{ s}} = \frac{60\text{ ft} - 80\text{ ft}}{1\text{ s}} = -20\,\frac{\text{ft}}{\text{s}}$$

NOTE ► We have used x_2 for the position at time $t = 2$ seconds and x_3 for the position at time $t = 3$ seconds. The subscripts serve the same role as before—identifying particular positions—but in this case the positions are identified by the time at which each position is reached. ◄

The two velocities have opposite signs because the bicycles are traveling in opposite directions. **Speed measures only how fast an object moves, but velocity tells us both an object's speed *and its direction.*** A positive velocity indicates motion to the right or, for vertical motion, upward. Similarly, an object moving to the left, or down, has a negative velocity.

NOTE ► Learning to distinguish between speed, which is always a positive number, and velocity, which can be either positive or negative, is one of the most important tasks in the analysis of motion. ◄

The velocity as defined by Equation 1.2 is actually what is called the *average* velocity. On average, over each 1 s interval bicycle 1 moves 20 ft, but we don't know if it was moving at exactly the same speed at every moment during this time interval. In Chapter 2, we'll develop the idea of *instantaneous* velocity, the velocity of an object at a particular instant in time. Since our goal in this chapter is to *visualize* motion with motion diagrams, we'll somewhat blur the distinction between average and instantaneous quantities, refining these definitions in Chapter 2, where our goal will be to develop the mathematics of motion.

EXAMPLE 1.2 Finding the speed of a seabird

Albatrosses are seabirds that spend most of their lives flying over the ocean looking for food. With a stiff tailwind, an albatross can fly at high speeds. Satellite data on one particularly speedy albatross showed it 60 miles east of its roost at 3:00 PM and then, at 3:20 PM, 86 miles east of its roost. What was its velocity?

PREPARE The statement of the problem provides us with a natural coordinate system: We can measure distances with respect to the roost, with distances to the east as positive. With this coordinate system, the motion of the albatross appears as in **FIGURE 1.18**. The motion takes place between 3:00 and 3:20, a time interval of 20 minutes, or 0.33 hour.

FIGURE 1.18 The motion of an albatross at sea.

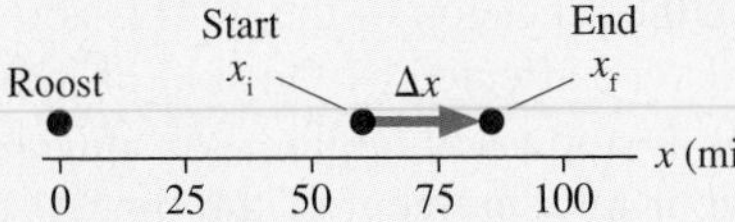

SOLVE We know the initial and final positions, and we know the time interval, so we can calculate the velocity:

$$v = \frac{\Delta x}{\Delta t} = \frac{x_f - x_i}{0.33\text{ h}} = \frac{26\text{ mi}}{0.33\text{ h}} = 79\text{ mph}$$

ASSESS The velocity is positive, which makes sense because Figure 1.18 shows that the motion is to the right. A speed of 79 mph is certainly fast, but the problem said it was a "particularly speedy" albatross, so our answer seems reasonable. (Indeed, albatrosses have been observed to fly at such speeds in the very fast winds of the Southern Ocean. This problem is based on real observations, as will be our general practice in this book.)

The "Per" in Meters Per Second

The units for speed and velocity are a unit of distance (feet, meters, miles) divided by a unit of time (seconds, hours). Thus we could measure velocity in units of m/s or mph, pronounced "meters *per* second" and "miles *per* hour." The word "per" will often arise in physics when we consider the ratio of two quantities. What do we mean, exactly, by "per"?

If a car moves with a speed of 23 m/s, we mean that it travels 23 meters *for each* 1 second of elapsed time. The word "per" thus associates the number of units in the numerator (23 m) with *one* unit of the denominator (1 s). We'll see many other examples of this idea as the book progresses. You may already know a bit about *density;* you can look up the density of gold and you'll find that it is 19.3 g/cm^3 ("grams *per* cubic centimeter"). This means that there are 19.3 grams of gold *for each* 1 cubic centimeter of the metal. Thinking about the word "per" in this way will help you better understand physical quantities whose units are the ratio of two other units.

1.4 A Sense of Scale: Significant Figures, Scientific Notation, and Units

Physics attempts to explain the natural world, from the very small to the exceedingly large. And in order to understand our world, we need to be able to *measure* quantities both minuscule and enormous. A properly reported measurement has three elements. First, we can measure our quantity with only a certain precision. To make this precision clear, we need to make sure that we report our measurement with the correct number of *significant figures.*

Second, writing down the really big and small numbers that often come up in physics can be awkward. To avoid writing all those zeros, scientists use *scientific notation* to express numbers both big and small.

Finally, we need to choose an agreed-upon set of *units* for the quantity. For speed, common units include meters per second and miles per hour. For mass, the kilogram is the most commonly used unit. Every physical quantity that we can measure has an associated set of units.

300 million light years $= 2.8 \times 10^{24}$ m

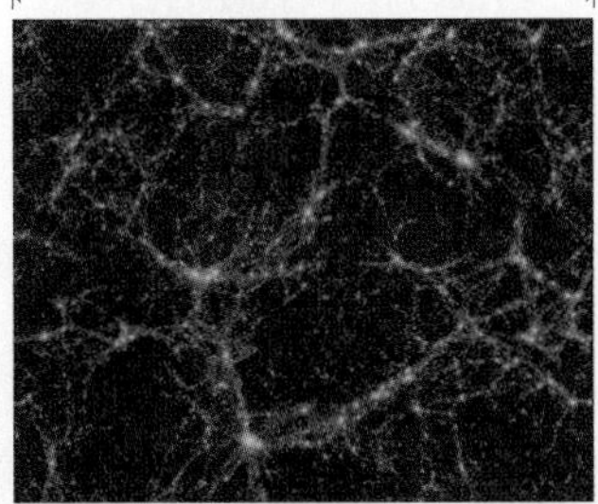

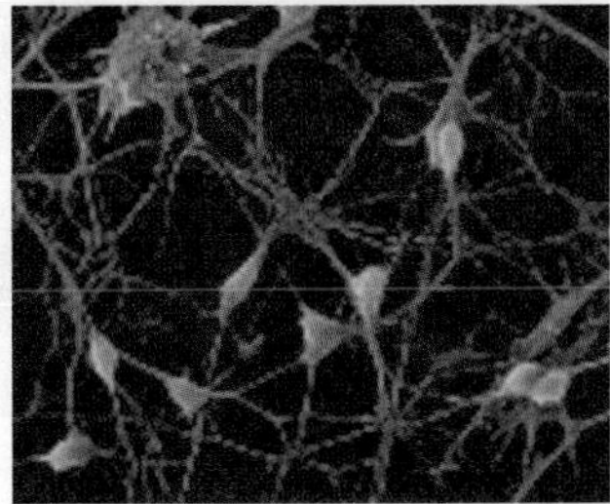

120 μm $= 1.2 \times 10^{-4}$ m

From galaxies to cells ... BIO In science, we need to express numbers both very large and very small. The top image is a computer simulation of the structure of the universe. Bright areas represent regions of clustered galaxies. The bottom image is cortical nerve cells. Nerve cells relay signals to each other through a complex web of dendrites. These images, though similar in appearance, differ in scale by a factor of about 2×10^{28}!

FIGURE 1.19 The precision of a measurement depends on the instrument used to make it.

These calipers have a precision of 0.01 mm.

Walter Davis's best long jump on this day was reported as 8.24 m. This implies that the actual length of the jump was between 8.235 m and 8.245 m, a spread of only 0.01 m, which is 1 cm. Does this claimed accuracy seem reasonable?

TRY IT YOURSELF

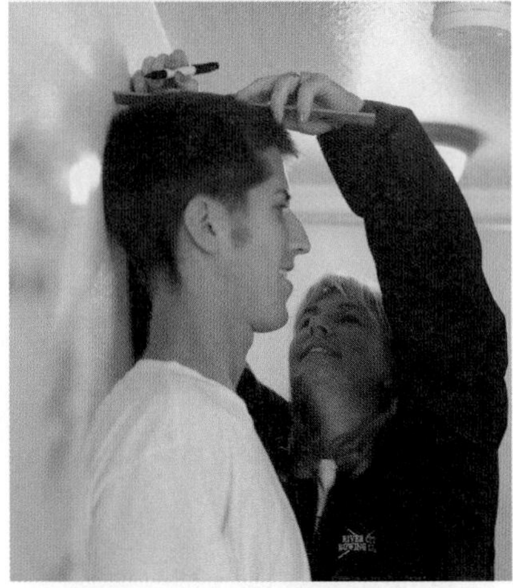

How tall are you really? If you measure your height in the morning, just after you wake up, and then in the evening, after a full day of activity, you'll find that your evening height is *shorter* by as much as 3/4 inch. Your height decreases over the course of the day as gravity compresses and reshapes your spine. If you give your height as 66 3/16 in, you are claiming more significant figures than are truly warranted; the 3/16 in isn't really reliably known because your height can vary by more than this. Expressing your height to the nearest inch is plenty!

Measurements and Significant Figures

When we measure any quantity, such as the length of a bone or the weight of a specimen, we can do so with only a certain *precision*. The digital calipers in FIGURE 1.19 can make a measurement to within ± 0.01 mm, so they have a precision of 0.01 mm. If you made the measurement with a ruler, you probably couldn't do better than about ± 1 mm, so the precision of the ruler is about 1 mm. The precision of a measurement can also be affected by the skill or judgment of the person performing the measurement. A stopwatch might have a precision of 0.001 s, but, due to your reaction time, your measurement of the time of a sprinter would be much less precise.

It is important that your measurement be reported in a way that reflects its actual precision. Suppose you use a ruler to measure the length of a particular specimen of a newly discovered species of frog. You judge that you can make this measurement with a precision of about 1 mm, or 0.1 cm. In this case, the frog's length should be reported as, say, 6.2 cm. We interpret this to mean that the actual value falls between 6.15 cm and 6.25 cm and thus rounds to 6.2 cm. Reporting the frog's length as simply 6 cm is saying less than you know; you are withholding information. On the other hand, to report the number as 6.213 cm is wrong. Any person reviewing your work would interpret the number 6.213 cm as meaning that the actual length falls between 6.2125 cm and 6.2135 cm, thus rounding to 6.213 cm. In this case, you are claiming to have knowledge and information that you do not really possess.

The way to state your knowledge precisely is through the proper use of **significant figures.** You can think of a significant figure as a digit that is reliably known. A measurement such as 6.2 cm has *two* significant figures, the 6 and the 2. The next decimal place—the hundredths—is not reliably known and is thus not a significant figure. Similarly, a time measurement of 34.62 s has four significant figures, implying that the 2 in the hundredths place is reliably known.

When we perform a calculation such as adding or multiplying two or more measured numbers, we can't claim more accuracy for the result than was present in the initial measurements. Nine out of ten numbers used in a calculation might be known with a precision of 0.01%, but if the tenth number is poorly known, with a precision of only 10%, then the result of the calculation cannot possibly be more precise than 10%.

Determining the proper number of significant figures is straightforward, but there are a few definite rules to follow. We will often spell out such technical details in what we call a "Tactics Box." A Tactics Box is designed to teach you particular skills and techniques. Each Tactics Box will use the icon to designate exercises in the *Student Workbook* that you can use to practice these skills.

TACTICS BOX 1.1 Using significant figures

❶ When you multiply or divide several numbers, or when you take roots, the number of significant figures in the answer should match the number of significant figures of the *least* precisely known number used in the calculation:

Three significant figures

$$3.73 \times 5.7 = 21$$

Two significant figures

Answer should have the *lower* of the two, or two significant figures.

Continued

❷ When you add or subtract several numbers, the number of decimal places in the answer should match the *smallest* number of decimal places of any number used in the calculation:

$$\begin{array}{r} 18.54 \text{ — Two decimal places} \\ +106.6 \text{ — One decimal place} \\ \hline 125.1 \end{array}$$

Answer should have the *lower* of the two, or one decimal place.

❸ **Exact numbers** have no uncertainty and, when used in calculations, do not change the number of significant figures of measured numbers. Examples of exact numbers are π and the number 2 in the relation $d = 2r$ between a circle's diameter and radius.

There is one notable exception to these rules:

- It is acceptable to keep one or two extra digits during *intermediate* steps of a calculation. The goal here is to minimize round-off errors in the calculation. But the *final* answer must be reported with the proper number of significant figures.

Exercise 15

EXAMPLE 1.3 Measuring the velocity of a car

To measure the velocity of a car, clocks A and B are set up at two points along the road, as shown in **FIGURE 1.20**. Clock A is precise to 0.01 s, while B is precise to only 0.1 s. The distance between these two clocks is carefully measured to be 124.5 m. The two clocks are automatically started when the car passes a trigger in the road; each clock stops automatically when the car passes that clock. After the car has passed both clocks, clock A is found to read $t_A = 1.22$ s, and clock B to read $t_B = 4.5$ s. The time from the less-precise clock B is correctly reported with fewer significant figures than that from A. What is the velocity of the car, and how should it be reported with the correct number of significant figures?

FIGURE 1.20 Measuring the velocity of a car.

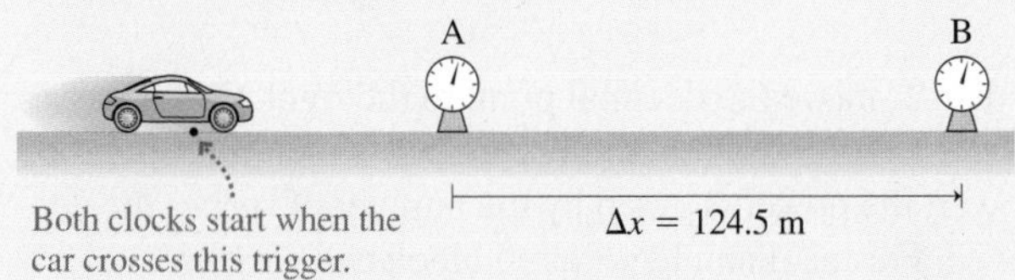

PREPARE To calculate the velocity, we need the displacement Δx and the time interval Δt as the car moves between the two clocks. The displacement is given as $\Delta x = 124.5$ m; we can calculate the time interval as the difference between the two measured times.

SOLVE The time interval is:

$$\Delta t = t_B - t_A = (4.5 \text{ s}) - (1.22 \text{ s}) = 3.3 \text{ s}$$

This number has one decimal place. This number has two decimal places. By rule 2 of Tactics Box 1.1, the result should have *one* decimal place.

We can now calculate the velocity with the displacement and the time interval:

$$v = \frac{\Delta x}{\Delta t} = \frac{124.5 \text{ m}}{3.3 \text{ s}} = 38 \text{ m/s}$$

The displacement has four significant figures. The time interval has two significant figures. By rule 1 of Tactics Box 1.1, the result should have *two* significant figures.

ASSESS Our final value has two significant figures. Suppose you had been hired to measure the speed of a car this way, and you reported 37.72 m/s. It would be reasonable for someone looking at your result to assume that the measurements you used to arrive at this value were correct to four significant figures and thus that you had measured time to the nearest 0.001 second. Our correct result of 38 m/s has all of the accuracy that you can claim, but no more!

Scientific Notation

It's easy to write down measurements of ordinary-sized objects: Your height might be 1.72 meters, the weight of an apple 0.34 pound. But the radius of a hydrogen atom is 0.000000000053 m, and the distance to the moon is 384000000 m. Keeping track of all those zeros is quite cumbersome.

Beyond requiring you to deal with all the zeros, writing quantities this way makes it unclear how many significant figures are involved. In the distance to the moon given above, how many of those digits are significant? Three? Four? All nine?

Writing numbers using scientific notation avoids both these problems. A value in scientific notation is a number with one digit to the left of the decimal point and zero or more to the right of it, multiplied by a power of ten. This solves the problem of all the zeros and makes the number of significant figures immediately apparent. In scientific notation, writing the distance to the sun as 1.50×10^{11} m implies that three digits are significant; writing it as 1.5×10^{11} m implies that only two digits are.

Even for smaller values, scientific notation can clarify the number of significant figures. Suppose a distance is reported as 1200 m. How many significant figures does this measurement have? It's ambiguous, but using scientific notation can remove any ambiguity. If this distance is known to within 1 m, we can write it as 1.200×10^3 m, showing that all four digits are significant; if it is accurate to only 100 m or so, we can report it as 1.2×10^3 m, indicating two significant figures.

Tactics Box 1.2 shows how to convert a number to scientific notation, and how to correctly indicate the number of significant figures.

TACTICS BOX 1.2 Using scientific notation

To convert a number into scientific notation:

❶ For a number greater than 10, move the decimal point to the left until only one digit remains to the left of the decimal point. The remaining number is then multiplied by 10 to a power; this power is given by the number of spaces the decimal point was moved. Here we convert the diameter of the earth to scientific notation:

We move the decimal point until there is only one digit to its left, counting the number of steps.

Since we moved the decimal point 6 steps, the power of ten is 6.

$$6\,370\,000 \text{ m} = 6.37 \times 10^6 \text{ m}$$

The number of digits here equals the number of significant figures.

❷ For a number less than 1, move the decimal point to the right until it passes the first digit that isn't a zero. The remaining number is then multiplied by 10 to a negative power; the power is given by the number of spaces the decimal point was moved. For the diameter of a red blood cell we have:

We move the decimal point until it passes the first digit that is not a zero, counting the number of steps.

Since we moved the decimal point 6 steps, the power of ten is −6.

$$0.000\,007\,5 \text{ m} = 7.5 \times 10^{-6} \text{ m}$$

The number of digits here equals the number of significant figures.

Exercise 16 

Proper use of significant figures is part of the "culture" of science. We will frequently emphasize these "cultural issues" because you must learn to speak the same language as the natives if you wish to communicate effectively! Most students know the rules of significant figures, having learned them in high school, but many fail to apply them. It is important that you understand the reasons for significant figures and that you get in the habit of using them properly.

The importance of units In 1999, the $125 million Mars Climate Orbiter burned up in the Martian atmosphere instead of entering a safe orbit from which it could perform observations. The problem was faulty units! An engineering team had provided critical data on spacecraft performance in English units, but the navigation team assumed these data were in metric units. As a consequence, the navigation team had the spacecraft fly too close to the planet, and it burned up in the atmosphere.

Units

As we have seen, in order to measure a quantity we need to give it a numerical value. But a measurement is more than just a number—it requires a *unit* to be given. You can't go to the deli and ask for "three quarters of cheese." You need to use a unit—here, one of weight, such as pounds—in addition to the number.

In your daily life, you probably use the English system of units, in which distances are measured in inches, feet, and miles. These units are well adapted for daily life, but they are rarely used in scientific work. Given that science is an international discipline, it is also important to have a system of units that is recognized around the world. For these reasons, scientists use a system of units called *le Système Internationale d'Unités,* commonly referred to as **SI units.** SI units were originally developed by the French in the late 1700s as a way of standardizing and regularizing numbers for commerce and science. We often refer to these as *metric units* because the meter is the basic standard of length.

The three basic SI quantities, shown in Table 1.1, are time, length (or distance), and mass. Other quantities needed to understand motion can be expressed as combinations of these basic units. For example, speed and velocity are expressed in meters per second or m/s. This combination is a ratio of the length unit (the meter) to the time unit (the second).

TABLE 1.1 Common SI units

Quantity	Unit	Abbreviation
time	second	s
length	meter	m
mass	kilogram	kg

Using Prefixes

We will have many occasions to use lengths, times, and masses that are either much less or much greater than the standards of 1 meter, 1 second, and 1 kilogram. We will do so by using *prefixes* to denote various powers of ten. For instance, the prefix "kilo" (abbreviation k) denotes 10^3, or a factor of 1000. Thus 1 km equals 1000 m, 1 MW equals 10^6 watts, and 1 μV equals 10^{-6} V. Table 1.2 lists the common prefixes that will be used frequently throughout this book. A more extensive list of prefixes is shown inside the cover of the book.

Although prefixes make it easier to talk about quantities, the proper SI units are meters, seconds, and kilograms. Quantities given with prefixed units must be converted to base SI units before any calculations are done. Thus 23.0 cm must be converted to 0.230 m before starting calculations. The exception is the kilogram, which is already the base SI unit.

TABLE 1.2 Common prefixes

Prefix	Abbreviation	Power of 10
mega-	M	10^6
kilo-	k	10^3
centi-	c	10^{-2}
milli-	m	10^{-3}
micro-	μ	10^{-6}
nano-	n	10^{-9}

Unit Conversions

Although SI units are our standard, we cannot entirely forget that the United States still uses English units. Even after repeated exposure to metric units in classes, most of us "think" in English units. Thus it remains important to be able to convert back and forth between SI units and English units. Table 1.3 shows some frequently used conversions that will come in handy.

TABLE 1.3 Useful unit conversions

1 inch (in) = 2.54 cm
1 foot (ft) = 0.305 m
1 mile (mi) = 1.609 km
1 mile per hour (mph) = 0.447 m/s
1 m = 39.37 in
1 km = 0.621 mi
1 m/s = 2.24 mph

One effective method of performing unit conversions begins by noticing that since, for example, 1 mi = 1.609 km, the ratio of these two distances—*including their units*—is equal to 1, so that

$$\frac{1 \text{ mi}}{1.609 \text{ km}} = \frac{1.609 \text{ km}}{1 \text{ mi}} = 1$$

A ratio of values equal to 1 is called a **conversion factor.** The following Tactics Box shows how to make a unit conversion.

TACTICS BOX 1.3 **Making a unit conversion**

❶ Start with the quantity you wish to convert.

❷ Multiply by the appropriate conversion factor. Because this conversion factor is equal to 1, multiplying by it does not change the value of the quantity—only its units.

❺ Remember to convert your final answer to the correct number of significant figures!

$$60 \text{ mi} = 60 \text{ mi} \times \frac{1.609 \text{ km}}{1 \text{ mi}} = 96.54 \text{ km} = 97 \text{ km}$$

❸ You can cancel the original unit (here, miles) because it appears in both the numerator and the denominator.

❹ Calculate the answer; it is in the desired units. Remember, 60 mi and 96.54 km are the same distance; they are simply in different units.

Note that we've rounded the answer to 97 kilometers because the distance we're converting, 60 miles, has only two significant figures.

More complicated conversions can be done with several successive multiplications of conversion factors, as we see in the next example.

EXAMPLE 1.4 **Can a bicycle go that fast?**

In Section 1.3, we calculated the speed of a bicycle to be 20 ft/s. Is this a reasonable speed for a bicycle?

PREPARE In order to determine whether or not this speed is reasonable, we will convert it to more familiar units. For speed, the unit you are most familiar with is likely miles per hour.

SOLVE We first collect the necessary unit conversions:

$$1 \text{ mi} = 5280 \text{ ft} \qquad 1 \text{ hour (1 h)} = 60 \text{ min} \qquad 1 \text{ min} = 60 \text{ s}$$

We then multiply our original value by successive factors of 1 in order to convert the units:

We want to cancel feet here in the numerator . . .

. . . so we multiply by $1 = \frac{1 \text{ mi}}{5280 \text{ ft}}$ to get the feet in the denominator.

$$20\frac{\text{ft}}{\text{s}} = 20\frac{\text{ft}}{\text{s}} \times \frac{1 \text{ mi}}{5280 \text{ ft}} \times \frac{60 \text{ s}}{1 \text{ min}} \times \frac{60 \text{ min}}{1 \text{ h}} = 14\frac{\text{mi}}{\text{h}} = 14 \text{ mph}$$

The unwanted units cancel in pairs, as indicated by the colors.

ASSESS Our final result of 14 miles per hour (14 mph) is a very reasonable speed for a bicycle, which gives us confidence in our answer. If we had calculated a speed of 140 miles per hour, we would have suspected that we had made an error because this is quite a bit faster than the average bicyclist can travel!

How many jellybeans are in the jar? Some reasoning about the size of one bean and the size of the jar can give you a one-significant-figure estimate.

Estimation

When scientists and engineers first approach a problem, they may do a quick measurement or calculation to establish the rough physical scale involved. This will help establish the procedures that should be used to make a more accurate measurement—or the estimate may well be all that is needed.

Suppose you see a rock fall off a cliff and would like to know how fast it was going when it hit the ground. By doing a mental comparison with the speeds of familiar objects, such as cars and bicycles, you might judge that the rock was traveling at about 20 mph. This is a one-significant-figure estimate. With some luck, you can probably distinguish 20 mph from either 10 mph or 30 mph, but you certainly cannot distinguish 20 mph from 21 mph just from a visual appearance. A one-significant-figure estimate or calculation, such as this estimate of speed, is called an **order-of-magnitude estimate.** An order-of-magnitude estimate is indicated by the symbol $\sim$, which indicates even less precision than the "approximately equal" symbol $\approx$. You would report your estimate of the speed of the falling rock as $v \sim 20$ mph.

A useful skill is to make reliable order-of-magnitude estimates on the basis of known information, simple reasoning, and common sense. This is a skill that is acquired by practice. Tables 1.4 and 1.5 have information that will be useful for doing estimates.

TABLE 1.4 Some approximate lengths

	Length (m)
Circumference of the earth	4×10^7
Distance from New York to Los Angeles	5×10^6
Distance you can drive in 1 hour	1×10^5
Altitude of jet planes	1×10^4
Distance across a college campus	1000
Length of a football field	100
Length of a classroom	10
Length of your arm	1
Width of a textbook	0.1
Length of your little fingernail	0.01
Diameter of a pencil lead	1×10^{-3}
Thickness of a sheet of paper	1×10^{-4}
Diameter of a dust particle	1×10^{-5}

EXAMPLE 1.5 How fast do you walk?

Estimate how fast you walk, in meters per second.

PREPARE In order to compute speed, we need a distance and a time. If you walked a mile to campus, how long would this take? You'd probably say 30 minutes or so—half an hour. Let's use this rough number in our estimate.

SOLVE Given this estimate, we compute speed as

$$\text{speed} = \frac{\text{distance}}{\text{time}} \sim \frac{1 \text{ mile}}{1/2 \text{ hour}} = 2\,\frac{\text{mi}}{\text{h}}$$

But we want the speed in meters per second. Since our calculation is only an estimate, we use an approximate form of the conversion factor from Table 1.3:

$$1\,\frac{\text{mi}}{\text{h}} \approx 0.5\,\frac{\text{m}}{\text{s}}$$

This gives an approximate walking speed of 1 m/s.

ASSESS Is this a reasonable value? Let's do another estimate. Your stride is probably about 1 yard long—about 1 meter. And you take about one step per second; next time you are walking, you can count and see. So a walking speed of 1 meter per second sounds pretty reasonable.

TABLE 1.5 Some approximate masses

	Mass (kg)
Large airliner	1×10^5
Small car	1000
Large human	100
Medium-size dog	10
Science textbook	1
Apple	0.1
Pencil	0.01
Raisin	1×10^{-3}
Fly	1×10^{-4}

This sort of estimation is very valuable. We will see many cases in which we need to know an approximate value for a quantity before we start a problem or after we finish a problem, in order to assess our results.

STOP TO THINK 1.4 Rank in order, from the most to the fewest, the number of significant figures in the following numbers. For example, if B has more than C, C has the same number as A, and A has more than D, give your answer as B > C = A > D.

A. 0.43 B. 0.0052 C. 0.430 D. 4.321×10^{-10}

1.5 Vectors and Motion: A First Look

Many physical quantities, such as time, temperature, and weight, can be described completely by a number with a unit. For example, the mass of an object might be 6 kg and its temperature 30° C. When a physical quantity is described by a single number (with a unit), we call it a **scalar quantity.** A scalar can be positive, negative, or zero.

Vectors and scalars

Scalars

Time, temperature and weight are all *scalar* quantities. To specify your weight, the temperature outside, or the current time, you only need a single number.

Vectors

The velocity of the race car is a *vector.* To fully specify a velocity, we need to give its magnitude (e.g., 120 mph) *and* its direction (e.g., west).

The force with which the boy pushes on his friend is another example of a vector. To completely specify this force, we must know not only how hard he pushes (the magnitude) but also in which direction.

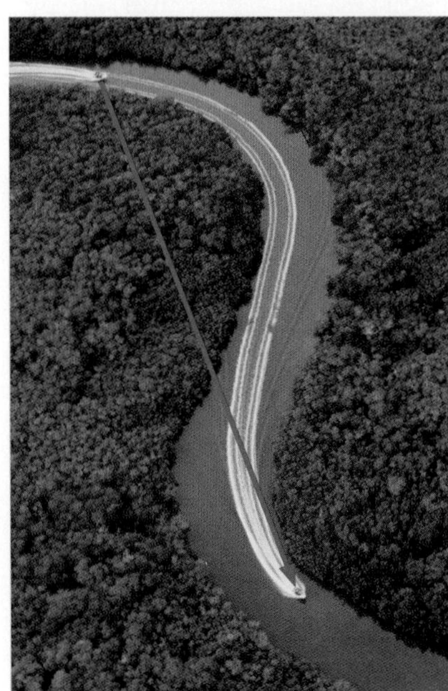

The boat's displacement is the straight-line connection from its initial to its final position.

Many other quantities, however, have a directional quality and cannot be described by a single number. To describe the motion of a car, for example, you must specify not only how fast it is moving, but also the *direction* in which it is moving. A **vector quantity** is a quantity that has both a *size* (How far? or How fast?) and a *direction* (Which way?). The size or length of a vector is called its **magnitude.** The magnitude of a vector can be positive or zero, but it cannot be negative.

Some examples of vector and scalar quantities are shown on the left.

We graphically represent a vector as an *arrow,* as illustrated for the velocity and force vectors. The arrow is drawn to point in the direction of the vector quantity, and the *length* of the arrow is proportional to the magnitude of the vector quantity.

When we want to represent a vector quantity with a *symbol,* we need somehow to indicate that the symbol is for a vector rather than for a scalar. We do this by drawing an arrow over the letter that represents the quantity. Thus $\vec{r}$ and $\vec{A}$ are symbols for vectors, whereas r and A, without the arrows, are symbols for scalars. In handwritten work you *must* draw arrows over all symbols that represent vectors. This may seem strange until you get used to it, but it is very important because we will often use both r and $\vec{r}$, or both A and $\vec{A}$, in the same problem, and they mean different things!

NOTE ▶ The arrow over the symbol always points to the right, regardless of which direction the actual vector points. Thus we write $\vec{r}$ or $\vec{A}$, never $\overleftarrow{r}$ or $\overleftarrow{A}$. ◀

Displacement Vectors

For motion along a line, we found in Section 1.2 that the displacement is a quantity that specifies not only how *far* an object moves but also the *direction*—to the left or to the right—that the object moves. Since displacement is a quantity that has both a magnitude (How far?) and a direction, it can be represented by a vector, the **displacement vector.** FIGURE 1.21 shows the displacement vector for Sam's trip that we discussed earlier. We've simply drawn an arrow—the vector—from his initial to his final position and assigned it the symbol $\vec{d}_S$. Because $\vec{d}_S$ has both a magnitude and a direction, it is convenient to write Sam's displacement as $\vec{d}_S = (100 \text{ ft, east})$. The first value in the parentheses is the magnitude of the vector (i.e., the size of the displacement), and the second value specifies its direction.

FIGURE 1.21 Two displacement vectors.

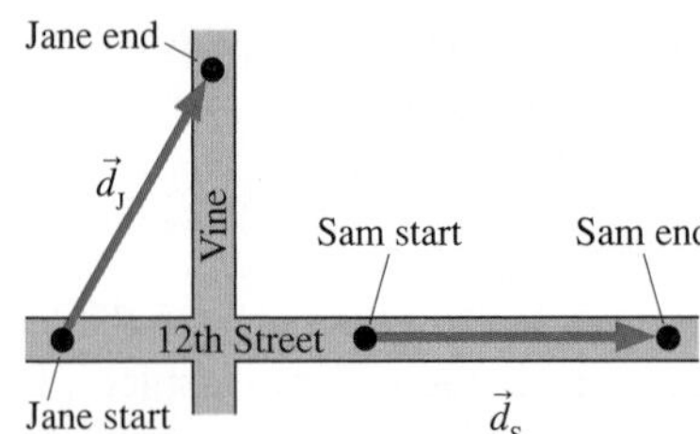

Also shown in Figure 1.21 is the displacement vector $\vec{d}_J$ for Jane, who started on 12th Street and ended up on Vine. As with Sam, we draw her displacement vector as an arrow from her initial to her final position. In this case, $\vec{d}_J = 100$ ft, 30° east of north).

Jane's trip illustrates an important point about displacement vectors. Jane started her trip on 12th Street and ended up on Vine, leading to the displacement vector shown. But to get from her initial to her final position, she needn't have walked along the straight-line path denoted by $\vec{d}_J$. If she walked east along 12th Street to the intersection and then headed north on Vine, her displacement would still be the vector shown. **An object's displacement vector is drawn from the object's initial position to its final position, regardless of the actual path followed between these two points.**

Vector Addition

Let's consider one more trip for the peripatetic Sam. In FIGURE 1.22, he starts at the intersection and walks east 50 ft; then he walks 100 ft to the northeast through a vacant lot. His displacement vectors for the two legs of his trip are labeled $\vec{d}_1$ and $\vec{d}_2$ in the figure.

Sam's trip consists of two legs that can be represented by the two vectors $\vec{d}_1$ and $\vec{d}_2$, but we can represent his trip as a whole, from his initial starting position to his overall final position, with the *net* displacement vector labeled $\vec{d}_{net}$. Sam's net displacement is in a sense the *sum* of the two displacements that made it up, so we can write

$$\vec{d}_{net} = \vec{d}_1 + \vec{d}_2$$

FIGURE 1.22 Sam undergoes two displacements.

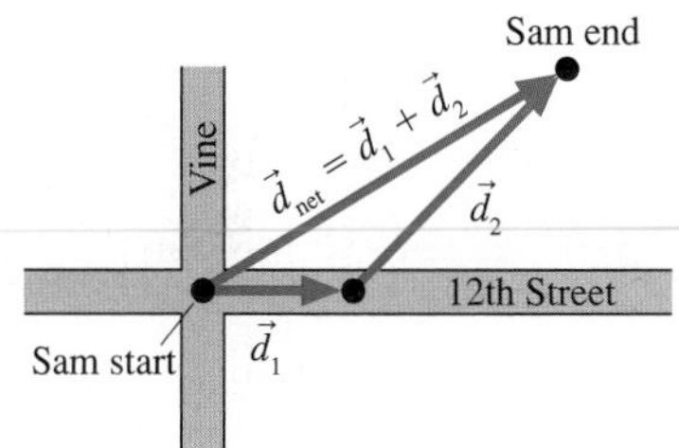

Sam's net displacement thus requires the *addition* of two vectors, but vector addition obeys different rules from the addition of two scalar quantities. The directions of the two vectors, as well as their magnitudes, must be taken into account. Sam's trip suggests that we can add vectors together by putting the "tail" of one vector at the tip of the other. This idea, which is reasonable for displacement vectors, in fact is how *any* two vectors are added. Tactics Box 1.4 shows how to add two vectors $\vec{A}$ and $\vec{B}$.

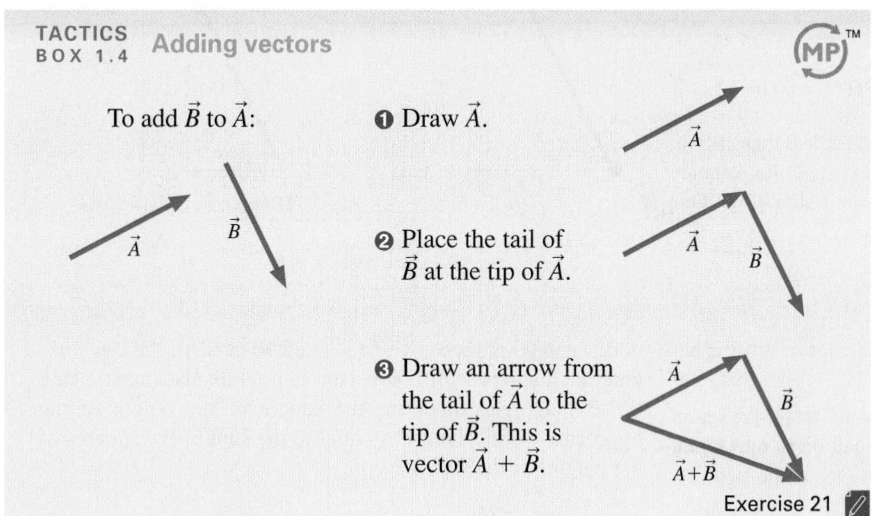

TACTICS BOX 1.4 Adding vectors
MP TM
To add $\vec{B}$ to $\vec{A}$:
$\vec{A}$
$\vec{B}$
❶ Draw $\vec{A}$.
$\vec{A}$
❷ Place the tail of $\vec{B}$ at the tip of $\vec{A}$.
$\vec{A}$
$\vec{B}$
❸ Draw an arrow from the tail of $\vec{A}$ to the tip of $\vec{B}$. This is vector $\vec{A} + \vec{B}$.
$\vec{A}$
$\vec{B}$
$\vec{A}+\vec{B}$
Exercise 21

Vectors and Trigonometry

When we need to add displacements or other vectors in more than one dimension, we'll end up computing lengths and angles of triangles. This is the job of trigonometry. Trigonometry will be our primary mathematical tool for vector addition; let's review the basic ideas.

Suppose we have a right triangle with hypotenuse H, angle θ, side opposite the angle O, and side adjacent to the angle A, as shown in FIGURE 1.23. The sine, cosine, and tangent (which we write as "sin," "cos," and "tan") of angle θ are defined as ratios of the sides of the triangle:

$$\sin\theta = \frac{O}{H} \qquad \cos\theta = \frac{A}{H} \qquad \tan\theta = \frac{O}{A} \tag{1.3}$$

FIGURE 1.23 A right triangle.

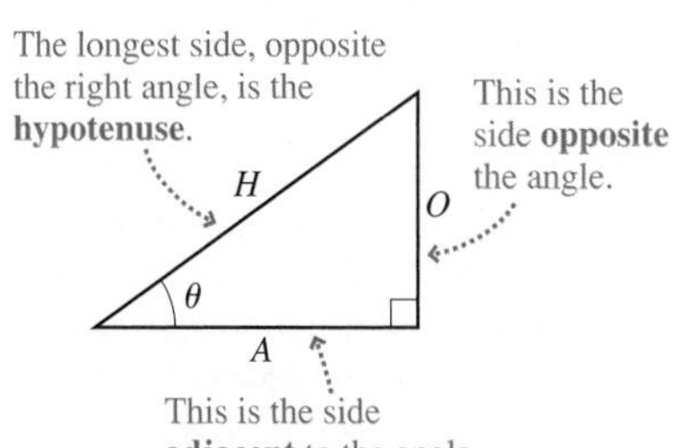

If you know the angle θ and the length of one side, you can use the sine, cosine, or tangent to find the lengths of the other sides. For example, if you know θ and the length A of the adjacent side, you can find the hypotenuse H by rearranging the middle Equation 1.3 to give $H = A/\cos\theta$.

Conversely, if you know two sides of the triangle, you can find the angle θ by using inverse trigonometric functions:

$$\theta = \sin^{-1}\left(\frac{O}{H}\right) \qquad \theta = \cos^{-1}\left(\frac{A}{H}\right) \qquad \theta = \tan^{-1}\left(\frac{O}{A}\right) \tag{1.4}$$

We will make regular use of these relationships in the following chapters.

EXAMPLE 1.6 How far north and east?

Suppose Alex is navigating using a compass. She starts walking at an angle 60° north of east and walks a total of 100 m. How far north is she from her starting point? How far east?

PREPARE A sketch of Alex's motion is shown in FIGURE 1.24a. We've shown north and east as they would be on a map, and we've noted Alex's displacement as a vector, giving its magnitude and direction. FIGURE 1.24b shows a triangle with this displacement as the hypotenuse. Alex's distance north of her starting point and her distance east of her starting point are the sides of this triangle.

SOLVE The sine and cosine functions are ratios of sides of right triangles, as we saw above. With the 60° angle as noted, the distance north of the starting point is the opposite side of the triangle; the distance east is the adjacent side. Thus:

$$\text{distance north of start} = (100\text{ m})\sin(60°) = 87\text{ m}$$

$$\text{distance east of start} = (100\text{ m})\cos(60°) = 50\text{ m}$$

ASSESS Both of the distances we calculated are less than 100 m, as they must be, and the distance east is less than the distance north, as our diagram in Figure 1.24b shows it should be. Our answers seem reasonable. In finding the solution to this problem, we "broke down" the displacement into two different distances, one north and one east. This hints at the idea of the *components* of a vector, something we'll explore in the next chapter.

FIGURE 1.24 An analysis of Alex's motion.

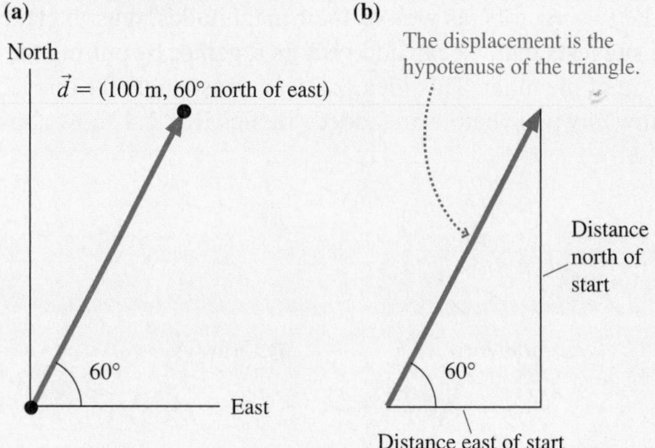

EXAMPLE 1.7 How far away is Anna?

Anna walks 90 m due east and then 50 m due north. What is her displacement from her starting point?

PREPARE Let's start with the sketch in FIGURE 1.25a. We set up a coordinate system with Anna's original position as the origin, and then we drew her two subsequent motions as the two displacement vectors $\vec{d}_1$ and $\vec{d}_2$.

FIGURE 1.25 Analyzing Anna's motion.

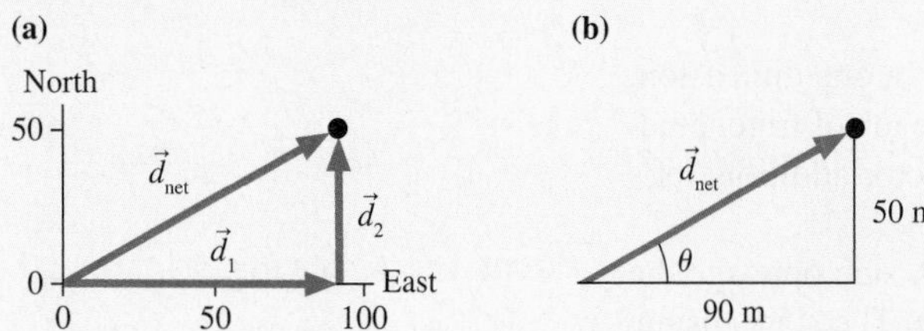

SOLVE We drew the two vector displacements with the tail of one vector starting at the head of the previous one—exactly what is needed to form a vector sum. The vector $\vec{d}_{net}$ in Figure 1.25a is the vector sum of the successive displacements and thus represents Anna's net displacement from the origin.

Anna's distance from the origin is the length of this vector $\vec{d}_{net}$. FIGURE 1.25b shows that this vector is the hypotenuse of a right triangle with sides 50 m (because Anna walked 50 m north) and 90 m (because she walked 90 m east). We can compute the magnitude of this vector, her net displacement, using the Pythagorean theorem (the square of the length of the hypotenuse of a triangle is equal to the sum of the squares of the lengths of the sides):

$$d_{net}^2 = (50\text{ m})^2 + (90\text{ m})^2$$

$$d_{net} = \sqrt{(50\text{ m})^2 + (90\text{ m})^2} = 103\text{ m} \approx 100\text{ m}$$

We have rounded off to the appropriate number of significant figures, giving us 100 m for the magnitude of the displacement vector. How about the direction? Figure 1.25b identifies the angle that gives the angle north of east of Anna's displacement. In the right triangle, 50 m is the opposite side and 90 m is the adjacent side, so the angle is given by

$$\theta = \tan^{-1}\left(\frac{50\text{ m}}{90\text{ m}}\right) = \tan^{-1}\left(\frac{5}{9}\right) = 29°$$

Putting it all together, we get a net displacement of

$$\vec{d}_{net} = (100\text{ m}, 29°\text{ north of east})$$

ASSESS We can use our drawing to assess our result. If the two sides of the triangle are 50 m and 90 m, a length of 100 m for the hypotenuse seems about right. The angle is certainly less than 45°, but not too much less, so 29° seems reasonable.

Velocity Vectors

We've seen that a basic quantity describing the motion of an object is its velocity. Velocity is a vector quantity because its specification involves not only how fast an object is moving (its speed) but also the direction in which the object is moving. We thus represent the velocity of an object by a **velocity vector** $\vec{v}$ that points in the direction of the object's motion, and whose magnitude is the object's speed.

FIGURE 1.26a shows the motion diagram of a car accelerating from rest. We've drawn vectors showing the car's displacement between successive positions of the motion diagram. How can we draw the velocity vectors on this diagram? First, note that the direction of the displacement vector is the direction of motion between successive points in the motion diagram. But the velocity of an object also points in the direction of motion, so an object's velocity vector points in the same direction as its displacement vector. Second, we've already noted that the magnitude of the velocity vector—How fast?—is the object's speed. Because higher speeds imply greater displacements in the same time interval, you can see that the length of the velocity vector should be proportional to the length of the displacement vector between successive points on a motion diagram. Consequently, the vectors connecting each dot of a motion diagram to the next, which we previously labeled as displacement vectors, could equally well be identified as velocity vectors. This is shown in FIGURE 1.26b. **From now on, we'll show and label velocity vectors on motion diagrams rather than displacement vectors.**

FIGURE 1.26 The motion diagram for a car starting from rest.

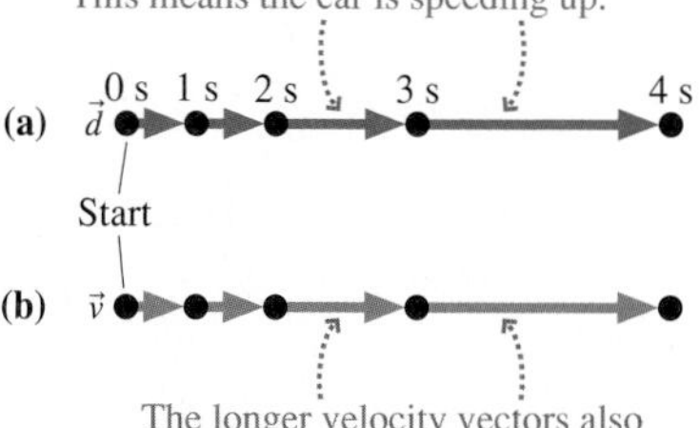

NOTE ▶ The velocity vectors shown in Figure 1.26b are actually *average* velocity vectors. Because the velocity is steadily increasing, it's a bit less than this average at the start of each time interval, and a bit more at the end. In Chapter 2 we'll refine these ideas as we develop the idea of instantaneous velocity. ◀

EXAMPLE 1.8 Drawing a ball's motion diagram

Jake hits a ball at a 60° angle from the horizontal. It is caught by Jim. Draw a motion diagram of the ball that shows velocity vectors rather than displacement vectors.

PREPARE This example is typical of how many problems in science and engineering are worded. The problem does not give a clear statement of where the motion begins or ends. Are we interested in the motion of the ball only during the time it is in the air between Jake and Jim? What about the motion *as* Jake hits it (ball rapidly speeding up) or *as* Jim catches it (ball rapidly slowing down)? Should we include Jim dropping the ball after he catches it? The point is that *you* will often be called on to make a *reasonable interpretation* of a problem statement. In this problem, the details of hitting and catching the ball are complex. The motion of the ball through the air is easier to describe, and it's a motion you might expect to learn about in a physics class. So our *interpretation* is that the motion diagram should start as the ball leaves Jake's bat (ball already moving) and should end the instant it touches Jim's hand (ball still moving). We will model the ball as a particle.

SOLVE With this interpretation in mind, FIGURE 1.27 shows the motion diagram of the ball. Notice how, in contrast to the car of Figure 1.26, the ball is already moving as the motion diagram movie begins. As before, the velocity vectors are shown by connecting the dots with arrows. You can see that the velocity vectors get shorter (ball slowing down), get longer (ball speeding up), and change direction. Each $\vec{v}$ is different, so this is *not* constant-velocity motion.

FIGURE 1.27 The motion diagram of a ball traveling from Jake to Jim.

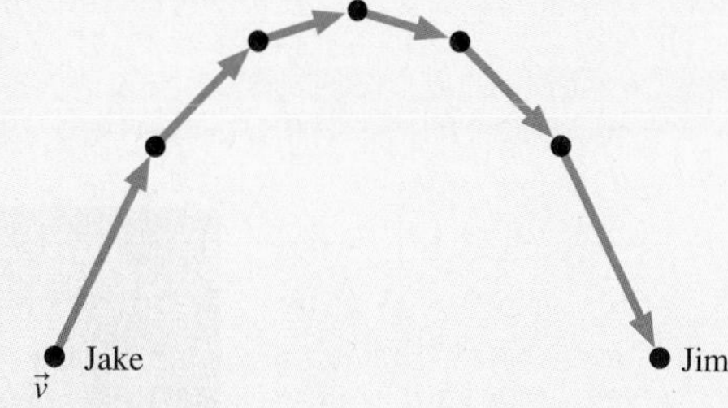

ASSESS We haven't learned enough to make a detailed analysis of the motion of the ball, but it's still worthwhile to do a quick assessment. Does our diagram make sense? Think about the velocity of the ball—we show it moving upward at the start and downward at the end. This does match what happens when you toss a ball back and forth, so our answer seems reasonable.

STOP TO THINK 1.5 $\vec{P}$ and $\vec{Q}$ are two vectors of equal length but different direction. Which vector shows the sum $\vec{P}+\vec{Q}$?

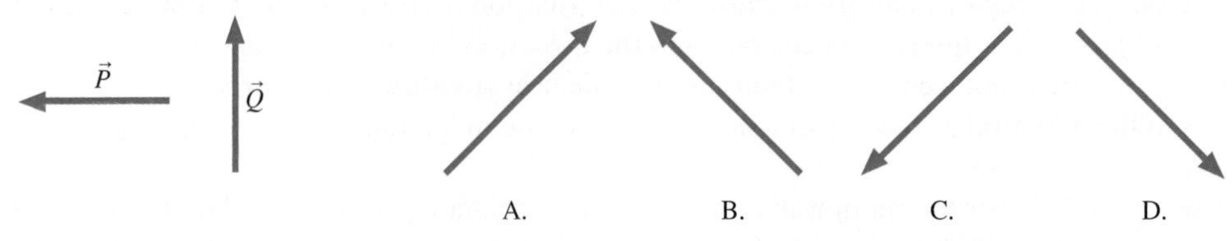

1.6 Where Do We Go from Here?

This first chapter has been an introduction to some of the fundamental ideas about motion and some of the basic techniques that you will use in the rest of the course. You have seen some examples of how to make *models* of a physical situation, thereby focusing in on the essential elements of the situation. You have learned some practical ideas, such as how to convert quantities from one kind of units to another. The rest of this book—and the rest of your course—will extend these themes. You will learn how to model many kinds of physical systems, and learn the technical skills needed to set up and solve problems using these models.

In each chapter of this book, you'll learn both new principles and more tools and techniques. We are starting with motion, but, by the end of the book, you'll have learned about more abstract concepts such as magnetic fields and the structure of the nucleus of the atom. As you proceed, you'll find that each new chapter depends on those that preceded it. The principles and the problem-solving strategies you learned in this chapter will still be needed in Chapter 30.

We'll give you some assistance integrating new ideas with the material of previous chapters. When you start a chapter, the **chapter preview** will let you know which topics are especially important to review. And the last element in each chapter will be an **integrated example** that brings together the principles and techniques you have just learned with those you learned previously. The integrated nature of these examples will also be a helpful reminder that the problems of the real world are similarly complex, and solving such problems requires you to do just this kind of integration.

Our first integrated example is reasonably straightforward because there's not much to integrate yet. The examples in future chapters will be much richer.

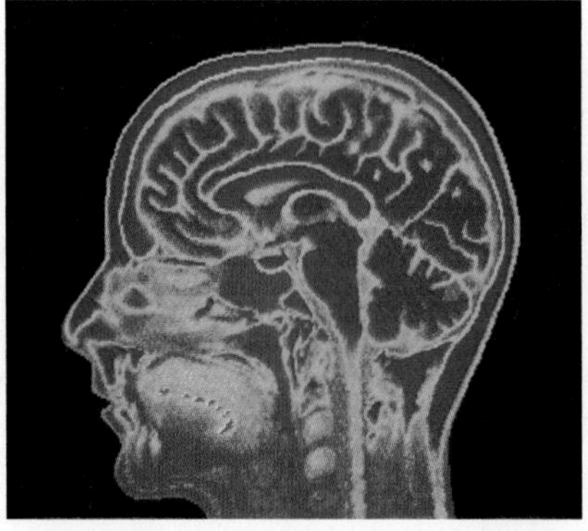

◀ Chapter 28 ends with an integrated example that explores the basic physics of magnetic resonance imaging (MRI), explaining how the interaction of magnetic fields with the nuclei of atoms in the body can be used to create an image of the body's interior.

INTEGRATED EXAMPLE 1.9 A goose gets its bearings

Migrating geese determine direction using many different tools: by noting local landmarks, by following rivers and roads, and by using the position of the sun in the sky. When the weather is overcast so that they can't use the sun's position to get their bearings, geese may start their day's flight in the wrong direction. **FIGURE 1.28** shows the path of a Canada goose that flew in a straight line for some time before making a corrective right-angle turn. One hour after beginning, the goose made a rest stop on a lake due east of its original position.

FIGURE 1.28 Trajectory of a misdirected goose.

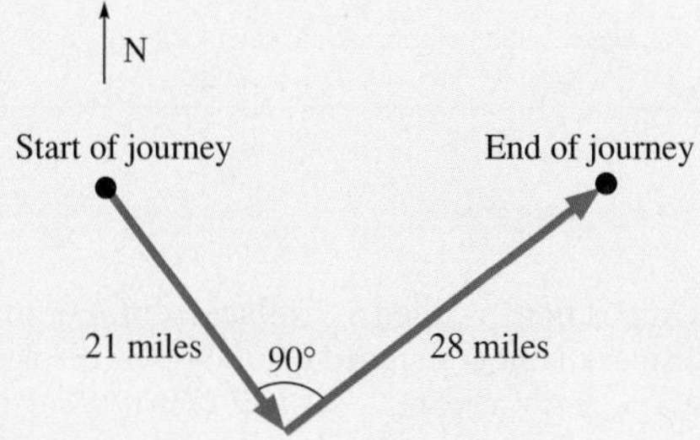

a. How much extra distance did the goose travel due to its initial error in flight direction? That is, how much farther did it fly than if it had simply flown directly to its final position on the lake?
b. What was the flight speed of the goose?
c. A typical flight speed for a migrating goose is 80 km/h. Given this, does your result seem reasonable?

PREPARE Figure 1.28 shows the trajectory of the goose, but it's worthwhile to redraw Figure 1.28 and note the displacement from the start to the end of the journey, the shortest distance the goose could have flown. (The examples in the chapter to this point have used professionally rendered drawings, but these are much more careful and detailed than you are likely to make. **FIGURE 1.29** shows a drawing that is more typical of what you might actually do when working problems yourself.) Drawing and labeling the displacement between the starting and ending points in Figure 1.29 show that it is the hypotenuse of a right triangle, so we can use our rules for triangles as we look for a solution.

FIGURE 1.29 A typical student sketch shows the motion and the displacement of the goose.

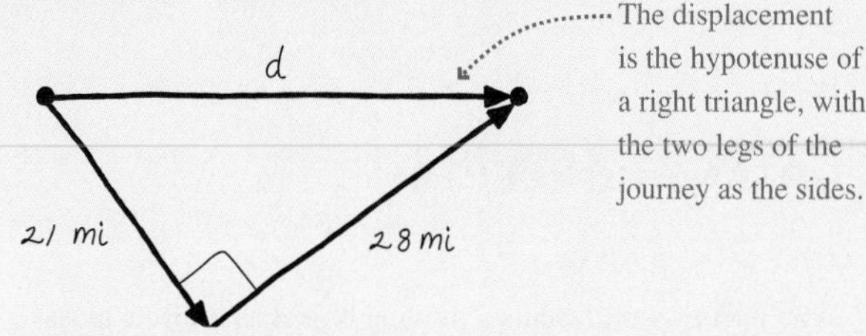

SOLVE

a. The minimum distance the goose *could* have flown, if it flew straight to the lake, is the hypotenuse of a triangle with sides 21 mi and 28 mi. This straight-line distance is

$$d = \sqrt{(21\text{ mi})^2 + (28\text{ mi})^2} = 35\text{ mi}$$

The actual distance the goose flew is the sum of the distances traveled for the two legs of the journey:

$$\text{distance traveled} = 21\text{ mi} + 28\text{ mi} = 49\text{ mi}$$

The extra distance flown is the difference between the actual distance flown and the straight-line distance—namely, 14 miles.

b. To compute the flight speed, we need to consider the distance that the bird actually flew. The flight speed is the total distance flown divided by the total time of the flight:

$$v = \frac{49\text{ mi}}{1.0\text{ h}} = 49\text{ mi/h}$$

c. To compare our calculated speed with a typical flight speed, we must convert our solution to km/h, rounding off to the correct number of significant digits:

$$49\ \frac{\text{mi}}{\text{h}} \times \frac{1.61\text{ km}}{1.00\text{ mi}} = 79\ \frac{\text{km}}{\text{h}}$$

A calculator will return many more digits, but the original data had only two significant figures, so we report the final result to this accuracy.

ASSESS In this case, an assessment was built into the solution of the problem. The calculated flight speed matches the expected value for a goose, which gives us confidence that our answer is correct. As a further check, our calculated net displacement of 35 mi seems about right for the hypotenuse of the triangle in Figure 1.29.

SUMMARY

The goals of Chapter 1 have been to introduce the fundamental concepts of motion and to review the related basic mathematical principles.

IMPORTANT CONCEPTS

Motion Diagrams

The particle model represents a moving object as if all its mass were concentrated at a single point. Using this model, we can represent motion with a **motion diagram,** where dots indicate the object's positions at successive times. In a motion diagram, the time interval between successive dots is always the same.

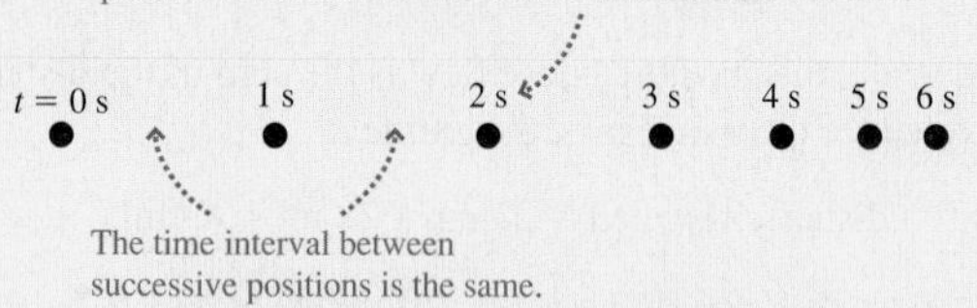

Each dot represents the position of the object. Each position is labeled with the time at which the dot was there.
t = 0 s
1 s
2 s
3 s
4 s
5 s
6 s
The time interval between successive positions is the same.

Scalars and Vectors

Scalar quantities have only a magnitude and can be represented by a single number. Temperature, time, and mass are scalars.

A vector is a quantity described by both a magnitude and a direction. Velocity and displacement are vectors.

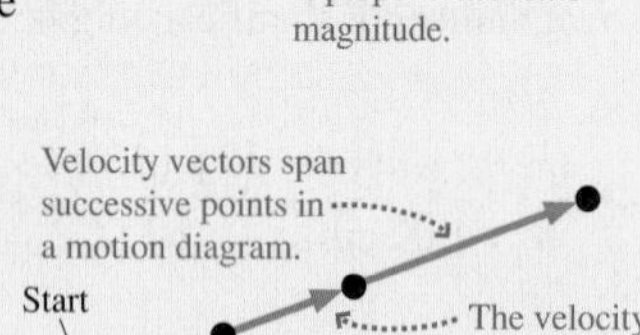

Velocity vectors can be drawn on a motion diagram by connecting successive points with a vector.

Velocity vectors span successive points in a motion diagram.

Start

$\vec{v}$

The velocity vectors are getting longer, so the object is speeding up.

Describing Motion

Position locates an object with respect to a chosen coordinate system. It is described by a **coordinate.**

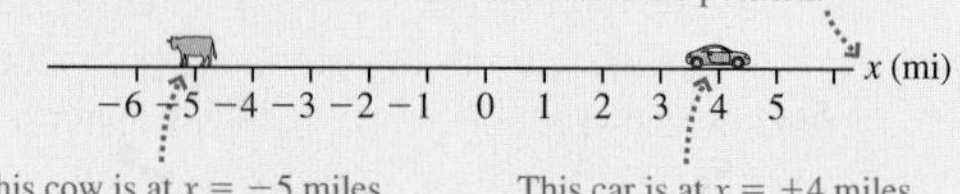

This cow is at x = −5 miles.
This car is at x = +4 miles.

A change in position is called a **displacement.** For motion along a line, a displacement is a signed quantity. The displacement from x_i to x_f is $\Delta x = x_f - x_i$.

Time is measured from a particular instant to which we assign $t = 0$. A **time interval** is the elapsed time between two specific instants t_i and t_f. It is given by $\Delta t = t_f - t_i$.

Velocity is the ratio of the displacement of an object to the time interval during which this displacement occurs:

$$v = \frac{\Delta x}{\Delta t}$$

Units

Every measurement of a quantity must include a unit.

The standard system of units used in science is the SI system. Common SI units include:

- Length: meters (m)
- Time: seconds (s)
- Mass: kilograms (kg)

APPLICATIONS

Working with Numbers

In scientific notation, a number is expressed as a decimal number between 1 and 10 multiplied by a power of ten. In scientific notation, the diameter of the earth is 1.27×10^7 m.

A prefix can be used before a unit to indicate a multiple of 10 or 1/10. Thus we can write the diameter of the earth as 12,700 km, where the k in km denotes 1000.

We can perform a unit conversion to convert the diameter of the earth to a different unit, such as miles. We do so by multiplying by a conversion factor equal to 1, such as 1 = 1 mi/1.61 km.

Significant figures are reliably known digits. The number of significant figures for:

- **Multiplication, division, and powers** is set by the value with the fewest significant figures.
- **Addition and subtraction** is set by the value with the smallest number of decimal places.

An order-of-magnitude estimate is an estimate that has an accuracy of about one significant figure. Such estimates are usually made using rough numbers from everyday experience.

For homework assigned on MasteringPhysics, go to www.masteringphysics.com

Problems labeled BIO are of biological or medical interest.

Problem difficulty is labeled as | (straightforward) to ||||| (challenging).

QUESTIONS

Conceptual Questions

1. a. Write a paragraph describing the *particle model.* What is it, and why is it important?
   b. Give two examples of situations, different from those described in the text, for which the particle model is appropriate.
   c. Give an example of a situation, different from those described in the text, for which it would be inappropriate.
2. A softball player slides into second base. Use the particle model to draw a motion diagram of the player from the time he begins to slide until he reaches the base. Number the dots in order, starting with zero.
3. A car travels to the left at a steady speed for a few seconds, then brakes for a stop sign. Use the particle model to draw a motion diagram of the car for the entire motion described here. Number the dots in order, starting with zero.
4. A ball is dropped from the roof of a tall building and students in a physics class are asked to sketch a motion diagram for this situation. A student submits the diagram shown in Figure Q1.4. Is the diagram correct? Explain.

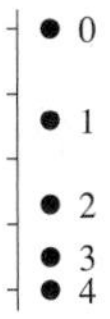

FIGURE Q1.4

5. Write a sentence or two describing the difference between position and displacement. Give one example of each.
6. Give an example of a trip you might take in your car for which the distance traveled as measured on your car's odometer is not equal to the displacement between your initial and final positions.
7. Write a sentence or two describing the difference between speed and velocity. Give one example of each.
8. The motion of a skateboard along a horizontal axis is observed for 5 s. The initial position of the skateboard is negative with respect to a chosen origin, and its velocity throughout the 5 s is also negative. At the end of the observation time, is the skateboard closer to or farther from the origin than initially? Explain.
9. Can the velocity of an object be positive during a time interval in which its position is always negative? Can its velocity be positive during a time interval in which its displacement is negative?
10. Two friends watch a jogger complete a 400 m lap around the track in 100 s. One of the friends states, "The jogger's velocity was 4 m/s during this lap." The second friend objects, saying, "No, the jogger's speed was 4 m/s." Who is correct? Justify your answer.
11. A softball player hits the ball and starts running toward first base. Draw a motion diagram, using the particle model, showing her velocity vectors during the first few seconds of her run.
12. A child is sledding on a smooth, level patch of snow. She encounters a rocky patch and slows to a stop. Draw a motion diagram, using the particle model, showing her velocity vectors.
13. A skydiver jumps out of an airplane. Her speed steadily increases until she deploys her parachute, at which point her speed quickly decreases. She subsequently falls to earth at a constant rate, stopping when she lands on the ground. Draw a motion diagram, using the particle model, that shows her position at successive times and includes velocity vectors.
14. Your roommate drops a tennis ball from a third-story balcony. It hits the sidewalk and bounces as high as the second story. Draw a motion diagram, using the particle model, showing the ball's velocity vectors from the time it is released until it reaches the maximum height on its bounce.
15. A car is driving north at a steady speed. It makes a gradual 90° left turn without losing speed, then continues driving to the west. Draw a motion diagram, using the particle model, showing the car's velocity vectors as seen from a helicopter hovering over the highway.
16. A toy car rolls down a ramp, then across a smooth, horizontal floor. Draw a motion diagram, using the particle model, showing the car's velocity vectors.
17. Estimate the average speed with which you go from home to campus (or another trip you commonly make) via whatever mode of transportation you use most commonly. Give your answer in both mph and m/s. Describe how you arrived at this estimate.
18. Estimate the number of times you sneezed during the past year. Describe how you arrived at this estimate.
19. Density is the ratio of an object's mass to its volume. Would you expect density to be a vector or a scalar quantity? Explain.

Multiple-Choice Questions

20. | A student walks 1.0 mi west and then 1.0 mi north. Afterward, how far is she from her starting point?
    A. 1.0 mi B. 1.4 mi
    C. 1.6 mi D. 2.0 mi
21. | Which of the following motions is described by the motion diagram of Figure Q1.21?
    A. An ice skater gliding across the ice.
    B. An airplane braking to a stop after landing.
    C. A car pulling away from a stop sign.
    D. A pool ball bouncing off a cushion and reversing direction.

FIGURE Q1.21

22. | A bird flies 3.0 km due west and then 2.0 km due north. What is the magnitude of the bird's displacement?
    A. 2.0 km B. 3.0 km
    C. 3.6 km D. 5.0 km

23. || A bird flies 3.0 km due west and then 2.0 km due north. Another bird flies 2.0 km due west and 3.0 km due north. What is the angle between the net displacement vectors for the two birds?
A. 23° B. 34° C. 56° D. 90°
24. | A woman walks briskly at 2.00 m/s. How much time will it take her to walk one mile?
A. 8.30 min B. 13.4 min C. 21.7 min D. 30.0 min
25. | Compute 3.24 m + 0.532 m to the correct number of significant figures.
A. 3.7 m B. 3.77 m C. 3.772 m D. 3.7720 m
26. | A rectangle has length 3.24 m and height 0.532 m. To the correct number of significant figures, what is its area?
A. 1.72 m^2 B. 1.723 m^2
C. 1.7236 m^2 D. 1.72368 m^2
27. | The earth formed 4.57×10^9 years ago. What is this time in seconds?
A. 1.67×10^{12} s B. 4.01×10^{13} s
C. 2.40×10^{15} s D. 1.44×10^{17} s
28. ||| An object's average density ρ is defined as the ratio of its mass to its volume: $\rho = M/V$. The earth's mass is 5.94×10^{24} kg, and its volume is 1.08×10^{12} km^3. What is the earth's average density?
A. 5.50×10^3 kg/m^3 B. 5.50×10^6 kg/m^3
C. 5.50×10^9 kg/m^3 D. 5.50×10^{12} kg/m^3

PROBLEMS

Section 1.1 Motion: A First Look

1. | You've made a video of a car as it skids to a halt to avoid hitting an object in the road. Use the images from the video to draw a motion diagram of the car from the time the skid begins until the car is stopped.
2. | A man rides a bike along a straight road for 5 min, then has a flat tire. He stops for 5 min to repair the flat, but then realizes he cannot fix it. He continues his journey by walking the rest of the way, which takes him another 10 min. Use the particle model to draw a motion diagram of the man for the entire motion described here. Number the dots in order, starting with zero.
3. | A jogger running east at a steady pace suddenly develops a cramp. He is lucky: A westbound bus is sitting at a bus stop just ahead. He gets on the bus and enjoys a quick ride home. Use the particle model to draw a motion diagram of the jogger for the entire motion described here. Number the dots in order, starting with zero.

Section 1.2 Position and Time: Putting Numbers on Nature

4. | Figure P1.4 shows Sue between her home and the cinema. What is Sue's position x if
a. Her home is the origin?
b. The cinema is the origin?

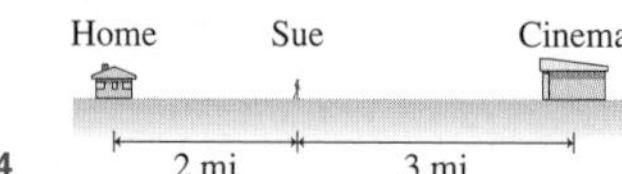

FIGURE P1.4

5. | Keira starts at position $x = 23$ m along a coordinate axis. She then undergoes a displacement of -45 m. What is her final position?
6. | A car travels along a straight east-west road. A coordinate system is established on the road, with x increasing to the east. The car ends up 14 mi west of the intersection with Mulberry Road. If its displacement was -23 mi, how far from and on which side of Mulberry Road did it start?
7. | Foraging bees often move in straight lines away from and
BIO toward their hives. Suppose a bee starts at its hive and flies 500 m due east, then flies 400 m west, then 700 m east. How far is the bee from the hive?

Section 1.3 Velocity

8. | A security guard walks 110 m in one trip around the perimeter of the building. It takes him 240 s to make this trip. What is his speed?
9. || List the following items in order of decreasing speed, from greatest to least: (i) A wind-up toy car that moves 0.15 m in 2.5 s. (ii) A soccer ball that rolls 2.3 m in 0.55 s. (iii) A bicycle that travels 0.60 m in 0.075 s. (iv) A cat that runs 8.0 m in 2.0 s.
10. || Figure P1.10 shows the motion diagram for a horse galloping in one direction along a straight path. Not every dot is labeled, but the dots are at equally spaced instants of time. What is the horse's velocity
a. During the first ten seconds of its gallop?
b. During the interval from 30 s to 40 s?
c. During the interval from 50 s to 70 s?

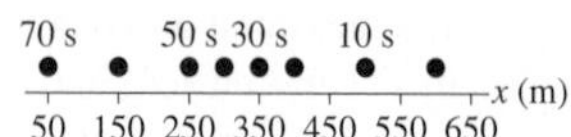

FIGURE P1.10

11. || It takes Harry 35 s to walk from $x = -12$ m to $x = -47$ m. What is his velocity?
12. | A dog trots from $x = -12$ m to $x = 3$ m in 10 s. What is its velocity?
13. | A ball rolling along a straight line with velocity 0.35 m/s goes from $x = 2.1$ m to $x = 7.3$ m. How much time does this take?

Section 1.4 A Sense of Scale: Significant Figures, Scientific Notation, and Units

14. || Convert the following to SI units:
a. 9.12 μs b. 3.42 km
c. 44 cm/ms d. 80 km/hour
15. | Convert the following to SI units:
a. 8.0 in b. 66 ft/s c. 60 mph
16. | Convert the following to SI units:
a. 1.0 hour b. 1.0 day c. 1.0 year
17. || List the following three speeds in order, from smallest to largest: 1 mm per μs, 1 km per ks, 1 cm per ms.

18. | How many significant figures does each of the following numbers have?
    a. 6.21 b. 62.1 c. 0.620 d. 0.062
19. | How many significant figures does each of the following numbers have?
    a. 0.621 b. 0.006200
    c. 1.0621 d. 6.21×10^3
20. | Compute the following numbers to 3 significant figures.
    a. 33.3×25.4 b. $33.3 - 25.4$
    c. $\sqrt{33.3}$ d. $333.3 \div 25.4$
21. ||||| The Empire State Building has a height of 1250 ft. Express this height in meters, giving your result in scientific notation with three significant figures.
22. | Estimate (don't measure!) the length of a typical car. Give your answer in both feet and meters. Briefly describe how you arrived at this estimate.
23. ||| Blades of grass grow from the bottom, so, as growth occurs, BIO the top of the blade moves upward. During the summer, when your lawn is growing quickly, estimate this speed in m/s. Explain how you made this estimate, and express your result in scientific notation.
24. || Estimate the average speed with which the hair on your head BIO grows. Give your answer in both m/s and μm/h. Briefly describe how you arrived at this estimate.
25. || Estimate the average speed at which your fingernails grow, in BIO both m/s and μm/h. Briefly describe how you arrived at this estimate.

Section 1.5 Vectors and Motion: A First Look

26. | Carol and Robin share a house. To get to work, Carol walks north 2.0 km while Robin drives west 7.5 km. How far apart are their workplaces?
27. | Joe and Max shake hands and say goodbye. Joe walks east 0.55 km to a coffee shop, and Max flags a cab and rides north 3.25 km to a bookstore. How far apart are their destinations?
28. || A city has streets laid out in a square grid, with each block 135 m long. If you drive north for three blocks, then west for two blocks, how far are you from your starting point?
29. || A butterfly flies from the top of a tree in the center of a garden to rest on top of a red flower at the garden's edge. The tree is 8.0 m taller than the flower, and the garden is 12 m wide. Determine the magnitude of the butterfly's displacement.
30. ||| A garden has a circular path of radius 50 m. John starts at the easternmost point on this path, then walks counterclockwise around the path until he is at its southernmost point. What is John's displacement? Use the (magnitude, direction) notation for your answer.
31. | Migrating geese tend to travel in straight-line paths at approx- BIO imately constant speed. A goose flies 32 km south, then turns to fly 20 km west. How far is the goose from its original position?
32. ||||| A ball on a porch rolls 60 cm to the porch's edge, drops 40 cm, continues rolling on the grass, and eventually stops 80 cm from the porch's edge. What is the magnitude of the ball's net displacement, in centimeters?
33. || A kicker punts a football from the very center of the field to the sideline 43 yards downfield. What is the net displacement of the ball? (A football field is 53 yards wide.)

General Problems

Problems 34 through 40 are motion problems similar to those you will learn to solve in Chapter 2. For now, simply *interpret* the problem by drawing a motion diagram showing the object's position and its velocity vectors. **Do *not* solve these problems** or do any mathematics.

34. || In a typical greyhound race, a dog accelerates to a speed of BIO 20 m/s over a distance of 30 m. It then maintains this speed. What would be a greyhound's time in the 100 m dash?
35. || Billy drops a watermelon from the top of a three-story building, 10 m above the sidewalk. How fast is the watermelon going when it hits?
36. || Sam is recklessly driving 60 mph in a 30 mph speed zone when he suddenly sees the police. He steps on the brakes and slows to 30 mph in three seconds, looking nonchalant as he passes the officer. How far does he travel while braking?
37. || A speed skater moving across frictionless ice at 8.0 m/s hits a 5.0-m-wide patch of rough ice. She slows steadily, then continues on at 6.0 m/s. What is her acceleration on the rough ice?
38. || The giant eland, an African antelope, is an exceptional BIO jumper, able to leap 1.5 m off the ground. To jump this high, with what speed must the eland leave the ground?
39. || A ball rolls along a smooth horizontal floor at 10 m/s, then starts up a 20° ramp. How high does it go before rolling back down?
40. || A motorist is traveling at 20 m/s. He is 60 m from a stop light when he sees it turn yellow. His reaction time, before stepping on the brake, is 0.50 s. What steady deceleration while braking will bring him to a stop right at the light?

Problems 41 through 46 show a motion diagram. For each of these problems, write a one or two sentence "story" about a *real object* that has this motion diagram. Your stories should talk about people or objects by name and say what they are doing. Problems 34 through 40 are examples of motion short stories.

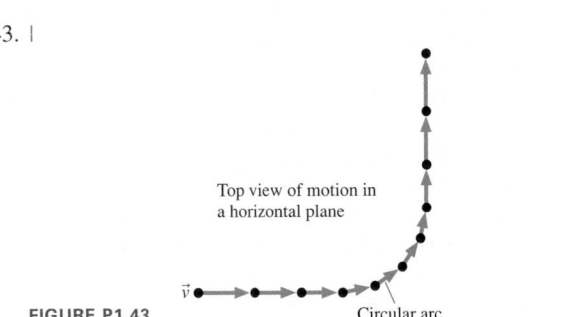

41. |

FIGURE P1.41

42. |

FIGURE P1.42

43. |

FIGURE P1.43

44. |

45. |

46. |

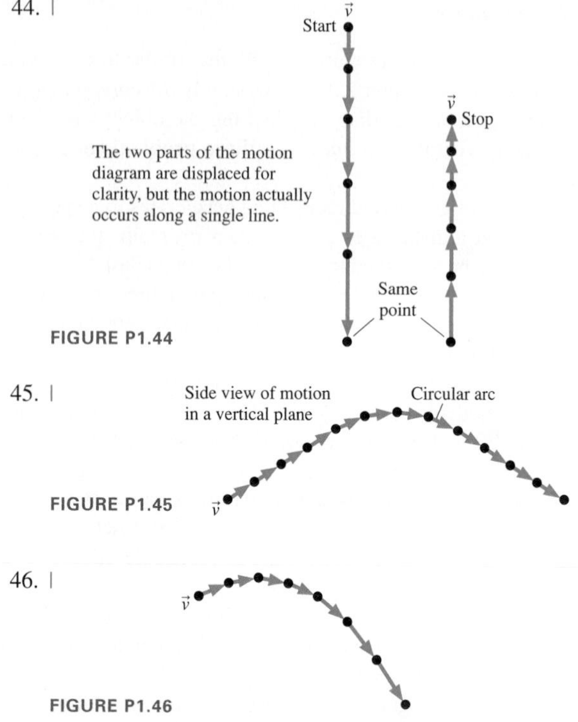

FIGURE P1.44

FIGURE P1.45

FIGURE P1.46

47. ||||| How many inches does light travel in one nanosecond? The speed of light is 3.0×10^8 m/s.

48. | Joseph watches the roadside mile markers during a long car trip on an interstate highway. He notices that at 10:45 A.M. they are passing a marker labeled 101, and at 11:00 A.M. the car reaches marker 119. What is the car's speed, in mph?

49. || Alberta is going to have dinner at her grandmother's house, but she is running a bit behind schedule. As she gets onto the highway, she knows that she must exit the highway within 45 min if she is not going to arrive late. Her exit is 32 mi away. What is the slowest speed at which she could drive and still arrive in time? Express your answer in miles per hour.

50. || The end of Hubbard Glacier in Alaska advances by an average of 105 feet per year. What is the speed of advance of the glacier in m/s?

51. | The earth completes a circular orbit around the sun in one year. The orbit has a radius of 93,000,000 miles. What is the speed of the earth around the sun in m/s? Report your result using scientific notation.

52. ||| Shannon decides to check the accuracy of her speedometer. She adjusts her speed to read exactly 70 mph on her speedometer and holds this steady, measuring the time between successive mile markers separated by exactly 1.00 mile. If she measures a time of 54 s, is her speedometer accurate? If not, is the speed it shows too high or too low?

53. || Motor neurons in mammals transmit signals from the brain to
BIO skeletal muscles at approximately 25 m/s. Estimate how much time in ms (10^{-3} s) it will take for a signal to get from your brain to your hand.

54. ||| Satellite data taken several times per hour on a particular
BIO albatross showed travel of 1200 km over a time of 1.4 days.
 a. Given these data, what was the bird's average speed in mph?
 b. Data on the bird's position were recorded only intermittently. Explain how this means that the bird's actual average speed was higher than what you calculated in part a.

55. | Your brain communicates with your body using *nerve*
BIO *impulses,* electrical signals propagated along axons. Axons come in two varieties: insulated axons with a sheath made of myelin, and uninsulated axons with no such sheath. Myelinated (sheathed) axons conduct nerve impulses much faster than unmyelinated (unsheathed) axons. The impulse speed depends on the diameter of the axons and the sheath, but a typical myelinated axon transmits nerve impulses at a speed of about 25 m/s, much faster than the typical 2.0 m/s for an unmyelinated axon. Figure P1.55 shows three equal-length nerve fibers consisting of eight axons in a row. Nerve impulses enter at the left side simultaneously and travel to the right.
 a. Draw motion diagrams for the nerve impulses traveling along fibers A, B, and C.
 b. Which nerve impulse arrives at the right side first?
 c. Which will be last?

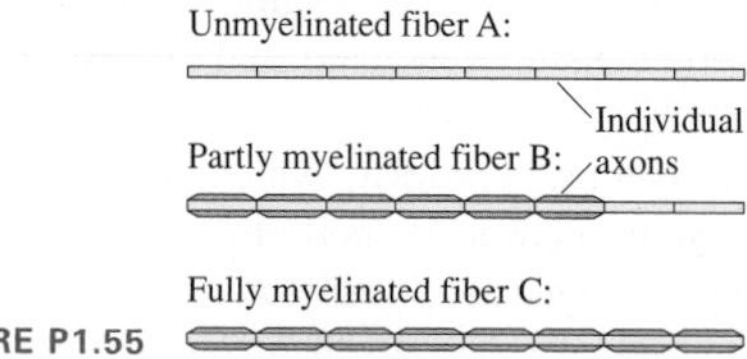

FIGURE P1.55

56. || The bacterium *Escherichia coli* (or *E. coli*) is a single-celled
BIO organism that lives in the gut of healthy humans and animals. Its body shape can be modeled as a 2-μm-long cylinder with a 1 μm diameter, and it has a mass of 1×10^{-12} g. Its chromosome consists of a single double-stranded chain of DNA 700 times longer than its body length. The bacterium moves at a constant speed of 20 μm/s, though not always in the same direction. Answer the following questions about *E. coli* using SI units (unless specifically requested otherwise) and correct significant figures. 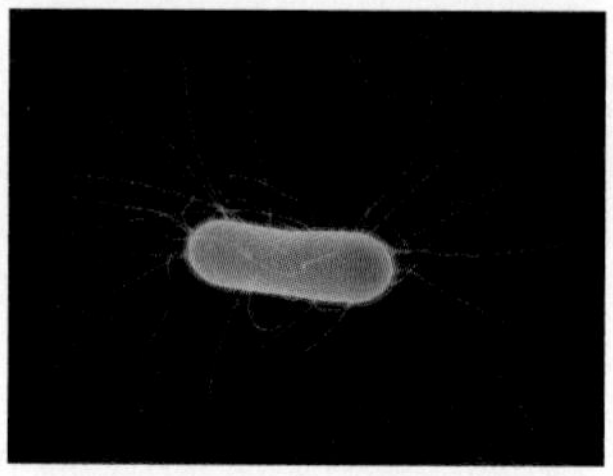
 a. What is its length?
 b. Diameter?
 c. Mass?
 d. What is the length of its DNA, in millimeters?
 e. If the organism were to move along a straight path, how many meters would it travel in one day?

57. || The bacterium *Escherichia*
BIO *coli* (or *E. coli*) is a single-celled organism that lives in the gut of healthy humans and animals. When grown in a uniform medium rich in salts and amino acids, it swims along zig-zag paths at a constant speed. Figure P1.57 shows the positions of an *E. coli* as it moves from point A to point J. Each segment of the motion can be identified by two letters, such as segment BC. During which segments, if any, does the bacterium have the same
 a. Displacement? b. Speed? c. Velocity?

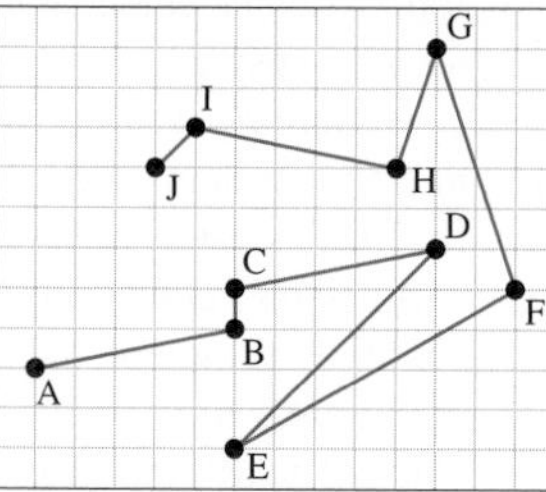

FIGURE P1.57

58. ||| In 2003, the population of the United States was 291 million people. The per-capita income was \$31,459. What was the total income of everyone in the United States? Express your answer in scientific notation, with the correct number of significant figures.
59. ||| The sun is 30° above the horizon. It makes a 52-m-long shadow of a tall tree. How high is the tree?
60. || A large passenger aircraft accelerates down the runway for a distance of 3000 m before leaving the ground. It then climbs at a steady 3.0° angle. After the plane has traveled 3000 m along this new trajectory, (a) how high is it, and (b) how far horizontally is it, from its initial position?
61. ||| Starting from its nest, an eagle flies at constant speed for 3.0 min due east, then 4.0 min due north. From there the eagle flies directly to its nest at the same speed. How long is the eagle in the air?
62. ||| John walks 1.00 km north, then turns right and walks 1.00 km east. His speed is 1.50 m/s during the entire stroll.
    a. What is the magnitude of his displacement, from beginning to end?
    b. If Jane starts at the same time and place as John, but walks in a straight line to the endpoint of John's stroll, at what speed should she walk to arrive at the endpoint just when John does?

Passage Problems

Growth Speed

The images of trees in Figure P1.63 come from a catalog advertising fast-growing trees. If we mark the position of the top of the tree in the successive years, as shown in the graph in the figure, we obtain a motion diagram much like ones we have seen for other kinds of motion. The motion isn't steady, of course. In some months the tree grows rapidly; in other months, quite slowly. We can see, though, that the average speed of growth is fairly constant for the first few years.

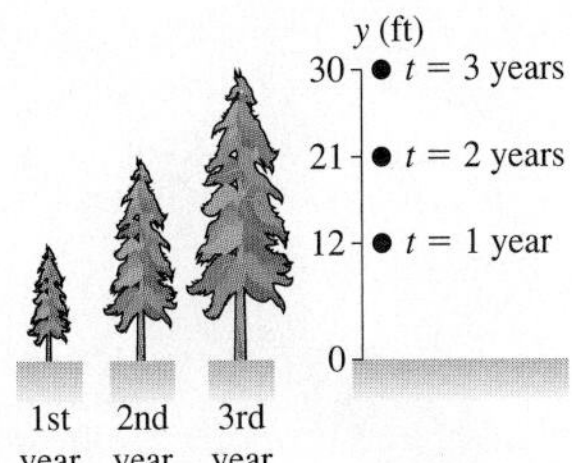

FIGURE P1.63

63. | What is the tree's speed of growth, in feet per year, from $t = 1$ yr to $t = 3$ yr?
    A. 12 ft/yr B. 9 ft/yr C. 6 ft/yr D. 3 ft/yr
64. | What is this speed in m/s?
    A. 9×10^{-8} m/s B. 3×10^{-9} m/s
    C. 5×10^{-6} m/s D. 2×10^{-6} m/s
65. | At the end of year 3, a rope is tied to the very top of the tree to steady it. This rope is staked into the ground 15 feet away from the tree. What angle does the rope make with the ground?
    A. 63° B. 60° C. 30° D. 27°

STOP TO THINK ANSWERS

Stop to Think 1.1: B. The images of B are farther apart, so B travels a greater distance than does A during the same intervals of time.

Stop to Think 1.2: A. Dropped ball. **B.** Dust particle. **C.** Descending rocket.

Stop to Think 1.3: C. Depending on her initial positive position and how far she moves in the negative direction, she could end up on either side of the origin.

Stop to Think 1.4: D > C > B = A.

Stop to Think 1.5: B. The vector sum is found by placing the tail of one vector at the head of the other.

2 Motion in One Dimension

A horse can run at 35 mph, much faster than a human. And yet, surprisingly, a man can win a race against a horse if the length of the course is right. When, and how, can a man outrun a horse?

LOOKING AHEAD ▸▸

The goal of Chapter 2 is to describe and analyze linear motion.

Describing Motion

We began discussing motion in Chapter 1. In Chapter 2, you'll learn more ways to represent motion. You will also learn general strategies for solving problems.

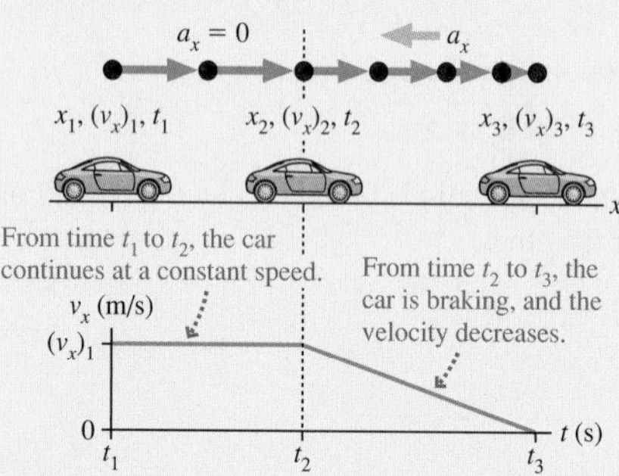

Motion diagrams and graphs are key parts of the problem-solving strategies that we will develop in this chapter.

Looking Back ◂◂
1.5 Velocity vectors and motion diagrams

Position

Position is defined in terms of a coordinate system and units of our choosing.

A game of football is really about motion in one dimension with a well-defined coordinate system.

Looking Back ◂◂
1.2 Position and displacement

Velocity

Velocity is the rate of change of position.

A small change in position during an interval of time means a small velocity; a larger change means a larger velocity.

Looking Back ◂◂
1.3 Velocity

Acceleration

Acceleration is the rate of change of velocity.

A cheetah is capable of a rapid change in velocity—that is a large acceleration. We'll see how to solve problems of changing velocity by using the concept of acceleration.

Analyzing Motion

Once you know how to describe motion, you'll be ready to do some analysis.

The main engines of a Saturn V fire for $2^1/_2$ minutes. How high will the rocket be and how fast will it be going when the engines shut off?

Constant Velocity

One important case we will consider is motion in a straight line at a constant velocity— uniform motion.

Each minute, the ship moves the same distance in the same direction.

Constant Acceleration

Another special case is motion with constant acceleration.

We think of acceleration as "speeding up," but braking to a stop involves a change in velocity—an acceleration—as well.

Free Fall

Free fall is a special case of constant acceleration.

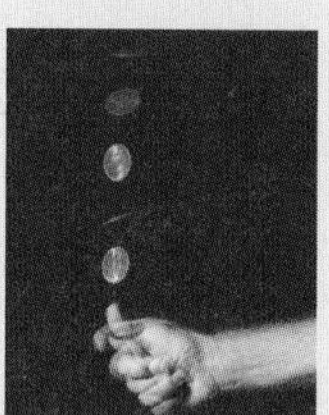

Once the coin leaves your hand, it's in free fall, and its motion is similar to that of a falling ball or a jumping gazelle.

2.1 Describing Motion

The modern name for the mathematical description of motion, without regard to causes, is **kinematics.** The term comes from the Greek word *kinema,* meaning "movement." You know this word through its English variation *cinema*—motion pictures! This chapter will focus on the kinematics of motion in one dimension, motion along a straight line.

Representing Position

As we saw in Chapter 1, kinematic variables such as position and velocity are measured with respect to a coordinate system, an axis that *you* impose on a system. We will use an x-axis to analyze both horizontal motion and motion on a ramp; a y-axis will be used for vertical motion. We will adopt the convention that the positive end of an x-axis is to the right and the positive end of a y-axis is up. This convention is illustrated in FIGURE 2.1.

FIGURE 2.1 Sign conventions for position.

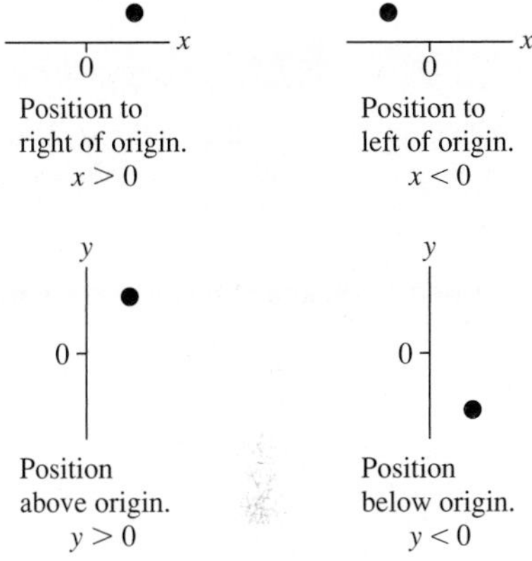

Now, let's look at a practical problem of the sort that we first saw in Chapter 1. FIGURE 2.2 is a motion diagram, made at 1 frame per minute, of a straightforward situation: A student walking to school. She is moving horizontally, so we use the variable x to describe her motion. We have set the origin of the coordinate system, $x = 0$, at her starting position, and we measure her position in meters. We have included velocity vectors connecting successive positions on the motion diagram, as we saw we could do in Chapter 1. The motion diagram shows that she leaves home at a time we choose to call $t = 0$ min, and then makes steady progress for a while. Beginning at $t = 3$ min there is a period in which the distance traveled during each time interval becomes shorter—perhaps she slowed down to speak with a friend. Then, at $t = 6$ min, the distances traveled within each interval are longer—perhaps, realizing she is running late, she begins walking more quickly.

FIGURE 2.2 The motion diagram of a student walking to school and a coordinate axis for making measurements.

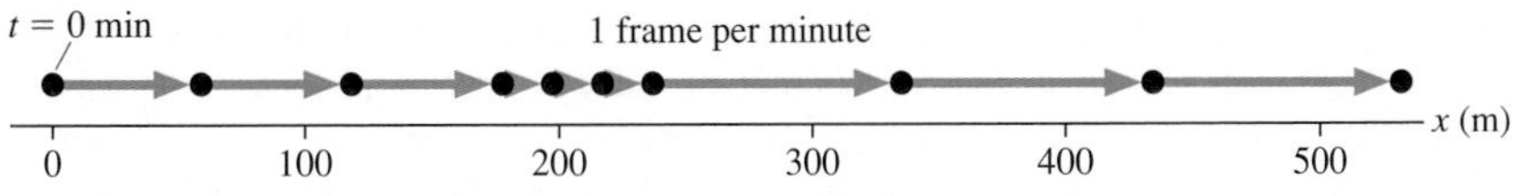

TABLE 2.1 Measured positions of a student walking to school

Time t (min)	Position x (m)	Time t (min)	Position x (m)
0	0	5	220
1	60	6	240
2	120	7	340
3	180	8	440
4	200	9	540

Every dot in the motion diagram of Figure 2.2 represents the student's position at a particular time. For example, the student is at position $x = 120$ m at $t = 2$ min. Table 2.1 lists her position for every point in the motion diagram.

The motion diagram of Figure 2.2 is one way to represent the student's motion. Presenting the data as in Table 2.1 is a second way to represent this motion. A third way to represent the motion is to make a graph. FIGURE 2.3 is a graph of the positions of the student at different times; we say it is a graph of x versus t for the student. We have merely taken the data from the table and plotted these particular points on the graph.

NOTE ► A graph of "a versus b" means that a is graphed on the vertical axis and b on the horizontal axis. ◄

We can flesh out the graph of Figure 2.3, though. Common sense tells us that the student was *somewhere specific* at all times: There was never a time when she failed to have a well-defined position, nor could she occupy two positions at one time. (As reasonable as this belief appears to be, we'll see that it's not entirely accurate when we get to quantum physics!) We also can assume that, from the start to the end of her motion, the student moved *continuously* through all intervening points of space, so

FIGURE 2.3 A graph of the student's motion.

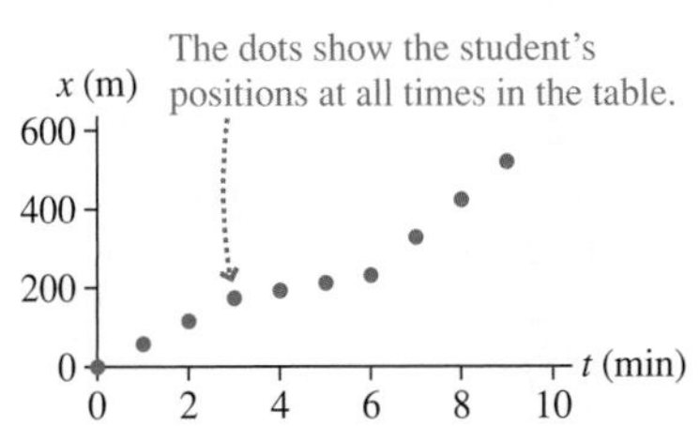

FIGURE 2.4 Extending the graph of Figure 2.3 to a position-versus-time graph.

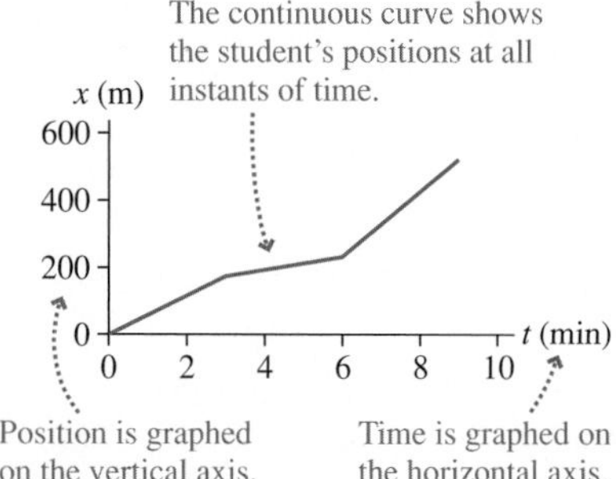

we can represent her motion as a continuous curve that passes through the measured points, as shown in **FIGURE 2.4**. Such a continuous curve that shows an object's position as a function of time is called a **position-versus-time graph** or, sometimes, just a *position graph.*

NOTE ► A graph is *not* a "picture" of the motion. The student is walking along a straight line, but the graph itself is not a straight line. Further, we've graphed her position on the vertical axis even though her motion is horizontal. A graph is an *abstract representation* of motion. We will place significant emphasis on the process of interpreting graphs, and many of the exercises and problems will give you a chance to practice these skills. ◄

CONCEPTUAL EXAMPLE 2.1 Interpreting a car's position-versus-time graph

The graph in **FIGURE 2.5** represents the motion of a car along a straight road. Describe (in words) the motion of the car.

FIGURE 2.5 Position-versus-time graph for the car.

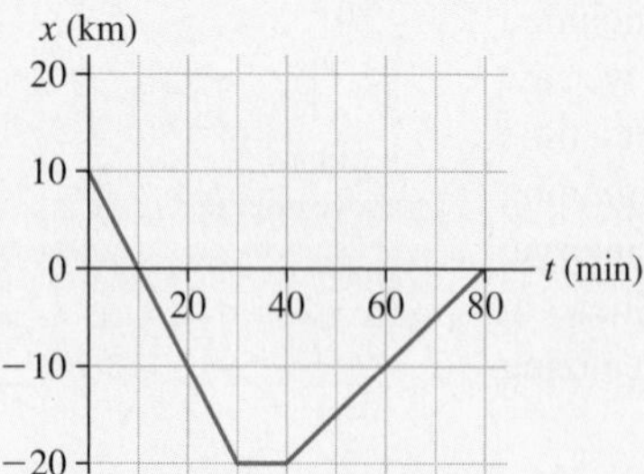

REASON The vertical axis in Figure 2.5 is labeled "x (km)"; position is measured in kilometers. Our convention for motion along the x-axis given in Figure 2.1 tells us that x increases as the car moves to the right and x decreases as the car moves to the left. The graph thus shows that the car travels to the left for 30 minutes, stops for 10 minutes, then travels to the right for 40 minutes. It ends up 10 km to the left of where it began. **FIGURE 2.6** gives a full explanation of the reasoning.

FIGURE 2.6 Looking at the position-versus-time graph in detail.

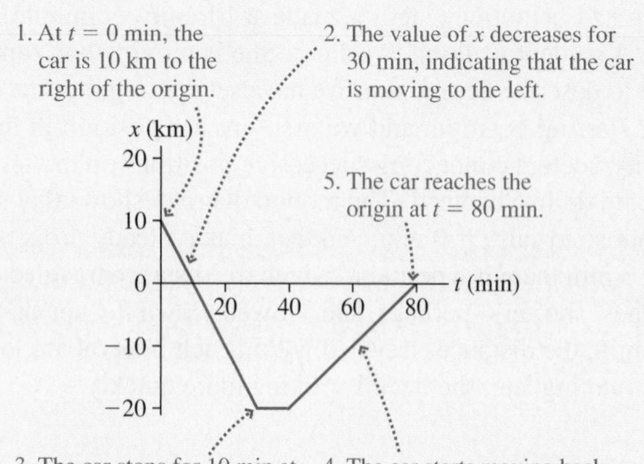

ASSESS The car travels to the left for 30 minutes and to the right for 40 minutes. Nonetheless, it ends up to the left of where it started. This means that the car was moving faster when it was moving to the left than when it was moving to the right. We can deduce this fact from the graph as well, as we will see in the next section.

Representing Velocity

FIGURE 2.7 Sign conventions for velocity.

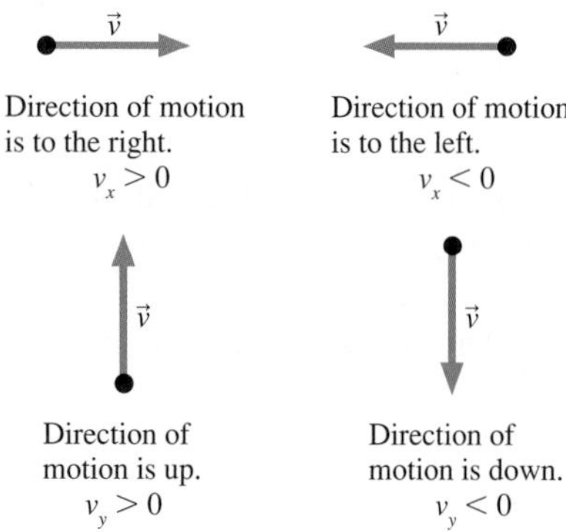

Velocity is a vector; it has both a magnitude and a direction. When we draw a velocity vector on a diagram, we use an arrow labeled with the symbol $\vec{v}$ to represent the magnitude and the direction. For motion in one dimension, vectors are restricted to point only "forward" or "backward" for horizontal motion (or "up" or "down" for vertical motion). This restriction lets us simplify our notation for vectors in one dimension. When we solve problems for motion along an x-axis, we will represent velocity with the symbol v_x. v_x will be positive or negative, corresponding to motion to the right or the left, as shown in **FIGURE 2.7**. For motion along a y-axis, we will use the symbol v_y to represent the velocity; the sign conventions are also illustrated in Figure 2.7. We will use the symbol v, with no subscript, to represent the speed of an object. **Speed is the *magnitude* of the velocity vector** and is always positive.

For motion along a line, the definition of velocity from Chapter 1 can be written as

$$v_x = \frac{\Delta x}{\Delta t} \tag{2.1}$$

This agrees with the sign conventions in Figure 2.7. If Δx is positive, x is increasing, the object is moving to the right, and Equation 2.1 gives a positive value for velocity. If Δx is negative, x is decreasing, the object is moving to the left, and Equation 2.1 gives a negative value for velocity.

Equation 2.1 is the first of many kinematic equations we'll see in this chapter. We'll often specify equations in terms of the coordinate x, but if the motion is vertical, in which case we use the coordinate y, the equations can be easily adapted. For example, Equation 2.1 for motion along a vertical axis becomes

$$v_y = \frac{\Delta y}{\Delta t} \tag{2.2}$$

From Position to Velocity

Let's take another look at the motion diagram of the student walking to school. As we see in **FIGURE 2.8**, where we have repeated the motion diagram of Figure 2.2, her motion has three clearly defined phases. In each phase her speed is constant (because the velocity vectors have the same length) but the speed varies from phase to phase.

FIGURE 2.8 Revisiting the motion diagram of the student walking to school.

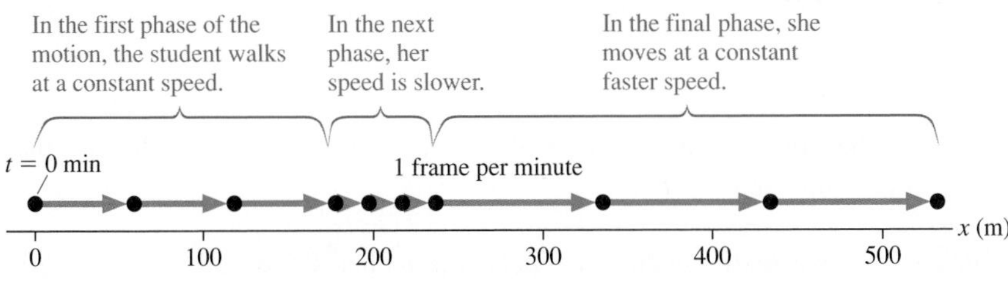

In the first phase of the motion, the student walks at a constant speed.
In the next phase, her speed is slower.
In the final phase, she moves at a constant faster speed.
t = 0 min
1 frame per minute
0
100
200
300
400
500
x (m)
Throughout the motion, she moves toward the right, in the direction of increasing x. Her velocity is always positive.

Her motion has three different phases; similarly, the position-versus-time graph redrawn in **FIGURE 2.9a** has three clearly defined segments with three different slopes. Looking at the different segments of the graph, we can see that there's a relationship between her speed and the slope of the graph: **A faster speed corresponds to a steeper slope.**

FIGURE 2.9 Revisiting the graph of the motion of the student walking to school.

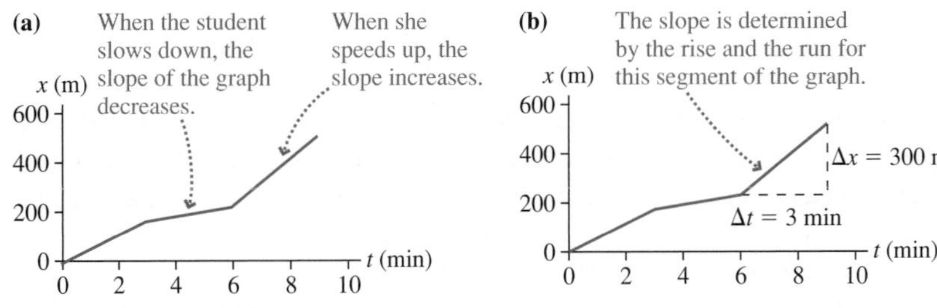

When she speeds up, the slope increases.

The correspondence is actually deeper than this. Let's look at the slope of the third segment of the position-versus-time graph, as shown in **FIGURE 2.9b**. The slope of a graph is defined as the ratio of the "rise," the vertical change, to the "run," the horizontal change. For the segment of the graph shown, the slope is:

$$\text{slope of graph} = \frac{\text{rise}}{\text{run}} = \frac{\Delta x}{\Delta t}$$

This ratio has a physical meaning—it's the velocity, exactly as we defined it in Equation 2.1. We've shown this correspondence for one particular graph, but it is a

Time lines BIO This section of the trunk of a pine tree shows the light bands of spring growth and the dark bands of summer and fall growth in successive years. If you focus on the spacing of successive dark bands, you can think of this picture as a motion diagram for the tree, representing its growth in diameter. The years of rapid growth (large distance between dark bands) during wet years and slow growth (small distance between dark bands) during years of drought are readily apparent.

general principle: **The slope of an object's position-versus-time graph is the object's velocity at that point in the motion.** This principle also holds for negative slopes, which correspond to negative velocities. We can associate the slope of a position-versus-time graph, a *geometrical* quantity, with velocity, a *physical* quantity. This is an important aspect of interpreting position-versus-time graphs, as outlined in Tactics Box 2.1.

TACTICS BOX 2.1 Interpreting position-versus-time graphs

Information about motion can be obtained from position-versus-time graphs as follows:

1. Determine an object's *position* at time t by reading the graph at that instant of time.
2. Determine the object's *velocity* at time t by finding the slope of the position graph at that point. Steeper slopes correspond to faster speeds.
3. Determine the *direction of motion* by noting the sign of the slope. Positive slopes correspond to positive velocities and, hence, to motion to the right (or up). Negative slopes correspond to negative velocities and, hence, to motion to the left (or down).

Exercises 2,3

NOTE ► The slope is a ratio of intervals, $\Delta x/\Delta t$, not a ratio of coordinates; that is, the slope is *not* simply x/t. ◄

NOTE ► We are distinguishing between the actual slope and the *physically meaningful* slope. If you were to use a ruler to measure the rise and the run of the graph, you could compute the actual slope of the line as drawn on the page. That is not the slope we are referring to when we equate the velocity with the slope of the line. Instead, we find the *physically meaningful* slope by measuring the rise and run using the scales along the axes. The "rise" Δx is some number of meters; the "run" Δt is some number of seconds. The physically meaningful rise and run include units, and the ratio of these units gives the units of the slope. ◄

We can now use the approach of Tactics Box 2.1 to analyze the student's position-versus-time graph that we saw in Figure 2.4. We can determine her velocity during the first phase of her motion by measuring the slope of the line:

$$v_x = \text{slope} = \frac{\Delta x}{\Delta t} = \frac{180\text{ m}}{3\text{ min}} = 60\frac{\text{m}}{\text{min}} \times \frac{1\text{ min}}{60\text{ s}} = 1.0\text{ m/s}$$

In completing this calculation, we've converted to more usual units for speed, m/s. During this phase of the motion, her velocity is constant, so a graph of velocity versus time appears as a horizontal line at 1.0 m/s, as shown in FIGURE 2.10. We can do similar calculations to show that her velocity during the second phase of her motion (i.e., the slope of the position graph) is +0.33 m/s, and then increases to +1.7 m/s during the final phase. We combine all of this information to create the **velocity-versus-time graph** shown in Figure 2.10.

FIGURE 2.10 Deducing the velocity-versus-time graph from the position-versus-time graph.

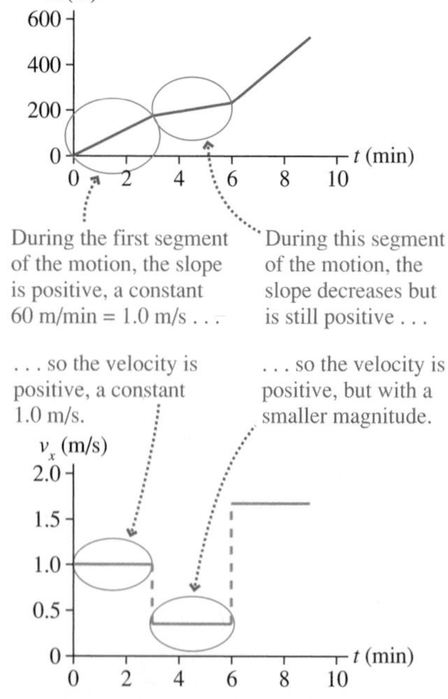
x (m)
600
400
200
0
t (min)
0 2 4 6 8 10
During the first segment of the motion, the slope is positive, a constant 60 m/min = 1.0 m/s . . .
During this segment of the motion, the slope decreases but is still positive . . .
. . . so the velocity is positive, a constant 1.0 m/s.
. . . so the velocity is positive, but with a smaller magnitude.
v_x (m/s)
2.0
1.5
1.0
0.5
0
t (min)
0 2 4 6 8 10

An inspection of the velocity-versus-time graph shows that it matches our understanding of the student's motion: There are three phases of the motion, each with constant speed. In each phase, the velocity is positive because she is always moving to the right. The second phase is slow (low velocity) and the third phase fast (high velocity.) All of this can be clearly seen on the velocity-versus-time graph, which is yet another way to represent her motion.

NOTE ► The velocity-versus-time graph in Figure 2.10 includes vertical segments in which the velocity changes instantaneously. Such rapid changes are an idealization; it actually takes a small amount of time to change velocity. ◄

EXAMPLE 2.2 **Analyzing a car's position graph**

FIGURE 2.11 gives the position-versus-time graph of a car.

a. Draw the car's velocity-versus-time graph.
b. Describe the car's motion in words.

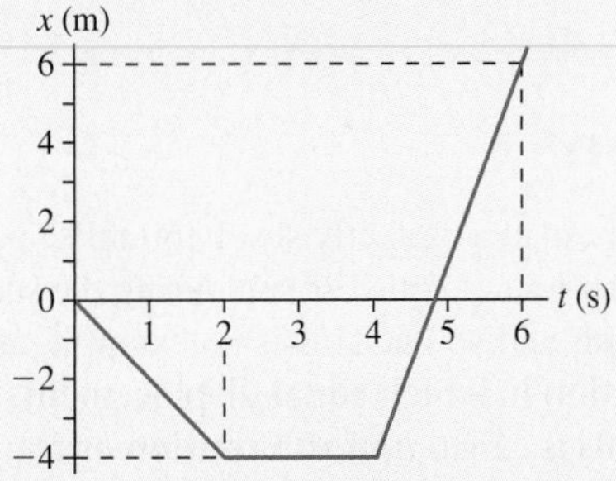

FIGURE 2.11 The position-versus-time graph of a car.

PREPARE Figure 2.11 is a graphical representation of the motion. The car's position-versus-time graph is a sequence of three straight lines. Each of these straight lines represents uniform motion at a constant velocity. We can determine the car's velocity during each interval of time by measuring the slope of the line.

SOLVE

a. From $t = 0$ s to $t = 2$ s ($\Delta t = 2$ s) the car's displacement is $\Delta x = -4\text{ m} - 0\text{ m} = -4$ m. The velocity during this interval is

$$v_x = \frac{\Delta x}{\Delta t} = \frac{-4\text{ m}}{2\text{ s}} = -2\text{ m/s}$$

The car's position does not change from $t = 2$ s to $t = 4$ s ($\Delta x = 0$ m), so $v_x = 0$ m/s. Finally, the displacement between $t = 4$ s and $t = 6$ s ($\Delta t = 2$ s) is $\Delta x = 10$ m. Thus the velocity during this interval is

$$v_x = \frac{10\text{ m}}{2\text{ s}} = 5\text{ m/s}$$

These velocities are represented graphically in FIGURE 2.12.

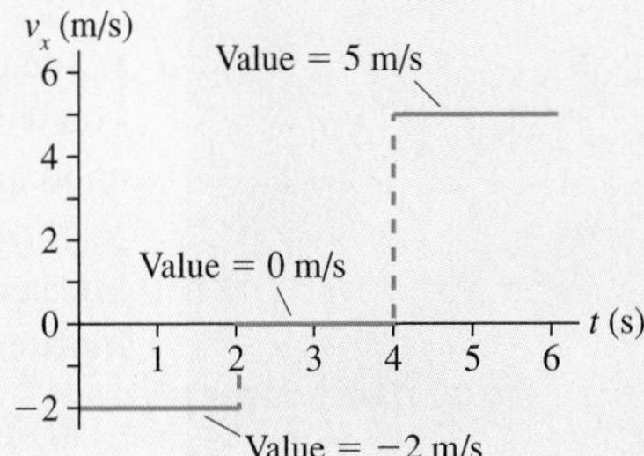

FIGURE 2.12 The velocity-versus-time graph for the car.

b. The velocity-versus-time graph of Figure 2.12 shows the motion in a way that we can describe in a straightforward manner: The car backs up for 2 s at 2 m/s, sits at rest for 2 s, then drives forward at 5 m/s for 2 s.

ASSESS Notice that the velocity graph and the position graph look completely different. They should! The value of the velocity graph at any instant of time equals the *slope* of the position graph. Since the position graph is made up of segments of constant slope, the velocity graph should be made up of segments of constant *value*, as it is. This gives us confidence that the graph we have drawn is correct.

From Velocity to Position

We've now seen how to move between different representations of uniform motion. There's one last issue to address: If you have a graph of velocity versus time, how can you determine the position graph?

Suppose you leave a lecture hall and begin walking toward your next class, which is down the hall to the west. You then realize that you left your textbook (which you always bring to class with you!) at your seat. You turn around and run back to the lecture hall to retrieve it. A velocity-versus-time graph for this motion appears as the top graph in FIGURE 2.13. There are two clear phases to the motion: walking away from class (velocity +1.0 m/s) and running back (velocity −3.0 m/s.) How can we deduce your position-versus-time graph?

As before, we can analyze the graph segment by segment. This process is shown in Figure 2.13, in which the upper velocity-versus-time graph is used to deduce the lower position-versus-time graph. For each of the two segments of the motion, the sign of the velocity tells us whether the slope of the graph is positive or negative; the magnitude of the velocity tells how steep the slope is. The final result makes sense; it shows 15 seconds of slowly increasing position (walking away) and then 5 seconds of rapidly decreasing position (running back.) And you end up back where you started.

There's one important detail that we didn't talk about in the preceding paragraph: How did we know that the position graph started at $x = 0$ m? The velocity graph tells us the *slope* of the position graph, but it doesn't tell us where the position graph should start. Although you're free to select any point you choose as the origin of the coordinate system, here it seems reasonable to set $x = 0$ m at your starting point in the lecture hall; as you walk away, your position increases.

FIGURE 2.13 Deducing a position graph from a velocity-versus-time graph.

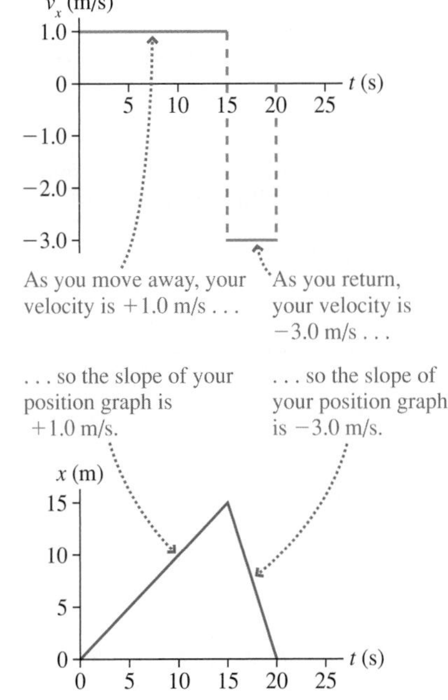

STOP TO THINK 2.1 Which position-versus-time graph best describes the motion diagram at left?

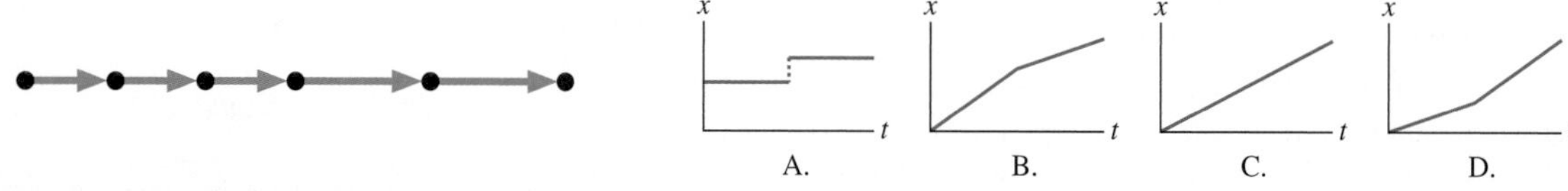

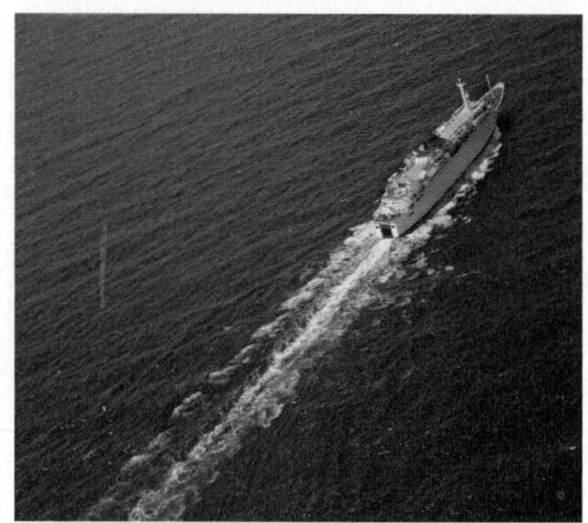

A ship on a constant heading at a steady speed is a practical example of uniform motion.

2.2 Uniform Motion

If you drive your car on a straight road at a perfectly steady 60 miles per hour (mph), you will cover 60 mi during the first hour, another 60 mi during the second hour, yet another 60 mi during the third hour, and so on. This is an example of what we call *uniform motion.* **Straight-line motion in which equal displacements occur during any successive equal-time intervals is called uniform motion or constant-velocity motion.**

NOTE ► The qualifier "any" is important. If during each hour you drive 120 mph for 30 min and stop for 30 min, you will cover 60 mi during each successive 1 hour interval. But you will *not* have equal displacements during successive 30 min intervals, so this motion is not uniform. Your constant 60 mph driving is uniform motion because you will find equal displacements no matter how you choose your successive time intervals. ◄

FIGURE 2.14 Motion diagram and position-versus-time graph for uniform motion.

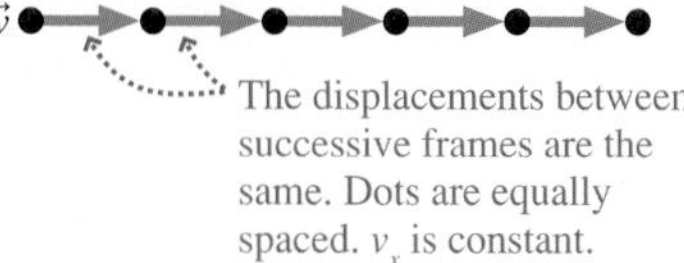

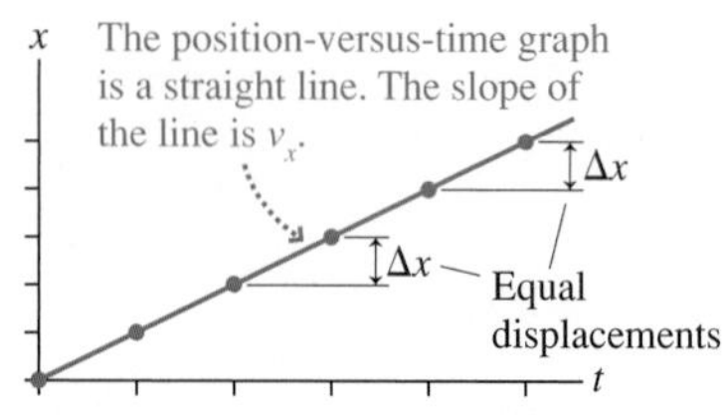

FIGURE 2.14 shows a motion diagram and a graph for an object in uniform motion. Notice that the position-versus-time graph for uniform motion is a straight line. This follows from the requirement that all values of Δx corresponding to the same value of Δt be equal. In fact, an alternative definition of uniform motion is: **An object's motion is uniform if and only if its position-versus-time graph is a straight line.**

Equations of Uniform Motion

An object is in uniform motion along the *x*-axis with the linear position-versus-time graph shown in FIGURE 2.15. Recall from Chapter 1 that we denote the object's initial position as x_i at time t_i. The term "initial" refers to the starting point of our analysis or the starting point in a problem. The object may or may not have been in motion prior to t_i. We use the term "final" for the ending point of our analysis or the ending point of a problem, and denote the object's final position x_f at the time t_f. As we've seen, the object's velocity v_x along the *x*-axis can be determined by finding the slope of the graph:

$$v_x = \frac{\text{rise}}{\text{run}} = \frac{\Delta x}{\Delta t} = \frac{x_f - x_i}{t_f - t_i} \tag{2.3}$$

FIGURE 2.15 Position-versus-time graph for an object in uniform motion.

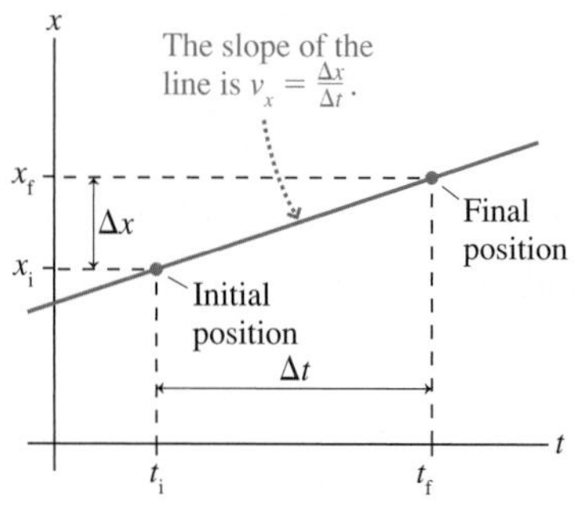

Equation 2.3 can be rearranged to give

$$x_f = x_i + v_x \Delta t \tag{2.4}$$

Position equation for an object in uniform motion (v_x is constant)

where $\Delta t = t_f - t_i$ is the interval of time in which the object moves from position x_i to position x_f. Equation 2.4 applies to any time interval Δt during which the velocity is constant. We can also write this in terms of the object's displacement, $\Delta x = x_f - x_i$:

$$\Delta x = v_x \Delta t \tag{2.5}$$

The velocity of an object in uniform motion tells us the amount by which its position changes during each second. An object with a velocity of 20 m/s *changes* its position by 20 m during every second of motion: by 20 m during the first second of its motion, by another 20 m during the next second, and so on. We say that position is changing at the *rate* of 20 m/s. If the object starts at $x_i = 10$ m, it will be at $x = 30$ m after 1 s of motion and at $x = 50$ m after 2 s of motion. Thinking of velocity like this will help you develop an intuitive understanding of the connection between velocity and position.

Physics may seem densely populated with equations, but most equations follow a few basic forms. The mathematical form of Equation 2.5 is a type that we will see again: The displacement Δx is *proportional* to the time interval Δt.

NOTE ► The important features of a proportional relationship are described below. In this text, the first time we use a particular mathematical form we will provide such an overview. In future chapters, when we see other examples of this type of relationship, we will refer back to this overview. ◄

Proportional relationships

We say that y is **proportional** to x if they are related by an equation of this form:

$$y = Cx$$

y is proportional to x

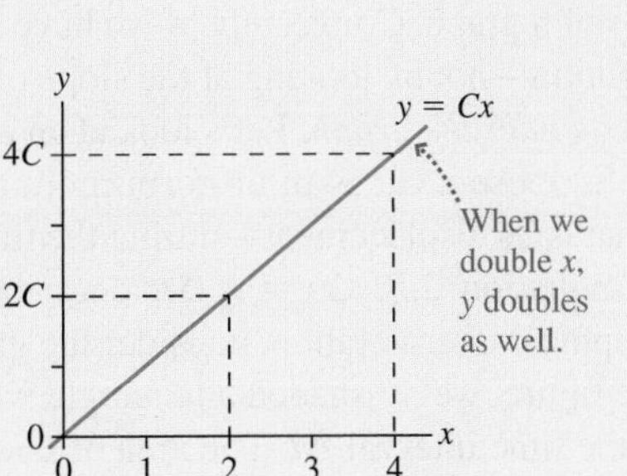

We call C the **proportionality constant.** A graph of y versus x is a straight line that passes through the origin.

SCALING If x has the initial value x_1, then y has the initial value $y_1 = Cx_1$. Changing x from x_1 to x_2 changes y from y_1 to y_2. The ratio of y_2 to y_1 is

$$\frac{y_2}{y_1} = \frac{Cx_2}{Cx_1} = \frac{x_2}{x_1}$$

The ratio of y_2 to y_1 is exactly the same as the ratio of x_2 to x_1. If y is proportional to x, which is often written $y \propto x$, then x and y change by the same factor:

- If you double x, you double y.
- If you decrease x by a factor of 3, you decrease y by a factor of 3.

If two variables have a proportional relationship, we can draw important conclusions from ratios without knowing the value of the proportionality constant C. We can often solve problems in a very straightforward manner by looking at such ratios. This is an important skill called *ratio reasoning*.

Exercise 11

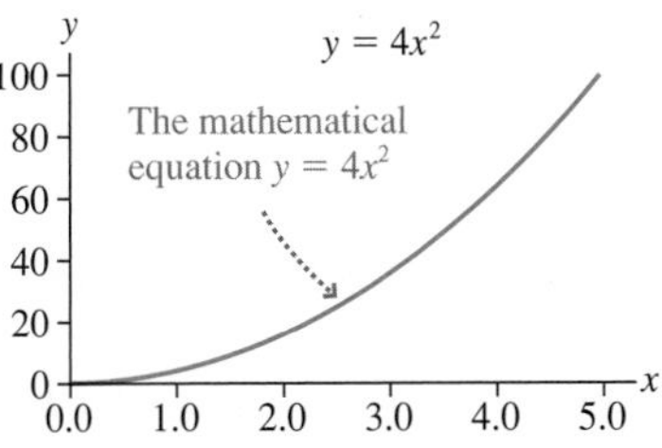

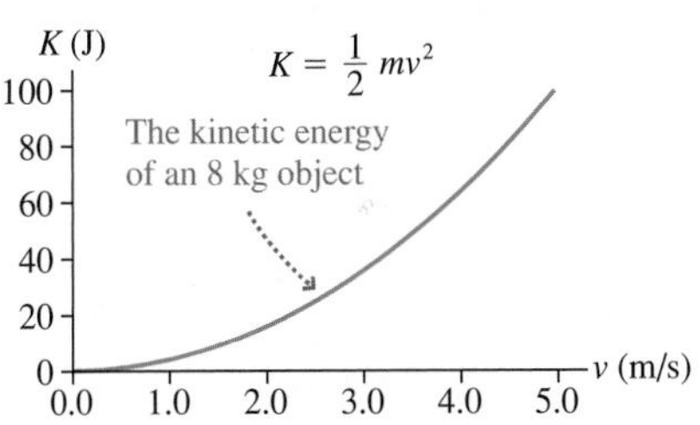

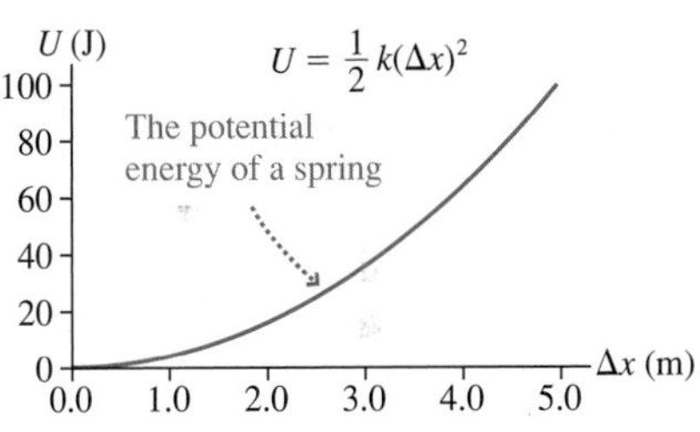

Mathematical Forms These three figures show graphs of a mathematical equation, the kinetic energy of a moving object versus its speed, and the potential energy of a spring versus the displacement of the end of the spring. All three graphs have the same overall appearance. The three expressions differ in their variables, but all three equations have the same **mathematical form.** There are only a handful of different mathematical forms that we'll use in this text. As we meet each form for the first time, we will give an overview. When you see it again, we'll insert an icon that refers back to the overview so that you can remind yourself of the key details.

EXAMPLE 2.3 If a train leaves Cleveland at 2:00 . . .

A train is moving due west at a constant speed. A passenger notes that it takes 10 minutes to travel 12 km. How long will it take the train to travel 60 km?

PREPARE For an object in uniform motion, Equation 2.5 shows that the distance traveled Δx is proportional to the time interval Δt, so this is a good problem to solve using ratio reasoning.

SOLVE We are comparing two cases: the time to travel 12 km and the time to travel 60 km. Because Δx is proportional to Δt, the ratio of the times will be equal to the ratio of the distances. The ratio of the distances is

$$\frac{\Delta x_2}{\Delta x_1} = \frac{60\text{ km}}{12\text{ km}} = 5$$

Continued

This is equal to the ratio of the times:

$$\frac{\Delta t_2}{\Delta t_1} = 5$$

$$\Delta t_2 = \text{time to travel 60 km} = 5\Delta t_1 = 5 \times (10\text{ min}) = 50\text{ min}$$

It takes 10 minutes to travel 12 km; it will take 50 minutes—5 times as long—to travel 60 km.

ASSESS For an object in steady motion, it makes sense that 5 times the distance requires 5 times the time. We can see that using ratio reasoning is a straightforward way to solve this problem. We don't need to know the proportionality constant (in this case, the velocity); we just used ratios of distances and times.

From Velocity to Position, One More Time

We've seen that we can deduce an object's velocity by measuring the slope of its position graph. Conversely, if we have a velocity graph, we can say something about position—not by looking at the slope of the graph, but by looking at what we call the *area under the graph.* Let's look at an example.

Suppose a car is in uniform motion at 12 m/s. How far does it travel—that is, what is its displacement—during the time interval between $t = 1.0$ s and $t = 3.0$ s?

Equation 2.5, $\Delta x = v_x \Delta t$, describes the displacement mathematically; for a graphical interpretation, consider the graph of velocity versus time in FIGURE 2.16. In the figure, we've shaded a rectangle whose height is the velocity v_x and whose base is the time interval Δt. The area of this rectangle is $v_x \Delta t$. Looking at Equation 2.5, we see that this quantity is also equal to the displacement of the car. The area of this rectangle is the area between the axis and the line representing the velocity; we call it the "area under the graph." We see that the **displacement Δx is equal to the area under the velocity graph during interval Δt.**

FIGURE 2.16 Displacement is the area under a velocity-versus-time graph.

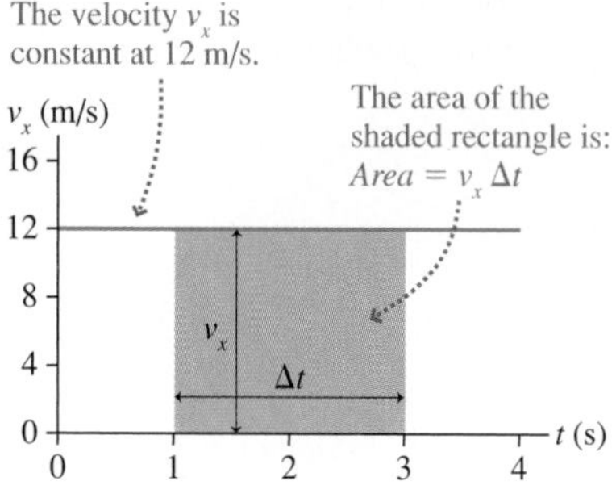

Whether we use Equation 2.5 or the area under the graph to compute the displacement, we get the same result:

$$\Delta x = v_x \Delta t = (12\text{ m/s})(2.0\text{ s}) = 24\text{ m}$$

Although we've shown that the displacement is the area under the graph only for uniform motion, where the velocity is constant, we'll soon see that this result applies to any one-dimensional motion.

NOTE ► Wait a minute! The displacement $\Delta x = x_f - x_i$ is a length. How can a length equal an area? Recall that earlier, when we found that the velocity is the slope of the position graph, we made a distinction between the *actual* slope and the *physically meaningful* slope? The same distinction applies here. The velocity graph does indeed bound a certain area on the page. That is the actual area, but it is *not* the area to which we are referring. Once again, we need to measure the quantities we are using, v_x and Δt, by referring to the scales on the axes. Δt is some number of seconds, while v_x is some number of meters per second. When these are multiplied together, the *physically meaningful* area has units of meters, appropriate for a displacement. ◄

STOP TO THINK 2.2 Four objects move with the velocity-versus-time graphs shown. Which object has the largest displacement between $t = 0$ s and $t = 2$ s?

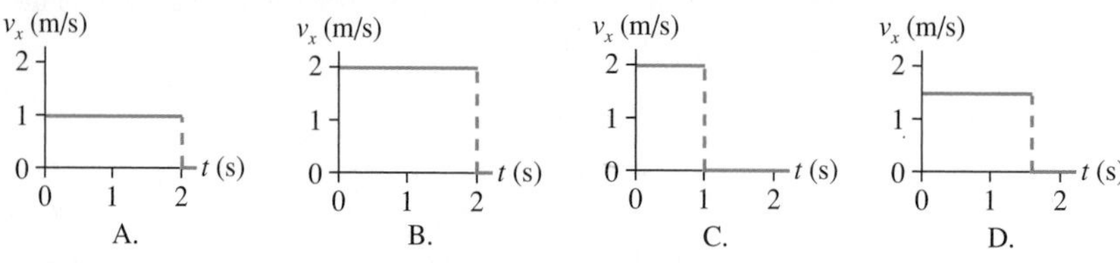

2.3 Instantaneous Velocity

The objects we've studied so far have moved with a constant, unchanging velocity or, like the car in Example 2.1, changed abruptly from one constant velocity to another. This is not very realistic. Real moving objects speed up and slow down, *changing* their velocity. As an extreme example, think about a drag racer. In a typical race, the car begins at rest but, 1 second later, is moving at over 25 miles per hour!

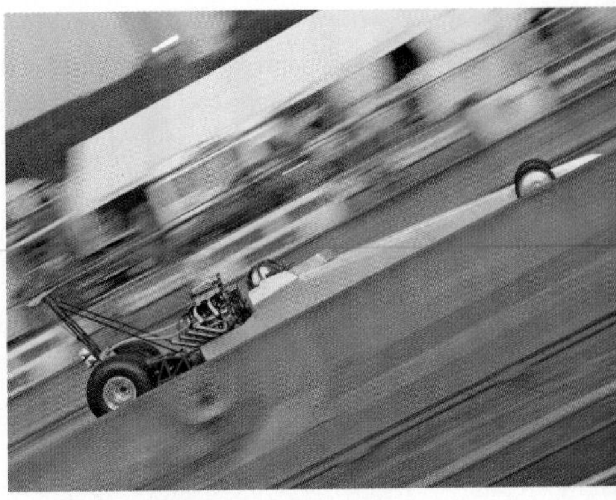

A drag racer moves with rapidly changing velocity.

For one-dimensional motion, an object changing its velocity is either speeding up or slowing down. When you drive your car, as you speed up or slow down—changing your velocity—a glance at your speedometer tells you how fast you're going *at that instant.* An object's velocity—a speed *and* a direction—at a specific *instant* of time t is called the object's **instantaneous velocity.**

But what does it mean to have a velocity "at an instant"? An instantaneous velocity of magnitude 60 mph means that the rate at which your car's position is changing—at that exact instant—is such that it would travel a distance of 60 miles in 1 hour *if* it continued at that rate without change. Said another way, if *just for an instant* your car matches the velocity of another car driving at a steady 60 mph, then your instantaneous velocity is 60 mph. **From now on, the word "velocity" will always mean instantaneous velocity.**

For uniform motion, we found that an object's position-versus-time graph is a straight line and the object's velocity is the slope of that line. In contrast, FIGURE 2.17 shows that the position-versus-time graph for a drag racer is a *curved* line. The displacement Δx during equal intervals of time gets greater as the car speeds up. Even so, we can use the slope of the position graph to measure the car's velocity. We can say that

$$\text{instantaneous velocity } v_x \text{ at time } t = \text{slope of position graph at time } t \qquad (2.6)$$

But how do we determine the slope of a curved line at a particular point? The following table gives the necessary details.

FIGURE 2.17 Position-versus-time graph for a drag racer.

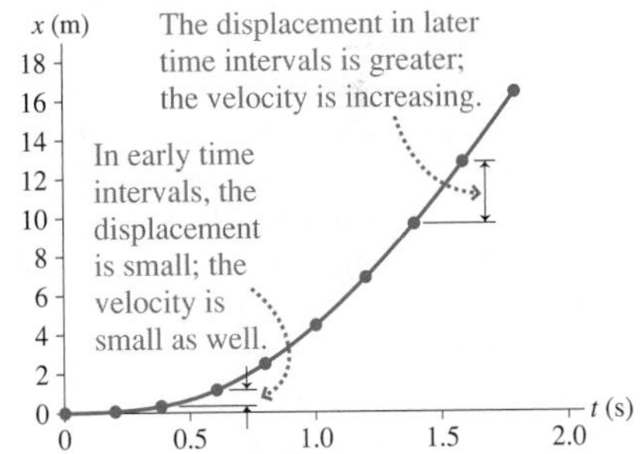

Finding the instantaneous velocity

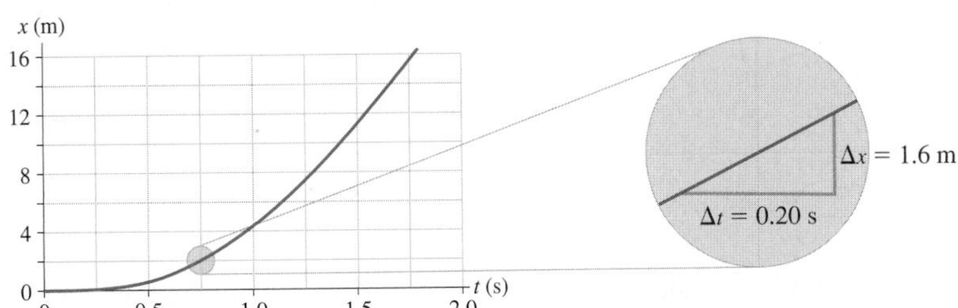

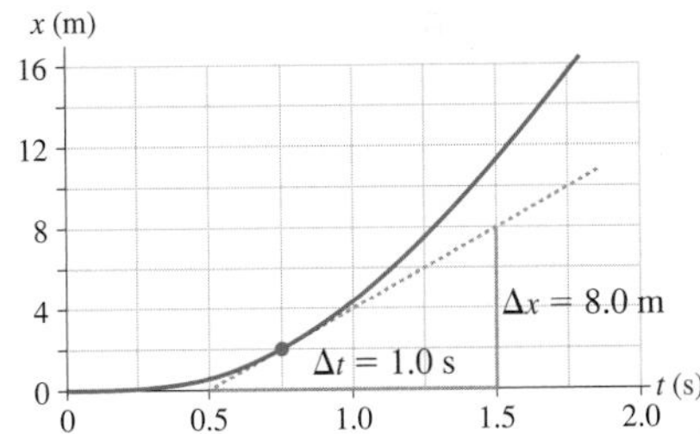

If the velocity changes, the position graph is a curved line. But we can still compute a slope by considering a small segment of the graph. Let's look at the motion in a very small time interval right around $t = 0.75$ s. This is highlighted with a circle, and we show a close-up in the next graph, at right.

Now that we have magnified a small part of the position graph, we see that the graph in this small part appears to have a constant slope. It is always possible to make the graph appear as a straight line by choosing a small enough time interval. We can find the slope of the line by calculating the rise over run, just as before:

$$v_x = \frac{1.6 \text{ m}}{0.20 \text{ s}} = 8.0 \text{ m/s}$$

This is the slope of the graph at $t = 0.75$ s and thus the velocity at this instant of time.

Graphically, the slope of the curve at a particular point is the same as the slope of a straight line drawn *tangent* to the curve at that point. **The slope of the tangent line is the instantaneous velocity at that instant of time.**

Calculating rise over run for the tangent line, we get

$$v_x = \frac{8.0 \text{ m}}{1.0 \text{ s}} = 8.0 \text{ m/s}$$

This is the same value we obtained from considering the close-up view.

CONCEPTUAL EXAMPLE 2.4 Analyzing an elevator's position graph

FIGURE 2.18 shows the position-versus-time graph of an elevator.

a. Sketch an approximate velocity-versus-time graph.
b. At which point or points is the elevator moving the fastest?
c. Is the elevator ever at rest? If so, at which point or points?

FIGURE 2.18 The position-versus-time graph for an elevator.

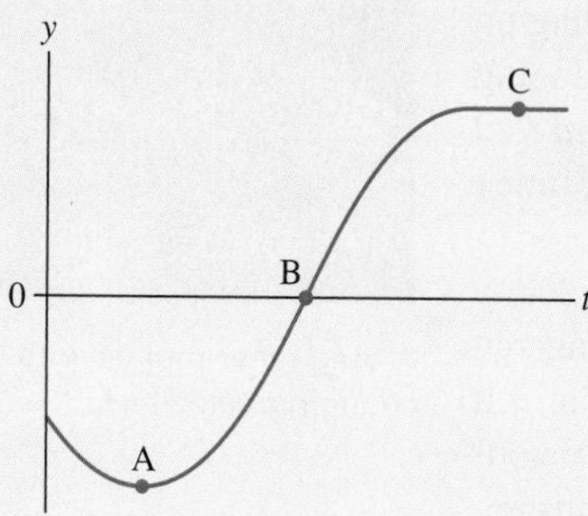

FIGURE 2.19 Finding a velocity graph from a position graph.

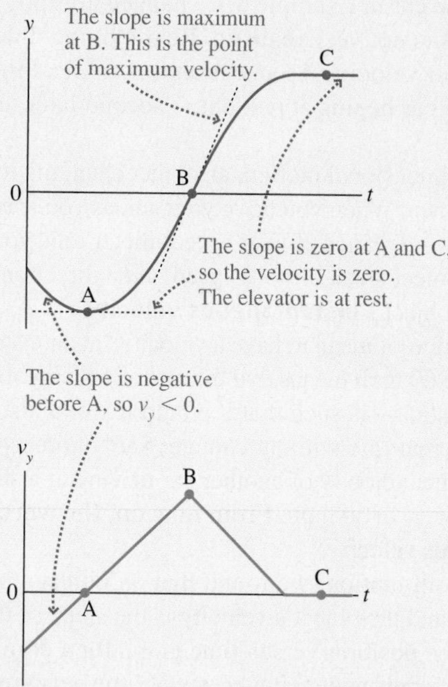

REASON a. Notice that the position graph shows *y* versus *t*, rather than *x* versus *t*, indicating that the motion is vertical rather than horizontal. Our analysis of one-dimensional motion has made no assumptions about the direction of motion, so it applies equally well to both horizontal and vertical motion. As we just found, the velocity at a particular instant of time is the slope of a tangent line to the position-versus-time graph at that time. We can move point-by-point along the position-versus-time graph, noting the slope of the tangent at each point. This will give us the velocity at that point.

Initially, to the left of point A, the slope is negative and thus the velocity is negative (i.e., the elevator is moving downward). But the slope decreases as the curve flattens out, and by the time the graph gets to point A, the slope is zero. The slope then increases to a maximum value at point B, decreases back to zero a little before point C, and remains at zero thereafter. This reasoning process is outlined in FIGURE 2.19a, and FIGURE 2.19b shows the approximate velocity-versus-time graph that results.

The other questions were answered during the construction of the graph:

b. The elevator moves the fastest at point B where the slope of the position graph is the steepest.

c. A particle at rest has $v_y = 0$. Graphically, this occurs at points where the tangent line to the position-versus-time graph is horizontal and thus has zero slope. Figure 2.19 shows that the slope is zero at points A and C. At point A, the velocity is only instantaneously zero as the particle reverses direction from downward motion (negative velocity) to upward motion (positive velocity). At point C, the elevator has actually stopped and remains at rest.

ASSESS The best way to check our work is to look at different segments of the motion and see if the velocity and position graphs match. Until point A, *y* is decreasing. The elevator is going down, so the velocity should be negative, which our graph shows. Between points A and C, *y* is increasing, so the velocity should be positive, which is also a feature of our graph. The steepest slope is at point B, so this should be the high point of our velocity graph, as it is.

For uniform motion we showed that the displacement Δx is the area under the velocity-versus-time graph during time interval Δt. We can generalize this idea to the case of an object whose velocity varies. FIGURE 2.20a on the next page is the velocity-versus-time graph for an object whose velocity changes with time. Suppose we know the object's position to be x_i at an initial time t_i. Our goal is to find its position x_f at a later time t_f.

Because we know how to handle constant velocities, let's *approximate* the velocity function of Figure 2.20a as a series of constant-velocity steps of width Δt as shown in FIGURE 2.21b. The velocity during each step is constant (uniform motion), so we can calculate the displacement during each step as the area of the rectangle under the curve. The total displacement of the object between t_i and t_f can be found as the sum of all the individual displacements during each of the constant-velocity steps. We can see in Figure 2.20b that the total displacement is approximately equal to the area under the graph, even in the case where the velocity varies. Although the approximation shown in the figure is rather rough, with only nine steps, we can imagine that it could be made as accurate as desired by having more and more ever-narrower steps.

FIGURE 2.20 Approximating a velocity-versus-time graph with a series of constant-velocity steps.

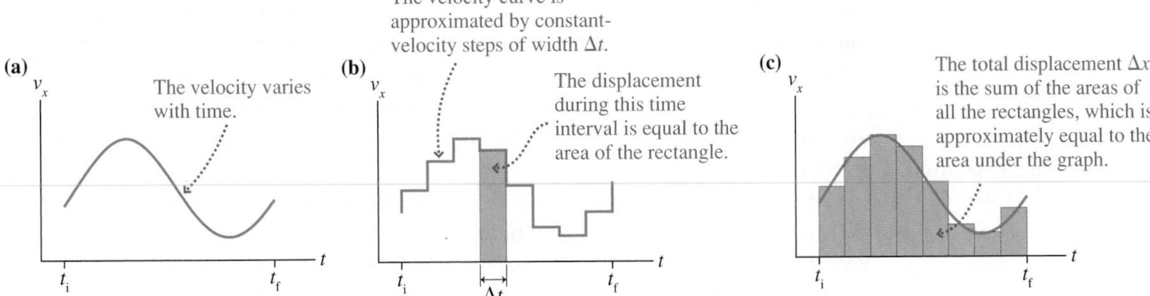

Consequently, an object's displacement is related to its velocity by

$$x_f - x_i = \Delta x = \text{area under the velocity graph } v_x \text{ between } t_i \text{ and } t_f \quad (2.7)$$

EXAMPLE 2.5 The displacement during a rapid start

FIGURE 2.21 shows the velocity-versus-time graph of a car pulling away from a stop. How far does the car move during the first 3.0 s?

PREPARE Figure 2.21 is a graphical representation of the motion. The question How far? indicates that we need to find a displacement Δx rather than a position x. According to Equation 2.7, the car's displacement $\Delta x = x_f - x_i$ between $t = 0$ s and $t = 3$ s is the area under the curve from $t = 0$ s to $t = 3$ s.

FIGURE 2.21 Velocity-versus-time graph for the car of Example 2.5.

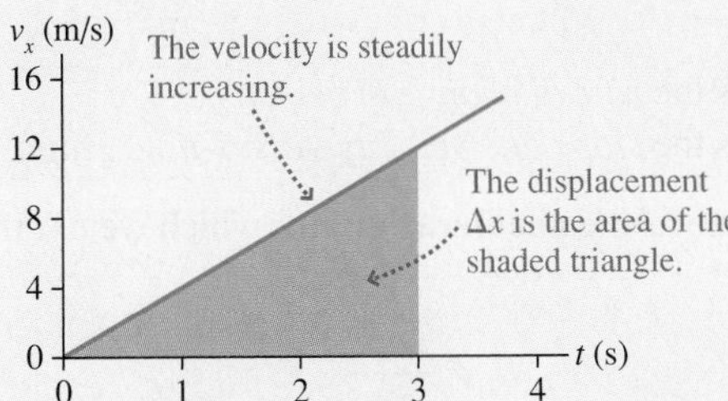

SOLVE The curve in this case is an angled line, so the area is that of a triangle:

$$\Delta x = \text{area of triangle between } t = 0 \text{ s and } t = 3 \text{ s}$$
$$= \tfrac{1}{2} \times \text{base} \times \text{height} = \tfrac{1}{2} \times 3 \text{ s} \times 12 \text{ m/s} = 18 \text{ m}$$

The car moves 18 m during the first 3 seconds as its velocity changes from 0 to 12 m/s.

ASSESS The physically meaningful area is a product of s and m/s, so Δx has the proper units of m. Let's check the numbers to see if they make physical sense. The final velocity, 12 m/s, is about 25 mph. Pulling away from a stop, you might expect to reach this speed in about 3 s—at least if you have a reasonably sporty vehicle! If the car had moved at a constant 12 m/s (the final velocity) during these 3 s, the distance would be 36 m. The actual distance traveled during the 3 s is 18 m—half of 36 m. This makes sense, as the velocity was 0 m/s at the start of the problem and increased steadily to 12 m/s.

STOP TO THINK 2.3 Which velocity-versus-time graph goes with the position-versus-time graph on the left?

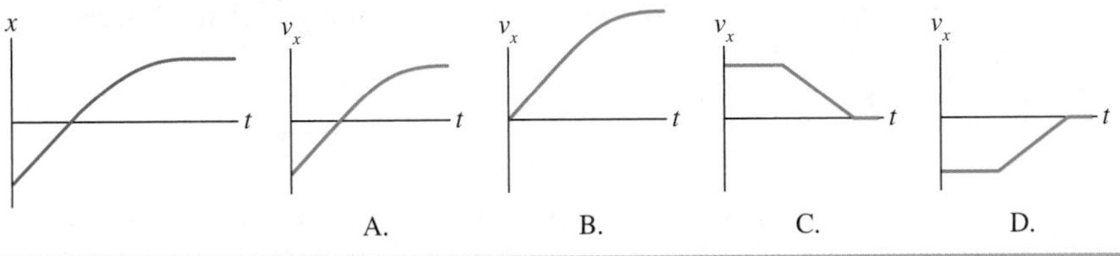

2.4 Acceleration

The goal of this chapter is to describe motion. We've seen that velocity describes the rate at which an object changes position. We need one more motion concept to complete the description, one that will describe an object whose velocity is changing.

As an example, let's look at a frequently quoted measurement of car performance, the time it takes the car to go from 0 to 60 mph. Table 2.2 shows this time for two different cars.

TABLE 2.2 Performance data for vehicles

Vehicle	Time to go from 0 to 60 mph
1997 Porsche 911 Turbo S	3.6 s
1973 Volkswagen Super Beetle Convertible	24 s

Let's look at motion diagrams for the Porsche and the Volkswagen in FIGURE 2.22. We can see two important facts about the motion. First, the lengths of the velocity vectors are increasing, showing that the speeds are increasing. Second, the velocity vectors for the Porsche are increasing in length more rapidly than those of the VW. The quantity we seek is one that measures how rapidly an object's velocity vectors change in length.

FIGURE 2.22 Motion diagrams for the Porsche and Volkswagen.

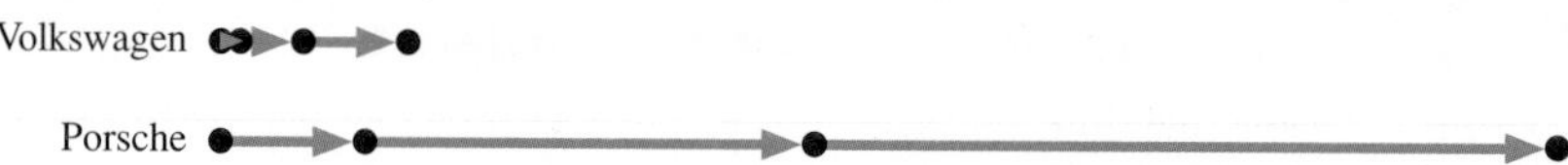

When we wanted to measure changes in position, the ratio $\Delta x/\Delta t$ was useful. This ratio, which we defined as the velocity, is the *rate of change of position.* Similarly, we can measure how rapidly an object's velocity changes with the ratio $\Delta v_x/\Delta t$. Given our experience with velocity, we can say a couple of things about this new ratio:

- The ratio $\Delta v_x/\Delta t$ is the *rate of change of velocity.*
- The ratio $\Delta v_x/\Delta t$ is the *slope of a velocity-versus-time graph.*

We will define this ratio as the **acceleration,** for which we use the symbol a_x:

$$a_x = \frac{\Delta v_x}{\Delta t} \tag{2.8}$$

Definition of acceleration as the rate of change of velocity

Cushion kinematics When a car hits an obstacle head-on, the damage to the car and its occupants can be reduced by making the acceleration as small as possible. As we can see from Equation 2.8, acceleration can be reduced by making the *time* for a change in velocity as long as possible. This is the purpose of the yellow crash cushion barrels you may have seen in work zones on highways—to lengthen the time of a collision with a barrier.

Similarly, $a_y = \Delta v_y/\Delta t$ for vertical motion.

As an example, let's calculate the acceleration for the Porsche and the Volkswagen. For both, the initial velocity $(v_x)_i$ is zero and the final velocity $(v_x)_f$ is 60 mph. Thus the *change* in velocity is $\Delta v_x = 60$ mph. In m/s, our SI unit of velocity, $\Delta v_x = 27$ m/s.

Now we can use Equation 2.8 to compute acceleration. Let's start with the Porsche, which speeds up to 27 m/s in $\Delta t = 3.6$ s:

$$a_{\text{Porsche}\,x} = \frac{\Delta v_x}{\Delta t} = \frac{27\text{ m/s}}{3.6\text{ s}} = 7.5\,\frac{\text{m/s}}{\text{s}}$$

Here's the meaning of this final figure: Every second, the Porsche's velocity changes by 7.5 m/s. In the first second of motion, the Porsche's velocity increases by 7.5 m/s; in the next second, it increases by another 7.5 m/s, and so on. After 1 second, the velocity is 7.5 m/s; after 2 seconds, it is 15 m/s. This increase continues as long as the Porsche has this acceleration. We thus interpret the units as 7.5 meters per second, per second—7.5 (m/s)/s.

The Volkswagen's acceleration is

$$a_{\text{VW}\,x} = \frac{\Delta v_x}{\Delta t} = \frac{27\text{ m/s}}{24\text{ s}} = 1.1\,\frac{\text{m/s}}{\text{s}}$$

In each second, the Volkswagen changes its speed by 1.1 m/s. This is only 1/7 the acceleration of the Porsche! The reasons the Porsche is capable of greater acceleration has to do with what *causes* the motion. We will explore the reasons for acceleration in Chapter 4. For now, we will simply note that the Porsche is capable of much greater acceleration, something you would have suspected.

NOTE ► It is customary to abbreviate the acceleration units (m/s)/s as m/s^2. For example, we'll write that the Volkswagen has an acceleration of 1.1 m/s^2. When you use this notation, keep in mind the *meaning* of the notation as "(meters per second) per second." ◄

TABLE 2.3 Velocity data for the Volkswagen and the Porsche

Time (s)	Velocity of VW (m/s)	Velocity of Porsche (m/s)
0	0	0
1	1.1	7.5
2	2.2	15.0
3	3.3	22.5
4	4.4	30.0

Representing Acceleration

Let's use the values we have computed for acceleration to make a table of velocities for the Porsche and the Volkswagen we considered earlier. Table 2.3 uses the idea that the VW's velocity increases by 1.1 m/s every second while the Porsche's velocity increases by 7.5 m/s every second. The data in Table 2.3 are the basis for the velocity-versus-time graphs in **FIGURE 2.23**. As you can see, an object undergoing constant acceleration has a straight–line velocity graph.

The slope of either of these lines—the rise over run—is $\Delta v_x/\Delta t$. Comparing this with Equation 2.8, we see that the equation for the slope is the same as that for the acceleration. That is, **an object's acceleration is the slope of its velocity-versus-time graph:**

$$\text{acceleration } a_x \text{ at time } t = \text{slope of velocity graph at time } t \qquad (2.9)$$

The VW has a smaller acceleration, so its velocity graph has a smaller slope.

FIGURE 2.23 Velocity-versus-time graphs for the two cars.

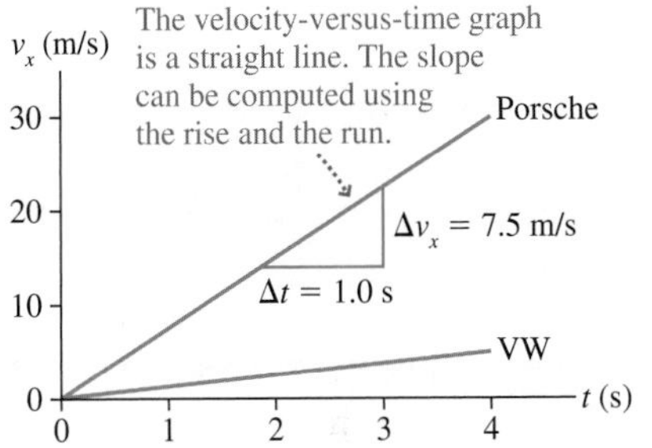

CONCEPTUAL EXAMPLE 2.6 Analyzing a car's velocity graph

FIGURE 2.24a is a graph of velocity versus time for a car. Sketch a graph of the car's acceleration versus time.

REASON The graph can be divided into three sections:

- An initial segment, in which the velocity increases at a steady rate.
- A middle segment, in which the velocity is constant.
- A final segment, in which the velocity decreases at a steady rate.

In each section, the acceleration is the slope of the velocity-versus-time graph. Thus the initial segment has constant, positive acceleration, the middle segment has zero acceleration, and the final segment has constant, *negative* acceleration. The acceleration graph appears in **FIGURE 2.24b**.

ASSESS This process is analogous to finding a velocity graph from the slope of a position graph. The middle segment having zero acceleration does *not* mean that the velocity is zero. The velocity is constant, which means it is *not changing* and thus the car is not accelerating. The car does accelerate during the initial and final segments. The magnitude of the acceleration is a measure of how quickly the velocity is changing. How about the sign? This is an issue we will address in the next section.

FIGURE 2.24 Finding an acceleration graph from a velocity graph.

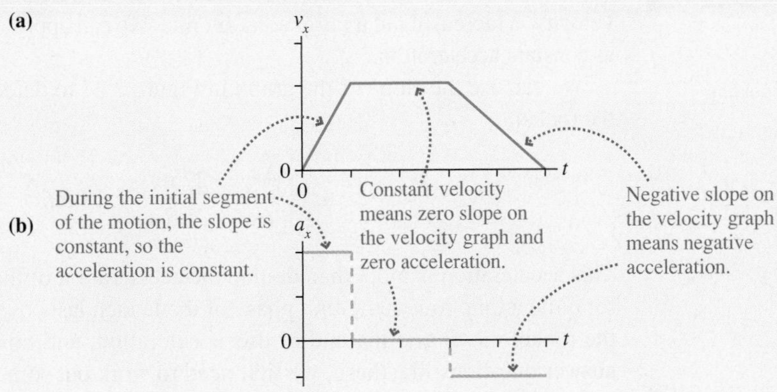

FIGURE 2.25 Determining the sign of the acceleration.

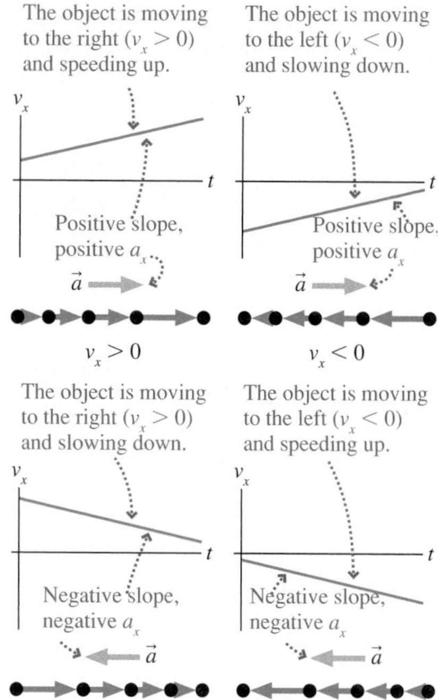

The Sign of the Acceleration

It's a natural tendency to think that a positive value of a_x or a_y describes an object that is speeding up while a negative value describes an object that is slowing down (decelerating). Unfortunately, this simple interpretation *does not work.*

Because an object can move right or left (or, equivalently, up and down) while either speeding up or slowing down, there are four situations to consider. FIGURE 2.25 shows a motion diagram and a velocity graph for each of these situations. As we've seen, an object's acceleration is the slope of its velocity graph, so a positive slope implies a positive acceleration and a negative slope implies a negative acceleration.

Acceleration, like velocity, is really a vector quantity, a concept that we will explore more fully in Chapter 3. Figure 2.25 shows the acceleration vectors for the four situations. The acceleration vector points in the same direction as the velocity vector $\vec{v}$ for an object that is speeding up and opposite to $\vec{v}$ for an object that is slowing down.

An object that speeds up as it moves to the right (positive v_x) has a positive acceleration, but an object that speeds up as it moves to the left (negative v_x) has a negative acceleration. Whether or not an object that is slowing down has a negative acceleration depends on whether the object is moving to the right or to the left. This is admittedly a bit more complex than thinking that negative acceleration always means slowing down, but our definition of acceleration as the slope of the velocity graph forces us to pay careful attention to the sign of the acceleration.

STOP TO THINK 2.4 A particle moves with the velocity-versus-time graph shown here. At which labeled point is the magnitude of the acceleration the greatest?

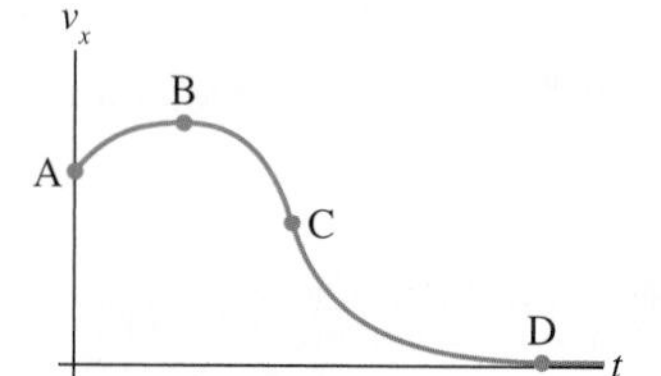

FIGURE 2.26 The red dots show the positions of the top of the Saturn V rocket at equally spaced intervals of time during liftoff.

2.5 Motion with Constant Acceleration

For uniform motion—motion with constant velocity—we found in Equation 2.3 a simple relationship between position and time. It's no surprise that there are also simple relationships that connect the various kinematic variables in constant-acceleration motion. We will start with a concrete example, the launch of a Saturn V rocket like the one that carried the Apollo astronauts to the moon in the 1960s and 1970s. FIGURE 2.26 shows one frame from a video of a rocket lifting off the launch pad. The red dots show the positions of the top of the rocket at equally spaced intervals of time in earlier frames of the video. This is a motion diagram for the rocket, and we can see that the velocity is increasing. The graph of velocity versus time in FIGURE 2.27 shows that the velocity is increasing at a fairly constant rate. We can approximate the rocket's motion as constant acceleration.

We can use the slope of the graph in Figure 2.27 to determine the acceleration of the rocket:

$$a_y = \frac{\Delta v_y}{\Delta t} = \frac{27 \text{ m/s}}{1.5 \text{ s}} = 18 \text{ m/s}^2$$

FIGURE 2.27 A graph of the rocket's velocity versus time.

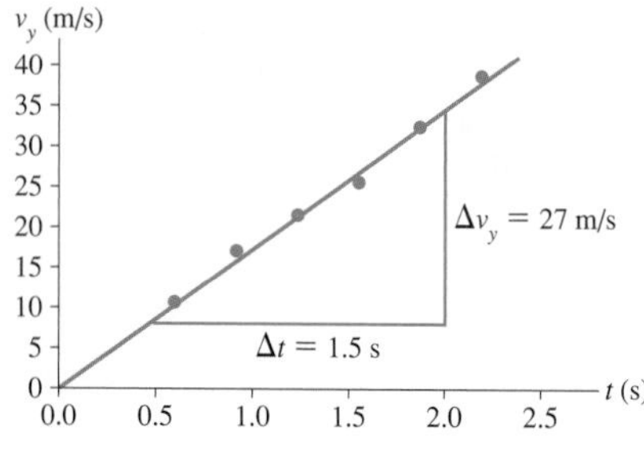

This acceleration is more than double the acceleration of the Porsche, and it goes on for quite a long time—the first phase of the launch lasts over 2 minutes! How fast is the rocket moving at the end of this acceleration, and how far has it traveled? To answer questions like these, we first need to work out some basic kinematic formulas for motion with constant acceleration.

▶ **Solar sailing** A rocket achieves a high speed by having a very high acceleration. A different approach is represented by a solar sail. A spacecraft with a solar sail accelerates due to the pressure of sunlight from the sun on a large, mirrored surface. The acceleration is minuscule, but it can continue for a long, long time. After an acceleration period of a few *years,* the spacecraft will reach a respectable speed!

Constant-Acceleration Equations

Consider an object whose acceleration a_x remains constant during the time interval $\Delta t = t_f - t_i$. At the beginning of this interval, the object has initial velocity $(v_x)_i$ and initial position x_i. Note that t_i is often zero, but it need not be. FIGURE 2.28a shows the acceleration-versus-time graph. It is a horizontal line between t_i and t_f, indicating a *constant* acceleration.

The object's velocity is changing because the object is accelerating. We can use the acceleration to find $(v_x)_f$ at a later time t_f. We defined acceleration as

$$a_x = \frac{\Delta v_x}{\Delta t} = \frac{(v_x)_f - (v_x)_i}{\Delta t} \tag{2.10}$$

which is rearranged to give

$$(v_x)_f = (v_x)_i + a_x \Delta t \tag{2.11}$$

Velocity equation for an object with constant acceleration

FIGURE 2.28 Acceleration and velocity graphs for motion with constant acceleration.

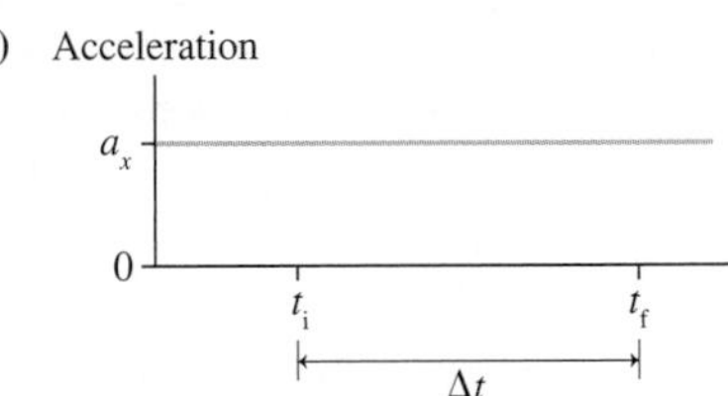

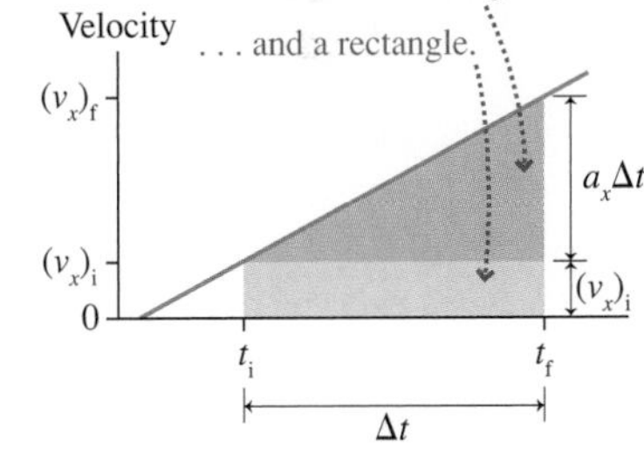

NOTE ▶ We have expressed this equation for motion along the *x*-axis, but it is a general result that will apply to any axis. ◀

The velocity-versus-time graph for this constant-acceleration motion, shown in FIGURE 2.28b, is a straight line with value $(v_x)_i$ at time t_i and with slope a_x.

We would also like to know the object's position x_f at time t_f. As you learned earlier, the displacement Δx during a time interval Δt is the area under the velocity-versus-time graph. The shaded area in Figure 2.28b can be subdivided into a rectangle of area $(v_x)_i \Delta t$ and a triangle of area $\frac{1}{2}(a_x \Delta t)(\Delta t) = \frac{1}{2}a_x(\Delta t)^2$. Adding these gives

$$x_f = x_i + (v_x)_i \Delta t + \tfrac{1}{2}a_x(\Delta t)^2 \tag{2.12}$$

Position equation for an object with constant acceleration

where $\Delta t = t_f - t_i$ is the elapsed time. The fact that the time interval Δt appears in the equation as $(\Delta t)^2$ causes the position-versus-time graph for constant-acceleration motion to have a parabolic shape. For the rocket launch of Figure 2.26, a graph of the position of the top of the rocket versus time appears as in FIGURE 2.29.

FIGURE 2.29 Position-versus-time graph for the Saturn V rocket launch.

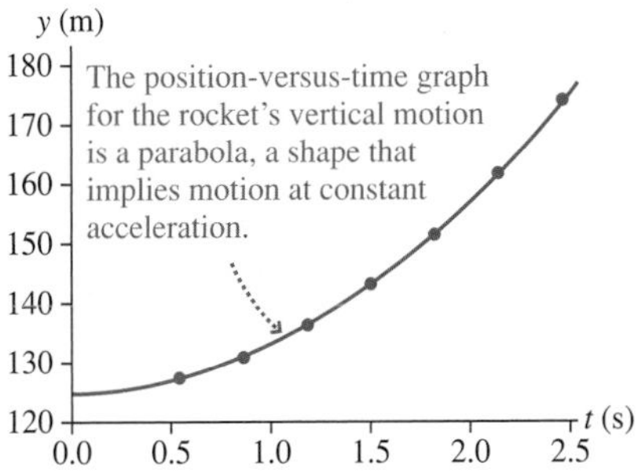

Equations 2.11 and 2.12 are two of the basic kinematic equations for motion with constant acceleration. They allow us to predict an object's position and velocity at a future instant of time. We need one more equation to complete our set, a direct relationship between displacement and velocity. To derive this relationship, we first use Equation 2.11 to write $\Delta t = ((v_x)_f - (v_x)_i)/a_x$. We can substitute this into Equation 2.12 to obtain

$$(v_x)_f^2 = (v_x)_i^2 + 2a_x \Delta x \tag{2.13}$$

Relating velocity and displacement for constant-acceleration motion

In Equation 2.13 $\Delta x = x_f - x_i$ is the *displacement* (not the distance!). Notice that Equation 2.13 does not require knowing the time interval Δt. This is an important equation in problems where you're not given information about times. Equations 2.11, 2.12, and 2.13 are the key equations for motion with constant acceleration. These results are summarized in Table 2.4.

TABLE 2.4 Kinematic equations for motion with constant acceleration

$(v_x)_f = (v_x)_i + a_x \Delta t$
$x_f = x_i + (v_x)_i \Delta t + \frac{1}{2}a_x(\Delta t)^2$
$(v_x)_f^2 = (v_x)_i^2 + 2a_x \Delta x$

EXAMPLE 2.7 Coming to a stop

As you drive in your car at 15 m/s (just a bit under 35 mph), you see a child's ball roll into the street ahead of you. You hit the brakes and stop as quickly as you can. In this case, you come to rest in 1.5 s. How far does your car travel as you brake to a stop?

PREPARE The problem statement gives us a description of motion in words. To help us visualize the situation, FIGURE 2.30 illustrates the key features of the motion with a motion diagram and a velocity graph. The graph is based on the car slowing from 15 m/s to 0 m/s in 1.5 s.

FIGURE 2.30 Motion diagram and velocity graph for a car coming to a stop.

We'll assume that the car moves to the right.

The velocity vectors get shorter, so the acceleration vector points to the left.

$\vec{a}$

$\vec{v}$

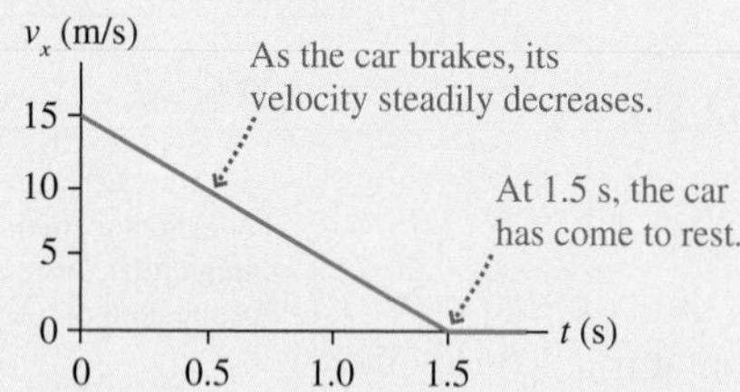

SOLVE We've assumed that your car is moving to the right, so its initial velocity is $(v_x)_i = +15$ m/s. After you come to rest, your final velocity is $(v_x)_f = 0$ m/s. The acceleration is given by Equation 2.10:

$$a_x = \frac{\Delta v_x}{\Delta t} = \frac{(v_x)_f - (v_x)_i}{\Delta t} = \frac{0\text{ m/s} - 15\text{ m/s}}{1.5\text{ s}} = -10\text{ m/s}^2$$

An acceleration of -10 m/s^2 (really -10 m/s per second) means the car slows by 10 m/s every second.

Now that we know the acceleration, we can compute the distance that the car moves as it comes to rest using Equation 2.12:

$$x_f - x_i = (v_x)_i\Delta t + \tfrac{1}{2}a_x(\Delta t)^2$$
$$= (15\text{ m/s})(1.5\text{ s}) + \tfrac{1}{2}(-10\text{ m/s}^2)(1.5\text{ s})^2 = 11\text{ m}$$

ASSESS 11 m is a little over 35 feet. That's a reasonable distance for a quick stop while traveling at about 35 mph. The purpose of the Assess step is not to prove that your solution is correct but to use common sense to recognize answers that are clearly wrong. Had you made a calculation error and ended up with an answer of 1.1 m—less than 4 feet—a moment's reflection should indicate that this couldn't possibly be correct.

1.1, 1.2, 1.3 ActivPhysics Online

Graphs will be an important component of our problem solutions, so we want to consider the types of graphs we are likely to encounter in more detail. FIGURE 2.31 is a graphical comparison of motion with constant velocity (uniform motion) and motion with constant acceleration. Notice that uniform motion is really a special case of constant-acceleration motion in which the acceleration happens to be zero.

FIGURE 2.31 Motion with constant velocity and constant acceleration. These graphs assume $x_i = 0$, $(v_x)_i > 0$, and (for constant acceleration) $a_x > 0$.

(a) Motion at constant velocity

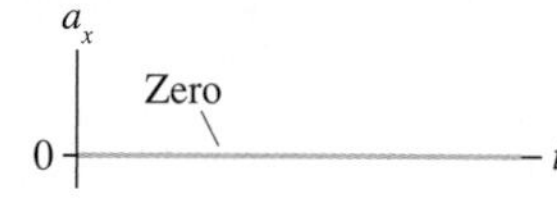

(b) Motion at constant acceleration

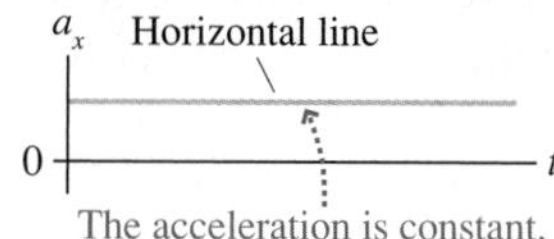

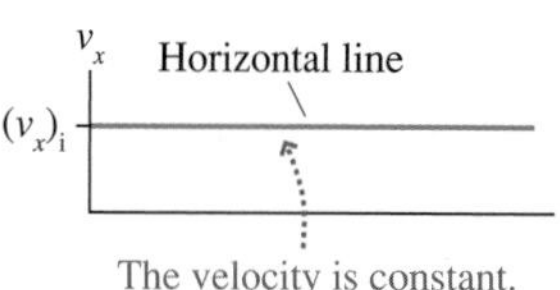

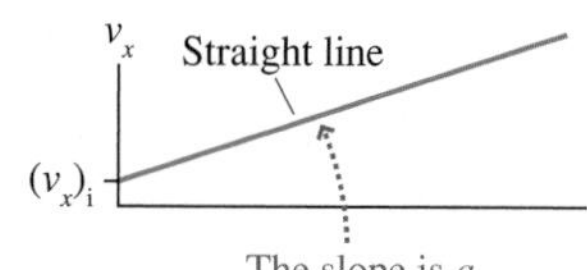

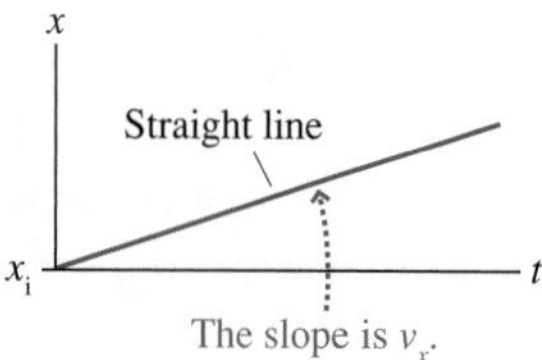

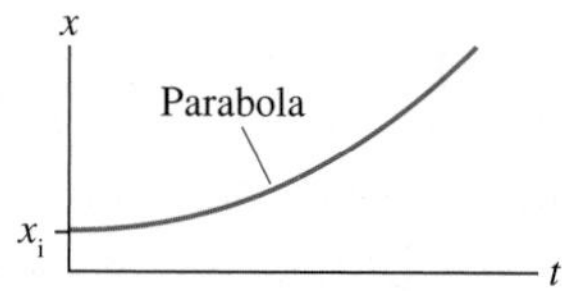

For motion at constant acceleration, a graph of position versus time is a *parabola*. This is a new mathematical form, one that we will see again. If $(v_x)_i = 0$, the second equation in Table 2.4 is simply

$$\Delta x = \tfrac{1}{2}a_x(\Delta t)^2 \tag{2.14}$$

Δx depends on the *square* of Δt; we call this a *quadratic relationship*.

Getting up to speed BIO A bird must have a minimum speed to fly. Generally, the larger the bird, the faster the takeoff speed. Small birds can get moving fast enough to fly with a vigorous jump, but larger birds may need a running start. This swan must accelerate for a long distance in order to achieve the high speed it needs to fly, so it makes a frenzied dash across the frozen surface of a pond. Swans require a long, clear stretch of water or land to become airborne. Airplanes require an even faster takeoff speed and thus an even longer runway, as we will see.

Quadratic relationships

MP™

Two quantities are said to have a **quadratic relationship** if y is proportional to the square of x. We write the mathematical relationship as

$$y = Ax^2$$

y is proportional to x^2

The graph of a quadratic relationship is a parabola.

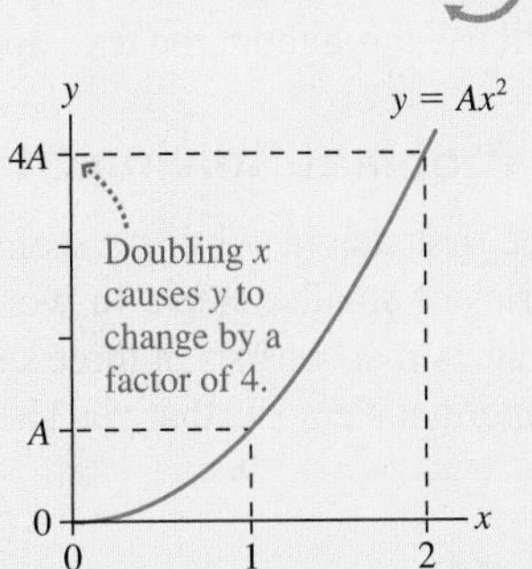

SCALING If x has the initial value x_1, then y has the initial value $y_1 = A(x_1)^2$. Changing x from x_1 to x_2 changes y from y_1 to y_2. The ratio of y_2 to y_1 is

$$\frac{y_2}{y_1} = \frac{A(x_2)^2}{A(x_1)^2} = \left(\frac{x_2}{x_1}\right)^2$$

The ratio of y_2 to y_1 is the square of the ratio of x_2 to x_1. If y is a quadratic function of x, a change in x by some factor changes y by the square of that factor:

- If you increase x by a factor of 2, you increase y by a factor of $2^2 = 4$.
- If you decrease x by a factor of 3, you decrease y by a factor of $3^2 = 9$.

Generally, we can say that:

Changing x by a factor of c changes y by a factor of c^2.

Exercise 19

EXAMPLE 2.8 Displacement of a drag racer

A drag racer, starting from rest, travels 6.0 m in 1.0 s. Suppose the car continues this acceleration for an additional 4.0 s. How far from the starting line will the car be?

PREPARE We assume that the acceleration is constant, so the displacement will follow Equation 2.14. This is a *quadratic relationship,* so the displacement will scale as the square of the time.

SOLVE After 1.0 s, the car has traveled 6.0 m; after another 4.0 s, a total of 5.0 s will have elapsed. The time has increased by a factor of 5, so the displacement will increase by a factor of 5^2, or 25. The total displacement is

$$\Delta x = 25(6.0 \text{ m}) = 150 \text{ m}$$

ASSESS This is a big distance in a short time, but drag racing is a fast sport, so our answer makes sense.

STOP TO THINK 2.5 A cyclist is at rest at a traffic light. When the light turns green, he begins accelerating at 1.2 m/s². How many seconds after the light turns green does he reach his cruising speed of 6.0 m/s?

A. 1.0 s B. 2.0 s C. 3.0 s D. 4.0 s E. 5.0 s

Dinner at a distance BIO A chameleon's tongue is a powerful tool for catching prey. Certain species can extend the tongue to a distance of over 1 ft in less than 0.1 s! A study of the kinematics of the motion of the chameleon tongue, using techniques like those in this chapter, reveals that the tongue has a period of rapid acceleration followed by a period of constant velocity. This knowledge is a very valuable clue in the analysis of the evolutionary relationships between chameleons and other animals.

2.6 Solving One-Dimensional Motion Problems

The big challenge when solving a physics problem is to translate the words into symbols that can be manipulated, calculated, and graphed. This translation from words to symbols is the heart of problem solving in physics. Ambiguous words and phrases must be clarified, the imprecise must be made precise, and you must arrive at an understanding of exactly what the question is asking.

In this section we will explore some general problem-solving strategies that we will use throughout the text, applying them to problems of motion along a line.

A Problem-Solving Strategy

The first step in solving a seemingly complicated problem is to break it down into a series of smaller steps. In worked examples in the text, we use a problem-solving strategy that consists of three steps: *prepare, solve,* and *assess.* Each of these steps has important elements that you should follow when you solve problems on your own.

Building a complex structure requires careful planning. The architect's visualization and drawings have to be complete before the detailed procedures of construction get under way. The same is true for solving problems in physics.

Problem-Solving Strategy

PREPARE The Prepare step of a solution is where you identify important elements of the problem and collect information you will need to solve it. It's tempting to jump right to the Solve step, but a skilled problem solver will spend the most time on this step, the preparation. Preparation includes:

- **Drawing a picture.** In many cases, this is the most important part of a problem. The picture lets you model the problem and identify the important elements. As you add information to your picture, the outline of the solution will take shape. For the problems in this chapter, a picture could be a motion diagram or a graph—or perhaps both.
- **Collecting necessary information.** The problem's statement may give you some values of variables. Other important information may be implied or must be looked up in a table. Gather everything you need to solve the problem and compile it in a list.
- **Doing preliminary calculations.** There are a few calculations, such as unit conversions, that are best done in advance of the main part of the solution.

SOLVE The Solve step of a solution is where you actually do the mathematics or reasoning necessary to arrive at the answer needed. This is the part of the problem-solving strategy that you likely think of when you think of "solving problems." But don't make the mistake of starting here! The Prepare step will help you be certain you understand the problem before you start putting numbers in equations.

ASSESS The Assess step of your solution is very important. When you have an answer, you should check to see whether it makes sense. Ask yourself:

- **Does my solution answer the question that was asked?** Make sure you have addressed all parts of the question and clearly written down your solutions.
- **Does my answer have the correct units and number of significant figures?**
- **Does the value I computed make physical sense?** In this book all calculations use physically reasonable numbers. You will not be given a problem to solve in which the final velocity of a bicycle is 100 miles per hour! If your answer seems unreasonable, go back and check your work.
- **Can I estimate what the answer should be to check my solution?**
- **Does my final solution make sense in the context of the material I am learning?**

The Pictorial Representation

Many physics problems, including 1-D motion problems, often have several variables and other pieces of information to keep track of. The best way to tackle such problems is to draw a picture, as we noted when we introduced a general problem-solving strategy. But what kind of picture should you draw?

In this section, we will begin to draw **pictorial representations** as an aid to solving problems. A pictorial representation shows all of the important details that we need to keep track of and will be very important in solving motion problems.

TACTICS BOX 2.2 Drawing a pictorial representation

❶ **Sketch the situation.** Not just any sketch: Show the object at the *beginning* of the motion, at the *end*, and at any point where the character of the motion changes. Very simple drawings are adequate.

❷ **Establish a coordinate system.** Select your axes and origin to match the motion.

❸ **Define symbols.** Use the sketch to define symbols representing quantities such as position, velocity, acceleration, and time. *Every* variable used later in the mathematical solution should be defined on the sketch.

We will generally combine the pictorial representation with a **list of values,** which will include:

- *Known information.* Make a table of the quantities whose values you can determine from the problem statement or that you can find quickly with simple geometry or unit conversions.
- *Desired unknowns.* What quantity or quantities will allow you to answer the question?

Exercise 21 

EXAMPLE 2.9 Drawing a pictorial representation

Complete a pictorial representation and a list of values for the following problem: A rocket sled accelerates at 50 m/s^2 for 5 s. What are the total distance traveled and the final velocity?

PREPARE FIGURE 2.32a shows a pictorial representation as drawn by an artist in the style of the figures in this book. This is certainly neater and more artistic than the sketches you will make when solving problems yourself! FIGURE 2.32b shows a sketch like one you might actually do. It's less formal, but it contains all of the important information you need to solve the problem.

FIGURE 2.32 Constructing a pictorial representation and a list of values.

(a) Artist's version
Pictorial representation

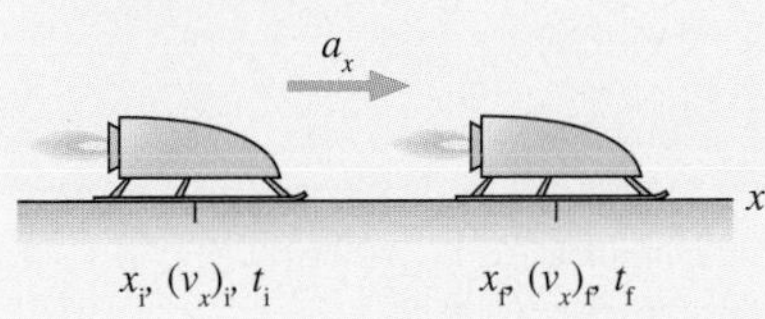

List of values

Known
$x_i = 0$ m
$(v_x)_i = 0$ m/s
$t_i = 0$ s
$a_x = 50$ m/s^2
$t_f = 5$ s

Find
x_f, $(v_x)_f$

(b) Student sketch

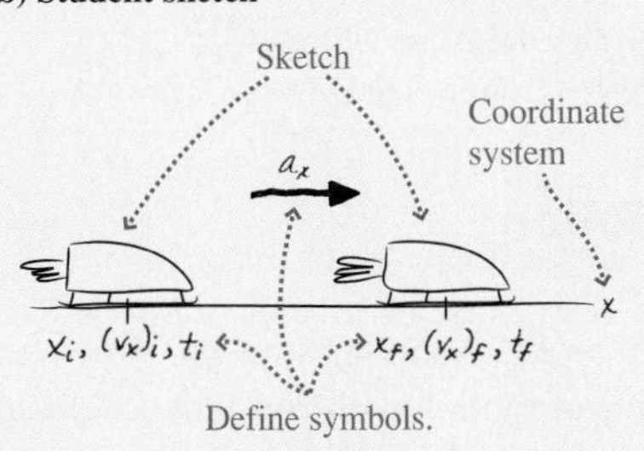

Known	Find
$x_i = 0$ m	x_f
$(v_x)_i = 0$ m/s	$(v_x)_f$
$t_i = 0$ s	
$a_x = 50$ m/s^2	
$t_f = 5$ s	

List known information.

Identify desired unknowns.

NOTE ► Throughout this book we will illustrate select examples with actual hand-drawn figures so that you have them to refer to as you work on your own pictures for homework and practice. ◄

Let's look at how these pictures were constructed. The motion has a clear beginning and end; these are the points sketched. A coordinate system has been chosen with the origin at the starting point. The quantities x, v_x, and t are needed at both points, so these have been defined on the sketch and distinguished by subscripts. The acceleration is associated with an interval between these points. Values for two of these quantities are given in the problem statement. Others, such as $x_i = 0$ m and $t_i = 0$ s, are inferred from our choice of coordinate system. The value $(v_x)_i = 0$ m/s is part of our *interpretation* of the problem. Finally, we identify x_f and $(v_x)_f$ as the quantities that will answer the question. We now understand quite a bit about the problem and would be ready to start a quantitative analysis.

ASSESS We didn't *solve* the problem; that was not our purpose. Constructing a pictorial representation and a list of values is part of a systematic approach to interpreting a problem and getting ready for a mathematical solution.

The Visual Overview

The pictorial representation and the list of values are a very good complement to the motion diagram and other ways of looking at a problem that we have seen. As we translate a problem into a form we can solve, we will combine these elements into what we will term a **visual overview.** The visual overview will consist of some or all of the following elements:

- A *motion diagram.* A good strategy for solving a motion problem is to start by drawing a motion diagram.
- A *pictorial representation*, as defined above.
- A *graphical representation.* For motion problems, it is often quite useful to include a graph of position and/or velocity.
- A *list of values.* This list should sum up all of the important values in the problem.

Future chapters will add other elements to this visual overview of the physics.

EXAMPLE 2.10 Kinematics of a rocket launch

A Saturn V rocket is launched straight up with a constant acceleration of 18 m/s^2. After 150 s, how fast is the rocket moving and how far has it traveled?

PREPARE FIGURE 2.33 shows a visual overview of the rocket launch that includes a motion diagram, a pictorial representation, and a list of values. The visual overview shows the whole problem in a nutshell. The motion diagram illustrates the motion of the rocket. The pictorial representation (produced according to Tactics Box 2.2) shows axes, identifies the important points of the motion, and defines variables. Finally, we have included a list of values that gives the known and unknown quantities. In the visual overview we have taken the statement of the problem in words and made it much more precise; it contains everything you need to know about the problem.

SOLVE Our first task is to find the final velocity. Our list of values includes the initial velocity, the acceleration, and the time interval, so we can use the first kinematic equation of Table 2.4 to find the final velocity:

$$(v_y)_f = (v_y)_i + a_y\,\Delta t = 0\text{ m/s} + (18\text{ m/s}^2)(150\text{ s}) = 2700\text{ m/s}$$

The distance traveled is found using the second equation in Table 2.4:

$$y_f = y_i + (v_y)_i\,\Delta t + \tfrac{1}{2}a_y(\Delta t)^2 = 0\text{ m} + (0\text{ m/s})(150\text{ s}) + \tfrac{1}{2}(18\text{ m/s}^2)(150\text{ s})^2 = 2.0 \times 10^5\text{ m} = 200\text{ km}$$

ASSESS The acceleration is very large, and it goes on for a long time, so the large final velocity and large distance traveled seem reasonable.

FIGURE 2.33 Visual overview of the rocket launch.

Motion diagram

The motion diagram for the rocket shows the full range of the motion.

Pictorial representation

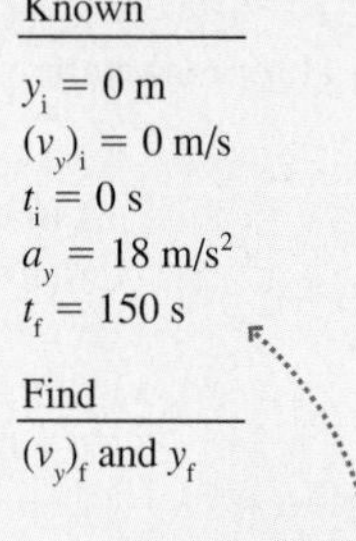

The pictorial representation identifies the two important points of the motion, the start and the end, and shows that the rocket accelerates between them.

List of values

Known
$y_i = 0$ m
$(v_y)_i = 0$ m/s
$t_i = 0$ s
$a_y = 18$ m/s^2
$t_f = 150$ s

Find
$(v_y)_f$ and y_f

The list of values makes everything concrete. We define the start of the problem to be at time 0 s, when the rocket has a position of 0 m and a velocity of 0 m/s. The end of the problem is at time 150 s. We are to find the position and velocity at this time.

Problem-Solving Strategy for Motion with Constant Acceleration

1.4–1.6, 1.8, 1.9, 1.11–1.14 ActivPhysics Online

Earlier in this section, we introduced a general problem-solving strategy. Now we will adapt this general strategy to solving problems of motion with constant acceleration. We will introduce such specific problem-solving strategies in future chapters as well.

PROBLEM-SOLVING STRATEGY 2.1 **Motion with constant acceleration**

PREPARE Draw a visual overview of the problem. This should include a motion diagram, a pictorial representation, and a list of values; a graphical representation may be useful for certain problems.

SOLVE The mathematical solution is based on the three equations in Table 2.4.

- Though the equations are phrased in terms of the variable x, it's customary to use y for motion in the vertical direction.
- Use the equation that best matches what you know and what you need to find. For example, if you know acceleration and time and are looking for a change in velocity, the first equation is the best one to use.
- Uniform motion with constant velocity has $a = 0$.

ASSESS Is your result believable? Does it have proper units? Does it make sense?

Exercise 25 

EXAMPLE 2.11 Calculating the minimum length of a runway

A fully loaded Boeing 747 with all engines at full thrust accelerates at 2.6 m/s^2. Its minimum takeoff speed is 70 m/s. How much time will the plane take to reach its takeoff speed? What minimum length of runway does the plane require for takeoff?

PREPARE The visual overview of FIGURE 2.34 summarizes the important details of the problem. We set x_i and t_i equal to zero at the starting point of the motion, when the plane is at rest and the acceleration begins. The final point of the motion is when the plane achieves the necessary takeoff speed of 70 m/s. The plane is accelerating to the right, so we will compute the time for the plane to reach a velocity of 70 m/s and the position of the plane at this time, giving us the minimum length of the runway.

FIGURE 2.34 Visual overview for an accelerating plane.

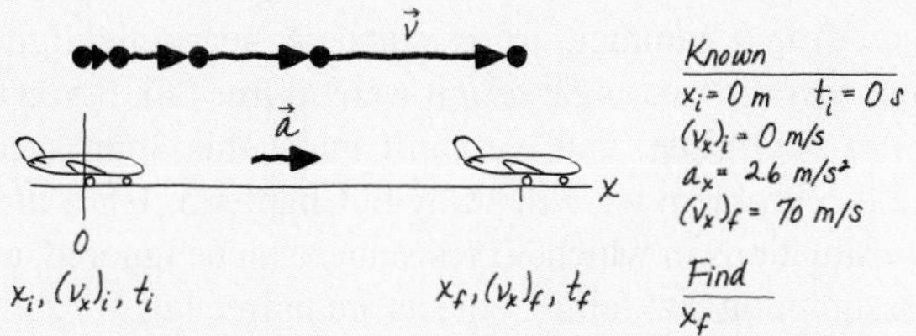

SOLVE First we solve for the time required for the plane to reach takeoff speed. We can use the first equation in Table 2.4 to compute this time:

$$(v_x)_f = (v_x)_i + a_x\,\Delta t$$

$$70\text{ m/s} = 0\text{ m/s} + (2.6\text{ m/s}^2)\,\Delta t$$

$$\Delta t = \frac{70\text{ m/s}}{2.6\text{ m/s}^2} = 26.9\text{ s}$$

We keep an extra significant figure here because we will use this result in the next step of the calculation.

Given the time that the plane takes to reach takeoff speed, we can compute the position of the plane when it reaches this speed using the second equation in Table 2.4:

$$x_f = x_i + (v_x)_i\,\Delta t + \tfrac{1}{2}a_x(\Delta t)^2$$
$$= 0\text{ m} + (0\text{ m/s})(26.9\text{ s}) + \tfrac{1}{2}(2.6\text{ m/s}^2)(26.9\text{ s})^2$$
$$= 940\text{ m}$$

Our final answers are thus that the plane will take 27 s to reach takeoff speed, with a minimum runway length of 940 m.

ASSESS Think about the last time you flew; 27 s seems like a reasonable time for a plane to accelerate on takeoff. Actual runway lengths at major airports are 3000 m or more, a few times greater than the minimum length, because they have to allow for emergency stops during an aborted takeoff. (If we had calculated a distance far greater than 3000 m, we would know we had done something wrong!)

EXAMPLE 2.12 Finding the braking distance

A car is traveling at a speed of 30 m/s, a typical highway speed, on wet pavement. The driver sees an obstacle ahead and decides to stop. From this instant, it takes him 0.75 s to begin applying the brakes. Once the brakes are applied, the car experiences an acceleration of -6.0 m/s^2. How far does the car travel from the instant the driver notices the obstacle until stopping?

PREPARE This problem is more involved than previous problems we have solved, so we will take more care with the visual overview in FIGURE 2.35. In addition to a motion diagram and a pictorial representation, we include a graphical representation. Notice that there are two different phases of the motion: a constant-velocity phase before braking begins, and a steady slowing

Continued

FIGURE 2.35 Visual overview for a car braking to a stop.

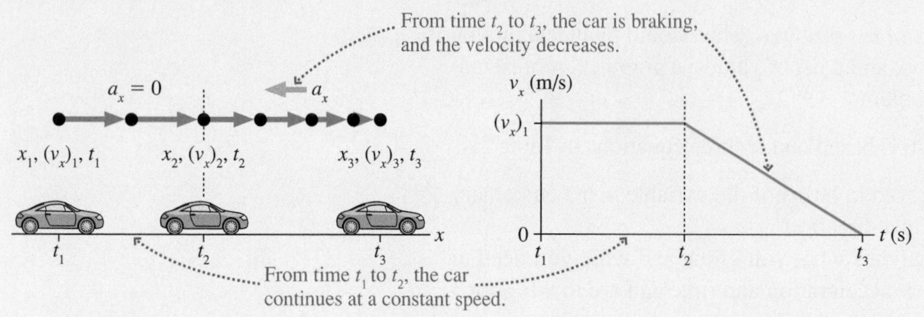

down once the brakes are applied. We will need to do two different calculations, one for each phase. Consequently, we've used numerical subscripts rather than a simple i and f.

SOLVE From t_1 to t_2 the velocity stays constant at 30 m/s. This is uniform motion, so the position at time t_2 is computed using Equation 2.4:

$$x_2 = x_1 + (v_x)_1(t_2 - t_1) = 0\text{ m} + (30\text{ m/s})(0.75\text{ s})$$
$$= 22.5\text{ m}$$

At t_2, the velocity begins to decrease at a steady -6.0 m/s^2 until the car comes to rest at t_3. This time interval can be computed using the first equation in Table 2.4, $(v_x)_3 = (v_x)_2 + a_x\,\Delta t$:

$$\Delta t = t_3 - t_2 = \frac{(v_x)_3 - (v_x)_2}{a_x} = \frac{0\text{ m/s} - 30\text{ m/s}}{-6.0\text{ m/s}^2} = 5.0\text{ s}$$

The position at time t_3 is computed using the second equation in Table 2.4; we take point 2 as the initial point and point 3 as the final point for this phase of the motion and use $\Delta t = t_3 - t_2$:

$$x_3 = x_2 + (v_x)_2\,\Delta t + \tfrac{1}{2}a_x(\Delta t)^2$$
$$= 22.5\text{ m} + (30\text{ m/s})(5.0\text{ s}) + \tfrac{1}{2}(-6.0\text{ m/s}^2)(5.0\text{ s})^2$$
$$= 98\text{ m}$$

x_3 is the position of the car at the end of the problem—and so the car travels 98 m before coming to rest.

ASSESS The numbers for the reaction time and the acceleration on wet pavement are reasonable ones for an alert driver in a car with good tires. The final distance is quite large—it is more than the length of a football field.

1.7, 1.10 ActivPhysics ONLINE

2.7 Free Fall

"Looks like Mr. Galileo was correct . . ." was the comment made by Apollo 15 astronaut David Scott, who dropped a hammer and a feather on the moon. The objects were dropped from the same height at the same time and hit the ground simultaneously—something that would not happen in the atmosphere of the earth!

A particularly important example of constant acceleration is the motion of an object moving under the influence of gravity only, and no other forces. This motion is called **free fall.** Strictly speaking, free fall occurs only in a vacuum, where there is no air resistance. But if you drop a hammer, air resistance is nearly negligible, so we'll make only a very slight error in treating it *as if* it were in free fall. If you drop a feather, air resistance is *not* negligible, and we can't make this approximation. Motion with air resistance is a problem we will study in Chapter 5. Until then, we will restrict our attention to situations in which air resistance can be ignored, and we will make the reasonable assumption that falling objects are in free fall.

As part of his early studies of motion, Galileo did the first careful experiments on free fall and made the surprising observation that two objects of different weight dropped from the same height will, if air resistance can be neglected, hit the ground at the same time and with the same speed. In fact—as Galileo surmised, and as a famous demonstration on the moon showed—in a vacuum, where there is no air resistance, this holds true for *any* two objects.

Galileo's discovery about free fall means that **any two objects in free fall, regardless of their mass, have the same acceleration.** This is an especially important conclusion. FIGURE 2.36a shows the motion diagram of an object that was released from rest and falls freely. The motion diagram and graph would be identical for a falling baseball or a falling boulder! FIGURE 2.36b shows the object's velocity graph. As we can see, the velocity changes at a steady rate. The slope of the velocity-versus-time graph is the free-fall acceleration $a_{\text{free fall}}$.

Instead of dropping the object, suppose we throw it upward. What happens then? You know that the object will move up and that its speed will decrease as it rises.

FIGURE 2.36 Motion of an object in free fall.

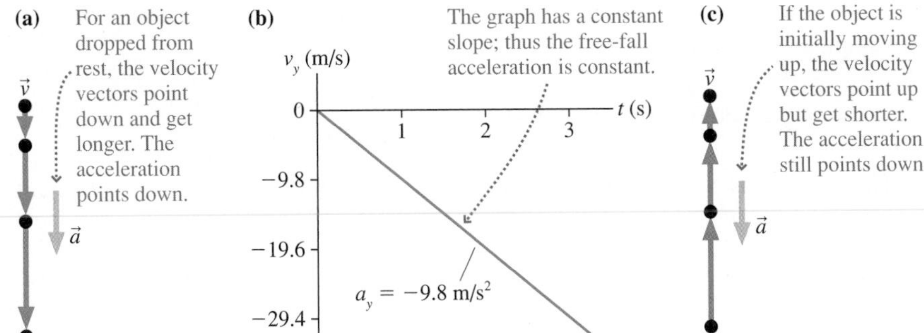

(a)
For an object dropped from rest, the velocity vectors point down and get longer. The acceleration points down.
(b)
v_y (m/s)
The graph has a constant slope; thus the free-fall acceleration is constant.
t (s)
0
−9.8
−19.6
−29.4
1
2
3
$a_y = -9.8$ m/s²
(c)
If the object is initially moving up, the velocity vectors point up but get shorter. The acceleration still points down.

This is illustrated in the motion diagram of FIGURE 2.36c, which shows a surprising result: Even though the object is moving up, its acceleration still points down. In fact, **the free-fall acceleration always points down,** no matter what direction an object is moving.

Some of the children are moving up and some are moving down, but all are in free fall—and so are accelerating downward at 9.8 m/s².

NOTE ▶ Despite the name, free fall is not restricted to objects that are literally falling. Any object moving under the influence of gravity only, and no other forces, is in free fall. This includes objects falling straight down, objects that have been tossed or shot straight up, objects in projectile motion (such as a passed football), and, as we will see, satellites in orbit. In this chapter we consider only objects that move up and down along a vertical line; projectile motion will be studied in Chapter 3. ◀

The free-fall acceleration is always in the same direction, and on earth, it always has approximately the same magnitude. Careful measurements show that the value of the free-fall acceleration varies slightly at different places on the earth, but for the calculations in this book we will use the the following average value:

$$\vec{a}_{\text{free fall}} = (9.80 \text{ m/s}^2, \text{vertically downward}) \quad (2.15)$$

Standard value for the acceleration of an object in free fall

The magnitude of the **free-fall acceleration** has the special symbol g:

$$g = 9.80 \text{ m/s}^2$$

We will generally work with two significant figures and so will use $g = 9.8 \text{ m/s}^2$.

Several points about free fall are worthy of note:

- g, by definition, is *always* positive. **There will never be a problem that uses a negative value for g.**
- The velocity graph in Figure 2.36b has a negative slope. Even though a falling object speeds up, it has *negative* acceleration. Alternatively, notice that the acceleration vector $\vec{a}_{\text{free fall}}$ points down. Thus g is *not* the object's acceleration, simply the magnitude of the acceleration. The one-dimensional acceleration is

$$a_y = a_{\text{free fall}} = -g$$

  It is a_y that is negative, not g.
- Because free fall is motion with constant acceleration, we can use the kinematic equations of Table 2.4 with the acceleration being due to gravity, $a_y = -g$.
- g is not called "gravity." Gravity is a force, not an acceleration. g is the *free-fall acceleration.*
- $g = 9.80 \text{ m/s}^2$ only on earth. Other planets have different values of g. You will learn in Chapter 6 how to determine g for other planets.

TRY IT YOURSELF

A reaction time challenge Hold a $1 (or larger!) bill by an upper corner. Have a friend prepare to grasp a lower corner, putting her fingers *near but not touching* the bill. Tell her to try to catch the bill when you drop it by simply closing her fingers without moving her hand downward—and that if she can catch it, she can keep it. Don't worry; the bill's free fall will keep your money safe. In the few tenths of a second that it takes your friend to react, free fall will take the bill beyond her grasp.

- We will sometimes compute acceleration in units of g. An acceleration of 9.8 m/s^2 is an acceleration of $1g$; an acceleration of 19.6 m/s^2 is $2g$. Generally, we can compute

$$\text{acceleration (in units of } g) = \frac{\text{acceleration (in units of m/s}^2)}{9.8 \text{ m/s}^2} \quad (2.16)$$

This allows us to express accelerations in units that have a definite physical reference.

EXAMPLE 2.13 Analyzing a rock's fall

A heavy rock is dropped from rest at the top of a cliff and falls 100 m before hitting the ground. How long does the rock take to fall to the ground, and what is its velocity when it hits?

PREPARE FIGURE 2.37 shows a visual overview with all necessary data. We have placed the origin at the ground, which makes $y_i = 100$ m.

FIGURE 2.37 Visual overview of a falling rock.

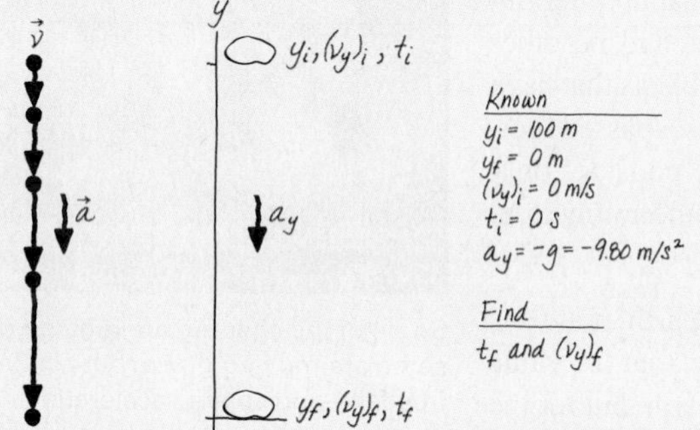

SOLVE Free fall is motion with the specific constant acceleration $a_y = -g$. The first question involves a relation between time and distance, a relation expressed by the second equation in Table 2.4. Using $(v_y)_i = 0$ m/s and $t_i = 0$ s, we find

$$y_f = y_i + (v_y)_i\,\Delta t + \tfrac{1}{2}a_y\,\Delta t^2 = y_i - \tfrac{1}{2}g\,\Delta t^2 = y_i - \tfrac{1}{2}gt_f^2$$

We can now solve for t_f:

$$t_f = \sqrt{\frac{2(y_i - y_f)}{g}} = \sqrt{\frac{2(100 \text{ m} - 0 \text{ m})}{9.80 \text{ m/s}^2}} = 4.52 \text{ s}$$

Now that we know the fall time, we can use the first kinematic equation to find $(v_y)_f$:

$$(v_y)_f = (v_y)_i - g\,\Delta t = -gt_f = -(9.80 \text{ m/s}^2)(4.52 \text{ s})$$
$$= -44.3 \text{ m/s}$$

ASSESS Are the answers reasonable? Well, 100 m is about 300 feet, which is about the height of a 30-floor building. How long does it take something to fall 30 floors? Four or five seconds seems pretty reasonable. How fast would it be going at the bottom? Using an approximate version of our conversion factor 1 m/s $\approx$ 2 mph, we find that 44.3 m/s $\approx$ 90 mph. That also seems like a pretty reasonable speed for something that has fallen 30 floors. Suppose we had made a mistake. If we misplaced a decimal point we could have calculated a speed of 443 m/s, or about 900 mph! This is clearly *not* reasonable. If we had misplaced the decimal point in the other direction, we would have calculated a speed of 4.3 m/s $\approx$ 9 mph. This is another unreasonable result, because this is slower than a typical bicycling speed.

CONCEPTUAL EXAMPLE 2.14 Analyzing the motion of a ball tossed upward

Draw a motion diagram and a velocity-versus-time graph for a ball tossed straight up in the air from the point that it leaves the hand until just before it is caught.

REASON You know what the motion of the ball looks like: The ball goes up, and then it comes back down again. This complicates the drawing of a motion diagram a bit, as the ball retraces its route as it falls. A literal motion diagram would show the upward motion and downward motion on top of each other, leading to confusion. We can avoid this difficulty by horizontally separating the upward motion and downward motion diagrams. This will not affect our conclusions because it does not change any of the vectors. The motion diagram and velocity-versus-time graph appear as in FIGURE 2.38 on the next page.

ASSESS The highest point in the ball's motion, where it reverses direction, is called a *turning point.* What are the velocity and the acceleration at this point? We can see from the motion diagram that the velocity vectors are pointing upward but getting shorter as the ball approaches the top. As it starts to fall, the velocity vectors are pointing downward and getting longer. There must be a moment—just an instant as $\vec{v}$ switches from pointing up to pointing down—when the velocity is zero. Indeed, the ball's velocity *is* zero for an instant at the precise top of the motion! We can also see on the velocity graph that there is one instant of time when $v_y = 0$. This is the turning point.

But what about the acceleration at the top? Many people expect the acceleration to be zero at the highest point. But recall that the velocity at the top point is changing—from up to down. If the velocity is changing, there *must* be an acceleration. The slope of the velocity graph at the instant when $v_y = 0$—that is, at the highest point—is no different than at any other point in the motion. The ball is still in free fall with acceleration $a_y = -g$!

Another way to think about this is to note that zero acceleration would mean no change of velocity. When the ball reached zero velocity at the top, it would hang there and not fall if the acceleration were also zero!

FIGURE 2.38 Motion diagram and velocity graph of a ball tossed straight up in the air.

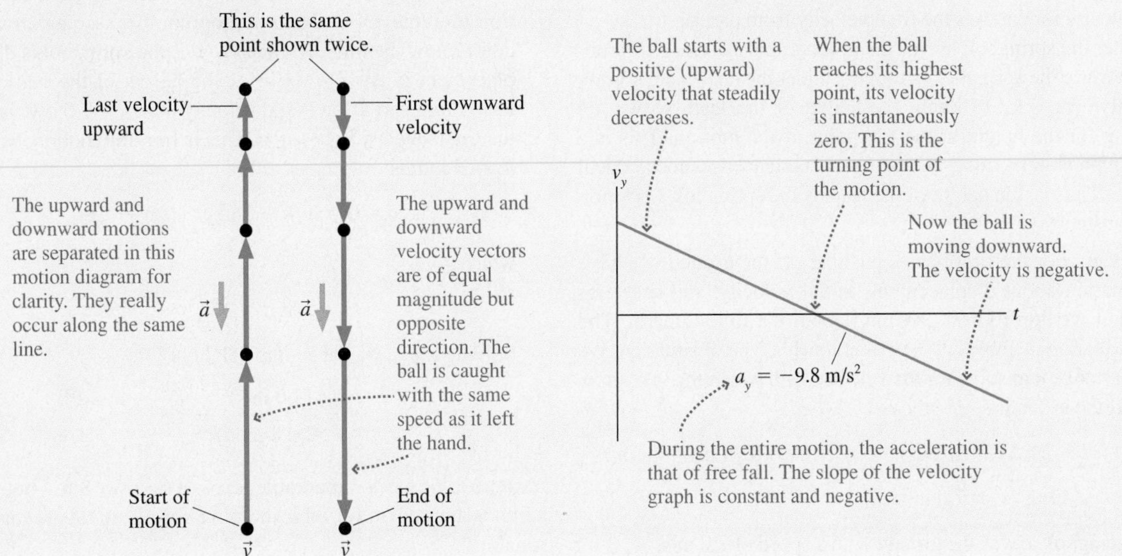

This is the same point shown twice.
Last velocity upward
First downward velocity
The upward and downward motions are separated in this motion diagram for clarity. They really occur along the same line.
The upward and downward velocity vectors are of equal magnitude but opposite direction. The ball is caught with the same speed as it left the hand.
Start of motion
End of motion
The ball starts with a positive (upward) velocity that steadily decreases.
When the ball reaches its highest point, its velocity is instantaneously zero. This is the turning point of the motion.
Now the ball is moving downward. The velocity is negative.
$a_y = -9.8$ m/s²
During the entire motion, the acceleration is that of free fall. The slope of the velocity graph is constant and negative.

EXAMPLE 2.15 Finding the height of a leap

A springbok is an antelope found in southern Africa that gets its name from its remarkable jumping ability. When a springbok is startled, it will leap straight up into the air—a maneuver called a "pronk." A springbok goes into a crouch to perform a pronk. It then extends its legs forcefully, accelerating at 35 m/s^2 for 0.70 m as its legs straighten. Legs fully extended, it leaves the ground and rises into the air.

a. At what speed does the springbok leave the ground?
b. How high does it go?

PREPARE We begin with the visual overview shown in **FIGURE 2.39**, where we've identified two different phases of the motion: the springbok pushing off the ground and the springbok rising into the air. We'll treat these as two separate problems that we solve in turn. We will "re-use" the variables y_i, y_f, $(v_y)_i$, and $(v_y)_f$ for the two phases of the motion.

For the first part of our solution, in Figure 2.39a we choose the origin of the y-axis at the position of the springbok deep in the crouch. The final position is the top extent of the push, at the instant the springbok leaves the ground. We want to find the velocity at this position because that's how fast the springbok is moving as it leaves the ground. Figure 2.39b essentially starts over—we have defined a new vertical axis with its origin at the ground, so the highest point of the springbok's motion is a

FIGURE 2.39 A visual overview of the springbok's leap.

(a) Pushing off the ground

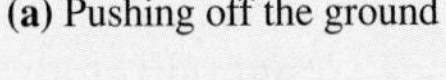

Known
$y_i = 0$ m
$y_f = 0.70$ m
$(v_y)_i = 0$ m/s
$a_y = 35$ m/s^2

Find
$(v_y)_f$

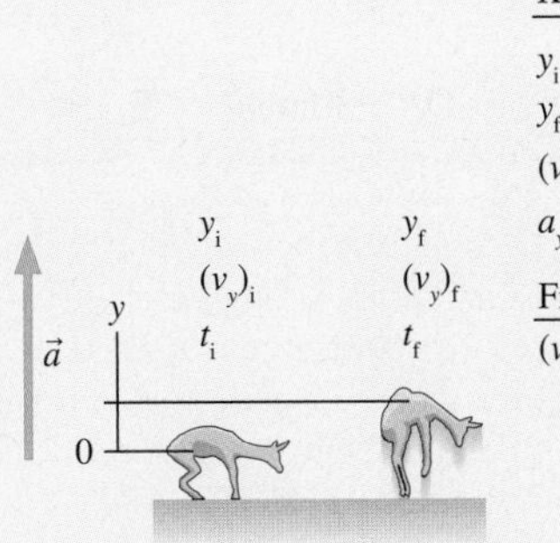

(b) Rising into the air

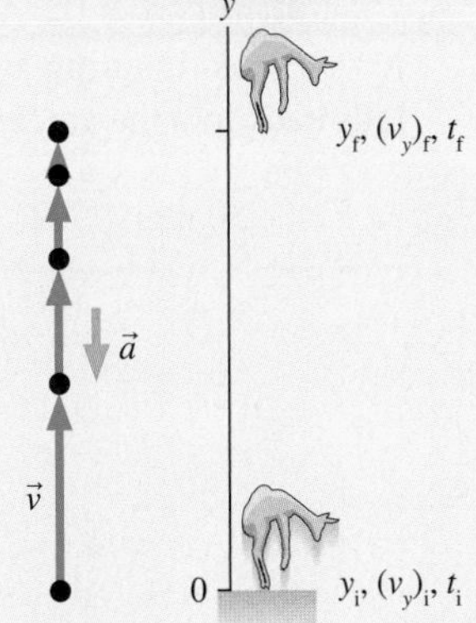

Known
$y_i = 0$ m
$(v_y)_i$ is equal to $(v_y)_f$ from part a
$(v_y)_f = 0$ m/s
$a_y = -9.8$ m/s^2

Find
y_f

Continued

distance above the ground. The table of values shows the key piece of information for this second part of the problem: The initial velocity for part b is the final velocity from part a.

After the springbok leaves the ground, this is a free-fall problem because the springbok is moving under the influence of gravity only. We want to know the height of the leap, so we are looking for the height at the top point of the motion. This is a turning point of the motion, with the instantaneous velocity equal to zero. Thus y_f, the height of the leap, is the springbok's position at the instant $(v_y)_f = 0$.

SOLVE a. For the first phase, pushing off the ground, we have information about displacement, initial velocity, and acceleration, but we don't know anything about the time interval. The third equation in Table 2.4 is perfect for this type of situation. We can rearrange it to solve for the velocity with which the springbok lifts off the ground:

$$(v_y)_f^2 = v_i^2 + 2a_y\Delta y = (0 \text{ m/s})^2 + 2(35 \text{ m/s}^2)(0.70 \text{ m}) = 49 \text{ m}^2/\text{s}^2$$

$$(v_y)_f = \sqrt{49 \text{ m}^2/\text{s}^2} = 7.0 \text{ m/s}$$

The springbok leaves the ground with a speed of 7.0 m/s.

b. Now we are ready for the second phase of the motion, the vertical motion after leaving the ground. The third equation in Table 2.4 is again appropriate because again we don't know the time. Because $y_i = 0$, the springbok's displacement is $\Delta y = y_f - y_i = y_f$, the height of the vertical leap. From part a, the initial velocity is $(v_y)_i = 7.0$ m/s, and the final velocity is $(v_y)_f = 0$. This is free-fall motion, with $a_y = -g$; thus

$$(v_y)_f^2 = 0 = (v_y)_i^2 - 2g\Delta y = (v_y)_i^2 - 2gy_f$$

which gives

$$(v_y)_i^2 = 2gy_f$$

Solving for y_f, we get a jump height of

$$y_f = \frac{(7.0 \text{ m/s})^2}{2(9.8 \text{ m/s}^2)} = 2.5 \text{ m}$$

ASSESS 2.5 m is a remarkable leap—a bit over 8 ft—but these animals are known for their jumping ability, so this seems reasonable.

FIGURE 2.40 Velocity-versus-time and position-versus-time graphs for a sprint between a man and a horse.

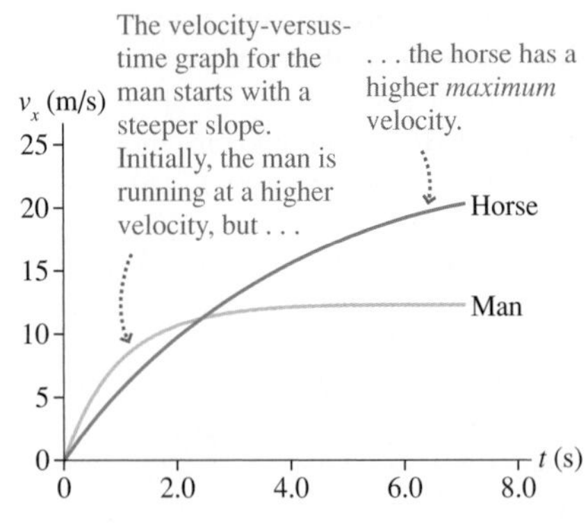
The velocity-versus-time graph for the man starts with a steeper slope. Initially, the man is running at a higher velocity, but . . .
. . . the horse has a higher *maximum* velocity.
v_x (m/s)
25
20
15
10
5
0
Horse
Man
0
2.0
4.0
6.0
8.0
t (s)

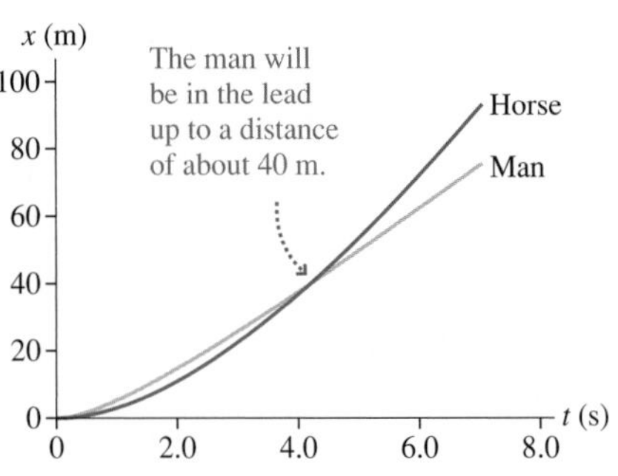
x (m)
100
80
60
40
20
0
The man will be in the lead up to a distance of about 40 m.
Horse
Man
0
2.0
4.0
6.0
8.0
t (s)

The caption accompanying the photo at the start of the chapter asked a question about animals and their athletic abilities: Who is the winner in a race between a horse and a man? The surprising answer is "It depends." Specifically, the winner depends on the length of the race.

Some animals are capable of high speed; others are capable of great acceleration. Horses can run much faster than humans, but, when starting from rest, humans are capable of much greater initial acceleration. FIGURE 2.40 shows velocity and position graphs for an elite male sprinter and a thoroughbred racehorse. The horse's maximum velocity is about twice that of the man, but the man's initial acceleration—the slope of the velocity graph at early times—is about twice that of the horse. As the second graph shows, a man could win a *very* short race. For a longer race, the horse's higher maximum velocity will put it in the lead; the men's world-record time for the mile is a bit under 4 min, but a horse can easily run this distance in less than 2 min.

For a race of many miles, another factor comes into play: energy. A very long race is less about velocity and acceleration than about endurance—the ability to continue expending energy for a long time. In such endurance trials, humans often win. We will explore such energy issues in Chapter 11.

STOP TO THINK 2.6 A volcano ejects a chunk of rock straight up at a velocity of $v_y = 30$ m/s. Ignoring air resistance, what will be the velocity v_y of the rock when it falls back into the volcano's crater?

A. > 30 m/s B. 30 m/s C. 0 m/s D. -30 m/s E. < -30 m/s

INTEGRATED EXAMPLE 2.16 Speed versus endurance

Cheetahs have the highest top speed of any land animal, but they usually fail in their attempts to catch their prey because their endurance is limited. They can maintain their maximum speed of 30 m/s for only about 15 s before they need to stop.

Thomson's gazelles, their preferred prey, have a lower top speed than cheetahs, but they can maintain this speed for a few minutes. When a cheetah goes after a gazelle, success or failure is a simple matter of kinematics: Is the cheetah's high speed enough to allow it to reach its prey before the cheetah runs out of steam? The following problem uses realistic data for such a chase.

A cheetah has spotted a gazelle. The cheetah leaps into action, reaching its top speed of 30 m/s in a few seconds. At this instant, the gazelle, 160 m from the running cheetah, notices the danger and heads directly away. The gazelle accelerates at 4.5 m/s² for 6.0 s, then continues running at a constant speed. After reaching its maximum speed, the cheetah can continue running for only 15 s. Does the cheetah catch the gazelle, or does the gazelle escape?

PREPARE The example asks, "Does the cheetah catch the gazelle?" Our most challenging task is to translate these words into a graphical and mathematical problem that we can solve using the techniques of the chapter.

There are two related problems: the motion of the cheetah and the motion of the gazelle, for which we'll use the subscripts "C" and "G". Let's take our starting time, $t_1 = 0$ s, as the instant that the gazelle notices the cheetah and begins to run. We'll take the position of the cheetah at this instant as the origin of our coordinate system, so $x_{1C} = 0$ m and $x_{1G} = 160$ m—the gazelle is 160 m away when it notices the cheetah. We've used this information to draw the visual overview in FIGURE 2.41, which includes motion diagrams and velocity graphs for the cheetah and the gazelle. The visual overview sums up everything we know about the problem.

With a clear picture of the situation, we can now rephrase the problem this way: Compute the position of the cheetah and the position of the gazelle at $t_3 = 15$ s, the time when the cheetah needs to break off the chase. If $x_{3G} \geq x_{3C}$, then the gazelle stays out in front and escapes. If $x_{3G} < x_{3C}$, the cheetah wins the race—and gets its dinner.

SOLVE The cheetah is in uniform motion for the entire duration of the problem, so we can use Equation 2.4 to solve for its position at $t_3 = 15$ s:

$$x_{3C} = x_{1C} + (v_x)_{1C}\Delta t = 0 \text{ m} + (30 \text{ m/s})(15 \text{ s}) = 450 \text{ m}$$

The gazelle's motion has two phases: one of constant acceleration and then one of constant velocity. We can solve for the position and the velocity at t_2, the end of the first phase, using the first two equations in Table 2.4. Let's find the velocity first:

$$(v_x)_{2G} = (v_x)_{1G} + (a_x)_G\Delta t = 0 \text{ m/s} + (4.5 \text{ m/s}^2)(6.0 \text{ s}) = 27 \text{ m/s}$$

The gazelle's position at t_2 is:

$$x_{2G} = x_{1G} + (v_x)_{1G}\,\Delta t + \tfrac{1}{2}(a_x)_G(\Delta t)^2$$
$$= 160 \text{ m} + 0 + \tfrac{1}{2}(4.5 \text{ m/s}^2)(6.0 \text{ s})^2 = 240 \text{ m}$$

Δt is the time for this phase of the motion, $t_2 - t_1 = 6.0$ s.

The gazelle has a head start; it begins at $x_{1G} = 160$ m.

From t_2 to t_3 the gazelle moves at a constant speed, so we can use the uniform motion equation, Equation 2.4, to find its final position:

$$x_{3G} = x_{2G} + (v_x)_{2G}\,\Delta t = 240 \text{ m} + (27 \text{ m/s})(9.0 \text{ s}) = 480 \text{ m}$$

The gazelle begins this phase of the motion at $x_{2G} = 240$ m.

Δt for this phase of the motion is $t_3 - t_2 = 9.0$ s.

x_{3C} is 450 m; x_{3G} is 480 m. The gazelle is 30 m ahead of the cheetah when the cheetah has to break off the chase, so the gazelle escapes.

ASSESS Does our solution make sense? Let's look at the final result. The numbers in the problem statement are realistic, so we expect our results to mirror real life. The speed for the gazelle is close to that of the cheetah, which seems reasonable for two animals known for their speed. And the result is the most common occurrence—the chase is close, but the gazelle gets away.

FIGURE 2.41 Visual overview for the cheetah and for the gazelle.

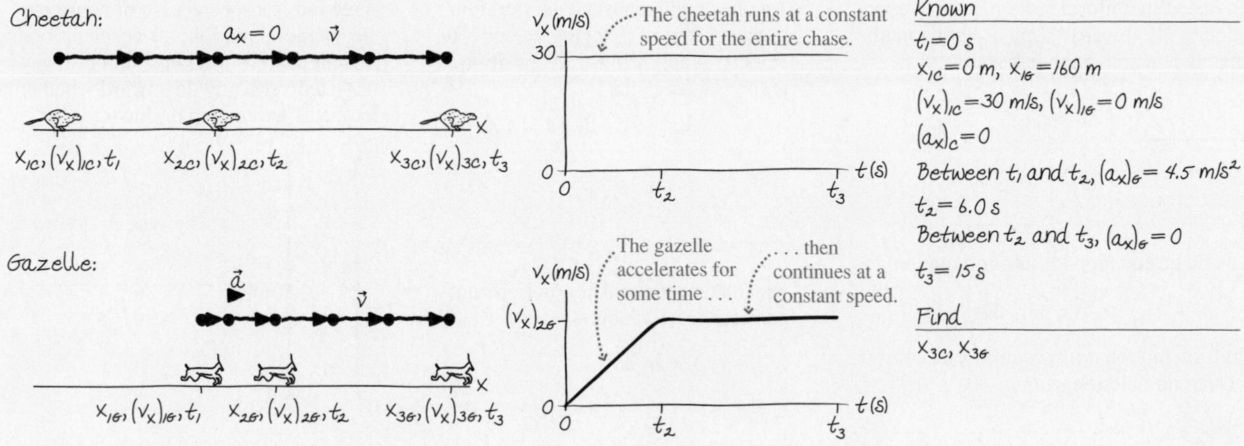

SUMMARY

The goal of Chapter 2 has been to describe and analyze linear motion.

GENERAL STRATEGIES

Problem-Solving Strategy

Our general problem-solving strategy has three parts:

PREPARE Set up the problem:

- Draw a picture.
- Collect necessary information.
- Do preliminary calculations.

SOLVE Do the necessary mathematics or reasoning.

ASSESS Check your answer to see if it is complete in all details and makes physical sense.

Visual Overview

A visual overview consists of several pieces that completely specify a problem. This may include any or all of the elements below:

Motion diagram **Pictorial representation** **Graphical representation** **List of values**

$\vec{v}$ $\vec{a}$

y $y_f, (v_y)_f, t_f$ $\vec{a}$ 0 $y_i, (v_y)_i, t_i$

v_y $(v_y)_f$ t t_f

Known
$y_i = 0$ m
$(v_y)_i = 0$ m/s
$t_i = 0$ s
$a_y = 18$ m/s^2
$t_f = 150$ s
Find
$(v_y)_f$ and y_f

IMPORTANT CONCEPTS

Velocity is the rate of change of position:

$$v_x = \frac{\Delta x}{\Delta t}$$

Acceleration is the rate of change of velocity:

$$a_x = \frac{\Delta v_x}{\Delta t}$$

The units of acceleration are m/s^2.

An object is speeding up if v_x and a_x have the same sign, slowing down if they have opposite signs.

A **position-versus-time graph** plots position on the vertical axis against time on the horizontal axis.

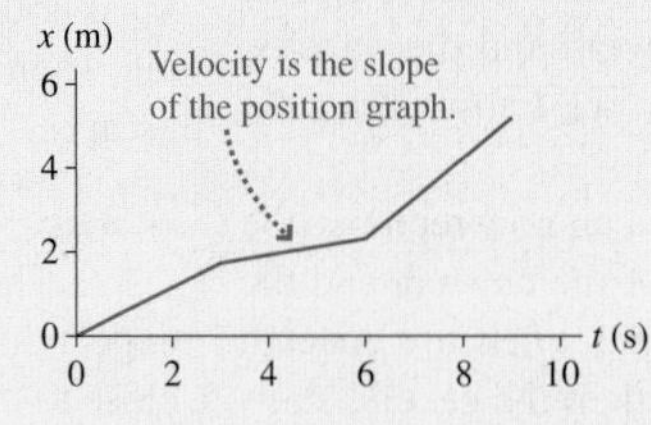

A **velocity-versus-time graph** plots velocity on the vertical axis against time on the horizontal axis.

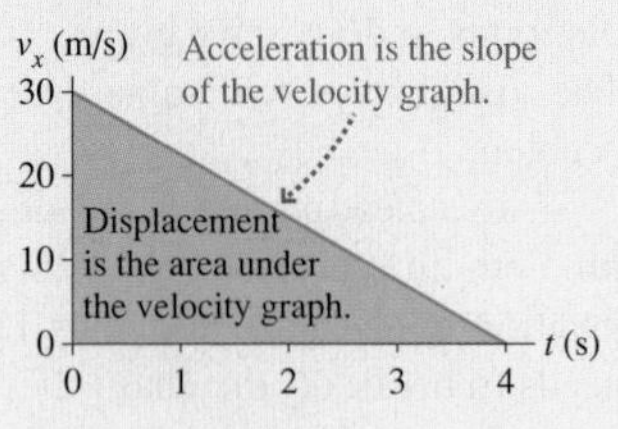

APPLICATIONS

Uniform motion

An object in uniform motion has a constant velocity. Its velocity graph is a horizontal line; its position graph is linear.

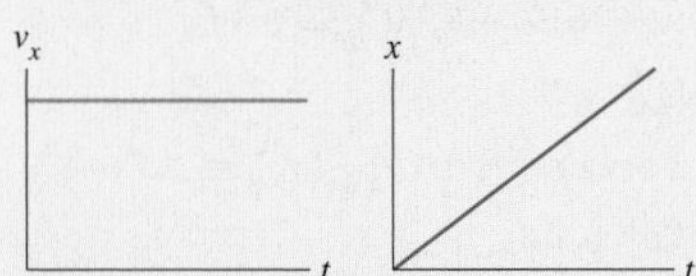

Kinematic equation for uniform motion:

$$x_f = x_i + v_x \, \Delta t$$

Uniform motion is a special case of constant-acceleration motion, with $a_x = 0$.

Motion with constant acceleration

An object with constant acceleration has a constantly changing velocity. Its velocity graph is linear; its position graph is a parabola.

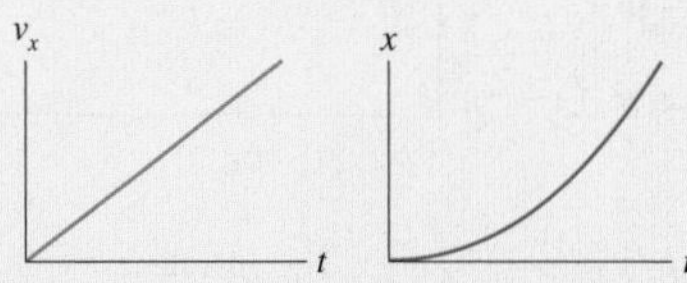

Kinematic equations for motion with constant acceleration:

$$(v_x)_f = (v_x)_i + a_x \, \Delta t$$

$$x_f = x_i + (v_x)_i \, \Delta t + \tfrac{1}{2} a_x (\Delta t)^2$$

$$(v_x)_f^2 = (v_x)_i^2 + 2a_x \, \Delta x$$

Free fall

Free fall is a special case of constant-acceleration motion; the acceleration has magnitude $g = 9.80$ m/s^2 and is always directed vertically downward whether an object is moving up or down.

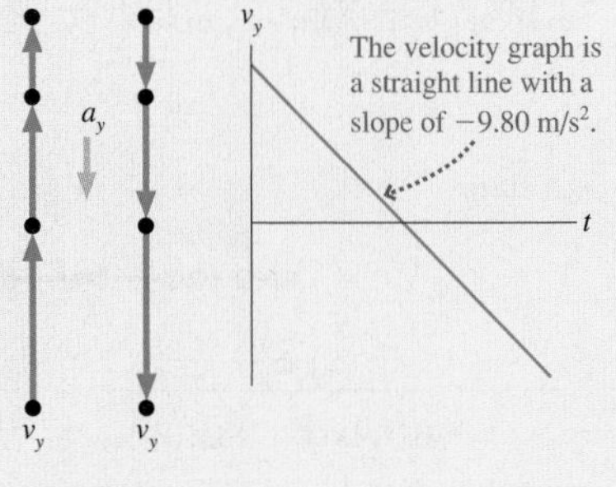

For homework assigned on MasteringPhysics, go to www.masteringphysics.com

Problems labeled INT integrate significant material from earlier chapters; BIO are of biological or medical interest.

Problem difficulty is labeled as | (straightforward) to ||||| (challenging).

QUESTIONS

Conceptual Questions

1. A person gets in an elevator on the ground floor and rides it to the top floor of a building. Sketch a velocity-versus-time graph for this motion.
2. a. Give an example of a vertical motion with a positive velocity and a negative acceleration.
   b. Give an example of a vertical motion with a negative velocity and a negative acceleration.
3. Sketch a velocity-versus-time graph for a rock that is thrown straight upward, from the instant it leaves the hand until the instant it hits the ground.
4. You are driving down the road at a constant speed. Another car going a bit faster catches up with you and passes you. Draw a position graph for both vehicles on the same set of axes, and note the point on the graph where the other vehicle passes you.
5. A car is traveling north. Can its acceleration vector ever point south? Explain.
6. BIO Certain animals are capable of running at great speeds; other animals are capable of tremendous accelerations. Speculate on which would be more beneficial to a predator—large maximum speed or large acceleration.
7. A ball is thrown straight up into the air. At each of the following instants, is the ball's acceleration a_y equal to g, $-g$, 0, $< g$, or $> g$?
   a. Just after leaving your hand?
   b. At the very top (maximum height)?
   c. Just before hitting the ground?
8. A rock is *thrown* (not dropped) straight down from a bridge into the river below.
   a. Immediately after being released, is the magnitude of the rock's acceleration greater than g, less than g, or equal to g? Explain.
   b. Immediately before hitting the water, is the magnitude of the the rock's acceleration greater than g, less than g, or equal to g? Explain.
9. Figure Q2.9 shows an object's position-versus-time graph. The letters A to E correspond to various segments of the motion in which the graph has constant slope.

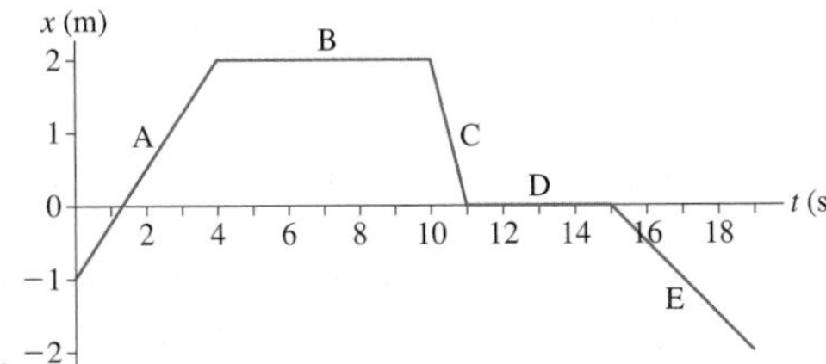

FIGURE Q2.9

   a. Write a realistic motion short story for an object that would have this position graph.
   b. In which segment(s) is the object at rest?
   c. In which segment(s) is the object moving to the right?
   d. Is the speed of the object during segment C greater than, equal to, or less than its speed during segment E? Explain.
10. Figure Q2.10 shows the position graph for an object moving along the horizontal axis.
    a. Write a realistic motion short story for an object that would have this position graph.
    b. Draw the corresponding velocity graph.

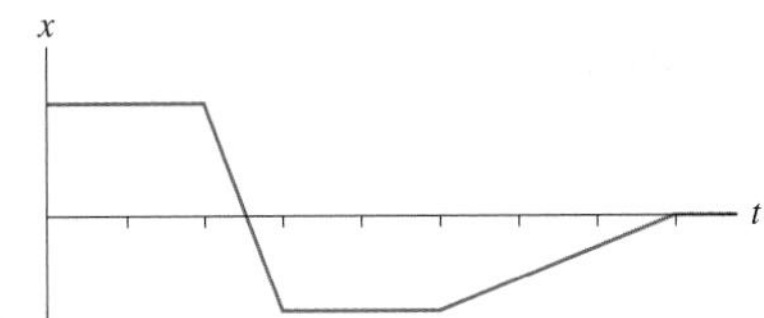

FIGURE Q2.10

11. Figure Q2.11 shows the position-versus-time graphs for two objects, A and B, that are moving along the same axis.
    a. At the instant $t = 1$ s, is the speed of A greater than, less than, or equal to the speed of B? Explain.
    b. Do objects A and B ever have the *same* speed? If so, at what time or times? Explain.

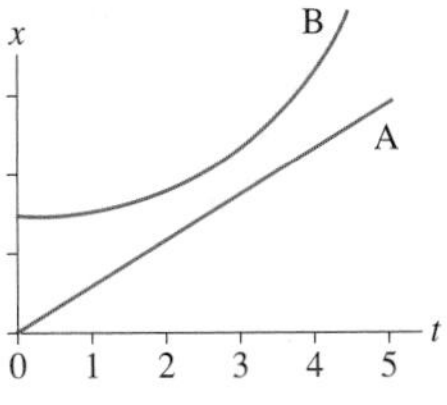

FIGURE Q2.11

12. Figure Q2.12 shows a position-versus-time graph. At which lettered point or points is the object
    a. Moving the fastest?
    b. Moving to the left?
    c. Speeding up?
    d. Slowing down?
    e. Turning around?

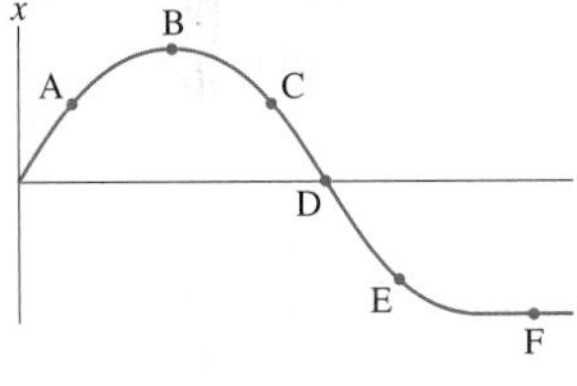

FIGURE Q2.12

13. Figure Q2.13 is the velocity-versus-time graph for an object moving along the x-axis.
    a. During which segment(s) is the velocity constant?
    b. During which segment(s) is the object speeding up?
    c. During which segment(s) is the object slowing down?
    d. During which segment(s) is the object standing still?
    e. During which segment(s) is the object moving to the right?

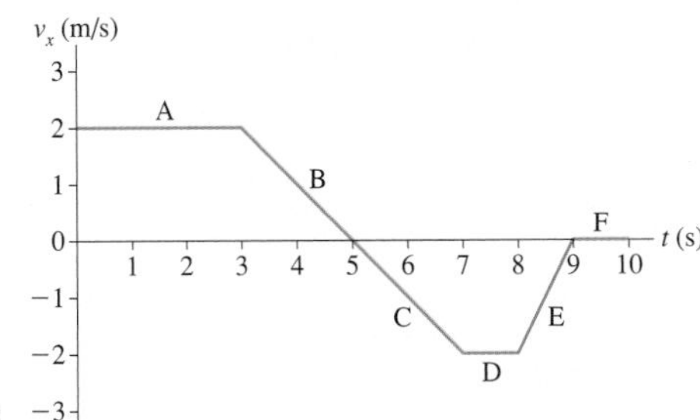

FIGURE Q2.13

14. A car traveling at velocity v takes distance d to stop after the brakes are applied. What is the stopping distance if the car is initially traveling at velocity $2v$? Assume that the acceleration due to the braking is the same in both cases.

Multiple-Choice Questions

15. | Figure Q2.15 shows the position graph of a car traveling on a straight road. At which labeled instant is the speed of the car greatest?

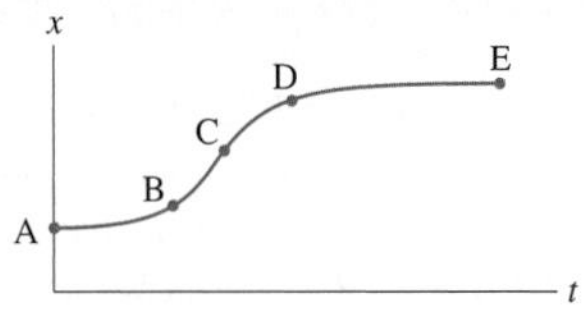

FIGURE Q2.15

16. | Figure Q2.16 shows the position graph of a car traveling on a straight road. The velocity at instant 1 is _________ and the velocity at instant 2 is _________.
    A. positive, negative
    B. positive, positive
    C. negative, negative
    D. negative, zero
    E. positive, zero

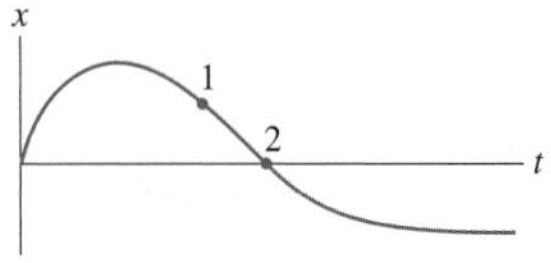

FIGURE Q2.16

17. | Figure Q2.17 shows an object's position-versus-time graph. What is the velocity of the object at $t = 6$ s?
    A. 0.67 m/s
    B. 0.83 m/s
    C. 3.3 m/s
    D. 4.2 m/s
    E. 25 m/s

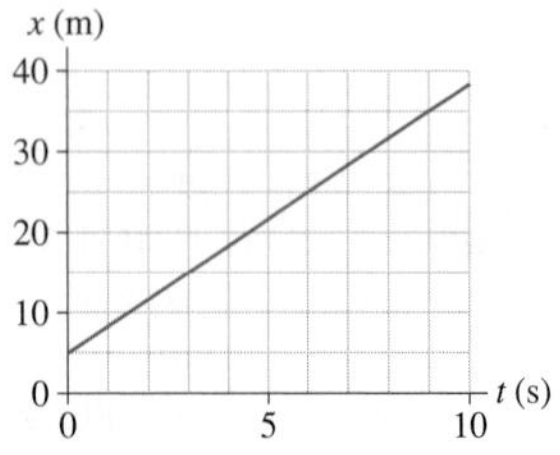

FIGURE Q2.17

18. | The following options describe the motion of four cars A–D. Which car has the largest acceleration?
    A. Goes from 0 m/s to 10 m/s in 5.0 s
    B. Goes from 0 m/s to 5.0 m/s in 2.0 s
    C. Goes from 0 m/s to 20 m/s in 7.0 s
    D. Goes from 0 m/s to 3.0 m/s in 1.0 s
19. | A car is traveling at $v_x = 20$ m/s. The driver applies the brakes, and the car slows with $a_x = -4.0$ m/s^2. What is the stopping distance?
    A. 5.0 m
    B. 25 m
    C. 40 m
    D. 50 m
20. || Velocity-versus-time graphs for three drag racers are shown in Figure Q2.20. At $t = 5.0$ s, which car has traveled the furthest?
    A. Andy
    B. Betty
    C. Carl
    D. All have traveled the same distance

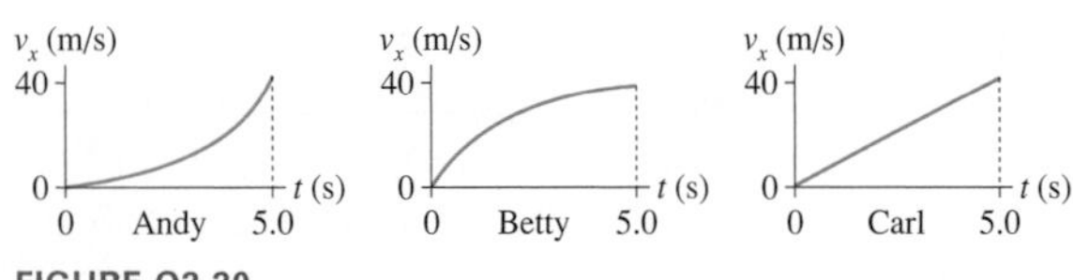

FIGURE Q2.20

21. | Which of the three drag racers in Question 20 had the greatest acceleration at $t = 0$ s?
    A. Andy
    B. Betty
    C. Carl
    D. All had the same acceleration
22. || Ball 1 is thrown straight up in the air and, at the same instant, ball 2 is released from rest and allowed to fall. Which velocity graph in Figure Q2.22 best represents the motion of the two balls?

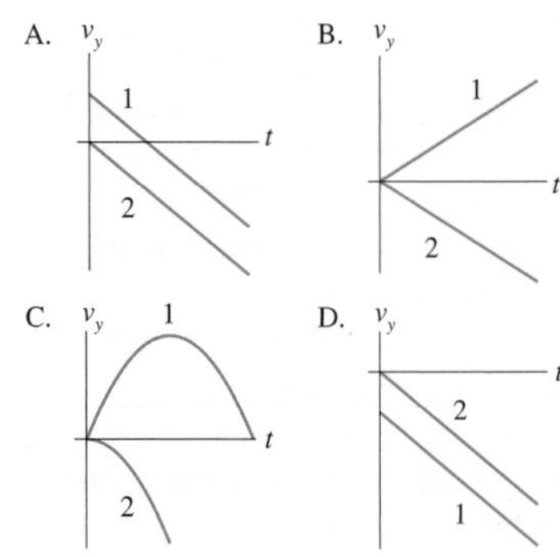

FIGURE Q2.22

23. || Figure Q2.23 shows a motion diagram with the clock reading (in seconds) shown at each position. From $t = 9$ s to $t = 15$ s the object is at the same position. After that, it returns along the same track. The positions of the dots for $t \geq 16$ s are offset for clarity. Which graph best represents the object's *velocity?*

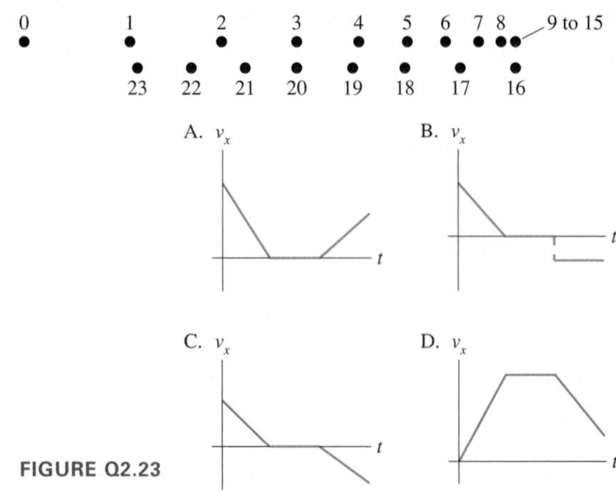

FIGURE Q2.23

24. || A car can go from 0 to 60 mph in 7.0 s. Assuming that it could maintain the same acceleration at higher speeds, how long would it take the car to go from 0 to 120 mph?
    A. 10 s
    B. 14 s
    C. 21 s
    D. 28 s
25. || A car can go from 0 to 60 mph in 12 s. A second car is capable of twice the acceleration of the first car. Assuming that it could maintain the same acceleration at higher speeds, how much time will this second car take to go from 0 to 120 mph?
    A. 12 s
    B. 9.0 s
    C. 6.0 s
    D. 3.0 s

PROBLEMS

Section 2.1 Describing Motion

1. ||| Figure P2.1 shows a motion diagram of a car traveling down a street. The camera took one frame every second. A distance scale is provided.
   a. Measure the x-value of the car at each dot. Place your data in a table, similar to Table 2.1, showing each position and the instant of time at which it occurred.
   b. Make a graph of x versus t, using the data in your table. Because you have data only at certain instants of time, your graph should consist of dots that are not connected together.

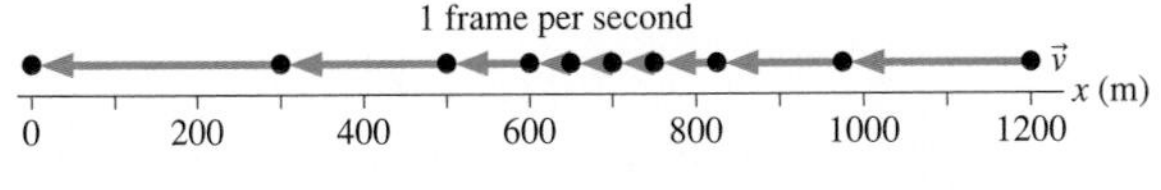

FIGURE P2.1

2. | For each motion diagram in Figure P2.2, determine the sign (positive or negative) of the position and the velocity.

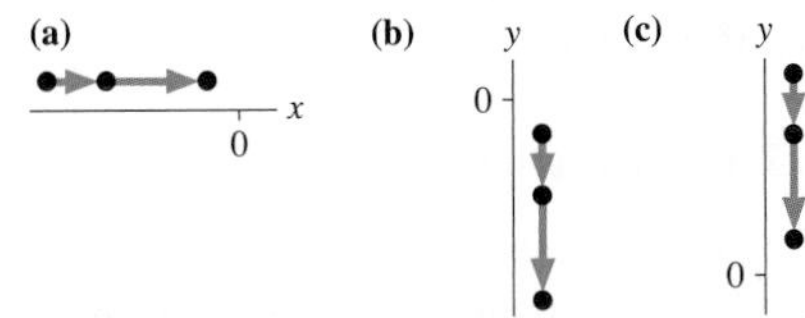

FIGURE P2.2

3. | Write a short description of the motion of a real object for which Figure P2.3 would be a realistic position-versus-time graph.

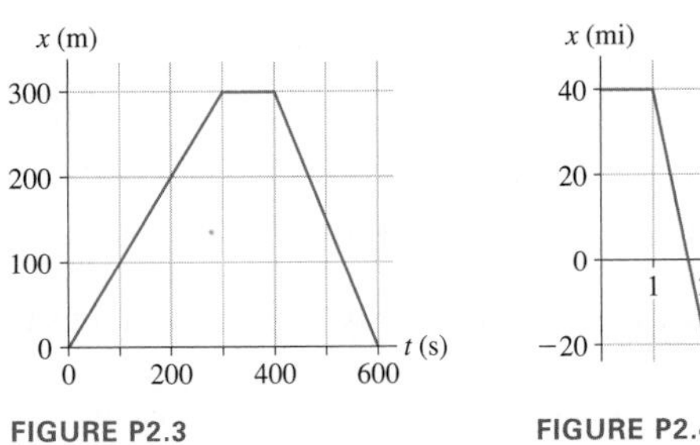

FIGURE P2.3

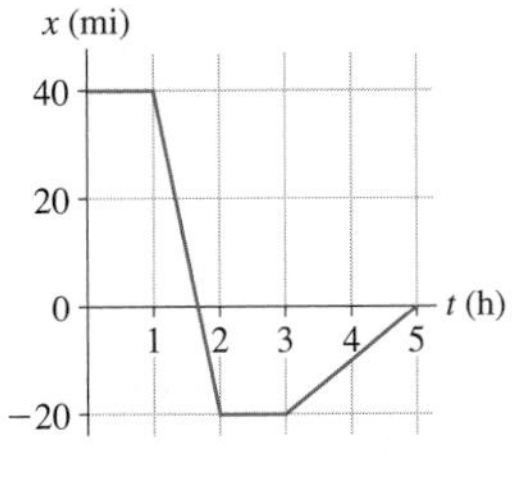

FIGURE P2.4

4. | Write a short description of the motion of a real object for which Figure P2.4 would be a realistic position-versus-time graph.
5. || The position graph of Figure P2.5 shows a dog slowly sneaking up on a squirrel, then putting on a burst of speed.
   a. For how many seconds does the dog move at the slower speed?
   b. Draw the dog's velocity-versus-time graph. Include a numerical scale on both axes.

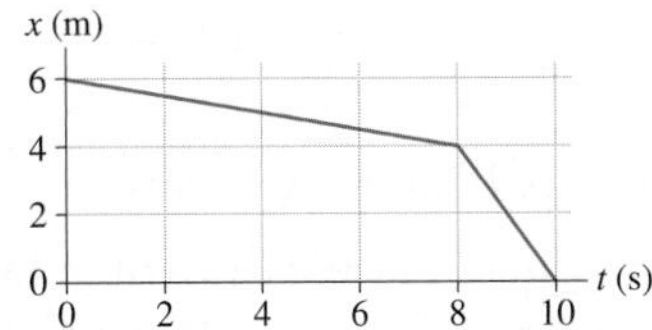

FIGURE P2.5

6. || The position graph of Figure P2.6 represents the motion of a ball being rolled back and forth by two children.
   a. At what positions are the two children sitting?
   b. Draw the ball's velocity-versus-time graph. Include a numerical scale on both axes.

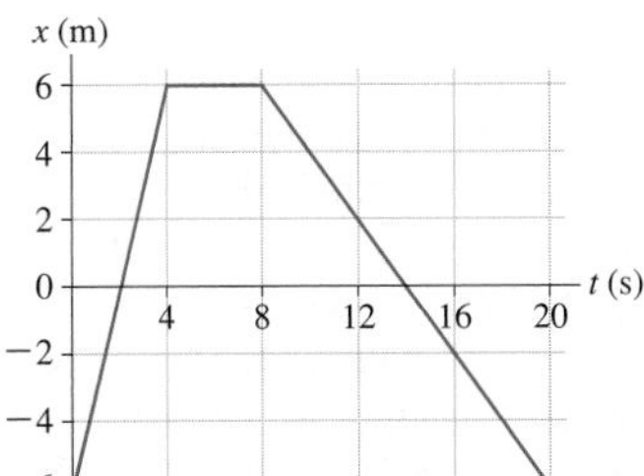

FIGURE P2.6

7. || A rural mail carrier is driving slowly, putting mail in mailboxes near the road. He overshoots one mailbox, stops, shifts into reverse, and then backs up until he is at the right spot. The velocity graph of Figure P2.7 represents his motion.
   a. Draw the mail carrier's position-versus-time graph. Assume that $x = 0$ m at $t = 0$ s.
   b. What is the position of the mailbox?

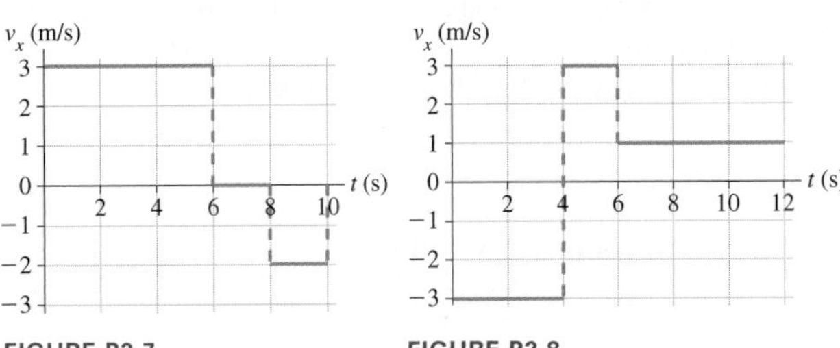

FIGURE P2.7

FIGURE P2.8

8. || For the velocity-versus-time graph of Figure P2.8:
   a. Draw the corresponding position-versus-time graph. Assume that $x = 0$ m at $t = 0$ s.
   b. What is the object's position at $t = 12$ s?
   c. Describe a moving object that could have these graphs.
9. || A bicyclist has the position-versus-time graph shown in Figure P2.9. What is the bicyclist's velocity at $t = 10$ s, at $t = 25$ s, and at $t = 35$ s?

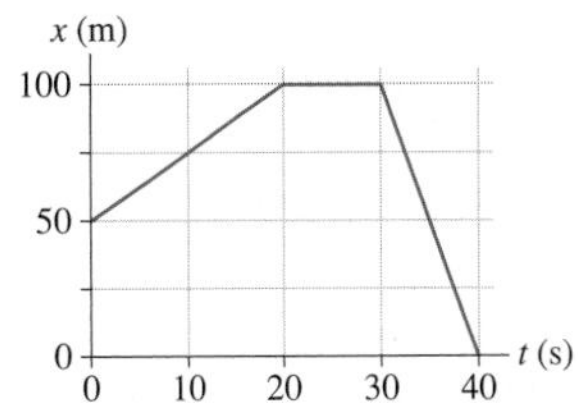

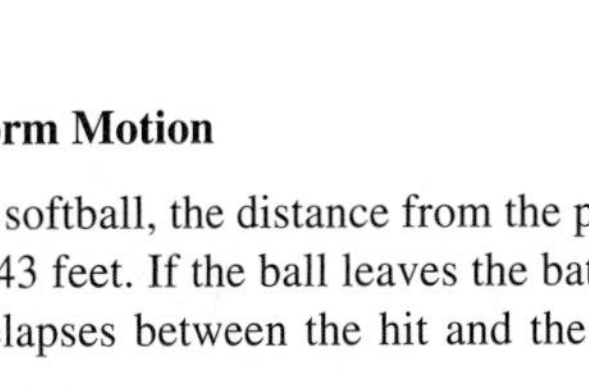

FIGURE P2.9

Section 2.2 Uniform Motion

10. | In college softball, the distance from the pitcher's mound to the batter is 43 feet. If the ball leaves the bat at 100 mph, how much time elapses between the hit and the ball reaching the pitcher?

11. ‖ Alan leaves Los Angeles at 8:00 A.M. to drive to San Francisco, 400 mi away. He travels at a steady 50 mph. Beth leaves Los Angeles at 9:00 A.M. and drives a steady 60 mph.
    a. Who gets to San Francisco first?
    b. How long does the first to arrive have to wait for the second?
12. ‖ Richard is driving home to visit his parents. 125 mi of the trip are on the interstate highway where the speed limit is 65 mph. Normally Richard drives at the speed limit, but today he is running late and decides to take his chances by driving at 70 mph. How many minutes does he save?
13. ||| In a 5.00 km race, one runner runs at a steady 12.0 km/h and another runs at 14.5 km/h. How long does the faster runner have to wait at the finish line to see the slower runner cross?
14. |||| In an 8.00 km race, one runner runs at a steady 11.0 km/h and another runs at 14.0 km/h. How far from the finish line is the slower runner when the faster runner finishes the race?
15. ‖ A car moves with constant velocity along a straight road. Its position is $x_1 = 0$ m at $t_1 = 0$ s and is $x_2 = 30$ m at $t_2 = 3.0$ s. Answer the following by considering ratios, without computing the car's velocity.
    a. What is the car's position at $t = 1.5$ s?
    b. What will be its position at $t = 9.0$ s?
16. ‖ While running a marathon, a long-distance runner uses a stopwatch to time herself over a distance of 100 m. She finds that she runs this distance in 18 s. Answer the following by considering ratios, without computing her velocity.
    a. If she maintains her speed, how much time will it take her to run the next 400 m?
    b. How long will it take her to run a mile at this speed?

Section 2.3 Instantaneous Velocity

17. | Figure P2.17 shows the position graph of a particle.
    a. Draw the particle's velocity graph for the interval $0 \text{ s} \le t \le 4$ s.
    b. Does this particle have a turning point or points? If so, at what time or times?

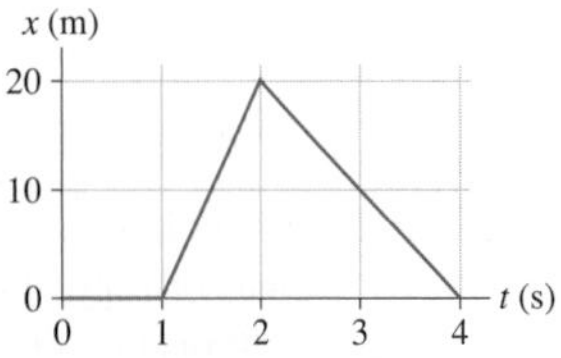

FIGURE P2.17

18. ‖ A somewhat idealized graph of the speed of the blood in the
BIO ascending aorta during one beat of the heart appears as in Figure P2.18.
    a. Approximately how far, in cm, does the blood move during one beat?
    b. Assume similar data for the motion of the blood in your aorta. Estimate how many beats of the heart will it take the blood to get from your heart to your brain.

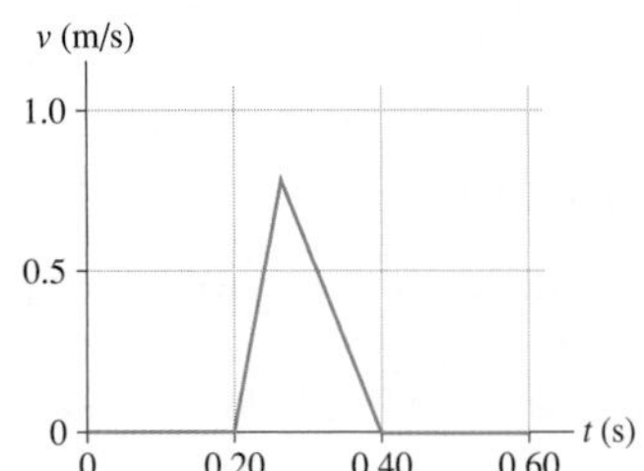

FIGURE P2.18

19. ||| A car starts from $x_i = 10$ m at $t_i = 0$ s and moves with the velocity graph shown in Figure P2.19.

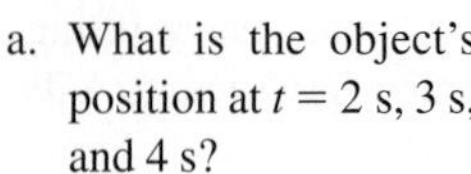

    a. What is the object's position at $t = 2$ s, 3 s, and 4 s?
    b. Does this car ever change direction? If so, at what time?

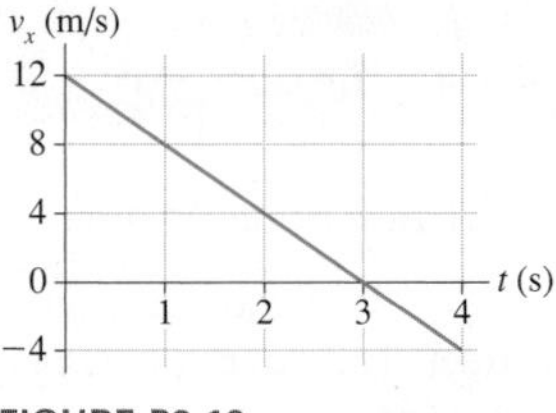

FIGURE P2.19

20. ‖ Figure P2.20 shows a graph of actual position-versus-time data for a particular type of drag racer known as a "funny car."
    a. Estimate the car's velocity at 2.0 s.
    b. Estimate the car's velocity at 4.0 s.

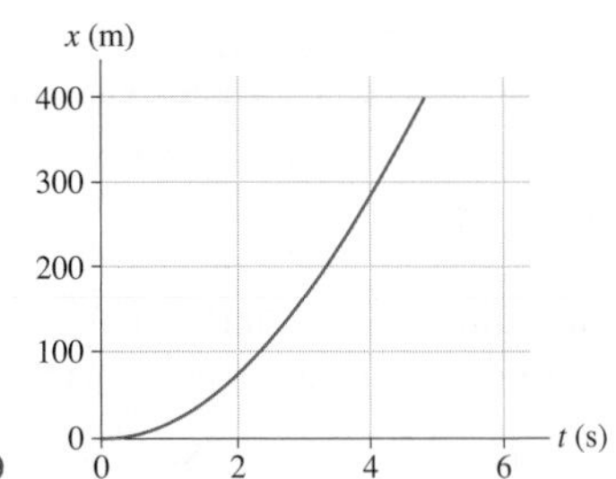

FIGURE P2.20

Section 2.4 Acceleration

21. ‖ Figure P2.21 shows the velocity graph of a bicycle. Draw the bicycle's acceleration graph for the interval $0 \text{ s} \le t \le 4$ s. Give both axes an appropriate numerical scale.

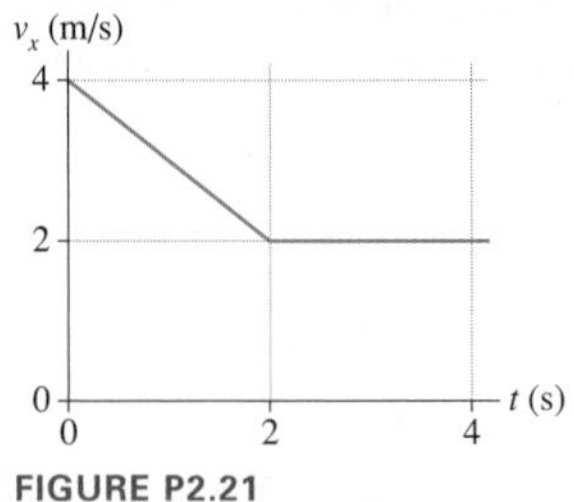

FIGURE P2.21

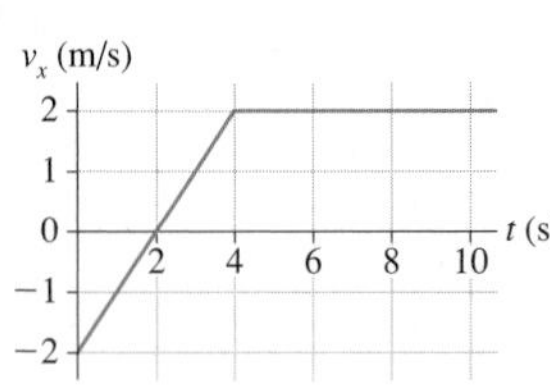

FIGURE P2.22

22. ||| Figure P2.22 shows the velocity graph of a train that starts from the origin at $t = 0$ s.
    a. Draw position and acceleration graphs for the train.
    b. Find the acceleration of the train at $t = 3.0$ s.
23. | For each motion diagram shown earlier in Figure P2.2, determine the sign (positive or negative) of the acceleration.
24. ‖ Figure P2.18 showed data for the speed of blood in the aorta.
BIO Determine the magnitude of the acceleration for both phases, speeding up and slowing down.
25. ‖ Figure P2.25 is a somewhat simplified velocity graph for Olympic sprinter Carl Lewis starting a 100 m dash. Estimate his acceleration during each of the intervals A, B, and C.

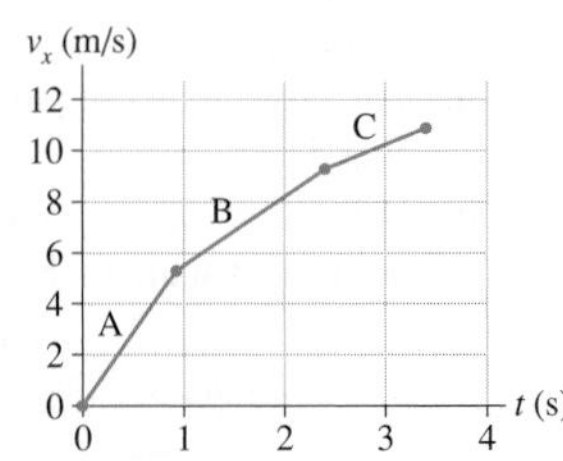

FIGURE P2.25

Section 2.5 Motion with Constant Acceleration

26. | A Thomson's gazelle can reach a speed of 13 m/s in 3.0 s. A
BIO lion can reach a speed of 9.5 m/s in 1.0 s. A trout can reach a speed of 2.8 m/s in 0.12 s. Which animal has the largest acceleration?

27. ||| When striking, the pike, a predatory fish, can accelerate from rest to a speed of 4.0 m/s in 0.11 s.
BIO

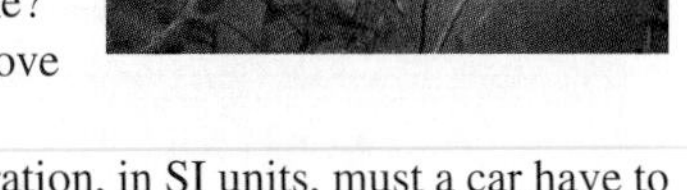

   a. What is the acceleration of the pike during this strike?
   b. How far does the pike move during this strike?

28. || a. What constant acceleration, in SI units, must a car have to go from zero to 60 mph in 10 s?
   b. What fraction of g is this?
   c. How far has the car traveled when it reaches 60 mph? Give your answer both in SI units and in feet.

29. || Light-rail passenger trains that provide transportation within and between cities are capable of modest accelerations. The magnitude of the maximum acceleration is typically 1.3 m/s^2, but the driver will usually maintain a constant acceleration that is less than the maximum. A train travels through a congested part of town at 5.0 m/s. Once free of this area, it speeds up to 12 m/s in 8.0 s. At the edge of town, the driver again accelerates, with the same acceleration, for another 16 s to reach a higher cruising speed. What is the final speed?

30. ||| A speed skater moving across frictionless ice at 8.0 m/s hits a 5.0-m-wide patch of rough ice. She slows steadily, then continues on at 6.0 m/s. What is her acceleration on the rough ice?

31. || A small propeller airplane can comfortably achieve a high enough speed to take off on a runway that is 1/4 mile long. A large, fully loaded passenger jet has about the same acceleration from rest, but it needs to achieve twice the speed to take off. What is the minimum runway length that will serve? **Hint:** You can solve this problem using ratios without having any additional information.

32. || Figure P2.32 shows a velocity-versus-time graph for a particle moving along the x-axis. At $t = 0$ s, assume that $x = 0$ m.
   a. What are the particle's position, velocity, and acceleration at $t = 1.0$ s?
   b. What are the particle's position, velocity, and acceleration at $t = 3.0$ s?

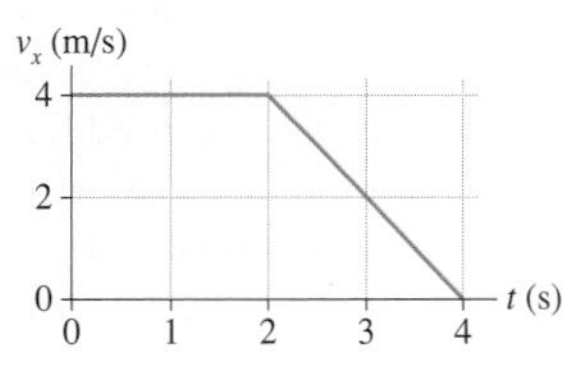

FIGURE P2.32

Section 2.6 Solving One-Dimensional Motion Problems

33. || A driver has a reaction time of 0.50 s, and the maximum deceleration of her car is 6.0 m/s^2. She is driving at 20 m/s when suddenly she sees an obstacle in the road 50 m in front of her. Can she stop the car in time to avoid a collision?

34. || Chameleons catch insects with their tongues, which they can rapidly extend to great lengths. In a typical strike, the chameleon's tongue accelerates at a remarkable 250 m/s^2 for 20 ms, then travels at constant speed for another 30 ms. During this total time of 50 ms, 1/20 of a second, how far does the tongue reach?
BIO

35. ||| You're driving down the highway late one night at 20 m/s when a deer steps onto the road 35 m in front of you. Your reaction time before stepping on the brakes is 0.50 s, and the maximum deceleration of your car is 10 m/s^2.
   a. How much distance is between you and the deer when you come to a stop?
   b. What is the maximum speed you could have and still not hit the deer?

36. ||| A light-rail train going from one station to the next on a straight section of track accelerates from rest at 1.1 m/s^2 for 20 s. It then proceeds at constant speed for 1100 m before slowing down at 2.2 m/s^2 until it stops at the station.
   a. What is the distance between the stations?
   b. How much time does it take the train to go between the stations?

37. ||| A simple model for a person running the 100 m dash is to assume the sprinter runs with constant acceleration until reaching top speed, then maintains that speed through the finish line. If a sprinter reaches his top speed of 11.2 m/s in 2.14 s, what will be his total time?

Section 2.7 Free Fall

38. ||| Ball bearings can be made by letting spherical drops of molten metal fall inside a tall tower—called a *shot tower*—and solidify as they fall.
   a. If a bearing needs 4.0 s to solidify enough for impact, how high must the tower be?
   b. What is the bearing's impact velocity?

39. ||| In the chapter, we saw that a person's reaction time is generally not quick enough to allow the person to catch a dollar bill dropped between the fingers. If a typical reaction time in this case is 0.25 s, how long would a bill need to be for a person to have a good chance of catching it?
BIO

40. || A ball is thrown vertically upward with a speed of 19.6 m/s.
   a. What are the ball's velocity and height after 1.00, 2.00, 3.00, and 4.00 s?
   b. Draw the ball's velocity-versus-time graph. Give both axes an appropriate numerical scale.

41. | A student at the top of a building of height h throws ball A straight upward with speed v_0 and throws ball B straight downward with the same initial speed.
   a. Compare the balls' accelerations, both direction and magnitude, immediately after they leave her hand. Is one acceleration larger than the other? Or are the magnitudes equal?
   b. Compare the final speeds of the balls as they reach the ground. Is one larger than the other? Or are they equal?

42. || Excellent human jumpers can leap straight up to a height of 110 cm off the ground. To reach this height, with what speed would a person need to leave the ground?

43. || A football is kicked straight up into the air; it hits the ground 5.2 s later.
   a. What was the greatest height reached by the ball? Assume it is kicked from ground level.
   b. With what speed did it leave the kicker's foot?

44. |||| In an action movie, the villain is rescued from the ocean by grabbing onto the ladder hanging from a helicopter. He is so intent on gripping the ladder that he lets go of his briefcase of counterfeit money when he is 130 m above the water. If the briefcase hits the water 6.0 s later, what was the speed at which the helicopter was ascending?

45. |||| A rock climber stands on top of a 50-m-high cliff overhanging a pool of water. He throws two stones vertically downward 1.0 s apart and observes that they cause a single splash. The initial speed of the first stone was 2.0 m/s.
   a. How long after the release of the first stone does the second stone hit the water?
   b. What was the initial speed of the second stone?
   c. What is the speed of each stone as they hit the water?

General Problems

46. ||| Actual velocity data for a lion pursuing prey are shown in BIO Figure P2.46. Estimate:
   a. The initial acceleration of the lion.
   b. The acceleration of the lion at 2 s and at 4 s.
   c. The distance traveled by the lion between 0 s and 8 s.

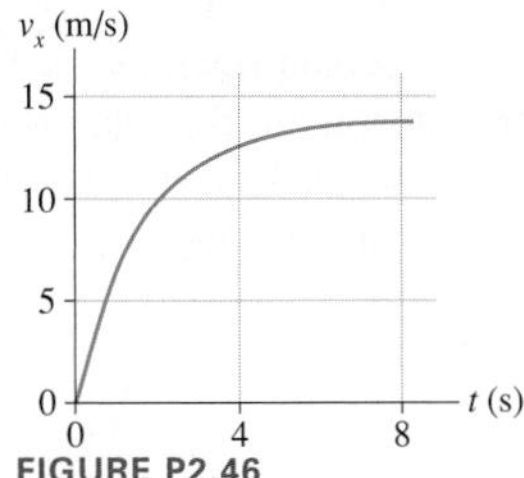

FIGURE P2.46

Problems 47 and 48 concern *nerve impulses*, electrical signals propagated along nerve fibers consisting of many *axons* (fiberlike extensions of nerve cells) connected end-to-end. Axons come in two varieties: insulated axons with a sheath made of myelin, and uninsulated axons with no such sheath. Myelinated (sheathed) axons conduct nerve impulses much faster than unmyelinated (unsheathed) axons. The impulse speed depends on the diameter of the axons and the sheath, but a typical myelinated axon transmits nerve impulses at a speed of about 25 m/s, much faster than the typical 2.0 m/s for an unmyelinated axon. Figure P2.47 shows small portions of three nerve fibers consisting of axons of equal size. Two-thirds of the axons in fiber B are myelinated.

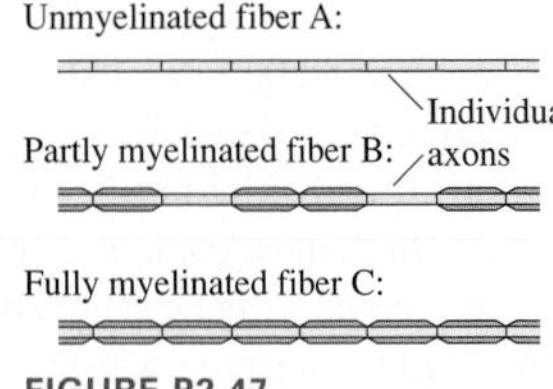

FIGURE P2.47

47. | Suppose nerve impulses simultaneously enter the left side of BIO the nerve fibers sketched in Figure P2.47, then propagate to the right. Draw qualitatively accurate position and velocity graphs for the nerve impulses in all three cases. A nerve fiber is made up of many axons, but show the propagation of the impulses only over the six axons shown here.
48. | Suppose that the nerve fibers in Figure P2.47 connect a finger to your brain, a distance of 1.2 m. BIO
   a. What are the travel times of a nerve impulse from finger to brain along fibers A and C?
   b. For fiber B, 2/3 of the length is composed of myelinated axons, 1/3 unmyelinated axons. Compute the travel time for a nerve impulse on this fiber.
   c. When you touch a hot stove with your finger, the sensation of pain must reach your brain as a nerve signal along a nerve fiber before your muscles can react. Which of the three fibers gives you the best protection against a burn? Are any of these fibers unsuitable for transmitting urgent sensory information?
49. || A truck driver has a shipment of apples to deliver to a destination 440 miles away. The trip usually takes him 8 hours. Today he finds himself daydreaming and realizes 120 miles into his trip that he is running 15 minutes later than his usual pace at this point. At what speed must he drive for the remainder of the trip to complete the trip in the usual amount of time?
50. || When you sneeze, the air in your lungs accelerates from rest BIO to approximately 150 km/h in about 0.50 seconds.
   a. What is the acceleration of the air in m/s^2?
   b. What is this acceleration, in units of g?
51. || Figure P2.51 shows the motion diagram, made at two frames of film per second, of a ball rolling along a track. The track has a 3.0-m-long sticky section.

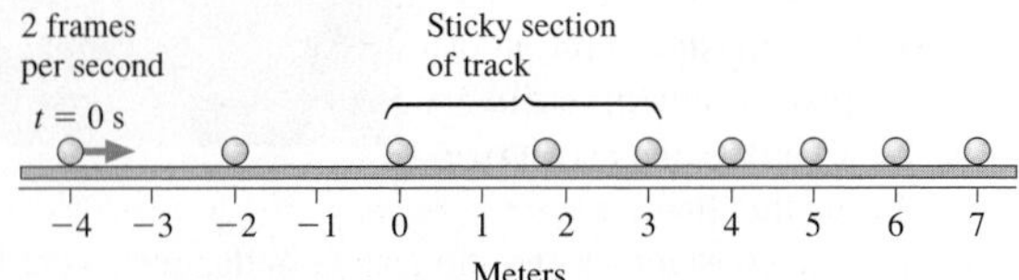

FIGURE P2.51

   a. Use the scale to determine the positions of the center of the ball. Place your data in a table, similar to Table 2.1, showing each position and the instant of time at which it occurred.
   b. Make a graph of x versus t for the ball. Because you have data only at certain instants of time, your graph should consist of dots that are not connected together.
   c. What is the *change* in the ball's position from $t = 0$ s to $t = 1.0$ s?
   d. What is the *change* in the ball's position from $t = 2.0$ s to $t = 4.0$ s?
   e. What is the ball's velocity before reaching the sticky section?
   f. What is the ball's velocity after passing the sticky section?
   g. Determine the ball's acceleration on the sticky section of the track.
52. || Julie drives 100 mi to Grandmother's house. On the way to Grandmother's, Julie drives half the *distance* at 40 mph and half the distance at 60 mph. On her return trip, she drives half the *time* at 40 mph and half the time at 60 mph.
   a. How long does it take Julie to complete the trip to Grandmother's house?
   b. How long does the return trip take?
53. || The takeoff speed for an Airbus A320 jetliner is 80 m/s. Velocity data measured during takeoff are as shown in the table.

t(s)	v_x(m/s)
0	0
10	23
20	46
30	69

   a. What is the takeoff speed in miles per hour?
   b. What is the jetliner's acceleration during takeoff?
   c. At what time do the wheels leave the ground?
   d. For safety reasons, in case of an aborted takeoff, the runway must be three times the takeoff distance. Can an A320 take off safely on a 2.5-mi-long runway?
54. |||| Does a real automobile have constant acceleration? Measured data for a Porsche 944 Turbo at maximum acceleration are as shown in the table.

t(s)	v_x(mph)
0	0
2	28
4	46
6	60
8	70
10	78

   a. Convert the velocities to m/s, then make a graph of velocity versus time. Based on your graph, is the acceleration constant? Explain.
   b. Draw a smooth curve through the points on your graph, then use your graph to *estimate* the car's acceleration at 2.0 s and 8.0 s. Give your answer in SI units. **Hint:** Remember that acceleration is the slope of the velocity graph.

55. || People hoping to travel to other worlds are faced with huge challenges. One of the biggest is the time required for a journey. The nearest star is 4.1×10^{16} m away. Suppose you had a spacecraft that could accelerate at $1.0g$ for half a year, then continue at a constant speed. (This is far beyond what can be achieved with any known technology.) How long would it take you to reach the nearest star to earth?

56. ||| You are driving to the grocery store at 20 m/s. You are 110 m from an intersection when the traffic light turns red. Assume that your reaction time is 0.70 s and that your car brakes with constant acceleration.
   a. How far are you from the intersection when you begin to apply the brakes?
   b. What acceleration will bring you to rest right at the intersection?
   c. How long does it take you to stop?

57. | When you blink your eye, the upper lid goes from rest with your
BIO eye open to completely covering your eye in a time of 0.024 s.
   a. Estimate the distance that the top lid of your eye moves during a blink.
   b. What is the acceleration of your eyelid? Assume it to be constant.
   c. What is your upper eyelid's final speed as it hits the bottom eyelid?

58. ||| A bush baby, an African
BIO primate, is capable of leaping vertically to the remarkable height of 2.3 m. To jump this high, the bush baby accelerates over a distance of 0.16 m while rapidly extending its legs. The acceleration during the jump is approximately constant. What is the acceleration in m/s^2 and in g's?

59. |||| When jumping, a flea reaches a takeoff speed of 1.0 m/s over
BIO a distance of 0.50 mm.
   a. What is the flea's acceleration during the jump phase?
   b. How long does the acceleration phase last?
   c. If the flea jumps straight up, how high will it go? (Ignore air resistance for this problem; in reality, air resistance plays a large role, and the flea will not reach this height.)

60. ||| Certain insects can achieve
BIO seemingly impossible accelerations while jumping. The click beetle accelerates at an astonishing $400g$ over a distance of 0.60 cm as it rapidly bends its thorax, making the "click" that gives it its name.

   a. Assuming the beetle jumps straight up, at what speed does it leave the ground?
   b. How much time is required for the beetle to reach this speed?
   c. Ignoring air resistance, how high would it go?

61. |||| Divers compete by diving into a 3.0-m-deep pool from a platform 10 m above the water. What is the magnitude of the minimum acceleration in the water needed to keep a diver from hitting the bottom of the pool? Assume the acceleration is constant.

62. |||| A student standing on the ground throws a ball straight up. The ball leaves the student's hand with a speed of 15 m/s when the hand is 2.0 m above the ground. How long is the ball in the air before it hits the ground? (The student moves her hand out of the way.)

63. |||| A rock is tossed straight up with a speed of 20 m/s. When it returns, it falls into a hole 10 m deep.
   a. What is the rock's velocity as it hits the bottom of the hole?
   b. How long is the rock in the air, from the instant it is released until it hits the bottom of the hole?

64. ||||| A 200 kg weather rocket is loaded with 100 kg of fuel and fired straight up. It accelerates upward at $30.0\ m/s^2$ for 30.0 s, then runs out of fuel. Ignore any air resistance effects.
   a. What is the rocket's maximum altitude?
   b. How long is the rocket in the air?
   c. Draw a velocity-versus-time graph for the rocket from liftoff until it hits the ground.

65. ||||| A juggler throws a ball straight up into the air with a speed of 10 m/s. With what speed would she need to throw a second ball half a second later, starting from the same position as the first, in order to hit the first ball at the top of its trajectory?

66. |||| A hotel elevator ascends 200 m with a maximum speed of 5.0 m/s. Its acceleration and deceleration both have a magnitude of $1.0\ m/s^2$.
   a. How far does the elevator move while accelerating to full speed from rest?
   b. How long does it take to make the complete trip from bottom to top?

67. |||| A car starts from rest at a stop sign. It accelerates at $2.0\ m/s^2$ for 6.0 seconds, coasts for 2.0 s, and then slows down at a rate of $1.5\ m/s^2$ for the next stop sign. How far apart are the stop signs?

68. |||| A toy train is pushed forward and released at $x_i = 2.0$ m with a speed of 2.0 m/s. It rolls at a steady speed for 2.0 s, then one wheel begins to stick. The train comes to a stop 6.0 m from the point at which it was released. What is the train's acceleration after its wheel begins to stick?

69. ||| Heather and Jerry are standing on a bridge 50 m above a river. Heather throws a rock straight down with a speed of 20 m/s. Jerry, at exactly the same instant of time, throws a rock straight up with the same speed. Ignore air resistance.
   a. How much time elapses between the first splash and the second splash?
   b. Which rock has the faster speed as it hits the water?

70. |||| A motorist is driving at 20 m/s when she sees that a traffic light 200 m ahead has just turned red. She knows that this light stays red for 15 s, and she wants to reach the light just as it turns green again. It takes her 1.0 s to step on the brakes and begin slowing at a constant deceleration. What is her speed as she reaches the light at the instant it turns green?

71. |||| A "rocket car" is launched along a long straight track at $t = 0$ s. It moves with constant acceleration $a_1 = 2.0\ m/s^2$. At $t = 2.0$ s, a second car is launched along a parallel track, from the same starting point, with constant acceleration $a_2 = 8.0\ m/s^2$.
   a. At what time does the second car catch up with the first one?
   b. How far have the cars traveled when the second passes the first?

72. ||| A Porsche challenges a Honda to a 400 m race. Because the Porsche's acceleration of $3.5\ m/s^2$ is larger than the Honda's $3.0\ m/s^2$, the Honda gets a 50-m head start. Assume, somewhat unrealistically, that both cars can maintain these accelerations the entire distance. Who wins, and by how much time?

73. ||||| The minimum stopping distance for a car traveling at a speed of 30 m/s is 60 m, including the distance traveled during the driver's reaction time of 0.50 s.
   a. What is the minimum stopping distance for the same car traveling at a speed of 40 m/s?
   b. Draw a position-versus-time graph for the motion of the car in part a. Assume the car is at $x_i = 0$ m when the driver first sees the emergency situation ahead that calls for a rapid halt.

74. ||||| A rocket is launched straight up with constant acceleration. Four seconds after liftoff, a bolt falls off the side of the rocket. The bolt hits the ground 6.0 s later. What was the rocket's acceleration?

Passage Problems

Free Fall on Different Worlds

Objects in free fall on the earth have acceleration $a_y = -9.8$ m/s^2. On the moon, free-fall acceleration is approximately 1/6 of the acceleration on earth. This changes the scale of problems involving free fall. For instance, suppose you jump straight upward, leaving the ground with velocity v_i and then steadily slowing until reaching zero velocity at your highest point. Because your initial velocity is determined mostly by the strength of your leg muscles, we can assume your initial velocity would be the same on the moon. But considering the final equation in Table 2.4 we can see that, with a smaller free-fall acceleration, your maximum height would be greater. The following questions ask you to think about how certain athletic feats might be performed in this reduced-gravity environment.

75. | If an astronaut can jump straight up to a height of 0.50 m on earth, how high could he jump on the moon?
   A. 1.2 m B. 3.0 m C. 3.6 m D. 18 m

76. | On the earth, an astronaut can safely jump to the ground from a height of 1.0 m; her velocity when reaching the ground is slow enough to not cause injury. From what height could the astronaut safely jump to the ground on the moon?
   A. 2.4 m B. 6.0 m C. 7.2 m D. 36 m

77. | On the earth, an astronaut throws a ball straight upward; it stays in the air for a total time of 3.0 s before reaching the ground again. If a ball were to be thrown upward with the same initial speed on the moon, how much time would pass before it hit the ground?
   A. 7.3 s B. 18 s C. 44 s D. 108 s

STOP TO THINK ANSWERS

Stop to Think 2.1: D. The motion consists of two constant-velocity phases; the second one has a greater velocity. The correct graph has two straight-line segments, with the second one having a steeper slope.

Stop to Think 2.2: B. The displacement is the area under a velocity-versus-time curve. In all four cases, the graph is a straight line, so the area under the curve is a rectangle. The area is the product of the length times the height, so the largest displacement belongs to the graph with the largest product of the length (the time interval, in s) times the height (the velocity, in m/s).

Stop to Think 2.3: C. Consider the slope of the position-versus-time graph; it starts out positive and constant, then decreases to zero. Thus the velocity graph must start with a constant positive value, then decrease to zero.

Stop to Think 2.4: C. Acceleration is the slope of the velocity-versus-time graph. The largest magnitude of the slope is at point C.

Stop to Think 2.5: E. An acceleration of 1.2 m/s^2 corresponds to an increase of 1.2 m/s every second. At this rate, the cruising speed of 6.0 m/s will be reached after 5.0 s.

Stop to Think 2.6: D. The final velocity will have the same *magnitude* as the initial velocity, but the velocity is negative because the rock will be moving downward.

3 Vectors and Motion in Two Dimensions

Once the leopard jumps, its trajectory is fixed by the initial speed and angle of the jump. How can we work out where the leopard will land?

LOOKING AHEAD ▸▸

The goals of Chapter 3 are to learn more about vectors and to use vectors as a tool to analyze motion in two dimensions.

Tools for Describing Motion

In the last chapter, we discussed motion along a line. In this chapter, we'll look at motion in which the direction changes. We'll need new tools.

Types of Motion

There are a few basic types of motion that we'll consider. In each case there is an acceleration due to a change in speed or direction—or both.

Looking Back ◂◂
1.1–1.2 Basic motion concepts

Vectors

We use vectors to describe quantities, like velocity, for which both the magnitude and direction are important.

Vectors specify a direction as well as a magnitude.

Looking Back ◂◂
1.5 Vectors

Vector Math

Our basic kinematic variables are vectors. Working with them means learning how to work with vectors.

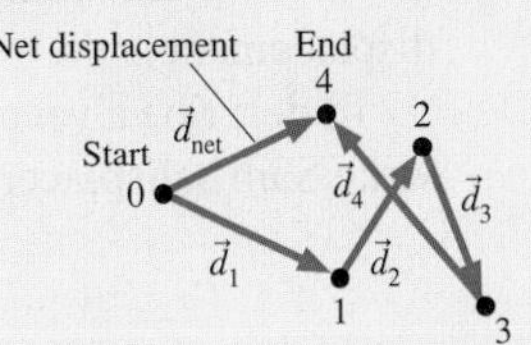

You'll learn how to add, subtract, and perform other mathematical operations on vectors.

Vector Components

We make measurements in a coordinate system. How do vectors fit in?

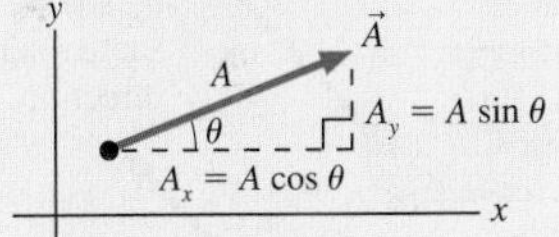

You'll learn how to find the components of vectors and how to add and subtract vectors using components.

Motion on a Ramp

Gravity causes the motion, but it's not straight down—it's at an angle.

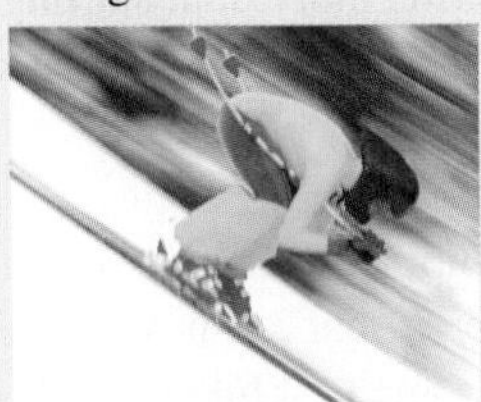

A speed skier is accelerating down a ramp. How fast is he moving at the end of his run?

Looking Back ◂◂
2.5 Motion with constant acceleration

Projectile Motion

Objects launched through the air follow a parabolic path. Water going over the falls, the leaping salmon, footballs, and jumping leopards all follow similar paths.

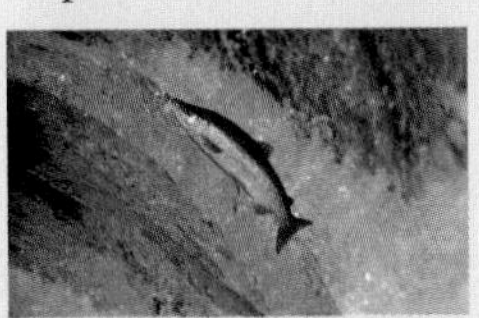

The salmon is moving horizontally and vertically—as do all objects undergoing projectile motion.

Looking Back ◂◂
2.7 Free fall

Circular Motion

Motion in a circle at a constant speed involves acceleration, but not the constant acceleration you studied in Chapter 2.

The riders are moving at a constant speed but with an ever-changing direction. It is the changing direction that causes the acceleration and makes the ride fun.

3.1 Using Vectors

FIGURE 3.1 The velocity vector $\vec{v}$ has both a magnitude and a direction.

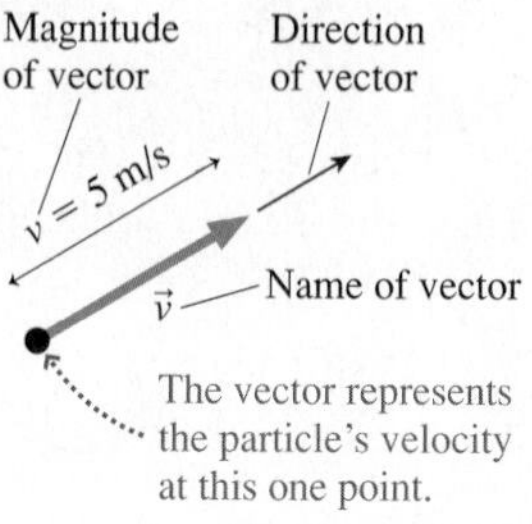

In the previous chapter, we solved many problems in which an object moved in a straight-line path. In this chapter, we will look at particles that take curving paths—motion in two dimensions. Because the direction of motion will be so important, we need to develop an appropriate mathematical language to describe it—the language of vectors.

We introduced the concept of a vector in Chapter 1, and in the next few sections we will develop that concept into a useful and powerful tool. We will practice using vectors by analyzing a problem of motion in one dimension (that of motion on a ramp) and by studying the interesting notion of relative velocity. We will then be ready to analyze the two-dimensional motion of projectiles and of objects moving in a circle.

Recall from Chapter 1 that a vector is a quantity with both a size (magnitude) and a direction. FIGURE 3.1 shows how to represent a particle's velocity as a vector $\vec{v}$. The particle's speed at this point is 5 m/s *and* it is moving in the direction indicated by the arrow. The magnitude of a vector is represented by the letter without an arrow. In this case, the particle's speed—the magnitude of the velocity vector $\vec{v}$—is $v = 5$ m/s. The magnitude of a vector, a *scalar* quantity, cannot be a negative number.

FIGURE 3.2 Displacement vectors.

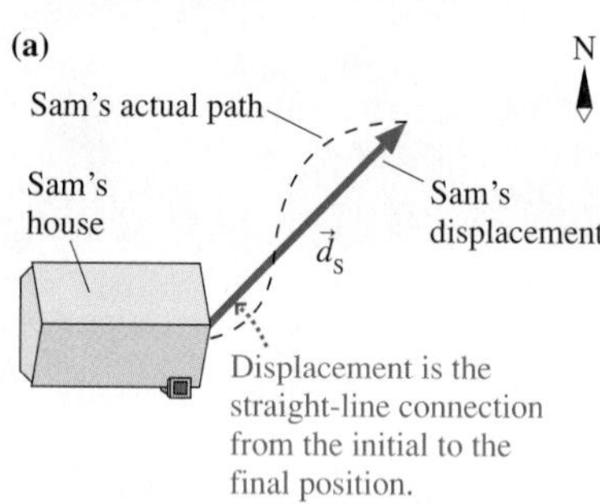

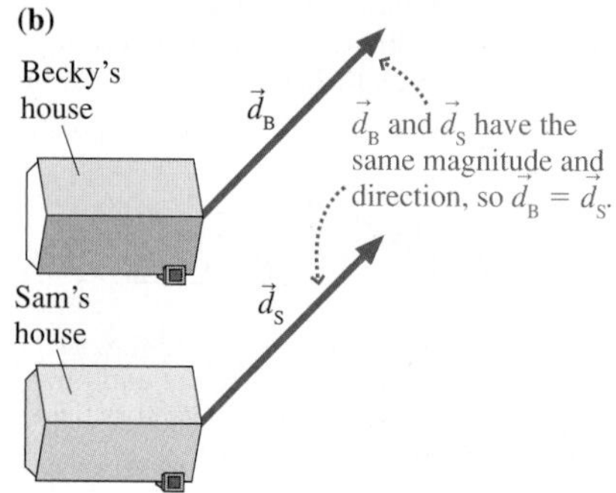

NOTE ▶ Although the vector arrow is drawn across the page, from its tail to its tip, this arrow does *not* indicate that the vector "stretches" across this distance. Instead, the arrow tells us the value of the vector quantity only at the one point where the tail of the vector is placed. ◀

We found in Chapter 1 that the displacement of an object is a vector drawn from its initial position to its position at some later time. Because displacement is an easy concept to think about, we can use it to introduce some of the properties of vectors. However, **all the properties we will discuss in this chapter (addition, subtraction, multiplication, components) apply to all types of vectors, not just to displacement.**

Suppose that Sam, our old friend from Chapter 1, starts from his front door, walks across the street, and ends up 200 ft to the northeast of where he started. Sam's displacement, which we will label $\vec{d}_S$, is shown in FIGURE 3.2a. The displacement vector is a straight-line connection from his initial to his final position, not necessarily his actual path. The dashed line indicates a possible route Sam might have taken, but his displacement is the vector $\vec{d}_S$.

To describe a vector we must specify both its magnitude and its direction. We can write Sam's displacement as

$$\vec{d}_S = (200 \text{ ft, northeast})$$

where the first number specifies the magnitude and the second item gives the direction. The magnitude of Sam's displacement is $d_S = 200$ ft, the distance between his initial and final points.

Sam's next-door neighbor Becky also walks 200 ft to the northeast, starting from her own front door. Becky's displacement $\vec{d}_B = (200 \text{ ft, northeast})$ has the same magnitude and direction as Sam's displacement $\vec{d}_S$. Because vectors are defined by their magnitude and direction, **two vectors are equal if they have the same magnitude and direction.** This is true regardless of the individual starting points of the vectors. Thus the two displacements in FIGURE 3.2b are equal to each other, and we can write $\vec{d}_B = \vec{d}_S$.

Vector Addition

FIGURE 3.3 The net displacement $\vec{C}$ resulting from two displacements $\vec{A}$ and $\vec{B}$.

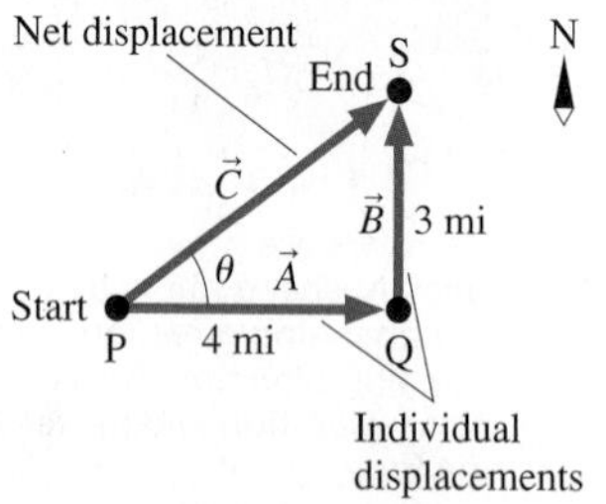

As we saw in Chapter 1, we can combine successive displacements by vector addition. Let's review and extend this concept. FIGURE 3.3 shows the displacement of a hiker who starts at point P and ends at point S. She first hikes 4 miles to the east, then 3 miles to the north. The first leg of the hike is described by the displacement vector $\vec{A} = (4 \text{ mi, east})$. The second leg of the hike has displacement $\vec{B} = (3 \text{ mi, north})$. By definition, a vector from her initial position P to her final position S is also a displacement. This is vector $\vec{C}$ on the figure. $\vec{C}$ is the *net displacement* because it describes the net result of the hiker's having first displacement $\vec{A}$, then displacement $\vec{B}$.

The word "net" implies addition. The net displacement $\vec{C}$ is an initial displacement $\vec{A}$ *plus* a second displacement $\vec{B}$, or

$$\vec{C} = \vec{A} + \vec{B} \tag{3.1}$$

The sum of two vectors is called the **resultant vector.** Vector addition is commutative: $\vec{A} + \vec{B} = \vec{B} + \vec{A}$. You can add vectors in any order you wish.

Look back at Tactics Box 1.4 on page 19 to review the three-step procedure for adding two vectors. This tip-to-tail method for adding vectors, which is used to find $\vec{C} = \vec{A} + \vec{B}$ in Figure 3.3, is called *graphical addition.* Any two vectors of the same type—two velocity vectors or two force vectors—can be added in exactly the same way.

When two vectors are to be added, it is often convenient to draw them with their tails together, as shown in FIGURE 3.4a. To evaluate $\vec{D} + \vec{E}$, you could move vector $\vec{E}$ over to where its tail is on the tip of $\vec{D}$, then use the tip-to-tail rule of graphical addition. This gives vector $\vec{F} = \vec{D} + \vec{E}$ in FIGURE 3.4b. Alternatively, FIGURE 3.4c shows that the vector sum $\vec{D} + \vec{E}$ can be found as the diagonal of the parallelogram defined by $\vec{D}$ and $\vec{E}$. This method is called the *parallelogram rule* of vector addition.

FIGURE 3.4 Two vectors can be added using the tip-to-tail rule or the parallelogram rule.

(a)

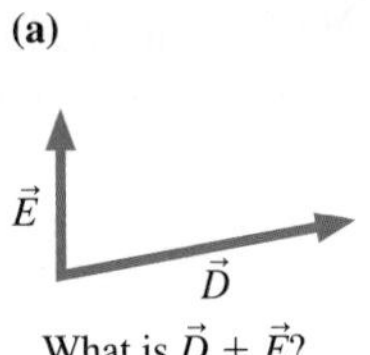

What is $\vec{D} + \vec{E}$?

(b)

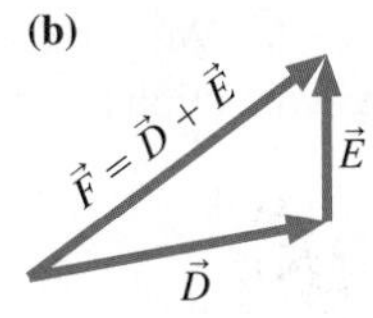

Tip-to-tail rule: Slide the tail of $\vec{E}$ to the tip of $\vec{D}$.

(c)

$\vec{E}$ $\vec{F} = \vec{D} + \vec{E}$ $\vec{D}$

Parallelogram rule: Find the diagonal of the parallelogram formed by $\vec{D}$ and $\vec{E}$.

Vector addition is easily extended to more than two vectors. FIGURE 3.5 shows the path of a hiker moving from initial position 0 to position 1, then position 2, then position 3, and finally arriving at position 4. These four segments are described by displacement vectors $\vec{d}_1$, $\vec{d}_2$, $\vec{d}_3$, and $\vec{d}_4$. The hiker's *net* displacement, an arrow from position 0 to position 4, is the vector $\vec{d}_{\text{net}}$. In this case,

$$\vec{d}_{\text{net}} = \vec{d}_1 + \vec{d}_2 + \vec{d}_3 + \vec{d}_4 \tag{3.2}$$

The vector sum is found by using the tip-to-tail method three times in succession.

FIGURE 3.5 The net displacement after four individual displacements.

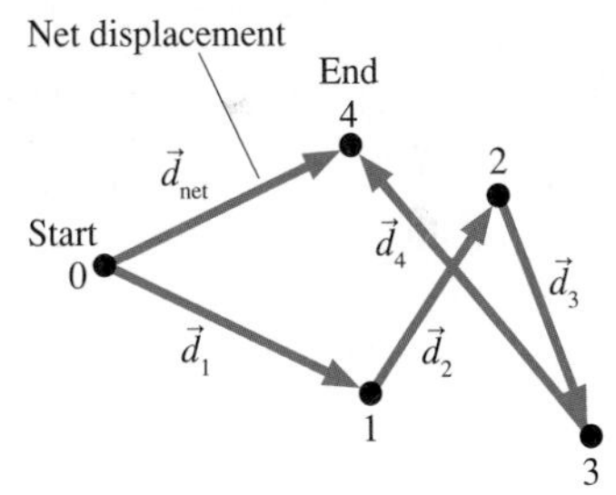

Multiplication by a Scalar

The hiker in Figure 3.3 started with displacement $\vec{A}_1 = (4 \text{ mi, east})$. Suppose a second hiker walks twice as far to the east. The second hiker's displacement will then certainly be $\vec{A}_2 = (8 \text{ mi, east})$. The words "twice as" indicate a multiplication, so we can say

$$\vec{A}_2 = 2\vec{A}_1$$

Multiplying a vector by a positive scalar gives another vector of *different magnitude* but pointing in the *same direction.*

Let the vector $\vec{A}$ be specified as a magnitude A and a direction θ_A; that is, $\vec{A} = (A, \theta_A)$. Now let $\vec{B} = c\vec{A}$, where c is a positive scalar constant. Then

$$\vec{B} = c\vec{A} \text{ means that } (B, \theta_B) = (cA, \theta_A) \tag{3.3}$$

The vector is stretched or compressed by the factor c (i.e., vector $\vec{B}$ has magnitude $B = cA$), but $\vec{B}$ points in the same direction as $\vec{A}$. This is illustrated in FIGURE 3.6.

Suppose we multiply $\vec{A}$ by zero. Using Equation 3.3, we get

$$0 \cdot \vec{A} = \vec{0} = (0 \text{ m, direction undefined}) \tag{3.4}$$

The product is a vector having zero length or magnitude. This vector is known as the **zero vector,** denoted $\vec{0}$. The direction of the zero vector is irrelevant; you cannot describe the direction of an arrow of zero length!

FIGURE 3.6 Multiplication of a vector by a positive scalar.

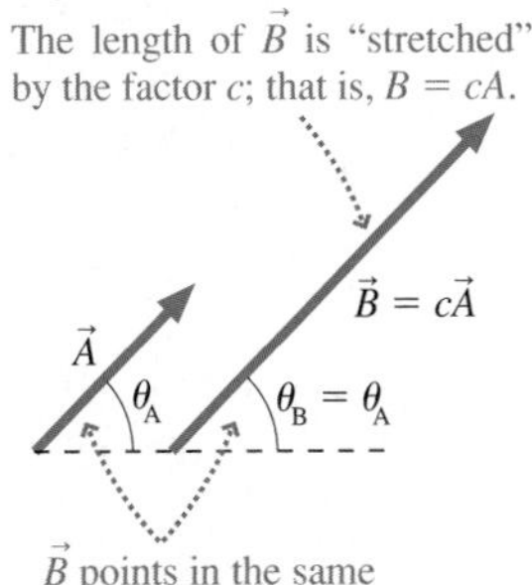

The length of $\vec{B}$ is "stretched" by the factor c; that is, $B = cA$.
$\vec{B} = c\vec{A}$
$\vec{A}$
θ_A
$\theta_B = \theta_A$
$\vec{B}$ points in the same direction as $\vec{A}$.

FIGURE 3.7 Vector $-\vec{A}$.

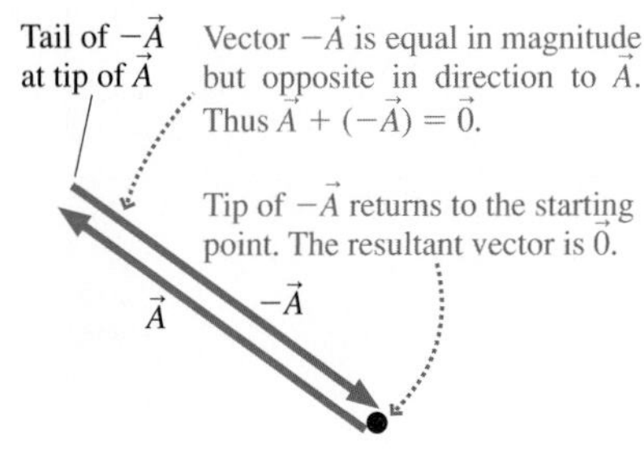

Tail of $-\vec{A}$ at tip of $\vec{A}$
Vector $-\vec{A}$ is equal in magnitude but opposite in direction to $\vec{A}$. Thus $\vec{A} + (-\vec{A}) = \vec{0}$.
Tip of $-\vec{A}$ returns to the starting point. The resultant vector is $\vec{0}$.
$\vec{A}$
$-\vec{A}$

FIGURE 3.8 Vectors $\vec{A}$, $2\vec{A}$, and $-3\vec{A}$.

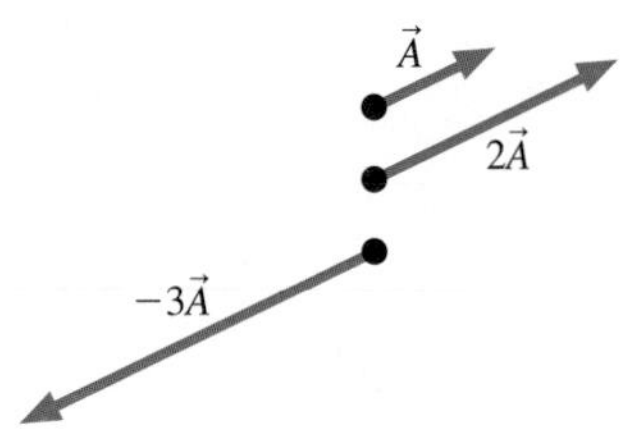

What happens if we multiply a vector by a negative number? Equation 3.3 does not apply if $c < 0$ because vector $\vec{B}$ cannot have a negative magnitude. Consider the vector $-\vec{A}$, which is equivalent to multiplying $\vec{A}$ by -1. Because

$$\vec{A} + (-\vec{A}) = \vec{0} \tag{3.5}$$

The vector $-\vec{A}$ must be such that, when it is added to $\vec{A}$, the resultant is the zero vector $\vec{0}$. In other words, the *tip* of $-\vec{A}$ must return to the *tail* of $\vec{A}$, as shown in FIGURE 3.7. This will be true only if $-\vec{A}$ is equal in magnitude to $\vec{A}$ but opposite in direction. Thus we can conclude that

$$-\vec{A} = (A, \text{direction opposite } \vec{A}) \tag{3.6}$$

Multiplying a vector by -1 reverses its direction without changing its length.

As an example, FIGURE 3.8 shows vectors $\vec{A}$, $2\vec{A}$, and $-3\vec{A}$. Multiplication by 2 doubles the length of the vector but does not change its direction. Multiplication by -3 stretches the length by a factor of 3 *and* reverses the direction.

Vector Subtraction

How might we *subtract* vector $\vec{B}$ from vector $\vec{A}$ to form the vector $\vec{A} - \vec{B}$? With numbers, subtraction is the same as the addition of a negative number. That is, $5 - 3$ is the same as $5 + (-3)$. Similarly, $\vec{A} - \vec{B} = \vec{A} + (-\vec{B})$. We can use the rules for vector addition and the fact that $-\vec{B}$ is a vector opposite in direction to $\vec{B}$ to form rules for vector subtraction.

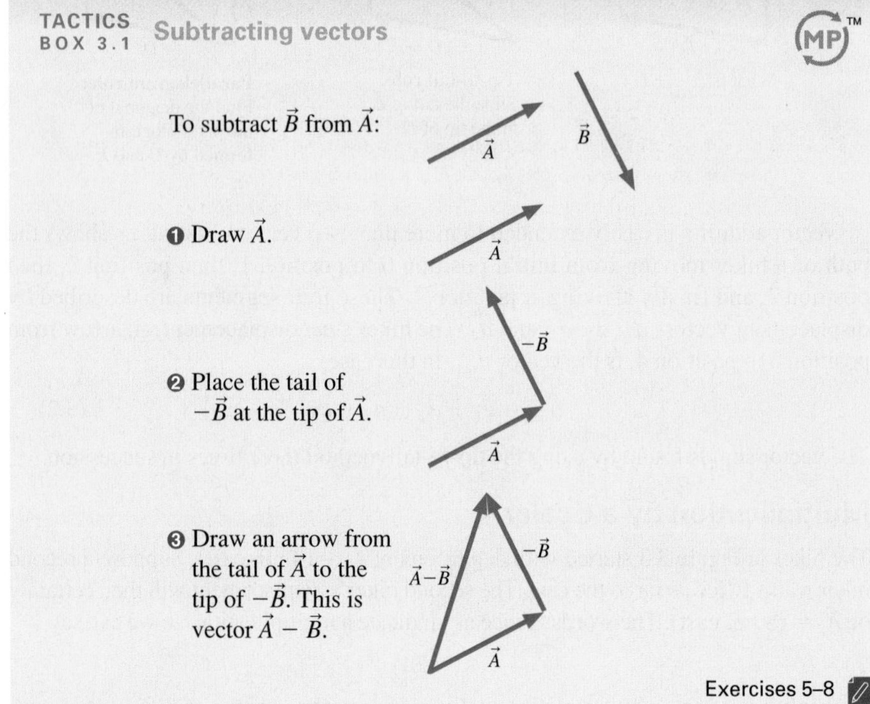

TACTICS BOX 3.1 Subtracting vectors
MP™
To subtract $\vec{B}$ from $\vec{A}$:
$\vec{A}$
$\vec{B}$
❶ Draw $\vec{A}$.
$\vec{A}$
❷ Place the tail of $-\vec{B}$ at the tip of $\vec{A}$.
$-\vec{B}$
$\vec{A}$
❸ Draw an arrow from the tail of $\vec{A}$ to the tip of $-\vec{B}$. This is vector $\vec{A} - \vec{B}$.
$-\vec{B}$
$\vec{A}-\vec{B}$
$\vec{A}$
Exercises 5–8

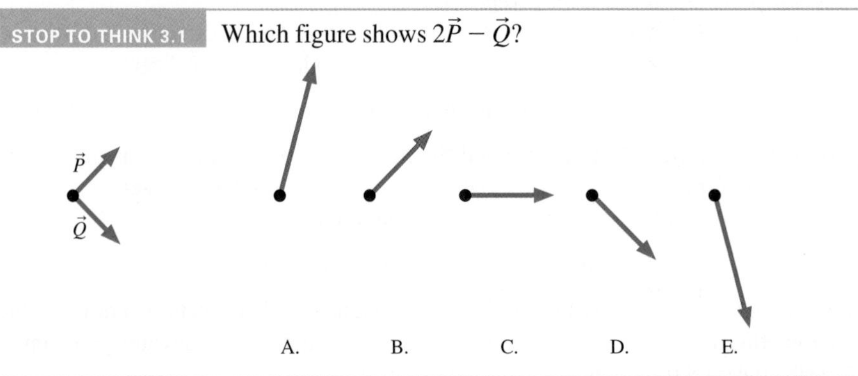

STOP TO THINK 3.1 Which figure shows $2\vec{P} - \vec{Q}$?

3.2 Using Vectors on Motion Diagrams

In Chapter 2, we defined velocity for one-dimensional motion as an object's displacement—the change in position—divided by the time interval in which the change occurs:

$$v_x = \frac{\Delta x}{\Delta t} = \frac{x_f - x_i}{\Delta t}$$

In two dimensions, an object's displacement is a vector. Suppose an object undergoes displacement $\vec{d}$ during the time interval Δt. Let's define an object's velocity *vector* to be

$$\vec{v} = \frac{\vec{d}}{\Delta t} = \left(\frac{d}{\Delta t}, \text{ same direction as } \vec{d}\right) \tag{3.7}$$

Definition of velocity in two or more dimensions

Notice that we've multiplied a vector by a scalar: The velocity vector is simply the displacement vector multiplied by the scalar $1/\Delta t$. Consequently, as we found in Chapter 1, **the velocity vector points in the direction of the displacement.** As a result, we can use the dot-to-dot vectors on a motion diagram to visualize the velocity.

NOTE ▶ Strictly speaking, the velocity defined in Equation 3.7 is the *average* velocity for the time interval Δt. This is adequate for using motion diagrams to visualize motion. As we did in Chapter 2, when we make Δt very small, we get an *instantaneous* velocity we can use in performing some calculations. ◀

EXAMPLE 3.1 Finding the velocity of an airplane

A small plane is 100 km due east of Denver. After 1 hour of flying at a constant speed in the same direction, it is 200 km due north of Denver. What is the plane's velocity?

PREPARE The initial and final positions of the plane are shown in **FIGURE 3.9**; the displacement $\vec{d}$ is the vector that points from the initial to the final position.

FIGURE 3.9 Displacement vector for an airplane.

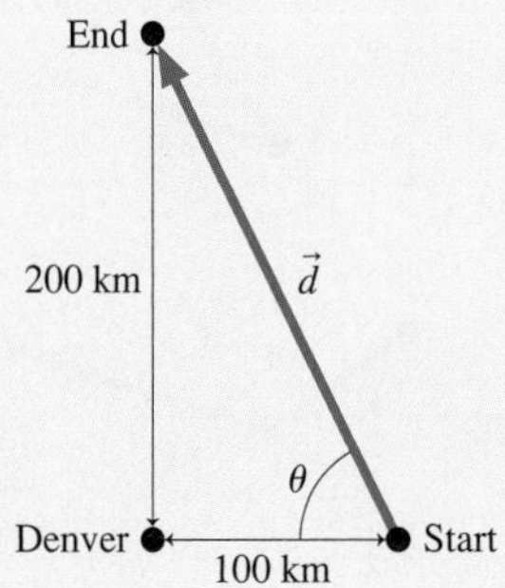

SOLVE The length of the displacement vector is the hypotenuse of a right triangle:

$$d = \sqrt{(100\text{ km})^2 + (200\text{ km})^2} = 224\text{ km}$$

The direction of the displacement vector is described by the angle θ in Figure 3.9. From trigonometry, this angle is

$$\theta = \tan^{-1}\left(\frac{200\text{ km}}{100\text{ km}}\right) = \tan^{-1}(2.00) = 63.4°$$

Thus the plane's displacement vector is

$$\vec{d} = (224\text{ km}, 63.4° \text{ north of west})$$

Because the plane undergoes this displacement during 1 hour, its velocity is

$$\vec{v} = \left(\frac{d}{\Delta t}, \text{ same direction as } \vec{d}\right) = \left(\frac{224\text{ km}}{1\text{ h}}, 63.4° \text{ north of west}\right)$$
$$= (224\text{ km/h}, 63.4° \text{ north of west})$$

ASSESS The plane's *speed* is the magnitude of the velocity, $v = 224$ km/h. This is approximately 140 mph, which is a reasonable speed for a small plane.

Lunging versus veering BIO The top photo shows a barracuda, a type of fish that catches prey with a rapid linear acceleration, a quick change in speed. The barracuda's body shape is optimized for such a straight-line strike. The butterfly fish in the bottom photo has a very different appearance. It can't rapidly change its speed, but its body shape lets it quickly change its direction. Once the barracuda gets up to speed, it can't change its direction very easily, so the butterfly fish can, by employing this other type of acceleration, avoid capture.

We defined an object's acceleration in one dimension as $a_x = \Delta v_x/\Delta t$. In two dimensions, we need to use a vector to describe acceleration. The vector definition of acceleration is a straightforward extension of the one-dimensional version:

$$\vec{a} = \frac{\vec{v}_f - \vec{v}_i}{t_f - t_i} = \frac{\Delta\vec{v}}{\Delta t} \tag{3.8}$$

Definition of acceleration in two or more dimensions

There is an acceleration whenever there is a *change* in velocity. Because velocity is a vector, it can change in either or both of two possible ways:

1. The magnitude can change, indicating a change in speed.
2. The direction of motion can change.

In Chapter 2 we saw how to compute an acceleration vector for the first case, in which an object speeds up or slows down while moving in a straight line. In this chapter we will examine the second case, in which an object changes its direction of motion.

Suppose an object has an initial velocity $\vec{v}_i$ at time t_i and later, at time t_f, has velocity $\vec{v}_f$. The fact that the velocity *changes* tells us the object undergoes an acceleration during the time interval $\Delta t = t_f - t_i$. We see from Equation 3.8 that the acceleration points in the same direction as the vector $\Delta\vec{v}$. This vector is the change in the velocity $\Delta\vec{v} = \vec{v}_f - \vec{v}_i$, so to know which way the acceleration vector points, we have to perform the vector subtraction $\vec{v}_f - \vec{v}_i$. Tactics Box 3.1 showed how to perform vector subtraction. Tactics Box 3.2 shows how to use vector subtraction to find the acceleration vector.

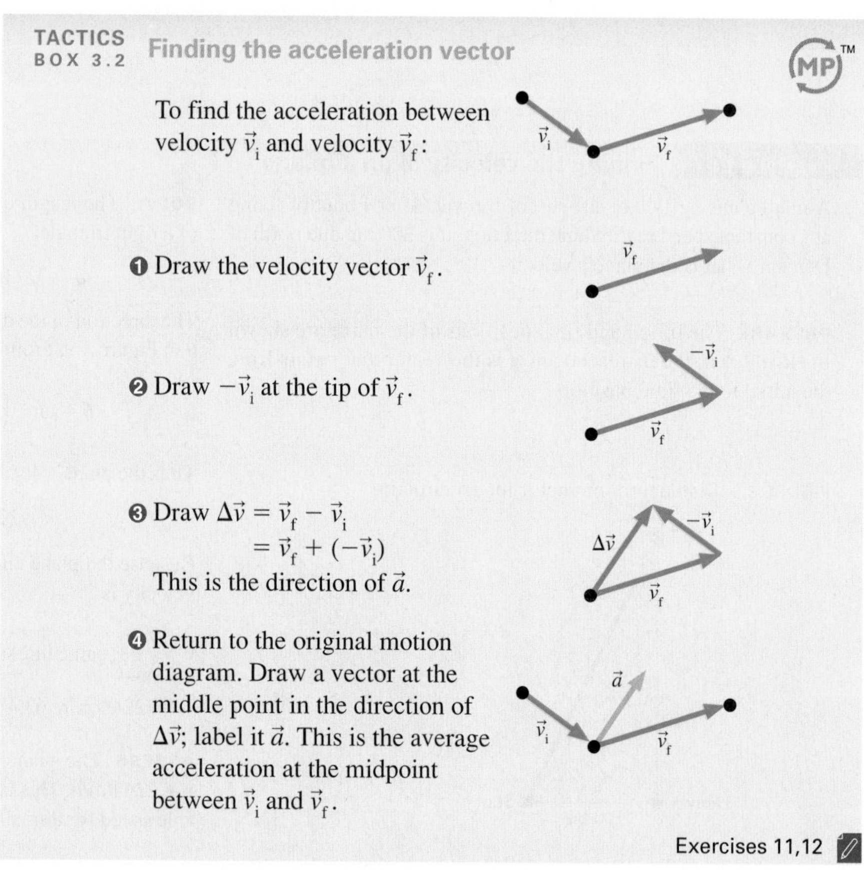

TACTICS BOX 3.2 Finding the acceleration vector

To find the acceleration between velocity $\vec{v}_i$ and velocity $\vec{v}_f$:

❶ Draw the velocity vector $\vec{v}_f$.

❷ Draw $-\vec{v}_i$ at the tip of $\vec{v}_f$.

❸ Draw $\Delta\vec{v} = \vec{v}_f - \vec{v}_i = \vec{v}_f + (-\vec{v}_i)$

This is the direction of $\vec{a}$.

❹ Return to the original motion diagram. Draw a vector at the middle point in the direction of $\Delta\vec{v}$; label it $\vec{a}$. This is the average acceleration at the midpoint between $\vec{v}_i$ and $\vec{v}_f$.

Exercises 11,12

Now that we know how to determine acceleration vectors, we can make a complete motion diagram with dots showing the position of the object, average velocity vectors found by connecting the dots with arrows, and acceleration vectors found using Tactics Box 3.2. Note that there is *one* acceleration vector linking each *two* velocity vectors, and $\vec{a}$ is drawn at the dot between the two velocity vectors it links. Let's look at two examples, one with changing speed and one with changing direction.

EXAMPLE 3.2 Drawing the acceleration for a Mars descent

A spacecraft slows as it safely descends to the surface of Mars. Draw a complete motion diagram for the last few seconds of the descent.

PREPARE FIGURE 3.10 shows two versions of a motion diagram: a professionally drawn version like you generally find in this text and a simpler version similar to what you might draw for a homework assignment. As the spacecraft slows in its descent, the dots get closer together and the velocity vectors get shorter.

SOLVE The inset in Figure 3.10 shows how Tactics Box 3.2 is used to determine the acceleration at one point. All the other acceleration vectors will be similar, because for each pair of velocity vectors the earlier one is longer than the later one.

ASSESS As the spacecraft slows, the acceleration vectors and velocity vectors point in opposite directions, consistent with what we learned about the sign of the acceleration in Chapter 2.

FIGURE 3.10 Motion diagram for a descending spacecraft.

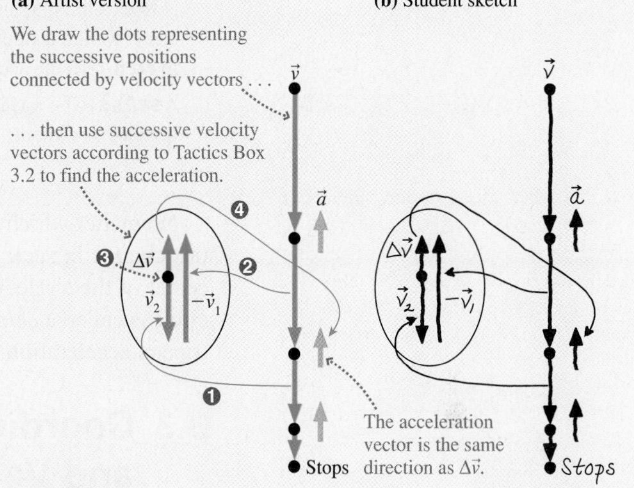

(a) Artist version
(b) Student sketch
We draw the dots representing the successive positions connected by velocity vectors . . .
. . . then use successive velocity vectors according to Tactics Box 3.2 to find the acceleration.
$\vec{v}$
$\vec{a}$
$\Delta\vec{v}$
$\vec{v}_2$
$-\vec{v}_1$
The acceleration vector is the same direction as $\Delta\vec{v}$.
Stops
Stops

EXAMPLE 3.3 Drawing the acceleration for a Ferris wheel ride

Anne rides a Ferris wheel at an amusement park. Draw a complete motion diagram for Anne's ride.

PREPARE FIGURE 3.11 shows 10 points of the motion during one complete revolution of the Ferris wheel. A person riding a Ferris wheel moves in a circle at a constant speed,

FIGURE 3.11 Motion diagram for Anne on a Ferris wheel.

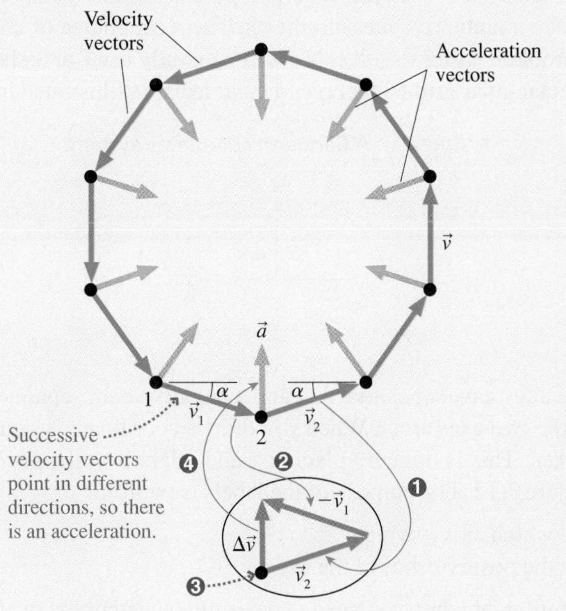

Velocity vectors
Acceleration vectors
$\vec{v}$
$\vec{a}$
1
2
α
α
$\vec{v}_1$
$\vec{v}_2$
Successive velocity vectors point in different directions, so there is an acceleration.
$-\vec{v}_1$
$\Delta\vec{v}$
$\vec{v}_2$

Continued

so we've shown equal distances between successive dots. As before, the velocity vectors are found by connecting each dot to the next. Note that the velocity vectors are *straight lines,* not curves.

We see that all the velocity vectors have the same length, but each has a different *direction,* and that means Anne is accelerating. This is not a "speeding up" or "slowing down" acceleration, but is, instead, a "change of direction" acceleration.

SOLVE The inset to Figure 3.11 shows how to use the steps of Tactics Box 3.2 to find the acceleration at one particular position, at the bottom of the circle. Vector $\vec{v}_1$ is the velocity vector that leads into this dot, while $\vec{v}_2$ moves away from it. From the circular geometry of the main figure, the two angles marked α are equal. Thus we see that $\vec{v}_2$ and $-\vec{v}_1$ form an isosceles triangle and vector $\Delta\vec{v} = \vec{v}_2 - \vec{v}_1$ is exactly vertical, toward the center of the circle. If we did a similar calculation for each point of the motion, we'd find a similar result: In each case, the acceleration points toward the center of the circle.

ASSESS The speed is constant but the direction is changing, so there is an acceleration, as we expect.

No matter which dot you select on the motion diagram of Figure 3.11, the velocities change in such a way that the acceleration vector $\vec{a}$ points directly toward the center of the circle. An acceleration vector that always points toward the center of a circle is called a *centripetal acceleration.* We will have much more to say about centripetal acceleration later in this chapter.

3.3 Coordinate Systems and Vector Components

Archaeologists establish a coordinate system so that they can precisely determine the positions of objects they excavate.

In the past two sections, we have seen how to add and subtract vectors graphically, using these operations to deduce important details of motion. But the graphical combination of vectors is not an especially good way to find quantitative results. In this section we will introduce a *coordinate description* of vectors that will be the basis for doing vector calculations.

Coordinate Systems

As we saw in Chapter 1, the world does not come with a coordinate system attached to it. A coordinate system is an artificially imposed grid that you place on a problem in order to make quantitative measurements. The right choice of coordinate system will make a problem easier to solve. We will generally use **Cartesian coordinates,** the familiar rectangular grid with perpendicular axes, as illustrated in FIGURE 3.12.

FIGURE 3.12 A Cartesian coordinate system.

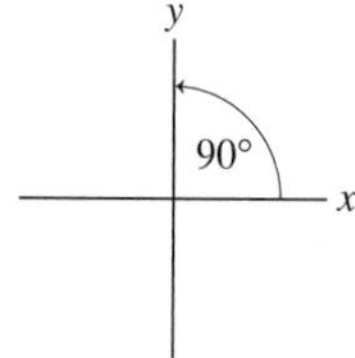

Coordinate axes have a positive end and a negative end, separated by zero at the origin where the two axes cross. When you draw a coordinate system, it is important to label the axes. This is done by placing x and y labels at the *positive* ends of the axes, as in Figure 3.12. The purpose of the labels is twofold:

- To identify which axis is which.
- To identify the positive ends of the axes.

This will be important when you need to determine whether the quantities in a problem should be assigned positive or negative values.

Component Vectors

FIGURE 3.13 shows a vector $\vec{A}$ and an xy-coordinate system that we've chosen. Once the directions of the axes are known, we can define two new vectors *parallel to the axes* that we call the **component vectors** of $\vec{A}$. Vector $\vec{A}_x$, called the *x-component vector,* is the projection of $\vec{A}$ along the x-axis. Vector $\vec{A}_y$, the *y-component vector,* is the projection of $\vec{A}$ along the y-axis. Notice that the component vectors are perpendicular to each other.

You can see, using the parallelogram rule, that $\vec{A}$ is the vector sum of the two component vectors:

$$\vec{A} = \vec{A}_x + \vec{A}_y \tag{3.9}$$

In essence, we have "broken" vector $\vec{A}$ into two perpendicular vectors that are parallel to the coordinate axes. We say that we have **decomposed** or **resolved** vector $\vec{A}$ into its component vectors.

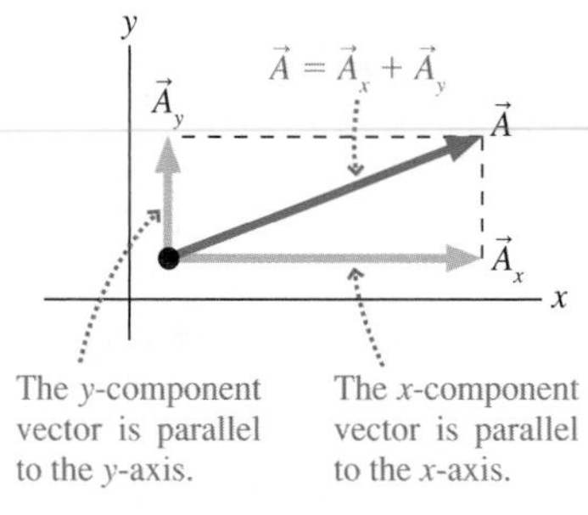

y
$\vec{A} = \vec{A}_x + \vec{A}_y$
$\vec{A}_y$
$\vec{A}$
$\vec{A}_x$
x
The y-component vector is parallel to the y-axis.
The x-component vector is parallel to the x-axis.

FIGURE 3.13 Component vectors $\vec{A}_x$ and $\vec{A}_y$ are drawn parallel to the coordinate axes such that $\vec{A} = \vec{A}_x + \vec{A}_y$.

NOTE ► It is not necessary for the tail of $\vec{A}$ to be at the origin. All we need to know is the *orientation* of the coordinate system so that we can draw $\vec{A}_x$ and $\vec{A}_y$ parallel to the axes. ◄

Components

You learned in Chapter 2 to give the one-dimensional kinematic variable v_x a positive sign if the velocity vector $\vec{v}$ points toward the positive end of the x-axis and a negative sign if $\vec{v}$ points in the negative x-direction. The basis of this rule is that v_x is the *x-component* of $\vec{v}$. We need to extend this idea to vectors in general.

Suppose we have a vector $\vec{A}$ that has been decomposed into component vectors $\vec{A}_x$ and $\vec{A}_y$ parallel to the coordinate axes. We can describe each component vector with a single number (a scalar) called the **component.** The *x-component* and *y-component* of vector $\vec{A}$, denoted A_x and A_y, are determined as follows:

TACTICS BOX 3.3 Determining the components of a vector

❶ The absolute value $|A_x|$ of the x-component A_x is the magnitude of the component vector $\vec{A}_x$.
❷ The *sign* of A_x is positive if $\vec{A}_x$ points in the positive x-direction, negative if $\vec{A}_x$ points in the negative x-direction.
❸ The y-component A_y is determined similarly.

Exercises 16–18

In other words, the component A_x tells us two things: how big $\vec{A}_x$ is and which end of the axis $\vec{A}_x$ points toward. FIGURE 3.14 shows three examples of determining the components of a vector.

FIGURE 3.14 Determining the components of a vector.

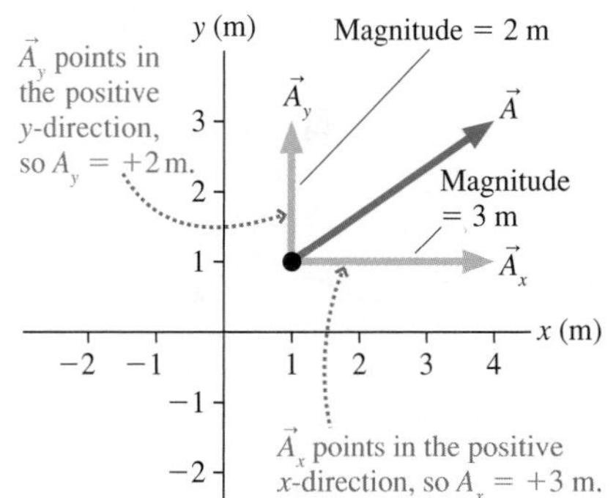

y (m)
Magnitude = 2 m
$\vec{A}_y$ points in the positive y-direction, so $A_y = +2$ m.
$\vec{A}_y$
$\vec{A}$
Magnitude = 3 m
$\vec{A}_x$
x (m)
$\vec{A}_x$ points in the positive x-direction, so $A_x = +3$ m.

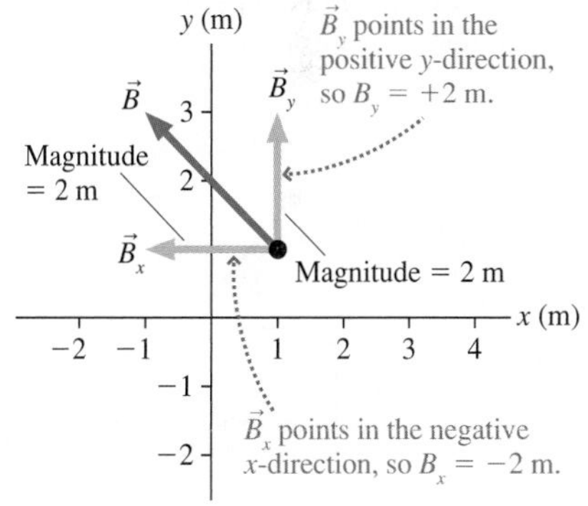

y (m)
$\vec{B}_y$ points in the positive y-direction, so $B_y = +2$ m.
$\vec{B}$
$\vec{B}_y$
Magnitude = 2 m
$\vec{B}_x$
Magnitude = 2 m
x (m)
$\vec{B}_x$ points in the negative x-direction, so $B_x = -2$ m.

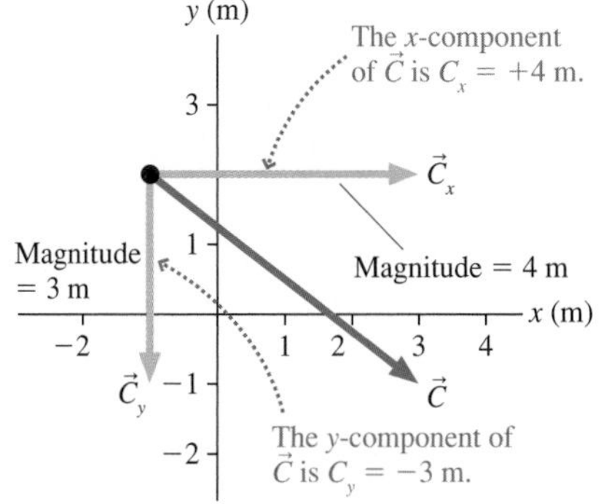

y (m)
The x-component of $\vec{C}$ is $C_x = +4$ m.
$\vec{C}_x$
Magnitude = 3 m
Magnitude = 4 m
x (m)
$\vec{C}_y$
$\vec{C}$
The y-component of $\vec{C}$ is $C_y = -3$ m.

NOTE ► $\vec{A}_x$ and $\vec{A}_y$ are *component vectors;* they have a magnitude and a direction. A_x and A_y are simply *components.* The components A_x and A_y are scalars—just numbers (with units) that can be positive or negative. ◄

Much of physics is expressed in the language of vectors. We will frequently need to decompose a vector into its components or to "reassemble" a vector from its components, moving back and forth between the graphical and the component representations of a vector.

Let's start with the problem of decomposing a vector into its x- and y-components. FIGURE 3.15a on the next page shows a vector $\vec{A}$ at an angle θ above horizontal. It is *essential* to use a picture or diagram such as this to define the angle you are using to describe a vector's direction. $\vec{A}$ points to the right and up, so Tactics Box 3.3 tells us that the components A_x and A_y are both positive.

FIGURE 3.15 Breaking a vector into components.

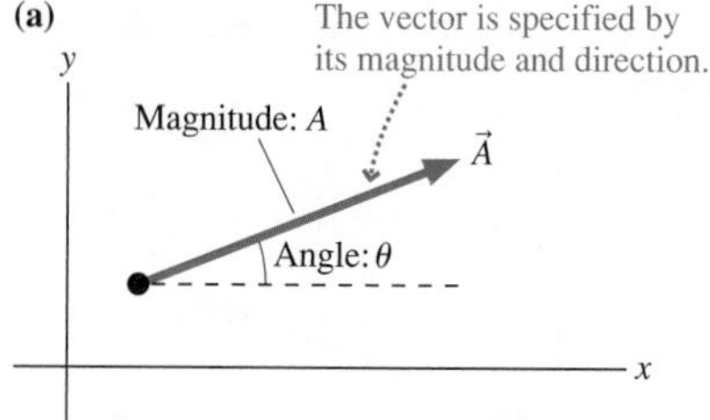

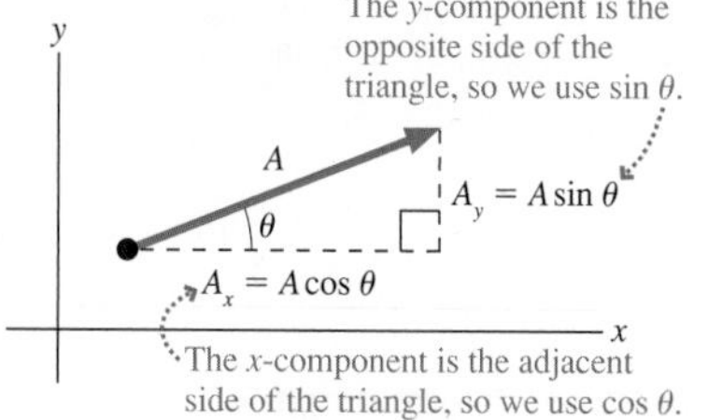

FIGURE 3.16 Specifying a vector from its components.

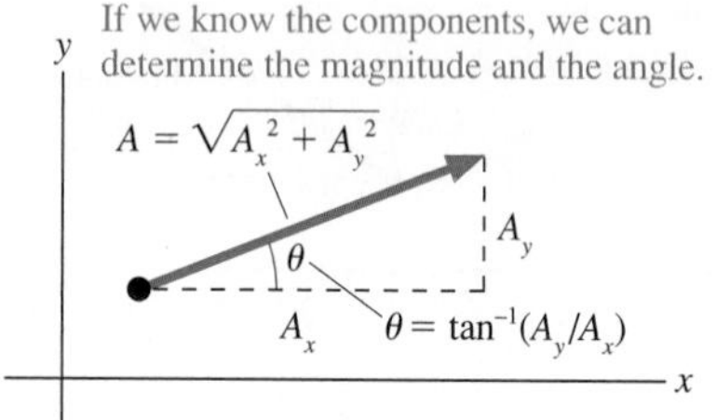

FIGURE 3.17 Relationships for a vector with a negative component.

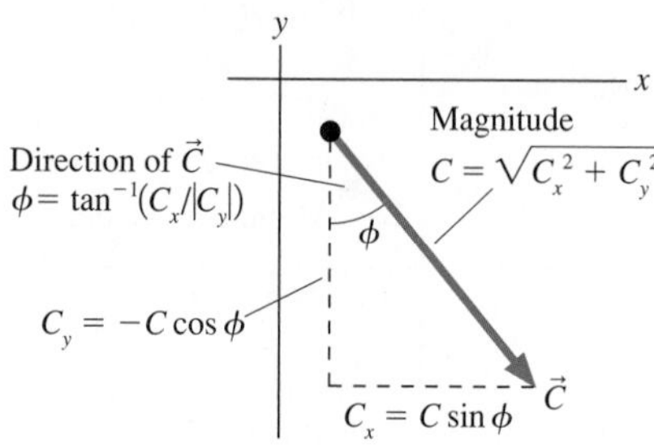

We can find the components using trigonometry, as illustrated in FIGURE 3.15b. For this case, we find that

$$A_x = A\cos\theta \qquad A_y = A\sin\theta \tag{3.10}$$

where A is the magnitude, or length, of $\vec{A}$. These equations convert the length and angle description of vector $\vec{A}$ into the vector's components, but they are correct *only* if θ is measured from horizontal.

Alternatively, if we are given the components of a vector, we can determine the length and angle of the vector from the x- and y-components, as shown in FIGURE 3.16. Because A in Figure 3.16 is the hypotenuse of a right triangle, its length is given by the Pythagorean theorem:

$$A = \sqrt{A_x^2 + A_y^2} \tag{3.11}$$

Similarly, the tangent of angle θ is the ratio of the opposite side to the adjacent side, so

$$\theta = \tan^{-1}\left(\frac{A_y}{A_x}\right) \tag{3.12}$$

Equations 3.11 and 3.12 can be thought of as the "inverse" of Equations 3.10.

How do things change if the vector isn't pointing to the right and up—that is, if one of the components is negative? FIGURE 3.17 shows vector $\vec{C}$ pointing to the right and down. In this case, the component vector $\vec{C}_y$ is pointing *down,* in the negative y-direction, so the y-component C_y is a *negative* number. The angle ϕ is drawn measured from the y-axis, so the components of $\vec{C}$ are

$$C_x = C\sin\phi \qquad C_y = -C\cos\phi \tag{3.13}$$

The roles of sine and cosine are reversed from those in Equations 3.10 because the angle ϕ is measured with respect to vertical, not horizontal.

NOTE ▶ Whether the x- and y-components use the sine or cosine depends on how you define the vector's angle. As noted above, you *must* draw a diagram to define the angle that you use, and you must be sure to refer to the diagram when computing components. Don't use Equations 3.10 or 3.13 as general rules—they aren't! They appear as they do because of how we defined the angles. ◀

Next, let's look at the "inverse" problem for this case: determining the length and direction of the vector given the components. The signs of the components don't matter for determining the length; the Pythagorean theorem always works to find the length or magnitude of a vector because the squares eliminate any concerns over the signs. The length of the vector in Figure 3.17 is simply

$$C = \sqrt{C_x^2 + C_y^2} \tag{3.14}$$

When we determine the direction of the vector from its components, we must consider the signs of the components. Finding the angle of vector $\vec{C}$ in Figure 3.17 requires the length of C_y *without* the minus sign, so vector $\vec{C}$ has direction

$$\phi = \tan^{-1}\left(\frac{C_x}{|C_y|}\right) \tag{3.15}$$

Notice that the roles of x and y differ from those in Equation 3.12.

EXAMPLE 3.4 **Finding the components of an acceleration vector**

Find the x- and y-components of the acceleration vector $\vec{a}$ shown in **FIGURE 3.18**.

FIGURE 3.18 Acceleration vector $\vec{a}$ of Example 3.4.

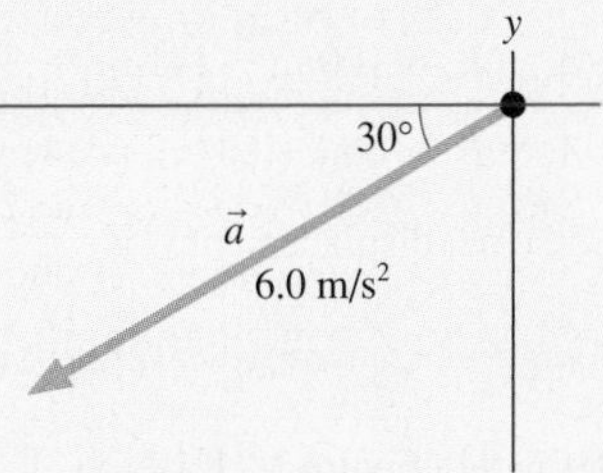

PREPARE It's important to *draw* the vectors. Making a sketch is crucial to setting up this problem. **FIGURE 3.19** shows the original vector $\vec{a}$ decomposed into component vectors parallel to the axes.

SOLVE The acceleration vector $\vec{a} = (6.0\text{ m/s}^2,\ 30°\text{ below the negative } x\text{-axis})$ points to the left (negative x-direction) and down (negative y-direction), so the components a_x and a_y are both negative:

FIGURE 3.19 The components of the acceleration vector.

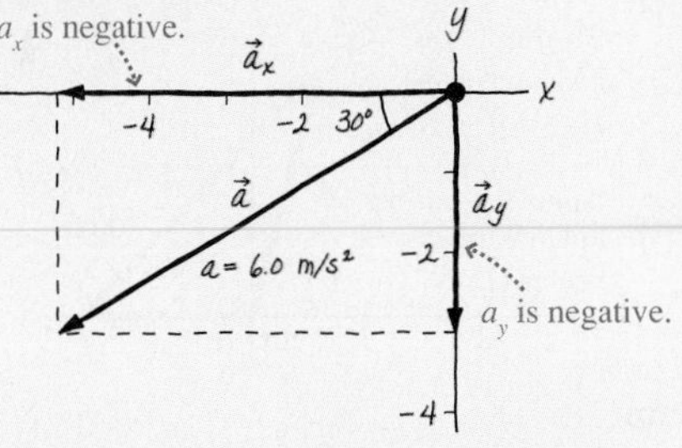

$$a_x = -a\cos 30° = -(6.0\text{ m/s}^2)\cos 30° = -5.2\text{ m/s}^2$$

$$a_y = -a\sin 30° = -(6.0\text{ m/s}^2)\sin 30° = -3.0\text{ m/s}^2$$

ASSESS The magnitude of the y-component is less than that of the x-component, as seems to be the case in Figure 3.19, a good check on our work. The units of a_x and a_y are the same as the units of vector $\vec{a}$. Notice that we had to insert the minus signs manually by observing that the vector points down and to the left.

STOP TO THINK 3.2 What are the x- and y-components C_x and C_y of vector $\vec{C}$?

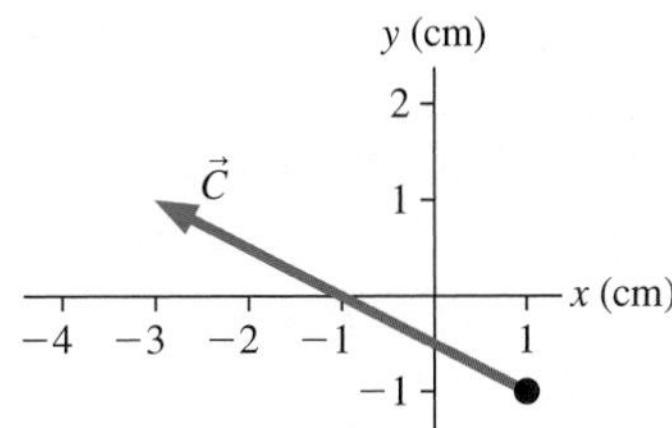

Working with Components

We've seen how to add vectors graphically, but there's an easier way: using components. To illustrate, let's look at the vector sum $\vec{C} = \vec{A} + \vec{B}$ for the vectors shown in **FIGURE 3.20**. You can see that the component vectors of $\vec{C}$ are the sums of the component vectors of $\vec{A}$ and $\vec{B}$. The same is true of the components: $C_x = A_x + B_x$ and $C_y = A_y + B_y$.

In general, if $\vec{D} = \vec{A} + \vec{B} + \vec{C} + \cdots$, then the x- and y-components of the resultant vector $\vec{D}$ are

$$\begin{aligned} D_x &= A_x + B_x + C_x + \cdots \\ D_y &= A_y + B_y + C_y + \cdots \end{aligned} \tag{3.16}$$

This method of vector addition is called *algebraic addition.*

FIGURE 3.20 Using components to add vectors.

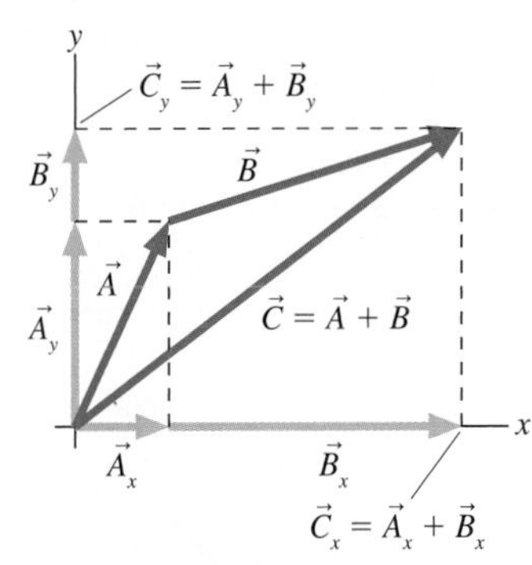

EXAMPLE 3.5 **Using algebraic addition to find a bird's displacement**

A bird flies 100 m due east from a tree, then 200 m northwest (that is, 45° north of west). What is the bird's net displacement?

PREPARE **FIGURE 3.21a** on the next page shows the displacement vectors $\vec{A} = (100\text{ m, east})$ and $\vec{B} = (200\text{ m, northwest})$ and also the net displacement $\vec{C}$. We draw vectors tip-to-tail if we are going to add them graphically, but it's usually easier to draw them all from the origin if we are going to use algebraic addition. **FIGURE 3.21b** redraws the vectors with their tails together.

Continued

FIGURE 3.21 Finding the net displacement.

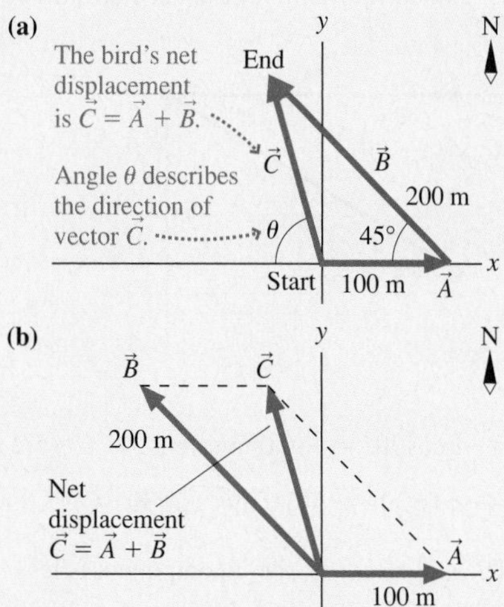

SOLVE To add the vectors algebraically we must know their components. From the figure these are seen to be

$$A_x = 100 \text{ m}$$
$$A_y = 0 \text{ m}$$
$$B_x = -(200 \text{ m})\cos 45° = -141 \text{ m}$$
$$B_y = (200 \text{ m})\sin 45° = 141 \text{ m}$$

We learned *from the figure* that $\vec{B}$ has a negative x-component. Adding $\vec{A}$ and $\vec{B}$ by components gives

$$C_x = A_x + B_x = 100 \text{ m} - 141 \text{ m} = -41 \text{ m}$$
$$C_y = A_y + B_y = 0 \text{ m} + 141 \text{ m} = 141 \text{ m}$$

The magnitude of the net displacement $\vec{C}$ is

$$C = \sqrt{C_x^2 + C_y^2} = \sqrt{(-41 \text{ m})^2 + (141 \text{ m})^2} = 147 \text{ m}$$

The angle θ, as defined in Figure 3.21, is

$$\theta = \tan^{-1}\left(\frac{C_y}{|C_x|}\right) = \tan^{-1}\left(\frac{141 \text{ m}}{41 \text{ m}}\right) = 74°$$

Thus the bird's net displacement is $\vec{C} = (147 \text{ m}, 74° \text{ north of west})$.

ASSESS The final values of C_x and C_y match what we would expect from the sketch in Figure 3.21. The geometric addition was a valuable check on the answer we found by algebraic addition.

When you look at a trail map for a hike in a mountainous region, it will give the length of a trail *and* the elevation gain—an important variable! The elevation gain is simply d_y, the vertical component of the displacement for the hike.

FIGURE 3.22 A coordinate system with tilted axes.

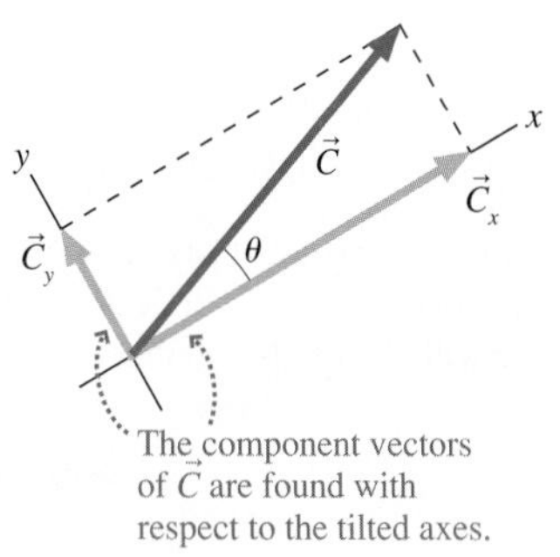

Vector subtraction and the multiplication of a vector by a scalar are also easily performed using components. To find $\vec{D} = \vec{P} - \vec{Q}$ we would compute

$$D_x = P_x - Q_x \qquad D_y = P_y - Q_y \tag{3.17}$$

Similarly, $\vec{T} = c\vec{S}$ is

$$T_x = cS_x \qquad T_y = cS_y \tag{3.18}$$

The next few chapters will make frequent use of *vector equations.* For example, you will learn that the equation to calculate the net force on a car skidding to a stop is

$$\vec{F} = \vec{n} + \vec{w} + \vec{f} \tag{3.19}$$

Equation 3.19 is really just a shorthand way of writing the two simultaneous equations:

$$F_x = n_x + w_x + f_x \qquad F_y = n_y + w_y + f_y \tag{3.20}$$

In other words, a vector equation is interpreted as meaning: Equate the x-components on both sides of the equals sign, then equate the y-components. Vector notation allows us to write these two equations in a more compact form.

Tilted Axes

Although we are used to having the x-axis horizontal, there is no requirement that it has to be that way. In Chapter 1, we saw that for motion on a slope, it is often most convenient to put the x-axis along the slope. When we add the y-axis, this gives us a tilted coordinate system such as that shown in FIGURE 3.22.

Finding components with tilted axes is no harder than what we have done so far. Vector $\vec{C}$ in Figure 3.22 can be decomposed into component vectors $\vec{C}_x$ and $\vec{C}_y$, with $C_x = C\cos\theta$ and $C_y = C\sin\theta$.

STOP TO THINK 3.3 Angle ϕ that specifies the direction of $\vec{C}$ is computed as

A. $\tan^{-1}(C_x/C_y)$.
B. $\tan^{-1}(C_x/|C_y|)$.
C. $\tan^{-1}(|C_x|/|C_y|)$.
D. $\tan^{-1}(C_y/C_x)$.
E. $\tan^{-1}(C_y/|C_x|)$.
F. $\tan^{-1}(|C_y|/|C_x|)$.

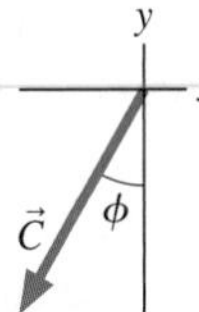

3.4 Motion on a Ramp

In this section, we will examine the problem of motion on a ramp or incline. There are three reasons to look at this problem. First, it will provide good practice at using vectors to analyze motion. Second, it is a simple problem for which we can find an exact solution. Third, this seemingly abstract problem has real and important applications.

We begin with a constant-velocity example to give us some practice with vectors and components before moving on to the more general case of accelerated motion.

EXAMPLE 3.6 Finding the height gained on a slope

A car drives up a steep 10° slope at a constant speed of 15 m/s. After 10 s, how much height has the car gained?

PREPARE FIGURE 3.23 is a visual overview, with x- and y-axes defined. The velocity vector $\vec{v}$ points up the slope. We are interested in the vertical motion of the car, so we decompose $\vec{v}$ into component vectors $\vec{v}_x$ and $\vec{v}_y$ as shown.

FIGURE 3.23 Visual overview of a car moving up a slope.

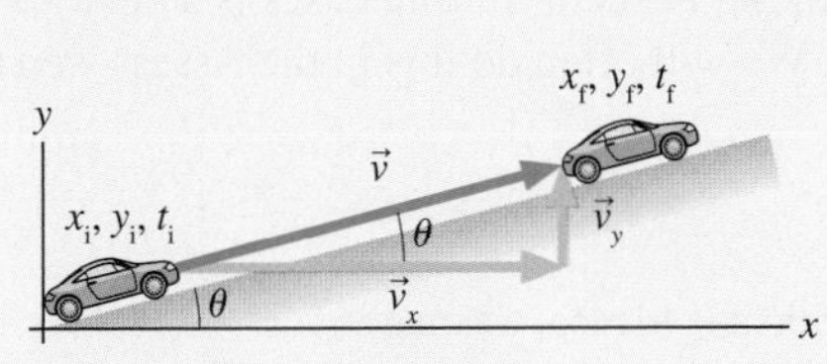

Known
$x_i = y_i = 0$ m
$t_i = 0$ s, $t_f = 10$ s
$v = 15$ m/s
$\theta = 10°$

Find
Δy

SOLVE The velocity component we need is v_y; this describes the vertical motion of the car. Using the rules for finding components outlined above, we find

$$v_y = v\sin\theta = (15 \text{ m/s})\sin(10°) = 2.6 \text{ m/s}$$

Because the velocity is constant, the car's vertical displacement (i.e., the height gained) during 10 s is

$$\Delta y = v_y\,\Delta t = (2.6 \text{ m/s})(10 \text{ s}) = 26 \text{ m}$$

ASSESS The car is traveling at a pretty good clip—15 m/s is a bit faster than 30 mph—up a steep slope, so it should climb a respectable height in 10 s. 26 m, or about 80 ft, seems reasonable.

Accelerated Motion on a Ramp

FIGURE 3.24a on the next page shows a crate sliding down a frictionless (i.e., smooth) ramp tilted at angle θ. The crate accelerates due to the action of gravity, but it is *constrained* to accelerate parallel to the surface. What is the acceleration?

A motion diagram for the crate is drawn in FIGURE 3.24b. There is an acceleration because the velocity is changing, with both the acceleration and velocity vectors parallel to the ramp. We can take advantage of the properties of vectors to find the crate's acceleration. To do so, FIGURE 3.24c sets up a coordinate system with the x-axis along the ramp and the y-axis perpendicular. All motion will be along the x-axis.

FIGURE 3.24 Acceleration on an inclined plane.

(a)

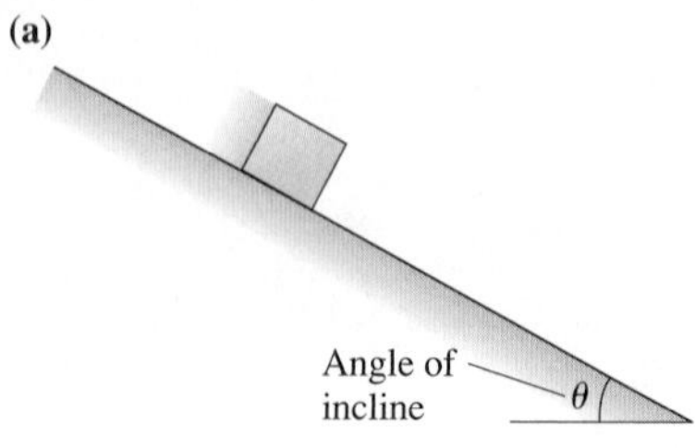
Angle of incline
θ

(b)

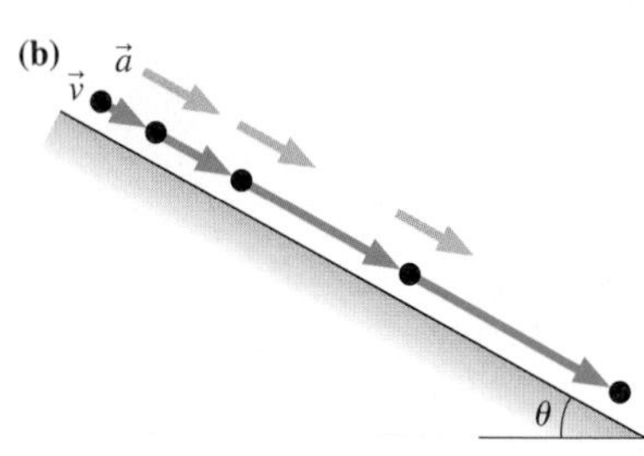
a
v
θ

(c)

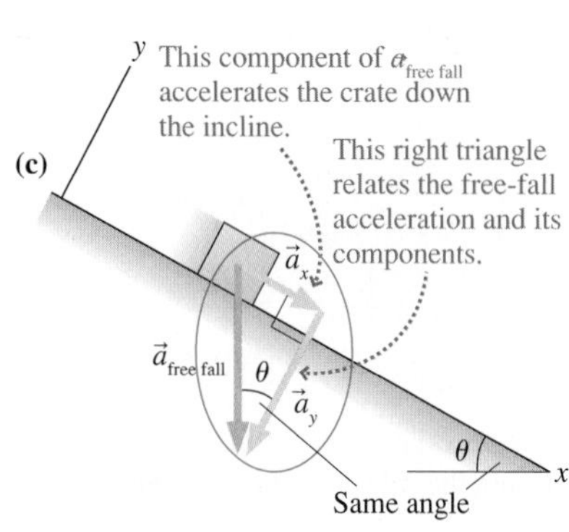
y
This component of $\vec{a}_{\text{free fall}}$ accelerates the crate down the incline.
This right triangle relates the free-fall acceleration and its components.
$\vec{a}_x$
$\vec{a}_{\text{free fall}}$
θ
$\vec{a}_y$
θ
x
Same angle

If the incline suddenly vanished, the object would have a free-fall acceleration $\vec{a}_{\text{free fall}}$ straight down. As Figure 3.24c shows, this acceleration vector can be decomposed into two component vectors: a vector $\vec{a}_x$ that is *parallel* to the incline and a vector $\vec{a}_y$ that is *perpendicular* to the incline. The vector addition rules studied earlier in this chapter tell us that $\vec{a}_{\text{free fall}} = \vec{a}_x + \vec{a}_y$.

The motion diagram shows that the object's actual acceleration $\vec{a}_x$ is parallel to the incline. The surface of the incline somehow "blocks" the other component of the acceleration $\vec{a}_y$, through a process we will examine in Chapter 5, but $\vec{a}_x$ is unhindered. It is this component of $\vec{a}_{\text{free fall}}$, parallel to the incline, that accelerates the object.

We can use trigonometry to work out the magnitude of this acceleration. Figure 3.24c shows that the three vectors $\vec{a}_{\text{free fall}}$, $\vec{a}_y$, and $\vec{a}_x$ form a right triangle with angle θ as shown; this angle is the same as the angle of the incline. By definition, the magnitude of $\vec{a}_{\text{free fall}}$ is g. This vector is the hypotenuse of the right triangle. The vector we are interested in, $\vec{a}_x$, is opposite angle θ. Thus the value of the acceleration along a frictionless slope is

$$a_x = \pm g \sin\theta \tag{3.21}$$

NOTE ► The correct sign depends on the direction in which the ramp is tilted. The acceleration in Figure 3.24 is $+g\sin\theta$, but upcoming examples will show situations in which the acceleration is $-g\sin\theta$. ◄

Let's look at Equation 3.21 to verify that it makes sense. A good way to do this is to consider some **limiting cases** in which the angle is at one end of its range. In these cases, the physics is clear and we can check our result. Let's look at two such possibilities:

1. Suppose the plane is perfectly horizontal, with $\theta = 0°$. If you place an object on a horizontal surface, you expect it to stay at rest with no acceleration. Equation 3.21 gives $a_x = 0$ when $\theta = 0°$, in agreement with our expectations.
2. Now suppose you tilt the plane until it becomes vertical, with $\theta = 90°$. You know what happens—the object will be in free fall, parallel to the vertical surface. Equation 3.21 gives $a_x = g$ when $\theta = 90°$, again in agreement with our expectations.

NOTE ► Checking your answer by looking at such limiting cases is a very good way to see if your answer makes sense. We will often do this in the Assess step of a solution. ◄

◄ **Extreme physics** A speed skier, on wide skis with little friction, wearing an aerodynamic helmet and crouched low to minimize air resistance, moves in a straight line down a steep slope—pretty much like an object sliding down a frictionless ramp. There is a maximum speed that a skier could possibly achieve at the end of the slope. Course designers set the starting point to keep this maximum speed within reasonable (for this sport!) limits.

EXAMPLE 3.7 Maximum possible speed for a skier

The Willamette Pass ski area in Oregon was the site of the 1993 U.S. National Speed Skiing Competition. The skiers started from rest and then accelerated down a stretch of the mountain with a reasonably constant slope, aiming for the highest possible speed at the end of this run. During this acceleration phase, the skiers traveled 360 m while dropping a vertical distance of 170 m. What is the fastest speed a skier could achieve at the end of this run? How much time would this fastest run take?

PREPARE We begin with the visual overview in **FIGURE 3.25**. The motion diagram shows the acceleration of the skier and the pictorial representation gives an overview of the problem including the dimensions of the slope. As before, we put the x-axis along the slope.

FIGURE 3.25 Visual overview of a skier accelerating down a slope.

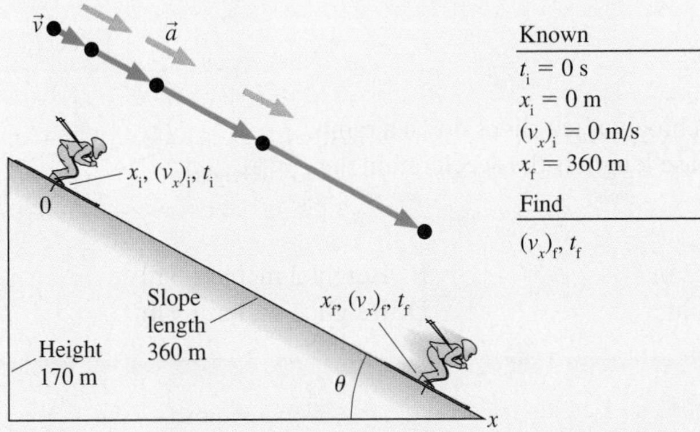

SOLVE The fastest possible run would be one without any friction or air resistance, meaning the acceleration down the slope is given by Equation 3.21. The acceleration is in the positive x-direction, so we use the positive sign. What is the angle in Equation 3.21? Figure 3.25 shows that the 360-m-long slope is the hypotenuse of a triangle of height 170 m, so we use trigonometry to find

$$\sin\theta = \frac{170 \text{ m}}{360 \text{ m}}$$

which gives $\theta = \sin^{-1}(170/360) = 28°$. Equation 3.21 then gives

$$a_x = +g\sin\theta = (9.8 \text{ m/s}^2)(\sin 28°) = 4.6 \text{ m/s}^2$$

For linear motion with constant acceleration, we can use the third of the kinematic equations in Table 2.4: $(v_x)_f^2 = (v_x)_i^2 + 2a_x\,\Delta x$. The initial velocity $(v_x)_i$ is zero; thus:

This is the distance along the slope, the length of the run.

$$(v_x)_f = \sqrt{2a_x\Delta x} = \sqrt{2(4.6 \text{ m/s}^2)(360 \text{ m})} = 58 \text{ m/s}$$

This is the fastest that any skier could hope to be moving at the end of the run. Any friction or air resistance would decrease this speed. Because the acceleration is constant and the initial velocity $(v_x)_i$ is zero, the time of the fastest-possible run is

$$\Delta t = \frac{(v_x)_f}{a_x} = \frac{58 \text{ m/s}}{4.6 \text{ m/s}^2} = 13 \text{ s}$$

A speed skiing event is a quick affair!

ASSESS The final speed we calculated is 58 m/s, which is about 130 mph, reasonable because we expect a high speed for this sport. In the competition noted, the actual winning speed was 111 mph, not much slower than the result we calculated. Obviously, the efforts to minimize friction and air resistance are working!

Skis on snow have very little friction, but there are other ways to reduce the friction between surfaces. For instance, a roller coaster car rolls along a track on low-friction wheels. No drive force is applied to the cars after they are released at the top of the first hill: the speed changes due to gravity alone. The cars speed up as they go down hills and slow down as they climb.

EXAMPLE 3.8 Speed of a roller coaster

A classic wooden coaster has cars that go down a big first hill, gaining speed. The cars then ascend a second hill with a slope of 30°. If the cars are going 25 m/s at the bottom and it takes them 2.0 s to climb this hill, how fast are they going at the top?

PREPARE We start with the visual overview in **FIGURE 3.26**, which includes a motion diagram, a pictorial representation, and a list of values. We've done this with a sketch such as you might draw for your homework. Notice how the motion diagram of Figure 3.26 differs from that of the previous example: The velocity decreases as the car moves up the hill, so the acceleration vector is opposite the direction of the velocity vector. The motion is along the x-axis, as before, but the acceleration vector points in the negative x-direction, so the component a_x is negative. In the motion diagram, notice that we drew only a single acceleration vector—a reasonable shortcut because we know that the acceleration is constant. One vector can represent the acceleration for the entire motion.

FIGURE 3.26 The coaster's speed decreases as it goes up the hill.

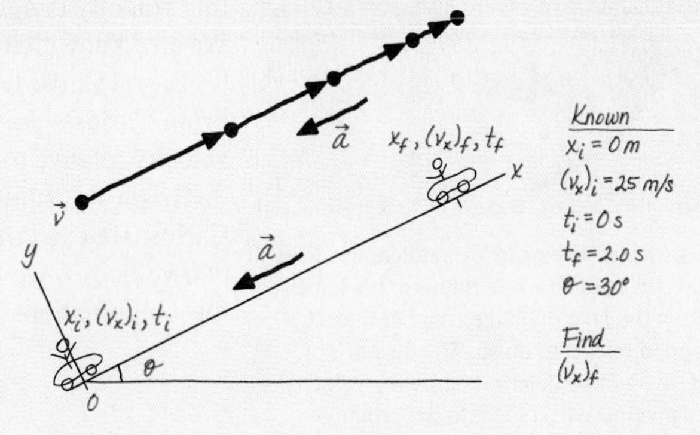

Continued

SOLVE To determine the final speed, we need to know the acceleration. We will assume that there is no friction or air resistance, so the magnitude of the roller coaster's acceleration is given by Equation 3.21 using the minus sign, as noted:

$$a_x = -g\sin\theta = -(9.8 \text{ m/s}^2)\sin 30° = -4.9 \text{ m/s}^2$$

The speed at the top of the hill can then be computed using our kinematic equation for velocity:

$$(v_x)_f = (v_x)_i + a_x\,\Delta t = 25 \text{ m/s} + (-4.9 \text{ m/s}^2)(2.0 \text{ s}) = 15 \text{ m/s}$$

ASSESS The speed is less at the top of the hill than at the bottom, as it should be, but the coaster is still moving at a pretty good clip at the top—almost 35 mph. This seems reasonable; a fast ride is a fun ride.

STOP TO THINK 3.4 A block of ice slides down a ramp. For which height and base length is the acceleration the greatest?

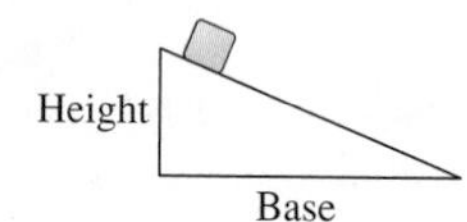

A. Height 4 m, base 12 m
B. Height 3 m, base 6 m
C. Height 2 m, base 5 m
D. Height 1 m, base 3 m

3.5 Relative Motion

FIGURE 3.27 Amy, Bill, and Carlos each measure the velocity of the runner. The velocities are shown relative to Amy.

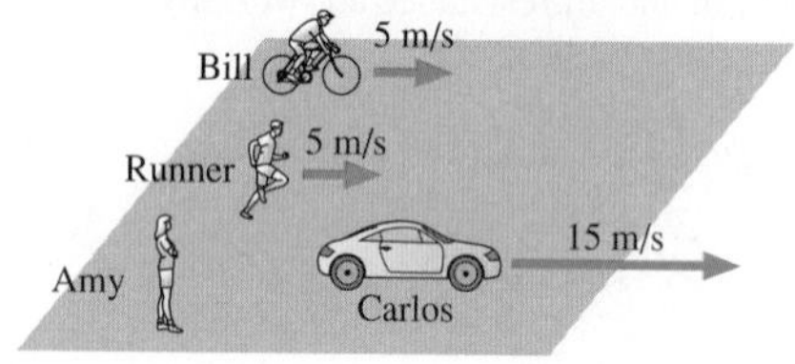

You've now dealt many times with problems that say something like "A car travels at 30 m/s" or "A plane travels at 300 m/s." But, as we will see, we may need to be a bit more specific.

In FIGURE 3.27, Amy, Bill, and Carlos are watching a runner. According to Amy, the runner's velocity is $v_x = 5$ m/s. But to Bill, who's riding alongside, the runner is lifting his legs up and down but going neither forward nor backward relative to Bill. As far as Bill is concerned, the runner's velocity is $v_x = 0$ m/s. Carlos sees the runner receding in his rearview mirror, in the *negative* x-direction, getting 10 m farther away from him every second. According to Carlos, the runner's velocity is $v_x = -10$ m/s. Which is the runner's *true* velocity?

Velocity is not a concept that can be true or false. The runner's velocity *relative to Amy* is 5 m/s; that is, his velocity is 5 m/s in a coordinate system attached to Amy and in which Amy is at rest. The runner's velocity relative to Bill is 0 m/s, and the velocity relative to Carlos is -10 m/s. These are all valid descriptions of the runner's motion.

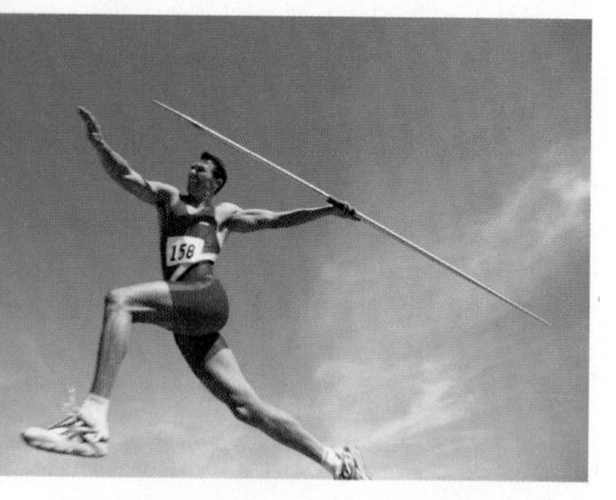

Throwing for the gold An athlete throwing the javelin does so while running. It's harder to throw the javelin on the run, but there's a very good reason to do so. The distance of the throw will be determined by the velocity of the javelin with respect to the ground—which is the sum of the velocity of the throw plus the velocity of the athlete. A faster run means a longer throw.

Relative Velocity

Suppose we know that the runner's velocity relative to Amy is 5 m/s; we will call this velocity $(v_x)_{RA}$. The second subscript "RA" means "**R**unner relative to **A**my." We also know that the velocity of **C**arlos relative to **A**my is 15 m/s; we write this as $(v_x)_{CA} = 15$ m/s. It is equally valid to compute Amy's velocity relative to Carlos. From Carlos's point of view, Amy is moving to the left at 15 m/s; we write Amy's velocity relative to Carlos as $(v_x)_{AC} = -15$ m/s; note that $(v_x)_{AC} = -(v_x)_{CA}$.

Given the runner's velocity relative to Amy and Amy's velocity relative to Carlos, we can compute the runner's velocity relative to Carlos by combining the two velocities we know. The subscripts as we have defined them are our guide for this combination:

$$(v_x)_{RC} = (v_x)_{RA} + (v_x)_{AC} \tag{3.22}$$

The "A" appears on the right of the first expression and on the left of the second; when we combine these velocities, we "cancel" the A to get $(v_x)_{RC}$.

Generally, you can add two relative velocities in this manner, by "canceling" subscripts as in Equation 3.22. In Chapter 27, when we learn about relativity, we will have a more rigorous scheme for computing relative velocities, but this technique will serve our purposes at present.

EXAMPLE 3.9 Speed of a seabird

Researchers doing satellite tracking of albatrosses in the Southern Ocean observed a bird maintaining sustained flight speeds of 35 m/s—nearly 80 mph! This seems surprisingly fast until you realize that this particular bird was flying with the wind, which was moving at 23 m/s. What was the bird's airspeed—its speed relative to the air? This is a truer measure of its flight speed.

PREPARE FIGURE 3.28 shows the wind and the albatross moving to the right, so all velocities will be positive. We've shown the velocity $(v_x)_{bw}$ of the **b**ird with respect to the **w**ater, which is the measured flight speed, and the velocity $(v_x)_{aw}$ of the **a**ir with respect to the **w**ater, which is the known wind speed. We want to find the bird's airspeed—the speed of the **b**ird with respect to the **a**ir.

FIGURE 3.28 Relative velocities for the albatross and the wind for Example 3.9.

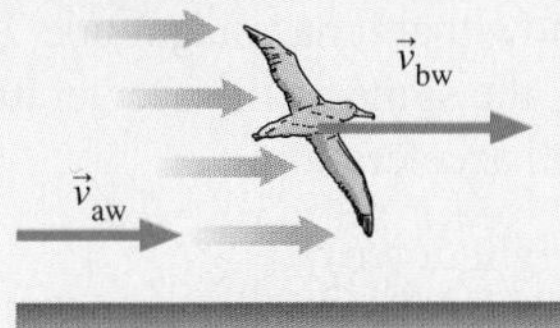

Known

$(v_x)_{bw} = 35$ m/s
$(v_x)_{aw} = 23$ m/s

Find

$(v_x)_{ba}$

SOLVE We need the subscript for the water to "cancel," so, according to Equation 3.22, we write

$$(v_x)_{ba} = (v_x)_{bw} + (v_x)_{wa}$$

The term $(v_x)_{wa}$, is the opposite of the second of our known values, so we use $(v_x)_{wa} = -(v_x)_{aw} = -23\text{ m/s}$ to find

$$(v_x)_{ba} = (35\text{ m/s}) + (-23\text{ m/s}) = 12\text{ m/s}$$

ASSESS 12 m/s—about 25 mph—is a reasonable airspeed for a bird. And it's slower than the observed flight speed, which makes sense because the bird is flying with the wind.

This technique for finding relative velocities also works for two-dimensional situations, as we see in the next example. Relative motion in two dimensions is another good exercise in working with vectors.

EXAMPLE 3.10 Finding the ground speed of an airplane

Cleveland is approximately 300 miles east of Chicago. A plane leaves Chicago flying due east at 500 mph. The pilot forgot to check the weather and doesn't know that the wind is blowing to the south at 100 mph. What is the plane's velocity relative to the ground?

PREPARE FIGURE 3.29 is a visual overview of the situation. We are given the speed of the **p**lane relative to the **a**ir ($\vec{v}_{pa}$) and the speed of the **a**ir relative to the **g**round ($\vec{v}_{ag}$); the speed of the **p**lane relative to the **g**round will be the vector sum of these velocities:

$$\vec{v}_{pg} = \vec{v}_{pa} + \vec{v}_{ag}$$

This vector sum is shown in Figure 3.29.

FIGURE 3.29 The wind causes a plane flying due east in the air to move to the southeast relative to the ground.

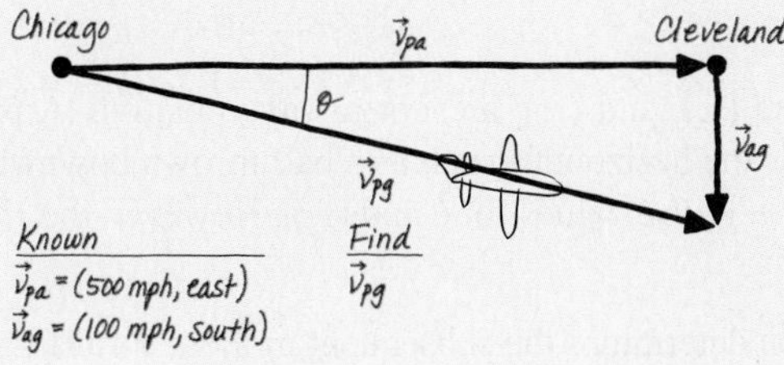

SOLVE The plane's speed relative to the ground is the hypotenuse of the right triangle in Figure 3.29; thus:

$$v_{pg} = \sqrt{v_{pa}^2 + v_{ag}^2} = \sqrt{(500\text{ mph})^2 + (100\text{ mph})^2} = 510\text{ mph}$$

The plane's direction can be specified by the angle θ measured from due east:

$$\theta = \tan^{-1}\left(\frac{100\text{ mph}}{500\text{ mph}}\right) = \tan^{-1}(0.20) = 11°$$

The velocity of the plane relative to the ground is thus

$$\vec{v}_{pg} = (510\text{ mph, } 11° \text{ south of east})$$

ASSESS The good news is that the wind is making the plane move a bit faster relative to the ground; the bad news is that the wind is making the plane move in the wrong direction!

FIGURE 3.30 The motion of a tossed ball. The inset shows how to find the direction of $\Delta\vec{v}$, the change in velocity. This is the direction in which the acceleration $\vec{a}$ points.

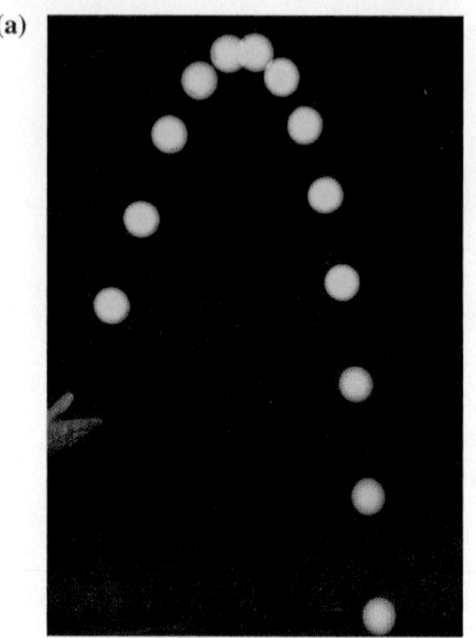

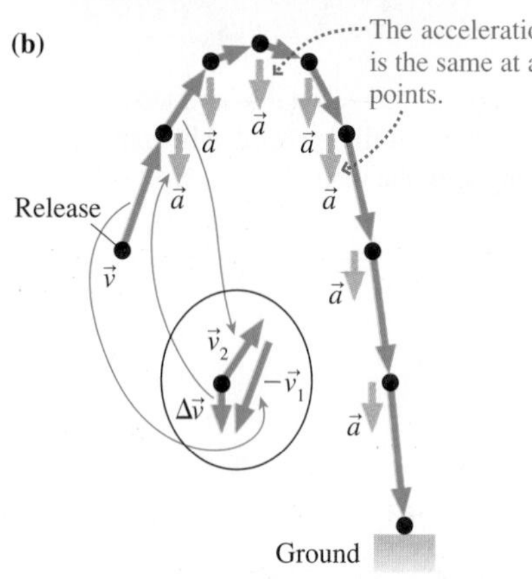

3.6 Motion in Two Dimensions: Projectile Motion

Balls flying through the air, long jumpers, and cars doing stunt jumps are all examples of the two-dimensional motion that we call *projectile motion.* Projectile motion is an extension to two dimensions of the free-fall motion we studied in Chapter 2. A **projectile is an object that moves in two dimensions under the influence of gravity and nothing else.** Although real objects are also influenced by air resistance, the effect of air resistance is small for reasonably dense objects moving at modest speeds, so we can ignore it for the cases we consider in this chapter. As long as we can neglect air resistance, any projectile will follow the same type of path: a trajectory with the mathematical form of a parabola. Because the form of the motion will always be the same, the strategies we develop to solve one projectile problem can be applied to others as well.

FIGURE 3.30a shows the parabolic arc of a ball tossed into the air; the camera has captured its position at equal intervals of time. In FIGURE 3.30b we show the motion diagram for this toss, with velocity vectors connecting the points. The acceleration vector points in the same direction as the change in velocity $\Delta\vec{v}$, which we can compute using the techniques of Tactics Box 3.2. You can see that the acceleration vector points straight down; a careful analysis would show that it has magnitude 9.80 m/s^2. Consequently, the acceleration of a projectile is the same as the acceleration of an object falling straight down—namely, the free-fall acceleration:

$$\vec{a}_{\text{free fall}} = (9.80\ \text{m/s}^2,\ \text{straight down})$$

Because the free-fall acceleration is the same for all objects, it is no wonder that the shape of the trajectory—a parabola—is the same as well.

As the projectile moves, the free-fall acceleration will change the vertical component of the velocity, but there will be no change to the horizontal component of the velocity. Therefore, the vertical and horizontal components of the acceleration are

$$\begin{aligned} a_x &= 0\ \text{m/s}^2 \\ a_y &= -g = -9.80\ \text{m/s}^2 \end{aligned} \tag{3.23}$$

The vertical component of acceleration a_y for all projectile motion is just the familiar $-g$ of free fall, while the horizontal component a_x is zero.

Analyzing Projectile Motion

FIGURE 3.31 The launch and motion of a projectile.

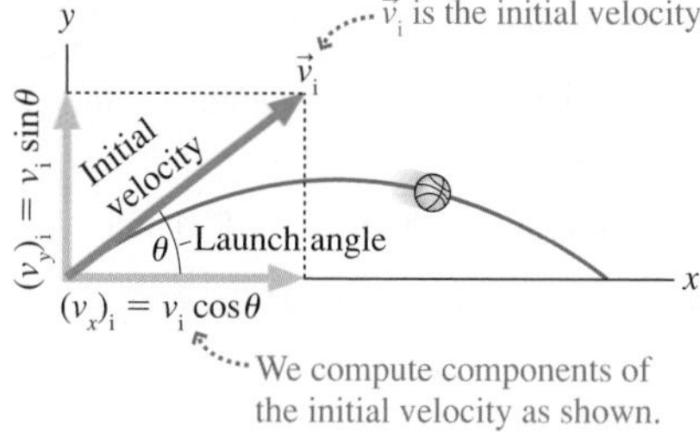

Suppose you toss a basketball down the court, as shown in FIGURE 3.31. To study this projectile motion, we've established a coordinate system with the x-axis horizontal and the y-axis vertical. The start of a projectile's motion is called the *launch,* and the angle θ of the initial velocity $\vec{v}_i$ above the horizontal (i.e., above the x-axis) is the **launch angle.** As you learned in Section 3.3, the initial velocity vector $\vec{v}_i$ can be expressed in terms of the x- and y-components $(v_x)_i$ and $(v_y)_i$. You can see from the figure that

$$\begin{aligned} (v_x)_i &= v_i \cos\theta \\ (v_y)_i &= v_i \sin\theta \end{aligned} \tag{3.24}$$

where v_i is the initial speed.

NOTE ▶ The components $(v_x)_i$ and $(v_y)_i$ are not always positive. A projectile launched at an angle *below* the horizontal (such as a ball thrown downward from the roof of a building) has *negative* values for θ and $(v_y)_i$. However, the *speed* v_i is always positive. ◀

To see how the acceleration determines the subsequent motion, FIGURE 3.32 shows a projectile launched at a speed of 22.0 m/s at an angle of 63° from the horizontal.

In Figure 3.32a, the initial velocity vector is broken into its horizontal and vertical components. In Figure 3.32b, the velocity vector and its component vectors are shown every subsequent 1.0 s. Because there is no horizontal acceleration ($a_x = 0$), the value of v_x never changes. In contrast, v_y decreases by 9.8 m/s every second. This is what it *means* to accelerate at $a_y = -9.8\ \text{m/s}^2 = (-9.8\ \text{m/s})$ per second. Nothing *pushes* the projectile along the curve. Instead, the downward acceleration changes the velocity vector as shown, causing it to increase downward as the motion proceeds. At the end of the motion, when the ball is at the same height as it started, v_y is -19.6 m/s, the negative of its initial value. **The ball finishes its motion moving downward at the same speed as it started moving upward,** just as we saw in the case of one-dimensional free fall in Chapter 2.

You can see from Figure 3.32 that **projectile motion is made up of two independent motions: uniform motion at constant velocity in the horizontal direction and free-fall motion in the vertical direction.** In Chapter 2, we saw kinematic equations for constant-velocity and constant-acceleration motion. We can adapt these general equations to this current case: The horizontal motion is constant-velocity motion at $(v_x)_i$; the vertical motion is constant-acceleration motion with initial velocity $(v_y)_i$ and an acceleration of $a_y = -g$.

$$x_f = x_i + (v_x)_i\,\Delta t \qquad y_f = y_i + (v_y)_i\,\Delta t - \tfrac{1}{2}g(\Delta t)^2$$
$$(v_x)_f = (v_x)_i = \text{constant} \qquad (v_y)_f = (v_y)_i - g\,\Delta t \tag{3.25}$$

Equations of motion for the parabolic trajectory of a projectile

FIGURE 3.32 The velocity and acceleration vectors of a projectile.

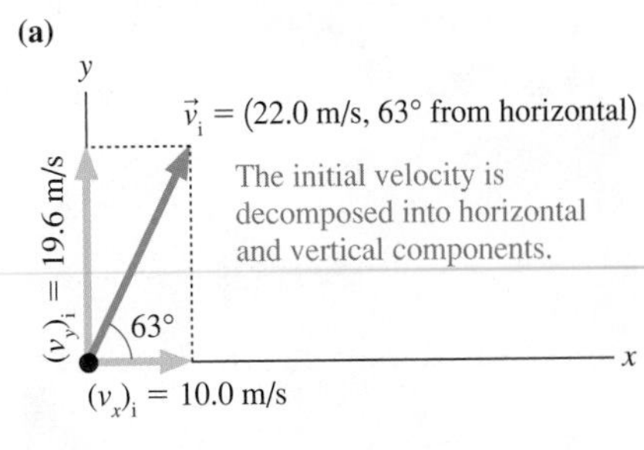
(a)
y
$\vec{v}_i$ = (22.0 m/s, 63° from horizontal)
The initial velocity is decomposed into horizontal and vertical components.
$(v_y)_i$ = 19.6 m/s
63°
x
$(v_x)_i$ = 10.0 m/s

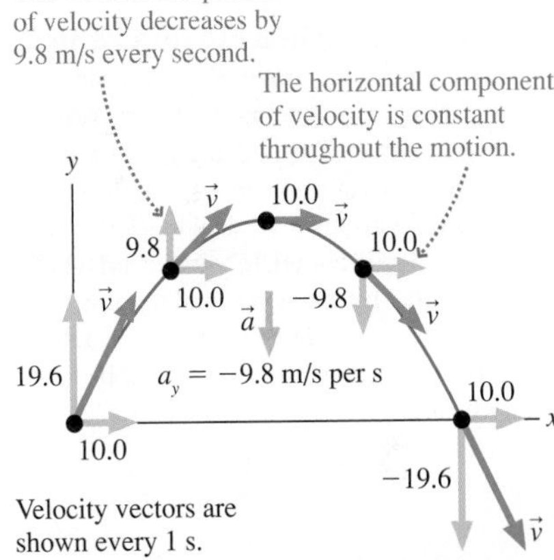
(b)
The vertical component of velocity decreases by 9.8 m/s every second.
The horizontal component of velocity is constant throughout the motion.
y
10.0
9.8
10.0
10.0
−9.8
a_y = −9.8 m/s per s
19.6
10.0
10.0
x
−19.6
Velocity vectors are shown every 1 s. Values are in m/s.
When the particle returns to its initial height, v_y is opposite its initial value.

A close look at these equations reveals a surprising fact: **The horizontal and vertical components of projectile motion are independent of each other.** The initial horizontal velocity has *no* influence over the vertical motion, and vice versa. This independence of the horizontal and vertical motions is illustrated in FIGURE 3.33, which shows a strobe photograph of two balls, one shot horizontally and the other released from rest at the same instant. The *vertical* motions of the two balls are identical, and they hit the floor simultaneously. Neither ball has any initial motion in the vertical direction, so both fall distance h in the same amount of time.

FIGURE 3.33 A projectile launched horizontally falls in the same time as a projectile that is released from rest.

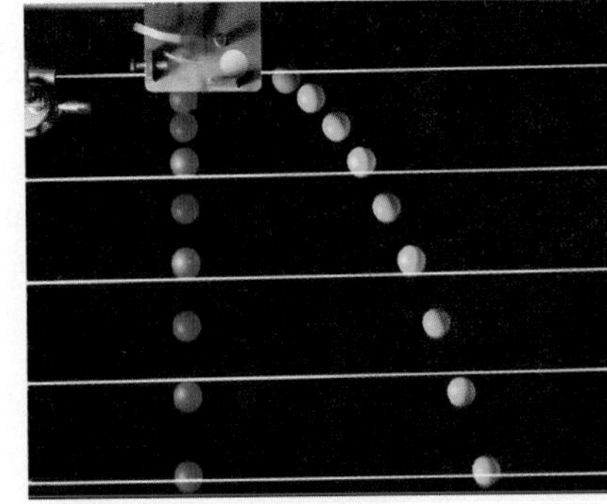

Let's extend these ideas to consider a "classic" problem in physics:

> A hungry hunter in the jungle wants to shoot down a coconut that is hanging from the branch of a tree. He aims the gun directly at the coconut, but as luck would have it the coconut falls from the branch at the exact instant the hunter pulls the trigger. Does the bullet hit the coconut?

FIGURE 3.34 shows a useful way to analyze this problem. Figure 3.34a shows the trajectory of a projectile. Without gravity, a projectile would follow a

FIGURE 3.34 A projectile follows a parabolic trajectory because it "falls" a distance $\frac{1}{2}gt^2$ below a straight-line trajectory.

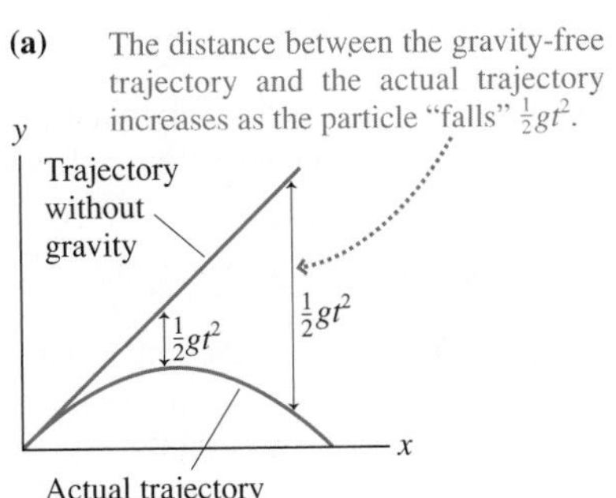
(a)
The distance between the gravity-free trajectory and the actual trajectory increases as the particle "falls" $\frac{1}{2}gt^2$.
y
Trajectory without gravity
$\frac{1}{2}gt^2$
$\frac{1}{2}gt^2$
x
Actual trajectory

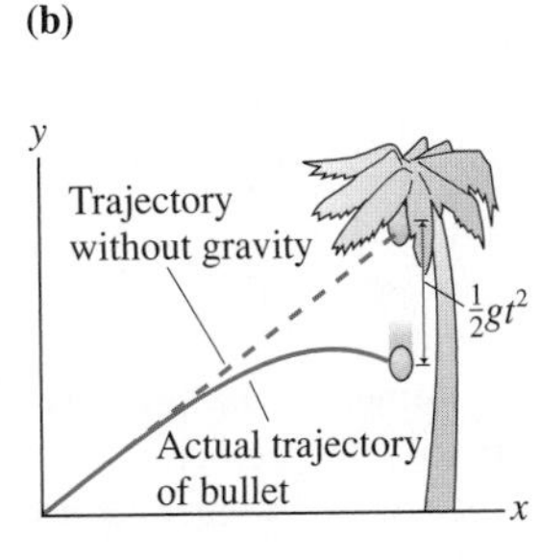
(b)
y
Trajectory without gravity
$\frac{1}{2}gt^2$
Actual trajectory of bullet
x

TRY IT YOURSELF

A game of catch in a moving vehicle While riding in a car moving at a constant speed, toss a ball or a coin into the air. You can easily catch it! The ball and you continue to move forward at a constant speed during the ball's up-and-down vertical motion. The vertical motion is completely independent of, and unaffected by, the horizontal motion. From the point of a view of a person watching you drive by, the ball's motion would be a parabolic arc.

straight line. Because of gravity, the particle at time t has "fallen" a distance $\frac{1}{2}gt^2$ below this line. The separation grows as $\frac{1}{2}gt^2$, giving the trajectory its parabolic shape. Figure 3.34b applies this reasoning to the bullet and coconut. Although the bullet travels very fast, it follows a slightly curved trajectory, not a straight line. Had the coconut stayed on the tree, the bullet would have curved under its target because gravity causes it to fall a distance $\frac{1}{2}gt^2$ below the straight line. But $\frac{1}{2}gt^2$ is also the distance the coconut falls while the bullet is in flight. Thus, as Figure 3.34b shows, the bullet and the coconut fall the same distance and meet at the same point!

STOP TO THINK 3.5 A 100 g ball rolls off a table and lands 2 m from the base of the table. A 200 g ball rolls off the same table with the same speed. How far does it land from the base of the table?

A. <1 m.
B. 1 m.
C. Between 1 m and 2 m.
D. 2 m.
E. Between 2 m and 4 m.
F. 4 m.

3.7 Projectile Motion: Solving Problems

Now that we have a good idea of how projectile motion works, we can use that knowledge to solve some true two-dimensional motion problems.

EXAMPLE 3.11 Planning a Hollywood stunt

To get the shots of cars flying through the air in movies, it is sometimes necessary to drive a car off a cliff and film it. Suppose a stunt man drives a car off a 10-m-high cliff at a speed of 20 m/s How far does the car land from the base of the cliff?

PREPARE We start with a visual overview of the situation in FIGURE 3.35. Note that we have chosen to put the origin of the coordinate system at the base of the cliff. We assume that the car is moving horizontally as it leaves the cliff. In this case, the x- and y-components of the initial velocity are

$$(v_x)_i = v_i = 20 \text{ m/s}$$
$$(v_y)_i = 0 \text{ m/s}$$

FIGURE 3.35 Visual overview for Example 3.11.

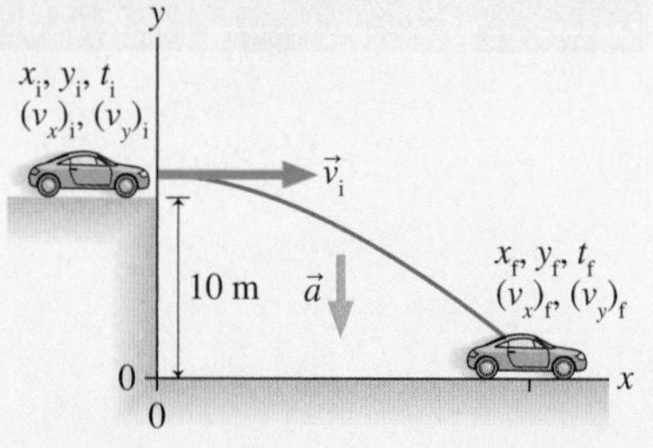

SOLVE Each point on the trajectory has x- and y-components of position, velocity, and acceleration but only *one* value of time. The time needed to move horizontally to the final position x_f is the *same* time needed to fall 10 m vertically. **Although the horizontal and vertical motions are independent, they are both related to the time t.** This is a critical observation for solving projectile motion problems. We will call the time interval between the car leaving the cliff and landing on the ground Δt. In this problem, we'll analyze the vertical motion first. We can solve the vertical-motion equations for the time interval Δt. We'll then use that value of Δt in the equation for the horizontal motion.

The vertical motion is just free fall. The initial vertical velocity is zero; the car falls from $y_i = 10$ m to $y_f = 0$ m. We can analyze this motion using the vertical-position equation from Equations 3.25:

$$y_f = y_i + (v_y)_i \,\Delta t - \tfrac{1}{2}g(\Delta t)^2$$

$$0 \text{ m} = 10 \text{ m} + (0 \text{ m/s})(\Delta t) - \tfrac{1}{2}(9.8 \text{ m/s}^2)(\Delta t)^2$$

Rearranging the terms and then solving for Δt give

$$-10 \text{ m} = -\tfrac{1}{2}(9.8 \text{ m/s}^2)(\Delta t)^2$$

$$\Delta t = \sqrt{\frac{2(10 \text{ m})}{9.8 \text{ m/s}^2}} = 1.43 \text{ s}$$

Now that we have the time, we can use the horizontal-position equation from Equations 3.25 to find out where the car lands:

$$x_f = x_i + (v_x)_i \,\Delta t$$
$$x_f = 0 \text{ m} + (20 \text{ m/s})(1.43 \text{ s}) = 29 \text{ m}$$

ASSESS The cliff height is $h \approx 33$ ft and the initial horizontal velocity is $(v_x)_i \approx 40$ mph. At this speed, a car moves faster than 60 feet per second, so traveling $x_f = 29 \text{ m} \approx 95$ ft before hitting the ground seems quite reasonable.

The approach of Example 3.11 is a general one. We can condense the relevant details into a problem-solving strategy.

PROBLEM-SOLVING STRATEGY 3.1 **Projectile motion problems** 

PREPARE There are a number of steps that you should go through in setting up the solution to a projectile motion problem:

- Make simplifying assumptions. Whether the projectile is a car or a basketball, the motion will be the same.
- Draw a visual overview including a pictorial representation showing the beginning and ending points of the motion.
- Establish a coordinate system with the x-axis horizontal and the y-axis vertical. In this case, you know that the horizontal acceleration will be zero and the vertical acceleration will be free fall: $a_x = 0$ and $a_y = -g$.
- Define symbols and write down a list of known values. Identify what the problem is trying to find.

SOLVE There are two sets of kinematic equations for projectile motion, one for the horizontal component and one for the vertical:

Horizontal	Vertical
$x_f = x_i + (v_x)_i \, \Delta t$	$y_f = y_i + (v_y)_i \, \Delta t - \frac{1}{2}g(\Delta t)^2$
$(v_x)_f = (v_x)_i = \text{constant}$	$(v_y)_f = (v_y)_i - g \, \Delta t$

Δt is the same for the horizontal and vertical components of the motion. Find Δt by solving for the vertical or the horizontal component of the motion; then use that value to complete the solution for the other component.

ASSESS Check that your result has the correct units, is reasonable, and answers the question.

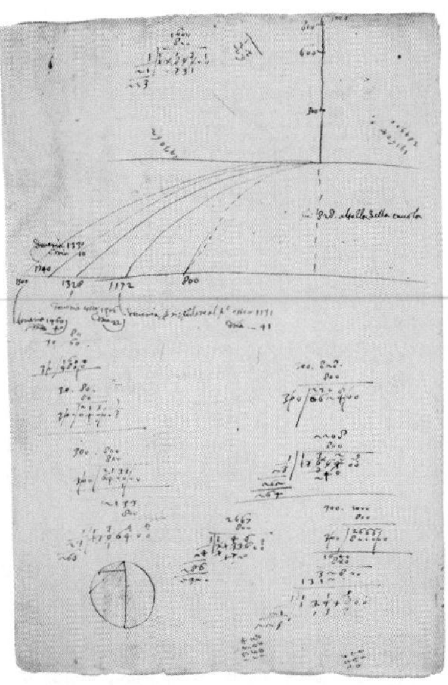

Galileo was the first person to make a serious study of projectile motion, deducing the independence of the horizontal and vertical components. This page from his notes shows his analysis of a projectile launched horizontally. In his day, this topic was cutting-edge science; now it is in Chapter 3 of a 30-chapter book!

ActivPhysics ONLINE 3.1, 3.2, 3.3, 3.4, 3.5, 3.6, 3.7

EXAMPLE 3.12 **Checking the feasibility of a Hollywood stunt**

The main characters in the movie *Speed* are on a bus that has been booby-trapped to explode if its speed drops below 50 mph. But there is a problem ahead: A 50 ft section of a freeway overpass is missing. They decide to jump the bus over the gap. The road leading up to the break has an angle of about 5°. A view of the speedometer just before the jump shows that the bus is traveling at 67 mph. The movie bus makes the jump and survives. Is this realistic, or movie fiction?

PREPARE We begin by converting speed and distance to SI units. The initial speed is $v_i = 30$ m/s and the size of the gap is $L = 15$ m. Next, following the problem-solving strategy, we make a sketch, the visual overview shown in FIGURE 3.36, and a list of values. In choosing our axes, we've placed the origin at the point where the bus starts its jump. The initial velocity vector is tilted 5° above horizontal, so the components of the initial velocity are

$$(v_x)_i = v_i \cos\theta = (30 \text{ m/s})(\cos 5°) = 30 \text{ m/s}$$

$$(v_y)_i = v_i \sin\theta = (30 \text{ m/s})(\sin 5°) = 2.6 \text{ m/s}$$

FIGURE 3.36 Visual overview of the bus jumping the gap.

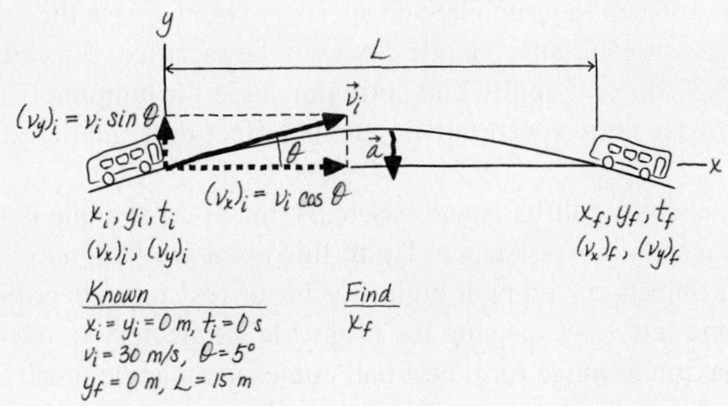

How do we specify the "end" of the problem? By setting $y_f = 0$ m, we'll solve for the horizontal distance x_f at which the bus returns to its initial height. If x_f exceeds 50 ft, the bus successfully clears the gap. We have optimistically drawn our diagram as if the bus makes the jump, but . . .

SOLVE Problem-Solving Strategy 3.1 suggests using one component of the motion to solve for Δt. We will begin with the vertical motion. The kinematic equation for the vertical position is

$$y_f = y_i + (v_y)_i \, \Delta t - \tfrac{1}{2}g(\Delta t)^2$$

We know that $y_f = y_i = 0$ m. If we factor out Δt, the position equation becomes

$$0 = \Delta t\left((v_y)_i - \tfrac{1}{2}g \, \Delta t\right)$$

One solution to this equation is $\Delta t = 0$ s. This is a legitimate solution, but it corresponds to the instant when $y = 0$ at the beginning

Continued

of the trajectory. We want the second solution, for $y = 0$ at the end of the trajectory, which is when

$$0 = (v_y)_i - \tfrac{1}{2}g\,\Delta t = (2.6\text{ m/s}) - \tfrac{1}{2}(9.8\text{ m/s}^2)\,\Delta t$$

which gives

$$\Delta t = \frac{2 \times (2.6\text{ m/s})}{9.8\text{ m/s}^2} = 0.53\text{ s}$$

During the 0.53 s that the bus is moving vertically it is also moving horizontally. The final horizontal position of the bus is $x_f = x_i + (v_x)_i\,\Delta t$, or

$$x_f = 0\text{ m} + (30\text{ m/s})(0.53\text{ s}) = 16\text{ m}$$

This is how far the bus has traveled horizontally when it returns to its original height. 16 m is a bit more than the width of the gap, so a bus coming off a 5° ramp at the noted speed would make it—just barely!

ASSESS We can do a quick check on our math by noting that the bus takes off and lands at the same height. This means, as we saw in Figure 3.32b, that the y-velocity at the landing should be the negative of its initial value. We can use the velocity equation for the vertical component of the motion to compute the final value and see that the final velocity value is as we predict:

$$(v_y)_f = (v_y)_i - g\,\Delta t$$
$$= (2.6\text{ m/s}) - (9.8\text{ m/s}^2)(0.53\text{ s}) = -2.6\text{ m/s}$$

During the filming of the movie, the filmmakers really did jump a bus over a gap in an overpass! The actual jump was a bit more complicated than our example because a real bus, being an extended object rather than a particle, will start rotating as the front end comes off the ramp. The actual stunt jump used an extra ramp to give a boost to the front end of the bus. Nonetheless, our example shows that the filmmakers did their homework and devised a situation in which the physics was correct.

The Range of a Projectile

When the quarterback throws a football down the field, how far will it go? What will be the **range** for this particular projectile motion, the horizontal distance traveled?

Example 3.12 was a range problem—for a given speed and a given angle, we wanted to know how far the bus would go. The speed and the angle are the two variables that determine the range. A higher speed means a greater range, of course. But how does angle figure in?

FIGURE 3.37 Trajectories of a projectile for different launch angles, assuming air resistance can be neglected.

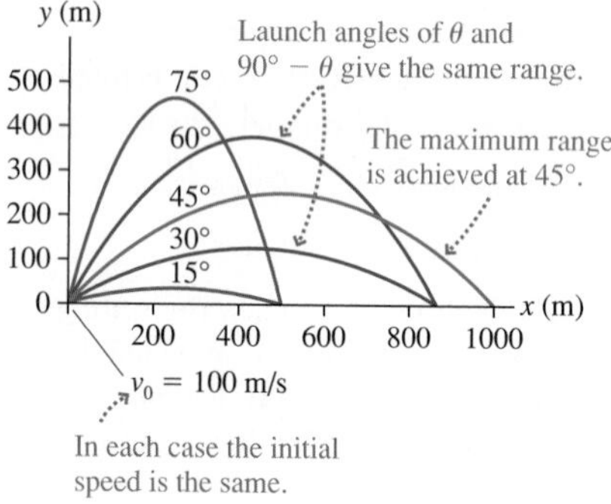

FIGURE 3.38 The trajectory of a long jumper.

FIGURE 3.37 shows the trajectory that a projectile launched at 100 m/s will follow for different launch angles. At very small or very large angles, the range is quite small. If you throw a ball at a 75° angle, it will do a great deal of up-and-down motion, but it won't achieve much horizontal travel. If you throw a ball at a 15° angle, the ball won't be in the air long enough to go very far. These cases both have the same range, as Figure 3.37 shows.

If the angle is too small or too large, the range is shorter than it could be. The "just right" case that gives the maximum range when landing at the same elevation as the launch is a launch angle of 45°, as Figure 3.37 shows.

If that's true, why does a long jumper take off at an angle that is so much less than 45°, as shown in FIGURE 3.38? One reason is that he changes the position of his legs as he jumps—he doesn't really land at the same height as that from which he took off, which changes things a bit. But there's a more important reason. In Figure 3.37 we looked at the range of projectiles that were launched at the *same speed* to see that a 45° angle gave the longest range. But the biomechanics of running and jumping don't allow you to keep the same launch speed as you increase the angle of your jump. Any increase in your launch angle comes at the sacrifice of speed, so the situation of Figure 3.37 doesn't apply. The optimum angle for jumping is less than 45° because your faster jump speed outweighs the effect of a smaller jump angle.

For other projectiles, such as golf balls and baseballs, the optimal angle is less than 45° for a different reason: air resistance. Up to this point we've ignored air resistance, but for small objects traveling at high speeds, air resistance is critical. Aerodynamic forces come into play, causing the projectile's trajectory to deviate from a parabola. The maximum range for a golf ball comes at an angle much less than 45°, as you no doubt know if you have ever played golf.

▶ **Physics of fielding** BIO The batter hits a high fly ball, and the fielder makes a graceful arc to the exact spot where it lands, catching it on the run. He didn't estimate velocity and calculate the ball's trajectory, so how did he do it? The key is that the fielder is in constant motion. He monitors the relative motion of the ball as he runs and makes adjustments in his velocity to keep the ball at a constant angle with respect to him. By doing this, he'll be at the right spot when the ball lands. He doesn't know where the ball will land—just how to be there when it does!

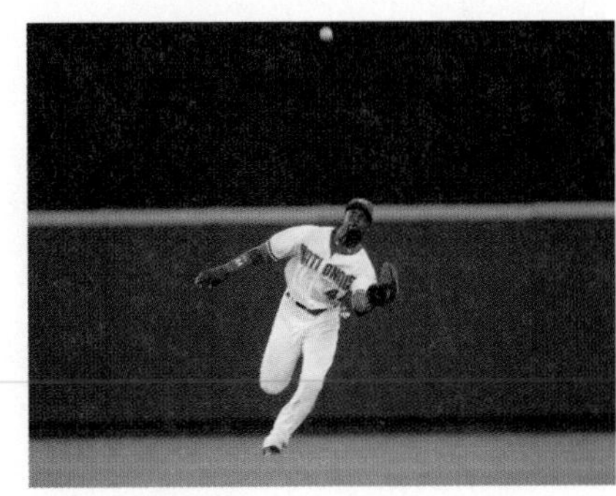

3.8 Motion in Two Dimensions: Circular Motion

The 32 cars on the London Eye Ferris wheel move at a constant speed of about 0.5 m/s in a vertical circle of radius 65 m. The cars may move at a constant speed, but they do *not* move with constant velocity. Velocity is a vector that depends on both an object's speed *and* its direction of motion, and the direction of circular motion is constantly changing. This is the hallmark of **uniform circular motion:** constant speed, but continuously changing direction. We will introduce some basic ideas about circular motion in this section, then return to treat it in considerably more detail in Chapter 6. For now, we will consider only objects that move around a circular trajectory at constant speed.

Period, Frequency, and Speed

The time interval it takes an object to go around a circle one time, completing one revolution (abbreviated rev), is called the **period** of the motion. Period is represented by the symbol T.

Rather than specify the time for one revolution, we can specify circular motion by its **frequency,** the number of revolutions per second, for which we use the symbol f. An object with a period of one-half second completes 2 revolutions each second. Similarly, an object can make 10 revolutions in 1 s if its period is one-tenth of a second. This shows that frequency is the inverse of the period:

$$f = \frac{1}{T} \tag{3.26}$$

The London Eye Ferris wheel.

Although frequency is often expressed as "revolutions per second," *revolutions* are not true units but merely the counting of events. Thus the SI unit of frequency is simply inverse seconds, or s^{-1}. Frequency may also be given in revolutions per minute (rpm) or another time interval, but these usually need to be converted to s^{-1} before doing calculations.

FIGURE 3.39 shows an object moving at a constant speed in a circular path of radius R. We know the time for one revolution—one period T—and we know the distance traveled, so we can write an equation relating the period, the radius, and the speed:

$$v = \frac{2\pi R}{T} \tag{3.27}$$

Given Equation 3.26 relating frequency and period, we can also write this equation as

$$v = 2\pi f R \tag{3.28}$$

FIGURE 3.39 Relating frequency and speed.

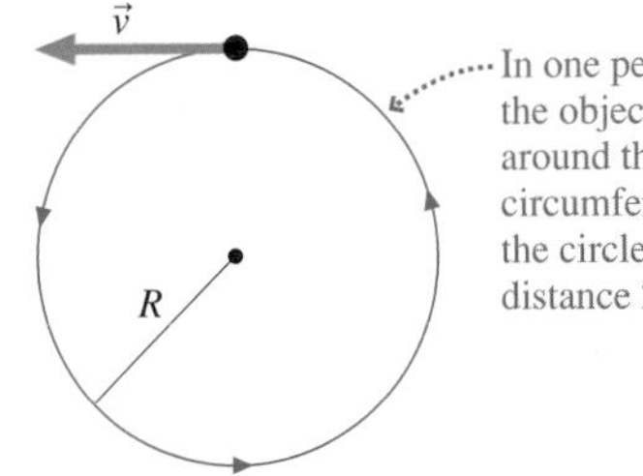

$\vec{v}$
In one period T, the object travels around the circumference of the circle, a distance $2\pi R$.
R

EXAMPLE 3.13 Spinning some tunes

An audio CD has a diameter of 120 mm and spins at up to 540 rpm. When a CD is spinning at its maximum rate, how much time is required for one revolution? If a speck of dust rides on the outside edge of the disk, how fast is it moving?

PREPARE Before we get started, we need to do some unit conversions. The diameter of a CD is given as 120 mm, which is 0.12 m. The radius is 0.060 m. The frequency is given in rpm; we need to convert this to s^{-1}:

$$f = 540\ \frac{\text{rev}}{\text{min}} \times \frac{1\ \text{min}}{60\ \text{s}} = 9.0\ \frac{\text{rev}}{\text{s}} = 9.0\ \text{s}^{-1}$$

Continued

SOLVE The time for one revolution is the period; this is given by Equation 3.26:

$$T = \frac{1}{f} = \frac{1}{9.0\ \text{s}^{-1}} = 0.11\ \text{s}$$

The dust speck is moving in a circle of radius 0.12 m at a frequency of $9.0\ \text{s}^{-1}$. We can use Equation 3.28 to find the speed:

$$v = 2\pi fR = 2\pi(9.0\ \text{s}^{-1})(0.060\ \text{m}) = 3.4\ \text{m/s}$$

ASSESS If you've watched a CD spin, you know that it takes much less than a second to go around, so the value for the period seems reasonable. The speed we calculate for the dust speck is nearly 8 mph, but for a point on the edge of the CD to go around so many times in a second, it must be moving pretty fast.

Acceleration in Circular Motion

FIGURE 3.40 The velocity and acceleration vectors for circular motion.

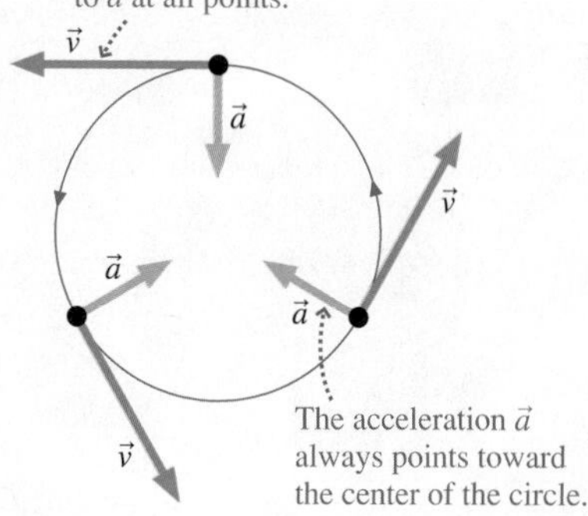

The velocity $\vec{v}$ is always tangent to the circle and perpendicular to $\vec{a}$ at all points.
$\vec{v}$
$\vec{a}$
The acceleration $\vec{a}$ always points toward the center of the circle.

It may seem strange to think that an object moving with constant speed can be accelerating, but that's exactly what an object in uniform circular motion is doing. It is accelerating because its velocity is changing as its direction of motion changes. What is the acceleration in this case? We saw in Example 3.3 that for circular motion at a constant speed, **the acceleration vector $\vec{a}$ points toward the center of the circle.** This is an idea that is worth reviewing. As you can see in FIGURE 3.40, the velocity is always tangent to the circle, so $\vec{v}$ and $\vec{a}$ are perpendicular to each other at all points on the circle.

An acceleration that always points directly toward the center of a circle is called a **centripetal acceleration.** The word "centripetal" comes from a Greek root meaning "center seeking."

NOTE ▶ Centripetal acceleration is not a new type of acceleration; all we are doing is *naming* an acceleration that corresponds to a particular type of motion. The magnitude of the centripetal acceleration is constant because each successive $\Delta\vec{v}$ in the motion diagram has the same length. ◀

To complete our description of circular motion, we need to find a quantitative relationship between the magnitude of the acceleration a and the speed v. Let's return to the case of the Ferris wheel. During a time Δt in which a car on the Ferris wheel moves around the circle from point 1 to point 2, the car moves through an angle θ and undergoes a displacement $\vec{d}$, as shown in FIGURE 3.41a. We've chosen a relatively large angle θ for our drawing so that angular relationships can be clearly seen, but for a small angle the displacement is essentially identical to the actual distance traveled, and we'll make this approximation.

FIGURE 3.41b shows how the velocity changes as the car moves, and FIGURE 3.41c shows the vector calculation of the change in velocity. The triangle we use to make this calculation is geometrically *similar* to the one that shows the displacement, as

FIGURE 3.41 Changing position and velocity for an object in circular motion.

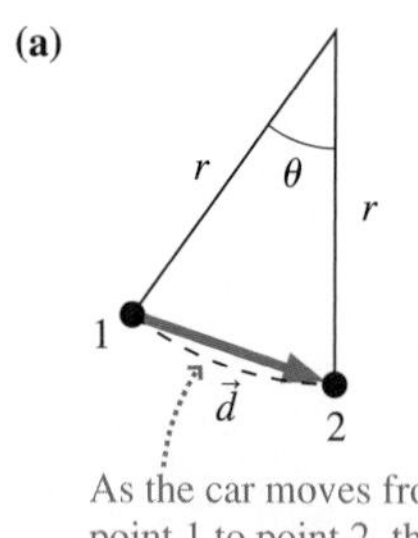

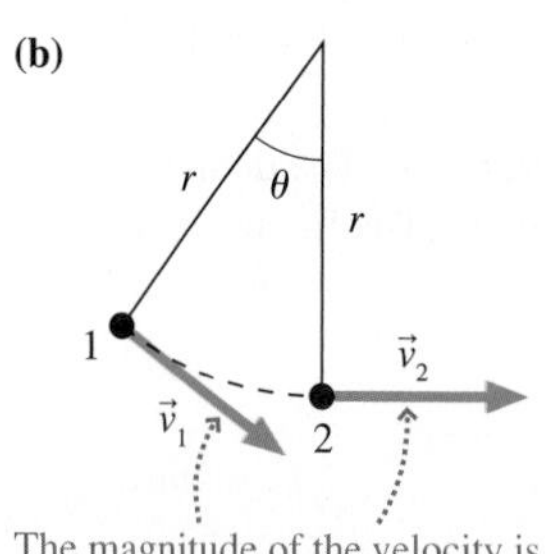

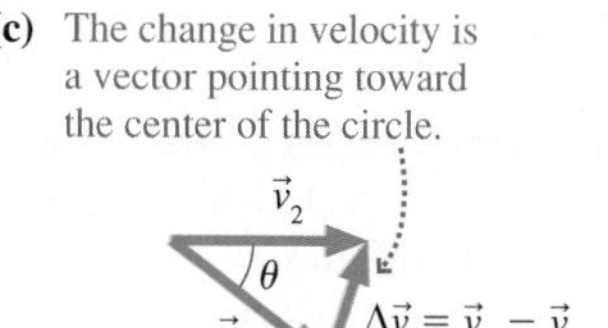

This triangle is the same as in part a, but rotated.

Figure 3.41c shows. This is a key piece of information: You'll remember from geometry that similar triangles have equal ratios of their sides, so we can write

$$\frac{\Delta v}{v} = \frac{d}{r} \tag{3.29}$$

where Δv is the magnitude of the velocity-change vector $\Delta\vec{v}$. We've used the unsubscripted speed v for the length of a side of the first triangle because it is the same for velocities $\vec{v}_1$ and $\vec{v}_2$.

Now we're ready to compute the acceleration. The displacement is just the speed v times the time interval Δt, so we can write

$$d = v\Delta t$$

We can substitute this for d in Equation 3.29 to obtain

$$\frac{\Delta v}{v} = \frac{v\Delta t}{r}$$

which we can rearrange like so:

$$\frac{\Delta v}{\Delta t} = \frac{v^2}{r}$$

We recognize the left-hand side of the equation as the acceleration, so this becomes

$$a = \frac{v^2}{r}$$

Combining this magnitude with the direction we noted above, we can write the centripetal acceleration as

$$\vec{a} = \left(\frac{v^2}{r}, \text{ toward center of circle}\right) \tag{3.30}$$

Centripetal acceleration of object moving in a circle of radius r at speed v

QUADRATIC p.47

CONCEPTUAL EXAMPLE 3.14 **Acceleration on a swing**

A child is riding a playground swing. The swing rotates in a circle around a central point where the rope or chain for the swing is attached. The speed isn't changing at the lowest point of the motion, but the direction is—this is circular motion, with an acceleration directed upward, as shown in **FIGURE 3.42**. More acceleration will mean a more exciting ride. What change could the child make to increase the acceleration she experiences?

FIGURE 3.42 A child at the lowest point of motion on a swing.

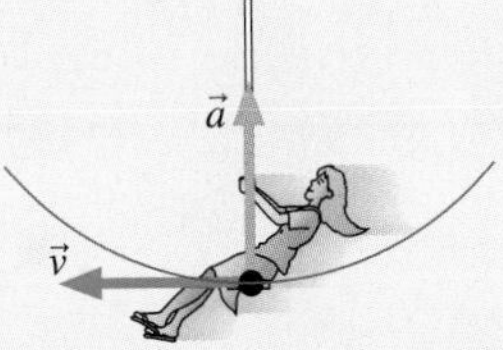

REASON The acceleration the child experiences is the "changing direction" acceleration of circular motion, given by Equation 3.30. The acceleration depends on the speed and the radius of the circle. The radius of the circle is determined by the length of the chain or rope, so the only easy way to change the acceleration is to change the speed, which she could do by swinging higher. Because the acceleration is proportional to the square of the speed, doubling the speed means a fourfold increase in the acceleration.

ASSESS If you have ever ridden a swing, you know that the acceleration you experience is greater the faster you go—so our answer makes sense.

EXAMPLE 3.15 Finding the acceleration of a Ferris wheel

A typical carnival Ferris wheel has a radius of 9.0 m and rotates 6.0 times per minute. What magnitude acceleration do the riders experience?

PREPARE The cars on a Ferris wheel move in a circle at constant speed; the acceleration the riders experience is a centripetal acceleration.

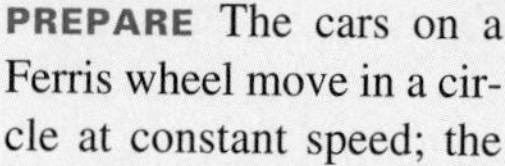

SOLVE In order to use Equation 3.30 to compute an acceleration, we need to know the speed v of a rider on the Ferris wheel. The wheel rotates 6.0 times per minute; therefore, the time for one rotation (i.e., the period) is 10 s. We can use Equation 3.27 to find the speed:

$$v = \frac{2\pi R}{T} = \frac{(2\pi)(9.0 \text{ m})}{10 \text{ s}} = 5.7 \text{ m/s}$$

Knowing the speed, we can use Equation 3.30 to find the magnitude of the acceleration:

$$a = \frac{v^2}{r} = \frac{(5.7 \text{ m/s})^2}{9.0 \text{ m}} = 3.6 \text{ m/s}^2$$

ASSESS This is about 1/3 of the free-fall acceleration; the acceleration, in units of g, is $0.37g$. This is enough to notice, but not enough to be scary! Our answer seems reasonable.

What Comes Next: Forces

So far we have been studying motion without saying too much about what actually *causes* motion. Kinematics, the mathematical description of motion, is a good place to start because motion is very visible and very familiar. And in our study of motion we have introduced many of the basic tools, such as vectors, that we will use in the rest of the book.

But now it's time to look at what causes motion: forces. By learning about forces, which you will do in the next several chapters, you will be able to explore a much wider range of problems in much more depth. As an example, think about the picture of a roller coaster with an inverted loop. How is it that riders can go through the loop and not fall out of their seats? This is just one of the problems that you will study once you know a bit about forces and the connection between forces and motion.

◀ **Amusement park kinematics** Acceleration is fun—at least that's what the designer of this roller coaster seems to think! The coaster has ramps that give linear acceleration, parabolic segments in which the coaster follows a projectile path with a free-fall acceleration, and circular arcs in which the centripetal acceleration is greater than g. All of this acceleration means there are forces on the riders—and the coaster must be carefully designed so that these forces are well within safe limits.

STOP TO THINK 3.6 Which of the following particles has the greatest centripetal acceleration?

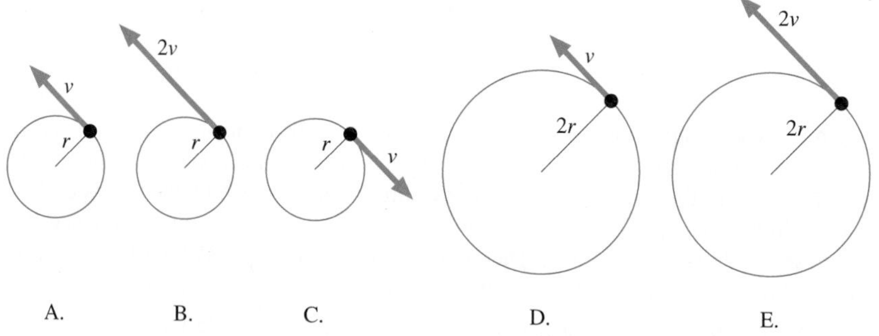

INTEGRATED EXAMPLE 3.16 World-record jumpers

Frogs, with their long, strong legs, are excellent jumpers. And thanks to the good folks of Calaveras County, California, who have a jumping frog contest every year in honor of a Mark Twain story, we have very good data as to just how far a determined frog can jump. The current record holder is Rosie the Ribeter, a bullfrog who made a leap of 6.5 m from a standing start. This compares favorably with the world record for a human, which is a mere 3.7 m.

Typical data for a serious leap by a bullfrog look like this: The frog goes into a crouch, then rapidly extends its legs by 15 cm as it pushes off, leaving the ground at an angle of 30° to the horizontal. It's in the air for 0.68 s before landing at the same height from which it took off. Given this leap, what is the acceleration while the frog is pushing off? How far does the frog jump?

PREPARE The problem really has two parts: the leap through the air and the acceleration required to produce this leap. We'll need to analyze the leap—the projectile motion—first, which will give us the frog's launch speed and the distance of the jump. Once we know the velocity with which the frog leaves the ground, we can calculate its acceleration while pushing off the ground. Let's start with a visual overview of the two parts, as shown in **FIGURE 3.43**. Notice that the second part of the problem uses a different x-axis, tilted as we did earlier for motion on a ramp.

SOLVE The "flying through the air" part of Figure 3.43a is projectile motion. The frog lifts off at a 30° angle with a speed v_i; the x- and y-components of the initial velocity are

$$(v_x)_i = v_i \cos(30°)$$
$$(v_y)_i = v_i \sin(30°)$$

The vertical motion can be analyzed as we did in Example 3.12. The kinematic equation is

$$y_f = y_i + (v_y)_i \Delta t + \tfrac{1}{2} a_y (\Delta t)^2$$

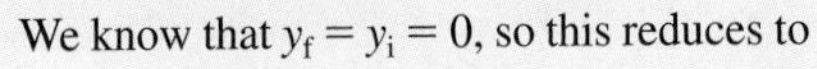

We know that $y_f = y_i = 0$, so this reduces to

$$(v_y)_i = -\tfrac{1}{2} a_y \Delta t = -\tfrac{1}{2}(-9.8 \text{ m/s}^2)(0.68 \text{ s}) = 3.3 \text{ m/s}$$

We know the y-component of the velocity and the angle, so we can find the magnitude of the velocity and the x-component:

$$v_i = \frac{(v_y)_i}{\sin 30°} = \frac{3.3 \text{ m/s}}{\sin 30°} = 6.6 \text{ m/s}$$

$$(v_x)_i = v_i \cos 30° = (6.6 \text{ m/s}) \cos 30° = 5.7 \text{ m/s}$$

The horizontal motion is uniform motion, so the frog's horizontal position when it returns to the ground is

$$x_f = x_i + (v_x)_i \Delta t = 0 + (5.7 \text{ m/s})(0.68 \text{ s}) = 3.9 \text{ m}$$

This is the length of the jump.

Now that we know how fast the frog is going when it leaves the ground, we can calculate the acceleration necessary to produce this jump—the "pushing off the ground" part of Figure 3.43b. We've drawn the x-axis along the direction of motion, as we did for problems of motion on a ramp. We know the displacement Δx of the jump but not the time, so we can use the third equation in Table 2.4:

$$(v_x)_f^2 = (v_x)_i^2 + 2a_x \Delta x$$

The initial velocity is zero, the final velocity is $(v_x)_f = 6.6$ m/s, and the displacement is the 15 cm (or 0.15 m) stretch of the legs during the jump. Thus the frog's acceleration while pushing off is

$$a_x = \frac{(v_x)_f^2}{2\Delta x} = \frac{(6.6 \text{ m/s})^2}{2(0.15 \text{ m})} = 150 \text{ m/s}^2$$

ASSESS A 3.9 m jump is more than a human can achieve, but it's less than the record for a frog, so the final result for the distance seems reasonable. Such a long jump must require a large acceleration during the pushing-off phase, which is what we found.

FIGURE 3.43 A visual overview for the leap of a frog.

(a) Flying through the air

Known
$x_i = 0$ m, $y_i = 0$ m, $t_i = 0$ s
$y_f = 0$ m, $\Delta t = 0.68$ s
$\theta = 30°$
$a_y = -9.8$ m/s^2

Find
v_i
x_f

(b) Pushing off the ground

Known
$(v_x)_i = 0$ m/s
$(v_x)_f$
$x_f = 0.15$ m

Find
a_x

The initial velocity for flying through the air is the final velocity for pushing off the ground.

SUMMARY

The goals of Chapter 3 have been to learn more about vectors and to use vectors as a tool to analyze motion in two dimensions.

GENERAL PRINCIPLES

Projectile Motion

A projectile is an object that moves through the air under the influence of gravity and nothing else.

The path of the motion is a parabola.

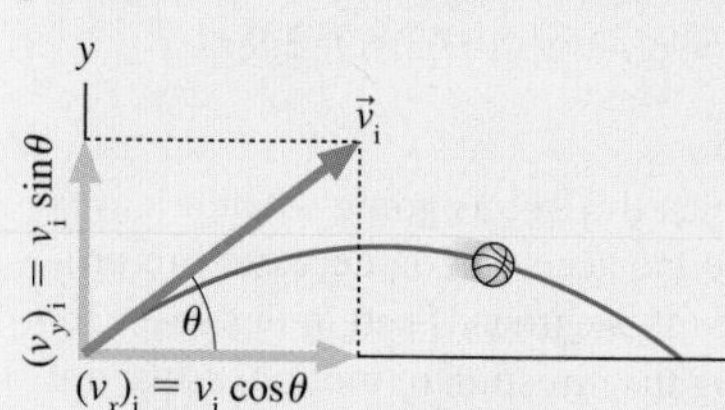

The motion consists of two pieces:

1. Vertical motion with free-fall acceleration, $a_y = -g$.
2. Horizontal motion with constant velocity.

Kinematic equations:

$$x_f = x_i + (v_x)_i \Delta t$$

$$(v_x)_f = (v_x)_i = \text{constant}$$

$$y_f = y_i + (v_y)_i \Delta t - \tfrac{1}{2}g(\Delta t)^2$$

$$(v_y)_f = (v_y)_i - g\,\Delta t$$

Circular Motion

For an object moving in a circle at a constant speed:

- The period T is the time for one rotation.
- The frequency $f = 1/T$ is the number of revolutions per second.
- The velocity is tangent to the circular path.
- The acceleration points toward the center of the circle and has magnitude

$$a = \frac{v^2}{r}$$

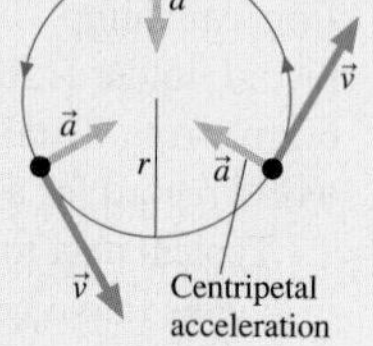

IMPORTANT CONCEPTS

Vectors and Components

A vector can be decomposed into *x*- and *y*-**components.**

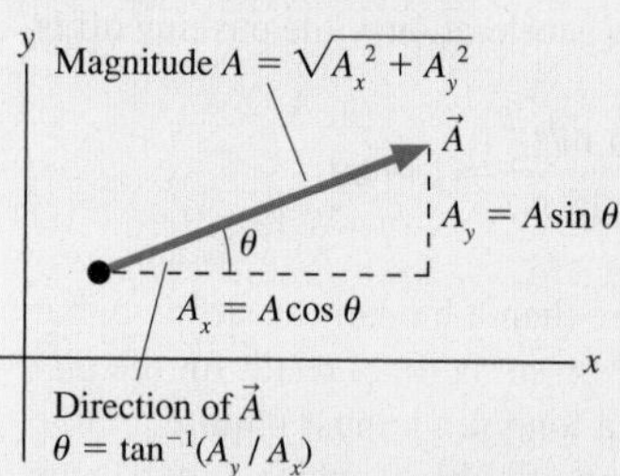

The magnitude and direction of a vector can be expressed in terms of its components.

The sign of the components depends on the direction of the vector:

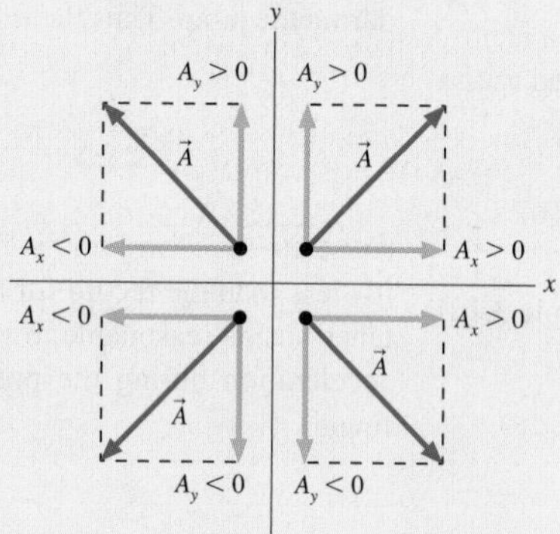

The Acceleration Vector

We define the acceleration vector as

$$\vec{a} = \frac{\vec{v}_f - \vec{v}_i}{t_f - t_i} = \frac{\Delta\vec{v}}{\Delta t}$$

We find the acceleration vector on a motion diagram as follows:

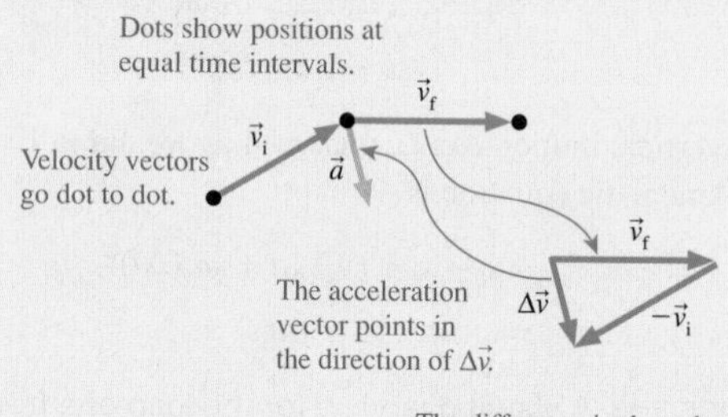

APPLICATIONS

Relative motion

Velocities can be expressed relative to an observer. We can add relative velocities to convert to another observer's point of view.

c = car, r = runner, g = ground

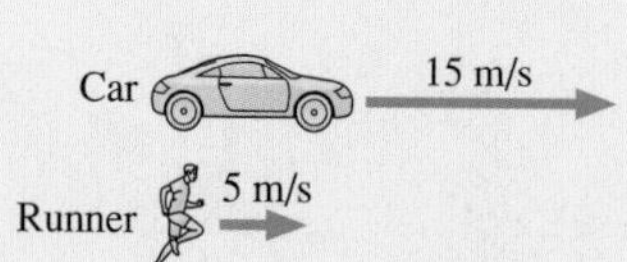

The speed of the car with respect to the runner is:

$$(v_x)_{cr} = (v_x)_{cg} + (v_x)_{gr}$$

Motion on a ramp

An object sliding down a ramp will accelerate parallel to the ramp:

$$a_x = \pm g\sin\theta$$

The correct sign depends on the direction in which the ramp is tilted.

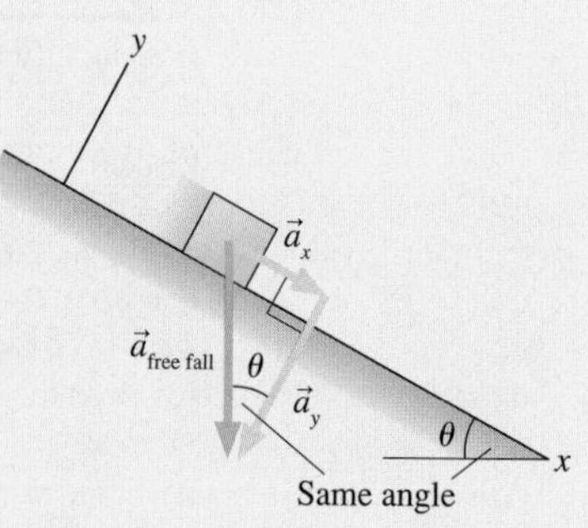

For homework assigned on MasteringPhysics, go to www.masteringphysics.com

Problems labeled INT integrate significant material from earlier chapters; BIO are of biological or medical interest.

Problem difficulty is labeled as | (straightforward) to ||||| (challenging).

QUESTIONS

Conceptual Questions

1. a. Can a vector have nonzero magnitude if a component is zero? If no, why not? If yes, give an example.
   b. Can a vector have zero magnitude and a nonzero component? If no, why not? If yes, give an example.
2. Is it possible to add a scalar to a vector? If so, demonstrate. If not, explain why not.
3. Suppose two vectors have unequal magnitudes. Can their sum be $\vec{0}$? Explain.
4. Suppose $\vec{C} = \vec{A} + \vec{B}$
   a. Under what circumstances does $C = A + B$?
   b. Could $C = A - B$? If so, how? If not, why not?
5. For a projectile, which of the following quantities are constant during the flight: x, y, v_x, v_y, v, a_x, a_y? Which of the quantities are zero throughout the flight?
6. A baseball player throws a ball at a 40° angle to the ground. The ball lands on the ground some distance away.
   a. Is there any point on the trajectory where $\vec{v}$ and $\vec{a}$ are parallel to each other? If so, where?
   b. Is there any point where $\vec{v}$ and $\vec{a}$ are perpendicular to each other? If so, where?
7. An athlete performing the long jump tries to achieve the maximum distance from the point of takeoff to the first point of touching the ground. After the jump, rather than land upright, she extends her legs forward as in the photo. How does this affect the time in the air? How does this give the jumper a longer range?

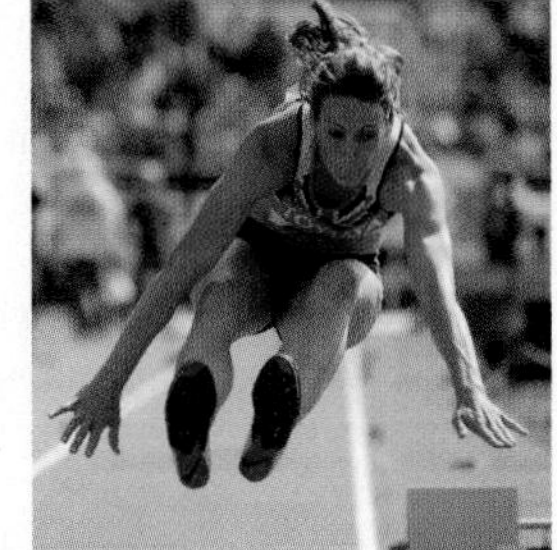

8. A person trying to throw a ball as far as possible will run forward during the throw. Explain why this increases the distance of the throw.
9. A passenger on a jet airplane claims to be able to walk at a speed in excess of 500 mph. Can this be true? Explain.
10. If you go to a ski area, you'll likely find that the beginner's slope has the smallest angle. Use the concept of acceleration on a ramp to explain why this is so.
11. In an amusement-park ride, cars rolling along at high speed suddenly head up a long, straight ramp. They roll up the ramp, reverse direction at the highest point, then roll backward back down the ramp. In each of the following segments of the motion, are the cars accelerating, or is their acceleration zero? If accelerating, which way does their acceleration vector point?
    a. As the cars roll up the ramp.
    b. At the highest point on the ramp.
    c. As the cars roll back down the ramp.
12. There are competitions in which pilots fly small planes low over the ground and drop weights, trying to hit a target. A pilot flying low and slow drops a weight; it takes 2.0 s to hit the ground, during which it travels a horizontal distance of 100 m. Now the pilot does a run at the same height but twice the speed. How much time does it take the weight to hit the ground? How far does it travel before it lands?
13. A cyclist goes around a level, circular track at constant speed. Do you agree or disagree with the following statement: "Because the cyclist's speed is constant, her acceleration is zero." Explain.
14. You are driving your car in a circular path on level ground at a constant speed of 20 mph. At the instant you are driving north, and turning left, are you accelerating? If so, toward what point of the compass (N, S, E, W) does your acceleration vector point? If not, why not?
15. An airplane has been directed to fly in a clockwise circle, as seen from above, at constant speed until another plane has landed. When the plane is going north, is it accelerating? If so, in what direction does the acceleration vector point? If not, why not?
16. When you go around a corner in your car, your car follows a path that is a segment of a circle. To turn safely, you should keep your car's acceleration below some safe upper limit. If you want to make a "tighter" turn—that is, turn in a circle with a smaller radius—how should you adjust your speed? Explain.

Multiple-Choice Questions

17. || Which combination of the vectors shown in Figure Q3.17 has the largest magnitude?
    A. $\vec{A} + \vec{B} + \vec{C}$
    B. $\vec{B} + \vec{A} - \vec{C}$
    C. $\vec{A} - \vec{B} + \vec{C}$
    D. $\vec{C} - \vec{A} - \vec{C}$

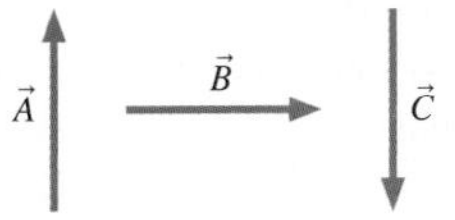

$\vec{A}$ $\vec{B}$ $\vec{C}$

FIGURE Q3.17

18. || Two vectors appear as in Figure Q3.18. Which combination points directly to the left?
    A. $\vec{P} + \vec{Q}$
    B. $\vec{P} - \vec{Q}$
    C. $\vec{Q} - \vec{P}$
    D. $-\vec{Q} - \vec{P}$

FIGURE Q3.18

19. | The gas pedal in a car is sometimes referred to as "the accelerator." Which other controls on the vehicle can be used to produce acceleration?
    A. The brakes.
    B. The steering wheel.
    C. The gear shift.
    D. All of the above.

20. | A car travels at constant speed along the curved path shown from above in Figure Q3.20. Five possible vectors are also shown in the figure; the letter E represents the zero vector. Which vector best represents
    a. The car's *velocity* at position 1?
    b. The car's *acceleration* at point 1?
    c. The car's *velocity* at position 2?
    d. The car's *acceleration* at point 2?
    e. The car's *velocity* at position 3?
    f. The car's *acceleration* at point 3?

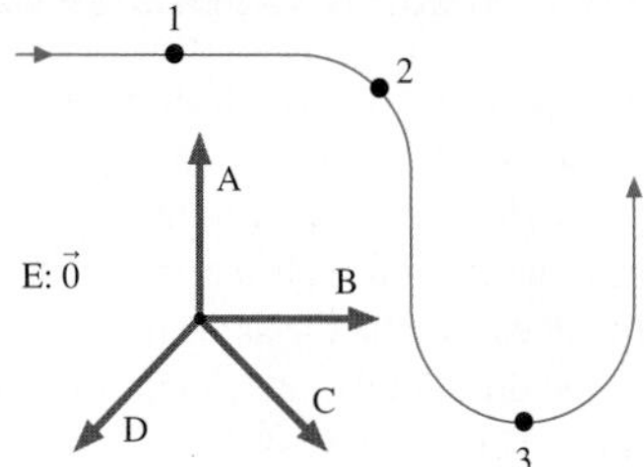

FIGURE Q3.20

21. | A ball is fired from a cannon at point 1 and follows the trajectory shown in Figure Q3.21. Air resistance may be neglected. Five possible vectors are also shown in the figure; the letter E represents the zero vector. Which vector best represents
    a. The ball's *velocity* at position 2?
    b. The ball's *acceleration* at point 2?
    c. The ball's *velocity* at position 3?
    d. The ball's *acceleration* at point 3?

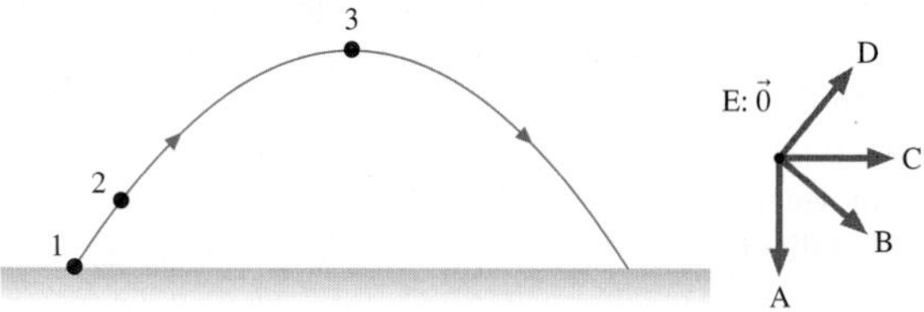

FIGURE Q3.21

22. | A ball thrown at an initial angle of 37.0° and initial velocity of 23.0 m/s reaches a maximum height h, as shown in Figure Q3.22. With what initial speed must a ball be thrown *straight up* to reach the same maximum height h?
    A. 13.8 m/s B. 17.3 m/s
    C. 18.4 m/s D. 23.0 m/s

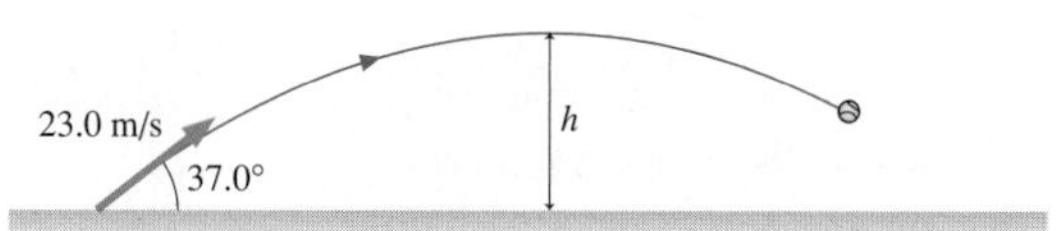

FIGURE Q3.22

23. | A cannon, elevated at 40° is fired at a wall 300 m away on level ground, as shown in Figure Q3.23. The initial speed of the cannonball is 89 m/s

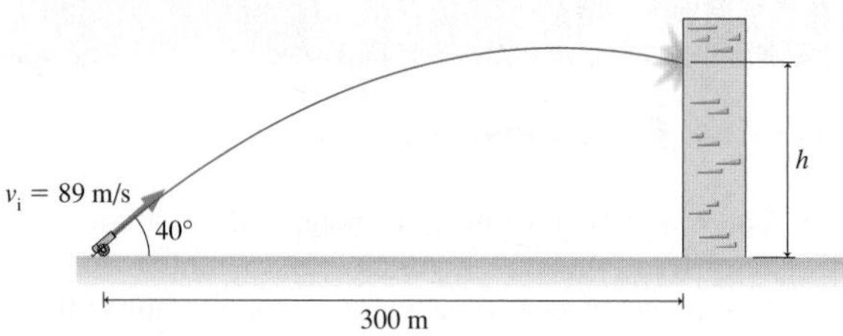

FIGURE Q3.23

a. How long does it take for the ball to hit the wall?
    A. 1.3 s B. 3.3 s C. 4.4 s
    D. 6.8 s E. 7.2 s
b. At what height h does the ball hit the wall?
    A. 39 m B. 47 m C. 74 m
    D. 160 m E. 210 m

24. | A car drives horizontally off a 73-m-high cliff at a speed of 27 m/s. Ignore air resistance.
    a. How long will it take the car to hit the ground?
    A. 2.0 s B. 3.2 s C. 3.9 s
    D. 4.9 s E. 5.0 s
    b. How far from the base of the cliff will the car hit?
    A. 74 m B. 88 m C. 100 m
    D. 170 m E. 280 m
25. | A football is kicked at an angle of 30° with a speed of 20 m/s. To the nearest second, how long will the ball stay in the air?
    A. 1 s B. 2 s C. 3 s D. 4 s
26. | A football is kicked at an angle of 30° with a speed of 20 m/s. To the nearest 5 m, how far will the ball travel?
    A. 15 m B. 25 m C. 35 m D. 45 m
27. | Riders on a Ferris wheel move in a circle with a speed of 4.0 m/s. As they go around, they experience a centripetal acceleration of 2.0 m/s^2. What is the diameter of this particular Ferris wheel?
    A. 4.0 m B. 6.0 m C. 8.0 m
    D. 16 m E. 24 m

PROBLEMS

Section 3.1 Using Vectors

1. || Trace the vectors in Figure P3.1 onto your paper. Then use graphical methods to draw the vectors (a) $\vec{A} + \vec{B}$ and (b) $\vec{A} - \vec{B}$.

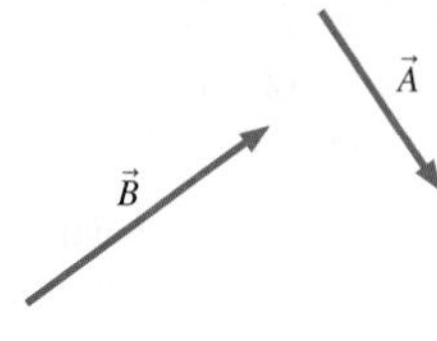

FIGURE P3.1

2. ||| Trace the vectors in Figure P3.2 onto your paper. Then use graphical methods to draw the vectors (a) $\vec{A} + \vec{B}$ and (b) $\vec{A} - \vec{B}$.

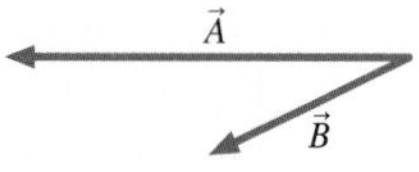

FIGURE P3.2

Section 3.2 Using Vectors on Motion Diagrams

3. | A car goes around a corner in a circular arc at constant speed. Draw a motion diagram including positions, velocity vectors, and acceleration vectors.
4. | a. Is the object's average speed between points 1 and 2 greater than, less than, or equal to its average speed between points 0 and 1? Explain how you can tell.
    b. Find the average acceleration vector at point 1 of the three-point motion diagram in Figure P3.4.

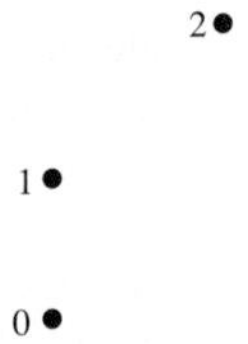

FIGURE P3.4

5. |||| Figure 3.11 showed the motion diagram for Anne as she rode a Ferris wheel that was turning at a constant speed. The inset to the figure showed how to find the acceleration vector at the lowest point in her motion. Use a similar analysis to find Anne's acceleration vector at the 12 o'clock, 4 o'clock, and 8 o'clock positions of the motion diagram. Use a ruler so that your analysis is accurate.

Section 3.3 Coordinate Systems and Vector Components

6. || A position vector with magnitude 10 m points to the right and up. Its x-component is 6.0 m. What is the value of its y-component?
7. ||| A velocity vector 40° above the positive x-axis has a y-component of 10 m/s. What is the value of its x-component?
8. || Jack and Jill ran up the hill at 3.0 m/s. The horizontal component of Jill's velocity vector was 2.5 m/s.
   a. What was the angle of the hill?
   b. What was the vertical component of Jill's velocity?
9. || A cannon tilted upward at 30° fires a cannonball with a speed of 100 m/s. At that instant, what is the component of the cannonball's velocity parallel to the ground?
10. || a. What are the x- and y-components of vector $\vec{E}$ of Figure P3.10 in terms of the angle θ and the magnitude E?
   b. For the same vector, what are the x- and y-components in terms of the angle ϕ and the magnitude E?

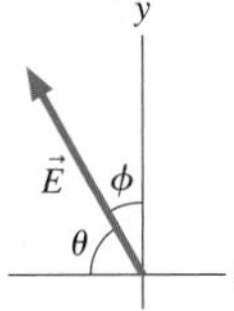

FIGURE P3.10

11. | Draw each of the following vectors, then find its x- and y-components.
   a. $\vec{d}$ = (100 m, 45° below +x-axis)
   b. $\vec{v}$ = (300 m/s, 20° above +x-axis)
   c. $\vec{a}$ = (5.0 m/s^2, −y-direction)
12. || Draw each of the following vectors, then find its x- and y-components.
   a. $\vec{d}$ = (2.0 km, 30° left of +y-axis)
   b. $\vec{v}$ = (5.0 cm/s, −x-direction)
   c. $\vec{a}$ = (10 m/s^2, 40° left of −y-axis)
13. | Each of the following vectors is given in terms of its x- and y-components. Draw the vector, label an angle that specifies the vector's direction, then find the vector's magnitude and direction.
   a. $v_x = 20$ m/s, $v_y = 40$ m/s
   b. $a_x = 2.0$ m/s^2, $a_y = -6.0$ m/s^2
14. | Each of the following vectors is given in terms of its x- and y-components. Draw the vector, label an angle that specifies the vector's direction, then find the vector's magnitude and direction.
   a. $v_x = 10$ m/s, $v_y = 30$ m/s
   b. $a_x = 20$ m/s^2, $a_y = 10$ m/s^2
15. |||| While visiting England, you decide to take a jog and find yourself in the neighborhood shown on the map in Figure P3.15. What is your displacement after running 2.0 km on Strawberry Fields, 1.0 km on Penny Lane, and 4.0 km on Abbey Road?

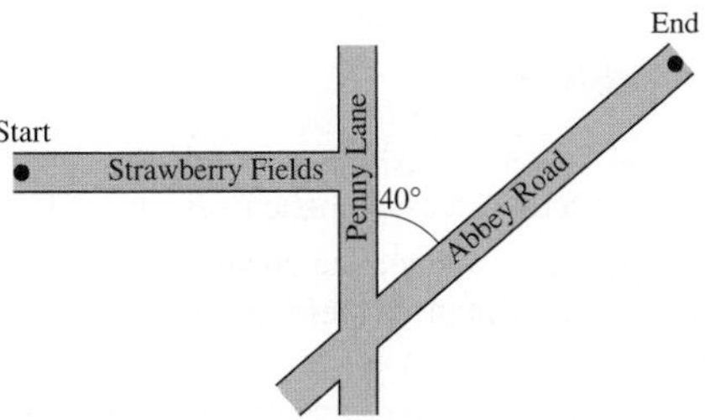

FIGURE P3.15

Section 3.4 Motion on a Ramp

16. ||| You begin sliding down a 15° ski slope. Ignoring friction and air resistance, how fast will you be moving after 10 s?
17. |||| A car traveling at 30 m/s runs out of gas while traveling up a 5.0° slope. How far will it coast before starting to roll back down?
18. || In the Soapbox Derby, young participants build non-motorized cars with very low-friction wheels. Cars race by rolling down a hill. The track at Akron's Derby Downs, where the national championship is held, begins with a 55-ft-long section tilted 13° below horizontal.

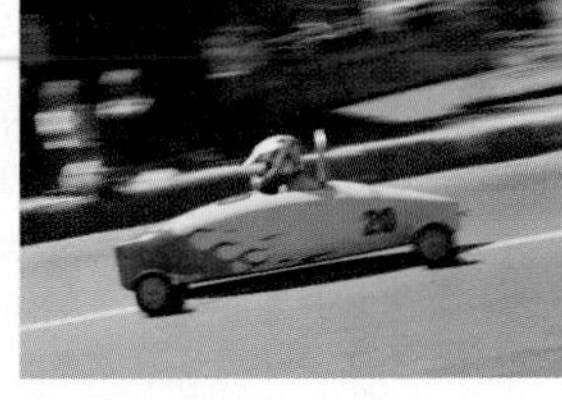

   a. What is the maximum possible acceleration of a car moving down this stretch of track?
   b. If a car starts from rest and undergoes this acceleration for the full 55 ft, what is its final speed in m/s?
19. ||| A piano has been pushed to the top of the ramp at the back of a moving van. The workers think it is safe, but as they walk away, it begins to roll down the ramp. If the back of the truck is 1.0 m above the ground and the ramp is inclined at 20°, how much time do the workers have to get to the piano before it reaches the bottom of the ramp?
20. || Starting from rest, several toy cars roll down ramps of differing lengths and angles. Rank them according to their speed at the bottom of the ramp, from slowest to fastest. Car A goes down a 10 m ramp inclined at 15°, car B goes down a 10 m ramp inclined at 20°, car C goes down an 8.0 m ramp inclined at 20°, and car D goes down a 12 m ramp inclined at 12°.

Section 3.5 Relative Motion

21. | Anita is running to the right at 5 m/s, as shown in Figure P3.21. Balls 1 and 2 are thrown toward her at 10 m/s by friends standing on the ground. According to Anita, what is the speed of each ball?

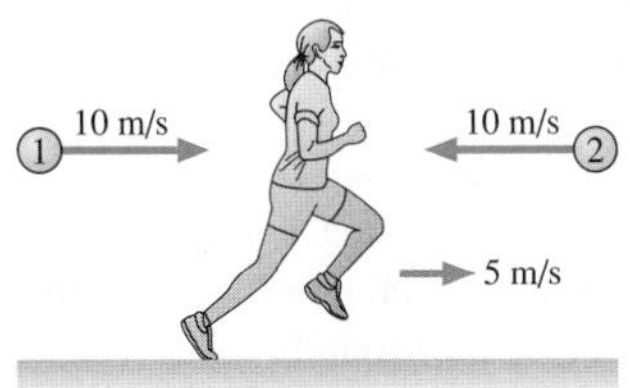

FIGURE P3.21

22. | Anita is running to the right at 5 m/s, as shown in Figure P3.22. Balls 1 and 2 are thrown toward her by friends standing on the ground. According to Anita, both balls are approaching her at 10 m/s. According to her friends, with what speeds were the balls thrown?

FIGURE P3.22

23. ||| A boat takes 3.0 h to travel 30 km down a river, then 5.0 h to return. How fast is the river flowing?

24. || Two children who are bored while waiting for their flight at the airport decide to race from one end of the 20-m-long moving sidewalk to the other and back. Phillippe runs on the sidewalk at 2.0 m/s (relative to the sidewalk). Renee runs on the floor at 2.0 m/s. The sidewalk moves at 1.5 m/s relative to the floor. Both make the turn instantly with no loss of speed.
   a. Who wins the race?
   b. By how much time does the winner win?

25. || A skydiver deploys his parachute when he is 1000 m directly above his desired landing spot. He then falls through the air at a steady 5.0 m/s. There is a breeze blowing to the west at 2.0 m/s.
   a. At what angle with respect to vertical does he fall?
   b. By what distance will he miss his desired landing spot?

Section 3.6 Motion in Two Dimensions: Projectile Motion

Section 3.7 Projectile Motion: Solving Problems

26. ||| An object is launched with an initial velocity of 50.0 m/s at a launch angle of 36.9° above the horizontal.
   a. Make a table showing values of x, y, v_x, v_y, and the speed v every 1 s from $t = 0$ s to $t = 6$ s.
   b. Plot a graph of the object's trajectory during the first 6 s of motion.

27. ||| A ball is thrown horizontally from a 20-m-high building with a speed of 5.0 m/s.
   a. Make a sketch of the ball's trajectory.
   b. Draw a graph of v_x, the horizontal velocity, as a function of time. Include units on both axes.
   c. Draw a graph of v_y, the vertical velocity, as a function of time. Include units on both axes.
   d. How far from the base of the building does the ball hit the ground?

28. || A ball with a horizontal speed of 1.25 m/s rolls off a bench 1.00 m above the floor.
   a. How long will it take the ball to hit the floor?
   b. How far from a point on the floor directly below the edge of the bench will the ball land?

29. |||| King Arthur's knights use a catapult to launch a rock from their vantage point on top of the castle wall, 12 m above the moat. The rock is launched at a speed of 25 m/s and an angle of 30° above the horizontal. How far from the castle wall does the launched rock hit the ground?

30. | Two spheres are launched horizontally from a 1.0-m-high table. Sphere A is launched with an initial speed of 5.0 m/s. Sphere B is launched with an initial speed of 2.5 m/s.
   a. What are the times for each sphere to hit the floor?
   b. What are the distances that each travels from the edge of the table?

31. ||| A rifle is aimed horizontally at a target 50 m away. The bullet hits the target 2.0 cm below the aim point.
   a. What was the bullet's flight time?
   b. What was the bullet's speed as it left the barrel?

32. ||| A gray kangaroo can bound across a flat stretch of ground BIO with each jump carrying it 10 m from the takeoff point. If the kangaroo leaves the ground at a 20° angle, what are its (a) takeoff speed and (b) horizontal speed?

33. ||| On the Apollo 14 mission to the moon, astronaut Alan Shepard hit a golf ball with a golf club improvised from a tool. The free-fall acceleration on the moon is 1/6 of its value on earth. Suppose he hit the ball with a speed of 25 m/s at an angle 30° above the horizontal.
   a. How long was the ball in flight?
   b. How far did it travel?
   c. Ignoring air resistance, how much farther would it travel on the moon than on earth?

Section 3.8 Motion in Two Dimensions: Circular Motion

34. | An old-fashioned LP record rotates at $33\frac{1}{3}$ rpm.
   a. What is its frequency, in rev/s?
   b. What is its period, in seconds?

35. | A typical hard disk in a computer spins at 5400 rpm.
   a. What is the frequency, in rev/s?
   b. What is the period, in seconds?

36. | Racing greyhounds are capable of rounding corners at very BIO high speeds. A typical greyhound track has turns that are 45-m-diameter semicircles. A greyhound can run around these turns at a constant speed of 15 m/s. What is its acceleration in m/s^2 and in units of g?

37. || A CD-ROM drive in a computer spins the 12-cm-diameter disks at 10,000 rpm.
   a. What are the disk's period (in s) and frequency (in rev/s)?
   b. What would be the speed of a speck of dust on the outside edge of this disk?
   c. What is the acceleration in units of g that this speck of dust experiences?

38. || To withstand "g-forces" of up to 10 g's, caused by suddenly BIO pulling out of a steep dive, fighter jet pilots train on a "human centrifuge." 10 g's is an acceleration of 98 m/s^2. If the length of the centrifuge arm is 12 m, at what speed is the rider moving when she experiences 10 g's?

39. || A particle rotates in a circle with centripetal acceleration $a = 8.0$ m/s^2. What is a if
   a. The radius is doubled without changing the particle's speed?
   b. The speed is doubled without changing the circle's radius?

40. || Entrance and exit ramps for freeways are often circular stretches of road. As you go around one at a constant speed, you will experience a constant acceleration. Suppose you drive through an entrance ramp at a modest speed and your acceleration is 3.0 m/s^2. What will be the acceleration if you double your speed?

41. || A peregrine falcon in a tight, circular turn can attain a cen- BIO tripetal acceleration 1.5 times the free-fall acceleration. If the falcon is flying at 20 m/s, what is the radius of the turn?

General Problems

42. | Suppose $\vec{C} = \vec{A} + \vec{B}$ where vector $\vec{A}$ has components $A_x = 5$, $A_y = 2$ and vector $\vec{B}$ has components $B_x = -3$, $B_y = -5$.
   a. What are the x- and y-components of vector $\vec{C}$?
   b. Draw a coordinate system and on it show vectors $\vec{A}$, $\vec{B}$, and $\vec{C}$.
   c. What are the magnitude and direction of vector $\vec{C}$?

43. | Suppose $\vec{D} = \vec{A} - \vec{B}$ where vector $\vec{A}$ has components $A_x = 5$, $A_y = 2$ and vector $\vec{B}$ has components $B_x = -3$, $B_y = -5$.
 a. What are the x- and y-components of vector $\vec{D}$?
 b. Draw a coordinate system and on it show vectors $\vec{A}$, $\vec{B}$, and $\vec{D}$.
 c. What are the magnitude and direction of vector $\vec{D}$?

44. || Suppose $\vec{E} = 2\vec{A} + 3\vec{B}$ where vector $\vec{A}$ has components $A_x = 5$, $A_y = 2$ and vector $\vec{B}$ has components $B_x = -3$, $B_y = -5$.
 a. What are the x- and y-components of vector $\vec{E}$?
 b. Draw a coordinate system and on it show vectors $\vec{A}$, $\vec{B}$, and $\vec{E}$.
 c. What are the magnitude and direction of vector $\vec{E}$?

45. || For the three vectors shown in Figure P3.45, the vector sum $\vec{D} = \vec{A} + \vec{B} + \vec{C}$ has components $D_x = 2$ and $D_y = 0$.
 a. What are the x- and y-components of vector $\vec{B}$?
 b. Write $\vec{B}$ as a magnitude and a direction.

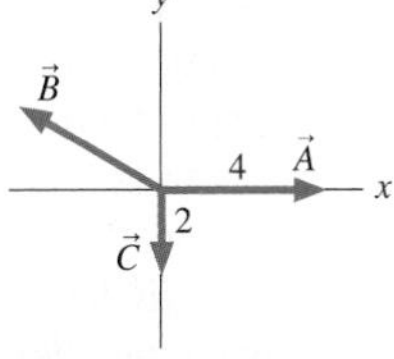

FIGURE P3.45

46. || Let $\vec{A} = (3.0\text{ m}, 20°\text{ south of east})$, $\vec{B} = (2.0\text{ m, north})$, and $\vec{C} = (5.0\text{ m}, 70°\text{ south of west})$.
 a. Draw and label $\vec{A}$, $\vec{B}$, and $\vec{C}$ with their tails at the origin. Use a coordinate system with the x-axis to the east.
 b. Write the x- and y-components of vectors $\vec{A}$, $\vec{B}$, and $\vec{C}$.
 c. Find the magnitude and the direction of $\vec{D} = \vec{A} + \vec{B} + \vec{C}$.

47. || A typical set of stairs is angled at 38°. You climb a set of stairs at a speed of 3.5 m/s.
 a. How much height will you gain in 2.0 s?
 b. How much horizontal distance will you cover in 2.0 s?

48. || The minute hand on a watch is 2.0 cm long. What is the displacement vector of the tip of the minute hand
 a. From 8:00 to 8:20 A.M.?
 b. From 8:00 to 9:00 A.M.?

49. ||||| A field mouse trying to escape a hawk runs east for 5.0 m, darts southeast for 3.0 m, then drops 1.0 m down a hole into its burrow. What is the magnitude of the net displacement of the mouse?

50. ||| A pilot in a small plane encounters shifting winds. He flies 26.0 km northeast, then 45.0 km due north. From this point, he flies an additional distance in an unknown direction, only to find himself at a small airstrip that his map shows to be 70.0 km directly north of his starting point. What were the length and direction of the third leg of his trip?

51. || A small plane is 100 km south of the equator. The plane is flying at 150 km/h at a heading of 30° to the west of north. In how many minutes will the plane cross the equator?

52. || The bacterium *Escherichia coli* (or *E. coli*) is a single-celled organism that lives in the gut of healthy humans and animals. When grown in a uniform medium rich in salts and amino acids, these bacteria swim along zig-zag paths at a constant speed of 20 μm/s. Figure P3.52 shows the trajectory of an *E. coli* as it moves from point A to point E. Each segment of the motion can be identified by two letters, such as segment BC.
BIO
 a. For each of the four segments in the bacterium's trajectory, calculate the x- and y-components of its displacement and of its velocity.
 b. Calculate both the total distance traveled and the magnitude of the net displacement for the entire motion.
 c. What are the magnitude and the direction of the bacterium's average velocity for the entire trip?

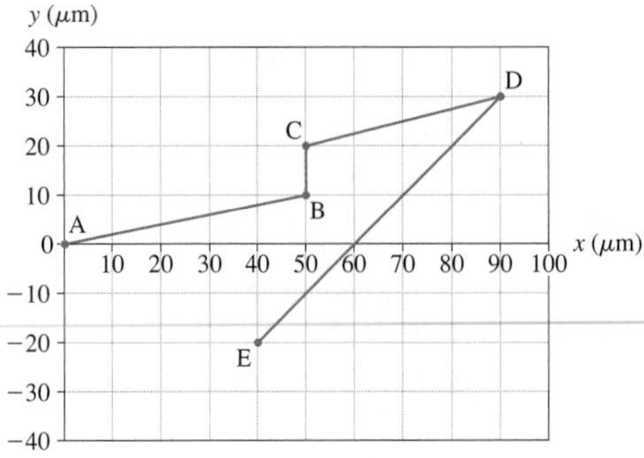

FIGURE P3.52

53. |||| A skier is gliding along at 3.0 m/s on horizontal, frictionless snow. He suddenly starts down a 10° incline. His speed at the bottom is 15 m/s.
 a. What is the length of the incline?
 b. How long does it take him to reach the bottom?

54. ||| A block slides along the frictionless track shown in Figure P3.54 with an initial speed of 5.0 m/s. Assume it turns all the corners smoothly, with no loss of speed.
 a. What is the block's speed as it goes over the top?
 b. What is its speed when it reaches the level track on the right side?
 c. By what percentage does the block's final speed differ from its initial speed? Is this surprising?

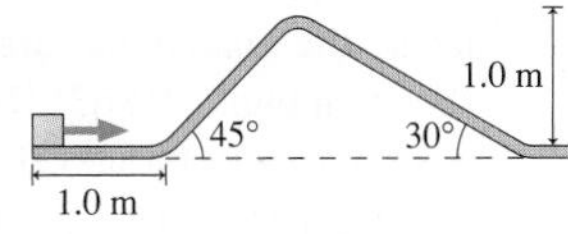

FIGURE P3.54

55. ||| One game at the amusement park has you push a puck up a long, frictionless ramp. You win a stuffed animal if the puck, at its highest point, comes to within 10 cm of the end of the ramp without going off. You give the puck a push, releasing it with a speed of 5.0 m/s when it is 8.5 m from the end of the ramp. The puck's speed after traveling 3.0 m is 4.0 m/s. Are you a winner?

56. |||| When the moving sidewalk at the airport is broken, as it often seems to be, it takes you 50 s to walk from your gate to the baggage claim. When it is working and you stand on the moving sidewalk the entire way, without walking, it takes 75 s to travel the same distance. How long will it take you to travel from the gate to baggage claim if you walk while riding on the moving sidewalk?

57. ||||| Ships A and B leave port together. For the next two hours, ship A travels at 20 mph in a direction 30° west of north while ship B travels 20° east of north at 25 mph.
 a. What is the distance between the two ships two hours after they depart?
 b. What is the speed of ship A as seen by ship B?

58. ||| Mary needs to row her boat across a 100-m-wide river that is flowing to the east at a speed of 1.0 m/s. Mary can row the boat with a speed of 2.0 m/s relative to the water.
 a. If Mary rows straight north, how far downstream will she land?
 b. Draw a picture showing Mary's displacement due to rowing, her displacement due to the river's motion, and her net displacement.

59. ||| A flock of ducks is trying to migrate south for the winter, but they keep being blown off course by a wind blowing from the west at 12 m/s. A wise elder duck finally realizes that the solution is to fly at an angle to the wind. If the ducks can fly at 16 m/s relative to the air, in what direction should they head in order to move directly south?

60. ||| A kayaker needs to paddle north across a 100-m-wide harbor. The tide is going out, creating a tidal current flowing east at 2.0 m/s. The kayaker can paddle with a speed of 3.0 m/s.
    a. In which direction should he paddle in order to travel straight across the harbor?
    b. How long will it take him to cross?
61. |||| A plane has an airspeed of 200 mph. The pilot wishes to reach a destination 600 mi due east, but a wind is blowing at 50 mph in the direction 30° north of east.
    a. In what direction must the pilot head the plane in order to reach her destination?
    b. How long will the trip take?
62. ||| The Gulf Stream off the east coast of the United States can flow at a rapid 3.6 m/s to the north. A ship in this current has a cruising speed of 10 m/s. The captain would like to reach land at a point due west from the current position.
    a. In what direction with respect to the water should the ship sail?
    b. At this heading, what is the ship's speed with respect to land?
63. || A physics student on Planet Exidor throws a ball, and it follows the parabolic trajectory shown in Figure P3.63. The ball's position is shown at 1.0 s intervals until $t = 3.0$ s. At $t = 1.0$ s, the ball's velocity has components $v_x = 2.0$ m/s, $v_y = 2.0$ m/s.

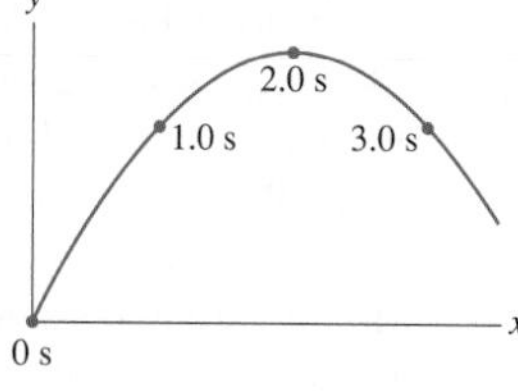

FIGURE P3.63

    a. Determine the x- and y-components of the ball's velocity at $t = 0.0$ s, 2.0 s, and 3.0 s.
    b. What is the value of g on Planet Exidor?
    c. What was the ball's launch angle?
64. ||| A ball thrown horizontally at 25 m/s travels a horizontal distance of 50 m before hitting the ground. From what height was the ball thrown?
65. ||| BIO In 1780, in what is now referred to as "Brady's Leap," Captain Sam Brady of the U.S. Continental Army escaped certain death from his enemies by running over the edge of the cliff above Ohio's Cuyahoga River, which is confined at that spot to a gorge. He landed safely on the far side of the river. It was reported that he leapt 22 ft across while falling 20 ft.

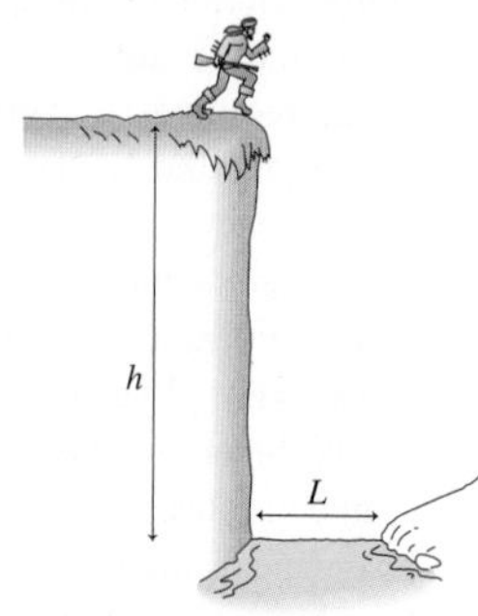

FIGURE P3.65

    a. Representing the distance jumped as L and the vertical drop as h, as shown in Figure P3.65, derive an expression for the minimum speed v he would need to make his leap if he ran straight off the cliff.
    b. Evaluate your expression for a 22 ft jump with a 20 ft drop to the other side.
    c. Is it reasonable that a person could make this leap? Use the fact that the world record for the 100 m dash is approximately 10 s to estimate the maximum speed such a runner would have.
66. ||| The longest recorded pass in an NFL game traveled 83 yards in the air from the quarterback to the receiver. Assuming that the pass was thrown at the optimal 45° angle, what was the speed at which the ball left the quarterback's hand?
67. ||| A spring-loaded gun, fired vertically, shoots a marble 6.0 m straight up in the air. What is the marble's range if it is fired horizontally from 1.5 m above the ground?
68. || In a shot-put event, an athlete throws the shot with an initial speed of 12.0 m/s at a 40.0° angle from the horizontal. The shot leaves her hand at a height of 1.80 m above the ground.
    a. How far does the shot travel?
    b. Repeat the calculation of part (a) for angles 42.5°, 45.0°, and 47.5°. Put all your results, including 40.0°, in a table. At what angle of release does she throw the farthest?
69. ||| A tennis player hits a ball 2.0 m above the ground. The ball leaves his racquet with a speed of 20 m/s at an angle 5.0° above the horizontal. The horizontal distance to the net is 7.0 m, and the net is 1.0 m high. Does the ball clear the net? If so, by how much? If not, by how much does it miss?
70. |||| Water at the top of Horseshoe Falls (part of Niagara Falls) is moving horizontally at 9.0 m/s as it goes off the edge and plunges 53 m to the pool below. If you ignore air resistance, at what angle is the falling water moving as it enters the pool?
71. ||| Figure 3.37 shows that the range of a projectile launched at a 60° angle has the same range as a projectile launched at a 30° angle—but they won't be in the air for the same amount of time. Suppose a projectile launched at a 30° angle is in the air for 2.0 s. How long will the projectile be in the air if it is launched with the same speed at a 60° angle?
72. || A supply plane needs to drop a package of food to scientists working on a glacier in Greenland. The plane flies 100 m above the glacier at a speed of 150 m/s. How far short of the target should it drop the package?
73. ||| A child slides down a frictionless 3.0-m-long playground slide tilted upward at an angle of 40°. At the end of the slide, there is an additional section that curves so that the child is launched off the end of the slide horizontally.
    a. How fast is the child moving at the bottom of the slide?
    b. If the end of the slide is 0.40 m above the ground, how far from the end does she land?
74. ||| INT A sports car is advertised to be able to "reach 60 mph in 5 seconds flat, corner at $0.85g$, and stop from 70 mph in only 168 feet."
    a. In which of those three situations is the magnitude of the car's acceleration the largest? In which is it the smallest?
    b. At 60 mph, what is the smallest turning radius that this car can navigate?
75. || INT A Ford Mustang can accelerate from 0 to 60 mph in a time of 5.6 s. A Mini Cooper isn't capable of such a rapid start, but it can turn in a very small circle 34 ft in diameter. How fast would you need to drive the Mini Cooper in this tight circle to match the magnitude of the Mustang's acceleration?
76. || The "Screaming Swing" is a carnival ride that is—not surprisingly—a giant swing. It's actually two swings moving in opposite directions. At the bottom of its arc, riders are moving at 30 m/s with respect to the ground in a 50-m-diameter circle.
    a. What is the acceleration, in m/s^2 and in units of g, that riders experience?
    b. At the bottom of the ride, as they pass each other, how fast do the riders move with respect to each other?
77. || On an otherwise straight stretch of road near Moffat, Colorado, the road suddenly turns. This bend in the road is a segment of a circle with radius 110 m. Drivers are cautioned to slow down to 40 mph as they navigate the curve.
    a. If you heed the sign and slow to 40 mph, what will be your acceleration going around the curve at this constant speed? Give your answer in m/s^2 and in units of g.
    b. At what speed would your acceleration be double that at the recommended speed?

Passage Problems

Riding the Water Slide

A rider on a water slide goes through three different kinds of motion, as illustrated in Figure P3.78. Use the data and details from the figure to answer the following questions.

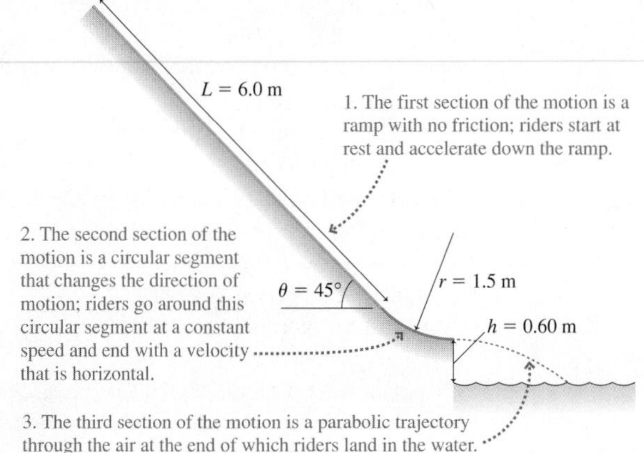

L = 6.0 m
1. The first section of the motion is a ramp with no friction; riders start at rest and accelerate down the ramp.
2. The second section of the motion is a circular segment that changes the direction of motion; riders go around this circular segment at a constant speed and end with a velocity that is horizontal.
θ = 45°
r = 1.5 m
h = 0.60 m
3. The third section of the motion is a parabolic trajectory through the air at the end of which riders land in the water.

FIGURE P3.78

78. | At the end of the first section of the motion, riders are moving at what approximate speed?
A. 3 m/s B. 6 m/s C. 9 m/s D. 12 m/s

79. | Suppose the acceleration during the second section of the motion is too large to be comfortable for riders. What change could be made to decrease the acceleration during this section?
A. Reduce the radius of the circular segment.
B. Increase the radius of the circular segment.
C. Increase the angle of the ramp.
D. Increase the length of the ramp.

80. | What is the vertical component of the velocity of a rider as he or she hits the water?
A. 2.4 m/s B. 3.4 m/s C. 5.2 m/s D. 9.1 m/s

81. | Suppose the designers of the water slide want to adjust the height h above the water so that riders land twice as far away from the bottom of the slide. What would be the necessary height above the water?
A. 1.2 m B. 1.8 m C. 2.4 m D. 3.0 m

82. | During which section of the motion is the magnitude of the acceleration experienced by a rider the greatest?
A. The first. B. The second.
C. The third. D. It is the same in all sections.

STOP TO THINK ANSWERS

Stop to Think 3.1: A. The graphical construction of $2\vec{P} - \vec{Q}$ is shown at right.

Stop to Think 3.2: From the axes on the graph, we can see that the x- and y-components are -4 cm and $+2$ cm, respectively.

Stop to Think 3.3: C. Vector $\vec{C}$ points to the left and down, so both C_x and C_y are negative. C_x is in the numerator because it is the side opposite ϕ.

Stop to Think 3.4: B. The angle of the slope is greatest in this case, leading to the greatest acceleration.

Stop to Think 3.5: D. Mass does not appear in the kinematic equations, so the mass has no effect; the balls will follow the same path.

Stop to Think 3.6: B. The magnitude of the acceleration is v^2/r. Acceleration is largest for the combination of highest speed and smallest radius.

4 Forces and Newton's Laws of Motion

These ice boats sail across the ice at great speeds. What gets the boats moving in the first place? What keeps them moving once they're going?

LOOKING AHEAD ▸▸

The goal of Chapter 4 is to establish a connection between force and motion.

What Causes Motion?

Galileo was the first to realize that objects in *uniform motion* require no "cause" for their motion. Only *changes* in motion—accelerations—require a cause: a *force*.

What is a Force?

We'll understand force by first examining the properties common to all forces, then by studying a number of forces we'll encounter often.

Forces are a *push* or a *pull*, act on an *object*, and have an identifiable *agent*. Forces are *vectors*.

Looking Back ◂◂
1.5 Vectors and motion

Newton's Third Law

When two objects interact, each exerts a force on the other. Newton's third law tells us that these two forces point in *opposite* directions but have the *same* magnitudes.

The force of the hammer on the nail has the same magnitude as the force of the nail on the hammer.

Some Important Forces

It's important to understand the characteristics of a number of important forces. Some of the forces you'll learn about in this chapter are . . .

Weight
The force of gravity acting on an object.

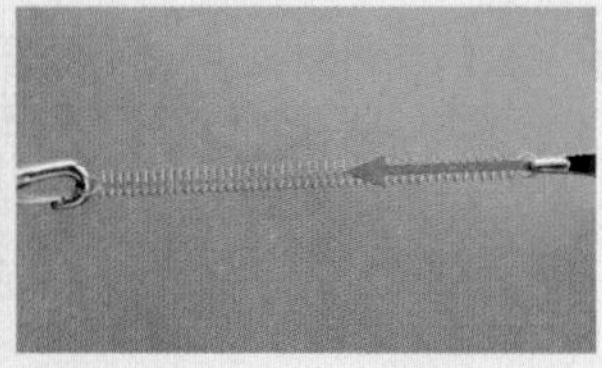

Spring force
The force exerted by a stretched or compressed spring.

Normal force
A force that a surface exerts on an object.

Newton's Second Law

Newton's second law tells us what forces *do* when applied to an object. We'll find that forces act to *accelerate* objects. We will use Newton's second law throughout this textbook to solve a wide variety of physics problems.

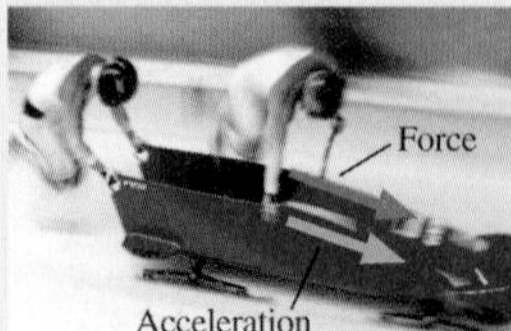

An object's acceleration vector is in the same direction as the net force acting on the object.

Looking Back ◂◂
2.4 Acceleration

Looking Back ◂◂
3.2–3.3 Vectors and coordinate systems

Identifying and Representing Forces

One of the most important skills you'll learn in this chapter is to properly identify the forces that act on an object. Then you'll learn to organize these forces in a *free-body diagram*.

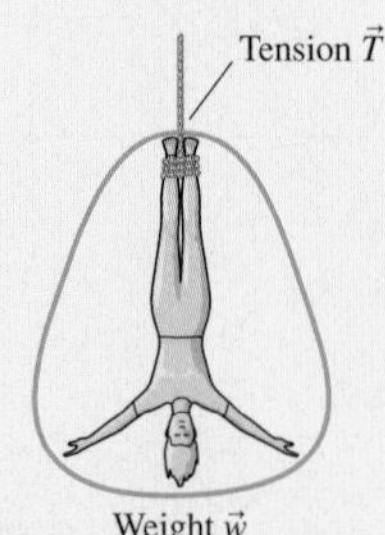

Other than the weight force, all forces acting on an object come from other objects that *touch* it.

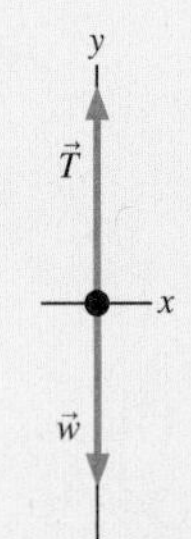

We can represent all the forces acting on an object in a free-body diagram.

4.1 What Causes Motion?

The ice boats shown in the chapter-opening photo fly across the frozen lake at some 60 mph. We could use kinematics to describe the boats' motion with pictures, graphs, and equations. Kinematics provides a language to describe *how* something moves, but tells us nothing about *why* the boats accelerate briskly before reaching their top speed. For the more fundamental task of understanding the *cause* of motion, we turn our attention to **dynamics.** Dynamics joins with kinematics to form **mechanics,** the general science of motion. We study dynamics qualitatively in this chapter, then develop it quantitatively in the next four chapters.

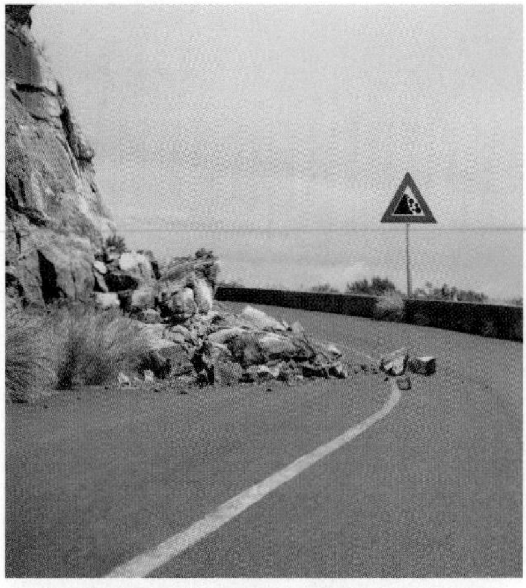

The rocks in this rockslide quickly came to rest. Is this the "natural state" of objects?

As we remarked in Chapter 1, Aristotle (384–322 BC) and his contemporaries in the world of ancient Greece were very interested in motion. One question they asked was: What is the "natural state" of an object if left to itself? It does not take an expensive research program to see that every moving object on earth, if left to itself, eventually comes to rest. You must push a shopping cart to keep it rolling, but when you stop pushing, the cart soon comes to rest; a boulder bounds downhill and then tumbles to a halt. Having observed many such examples himself, Aristotle concluded that the natural state of an earthly object is to be *at rest.* An object at rest requires no explanation; it is doing precisely what comes naturally to it. We'll soon see, however, that this simple viewpoint is *incomplete.*

Aristotle further pondered moving objects. A moving object is *not* in its natural state and thus requires an explanation: Why is this object moving? What keeps it going and prevents it from being in its natural state? When a puck is sliding across the ice, what keeps it going? Why does an arrow fly through the air once it is no longer being pushed by the bowstring? Although these questions seem like reasonable ones to pose, it was Galileo who first showed that the questions being asked were, in fact, the wrong ones.

Galileo reopened the question of the "natural state" of objects. He suggested focusing on the *idealized case* in which resistance to the motion (e.g., friction or air resistance) is zero. He performed many experiments to study motion. Let's imagine a modern experiment of this kind, as shown in FIGURE 4.1.

FIGURE 4.1 Sleds sliding on increasingly smooth surfaces.

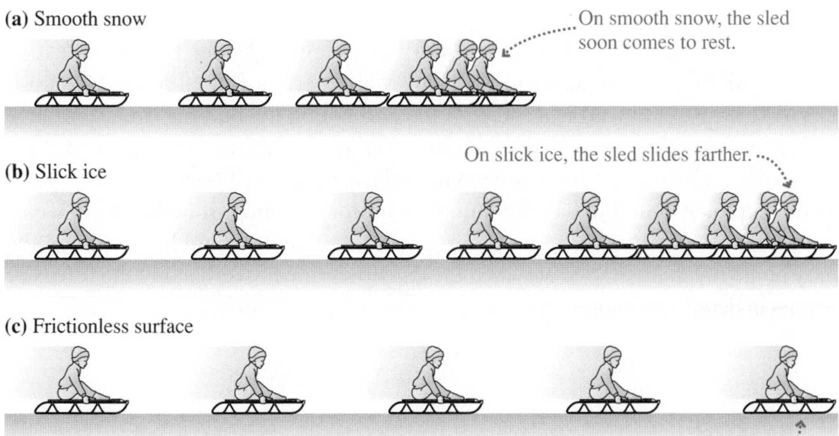

(a) Smooth snow
On smooth snow, the sled soon comes to rest.
(b) Slick ice
On slick ice, the sled slides farther.
(c) Frictionless surface
If friction could be reduced to zero, the sled would *never* stop.

▶ **Interstellar coasting** A nearly perfect example of Newton's first law is the pair of Voyager space probes launched in 1977. Both spacecraft long ago ran out of fuel and are now coasting through the frictionless vacuum of space. Although not entirely free of influence from the sun's gravity, they are now so far from the sun and other stars that gravitational influences are very nearly zero. Thus, according to the first law, they will continue at their current speed of about 40,000 miles per hour essentially forever. Billions of years from now, long after our solar system is dead, the Voyagers will still be drifting through the stars.

TRY IT YOURSELF

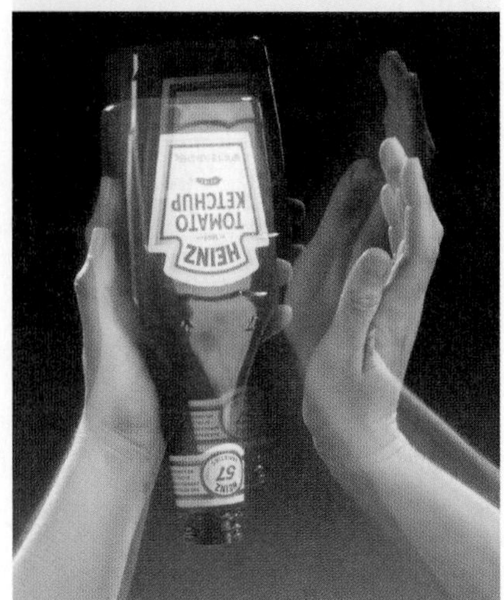

Getting the ketchup out The ketchup stuck at the bottom of the bottle is initially at rest. If you hit the bottom of the bottle, the bottle suddenly moves down, taking the ketchup on the bottom of the bottle with it, so that the ketchup just stays stuck to the bottom. But if instead you hit *up* on the bottle, as shown, you force the bottle rapidly upward. By the first law, the ketchup that was stuck to the bottom stays at rest, so it separates from the upward-moving bottle: The ketchup has moved forward with respect to the bottle!

Tyler slides down a hill on his sled, then out onto a horizontal patch of smooth snow, which is shown in Figure 4.1a. Even if the snow is quite smooth, the friction between the sled and the snow will soon cause the sled to come to rest. What if Tyler slides down the hill onto some very slick ice, as in Figure 4.1b? This gives very low friction, and the sled could slide for quite a distance before stopping. Galileo's genius was to imagine the case where *all* sources of friction, air resistance, and other retarding influences were removed, as for the sled in Figure 4.1c sliding on idealized *frictionless* ice. We can imagine in that case that the sled, once started in its motion, would continue in its motion *forever,* moving in a straight line with no loss of speed. In other words, **the natural state of an object—its behavior if free of external influences—is *uniform motion* with constant velocity!** Further, "at rest" has no special significance in Galileo's view of motion; it is simply uniform motion that happens to have a velocity of zero. This implies that an object at rest, in the absence of external influences, will remain at rest forever.

Galileo's ideas were completely counter to those of the ancient Greeks. We no longer need to explain why a sled continues to slide across the ice; that motion is its "natural" state. What needs explanation, in this new viewpoint, is why objects *don't* continue in uniform motion. Why does a sliding puck eventually slow to a stop? Why does a stone, thrown upward, slow and eventually fall back down? Galileo's new viewpoint was that the stone and the puck are *not* free of "influences": The stone is somehow pulled toward the earth, and some sort of retarding influence acted to slow the sled down. Today, we call such influences that lead to deviations from uniform motion **forces.**

Galileo's experiments were limited to motion along horizontal surfaces. It was left to Newton to generalize Galileo's conclusions, and today we call this generalization Newton's first law of motion.

Newton's first law Consider an object with no force acting on it. If it is at rest, it will remain at rest; if it is moving, it will continue to move in a straight line at a constant speed.

FIGURE 4.2 Newton's first law tells us: Wear your seatbelts!

At the instant of impact, the car and driver are moving at the same speed.

The car slows as it hits, but the driver continues at the same speed . . .

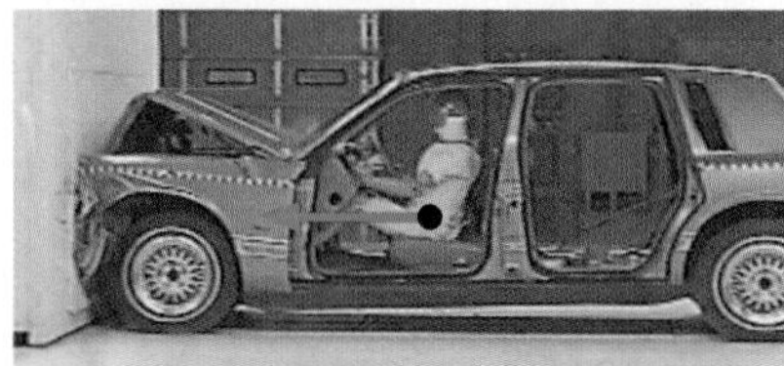

. . . until he hits the now-stationary dashboard. Ouch!

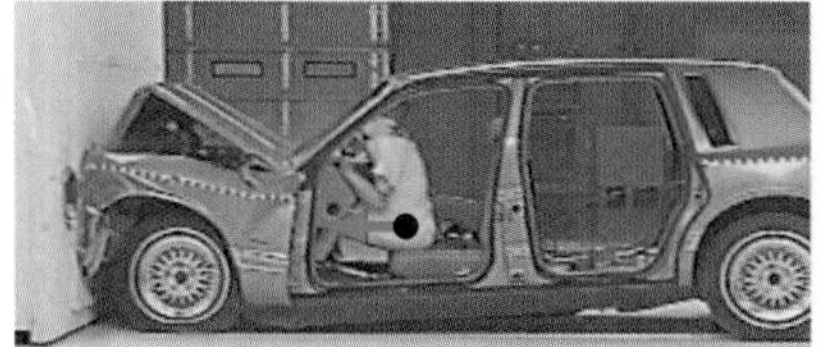

As an important application of Newton's first law, consider the crash test of FIGURE 4.2. As the car contacts the wall, the wall exerts a force on the car and it begins to slow. But the wall is a force on the *car,* not on the dummy. In accordance with Newton's first law, the unbelted dummy continues to move straight ahead at his original speed. Only when he collides violently with the dashboard of the stopped car is there a force acting to halt the dummy's uniform motion. If he had been wearing a seatbelt, the influence (i.e., the force) of the seatbelt would have slowed the dummy at the much lower rate at which the car slows down. We'll study the forces of collisions in detail in Chapter 10.

4.2 Force

Newton's first law tells us that an object in motion subject to no forces will continue to move in a straight line forever. But this law does not explain in any detail exactly what a force *is.* Unfortunately, there is no simple one-sentence definition of force. The concept of force is best introduced by looking at examples of some common forces and considering the basic properties shared by all forces. This will be our task in the next two sections. Let's begin by examining the properties that all forces have in common, as presented in the table on the next page.

What is a force?

A force is a push or a pull.

Our commonsense idea of a **force** is that it is a *push* or a *pull.* We will refine this idea as we go along, but it is an adequate starting point. Notice our careful choice of words: We refer to "*a* force" rather than simply "force." We want to think of a force as a very specific *action,* so that we can talk about a single force or perhaps about two or three individual forces that we can clearly distinguish—hence the concrete idea of "a force" acting on an object.

A force acts on an object.

Implicit in our concept of force is that **a force acts on an object.** In other words, pushes and pulls are applied *to* something—an object. From the object's perspective, it has a force *exerted* on it. Forces do not exist in isolation from the object that experiences them.

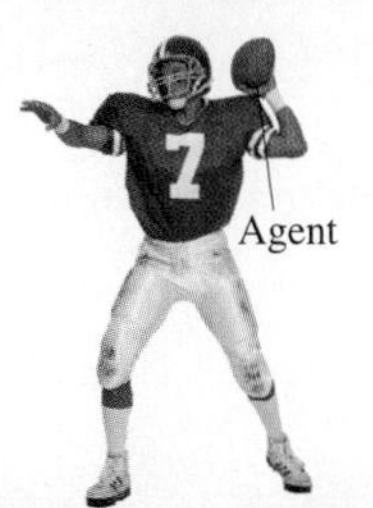

A force requires an agent.

Every force has an **agent,** something that acts or pushes or pulls; that is, a force has a specific, identifiable *cause.* As you throw a ball, it is your hand, while in contact with the ball, that is the agent or the cause of the force exerted on the ball. *If* a force is being exerted on an object, you must be able to identify a specific cause (i.e., the agent) of that force. Conversely, a force is not exerted on an object *unless* you can identify a specific cause or agent. Note that an agent can be an inert object such as a tabletop or a wall. Such agents are the cause of many common forces.

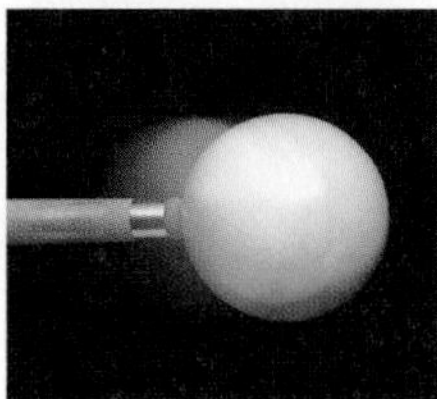

A force is a vector.

If you push an object, you can push either gently or very hard. Similarly, you can push either left or right, up or down. To quantify a push, we need to specify both a magnitude *and* a direction. It should thus come as no surprise that a force is a vector quantity. The general symbol for a force is the vector symbol $\vec{F}$. The size or strength of such a force is its magnitude F.

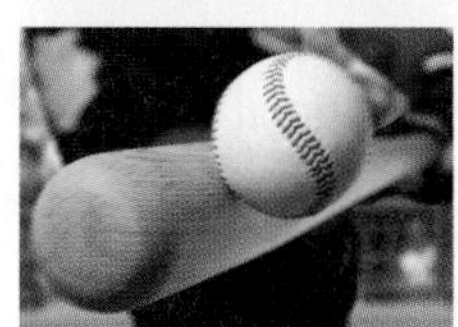

A force can be either a contact force . . .

There are two basic classes of forces, depending on whether the agent touches the object or not. **Contact forces** are forces that act on an object by touching it at a point of contact. The bat must touch the ball to hit it. A string must be tied to an object to pull it. The majority of forces that we will examine are contact forces.

. . . or a long-range force.

Long-range forces are forces that act on an object without physical contact. Magnetism is an example of a long-range force. You have undoubtedly held a magnet over a paper clip and seen the paper clip leap up to the magnet. A coffee cup released from your hand is pulled to the earth by the long-range force of gravity.

Let's summarize these ideas as our definition of force:

- A force is a push or a pull on an object.
- A force is a vector. It has both a magnitude and a direction.
- A force requires an agent. Something does the pushing or pulling. The agent can be an inert object such as a tabletop or a wall.
- A force is either a contact force or a long-range force. Gravity is the only long-range force we will deal with until much later in the book.

There's one more important aspect of forces. If you push against a door (the object) to close it, the door pushes back against your hand (the agent). If a tow rope pulls on a car (the object), the car pulls back on the rope (the agent). In general, if an agent exerts a force on an object, the object exerts a force on the agent. We really need to think of a force as an *interaction* between two objects. Although the interaction perspective is a more exact way to view forces, it adds complications that we would like to avoid for now. Our approach will be to start by focusing on how a single object responds to forces exerted on it. Later in this chapter, we'll return to the larger issue of how two or more objects interact with each other.

Force Vectors

We can use a simple diagram to visualize how forces are exerted on objects. Because we are using the particle model, in which objects are treated as points, the process of drawing a force vector is straightforward. Here is how it goes:

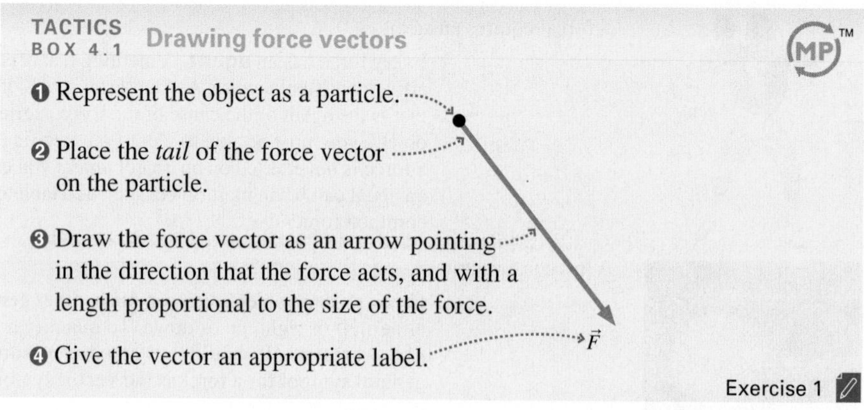

TACTICS BOX 4.1 Drawing force vectors
MP™
❶ Represent the object as a particle.
❷ Place the *tail* of the force vector on the particle.
❸ Draw the force vector as an arrow pointing in the direction that the force acts, and with a length proportional to the size of the force.
❹ Give the vector an appropriate label.
$\vec{F}$
Exercise 1

Step 2 may seem contrary to what a "push" should do (it may look as if the force arrow is *pulling* the object rather than *pushing* it), but recall that moving a vector does not change it as long as the length and angle do not change. The vector $\vec{F}$ is the same regardless of whether the tail or the tip is placed on the particle. Our reason for using the tail will become clear when we consider how to combine several forces.

FIGURE 4.3 shows three examples of force vectors. One is a pull, one a push, and one a long-range force, but in all three the *tail* of the force vector is placed on the particle representing the object.

FIGURE 4.3 Three force vectors.

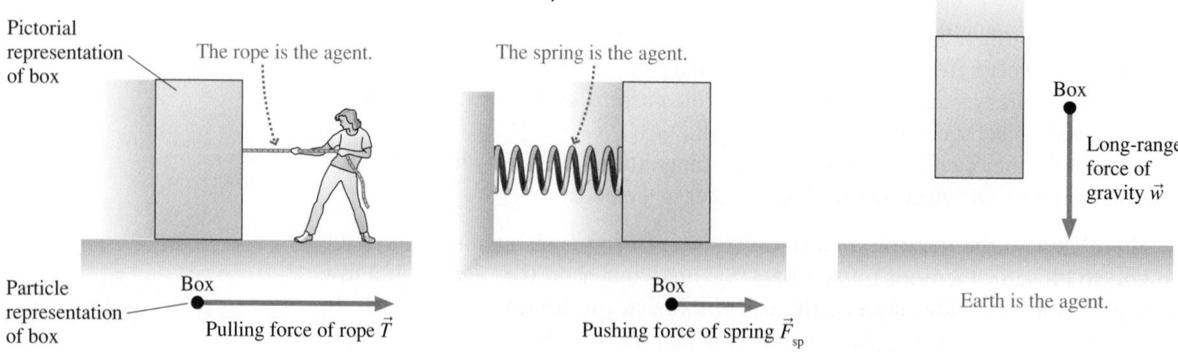

Combining Forces

FIGURE 4.4a shows a top view of a box being pulled by two ropes, each exerting a force on the box. How will the box respond? Experimentally, we find that when several forces $\vec{F}_1, \vec{F}_2, \vec{F}_3, \ldots$ are exerted on an object, they combine to form a **net force** that is the *vector* sum of all the forces:

$$\vec{F}_{net} = \vec{F}_1 + \vec{F}_2 + \vec{F}_3 + \cdots \quad (4.1)$$

That is, the single force $\vec{F}_{net}$ causes the exact same motion of the object as the combination of original forces $\vec{F}_1, \vec{F}_2, \vec{F}_3, \ldots$. Mathematically, this summation is called a *superposition* of forces. The net force is sometimes called the *resultant force.* FIGURE 4.4b shows the net force on the box.

NOTE ▶ It is important to realize that the net force $\vec{F}_{net}$ is not a new force acting *in addition* to the original forces $\vec{F}_1, \vec{F}_2, \vec{F}_3, \ldots$. Instead, we should think of the original forces being *replaced* by $\vec{F}_{net}$. ◀

FIGURE 4.4 Two forces applied to a box.

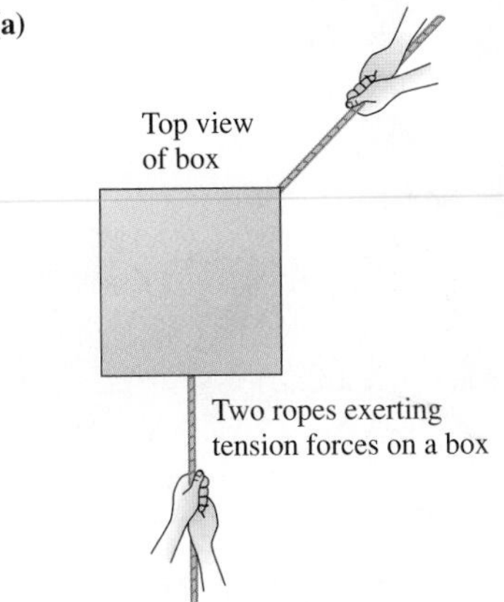

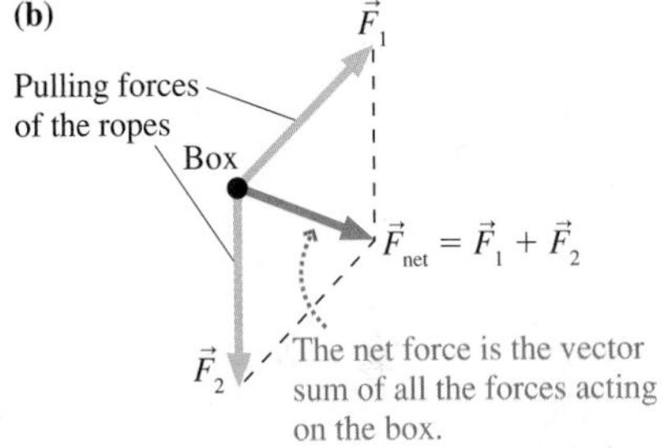

STOP TO THINK 4.1 Two of the three forces exerted on an object are shown. The net force points directly to the left. Which is the missing third force?

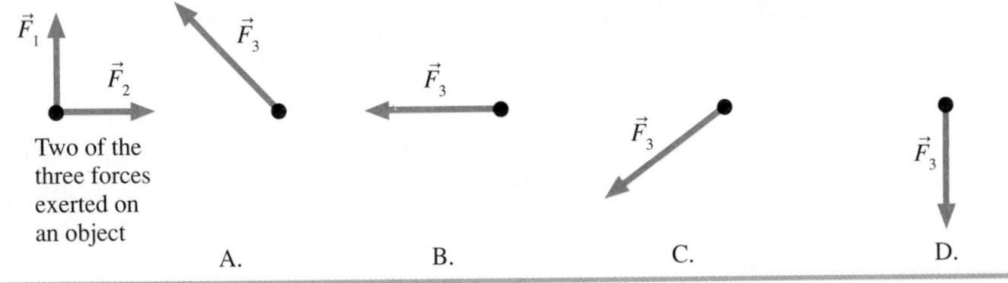

4.3 A Short Catalog of Forces

There are many forces we will deal with over and over. This section will introduce you to some of them and to the symbols we use to represent them.

Weight

A falling rock is pulled toward the earth by the long-range force of gravity. Gravity is what keeps you in your chair, keeps the planets in their orbits around the sun, and shapes the large-scale structure of the universe. We'll have a thorough look at gravity in Chapter 6. For now we'll concentrate on objects on or near the surface of the earth (or other planet).

The gravitational pull of the earth on an object on or near the surface of the earth is called **weight.** The symbol for weight is $\vec{w}$. Weight is the only long-range force we will encounter in the next few chapters. The agent for the weight force is the *entire earth* pulling on an object. The weight force is in some ways the simplest force we'll study. As FIGURE 4.5 shows, **an object's weight vector always points vertically downward,** no matter how the object is moving.

NOTE ▶ We often refer to "the weight" of an object. This is an informal expression for w, the magnitude of the weight force exerted on the object. Note that **weight is not the same thing as mass.** We will briefly examine mass later in the chapter and explore the connection between weight and mass in Chapter 5. ◀

FIGURE 4.5 Weight always points vertically downward.

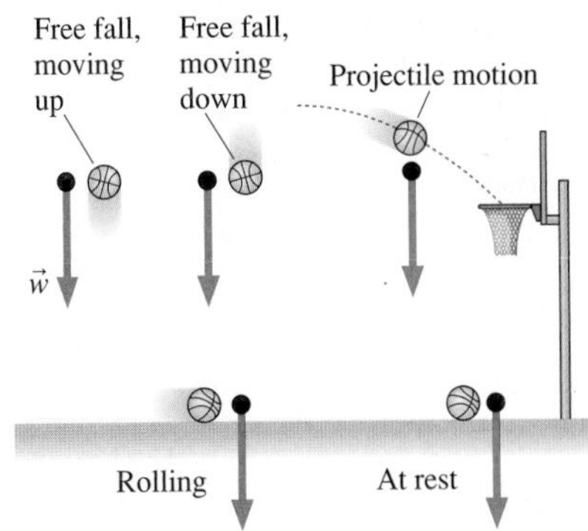

Springs come in many forms. When deflected, they push or pull with a spring force.

Spring Force

Springs exert one of the most basic contact forces. A spring can either push (when compressed) or pull (when stretched). FIGURE 4.6 shows the **spring force.** In both cases, pushing and pulling, the tail of the force vector is placed on the particle in the force diagram. There is no special symbol for a spring force, so we simply use a subscript label: $\vec{F}_{sp}$.

FIGURE 4.6 The spring force is parallel to the spring.

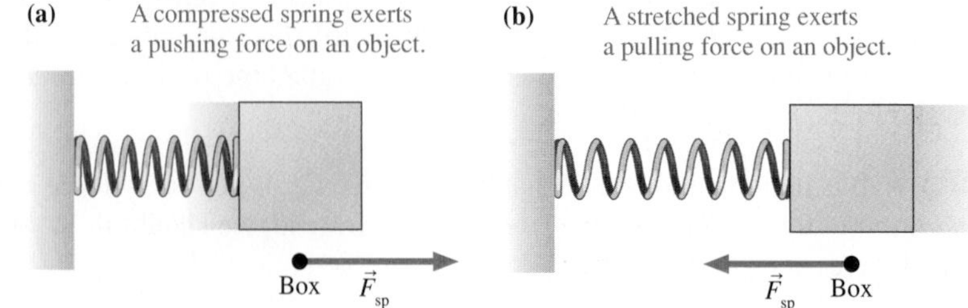

Although you may think of a spring as a metal coil that can be stretched or compressed, this is only one type of spring. Hold a ruler, or any other thin piece of wood or metal, by the ends and bend it slightly. It flexes. When you let go, it "springs" back to its original shape. This is just as much a spring as is a metal coil.

Tension Force

FIGURE 4.7 Tension is parallel to the rope.

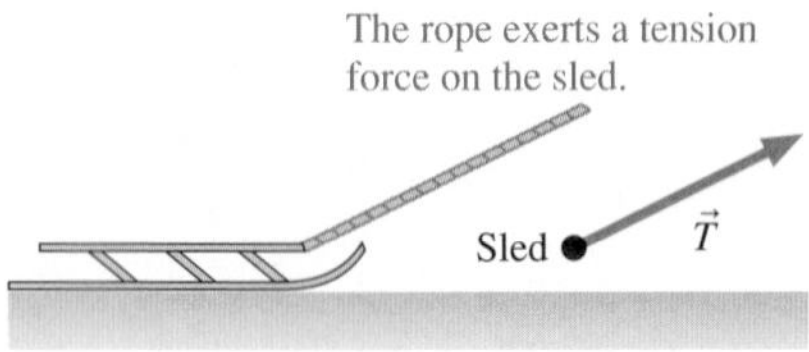

When a string or rope or wire pulls on an object, it exerts a contact force that we call the **tension force,** represented by $\vec{T}$. **The direction of the tension force is always in the direction of the string or rope,** as you can see in FIGURE 4.7. When we speak of "the tension" in a string, this is an informal expression for T, the size or magnitude of the tension force. Note that the tension force can only *pull* in the direction of the string; if you try to *push* with a string, it will go slack and be unable to exert a force.

We can think about the tension force using a microscopic picture. If you were to use a very powerful microscope to look inside a rope, you would "see" that it is made of *atoms* joined together by *molecular bonds.* Molecular bonds are not rigid connections between the atoms. They are more accurately thought of as tiny *springs* holding the atoms together, as in FIGURE 4.8. Pulling on the ends of a string or rope stretches the molecular springs ever so slightly. The tension within a rope and the tension force experienced by an object at the end of the rope are really the net spring force exerted by billions and billions of microscopic springs.

FIGURE 4.8 An atomic model of tension.

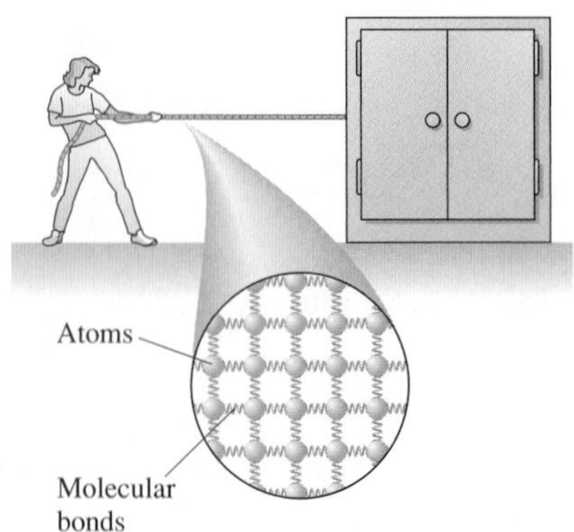

This atomic-level view of tension introduces a new idea: a microscopic **atomic model** for understanding the behavior and properties of **macroscopic** (i.e., containing many atoms) objects. We will frequently use atomic models to obtain a deeper understanding of our observations.

The atomic model of tension also helps to explain one of the basic properties of ropes and strings. When you pull on a rope tied to a heavy box, the rope in turn exerts a tension force on the box. If you pull harder, the tension force on the box becomes greater. How does the box "know" that you are pulling harder on the other end of the rope? According to our atomic model, when you pull harder on the rope, its microscopic springs stretch a bit more, increasing the spring force they exert on each other—and on the box they're attached to.

Normal Force

If you sit on a bed, the springs in the mattress compress and, as a consequence of the compression, exert an upward force on you. Stiffer springs would show less compression but would still exert an upward force. The compression of extremely stiff springs might be measurable only by sensitive instruments. Nonetheless, the springs would compress ever so slightly and exert an upward spring force on you.

FIGURE 4.9 shows a book resting on top of a sturdy table. The table may not visibly flex or sag, but—just as you do to the bed—the book compresses the molecular springs in the table. The compression is very small, but it is not zero. As a consequence, the compressed molecular springs *push upward* on the book. We say that "the table" exerts the upward force, but it is important to understand that the pushing is *really* done by molecular springs. Similarly, an object resting on the ground compresses the molecular springs holding the ground together and, as a consequence, the ground pushes up on the object.

FIGURE 4.9 An atomic model of the force exerted by a table.

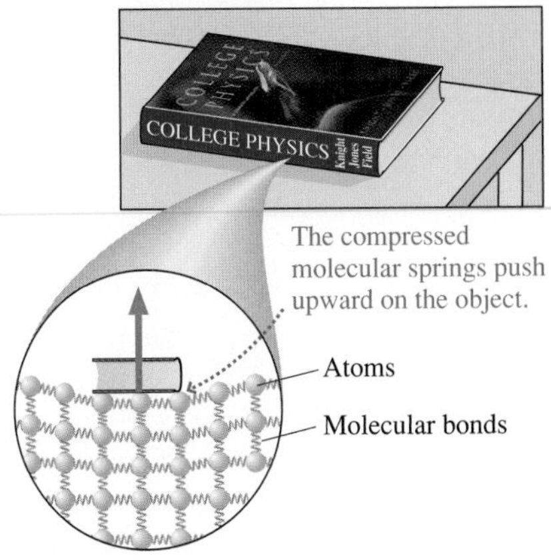
COLLEGE PHYSICS
The compressed molecular springs push upward on the object.
Atoms
Molecular bonds

We can extend this idea. Suppose you place your hand on a wall and lean against it, as shown in FIGURE 4.10. Does the wall exert a force on your hand? As you lean, you compress the molecular springs in the wall and, as a consequence, they push outward *against* your hand. So the answer is Yes, the wall does exert a force on you. It's not hard to see this if you examine your hand as you lean: You can see that your hand is slightly deformed, and becomes more so the harder you lean. This deformation is direct evidence of the force that the wall exerts on your hand. Consider also what would happen if the wall suddenly vanished. Without the wall there to push against you, you would topple forward.

FIGURE 4.10 The wall pushes outward against your hand.

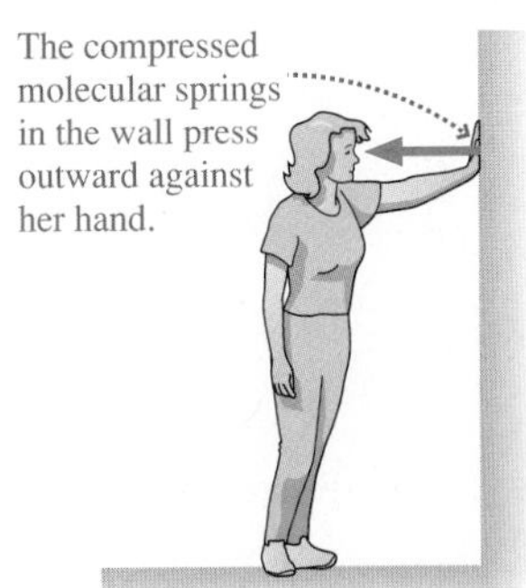
The compressed molecular springs in the wall press outward against her hand.

The force the table surface exerts is vertical, while the force the wall exerts is horizontal. In all cases, the force exerted on an object that is pressing against a surface is in a direction *perpendicular* to the surface. Mathematicians refer to a line that is perpendicular to a surface as being *normal* to the surface. In keeping with this terminology, we define the **normal force** as the force exerted by a surface (the agent) against an object that is pressing against the surface. The symbol for the normal force is $\vec{n}$.

We're not using the word "normal" to imply that the force is an "ordinary" force or to distinguish it from an "abnormal force." A surface exerts a force *perpendicular* (i.e., normal) to itself as the molecular springs press *outward.* FIGURE 4.11 shows an object on an inclined surface, a common situation. Notice how the normal force $\vec{n}$ is perpendicular to the surface.

The normal force is a very real force arising from the very real compression of molecular bonds. It is in essence just a spring force, but one exerted by a vast number of microscopic springs acting at once. The normal force is responsible for the "solidness" of solids. It is what prevents you from passing right through the chair you are sitting in and what causes the pain and the lump if you bang your head into a door. Your head can then tell you that the force exerted on it by the door was very real!

FIGURE 4.11 The normal force is perpendicular to the surface.

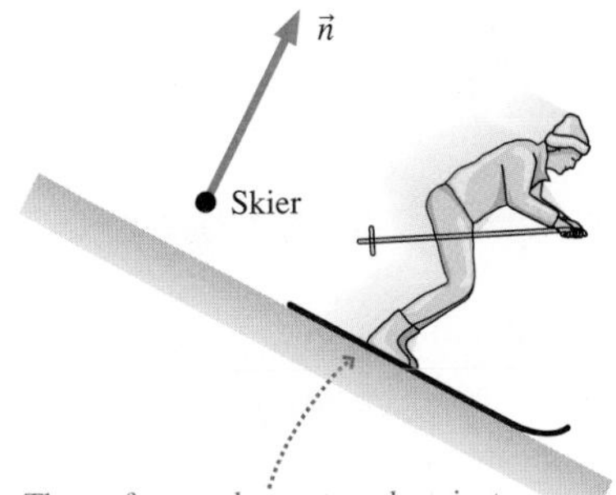
$\vec{n}$
Skier
The surface pushes outward against the bottom of the skis. The force is perpendicular to the surface.

Friction

You've certainly observed that a rolling or sliding object, if not pushed or propelled, slows down and eventually stops. You've probably discovered that you can slide better across a sheet of ice than across asphalt. And you also know that most objects stay in place on a table without sliding off even if the table is tilted a bit. The force responsible for these sorts of behavior is **friction.** The symbol for friction is $\vec{f}$.

Friction, like the normal force, is exerted by a surface. Unlike the normal force, however, **the frictional force is always *parallel* to the surface,** not perpendicular to it. (In many cases, a surface will exert *both* a normal and a frictional force.) On a microscopic level, friction arises as atoms from the object and atoms on the surface run into each other. The rougher the surface is, the more these atoms are forced into close proximity and, as a result, the larger the friction force. We will develop a simple model of friction in the next chapter that will be sufficient for our needs. For now, it is useful to distinguish between two kinds of friction:

- *Kinetic friction,* denoted $\vec{f}_k$, acts as an object slides across a surface. Kinetic friction is a force that always "opposes the motion," meaning that the friction force $\vec{f}_k$ on a sliding object points in the direction opposite the direction of the object's motion.
- *Static friction,* denoted $\vec{f}_s$, is the force that keeps an object "stuck" on a surface and prevents its motion relative to the surface. Finding the direction of $\vec{f}_s$ is a little trickier than finding it for $\vec{f}_k$. Static friction points opposite the direction in which the object *would* move if there were no friction. That is, it points in the direction necessary to *prevent* motion.

FIGURE 4.12 shows examples of kinetic and static friction.

FIGURE 4.12 Kinetic and static friction are parallel to the surface.

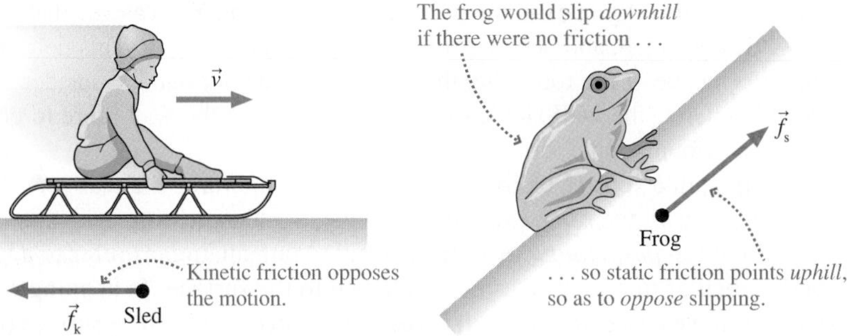

FIGURE 4.13 Air resistance is an example of drag.

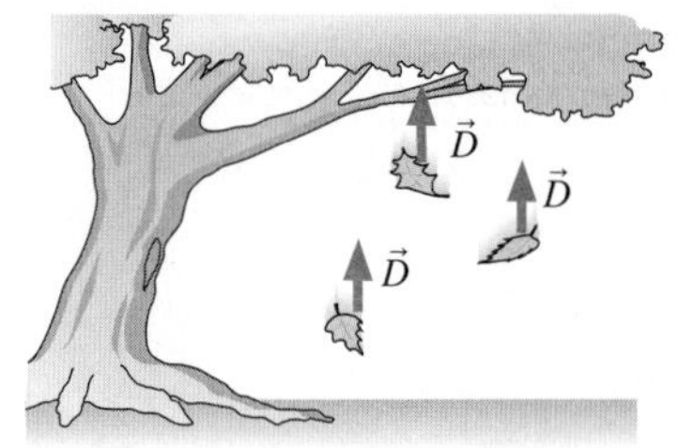

Drag

Friction at a surface is one example of a *resistive force,* a force that opposes or resists motion. Resistive forces are also experienced by objects moving through *fluids*—gases (like air) and liquids (like water). This kind of resistive force—the force of a fluid on a moving object—is called **drag** and is symbolized as $\vec{D}$. Like kinetic friction, **drag points opposite the direction of motion.** FIGURE 4.13 shows an example of drag.

Drag can be a large force for objects moving at high speeds or in dense fluids. Hold your arm out the window as you ride in a car and feel how hard the air pushes against your arm; note also how the air resistance against your arm increases rapidly as the car's speed increases. Drop a lightweight bead into a beaker of water and watch how slowly it settles to the bottom. The drag force of the water on the bead is significant.

On the other hand, for objects that are heavy and compact, moving in air, and with a speed that is not too great, the drag force of air resistance is fairly small. To keep things as simple as possible, **you can neglect air resistance in all problems unless a problem explicitly asks you to include it.** The error introduced into calculations by this approximation is generally pretty small.

FIGURE 4.14 The thrust force on a rocket is opposite the direction of the expelled gases.

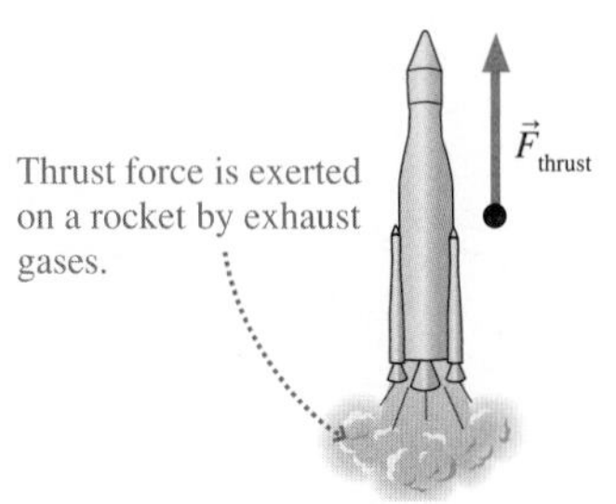

Thrust

A jet airplane obviously has a force that propels it forward; likewise for the rocket in FIGURE 4.14. This force, called **thrust,** occurs when a jet or rocket engine expels gas molecules at high speed. Thrust is a contact force, with the exhaust gas being the agent that pushes on the engine. The process by which thrust is generated is rather subtle and requires an appreciation of Newton's third law, introduced later in this chapter. For now, we need only consider that **thrust is a force opposite the direction in which the exhaust gas is expelled.** There's no special symbol for thrust, so we will call it $\vec{F}_{\text{thrust}}$.

It's a drag At the high speeds attained by racing cyclists, air drag can become very significant. The world record for the longest distance traveled in one hour on an ordinary bicycle is 56.38 km, set by Chris Boardman in 1996. But a bicycle with an aerodynamic shell has a much lower drag force, allowing it to attain significantly higher speeds. The bike shown here was pedaled 84.22 km in one hour by Sam Whittingham in 2004, for an amazing average speed of 52.3 mph!

Electric and Magnetic Forces

Electricity and magnetism, like gravity, exert long-range forces. The forces of electricity and magnetism act on charged particles. We will study electric and magnetic forces in detail in Part VI of this book. For now, it is worth noting that the forces holding molecules together—the molecular bonds—are not actually tiny springs. Atoms and molecules are made of charged particles—electrons and protons—and what we call a molecular bond is really an electric force between these particles. So when we say that the normal force and the tension force are due to "molecular springs," or that friction is due to atoms running into each other, what we're really saying is that these forces, at the most fundamental level, are actually electric forces between the charged particles in the atoms.

4.4 Identifying Forces

Force and motion problems generally have two basic steps:

1. Identify all of the forces acting on an object.
2. Use Newton's laws and kinematics to determine the motion.

Understanding the first step is the primary goal of this chapter. We'll turn our attention to step 2 in the next chapter.

A typical physics problem describes an object that is being pushed and pulled in various directions. Some forces are given explicitly, while others are only implied. In order to proceed, it is necessary to determine all the forces that act on the object. It is also necessary to avoid including forces that do not really exist. Now that you have learned the properties of forces and seen a catalog of typical forces, we can develop a step-by-step method for identifying each force in a problem. A list of the most common forces we'll come across in the next few chapters is given in Table 4.1.

TABLE 4.1 Common forces and their notation

Force	Notation
General force	$\vec{F}$
Weight	$\vec{w}$
Spring force	$\vec{F}_{\text{sp}}$
Tension	$\vec{T}$
Normal force	$\vec{n}$
Static friction	$\vec{f}_{\text{s}}$
Kinetic friction	$\vec{f}_{\text{k}}$
Drag	$\vec{D}$
Thrust	$\vec{F}_{\text{thrust}}$

TACTICS BOX 4.2 Identifying forces

❶ **Identify the object of interest.** This is the object whose motion you wish to study.

❷ **Draw a picture of the situation.** Show the object of interest and all other objects—such as ropes, springs, and surfaces—that touch it.

❸ **Draw a closed curve around the object.** Only the object of interest is inside the curve; everything else is outside.

❹ **Locate every point on the boundary of this curve where other objects touch the object of interest.** These are the points where *contact forces* are exerted on the object.

❺ **Name and label each contact force acting on the object.** There is at least one force at each point of contact; there may be more than one. When necessary, use subscripts to distinguish forces of the same type.

❻ **Name and label each long-range force acting on the object.** For now, the only long-range force is weight.

Exercises 4–8 

CONCEPTUAL EXAMPLE 4.1 **Identifying forces on a bungee jumper**

A bungee jumper has leapt off a bridge and is nearing the bottom of her fall. What forces are being exerted on the bungee jumper?

REASON FIGURE 4.15 Forces on a bungee jumper.

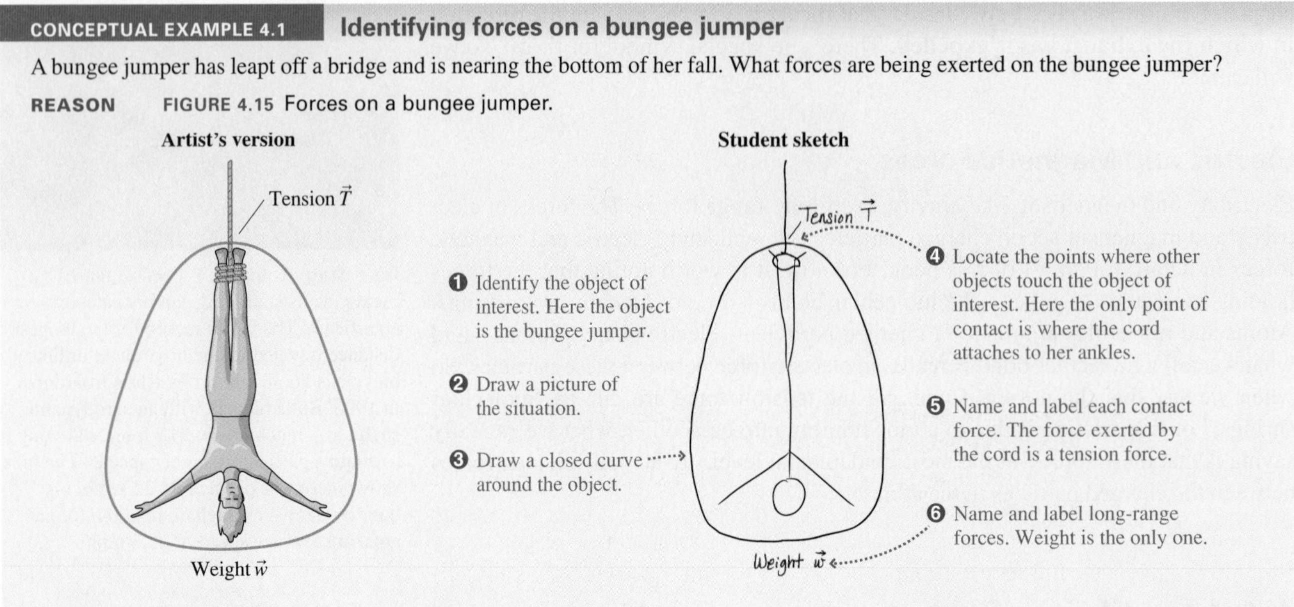

CONCEPTUAL EXAMPLE 4.2 **Identifying forces on a skier**

A skier is being towed up a snow-covered hill by a tow rope. What forces are being exerted on the skier?

REASON FIGURE 4.16 Forces on a skier.

Tension $\vec{T}$

Normal force $\vec{n}$

Kinetic friction $\vec{f}_k$

Weight $\vec{w}$

❶ Identify the object of interest. Here the object is the skier.

❷ Draw a picture of the situation.

❸ Draw a closed curve around the object.

❹ Locate the points where other objects touch the object of interest. Here the rope and the ground touch the skier.

❺ Name and label each contact force. The rope exerts a tension force, and the ground exerts both a normal and a kinetic friction force.

❻ Name and label long-range forces. Weight is the only one.

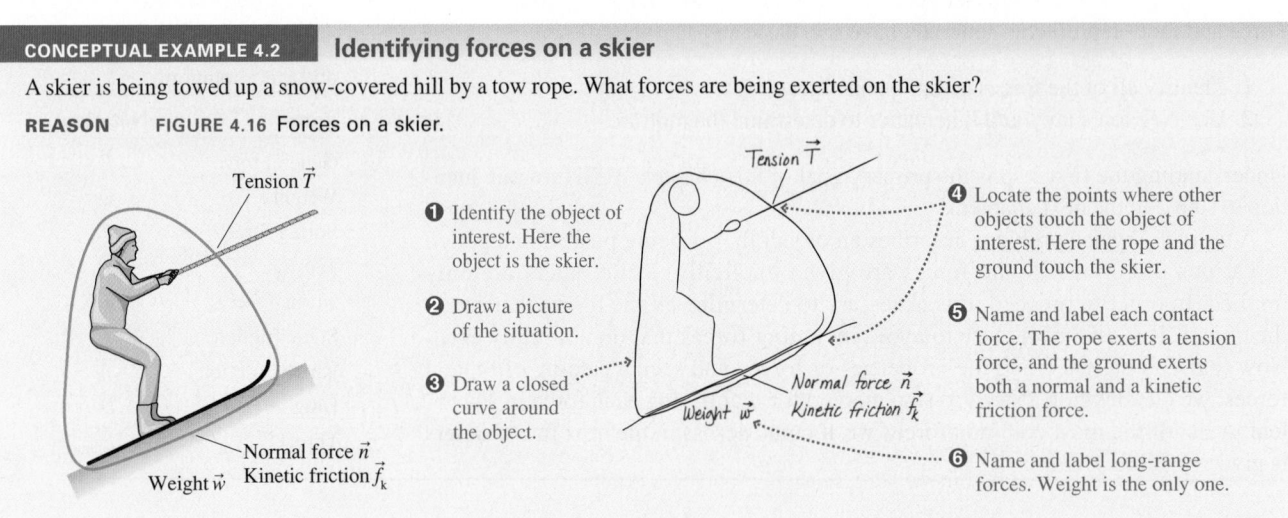

NOTE ▶ You might have expected two friction forces and two normal forces in Example 4.2, one on each ski. Keep in mind, however, that we're working within the particle model, which represents the skier by a single point. A particle has only one contact with the ground, so there is a single normal force and a single friction force. The particle model is valid if we want to analyze the motion of the skier as a whole, but we would have to go beyond the particle model to find out what happens to each ski. ◀

CONCEPTUAL EXAMPLE 4.3 **Identifying forces on a rocket**

A rocket is being launched to place a new satellite in orbit. Air resistance is not negligible. What forces are being exerted on the rocket?

REASON

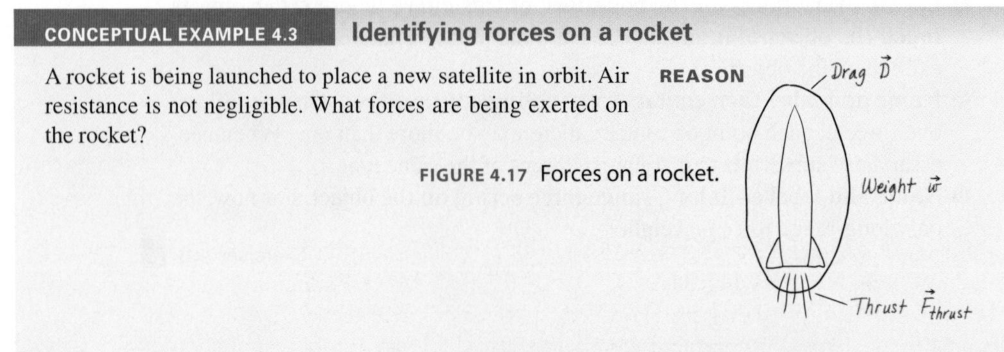

FIGURE 4.17 Forces on a rocket.

STOP TO THINK 4.2 You've just kicked a rock, and it is now sliding across the ground about 2 meters in front of you. Which of these are forces acting on the rock? List all that apply.

A. Gravity, acting downward
B. The normal force, acting upward
C. The force of the kick, acting in the direction of motion
D. Friction, acting opposite the direction of motion
E. Air resistance, acting opposite the direction of motion

4.5 What Do Forces Do?

The fundamental question is: How does an object move when a force is exerted on it? The only way to answer this question is to do experiments. To do experiments, however, we need a way to reproduce the same force again and again, and we need a standard object so that our experiments are repeatable.

FIGURE 4.18 shows how you can use your fingers to stretch a rubber band to a certain length—say, 10 centimeters—that you can measure with a ruler. We'll call this the *standard length.* You know that a stretched rubber band exerts a force because your fingers *feel* the pull. Furthermore, this is a reproducible force. The rubber band exerts the same force every time you stretch it to the standard length. We'll call the magnitude of this force the *standard force F*. Not surprisingly, two identical rubber bands, each stretched to the standard length, exert twice the force of one rubber band; three rubber bands exert three times the force; and so on.

FIGURE 4.18 A reproducible force.

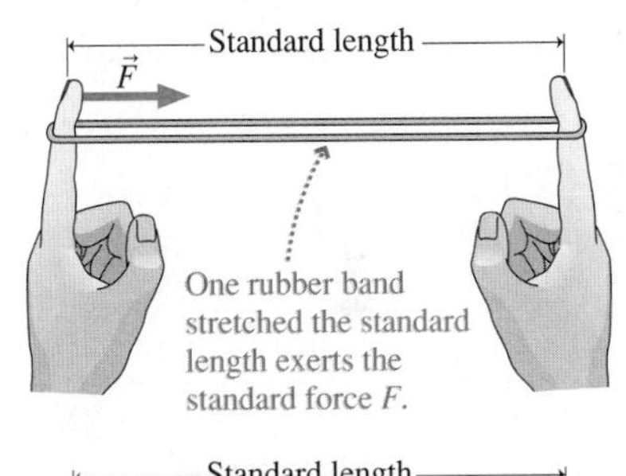
Standard length
$\vec{F}$
One rubber band stretched the standard length exerts the standard force F.

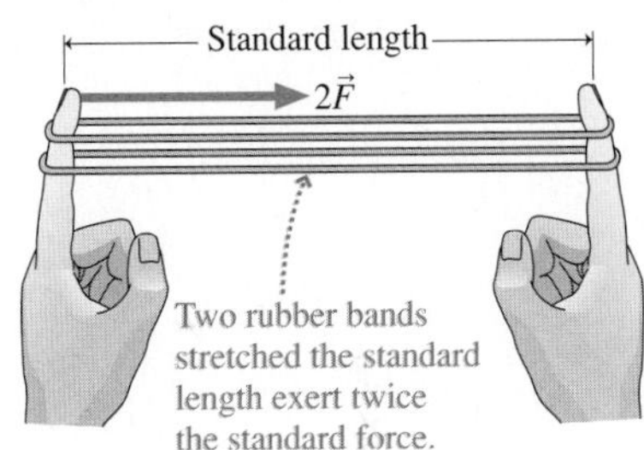
Standard length
$2\vec{F}$
Two rubber bands stretched the standard length exert twice the standard force.

We'll also need several identical standard objects to which the force will be applied. As we learned in Chapter 1, the SI unit of mass is the kilogram (kg). The kilogram is defined in terms of a particular metal block kept in a vault in Paris. For our standard objects, we will make ourselves several identical copies, each with, by definition, a mass of 1 kg. At this point, you can think of mass as the "quantity of matter" in an object. This idea will suffice for now, but by the end of this section, we'll be able to give a more precise meaning to the concept of mass.

Now we're ready to start the virtual experiment. First, place one of the 1 kg blocks on a frictionless surface. (In a real experiment, we can nearly eliminate friction by floating the block on a cushion of air.) Second, attach a rubber band to the block and stretch the band to the standard length. Then the block experiences the same force F as your finger did. As the block starts to move, in order to keep the pulling force constant you must *move your hand* in just the right way to keep the length of the rubber band—and thus the force—*constant.* **FIGURE 4.19** shows the experiment being carried out. Once the motion is complete, you can use motion diagrams and kinematics to analyze the block's motion.

FIGURE 4.19 Measuring the motion of a 1 kg block that is pulled with a constant force.

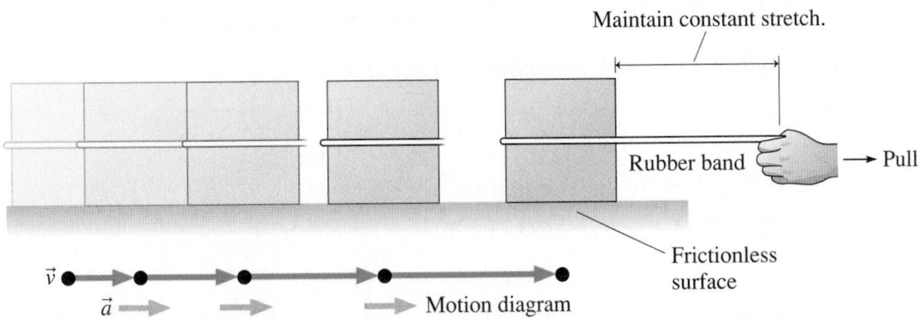
Maintain constant stretch.
Rubber band
Pull
Frictionless surface
$\vec{v}$
$\vec{a}$
Motion diagram

The motion diagram in Figure 4.19 shows that the velocity vectors are getting longer, so the velocity is increasing: The block is *accelerating*. Furthermore, a close inspection of the motion diagram shows that the acceleration vectors are all the same length. This is the first important finding of this experiment: **An object pulled with**

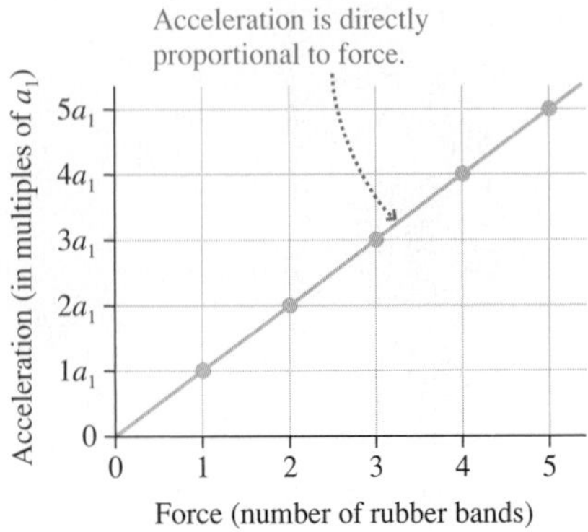

FIGURE 4.20 Graph of acceleration versus force.

FIGURE 4.21 Graph of acceleration versus number of blocks.

a constant force moves with a constant acceleration. This finding could not have been anticipated in advance. It's conceivable that the object would speed up for a while and then move with a steady speed. Or that it would continue to speed up, but that the *rate* of increase, the acceleration, would steadily decline. But these descriptions do not match what happens. Instead, the object continues *with a constant acceleration* for as long as you pull it with a constant force. We'll call this constant acceleration of *one* block pulled by *one* band a_1.

What happens if you increase the force by using several rubber bands? To find out, use two rubber bands. Stretch both to the standard length to double the force to $2F$, then measure the acceleration. Measure the acceleration due to three rubber bands, then four, and so on. **FIGURE 4.20** is a graph of the results. Force is the independent variable, the one you can control, so we've placed force on the horizontal axis to make an acceleration-versus-force graph. The graph reveals our second important finding: **Acceleration is directly proportional to force.**

The final question for our virtual experiment is: How does the acceleration of an object depend on the mass of the object, the "quantity of matter" that it contains? To find out, we'll glue two of our 1 kg blocks together, so that we have a block with twice as much matter as a 1 kg block—that is, a 2 kg block. Now apply the same force—a single rubber band—as you applied to the single 1 kg block. **FIGURE 4.21** shows that the acceleration is *one-half* as great as that of the single block. If we glue three blocks together, making a 3 kg object, we find that the acceleration is only *one-third* of the 1 kg block's acceleration. In general, we find that the acceleration is proportional to the *inverse* of the mass of the object. So our third important result is: **Acceleration is *inversely proportional* to an object's mass.**

Inversely proportional relationships

Two quantities are said to be **inversely proportional** to each other if one quantity is proportional to the *inverse* of the other. Mathematically, this means that

$$y = \frac{A}{x}$$

y is inversely proportional to x

Here, A is a proportionality constant. This relationship is sometimes written as $y \propto 1/x$.

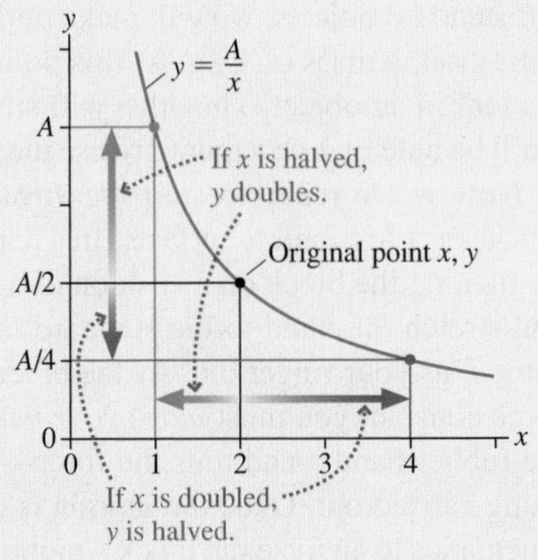

SCALING
- If you double x, you halve y.
- If you triple x, y is reduced by a factor of 3.
- If you halve x, y doubles.
- If you reduce x by a factor of 3, y becomes 3 times as large.

RATIOS For any two values of x—say, x_1 and x_2—we have

$$y_1 = \frac{A}{x_1} \quad \text{and} \quad y_2 = \frac{A}{x_2}$$

Dividing the y_1 equation by the y_2 equation, we find

$$\frac{y_1}{y_2} = \frac{A/x_1}{A/x_2} = \frac{A}{x_1}\frac{x_2}{A} = \frac{x_2}{x_1}$$

That is, the ratio of y-values is the inverse of the ratio of the corresponding values of x.

LIMITS
- As x gets very large, y approaches zero.
- As x approaches zero, y gets very large.

Exercise 10

Our original idea of mass was that it was a measure of the "quantity of matter" that an object contains. Now we see that a more precise way of defining the mass of an object is in terms of its *acceleration*. You're familiar with this idea: It's much harder to get your car rolling by pushing it than to get your bicycle rolling; it's harder to stop a heavily loaded grocery cart than to stop a skateboard. This tendency to resist a change in velocity (i.e., to resist speeding up or slowing down) is called **inertia.** Thus we can say that more massive objects have more inertia.

These considerations allow us to unambiguously determine the mass of an object by measuring its acceleration, as the next example shows.

EXAMPLE 4.4 Finding the mass of an unknown block

When a rubber band is stretched to pull on a 1.0 kg block with a constant force, the acceleration of the block is measured to be 3.0 m/s^2. When a block with an unknown mass is pulled with the same rubber band, using the same force, its acceleration is 5.0 m/s^2. What is the mass of the unknown block?

PREPARE Each block's acceleration is inversely proportional to its mass.

SOLVE We can use the result of the Inversely Proportional Relationships box to write

$$\frac{3.0\ \text{m/s}^2}{5.0\ \text{m/s}^2} = \frac{m}{1.0\ \text{kg}}$$

or

$$m = \frac{3.0\ \text{m/s}^2}{5.0\ \text{m/s}^2}(1.0\ \text{kg}) = 0.60\ \text{kg}$$

ASSESS With the same force applied, the unknown block had a *larger* acceleration than the 1.0 kg block. It makes sense, then, that its mass—its resistance to acceleration—is *less* than 1.0 kg.

STOP TO THINK 4.3 Two rubber bands stretched to the standard length cause an object to accelerate at 2 m/s^2. Suppose another object with twice the mass is pulled by four rubber bands stretched to the standard length. What is the acceleration of this second object?

A. 1 m/s^2 B. 2 m/s^2 C. 4 m/s^2 D. 8 m/s^2 E. 16 m/s^2

4.6 Newton's Second Law

We can now summarize the results of our experiments. We've seen that **a force causes an object to accelerate. The acceleration a is directly proportional to the force F and inversely proportional to the mass m.** We can express both these relationships in equation form as

$$a = \frac{F}{m} \tag{4.2}$$

Note that if we double the size of the force F, the acceleration a will double, as we found experimentally. And if we triple the mass m, the acceleration will be only one-third as great, again agreeing with experiment.

Equation 4.2 tells us the magnitude of an object's acceleration in terms of its mass and the force applied. But our experiments also had another important finding: The *direction* of the acceleration was the same as the direction of the force. We can express this fact by writing Equation 4.2 in *vector* form as

$$\vec{a} = \frac{\vec{F}}{m} \tag{4.3}$$

TRY IT YOURSELF

Feel the difference Because of its high sugar content, a can of regular soda has a mass about 4% greater than that of a can of diet soda. If you try to judge which can is more massive by simply holding one in each hand, this small difference is almost impossible to detect. If you *move* the cans up and down, however, the difference becomes subtly but noticeably apparent: People evidently are more sensitive to how the mass of each can resists acceleration than they are to the cans' weights alone.

Finally, our experiment was limited to looking at an object's response to a *single* applied force acting in a single direction. Realistically, an object is likely to be subjected to several distinct forces $\vec{F}_1, \vec{F}_2, \vec{F}_3, \ldots$ that may point in different directions. What happens then? Experiments show that the acceleration of the object is determined by the *net force* acting on it. Recall from Figure 4.4 and Equation 4.1 that the net force is the *vector sum* of all forces acting on the object. So if several forces are acting, we use the *net* force in Equation 4.4.

Newton was the first to recognize these connections between force and motion. This relationship is known today as Newton's second law.

Newton's second law An object of mass m subjected to forces $\vec{F}_1, \vec{F}_2, \vec{F}_3, \ldots$ will undergo an acceleration $\vec{a}$ given by

$$\vec{a} = \frac{\vec{F}_{\text{net}}}{m} \tag{4.4}$$

INVERSE p.114

where the net force $\vec{F}_{\text{net}} = \vec{F}_1 + \vec{F}_2 + \vec{F}_3 + \cdots$ is the vector sum of all forces acting on the object. **The acceleration vector $\vec{a}$ points in the same direction as the net force vector $\vec{F}_{\text{net}}$.**

An unfair advantage? Race car driver Danica Patrick was the subject of controversial comments by other drivers who thought her small mass of 45 kg gave her an advantage over heavier drivers; the next-lightest driver's mass was 61 kg. Because every driver's car must have the same mass, Patrick's overall racing mass was lower than any other driver's. Because a car's acceleration is inversely proportional to its mass, her car could be expected to have a slightly greater acceleration.

We'll use Newton's second law in Chapter 5 to solve many kinds of motion problems; for the moment, however, the critical idea is that an object accelerates in the direction of the net force acting on it.

The significance of Newton's second law cannot be overstated. There was no reason to suspect that there should be any simple relationship between force and acceleration. Yet a simple but exceedingly powerful equation relates the two. Newton's work, preceded to some extent by Galileo's, marks the beginning of a highly successful period in the history of science during which it was learned that the behavior of physical objects can often be described and predicted by mathematical relationships. While some relationships are found to apply only in special circumstances, others seem to have universal applicability. Those equations that appear to apply at all times and under all conditions have come to be called "laws of nature." Newton's second law is a law of nature; you will meet others as we go through this book.

We can rewrite Newton's second law in the form

$$\vec{F}_{\text{net}} = m\vec{a} \tag{4.5}$$

which is how you'll see it presented in many textbooks and how, in practice, we'll often use the second law. Equations 4.4 and 4.5 are mathematically equivalent, but Equation 4.4 better describes the central idea of Newtonian mechanics: A force applied to an object causes the object to accelerate.

NOTE ► When several forces act on an object, be careful not to think that the strongest force "overcomes" the others to determine the motion on its own. It is $\vec{F}_{\text{net}}$, the sum of *all* the forces, that determines the acceleration $\vec{a}$. ◄

CONCEPTUAL EXAMPLE 4.5 Acceleration of a wind-blown basketball

A basketball is released from rest in a stiff breeze directed to the right. In what direction does the ball accelerate?

REASON As shown in **FIGURE 4.22a**, two forces are acting on the ball: its weight $\vec{w}$ directed downward and a wind force $\vec{F}_{\text{wind}}$ pushing the ball to the right. Newton's second law tells us that the direction of the acceleration is the same as the direction of the net force $\vec{F}_{\text{net}}$. In **FIGURE 4.22b** we find $\vec{F}_{\text{net}}$ by graphical vector addition of $\vec{w}$ and $\vec{F}_{\text{wind}}$. We see that $\vec{F}_{\text{net}}$ and therefore $\vec{a}$ point down and to the right.

FIGURE 4.22 A basketball falling in a strong breeze.

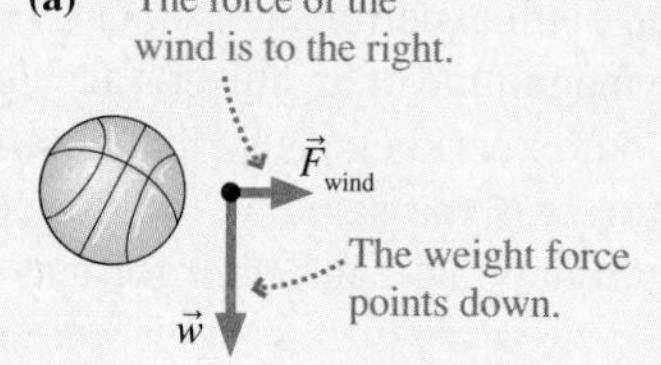

The force of the wind is to the right.

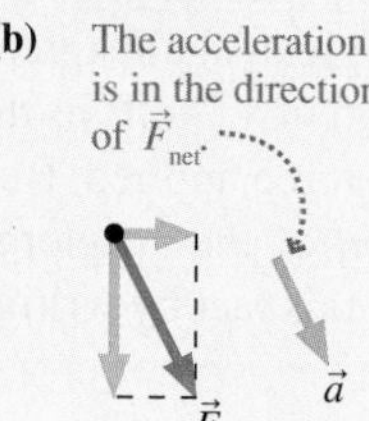

(b)
The acceleration is in the direction of $\vec{F}_{\text{net}}$.
$\vec{F}_{\text{net}}$
$\vec{a}$

Units of Force

Because $\vec{F}_{net} = m\vec{a}$, the unit of force must be mass units multiplied by acceleration units. We've previously specified the SI unit of mass as the kilogram. We can now define the basic unit of force as "the force that causes a 1 kg mass to accelerate at 1 m/s^2." From Newton's second law, this force is

$$1 \text{ basic unit of force} = (1 \text{ kg}) \times (1 \text{ m/s}^2) = 1 \frac{\text{kg} \cdot \text{m}}{\text{s}^2}$$

This basic unit of force is called a *newton:* One **newton** is the force that causes a 1 kg mass to accelerate at 1 m/s^2. The abbreviation for newton is N. Mathematically, $1 \text{ N} = 1 \text{ kg} \cdot \text{m/s}^2$.

The newton is a *secondary unit,* meaning that it is defined in terms of the *primary units* of kilograms, meters, and seconds. We will introduce other secondary units as needed.

It is important to develop a feeling for what the size of forces should be. Table 4.2 lists some typical forces. As you can see, "typical" forces on "typical" objects are likely to be in the range 0.01–10,000 N. Forces less than 0.01 N are too small to consider unless you are dealing with very small objects. Forces greater than 10,000 N would make sense only if applied to very massive objects.

The unit of force in the English system is the *pound* (abbreviated lb). Although the definition of the pound has varied throughout history, it is now defined in terms of the newton:

$$1 \text{ pound} = 1 \text{ lb} = 4.45 \text{ N}$$

You very likely associate pounds with kilograms rather than with newtons. Everyday language often confuses the ideas of mass and weight, but we're going to need to make a clear distinction between them. We'll have more to say about this in the next chapter.

TABLE 4.2 Approximate magnitude of some typical forces

Force	Approximate magnitude (newtons)
Weight of a U.S. nickel	0.05
Weight of a 1-pound object	5
Weight of a 110-pound person	500
Propulsion force of a car	5000
Thrust force of a rocket motor	5,000,000

EXAMPLE 4.6 **Pulling an airplane**

In 2000, a team of 60 British police officers set a world record by pulling a Boeing 747, with a mass of 205,000 kg, a distance of 100 m in 53.3 s. Estimate the force with which each officer pulled on the plane.

PREPARE If we assume that the plane undergoes a constant acceleration, we can use kinematics to find the magnitude of that acceleration. Then we can use Newton's second law to find the force applied to the airplane. **FIGURE 4.23** shows the visual overview of the airplane.

FIGURE 4.23 Visual overview of the airplane accelerating.

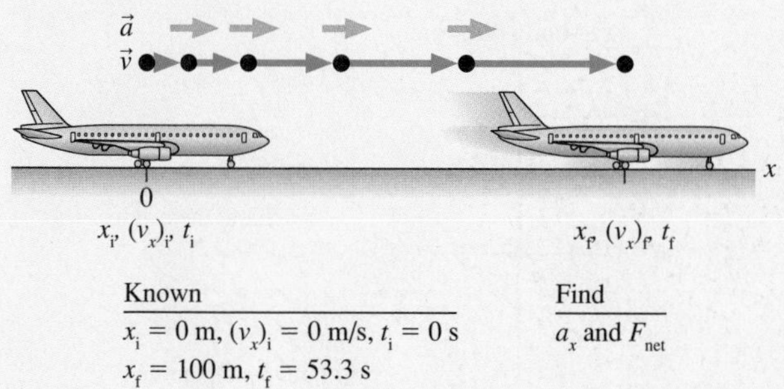

SOLVE Because we know the net displacement of the plane and the time it took to move, we can use the kinematic equation

$$x_f = x_i + (v_x)_i \, \Delta t + \tfrac{1}{2} a_x (\Delta t)^2$$

to find the airplane's acceleration a_x. Using the known values $x_i = 0$ m and $(v_x)_i = 0$ m/s, we can solve for the acceleration:

$$a_x = \frac{2x_f}{(\Delta t)^2} = \frac{2(100 \text{ m})}{(53.3 \text{ s})^2} = 0.0704 \text{ m/s}^2$$

Now we apply Newton's second law. The net force is

$$F_{net} = ma_x = (205{,}000 \text{ kg})(0.0704 \text{ m/s}^2) = 1.44 \times 10^4 \text{ N}$$

This is the force applied by all 60 men. Each man thus applies about 1/60th of this force, or around 240 N.

ASSESS Converting this force to pounds, we have

$$F = 240 \text{ N} \times \frac{1 \text{ lb}}{4.45 \text{ N}} = 54 \text{ lb}$$

Our answer is suspiciously low. Burly policemen can certainly apply a force greater than 54 lb. The fact that our calculation ended with a force that appears too small suggests we've overlooked something. In fact, we have. We've neglected the rolling friction of the plane's tires. We'll learn how to deal with friction in the next chapter, where we'll find that, because of the opposing friction force, the men have to pull harder than our estimate, in which we've ignored friction.

STOP TO THINK 4.4 Three forces act on an object. In which direction does the object accelerate?

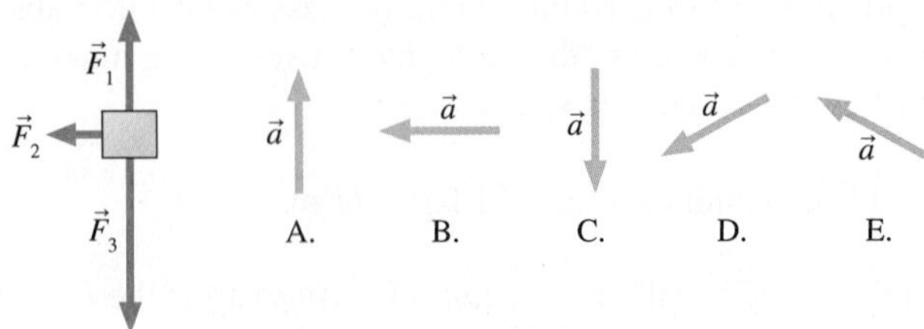

4.7 Free-Body Diagrams

Having discussed at length what is and is not a force, and what forces do to an object, we are ready to assemble our knowledge about force and motion into a single diagram called a **free-body diagram.** A free-body diagram represents the object as a particle and shows *all* of the forces acting on the object. Learning how to draw a correct free-body diagram is a very important skill, one that in the next chapter will become a critical part of our strategy for solving motion problems. For now, let's concentrate on the basic skill of constructing a correct free-body diagram.

TACTICS BOX 4.3 Drawing a free-body diagram MP™

1. **Identify all forces acting on the object.** This step was described in Tactics Box 4.2.
2. **Draw a coordinate system.** Use the axes defined in your pictorial representation (Tactics Box 2.2). If those axes are tilted, for motion along an incline, then the axes of the free-body diagram should be similarly tilted.
3. **Represent the object as a dot at the origin of the coordinate axes.** This is the particle model.
4. **Draw vectors representing each of the identified forces.** This was described in Tactics Box 4.1. Be sure to label each force vector.
5. **Draw and label the *net force* vector $\vec{F}_{net}$.** Draw this vector beside the diagram, not on the particle. Then check that $\vec{F}_{net}$ points in the same direction as the acceleration vector $\vec{a}$ on your motion diagram. Or, if appropriate, write $\vec{F}_{net} = \vec{0}$.

Exercises 17–22 

EXAMPLE 4.7 Forces on an upward-accelerating elevator

An elevator, suspended by a cable, speeds up as it moves upward from the ground floor. Draw a free-body diagram of the elevator.

PREPARE FIGURE 4.24 illustrates the steps listed in Tactics Box 4.3.

FIGURE 4.24 Free-body diagram of an elevator accelerating upward.

Force identification

Tension $\vec{T}$

Weight $\vec{w}$

1. Identify all forces acting on the object.

Free-body diagram

y
x
$\vec{T}$
$\vec{w}$
$\vec{F}_{net}$

2. Draw a coordinate system.
3. Represent the object as a dot at the origin.
4. Draw vectors for the identified forces.
5. Draw and label $\vec{F}_{net}$ beside the diagram.

ASSESS The coordinate axes, with a vertical y-axis, are the ones we use in a pictorial representation of the motion. The elevator is accelerating upward, so $\vec{F}_{net}$ must point upward. For this to be true, the magnitude of $\vec{T}$ must be greater than the magnitude of $\vec{w}$. The diagram has been drawn accordingly.

EXAMPLE 4.8 Forces on a rocket-propelled ice block

Bobby straps a small model rocket to a block of ice and shoots it across the smooth surface of a frozen lake. Friction is negligible. Draw a visual overview—a motion diagram, force identification diagram, and free-body diagram—of the block of ice.

PREPARE We treat the block of ice as a particle. The visual overview consists of a motion diagram to determine $\vec{a}$, a force identification picture, and a free-body diagram. The statement of the situation tells us that friction is negligible. We can draw these three pictures using Problem-Solving Strategy 1.1 for the motion diagram, Tactics Box 4.2 to identify the forces, and Tactics Box 4.3 to draw the free-body diagrams. These pictures are shown in FIGURE 4.25.

ASSESS The motion diagram tells us that the acceleration is in the positive x-direction. According to the rules of vector addition, this can be true only if the upward-pointing $\vec{n}$ and the downward-pointing $\vec{w}$ are equal in magnitude and thus cancel each other. The vectors have been drawn accordingly, and this leaves the net force vector pointing toward the right, in agreement with $\vec{a}$ from the motion diagram.

FIGURE 4.25 Visual overview for a block of ice shooting across a frictionless frozen lake.

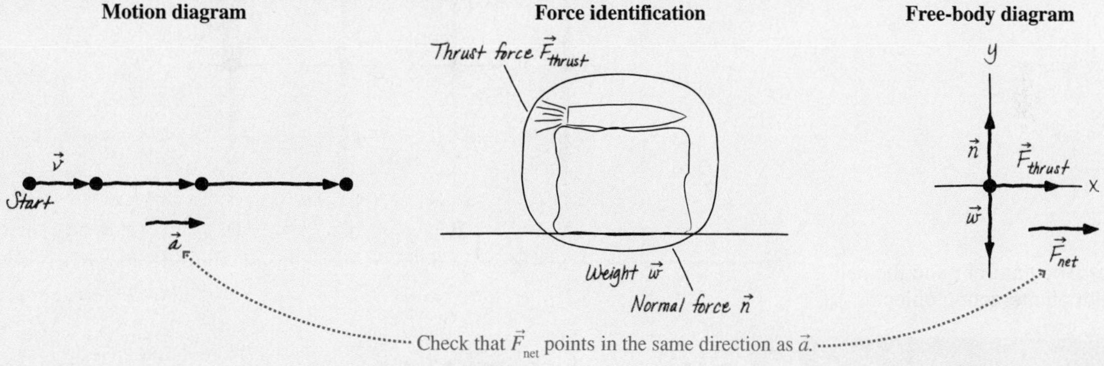

EXAMPLE 4.9 Forces on a towed skier

A tow rope pulls a skier up a snow-covered hill at a constant speed. Draw a full visual overview of the skier.

PREPARE This is Example 4.2 again with the additional information that the skier is moving at a constant speed. If we were doing a kinematics problem, the pictorial representation would use a tilted coordinate system with the x-axis parallel to the slope, so we use these same tilted coordinate axes for the free-body diagram. The motion diagram, force identification diagram, and free-body diagram are shown in FIGURE 4.26.

FIGURE 4.26 Visual overview for a skier being towed at a constant speed.

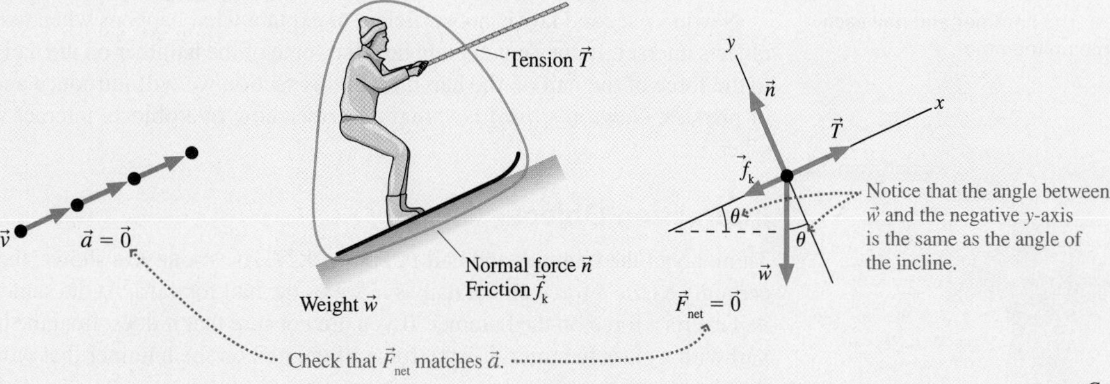

Continued

ASSESS We have shown $\vec{T}$ pulling parallel to the slope and $\vec{f}_k$, which opposes the direction of motion, pointing down the slope. The normal force $\vec{n}$ is perpendicular to the surface and thus along the y-axis. Finally, and this is important, the weight $\vec{w}$ is *vertically* downward, *not* along the negative y-axis.

The skier moves in a straight line with constant speed, so $\vec{a} = \vec{0}$. Newton's second law then tells us that $\vec{F}_{net} = m\vec{a} = \vec{0}$. Thus we have drawn the vectors such that the forces add to zero. We'll learn more about how to do this in Chapter 5.

Free-body diagrams will be our major tool for the next several chapters. Careful practice with the workbook exercises and homework in this chapter will pay immediate benefits in the next chapter. Indeed, it is not too much to assert that a problem is more than half solved when you correctly complete the free-body diagram.

STOP TO THINK 4.5 An elevator suspended by a cable is moving upward and slowing to a stop. Which free-body diagram is correct?

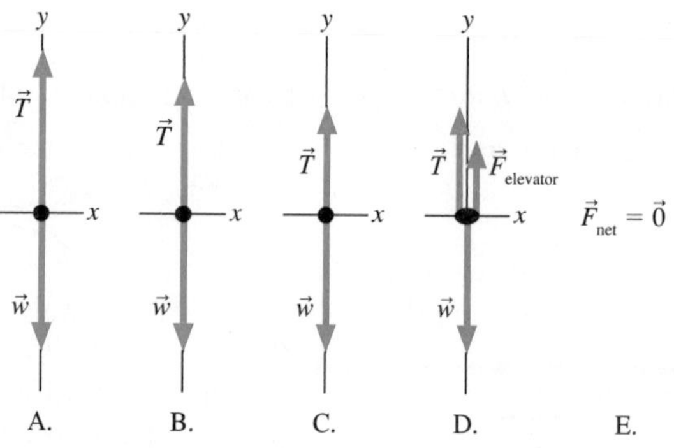

FIGURE 4.27 The hammer and the nail are a system of interacting objects.

FIGURE 4.28 The hammer and nail each exert a force on the other.

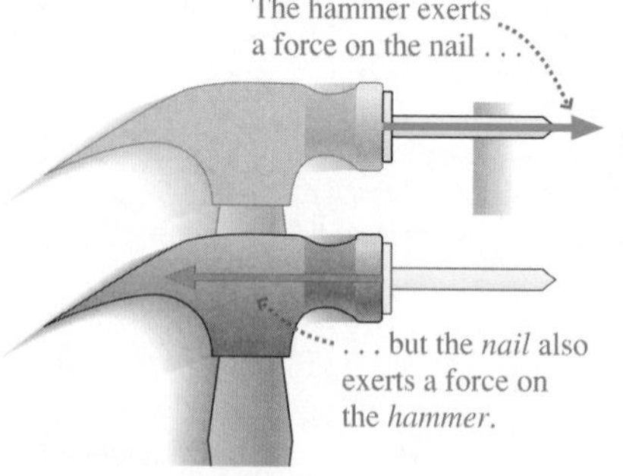

The hammer exerts a force on the nail . . .
. . . but the *nail* also exerts a force on the *hammer*.

4.8 Newton's Third Law

Thus far, we've focused on the motion of a single particle responding to well-defined forces exerted by other objects, or to long-range forces. A skier sliding downhill, for instance, is subject to frictional and normal forces from the slope, and the pull of gravity on his body. Once we have identified these forces, we can use Newton's second law to calculate the acceleration, and hence the overall motion, of the skier.

But motion in the real world often involves two or more objects *interacting* with each other. Consider the hammer and nail in FIGURE 4.27. As the hammer hits the nail, the nail pushes back on the hammer. A bat and a ball, your foot and a soccer ball, and the earth–moon system are other examples of interacting objects.

Newton's second law is not sufficient to explain what happens when two or more objects interact. It does not explain how the force of the hammer on the nail is related to the force of the nail on the hammer. In this section we will introduce another law of physics, Newton's *third* law, that describes how two objects interact with each other.

Interacting Objects

Think about the hammer and nail in Figure 4.27. As FIGURE 4.28 shows, the hammer certainly exerts a force on the nail as it drives the nail forward. At the same time, the nail exerts a force on the hammer. If you are not sure that it does, imagine hitting the nail with a glass hammer. It's the force of the nail on the hammer that would cause the glass to shatter.

Indeed, if you stop to think about it, any time that object A pushes or pulls on object B, object B pushes or pulls back on object A. As you push on a filing cabinet to move it, the cabinet pushes back on you. (If you pushed forward without the cabinet pushing back, you would fall forward in the same way you do if someone suddenly opens a door you're leaning against.) Your chair pushes upward on you (the normal force that keeps you from falling) while, at the same time, you push down on the chair, compressing the cushion. These are examples of what we call an *interaction*. An **interaction** is the mutual influence of two objects on each other.

These examples illustrate a key aspect of interactions: The forces involved in an interaction between two objects always occur as a *pair*. To be more specific, if object A exerts a force $\vec{F}_{\text{A on B}}$ on object B, then object B exerts a force $\vec{F}_{\text{B on A}}$ on object A. This pair of forces, shown in FIGURE 4.29, is called an **action/reaction pair.** Two objects interact by exerting an action/reaction pair of forces on each other. Notice the very explicit subscripts on the force vectors. The first letter is the *agent*—the source of the force—and the second letter is the *object* on which the force acts. $\vec{F}_{\text{A on B}}$ is thus the force exerted *by* A *on* B.

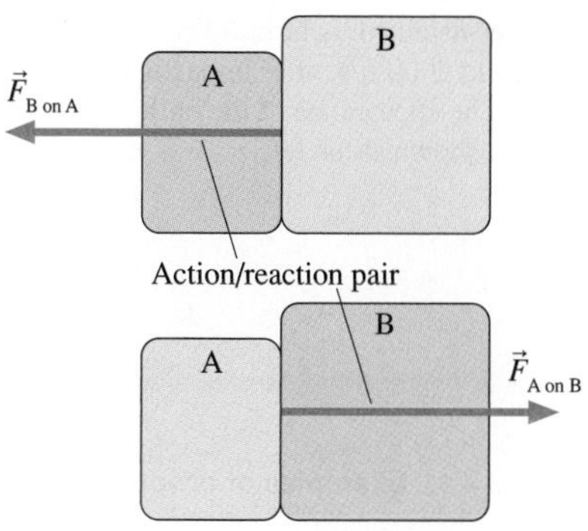
B
A
$\vec{F}_{\text{B on A}}$
Action/reaction pair
B
A
$\vec{F}_{\text{A on B}}$

FIGURE 4.29 An action/reaction pair of forces.

NOTE ► The name "action/reaction pair" is somewhat misleading. The forces occur simultaneously, and we cannot say which is the "action" and which the "reaction." Neither is there any implication about cause and effect; the action does not cause the reaction. **An action/reaction pair of forces exists as a pair, or not at all.** For action/reaction pairs, the labels are the key: Force $\vec{F}_{\text{A on B}}$ is paired with force $\vec{F}_{\text{B on A}}$. ◄

Reasoning with Newton's Third Law

We've discovered that two objects always interact via an action/reaction pair of forces. Newton was the first to recognize how the two members of an action/reaction pair of forces are related to each other. Today we know this as Newton's third law:

> **Newton's third law** Every force occurs as one member of an action/reaction pair of forces.
>
> - The two members of an action/reaction pair act on two *different* objects.
> - The two members of an action/reaction pair point in *opposite* directions and are *equal in magnitude.*

Newton's third law is often stated: "For every action there is an equal but opposite reaction." While this is a catchy phrase, it lacks the preciseness of our preferred version. In particular, it fails to capture an essential feature of the two members of an action/reaction pair—that each acts on a *different* object. This is shown in FIGURE 4.30, where a hammer hitting a nail exerts a force $\vec{F}_{\text{hammer on nail}}$ on the nail; by the third law, the nail must exert a force $\vec{F}_{\text{nail on hammer}}$ to complete the action/reaction pair.

Figure 4.30 also illustrates that these two forces point in *opposite directions.* This feature of the third law is also in accord with our experience. If the hammer hits the nail with a force directed to the right, the force of the nail on the hammer is directed to the left; if the force of my chair on me pushes up, the force of me on the chair pushes down.

Finally, Figure 4.30 shows that, according to Newton's third law, the two members of an action/reaction pair have *equal* magnitudes, so that $F_{\text{hammer on nail}} = F_{\text{nail on hammer}}$. This is something new, and it is by no means obvious. Indeed, this statement causes students the most trouble when applying the third law because it seems so counter to our intuition, as the following example shows.

FIGURE 4.30 Newton's third law.

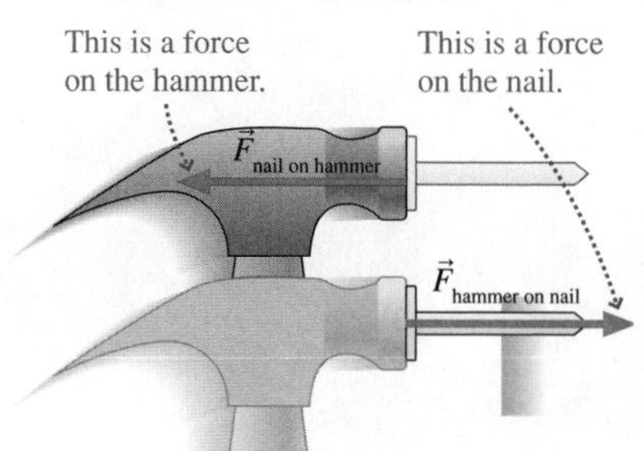
Each force in an action/reaction pair acts on a *different* object.
This is a force on the hammer.
This is a force on the nail.
$\vec{F}_{\text{nail on hammer}}$
$\vec{F}_{\text{hammer on nail}}$
The members of the pair point in *opposite directions*, but are of *equal magnitude*.

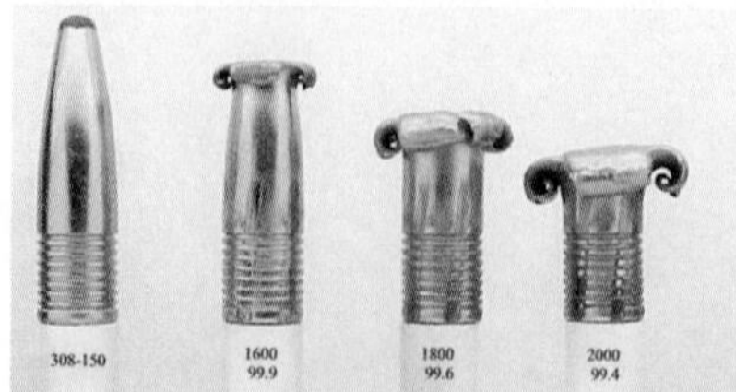
308-150
1600
99.9
1800
99.6
2000
99.4

Revenge of the target We normally think of the damage that the force of a bullet inflicts on its target. But according to Newton's third law, the target exerts an equal force on the bullet. The photo shows the damage sustained by bullets fired at 1600, 1800, and 2000 ft/s, after impacting a test target. The appearance of the bullet before firing is shown at the left.

CONCEPTUAL EXAMPLE 4.10 The bug versus the windshield

During the collision between a bug and the windshield of a fast-moving truck, which force has greater magnitude: the force of the windshield on the bug or the force of the bug on the windshield?

REASON The third law tells us that the magnitude of the force of the windshield on the bug must be *equal* to that of the bug on the windshield! How can this be, when the bug is so small compared to the truck? The source of puzzlement in problems like this is that Newton's third law equates the size of the *forces* acting on the two objects, not their *accelerations.* The acceleration of each object depends not only on the force applied to it, but also, according to Newton's second law, on its mass. The bug and the truck do in fact feel forces of equal strength from the other, but the bug, with its very small mass, undergoes an extreme acceleration from this force while the acceleration of the heavy truck is negligible.

ASSESS It is important to separate the *effects* of the forces (the accelerations) from the causes (the forces themselves). Because two interacting objects can have very different masses, their accelerations can be very different even though the interaction forces are of the same strength.

We'll return to Newton's third law in Chapter 5, where we'll use it to solve problems involving two or more interacting objects.

FIGURE 4.31 Examples of propulsion.

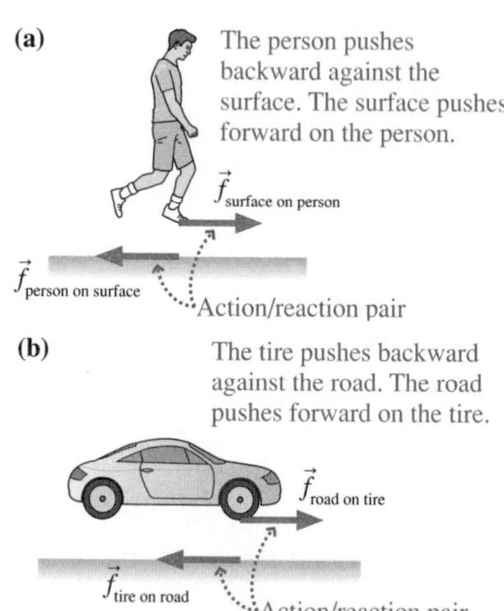
(a)
The person pushes backward against the surface. The surface pushes forward on the person.
$\vec{f}_{\text{surface on person}}$
$\vec{f}_{\text{person on surface}}$
Action/reaction pair
(b)
The tire pushes backward against the road. The road pushes forward on the tire.
$\vec{f}_{\text{road on tire}}$
$\vec{f}_{\text{tire on road}}$
Action/reaction pair

FIGURE 4.32 When the driver hits the gas, the force of the track on the tire is so great that the tire deforms.

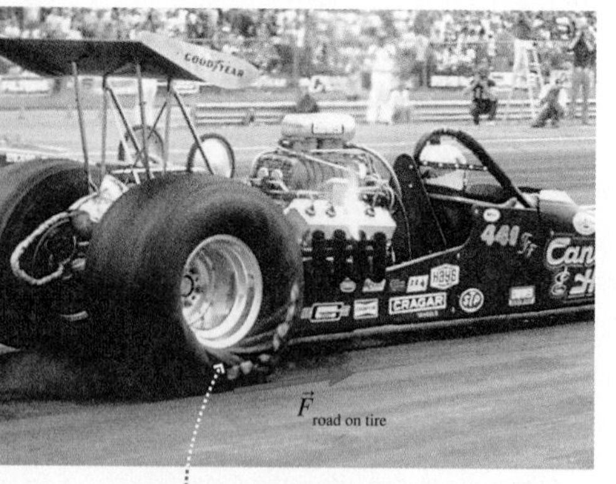
$\vec{F}_{\text{road on tire}}$

You can *see* that the force of the road on the tire points forward by the way it twists the rubber of the tire.

Propulsion

A sprinter accelerates out of the blocks. Because he's accelerating, there must be a force on him in the forward direction. For a system with an internal source of energy, a force that drives the system is a force of **propulsion.** Propulsion is an important feature not only of walking or running but also of the forward motion of cars, jets, and rockets. Propulsion is somewhat counterintuitive, so it is worth a closer look.

If you tried to walk across a frictionless floor, your foot would slip and slide *backward.* In order for you to walk, the floor needs to have friction so that your foot *sticks* to the floor as you straighten your leg, moving your body forward. The friction that prevents slipping is *static* friction. Static friction, you will recall, acts in the direction that prevents slipping, so the static friction force $\vec{f}_{\text{SonP}}$ (for **S**urface on **P**erson) has to point in the *forward* direction to prevent your foot from slipping backward. As shown in FIGURE 4.31a, it is this forward-directed static friction force that propels you forward! The force of your foot on the floor, $\vec{f}_{\text{PonS}}$, is the other half of the action/reaction pair, and it points in the opposite direction as you push backward against the floor.

Similarly, the car in FIGURE 4.31b uses static friction to propel itself. The car uses its motor to turn the tires, causing the tires to push backward against the road ($\vec{f}_{\text{tire on road}}$). The road surface responds by pushing the car forward ($\vec{f}_{\text{road on tire}}$). This force of the road on the tire can be seen in photos of drag racers, where the forces are very great (FIGURE 4.32). Again, the forces involved are *static* friction forces. The tire is rolling, but the bottom of the tire, where it contacts the road, is instantaneously at rest. If it weren't, you would leave one giant skid mark as you drove and would burn off the tread within a few miles.

Rocket motors are somewhat different because they are not pushing *against* anything external. That's why rocket propulsion works in the vacuum of space. Instead, the rocket engine pushes hot, expanding gases out of the back of the rocket, as shown in FIGURE 4.33. In response, the exhaust gases push the rocket forward with the force we've called *thrust.*

Now we've assembled all the pieces we need in order to start solving problems in dynamics. We have seen what forces are and how to identify them, and we've learned how forces cause objects to accelerate according to Newton's second law. We've also found how Newton's third law governs the interaction forces between two objects. Our goal in the next several chapters is to apply Newton's laws to a variety of problems involving straight-line and circular motion.

FIGURE 4.33 Rocket propulsion.

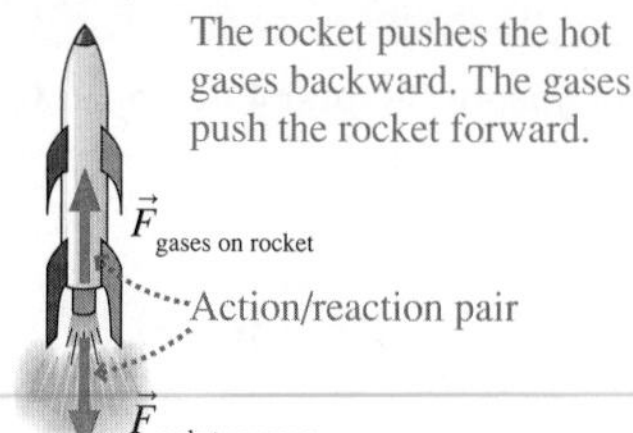

STOP TO THINK 4.6 A small car is pushing a larger truck that has a dead battery. The mass of the truck is greater than the mass of the car. Which of the following statements is true?

A. The car exerts a force on the truck, but the truck doesn't exert a force on the car.
B. The car exerts a larger force on the truck than the truck exerts on the car.
C. The car exerts the same amount of force on the truck as the truck exerts on the car.
D. The truck exerts a larger force on the car than the car exerts on the truck.
E. The truck exerts a force on the car, but the car doesn't exert a force on the truck.

INTEGRATED EXAMPLE 4.11 Pulling an excursion train

An engine slows as it pulls two cars of an excursion train up a mountain. Draw a visual overview (motion diagram, a force identification diagram, and free-body diagram) for the car just behind the engine. Ignore friction.

PREPARE Because the train is slowing down, the motion diagram consists of a series of particle positions that become closer together at successive times; the corresponding velocity vectors become shorter and shorter. To identify the forces acting on the car we use the steps of Tactics Box 4.2. Finally, we can draw a free-body diagram using Tactics Box 4.3.

SOLVE Finding the forces acting on car 1 can be tricky. The engine exerts a forward force $\vec{F}_{\text{engine on 1}}$ on car 1 where the engine touches the front of car 1. At its back, car 1 touches car 2, so car 2 must also exert a force on car 1. The direction of this force can be understood from Newton's third law. Car 1 exerts an uphill force on car 2 in order to pull it up the mountain. Thus, by Newton's third law, car 2 must exert an oppositely directed *downhill* force on car 1. This is the force we label $\vec{F}_{2\text{ on }1}$. The three diagrams that make up the full visual overview are shown in FIGURE 4.34.

ASSESS Correctly preparing the three diagrams illustrated in this example is critical for solving problems using Newton's laws. The motion diagram allows you to determine the direction of the acceleration and hence of $\vec{F}_{\text{net}}$. Using the force identification diagram, you will correctly identify all the forces acting on the object and, just as important, not add any extraneous forces. And by properly drawing these force vectors in a free-body diagram, you'll be ready for the quantitative application of Newton's laws that is the focus of Chapter 5.

FIGURE 4.34 Visual overview for a slowing train car being pulled up a mountain.

Motion diagram

Because the train is slowing down, its acceleration vector points in the direction opposite its motion.

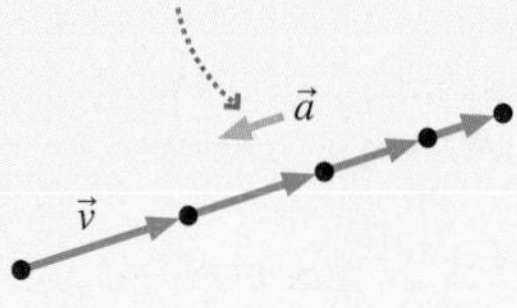

Force identification
(Numbered steps from Tactics Box 4.2)

1. The object of interest is car 1.
2. Draw a picture.
3. Draw a closed curve around the object.
4. Locate the points where the object touches other objects.
5. Name and label each contact force.
6. Weight is the only long-range force.

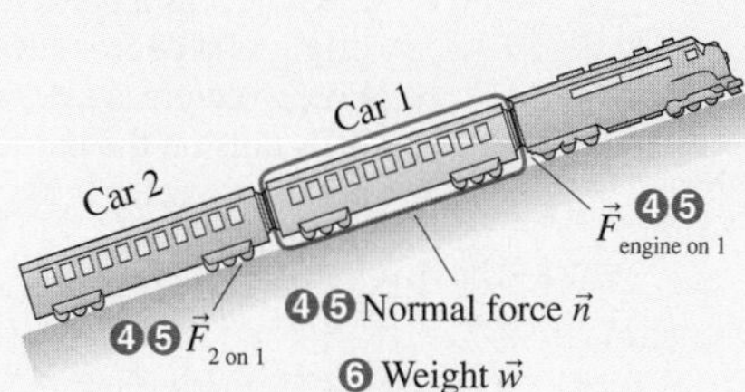

Free-body diagram
(Numbered steps from Tactics Box 4.3)

1. Identify all forces (already done).
2. Draw a coordinate system. Because the motion here is along an incline, we tilt our x-axis to match.
3. Represent the object as a dot at the origin.
4. Draw vectors representing each identified force.
5. Draw the net force vector. Check that it points in the same direction as $\vec{a}$.

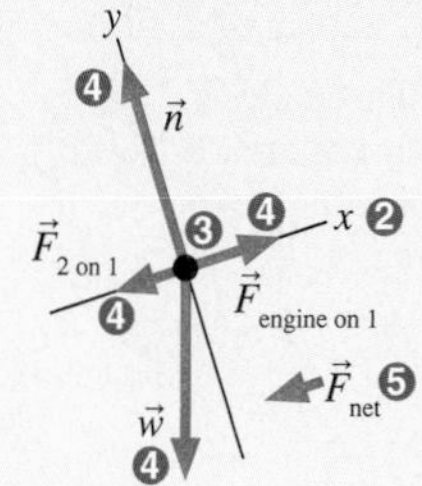

SUMMARY

The goal of Chapter 4 has been to establish a connection between force and motion.

GENERAL PRINCIPLES

Newton's First Law

Consider an object with no force acting on it. If it is at rest, it will remain at rest. If it is in motion, then it will continue to move in a straight line at a constant speed.

$\vec{F} = \vec{0}$

$\vec{v}$ $\vec{v}$ $\vec{v}$ $\vec{v}$ $\vec{v}$

$\vec{a} = \vec{0}$

The first law tells us that no "cause" is needed for motion. Uniform motion is the "natural state" of an object.

Newton's Second Law

An object with mass m will undergo acceleration

$$\vec{a} = \frac{\vec{F}_{\text{net}}}{m}$$

where the net force $\vec{F}_{\text{net}} = \vec{F}_1 + \vec{F}_2 + \vec{F}_3 + \cdots$ is the vector sum of all the individual forces acting on the object.

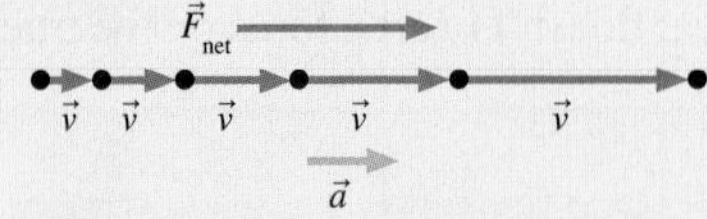

The second law tells us that a net force causes an object to accelerate. This is the connection between force and motion. The acceleration points in the direction of $\vec{F}_{\text{net}}$.

Newton's Third Law

Every force occurs as one member of an **action/reaction** pair of forces. The two members of an action/reaction pair:

- act on two *different* objects.
- point in opposite directions and are equal in magnitude:

$$\vec{F}_{\text{A on B}} = -\vec{F}_{\text{B on A}}$$

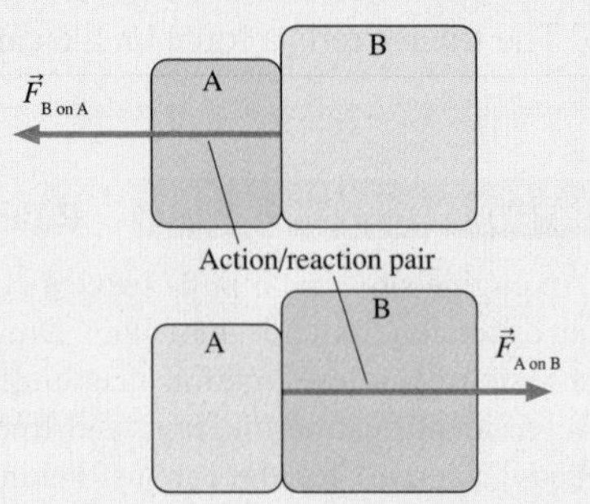

IMPORTANT CONCEPTS

Force is a push or pull on an object.

- Force is a vector, with a magnitude and a direction.
- A force requires an agent.
- A force is either a contact force or a long-range force.

The SI unit of force is the **newton** (N). A 1 N force will cause a 1 kg mass to accelerate at 1 m/s^2.

Net force is the vector sum of all the forces acting on an object.

$\vec{F}_1$, $\vec{F}_2$, $\vec{F}_3$ = $\vec{F}_{\text{net}} = \vec{F}_1 + \vec{F}_2 + \vec{F}_3$

Mass is the property of an object that determines its resistance to acceleration.

If the same force is applied to objects A and B, then the ratio of their accelerations is related to the ratio of their masses as

$$\frac{a_{\text{A}}}{a_{\text{B}}} = \frac{m_{\text{B}}}{m_{\text{A}}}$$

The mass of objects can be determined in terms of their accelerations.

APPLICATIONS

Identifying Forces

Forces are identified by locating the points where other objects touch the object of interest. These are points where contact forces are exerted. In addition, objects feel a long-range weight force.

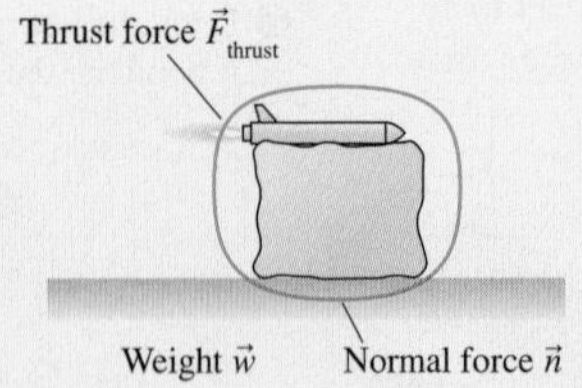

Free-Body Diagrams

A free-body diagram represents the object as a particle at the origin of a coordinate system. Force vectors are drawn with their tails on the particle. The net force vector is drawn beside the diagram.

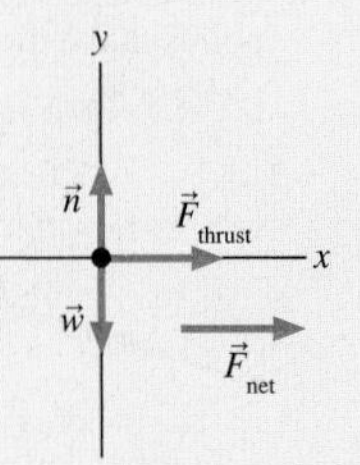

For homework assigned on MasteringPhysics, go to www.masteringphysics.com

Problem difficulty is labeled as | (straightforward) to ||||| (challenging).

Problems labeled INT integrate significant material from earlier chapters; BIO are of biological or medical interest.

QUESTIONS

Conceptual Questions

1. A hockey puck slides along the surface of the ice. If friction and air resistance are negligible, what force is required to keep the puck moving?
2. If an object is not moving, does that mean that there are no forces acting on it? Explain.
3. An object moves in a straight line at a constant speed. Is it true that there must be no forces of any kind acting on this object? Explain.
4. A ball sits near the front of a child's wagon. As she pulls on the wagon and it begins to move forward, the ball rolls toward the back of the wagon. Explain why the ball rolls in this direction.
5. If you know all of the forces acting on a moving object, can you tell in which direction the object is moving? If the answer is Yes, explain how. If the answer is No, give an example.
6. Three arrows are shot horizontally. They have left the bow and are traveling parallel to the ground as shown in Figure Q4.6. Air resistance is negligible. Rank in order, from largest to smallest, the magnitudes of the *horizontal* forces F_1, F_2, and F_3 acting on the arrows. Some may be equal. State your reasoning.

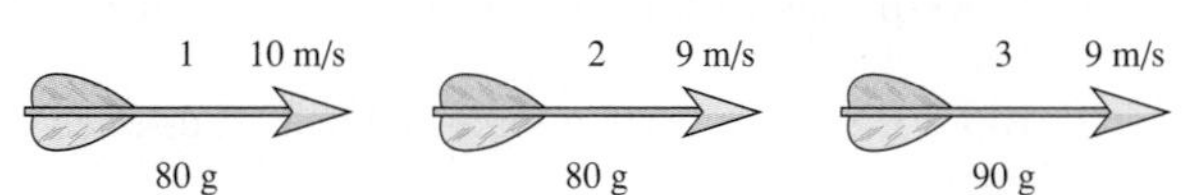

FIGURE Q4.6

7. A carpenter wishes to tighten the heavy head of his hammer onto its light handle. Which method shown in Figure Q4.7 will better tighten the head? Explain.

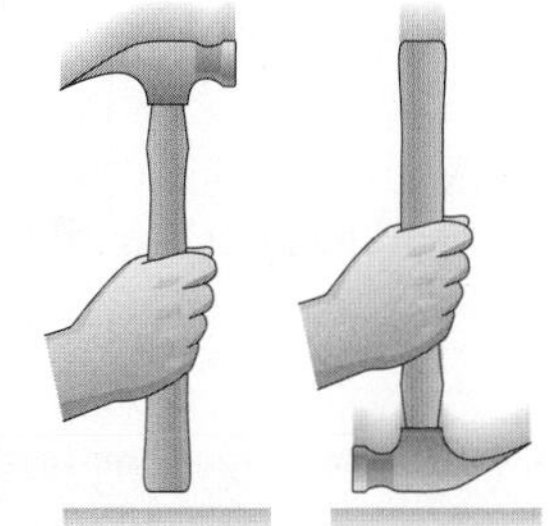
FIGURE Q4.7

8. BIO Internal injuries in vehicular accidents may be due to what is called the "third collision." The first collision is the vehicle hitting the external object. The second collision is the person hitting something on the inside of the car, such as the dashboard or windshield. This may cause external lacerations. The third collision, possibly the most damaging to the body, is when organs, such as the heart or brain, hit the ribcage, skull, or other confines of the body, bruising the tissues on the leading edge and tearing the organ from its supporting structures on the trailing edge.
   a. Why is there a third collision? In other words, why are the organs still moving after the second collision?
   b. If the vehicle was traveling at 60 mph before the first collision, would the organs be traveling more than, equal to, or less than 60 mph just before the third collision?
9. a. Give an example of the motion of an object in which the frictional force on the object is directed opposite to the motion.
   b. Give an example of the motion of an object in which the frictional force on the object is in the same direction as the motion.
10. Suppose you are an astronaut in deep space, far from any source of gravity. You have two objects that look identical, but one has a large mass and the other a small mass. How can you tell the difference between the two?
11. Jonathan accelerates away from a stop sign. His eight-year-old daughter sits in the passenger seat. On whom does the back of the seat exert a greater force?
12. The weight of a box sitting on the floor points directly down. The normal force of the floor on the box points directly up. Need these two forces have the same magnitude? Explain.
13. A ball weighs 2.0 N when placed on a scale. It is then thrown straight up. What is its weight at the very top of its motion? Explain.
14. Josh and Taylor, standing face-to-face on frictionless ice, push off each other, causing each to slide backward. Josh is much bigger than Taylor. After the push, which of the two is moving faster?
15. A person sits on a sloped hillside. Is it ever possible to have the static friction force on this person point down the hill? Explain.
16. BIO Walking without slipping requires a static friction force between your feet (or footwear) and the floor. As described in this chapter, the force on your foot as you push off the floor is forward while the force exerted by your foot on the floor is backward. But what about your *other* foot, the one moved during a stride? What is the direction of the force on that foot as it comes into contact with the floor? Explain.
17. Figure 4.31b showed a situation in which the force of the road on the car's tire points forward. In other situations, the force points backward. Give an example of such a situation.
18. Alyssa pushes to the right on a filing cabinet; the friction force from the floor pushes on it to the left. Because the cabinet doesn't move, these forces have the same magnitude. Do they form an action/reaction pair? Explain.
19. A very smart three-year-old child is given a wagon for her birthday. She refuses to use it. "After all," she says, "Newton's third law says that no matter how hard I pull, the wagon will exert an equal but opposite force on me. So I will never be able to get it to move forward." What would you say to her in reply?
20. Will hanging a magnet in front of an iron cart, as shown in Figure Q4.20, make it go? Explain why or why not.

FIGURE Q4.20

Multiple-Choice Questions

21. | Figure Q4.21 shows the view looking down onto a frictionless sheet of ice. A puck, tied with a string to point P, slides on the surface of the ice in the circular path shown. If the string suddenly snaps when the puck is in the position shown, which path best represents the puck's subsequent motion?

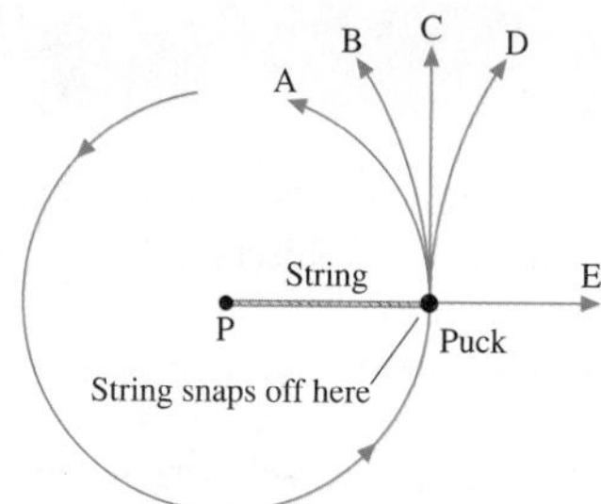

FIGURE Q4.21

22. | A block has acceleration a when pulled by a string. If two identical blocks are glued together and pulled with twice the original force, their acceleration will be
A. $(1/4)a$ B. $(1/2)a$
C. a D. $2a$
E. $4a$
23. | A 5.0 kg block has an acceleration of 0.20 m/s² when a force is exerted on it. A second block has an acceleration of 0.10 m/s² when subject to the same force. What is the mass of the second block?
A. 10 kg B. 5.0 kg C. 2.5 kg D. 7.5 kg
24. | Tennis balls experience a large drag force. A tennis ball is hit so that it goes straight up and then comes back down. The direction of the drag force is
A. Always up.
B. Up and then down.
C. Always down.
D. Down and then up.
25. | A person gives a box a shove so that it slides up a ramp, then reverses its motion and slides down. The direction of the force of friction is
A. Always down the ramp.
B. Up the ramp and then down the ramp.
C. Always down the ramp.
D. Down the ramp and then up the ramp.
26. | A person is pushing horizontally on a box with a constant force, causing it to slide across the floor with a constant speed. If the person suddenly stops pushing on the box, the box will
A. Immediately come to a stop.
B. Continue moving at a constant speed for a while, then gradually slow down to a stop.
C. Immediately change to a slower but constant speed.
D. Immediately begin slowing down and eventually stop.
27. | Rachel is pushing a box across the floor while Jon, at the same time, is hoping to stop the box by pushing in the opposite direction. There is friction between the box and floor. If the box is moving at constant speed, then the magnitude of Rachel's pushing force is
A. Greater than the magnitude of Jon's force.
B. Equal to the magnitude of Jon's force.
C. Less than the magnitude of Jon's force.
D. The question can't be answered without knowing how large the friction force is.
28. || Dave pushes his four-year-old son Thomas across the snow on a sled. As Dave pushes, Thomas speeds up. Which statement is true?
A. The force of Dave on Thomas is larger than the force of Thomas on Dave.
B. The force of Thomas on Dave is larger than the force of Dave on Thomas.
C. Both forces have the same magnitude.
D. It depends on how hard Dave pushes on Thomas.
29. | Figure Q4.29 shows block A sitting on top of block B. A constant force $\vec{F}$ is exerted on block B, causing block B to accelerate to the right. Block A rides on block B without slipping. Which statement is true?
A. Block B exerts a friction force on block A, directed to the left.
B. Block B exerts a friction force on block A, directed to the right.
C. Block B does not exert a friction force on block A.

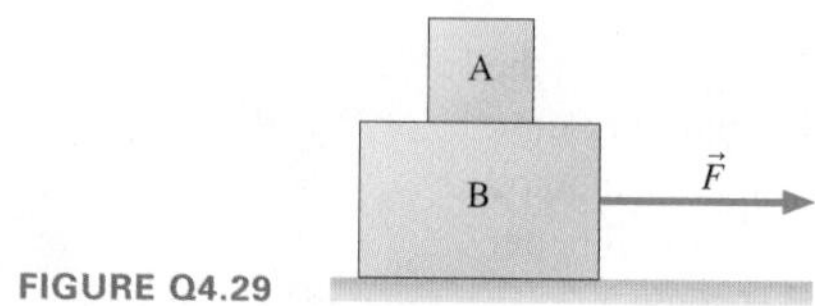

FIGURE Q4.29

PROBLEMS

Section 4.1 What Causes Motion?

1. | Whiplash injuries during an automobile accident are caused
BIO by the inertia of the head. If someone is wearing a seatbelt, her body will tend to move with the car seat. However, her head is free to move until the neck restrains it, causing damage to the neck. Brain damage can also occur.

Figure P4.1 shows two sequences of head and neck motion for a passenger in an auto accident. One corresponds to a head-on collision, the other to a rear-end collision. Which is which? Explain.

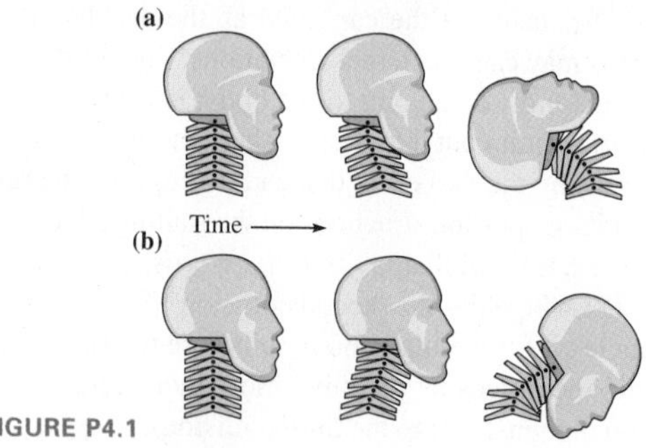

FIGURE P4.1

2. | An automobile has a head-on BIO collision. A passenger in the car experiences a compression injury to the brain. Is this injury most likely to be in the front or rear portion of the brain? Explain.

3. | In a head-on collision, an infant is much safer in a child safety seat when the seat is installed facing the rear of the car. Explain.

Section 4.2 Force

Problems 4 through 6 show two forces acting on an object at rest. Redraw the diagram, then add a third force that will allow the object to remain at rest. Label the new force $\vec{F}_3$.

4. || 5. || 6. ||

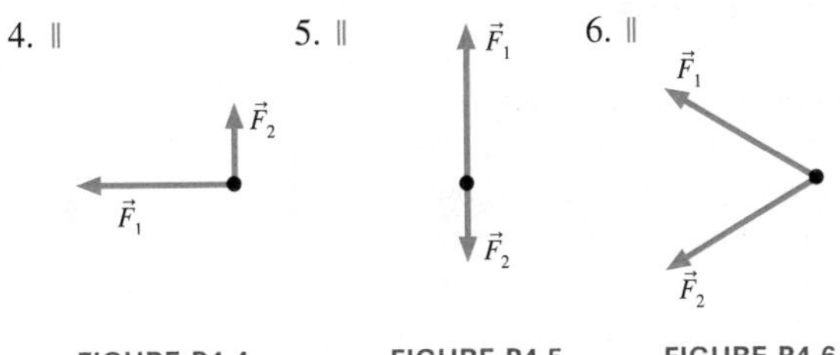

FIGURE P4.4 FIGURE P4.5 FIGURE P4.6

Section 4.3 A Short Catalog of Forces

Section 4.4 Identifying Forces

7. || A mountain climber is hanging from a rope in the middle of a crevasse. The rope is vertical. Identify the forces on the mountain climber.
8. ||| A circus clown hangs from one end of a large spring. The other end is anchored to the ceiling. Identify the forces on the clown.
9. |||| A baseball player is sliding into second base. Identify the forces on the baseball player.
10. ||||| A jet plane is speeding down the runway during takeoff. Air resistance is not negligible. Identify the forces on the jet.
11. | A skier is sliding down a 15° slope. Friction is not negligible. Identify the forces on the skier.
12. || A tennis ball is flying horizontally across the net. Air resistance is not negligible. Identify the forces on the ball.

Section 4.5 What Do Forces Do?

13. |||| Figure P4.13 shows an acceleration-versus-force graph for three objects pulled by rubber bands. The mass of object 2 is 0.20 kg. What are the masses of objects 1 and 3? Explain your reasoning.

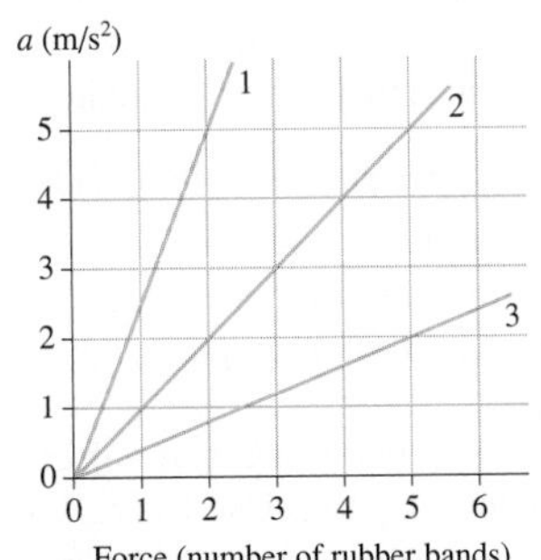

FIGURE P4.13

14. | A constant force applied to object A causes it to accelerate at 5 m/s^2. The same force applied to object B causes an acceleration of 3 m/s^2. Applied to object C, it causes an acceleration of 8 m/s^2.
    a. Which object has the largest mass?
    b. Which object has the smallest mass?
    c. What is the ratio of mass A to mass B (m_A/m_B)?
15. | Two rubber bands pulling on an object cause it to accelerate at 1.2 m/s^2.
    a. What will be the object's acceleration if it is pulled by four rubber bands?
    b. What will be the acceleration of two of these objects glued together if they are pulled by two rubber bands?
16. | A constant force is applied to an object, causing the object to accelerate at 10 m/s^2. What will the acceleration be if
    a. The force is halved?
    b. The object's mass is halved?
    c. The force and the object's mass are both halved?
    d. The force is halved and the object's mass is doubled?
17. | A constant force is applied to an object, causing the object to accelerate at 8.0 m/s^2. What will the acceleration be if
    a. The force is doubled?
    b. The object's mass is doubled?
    c. The force and the object's mass are both doubled?
    d. The force is doubled and the object's mass is halved?
18. ||| A man pulling an empty wagon causes it to accelerate at 1.4 m/s^2. What will the acceleration be if he pulls with the same force when the wagon contains a child whose mass is three times that of the wagon?
19. | A car has a maximum acceleration of 5.0 m/s^2. What will the maximum acceleration be if the car is towing another car of the same mass?

Section 4.6 Newton's Second Law

20. || Figure P4.20 shows an acceleration-versus-force graph for a 500 g object. Redraw this graph and add appropriate acceleration values on the vertical scale.

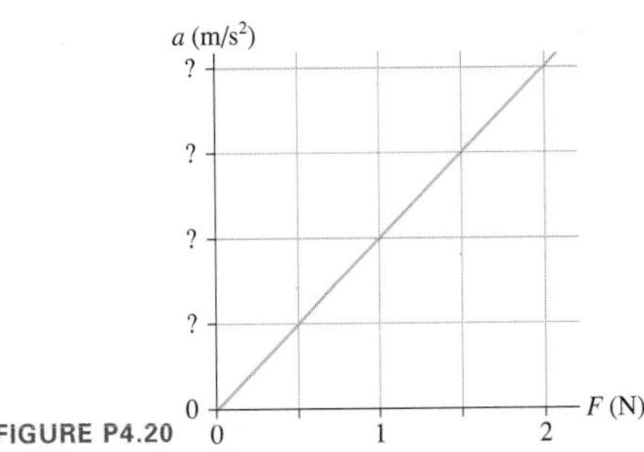

FIGURE P4.20

21. | Figure P4.21 shows an object's acceleration-versus-force graph. What is the object's mass?

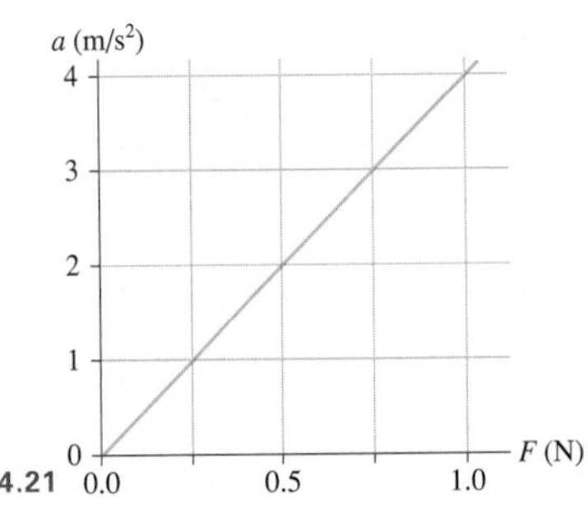

FIGURE P4.21

22. || Two children fight over a 200 g stuffed bear. The 25 kg boy pulls to the right with a 15 N force and the 20 kg girl pulls to the left with a 17 N force. Ignore all other forces on the bear (such as its weight).
   a. At this instant, can you say what the velocity of the bear is? If so, what are the magnitude and direction of the velocity?
   b. At this instant, can you say what the acceleration of the bear is? If so, what are the magnitude and direction of the acceleration?

23. || A 1500 kg car is traveling along a straight road at 20 m/s. INT Two seconds later its speed is 21 m/s. What is the magnitude of the net force acting on the car during this time?

24. || Very small forces can have tremendous effects on the motion of very small objects. Consider a single electron, with a mass of 9.1×10^{-31} kg, subject to a single force equal to the weight of a penny, 2.5×10^{-2} N. What is the acceleration of the electron?

25. || The motion of a very massive object is hardly affected by what would seem to be a substantial force. Consider a supertanker, with a mass of 3.0×10^{8} kg. If it is pushed by a rocket motor (see Table 4.2) and is subject to no other forces, what will be the magnitude of its acceleration?

Section 4.7 Free-Body Diagrams

Problems 26 through 28 show a free-body diagram. For each, (a) Redraw the free-body diagram and (b) Write a short description of a real object for which this is the correct free-body diagram. Use Examples 4.3, 4.4, and 4.5 as models of what a description should be like.

26. |

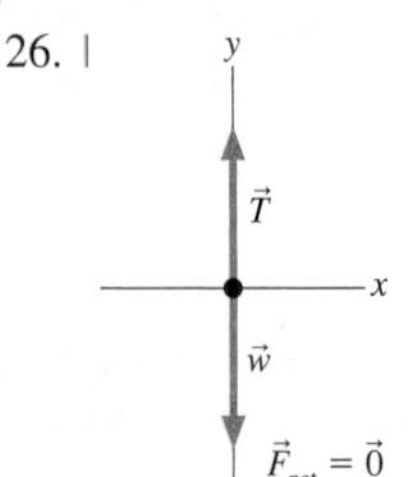

FIGURE P4.26

27. |

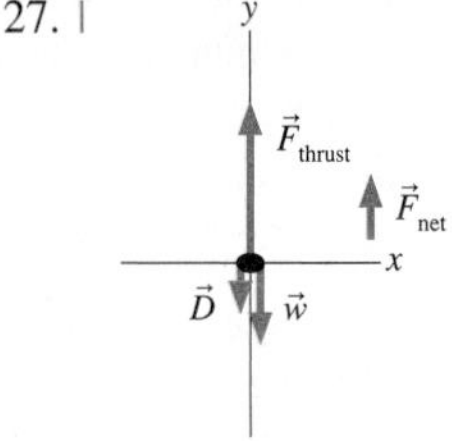

FIGURE P4.27

28. |

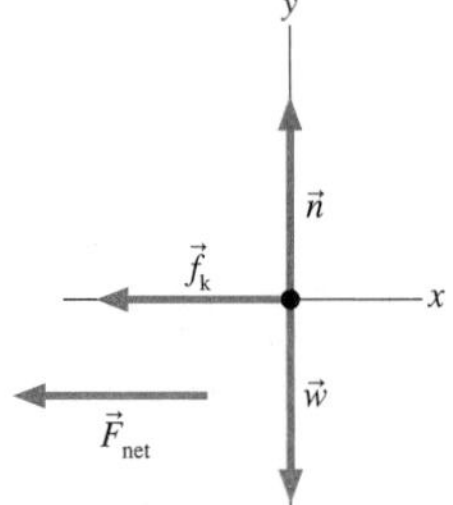

FIGURE P4.28

Problems 29 through 35 describe a situation. For each, identify all forces acting on the object and draw a free-body diagram of the object.

29. || Your car is sitting in the parking lot.
30. || Your car is accelerating from a stop.
31. || Your car is slowing to a stop from a high speed.
32. || Your physics textbook is sliding across the table.
33. || An ascending elevator, hanging from a cable, is coming to a stop.
34. || A skier slides down a slope at a constant speed.
35. ||| You hold a picture motionless against a wall by pressing on it, as shown in Figure P4.35.

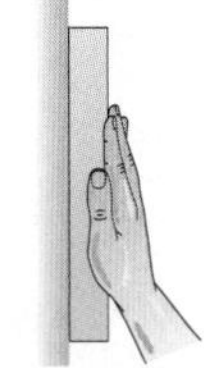

FIGURE P4.35

Section 4.8 Newton's Third Law

36. || A weightlifter stands up from a squatting position while holding a heavy barbell across his shoulders. Identify all the action/reaction pairs of forces between the weight lifter and the barbell.

37. || Three ice skaters, numbered 1, 2, and 3, stand in a line, each with her hands on the shoulders of the skater in front. Skater 3, at the rear, pushes on skater 2. Identify all the action/reaction pairs of forces between the three skaters. Draw a free-body diagram for skater 2, in the middle. Assume the ice is frictionless.

38. | A girl stands on a sofa. Identify all the action/reaction pairs of forces between the girl and the sofa.

General Problems

39. | Redraw the motion diagram INT shown in Figure P4.39, then draw a vector beside it to show the direction of the net force acting on the object. Explain your reasoning.

40. | Redraw the motion diagram INT shown in Figure P4.40, then draw a vector beside it to show the direction of the net force acting on the object. Explain your reasoning.

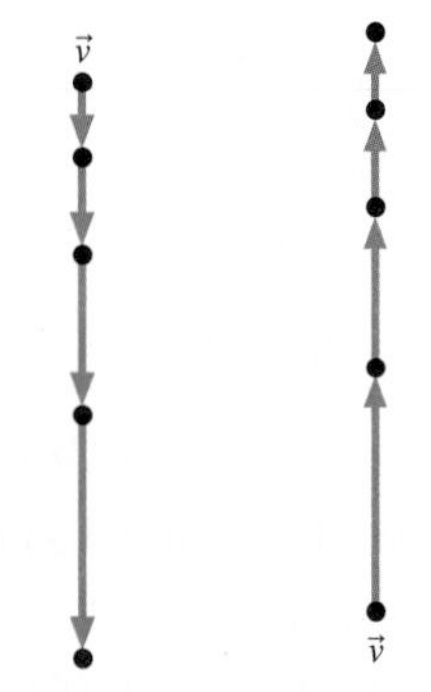

FIGURE P4.39 FIGURE P4.40

41. | Redraw the motion diagram INT shown in Figure P4.41, then draw a vector beside it to show the direction of the net force acting on the object. Explain your reasoning.

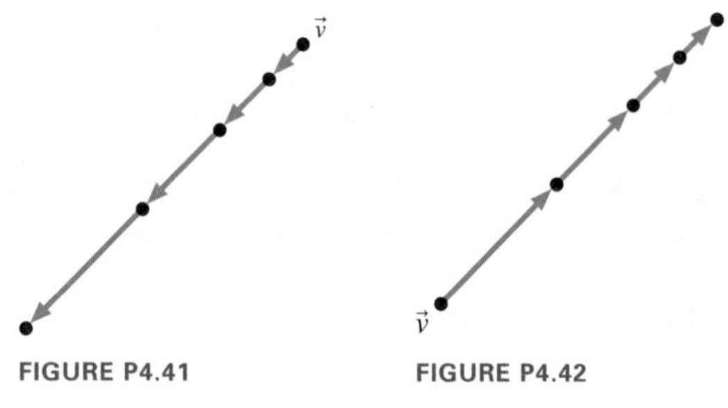

FIGURE P4.41 FIGURE P4.42

42. | Redraw the motion diagram shown in Figure P4.42, then INT draw a vector beside it to show the direction of the net force acting on the object. Explain your reasoning.

Problems 43 through 49 show a free-body diagram. For each:

a. Redraw the diagram.
b. Identify the direction of the acceleration vector $\vec{a}$ and show it as a vector next to your diagram. Or, if appropriate, write $\vec{a} = \vec{0}$.
c. Write a short description of a real object for which this is the correct free-body diagram. Use Examples 4.7, 4.8, and 4.9 as models of what a description should be like.

43. |

44. |

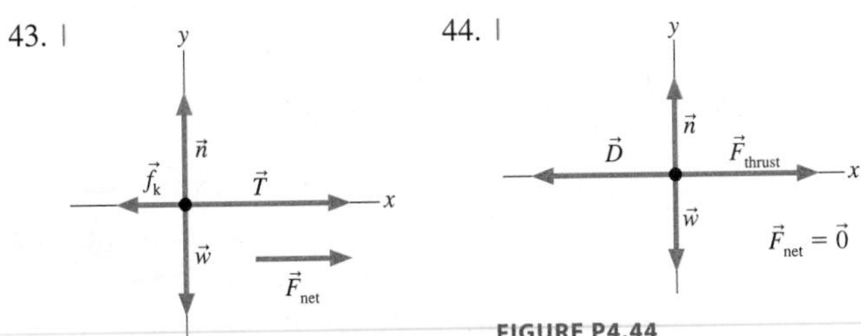

FIGURE P4.43

FIGURE P4.44

45. |

46. |

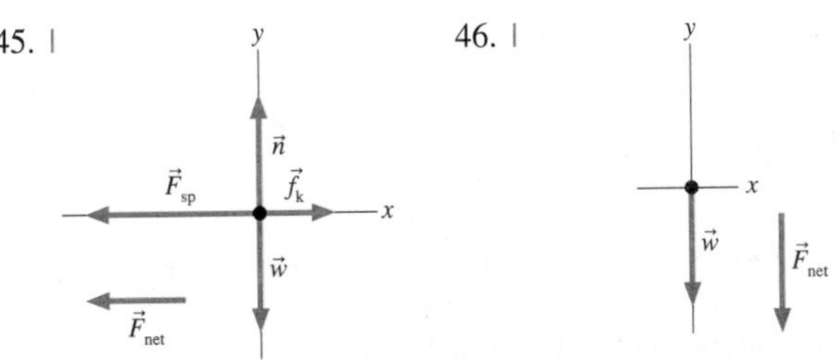

FIGURE P4.45

FIGURE P4.46

47. |

48. |

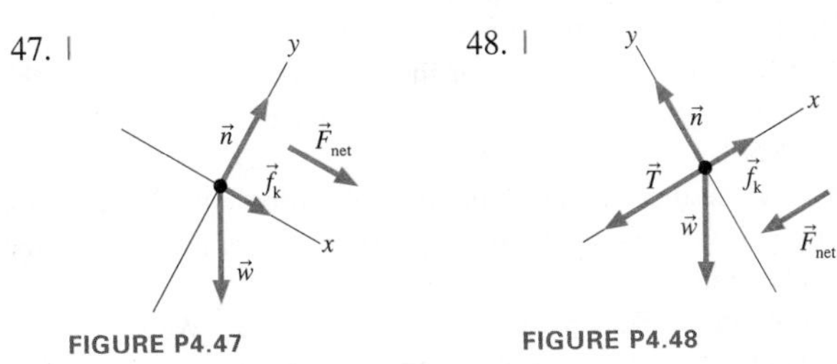

FIGURE P4.47

FIGURE P4.48

49. |

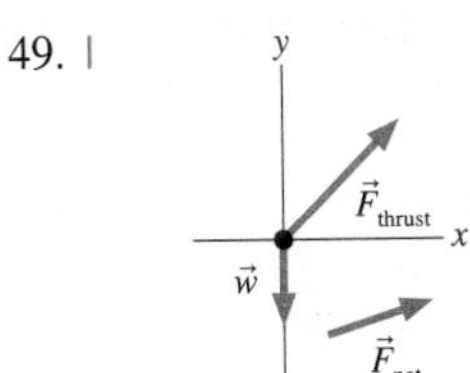

FIGURE P4.49

50. |||| A student draws the flawed free-body diagram shown in Figure P4.50 to represent the forces acting on a car traveling at constant speed on a level road. Identify the errors in the diagram, then draw a correct free-body diagram for this situation.

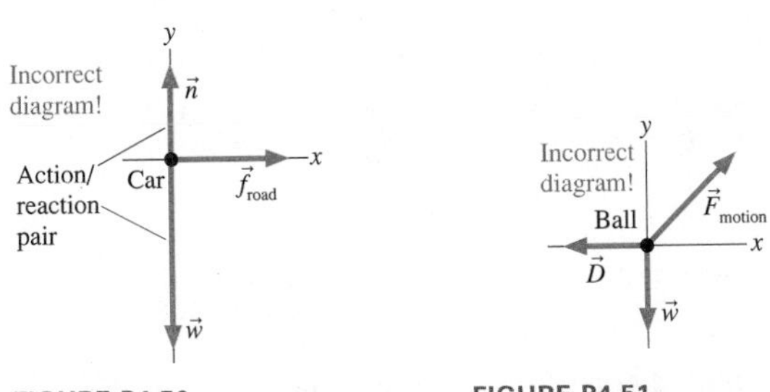

FIGURE P4.50

FIGURE P4.51

51. ||| A student draws the flawed free-body diagram shown in Figure P4.51 to represent the forces acting on a golf ball that is traveling upward and to the right a very short time after being hit off the tee. Air resistance is assumed to be relevant. Identify the errors in the diagram, then draw a correct free-body diagram for this situation.

Problems 52 through 63 describe a situation. For each, draw a motion diagram, a force identification diagram, and a free-body diagram.

52. || An elevator, suspended by a single cable, has just left the tenth floor and is speeding up as it descends toward the ground floor.
53. ||| A rocket is being launched straight up. Air resistance is not negligible.
54. ||| A jet plane is speeding down the runway during takeoff. Air resistance is not negligible.
55. || You've slammed on the brakes and your car is skidding to a stop while going down a 20° hill.
56. || A skier is going down a 20° slope. A *horizontal* headwind is blowing in the skier's face. Friction is small, but not zero.
57. || A bale of hay sits on the bed of a trailer. The trailer is starting to accelerate forward, and the bale is slipping toward the back of the trailer.
58. || A Styrofoam ball has just been shot straight up. Air resistance is not negligible.
59. ||| A spring-loaded gun shoots a plastic ball. The trigger has just been pulled and the ball is starting to move down the barrel. The barrel is horizontal.
60. || A person on a bridge throws a rock straight down toward the water. The rock has just been released.
61. ||| A gymnast has just landed on a trampoline. She's still moving downward as the trampoline stretches.
62. ||| A heavy box is in the back of a truck. The truck is accelerating to the right. Apply your analysis to the box.
63. || A bag of groceries is on the back seat of your car as you stop for a stop light. The bag does not slide. Apply your analysis to the bag.
64. || A rubber ball bounces. We'd like to understand *how* the ball bounces.
   a. A rubber ball has been dropped and is bouncing off the floor. Draw a motion diagram of the ball during the brief time interval that it is in contact with the floor. Show 4 or 5 frames as the ball compresses, then another 4 or 5 frames as it expands. What is the direction of $\vec{a}$ during each of these parts of the motion?
   b. Draw a picture of the ball in contact with the floor and identify all forces acting on the ball.
   c. Draw a free-body diagram of the ball during its contact with the ground. Is there a net force acting on the ball? If so, in which direction?
   d. During contact, is the force of the ground on the ball larger, smaller, or equal to the weight of the ball? Use your answers to parts a–c to explain your reasoning.
65. || If a car stops suddenly, you feel "thrown forward." We'd like to understand what happens to the passengers as a car stops. Imagine yourself sitting on a *very* slippery bench inside a car. This bench has no friction, no seat back, and there's nothing for you to hold on to.
   a. Draw a picture and identify all of the forces acting on you as the car travels in a straight line at a perfectly steady speed on level ground.
   b. Draw your free-body diagram. Is there a net force on you? If so, in which direction?
   c. Repeat parts a and b with the car slowing down.
   d. Describe what happens to you as the car slows down.
   e. Use Newton's laws to explain why you seem to be "thrown forward" as the car stops. Is there really a force pushing you forward?

66. The fastest pitched baseball was clocked at 46 m/s. If the pitcher exerted his force (assumed to be horizontal and constant) over a distance of 1.0 m, and a baseball has a mass of 145 g,
BIO
    a. Draw a free-body diagram of the ball during the pitch.
    b. What force did the pitcher exert on the ball during this record-setting pitch?
    c. Estimate the force in part b as a fraction of the pitcher's weight.
67. The froghopper, champion leaper of the insect world, can jump straight up at 4.0 m/s. The jump itself lasts a mere 1.0 ms before the insect is clear of the ground.
BIO
    a. Draw a free-body diagram of this mighty leaper while the jump is taking place.
    b. While the jump is taking place, is the force that the ground exerts on the froghopper greater than, less than, or equal to the insect's weight? Explain.
68. A beach ball is thrown straight up, and some time later it lands on the sand. Is the magnitude of the net force on the ball greatest when it is going up or when it is on the way down? Or is it the same in both cases? Explain. Air resistance should not be neglected for a large, light object.

Passage Problems

A Simple Solution for a Stuck Car

If your car is stuck in the mud and you don't have a winch to pull it out, you can use a piece of rope and a tree to do the trick. First, you tie one end of the rope to your car and the other to a tree, then pull as hard as you can on the middle of the rope, as shown in Figure P4.69a. This technique applies a force to the car much larger than the force that you can apply directly. To see why the car experiences such a large force, look at the forces acting on the center point of the rope, as shown in Figure P4.69b. The sum of the forces is zero, thus the tension is much greater than the force you apply. It is this tension force that acts on the car and, with luck, pulls it free.

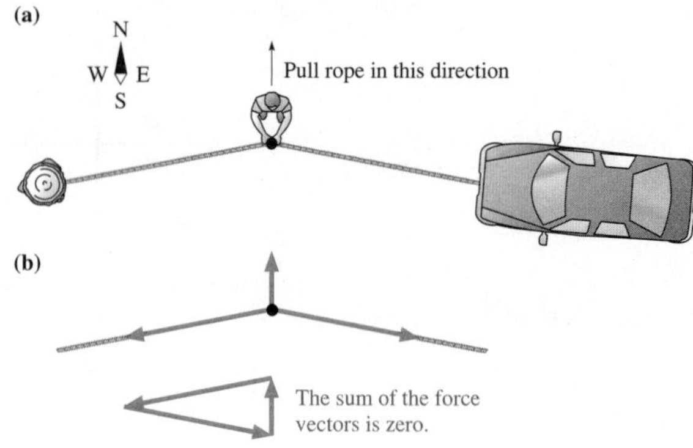
(a)
N
W
E
S
Pull rope in this direction
(b)
The sum of the force vectors is zero.

FIGURE P4.69

69. The sum of the three forces acting on the center point of the rope is assumed to be zero because
    A. This point has a very small mass.
    B. Tension forces in a rope always cancel.
    C. This point is not accelerating.
    D. The angle of deflection is very small.
70. When you are pulling on the rope as shown, what is the approximate direction of the tension force on the tree?
    A. North B. South C. East D. West
71. Assume that you are pulling on the rope but the car is not moving. What is the approximate direction of the force of the mud on the car?
    A. North B. South C. East D. West
72. Suppose your efforts work, and the car begins to move forward out of the mud. As it does so, the force of the car on the rope is
    A. Zero.
    B. Less than the force of the rope on the car.
    C. Equal to the force of the rope on the car.
    D. Greater than the force of the rope on the car.

STOP TO THINK ANSWERS

Stop to Think 4.1: C.

$\vec{F}_1$ $\vec{F}_2$ $\vec{F}_3$

y-component of $\vec{F}_3$ cancels y-component of $\vec{F}_1$.

x-component of $\vec{F}_3$ is to the left and larger than the x-component of $\vec{F}_2$.

Stop to Think 4.2: A, B, and D. Friction and the normal force are the only contact forces. Nothing is touching the rock to provide a "force of the kick." We've agreed to ignore air resistance unless a problem specifically calls for it.

Stop to Think 4.3: B. Acceleration is proportional to force, so doubling the number of rubber bands doubles the acceleration of the original object from $2\ \text{m/s}^2$ to $4\ \text{m/s}^2$. But acceleration is also inversely proportional to mass. Doubling the mass cuts the acceleration in half, back to $2\ \text{m/s}^2$.

Stop to Think 4.4: D

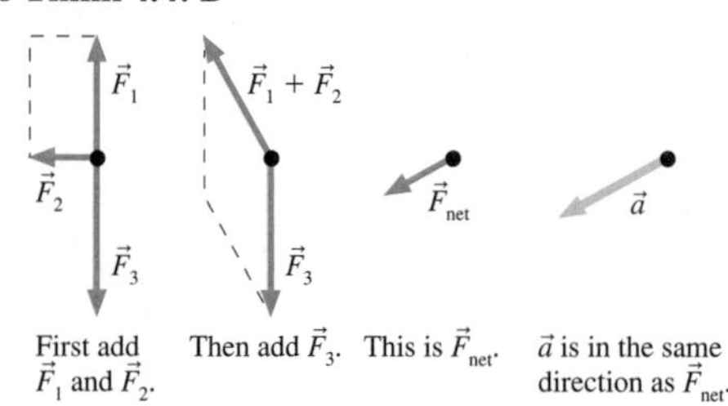
$\vec{F}_1$
$\vec{F}_1 + \vec{F}_2$
$\vec{F}_2$
$\vec{F}_3$
$\vec{F}_3$
$\vec{F}_{net}$
$\vec{a}$
First add $\vec{F}_1$ and $\vec{F}_2$.
Then add $\vec{F}_3$.
This is $\vec{F}_{net}$.
$\vec{a}$ is in the same direction as $\vec{F}_{net}$.

Stop to Think 4.5: C. The acceleration vector points downward as the elevator slows. $\vec{F}_{net}$ points in the same direction as $\vec{a}$, so $\vec{F}_{net}$ also points downward. This will be true if the tension is less than the weight: $T < w$.

Stop to Think 4.6: C. Newton's third law says that the force of A on B is *equal* and opposite to the force of B on A. This is always true. The mass of the objects isn't relevant.

5 Applying Newton's Laws

Why does this sky surfer fall at a constant speed? And why does he suddenly slow down when his parachute opens?

LOOKING AHEAD ▸▸

The goal of Chapter 5 is to learn how to solve problems about motion along a straight line.

Equilibrium Problems

An object at rest or moving at a constant velocity has zero acceleration. According to Newton's second law, this means that the net force on it must also be zero.

This boulder is in *static equilibrium:* It remains at rest.

A ski lift, moving at a constant velocity, is in *dynamic equilibrium.*

Looking Back ◂◂
4.4, 4.7 Identifying forces, free-body diagrams
4.6 Newton's second law

Interacting Objects

When two objects interact with each other, each exerts a force on the other. By Newton's third law, these forces are equal in magnitude but oppositely directed.

The barge pushes back on the tug just as hard as the tug pushes on the barge. Newton's third law will help you understand the motion of objects that are in contact with each other.

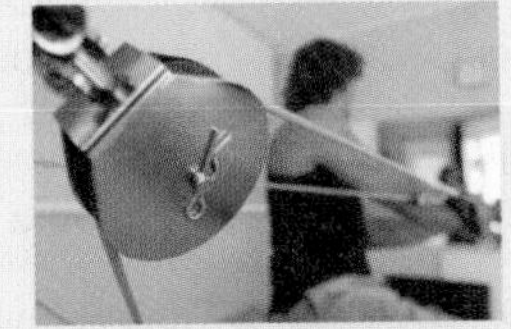

A common way for two objects to interact is with ropes or strings under tension. You'll learn how to solve problems involving tension and how pulleys act to change the direction of the tension force.

Applying Newton's Second Law

In this chapter, you'll learn how to use Newton's second law in component form to solve a variety of problems in mechanics.

What are the blocks' accelerations?

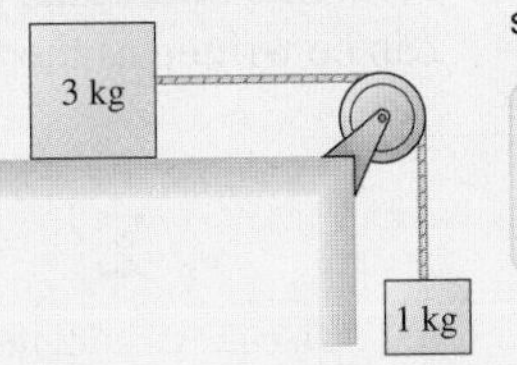

Friction, tension, gravity, and the pulley all act in this problem. You'll learn explicit strategies to solve problems like this.

Looking Back ◂◂
2.5, 2.7 Constant acceleration and free fall
3.2–3.3 Vectors and components

Forces and Newton's Second Law

Chapter 4 introduced several important forces. Now you'll need to understand these forces in more detail, so that you can use them in Newton's second law. For example, you'll learn . . .

. . . that **mass and weight are not the same thing.** However, you'll find that there is a simple relationship between the two.

This astronaut on the moon weighs only 1/6 of what he does on earth. His *mass*, however, is the same on both the moon and the earth.

. . . a simple model for friction that provides a reasonably accurate description of how static and kinetic friction behave.

Static friction adjusts its magnitude as needed to keep the sofa from slipping.

Kinetic (sliding) friction does not depend on an object's speed.

This human tower is in equilibrium because the net force on each man is zero.

5.1 Equilibrium

Chapter 4 introduced Newton's three laws of motion. Now, in Chapter 5, we want to use these laws to solve force and motion problems. This chapter focuses on objects that are at rest or that move in a straight line, such as runners, bicycles, cars, planes, and rockets. Circular motion and rotational motion will be treated in Chapters 6 and 7.

The simplest applications of Newton's second law, $\vec{F}_{net} = m\vec{a}$, are those for which the acceleration $\vec{a}$ is *zero.* In such cases, the net force acting on the object must be zero as well. One way an object can have $\vec{a} = \vec{0}$ is to be at rest. An object that remains at rest is said to be in **static equilibrium**. A second way for an object to have $\vec{a} = \vec{0}$ is to move in a straight line at a constant speed. Such an object is in **dynamic equilibrium.** The key property of both these cases of **equilibrium** is that the net force acting on the object is $\vec{F}_{net} = \vec{0}$.

To use Newton's laws, we have to identify all the forces acting on an object and then evaluate $\vec{F}_{net}$. Recall that $\vec{F}_{net}$ is the vector sum

$$\vec{F}_{net} = \vec{F}_1 + \vec{F}_2 + \vec{F}_3 + \cdots$$

where $\vec{F}_1$, $\vec{F}_2$, and so on are the individual forces, such as tension or friction, acting on the object. We found in Chapter 3 that vector sums can be evaluated in terms of the x- and y-components of the vectors; that is, the x-component of the net force is $(F_{net})_x = F_{1x} + F_{2x} + F_{3x} + \cdots$. If we restrict ourselves to problems where all the forces are in the xy-plane, then the equilibrium requirement $\vec{F}_{net} = \vec{a} = \vec{0}$ is a shorthand way of writing two simultaneous equations:

$$(F_{net})_x = F_{1x} + F_{2x} + F_{3x} + \cdots = 0$$

$$(F_{net})_y = F_{1y} + F_{2y} + F_{3y} + \cdots = 0$$

Recall from your math classes that the Greek letter Σ (sigma) stands for "the sum of." It will be convenient to abbreviate the sum of the x-components of all forces as

$$F_{1x} + F_{2x} + F_{3x} + \cdots = \sum F_x$$

With this notation, Newton's second law for an object in equilibrium, with $\vec{a} = \vec{0}$, can be written as the two equations

$$\sum F_x = ma_x = 0 \qquad \text{and} \qquad \sum F_y = ma_y = 0 \tag{5.1}$$

In equilibrium, the sums of the x- and y-components of the force are zero

Although this may look a bit forbidding, we'll soon see how to use a free-body diagram of the forces to help evaluate these sums.

When an object is in equilibrium, we are usually interested in finding the forces that keep it in equilibrium. Newton's second law is the basis for a strategy for solving equilibrium problems.

PROBLEM-SOLVING STRATEGY 5.1 Equilibrium problems

PREPARE First check that the object is in equilibrium: Does $\vec{a} = \vec{0}$?

- An object at rest is in static equilibrium.
- An object moving at a constant velocity is in dynamic equilibrium.

Then identify all forces acting on the object and show them on a free-body diagram. Determine which forces you know and which you need to solve for.

Continued

SOLVE An object in equilibrium must satisfy Newton's second law for the case where $\vec{a} = \vec{0}$. In component form, the requirement is

$$\sum F_x = ma_x = 0 \quad \text{and} \quad \sum F_y = ma_y = 0$$

You can find the force components that go into these sums directly from your free-body diagram. From these two equations, solve for the unknown forces in the problem.

ASSESS Check that your result has the correct units, is reasonable, and answers the question.

Static Equilibrium

EXAMPLE 5.1 Forces supporting an orangutan

An orangutan weighing 500 N hangs from a vertical vine. What is the tension in the vine?

PREPARE The orangutan is at rest, so it is in static equilibrium. The net force on it must then be zero. FIGURE 5.1 first identifies the forces acting on the orangutan: the upward force of the tension in the vine and the downward, long-range force of gravity. These forces are then shown on a free-body diagram, where it's noted that equilibrium requires $\vec{F}_{net} = \vec{0}$.

FIGURE 5.1 The forces on an orangutan.

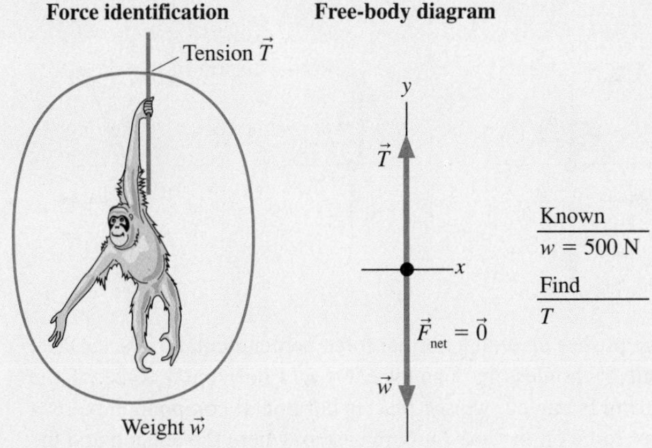

SOLVE Neither force has an x-component, so we need to examine only the y-components of the forces. In this case, the y-component of Newton's second law is

$$\sum F_y = T_y + w_y = ma_y = 0$$

You might have been tempted to write $T_y - w_y$ because the weight force points down. But remember that T_y and w_y are *components* of vectors, and can thus be positive (for a vector such as $\vec{T}$ that points up) or negative (for a vector such as $\vec{w}$ that points down). The fact that $\vec{w}$ points down is taken into account when we *evaluate* the components—that is, when we write them in terms of the *magnitudes* T and w of the vectors $\vec{T}$ and $\vec{w}$.

Because the tension vector $\vec{T}$ points straight up, in the positive y-direction, its y-component is $T_y = T$. Because the weight vector $\vec{w}$ points straight down, in the negative y-direction, its y-component is $w_y = -w$. This is where the signs enter. With these components, Newton's second law becomes

$$T - w = 0$$

This equation is easily solved for the tension in the vine:

$$T = w = 500 \text{ N}$$

ASSESS It's not surprising that the tension in the vine equals the weight of the orangutan. However, we'll soon see that this is *not* the case if the object is accelerating.

EXAMPLE 5.2 Readying a wrecking ball

A wrecking ball weighing 2500 N hangs from a cable. Prior to swinging, it is pulled back to a 20° angle by a second, horizontal cable. What is the tension in the horizontal cable?

PREPARE Because the ball is not moving, it hangs in static equilibrium, with $\vec{a} = \vec{0}$, until it is released. In FIGURE 5.2, we start by identifying all the forces acting on the ball: a tension force from each cable and the ball's weight. We've used different symbols $\vec{T}_1$ and $\vec{T}_2$ for the two different tension forces. We then construct a free-body diagram for these three forces, noting that $\vec{F}_{net} = m\vec{a} = \vec{0}$. We're looking for the magnitude T_1 of the tension force $\vec{T}_1$ in the horizontal cable.

FIGURE 5.2 Visual overview of a wrecking ball just before release.

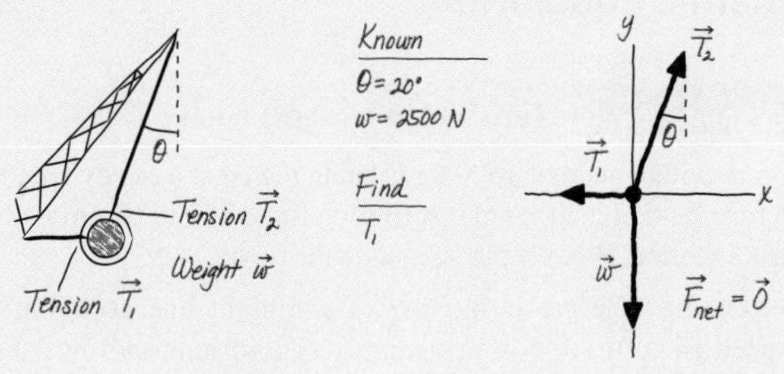

Continued

SOLVE The requirement of equilibrium is $\vec{F}_{net} = m\vec{a} = \vec{0}$. In component form, we have the two equations:

$$\sum F_x = T_{1x} + T_{2x} + w_x = ma_x = 0$$
$$\sum F_y = T_{1y} + T_{2y} + w_y = ma_y = 0$$

As always, we *add* the force components together. Now we're ready to write the components of each force vector in terms of the magnitudes and directions of those vectors. We learned how to do this in Section 3.3 of Chapter 3. With practice you'll learn to read the components directly off the free-body diagram, but to begin it's worthwhile to organize the components into a table.

Force	Name of x-component	Value of x-component	Name of y-component	Value of y-component
$\vec{T}_1$	T_{1x}	$-T_1$	T_{1y}	0
$\vec{T}_2$	T_{2x}	$T_2\sin\theta$	T_{2y}	$T_2\cos\theta$
$\vec{w}$	w_x	0	w_y	$-w$

We see from the free-body diagram that $\vec{T}_1$ points along the negative x-axis, so $T_{1x} = -T_1$ and $T_{1y} = 0$. We need to be careful with our trigonometry as we find the components of $\vec{T}_2$. Remembering that the side adjacent to the angle is related to the cosine, we see that the vertical (y) component of $\vec{T}_2$ is $T_2\cos\theta$. Similarly, the horizontal (x) component is $T_2\sin\theta$. The weight vector points straight down, so its y-component is $-w$. Notice that negative signs enter as we evaluate the components of the vectors, *not* when we write Newton's second law. This is a critical aspect of solving force and motion problems. With these components, Newton's second law now becomes

$$-T_1 + T_2\sin\theta + 0 = 0 \quad \text{and} \quad 0 + T_2\cos\theta - w = 0$$

We can rewrite these equations as

$$T_2\sin\theta = T_1 \quad \text{and} \quad T_2\cos\theta = w$$

These are two simultaneous equations with two unknowns: T_1 and T_2. To eliminate T_2 from the two equations, we solve the second equation for T_2, giving $T_2 = w/\cos\theta$. Then we insert this expression for T_2 into the first equation to get

$$T_1 = \frac{w}{\cos\theta}\sin\theta = \frac{\sin\theta}{\cos\theta}w = w\tan\theta = (2500\text{ N})\tan 20° = 910\text{ N}$$

where we made use of the fact that $\tan\theta = \sin\theta/\cos\theta$.

ASSESS It seems reasonable that to pull the ball back to this modest angle, a force substantially less than the ball's weight will be required.

CONCEPTUAL EXAMPLE 5.3 Forces in static equilibrium

A rod is free to slide on a frictionless sheet of ice. One end of the rod is lifted by a string. If the rod is at rest, which diagram in FIGURE 5.3 shows the correct angle of the string?

FIGURE 5.3 Which is the correct angle of the string?

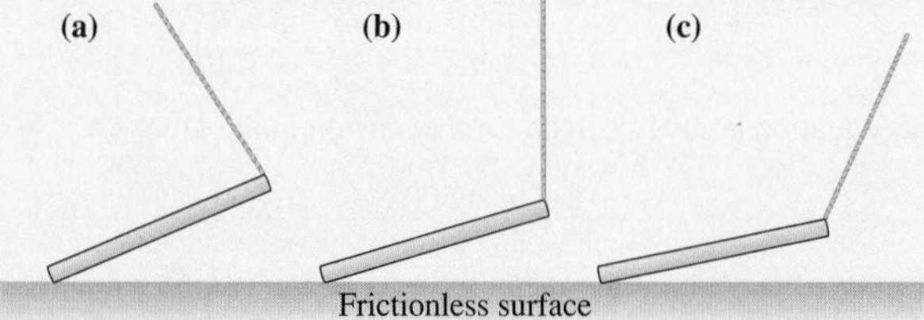

REASON If the rod is to hang motionless, it must be in static equilibrium with $\sum F_x = ma_x = 0$ and $\sum F_y = ma_y = 0$. FIGURE 5.4 shows free-body diagrams for the three string orientations. Remember that tension always acts along the direction of the string and that the weight force always points straight down. The ice pushes up with a normal force perpendicular to the surface, but frictionless ice cannot exert any horizontal force. If the string is angled, we see that its horizontal component exerts a net force on the rod. Only in case b, where the tension and the string are vertical, can the net force be zero.

FIGURE 5.4 Free-body diagrams for three angles of the string.

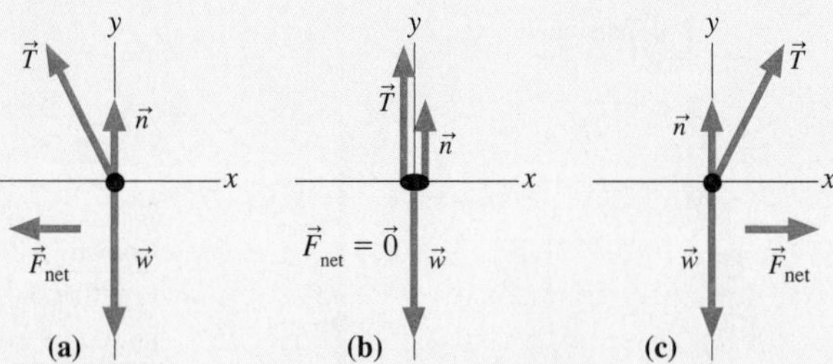

ASSESS If friction were present, the rod could in fact hang as in cases a or c. But without friction, the rods in these cases would slide until they came to rest as in case b.

Dynamic Equilibrium

EXAMPLE 5.4 Tension in towing a car

A car with a mass of 1500 kg is being towed at a steady speed by a rope held at a 20° angle. A friction force of 320 N opposes the car's motion. What is the tension in the rope?

PREPARE The car is moving in a straight line at a constant speed ($\vec{a} = \vec{0}$) so it is in dynamic equilibrium and must have $\vec{F}_{net} = m\vec{a} = \vec{0}$. FIGURE 5.5 shows three contact forces acting on the car—the tension force $\vec{T}$, friction $\vec{f}$, and the normal force $\vec{n}$—and the long-range force of gravity $\vec{w}$. These four forces are shown on the free-body diagram.

FIGURE 5.5 Visual overview of a car being towed.

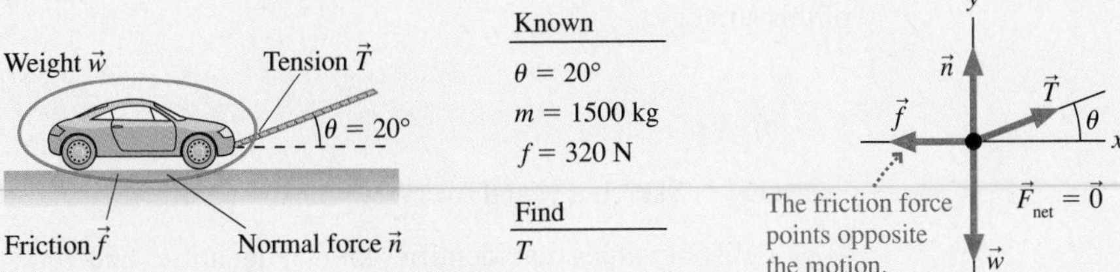

SOLVE This is still an equilibrium problem, even though the car is moving, so our problem-solving procedure is unchanged. With four forces, the requirement of equilibrium is

$$\sum F_x = n_x + T_x + f_x + w_x = ma_x = 0$$

$$\sum F_y = n_y + T_y + f_y + w_y = ma_y = 0$$

We can again determine the horizontal and vertical components of the forces by "reading" the free-body diagram. The results are shown in the table.

Force	Name of *x*-component	Value of *x*-component	Name of *y*-component	Value of *y*-component
$\vec{n}$	n_x	0	n_y	n
$\vec{T}$	T_x	$T\cos\theta$	T_y	$T\sin\theta$
$\vec{f}$	f_x	$-f$	f_y	0
$\vec{w}$	w_x	0	w_y	$-w$

With these components, Newton's second law becomes

$$T\cos\theta - f = 0$$

$$n + T\sin\theta - w = 0$$

The first equation can be used to solve for the tension in the rope:

$$T = \frac{f}{\cos\theta} = \frac{320\text{ N}}{\cos 20°} = 340\text{ N}$$

to two significant figures. It turned out that we did not need the y-component equation in this problem. We would need it if we wanted to find the normal force $\vec{n}$.

ASSESS Had we pulled the car with a horizontal rope, the tension would need to exactly balance the friction force of 320 N. Because we are pulling at an angle, however, part of the tension in the rope pulls *up* on the car instead of in the forward direction. Thus we need a little more tension in the rope when it's at an angle.

5.2 Dynamics and Newton's Second Law

2.1–2.4

Newton's second law is the essential link between force and motion. The essence of Newtonian mechanics can be expressed in two steps:

- The forces acting on an object determine its acceleration $\vec{a} = \vec{F}_{net}/m$.
- The object's motion can be found by using $\vec{a}$ in the equations of kinematics.

We want to develop a strategy to solve a variety of problems in mechanics, but first we need to write the second law in terms of its components. To do so, let's first rewrite Newton's second law in the form

$$\vec{F}_{net} = \vec{F}_1 + \vec{F}_2 + \vec{F}_3 + \cdots = m\vec{a}$$

where $\vec{F}_1, \vec{F}_2, \vec{F}_3$, and so on are the forces acting on an object. To write the second law in component form merely requires that we use the x- and y-components of the acceleration. Thus Newton's second law, $\vec{F}_{net} = m\vec{a}$, is

$$\sum F_x = ma_x \quad \text{and} \quad \sum F_y = ma_y \qquad (5.2)$$

Newton's second law in component form

The first equation says that **the component of the acceleration in the *x*-direction is determined by the sum of the *x*-components of the forces acting on the object.** A similar statement applies to the y-direction.

There are two basic types of problems in mechanics. In the first, you use information about forces to find an object's acceleration, then use kinematics to determine the object's motion. In the second, you use information about the object's motion to determine its acceleration, then solve for unknown forces. Either way, the two equations of Equation 5.2 are the link between force and motion, and they form the basis of a problem-solving strategy. The primary goal of this chapter is to illustrate the use of this strategy.

PROBLEM-SOLVING STRATEGY 5.2 Dynamics problems

PREPARE Sketch a visual overview consisting of:

- A list of values that identifies known quantities and what the problem is trying to find.
- A force identification diagram to help you identify all the forces acting on the object.
- A free-body diagram that shows all the forces acting on the object.

If you'll need to use kinematics to find velocities or positions, you'll also need to sketch:

- A motion diagram to determine the direction of the acceleration.
- A pictorial representation that establishes a coordinate system, shows important points in the motion, and defines symbols.

It's OK to go back and forth between these steps as you visualize the situation.

SOLVE Write Newton's second law in component form as

$$\sum F_x = ma_x \qquad \text{and} \qquad \sum F_y = ma_y$$

You can find the components of the forces directly from your free-body diagram. Depending on the problem, either:

- Solve for the acceleration, then use kinematics to find velocities and positions.
- Use kinematics to determine the acceleration, then solve for unknown forces.

ASSESS Check that your result has the correct units, is reasonable, and answers the question.

Exercise 24

EXAMPLE 5.5 Putting a golf ball

A golfer putts a 46 g ball with a speed of 3.0 m/s. Friction exerts a 0.020 N retarding force on the ball, slowing it down. Will her putt reach the hole, 10 m away?

PREPARE FIGURE 5.6 is a visual overview of the problem. We've collected the known information, drawn a sketch, and identified what we want to find. The motion diagram shows that the ball is slowing down as it rolls to the right, so the acceleration vector points to the left. Next, we identify the forces acting on the ball and show them on a free-body diagram. Note that the net force points to the left, as it must because the acceleration points to the left.

FIGURE 5.6 Visual overview of a golf putt.

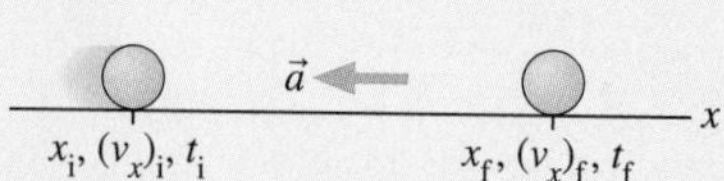

Known

$x_i = 0$ m $f = 0.020$ N
$(v_x)_i = 3.0$ m/s $m = 0.046$ kg
$(v_x)_f = 0$ m/s

Find

x_f

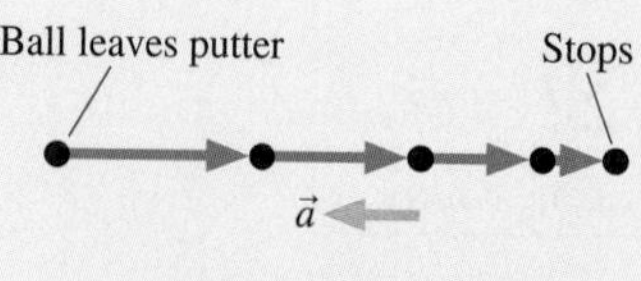

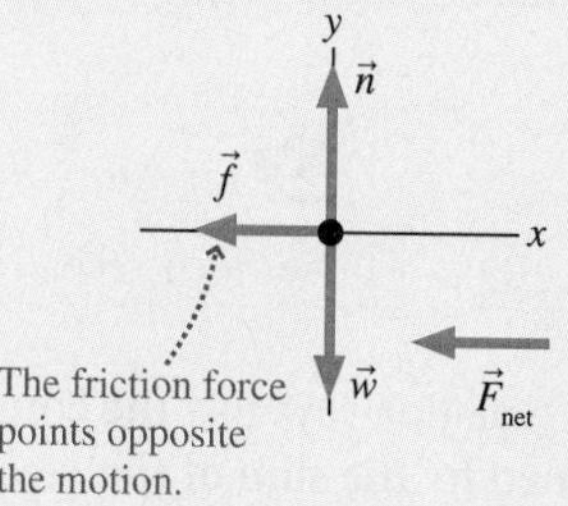

SOLVE Newton's second law in component form is

$$\sum F_x = n_x + f_x + w_x = 0 - f + 0 = ma_x$$

$$\sum F_y = n_y + f_y + w_y = n + 0 - w = ma_y = 0$$

We've written the equations as sums, as we did with equilibrium problems, then "read" the values of the force components from the free-body diagram. The components are simple enough in this problem that we don't really need to show them in a table. It is particularly important to notice that we set $a_y = 0$ in the second equation. This is because the ball does not move in the y-direction, so it can't have any acceleration in the y-direction. This will be an important step in many problems.

The first equation is $-f = ma_x$, from which we find

$$a_x = -\frac{f}{m} = \frac{-(0.020\text{ N})}{0.046\text{ kg}} = -0.43\text{ m/s}^2$$

The negative sign shows that the acceleration is directed to the left, as expected.

Now that we know the acceleration, we can use kinematics to find how far the ball will roll before stopping. We don't have any information about the time it takes for the ball to stop, so we'll use the kinematic equation $(v_x)_f^2 = (v_x)_i^2 + 2a_x(x_f - x_i)$. This gives

$$x_f = x_i + \frac{(v_x)_f^2 - (v_x)_i^2}{2a_x} = 0\text{ m} + \frac{(0\text{ m/s})^2 - (3.0\text{ m/s})^2}{2(-0.43\text{ m/s}^2)} = 10.5\text{ m}$$

If her aim is true, the ball will just make it into the hole.

ASSESS It seems reasonable that a ball putted on grass with an initial speed of 3 m/s—about jogging speed—would travel roughly 10 m.

EXAMPLE 5.6 Finding a rocket cruiser's acceleration

A rocket cruiser with a mass of 2200 kg and weighing 5000 N is flying horizontally over the surface of a distant planet. At its present speed, a 3000 N drag force acts on the cruiser. The cruiser's engines can be tilted so as to provide a thrust angled up or down. The pilot turns the thrust up to 14,000 N while pivoting the engines to continue flying horizontally. What is the cruiser's acceleration?

PREPARE FIGURE 5.7 is a visual overview in which we've listed the known information, identified the forces on the cruiser, and drawn a free-body diagram. (Because kinematics is not needed to find the acceleration, we don't need a pictorial diagram.) As discussed in Chapter 4, the thrust force points *opposite* the direction of the rocket exhaust, which we've shown at angle θ. The thrust must have an upward vertical component to balance the weight force; otherwise, the cruiser would fall. To continue flying horizontally requires the net force to be directed forward.

SOLVE Newton's second law in component form is

$$\sum F_x = (F_{\text{thrust}})_x + D_x + w_x = ma_x$$

$$\sum F_y = (F_{\text{thrust}})_y + D_y + w_y = ma_y$$

From the free-body diagram, we see that $(F_{\text{thrust}})_x = F_{\text{thrust}}\cos\theta$, $(F_{\text{thrust}})_y = F_{\text{thrust}}\sin\theta$, $D_x = -D$, $D_y = 0$, $w_x = 0$, and $w_y = -w$. We know that a_y must be zero because the cruiser is to accelerate *horizontally*. Thus the second law becomes

$$F_{\text{thrust}}\cos\theta - D = ma_x$$

$$F_{\text{thrust}}\sin\theta - w = 0$$

The first of these equations contains a_x, the quantity we want to find, but we can't solve for a_x without knowing what θ is. Fortunately, we can use the second equation to find θ, then use this value of θ in the first equation to find a_x.

The second equation gives

$$\sin\theta = \frac{w}{F_{\text{thrust}}} = \frac{5000\text{ N}}{14{,}000\text{ N}} = 0.357$$

$$\theta = \sin^{-1}(0.357) = 20.9°$$

Now we can use this value in the first equation to get

$$a_x = \frac{1}{m}(F_{\text{thrust}}\cos\theta - D)$$
$$= \frac{1}{2200\text{ kg}}[(14{,}000\text{ N})\cos(20.9°) - 3000\text{ N}] = 4.6\text{ m/s}^2$$

ASSESS An important key to solving this problem was to use the information that the cruiser accelerates only in the horizontal direction. Mathematically, this means that $a_y = 0$. Because the thrust is much greater than the weight, we need only a modest downward component of the thrust to cancel the weight and let the cruiser accelerate horizontally. So our engine tilt seems reasonable.

FIGURE 5.7 Visual overview of a rocket cruiser.

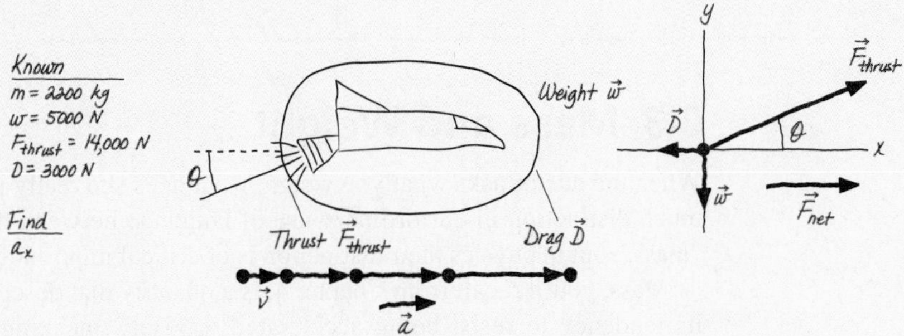

EXAMPLE 5.7 Towing a car with acceleration

A car with a mass of 1500 kg is being towed by a rope held at a 20° angle. A friction force of 320 N opposes the car's motion. What is the tension in the rope if the car goes from rest to 12 m/s in 10 s?

PREPARE You should recognize that this problem is almost identical to Example 5.4. The difference is that the car is now accelerating, so it is no longer in equilibrium. This means, as shown in FIGURE 5.8, that the net force is not zero. We've already identified all the forces in Example 5.4.

SOLVE Newton's second law in component form is

$$\sum F_x = n_x + T_x + f_x + w_x = ma_x$$
$$\sum F_y = n_y + T_y + f_y + w_y = ma_y = 0$$

We've again used the fact that $a_y = 0$ for motion that is purely along the x-axis. The components of the forces were worked out in Example 5.4. With that information, Newton's second law in component form is

$$T\cos\theta - f = ma_x$$
$$n + T\sin\theta - w = 0$$

Because the car speeds up from rest to 12 m/s in 10 s, we can use kinematics to find the acceleration:

$$a_x = \frac{\Delta v_x}{\Delta t} = \frac{(v_x)_\text{f} - (v_x)_\text{i}}{t_\text{f} - t_\text{i}} = \frac{(12\text{ m/s}) - (0\text{ m/s})}{(10\text{ s}) - (0\text{ s})} = 1.2\text{ m/s}^2$$

We can now use the first Newton's-law equation above to solve for the tension. We have

$$T = \frac{ma_x + f}{\cos\theta} = \frac{(1500\text{ kg})(1.2\text{ m/s}^2) + 320\text{ N}}{\cos 20°} = 2300\text{ N}$$

ASSESS The tension is substantially more than the 340 N found in Example 5.4. It takes a much more force to accelerate the car than to keep it rolling at a constant speed.

FIGURE 5.8 Visual overview of a car being towed.

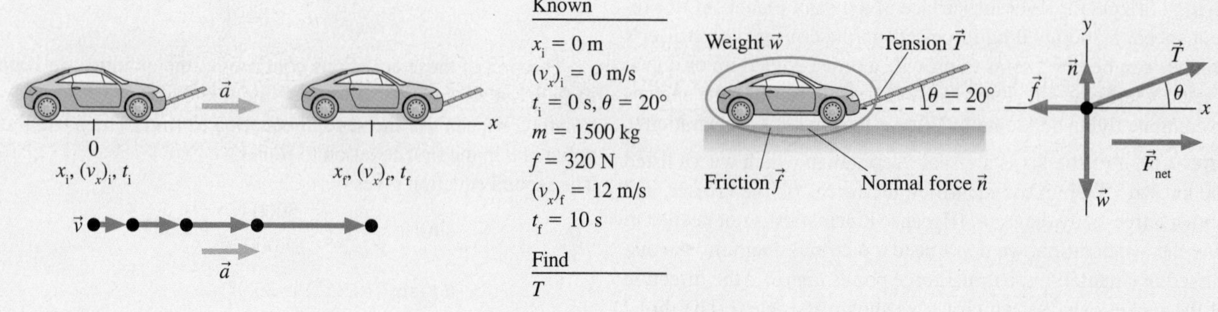

These first examples have shown all the details of our problem-solving strategy. Our purpose has been to demonstrate how the strategy is put into practice. Future examples will be briefer, but the basic *procedure* will remain the same.

STOP TO THINK 5.1 A Martian lander is approaching the surface. It is slowing its descent by firing its rocket motor. Which is the correct free-body diagram for the lander?

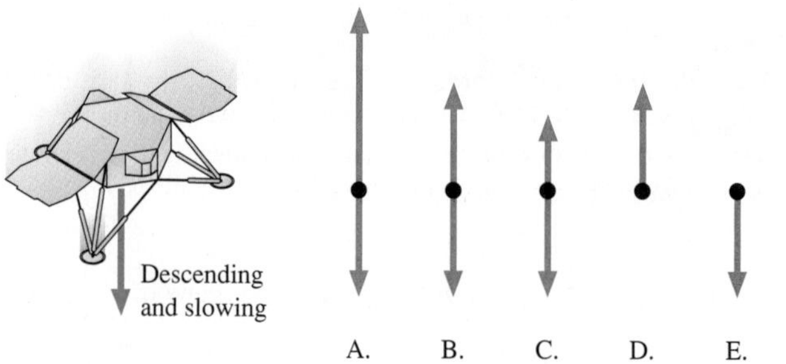

5.3 Mass and Weight

When the doctor asks what you weigh, what does she really mean? We do not make much distinction in our ordinary use of language between the terms "weight" and "mass," but in physics their distinction is of critical importance.

Mass, you'll recall from Chapter 4, is a quantity that describes an object's inertia, its tendency to resist being accelerated. Loosely speaking, it also describes the amount of matter in an object. Mass, measured in kilograms, is an intrinsic property of an object; it has the same value wherever the object may be and whatever forces might be acting on it.

Weight, on the other hand, is a *force*. Specifically, it is the gravitational force exerted on an object by a planet. Weight is a vector, not a scalar, and the vector's direction is always straight down. Weight is measured in newtons.

Mass and weight are not the same thing, but they are related. We can use Galileo's discovery about free fall to make the connection. FIGURE 5.9 shows the free-body diagram of an object in free fall. The *only* force acting on this object is its weight $\vec{w}$ the downward pull of gravity. Newton's second law for this object is

$$\vec{F}_{\text{net}} = \vec{w} = m\vec{a} \tag{5.3}$$

FIGURE 5.9 The free-body diagram of an object in free fall.

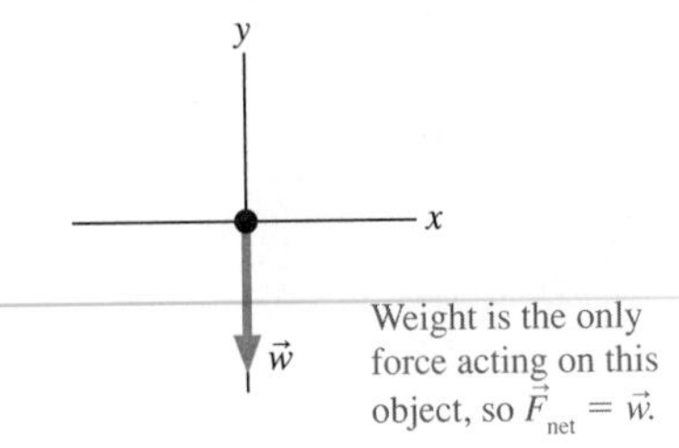

y
x
$\vec{w}$
Weight is the only force acting on this object, so $\vec{F}_{\text{net}} = \vec{w}$.

Recall Galileo's discovery that *any* object in free fall, regardless of its mass, has the same acceleration:

$$\vec{a}_{\text{free fall}} = (g, \text{downward}) \tag{5.4}$$

where $g = 9.80\ \text{m/s}^2$ is the free-fall acceleration at the earth's surface. So a_y in Equation 5.3 is equal to $-g$, and we have $-w = -mg$, or

$$w = mg \tag{5.5}$$

The magnitude of the weight force, which we call simply "the weight," is directly proportional to the mass, with g as the constant of proportionality. Thus, for example, the weight of a 3.6 kg book is $w = (3.6\ \text{kg})(9.8\ \text{m/s}^2) = 35\ \text{N}$.

NOTE ▶ Although we derived the relationship between mass and weight for an object in free fall, the weight of an object is *independent* of its state of motion. Equation 5.5 holds for an object at rest on a table, sliding horizontally, or moving in any other way. ◀

On the moon, astronaut John Young jumps 2 feet straight up, despite his spacesuit that weighs 370 pounds on earth. On the moon, where $g = 1.6\ \text{m/s}^2$, he and his suit together weighed only 90 pounds.

Because an object's weight depends on g, and the value of g varies from planet to planet, weight is not a fixed, constant property of an object. The value of g at the surface of the moon is about one-sixth its earthly value, so an object on the moon would have only one-sixth its weight on earth. The object's weight on Jupiter would be greater than its weight on earth. Its mass, however, would be the same. The amount of matter has not changed, only the gravitational force exerted on that matter.

So, when the doctor asks what you weigh, she really wants to know your *mass*. That's the amount of matter in your body. You can't really "lose weight" by going to the moon, even though you would weigh less there!

Measuring Mass and Weight

A *pan balance,* shown in FIGURE 5.10, is a device for measuring *mass*. You may have used a pan balance to "weigh" chemicals in a chemistry lab. An unknown mass is placed in one pan, then known masses are added to the other until the pans balance. Gravity pulls down on both sides, effectively *comparing* the masses, and the unknown mass equals the sum of the known masses that balance it. Although a pan balance requires gravity in order to function, it does not depend on the value of g. Consequently, the pan balance would give the same result on another planet.

A *spring scale* measures weight, not mass. A spring scale can be understood on the basis of Newton's second law. The object being weighed compresses the springs, as shown in FIGURE 5.11 on the following page, which then push up with force $\vec{F}_{\text{sp}}$. But because the object is at rest, in static equilibrium, the net force on it must be zero. Thus the upward spring force must exactly balance the downward weight force:

$$F_{\text{sp}} = w = mg \tag{5.6}$$

The *reading* of a spring scale is F_{sp}, the magnitude of the force that the spring is exerting. If the object is in equilibrium, then F_{sp} is exactly equal to the object's weight w. The scale does not "know" the weight of the object. All it can do is measure how much its spring is stretched or compressed. On a different planet, with a

FIGURE 5.10 A pan balance measures mass.

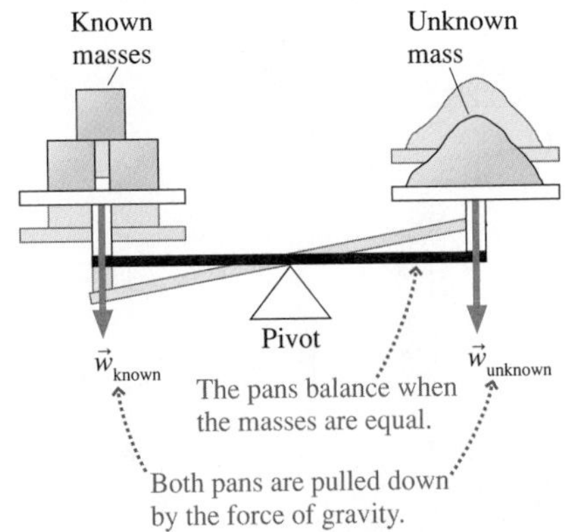

FIGURE 5.11 A spring scale measures weight.

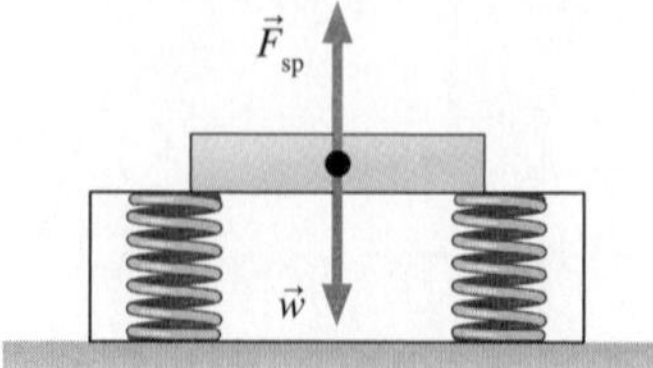

different value for g, the expansion or compression of the spring would be different and the scale's reading would be different.

The unit of force in the English system is the *pound.* We noted in Chapter 4 that the pound is defined as 1 lb = 4.45 N. An object whose weight $w = mg$ is 4.45 N has a mass

$$m = \frac{w}{g} = \frac{4.45\text{ N}}{9.80\text{ m/s}^2} = 0.454\text{ kg} = 454\text{ g}$$

You may have learned in previous science classes that "1 pound = 454 grams" or, equivalently, "1 kg = 2.2 lb." Strictly speaking, these well-known "conversion factors" are not true. They are comparing a weight (pounds) to a mass (kilograms). The correct statement is: "A mass of 1 kg has a weight on *earth* of 2.2 pounds." On another planet, the weight of a 1 kg mass would be something other than 2.2 pounds.

EXAMPLE 5.8 Masses of people

What is the mass, in kilograms, of a 90 pound gymnast, a 160 pound professor, and a 240 pound football player?

SOLVE We must convert their weights into newtons; then we can find their masses from $m = w/g$:

$$w_{\text{gymnast}} = 90\text{ lb} \times \frac{4.45\text{ N}}{1\text{ lb}} = 400\text{ N} \qquad m_{\text{gymnast}} = \frac{w_{\text{gymnast}}}{g} = \frac{400\text{ N}}{9.80\text{ m/s}^2} = 41\text{ kg}$$

$$w_{\text{prof}} = 160\text{ lb} \times \frac{4.45\text{ N}}{1\text{ lb}} = 710\text{ N} \qquad m_{\text{prof}} = \frac{w_{\text{prof}}}{g} = \frac{710\text{ N}}{9.80\text{ m/s}^2} = 72\text{ kg}$$

$$w_{\text{football}} = 240\text{ lb} \times \frac{4.45\text{ N}}{1\text{ lb}} = 1070\text{ N} \qquad m_{\text{football}} = \frac{w_{\text{football}}}{g} = \frac{1070\text{ N}}{9.80\text{ m/s}^2} = 110\text{ kg}$$

ASSESS It's worth remembering that a *typical* adult has a mass in the range of 60 to 80 kg.

This popular amusement park ride shoots you straight up with an acceleration of $4g$. As a result, you feel five times as heavy as usual.

Apparent Weight

The weight of an object is the force of gravity on that object. You may never have thought about it, but gravity is not a force that you can feel or sense directly. Your *sensation* of weight—how heavy you *feel*—is due to *contact forces* pressing against you. Surfaces touch you and activate nerve endings in your skin. As you read this, your sensation of weight is due to the normal force exerted on you by the chair in which you are sitting. When you stand, you feel the contact force of the floor pushing against your feet.

When you stand on a scale, the contact force is the upward spring force F_{sp} acting on your feet. If you and the scale are in equilibrium, with $\vec{a} = \vec{0}$, this spring force and thus the scale reading are equal to your weight.

What would happen, however, if you stood on a scale while *accelerating* upward in an elevator? What would the scale read then? Recall the sensations you feel while being accelerated. You feel "heavy" when an elevator suddenly accelerates upward, and you feel lighter than normal as the upward-moving elevator brakes to a halt. Your true weight $w = mg$ has not changed during these events, but your *sensation* of your weight has.

To investigate this, imagine a man weighing himself by standing on a spring scale in an elevator as it accelerates upward. What does the scale read? As FIGURE 5.12 shows, the only forces acting on the man are the upward spring force of the scale and the downward weight force. Because the man now has an acceleration $\vec{a}$, according to Newton's second law there must be a net force acting on the man in the direction of $\vec{a}$.

FIGURE 5.12 A man weighing himself in an accelerating elevator.

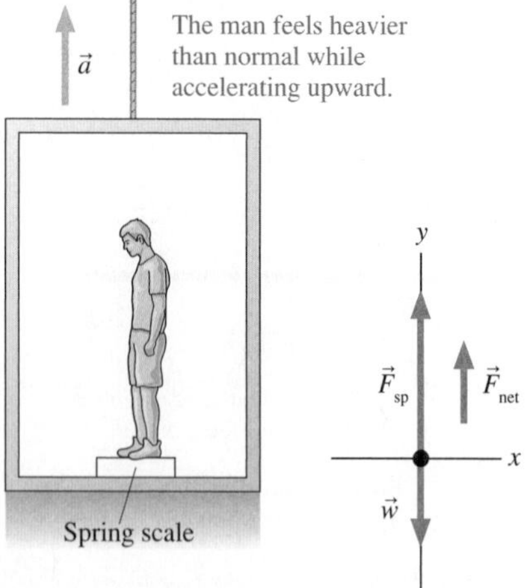

Looking at the free-body diagram in Figure 5.12, we see that the y-component of Newton's second law is

$$\sum F_y = (F_{sp})_y + w_y = F_{sp} - w = ma_y \qquad (5.7)$$

where m is the man's mass. Solving Equation 5.7 for F_{sp} gives

$$F_{sp} = w + ma_y \qquad (5.8)$$

If the elevator is either at rest or moving with constant velocity, then $a_y = 0$ and the man is in equilibrium. In that case, $F_{sp} = w$ and the scale correctly reads his weight. But if $a_y \neq 0$, the scale's reading is *not* the man's true weight. If the elevator is accelerating upward, then $a_y = +a$, and Equation 5.8 reads $F_{sp} = w + a$. Thus $F_{sp} > w$ and the man *feels* heavier than normal. If the elevator is accelerating downward, the acceleration vector $\vec{a}$ points downward and $a_y = -a$. Thus $F_{sp} < w$ and the man feels lighter.

Let's define a person's **apparent weight** w_{app} as the magnitude of the contact force that supports him; in this case, this is the spring force F_{sp}. From Equation 5.8, the apparent weight is

$$w_{app} = w + ma_y = mg + ma_y = m(g + a_y) \qquad (5.9)$$

Thus, when the man accelerates upward, his apparent weight is greater than his true weight; when accelerating downward, his apparent weight is less than his true weight.

An object doesn't have to be on a scale for its apparent weight to differ from its true weight. An object's apparent weight is the magnitude of the contact force supporting it. It makes no difference whether this is the spring force of the scale or simply the normal force of the floor.

The idea of apparent weight has important applications. Astronauts are nearly crushed by their apparent weight during a rocket launch when a is much greater than g. Much of the thrill of amusement park rides, such as roller coasters, comes from rapid changes in your apparent weight.

TRY IT YOURSELF

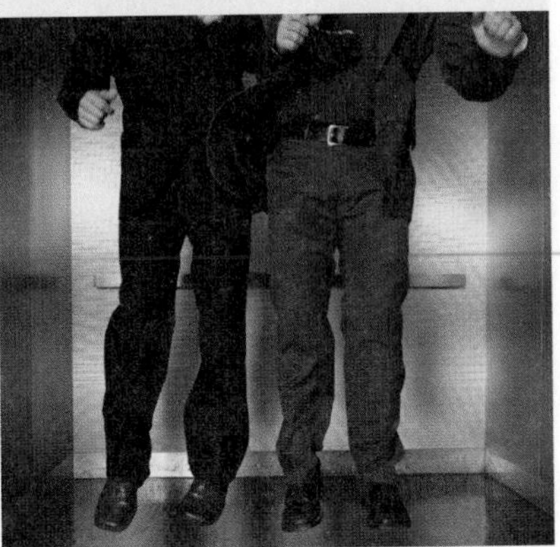

Physics students can't jump The next time you ride up in an elevator, try jumping in the air just as the elevator starts to rise. You'll feel like you can hardly get off the ground. This is because with $a_y > 0$ your apparent weight is *greater* than your actual weight; for an elevator with a large acceleration it's like trying to jump while carrying an extra 20 pounds. What will happen if you jump as the elevator slows at the top?

EXAMPLE 5.9 Apparent weight in an elevator

Anjay's mass is 70 kg. He's standing on a scale in an elevator. As the elevator stops, the scale reads 750 N. Had the elevator been moving up or down? If the elevator had been moving at 5.0 m/s, how long does it take to stop?

PREPARE The scale reading as he stops is his apparent weight, so $w_{app} = 750$ N. Because we know his mass m, we can use Equation 5.9 to find the elevator's acceleration a_y. Then we can use kinematics to find the time it takes to stop the elevator.

SOLVE From Equation 5.9 we have $w_{app} = m(g + a_y)$, so that

$$a_y = \frac{w_{app}}{m} - g = \frac{750 \text{ N}}{70 \text{ kg}} - 9.80 \text{ m/s}^2 = 0.91 \text{ m/s}^2$$

This is a *positive* acceleration. If the elevator is stopping with a positive acceleration, it must have been moving *down,* with a negative velocity.

To find the stopping time, we can use the kinematic equation

$$(v_y)_f = (v_y)_i + a_y \Delta t$$

to get

$$\Delta t = \frac{(v_y)_f - (v_y)_i}{a_y} = \frac{(0 \text{ m/s}) - (-5.0 \text{ m/s})}{0.91 \text{ m/s}^2} = 5.5 \text{ s}$$

Notice that we used -5.0 m/s as the initial velocity because the elevator was moving down before it stopped.

ASSESS Anjay's true weight is $mg = (70 \text{ kg})(9.8 \text{ m/s}^2) = 670$ N. Thus his apparent weight is *greater* than his true weight. You have no doubt experienced this sensation in an elevator that is stopping as it reaches the ground floor. If it had stopped while going up, you'd feel *lighter* than your true weight.

Weightlessness

One last issue before leaving this topic: Suppose the elevator cable breaks and the elevator, along with the man and his scale, plunges straight down in free fall! What will the scale read? The acceleration in free fall is $a_y = -g$. When this acceleration is used in Equation 5.9, we find that $w_{app} = 0$! In other words, the man has *no sensation* of weight.

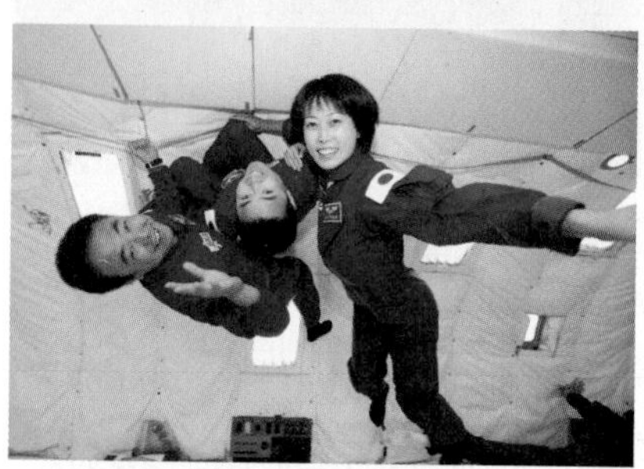

A weightless experience You probably wouldn't want to experience weightlessness in a falling elevator. But, as we learned in Chapter 3, objects undergoing projectile motion are in free fall as well. The special plane shown flies in the same parabolic trajectory as would a projectile with no air resistance. Objects inside, such as these passengers, are then moving along a perfect free-fall trajectory. Just as for the man in the elevator, they then float with respect to the plane's interior. Such flights can last up to 30 seconds.

Think about this carefully. Suppose, as the elevator falls, the man inside releases a ball from his hand. In the absence of air resistance, as Galileo discovered, both the man and the ball would fall at the same rate. From the man's perspective, the ball would appear to "float" beside him. Similarly, the scale would float beneath him and not press against his feet. He is what we call *weightless*.

Surprisingly, "weightless" does *not* mean "no weight." An object that is **weightless** has no *apparent* weight. The distinction is significant. The man's weight is still mg because gravity is still pulling down on him, but he has no *sensation* of weight as he free falls. The term "weightless" is a very poor one, likely to cause confusion because it implies that objects have no weight. As we see, that is not the case.

But isn't this exactly what happens to astronauts orbiting the earth? You've seen films of astronauts and various objects floating inside the Space Shuttle. If an astronaut tries to stand on a scale, it does not exert any force against her feet and reads zero. She is said to be weightless. But if the criterion to be weightless is to be in free fall, and if astronauts orbiting the earth are weightless, does this mean that they are in free fall? This is a very interesting question to which we shall return in Chapter 6.

STOP TO THINK 5.2 You're bouncing up and down on a trampoline. At the very highest point of your motion, your apparent weight is

A. More than your true weight.
B. Less than your true weight.
C. Equal to your true weight.
D. Zero.

5.4 Normal Forces

In Chapter 4 we saw that an object at rest on a table is subject to an upward force due to the table. This force is called the *normal force* because it is always directed normal, or perpendicular, to the surface of contact. As we saw, the normal force has its origin in the atomic "springs" that make up the surface. The harder the object bears down on the surface, the more these springs are compressed and the harder they push back. Thus the normal force *adjusts* itself so that the object stays on the surface without penetrating it. This fact is key in solving for the normal force.

EXAMPLE 5.10 Normal force on a pressed book

A 1.2 kg book lies on a table. The book is pressed down from above with a force of 15 N. What is the normal force acting on the book from the table below?

PREPARE The book is not moving and is thus in static equilibrium. We need to identify the forces acting on the book, and prepare a free-body diagram showing these forces. These steps are illustrated in **FIGURE 5.13**.

FIGURE 5.13 Finding the normal force on a book pressed from above.

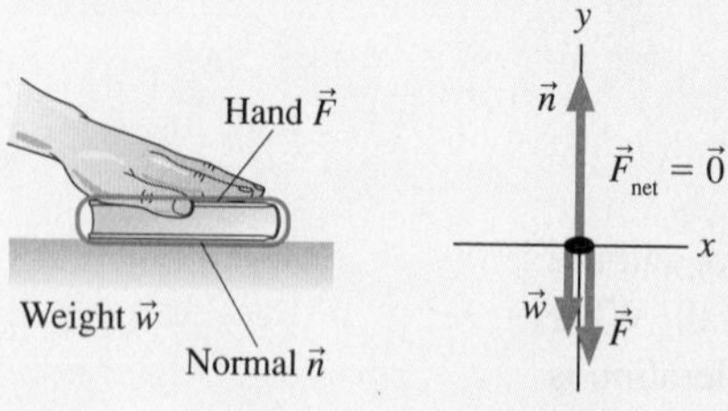

SOLVE Because the book is in static equilibrium, the net force on it must be zero. The only forces acting are in the y-direction, so Newton's second law is

$$\sum F_y = n_y + w_y + F_y = n - w - F = ma_y = 0$$

We learned in the last section that the weight force is $w = mg$. The weight of the book is thus

$$w = mg = (1.2\text{ kg})(9.8\text{ m/s}^2) = 12\text{ N}$$

With this information, we see that the normal force exerted by the table is

$$n = F + w = 15\text{ N} + 12\text{ N} = 27\text{ N}$$

ASSESS The magnitude of the normal force is *larger* than the weight of the book. From the table's perspective, the extra force from the hand pushes the book further into the atomic springs of the table. These springs then push back harder, giving a normal force that is greater than the weight of the book.

A common situation is an object on a ramp or incline. If friction is neglected, there are only two forces acting on the object: gravity and the normal force. However, we need to carefully work out the components of these two forces in order to solve dynamics problems. **FIGURE 5.14a** shows how. Be sure you avoid the two common errors shown in **FIGURE 5.14b**.

FIGURE 5.14 The forces on an object on an incline.

(a) Analyzing forces on an incline

The normal force always points perpendicular to the surface.

When we rotate the x-axis to match the surface, the angle between $\vec{w}$ and the negative y-axis is the same as the angle θ of the slope.

The weight force always points straight down.

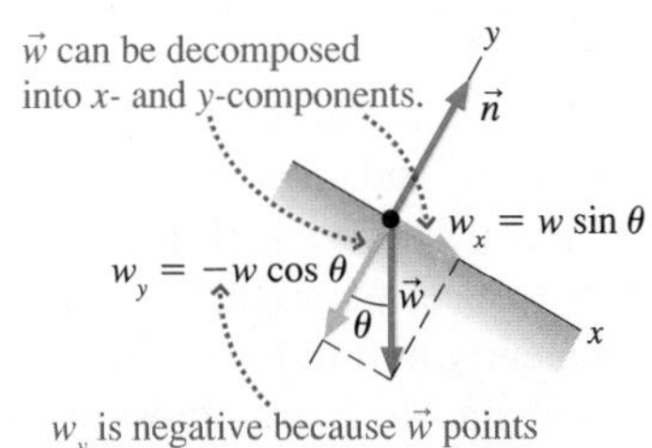

w can be decomposed into x- and y-components.
y
n
wx = w sin θ
wy = −w cos θ
w
θ
x
wy is negative because w points in the negative y-direction.

(b) Two common mistakes to avoid

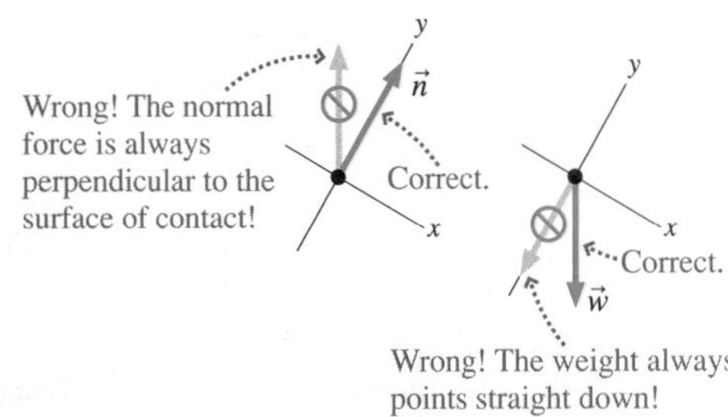

Wrong! The normal force is always perpendicular to the surface of contact!
y
n
Correct.
x
y
x
Correct.
w
Wrong! The weight always points straight down!

EXAMPLE 5.11 Acceleration of a downhill skier

A skier slides down a steep slope of 27° on ideal, frictionless snow. What is his acceleration?

PREPARE **FIGURE 5.15** is a visual overview. We choose a coordinate system tilted so that the x-axis points down the slope. This greatly simplifies the analysis, because with this choice $a_y = 0$ (the skier does not move in the y-direction at all). The free-body diagram is based on the information in Figure 5.14.

FIGURE 5.15 Visual overview of a downhill skier.

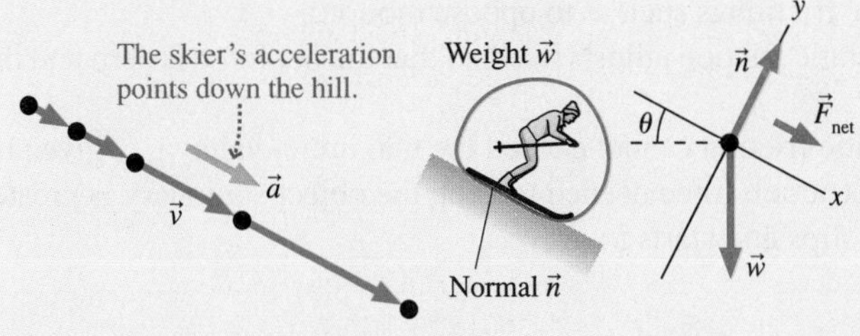

The skier's acceleration points down the hill.
a
v
Weight w
Normal n
y
n
θ
F net
x
w

SOLVE We can now use Newton's second law in component form to find the skier's acceleration:

$$\sum F_x = w_x + n_x = ma_x$$
$$\sum F_y = w_y + n_y = ma_y$$

Because $\vec{n}$ points directly in the positive y-direction, $n_y = n$ and $n_x = 0$. Figure 5.14a showed the important fact that the angle between $\vec{w}$ and the negative y-axis is the *same* as the slope angle θ. With this information, the components of $\vec{w}$ are $w_x = w\sin\theta = mg\sin\theta$ and $w_y = -w\cos\theta = -mg\cos\theta$, where we used the fact that $w = mg$. With these components in hand, Newton's second law becomes

$$\sum F_x = w_x + n_x = mg\sin\theta = ma_x$$
$$\sum F_y = w_y + n_y = -mg\cos\theta + n = ma_y = 0$$

In the second equation we used the fact that $a_y = 0$. The m cancels in the first of these equations, leaving us with

$$a_x = g\sin\theta$$

This is the expression for acceleration on a frictionless surface that we presented, without proof, in Chapter 3. Now we've justified our earlier assertion. We can use this to calculate the skier's acceleration:

$$a_x = g\sin\theta = (9.8\ \text{m/s}^2)\sin(27°) = 4.4\ \text{m/s}^2$$

ASSESS Our result shows that when $\theta = 0$, so that the slope is horizontal, the skier's acceleration is zero, as it should be. Further, when $\theta = 90°$ (a vertical slope), his acceleration is g, which makes sense because he's in free fall when $\theta = 90°$. Notice that the mass canceled out, so we didn't need to know the skier's mass.

5.5 Friction

In everyday life, friction is everywhere. Friction is absolutely essential for many things we do. Without friction you could not walk, drive, or even sit down (you would slide right off the chair!). It is sometimes useful to think about idealized frictionless situations, but it is equally necessary to understand a real world where friction is present. Although friction is a complicated force, many aspects of friction can be described with a simple model.

FIGURE 5.16 Static friction keeps an object from slipping.

(a) Force identification

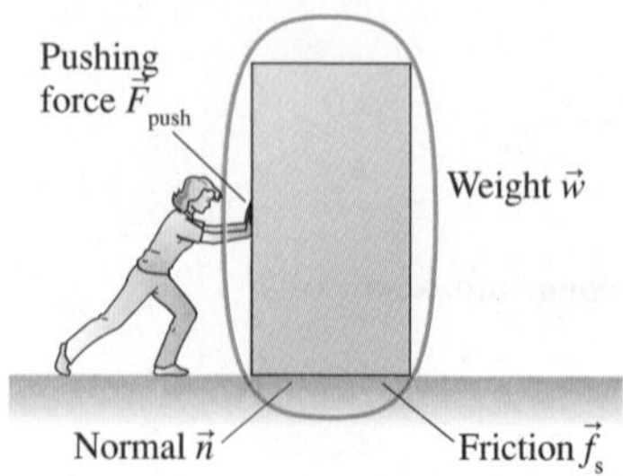

(b) Free-body diagram

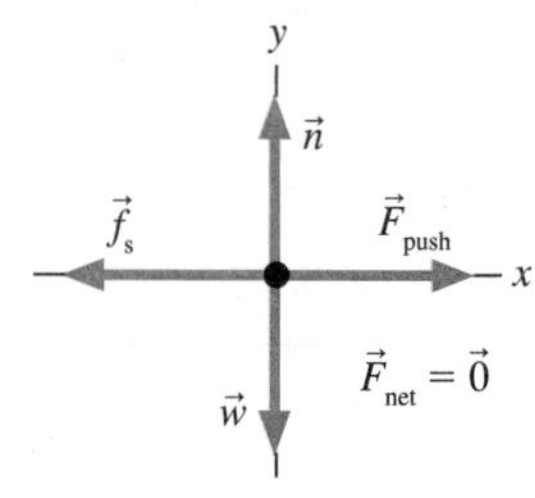

FIGURE 5.17 Static friction acts in *response* to an applied force.

(a) Pushing gently: friction pushes back gently.

$\vec{f}_s$ balances $\vec{F}_{push}$ and the box does not move.

(b) Pushing harder: friction pushes back harder.

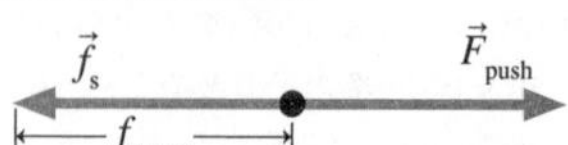

$\vec{f}_s$ grows as $\vec{F}_{push}$ increases, but the two still cancel and the box remains at rest.

(c) Pushing harder still: $\vec{f}_s$ is now pushing back as hard as it can.

$\vec{f}_s$ $\vec{F}_{push}$

$f_{s\,max}$

Now the magnitude of f_s has reached its maximum value $f_{s\,max}$. If $\vec{F}_{push}$ gets any bigger, the forces will *not* cancel and the box will start to accelerate.

TABLE 5.1 Coefficients of friction

Materials	Static μ_s	Kinetic μ_k	Rolling μ_r
Rubber on concrete	1.00	0.80	0.02
Steel on steel (dry)	0.80	0.60	0.002
Steel on steel (lubricated)	0.10	0.05	
Wood on wood	0.50	0.20	
Wood on snow	0.12	0.06	
Ice on ice	0.10	0.03	

Static Friction

Chapter 4 defined static friction $\vec{f}_s$ as the force that a surface exerts on an object to keep it from slipping across that surface. Consider the woman pushing on the box in FIGURE 5.16a. Because the box is not moving with respect to the floor, the woman's push to the right must be balanced by a static friction force $\vec{f}_s$ pointing to the left. This is the general rule for finding the *direction* of $\vec{f}_s$: Decide which way the object *would* move if there were no friction. The static friction force $\vec{f}_s$ then points in the opposite direction, to prevent motion relative to the surface.

Determining the *magnitude* of $\vec{f}_s$ is a bit trickier. Because the box is at rest, it's in static equilibrium. From the free-body diagram of FIGURE 5.16b, this means that the static friction force must exactly balance the pushing force, so that $f_s = F_{push}$. As shown in FIGURES 5.17a and 5.17b, the harder the woman pushes, the harder the friction force from the floor pushes back. If she reduces her pushing force, the friction force will automatically be reduced to match. Static friction acts in *response* to an applied force.

But there's clearly a limit to how big $\vec{f}_s$ can get. If she pushes hard enough, the box will slip and start to move across the floor. In other words, the static friction force has a *maximum* possible magnitude $f_{s\,max}$, as illustrated in FIGURE 5.17c. Experiments with friction (first done by Leonardo da Vinci) show that $f_{s\,max}$ is proportional to the magnitude of the normal force between the surface and the object; that is,

$$f_{s\,max} = \mu_s n \tag{5.10}$$

where μ_s is called the **coefficient of static friction.** The coefficient is a number that depends on the materials from which the object and the surface are made. The higher the coefficient of static friction, the greater the "stickiness" between the object and the surface, and the harder it is to make the object slip. Table 5.1 lists some approximate values of coefficients of friction.

NOTE ► Equation 5.10 does *not* say $f_s = \mu_s n$. The value of f_s depends on the force or forces that static friction has to balance to keep the object from moving. It can have any value from zero up to, but not exceeding, $\mu_s n$. ◄

So our rules for static friction are:

- The direction of static friction is such as to oppose motion.
- The magnitude f_s of static friction adjusts itself so that the net force is zero and the object doesn't move.
- The magnitude of static friction cannot exceed the maximum value $f_{s\,max}$ given by Equation 5.10. If the friction force needed to keep the object stationary is greater than $f_{s\,max}$, the object slips and starts to move.

Kinetic Friction

Once the box starts to slide, as in FIGURE 5.18, the static friction force is replaced by a kinetic (or sliding) friction force $\vec{f}_k$. Kinetic friction is in some ways simpler than static friction: The direction of $\vec{f}_k$ is always opposite the direction in which an object

FIGURE 5.18 The kinetic friction force is *opposite* the direction of motion.

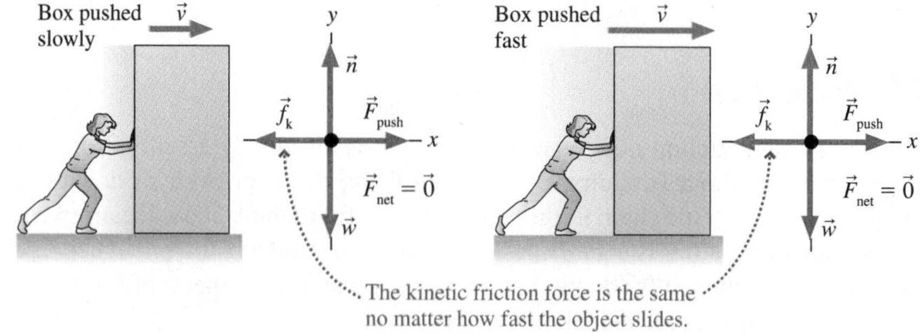

slides across the surface, and experiments show that kinetic friction, unlike static friction, has a nearly *constant* magnitude, given by

$$f_k = \mu_k n \tag{5.11}$$

where μ_k is called the **coefficient of kinetic friction.** Equation 5.11 also shows that kinetic friction, like static friction, is proportional to the magnitude of the normal force n. Notice that **the magnitude of the kinetic friction force does not depend on how fast the object is sliding.**

Table 5.1 includes approximate values of μ_k. You can see that $\mu_k < \mu_s$, which explains why it is easier to keep a box moving than it was to start it moving.

FIGURE 5.19 The bottom of the wheel is stationary.

Rolling Friction

If you slam on the brakes hard enough, your car tires slide against the road surface and leave skid marks. This is kinetic friction because the tire and the road are *sliding* against each other. A wheel *rolling* on a surface also experiences friction, but not kinetic friction: The portion of the wheel that contacts the surface is stationary with respect to the surface, not sliding. The photo in FIGURE 5.19 was taken with a stationary camera. Note how the part of the wheel touching the ground is not blurred, indicating that this part of the wheel is not moving with respect to the ground.

Textbooks draw wheels as circles, but no wheel is perfectly round. The weight of the wheel, and of any object supported by the wheel, causes the bottom of the wheel to flatten where it touches the surface, as FIGURE 5.20 shows. As a wheel rolls forward, the leading part of the tire must become deformed. This requires that the road push *backward* on the tire. In this way the road causes a backward force, even without slipping between the tire and the road.

The force of this *rolling friction* can be calculated in terms of a **coefficient of rolling friction** μ_r:

$$f_r = \mu_r n \tag{5.12}$$

with the *direction* of the force opposing the direction of motion. Thus rolling friction acts very much like kinetic friction, but values of μ_r (see Table 5.1) are much lower than values of μ_k. This is why it is easier to roll an object on wheels than to slide it.

FIGURE 5.20 Rolling friction is due to deformation of a wheel.

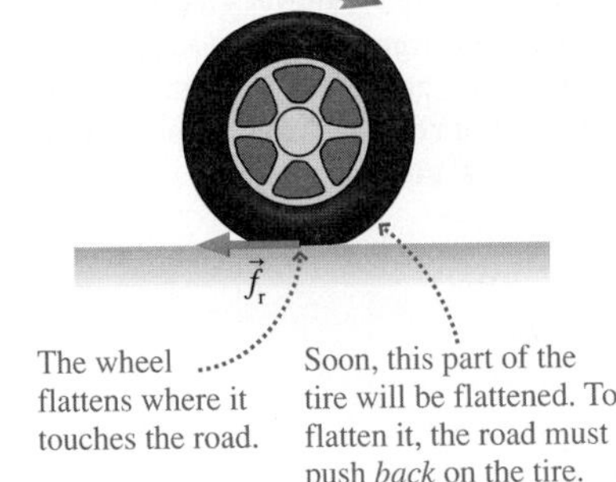

STOP TO THINK 5.3 Rank in order, from largest to smallest, the size of the friction forces $\vec{f}_A$ to $\vec{f}_E$ in the five different situations (one or more friction forces could be zero). The box and the floor are made of the same materials in all situations.

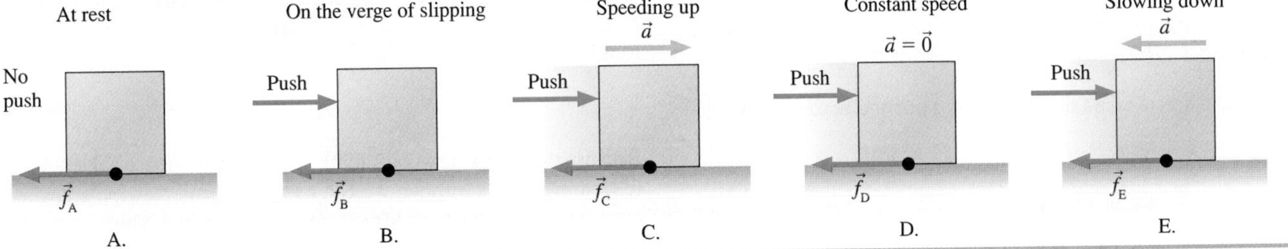

Working with Friction Forces

2.5, 2.6

These ideas can be summarized in a *model* of friction:

$$\begin{aligned}
&\text{Static: } \vec{f}_s = (\text{magnitude} \le f_{s\,max} = \mu_s n, \\
&\qquad \text{direction as necessary to prevent motion}) \\
&\text{Kinetic: } \vec{f}_k = (\mu_k n, \text{direction opposite the motion}) \\
&\text{Rolling: } \vec{f}_r = (\mu_r n, \text{direction opposite the motion})
\end{aligned} \tag{5.13}$$

f, n, p.37, PROPORTIONAL

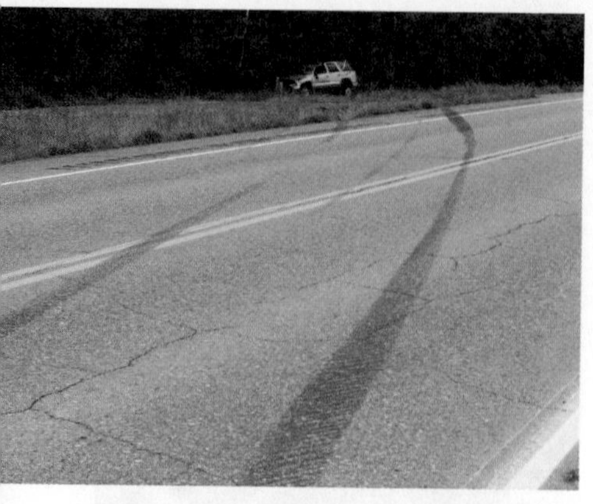

Optimized braking If you slam on your brakes, your wheels will lock up and you'll go into a skid. Then it is the *kinetic* friction force between the road and your tires that slows your car to a halt. If, however, you apply the brakes such that you don't quite skid and your tires continue to roll, the force stopping you is the *static* friction force between the road and your tires. This is a better way to brake, because the maximum static friction force is always greater than the kinetic friction force. *Antilock braking systems* (ABS) automatically do this for you when you slam on the brakes, stopping you in the shortest possible distance.

Here "motion" means "motion relative to the surface." The maximum value of static friction $f_{s\,max} = \mu_s n$ occurs at the point where the object slips and begins to move. Note that only one kind of friction force at a time can act on an object.

NOTE ► Equations 5.13 are a "model" of friction, not a "law" of friction. These equations provide a reasonably accurate, but not perfect, description of how friction forces act. They are a simplification of reality that works reasonably well, which is what we mean by a "model." They are not a "law of nature" on a level with Newton's laws. ◄

TACTICS BOX 5.1 Working with friction forces

1. If the object is *not moving* relative to the surface it's in contact with, then the friction force is **static friction.** Draw a free-body diagram of the object. The *direction* of the friction force is such as to oppose sliding of the object relative to the surface. Then use Problem-Solving Strategy 5.1 to solve for f_s. If f_s is greater than $f_{s\,max} = \mu_s n$, then static friction cannot hold the object in place. The assumption that the object is at rest is not valid, and you need to redo the problem using kinetic friction.
2. If the object is *sliding* relative to the surface, then **kinetic friction** is acting. From Newton's second law, find the normal force n. Equation 5.13 then gives the magnitude and direction of the friction force.
3. If the object is *rolling* along the surface, then **rolling friction** is acting. From Newton's second law, find the normal force n. Equation 5.13 then gives the magnitude and direction of the friction force.

Exercises 20, 21

EXAMPLE 5.12 Finding the force to push a box

Carol pushes a 10.0 kg wood box across a wood floor at a steady speed of 2.0 m/s. How much force does Carol exert on the box?

PREPARE Let's assume the box slides to the right. In this case, a kinetic friction force $\vec{f}_k$, opposes the motion by pointing to the left. In FIGURE 5.21 we identify the forces acting on the box and construct a free-body diagram.

FIGURE 5.21 Forces on a box being pushed across a floor.

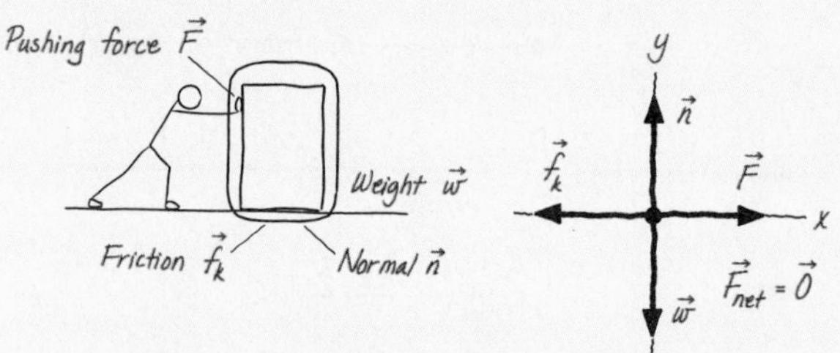

SOLVE The box is moving at a constant speed, so it is in dynamic equilibrium with $\vec{F}_{net} = \vec{0}$. This means that the x- and y-components of the net force must be zero:

$$\sum F_x = n_x + w_x + F_x + (f_k)_x = 0 + 0 + F - f_k = 0$$

$$\sum F_y = n_y + w_y + F_y + (f_k)_y = n - w + 0 + 0 = 0$$

In the first equation, the x-component of $\vec{f}_k$ is equal to $-f_k$ because $\vec{f}_k$ is directed to the left. Similarly, $w_y = -w$ because the weight force points down.

From the first equation, we see that Carol's pushing force is $F = f_k$. To evaluate this, we need f_k. Here we can use our model for kinetic friction:

$$f_k = \mu_k n$$

Because the friction is wood sliding on wood, we can use Table 5.1 to find $\mu_k = 0.20$. Further, we can use the second Newton's-law equation to find that the normal force is $n = w = mg$. Thus

$$F = f_k = \mu_k n = \mu_k mg$$

$$= (0.20)(10.0\ \text{kg})(9.80\ \text{m/s}^2) = 20\ \text{N}$$

This is the force that Carol needs to apply to the box to keep it moving at a steady speed.

ASSESS The speed of 2.0 m/s with which Carol pushes the box does not enter into the answer. This is because our model of kinetic friction does not depend on the speed of the sliding object.

CONCEPTUAL EXAMPLE 5.13 To push or pull a lawn roller?

A lawn roller is a heavy cylinder used to flatten a bumpy lawn, as shown in FIGURE 5.22. Is it easier to push or pull such a roller? Which is more effective for flattening the lawn: pushing or pulling? Assume that the pushing or pulling force is directed along the handle of the roller.

FIGURE 5.22 Pushing and pulling a lawn roller.

REASON FIGURE 5.23 shows free-body diagrams for the two cases. We assume that the roller is pushed at a constant speed so that it is in dynamic equilibrium with $\vec{F}_{net} = \vec{0}$. Because the roller does not move in the y-direction, the y-component of the net force must be zero. According to our model, the magnitude f_r of rolling friction is proportional to the magnitude n of the normal force. If we *push* on the roller, our pushing force $\vec{F}$ will have a downward y-component. To compensate for this, the normal force must increase and, because $f_r = \mu_r n$, the rolling friction will increase as well. This makes the roller harder to move. If we *pull* on the roller, the now upward y-component of $\vec{F}$ will lead to a *reduced* value of n and hence of f_r. Thus the roller is easier to pull than to push.

FIGURE 5.23 Free-body diagrams for the lawn roller.

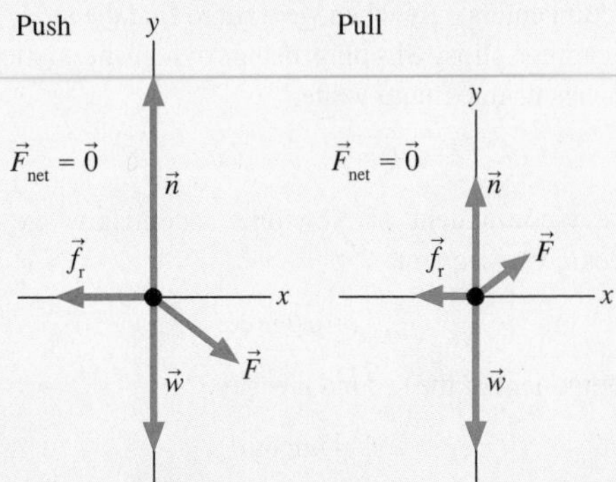

However, the purpose of the roller is to flatten the soil. If the normal force $\vec{n}$ of the ground on the roller is greater, then by Newton's third law the force of the roller on the ground will be greater as well. So for smoothing your lawn, it's better to push.

ASSESS You've probably experienced this effect while using an upright vacuum cleaner. The vacuum is harder to push on the forward stroke than when drawing it back.

EXAMPLE 5.14 How to dump a file cabinet

A 50.0 kg steel file cabinet is in the back of a dump truck. The truck's bed, also made of steel, is slowly tilted. What is the magnitude of the static friction force on the cabinet when the bed is tilted 20°? At what angle will the file cabinet begin to slide?

PREPARE We'll use our model of static friction. The file cabinet will slip when the static friction force reaches its maximum possible value $f_{s\,max}$. FIGURE 5.24 shows the visual overview when the truck bed is tilted at angle θ. We can make the analysis easier if we tilt the coordinate system to match the bed of the truck. To prevent the file cabinet from slipping, the static friction force must point *up* the slope.

SOLVE Before it slips, the file cabinet is in static equilibrium. Newton's second law gives

$$\sum F_x = n_x + w_x + (f_s)_x = 0$$

$$\sum F_y = n_y + w_y + (f_s)_y = 0$$

From the free-body diagram we see that f_s has only a negative x-component and that n has only a positive y-component. We also have $w_x = w\sin\theta$ and $w_y = -w\cos\theta$. Thus the second law becomes

$$\sum F_x = w\sin\theta - f_s = mg\sin\theta - f_s = 0$$

$$\sum F_y = n - w\cos\theta = n - mg\cos\theta = 0$$

FIGURE 5.24 Visual overview of a file cabinet in a tilted dump truck.

Known
$\mu_s = 0.80$ $m = 50.0$ kg
$\mu_k = 0.60$

Find
f_s when $\theta = 20°$
θ at which cabinet slips

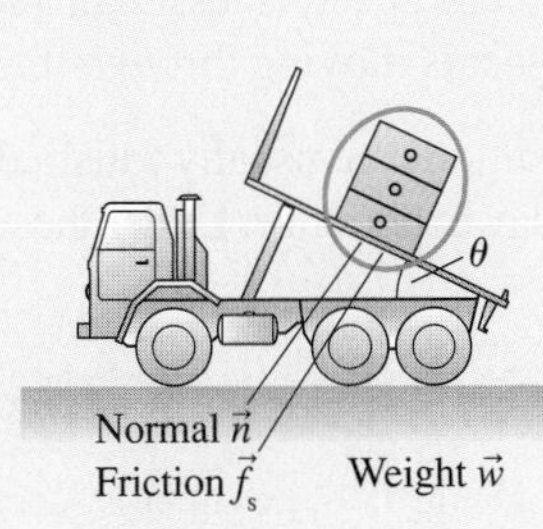

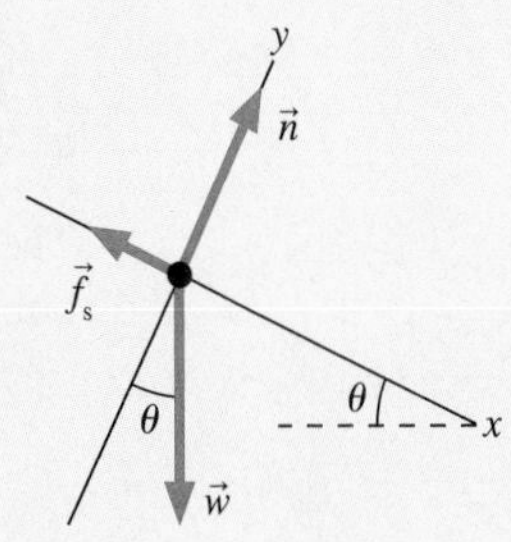

Continued

The x-component equation allows us to determine the magnitude of the static friction force when $\theta = 20°$:

$$f_s = mg\sin\theta = (50.0\text{ kg})(9.80\text{ m/s}^2)\sin 20° = 168\text{ N}$$

This value does not require that we know μ_s. The coefficient of static friction enters only when we want to find the angle at which the file cabinet slips. Slipping occurs when the static friction force reaches its maximum value:

$$f_s = f_{s\max} = \mu_s n$$

From the y-component of Newton's second law we see that $n = mg\cos\theta$. Consequently,

$$f_{s\max} = \mu_s mg\cos\theta$$

The x-component of the second law gave

$$f_s = mg\sin\theta$$

Setting $f_s = f_{s\max}$ then gives

$$mg\sin\theta = \mu_s mg\cos\theta$$

The mg in both terms cancels, and we find

$$\frac{\sin\theta}{\cos\theta} = \tan\theta = \mu_s$$

$$\theta = \tan^{-1}\mu_s = \tan^{-1}(0.80) = 39°$$

ASSESS Steel doesn't slide all that well on unlubricated steel, so a fairly large angle is not surprising. The answer seems reasonable. It is worth noting that $n = mg\cos\theta$ in this example. A common error is to use simply $n = mg$. Be sure to evaluate the normal force within the context of each particular problem.

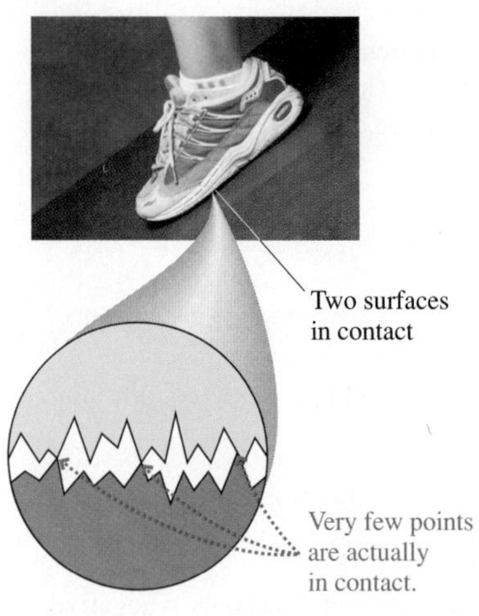
Two surfaces
in contact
Very few points
are actually
in contact.

FIGURE 5.25 A microscopic view of friction.

Causes of Friction

It is worth a brief pause to look at the *causes* of friction. All surfaces, even those quite smooth to the touch, are very rough on a microscopic scale. When two objects are placed in contact, they do not make a smooth fit. Instead, as FIGURE 5.25 shows, the high points on one surface become jammed against the high points on the other surface, while the low points are not in contact at all. Only a very small fraction (typically 10^{-4}) of the surface area is in actual contact. The amount of contact depends on how hard the surfaces are pushed together, which is why friction forces are proportional to n.

For an object to slip, you must push it hard enough to overcome the forces exerted at these contact points. Once the two surfaces are sliding against each other, their high points undergo constant collisions, deformations, and even brief bonding that lead to the resistive force of kinetic friction.

5.6 Drag

The air exerts a drag force on objects as they move through it. You experience drag forces every day as you jog, bicycle, ski, or drive your car. The drag force $\vec{D}$:

- Is opposite in direction to the velocity $\vec{v}$.
- Increases in magnitude as the object's speed increases.

At relatively low speeds, the drag force in air is small and can usually be neglected, but drag plays an important role as speeds increase. Fortunately, we can use a fairly simple *model* of drag if the following three conditions are met:

- The object's size (diameter) is between a few millimeters and a few meters.
- The object's speed is less than a few hundred meters per second.
- The object is moving through the air near the earth's surface.

These conditions are usually satisfied for balls, people, cars, and many other objects in our everyday experience. Under these conditions, the drag force can be written:

$$\vec{D} = \left(\tfrac{1}{2}C_D\rho A v^2,\text{ direction opposite the motion}\right) \tag{5.14}$$

Drag force on an object of cross-section area A moving at speed v

D v p.47 QUADRATIC

Here, ρ is the density of air ($\rho = 1.22\ \text{kg/m}^3$ at sea level), A is the cross-section area of the object (in m^2), and the **drag coefficient** C_D depends on the details of the object's shape. However, the value of C_D for everyday moving objects is roughly 1/2, so a good approximation to the drag force is

$$D \approx \tfrac{1}{4}\rho A v^2 \tag{5.15}$$

This is the expression for the magnitude of the drag force that we'll use in this chapter.

The size of the drag force in air is proportional to the *square* of the object's speed: If the speed doubles, the drag increases by a factor of 4. This model of drag fails for objects that are very small (such as dust particles) or very fast (such as jet planes) or that move in other media (such as water).

FIGURE 5.26 shows that the area A in Equation 5.14 is the cross section of the object as it "faces into the wind." It's interesting to note that the magnitude of the drag force depends on the object's *size and shape* but not on its *mass.* This has important consequences for the motion of falling objects.

FIGURE 5.26 How to calculate the cross-section area A.

The cross-section area of a sphere is a circle. For this soccer ball, $A = \pi r^2$.

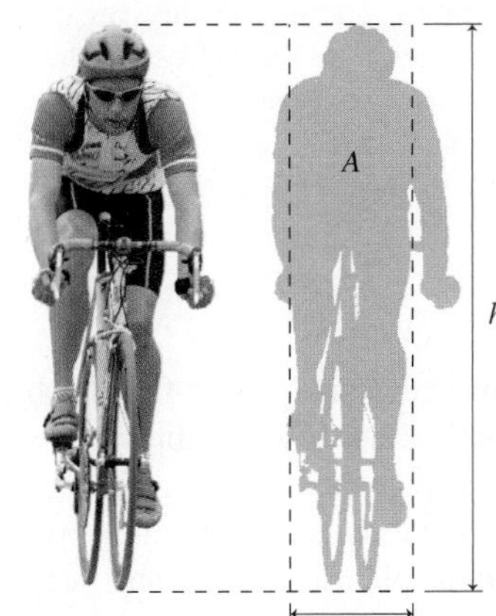

A is the cross-section area of the cyclist as seen from the front. This area is approximated by the rectangle shown, with area $A = h \times w$.

Terminal Speed

Just after an object is released from rest, its speed is low and the drag force is small (as shown in FIGURE 5.27a). Because the net force is nearly equal to the weight, the object will fall with an acceleration only a little less than g. As it falls farther, its speed and hence the drag force increase. Now the net force is smaller, so the acceleration is smaller (as shown in FIGURE 5.27b). It's still speeding up, but at a lower *rate.* Eventually the speed will increase to a point such that the magnitude of the drag force *equals* the weight (as shown in FIGURE 5.27c). The net force and hence the acceleration at this speed are then *zero,* and the object falls with a *constant* speed. The speed at which the exact balance between the upward drag force and the downward weight force causes an object to fall without acceleration is called the **terminal speed** v_{term}. **Once an object has reached terminal speed, it will continue falling at that speed until it hits the ground.**

It's straightforward to compute the terminal speed. It is the speed, by definition, at which $D = w$ or, equivalently, $\frac{1}{4}\rho A v^2 = mg$. This speed is then

$$v_{\text{term}} \approx \sqrt{\frac{4mg}{\rho A}} \tag{5.16}$$

This equation shows that a more massive object has a greater terminal speed than a less massive object of equal size. A 10-cm-diameter lead ball, with a mass of 6 kg, has a terminal speed of 150 m/s, while a 10-cm-diameter Styrofoam ball, with a mass of 50 g, has a terminal speed of only 14 m/s.

FIGURE 5.27 A falling object eventually reaches terminal speed.

(a) At low speeds, D is small and the ball falls with $a \approx g$.

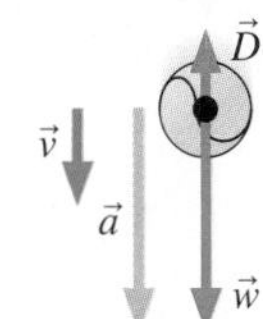

(b) As v increases, so does D. The net force and hence a get smaller.

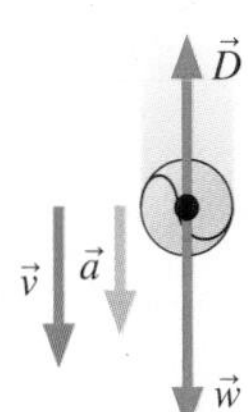

(c) Eventually, v reaches a value such that $D = w$. Then the net force is zero and the ball falls at a constant speed.

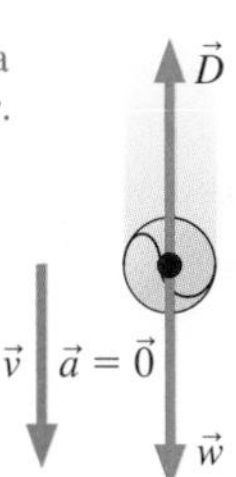

EXAMPLE 5.15 **Terminal speeds of a skydiver and a mouse**

A skydiver and his pet mouse jump from a plane. Estimate their terminal speeds.

PREPARE To use Equation 5.16 we need to estimate the mass m and cross-section area A of both man and mouse. **FIGURE 5.28** shows how. A typical skydiver might be 1.8 m long and 0.4 m wide ($A = 0.72\text{ m}^2$) with a mass of 75 kg, while a mouse has a mass of perhaps 20 g (0.020 kg) and is 7 cm long and 3 cm wide ($A = 0.07\text{ m} \times 0.03\text{ m} = 0.0021\text{ m}^2$).

FIGURE 5.28 The cross-section areas of a skydiver and a mouse.

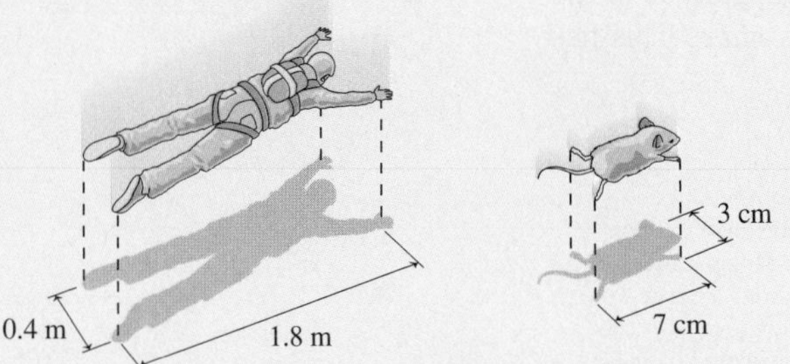

SOLVE We can use Equation 5.16 to find that for the skydiver

$$v_{\text{term}} \approx \sqrt{\frac{4mg}{\rho A}} = \sqrt{\frac{4(75\text{ kg})(9.8\text{ m/s}^2)}{(1.22\text{ kg/m}^3)(0.72\text{ m}^2)}} = 58\text{ m/s}$$

This is roughly 130 mph. A higher speed can be reached by falling feet first or head first, which reduces the area A. Fortunately the skydiver can open his parachute, greatly increasing A. This brings his terminal speed down to a safe value.

For the mouse we have

$$v_{\text{term}} \approx \sqrt{\frac{4mg}{\rho A}} = \sqrt{\frac{4(0.020\text{ kg})(9.8\text{ m/s}^2)}{(1.22\text{ kg/m}^2)(0.0021\text{ m}^2)}} = 17\text{ m/s}$$

The mouse has no parachute—nor does he need one! A mouse's terminal speed is slow enough that he can fall from any height, even out of an airplane, and survive. Cats, too, have relatively low terminal speeds. In a study of cats that fell from high rises, over 90% survived—including one that fell 45 stories!

ASSESS The mouse survives the fall not only because of its lower terminal speed. The smaller an animal's body, the proportionally more robust it is. Further, a small animal's low mass and terminal speed mean that it has a very small *kinetic energy,* an idea we'll study in Chapter 10.

Although we've focused our analysis on falling objects, the same ideas apply to objects moving horizontally. If an object is thrown or shot horizontally, $\vec{D}$ causes the object to slow down. An airplane reaches its maximum speed, which is analogous to the terminal speed, when the drag is equal to and opposite the thrust: $D = F_{\text{thrust}}$. The net force is then zero and the plane cannot go any faster.

We will continue to neglect drag unless a problem specifically calls for drag to be considered.

STOP TO THINK 5.4 The terminal speed of a Styrofoam ball is 15 m/s. Suppose a Styrofoam ball is shot straight down with an initial speed of 30 m/s. Which velocity graph is correct?

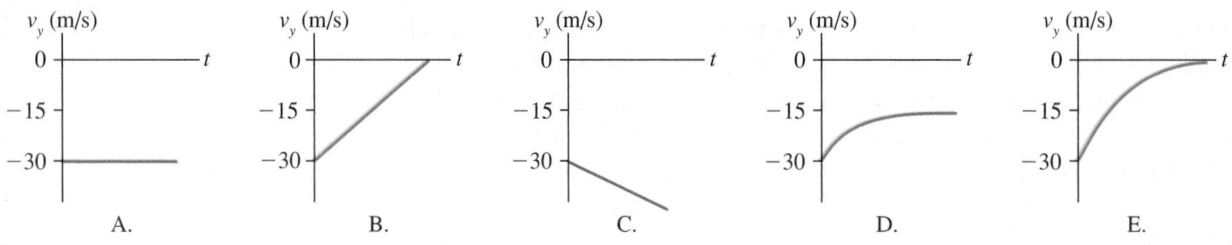

5.7 Interacting Objects

2.7–2.9 ActivPhysics ONLINE

Up to this point we have studied the dynamics of a single object subject to forces exerted on it by other objects. In Example 5.11, for instance, the box was acted upon by friction, normal, weight, and pushing forces that came from the floor, the earth, and the person pushing. As we've seen, such problems can be solved by an application of Newton's second law after all the forces have been identified.

But in Chapter 4 we found that real-world motion often involves two or more objects interacting with each other. We further found that forces always come in action/reaction *pairs* that are related by Newton's third law. To remind you, Newton's third law states:

- Every force occurs as one member of an action/reaction pair of forces. The two members of the pair always act on *different* objects.
- The two members of an action/reaction pair point in *opposite* directions and are *equal* in magnitude.

Our goal in this section is to learn how to apply the second *and* third laws to interacting objects.

Objects in Contact

One common way that two objects interact is via direct contact forces between them. Consider, for example, the two blocks being pushed across a frictionless table in FIGURE 5.29. To analyze block A's motion, we need to identify all the forces acting on it and then draw its free-body diagram. We repeat the same steps to analyze the motion of block B. However, the forces on A and B are *not* independent: Forces $\vec{F}_{\text{B on A}}$ acting on block A and $\vec{F}_{\text{A on B}}$ acting on block B are an action/reaction pair and thus have the same magnitude. Furthermore, because the two blocks are in contact, their *accelerations* must be the same, so that $a_{Ax} = a_{Bx} = a_x$. Because the accelerations of both blocks are equal, we can drop the subscripts A and B and call both accelerations a_x.

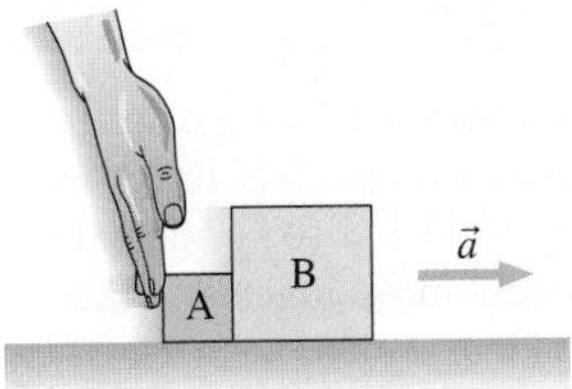

FIGURE 5.29 Two boxes moving together have the same acceleration.

These observations suggest that we can't solve for the motion of one block without considering the motion of the other. The following revised version of our basic Problem-Solving Strategy 5.2 that was developed earlier in this chapter shows how to do this.

PROBLEM-SOLVING STRATEGY 5.3 **Objects-in-contact problems** 

PREPARE Identify those objects whose motion you wish to study. Make simplifying assumptions.

Prepare a visual overview:

- Make a sketch of the situation. Define symbols and identify what the problem is trying to find. You may want to give each object its own coordinate system.
- Draw each object separately and prepare a *separate* force identification diagram for each object.
- Identify the action/reaction pairs of forces. If object A acts on object B with force $\vec{F}_{\text{A on B}}$, then identify the force $\vec{F}_{\text{B on A}}$ that B exerts on A.
- Draw a *separate* free-body diagram for each object. Use subscript labels to distinguish forces, such as $\vec{n}$ and $\vec{w}$, that act independently on more than one object.

SOLVE Use Newton's second and third laws:

- Write Newton's second law in component form for each object. Find the force components from the free-body diagrams.
- Equate the magnitudes of the two forces in each action/reaction pair.
- Determine how the accelerations of the objects are related to each other. Objects in contact will have the *same* acceleration; Section 5.8 shows how accelerations are related for objects connected by ropes or strings.
- Solve for the unknown forces or acceleration.

ASSESS Check that your result has the correct units, is reasonable, and answers the question.

NOTE ► Two steps are especially important when drawing the free-body diagrams. First, draw a *separate* diagram for each object. They need not have the same coordinate system. Second, show only the forces acting *on* that object. The force $\vec{F}_{\text{A on B}}$ goes on the free-body diagram of object B, but $\vec{F}_{\text{B on A}}$ goes on the diagram of object A. The two members of an action/reaction pair *always* appear on two different free-body diagrams—*never* on the same diagram. ◄

You might be puzzled that the Solve step calls for the use of the third law to equate just the *magnitudes* of action/reaction forces. What about the "opposite in direction" part of the third law? You have already used it! Your free-body diagrams should show the two members of an action/reaction pair to be opposite in direction, and that information will have been utilized in writing the second-law equations. Because the directional information has already been used, all that is left is the magnitude information.

EXAMPLE 5.16 Pushing two blocks

FIGURE 5.30 shows a 5.0 kg block A being pushed with a 3.0 N force. In front of this block is a 10 kg block B; the two blocks move together. What force does block A exert on block B?

FIGURE 5.30 Two blocks are pushed by a hand.

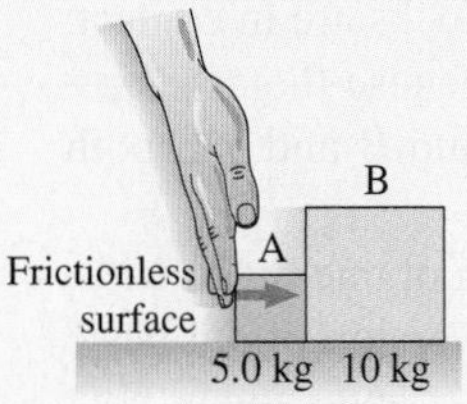

PREPARE The visual overview of FIGURE 5.31 lists the known information and identifies $F_{\text{A on B}}$ as what we're trying to find. Then, following the steps of Problem-Solving Strategy 5.3, we've drawn *separate* force identification diagrams and *separate* free-body diagrams for the two blocks. Both blocks have a weight force and a normal force, so we've used subscripts A and B to distinguish between them.

The force $\vec{F}_{\text{A on B}}$ is the contact force that block A exerts on B; it forms an action/reaction pair with the force $\vec{F}_{\text{B on A}}$ that block B exerts on A. Notice that force $\vec{F}_{\text{A on B}}$ is drawn acting on block B; it is the force *of* A *on* B. **Force vectors are always drawn on the free-body diagram of the object that *experiences* the force,** not the object exerting the force. Because action/reaction pairs act in opposite directions, force $\vec{F}_{\text{B on A}}$ pushes backward on block A and appears on A's free-body diagram.

SOLVE We begin by writing Newton's second law in component form for each block. Because the motion is only in the x-direction, we need only the x-component of the second law. For block A,

$$\sum F_x = (F_{\text{H}})_x + (F_{\text{B on A}})_x = m_{\text{A}} a_{\text{A}x}$$

The force components can be "read" from the free-body diagram, where we see $\vec{F}_{\text{H}}$ pointing to the right and $\vec{F}_{\text{B on A}}$ pointing to the left. Thus

$$F_{\text{H}} - F_{\text{B on A}} = m_{\text{A}} a_{\text{A}x}$$

For B, we have

$$\sum F_x = (F_{\text{A on B}})_x = F_{\text{A on B}} = m_{\text{B}} a_{\text{B}x}$$

We have two additional pieces of information: First, Newton's third law tells us that $F_{\text{B on A}} = F_{\text{A on B}}$. Second, the boxes are in contact and must have the same acceleration a_x: that is, $a_{\text{A}x} = a_{\text{B}x} = a_x$. With this information, the two x-component equations become

$$F_{\text{H}} - F_{\text{A on B}} = m_{\text{A}} a_x$$

$$F_{\text{A on B}} = m_{\text{B}} a_x$$

Our goal is to find $F_{\text{A on B}}$, so we need to eliminate the unknown acceleration a_x. From the first equation, $a_x = F_{\text{A on B}}/m_{\text{B}}$. Substituting this into the second equation gives

$$F_{\text{H}} - F_{\text{A on B}} = \frac{m_{\text{A}}}{m_{\text{B}}} F_{\text{A on B}}$$

This can be solved for the force of block A on block B, giving

$$F_{\text{A on B}} = \frac{F_{\text{H}}}{1 + m_{\text{A}}/m_{\text{B}}} = \frac{3.0\text{ N}}{1 + (5.0\text{ kg})/(10\text{ kg})} = \frac{3.0\text{ N}}{1.5} = 2.0\text{ N}$$

ASSESS Force F_{H} accelerates both blocks, a total mass of 15 kg, but force $F_{\text{A on B}}$ on B accelerates only block B, with a mass of 10 kg. Thus it makes sense that $F_{\text{A on B}} < F_{\text{H}}$.

FIGURE 5.31 A visual overview of the two blocks.

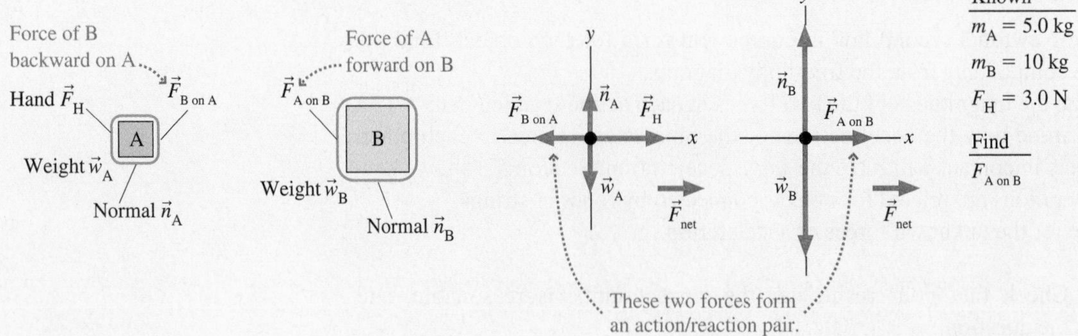

STOP TO THINK 5.5 Boxes P and Q are sliding to the right across a frictionless table. The hand H is slowing them down. The mass of P is larger than the mass of Q. Rank in order, from largest to smallest, the *horizontal* forces on P, Q, and H.

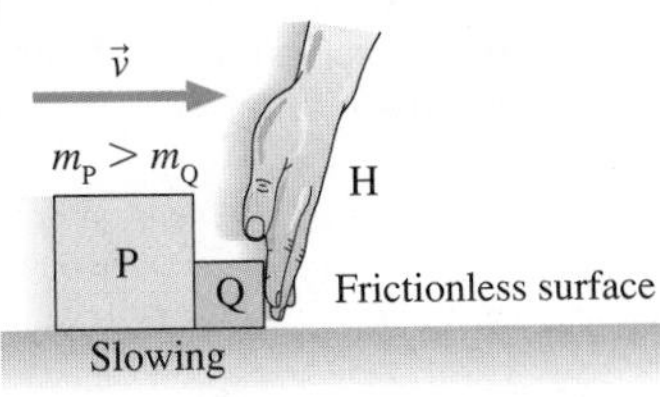

A. $F_{QonH} = F_{HonQ} = F_{PonQ} = F_{QonP}$
B. $F_{QonH} = F_{HonQ} > F_{PonQ} = F_{QonP}$
C. $F_{QonH} = F_{HonQ} < F_{PonQ} = F_{QonP}$
D. $F_{HonQ} = F_{HonP} > F_{PonQ}$

5.8 Ropes and Pulleys

Many objects are connected by strings, ropes, cables, and so on. In single-particle dynamics, we defined *tension* as the force exerted on an object by a rope or string. We can learn several important facts about ropes and tension by considering the box being pulled by a rope in **FIGURE 5.32**; the rope in turn is being pulled by a hand that exerts a force $\vec{F}$ on the rope.

FIGURE 5.32 A box being pulled by a rope.

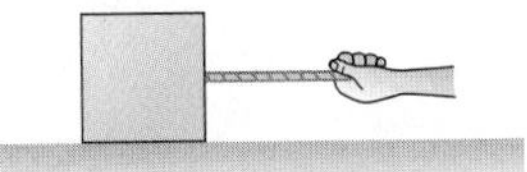

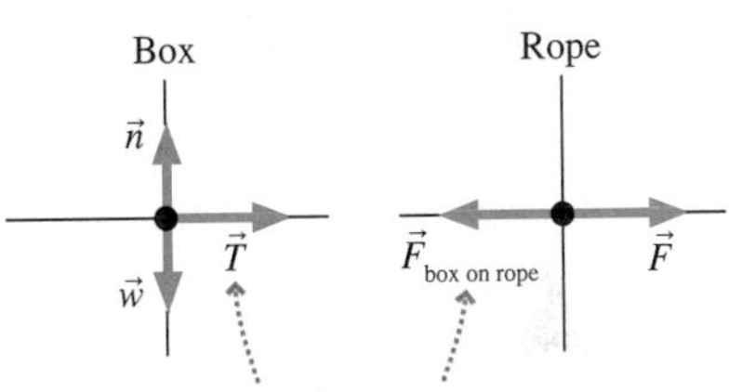

The tension $\vec{T}$ is the force that the rope exerts on the box. Thus $\vec{T}$ and $\vec{F}_{\text{box on rope}}$ are an action/reaction pair and have the same magnitude.

The box is pulled by the rope, so the box's free-body diagram shows a tension force $\vec{T}$. The *rope* is subject to two horizontal forces: the force $\vec{F}$ of the hand on the rope, and the force $\vec{F}_{\text{box on rope}}$ with which the box pulls back on the rope. $\vec{T}$ and $\vec{F}_{\text{box on rope}}$ form an action/reaction pair, so their magnitudes are equal: $F_{\text{box on rope}} = T$. Newton's second law *for the rope* is thus

$$\sum F_x = F - F_{\text{box on rope}} = F - T = m_{\text{rope}} a_x \tag{5.17}$$

where m_{rope} is the rope's mass.

In many problems, the mass of a string or rope is significantly less than the mass of the objects it pulls on. In that case, it's reasonable to make the approximation—called the **massless string approximation**—that $m_{\text{rope}} = 0$. If $m_{\text{rope}} = 0$, then the right side of Equation 5.17 is zero and so $T = F$. In other words, **the tension in a massless string or rope equals the magnitude of the force pulling on the end of the string or rope.** As a result:

- A massless string or rope "transmits" a force undiminished from one end to the other: If you pull on one end of a rope with force F, the other end of the rope pulls on what it's attached to with a force of the same magnitude F.
- The tension in a massless string or rope is the same from one end to the other.

CONCEPTUAL EXAMPLE 5.17 Pulling a rope

FIGURE 5.33a shows a student pulling horizontally with a 100 N force on a rope that is attached to a wall. In **FIGURE 5.33b**, two students in a tug-of-war pull on opposite ends of a rope with 100 N each. Is the tension in the second rope larger, smaller, or the same as that in the first?

FIGURE 5.33 Pulling on a rope. Which produces a larger tension?

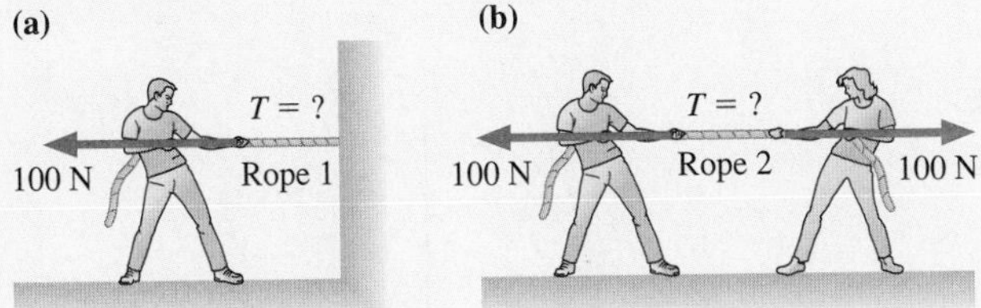

REASON Surely pulling on a rope from both ends causes more tension than pulling on one end. Right? Before jumping to conclusions, let's analyze the situation carefully. We found above that the force pulling on the end of a rope—here, the 100 N force exerted by the student—and the tension in the rope have the same magnitude. Thus, the tension in rope 1 is 100 N, the force with which the student pulls on the rope.

To find the tension in the second rope, consider the force that the *wall* exerts on the *first* rope. The first rope is in equilibrium, so the 100 N force exerted by the student must be balanced by a 100 N force on the rope from the wall. The first rope is being pulled from *both* ends by a 100 N force—the exact same situation as for the second rope, pulled by the students. A rope doesn't care whether it's being pulled on by a wall or by a person, so the tension in the second rope is the *same* as that in the first, or 100 N.

ASSESS This example reinforces what we just learned about ropes: A rope pulls on the objects at each of its ends with a force equal in magnitude to the tension, and the external force applied to each end of the rope and the rope's tension have equal magnitude.

FIGURE 5.34 An ideal pulley changes the direction in which a tension force acts, but not its magnitude.

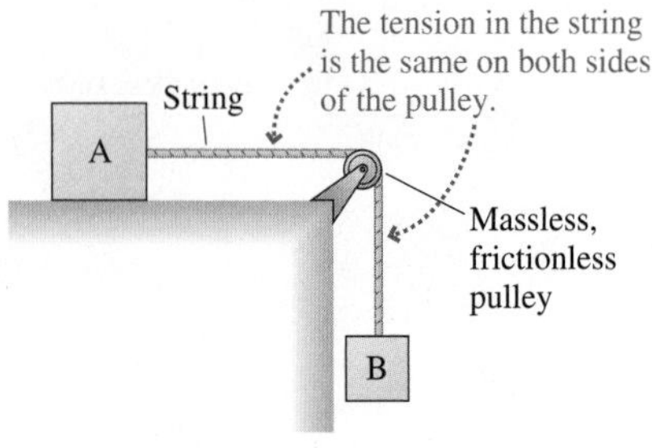

The tension in the string is the same on both sides of the pulley.
String
A
Massless, frictionless pulley
B

FIGURE 5.35 Tension forces on a pulley.

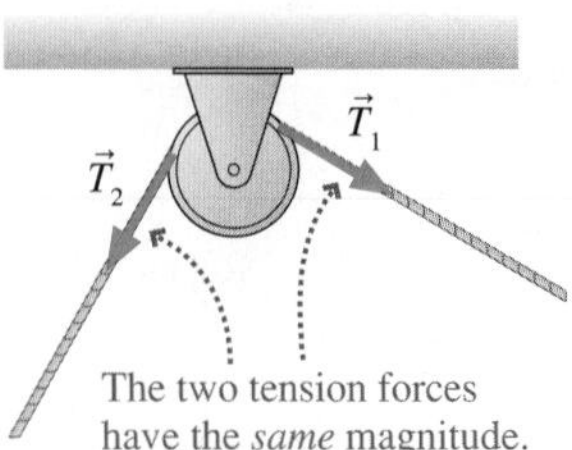

$\vec{T}_1$
$\vec{T}_2$
The two tension forces have the *same* magnitude.

Pulleys

Strings and ropes often pass over pulleys. FIGURE 5.34 shows a simple situation in which block B drags block A across a table as it falls. As the string moves, static friction between the string and the pulley causes the pulley to turn. If we assume that

- The string *and* the pulley are both massless, and
- There is no friction where the pulley turns on its axle,

then no net force is needed to accelerate the string or turn the pulley. In this case, **the tension in a massless string is unchanged by passing over a massless, frictionless pulley.** We'll assume such an ideal pulley for problems in this chapter. Later, when we study rotational motion in Chapter 8, we'll consider the effect of the pulley's mass.

In some situations we are interested in the force exerted on the pulley by the rope. Even though the pulley in FIGURE 5.35 is frictionless, the tension force still pulls on the pulley. Because the rope is under tension on *both* sides of the pulley, the pulley is subject to *two* tension forces, one from each side of the rope. The net force on the pulley due to the rope is then the vector sum of these two tension forces, both of which have the same magnitude, equal to the rope's tension.

We can collect all these observations about ropes, pulleys, and tension into a Tactics Box. We'll use these three rules extensively in solving problems with ropes, strings, and pulleys.

TACTICS BOX 5.2 Working with ropes and pulleys

For massless ropes or strings and massless, frictionless pulleys:

- If a force pulls on one end of a rope, the tension in the rope equals the magnitude of the pulling force.
- If two objects are connected by a rope, the tension is the same at both ends.
- If the rope passes over a pulley, the tension in the rope is unaffected.

Exercises 29–32

EXAMPLE 5.18 Placing a leg in traction

For serious fractures of the leg, the leg may need to have a stretching force applied to it to keep contracting leg muscles from forcing the broken bones together too hard. This is often done using *traction*, an arrangement of a rope, a weight, and pulleys as shown in FIGURE 5.36. The rope must make the same angle θ on both sides of the pulley so that the net force of the rope on the pulley is horizontally to the right, but θ can be adjusted to control the amount of traction. The doctor has specified 50 N of traction for this patient, with a 4.2 kg hanging mass. What is the proper angle θ?

FIGURE 5.36 A leg in traction.

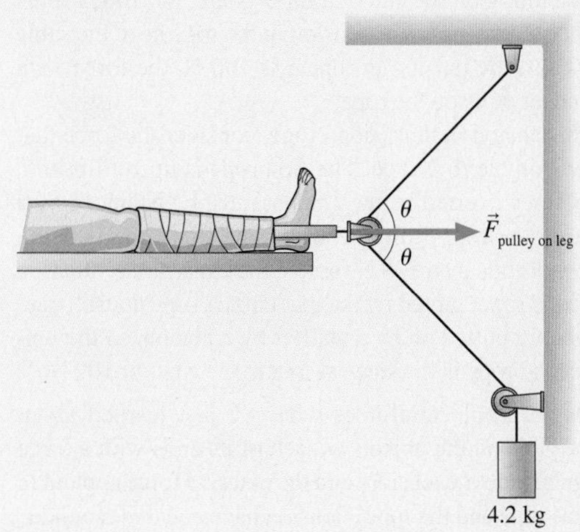

θ
θ
$\vec{F}_{\text{pulley on leg}}$
4.2 kg

PREPARE The pulley attached to the patient's leg is in static equilibrium, so the net force on it must be zero. FIGURE 5.37 shows a free-body diagram for the pulley, which we'll assume to be frictionless. Forces $\vec{T}_1$ and $\vec{T}_2$ are the tension forces of the rope as it

FIGURE 5.37 Free-body diagram for the pulley.

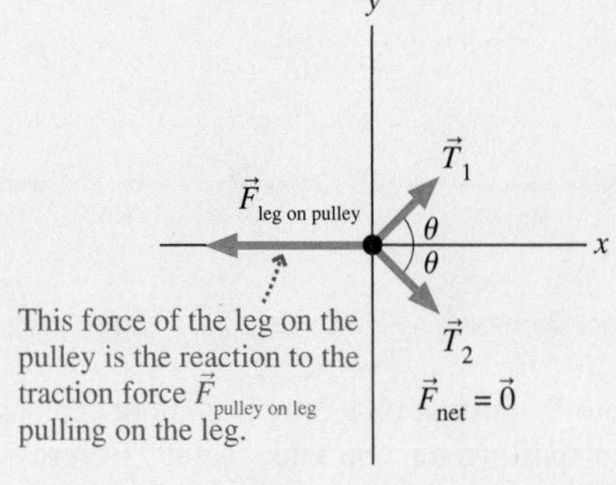

y
x
$\vec{F}_{\text{leg on pulley}}$
$\vec{T}_1$
θ
θ
$\vec{T}_2$
This force of the leg on the pulley is the reaction to the traction force $\vec{F}_{\text{pulley on leg}}$ pulling on the leg.
$\vec{F}_{\text{net}} = \vec{0}$

pulls on the pulley. These forces are equal in magnitude for a frictionless pulley, and their combined pull is to the right. This force is balanced by the force $\vec{F}_{\text{leg on pulley}}$ of the patient's leg pulling to the left. The traction force $\vec{F}_{\text{pulley on leg}}$ forms an action/reaction pair with $\vec{F}_{\text{leg on pulley}}$, so 50 N of traction means that $\vec{F}_{\text{leg on pulley}}$ also has a magnitude of 50 N.

SOLVE Two important properties of ropes, given in Tactics Box 5.2, are that (1) the tension equals the magnitude of the force pulling on its end and (2) the tension is the same throughout the rope. Thus, if a hanging mass m pulls on the rope with its weight mg, the tension along the entire rope is $T = mg$. For a 4.2 kg hanging mass, the tension is then $T = mg = 41$ N.

The pulley, in equilibrium, must satisfy Newton's second law for the case where $\vec{a} = \vec{0}$. Thus

$$\sum F_x = T_{1x} + T_{2x} + (F_{\text{leg on pulley}})_x = ma_x = 0$$

The tension forces both have the same magnitude T, and both are at angle θ from horizontal. The x-component of the leg force is negative because it's directed to the left. Then Newton's law becomes

$$2T\cos\theta - F_{\text{leg on pulley}} = 0$$

so that

$$\cos\theta = \frac{F_{\text{leg on pulley}}}{2T} = \frac{50\text{ N}}{82\text{ N}} = 0.61$$

$$\theta = \cos^{-1}(0.61) = 52°$$

ASSESS The traction force would approach $2mg = 82$ N if angle θ approached zero because the two tensions would pull in parallel. Conversely, the traction force would approach 0 N if θ approached 90°. Because the desired traction force is roughly halfway between 0 N and 82 N, an angle near 45° is reasonable.

EXAMPLE 5.19 Lifting a stage set

A 200 kg set used in a play is stored in the loft above the stage. The rope holding the set passes up and over a pulley, then is tied backstage. The director tells a 100 kg stagehand to lower the set. When he unties the rope, the set falls and the unfortunate man is hoisted into the loft. What is the stagehand's acceleration?

PREPARE FIGURE 5.38 shows the visual overview. The objects of interest are the stagehand M and the set S, for which we've drawn separate free-body diagrams. Assume a massless rope and a massless, frictionless pulley. Tension forces $\vec{T}_S$ and $\vec{T}_M$ are due to a massless rope going over an ideal pulley, so their magnitudes are the same.

FIGURE 5.38 Visual overview for the stagehand and set.

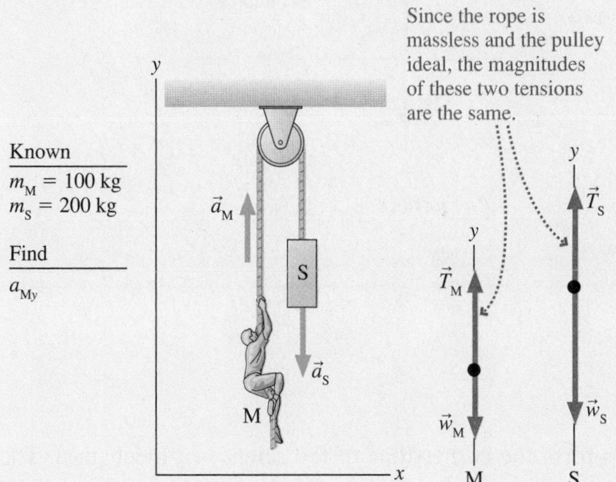

SOLVE From the two free-body diagrams, we can write Newton's second law in component form. For the man we have

$$\sum F_{My} = T_M - w_M = T_M - m_M g = m_M a_{My}$$

For the set we have

$$\sum F_{Sy} = T_S - w_S = T_S - m_S g = m_S a_{Sy}$$

Only the y-equations are needed. Because the stagehand and the set are connected by a rope, the upward distance traveled by one is the *same* as the downward distance traveled by the other. Thus the *magnitudes* of their accelerations must be the same, but, as Figure 5.38 shows, their *directions* are opposite. We can express this mathematically as $a_{Sy} = -a_{My}$. We also know that the two tension forces have equal magnitudes, which we'll call T. Inserting this information into the above equations gives

$$T - m_M g = m_M a_{My}$$

$$T - m_S g = -m_S a_{My}$$

These are simultaneous equations in the two unknowns T and a_{My}. We can solve for T in the first equation to get

$$T = m_M a_{My} + m_M g$$

Inserting this value of T into the second equation then gives

$$m_M a_{My} + m_M g - m_S g = -m_S a_{My}$$

which we can rewrite as

$$(m_S - m_M)g = (m_S + m_M)a_{My}$$

Finally, we can solve for the hapless stagehand's acceleration:

$$a_{My} = \frac{m_S - m_M}{m_S + m_M}g = \left(\frac{100\text{ kg}}{300\text{ kg}}\right) \times 9.80\text{ m/s}^2 = 3.3\text{ m/s}^2$$

This is also the acceleration with which the set falls. If the rope's tension was needed, we could now find it from $T = m_M a_{My} + m_M g$.

ASSESS If the stagehand weren't holding on, the set would fall with free-fall acceleration g. The stagehand acts as a *counterweight* to reduce the acceleration.

EXAMPLE 5.20 A not-so-clever bank robbery

Bank robbers have pushed a 1000 kg safe to a second-story floor-to-ceiling window. They plan to break the window, then lower the safe 3.0 m to their truck. Not being too clever, they stack up 500 kg of furniture, tie a rope between the safe and the furniture, and place the rope over a pulley. Then they push the safe out the window. What is the safe's speed when it hits the truck? The coefficient of kinetic friction between the furniture and the floor is 0.50.

PREPARE The visual overview in **FIGURE 5.39** establishes a coordinate system and defines the symbols that will be needed to calculate the safe's motion. The objects of interest are the safe S and the furniture F, which we will model as particles. We will assume a massless rope and a massless, frictionless pulley; the tension is then the same everywhere in the rope.

SOLVE We can write Newton's second law directly from the free-body diagrams. For the furniture,

$$\sum F_{\mathrm{F}x} = T_{\mathrm{F}} - f_{\mathrm{k}} = T - f_{\mathrm{k}} = m_{\mathrm{F}} a_{\mathrm{F}x}$$

$$\sum F_{\mathrm{F}y} = n - w_{\mathrm{F}} = n - m_{\mathrm{F}} g = 0$$

And for the safe,

$$\sum F_{\mathrm{S}y} = T_{\mathrm{S}} - w_{\mathrm{S}} = T - m_{\mathrm{S}} g = m_{\mathrm{S}} a_{\mathrm{S}y}$$

The safe and the furniture are tied together, so their accelerations have the same magnitude. But as the furniture slides to the right with positive acceleration $a_{\mathrm{F}x}$, the safe falls in the negative y-direction, so its acceleration $a_{\mathrm{S}y}$ is negative; we can express this mathematically as $a_{\mathrm{F}x} = -a_{\mathrm{S}y}$. We also have made use of the fact that $T_{\mathrm{S}} = T_{\mathrm{F}} = T$. We have one additional piece of information, the model of kinetic friction:

$$f_{\mathrm{k}} = \mu_{\mathrm{k}} n = \mu_{\mathrm{k}} m_{\mathrm{F}} g$$

where we used the y-equation of the furniture to deduce that $n = m_{\mathrm{F}} g$. Substitute this result for f_{k} into the x-equation of the furniture, then rewrite the furniture's x-equation and the safe's y-equation:

$$T - \mu_{\mathrm{k}} m_{\mathrm{F}} g = -m_{\mathrm{F}} a_{\mathrm{S}y}$$

$$T - m_{\mathrm{S}} g = m_{\mathrm{S}} a_{\mathrm{S}y}$$

We have succeeded in reducing our knowledge to two simultaneous equations in the two unknowns $a_{\mathrm{S}y}$ and T. We subtract the second equation from the first to eliminate T:

$$(m_{\mathrm{S}} - \mu_{\mathrm{k}} m_{\mathrm{F}}) g = -(m_{\mathrm{S}} + m_{\mathrm{F}}) a_{\mathrm{S}y}$$

Finally, we can solve for the safe's acceleration:

$$a_{\mathrm{S}y} = -\left(\frac{m_{\mathrm{S}} - \mu_{\mathrm{k}} m_{\mathrm{F}}}{m_{\mathrm{S}} + m_{\mathrm{F}}}\right) g$$

$$= -\frac{1000\text{ kg} - 0.5(500\text{ kg})}{1000\text{ kg} + 500\text{ kg}} \times 9.80\text{ m/s}^2 = -4.9\text{ m/s}^2$$

Now we need to calculate the kinematics of the falling safe. Because the time of the fall is not known or needed, we can use

$$(v_y)_{\mathrm{f}}^2 = (v_y)_{\mathrm{i}}^2 + 2a_{\mathrm{S}y}\,\Delta y = 0 + 2a_{\mathrm{S}y}(y_{\mathrm{f}} - y_{\mathrm{i}}) = -2a_{\mathrm{S}y} y_{\mathrm{i}}$$

$$(v_y)_{\mathrm{f}} = \sqrt{-2a_{\mathrm{S}y} y_{\mathrm{i}}} = \sqrt{-2(-4.9\text{ m/s}^2)(3.0\text{ m})} = 5.4\text{ m/s}$$

The value of $(v_y)_{\mathrm{f}}$ is negative, but we only needed to find the speed, so we took the absolute value. It seems unlikely that the truck will survive the impact of the 1000 kg safe!

FIGURE 5.39 Visual overview of the furniture and falling safe.

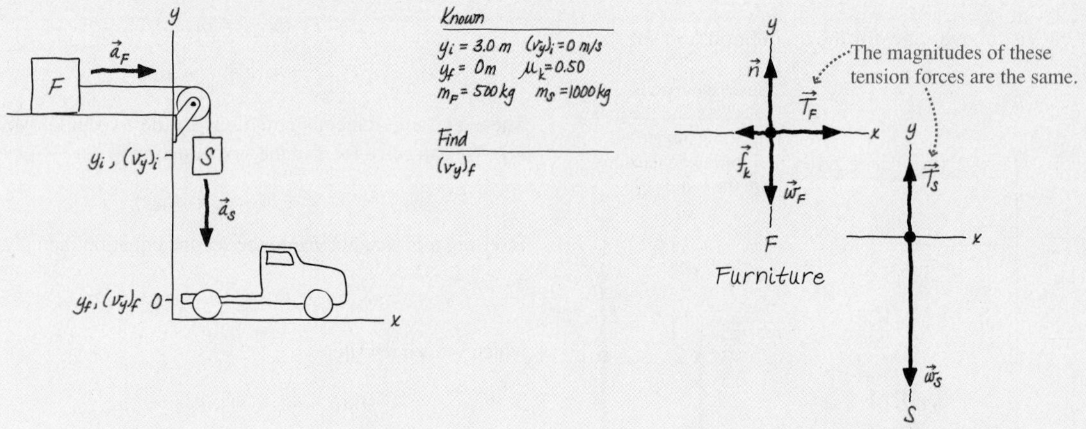

Newton's three laws form the cornerstone of the science of mechanics. These laws allowed scientists to understand many diverse phenomena, from the motion of a raindrop to the orbits of the planets. These laws were so precise at predicting motion that they went unchallenged for well over two hundred years. At the beginning of the twentieth century, however, it began to be apparent that the laws of mechanics and the laws of electricity and magnetism were somehow inconsistent. Bringing these two apparently disconnected theories into harmony required the genius of a young patent clerk named Albert Einstein. In doing so, he shook the foundations not only of Newtonian mechanics but also of our very notions of space and time. We will continue to develop Newtonian mechanics for the next few chapters because of its tremendous importance to the physics of everyday life. But it's worth keeping in the back of your mind that Newton's laws aren't the ultimate statement about motion. Later in this textbook we'll reexamine motion and mechanics from the perspective of Einstein's theory of relativity.

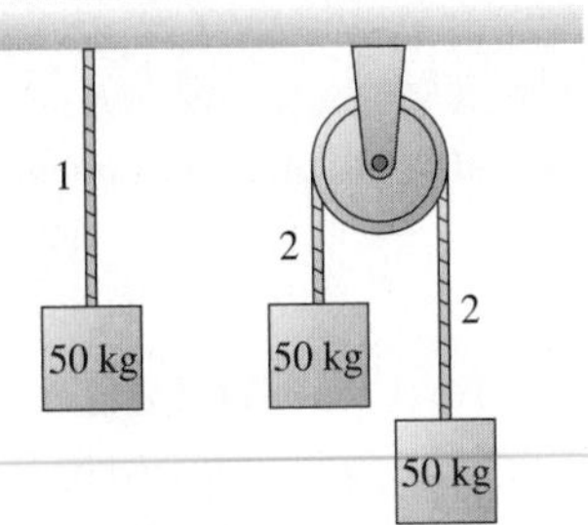

STOP TO THINK 5.6 All three 50 kg blocks are at rest. Is the tension in rope 2 greater than, less than, or equal to the tension in rope 1?

INTEGRATED EXAMPLE 5.21 Stopping distances

A 1500 kg car is traveling at a speed of 30 m/s when the driver slams on the brakes and skids to a halt. Determine the stopping distance if the car is traveling up a 10° slope, down a 10° slope, or on a level road.

PREPARE We'll represent the car as a particle and we'll use the model of kinetic friction. We want to solve the problem only once, not three separate times, so we'll leave the slope angle θ unspecified until the end.

FIGURE 5.40 shows the visual overview. We've shown the car sliding uphill, but these representations work equally well for a level or downhill slide if we let θ be zero or negative, respectively. We've used a tilted coordinate system so that the motion is along the x-axis. The car *skids* to a halt, so we've taken the coefficient of *kinetic* friction for rubber on concrete from Table 5.1.

SOLVE Newton's second law and the model of kinetic friction are

$$\sum F_x = n_x + w_x + (f_k)_x$$
$$= 0 - mg\sin\theta - f_k = ma_x$$
$$\sum F_y = n_y + w_y + (f_k)_y$$
$$= n - mg\cos\theta + 0 = ma_y = 0$$

We've written these equations by "reading" the motion diagram and the free-body diagram. Notice that both components of the weight vector $\vec{w}$ are negative. $a_y = 0$ because the motion is entirely along the x-axis.

The second equation gives $n = mg\cos\theta$. Using this in the friction model, we find $f_k = \mu_k mg\cos\theta$. Inserting this result back into the first equation then gives

$$ma_x = -mg\sin\theta - \mu_k mg\cos\theta$$
$$= -mg(\sin\theta + \mu_k\cos\theta)$$
$$a_x = -g(\sin\theta + \mu_k\cos\theta)$$

This is a constant acceleration. Constant-acceleration kinematics gives

$$(v_x)_f^2 = 0 = (v_x)_i^2 + 2a_x(x_f - x_i) = (v_x)_i^2 + 2a_x x_f$$

which we can solve for the stopping distance x_f:

$$x_f = -\frac{(v_x)_i^2}{2a_x} = \frac{(v_x)_i^2}{2g(\sin\theta + \mu_k\cos\theta)}$$

Notice how the minus sign in the expression for a_x canceled the minus sign in the expression for x_f. Evaluating our result at the three different angles gives the stopping distances:

$$x_f = \begin{cases} 48\text{ m} & \theta = 10° & \text{uphill} \\ 57\text{ m} & \theta = 0° & \text{level} \\ 75\text{ m} & \theta = -10° & \text{downhill} \end{cases}$$

The implications are clear about the danger of driving downhill too fast!

ASSESS 30m/s $\approx$ 60 mph and 57m $\approx$ 180 feet on a level surface. These are similar to the stopping distances you learned when you got your driver's license, so the results seem reasonable. Additional confirmation comes from noting that the expression for a_x becomes $-g\sin\theta$ if $\mu_k = 0$. This is what you learned in Chapter 3 for the acceleration on a frictionless inclined plane.

FIGURE 5.40 Visual overview for a skidding car.

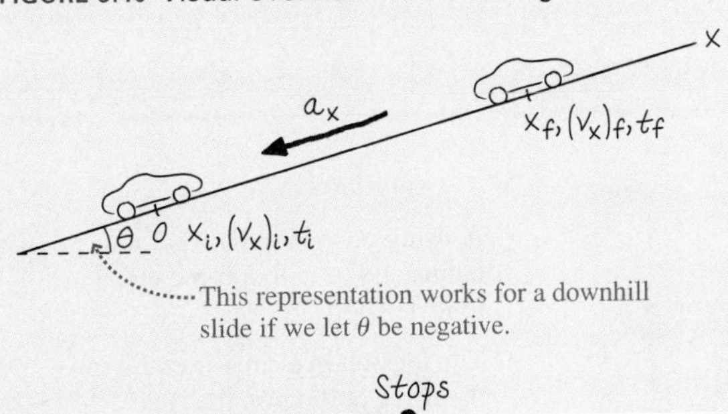

Known
$x_i = 0$ m, $t_i = 0$ s $(v_x)_i = 30$ m/s
$m = 1500$ kg $(v_x)_f = 0$ m/s
$\mu_k = 0.80$
$\theta = -10°, 0, 10$
Find
$\Delta x = x_f - x_i = x_f$

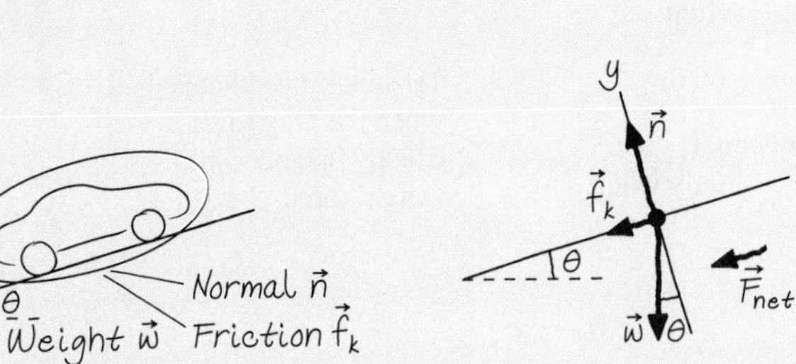

SUMMARY

The goal of Chapter 5 has been to learn how to solve problems about motion in a straight line.

GENERAL STRATEGY

All examples in this chapter follow a three-part strategy. You'll become a better problem solver if you adhere to it as you do the homework problems. The *Dynamics Worksheets* in the *Student Workbook* will help you structure your work in this way.

Equilibrium Problems

Object at rest or moving at constant velocity.

PREPARE Make simplifying assumptions.

- Check that the object is either at rest or moving with constant velocity ($\vec{a} = \vec{0}$).
- Identify forces and show them on a free-body diagram.

SOLVE Use Newton's second law in component form:

$$\sum F_x = ma_x = 0$$
$$\sum F_y = ma_y = 0$$

"Read" the components from the free-body diagram.

ASSESS Is your result reasonable?

Dynamics Problems

Object accelerating.

PREPARE Make simplifying assumptions. Make a **visual overview:**

- Sketch a pictorial representation.
- Identify known quantities and what the problem is trying to find.
- Identify all forces and show them on a free-body diagram.

SOLVE Use Newton's second law in component form:

$$\sum F_x = ma_x \text{ and } \sum F_y = ma_y$$

"Read" the components of the vectors from the free-body diagram. If needed, use kinematics to find positions and velocities.

ASSESS Is your result reasonable?

Objects in Contact

Two or more objects interacting.

PREPARE Make a **visual overview:**

- Sketch a pictorial representation.
- Identify all forces acting on *each* object.
- Identify action/reaction pairs of forces acting on objects in the system.
- Draw a *separate* free-body diagram for each object.

SOLVE Write Newton's second law for each object. Use Newton's third law to equate the magnitudes of action/reaction pairs. Determine how the accelerations of the objects are related to each other.

ASSESS Is your result reasonable?

IMPORTANT CONCEPTS

Specific information about three important forces:

Weight $\vec{w} = (mg, \text{downward})$

Friction $\vec{f}_s = (0 \text{ to } \mu_s n, \text{direction as necessary to prevent motion})$

$\vec{f}_k = (\mu_k n, \text{direction opposite the motion})$

$\vec{f}_r = (\mu_r n, \text{direction opposite the motion})$

Drag $\vec{D} \approx (\frac{1}{4}\rho A v^2, \text{direction opposite the motion})$ for motion in air

Newton's laws are vector expressions. You must write them out by **components:**

$$(F_{net})_x = \sum F_x = ma_x$$
$$(F_{net})_y = \sum F_y = ma_y$$

For equilibrium problems, $a_x = 0$ and $a_y = 0$.

APPLICATIONS

Apparent weight is the magnitude of the contact force supporting an object. It is what a scale would read, and it is your sensation of weight:

$$w_{app} = m(g + a_y)$$

Apparent weight equals your true weight $w = mg$ only when $a_y = 0$.

A falling object reaches **terminal speed**

$$v_{term} \approx \sqrt{\frac{4mg}{\rho A}}$$

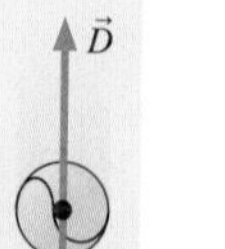

Terminal speed is reached when the drag force exactly balances the weight force: $\vec{a} = \vec{0}$.

Strings and pulleys

- A string or rope pulls what it's connected to with a force equal to its tension.
- The tension in a rope is equal to the force pulling on the rope.
- The tension in a massless rope is the same at all points in the rope.
- Tension does not change when a rope passes over a massless, frictionless pulley.

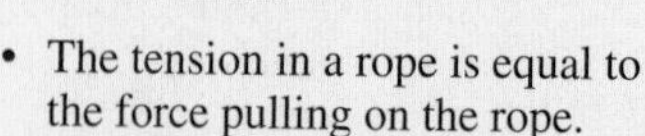

For homework assigned on MasteringPhysics, go to www.masteringphysics.com

Problem difficulty is labeled as | (straightforward) to ||||| (challenging).

Problems labeled [workbook icon] can be done on a Workbook Dynamics Worksheet; INT integrate significant material from earlier chapters; BIO are of biological or medical interest.

QUESTIONS

Conceptual Questions

1. An object is subject to two forces that do not point in opposite directions. Is it possible to choose their magnitudes so that the object is in equilibrium? Explain.
2. Are the objects described here in static equilibrium, dynamic equilibrium, or not in equilibrium at all?
   a. A girder is lifted at constant speed by a crane.
   b. A girder is lowered by a crane. It is slowing down.
   c. You're straining to hold a 200 lb barbell over your head.
   d. A jet plane has reached its cruising speed and altitude.
   e. A rock is falling into the Grand Canyon.
   f. A box in the back of a truck doesn't slide as the truck stops.
3. What forces are acting on you right now? What net force is acting on you right now?
4. Decide whether each of the following is true or false. Give a reason!
   a. The mass of an object depends on its location.
   b. The weight of an object depends on its location.
   c. Mass and weight describe the same thing in different units.
5. An astronaut takes his bathroom scale to the moon and then stands on it. Is the reading of the scale his true weight? Explain.
6. A light block of mass m and a heavy block of mass M are attached to the ends of a rope. A student holds the heavier block and lets the lighter block hang below it, as shown in Figure Q5.6. Then she lets go. Air resistance can be neglected.
   a. What is the tension in the rope while the blocks are falling, before either hits the ground?
   b. Would your answer be different if she had been holding the lighter block initially?

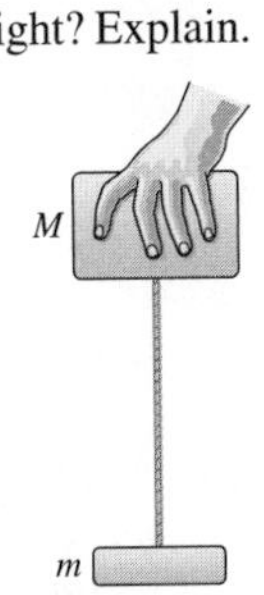

FIGURE Q5.6

7. Four balls are thrown straight up. Figure Q5.7 is a "snapshot" showing their velocities. They have the same size but different mass. Air resistance is negligible. Rank in order, from largest to smallest, the magnitudes of the net forces, F_{net1}, F_{net2}, F_{net3}, F_{net4}, acting on the balls. Some may be equal. Give your answer in the form A > B = C > D, and state your reasoning.

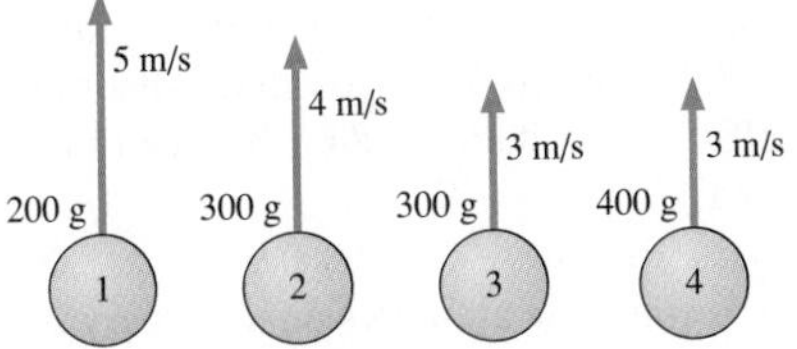

FIGURE Q5.7

8. Suppose you attempt to pour out 100 g of salt, using a pan balance for measurements, while in an elevator that is accelerating upward. Will the quantity of salt be too much, too little, or the correct amount? Explain.
9. a. Can the normal force on an object be directed horizontally? If not, why not? If so, provide an example.
   b. Can the normal force on an object be directed downward? If not, why not? If so, provide an example.
10. A ball is thrown straight up. Taking the drag force of air into account, does it take longer for the ball to travel to the top of its motion or for it to fall back down again?
11. Three objects move through the air as shown in Figure Q5.11. Rank in order, from largest to smallest, the three drag forces D_1, D_2, and D_3. Some may be equal. Give your answer in the form A > B = C and state your reasoning.

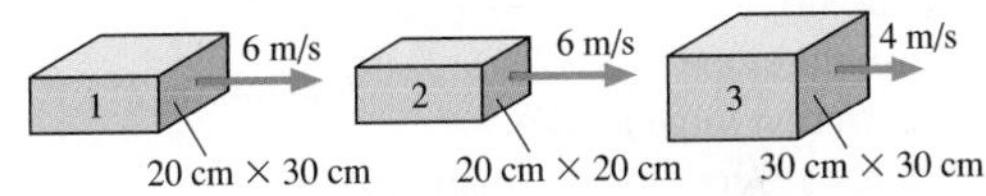

FIGURE Q5.11

12. A skydiver is falling at her terminal speed. Right after she opens her parachute, which has a very large area, what is the direction of the net force on her?
13. Raindrops can fall at different speeds; some fall quite quickly, others quite slowly. Why might this be true?
14. An airplane moves through the air at a constant speed. The jet engine's thrust applies a force in the direction of motion. Reducing thrust will cause the plane to fly at a slower—but still constant—speed. Explain why this is so.
15. Is it possible for an object to travel in air faster than its terminal speed? If not, why not? If so, explain how this might happen.

For Questions 16 through 19, determine the tension in the rope at the point indicated with a dot.

- All objects are at rest.
- The strings and pulleys are massless, and the pulleys are frictionless.

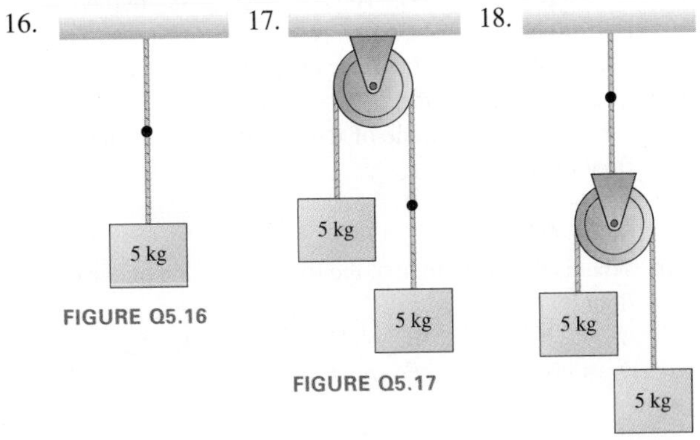

FIGURE Q5.16

FIGURE Q5.17

FIGURE Q5.18

19.

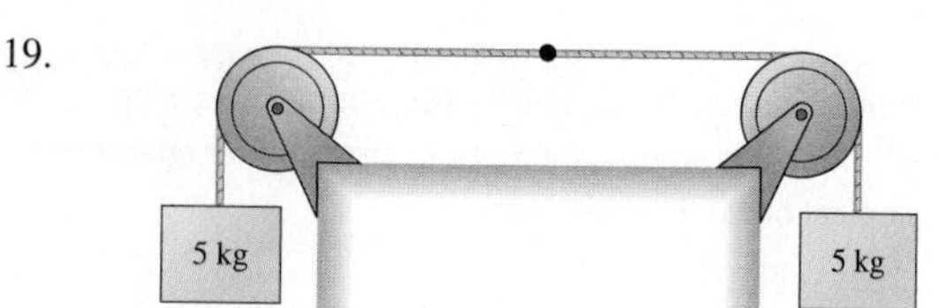

FIGURE Q5.19

20. The floor is frictionless. In which direction is the kinetic friction force on block 1 in Figure Q5.20? On block 2? Explain.

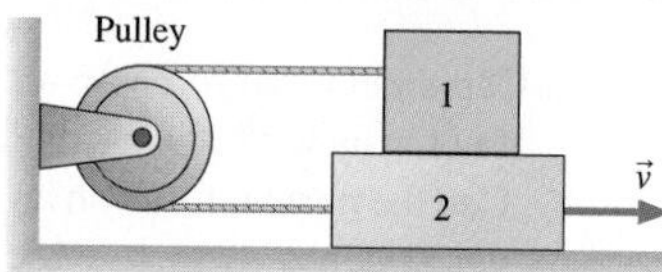

FIGURE Q5.20

Multiple-Choice Questions

21. || The wood block in Figure Q5.21 is at rest on a wood ramp. In which direction is the static friction force on block 1?
A. Up the slope.
B. Down the slope.
C. The friction force is zero.
D. There's not enough information to tell.

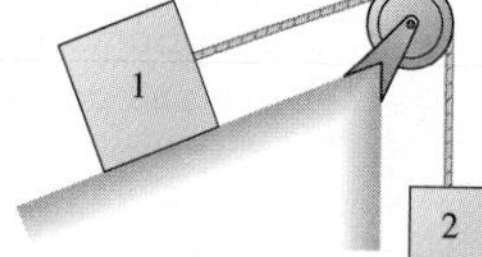

FIGURE Q5.21

22. || A 2.0 kg ball is suspended by two light strings as shown in Figure Q5.22. What is the tension T in the angled string?
A. 9.5 N B. 15 N C. 20 N
D. 26 N E. 30 N

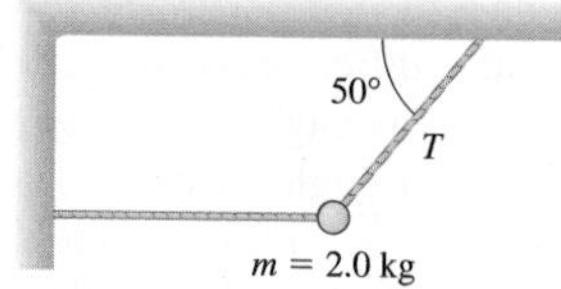

FIGURE Q5.22

23. | While standing in a low tunnel, you raise your arms and push against the ceiling with a force of 100 N. Your mass is 70 kg.
a. What force does the ceiling exert on you?
A. 10 N B. 100 N C. 690 N
D. 790 N E. 980 N
b. What force does the floor exert on you?
A. 10 N B. 100 N C. 690 N
D. 790 N E. 980 N

24. | A 5.0 kg dog sits on the floor of an elevator that is accelerating *downward* at 1.20 m/s^2.
a. What is the magnitude of the normal force of the elevator floor on the dog?
A. 34 N B. 43 N C. 49 N
D. 55 N E. 74 N
b. What is the magnitude of the force of the dog on the elevator floor?
A. 4.2 N B. 49 N C. 55 N
D. 43 N E. 74 N

25. | A 3.0 kg puck slides due east on a horizontal frictionless surface at a constant speed of 4.5 m/s. Then a force of magnitude 6.0 N, directed due north, is applied for 1.5 s. Afterward,
a. What is the northward component of the puck's velocity?
A. 0.50 m/s B. 2.0 m/s C. 3.0 m/s
D. 4.0 m/s E. 4.5 m/s
b. What is the speed of the puck?
A. 4.9 m/s B. 5.4 m/s C. 6.2 m/s
D. 7.5 m/s E. 11 m/s

26. | A rocket in space, initially at rest, fires its main engines at a constant thrust. As it burns fuel, the mass of the rocket decreases. Which of the graphs in Figure Q5.26 best represents the velocity of the rocket as a function of time?

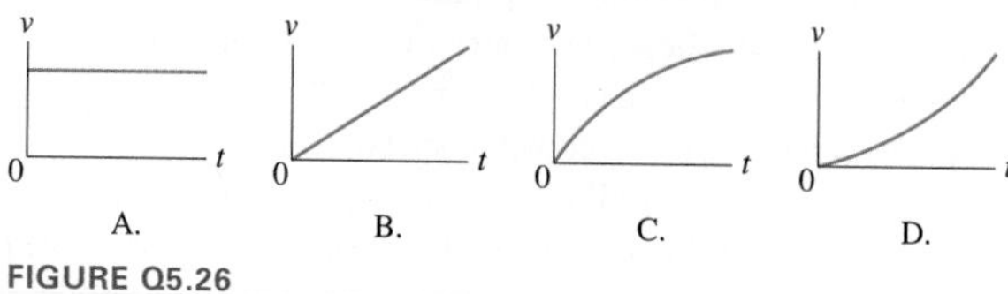

FIGURE Q5.26

27. | Eric has a mass of 60 kg. He is standing on a scale in an elevator that is accelerating downward at 1.7 m/s^2. What is the approximate reading on the scale?
A. 0 N B. 400 N C. 500 N D. 600 N

28. | The two blocks in Figure Q5.28 are at rest on frictionless surfaces. What must be the mass of the right block in order that the two blocks remain stationary?
A. 4.9 kg B. 6.1 kg C. 7.9 kg
D. 9.8 kg E. 12 kg

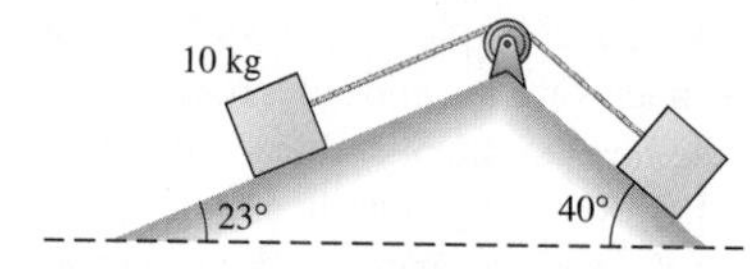

FIGURE Q5.28

29. | A football player at practice pushes a 60 kg blocking sled across the field at a constant speed. The coefficient of kinetic friction between the grass and the sled is 0.30. How much force must he apply to the sled?
A. 18 N B. 60 N C. 180 N D. 600 N

30. | Two football players are pushing a 60 kg blocking sled across the field at a constant speed of 2.0 m/s. The coefficient of kinetic friction between the grass and the sled is 0.30. Once they stop pushing, how far will the sled slide before coming to rest?
A. 0.20 m B. 0.68 m C. 1.0 m D. 6.6 m

31. || Land Rover ads used to claim that their vehicles could climb a slope of 45°. For this to be possible, what must be the minimum coefficient of static friction between the vehicle's tires and the road?
A. 0.5 B. 0.7 C. 0.9 D. 1.0

32. || A truck is traveling at 30 m/s on a slippery road. The driver slams on the brakes and the truck starts to skid. If the coefficient of kinetic friction between the tires and the road is 0.20, how far will the truck skid before stopping?
A. 230 m B. 300 m C. 450 m D. 680 m

PROBLEMS

Section 5.1 Equilibrium

1. | The three ropes in Figure P5.1 are tied to a small, very light ring. Two of the ropes are anchored to walls at right angles, and the third rope pulls as shown. What are T_1 and T_2, the magnitudes of the tension forces in the first two ropes?

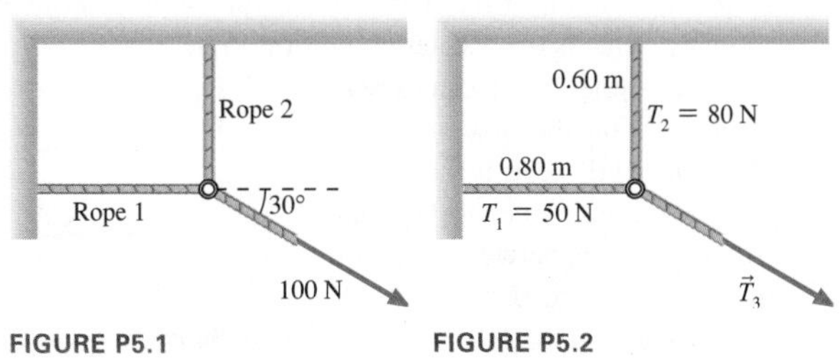

FIGURE P5.1 FIGURE P5.2

2. ||| The three ropes in Figure P5.2 are tied to a small, very light ring. Two of these ropes are anchored to walls at right angles with the tensions shown in the figure. What are the magnitude and direction of the tension $\vec{T}_3$ in the third rope?

3. |||| A 20 kg loudspeaker is suspended 2.0 m below the ceiling by two cables that are each 30° from vertical. What is the tension in the cables?

4. || A 1000 kg steel beam is supported by the two ropes shown in Figure P5.4. Each rope can support a maximum sustained tension of 5600 N. Do the ropes break?

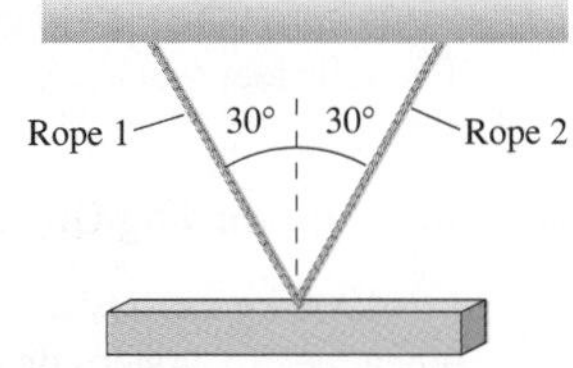

FIGURE P5.4

5. | A cable is used to raise a 25 kg urn from an underwater archeological site. There is a 25 N drag force from the water as the urn is raised at a constant speed. What is the tension in the cable?

6. |||| BIO When you bend your knee, the quadriceps muscle is stretched. This increases the tension in the quadriceps tendon attached to your kneecap (patella), which, in turn, increases the tension in the patella tendon that attaches your kneecap to your lower leg bone (tibia). Simultaneously, the end of your upper leg bone (femur) pushes outward on the patella. Figure P5.6 shows how these parts of a knee joint are arranged. What size force does the femur exert on the kneecap if the tendons are oriented as in the figure and the tension in each tendon is 60 N?

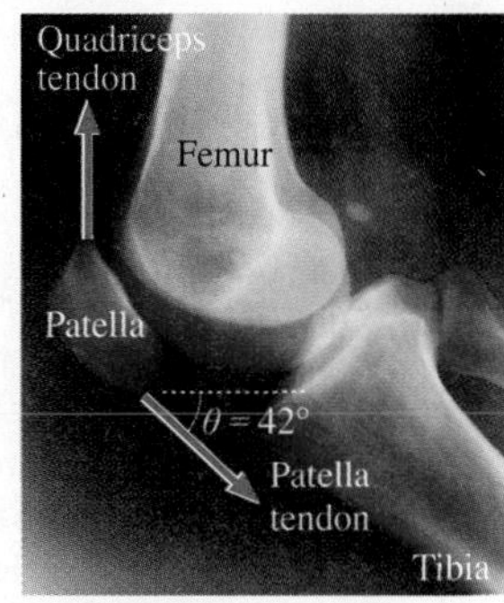

FIGURE P5.6

7. || The two angled ropes used to support the crate in Figure P5.7 can withstand a maximum tension of 1500 N before they break. What is the largest mass the ropes can support?

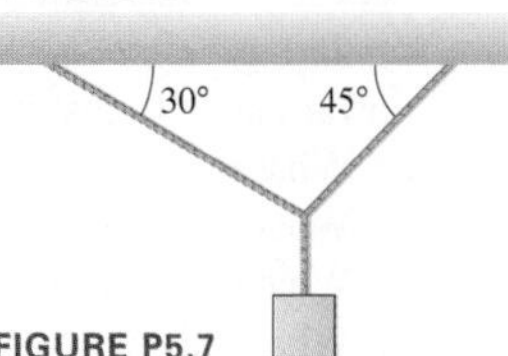

FIGURE P5.7

Section 5.2 Dynamics and Newton's Second Law

8. || A force with x-component F_x acts on a 500 g object as it moves along the x-axis. The object's acceleration graph (a_x versus t) is shown in Figure P5.8. Draw a graph of F_x versus t.

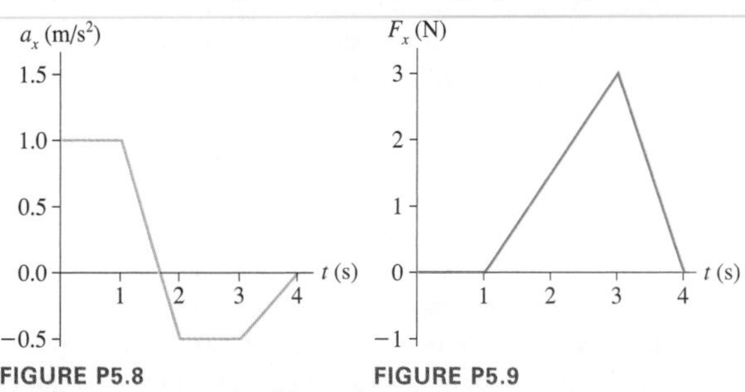

FIGURE P5.8 FIGURE P5.9

9. || A force with x-component F_x acts on a 2.0 kg object as it moves along the x-axis. A graph of F_x versus t is shown in Figure P5.9. Draw an acceleration graph (a_x versus t) for this object.

10. | A force with x-component F_x acts on a 500 g object as it moves along the x-axis. A graph of F_x versus t is shown in Figure P5.10. Draw an acceleration graph (a_x versus t) for this object.

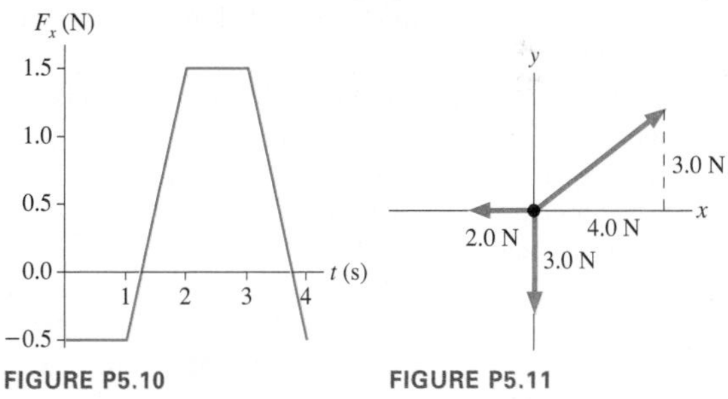

FIGURE P5.10 FIGURE P5.11

11. || The forces in Figure P5.11 are acting on a 2.0 kg object. Find the values of a_x and a_y, the x- and y-components of the object's acceleration.

12. | The forces in Figure P5.12 are acting on a 2.0 kg object. Find the values of a_x and a_y, the x- and y-components of the object's acceleration.

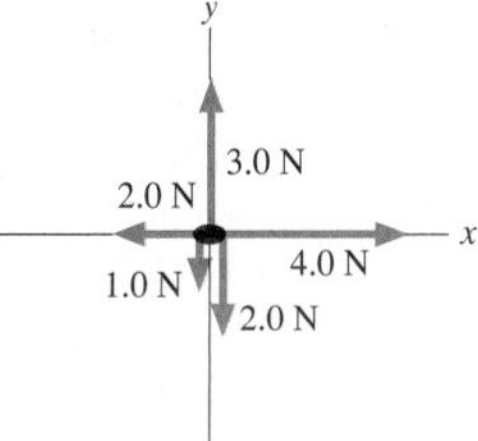

FIGURE P5.12

13. | A horizontal rope is tied to a 50 kg box on frictionless ice. What is the tension in the rope if
    a. The box is at rest?
    b. The box moves at a steady 5.0 m/s?
    c. The box has $v_x = 5.0$ m/s and $a_x = 5.0$ m/s²?

14. |||| A crate pushed along the floor with velocity $\vec{v}_i$ slides a distance d after the pushing force is removed.
    a. If the mass of the crate is doubled but the initial velocity is not changed, what distance does the crate slide before stopping? Explain.
    b. If the initial velocity of the crate is doubled to $2\vec{v}_i$ but the mass is not changed, what distance does the crate slide before stopping? Explain.

15. || In a head-on collision, a car stops in 0.10 s from a speed of 14 m/s. The driver has a mass of 70 kg, and is, fortunately, tightly strapped into his seat. What force is applied to the driver by his seat belt during that fraction of a second?

Section 5.3 Mass and Weight

16. | An astronaut's weight on earth is 800 N. What is his weight on Mars, where $g = 3.76\ \text{m/s}^2$?
17. | A woman has a mass of 55.0 kg.
    a. What is her weight on earth?
    b. What are her mass and her weight on the moon, where $g = 1.62\ \text{m/s}^2$?
18. ||| A box with a 75 kg passenger inside is launched straight up into the air by a giant rubber band. After the box has left the rubber band but is still moving *upward,*
    a. What is the passenger's true weight?
    b. What is the passenger's apparent weight?
19. || a. How much force does an 80 kg astronaut exert on his chair while sitting at rest on the launch pad?
    b. How much force does the astronaut exert on his chair while accelerating straight up at $10\ \text{m/s}^2$?
20. | It takes the elevator in a skyscraper 4.0 s to reach its cruising speed of 10 m/s. A 60 kg passenger gets aboard on the ground floor. What is the passenger's apparent weight
    a. Before the elevator starts moving?
    b. While the elevator is speeding up?
    c. After the elevator reaches its cruising speed?
21. || Zach, whose mass is 80 kg, is in an elevator descending at 10 m/s. The elevator takes 3.0 s to brake to a stop at the first floor.
    a. What is Zach's apparent weight before the elevator starts braking?
    b. What is Zach's apparent weight while the elevator is braking?
22. ||| INT Figure P5.22 shows the velocity graph of a 75 kg passenger in an elevator. What is the passenger's apparent weight at $t = 1.0$ s? At 5.0 s? At 9.0 s?

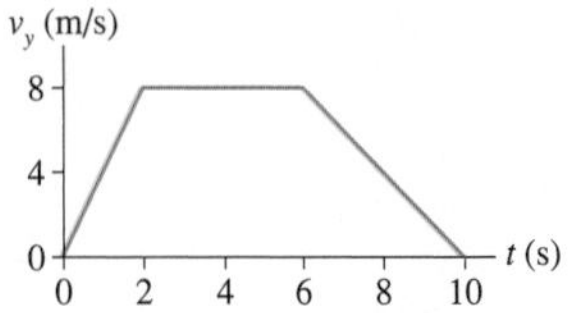

FIGURE P5.22

Section 5.4 Normal Forces

23. || a. A 0.60 kg bullfrog is sitting at rest on a level log. How large is the normal force of the log on the bullfrog?
    b. A second 0.60 kg bullfrog is on a log tilted 30° above horizontal. How large is the normal force of the log on this bullfrog?
24. ||| A 23 kg child goes down a straight slide inclined 38° above horizontal. The child is acted on by his weight, the normal force from the slide, and kinetic friction.
    a. Draw a free-body diagram of the child.
    b. How large is the normal force of the slide on the child?

Section 5.5 Friction

25. ||| Bonnie and Clyde are sliding a 300 kg bank safe across the floor to their getaway car. The safe slides with a constant speed if Clyde pushes from behind with 385 N of force while Bonnie pulls forward on a rope with 350 N of force. What is the safe's coefficient of kinetic friction on the bank floor?
26. ||| A 4000 kg truck is parked on a 15° slope. How big is the friction force on the truck?
27. ||| A 1000 kg car traveling at a speed of 40 m/s skids to a halt on wet concrete where $\mu_k = 0.60$. How long are the skid marks?
28. | A stubborn 120 kg mule sits down and refuses to move. To drag the mule to the barn, the exasperated farmer ties a rope around the mule and pulls with his maximum force of 800 N. The coefficients of friction between the mule and the ground are $\mu_s = 0.80$ and $\mu_k = 0.50$. Is the farmer able to move the mule?
29. ||| A 10 kg crate is placed on a horizontal conveyor belt. The materials are such that $\mu_s = 0.50$ and $\mu_k = 0.30$.
    a. Draw a free-body diagram showing all the forces on the crate if the conveyer belt runs at constant speed.
    b. Draw a free-body diagram showing all the forces on the crate if the conveyer belt is speeding up.
    c. What is the maximum acceleration the belt can have without the crate slipping?
    d. If acceleration of the belt exceeds the value determined in part c, what is the acceleration of the crate?
30. || What is the minimum downward force on the box in Figure P5.30 that will keep it from slipping? The coefficients of static and kinetic friction between the box and the floor are 0.35 and 0.25, respectively.

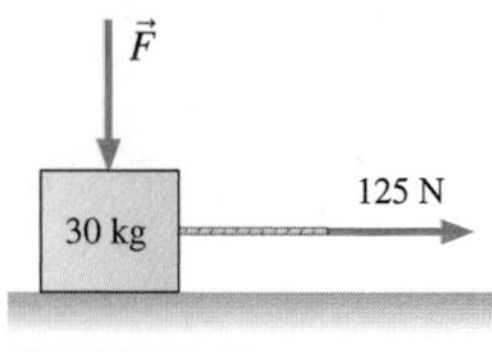

FIGURE P5.30

Section 5.6 Drag

31. || What is the drag force on a 1.6-m-wide, 1.4-m-high car traveling at
    a. 10 m/s (≈22 mph)? b. 30 m/s (≈65 mph)?
32. ||| A 22-cm-diameter bowling ball has a terminal speed of 77 m/s. What is the ball's mass?
33. |||| A 75 kg skydiver can be modeled as a rectangular "box" with dimensions 20 cm × 40 cm × 1.8 m. What is his terminal speed if he falls feet first?

Section 5.7 Interacting Objects

34. ||| A 1000 kg car pushes a 2000 kg truck that has a dead battery. When the driver steps on the accelerator, the drive wheels of the car push backward against the ground with a force of 4500 N.
    a. What is the magnitude of the force of the car on the truck?
    b. What is the magnitude of the force of the truck on the car?
35. |||| Blocks with masses of 1.0 kg, 2.0 kg, and 3.0 kg are lined up in a row on a frictionless table. All three are pushed forward by a 12 N force applied to the 1.0 kg block. How much force does the 2.0 kg block exert on (a) the 3.0 kg block and (b) the 1.0 kg block?

Section 5.8 Ropes and Pulleys

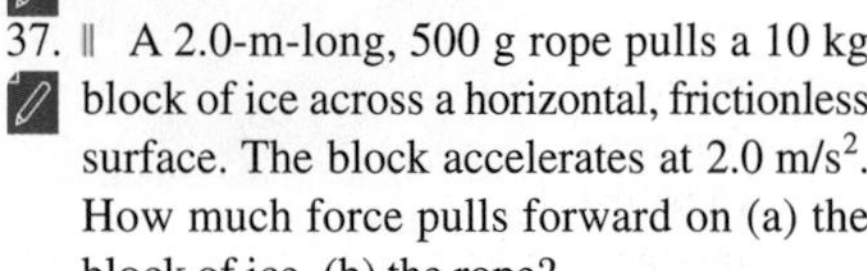

36. ||| What is the tension in the rope of Figure P5.36?
37. | A 2.0-m-long, 500 g rope pulls a 10 kg block of ice across a horizontal, frictionless surface. The block accelerates at $2.0\ \text{m/s}^2$. How much force pulls forward on (a) the block of ice, (b) the rope?

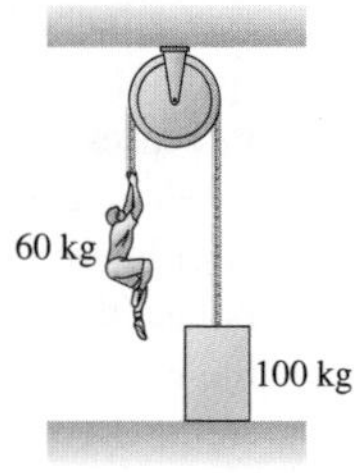

FIGURE P5.36

38. ||| Figure P5.38 shows two 1.00 kg blocks connected by a rope. A second rope hangs beneath the lower block. Both ropes have a mass of 250 g. The entire assembly is accelerated upward at $3.00\ \text{m/s}^2$ by force $\vec{F}$.
    a. What is F?
    b. What is the tension at the top end of rope 1?
    c. What is the tension at the bottom end of rope 1?
    d. What is the tension at the top end of rope 2?

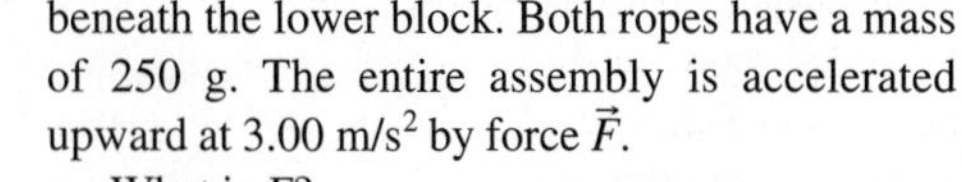

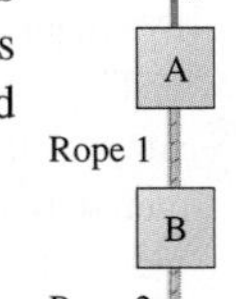

FIGURE P5.38

39. || Each of 100 identical blocks sitting on a frictionless surface is connected to the next block by a massless string. The first block is pulled with a force of 100 N.
  a. What is the tension in the string connecting block 100 to block 99?
  b. What is the tension in the string connecting block 50 to block 51?

40. || Two blocks on a frictionless table, A and B, are connected by a massless string. When block A is pulled with a certain force, dragging block B, the tension in the string is 24 N. When block B is pulled by the same force, dragging block A, the tension is 18 N. What is the ratio m_A/m_B of the blocks' masses?

General Problems

41. ||| A 500 kg piano is being lowered into position by a crane while two people steady it with ropes pulling to the sides. Bob's rope pulls to the left, 15° below horizontal, with 500 N of tension. Ellen's rope pulls toward the right, 25° below horizontal.
  a. What tension must Ellen maintain in her rope to keep the piano descending vertically at constant speed?
  b. What is the tension in the vertical main cable supporting the piano?

42. || Dana has a sports medal suspended by a long ribbon from her rearview mirror. As she accelerates onto the highway, she notices that the medal is hanging at an angle of 10° from the vertical.
  a. Does the medal lean toward or away from the windshield? Explain.
  b. What is her acceleration?

43. || Figure P5.43 shows the velocity graph of a 2.0 kg object as it moves along the x-axis. What is the net force acting on this object at $t = 1$ s? At 4 s? At 7 s?

INT

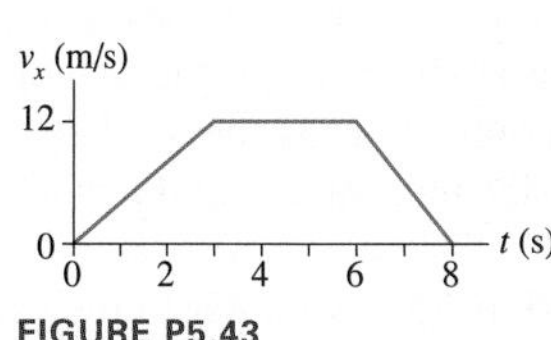

FIGURE P5.43

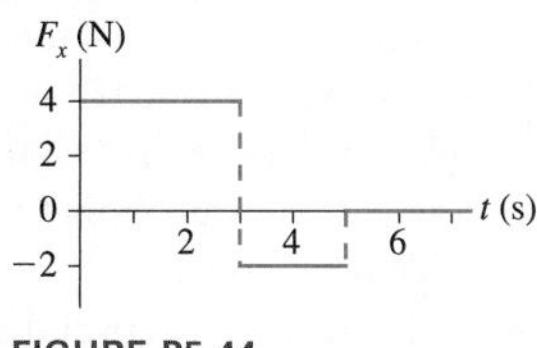

FIGURE P5.44

44. || Figure P5.44 shows the net force acting on a 2.0 kg object as it moves along the x-axis. The object is at rest at the origin at $t = 0$ s. What are its acceleration and velocity at $t = 6.0$ s?

INT

45. || A 50 kg box hangs from a rope. What is the tension in the rope if
  a. The box is at rest?
  b. The box has $v_y = 5.0$ m/s and is speeding up at 5.0 m/s^2?

46. || A 50 kg box hangs from a rope. What is the tension in the rope if
  a. The box moves up at a steady 5.0 m/s?
  b. The box has $v_y = 5.0$ m/s and is slowing down at 5.0 m/s^2?

47. | Your forehead can withstand a force of about 6.0 kN before fracturing, while your cheekbone can only withstand about 1.3 kN.

BIO
  a. If a 140 g baseball strikes your head at 30 m/s and stops in 0.0015 s, what is the magnitude of the ball's acceleration?
  b. What is the magnitude of the force that stops the baseball?
  c. What force does the baseball apply to your head? Explain.
  d. Are you in danger of a fracture if the ball hits you in the forehead? In the cheek?

48. ||| Seat belts and air bags save lives by reducing the forces exerted on the driver and passengers in an automobile collision. Cars are designed with a "crumple zone" in the front of the car. In the event of an impact, the passenger compartment decelerates over a distance of about 1 m as the front of the car crumples. An occupant restrained by seat belts and air bags decelerates with the car. By contrast, an unrestrained occupant keeps moving forward with no loss of speed (Newton's first law!) until hitting the dashboard or windshield, as we saw in Figure 4.2. These are unyielding surfaces, and the unfortunate occupant then decelerates over a distance of only about 5 mm.

BIO
  a. A 60 kg person is in a head-on collision. The car's speed at impact is 15 m/s. Estimate the net force on the person if he or she is wearing a seat belt and if the air bag deploys.
  b. Estimate the net force that ultimately stops the person if he or she is not restrained by a seat belt or air bag.
  c. How do these two forces compare to the person's weight?

49. |||| Bob, who has a mass of 75 kg, can throw a 500 g rock with a speed of 30 m/s. The distance through which his hand moves as he accelerates the rock forward from rest until he releases it is 1.0 m.

INT
  a. What constant force must Bob exert on the rock to throw it with this speed?
  b. If Bob is standing on frictionless ice, what is his recoil speed after releasing the rock?

50. |||| An 80 kg spacewalking astronaut pushes off a 640 kg satellite, exerting a 100 N force for the 0.50 s it takes him to straighten his arms. How far apart are the astronaut and the satellite after 1.0 min?

INT

51. || What thrust does a 200 g model rocket need in order to have a vertical acceleration of 10.0 m/s^2
  a. On earth?
  b. On the moon, where $g = 1.62$ m/s^2?

52. ||| A 20,000 kg rocket has a rocket motor that generates 3.0×10^5 N of thrust.
  a. What is the rocket's initial upward acceleration?
  b. At an altitude of 5.0 km the rocket's acceleration has increased to 6.0 m/s^2. What mass of fuel has it burned?

53. ||| You've always wondered about the acceleration of the elevators in the 101-story-tall Empire State Building. One day, while visiting New York, you take your bathroom scales into the elevator and stand on them. The scales read 150 lb as the door closes. The reading varies between 120 lb and 170 lb as the elevator travels 101 floors.
  a. What is the magnitude of the acceleration as the elevator starts upward?
  b. What is the magnitude of the acceleration as the elevator brakes to a stop?

54. |||| A 23 kg child goes down a straight slide inclined 38° above horizontal. The child is acted on by his weight, the normal force from the slide, kinetic friction, and a horizontal rope exerting a 30 N force as shown in Figure P5.54. How large is the normal force of the slide on the child?

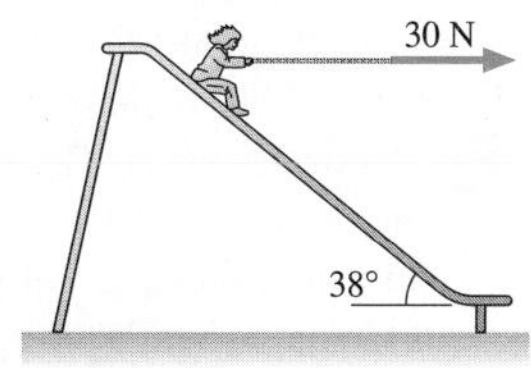

FIGURE P5.54

55. || Josh starts his sled at the top of a 3.0-m-high hill that has a constant slope of 25°. After reaching the bottom, he slides across a horizontal patch of snow. The hill is frictionless, but the coefficient of kinetic friction between his sled and the snow is 0.05. How far from the base of the hill does he end up?

INT

56. || A wood block, after being given a starting push, slides down a wood ramp at a constant speed. What is the angle of the ramp above horizontal?

57. || Researchers often use *force plates* to measure the forces BIO INT that people exert against the floor during movement. A force plate works like a bathroom scale, but it keeps a record of how the reading changes with time. Figure P5.57 shows the data from a force plate as a woman jumps straight up and then lands.
   a. What was the vertical component of her acceleration during push-off?
   b. What was the vertical component of her acceleration while in the air?
   c. What was the vertical component of her acceleration during the landing?
   d. What was her speed as her feet left the force plate?
   e. How high did she jump?

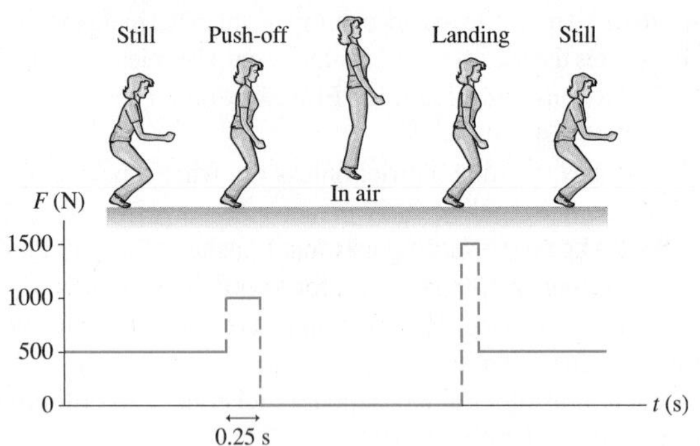

FIGURE P5.57

58. |||| A 77 kg sprinter is running the 100 m dash. At one instant, BIO early in the race, his acceleration is 4.7 m/s^2.
   a. What *total* force does the track surface exert on the sprinter? Assume his acceleration is parallel to the ground. Give your answer as a magnitude and an angle with respect to the horizontal.
   b. This force is applied to one foot (the other foot is in the air), which for a fraction of a second is stationary with respect to the track surface. Because the foot is stationary, the net force on it must be zero. Thus the force of the lower leg bone on the foot is equal but opposite to the force of the track on the foot. If the lower leg bone is 60° from horizontal, what are the components of the leg's force on the foot in the directions parallel and perpendicular to the leg? (Force components perpendicular to the leg can cause dislocation of the ankle joint.)

59. |||| Sam, whose mass is 75 kg, takes off across level snow on his jet-powered skis. The skis have a thrust of 200 N and a coefficient of kinetic friction on snow of 0.10. Unfortunately, the skis run out of fuel after only 10 s.
   a. What is Sam's top speed?
   b. How far has Sam traveled when he finally coasts to a stop?

60. ||| A person with compromised pinch strength in his fingers can only exert a normal force of 6.0 N to either side of a pinch-BIO held object, such as the book shown in Figure P5.60. What is the heaviest book he can hold onto vertically before it slips out of his fingers? The coefficient of static friction of the surface between the fingers and the book cover is 0.80.

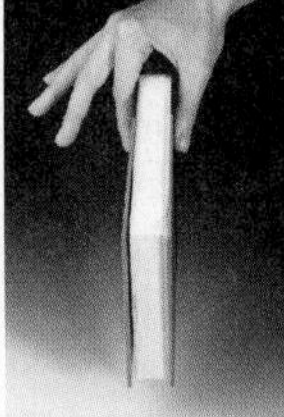
FIGURE P5.60

61. ||| A 1.0 kg wood block is pressed against a vertical wood wall by a 12 N force as shown in Figure P5.61. If the block is initially at rest, will it move upward, move downward, or stay at rest?

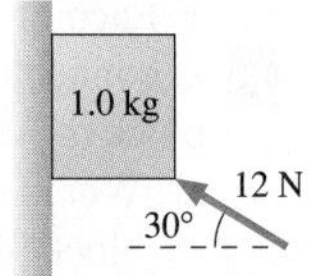

FIGURE P5.61

62. |||| A 50,000 kg locomotive, with steel wheels, is traveling at 10 m/s on steel rails when its engine and brakes both fail. How far will the locomotive roll before it comes to a stop?

63. || An Airbus A320 jetliner has a takeoff mass of 75,000 kg. It reaches its takeoff speed of 82 m/s (180 mph) in 35 s. What is the thrust of the engines? You can neglect air resistance but not rolling friction.

64. |||| A 2.0 kg wood block is launched up a wooden ramp that is inclined at a 35° angle. The block's initial speed is 10 m/s.
   a. What vertical height does the block reach above its starting point?
   b. What speed does it have when it slides back down to its starting point?

65. ||| Two blocks are at rest on a frictionless incline, as shown in Figure P5.65. What are the tensions in the two strings?

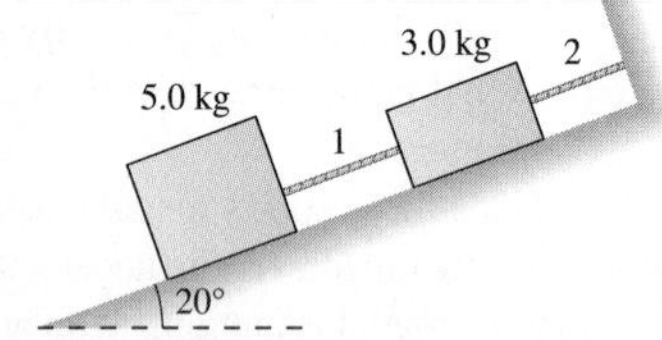

FIGURE P5.65

66. || Two identical blocks are stacked one on top of the other. The bottom block is free to slide on a frictionless surface. The coefficient of static friction between the blocks is 0.35. What is the maximum horizontal force that can be applied to the lower block without the upper block slipping?

67. ||| A wood block is sliding up a wood ramp. If the ramp is very steep, the block will reverse direction at its highest point and slide back down. If the ramp is shallow, the block will stop when it reaches its highest point. What is the smallest ramp angle, measured from the horizontal, for which the block will slide back down?

68. |||| The fastest recorded skydive was by an Air Force officer who jumped from a helium balloon at an elevation of 103,000 ft, three times higher than airliners fly. Because the density of air is so low at these altitudes, he reached a speed of 614 mph at an elevation of 90,000 ft, then gradually slowed as the air became more dense. Assume that he fell in the spread-eagle position of Example 5.15 and that his low-altitude terminal speed is 125 mph. Use this information to determine the density of air at 90,000 ft.

69. ||| A 2.7 g Ping-Pong ball has a diameter of 4.0 cm.
   a. The ball is shot straight up at twice its terminal speed. What is its initial acceleration?
   b. The ball is shot straight down at twice its terminal speed. What is its initial acceleration?

70. ||| Two blocks are connected by a string as in Figure P5.70. What is the upper block's acceleration if the coefficient of kinetic friction between the block and the table is 0.20?

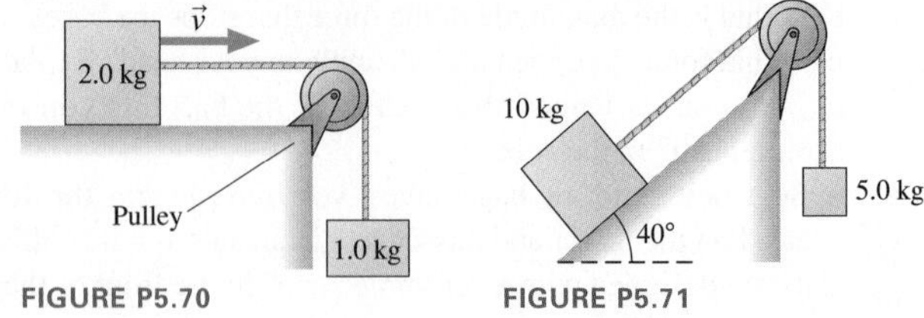

FIGURE P5.70
FIGURE P5.71

71. |||| The 10 kg block in Figure P5.71 slides down a frictionless ramp. What is its acceleration?

72. || A 2.0 kg wood block is pulled along a wood floor at a steady speed. A second wood block, with mass 3.0 kg, is attached to the first by a horizontal string. What is the magnitude of the force pulling on the first block?

73. |||| A magician pulls a tablecloth out from under some dishes. How far do the dishes move during the 0.25 s it takes to pull out the tablecloth? The coefficient of kinetic friction between the cloth and the dishes is $\mu_k = 0.12$.

74. ||| The 100 kg block in Figure P5.74 takes 6.0 s to reach the floor after being released from rest. What is the mass of the block on the left?

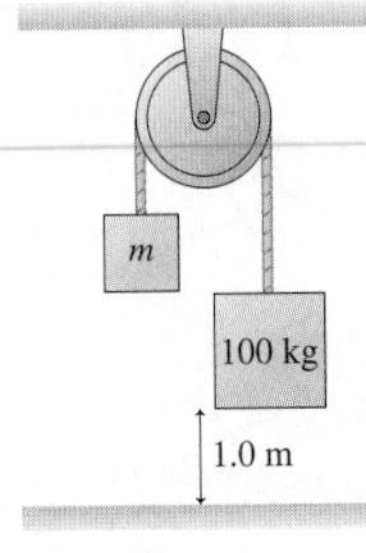

FIGURE P5.74

Problems 75 and 76 show free-body diagrams. For each,

a. Write a realistic dynamics problem for which this is the correct free-body diagram. Your problem should ask a question that can be answered with a value of position or velocity (such as "How far?" or "How fast?"), and should give sufficient information to allow a solution.
b. Solve your problem!

75. |

76. |||

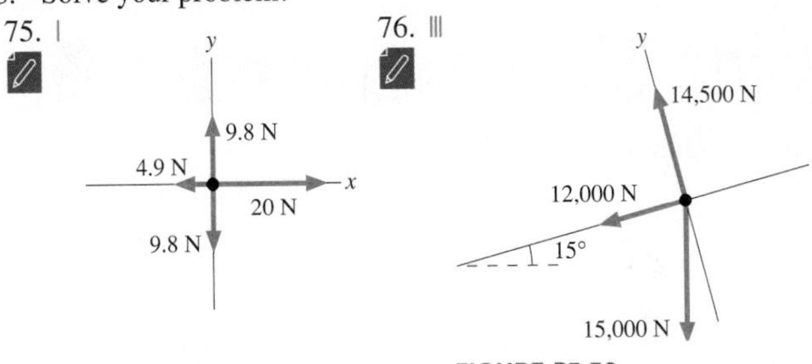

FIGURE P5.75

FIGURE P5.76

In Problems 77 through 79 you are given the dynamics equations that are used to solve a problem. For each of these, you are to

a. Write a realistic problem for which these are the correct equations.
b. Draw the free-body diagram and the pictorial representation for your problem.
c. Finish the solution of the problem.

77. || $-0.80n = (1500\text{ kg})a_x$
$n - (1500\text{ kg})(9.8\text{ m/s}^2) = 0$

78. || $T - 0.2n - (20\text{ kg})(9.8\text{ m/s}^2)\sin 20° = (20\text{ kg})(2.0\text{ m/s}^2)$
$n - (20\text{ kg})(9.8\text{ m/s}^2)\cos 20° = 0$

79. || $(100\text{ N})\cos 30° - f_k = (20\text{ kg})a_x$
$n + (100\text{ N})\sin 30° - (20\text{ kg})(9.8\text{ m/s}^2) = 0$
$f_k = 0.20n$

Passage Problems

Sliding on the Ice

In the winter sport of curling, players give a 20 kg stone a push across a sheet of ice. The stone moves approximately 40 m before coming to rest. The final position of the stone, in principle, only depends on the initial speed at which it is launched and the force of friction between the ice and the stone, but team members can use brooms to sweep the ice in front of the stone to adjust its speed and trajectory a bit; they must do this without touching the stone. Judicious sweeping can lengthen the travel of the stone by 3 m.

80. | A curler pushes a stone to a speed of 3.0 m/s over a time of 2.0 s. Ignoring the force of friction, how much force must the curler apply to the stone to bring it up to speed?
A. 3.0 N B. 15 N C. 30 N D. 150 N

81. | The sweepers in a curling competition adjust the trajectory of the stone by
A. Decreasing the coefficient of friction between the stone and the ice.
B. Increasing the coefficient of friction between the stone and the ice.
C. Changing friction from kinetic to static.
D. Changing friction from static to kinetic.

82. | Suppose the stone is launched with a speed of 3 m/s and travels 40 m before coming to rest. What is the *approximate* magnitude of the friction force on the stone?
A. 0 N B. 2 N C. 20 N D. 200 N

83. | Suppose the stone's mass is increased to 40 kg, but it is launched at the same 3 m/s. Which one of the following is true?
A. The stone would now travel a longer distance before coming to rest.
B. The stone would now travel a shorter distance before coming to rest.
C. The coefficient of friction would now be greater.
D. The force of friction would now be greater.

STOP TO THINK ANSWERS

Stop to Think 5.1: A. The lander is descending and slowing. The acceleration vector points upward, and so $\vec{F}_{net}$ points upward. This can be true only if the thrust has a larger magnitude than the weight.

Stop to Think 5.2: D. When you are in the air, there is *no* contact force supporting you, so your apparent weight is zero: You are weightless.

Stop to Think 5.3: $f_B > f_C = f_D = f_E > f_A$. Situations C, D, and E are all kinetic friction, which does not depend on either velocity or acceleration. Kinetic friction is less than the maximum static friction that is exerted in B. $f_A = 0$ because no friction is needed to keep the object at rest.

Stop to Think 5.4: D. The ball is shot *down* at 30 m/s, so $v_{0y} = -30$ m/s. This exceeds the terminal speed, so the upward drag force is *greater* than the downward weight force. Thus the ball *slows down* even though it is "falling." It will slow until $v_y = -15$ m/s, the terminal velocity, then maintain that velocity.

Stop to Think 5.5: B. $F_{Q\text{on}H} = F_{H\text{on}Q}$ and $F_{P\text{on}Q} = F_{Q\text{on}P}$ because these are action/reaction pairs. Box Q is slowing down and therefore must have a net force to the left. So from Newton's second law we also know that $F_{H\text{on}Q} > F_{P\text{on}Q}$.

Stop to Think 5.6: Equal to. Each block is hanging in equilibrium, with no net force, so the upward tension force is mg.

6 Circular Motion, Orbits, and Gravity

Motorcyclists in the "Globe of Death" ride their bikes on the inside of a spherical steel frame, seeming to defy gravity as they ride up the sides and upside down over the top. What prevents them from falling?

LOOKING AHEAD ▸▸

The goal of Chapter 6 is to learn about motion in a circle, including orbital motion under the influence of a gravitational force.

Uniform Circular Motion

A particle moving in a circle at a constant speed undergoes **uniform circular motion**. In this chapter you'll learn how to describe a particle's motion in terms of its *angular* position and *angular* velocity.

We'll also review the important idea that a particle moving in a circle has an acceleration directed toward the center of the circle.

Looking Back ◂◂
2.2 Uniform motion
3.8 Circular motion

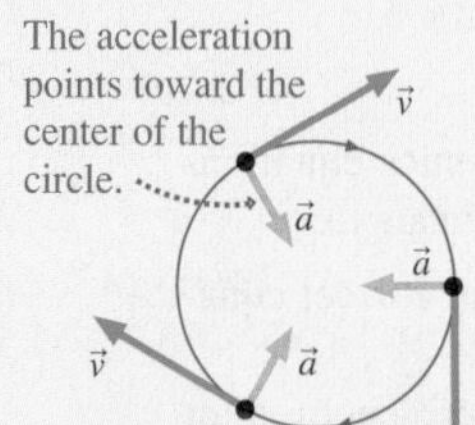

Dynamics of Uniform Circular Motion

Because a particle moving in a circle has an acceleration that points toward the center of the circle, there must be a net force toward the center to cause this acceleration.

Looking Back ◂◂
5.2 Using Newton's second law

The net force on the girl is directed toward the center of the circle.

The normal force of the track and the car's weight combine to provide a net force toward the circle's center.

Apparent Forces in Circular Motion

An object moving in a circle appears to experience a force that "flings" it outward. You'll learn that these apparent forces are not real forces, but are in fact a consequence of Newton's first law.

Looking Back ◂◂
5.3 Weight and apparent weight

What holds these riders in this carnival ride against the wall?

Newton's Law of Gravity

Newton discovered the law that governs gravity. You'll learn how it applies to an apple falling to earth or a rock falling on the moon, and how this law governs the motions of the moon, the planets, and even distant galaxies.

Newton's great insight was that the law of gravity described not only falling objects but also the orbits of the moon and planets.

Even the structure of distant galaxies is determined by Newton's law of gravity.

Gravity and Orbits

If an object moves fast enough, it can orbit the earth, the sun, or another planet.

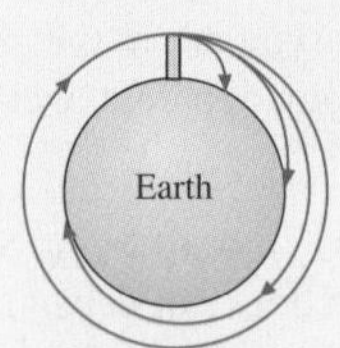

An orbit can be thought of as projectile motion where the ground curves away just as fast as the object falls.

The space station appears weightless, but gravity still acts strongly on it; only its *apparent weight* is zero.

6.1 Uniform Circular Motion

We began our study of circular motion in Section 3.8. There, we learned how to describe the circular motion of a particle in terms of its period and frequency. We also learned that a particle moving in a circle has an acceleration—even if the particle's speed is constant—because the *direction* of its velocity is constantly changing.

In this chapter, we'll study the simplest kind of circular motion, in which a particle moves at a *constant* speed around its circular path. **FIGURE 6.1** shows a particle undergoing this uniform circular motion. The particle might be a satellite moving in an orbit, a ball on the end of a string, or even just a dot painted on the side of a wheel. Regardless of what the particle represents, its velocity vector is always tangent to the circular path. The particle's speed v is constant, so the vector's length stays constant as the particle moves around the circle.

FIGURE 6.1 A particle in uniform circular motion.

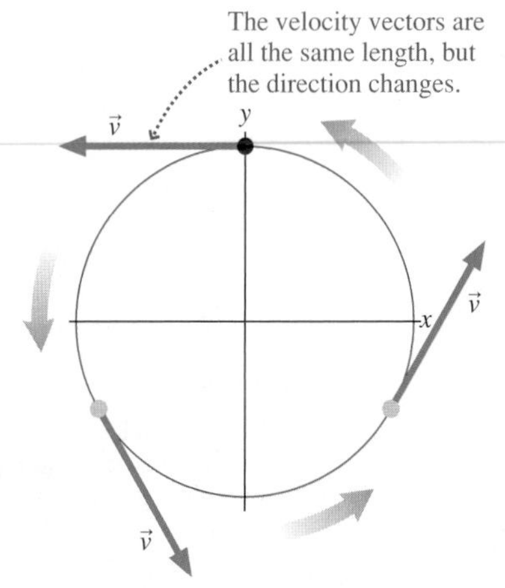

Angular Position

In order to describe the position of a particle as it moves around the circle, it is convenient to use the angle θ from the positive x-axis. This is shown in **FIGURE 6.2**. Because the particle travels in a circle with a fixed radius r, specifying θ completely locates the position of the particle. Thus we call angle θ the **angular position** of the particle.

We define θ to be positive when measured *counterclockwise* from the positive x-axis. An angle measured *clockwise* from the positive x-axis has a negative value. "Clockwise" and "counterclockwise" in circular motion are analogous, respectively, to "left of the origin" and "right of the origin" in linear motion, which we associated with negative and positive values of x.

FIGURE 6.2 A particle's angular position is described by angle θ.

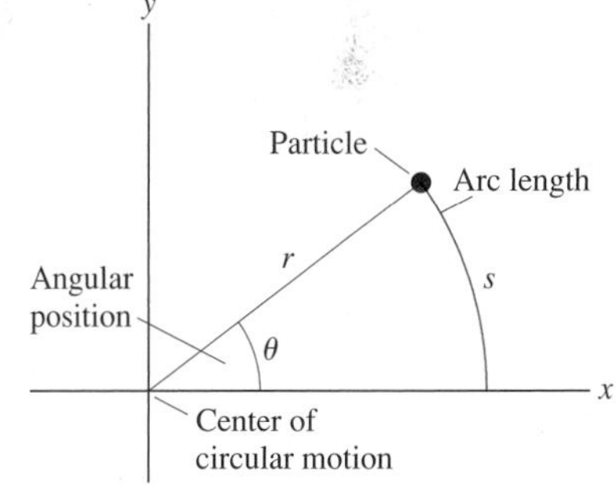

Rather than measure angles in degrees, mathematicians and scientists usually measure angle θ in the angular unit of *radians.* In Figure 6.2, we also show the **arc length** s, the distance that the particle has traveled along its circular path. We define the particle's angle θ in **radians** in terms of this arc length and the radius of the circle:

$$\theta \text{ (radians)} = \frac{s}{r} \tag{6.1}$$

This is a sensible definition of an angle: The farther the particle has traveled around the circle (i.e., the greater s is), the larger the angle θ in radians. The radian, abbreviated rad, is the SI unit of angle. An angle of 1 rad has an arc length s exactly equal to the radius r. An important consequence of Equation 6.1 is that the arc length spanning the angle θ is

$$s = r\theta \tag{6.2}$$

NOTE ► Equation 6.2 is valid only if θ is measured in radians, not degrees. This very simple relationship between angle and arc length is one of the primary motivations for using radians. ◄

When a particle travels all the way around the circle—completing one *revolution,* abbreviated rev—the arc length it travels is the circle's circumference $2\pi r$. Thus the angle of a full circle is

$$\theta_{\text{full circle}} = \frac{s}{r} = \frac{2\pi r}{r} = 2\pi \text{ rad}$$

We can use this fact to define conversion factors among revolutions, radians, and degrees:

$$1 \text{ rev} = 360° = 2\pi \text{ rad}$$

$$1 \text{ rad} = 1 \text{ rad} \times \frac{360°}{2\pi \text{ rad}} = 57.3°$$

We will often specify angles in degrees, but keep in mind that the SI unit is the radian. You can visualize angles in radians by remembering that 1 rad is just about 60°.

Angular Displacement and Angular Velocity

For the *linear* motion you studied in Chapters 1 and 2, a particle with a larger velocity undergoes a greater displacement in each second than one with a smaller velocity, as **FIGURE 6.3a** shows. **FIGURE 6.3b** shows two particles undergoing uniform *circular* motion. The particle on the left is moving slowly around the circle; it has gone only one-quarter of the way around after 5 seconds. The particle on the right is moving much faster around the circle, covering half of the circle in the same 5 seconds. You can see that the particle to the right undergoes twice the **angular displacement** $\Delta\theta$ during each interval as the particle to the left. Its **angular velocity,** the angular displacement through which the particle moves each second, is twice as large.

FIGURE 6.3 Comparing uniform linear and circular motion.

(a) Uniform linear motion

A particle with a small velocity v undergoes a small displacement Δx each second.

0 s 1 s 2 s 3 s 4 s 5 s

A particle with a large velocity v undergoes a large displacement Δx each second.

(b) Uniform circular motion

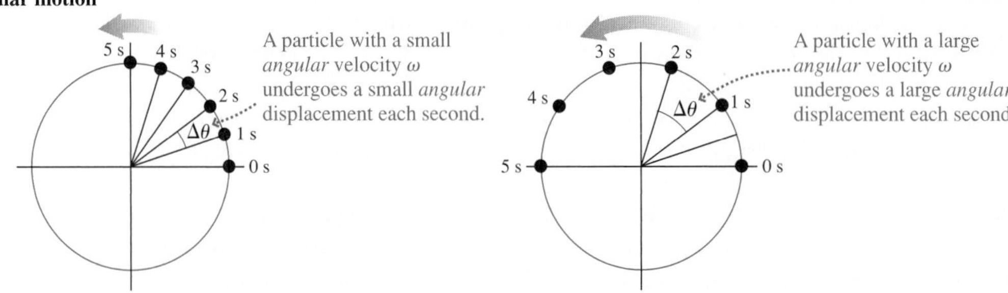

5 s
4 s
3 s
2 s
1 s
0 s
Δθ
A particle with a small angular velocity ω undergoes a small angular displacement each second.
3 s
2 s
4 s
1 s
5 s
0 s
Δθ
A particle with a large angular velocity ω undergoes a large angular displacement each second.

In analogy with linear motion, where $v_x = \Delta x/\Delta t$, we thus define the angular velocity as

$$\omega = \frac{\text{angular displacement}}{\text{time interval}} = \frac{\Delta\theta}{\Delta t} \tag{6.3}$$

Angular velocity of a particle in uniform circular motion

The symbol ω is a lowercase Greek omega, *not* an ordinary w. The SI unit of angular velocity is rad/s.

Figure 6.3a shows that the displacement Δx of a particle in uniform linear motion changes by the same amount each second. Similarly, as Figure 6.3b shows, the *angular* displacement $\Delta\theta$ of a particle in uniform *circular* motion changes by the same amount each second. This means that **the angular velocity $\omega = \Delta\theta/\Delta t$ is constant for a particle moving with uniform circular motion.**

EXAMPLE 6.1 Comparing angular velocities

Find the angular velocities of the two particles in Figure 6.3b.

PREPARE For uniform circular motion, we can use any angular displacement $\Delta\theta$, as long as we use the corresponding time interval Δt. For each particle, we'll choose the angular displacement corresponding to the motion from $t = 0$ s to $t = 5$ s.

SOLVE The particle on the left travels one-quarter of a full circle during the 5 s time interval. We learned earlier that a full circle corresponds to an angle of 2π rad, so the angular displacement for this particle is $\Delta\theta = (2\pi \text{ rad})/4 = \pi/2$ rad. Thus its angular velocity is

$$\omega = \frac{\Delta\theta}{\Delta t} = \frac{\pi/2 \text{ rad}}{5 \text{ s}} = 0.31 \text{ rad/s}$$

The particle on the right travels halfway around the circle, or π rad, in the 5 s interval. Its angular velocity is

$$\omega = \frac{\Delta\theta}{\Delta t} = \frac{\pi \text{ rad}}{5 \text{ s}} = 0.63 \text{ rad/s}$$

ASSESS The angular velocity of the particle on the right is 0.63 rad/s, meaning that the particle travels through an angle of 0.63 rad each second. Because 1 rad $\approx 60°$, 0.63 rad is roughly 35°. In Figure 6.3b, the particle on the right appears to move through an angle of about this size during each 1 s time interval, so our answer is reasonable.

Angular velocity, like the velocity v_x of one-dimensional motion, can be positive or negative. The signs for ω noted in **FIGURE 6.4** are based on the convention that angles are positive when measured counterclockwise from the positive x-axis.

We've already noted how circular motion is analogous to linear motion, with angular variables replacing linear variables. Thus much of what you learned about linear kinematics and dynamics carries over to circular motion. For example, Equation 2.4 gave us a formula for computing a linear displacement during a time interval:

$$x_f - x_i = \Delta x = v_x \, \Delta t$$

You can see from Equation 6.3 that we can write a similar equation for the angular displacement:

$$\theta_f - \theta_i = \Delta\theta = \omega \, \Delta t \quad (6.4)$$

Angular displacement for uniform circular motion

FIGURE 6.4 Positive and negative angular velocities.

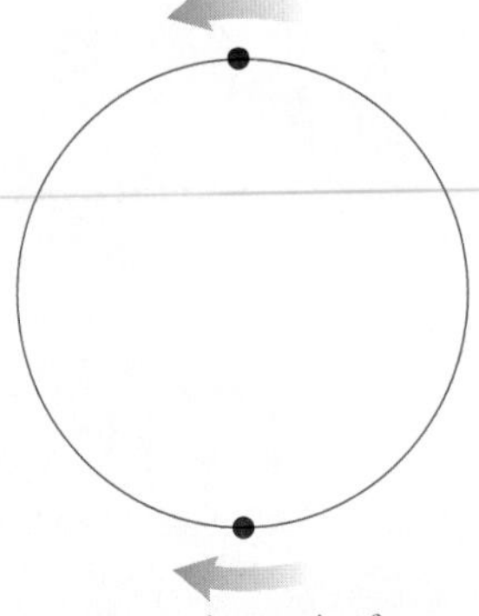
ω is positive for a counterclockwise rotation.
ω is negative for a clockwise rotation.

For linear motion, we use the term *speed v* when we are not concerned with the direction of motion, *velocity* v_x when we are. For circular motion, we define the **angular speed** to be the absolute value of the angular velocity, so that it's a positive quantity irrespective of the particle's direction of rotation. Although potentially confusing, it is customary to use the symbol ω for angular speed *and* for angular velocity. If the direction of rotation is not important, we will interpret ω to mean angular speed. In kinematic equations, such as Equation 6.4, ω is always the angular velocity, and you need to use a negative value for clockwise rotation.

EXAMPLE 6.2 **Kinematics at the roulette wheel**

A small steel ball rolls counterclockwise around the inside of a 30.0-cm-diameter roulette wheel. The ball completes exactly 2 rev in 1.20 s.

a. What is the ball's angular velocity?
b. What is the ball's angular position at $t = 2.00$ s? Assume $\theta_i = 0$.

PREPARE Treat the ball as a particle in uniform circular motion.

SOLVE

a. The ball's angular velocity is $\omega = \Delta\theta/\Delta t$. We know that the ball completes 2 revolutions in 1.20 s, and that each revolution corresponds to an angular displacement $\Delta\theta = 2\pi$ rad. Thus

$$\omega = \frac{2(2\pi \text{ rad})}{1.20 \text{ s}} = 10.5 \text{ rad/s}$$

Because the rotation direction is counterclockwise, the angular velocity is positive.

b. The ball moves with constant angular velocity, so its angular position is given by Equation 6.4. Thus the ball's angular position at $t = 2.00$ s is

$$\theta_f = \theta_i + \omega \, \Delta t = 0 \text{ rad} + (10.5 \text{ rad/s})(2.00 \text{ s}) = 21.0 \text{ rad}$$

If we're interested in where the ball is in the wheel at $t = 2.00$ s, we can write its angular position as an integer multiple of 2π (representing the number of complete revolutions the ball has made) plus a remainder:

$$\theta_f = 21.0 \text{ rad} = 3.34 \times 2\pi \text{ rad}$$
$$= 3 \times 2\pi \text{ rad} + 0.34 \times 2\pi \text{ rad}$$
$$= 3 \times 2\pi \text{ rad} + 2.1 \text{ rad}$$

In other words, at $t = 2.00$ s, the ball has completed 3 rev and is 2.1 rad = 120° into its fourth revolution. An observer would say that the ball's angular position is $\theta = 120°$.

ASSESS Since the ball completes 2 revolutions in 1.20 s, it seems reasonable that it completes 3.34 revolutions in 2.00 s.

The angular speed ω is closely related to the period T and the frequency f of the motion. If a particle in uniform circular motion moves around a circle once, which by definition takes time T, its angular displacement is $\Delta\theta = 2\pi$ rad. The angular speed is thus

$$\omega = \frac{2\pi \text{ rad}}{T} \quad (6.5)$$

◀ **Why do clocks go clockwise?** In the northern hemisphere, the rotation of the earth causes the sun to follow a circular arc through the southern sky, rising in the east and setting in the west. For millennia, humans have marked passing time by noting the position of shadows cast by the sun, which sweep in an arc from west to east—eventually leading to the development of the sundial, the first practical timekeeping device. In the northern hemisphere, sundials point north, and the shadow sweeps around the dial from left to right. Early clockmakers used the same convention, which is how it came to be clockwise.

We can also write the angular speed in terms of the frequency $f = 1/T$:

$$\omega = (2\pi \text{ rad})f \tag{6.6}$$

where f must be in rev/s. For example, a particle in circular motion with frequency 10 rev/s would have angular speed $\omega = 20\pi$ rad/s $= 62.8$ rad/s.

EXAMPLE 6.3 Rotations in a car engine

The crankshaft in your car engine is turning at 3000 rpm. What is the shaft's angular velocity?

PREPARE We'll need to convert rpm to rev/s; then we can use Equation 6.6.

SOLVE We convert rpm to rev/s by

$$\left(3000 \frac{\text{rev}}{\text{min}}\right)\left(\frac{1 \text{ min}}{60 \text{ s}}\right) = 50 \text{ rev/s}$$

Thus the crankshaft's angular velocity is

$$\omega = (2\pi \text{ rad})f = (2\pi \text{ rad})(50 \text{ rev/s}) = 314 \text{ rad/s}$$

FIGURE 6.5 Angular position for the ball on the roulette wheel.

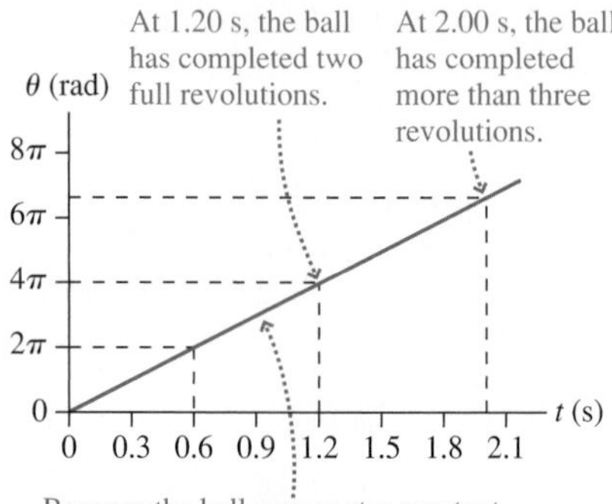

Angular-Position and Angular-Velocity Graphs

For the one-dimensional motion you studied in Chapter 3, we found that position- and velocity-versus-time graphs were important and useful representations of motion. We can use the same kinds of graphs to represent angular motion. Let's begin by considering the motion of the roulette ball of Example 6.2. We found that it had angular velocity $\omega = 10.5$ rad/s, meaning that its angular *position* changed by $+10.5$ rad every second. This is exactly analogous to the one-dimensional motion problem of a car driving in a straight line with a velocity of 10.5 m/s, so that its position increases by 10.5 m each second. Using this analogy, we can construct the **angular position-versus-time graph** for the roulette ball shown in FIGURE 6.5.

The angular velocity is given by $\omega = \Delta\theta/\Delta t$. Graphically, this is the *slope* of the angular position-versus-time graph, just as the ordinary velocity is the slope of the position-versus-time graph. Thus we can create an **angular velocity-versus-time graph** by finding the slope of the corresponding angular position-versus-time graph.

EXAMPLE 6.4 Graphing a bike ride

Jake rides his bicycle home from campus. FIGURE 6.6 is the angular position-versus-time graph for a small rock stuck in the tread of his tire. First, draw the rock's angular velocity-versus-time graph, using rpm on the vertical axis. Then interpret the graphs with a story about Jake's ride.

PREPARE Angular velocity ω is the slope of the angular position-versus-time graph.

SOLVE We can see that $\omega = 0$ rad/s during the first and last 30 s of Jake's ride because the horizontal segments of the graph have zero slope. Between $t = 30$ s and $t = 150$ s, an interval of 120 s,

FIGURE 6.6 Angular position-versus-time graph for Jake's bike ride.

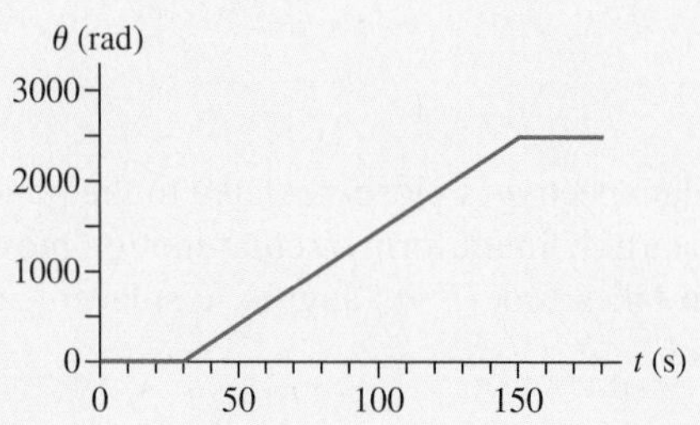

the rock's angular velocity (the slope of the angular position-versus-time graph) is

$$\omega = \text{slope} = \frac{2500 \text{ rad} - 0 \text{ rad}}{120 \text{ s}} = 20.8 \text{ rad/s}$$

We need to convert this to rpm:

$$\omega = \left(\frac{20.8 \text{ rad}}{1 \text{ s}}\right)\left(\frac{1 \text{ rev}}{2\pi \text{ rad}}\right)\left(\frac{60 \text{ s}}{1 \text{ min}}\right) = 200 \text{ rpm}$$

These values have been used to draw the angular velocity-versus-time graph of FIGURE 6.7. It looks like Jake waited 30 s for the light to change, then pedaled so that the bike wheel turned at a constant angular velocity of 200 rpm. 2.0 min later, he quickly braked to a stop for another 30-s-long red light.

FIGURE 6.7 Angular velocity-versus-time graph for Jake's bike ride.

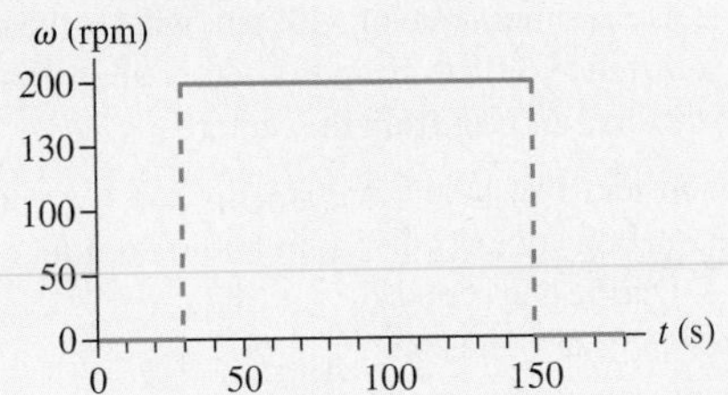

ASSESS At 200 rpm for 2 minutes, the wheel would turn roughly 400 times or, at ≈6 rad/rev, through about 2400 rad. Our answer seems reasonable.

STOP TO THINK 6.1 Which particle has angular position $5\pi/2$?

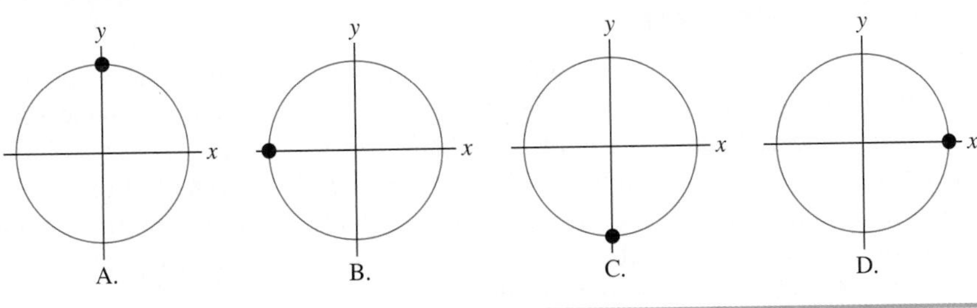

6.2 Speed, Velocity, and Acceleration in Uniform Circular Motion

The preceding section described uniform circular motion in terms of angular variables. In Chapter 3, we introduced a description of uniform circular motion in terms of velocity and acceleration vectors. We will now unite these two different descriptions, which will enable us to consider a much wider range of problems.

Speed

In Chapter 3, we found that the speed of a particle moving with frequency f around a circular path of radius r is $v = 2\pi fr$. If we combine this result with Equation 6.5 for the angular speed, we find that speed v and angular speed ω are related by

$$v = \omega r \tag{6.7}$$

Relationship between speed and angular speed

NOTE ▶ In Equation 6.7, **ω must be in units of rad/s.** If you are given a frequency in rev/s or rpm, you should convert it to an angular speed in rad/s. ◀

EXAMPLE 6.5 **Finding the speed at two points on a CD**

The diameter of an audio compact disc is 12.0 cm. When the disc is spinning at its maximum rate of 540 rpm, what is the speed of a point (a) at a distance 3.0 cm from the center and (b) at the outside edge of the disc, 6.0 cm from the center?

PREPARE Consider two points A and B on the rotating compact disc in **FIGURE 6.8**. During one period T, the disc rotates once, and both points rotate through the same angle, 2π rad. Thus the angular speed, $\omega = 2\pi/T$, is the same for these two points; in fact, it is the same for all points on the disc. But as they go around one time, the two points move different *distances;* the outer point B goes around a larger circle. The two points thus have different *speeds.* We can solve this problem by first finding the angular speed of the disc and then computing the speeds at the two points.

FIGURE 6.8 The rotation of an audio compact disc.

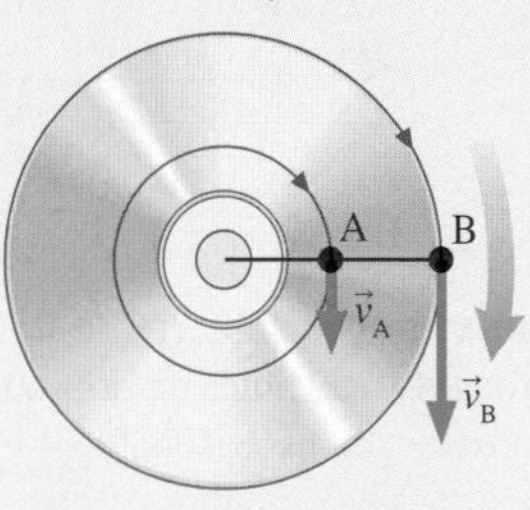
A
B
$\vec{v}_A$
$\vec{v}_B$

SOLVE We first convert the frequency of the disc from rpm to rev/s:

$$f = \left(540\,\frac{\text{rev}}{\text{min}}\right) \times \left(\frac{1\text{ min}}{60\text{ s}}\right) = 9.00\text{ rev/s}$$

We can compute the angular speed using Equation 6.6:

$$\omega = (2\pi\text{ rad})(9.00\text{ rev/s}) = 56.5\text{ rad/s}$$

We can now use Equation 6.7 to compute the speeds of points on the disc. At point A, $r = 3.0\text{ cm} = 0.030\text{ m}$, so the speed is

$$v_B = \omega r = (56.5\text{ rad/s})(0.030\text{ m}) = 1.7\text{ m/s}$$

At point B, $r = 6.0\text{ cm} = 0.060\text{ m}$, so the speed at the outside edge is

$$v_B = \omega r = (56.5\text{ rad/s})(0.060\text{ m}) = 3.4\text{ m/s}$$

ASSESS The speeds are a few meters per second, which seems reasonable. The point farther from the center is moving at a higher speed, as we expected.

Velocity and Acceleration

FIGURE 6.9 Velocity and acceleration for uniform circular motion.

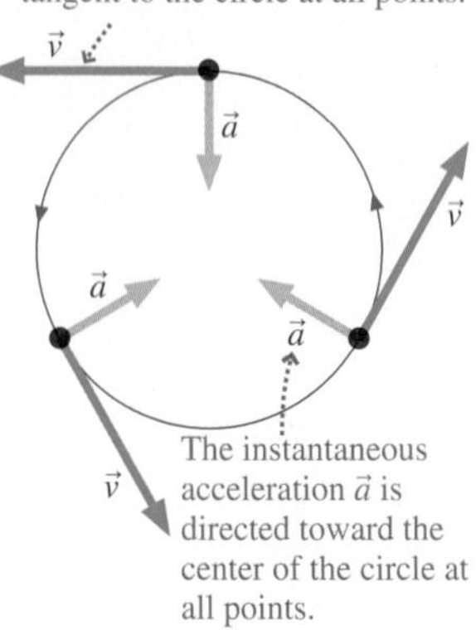
The instantaneous velocity $\vec{v}$ is tangent to the circle at all points.
$\vec{v}$
$\vec{a}$
The instantaneous acceleration $\vec{a}$ is directed toward the center of the circle at all points.

Although the *speed* of a particle in uniform circular motion is constant, its *velocity* is not constant because the *direction* of the motion is always changing. As you learned in Chapter 3, and as **FIGURE 6.9** reminds you, there is an acceleration at every point in the motion, with the acceleration vector $\vec{a}$ pointing toward the center of the circle. We called this the *centripetal acceleration,* and we showed that for uniform circular motion the acceleration was given by $a = v^2/r$. Because $v = \omega r$, we can also write this relationship in terms of the angular speed:

$$a = \frac{v^2}{r} = \omega^2 r \tag{6.8}$$

Centripetal acceleration for uniform circular motion

a — ω — p.47 — QUADRATIC

Acceleration depends on speed, but also distance from the center of the circle.

CONCEPTUAL EXAMPLE 6.6 **Who has the larger acceleration?**

Two children are riding in circles on a merry-go-round, as shown in **FIGURE 6.10**. Which child experiences the larger acceleration?

FIGURE 6.10 Top view of a merry-go-round.

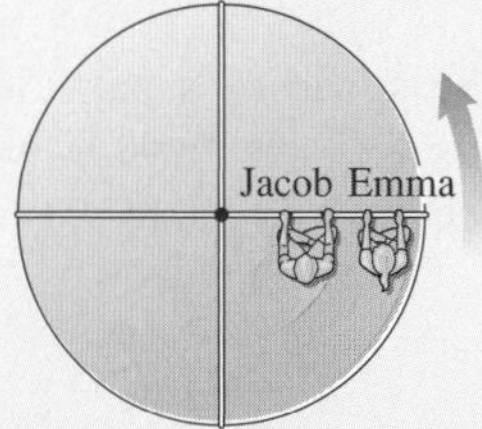
Jacob Emma

REASON As Example 6.2 showed, all points on the merry-go-round move at the same angular speed. The second expression for the acceleration in Equation 6.8 tells us that $a = \omega^2 r$. As the two children are moving with the same angular speed, Emma, with a larger value of r, experiences a larger acceleration.

ASSESS In Example 6.5, we saw that points farther from the center move at a higher speed. This would imply a higher acceleration as well, so our answer makes sense.

EXAMPLE 6.7 Finding the period of a carnival ride

In the Quasar carnival ride, passengers travel in a horizontal 5.0-m-radius circle. For safe operation, the maximum sustained acceleration that riders may experience is 20 m/s^2, approximately twice the free-fall acceleration. What is the period of the ride when it is being operated at the maximum acceleration?

FIGURE 6.11 Visual overview for the Quasar carnival ride.

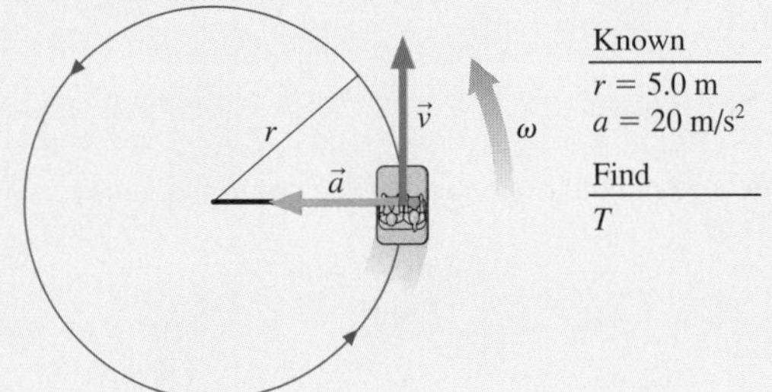

PREPARE We will assume that the cars on the ride are in uniform circular motion. The visual overview of **FIGURE 6.11** shows a top view of the motion of the ride.

SOLVE The angular speed can be computed from the acceleration by rearranging Equation 6.8:

$$\omega = \sqrt{\frac{a}{r}} = \sqrt{\frac{20\text{ m/s}^2}{5.0\text{ m}}} = 2.0\text{ rad/s}$$

At this angular speed, the period is $T = 2\pi/\omega = 3.1$ s.

ASSESS One rotation in just over 3 seconds seems reasonable for a pretty zippy carnival ride. The period for this particular ride is actually 3.7 s, so it runs a bit slower than the maximum safe speed.

STOP TO THINK 6.2 Rank in order, from largest to smallest, the centripetal accelerations of particles A to D.

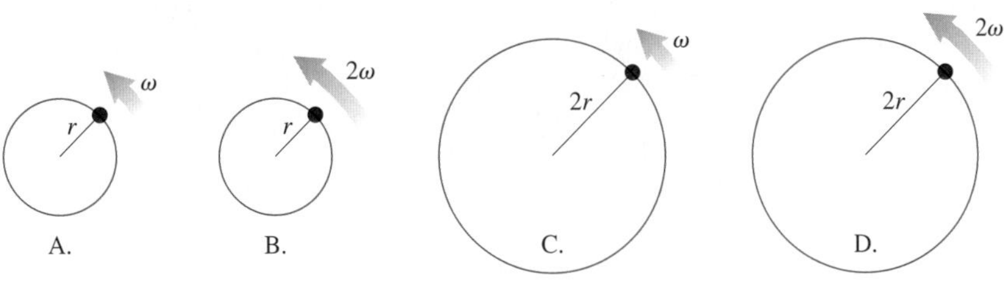

Hurling the heavy hammer This man is throwing a hammer that weighs over 30 pounds as far as he can by spinning the hammer around in a circle and then letting go. While he holds the handle, the hammer follows a circular path. He must provide a very large force directed toward the center of the circle to produce the centripetal acceleration, as you can see by how he is leaning away from the hammer. When he lets go, there is no longer a force directed toward the center, and the hammer will stop going in a circle and fly across the field.

6.3 Dynamics of Uniform Circular Motion

Riders traveling around on a circular carnival ride are accelerating, as we have just seen. Consequently, according to Newton's second law, the riders must have a net *force* acting on them. In this section, we'll look at the forces that cause uniform circular motion.

We've already determined the acceleration of a particle in uniform circular motion—the centripetal acceleration of Equation 6.8. Newton's second law tells us what the net force must be to cause this acceleration:

$$\vec{F}_{\text{net}} = m\vec{a} = \left(\frac{mv^2}{r} = m\omega^2 r, \text{ toward center of circle}\right) \quad (6.9)$$

Net force producing the centripetal acceleration of uniform circular motion

F_{net} vs v: QUADRATIC p.47; F_{net} vs r: INVERSE p.114

In other words, **a particle of mass m moving at constant speed v around a circle of radius r must always have a net force of magnitude $mv^2/r = m\omega^2 r$ pointing toward the center of the circle,** as in **FIGURE 6.12**. It is this net force that causes the centripetal acceleration of circular motion. Without such a net force, the particle would move off in a straight line tangent to the circle.

FIGURE 6.12 Net force for circular motion.

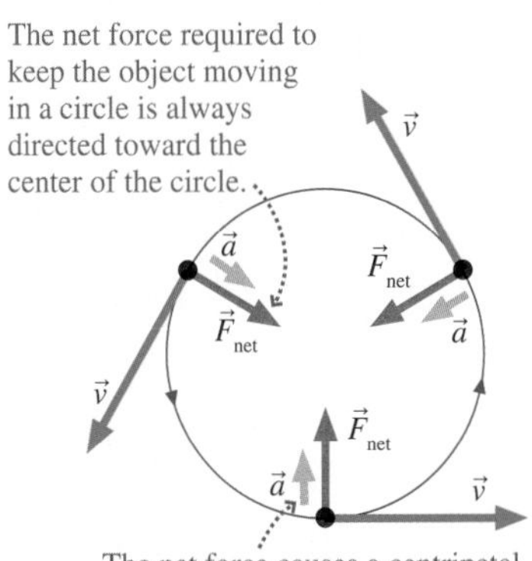

The net force required to keep the object moving in a circle is always directed toward the center of the circle.
$\vec{v}$
$\vec{a}$
$\vec{F}_{net}$
The net force causes a centripetal acceleration.

The force described by Equation 6.9 is not a *new* kind of force. The net force will be due to one or more of our familiar forces, such as tension, friction, or the normal

force. Equation 6.9 simply tells us how the net force needs to act—how strongly and in which direction—to cause the particle to move with speed v in a circle of radius r.

In each example of circular motion that we will consider in this chapter, a physical force or a combination of forces directed toward the center produces the necessary acceleration. In some cases, the circular motion and the force are obvious, as in the hammer throw. Other cases are more subtle. For instance, for a car following a circular path on a level road, the necessary force is provided by the friction force between the tires and the road.

CONCEPTUAL EXAMPLE 6.8 Forces on a car

A car drives through a circularly shaped valley at a constant speed. At the very bottom of the valley, is the normal force of the road on the car greater than, less than, or equal to the car's weight?

REASON FIGURE 6.13 shows a visual overview of the situation. The car is accelerating, even though it is moving at a constant speed, because its direction is changing. When the car is at the bottom of the valley, the center of its circular path is directly above it and so its acceleration vector points straight up. The free-body diagram of Figure 6.13 identifies the only two forces acting on the car as the normal force, pointing upward, and its weight, pointing downward. Which is larger: n or w?

Because $\vec{a}$ points upward, by Newton's second law there must be a net force on the car that also points upward. In order for this to be the case, the free-body diagram shows that the magnitude of the normal force must be *greater* than the weight.

FIGURE 6.13 Visual overview for the car in the valley.

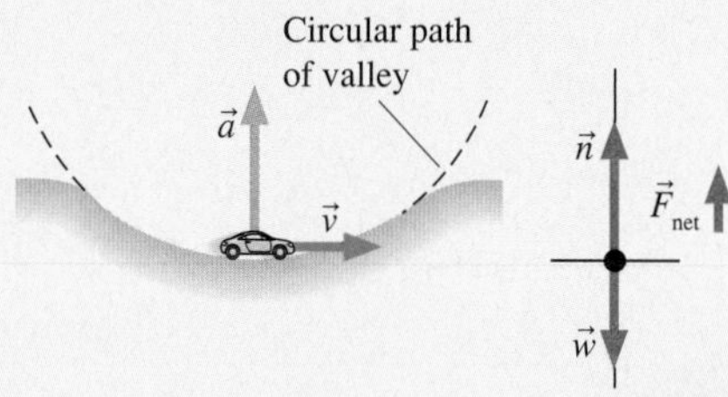

ASSESS You have probably experienced this situation. As you drive through a dip in the road, you feel "heavier" than normal. As discussed in Section 5.3, this is because your apparent weight—the normal force that supports you—is greater than your true weight.

Solving Circular Dynamics Problems

4.2, 4.3, 4.4, 4.5 

We have one basic equation for circular dynamics problems, Equation 6.9, which is just a version of Newton's second law. The techniques for solving circular dynamics problems are thus quite similar to those we have used for solving other Newton's second-law problems.

PROBLEM-SOLVING STRATEGY 6.1 Circular dynamics problems

PREPARE Begin your visual overview with a pictorial representation in which you sketch the motion, define symbols, define axes, and identify what the problem is trying to find. There are two common situations:

- If the motion is in a horizontal plane, like a tabletop, draw the free-body diagram with the circle viewed edge-on, the x-axis pointing toward the center of the circle, and the y-axis perpendicular to the plane of the circle.
- If the motion is in a vertical plane, like a Ferris wheel, draw the free-body diagram with the circle viewed face-on, the x-axis pointing toward the center of the circle, and the y-axis tangent to the circle.

SOLVE Newton's second law for uniform circular motion, $\vec{F}_{net} = (mv^2/r$, toward center of circle), is a vector equation. Some forces act in the plane of the circle, some act perpendicular to the circle, and some may have components in both directions. In the coordinate system described above, with the x-axis pointing toward the center of the circle, Newton's second law is

$$\sum F_x = \frac{mv^2}{r} = m\omega^2 r \quad \text{and} \quad \sum F_y = 0$$

Continued

That is, the net force toward the center of the circle has magnitude $mv^2/r = m\omega^2 r$ while the net force perpendicular to the circle is zero. The components of the forces are found directly from the free-body diagram. Depending on the problem, either:

- Use the net force to determine the speed v, then use circular kinematics to find frequencies or angular velocities.
- Use circular kinematics to determine the speed v, then solve for unknown forces.

ASSESS Make sure your net force points toward the center of the circle. Check that your result has the correct units, is reasonable, and answers the question.

Exercise 13 

EXAMPLE 6.9 Analyzing the motion of a cart

An energetic father places his 20 kg child on a 5.0 kg cart to which a 2.0-m-long rope is attached. He then holds the end of the rope and spins the cart and child around in a circle, keeping the rope parallel to the ground. If the tension in the rope is 100 N, how much time does it take for the cart to make one rotation?

PREPARE We proceed according to the steps of Problem-Solving Strategy 6.1. **FIGURE 6.14** shows a visual overview of the problem. The main reason for the pictorial representation on the left is to illustrate the relevant geometry and to define the symbols that will be used. A circular dynamics problem usually does not have starting and ending points like a projectile problem, so subscripts such as x_i or y_f are usually not needed. Here we need to define the cart's speed v and the radius r of the circle.

The object moving in the circle is the cart plus the child, a total mass of 25 kg; the free-body diagram shows the forces. Because the motion is in a horizontal plane, Problem-Solving Strategy 6.1 tells us to draw the free-body diagram looking at the edge of the circle, with the x-axis pointing toward the center of the circle and the y-axis perpendicular to the plane of the circle. Three forces are acting on the cart: the weight force $\vec{w}$, the normal force of the ground $\vec{n}$, and the tension force of the rope $\vec{T}$.

Notice that there are two quantities for which we use the symbol T: the tension and the period. We will include additional information when necessary to distinguish the two.

SOLVE There is no net force in the y-direction, perpendicular to the circle, so $\vec{w}$ and $\vec{n}$ must be equal and opposite. There is a net force in the x-direction, toward the center of the circle, as there must be to cause the centripetal acceleration of circular motion. Only the tension force has an x-component, so Newton's second law is

$$\sum F_x = T = \frac{mv^2}{r}$$

We know the mass, the radius of the circle, and the tension, so we can solve for v:

$$v = \sqrt{\frac{Tr}{m}} = \sqrt{\frac{(100\ \text{N})(2.0\ \text{m})}{25\ \text{kg}}} = 2.83\ \text{m/s}$$

From this, we can compute the period with a slight rearrangement of Equation 3.27:

$$T = \frac{2\pi r}{v} = \frac{(2\pi)(2.0\ \text{m})}{2.83\ \text{m/s}} = 4.4\ \text{s}$$

ASSESS The speed is about 3 m/s. Because 1 m/s ≈ 2 mph, the child is going about 6 mph. A trip around the circle in just over 4 s at a speed of about 6 mph sounds reasonable; it's a fast ride, but not so fast as to be scary!

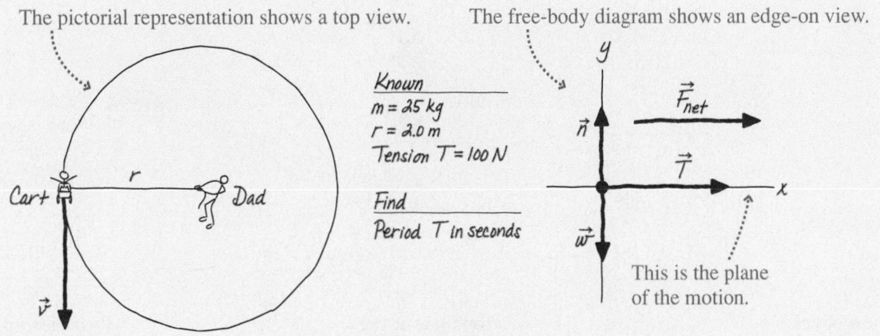

FIGURE 6.14 A visual overview of the cart spinning in a circle.

The pictorial representation shows a top view.

The free-body diagram shows an edge-on view.

Known
$m = 25$ kg
$r = 2.0$ m
Tension $T = 100$ N

Find
Period T in seconds

This is the plane of the motion.

EXAMPLE 6.10 Finding the maximum speed to turn a corner

What is the maximum speed with which a 1500 kg car can make a turn around a curve of radius 20 m on a level (unbanked) road without sliding? (This radius turn is about what you might expect at a major intersection in a city.)

PREPARE We start with the visual overview in **FIGURE 6.15**. The car moves along a circular arc at a constant speed—uniform circular motion—for the quarter-circle necessary to complete the turn. The motion before and after the turn is not relevant to the problem. The more interesting issue is *how* a car turns a corner. What force or forces can we identify that cause the direction of the velocity vector to change? Imagine you are driving a car on a frictionless road, such as a very icy road. You would not be able to turn a corner. Turning the steering wheel would be of no use; the car would slide straight ahead, in accordance with both Newton's first law and the experience of anyone who has ever driven on ice! So it must be *friction* that causes the car to turn.

The top view of the tire in Figure 6.15 shows the force on one of the car's tires as it turns a corner. If the road surface were frictionless, the tire would slide straight ahead. The force that prevents an object from sliding across a surface is *static friction*. Static friction $\vec{f}_s$ pushes *sideways* on the tire, toward the center of the circle. How do we know the direction is sideways? If $\vec{f}_s$ had a component either parallel to $\vec{v}$ or opposite $\vec{v}$, it would cause the car to speed up or slow down. Because the car changes direction but not speed, static friction must be perpendicular to $\vec{v}$. Thus $\vec{f}_s$ causes the centripetal acceleration of circular motion around the curve. With this in mind, the free-body diagram, drawn from behind the car, shows the static friction force pointing toward the center of the circle. Because the motion is in a horizontal plane, we've again chosen an x-axis toward the center of the circle and a y-axis perpendicular to the plane of motion.

SOLVE The only force in the x-direction, toward the center of the circle, is static friction. Newton's second law along the x-axis is

$$\sum F_x = f_s = \frac{mv^2}{r}$$

The only difference between this example and the preceding one is that the tension force toward the center has been replaced by a static friction force toward the center.

Newton's second law in the y-direction is

$$\sum F_y = n - w = ma_y = 0$$

so that $n = w = mg$.

The net force toward the center of the circle is the force of static friction. Recall from Equation 5.10 in Chapter 5 that static friction has a maximum possible value:

$$f_{s\,max} = \mu_s n = \mu_s mg$$

Because the static friction force has a maximum value, there will be a maximum speed at which a car can turn without sliding. This speed is reached when the static friction force reaches its maximum value $f_{s\,max} = \mu_s mg$. If the car enters the curve at a speed higher than the maximum, static friction cannot provide the necessary centripetal acceleration and the car will slide.

Thus the maximum speed occurs at the maximum value of the force of static friction, or when

$$f_{s\,max} = \frac{mv_{max}^2}{r}$$

Using the known value of $f_{s\,max}$, we find

$$\frac{mv_{max}^2}{r} = f_{s\,max} = \mu_s mg$$

Rearranging, we get

$$v_{max}^2 = \mu_s gr$$

For rubber tires on pavement, we find from Table 5.1 that $\mu_s = 1.0$. We then have

$$v_{max} = \sqrt{\mu_s gr} = \sqrt{(1.0)(9.8\ \text{m/s}^2)(20\ \text{m})} = 14\ \text{m/s}$$

ASSESS 14 m/s ≈ 30 mph, which seems like a reasonable upper limit for the speed at which a car can go around a curve without sliding. There are two other things to note about the solution:

- The car's mass canceled out. The maximum speed *does not* depend on the mass of the vehicle, though this may seem surprising.
- The final expression for v_{max} *does* depend on the coefficient of friction and the radius of the turn. v_{max} decreases if μ is less (a slipperier road) or if r is smaller (a tighter turn). Both make sense.

FIGURE 6.15 Visual overview of a car turning a corner.

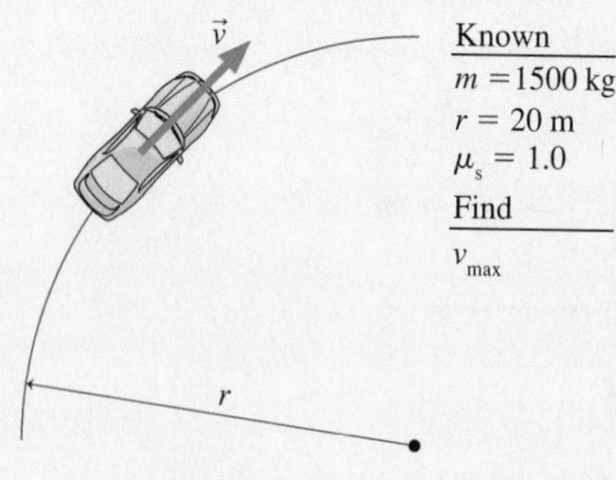

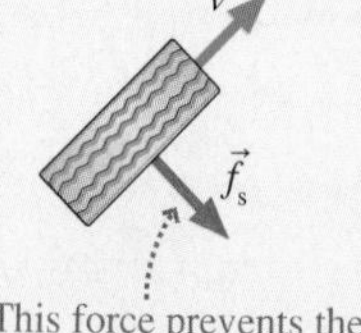

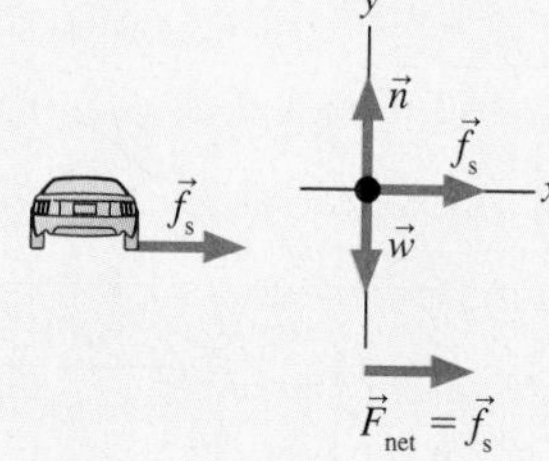

Because v_{max} depends on μ_s and because μ_s depends on road conditions, the maximum safe speed through turns can vary dramatically. Wet or icy roads lower the value of μ_s and thus lower the maximum speed of turns. A car that easily handles a curve in dry weather can suddenly slide out of control when the pavement is wet. Icy conditions are even worse. If you lower the value of the coefficient of friction in Example 6.10 from 1.0 (dry pavement) to 0.1 (icy pavement), the maximum speed for the turn goes down to 4.4 m/s—about 10 mph!

Race cars turn corners at much higher speeds than normal passenger vehicles. One design modification of the *cars* to allow this is the addition of wings, as on the car in **FIGURE 6.16**. The wings provide an additional force pushing the car *down* onto the pavement by deflecting air upward. This extra downward force increases the normal force, thus increasing the maximum static friction force and making faster turns possible.

There are also design modifications of the *track* to allow race cars to take corners at high speeds. If the track is banked by raising the outside edge of curved sections, the normal force can provide some of the force necessary to produce the centripetal acceleration, as we will see in the next example. The curves on racetracks may be quite sharply banked. Curves on ordinary highways are often banked as well, though at more modest angles suiting the lower speeds.

FIGURE 6.16 Wings on an Indy racer.

A banked turn on a racetrack.

EXAMPLE 6.11 **Finding speed on a banked turn**

A curve on a racetrack of radius 70 m is banked at a 15° angle. At what speed can a car take this curve without assistance from friction?

PREPARE After drawing the pictorial representation in **FIGURE 6.17**, we use the force identification diagram to find that, given that there is no friction acting, the only two forces are the normal force and the car's weight. We can then construct the free-body diagram, making sure that we draw the normal force perpendicular to the road's surface.

Even though the car is tilted, it is still moving in a *horizontal* circle. Thus, following Problem-Solving Strategy 6.1, we choose the x-axis to be horizontal and pointing toward the center of the circle.

FIGURE 6.17 Visual overview for the car on a banked turn.

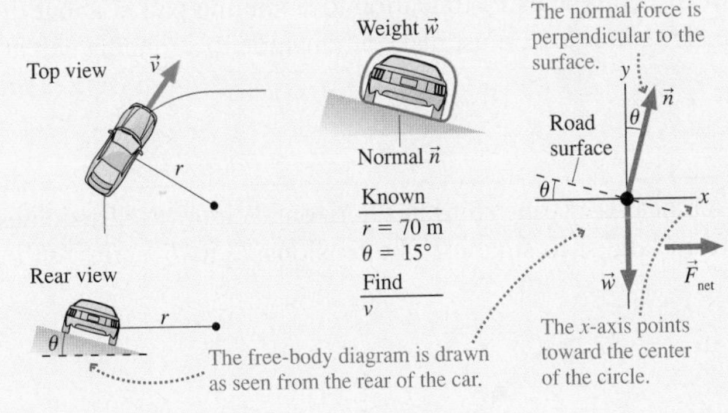
Top view
Weight $\vec{w}$
The normal force is perpendicular to the surface.
Road surface
Normal $\vec{n}$
Known
$r = 70$ m
$\theta = 15°$
Find
v
Rear view
The free-body diagram is drawn as seen from the rear of the car.
The x-axis points toward the center of the circle.
$\vec{F}_{net}$

SOLVE Without friction, $n_x = n\sin\theta$ is the only component of force toward the center of the circle. It is this inward component of the normal force on the car that causes it to turn the corner. Newton's second law is

$$\sum F_x = n\sin\theta = \frac{mv^2}{r}$$

$$\sum F_y = n\cos\theta - w = 0$$

where θ is the angle at which the road is banked, and we've assumed that the car is traveling at the correct speed v. From the y-equation,

$$n = \frac{w}{\cos\theta} = \frac{mg}{\cos\theta}$$

Substituting this into the x-equation and solving for v give

$$\left(\frac{mg}{\cos\theta}\right)\sin\theta = mg\tan\theta = \frac{mv^2}{r}$$

$$v = \sqrt{rg\tan\theta} = 14 \text{ m/s}$$

ASSESS This is $\approx$30 mph, a reasonable speed. Only at this exact speed can the turn be negotiated without reliance on friction forces.

Maximum Walking Speed

Humans and other two-legged animals have two basic gaits: walking and running. At slow speeds, you walk. When you need to go faster, you run. Why don't you just walk faster? There is an upper limit to the speed of walking, and this limit is set by the physics of circular motion.

FIGURE 6.18 Analysis of a walking stride.

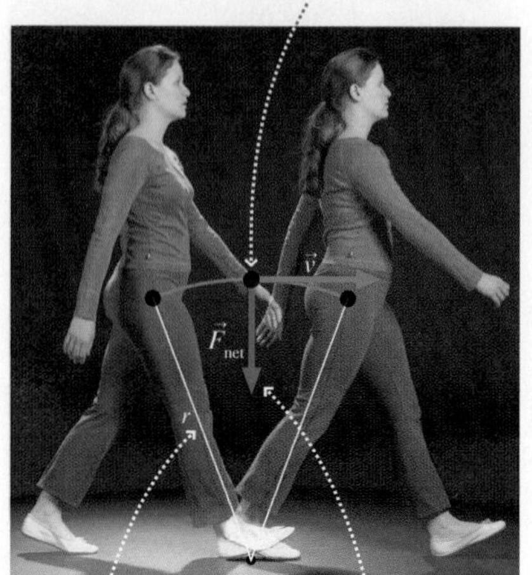

(a) Walking stride
During each stride, her hip undergoes circular motion.
The radius of the circular motion is the length of the leg from the foot to the hip.
The circular motion requires a force directed toward the center of the circle.

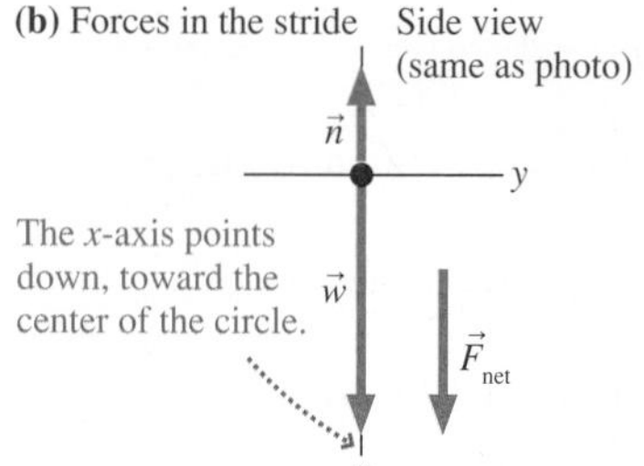

(b) Forces in the stride
Side view (same as photo)
The x-axis points down, toward the center of the circle.

Think about the motion of your body as you take a walking stride. You put one foot forward, then push off with your rear foot. Your body pivots over your front foot, and you bring your rear foot forward to take the next stride. As you can see in FIGURE 6.18a, the path that your body takes during this stride is the arc of a circle. **In a walking gait, your body is in circular motion as you pivot on your forward foot.**

A force toward the center of the circle is required for this circular motion, as shown in Figure 6.18. FIGURE 6.18b shows the forces acting on the woman's body during the midpoint of the stride: her weight, directed down, and the normal force of the ground, directed up. Newton's second law for the x-axis is

$$\sum F_x = w - n = \frac{mv^2}{r}$$

Because of her circular motion, the net force must point toward the center of the circle, or, in this case, down. In order for the net force to point down, the normal force must be *less* than her weight. Your body tries to "lift off" as it pivots over your foot, decreasing the normal force exerted on you by the ground. The normal force becomes smaller as you walk faster, but n cannot be less than zero. Thus the maximum possible walking speed v_{max} occurs when $n = 0$. Setting $n = 0$ in Newton's second law gives

$$w = mg = \frac{m{v_{max}}^2}{r}$$

Thus

$$v_{max} = \sqrt{gr} \tag{6.10}$$

The maximum possible walking speed is limited by r, the length of the leg, and g, the free-fall acceleration. This formula is a good approximation of the maximum walking speed for humans and other animals. The maximum walking speed is higher for animals with longer legs. Giraffes, with their very long legs, can walk at high speeds. Animals such as mice with very short legs have such a low maximum walking speed that they rarely use this gait. Mice generally run to get from one place to another.

For humans, the length of the leg is approximately 0.7 m, so we calculate a maximum speed of

$$v_{max} \approx 2.6 \text{ m/s} \approx 6 \text{ mph}$$

You *can* walk this fast, though it becomes energetically unfavorable to do so at speeds above 4 mph. Most people make a transition to a running gait at about this speed. Children, with their shorter legs, must make a transition to a running gait at a much lower speed.

STOP TO THINK 6.3 A block on a string spins in a horizontal circle on a frictionless table. Rank in order, from largest to smallest, the tensions T_A to T_E acting on the blocks A to E.

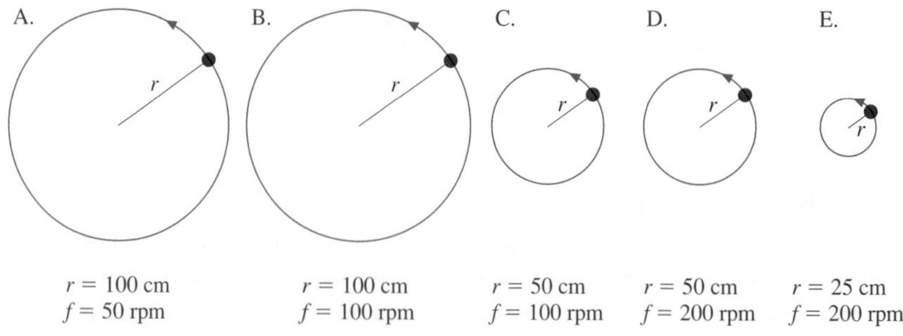

6.4 Apparent Forces in Circular Motion

FIGURE 6.19 shows a carnival ride that spins the riders around inside a large cylinder. The people are "stuck" to the inside wall of the cylinder! As you probably know from experience, the riders *feel* that they are being pushed outward, into the wall. But our analysis has found that an object in circular motion must have an *inward* force to create the centripetal acceleration. How can we explain this apparent difference?

FIGURE 6.19 Inside the Gravitron, a rotating circular room.

Centrifugal Force?

If you are a passenger in a car that turns a corner quickly, you may feel "thrown" by some mysterious force against the door. But is there really such a force? FIGURE 6.20 shows a bird's-eye view of you riding in a car as it makes a left turn. You try to continue moving in a straight line, obeying Newton's first law, when—without having been provoked—the door starts to turn in toward you and so runs into you! You do then feel the force of the door because it is now the force of the door, pushing *inward* toward the center of the curve, that is causing you to turn the corner. But you were not "thrown" into the door; the door ran into you.

A "force" that *seems* to push an object to the outside of a circle is called a *centrifugal force*. Despite having a name, there really is no such force. What you feel is your body trying to move ahead in a straight line (which would take you away from the center of the circle) as outside forces act to turn you in a circle. The only real forces, those that appear on free-body diagrams, are the ones pushing inward toward the center. **A centrifugal force will never appear on a free-body diagram and never be included in Newton's laws.**

FIGURE 6.20 Bird's-eye view of a passenger in a car turning a corner.

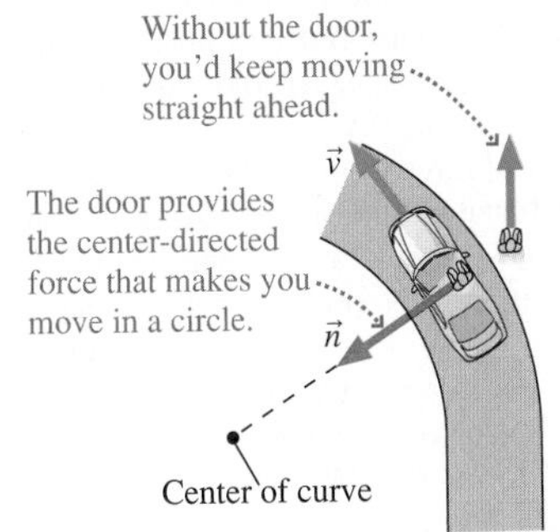

With this in mind, let's revisit the rotating carnival ride. A person watching from above would see the riders in the cylinder moving in a circle with the walls providing the inward force that causes their centripetal acceleration. The riders *feel* as if they're being pushed outward because their natural tendency to move in a straight line is being resisted by the wall of the cylinder, which keeps getting in the way. But feelings aren't forces. The only actual force is the contact force of the cylinder wall pushing *inward*.

Apparent Weight in Circular Motion

Imagine swinging a bucket of water over your head. If you swing the bucket quickly, the water stays in. But you'll get a shower if you swing too slowly. Why does the water stay in the bucket? Or think about a roller coaster that does a loop-the-loop. How does the car stay on the track when it's upside down? You might have said that there was a centrifugal force holding the water in the bucket and the car on the track, but we have seen that there really isn't a centrifugal force. Analyzing these questions will tell us a lot about forces in general and circular motion in particular.

FIGURE 6.21a shows a roller coaster car going around a vertical loop-the-loop of radius r. If you've ever ridden a roller coaster, you know that your sensation of weight changes as you go over the crests and through the dips. To understand why, let's look at the forces on passengers going through the loop. To simplify our analysis, we will assume that the speed of the car stays constant as it moves through the loop.

FIGURE 6.21b shows a passenger's free-body diagram at the top and the bottom of the loop. Let's start by examining the forces on the passenger at the bottom of the loop. The only forces acting on her are her weight $\vec{w}$ and the normal force $\vec{n}$ of the seat pushing up on her. Recall from Chapter 5 that a person's apparent weight is the magnitude of the force that supports her. Here the seat is supporting the passenger with the normal force $\vec{n}$, so her apparent weight is $w_{app} = n$. Based on our understanding of circular motion, we can say:

- She's moving in a circle, so there must be a net force directed toward the center of the circle—currently directly above her head—to provide the centripetal acceleration.

FIGURE 6.21 A roller coaster car going around a loop-the-loop.

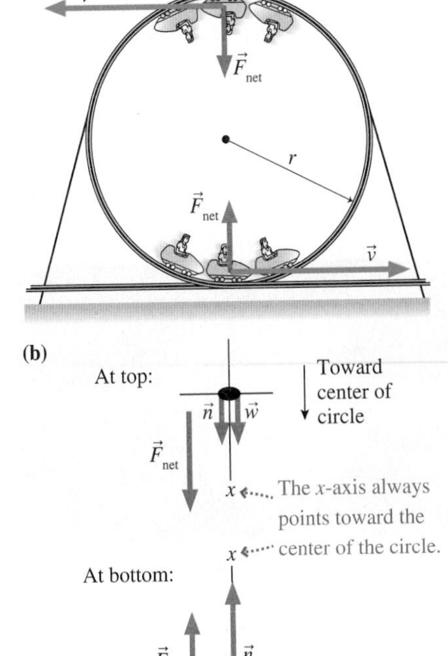

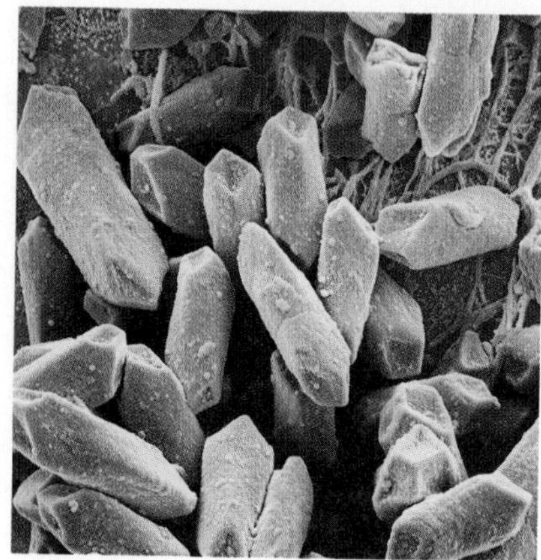
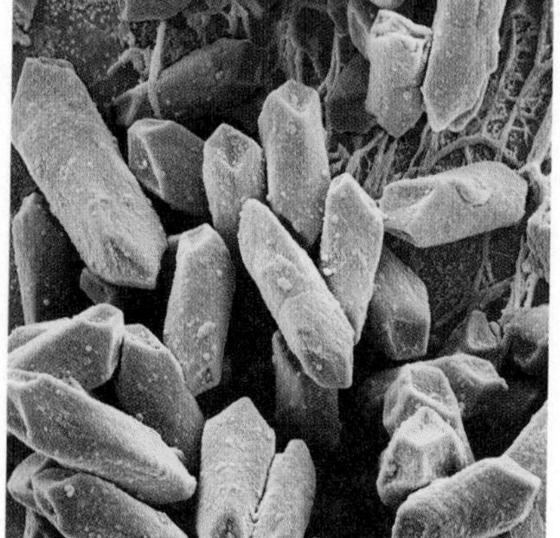

When "down" is up BIO You can tell, even with your eyes closed, what direction is down. This sense is due to small crystals of calcium carbonate, called *otoliths,* in your inner ears. Gravity pulls the otoliths down, so a normal force from a sensitive supporting membrane must push them up. Your brain interprets "down" as the opposite of this normal force. Normally, what your ears tell you is "down" is really down. But at the top of a loop in a roller coaster, the normal force is directed down, so your inner ear tells you that "down" is up! If your ears tell you one thing and your eyes another, it can be disorienting.

- The net force points *upward,* so it must be the case that $n > w$.
- Her apparent weight is $w_{app} = n$, so her apparent weight is greater than her true weight ($w_{app} > w$). Thus she "feels heavy" at the bottom of the circle.

This situation is the same as for the car driving through a valley in Conceptual Example 6.8. To analyze the situation quantitatively, we'll apply the steps of Problem-Solving Strategy 6.1. As always, we choose the x-axis to point toward the center of the circle or, in this case, vertically upward. Then Newton's second law is

$$\sum F_x = n_x + w_x = n - w = \frac{mv^2}{r}$$

From this equation, her apparent weight is

$$w_{app} = n = w + \frac{mv^2}{r} \tag{6.11}$$

The passenger's apparent weight at the bottom is *greater* than her true weight w, which agrees with your experience when you go through a dip or a valley.

Now let's look at the roller coaster car as it crosses the top of the loop. Things are a little trickier here. As Figure 6.21b shows, whereas the normal force of the seat pushes up when the passenger is at the bottom of the circle, it pushes *down* when she is at the top and the seat is above her. It's worth thinking carefully about this diagram to make sure you understand what it is showing.

The passenger is still moving in a circle, so there must be a net force *downward,* toward the center of the circle, to provide her centripetal acceleration. As always, we define the x-axis to be toward the center of the circle, so here the x-axis points vertically downward. Newton's second law gives

$$\sum F_x = n_x + w_x = n + w = \frac{mv^2}{r}$$

Note that w_x is now *positive* because the x-axis is directed downward. We can solve for her apparent weight:

$$w_{app} = n = \frac{mv^2}{r} - w \tag{6.12}$$

If v is sufficiently large, her apparent weight can exceed the true weight, just as it did at the bottom of the track.

But let's look at what happens if the car goes slower. Notice from Equation 6.12 that, as v decreases, there comes a point when $mv^2/r = w$ and n becomes zero. At that point, the seat is *not* pushing against the passenger at all! Instead, she is able to complete the circle because her weight force alone provides sufficient centripetal acceleration.

The speed for which $n = 0$ is called the *critical speed* v_c. Because for n to be zero we must have $mv_c^2/r = w$, the critical speed is

$$v_c = \sqrt{\frac{rw}{m}} = \sqrt{\frac{rmg}{m}} = \sqrt{gr} \tag{6.13}$$

What happens if the speed is slower than the critical speed? In this case, Equation 6.12 gives a *negative* value for n if $v < v_c$. But that is physically impossible. The seat can push against the passenger ($n > 0$), but it can't *pull* on her, so the slowest possible speed is the speed for which $n = 0$ at the top. Thus, **the critical speed is the slowest speed at which the car can complete the circle.** If $v < v_c$, the passenger cannot turn the full loop but, instead, will fall from the car as a projectile! (This is why you're always strapped into a roller coaster.)

Water stays in a bucket swung over your head for the same reason. The bottom of the bucket pushes against the water to provide the inward force that causes circular motion. If you swing the bucket too slowly, the force of the bucket on the water drops to zero. At that point, the water leaves the bucket and becomes a projectile following a parabolic trajectory onto your head!

A fast-spinning world Saturn, a gas giant planet composed largely of fluid matter, is quite a bit larger than the earth. It also rotates much more quickly, completing one rotation in just under 11 hours. The rapid rotation decreases the apparent weight at the equator enough to distort the fluid surface; the planet is noticeably out of round, as the red circle shows. The diameter at the equator is 11% greater than the diameter at the poles.

EXAMPLE 6.12 How slow can you go?

A motorcyclist in the Globe of Death, pictured at the start of the chapter, rides in a 2.2-m-radius vertical loop. To keep control of the bike, the rider wants the normal force on his tires at the top of the loop to equal or exceed his and the bike's combined weight. What is the minimum speed at which the rider can take the loop?

PREPARE The visual overview for this problem is shown in FIGURE 6.22. At the top of the loop, the normal force of the cage on the tires is a *downward* force. In accordance with Problem-Solving Strategy 6.1, we've chosen the x-axis to point toward the center of the circle.

FIGURE 6.22 Riding in a vertical loop around the Globe of Death.

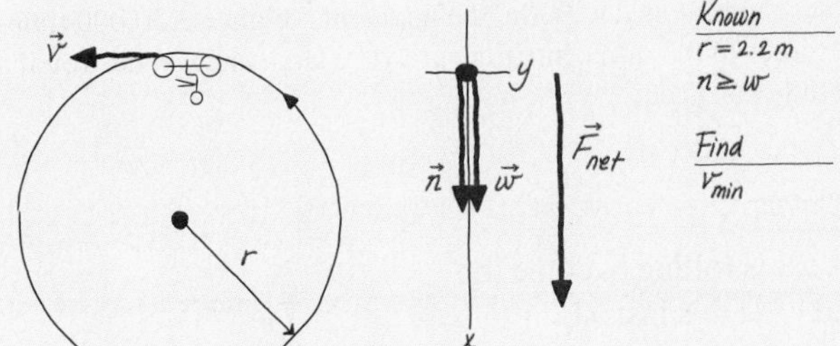

SOLVE We will consider the forces at the top point of the loop. Because the x-axis points downward, Newton's second law is

$$\sum F_x = w + n = \frac{mv^2}{r}$$

The minimum acceptable speed occurs when $n = w$; thus

$$2w = 2mg = \frac{mv_{min}^2}{r}$$

Solving for the speed, we find

$$v_{min} = \sqrt{2gr} = \sqrt{2(9.8 \text{ m/s}^2)(2.2 \text{ m})} = 6.6 \text{ m/s}$$

ASSESS The minimum speed is $\approx$15 mph, which isn't all that fast; the bikes can easily reach this speed. But normally several bikes are in the globe at one time. The big challenge is to keep all of the riders in the cage moving at this speed in synchrony. The period for the circular motion at this speed is $T = 2\pi r/v \approx 2$ s, leaving little room for error!

Centrifuges

The *centrifuge,* an important biological application of circular motion, is used to separate the components of a liquid with different densities. Typically these are different types of cells, or the components of cells, suspended in water. You probably know that small particles suspended in water will eventually settle to the bottom. However, the downward motion due to gravity for extremely small objects such as cells is so slow that it could take days or even months for the cells to settle out. It's not practical to wait for biological samples to separate due to gravity alone.

The separation would go faster if the force of gravity could be increased. Although we can't change gravity, we can increase the apparent weight of objects in the sample by spinning them very fast, and that is what the centrifuge in FIGURE 6.23 does. By using very high angular velocities, the centrifuge produces centripetal accelerations that are thousands of times greater than free-fall acceleration. As the centrifuge effectively increases gravity to thousands of times its normal value, the cells or cell components settle out and separate by density in a matter of minutes or hours.

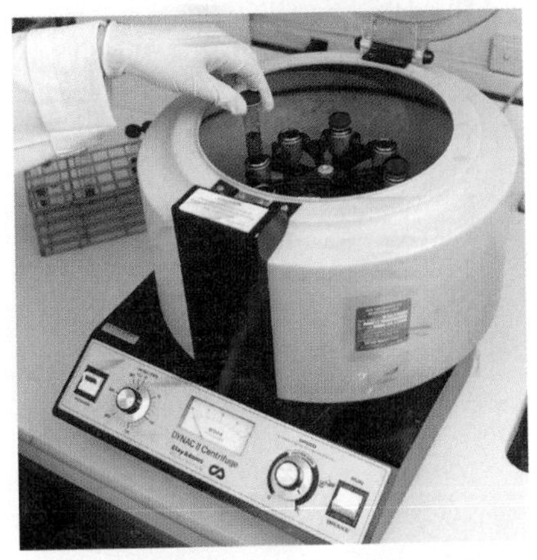

A centrifuge.

EXAMPLE 6.13 Analyzing the ultracentrifuge

An 18-cm-diameter ultracentrifuge produces an extraordinarily large centripetal acceleration of 250,000g, where g is the free-fall acceleration due to gravity. What is its frequency in rpm? What is the apparent weight of a sample with a mass of 0.0030 kg?

PREPARE The acceleration in SI units is

$$a = 250{,}000(9.80 \text{ m/s}^2) = 2.45 \times 10^6 \text{ m/s}^2$$

The radius is half the diameter, or $r = 9.0 \text{ cm} = 0.090$ m.

SOLVE The centripetal acceleration is related to the angular speed by $a = \omega^2 r$. Thus

$$\omega = \sqrt{\frac{a}{r}} = \sqrt{\frac{2.45 \times 10^6 \text{ m/s}^2}{0.090 \text{ m}}} = 5.22 \times 10^3 \text{ rad/s}$$

Continued

FIGURE 6.23 The operation of a centrifuge.

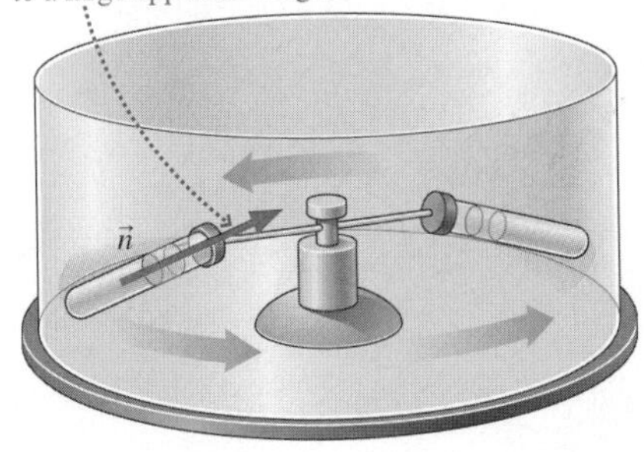

The high angular velocity requires a large normal force, which leads to a large apparent weight.
$\vec{n}$

TRY IT YOURSELF

Human centrifuge BIO If you spin your arm rapidly in a vertical circle, the motion will produce an effect like that in a centrifuge. The motion will assist outbound blood flow in your arteries and retard inbound blood flow in your veins. There will be a buildup of fluid in your hand that you will be able to see (and feel!) quite easily.

By Equation 6.6, this corresponds to a frequency

$$f = \frac{\omega}{2\pi} = \frac{5.22 \times 10^3 \text{ rad/s}}{2\pi} = 830 \text{ rev/s}$$

Converting to rpm, we find

$$830 \frac{\text{rev}}{\text{s}} \times \frac{60 \text{ s}}{1 \text{ min}} = 50{,}000 \text{ rpm}$$

At this rotation rate, the 0.0030 kg mass has an apparent weight

$$w_{\text{app}} = ma = (3.0 \times 10^{-3} \text{ kg})(2.45 \times 10^6 \text{ m/s}^2) = 7.4 \times 10^3 \text{ N}$$

The three gram sample has an effective weight of about 1700 pounds!

ASSESS Because the acceleration is 250,000g, the apparent weight is 250,000 times the actual weight. The forces in the ultracentrifuge are very large and can destroy the machine if it is not carefully balanced.

STOP TO THINK 6.4 A car is rolling over the top of a hill at constant speed v. At this instant,

A. $n > w$.
B. $n < w$.
C. $n = w$.
D. We can't tell about n without knowing v.

6.5 Circular Orbits and Weightlessness

The Space Shuttle orbits the earth in a circular path at a speed of over 15,000 miles per hour. What forces act on it? Why does it move in a circle? Before we start considering the physics of orbital motion, let's return, for a moment, to projectile motion. Projectile motion occurs when the only force on an object is gravity. Our analysis of projectiles made an implicit assumption that the earth is flat and that the free-fall acceleration, due to gravity, is everywhere straight down. This is an acceptable approximation for projectiles of limited range, such as baseballs or cannon balls, but there comes a point where we can no longer ignore the curvature of the earth.

Orbital Motion

FIGURE 6.24 Projectiles being launched at increasing speeds from height h on a smooth, airless planet.

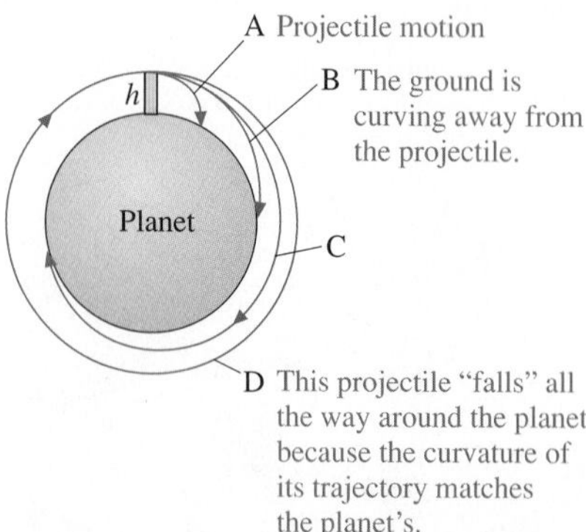
A Projectile motion
h
B The ground is curving away from the projectile.
Planet
C
D This projectile "falls" all the way around the planet because the curvature of its trajectory matches the planet's.

FIGURE 6.24 shows a perfectly smooth, spherical, airless planet with a vertical tower of height h. A projectile is launched from this tower with initial speed v_i parallel to the ground. If v_i is very small, as in trajectory A, the "flat-earth approximation" is valid and the problem is identical to Example 3.11 in which a car drove off a cliff. The projectile simply falls to the ground along a parabolic trajectory.

As the initial speed v_i is increased, it seems to the projectile that the ground is curving out from beneath it. It is still falling the entire time, always getting closer to the ground, but the distance that the projectile travels before finally reaching the ground—that is, its range—increases because the projectile must "catch up" with the ground that is curving away from it. Trajectories B and C are like this.

If the launch speed v_i is sufficiently large, there comes a point at which the curve of the trajectory and the curve of the earth are parallel. In this case, the projectile "falls" but it never gets any closer to the ground! This is the situation for trajectory D. The projectile returns to the point from which it was launched, at the same speed at which it was launched, making a closed trajectory. Such a closed trajectory around a planet or star is called an **orbit.**

The most important point of this qualitative analysis is that, in the absence of air resistance, **an orbiting projectile is in free fall.** This is, admittedly, a strange idea, but one worth careful thought. An orbiting projectile is really no different from a thrown baseball or a car driving off a cliff. The only force acting on it is gravity, but its tangential velocity is so great that the curvature of its trajectory matches the curvature of the earth. When this happens, the projectile "falls" under the influence of gravity but never gets any closer to the surface, which curves away beneath it.

When we first studied free fall in Chapter 2, we said that free-fall acceleration is always directed vertically downward. As we see in FIGURE 6.25, "downward" really means "toward the center of the earth." For a projectile in orbit, the direction of the force of gravity changes, always pointing toward the center of the earth.

FIGURE 6.25 The force of gravity is really directed toward the center of the earth.

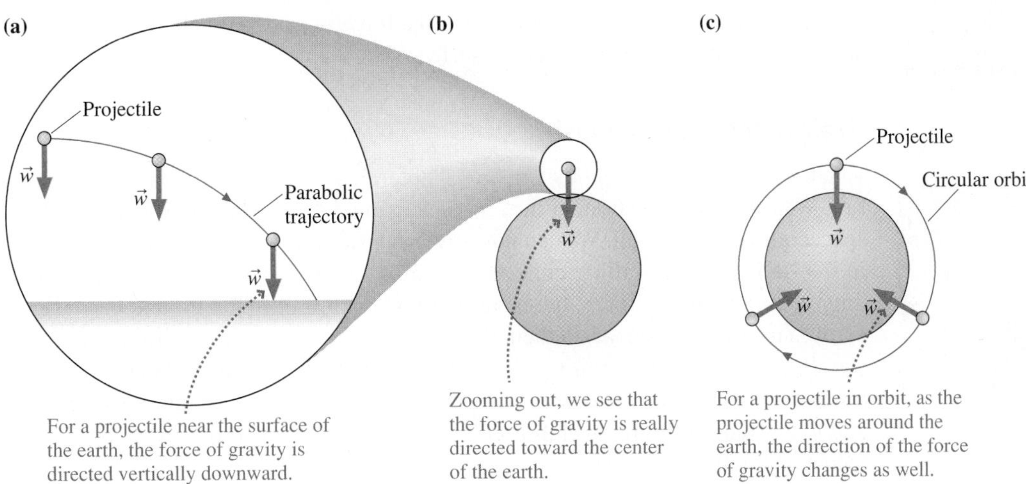

(a)
(b)
(c)
Projectile
Parabolic trajectory
Projectile
Circular orbit
For a projectile near the surface of the earth, the force of gravity is directed vertically downward.
Zooming out, we see that the force of gravity is really directed toward the center of the earth.
For a projectile in orbit, as the projectile moves around the earth, the direction of the force of gravity changes as well.

As you have learned, a force of constant magnitude that always points toward the center of a circle causes the centripetal acceleration of uniform circular motion. Because the only force acting on the orbiting projectile in Figure 6.25 is gravity, and we're assuming the projectile is very near the surface of the earth, we can write

$$a = \frac{F_{\text{net}}}{m} = \frac{w}{m} = \frac{mg}{m} = g \tag{6.14}$$

An object moving in a circle of radius r at speed v_{orbit} will have this centripetal acceleration if

$$a = \frac{(v_{\text{orbit}})^2}{r} = g \tag{6.15}$$

That is, if an object moves parallel to the surface with the speed

$$v_{\text{orbit}} = \sqrt{gr} \tag{6.16}$$

then the free-fall acceleration provides exactly the centripetal acceleration needed for a circular orbit of radius r. An object with any other speed will not follow a circular orbit.

The earth's radius is $r = R_e = 6.37 \times 10^6$ m. The orbital speed of a projectile just skimming the surface of a smooth, airless earth is

$$v_{\text{orbit}} = \sqrt{gR_e} = \sqrt{(9.80\ \text{m/s}^2)(6.37 \times 10^6\ \text{m})} = 7900\ \text{m/s} \approx 18{,}000\ \text{mph}$$

We can use v_{orbit} to calculate the period of the satellite's orbit:

$$T = \frac{2\pi r}{v_{\text{orbit}}} = 2\pi\sqrt{\frac{r}{g}} \tag{6.17}$$

For this earth-skimming orbit, $T = 5065\text{ s} = 84.4\text{ min}$.

Of course, this orbit is unrealistic; even if there were no trees and mountains, a real projectile moving at this speed would burn up from the friction of air resistance. Suppose, however, that we launched the projectile from a tower of height $h = 200\text{ mi} \approx 3.2 \times 10^5\text{ m}$, above most of the earth's atmosphere. This is approximately the height of low-earth-orbit satellites, such as the Space Shuttle. Note that $h \ll R_e$, so the radius of the orbit $r = R_e + h = 6.69 \times 10^6\text{ m}$ is only 5% larger than the earth's radius. Many people have a mental image that satellites orbit far above the earth, but in fact most satellites come pretty close to skimming the surface.

Zero apparent weight in the Space Shuttle.

At this slightly larger value of r, Equation 6.17 gives $T = 87\text{ min}$. The actual period of the Space Shuttle at an altitude of 200 mi is about 91 minutes, so our calculation is very good—but not perfect. As we'll see in the next section, a correct calculation must take into account the fact that the force of gravity gradually gets weaker at higher elevations above the earth's surface.

Weightlessness in Orbit

When we discussed *weightlessness* in Chapter 5, we saw that it occurs during free fall. We asked the question, at the end of Section 5.4, whether astronauts and their spacecraft are in free fall. We can now give an affirmative answer: They are, indeed, in free fall. They are falling continuously around the earth, under the influence of only the gravitational force, but never getting any closer to the ground because the earth's surface curves beneath them. Weightlessness in space is no different from the weightlessness in a free-falling elevator. **Weightlessness does *not* occur from an absence of weight or an absence of gravity.** Instead, the astronaut, the spacecraft, and everything in it are "weightless" (i.e., their *apparent* weight is zero) because they are all falling together.

Rotating space stations BIO The weightlessness astronauts experience in orbit has serious physiological consequences. Astronauts who spend time in weightless environments lose bone and muscle mass and suffer other adverse effects. One solution is to introduce "artificial gravity." On a space station, the easiest way to do this would be to make the station rotate, producing an apparent weight. The designers of this space station model for the movie *2001: A Space Odyssey* made it rotate for just that reason.

The Orbit of the Moon

If a satellite is simply "falling" around the earth, with the gravitational force causing a centripetal acceleration, then what about the moon? Is it obeying the same laws of physics? Or do celestial objects obey laws that we cannot discover by experiments here on earth?

The radius of the moon's orbit around the earth is $3.84 \times 10^8\text{ m}$. If we use Equation 6.17 to calculate the period of the moon's orbit, the time the moon takes to circle the earth once, we get

$$T = 2\pi\sqrt{\frac{r}{g}} = 2\pi\sqrt{\frac{3.84 \times 10^8\text{ m}}{9.80\text{ m/s}^2}} = 655\text{ min} \approx 11\text{ h} \tag{6.18}$$

This is clearly wrong; the period of the moon's orbit is approximately one month.

Newton believed that the laws of motion he had discovered were *universal* and so should apply to the motion of the moon as well as to the motion of objects in the laboratory. But why should we assume that the free-fall acceleration g is the same at the distance of the moon as it is on or near the earth's surface? If gravity is the force of the earth pulling on an object, it seems plausible that the size of that force, and thus the size of g, should diminish with increasing distance from the earth.

Newton proposed the idea that the earth's force of gravity decreases with the square of the distance from the earth. This is the basis of *Newton's law of gravity,* a topic we will study in the next section. The force of gravity is less at the distance of the moon—exactly the strength needed to make the moon orbit at the observed rate. The moon, just like the Space Shuttle, is simply "falling" around the earth!

6.6 Newton's Law of Gravity

A popular image has Newton thinking of the idea of gravity after an apple fell on his head. This amusing story is at least close to the truth. Newton himself said that the "notion of gravitation" came to him as he "sat in a contemplative mood" and "was occasioned by the fall of an apple." It occurred to him that, perhaps, the apple was attracted to the center of the earth but was prevented from getting there by the earth's surface. And if the apple was so attracted, why not the moon? Newton's genius was his sudden realization that **the force that attracts the moon to the earth (and the planets to the sun) was identical to the force that attracts an apple to the earth.** In other words, gravitation is a *universal* force between all objects in the universe! This is not shocking today, but no one before Newton had ever thought that the mundane motion of objects on earth had any connection at all with the stately motion of the planets around the sun.

Isaac Newton was born to a poor farming family in 1642, the year of Galileo's death. He entered Trinity College at Cambridge University at age 19 as a "subsizar," a poor student who had to work his way through school. Newton graduated in 1665, at age 23, just as an outbreak of the plague in England forced the universities to close for two years. He returned to his family farm for that period, during which he made important experimental discoveries in optics, laid the foundations for his theories of mechanics and gravitation, and made major progress toward his invention of calculus as a whole new branch of mathematics.

Gravity Obeys an Inverse-Square Law

Newton also recognized that the strength of gravity must decrease with distance. These two notions about gravity—that it is universal and that it decreases with distance—form the basis for Newton's law of gravity.

Newton proposed that *every* object in the universe attracts *every other* object with a force that has the following properties:

1. The force is inversely proportional to the square of the distance between the objects.
2. The force is directly proportional to the product of the masses of the two objects.

FIGURE 6.26 shows two spherical objects with masses m_1 and m_2 separated by distance r. Each object exerts an attractive force on the other, a force that we call the **gravitational force.** These two forces form an action/reaction pair, so $\vec{F}_{1\text{on}2}$ is equal in magnitude and opposite in direction to $\vec{F}_{2\text{on}1}$. The magnitude of the forces is given by Newton's law of gravity.

FIGURE 6.26 The gravitational forces on masses m_1 and m_2.

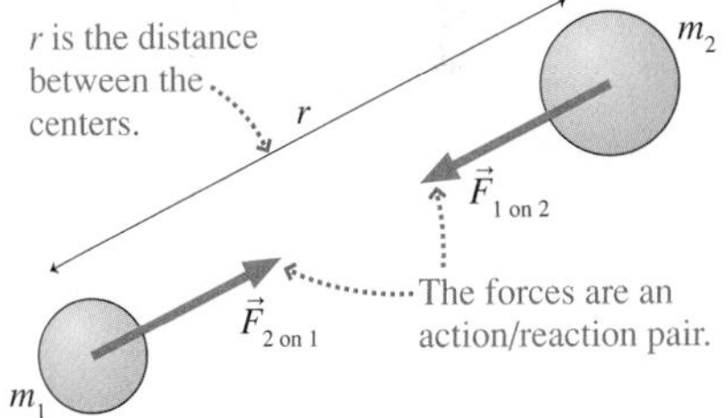

Newton's law of gravity If two objects with masses m_1 and m_2 are a distance r apart, the objects exert attractive forces on each other of magnitude

$$F_{1\text{on}2} = F_{2\text{on}1} = \frac{Gm_1m_2}{r^2} \quad (6.19)$$

p.186 INVERSE-SQUARE

The forces are directed along the line joining the two objects.

The constant G is called the **gravitational constant.** In the SI system of units,

$$G = 6.67 \times 10^{-11}\ \text{N} \cdot \text{m}^2/\text{kg}^2$$

NOTE ► Strictly speaking, Newton's law of gravity applies to *particles* with masses m_1 and m_2. However, it can be shown that the law also applies to the force between two spherical objects if r is the distance between their centers. ◄

As the distance r between two objects increases, the gravitational force between them decreases. Because the distance appears squared in the denominator, Newton's law of gravity is what we call an **inverse-square** law. Doubling the distance between two masses causes the force between them to decrease by a factor of 4. This mathematical form is one we will see again, so it is worth our time to explore it in more detail.

Inverse-square relationships

Two quantities have an **inverse-square relationship** if y is inversely proportional to the *square* of x. We write the mathematical relationship as

$$y = \frac{A}{x^2}$$

y is inversely proportional to x^2

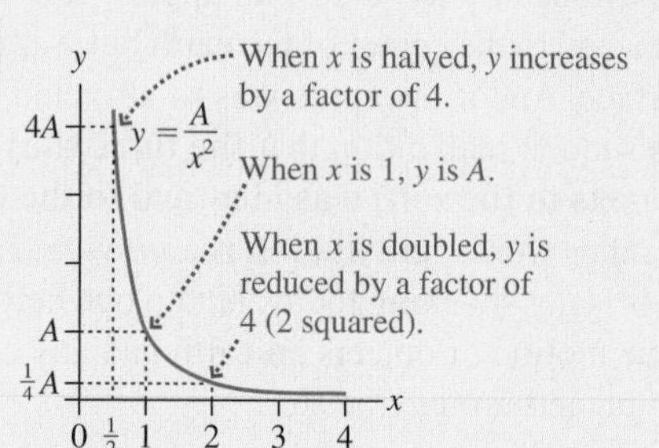

y
4A
A
$\frac{1}{4}$A
$y = \frac{A}{x^2}$
When x is halved, y increases by a factor of 4.
When x is 1, y is A.
When x is doubled, y is reduced by a factor of 4 (2 squared).
0 $\frac{1}{2}$ 1 2 3 4
x

Here, A is a constant. This relationship is sometimes written as $y \propto 1/x^2$.

SCALING As the graph shows, inverse-square scaling means, for example:

- If you double x, you decrease y by a factor of 4.
- If you halve x, you increase y by a factor of 4.
- If you increase x by a factor of 3, you decrease y by a factor of 9.
- If you decrease x by a factor of 3, you increase y by a factor of 9.

Generally, **if x *increases* by a factor of C, y *decreases* by a factor of C^2.** If x *decreases* by a factor of C, y *increases* by a factor of C^2.

RATIOS For any two values of x—say, x_1 and x_2—we have

$$y_1 = \frac{A}{x_1^2} \quad \text{and} \quad y_2 = \frac{A}{x_2^2}$$

Dividing the y_1-equation by the y_2-equation, we find

$$\frac{y_1}{y_2} = \frac{A/x_1^2}{A/x_2^2} = \frac{A}{x_1^2}\frac{x_2^2}{A} = \frac{x_2^2}{x_1^2}$$

That is, the ratio of y-values is the inverse of the ratio of the squares of the corresponding values of x.

LIMITS As x becomes large, y becomes very small; as x becomes small, y becomes very large.

Exercises 23, 24 

CONCEPTUAL EXAMPLE 6.14 Varying gravitational force

The gravitational force between two giant lead spheres is 0.010 N when the centers of the spheres are 20 m apart. What is the distance between their centers when the gravitational force between them is 0.160 N?

REASON We can solve this problem without knowing the masses of the two spheres. The key is to consider the ratios of forces and distances. Gravity is an inverse-square relationship; the force is related to the inverse square of the distance. The force *increases* by a factor of (0.160 N)/(0.010 N) = 16, so the distance must *decrease* by a factor of $\sqrt{16} = 4$. The distance is thus (20 m)/4 = 5.0 m.

ASSESS This type of ratio reasoning is a very good way to get a quick handle on the solution to a problem.

EXAMPLE 6.15 Gravitational force between two people

You are seated in your physics class next to another student 0.60 m away. Estimate the magnitude of the gravitational force between you. Assume that you each have a mass of 65 kg.

PREPARE We will model each of you as a sphere; this is not a particularly good model, but it will do for making an estimate. We will take the 0.60 m as the distance between your centers.

SOLVE The gravitational force is given by Equation 6.19:

$$F_{\text{(you) on (other student)}} = \frac{Gm_{\text{you}}m_{\text{other student}}}{r^2} = \frac{(6.67 \times 10^{-11}\ \text{N} \cdot \text{m}^2/\text{kg}^2)(65\ \text{kg})(65\ \text{kg})}{(0.60\ \text{m})^2} = 7.8 \times 10^{-7}\ \text{N}$$

ASSESS The force is quite small, roughly the weight of one hair on your head.

There is a gravitational force between all objects in the universe, but the gravitational force between two ordinary-sized objects is extremely small. Only when one (or both) of the masses is exceptionally large does the force of gravity become important. The downward force of the earth on you—your weight—is large because the earth has an enormous mass. And the attraction is mutual; by Newton's third law, you exert an upward force on the earth that is equal to your weight. However, the large mass of the earth makes the effect of this force on the earth negligible.

EXAMPLE 6.16 Gravitational force of the earth on a person

What is the magnitude of the gravitational force of the earth on a 60 kg person? The earth has mass 5.98×10^{24} kg and radius 6.37×10^6 m.

PREPARE We'll again model the person as a sphere. The distance r in Newton's law of gravity is the distance between the *centers* of the two spheres. The size of the person is negligible compared to the size of the earth, so we can use the earth's radius as r.

SOLVE The force of gravity on the person due to the earth can be computed using Equation 6.19:

$$F_{\text{earth on person}} = \frac{GM_e m}{R_e^2} = \frac{(6.67 \times 10^{-11}\ \text{N} \cdot \text{m}^2/\text{kg}^2)(5.98 \times 10^{24}\ \text{kg})(60\ \text{kg})}{(6.37 \times 10^6\ \text{m})^2} = 590\ \text{N}$$

ASSESS This force is exactly the same as we would calculate using the formula for the weight force, $w = mg$. This isn't surprising, though. Chapter 5 introduced the weight of an object as simply the "force of gravity" acting on it. Newton's law of gravity is a more fundamental law for calculating the force of gravity, but it's still the same force that we earlier called "weight."

NOTE ► We will use uppercase R and M to represent the large mass and radius of a star or planet, as we did in Example 6.16. ◄

The force of gravitational attraction between the earth and you is responsible for your weight. If you were to venture to another planet, your *mass* would be the same but your *weight* would vary, as we discussed in Chapter 5. We will now explore this concept in more detail.

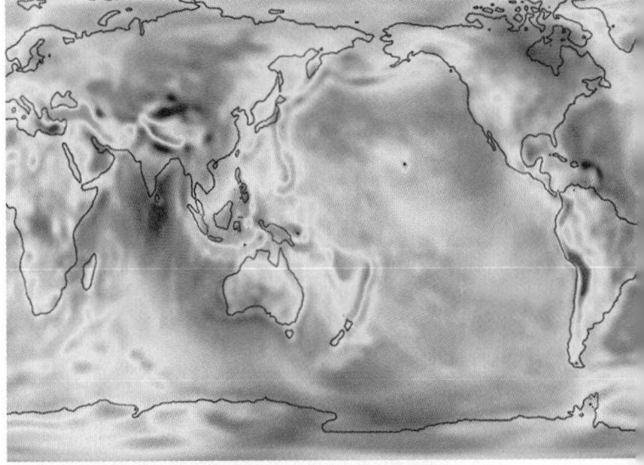

Variable gravity When we calculated the force of the earth's gravity, we assumed that the earth's shape and composition are uniform. Neither is quite true, so there is a very small variation in gravity at the surface of the earth, as shown in this image. Red means slightly stronger surface gravity; blue, slightly weaker. These variations are caused by differing distances from the earth's center and by unevenness in the density of the earth's crust. Though these variations are important for scientists studying the earth, they are small enough that we can ignore them for the computations we'll do in this textbook.

Gravity on Other Worlds

When astronauts ventured to the moon, television images showed them walking—and even jumping and skipping—with some ease, even though they were wearing life support systems with a mass of over 80 kg. This was a visible reminder that the weight of objects is less on the moon. Let's consider why this is so.

FIGURE 6.27 on the next page shows an astronaut on the moon weighing a rock of mass m. When we compute the weight of an object on the surface of the earth, we

FIGURE 6.27 An astronaut weighing a mass on the moon.

"Little g" perspective:
$F = mg_{moon}$

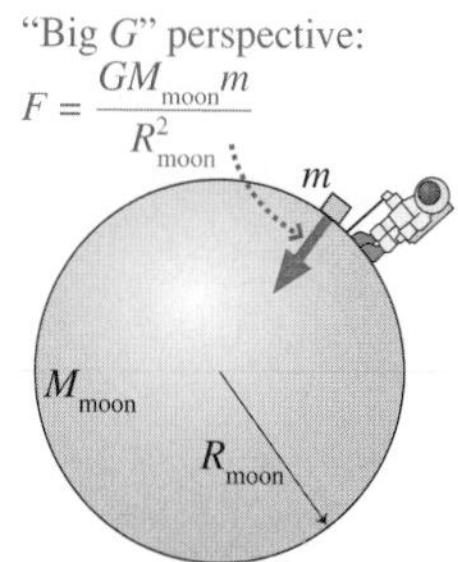

use the formula $w = mg$. We can do the same calculation for a mass on the moon, as long as we use the value of g on the moon:

$$w = mg_{moon} \tag{6.20}$$

This is the "little g" perspective. Falling-body experiments on the moon would give the value of g_{moon} as 1.62 m/s^2.

But we can also take a "big G" perspective. The weight of the rock comes from the gravitational attraction of the moon, and we can compute this weight using Equation 6.19. The distance r is the radius of the moon, which we'll call R_{moon}. Thus

$$F_{moon\,on\,m} = \frac{GM_{moon}m}{R_{moon}^2} \tag{6.21}$$

Because Equations 6.20 and 6.21 are two names and two expressions for the same force, we can equate the right-hand sides to find that

$$g_{moon} = \frac{GM_{moon}}{R_{moon}^2}$$

We have done this calculation for an object on the moon, but the result is completely general. At the surface of a planet (or a star), the free-fall acceleration g, a consequence of gravity, can be computed as

$$g_{planet} = \frac{GM_{planet}}{R_{planet}^2} \tag{6.22}$$

Free-fall acceleration on the surface of a planet

PROPORTIONAL p. 37 INVERSE-SQUARE p. 186

If we use values for the mass and the radius of the moon from the table inside the cover of the book, we can compute $g_{moon} = 1.62$ m/s^2. This means that an object would weigh less on the moon than it would on the earth, where g is 9.80 m/s^2. A 70 kg astronaut wearing an 80 kg spacesuit would weigh over 330 lb on the earth but only 54 lb on the moon.

The low lunar gravity makes walking very easy, but a walking pace on the moon would be very slow. Earlier in the chapter we found that the maximum walking speed is $v_{max} = \sqrt{gr}$, where r is the length of the leg. For a typical leg length of 0.7 m and the gravity of the moon, the *maximum* walking speed would be about 1 m/s, just over 2 mph—a very gentle stroll!

Equation 6.22 gives g at the surface of a planet. More generally, imagine an object at distance $r > R$ from the center of a planet. Its free-fall acceleration at this distance is

$$g = \frac{GM}{r^2} \tag{6.23}$$

This more general result agrees with Equation 6.22 if $r = R$, but it allows us to determine the "local" free-fall acceleration at distances $r > R$. Equation 6.23 expresses Newton's idea that the size of g should decrease as you get farther from the earth.

As you're flying in a jet airplane at a height of about 10 km, the free-fall acceleration is about 0.3% less than on the ground. At the height of the Space Shuttle, about 300 km, Equation 6.23 gives $g = 8.9$ m/s^2, about 10% less than the free-fall acceleration on the earth's surface. If you use this slightly smaller value of g in Equation 6.17 for the period of a satellite's orbit, you'll get the correct period of about 90 minutes. This value of g, only slightly less than the ground-level value, emphasizes the

◀ **Walking on the moon** BIO The low lunar gravity made walking at a reasonable pace difficult for the Apollo astronauts, but the reduced weight made jumping quite easy. Videos from the surface of the moon often show the astronauts getting from place to place by hopping or skipping—not for fun, but for speed and efficiency.

point that an object in orbit is not "weightless" due to the absence of gravity, but rather because it is in free fall.

EXAMPLE 6.17 Gravity on Saturn

Saturn, at 5.68×10^{26} kg, has nearly 100 times the mass of the earth. It is also much larger, with a radius of 5.85×10^7 m. What is the value of g on the surface of Saturn?

SOLVE We can use Equation 6.22 to compute the value of g_{Saturn}:

$$g_{\text{Saturn}} = \frac{GM_{\text{Saturn}}}{R_{\text{Saturn}}^2} = \frac{(6.67 \times 10^{-11}\ \text{N} \cdot \text{m}^2/\text{kg}^2)(5.68 \times 10^{26}\ \text{kg})}{(5.85 \times 10^7\ \text{m})^2} = 11.1\ \text{m/s}^2$$

ASSESS Even though Saturn is much more massive than the earth, its larger radius gives it a surface gravity that is not markedly different from that of the earth. If Saturn had a solid surface, you could walk and move around quite normally.

EXAMPLE 6.18 Finding the speed to orbit Deimos

Mars has two moons, each much smaller than the earth's moon. The smaller of these two bodies, Deimos, has an average radius of only 6.3 km and a mass of 1.8×10^{15} kg. At what speed would a projectile move in a very low orbit around Deimos?

SOLVE The free-fall acceleration at the surface of Deimos is quite small:

$$g_{\text{Deimos}} = \frac{GM_{\text{Deimos}}}{R_{\text{Deimos}}^2} = \frac{(6.67 \times 10^{-11}\ \text{N} \cdot \text{m}^2/\text{kg}^2)(1.8 \times 10^{15}\ \text{kg})}{(6.3 \times 10^3\ \text{m})^2}$$

$$= 0.0030\ \text{m/s}^2$$

Given this, we can use Equation 6.16 to calculate the orbital speed:

$$v_{\text{orbit}} = \sqrt{gr} = \sqrt{(0.0030\ \text{m/s}^2)(6.3 \times 10^3\ \text{m})} = 4.3\ \text{m/s} \approx 10\ \text{mph}$$

ASSESS This is quite slow. With a good jump, you could easily launch yourself into an orbit around Deimos!

STOP TO THINK 6.5 Rank in order, from largest to smallest, the free-fall accelerations on the surfaces of the following planets.

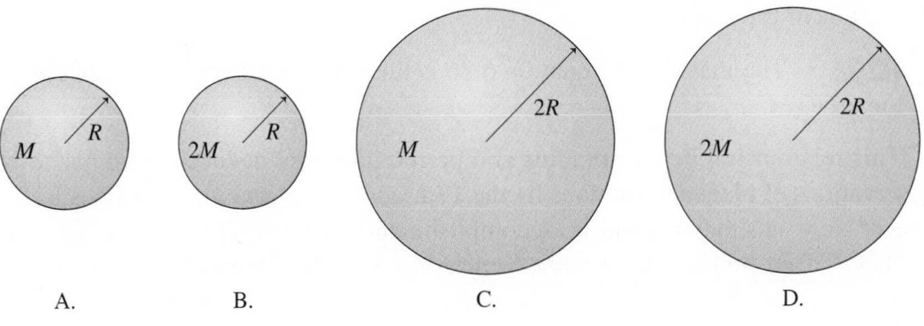

6.7 Gravity and Orbits

The planets of the solar system orbit the sun because the sun's gravitational pull, a force that points toward the center, causes the centripetal acceleration of circular motion. Mercury, the closest planet, experiences the largest acceleration, while Pluto, the most distant, has the smallest.

4.6

FIGURE 6.28 on the following page shows a large body of mass M, such as the earth or the sun, with a much smaller body of mass m orbiting it. The smaller body is called a **satellite**, even though it may be a planet orbiting the sun. Newton's second law tells us that $F_{M \text{ on } m} = ma$, where $F_{M \text{ on } m}$ is the gravitational force of the large body

FIGURE 6.28 The orbital motion of a satellite is due to the force of gravity.

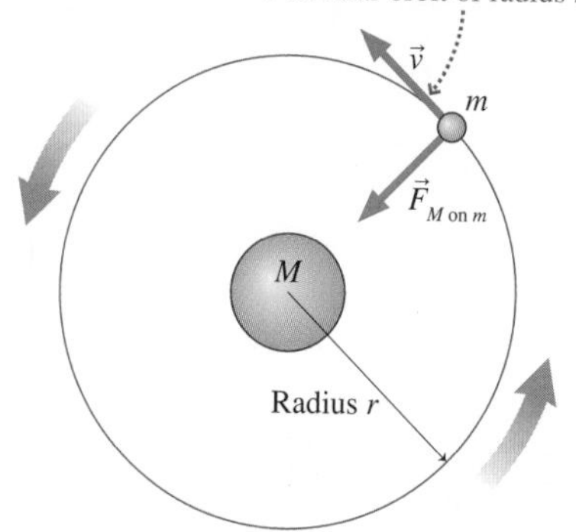
The satellite must have speed $\sqrt{GM/r}$ to maintain a circular orbit of radius r.
$\vec{v}$
m
$\vec{F}_{M \text{ on } m}$
M
Radius r

on the satellite and a is the satellite's acceleration. $F_{M \text{on} m}$ is given by Equation 6.19, and, because it's moving in a circular orbit, the satellite's acceleration is its centripetal acceleration, mv^2/r. Thus Newton's second law gives

$$F_{M \text{on} m} = \frac{GMm}{r^2} = ma = \frac{mv^2}{r} \qquad (6.24)$$

Solving for v, we find that the speed of a satellite in a circular orbit is

$$v = \sqrt{\frac{GM}{r}} \qquad (6.25)$$

Speed of a satellite in a circular orbit of radius r about a star or planet of mass M

A satellite must have this specific speed in order to maintain a circular orbit of radius r about the larger mass M. If the velocity differs from this value, the orbit will become elliptical rather than circular. Notice that the orbital speed does not depend on the satellite's mass m. This is consistent with our previous discoveries that free-fall motion and projectile motion due to gravity are independent of the mass.

For a planet orbiting the sun, the period T is the time to complete one full orbit. The relationship among speed, radius, and period is the same as for any circular motion, $v = 2\pi r/T$. Combining this with the value of v for a circular orbit from Equation 6.25 gives

$$\sqrt{\frac{GM}{r}} = \frac{2\pi r}{T}$$

If we square both sides and rearrange, we find the period of a satellite:

$$T^2 = \left(\frac{4\pi^2}{GM}\right) r^3 \qquad (6.26)$$

Relationship between the orbital period T and radius r for a satellite in a circular orbit around an object of mass M

In other words, **the square of the period of the orbit is proportional to the cube of the radius of the orbit.**

NOTE ► The mass M in Equation 6.26 is the mass of the object at the center of the orbit. ◄

This relationship between radius and period had been deduced from naked-eye observations of planetary motions by the 17th-century astronomer Johannes Kepler. One of Newton's major scientific accomplishments was to use his law of gravity and his laws of motion to prove what Kepler had deduced from observations. Even today, Newton's law of gravity and equations such as Equation 6.26 are essential tools for the NASA engineers who launch probes to other planets in the solar system.

The table inside the back cover of this book contains astronomical information about the sun and the planets that will be useful for many of the end-of-chapter problems. Note that planets farther from the sun have longer periods, in agreement with Equation 6.26.

EXAMPLE 6.19 Locating a geostationary satellite

Communication satellites appear to "hover" over one point on the earth's equator. A satellite that appears to remain stationary as the earth rotates is said to be in a *geostationary orbit*. What is the radius of the orbit of such a satellite?

PREPARE For the satellite to remain stationary with respect to the earth, the satellite's orbital period must be 24 hours; in seconds this is $T = 8.64 \times 10^4$ s.

SOLVE We solve for the radius of the orbit by rearranging Equation 6.26. The mass at the center of the orbit is the earth:

$$r = \left(\frac{GM_e T^2}{4\pi^2}\right)^{\frac{1}{3}} = \left(\frac{(6.67 \times 10^{-11}\ \text{N} \cdot \text{m}^2/\text{kg}^2)(5.98 \times 10^{24}\ \text{kg})(8.64 \times 10^4\ \text{s})^2}{4\pi^2}\right)^{\frac{1}{3}}$$

$$= 4.22 \times 10^7\ \text{m}$$

ASSESS This is a high orbit; the radius is about 7 times the radius of the earth. Recall that the radius of the Space Shuttle's orbit is only about 5% larger than that of the earth.

Gravity on a Grand Scale

Although relatively weak, gravity is a long-range force. No matter how far apart two objects may be, there is a gravitational attraction between them. Consequently, gravity is the most ubiquitous force in the universe. It not only keeps your feet on the ground, but also is at work on a much larger scale. The Milky Way galaxy, the collection of stars of which our sun is a part, is held together by gravity. But why doesn't the attractive force of gravity simply pull all of the stars together?

A spiral galaxy, similar to our Milky Way galaxy.

The reason is that all of the stars in the galaxy are in orbit around the center of the galaxy. The gravitational attraction keeps the stars moving in orbits around the center of the galaxy rather than falling inward, much as the planets orbit the sun rather than falling into the sun. In the nearly 5 billion years that our solar system has existed, it has orbited the center of the galaxy approximately 20 times.

The galaxy as a whole doesn't rotate at a fixed angular speed, though. All of the stars in the galaxy are different distances from the galaxy's center, and so orbit with different periods. Stars closer to the center complete their orbits in less time, as we would expect from Equation 6.26. As the stars orbit, their relative positions shift. Stars that are relatively near neighbors now could be on opposite sides of the galaxy at some later time.

The rotation of a *rigid body* like a wheel is much simpler. As a wheel rotates, all of the points keep the same relationship to each other; every point on the wheel moves with the same angular velocity. The rotational dynamics of such rigid bodies is a topic we will take up in the next chapter.

STOP TO THINK 6.6 If the mass of the moon were doubled but it stayed in its present orbit, how would its orbital period change?

A. The period would increase.
B. The period would decrease.
C. The period would stay the same.

INTEGRATED EXAMPLE 6.20 **A hunter and his sling**

A Stone Age hunter stands on a cliff overlooking a flat plain. He places a 1.0 kg rock in a sling, ties the sling to a 1.0-m-long vine, then swings the rock in a horizontal circle around his head. The plane of the motion is 25 m above the plain below. The tension in the vine increases as the rock goes faster and faster. Suddenly, just as the tension reaches 200 N, the vine snaps. If the rock is moving toward the cliff at this instant, how far out on the plain (from the base of the cliff) will it land?

PREPARE We model the rock as a particle in uniform circular motion. We can use Problem-Solving Strategy 6.1 to analyze this part of the motion. Once the vine breaks, the rock undergoes projectile motion with an initial velocity that is horizontal.

The force identification diagram of **FIGURE 6.29a** on the next page shows that the only contact force acting on the rock is the tension in the vine. Because the rock moves in a horizontal circle, you may be tempted to draw a free-body diagram like **FIGURE 6.29b**, on the next page where $\vec{T}$ is directed along the x-axis. You will quickly run into trouble, however, because in this diagram the net force has a downward y-component that would cause the rock to rapidly accelerate downward. But we know that it moves in a horizontal circle and that the net force must point toward the center of the circle. In this free-body diagram, the weight force $\vec{w}$ points straight down and is certainly correct, so the difficulty must be with $\vec{T}$.

Continued

FIGURE 6.29 Visual overview of a hunter swinging a rock.

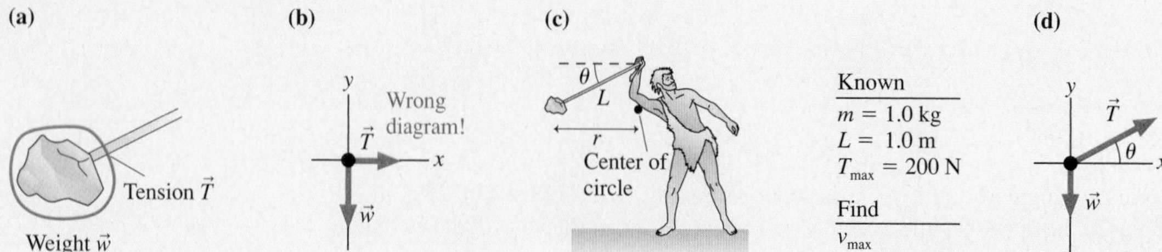

As an experiment, tie a small weight to a string, swing it over your head, and check the angle of the string. You will discover that the string is not horizontal but, instead, is angled downward. The sketch of **FIGURE 6.29c** labels this angle θ. Notice that the rock moves in a *horizontal* circle, so the center of the circle is not at his hand. The x-axis points horizontally, to the center of the circle, but the tension force is directed along the vine. Thus the correct free-body diagram is the one in **FIGURE 6.29d**.

Once the vine breaks, the visual overview of the situation is shown in **FIGURE 6.30**. The important thing to note here is that the initial x-component of velocity is the speed the rock had an instant before the vine broke.

SOLVE From the free-body diagram of Figure 6.29d, Newton's second law for circular motion is

$$\sum F_x = T\cos\theta = \frac{mv^2}{r}$$
$$\sum F_y = T\sin\theta - mg = 0$$

where θ is the angle of the vine below the horizontal. We can use the y-equation to find the angle of the vine:

$$\sin\theta = \frac{mg}{T}$$
$$\theta = \sin^{-1}\left(\frac{mg}{T}\right) = \sin^{-1}\left(\frac{(1.0\text{ kg})(9.8\text{ m/s}^2)}{200\text{ N}}\right) = 2.81°$$

where we've evaluated the angle at the maximum tension of 200 N. The vine's angle of inclination is small but not zero.

Turning now to the x-equation, we find the rock's speed around the circle is

$$v = \sqrt{\frac{rT\cos\theta}{m}}$$

Be careful! The radius r of the circle is not the length L of the vine. You can see in Figure 6.29c that $r = L\cos\theta$. Thus

$$v = \sqrt{\frac{LT\cos^2\theta}{m}} = \sqrt{\frac{(1.0\text{ m})(200\text{ N})(\cos 2.81°)^2}{1.0\text{ kg}}} = 14\text{ m/s}$$

Because this is the horizontal speed of the rock just when the vine breaks, the initial velocity $(v_x)_i$ in the visual overview of the projectile motion, Figure 6.30, must be $(v_x)_i = 14$ m/s. Recall that a projectile has no horizontal acceleration, so the rock's final position is

$$x_f = x_i + (v_x)_i\,\Delta t = 0\text{ m} + (14\text{ m/s})\Delta t$$

where Δt is the time the projectile is in the air. We're not given Δt, but we can find it from the vertical motion. For a projectile, the vertical motion is just free-fall motion, so we have

$$y_f = y_i + (v_y)_i\Delta t - \frac{1}{2}g(\Delta t)^2$$

The initial height is $y_i = 25$ m, the final height is $y_f = 0$ m, and the initial vertical velocity is $(v_y)_i = 0$ m/s. With these values, we have

$$0\text{ m} = 25\text{ m} + (0\text{ m/s})\Delta t - \frac{1}{2}(9.8\text{ m/s}^2)(\Delta t)^2$$

Solving this for Δt gives

$$\Delta t = \sqrt{\frac{2(25\text{ m})}{9.8\text{ m/s}^2}} = 2.3\text{ s}$$

Now we can use this time to find

$$x_f = 0\text{ m} + (14\text{ m/s})(2.3\text{ s}) = 32\text{ m}$$

The rock lands 32 m from the base of the cliff.

ASSESS The circumference of the rock's circle is $2\pi r$, or about 6 m. At a speed of 14 m/s, the rock takes roughly half a second to go around once. This seems reasonable. The 32 m distance is about 100 ft, which seems easily attainable from a cliff over 75 feet high.

FIGURE 6.30 Visual overview of the rock in projectile motion.

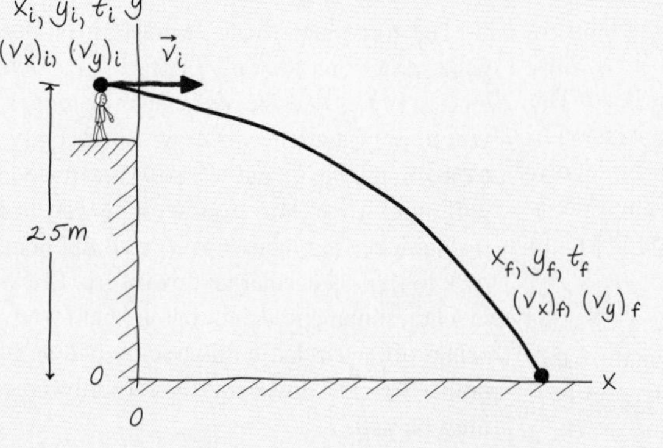

Known
$y_i = 25$ m, $y_f = 0$ m
$(v_y)_i = 0$ m/s
$t_i = 0$ s
$x_i = 0$ m
$(v_x)_i$ = speed of circular motion when vine breaks
Find
x_f

SUMMARY

The goal of Chapter 6 has been to learn about motion in a circle, including orbital motion under the influence of a gravitational force.

GENERAL PRINCIPLES

Uniform Circular Motion

An object moving in a circular path is in uniform circular motion if v is constant.

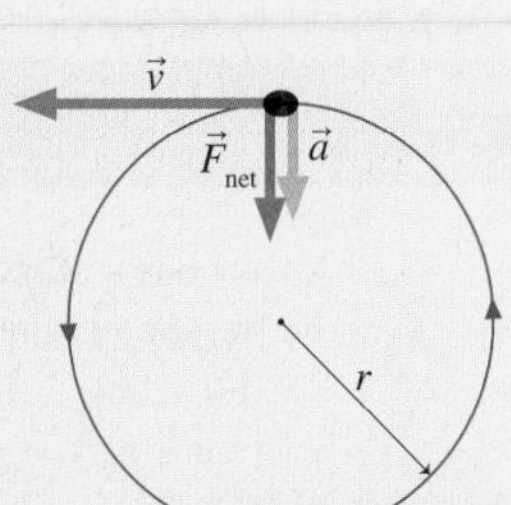

- The speed is constant, but the direction of motion is constantly changing.
- The **centripetal acceleration** is directed toward the center of the circle and has magnitude

$$a = \frac{v^2}{r}$$

- This acceleration requires a net force directed toward the center of the circle. Newton's second law for circular motion is

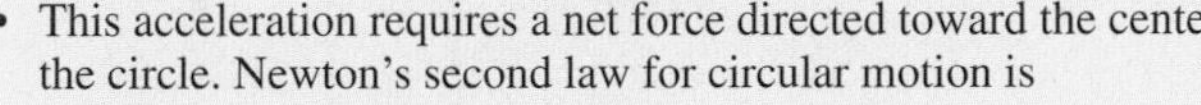

$$\vec{F}_{net} = m\vec{a} = \left(\frac{mv^2}{r}, \text{ toward center of circle}\right)$$

Universal Gravitation

Two objects with masses m_1 and m_2 that are distance r apart exert attractive gravitational forces on each other of magnitude

$$F_{1 on 2} = F_{2 on 1} = \frac{Gm_1m_2}{r^2}$$

where the gravitational constant is

$$G = 6.67 \times 10^{-11}\ \text{N}\cdot\text{m}^2/\text{kg}^2$$

This is **Newton's law of gravity.** Gravity is an inverse-square law.

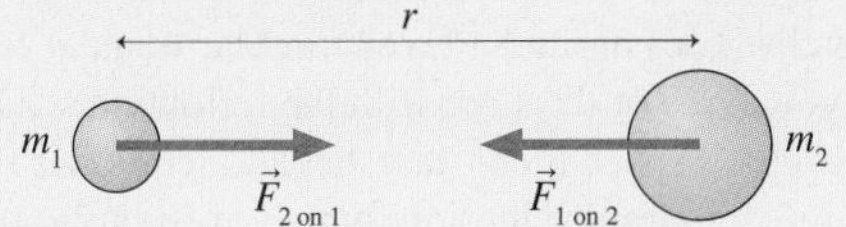

IMPORTANT CONCEPTS

Describing circular motion

We define new variables for circular motion. By convention, counterclockwise is positive.

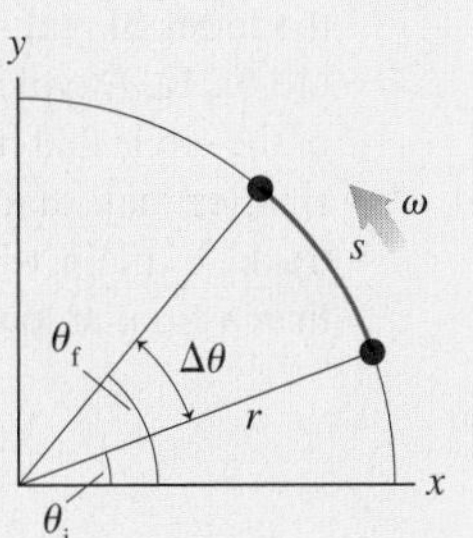

Angular position: θ

Angular displacement: $\Delta\theta = \theta_f - \theta_i$

Angular velocity: $\omega = \dfrac{\Delta\theta}{\Delta t}$

Angles are measured in radians, where 1 rev = 360° = 2π rad. The SI units of angular velocity are rad/s.

Period: $T =$ time for one complete circle.

Frequency: $f = \dfrac{1}{T}$

Uniform circular motion kinematics

For uniform circular motion:

$$\omega = 2\pi f \qquad \theta_f - \theta_i = \Delta\theta = \omega\,\Delta t$$

The velocity, acceleration, and circular motion variables are related as follows:

$$v = \frac{2\pi r}{T}$$

$$v = \omega r$$

$$a = \frac{v^2}{r} = \omega^2 r$$

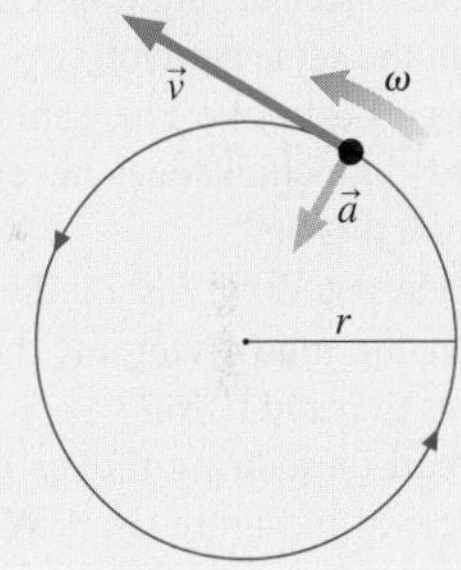

APPLICATIONS

Apparent weight and weightlessness

Circular motion requires a net force pointing to the center. The apparent weight $w_{app} = n$ is usually not the same as the true weight w. n must be > 0 for the object to be in contact with a surface.

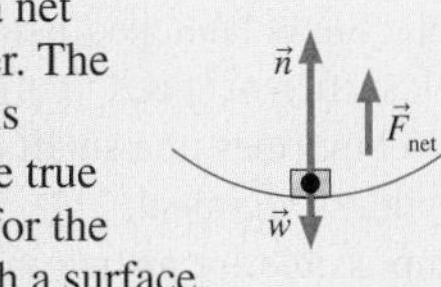

In orbital motion, the net force is provided by gravity. An astronaut and his spacecraft are both in free fall, so he feels weightless.

Planetary gravity and orbital motion

For a planet of mass M and radius R, the free-fall acceleration on the surface is

$$g = \frac{GM}{R^2}$$

The speed of a satellite in a low orbit is

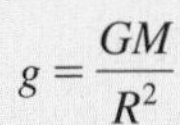

$$v = \sqrt{gr}$$

A **satellite** in a circular orbit of radius r around an object of mass M moves at a speed v given by

$$v = \sqrt{\frac{GM}{r}}$$

The period and radius are related as follows:

$$T^2 = \left(\frac{4\pi^2}{GM}\right)r^3$$

For homework assigned on MasteringPhysics, go to www.masteringphysics.com

Problem difficulty is labeled as | (straightforward) to ||||| (challenging).

Problems labeled can be done on a Workbook Dynamics Worksheet; INT integrate significant material from earlier chapters; BIO are of biological or medical interest.

QUESTIONS

Conceptual Questions

1. The batter in a baseball game hits a home run. As he circles the bases, is his angular velocity positive or negative?
2. Viewed from somewhere in space above the north pole, would a point on the earth's equator have a positive or negative angular velocity due to the earth's rotation?
3. A cyclist goes around a level, circular track at constant speed. Do you agree or disagree with the following statement? "Since the cyclist's speed is constant, her acceleration is zero." Explain.
4. In uniform circular motion, which of the following quantities are constant: speed, instantaneous velocity, angular velocity, centripetal acceleration, the magnitude of the net force?
5. A particle moving along a straight line can have nonzero acceleration even when its speed is zero (for instance, a ball in free fall at the top of its path). Can a particle moving in a circle have nonzero *centripetal* acceleration when its speed is zero? If so, give an example. If not, why not?
6. Would having four-wheel drive on a car make it possible to drive faster around corners on an icy road, without slipping, than the same car with two-wheel drive? Explain.
7. BIO Large birds like pheasants often walk short distances. Small birds like chickadees never walk. They either hop or fly. Why might this be?
8. When you drive fast on the highway with muddy tires, you can hear the mud flying off the tires into your wheel wells. Why does the mud fly off?
9. A ball on a string moves in a vertical circle as in Figure Q6.9. When the ball is at its lowest point, is the tension in the string greater than, less than, or equal to the ball's weight? Explain. (You may want to include a free-body diagram as part of your explanation.)

FIGURE Q6.9

10. Give an everyday example of circular motion for which the centripetal acceleration is mostly or completely due to a force of the type specified: (a) Static friction. (b) Tension.
11. Give an everyday example of circular motion for which the centripetal acceleration is mostly or completely due to a force of the type specified: (a) Gravity. (b) Normal force.
12. It's been proposed that future space stations create "artificial gravity" by rotating around an axis. (The space station would have to be much larger than the present space station for this to be feasible.)
    a. How would this work? Explain.
    b. Would the artificial gravity be equally effective throughout the space station? If not, where in the space station would the residents want to live and work?
13. A car coasts at a constant speed over a circular hill. Which of the free-body diagrams in Figure Q6.13 is correct? Explain.

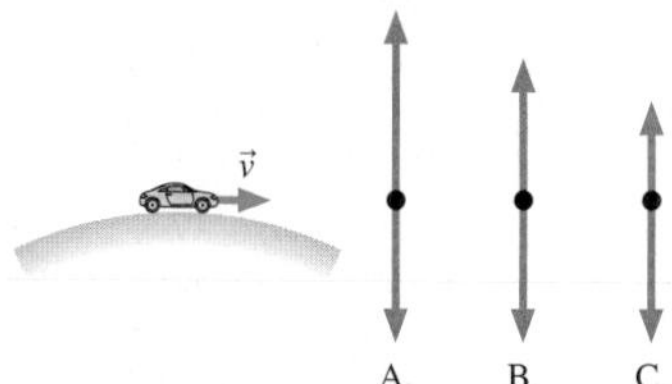

FIGURE Q6.13

14. Riding in the back of a pickup truck can be very dangerous. If the truck turns suddenly, the riders can be thrown from the truck bed. Why are the riders ejected from the bed?
15. Variation in your apparent weight is desirable when you ride a roller coaster; it makes the ride fun. However, too much variation over a short period of time can be painful. For this reason, the loops of real roller coasters are not simply circles like Figure 6.21a. A typical loop is shown in Figure Q6.15. The radius of the circle that matches the track at the top of the loop is much smaller than that of a matching circle at other places on the track. Explain why this shape gives a more comfortable ride than a circular loop.

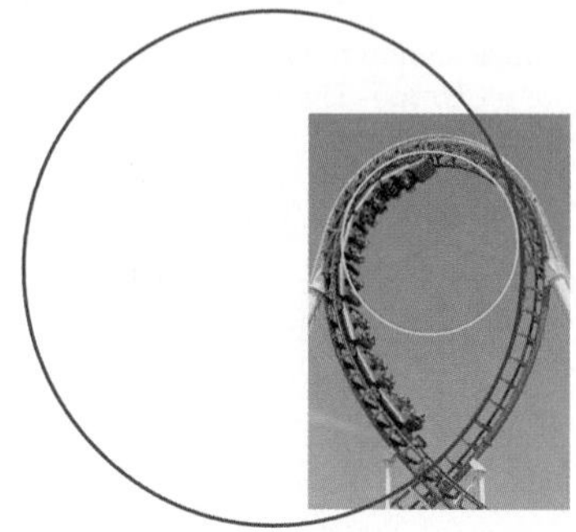
FIGURE Q6.15

16. A small projectile is launched parallel to the ground at height $h = 1$ m with sufficient speed to orbit a completely smooth, airless planet. A bug rides in a small hole inside the projectile. Is the bug weightless? Explain.
17. Why is it impossible for an astronaut inside an orbiting space shuttle to go from one end to the other by walking normally?
18. If every object in the universe feels an attractive gravitational force due to every other object, why don't you feel a pull from someone seated next to you?
19. A mountain climber's weight is less on the top of a tall mountain than at the base, though his mass is the same. Why?
20. Is the earth's gravitational force on the sun larger, smaller, or equal to the sun's gravitational force on the earth? Explain.

Multiple-Choice Questions

21. | A ball on a string moves around a complete circle, once a second, on a frictionless, horizontal table. The tension in the string is measured to be 6.0 N. What would the tension be if the ball went around in only half a second?
A. 1.5 N B. 3.0 N C. 12 N D. 24 N
22. | As seen from above, a car rounds the curved path shown in Figure Q6.22 at a constant speed. Which vector best represents the net force acting on the car?

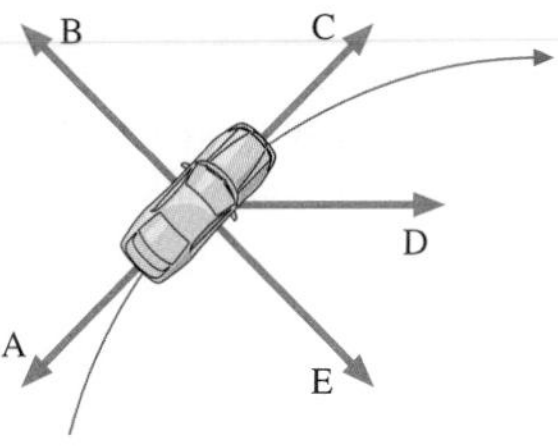

FIGURE Q6.22

23. | Suppose you and a friend, each of mass 60 kg, go to the park and get on a 4.0-m-diameter merry-go-round. You stand on the outside edge of the merry-go-round, while your friend pushes so that it rotates once every 6.0 s. What is the magnitude of the (apparent) outward force that you feel?
A. 7 N B. 63 N C. 130 N D. 260 N
24. | The cylindrical space station in Figure Q6.24, 200 m in diameter, rotates in order to provide artificial gravity of g for the occupants. How much time does the station take to complete one rotation?
A. 3 s B. 20 s C. 28 s D. 32 s

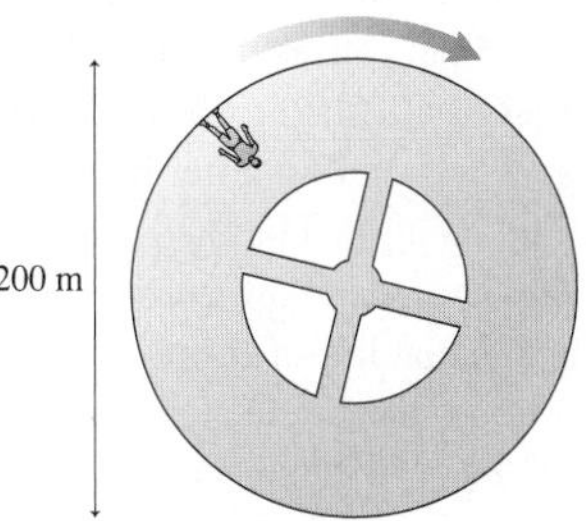

FIGURE Q6.24

25. || Two cylindrical space stations, the second four times the diameter of the first, rotate so as to provide the same amount of artificial gravity. If the first station makes one rotation in the time T, then the second station makes one rotation in time
A. $T/4$ B. $2T$ C. $4T$ D. $16T$
26. | A newly discovered planet has twice the mass and three times the radius of the earth. What is the free-fall acceleration at its surface, in terms of the free-fall acceleration g at the surface of the earth?
A. $\frac{2}{9}g$ B. $\frac{2}{3}g$ C. $\frac{3}{4}g$ D. $\frac{4}{3}g$
27. || Suppose one night the radius of the earth doubled but its mass stayed the same. What would be an approximate new value for the free-fall acceleration at the surface of the earth?
A. 2.5 m/s^2 B. 5.0 m/s^2 C. 10 m/s^2 D. 20 m/s^2
28. | Currently, the moon goes around the earth once every 27.3 days. If the moon could be brought into a new circular orbit with a smaller radius, its orbital period would be
A. More than 27.3 days.
B. 27.3 days.
C. Less than 27.3 days.
29. || Two planets orbit a star. Planet 1 has orbital radius r_1 and planet 2 has $r_2 = 4r_1$. Planet 1 orbits with period T_1. Planet 2 orbits with period
A. $T_2 = \frac{1}{2}T_1$ B. $T_2 = 2T_1$ C. $T_2 = 4T_1$ D. $T_2 = 8T_1$
30. | A particle undergoing circular motion in the xy-plane stops on the positive y-axis. Which of the following does *not* describe its angular position?
A. $\pi/2$ rad B. π rad C. $5\pi/2$ rad D. $-3\pi/2$ rad

Questions 31 through 33 concern a classic figure-skating jump called the axel. A skater starts the jump moving forward as shown in Figure Q6.31, leaps into the air, and turns one-and-a-half revolutions before landing. The typical skater is in the air for about 0.5 s, and the skater's hands are located about 0.8 m from the rotation axis.

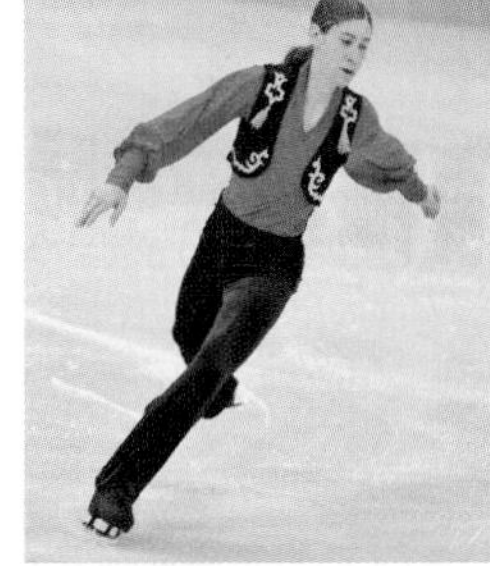
FIGURE Q6.31

31. || What is the approximate angular speed of the skater during the leap?
A. 2 rad/s B. 6 rad/s C. 9 rad/s D. 20 rad/s
32. | The skater's arms are fully extended during the jump. What is the approximate centripetal acceleration of the skater's hand?
A. 10 m/s^2 B. 30 m/s^2 C. 300 m/s^2 D. 450 m/s^2
33. | What is the approximate speed of the skater's hand?
A. 1 m/s B. 3 m/s C. 9 m/s D. 15 m/s

PROBLEMS

Section 6.1 Uniform Circular Motion

1. || What is the angular position in radians of the minute hand of a clock at (a) 5:00, (b) 7:15, and (c) 3:35?
2. | A child on a merry-go-round takes 3.0 s to go around once. What is his angular displacement during a 1.0 s time interval?
3. ||| What is the angular speed of the tip of the minute hand on a clock, in rad/s?
4. || An old-fashioned vinyl record rotates on a turntable at 45 rpm. What are (a) the angular speed in rad/s and (b) the period of the motion?
5. ||| The earth's radius is about 4000 miles. Kampala, the capital of Uganda, and Singapore are both nearly on the equator. The distance between them is 5000 miles.
   a. Through what angle do you turn, relative to the earth, if you fly from Kampala to Singapore? Give your answer in both radians and degrees.
   b. The flight from Kampala to Singapore takes 9 hours. What is the plane's angular speed relative to the earth?

6. ||| A Ferris wheel rotates at an angular velocity of 0.036 rad/s. At $t = 0$ min, your friend Seth is at the very top of the ride. What is Seth's angular position at $t = 3.0$ min, measured counterclockwise from the top? Give your answer as an angle in degrees between 0° and 360°.

7. ||||| A turntable rotates counterclockwise at 78 rpm. A speck of dust on the turntable is at $\theta = 0.45$ rad at $t = 0$ s. What is the angle of the speck at $t = 8.0$ s? Your answer should be between 0 and 2π rad.

8. || A fast-moving superhero in a comic book runs around a circular, 70-m-diameter track five and a half times (ending up directly opposite her starting point) in 3.0 s. What is her angular speed, in rad/s?

9. || Figure P6.9 shows the angular position of a potter's wheel.
   a. What is the angular displacement of the wheel between $t = 5$ s and $t = 15$ s?
   b. What is the angular velocity of the wheel at $t = 15$ s?

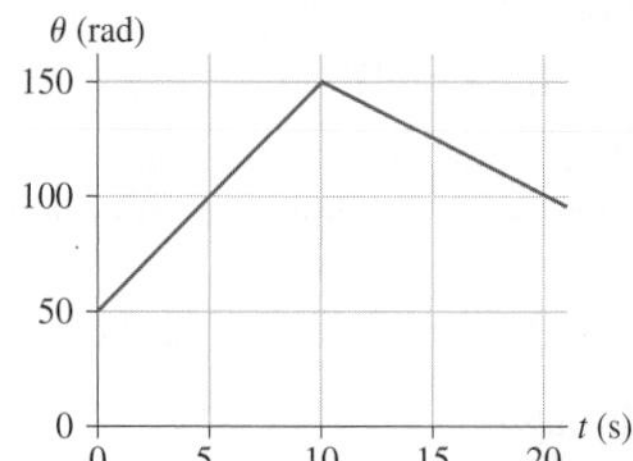

FIGURE P6.9

10. || The angular velocity (in rpm) of the blade of a blender is given in Figure P6.10.
    a. If $\theta = 0$ rad at $t = 0$ s, what is the blade's angular position at $t = 20$ s?
    b. At what time has the blade completed 10 full revolutions?

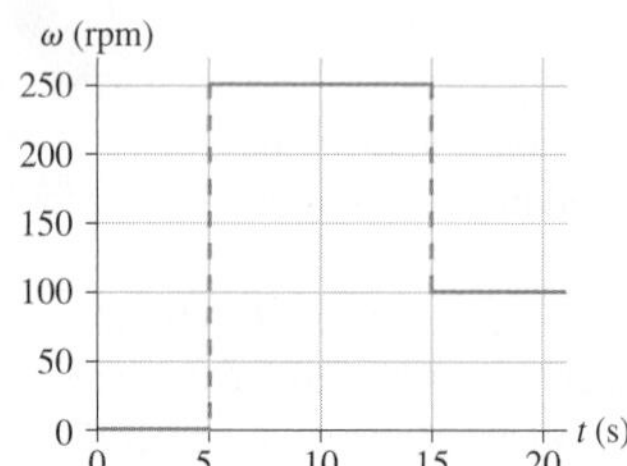

FIGURE P6.10

Section 6.2 Speed, Velocity, and Acceleration in Uniform Circular Motion

11. || A 5.0-m-diameter merry-go-round is turning with a 4.0 s period. What is the speed of a child on the rim?

12. | The blade on a table saw spins at 3450 rpm. Its diameter is 25.0 cm. What is the speed of a tooth on the edge of the blade, in both m/s and mph?

13. || The horse on a carousel is 4.0 m from the central axis.
    a. If the carousel rotates at 0.10 rev/s, how long does it take the horse to go around twice?
    b. How fast is a child on the horse going (in m/s)?

14. ||| The radius of the earth's very nearly circular orbit around the sun is 1.50×10^{11} m. Find the magnitude of the earth's (a) velocity, (b) angular velocity, and (c) centripetal acceleration as it travels around the sun. Assume a year of 365 days.

15. || Your roommate is working on his bicycle and has the bike upside down. He spins the 60-cm-diameter wheel, and you notice that a pebble stuck in the tread goes by three times every second. What are the pebble's speed and acceleration?

16. || To withstand "g-forces" of up to $10g$, caused by suddenly pulling out of a steep dive, fighter jet pilots train on a "human centrifuge." $10g$ is an acceleration of 98 m/s^2. If the length of the centrifuge arm is 12 m, at what speed is the rider moving when she experiences $10g$?

Section 6.3 Dynamics of Uniform Circular Motion

17. ||||| Figure P6.17 is a bird's-eye view of particles on a string moving in horizontal circles on a tabletop. All are moving at the same speed. Rank in order, from largest to smallest, the tensions T_1 to T_4.

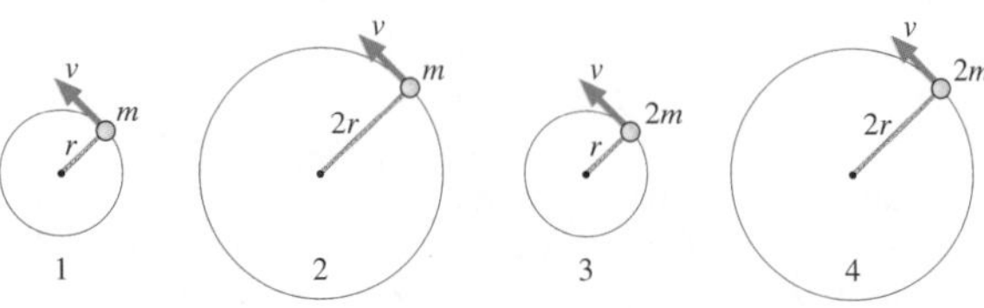

FIGURE P6.17

18. ||| A 200 g block on a 50-cm-long string swings in a circle on a horizontal, frictionless table at 75 rpm.
    a. What is the speed of the block?
    b. What is the tension in the string?

19. || A 1500 kg car drives around a flat 200-m-diameter circular track at 25 m/s. What are the magnitude and direction of the net force on the car? What causes this force?

20. || A fast pitch softball player does a "windmill" pitch, illustrated in Figure P6.20, moving her hand through a circular arc to pitch a ball at 70 mph. The 0.19 kg ball is 50 cm from the pivot point at her shoulder. At the lowest point of the circle, the ball has reached its maximum speed. BIO
    a. At the bottom of the circle, just before the ball leaves her hand, what is its centripetal acceleration?
    b. What are the magnitude and direction of the force her hand exerts on the ball at this point?

FIGURE P6.20

21. || A baseball pitching machine works by rotating a light and stiff rigid rod about a horizontal axis until the ball is moving toward the target. Suppose a 144 g baseball is held 85 cm from the axis of rotation and released at the major league pitching speed of 85 mph.
    a. What is the ball's centripetal acceleration just before it is released?
    b. What is the magnitude of the net force that is acting on the ball just before it is released?

22. || You're driving your pickup truck around a curve with a radius of 20 m. A box in the back of the truck is pressed up against the wall of the truck. How fast must you drive so that the force of the wall on the box equals the weight of the box?

Section 6.4 Apparent Forces in Circular Motion

23. ||| The passengers in a roller coaster car feel 50% heavier than their true weight as the car goes through a dip with a 30 m radius of curvature. What is the car's speed at the bottom of the dip?
24. || You hold a bucket in one hand. In the bucket is a 500 g rock. You swing the bucket so the rock moves in a vertical circle 2.2 m in diameter. What is the minimum speed the rock must have at the top of the circle if it is to always stay in contact with the bottom of the bucket?
25. ||| As a roller coaster car crosses the top of a 40-m-diameter loop-the-loop, its apparent weight is the same as its true weight. What is the car's speed at the top?
26. ||| A typical laboratory centrifuge rotates at 4000 rpm. Test tubes have to be placed into a centrifuge very carefully because of the very large accelerations.
BIO INT
    a. What is the acceleration at the end of a test tube that is 10 cm from the axis of rotation?
    b. For comparison, what is the magnitude of the acceleration a test tube would experience if stopped in a 1.0-ms-long encounter with a hard floor after falling from a height of 1.0 m?

Section 6.5 Circular Orbits and Weightlessness

27. ||| A satellite orbiting the moon very near the surface has a period of 110 min. Use this information, together with the radius of the moon from the table on the inside of the back cover, to calculate the free-fall acceleration on the moon's surface.

Section 6.6 Newton's Law of Gravity

28. |||| The centers of a 10 kg lead ball and a 100 g lead ball are separated by 10 cm.
    a. What gravitational force does each exert on the other?
    b. What is the ratio of this gravitational force to the weight of the 100 g ball?
29. || The gravitational force of a star on an orbiting planet 1 is F_1. Planet 2, which is twice as massive as planet 1 and orbits at twice the distance from the star, experiences gravitational force F_2. What is the ratio F_2/F_1?
30. || The free-fall acceleration at the surface of planet 1 is 20 m/s^2. The radius and the mass of planet 2 are twice those of planet 1. What is the free-fall acceleration on planet 2?
31. |||| What is the ratio of the sun's gravitational force on you to the earth's gravitational force on you?
32. ||| Suppose the free-fall acceleration at some location on earth was exactly 9.8000 m/s^2. What would it be at the top of a 1000-m-tall tower at this location? (Give your answer to five significant figures.)
33. || a. What is the gravitational force of the sun on the earth?
    b. What is the gravitational force of the moon on the earth?
    c. The moon's force is what percent of the sun's force?
34. || What is the free-fall acceleration at the surface of (a) Mars and (b) Jupiter?

Section 6.7 Gravity and Orbits

35. ||| Planet X orbits the star Omega with a "year" that is 200 earth days long. Planet Y circles Omega at four times the distance of Planet X. How long is a year on Planet Y?
36. |||| Satellite A orbits a planet with a speed of 10,000 m/s. Satellite B is twice as massive as satellite A and orbits at twice the distance from the center of the planet. What is the speed of satellite B?
37. ||| The Space Shuttle is in a 250-mile-high orbit. What are the shuttle's orbital period, in minutes, and its speed?
38. || The *asteroid belt* circles the sun between the orbits of Mars and Jupiter. One asteroid has a period of 5.0 earth years. What are the asteroid's orbital radius and speed?
39. |||| An earth satellite moves in a circular orbit at a speed of 5500 m/s. What is its orbital period?

General Problems

40. || How fast must a plane fly along the earth's equator so that the sun stands still relative to the passengers? In which direction must the plane fly, east to west or west to east? Give your answer in both km/h and mph. The radius of the earth is 6400 km.
41. ||| The car in Figure P6.41 travels at a constant speed along the road shown. Draw vectors showing its acceleration at the three points A, B, and C, or write $\vec{a} = \vec{0}$. The lengths of your vectors should correspond to the magnitudes of the accelerations.

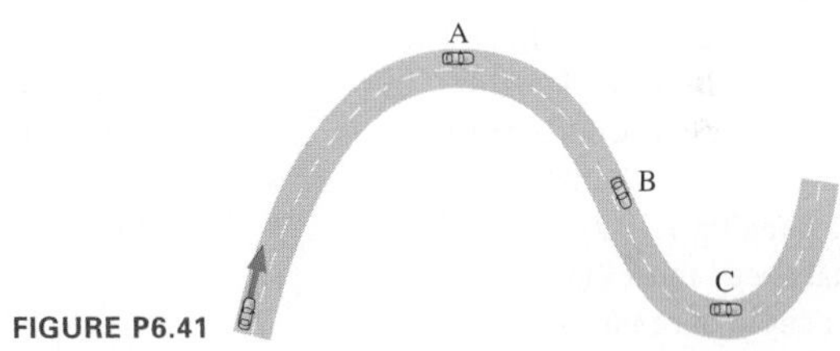

FIGURE P6.41

42. ||| In the Bohr model of the hydrogen atom, an electron (mass $m = 9.1 \times 10^{-31}$ kg) orbits a proton at a distance of 5.3×10^{-11} m. The proton pulls on the electron with an electric force of 8.2×10^{-8} N. How many revolutions per second does the electron make?
43. || A 75 kg man weighs himself at the north pole and at the equator. Which scale reading is higher? By how much? Assume the earth is a perfect sphere. Explain why the readings differ.
44. | A 1500 kg car takes a 50-m-radius unbanked curve at 15 m/s. What is the size of the friction force on the car?
45. |||| A 500 g ball swings in a vertical circle at the end of a 1.5-m-long string. When the ball is at the bottom of the circle, the tension in the string is 15 N. What is the speed of the ball at that point?
46. ||| Suppose the moon were held in its orbit not by gravity but by a massless cable attached to the center of the earth. What would be the tension in the cable? See the inside of the back cover for astronomical data.
47. ||| A 30 g ball rolls around a 40-cm-diameter L-shaped track, shown in Figure P6.47, at 60 rpm. Rolling friction can be neglected.

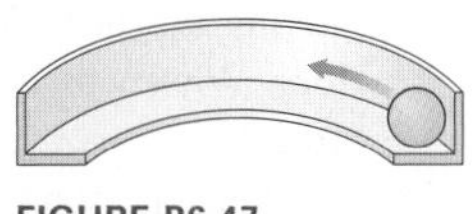
FIGURE P6.47

    a. How many different contact forces does the track exert on the ball? Name them.
    b. What is the magnitude of the net force on the ball?
48. || A 5.0 g coin is placed 15 cm from the center of a turntable. The coin has static and kinetic coefficients of friction with the turntable surface of $\mu_s = 0.80$ and $\mu_k = 0.50$. The turntable very slowly speeds up to 60 rpm. Does the coin slide off?
INT

49. ||| A *conical pendulum* is formed by attaching a 500 g ball to a 1.0-m-long string, then allowing the mass to move in a horizontal circle of radius 20 cm. Figure P6.49 shows that the string traces out the surface of a cone, hence the name.
 a. What is the tension in the string?
 b. What is the ball's angular velocity, in rpm?

Hint: Determine the horizontal and vertical components of the forces acting on the ball, and use the fact that the vertical component of acceleration is zero since there is no vertical motion.

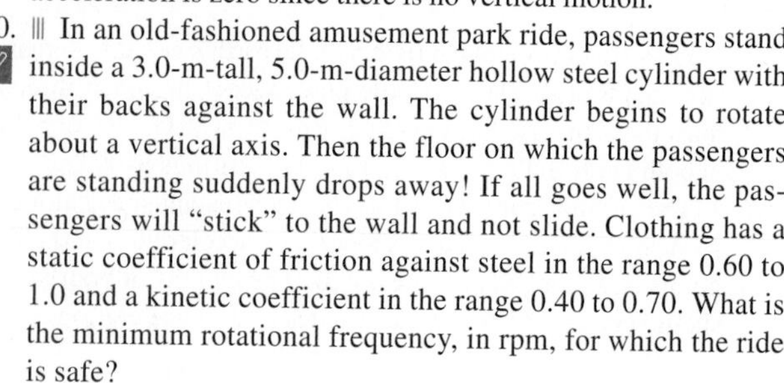

FIGURE P6.49

50. ||| In an old-fashioned amusement park ride, passengers stand inside a 3.0-m-tall, 5.0-m-diameter hollow steel cylinder with their backs against the wall. The cylinder begins to rotate about a vertical axis. Then the floor on which the passengers are standing suddenly drops away! If all goes well, the passengers will "stick" to the wall and not slide. Clothing has a static coefficient of friction against steel in the range 0.60 to 1.0 and a kinetic coefficient in the range 0.40 to 0.70. What is the minimum rotational frequency, in rpm, for which the ride is safe?

51. ||| The 0.20 kg puck on the frictionless, horizontal table in Figure P6.51 is connected by a string through a hole in the table to a hanging 1.20 kg block. With what speed must the puck rotate in a circle of radius 0.50 m if the block is to remain hanging at rest?

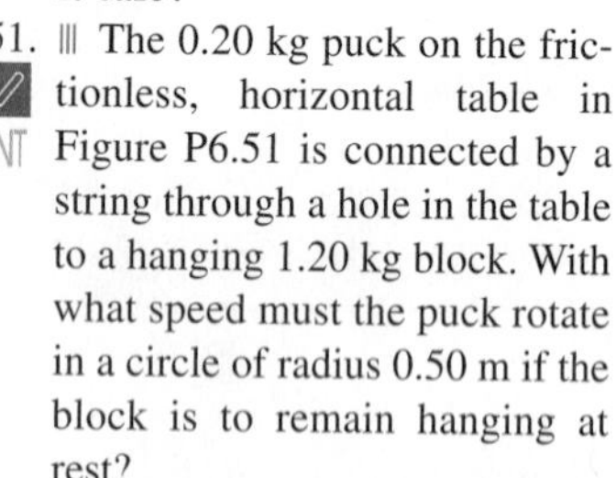

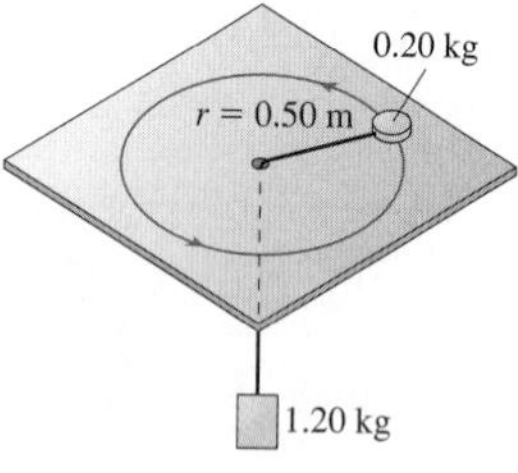

FIGURE P6.51

52. || While at the county fair, you decide to ride the Ferris wheel. Having eaten too many candy apples and elephant ears, you find the motion somewhat unpleasant. To take your mind off your stomach, you wonder about the motion of the ride. You estimate the radius of the big wheel to be 15 m, and you use your watch to find that each loop around takes 25 s.
 a. What are your speed and magnitude of your acceleration?
 b. What is the ratio of your apparent weight to your true weight at the top of the ride?
 c. What is the ratio of your apparent weight to your true weight at the bottom?

53. || A car drives over the top of a hill that has a radius of 50 m. What maximum speed can the car have without flying off the road at the top of the hill?

54. ||| A 100 g ball on a 60-cm-long string is swung in a vertical circle whose center is 200 cm above the floor. The string suddenly breaks when it is parallel to the ground and the ball is moving upward. The ball reaches a height 600 cm above the floor. What was the tension in the string an instant before it broke?

55. |||| While a person is walking, his arms (each with typical length 70 cm measured from the shoulder joint) swing through approximately a 45° angle in 0.50 s. As a reasonable approximation, we can assume that the arm moves with constant speed during each swing.
 a. What is the acceleration of a 1.0 g drop of blood in the fingertips at the bottom of the swing?
 b. Draw a free-body diagram for the drop of blood in part a.
 c. Find the magnitude and direction of the force that the blood vessel must exert on the drop of blood.
 d. What force would the blood vessel exert if the arm were not swinging?

56. ||| The two identical pucks in Figure P6.56 rotate together on a frictionless, horizontal table. They are tied together by strings 1 and 2, each of length l. If their common angular speed is ω, what are the tensions in the two strings?

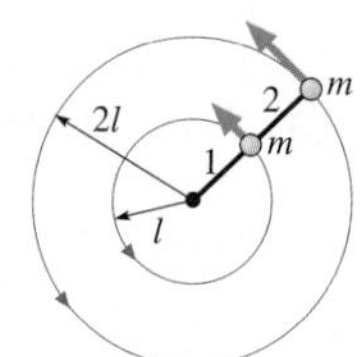

FIGURE P6.56

57. ||| The ultracentrifuge is an important tool for separating and analyzing proteins in biological research. Because of the enormous centripetal accelerations that can be achieved, the apparatus (see Figure 6.23) must be carefully balanced so that each sample is matched by another on the opposite side of the rotor shaft. Failure to do so is a costly mistake, as seen in Figure P6.57. Any difference in mass of the opposing samples will cause a net force in the horizontal plane on the shaft of the rotor. Suppose that a scientist makes a slight error in sample preparation, and one sample has a mass 10 mg greater than the opposing sample. If the samples are 10 cm from the axis of the rotor and the ultracentrifuge spins at 70,000 rpm, what is the magnitude of the net force on the rotor due to the unbalanced samples?

FIGURE P6.57

58. || The Space Shuttle orbits 300 km above the surface of the earth.
 a. What is the force of gravity on a 1.0 kg sphere inside the Space Shuttle?
 b. The sphere floats around inside the Space Shuttle, apparently "weightless." How is this possible?

59. |||| A sensitive gravimeter at a mountain observatory finds that the free-fall acceleration is 0.0075 m/s^2 less than that at sea level. What is the observatory's altitude?

60. || Suppose we could shrink the earth without changing its mass. At what fraction of its current radius would the free-fall acceleration at the surface be three times its present value?

61. ||| Planet Z is 10,000 km in diameter. The free-fall acceleration on Planet Z is 8.0 m/s^2.
 a. What is the mass of Planet Z?
 b. What is the free-fall acceleration 10,000 km above Planet Z's north pole?

62. ||| What are the speed and altitude of a geostationary satellite (see Example 6.19) orbiting Mars? Mars rotates on its axis once every 24.8 hours.

63. |||| a. What is the free-fall acceleration on Mars?
 b. Estimate the maximum speed at which an astronaut can walk on the surface of Mars.

64. ||| How long will it take a rock dropped from 2.0 m above the surface of Mars to reach the ground?

65. ||| A 20 kg sphere is at the origin and a 10 kg sphere is at
INT $(x, y) = (20\text{ cm}, 0\text{ cm})$. At what point or points could you place a small mass such that the net gravitational force on it due to the spheres is zero?

66. ||| a. At what height above the earth is the free-fall acceleration 10% of its value at the surface?
b. What is the speed of a satellite orbiting at that height?

67. | Mars has a small moon, Phobos, that orbits with a period of 7 h 39 min. The radius of Phobos' orbit is 9.4×10^6 m. Use only this information (and the value of G) to calculate the mass of Mars.

68. || You are the science officer on a visit to a distant solar system. Prior to landing on a planet you measure its diameter to be 1.80×10^7 m and its rotation period to be 22.3 h. You have previously determined that the planet orbits 2.20×10^{11} m from its star with a period of 402 earth days. Once on the surface you find that the free-fall acceleration is 12.2 m/s^2. What are the masses of (a) the planet and (b) the star?

69. ||||| Europa, a satellite of Jupiter,
BIO is believed to have a liquid ocean of water (with a possibility of life) beneath its icy surface. In planning a future mission to Europa, what is the fastest that an astronaut with legs of length 0.70 m could walk on the surface of Europa? Europa is 3100 km in diameter and has a mass of 4.8×10^{22} kg.

In Problems 70 through 73 you are given the equation (or equations) used to solve a problem. For each of these, you are to

a. Write a realistic problem for which this is the correct equation. The last two questions should involve real planets. Be sure that the answer your problem requests is consistent with the equation given.
b. Finish the solution of the problem.

70. || $60\text{ N} = (0.30\text{ kg})\omega^2(0.50\text{ m})$

71. || $(1500\text{ kg})(9.80\text{ m/s}^2) - 11{,}760\text{ N} = (1500\text{ kg})v^2/(200\text{ m})$

72. || $$\frac{(6.67 \times 10^{-11}\text{ N}\cdot\text{m}^2/\text{kg}^2)(1.90 \times 10^{27}\text{ kg})}{r^2} = \frac{(6.67 \times 10^{-11}\text{ N}\cdot\text{m}^2/\text{kg}^2)(5.98 \times 10^{24}\text{ kg})}{(6.37 \times 10^6\text{ m})^2}$$

73. || $$\frac{(6.67 \times 10^{-11}\text{ N}\cdot\text{m}^2/\text{kg}^2)(5.98 \times 10^{24}\text{ kg})(1000\text{ kg})}{r^2} = \frac{(1000\text{ kg})(1997\text{ m/s})^2}{r}$$

Passage Problems

Orbiting the Moon

Suppose a spacecraft orbits the moon in a very low, circular orbit, just a few hundred meters above the lunar surface. The moon has a diameter of 3500 km, and the free-fall acceleration at the surface is 1.6 m/s^2.

74. | The direction of the net force on the craft is
A. Away from the surface of the moon.
B. In the direction of motion.
C. Toward the center of the moon.
D. Nonexistent, because the net force is zero.

75. | How fast is this spacecraft moving?
A. 53 m/s B. 75 m/s C. 1700 m/s D. 2400 m/s

76. | How much time does it take for the spacecraft to complete one orbit?
A. 38 min B. 76 min C. 110 min D. 220 min

77. | The material that comprises the side of the moon facing the earth is actually slightly more dense than the material on the far side. When the spacecraft is above a more dense area of the surface, the moon's gravitational force on the craft is a bit stronger. In order to stay in a circular orbit of constant height and speed, the spacecraft could fire its rockets while passing over the denser area. The rockets should be fired so as to generate a force on the craft
A. Away from the surface of the moon.
B. In the direction of motion.
C. Toward the center of the moon.
D. Opposite the direction of motion.

STOP TO THINK ANSWERS

Stop to Think 6.1: A. Because $5\pi/2$ rad $= 2\pi$ rad $+ \pi/2$ rad, the particle's position is one complete revolution (2π rad) plus an extra $\pi/2$ rad. This extra $\pi/2$ rad puts the particle at position A.

Stop to Think 6.2: D > B > C > A. The centripetal acceleration is $\omega^2 r$. Changing r by a factor of 2 changes the centripetal acceleration by a factor of 2, but changing ω by a factor of 2 changes the centripetal acceleration by a factor of 4.

Stop to Think 6.3: $T_D > T_B = T_E > T_C > T_A$. The center-directed force is $m\omega^2 r$. Changing r by a factor of 2 changes the tension by a factor of 2, but changing f (and thus ω) by a factor of 2 changes the tension by a factor of 4.

Stop to Think 6.4: B. The car is moving in a circle, so there must be a net force toward the center of the circle. The center of the circle is below the car, so the net force must point downward. This can be true only if $w > n$.

Stop to Think 6.5: B > A > D > C. The free-fall acceleration is proportional to the mass, but inversely proportional to the square of the radius.

Stop to Think 6.6: C. The period of the orbit does not depend on the mass of the orbiting object.

7 Rotational Motion

To get the roulette wheel spinning, the croupier must give the wheel a push in the direction of its motion. Does it matter if she pushes closer to the rim or nearer the center?

LOOKING AHEAD ▸▸

The goal of Chapter 7 is to understand the physics of rotating objects.

The Rotation of a Rigid Body

A **rigid body** is an extended object whose size and shape do not change as it moves.

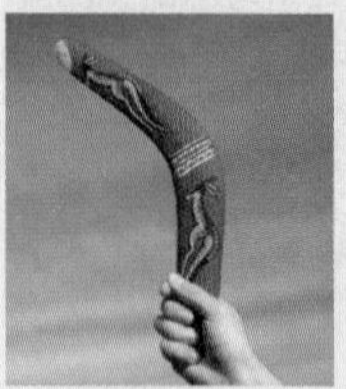

Boomerangs and bicycle wheels are examples of rigid bodies.

A rigid body whose angular velocity is changing has an **angular acceleration.**

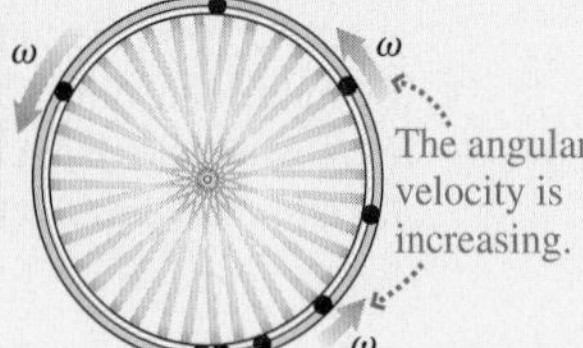

Looking Back ◂◂
6.1–6.2 Uniform circular motion

Newton's Second Law for Rotational Motion

You've learned Newton's second law of motion for *translational* motion: **A net force causes an object to accelerate.** In this chapter, we'll study Newton's second law for *rotational* motion: **A net torque causes an object to have an *angular* acceleration.**

To make the merry-go-round speed up, the girl has to apply a torque to it by pushing at its edge.

Looking Back ◂◂
4.6 Newton's second law

Torque

Torque is the rotational equivalent of force. To get an object rotating, you need to apply a torque to it. The farther from the axis of rotation a force is applied, the greater the torque.

By applying forces at the edge of the large wheel, the sailor can exert a large torque upon it.

Gravitational Torque

For the purpose of calculating torque, the entire weight of an object can be considered as acting at a single point, the **center of gravity.**

The weight of this tree, acting at its center of gravity, tries to rotate the tree about its base.

Moment of Inertia

We have learned that *mass* is the property of an object that resists acceleration. The property of an object that resists angular acceleration is its *moment of inertia.* The moment of inertia of an object depends not only on its mass but also on how that mass is distributed.

By extending its tail, this cat increases its moment of inertia. This increases its resistance to angular acceleration, making it harder for it to fall.

7.1 The Rotation of a Rigid Body

7.7

So far, our study of physics has focused almost exclusively on the *particle model* in which an entire object is represented as a single point in space. The particle model is entirely adequate for understanding motion in a wide variety of situations, but there are also cases for which we need to consider the motion of an *extended object*—a system of particles for which the size and shape *do* make a difference and cannot be neglected.

A **rigid body** is an extended object whose size and shape do not change as it moves. For example, a bicycle wheel can be thought of as a rigid body. FIGURE 7.1 shows a rigid body as a collection of atoms held together by the rigid "massless rods" of molecular bonds.

FIGURE 7.1 The rigid-body model of an extended object.

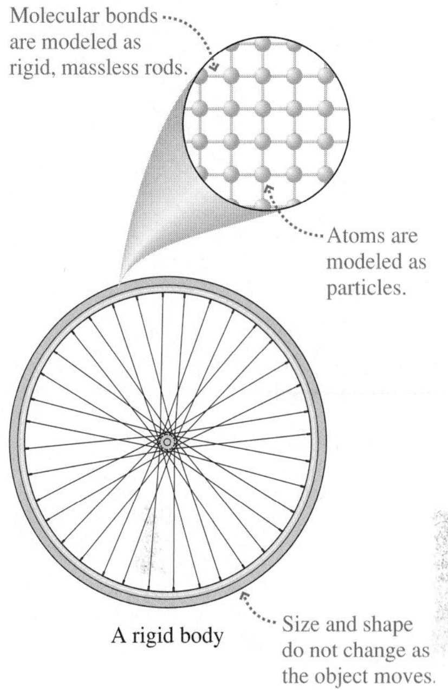
Molecular bonds are modeled as rigid, massless rods.
Atoms are modeled as particles.
A rigid body
Size and shape do not change as the object moves.

Real molecular bonds are, of course, not perfectly rigid. That's why an object seemingly as rigid as a bicycle wheel can flex and bend. Thus Figure 7.1 is really a simplified *model* of an extended object, the **rigid-body model.** The rigid-body model is a very good approximation for many real objects of practical interest, such as wheels and axles. Even nonrigid objects can often be modeled as rigid bodies during segments of their motion. For example, a diver is well described as a rotating rigid body while she's in the tuck position.

FIGURE 7.2 illustrates the three basic types of motion of a rigid body: **translational motion, rotational motion,** and **combination motion.** We've already studied translational motion of a rigid body using the particle model. If a rigid body doesn't rotate, this model is often adequate for describing its motion. The rotational motion of a rigid body will be the main focus of this chapter. We'll also discuss an important case of combination motion—that of a *rolling* object—later in this chapter.

FIGURE 7.2 Three basic types of motion of a rigid body.

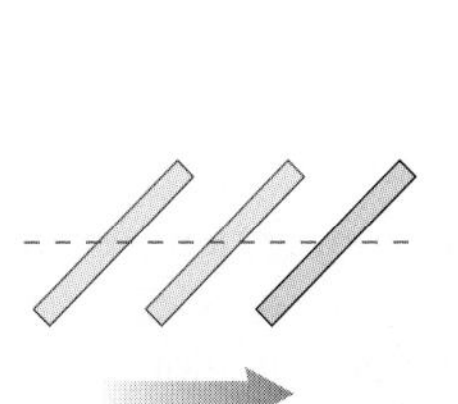

Translational motion: The object as a whole moves along a trajectory but does not rotate.

Rotational motion: The object rotates about a fixed point. Every point on the object moves in a circle.

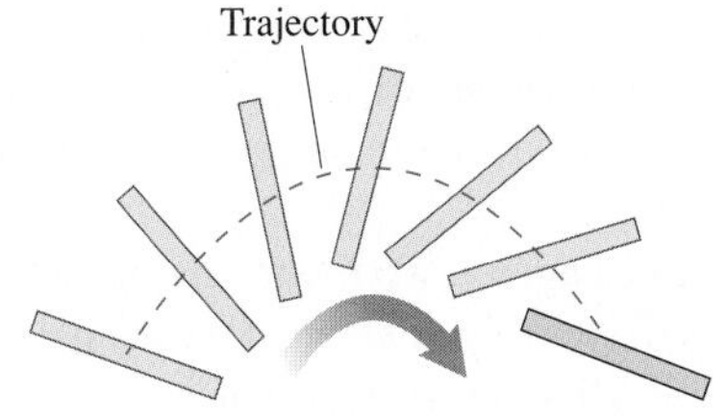
Trajectory

Combination motion: An object rotates as it moves along a trajectory.

Rotational Motion of a Rigid Body

FIGURE 7.3 shows a wheel rotating on an axle. Notice that as the wheel rotates for a time interval Δt, two points 1 and 2 on the wheel, marked with dots, turn through the *same angle,* even through their distances r from the axis of rotation may be different; that is, $\Delta\theta_1 = \Delta\theta_2$ during the time interval Δt. As a consequence, the two points have equal angular velocities: $\omega_1 = \omega_2$. In general, **every point on a rotating rigid body has the same angular velocity.** Because of this, we can refer to the angular velocity ω *of the wheel.*

FIGURE 7.3 All points on a wheel rotate with the same angular velocity.

Every point on the wheel undergoes circular motion with the same angular velocity ω.

Recall from Chapter 6 that the speed of a particle moving in a circle is $v = \omega r$, so two points of a rotating object will have different *speeds* if they have different distances from the axis of rotation, but *all* points have the *same* angular velocity ω. Thus angular velocity is one of the most important parameters of a rotating object.

Because every point on a rotating object moves in a circle, we can carry forward all the results for circular motion from Chapter 6. Thus the angular displacement of any point on the wheel shown in Figure 7.3 is found from Equation 6.4 as $\Delta\theta = \omega\,\Delta t$; the speed of any particle in the wheel is $v = \omega r$, where r is the particle's distance from the axis; and the particle's centripetal acceleration is $a = \omega^2 r$.

FIGURE 7.4 A rotating wheel with a changing angular velocity.

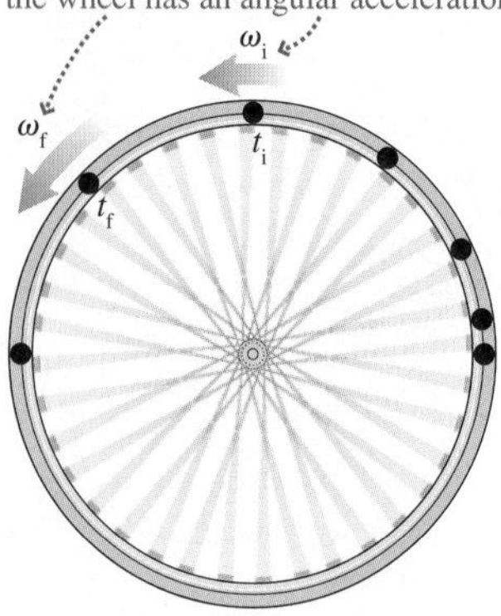

Angular Acceleration

If you push on the edge of a bicycle wheel, it begins to rotate. If you continue to push, it rotates ever faster. Its angular velocity is *changing*. To understand the dynamics of rotating objects, we'll need to be able to describe this case of changing angular velocity—that is, the case of *nonuniform* circular motion.

FIGURE 7.4 shows a bicycle wheel whose angular velocity is changing. The dot represents a particular point on the wheel at successive times. At time t_i the angular velocity is ω_i; at a later time $t_f = t_i + \Delta t$ the angular velocity has changed to ω_f. The change in angular velocity during this time interval is

$$\Delta\omega = \omega_f - \omega_i$$

Recall that in Chapter 2 we defined the *linear* acceleration as

$$a_x = \frac{\Delta v_x}{\Delta t} = \frac{(v_x)_f - (v_x)_i}{\Delta t}$$

By analogy, we now define the **angular acceleration** as

$$\alpha = \frac{\text{change in angular velocity}}{\text{time interval}} = \frac{\Delta\omega}{\Delta t} \tag{7.1}$$

Angular acceleration for a particle in nonuniform circular motion

FIGURE 7.5 Determining the sign of the angular acceleration.

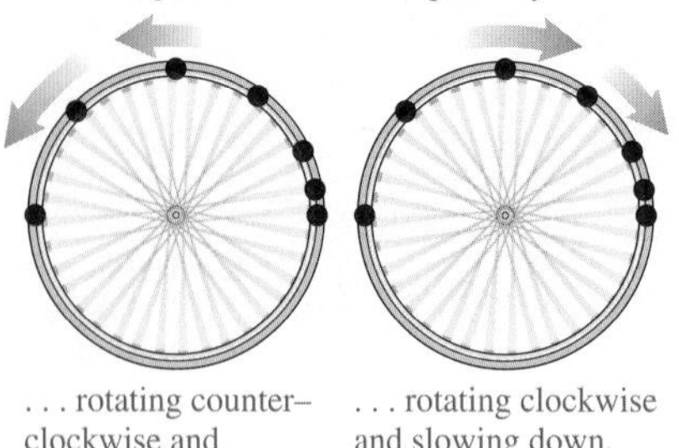

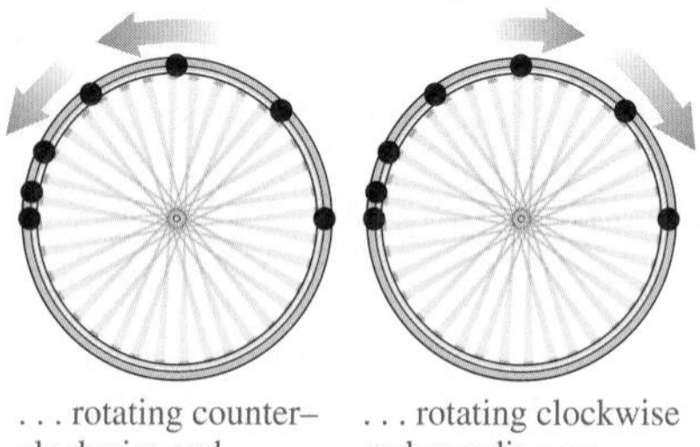

α is *negative* when the rigid body is . . .
. . . rotating counter–clockwise and slowing down.
. . . rotating clockwise and speeding up.

We use the symbol α (Greek alpha) for angular acceleration. Because the units of ω are rad/s, the units of angular acceleration are (rad/s)/s, or rad/s^2. From Equation 7.1, the sign of α is the same as the sign of $\Delta\omega$. FIGURE 7.5 shows how to determine the sign of α. Be careful with the sign of α; just as with linear acceleration, positive and negative values of α can't be interpreted as simply "speeding up" and "slowing down." Like ω, the angular acceleration α is the same for every point on a rotating rigid body.

In Chapter 6 we found analogies between linear and angular positions and velocities. Here we've extended those analogies to include linear and angular accelerations. Table 7.1 summarizes all of these analogies between linear and circular motion.

NOTE ► Don't confuse the angular acceleration with the centripetal acceleration introduced in Chapter 6. The angular acceleration indicates how rapidly the *angular* velocity is changing. The centripetal acceleration is a vector quantity that points toward the center of a particle's circular path; it is nonzero even if the angular velocity is constant. ◄

In addition, the various equations of one-dimensional kinematics have analogs for rotational or circular motion. Table 7.2 lists the equations for one-dimensional motion and the analogous equations for the kinematics of circular motion.

TABLE 7.1 Linear and circular motion variables

Linear motion	Circular motion
Position x	Angular position θ
Velocity $v_x = \Delta x/\Delta t$	Angular velocity $\omega = \Delta\theta/\Delta t$
Acceleration $a_x = \Delta v_x/\Delta t$	Angular acceleration $\alpha = \Delta\omega/\Delta t$

TABLE 7.2 Linear and circular motion equations

Linear motion	Circular motion
Displacement at constant speed: $\Delta x = v\,\Delta t$	Angular displacement at constant angular speed: $\Delta\theta = \omega\,\Delta t$
Change in velocity at constant acceleration: $\Delta v = a\,\Delta t$	Change in angular velocity at constant angular acceleration: $\Delta\omega = \alpha\,\Delta t$
Displacement at constant acceleration: $\Delta x = v_i\,\Delta t + \frac{1}{2}a\,\Delta t^2$	Angular displacement at constant angular acceleration: $\Delta\theta = \omega_i\,\Delta t + \frac{1}{2}\alpha\,\Delta t^2$

EXAMPLE 7.1 Spinning up a computer disk

The disk in a computer disk drive spins up to 5400 rpm in 2.00 s. What is the angular acceleration of the disk? At the end of 2.00 s, how many revolutions has the disk made?

PREPARE The initial angular velocity is $\omega_i = 0$ rad/s. The final angular velocity is $\omega_f = 5400$ rpm. However, this value is not in the correct SI units of rad/s. The conversion is

$$\omega_f = \frac{5400\ \text{rev}}{\text{min}} \times \frac{1\ \text{min}}{60\ \text{s}} \times \frac{2\pi\ \text{rad}}{1\ \text{rev}} = 565\ \text{rad/s}$$

SOLVE From the definition of angular acceleration, we have

$$\alpha = \frac{\Delta\omega}{\Delta t} = \frac{565\ \text{rad/s} - 0\ \text{rad/s}}{2.00\ \text{s}} = 283\ \text{rad/s}^2$$

We can compute the angular displacement during this acceleration by using the angular displacement equation from Table 7.2:

$$\begin{aligned}\Delta\theta &= \omega_i \Delta t + \tfrac{1}{2}\alpha\, \Delta t^2 \\ &= (0\ \text{rad/s})(2.00\ \text{s}) + \tfrac{1}{2}(283\ \text{rad/s}^2)(2.00\ \text{s})^2 \\ &= 566\ \text{rad}\end{aligned}$$

Each revolution corresponds to an angular displacement of 2π, so we have

$$\begin{aligned}\text{number of revolutions} &= \frac{566\ \text{rad}}{2\pi\ \text{rad/revolution}} \\ &= 90\ \text{revolutions}\end{aligned}$$

The disk completes 90 revolutions during the first 2 seconds.

ASSESS It seems reasonable that a fast-spinning disk would turn 90 times in a few seconds.

Graphs for Rotational Motion with Constant Angular Acceleration

In Chapter 2 we studied position, velocity, and acceleration graphs for motion with constant acceleration. A review of Section 2.5 is highly recommended. Because of the analogies between linear and angular quantities in Table 7.1, the rules for graphing angular variables are identical with those for linear variables. In particular, **the angular velocity is the slope of the angular position-versus-time graph** (as we discussed in Chapter 6), and **the angular acceleration is the slope of the angular velocity-versus-time graph.** When the angular acceleration is constant, the equations for circular motion in Table 7.2 show that the angular velocity graph is linear while the angular position graph is parabolic.

EXAMPLE 7.2 Graphing angular quantities

FIGURE 7.6 shows the angular velocity-versus-time graph for the propeller of a ship.

a. Describe the motion of the propeller.
b. Draw the angular acceleration graph for the propeller.

FIGURE 7.6 The propeller's angular velocity.

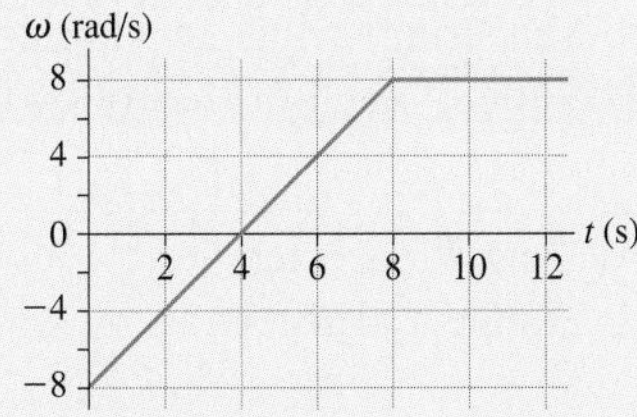

PREPARE The angular acceleration graph is the slope of the angular velocity graph.

SOLVE

a. Initially the propeller has a negative angular velocity, so it is turning clockwise. It slows down until, at $t = 4$ s, it is instantaneously stopped. It then speeds up in the opposite direction until it is turning counterclockwise at a constant angular velocity.
b. The angular acceleration graph is the slope of the angular velocity graph. From $t = 0$ s, to $t = 8$ s, the slope is

$$\frac{\Delta\omega}{\Delta t} = \frac{\omega_f - \omega_i}{\Delta t} = \frac{(8.0\ \text{rad/s}) - (-8.0\ \text{rad/s})}{8.0\ \text{s}} = 2.0\ \text{rad/s}^2$$

After $t = 8$ s, the slope is zero, so the angular acceleration is zero. This graph is plotted in **FIGURE 7.7**.

FIGURE 7.7 Angular acceleration graph for a propeller.

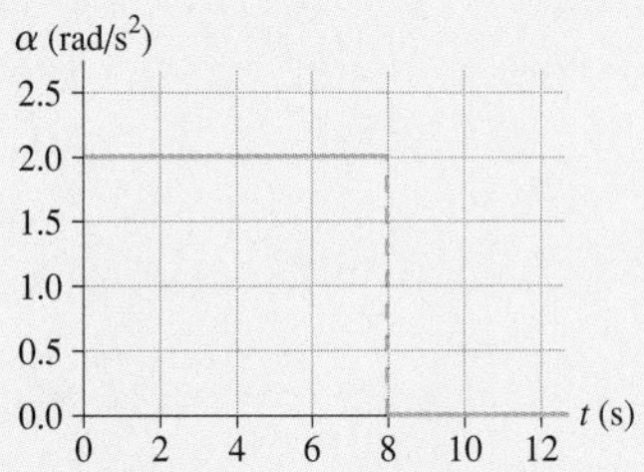

ASSESS A comparison of these graphs with their linear analogs in Figure 2.24 suggests that we're on the right track.

FIGURE 7.8 Uniform and nonuniform circular motion.

(a) Uniform circular motion

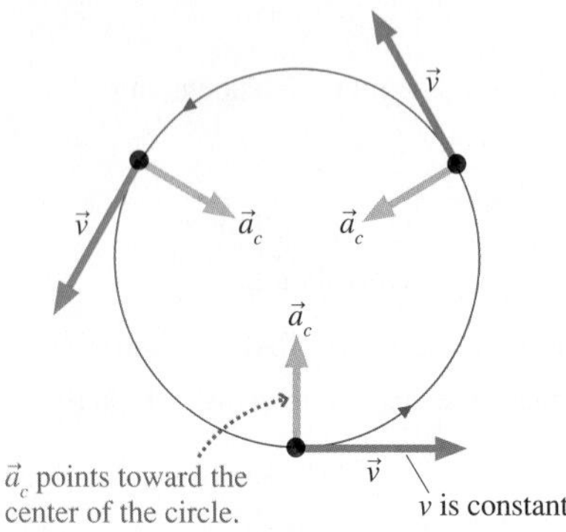
$\vec{v}$
$\vec{v}$
$\vec{a}_c$
$\vec{a}_c$
$\vec{a}_c$
$\vec{v}$
$\vec{a}_c$ points toward the center of the circle.
v is constant.

(b) Nonuniform circular motion

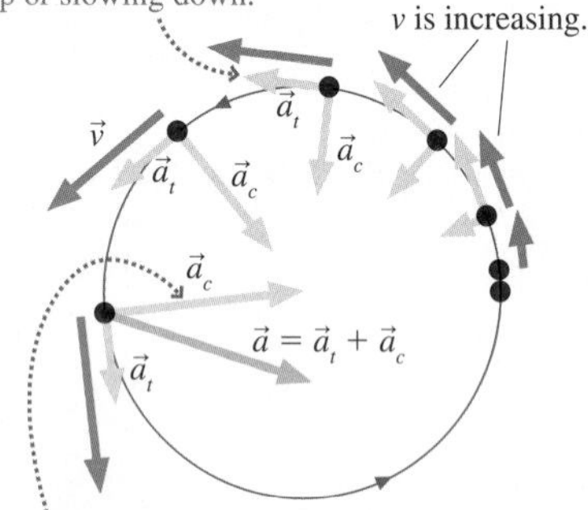
The tangential acceleration $\vec{a}_t$ causes the particle's *speed* to change. There's a tangential acceleration *only* when the particle is speeding up or slowing down.
v is increasing.
$\vec{v}$
$\vec{a}_t$
$\vec{a}_c$
$\vec{a}_t$
$\vec{a}_c$
$\vec{a}_c$
$\vec{a} = \vec{a}_t + \vec{a}_c$
$\vec{a}_t$
The centripetal acceleration $\vec{a}_c$ causes the particle's *direction* to change. As the particle speeds up, a_c gets larger. Circular motion *always* has a centripetal acceleration.

Tangential Acceleration

As you learned in Chapter 6, and as FIGURE 7.8a reminds you, a particle undergoing uniform circular motion has an acceleration directed inward toward the center of the circle. This centripetal acceleration $\vec{a}_c$ is due to the change in the *direction* of the particle's velocity. Recall that the magnitude of the centripetal acceleration is $a_c = v^2/r = \omega^2 r$.

NOTE ► Centripetal acceleration will now be denoted a_c to distinguish it from tangential acceleration a_t, discussed below. ◄

If the particle's circular motion is *nonuniform*, so that the particle's speed is changing, then the particle will have another component to its acceleration. FIGURE 7.8b shows a particle whose speed is increasing as it moves around its circular path. Because the *magnitude* of the velocity is increasing, this second component of the acceleration is directed *tangentially* to the circle, in the same direction as the velocity. This component of acceleration is called the **tangential acceleration.** As shown in Figure 7.8b, **the full acceleration $\vec{a}$ is then the vector sum of these two components:** the centripetal acceleration $\vec{a}_c$ and the tangential acceleration $\vec{a}_t$.

The tangential acceleration measures the rate at which the particle's speed around the circle increases. Thus its magnitude is

$$a_t = \frac{\Delta v}{\Delta t}$$

We can relate the tangential acceleration to the *angular* acceleration by using the relation $v = \omega r$ between the speed of a particle moving in a circle of radius r and its angular velocity ω. We have

$$a_t = \frac{\Delta v}{\Delta t} = \frac{\Delta(\omega r)}{\Delta t} = \frac{\Delta \omega}{\Delta t} r$$

or, because $\alpha = \Delta\omega/\Delta t$ from Equation 7.1,

$$a_t = \alpha r \qquad (7.2)$$

Relationship between tangential and angular acceleration

We've seen that all points on a rotating rigid body have the same angular acceleration. From Equation 7.2, however, the centripetal and tangential accelerations of a point on a rotating object depends on the point's distance r from the axis, so that these accelerations are *not* the same for all points.

STOP TO THINK 7.1 A ball on the end of a string swings in a horizontal circle once every second. State whether the magnitude of each of the following quantities is zero, constant (but not zero), or changing.

a. Velocity
b. Angular velocity
c. Centripetal acceleration
d. Angular acceleration
e. Tangential acceleration

7.2 Torque

Newton's genius, summarized in his second law of motion, was to recognize force as the cause of acceleration. But what about *angular* acceleration? What do Newton's laws tell us about rotational motion? To begin our study of rotational motion, we'll need to find a rotational equivalent of force.

Consider the common experience of pushing open a heavy door. **FIGURE 7.9** is a top view of a door that is hinged on the left. Four forces are shown, all of equal strength. Which of these will be most effective at opening the door?

Force $\vec{F}_1$ will open the door, but force $\vec{F}_2$, which pushes straight at the hinge, will not. Force $\vec{F}_3$ will open the door, but not as easily as $\vec{F}_1$. What about $\vec{F}_4$? It is perpendicular to the door and it has the same magnitude as $\vec{F}_1$, but you know from experience that pushing close to the hinge is not as effective as pushing at the outer edge of the door.

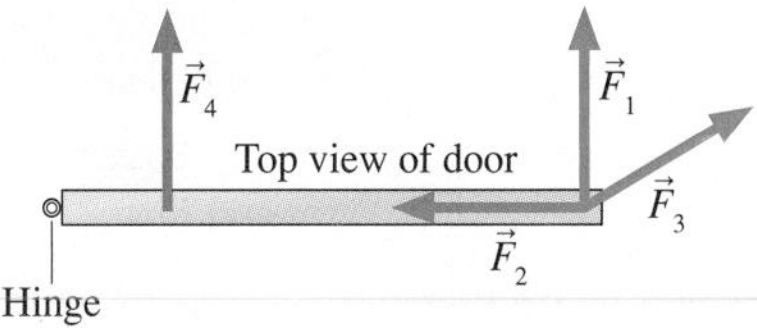

FIGURE 7.9 The four forces are the same strength, but they have different effects on the swinging door.

The ability of a force to cause a rotation thus depends on three factors:

1. The magnitude F of the force
2. The distance r from the pivot—the axis about which the object can rotate—to the point at which the force is applied
3. The angle at which the force is applied

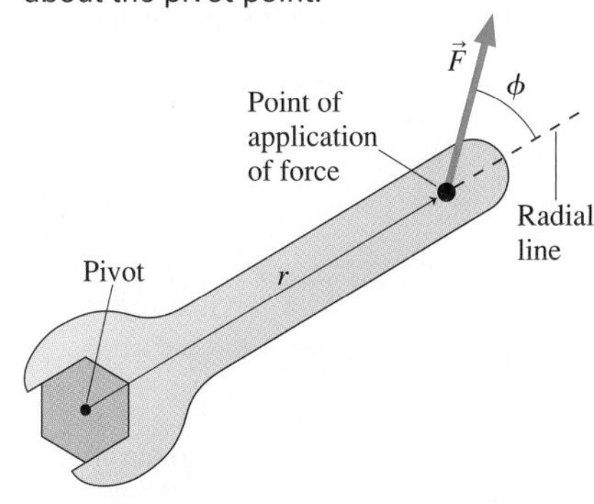

FIGURE 7.10 Force $\vec{F}$ exerts a torque about the pivot point.

We can incorporate these three observations into a single quantity called the **torque τ** (Greek tau). Loosely speaking, τ measures the "effectiveness" of a force at causing an object to rotate about a pivot. **Torque is the rotational equivalent of force.** In Figure 7.9, for instance, the torque τ_1 due to $\vec{F}_1$ is greater than τ_4 due to $\vec{F}_4$.

To make these ideas specific, **FIGURE 7.10** shows a force $\vec{F}$ applied at one point of a wrench that's loosening a nut. Figure 7.10 defines the distance r from the pivot to the point at which the force is applied; the **radial line,** the line starting at the pivot and extending through this point; and the angle ϕ (Greek phi) measured from the radial line to the direction of the force.

We saw in Figure 7.9 that force $\vec{F}_1$, which was directed perpendicular to the door, was effective in opening it, but force $\vec{F}_2$, directed toward the hinges, had no effect on its rotation. As shown in **FIGURE 7.11**, this suggests breaking the force $\vec{F}$ applied to the wrench into two component vectors: $\vec{F}_\perp$ directed perpendicular to the radial line, and $\vec{F}_\parallel$ directed parallel to it. Because $\vec{F}_\parallel$ points either directly toward or away from the pivot, it has no effect on the wrench's rotation, and thus contributes nothing to the torque. Only $\vec{F}_\perp$ tends to cause rotation of the wrench, so it is this component of the force that determines the torque.

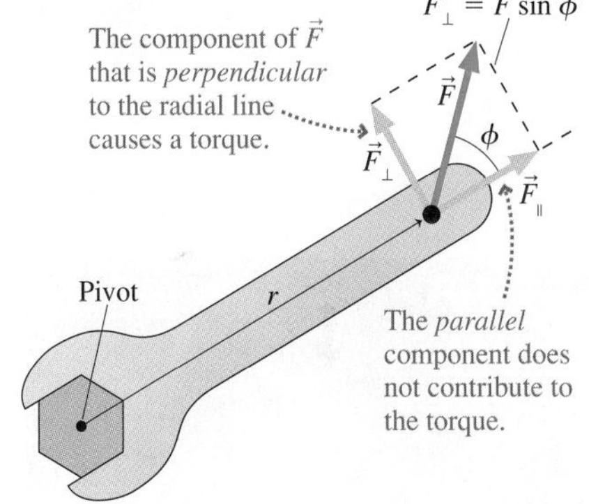

FIGURE 7.11 Torque is due to the component of the force perpendicular to the radial line.

NOTE ► The perpendicular component $\vec{F}_\perp$ is pronounced "F perpendicular" and the parallel component $\vec{F}_\parallel$ is "F parallel." ◄

We've seen that a force applied at a greater distance r from the pivot has a greater effect on rotation, so we expect a larger value of r to give a greater torque. We also saw that only $\vec{F}_\perp$ contributes to the torque. Both these observations are contained in our first expression for torque:

$$\tau = rF_\perp \tag{7.3}$$

Torque due to a force with perpendicular component $F_\perp$ acting at a distance r from the pivot

From this equation, we see that the SI units of torque are newton-meters, abbreviated N · m.

EXAMPLE 7.3 **Torque in opening a door**

In trying to open a stuck door, Ryan pushes it at a point 0.75 m from the hinges with a 240 N force directed 20° away from being perpendicular to the door. What torque does Ryan exert on the door?

PREPARE In **FIGURE 7.12** on the next page the radial line is shown drawn from the pivot—the hinge—through the point at which the force $\vec{F}$ is applied. We see that the component of $\vec{F}$ that is perpendicular to the radial line is $F_\perp = F \cos 20° = 226$ N. The distance from the hinge to the point at which the force is applied is $r = 0.75$ m.

Continued

FIGURE 7.12 Ryan's force exerts a torque on the door.

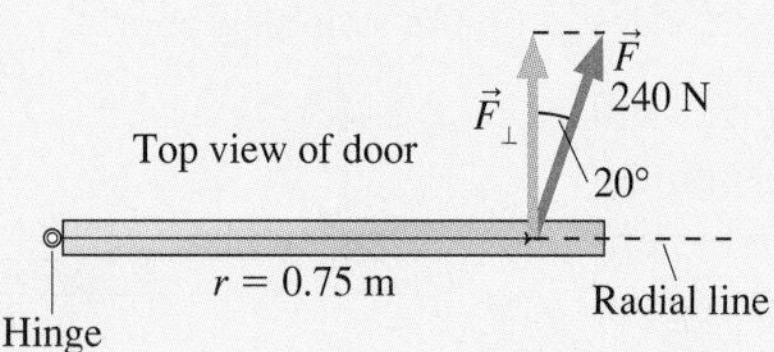

Top view of door
$\vec{F}_\perp$
$\vec{F}$
240 N
20°
$r = 0.75$ m
Hinge
Radial line

SOLVE We can find the torque on the door from Equation 7.3:

$$\tau = rF_\perp = (0.75 \text{ m})(226 \text{ N}) = 170 \text{ N} \cdot \text{m}$$

ASSESS Ryan could slightly increase the torque he exerts by pushing with the same force but exactly perpendicular to the door.

FIGURE 7.13 You can also calculate torque in terms of the moment arm between the pivot and the line of action.

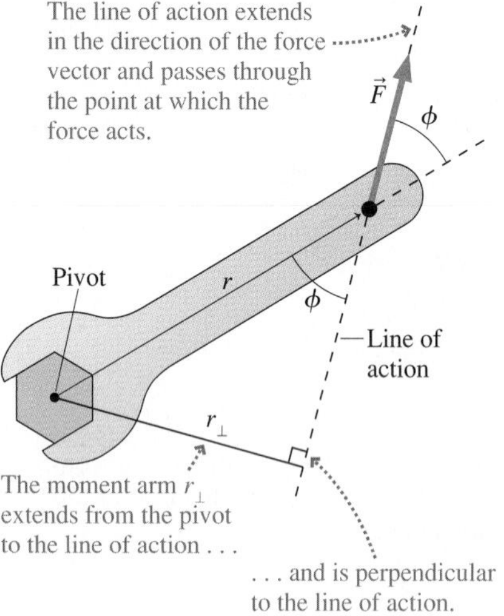

The line of action extends in the direction of the force vector and passes through the point at which the force acts.
$\vec{F}$
ϕ
Pivot
r
ϕ
Line of action
$r_\perp$
The moment arm $r_\perp$ extends from the pivot to the line of action . . .
. . . and is perpendicular to the line of action.

FIGURE 7.13 shows an alternative way to calculate torque. The line that is in the direction of the force, and passes through the point at which the force acts, is called the *line of action*. The perpendicular distance from this line to the pivot is called the **moment arm** (or *lever arm*) $r_\perp$. You can see from the figure that $r_\perp = r\sin\phi$. Further, Figure 7.11 showed that $F_\perp = F\sin\phi$. We can then write Equation 7.3 as $\tau = rF\sin\phi = F(r\sin\phi) = Fr_\perp$. Thus an equivalent expression for the torque is

$$\tau = r_\perp F \tag{7.4}$$

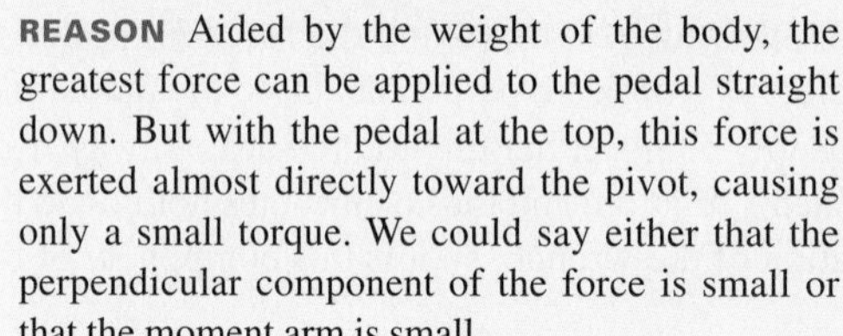

Torque due to a force F with moment arm $r_\perp$

CONCEPTUAL EXAMPLE 7.4 Starting a bike

It is hard to get going if you try to start your bike with the pedal at the highest point. Why is this?

REASON Aided by the weight of the body, the greatest force can be applied to the pedal straight down. But with the pedal at the top, this force is exerted almost directly toward the pivot, causing only a small torque. We could say either that the perpendicular component of the force is small or that the moment arm is small.

ASSESS If you've ever climbed a steep hill while standing on the pedals, you know that you get the greatest forward motion when one pedal is completely forward with the crank parallel to the ground. This gives the maximum possible torque because the force you apply is entirely perpendicular to the radial line, and the moment arm is as long as it can be.

We've seen that Equation 7.3 can be written as $\tau = rF_\perp = r(F\sin\phi)$, and Equation 7.4 as $\tau = r_\perp F = (r\sin\phi)F$. This shows that both methods of calculating torque lead to the same expression for torque—namely:

$$\tau = rF\sin\phi \tag{7.5}$$

Torque due to a force F applied at a distance r from the pivot, at an angle ϕ to the radial line

MEXICO
GREAT BRITAIN
71

◀ **Torque versus speed** To start and stop quickly, the basketball player needs to apply a large torque to her wheel. To make the torque as large as possible, the handrim—the outside wheel that she actually grabs—is almost as big as the wheel itself. The racer needs to move continuously at high speed, so his wheel spins much faster. To allow his hands to keep up, his handrim is much smaller than his chair's wheel, making its linear velocity correspondingly lower. The smaller radius means, however, that the torque he can apply is lower as well.

NOTE ► Torque differs from force in a very important way. Torque is calculated or measured *about a particular point.* To say that a torque is 20 N · m is meaningless without specifying the point about which the torque is calculated. Torque can be calculated about any point, but its value depends on the point chosen because this choice determines r and ϕ. In practice, we usually calculate torques about a hinge, pivot, or axle. ◄

Equations 7.3–7.5 are three different ways of thinking about—and calculating—the torque due to a force. Depending on the problem at hand, one might be easier to use than the others. But they all calculate the *same* torque, and all will give the same value for the torque.

These equations give only the magnitude of the torque. But torque, like a force component, has a sign. **A torque that tends to rotate the object in a counterclockwise direction is positive, while a torque that tends to rotate the object in a clockwise direction is negative.** FIGURE 7.14 summarizes the signs. Notice that a force pushing straight toward the pivot or pulling straight out from the pivot exerts *no* torque.

NOTE ► When calculating a torque, you must supply the appropriate sign by observing the direction in which the torque acts. ◄

FIGURE 7.14 Signs and strengths of the torque.

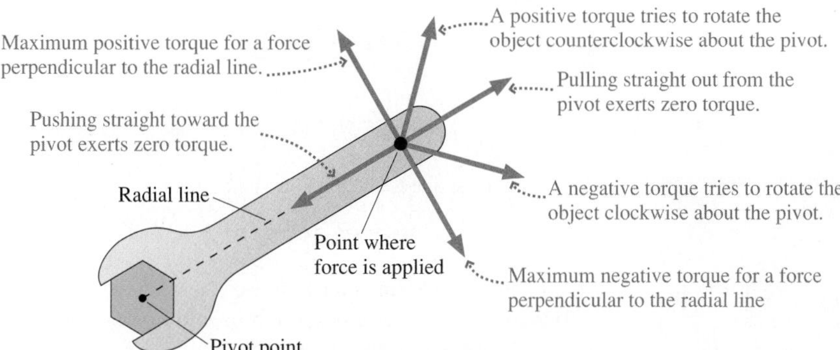
Maximum positive torque for a force perpendicular to the radial line.
A positive torque tries to rotate the object counterclockwise about the pivot.
Pulling straight out from the pivot exerts zero torque.
Pushing straight toward the pivot exerts zero torque.
Radial line
A negative torque tries to rotate the object clockwise about the pivot.
Point where force is applied
Maximum negative torque for a force perpendicular to the radial line
Pivot point

EXAMPLE 7.5 Calculating the torque on a nut

Luis uses a 20-cm-long wrench to turn a nut. The wrench handle is tilted 30° above the horizontal, and Luis pulls straight down on the end with a force of 100 N. How much torque does Luis exert on the nut?

PREPARE FIGURE 7.15 shows the situation. The two illustrations correspond to two methods of calculating torque, corresponding to Equations 7.3 and 7.4.

FIGURE 7.15 A wrench being used to turn a nut.

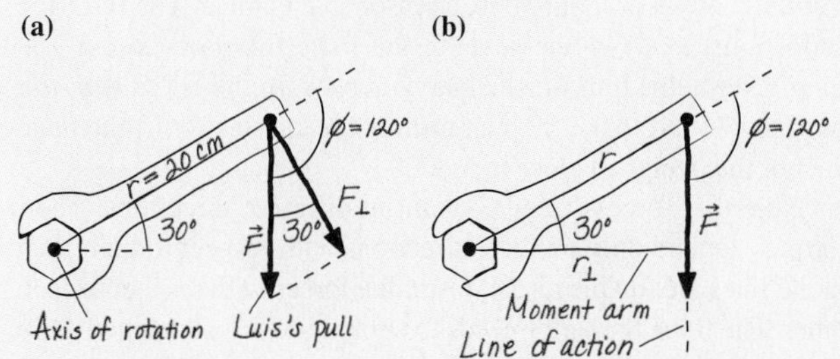
(a)
(b)
r = 20 cm
φ = 120°
F⊥
30°
F
30°
Axis of rotation
Luis's pull
r
φ = 120°
30°
F
r⊥
Moment arm
Line of action

SOLVE According to Equation 7.3 and 7.5, the torque can be calculated as $\tau = rF_\perp = rF\sin\phi$. From Figure 7.15a we see that the angle between the force and the radial line is $\phi = 30° + 90° = 120°$. The torque is then

$$\tau = -rF\sin\phi = -(0.20\text{ m})(100\text{ N})(\sin 120°) = -17\text{ N}\cdot\text{m}$$

We put in the minus sign because the torque is negative—it tries to rotate the nut in a *clockwise* direction.

Alternatively, we can use Equation 7.4 to find the torque. Figure 7.15b shows the moment arm $r_\perp$, the perpendicular distance from the pivot to the line of action. From the figure we see that

$$r_\perp = r\cos 30° = (0.20\text{ m})(\cos 30°) = 0.17\text{ m}$$

Then the torque is

$$\tau = -r_\perp F = -(0.17\text{ m})(100\text{ N}) = -17\text{ N}\cdot\text{m}$$

Again, we insert the minus sign because the torque acts to give a clockwise rotation.

ASSESS Both methods give the same answer for the torque, as expected. In general, however, you need use only one of Equations 7.3–7.5 to find the torque in any given situation. In using any of these methods to find the torque, remember to include the minus sign if the torque acts to rotate the object in a clockwise direction.

STOP TO THINK 7.2 Rank in order, from largest to smallest, the five torques τ_A to τ_E. The rods all have the same length and are pivoted at the dot.

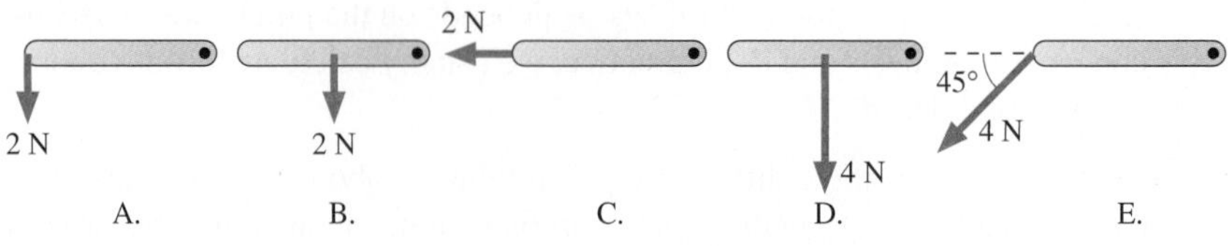
2 N
2 N
2 N
4 N
45°
4 N
A.
B.
C.
D.
E.

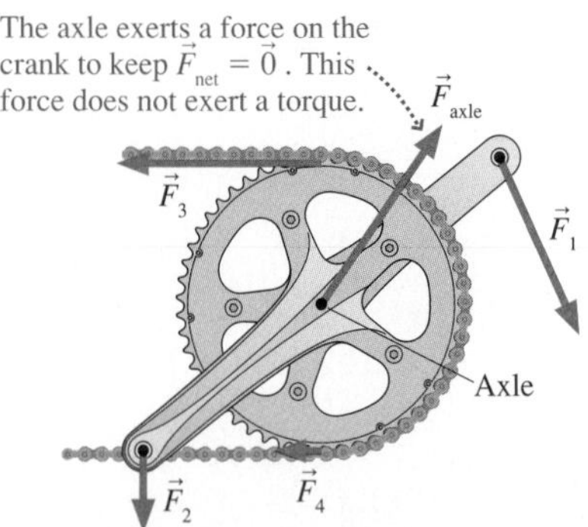
The axle exerts a force on the crank to keep $\vec{F}_{net} = \vec{0}$. This force does not exert a torque.
$\vec{F}_{axle}$
$\vec{F}_3$
$\vec{F}_1$
Axle
$\vec{F}_2$
$\vec{F}_4$

FIGURE 7.16 The forces exert a net torque about the pivot point.

Net Torque

FIGURE 7.16 shows the forces acting on the crankset of a bicycle. Forces $\vec{F}_1$ and $\vec{F}_2$ are due to the rider pushing on the pedals, and $\vec{F}_3$ and $\vec{F}_4$ are tension forces from the chain. The crankset is free to rotate about a fixed axle, but the axle prevents it from having any translational motion with respect to the bike frame. It does so by exerting force $\vec{F}_{axle}$ on the object to balance the other forces and keep $\vec{F}_{net} = \vec{0}$.

Forces $\vec{F}_1$, $\vec{F}_2$, $\vec{F}_3$, and $\vec{F}_4$ exert torques τ_1, τ_2, τ_3, and τ_4 on the crank (measured about the axle), but $\vec{F}_{axle}$ does *not* exert a torque because it is applied at the pivot point—the axle—and so has zero moment arm. Thus the *net* torque about the axle is the sum of the torques due to the *applied* forces:

$$\tau_{net} = \tau_1 + \tau_2 + \tau_3 + \tau_4 + \cdots = \sum \tau \qquad (7.6)$$

EXAMPLE 7.6 Force in turning a capstan

A capstan is a device used on old sailing ships to raise the anchor. A sailor pushes the long lever, turning the capstan and winding up the anchor rope. If the capstan turns at a constant speed, the net torque on it, as we'll learn later in the chapter, is zero.

Suppose the rope tension due to the weight of the anchor is 1500 N. If the distance from the axis to the point on the lever where the sailor pushes is exactly seven times the radius of the capstan around which the rope is wound, with what force must the sailor push if the net torque on the capstan is to be zero?

PREPARE Shown in **FIGURE 7.17** is a view looking down from above the capstan. The rope pulls with a tension force $\vec{T}$ at distance R from the axis of rotation. The sailor pushes with a force $\vec{F}$ at distance $7R$ from the axis. Both forces are perpendicular to their radial lines, so ϕ in Equation 7.5 is 90°.

FIGURE 7.17 Top view of a sailor turning a capstan.

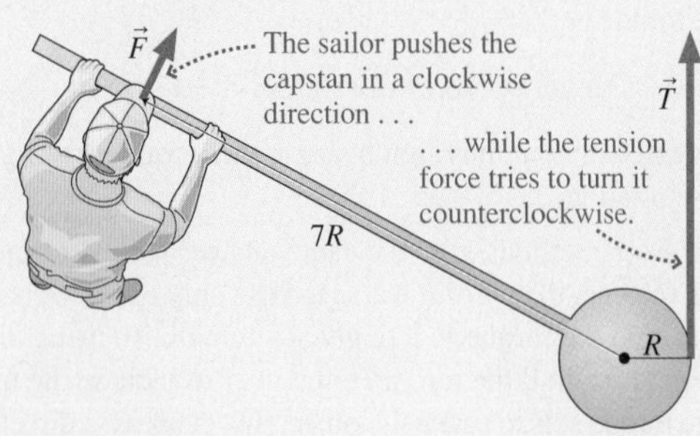
$\vec{F}$
The sailor pushes the capstan in a clockwise direction . . .
$\vec{T}$
. . . while the tension force tries to turn it counterclockwise.
$7R$
R

SOLVE The torque due to the tension in the rope is

$$\tau_T = RT\sin 90° = RT$$

We don't know the capstan radius, so we'll just leave it as R for now. This torque is positive because it tries to turn the capstan counterclockwise. The torque due to the sailor is

$$\tau_S = -(7R)F\sin 90° = -7RF$$

We put the minus sign in because this torque acts in the clockwise (negative) direction. The net torque is zero, so we have $\tau_T + \tau_S = 0$, or

$$RT - 7RF = 0$$

Note that the radius R cancels, leaving

$$F = \frac{T}{7} = \frac{1500\text{ N}}{7} = 210\text{ N}$$

ASSESS 210 N is about 50 lb, a reasonable number. The force the sailor must exert is one-seventh the force the rope exerts: The long lever helps him lift the heavy anchor. In the HMS *Warrior,* built in 1860, it took 200 men turning the capstan to lift the huge anchor that weighed close to 55,000 N!

Note that forces $\vec{F}$ and $\vec{T}$ point in different directions. Their torques depend only on their directions with respect to their own radial lines, not on the directions of the forces with respect to each other. The force the sailor needs to apply remains unchanged as he circles the capstan.

STOP TO THINK 7.3 Two forces act on the wheel shown. What third force, acting at point P, will make the net torque on the wheel zero?

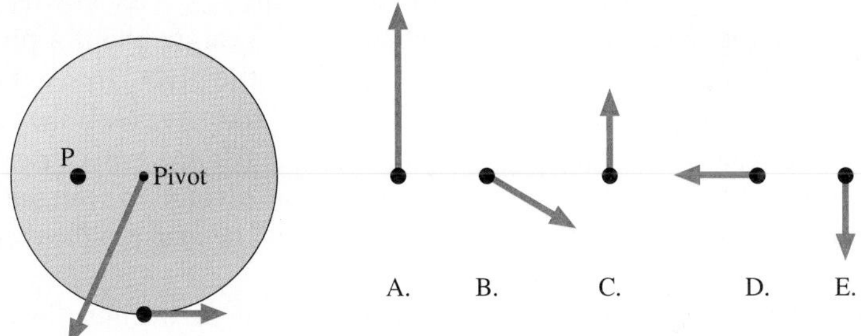

7.3 Gravitational Torque and the Center of Gravity

As the gymnast in FIGURE 7.18 pivots around the bar, a torque due to the force of gravity causes her to rotate toward a vertical position. A falling tree and a car hood slamming shut are other examples where gravity exerts a torque on an object. Stationary objects can also experience a torque due to gravity. A diving board experiences a gravitational torque about its fixed end. It doesn't rotate because of a counteracting torque provided by forces from the base at its fixed end.

We've learned how to calculate the torque due to a single force acting on an object. But gravity doesn't act at a single point on an object. It pulls downward on *every particle* that makes up the object, as shown for the gymnast in Figure 7.18a, and so each particle experiences a small torque due to the force of gravity that acts upon it. The gravitational torque on the object as a whole is then the *net* torque exerted on all the particles. We won't prove it, but the gravitational torque can be calculated by assuming that the net force of gravity—that is, the object's weight $\vec{w}$—acts at a single special point on the object called its **center of gravity** (symbol ⊕). Then we can calculate the torque due to gravity by the methods learned earlier for a single force ($\vec{w}$) acting at a single point (the center of gravity). Figure 7.18b shows how we can consider the gymnast's weight as acting at her center of gravity.

FIGURE 7.18 The center of gravity is the point where the weight appears to act.

(a) Gravity exerts a force and a torque on each particle that makes up the gymnast.

(b) The weight force provides a torque about the rotation axis.

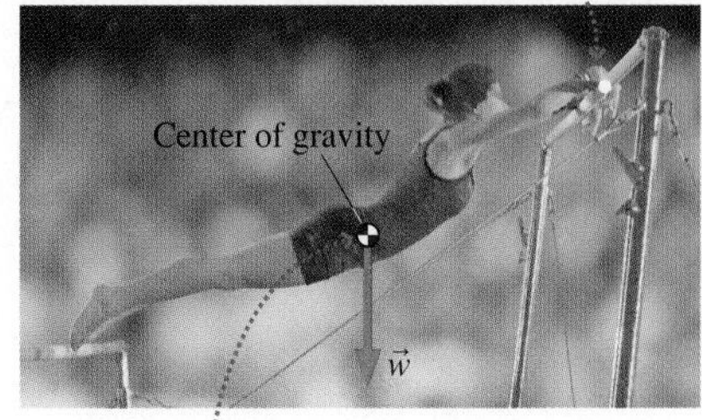

The gymnast responds *as if* her entire weight acts at her center of gravity.

EXAMPLE 7.7 The torque on a flagpole

A 3.2 kg flagpole extends from a wall at an angle of 25° from the horizontal. Its center of gravity is 1.6 m from the point where the pole is attached to the wall. What is the gravitational torque on the flagpole about the point of attachment?

PREPARE FIGURE 7.19 shows the situation. For the purpose of calculating torque, we can consider the entire weight of the pole as acting at the center of gravity. We can use any of the three methods discussed in Section 7.2 to calculate the torque. Because the moment arm $r_\perp$ is simple to visualize here, we'll use Equation 7.4 for the torque.

FIGURE 7.19 Visual overview of the flagpole.

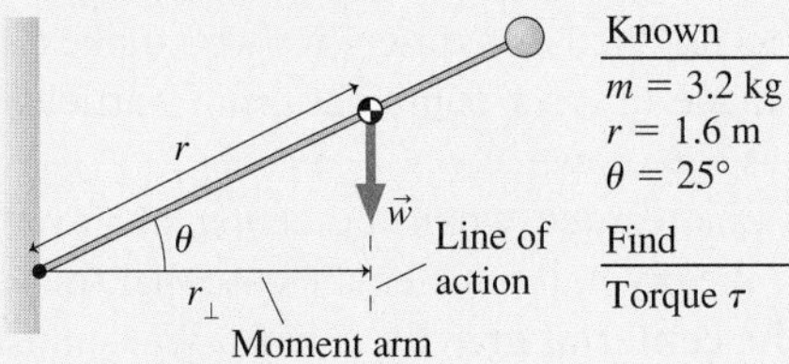

SOLVE From Figure 7.19, we see that the moment arm is $r_\perp = (1.6\text{ m})\cos 25° = 1.5\text{ m}$. Thus the gravitational torque on the flagpole, about the point where it attaches to the wall, is

$$\tau = -r_\perp w = -r_\perp mg = -(1.5\text{ m})(3.2\text{ kg})(9.8\text{ m/s}^2) = -47\text{ N}\cdot\text{m}$$

We inserted the minus sign because the torque tries to rotate the pole in a clockwise direction.

ASSESS If the pole were attached to the wall by a hinge, the gravitational torque would cause the pole to fall. However, the actual rigid connection provides a counteracting (positive) torque to the pole that prevents this.

FIGURE 7.20 Method for finding the center of gravity of an object.

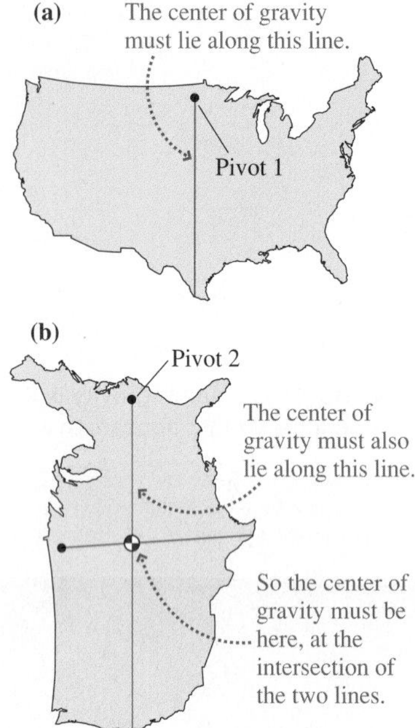

Finding the Center of Gravity

To calculate the gravitational torque, we need to locate the object's center of gravity. There is a simple experimental method for finding the center of gravity of any object, based on the observation that **any object free to rotate about a pivot will come to rest with its center of gravity directly below the pivot.** To see this, consider the cutout map of the continental United States in FIGURE 7.20a. If the center of gravity is to the right or left of the blue line, a gravitational torque will cause the map to swing. If the center of gravity lies directly *below* the pivot, however, the weight force lies along the line of action and the torque is zero. The map can then remain at rest with no tendency to swing.

We know that the center of gravity lies somewhere along the blue line, but we don't yet know where. To find out, we need to suspend the map from a second pivot, as shown in FIGURE 7.20b. Then the center of gravity will fall somewhere along the red line shown. Because the center of gravity must lie on both the blue and red lines, it must be at their *intersection.* Interestingly, the geographical center of the continental United States is defined in just this way, as the center of gravity of a map of the contiguous United States. This point is one mile northwest of Lebanon, Kansas.

For a simple symmetrical object, such as a rod, sphere, or cube made of a uniform material, FIGURE 7.21 shows that **the center of gravity of a symmetrical object lies at its geometrical center.** A particularly simple case of this is a point particle, whose center of gravity lies at the position of the particle.

FIGURE 7.21 The center of gravity of a symmetrical object lies at its center.

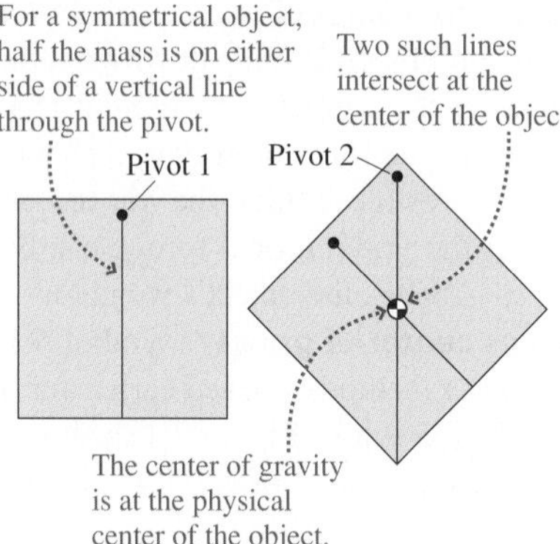

For a symmetrical object, half the mass is on either side of a vertical line through the pivot.
Two such lines intersect at the center of the object.
Pivot 1
Pivot 2
The center of gravity is at the physical center of the object.

FIGURE 7.22 Balancing a ruler.

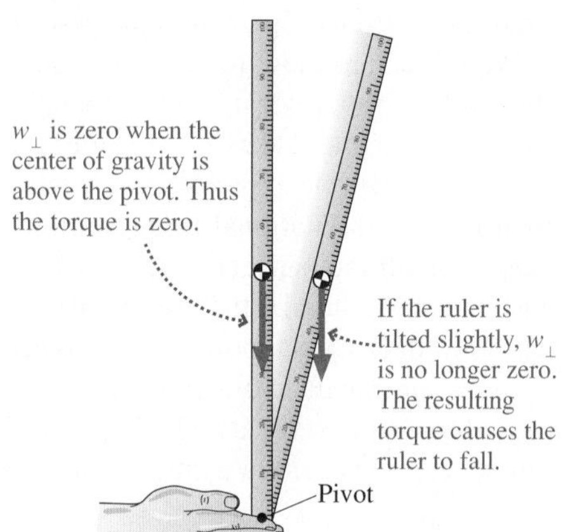

If the ruler is tilted slightly, $w_\perp$ is no longer zero. The resulting torque causes the ruler to fall.

As we've seen, an object free to pivot will rotate until its center of gravity is directly below the pivot. If the center of gravity lies directly *above* the pivot, as in FIGURE 7.22, there is no torque due to the object's weight and it can remain balanced. However, if the object is even slightly displaced to either side, the gravitational torque will no longer be zero and the object will begin to rotate. This question of *balance*—the behavior of an object whose center of gravity lies above the pivot—will be explored in depth in Chapter 8.

Calculating the Position of the Center of Gravity

It is nice to know you can locate an object's center of gravity by suspending it from a pivot, but it is rarely a practical technique. More often, we would like to calculate the center of gravity of an object made up of a combination of particles and objects whose center-of-gravity positions are known.

Because there's no gravitational torque when the center of gravity lies either directly above or directly below the pivot, it must be the case that **the torque due to gravity when the pivot is *at* the center of gravity is zero.** We can use this fact to find a general expression for the position of the center of gravity.

Consider the dumbbell shown in FIGURE 7.23. If we slide the triangular pivot back and forth until the dumbbell balances, the pivot must then be at the center of gravity (at position x_{cg}), and the torque due to gravity must therefore be zero. But we can calculate the gravitational torque directly by calculating and summing the torques about this point due to the two individual weights. Gravity acts on weight 1 with moment arm r_1, so the torque about the pivot at position x_{cg} is

$$\tau_1 = r_1 w_1 = (x_{cg} - x_1)m_1 g$$

Similarly, the torque due to weight 2 is

$$\tau_2 = -r_2 w_2 = -(x_2 - x_{cg})m_2 g$$

This torque is negative because it tends to rotate the dumbbell in a clockwise direction. We've just argued that the net torque must be zero because the pivot is directly under the center of gravity, so

$$\tau_{net} = 0 = \tau_1 + \tau_2 = (x_{cg} - x_1)m_1 g - (x_2 - x_{cg})m_2 g$$

We can solve this equation for the position of the center of gravity x_{cg}:

$$x_{cg} = \frac{x_1 m_1 + x_2 m_2}{m_1 + m_2} \qquad (7.7)$$

The following Tactics Box shows how Equation 7.7 can be generalized to find the center of gravity of *any* number of particles. If the particles don't all lie along the x-axis, then we'll also need to find the y-coordinate of the center of gravity.

FIGURE 7.23 Finding the center of gravity of a dumbbell.

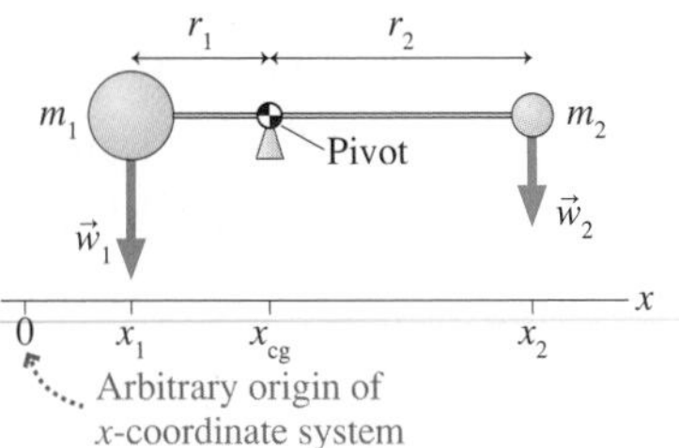

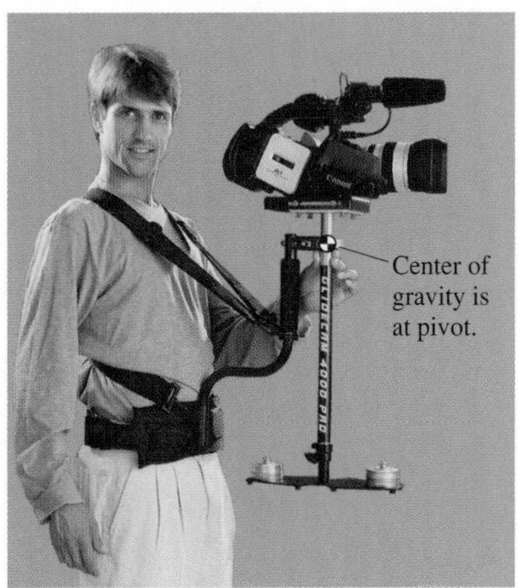

Holding steady Many movie-making scenes require handheld shots for which the cameraman must walk along, following the action. The stabilizer shown reduces unwanted camera motion as the cameraman walks. The center of gravity of the camera and its hanging weight arm is located exactly at a pivot that can swing freely in any direction. The frictionless pivot exerts no torque on the camera and arm. Neither does the weight, because it acts at the pivot. With no torque acting on it, the camera has no tendency to rotate. The long arm also increases the system's *moment of inertia,* further decreasing unwanted rotations. More about this later!

TACTICS BOX 7.1 Finding the center of gravity (MP)™

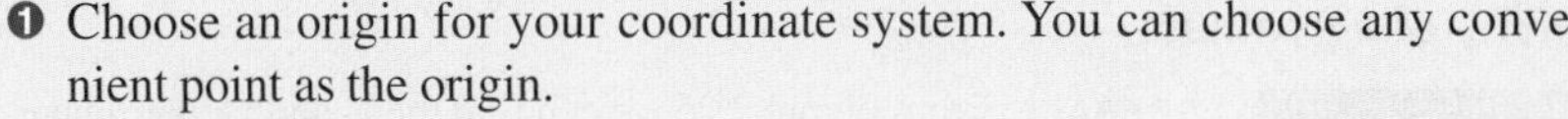

1. Choose an origin for your coordinate system. You can choose any convenient point as the origin.
2. Determine the coordinates $(x_1, y_1), (x_2, y_2), (x_3, y_3), \ldots$ for the particles of masses $m_1, m_2, m_3, \ldots$, respectively.
3. The x-coordinate of the center of gravity is

$$x_{cg} = \frac{x_1 m_1 + x_2 m_2 + x_3 m_3 + \cdots}{m_1 + m_2 + m_3 + \cdots} \qquad (7.8)$$

4. Similarly, the y-coordinate of the center of gravity is

$$y_{cg} = \frac{y_1 m_1 + y_2 m_2 + y_3 m_3 + \cdots}{m_1 + m_2 + m_3 + \cdots} \qquad (7.9)$$

Exercises 12–15

Because the center of gravity depends on products such as $x_1 m_1$, objects with large masses count more heavily than objects with small masses. Consequently, **the center of gravity tends to lie closer to the heavier objects or particles** that make up the entire object.

EXAMPLE 7.8 Where should the dumbbell be lifted?

A 1.0-m-long dumbbell has a 10 kg mass on the left and a 5.0 kg mass on the right. Find the position of the center of gravity, the point where the dumbbell should be lifted in order to remain balanced.

PREPARE We'll treat the two masses as point particles separated by a massless rod. Then we can use the steps from Tactics Box 7.1 to find the center of gravity. Let's choose the origin to be at the position of the 10 kg mass on the left, making $x_1 = 0$ m and $x_2 = 1.0$ m. Because the dumbbell masses lie on the x-axis, the y-coordinate of the center of gravity must also lie on the x-axis. Thus we only need to solve for the x-coordinate of the center of gravity.

Continued

SOLVE The x-coordinate of the center of gravity is found from Equation 7.8:

$$x_{cg} = \frac{x_1 m_1 + x_2 m_2}{m_1 + m_2} = \frac{(0\text{ m})(10\text{ kg}) + (1.0\text{ m})(5.0\text{ kg})}{(10\text{ kg}) + (5.0\text{ kg})}$$
$$= 0.33\text{ m}$$

The center of gravity is 0.33 m from the 10 kg mass or, equivalently, 0.17 m left of the center of the bar.

ASSESS The position of the center of gravity is closer to the larger mass. This agrees with our general statement that the center of gravity tends to lie closer to the heavier particles.

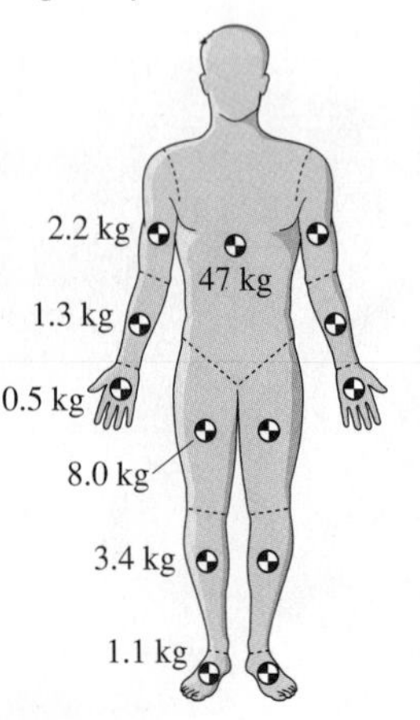

FIGURE 7.24 Body segment masses and centers of gravity.

The center of gravity of an extended object can often be found by considering the object as made up of pieces, each with mass and center of gravity that are known or can be found. Then the coordinates of the entire object's center of gravity are given by Equations 7.8 and 7.9, with (x_1, y_1), (x_2, y_2), (x_3, y_3), ... the coordinates of the center of gravity of each piece and $m_1, m_2, m_3, \ldots$ their masses.

This method is widely used in biomechanics and kinesiology to calculate the center of gravity of the human body. FIGURE 7.24 shows how the body can be considered to be made of segments, each of whose mass and center of gravity have been measured. The numbers shown are appropriate for a man with a total mass of 80 kg. For a given posture the positions of the segments and their centers of gravity can be found, and thus the whole-body center of gravity from Equations 7.8 and 7.9 (and a third equation for the z-coordinate). Example 7.9 explores a simplified version of this method.

EXAMPLE 7.9 Finding the center of gravity of a gymnast

A gymnast performing on the rings holds himself in the pike position. FIGURE 7.25 shows how we can consider his body to be made up of two segments whose masses and center-of-gravity positions are shown. The upper segment includes his head, trunk, and arms, while the lower segment consists of his legs. Locate the overall center of gravity of the gymnast.

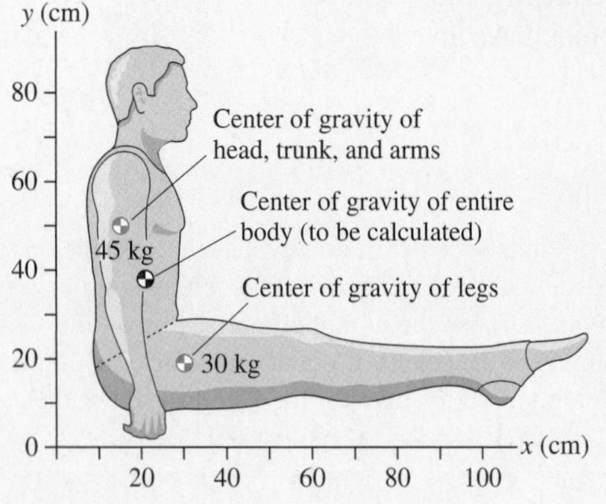

FIGURE 7.25 Centers of gravity of two segments of a gymnast.

PREPARE From Figure 7.25 we can find the x- and y-coordinates of the segment centers of gravity:

$$x_{\text{trunk}} = 15\text{ cm} \qquad y_{\text{trunk}} = 50\text{ cm}$$
$$x_{\text{legs}} = 30\text{ cm} \qquad y_{\text{legs}} = 20\text{ cm}$$

SOLVE The x- and y-coordinates of the center of gravity are given by Equations 7.8 and 7.9:

$$x_{cg} = \frac{x_{\text{trunk}} m_{\text{trunk}} + x_{\text{legs}} m_{\text{legs}}}{m_{\text{trunk}} + m_{\text{legs}}}$$
$$= \frac{(15\text{ cm})(45\text{ kg}) + (30\text{ cm})(30\text{ kg})}{45\text{ kg} + 30\text{ kg}} = 21\text{ cm}$$

and

$$y_{cg} = \frac{y_{\text{trunk}} m_{\text{trunk}} + y_{\text{legs}} m_{\text{legs}}}{m_{\text{trunk}} + m_{\text{legs}}}$$
$$= \frac{(50\text{ cm})(45\text{ kg}) + (20\text{ cm})(30\text{ kg})}{45\text{ kg} + 30\text{ kg}} = 38\text{ cm}$$

ASSESS The center-of-gravity position of the entire body, shown in Figure 7.25, is closer to that of the heavier trunk segment than to that of the lighter legs. It also lies along a line connecting the two segment centers of gravity, just as it would for the center of gravity of two point particles. Note also that the gymnast's hands—the pivot point—must lie directly below his center of gravity. Otherwise he would rotate forward or backward.

STOP TO THINK 7.4 The balls are connected by very lightweight rods pivoted at the point indicated by a dot. The rod lengths are all equal except for A, which is twice as long. Rank in order, from least to greatest, the magnitudes of the gravitational torques about the pivots for arrangements A to E.

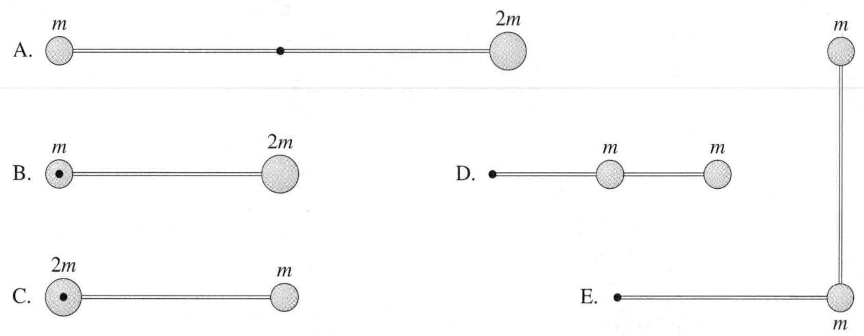

7.4 Rotational Dynamics and Moment of Inertia

In Section 7.2 we asked: What do Newton's laws tell us about rotational motion? We can now answer that question: **A torque causes an angular acceleration.** This is the rotational equivalent of our earlier discovery, for motion along a line, that a force causes an acceleration.

To see where this connection between torque and angular acceleration comes from, let's start by examining a *single particle* subject to a torque. FIGURE 7.26 shows a particle of mass m attached to a lightweight, rigid rod of length r that constrains the particle to move in a circle. The particle is subject to two forces. Because it's moving in a circle, there must be a force—here, the tension $\vec{T}$ from the rod—directed toward the center of the circle. As we learned in Chapter 6, this is the force responsible for changing the *direction* of the particle's velocity. The acceleration associated with this change in the particle's velocity is the centripetal acceleration $\vec{a}_c$.

But the particle in Figure 7.26 is also subject to the force $\vec{F}$ that changes the *speed* of the particle. This force causes a tangential acceleration $\vec{a}_t$. Applying Newton's second law in the direction tangent to the circle gives

$$a_t = \frac{F}{m} \tag{7.10}$$

Now the tangential and angular accelerations are related by $a_t = \alpha r$, so we can rewrite Equation 7.10 as $\alpha r = F/m$, or

$$\alpha = \frac{F}{mr} \tag{7.11}$$

We can now connect this angular acceleration to the torque because force $\vec{F}$, which is perpendicular to the radial line, exerts torque

$$\tau = rF$$

With this relation between F and τ, we can write Equation 7.11 as

$$\alpha = \frac{\tau}{mr^2} \tag{7.12}$$

Equation 7.12 gives a relationship between the torque on a single particle and its angular acceleration. Now all that remains is to expand this idea from a single particle to an extended object.

FIGURE 7.26 A tangential force $\vec{F}$ exerts a torque on the particle and causes an angular acceleration.

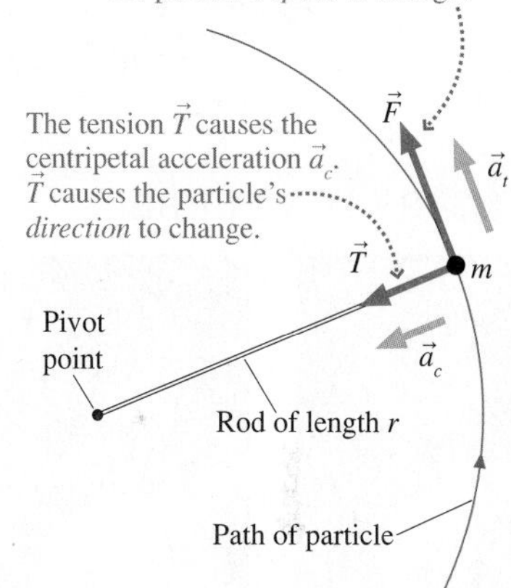

This large granite ball, with a mass of 26,400 kg, floats with nearly zero friction on a thin layer of pressurized water. Even though the girl exerts a large torque on the ball, its angular acceleration is small because of its large moment of inertia.

Newton's Second Law for Rotational Motion

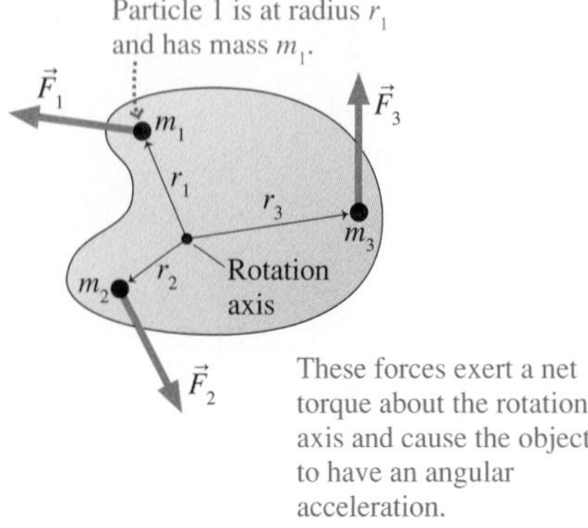
Particle 1 is at radius r_1 and has mass m_1.
$\vec{F}_1$
m_1
$\vec{F}_3$
r_1
r_3
m_3
m_2
r_2
Rotation axis
$\vec{F}_2$
These forces exert a net torque about the rotation axis and cause the object to have an angular acceleration.

FIGURE 7.27 The forces on a rigid body exert a torque about the rotation axis.

FIGURE 7.27 shows a rigid body that undergoes rotation about a fixed and unmoving axis. According to the rigid-body model, we can think of the object as consisting of particles with masses $m_1, m_2, m_3, \ldots$ at fixed distances $r_1, r_2, r_3, \ldots$ from the axis. Suppose forces $\vec{F}_1, \vec{F}_2, \vec{F}_3, \ldots$ act on these particles. These forces exert torques around the rotation axis, so the object will undergo an angular acceleration α. Because all the particles that make up the object rotate together, each particle has this *same* angular acceleration α. Rearranging Equation 7.12 slightly, we can write the torques on the particles as

$$\tau_1 = m_1 r_1^2 \alpha \qquad \tau_2 = m_2 r_2^2 \alpha \qquad \tau_3 = m_3 r_3^2 \alpha$$

and so on for every particle in the object. If we add up all these torques, the *net* torque on the object is

$$\begin{aligned}\tau_{net} &= \tau_1 + \tau_2 + \tau_3 + \cdots = m_1 r_1^2 \alpha + m_2 r_2^2 \alpha + m_3 r_3^2 \alpha + \cdots \\ &= \alpha(m_1 r_1^2 + m_2 r_2^2 + m_3 r_3^2 + \cdots) = \alpha \sum m_i r_i^2 \qquad (7.13)\end{aligned}$$

By factoring α out of the sum, we're making explicit use of the fact that every particle in a rotating rigid body has the *same* angular acceleration α.

7.6 Activ Online Physics The quantity $\Sigma m r^2$ in Equation 7.13, which is the proportionality constant between angular acceleration and net torque, is called the object's **moment of inertia** I:

$$I = m_1 r_1^2 + m_2 r_2^2 + m_3 r_3^2 + \cdots = \sum m_i r_i^2 \qquad (7.14)$$

Moment of inertia of a collection of particles

The units of moment of inertia are mass times distance squared, or kg · m². An object's moment of inertia, like torque, *depends on the axis of rotation.* Once the axis is specified, allowing the values of $r_1, r_2, r_3, \ldots$ to be determined, the moment of inertia *about that axis* can be calculated from Equation 7.14.

NOTE ► The word "moment" in "moment of inertia" and "moment arm" has nothing to do with time. It stems from the Latin *momentum,* meaning "motion." ◄

Substituting the moment of inertia into Equation 7.13 puts the final piece of the puzzle into place, giving us the fundamental equation for rigid-body dynamics:

Newton's second law for rotation An object that experiences a net torque τ_{net} about the axis of rotation undergoes an angular acceleration

$$\alpha = \frac{\tau_{net}}{I} \qquad (7.15)$$

where I is the moment of inertia of the object *about the rotation axis.*

In practice we often write $\tau_{net} = I\alpha$, but Equation 7.15 better conveys the idea that **a net torque is the cause of angular acceleration.** In the absence of a net torque ($\tau_{net} = 0$), the object has zero angular acceleration α, so it either does not rotate ($\omega = 0$) or rotates with *constant* angular velocity (ω = constant).

TRY IT YOURSELF

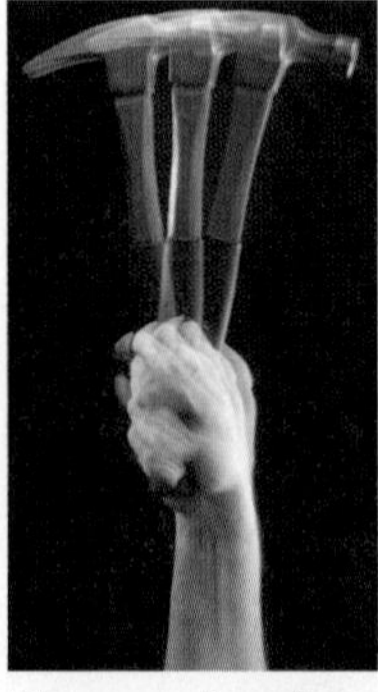

Hammering home inertia Most of the mass of a hammer is in its head, so the hammer's moment of inertia is large when calculated about an axis passing through the end of the handle (far from the head), but small when calculated about an axis passing through the head itself. You can *feel* this difference by attempting to wave a hammer back and forth about the handle end and the head end. It's much harder to do about the handle end because the large moment of inertia keeps the angular acceleration small.

Interpreting the Moment of Inertia

Before rushing to calculate moments of inertia, let's get a better understanding of its meaning. First, notice that **moment of inertia is the rotational equivalent of mass.** It plays the same role in Equation 7.15 as does mass m in the now-familiar $\vec{a} = \vec{F}_{net}/m$. Recall that objects with larger mass have a larger *inertia,* meaning that they're harder to accelerate. Similarly, an object with a larger moment of inertia is harder to get rotating: It takes a larger torque to spin up an object with a larger moment of inertia than an object with a smaller moment of inertia. The fact that "moment of inertia" retains the word "inertia" reminds us of this.

But why does the moment of inertia depend on the distances r from the rotation axis? Think about trying to start a merry-go-round from rest, as shown in FIGURE 7.28. By pushing on the rim of the merry-go-round, you exert a torque on it, and its angular velocity begins to increase. If your friends sit at the rim of the merry-go-round, as in Figure 7.28a, their distances r from the axle are large. The moment of inertia is large, according to Equation 7.14, and it will be difficult to get the merry-go-round rotating. If, however, your friends sit near the axle, as in Figure 7.28b, then r and the moment of inertia are small. You'll find it's much easier to get the merry-go-round going.

Thus an object's moment of inertia depends not only on the object's mass but also on *how the mass is distributed* around the rotation axis. This is well known to bicycle racers. Every time a cyclist accelerates, she has to "spin up" the wheels and tires. The larger the moment of inertia, the more effort it takes and the slower her acceleration. For this reason, racers use the lightest possible tires, and they put those tires on wheels that have been designed to keep the mass as close as possible to the center without sacrificing the necessary strength and rigidity.

Table 7.3 summarizes the analogies between linear and rotational dynamics.

FIGURE 7.28 Moment of inertia depends on both the mass and how the mass is distributed.

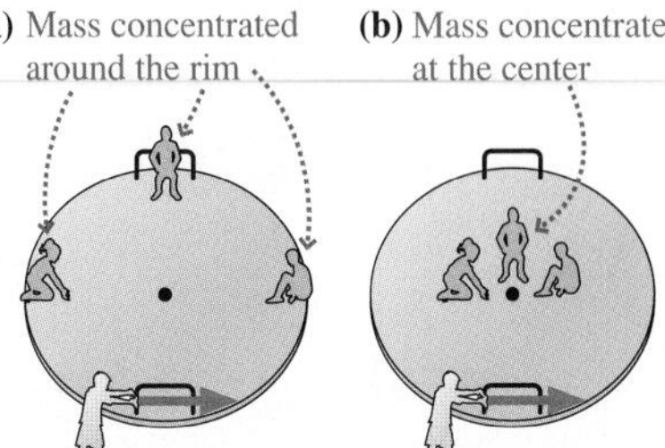

TABLE 7.3 Rotational and linear dynamics

Rotational dynamics		Linear dynamics	
Torque	τ_{net}	Force	$\vec{F}_{net}$
Moment of inertia	I	Mass	m
Angular acceleration	α	Acceleration	$\vec{a}$
Second law	$\alpha = \tau_{net}/I$	Second law	$\vec{a} = \vec{F}_{net}/m$

EXAMPLE 7.10 Calculating the moment of inertia

Your friend is creating an abstract sculpture that consists of three small, heavy spheres attached by very lightweight 10-cm-long rods as shown in FIGURE 7.29. The spheres have masses $m_1 = 1.0$ kg, $m_2 = 1.5$ kg, and $m_3 = 1.0$ kg. What is the object's moment of inertia if it is rotated about axis a? About axis b?

FIGURE 7.29 Three point particles separated by lightweight rods.

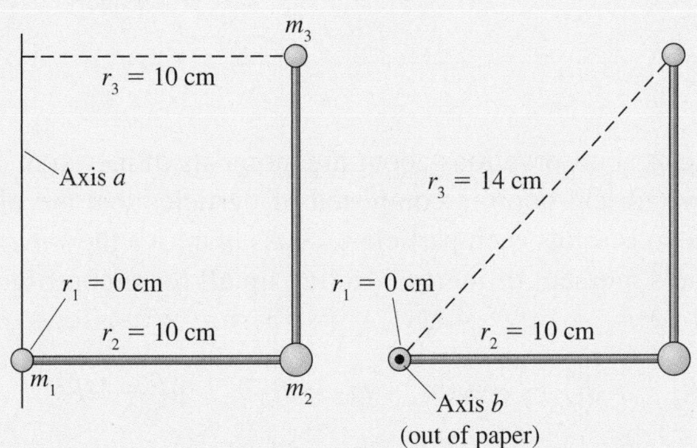

PREPARE We'll use Equation 7.14 for the moment of inertia:

$$I = m_1 r_1^2 + m_2 r_2^2 + m_3 r_3^2$$

In this expression, r_1, r_2, and r_3 are the distances of each particle from the axis of rotation, so they depend on the axis chosen. Particle 1 lies on both axes, so $r_1 = 0$ cm in both cases. Particle 2 lies 10 cm (0.10 m) from both axes. Particle 3 is 10 cm from axis a, but farther from axis b. We can find r_3 for axis b by using the Pythagorean theorem, which gives $r_3 = 14$ cm. These distances are indicated in the figure.

SOLVE For each axis, we can prepare a table of the values of r, m, and mr^2 for each particle, then add the values of mr^2. For axis a we have

Particle	r	m	mr^2
1	0 m	1.0 kg	$0\ \text{kg} \cdot \text{m}^2$
2	0.10 m	1.5 kg	$0.015\ \text{kg} \cdot \text{m}^2$
3	0.10 m	1.0 kg	$0.010\ \text{kg} \cdot \text{m}^2$
			$I_a = 0.025\ \text{kg} \cdot \text{m}^2$

For axis b we have

Particle	r	m	mr^2
1	0 m	1.0 kg	$0\ \text{kg} \cdot \text{m}^2$
2	0.10 m	1.5 kg	$0.015\ \text{kg} \cdot \text{m}^2$
3	0.14 m	1.0 kg	$0.020\ \text{kg} \cdot \text{m}^2$
			$I_b = 0.035\ \text{kg} \cdot \text{m}^2$

ASSESS We've already noted that the moment of inertia of an object is higher when its mass is distributed farther from the axis of rotation. Here, m_3 is farther from axis b than from axis a, leading to a higher moment of inertia about that axis.

The Moments of Inertia of Common Shapes

Newton's second law for rotational motion is easy to write, but we can't make use of it without knowing an object's moment of inertia. Unlike mass, we can't measure moment of inertia by putting an object on a scale. And although we can guess that the center of gravity of a symmetrical object is at the physical center of the object, we can *not* guess the moment of inertia of even a simple object.

For an object consisting of only a few point particles connected by massless rods, we can use Equation 7.14 to directly calculate I. But such an object is pretty unrealistic. All real objects are made up of solid material that is itself composed of countless atoms. To calculate the moment of inertia of even a simple object requires integral calculus and is beyond the scope of this text. A short list of common moments of inertia is given in Table 7.4. We use a capital M for the total mass of an extended object.

TABLE 7.4 Moments of inertia of objects with uniform density and total mass M

Object and axis	Picture	I	Object and axis	Picture	I
Thin rod (of any cross section), about center	L	$\frac{1}{12}ML^2$	Cylinder or disk, about center	R	$\frac{1}{2}MR^2$
Thin rod (of any cross section), about end	L	$\frac{1}{3}ML^2$	Cylindrical hoop, about center	R	MR^2
Plane or slab, about center	a, b	$\frac{1}{12}Ma^2$	Solid sphere, about diameter	R	$\frac{2}{5}MR^2$
Plane or slab, about edge	a, b	$\frac{1}{3}Ma^2$	Spherical shell, about diameter	R	$\frac{2}{3}MR^2$

We can make some general observations about the moments of inertia in Table 7.4. For instance, the cylindrical hoop is composed of particles that are all the same distance R from the axis. Thus each particle of mass m makes the *same* contribution mR^2 to the hoop's moment of inertia. Adding up all these contributions gives

$$I = m_1R^2 + m_2R^2 + m_3R^2 + \cdots = (m_1 + m_2 + m_3 + \cdots)R^2 = MR^2$$

as given in the table. The solid cylinder of the same mass and radius has a *lower* moment of inertia than the hoop because much of the cylinder's mass is nearer its center. In the same way we can see why a slab rotated about its center has a lower moment of inertia than the same slab rotated about its edge: In the latter case, some of the mass is twice as far from the axis as the farthest mass in the former case. Those particles contribute *four times* as much to the moment of inertia, leading to an overall larger moment of inertia for the slab rotated about its edge.

▶ **Novel golf clubs** The latest craze in golf putters is heads with high moments of inertia. When the putter hits the ball, the ball—by Newton's third law—exerts a force on the putter and thus exerts a torque that causes the head of the putter to rotate around the shaft. Any rotation while the putter is still in contact with the ball will affect the ball's direction. If the putter's mass is largely placed rather far from the shaft (the rotation axis), the moment of inertia about the shaft can be greatly increased. The large moment of inertia of the head will keep its angular acceleration small—reducing unwanted rotation and allowing a truer putt.

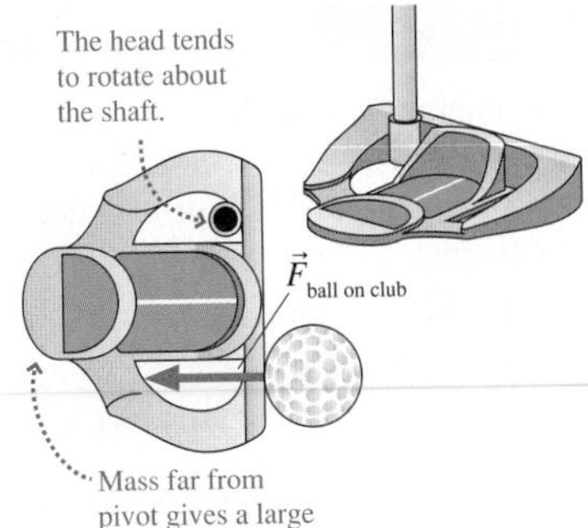
The head tends to rotate about the shaft.
$\vec{F}_{\text{ball on club}}$
Mass far from pivot gives a large moment of inertia.

STOP TO THINK 7.5 Four very lightweight disks of equal radii each have three identical heavy marbles glued to them as shown. Rank in order, from largest to smallest, the moments of inertia of the disks about the indicated axis.

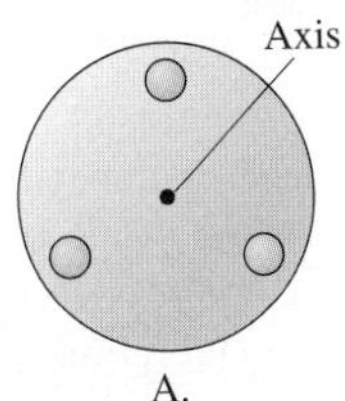
Axis
A.

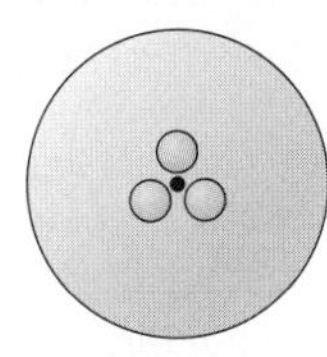
B.

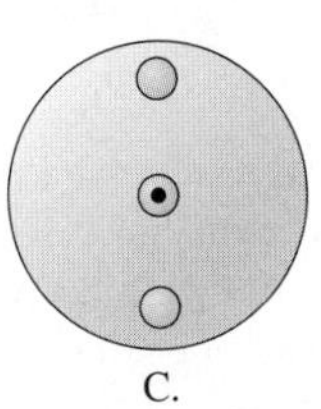
C.

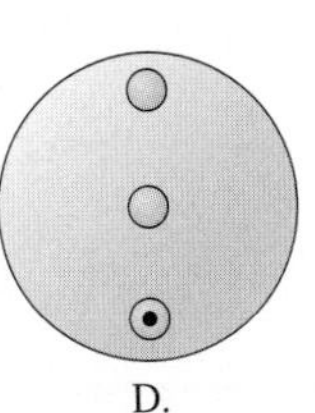
D.

7.5 Using Newton's Second Law for Rotation

In this section we'll look at several examples of rotational dynamics for rigid bodies that rotate about a *fixed axis*. The restriction to a fixed axis avoids complications that arise for an object undergoing a combination of rotational and translational motion. The problem-solving strategy for rotational dynamics is very similar to that for linear dynamics in Chapter 5.

7.8, 7.9, 7.10

PROBLEM-SOLVING STRATEGY 7.1 **Rotational dynamics problems**

PREPARE Model the object as a simple shape. Draw a pictorial representation to clarify the situation, define coordinates and symbols, and list known information.

- Identify the axis about which the object rotates.
- Identify the forces and determine their distance from the axis.
- Calculate the torques caused by the forces, and find the signs of the torques.

SOLVE The mathematical representation is based on Newton's second law for rotational motion:

$$\tau_{\text{net}} = I\alpha \quad \text{or} \quad \alpha = \frac{\tau_{\text{net}}}{I}$$

- Find the moment of inertia either by direct calculation using Equation 7.14 or from Table 7.4 for common shapes of objects.
- Use rotational kinematics to find angular positions and velocities.

ASSESS Check that your result has the correct units, is reasonable, and answers the question.

Exercise 25

EXAMPLE 7.11 Starting an airplane engine

The engine in a small airplane is specified to have a torque of 500 N · m. This engine drives a 2.0-m-long, 40 kg single-blade propeller. On start-up, how long does it take the propeller to reach 2000 rpm?

PREPARE The propeller can be modeled as a rod that rotates about its center. The engine exerts a torque on the propeller. **FIGURE 7.30** shows the propeller and the rotation axis.

FIGURE 7.30 A rotating airplane propeller.

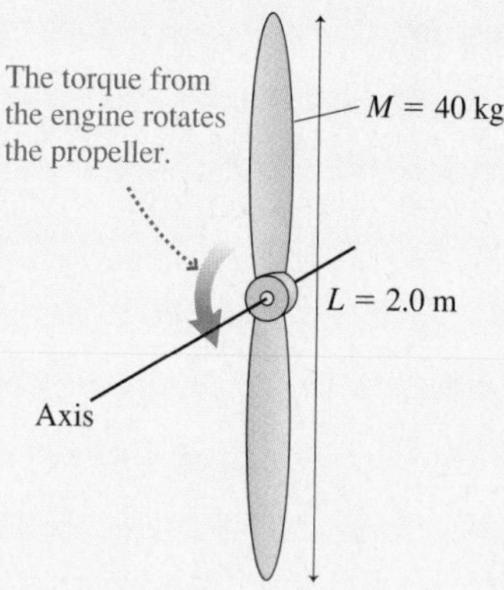

SOLVE The moment of inertia of a rod rotating about its center is found from Table 7.4:

$$I = \frac{1}{12}ML^2 = \frac{1}{12}(40\text{ kg})(2.0\text{ m})^2 = 13.3\text{ kg}\cdot\text{m}^2$$

The 500 N · m torque of the engine causes an angular acceleration of

$$\alpha = \frac{\tau}{I} = \frac{500\text{ N}\cdot\text{m}}{13.3\text{ kg}\cdot\text{m}^2} = 37.5\text{ rad/s}^2$$

The time needed to reach $\omega_f = 2000$ rpm $= 33.3$ rev/s $=$ 209 rad/s is

$$\Delta t = \frac{\Delta\omega}{\alpha} = \frac{\omega_f - \omega_i}{\alpha} = \frac{209\text{ rad/s} - 0\text{ rad/s}}{37.5\text{ rad/s}^2} = 5.6\text{ s}$$

ASSESS We've assumed a constant angular acceleration, which is reasonable for the first few seconds while the propeller is still turning slowly. Eventually, air resistance and friction will cause opposing torques and the angular acceleration will decrease. At full speed, the negative torque due to air resistance and friction cancels the torque of the engine. Then $\tau_{net} = 0$ and the propeller turns at *constant* angular velocity with no angular acceleration.

EXAMPLE 7.12 Angular acceleration of a falling pole

A 7.0-m-tall telephone pole with a mass of 260 kg has just been placed in the ground. Before the wires can be connected, the pole is hit by lightning, nearly severing the pole at its base. The pole begins to fall, rotating about the part still connected to the base. Estimate the pole's angular acceleration when it has fallen by 25° from the vertical.

PREPARE The situation is shown in **FIGURE 7.31**, where we define our symbols and list the known information. Two forces are acting on the pole: the pole's weight $\vec{w}$, which acts at the center of gravity, and the force of the base on the pole (not shown).

FIGURE 7.31 A falling telephone pole undergoes an angular acceleration due to a gravitational torque.

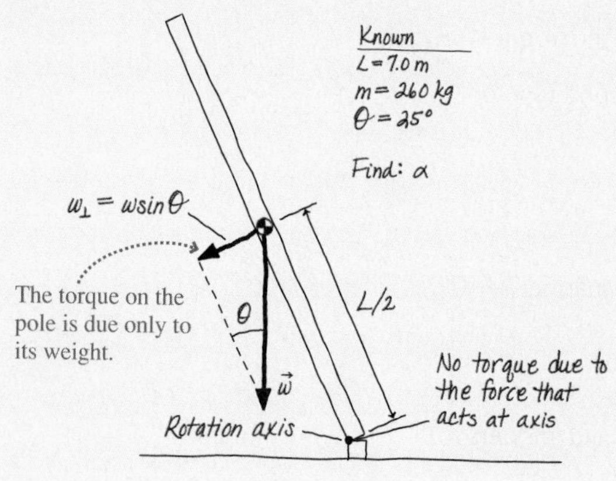

This second force exerts no torque because it acts at the axis of rotation. The torque on the pole is thus due only to gravity. From the figure we see that this torque tends to rotate the pole in a counterclockwise direction, so the torque is positive.

SOLVE We'll model the pole as a uniform thin rod rotating about one end. Its center of gravity is at its center, a distance $L/2$ from the axis. You can see from the figure that the perpendicular component of $\vec{w}$ is $w_\perp = w\sin\theta$. Thus the torque due to gravity is

$$\tau_{net} = \left(\frac{L}{2}\right)w_\perp = \left(\frac{L}{2}\right)w\sin\theta = \frac{mgL}{2}\sin\theta$$

From Table 7.4, the moment of inertia of a thin rod rotated about its end is $I = \frac{1}{3}mL^2$. Thus, from Newton's second law for rotational motion, the angular acceleration is

$$\alpha = \frac{\tau_{net}}{I} = \frac{\frac{1}{2}mgL\sin\theta}{\frac{1}{3}mL^2} = \frac{3g\sin\theta}{2L}$$

$$= \frac{3(9.8\text{ m/s}^2)\sin 25°}{2(7.0\text{ m})} = 0.9\text{ rad/s}^2$$

ASSESS The answer is given to only one significant figure because the problem asked for an *estimate* of the angular acceleration. This is usually a hint that you should make some simplifying assumptions, as we did here in modeling the pole as a thin rod.

CONCEPTUAL EXAMPLE 7.13 Balancing a meter stick

You've probably tried balancing a rod-shaped object vertically on your fingertip. If the object is very long, like a meter stick or a baseball bat, it's not too hard. But if it's short, like a pencil, it's almost impossible. Why is this?

REASON Suppose you've managed to balance a vertical stick on your fingertip, but then it starts to fall. You'll need to quickly adjust your finger to bring the stick back into balance. As Example 7.12 showed, the angular acceleration α of a thin rod is *inversely proportional* to L. Thus a long object like a meter stick topples much more slowly than a short one like a pencil. Your reaction time is fast enough to correct for a slowly falling meter stick but not for a rapidly falling pencil.

ASSESS If we double the length of a rod, its mass doubles and its center of gravity is twice as high, so the gravitational torque τ on it is four times as much. But because a rod's moment of inertia is $I = \frac{1}{3}ML^2$, the longer rod's moment of inertia will be *eight* times greater, so the angular acceleration will be only half as large.

Constraints Due to Ropes and Pulleys

Many important applications of rotational dynamics involve objects that are attached to ropes that pass over pulleys. FIGURE 7.32 shows a rope passing over a pulley and connected to an object in linear motion. If the pulley turns *without the rope slipping on it,* then the rope's speed v_{rope} must exactly match the speed of the rim of the pulley, which is $v_{rim} = \omega R$. If the pulley has an angular acceleration, the rope's acceleration a_{rope} must match the *tangential* acceleration of the rim of the pulley, $a_t = \alpha R$.

The object attached to the other end of the rope has the same speed and acceleration as the rope. Consequently, the object must obey the constraints

$$v_{obj} = \omega R$$
$$a_{obj} = \alpha R \tag{7.16}$$

Motion constraints for an object connected to a pulley of radius R by a nonslipping rope

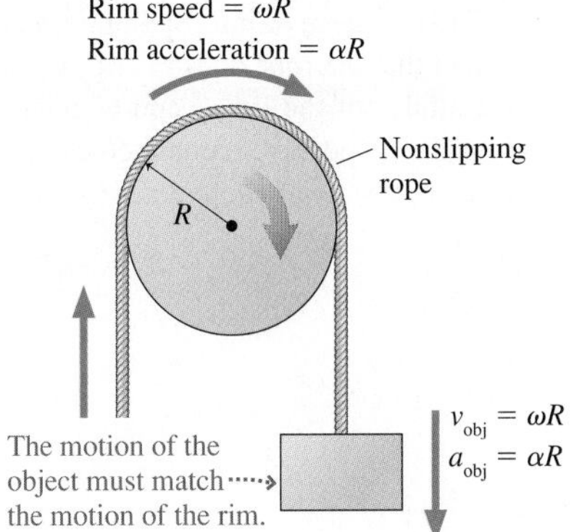

FIGURE 7.32 The rope's motion must match the motion of the rim of the pulley.

These constraints are similar to the acceleration constraints introduced in Chapter 5 for two objects connected by a string or rope.

NOTE ▶ The constraints are given as magnitudes. Specific problems will require you to specify signs that depend on the direction of motion and on the choice of coordinate system. ◀

EXAMPLE 7.14 Time for a bucket to fall

Josh has just raised a 2.5 kg bucket of water using a well's winch when he accidentally lets go of the handle. The winch consists of a rope wrapped around a 3.0 kg, 4.0-cm-diameter cylinder, which rotates on an axle through the center. The bucket is released from rest 4.0 m above the water level of the well. How long does it take to reach the water?

PREPARE Assume the rope is massless and does not slip. FIGURE 7.33a gives a visual overview of the falling bucket. FIGURE 7.33b shows the free-body diagrams for the cylinder and the bucket. The rope tension exerts an upward force on the bucket and a downward force on the outer edge of the cylinder. The rope is massless, so these two tension forces have equal magnitudes, which we'll call T.

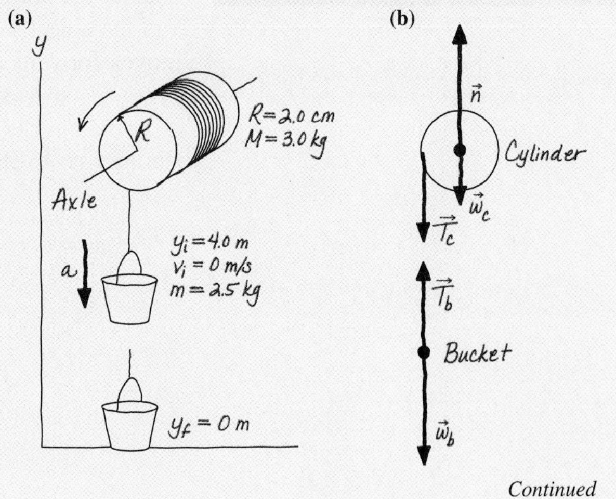

FIGURE 7.33 Visual overview of a falling bucket.

Continued

SOLVE Newton's second law applied to the linear motion of the bucket is

$$ma_y = T - mg$$

where, as usual, the y-axis points upward. What about the cylinder? There is a normal force $\vec{n}$ on the cylinder due to the axle and the weight of the cylinder $\vec{w}_c$. However, neither of these forces exerts a torque because each passes through the rotation axis. The only torque comes from the rope tension. The moment arm for the tension is $r_\perp = R$, and the torque is positive because the rope turns the cylinder counterclockwise. Thus $\tau_{rope} = TR$, and Newton's second law for the rotational motion is

$$\alpha = \frac{\tau_{net}}{I} = \frac{TR}{\frac{1}{2}MR^2} = \frac{2T}{MR}$$

The moment of inertia of a cylinder rotating about a center axis was taken from Table 7.4.

The last piece of information we need is the constraint due to the fact that the rope doesn't slip. Equation 7.16 relates only the magnitudes of the linear and angular accelerations, but in this problem α is positive (counterclockwise acceleration), while a_y is negative (downward acceleration). Hence

$$a_y = -\alpha R$$

Using α from the cylinder's equation in the constraint, we find

$$a_y = -\alpha R = -\frac{2T}{MR}R = -\frac{2T}{M}$$

Thus the tension is $T = -\frac{1}{2}Ma_y$. If we use this value of the tension in the bucket's equation, we can solve for the acceleration:

$$ma_y = -\tfrac{1}{2}Ma_y - mg$$

$$a_y = -\frac{g}{(1 + M/2m)} = -6.1 \text{ m/s}^2$$

The time to fall through $\Delta y = y_f - y_i = -4.0$ m is found from kinematics:

$$\Delta y = \tfrac{1}{2}a_y(\Delta t)^2$$

$$\Delta t = \sqrt{\frac{2\Delta y}{a_y}} = \sqrt{\frac{2(-4.0 \text{ m})}{-6.1 \text{ m/s}^2}} = 1.1 \text{ s}$$

ASSESS The expression for the acceleration gives $a_y = -g$ if $M = 0$. This makes sense because the bucket would be in free fall if there were no cylinder. When the cylinder has mass, the downward force of gravity on the bucket has to accelerate the bucket *and* spin the cylinder. Consequently, the acceleration is reduced and the bucket takes longer to fall.

7.6 Rolling Motion

FIGURE 7.34 The trajectories of the center of a wheel and of a point on the rim are seen in a time-exposure photograph.

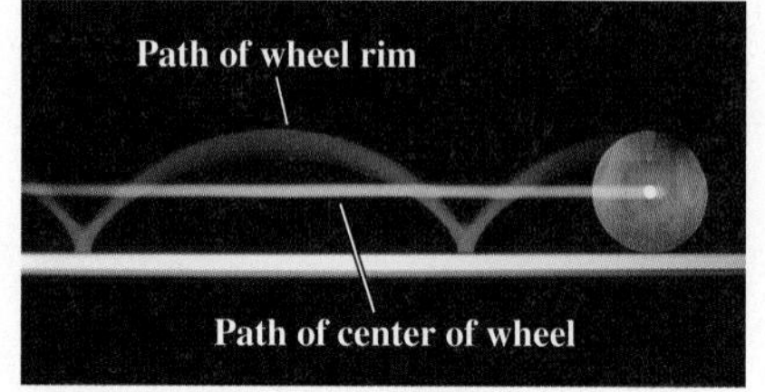

Rolling is a *combination motion* in which an object rotates about an axis that is moving along a straight-line trajectory. For example, FIGURE 7.34 is a time-exposure photo of a rolling wheel with one lightbulb on the axis and a second lightbulb at the edge. The axis light moves straight ahead, but the edge light follows a curve called a *cycloid.* Let's see if we can understand this interesting motion. We'll consider only objects that roll without slipping.

To understand rolling motion, consider FIGURE 7.35, which shows a round object—a wheel or a sphere—that rolls forward, *without slipping,* exactly one revolution. The point initially at the bottom follows the blue curve to the top and then back to the bottom. The overall position of the object is measured by the position x of the object's center. Because the object doesn't slip, in one revolution the center moves forward exactly one circumference, so that $\Delta x = 2\pi R$. The time for the

FIGURE 7.35 An object rolling through one revolution.

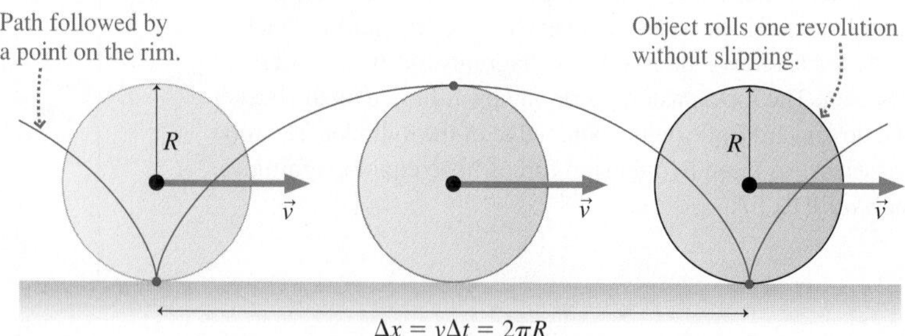

object to turn one revolution is its period T, so we can compute the speed of the object's center as

$$v = \frac{\Delta x}{T} = \frac{2\pi R}{T} \tag{7.17}$$

But $2\pi/T$ is the angular velocity ω, as you learned in Chapter 6, which leads to

$$v = \omega R \tag{7.18}$$

Equation 7.18 is the **rolling constraint,** the basic link between translation and rotation for objects that roll without slipping.

We can find the velocity for any point on a rolling object by adding the velocity of that point when the object is in pure translation, without rolling, to the velocity of the point when the object is in pure rotation, without translating. **FIGURE 7.36** shows how the velocity vectors at the top, center, and bottom of a rotating wheel are found in this way.

Ancient movers The great stone *moai* of Easter Island were moved as far as 16 km from a quarry to their final positions. Archeologists believe that one possible method of moving these 14 ton statues was to place them on rollers. One disadvantage of this method is that the statues, placed on top of the rollers, move twice as fast as the rollers themselves. Thus rollers are continuously left behind and have to be carried back to the front and reinserted. Sadly, the indiscriminate cutting of trees for moving *moai* may have hastened the demise of this island civilization.

FIGURE 7.36 Rolling is a combination of translation and rotation.

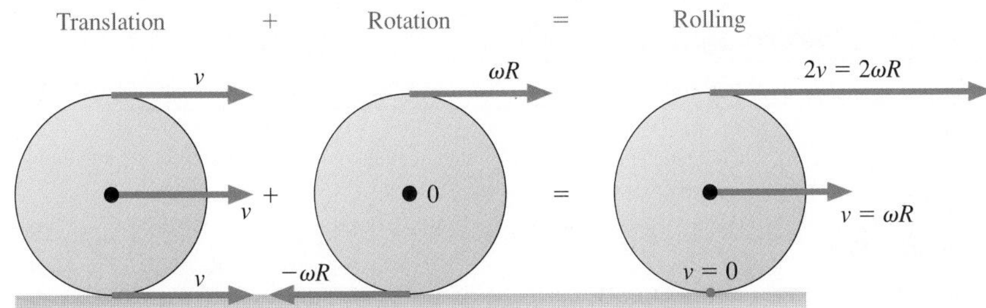

Thus the point at the top of the wheel has a forward speed of v due to its translational motion plus a forward speed of $\omega R = v$ due to its rotational motion. The speed of a point at the top of a wheel is then $2v = 2\omega R$, or *twice* the speed of its center of mass. On the other hand, the point at the bottom of the wheel, where it touches the ground, still has a forward speed of v due to its translational motion. But its velocity due to rotation points *backward* with a magnitude of $\omega R = v$. Adding these, we find that the velocity of this lowest point is *zero*. In other words, **the point on the bottom of a rolling object is instantaneously at rest.**

Although this seems surprising, it is really what we mean by "rolling without slipping." If the bottom point had a velocity, it would be moving horizontally relative to the surface. In other words, it would be slipping or sliding across the surface. To roll without slipping, the bottom point, the point touching the surface, must be at rest.

EXAMPLE 7.15 Rotating your tires

The diameter of your tires is 0.60 m. You take a 60 mile trip at a speed of 45 mph.

a. During this trip, what was your tires' angular speed?
b. How many times did they revolve?

PREPARE The angular speed is related to the speed of a wheel's center by Equation 7.18: $v = \omega R$. Because the center of the wheel turns on an axle fixed to the car, the speed v of the wheel's center is the same as that of the car. We prepare by converting the car's speed to SI units:

$$v = (45\text{ mph}) \times \left(0.447\,\frac{\text{m/s}}{\text{mph}}\right) = 20\text{ m/s}$$

Once we know the angular speed, we can find the number of times the tires turned from the rotational-kinematic equation $\Delta\theta = \omega\,\Delta t$. We'll need to find the time traveled Δt from $v = \Delta x/\Delta t$.

SOLVE a. From Equation 7.18 we have

$$\omega = \frac{v}{R} = \frac{20\text{ m/s}}{0.30\text{ m}} = 67\text{ rad/s}$$

b. The time of the trip is

$$\Delta t = \frac{\Delta x}{v} = \frac{60\text{ mi}}{45\text{ mi/h}} = 1.33\text{ h} \times \frac{3600\text{ s}}{1\text{ h}} = 4800\text{ s}$$

Continued

Thus the total angle through which the tires turn is

$$\Delta\theta = \omega\,\Delta t = (67\text{ rad/s})(4800\text{ s}) = 3.2 \times 10^5\text{ rad}$$

Because each turn of the wheel is 2π rad, the number of turns is

$$\frac{3.2 \times 10^5\text{ rad}}{2\pi\text{ rad}} = 51{,}000\text{ turns}$$

ASSESS You probably know from seeing tires on passing cars that a tire rotates several times a second at 45 mph. Because there are 3600 s in an hour, and your 60 mile trip at 45 mph is going to take over an hour—say, ≈5000 s—you would expect the tire to make many thousands of revolutions. So 51,000 turns seems to be a reasonable answer. You can see that your tires rotate roughly a thousand times per mile. During the lifetime of a tire, about 50,000 miles, it will rotate about 50 million times!

STOP TO THINK 7.6 A wheel rolls without slipping. Which is the correct velocity vector for point P on the wheel?

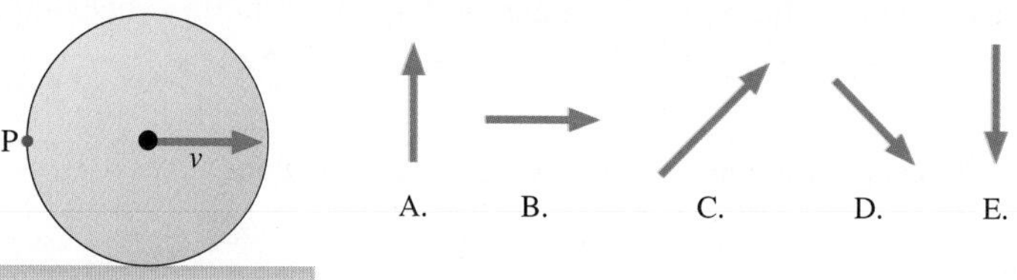

INTEGRATED EXAMPLE 7.16 **Spinning a gyroscope**

A gyroscope is a top-like toy consisting of a heavy ring attached by light spokes to a central axle. The axle and ring are free to turn on bearings. To get the gyroscope spinning, a 30-cm-long string is wrapped around the 2.0-mm-diameter axle, then pulled with a constant force of 5.0 N. If the ring's diameter is 5.0 cm and its mass is 30 g, at what rate is it spinning, in rpm, once the string is completely unwound?

PREPARE Because the ring is heavy compared to the spokes and the axle, we'll model it as a cylindrical hoop, taking its moment of inertia from Table 7.4 to be $I = MR^2$. **FIGURE 7.37** shows a visual overview of the problem. Two points are worth noting. First, rule 2 of Tactics Box 5.2 tells us that the tension in the string has the same magnitude as the force that pulls on the string, so the tension is $T = 5.0$ N. Second, it is a good idea to convert all the known quantities in the problem statement to SI units, and to collect them all in one place as we have done in the visual overview of Figure 7.37. Here, radius R is half the 5.0 cm ring diameter, and radius r is half the 2.0 mm axle diameter.

We are asked at what rate the ring is spinning when the string is unwound. This is a question about the ring's final *angular velocity*, which we've labeled ω_f. We've assumed that the initial angular velocity is $\omega_i = 0$ rad/s. Because the angular velocity is changing, the ring must have an angular acceleration that, as we know, is caused by a torque. So a good strategy will be to find the torque on the ring, from which we can find its angular acceleration and, using kinematics, the final angular velocity.

FIGURE 7.37 Visual overview of a gyroscope being spun.

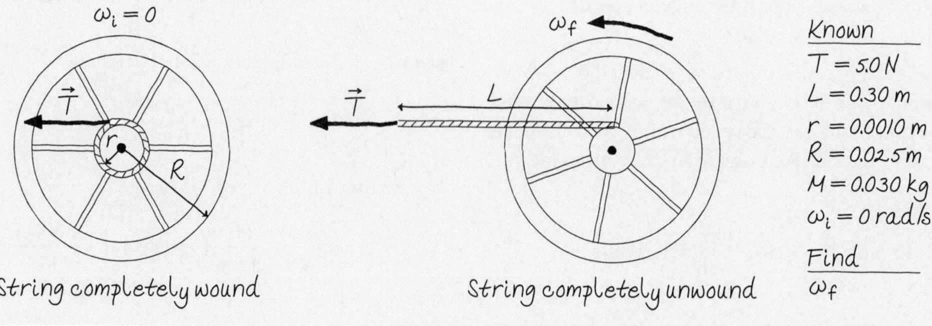

SOLVE The torque on the ring is due to the tension in the string. Because the string—and the line of action of the tension—is tangent to the axle, the moment arm of the tension force is the radius r of the axle. Thus $\tau = r_\perp T = rT$. Now we can apply Newton's second law for rotational motion, Equation 7.15, to find the angular acceleration:

$$\alpha = \frac{\tau_{\text{net}}}{I} = \frac{rT}{MR^2} = \frac{(0.0010\ \text{m})(5.0\ \text{N})}{(0.030\ \text{kg})(0.025\ \text{m})^2} = 267\ \text{rad/s}^2$$

We next use constant-angular-acceleration kinematics to find the final angular velocity. For the equation $\Delta\theta = \omega_i \Delta t + \frac{1}{2}\alpha \Delta t^2$ of Table 7.2, we know α and ω_i, and we should be able to find $\Delta\theta$ from the length of string unwound, but we don't know Δt. For the equation $\Delta\omega = \omega_f - \omega_i = \alpha \Delta t$, we know α and ω_i, and ω_f is what we want to find, but again we don't know Δt. To find an equation that doesn't contain Δt, we first write

$$\Delta t = \frac{\omega_f - \omega_i}{\alpha}$$

from the second kinematic equation. Inserting this value for Δt into the first equation gives

$$\Delta\theta = \omega_i \frac{\omega_f - \omega_i}{\alpha} + \frac{1}{2}\alpha\left(\frac{\omega_f - \omega_i}{\alpha}\right)^2$$

which can be simplified to

$$\omega_f^2 = \omega_i^2 + 2\alpha\,\Delta\theta$$

This equation, which is the rotational analog of the linear motion Equation 2.13, will allow us to find ω_f once $\Delta\theta$ is known.

FIGURE 7.38 shows how to find $\Delta\theta$. As a segment of string of length s unwinds, the axle turns through an angle (based on the definition of radian measure) $\theta = s/r$. Thus as the whole string, of length L, unwinds, the axle (and the ring) turns through an angular displacement

FIGURE 7.38 Relating the angle turned to the length of string unwound.

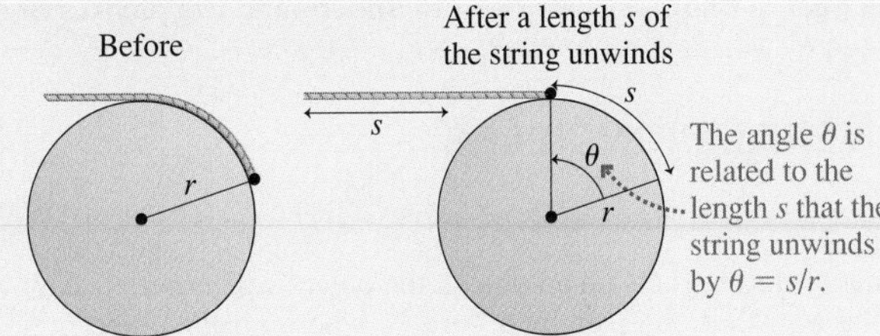

$$\Delta\theta = \frac{L}{r} = \frac{0.30\ \text{m}}{0.0010\ \text{m}} = 300\ \text{rad}$$

Now we can use our kinematic equation to find that

$$\omega_f^2 = \omega_i^2 + 2\alpha\Delta\theta = (0\ \text{rad/s})^2 + 2(267\ \text{rad/s}^2)(300\ \text{rad})$$
$$= 160{,}000\ (\text{rad/s})^2$$

from which we find that $\omega_f = 400$ rad/s. Converting rad/s to rpm, we find that the gyroscope ring is spinning at

$$400\ \text{rad/s} = \left(\frac{400\ \text{rad}}{\text{s}}\right)\left(\frac{60\ \text{s}}{1\ \text{min}}\right)\left(\frac{1\ \text{rev}}{2\pi\ \text{rad}}\right) = 3800\ \text{rpm}$$

ASSESS This is fast, about the speed of your car engine when its on the highway, but if you've ever played with a gyroscope or a string-wound top, you know you can really get it spinning fast.

SUMMARY

The goal of Chapter 7 has been to understand the physics of rotating objects.

GENERAL PRINCIPLES

Newton's Second Law for Rotational Motion

If a net torque τ_{net} acts on an object, the object will experience an angular acceleration given by $\alpha = \tau_{net}/I$, where I is the object's moment of inertia about the rotation axis.

This law is analogous to Newton's second law for linear motion, $\vec{a} = \vec{F}_{net}/m$.

IMPORTANT CONCEPTS

Torque is the rotational analog of force. Just as a force causes an object to undergo a linear acceleration, a torque causes an object to undergo an angular acceleration.

There are two interpretations of torque:

Interpretation 1: $\tau = rF_\perp$ Interpretation 2: $\tau = r_\perp F$

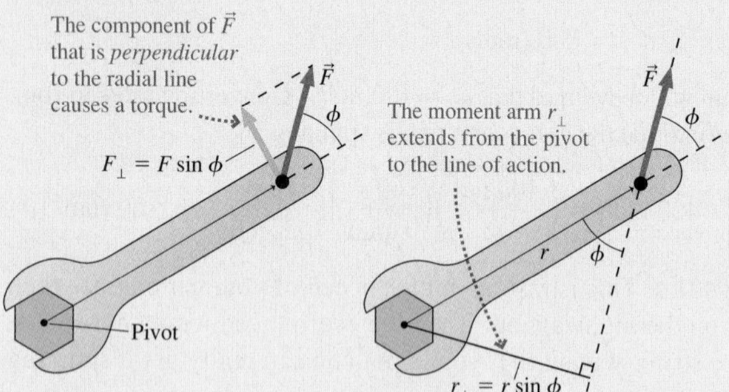

Both interpretations give the same expression for the magnitude of the torque: $\tau = rF\sin\phi$.

A torque is positive if it tends to rotate the object counterclockwise; negative if it tends to rotate the object clockwise.

Angular and tangential acceleration

A particle moving in a circle has

- A velocity tangent to the circle.
- A centripetal acceleration $\vec{a}_c$ directed toward the center of the circle.

If the particle's speed is increasing, it will also have

- A tangential acceleration $\vec{a}_t$ directed tangent to the circle.
- An angular acceleration α.

Centripetal acceleration
Angular velocity
Tangential acceleration
Velocity

The angular and tangential accelerations are related by $a_t = \alpha r$.

For a **rigid body,** the angular velocity and angular acceleration are the same for every point on the object.

The **moment of inertia** is the rotational equivalent of mass. The larger an object's moment of inertia, the more difficult it is to get the object rotating. For an object made up of particles of masses $m_1, m_2, \ldots$ at distances $r_1, r_2, \ldots$ from the axis, the moment of inertia is

$$I = m_1r_1^2 + m_2r_2^2 + m_3r_3^2 + \cdots = \sum mr^2$$

Center of gravity

The **center of gravity** of an object is the point at which gravity can be considered as acting.

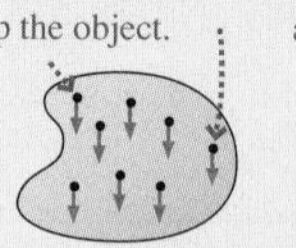

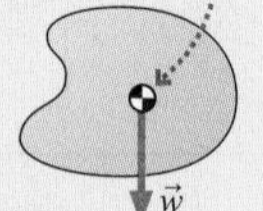

The **position of the center of gravity** depends on the distance $x_1, x_2, \ldots$ of each particle of mass $m_1, m_2, \ldots$ from the origin:

$$x_{cg} = \frac{x_1m_1 + x_2m_2 + x_3m_3 + \cdots}{m_1 + m_2 + m_3 + \cdots}$$

APPLICATIONS

Moments of inertia of common shapes

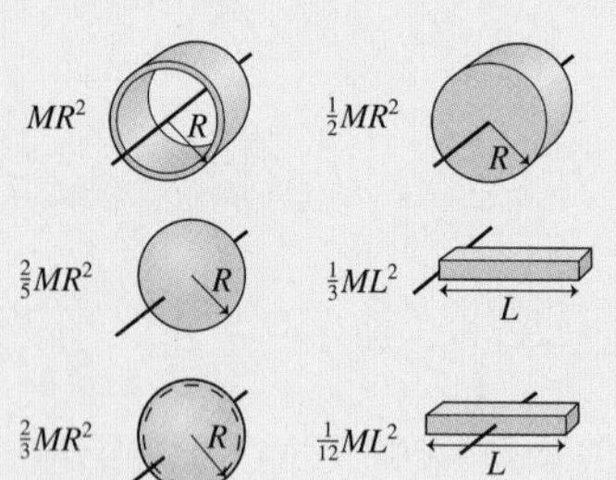

Rotation about a fixed axis

When a net torque is applied to an object that rotates about a fixed axis, the object will undergo an **angular acceleration** given by

$$\alpha = \frac{\tau_{net}}{I}$$

If a rope unwinds from a pulley of radius R, the linear motion of an object tied to the rope is related to the angular motion of the pulley by

$$a_{obj} = \alpha R \qquad v_{obj} = \omega R$$

Rolling motion

For an object that rolls without slipping,

$$v = \omega R$$

The velocity of a point at the top of the object is twice that of the center.

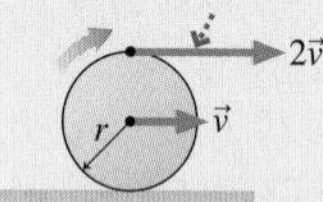

For homework assigned on MasteringPhysics, go to www.masteringphysics.com

Problems labeled INT integrate significant material from earlier chapters; BIO are of biological or medical interest.

Problem difficulty is labeled as | (straightforward) to ||||| (challenging).

QUESTIONS

Conceptual Questions

1. Figure Q7.1 shows four pulleys, each with a heavy and a light block strung over them. The blocks' velocities are shown. What are the signs (+ or −) of the angular velocity and angular acceleration of the pulley in each case?

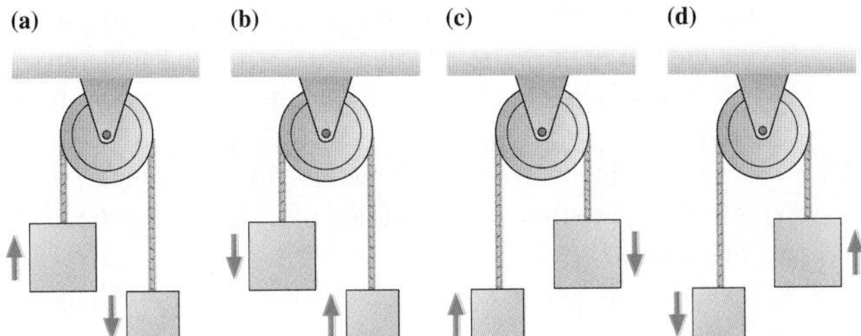

FIGURE Q7.1

2. If you are using a wrench to loosen a very stubborn nut, you can make the job easier by using a "cheater pipe." This is a piece of pipe that slides over the handle of the wrench, as shown in Figure Q7.2, making it effectively much longer. Explain why this would help you loosen the nut.

FIGURE Q7.2

3. Five forces are applied to a door, as seen from above in Figure Q7.3. For each force, is the torque about the hinge positive, negative, or zero?

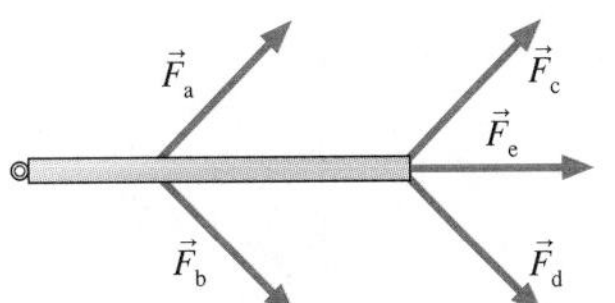

FIGURE Q7.3

4. A screwdriver with a very thick handle requires less force to operate than one with a very skinny handle. Explain why this is so.
5. If you have ever driven a truck, you likely found that it had a steering wheel with a larger diameter than that of a passenger car. Why is this?
6. A common type of door stop is a wedge made of rubber. Is such a stop more effective when jammed under the door near or far from the hinges? Why?
7. Suppose you are hanging from a tree branch. If you move out along the branch, farther away from the trunk, the branch will be more likely to break. Explain why this is so.
8. A student gives a quick push to a ball at the end of a massless, rigid rod, causing the ball to rotate clockwise in a *horizontal* circle as shown in Figure Q7.8. The rod's pivot is frictionless.
   a. As the student is pushing, is the torque about the pivot positive, negative, or zero?
   b. After the push has ended, what does the ball's angular velocity do? Steadily increase? Increase for a while, then hold steady? Hold steady? Decrease for a while, then hold steady? Steadily decrease? Explain.
   c. Right after the push has ended, is the torque positive, negative, or zero?

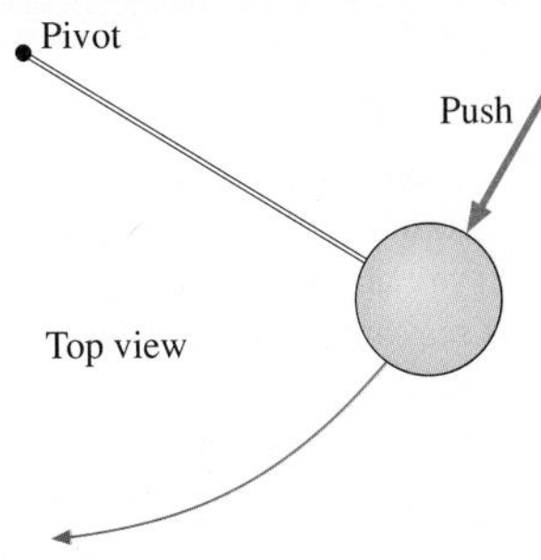

FIGURE Q7.8

9. The two ends of the dumbbell shown in Figure Q7.9 are made of the same material. Is the dumbbell's center of gravity at point 1, 2, or 3? Explain.

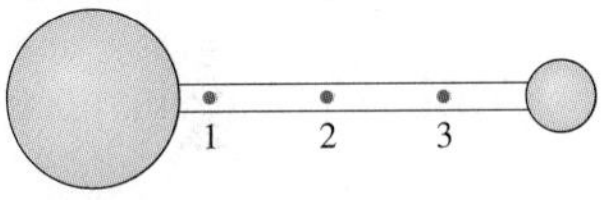

FIGURE Q7.9

10. When you rise from a chair, you have to lean quite far forward (try it!). Why is this?
11. Suppose you have two identical-looking metal spheres of the same size and the same mass. One of them is solid, the other is hollow. How can you tell which is which?
12. The moment of inertia of a uniform rod about an axis through its center is $ML^2/12$. The moment of inertia about an axis at one end is $ML^2/3$. Explain *why* the moment of inertia is larger about the end than about the center.
13. A heavy steel rod, 1.0 m long, and a light pencil, 0.15 m long, are held 15° from the vertical with one end on a table, then released simultaneously. Which will hit the table first? Or will it be a tie? Explain.
14. The wheel in Figure Q7.14 is rolling to the right without slipping. Rank in order, from fastest to slowest, the *speeds* of the points labeled 1 through 5. Explain your reasoning.

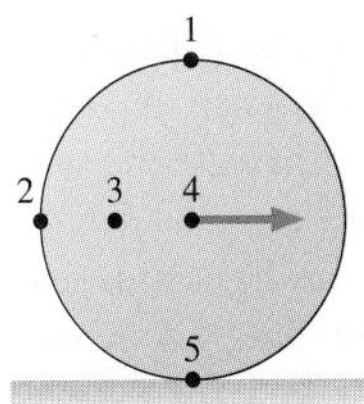

FIGURE Q7.14

15. A car traveling at 60 mph has a pebble stuck in one of its tires. Eventually the pebble works loose, and at the instant of release it is at the top of the tire. Explain why the pebble then slams *hard* into the *front* of the wheel well.

Multiple-Choice Questions

16. | A nut needs to be tightened with a wrench. Which force shown in Figure Q7.16 will apply the greatest torque to the nut?

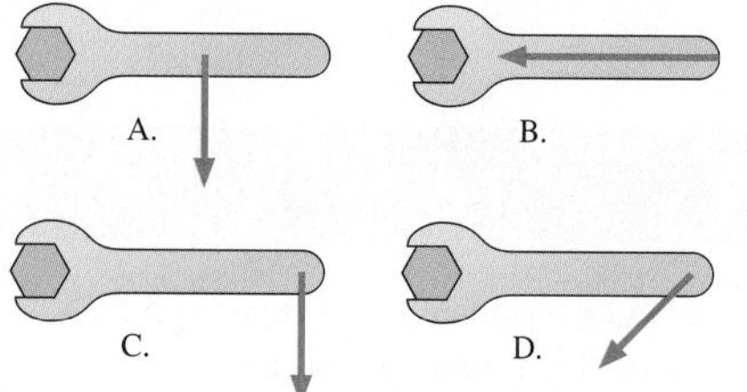

FIGURE Q7.16

17. | Suppose a bolt on your car engine needs to be tightened to a torque of 20 N · m. You are using a 15-cm-long wrench, and you apply a force at the very end in the direction that produces maximum torque. What force should you apply?
A. 1300 N B. 260 N C. 130 N D. 26 N

18. | A machine part is made up of two pieces, with centers of gravity shown in Figure Q7.18. Which point could be the center of gravity of the entire part?

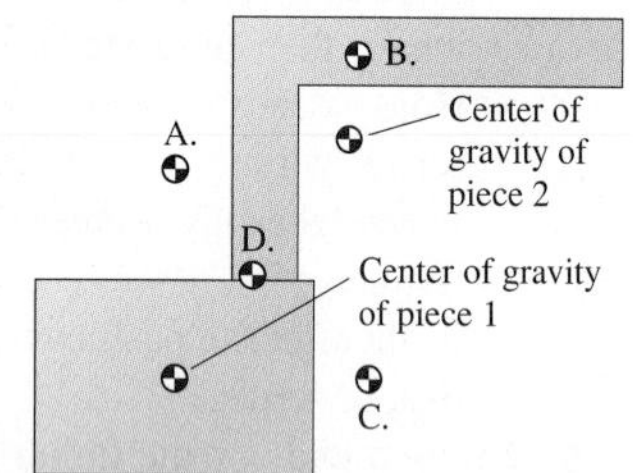

FIGURE Q7.18

19. || A typical compact disk has a mass of 15 g and a diameter of 120 mm. What is its moment of inertia about an axis through its center, perpendicular to the disk?
A. 2.7×10^{-5} kg · m^2 B. 5.4×10^{-5} kg · m^2
C. 1.1×10^{-4} kg · m^2 D. 2.2×10^{-4} kg · m^2

20. ||| Suppose you make a new kind of compact disk that is the same thickness as a current disk but twice the diameter. By what factor will the moment of inertia increase?
A. 2 B. 4 C. 8 D. 16

21. | Doors 1 and 2 have the same mass, height, and thickness. Door 2 is twice as wide as door 1. Bob pushes straight against the outer edge of door 2 with force F, and Barb pushes straight against the outer edge of door 1 with force $2F$. How do the angular accelerations α_1 and α_2 of the two doors compare?
A. $\alpha_1 > \alpha_2$ B. $\alpha_1 = \alpha_2$ C. $\alpha_1 < \alpha_2$

22. | A baseball bat has a heavy barrel and a thin handle. If you want to hold a baseball bat on your palm so that it balances vertically, you should
A. Put the end of the handle in your palm, with the barrel up.
B. Put the end of the barrel in your palm, with the handle up.
C. The bat will be equally easy to balance in either configuration.

23. | A car traveling at a steady 30 m/s has 74-cm-diameter tires. What is the approximate acceleration of a piece of the tread on any of the tires?
A. 24 m/s^2 B. 48 m/s^2 C. 2400 m/s^2 D. 4800 m/s^2

PROBLEMS

Section 7.1 The Rotation of a Rigid Body

1. ||| To throw a discus, the thrower holds it with a fully outstretched arm. Starting from rest, he begins to turn with a constant angular acceleration, releasing the discus after making one complete revolution. The diameter of the circle in which the discus moves is about 1.8 m. If the thrower takes 1.0 s to complete one revolution, starting from rest, what will be the speed of the discus at release?

2. ||| A computer hard disk starts from rest, then speeds up with an angular acceleration of 190 rad/s^2 until it reaches its final angular speed of 7200 rpm. How many revolutions has the disk made 10.0 s after it starts up?

3. ||| The crankshaft in a race car goes from rest to 3000 rpm in 2.0 s.
a. What is the crankshaft's angular acceleration?
b. How many revolutions does it make while reaching 3000 rpm?

Section 7.2 Torque

4. | Reconsider the situation in Example 7.5. If Luis pulls straight down on the end of a wrench that is in the same orientation but is 35 cm long, rather than 20 cm, what force must he apply to exert the same torque?

5. ||| Balls are attached to light rods and can move in horizontal circles as shown in Figure P7.5. Rank in order, from smallest to largest, the torques τ_1 to τ_4 about the centers of the circles. Explain.

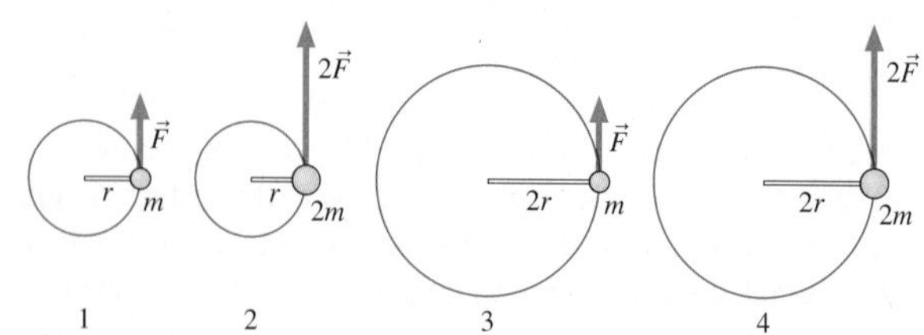

FIGURE P7.5

6. |||| Six forces, each of magnitude either F or $2F$, are applied to a door as seen from above in Figure P7.6. Rank in order, from smallest to largest, the six torques τ_1 to τ_6 about the hinge.

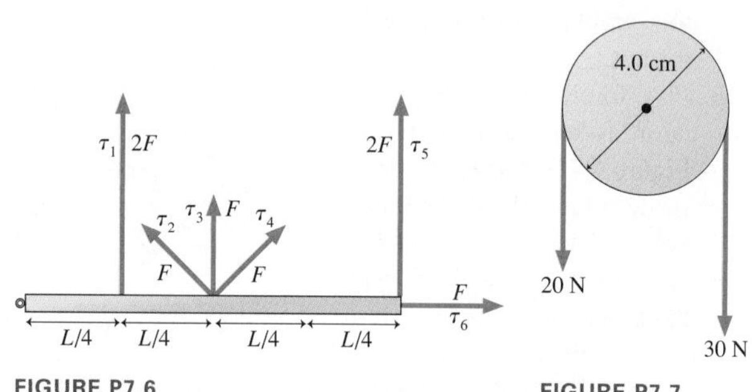

FIGURE P7.6

FIGURE P7.7

7. | What is the net torque about the axle on the pulley in Figure P7.7?

8. ||| The tune-up specifications of a car call for the spark plugs to be tightened to a torque of 38 N · m. You plan to tighten the plugs by pulling on the end of a 25-cm-long wrench. Because of the cramped space under the hood, you'll need to pull at an angle of 120° with respect to the wrench shaft. With what force must you pull?

9. ||| A professor's office door is 0.91 m wide, 2.0 m high, and 4.0 cm thick; has a mass of 25 kg; and pivots on frictionless hinges. A "door closer" is attached to the door and the top of the door frame. When the door is open and at rest, the door closer exerts a torque of 5.2 N · m. What is the least force that you need to apply to the door to hold it open?

10. || In Figure P7.10, force $\vec{F}_2$ acts half as far from the pivot as $\vec{F}_1$. What magnitude of $\vec{F}_2$ causes the net torque on the rod to be zero?

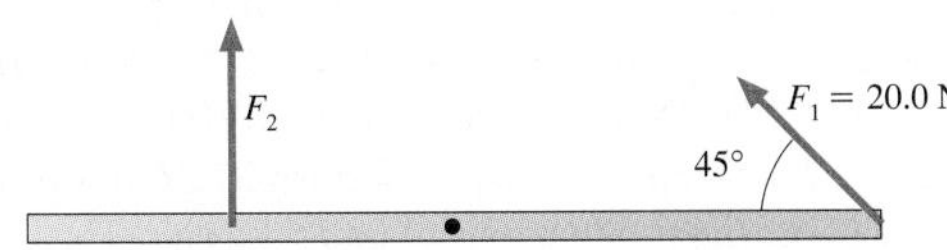

FIGURE P7.10

11. ||| Tom and Jerry both push on the 3.00-m-diameter merry-go-round shown in Figure P7.11.
   a. If Tom pushes with a force of 40.0 N and Jerry pushes with a force of 35.2 N, what is the net torque on the merry-go-round?
   b. What is the net torque if Jerry reverses the direction he pushes by 180° without changing the magnitude of his force?

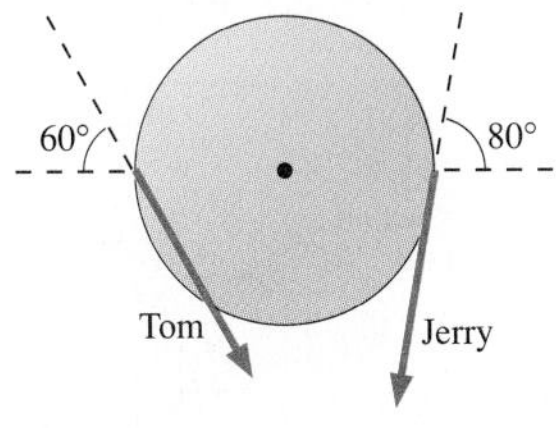

FIGURE P7.11

12. || What is the net torque of the bar shown in Figure P7.12, about the axis indicated by the dot?

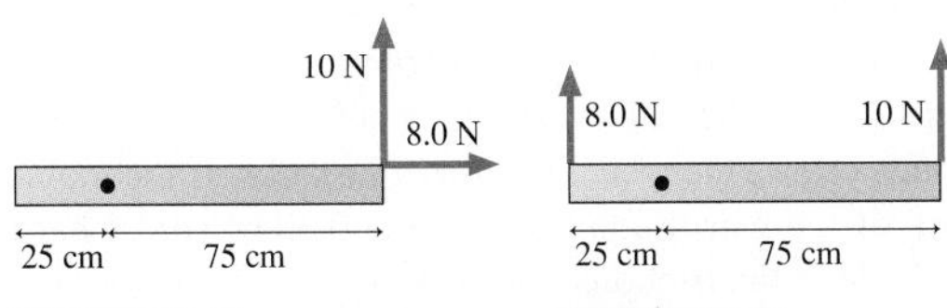

FIGURE P7.12 FIGURE P7.13

13. || What is the net torque of the bar shown in Figure P7.13, about the axis indicated by the dot?

14. || What is the net torque of the bar shown in Figure P7.14, about the axis indicated by the dot?

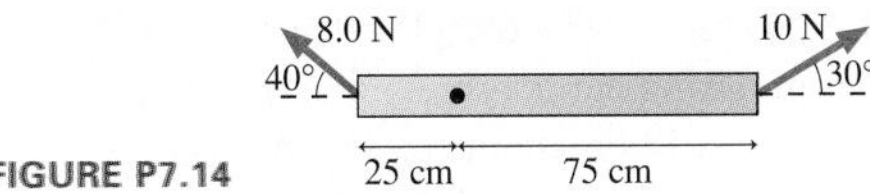

FIGURE P7.14

15. || A 1.7-m-long barbell has a 20 kg weight on its left end and a 35 kg weight on its right end.
   a. If you ignore the weight of the bar itself, how far from the left end of the barbell is the center of gravity?
   b. Where is the center of gravity if the 8.0 kg mass of the barbell itself is taken into account?

Section 7.3 Gravitational Torque and the Center of Gravity

16. || Three identical coins lie on three corners of a square 10.0 cm on a side, as shown in Figure P7.16. Determine the x and y coordinates of the center of gravity of the three coins.

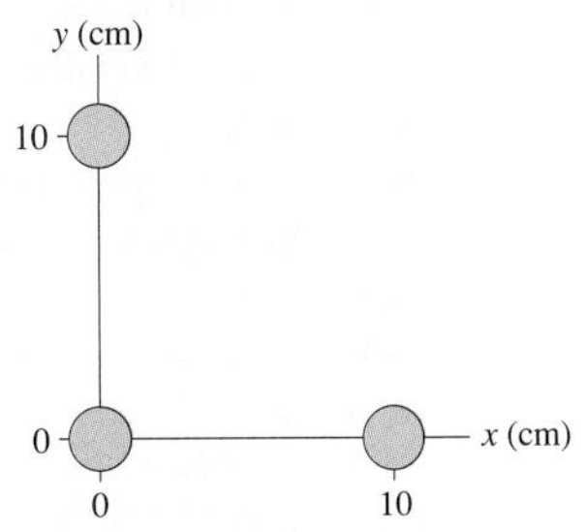

FIGURE P7.16

17. | Hold your arm outstretched so that it is horizontal. Estimate the mass of your arm and the position of its center of gravity. What is the gravitational torque on your arm in this position, computed around the shoulder joint?
BIO

18. ||| A solid cylinder sits on top of a solid cube as shown in Figure P7.18. How far above the table's surface is the center of gravity of the combined object?

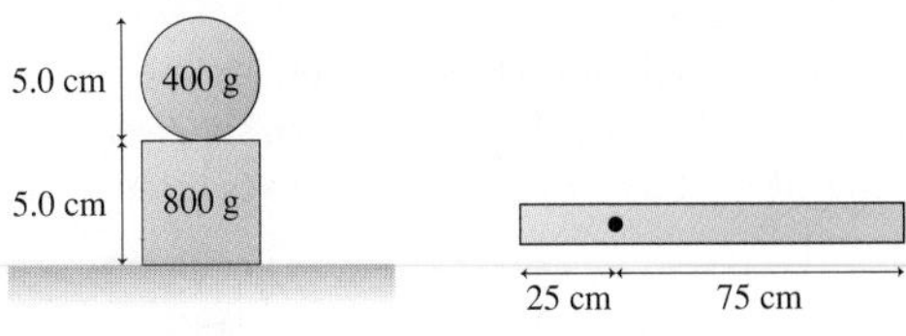

FIGURE P7.18 FIGURE P7.19

19. || The 2.0 kg, uniform, horizontal rod in Figure P7.19 is seen from the side. What is the gravitational torque about the point shown?

20. |||| A 4.00-m-long, 500 kg steel beam extends horizontally from the point where it has been bolted to the framework of a new building under construction. A 70.0 kg construction worker stands at the far end of the beam. What is the magnitude of the torque about the point where the beam is bolted into place?

21. ||| An athlete at the gym holds a 3.0 kg steel ball in his hand. His arm is 70 cm long and has a mass of 4.0 kg. What is the magnitude of the torque about his shoulder if he holds his arm
BIO
   a. Straight out to his side, parallel to the floor?
   b. Straight, but 45° below horizontal?

22. ||| The 2.0-m-long, 15 kg beam in Figure P7.22 is hinged at its left end. It is "falling" (rotating clockwise, under the influence of gravity), and the figure shows its position at three different times. What is the gravitational torque on the beam about an axis through the hinged end when the beam is at the
   a. Upper position?
   b. Middle position?
   c. Lower position?

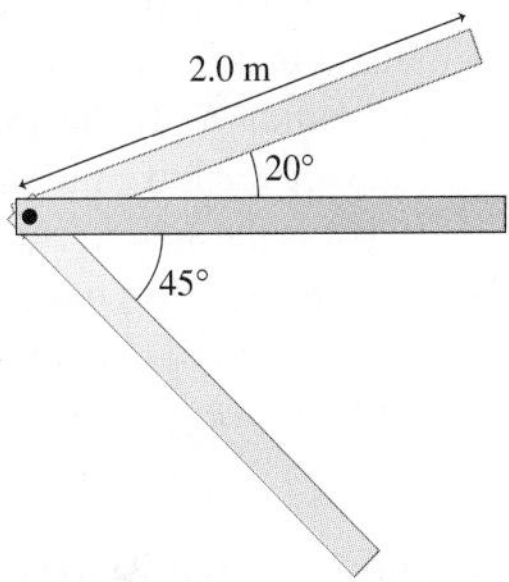

FIGURE P7.22

23. |||| Two thin beams are joined end-to-end as shown in Figure P7.23 to make a single object. The left beam is 10.0 kg and 1.00 m long and the right one is 40.0 kg and 2.00 m long.
   a. How far from the left end of the left beam is the center of gravity of the object?
   b. What is the gravitational torque on the object about an axis through its left end? The object is seen from the side.

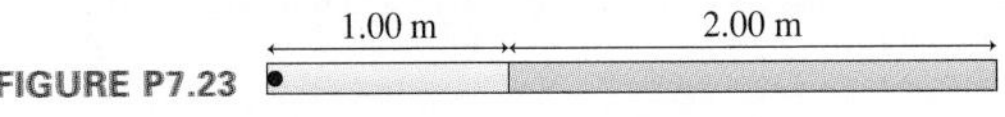

FIGURE P7.23

24. ||| Figure P7.24 shows two thin beams joined at right angles. The vertical beam is 15.0 kg and 1.00 m long and the horizontal beam is 25.0 kg and 2.00 m long.
   a. Find the center of gravity of the two joined beams. Express your answer in the form (x, y), taking the origin at the corner where the beams join.
   b. Calculate the gravitational torque on the joined beams about an axis through the corner. The beams are seen from the side.

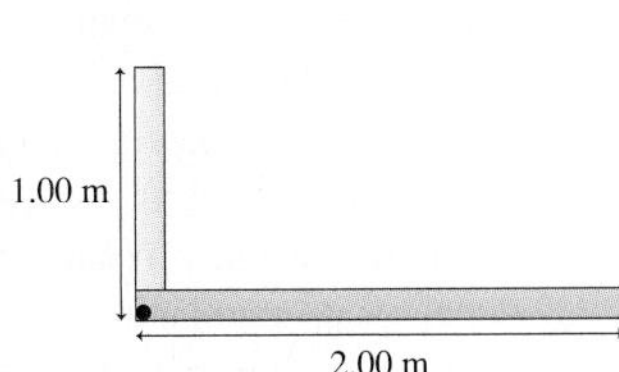

FIGURE P7.24

Section 7.4 Rotational Dynamics and Moment of Inertia

25. ||| A regulation table tennis ball has a mass of 2.7 g and is 40 mm in diameter. What is its moment of inertia about an axis that passes through its center?

26. || Three pairs of balls are connected by very light rods as shown in Figure P7.26. Rank in order, from smallest to largest, the moments of inertia I_1, I_2, and I_3 about axes through the centers of the rods.

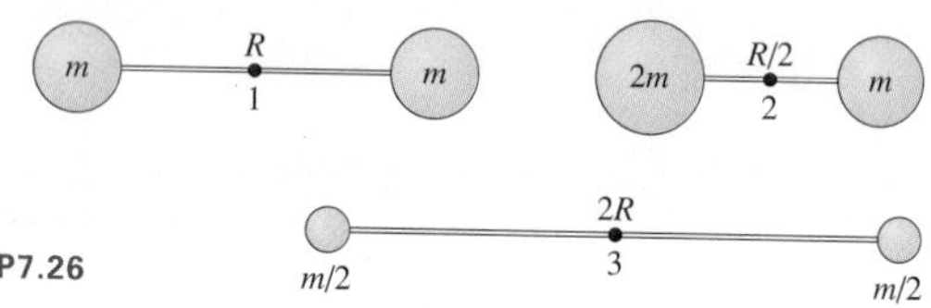

FIGURE P7.26

27. || A playground toy has four seats, each 5.0 kg, attached to very light, 1.5-m-long rods, as seen from above in Figure P7.27. If two children, with masses of 15 kg and 20 kg, sit in seats opposite one another, what is the moment of inertia about the rotation axis?

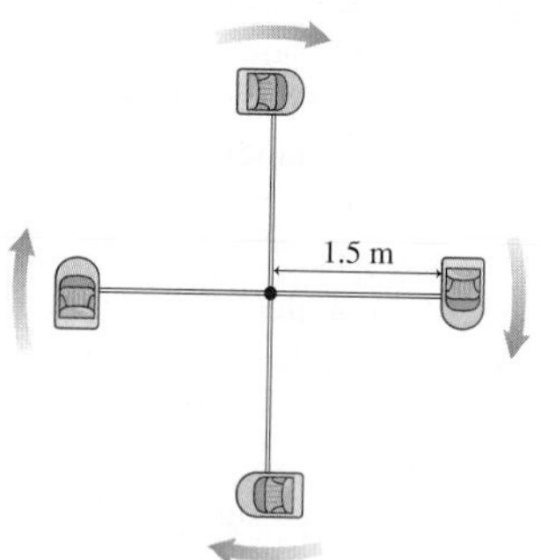

FIGURE P7.27

28. |||| A solid cylinder with a radius of 4.0 cm has the same mass as a solid sphere of radius R. If the cylinder and sphere have the same moment of inertia about their centers, what is the sphere's radius?

29. || A bicycle rim has a diameter of 0.65 m and a moment of inertia, measured about its center, of 0.19 kg · m². What is the mass of the rim?

Section 7.5 Using Newton's Second Law for Rotation

30. || The left part of Figure P7.30 shows a bird's-eye view of two identical balls connected by a light rod that rotates about a vertical axis through its center. The right part shows a ball of twice the mass connected to a light rod of half the length, that rotates about its left end. If equal forces are applied as shown in the figure, which of the two rods will have the greater angular acceleration?

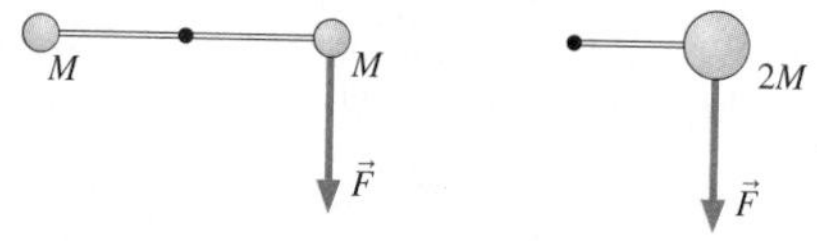

FIGURE P7.30

31. |||| a. What is the moment of inertia of the door in Problem 9?
    b. If you let go of the open door, what is its angular acceleration immediately afterward?

32. | A small grinding wheel has a moment of inertia of 4.0×10^{-5} kg · m². What net torque must be applied to the wheel for its angular acceleration to be 150 rad/s²?

33. || While sitting in a swivel chair, you push against the floor with your heel to make the chair spin. The 7.0 N frictional force is applied at a point 40 cm from the chair's rotation axis, in the direction that causes the greatest angular acceleration. If that angular acceleration is 1.8 rad/s², what is the total moment of inertia about the axis of you and the chair?

34. | An object's moment of inertia is 2.0 kg · m². Its angular velocity is increasing at the rate of 4.0 rad/s per second. What is the net torque on the object?

35. |||| A 200 g, 20-cm-diameter plastic disk is spun on an axle through its center by an electric motor. What torque must the motor supply to take the disk from 0 to 1800 rpm in 4.0 s?

36. |||| The 2.5 kg object shown in Figure P7.36 has a moment of inertia about the rotation axis of 0.085 kg · m². The rotation axis is horizontal. When released, what will be the object's initial angular acceleration?

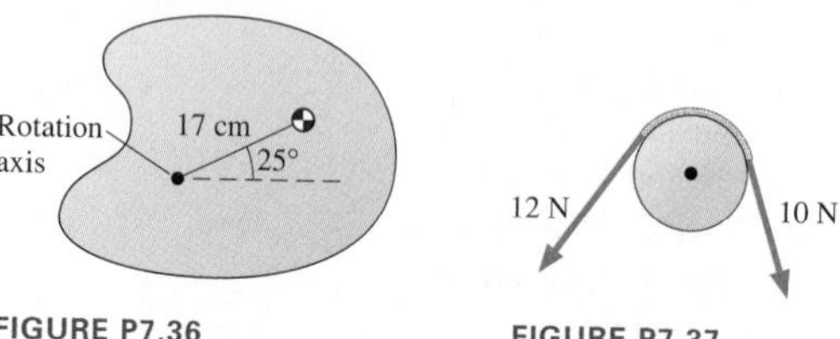

FIGURE P7.36 FIGURE P7.37

37. ||| A frictionless pulley, which can be modeled as a 0.80 kg solid cylinder with a 0.30 m radius, has a rope going over it, as shown in Figure P7.37. The tension in the rope is 10 N on one side and 12 N on the other. What is the angular acceleration of the pulley?

38. ||| If you lift the front wheel of a poorly maintained bicycle off the ground and then start it spinning at 0.72 rev/s, friction in the bearings causes the wheel to stop in just 12 s. If the moment of inertia of the wheel about its axle is 0.30 kg · m², what is the magnitude of the frictional torque?

39. ||||| A toy top with a spool of diameter 5.0 cm has a moment of inertia of 3.0×10^{-5} kg · m² about its rotation axis. To get the top spinning, its string is pulled with a tension of 0.30 N. How long does it take for the top to complete the first five revolutions? The string is long enough that it is wrapped around the top more than five turns.

40. ||| A 34-cm-diameter potter's wheel with a mass of 20 kg is spinning at 180 rpm. Using her hands, a potter forms a pot, centered on the wheel, with a 14 cm diameter. Her hands apply a net friction force of 1.3 N to the edge of the pot. If the power goes out, so that the wheel's motor no longer provides any torque, how long will it take for the wheel to come to a stop in her hands?

41. ||||| A 1.5 kg block and a 2.5 kg block are attached to opposite ends of a light rope. The rope hangs over a solid, frictionless pulley that is 30 cm in diameter and has a mass of 0.75 kg. When the blocks are released, what is the acceleration of the lighter block?

Section 7.6 Rolling Motion

42. ||| A bicycle with 0.80-m-diameter tires is coasting on a level road at 5.6 m/s. A small blue dot has been painted on the tread of the rear tire.
    a. What is the angular speed of the tires?
    b. What is the speed of the blue dot when it is 0.80 m above the road?
    c. What is the speed of the blue dot when it is 0.40 m above the road?

43. |||| A 1.2 g pebble is stuck in a tread of a 0.76-m-diameter automobile tire, held in place by static friction that can be at most 3.6 N. The car starts from rest and gradually accelerates on a straight road. How fast is the car moving when the pebble flies out of the tire tread?

General Problems

44. | Figure P7.44 shows the angular position-versus-time graph
INT for a particle moving in a circle.
    a. Write a description of the particle's motion.
    b. Draw the angular velocity-versus-time graph.

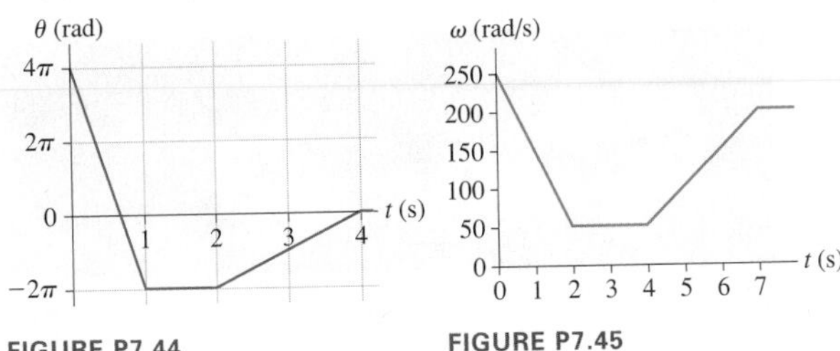

FIGURE P7.44 FIGURE P7.45

45. | The graph in Figure P7.45 shows the angular velocity of the
INT crankshaft in a car. Draw a graph of the angular acceleration versus time. Include appropriate numerical scales on both axes.

46. ||| A computer disk is 8.0 cm in diameter. A reference dot on the edge of the disk is initially located at $\theta = 45°$. The disk accelerates steadily for 0.50 s, reaching 2000 rpm, then coasts at steady angular velocity for another 0.50 s.
    a. What is the tangential acceleration of the reference dot at $t = 0.25$ s?
    b. What is the centripetal acceleration of the reference dot at $t = 0.25$ s?
    c. What is the angular position of the reference dot at $t = 1.0$ s?
    d. What is the speed of the reference dot at $t = 1.0$ s?

47. || A car with 58-cm-diameter tires accelerates uniformly from
INT rest to 20 m/s in 10 s. How many times does each tire rotate?

48. ||| The cable lifting an elevator is wrapped around a 1.0-m-diameter cylinder that is turned by the elevator's motor. The elevator is moving upward at a speed of 1.6 m/s. It then slows to a stop, while the cylinder turns one complete revolution. How long does it take for the elevator to stop?

49. |||| The 20-cm-diameter disk in Figure P7.49 can rotate on an axle through its center. What is the net torque about the axle?

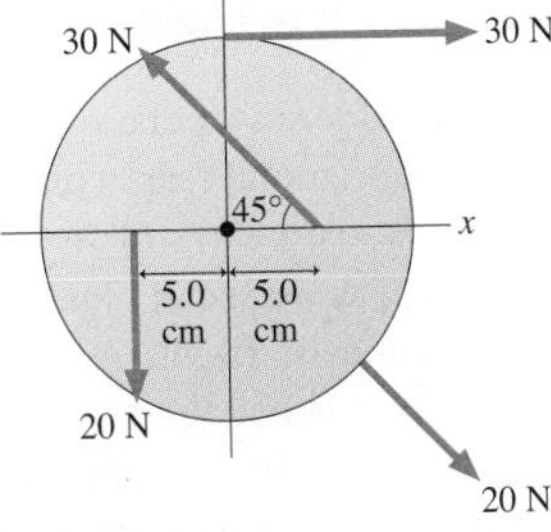

FIGURE P7.49

50. |||| A combination lock has a
INT 1.0-cm-diameter knob that is part of the dial you turn to unlock the lock. To turn that knob, you grip it between your thumb and forefinger with a force of 0.60 N as you twist your wrist. Suppose the coefficient of static friction between the knob and your fingers is only 0.12 because some oil accidentally got onto the knob. What is the most torque you can exert on the knob without having it slip between your fingers?

51. ||||| A 70 kg man's arm, including the hand, can be modeled as a
BIO 75-cm-long uniform cylinder with a mass of 3.5 kg. In raising both his arms, from hanging down to straight up, by how much does he raise his center of gravity?

52. ||| A penny has a mass of 2.5 g and is 1.5 mm thick; a nickel has a mass of 5.7 g and is 1.9 mm thick. If you make a stack of coins on a table, starting with five nickels and finishing with four pennies, how far above the tabletop is the center of gravity of the stack?

53. |||| The machinist's square shown in Figure P7.53 consists of a thin, rectangular blade connected to a rectangular handle.

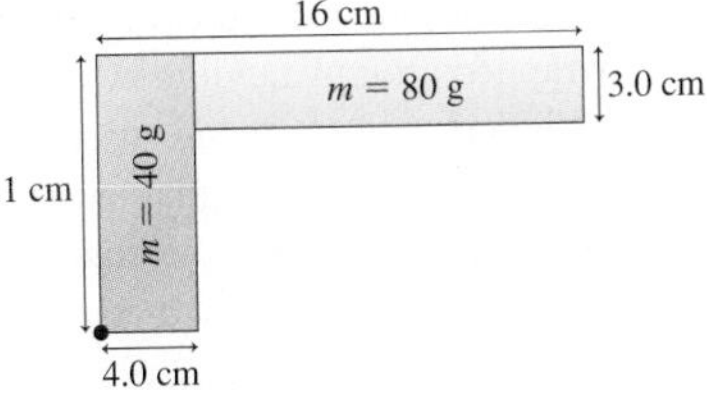

FIGURE P7.53

    a. Determine the x and y coordinates of the center of gravity. Let the lower left corner be $x = 0, y = 0$.
    b. Sketch how the tool would hang if it were allowed to freely pivot about the point $x = 0, y = 0$.
    c. When hanging from that point, what angle would the long side of the blade make with the vertical?

54. |||| The four masses shown in Figure P7.54 are connected by massless, rigid rods.
    a. Find the coordinates of the center of gravity.
    b. Find the moment of inertia about an axis that passes through mass A and is perpendicular to the page.
    c. Find the moment of inertia about a diagonal axis that passes through masses B and D.

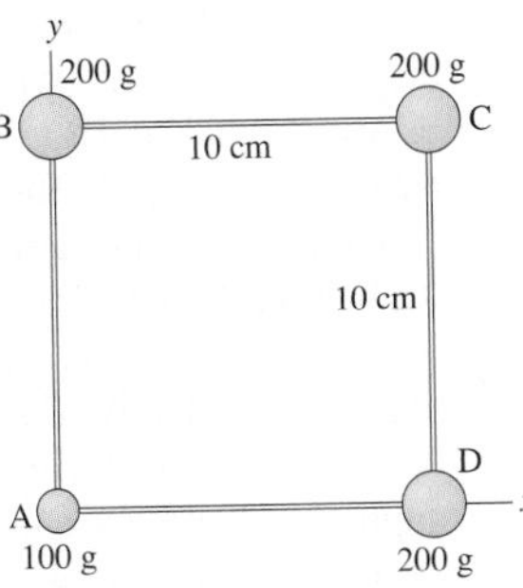

FIGURE P7.54

55. |||| Three 0.10 kg balls are connected by light rods to form an equilateral triangle with a side length of 0.30 m. What is the moment of inertia of this triangle about an axis perpendicular to its plane and passing through one of the balls?

56. |||| The three masses shown in Figure P7.56 are connected by massless, rigid rods.
    a. Find the coordinates of the center of gravity.
    b. Find the moment of inertia about an axis that passes through mass A and is perpendicular to the page.
    c. Find the moment of inertia about an axis that passes through masses B and C.

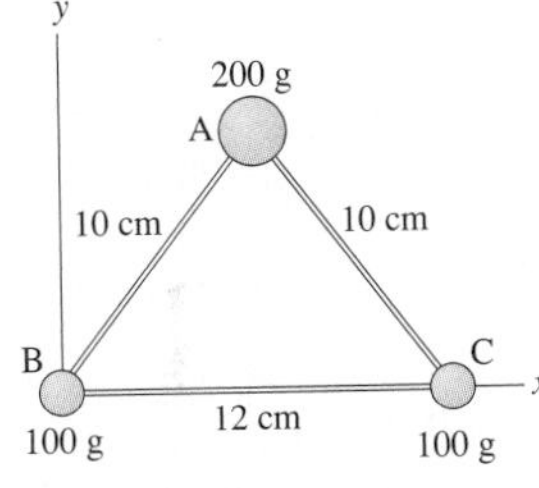

FIGURE P7.56

57. || A reasonable estimate of the moment of inertia of an ice
BIO skater spinning with her arms at her sides can be made by modeling most of her body as a uniform cylinder. Suppose the skater has a mass of 64 kg. One-eighth of that mass is in her arms, which are 60 cm long and 20 cm from the vertical axis about which she rotates. The rest of her mass is approximately in the form of a 20-cm-radius cylinder.
    a. Estimate the skater's moment of inertia to two significant figures.
    b. If she were to hold her arms outward, rather than at her sides, would her moment of inertia increase, decrease, or remain unchanged? Explain.

58. || Starting from rest, a 12-cm-diameter compact disk takes 3.0 s to reach its operating angular velocity of 2000 rpm. Assume that the angular acceleration is constant. The disk's moment of inertia is 2.5×10^{-5} kg $\cdot$ m^2.
    a. How much torque is applied to the disk?
    b. How many revolutions does it make before reaching full speed?

59. ⅡⅡ The ropes in Figure P7.59 are each wrapped around a cylinder, and the two cylinders are fastened together. The smaller cylinder has a diameter of 10 cm and a mass of 5.0 kg; the larger cylinder has a diameter of 20 cm and a mass of 20 kg. What is the angular acceleration of the cylinders? Assume that the cylinders turn on a frictionless axle.

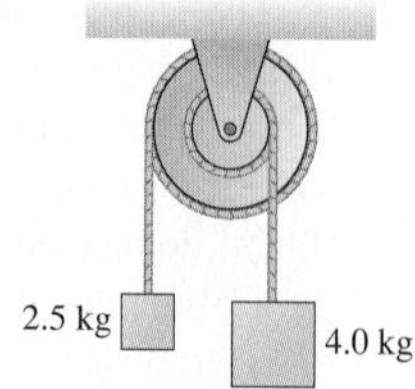

FIGURE P7.59

60. ⅡⅡ Flywheels are large, massive wheels used to store energy. They can be spun up slowly, then the wheel's energy can be released quickly to accomplish a task that demands high power. An industrial flywheel has a 1.5 m diameter and a mass of 250 kg. A motor spins up the flywheel with a constant torque of 50 N · m. How long does it take the flywheel to reach top angular speed of 1200 rpm?

61. ⅡⅡ A 1.0 kg ball and a 2.0 kg ball are connected by a 1.0-m-long rigid, massless rod. The rod and balls are rotating clockwise about their center of gravity at 20 rpm. What torque will bring the balls to a halt in 5.0 s?

62. ⅡⅡ A 1.5 kg block is connected by a rope across a 50-cm-diameter, 2.0 kg, frictionless pulley, as shown in Figure P7.62. A constant 10 N tension is applied to the other end of the rope. Starting from rest, how long does it take the block to move 30 cm?
INT

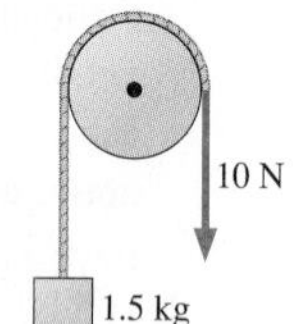

FIGURE P7.62

63. ⅡⅡⅡ The two blocks in Figure P7.63 are connected by a massless rope that passes over a pulley. The pulley is 12 cm in diameter and has a mass of 2.0 kg. As the pulley turns, friction at the axle exerts a torque of magnitude 0.50 N · m. If the blocks are released from rest, how long does it take the 4.0 kg block to reach the floor?
INT

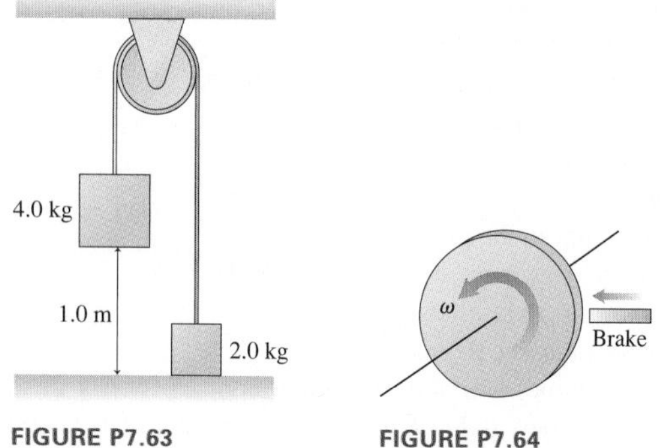

FIGURE P7.63 FIGURE P7.64

64. ⅡⅡ The 2.0 kg, 30-cm-diameter disk in Figure P7.64 is spinning at 300 rpm. How much friction force must the brake apply to the rim to bring the disk to a halt in 3.0 s?

65. ⅡⅡ A tradesman sharpens a knife by pushing it against the rim of a grindstone. The 30-cm-diameter stone is spinning at 200 rpm and has a mass of 28 kg. The coefficient of kinetic friction between the knife and the stone is 0.20. If the stone loses 10% of its speed in 10 s of grinding, what is the force with which the man presses the knife against the stone?
INT

66. Ⅱ The bunchberry flower has the fastest-moving parts ever seen in a plant. Initially, the stamens are held by the petals in a bent position, storing elastic energy like a coiled spring. As the petals release, the tips of the stamens act like medieval catapults, flipping through a 60° angle in just 0.30 ms to launch pollen from the anther sacs at their ends. The human eye just sees a burst of pollen; careful photography (see Figure P7.66a) reveals the details. As shown in Figure P7.66b, we can model a stamen tip as a 1.0-mm-long, 10 μg rigid rod with a 10 μg anther sac at one end and a pivot point at the opposite end. Although oversimplifying, we will assume that the angular acceleration is constant throughout the motion.
BIO

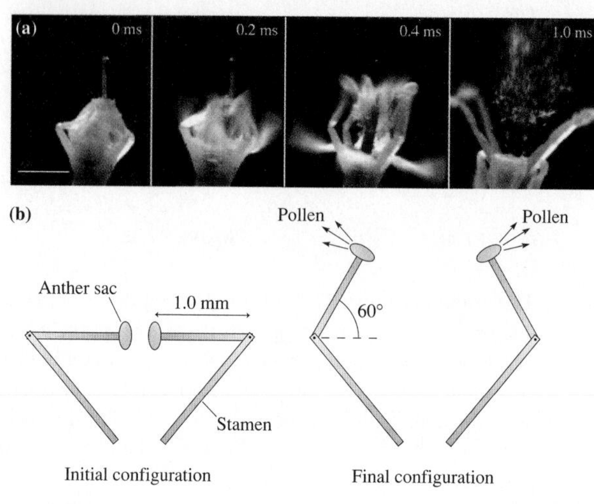

FIGURE P7.66

a. What is the tangential acceleration of the anther sac during the motion?
b. What is the speed of the anther sac as it releases its pollen?
c. How large is the "straightening torque"? Neglect gravitational forces in your calculation.
d. Compute the gravitational torque on the stamen tip (including the anther sac) in its initial orientation. Was it reasonable to neglect the gravitational torque in part c?

Passage Problems

The Illusion of Flight

The grand jeté is a classic ballet maneuver in which a dancer executes a horizontal leap while moving her arms and legs up and then down. At the center of the leap, the arms and legs are gracefully extended, as we see in Figure P7.67a. The goal of the leap is to create the illusion of flight. As discussed in Section 7.3, the center of mass—and hence the center of gravity—of an extended object follows a parabolic trajectory when undergoing projectile motion. But when you watch a dancer leap through the air, you don't watch her center of gravity, you watch her head. If the translational motion of her head is horizontal—not parabolic—this creates the illusion that she is flying through the air, held up by unseen forces.

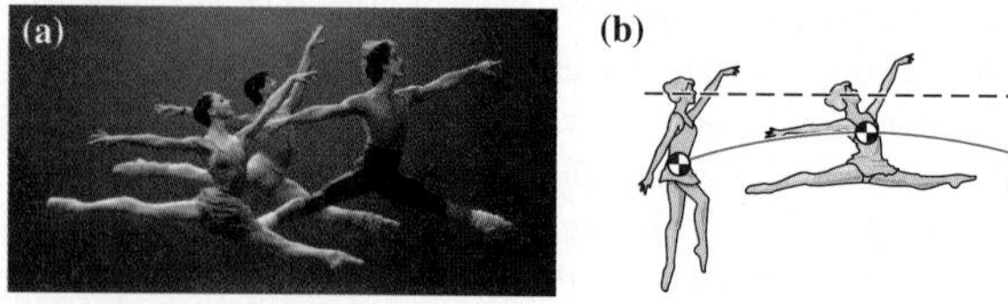

FIGURE P7.67

Figure P7.67b illustrates how the dancer creates this illusion. While in the air, she changes the position of her center of gravity relative to her body by moving her arms and legs up, then down. Her center of gravity moves in a parabolic path, but her head moves in a straight line. It's not flight, but it will appear that way, at least for a moment.

67. | To perform this maneuver, the dancer relies on the fact that the position of her center of gravity
    A. Is near the center of the torso.
    B. Is determined by the positions of her arms and legs.
    C. Moves in a horizontal path.
    D. Is outside of her body.

68. | Suppose you wish to make a vertical leap with the goal of getting your head as high as possible above the ground. At the top of your leap, your arms should be
    A. Held at your sides.
    B. Raised above your head.
    C. Outstretched, away from your body.

69. | When the dancer is in the air, is there a gravitational torque on her? Take the dancer's rotation axis to be through her center of gravity.
    A. Yes, there is a gravitational torque.
    B. No, there is not a gravitational torque.
    C. It depends on the positions of her arms and legs.

70. | In addition to changing her center of gravity, a dancer may change her moment of inertia. Consider her moment of inertia about a vertical axis through the center of her body. When she raises her arms and legs, this
    A. Increases her moment of inertia.
    B. Decreases her moment of inertia.
    C. Does not change her moment of inertia.

STOP TO THINK ANSWERS

Stop to Think 7.1: a. constant (but not zero), b. constant (but not zero), c. constant (but not zero), d. zero, e. zero. The angular velocity ω is constant. Thus the magnitude of the velocity $v = \omega r$ and the centripetal acceleration $a_c = \omega^2 r$ are constant. This also means that the ball's angular acceleration α and tangential acceleration $a_t = \alpha r$ are both zero.

Stop to Think 7.2: $\tau_E > \tau_A = \tau_D > \tau_B > \tau_C$. The perpendicular component in E is larger than 2 N.

Stop to Think 7.3: A. The force acting at the axis exerts no torque. Thus the third force needs to exert an equal but opposite torque to that exerted by the force acting at the rim. Force A, which has twice the magnitude but acts at half the distance from the axis, does so.

Stop to Think 7.4: $\tau_E = \tau_B > \tau_D > \tau_A = \tau_C$. The torques are $\tau_B = \tau_E = 2mgL$, $\tau_D = \frac{3}{2}mgL$, and $\tau_A = \tau_C = mgL$, where L is the length of the rod in B.

Stop to Think 7.5: $I_D > I_A > I_C > I_B$. The moments of inertia are $I_B \approx 0$, $I_C = 2mr^2$, $I_A = 3mr^2$, and $I_D = mr^2 + m(2r)^2 = 5mr^2$.

Stop to Think 7.6: C. The velocity of P is the vector sum of $\vec{v}$ directed to the right and an upward velocity of the same magnitude due to the rotation of the wheel.

8 Equilibrium and Elasticity

How does a dancer balance so gracefully *en pointe*? And how does her foot withstand the great stresses concentrated on her toes? In this chapter we'll find answers to both these questions.

LOOKING AHEAD ▸▸

The goals of Chapter 8 are to learn about the static equilibrium of extended objects and to understand the basic properties of springs and elastic materials.

Torque and Static Equilibrium

In Chapter 5 we found that a particle can be in static equilibrium only if the net force acting on it is zero. For an extended object such as the dancer shown above, we'll learn that the net torque on the object must also be zero for it to be in equilibrium.

For this cyclist and his bike to stand motionless, the net force *and* net torque on them must be zero.

Looking Back ◂◂
5.1 Equilibrium

Springs

We'll study springs and similar elastic objects, and we'll find the important result that the spring force is proportional to the distance the spring is stretched or compressed from its natural length.

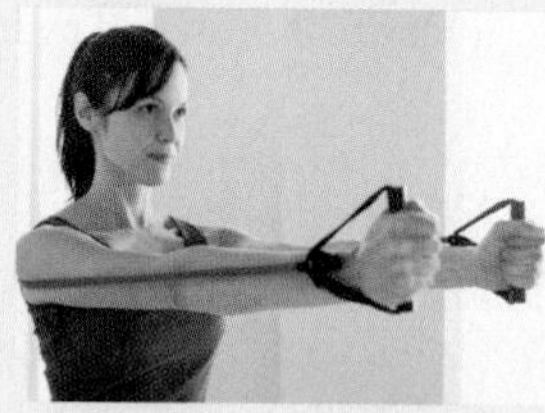

You can get a better workout by stretching the band farther, because the farther it stretches, the harder it is to pull.

Looking Back ◂◂
4.3 Spring forces

Stability and Balance

Why are some objects more *stable* than others; that is, why are some easy to tip over while others are more solidly planted? We'll learn that an object is more stable when its center of gravity is low and its contact points with the ground are widely separated.

For maximal stability, this football player keeps his center of gravity low and his stance wide.

The girls on this tree trunk have a hard time balancing because of their tall stance and narrow footprint.

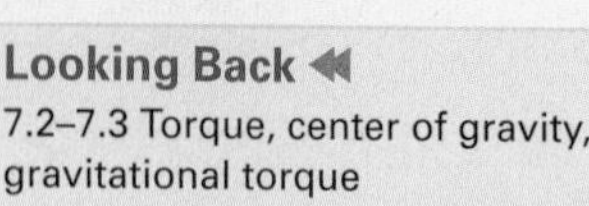

Looking Back ◂◂
7.2–7.3 Torque, center of gravity, gravitational torque

Elastic Materials

All materials, even seemingly rigid ones like glass or steel, stretch slightly when you pull on them—they act like very stiff springs. We'll learn how to calculate the amount a solid object stretches or compresses as forces are applied to it.

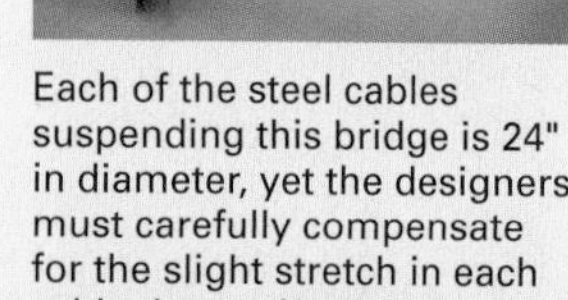

Each of the steel cables suspending this bridge is 24" in diameter, yet the designers must carefully compensate for the slight stretch in each cable due to the enormous load of the bridge.

Biological Materials

The elastic properties of biological materials such as bone, tendon, and even spider silk play an important role in the world of living things. Bone, of course, is quite rigid, but did you know that spider silk is as strong as steel?

8.1 Torque and Static Equilibrium

We have now spent several chapters studying motion and its causes. In many disciplines, it is just as important to understand the conditions under which objects do *not* move. In structural engineering, buildings and dams must be designed such that they remain motionless, even when huge forces act on them. In sports science, a correct stationary position is often the starting point for a successful athletic event. And joints in the body must sustain large forces when the body is supporting heavy loads, as in holding or carrying heavy objects.

Recall from Section 5.1 that an object at rest is in *static equilibrium.* As long as an object can be modeled as a *particle,* the condition necessary for static equilibrium is that the net force $\vec{F}_{net}$ on the particle is zero. Such a situation is shown in **FIGURE 8.1a**, where the two forces applied to the particle balance and the particle can remain at rest.

But in Chapter 7 we moved beyond the particle model to study extended objects that can rotate. Consider, for example, the block in **FIGURE 8.1b**. In this case the two forces act along the same line, the net force is zero, and the block is in equilibrium. But what about the block in **FIGURE 8.1c**? The net force is still zero, but this time the block begins to rotate because the two forces exert a net *torque.* For an extended object, $\vec{F}_{net} = \vec{0}$ is not by itself enough to ensure static equilibrium. There is a second condition for static equilibrium of an extended object: The net torque τ_{net} on the object must also be zero.

FIGURE 8.1 A block with no net force acting on it may still be out of equilibrium.

(a) When the net force on a particle is zero, the particle is in static equilibrium.

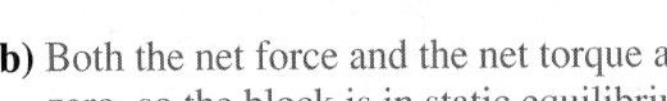

(b) Both the net force and the net torque are zero, so the block is in static equilibrium.

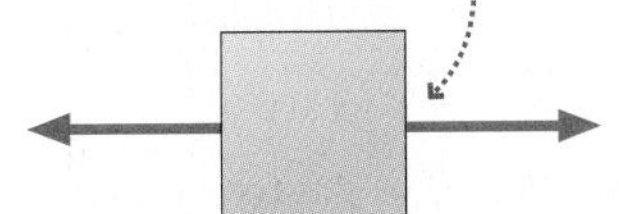

(c) The net force is still zero, but the net torque is *not* zero. The block is not in equilibrium.

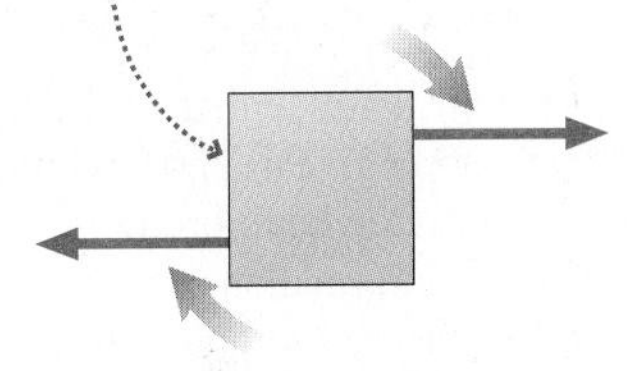

If we write the net force in component form, the conditions for static equilibrium of an extended object are

$$\left.\begin{aligned}\sum F_x &= 0\\ \sum F_y &= 0\end{aligned}\right\}\ \text{No net force}$$
$$\sum \tau = 0 \ \}\ \text{No net torque} \qquad (8.1)$$

Conditions for static equilibrium of an extended object

If motion is possible in the z-direction, we'd also require that $\sum F_z = 0$. In this chapter, however, we'll consider only motion restricted to the xy-plane.

EXAMPLE 8.1 **Finding the force from the biceps tendon**

Weightlifting can exert extremely large forces on the body's joints and tendons. In the *strict curl* event, a standing athlete lifts a barbell by moving only his forearms, which pivot at the elbow. The record weight lifted in the strict curl is over 200 pounds (about 900 N). **FIGURE 8.2** shows the arm bones and the main lifting muscle when the forearm is horizontal. The distance from the tendon to the elbow joint is 4.0 cm, and from the barbell to the elbow 35 cm.

FIGURE 8.2 An arm holding a barbell.

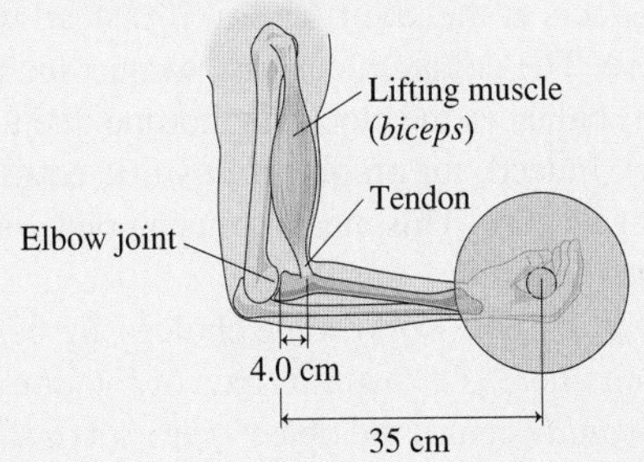

a. What is the tension in the tendon connecting the biceps muscle to the bone while a 900 N barbell is held stationary in this position?

b. What is the force exerted by the elbow on the forearm bones?

PREPARE **FIGURE 8.3** shows a simplified model of the arm and the forces acting on the forearm. $\vec{F}_t$ is the tension force due to the muscle, $\vec{F}_b$ is the downward force of the barbell, and $\vec{F}_e$ is the force of the elbow joint on the forearm. As a simplification, we've neglected the weight of the arm itself because it is so much less than the weight of the barbell. Because $\vec{F}_t$ and $\vec{F}_b$ have no x-component, neither can $\vec{F}_e$. If it did, the net force in the x-direction would not be zero, and the forearm could not be in equilibrium. Because each arm supports half the weight of the barbell, the magnitude of the barbell force is $F_b = 450$ N.

Continued

FIGURE 8.3 Visual overview of holding a barbell.

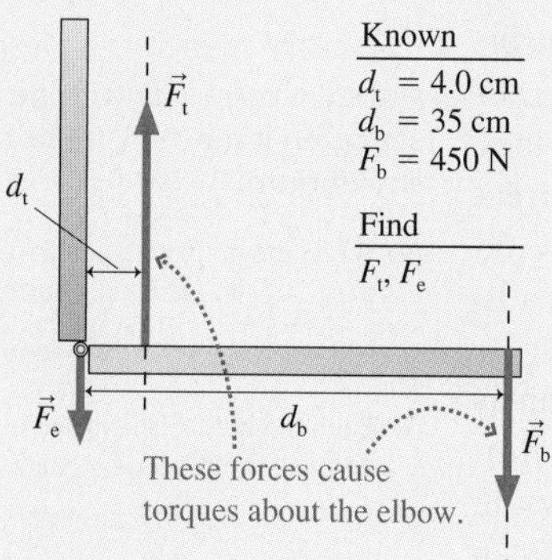

SOLVE a. For the forearm to be in static equilibrium, the net force and net torque on it must both be zero. Setting the net force to zero gives

$$\sum F_y = F_t - F_e - F_b = 0$$

We don't know either of the forces F_t and F_e, nor does the force equation give us enough information to find them. But the fact that in static equilibrium the torque also must be zero gives us the extra information that we need.

Recall that the torque must be calculated about a particular point. Here, a natural choice is the elbow joint, about which the forearm can pivot. Given this pivot, we can calculate the torque due to each of the three forces in terms of their magnitudes F and moment arms $r_\perp$ as $\tau = r_\perp F$. The moment arm is the perpendicular distance between the pivot and the "line of action" along which the force is applied. Figure 8.3 shows that the moment arms for $\vec{F}_t$ and $\vec{F}_b$ are the distances d_t and d_b, respectively, measured along the beam representing the forearm. The moment arm for $\vec{F}_e$ is zero, because this force acts directly at the pivot. Thus we have

$$\tau_{net} = F_e \times 0 + F_t d_t - F_b d_b = 0$$

The tension in the tendon tries to rotate the arm counterclockwise, so it produces a positive torque; the torque due to the barbell, which tries to rotate the arm in a clockwise direction, is negative. We can solve the torque equation for F_t:

$$F_t = F_b \frac{d_b}{d_t} = (450 \text{ N}) \frac{35 \text{ cm}}{4.0 \text{ cm}} = 3900 \text{ N}$$

b. We now need to make use of the force equation:

$$F_e = F_t - F_b = 3900 \text{ N} - 450 \text{ N} = 3450 \text{ N}$$

ASSESS This large value for F_t makes sense: The short distance d_t from the tendon to the elbow joint means that the force supplied by the biceps has to be very large to counter the torque generated by a force applied at the opposite end of the forearm.

STOP TO THINK 8.1 Which of these objects is in static equilibrium?

A. B. C. D.

Choosing the Pivot Point

In Example 8.1, we calculated the net torque using the elbow joint as the axis of rotation or pivot point. But we learned in Chapter 7 that the torque depends on which point is chosen as the pivot point. Was there something special about our choice of the elbow joint?

FIGURE 8.4 A hammer resting on two pegs.

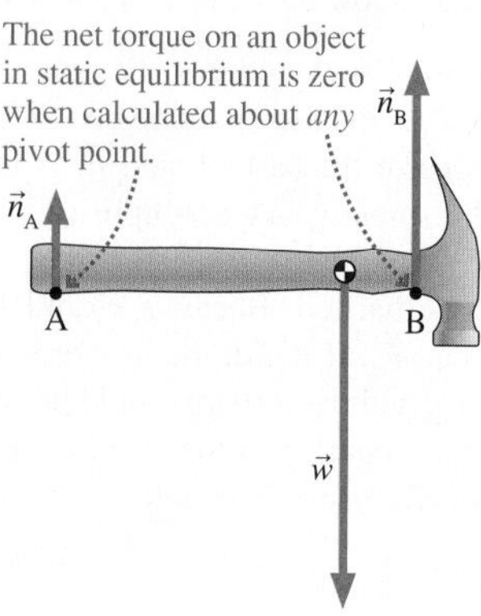

Consider the hammer shown in FIGURE 8.4, supported on a pegboard by two pegs A and B. Because the hammer is in static equilibrium, the net torque around the pivot at peg A must be zero: The clockwise torque due to the weight $\vec{w}$ is exactly balanced by the counterclockwise torque due to the force $\vec{n}_B$ of peg B. (Recall that the torque due to $\vec{n}_A$ is zero, because here $\vec{n}_A$ acts at the pivot A.) But if instead we take B as the pivot, the net torque is still zero. The counterclockwise torque due to $\vec{w}$ (with a large force but small moment arm) balances the clockwise torque due to $\vec{n}_A$ (with a small force but large moment arm). Indeed, **for an object in static equilibrium, the net torque about *every* point must be zero.** This means you can pick *any* point you wish as a pivot point for calculating the torque.

Although any choice of a pivot point will work, some choices are better because they simplify the calculations. Often, there is a "natural" axis of rotation in the problem, an axis about which rotation *would* occur if the object were not in static equilibrium. Example 8.1 is of this type, with the elbow joint as a natural axis of rotation.

If no point naturally suggests itself as an axis, look for a point on the object at which several forces act, or at which a force acts whose magnitude you don't know. Such a point is a good choice because any force acting at that point does not contribute to the torque. For instance, the woman in FIGURE 8.5 is in equilibrium as she rests on the rock wall. A good choice of pivot point would be where her foot contacts the wall because this choice eliminates the torque due to the force $\vec{F}$ of the wall on her foot. But don't agonize over the choice of a pivot point! You can still solve the problem no matter which point you choose.

FIGURE 8.5 Choosing the pivot for a woman rappelling down a rock wall.

The torque due to $\vec{F}$ about this point is zero. This makes this point a good choice as the pivot.

PROBLEM-SOLVING STRATEGY 8.1 Static equilibrium problems

PREPARE Model the object as a simple shape. Draw a visual overview that shows all forces and distances. List known information.

- Pick an axis or pivot about which the torques will be calculated.
- Determine the torque about this pivot point due to each force acting on the object. The torques due to any forces acting *at* the pivot are zero.
- Determine the sign of each torque about this pivot point.

SOLVE The mathematical steps are based on the fact that an object in static equilibrium has no net force and no net torque:

$$\vec{F}_{\text{net}} = \vec{0} \quad \text{and} \quad \tau_{\text{net}} = 0$$

- Write equations for $\Sigma F_x = 0$, $\Sigma F_y = 0$, and $\Sigma \tau = 0$.
- Solve the resulting equations.

ASSESS Check that your result is reasonable and answers the question.

ActivPhysics Online 7.2, 7.3, 7.4, 7.5

EXAMPLE 8.2 Forces on a board on sawhorses

A board weighing 100 N sits across two sawhorses, as shown in FIGURE 8.6. What are the magnitudes of the normal forces of the sawhorses acting on the board?

FIGURE 8.6 A board sitting on two sawhorses.

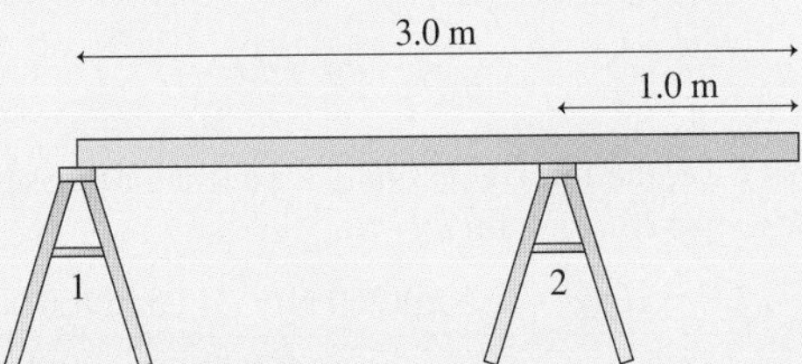

PREPARE The board and the forces acting on it are shown in FIGURE 8.7. $\vec{n}_1$ and $\vec{n}_2$ are the normal forces on the board due to the sawhorses, and $\vec{w}$ is the weight of the board acting at the center of gravity. The distance d_1 to the center of the board is half the board's length, or 1.5 m. Then d_2 is $d_1 - 1.0$ m, or 0.5 m.

As discussed above, a good choice for the pivot is a point at which an unknown force acts, because that force contributes nothing to the torque. Either the point where $\vec{n}_1$ acts or the point where $\vec{n}_2$ acts will work; let's choose the left end of the board, where $\vec{n}_1$ acts, for this example. With this choice of pivot point, the moment arm for $\vec{w}$ is $d_1 = 1.5$ m. Because $\vec{w}$ tends to rotate the board clockwise, its torque is negative. The moment arm for $\vec{n}_2$ is the distance $d_1 + d_2 = 2.0$ m, and its torque is positive.

FIGURE 8.7 Visual overview of a board on two sawhorses.

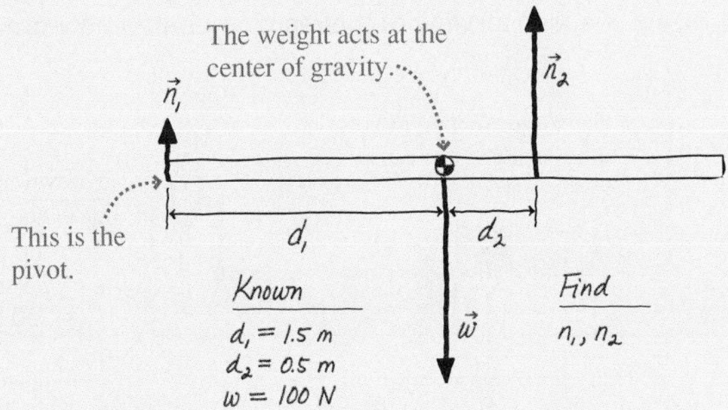

SOLVE The board is in static equilibrium, so the net force $\vec{F}_{\text{net}}$ and the net torque τ_{net} must both be zero. The forces have only y-components, so the force equation is

$$\sum F_y = n_1 - w + n_2 = 0$$

The torque equation, computed around the left end of the board, is

$$\tau_{\text{net}} = -d_1 w + (d_1 + d_2) n_2 = 0$$

Continued

We now have two simultaneous equations with the two unknowns n_1 and n_2. To solve these, let's solve for n_2 in the torque equation and then substitute that result into the force equation. From the torque equation,

$$n_2 = \frac{d_1 w}{d_1 + d_2} = \frac{(1.5\text{ m})(100\text{ N})}{2.0\text{ m}} = 75\text{ N}$$

The force equation is then $n_1 - 100\text{ N} + 75\text{ N} = 0$, which we can solve for n_1:

$$n_1 = w - n_2 = 100\text{ N} - 75\text{ N} = 25\text{ N}$$

ASSESS It seems reasonable that $n_2 > n_1$ because more of the board sits over the right sawhorse.

STOP TO THINK 8.2 A beam with a pivot on its left end is suspended from a rope. In which direction is the force of the pivot on the beam?

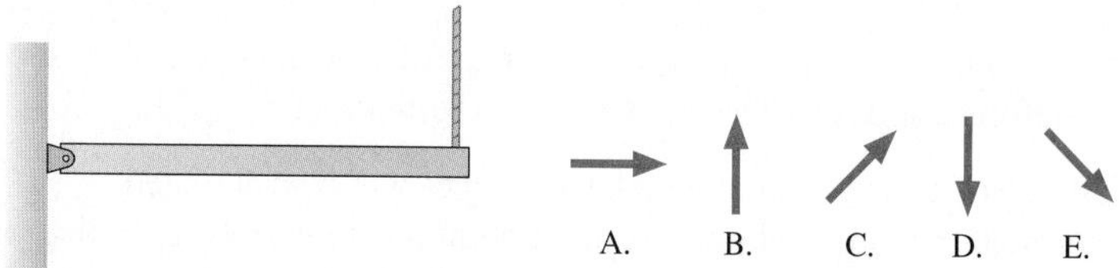

An interesting application of static equilibrium is to find the center of gravity of the human body. Because the human body is highly flexible, the position of the center of gravity is quite variable and depends on just how the body is posed. The horizontal position of the body's center of gravity can be located accurately from simple measurements with a *reaction board* and a scale. The following example shows how this is done.

EXAMPLE 8.3 Finding the center of gravity of the human body

A woman weighing 600 N lies on a 2.5-m-long, 60 N reaction board with her feet over the pivot. The scale on the right reads 250 N. What is the distance d from the woman's feet to her center of gravity?

PREPARE The forces and distances in the problem are shown in **FIGURE 8.8**. We'll consider the board and woman as a single object. We've assumed that the board is uniform, so its center of gravity is at its midpoint. To eliminate the unknown magnitude of $\vec{n}$ from the torque equation, we'll choose the pivot to be the left end of the board. The torque due to $\vec{F}$ is positive, and those due to $\vec{w}$ and $\vec{w}_b$ are negative.

FIGURE 8.8 Visual overview of the reaction board and woman.

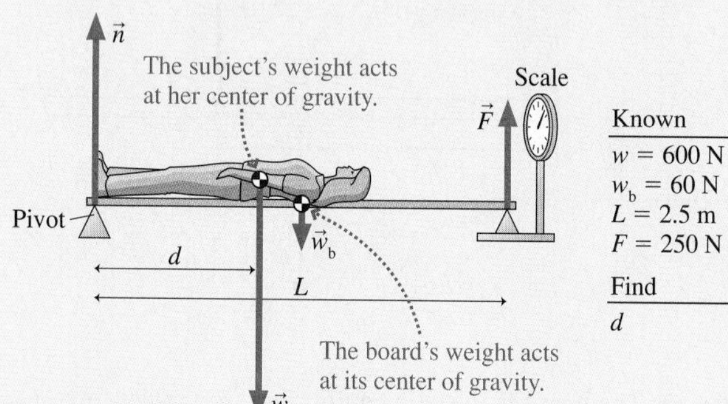

SOLVE Because the board and woman are in static equilibrium, the net force and net torque on them must be zero. The force equation reads

$$\sum F_y = n - w_b - w + F = 0$$

and the torque equation gives

$$\sum \tau = -\frac{L}{2} w_b - dw + LF = 0$$

In this case, the force equation isn't needed because we can solve the torque equation for d:

$$d = \frac{LF - \frac{1}{2}Lw_b}{w} = \frac{(2.5\text{ m})(250\text{ N}) - \frac{1}{2}(2.5\text{ m})(60\text{ N})}{600\text{ N}}$$

$$= 0.92\text{ m}$$

ASSESS If the woman is 5′ 6″ (1.68 m) tall, her center of gravity is (0.92 m)/(1.68 m) = 55% of her height, or a little more than halfway up her body. This seems reasonable.

EXAMPLE 8.4 Will the ladder slip?

A 3.0-m-long ladder leans against a frictionless wall at an angle of 60° with respect to the floor. What is the minimum value of μ_s, the coefficient of static friction with the ground, that will prevent the ladder from slipping?

PREPARE The ladder is a rigid rod of length L. To not slip, both the net force and net torque on the ladder must be zero. **FIGURE 8.9** shows the ladder and the forces acting on it. The bottom corner of the ladder is a good choice of a pivot point because two of the

FIGURE 8.9 Visual overview of a ladder in static equilibrium.

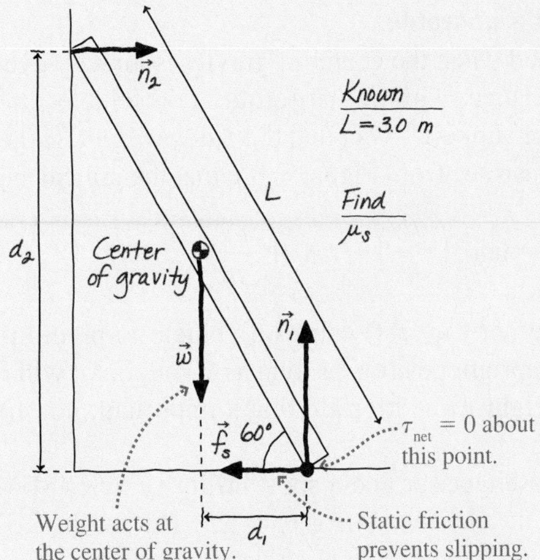

forces pass through this point and thus produce no torque about it. With this choice, the weight of the ladder, acting at the center of gravity, exerts torque $d_1 w$ and the force of the wall exerts torque $-d_2 n_2$. The signs are based on the observation that $\vec{w}$ would cause the ladder to rotate counterclockwise, while $\vec{n}_2$ would cause it to rotate clockwise.

SOLVE The x- and y-components of $\vec{F}_{net} = \vec{0}$ are

$$\sum F_x = n_2 - f_s = 0$$
$$\sum F_y = n_1 - w = n_1 - Mg = 0$$

The torque about the bottom corner is

$$\tau_{net} = d_1 w - d_2 n_2 = \frac{1}{2}(L\cos 60°)Mg - (L\sin 60°)n_2 = 0$$

Altogether, we have three equations with the three unknowns n_1, n_2, and f_s. If we solve the third equation for n_2,

$$n_2 = \frac{\frac{1}{2}(L\cos 60°)Mg}{L\sin 60°} = \frac{Mg}{2\tan 60°}$$

we can then substitute this into the first equation to find

$$f_s = \frac{Mg}{2\tan 60°}$$

Our model of static friction is $f_s \le f_{s\,max} = \mu_s n_1$. We can find n_1 from the second equation: $n_1 = Mg$. From this, the model of friction tells us that

$$f_s \le \mu_s Mg$$

Comparing these two expressions for f_s, we see that μ_s must obey

$$\mu_s \ge \frac{1}{2\tan 60°} = 0.29$$

Thus the minimum value of the coefficient of static friction is 0.29.

ASSESS You know from experience that you can lean a ladder or other object against a wall if the ground is "rough," but it slips if the surface is too smooth. 0.29 is a "medium" value for the coefficient of static friction, which is reasonable.

8.2 Stability and Balance

If you tilt a box up on one edge by a small amount and let go, it falls back down. If you tilt it too much, it falls over. And if you tilt it "just right," you can get the box to balance on its edge. What determines these three possible outcomes?

FIGURE 8.10 illustrates the idea with a car, but the results are general and apply in many situations. An extended object, whether it's a car, a box, or a person, has a *base of support* on which it rests when in static equilibrium. If you tilt the object, one edge of the base of support becomes a pivot point. As long as the object's center of gravity remains over the base of support, torque due to gravity will rotate the object back toward its stable equilibrium position; we say that the object is **stable.** This is the

FIGURE 8.10 A car—or any object—will fall over when tilted too far.

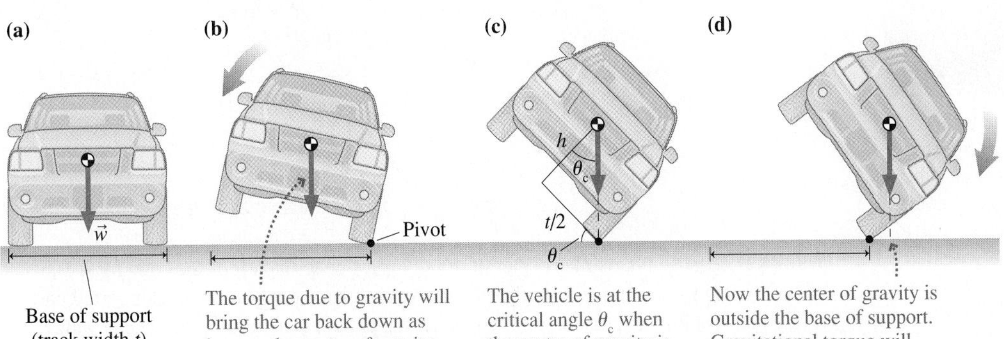

(a)
(b)
(c)
(d)
$\vec{w}$
Pivot
h
θ_c
t/2
θ_c
Base of support (track width t)
The torque due to gravity will bring the car back down as long as the center of gravity is over the base of support.
The vehicle is at the critical angle θ_c when the center of gravity is directly over the pivot.
Now the center of gravity is outside the base of support. Gravitational torque will cause it to roll over.

FIGURE 8.11 Compared to a passenger car, an SUV has a high center of gravity relative to its width.

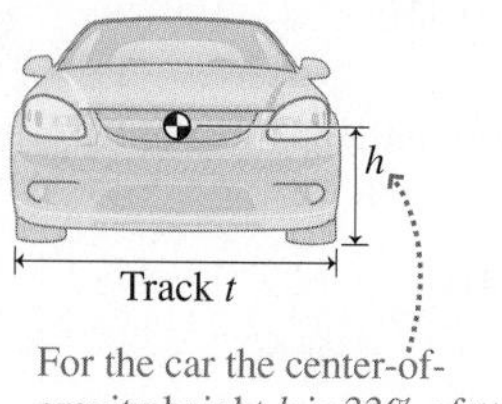
h
Track t

For the car the center-of-gravity height h is 33% of t.

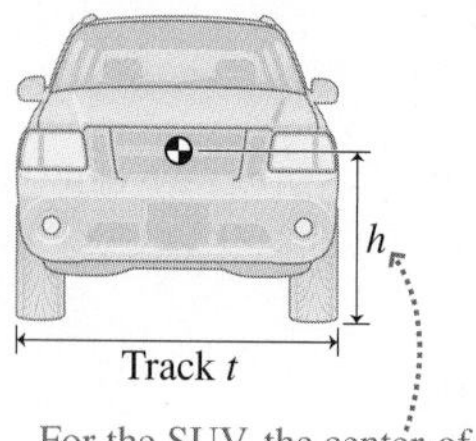
h
Track t

For the SUV, the center-of-gravity height h is 47% of t.

situation in Figure 8.10b. But if the center of gravity gets outside the base of support, as in Figure 8.10d, the gravitational torque causes a rotation in the opposite direction. Now the car rolls over; it is **unstable.**

A *critical angle* θ_c is reached when the center of gravity is directly over the pivot point. This is the point of balance, with no net torque. For vehicles, the distance between the tires—the base of support—is called the *track width t*. If the height of the center of gravity is h, you can see from Figure 8.10c that the critical angle is

$$\theta_c = \tan^{-1}\left(\frac{t/2}{h}\right) = \tan^{-1}\left(\frac{t}{2h}\right) \tag{8.2}$$

If an accident (or taking a corner too fast) causes a vehicle to pivot up onto two wheels, it will roll back to an upright position as long as $\theta < \theta_c$ but it will roll over if $\theta > \theta_c$. Notice that it's the height-to-width ratio that's important, not the absolute height of the center of gravity.

FIGURE 8.11 compares a passenger car and a sport utility vehicle (SUV). For the passenger car, with $h \approx 0.33t$, the critical angle is $\theta_c \approx 57°$. But for the SUV, with its higher center of gravity ($h \approx 0.47t$), the critical angle is only $\theta_c \approx 47°$. Loading an SUV with cargo further raises the center of gravity, especially if the roof rack is used, thus reducing θ_c even more. Various automobile safety groups have determined that a vehicle with $\theta_c > 50°$ is unlikely to roll over in an accident. A rollover becomes increasingly likely when θ_c is less than 50°. The same argument that leads to Equation 8.2 for tilted vehicles can be made for any object, leading to the general rule that **a wider base of support and/or a lower center of gravity improve stability.**

CONCEPTUAL EXAMPLE 8.5 How far to walk the plank?

A cat walks along a plank that extends out from a table. If the cat walks too far out on the plank, the plank will begin to tilt. What determines when this happens?

REASON An object is stable if its center of gravity lies over its base of support, and unstable otherwise. Let's take the cat and the plank to be one combined object whose center of gravity lies along a line between the cat's center of gravity and that of the plank.

In FIGURE 8.12a, when the cat is near the left end of the plank, the combined center of gravity is over the base of support and the plank is stable. As the cat moves to the right, he reaches a point where the combined center of gravity is directly over the edge of the table, as shown in FIGURE 8.12b. If the cat takes one more step, the cat and plank will become unstable and the plank will begin to tilt.

FIGURE 8.12 Changing stability as a cat walks on a plank.

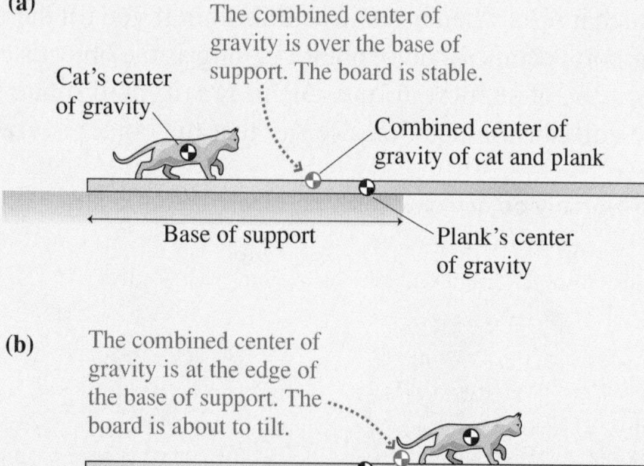
(a)
The combined center of gravity is over the base of support. The board is stable.
Cat's center of gravity
Combined center of gravity of cat and plank
Base of support
Plank's center of gravity
(b)
The combined center of gravity is at the edge of the base of support. The board is about to tilt.

ASSESS Because the plank's center of gravity must be to the left of the edge for it to be stable by itself, the cat can actually walk a short distance out onto the unsupported part of the plank before it starts to tilt. The heavier the plank is, the farther the cat can walk.

TRY IT YOURSELF

diet Sprite

Balancing a soda can Try to balance a soda can—full or empty—on the narrow bevel at the bottom. It can't be done because, either full or empty, the center of gravity is near the center of the can. If the can is tilted enough to sit on the bevel, the center of gravity lies far outside this small base of support. But if you put about 2 ounces (60 ml) of water in an empty can, the center of gravity will be right over the bevel and the can will balance.

Stability and Balance of the Human Body

The human body is remarkable for its ability to constantly adjust its stance to remain stable on just two points of support. In walking, running, or even in the simple act of rising from a chair, the position of the body's center of gravity is constantly changing. To maintain stability, we unconsciously adjust the positions of our arms and legs to keep our center of gravity over our base of support.

A simple example of how the body naturally realigns its center of gravity is found in the act of standing up on tiptoes. **FIGURE 8.13a** shows the body in its normal standing position. Notice that the center of gravity is well centered over the base of support (the feet), ensuring stability. If the subject were now to stand on tiptoes *without* otherwise adjusting the body position, her center of gravity would fall behind the base of support, which is now the balls of the feet, and she would fall backward. To prevent this, as shown in **FIGURE 8.13b**, the body naturally leans forward, regaining stability by moving the center of gravity over the balls of the feet.

FIGURE 8.13 Standing on tiptoes.

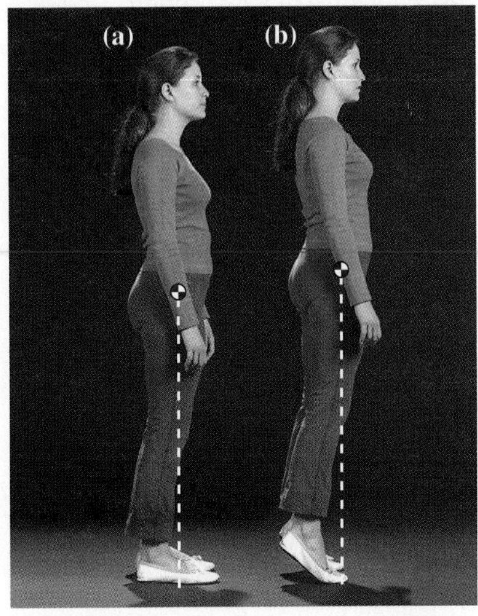

STOP TO THINK 8.3 Rank in order, from least stable to most stable, the three objects shown in the figure. The positions of their centers of gravity are marked. (For the centers of gravity to be positioned like this, the objects must have a nonuniform composition.)

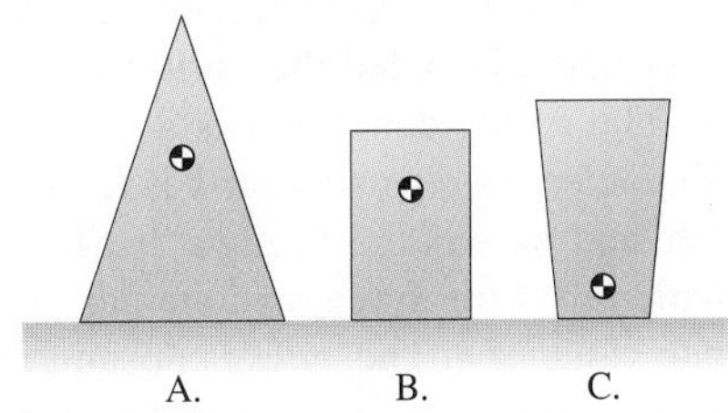

TRY IT YOURSELF

Impossible balance Stand facing a wall with your toes touching the base of the wall. Now rise onto your tiptoes. You will not be able to do so without falling backward. As we see from Figure 8.13b, your body has to lean forward to stand on tiptoes. With the wall in your way, you cannot lean enough to maintain your balance, and you will begin to topple backward.

8.3 Springs and Hooke's Law

We have assumed that objects in equilibrium maintain their shape as forces and torques are applied to them. In reality this is an oversimplification. Every solid object stretches, compresses, or deforms when a force acts upon it. This change is easy to see when you press on a green twig on a tree, but even the largest branch on the tree will bend slightly under your weight.

If you stretch a rubber band, there is a force that tries to pull the rubber band back to its equilibrium, or unstretched, length. A force that restores a system to an equilibrium position is called a **restoring force.** Systems that exhibit such restoring forces are called **elastic.** The most basic examples of **elasticity** are things like springs and rubber bands. If you stretch a spring, a tension-like force pulls back. Similarly, a compressed spring tries to re-expand to its equilibrium length. Elasticity and restoring forces are properties of much stiffer systems as well. The steel beams of a bridge bend slightly as you drive your car over it, but they are restored to equilibrium after your car passes by. Your leg bones flex a bit during each step you take. Nearly everything that stretches, compresses, bends, or twists exhibits a restoring force and can be called elastic.

The behavior of a simple spring illustrates the basic ideas of elasticity. When no forces act on a spring to compress or extend it, it will relax to its **equilibrium length.** If we now stretch the spring by a displacement Δx, how hard does it pull back? **FIGURE 8.14** shows what happens: The farther we stretch the spring, the harder the restoring force of the spring pulls back.

FIGURE 8.14 The spring force depends on how far the spring is stretched.

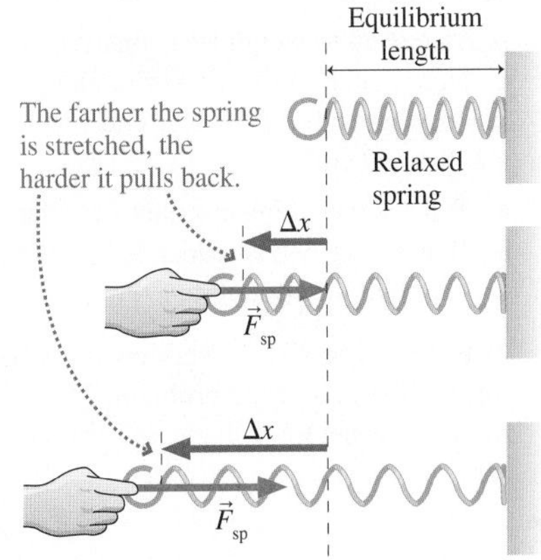

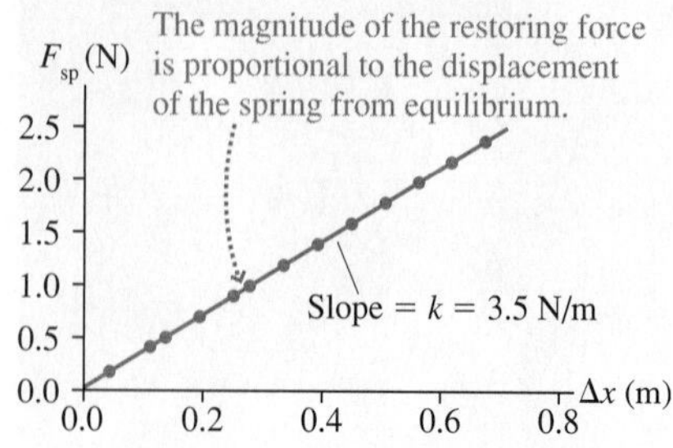

FIGURE 8.15 Measured data for the restoring force of a real spring.

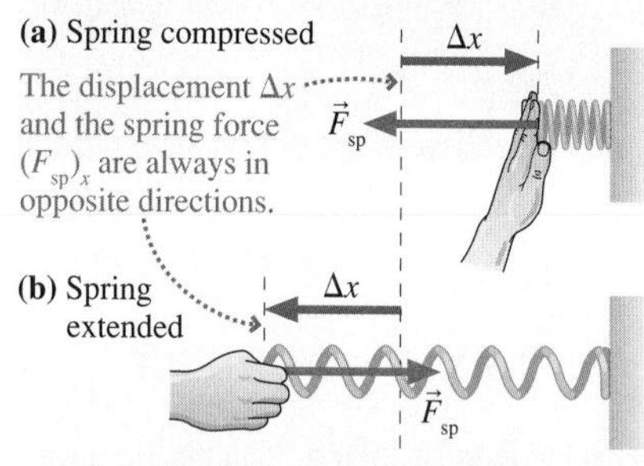

FIGURE 8.16 The spring force is always directed opposite the displacement.

Elasticity in action A golf ball compresses quite a bit when struck. The restoring force that pushes the ball back into its original shape helps launch the ball off the face of the club, making for a longer drive.

In FIGURE 8.15, data for the magnitude of the restoring force of a real spring show that **the force of the spring is *proportional* to the displacement of the end of the spring.** That is, compressing or stretching the spring twice as far results in a restoring force that is twice as large. This is a *linear relationship,* and the slope k of the line is the proportionality constant:

$$F_{sp} = k\,\Delta x \tag{8.3}$$

A second important fact about spring forces is illustrated in FIGURE 8.16. If the spring is compressed, as in Figure 8.16a, Δx is positive and, because $\vec{F}_{sp}$ points to the left, its component $(F_{sp})_x$ is negative. If, however, the spring is stretched, as in Figure 8.16b, Δx is negative and, because $\vec{F}_{sp}$ points to the right, its component $(F_{sp})_x$ is positive. In general, **the spring force always points in the opposite direction to the displacement from equilibrium.** We can express this fact, along with what we've learned about the magnitude of the spring force, by rewriting Equation 8.3 in terms of the *component* of the spring force:

$$(F_{sp})_x = -k\,\Delta x \tag{8.4}$$

Hooke's law for the force due to a spring

$(F_{sp})_x$ vs. Δx — p.37 PROPORTIONAL

The minus sign in Equation 8.4 reflects the fact that $(F_{sp})_x$ and Δx are always of opposite sign. (For motion in the vertical (y) direction, Hooke's law is $(F_{sp})_y = -k\,\Delta y$.)

The proportionality constant k is called the **spring constant.** The units of the spring constant are N/m. The spring constant k is a property that characterizes a spring, just as mass m characterizes a particle. If k is large, it takes a large pull to cause a significant stretch, and we call the spring a "stiff" spring. If k is small, we can stretch the spring with very little force, and we call it a "soft" spring. Every spring has its own, unique value of k. The spring constant for the spring in Figure 8.15 can be determined from the slope of the straight line to be $k = 3.5$ N/m.

Equation 8.4 for the restoring force of a spring was first suggested by Robert Hooke, a contemporary (and sometimes bitter rival) of Newton. Hooke's law is not a true "law of nature," in the sense that Newton's laws are, but is actually just a *model* of a restoring force. It works extremely well for some springs, as in Figure 8.15, but less well for others. Hooke's law will fail for any spring if it is compressed or stretched too far.

NOTE ► Just as we used massless strings, we will adopt the idealization of a *massless spring.* Though not a perfect description, it is a good approximation if the mass attached to a spring is much greater than the mass of the spring itself. ◄

EXAMPLE 8.6 Weighing a fish

A scale used to weigh fish consists of a spring connected to the ceiling. The spring's equilibrium length is 30 cm. When a 4.0 kg fish is suspended from the end of the spring, it stretches to a length of 42 cm.

a. What is the spring constant k for this spring?
b. If an 8.0 kg fish is suspended from the spring, what will be the length of the spring?

PREPARE The visual overview in FIGURE 8.17 shows the details for the first part of the problem. The fish hangs in static equilibrium, so the net force in the y-direction and the net torque must be zero.

FIGURE 8.17 Visual overview of a mass suspended from a spring.

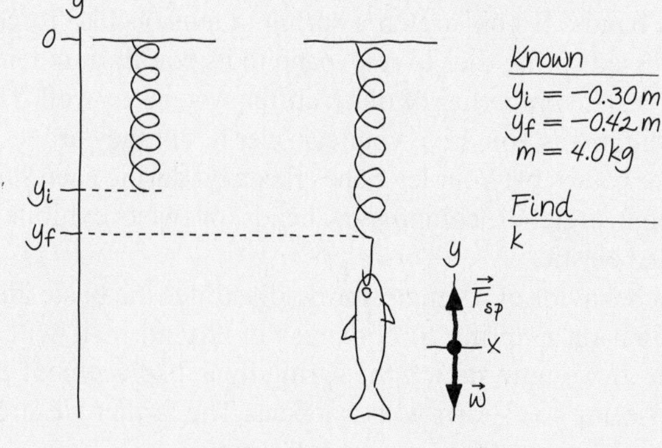

SOLVE a. Because the fish is in static equilibrium, we have

$$\sum F_y = (F_{sp})_y + w_y = -k\,\Delta y - mg = 0$$

so that $k = -mg/\Delta y$. (The net torque is zero because the fish's center of gravity comes to rest directly under the pivot point of the hook.) From Figure 8.17, the displacement of the spring from equilibrium is $\Delta y = y_f - y_i = (-0.42\text{ m}) - (-0.30\text{ m}) = -0.12\text{ m}$. This displacement is *negative* because the fish moves in the $-y$-direction. We can now solve for the spring constant:

$$k = -\frac{mg}{\Delta y} = -\frac{(4.0\text{ kg})(9.8\text{ m/s}^2)}{-0.12\text{ m}} = 330\text{ N/m}$$

b. The restoring force is proportional to the displacement of the spring from its equilibrium length. If we double the mass (and thus the weight) of the fish, the displacement of the end of the spring will double as well, to $\Delta y = -0.24\text{ m}$. Thus the spring will be 0.24 m longer, so its new length is $0.30\text{ m} + 0.24\text{ m} = 0.54\text{ m}$.

ASSESS A spring constant of 330 N/m means that when the spring is stretched by 1.0 m it will exert a force of 330 N (about 75 lb). This seems reasonable for a spring used to weigh objects of 10 or 20 lb.

EXAMPLE 8.7 When does the block slip?

FIGURE 8.18 shows a spring attached to a 2.0 kg block. The other end of the spring is pulled by a motorized toy train that moves forward at 5.0 cm/s. The spring constant is 50 N/m, and the coefficient of static friction between the block and the surface is 0.60. The spring is at its equilibrium length at $t = 0$ s when the train starts to move. When does the block slip?

FIGURE 8.18 A toy train stretches the spring until the block slips.

PREPARE We model the block as a particle and the spring as a massless spring. **FIGURE 8.19** is a free-body diagram for the block. We convert the speed of the train into m/s: $v = 0.050$ m/s.

FIGURE 8.19 Free-body diagram for the block.

When the spring force exceeds the maximum force of static friction, the block will slip.

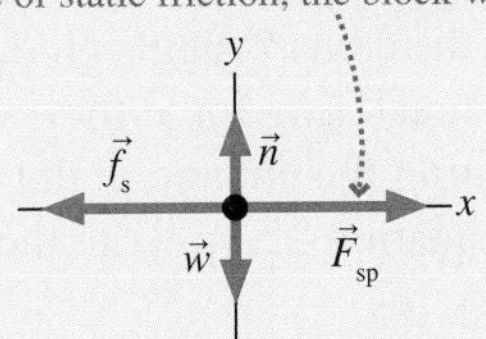

SOLVE Recall that the tension in a massless string pulls equally at *both* ends of the string. The same is true for the spring force: It pulls (or pushes) equally at *both* ends. Imagine holding a rubber band with your left hand and stretching it with your right hand. Your left hand feels the pulling force, even though it was the right end of the rubber band that moved.

This is the key to solving the problem. As the right end of the spring moves, stretching the spring, the spring pulls backward on the train *and* forward on the block with equal strength. The train is moving to the right, and so the spring force pulls to the left on the train—as we would expect. But the block is at the other end of the spring; the spring force pulls to the right on the block, as shown in Figure 8.19. As the spring stretches, the static friction force on the block increases in magnitude to keep the block at rest. The block is in static equilibrium, so

$$\sum F_x = (F_{sp})_x + (f_s)_x = F_{sp} - f_s = 0$$

where F_{sp} is the magnitude of the spring force. This magnitude is $F_{sp} = k\,\Delta x$, where $\Delta x = vt$ is the distance the train has moved. Thus

$$f_s = F_{sp} = k\,\Delta x$$

The block slips when the static friction force reaches its maximum value $f_{s\,max} = \mu_s n = \mu_s mg$. This occurs when the train has moved a distance

$$\Delta x = \frac{f_{s\,max}}{k} = \frac{\mu_s mg}{k} = \frac{(0.60)(2.0\text{ kg})(9.8\text{ m/s}^2)}{50\text{ N/m}} = 0.235\text{ m}$$

The time at which the block slips is

$$t = \frac{\Delta x}{v} = \frac{0.235\text{ m}}{0.050\text{ m/s}} = 4.7\text{ s}$$

ASSESS The result of about 5 s seems reasonable for a slowly moving toy train to stretch the spring enough for the block to slip.

STOP TO THINK 8.4 A 1.0 kg weight is suspended from a spring, stretching it by 5.0 cm. How much does the spring stretch if the 1.0 kg weight is replaced by a 3.0 kg weight?

A. 5.0 cm B. 10.0 cm C. 15.0 cm D. 20.0 cm

FIGURE 8.20 Stretching a steel rod.

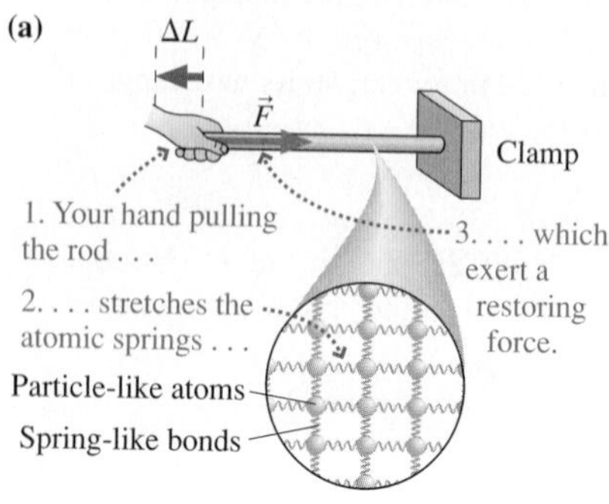

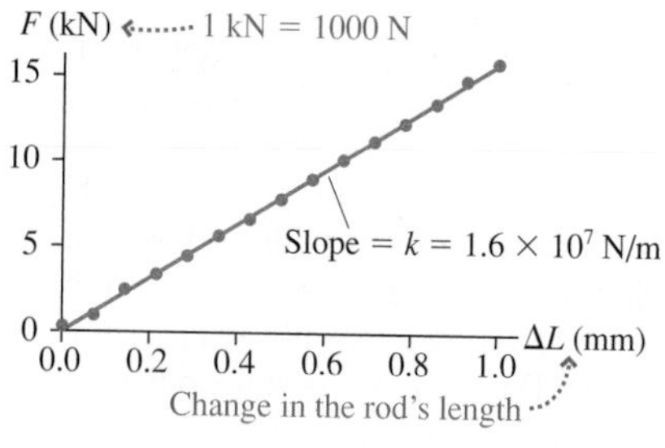

FIGURE 8.21 A rod stretched by length ΔL.

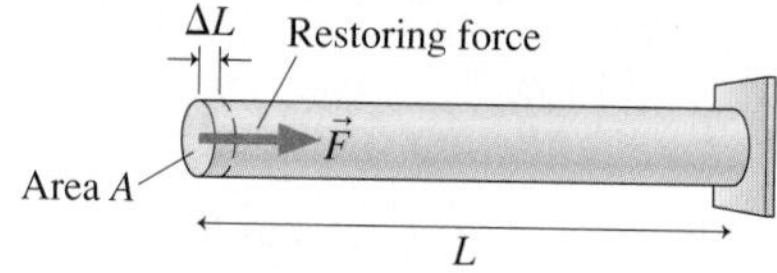

TABLE 8.1 Young's modulus for rigid materials

Material	Young's modulus (10^{10} N/m^2)
Cast iron	20
Steel	20
Silicon	13
Copper	11
Aluminum	7
Glass	7
Concrete	3
Wood (Douglas Fir)	1

8.4 Stretching and Compressing Materials

In Chapter 4 we noted that we could model most solid materials as being made of particle-like atoms connected by spring-like bonds. We can model a steel rod this way, as illustrated in FIGURE 8.20a. The spring-like bonds between the atoms in steel are quite stiff, but they can be stretched or compressed, meaning that even a steel rod is elastic. If you pull on the end of a steel rod, as in Figure 8.20a, you will slightly stretch the bonds between the particles that make it up, and the rod itself will stretch. The stretched bonds pull back on your hand with a restoring force that causes the rod to return to its original length when released. In this sense, the entire rod acts like a very stiff spring. As is the case for a spring, a restoring force is also produced by compressing the rod.

In FIGURE 8.20b, real data for a 1.0-m-long, 1.0-cm-diameter steel rod show that, just as for a spring, the restoring force is proportional to the change in length. However, the *scale* of the stretch of the rod and the restoring force is much different from that for a spring. It would take a force of 16,000 N to stretch the rod by only 1 mm, corresponding to a spring constant of 1.6×10^7 N/m! Steel is elastic, but under normal forces, it experiences only very small changes in dimension. Materials of this sort are called **rigid.**

The behavior of other materials, such as the rubber in a rubber band, can be quite different. A rubber band can be stretched quite far—several times its equilibrium length—with a very small force, and then snaps back to its original shape when released. Materials that show large deformations with small forces are called **pliant.**

A rod's spring constant depends on several factors, as shown in FIGURE 8.21. First, we expect that a thick rod, with a large cross-section area A, will be more difficult to stretch than a thinner rod. Second, a rod with a long length L will be easier to stretch by a given amount than a short rod (think of trying to stretch a rope by 1 cm—this would be easy to do for a 10-m-long rope, but it would be pretty hard for a piece of rope only 10 cm long). Finally, the stiffness of the rod will depend on the material that it's made of. Experiments bear out these observations, and it is found that the spring constant of the rod can be written as

$$k = \frac{YA}{L} \tag{8.5}$$

where the constant Y is called **Young's modulus.** Young's modulus is a property of the *material* from which the rod is made—it does not depend on the object's shape or size. All rods made from steel have the same Young's modulus, regardless of their length or area, while aluminum rods have a different Young's modulus.

From Equation 8.3, the magnitude of the restoring force for a spring is related to the change in its length as $F_{sp} = k\,\Delta x$. Writing the change in the length of a rod as ΔL, as shown in Figure 8.21, we can use Equation 8.5 to write the restoring force F of a rod as

$$F = \frac{YA}{L}\Delta L \tag{8.6}$$

Equation 8.6 applies both to elongation (stretching) and to compression.

It's useful to rearrange Equation 8.6 in terms of two new ratios, the *stress* and the *strain*:

$$\frac{F}{A} = Y\left(\frac{\Delta L}{L}\right) \tag{8.7}$$

The ratio of force to cross-section area is called **stress**. The ratio of the change in length to the original length is called **strain**.

The unit of stress is N/m^2. If the stress is due to stretching, we call it a **tensile stress.** The strain is the fractional change in the rod's length. If the rod's length changes by 1%, the strain is 0.01. Because strain is dimensionless, Young's modulus Y has the same units as stress. Table 8.1 gives values of Young's modulus for several rigid materials. Large values of Y characterize materials that are stiff. "Softer" materials have smaller values of Y. Because the values of Young's modulus for materials such as steel or aluminum are very large, it takes a significant stress to produce even a small strain.

EXAMPLE 8.8 **Finding the stretch of a wire**

A *Foucault pendulum* in a physics department (used to prove that the earth rotates) consists of a 120 kg steel ball that swings at the end of a 6.0-m-long steel cable. The cable has a diameter of 2.5 mm. When the ball was first hung from the cable, by how much did the cable stretch?

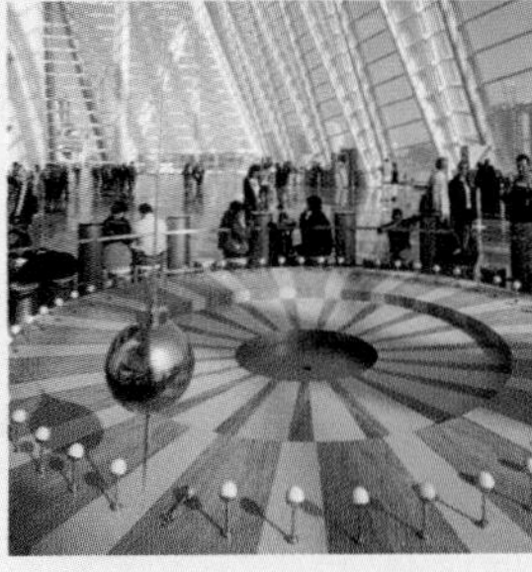

PREPARE The amount by which the cable stretches depends on the elasticity of the steel cable. Young's modulus for steel is given in Table 8.1 as $Y = 20 \times 10^{10}\ \text{N/m}^2$.

SOLVE Equation 8.7 relates the stretch of the cable ΔL to the restoring force F and to the properties of the cable. Rearranging terms, we find that the cable stretches by

$$\Delta L = \frac{LF}{AY}$$

The cross-section area of the cable is

$$A = \pi r^2 = \pi(0.00125\ \text{m})^2 = 4.91 \times 10^{-6}\ \text{m}^2$$

The restoring force of the cable is equal to the ball's weight:

$$F = w = mg = (120\ \text{kg})(9.8\ \text{m/s}^2) = 1180\ \text{N}$$

The change in length is thus

$$\Delta L = \frac{(6.0\ \text{m})(1180\ \text{N})}{(4.91 \times 10^{-6}\ \text{m}^2)(20 \times 10^{10}\ \text{N/m}^2)} = 0.0072\ \text{m} = 7.2\ \text{mm}$$

ASSESS If you've ever strung a guitar with steel strings, you know that the strings stretch several millimeters with the force you can apply by turning the tuning pegs. So a stretch of 7 mm under a 120 kg load seems reasonable.

Beyond the Elastic Limit

In the previous section, we found that if we stretch a rod by a small amount ΔL, it will pull back with a restoring force F, according to Equation 8.6. But if we continue to stretch the rod, this simple linear relationship between ΔL and F will eventually break down. **FIGURE 8.22** is a graph of the rod's restoring force from the start of the stretch until the rod finally breaks.

As you can see, the graph has a *linear region,* the region where F and ΔL are proportional to each other, obeying Hooke's law: $F = k\,\Delta L$. **As long as the stretch stays within the linear region, a solid rod acts like a spring and obeys Hooke's law.**

How far can you stretch the rod before damaging it? As long as the stretch is less than the **elastic limit,** the rod will return to its initial length L when the force is removed. The elastic limit is the end of the **elastic region.** Stretching the rod beyond the elastic limit will permanently deform it, and the rod won't return to its original length. Finally, at a certain point the rod will reach a breaking point, where it will snap in two. The maximum stress that a material can be subjected to before failing is called the **tensile strength.** Table 8.2 lists values of tensile strength for rigid materials. When we speak of the *strength* of a material, we are referring to its tensile strength.

FIGURE 8.22 Stretch data for a steel rod.

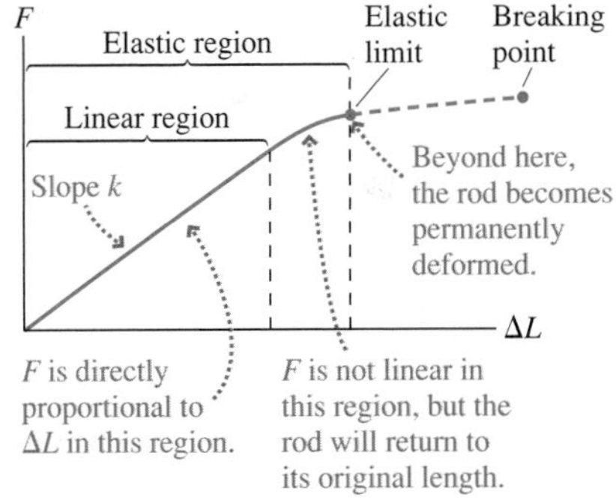

EXAMPLE 8.9 **Breaking a pendulum cable**

After a late night of studying physics, several 80 kg students decide it would be fun to swing on the Foucault pendulum of Example 8.8. What's the maximum number of students that the pendulum cable could support?

PREPARE The tensile strength, given for steel in Table 8.2 as $1000 \times 10^6\ \text{N/m}^2$, or $1.0 \times 10^9\ \text{N/m}^2$, is the largest stress the cable can sustain. Because the stress in the cable is F/A, we can find the maximum force F_{max} that can be applied to the cable before it fails.

Continued

TABLE 8.2 Tensile strengths of rigid materials

Material	Tensile strength (N/m²)
Polypropylene	20×10^6
Glass	60×10^6
Cast iron	150×10^6
Aluminum	400×10^6
Steel	1000×10^6

Spider silk BIO The glands on the abdomen of a spider produce different kinds of silk. The silk that is used in webs can be quite stretchy; that used to subdue prey is generally not. An individual strand of silk may be a mix of fibers of different types, allowing spiders great flexibility in their material.

SOLVE We have

$$F_{\text{max}} = A(1.0 \times 10^9\ \text{N/m}^2)$$

From Example 8.8, the diameter of the cable is 2.5 mm, so its radius is 0.00125 m. Thus

$$F_{\text{max}} = \left(\pi(0.00125\ \text{m})^2\right)(1.0 \times 10^9\ \text{N/m}^2) = 4.9 \times 10^3\ \text{N}$$

This force is the weight of the heaviest mass the cable can support: $w = m_{\text{max}}g$. The maximum mass that can be supported is

$$m_{\text{max}} = \frac{F_{\text{max}}}{g} = 500\ \text{kg}$$

The ball has a mass of 120 kg, leaving 380 kg for the students. Four students have a mass of 320 kg, which is less than this value. But five students, totaling 400 kg, would cause the cable to break.

ASSESS Steel has a very large tensile strength. This very narrow wire can still support $4900\ \text{N} \approx 1100\ \text{lb}$.

Biological Materials

FIGURE 8.23 Stress-versus-strain graphs for steel and spider silk.

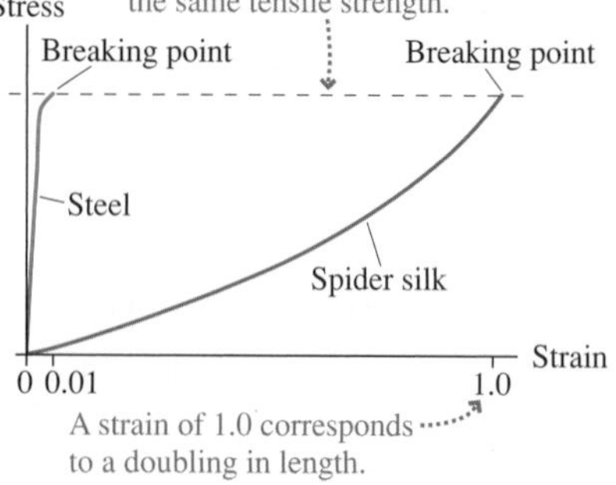
Both materials fail at approximately the same stress, so both have about the same tensile strength.
Stress
Breaking point
Breaking point
Steel
Spider silk
Strain
0 0.01
1.0
A strain of 1.0 corresponds to a doubling in length.

Suppose we take equal lengths of spider silk and steel wire, stretch each, and measure the restoring force of each until it breaks. The graph of stress versus strain might appear as in FIGURE 8.23.

The spider silk is certainly less stiff: For a given stress, the silk will stretch about 100 times farther than steel. Interestingly, though, spider silk and steel eventually fail at approximately the same stress. In this sense, spider silk is "as strong as steel." Many pliant biological materials share this combination of low stiffness and large tensile strength. These materials can undergo significant deformations without failing. Tendons, the walls of arteries, and the web of a spider are all quite strong but nonetheless capable of significant stretch.

Bone is an interesting example of a rigid biological material. Most bones in your body are made of two different kinds of bony material: dense and rigid cortical (or compact) bone on the outside, and porous, flexible cancellous (or spongy) bone on the inside. FIGURE 8.24 shows a cross section of a typical bone. Cortical and cancellous bones have very different values of Young's modulus. Young's modulus for cortical bone approaches that of concrete, so it is very rigid with little ability to stretch or compress. In contrast, cancellous bone has a much lower Young's modulus. Consequently, the elastic properties of bones can be well modeled as those of a hollow cylinder.

FIGURE 8.24 Cross section of a long bone.

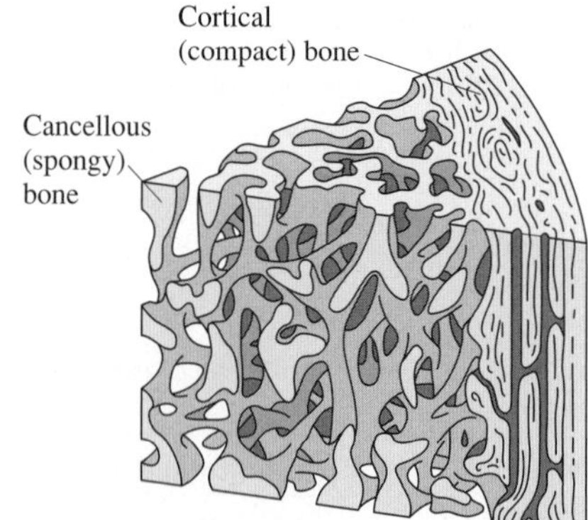
Cortical (compact) bone
Cancellous (spongy) bone

The structure of bones in birds actually approximates a hollow cylinder quite well. FIGURE 8.25 shows that a typical bone is a thin-walled tube of cortical bone with a tenuous structure of cancellous bone inside. Most of a cylinder's rigidity comes from the material near its surface. A hollow cylinder retains most of the rigidity of a solid one, but it is much lighter. Bird bones carry this idea to its extreme.

FIGURE 8.25 Section of a bone from a bird.

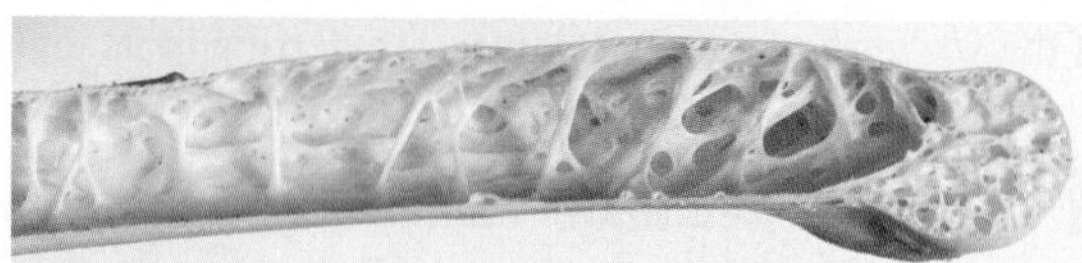

Table 8.3 gives values of Young's modulus for biological materials. Note the large difference between pliant and rigid materials. Table 8.4 shows the tensile strengths for biological materials. Interestingly, spider silk, a pliant material, has a greater tensile strength than bone!

The values in Table 8.4 are for static forces—forces applied for a long time in a testing machine. Bone can withstand significantly greater stresses if the forces are applied for only a very short period of time.

EXAMPLE 8.10 Finding the compression of a bone

The femur, the long bone in the thigh, can be modeled as a tube of cortical bone for most of its length. A 70 kg person has a femur with a cross-section area (of the cortical bone) of $4.8 \times 10^{-4}\ \text{m}^2$, a typical value.

a. If this person supports his entire weight on one leg, what fraction of the tensile strength of the bone does this stress represent?
b. By what fraction of its length does the femur shorten?

PREPARE The stress on the femur is F/A. Here F, the force compressing the femur, is the person's weight, so $F = mg$. The fractional change $\Delta L/L$ in the femur is the strain, which we can find using Equation 8.7, taking the value of Young's modulus for cortical bone from Table 8.3.

SOLVE

a. The person's weight is $mg = (70\ \text{kg})(9.8\ \text{m/s}^2) = 690\ \text{N}$. The resulting stress on the femur is

$$\frac{F}{A} = \frac{690\ \text{N}}{4.8 \times 10^{-4}\ \text{m}^2} = 1.4 \times 10^6\ \text{N/m}^2$$

A stress of $1.4 \times 10^6\ \text{N/m}^2$ is 1.4% of the tensile strength of cortical bone given in Table 8.4.

b. We can compute the strain as

$$\frac{\Delta L}{L} = \left(\frac{1}{Y}\right)\frac{F}{A} = \left(\frac{1}{1.6 \times 10^{10}\ \text{N/m}^2}\right)(1.4 \times 10^6\ \text{N/m}^2) = 8.8 \times 10^{-5} \approx 0.0001$$

The femur compression is $\Delta L \approx 0.0001L$, or $\approx 0.01\%$ of its length.

ASSESS It makes sense that, under ordinary standing conditions, the stress on the femur is only a percent or so of the maximum value it can sustain.

TABLE 8.3 Young's modulus for biological materials

Material	Young's modulus ($10^{10}\ \text{N/m}^2$)
Tooth enamel	6
Cortical bone	1.6
Cancellous bone	0.02–0.3
Spider silk	0.2
Tendon	0.15
Cartilage	0.0001
Blood vessel (aorta)	0.00005

TABLE 8.4 Tensile strength of biological materials

Material	Tensile strength (N/m^2)
Cancellous bone	5×10^6
Cortical bone	100×10^6
Tendon	100×10^6
Spider silk	1000×10^6

The dancer in the chapter-opening photo stands *en pointe,* balanced delicately on the tip of her shoe with her entire weight supported on a very small area. The stress on the bones in her toes is very large, but it is still much less than the tensile strength of bone.

STOP TO THINK 8.5 A 10 kg mass is hung from a 1-m-long cable, causing the cable to stretch by 2 mm. Suppose a 10 kg mass is hung from a 2 m length of the same cable. By how much does the cable stretch?

A. 0.5 mm B. 1 mm C. 2 mm D. 3 mm E. 4 mm

INTEGRATED EXAMPLE 8.11 Holding a barrel on a hill

FIGURE 8.26 shows a 60-cm-diameter barrel of sand, with a mass of 600 kg, being held in place on a hill by a polypropylene rope wrapped around the barrel. The coefficient of static friction between the barrel and the hill is 0.25.

a. What is the tension in the rope?
b. What is the steepest hill that the barrel could rest on without slipping?
c. What is the smallest-diameter rope that can be used without the rope breaking?

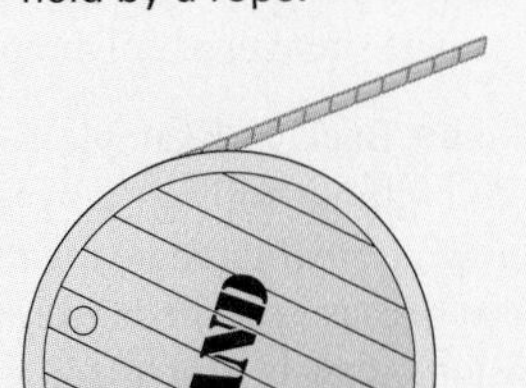

FIGURE 8.26 A barrel being held by a rope.

PREPARE We'll follow Problem-Solving Strategy 8.1: For an object in static equilibrium, the net torque is zero, $\Sigma\tau = 0$, and the net force is zero, $\Sigma F_x = 0$ and $\Sigma F_y = 0$. To find the net torque on the barrel, we'll redraw it in **FIGURE 8.27** on the next page with all the forces shown at the points at which they act. To find the components of the net force, we'll draw the free-body diagram of Figure 8.27. As usual, we tilt our x-axis so that it's parallel to the surface of the hill. Recall from Figure 5.14 that the angle between the weight vector and the $-y$-axis is the same as the angle of the slope.

The direction of the static friction force is chosen to keep the bottom of the barrel from slipping down the hill. Imagine what

Continued

FIGURE 8.27 Visual overview of the barrel.

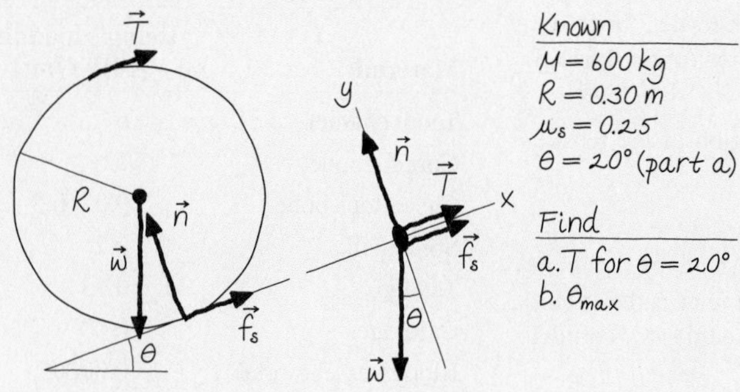

would happen if friction suddenly vanished while the rope was holding the top of the barrel in place.

We need to pick a pivot about which the torque will be calculated. If we choose the point of contact of the barrel with the hill as our pivot, then the unknown forces $\vec{n}$ and $\vec{f}_s$, which act at that point, make no contribution to the torque. Only the weight, whose magnitude we know, and the tension, which is what we want to find, contribute to the torque.

For part c, the rope will fail if the stress exceeds the tensile strength of polypropylene, given in Table 8.2 as 2.0×10^7 N/m^2.

SOLVE

a. FIGURE 8.28 shows how to calculate the torque. Forces $\vec{n}$ and $\vec{f}_s$ are not shown because, as just discussed, they act at the pivot and do not contribute to the torque. In Figure 8.28a, the tension $\vec{T}$ acts perpendicular to the radial line, at a distance $2R$ from the pivot, so the torque due to the tension is $\tau_T = -2RT$. It's negative because the tension tries to rotate the barrel clockwise.

FIGURE 8.28 Calculating the torque.

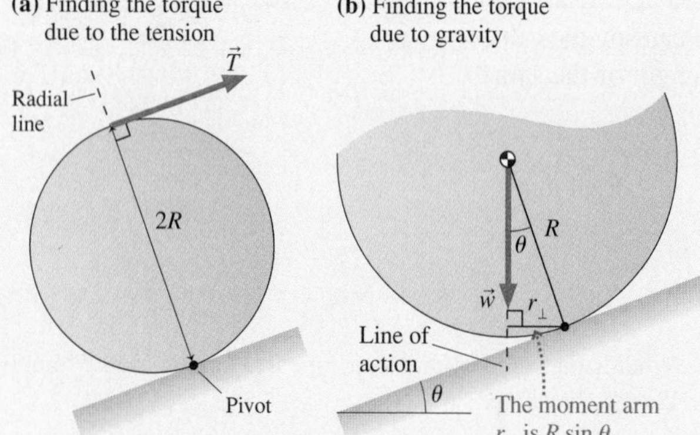

We'll use Equation 7.4, $\tau = r_\perp F$, to find the torque due to the weight, which acts at the center of gravity. From Figure 8.28b we see that the moment arm $r_\perp$, the perpendicular distance from the line of action to the pivot, is $r_\perp = R \sin\theta$. The magnitude of the weight force is Mg, so the torque due to the weight is $\tau_w = MgR \sin\theta$.

We can now write the condition that the net torque is zero as

$$\tau_{net} = \tau_w + \tau_T = MgR\sin\theta - 2RT = 0$$

Solving this equation for the tension gives

$$T = \tfrac{1}{2}Mg\sin\theta = \frac{(600\text{ kg})(9.8\text{ m/s}^2)}{2}\sin 20° = 1000\text{ N}$$

Note that R cancels, so the tension does not depend on the radius of the barrel.

b. Part a was solved using only the net torque equation. For this part, we'll need the two force equations as well. From the free-body diagram of Figure 8.27, we can write

$$\sum F_x = T + f_s - Mg\sin\theta = 0$$

$$\sum F_y = n - Mg\cos\theta = 0$$

We can solve the first of these equations for the friction force:

$$f_s = Mg\sin\theta - T = Mg\sin\theta - \tfrac{1}{2}Mg\sin\theta = \tfrac{1}{2}Mg\sin\theta$$

Here we used the result for the tension T found in part a.

The static friction must have this value to keep the barrel from slipping. But static friction can't exceed the maximum possible value, $f_{s\,max} = \mu_s n$. From the y force equation, the normal force is $n = Mg\cos\theta$, so $f_{s\,max} = \mu_s Mg\cos\theta$. The barrel will slip when the friction force equals its maximum possible value, or when

$$f_s = \tfrac{1}{2}Mg\sin\theta = f_{s\,max} = \mu_s Mg\cos\theta$$

The factor Mg cancels from both sides of this equation, giving $\tfrac{1}{2}\sin\theta = \mu_s\cos\theta$, which, after dividing both sides by $\cos\theta$, can be written as $\tfrac{1}{2}\tan\theta = \mu_s$. Thus the angle at which the barrel will slip is given by $\tan\theta = 2\mu_s$, or

$$\theta = \tan^{-1}(2\mu_s) = \tan^{-1}(2\cdot 0.25) = 27°$$

c. The maximum possible stress in the rope, when it is at its breaking point, is

$$\frac{F}{A} = \frac{T}{\pi r^2} = \frac{1000\text{ N}}{\pi r^2} = 2.0 \times 10^7\text{ N/m}^2$$

The radius of the rope at this level of stress is

$$r = \sqrt{\frac{1000\text{ N}}{\pi(2.0\times 10^7\text{ N/m}^2)}} = 4.0 \times 10^{-3}\text{ m}$$

The minimum rope diameter is twice this radius, or 8.0 mm. If the rope were any smaller, the stress would exceed the tensile strength of polypropylene and the rope would break.

ASSESS Back in Example 5.14, we found that an object will *slide* (without rolling) down a slope when the angle exceeds $\tan^{-1}\mu_s$, a *smaller* angle than for our rolling object. This makes sense because for the barrel there is an extra uphill force—the tension—that is absent for a sliding object. This uphill force allows the slope to be steeper before a round object begins to slip.

SUMMARY

The goals of Chapter 8 have been to learn about the static equilibrium of extended objects and to understand the basic properties of springs and elastic materials.

GENERAL PRINCIPLES

Static Equilibrium

An object in **static equilibrium** must have no net force on it and no net torque. Mathematically, we express this as

$$\sum F_x = 0$$
$$\sum F_y = 0$$
$$\sum \tau = 0$$

Since the net torque is zero about *any* point, the pivot point for calculating the torque can be chosen at any convenient location.

Springs and Hooke's Law

When a spring is stretched or compressed, it exerts a force proportional to the change Δx in its length but in the opposite direction. This is known as **Hooke's law:**

$$(F_{sp})_x = -k\,\Delta x$$

The constant of proportionality k is called the **spring constant.** It is larger for a "stiff" spring.

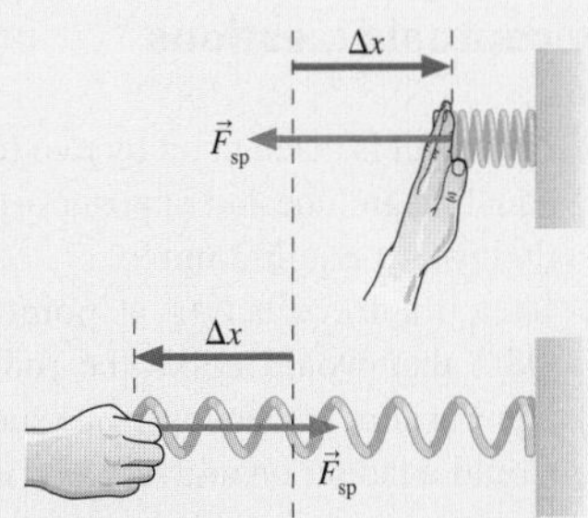

IMPORTANT CONCEPTS

Stability

An object is **stable** if its center of gravity is over its base of support; otherwise, it is **unstable.**

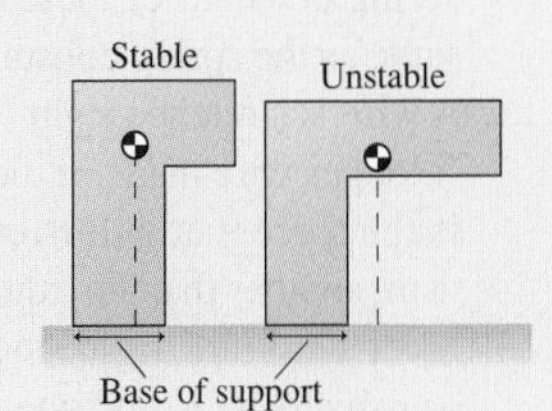

If an object is tipped, it will reach the limit of its stability when its center of gravity is over the edge of the base. This defines the **critical angle** θ_c.

This object is at its critical angle.

This object has a wider base of support and hence a larger critical angle.

This object has a lower center of gravity, so its critical angle is larger too.

θ_c θ_c θ_c

Greater stability is possible with a lower center of gravity or a broader base of support.

Elastic materials and Young's modulus

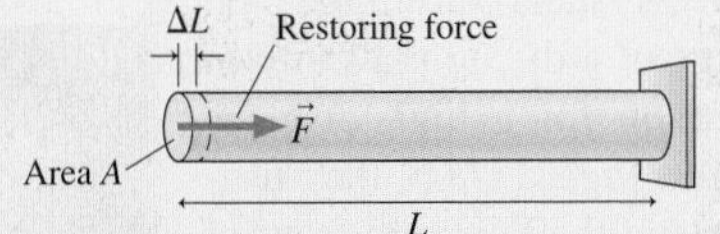

A solid rod illustrates how materials respond when stretched or compressed.

$$\left(\frac{F}{A}\right) = Y\left(\frac{\Delta L}{L}\right)$$

Stress is the restoring force of the rod divided by its cross-section area. **Strain** is the fractional change in the rod's length. Y: Young's modulus

This equation can also be written as

$$F = \left(\frac{YA}{L}\right)\Delta L$$

This is the "spring constant" k for the rod.

showing that a rod obeys Hooke's law and acts like a very stiff spring.

APPLICATIONS

Forces in the body

Muscles and tendons apply the forces and torques needed to maintain static equilibrium. These forces may be quite large.

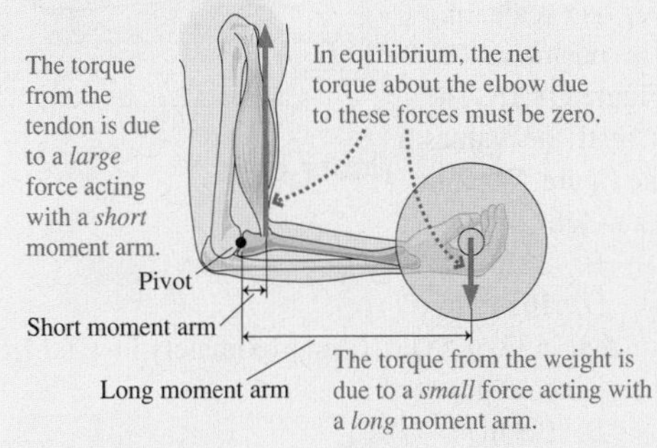

The torque from the tendon is due to a *large* force acting with a *short* moment arm.
In equilibrium, the net torque about the elbow due to these forces must be zero.
Pivot
Short moment arm
Long moment arm
The torque from the weight is due to a *small* force acting with a *long* moment arm.

The elastic limit and beyond

If a rod or other object is not stretched too far, when released it will return to its original shape.

If stretched too far, an object will permanently deform, and finally break. The stress at which an object breaks is its **tensile stress.**

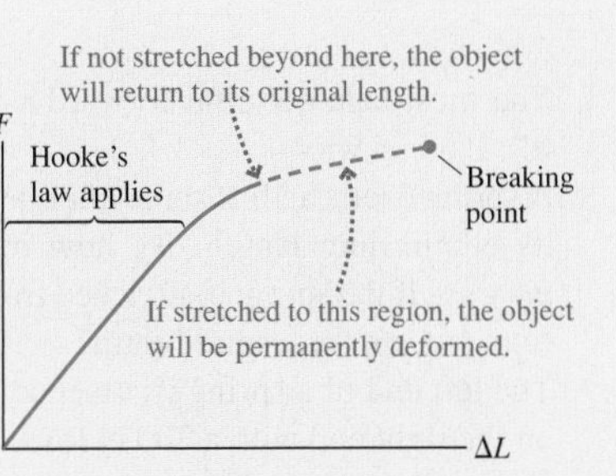

For homework assigned on MasteringPhysics, go to www.masteringphysics.com

Problem difficulty is labeled as | (straightforward) to ||||| (challenging).

Problems labeled INT integrate significant material from earlier chapters; BIO are of biological or medical interest.

QUESTIONS

Conceptual Questions

1. An object is acted upon by two (and only two) forces that are of equal magnitude and oppositely directed. Is the object necessarily in static equilibrium?
2. Sketch a force acting at point P in Figure Q8.2 that would make the rod be in static equilibrium. Is there only one such force?
3. Could a ladder on a level floor lean against a wall in static equilibrium if there were no friction forces? Explain.
4. Suppose you are hanging from a tree branch. If you move out the branch, farther away from the trunk, the branch will be more likely to break. Explain why this is so.

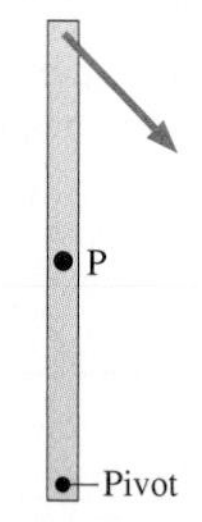

FIGURE Q8.2

5. As divers stand on tiptoes on the edge of a diving platform, in preparation for a high dive, as shown in Figure Q8.5, they usually extend their arms in front of them. Why do they do this?
6. Where are the centers of gravity of the two people doing the classic yoga poses shown in Figure Q8.6?

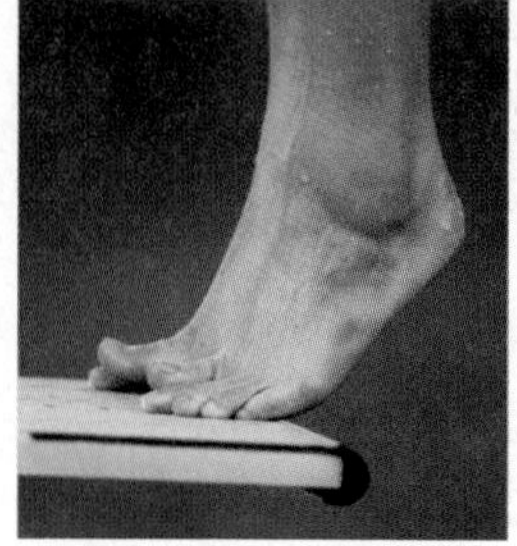

FIGURE Q8.5

FIGURE Q8.6

7. You must lean quite far forward as you rise from a chair (try it!). Explain why.
8. A spring exerts a 10 N force after being stretched by 1 cm from its equilibrium length. By how much will the spring force *increase* if the spring is stretched from 4 cm away from equilibrium to 5 cm from equilibrium?
9. The left end of a spring is attached to a wall. When Bob pulls on the right end with a 200 N force, he stretches the spring by 20 cm. The same spring is then used for a tug-of-war between Bob and Carlos. Each pulls on his end of the spring with a 200 N force.
   a. How far does Bob's end of the spring move? Explain.
   b. How far does Carlos's end of the spring move? Explain.
10. A spring is attached to the floor and pulled straight up by a string. The string's tension is measured. The graph in Figure Q8.10 shows the tension in the spring as a function of the spring's length L.
    a. Does this spring obey Hooke's law? Explain.
    b. If it does, what is the spring constant?

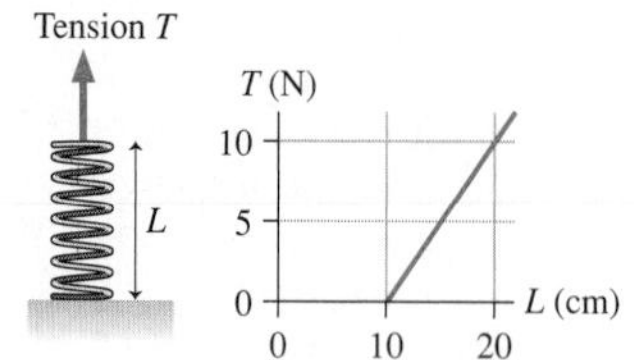

FIGURE Q8.10

11. Take a spring and cut it in half to make two springs. Is the spring constant of these smaller springs larger, smaller, or the same as the spring constant of the original spring? Explain.
12. A wire is stretched right to its breaking point by a 5000 N force. A longer wire made of the same material has the same diameter. Is the force that will stretch it right to its breaking point larger than, smaller than, or equal to 5000 N? Explain.
13. Steel nails are rigid and unbending. Steel wool is soft and squishy. How would you account for this difference?

Multiple-Choice Questions

14. || Two children carry a lightweight 1.8-m-long horizontal pole with a water bucket hanging from it. The older child supports twice as much weight as the younger child. How far is the bucket from the older child?
    A. 0.3 m B. 0.6 m
    C. 0.9 m D. 1.2 m
15. || The uniform rod in Figure Q8.15 has a weight of 14.0 N. What is the magnitude of the normal force exerted on the rod by the surface?
    A. 7 N B. 14 N
    C. 20 N D. 28 N

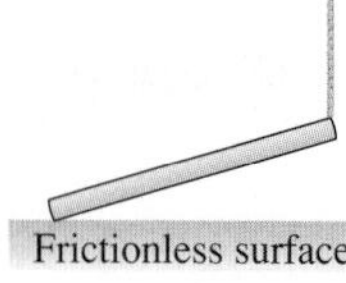

FIGURE Q8.15

16. | A student lies on a very light, rigid board with a scale under each end. Her feet are directly over one scale, and her body is positioned as shown in Figure Q8.16. The two scales read the values shown in the figure. What is the student's weight?
    A. 65 lb B. 75 lb
    C. 100 lb D. 165 lb

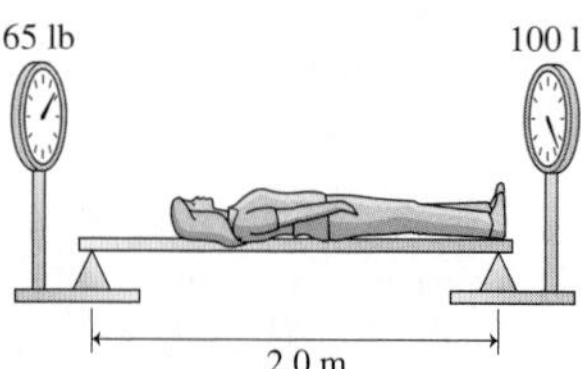

FIGURE Q8.16

17. | For the student in Figure Q8.16, approximately how far from her feet is her center of gravity?
    A. 0.6 m B. 0.8 m
    C. 1.0 m D. 1.2 m

Questions 18 through 20 use the information in the following paragraph and figure.

Suppose you stand on one foot while holding your other leg up behind you. Your muscles will have to apply a force to hold your leg in this raised position. We can model this situation as in Figure Q8.18. The leg pivots at the knee joint, and the force to hold the leg up is provided by a tendon attached to the lower leg as shown. Assume that the lower leg and the foot together have a combined mass of 4.0 kg, and that their combined center of gravity is at the center of the lower leg.

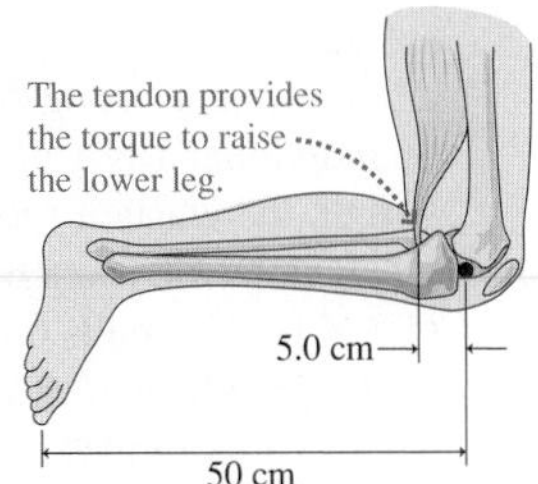

The tendon provides the torque to raise the lower leg.
5.0 cm
50 cm

FIGURE Q8.18

18. | How much force must the tendon exert to keep the leg in this BIO position?
A. 40 N B. 200 N C. 400 N D. 1000 N

19. | As you hold your leg in this position, the upper leg exerts a BIO force on the lower leg at the knee joint. What is the direction of this force?
A. Up B. Down C. Right D. Left

20. | What is the magnitude of the force of the upper leg on the BIO lower leg at the knee joint?
A. 40 N B. 160 N C. 200 N D. 240 N

21. ||| You have a heavy piece of equipment hanging from a 1.0-mm-diameter wire. Your supervisor asks that the length of the wire be doubled without changing how far the wire stretches. What diameter must the new wire have?
A. 1.0 mm B. 1.4 mm C. 2.0 mm D. 4.0 mm

22. ||| A 30.0-cm-long board is placed on a table such that its right end hangs over the edge by 8.0 cm. A second identical board is stacked on top of the first, as shown in Figure Q8.22. What is the largest that the distance x can be before both boards topple over?
A. 4.0 cm B. 8.0 cm
C. 14 cm D. 15 cm

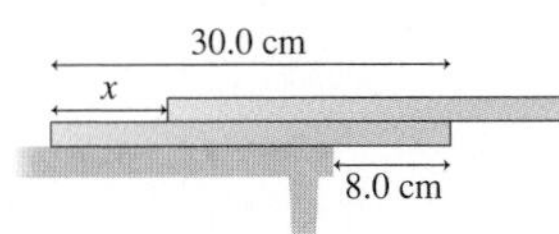

FIGURE Q8.22

23. || Two 20 kg blocks are connected by a 2.0-m-long, 5.0-mm-diameter rope. Young's modulus for this rope is 1.5×10^9 N/m^2. The rope is then hung over a pulley, so that the blocks, hanging from each side of the pulley, are in static equilibrium. By how much does the rope stretch?
A. 3.0 mm B. 6.3 mm
C. 9.3 mm D. 13 mm

PROBLEMS

Section 8.1 Torque and Static Equilibrium

1. || A 64 kg woman stands on a very light, rigid board that rests on a bathroom scale at each end, as shown in Figure P8.1. What is the reading on each of the scales?

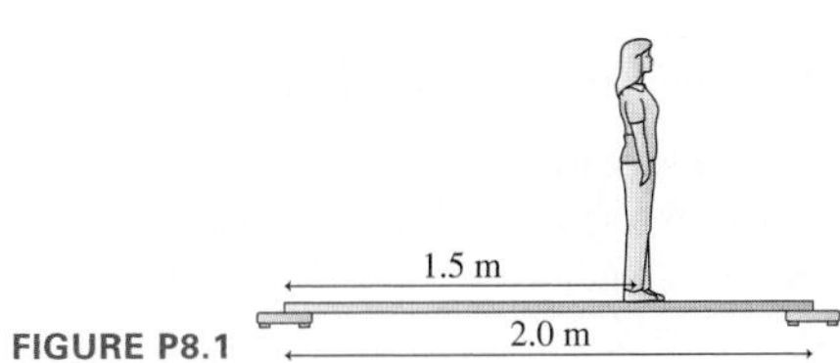

FIGURE P8.1

2. ||| Suppose the woman in Figure P8.1 is 54 kg, and the board she is standing on has a 10 kg mass. What is the reading on each of the scales?

3. ||| How close to the right edge of the 56 kg picnic table shown in Figure P8.3 can a 70 kg man stand without the table tipping over?

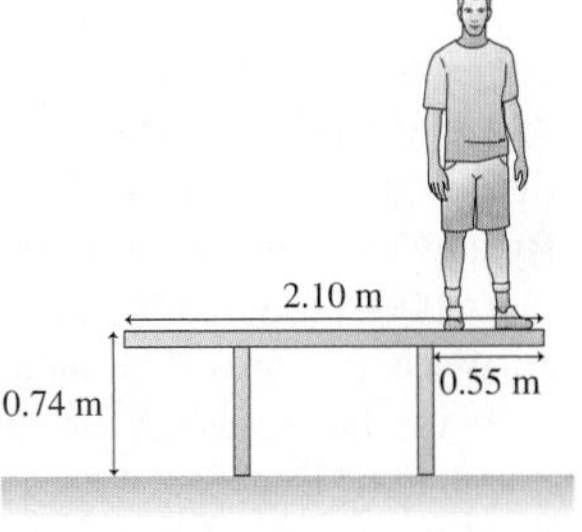

FIGURE P8.3

4. || In Figure P8.4, a 70 kg man walks out on a 10 kg beam that rests on, but is not attached to, two supports. When the beam just starts to tip, what is the force exerted on the beam by the right support?

FIGURE P8.4

5. ||| You're carrying a 3.6-m-long, 25 kg pole to a construction site when you decide to stop for a rest. You place one end of the pole on a fence post and hold the other end of the pole 35 cm from its tip. How much force must you exert to keep the pole motionless in a horizontal position?

6. ||| How much torque must the pin exert to keep the rod in Figure P8.6 from rotating? Calculate this torque about an axis that passes through the point where the pin enters the rod and is perpendicular to the plane of the figure.

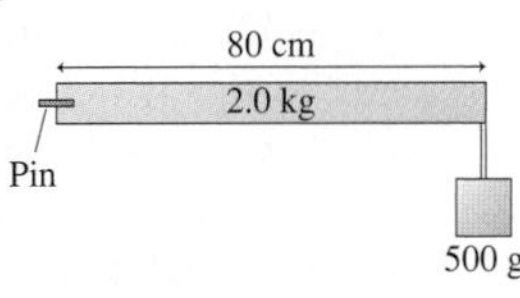

FIGURE P8.6

7. ||| Is the object in Figure P8.7 in equilibrium? Explain.

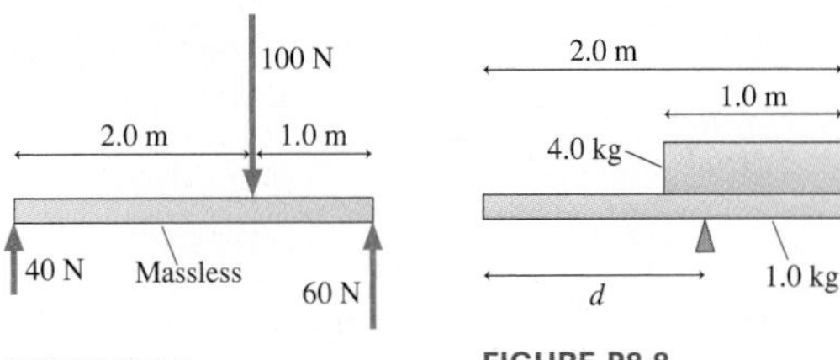

FIGURE P8.7

FIGURE P8.8

8. ||| The two objects in Figure P8.8 are balanced on the pivot. What is distance d?

9. ||| A 60 kg diver stands at the end of a 30 kg springboard, as shown in Figure P8.9. The board is attached to a hinge at the left end but simply rests on the right support. What is the magnitude of the vertical force exerted by the hinge on the board?

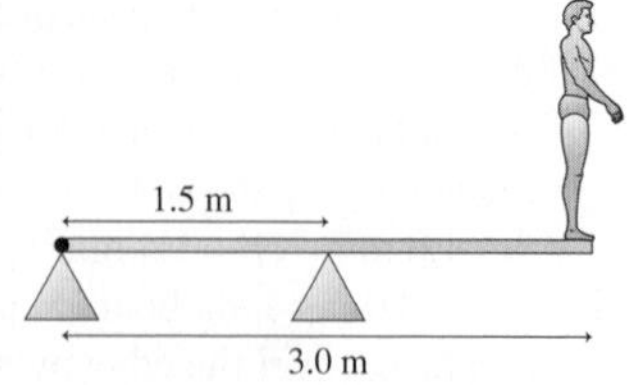

FIGURE P8.9

10. ||| A uniform beam of length 1.0 m and mass 10 kg is attached to a wall by a cable, as shown in Figure P8.10. The beam is free to pivot at the point where it attaches to the wall. What is the tension in the cable?

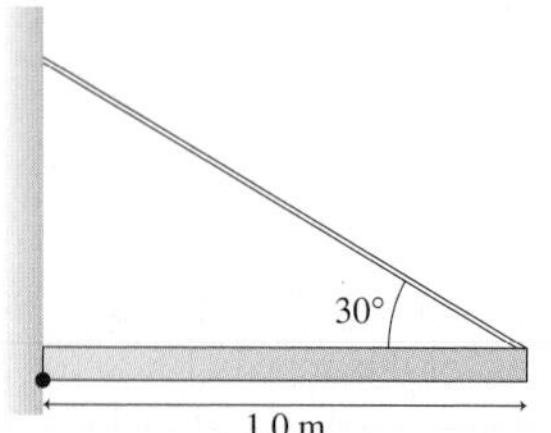

FIGURE P8.10

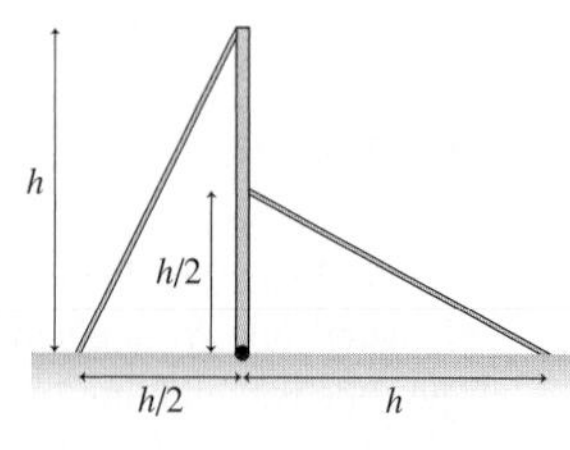

FIGURE P8.11

11. || Figure P8.11 shows a vertical pole of height h that can rotate about a hinge at the bottom. The pole is held in position by two wires under tension. What is the ratio of the tension in the left wire to the tension in the right wire?

Section 8.2 Stability and Balance

12. || You want to slowly push a stiff board across a 20 cm gap between two tabletops that are at the same height. If you apply only a horizontal force, how long must the board be so that it doesn't tilt down into the gap before reaching the other side?

13. | A magazine rack has a center of gravity 16 cm above the floor, as shown in Figure P8.13. Through what maximum angle, in degrees, can the rack be tilted without falling over?

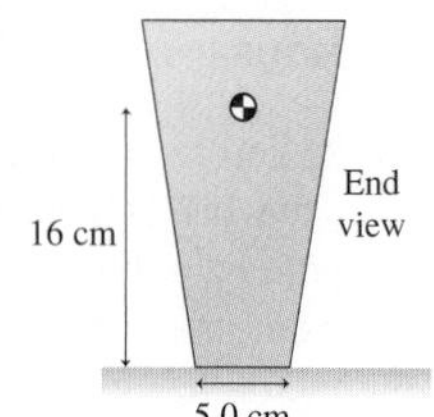

FIGURE P8.13

14. || A car manufacturer claims that you can drive its new vehicle across a hill with a 47° slope before the vehicle starts to tip. If the vehicle is 2.0 m wide, how high is its center of gravity?

15. || A thin 2.00 kg box rests on a 6.00 kg board that hangs over the end of a table, as shown in Figure P8.15. How far can the center of the box be from the end of the table before the board begins to tilt?

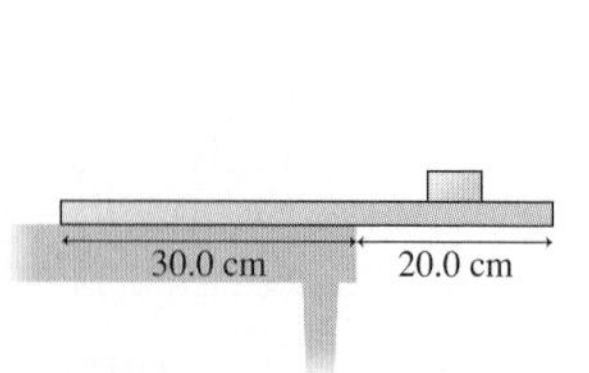

FIGURE P8.15

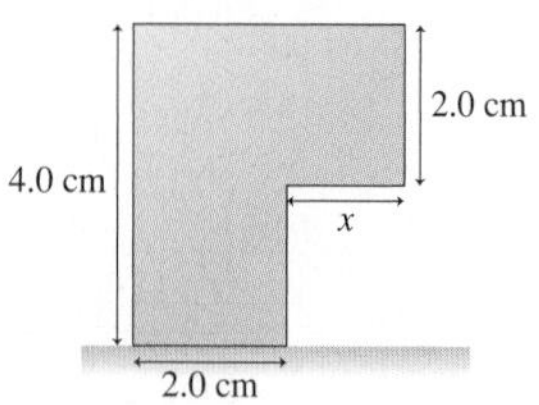

FIGURE P8.16

16. ||| The object shown in Figure P8.16 is made of a uniform material. What is the greatest that x can be without the object tipping over?

Section 8.3 Springs and Hooke's Law

17. | One end of a spring is attached to a wall. A 25 N pull on the other end causes the spring to stretch by 3.0 cm. What is the spring constant?

18. | Experiments using "optical tweezers" measure the elasticity
BIO of individual DNA molecules. For small enough changes in length, the elasticity has the same form as that of a spring. A DNA molecule is anchored at one end, then a force of 1.5 nN (1.5×10^{-9} N) pulls on the other end, causing the molecule to stretch by 5.0 nm (5.0×10^{-9} m). What is the spring constant of that DNA molecule?

19. || A spring has an unstretched length of 10 cm. It exerts a restoring force F when stretched to a length of 11 cm.
    a. For what total stretched length of the spring is its restoring force $3F$?
    b. At what compressed length is the restoring force $2F$?

20. || A 10-cm-long spring is attached to the ceiling. When a 2.0 kg mass is hung from it, the spring stretches to a length of 15 cm.
    a. What is the spring constant?
    b. How long is the spring when a 3.0 kg mass is suspended from it?

21. || A spring stretches 5.0 cm when a 0.20 kg block is hung from it. If a 0.70 kg block replaces the 0.20 kg block, how far does the spring stretch?

22. || A 1.2 kg block is hung from a vertical spring, causing the spring to stretch by 2.4 cm. How much farther will the spring stretch if a 0.60 kg block is added to the 1.2 kg block?

23. || A runner wearing spiked shoes pulls a 20 kg sled across frictionless ice using a horizontal spring with spring constant 1.5×10^2 N/m. The spring is stretched 20 cm from its equilibrium length. What is the acceleration of the sled?

24. | You need to make a spring scale to measure the mass of objects hung from it. You want each 1.0 cm length along the scale to correspond to a mass difference of 0.10 kg. What should be the value of the spring constant?

Section 8.4 Stretching and Compressing Materials

25. || A force stretches a wire by 1.0 mm.
    a. A second wire of the same material has the same cross section and twice the length. How far will it be stretched by the same force?
    b. A third wire of the same material has the same length and twice the diameter as the first. How far will it be stretched by the same force?

26. |||| What hanging mass will stretch a 2.0-m-long, 0.50-mm-diameter steel wire by 1.0 mm?

27. ||| How much force does it take to stretch a 10-m-long, 1.0-cm-diameter steel cable by 5.0 mm?

28. ||| An 80-cm-long, 1.0-mm-diameter steel guitar string must be tightened to a tension of 2.0 kN by turning the tuning screws. By how much is the string stretched?

29. ||| A 2000 N force stretches a wire by 1.0 mm.
    a. A second wire of the same material is twice as long and has twice the diameter. How much force is needed to stretch it by 1.0 mm? Explain.
    b. A third wire of the same material is twice as long as the first and has the same diameter. How far is it stretched by a 4000 N force?

30. ||| A 1.2-m-long steel rod with a diameter of 0.50 cm hangs vertically from the ceiling. An auto engine weighing 4.7 kN is hung from the rod. By how much does the rod stretch?

31. ||| A mine shaft has an elevator hung from a single steel-wire cable of diameter 2.5 cm. When the cable is fully extended, the end of the cable is 500 m below the support. How much does the fully extended cable stretch when 3000 kg of ore is loaded into the elevator?

32. || The normal force of the ground on the foot can reach three BIO times a runner's body weight when the foot strikes the pavement. By what amount does the 52-cm-long femur of an 80 kg runner compress at this moment? The cross-section area of the bone of the femur can be taken as $5.2 \times 10^{-4}\ \text{m}^2$.

33. |||| A three-legged wooden bar stool made out of solid Douglas fir has legs that are 2.0 cm in diameter. When a 75 kg man sits on the stool, by what percent does the length of the legs decrease? Assume, for simplicity, that the stool's legs are vertical and that each bears the same load.

34. ||| A 3.0-m-tall, 50-cm-diameter concrete column supports a 200,000 kg load. By how much is the column compressed?

General Problems

35. |||| A 3.0-m-long rigid beam with a mass of 100 kg is supported at each end, as shown in Figure P8.35. An 80 kg student stands 2.0 m from support 1. How much upward force does each support exert on the beam?

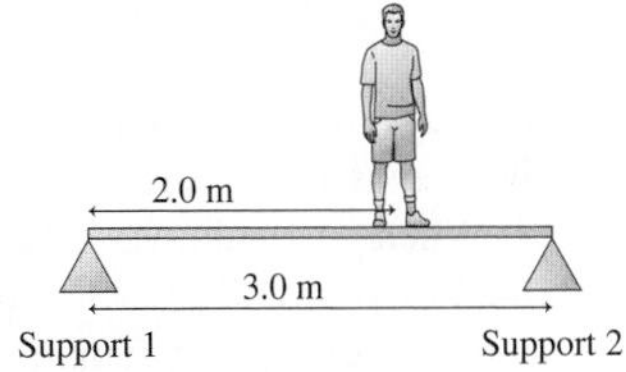

FIGURE P8.35

36. ||| An 80 kg construction worker sits down 2.0 m from the end of a 1450 kg steel beam to eat his lunch, as shown in Figure P8.36. The cable supporting the beam is rated at 15,000 N. Should the worker be worried?

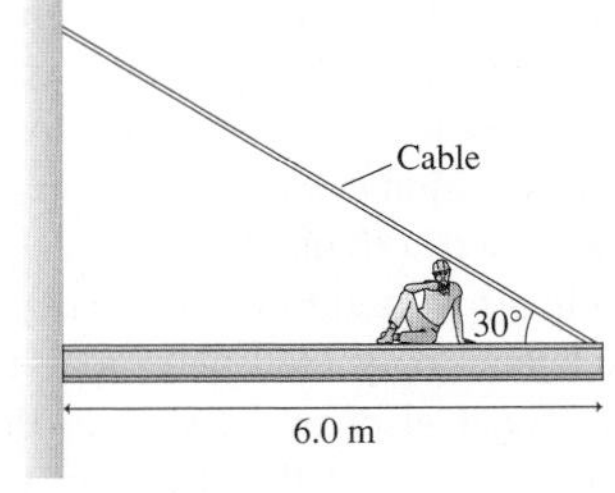

FIGURE P8.36

37. |||| Using the information in Figure 8.2, calculate the tension in BIO the biceps tendon if the hand is holding a 10 kg ball while the forearm is held 45° below horizontal.

38. ||| A woman weighing 580 N does a BIO pushup from her knees, as shown in Figure P8.38. What are the normal forces of the floor on (a) each of her hands and (b) each of her knees?

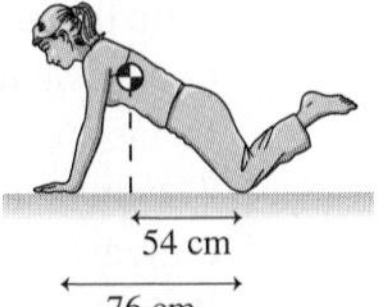

FIGURE P8.38

39. |||| When you bend over, a series of large BIO muscles, the erector spinae, pull on your spine to hold you up. Figure P8.39 shows a simplified model of the spine as a rod of length L that pivots at its lower end. In this model, the center of gravity of the 320 N weight of the upper torso is at the center of the spine. The 160 N weight of the head and arms acts at the top of the spine. The erector spinae muscles are modeled as a single muscle that acts at an 12° angle to the spine. Suppose the person in Figure P8.39 bends over to an angle of 30° from the horizontal.

a. What is the tension in the erector muscle?

Hint: Align your x-axis with the axis of the spine.

b. A force from the pelvic girdle acts on the base of the spine. What is the component of this force in the direction of the spine? (This large force is the cause of many back injuries).

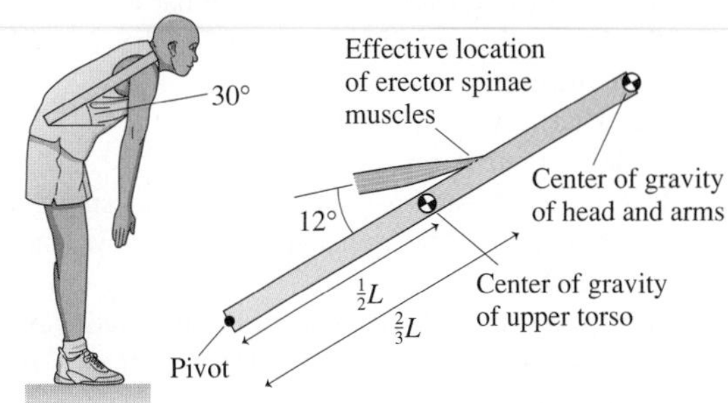

FIGURE P8.39

40. | The woman lying on the reaction board in Example 8.3 spreads her arms out in the plane of the board. The reading of the scale is observed to increase by 10 N. By how much does the distance from her feet to her center of gravity change?

41. ||| A man is attempting to raise a 7.5-m-long, 28 kg flagpole that has a hinge at the base by pulling on a rope attached to the top of the pole, as shown in Figure P8.41. With what force does the man have to pull on the rope to hold the pole motionless in this position?

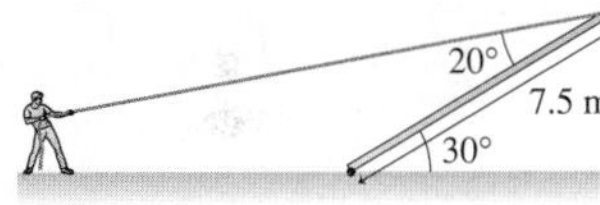

FIGURE P8.41

42. ||||| A library ladder of length L rolls on wheels as shown in Figure P8.42. The two legs of the ladder freely pivot at the hinge at the top. The legs are kept from splaying apart by a lightweight chain that is attached halfway up the ladder. If the ladder weighs 200 N, and the angle between each leg and the vertical is 25°, what is the tension in the chain?

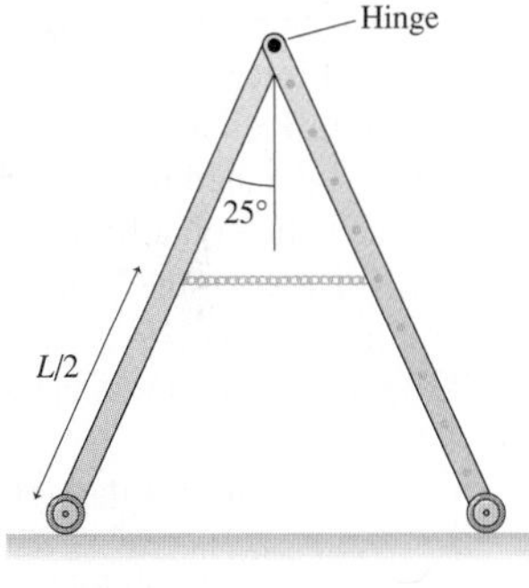

FIGURE P8.42

43. |||| A 40 kg, 5.0-m-long beam is supported by, but not attached to, the two posts in Figure P8.43. A 20 kg boy starts walking along the beam. How close can he get to the right end of the beam without it tipping?

FIGURE P8.43

FIGURE P8.44

44. ||| The wheel of mass m in Figure P8.44 is pulled on by a horizontal force applied at its center. The wheel is touching a curb whose height is half the wheel's radius. What is the minimum force required to just raise the wheel off the ground?

45. || A 5.0 kg mass hanging from a spring scale is slowly lowered onto a vertical spring, as shown in Figure P8.45. The scale reads in newtons.
   a. What does the spring scale read just before the mass touches the lower spring?
   b. The scale reads 20 N when the lower spring has been compressed by 2.0 cm. What is the value of the spring constant for the lower spring?
   c. At what compression distance will the scale read zero?

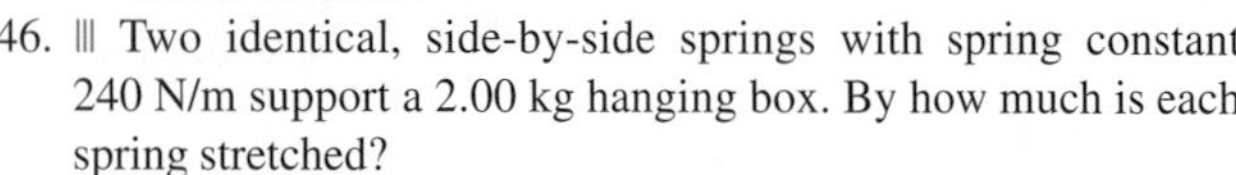

FIGURE P8.45

46. ||| Two identical, side-by-side springs with spring constant 240 N/m support a 2.00 kg hanging box. By how much is each spring stretched?

47. | Two springs have the same equilibrium length but different spring constants. They are arranged as shown in Figure P8.47, then a block is pushed against them, compressing both by 1.00 cm. With what net force do they push back on the block?

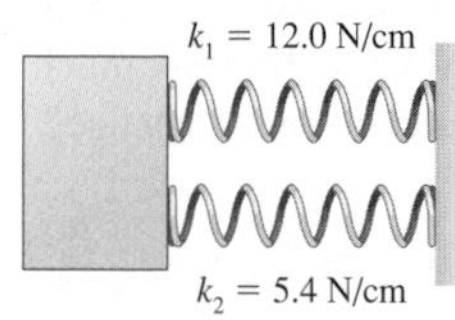

FIGURE P8.47

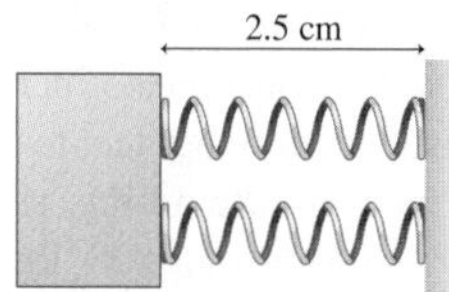

FIGURE P8.48

48. || Two springs have the same spring constant $k = 130$ N/m but different equilibrium lengths, 3.0 cm and 5.0 cm. They are arranged as shown in Figure P8.48, then a block is pushed against them, compressing both to a length of 2.5 cm. With what net force do they push back on the block?

49. || Figure P8.49 shows two springs attached to a block that can slide on a frictionless surface. In the block's equilibrium position, the left spring is compressed by 2.0 cm.
   a. By how much is the right spring compressed?
   b. What is the net force on the block if it is moved 15 cm to the right of its equilibrium position?

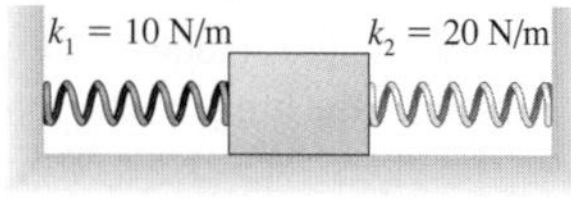

FIGURE P8.49

50. ||| Figure P8.50 shows two springs attached to each other, and also attached to a box that can slide on a frictionless surface. In the block's equilibrium position, neither spring is stretched. What is the net force on the block if it is moved 15 cm to the right of its equilibrium position?
   Hint: There is zero net force on the point where the two springs meet. This implies a relationship between the amounts the two springs stretch.

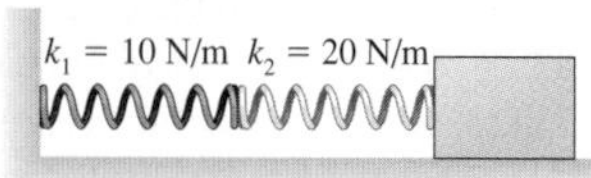

FIGURE P8.50

51. || A 60 kg student is standing atop a spring in an elevator that is
INT accelerating upward at 3.0 m/s^2. The spring constant is 2.5×10^3 N/m. By how much is the spring compressed?

52. ||| A 25 kg child bounces on a pogo stick. The pogo stick has a
INT spring with spring constant 2.0×10^4 N/m. When the child makes a nice big bounce, she finds that at the bottom of the bounce she is accelerating *upward* at 9.8 m/s^2. How much is the spring compressed?

53. |||| Two 3.0 kg blocks on a
INT level, frictionless surface are connected by a spring with spring constant 1000 N/m, as shown in Figure P8.53. The left block is pushed by a horizontal force $\vec{F}$. At $t = 0$ s, both blocks have velocity 3.2 m/s to the right. For the next second, the spring's compression is a constant 1.5 cm.
   a. What is the velocity of the right block at $t = 1.0$ s?
   b. What is the magnitude of $\vec{F}$ during that 1.0 s interval?

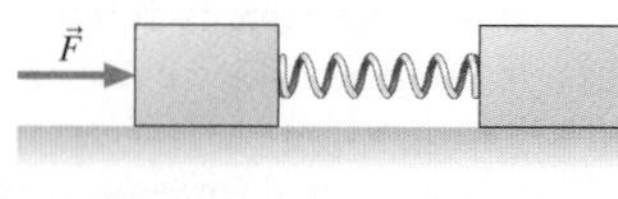

FIGURE P8.53

54. || What is the effective spring constant (that is, the ratio of force to change in length) of a copper cable that is 5.0 mm in diameter and 5.0 m long?

55. ||| Figure P8.55 shows a 100 kg plank supported at its right end by a 7.0-mm-diameter rope with a tensile strength of 6.0×10^7 N/m^2. How far along the plank, measured from the pivot, can the center of gravity of an 800 kg piece of heavy machinery be placed before the rope snaps?

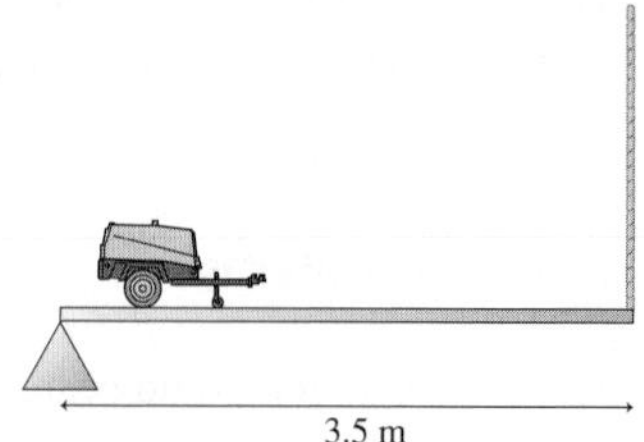

FIGURE P8.55

56. ||| When you walk, your Achilles tendon, which connects your
BIO heel to your calf muscles, repeatedly stretches and contracts, much like a spring. This helps make walking more efficient. Suppose your Achilles tendon is 15 cm long and has a cross-section area of 110 mm^2, typical values. If you model the Achilles tendon as a spring, what is its spring constant?

57. ||| There is a disk of cartilage between each pair of vertebrae in
BIO your spine. Suppose a disk is 0.50 cm thick and 4.0 cm in diameter. If this disk supports half the weight of a 65 kg person, by what fraction of its thickness does the disk compress?

58. || In Example 8.1, the tension in the biceps tendon for a person
BIO doing a strict curl of a 900 N barbell was found to be 3900 N. What fraction does this represent of the maximum possible tension the biceps tendon can support? You can assume a typical cross-section area of 130 mm^2.

59. ||| Larger animals have sturdier bones than smaller animals. A
BIO mouse's skeleton is only a few percent of its body weight, compared to 16% for an elephant. To see why this must be so, recall, from Example 8.10, that the stress on the femur for a man standing on one leg is 1.4% of the bone's tensile strength. Suppose we scale this man up by a factor of 10 in all dimensions, keeping the same body proportions. Use the data for Example 8.10 to compute the following.
   a. Both the inside and outside diameter of the femur, the region of cortical bone, will increase by a factor of 10. What will be the new cross-section area?
   b. The man's body will increase by a factor of 10 in each dimension. What will be his new mass?
   c. If the scaled-up man now stands on one leg, what fraction of the tensile strength is the stress on the femur?

60. || Orb spiders make silk with a typical diameter of 0.15 mm.
BIO a. A typical large orb spider has a mass of 0.50 g. If this spider suspends itself from a single 12-cm-long strand of silk, by how much will the silk stretch?
   b. What is the maximum weight that a single thread of this silk could support?

Passage Problems

Standing on Tiptoes BIO

When you stand on your tiptoes, your feet pivot about your ankle. As shown in Figure P8.61, the forces on your foot are an upward force on your toes from the floor, a downward force on your ankle from the lower leg bone, and an upward force on the heel of your foot from your Achilles tendon. Suppose a 60 kg woman stands on tiptoes with the sole of her foot making a 25° angle with the floor. Assume that each foot supports half her weight.

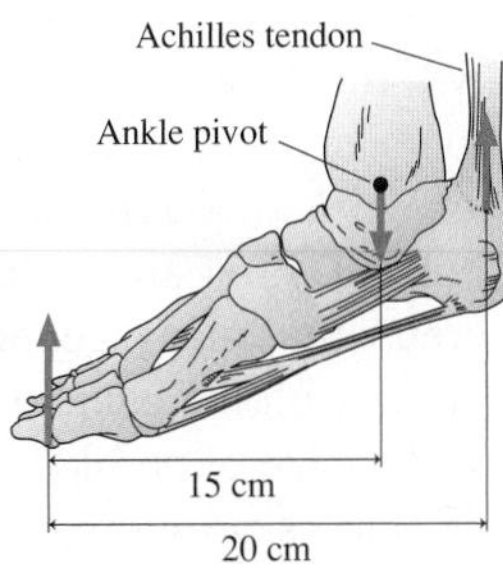

FIGURE P8.61

61. ||| What is the upward force of the floor on the toes of one foot?
A. 140 N B. 290 N
C. 420 N D. 590 N

62. || What upward force does the Achilles tendon exert on the heel of her foot?
A. 290 N B. 420 N
C. 590 N D. 880 N

63. ||| The tension in the Achilles tendon will cause it to stretch. If the Achilles tendon is 15 cm long and has a cross-section area of $110\ \text{mm}^2$, by how much will it stretch under this force?
A. 0.2 mm B. 0.8 mm
C. 2.3 mm D. 5.2 mm

STOP TO THINK ANSWERS

Stop to Think 8.1: D. Only object D has both zero net force and zero net torque.

Stop to Think 8.2: B. The tension in the rope and the weight have no horizontal component. To make the net force zero, the force due to the pivot must also have no horizontal component, so we know it points either up or down. Now consider the torque about the point where the rope is attached. The tension provides no torque. The weight exerts a counterclockwise torque. To make the net torque zero, the pivot force must exert a *clockwise* torque, which it can do only if it points *up*.

Stop to Think 8.3: B, A, C. The critical angle θ_c, shown in the figure, measures how far the object can be tipped before falling. B has the smallest critical angle, followed by A, then C.

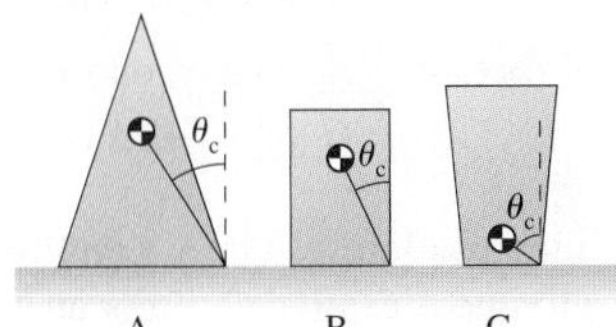

Stop to Think 8.4: C. The restoring force of the spring is proportional to the stretch. Increasing the restoring force by a factor of 3 requires increasing the stretch by a factor of 3.

Stop to Think 8.5: E. The cables have the same diameter, and the force is the same, so the stress is the same in both cases. This means that the strain, $\Delta L/L$, is the same. The 2 m cable will experience twice the change in length of the 1 m cable.

PART I SUMMARY

Force and Motion

The goal of Part I has been to discover the connection between force and motion. We started with kinematics, the mathematical description of motion; then we proceeded to dynamics, the explanation of motion in terms of forces. We then used these descriptions to analyze and explain motions ranging from the motion of the moon about the earth to the forces in your elbow when you lift a weight. Newton's three laws of motion formed the basis of all of our explanations.

The table below is a *knowledge structure* for force and motion. The knowledge structure does not represent everything you have learned over the past eight chapters. It's a summary of the "big picture," outlining the basic goals, the general principles, and the primary applications of the part of the book we have just finished. When you are immersed in a chapter, it may be hard to see the connections among all of the different topics. Before we move on to new topics, we will finish each part of the book with a knowledge structure to make these connections clear.

Work through the knowledge structure from top to bottom. First are the goals and general principles. There aren't that many general principles, but we can use them along with the general problem-solving strategy to solve a wide range of problems. Once you recognize a problem as a dynamics problem, you immediately know to start with Newton's laws. You can then determine the category of motion and apply Newton's second law in the appropriate form. The kinematic equations for that category of motion then allow you to reach the solution you seek. These equations and other detailed information from the chapters are summarized in the bottom section.

KNOWLEDGE STRUCTURE I Force and Motion

BASIC GOALS

How can we describe motion?
How does an object respond to a force?
How do systems interact?
What is the nature of the force of gravity?
How can we analyze the motion and deformation of extended objects?

GENERAL PRINCIPLES

Newton's first law	An object with no forces acting on it will remain at rest or move in a straight line at a constant speed.
Newton's second law	$\vec{F}_{\text{net}} = m\vec{a}$
Newton's third law	$\vec{F}_{\text{A on B}} = -\vec{F}_{\text{B on A}}$
Newton's law of gravity	$F_{1\text{ on }2} = F_{2\text{ on }1} = \dfrac{Gm_1m_2}{r^2}$

BASIC PROBLEM-SOLVING STRATEGY

Use Newton's second law for each particle or system. Use Newton's third law to equate the magnitudes of the two members of an action/reaction pair.

Types of forces:

$\vec{w} = (mg, \text{downward})$

$\vec{f}_k = (\mu_k n, \text{opposite motion})$

$(F_{sp})_x = -k\,\Delta x$

Linear and projectile motion:

$$\left.\begin{aligned}\textstyle\sum F_x &= ma_x\\ \textstyle\sum F_y &= 0\end{aligned}\right\} \quad \text{or} \quad \left\{\begin{aligned}\textstyle\sum F_x &= 0\\ \textstyle\sum F_y &= ma_y\end{aligned}\right.$$

Circular motion:

The force is directed to the center:

$$\vec{F}_{\text{net}} = \left(\frac{mv^2}{r}, \text{ toward center of circle}\right)$$

Rigid-body motion:

When a torque is exerted on an object with moment of inertia I,

$$\tau_{\text{net}} = I\alpha$$

Equilibrium:

For an object at rest,

$$\textstyle\sum F_x = 0 \quad \sum \tau = 0$$
$$\textstyle\sum F_y = 0$$

Linear and projectile kinematics

Uniform motion: $x_f = x_i + v_x\,\Delta t$

($a_x = 0$, $v_x =$ constant)

Constant acceleration: $(v_x)_f = (v_x)_i + a_x\,\Delta t$

($a_x =$ constant)

$x_f = x_i + (v_x)_i\,\Delta t + \frac{1}{2}a_x(\Delta t)^2$

$(v_x)_f^2 = (v_x)_i^2 + 2a_x\,\Delta x$

Projectile motion:
Projectile motion is uniform horizontal motion and constant-acceleration vertical motion with $a_y = -g$.

Velocity is the slope of the position-versus-time graph.
Acceleration is the slope of the velocity-versus-time graph.

Circular kinematics

Uniform circular motion:

$f = \dfrac{1}{T}$ $\quad \omega = 2\pi f$

$v = \dfrac{2\pi r}{T} = \omega r$ $\quad a = \dfrac{v^2}{r} = \omega^2 r$

Rigid bodies

Torque $\tau = rF_\perp = r_\perp F$

Center of gravity $x_{cg} = \dfrac{x_1m_1 + x_2m_2 + \cdots}{m_1 + m_2 + \cdots}$

Moment of inertia $I = \sum mr^2$

Dark Matter and the Structure of the Universe

The idea that the earth exerts a gravitational force on us is something we now accept without questioning. But when Isaac Newton developed this idea to show that the gravitational force also holds the moon in its orbit, it was a remarkable, ground-breaking insight. It changed the way that we look at the universe we live in.

Newton's laws of motion and gravity are tools that allow us to continue Newton's quest to better understand our place in the cosmos. But it sometimes seems that the more we learn, the more we realize how little we actually know and understand.

Here's an example. Advances in astronomy over the past 100 years have given us great insight into the structure of the universe. But everything our telescopes can see appears to be only a small fraction of what is out there. As much as 90% of the mass in the universe is *dark matter*—matter that gives off no light or other radiation that we can detect. Everything that we have ever seen through a telescope is merely the tip of the cosmic iceberg.

What is this dark matter? Black holes? Neutrinos? Some form of exotic particle? No one knows. It could be any of these, or all of them—or something entirely different that no one has yet dreamed of. You might wonder how we know that such matter exists if no one has seen it. Even though we can't directly observe dark matter, we see its effects. And you now know enough physics to understand why.

Whatever dark matter is, it has mass, and so it has gravity. This picture of the Andromeda galaxy shows a typical spiral galaxy structure: a dense collection of stars in the center surrounded by a disk of stars and other matter. This is the shape of our own Milky Way galaxy.

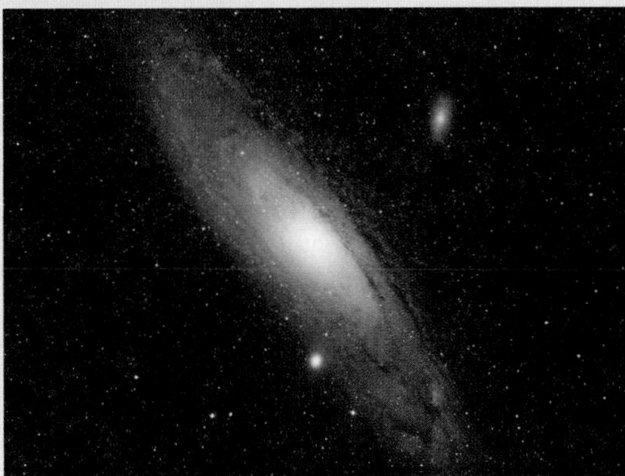

The spiral Andromeda galaxy.

This structure is reminiscent of the structure of the solar system: a dense mass (the sun) in the center surrounded by a disk of other matter (the planets, asteroids, and comets). The sun's gravity keeps the planets in their orbits, but the planets would fall into the sun unless they were in constant motion around it. The same is true of a spiral galaxy; everything in the galaxy orbits its center. Our solar system orbits the center of our galaxy with a period of about 200 million years.

The orbital speed of an object depends on the mass that pulls on it. If you analyze our sun's motion about the center of the Milky Way, or the motion of stars in the Andromeda galaxy about its center, you find that the orbits are much faster than they should be, based on how many stars we see. There must be some other mass present.

There's another problem with the orbital motion of stars around the center of their galaxies. We know that the orbital speeds of planets decrease with distance from the sun; Neptune orbits at a much slower speed than the earth. We might expect something similar for galaxies: Stars farther from the center should orbit at reduced speeds. But they don't. As we measure outward from the center of the galaxy, the orbital speed stays about the same—even as we get to the edge of the visible disk. There must be some other mass—the invisible dark matter—exerting a gravitational force on the stars. This dark matter, which far outweighs the matter we can see, seems to form a halo around the centers of galaxies, providing the gravitational force necessary to produce the observed rotation. Other observations of the motions of galaxies with respect to each other verify this basic idea.

On a cosmic scale, the picture is even stranger. The universe is currently expanding. The mutual gravitational attraction of all matter—regular and dark—in the universe should slow this expansion. But recent observations of the speeds of distant galaxies imply that the expansion of the universe is accelerating, so there must be yet another component to the universe, something that "pushes out". The best explanation at present is that the acceleration is caused by *dark energy*. The nature of dark matter isn't known, but the nature of dark energy is even more mysterious. If current theories hold, it's the most abundant stuff in the universe. And we don't know what it is.

This sort of mystery is what drives scientific investigation. It's what drove Newton to wonder about the connection between the fall of an apple and the motion of the moon, and what drove investigators to develop all of the techniques and theories you will learn about in the coming chapters.

PART I PROBLEMS

The following questions are related to the passage "Dark Matter and the Structure of the Universe" on the previous page.

1. As noted in the passage, our solar system orbits the center of the Milky Way galaxy in about 200 million years. If there were no dark matter in our galaxy, this period would be
   A. Longer.
   B. The same.
   C. Shorter.
2. Saturn is approximately 10 times as far away from the sun as the earth. This means that its orbital acceleration is ______ that of the earth.
   A. 1/10
   B. 1/100
   C. 1/1000
   D. 1/10,000
3. Saturn is approximately 10 times as far away from the sun as the earth. If dark matter changed the orbital properties of the planets so that Saturn had the same orbital speed as the earth, Saturn's orbital acceleration would be ______ that of the earth.
   A. 1/10
   B. 1/100
   C. 1/1000
   D. 1/10,000
4. Which of the following might you expect to be an additional consequence of the fact that galaxies contain more mass than expected?
   A. The gravitational force between galaxies is greater than expected.
   B. Galaxies appear less bright than expected.
   C. Galaxies are farther away than expected.
   D. There are more galaxies than expected.

The following passages and associated questions are based on the material of Part I.

Animal Athletes BIO

Different animals have very different capacities for running. A horse can maintain a top speed of 20 m/s for a long distance but has a maximum acceleration of only 6.0 m/s^2, half what a good human sprinter can achieve with a block to push against. Greyhounds, dogs especially bred for feats of running, have a top speed of 17 m/s, but their acceleration is much greater than that of the horse. Greyhounds are particularly adept at turning corners at a run.

FIGURE I.1

5. If a horse starts from rest and accelerates at the maximum value until reaching its top speed, how much time elapses, to the nearest second?
   A. 1 s B. 2 s
   C. 3 s D. 4 s
6. If a horse starts from rest and accelerates at the maximum value until reaching its top speed, how far does it run, to the nearest 10 m?
   A. 40 m B. 30 m
   C. 20 m D. 10 m
7. A greyhound on a racetrack turns a corner at a constant speed of 15 m/s with an acceleration of 7.1 m/s^2. What is the radius of the turn?
   A. 40 m B. 30 m
   C. 20 m D. 10 m
8. A human sprinter of mass 70 kg starts a run at the maximum possible acceleration, pushing backward against a block set in the track. What is the force of his foot on the block?
   A. 1500 N B. 840 N
   C. 690 N D. 420 N
9. In the photograph of the greyhounds in Figure I.1, what is the direction of the net force on each dog?
   A. Up
   B. Down
   C. Left, toward the outside of the turn
   D. Right, toward the inside of the turn

Sticky Liquids BIO

The drag force on an object moving in a liquid is quite different from that in air. Drag forces in air are largely the result of the object having to push the air out of its way as it moves. For an object moving slowly through a liquid, however, the drag force is mostly due to the *viscosity* of the liquid, a measure of how much resistance to flow the fluid has. Honey, which drizzles slowly out of its container, has a much higher viscosity than water, which flows fairly freely.

The *viscous drag* force in a liquid depends on the shape of the object, but there is a simple result called *Stokes's law* for the drag on a sphere. The drag force on a sphere of radius r moving at speed v through a fluid with viscosity η is

$$\vec{D} = (6\pi\eta r v,\ \text{direction opposite motion})$$

At small scales, viscous drag becomes very important. To a paramecium (Figure I.2), a single-celled animal that can propel itself through water with fine hairs on its body, swimming through water feels like swimming through honey would to you. We can model a paramecium as a sphere of diameter 50 μm, with a mass of 6.5×10^{-11} kg. Water has a viscosity of 0.0010 N · s/m^2.

FIGURE I.2

10. A paramecium swimming at a constant speed of 0.25 mm/s ceases propelling itself and slows to a stop. At the instant it stops swimming, what is the magnitude of its acceleration?
    A. $0.2g$ B. $0.5g$
    C. $2g$ D. $5g$
11. If the acceleration of the paramecium in Problem 10 were to stay constant as it comes to rest, approximately how far would it travel before stopping?
    A. 0.02 μm B. 0.2 μm
    C. 2 μm D. 20 μm
12. If the paramecium doubles its swimming speed, how does this change the drag force?
    A. The drag force decreases by a factor of 2.
    B. The drag force is unaffected.
    C. The drag force increases by a factor of 2.
    D. The drag force increases by a factor of 4.

13. You can test the viscosity of a liquid by dropping a steel sphere into it and measuring the speed at which it sinks. For viscous fluids, the sphere will rapidly reach a terminal speed. At this terminal speed, the net force on the sphere is
 A. Directed downward.
 B. Zero.
 C. Directed upward.

Pulling Out of a Dive BIO

Falcons are excellent fliers that can reach very high speeds by diving nearly straight down. To pull out of such a dive, a falcon extends its wings and flies through a circular arc that redirects its motion. The forces on the falcon that control its motion are its weight and an upward lift force—like an airplane—due to the air flowing over its wings. At the bottom of the arc, as in Figure I.3, a falcon can easily achieve an acceleration of 15 m/s^2.

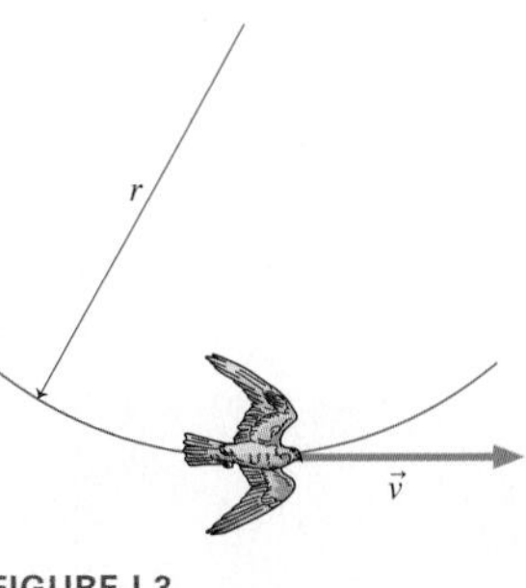

FIGURE I.3

14. At the bottom of the arc, as in Figure I.3, what is the direction of the net force on the falcon?
 A. To the left, opposite the motion
 B. To the right, in the direction of the motion
 C. Up
 D. Down
 E. The net force is zero.
15. Suppose the falcon weighs 8.0 N and is turning with an acceleration of 15 m/s^2 at the lowest point of the arc. What is the magnitude of the upward lift force at this instant?
 A. 8.0 N B. 12 N
 C. 16 N D. 20 N
16. A falcon starts from rest, does a free-fall dive from a height of 30 m, and then pulls out by flying in a circular arc of radius 50 m. Which segment of the motion has a higher acceleration?
 A. The free-fall dive
 B. The circular arc
 C. The two accelerations are equal.

Bending Beams

If you bend a rod down, it compresses the lower side of the rod and stretches the top, resulting in a restoring force. Figure I.4 shows a beam of length L, width w, and thickness t fixed at one end and free to move at the other. Deflecting the end of the beam causes a restoring force F at the end of the beam. The magnitude of the restoring force F depends on the dimensions of the beam, the Young's modulus Y for the material, and the deflection d. For small values of the deflection, the restoring force is

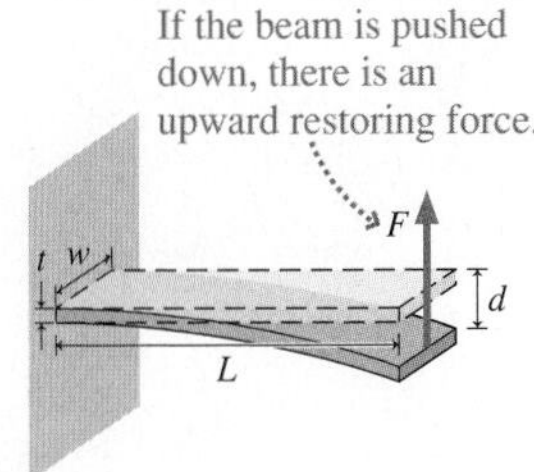

If the beam is pushed down, there is an upward restoring force.
F
t
w
d
L

FIGURE I.4

$$F = \left[\frac{Ywt^3}{4L^3}\right]d$$

This is similar to the formula for the restoring force of a spring, with the quantity in brackets playing the role of the spring constant k.

When a 70 kg man stands on the end of a springboard (a type of diving board), the board deflects by 4.0 cm.

17. If a 35 kg child stands at the end of the board, the deflection is
 A. 1.0 cm. B. 2.0 cm.
 C. 3.0 cm. D. 4.0 cm.
18. A 70 kg man jumps up and lands on the end of the board, deflecting it by 12 cm. At this instant, what is the approximate magnitude of the upward force the board exerts on his feet?
 A. 700 N B. 1400 N
 C. 2100 N D. 2800 N
19. If the board is replaced by one that is half the length but otherwise identical, how much will it deflect when a 70 kg man stands on the end?
 A. 0.50 cm B. 1.0 cm
 C. 2.0 cm D. 4.0 cm

Additional Integrated Problems

20. You go to the playground and slide down the slide, a 3.0-m-long ramp at an angle of 40° with respect to horizontal. The pants that you've worn aren't very slippery; the coefficient of kinetic friction between your pants and the slide is $\mu_k = 0.45$. A friend gives you a very slight push to get you started. How long does it take you to reach the bottom of the slide?
21. If you stand on a scale at the equator, the scale will read slightly less than your true weight due to your circular motion with the rotation of the earth.
 a. Draw a free-body diagram to show why this is so.
 b. By how much is the scale reading reduced for a person with a true weight of 800 N?
22. Dolphins and other sea creatures can leap to great heights by swimming straight up and exiting the water at a high speed. A 210 kg dolphin leaps straight up to a height of 7.0 m. When the dolphin reenters the water, drag from the water brings it to a stop in 1.5 m. Assuming that the force of the water on the dolphin stays constant as it slows down,
 a. How much time does it take for the dolphin to come to rest?
 b. What is the force of the water on the dolphin as it is coming to rest?

PART II

Conservation Laws

The kestrel is pulling in its wings to begin a steep dive, in which it can achieve a speed of 60 mph. How does the bird achieve such a speed, and why does this speed help the kestrel catch its prey? Such questions are best answered by considering the conservation of energy and momentum.

OVERVIEW

Why Some Things Stay the Same

Part I of this textbook was about *change*. Simple observations show us that most things in the world around us are changing. Even so, there are some things that *don't* change even as everything else is changing around them. Our emphasis in Part II will be on things that stay the same.

Consider, for example, a strong, sealed box in which you have replaced all the air with a mixture of hydrogen and oxygen. The mass of the box plus the gases inside is 600.0 g. Now, suppose you use a spark to ignite the hydrogen and oxygen. As you know, this is an explosive reaction, with the hydrogen and oxygen combining to create water—and quite a bang. But the strong box contains the explosion and all of its products.

What is the mass of the box after the reaction? The gas inside the box is different now, but a careful measurement would reveal that the mass hasn't changed—it's still 600.0 g! We say that the mass is *conserved*. Of course, this is true only if the box has stayed sealed. For conservation of mass to apply, the system must be *closed*.

Conservation Laws

A closed system of interacting particles has another remarkable property. Each system is characterized by a certain number, and no matter how complex the interactions, the value of this number never changes. This number is called the *energy* of the system, and the fact that it never changes is called the *law of conservation of energy*. It is, perhaps, the single most important physical law ever discovered.

The law of conservation of energy is much more general than Newton's laws. Energy can be converted to many different forms, and, in all cases, the total energy stays the same:

- Gasoline, diesel, and jet engines convert the energy of a fuel into the mechanical energy of moving pistons, wheels, and gears.
- A solar cell converts the electromagnetic energy of light into electrical energy.
- An organism converts the chemical energy of food into a variety of other forms of energy, including kinetic energy, sound energy, and thermal energy.

Energy will be *the* most important concept throughout the remainder of this textbook, and much of Part II will focus on understanding what energy is and how it is used.

But energy is not the only conserved quantity. We will begin Part II with the study of two other quantities that are conserved in a closed system: *momentum* and *angular momentum*. Their conservation will help us understand a wide range of physical processes, from the forces when two rams butt heads to the graceful spins of ice skaters.

Conservation laws will give us a new and different *perspective* on motion. Some situations are most easily analyzed from the perspective of Newton's laws, but others make much more sense when analyzed from a conservation-law perspective. An important goal of Part II is to learn which perspective is best for a given problem.

9 Momentum

Male rams butt heads at high speeds in a ritual to assert their dominance. How can the force of this collision be minimized so as to avoid damage to their brains?

LOOKING AHEAD ▸▸

The goals of Chapter 9 are to introduce the ideas of impulse, momentum, and angular momentum and to learn a new problem-solving strategy based on conservation laws.

Impulse

We'll begin by studying what happens to an object subject to a strong but short-duration **impulsive force.**

We say that the club has delivered an **impulse** to the ball.

A golf ball being struck by a club is subject to a brief but very large force.

Looking Back ◂◂
5.2 Newton's second law

Momentum

The **momentum** of an object is the product of its mass and its velocity. A heavy, fast-moving object such as a car has much greater momentum than a light, slow-moving one such as a falling raindrop.

The momentum of a moving car is a billion times greater than that of a falling raindrop.

Momentum and Impulse

We'll discover how an impulse delivered to an object *changes* the object's momentum.

The player's head delivers an impulse to the ball, changing its momentum and thus its direction.

Conservation of Momentum

When two or more objects interact only with each other, the total momentum of these objects is conserved—it is the same after the interaction as it was before. This **law of conservation of momentum** will lead us to a powerful new *before-and-after* problem-solving strategy.

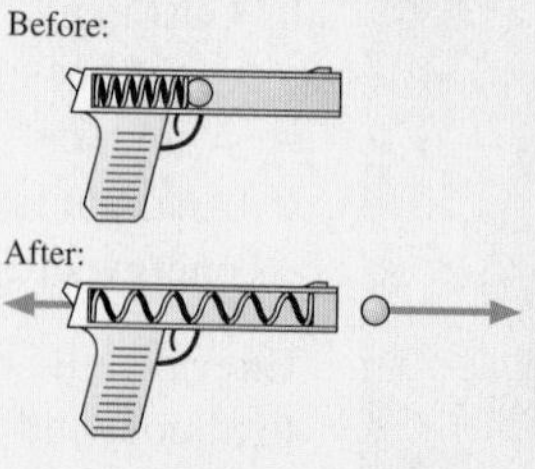

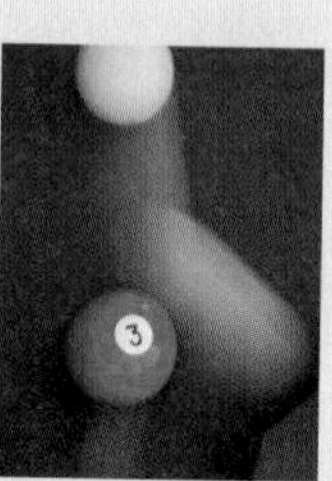

Conservation of momentum is an important tool for analyzing *explosions*, like the ball shot from this toy gun, or *collisions*, like two pool balls hitting and flying apart.

Looking Back ◂◂
4.8 Newton's third law
5.7 Interacting objects

Angular Momentum

Rotating objects have *angular* momentum. Angular momentum is changed by an applied torque. When no torques act on a rotating object, its angular momentum is conserved.

Conservation of angular momentum causes this skater's spin to increase as she pulls her arms in toward her body.

Looking Back ◂◂
7.2, 7.4 Torque and moment of inertia

9.1 Impulse

Suppose that two or more objects have an intense and perhaps complex interaction, such as a collision or an explosion. Our goal is to find a relationship between the velocities of the objects before the interaction and their velocities after the interaction. We'll start by looking at collisions.

A **collision** is a short-duration interaction between two objects. The collision between a tennis ball and a racket, or a baseball and a bat, may seem instantaneous to your eye, but that is a limitation of your perception. A careful look at the tennis ball/racket collision in **FIGURE 9.1** reveals that the right side of the ball is flattened and pressed up against the strings of the racket. It takes time to compress the ball, and more time for the ball to re-expand as it leaves the racket.

The duration of a collision depends on the materials from which the objects are made, but 1 to 10 ms (0.001 to 0.010 s) is typical. This is the time during which the two objects are in contact with each other. The harder the objects, the shorter the contact time. A collision between two steel balls lasts less than 1 ms, while that between a tennis ball and racket might last 10 ms.

FIGURE 9.1 A tennis ball collides with a racket. Notice that the right side of the ball is flattened.

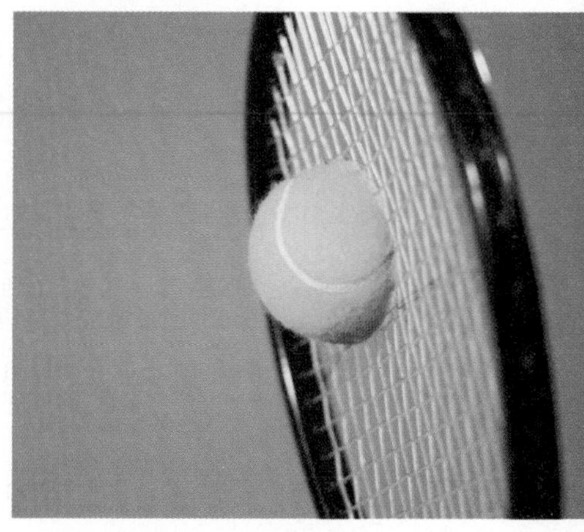

FIGURE 9.2 A sequence of high-speed photos of a soccer ball being kicked.

Let's begin our discussion by considering a collision that most of us have experienced: kicking a soccer ball. A sequence of high-speed photos of a soccer kick is shown in **FIGURE 9.2**. As the foot and the ball just come into contact, as shown in the left frame, the ball is just beginning to compress. By the middle frame of Figure 9.2, the ball has sped up and become greatly compressed. Finally, as shown in the right frame, the ball, now moving very fast, is again only slightly compressed.

The amount by which the ball is compressed is a measure of the magnitude of the force the foot exerts on the ball; more compression indicates a greater force. If we were to graph this force versus time, it would look something like **FIGURE 9.3**. The force is zero until the foot first contacts the ball, rises quickly to a maximum value, and then falls back to zero as the ball leaves the foot. Thus there is a well-defined duration Δt of the force. A large force like this exerted during a short interval of time is called an **impulsive force.** The forces of a hammer on a nail and of a bat on a baseball are other examples of impulsive forces.

FIGURE 9.3 The force on a soccer ball changes rapidly.

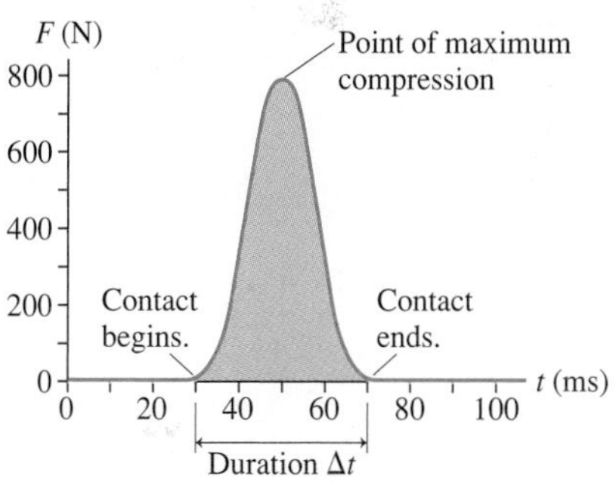

A harder kick (i.e., a taller force curve) or a kick of longer duration (a wider force curve) causes the ball to leave the kicker's foot with a higher speed; that is, the *effect* of the kick is larger. Now a taller or wider force-versus-time curve has a larger *area* between the curve and the axis (i.e., the area "under" the force curve is larger), so we can say that **the effect of an impulsive force is proportional to the area under the force-versus-time curve.** This area, shown in **FIGURE 9.4a**, is called the **impulse** J of the force.

Impulsive forces can be complex, and the shape of the force-versus-time graph often changes in a complicated way. Consequently, it is often useful to think of the collision in terms of an *average* force F_{avg}. As **FIGURE 9.4b** shows, F_{avg} is defined to be the constant force that has the same duration Δt and the same area under the force curve as the real force. You can see from the figure that the area under the force curve can be written simply as $F_{avg}\,\Delta t$. Thus

$$\text{impulse } J = \text{area under the force curve} = F_{avg}\,\Delta t \tag{9.1}$$

Impulse due to a force acting for a duration Δt

FIGURE 9.4 Looking at the impulse graphically.

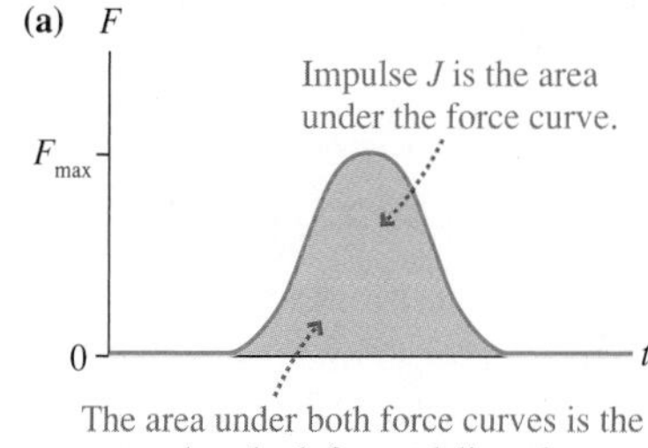

The area under both force curves is the same; thus, both forces deliver the same impulse. This means that they have the same effect on the object.

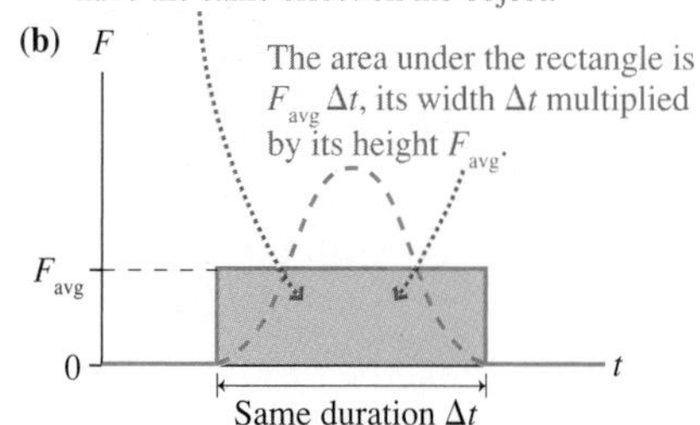

From Equation 9.1 we see that impulse has units of N · s, but you should be able to show that N · s are equivalent to kg · m/s. We'll see shortly why the latter are the preferred units for impulse.

So far, we've been assuming the force is directed along a coordinate axis, such as the x-axis. In this case impulse is a *signed* quantity—it can be positive or negative. A positive impulse results from an average force directed in the positive x-direction (that is, F_{avg} is positive), while a negative impulse is due to a force directed in the negative x-direction (F_{avg} is negative). More generally, the impulse is a *vector* quantity pointing in the direction of the average force vector:

$$\vec{J} = \vec{F}_{avg}\,\Delta t \tag{9.2}$$

EXAMPLE 9.1 Finding the impulse on a bouncing ball

A rubber ball experiences the force shown in FIGURE 9.5 as it bounces off the floor.

a. What is the impulse on the ball?
b. What is the average force on the ball?

FIGURE 9.5 The force of the floor on a bouncing ball.

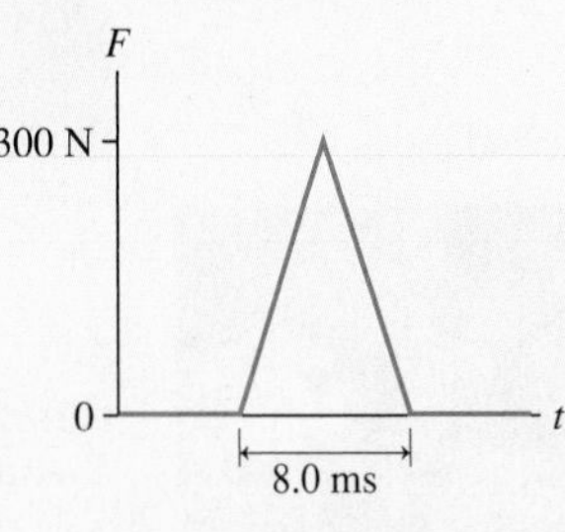
F
300 N
0
t
8.0 ms

PREPARE The impulse is the area under the force curve. Here the shape of the graph is triangular, so we'll need to use the fact that the area of a triangle is $\frac{1}{2}$ × height × base.

SOLVE a. The impulse is

$$J = \tfrac{1}{2}(300\text{ N})(0.0080\text{ s}) = 1.2\text{ N}\cdot\text{s} = 1.2\text{ kg}\cdot\text{m/s}$$

b. From Equation 9.1, $J = F_{avg}\,\Delta t$, we can find the average force that would give this same impulse:

$$F_{avg} = \frac{J}{\Delta t} = \frac{1.2\text{ N}\cdot\text{s}}{0.0080\text{ s}} = 150\text{ N}$$

ASSESS In this particular example, the average value of the force is half the maximum value. This is not surprising for a triangular force because the area of a triangle is *half* the base times the height.

9.2 Momentum and the Impulse-Momentum Theorem

We've noted that the effect of an impulsive force depends on the impulse delivered to the object. The effect also depends on the object's mass. Our experience tells us that giving a kick to a heavy object will change its velocity much less than giving the same kick to a light object. We want now to find a quantitative relationship for impulse, mass, and velocity change.

FIGURE 9.6 The stick exerts an impulse on the puck, changing its speed.

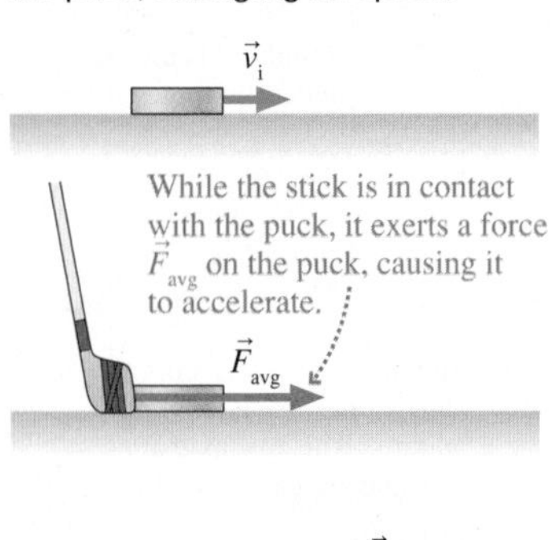
$\vec{v}_i$
While the stick is in contact with the puck, it exerts a force $\vec{F}_{avg}$ on the puck, causing it to accelerate.
$\vec{F}_{avg}$
$\vec{v}_f$

Consider the puck of mass m in FIGURE 9.6, sliding with an initial velocity $\vec{v}_i$. It is struck by a hockey stick that delivers an impulse $\vec{J} = \vec{F}_{avg}\,\Delta t$ to the puck. After the impulse, the puck leaves the stick with a final velocity $\vec{v}_f$. How is this final velocity related to the initial velocity?

From Newton's second law, the average acceleration of the puck during the time the stick is in contact with it is

$$\vec{a}_{avg} = \frac{\vec{F}_{avg}}{m} \tag{9.3}$$

The average acceleration is related to the change in the velocity by

$$\vec{a}_{avg} = \frac{\Delta\vec{v}}{\Delta t} = \frac{\vec{v}_f - \vec{v}_i}{\Delta t} \tag{9.4}$$

Combining Equations 9.3 and 9.4, we have

$$\frac{\vec{F}_{avg}}{m} = \vec{a}_{avg} = \frac{\vec{v}_f - \vec{v}_i}{\Delta t}$$

or, rearranging,

$$\vec{F}_{avg}\,\Delta t = m\vec{v}_f - m\vec{v}_i \tag{9.5}$$

We recognize the left side of this equation as the impulse $\vec{J}$. The right side is the *change* in the quantity $m\vec{v}$. This quantity, the product of the object's mass and velocity, is called the **momentum** of the object. The symbol for momentum is $\vec{p}$:

$$\vec{p} = m\vec{v} \tag{9.6}$$

Momentum of an object of mass m and velocity $\vec{v}$

From Equation 9.6, the units of momentum are those of mass times velocity, or kg · m/s. We noted above that kg · m/s are the preferred units of impulse. Now we see that the reason for that preference is to match the units of momentum.

FIGURE 9.7 shows that the momentum $\vec{p}$ is a *vector* quantity that points in the same direction as the velocity vector $\vec{v}$. Like any vector, $\vec{p}$ can be decomposed into x- and y-components. Equation 9.6, which is a vector equation, is a shorthand way to write the two equations

$$\begin{aligned} p_x &= mv_x \\ p_y &= mv_y \end{aligned} \tag{9.7}$$

NOTE ► One of the most common errors in momentum problems is failure to use the correct signs. The momentum component p_x has the same sign as v_x. Just like velocity, momentum is positive for a particle moving to the right (on the x-axis) or up (on the y-axis), but *negative* for a particle moving to the left or down. ◄

The *magnitude* of an object's momentum is simply the product of the object's mass and speed, or $p = mv$. A heavy, fast-moving object will have a great deal of momentum, while a light, slow-moving object will have very little. Two objects with very different masses can have similar momenta if their speeds are very different as well. Table 9.1 gives some typical values of the momenta (the plural of *momentum*) of various moving objects. You can see that the momenta of a bullet and a fastball are similar. The momentum of a moving car is almost a billion times greater than that of a falling raindrop.

FIGURE 9.7 A particle's momentum vector $\vec{p}$ can be decomposed into x- and y-components.

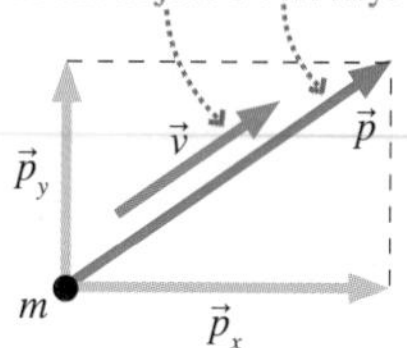
Momentum is a vector that points in the same direction as the object's velocity.
$\vec{v}$
$\vec{p}$
$\vec{p}_y$
m
$\vec{p}_x$

TABLE 9.1 Some typical momenta (approximate)

Object	Mass (kg)	Speed (m/s)	Momentum (kg · m/s)
Falling raindrop	2×10^{-5}	5	10^{-4}
Bullet	0.004	500	2
Pitched baseball	0.15	40	6
Running person	70	3	200
Car on highway	1000	30	3×10^4

The Impulse-Momentum Theorem

We can now write Equation 9.5 in terms of impulse and momentum:

$$\vec{J} = \vec{p}_f - \vec{p}_i = \Delta\vec{p} \tag{9.8}$$

Impulse-momentum theorem

where $\vec{p}_i = m\vec{v}_i$ is the object's initial momentum, $\vec{p}_f = m\vec{v}_f$ is its final momentum after the impulse, and $\Delta\vec{p} = \vec{p}_f - \vec{p}_i$ is the *change* in its momentum. This expression is known as the **impulse-momentum theorem.** It states that **an impulse delivered to an object causes the object's momentum to change.** That is, the *effect* of an impulsive force is to change the object's momentum from $\vec{p}_i$ to

$$\vec{p}_f = \vec{p}_i + \vec{J} \tag{9.9}$$

Equation 9.8 can also be written in terms of its x- and y-components as

$$\begin{aligned} J_x &= \Delta p_x = (p_x)_f - (p_x)_i = m(v_x)_f - m(v_x)_i \\ J_y &= \Delta p_y = (p_y)_f - (p_y)_i = m(v_y)_f - m(v_y)_i \end{aligned} \tag{9.10}$$

The impulse-momentum theorem is illustrated by two examples in FIGURE 9.8 on the next page. In the first, the putter strikes the ball, exerting a force on it and delivering an impulse $\vec{J} = \vec{F}_{avg}\,\Delta t$. Notice that the direction of the impulse is the same as

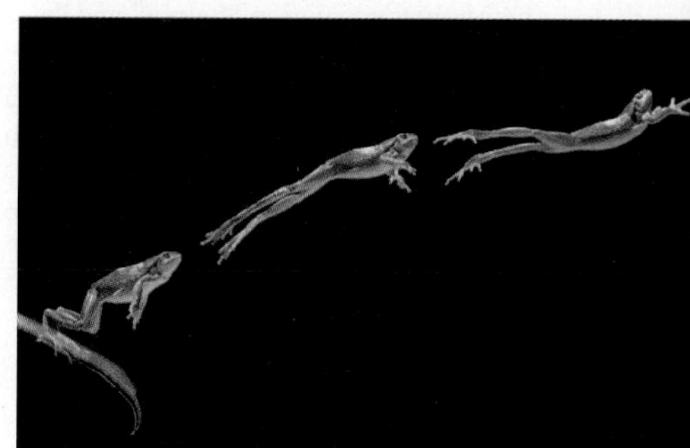

Legging it BIO A frog making a jump wants to gain as much momentum as possible before leaving the ground. This means that he wants the greatest impulse $J = F_{avg}\,\Delta t$ delivered to him by the ground. There is a maximum force that muscles can exert, limiting F_{avg}. But the time interval Δt over which the force is exerted can be greatly increased by having long legs. Many animals that are good jumpers have particularly long legs.

FIGURE 9.8 Impulse causes a *change* in momentum.

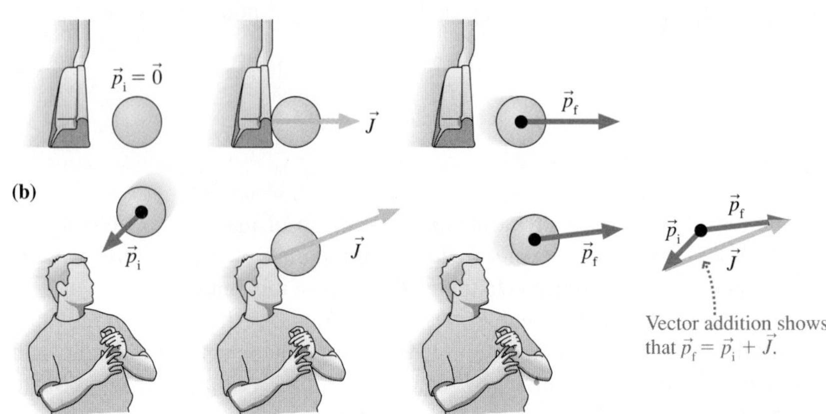

that of the force. Because $\vec{p}_i = \vec{0}$ in this situation, we can use the impulse-momentum theorem to find that the ball leaves the putter with momentum $\vec{p}_f = \vec{p}_i + \vec{J} = \vec{J}$.

NOTE ► You can think of the putter as changing the ball's momentum by transferring momentum to it as an impulse. Thus we say the putter *delivers* an impulse to the ball, and the ball *receives* an impulse from the putter. ◄

The soccer player in Figure 9.8b presents a more complicated case. Here, the initial momentum of the ball is directed downward to the left. The impulse delivered to it by the player's head, upward to the right, is strong enough to reverse the ball's motion and send it off in a new direction. The graphical addition of vectors in Figure 9.8b again shows that $\vec{p}_f = \vec{p}_i + \vec{J}$.

EXAMPLE 9.2 Calculating the change in momentum

A ball of mass $m = 0.25$ kg rolling to the right at 1.3 m/s strikes a wall and rebounds to the left at 1.1 m/s. What is the change in the ball's momentum? What is the impulse delivered to it by the wall?

PREPARE A visual overview of the ball bouncing is shown in FIGURE 9.9. This is a new kind of visual overview, one in which we show the situation "before" and "after" the interaction. We'll have more to say about before-and-after pictures in the next section. The ball is moving along the x-axis, so we'll write the momentum in component form, as in Equation 9.7. The change in momentum is then the difference between the final and initial values of the momentum. By the impulse-momentum theorem, the impulse is equal to this change in momentum.

FIGURE 9.9 Visual overview for a ball bouncing off a wall.

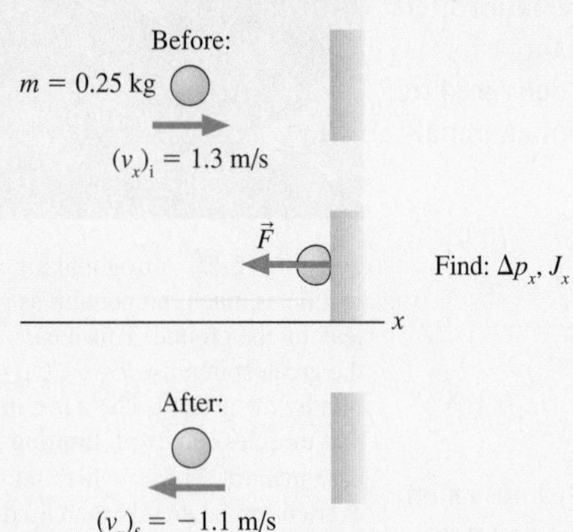

SOLVE The x-component of the initial momentum is

$$(p_x)_i = m(v_x)_i = (0.25 \text{ kg})(1.3 \text{ m/s}) = 0.33 \text{ kg} \cdot \text{m/s}$$

The y-component of the momentum is zero both before and after the bounce. After the ball rebounds, the x-component of its momentum is

$$(p_x)_f = m(v_x)_f = (0.25 \text{ kg})(-1.1 \text{ m/s}) = -0.28 \text{ kg} \cdot \text{m/s}$$

It is particularly important to notice that the x-component of the momentum, like that of the velocity, is negative. This indicates that the ball is moving to the *left*. The change in momentum is

$$\Delta p_x = (p_x)_f - (p_x)_i = (-0.28 \text{ kg} \cdot \text{m/s}) - (0.33 \text{ kg} \cdot \text{m/s})$$
$$= -0.61 \text{ kg} \cdot \text{m/s}$$

The change in the momentum is negative. By the impulse-momentum theorem, the impulse delivered to the ball by the wall is equal to this change, so

$$J_x = \Delta p_x = -0.61 \text{ kg} \cdot \text{m/s}$$

ASSESS The impulse is negative, indicating that the force causing the impulse is pointing to the left, which makes sense.

TRY IT YOURSELF

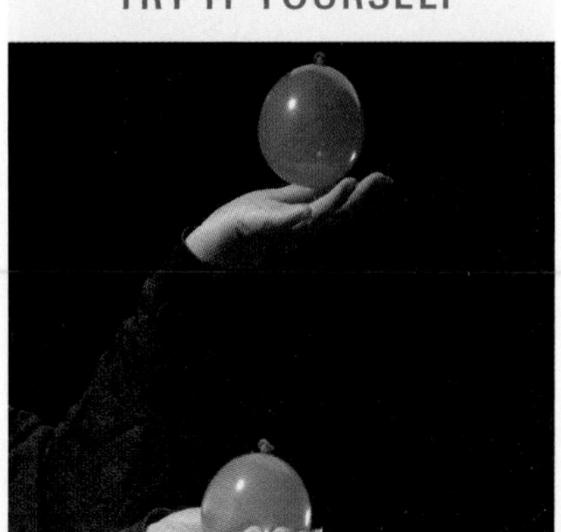

▶ Water balloon catch If you've ever tried to catch a water balloon, you may have learned the hard way not to catch it with your arms rigidly extended. The brief collision time implies a large, balloon-bursting force. A better way to catch a water balloon is to pull your arms in toward your body as you catch it, lengthening the collision time and hence reducing the force on the balloon.

An interesting application of the impulse-momentum theorem is to the question of how to slow down a fast-moving object in the gentlest possible way. For instance, a car is headed for a collision with a bridge abutment. How can this crash be made survivable? How do the rams in the chapter-opening photo avoid injury when they collide?

In these examples, the object has momentum $\vec{p}_\mathrm{i}$ just before impact and zero momentum after (i.e., $\vec{p}_\mathrm{f} = \vec{0}$). The impulse-momentum theorem tells us that

$$\vec{J} = \vec{F}_\mathrm{avg}\,\Delta t = \Delta\vec{p} = \vec{p}_\mathrm{f} - \vec{p}_\mathrm{i} = -\vec{p}_\mathrm{i}$$

or

$$\vec{F}_\mathrm{avg} = -\frac{\vec{p}_\mathrm{i}}{\Delta t} \tag{9.11}$$

F
p.114
Δt
INVERSE

That is, the average force needed to stop an object is *inversely proportional* to the duration Δt of the collision. **If the duration of the collision can be increased, the force of the impact will be decreased.** This is the principle used in most impact-lessening techniques.

For example, obstacles such as bridge abutments are made safer by placing a line of water-filled barrels in front of them. Water is heavy but deformable. In case of a collision, the time it takes for the car to plow through these barrels is much longer than the time it would take it to stop if it hit the abutment head-on. The force on the car (and on the driver from his or her seat belt) is greatly reduced by the longer-duration collision with the barrels.

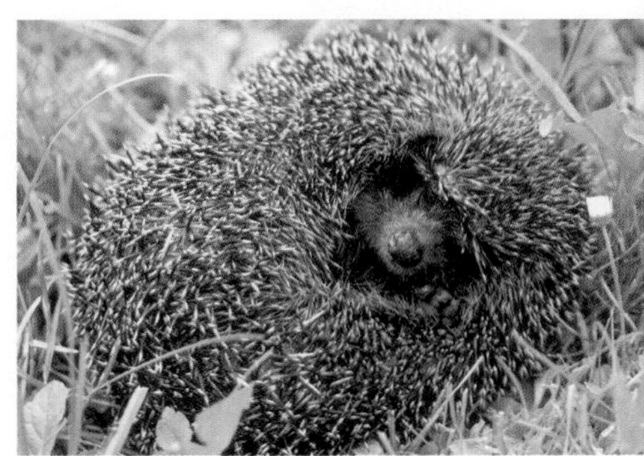

A hedgehog is its own crash cushion!

The spines of a hedgehog obviously help protect it from predators. But they serve another function as well. If a hedgehog falls from a tree—a not uncommon occurrence—it simply rolls itself into a ball before it lands. Indeed, hedgehogs have been observed to purposely descend to the ground by simply dropping from the tree. Its thick spines then cushion the blow by increasing the time it takes for the animal to come to rest. Along with its small size, this adaptation allows the hedgehog to easily survive long falls unhurt.

The butting rams shown in the photo at the beginning of this chapter also have adaptations that allow them to collide at high speeds without injury to their brains. The cranium has a double wall to prevent skull injuries, and there is a thick spongy mass that increases the time it takes for the brain to come to rest upon impact, again reducing the magnitude of the force on the brain.

Total Momentum

If we have more than one object moving—a *system* of particles—then the system as a whole has an overall momentum. The **total momentum** $\vec{P}$ (note the capital P) of a system of particles is the vector sum of the momenta of the individual particles:

$$\vec{P} = \vec{p}_1 + \vec{p}_2 + \vec{p}_3 + \cdots$$

FIGURE 9.10 shows how the momentum vectors of three moving pool balls are graphically added to find the total momentum. The concept of total momentum will be of key importance when we discuss the conservation law for momentum in Section 9.4.

FIGURE 9.10 The total momentum of three pool balls.

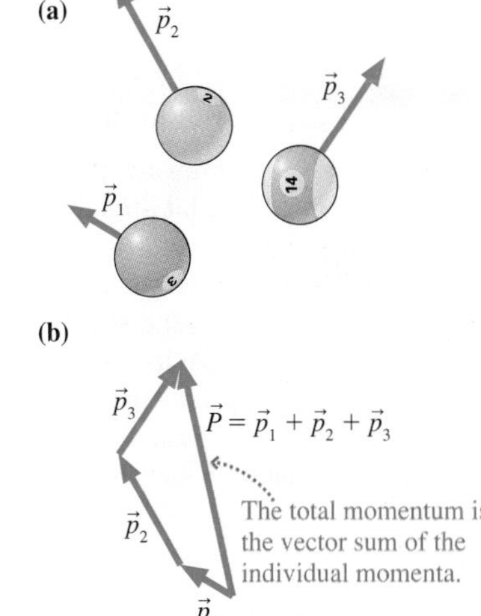

STOP TO THINK 9.1 The cart's change of momentum is

A. $-30\text{ kg} \cdot \text{m/s}$.
B. $-20\text{ kg} \cdot \text{m/s}$.
C. $-10\text{ kg} \cdot \text{m/s}$.
D. $10\text{ kg} \cdot \text{m/s}$.
E. $20\text{ kg} \cdot \text{m/s}$.
F. $30\text{ kg} \cdot \text{m/s}$.

Before:

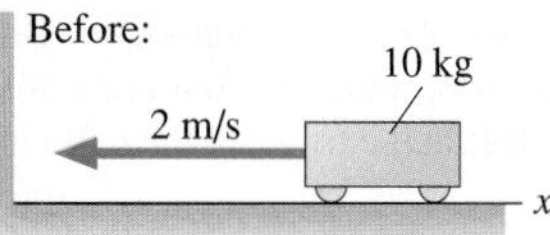

After:

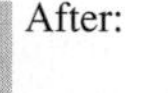

9.3 Solving Impulse and Momentum Problems

Visual overviews have become an important problem-solving tool. The visual overviews and free-body diagrams that you learned to draw in Chapters 1–8 were oriented toward the use of Newton's laws and a subsequent kinematical analysis. Now we are interested in making a connection between "before" and "after."

TACTICS BOX 9.1 Drawing a before-and-after visual overview

❶ **Sketch the situation.** Use two drawings, labeled "Before" and "After," to show the objects *immediately before* they interact and again *immediately after* they interact.
❷ **Establish a coordinate system.** Select your axes to match the motion.
❸ **Define symbols.** Define symbols for the masses and for the velocities before and after the interaction. Position and time are not needed.
❹ **List known information.** List the values of quantities known from the problem statement or that can be found quickly with simple geometry or unit conversions. Before-and-after pictures are usually simpler than the pictures you used for dynamics problems, so listing known information on the sketch is often adequate.
❺ **Identify the desired unknowns.** What quantity or quantities will allow you to answer the question? These should have been defined as symbols in step 3.

Exercises 9–11

EXAMPLE 9.3 Force in hitting a baseball

A 150 g baseball is thrown with a speed of 20 m/s. It is hit straight back toward the pitcher at a speed of 40 m/s. The impulsive force of the bat on the ball has the shape shown in FIGURE 9.11. What is the *maximum* force F_{max} that the bat exerts on the ball? What is the *average* force that the bat exerts on the ball?

FIGURE 9.11 The interaction force between the baseball and the bat.

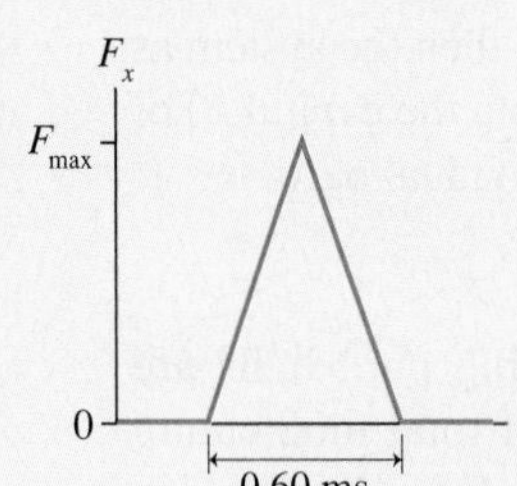

PREPARE We can model the interaction as a collision. FIGURE 9.12 is a before-and-after visual overview in which the steps from Tactics Box 9.1 are explicitly noted. Because F_x is positive (a force to the right), we know the ball was initially moving toward the left and is hit back toward the right. Thus we converted the statements about *speeds* into information about *velocities,* with $(v_x)_i$ negative.

SOLVE In the last several chapters we've started the mathematical solution with Newton's second law. Now we want to use the impulse-momentum theorem:

$$\Delta p_x = J_x = \text{area under the force curve}$$

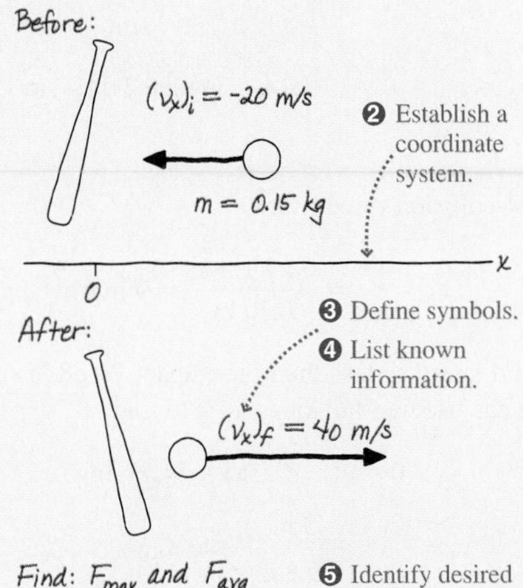

FIGURE 9.12 A before-and-after visual overview.

We know the velocities before and after the collision, so we can find the change in the ball's momentum:

$$\Delta p_x = m(v_x)_f - m(v_x)_i = (0.15 \text{ kg})(40 \text{ m/s} - (-20 \text{ m/s}))$$
$$= 9.0 \text{ kg} \cdot \text{m/s}$$

The force curve is a triangle with height F_{max} and width 0.60 ms. As in Example 9.1, the area under the curve is

$$J_x = \text{area} = \tfrac{1}{2} \times F_{max} \times (6.0 \times 10^{-4} \text{ s})$$
$$= (F_{max})(3.0 \times 10^{-4})$$

According to the impulse-momentum theorem, $\Delta p_x = J_x$, so we have

$$9.0 \text{ kg} \cdot \text{m/s} = (F_{max})(3.0 \times 10^{-4} \text{ s})$$

Thus the *maximum* force is

$$F_{max} = \frac{9.0 \text{ kg} \cdot \text{m/s}}{3.0 \times 10^{-4} \text{ s}} = 30{,}000 \text{ N}$$

Using Equation 9.1, we find that the *average* force, which depends on the collision duration $\Delta t = 6.0 \times 10^{-4}$ s, has the smaller value:

$$F_{avg} = \frac{J_x}{\Delta t} = \frac{\Delta p_x}{\Delta t} = \frac{9.0 \text{ kg} \cdot \text{m/s}}{6.0 \times 10^{-4} \text{ s}} = 15{,}000 \text{ N}$$

ASSESS F_{max} is a large force, but quite typical of the impulsive forces during collisions.

The Impulse Approximation

When two objects interact during a collision or other brief interaction, such as that between the bat and ball of Example 9.3, the forces *between* them are generally quite large. Other forces may also act on the interacting objects, but usually these forces are *much* smaller than the interaction forces. In Example 9.3, for example, the 1.5 N weight of the ball is vastly less than the 30,000 N force of the bat on the ball. We can reasonably neglect these small forces *during* the brief time of the impulsive force. Doing so is called the **impulse approximation.**

When we use the impulse approximation, $(p_x)_i$ and $(p_x)_f$—and $(v_x)_i$ and $(v_x)_f$—are then the momenta (and velocities) *immediately* before and *immediately* after the collision. For example, the velocities in Example 9.3 are those of the ball just before and after it collides with the bat. We could then do a follow-up problem, including weight and drag, to find the ball's speed a second later as the second baseman catches it.

EXAMPLE 9.4 Height of a bouncing ball

A 100 g rubber ball is thrown straight down onto a hard floor so that it strikes the floor with a speed of 11 m/s. **FIGURE 9.13** shows the force that the floor exerts on the ball. Estimate the height of the ball's bounce.

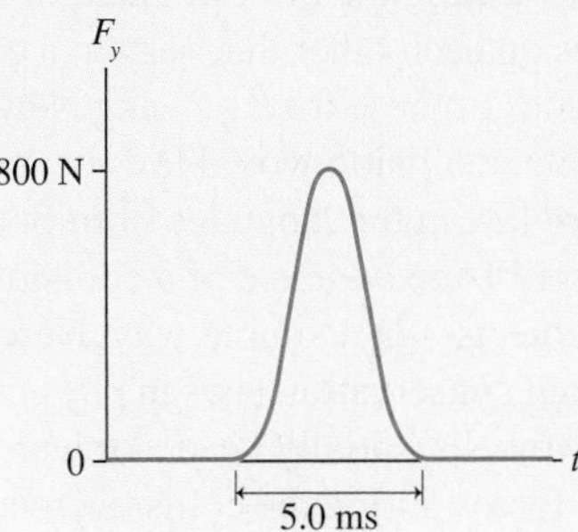

FIGURE 9.13 The force of the floor on a bouncing rubber ball.

PREPARE The ball experiences an impulsive force while in contact with the floor. Using the impulse approximation, we'll neglect the ball's weight during these 5.0 ms. The ball's rise after the bounce is free-fall motion—that is, motion subject only to the force of gravity. We'll use free-fall kinematics to describe the motion after the bounce.

FIGURE 9.14 on the next page is a visual overview. Here we have a two-part problem, an impulsive collision followed by upward free fall. The overview thus shows the ball just before the collision, where we label its velocity as v_{1y}; just after the collision, where its velocity is v_{2y}; and at the highest point of its rising free fall, where its velocity is $v_{3y} = 0$.

Continued

FIGURE 9.14 Before-and-after visual overview for a bouncing ball.

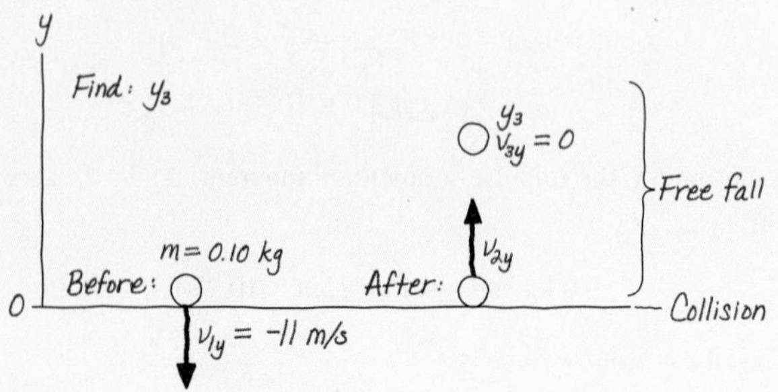

SOLVE The impulse-momentum theorem tells us that $J_y = \Delta p_y = p_{2y} - p_{1y}$, so that $p_{2y} = p_{1y} + J_y$. The initial momentum, just before the collision, is $p_{1y} = mv_{1y} = (0.10\text{ kg})(-11\text{ m/s}) = -1.1\text{ kg}\cdot\text{m/s}$.

Next, we need to find the impulse J_y, which is the area under the curve in Figure 9.13. Because the force is given as a smooth curve, we'll have to *estimate* this area. Recall that the area can be written as $F_{avg}\,\Delta t$. From the curve, we might estimate F_{avg} to be about 400 N, or half the maximum value of the force. With this estimate we have

$$J_y = \text{area under the force curve} \approx (400\text{ N}) \times (0.0050\text{ s})$$
$$= 2.0\text{ N}\cdot\text{s} = 2.0\text{ kg}\cdot\text{m/s}$$

Thus

$$p_{2y} = p_{1y} + J_y = (-1.1\text{ kg}\cdot\text{m/s}) + 2.0\text{ kg}\cdot\text{m/s}$$
$$= 0.9\text{ kg}\cdot\text{m/s}$$

and the post-collision velocity is

$$v_{2y} = \frac{p_{2y}}{m} = \frac{0.9\text{ kg}\cdot\text{m/s}}{0.10\text{ kg}} = 9\text{ m/s}$$

The rebound speed is less than the impact speed, as expected. Finally, we can use free-fall kinematics to find

$$v_{3y}^2 = 0 = v_{2y}^2 - 2g\,\Delta y = v_{2y}^2 - 2gy_3$$
$$y_3 = \frac{v_{2y}^2}{2g} = \frac{(9\text{ m/s})^2}{2(9.8\text{ m/s}^2)} = 4\text{ m}$$

We estimate that the ball bounces to a height of 4 m.

ASSESS This is a reasonable height for a rubber ball thrown down quite hard.

STOP TO THINK 9.2 A 10 g rubber ball and a 10 g clay ball are each thrown at a wall with equal speeds. The rubber ball bounces; the clay ball sticks. Which ball receives the greater impulse from the wall?

A. The clay ball receives a greater impulse because it sticks.
B. The rubber ball receives a greater impulse because it bounces.
C. They receive equal impulses because they have equal momenta.
D. Neither receives an impulse because the wall doesn't move.

FIGURE 9.15 A collision between two balls.

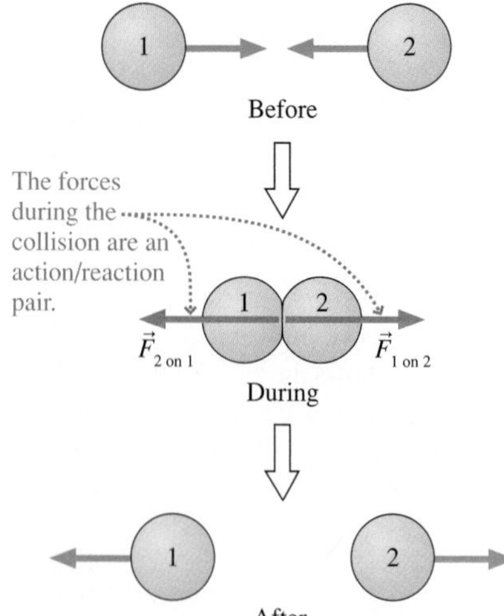

9.4 Conservation of Momentum

The impulse-momentum theorem was derived from Newton's second law and is really just an alternative way of looking at that law. It is used in the context of single-particle dynamics, much as we used Newton's law in Chapters 4–7.

However, consider two objects, such as the rams shown in the opening photo of this chapter, that interact during the brief moment of a collision. During a collision, two objects exert forces on each other that vary in a complex way. We usually don't even know the magnitudes of these forces. Using Newton's second law alone to predict the outcome of such a collision would thus be a daunting challenge. However, by using Newton's *third* law in the language of impulse and momentum, we'll find that it's possible to describe the *outcome* of a collision—the final speeds and directions of the colliding objects—in a simple way. Newton's third law will lead us to one of the most important conservation laws in physics.

FIGURE 9.15 shows two balls initially headed toward each other. The balls collide, then bounce apart. The forces during the collision, when the balls are interacting, are the action/reaction pair $\vec{F}_{1\text{on}2}$ and $\vec{F}_{2\text{on}1}$. For now, we'll continue to assume that the motion is one dimensional along the x-axis.

During the collision, the impulse J_{2x} delivered to ball 2 by ball 1 is the average value of $\vec{F}_{1\text{ on }2}$ multiplied by the collision time Δt. Likewise, the impulse J_{1x} delivered to ball 1 by ball 2 is the average value of $\vec{F}_{2\text{ on }1}$ multiplied by Δt. Because $\vec{F}_{1\text{ on }2}$ and $\vec{F}_{2\text{ on }1}$ form an action/reaction pair, they have equal magnitudes but opposite directions. As a result, the two impulses J_{1x} and J_{2x} are also equal in magnitude but opposite in sign, so that $J_{1x} = -J_{2x}$.

According to the impulse-momentum theorem, the change in the momentum of ball 1 is $\Delta p_{1x} = J_{1x}$ and the change in the momentum of ball 2 is $\Delta p_{2x} = J_{2x}$. Because $J_{1x} = -J_{2x}$, the change in the momentum of ball 1 is equal in magnitude but opposite in sign to the change in momentum of ball 2. If ball 1's momentum increases by a certain amount during the collision, ball 2's momentum will *decrease* by exactly the same amount. This implies that **the total momentum $P_x = p_{1x} + p_{2x}$ of the two balls is *unchanged* by the collision;** that is,

$$(P_x)_\text{f} = (P_x)_\text{i} \tag{9.12}$$

Because it doesn't change during the collision, we say that the x-component of total momentum is *conserved*. Equation 9.12 is our first example of a *conservation law*.

Law of Conservation of Momentum

The same arguments just presented for the two colliding balls can be extended to systems containing any number of objects. FIGURE 9.16 shows the idea. Each pair of particles in the system (the boundary of which is denoted by the red line) interacts via forces that are an action/reaction pair. Exactly as for the two-particle collision, the change in momentum of particle 2 due to the force from particle 3 is equal in magnitude, but opposite in direction, to the change in particle 3's momentum due to particle 2. The *net* change in the momentum of these two particles due to their interaction forces is thus zero. The same argument holds for every pair, with the result that, no matter how complicated the forces between the particles, **there is no change in the *total* momentum $\vec{P}$ of the system.** The total momentum of the system remains constant: It is *conserved*.

FIGURE 9.16 A system of three particles.

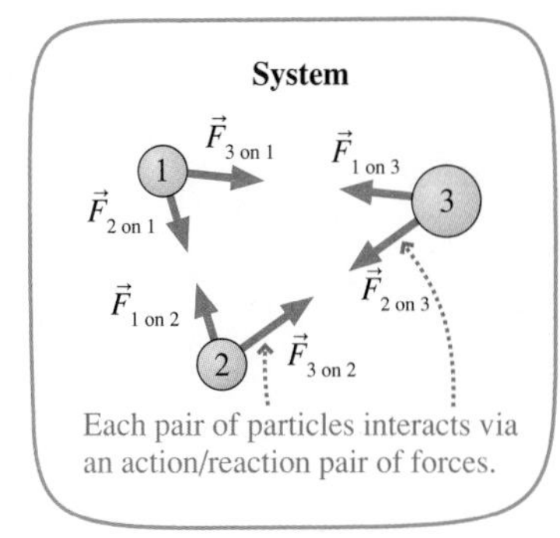

Figure 9.16 showed particles interacting only with other particles inside the system. Forces that act only between particles within the system are called **internal forces.** As we've just seen, **the total momentum of a system subject only to internal forces is conserved.**

Most systems are also subject to forces from agents outside the system. These forces are called **external forces.** For example, the system consisting of a student on a skateboard is subject to three external forces—the normal force of the ground on the skateboard, the force of gravity on the student, and the force of gravity on the board. How do external forces affect the momentum of a system of particles?

In FIGURE 9.17 we show the same three-particle system of Figure 9.16, but now with *external* forces acting on the three particles. These external forces *can* change the momentum of the system. During a time interval Δt, for instance, the external force $\vec{F}_{\text{ext on }1}$ acting on particle 1 changes its momentum, according to the impulse-momentum theorem, by $\Delta\vec{p}_1 = (\vec{F}_{\text{ext on }1})\Delta t$. The momenta of the other two particles change similarly. Thus the change in the total momentum is

$$\begin{aligned}\Delta\vec{P} &= \Delta\vec{p}_1 + \Delta\vec{p}_2 + \Delta\vec{p}_3\\ &= (\vec{F}_{\text{ext on }1}\Delta t) + (\vec{F}_{\text{ext on }2}\Delta t) + (\vec{F}_{\text{ext on }3}\Delta t)\\ &= (\vec{F}_{\text{ext on }1} + \vec{F}_{\text{ext on }2} + \vec{F}_{\text{ext on }3})\Delta t\\ &= \vec{F}_\text{net}\Delta t\end{aligned} \tag{9.13}$$

where $\vec{F}_\text{net}$ is the net force due to *external forces*.

FIGURE 9.17 A system of particles subject to external forces.

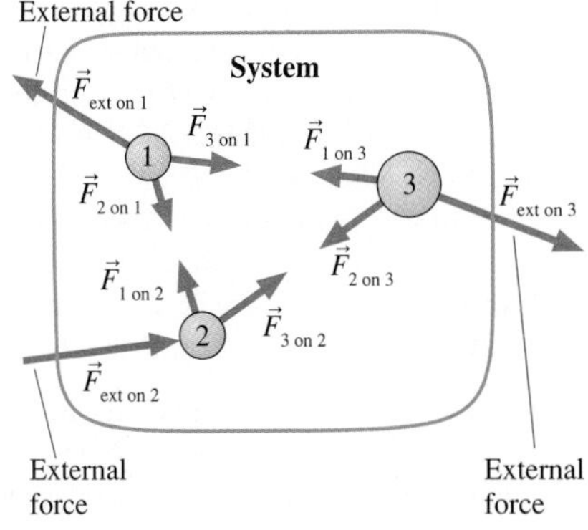

Equation 9.13 has a very important implication in the case where the net force on a system is zero: **If $\vec{F}_\text{net} = \vec{0}$ the *total* momentum $\vec{P}$ of the system does not change.** The total momentum remains constant, *regardless* of whatever interactions are going on *inside* the system.

Earlier, we found that a system's total momentum is conserved when the system has no external forces acting on it. Now we've found that the system's total momentum is also conserved when the net external force acting on it is zero. With no external forces acting that can change its momentum, we call a system with $\vec{F}_{\text{net}} = \vec{0}$ an **isolated system.**

The importance of these results is sufficient to elevate them to a law of nature, alongside Newton's laws.

Law of conservation of momentum The total momentum $\vec{P}$ of an isolated system is a constant. Interactions within the system do not change the system's total momentum.

NOTE ► It is worth emphasizing the critical role of Newton's third law in the derivation of Equation 9.13. The law of conservation of momentum is a direct consequence of the fact that interactions within an isolated system are action/reaction pairs. ◄

Mathematically, the law of conservation of momentum for an isolated system is

$$\vec{P}_{\text{f}} = \vec{P}_{\text{i}} \tag{9.14}$$

Law of conservation of momentum for an isolated system

The total momentum after an interaction is equal to the total momentum before the interaction. Because Equation 9.14 is a vector equation, the equality is true for each of the components of the momentum vector; that is,

$$\begin{aligned} &\text{x-component:}\quad \underbrace{(p_{1x})_{\text{f}} + (p_{2x})_{\text{f}} + (p_{3x})_{\text{f}} + \cdots}_{\text{Final momentum}} = \underbrace{(p_{1x})_{\text{i}} + (p_{2x})_{\text{i}} + (p_{3x})_{\text{i}} + \cdots}_{\text{Initial momentum}} \\ &\text{y-component:}\quad (p_{1y})_{\text{f}} + (p_{2y})_{\text{f}} + (p_{3y})_{\text{f}} + \cdots = (p_{1y})_{\text{i}} + (p_{2y})_{\text{i}} + (p_{3y})_{\text{i}} + \cdots \end{aligned} \tag{9.15}$$

(Terms 1, 2, 3 refer to Particle 1, Particle 2, Particle 3.)

EXAMPLE 9.5 **Speed of ice skaters pushing off**

Two ice skaters, Sandra and David, stand facing each other on frictionless ice. Sandra has a mass of 45 kg, David a mass of 80 kg. They then push off from each other. After the push, Sandra moves off at a speed of 2.2 m/s. What is David's speed?

PREPARE The two skaters interact with each other, but they form an isolated system because, for each skater, the upward normal force of the ice balances their downward weight force to make $\vec{F}_{\text{net}} = \vec{0}$. Thus the total momentum of the system of the two skaters is conserved.

FIGURE 9.18 shows a before-and-after visual overview for the two skaters. The total momentum before they push off is $\vec{P}_{\text{i}} = \vec{0}$ because both skaters are at rest. Consequently, the total momentum will still be $\vec{0}$ *after* they push off.

SOLVE Since the motion is only in the x-direction, we'll only need to consider x-components of momentum. We write Sandra's initial momentum as $(p_{\text{S}x})_{\text{i}} = m_{\text{S}}(v_{\text{S}x})_{\text{i}}$, where m_{S} is her mass and

FIGURE 9.18 Before-and-after visual overview for two skaters pushing off from each other.

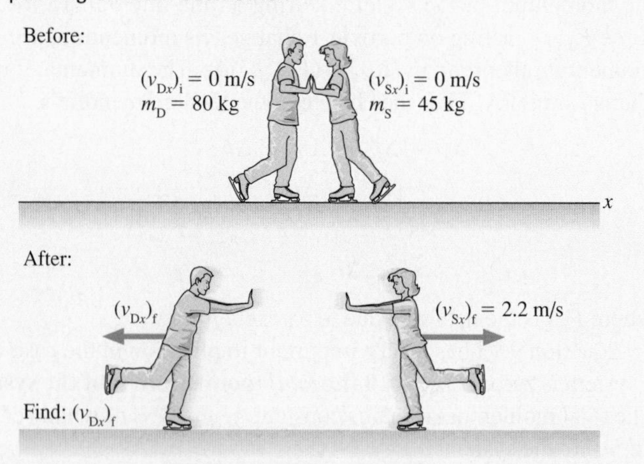

$(v_{Sx})_i$ her initial velocity. Similarly, we write David's initial momentum as $(p_{Dx})_i = m_D(v_{Dx})_i$. Both these momenta are zero because both skaters are initially at rest.

We can now apply the mathematical statement of momentum conservation, Equation 9.15. Writing the final momentum of Sandra as $m_S(v_{Sx})_f$ and that of David as $m_D(v_{Dx})_f$, we have

$$\underbrace{m_S(v_{Sx})_f + m_D(v_{Dx})_f}_{\text{The skaters' final momentum . . .}} = \underbrace{m_S(v_{Sx})_i + m_D(v_{Dx})_i}_{\text{. . . equals their initial momentum . . .}} = \underbrace{0}_{\text{. . . which was zero.}}$$

Solving for $(v_{Dx})_f$, we find

$$(v_{Dx})_f = -\frac{m_S}{m_D}(v_{Sx})_f = -\frac{45 \text{ kg}}{80 \text{ kg}} \times 2.2 \text{ m/s} = -1.2 \text{ m/s}$$

David moves backward with a *speed* of 1.2 m/s.

Notice that we didn't need to know any details about the force between David and Sandra in order to find David's final speed. Conservation of momentum *mandates* this result.

ASSESS The *total* momentum of the system is zero both before and after they push off, but the individual momenta are not zero. Because $(p_{Sx})_f$ is positive (Sandra moves to the right), $(p_{Dx})_f$ must have the same magnitude but the opposite sign (David moves to the left).

A Strategy for Conservation of Momentum Problems

Our derivation of the law of conservation of momentum, and the conditions under which it holds, suggests a problem-solving strategy.

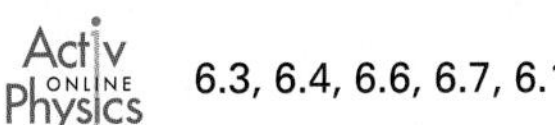

PROBLEM-SOLVING STRATEGY 9.1 **Conservation of momentum problems**

PREPARE Clearly define *the system.*

- If possible, choose a system that is isolated ($\vec{F}_{net} = \vec{0}$) or within which the interactions are sufficiently short and intense that you can ignore external forces for the duration of the interaction (the impulse approximation). Momentum is then conserved.
- If it's not possible to choose an isolated system, try to divide the problem into parts such that momentum is conserved during one segment of the motion. Other segments of the motion can be analyzed using Newton's laws or, as you'll learn in Chapter 10, conservation of energy.

Following Tactics Box 9.1, draw a before-and-after visual overview. Define symbols that will be used in the problem, list known values, and identify what you're trying to find.

SOLVE The mathematical representation is based on the law of conservation of momentum: $\vec{P}_f = \vec{P}_i$. In component form, this is

$$(p_{1x})_f + (p_{2x})_f + (p_{3x})_f + \cdots = (p_{1x})_i + (p_{2x})_i + (p_{3x})_i + \cdots$$
$$(p_{1y})_f + (p_{2y})_f + (p_{3y})_f + \cdots = (p_{1y})_i + (p_{2y})_i + (p_{3y})_i + \cdots$$

ASSESS Check that your result has the correct units, is reasonable, and answers the question.

Exercise 17 

EXAMPLE 9.6 **Getaway speed of a cart**

Bob is running from the police and thinks he can make a faster getaway by jumping on a stationary cart in front of him. He runs toward the cart, jumps on, and rolls along the horizontal street. Bob has a mass of 75 kg and the cart's mass is 25 kg. If Bob's speed is 4.0 m/s when he jumps onto the cart, what is the cart's speed after Bob jumps on?

PREPARE When Bob lands on and sticks to the cart, a "collision" occurs between Bob and the cart. If we take Bob and the cart together to be the system, the forces involved in this collision—friction forces between Bob's feet and the cart—are internal forces. Because the normal force balances the weight of both Bob and the cart, the net external force on the system is zero, so the

Continued

total momentum of Bob + cart is conserved: It is the same before and after the collision.

The visual overview in FIGURE 9.19 shows the important point that Bob and the cart move together after he lands on the cart, so $(v_x)_f$ is their common final velocity.

FIGURE 9.19 Before-and-after visual overview of Bob and the cart.

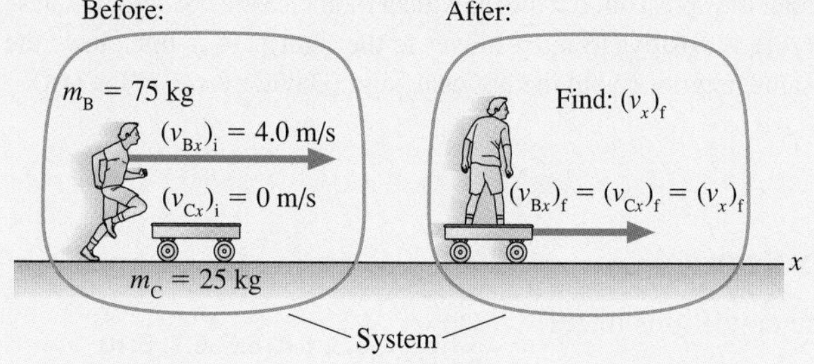

SOLVE To solve for the final velocity of Bob and the cart, we'll use conservation of momentum: $(P_x)_f = (P_x)_i$. Written in terms of the individual momenta, we have

$$(P_x)_i = m_B(v_{Bx})_i + m_C\underbrace{(v_{Cx})_i}_{0\text{ m/s}} = m_B(v_{Bx})_i$$

$$(P_x)_f = m_B(v_x)_f + m_C(v_x)_f = (m_B + m_C)(v_x)_f$$

In the second equation, we've used the fact that both Bob and the cart travel at the common velocity of $(v_x)_f$. Equating the final and initial total momenta gives

$$(m_B + m_C)(v_x)_f = m_B(v_{Bx})_i$$

Solving this for $(v_x)_f$, we find

$$(v_x)_f = \frac{m_B}{m_B + m_C}(v_{Bx})_i = \frac{75\text{ kg}}{100\text{ kg}} \times 4.0\text{ m/s} = 3.0\text{ m/s}$$

The cart's speed is 3.0 m/s immediately after Bob jumps on.

ASSESS It makes sense that Bob has *lost* speed because he had to share his initial momentum with the cart. Not a good way to make a getaway!

Notice how easy this was! No forces, no kinematic equations, no simultaneous equations. Why didn't we think of this before? Although conservation laws are indeed powerful, they can answer only certain questions. Had we wanted to know how far Bob slid across the cart before sticking to it, how long the slide took, or what the cart's acceleration was during the collision, we would not have been able to answer such questions on the basis of the conservation law. There is a price to pay for finding a simple connection between before and after, and that price is the loss of information about the details of the interaction. If we are satisfied with knowing only about before and after, then conservation laws are a simple and straightforward way to proceed. But many problems *do* require us to understand the interaction, and for these there is no avoiding Newton's laws and all they entail.

It Depends on the System

The first step in the problem-solving strategy asks you to clearly define *the system.* This is worth emphasizing, because many problem-solving errors arise from trying to apply momentum conservation to an inappropriate system. **The goal is to choose a system whose momentum will be conserved.** Even then, it is the *total* momentum of the system that is conserved, not the momenta of the individual particles within the system.

FIGURE 9.20 An analysis of the system consisting of Bob alone.

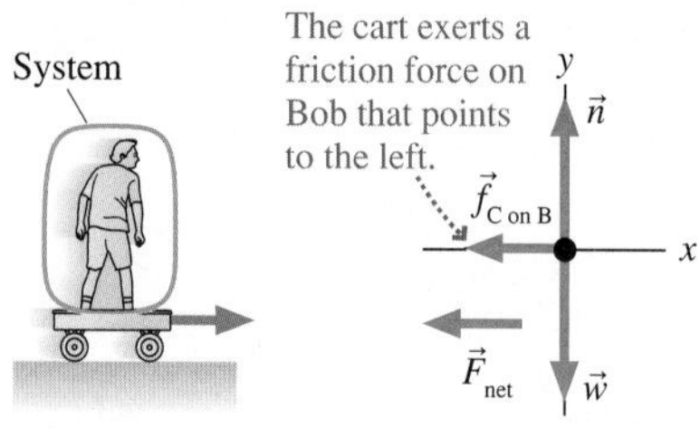

In Example 9.6, we chose the system to be Bob and the cart. Why this choice? Let's see what would happen if we had chosen the system to be Bob alone, as shown in FIGURE 9.20. As the free-body diagram shows, as Bob lands on the cart, there are three forces acting on him: the normal force $\vec{n}$ of the cart on Bob, his weight $\vec{w}$, and a friction force $\vec{f}_{C\text{on}B}$ of the cart on Bob. This last force is subtle. We know that Bob's feet must exert a rightward-directed friction force $\vec{f}_{B\text{on}C}$ *on the cart* as he lands; it is this friction force that causes the cart to speed up. By Newton's third law, then, the cart exerts a leftward directed force $\vec{f}_{C\text{on}B}$ on Bob.

The free-body diagram of Figure 9.20 then shows that there is a net force on Bob directed to the left. Thus the system consisting of Bob alone is *not* isolated, and Bob's momentum will not be conserved. Indeed, we know that Bob slows down after landing on the cart, so that his momentum clearly *decreases*.

If we had chosen the cart to be the system, the unbalanced rightward force $\vec{f}_{B\text{on}C}$ of Bob on the cart would also lead to a nonzero net force. Thus the cart's momentum would not be conserved; in fact, we know it *increases* because the cart speeds up.

Only by choosing the system to be Bob and the cart *together* is the net force on the system zero and the total momentum conserved. The momentum lost by Bob is gained by the cart, so the total momentum of the two is unchanged.

Explosions

An **explosion,** where the particles of the system move apart after a brief, intense interaction, is the opposite of a collision. The explosive forces, which could be from an expanding spring or from expanding hot gases, are *internal* forces. If the system is isolated, its total momentum during the explosion will be conserved.

EXAMPLE 9.7 Recoil speed of a rifle

A 30 g ball is fired from a 1.2 kg spring-loaded toy rifle with a speed of 15 m/s. What is the recoil speed of the rifle?

PREPARE As the ball moves down the barrel, there are complicated forces exerted on the ball and on the rifle. However, if we take the system to be the ball + rifle, these are *internal* forces that do not change the total momentum.

The *external* forces of the rifle's and ball's weights are balanced by the external force exerted by the person holding the rifle, so $\vec{F}_{net} = \vec{0}$. This is an isolated system and the law of conservation of momentum applies.

FIGURE 9.21 Before-and-after visual overview for a toy rifle.

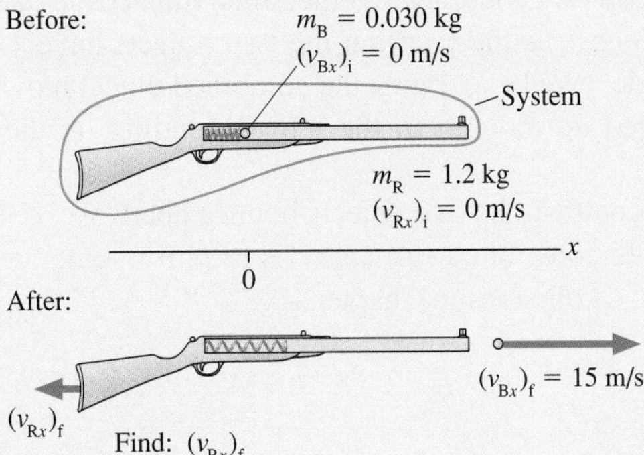

FIGURE 9.21 shows a visual overview before and after the ball is fired. We'll assume the ball is fired in the $+x$-direction.

SOLVE The x-component of the total momentum is $P_x = p_{Bx} + p_{Rx}$. Everything is at rest before the trigger is pulled, so the initial momentum is zero. After the trigger is pulled, the internal force of the spring pushes the ball down the barrel *and* pushes the rifle backward. Conservation of momentum gives

$$(P_x)_f = m_B(v_{Bx})_f + m_R(v_{Rx})_f = (P_x)_i = 0$$

Solving for the rifle's velocity, we find

$$(v_{Rx})_f = -\frac{m_B}{m_R}(v_{Bx})_f = -\frac{0.030 \text{ kg}}{1.2 \text{ kg}} \times 15 \text{ m/s} = -0.38 \text{ m/s}$$

The minus sign indicates that the rifle's recoil is to the left. The recoil *speed* is 0.38 m/s.

ASSESS Real rifles fire their bullets at much higher velocities, and their recoil is correspondingly higher. Shooters need to brace themselves against the "kick" of the rifle back against their shoulder.

We would not know where to begin to solve a problem such as this using Newton's laws. But Example 9.7 is a simple problem when approached from the before-and-after perspective of a conservation law. The selection of ball + rifle as "the system" was the critical step. For momentum conservation to be a useful principle, we had to select a system in which the complicated forces due to the spring and to friction were all internal forces. The rifle by itself is *not* an isolated system, so its momentum is *not* conserved.

Much the same reasoning explains how a rocket or jet aircraft accelerates. **FIGURE 9.22** shows a rocket with a parcel of fuel on board. Burning converts the fuel to hot gases that are expelled from the rocket motor. If we choose rocket + gases to be the system, then the burning and expulsion are internal forces. In deep space there are no other forces, so the total momentum of the rocket + gases system must be conserved. The rocket gains forward velocity and momentum as the exhaust gases are shot out the back, but the *total* momentum of the system remains zero.

FIGURE 9.22 Rocket propulsion is an example of conservation of momentum.

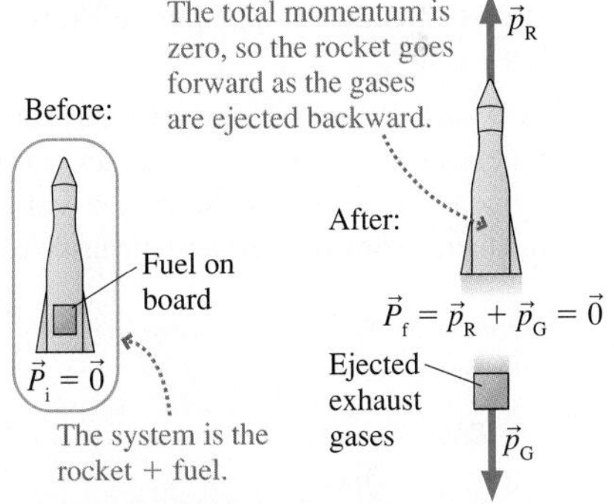

Many people find it hard to understand how a rocket can accelerate in the vacuum of space because there is nothing to "push against." Thinking in terms of momentum, you can see that the rocket does not push against anything *external,* but only against the gases that it pushes out the back. In return, in accordance with Newton's third law, the gases push forward on the rocket.

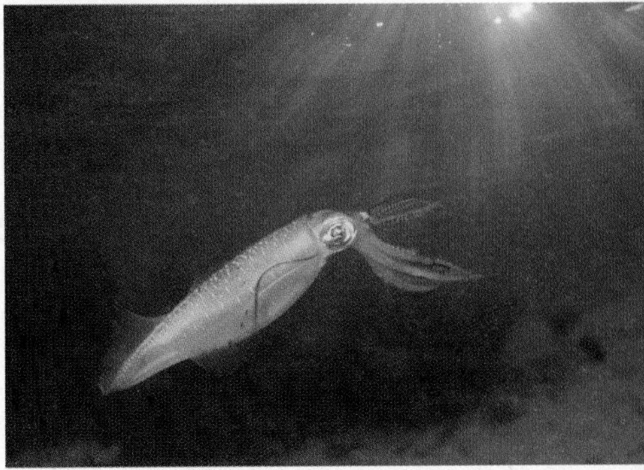

► **Squid propulsion** BIO Squids use a form of "rocket propulsion" to make quick movements to escape enemies or catch prey. The squid draws in water through a pair of valves in its outer sheath, or mantle, and then quickly expels the water through a funnel, propelling the squid backward. The funnel's direction is adjustable, allowing the squid to move in any backward direction.

STOP TO THINK 9.3 An explosion in a rigid pipe shoots three balls out of its ends. A 6 g ball comes out the right end. A 4 g ball comes out the left end with twice the speed of the 6 g ball. From which end, left or right, does the third ball emerge?

9.5 Inelastic Collisions

Collisions can have different possible outcomes. A rubber ball dropped on the floor bounces—it's *elastic*—but a ball of clay sticks to the floor without bouncing; we call such a collision *inelastic*. A golf club hitting a golf ball causes the ball to rebound away from the club (elastic), but a bullet striking a block of wood becomes embedded in the block (inelastic).

FIGURE 9.23 A perfectly inelastic collision.

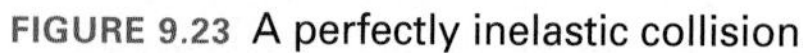

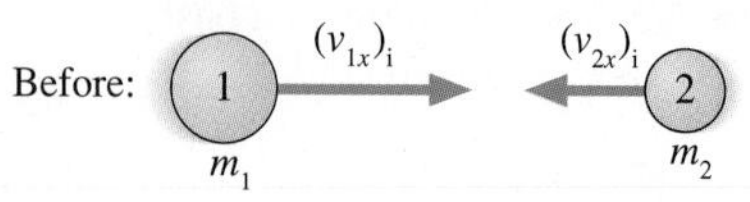

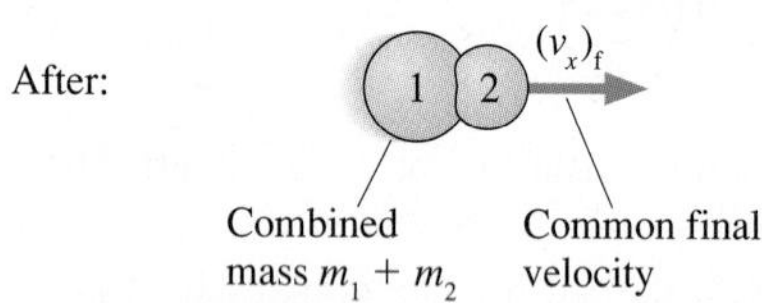

A collision in which the two objects stick together and move with a common final velocity is called a **perfectly inelastic collision.** The clay hitting the floor and the bullet embedding itself in the wood are examples of perfectly inelastic collisions. Other examples include railroad cars coupling together upon impact and darts hitting a dart board. FIGURE 9.23 emphasizes the fact that the two objects have a common final velocity after they collide. (We have drawn the combined object moving to the right, but it could have ended up moving to the left, depending on the objects' masses and initial velocities.)

In an *elastic collision,* by contrast, the two objects bounce apart. We've looked at some examples of elastic collisions, but a full analysis requires some ideas about energy. We will return to elastic collisions in Chapter 10.

EXAMPLE 9.8 Speeds in a perfectly inelastic glider collision

In a laboratory experiment, a 200 g air-track glider and a 400 g air-track glider are pushed toward each other from opposite ends of the track. The gliders have Velcro tabs on their fronts so that they will stick together when they collide. The 200 g glider is pushed with an initial speed of 3.0 m/s. The collision causes it to reverse direction at 0.50 m/s. What was the initial speed of the 400 g glider?

PREPARE We model the gliders as particles and define the two gliders as the system. This is an isolated system, so its total momentum is conserved in the collision. The gliders stick together, so this is a perfectly inelastic collision.

FIGURE 9.24 Before-and-after visual overview for two gliders colliding on an air track.

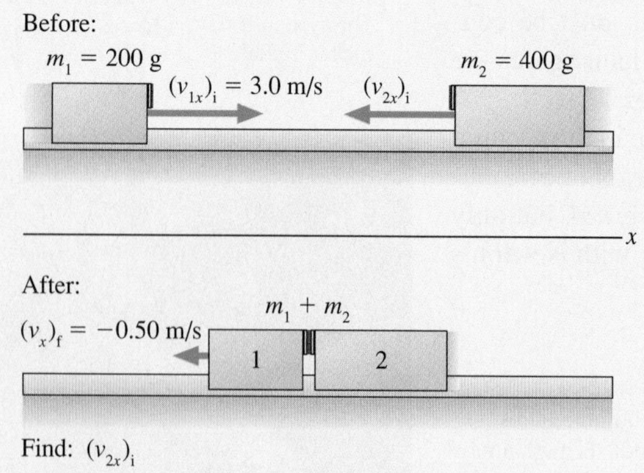

FIGURE 9.24 shows a visual overview. We've chosen to let the 200 g glider (glider 1) start out moving to the right, so $(v_{1x})_i$ is a positive 3.0 m/s. The gliders move to the left after the collision, so their common final velocity is $(v_x)_f = -0.50$ m/s. You can see that velocity $(v_{2x})_i$ must be negative in order to "turn around" both gliders.

SOLVE The law of conservation of momentum, $(P_x)_f = (P_x)_i$, is

$$(m_1 + m_2)(v_x)_f = m_1(v_{1x})_i + m_2(v_{2x})_i$$

where we made use of the fact that the combined mass $m_1 + m_2$ moves together after the collision. We can easily solve for the initial velocity of the 400 g glider:

$$(v_{2x})_i = \frac{(m_1 + m_2)(v_x)_f - m_1(v_{1x})_i}{m_2}$$
$$= \frac{(0.60\text{ kg})(-0.50\text{ m/s}) - (0.20\text{ kg})(3.0\text{ m/s})}{0.40\text{ kg}}$$
$$= -2.3\text{ m/s}$$

The negative sign, which we anticipated, indicates that the 400 g glider started out moving to the left. The initial *speed* of the glider, which we were asked to find, is 2.3 m/s.

ASSESS The key step in solving inelastic collision problems is that both objects move after the collision with the same velocity. You should thus choose a single symbol (here, $(v_x)_f$) for this common velocity.

STOP TO THINK 9.4 The two particles shown collide and stick together. After the collision, the combined particles

A. Move to the right as shown.
B. Move to the left.
C. Are at rest.

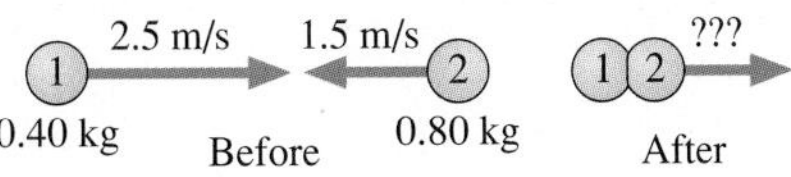

9.6 Momentum and Collisions in Two Dimensions

Our examples thus far have been confined to motion along a one-dimensional axis. Many practical examples of momentum conservation involve motion in a plane. The total momentum $\vec{P}$ is the *vector* sum of the momenta $\vec{p} = m\vec{v}$ of the individual particles. Consequently, as Equation 9.15 showed, momentum is conserved only if each component of $\vec{P}$ is conserved:

$$\begin{aligned} (p_{1x})_f + (p_{2x})_f + (p_{3x})_f + \cdots &= (p_{1x})_i + (p_{2x})_i + (p_{3x})_i + \cdots \\ (p_{1y})_f + (p_{2y})_f + (p_{3y})_f + \cdots &= (p_{1y})_i + (p_{2y})_i + (p_{3y})_i + \cdots \end{aligned} \quad (9.16)$$

In this section we'll apply momentum conservation to motion in two dimensions.

Collisions and explosions often involve motion in two dimensions.

EXAMPLE 9.9 Analyzing a peregrine falcon strike

Peregrine falcons often grab their prey from above while both falcon and prey are in flight. A falcon, flying at 18 m/s, swoops down at a 45° angle from behind a pigeon flying horizontally at 9.0 m/s. The falcon has a mass of 0.80 kg and the pigeon a mass of 0.36 kg. What are the speed and direction of the falcon (now holding the pigeon) immediately after impact?

PREPARE This is a perfectly inelastic collision because after the collision the falcon and pigeon move at a common velocity. The total momentum of the falcon + pigeon system is conserved. For a two-dimensional collision, this means that the x-component of the total momentum before the collision must equal the x-component of the total momentum after the collision, and similarly for the y-components. **FIGURE 9.25** is a before-and-after visual overview.

FIGURE 9.25 Before-and-after visual overview for a falcon catching a pigeon.

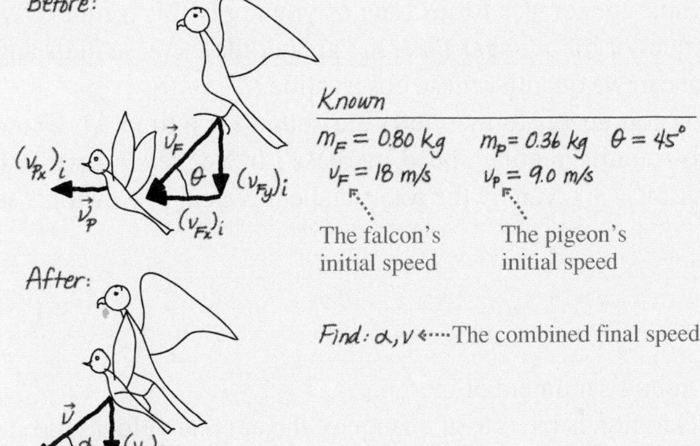

SOLVE We'll start by finding the x- and y-components of the momentum before the collision. For the x-component we have

The x-component of the initial momentum . . . (Both velocity components are negative, since they point to the left.)

$$(P_x)_i = m_F(v_{Fx})_i + m_P(v_{Px})_i = m_F(-v_F \cos\theta) + m_P(-v_P)$$

. . . equals the x-component of the initial momentum of the falcon . . . . . . plus the x-component of the initial momentum of the pigeon.

$$= (0.80 \text{ kg})(-18 \text{ m/s})(\cos 45°) + (0.36 \text{ kg})(-9.0 \text{ m/s})$$

$$= -13.4 \text{ kg} \cdot \text{m/s}$$

Similarly, for the y-component of the initial momentum we have

$$(P_y)_i = m_F(v_{Fy})_i + m_P(v_{Py})_i = m_F(-v_F \sin\theta) + 0$$

$$= (0.80 \text{ kg})(-18.0 \text{ m/s})(\sin 45°) = -10.2 \text{ kg} \cdot \text{m/s}$$

After the collision, the two birds move with a common velocity $\vec{v}$ that is directed at an angle α from the horizontal. The x-component of the final momentum is then

$$(P_x)_f = (m_F + m_P)(v_x)_f$$

Continued

Momentum conservation requires $(P_x)_f = (P_x)_i$, so

$$(v_x)_f = \frac{(P_x)_i}{m_F + m_P} = \frac{-13.4\ \text{kg}\cdot\text{m/s}}{(0.80\ \text{kg}) + (0.36\ \text{kg})} = -11.6\ \text{m/s}$$

Similarly, $(P_y)_f = (P_y)_i$ gives

$$(v_y)_f = \frac{(P_y)_i}{m_F + m_P} = \frac{-10.2\ \text{kg}\cdot\text{m/s}}{(0.80\ \text{kg}) + (0.36\ \text{kg})} = -8.79\ \text{m/s}$$

From the figure we see that $\tan\alpha = (v_y)_f/(v_x)_f$, so that

$$\alpha = \tan^{-1}\left(\frac{(v_y)_f}{(v_x)_f}\right) = \tan^{-1}\left(\frac{-8.79\ \text{m/s}}{-11.6\ \text{m/s}}\right) = 37^\circ$$

The magnitude of the final velocity (i.e., the speed) can be found from the Pythagorean theorem as

$$v = \sqrt{(v_x)_f^2 + (v_y)_f^2}$$
$$= \sqrt{(-11.6\ \text{m/s})^2 + (-8.79\ \text{m/s})^2} = 15\ \text{m/s}$$

Thus immediately after impact the falcon, with its meal, is moving 37° below horizontal at a speed of 15 m/s.

ASSESS It makes sense that the falcon slows down after catching the slower-moving pigeon. Also, the final angle is closer to the horizontal than the falcon's initial angle. This seems reasonable because the pigeon was initially flying horizontally, making the total momentum vector more horizontal than the direction of the falcon's initial momentum.

FIGURE 9.26 The momentum vectors of the falcon strike.

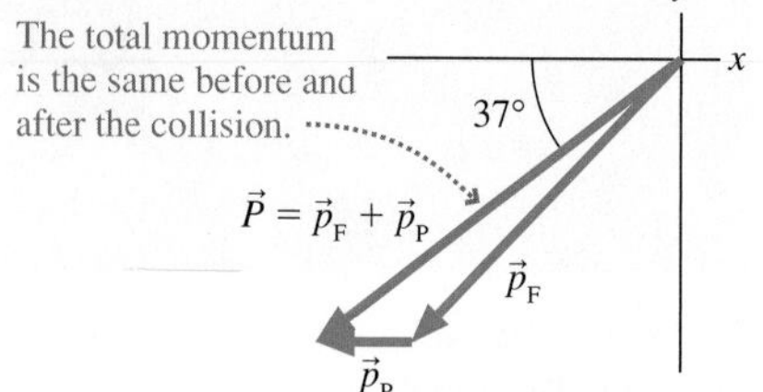

It's instructive to examine this collision with a picture of the momentum vectors. The vectors $\vec{p}_F$ and $\vec{p}_P$ before the collision, and their sum $\vec{P} = \vec{p}_F + \vec{p}_P$, are shown in **FIGURE 9.26**. You can see that the total momentum vector makes a 53° angle with the negative y-axis. The individual momenta change in the collision, *but the total momentum does not.*

9.7 Angular Momentum

For a single particle, we can think of the law of conservation of momentum as an alternative way of stating Newton's first law. Rather than saying that a particle will continue to move in a straight line at constant velocity unless acted on by a net force, we can say that the momentum of an isolated particle is conserved. Both express the idea that a particle moving in a straight line tends to "keep going" unless something acts on it to change its motion.

Another important motion you've studied is motion in a circle. The momentum $\vec{p}$ is *not* conserved for a particle undergoing circular motion. Momentum is a vector, and the momentum of a particle in circular motion changes as the direction of motion changes.

Nonetheless, a spinning bicycle wheel would keep turning if it were not for friction, and a ball moving in a circle at the end of a string tends to "keep going" in a circular path. The quantity that expresses this idea for circular motion is called *angular momentum*.

FIGURE 9.27 By applying a torque to the merry-go-round, the girl is increasing its angular momentum.

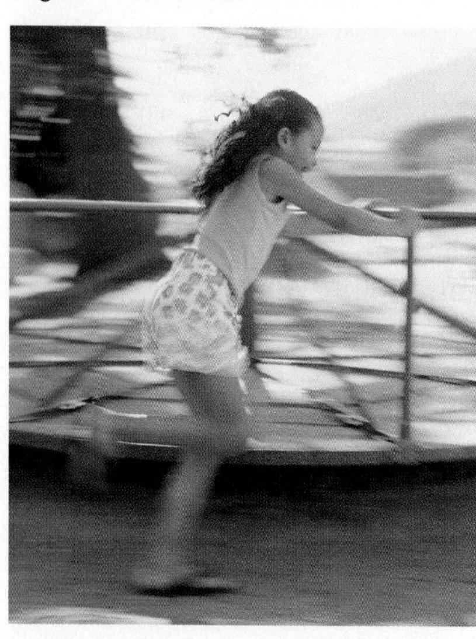

Let's start by looking at an example from everyday life: pushing a merry-go-round, as in **FIGURE 9.27**. If you push tangentially to the rim, you are applying a *torque* to the merry-go-round. As we learned in Chapter 7, the merry-go-round's angular speed will continue to increase for as long as you apply this torque. If you push *harder* (greater torque) or for a *longer time,* the greater the increase in its angular velocity will be. How can we quantify these observations?

Let's apply a constant torque τ_{net} to the merry-go-round for a time Δt. By how much will the merry-go-round's angular speed increase? In Section 7.4 we found that the angular acceleration α is given by the rotational equivalent of Newton's second law, or

$$\alpha = \frac{\tau_{net}}{I} \tag{9.17}$$

where I is the merry-go-round's moment of inertia.

Now the angular acceleration is the rate of change of the angular velocity, so

$$\alpha = \frac{\Delta\omega}{\Delta t} \tag{9.18}$$

Setting Equations 9.17 and 9.18 equal to each other gives

$$\frac{\Delta\omega}{\Delta t} = \frac{\tau_{net}}{I}$$

or, rearranging,

$$\tau_{net}\,\Delta t = I\,\Delta\omega \tag{9.19}$$

If you recall the impulse-momentum theorem for *linear* motion, which is

$$\vec{F}_{net}\,\Delta t = m\,\Delta\vec{v} = \Delta\vec{p} \tag{9.20}$$

you can see that Equation 9.19 is an analogous statement about rotational motion. Because the quantity $I\omega$ is evidently the rotational equivalent of $m\vec{v}$, the linear momentum $\vec{p}$, it seems reasonable to define the **angular momentum** L to be

$$L = I\omega \tag{9.21}$$

Angular momentum of an object with moment of inertia I rotating at angular velocity ω

The SI units of angular momentum are those of moment of inertia times angular velocity, or kg · m²/s.

Just as an object in linear motion can have a large momentum by having either a large mass or a high speed, a rotating object can have a large angular momentum by having a large moment of inertia or a large angular velocity. The merry-go-round in Figure 9.27 has a larger angular momentum if it's spinning fast than if it's spinning slowly. Also, the merry-go-round (large I) has a much larger angular momentum than a toy top (small I) spinning with the same angular velocity.

Table 9.2 summarizes the analogies between linear and rotational quantities that you learned in Chapter 7 and adds the analogy between linear momentum and angular momentum.

TABLE 9.2 Rotational and linear dynamics

Rotational dynamics	Linear dynamics
Torque τ_{net}	Force $\vec{F}_{net}$
Moment of inertia I	Mass m
Angular velocity ω	Velocity $\vec{v}$
Angular momentum $L = I\omega$	Linear momentum $\vec{p} = m\vec{v}$

Conservation of Angular Momentum

Having now defined angular momentum, we can write Equation 9.19 as

$$\tau_{net}\,\Delta t = \Delta L \tag{9.22}$$

in exact analogy with its linear dynamics equivalent, Equation 9.20. This equation states that the change in the angular momentum of an object is proportional to the net torque applied to the object. If the net external torque on an object is *zero,* the rotational analog of an isolated system, then the change in the angular momentum is zero as well. That is, a rotating object will continue to rotate with *constant* angular velocity—to "keep going"—unless acted upon by an external torque. We can state this conclusion as the *law of conservation of angular momentum*:

ActivPhysics ONLINE 7.14

Law of conservation of angular momentum The angular momentum of a rotating object subject to no net external torque ($\tau_{net} = 0$) is a constant. The final angular momentum L_f is equal to the initial angular momentum L_i.

This law is analogous to that for the conservation of linear momentum; there, linear momentum is conserved if the net *force* is zero. Because the angular momentum is $L = I\omega$, the mathematical statement of the law of conservation of angular momentum is

$$I_f\omega_f = I_i\,\omega_i \tag{9.23}$$

EXAMPLE 9.10 **Period of a merry-go-round**

Joey, whose mass is 36 kg, stands at the center of a 200 kg merry-go-round that is rotating once every 2.5 s. While it is rotating, Joey walks out to the edge of the merry-go-round, 2.0 m from its center. What is the rotational period of the merry-go-round when Joey gets to the edge?

PREPARE Take the system to be Joey + merry-go-round and assume frictionless bearings. There is no external torque on this system, so the angular momentum of the system will be conserved. As shown in the visual overview of FIGURE 9.28, we model the merry-go-round as a uniform disk of radius $R = 2.0$ m. From Table 7.2, the moment of inertia of a disk is $I_{disk} = \frac{1}{2}MR^2$. If we model Joey as a particle of mass m, his moment of inertia is zero when he is at the center, but it increases to mR^2 when he reaches the edge.

FIGURE 9.28 Visual overview of the merry-go-round.

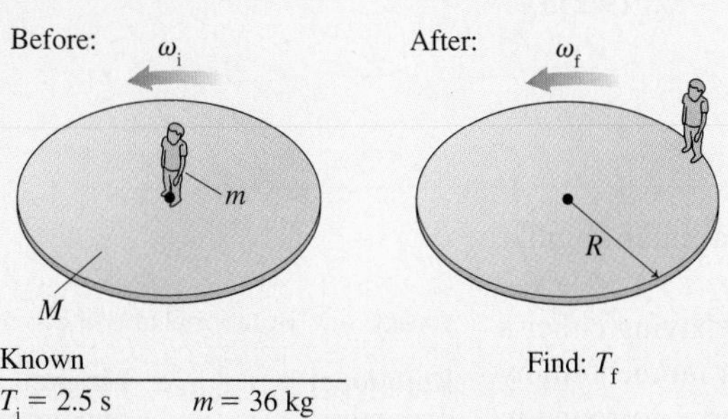

Known	
$T_i = 2.5$ s	$m = 36$ kg
$M = 200$ kg	$R = 2.0$ m

Find: T_f

SOLVE The mathematical statement of the law of conservation of momentum is $L_i = L_f$ or, from Equation 9.23, $I_f\omega_f = I_i\omega_i$, which we can rewrite as

$$\omega_f = \frac{I_i}{I_f}\omega_i$$

As Joey moves out to the edge, the moment of inertia of the system increases and, as a result, the angular velocity decreases. Initially, the moment of inertia of the system is just that of the merry-go-round because Joey's contribution is zero. Thus

$$I_i = I_{disk} = \tfrac{1}{2}MR^2 = \tfrac{1}{2}(200 \text{ kg})(2.0 \text{ m})^2 = 400 \text{ kg} \cdot \text{m}^2$$

When Joey reaches the edge, the total moment of inertia becomes

$$I_f = I_{disk} + mR^2 = 400 \text{ kg} \cdot \text{m}^2 + (36 \text{ kg})(2.0 \text{ m})^2$$
$$= 540 \text{ kg} \cdot \text{m}^2$$

The initial angular velocity is related to the initial period of rotation T_i by

$$\omega_i = \frac{2\pi}{T_i} = \frac{2\pi}{2.5 \text{ s}} = 2.5 \text{ rad/s}$$

Thus the final angular velocity is

$$\omega_f = \frac{I_i}{I_f}\omega_i = \frac{400 \text{ kg} \cdot \text{m}^2}{540 \text{ kg} \cdot \text{m}^2}(2.5 \text{ rad/s}) = 1.9 \text{ rad/s}$$

When Joey reaches the edge, the period of the merry-go-round has increased to

$$T_f = \frac{2\pi}{\omega_f} = \frac{2\pi}{1.9 \text{ rad/s}} = 3.3 \text{ s}$$

ASSESS The merry-go-round rotates *more slowly* after Joey moves out to the edge. This makes sense because if the system's moment of inertia increases, as it does when Joey moves out, the angular velocity must decrease to keep the angular momentum constant.

FIGURE 9.29 A spinning figure skater.

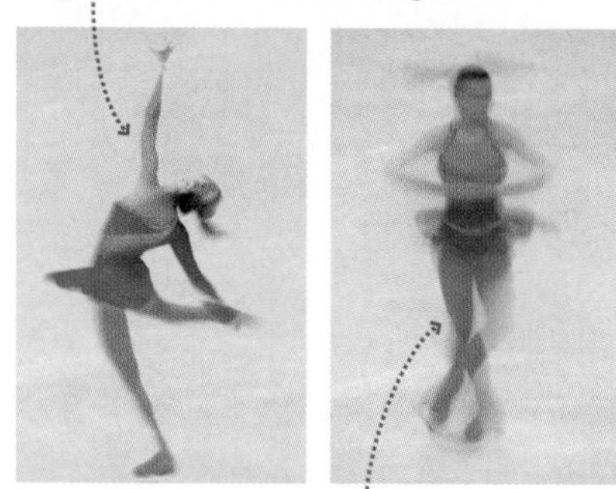

Conservation of momentum enters into many aspects of sports. Because no external torques act, the angular momentum of a platform diver is conserved while she's in the air. Just as for Joey and the merry-go-round of Example 9.10, she spins slowly when her moment of inertia is large; by decreasing her moment of inertia, she increases her rate of spin. Divers can thus markedly increase their spin rate by changing their body from an extended posture to a tuck position. Figure skaters also increase their spin rate by decreasing their moment of inertia, as shown in FIGURE 9.29. The following example gives a simplified treatment of this process.

EXAMPLE 9.11 **Analyzing a spinning ice skater**

An ice skater spins around on the tips of his blades while holding a 5.0 kg weight in each hand. He begins with his arms straight out from his body and his hands 140 cm apart. While spinning at 2.0 rev/s, he pulls the weights in and holds them 50 cm apart against his shoulders. If we neglect the mass of the skater, how fast is he spinning after pulling the weights in?

PREPARE Although the mass of the skater is larger than the mass of the weights, neglecting the skater's mass is not a bad approximation. Moment of inertia depends on the *square* of the distance of the mass from the axis of rotation. The skater's mass is concentrated in his torso, which has an effective radius (i.e., where most of the mass is concentrated) of only 9 or 10 cm. The weights move in much larger circles and have a disproportionate influence on his motion. The skater's arms exert radial forces on the weights just to keep them moving in circles, and even larger radial forces as he pulls them in. But there is no external torque on the weights, so their total angular momentum is conserved. **FIGURE 9.30** shows a before-and-after visual overview, as seen from above.

FIGURE 9.30 Top view visual overview of the spinning ice skater.

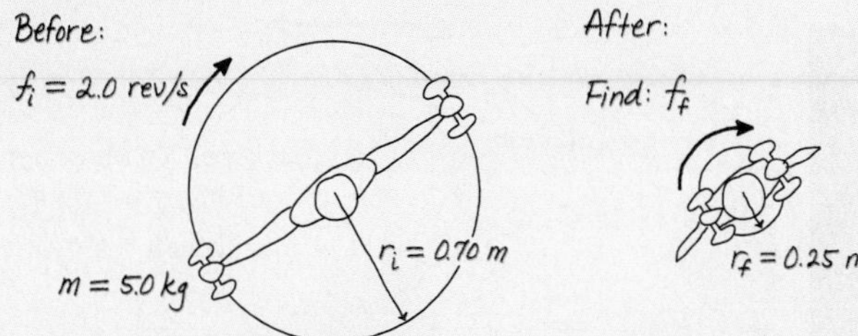

SOLVE The two weights have the same mass, move in circles with the same radius, and have the same angular velocity. Thus the total angular momentum is twice that of one weight. The mathematical statement of angular momentum conservation, $I_f\omega_f = I_i\omega_i$, is

There are two weights.

$$\underbrace{(2\,mr_f^2)}_{I_f}\omega_f = \underbrace{(2\,mr_i^2)}_{I_i}\omega_i$$

Because the angular velocity is related to the rotation frequency f by $\omega = 2\pi f$, this equation simplifies to

$$f_f = \left(\frac{r_i}{r_f}\right)^2 f_i$$

When he pulls the weights in, his rotation frequency increases to

$$f_f = \left(\frac{0.70\text{ m}}{0.25\text{ m}}\right)^2 \times 2.0\text{ rev/s} = 16\text{ rev/s}$$

ASSESS Pulling in the weights increases the skater's spin from 2 rev/s to 16 rev/s. This is somewhat high, because we neglected the mass of the skater, but it illustrates how skaters do "spin up" by pulling their mass in toward the rotation axis.

Solving either of these two examples using Newton's laws would be quite difficult. We would have to deal with internal forces, such as Joey's feet against the merry-go-round, and other complications. For problems like these, where we're interested only in the before-and-after aspects of the motion, using a conservation law makes the solution much simpler.

▶ **The eye of a hurricane** As air masses from the slowing rotating outer zones are drawn toward the low-pressure center, their moment of inertia decreases. Because the angular momentum of these air masses is conserved, their speed must *increase* as they approach the center, leading to the high wind speeds near the center of the storm.

STOP TO THINK 9.5 The left figure shows two boys of equal mass standing halfway to the edge on a turntable that is freely rotating at angular speed ω_i. They then walk to the positions shown in the right figure. The final angular speed ω_f is

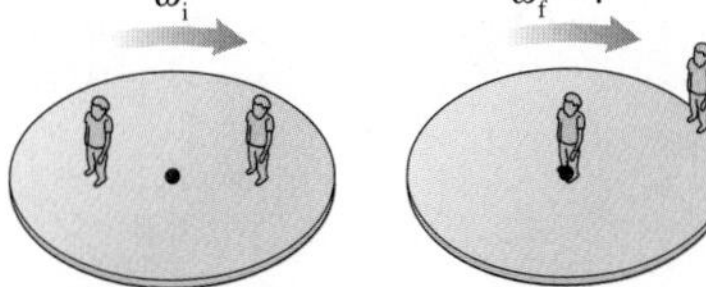

A. Greater than ω_i. B. Less than ω_i. C. Equal to ω_i.

INTEGRATED EXAMPLE 9.12 Aerial firefighting

A forest fire is easiest to attack when it's just getting started. In remote locations, this often means using airplanes to rapidly deliver large quantities of water and fire suppressant to the blaze.

The "Superscooper" is an amphibious aircraft that can pick up a 6000 kg load of water by skimming over the surface of a river or lake and scooping water directly into its storage tanks. As it approaches the water's surface at a speed of 35 m/s, an empty Superscooper has a mass of 13,000 kg.

a. It takes the plane 12 s to pick up a full load of water. If we ignore the force on the plane due to the thrust of its propellers, what is its speed immediately after picking up the water?
b. What is the impulse delivered to the plane by the water?
c. What is the average force of the water on the plane?
d. The plane then flies over the fire zone at 40 m/s. It releases water by opening doors in the belly of the plane, allowing the water to fall straight down with respect to the plane. What is the plane's speed after dropping the water if it takes 5.0 s to do so?

PREPARE We can solve part a, and later part d, using conservation of momentum, following Problem-Solving Strategy 9.1. We'll need to choose the system with care, so that $\vec{F}_{\text{net}} = \vec{0}$. The plane alone is not an appropriate system for using conservation of momentum: As the plane scoops up the water, the water exerts a large external drag force on the plane, so $\vec{F}_{\text{net}}$ is definitely not zero. Instead, we should choose the plane *and* the water it is going to scoop up as the system. Then there are no external forces in the x-direction, and the net force in the y-direction is zero, since neither plane nor water accelerates appreciably in this direction during the scooping process. The complicated forces between plane and water are now *internal* forces that do not change the total momentum of the plane + water system.

With the system chosen, we follow the steps of Tactics Box 9.1 to prepare the before-and-after visual overview shown in FIGURE 9.31.

Parts b and c are impulse-and-momentum problems, so to solve them we'll use the impulse-momentum theorem, Equation 9.8. The impulse-momentum theorem considers the dynamics of a *single* object—here, the plane—subject to external forces—in this case, from the water.

FIGURE 9.31 Visual overview of the plane and water.

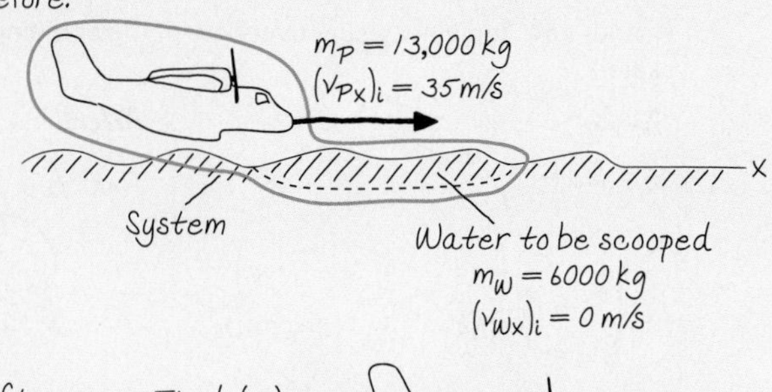

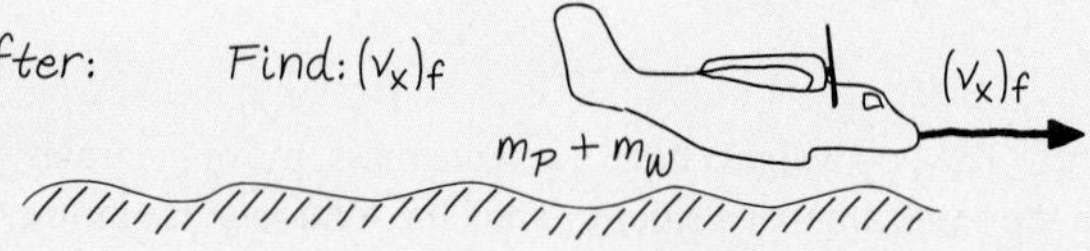

SOLVE a. Conservation of momentum in the x-direction is

$$(P_x)_f = (P_x)_i$$

or

$$(m_P + m_W)(v_x)_f = m_P(v_{Px})_i + m_W(v_{Wx})_i = m_P(v_{Px})_i + 0$$

Here we've used the facts that the initial velocity of the water is zero and that the final situation, as in an inelastic collision, has the combined mass of the plane and water moving with the same velocity $(v_x)_f$. Solving for $(v_x)_f$, we find

$$(v_x)_f = \frac{m_P(v_{Px})_i}{m_P + m_W} = \frac{(13{,}000\text{ kg})(35\text{ m/s})}{(13{,}000\text{ kg}) + (6000\text{ kg})} = 24\text{ m/s}$$

b. The impulse-momentum theorem is $J_x = \Delta p_x$, where $\Delta p_x = m_P \Delta v_x$ is the change in the plane's momentum. Thus

$$J_x = m_P \Delta v_x = m_P[(v_x)_f - (v_{Px})_i]$$
$$= (13{,}000\text{ kg})(24\text{ m/s} - 35\text{ m/s}) = -1.4 \times 10^5\text{ kg}\cdot\text{m/s}$$

c. From Equation 9.1, the definition of impulse, we have

$$(F_{\text{avg}})_x = \frac{J_x}{\Delta t} = \frac{-1.4 \times 10^5\text{ kg}\cdot\text{m/s}}{12\text{ s}} = -12{,}000\text{ N}$$

d. Because the water drops straight down *relative to the plane*, it has the same x-component of velocity immediately after being dropped as before being dropped. That is, simply opening the doors doesn't cause the water to speed up or slow down horizontally, so the water's horizontal momentum doesn't change upon being dropped. Because the total momentum of the plane + water system is conserved, the momentum of the plane doesn't change either. The plane's speed after the drop is still 40 m/s.

ASSESS The mass of the water is nearly half that of the plane, so the significant decrease in the plane's velocity as it scoops up the water is reasonable. The force of the water on the plane is large, but is still only about 10% of the plane's weight, $mg = 130{,}000$ N, so the answer seems to be reasonable.

SUMMARY

The goals of Chapter 9 have been to introduce the ideas of impulse, momentum, and angular momentum and to learn a new problem-solving strategy based on conservation laws.

GENERAL PRINCIPLES

Law of Conservation of Momentum

The total momentum $\vec{P} = \vec{p}_1 + \vec{p}_2 + \cdots$ of an isolated system is a constant. Thus

$$\vec{P}_f = \vec{P}_i$$

Conservation of Angular Momentum

The angular momentum L of a rotating object subject to zero external torque does not change. Thus

$$L_f = L_i$$

This can be written in terms of the moment of inertia and angular velocity as

$$I_f\omega_f = I_i\omega_i$$

Solving Momentum Conservation Problems

PREPARE Choose an isolated system or a system that is isolated during at least part of the problem. Draw a visual overview of the system before and after the interaction.

SOLVE Write the law of conservation of momentum in terms of vector components:

$$(p_{1x})_f + (p_{2x})_f + \cdots = (p_{1x})_i + (p_{2x})_i + \cdots$$

$$(p_{1y})_f + (p_{2y})_f + \cdots = (p_{1y})_i + (p_{2y})_i + \cdots$$

In terms of masses and velocities, this is

$$m_1(v_{1x})_f + m_2(v_{2x})_f + \cdots = m_1(v_{1x})_i + m_2(v_{2x})_i + \cdots$$

$$m_1(v_{1y})_f + m_2(v_{2y})_f + \cdots = m_1(v_{1y})_i + m_2(v_{2y})_i + \cdots$$

ASSESS Is the result reasonable?

IMPORTANT CONCEPTS

Momentum $\vec{p} = m\vec{v}$

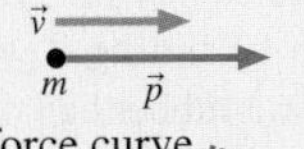

Impulse J_x = area under force curve

Impulse and momentum are related by the **impulse-momentum theorem**

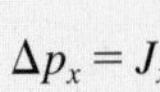

$$\Delta p_x = J_x$$

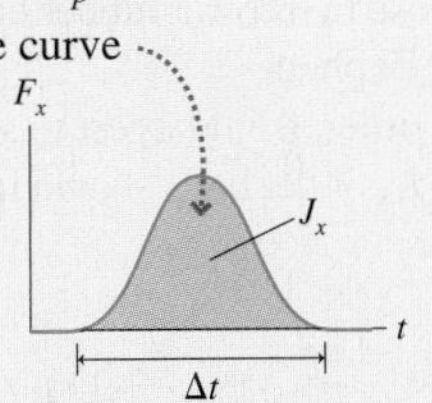

This is an alternative statement of Newton's second law.

Angular momentum $L = I\omega$ is the rotational analog of linear momentum $\vec{p} = m\vec{v}$.

System A group of interacting particles.

Isolated system A system on which the net external force is zero.

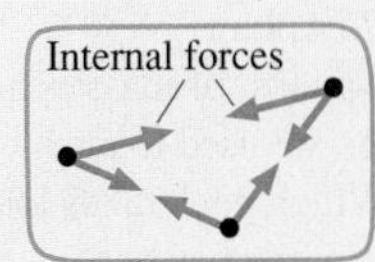

Before-and-after visual overview

- Define the system.
- Use two drawings to show the system *before* and *after* the interaction.
- List known information and identify what you are trying to find.

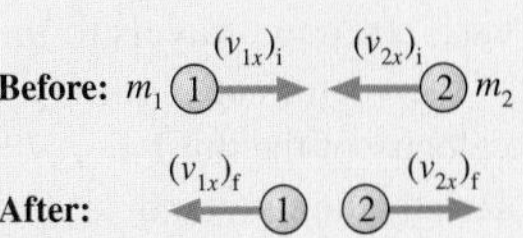

APPLICATIONS

Collisions Two or more particles come together. In a perfectly inelastic collision, they stick together and move with a common final velocity.

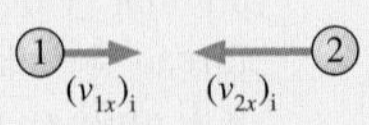

Explosions Two or more particles move away from each other.

(1) $(v_{1x})_f$ (2) $(v_{2x})_f$

Two dimensions Both the x- and y-components of the total momentum P must be conserved, giving two simultaneous equations.

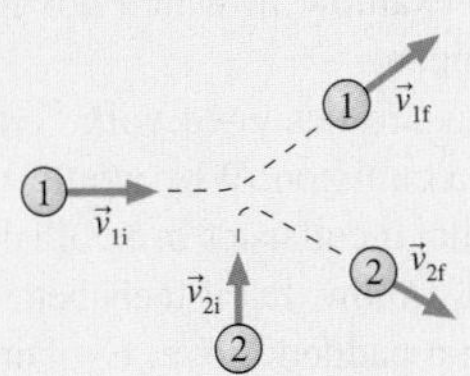

For homework assigned on MasteringPhysics, go to www.masteringphysics.com

Problem difficulty is labeled as | (straightforward) to ||||| (challenging).

Problems labeled can be done on a Workbook Momentum Worksheet; INT integrate significant material from earlier chapters; BIO are of biological or medical interest.

QUESTIONS

Conceptual Questions

1. Rank in order, from largest to smallest, the momenta p_{1x} through p_{5x} of the objects presented in Figure Q9.1. Explain.

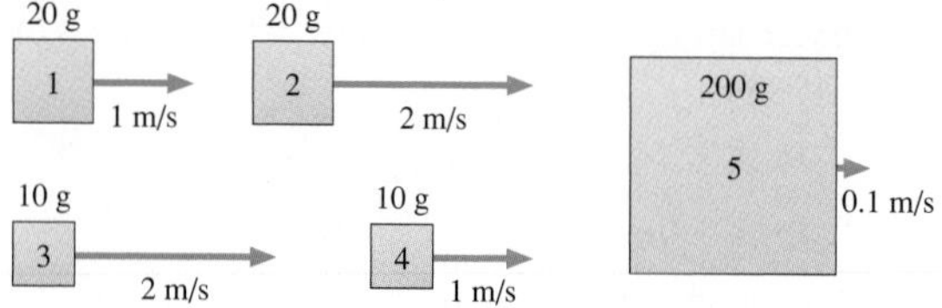

FIGURE Q9.1

2. Starting from rest, object 1 is subject to a 12 N force for 2.0 s. Object 2, with twice the mass, is subject to a 15 N force for 3.0 s. Which object has the greater final speed? Explain.
3. A 0.2 kg plastic cart and a 20 kg lead cart can roll without friction on a horizontal surface. Equal forces are used to push both carts forward for a time of 1 s, starting from rest. After the force is removed at $t = 1$ s, is the momentum of the plastic cart greater than, less than, or equal to the momentum of the lead cart? Explain.
4. Two pucks, of mass m and $4m$, lie on a frictionless table. Equal forces are used to push both pucks forward a distance of 1 m.
   a. Which puck takes longer to travel the distance? Explain.
   b. Which puck has the greater momentum upon completing the distance? Explain.
5. A stationary firecracker explodes into three pieces. One piece travels off to the east; a second travels to the north. Which of the vectors of Figure Q9.5 could be the velocity of the third piece? Explain.

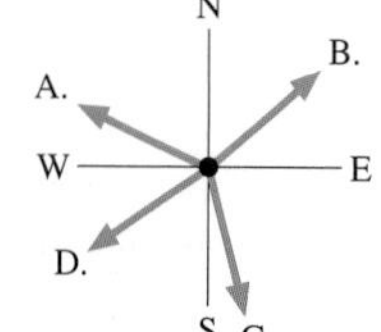

FIGURE Q9.5

6. Two students stand at rest, facing each other on frictionless skates. They then start tossing a heavy ball back and forth between them. Describe their subsequent motion.
7. Two particles collide, one of which was initially moving and the other initially at rest.
   a. Is it possible for *both* particles to be at rest after the collision? Give an example in which this happens, or explain why it can't happen.
   b. Is it possible for *one* particle to be at rest after the collision? Give an example in which this happens, or explain why it can't happen.
8. Automobiles are designed with "crumple zones" intended to collapse in a collision. Why would a manufacturer design part of a car so that it collapses in a collision?
9. You probably know that it feels better to catch a baseball if you are wearing a padded glove. Explain why this is so, using the ideas of momentum and impulse.
10. In the early days of rocketry, some people claimed that rockets couldn't fly in outer space as there was no air for the rockets to push against. Suppose you were an early investigator in the field of rocketry and met someone who made this argument. How would you convince the person that rockets could travel in space?

11. Two ice skaters, Megan and Jason, push off from each other on frictionless ice. Jason's mass is twice that of Megan.
    a. Which skater, if either, experiences the greater impulse during the push? Explain.
    b. Which skater, if either, has the greater speed after the push-off? Explain.
12. Suppose a rubber ball and a steel ball collide. Which, if either, receives the larger impulse? Explain.
13. While standing still on a basketball court, you throw the ball to a teammate. Why do you not move backward as a result? Is the law of conservation of momentum violated?
14. To win a prize at the county fair, you're trying to knock down a heavy bowling pin by hitting it with a thrown object. Should you choose to throw a rubber ball or a beanbag of equal size and weight? Explain.
15. Rank in order, from largest to smallest, the angular momenta L_1 through L_5 of the balls shown in Figure Q9.15. Explain.

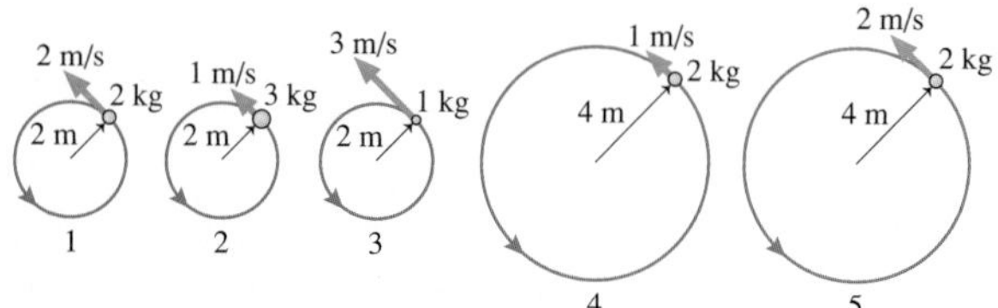

FIGURE Q9.15

16. Figure Q9.16 shows two masses held together by a thread on a rod that is rotating about its center with angular velocity ω. If the thread breaks, the masses will slide out to the ends of the rod. If that happens, will the rod's angular velocity increase, decrease, or remain unchanged? Explain.

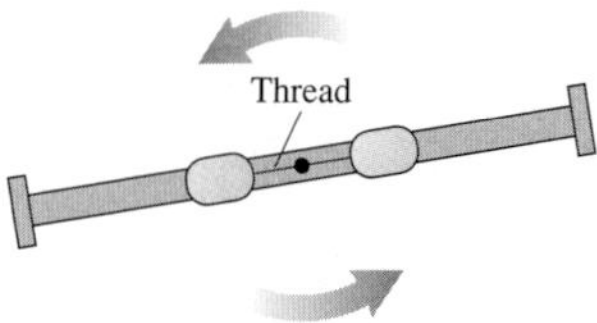

FIGURE Q9.16

17. If the earth warms significantly, the polar ice caps will melt. Water will move from the poles, near the earth's rotation axis, and will spread out around the globe. In principle, this will change the length of the day. Why? Will the length of the day increase or decrease?

18. The disks shown in Figure Q9.18 have equal mass. Is the angular momentum of disk 2, on the right, larger than, smaller than, or equal to the angular momentum of disk 1? Explain.

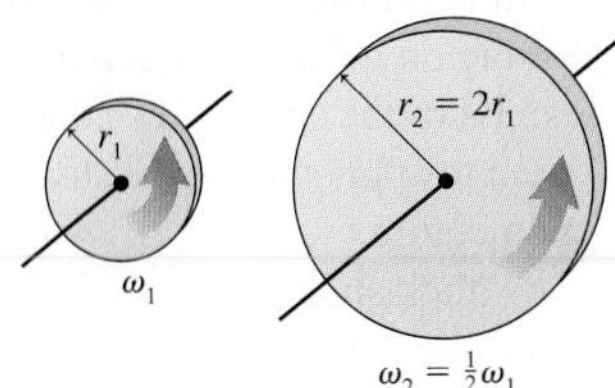

FIGURE Q9.18

Multiple-Choice Questions

19. | Curling is a sport played with 20 kg stones that slide across an ice surface. Suppose a curling stone sliding at 1 m/s strikes another stone and comes to rest in 2 ms. Approximately how much force is there on the stone during the impact?
A. 200 N B. 1000 N C. 2000 N D. 10,000 N

20. | Two balls are hung from cords. The first ball, of mass 1.0 kg, is pulled to the side and released, reaching a speed of 2.0 m/s at the bottom of its arc. Then, as shown in Figure Q9.20, it hits and sticks to another ball. The speed of the pair just after the collision is 1.2 m/s. What is the mass of the second ball?
A. 0.67 kg
B. 2.0 kg
C. 1.7 kg
D. 1.0 kg

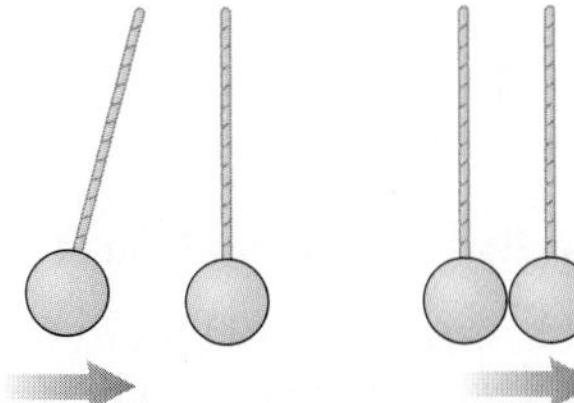

FIGURE Q9.20

21. | Figure Q9.21 shows two blocks sliding on a frictionless surface. Eventually the smaller block overtakes the larger one, collides with it, and sticks. What is the speed of the two blocks after the collision?
A. $v_i/2$ B. $4v_i/5$ C. v_i D. $5v_i/4$ E. $2v_i$

FIGURE Q9.21

22. | Two friends are sitting in a stationary canoe. At $t = 3.0$ s the person at the front tosses a sack to the person in the rear, who catches the sack 0.2 s later. Which plot in Figure Q9.22 shows the velocity of the boat as a function of time? Positive velocity is forward, negative velocity is backward. Neglect any drag force on the canoe from the water.

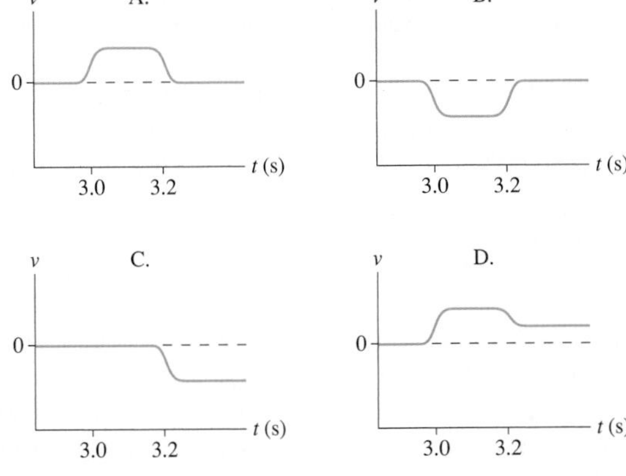

FIGURE Q9.22

23. || Two blocks, with masses $m_1 = 2.5$ kg and $m_2 = 14$ kg, approach each other along a horizontal, frictionless track. The initial velocities of the blocks are $v_1 = 12.0$ m/s to the right and $v_2 = 3.4$ m/s to the left. The two blocks then collide and stick together. Which of the graphs could represent the force of block 1 on block 2 during the collision?

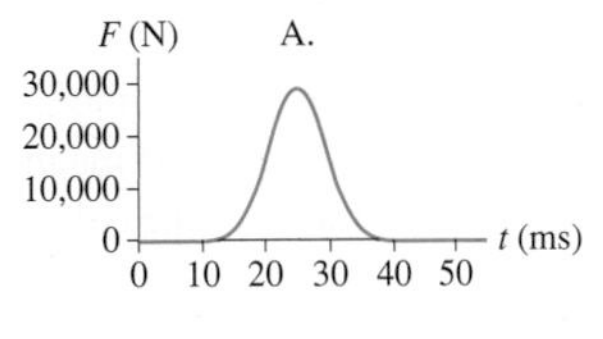

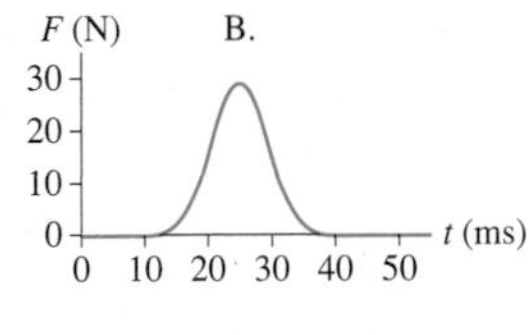

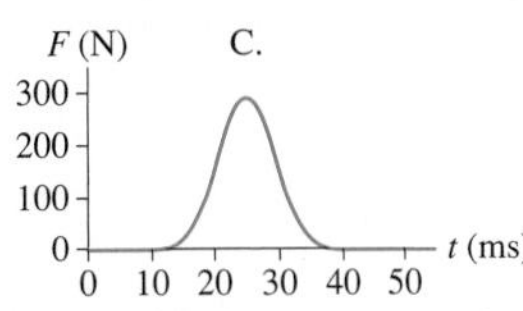

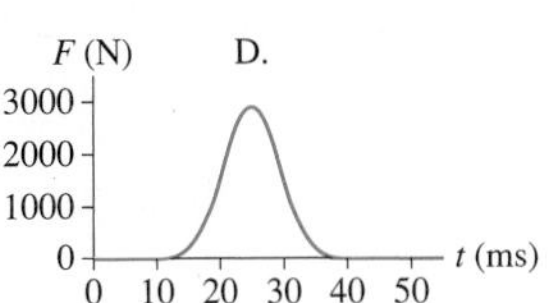

FIGURE Q9.23

24. | A small puck is sliding to the right with momentum $\vec{p}_i$ on a horizontal, frictionless surface, as shown in Figure Q9.24. A force is applied to the puck for a short time and its momentum afterward is $\vec{p}_f$. Which lettered arrow shows the direction of the impulse that was delivered to the puck?

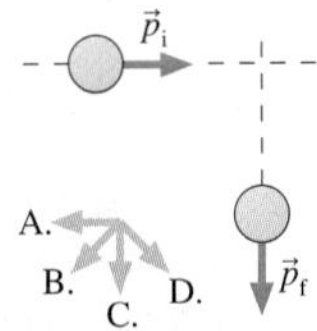

FIGURE Q9.24

25. | A red ball, initially at rest, is simultaneously hit by a blue ball traveling from west to east at 3 m/s and a green ball traveling east to west at 3 m/s. All three balls have equal mass. Afterward, the red ball is traveling south and the green ball is moving to the east. In which direction is the blue ball traveling?
A. West
B. North
C. Between north and west
D. Between north and east
E. Between south and west

26. | A 24 g, 3-cm-diameter thin, hollow sphere rotates at 30 rpm about a vertical, frictionless axis through its center. A 4 g bug stands at the top of the sphere. He then walks along the surface of the sphere until he reaches its "equator." When he reaches the equator, the sphere is rotating at
A. 15 rpm
B. 24 rpm
C. 30 rpm
D. 37 rpm
E. 45 rpm

27. | A 5.0 kg solid cylinder of radius 12 cm rotates with $\omega_i = 3.7$ rad/s about an axis through its center. A torque of 0.040 N · m is applied to the cylinder for 5.0 s. By how much does the cylinder's angular momentum change?
A. 0.12 kg · m²/s
B. 0.20 kg · m²/s
C. 0.38 kg · m²/s
D. 0.52 kg · m²/s
E. 0.88 kg · m²/s

PROBLEMS

Section 9.1 Impulse

Section 9.2 Momentum and the Impulse-Momentum Theorem

1. | At what speed do a bicycle and its rider, with a combined mass of 100 kg, have the same momentum as a 1500 kg car traveling at 5.0 m/s?
2. | A 57 g tennis ball is served at 45 m/s. If the ball started from rest, what impulse was applied to the ball by the racket?
3. || A student throws a 120 g snowball at 7.5 m/s at the side of the schoolhouse, where it hits and sticks. What is the magnitude of the average force on the wall if the duration of the collision is 0.15 s?
4. |||| In Figure P9.4, what value of F_{max} gives an impulse of 6.0 N · s?

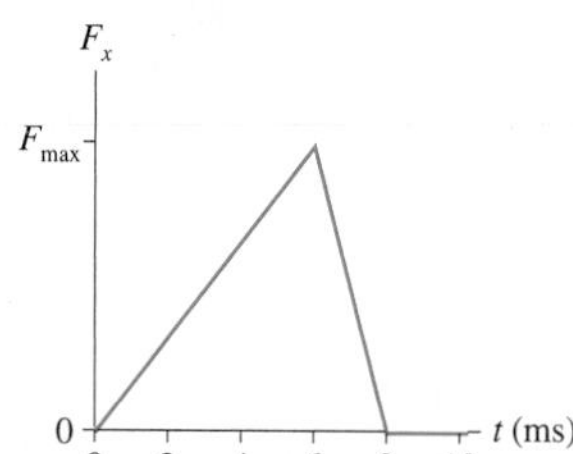

FIGURE P9.4

5. | A sled and rider, gliding over horizontal, frictionless ice at 4.0 m/s, have a combined mass of 80 kg. The sled then slides over a rough spot in the ice, slowing down to 3.0 m/s. What impulse was delivered to the sled by the friction force from the rough spot?

Section 9.3 Solving Impulse and Momentum Problems

6. || Use the impulse-momentum theorem to find how long a stone falling straight down takes to increase its speed from 5.5 m/s to 10.4 m/s.
7. || a. A 2.0 kg object is moving to the right with a speed of 1.0 m/s when it experiences the force shown in Figure P9.7a. What are the object's speed and direction after the force ends?
   b. Answer this question for the force shown in Figure P9.7b.

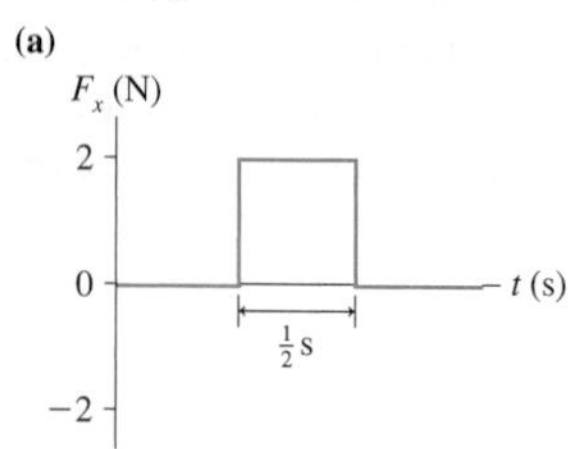

FIGURE P9.7

8. |||| A 60 g tennis ball with an initial speed of 32 m/s hits a wall and rebounds with the same speed. Figure P9.8 shows the force of the wall on the ball during the collision. What is the value of F_{max}, the maximum value of the contact force during the collision?

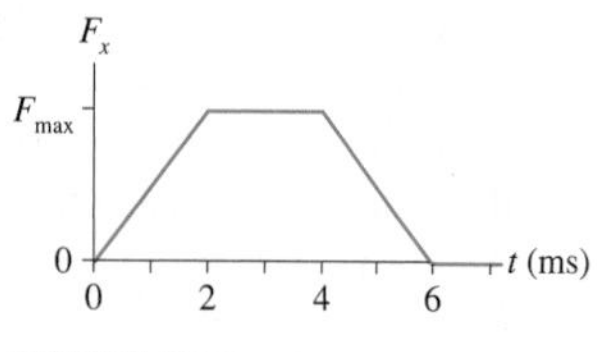

FIGURE P9.8

9. || A child is sliding on a sled at 1.5 m/s to the right. You stop the sled by pushing on it for 0.50 s in a direction opposite to its motion. If the mass of the child and sled is 35 kg, what average force do you need to apply to stop the sled? Use the concepts of impulse and momentum.
10. ||| An ice hockey puck slides along the ice at 12 m/s. A hockey stick delivers an impulse of 4.0 kg · m/s, causing the puck to move off in the opposite direction with the same speed. What is the mass of the puck?
11. | As part of a safety investigation, two 1400 kg cars traveling at 20 m/s are crashed into different barriers. Find the average forces exerted on (a) the car that hits a line of water barrels and takes 1.5 s to stop, and (b) the car that hits a concrete barrier and takes 0.1 s to stop.
12. || In a Little League baseball game, the 145 g ball enters the strike zone with a speed of 15.0 m/s. The batter hits the ball, and it leaves his bat with a speed of 20.0 m/s in exactly the opposite direction.
    a. What is the magnitude of the impulse delivered by the bat to the ball?
    b. If the bat is in contact with the ball for 1.5 ms, what is the magnitude of the average force exerted by the bat on the ball?

Section 9.4 Conservation of Momentum

13. |||| A small, 100 g cart is moving at 1.20 m/s on an air track when it collides with a larger, 1.00 kg cart at rest. After the collision, the small cart recoils at 0.850 m/s. What is the speed of the large cart after the collision?
14. || A man standing on very slick ice fires a rifle horizontally. The mass of the man together with the rifle is 70 kg, and the mass of the bullet is 10 g. If the bullet leaves the muzzle at a speed of 500 m/s, what is the final speed of the man?
15. ||| A 2.7 kg block of wood sits on a table. A 3.0 g bullet, fired horizontally at a speed of 500 m/s, goes completely through the block, emerging at a speed of 220 m/s. What is the speed of the block immediately after the bullet exits?
16. | A strong man is compressing a lightweight spring between two weights. One weight has a mass of 2.3 kg, the other a mass of 5.3 kg. He is holding the weights stationary, but then he loses his grip and the weights fly off in opposite directions. The lighter of the two is shot out at a speed of 6.0 m/s. What is the speed of the heavier weight?
17. || A 10,000 kg railroad car is rolling at 2.00 m/s when a 4000 kg load of gravel is suddenly dropped in. What is the car's speed just after the gravel is loaded?
18. | A 5000 kg open train car is rolling on frictionless rails at 22.0 m/s when it starts pouring rain. A few minutes later, the car's speed is 20.0 m/s. What mass of water has collected in the car?
19. || A 50.0 kg archer, standing on frictionless ice, shoots a 40 g arrow at a speed of 60 m/s. What is the recoil speed of the archer?
20. || A 9.5 kg dog takes a nap in a canoe and wakes up to find the canoe has drifted out onto the lake but now is stationary. He walks along the length of the canoe at 0.50 m/s, relative to the water, and the canoe simultaneously moves in the opposite direction at 0.15 m/s. What is the mass of the canoe?

Section 9.5 Inelastic Collisions

21. || A 300 g bird flying along at 6.0 m/s sees a 10 g insect heading straight toward it with a speed of 30 m/s. The bird opens its mouth wide and enjoys a nice lunch. What is the bird's speed immediately after swallowing?
22. || A 71 kg baseball player jumps straight up to catch a line drive. If the 140 g ball is moving horizontally at 28 m/s, and the catch is made when the ballplayer is at the highest point of his leap, what is his speed immediately after stopping the ball?
23. ||| A kid at the junior high cafeteria wants to propel an empty milk carton along a lunch table by hitting it with a 3.0 g spit ball. If he wants the speed of the 20 g carton just after the spit ball hits it to be 0.30 m/s, at what speed should his spit ball hit the carton?
24. | The parking brake on a 2000 kg Cadillac has failed, and it is rolling slowly, at 1 mph, toward a group of small children. Seeing the situation, you realize you have just enough time to drive your 1000 kg Volkswagen head-on into the Cadillac and save the children. With what speed should you impact the Cadillac to bring it to a halt?
25. | A 2.0 kg block slides along a frictionless surface at 1.0 m/s. A second block, sliding at a faster 4.0 m/s, collides with the first from behind and sticks to it. The final velocity of the combined blocks is 2.0 m/s. What was the mass of the second block?

Section 9.6 Momentum and Collisions in Two Dimensions

26. |||| A 20 g ball of clay traveling east at 3.0 m/s collides with a 30 g ball of clay traveling north at 2.0 m/s. What are the speed and the direction of the resulting 50 g ball of clay?
27. || Two particles collide and bounce apart. Figure P9.27 shows the initial momenta of both and the final momentum of particle 2. What is the final momentum of particle 1? Show your answer by copying the figure and drawing the final momentum vector on the figure.

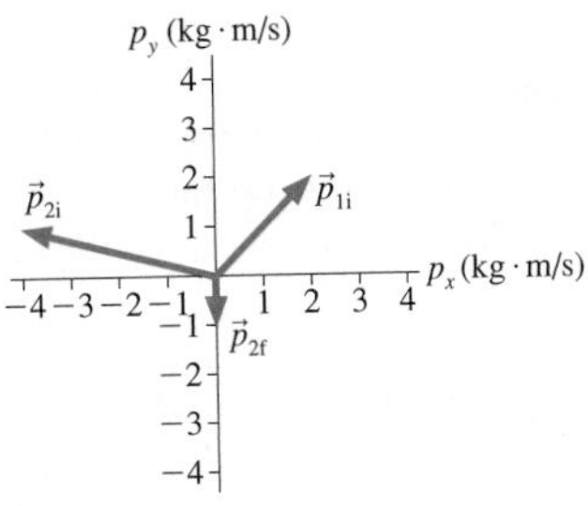

FIGURE P9.27

28. || A 20 g ball of clay traveling east at 2.0 m/s collides with a 30 g ball of clay traveling 30° south of west at 1.0 m/s. What are the speed and direction of the resulting 50 g blob of clay?
29. || A firecracker in a coconut blows the coconut into three pieces. Two pieces of equal mass fly off south and west, perpendicular to each other, at 20 m/s. The third piece has twice the mass as the other two. What are the speed and direction of the third piece?

Section 9.7 Angular Momentum

30. ||| What is the angular momentum of the moon around the INT earth? The moon's mass is 7.4×10^{22} kg and it orbits 3.8×10^{8} m from the earth.
31. ||| A little girl is going on the merry-go-round for the first time, and wants her 47 kg mother to stand next to her on the ride, 2.6 m from the merry-go-round's center. If her mother's speed is 4.2 m/s when the ride is in motion, what is her angular momentum around the center of the merry-go-round?
32. || What is the angular momentum about the axle of the 500 g rotating bar in Figure P9.32?

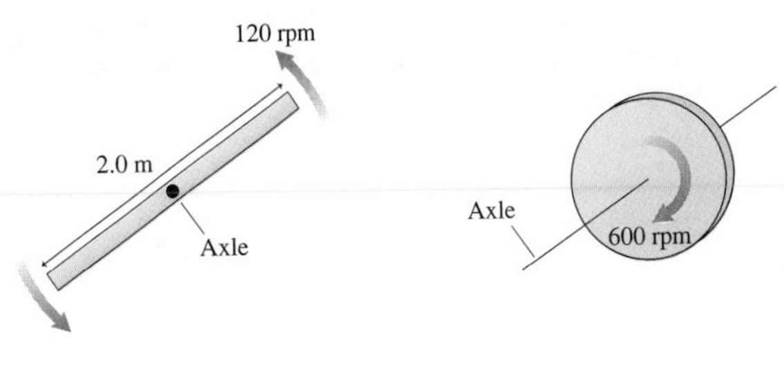

FIGURE P9.32 FIGURE P9.33

33. |||| What is the angular momentum about the axle of the 2.0 kg, 4.0-cm-diameter rotating disk in Figure P9.33?
34. | Divers change their body position in midair while rotating BIO about their center of mass. In one dive, the diver leaves the board with her body nearly straight, then tucks into a somersault position. If the moment of inertia of the diver in a straight position is 14 kg · m² and in a tucked position is 4.0 kg · m², by what factor is her angular velocity when tucked greater than when straight?
35. || Ice skaters often end their performances with spin turns, BIO where they spin very fast about their center of mass with their arms folded in and legs together. Upon ending, their arms extend outward, proclaiming their finish. Not quite as noticeably, one leg goes out as well. Suppose that the moment of inertia of a skater with arms out and one leg extended is 3.2 kg · m² and for arms and legs in is 0.80 kg · m². If she starts out spinning at 5.0 rev/s, what is her angular speed (in rev/s) when her arms and one leg open outward?

General Problems

36. ||| What is the impulse on a 3.0 kg particle that experiences the force described by the graph in Figure P9.36?

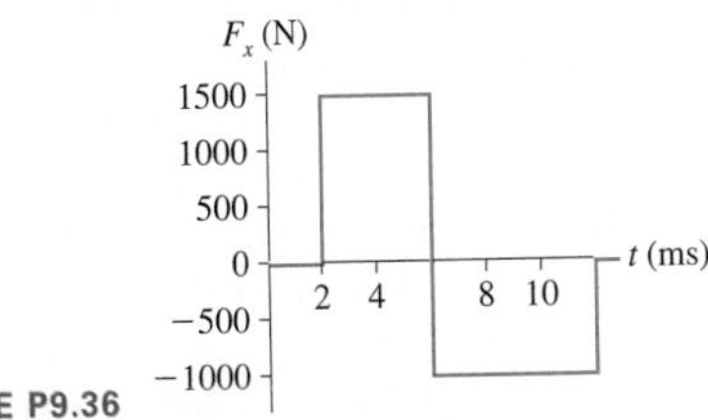

FIGURE P9.36

37. ||| A 600 g air-track glider collides with a spring at one end of the track. Figure P9.37 shows the glider's velocity and the force exerted on the glider by the spring. How long is the glider in contact with the spring?

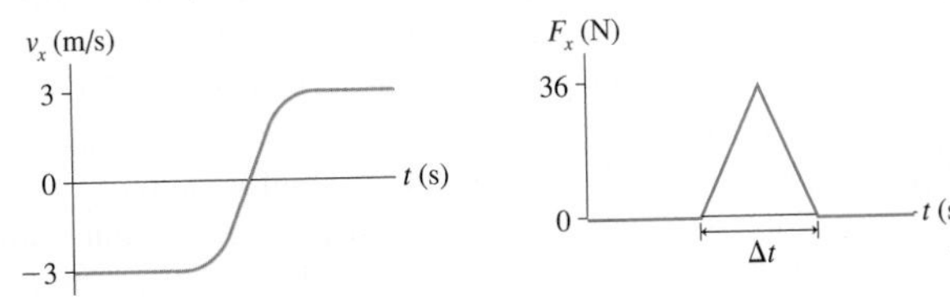

FIGURE P9.37

38. || Far in space, where gravity is negligible, a 425 kg rocket traveling at 75.0 m/s in the positive x-direction fires its engines. Figure P9.38 shows the thrust force as a function of time. The mass lost by the rocket during these 30.0 s is negligible.
   a. What impulse does the engine impart to the rocket?
   b. At what time does the rocket reach its maximum speed? What is the maximum speed?

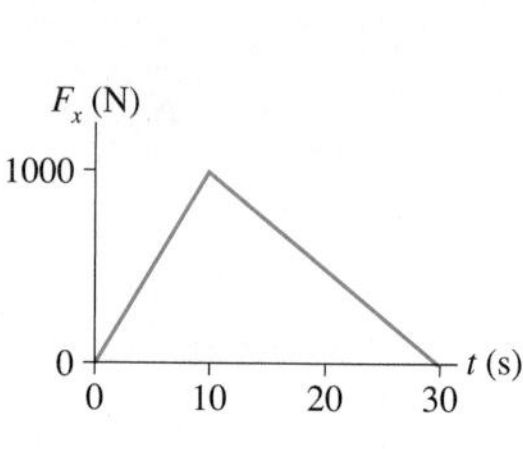

FIGURE P9.38

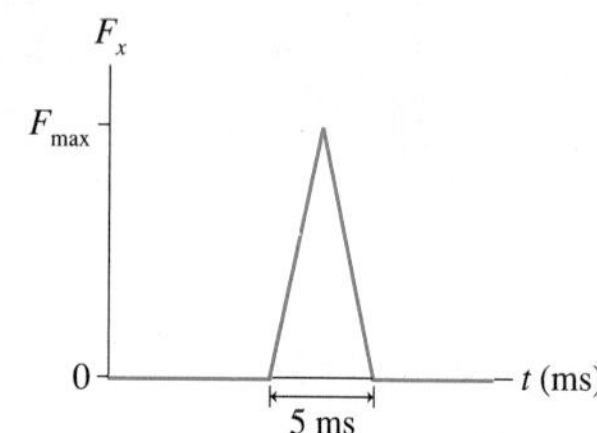

FIGURE P9.39

39. |||| A 200 g ball is dropped from a height of 2.0 m, bounces on a hard floor, and rebounds to a height of 1.5 m. Figure P9.39 shows the impulse received from the floor. What maximum force does the floor exert on the ball?

INT

40. |||| A 200 g ball is dropped from a height of 2.0 m and bounces on a hard floor. The force on the ball from the floor is shown in Figure P9.40. How high does the ball rebound?

INT

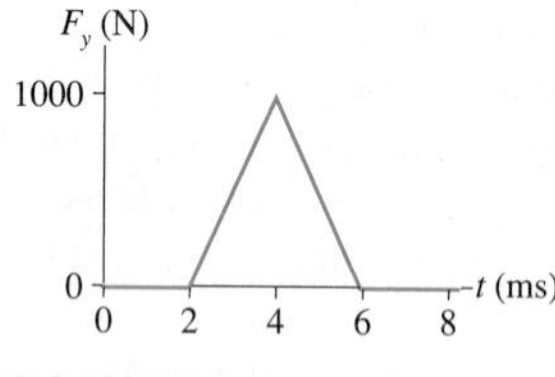

FIGURE P9.40

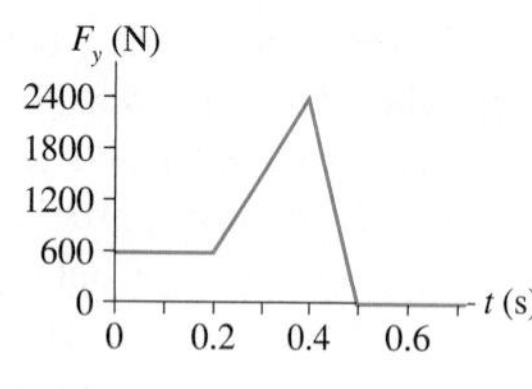

FIGURE P9.41

41. ||| Figure P9.41 is a graph of the force exerted by the floor on a woman making a vertical jump. At what speed does she leave the ground?

INT

Hint: The force of the floor is not the only force acting on the woman.

42. || A sled slides along a horizontal surface for which the coefficient of kinetic friction is 0.25. Its velocity at point A is 8.0 m/s and at point B is 5.0 m/s. Use the impulse-momentum theorem to find how long the sled takes to travel from A to B.

INT

43. ||| A 140 g baseball is moving horizontally to the right at 35 m/s when it is hit by the bat. The ball flies off to the left at 55 m/s, at an angle of 25° above the horizontal. What are the magnitude and direction of the impulse that the bat delivers to the ball?

44. || Squids rely on jet propulsion, a versatile technique to move around in water. A 1.5 kg squid at rest suddenly expels 0.10 kg of water backward to quickly get itself moving forward at 3.0 m/s. If other forces (such as the drag force on the squid) are ignored, what is the speed with which the squid expels the water?

BIO

45. ||| The flowers of the bunchberry plant open with astonishing force and speed, causing the pollen grains to be ejected out of the flower in a mere 0.30 ms at an acceleration of 2.5×10^4 m/s^2. If the acceleration is constant, what impulse is delivered to a pollen grain with a mass of 1.0×10^{-7} g?

BIO

46. || a. With what speed are pollen grains ejected from a bunchberry flower? See Problem 45 for information.

BIO

   b. Suppose that 1000 ejected pollen grains slam into the abdomen of a 5.0 g bee that is hovering just above the flower. If the collision is perfectly inelastic, what is the bee's speed immediately afterward? Is the bee likely to notice?

47. ||| A tennis player swings her 1000 g racket with a speed of 10 m/s. She hits a 60 g tennis ball that was approaching her at a speed of 20 m/s. The ball rebounds at 40 m/s.
   a. How fast is her racket moving immediately after the impact? You can ignore the interaction of the racket with her hand for the brief duration of the collision.
   b. If the tennis ball and racket are in contact for 10 ms, what is the average force that the racket exerts on the ball?

48. || A 20 g ball of clay is thrown horizontally at 30 m/s toward a 1.0 kg block sitting at rest on a frictionless surface. The clay hits and sticks to the block.
   a. What is the speed of the block and clay right after the collision?
   b. Use the block's initial and final speeds to calculate the impulse the clay exerts on the block.
   c. Use the clay's initial and final speeds to calculate the impulse the block exerts on the clay.
   d. Does $\vec{J}_{\text{block on clay}} = -\vec{J}_{\text{clay on block}}$?

49. || Dan is gliding on his skateboard at 4.0 m/s. He suddenly jumps backward off the skateboard, kicking the skateboard forward at 8.0 m/s. How fast is Dan going as his feet hit the ground? Dan's mass is 50 kg and the skateboard's mass is 5.0 kg.

50. | James and Sarah stand on a stationary cart with frictionless wheels. The total mass of the cart and riders is 130 kg. At the same instant, James throws a 1.0 kg ball to Sarah at 4.5 m/s, while Sarah throws a 0.50 kg ball to James at 1.0 m/s. James's throw is to the right and Sarah's is to the left.
   a. While the two balls are in the air, what are the speed and direction of the cart and its riders?
   b. After the balls are caught, what are the speed and direction of the cart and riders?

51. ||| Ethan, whose mass is 80 kg, stands at one end of a very long, stationary wheeled cart that has a mass of 500 kg. He then starts sprinting toward the other end of the cart. He soon reaches his top speed of 8.0 m/s, measured relative to the cart. What is the speed of the cart when Ethan has reached his top speed?

52. || The cars of a long coal train are filled by pulling them under a hopper, from which coal falls into the cars at a rate of 10,000 kg/s. Ignoring friction due to the rails, what is the average force that the engine must exert on the coal train to keep it moving under the hopper at a speed of 0.50 m/s?

53. || Three identical train cars, coupled together, are rolling east at 2.0 m/s. A fourth car traveling east at 4.0 m/s catches up with the three and couples to make a four-car train. A moment later, the train cars hit a fifth car that was at rest on the tracks, and it couples to make a five-car train. What is the speed of the five-car train?

54. | A 110 kg linebacker running at 2.0 m/s and an 82 kg quarterback running at 3.0 m/s have a head-on collision in midair. The linebacker grabs and holds onto the quarterback. Who ends up moving forward after they hit?

55. || Most geologists believe that the dinosaurs became extinct 65 million years ago when a large comet or asteroid struck the earth, throwing up so much dust that the sun was blocked out for a period of many months. Suppose an asteroid with a diameter of 2.0 km and a mass of 1.0×10^{13} kg hits the earth with an impact speed of 4.0×10^4 m/s.

INT

   a. What is the earth's recoil speed after such a collision? (Use a reference frame in which the earth was initially at rest.)
   b. What percentage is this of the earth's speed around the sun? (Use the astronomical data inside the back cover.)

56. ||| At the center of a 50-m-diameter circular ice rink, a 75 kg skater traveling north at 2.5 m/s collides with and holds onto a 60 kg skater who had been heading west at 3.5 m/s.
    a. How long will it take them to glide to the edge of the rink?
    b. Where will they reach it? Give your answer as an angle north of west.
57. || Two ice skaters, with masses of 50 kg and 75 kg, are at the center of a 60-m-diameter circular rink. The skaters push off against each other and glide to opposite edges of the rink. If the heavier skater reaches the edge in 20 s, how long does the lighter skater take to reach the edge?
58. ||| One billiard ball is shot east at 2.00 m/s. A second, identical billiard ball is shot west at 1.00 m/s. The balls have a glancing collision, not a head-on collision, deflecting the second ball by 90° and sending it north at 1.41 m/s. What are the speed and direction of the first ball after the collision?
59. |||| A 10 g bullet is fired into a 10 kg wood block that is at rest on a wood table. The block, with the bullet embedded, slides 5.0 cm across the table. What was the speed of the bullet?
60. ||| You are part of a search-and-rescue mission that has been called out to look for a lost explorer. You've found the missing explorer, but you're separated from him by a 200-m-high cliff and a 30-m-wide raging river, as shown in Figure P9.60. To save his life, you need to get a 5.0 kg package of emergency supplies across the river. Unfortunately, you can't throw the package hard enough to make it across. Fortunately, you happen to have a 1.0 kg rocket intended for launching flares. Improvising quickly, you attach a sharpened stick to the front of the rocket, so that it will impale itself into the package of supplies, then fire the rocket at ground level toward the supplies. What minimum speed must the rocket have just before impact in order to save the explorer's life?

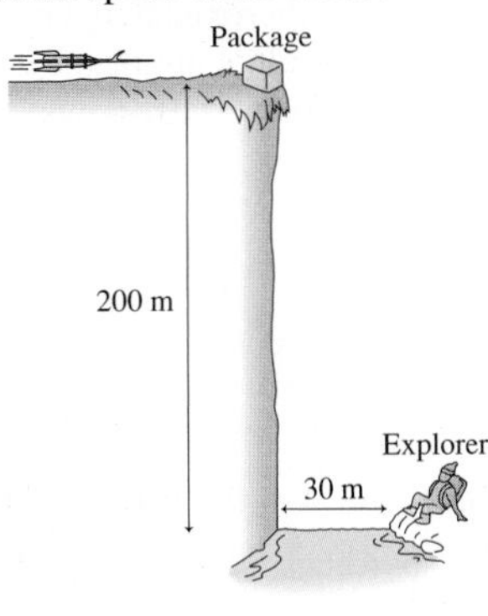

FIGURE P9.60

61. |||| A 1500 kg weather rocket accelerates upward at 10.0 m/s^2. It explodes 2.00 s after liftoff and breaks into two fragments, one twice as massive as the other. Photos reveal that the lighter fragment traveled straight up and reached a maximum height of 530 m. What were the speed and direction of the heavier fragment just after the explosion?
62. || Two 500 g blocks of wood are 2.0 m apart on a frictionless table. A 10 g bullet is fired at 400 m/s toward the blocks. It passes all the way through the first block, then embeds itself in the second block. The speed of the first block immediately afterward is 6.0 m/s. What is the speed of the second block after the bullet stops?
63. | A 500 kg cannon fires a 10 kg cannonball with a speed of 200 m/s relative to the muzzle. The cannon is on wheels that roll without friction. When the cannon fires, what is the speed of the cannonball relative to the earth?
64. ||| Laura, whose mass is 35 kg, jumps horizontally off a 55 kg canoe at 1.5 m/s relative to the canoe. What is the canoe's speed just after she jumps?
65. || A spaceship of mass 2.0×10^6 kg is cruising at a speed of 5.0×10^6 m/s when the antimatter reactor fails, blowing the ship into three pieces. One section, having a mass of 5.0×10^5 kg, is blown straight backward with a speed of 2.0×10^6 m/s. A second piece, with mass 8.0×10^5 kg, continues forward at 1.0×10^6 m/s. What are the direction and speed of the third piece?
66. ||| A proton is shot at 5.0×10^7 m/s toward a gold target. The nucleus of a gold atom, with a mass 197 times that of the proton, repels the proton and deflects it straight back with 90% of its initial speed. What is the recoil speed of the gold nucleus?
67. ||| Figure P9.67 shows a collision between three balls of clay. The three hit simultaneously and stick together. What are the speed and direction of the resulting blob of clay?

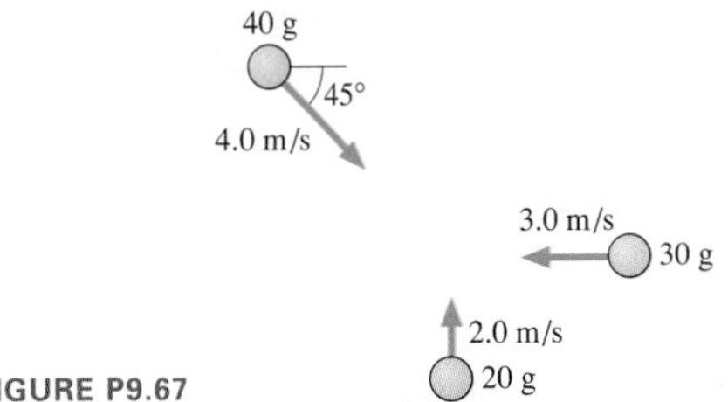

FIGURE P9.67

68. ||| The carbon isotope ^{14}C is used for carbon dating of archeological artifacts. ^{14}C (mass 2.34×10^{-26} kg) decays by the process known as *beta decay* in which the nucleus emits an electron (the beta particle) and a subatomic particle called a neutrino. In one such decay, the electron and the neutrino are emitted at right angles to each other. The electron (mass 9.11×10^{-31} kg) has a speed of 5.00×10^7 m/s and the neutrino has a momentum of 8.00×10^{-24} kg · m/s. What is the recoil speed of the nucleus?
69. ||| A 1.0-m-long massless rod is pivoted at one end and swings around in a circle on a frictionless table. A block with a hole through the center can slide in and out along the rod. Initially, a small piece of wax holds the block 30 cm from the pivot. The block is spun at 50 rpm, then the temperature of the rod is slowly increased. When the wax melts, the block slides out to the end of the rod. What is the final angular speed? Give your answer in rpm.
70. || A 200 g puck revolves in a circle on a frictionless table at the end of a 50.0-cm-long string. The puck's angular momentum about the center of the circle is 3.00 kg · m^2/s. What is the tension in the string?
71. ||| Figure P9.71 shows a 100 g puck revolving in a 20-cm-radius circle on a frictionless table. The string passes through a hole in the center of the table and is tied to two 200 g weights.
    a. What speed does the puck need to support the two weights?
    b. The lower weight is a light bag filled with sand. Suppose a pin pokes a hole in the bag and the sand slowly leaks out while the puck is revolving. What will be the puck's speed and the radius of its trajectory after all of the sand is gone?

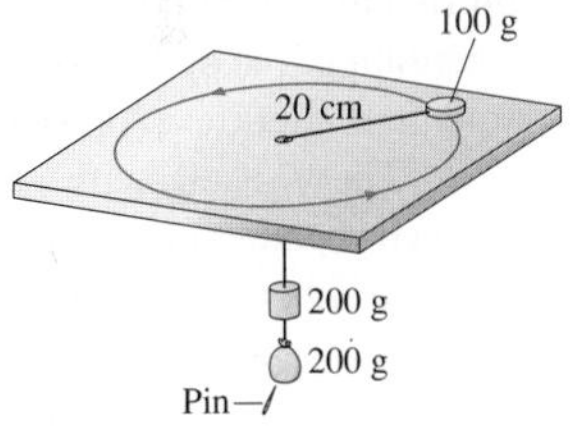

FIGURE P9.71

72. || A 2.0 kg, 20-cm-diameter turntable rotates at 100 rpm on frictionless bearings. Two 500 g blocks fall from above, hit the turntable simultaneously at opposite ends of a diagonal, and stick. What is the turntable's angular speed, in rpm, just after this event?
73. || Joey, from Example 9.10, stands at rest at the outer edge of the frictionless merry-go-round of Figure 9.28. The merry-go-round is also at rest. Joey then begins to run around the perimeter of the merry-go-round, finally reaching a constant speed, measured relative to the ground, of 5.0 m/s. What is the final angular speed of the merry-go-round?

74. ||| A 3.0-m-diameter merry-go-round with a mass of 250 kg is spinning at 20 rpm. John runs around the merry-go-round at 5.0 m/s, in the same direction that it is turning, and jumps onto the outer edge. John's mass is 30 kg. What is the merry-go-round's angular speed, in rpm, after John jumps on?

75. ||| Disk A, with a mass of 2.0 kg and a radius of 40 cm, rotates clockwise about a frictionless vertical axle at 30 rev/s. Disk B, also 2.0 kg but with a radius of 20 cm, rotates counterclockwise about that same axle, but at a greater height than disk A, at 30 rev/s. Disk B slides down the axle until it lands on top of disk A, after which they rotate together. After the collision, what is their common angular speed (in rev/s) and in which direction do they rotate?

Passage Problems

Hitting a Golf Ball

Consider a golf club hitting a golf ball. To a good approximation, we can model this as a collision between the rapidly moving head of the golf club and the stationary golf ball, ignoring the shaft of the club and the golfer.

A golf ball has a mass of 46 g. Suppose a 200 g club head is moving at a speed of 40 m/s just before striking the golf ball. After the collision, the golf ball's speed is 60 m/s.

76. | What is the momentum of the club + ball system right before the collision?
A. 1.8 kg · m/s
B. 8.0 kg · m/s
C. 3220 kg · m/s
D. 8000 kg · m/s

77. | Immediately after the collision, the momentum of the club + ball system will be
A. Less than before the collision.
B. The same as before the collision.
C. More than before the collision.

78. | A manufacturer makes a golf ball that compresses more than a traditional golf ball when struck by a club. How will this affect the average force during the collision?
A. The force will decrease.
B. The force will not be affected.
C. The force will increase.

79. | By approximately how much does the club head slow down as a result of hitting the ball?
A. 4 m/s B. 6 m/s C. 14 m/s D. 26 m/s

STOP TO THINK ANSWERS

Stop to Think 9.1: F. The cart is initially moving in the negative x-direction, so $(p_x)_i = -20$ kg · m/s. After it bounces, $(p_x)_f =$ 10 kg · m/s. Thus $\Delta p =$ (10 kg · m/s) − (−20 kg · m/s) = 30 kg · m/s.

Stop to Think 9.2: B. The clay ball goes from $(v_x)_i = v$ to $(v_x)_f = 0$, so $J_{clay} = \Delta p_x = -mv$. The rubber ball rebounds, going from $(v_x)_i = v$ to $(v_x)_f = -v$ (same speed, opposite direction). Thus $J_{rubber} = \Delta p_x = -2mv$. The rubber ball has a greater momentum change, and this requires a greater impulse.

Stop to Think 9.3: Right end. The balls started at rest, so the total momentum of the system is zero. It's an isolated system, so the total momentum after the explosion is still zero. The 6 g ball has momentum $6v$. The 4 g ball, with velocity $-2v$, has momentum $-8v$. The combined momentum of these two balls is $-2v$. In order for P to be zero, the third ball must have a *positive* momentum $(+2v)$ and thus a positive velocity.

Stop to Think 9.4: B. The momentum of particle 1 is (0.40 kg)(2.5 m/s) = 1.0 kg · m/s, while that of particle 2 is (0.80 kg)(−1.5 m/s) = −1.2 kg · m/s. The total momentum is then 1.0 kg · m/s − 1.2 kg · m/s = −0.2 kg · m/s. Because it's negative, the total momentum, and hence the final velocity of the particles, is directed to the left.

Stop to Think 9.5: B. Angular momentum $L = I\omega$ is conserved. Both boys have mass m and initially stand distance $R/2$ from the axis. Thus the initial moment of inertia is $I_i = I_{disk} + 2 \times m(R/2)^2 = I_{disk} + \frac{1}{2}mR^2$. The final moment of inertia is $I_f = I_{disk} + 0 + mR^2$, because the boy standing at the axis contributes nothing to the moment of inertia. Because $I_f > I_i$ we must have $\omega_f < \omega_i$.

10 Energy and Work

As this bungee jumper falls, he gains kinetic energy, the energy of motion. Where does this energy come from? And where does it go as he slows at the bottom of his fall?

LOOKING AHEAD ▸▸

The goals of Chapter 10 are to introduce the concept of energy and to learn a new problem-solving strategy based on conservation of energy.

Forms of Energy

A principal goal of this chapter is to learn about several important forms of energy.

Kinetic energy is the energy of motion. This heavy, fast-moving rhinoceros has lots of kinetic energy.

These passengers gain **potential energy,** the energy of position, as they ride up the escalator.

The **thermal energy** of this red-hot horseshoe is associated with the microscopic motion of its molecules.

Looking Back ◂◂
2.5 Motion with constant acceleration
7.1, 7.4 Rotation and moment of inertia
8.3 Hooke's law

The Law of Conservation of Energy

One of the most fundamental laws of physics, the **law of conservation of energy** states that the total energy of an isolated system is a constant.

How fast are these water sliders moving at the bottom? How fast does the rock fly out of the slingshot? We'll use conservation of energy and the before-and-after analysis introduced in Chapter 9 to solve these kinds of problems.

Looking Back ◂◂
9.2–9.3 Before-and after visual overviews

Transferring Energy

Energy can be *transferred* into a system by pushing on it, a process called **work.**

The bobsledders do work on the sled, *transferring* energy to it and causing it to speed up.

Transforming Energy

Energy of one kind can change into energy of a different kind. These **energy transformations** are what make the world an interesting place.

As this race car skids to a stop, its kinetic energy is being *transformed* into thermal energy, making the tires hot enough to smoke.

Power

We're very often interested in **power,** the *rate* at which energy is transformed from one kind into another.

As they climb, this truck and these jets both transform the chemical energy of their fuel into potential energy. But the jet engines transform energy at a rate 70 times that of the truck's engine—their *power* is much greater.

10.1 The Basic Energy Model

Energy. It's a word you hear all the time. We use chemical energy to heat our homes and bodies, electric energy to run our lights and computers, and solar energy to grow our crops and forests. We're told to use energy wisely and not to waste it. Athletes and weary students consume "energy bars" and "energy drinks."

But just what is energy? The concept of energy has grown and changed over time, and it is not easy to define in a general way just what energy is. Rather than starting with a formal definition, we'll let the concept of energy expand slowly over the course of several chapters. In this chapter we introduce several fundamental forms of energy, including kinetic energy, potential energy, and thermal energy. Our goal is to understand the characteristics of energy, how energy is used, and, especially important, how energy is transformed from one form into another. Much of modern technology is concerned with transforming energy, such as changing the chemical energy of oil molecules into electric energy or into the kinetic energy of your car.

We'll also learn how energy can be transferred to or from a system by the application of mechanical forces. By pushing on a sled, you increase its speed, and hence its energy of motion. By lifting a heavy object, you increase its gravitational potential energy.

These observations will lead us to discover a very powerful conservation law for energy. Energy is neither created nor destroyed: If one form of energy in a system decreases, it must appear in an equal amount in another form. Many scientists consider the law of conservation of energy to be the most important of all the laws of nature. This law will have implications throughout the rest of this book.

Systems and Energy

In Chapter 9 we introduced the idea of a *system* of interacting objects. A system can be as simple as a falling acorn or as complex as a city. But whether simple or complex, every system in nature has associated with it a quantity we call its **total energy** E. The total energy is the sum of the different kinds of energies present in the system. In the table below, we give a brief overview of some of the more important forms of energy; in the rest of the chapter, we'll look at several of these forms of energy in much greater detail.

A system may have many of these kinds of energy at one time. For instance, a moving car has kinetic energy of motion, chemical energy stored in its gasoline, thermal energy in its hot engine, and many other forms of energy. The total energy of the system, E, is the *sum* of all the different energies present in the system:

$$E = K + U_g + U_s + E_{th} + E_{chem} + \cdots \tag{10.1}$$

The energies shown in this sum are the forms of energy in which we'll be most interested in this and the next chapter. The ellipses ($\cdots$) stand for other forms of energy, such as nuclear or electric, that also might be present. We'll treat these and others in later chapters.

Some important forms of energy

Kinetic energy K

Kinetic energy is the energy of *motion.* All moving objects have kinetic energy. The heavier an object and the faster it moves, the more kinetic energy it has. The wrecking ball in this picture is effective in part because of its large kinetic energy.

Gravitational potential energy U_g

Gravitational potential energy is *stored* energy associated with an object's *height above the ground.* As this coaster ascends, energy is stored as gravitational potential energy. As it descends, this stored energy is converted into kinetic energy.

Elastic or spring potential energy U_s

Elastic potential energy is energy stored when a spring or other elastic object, such as this archer's bow, is *stretched.* This energy can later be transformed into the kinetic energy of the arrow.

Continued

Thermal energy E_{th}

Hot objects have more *thermal energy* than cold ones because the molecules in a hot object jiggle around more than those in a cold object. Thermal energy is the sum of the microscopic kinetic and potential energies of all the molecules in an object. In boiling water, some molecules have enough energy to escape the water as steam.

Chemical energy E_{chem}

Electric forces cause atoms to bind together to make molecules. Energy can be stored in these bonds, energy that can later be released as the bonds are rearranged during chemical reactions. When we burn fuel to run our car or eat food to power our bodies, we are using *chemical energy.*

Nuclear energy $E_{nuclear}$

An enormous amount of energy is stored in the *nucleus,* the tiny core of an atom. Certain nuclei can be made to break apart, releasing some of this *nuclear energy,* which is transformed into the kinetic energy of the fragments and then into thermal energy. The ghostly blue glow of a nuclear reactor results from high-energy fragments as they travel through water.

Energy Transformations

We've seen that all systems contain energy in many different forms. But if the amounts of each form of energy never changed, the world would be a very dull place. What makes the world interesting is that **energy of one kind can be *transformed* into energy of another kind.** The gravitational potential energy of the roller coaster at the top of the track is rapidly transformed into kinetic energy as the coaster descends; the chemical energy of gasoline is transformed into the kinetic energy of your moving car. The following table illustrates a few common energy transformations. In this table, we use an arrow $\rightarrow$ as a shorthand way of representing an energy transformation.

Some energy transformations

A weightlifter lifts a barbell over her head

The barbell has much more gravitational potential energy when high above her head than when on the floor. To lift the barbell, she is transforming chemical energy in her body into gravitational potential energy of the barbell.

$$E_{chem} \rightarrow U_g$$

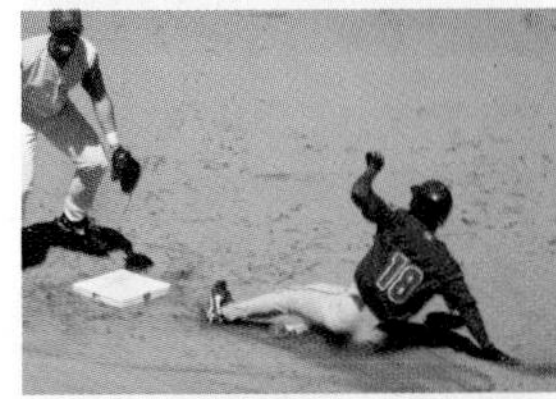

A base runner slides into the base

When running, he has lots of kinetic energy. After sliding, he has none. His kinetic energy is transformed mainly into thermal energy: The ground and his legs are slightly warmer.

$$K \rightarrow E_{th}$$

A burning campfire

The wood contains considerable chemical energy. When the carbon in the wood combines chemically with oxygen in the air, this chemical energy is transformed largely into thermal energy of the hot gases and embers.

$$E_{chem} \rightarrow E_{th}$$

A springboard diver

Here's a two-step energy transformation. At the instant shown, the board is flexed to its maximum extent, so that elastic potential energy stored in the board. Soon this energy will begin to be transformed into kinetic energy; as the diver rises into the air and slows, this kinetic energy will be transformed into gravitational potential energy.

$$U_s \rightarrow K \rightarrow U_g$$

FIGURE 10.1 Energy transformations occur within the system.

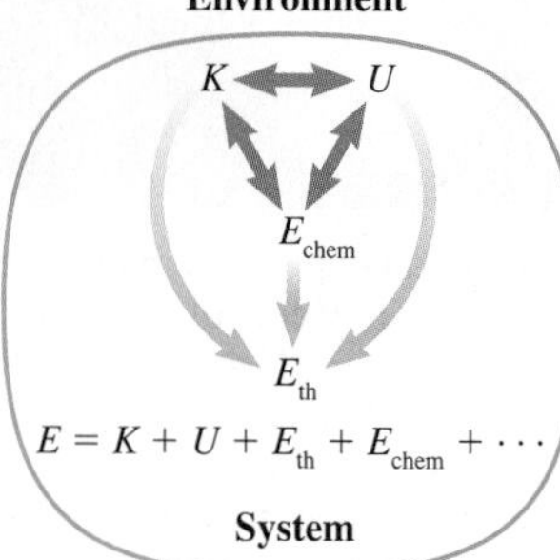
The *environment* is everything that is *not* part of the system.
Environment
K
U
E_{chem}
E_{th}
$E = K + U + E_{th} + E_{chem} + \cdots$
System

FIGURE 10.1 reinforces the idea that **energy transformations are changes of energy *within* the system from one form to another.** (The U in this figure is a generic potential energy; it could be gravitational potential energy U_g, spring potential energy U_s, or some other form of potential energy.) Note that it is easy to convert kinetic, potential, and chemical energies into thermal energy, but converting thermal energy back into these other forms is not so easy. How it can be done, and what possible limitations there might be in doing so, will form a large part of the next chapter.

Energy Transfers and Work

We've just seen that energy *transformations* occur between forms of energy *within* a system. But every physical system also interacts with the world around it—that is, with its *environment.* In the course of these interactions, the system can exchange energy with the environment. **An exchange of energy between system and environment is called an energy *transfer*.** There are two primary energy-transfer processes: **work,** the *mechanical* transfer of energy to or from a system by pushing or pulling on it, and **heat,** the *nonmechanical* transfer of energy from the environment to the system (or vice versa) because of a temperature difference between the two.

FIGURE 10.2 The basic energy model shows that work and heat are energy transfers into and out of the system, while energy transformations occur within the system.

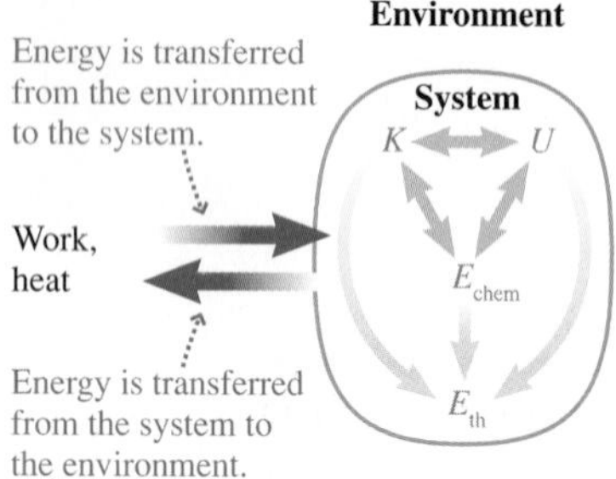
Environment
Energy is transferred from the environment to the system.
System
K
U
Work, heat
E_{chem}
Energy is transferred from the system to the environment.
E_{th}

FIGURE 10.2, which we call the **basic energy model,** shows how our energy model is modified to include energy transfers into and out of the system as well as energy transformations within the system. In this chapter we'll consider only energy transfers by means of work; the concept of heat will be developed much further in Chapters 11 and 12.

"Work" is a common word in the English language, with many meanings. When you first think of work, you probably think of physical effort or the job you do to make a living. After all, we talk about "working out," or we say, "I just got home from work." But that is not what work means in physics.

In physics, "work" is the process of *transferring* energy from the environment to a system, or from a system to the environment, by the application of mechanical forces—pushes and pulls—to the system. Once the energy has been transferred to the system, it can appear in many forms. Exactly what form it takes depends on the details of the system and how the forces are applied. The table below gives three examples of energy transfers due to work. We use W as the symbol for work.

Energy transfers: work

Putting a shot

The system: The shot

The environment: The athlete

As the athlete pushes on the shot to get it moving, he is doing work on the system; that is, he is transferring energy from himself to the ball. The energy transferred to the system appears as kinetic energy.

The transfer: $W \rightarrow K$

Striking a match

The system: The match and matchbox

The environment: The hand

As the hand quickly pulls the match across the box, the hand does work on the system, increasing its thermal energy. The match head becomes hot enough to ignite.

The transfer: $W \rightarrow E_{th}$

Firing a slingshot

The system: The slingshot

The environment: The boy

As the boy pulls back on the elastic bands, he does work on the system, increasing its elastic potential energy.

The transfer: $W \rightarrow U_s$

Notice that in each example on the previous page, the environment applies a force while the system undergoes a *displacement.* Energy is transferred as work only when the system *moves* while the force acts. A force applied to a stationary object, such as when you push against a wall, transfers no energy to the object and thus does no work.

NOTE ► In the table on the previous page, energy is being transferred *from* the athlete *to* the shot by the force of his hand. We say he "does work" on the shot, or "work is done" by the force of his hand. ◄

The Law of Conservation of Energy

Work done on a system represents energy that is transferred into or out of the system. This transferred energy *changes* the system's energy by exactly the amount of work W that was done. Writing the change in the system's energy as ΔE, we can represent this idea mathematically as

$$\Delta E = W \tag{10.2}$$

Now the total energy E of a system is, according to Equation 10.1, the sum of the different energies present in the system. Thus the change in E is the sum of the *changes* of the different energies present. Then Equation 10.2 gives what is called the *work-energy equation:*

The work-energy equation The total energy of a system changes by the amount of work done on it:

$$\Delta E = \Delta K + \Delta U_g + \Delta U_s + \Delta E_{th} + \Delta E_{chem} + \cdots = W \tag{10.3}$$

NOTE ► Equation 10.3, the work-energy equation, is the mathematical representation of the basic energy model of Figure 10.2. Together, they are the heart of what the subject of energy is all about. ◄

Suppose now we have an **isolated system,** one that is separated from its surrounding environment in such a way that no energy is transferred into or out of the system. This means that *no work is done on the system.* The energy within the system may be transformed from one form into another, but it is a deep and remarkable fact of nature that, during these transformations, the total energy of an isolated system—the *sum* of all the individual kinds of energy—remains *constant,* as shown in FIGURE 10.3. We say that **the total energy of an isolated system is *conserved*.**

FIGURE 10.3 An isolated system.

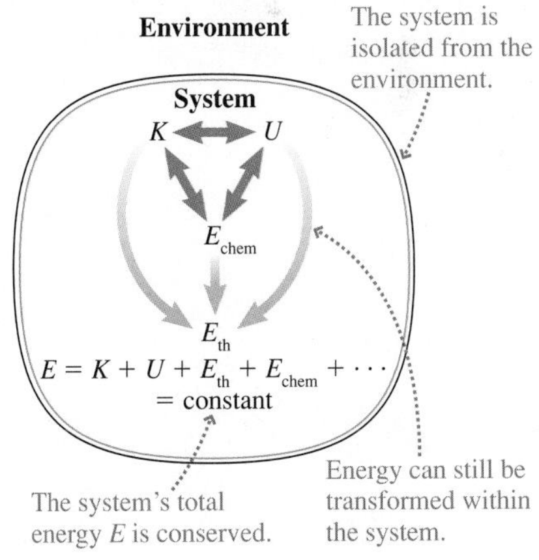
Environment
The system is isolated from the environment.
System
K
U
E_{chem}
E_{th}
$E = K + U + E_{th} + E_{chem} + \cdots$
= constant
The system's total energy E is conserved.
Energy can still be transformed within the system.

For an isolated system, we must set $W = 0$ in Equation 10.3, leading to the following *law of conservation of energy:*

Law of conservation of energy The total energy of an isolated system remains constant:

$$\Delta E = \Delta K + \Delta U_g + \Delta U_s + \Delta E_{th} + \Delta E_{chem} + \cdots = 0 \tag{10.4}$$

The law of conservation of energy is similar to the law of conservation of momentum. Momentum changes when an impulse acts on a system; the total momentum of an isolated system doesn't change. Similarly, energy changes when external forces do work on a system; the total energy of an isolated system doesn't change.

In solving momentum problems, we adopted a new before-and-after perspective: The momentum *after* an interaction was the same as the momentum *before* the interaction. We will introduce a similar before-and-after perspective for energy that will lead to an extremely powerful problem-solving strategy.

Before using energy ideas to solve problems, however, we first need to develop quantitative expressions for work, kinetic energy, potential energy, and thermal energy. This will be our task in the next several sections.

STOP TO THINK 10.1 A child slides down a playground slide at constant speed. The energy transformation is

A. $U_g \rightarrow K$ B. $K \rightarrow U_g$ C. $W \rightarrow K$ D. $U_g \rightarrow E_{th}$ E. $K \rightarrow E_{th}$

10.2 Work

Our first task is to learn how work is calculated. We've just seen that work is the transfer of energy to or from a system by the application of forces exerted on the system by the environment. Thus work is done on a system by forces from *outside* the system; we call such forces *external forces*. Only external forces can change the energy of a system. *Internal forces*—forces between objects *within* the system—cause energy transformations within the system but don't change the system's total energy.

We also learned that in order for energy to be transferred as work, the system must undergo a displacement—it must *move*—during the time that the force is applied. Let's further investigate the relationship among work, force, and displacement.

Consider a system consisting of a windsurfer at rest, as shown on the left in FIGURE 10.4. Let's assume that there is no friction between his board and the water. Initially the system has no kinetic energy. But if a force from outside the system, such as the force due to the wind, begins to act on the system, the surfer will begin to speed up, and his kinetic energy will increase. In terms of energy transfers, we would say that the energy of the system has increased because of the work done on the system by the force of the wind.

FIGURE 10.4 The force of the wind does work on the system, increasing its kinetic energy K.

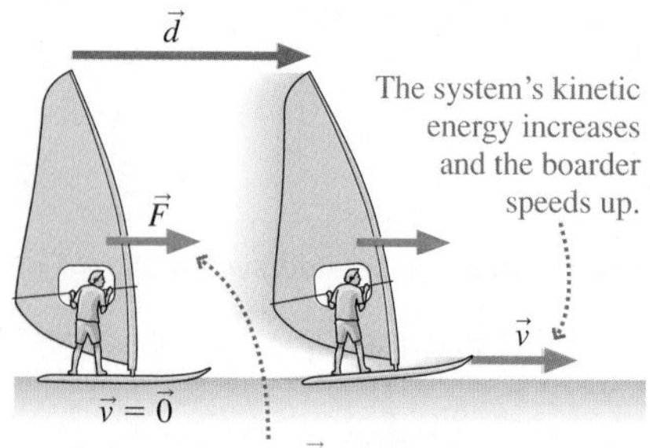
$\vec{d}$
The system's kinetic energy increases and the boarder speeds up.
$\vec{F}$
$\vec{v}$
$\vec{v} = \vec{0}$
The force of the wind $\vec{F}$ does work on the system.

What determines how much work is done by the force of the wind? First, we note that the greater the distance over which the wind pushes the surfer, the faster the surfer goes, and the more his kinetic energy increases. This implies a greater transfer of energy. So, **the larger the displacement, the greater the work done.** Second, if the wind pushes with a stronger force, the surfer speeds up more rapidly, and the change in his kinetic energy is greater than with a weaker force. **The stronger the force, the greater the work done.**

This experiment suggests that the amount of energy transferred to a system by a force $\vec{F}$—that is, the amount of work done by $\vec{F}$—depends on both the magnitude F of the force *and* the displacement d of the system. Many experiments of this kind have established that the amount of work done by $\vec{F}$ is *proportional* to both F and d. For the simplest case described above, where the force $\vec{F}$ is constant and points in the direction of the object's displacement, the expression for the work done is found to be

$$W = Fd \tag{10.5}$$

Work done by a constant force $\vec{F}$ in the direction of a displacement $\vec{d}$

The unit of work, that of force multiplied by distance, is N · m. This unit is so important that it has been given its own name, the **joule** (rhymes with *tool*). We define:

$$1 \text{ joule} = 1 \text{ J} = 1 \text{ N} \cdot \text{m}$$

Because work is simply energy being transferred, **the joule is the unit of *all* forms of energy.** Note that work is a *scalar* quantity.

EXAMPLE 10.1 Work done in pushing a crate

Sarah pushes a heavy crate 3.0 m along the floor at a constant speed. She pushes with a constant horizontal force of magnitude 70 N. How much work does Sarah do on the crate?

PREPARE We begin with the visual overview in **FIGURE 10.5**. Sarah pushes with a constant force in the direction of the crate's motion, so we can use Equation 10.5 to find the work done.

FIGURE 10.5 Sarah pushing a crate.

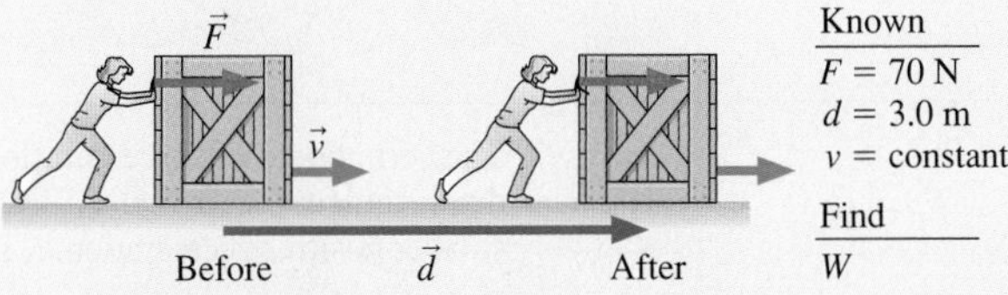

SOLVE The work done by Sarah is

$$W = Fd = (70 \text{ N})(3.0 \text{ m}) = 210 \text{ J}$$

ASSESS Work represents a transfer of energy into a system, so here the energy of the system—the box and the floor—increases. Unlike the windsurfer, the box doesn't speed up, so its kinetic energy doesn't increase. Instead, the work increases the thermal energy in the crate and the part of the floor along which it slides, increasing the temperature of both. Using the notation of Equation 10.3, we can write this energy transfer as $\Delta E_{th} = W$.

Force at an Angle to the Displacement

A force does the greatest possible amount of work on an object when the force points in the same direction as the object's displacement. Less work is done when the force acts at an angle to the displacement. To see this, consider the kite buggy of **FIGURE 10.6a**, pulled along a horizontal path by the angled force of the kite string $\vec{F}$. As shown in **FIGURE 10.6b**, we can divide $\vec{F}$ into a component $F_\perp$ perpendicular to the motion, and a component $F_\parallel$ parallel to the motion. Only the parallel component acts to accelerate the rider and increase his kinetic energy, so only the parallel component does work on the rider. From Figure 10.6b, we see that if the angle between $\vec{F}$ and the displacement is θ, then the parallel component is $F_\parallel = F\cos\theta$. So, when the force acts at an angle θ to the direction of the displacement, we have

FIGURE 10.6 Finding the work done when the force is at an angle to the displacement.

(a)

(b) The rider undergoes a displacement $\vec{d}$.

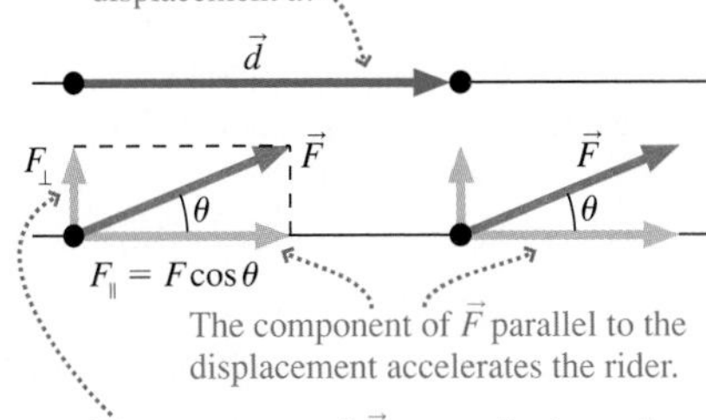

The component of $\vec{F}$ perpendicular to the displacement only pulls up on the rider. It doesn't accelerate him.

$$W = F_\parallel d = Fd\cos\theta \quad (10.6)$$

Work done by a constant force $\vec{F}$ at an angle θ to the displacement $\vec{d}$

Notice that this more general definition of work agrees with Equation 10.5 if $\theta = 0°$.

Tactics Box 10.1 shows how to calculate the work done by a force at any angle to the direction of motion. The system illustrated is a block sliding on a frictionless, horizontal surface, so that only the kinetic energy is changing. However, the same relationships hold for any object undergoing a displacement.

The quantities F and d are always positive, so **the sign of W is determined entirely by the angle θ between the force and the displacement.** Note that Equation 10.6, $W = Fd\cos\theta$, is valid for any angle θ. In three special cases, $\theta = 0°$, $\theta = 90°$, and $\theta = 180°$, however, there are simple versions of Equation 10.6 that you can use. These are noted in Tactics Box 10.1.

5.1

TACTICS BOX 10.1 Calculating the work done by a constant force

Direction of force relative to displacement	Angles and work done	Sign of W	Energy transfer
Before: $\vec{v}_i$ After: $\vec{v}_f$ $\vec{d}$ $\vec{F}$ $\theta = 0°$	$\theta = 0°$ $\cos\theta = 1$ $W = Fd$	+	The force is in the direction of motion. The block has its greatest positive acceleration. K increases the most: **Maximum energy transfer to system.**
$\theta < 90°$ $\vec{d}$ $\vec{F}$	$\theta < 90°$ $W = Fd\cos\theta$	+	The component of force parallel to the displacement is less than F. The block has a smaller positive acceleration. K increases less: **Decreased energy transfer to system.**
$\theta = 90°$ $\vec{d}$ $\vec{F}$	$\theta = 90°$ $\cos\theta = 0$ $W = 0$	0	There is no component of force in the direction of motion. The block moves at constant speed. No change in K: **No energy transferred.**
$\theta > 90°$ $\vec{F}$ $\vec{d}$	$\theta > 90°$ $W = Fd\cos\theta$	−	The component of force parallel to the displacement is opposite the motion. The block slows down, and K decreases: **Decreased energy transfer *out* of system.**
$\theta = 180°$ $\vec{F}$ $\vec{d}$	$\theta = 180°$ $\cos\theta = -1$ $W = -Fd$	−	The force is directly opposite the motion. The block has its greatest deceleration. K decreases the most: **Maximum energy transfer *out* of system.**

Exercises 5–6

EXAMPLE 10.2 Work done in pulling a suitcase

A strap inclined upward at a 45° angle pulls a suitcase through the airport. The tension in the strap is 20 N. How much work does the tension do if the suitcase is pulled 100 m at a constant speed?

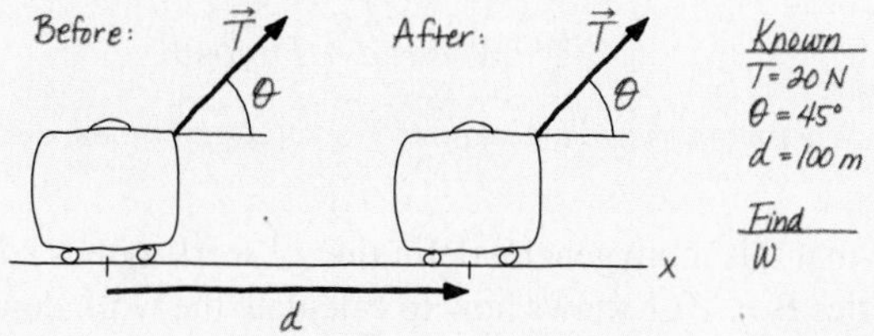

FIGURE 10.7 A suitcase pulled by a strap.

PREPARE FIGURE 10.7 shows a visual overview. Since the suitcase moves at a constant speed, there must be a rolling friction force acting to the left.

SOLVE We can use Equation 10.6, with force $F = T$, to find that the tension does work:

$$W = Td\cos\theta = (20\text{ N})(100\text{ m})\cos 45° = 1400\text{ J}$$

ASSESS Because a person is pulling on the other end of the strap, causing the tension, we would say informally that the person does 1400 J of work on the suitcase. This work represents energy transferred into the suitcase + floor system. Since the suitcase moves at a constant speed, the system's kinetic energy doesn't change. Thus, just as for Sarah pushing the crate in Example 10.1, the work done goes entirely into increasing the thermal energy E_{th} of the suitcase and the floor.

CONCEPTUAL EXAMPLE 10.3 **Work done by a parachute**

A drag racer is slowed by a parachute. What is the sign of the work done?

REASON The drag force on the drag racer is shown in **FIGURE 10.8**, along with the dragster's displacement as it slows. The force points in the direction opposite the displacement, so that the angle θ in Equation 10.6 is 180°. Then $\cos\theta = \cos(180°) = -1$. Because F and d in Equation 10.6 are magnitudes, and hence positive, the work $W = Fd\cos\theta = -Fd$ done by the drag force is *negative.*

FIGURE 10.8 The force acting on a drag racer.

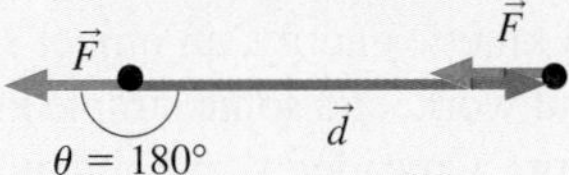

ASSESS Applying Equation 10.3 to this situation, we have

$$\Delta K = W$$

because the only system energy that changes is the racer's kinetic energy K. Because the kinetic energy is decreasing, its change ΔK is negative. This agrees with the sign of W. This example illustrates the general principle that **negative work represents a transfer of energy out of the system.**

If several forces act on an object that undergoes a displacement, each does work on the object. The **total** (or **net**) **work** W_{total} is the sum of the work done by each force. The total work represents the total energy transfer *to* the system from the environment (if $W_{total} > 0$) or *from* the system to the environment (if $W_{total} < 0$).

Forces That Do No Work

The fact that a force acts on an object doesn't mean that the force will do work on the object. The table below shows three common cases where a force does no work.

Forces that do no work

If the object undergoes no displacement while the force acts, no work is done.

This can sometimes seem counterintuitive. The weightlifter struggles mightily to hold the barbell over his head. But during the time the barbell remains stationary, he does no work on it because its displacement is zero. Why then is it so hard for him to hold it there? We'll see in Chapter 11 that it takes a rapid conversion of his internal chemical energy to keep his arms extended under this great load.

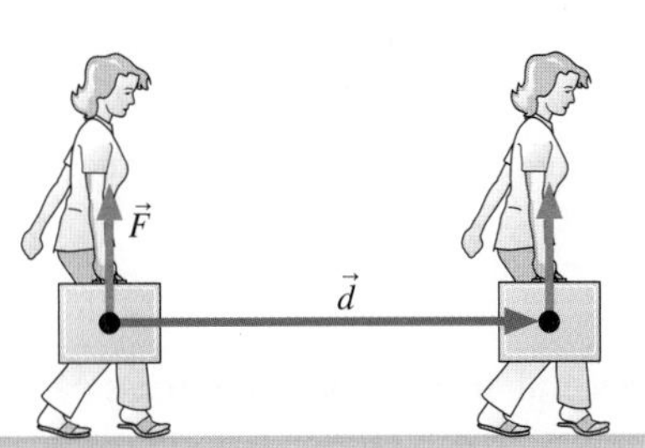

A force perpendicular to the displacement does no work.

The woman exerts only a vertical force on the briefcase she's carrying. This force has no component in the direction of the displacement, so the briefcase moves at a constant velocity and its kinetic energy remains constant. Since the energy of the briefcase doesn't change, it must be that no energy is being transferred to it as work.

(This is the case where $\theta = 90°$ in Tactics Box 10.1.)

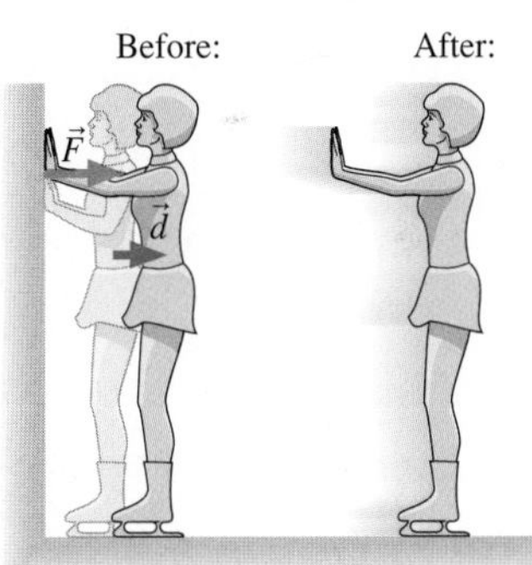

If the part of the object on which the force acts undergoes no displacement, no work is done.

Even though the wall pushes on the skater with a normal force $\vec{n}$ and she undergoes a displacement $\vec{d}$, the wall does no work on her, because the point of her body on which $\vec{n}$ acts—her hands—undergoes no displacement. This makes sense: How could energy be transferred as work from an inert, stationary object? So where does her kinetic energy come from? This will be the subject of much of Chapter 11. Can you guess?

STOP TO THINK 10.2 Which force does the most work?

A. The 10 N force.
B. The 8 N force.
C. The 6 N force.
D. They all do the same amount of work.

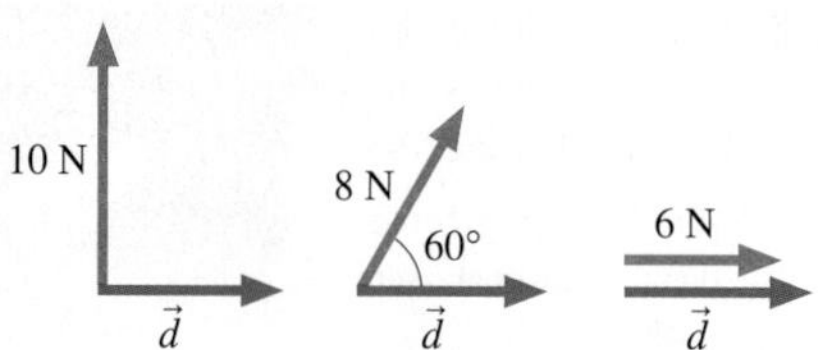

10.3 Kinetic Energy

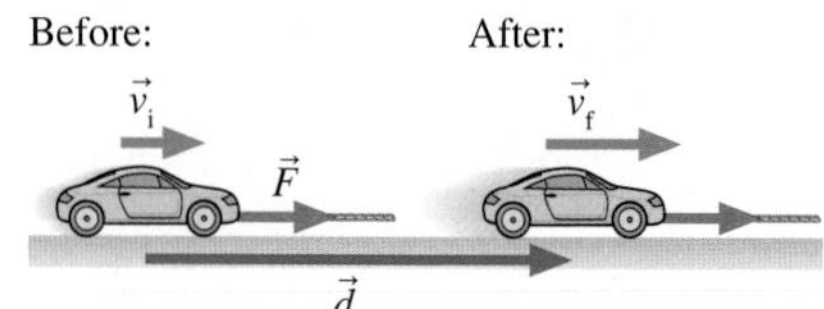

FIGURE 10.9 The work done by the tow rope increases the car's kinetic energy.

We've already qualitatively discussed kinetic energy, an object's energy of motion. Let's now use what we've learned about work, and some simple kinematics, to find a quantitative expression for kinetic energy. Consider a car being pulled by a tow rope, as in FIGURE 10.9. The rope pulls with a constant force $\vec{F}$ while the car undergoes a displacement $\vec{d}$, so the force does work $W = Fd$ on the car. If we ignore friction and drag, the work done by $\vec{F}$ is transferred entirely into the car's energy of motion—its kinetic energy. In this case, the change in the car's kinetic energy is given by the work-energy equation, Equation 10.3, as

$$W = \Delta K = K_f - K_i \tag{10.7}$$

Using kinematics, we can find another expression for the work done, in terms of the car's initial and final speeds. Recall from Chapter 2 the kinematic equation

$$v_f^2 = v_i^2 + 2a\Delta x$$

Applied to the motion of our car, $\Delta x = d$ is the car's displacement and, from Newton's second law, the acceleration is $a = F/m$. Thus we can write

$$v_f^2 = v_i^2 + \frac{2Fd}{m} = v_i^2 + \frac{2W}{m}$$

where we have replaced Fd with the work W. If we now solve for the work, we find

$$W = \frac{1}{2}m\left(v_f^2 - v_i^2\right) = \frac{1}{2}mv_f^2 - \frac{1}{2}mv_i^2$$

If we compare this result with Equation 10.7, we see that

$$K_f = \frac{1}{2}mv_f^2 \qquad \text{and} \qquad K_i = \frac{1}{2}mv_i^2$$

6.1 Activ ONLINE Physics

In general, then, an object of mass m moving with speed v has kinetic energy

$$K = \frac{1}{2}mv^2 \tag{10.8}$$

Kinetic energy of an object of mass m moving with speed v

K v p.47 QUADRATIC

TABLE 10.1 Some approximate kinetic energies

Object	Kinetic energy
Ant walking	1×10^{-8} J
Penny dropped 1 m	2.5×10^{-3} J
Person walking	70 J
Fastball, 100 mph	150 J
Bullet	5000 J
Car, 60 mph	5×10^{5} J
Supertanker, 20 mph	2×10^{10} J

From Equation 10.8, the units of kinetic energy are those of mass times speed squared, or kg · (m/s)^2. But

$$1 \text{ kg} \cdot (\text{m/s})^2 = \underbrace{1 \text{ kg} \cdot (\text{m/s}^2)}_{1\text{ N}} \cdot \text{m} = 1 \text{ N} \cdot \text{m} = 1 \text{ J}$$

We see that the units of kinetic energy are the same as those of work, as they must be. Table 10.1 gives some approximate kinetic energies. Everyday kinetic energies range from a tiny fraction of a fraction of a joule to nearly a million joules for a speeding car.

CONCEPTUAL EXAMPLE 10.4 Kinetic energy changes for a car

Compare the increase in a 1000 kg car's kinetic energy as it speeds up by 5.0 m/s, starting from 5.0 m/s, to its increase in kinetic energy as it speeds up by 5.0 m/s, starting from 10 m/s.

REASON The change in the car's kinetic energy in going from 5.0 m/s to 10 m/s is

$$\Delta K_{5\to10} = \frac{1}{2}mv_f^2 - \frac{1}{2}mv_i^2$$

This gives

$$\Delta K_{5\to10} = \frac{1}{2}(1000\ \text{kg})(10\ \text{m/s})^2 - \frac{1}{2}(1000\ \text{kg})(5.0\ \text{m/s})^2$$

$$= 3.8 \times 10^4\ \text{J}$$

Similarly, increasing from 10 m/s to 15 m/s requires

$$\Delta K_{10\to15} = \frac{1}{2}(1000\ \text{kg})(15\ \text{m/s})^2 - \frac{1}{2}(1000\ \text{kg})(10\ \text{m/s})^2$$

$$= 6.3 \times 10^4\ \text{J}$$

Even though the increase in the car's *speed* is the same in both cases, the increase in kinetic energy is substantially greater in the second case.

ASSESS Kinetic energy depends on the *square* of the speed v. In **FIGURE 10.10**, which plots kinetic energy versus speed, we see that the energy of the car increases rapidly with speed. We can also see graphically why the change in K for a 5 m/s change in v is greater at high speeds than at low speeds. In part this is why it's harder to accelerate your car at high speeds than at low speeds.

FIGURE 10.10 The kinetic energy increases as the *square* of the speed.

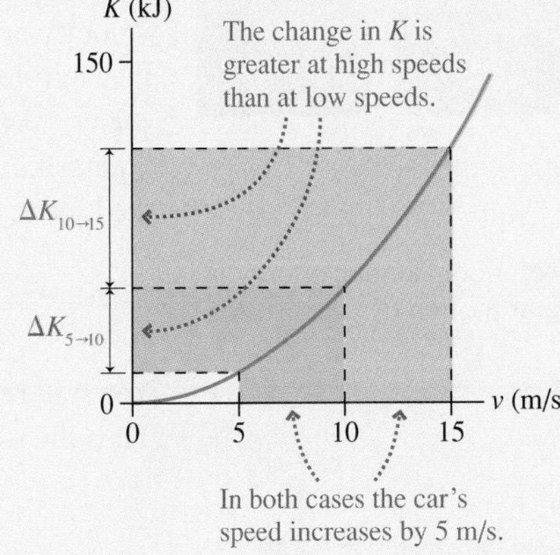

EXAMPLE 10.5 Speed of a bobsled after pushing

A two-man bobsled has a mass of 390 kg. Starting from rest, the two racers push the sled for the first 50 m with a net force of 270 N. Neglecting friction, what is the sled's speed at the end of the 50 m?

PREPARE We can find the sled's final speed if we can find its final kinetic energy. We can do so by equating the work done by the racers as they push on the sled to the change in its kinetic energy. **FIGURE 10.11** lists the known quantities and the quantity (v_f) that we want to find.

FIGURE 10.11 The work done by the pushers increases the sled's kinetic energy.

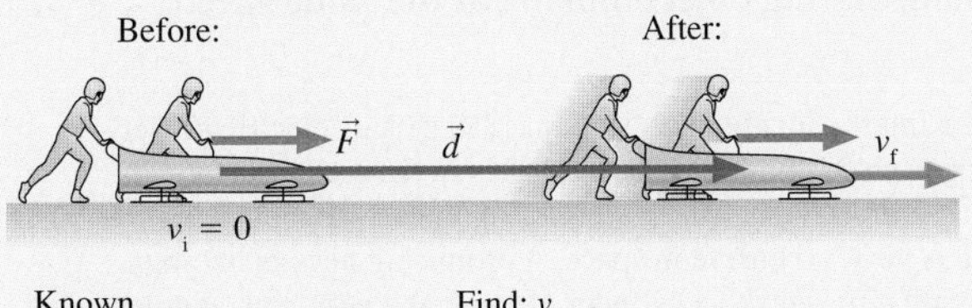

Known

$m = 390$ kg $F = 270$ N
$d = 50$ m $v_i = 0$ m/s

Find: v_f

SOLVE From Equation 10.3, the work-energy equation, the change in the sled's kinetic energy is $\Delta K = K_f - K_i = W$. The sled's final kinetic energy is thus

$$K_f = K_i + W$$

Using our expressions for kinetic energy and work, we get

$$\frac{1}{2}mv_f^2 = \frac{1}{2}mv_i^2 + Fd$$

Because $v_i = 0$, the work-energy equation reduces to

$$\frac{1}{2}mv_f^2 = Fd$$

We can solve for the final speed to get

$$v_f = \sqrt{\frac{2Fd}{m}} = \sqrt{\frac{2(270\ \text{N})(50\ \text{m})}{390\ \text{kg}}} = 8.3\ \text{m/s}$$

ASSESS 8.3 m/s, about 18 mph, seems a reasonable speed for two fast pushers to attain.

STOP TO THINK 10.3 Rank in order, from greatest to least, the kinetic energies of the sliding pucks.

A. 1 kg, 2 m/s
B. 1 kg, 3 m/s
C. 1 kg, −2 m/s
D. 2 kg, 2 m/s

FIGURE 10.12 The large rotating blades of a windmill have a great deal of kinetic energy.

Rotational Kinetic Energy

We've just found an expression for the kinetic energy of an object moving along a line or some other path. This energy is called **translational kinetic energy.** Consider now an object rotating about a fixed axis, such as the windmill blades in FIGURE 10.12. Although the blades have no overall translational motion, each particle in the blade is moving and hence has kinetic energy. Adding up the kinetic energy for each particle that makes up the blades, we find that the blades have **rotational kinetic energy,** the kinetic energy due to rotation.

FIGURE 10.13 shows two of the particles making up a windmill blade that rotates with angular velocity ω. Recall from Section 6.2 that a particle moving with angular velocity ω in a circle of radius r has a speed $v = \omega r$. Thus particle 1, which rotates in a circle of radius r_1, moves with speed $v_1 = r_1\omega$ and so has kinetic energy $\frac{1}{2}m_1v_1^2 = \frac{1}{2}m_1r_1^2\omega^2$. Similarly, particle 2, which rotates in a circle with a larger radius r_2, has kinetic energy $\frac{1}{2}m_2r_2^2\omega^2$. The object's rotational kinetic energy is the sum of the kinetic energies of *all* the particles:

FIGURE 10.13 Rotational kinetic energy is due to the circular motion of the particles.

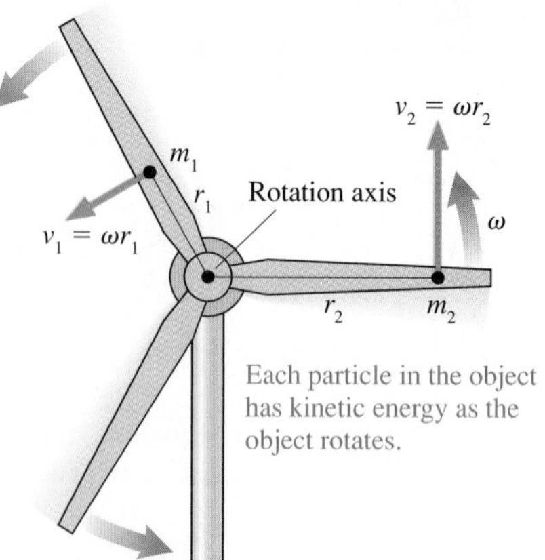

$$K_{\text{rot}} = \frac{1}{2}m_1r_1^2\omega^2 + \frac{1}{2}m_2r_2^2\omega^2 + \cdots = \frac{1}{2}\left(\sum mr^2\right)\omega^2$$

You will recognize the term in parentheses as our old friend, the moment of inertia I. Thus the rotational kinetic energy is

$$K_{\text{rot}} = \frac{1}{2}I\omega^2 \qquad (10.9)$$

Rotational kinetic energy of an object with moment of inertia I and angular velocity ω

K vs. ω: QUADRATIC p.47

NOTE ▶ Rotational kinetic energy is *not* a new form of energy. It is the ordinary kinetic energy of motion, only now expressed in a form that is especially convenient for rotational motion. Comparison with the familiar $\frac{1}{2}mv^2$ shows again that the moment of inertia I is the rotational equivalent of mass. ◀

A rolling object, such as a wheel, is undergoing both rotational *and* translational motions. Consequently, its total kinetic energy is the sum of its rotational and translational kinetic energies:

$$K = K_{\text{trans}} + K_{\text{rot}} = \frac{1}{2}mv^2 + \frac{1}{2}I\omega^2 \qquad (10.10)$$

This illustrates an important fact: **The kinetic energy of a rolling object is always greater than that of a nonrotating object moving at the same speed.**

◀ **Rotational recharge** The International Space Station (ISS) gets its electric power from solar panels. But during each 92-minute orbit, the ISS is in the earth's shadow for 30 minutes. The batteries that currently provide power during these blackouts need periodic replacement, which is very expensive in space. A promising new technology would replace the batteries with a *flywheel*—a cylinder rotating at a very high angular speed. Energy from the solar panels is used to speed up the flywheel, storing energy as rotational kinetic energy, which can then be converted back into electric energy when the ISS is in shadow.

EXAMPLE 10.6 Kinetic energy of a bicycle

Bike 1 has a 10.0 kg frame and 1.00 kg wheels; bike 2 has a 9.00 kg frame and 1.50 kg wheels. Both bikes thus have the same 12.0 kg total mass. What is the kinetic energy of each bike when they are ridden at 12.0 m/s? Model each wheel as a hoop of radius 35.0 cm.

PREPARE Each bike's frame has only translational kinetic energy $K_{\text{frame}} = \frac{1}{2}mv^2$, where m is the mass of the frame. The kinetic energy of each rolling wheel is given by Equation 10.10. From Table 7.4, we find that I for a hoop is MR^2, where M is the mass of one wheel.

SOLVE From Equation 10.10 the kinetic energy of each rolling wheel is

$$K_{\text{wheel}} = \frac{1}{2}Mv^2 + \frac{1}{2}I\omega^2 = \frac{1}{2}Mv^2 + \frac{1}{2}\underbrace{(MR^2)}_{I}\underbrace{\left(\frac{v}{R}\right)^2}_{\omega^2} = Mv^2$$

Then the total kinetic energy of a bike is

$$K = K_{\text{frame}} + 2K_{\text{wheel}} = \frac{1}{2}mv^2 + 2Mv^2$$

The factor of 2 in the second term occurs because each bike has two wheels. Thus the kinetic energies of the two bikes are

$$K_1 = \frac{1}{2}(10.0\text{ kg})(12.0\text{ m/s})^2 + 2(1.00\text{ kg})(12.0\text{ m/s})^2 = 1010\text{ J}$$

$$K_2 = \frac{1}{2}(9.00\text{ kg})(12.0\text{ m/s})^2 + 2(1.50\text{ kg})(12.0\text{ m/s})^2 = 1080\text{ J}$$

The kinetic energy of bike 2 is about 7% higher than that of bike 1. Note that the radius of the wheels was not needed in this calculation.

ASSESS As the cyclists on these bikes accelerate from rest to 12 m/s, they must convert some of their internal chemical energy into the kinetic energy of the bikes. Racing cyclists want to use as little of their own energy as possible. Although both bikes have the same total mass, the one with the lighter wheels will take less energy to get it moving. Shaving a little extra weight off your wheels is more useful than taking that same weight off your frame.

It's important that racing bike wheels are as light as possible.

10.4 Potential Energy

When two or more objects in a system interact, it is sometimes possible to *store* energy in the system in a way that the energy can be easily recovered. For instance, the earth and a ball interact by the gravitational force between them. If the ball is lifted up into the air, energy is stored in the ball + earth system, energy that can later be recovered as kinetic energy when the ball is released and falls. Similarly, a spring is a system made up of countless atoms that interact via their atomic "springs." If we push a box against a spring, energy is stored that can be recovered when the spring later pushes the box across the table. This sort of stored energy is called **potential energy,** since it has the *potential* to be converted into other forms of energy, such as kinetic or thermal energy.

The forces due to gravity and springs are special in that they allow for the storage of energy. Other interaction forces do not. When a crate is pushed across the floor, the crate and the floor interact via the force of friction, and the work done on the system is converted into thermal energy. But this energy is *not* stored up for later recovery—it slowly diffuses into the environment and cannot be recovered.

Interaction forces that can store useful energy are called **conservative forces.** The name comes from the important fact that, as we'll see, the mechanical energy of a system is *conserved* when only conservative forces act. Gravity and elastic forces are conservative forces, and later we'll find that the electric force is a conservative force as well. Friction, on the other hand, is a **nonconservative force.** When two objects interact via a friction force, energy is not stored. It is usually transformed into thermal energy.

Let's look more closely at the potential energies associated with the two conservative forces—gravity and springs—that we'll study in this chapter.

Gravitational Potential Energy

To find an expression for gravitational potential energy, let's consider the system of the book and the earth shown in FIGURE 10.14a on the next page. The book is lifted at a constant speed from its initial position at y_i to a final height y_f. The lifting force of the hand is external to the system and so does work W on the system, increasing its energy. The book is lifted at a constant speed, so its kinetic energy doesn't change. Because there's no friction, the book's thermal energy doesn't change either. Thus

FIGURE 10.14 Lifting a book increases the system's gravitational potential energy.

(a)

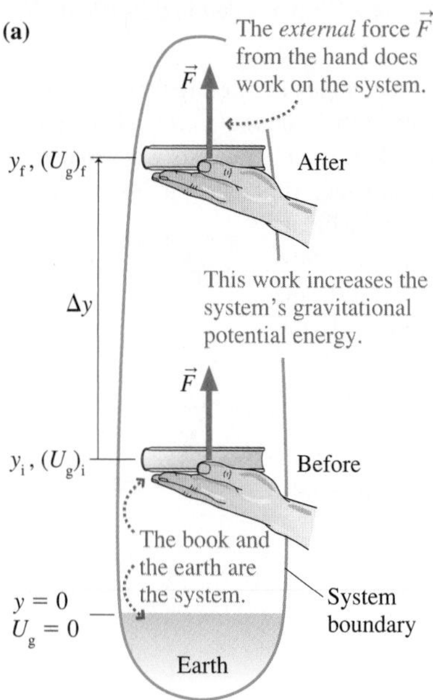
The *external* force $\vec{F}$ from the hand does work on the system.
$\vec{F}$
$y_f, (U_g)_f$
After
Δy
This work increases the system's gravitational potential energy.
$\vec{F}$
$y_i, (U_g)_i$
Before
The book and the earth are the system.
$y = 0$
$U_g = 0$
System boundary
Earth

(b) Because the book is being lifted at a constant speed, it is in dynamic equilibrium with $\vec{F}_{net} = \vec{0}$. Thus $F = w = mg$.

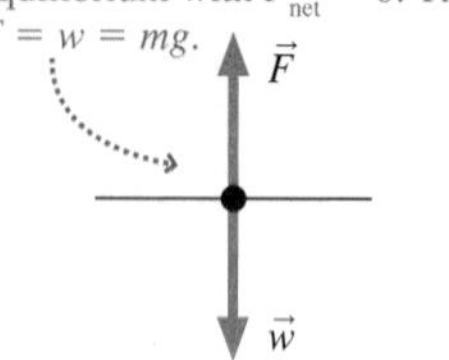
$\vec{F}$
$\vec{w}$

the work done goes entirely into increasing the gravitational potential energy of the system. According to Equation 10.3, the work-energy equation, $\Delta U_g = W$. Because $\Delta U_g = (U_g)_f - (U_g)_i$, Equation 10.3 can be written

$$(U_g)_f = (U_g)_i + W \tag{10.11}$$

The work done is $W = Fd$, where $d = \Delta y = y_f - y_i$ is the vertical distance that the book is lifted. From the free-body diagram of FIGURE 10.14b, we see that $F = mg$. Thus $W = mg\,\Delta y$, and so

$$(U_g)_f = (U_g)_i + mg\Delta y \tag{10.12}$$

Because our final height was greater than our initial height, Δy is positive and $(U_g)_f > (U_g)_i$. **The higher the object is lifted, the greater the gravitational potential energy in the object + earth system.**

Equation 10.12 gives the final gravitational potential energy $(U_g)_f$ in terms of its initial value $(U_g)_i$. But what is the value of $(U_g)_i$? We can gain some insight by writing Equation 10.12 in terms of energy *changes:*

$$(U_g)_f - (U_g)_i = \Delta U_g = mg\Delta y$$

For example, if we lift a 1.5 kg book up by $\Delta y = 2.0$ m, we increase the system's gravitational potential energy by $\Delta U_g = (1.5\text{ kg})(9.8\text{ m/s}^2)(2.0\text{ m}) = 29.4$ J. This increase is *independent* of the book's starting height: The gravitational potential energy increases by 29.4 J whether we lift the book 2.0 m starting at sea level or starting at the top of the Washington Monument. This illustrates an important general fact about *every* form of potential energy: **Only *changes* in potential energy are significant.**

Because of this fact, we are free to choose a *reference level* where we define U_g to be zero. Our expression for U_g is particularly simple if we choose this reference level to be at $y = 0$. We then have

$$U_g = mgy \tag{10.13}$$

Gravitational potential energy of an object of mass m at height y
(assuming $U_g = 0$ when the object is at $y = 0$)

NOTE ► We've emphasized that gravitational potential energy is an energy of the earth + object *system.* In solving problems using the law of conservation of energy, you'll need to include the earth as part of your system. For simplicity, we'll usually speak of "the gravitational potential energy of the ball," but what we really mean is the potential energy of the earth + ball system. ◄

EXAMPLE 10.7 Racing up a skyscraper

In the Empire State Building Run-Up, competitors race up the 1576 steps of the Empire State Building, climbing a total vertical distance of 320 m. How much gravitational potential energy does a 70 kg racer gain during this race?

Racers head up the staircase in the Empire State Building Run-Up.

PREPARE We choose $y = 0$ m and $U_g = 0$ J at the ground floor of the building.

SOLVE At the top, the racer's gravitational potential energy is

$$U_g = mgy = (70\text{ kg})(9.8\text{ m/s}^2)(320\text{ m}) = 2.2 \times 10^5\text{ J}$$

Because the racer's gravitational potential energy was 0 J at the ground floor, the change in his potential energy is 2.2×10^5 J.

ASSESS This is a large amount of energy. According to Table 10.1, it's comparable to the energy of a speeding car. But if you think how hard it would be to climb the Empire State Building, it seems like a plausible result.

An important conclusion from Equation 10.13 is that gravitational potential energy depends only on the height of the object above the reference level $y = 0$, not on the object's horizontal position. To understand why, consider carrying a briefcase while walking on level ground at a constant speed. As shown in the table on page 297, the vertical force of your hand on the briefcase is *perpendicular* to the displacement. *No work* is done on the briefcase, so its gravitational potential energy remains constant as long as its height above the ground doesn't change.

This idea can be applied to more complicated cases, such as the 82 kg hiker in FIGURE 10.15. His gravitational potential energy depends *only* on his height y above the reference level. Along path A, it's the same value $U_g = mgy = 80$ kJ at any point where he is at height $y = 100$ m above the reference level. If he had instead taken path B, his gravitational potential energy at $y = 100$ m would be the same 80 kJ. It doesn't matter *how* he gets to the 100 m elevation; his potential energy at that height is always the same. **Gravitational potential energy depends only on the *height* of an object and not on the path the object took to get to that position.** This fact will allow us to use the law of conservation of energy to easily solve a variety of problems that would be very difficult to solve using Newton's laws alone.

FIGURE 10.15 The hiker's gravitational potential energy depends only on his height above the $y = 0$ m reference level.

The hiker's potential energy at the top is 160 kJ regardless of whether he took path A or path B. $U_g = 160$ kJ

His potential energy is the same at any point where his elevation is 100 m.

200 m, 100 m, 0 m, Path A, Path B, $U_g = 80$ kJ

The reference level $y = 0$ m is where $U_g = 0$ J.

STOP TO THINK 10.4 Rank in order, from largest to smallest, the gravitational potential energies of identical balls 1 through 4.

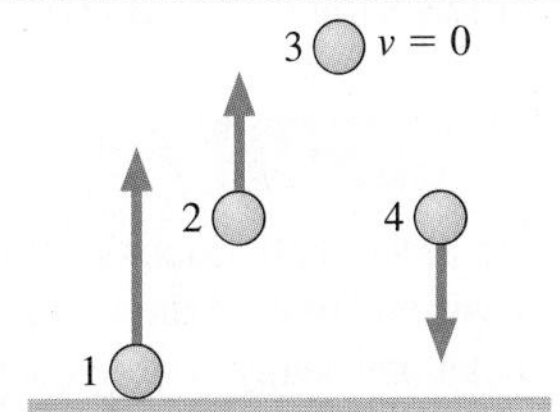

Elastic Potential Energy

Energy can also be stored in a compressed or extended spring as **elastic** (or **spring**) **potential energy** U_s. We can find out how much energy is stored in a spring by using an external force to slowly compress the spring. This external force does work on the spring, transferring energy to the spring. Since only the elastic potential energy of the spring is changing, Equation 10.3 reads

$$\Delta U_s = W \qquad (10.14)$$

That is, we can find out how much elastic potential energy is stored in the spring by calculating the amount of work needed to compress the spring.

FIGURE 10.16 shows a spring being compressed by a hand. In Section 8.3 we found that the force the spring exerts on the hand is $F_s = -k\Delta x$ (Hooke's law), where Δx is the displacement of the end of the spring from its equilibrium position and k is the spring constant. In Figure 10.16 we have set the origin of our coordinate system at the equilibrium position. The displacement from equilibrium Δx is therefore equal to x, and the spring force is then $-kx$. By Newton's third law, the force that the hand exerts on the spring is thus $F = +kx$.

As the hand pushes the end of the spring from its equilibrium position to a final position x, the applied force increases from 0 to kx. This is not a constant force, so we can't use Equation 10.5, $W = Fd$, to find the work done. However, it seems reasonable to calculate the work by using the *average* force in Equation 10.5. Because the force varies from $F_i = 0$ to $F_f = kx$, the average force used to compress the spring is $F_{avg} = \frac{1}{2}kx$. Thus the work done by the hand is

$$W = F_{avg}d = F_{avg}x = \left(\frac{1}{2}kx\right)x = \frac{1}{2}kx^2$$

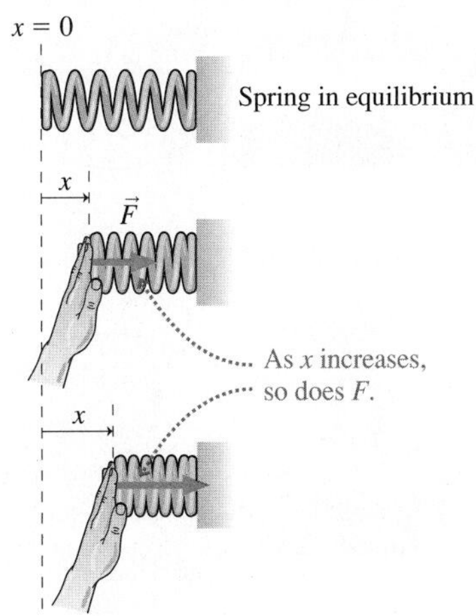

FIGURE 10.16 The force required to compress a spring is not constant.

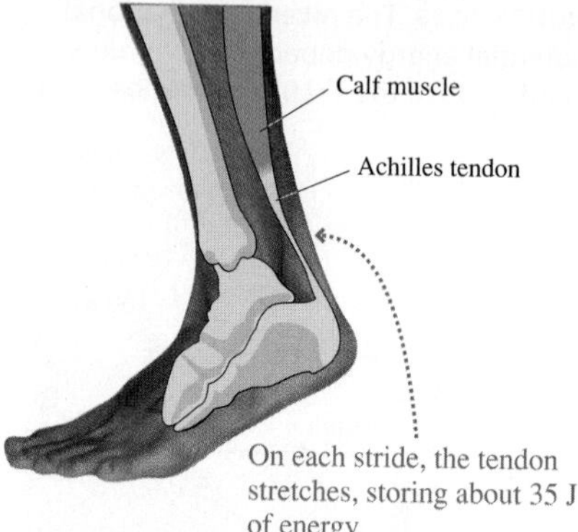

Calf muscle
Achilles tendon
On each stride, the tendon stretches, storing about 35 J of energy.

Spring in your step BIO As you run, you lose some of your mechanical energy each time your foot strikes the ground; this energy is transformed into unrecoverable thermal energy. Luckily, about 35% of the decrease of your mechanical energy when your foot lands is stored as elastic potential energy in the stretchable Achilles tendon of the lower leg. On each plant of the foot, the tendon is stretched, storing some energy. The tendon springs back as you push off the ground again, helping to propel you forward. This recovered energy reduces the amount of internal chemical energy you use, increasing your efficiency.

This work is stored as potential energy in the spring, so we can use Equation 10.14 to find that as the spring is compressed, the elastic potential energy increases by

$$\Delta U_s = \frac{1}{2}kx^2$$

Just as in the case of gravitational potential energy, we have found an expression for the *change* in U_s, not U_s itself. Again, we are free to set $U_s = 0$ at any convenient spring extension. An obvious choice is to set $U_s = 0$ at the point where the spring is in equilibrium, neither compressed nor stretched—that is, at $x = 0$. With this choice we have

$$U_s = \frac{1}{2}kx^2 \quad (10.15)$$

Elastic potential energy of a spring displaced a distance x from equilibrium (assuming $U_s = 0$ when the end of the spring is at $x = 0$)

U_s — x — p.47 QUADRATIC

NOTE ▶ Because U_s depends on the *square* of the displacement x, U_s is the same whether x is positive (the spring is compressed as in Figure 10.16) or negative (the spring is stretched). ◀

EXAMPLE 10.8 Pulling back on a bow

An archer pulls back the string on her bow to a distance of 70 cm from its equilibrium position. To hold the string at this position takes a force of 140 N. How much elastic potential energy is stored in the bow?

PREPARE A bow is an elastic material, so we will model it as obeying Hooke's law, $F_s = -kx$, where x is the distance the string is pulled back. We can use the force required to hold the string, and the distance it is pulled back, to find the bow's spring constant k. Then we can use Equation 10.15 to find the elastic potential energy.

SOLVE From Hooke's law, the spring constant is

$$k = \frac{F}{x} = \frac{140\ \text{N}}{0.70\ \text{m}} = 200\ \text{N/m}$$

Then the elastic potential energy of the flexed bow is

$$U_s = \frac{1}{2}kx^2 = \frac{1}{2}(200\ \text{N/m})(0.70\ \text{m})^2 = 49\ \text{J}$$

ASSESS When the arrow is released, this elastic potential energy will be transformed into the kinetic energy of the arrow. Because arrows are quite light, 49 J of kinetic energy will correspond to a very high speed.

STOP TO THINK 10.5 When a spring is stretched by 5 cm, its elastic potential energy is 1 J. What will its elastic potential energy be if it is *compressed* by 10 cm?

A. −4 J B. −2 J C. 2 J D. 4 J

FIGURE 10.17 A molecular view of thermal energy.

Hot object: Fast-moving molecules have lots of kinetic and elastic potential energy.

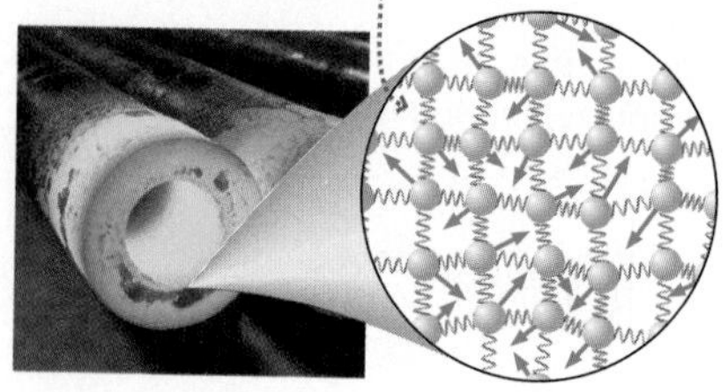

Cold object: Slow-moving molecules have little kinetic and elastic potential energy.

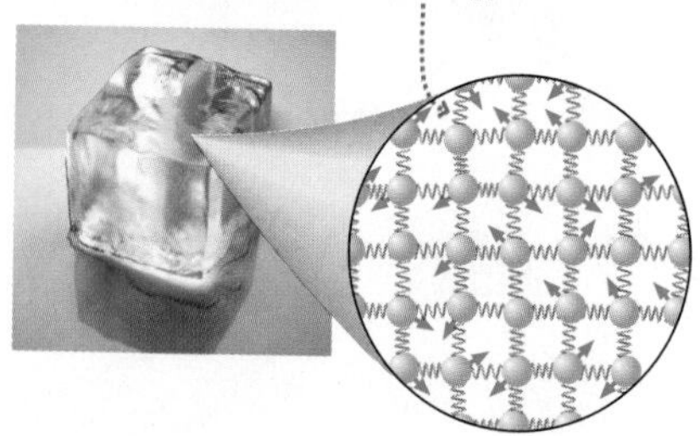

10.5 Thermal Energy

We noted earlier that thermal energy is related to the microscopic motion of the molecules of an object. As **FIGURE 10.17** shows, the molecules in a hot object jiggle around their average positions more than the molecules in a cold object. This has two consequences. First, each atom is on average moving faster in the hot object. This means that each atom has a higher *kinetic energy*. Second, each atom in the hot object tends to stray farther from its equilibrium position, leading to a greater stretching or compressing of the spring-like molecular bonds. This means that each atom has on average a higher *potential energy*. The potential energy stored in any one bond and the kinetic energy of any one atom are both exceedingly small, but there are incredibly many bonds and atoms. The sum of all these microscopic potential and kinetic energies is what we call **thermal energy.** Increasing an object's thermal energy corresponds to increasing its temperature.

Creating Thermal Energy

FIGURE 10.18 shows a thermogram of a heavy box and the floor across which it has just been dragged. In this image, warmer areas appear light blue or green. You can see that the bottom of the box and the region of the floor that the box moved over are noticeably warmer than their surroundings. In the process of dragging the box, thermal energy has appeared in the box and the floor.

FIGURE 10.18 A thermograph of a box that's been dragged across the floor.

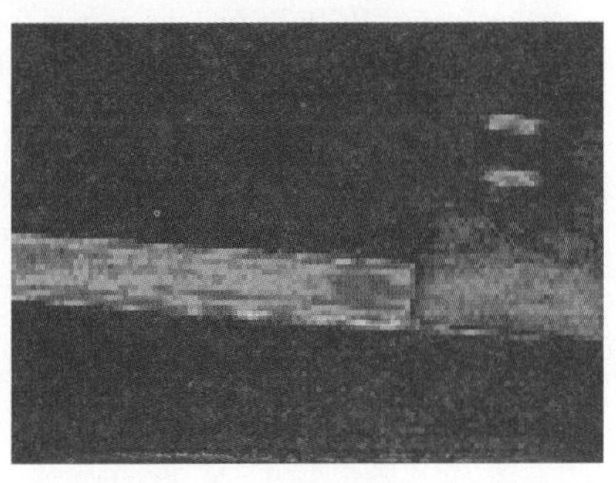

We can find a quantitative expression for the change in thermal energy by considering such a box pulled by a rope at a constant speed. As the box is pulled across the floor, the rope exerts a constant forward force $\vec{F}$ on the box, while the friction force $\vec{f}_k$ exerts a constant force on the box that is directed backward. Because the box moves at a constant speed, the magnitudes of these two forces are equal: $F = f_k$.

As the box moves through a displacement $d = \Delta x$, the rope does work $W = F\Delta x$ on the box. This work represents energy transferred into the system, so the system's energy must *increase*. In what form is this increased energy? The box's speed remains constant, so there is no change in its kinetic energy ($\Delta K = 0$). And its height doesn't change, so its gravitational potential energy is unchanged as well ($\Delta U_g = 0$). Instead, the increased energy must be in the form of *thermal* energy E_{th}. As Figure 10.18 shows, this energy appears as an increased temperature of both the box *and* the floor across which it was dragged.

We can write the work-energy equation, Equation 10.3, for the case where only thermal energy changes:

$$\Delta E_{th} = W$$

or, because the work is $W = F\Delta x = f_k\Delta x$,

$$\Delta E_{th} = f_k\Delta x \qquad (10.16)$$

This increase in thermal energy is a general feature of any system where friction between sliding objects is present. An atomic-level explanation is shown in FIGURE 10.19. Although we arrived at Equation 10.16 by considering energy transferred into the system via work done by an external force, the equation is equally valid for the transformation of mechanical energy into thermal energy when, for instance, an object slides to a halt on a rough surface. Equation 10.16 also applies to rolling friction; we need only replace f_k by f_r.

FIGURE 10.19 How friction causes an increase in thermal energy.

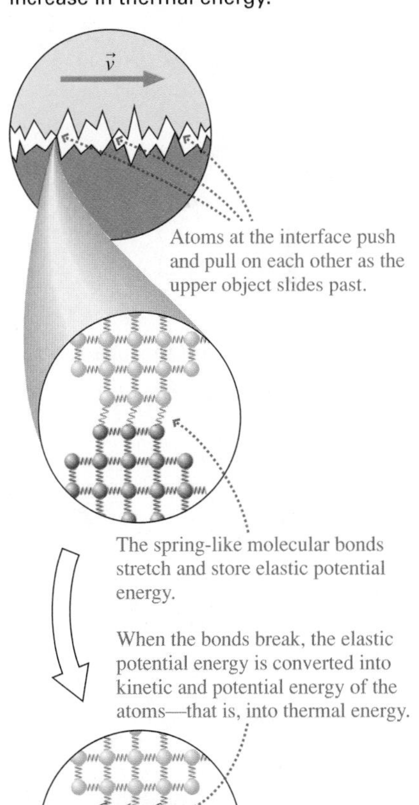

$\vec{v}$
Atoms at the interface push and pull on each other as the upper object slides past.
The spring-like molecular bonds stretch and store elastic potential energy.
When the bonds break, the elastic potential energy is converted into kinetic and potential energy of the atoms—that is, into thermal energy.

TRY IT YOURSELF

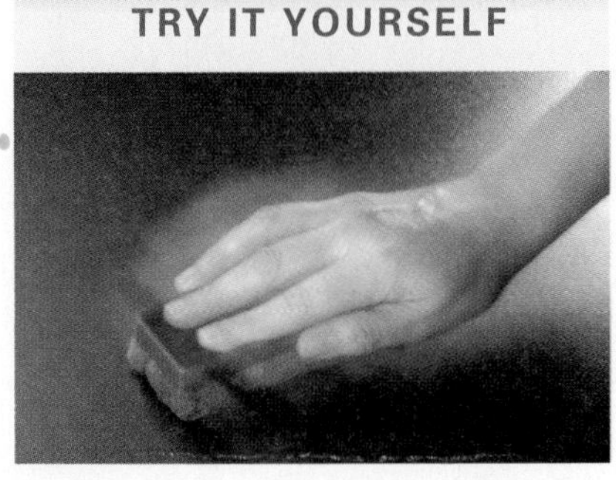

◀ **Agitating atoms** Vigorously rub a somewhat soft object such as a blackboard eraser on your desktop for about 10 seconds. If you then pass your fingers over the spot where you rubbed, you'll feel a distinct warm area. Congratulations: You've just set some 100,000,000,000,000,000,000,000 atoms into motion!

EXAMPLE 10.9 Creating thermal energy by rubbing

A 0.30 kg block of wood is rubbed back and forth against a wood table 30 times in each direction. The block is moved 8.0 cm during each stroke and pressed against the table with a force of 22 N. How much thermal energy is created in this process?

PREPARE The hand holding the block does work to push the block back and forth. Work transfers energy into the block + table system, where it appears as thermal energy according to Equation 10.16. The force of friction can be found from the model of kinetic friction introduced in Chapter 5, $f_k = \mu_k n$; from Table 5.1 the coefficient of kinetic friction for wood sliding on wood is $\mu_k = 0.20$. To find the normal force n acting on the block, we draw the free-body diagram of **FIGURE 10.20**, which shows only the *vertical* forces acting on the block.

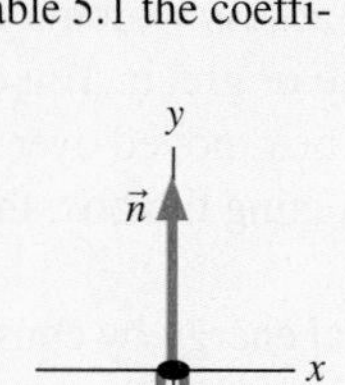

FIGURE 10.20 Free-body diagram (vertical forces only) for a block being rubbed against a table.

SOLVE From Equation 10.16 we have $\Delta E_{th} = f_k \Delta x$, where $f_k = \mu_k n$. The block is not accelerating in the y-direction, so from the free-body diagram Newton's second law gives

$$\sum F_y = n - w - F = ma_y = 0$$

or

$$n = w + F = mg + F = (0.30\text{ kg})(9.8\text{ m/s}^2) + 22\text{ N} = 25\text{ N}$$

The friction force is then $f_k = \mu_k n = (0.20)(25\text{ N}) = 5.0\text{ N}$. The total displacement of the block is $2 \times 30 \times 8.0\text{ cm} = 4.8\text{ m}$. Thus the thermal energy created is

$$\Delta E_{th} = f_k \Delta x = (5.0\text{ N})(4.8\text{ m}) = 24\text{ J}$$

ASSESS This modest amount of thermal energy seems reasonable for a person to create by rubbing.

10.6 Using the Law of Conservation of Energy

The law of conservation of energy, Equation 10.4, states that **the total energy of an *isolated system* is conserved** so that its change is zero:

$$\Delta E = \Delta K + \Delta U_g + \Delta U_s + \Delta E_{th} + \Delta E_{chem} + \cdots = 0 \tag{10.17}$$

This law applies to every form of energy, from kinetic to chemical to nuclear. For the rest of this chapter, however, we'll narrow our focus and concern ourselves with only the forms of energy typically transformed during the motion of ordinary objects. These forms are kinetic energy K, gravitational and elastic potential energies U_g and U_s, and thermal energy E_{th}.

We defined an isolated system as one on which external forces do no work, so that no energy is transferred into or out of the system. The following table shows how to choose an isolated system for four common situations.

TABLE 10.2 Choosing an isolated system

An object in free fall	An object sliding down a frictionless ramp	An object compressing a spring	An object sliding along a surface with friction
$\vec{w} = \vec{F}_{\text{earth on ball}}$ $\vec{F}_{\text{ball on earth}}$ Earth	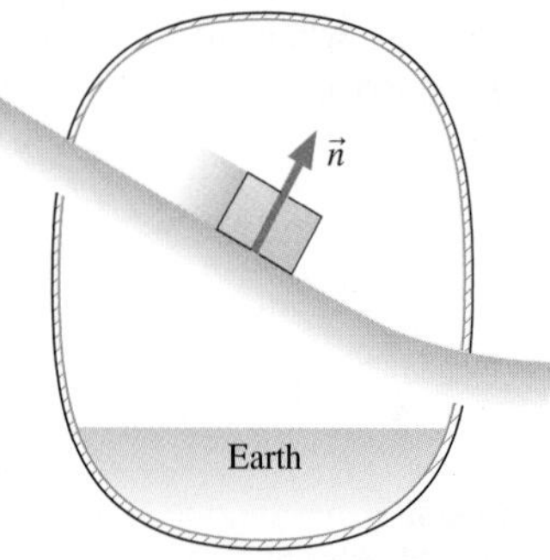	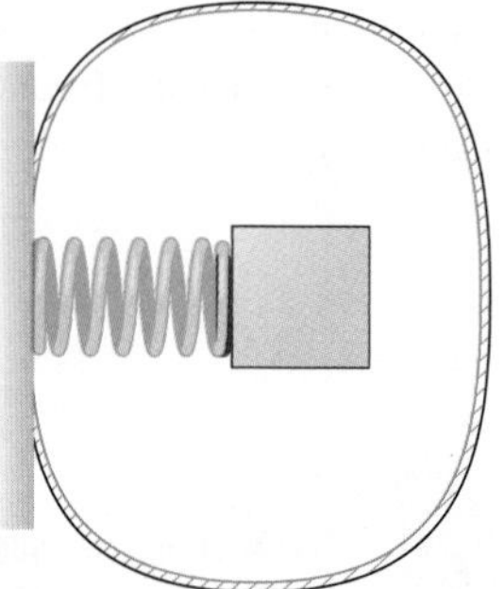	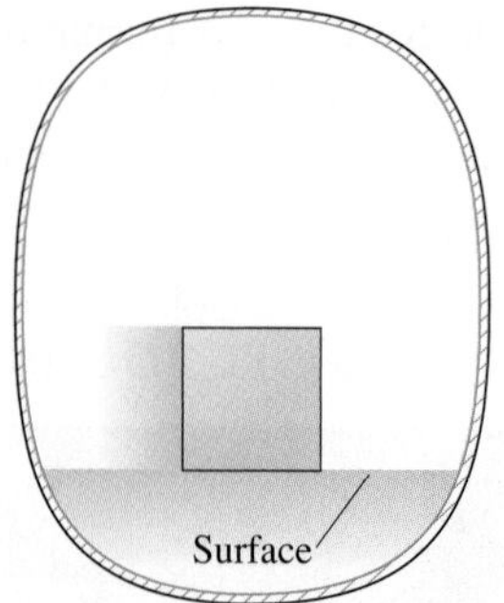
We choose the ball *and* the earth as the system, so that the forces between them are *internal* forces. There are no external forces to do work, so the system is isolated.	The external force the ramp exerts on the object is perpendicular to the motion, and so does no work. The object and the earth together form an isolated system.	We choose the object and the spring to be the system. The forces between them are internal forces, so no work is done.	The block and the surface interact via kinetic friction forces, but these forces are internal to the system. There are no external forces to do work, so the system is isolated.

Just as for momentum conservation, we wish to develop a before-and-after perspective for energy conservation. We can do so by noting that $\Delta K = K_f - K_i$, $\Delta U_g = (U_g)_f - (U_g)_i$, and so on. Then Equation 10.17 can be written as

$$K_f + (U_g)_f + (U_s)_f + \Delta E_{th} = K_i + (U_g)_i + (U_s)_i \quad (10.18)$$

Equation 10.18 is the before-and-after version of the law of conservation of energy: It equates the final value of an isolated system's energy to its initial energy. This equation will be the basis for a powerful problem-solving strategy.

NOTE ► We don't write ΔE_{th} as $(E_{th})_f - (E_{th})_i$ because the initial and final values of the thermal energy are typically unknown; only their *difference* ΔE_{th} can be measured. ◄

Conservation of Mechanical Energy

If we further restrict ourselves to cases where friction can be neglected, so that $\Delta E_{th} = 0$, the law of conservation of energy, Equation 10.18, becomes

$$K_f + (U_g)_f + (U_s)_f = K_i + (U_g)_i + (U_s)_i \quad (10.19)$$

Activ Physics ONLINE 5.2, 5.3, 5.4, 5.5, 5.6, 5.7, 7.11, 7.12, 7.13

The sum of the kinetic and potential energies, $K + U_g + U_s$, is called the **mechanical energy** of the system, so Equation 10.19 says that **the mechanical energy is conserved for an isolated system without friction.**

These observations about the conservation of energy suggest the following problem-solving strategy.

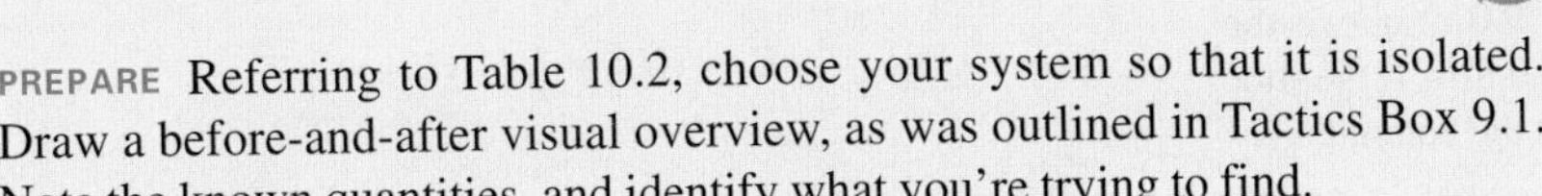

PROBLEM-SOLVING STRATEGY 10.1 Conservation of energy problems MP™

PREPARE Referring to Table 10.2, choose your system so that it is isolated. Draw a before-and-after visual overview, as was outlined in Tactics Box 9.1. Note the known quantities, and identify what you're trying to find.

SOLVE There are two important situations:

- If the system is isolated *and* there's no friction, the mechanical energy is conserved:

$$K_f + (U_g)_f + (U_s)_f = K_i + (U_g)_i + (U_s)_i$$

- If the system is isolated but there is friction within the system, the total energy is conserved:

$$K_f + (U_g)_f + (U_s)_f + \Delta E_{th} = K_i + (U_g)_i + (U_s)_i$$

Depending on the problem, you'll need to calculate the initial and/or final values of these energies; you can then solve for the unknown energies, and from these any unknown speeds (from K), heights (from U_g and U_s), or displacements or friction forces (from $\Delta E_{th} = f_k \Delta x$).

ASSESS Check the signs of your energies. Kinetic energy is always positive, as is the change in thermal energy. Check that your result has the correct units, is reasonable, and answers the question.

Exercise 23

Spring into action BIO A locust can jump as far as 1 meter, an impressive distance for such a small animal. To make such a jump, its legs must extend much more rapidly than muscles can ordinarily contract. Thus, instead of using its muscles to make the jump directly, the locust uses them to more slowly stretch an internal "spring" near its knee joint. This stores elastic potential energy in the spring. When the muscles relax, the spring is suddenly released, and its energy is rapidly converted into kinetic energy of the insect.

EXAMPLE 10.10 Hitting the bell

At the county fair, Katie tries her hand at the ring-the-bell attraction, as shown in FIGURE 10.21 on the next page. She swings the mallet hard enough to give the ball an initial upward speed of 8.0 m/s. Will the ball ring the bell, 3.0 m from the bottom?

PREPARE We'll follow the steps of Problem-Solving Strategy 10.1. From Table 10.2, we see that once the ball is in the air, the system consisting of the ball and the earth is isolated. If we assume that the track along which the ball moves is frictionless, then the

Continued

system's mechanical energy is conserved. Figure 10.21 shows a before-and-after visual overview in which we've chosen $y = 0$ m to be at the ball's starting point. We can then use conservation of mechanical energy, Equation 10.19.

FIGURE 10.21 Before-and-after visual overview of the ring-the-bell attraction.

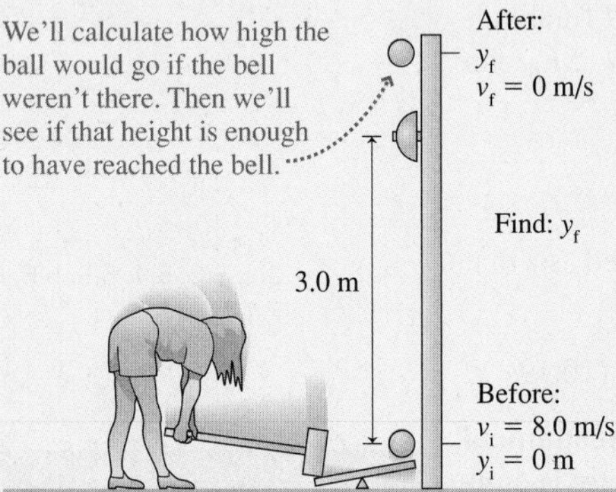

SOLVE Equation 10.19 tells us that $K_f + (U_g)_f = K_i + (U_g)_i$. We can use our expressions for kinetic and potential energy to write this as

$$\frac{1}{2}mv_f^2 + mgy_f = \frac{1}{2}mv_i^2 + mgy_i$$

Let's ignore the bell for the moment and figure out how far the ball would rise if there were nothing in its way. We know that the ball starts at $y_i = 0$ m and that its speed v_f at the highest point is 0 m/s. Thus the energy equation simplifies to

$$mgy_f = \frac{1}{2}mv_i^2$$

This is easily solved for the height y_f:

$$y_f = \frac{v_i^2}{2g} = \frac{(8.0 \text{ m/s})^2}{2(9.8 \text{ m/s}^2)} = 3.3 \text{ m}$$

This is higher than the point where the bell sits, so the ball would actually hit it on the way up.

ASSESS It seems reasonable that Katie could swing the mallet hard enough to make the ball rise by about 3 m.

EXAMPLE 10.11 Speed at the bottom of a water slide

Still at the county fair, Katie tries the water slide, whose shape is shown in **FIGURE 10.22**. The starting point is 9.0 m above the ground. She pushes off with an initial speed of 2.0 m/s. If the slide is frictionless, how fast will Katie be traveling at the bottom?

PREPARE Table 10.2 showed that the system consisting of Katie and the earth is isolated because the normal force of the slide is perpendicular to Katie's motion and does no work. If we assume the slide is frictionless, we can use the conservation of mechanical energy equation. Figure 10.22 is a visual overview of the problem.

FIGURE 10.22 Before-and-after visual overview of Katie on the water slide.

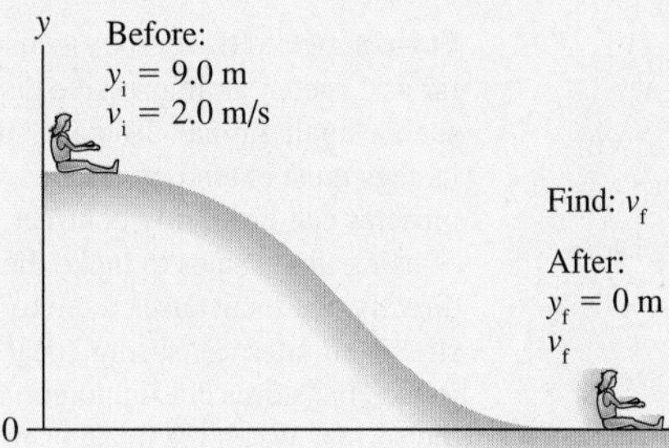

SOLVE Conservation of mechanical energy gives

$$K_f + (U_g)_f = K_i + (U_g)_i$$

or

$$\frac{1}{2}mv_f^2 + mgy_f = \frac{1}{2}mv_i^2 + mgy_i$$

Taking $y_f = 0$ m, we have

$$\frac{1}{2}mv_f^2 = \frac{1}{2}mv_i^2 + mgy_i$$

which we can solve to get

$$v_f = \sqrt{v_i^2 + 2gy_i}$$

$$= \sqrt{(2.0 \text{ m/s})^2 + 2(9.8 \text{ m/s}^2)(9.0 \text{ m})} = 13 \text{ m/s}$$

ASSESS This speed is about 30 mph. This is probably faster than you really would go on a water slide but, because we have ignored friction, our answer is reasonable. It is important to realize that the *shape* of the slide does not matter because gravitational potential energy depends only on the *height* above a reference level. **If you slide down any (frictionless) slide of the same height, your speed at the bottom is the same.**

EXAMPLE 10.12 Speed of a spring-launched ball

A spring-loaded toy gun is used to launch a 10 g plastic ball. The spring, which has a spring constant of 10 N/m, is compressed by 10 cm as the ball is pushed into the barrel. When the trigger is pulled, the spring is released and shoots the ball back out. What is the ball's speed as it leaves the barrel? Assume that friction is negligible.

PREPARE Assume the spring obeys Hooke's law, $F_s = -kx$, and is massless so that it has no kinetic energy of its own. Using Table 10.2, we choose the isolated system to be the spring and the ball. There's no friction; hence the system's mechanical energy $K + U_s$ is conserved.

FIGURE 10.23 Before-and-after visual overview of a ball being shot out of a spring-loaded toy gun.

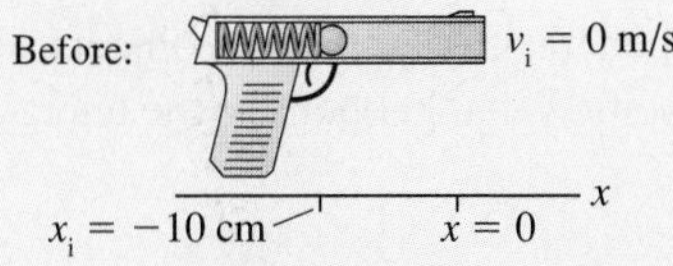

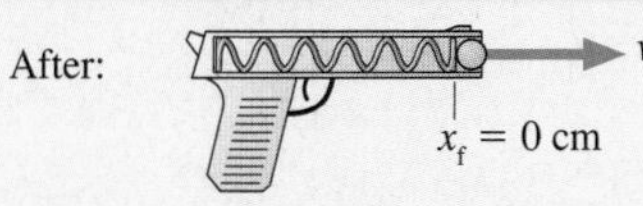

FIGURE 10.23 shows a before-and-after visual overview. The compressed spring will push on the ball until the spring has returned to its equilibrium length. We have chosen the origin of the coordinate system at the equilibrium position of the free end of the spring, making $x_i = -10$ cm and $x_f = 0$ cm.

SOLVE The energy conservation equation is $K_f + (U_s)_f = K_i + (U_s)_i$. We can use the elastic potential energy of the spring, Equation 10.15, to write this as

$$\tfrac{1}{2}mv_f^2 + \tfrac{1}{2}kx_f^2 = \tfrac{1}{2}mv_i^2 + \tfrac{1}{2}kx_i^2$$

We know that $x_f = 0$ m and $v_i = 0$ m/s, so this simplifies to

$$\tfrac{1}{2}mv_f^2 = \tfrac{1}{2}kx_i^2$$

It is now straightforward to solve for the ball's speed:

$$v_f = \sqrt{\frac{kx_i^2}{m}} = \sqrt{\frac{(10\text{ N/m})(-0.10\text{ m})^2}{0.010\text{ kg}}} = 3.2\text{ m/s}$$

ASSESS This is *not* a problem that we could have easily solved with Newton's laws. The acceleration is not constant, and we have not learned how to handle the kinematics of nonconstant acceleration. But with conservation of energy—it's easy!

Friction and Thermal Energy

Thermal energy is always created when kinetic friction is present, so we must use the more general conservation of energy equation, Equation 10.18, which includes thermal-energy changes ΔE_{th}. Furthermore, we know from Section 10.5 that the change in the thermal energy when an object slides a distance Δx while subject to a friction force f_k is $\Delta E_{th} = f_k \Delta x$.

EXAMPLE 10.13 Where will the sled stop?

A sledder, starting from rest, slides down a 10-m-high hill. At the bottom of the hill is a long horizontal patch of rough snow. The hill is nearly frictionless, but the coefficient of friction between the sled and the rough snow at the bottom is $\mu_k = 0.30$. How far will the sled slide along the rough patch?

PREPARE In order to be isolated, the system must include the sled, the earth, *and* the rough snow. As Table 10.2 shows, this makes the friction force an internal force so that no work is done on the system. We can use conservation of energy, but we will need to include thermal energy. A visual overview of the problem is shown in **FIGURE 10.24**.

FIGURE 10.24 Visual overview of a sledder sliding downhill.

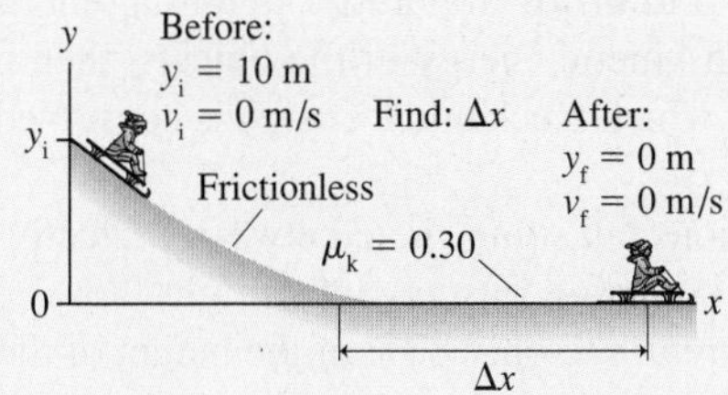

SOLVE At the top of the hill the sled has only gravitational potential energy $(U_g)_i = mgy_i$. It has no kinetic or potential energy after stopping at the bottom of the hill, so $K_f = (U_g)_f = 0$. However, friction in the rough patch causes an increase in thermal energy. Thus our conservation of energy equation $K_f + (U_g)_f + \Delta E_{th} = K_i + (U_g)_i$ is

$$\Delta E_{th} = (U_g)_i = mgy_i$$

The change in thermal energy is $\Delta E_{th} = f_k \Delta x = \mu_k n \Delta x$. The normal force $\vec{n}$ balances the sled's weight $\vec{w}$ as it crosses the rough patch, so $n = w = mg$. Thus

$$\Delta E_{th} = \mu_k n \Delta x = \mu_k (mg) \Delta x = mgy_i$$

from which we find

$$\Delta x = \frac{y_i}{\mu_k} = \frac{10\text{ m}}{0.30} = 33\text{ m}$$

ASSESS It seems reasonable that the sledder would slide a distance that is greater than the height of the hill he started down.

10.7 Energy in Collisions

6.2, 6.5, 6.8, 6.9

In Chapter 9 we studied collisions between two objects. We found that if no external forces are acting on the objects, the total *momentum* of the objects will be conserved. Now we wish to study what happens to *energy* in collisions. The energetics of collisions are important in many applications in biokinetics, such as designing safer automobiles and bicycle helmets.

Let's first re-examine a perfectly inelastic collision. We studied just such a collision in Example 9.8. Recall that in such a collision the two objects stick together and then move with a common final velocity. What happens to the energy?

EXAMPLE 10.14 Energy transformations in a perfectly inelastic collision

FIGURE 10.25 shows two air-track gliders that are pushed toward each other, collide, and stick together. In Example 9.8, we used conservation of momentum to find the final velocity shown in Figure 10.25 from the given initial velocities. How much thermal energy is created in this collision?

FIGURE 10.25 Before-and-after visual overview of a completely inelastic collision.

PREPARE We'll choose our system to be the two gliders. Because the track is horizontal, there is no change in potential energy. Thus the law of conservation of energy, Equation 10.18, is $K_f + \Delta E_{th} = K_i$. The total energy before the collision must equal the total energy afterward, but the *mechanical* energies need not be equal.

SOLVE The initial kinetic energy is

$$K_i = \frac{1}{2}m_1(v_{1x})_i^2 + \frac{1}{2}m_2(v_{2x})_i^2$$

$$= \frac{1}{2}(0.200\text{ kg})(3.00\text{ m/s})^2 + \frac{1}{2}(0.400\text{ kg})(-2.25\text{ m/s})^2$$

$$= 1.91\text{ J}$$

Because the gliders stick together and move as a single object with mass $m_1 + m_2$, the final kinetic energy is

$$K_f = \frac{1}{2}(m_1 + m_2)(v_x)_f^2$$

$$= \frac{1}{2}(0.600\text{ kg})(-0.500\text{ m/s})^2 = 0.0750\text{ J}$$

From the conservation of energy equation above, we find that the thermal energy increases by

$$\Delta E_{th} = K_i - K_f = 1.91\text{ J} - 0.075\text{ J} = 1.84\text{ J}$$

This amount of the initial kinetic energy is transformed into thermal energy during the impact of the collision.

ASSESS About 96% of the initial kinetic energy is transformed into thermal energy. This is typical of many real-world collisions.

Elastic Collisions

Figure 9.1 showed a collision of a tennis ball with a racket. The ball is compressed and the racket strings stretch as the two collide, then the ball expands and the strings relax as the two are pushed apart. In the language of energy, the kinetic energy of the objects is transformed into the elastic potential energy of the ball and strings, then back into kinetic energy as the two objects spring apart. If *all* of the kinetic energy is stored as elastic potential energy, and *all* of the elastic potential energy is transformed back into the post-collision kinetic energy of the objects, then mechanical energy is conserved. A collision for which mechanical energy is conserved is called a **perfectly elastic collision.**

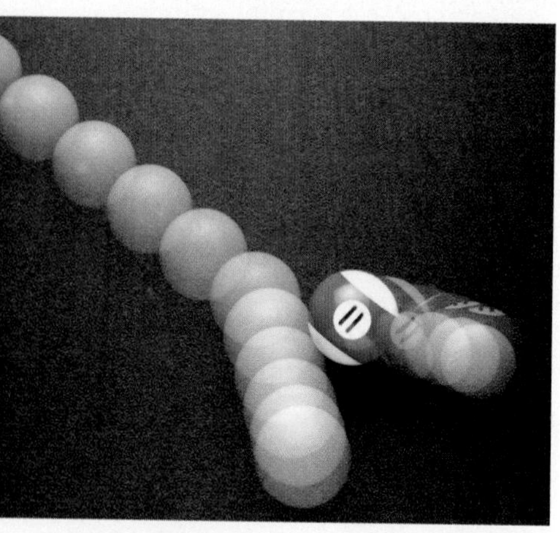

In a collision between a cue ball and a stationary ball, the mechanical energy of the balls is almost perfectly conserved.

Needless to say, most real collisions fall somewhere between perfectly elastic and perfectly inelastic. A rubber ball bouncing on the floor might "lose" 20% of its kinetic energy on each bounce and return to only 80% of the height of the preceding bounce. But collisions between two very hard objects, such as two pool balls or two steel balls, come close to being perfectly elastic. And collisions between microscopic particles, such as atoms or electrons, can be perfectly elastic.

FIGURE 10.26 on the next page shows a head-on, perfectly elastic collision of a ball of mass m_1, having initial velocity $(v_{1x})_i$, with a ball of mass m_2 that is initially at rest. The balls' velocities after the collision are $(v_{1x})_f$ and $(v_{2x})_f$. These are velocities, not speeds, and have signs. Ball 1, in particular, might bounce backward and have a negative value for $(v_{1x})_f$.

The collision must obey two conservation laws: conservation of momentum (obeyed in any collision) and conservation of mechanical energy (because the collision is perfectly elastic). Although the energy is transformed into potential energy during the collision, the mechanical energy before and after the collision is purely kinetic energy. Thus,

$$\text{momentum conservation:} \quad m_1(v_{1x})_i = m_1(v_{1x})_f + m_2(v_{2x})_f$$

$$\text{energy conservation:} \quad \frac{1}{2}m_1(v_{1x})_i^2 = \frac{1}{2}m_1(v_{1x})_f^2 + \frac{1}{2}m_2(v_{2x})_f^2$$

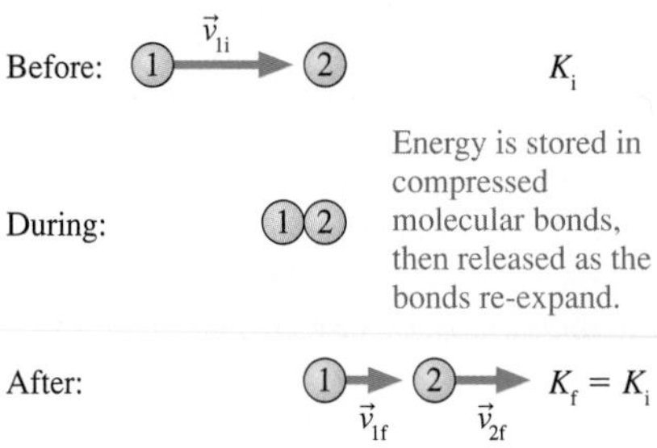

Momentum conservation alone is not sufficient to analyze the collision because there are two unknowns: the two final velocities. That is why we did not consider perfectly elastic collisions in Chapter 9. Energy conservation gives us another condition. The complete solution of these two equations involves straightforward but rather lengthy algebra. We'll just give the solution here:

$$(v_{1x})_f = \frac{m_1 - m_2}{m_1 + m_2}(v_{1x})_i \qquad (v_{2x})_f = \frac{2m_1}{m_1 + m_2}(v_{1x})_i \tag{10.20}$$

Perfectly elastic collision with object 2 initially at rest

Equations 10.20 allow us to compute the final velocity of each object. Let's look at a common and important example: a perfectly elastic collision between two objects of equal mass.

FIGURE 10.26 A perfectly elastic collision.

Before: ① $\vec{v}_{1i}$ ② K_i

During:

Energy is stored in compressed molecular bonds, then released as the bonds re-expand.

After: ① $\vec{v}_{1f}$ ② $\vec{v}_{2f}$ $K_f = K_i$

EXAMPLE 10.15 Velocities in an air hockey collision

On an air hockey table, a moving puck, traveling to the right at 2.3 m/s, makes a head-on collision with an identical puck at rest. What is the final velocity of each puck?

PREPARE The before-and-after visual overview is shown in **FIGURE 10.27**. We've shown the final velocities in the picture, but we don't really know yet which way the pucks will move. Because one puck was initially at rest, we can use Equation 10.20 to find the final velocities of the pucks. The pucks are identical, so we have $m_1 = m_2 = m$.

FIGURE 10.27 A moving puck collides with a stationary puck.

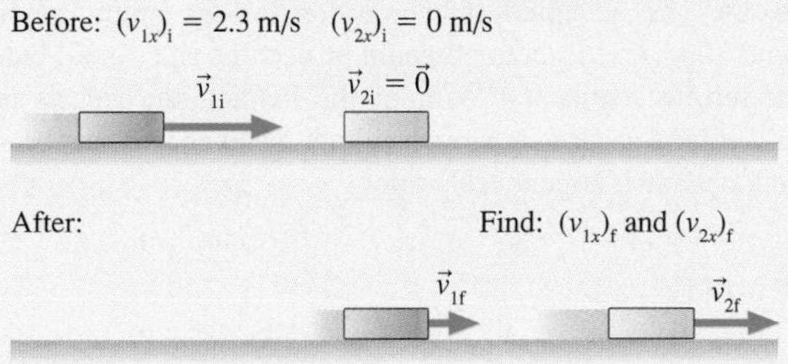

Before: $(v_{1x})_i = 2.3$ m/s $\quad (v_{2x})_i = 0$ m/s

$\vec{v}_{1i} \quad \vec{v}_{2i} = \vec{0}$

After: $\quad$ Find: $(v_{1x})_f$ and $(v_{2x})_f$

$\vec{v}_{1f} \quad \vec{v}_{2f}$

SOLVE We use Equation 10.20 with $m_1 = m_2 = m$ to get

$$(v_{1x})_f = \frac{m - m}{m + m}(v_{1x})_i = 0 \text{ m/s}$$

$$(v_{2x})_f = \frac{2m}{m + m}(v_{1x})_i = (v_{1x})_i = 2.3 \text{ m/s}$$

The incoming puck stops dead, and the initially stationary puck goes off with the same velocity that the incoming one had.

ASSESS You can see that momentum and energy are conserved: The incoming puck's momentum and energy are completely transferred to the outgoing puck. If you've ever played pool, you've probably seen this sort of collision when you hit a ball head-on with the cue ball. The cue ball stops and the other ball picks up the cue ball's velocity.

Other cases where the colliding objects have unequal masses will be treated in the end-of-chapter problems.

Forces in Collisions

The collision between two pool balls occurs very quickly, and the forces are typically very large and difficult to calculate. Fortunately, by using the concepts of momentum and energy conservation, we can often calculate the final velocities of the balls without having to know the forces between them. There are collisions, however, where knowing the forces involved is of critical importance. The following example shows how a helmet helps protect the head from the large forces involved in a bicycle accident.

EXAMPLE 10.16 Protecting your head

A bike helmet—basically a shell of hard, crushable foam—is tested by being strapped onto a 5.0 kg headform and dropped from a height of 2.0 m onto a hard anvil. What force is encountered by the headform if the impact crushes the foam by 3.0 cm?

The foam inside a bike helmet is designed to crush upon impact.

PREPARE A before-and-after visual overview of the test is shown in FIGURE 10.28. We've chosen the endpoint of the problem to be when the headform comes to rest with the foam crushed. We can use the work-energy equation, Equation 10.3, to calculate the force on the headform. We'll choose the headform and the earth to be the system; the foam in the helmet is part of the environment. We make this choice so that the force on the headform due to the foam is an *external* force that does work W on the headform.

FIGURE 10.28 Before-and-after visual overview of the bike helmet test.

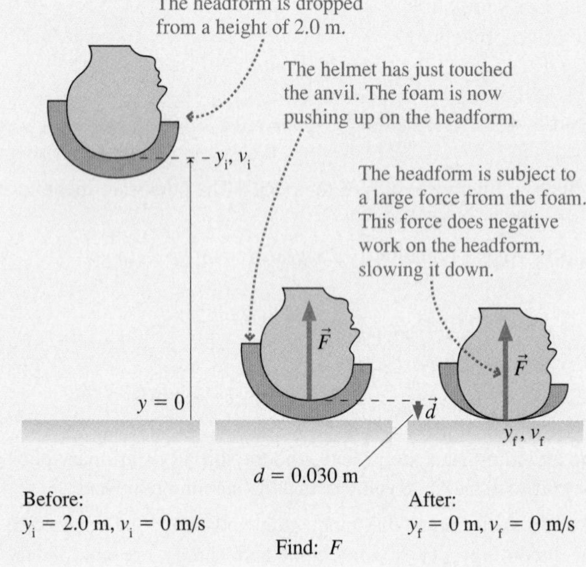

The headform is dropped from a height of 2.0 m.
The helmet has just touched the anvil. The foam is now pushing up on the headform.
y_i, v_i
The headform is subject to a large force from the foam. This force does negative work on the headform, slowing it down.
$\vec{F}$
$\vec{F}$
$y = 0$
$\vec{d}$
y_f, v_f
$d = 0.030$ m
Before:
$y_i = 2.0$ m, $v_i = 0$ m/s
After:
$y_f = 0$ m, $v_f = 0$ m/s
Find: F

SOLVE The work-energy equation $\Delta K + \Delta U_g + \Delta E_{th} = W$ tells us that the work done by external forces—in this case, the force of the foam on the headform—changes the energy of the system. The headform starts at rest, speeds up as it falls, then returns to rest during the impact. Overall, then, $\Delta K = 0$. Furthermore, $\Delta E_{th} = 0$ because there's no friction to increase the thermal energy. Only the gravitational potential energy changes, giving

$$\Delta U_g = (U_g)_f - (U_g)_i = W$$

The upward force of the foam on the headform is opposite the downward displacement of the headform. Referring to Tactics Box 10.1, we see that the work done is negative: $W = -Fd$, where we've assumed that the force is relatively constant. Using this result in the work-energy equation and solving for F, we find

$$F = -\frac{(U_g)_f - (U_g)_i}{d} = \frac{(U_g)_i - (U_g)_f}{d}$$

Taking our reference height to be $y = 0$ m at the anvil, we have $(U_g)_f = 0$. We're left with $(U_g)_i = mgy_i$, so

$$F = \frac{mgy_i}{d} = \frac{(5.0 \text{ kg})(9.8 \text{ m/s})(2.0 \text{ m})}{0.030 \text{ m}} = 3300 \text{ N}$$

This is the force that acts on the head to bring it to a halt in 3.0 cm. More important from the perspective of possible brain injury is the head's *acceleration:*

$$a = \frac{F}{m} = \frac{3300 \text{ N}}{5.0 \text{ kg}} = 660 \text{ m/s}^2 = 67g$$

ASSESS The accepted threshold for serious brain injury is around 300g, so this helmet would protect the rider in all but the most serious accidents. Without the helmet, the rider's head would come to a stop in a much shorter distance and thus be subjected to a much larger acceleration.

10.8 Power

We've now studied how energy can be transformed from one kind into another and how it can be transferred between the environment and the system as work. In many situations we would like to know *how quickly* the energy is transformed or transferred. Is a transfer of energy very rapid, or does it take place over a long time? In passing a truck, your car needs to transform a certain amount of the chemical energy in its fuel into kinetic energy. It makes a *big* difference whether your engine can do this in 20 s or 60 s!

The question How quickly? implies that we are talking about a *rate*. For example, the velocity of an object—how fast it is going—is the *rate of change* of position. So, when we raise the issue of how fast the energy is transformed, we are talking about the *rate of transformation* of energy. Suppose in a time interval Δt an amount of energy ΔE is transformed from one form to another. The rate at which this energy is transformed is called the **power** P and is defined as

$$P = \frac{\Delta E}{\Delta t} \qquad (10.21)$$

Power when an amount of energy ΔE is transformed in a time interval Δt

The unit of power is the **watt,** which is defined as 1 watt = 1 W = 1 J/s.

Power also measures the rate at which energy is transferred into or out of a system as work W. If work W is done in time interval Δt, the rate of energy *transfer* is

$$P = \frac{W}{\Delta t} \qquad (10.22)$$

Power when an amount of work W is done in a time interval Δt

A force that is doing work (i.e., transferring energy) at a rate of 3 J/s has an "output power" of 3 W. A system that is gaining energy at the rate of 3 J/s is said to "consume" 3 W of power. Common prefixes used for power are mW (milliwatts), kW (kilowatts), and MW (megawatts).

We can express Equation 10.22 in a different form. If in the time interval Δt an object undergoes a displacement Δx, the work done by a force acting on the object is $W = F\Delta x$. Then Equation 10.22 can be written as

$$P = \frac{W}{\Delta t} = \frac{F\Delta x}{\Delta t} = F\frac{\Delta x}{\Delta t} = Fv$$

The rate at which energy is transferred to an object as work—the power—is the product of the force that does the work and the velocity of the object:

$$P = Fv \qquad (10.23)$$

Rate of energy transfer due to a force F acting on an object moving at velocity v

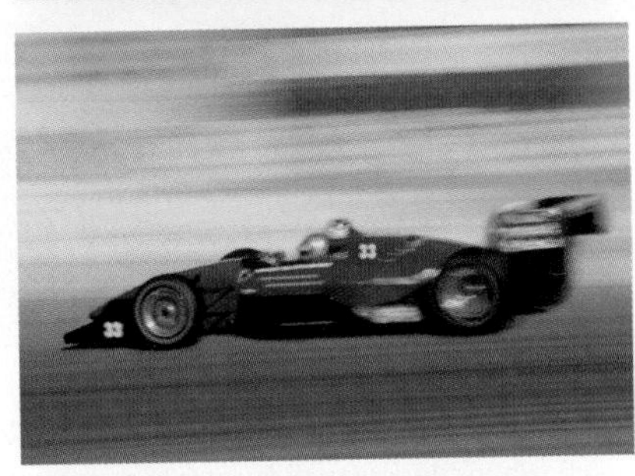

Both these cars take about the same energy to reach 60 mph, but the race car gets there in a much shorter time, so its *power* is much greater.

The English unit of power is the *horsepower.* The conversion factor to watts is

$$1 \text{ horsepower} = 1 \text{ hp} = 746 \text{ W}$$

Many common appliances, such as motors, are rated in hp.

EXAMPLE 10.17 Power to pass a truck

Your 1500 kg car is behind a truck traveling at 60 mph (27 m/s). To pass it, you speed up to 75 mph (34 m/s) in 6.0 s. What engine power is required to do this?

PREPARE Your engine is transforming the chemical energy of its fuel into the kinetic energy of the car. We can calculate the rate of transformation by finding the change ΔK in the kinetic energy and using the known time interval.

SOLVE We have

$$K_i = \frac{1}{2}mv_i^2 = \frac{1}{2}(1500 \text{ kg})(27 \text{ m/s})^2 = 5.47 \times 10^5 \text{ J}$$

$$K_f = \frac{1}{2}mv_f^2 = \frac{1}{2}(1500 \text{ kg})(34 \text{ m/s})^2 = 8.67 \times 10^5 \text{ J}$$

so that

$$\Delta K = K_f - K_i = (8.67 \times 10^5 \text{ J}) - (5.47 \times 10^5 \text{ J}) = 3.20 \times 10^5 \text{ J}$$

To transform this amount of energy in 6 s, the power required is

$$P = \frac{\Delta K}{\Delta t} = \frac{3.20 \times 10^5 \text{ J}}{6.0 \text{ s}} = 53{,}000 \text{ W} = 53 \text{ kW}$$

This is about 71 hp. This power is in addition to the power needed to overcome drag and friction and cruise at 60 mph, so the total power required from the engine will be even greater than this.

ASSESS You use a large amount of energy to perform a simple driving maneuver such as this. 3.20×10^5 J is enough energy to lift an 80 kg person 410 m in the air—the height of a tall skyscraper. And 53 kW would lift him there in only 6 s!

STOP TO THINK 10.6 Four students run up the stairs in the times shown. Rank in order, from largest to smallest, their power outputs P_A through P_D.

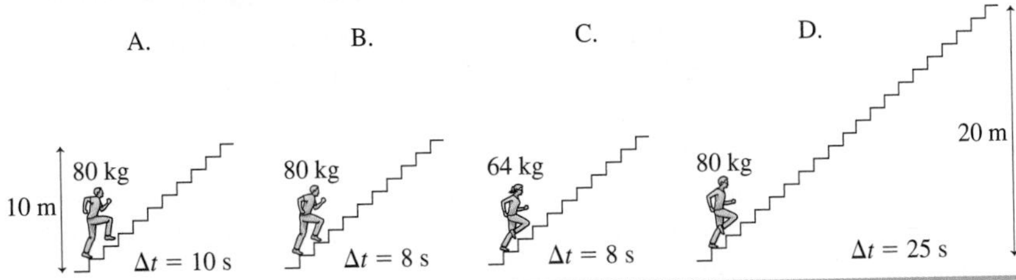

INTEGRATED EXAMPLE 10.18 Stopping a runaway truck

A truck's brakes can overheat and fail while descending mountain highways, leading to an extremely dangerous runaway truck. Some highways have *runaway-truck ramps* to safely bring out-of-control trucks to a stop. These uphill ramps are covered with a deep bed of gravel. The uphill slope and the large coefficient of rolling friction as the tires sink into the gravel bring the truck to a safe halt.

A runaway-truck ramp along Interstate 70 in Colorado.

A 22,000 kg truck heading down a 3.5° slope at 20 m/s ($\approx$45 mph) suddenly has its brakes fail. Fortunately, there's a runaway-truck ramp 600 m ahead. The ramp slopes upward at an angle of 10°, and the coefficient of rolling friction between the truck's tires and the loose gravel is $\mu_r = 0.40$. Ignore air resistance and rolling friction as the truck rolls down the highway.

a. Use conservation of energy to find how far along the ramp the truck travels before stopping.
b. By how much does the thermal energy of the truck and ramp increase as the truck stops?

PREPARE Parts a and b can be solved using energy conservation by following Problem-Solving Strategy 10.1. **FIGURE 10.29** shows a before-and-after visual overview. Because we're going to need to determine friction forces to calculate the increase in thermal energy, we've also drawn a free-body diagram for the truck as it moves up the ramp. One slight complication is that the *y*-axis of free-body diagrams is drawn perpendicular to the slope, whereas the calculation of gravitational potential energy needs a vertical *y*-axis to measure height. We've dealt with this by labeling the free-body diagram axis the y'-axis.

FIGURE 10.29 Visual overview of the runaway truck.

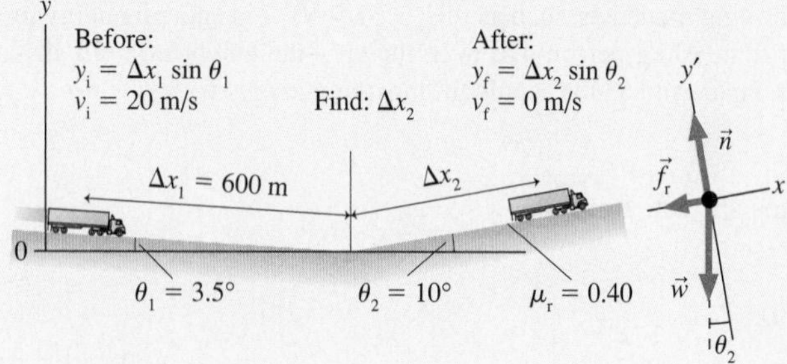

SOLVE a. The law of conservation of energy for the motion of the truck, from the moment its brakes fail to when it finally stops, is

$$K_f + (U_g)_f + \Delta E_{th} = K_i + (U_g)_i$$

Because friction is present only along the ramp, thermal energy will be created only as the truck moves up the ramp. This thermal energy is then given by $\Delta E_{th} = f_r \Delta x_2$, because Δx_2 is the length of the ramp. The conservation of energy equation then is

$$\frac{1}{2}mv_f^2 + mgy_f + f_r\Delta x_2 = \frac{1}{2}mv_i^2 + mgy_i$$

From Figure 10.29 we have $y_i = \Delta x_1 \sin\theta_1$, $y_f = \Delta x_2 \sin\theta_2$, and $v_f = 0$, so the equation becomes

$$mg\Delta x_2 \sin\theta_2 + f_r\Delta x_2 = \frac{1}{2}mv_i^2 + mg\Delta x_1 \sin\theta_1$$

To find $f_r = \mu_r n$ we need to find the normal force n. The free-body diagram shows that

$$\sum F_{y'} = n - mg\cos\theta_2 = a_{y'} = 0$$

from which $f_r = \mu_r n = \mu_r mg\cos\theta_2$. With this result for f_r, our conservation of energy equation is

$$mg\Delta x_2 \sin\theta_2 + \mu_r mg\cos\theta_2 \Delta x_2 = \frac{1}{2}mv_i^2 + mg\Delta x_1 \sin\theta_1$$

which, after we divide both sides by mg, simplifies to

$$\Delta x_2 \sin\theta_2 + \mu_r \cos\theta_2 \Delta x_2 = \frac{v_i^2}{2g} + \Delta x_1 \sin\theta_1$$

Solving this for Δx_2 gives

$$\Delta x_2 = \frac{\dfrac{v_i^2}{2g} + \Delta x_1 \sin\theta_1}{\sin\theta_2 + \mu_r\cos\theta_2}$$

$$= \frac{\dfrac{(20\text{ m/s})^2}{2(9.8\text{ m/s}^2)} + (600\text{ m})(\sin 3.5°)}{\sin 10° + 0.40(\cos 10°)} = 100\text{ m}$$

b. We know that $\Delta E_{th} = f_r\Delta x_2 = (\mu_r mg\cos\theta_2)\Delta x_2$, so that

$$\Delta E_{th} = (0.40)(22{,}000\text{ kg})(9.8\text{ m/s}^2)(\cos 10°)(100\text{ m})$$
$$= 8.5 \times 10^6\text{ J}$$

ASSESS It seems reasonable that a truck that speeds up as it rolls 600 m downhill takes only 100 m to stop on a steeper, high-friction ramp. We also expect the thermal energy to be roughly comparable to the kinetic energy of the truck, since it's largely the kinetic energy that is transformed into thermal energy. At the top of the hill the truck's kinetic energy is $K_i = \frac{1}{2}mv_i^2 = \frac{1}{2}(22{,}000\text{ kg})(20\text{ m/s})^2 = 4.4 \times 10^6$ J, which is of the same order of magnitude as ΔE_{th}. Our answer is reasonable.

SUMMARY

The goals of Chapter 10 are to introduce the concept of energy and to learn a new problem-solving strategy based on conservation of energy.

GENERAL PRINCIPLES

Basic Energy Model

Within a system, energy can be **transformed** between various forms.

Energy can be **transferred** into or out of a system in two basic ways:

- **Work:** The transfer of energy by mechanical forces.
- **Heat:** The nonmechanical transfer of energy from a hotter to a colder object.

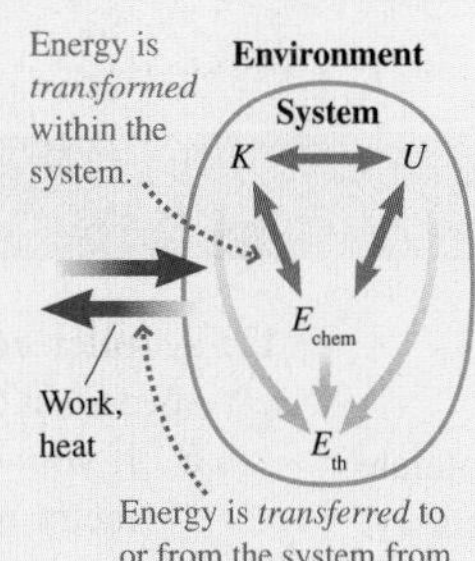

Conservation of Energy

When work W is done on a system, the system's total energy changes by the amount of work done. In mathematical form, this is the **work-energy equation:**

$$\Delta E = \Delta K + \Delta U_g + \Delta U_s + \Delta E_{th} + \Delta E_{chem} + \cdots = W$$

A system is **isolated** when no energy is transferred into or out of the system. This means the work is zero, giving the **law of conservation of energy:**

$$\Delta K + \Delta U_g + \Delta U_s + \Delta E_{th} + \Delta E_{chem} + \cdots = 0$$

Solving Energy Conservation Problems

PREPARE Choose your system so that it's isolated. Draw a before-and-after visual overview.

SOLVE

- If the system is isolated and there's no friction, then mechanical energy is conserved:

$$K_f + (U_g)_f + (U_s)_f = K_i + (U_g)_i + (U_s)_i$$

- If the system is isolated but there's friction present, then the total energy is conserved:

$$K_f + (U_g)_f + (U_s)_f + \Delta E_{th} = K_i + (U_g)_i + (U_s)_i$$

ASSESS Kinetic energy is always positive, as is the change in thermal energy.

IMPORTANT CONCEPTS

Kinetic energy is an energy of motion:

$$K = \frac{1}{2}mv^2 + \frac{1}{2}I\omega^2$$

Translational ... Rotational

Potential energy is energy stored in a system of interacting objects.

- **Gravitational potential energy:** $U_g = mgy$
- **Elastic potential energy:** $U_s = \frac{1}{2}kx^2$

Mechanical energy is the sum of a system's kinetic and potential energies:

$$\text{Mechanical energy} = K + U = K + U_g + U_s$$

Thermal energy is the sum of the microscopic kinetic and potential energies of all the molecules in an object. The hotter an object, the more thermal energy it has. When kinetic (sliding) friction is present, the increase in the thermal energy is $\Delta E_{th} = f_k \Delta x$.

Work is the process by which energy is transferred to or from a system by the application of mechanical forces.

If a particle moves through a displacement $\vec{d}$ while acted upon by a constant force $\vec{F}$, the force does work

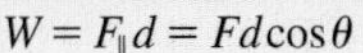

$$W = F_\parallel d = Fd\cos\theta$$

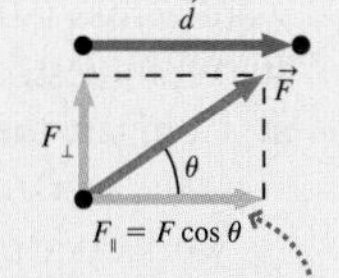

Only the component of the force parallel to the displacement does work.

APPLICATIONS

Perfectly elastic collisions
Both mechanical energy and momentum are conserved.

$$(v_{1x})_f = \frac{m_1 - m_2}{m_1 + m_2}(v_{1x})_i$$

$$(v_{2x})_f = \frac{2m_1}{m_1 + m_2}(v_{1x})_i$$

Object 2 initially at rest

Before: 1 $(v_{1x})_i$ → 2 K_i

After: $K_f = K_i$ 1 $(v_{1x})_f$ 2 $(v_{2x})_f$

Power is the rate at which energy is transformed . . .

$$P = \frac{\Delta E}{\Delta t}$$

ΔE: Amount of energy transformed; Δt: Time required to transform it

. . . or at which work is done.

$$P = \frac{W}{\Delta t}$$

W: Amount of work done; Δt: Time required to do work

For homework assigned on MasteringPhysics, go to www.masteringphysics.com

Problem difficulty is labeled as | (straightforward) to ||||| (challenging).

Problems labeled ✎ can be done on a Workbook Energy Worksheet; INT integrate significant material from earlier chapters; BIO are of biological or medical interest.

QUESTIONS

Conceptual Questions

1. The brake shoes of your car are made of a material that can tolerate very high temperatures without being damaged. Why is this so?
2. When you pound a nail with a hammer, the nail gets quite warm. Describe the energy transformations that lead to the addition of thermal energy in the nail.

For Questions 3 through 10, give a specific example of a system with the energy transformation shown. In these questions, W is the work done on the system, and K, U, and E_{th} are the kinetic, potential, and thermal energies of the system, respectively. Any energy not mentioned in the transformation is assumed to remain constant; if work is not mentioned, it is assumed to be zero.

3. $W \rightarrow K$
4. $W \rightarrow U$
5. $K \rightarrow U$
6. $K \rightarrow W$
7. $U \rightarrow K$
8. $W \rightarrow \Delta E_{th}$
9. $U \rightarrow \Delta E_{th}$
10. $K \rightarrow \Delta E_{th}$
11. A ball of putty is dropped from a height of 2 m onto a hard floor, where it sticks. What object or objects need to be included within the system if the system is to be isolated during this process?
12. A 0.5 kg mass on a 1-m-long string swings in a circle on a horizontal, frictionless table at a steady speed of 2 m/s. How much work does the tension in the string do on the mass during one revolution? Explain.
13. Particle A has less mass than particle B. Both are pushed forward across a frictionless surface by equal forces for 1 s. Both start from rest.
    a. Compare the amount of work done on each particle. That is, is the work done on A greater than, less than, or equal to the work done on B? Explain.
    b. Compare the impulses delivered to particles A and B. Explain.
    c. Compare the final speeds of particles A and B. Explain.
14. The meaning of the word "work" is quite different in physics from its everyday usage. Give an example of an action a person could do that "feels like work" but that does not involve any work as we've defined it in this chapter.
15. To change a tire, you need to use a jack to raise one corner of your car. While doing so, you happen to notice that pushing the jack handle down 20 cm raises the car only 0.2 cm. Use energy concepts to explain why the handle must be moved so far to raise the car by such a small amount.
16. You drop two balls from a tower, one of mass m and the other of mass $2m$. Just before they hit the ground, which ball, if either, has the larger kinetic energy? Explain.
17. A roller coaster car rolls down a frictionless track, reaching speed v at the bottom.
    a. If you want the car to go twice as fast at the bottom, by what factor must you increase the height of the track?
    b. Does your answer to part a depend on whether the track is straight or not? Explain.
18. A spring gun shoots out a plastic ball at speed v. The spring is then compressed twice the distance it was on the first shot.
    a. By what factor is the spring's potential energy increased?
    b. By what factor is the ball's speed increased? Explain.
19. Sandy and Chris stand on the edge of a cliff and throw identical mass rocks at the same speed. Sandy throws her rock horizontally while Chris throws his upward at an angle of 45° to the horizontal. Are the rocks moving at the same speed when they hit the ground, or is one moving faster than the other? If one is moving faster, which one? Explain.
20. A solid cylinder and a cylindrical shell have the same mass, same radius, and turn on frictionless, horizontal axles. (The cylindrical shell has lightweight spokes connecting the shell to the axle.) A rope is wrapped around each cylinder and tied to a block. The blocks have the same mass and are held the same height above the ground as shown in Figure Q10.20. Both blocks are released simultaneously. The ropes do not slip. Which block hits the ground first? Or is it a tie? Explain.

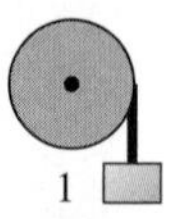

FIGURE Q10.20

21. BIO You are much more likely to be injured if you fall and your head strikes the ground than if your head strikes a gymnastics pad. Use energy and work concepts to explain why this is so.

Multiple-Choice Questions

22. || If you walk up a flight of stairs at constant speed, gaining vertical height h, the work done on you (the system, of mass m) is
    A. $+mgh$, by the normal force of the stairs.
    B. $-mgh$, by the normal force of the stairs.
    C. $+mgh$, by the gravitational force of the earth.
    D. $-mgh$, by the gravitational force of the earth.
23. | You and a friend each carry a 15 kg suitcase up two flights of stairs, walking at a constant speed. Take each suitcase to be the system. Suppose you carry your suitcase up the stairs in 30 s while your friend takes 60 s. Which of the following is true?
    A. You did more work, but both of you expended the same power.
    B. You did more work and expended more power.
    C. Both of you did equal work, but you expended more power.
    D. Both of you did equal work, but you expended less power.

24. | A woman uses a pulley and a rope to raise a 20 kg weight to a height of 2 m. If it takes 4 s to do this, about how much power is she supplying?
A. 100 W B. 200 W C. 300 W D. 400 W

25. | A hockey puck sliding along frictionless ice with speed v to the right collides with a horizontal spring and compresses it by 2.0 cm before coming to a momentary stop. What will be the spring's maximum compression if the same puck hits it at a speed of $2v$?
A. 2.0 cm B. 2.8 cm C. 4.0 cm
D. 5.6 cm E. 8.0 cm

26. || A block slides down a smooth ramp, starting from rest at a height h. When it reaches the bottom it's moving at speed v. It then continues to slide up a second smooth ramp. At what height is its speed equal to $v/2$?
A. $h/4$ B. $h/2$ C. $3h/4$ D. $2h$

27. | A wrecking ball is suspended from a 5.0-m-long cable that makes a 30° angle with the vertical. The ball is released and swings down. What is the ball's speed at the lowest point?
A. 7.7 m/s B. 4.4 m/s C. 3.6 m/s D. 3.1 m/s

PROBLEMS

Section 10.2 Work

1. || During an etiquette class, you walk slowly and steadily at 0.20 m/s for 2.5 m with a 0.75 kg book balanced on top of your head. How much work does your head do on the book?
2. || A 2.0 kg book is lying on a 0.75-m-high table. You pick it up and place it on a bookshelf 2.3 m above the floor.
   a. How much work does gravity do on the book?
   b. How much work does your hand do on the book?
3. || The two ropes seen in Figure P10.3 are used to lower a 255 kg piano exactly 5 m from a second-story window to the ground. How much work is done by each of the three forces?

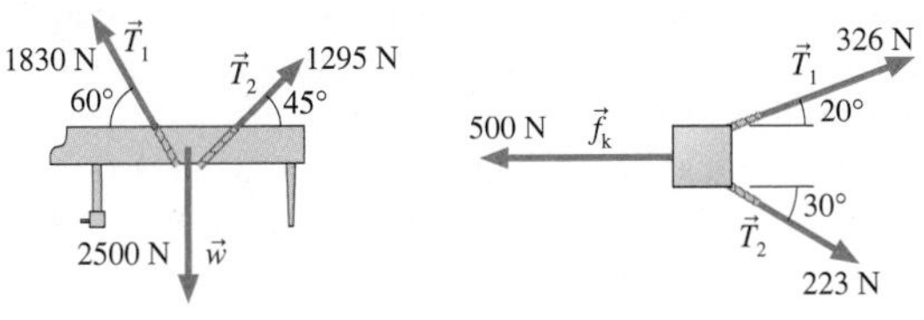

FIGURE P10.3 FIGURE P10.4

4. | The two ropes shown in the bird's-eye view of Figure P10.4 are used to drag a crate exactly 3 m across the floor. How much work is done by each of the ropes on the crate?
5. || a. At the airport, you ride a "moving sidewalk" that carries you horizontally for 25 m at 0.70 m/s. Assuming that you were moving at 0.70 m/s before stepping onto the moving sidewalk and continue at 0.70 m/s afterward, how much work does the moving sidewalk do on you? Your mass is 60 kg.
   b. An escalator carries you from one level to the next in the airport terminal. The upper level is 4.5 m above the lower level, and the length of the escalator is 7.0 m. How much work does the up escalator do on you when you ride it from the lower level to the upper level?
   c. How much work does the down escalator do on you when you ride it from the upper level to the lower level?
6. | A boy flies a kite with the string at a 30° angle to the horizontal. The tension in the string is 4.5 N. How much work does the string do on the boy if the boy
   a. Stands still?
   b. Walks a horizontal distance of 11 m away from the kite?
   c. Walks a horizontal distance of 11 m toward the kite?

Section 10.3 Kinetic Energy

7. | Which has the larger kinetic energy, a 10 g bullet fired at 500 m/s or a 10 kg bowling ball sliding at 10 m/s?
8. || At what speed does a 1000 kg compact car have the same kinetic energy as a 20,000 kg truck going 25 km/hr?
9. | A car is traveling at 10 m/s.
   a. How fast would the car need to go to double its kinetic energy?
   b. By what factor does the car's kinetic energy increase if its speed is doubled to 20 m/s?
10. ||| Sam's job at the amusement park is to slow down and bring to a stop the boats in the log ride. If a boat and its riders have a mass of 1200 kg and the boat drifts in at 1.2 m/s, how much work does Sam do to stop it?
11. ||| A 20 g plastic ball is moving to the left at 30 m/s. How much work must be done on the ball to cause it to move to the right at 30 m/s?
12. ||| The turntable in a microwave oven has a moment of inertia of $0.040 \text{ kg}\cdot\text{m}^2$ and is rotating once every 4.0 s. What is its kinetic energy?
13. |||| An energy storage system based on a flywheel (a rotating disk) can store a maximum of 4.0 MJ when the flywheel is rotating at 20,000 revolutions per minute. What is the moment of inertia of the flywheel?

Section 10.4 Potential Energy

14. || The lowest point in Death Valley is 85.0 m below sea level. The summit of nearby Mt. Whitney has an elevation of 4420 m. What is the change in gravitational potential energy of an energetic 65.0 kg hiker who makes it from the floor of Death Valley to the top of Mt. Whitney?
15. | a. What is the kinetic energy of a 1500 kg car traveling at a speed of 30 m/s ($\approx$65 mph)?
   b. From what height should the car be dropped to have this same amount of kinetic energy just before impact?
   c. Does your answer to part b depend on the car's mass?
16. | The world's fastest humans can reach speeds of about 11 m/s. In order to increase his gravitational potential energy by an amount equal to his kinetic energy at full speed, how high would such a sprinter need to climb?
17. | A 72 kg bike racer climbs a 1200-m-long section of road that has a slope of 4.3°. By how much does his gravitational potential energy change during this climb?
18. || A 1000 kg wrecking ball hangs from a 15-m-long cable. The ball is pulled back until the cable makes an angle of 25° with the vertical. By how much has the gravitational potential energy of the ball changed?
19. || How far must you stretch a spring with $k = 1000$ N/m to store 200 J of energy?

20. || How much energy can be stored in a spring with a spring constant of 500 N/m if its maximum possible stretch is 20 cm?

21. ||||| BIO The elastic energy stored in your tendons can contribute up to 35% of your energy needs when running. Sports scientists have studied the change in length of the knee extensor tendon in sprinters and nonathletes. They find (on average) that the sprinters' tendons stretch 41 mm, while nonathletes' stretch only 33 mm. The spring constant for the tendon is the same for both groups, 33 N/mm. What is the difference in maximum stored energy between the sprinters and the nonathletes?

Section 10.5 Thermal Energy

22. | Marissa drags a 23 kg duffel bag 14 m across the gym floor. If the coefficient of kinetic friction between the floor and bag is 0.15, how much thermal energy does Marissa create?

23. || Mark pushes his broken car 150 m down the block to his friend's house. He has to exert a 110 N horizontal force to push the car at a constant speed. How much thermal energy is created in the tires and road during this short trip?

24. ||| A 900 N crate slides 12 m down a ramp that makes an angle of 35° with the horizontal. If the crate slides at a constant speed, how much thermal energy is created?

25. ||| A 25 kg child slides down a playground slide at a *constant speed.* The slide has a height of 3.0 m and is 7.0 m long. Using energy considerations, find the magnitude of the kinetic friction force acting on the child.

Section 10.6 Using the Law of Conservation of Energy

26. || A boy reaches out of a window and tosses a ball straight up with a speed of 10 m/s. The ball is 20 m above the ground as he releases it. Use conservation of energy to find
   a. The ball's maximum height above the ground.
   b. The ball's speed as it passes the window on its way down.
   c. The speed of impact on the ground.

27. || a. With what minimum speed must you toss a 100 g ball straight up to just barely hit the 10-m-high ceiling of the gymnasium if you release the ball 1.5 m above the floor? Solve this problem using energy.
   b. With what speed does the ball hit the floor?

28. ||| What minimum speed does a 100 g puck need to make it to the top of a frictionless ramp that is 3.0 m long and inclined at 20°?

29. || A car is parked at the top of a 50-m-high hill. It slips out of gear and rolls down the hill. How fast will it be going at the bottom? (Ignore friction.)

30. ||| A 1500 kg car is approaching the hill shown in Figure P10.30 at 10 m/s when it suddenly runs out of gas.
   a. Can the car make it to the top of the hill by coasting?
   b. If your answer to part a is yes, what is the car's speed after coasting down the other side?

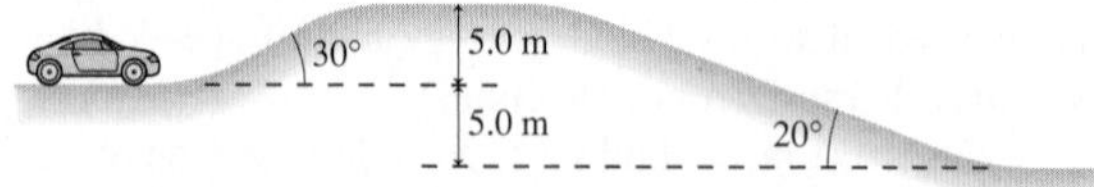

FIGURE P10.30

31. || A 10 kg runaway grocery cart runs into a spring with spring constant 250 N/m and compresses it by 60 cm. What was the speed of the cart just before it hit the spring?

32. || As a 15,000 kg jet lands on an aircraft carrier, its tail hook snags a cable to slow it down. The cable is attached to a spring with spring constant 60,000 N/m. If the spring stretches 30 m to stop the plane, what was the plane's landing speed?

33. || Your friend's Frisbee has become stuck 16 m above the ground in a tree. You want to dislodge the Frisbee by throwing a rock at it. The Frisbee is stuck pretty tight, so you figure the rock needs to be traveling at least 5.0 m/s when it hits the Frisbee. If you release the rock 2.0 m above the ground, with what minimum speed must you throw it?

34. || A fireman of mass 80 kg slides down a pole. When he reaches the bottom, 4.2 m below his starting point, his speed is 2.2 m/s. By how much has thermal energy increased during his slide?

35. || A 20 kg child slides down a 3.0-m-high playground slide. She starts from rest, and her speed at the bottom is 2.0 m/s.
   a. What energy transfers and transformations occur during the slide?
   b. What is the total change in the thermal energy of the slide and the seat of her pants?

36. || A hockey puck is given an initial speed of 5.0 m/s. If the coefficient of kinetic friction between the puck and the ice is 0.05, how far does the puck slide before coming to rest? Solve this problem using conservation of energy.

Section 10.7 Energy in Collisions

37. || A 50 g marble moving at 2.0 m/s strikes a 20 g marble at rest. What is the speed of each marble immediately after the collision? Assume the collision is perfectly elastic and the marbles collide head-on.

38. || Ball 1, with a mass of 100 g and traveling at 10 m/s, collides head-on with ball 2, which has a mass of 300 g and is initially at rest. What are the final velocities of each ball if the collision is (a) perfectly elastic? (b) perfectly inelastic?

39. | An air-track glider undergoes a perfectly inelastic collision with an identical glider that is initially at rest. What fraction of the first glider's initial kinetic energy is transformed into thermal energy in this collision?

40. | Two balls undergo a perfectly elastic head-on collision, with one ball initially at rest. If the incoming ball has a speed of 200 m/s, what are the final speed and direction of each ball if
   a. The incoming ball is *much* more massive than the stationary ball?
   b. The stationary ball is *much* more massive than the incoming ball?

Section 10.8 Power

41. || a. How much work must you do to push a 10 kg block of steel across a steel table at a steady speed of 1.0 m/s for 3.0 s? The coefficient of kinetic friction for steel on steel is 0.60.
   b. What is your power output while doing so?

42. | a. How much work does an elevator motor do to lift a 1000 kg elevator a height of 100 m?
   b. How much power must the motor supply to do this in 50 s at constant speed?

43. ||| A 1000 kg sports car accelerates from 0 to 30 m/s in 10 s. What is the average power of the engine?

44. ||| In just 0.30 s, you compress a spring (spring constant 5000 N/m), which is initially at its equilibrium length, by 4.0 cm. What is your average power output?

45. ||| In the winter sport of curling, players give a 20 kg stone a push across a sheet of ice. A curler accelerates a stone to a speed of 3.0 m/s over a time of 2.0 s.
   a. How much force does the curler exert on the stone?
   b. What average power does the curler use to bring the stone up to speed?

46. || A 710 kg car drives at a constant speed of 23 m/s. It is subject to a drag force of 500 N. What power is required from the car's engine to drive the car
 a. On level ground?
 b. Up a hill with a slope of 2.0°?

47. ||| An elevator weighing 2500 N ascends at a constant speed of 8.0 m/s. How much power must the motor supply to do this?

General Problems

48. || A 2.3 kg box, starting from rest, is pushed up a ramp by a 10 N force parallel to the ramp. The ramp is 2.0 m long and tilted at 17°. The speed of the box at the top of the ramp is 0.80 m/s. Consider the system to be the box + ramp + earth.
 a. How much work W does the force do on the system?
 b. What is the change ΔK in the kinetic energy of the system?
 c. What is the change ΔU_g in the gravitational potential energy of the system?
 d. What is the change ΔE_{th} in the thermal energy of the system?

49. | A 55 kg skateboarder wants to just make it to the upper edge of a "half-pipe" with a radius of 3.0 m, as shown in Figure P10.49. What speed v_i does he need at the bottom if he is to coast all the way up?

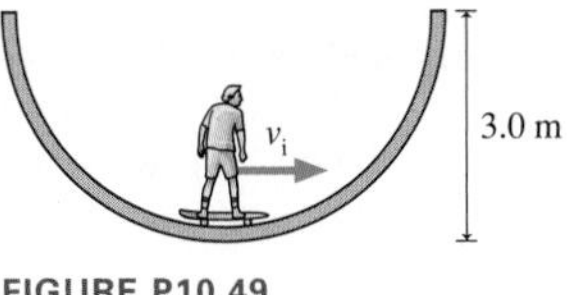

FIGURE P10.49

 a. First do the calculation treating the skateboarder and board as a point particle, with the entire mass nearly in contact with the half-pipe.
 b. More realistically, the mass of the skateboarder in a deep crouch might be thought of as concentrated 0.75 m from the half-pipe. Assuming he remains in that position all the way up, what v_i is needed to reach the upper edge?

50. ||| Fleas have remarkable jumping ability. A 0.50 mg flea, jumping straight up, would reach a height of 40 cm if there were no
BIO air resistance. In reality, air resistance limits the height to 20 cm.
 a. What is the flea's kinetic energy as it leaves the ground?
 b. At its highest point, what fraction of the initial kinetic energy has been converted to potential energy?

51. ||| A marble slides without friction in a *vertical* plane around the inside of a smooth, 20-cm-diameter horizontal pipe. The mar-
INT ble's speed at the bottom is 3.0 m/s; this is fast enough so that the marble makes a complete loop, never losing contact with the pipe. What is its speed at the top?

52. || A 20 kg child is on a swing that hangs from 3.0-m-long chains, as shown in Figure P10.52. What is her speed v_i at the bottom of the arc if she swings out to a 45° angle before reversing direction?

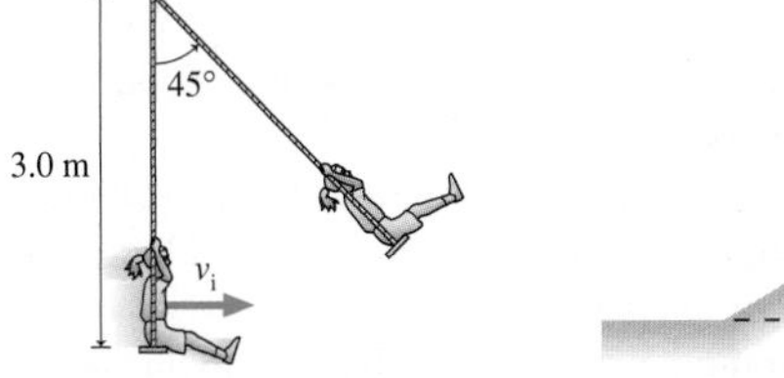

FIGURE P10.52

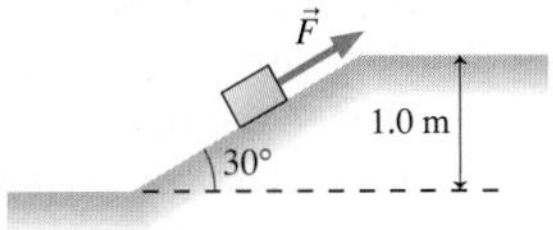

FIGURE P10.53

53. || Suppose you lift a 20 kg box by a height of 1.0 m.
 a. How much work do you do in lifting the box?
 Instead of lifting the box straight up, suppose you push it up a 1.0-m-high ramp that makes a 30° degree angle with the horizontal, as shown in Figure P10.53. Being clever, you choose a ramp with no friction.
 b. How much force F is required to push the box straight up the slope at a constant speed?
 c. How long is the ramp?
 d. Use your force and distance results to calculate the work you do in pushing the box up the ramp. How does this compare to your answer to part a?

54. || A cannon tilted up at a 30° angle fires a cannon ball at 80 m/s from atop a 10-m-high fortress wall. What is the ball's impact speed on the ground below? Ignore air resistance.

55. | The sledder shown in Figure P10.55 starts from the top of a frictionless hill and slides down into the valley. What initial speed v_i does the sledder need to just make it over the next hill?

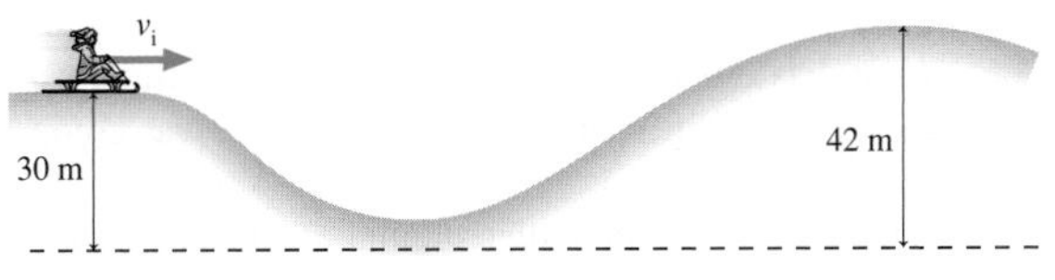

FIGURE P10.55

56. |||| In a physics lab experiment, a spring clamped to the table shoots a 20 g ball horizontally. When the spring is compressed
INT 20 cm, the ball travels horizontally 5.0 m and lands on the floor 1.5 m below the point at which it left the spring. What is the spring constant?

57. ||| A 50 g ice cube can slide without friction up and down a 30° slope. The ice cube is pressed against a spring at the bottom of the slope, compressing the spring 10 cm. The spring constant is 25 N/m. When the ice cube is released, what distance will it travel up the slope before reversing direction?

58. |||| The maximum energy a bone can absorb without breaking is
BIO surprisingly small. For a healthy human of mass 60 kg, experimental data show that the leg bones can absorb about 200 J.
 a. From what maximum height could a person jump and land rigidly upright on both feet without breaking his legs? Assume that all the energy is absorbed in the leg bones in a rigid landing.
 b. People jump from much greater heights than this; explain how this is possible.
 Hint: Think about how people land when they jump from greater heights.

59. || In an amusement park water slide, people slide down an essentially frictionless tube. They drop 3.0 m and exit the slide,
INT moving horizontally, 1.2 m above a swimming pool. What horizontal distance do they travel from the exit point before hitting the water? Does the mass of the person make any difference?

60. || The 5.0-m-long rope in Figure P10.60 hangs vertically from a tree right at the edge of a ravine. A woman wants to use the rope to swing to the other side of the ravine. She runs as fast as she can, grabs the rope, and swings out over the ravine.

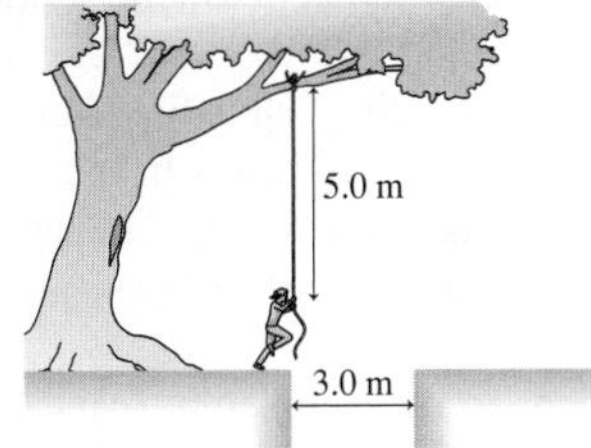

FIGURE P10.60

 a. As she swings, what energy conversion is taking place?
 b. When she's directly over the far edge of the ravine, how much higher is she than when she started?
 c. Given your answers to parts a and b, how fast must she be running when she grabs the rope in order to swing all the way across the ravine?

61. ||| You have been asked to design a "ballistic spring system" to measure the speed of bullets. A bullet of mass m is fired into a block of mass M. The block, with the embedded bullet, then slides across a frictionless table and collides with a horizontal spring whose spring constant is k. The opposite end of the spring is anchored to a wall. The spring's maximum compression d is measured.
INT
   a. Find an expression for the bullet's initial speed v_B in terms of m, M, k, and d.

   Hint: This is a two-part problem. The bullet's collision with the block is an inelastic collision. What quantity is conserved in an inelastic collision? Subsequently the block hits a spring on a frictionless surface. What quantity is conserved in this collision?

   b. What was the speed of a 5.0 g bullet if the block's mass is 2.0 kg and if the spring, with $k = 50$ N/m, was compressed by 10 cm?
   c. What fraction of the bullet's initial kinetic energy is "lost"? Where did it go?

62. ||| A new event, shown in Figure P10.62, has been proposed for the Winter Olympics. An athlete will sprint 100 m, starting from rest, then leap onto a 20 kg bobsled. The person and bobsled will then slide down a 50-m-long ice-covered ramp, sloped at 20°, and into a spring with a carefully calibrated spring constant of 2000 N/m. The athlete who compresses the spring the farthest wins the gold medal. Lisa, whose mass is 40 kg, has been training for this event. She can reach a maximum speed of 12 m/s in the 100 m dash.
INT

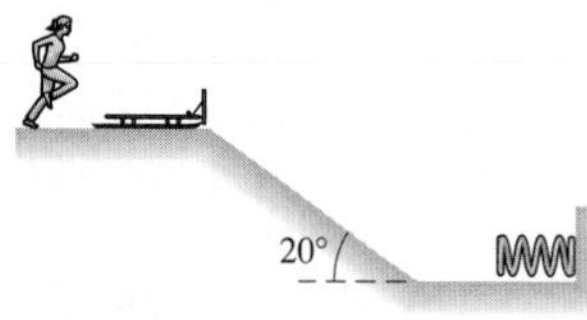

FIGURE P10.62

   a. How far will Lisa compress the spring?
   b. The Olympic committee has very exact specifications about the shape and angle of the ramp. Is this necessary? If the committee asks your opinion, what factors about the ramp will you tell them are important?

63. || Boxes A and B in Figure P10.63 have masses of 12.0 kg and 4.0 kg, respectively. The two boxes are released from rest. Use conservation of energy to find the boxes' speed when box B has fallen a distance of 0.50 m. Assume a frictionless upper surface.

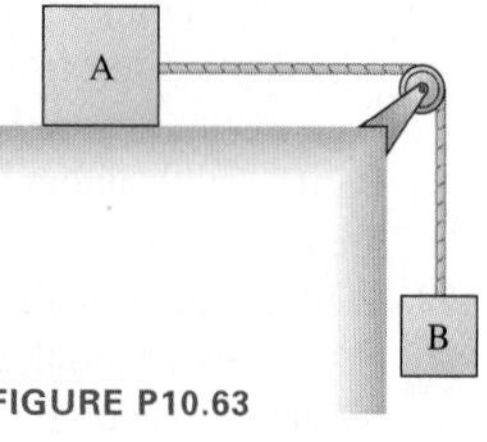

FIGURE P10.63

64. |||| What would be the speed of the boxes in Problem 63 if the coefficient of kinetic friction between box A and the surface it slides on were 0.20? Use conservation of energy.

65. ||||| A 20 g ball is fired horizontally with initial speed v_i toward a 100 g ball that is hanging motionless from a 1.0-m-long string. The balls undergo a head-on, perfectly elastic collision, after which the 100 g ball swings out to a maximum angle $\theta_{max} = 50°$. What was v_i?

66. || Two coupled boxcars are rolling along at 2.5 m/s when they collide with and couple to a third, stationary boxcar.
INT
   a. What is the final speed of the three coupled boxcars?
   b. What fraction of the cars' initial kinetic energy is transformed into thermal energy?

67. || A fish scale, consisting of a spring with spring constant $k =$ 200 N/m, is hung vertically from the ceiling. A 5.0 kg fish is attached to the end of the unstretched spring and then released. The fish moves downward until the spring is fully stretched, then starts to move back up as the spring begins to contract. What is the maximum distance through which the fish falls?

68. | A 70 kg human sprinter can accelerate from rest to 10 m/s in 3.0 s. During the same interval, a 30 kg greyhound can accelerate from rest to 20 m/s. Compute (a) the change in kinetic energy and (b) the average power output for each.
BIO

69. ||| A 50 g ball of clay traveling at speed v_i hits and sticks to a 1.0 kg block sitting at rest on a frictionless surface.
INT
   a. What is the speed of the block after the collision?
   b. Show that the mechanical energy is *not* conserved in this collision. What percentage of the ball's initial kinetic energy is "lost"? Where did this kinetic energy go?

70. || A package of mass m is released from rest at a warehouse loading dock and slides down a 3.0-m-high frictionless chute to a waiting truck. Unfortunately, the truck driver went on a break without having removed the previous package, of mass $2m$, from the bottom of the chute as shown in Figure P10.70.
INT
   a. Suppose the packages stick together. What is their common speed after the collision?
   b. Suppose the collision between the packages is perfectly elastic. To what height does the package of mass m rebound?

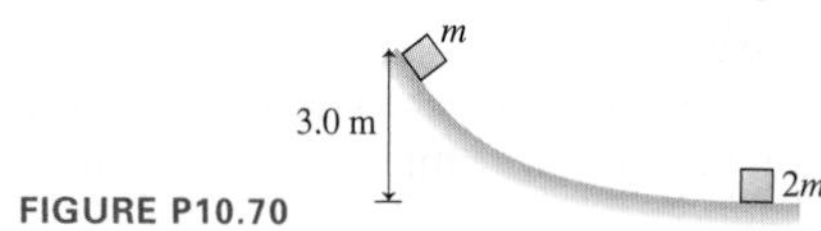

FIGURE P10.70

71. |||| A 50 kg sprinter, starting from rest, runs 50 m in 7.0 s at constant acceleration.
INT
   a. What is the magnitude of the horizontal force acting on the sprinter?
   b. What is the sprinter's average power output during the first 2.0 s of his run?
   c. What is the sprinter's average power output during the final 2.0 s?

72. || Bob can throw a 500 g rock with a speed of 30 m/s. He moves his hand forward 1.0 m while doing so.
INT
   a. How much force, assumed to be constant, does Bob apply to the rock?
   b. How much work does Bob do on the rock?

73. ||| A 2.0 hp electric motor on a water well pumps water from 10 m below the surface. The density of water is 1.0 kg per L. How many liters of water can the motor pump in 1 h?

74. || The human heart has to pump the average adult's 6.0 L of blood through the body every minute. The heart must do work to overcome frictional forces that resist the blood flow. The average blood pressure is 1.3×10^4 N/m^2.
BIO
   a. Compute the work done moving the 6.0 L of blood completely through the body, assuming the blood pressure always takes its average value.
   b. What power output must the heart have to do this task once a minute?

   Hint: When the heart contracts, it applies force to the blood. Pressure is just force/area, so we can write work = (pressure)(area)(distance). But (area)(distance) is just the blood volume passing through the heart.

Passage Problems

Tennis Ball Testing

A tennis ball bouncing on a hard surface compresses and then rebounds. The details of the rebound are specified in tennis regulations. Tennis balls, to be acceptable for tournament play, must have a mass of 57.5 g. When dropped from a height of 2.5 m onto a concrete surface, a ball must rebound to a height of 1.4 m. During impact, the ball compresses by approximately 6 mm.

75. | How fast is the ball moving when it hits the concrete surface? (Ignore air resistance.)
A. 5 m/s B. 7 m/s C. 25 m/s D. 50 m/s
76. | If the ball accelerates uniformly when it hits the floor, what is its approximate acceleration as it comes to rest before rebounding?
A. 1000 m/s^2 B. 2000 m/s^2 C. 3000 m/s^2 D. 4000 m/s^2
77. | The ball's kinetic energy just after the bounce is less than just before the bounce. In what form does this lost energy end up?
A. Elastic potential energy
B. Gravitational potential energy
C. Thermal energy
D. Rotational kinetic energy
78. | By approximately what percent does the kinetic energy decrease?
A. 35% B. 45% C. 55% D. 65%
79. | When a tennis ball bounces from a racket, the ball loses approximately 30% of its kinetic energy to thermal energy. A ball that hits a racket at a speed of 10 m/s will rebound with approximately what speed?
A. 8.5 m/s B. 7.0 m/s C. 4.5 m/s D. 3.0 m/s

Work and Power in Cycling

When you ride a bicycle at constant speed, almost all of the energy you expend goes into the work you do against the drag force of the air. In this problem, assume that *all* of the energy expended goes into working against drag. As we saw in Section 5.6, the drag force on an object is approximately proportional to the square of its speed with respect to the air. For this problem, assume that $F \propto v^2$ exactly and that the air is motionless with respect to the ground unless noted otherwise. Suppose a cyclist and her bicycle have a combined mass of 60 kg and she is cycling along at a speed of 5 m/s.

80. | If the drag force on the cyclist is 10 N, how much energy does she use in cycling 1 km?
A. 6 kJ B. 10 kJ C. 50 kJ D. 100 kJ
81. | Under these conditions, how much power does she expend as she cycles?
A. 10 W B. 50 W C. 100 W D. 200 W
82. | If she doubles her speed to 10 m/s, how much energy does she use in cycling 1 km?
A. 20 kJ B. 40 kJ C. 200 kJ D. 400 kJ
83. | How much power does she expend when cycling at that speed?
A. 100 W B. 200 W C. 400 W D. 1000 W
84. | Upon reducing her speed back down to 5 m/s, she hits a headwind of 5 m/s. How much power is she expending now?
A. 100 W B. 200 W C. 500 W D. 1000 W

STOP TO THINK ANSWERS

Stop to Think 10.1: D. Since the child slides at a constant speed, his kinetic energy doesn't change. But his gravitational potential energy decreases as he descends. It is transformed into thermal energy in the slide and his bottom.

Stop to Think 10.2: C. $W = Fd\cos\theta$. The 10 N force at 90° does no work at all. $\cos 60° = \frac{1}{2}$, so the 8 N force does less work than the 6 N force.

Stop to Think 10.3: B > D > A = C. $K = (1/2)mv^2$. Using the given masses and velocities, we find $K_A = 2.0$ J, $K_B = 4.5$ J, $K_C = 2.0$ J, $K_D = 4.0$ J.

Stop to Think 10.4: $(U_g)_3 > (U_g)_2 = (U_g)_4 > (U_g)_1$. Gravitational potential energy depends only on height, not speed.

Stop to Think 10.5: D. The potential energy of a spring depends on the *square* of the displacement x, so the energy is positive whether the spring is compressed or extended. Furthermore, if the spring is compressed by twice the amount it had been stretched, the energy will increase by a factor of $2^2 = 4$. So the energy will be $4 \times 1\text{ J} = 4\text{ J}$.

Stop to Think 10.6: $P_B > P_A = P_C > P_D$. The power here is the rate at which each runner's internal chemical energy is converted into gravitational potential energy. The change in gravitational potential energy is $mg\Delta y$, so the power is $mg\Delta y/\Delta t$. For runner A, the ratio $m\Delta y/\Delta t$ equals (80 kg)(10 m)/(10 s) = 80 kg · m/s. For C, it's the same. For B, it's 100 kg · m/s, while for D the ratio is 64 kg · m/s.

11 Using Energy

The odd hopping gait of a kangaroo has a very practical purpose: It lets the kangaroo cover great distances with minimal energy input. How is this efficiency possible?

LOOKING AHEAD ▸▸

The goals of Chapter 11 are to learn more about energy transformations and transfers, the laws of thermodynamics, and theoretical and practical limitations on energy use.

Efficiency

In this chapter, we'll look at practical limits on energy transfers and transformations.

Looking Back ◂◂
10.1 Basic energy concepts; forms of energy
10.6 The law of conservation of energy

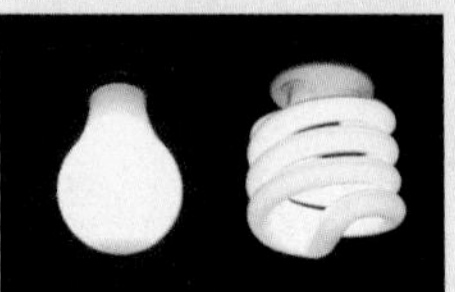

Both bulbs put out the same amount of light, but the one on the right uses 1/4 the electric power. Both bulbs perform the same transformation, but one is much more efficient.

Energy in the Body

All the energy that your body uses for all of the tasks you complete during the day comes from food. How efficient is your body at converting this energy? How much energy does your body actually use to run, climb, and move?

Climbing stairs requires a change in potential energy. How much energy does your body use to make this climb?

These fighters wear masks that allow a direct measurement of the energy used by their bodies as they exercise.

Thermal Energy and Temperature

An object's temperature is related to its thermal energy, the energy of motion on an atomic scale.

Before

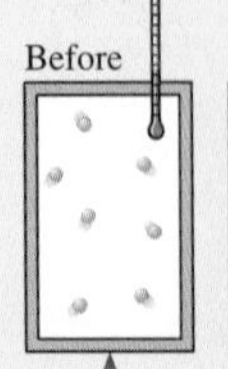

After

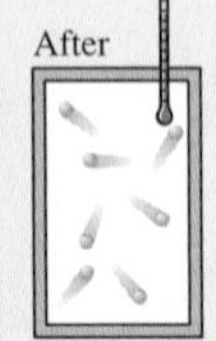

We'll use the ideal gas model to help us understand the nature of thermal energy and temperature.

Looking Back ◂◂
10.7 Thermal energy

Heat and Thermodynamics

Processes in which only thermal energy changes are the domain of **thermodynamics.**

The thermal energy of the kettle is increased by the heat from the burner, which is at a higher temperature.

Heat Engines and Heat Pumps

A **heat engine** can convert thermal energy into other forms of energy. A **heat pump** moves thermal energy from one place to another.

This geothermal plant uses volcanic thermal energy to generate electricity. "Waste" heat warms the lagoon. Why must any energy be "wasted"?

The inside of a refrigerator is cold because heat has been "pumped" out. What's the energy cost to move this heat?

Entropy

Entropy is a measure of disorder at an atomic level that helps us explain some basic observations about the world.

When you stir a cup of coffee, the cream mixes in. This is irreversible; stirring backward won't cause it to unmix. The concept of entropy explains why the future is different from the past and why there are theoretical limits on energy use.

chemical energy is converted to thermal energy by burning, and the resulting thermal energy is converted to electric energy. A typical power-plant cycle is shown in FIGURE 11.2. "What you get" is the energy output, the 35 J of electric energy. "What you had to pay" is the energy input, the 100 J of chemical energy, giving an efficiency of

$$e = \frac{35\text{ J}}{100\text{ J}} = 0.35 = 35\%$$

Unlike your stair-climbing efficiency, the rather modest power-plant efficiency turns out to be largely due to a *fundamental limitation:* Thermal energy cannot be transformed into other forms of energy with 100% efficiency. A 35% efficiency is close to the theoretical maximum, and no power plant could be designed that would do better than this maximum. In later sections, we will explore the fundamental properties of thermal energy that make this so.

FIGURE 11.2 Energy transformations in a coal-fired power plant.

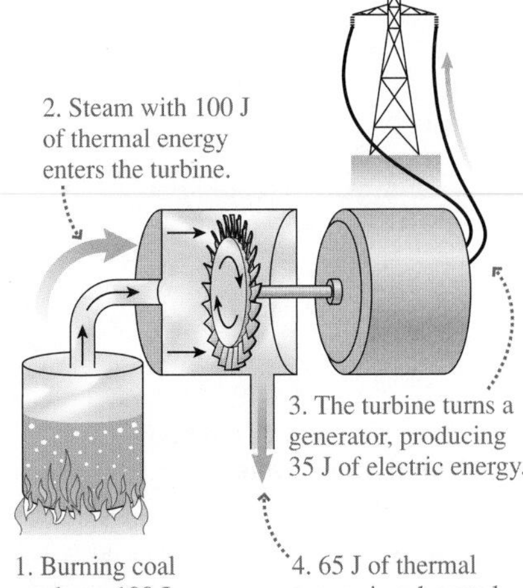

2. Steam with 100 J of thermal energy enters the turbine.
3. The turbine turns a generator, producing 35 J of electric energy.
1. Burning coal produces 100 J of thermal energy.
4. 65 J of thermal energy is exhausted into the environment.

PROBLEM-SOLVING STRATEGY 11.1 **Energy efficiency problems**

PREPARE There are two key components to define before we compute efficiency:

❶ Choose what energy to count as "what you get." This could be the useful energy output of an engine or process or the work that is done in completing a process. For example, when you climb a flight of stairs, "what you get" is your change in potential energy.

❷ "What you had to pay" will generally be the total energy input needed for an engine, task, or process. For example, when you run your air conditioner, "what you had to pay" is the electric energy input.

SOLVE You may need to do additional calculations:

- Compute values for "what you get" and "what you had to pay."
- Be certain that all energy values are in the same units.

After this, compute the efficiency using $e = \dfrac{\text{what you get}}{\text{what you had to pay}}$.

ASSESS Check your answer to see if it is reasonable, given what you know about typical efficiencies for the process under consideration.

EXAMPLE 11.1 **Lightbulb efficiency**

A 15 W compact fluorescent bulb and a 75 W incandescent bulb each produce 3.0 W of visible-light energy. What are the efficiencies of these two types of bulbs for converting electric energy into light?

PREPARE The problem statement doesn't give us values for energy; we are given values for power. But 15 W is 15 J/s, so we will consider the value for the power to be the energy in 1 second.

FIGURE 11.3 Incandescent and compact fluorescent bulbs.

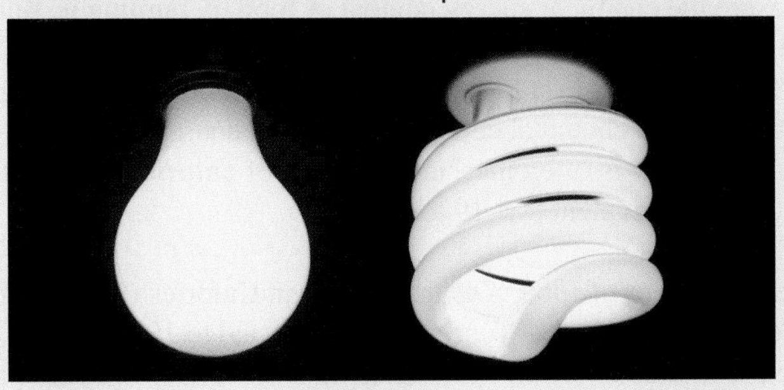

For each of the bulbs, "what you get" is the visible-light output—3.0 J of light every second for each bulb. "What you had to pay" is the electric energy to run the bulb. This is how the bulbs are rated. A bulb labeled "15 W" uses 15 J of electric energy each second. A 75 W bulb uses 75 J each second.

SOLVE The efficiencies of the two bulbs are computed using the energy in 1 second:

$$e(\text{compact fluorescent bulb}) = \frac{3.0\text{ J}}{15\text{ J}} = 0.20 = 20\%$$

$$e(\text{incandescent bulb}) = \frac{3.0\text{ J}}{75\text{ J}} = 0.040 = 4\%$$

ASSESS Both bulbs produce the same visible-light output, but the compact fluorescent bulb does so with a significantly lower energy input, so it is more efficient. Compact fluorescent bulbs are more efficient than incandescent bulbs, but their efficiency is still relatively low—only 20%.

STOP TO THINK 11.1 Crane 1 uses 10 kJ of energy to lift a 50 kg box to the roof of a building. Crane 2 uses 20 kJ to lift a 100 kg box the same distance. Which crane is more efficient?

A. Crane 1.
B. Crane 2.
C. Both cranes have the same efficiency.

11.2 Energy in the Body: Energy Inputs

In this section, we will look at energy in the body, which will give us the opportunity to explore a number of different energy transformations and transfers in a practical context. **FIGURE 11.4** shows the body considered as the system for energy analysis. The chemical energy in food provides the necessary energy input for your body to function. It is this energy that is used for energy transfers with the environment.

FIGURE 11.4 Energy of the body, considered as the system.

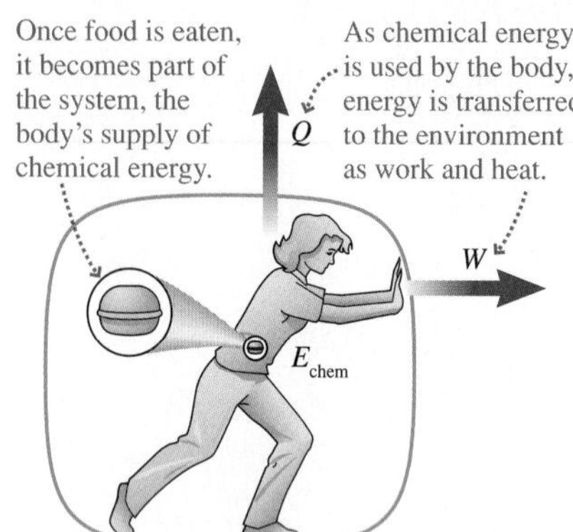

Once food is eaten, it becomes part of the system, the body's supply of chemical energy.
As chemical energy is used by the body, energy is transferred to the environment as work and heat.
Q
W
E_{chem}

Getting Energy from Food

When you walk up a flight of stairs, where does the energy come from to increase your body's potential energy? At some point, the energy came from the food you ate, but what were the intermediate steps? The chemical energy in food is made available to the cells in the body by a two-step process. First, the digestive system breaks down food into simpler molecules such as glucose, a simple sugar, or long chains of glucose molecules called glycogen. These molecules are delivered via the bloodstream to cells in the body, where they are metabolized by combining with oxygen, as in Equation 11.3:

Glucose from the digestion of food combines with oxygen that is breathed in to produce ... ... carbon dioxide, which is exhaled; water, which can be used by the body; and energy.

$$\underset{\text{Glucose}}{C_6H_{12}O_6} + \underset{\text{Oxygen}}{6O_2} \longrightarrow \underset{\text{Carbon dioxide}}{6CO_2} + \underset{\text{Water}}{6H_2O} + \text{energy} \tag{11.3}$$

This metabolism releases energy, much of which is stored in a molecule called ATP, or adenosine triphosphate. Cells in the body use this ATP to do all the work of life: Muscle cells use it to contract, nerve cells use it to produce electrical signals, and so on.

Oxidation reactions like those in Equation 11.3 "burn" the fuel that you obtain by eating. The oxidation of 1 g of glucose (or any other carbohydrate) will release approximately 17 kJ of energy. Table 11.1 compares the energy content of carbohydrates and other foods to other familiar sources of chemical energy.

TABLE 11.1 Energy in fuels

Fuel	Energy in 1 g of fuel (in kJ)
Hydrogen	121
Gasoline	44
Fat (in food)	38
Coal	27
Carbohydrates (in food)	17
Wood chips	15

It is possible to measure the chemical energy content of food by burning it. Burning food may seem quite different from metabolizing it, but if glucose is burned, the chemical formula for the reaction is Equation 11.3; the two reactions are the same. Burning food transforms all of its chemical energy into thermal energy, which can be easily measured. Thermal energy is often measured in units of **calories (cal)** rather than in joules; 1.00 calorie is equivalent to 4.19 joules.

NOTE ► The energy content of food is usually given in Calories (Cal) (with a capital "C"); one Calorie (also called a "food calorie") is equal to 1000 calories or 1 kcal. If a candy bar contains 230 Cal, this means that, if burned, it would produce 230,000 cal (or 964 kJ) of thermal energy. ◄

EXAMPLE 11.2 **Energy in food**

A 12 oz can of soda contains approximately 40 g (or a bit less than 1/4 cup) of sugar, a simple carbohydrate. What is the chemical energy in joules? How many Calories is this?

SOLVE 1 g of sugar contains 17 kJ of energy; 40 g contains

$$40\text{ g} \times \frac{17 \times 10^3\text{ J}}{1\text{ g}} = 68 \times 10^4\text{ J} = 680\text{ kJ}$$

Converting to Calories, we get

$$680\text{ kJ} = 6.8 \times 10^5\text{ J} = (6.8 \times 10^5\text{ J})\frac{1.00\text{ cal}}{4.19\text{ J}}$$

$$= 1.6 \times 10^5\text{ cal} = 160\text{ Cal}$$

ASSESS 160 Calories is a typical value for the energy content of a 12 oz can of soda (check the nutrition label on one to see), so this result seems reasonable.

Counting calories BIO Most foods burn quite well, as this photo of corn chips illustrates. You could set food on fire to measure its energy content, but this isn't really necessary. The chemical energies of the basic components of food (carbohydrates, proteins, fats) have been carefully measured—by burning—in a device called a *calorimeter.* Foods are analyzed to determine their composition, and their chemical energy can then be calculated.

The first item on the nutrition label on packaged foods is Calories—a measure of the chemical energy in the food. (In Europe, where SI units are standard, you will find the energy content listed in kJ.) The energy content of some common foods is given in Table 11.2.

TABLE 11.2 Energy content of foods

Food	Energy content in Cal	Energy content in kJ	Food	Energy content in Cal	Energy content in kJ
Carrot (large)	30	125	Slice of pizza	300	1260
Fried egg	100	420	Frozen burrito	350	1470
Apple (large)	125	525	Apple pie slice	400	1680
Beer (can)	150	630	Fast-food meal:		
BBQ chicken wing	180	750	burger, fries,		
Latte (whole milk)	260	1090	drink (large)	1350	5660

11.3 Energy in the Body: Energy Outputs

Your body uses energy in many ways. Even at rest, your body uses energy for tasks such as building and repairing tissue, digesting food, and keeping warm. The numbers of joules used per second (that is, the power in watts) by different tissues in the resting body are listed in Table 11.3.

Your body uses energy at the rate of approximately 100 W when at rest. This energy, from chemical energy in your body's stores, is ultimately converted entirely to thermal energy, which is then transferred as heat to the environment. A hundred people in a lecture hall add thermal energy to the room at a rate of 10,000 W, and the air conditioning must be designed to take account of this.

TABLE 11.3 Energy use at rest

Organ	Resting power (W) of 68 kg individual
Liver	26
Brain	19
Heart	7
Kidneys	11
Skeletal muscle	18
Remainder of body	19
Total	**100**

Energy Use in Activities

Your body stores very little energy as ATP. As your body uses energy, your cells must continuously metabolize carbohydrates, which requires oxygen, as we saw in Equation 11.3. Physiologists can precisely measure the body's energy use by measuring how much oxygen the body is taking up with a respiratory apparatus, as

FIGURE 11.5 The odd masks worn by the fighters measure respiration—the oxygen used by their bodies.

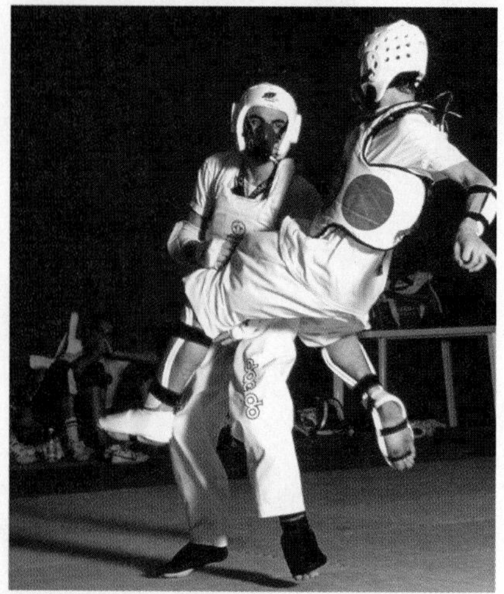

seen in FIGURE 11.5. The device determines the body's *total metabolic energy use*—all of the energy used by the body while performing an activity. This total will include all of the body's basic processes such as breathing plus whatever additional energy is needed to perform the activity. This corresponds to measuring "what you had to pay."

The metabolic energy used in an activity depends on an individual's size, level of fitness, and other variables. But we can make reasonable estimates for the power used in various activities for a typical individual. Some values are given in Table 11.4.

TABLE 11.4 Metabolic power use during activities

Activity	Metabolic power (W) of 68 kg individual
Typing	125
Ballroom dancing	250
Walking at 5 km/h	380
Cycling at 15 km/h	480
Swimming at a fast crawl	800
Running at 15 km/h	1150

Efficiency of the Human Body

Suppose you climb a set of stairs at a constant speed, as in FIGURE 11.6. What is your body's efficiency for this process? In this case, we can do a conservation of energy calculation similar to those in Chapter 10; when you climb stairs, "what you get" is a change in potential energy.

For such problems, we can adapt the work-energy equation to include only those energy changes that we need to consider. If you climb at a constant speed, there's no change in kinetic energy. And given how we define the system, there is no work—there is no external input of energy, as there would be if you took the elevator. As you climb the stairs, your body uses chemical energy to run your muscles for the climb; chemical energy decreases and potential energy increases. As with other cases we've seen, thermal energy increases as well. This is something you know well: If you climb several sets of stairs, you certainly warm up in the process!

For this case of climbing stairs, the work-energy equation reduces to

$$\Delta E_{\text{chem}} + \Delta U_{\text{g}} + \Delta E_{\text{th}} = 0 \qquad (11.4)$$

Thermal energy and gravitational potential energy are increasing, so ΔE_{th} and ΔU_{g} are positive; chemical energy is being used, so ΔE_{chem} is a negative number. We can get a better feeling about what is happening by rewriting Equation 11.4 as

$$|\Delta E_{\text{chem}}| = \Delta U_{\text{g}} + \Delta E_{\text{th}}$$

The *magnitude* of the change in the chemical energy is equal to the sum of the changes in the gravitational potential and thermal energies. Chemical energy from your body is converted into potential energy and thermal energy; in the final position, you are at a greater height and your body is slightly warmer.

FIGURE 11.6 Climbing a set of stairs.

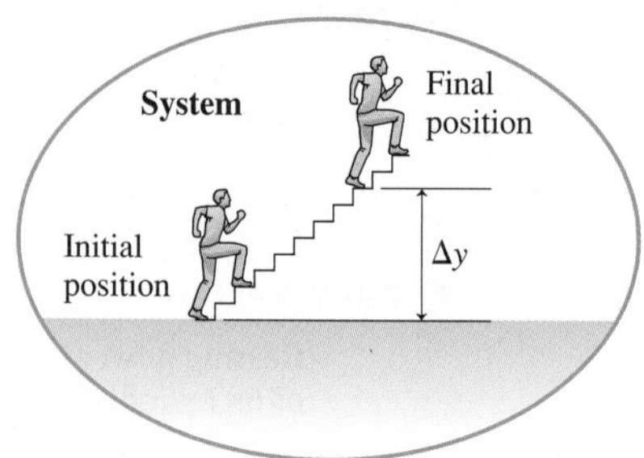

Earlier in the chapter, we noted that the efficiency for stair climbing is about 25%. Let's see where that number comes from.

1. *What you get.* What you get is the change in potential energy: You have raised your body to the top of the stairs. If you climb a flight of stairs of vertical height Δy, we can easily compute the increase in potential energy $\Delta U_{\text{g}} = mg\Delta y$. Assuming a mass of 68 kg and a change in height of 2.7 m (about 9 ft, a reasonable value for a flight of stairs), we compute (to two significant figures)

$$\Delta U_{\text{g}} = (68\text{ kg})(9.8\text{ m/s}^2)(2.7\text{ m}) = 1800\text{ J}$$

2. *What you had to pay.* The cost is the metabolic energy your body used in completing the task. As we've seen, physiologists can measure directly how much energy $|\Delta E_{\text{chem}}|$ your body uses to perform a task. A typical value for climbing a flight of stairs is

$$|E_{\text{chem}}| = 7200\text{ J}$$

Given the definition of efficiency in Equation 11.2, we can compute an efficiency for climbing the stairs:

$$e = \frac{\Delta U_{\text{g}}}{|\Delta E_{\text{chem}}|} = \frac{1800\text{ J}}{7200\text{ J}} = 0.25 = 25\%$$

▶ **High heating costs?** BIO The daily energy use of mammals is much higher than that of reptiles, largely because mammals use energy to maintain a constant body temperature. A 40 kg timber wolf uses approximately 19,000 kJ during the course of a day. A Komodo dragon, a reptilian predator of the same size, uses only 2100 kJ.

For the types of activities we will consider in this chapter, such as running, walking, and cycling, the body's efficiency is typically in the range of 20–30%. **We will generally use a value of 25% for the body's efficiency for our calculations.** Efficiency varies from individual to individual and from activity to activity, but this rough approximation will be sufficient for our purposes in this chapter.

The metabolic power values given in Table 11.4 represent the energy *used by the body* while these activities are being performed. Given that we assume an efficiency of 25%, the body's actual *useful power output* is quite a bit less than this. The table's value for cycling at 15 km/h (a bit less than 10 mph) is 480 W. If we assume that the efficiency for cycling is 25%, the actual power going to forward propulsion will only be 120 W. An elite racing cyclist whizzing along at 35 km/h is using about 300 W for forward propulsion. This is a surprisingly low figure, as noted in FIGURE 11.7.

The energy you use per second while running is proportional to your speed; running twice as fast takes approximately twice as much power. But running twice as fast takes you twice as far in the same time, so the energy you use to run a certain distance doesn't depend on how fast you run! Running a marathon takes approximately the same amount of energy whether you complete it in 2 hours, 3 hours, or 4 hours; it is only the power that varies.

FIGURE 11.7 Amazingly, a racing cyclist moving at 35 km/h uses about the same power as an electric scooter moving at 5 km/h.

NOTE ▶ It is important to remember the distinction between the metabolic energy used to perform a task and the work done in a physics sense; these values can be quite different. Your muscles use power when applying a force, even when there is no motion. Holding a weight above your head involves no external work, but it clearly takes metabolic power to keep the weight in place! ◀

This distinction—between the work done (what you get) and the energy used by the body (what you had to pay)—is important to keep in mind when you do calculations on energy used by the body. We'll consider two different cases:

- For some tasks, such as climbing stairs, we compute the energy change that is the outcome of the task; that is, we compute what you get. If we assume an efficiency of 25%, the energy used by the body (what you had to pay) is 4 times this amount.
- For other tasks, such as cycling, we use data from metabolic studies (such as data in Table 11.4). This is the actual power used by the body to complete a task—in other words, what you had to pay. If we assume an efficiency of 25%, the useful power output (what you get) is 1/4 of this value.

CONCEPTUAL EXAMPLE 11.3 Energy in weightlifting

A weightlifter lifts a 50 kg bar from the floor to a position over his head and back to the floor again 10 times in succession. At the end of this exercise, what energy transformations have taken place?

REASON We will take the system to be the weightlifter plus the bar. The environment does no work on the system, and we will assume that the time is short enough that no heat is transferred from the system to the environment. The bar has returned to its starting position and is not moving, so there has been no change in potential or kinetic energy. The equation for energy conservation is thus

$$\Delta E_{\text{chem}} + \Delta E_{\text{th}} = 0$$

This equation tells us that ΔE_{chem} must be negative. This makes sense because the muscles *use* chemical energy—depleting the body's store—each time the bar is raised or lowered. Ultimately, all of this energy is transformed into thermal energy.

ASSESS Most exercises in the gym—lifting weights, running on a treadmill—involve only the transformation of chemical energy into thermal energy.

EXAMPLE 11.4 **Energy usage for a cyclist**

A cyclist pedals for 20 min at a speed of 15 km/h. How much metabolic energy is required? How much energy is used for forward propulsion?

PREPARE Table 11.4 gives a value of 480 W for the power used in cycling at a speed of 15 km/h. 480 W is the power used by the body; the power going into forward propulsion is much less than this. The cyclist uses energy at this rate for 20 min, or 1200 s.

SOLVE We know the power and the time, so we can compute the energy needed by the body as follows:

$$\Delta E = P\Delta t = (480 \text{ J/s})(1200 \text{ s}) = 580 \text{ kJ}$$

If we assume an efficiency of 25%, only 25%, or 140 kJ, of this energy is used for forward propulsion. The remainder goes into thermal energy.

ASSESS How much energy is 580 kJ? A look at Table 11.2 shows that this is slightly more than the amount of energy available in a large apple, and only 10% of the energy available in a large fast-food meal. If you eat such a meal and plan to "work it off" by cycling, you should plan on cycling at a pretty good clip for a bit over 3 h.

EXAMPLE 11.5 **How many flights?**

How many flights of stairs could you climb on the energy contained in a 12 oz can of soda? Assume that your mass is 68 kg and that a flight of stairs has a vertical height of 2.7 m (9 ft).

PREPARE In Example 11.2 we found that the soda contains 680 kJ of chemical energy. We'll assume that all of this added energy is available to be transformed into the mechanical energy of climbing stairs, at the typical 25% efficiency. We'll also assume you ascend the stairs at constant speed, so your kinetic energy doesn't change. What you get in this case is increased gravitational potential energy, and what you had to pay is the 680 kJ obtained by "burning" the chemical energy of the soda.

SOLVE At 25% efficiency, the amount of chemical energy transformed into increased potential energy is

$$\Delta U_g = (0.25)(680 \times 10^3 \text{ J}) = 1.7 \times 10^5 \text{ J}$$

Because $\Delta U_g = mg\Delta y$, the height gained is

$$\Delta y = \frac{\Delta U_g}{mg} = \frac{1.7 \times 10^5 \text{ J}}{(68 \text{ kg})(9.8 \text{ m/s}^2)} = 255 \text{ m}$$

With each flight of stairs having a height of 2.7 m, the number of flights climbed is

$$\frac{255 \text{ m}}{2.7 \text{ m}} \cong 94 \text{ flights}$$

ASSESS This is almost enough to get to the top of the Empire State Building—all fueled by one can of soda! This is a remarkable result.

Energy Storage

The body gets energy from food; if this energy is not used, it will be stored. A small amount of energy needed for immediate use is stored as ATP. A larger amount of energy is stored as chemical energy of glycogen and glucose in muscle tissue and the liver. A healthy adult might store 400 g of these carbohydrates, which is a little more carbohydrate than is typically consumed in one day.

If the energy input from food continuously exceeds the energy outputs of the body, this energy will be stored in the form of fat under the skin and around the organs. From an energy point of view, gaining weight is simply explained!

EXAMPLE 11.6 **Running out of fuel**

The body stores about 400 g of carbohydrates. Approximately how far could a 68 kg runner travel on this stored energy?

PREPARE Table 11.1 gives a value of 17 kJ per g of carbohydrate. The 400 g of carbohydrates in the body contain an energy of

$$E_{\text{chem}} = (400 \text{ g})(17 \times 10^3 \text{ J/g}) = 6.8 \times 10^6 \text{ J}$$

SOLVE Table 11.4 gives the power used in running at 15 km/hr as 1150 W. The time that the stored chemical energy will last at this rate is

$$\Delta t = \frac{\Delta E_{\text{chem}}}{P} = \frac{6.8 \times 10^6 \text{ J}}{1150 \text{ W}} = 5.91 \times 10^3 \text{ s} = 1.64 \text{ h}$$

And the distance that can be covered during this time at 15 km/h is

$$\Delta x = v\Delta t = (15 \text{ km/h})(1.64 \text{ h}) = 25 \text{ km}$$

to two significant figures.

ASSESS A marathon is longer than this—just over 42 km. Even with "carbo loading" before the event (eating high-carbohydrate meals), many marathon runners "hit the wall" before the end of the race as they reach the point where they have exhausted their store of carbohydrates. The body has other energy stores (in fats, for instance), but the rate that they can be drawn on is much lower.

Energy and Locomotion

When you walk at a constant speed on level ground, your kinetic energy is constant. Your potential energy is also constant. So why does your body need energy to walk? Where does this energy go?

We use energy to walk because of mechanical inefficiencies in our gait. FIGURE 11.8 shows how the speed of your foot typically changes during each stride. The kinetic energy of your leg and foot increases, only to go to zero at the end of the stride. The kinetic energy is mostly transformed into thermal energy in your muscles and in your shoe. This thermal energy is lost; it can't be used for making more strides.

Footwear can be designed to minimize the loss of kinetic energy to thermal energy. A spring in the sole of the shoe can store potential energy, which can be returned to kinetic energy during the next stride. Such a spring will make the collision with the ground more elastic. We saw in Chapter 10 that the tendons in the ankle do store a certain amount of energy during a stride; very stout tendons in the legs of kangaroos store energy even more efficiently. Their peculiar hopping gait is efficient at high speeds.

FIGURE 11.8 Human locomotion analysis.

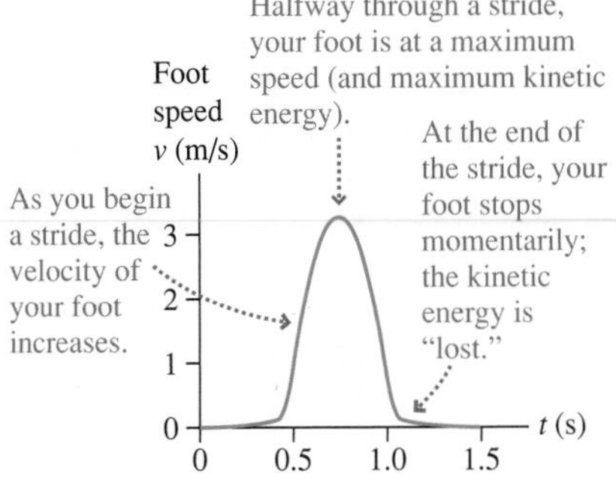

STOP TO THINK 11.2 A runner is moving at a constant speed on level ground. Chemical energy in the runner's body is being transformed into other forms of energy; most of the chemical energy is transformed into

A. Kinetic energy. B. Potential energy. C. Thermal energy.

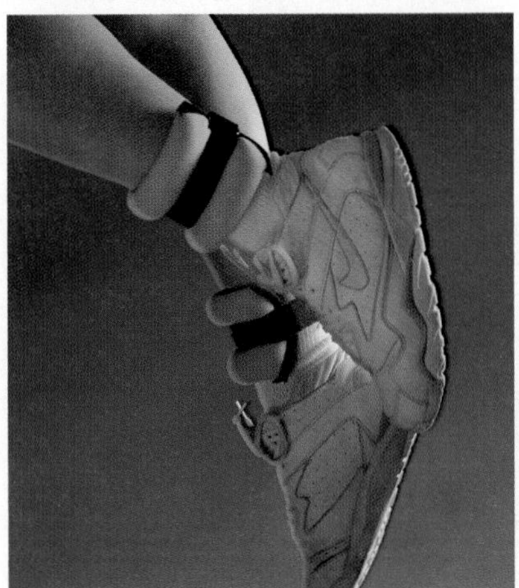

Where do you wear the weights? BIO If you wear a backpack with a mass equal to 1% of your body mass, your energy expenditure for walking will increase by 1%. But if you wear ankle weights with a combined mass of 1% of your body mass, the increase in energy expenditure is 6%, because you must repeatedly accelerate this extra mass. If you want to "burn more fat," wear the weights on your ankles, not on your back! If you are a runner who wants to shave seconds off your time in the mile, you might try lighter shoes.

11.4 Thermal Energy and Temperature

We have frequently spoken of the energy that is transformed into thermal energy as being "lost." Regardless of whether "the system" is a car, a power plant, or your body, this thermal energy is simply exhausted into the environment. But why isn't thermal energy in your body and other systems converted to other forms of energy and used for practical purposes? To continue our study of how energy is used, we need to understand a bit more about one particular kind of energy—thermal energy.

What do you mean when you say something is "hot"? Do you mean that it has a high temperature? Or do you mean that it has a lot of thermal energy? Are both of these definitions the same? Let's give some thought to the definitions of temperature and thermal energy, and the relationship between them.

An Atomic View of Thermal Energy and Temperature

Consider the simplest possible atomic system, the **ideal gas** of atoms seen in FIGURE 11.9. In Chapter 10, we defined thermal energy to be the energy associated with the motion of the atoms and molecules that make up an object. Because there are no molecular bonds, the atoms in an ideal gas have only the kinetic energy of their motion. Thus, **the thermal energy of an ideal gas is equal to the total kinetic energy of the moving atoms in the gas.**

If you take a container of an ideal gas and place it over a flame, as in FIGURE 11.10 on the next page, energy will be transferred from the hot flame to the cooler gas. This is a new type of energy transfer, one we'll call **heat**. We will define the nature of heat later in the chapter; for now, we'll focus on the changes in the gas when it is heated.

FIGURE 11.9 Motion of atoms in an ideal gas.

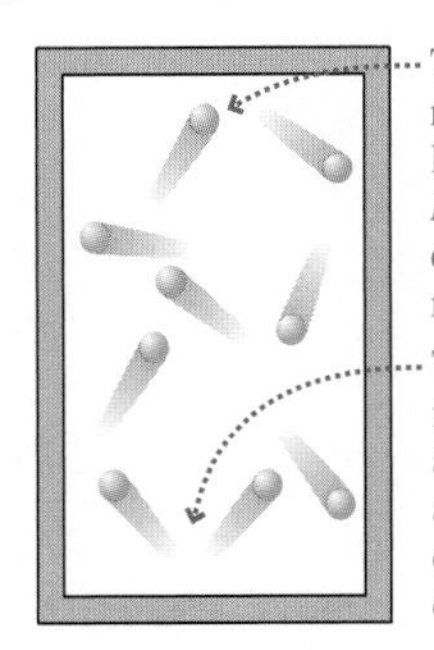

The gas is made up of a large number N of atoms, each moving randomly.
The only interactions among the atoms are elastic collisions.

FIGURE 11.10 Heating an ideal gas.

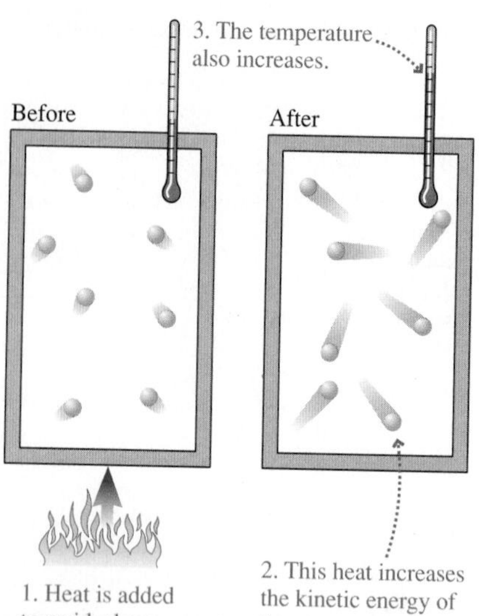
3. The temperature also increases.
Before
After
1. Heat is added to an ideal gas.
2. This heat increases the kinetic energy of the gas atoms.

As Figure 11.10 shows, heating the gas causes the atoms to move faster and thus increases the thermal energy of the gas. Heating the gas also increases its temperature. The observation that both increase suggests that temperature must be related to the kinetic energy of the gas atoms, but how? We can get an important hint from the following fact: The temperature of a system does not depend on the size of the system. If you mix together two glasses of water, each at a temperature of 20°C, you will have a larger volume of water at the same temperature of 20°C. The combined volume has more atoms, and therefore more *total* thermal energy, but each atom is moving about just as it was before, and so the *average* kinetic energy per atom is unchanged. It is this average kinetic energy of the atoms that is related to the temperature. **The temperature of an ideal gas is a measure of the *average* kinetic energy of the atoms that make up the gas.**

We'll formalize and extend these ideas in Chapter 12, but, for now, we'll be able to say quite a bit about temperature and thermal energy by referring to the properties of this simple model.

Temperature Scales

Temperature, in an ideal gas, is related to the average kinetic energy of atoms. We can make a similar statement about other materials, be they solid, liquid, or gas: The higher the temperature, the faster the atoms move. But how do you measure temperature? The common glass-tube thermometer works by allowing a small volume of mercury or alcohol to expand or contract when placed in contact with a "hot" or "cold" object. Other thermometers work in different ways, but all—at a microscopic level—depend on the speed of the object's atoms as they collide with the atoms in the thermometer. That is, all thermometers are sampling the average kinetic energy of the atoms in the object.

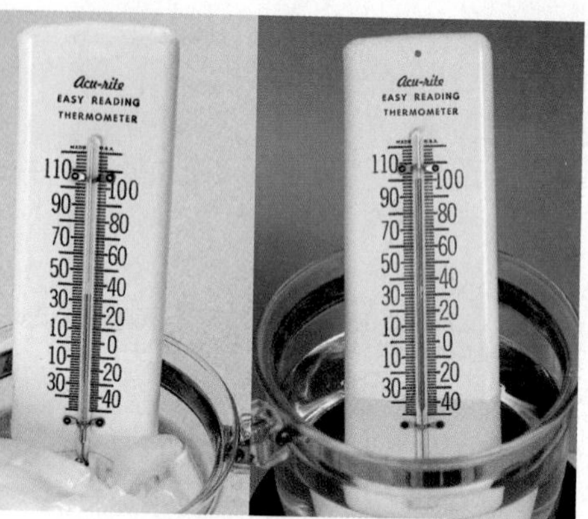
Acu-rite
EASY READING THERMOMETER
Acu-rite
EASY READING THERMOMETER

Thermal expansion of the liquid in the thermometer pushes it higher when immersed in hot water than in ice water.

A thermometer needs a temperature scale to be a useful measuring device. The scale used in scientific work (and in almost every country in the world) is the *Celsius scale*. As you likely know, the Celsius scale is defined so that the freezing point of water is 0°C and the boiling point is 100°C. The units of the Celsius temperature scale are "degrees Celsius," which we abbreviate as °C. The Fahrenheit scale, still widely used in the United States, is related to the Celsius scale by

$$T(°\text{C}) = \frac{5}{9}(T(°\text{F}) - 32°) \qquad T(°\text{F}) = \frac{9}{5}T(°\text{C}) + 32° \qquad (11.5)$$

Both the Celsius and the Fahrenheit scales have a zero point that is arbitrary—simply an agreed-upon convention—and both allow negative temperatures. If, instead, we use the average kinetic energy of ideal-gas atoms as our basis for the definition of temperature, our temperature scale will have a natural zero—the point at which kinetic energy is zero. Kinetic energy is always positive, so the zero on our temperature scale will be an **absolute zero;** no temperature below this is possible.

This is how zero is defined on the temperature scale called the *Kelvin scale:* **Zero degrees is the point at which the kinetic energy of atoms is zero.** All temperatures on the Kelvin scale are positive, so it is often called an *absolute temperature scale.* The units of the Kelvin temperature scale are "kelvin" (not degrees kelvin!), abbreviated K.

The spacing between divisions on the Kelvin scale is the same as that of the Celsius scale; the only difference is the position of the zero point. Absolute zero—the temperature at which atoms would cease moving—is −273°C. The conversion between Celsius and Kelvin temperatures is therefore quite straightforward:

$$T(\text{K}) = T(°\text{C}) + 273 \qquad T(°\text{C}) = T(\text{K}) - 273 \qquad (11.6)$$

The size of a degree on the Kelvin and Celsius scales is the same. This means that a temperature *difference* is the same on both scales:

$$\Delta T(\text{K}) = \Delta T(°\text{C})$$

On the Kelvin scale, the freezing point of water at 0°C is $T = 0 + 273 = 273$ K. A 30°C warm summer day is $T = 303$ K on the Kelvin scale. **FIGURE 11.11** gives a side-by-side comparison of these scales.

NOTE ▶ From now on, we will use the symbol T for temperature in kelvin. We will denote other scales by showing the units in parentheses. In the equations in this chapter and the rest of the text, **T must be interpreted as a temperature in kelvin.** ◀

FIGURE 11.11 Celsius and Kelvin temperature scales.

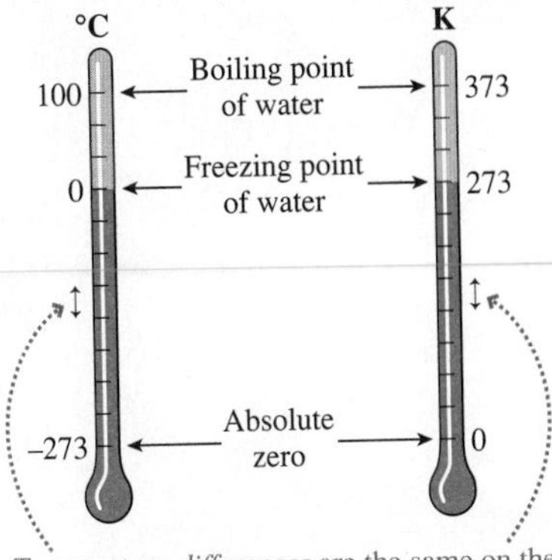

Temperature *differences* are the same on the Celsius and Kelvin scales. The temperature difference between the freezing point and boiling point of water is 100°C or 100 K.

EXAMPLE 11.7 **Temperature scales**

The coldest temperature ever measured on earth was −129°F, in Antarctica. What is this in °C and K?

SOLVE We use Equation 11.5 to convert the temperature to the Celsius scale:

$$T(°\text{C}) = \frac{5}{9}(-129° - 32°) = -89°\text{C}$$

We can then use Equation 11.6 to convert this to kelvin:

$$T = -89 + 273 = 184 \text{ K}$$

ASSESS This is cold, but quite a bit warmer than the coldest temperatures achieved in the laboratory.

Relating Temperature and Thermal Energy

We noted above that the temperature of an ideal gas is a measure of the average kinetic energy of the atoms. It can be shown that temperature on the Kelvin scale is related to the average kinetic energy per atom by

$$T = \frac{2}{3}\frac{K_{\text{avg}}}{k_{\text{B}}} \tag{11.7}$$

where k_{B} is a constant known as **Boltzmann's constant.** Its value is

$$k_{\text{B}} = 1.38 \times 10^{-23} \text{ J/K}$$

We can rearrange Equation 11.7 to give the average kinetic energy in terms of the temperature:

$$K_{\text{avg}} = \frac{3}{2}k_{\text{B}}T \tag{11.8}$$

The thermal energy of an ideal gas consisting of N atoms is the sum of the kinetic energies of the individual atoms:

$$E_{\text{th}} = NK_{\text{avg}} = \frac{3}{2}Nk_{\text{B}}T \tag{11.9}$$

Thermal energy of an ideal gas of N atoms

E_{th} vs. N: PROPORTIONAL p.37; E_{th} vs. T: PROPORTIONAL p.37

For an ideal gas, **thermal energy is directly proportional to temperature.** Consequently, a change in the thermal energy of an ideal gas is proportional to a change in temperature:

$$\Delta E_{\text{th}} = \frac{3}{2}Nk_{\text{B}}\Delta T \tag{11.10}$$

Optical molasses It isn't possible to reach absolute zero, where the atoms would be still, but it is possible to get quite close by slowing the atoms down directly. These crossed laser beams produce what is known as "optical molasses." As we will see in Chapter 28, light is made of photons, which carry energy and momentum. Interactions of the atoms of a diffuse gas with the photons cause the atoms to slow down. In this manner atoms can be slowed to speeds that correspond to a temperature as cold as 5×10^{-10} K!

ActivPhysics ONLINE 8.1, 8.2, 8.3

Is it cold in space? The Space Shuttle orbits in the upper thermosphere, about 300 km above the surface of the earth. There is still a trace of atmosphere left at this altitude, and it has quite a high temperature—over 1000°C. Although the average speed of the air molecules here is high, there are so few air molecules present that the thermal energy is extremely low.

This relationship between a change in temperature and a change in thermal energy is for an ideal gas, but solids, liquids, and other gases all follow similar rules, as we will see in the next chapter. For now, we will simply note two important conclusions that apply to any substance:

1. The thermal energy of a substance is proportional to the number of atoms. A gas with more atoms has more thermal energy than a gas at the same temperature with fewer atoms.
2. A change in temperature causes a proportional change in the substance's thermal energy. A larger temperature change causes a larger change in thermal energy.

EXAMPLE 11.8 Energy needed to warm up a room

A large bedroom contains about 1×10^{27} molecules of air. (In the next chapter, we'll see how to calculate this.) Estimate the energy required to raise the temperature of the air in the room by 5°C.

PREPARE We'll model the air as an ideal gas. Equation 11.10 relates the change in thermal energy of an ideal gas to a change in temperature. The actual temperature of the gas doesn't matter—only the change. The temperature increase is given as 5°C, implying a change in the absolute temperature by the same amount: $\Delta T = 5\text{ K}$.

SOLVE We can use Equation 11.10 to calculate the amount by which the room's thermal energy must be increased:

$$\Delta E_{\text{th}} = \frac{3}{2}Nk_{\text{B}}\Delta T = \frac{3}{2}(1 \times 10^{27})(1.38 \times 10^{-23}\text{ J/K})(5\text{ K}) = 1 \times 10^{5}\text{ J} = 100\text{ kJ}$$

This is the energy we would have to supply—probably in the form of heat from a furnace—to raise the temperature.

ASSESS 100 kJ isn't that much energy. Table 11.2 shows it to be less than the food energy in a carrot! This seems reasonable because you know that your furnace can quickly warm up the air in a room. Heating up the walls and furnishings is another story.

STOP TO THINK 11.3 Two samples of ideal gas, sample 1 and sample 2, have the same thermal energy. Sample 1 has twice as many atoms as sample 2. What can we say about the temperatures of the two samples?

A. $T_1 > T_2$ B. $T_1 = T_2$ C. $T_1 < T_2$

11.5 Heat and the First Law of Thermodynamics

In Chapter 10, we saw that a system could exchange energy with the environment through two different means: work and heat. Work was treated in some detail in Chapter 10; now it is time to look at the transfer of energy by heat. This will begin our study of a topic called *thermodynamics,* the study of thermal energy and heat and their relationships to other forms of energy and energy transfer.

What Is Heat?

Heat is a more elusive concept than work. We use the word "heat" very loosely in the English language, often as synonymous with "hot." We might say, on a very hot day, "This heat is oppressive." If your apartment is cold, you may say, "Turn up the heat." It's time to develop more precise language to discuss these concepts.

Suppose you put a pan of cold water on the stove. If you light the burner so that there is a hot flame under the pan, as in FIGURE 11.12a, the temperature of the water increases. We've added energy to the system—the pan of water—as heat. You know, from everyday experience, that heat always flows "downhill" in the sense that energy is transferred from a hotter object to a colder object. If there is no temperature difference, no energy is transferred. We'll explore this idea further, but, for now, we can use this familiar observation as a definition of heat: **Heat is energy transferred between two objects at different temperatures.**

FIGURE 11.12 Two different means of raising the temperature of a pan of water.

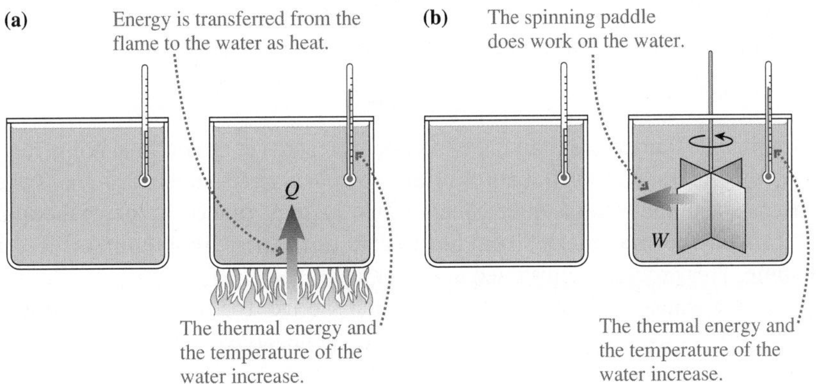

(a)
Energy is transferred from the flame to the water as heat.
Q
The thermal energy and the temperature of the water increase.
(b)
The spinning paddle does work on the water.
W
The thermal energy and the temperature of the water increase.

FIGURE 11.12b shows the same system, but this time work is being done on the system by means of a rapidly spinning paddle. Careful experiments by the British physicist James Joule in the 1840s found that doing work on the system also increases the temperature. In fact, the temperature increase and the final state of the system are exactly the same regardless of whether energy is added as heat or an equal amount of energy is added by doing work. This implies that heat and work are in some sense equivalent: **Heat and work are simply two different ways of transferring energy to or from a system.**

An Atomic Model of Heat

Let's consider an atomic model to explain why thermal energy is transferred from higher temperatures to lower. FIGURE 11.13 shows a rigid, insulated container that is divided into two sections by a very thin membrane. Each side is filled with a gas of the same kind of atoms. The left side, which we'll call system 1, is at an initial temperature T_{1i}. System 2 on the right is at an initial temperature T_{2i}. We imagine the membrane to be so thin that atoms can collide at the boundary as if the membrane were not there, yet it is a barrier that prevents atoms from moving from one side to the other.

Suppose that system 1 is initially at a higher temperature: $T_{1i} > T_{2i}$. This means that the atoms in system 1 have a higher average kinetic energy. Figure 11.13 shows a fast atom and a slow atom approaching the barrier from opposite sides. They undergo a perfectly elastic collision at the barrier. Although no net energy is lost in a perfectly elastic collision, the faster atom loses energy while the slower one gains energy. In other words, there is an energy *transfer* from the faster atom's side to the slower atom's side.

Because the atoms in system 1 are, on average, more energetic than the atoms in system 2, *on average* the collisions transfer energy from system 1 to system 2. This is not true for every collision. Sometimes a fast atom in system 2 collides with a slow atom in system 1, transferring energy from 2 to 1. But the net energy transfer, from all collisions, is from the warmer system 1 to the cooler system 2. This transfer of energy is heat; **thermal energy is transferred from the faster moving atoms on the warmer side to the slower moving atoms on the cooler side.**

TRY IT YOURSELF

Energetic cooking You can do a modern version of Joule's experiment in the kitchen. Next time you mix food in a blender, notice that it actually warms the food. As the blades rotate, friction increases the thermal energy of the food in the blender. A blender uses about as much power as a microwave, and most of this energy ends up as thermal energy in the food, so the temperature rises as the food is chopped and blended.

FIGURE 11.13 Collisions at a barrier transfer energy from faster molecules to slower molecules.

Insulation prevents heat from entering or leaving the container.

A thin barrier prevents atoms from moving from system 1 to 2 but still allows them to collide.

System 1 T_{1i} Fast

System 2 T_{2i} Slow

Elastic collision

Loses energy

Gains energy

System 1 is initially at a higher temperature, and so has atoms with a higher average speed.

Collisions transfer energy to the atoms in system 2.

FIGURE 11.14 Two systems in thermal contact exchange thermal energy.

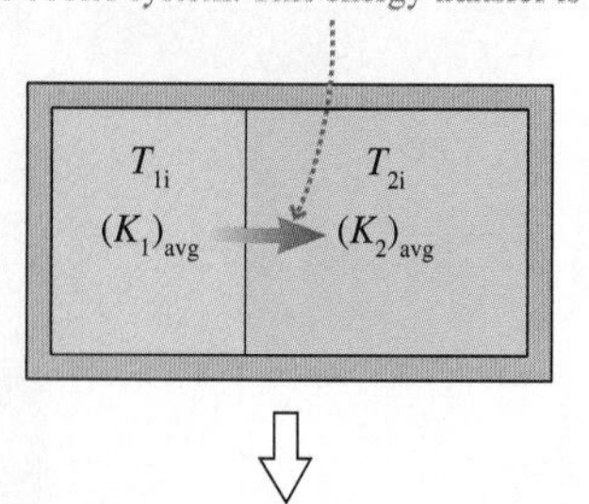
Collisions transfer energy from the warmer system to the cooler system. This energy transfer is heat.
T_{1i}
T_{2i}
$(K_1)_{avg}$
$(K_2)_{avg}$

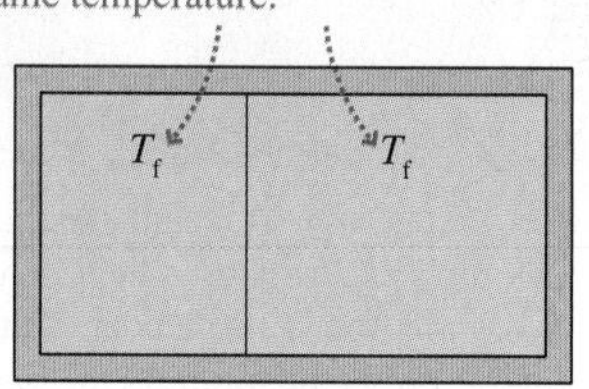
Thermal equilibrium occurs when the systems have the same average kinetic energy and thus the same temperature.
T_f
T_f

FIGURE 11.15 The sign of Q.

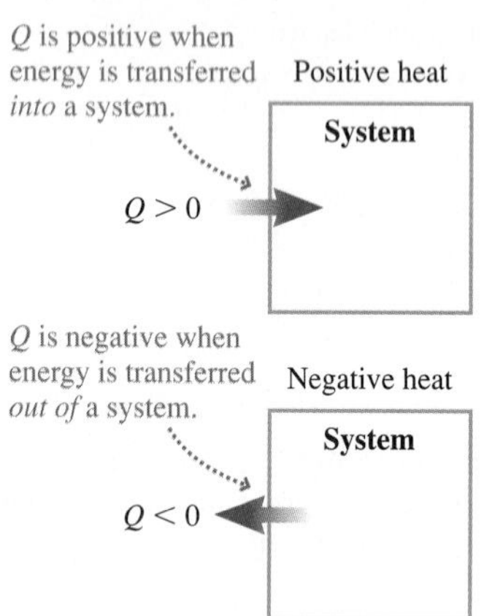
Q is positive when energy is transferred *into* a system.
Positive heat
System
$Q > 0$
Q is negative when energy is transferred *out of* a system.
Negative heat
System
$Q < 0$

This transfer will continue until a stable situation is reached. This is a situation we call **thermal equilibrium.** How do the systems "know" when they've reached thermal equilibrium? Energy transfer continues until the atoms on both sides of the barrier have the *same average kinetic energy.* Once the average kinetic energies are the same, individual collisions will still transfer energy from one side to the other. But since both sides have atoms with the same average kinetic energies, the amount of energy transferred from 1 to 2 will equal that transferred from 2 to 1. Once the average kinetic energies are the same, there will be no more net energy transfer.

As we've seen, the average kinetic energy of the atoms in a system is directly proportional to the system's temperature. If two systems exchange energy until their atoms have the same average kinetic energy, we can say that

$$T_{1f} = T_{2f} = T_f$$

That is, heat is transferred until the two systems reach a common final temperature; we call this final state **thermal equilibrium.** We considered a rather artificial system in this case, but the result is quite general: **Two systems placed in thermal contact will transfer thermal energy from hot to cold until their final temperatures are the same.** This process is illustrated in FIGURE 11.14.

Heat is a transfer of energy. The sign that we use for transfers is defined in FIGURE 11.15. In the process of Figure 11.14, Q_1 is negative because system 1 loses energy; Q_2 is positive because system 2 gains energy. No energy escapes from the container, so all of the energy that was lost by system 1 was gained by system 2. We can write that as:

$$Q_2 = -Q_1$$

The heat energy lost by one system is gained by the other.

The First Law of Thermodynamics

We need to broaden our work-energy equation, Equation 11.1, to include energy transfers in the form of heat. The sign conventions for heat, as shown in Figure 11.15, have the same form as those for work—a positive value means a transfer into the system, a negative value means a transfer out of the system. Thus, when we include heat Q in the work-energy equation, it appears on the right side of the equation along with work W:

$$\Delta K + \Delta U + \Delta E_{th} + \Delta E_{chem} + \cdots = W + Q \qquad (11.11)$$

This equation is our most general statement to date about energy and the conservation of energy; it includes all of the energy transfers and transformations we have discussed.

In Chapter 10, we focused on systems where the potential and kinetic energies could change, such as a sled moving down a hill. Earlier in this chapter, we looked at the body, where the chemical energy changes. Now let's consider systems in which only the thermal energy changes. That is, we will consider systems that aren't moving, that aren't changing chemically, but whose temperatures can change. Such systems are the province of what is called **thermodynamics.** The question of how to keep your house cool in the summer is a question of thermodynamics. If energy is transferred into your house, the thermal energy increases and the temperature rises. To reduce the temperature, you must transfer energy out of the house. This is the purpose of an air conditioner, as we'll see in a later section.

If we consider cases in which only thermal energy changes, Equation 11.11 can be simplified. This simpler version is a statement of conservation of energy for systems in which only the thermal energy changes; we call it the **first law of thermodynamics:**

First law of thermodynamics For systems in which only the thermal energy changes, the change in thermal energy is equal to the energy transferred into or out of the system as work W, heat Q, or both:

$$\Delta E_{th} = W + Q \qquad (11.12)$$

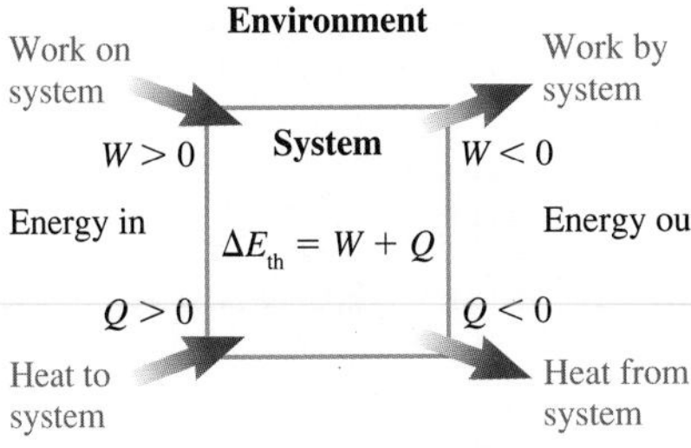

FIGURE 11.16 Energy transfers in a thermodynamic system.

Only work and heat, two ways of transferring energy between a system and the environment, cause the system's energy to change. In thermodynamic systems, the only energy change will be a change in thermal energy. Whether this energy increases or decreases depends on the signs of W and Q, as we've seen. The possible energy transfers between a system and the environment are illustrated in FIGURE 11.16.

CONCEPTUAL EXAMPLE 11.9 Compressing a gas

Suppose a gas is in an insulated container, so that no heat energy can escape. If a piston is used to compress the gas, what happens to the temperature of the gas?

REASON The piston applies a force to the gas, and there is a displacement. This means that work is done on the gas by the piston ($W > 0$). No thermal energy can be exchanged with the environment, meaning $Q = 0$. Since energy is transferred into the system, the thermal energy of the gas must increase. This means that the temperature must increase as well.

ASSESS This result makes sense in terms of everyday observations you may have made. When you use a bike pump to inflate a tire, the pump and the tire get warm. This temperature increase is largely due to the warming of the air by the compression.

EXAMPLE 11.10 Work and heat in an ideal gas

Suppose an uninsulated container holds 5.0×10^{22} molecules of an ideal gas. 50 J of work are done on the gas by a piston that compresses it. The temperature of the gas increases by 30°C during this process. How much heat is transferred to or from the environment?

PREPARE This is an energy conservation problem. We can first use the temperature increase to determine the change in thermal energy of the gas; then use this to determine how much heat energy goes in or out.

SOLVE For an ideal gas, the change in thermal energy is given by Equation 11.10:

$$\Delta E_{th} = \frac{3}{2}Nk_B\Delta T$$
$$= \left(\frac{3}{2}\right)(5.0 \times 10^{22})(1.38 \times 10^{-23}\ \text{J/K})(30\ \text{K}) = 31\ \text{J}$$

The first law of thermodynamics, Equation 11.12, tells us that the change in thermal energy ΔE_{th} of the gas is the sum of W and Q, the total energy added to the system. W is known, and we have just calculated ΔE_{th}. Combining, we get

$$Q = \Delta E_{th} - W = 31\ \text{J} - 50\ \text{J} = -19\ \text{J}$$

ASSESS Q is negative, meaning that energy is transferred outward from the hot gas to the cooler environment in this process.

Energy-Transfer Diagrams

Suppose you drop a hot rock into the ocean. Heat is transferred from the rock to the ocean until the rock and ocean are the same temperature. Although the ocean warms up ever so slightly, ΔT_{ocean} is so small as to be completely insignificant.

An **energy reservoir** is an object or a part of the environment so large that, like the ocean, its temperature does not noticeably change when heat is transferred between the system and the reservoir. A reservoir at a higher temperature than the system is called a *hot reservoir.* A vigorously burning flame is a hot reservoir for small objects placed in the flame. A reservoir at a lower temperature than the system is called a *cold reservoir.* The ocean is a cold reservoir for the hot rock. We will use T_H and T_C to designate the temperatures of the hot and cold reservoirs.

Heat energy is transferred between a system and a reservoir if they have different temperatures. We will define

$$Q_H = \text{amount of heat transferred to or from a hot reservoir}$$

$$Q_C = \text{amount of heat transferred to or from a cold reservoir}$$

By definition, Q_H and Q_C are *positive* quantities.

FIGURE 11.17a shows a heavy copper bar placed between a hot reservoir (at temperature T_H) and a cold reservoir (at temperature T_C). Heat Q_H is transferred from the hot

FIGURE 11.17 Energy-transfer diagrams.

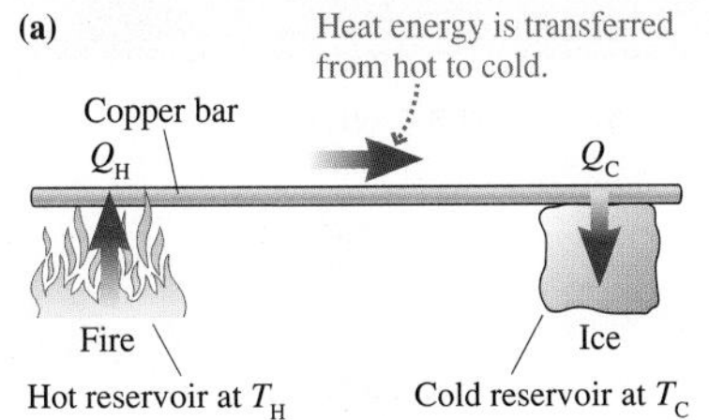

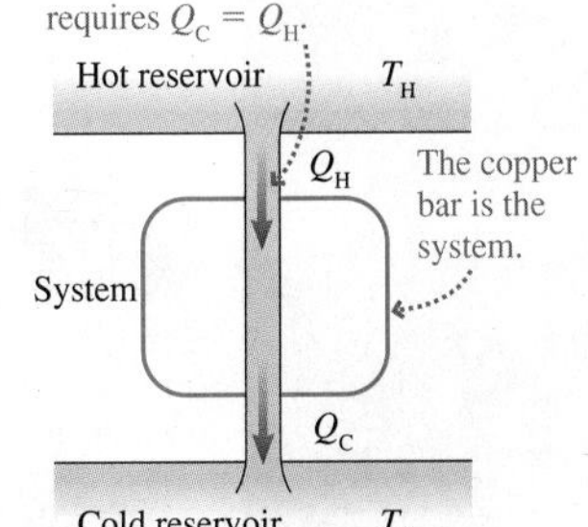

FIGURE 11.18 An impossible energy transfer.

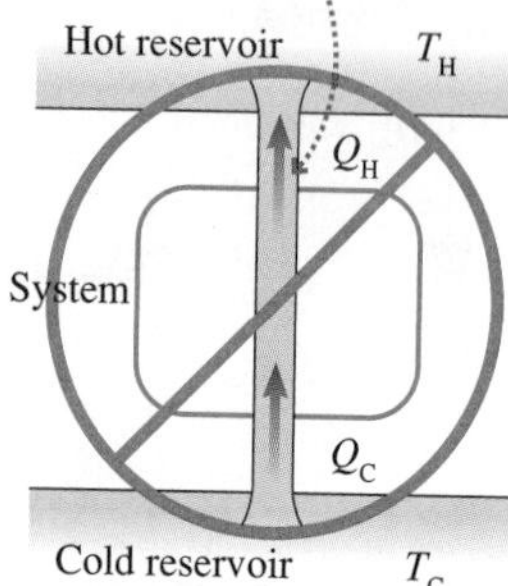
Heat is never spontaneously transferred from a colder object to a hotter object.
Hot reservoir
T_H
Q_H
System
Q_C
Cold reservoir
T_C

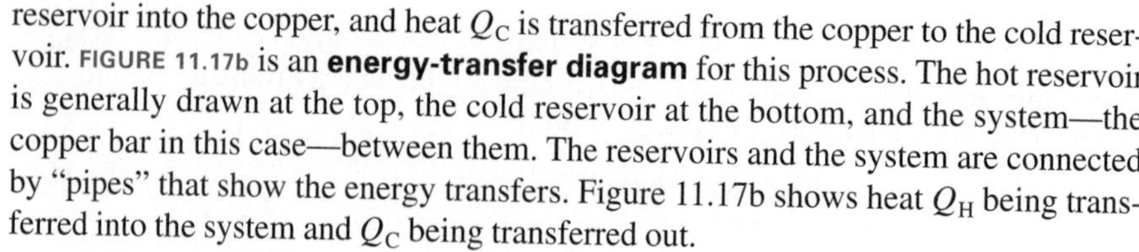

reservoir into the copper, and heat Q_C is transferred from the copper to the cold reservoir. FIGURE 11.17b is an **energy-transfer diagram** for this process. The hot reservoir is generally drawn at the top, the cold reservoir at the bottom, and the system—the copper bar in this case—between them. The reservoirs and the system are connected by "pipes" that show the energy transfers. Figure 11.17b shows heat Q_H being transferred into the system and Q_C being transferred out.

FIGURE 11.18 illustrates an important fact about heat transfers that we have discussed: Spontaneous transfers go in one direction only, from hot to cold. This is an important result that has significant practical implications.

CONCEPTUAL EXAMPLE 11.11 Energy transfers and the body

Why—in physics terms—is it more taxing on the body to exercise in very hot weather?

REASON Your body continuously converts chemical energy to thermal energy, as we have seen. In order to maintain a constant body temperature, your body must continuously transfer heat to the environment. This is a simple matter in cool weather when heat is spontaneously transferred to the environment, but when the air temperature is higher than your body temperature, your body cannot cool itself this way and must use other mechanisms to transfer this energy, such as perspiring. These mechanisms require additional energy expenditure.

ASSESS Strenuous exercise in hot weather can easily lead to a rise in body temperature if the body cannot exhaust heat quickly enough.

Falling water turns a waterwheel.

STOP TO THINK 11.4 You have driven your car for a while and now turn off the engine. Your car's radiator is at a higher temperature than the air around it. Considering the radiator as the system, we can say that

A. $Q > 0$ B. $Q = 0$ C. $Q < 0$

11.6 Heat Engines

In the early stages of the industrial revolution, most of the energy needed to run mills and factories came from water power. Water in a high reservoir will naturally flow downhill. A waterwheel can be used to harness this natural flow of water to produce some useful energy because some of the potential energy lost by the water as it flows downhill can be converted into other forms.

It is possible to do something similar with heat. Thermal energy is naturally transferred from a hot reservoir to a cold reservoir; it is possible to take some of this energy as it is transferred and convert it to other forms. This is the job of a device known as a **heat engine.**

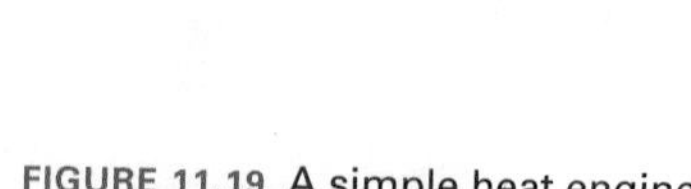

FIGURE 11.19 A simple heat engine.

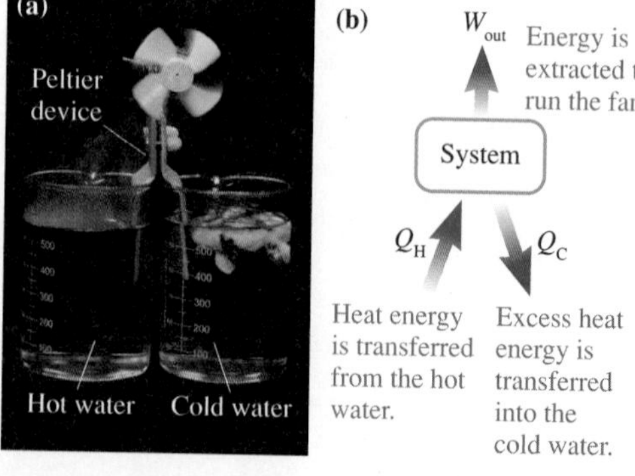
(a)
Peltier device
Hot water
Cold water
(b)
W_{out}
Energy is extracted to run the fan.
System
Q_H
Q_C
Heat energy is transferred from the hot water.
Excess heat energy is transferred into the cold water.

A simple example of a heat engine is shown in FIGURE 11.19a. Here a *Peltier device*—a device that produces a voltage if there is a temperature difference between its two sides—is sandwiched between two aluminum vanes that sit in cups of water. The cup of water on the left is hot; the cup on the right is cold. The temperature difference produces a voltage that drives the fan.

FIGURE 11.19b is a diagram that illustrates the operation of this device in thermodynamic terms. The Peltier device is the system. Heat energy is transferred through the device from the hot water to the cold water. As the transfer occurs, some of this energy is transformed into electric energy to run the fan.

The energy-transfer diagram of FIGURE 11.20 on the next page illustrates the basic physics of a heat engine. It takes in energy as heat from the hot reservoir, turns some of it into useful work, and exhausts the balance as waste heat in the cold reservoir. Any heat engine has exactly the same schematic.

Efficiency of a Heat Engine

We assume, for a heat engine, that the engine's thermal energy doesn't change. This means that there is no net energy transfer into or out of the heat engine. Because energy is conserved, we can say that the useful work extracted is equal to the difference between the heat energy transferred from the hot reservoir and the heat exhausted into the cold reservoir:

$$W_{out} = Q_H - Q_C$$

The energy input to the engine is Q_H and the energy output is W_{out}.

NOTE ▶ We earlier defined Q_H and Q_C to be positive quantities. We also define the energy output of a heat engine W_{out} to be a positive quantity. For heat engines, the directions of the transfers will always be clear, and we will take all of these basic quantities to be positive. ◀

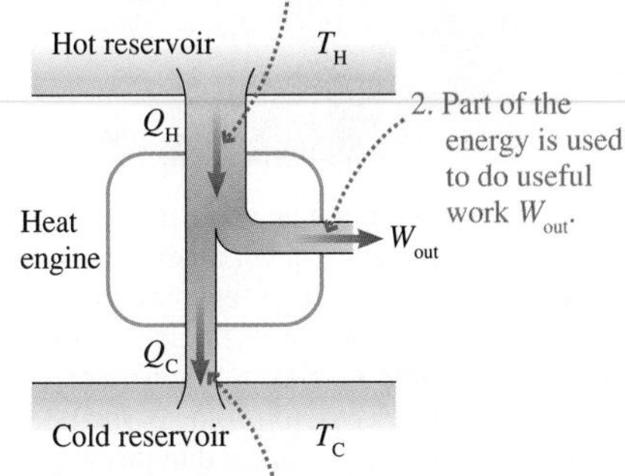
1. Heat energy Q_H is transferred from the hot reservoir to the system.
Hot reservoir
T_H
Q_H
2. Part of the energy is used to do useful work W_{out}.
Heat engine
W_{out}
Q_C
Cold reservoir
T_C
3. The remaining energy $Q_C = Q_H - W_{out}$ is exhausted to the cold reservoir as waste heat.

FIGURE 11.20 The operation of a heat engine.

We can use the definition of efficiency, from earlier in the chapter, to compute the heat engine's efficiency:

$$e = \frac{\text{what you get}}{\text{what you had to pay}} = \frac{W_{out}}{Q_H} = \frac{Q_H - Q_C}{Q_H} \tag{11.13}$$

Q_H is what you had to pay because this is the energy of the fuel burned—and, quite literally, paid for—to provide the high temperature of the hot reservoir. The heat energy that is not converted to work ends up in the cold reservoir as waste heat.

Why should we waste energy this way? Why don't we make a heat engine like the one shown in **FIGURE 11.21** that converts 100% of the heat into useful work? The surprising answer is that we can't. **No heat engine can operate without exhausting some fraction of the heat into a cold reservoir**. This isn't a limitation on our engineering abilities. As we'll see, it's a fundamental law of nature.

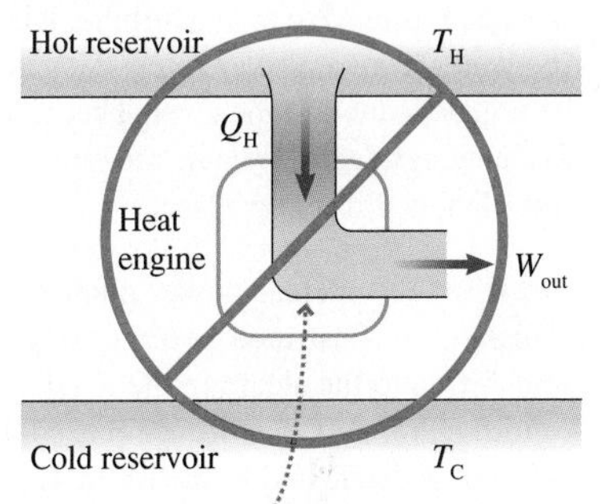
Hot reservoir
T_H
Q_H
Heat engine
W_{out}
Cold reservoir
T_C
This is impossible! No heat engine can convert 100% of heat into useful work.

FIGURE 11.21 A perfect (and impossible!) heat engine.

The maximum possible efficiency of a heat engine is fixed by the *second law of thermodynamics,* which we will explore in detail in Section 11.8. We will not do a detailed derivation, but simply note that the second law gives the maximum efficiency of a heat engine as

$$e_{max} = 1 - \frac{T_C}{T_H} \tag{11.14}$$

Theoretical maximum efficiency of a heat engine

The maximum efficiency of any heat engine is therefore fixed by the ratio of the temperatures of the hot and cold reservoirs. It is possible to increase the efficiency of a heat engine by increasing the temperature of the hot reservoir or decreasing the temperature of the cold reservoir. The efficiency of Equation 11.14 is also called the *Carnot efficiency,* after a particular heat engine that achieves this maximum possible efficiency. The actual efficiency of real heat engines is usually much less than the theoretical maximum.

NOTE ▶ The temperatures in Equation 11.14 *must* be in K, not °C. ◀

Because $e_{max} < 1$, the work done is always less than the heat input ($W_{out} < Q_H$). Consequently, there *must* be heat Q_C exhausted to the cold reservoir. That is why the heat engine of Figure 11.21 is impossible. Think of a heat engine as analogous to a water wheel: The engine "siphons off" some of the energy that is spontaneously flowing from hot to cold, but it can't completely shut off the flow.

The fact that heat engine efficiency is limited—and that heat must be exhausted into a cold reservoir—is of tremendous practical importance. Most of the energy that you use daily comes from the conversion of chemical energy into thermal energy and the subsequent conversion of that energy into other forms. Let's look at some common examples of heat engines.

Heat engines

Most of the electricity that you use was generated by heat engines. Coal or other fossil fuels are burned to produce high-temperature, high-pressure steam. The steam does work by spinning a turbine attached to a generator, which produces electricity. Some of the energy of the steam is extracted in this way, but more than half simply flows "downhill" and is deposited in a cold reservoir, often a lake or a river.

Your car gets the energy it needs to run from the chemical energy in gasoline. The gasoline is burned; the resulting hot gases are the hot reservoir. Some of the thermal energy is converted into the kinetic energy of the moving vehicle, but more than 90% is lost as heat to the surrounding air via the radiator and the exhaust, as shown in this thermogram.

There are many small, simple heat engines that are part of things you use daily. This fan, which can be put on top of a wood stove, uses the thermal energy of the stove to provide power to drive air around the room. Where are the hot and cold reservoirs in this device?

EXAMPLE 11.12 The efficiency of a nuclear power plant

Energy from nuclear reactions in the core of a nuclear reactor produces high-pressure steam at a temperature of 290°C. After the steam is used to spin a turbine, it is condensed (by using cooling water from a nearby river) back to water at 20°C. The excess heat is deposited in the river. The water is then reheated, and the cycle beings again. What is the maximum possible efficiency that this plant could achieve?

PREPARE A nuclear power plant is a heat engine, with energy transfers as illustrated in Figure 11.20. Q_H is the heat energy transferred to the steam in the reactor core. T_H is the temperature of the steam, 290°C. The steam is cooled and condensed, and the heat Q_C is exhausted to the river. The river is the cold reservoir, so T_C is 20°C.

In kelvin, these temperatures are

$$T_H = 290°C = 563 \text{ K} \qquad T_C = 20°C = 293 \text{ K}$$

SOLVE We use Equation 11.14 to compute the maximum possible efficiency:

$$e_{max} = 1 - \frac{T_C}{T_H} = 1 - \frac{293 \text{ K}}{563 \text{ K}} = 0.479 \approx 48\%$$

ASSESS This is the maximum possible efficiency. There are practical limitations as well that limit real power plants, whether nuclear or coal- or gas-fired, to an efficiency $e \approx 0.35$. This means that 65% of the energy from the fuel is exhausted as waste heat into a river or lake, where it may cause problematic warming in the local environment.

Heat engines take energy from a hot reservoir, transform some into useful forms of energy, such as mechanical or electric energy, and deposit the rest as thermal energy in a cold reservoir. The laws of physics say that some energy must be deposited in the cold reservoir, but this energy need not be "wasted." Those of you who live in cold climates use the heat energy from your car's engine to warm the car's interior—heat that would otherwise simply be deposited in the environment. There is no cost to warm your car's interior on a chilly day; you are simply putting this "waste" heat to good use. Many college campuses have combined heat and power systems in which the exhausted heat from an electric power plant is used to warm campus buildings. This scheme is quite common in cities in Europe and, increasingly, the United States.

STOP TO THINK 11.5 Which of the following changes (there may be more than one) would increase the maximum theoretical efficiency of a heat engine?

A. Increase T_H B. Increase T_C C. Decrease T_H D. Decrease T_C

11.7 Heat Pumps

The inside of your refrigerator is colder than the air in your kitchen, so heat energy will be transferred from the room to the inside of the refrigerator, warming it. This happens every time you open the door and as heat "leaks" through the walls. But you want the inside of the refrigerator to stay cool. To keep it cool, you need some way to move this heat back out to the warmer room. Transferring heat energy from a cold reservoir to a hot reservoir—the opposite of the natural direction—is the job of a **heat pump.**

The Peltier device that we saw in the last section can also be used as a heat pump. If an electric current is put through the device, a temperature difference develops between the two sides. As time goes by, the cold side gets colder and the hot side gets hotter. A practical application of such a heat pump is the water cooler/heater shown in FIGURE 11.22. The energy that is removed from the cold water ends up as increased thermal energy in the hot water. A single process cools the cold water and heats the hot water.

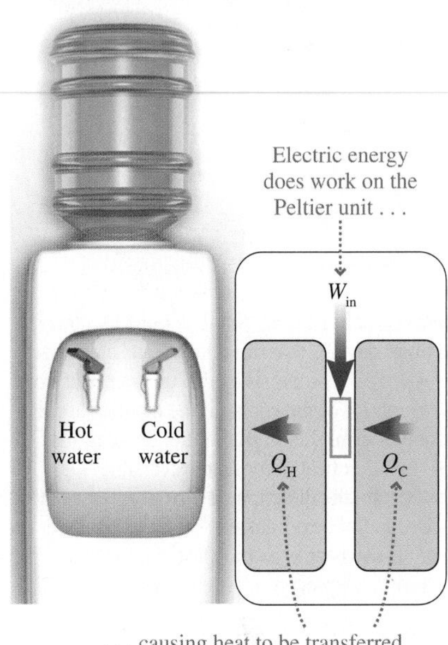
Electric energy does work on the Peltier unit . . .
W_{in}
Hot water
Cold water
Q_H
Q_C
. . . causing heat to be transferred from the cold water to the hot water.

FIGURE 11.22 A heat pump provides hot and cold water.

The heat pump in a refrigerator transfers heat from the cold air *inside* the refrigerator to the warmer air in the room. Coils inside the refrigerator that are colder than the inside air take in thermal energy. This energy is transferred to warm coils *outside* the refrigerator that transfer heat to the room. In the antique refrigerator shown on the next page, the cold coils are in the metal bracket at the top of the cabinet, the warm coils are outside the cabinet in the cylindrical unit on top. When the refrigerator is running, these top coils are quite warm. If you look closely at your refrigerator, you will find warm coils or a warm air exhaust that transfers heat to the room, heat that was removed from the inside.

An air conditioner works similarly, transferring heat from the cool air of a house (or a car) to the warmer air outside. You have probably seen room air conditioners that are single units meant to fit in a window. The side of the air conditioner that faces the room is the cool side; the other side of the air conditioner is warm. Heat is pumped from the cool side to the warm side, cooling the room.

You can also use a heat pump to *warm* your house in the winter by moving thermal energy from outside your house to inside. A unit outside the house takes in heat energy that is pumped to a unit inside the house, warming the inside air.

In all of these cases, we are moving energy against the natural direction it would flow. This requires an energy input—work must be done—as shown in the energy-transfer diagram of a heat pump in FIGURE 11.23. Energy must be conserved, so the heat deposited in the hot side must equal the sum of the heat removed from the cold side and the work input:

$$Q_H = Q_C + W_{in}$$

For heat pumps, rather than compute efficiency, we compute an analogous quantity called the **coefficient of performance (COP).** There are two different ways that one can use a heat pump, as noted above. A refrigerator uses a heat pump for cooling, removing heat from a cold reservoir to keep it cold. As Figure 11.23 shows, we must do work to make this happen.

If we use a heat pump for cooling, we define the coefficient of performance as

$$\text{COP} = \frac{\text{what you get}}{\text{what you had to pay}} = \frac{\text{energy removed from the cold reservoir}}{\text{work required to perform the transfer}} = \frac{Q_C}{W_{in}}$$

The second law of thermodynamics limits the efficiency of a heat pump just as it limits the efficiency of a heat engine. The maximum possible coefficient of performance is related to the temperatures of the hot and cold reservoirs:

$$\text{COP}_{\text{max}} = \frac{T_C}{T_H - T_C} \quad (11.15)$$

Theoretical maximum coefficient of performance of a heat pump used for cooling

FIGURE 11.23 The operation of a heat pump.

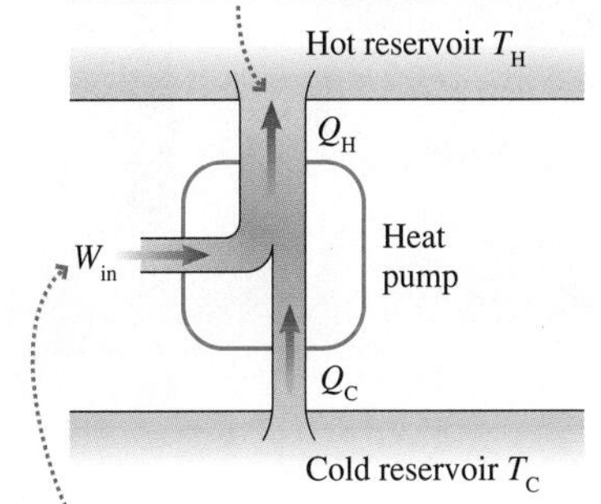
The amount of heat exhausted to the hot reservoir is larger than the amount of heat extracted from the cold reservoir.
Hot reservoir T_H
Q_H
Heat pump
W_{in}
Q_C
Cold reservoir T_C
External work is used to remove heat from a cold reservoir and exhaust heat to a hot reservoir.

Hot or cold lunch? Small coolers like this use Peltier devices that run off the 12 V electrical system of your car. They can transfer heat from the interior of the cooler to the outside, keeping food or drinks cool. But Peltier devices, like some other heat pumps, are reversible; switching the direction of current reverses the direction of heat transfer. A flick of a switch will cause heat to be transferred into the interior, keeping your lunch warm!

A refrigerator from 1934.

We can also use a heat pump for heating, moving heat from a cold reservoir to a hot reservoir to keep it warm. In that case, we define the coefficient of performance as

$$\text{COP} = \frac{\text{what you get}}{\text{what you had to pay}} = \frac{\text{energy added to the hot reservoir}}{\text{work required to perform the transfer}} = \frac{Q_H}{W_{in}}$$

In this case, the maximum possible coefficient of performance is

$$\text{COP}_{max} = \frac{T_H}{T_H - T_C} \tag{11.16}$$

Theoretical maximum coefficient of performance of a heat pump used for heating

In both cases, **a larger coefficient of performance means a more efficient heat pump.** Unlike the efficiency of a heat engine, which must be less than 1, the COP of a heat pump can be—and usually is—greater than 1. The following example shows that the COP can be quite high for typical temperatures.

EXAMPLE 11.13 Coefficient of performance of a refrigerator

The inside of your refrigerator is approximately 0°C. Heat from the inside of your refrigerator is deposited into the air in your kitchen, which has a temperature of approximately 20°C. At these operating temperatures, what is the maximum possible coefficient of performance of your refrigerator?

PREPARE The temperatures of the hot side and the cold side must be expressed in kelvin:

$$T_H = 20°\text{C} = 293\text{ K} \qquad T_C = 0°\text{C} = 273\text{ K}$$

SOLVE We use Equation 11.15 to compute the maximum coefficient of performance:

$$\text{COP}_{max} = \frac{T_C}{T_H - T_C} = \frac{273\text{ K}}{293\text{ K} - 273\text{ K}} = 13.6$$

ASSESS A coefficient of performance of 13.6 means that we pump 13.6 J of heat for an energy cost of 1 J. Due to practical limitations, the coefficient of performance of an actual refrigerator is typically ≈ 5. Other factors affect the overall efficiency of the appliance, including how well insulated it is.

CONCEPTUAL EXAMPLE 11.14 Keeping your cool?

It's a hot day, and your apartment is rather warm. Your roommate suggests cooling off the apartment by keeping the door of the refrigerator open. Will this help the situation?

REASON It's time for a physics lesson for your roommate! What you want to do is remove heat from the room. Your refrigerator, a heat pump, is designed to transfer heat from its inside to the outside. If you leave the door open, all parts of the refrigerator are exposed to the room air. It simply transfers heat from one part of the room to another—it won't make the room cooler.

ASSESS An air conditioner is a heat pump too. It must have a hot side that is outside the house, so that there is a net transfer of heat from inside to outside.

STOP TO THINK 11.6 Which of the following changes would allow your refrigerator to use less energy to run? (There may be more than one correct answer.)

A. Increasing the temperature inside the refrigerator
B. Increasing the temperature of the kitchen
C. Decreasing the temperature inside the refrigerator
D. Decreasing the temperature of the kitchen

11.8 Entropy and the Second Law of Thermodynamics

Throughout the chapter, we have noticed certain trends and certain limitations in energy transformations and transfers. Heat is transferred spontaneously from hot to cold, not from cold to hot. Once energy is transformed to thermal energy, it is (in some sense) "lost." The spontaneous transfer of heat from hot to cold is an example of an **irreversible** process, a process that can happen in only one direction. Why are some processes irreversible? The spontaneous transfer of heat from cold to hot would not violate any law of physics that we have seen to this point, but it is never observed. There must be another law of physics that prevents it. In this section we will explore the basis for this law, which we will call the **second law of thermodynamics.**

FIGURE 11.24 Molecular collisions are reversible.

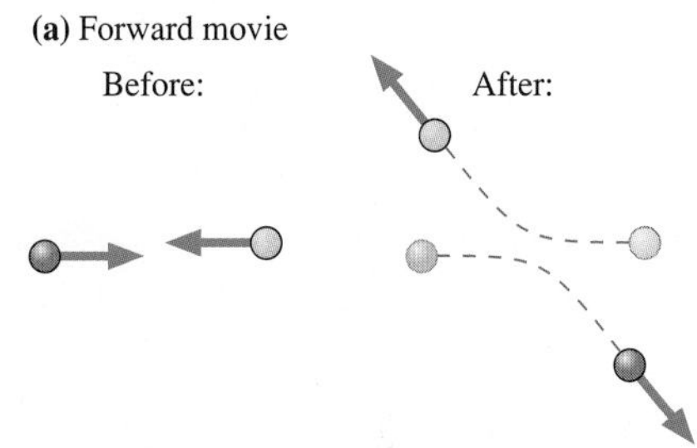

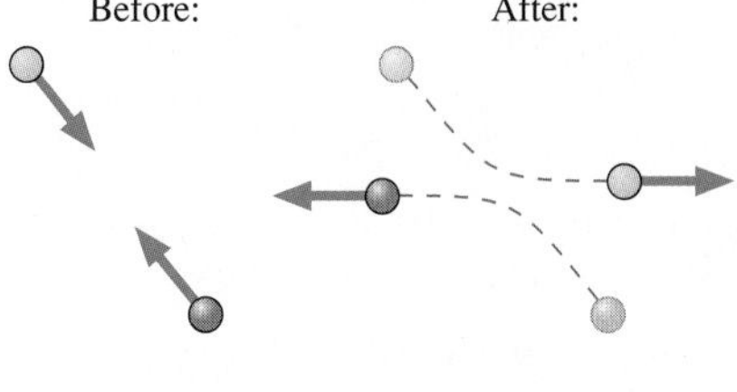

Reversible and Irreversible Processes

Stirring the cream in your coffee mixes the cream and coffee together. No amount of stirring ever unmixes them. If you watched a movie of someone stirring a cup of coffee and unmixing the cream, you'd be quite certain that the movie was running backward. In fact, a reasonable definition of an irreversible process is one for which a backward-running movie shows a physically impossible process.

At a microscopic level, collisions between molecules are completely reversible. In FIGURE 11.24 we see two possible movies of a collision between two gas molecules, one forward and the other backward. You can't tell by looking which is really going forward and which is being played backward. Nothing in either collision looks wrong, and no measurements you might make on either would reveal any violations of Newton's laws. Interactions at the molecular level are reversible processes.

At a macroscopic level, it's a different story. FIGURE 11.25 shows two possible movies of the collision of a car with a barrier. One movie is being run forward, the other backward. The backward movie of Figure 11.25b is obviously wrong. But what has been violated in the backward movie? To have the car return to its original shape and spring away from the wall would not violate any laws of physics we have so far discussed.

If microscopic motions are all reversible, how can macroscopic phenomena such as the car crash end up being irreversible? If reversible collisions can cause heat to be transferred from hot to cold, why do they never cause heat to be transferred from cold to hot? There must be something at work that can distinguish the past from the future.

FIGURE 11.25 Macroscopic collisions are not reversible.

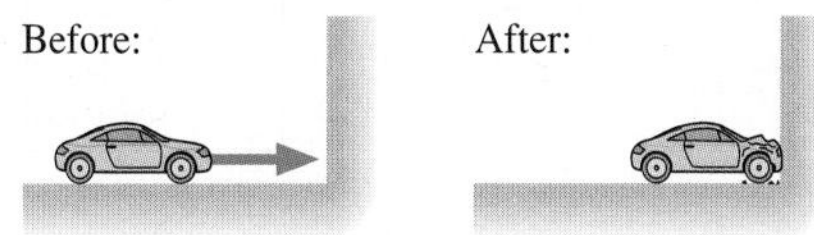

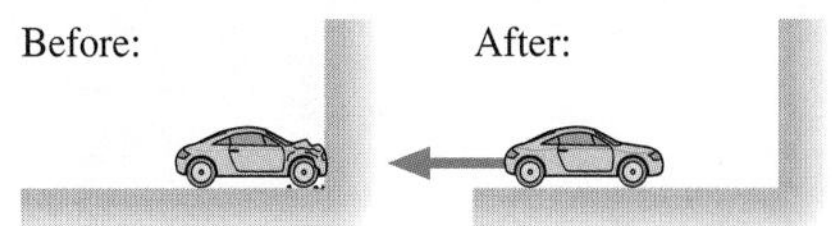

Which Way to Equilibrium?

How do two systems initially at different temperatures "know" which way to go to reach equilibrium? Perhaps an analogy will help.

FIGURE 11.26 on the next page shows two boxes, numbered 1 and 2, containing identical balls. Box 1 starts with more balls than box 2, so $N_{1i} > N_{2i}$. Once every second, a ball in one of the two boxes is chosen at random and moved to the other box. This is a reversible process because a ball can move from box 2 to box 1 just as easily as from box 1 to box 2. What do you expect to see if you return several hours later?

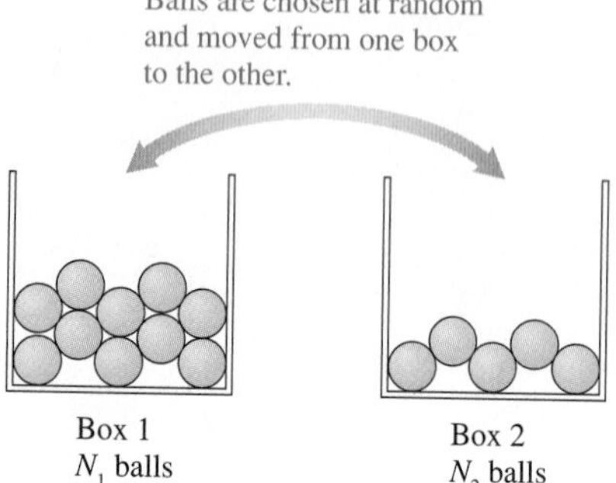

FIGURE 11.26 Moving balls between boxes.

Because balls are chosen at random, and because $N_{1i} > N_{2i}$, it's initially more likely that a ball will move from box 1 to box 2 than from box 2 to box 1. Sometimes a ball will move "backward" from box 2 to box 1, but overall there's a net movement of balls from box 1 to box 2. The system will evolve until $N_1 \approx N_2$. We have reached a stable situation—equilibrium!—with an equal number of balls moving in both directions.

But couldn't it go the other way, with N_1 getting even larger while N_2 decreases? In principle, any arrangement of the balls is possible. But certain arrangements are more likely. Each ball is equally likely to be in either box. With four balls, the odds are 1 in 2^4, or 1 in 16, that, at a randomly chosen instant of time, you would find all the balls in box 1. Were you to do so, you wouldn't find that to be terribly surprising. But with 100 balls, the probability has dropped to about 1 in 1,000,000,000,000,000,000,000,000,000,000, or 1 in 10^{30}, that all of the balls will be in box 1.

Although each transfer is reversible, **the statistics of large numbers make it overwhelmingly likely that the system will evolve toward a state in which $N_1 \approx N_2$.** For the 10^{23} or so particles in a realistic system of atoms or molecules, we will never see a state that deviates appreciably from the equilibrium case.

So imagine: Is it possible that all of the air molecules in the room in which you are sitting could, in the next second, end up moving to one side of the room, leaving the other side empty? In principle it's possible, but this situation is so extraordinarily unlikely that we can, in good conscience, call it impossible.

A system reaches thermal equilibrium not by any plan or by outside intervention, but simply because **equilibrium is the most probable state in which to be.** The consequence of having a vast number of random events is that the system evolves in one direction, toward equilibrium, and not the other. Reversible microscopic events lead to irreversible macroscopic behavior because some macroscopic states are vastly more probable than others.

Order, Disorder, and Entropy

FIGURE 11.27 shows microscopic views of three containers of gas. The top diagram shows a group of atoms arranged in a regular pattern. This is a highly ordered and nonrandom system, with each atom's position precisely specified. Contrast this with the system on the bottom, in which there is no order at all. It is extremely improbable that the atoms in a container would *spontaneously* arrange themselves into the ordered pattern of the top picture; even a small change in the pattern is quite noticeable. By contrast, there are a vast number of arrangements like the one on the bottom that randomly fill the container. A small change in the pattern is hard to detect.

FIGURE 11.27 Ordered and disordered arrangements of atoms in a gas.

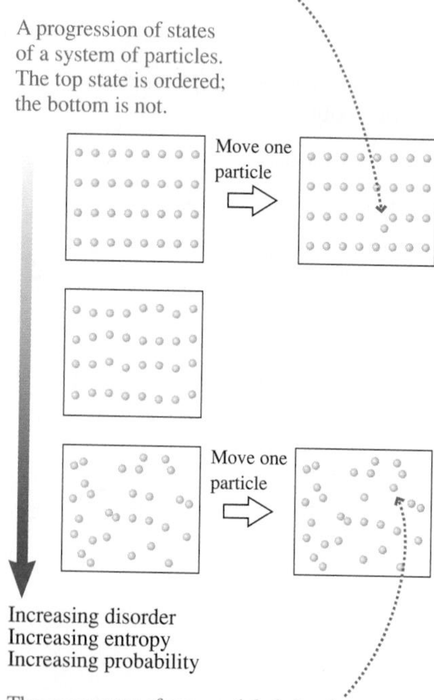

The arrangement is unlikely; all of the particles must be precisely arranged. The movement of one particle is easy to notice.
A progression of states of a system of particles. The top state is ordered; the bottom is not.
Move one particle
Move one particle
Increasing disorder
Increasing entropy
Increasing probability
The movement of one particle is hard to spot. Many arrangements have a similar appearance, so an arrangement like this is quite likely.

Scientists and engineers use the term **entropy** to quantify the probability that a certain state of a system will occur. The ordered arrangement of the top system, which has a very small probability of spontaneous occurrence, has a very low entropy. The entropy of the randomly filled container is high; it has a large probability of occurrence. The entropy in Figure 11.27 increases as you move from the ordered system on the top to the disordered system on the bottom.

Suppose you ordered the atoms in Figure 11.27 as they are in the top diagram, and then you let them go. After some time, you would expect their arrangement to appear as in the bottom diagram. In fact, the series of diagrams from top to bottom may be thought of as a series of movie frames showing the evolution of the positions of the atoms after they are released, moving toward a state of higher entropy. You would correctly expect this process to be irreversible—the particles will never spontaneously recreate their initial ordered state.

Two thermally interacting systems with different temperatures have a low entropy. These systems are ordered: The faster atoms are on one side of the barrier, the slower atoms on the other. The most random possible distribution of energy, and hence the least ordered system, corresponds to the situation where the two systems are in thermal equilibrium with equal temperatures. **Entropy increases as two systems with initially different temperatures move toward thermal equilibrium.**

TRY IT YOURSELF

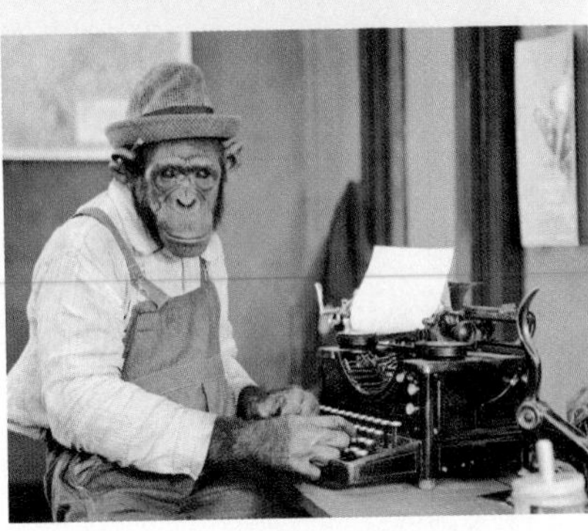

▶ **Typing Shakespeare** Make a new document in your word processor. Close your eyes and type randomly for a while. Now open your eyes. Did you type any recognizable words? There is a chance that you did, but you probably didn't. One thousand chimps in a room, typing away randomly, *could* type the works of Shakespeare. Molecular collisions *could* transfer energy from a cold object to a hot object. But, the probability is so tiny that the outcome is never seen in the real world.

The fact that macroscopic systems evolve irreversibly toward equilibrium is a new law of physics, **the second law of thermodynamics:**

Second law of thermodynamics The entropy of an isolated system never decreases. The entropy either increases, until the system reaches equilibrium, or, if the system began in equilibrium, stays the same.

NOTE ▶ The qualifier "isolated" is crucial. We can order the system by reaching in from the outside, perhaps using little tweezers to place atoms in a lattice. Similarly, we can transfer heat from cold to hot by using a refrigerator. The second law is about what a system can or cannot do *spontaneously,* on its own, without outside intervention. ◀

The second law of thermodynamics tells us that an isolated system evolves such that:

- Order turns into disorder and randomness.
- Information is lost rather than gained.
- The system "runs down" as other forms of energy are transformed into thermal energy.

An isolated system never spontaneously generates order out of randomness. It is not that the system "knows" about order or randomness, but rather that there are vastly more states corresponding to randomness than there are corresponding to order. As collisions occur at the microscopic level, the laws of probability dictate that the system will, on average, move inexorably toward the most probable and thus most random macroscopic state.

FIGURE 11.28 Kinetic energy and thermal energy compared.

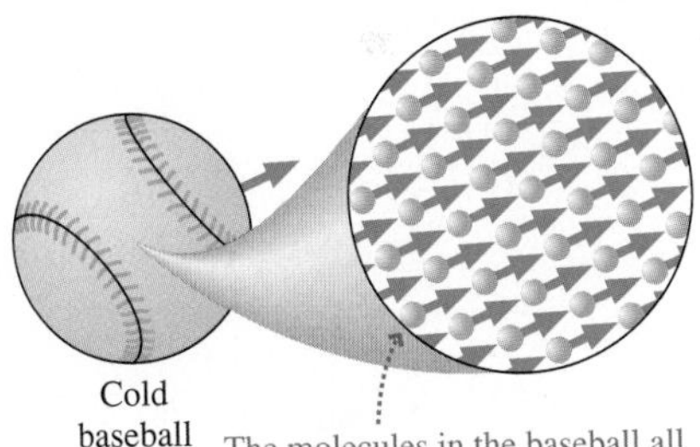

Cold baseball
The molecules in the baseball all move in the same direction at the same speed. This ordered motion is the ball's kinetic energy.

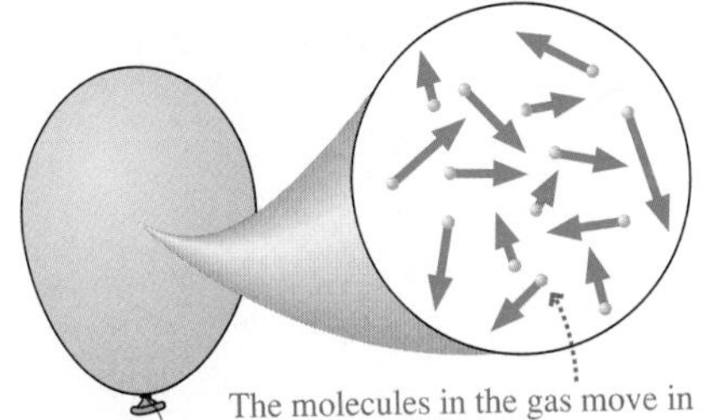

Helium balloon
The molecules in the gas move in different directions at different speeds. This random motion is the thermal energy of the gas.

Entropy and Thermal Energy

Suppose we have a very cold, moving baseball, with essentially no thermal energy, and a stationary helium balloon at room temperature, as shown in FIGURE 11.28. The atoms in the baseball and the atoms in the balloon are all moving, but there is a big difference in their motions. The atoms in the baseball are all moving in the same direction at the same speed, but the atoms in the balloon are moving in random directions. The ordered, organized motion of the baseball has low entropy, while the disorganized, random motion of the gas atoms—what we have called thermal energy—has high entropy. You can see that a conversion of macroscopic kinetic energy into thermal energy means an increase in entropy. We saw, in Section 11.5, that the conversion of other forms of energy into thermal energy was irreversible. Now, we can explain why: **When another form of energy is converted into thermal energy, there is an increase in entropy.**

This is why converting thermal energy into other forms of energy can't be done with 100% efficiency. We have redrawn our schematic diagram of a heat engine in FIGURE 11.29 on the next page, adding arrows representing entropy. Now we can see why heat must be exhausted into the cold reservoir: We need to get rid of entropy! Efficiencies or coefficients of performance higher than those in Equations 11.14, 11.15, and 11.16 would reduce entropy and thus violate the second law of thermodynamics.

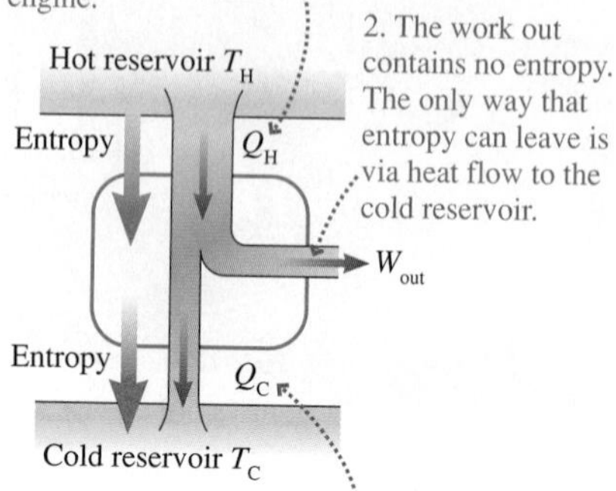

1. Heat energy Q_H comes into the system from the hot reservoir. Entropy is added to the heat engine.
2. The work out contains no entropy. The only way that entropy can leave is via heat flow to the cold reservoir.
Hot reservoir T_H
Entropy
Q_H
W_{out}
Entropy
Q_C
Cold reservoir T_C
3. Heat energy Q_C goes out of the heat engine into the cold reservoir. Entropy is removed from the heat engine.

FIGURE 11.29 A diagram of a heat engine, with entropy changes noted.

Of all of the forms of energy we've seen, only thermal energy has entropy; the others are all ordered. As long as thermal energy isn't involved, you can freely and reversibly convert between different forms of energy, as we've seen. When you toss a ball into the air, the kinetic energy is converted to potential energy as the ball rises, then back to kinetic energy as the ball falls. But when some form of energy is transformed into thermal energy, entropy has increased, and the second law of thermodynamics tells us that this change is irreversible. When you drop a ball on the floor, it bounces several times and then stops moving. The kinetic energy of the ball in motion has become thermal energy of the slightly warmer ball at rest. This process won't reverse—the ball won't suddenly cool and jump into the air! This process would conserve energy, but it would reduce entropy, and so would violate this new law of physics.

This property of thermal energy has consequences for the efficiencies of heat engines, as we've seen, but it has consequences for the efficiencies of other devices as well.

CONCEPTUAL EXAMPLE 11.15 **Efficiency of hybrid vehicles**

Hybrid vehicles are powered by a gasoline engine paired with an electric motor and batteries. They get much better mileage in the stop and go of city driving than conventional vehicles do. When you brake to a stop in a conventional car, friction converts the kinetic energy of the car's motion into thermal energy in the brakes. In a typical hybrid car, some of this energy is converted into chemical energy in a battery. Explain how this makes a hybrid vehicle more efficient.

REASON When energy is transformed into thermal energy, the increase in entropy makes this change irreversible. When you brake a conventional car, the kinetic energy is transformed into the thermal energy of hot brakes and is lost to your use. In a hybrid vehicle, the kinetic energy is converted into chemical energy in a battery. This change is reversible; when the car starts again, the energy can be transformed back into kinetic energy.

ASSESS Whenever energy is converted to thermal energy, it is in some sense "lost," which reduces efficiency. The hybrid vehicle avoids this transformation, and so is more efficient.

STOP TO THINK 11.7 Which of the following processes does not involve a change in entropy?

A. An electric heater raises the temperature of a cup of water by 20°C.
B. A ball rolls up a ramp, decreasing in speed as it rolls higher.
C. A basketball is dropped from 2 m and bounces until it comes to rest.
D. The sun shines on a black surface and warms it.

11.9 Systems, Energy, and Entropy

Over the past two chapters, we have learned a good deal about energy and how it is used. By introducing the concept of entropy, we were able to see how limits on our ability to use energy come about. We will close this chapter, and this part of the book, by considering a few final questions that bring these different pieces together.

The Conservation of Energy and Energy Conservation

We have all heard for many years that it is important to "conserve energy." We are asked to turn off lights when we leave rooms, to drive our cars less, to turn down our thermostats. But this brings up an interesting question: If we have a law of conservation of energy, which states that energy can't be created or destroyed, what do we really mean by "conserving energy"? If energy can't be created or destroyed, how can there be an "energy crisis"?

We started this chapter looking at energy transformations. We saw that whenever energy is transformed, some of it is "lost." And now we know what this means: The energy isn't really lost, but it is converted into thermal energy. This change is irreversible; thermal energy can't be efficiently converted back into other forms of energy.

And that's the problem. We aren't, as a society or as a planet, running out of energy. We can't! What we can run out of is high-quality sources of energy. Oil is a good example. A gallon of gasoline contains a great deal of chemical energy. It is a liquid, so is easily transported, and it is easily burned in a host of devices to generate heat, electricity, or motion. When you burn gasoline in your car, you don't use up its energy—you simply convert its chemical energy into thermal energy. As you do this, you decrease the amount of high-quality chemical energy in the world and increase the supply of thermal energy. The amount of energy in the world is still the same; it's just in a less useful form.

Perhaps the best way to "conserve energy" is to concentrate on efficiency, to reduce "what you had to pay." More efficient lightbulbs, more efficient cars—all of these use less energy to produce the same final result.

Sealed, but not isolated BIO This glass container is a completely sealed system containing living organisms, shrimp and algae. But the organisms will live and grow for many years. The reason this is possible is that the glass sphere, though sealed, is not an *isolated* system. Energy can be transferred in and out as light and heat. If the container were placed in a darkened room, the organisms would quickly perish.

Entropy and Life

The second law of thermodynamics predicts that systems will "run down," that ordered states will evolve toward disorder and randomness. But living organisms seem to violate this rule:

- Plants grow from simple seeds to complex entities.
- Single-celled fertilized eggs grow into complex adult organisms.
- Over the last billion years or so, life has evolved from simple unicellular organisms to very complex forms.

How can this be?

There is an important qualification in the second law of thermodynamics: It applies only to isolated systems, systems that do not exchange energy with their environment. The situation is entirely different if energy is transferred into or out of the system.

Your body is not an isolated system. Every day, you take in chemical energy in the food you eat. As you use this energy, most of it ends up as thermal energy that you exhaust as heat into the environment, taking away the entropy created by natural processes in your body. An energy and entropy diagram of this situation is given in FIGURE 11.30. Each second, as you sit quietly and read this book, your body is using 100 J of chemical energy and exhausting 100 J of thermal energy to the environment. The entropy of your body is staying approximately constant, but the entropy of the environment is increasing due to the thermal energy from your body. To grow and develop, organisms must take in high-quality forms of energy and exhaust thermal energy. This continuous exchange of energy (and entropy) with the environment makes your life—and all life—possible without violating any laws of physics.

FIGURE 11.30 Thermodynamic view of the body.

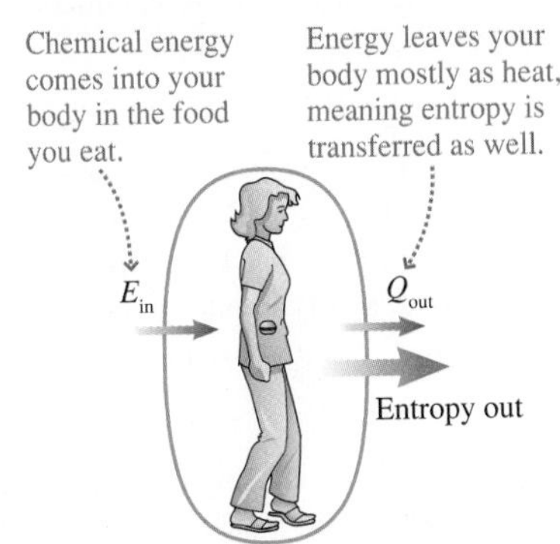
Chemical energy comes into your body in the food you eat.
Energy leaves your body mostly as heat, meaning entropy is transferred as well.
E_{in}
Q_{out}
Entropy out

INTEGRATED EXAMPLE 11.16 Efficiency of an automobile

In the absence of external forces, Newton's first law tells us that a car would continue at a constant speed once it was moving. So why can't you get your car up to speed and then take your foot off the gas? Because there *are* external forces. At highway speeds, nearly all of the force opposing your car's motion is the drag force of the air. (Rolling friction is much smaller than drag at highway speeds, so we'll ignore it.) FIGURE 11.31 shows that as your car moves through the air it pushes the air aside. Doing so takes energy, and your car's engine must keep running to replace this lost energy.

FIGURE 11.31 A wind-tunnel test shows airflow around a car.

Sports cars, with their aerodynamic shape, are often those that lose the least energy to drag. A typical sports car has a 350 hp engine, a drag coefficient of 0.30, and a low profile with the area of the front of the car being a modest 1.8 m^2. Such a car gets about 25 miles per gallon of gasoline at a highway speed of 30 m/s (just over 65 mph).

Suppose this car is driven 25 miles at 30 m/s. It will consume 1 gallon of gasoline, containing 1.4×10^8 J of chemical energy. What is the car's efficiency for this trip?

PREPARE We can use what we learned about the drag force in Chapter 5 to compute the amount of energy needed to move the car forward through the air. We can use this value in Equation 11.2 as what you get; it is the minimum amount of energy that *could* be used to move the car forward at this speed. We can then use the mileage data to calculate what you had to pay; this is the energy actually used by the car's engine. Once we have these two pieces of information, we can calculate efficiency.

SOLVE Heat and work both play a role in this process, so we need to use our most general equation about energy and energy conservation, Equation 11.11:

$$\Delta K + \Delta U + \Delta E_{\text{th}} + \Delta E_{\text{chem}} + \cdots = W + Q$$

The car is moving at a constant speed, so its kinetic energy is not changing. The road is assumed to be level, so there is no change in potential energy. Once the car is warmed up, its temperature will be constant, so its thermal energy isn't changing. Equation 11.11 reduces to

$$\Delta E_{\text{chem}} = W + Q$$

The car's engine is using chemical energy as it burns fuel. The engine transforms some of this energy to work—the propulsion force pushing the car forward against the opposing force of air drag. But the engine also transforms much of the chemical energy to "waste heat" that is transferred to the environment through the radiator and the exhaust. ΔE_{chem} is negative because the amount of energy stored in the gas tank is decreasing as gasoline is burned. W and Q are also negative, according to the sign convention of Figure 11.16, because these energies are being transferred *out* of the system to the environment.

In Chapter 5 we saw that the drag force on an object moving at a speed v is given by

$$\vec{D} = \left(\tfrac{1}{2}C_{\text{D}}\rho A v^2, \text{ direction opposite the motion}\right)$$

where C_{D} is the drag coefficient, ρ is the density of air (approximately 1.2 kg/m^3), and A is the area of the front of the car. The drag force on the car moving at 30 m/s is

$$D = \tfrac{1}{2}(0.30)(1.2\text{ kg/m}^3)(1.8\text{ m}^2)(30\text{ m/s})^2 = 290\text{ N}$$

The drag force opposes the motion. To move the car forward at constant speed, and thus with no *net* force, requires a propulsion force $F = 290$ N in the forward direction. In Chapter 10 we found that the power—the rate of energy expenditure—to move an object at speed v with force F is $P = Fv$. Thus the power the car must supply—at the wheels—to keep the car moving down the highway is

$$P = Fv = (290\text{ N})(30\text{ m/s}) = 8700\text{ W}$$

It's interesting to compare this to the engine power by converting to horsepower:

$$P = 8700\text{ W}\left(\frac{1\text{ hp}}{746\text{ W}}\right) = 12\text{ hp}$$

Only a small fraction of the engine's 350 horsepower is needed to keep the car moving at highway speeds.

The distance traveled is

$$\Delta x = 25\text{ mi} \times \frac{1.6\text{ km}}{1\text{ mi}} \times \frac{1000\text{ m}}{1\text{ km}} = 40{,}000\text{ m}$$

and the time required to travel it is $\Delta t = (40{,}000\text{ m})/(30\text{ m/s}) = 1333$ s. Thus the minimum energy needed to travel 40 km—the energy needed simply to push the air aside—is

$$E_{\text{min}} = P\Delta t = (8700\text{ W})(1333\text{ s}) = 1.2 \times 10^7\text{ J}$$

In traveling this distance, the car uses 1 gallon of gas, with 1.4×10^8 J of chemical energy. This is, quite literally, what you had to pay to drive this distance. Thus the car's efficiency is

$$e = \frac{\text{what you get}}{\text{what you had to pay}} = \frac{1.2 \times 10^7\text{ J}}{1.4 \times 10^8\text{ J}} = 0.086 = 8.6\%$$

ASSESS The efficiency of the car is quite low, even compared to other engines that we've seen. Nonetheless, our calculation agrees reasonably well with actual measurements. Gasoline-powered vehicles simply are inefficient, which is one factor favoring more efficient alternative vehicles. Smaller mass and better aerodynamic design would improve the efficiency of vehicles, but a large part of the inefficiency of a gasoline-powered vehicle is inherent in the thermodynamics of the engine itself and in the complex drive train needed to transfer the engine's power to the wheels.

SUMMARY

The goals of Chapter 11 have been to learn more about energy transformations and transfers, the laws of thermodynamics, and theoretical and practical limitations on energy use.

GENERAL PRINCIPLES

Energy and Efficiency

When energy is transformed from one form into another, some may be "lost," usually to thermal energy, due to practical or theoretical constraints. This limits the efficiency of processes. We define **efficiency** as

$$e = \frac{\text{what you get}}{\text{what you had to pay}}$$

Entropy and Irreversibility

Systems move toward more probable states. These states have higher **entropy**—more disorder. This change is irreversible. Changing other forms of energy to thermal energy is irreversible.

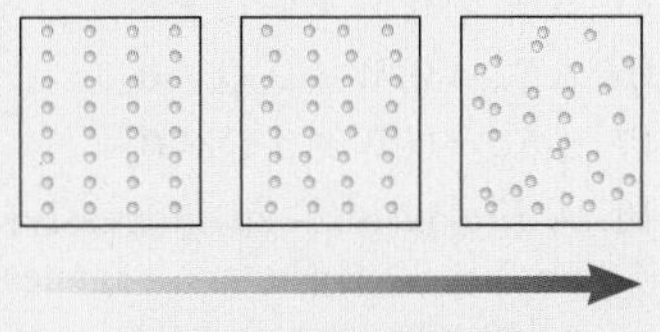

Increasing probability
Increasing entropy

The Laws of Thermodynamics

The **first law of thermodynamics** is a statement of conservation of energy for systems in which only thermal energy changes:

$$\Delta E_{th} = W + Q$$

The **second law of thermodynamics** specifies the way that isolated systems can evolve:

The entropy of an isolated system always increases.

This law has practical consequences:

- Heat energy spontaneously flows only from hot to cold.
- A transformation of energy into thermal energy is irreversible.
- No heat engine can be 100% efficient.

Heat is energy transferred between two objects at different temperatures. Energy will be transferred until thermal equilibrium is reached.

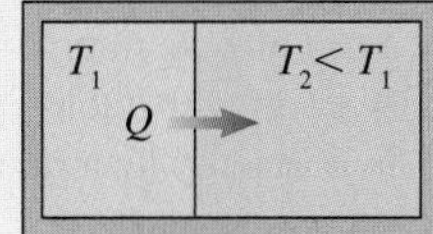

IMPORTANT CONCEPTS

Thermal energy

- For a gas, the thermal energy is the **total kinetic energy** of motion of the atoms.
- Thermal energy is random kinetic energy and so has entropy.

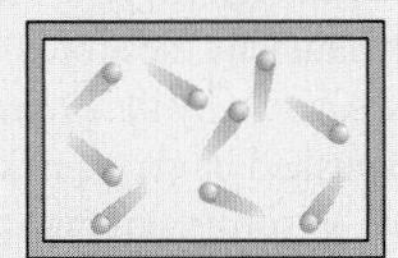

$$E_{th} = NK_{avg} = \frac{3}{2}Nk_BT$$

Temperature

- For a gas, temperature is proportional to the **average kinetic energy** of the motion of the atoms.

$$T = \frac{2}{3}\frac{K_{avg}}{k_B}$$

- Two systems are in **thermal equilibrium** if they are at the same temperature. No heat energy is transferred at thermal equilibrium.

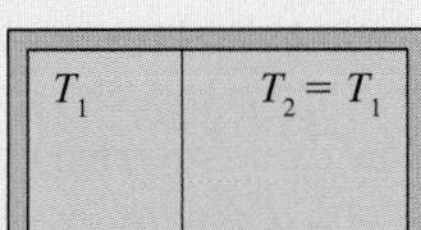

A **heat engine** converts thermal energy from a hot reservoir into useful work. Some heat is exhausted into a cold reservoir, limiting efficiency.

$$e_{max} = 1 - \frac{T_C}{T_H}$$

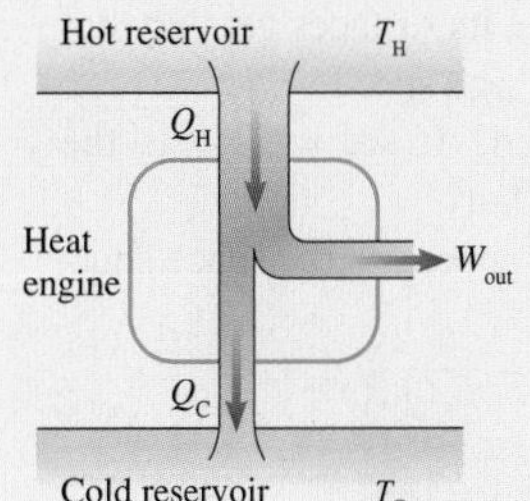

A **heat pump** uses an energy input to transfer heat from a cold side to a hot side. The **coefficient of performance** is analogous to efficiency. For cooling, it is:

$$COP_{max} = \frac{T_C}{T_H - T_C}$$

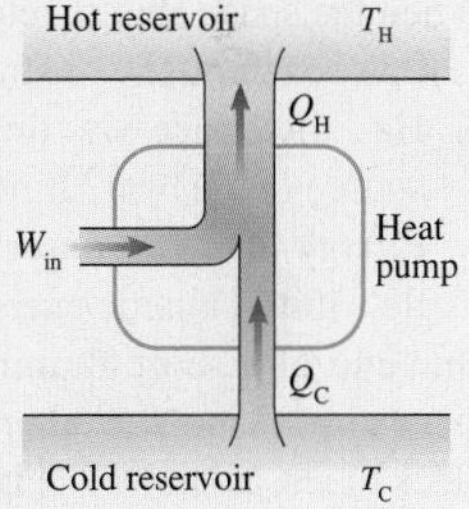

APPLICATIONS

Efficiencies

Energy in the body
Cells in the body metabolize chemical energy in food. Efficiency for most actions is about 25%.

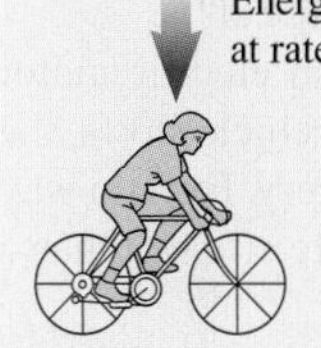

Power plants
A typical power plant converts about 1/3 of the energy input into useful work. The rest is exhausted as waste heat.

Temperature scales

Zero on the **Kelvin temperature scale** is the temperature at which the kinetic energy of atoms is zero. This is **absolute zero.** The conversion from °C to K is

$$T(\text{K}) = T(°\text{C}) + 273$$

► All temperatures in equations must be in kelvin. ◄

For homework assigned on MasteringPhysics, go to www.masteringphysics.com

Problems labeled INT integrate significant material from earlier chapters; BIO are of biological or medical interest.

Problem difficulty is labeled as | (straightforward) to ||||| (challenging).

QUESTIONS

Conceptual Questions

1. Rub your hands together vigorously. What happens? Discuss the energy transfers and transformations that take place.
2. Write a few sentences describing the energy transformations that occur from the time moving water enters a hydroelectric plant until you see some water being pumped out of a nozzle in a public fountain. Use the "Energy transformations" table on page 324 as an example.
3. BIO Describe the energy transfers and transformations that occur from the time you sit down to breakfast until you've completed a fast bicycle ride.
4. BIO According to Table 11.4, cycling at 15 km/h requires less metabolic energy than running at 15 km/h. Suggest reasons why this is the case.
5. You're stranded on a remote desert island with only a chicken, a bag of corn, and a shade tree. To survive as long as possible in hopes of being rescued, should you eat the chicken at once and then the corn? Or eat the corn, feeding enough to the chicken to keep it alive, and then eat the chicken when the corn is gone? Or are your survival chances the same either way? Explain.
6. For most automobiles, the number of miles per gallon decreases as highway speed increases. Fuel economy drops as speeds increase from 55 to 65 mph, then decreases further as speeds increase to 75 mph. Explain why this is the case.
7. When the space shuttle returns to earth, its surfaces get very hot as it passes through the atmosphere at high speed.
   a. Has the space shuttle been heated? If so, what was the source of the heat? If not, why is it hot?
   b. Energy must be conserved. What happens to the space shuttle's initial kinetic energy?
8. One end of a short aluminum rod is in a campfire and the other end is in a block of ice, as shown in Figure Q11.8. If 100 J of energy are transferred from the fire to the rod, and if the temperature at every point in the rod has reached a steady value, how much energy goes from the rod into the ice?

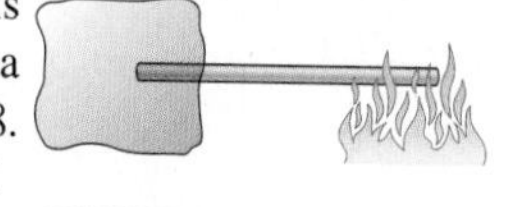
FIGURE Q11.8

9. Two blocks of copper, one of mass 1 kg and the second of mass 3 kg, are at the same temperature. Which block has more thermal energy? If the blocks are placed in thermal contact, will the thermal energy of the blocks change? If so, how?
10. If the temperature T of an ideal gas doubles, by what factor does the average kinetic energy of the atoms change?
11. A bottle of helium gas and a bottle of argon gas contain equal numbers of atoms at the same temperature. Which bottle, if either, has the greater total thermal energy?

For Questions 12 through 17, give a specific example of a process that has the energy changes and transfers described. (For example, if the question states "$\Delta E_{th} > 0$, $W = 0$," you are to describe a process that has an increase in thermal energy and no transfer of energy by work. You could write "Heating a pan of water on the stove.")

12. $\Delta E_{th} < 0$, $W = 0$
13. $\Delta E_{th} > 0$, $Q = 0$
14. $\Delta E_{th} < 0$, $Q = 0$
15. $\Delta E_{th} > 0$, $W \neq 0$, $Q \neq 0$
16. $\Delta E_{th} < 0$, $W \neq 0$, $Q \neq 0$
17. $\Delta E_{th} = 0$, $W \neq 0$, $Q \neq 0$
18. A fire piston—an impressive physics demonstration—ignites a fire without matches. The operation is shown in Figure Q11.18. A wad of cotton is placed at the bottom of a sealed syringe with a tight-fitting plunger. When the plunger is rapidly depressed, the air temperature in the syringe rises enough to ignite the cotton. Explain why the air temperature rises, and why the plunger must be pushed in very quickly.

FIGURE Q11.18

19. In a gasoline engine, fuel vapors are ignited by a spark. In a diesel engine, a fuel–air mixture is drawn in, then rapidly compressed to as little as 1/20 the original volume, in the process increasing the temperature enough to ignite the fuel–air mixture. Explain why the temperature rises during the compression.
20. A drop of green ink falls into a beaker of clear water. First *describe* what happens. Then *explain* the outcome in terms of entropy.
21. If you hold a rubber band loosely between two fingers and then stretch it, you can tell by touching it to the sensitive skin of your forehead that stretching the rubber band has increased its temperature. If you then let the rubber band rest against your forehead, it soon returns to its original temperature. What are the signs of W and Q for the entire process?
22. In areas in which the air temperature drops very low in the winter, the exterior unit of a heat pump designed for heating is sometimes buried underground in order to use the earth as a thermal reservoir. Why is it worthwhile to bury the heat exchanger, even if the underground unit costs more to purchase and install than one above ground?
23. Assuming improved materials and better processes, can engineers ever design a heat engine that exceeds the maximum efficiency indicated by Equation 11.14? If not, why not?
24. Electric vehicles increase speed by using an electric motor that draws energy from a battery. When the vehicle slows, the motor runs as a generator, recharging the battery. Explain why this means that an electric vehicle can be more efficient than a gasoline-fueled vehicle.
25. When the sun's light hits the earth, the temperature rises. Is there an entropy change to accompany this transformation? Explain.

26. When you put an ice cube tray filled with liquid water in your freezer, the water eventually becomes solid ice. The solid is more ordered than the liquid—it has less entropy. Explain how this transformation is possible without violating the second law of thermodynamics.
27. A company markets an electric heater that is described as 100% efficient at converting electric energy to thermal energy. Does this violate the second law of thermodynamics?

Multiple-Choice Questions

28. ||| A person is walking on level ground at constant speed. What energy transformation is taking place?
   A. Chemical energy is being transformed to thermal energy.
   B. Chemical energy is being transformed to kinetic energy.
   C. Chemical energy is being transformed to kinetic energy and thermal energy.
   D. Chemical energy and thermal energy are being transformed to kinetic energy.
29. | A person walks 1 km, turns around, and runs back to where he started. Compare the energy used and the power during the two segments.
   A. The energy used and the power are the same for both.
   B. The energy used while walking is greater, the power while running is greater.
   C. The energy used while running is greater, the power while running is greater.
   D. The energy used is the same for both segments, the power while running is greater.
30. | The temperature of the air in a basketball increases as it is pumped up. This means that
   A. The total kinetic energy of the air is increasing and the average kinetic energy of the molecules is decreasing.
   B. The total kinetic energy of the air is increasing and the average kinetic energy of the molecules is increasing.
   C. The total kinetic energy of the air is decreasing and the average kinetic energy of the molecules is decreasing.
   D. The total kinetic energy of the air is decreasing and the average kinetic energy of the molecules is increasing.
31. | The thermal energy of a container of helium gas is halved. What happens to the temperature, in kelvin?
   A. It decreases to one-fourth its initial value.
   B. It decreases to one-half its initial value.
   C. It stays the same.
   D. It increases to twice its initial value.
32. | An inventor approaches you with a device that he claims will take 100 J of thermal energy input and produce 200 J of electricity. You decide not to invest your money because this device would violate
   A. The first law of thermodynamics.
   B. The second law of thermodynamics.
   C. Both the first and second laws of thermodynamics.
33. | While keeping your food cold, your refrigerator transfers energy from the inside to the surroundings. Thus thermal energy goes from a colder object to a warmer one. What can you say about this?
   A. It is a violation of the second law of thermodynamics.
   B. It is not a violation of the second law of thermodynamics because refrigerators can have efficiency of 100%.
   C. It is not a violation of the second law of thermodynamics because the second law doesn't apply to refrigerators.
   D. The second law of thermodynamics applies in this situation, but it is not violated because the energy did not spontaneously go from cold to hot.
34. || An electric power plant uses energy from burning coal to generate steam at 450°C. The plant is cooled by 20°C water from a nearby river. If burning coal provides 100 MJ of heat, what is the theoretical minimum amount of heat that must be transferred to the river during the conversion of heat to electric energy?
   A. 100 MJ B. 90 MJ
   C. 60 MJ D. 40 MJ
35. || A refrigerator's freezer compartment is set at −10°C; the kitchen is 24°C. What is the theoretical minimum amount of electric energy necessary to pump 1.0 J of energy out of the freezer compartment?
   A. 0.89 J B. 0.87 J
   C. 0.13 J D. 0.11 J

PROBLEMS

Section 11.1 Transforming Energy

1. || A 10% efficient engine accelerates a 1500 kg car from rest to 15 m/s. How much energy is transferred to the engine by burning gasoline?
2. || A 60% efficient device uses chemical energy to generate 600 J of electric energy.
   a. How much chemical energy is used?
   b. A second device uses twice as much chemical energy to generate half as much electric energy. What is its efficiency?
3. | A typical photovoltaic cell delivers 4.0×10^{-3} W of electric energy when illuminated with 1.2×10^{-1} W of light energy. What is the efficiency of the cell?
4. || An individual white LED (light-emitting diode) has an efficiency of 20% and uses 1.0 W of electric power. How many LEDs must be combined into one light source to give a total of 1.6 W of visible-light output (comparable to the light output of a 40 W incandescent bulb)? What total power is necessary to run this LED light source?

Section 11.2 Energy in the Body: Energy Inputs

5. || A fast-food hamburger (with cheese and bacon) contains 1000 Calories. What is the burger's energy in joules? BIO
6. ||| In an average human, basic life processes require energy to be supplied at a steady rate of 100 W. What daily energy intake, in Calories, is required to maintain these basic processes? BIO
7. | An "energy bar" contains 6.0 g of fat. How much energy is this in joules? In calories? In Calories? BIO
8. | An "energy bar" contains 22 g of carbohydrates. How much energy is this in joules? In calories? In Calories? BIO

Section 11.3 Energy in the Body: Energy Outputs

9. ||| An "energy bar" contains 22 g of carbohydrates. If the energy bar was his only fuel, how far could a 68 kg person walk at 5.0 km/h? BIO
10. ||| Suppose your body was able to use the chemical energy in gasoline. How far could you pedal a bicycle at 15 km/h on the energy in 1 gal of gas? (1 gal of gas has a mass of 3.2 kg.) BIO

11. ||||| The label on a candy bar says 400 Calories. Assuming a typical efficiency for energy use by the body, if a 60 kg person were to use the energy in this candy bar to climb stairs, how high could she go?
BIO

12. ||| A weightlifter curls a 30 kg bar, raising it each time a distance of 0.60 m. How many times must he repeat this exercise to burn off the energy in one slice of pizza?
BIO

13. ||| A weightlifter works out at the gym each day. Part of her routine is to lie on her back and lift a 40 kg barbell straight up from chest height to full arm extension, a distance of 0.50 m.
BIO
    a. How much work does the weightlifter do to lift the barbell one time?
    b. If the weightlifter does 20 repetitions a day, what total energy does she expend on lifting, assuming a typical efficiency for energy use by the body.
    c. How many 400 Calorie donuts can she eat a day to supply that energy?

Section 11.4 Thermal Energy and Temperature

14. | The planet Mercury's surface temperature varies from 700 K during the day to 90 K at night. What are these values in °C and °F?
15. || An ideal gas is at 20°C. If we double the average kinetic energy of the gas atoms, what is the new temperature in °C?
16. || An ideal gas is at 20°C. The gas is cooled, reducing the thermal energy by 10%. What is the new temperature in °C?
17. || An ideal gas at 0°C consists of 1.0×10^{23} atoms. 10 J of thermal energy are added to the gas. What is the new temperature in °C?
18. || An ideal gas at 20°C consists of 2.2×10^{22} atoms. 4.3 J of thermal energy are removed from the gas. What is the new temperature in °C?

Section 11.5 Heat and the First Law of Thermodynamics

19. || 500 J of work are done on a system in a process that decreases the system's thermal energy by 200 J. How much energy is transferred to or from the system as heat?
20. | 600 J of heat energy are transferred to a system that does 400 J of work. By how much does the system's thermal energy change?
21. | 300 J of energy are transferred to a system in the form of heat while the thermal energy increases by 150 J. How much work is done on or by the system?
22. | 10 J of heat are removed from a gas sample while it is being compressed by a piston that does 20 J of work. What is the change in the thermal energy of the gas? Does the temperature of the gas increase or decrease?

Section 11.6 Heat Engines

23. | A heat engine extracts 55 kJ from the hot reservoir and exhausts 40 kJ into the cold reservoir. What are (a) the work done and (b) the efficiency?
24. || A heat engine does 20 J of work while exhausting 30 J of waste heat. What is the engine's efficiency?
25. || A heat engine does 200 J of work while exhausting 600 J of heat to the cold reservoir. What is the engine's efficiency?
26. | A heat engine with an efficiency of 40% does 100 J of work. How much heat is (a) extracted from the hot reservoir and (b) exhausted into the cold reservoir?
27. || a. At what cold-reservoir temperature (in °C) would an engine operating at maximum theoretical efficiency with a hot-reservoir temperature of 427°C have an efficiency of 60%?
    b. If another engine, operating at maximum theoretical efficiency with a hot-reservoir temperature of 400°C, has the same efficiency, what is its cold-reservoir temperature?
28. ||| A heat engine operating between energy reservoirs at 20°C and 600°C has 30% of the maximum possible efficiency. How much energy does this engine extract from the hot reservoir to do 1000 J of work?
29. | A newly proposed device for generating electricity from the sun is a heat engine in which the hot reservoir is created by focusing sunlight on a small spot on one side of the engine. The cold reservoir is ambient air at 20°C. The designer claims that the efficiency will be 60%. What minimum hot-reservoir temperature, in °C, would be required to produce this efficiency?
30. | Converting sunlight to electricity with solar cells has an efficiency of ≈15%. It's possible to achieve a higher efficiency (though currently at higher cost) by using concentrated sunlight as the hot reservoir of a heat engine. Each dish in Figure P11.30 concentrates sunlight on one side of a heat engine, producing a hot-reservoir temperature of 650°C. The cold reservoir, ambient air, is approximately 30°C. The actual working efficiency of this device is ≈30%. What is the theoretical maximum efficiency?

FIGURE P11.30

Section 11.7 Heat Pumps

31. || A refrigerator takes in 20 J of work and exhausts 50 J of heat. What is the refrigerator's coefficient of performance?
32. || Air conditioners are rated by their coefficient of performance at 80°F inside temperature and 95°F outside temperature. An efficient but realistic air conditioner has a coefficient of performance of 3.2. What is the maximum possible coefficient of performance?
33. || 50 J of work are done on a refrigerator with a coefficient of performance of 4.0. How much heat is (a) extracted from the cold reservoir and (b) exhausted to the hot reservoir?
34. || Find the maximum possible coefficient of performance for a heat pump used to heat a house in a northerly climate in winter. The inside is kept at 20°C while the outside is −20°C.

Section 11.8 Entropy and the Second Law of Thermodynamics

35. | Which, if any, of the heat engines in Figure P11.35 below violate (a) the first law of thermodynamics or (b) the second law of thermodynamics? Explain.

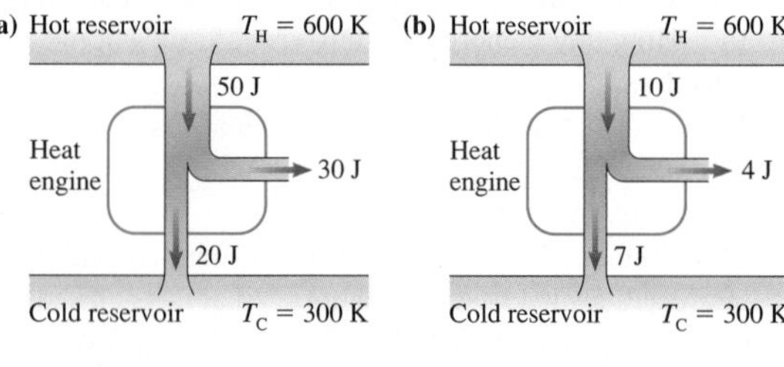

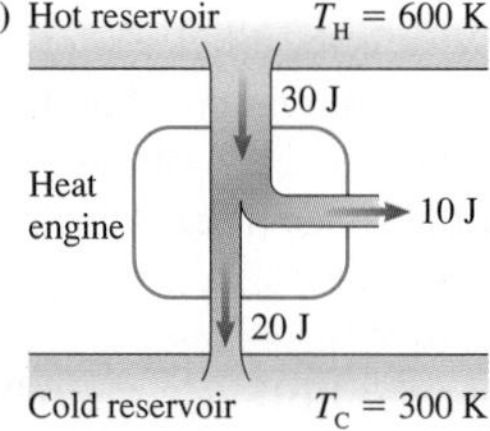

FIGURE P11.35

36. | Which, if any, of the refrigerators in Figure P11.36 below violate (a) the first law of thermodynamics or (b) the second law of thermodynamics? Explain.

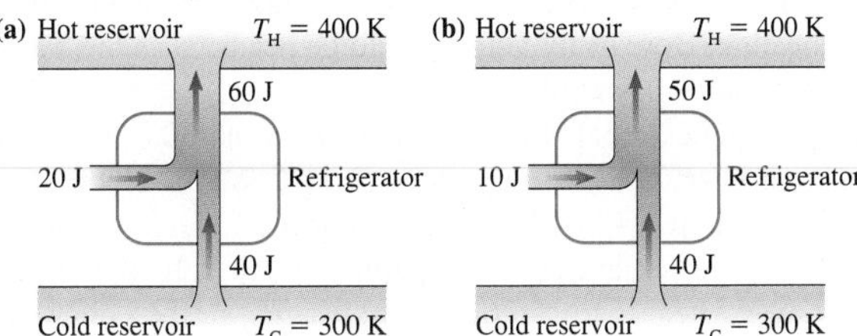

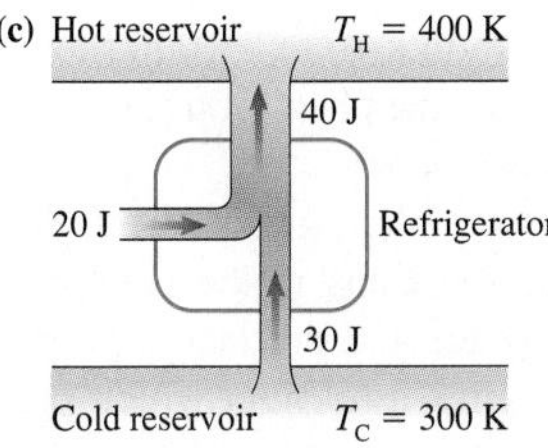

FIGURE P11.36

37. || Draw all possible distinct arrangements in which three balls (labeled A, B, C) are placed into two different boxes (1 and 2), as in Figure 11.26. If all arrangements are equally likely, what is the probability that all three will be in box 1?

General Problems

38. ||||| How many slices of pizza must you eat to walk for 1.0 h at a
BIO speed of 5.0 km/h? (Assume your mass is 68 kg.)

39. || A 60 kg hiker climbs to the top of a 500-m-high hill. Ignoring the energy needed for horizontal motion and assuming a typical efficiency for energy use by the body, how many frozen burritos would be needed to fuel this climb?

40. ||||| For how long would a 68 kg athlete have to swim at a fast
BIO crawl to use all the energy available in a typical fast-food meal of burger, fries, and a drink?

41. || a. How much metabolic energy is required for a 68 kg
BIO runner to run at a speed of 15 km/h for 20 min?
b. How much metabolic energy is required for this runner to walk at a speed of 5.0 km/h for 60 min? Compare your result to your answer to part a.
c. Compare your results of parts a and b to the result of Example 11.4. Of these three modes of human motion, which is the most efficient?

42. | To a good approximation, the only external force that does
BIO work on a cyclist moving on level ground is the force of air resistance. Suppose a cyclist is traveling at 15 km/h on level ground. Assume he is using 480 W of metabolic power.
a. Estimate the amount of power he uses for forward motion.
b. How much force must he exert to overcome the force of air resistance?

43. | The winning time for the 2005 annual race up 86 floors of the
BIO Empire State Building was 10 min and 49 s. The winner's mass was 60 kg.
a. If each floor was 3.7 m high, what was the winner's change in gravitational potential energy?
b. If the efficiency in climbing stairs is 25%, what total energy did the winner expend during the race?
c. How many food Calories did the winner "burn" in the race?
d. Of those Calories, how many were converted to thermal energy?
e. What was the winner's metabolic power in watts during the race up the stairs?

44. ||| Championship swimmers take about 22 s and about 30 arm
BIO strokes to move through the water in a 50 m freestyle race.
INT a. From Table 11.4, a swimmer's metabolic power is 800 W. If the efficiency for swimming is 25%, how much energy is expended moving through the water in a 50 m race?
b. If half the energy is used in arm motion and half in leg motion, what is the energy expenditure per arm stroke?
c. Model the swimmer's hand as a paddle. During one arm stroke, the paddle moves halfway around a 90-cm-radius circle. If all the swimmer's forward propulsion during an arm stroke comes from the hand pushing on the water and none from the arm (somewhat of an oversimplification), what is the average force of the hand on the water?

45. |||| A 68 kg hiker walks at 5.0 km/h up a 7% slope. What is the
BIO necessary metabolic power? **Hint:** You can model her power needs as the sum of the power to walk on level ground plus the power needed to raise her body by the appropriate amount.

46. ||| A 70 kg student consumes 2500 Cal each day and stays the
BIO same weight. One day, he eats 3500 Cal and, wanting to keep
INT from gaining weight, decides to "work off" the excess by jumping up and down. With each jump, he accelerates to a speed of 3.3 m/s before leaving the ground. How many jumps must he make? Assume that the efficiency of the body in using energy is 25%

47. || To make your workouts more productive, you can get a gen-
BIO erator that you drive with the rear wheel of your bicycle when it is mounted in a stand.
a. Your laptop charger uses 75 W. What is your body's metabolic power use while running the generator to power your laptop charger, given the typical efficiency for such tasks? Assume 100% efficiency for the generator.
b. Your laptop takes 1 hour to recharge. If you run the generator for 1 hour, how much energy does your body use? Express your result in joules and in Calories.

48. || The resistance of an exercise bike is often provided by a gen-
BIO erator; that is, the energy that you expend is used to generate electric energy, which is then dissipated. Rather than dissipate the energy, it could be used for practical purposes.
a. A typical person can maintain a steady energy expenditure of 400 W on a bicycle. Assuming a typical efficiency for the body and a generator that is 80% efficient, what useful electric power could you produce with a bicycle-powered generator?
b. How many people would need to ride bicycle generators simultaneously to power a 400 W TV in the gym?

49. || Smaller mammals use proportionately more energy than
BIO larger mammals; that is, it takes more energy per gram to power a mouse than a human. A typical mouse has a mass of 20 g and, at rest, needs to consume 3.0 Cal each day for basic body processes.
a. If a 68 kg human used the same energy per kg of body mass as a mouse, how much energy would be needed each day?
b. What resting power does this correspond to? How much greater is this than the resting power noted in the chapter?

50. || Larger animals use proportionately less energy than smaller animals; that is, it takes less energy per kg to power an elephant than to power a human. A 5000 kg African elephant requires about 70,000 Cal for basic needs for one day.
BIO

a. If a 68 kg human required the same energy per kg of body mass as an elephant, how much energy would be required each day?
b. What resting power does this correspond to? How much less is this than the resting power noted in the chapter?

51. || A large horse can perform work at a steady rate of about 1 horsepower, as you might expect.
BIO
a. Assuming a 25% efficiency, how many Calories would a horse need to consume to work at 1.0 hp for 1.0 h?
b. Dry hay contains about 10 MJ per kg. How many kilograms of hay would the horse need to eat to perform this work?

52. || A college student is working on her physics homework in her dorm room. Her room contains a total of 6.0×10^{26} gas molecules. As she works, her body is converting chemical energy into thermal energy at a rate of 125 W. If her dorm room were an isolated system (dorm rooms can certainly feel like that) and if all of this thermal energy were transferred to the air in the room, by how much would the temperature increase in 10 min?

53. || A container holding argon atoms changes temperature by 20°C when 30 J of heat are removed. How many atoms are in the container?

54. || A heat engine with a high-temperature reservoir at 400 K has an efficiency of 0.20. What is the maximum possible temperature of the cold reservoir?

55. ||| An engine does 10 J of work and exhausts 15 J of waste heat.
a. What is the engine's efficiency?
b. If the cold-reservoir temperature is 20°C, what is the minimum possible temperature in °C of the hot reservoir?

56. | The heat exhausted to the cold reservoir of an engine operating at maximum theoretical efficiency is two-thirds the heat extracted from the hot reservoir. What is the temperature ratio T_C/T_H?

57. ||| An engine operating at maximum theoretical efficiency whose cold-reservoir temperature is 7°C is 40% efficient. By how much should the temperature of the hot reservoir be increased to raise the efficiency to 60%?

58. || Some heat engines can run on very small temperature differences. One manufacturer claims to have a very small heat engine that can run on the temperature difference between your hand and the air in the room. Estimate the theoretical maximum efficiency of this heat engine.

59. ||| The coefficient of performance of a refrigerator is 5.0.
a. If the compressor uses 10 J of energy, how much heat is exhausted to the hot reservoir?
b. If the hot-reservoir temperature is 27°C, what is the lowest possible temperature in °C of the cold reservoir?

60. ||| An engineer claims to have measured the characteristics of a heat engine that takes in 100 J of thermal energy and produces 50 J of useful work. Is this engine possible? If so, what is the smallest possible ratio of the temperatures (in kelvin) of the hot and cold reservoirs?

61. ||| A 32% efficient electric power plant produces 900 MJ of electric energy per second and discharges waste heat into 20°C ocean water. Suppose the waste heat could be used to heat homes during the winter instead of being discharged into the ocean. A typical American house requires an average 20 kW for heating. How many homes could be heated with the waste heat of this one power plant?

62. |||| A typical coal-fired power plant burns 300 metric tons of coal *every hour* to generate 2.7×10^6 MJ of electric energy. 1 metric ton = 1000 kg; 1 metric ton of coal has a volume of 1.5 m^3. The heat of combustion of coal is 28 MJ/kg. Assume that *all* heat is transferred from the fuel to the boiler and that *all* the work done in spinning the turbine is transformed into electric energy.
a. Suppose the coal is piled up in a 10 m × 10 m room. How tall must the pile be to operate the plant for one day?
b. What is the power plant's efficiency?

63. ||| Each second, a nuclear power plant generates 2000 MJ of thermal energy from nuclear reactions in the reactor's core. This energy is used to boil water and produce high-pressure steam at 300°C. The steam spins a turbine, which produces 700 MJ of electric power, then the steam is condensed and the water is cooled to 30°C before starting the cycle again.
a. What is the maximum possible efficiency of the plant?
b. What is the plant's actual efficiency?

64. ||| 250 students sit in an auditorium listening to a physics lecture. Because they are thinking hard, each is using 125 W of metabolic power, slightly more than they would use at rest. An air conditioner with a COP of 5.0 is being used to keep the room at a constant temperature. What minimum electric power must be used to operate the air conditioner?

65. || Driving on asphalt roads entails very little rolling resistance, so most of the energy of the engine goes to overcoming air resistance. But driving slowly in dry sand is another story. If a 1500 kg car is driven in sand at 5.0 m/s, the coefficient of rolling friction is 0.06. In this case, nearly all of the energy that the car uses to move goes to overcoming rolling friction, so you can ignore air drag in this problem.
INT
a. What propulsion force is needed to keep the car moving forward at a constant speed?
b. What power is required for propulsion at 5.0 m/s?
c. If the car gets 15 mpg when driving on sand, what is the car's efficiency?

66. ||| Air conditioners sold in the United States are given a seasonal energy-efficiency ratio (SEER) rating that consumers can use to compare different models. A SEER rating is the ratio of heat pumped to energy input, similar to a COP but using English units, so a higher SEER rating means a more efficient model. You can determine the COP of an air conditioner by dividing the SEER rating by 3.4. For typical inside and outside temperatures when you'd be using air conditioning, estimate the theoretical maximum SEER rating of an air conditioner. (New air conditioners must have a SEER rating that exceeds 13, quite a bit less than the theoretical maximum, but there are practical issues that reduce efficiency.)

67. || The surface waters of tropical oceans are at a temperature of 27°C while water at a depth of 1200 m is at 3°C. It has been suggested these warm and cold waters could be the energy reservoirs for a heat engine, allowing us to do work or generate electricity from the thermal energy of the ocean. What is the maximum efficiency possible of such a heat engine?

68. || The light energy that falls on a square meter of ground over the course of a typical sunny day is about 20 MJ. The average rate of electric energy consumption in one house is 1.0 kW.
   a. On average, how much energy does one house use during each 24 h day?
   b. If light energy to electric energy conversion using solar cells is 15% efficient, how many square miles of land must be covered with solar cells to supply the electrical energy for 250,000 houses? Assume there is no cloud cover.

Passage Problems

Kangaroo Locomotion BIO

Kangaroos have very stout tendons in their legs that can be used to store energy. When a kangaroo lands on its feet, the tendons stretch, transforming kinetic energy of motion to elastic potential energy. Much of this energy can be transformed back into kinetic energy as the kangaroo takes another hop. The kangaroo's peculiar hopping gait is not very efficient at low speeds but is quite efficient at high speeds.

Figure P11.69 shows the energy cost of human and kangaroo locomotion. The graph shows oxygen uptake (in mL/s) per kg of body mass, allowing a direct comparison between the two species.

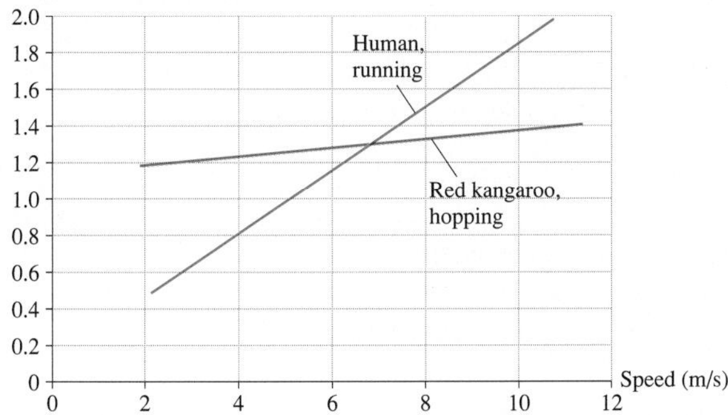

FIGURE P11.69 Oxygen uptake (a measure of energy use per second) for a running human and a hopping kangaroo.

For humans, the energy used per second (i.e., power) is proportional to the speed. That is, the human curve nearly passes through the origin, so running twice as fast takes approximately twice as much power. For a hopping kangaroo, the graph of energy use has only a very small slope. In other words, the energy used per second changes very little with speed. Going faster requires very little additional power. Treadmill tests on kangaroos and observations in the wild have shown that they do not become winded at any speed at which they are able to hop. No matter how fast they hop, the necessary power is approximately the same.

69. | A person runs 1 km. How does his speed affect the total energy needed to cover this distance?
   A. A faster speed requires less total energy.
   B. A faster speed requires more total energy.
   C. The total energy is about the same for a fast speed and a slow speed.

70. | A kangaroo hops 1 km. How does its speed affect the total energy needed to cover this distance?
   A. A faster speed requires less total energy.
   B. A faster speed requires more total energy.
   C. The total energy is about the same for a fast speed and a slow speed.

71. | At a speed of 4 m/s,
   A. A running human is more efficient than an equal-mass hopping kangaroo.
   B. A running human is less efficient than an equal-mass hopping kangaroo.
   C. A running human and an equal-mass hopping kangaroo have about the same efficiency.

72. | At approximately what speed would a human use half the power of an equal-mass kangaroo moving at the same speed?
   A. 3 m/s B. 4 m/s C. 5 m/s D. 6 m/s

73. | At what speed does the hopping motion of the kangaroo become more efficient than the running gait of a human?
   A. 3 m/s B. 5 m/s C. 7 m/s D. 9 m/s

STOP TO THINK ANSWERS

Stop to Think 11.1: C. In each case, what you get is the potential-energy change of the box. Crane 2 lifts a box with twice the mass the same distance as crane 1, so you get twice as much energy with crane 2. How about what you have to pay? Crane 2 uses 20 kJ, crane 1 only 10 kJ. Comparing crane 1 and crane 2, we find crane 2 has twice the energy out for twice the energy in, so the efficiencies are the same.

Stop to Think 11.2: C. As the body uses chemical energy from food, approximately 75% is transformed into thermal energy. Also, kinetic energy of motion of the legs and feet is transformed into thermal energy with each stride. Most of the chemical energy is transformed into thermal energy.

Stop to Think 11.3: C. Samples 1 and 2 have the same thermal energy, which is the total kinetic energy of all the atoms. Sample 1 has twice as many atoms, so the average energy per atom, and thus the temperature, must be less.

Stop to Think 11.4: C. The radiator is at a higher temperature than the surrounding air. Thermal energy is transferred out of the system to the environment, so $Q < 0$.

Stop to Think 11.5: A, D. The efficiency is fixed by the ratio of T_C to T_H. Decreasing this ratio increases efficiency; the heat engine will be more efficient with a hotter hot reservoir or a colder cold reservoir.

Stop to Think 11.6: A, D. The closer the temperatures of the hot and cold reservoirs, the more efficient the heat pump can be. (It is also true that having the two temperatures be closer will cause less thermal energy to "leak" out.) Any change that makes the two temperatures closer will allow the refrigerator to use less energy to run.

Stop to Think 11.7: B. In this case, kinetic energy is transformed into potential energy; there is no entropy change. In the other cases, energy is transformed into thermal energy, meaning entropy increases.

PART II SUMMARY

Conservation Laws

In Part II we have discovered that we don't need to know all the details of an interaction to relate the properties of a system "before" the interaction to those "after" the interaction. We also found two important quantities, momentum and energy, that are often conserved. Momentum and energy are characteristics of a system.

Momentum and energy have conditions under which they are conserved. The total momentum $\vec{P}$ and the total energy E are conserved for an *isolated system.* Of course, not all systems are isolated. For both momentum and energy, it was useful to develop a *model* of a system interacting with its environment. Interactions within the system do not change $\vec{P}$ or E. The kinetic, potential, and thermal energies *within* the system can be transformed without changing E. Interactions between the system and the environment *do* change the system's momentum and energy. In particular:

- Impulse is the transfer of momentum to or from the system: $\Delta\vec{p} = \vec{J}$.
- Work is the transfer of energy to or from the system in a mechanical interaction: $\Delta E = W$.
- Heat is the energy transferred to or from the system in a thermal interaction: $\Delta E = Q$.

The laws of conservation of momentum and energy, when coupled with the laws of Newtonian mechanics of Part I, form a powerful set of tools for analyzing motion. But energy is a concept that can be used for much more than the study of motion; it can be used to analyze how your body uses food, how the sun shines, and a host of other problems.

The study of the transformation of energy from one form into another reveals certain limits. Thermal energy is different from other forms of energy. A transformation of energy from another form into thermal energy is *irreversible.* The study of heat and thermal energy thus led us to the discipline of thermodynamics, a subject we will take up further in Part III as we look at properties of matter.

KNOWLEDGE STRUCTURE II Conservation Laws

BASIC GOALS

How is the system "after" an interaction related to the system "before"?
What quantities are conserved, and under what conditions?
Why are some energy changes more efficient than others?

GENERAL PRINCIPLES

Law of conservation of momentum For an isolated system, $\vec{P}_f = \vec{P}_i$.

Law of conservation of energy For an isolated system, there is no change in the system's energy:

$$\Delta K + \Delta U_g + \Delta U_s + \Delta E_{th} + \Delta E_{chem} + \cdots = 0$$

Energy can be exchanged with the environment as work or heat:

$$\Delta K + \Delta U_g + \Delta U_s + \Delta E_{th} + \Delta E_{chem} + \cdots = W + Q$$

Laws of thermodynamics First law: If only thermal energy changes, $\Delta E_{th} = W + Q$.
Second law: The entropy of an isolated system always increases.

BASIC PROBLEM-SOLVING STRATEGY

Draw a visual overview for the system "before" and "after"; then use the conservation of momentum or energy equations to relate the two. If necessary, calculate impulse and/or work.

Momentum and impulse

In a collision, the total momentum

$$\vec{P} = \vec{p}_1 + \vec{p}_2 = m_1\vec{v}_1 + m_2\vec{v}_2$$

is the same before and after.

Before: m_1 (1) $(v_{1x})_i$ → ← $(v_{2x})_i$ (2) m_2

After: ← $(v_{1x})_f$ (1) (2) $(v_{2x})_f$ →

A force can change the momentum of an object. The change is the **impulse:** $\Delta p_x = J_x$.

J_x = area under force curve

F_x, J_x, t, Δt

Basic model of energy

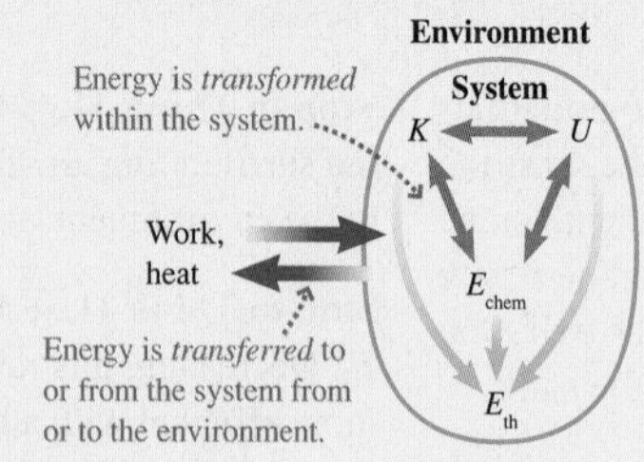

Work $W = F_{\parallel}d$ is done by the component of a force parallel to a displacement.

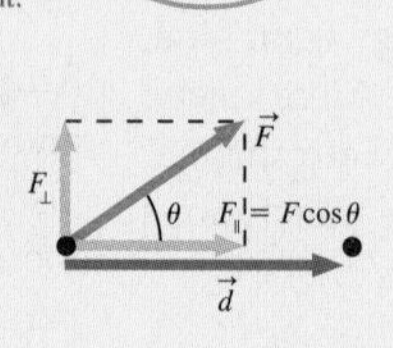

Limitations on energy transfers and transformations

Thermal energy is random kinetic energy. Changing other forms of energy to thermal energy is **irreversible.** When transforming energy from one form into another, some may be "lost" as thermal energy. This limits efficiency:

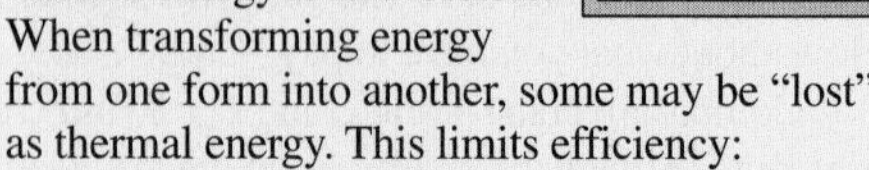

$$\text{efficiency: } e = \frac{\text{what you get}}{\text{what you had to pay}}$$

A heat engine can convert thermal energy to useful work. The efficiency must be less than 100%.

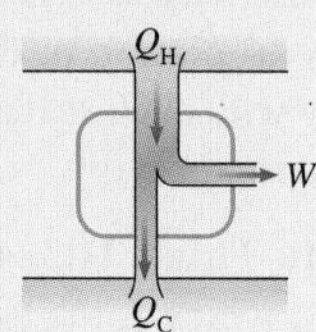

Order Out of Chaos

The second law of thermodynamics specifies that "the future" is the direction of entropy increase. But, as we have seen, this doesn't mean that systems must invariably become more random. You don't need to look far to find examples of systems that spontaneously evolve to a state of greater order.

A snowflake is a perfect example. As water freezes, the random motion of water molecules is transformed into the orderly arrangement of a crystal. The entropy of the snowflake is less than that of the water vapor from which it formed. Has the second law of thermodynamics been turned on its head?

The entropy of the water molecules in the snowflake certainly decreases, but the water doesn't freeze as an isolated system. For it to freeze, heat energy must be transferred from the water to the surrounding air. The entropy of the air increases by *more* than the entropy of the water decreases. Thus the *total* entropy of the water + air system increases when a snowflake is formed, just as the second law predicts. If the system isn't isolated, its entropy can decrease without violating the second law as long as the entropy increases somewhere else.

Systems that become *more* ordered as time passes, and in which the entropy decreases, are called *self-organizing systems.* These systems can't be isolated. It is common in self-organizing systems to find a substantial flow of energy *through* the system. Your body takes in chemical energy from food, makes use of that energy, and then gives waste heat back to the environment. It is this energy flow that allows systems to develop a high degree of order and a very low entropy. The entropy of the environment undergoes a significant *increase* so as to let selected subsystems decrease their entropy and become more ordered.

Self-organizing systems don't violate the second law of thermodynamics, but this fact doesn't really explain their existence. If you toss a coin, no law of physics says that you can't get heads 100 times in a row—but you don't expect this to happen. Can we show that self-organization isn't just possible, but likely?

Let's look at a simple example. Suppose you heat a shallow dish of oil at the bottom, while holding the temperature of the top constant. When the temperature difference between the top and the bottom of the dish is small, heat is transferred from the bottom to the top by conduction. But convection begins when the temperature difference becomes large enough. The pattern of convection needn't be random, though; it can develop in a stable, highly ordered pattern, as we see in the figure. Convection is a much more efficient means of transferring energy than conduction, so the rate of transfer is *increased* as a result of the development of these ordered *convection cells.*

The development of the convection cells is an example of self-organization. The roughly 10^{23} molecules in the fluid had been moving randomly but now have begun behaving in a very orderly fashion. But there is more to the story. The convection cells transfer energy from the hot lower side of the dish to the cold upper side. This hot-to-cold energy transfer increases the entropy of the surrounding environment, as we have seen. In becoming more organized, the system has become more effective at transferring heat, resulting in a greater rate of entropy increase! Order has arisen out of disorder in the system, but the net result is a more rapid increase of the disorder of the universe.

Convection cells are thus a thermodynamically favorable form of order. We should expect this, because convection cells aren't confined to the laboratory. We see them in the sun, where they transfer energy from lower levels to the surface, and in the atmosphere of the earth, where they give rise to some of our most dramatic weather.

Self-organizing systems are a very active field of research in physical and biological sciences. The 1977 Nobel Prize in chemistry was awarded to the Belgian scientist Ilya Prigogine for his studies of *nonequilibrium thermodynamics,* the basic science underlying self-organizing systems. Prigogine and others have shown how energy flow through a system can, when the conditions are right, "bring order out of chaos." And this spontaneous ordering is not just possible—it can be probable. The existence and evolution of self-organizing systems, from thunderstorms to life on earth, might just be nature's preferred way of increasing entropy in the universe.

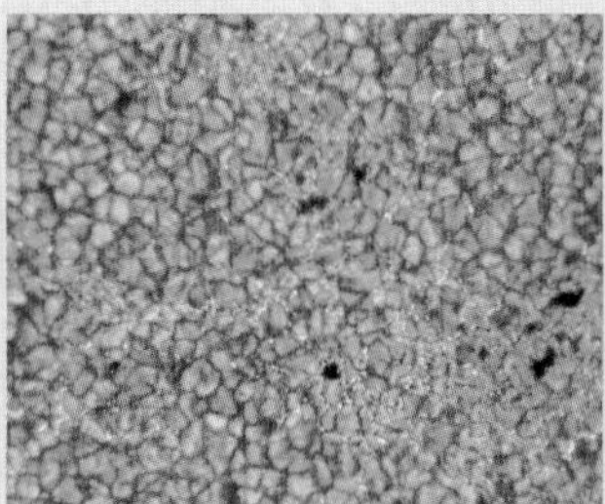

Convection cells in a shallow dish of oil heated from below (left) and in the sun (right). In both, warmer fluid is rising (lighter color) and cooler fluid is sinking (darker color).

PART II PROBLEMS

The following questions are related to the passage "Order Out of Chaos" on the previous page.

1. When water freezes to make a snowflake crystal, the entropy of the water
   A. Decreases.
   B. Increases.
   C. Does not change.
2. When thermal energy is transferred from a hot object to a cold object, the overall entropy
   A. Decreases.
   B. Increases.
   C. Does not change.
3. Do convection cells represent a reversible process?
   A. Yes, because they are orderly.
   B. No, because they transfer thermal energy from hot to cold.
   C. It depends on the type of convection cell.
4. In an isolated system far from thermal equilibrium, as time passes,
   A. The total energy stays the same; the total entropy stays the same.
   B. The total energy decreases; the total entropy increases.
   C. The total energy stays the same; the total entropy increases.
   D. The total energy decreases; the total entropy stays the same.

The following passages and associated questions are based on the material of Part II.

Big Air

A new generation of pogo sticks lets a rider bounce more than 2 meters off the ground by using elastic bands to store energy. When the pogo's plunger hits the ground, the elastic bands stretch as the pogo and rider come to rest. At the low point of the bounce, the stretched bands start to contract, pushing out the plunger and launching the rider into the air. For a total mass of 80 kg (rider plus pogo), a stretch of 0.40 m launches a rider 2.0 m above the starting point.

5. If you were to jump to the ground from a height of 2 meters, you'd likely injure yourself. But a pogo rider can do this repeatedly, bounce after bounce. How does the pogo stick make this possible?
   A. The elastic bands absorb the energy of the bounce, keeping it from hurting the rider.
   B. The elastic bands warm up as the rider bounces, absorbing dangerous thermal energy.
   C. The elastic bands simply convert the rider's kinetic energy to potential energy.
   D. The elastic bands let the rider come to rest over a longer time, meaning less force.
6. Assuming that the elastic bands stretch and store energy like a spring, how high would the 80 kg pogo and rider go for a stretch of 0.20 m?
   A. 2.0 m B. 1.5 m C. 1.0 m D. 0.50 m
7. Suppose a much smaller rider (total mass of rider plus pogo of 40 kg) mechanically stretched the elastic bands of the pogo by 0.40 m, then got on the pogo and released the bands. How high would this unwise rider go?
   A. 8.0 m B. 6.0 m C. 4.0 m D. 3.0 m
8. A pogo and rider of 80 kg total mass at the high point of a 2.0 m jump will drop 1.6 m before the pogo plunger touches the ground, slowing to a stop over an additional 0.40 m as the elastic bands stretch. What approximate average force does the pogo stick exert on the ground during the landing?
   A. 4000 N B. 3200 N C. 1600 N D. 800 N
9. Riders can use fewer elastic bands, reducing the effective spring constant of the pogo. The maximum stretch of the bands is still 0.40 m. Reducing the number of bands will
   A. Reduce the force on the rider and give a lower jump height.
   B. Not change the force on the rider but give a lower jump height.
   C. Reduce the force on the rider but give the same jump height.
   D. Make no difference to the force on the rider or the jump height.

Testing Tennis Balls

Tennis balls are tested by being dropped from a height of 2.5 m onto a concrete floor. The 57 g ball hits the ground, compresses, then rebounds. A ball will be accepted for play if it rebounds to a height of about 1.4 m; it will be rejected if the bounce height is much more or much less than this.

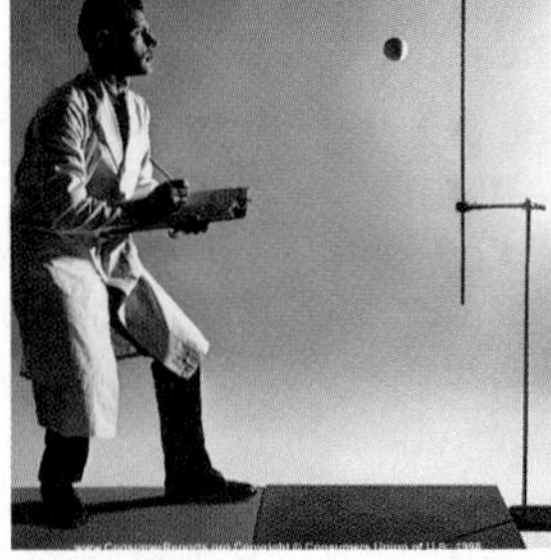

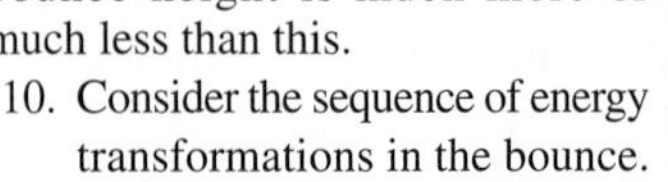

10. Consider the sequence of energy transformations in the bounce. When the dropped ball is motionless on the floor, compressed, and ready to rebound, most of the energy is in the form of
    A. Kinetic energy.
    B. Gravitational potential energy.
    C. Thermal energy.
    D. Elastic potential energy.
11. If a ball is "soft," it will spend more time in contact with the floor and won't rebound as high as it is supposed to. The force on the floor of the "soft" ball is ________ the force on the floor of a "normal" ball.
    A. Greater than
    B. The same as
    C. Less than
12. Suppose a ball is dropped from 2.5 m and rebounds to 1.4 m.
    a. How fast is the ball moving just before it hits the floor?
    b. What is the ball's speed just after leaving the floor?
    c. What happens to the "lost" energy?
    d. If the time of the collision with the floor is 6.0 ms, what is the average force on the ball during the impact?

Squid Propulsion BIO

Squid usually move by using their fins, but they can utilize a form of "jet propulsion," ejecting water at high speed to rocket them backward, as shown in Figure II.1. A 4.0 kg squid can slowly draw in and then quickly eject 0.30 kg of water. The water is ejected in 0.10 s at a speed of 10 m/s. This gives the squid a quick burst of speed to evade predators or catch prey.

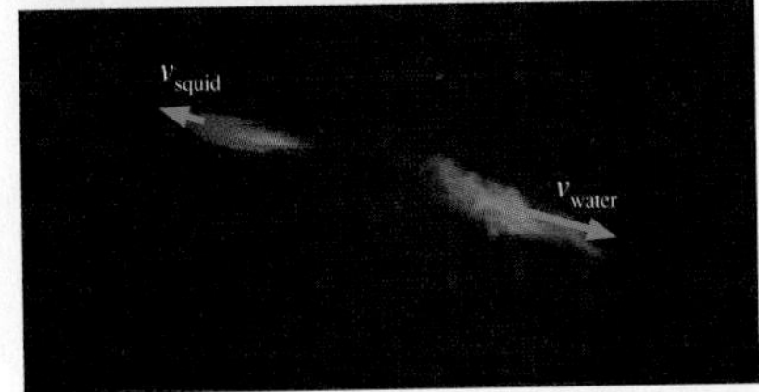

FIGURE II.1

13. What is the speed of the squid immediately after the water is ejected?
A. 10 m/s
B. 7.5 m/s
C. 1.3 m/s
D. 0.75 m/s
14. What is the squid's approximate acceleration in g?
A. $10g$ B. $7.5g$ C. $1.0g$ D. $0.75g$
15. What is the average force on the water during the jet?
A. 100 N B. 30 N C. 10 N D. 3.0 N
16. This form of locomotion is speedy, but is it efficient? The energy that the squid expends goes two places: the kinetic energy of the squid and the kinetic energy of the water. Think about how to define "what you get" and "what you had to pay"; then calculate an efficiency for this particular form of locomotion. (You can ignore biomechanical efficiency for this problem.)

Teeing Off

A golf club has a lightweight flexible shaft with a heavy block of wood or metal (called the head of the club) at the end. A golfer making a long shot off the tee uses a driver, a club whose 300 g head is much more massive than the 46 g ball it will hit. The golfer swings the driver so that the club head is moving at 40 m/s just before it collides with the ball. The collision is so rapid that it can be treated as the collision of a moving 300 g mass (the club head) with a stationary 46 g mass (the ball); the shaft of the club and the golfer can be ignored. The collision takes 5.0 ms, and the ball leaves the tee with a speed of 69 m/s.

17. What is the change in momentum of the ball during the collision?
A. 1.4 kg · m/s B. 1.8 kg · m/s
C. 3.2 kg · m/s D. 5.1 kg · m/s
18. What is the speed of the club head immediately after the collision?
A. 29 m/s B. 25 m/s C. 19 m/s D. 11 m/s
19. Is this a perfectly elastic collision?
A. Yes
B. No
C. There is insufficient information to make this determination.
20. If we define the kinetic energy of the club head before the collision as "what you paid" and the kinetic energy of the ball immediately after as "what you get," what is the efficiency of this energy transfer?
A. 0.54 B. 0.46 C. 0.37 D. 0.27

Additional Integrated Problems

21. Football players measure their acceleration by seeing how fast they can sprint 40 yards (37 m). A zippy player can, from a standing start, run 40 yards in 4.1 s, reaching a top speed of about 11 m/s. For an 80 kg player, what is the average power output for this sprint?
A. 300 W B. 600 W C. 900 W D. 1200 W
22. The unit of horsepower was defined by considering the power output of a typical horse. Working-horse guidelines in the 1900s called for them to pull with a force equal to 10% of their body weight at a speed of 3.0 mph. For a typical working horse of 1200 lb, what power does this represent in W and in hp?
23. A 100 kg football player is moving at 6.0 m/s to the east; a 130 kg player is moving at 5.0 m/s to the west. They meet, each jumping into the air and grabbing the other player. While they are still in the air, which way is the pair moving, and how fast?
24. A swift blow with the hand can break a pine board. As the hand hits the board, the kinetic energy of the hand is transformed into elastic potential energy of the bending board; if the board bends far enough, it breaks. Applying a force to the center of a particular pine board deflects the center of the board by a distance that increases in proportion to the force. Ultimately the board breaks at an applied force of 800 N and a deflection of 1.2 cm.
   a. To break the board with a blow from the hand, how fast must the hand be moving? Use 0.50 kg for the mass of the hand.
   b. If the hand is moving this fast and comes to rest in a distance of 1.2 cm, what is the average force on the hand?
25. A child's sled has rails that slide with little friction across the snow. Logan has an old wooden sled with heavy iron rails that has a mass of 10 kg—quite a bit for a 30 kg child! Logan runs at 4.0 m/s and leaps onto the stationary sled and holds on tight as it slides forward. The impact time with the sled is 0.25 s.
   a. Immediately after Logan jumps on the sled, how fast is it moving?
   b. What was the force on the sled during the impact?
   c. How much energy was "lost" in the impact? Where did this energy go?

PART III

Properties of Matter

Individual bees do not have the ability to regulate their body temperature. But a colony of bees, working together, can very precisely regulate the temperature of their hive. How can the bees use the heat generated by their muscles, the structure of the hive, and the evaporation of water to achieve this control? In Part III, we will learn about the flows of matter and energy that drive such processes.

Beyond the Particle Model

The first 11 chapters of this book have made extensive use of the *particle model* in which we represent objects as point masses. The particle model is especially useful for describing how discrete objects move through space and how they interact with each other. Whether a ball is made of metal or wood is irrelevant to calculating its trajectory.

But there are many situations where the distinction between metal and wood is crucial. If you toss a metal ball and a wood ball into a pond, one sinks and the other floats. If you stir a pan on the stove with a metal spoon, it can quickly get too hot to hold unless it has a wooden handle.

Wood and metal have different physical properties. So do air and water. Our goal in Part III is to describe and understand the similarities and differences of different materials. To do so, we must go beyond the particle model and dig deeper into the nature of matter.

Macroscopic Physics

In Part III, we will be concerned with systems that are solids, liquids, or gases. Properties such as pressure, temperature, specific heat, and viscosity are characteristics of the system as a whole, not of the individual particles. Solids, liquids, and gases are often called *macroscopic* systems—the prefix "macro" (the opposite of "micro") meaning "large." We'll make sense of the behavior of these macroscopic properties by considering a microscopic view in which we think of these systems as collections of particle-like atoms. This "micro-to-macro" development will be a key piece of the following chapters.

In the coming chapters, we'll consider a wide range of practical questions, such as:

- How do the temperature and pressure of a system change when you heat it? Why do some materials respond quickly, others slowly?
- What are the mechanisms by which a system exchanges heat energy with its environment? Why does blowing on a cup of hot coffee cause it to cool off?
- Why are there three phases of matter—solids, liquids, gases? What happens during a phase change?
- Why do some objects float while others, with the same mass, sink? What keeps a massive steel ship afloat?
- What are the laws of motion of a flowing liquid? How do they differ from the laws governing the motion of a particle?

Both Newton's laws and the law of conservation of energy will remain important tools—they are, after all, the basic laws of physics—but we'll have to learn how they apply to macroscopic systems.

It should come as no surprise that an understanding of macroscopic systems and their properties is essential for understanding the world around us. Biological systems, from cells to ecosystems, are macroscopic systems exchanging energy with their environment. On a larger scale, energy transport on earth gives us weather, and the exchange of energy between the earth and space determines our climate.

12 Thermal Properties of Matter

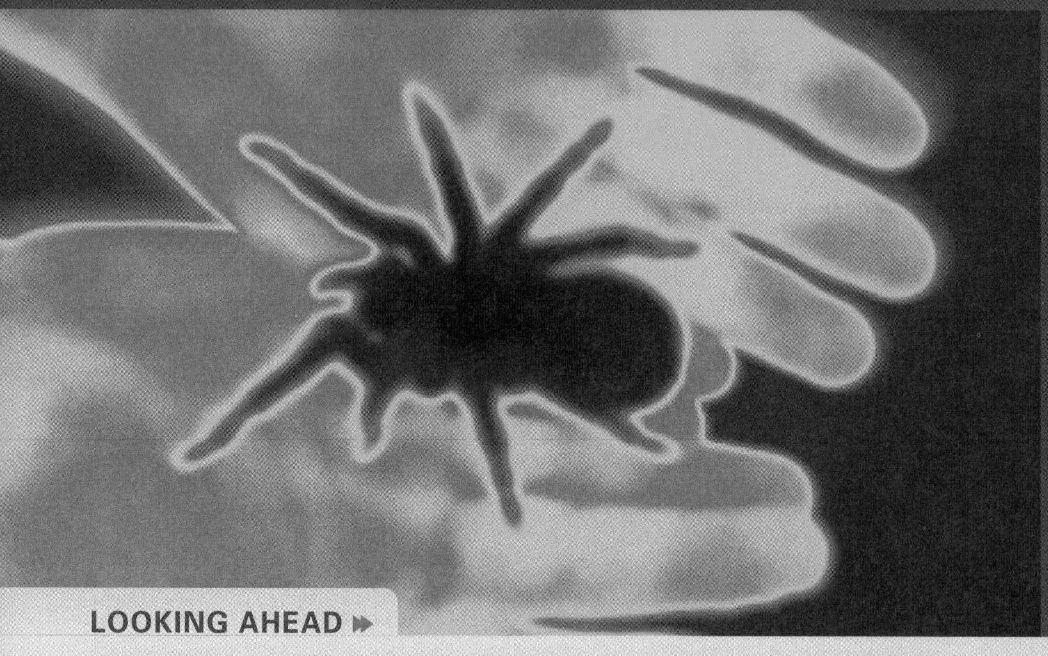

This thermal image shows a person with warm hands holding a much cooler tarantula. Why do the hands radiate energy? And why does the tarantula radiate less?

LOOKING AHEAD ▸▸

The goal of Chapter 12 is to use the atomic model of matter to explain and explore many macroscopic phenomena associated with heat, temperature, and the properties of matter.

The Ideal Gas

We'll continue to develop a model of the ideal gas based on the motion of individual atoms or molecules.

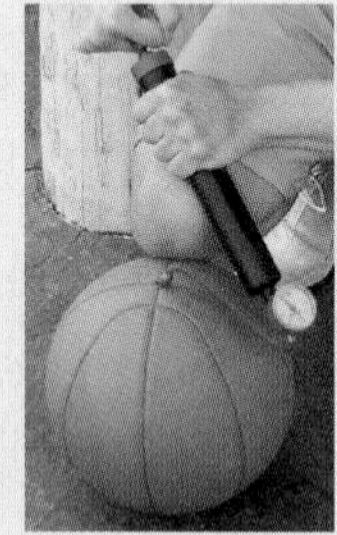

Pressure is due to collisions of molecules with the inside surface of the ball. More molecules or faster molecules mean more pressure.

Looking Back ◂◂
10.5 The nature of thermal energy
11.4 The ideal-gas model, the meaning of temperature

The Atomic Model

The central theme that runs through this chapter is the atomic model. We will explain the thermal properties of matter by modeling solids, liquids, and gases in terms of atoms and the interactions between atoms.

We can explain a wide range of physical phenomena in terms of this simple underlying model.

Thermal Expansion

When liquids and solids are heated, they expand because of the increased motion of their atoms or molecules.

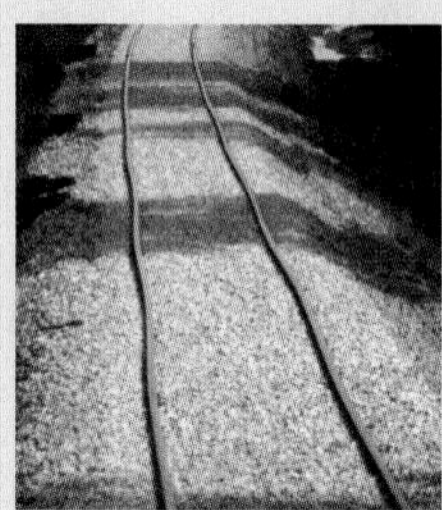

In Alaska, the heating of normally cool railroad tracks in the summer sun makes them expand and may create "sun kinks."

Ideal-Gas Processes

The pressure, volume, and temperature of a gas are related. Changing one means changing one or both of the others.

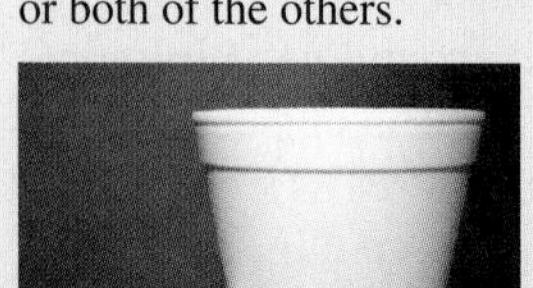

A Styrofoam cup is made of gas bubbles in plastic. The cup on the left was lowered deep into the ocean, where the huge pressure compressed the bubbles and deformed the cup.

Changes on Heating and Cooling

Heat causes both changes in temperature and changes of **phase**— solids change to liquid, liquids to gas.

Molten rock cools and solidifies when it hits the ocean. This transfer of heat causes the ocean water to boil.

Looking Back ◂◂
11.5 Heat and the first law of thermodynamics

Heat Transfer

We'll explore the mechanisms by which heat is transferred from one object to another.

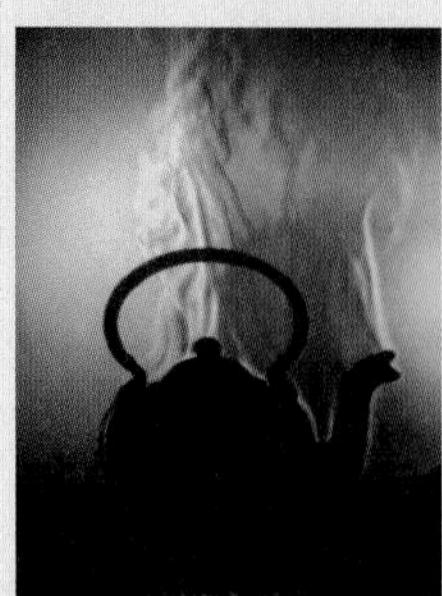

Air heated by the kettle streams upward, an example of heat transfer by **convection.**

12.1 The Atomic Model of Matter

We began exploring the concepts of thermal energy, temperature, and heat in Chapter 11, but many unanswered questions remain. How do the properties of matter depend on temperature? When you add heat to a system, by how much does its temperature change? And how is heat transferred to or from a system?

These are questions about the *macroscopic* state of systems, but we'll start our exploration by looking at a *microscopic* view, the atomic model that we've used to explain friction, elastic forces, and the nature of thermal energy. In this chapter, we'll use the atomic model to understand and explain the thermal properties of matter.

As you know, each element and most compounds can exist as a solid, liquid, or gas. These three **phases** of matter are familiar from everyday experience. An atomic view of the three phases is shown in **FIGURE 12.1**.

- A **gas** is a system in which each particle moves freely through space until, on occasion, it collides with another particle or the wall of its container.
- In a **liquid,** weak bonds permit motion while keeping the particles close together.
- A rigid **solid** has a definite shape and can be compressed or deformed only slightly, as we saw in Chapter 8. It consists of atoms connected by spring-like molecular bonds.

FIGURE 12.1 Atomic models of the three phases of matter: solid, liquid, and gas.

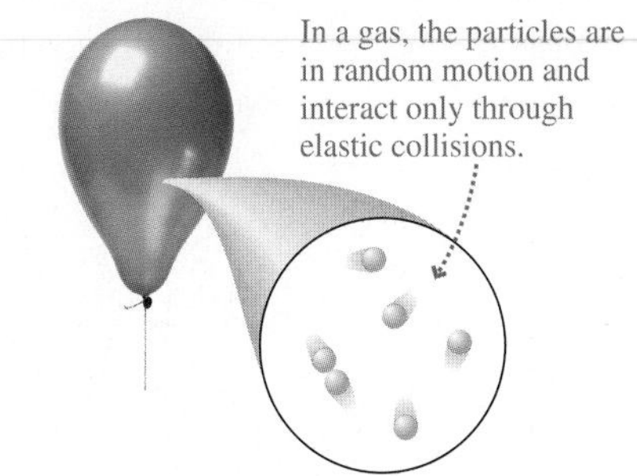

In a gas, the particles are in random motion and interact only through elastic collisions.

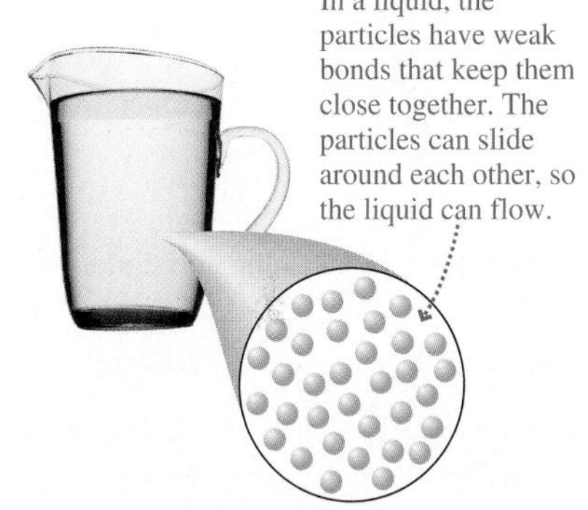

In a liquid, the particles have weak bonds that keep them close together. The particles can slide around each other, so the liquid can flow.

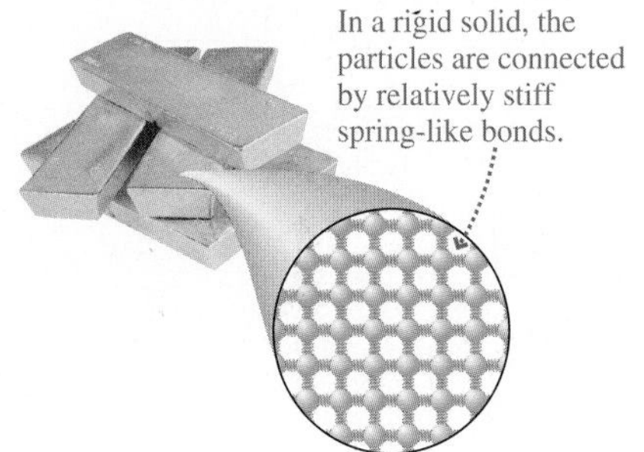

In a rigid solid, the particles are connected by relatively stiff spring-like bonds.

Our atomic model makes some simplifications that are worth noting. The basic particles of the gas in Figure 12.1 are drawn as simple spheres; no mention is made of the nature of the particles. The balloon might contain either helium (in which the basic particles are helium atoms) or air (in which the basic particles are nitrogen and oxygen molecules). A helium atom and a nitrogen molecule are quite different from each other, but many of the properties of the gas as a whole do not depend on the nature of the particles—a gas of helium atoms or oxygen molecules may behave identically. In such cases, we will simply refer to gas *particles,* which may be either atoms or molecules.

The basic particles in the liquid water in Figure 12.1 are water molecules. We ignore the structure of the molecules, so when we speak of the bonds that hold the particles in the liquid state, we are referring to the relatively weak intermolecular bonds among the water molecules, not the strong bonds between the hydrogen and oxygen atoms that form the molecules. The basic particles of the gold bars in Figure 12.1 are gold atoms. The bonds that hold these atoms together are the bonds that give the solid its structure. But there are solids composed of molecules that are held together by bonds between the molecules; water ice is one such example. In this case, the particles are the molecules, and the bonds that form the solid are the bonds between the molecules.

Atomic Mass and Atomic Mass Number

Before we see how the atomic model explains the thermal properties of matter, we need to remind you of some "atomic accounting." Recall that atoms of different elements have different masses. The mass of an atom is determined primarily by its most massive constituents: the protons and neutrons in its nucleus. The *sum* of the number of protons and the number of neutrons is the **atomic mass number** A:

$$A = \text{number of protons} + \text{number of neutrons}$$

A, which by definition is an integer, is written as a leading superscript on the atomic symbol. For example, the primary isotope of carbon, with six protons (which makes it carbon) and six neutrons, has $A = 12$ and is written ^{12}C. The radioactive isotope ^{14}C, used for carbon dating of archeological finds, contains six protons and eight neutrons.

TABLE 12.1 Some atomic mass numbers

Element	Symbol	A
Hydrogen	^{1}H	1
Helium	^{4}He	4
Carbon	^{12}C	12
Nitrogen	^{14}N	14
Oxygen	^{16}O	16
Neon	^{20}Ne	20
Aluminum	^{27}Al	27
Argon	^{40}Ar	40
Lead	^{207}Pb	207

The **atomic mass** scale is established by defining the mass of ^{12}C to be exactly 12 u, where u is the symbol for the *atomic mass unit.* That is, $m(^{12}C) = 12$ u. In kg, the atomic mass unit is

$$1 \text{ u} = 1.66 \times 10^{-27} \text{ kg}$$

Atomic masses are all very nearly equal to the integer atomic mass number A. For example, the mass of ^{1}H, with $A = 1$, is $m = 1.0078$ u. For our present purposes, it will be sufficient to use the integer atomic mass numbers as the values of the atomic mass. That is, we'll use $m(^{1}H) = 1$ u, $m(^{4}He) = 4$ u, and $m(^{16}O) = 16$ u. For molecules, the **molecular mass** is the sum of the atomic masses of the atoms that form the molecule. Thus the molecular mass of the diatomic molecule O_2, the constituent of oxygen gas, is $m(O_2) = 2m(^{16}O) = 32$ u.

NOTE ► An element's atomic mass number is *not* the same as its atomic number. The *atomic number,* which gives the element's position in the periodic table, is the number of protons. ◄

Table 12.1 lists the atomic mass numbers of some of the elements that we'll use for examples and homework problems. A complete periodic table, including atomic masses, is found in Appendix B.

The Definition of the Mole

TABLE 12.2 Monatomic and diatomic gases

Monatomic	Diatomic
Helium (He)	Hydrogen (H_2)
Neon (Ne)	Nitrogen (N_2)
Argon (Ar)	Oxygen (O_2)

One way to specify the amount of substance in a system is to give its mass. Another way, one connected to the number of atoms, is to measure the amount of substance in *moles.* **1 mole of substance, abbreviated 1 mol, is 6.02×10^{23} basic particles.**

The basic particle depends on the substance. Helium is a **monatomic gas,** meaning that the basic particle is the helium atom. Thus 6.02×10^{23} helium atoms are 1 mol of helium. But oxygen gas is a **diatomic gas** because the basic particle is the two-atom diatomic molecule O_2. 1 mol of oxygen gas contains 6.02×10^{23} *molecules* of O_2 and thus $2 \times 6.02 \times 10^{23}$ oxygen atoms. Table 12.2 lists the monatomic and diatomic gases that we will use for examples and problems.

The number of basic particles per mole of substance is called **Avogadro's number** N_A. The value of Avogadro's number is thus

$$N_A = 6.02 \times 10^{23} \text{ mol}^{-1}$$

The number n of moles in a substance containing N basic particles is

$$n = \frac{N}{N_A} \qquad (12.1)$$

Moles of a substance in terms of the number of basic particles

One mole of helium, sulfur, copper, and mercury.

The **molar mass** of a substance, M_{mol}, is the mass *in grams* of 1 mol of substance. To a good approximation, the numerical value of the molar mass equals the numerical value of the atomic or molecular mass. That is, the molar mass of He, with $m = 4$ u, is M_{mol} (He) = 4 g/mol, and the molar mass of diatomic O_2 is $M_{mol}(O_2) = 32$ g/mol.

You can use the molar mass to determine the number of moles. In one of the few instances where the proper units are *grams* rather than kilograms, the number of moles contained in a system of mass M consisting of atoms or molecules with molar mass M_{mol} is

$$n = \frac{M \text{ (in grams)}}{M_{mol}} \qquad (12.2)$$

Moles of a substance in terms of its mass

EXAMPLE 12.1 Determining quantities of oxygen

A system contains 100 g of oxygen. How many moles does it contain? How many molecules?

SOLVE The diatomic oxygen molecule O_2 has molar mass $M_{mol} = 32$ g/mol. From Equation 12.2,

$$n = \frac{100 \text{ g}}{32 \text{ g/mol}} = 3.1 \text{ mol}$$

Each mole contains N_A molecules, so the total number is $N = nN_A = 1.9 \times 10^{24}$ molecules.

Volume

An important property that characterizes a macroscopic system is its volume V, the amount of space the system occupies. The SI unit of volume is m^3. Nonetheless, both cm^3 and, to some extent, liters (L) are widely used metric units of volume. In most cases, you *must* convert these to m^3 before doing calculations.

While it is true that 1 m = 100 cm, it is *not* true that $1 \text{ m}^3 = 100 \text{ cm}^3$. **FIGURE 12.2** shows that the volume conversion factor is $1 \text{ m}^3 = 10^6 \text{ cm}^3$. A liter is 1000 cm^3, so $1 \text{ m}^3 = 10^3$ L. A milliliter (1 mL) is the same as 1 cm^3.

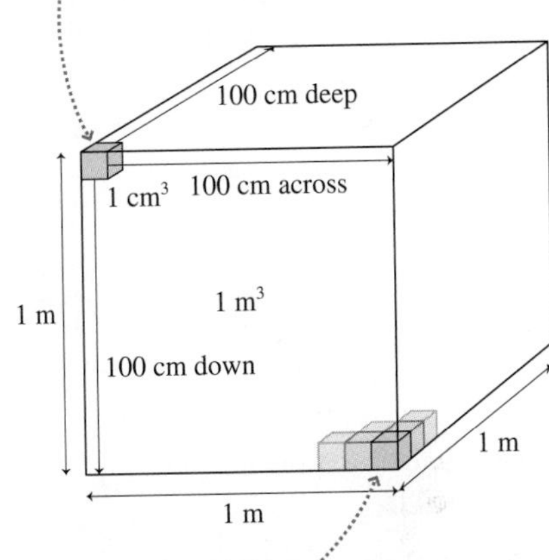

Subdivide the 1 m × 1 m × 1 m cube into little cubes 1 cm on a side. You will get 100 subdivisions along each edge.

100 cm across

100 cm down

FIGURE 12.2 There are 10^6 cm^3 in 1 m^3.

STOP TO THINK 12.1 Which system contains more atoms: 5 mol of helium ($A = 4$) or 1 mol of neon ($A = 20$)?

A. Helium. B. Neon. C. They have the same number of atoms.

12.2 The Atomic Model of an Ideal Gas

Solids and liquids are nearly incompressible because the atomic particles are in close contact with each other. Gases, in contrast, are highly compressible because the atomic particles are far apart. In Chapter 11, we introduced an atomic-level model of an *ideal gas,* reviewed in **FIGURE 12.3**. Our goal in this section is to further develop this model to understand the pressure of an ideal gas.

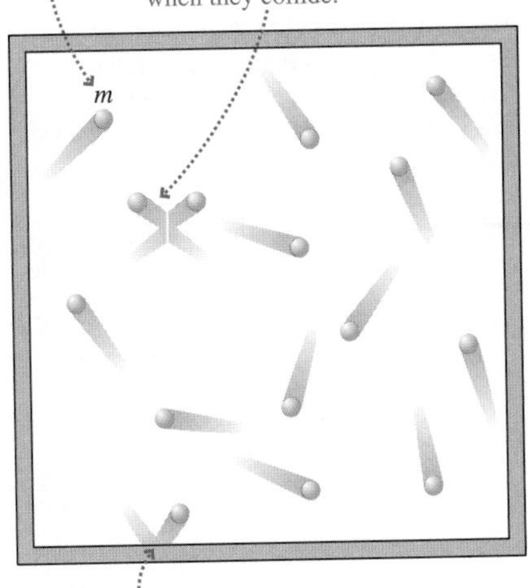

1. The gas is made up of a large number N of particles of mass m, each moving randomly.
2. The particles are quite far from each other and interact only rarely when they collide.
m
3. The collisions of the particles with each other (and with walls of the container) are elastic; no energy is lost in these collisions.

FIGURE 12.3 The ideal-gas model.

Molecular Speeds and Temperature

The atomic model of an ideal gas is based on random motion, so it's no surprise that the individual atoms in a gas are moving at different speeds. **FIGURE 12.4** on the next page shows data from an experiment to measure the molecular speeds in nitrogen gas at 20°C. The results are presented as a histogram, a bar chart in which the height of the bar indicates what percentage of the molecules have a speed in the range of speeds shown below the bar. For example, 16% of the molecules have speeds in the range from 600 m/s to 700 m/s. The most probable speed, as judged from the tallest bar, is ≈ 500 m/s. This is quite fast: ≈ 1200 mph!

Because temperature is related to the average kinetic energy of the atoms, as we learned in Chapter 11, it will be useful to calculate the average kinetic energy for this distribution. An individual atom of mass m and velocity v has kinetic energy $K = \frac{1}{2}mv^2$. Recall that the average of a series of measurements is found by adding all the values and then dividing by the number of data points. Similarly, we can find the average kinetic energy by adding up all the kinetic energies of all the atoms and then dividing by the number of atoms:

$$K_{avg} = \frac{\Sigma \frac{1}{2}mv^2}{N} = \frac{1}{2}m\frac{\Sigma v^2}{N} = \frac{1}{2}m(v^2)_{avg} \qquad (12.3)$$

FIGURE 12.4 The distribution of molecular speeds in nitrogen gas at 20°C.

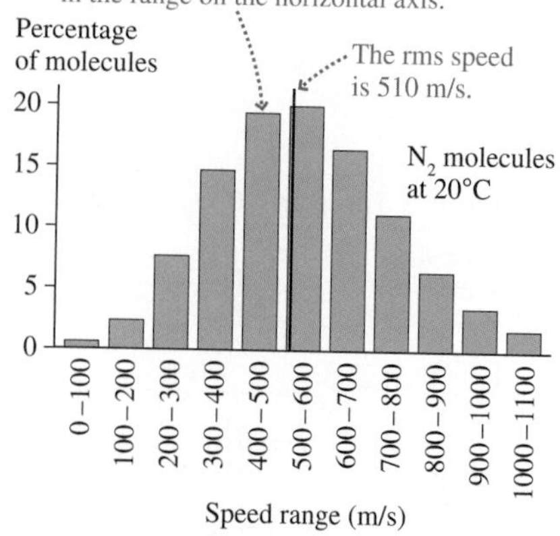

The quantity $\Sigma v^2/N$ is the sum of the values of v^2 for all the atoms divided by the number of atoms. By definition, this is average of the *squares* of all the individual speeds, which we've written $(v^2)_{avg}$.

The square root of this average is about how fast a typical atom in the gas is moving. Because we'll be taking the square root of the average, or mean, of the square of the speeds, we define the **root-mean-square speed** as

$$v_{rms} = \sqrt{(v^2)_{avg}} = \text{speed of a typical atom} \qquad (12.4)$$

The root-mean-square speed is often referred to as the *rms speed*. The rms speed isn't the average speed of atoms in the gas; it's the speed of an atom with the average kinetic energy. But the average speed and the rms speed are very nearly equal, so we'll interpret an rms speed as telling us the speed of a typical atom in the gas.

Rewriting Equation 12.3 in terms of v_{rms} gives the average kinetic energy per atom:

$$K_{avg} = \frac{1}{2}mv_{rms}^2 \qquad (12.5)$$

We learned in Chapter 11 that the temperature of a gas is related to the average kinetic energy of the atoms in the gas by

$$T = \frac{2}{3}\frac{K_{avg}}{k_B} \qquad (12.6)$$

where $k_B = 1.38 \times 10^{-23}$ J/K is Boltzmann's constant. Indeed, Equation 12.4 is really the definition of temperature for an ideal gas—temperature measures the average kinetic energy of the particles in a system. We can now relate the temperature to the speeds of the atoms if we substitute Equation 12.5 into Equation 12.6:

$$T = \frac{1}{3}\frac{mv_{rms}^2}{k_B} \qquad (12.7)$$

Solving Equation 12.7 for the rms speed of the atoms, we find that

$$v_{rms} = \sqrt{\frac{3k_BT}{m}} \qquad (12.8)$$

NOTE ► You must use absolute temperature, in kelvin, to compute rms speeds. ◄

We have been considering ideal gases made of atoms, but our results are equally valid for real gases made of either atoms (such as helium, He) or molecules (such as oxygen, O_2). At a given temperature, Equation 12.8 shows that the speed of atoms or molecules in a gas varies with the atomic or molecular mass. A gas with lighter atoms will have faster atoms, on average, than a gas with heavier atoms. The equation also shows that higher temperatures correspond to faster atomic or molecular speeds. The rms speed is proportional to the *square root* of the temperature. This is a new mathematical form that we will see again, so we will take a look at its properties.

Martian airsicles The atmosphere of Mars is mostly carbon dioxide. At night, the temperature may drop so low that the molecules in the atmosphere will slow down enough to stick together—the atmosphere actually freezes. The frost on the surface in this image from the Viking 2 lander is composed partially of frozen carbon dioxide.

Square-root relationships

MP™

Two quantities are said to have a **square-root relationship** if y is proportional to the square root of x. We write the mathematical relationship as

$$y = A\sqrt{x}$$

y is proportional to the square root of x

The graph of a square-root relationship is a parabola that has been rotated by 90°.

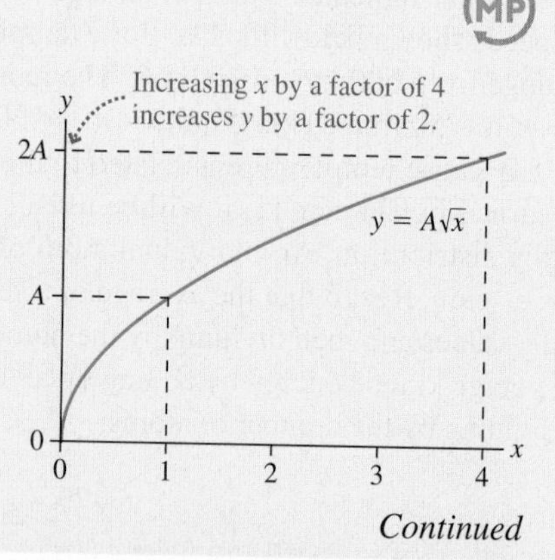

Continued

SCALING If x has the initial value x_1, then y has the initial value y_1. Changing x from x_1 to x_2 changes y from y_1 to y_2. The ratio of y_2 to y_1 is

$$\frac{y_2}{y_1} = \frac{A\sqrt{x_2}}{A\sqrt{x_1}} = \sqrt{\frac{x_2}{x_1}}$$

which is the square root of the ratio of x_2 to x_1.

- If you increase x by a factor of 4, you increase y by a factor of $\sqrt{4} = 2$.
- If you decrease x by a factor of 9, you decrease y by a factor of $\sqrt{9} = 3$.

These examples illustrate a general rule:

Changing x by a factor of c changes y by a factor of $\sqrt{c}$.

Exercise 14 

EXAMPLE 12.2 Speeds of air molecules

Most of the earth's atmosphere is the gas nitrogen, which consists of molecules, N_2. At the coldest temperature ever observed on earth, −129°C, what is the root-mean-square speed of the nitrogen molecules? Does the temperature at the earth's surface ever get high enough that a typical molecule is moving at twice this speed? (The highest temperature ever observed on earth was 58°C.)

PREPARE You can use the periodic table to determine that the mass of a nitrogen atom is 14 u. A molecule consists of two atoms, so its mass is 28 u. Thus the molecular mass in SI units (i.e., kg) is

$$m = 28 \text{ u} \times \frac{1.66 \times 10^{-27} \text{ kg}}{1 \text{ u}} = 4.6 \times 10^{-26} \text{ kg}$$

The problem statement gives two temperatures we'll call T_1 and T_2; we need to express these in kelvin. The lowest temperature ever observed on earth is $T_1 = -129 + 273 = 144$ K; the highest temperature is $T_2 = 58 + 273 = 331$ K.

SOLVE We use Equation 12.8 to find v_{rms} for the nitrogen molecules at T_1:

$$v_{rms} = \sqrt{\frac{3k_B T_1}{m}} = \sqrt{\frac{3(1.38 \times 10^{-23} \text{ J/K})(144 \text{ K})}{4.6 \times 10^{-26} \text{ kg}}} = 360 \text{ m/s}$$

Because the rms speed is proportional to the square root of the temperature, doubling the rms speed would require increasing the temperature by a factor of 4. The ratio of the highest temperature ever recorded to the lowest temperature ever recorded is less than this:

$$\frac{T_2}{T_1} = \frac{331 \text{ K}}{144 \text{ K}} = 2.3$$

The temperature at the earth's surface is never high enough that nitrogen molecules move at twice the computed speed.

ASSESS We can use the square-root relationship to assess our computed result for the molecular speed. Figure 12.4 shows an rms speed of 510 m/s for nitrogen molecules at 20°C, or 293 K. Temperature T_1 is approximately half of this, so we'd expect to compute a speed that is lower by about $1/\sqrt{2}$, which is what we found.

Pressure

Everyone has some sense of the concept of *pressure.* If you get a hole in your bicycle tire, the higher-pressure air inside comes squirting out. It's hard to get the lid off a vacuum-sealed jar because of the low pressure inside. But just what is pressure?

Let's take an atomic-scale view of pressure, defining it in terms of the motion of particles of a gas. Suppose we have a sample of gas in a container with rigid walls. As particles in the gas move around, they sometimes collide with and bounce off the walls, creating a force on the walls, as illustrated in FIGURE 12.5.

The force on a surface is proportional to the area. Doubling the surface area will double the number of collisions and thus double the force. Rather than the force itself, a more useful quantity is the force-to-area ratio F/A. This ratio is a property of the gas itself, independent of the surface, and it's what we call the gas **pressure:**

$$p = \frac{F}{A} \qquad (12.9)$$

Definition of pressure in a gas

You can see from Equation 12.9 that a gas exerts a force of magnitude

$$F = pA \qquad (12.10)$$

on a surface of area A. The force is *perpendicular* to the surface.

FIGURE 12.5 The pressure in a gas is due to the net force of the particles colliding with the walls.

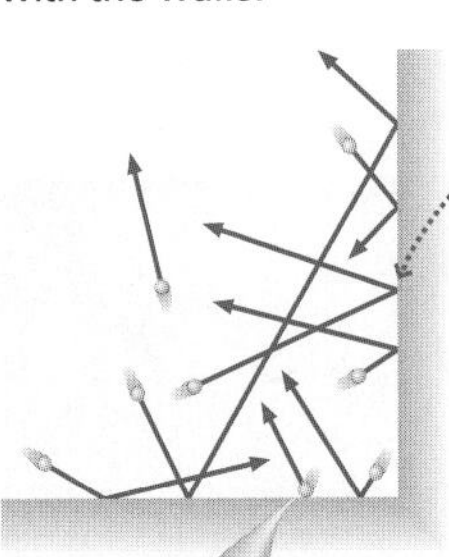

There are an enormous number of collisions of particles against the wall every second.

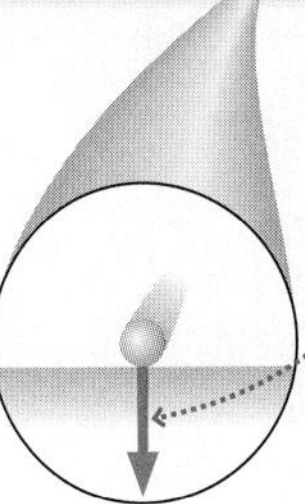

Each collision exerts a force on the wall. The net force due to all the collisions causes the gas to have a pressure. The force of any one collision is incredibly small, but the number of collisions each second is exceedingly large, so the pressure can be considerable.

NOTE ► Pressure itself is *not* a force, even though we sometimes talk informally about "the force exerted by the pressure." The correct statement is that the *gas* exerts a force on a surface. ◄

From its definition, you can see that pressure has units of N/m^2. The SI unit of pressure is the **pascal,** defined as

$$1 \text{ pascal} = 1 \text{ Pa} = 1 \frac{\text{N}}{\text{m}^2}$$

This unit is named for the 17th-century French scientist Blaise Pascal, who was one of the first to study gases. A pascal is a very small pressure, so we usually see pressures in kilopascals, where 1 kPa = 1000 Pa.

The total force on the surface of your body due to the pressure of the atmosphere is over 40,000 pounds. Why doesn't this enormous force pushing in simply crush you? The key is that there is also a force pushing out. **FIGURE 12.6a** shows an empty plastic soda bottle. The bottle isn't a very sturdy structure, but the inward force due to the pressure of the atmosphere outside the bottle doesn't crush it because there is an equal pressure due to air inside the bottle that pushes out. The forces pushing on both sides are quite large, but they exactly balance and so there is no net force.

FIGURE 12.6 The net force depends on the pressure difference.

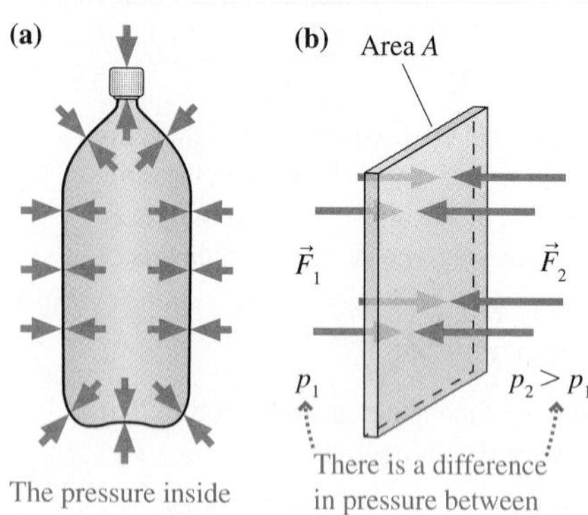
(a)
(b)
Area A
$\vec{F}_1$
$\vec{F}_2$
p_1
$p_2 > p_1$
The pressure inside the bottle equals the pressure outside, so there is no net force.
There is a difference in pressure between the two sides.

A net pressure force is exerted only where there's a pressure *difference* between the two sides of a surface. **FIGURE 12.6b** shows a surface of area A with a pressure difference $\Delta p = p_2 - p_1$ between the two sides. The net pressure force is

$$F_{\text{net}} = F_2 - F_1 = p_2 A - p_1 A = A(p_2 - p_1) = A\,\Delta p$$

This is the force that holds the lid on a vacuum-sealed jar, where the pressure inside is less than the pressure outside. To remove the lid, you have to exert a force greater than the force due to the pressure difference.

Decreasing the number of molecules in a container decreases the pressure because there are fewer collisions with the walls. The pressure in a completely empty container would be $p = 0$ Pa. This is called a *perfect vacuum.* A perfect vacuum cannot be achieved because it's impossible to remove every molecule from a region of space. In practice, a **vacuum** is an enclosed space in which $p \ll 1$ atm. Using $p = 0$ Pa is then a good approximation.

Measuring Pressure

The most important pressure for our daily lives is the pressure of the atmosphere. The pressure of the atmosphere varies with altitude and the weather, but the global average pressure at sea level, called the *standard atmosphere,* is

$$1 \text{ standard atmosphere} = 1 \text{ atm} = 101{,}300 \text{ Pa} = 101.3 \text{ kPa}$$

Just as we measured acceleration in units of g, we will often measure pressure in units of atm.

In the United States, pressure is often expressed in pounds per square inch, or psi. When you measure pressure in the tires on your car or bike, you probably use a gauge that reads in psi. The conversion factor is

$$1 \text{ atm} = 14.7 \text{ psi}$$

◄ **Too little pressure too fast** BIO This photo shows a rockfish, a popular game fish that can be caught at ocean depths up to 1000 ft. The great pressure at these depths is balanced by an equally large pressure inside the fish's swim bladder, a gas-filled organ that we will learn more about in Chapter 13. If the fish is hooked and rapidly raised to the lower pressure of the surface, the pressure inside the swim bladder is suddenly much greater than the pressure outside, causing the swim bladder to expand dramatically, damaging nearby organs. When reeled in from a great depth, these fish seldom survive.

Because the effects of pressure depend on pressure differences, most gauges measure not the actual or *absolute pressure p* but what is called the **gauge pressure.** The gauge pressure, denoted p_g, is the pressure *in excess* of atmospheric pressure; that is,

$$p_g = p - 1 \text{ atm}$$

You need to add 1 atm to the reading of a pressure gauge to find the absolute pressure p you need for doing most calculations in this chapter.

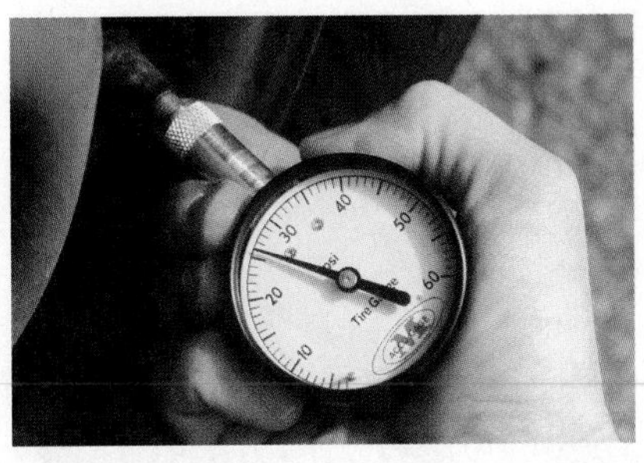

A tire-pressure gauge reads the gauge pressure p_g, not the absolute pressure p. Zero gauge pressure doesn't mean that there's no pressure inside the tire. It means that the pressure inside the tire is equal to atmospheric pressure. There's no pressure difference between the inside and the outside, and so the tire is flat.

EXAMPLE 12.3 Finding the force due to a pressure difference

Patients suffering from decompression sickness may be treated in a hyperbaric oxygen chamber filled with oxygen at greater than atmospheric pressure. A cylindrical chamber with flat end plates of diameter 0.75 m is filled with oxygen to a gauge pressure of 27 kPa. What is the resulting force on the end plate of the cylinder?

PREPARE There is a force on the end plate because of the pressure *difference* between the inside and outside. 27 kPa is the pressure in excess of 1 atm. If we assume the pressure outside is 1 atm, then 27 kPa is Δp, the pressure difference across the surface.

SOLVE The end plate has area $A = \pi(0.75 \text{ m}/2)^2 = 0.442 \text{ m}^2$. The pressure difference results in a net force

$$F_{\text{net}} = A\,\Delta p = (0.442 \text{ m}^2)(27{,}000 \text{ Pa}) = 12 \text{ kN}$$

ASSESS The area of the end plate is large, so we expect a large force. Our answer makes sense, although it is remarkable to think that this force results from the collisions of individual molecules with the plate. The large pressure force must be offset with an equally large force to keep the plate in place, so the end plate is fastened with stout bolts.

From Collisions to Pressure and the Ideal-Gas Law

We can use the fact that the pressure in a gas is due to the collisions of particles with the walls to make some qualitative predictions. **FIGURE 12.7** presents a few such predictions.

FIGURE 12.7 Relating gas pressure to other variables.

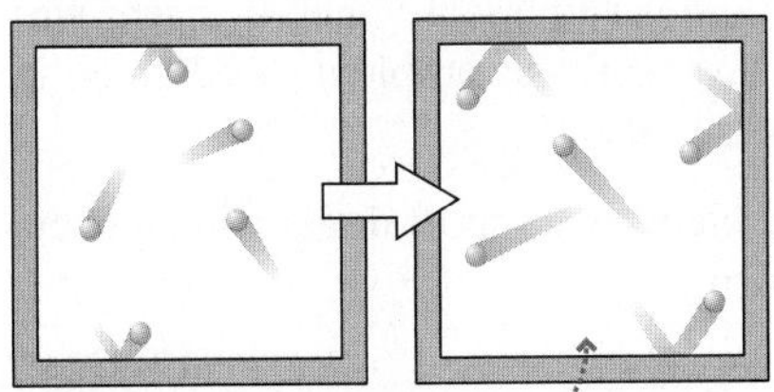

Increasing the temperature of the gas means the particles move at higher speeds. They hit the walls more often and with more force, so there is more pressure.

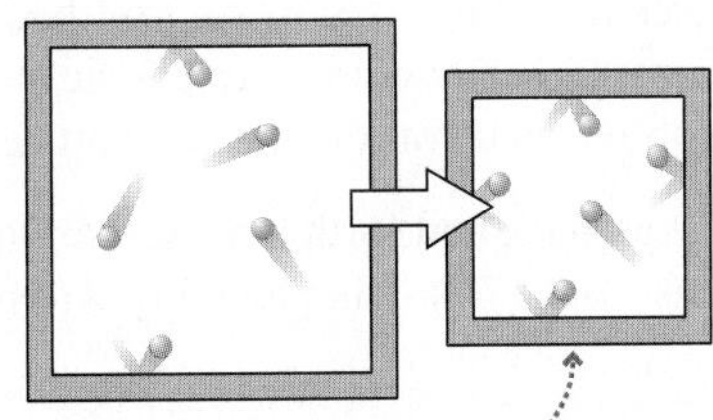

Decreasing the volume of the container means more frequent collisions with the walls of the container, and thus more pressure.

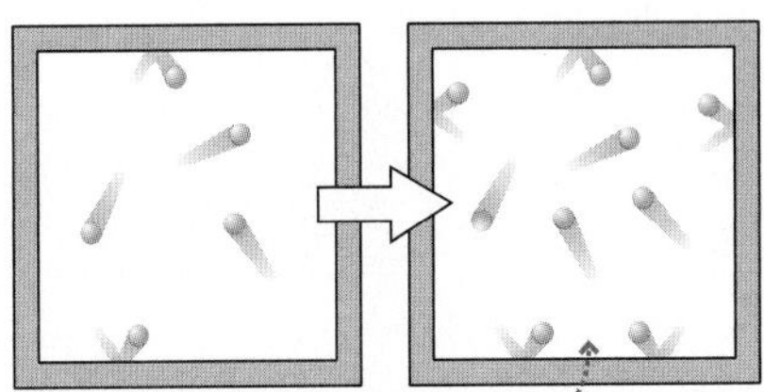

Increasing the number of particles in the container means more frequent collisions with the walls of the container, and thus more pressure.

Based on the reasoning in Figure 12.7, we expect the following proportionalities:

- Pressure should be proportional to the temperature of the gas: $p \propto T$.
- Pressure should be inversely proportional to the volume of the container: $p \propto 1/V$.
- Pressure should be proportional to the number of gas particles: $p \propto N$.

In fact, careful experiments back up each of these predictions. It's not hard to do a quantitative extension of the qualitative treatment above that looks at the collisions with the walls and the resulting forces. This theoretical analysis leads to the same

Pumping up a bicycle tire adds more particles to the fixed volume of the tire, increasing the pressure.

proportionalities noted above, and ultimately produces a single equation that expresses these proportionalities:

$$p = C\frac{NT}{V}$$

The proportionality constant C is none other than Boltzmann's constant k_B, which allows us to write

$$pV = Nk_BT \qquad (12.11)$$

Ideal-gas law, version 1

Equation 12.11 is known as the **ideal-gas law.**

NOTE ► When using the ideal-gas law, make sure that pressure p is the *absolute* pressure (in Pa) and temperature T is the *absolute* temperature (in K). ◄

The important point of the above discussion is that the ideal-gas law is more than just an empirical finding; it can be understood on the basis of the atomic model and an analysis of the forces in the collisions of particles in a gas with the walls of their container.

Equation 12.11 is written in terms of the number N of particles in the gas, whereas the ideal-gas law is stated in chemistry in terms of the number n of moles. But the change is easy to make. The number of particles is $N = nN_A$, so we can rewrite Equation 12.11 as

$$pV = nN_Ak_BT = nRT \qquad (12.12)$$

Ideal-gas law, version 2

In this version of the equation, the proportionality constant—known as the *gas constant*—is

$$R = N_Ak_B = 8.31\ \text{J/mol}\cdot\text{K}$$

The units may seem unusual, but the product of Pa and m^3, the units of pV, is equivalent to J.

NOTE ► You may have learned to work gas problems using units of atmospheres and liters. To do so, you had a different numerical value of R that was expressed in those units. In physics we always work gas problems in SI units, so the above value of R is the one you need to use. ◄

We will go back and forth between solving problems in terms of moles and numbers of atoms, so the following equality will be useful:

$$Nk_B = nN_Ak_B = nR \qquad (12.13)$$

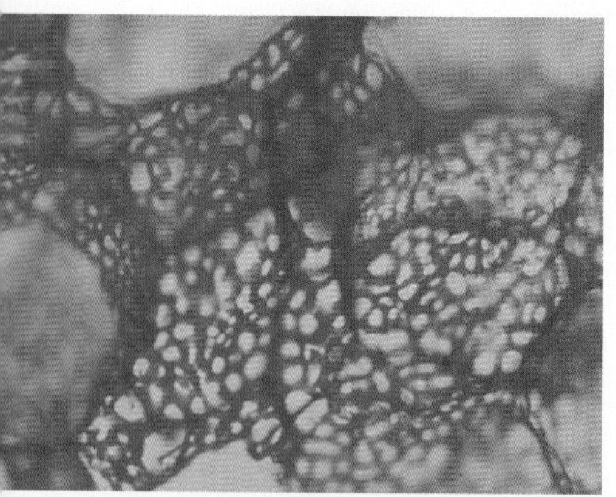

◄ **Gas exchange in the lungs** BIO When you draw a breath, how does the oxygen get into your bloodstream? The atomic model provides some insight. The picture is a highly magnified view of the alveoli, air sacs in the lungs, surrounded by capillaries, fine blood vessels. The thin membranes of the alveoli and the capillaries are permeable—small molecules can move across them. If a permeable membrane separates two regions of space having different concentrations of a molecule, the rapid motion of the molecules causes a net transport of molecules in the direction of lower concentration, as we might expect from the discussion of why heat flows from hot to cold in Chapter 11. This transport—entirely due to the motion of the molecules—is known as **diffusion.** In the lungs, the higher concentration of oxygen in the alveoli drives the diffusion of oxygen into the blood in the capillaries. At the same time, carbon dioxide diffuses from the blood into the alveoli. Diffusion is a rapid and effective means of transport in this case because the membranes are thin. Diffusion won't work to get oxygen from the lungs to other parts of the body, because diffusion is slow over large distances, so the oxygenated blood must be pumped throughout the body.

Let's review the meanings and the units of the various quantities in the ideal-gas law:

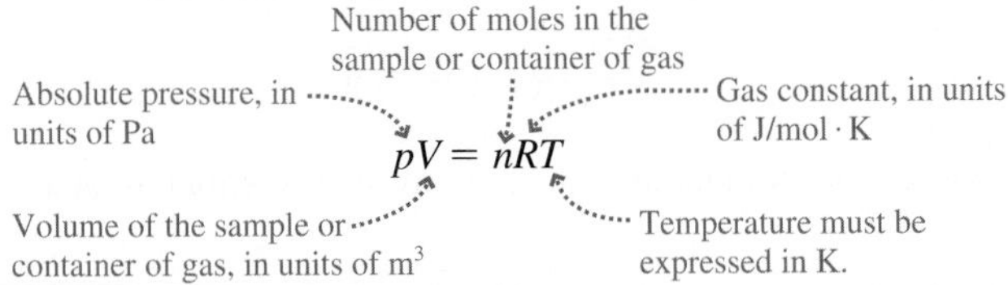

EXAMPLE 12.4 Finding the volume of a mole

What volume is occupied by 1 mole of an ideal gas at a pressure of 1.00 atm and a temperature of 0°C?

PREPARE The first step in ideal-gas law calculations is to convert all quantities to SI units:

$$p = 1.00 \text{ atm} = 101.3 \times 10^3 \text{ Pa}$$

$$T = 0 + 273 = 273 \text{ K}$$

SOLVE We use the ideal-gas law equation to compute

$$V = \frac{nRT}{p} = \frac{(1.00 \text{ mol})(8.31 \text{ J/mol} \cdot \text{K})(273 \text{ K})}{101.3 \times 10^3 \text{ Pa}} = 0.0224 \text{ m}^3$$

We recall from earlier in the chapter that $1.00 \text{ m}^3 = 1000 \text{ L}$, so we can write

$$V = 22.4 \text{ L}$$

ASSESS At this temperature and pressure, we find that the volume of 1 mole of a gas is 22.4 L, a result you might recall from chemistry. When we do calculations using gases, it will be useful to keep this volume in mind to see if our answers make physical sense.

NOTE ▶ The conditions of this example, 1 atm pressure and 0°C, are called *standard temperature and pressure,* abbreviated **STP.** They are a common reference point for many gas processes. ◀

STOP TO THINK 12.2 A sample of ideal gas is in a sealed container. The temperature of the gas and the volume of the container are both increased. What other properties of the gas necessarily change? (More than one answer may be correct.)

A. The rms speed of the gas atoms
B. The thermal energy of the gas
C. The pressure of the gas
D. The number of molecules of gas

12.3 Ideal-Gas Processes

8.4, 8.5, 8.6, 8.8, 8.9, 8.10, 8.11

This chapter is about the thermal properties of matter. What changes occur to matter as you change the temperature? We can use the ideal-gas law to make such deductions for gases because it provides a connection among the pressure, volume, and temperature of a gas. For example, suppose you measure the pressure in the tires on your car on a cold morning. How much will the tire pressure increase after the tires warm up in the sun and the temperature of the air in the tires has increased? We will solve this problem later in the chapter, but, for now, note these properties of this process:

- The quantity of gas is fixed. No air is added to or removed from the tire. All of the processes we will consider will involve a fixed quantity of gas.
- There is a well-defined initial state. The initial values of pressure, volume, and temperature will be designated p_i, V_i, and T_i.
- There is a well-defined final state in which the pressure, volume, and temperature have values p_f, V_f, and T_f.

For gases in sealed containers, the number of moles (and the number of molecules) does not change. In that case, the ideal-gas law can be written as

$$\frac{pV}{T} = nR = \text{constant}$$

The values of the variables in the initial and final states are then related by

$$\frac{p_f V_f}{T_f} = \frac{p_i V_i}{T_i} \tag{12.14}$$

Initial and final states for an ideal gas in a sealed container

This before-and-after relationship between the two states, reminiscent of a conservation law, will be valuable for many problems.

NOTE ► Because pressure and volume appear on both sides of the equation, this is a rare case of an equation for which we may use any units we wish, not just SI units. However, temperature *must* be in K. Unit-conversion factors for pressure and volume are multiplicative factors, and the same factor on both sides of the equation cancels. But the conversion from K to °C is an *additive* factor, and additive factors in the denominator don't cancel. ◄

FIGURE 12.8 The state of the gas and ideal-gas processes can be shown on a *pV* diagram.

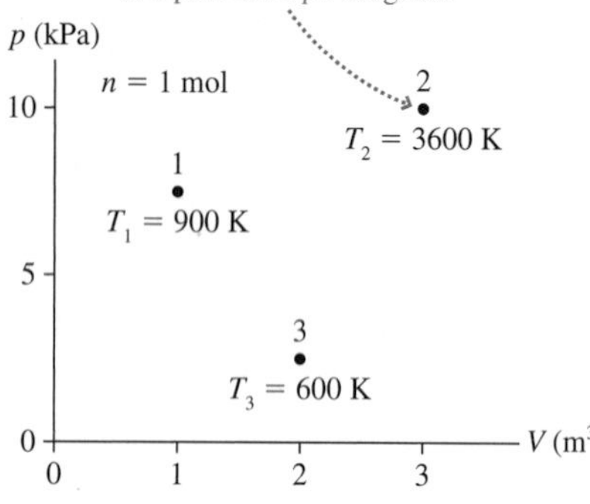

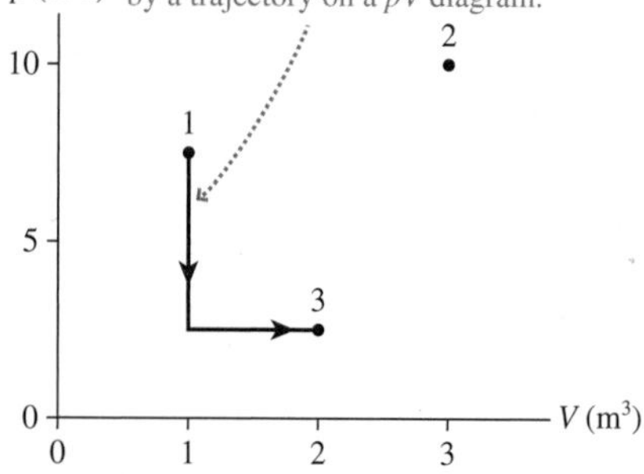

pV Diagrams

It's useful to represent ideal-gas processes on a graph called a ***pV* diagram.** The important idea behind a *pV* diagram is that each point on the graph represents a single, unique state of the gas. This may seem surprising because a point on a graph specifies only the value of pressure and volume. But knowing p and V, and assuming that n is known for a sealed container, we can find the temperature from the ideal-gas law. Thus each point on a *pV* diagram actually represents a triplet of values (p, V, T) specifying the state of the gas.

For example, FIGURE 12.8a is a *pV* diagram showing three states of a system consisting of 1 mol of gas. The values of p and V can be read from the axes, then the temperature at that point calculated from the ideal-gas law. An ideal-gas process—a process that changes the state of the gas by, for example, heating it or compressing it—can be represented as a "trajectory" in the *pV* diagram. FIGURE 12.8b shows one possible process by which the gas of Figure 12.8a is changed from state 1 to state 3.

Constant-Volume Processes

Suppose you have a gas in the closed, rigid container shown in FIGURE 12.9a. Warming the gas will raise its pressure without changing its volume. This is an example of a **constant-volume process.** $V_f = V_i$ for a constant-volume process.

Because the value of V doesn't change, this process is shown as the vertical line $i \rightarrow f$ on the *pV* diagram of FIGURE 12.9b. **A constant-volume process appears on a *pV* diagram as a vertical line.**

FIGURE 12.9 A constant-volume process.

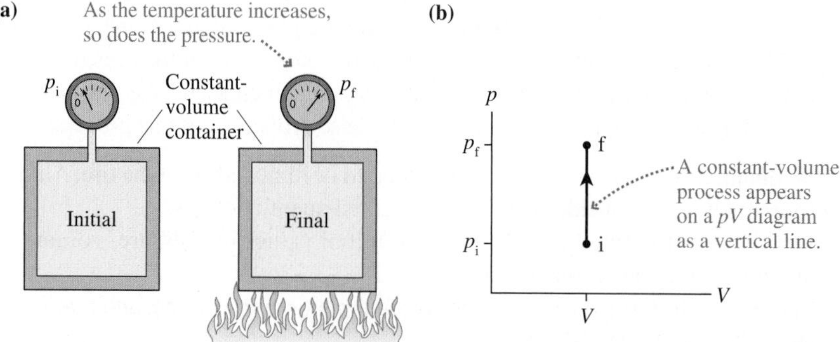

EXAMPLE 12.5 Computing tire pressure on a hot day

The pressure in a car tire is 30.0 psi on a cool morning when the air temperature is 0°C. After the day warms up and bright sun shines on the black tire, the temperature of the air inside the tire reaches 30°C. What is the tire pressure at this temperature?

PREPARE A tire is (to a good approximation) a sealed container with constant volume; this is a constant-volume process. The measured tire pressure is a gauge pressure, but the ideal-gas law requires an absolute pressure. We must correct for this. The initial pressure is

$$p_i = (p_g)_i + 1.00 \text{ atm} = 30.0 \text{ psi} + 14.7 \text{ psi} = 44.7 \text{ psi}$$

Temperatures must be in kelvin, so we convert:

$$T_i = 0°\text{C} + 273 = 273 \text{ K}$$

$$T_f = 30°\text{C} + 273 = 303 \text{ K}$$

SOLVE The gas is in a sealed container, so we can use the ideal-gas law as given in Equation 12.14 to solve for the final pressure. In this equation, we divide both sides by V_f, and then cancel the ratio of the two volumes, which is equal to 1 for this constant-volume process:

$$p_f = p_i \frac{V_i}{V_f}\frac{T_f}{T_i} = p_i \frac{T_f}{T_i}$$

The units for p_f will be the same as those for p_i, so we can keep the initial pressure in psi. The pressure at the higher temperature is

$$p_f = 44.7 \text{ psi} \times \frac{303 \text{ K}}{273 \text{ K}} = 49.6 \text{ psi}$$

This is an absolute pressure, but the problem asks for the measured pressure in the tire—a gauge pressure. Converting to gauge pressure gives

$$(p_g)_f = p_f - 1.00 \text{ atm} = 49.6 \text{ psi} - 14.7 \text{ psi} = 34.9 \text{ psi}$$

ASSESS The temperature has changed by 30 K, which is a bit more than 10% of the initial temperature, so we expect a large change in pressure. Our result seems reasonable, and it has practical implications: If you check the pressure in your tires when they are at a particular temperature, don't expect the pressure to be the same when conditions change!

Constant-Pressure Processes

Many gas processes take place at a constant, unchanging pressure. A constant-pressure process is also called an **isobaric process.** $p_f = p_i$ for an isobaric process.

One way to produce an isobaric process is shown in FIGURE 12.10a, where a gas is sealed in a cylinder by a lightweight, tight-fitting cap—a *piston*—that is free to slide up and down. In fact, the piston *will* slide up or down, compressing or expanding the gas inside, until it reaches the position at which $p_{gas} = p_{ext}$. That's the equilibrium position for the piston, the position at which the upward force $F_{gas} = p_{gas}A$, where A is the area of the face of the piston, exactly balances the downward force $F_{ext} = p_{ext}A$ due to the external pressure p_{ext}. The gas pressure in this situation is controlled by the external pressure. As long as the external pressure doesn't change, neither can the gas pressure inside the cylinder.

FIGURE 12.10 A constant-pressure (isobaric) process.

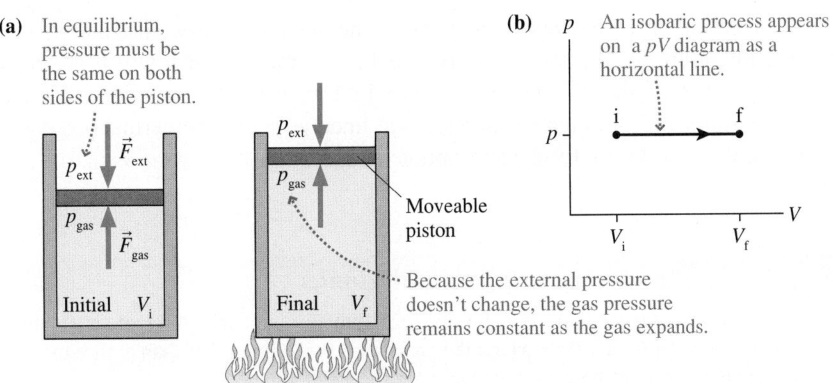

Suppose we heat the gas in the cylinder. The gas pressure doesn't change because the pressure is controlled by the unchanging external pressure, not by the temperature. But as the temperature rises, the faster-moving atoms cause the gas to expand, pushing the piston outward. Because the pressure is always the same, this is an isobaric process, with a trajectory as shown in FIGURE 12.10b. **A constant-pressure process appears on a pV diagram as a horizontal line.**

EXAMPLE 12.6 A constant-pressure compression

A gas in a cylinder with a movable piston occupies 50.0 cm^3 at 50°C. The gas is cooled at constant pressure until the temperature is 10°C. What is the final volume?

PREPARE This is a sealed container, so we can use Equation 12.14. The pressure of the gas doesn't change, so this is an isobaric process with $p_i/p_f = 1$.

The temperatures must be in kelvin, so we convert:

$$T_i = 50°C + 273 = 323\ K$$

$$T_f = 10°C + 273 = 283\ K$$

SOLVE We can use the ideal-gas law for a sealed container to solve for V_f:

$$V_f = V_i \frac{p_i}{p_f}\frac{T_f}{T_i} = 50.0\ cm^3 \times 1 \times \frac{283\ K}{323\ K} = 43.8\ cm^3$$

ASSESS In this example and the previous one, we have not converted pressure and volume units because these multiplicative factors cancel. But we did convert temperature to kelvin because this *additive* factor does *not* cancel.

FIGURE 12.11 A constant-temperature (isothermal) process.

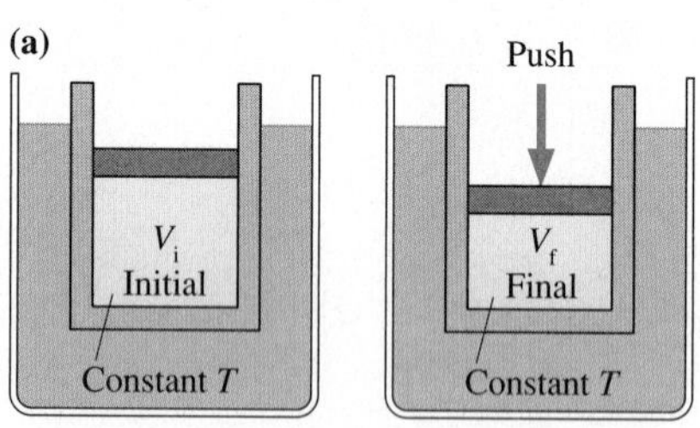

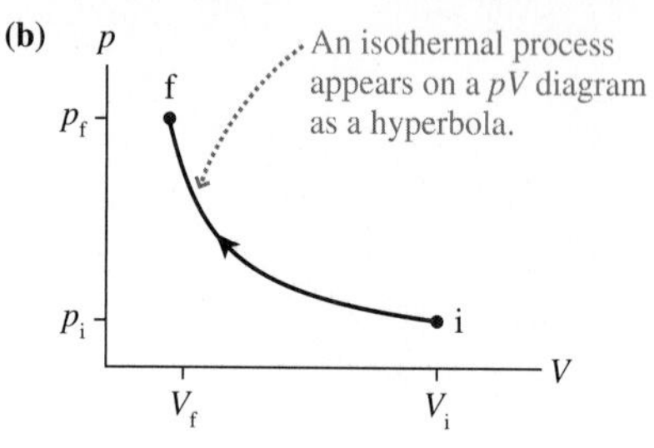

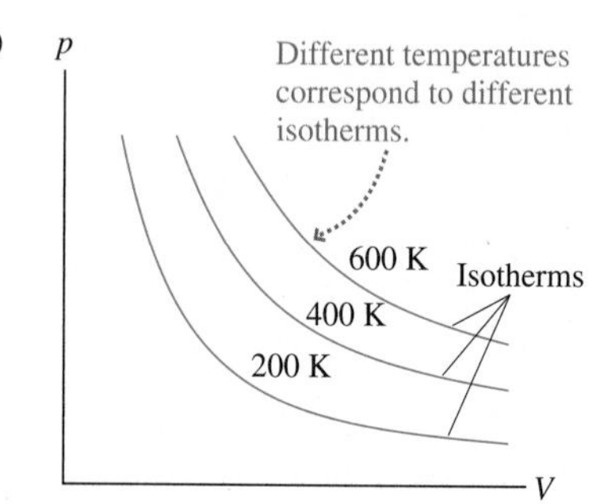

Constant-Temperature Processes

A constant-temperature process is also called an **isothermal process.** For an isothermal process $T_f = T_i$. One possible isothermal process is illustrated in FIGURE 12.11. A piston is being pushed down to compress a gas, but the gas cylinder is submerged in a large container of liquid that is held at a constant temperature. If the piston is pushed *slowly,* then heat-energy transfer through the walls of the cylinder will keep the gas at the same temperature as the surrounding liquid. This would be an *isothermal compression.* The reverse process, with the piston slowly pulled out, would be an *isothermal expansion.*

Representing an isothermal process on the pV diagram is a little more complicated than the two preceding processes because both p and V change. As long as T remains fixed, we have the relationship

$$p = \frac{nRT}{V} = \frac{\text{constant}}{V} \qquad (12.15)$$

p.114 INVERSE

Because there is an inverse relationship between p and V, the graph of an isothermal process is a *hyperbola*.

The process shown as i → f in Figure 12.11b represents the *isothermal compression* shown in Figure 12.11a. An *isothermal expansion* would move in the opposite direction along the hyperbola. The graph of an isothermal process is known as an **isotherm.**

The location of the hyperbola depends on the value of T. If we use a higher constant temperature for the process in Figure 12.11a, the isotherm will move farther from the origin of the pV diagram. Figure 12.11c shows three isotherms for this process at three different temperatures. A gas undergoing an isothermal process will move along the isotherm for the appropriate temperature.

EXAMPLE 12.7 Compressing air in the lungs

A snorkeler takes a deep breath at the surface, filling his lungs with 4.0 L of air. He then descends to a depth of 5.0 m, where the pressure is 0.50 atm higher than at the surface. At this depth, what is the volume of air in the snorkeler's lungs?

PREPARE At the surface, the pressure of the air inside the snorkeler's lungs is 1.0 atm—it's atmospheric pressure at sea level. As he descends, the pressure inside his lungs must rise to match the pressure of the surrounding water, because the body can't sustain large pressure differences between inside and out. Further, the air stays at body temperature, making this an isothermal process with $T_f = T_i$.

SOLVE The ideal-gas law for a sealed container (the lungs) gives

$$V_f = V_i \frac{p_i}{p_f}\frac{T_f}{T_i} = 4.0\text{ L} \times \frac{1.0\text{ atm}}{1.5\text{ atm}} = 2.7\text{ L}$$

Notice that we didn't need to convert pressure to SI units. As long as the units are the same in the numerator and the denominator, they cancel.

ASSESS The air has a smaller volume at the greater pressure, as we would expect. The air inside your lungs does compress—significantly!—when you dive below the surface.

Thermodynamics of Ideal-Gas Processes

Chapter 11 introduced the first law of thermodynamics, and we saw that heat and work are just two different ways to add energy to a system. We've been considering the changes when we heat gases, but now we want to consider the other form of energy transfer—work.

When gases expand, they can do work by pushing against a piston. This is how the engine under the hood of your car works: When the spark plug fires in a cylinder in the engine, it ignites the gaseous fuel-air mixture inside. The hot gas expands, pushing the piston out and, through various mechanical linkages, turning the wheels of your car. Energy is transferred out of the gas as work; we say that the gas does work on the piston. Similarly, the gas in Figure 12.10a does work as it pushes on and moves the piston.

You learned in Chapter 10 that the work done by a constant force F in pushing an object a distance d is $W = Fd$. Let's apply this idea to a gas. FIGURE 12.12a shows a gas cylinder sealed at one end by a movable piston. Force $\vec{F}_{gas}$ is due to the gas pressure and has magnitude $F_{gas} = pA$. Force $\vec{F}_{ext}$, perhaps a force applied by a piston rod, is equal in magnitude and opposite in direction to $\vec{F}_{gas}$. The gas pressure would blow the piston out if the external force weren't there!

Suppose the gas expands at constant pressure, pushing the piston outward from x_i to x_f, a distance $d = x_f - x_i$, as shown in FIGURE 12.12b. As it does, the force due to the gas pressure does work

$$W_{gas} = F_{gas}d = (pA)(x_f - x_i) = p(x_f A - x_i A)$$

But $x_i A$ is the cylinder's initial volume V_i (recall that the volume of a cylinder is length times the area of the base), and $x_f A$ is the final volume V_f. Thus the work done is

$$W_{gas} = p(V_f - V_i) = p\,\Delta V \qquad (12.16)$$

Work done by a gas in an isobaric process

where ΔV is the *change* in volume.

Equation 12.16 has a particularly simple interpretation on a pV diagram. As FIGURE 12.13a shows, $p\,\Delta V$ is the "area under the pV graph" between V_i and V_f.

FIGURE 12.12 The expanding gas does work on the piston.

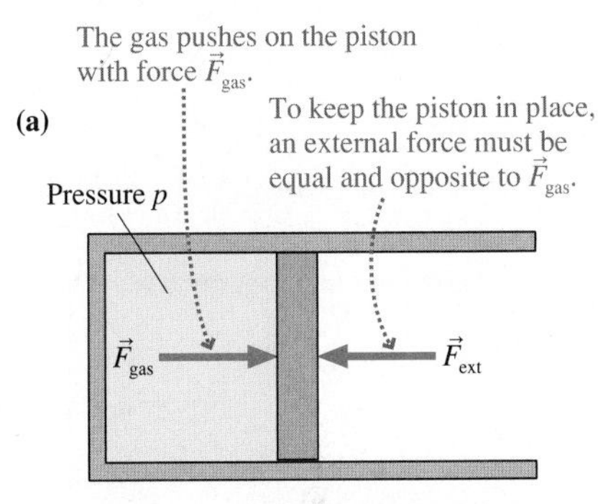

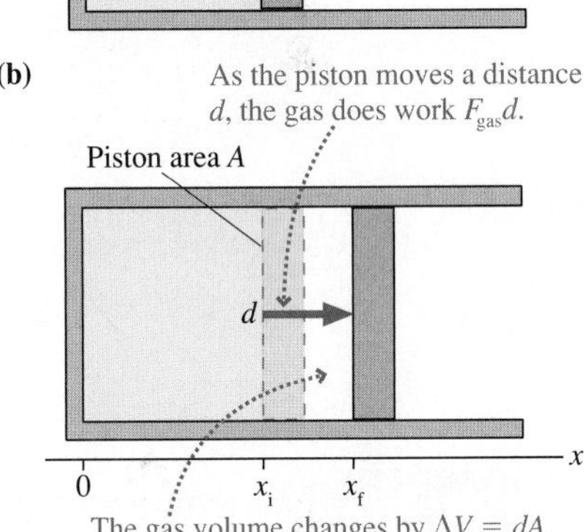

FIGURE 12.13 Calculating the work done in an ideal-gas process.

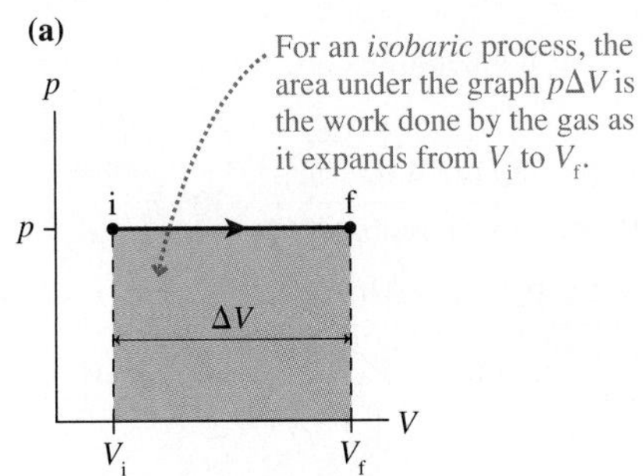

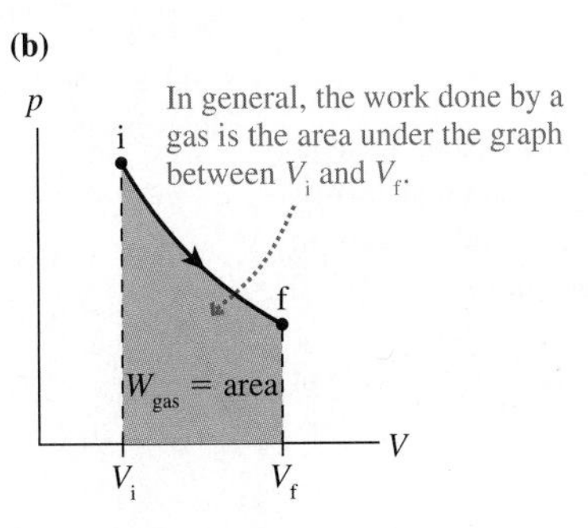

Although we've shown this result only for an isobaric process, it turns out to be true for all ideal-gas processes. That is, as FIGURE 12.13b shows,

$$W_{gas} = \text{area under the } pV \text{ graph between } V_i \text{ and } V_f$$

There are a few things to clarify:

- In order for the gas to do work, its volume must change. No work is done in a constant-volume process.
- The simple relationship of Equation 12.16 applies only to constant-pressure processes. For any other ideal-gas process, you must use the geometry of the pV diagram to calculate the area under the graph.
- To calculate work, pressure must be in Pa and volume in m^3. The product of Pa (which is N/m^2) and m^3 is $N \cdot m$. But $1\ N \cdot m$ is 1 J—the unit of work and energy.
- W_{gas} is positive if the gas expands ($\Delta V > 0$). The gas does work by pushing against the piston. In this case, the work done is energy transferred out of the system, and the energy of the gas decreases. W_{gas} is negative if the piston compresses the gas ($\Delta V < 0$) because the force $\vec{F}_{gas}$ is opposite the displacement of the piston. Energy is transferred into the system as work, and the energy of the gas increases. We often say "work is done *on* the gas," but this just means that W_{gas} is negative.

In the first law of thermodynamics, $\Delta E_{th} = Q + W$, W is the work done by the environment—that is, by force $\vec{F}_{ext}$ acting on the system. But $\vec{F}_{ext}$ and $\vec{F}_{gas}$ are equal and opposite forces, as we noted above, so the work done by the environment is the negative of the work done by the gas: $W = -W_{gas}$. Consequently, the first law of thermodynamics can be written as

$$\Delta E_{th} = Q - W_{gas} \tag{12.17}$$

In Chapter 11 we learned that the thermal energy of an ideal gas depends only on its temperature as $E_{th} = \frac{3}{2}Nk_B T$. Equation 12.13 tells us that Nk_B is equal to nR, so we can write the change in thermal energy of an ideal gas as

$$\Delta E_{th} = \frac{3}{2}Nk_B\,\Delta T = \frac{3}{2}nR\,\Delta T \tag{12.18}$$

EXAMPLE 12.8 Thermodynamics of an expanding gas

A cylinder with a movable piston contains 0.016 mol of helium. A researcher expands the gas via the process illustrated in FIGURE 12.14. To achieve this, does she need to heat the gas? If so, how much heat energy must be added or removed?

FIGURE 12.14 pV diagram for Example 12.8.

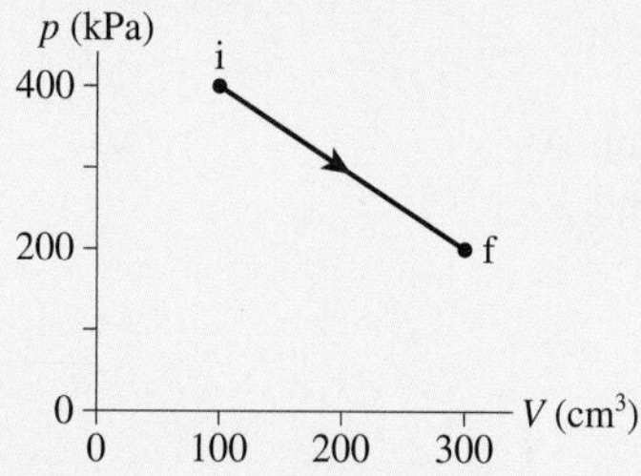

PREPARE As the gas expands, it does work on the piston. Its temperature may change as well, implying a change in thermal energy. We can use the version of the first law of thermodynamics in Equation 12.17 to describe the energy changes in the gas. We'll first find the change in thermal energy by computing the temperature change, then calculate how much work is done by looking at the area under the graph. Once we know W_{gas} and ΔE_{th}, we can use the first law to determine the sign and the magnitude of the heat—telling us whether heat energy goes in or out, and how much.

The graph tells us the pressure and the volume, so we can use the ideal-gas law to compute the temperature at the initial and final points. To do this we'll need the volumes in SI units. Reading the initial and final volumes on the graph and converting, we find

$$V_i = 100\ \text{cm}^3 \times \frac{1\ \text{m}^3}{10^6\ \text{cm}^3} = 1.0 \times 10^{-4}\ \text{m}^3$$

$$V_f = 300\ \text{cm}^3 \times \frac{1\ \text{m}^3}{10^6\ \text{cm}^3} = 3.0 \times 10^{-4}\ \text{m}^3$$

SOLVE The initial and final temperatures are found using the ideal-gas law:

$$T_i = \frac{p_i V_i}{nR} = \frac{(4.0 \times 10^5\ \text{Pa})(1.0 \times 10^{-4}\ \text{m}^3)}{(0.016\ \text{mol})(8.31\ \text{J/mol} \cdot \text{K})} = 300\ \text{K}$$

$$T_f = \frac{p_f V_f}{nR} = \frac{(2.0 \times 10^5\ \text{Pa})(3.0 \times 10^{-4}\ \text{m}^3)}{(0.016\ \text{mol})(8.31\ \text{J/mol} \cdot \text{K})} = 450\ \text{K}$$

The temperature increases, and so must the thermal energy. We can use Equation 12.18 to compute this change:

$$\Delta E_{th} = \frac{3}{2}(0.016 \text{ mol})(8.31 \text{ J/mol} \cdot \text{K})(450 \text{ K} - 300 \text{ K}) = 30 \text{ J}$$

The other piece of the puzzle is to compute the work done. We do this by finding the area under the graph for the process. **FIGURE 12.15** shows that we can do this calculation by viewing the area as a triangle on top of a rectangle. Notice that the areas are in joules because they are the product of Pa and m^3. The total work is

$$W_{gas} = \text{area of triangle} + \text{area of rectangle}$$
$$= 20 \text{ J} + 40 \text{ J} = 60 \text{ J}$$

Now we can use the first law as written in Equation 12.17 to find the heat:

$$Q = \Delta E_{th} + W_{gas} = 30 \text{ J} + 60 \text{ J} = 90 \text{ J}$$

FIGURE 12.15 The work done by the expanding gas is the total area under the graph.

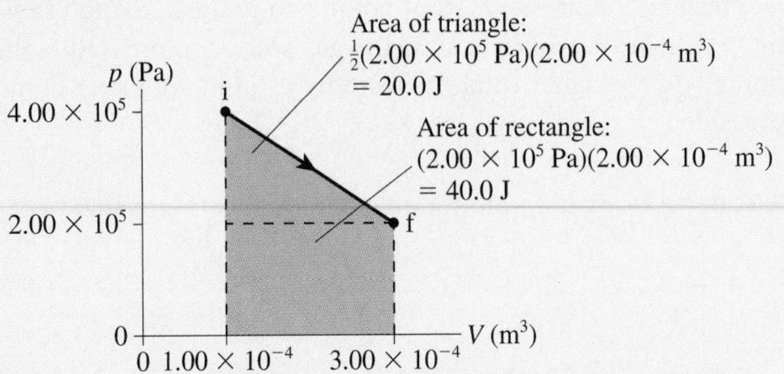

This is a positive number, so—using the conventions introduced in Chapter 11—we see that 90 J of heat energy must be added to the gas.

ASSESS The gas does work—a loss of energy—but its temperature increases, so it makes sense that heat energy must be added.

Adiabatic Processes

You may have noticed that when you pump up a bicycle tire with a hand pump, the pump gets warm. We noted the reason for this in Chapter 11: When you press down on the handle of the pump, a piston in the pump's chamber compresses the gas. According to the first law of thermodynamics, doing work on the gas increases its thermal energy. So the gas temperature goes up, and heat is then transferred through the walls of the pump to your hand.

Now suppose you compress a gas in an insulated container, so that no heat is exchanged with the environment, or you compress a gas so quickly that there is no time for heat to be transferred. In either case, $Q = 0$. If a gas process has $Q = 0$, for either a compression or an expansion, we call this an **adiabatic process.**

An expanding gas does work, so $W_{gas} > 0$. If the expansion is adiabatic, meaning $Q = 0$, then the first law of thermodynamics as written in Equation 12.17 tells us that $\Delta E_{th} < 0$. Temperature is proportional to thermal energy, so the temperature will decrease as well. **An adiabatic expansion lowers the temperature of a gas.** If the gas is compressed, work is done on the gas ($W_{gas} < 0$). If the compression is adiabatic, the first law of thermodynamics implies that $\Delta E_{th} > 0$ and thus that the temperature increases. **An adiabatic compression raises the temperature of a gas.** Adiabatic processes allow you to use work, rather than heat, to change the temperature of a gas.

Adiabatic processes have many important applications. In the atmosphere, large masses of air moving across the planet exchange heat energy with their surroundings very slowly, so adiabatic expansions and compressions can lead to large and abrupt changes in temperature, as we see in the discussion of the Chinook wind.

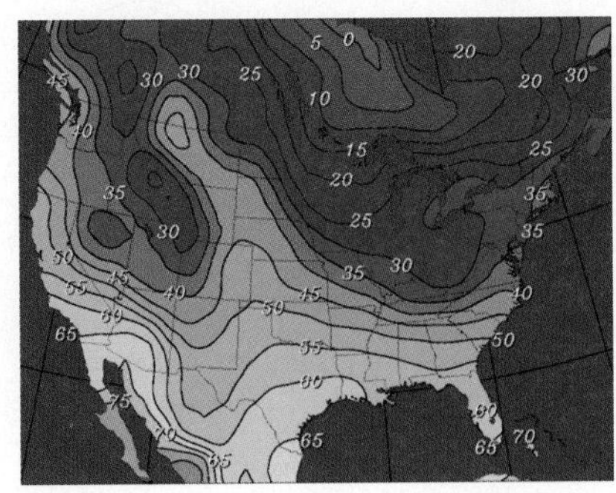

▶ **Warm mountain winds** This image shows surface temperatures (in °F) in North America on a winter day. Notice the bright green area of unseasonably warm temperatures extending north and west from the center of the continent. On this day, a strong westerly wind, known as a Chinook wind, was blowing down off the Rocky Mountains, rapidly moving from high elevations (and low pressures) to low elevations (and higher pressures). The air was rapidly compressed as it descended. The compression was so rapid that no heat was exchanged with the environment, so this was an adiabatic process that significantly increased the air temperature.

CONCEPTUAL EXAMPLE 12.9 Adiabatic curves on a pV diagram

FIGURE 12.16 shows the pV diagram of a gas undergoing an isothermal compression from point 1 to point 2. Sketch how the pV diagram would look if the gas were compressed from point 1 to the same final pressure by a rapid adiabatic compression.

FIGURE 12.16 pV diagram for an isothermal compression.

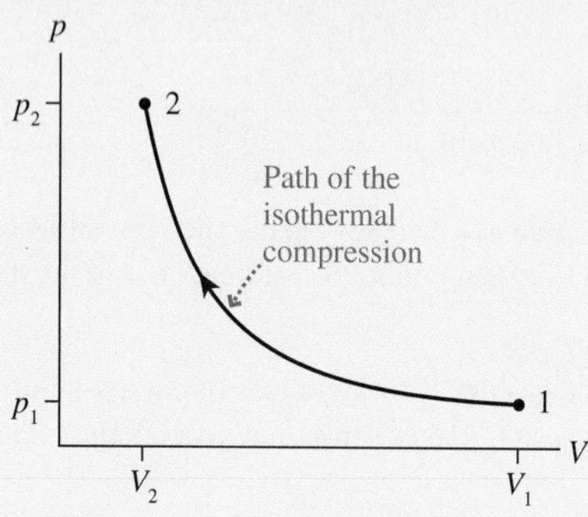

REASON An adiabatic compression increases the temperature of the gas as the work done on the gas is transformed into thermal energy. Consequently, as seen in **FIGURE 12.17**, the curve of the adiabatic compression cuts across the isotherms to end on a higher-temperature isotherm when the gas pressure reaches p_2.

FIGURE 12.17 pV diagram for an adiabatic compression.

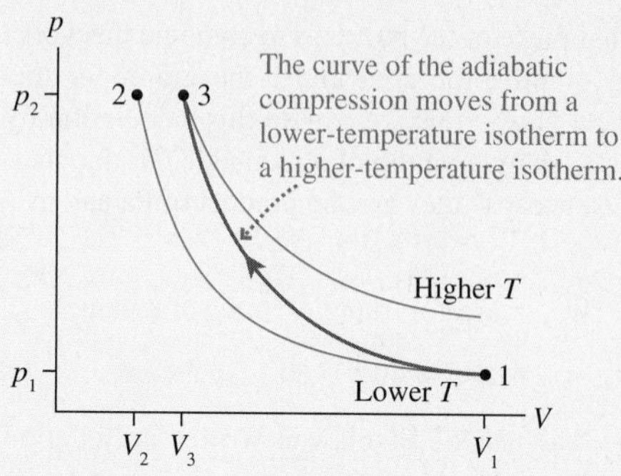

ASSESS In an isothermal compression, heat energy is transferred out of the gas so that the gas temperature stays the same. This heat transfer doesn't happen in an adiabatic compression, so we'd expect the gas to have a higher final temperature. In general, the temperature at the final point of an adiabatic compression is higher than at the starting point. Similarly, an adiabatic expansion ends on a lower-temperature isotherm.

STOP TO THINK 12.3 What is the ratio T_f/T_i for this process?

A. 1/4
B. 1/2
C. 1 (no change)
D. 2
E. 4
F. There is not enough information to decide.

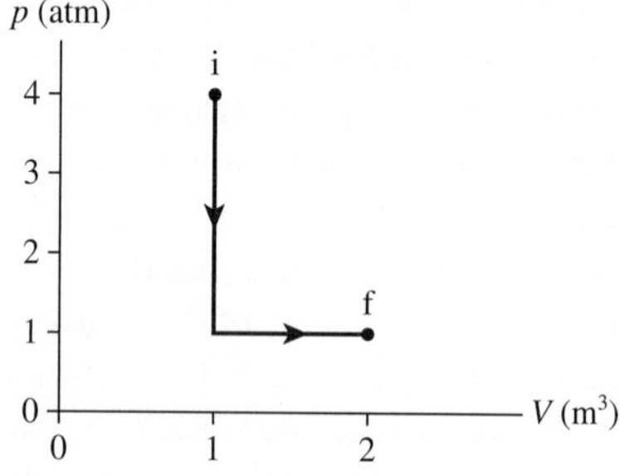

FIGURE 12.18 Increasing the temperature of a solid causes it to expand.

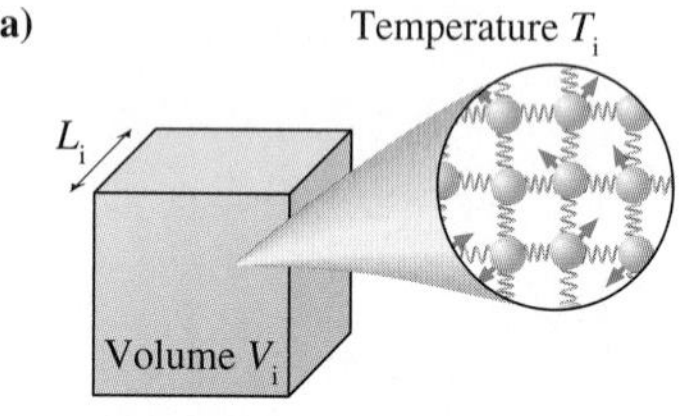

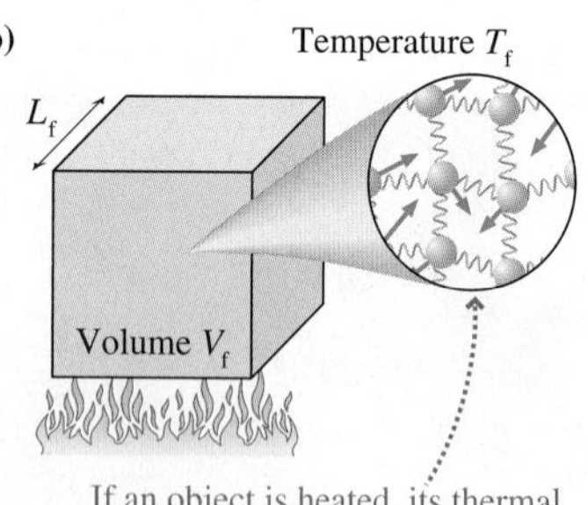

12.4 Thermal Expansion

The bonds between atoms in solids and liquids mean that solids and liquids are much less compressible than gases, as we've noted. But raising the temperature of a solid or a liquid does produce a small, measurable change in volume. You've seen this principle at work if you've ever used a thermometer that contains mercury or alcohol (the kind with the red liquid) to measure a temperature. When you immerse the bulb of the thermometer in something warm, the liquid moves up the column for a simple reason: It expands. This **thermal expansion** underlies many practical phenomena.

FIGURE 12.18a shows a cube of material, initially at temperature T_i. The edge of the cube has length L_i, and the cube's volume is V_i. We then heat the cube, increasing its temperature to T_f, as shown in **FIGURE 12.18b**. The atoms now jiggle around faster, so the bonds are stretched. This increases the average distance between atoms, and so the cube's volume increases to V_f.

For most substances, the change in volume $\Delta V = V_f - V_i$ is linearly related to the change in temperature $\Delta T = T_f - T_i$ by

$$\Delta V = \beta V_i \, \Delta T \qquad (12.19)$$

Volume thermal expansion

ΔV ΔT p.37 PROPORTIONAL

The constant β is known as the **coefficient of volume expansion.** Its value depends on the nature of the bonds in a material; it can vary quite a bit from one substance to another. Because ΔT is measured in K, the units of β are K^{-1}.

As a solid's volume increases, each of its linear dimensions increases as well. We can write a similar expression for this linear thermal expansion. If an object of initial length L_i undergoes a temperature change ΔT, its length changes to L_f. The change in length, $\Delta L = L_f - L_i$, is given by

$$\Delta L = \alpha L_i \, \Delta T \qquad (12.20)$$

Linear thermal expansion

Expanding spans A long steel bridge will slightly increase in length on a hot day and decrease on a cold day. Thermal expansion joints let the bridge's length change without causing the roadway to buckle.

The constant α is the **coefficient of linear expansion.** Note that Equations 12.19 and 12.20 apply equally well to thermal *contractions,* in which case both ΔT and ΔV (or ΔL) are negative.

NOTE ▶ The expressions for thermal expansion are only approximate expressions that apply over a limited range of temperatures. They are what we call **empirical formulas;** they are a good fit to measured data, but they do not represent any underlying fundamental law. There are materials that do *not* closely follow Equations 12.19 and 12.20; water at low temperatures is one example, as we will see. ◀

Values of α and β for some common materials are listed in Table 12.3. The volume expansion of a liquid can be measured, but, because a liquid can change shape, we don't assign a coefficient of linear expansion to a liquid. Although α and β do vary slightly with temperature, the room-temperature values in Table 12.3 are adequate for all problems in this book. The values of α and β are very small, so the changes in length and volume, ΔL and ΔV, are a small fraction of the original length.

TABLE 12.3 Coefficients of linear and volume thermal expansion at 20°C

Substance	Linear α (K^{-1})	Volume β (K^{-1})
Aluminum	23×10^{-6}	69×10^{-6}
Glass	9×10^{-6}	27×10^{-6}
Iron or steel	12×10^{-6}	36×10^{-6}
Concrete	12×10^{-6}	36×10^{-6}
Ethyl alcohol		1100×10^{-6}
Water		210×10^{-6}
Air (and other gases)		3400×10^{-6}

EXAMPLE 12.10 How much closer to space?

The height of the Space Needle, a steel observation tower in Seattle, is 180 meters on a 0°C winter day. How much taller is it on a hot summer day when the temperature is 30°C?

PREPARE The steel expands because of an increase in temperature, which is

$$\Delta T = T_f - T_i = 30°C - 0°C = 30°C = 30 \text{ K}$$

SOLVE The coefficient of linear expansion is given in Table 12.3; we can use this value in Equation 12.20 to compute the increase in height:

$$\Delta L = \alpha L_i \, \Delta T = (12 \times 10^{-6} \text{ K}^{-1})(180 \text{ m})(30 \text{ K}) = 0.065 \text{ m}$$

ASSESS You don't notice buildings getting taller on hot days, so we expect the final answer to be small. The change is a small fraction of the height of the tower, as we expect. Our answer makes physical sense. Compared to 180 m, an expansion of 6.5 cm is not something you would easily notice—but it isn't negligible. The thermal expansion of structural elements in towers and bridges must be accounted for in the design to avoid damaging stresses. When designers failed to properly account for thermal stresses in the marble panels cladding the Amoco Building in Chicago, all 43,000 panels had to be replaced, at great cost.

TRY IT YOURSELF

Thermal expansion to the rescue If you have a stubborn lid on a glass jar, try this: Put the lid under very hot water for a short time. Heating the lid and the jar makes them both expand, but, as you can see from the data in Table 12.3, the steel lid—and the opening in the lid that fits over the glass jar—expands by more than the glass jar. The jar lid is now looser and can be more easily removed.

CONCEPTUAL EXAMPLE 12.11 What happens to the hole?

A metal plate has a circular hole in it. As the plate is heated, does the hole get larger or smaller?

REASON As the plate expands, you might think the hole would shrink (with the metal expanding into the hole). But suppose we took a metal plate and simply drew a circle where a hole could be cut. On heating, the plate and the marked area both expand, as we see in FIGURE 12.19. We could cut a hole on the marked line before or after heating; the size of the hole would be larger in the latter case. Therefore, the size of the hole must expand as the plate expands.

FIGURE 12.19 The thermal expansion of a hole.

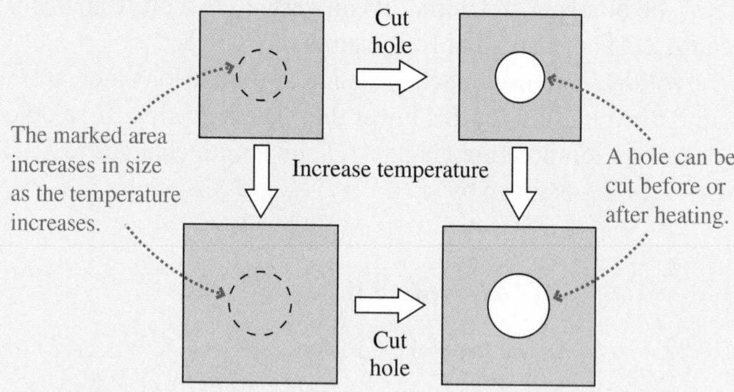

ASSESS This somewhat counterintuitive result has many practical applications, as you can see from the "Try It Yourself" exercise on loosening a jar lid.

Special Properties of Water and Ice

Water, the most important molecule for life, differs from other liquids in many important ways. If you cool a sample of water toward its freezing point of 0°C, you would expect its volume to decrease. FIGURE 12.20a shows a graph of the volume of a mole of water versus temperature. As the water is cooled, you can see that the volume indeed decreases—to a point. Once the water is cooled below 4°C, further cooling results in an *increase* in volume.

FIGURE 12.20 Variation of the volume of a mole of water with temperature.

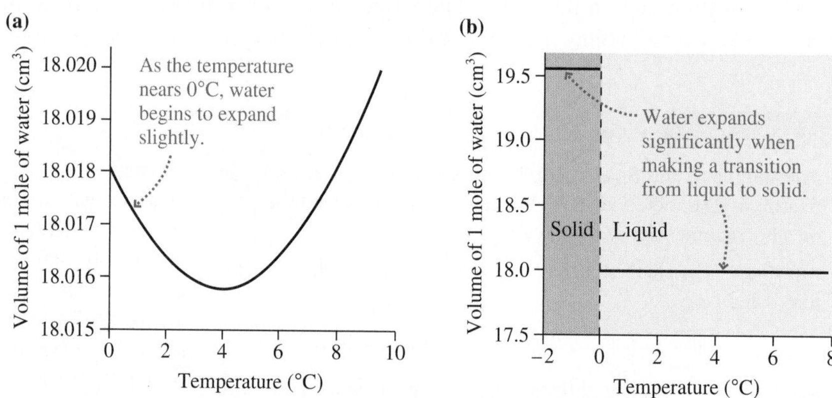

This anomalous behavior of water is a result of the charge distribution on the water molecules, which determines how nearby molecules interact with each other. As water approaches the freezing point, molecules begin to form clusters that are more strongly bound. Water molecules must actually get a bit farther apart to form such clusters, meaning the volume increases. Freezing the water results in an even greater increase in volume, as illustrated in FIGURE 12.20b.

This expansion on freezing has important consequences. In most materials, the solid phase is denser than the liquid phase, and so the solid material sinks. Because water *expands* as it freezes, ice is less dense than liquid water and floats. Not only do ice cubes float in your cold drink, but a lake freezes by forming ice on the top rather than at the bottom. This layer of ice then insulates the water below from much colder air above. Thus most lakes do not freeze solid, allowing aquatic life to survive even the harshest winters.

STOP TO THINK 12.4 An aluminum ring is tight around a solid iron rod. If we wish to loosen the ring to remove it from the rod, we should

A. Increase the temperature of the ring and rod.
B. Decrease the temperature of the ring and rod.

12.5 Specific Heat and Heat of Transformation

If you hold a glass of cold water in your hand, the heat from your hand will raise the temperature of the water. Similarly, the heat from your hand will melt an ice cube; this is what we will call a *phase change*. Increasing the temperature and changing the phase are two possible consequences of heating a system, and they form an important part of our understanding of the thermal properties of matter.

Specific Heat

Adding 4190 J of heat energy to 1 kg of water raises its temperature by 1 K. If you are fortunate enough to have 1 kg of gold, you need only 129 J of heat to raise its temperature by 1 K. The amount of heat that raises the temperature of 1 kg of a substance by 1 K is called the **specific heat** of that substance. The symbol for specific heat is c. Water has specific heat $c_{water} = 4190$ J/kg · K, and the specific heat of gold is $c_{gold} = 129$ J/kg · K. Specific heat depends only on the material from which an object is made. Table 12.4 lists the specific heats for some common liquids and solids.

TABLE 12.4 Specific heats of solids and liquids

Substance	c (J/kg · K)
Solids	
Lead	128
Gold	129
Copper	385
Iron	449
Aluminum	900
Water ice	2090
Mammalian body	3400
Liquids	
Mercury	140
Ethyl alcohol	2400
Water	4190

If heat c is required to raise the temperature of 1 kg of a substance by 1 K, then heat Mc is needed to raise the temperature of mass M by 1 K and $Mc\,\Delta T$ is needed to raise the temperature of mass M by ΔT. In general, the heat needed to bring about a temperature change ΔT is

$$Q = Mc\,\Delta T \qquad (12.21)$$

Heat needed to produce a temperature change ΔT for mass M with specific heat c

Q can be either positive (temperature goes up) or negative (temperature goes down).

Because $\Delta T = Q/Mc$, it takes more heat energy to change the temperature of a substance with a large specific heat than to change the temperature of a substance with a small specific heat. Water, with a very large specific heat, is slow to warm up and slow to cool down. This is fortunate for us. The large "thermal inertia" of water is essential for the biological processes of life.

Hurricane season At night, the large specific heat of water prevents the temperature of a body of water from dropping nearly as much as that of the surrounding air. Early in the morning, water vapor evaporating from a warm lake quickly condenses in the colder air above, forming mist. During the day, the opposite happens: The air becomes much warmer than the water. This lag between air and water temperatures is also at work at larger scales. Ocean temperatures in the northern hemisphere don't reach their maximum values until the late summer or early fall. This is why hurricanes, which are fueled by the energy in warm ocean water, happen in the fall but not the spring.

EXAMPLE 12.12 Energy to run a fever

A 70 kg student catches the flu, and his body temperature increases from 37.0°C (98.6°F) to 39.0°C (102.2°F). How much energy is required to raise his body's temperature?

PREPARE The increase in temperature requires the addition of energy. The change in temperature ΔT is 2.0°C, or 2.0 K.

SOLVE Raising the temperature of the body uses energy supplied internally from the chemical reactions of the body's metabolism, which transfer heat to the body. The specific heat of the body is given in Table 12.4 as 3400 J/kg · K. We can use Equation 12.21 to find the necessary heat energy:

$$Q = Mc\,\Delta T = (70\text{ kg})(3400\text{ J/kg}\cdot\text{K})(2.0\text{ K}) = 4.8 \times 10^5\text{ J}$$

ASSESS The body is mostly water, with a large specific heat, and the mass of the body is large, so we'd expect a large amount of energy to be necessary. Looking back to Chapter 11, we see that this is approximately the energy in a large apple, or the amount of energy required to walk 1 mile.

Phase Changes

The temperature inside the freezer compartment of a refrigerator is typically about −20°C. Suppose you remove a few ice cubes from the freezer and place them in a sealed container with a thermometer. Then, as in FIGURE 12.21a, you put a steady flame under the container. We'll assume that the heating is done slowly so that the inside of the container always has a single, uniform temperature.

FIGURE 12.21b shows the temperature as a function of time. After steadily rising from the initial −20°C, the temperature remains fixed at 0°C for an extended period of time. This is the interval of time during which the ice melts. As it's melting, the ice temperature is 0°C and the liquid water temperature is 0°C. Even though the system is being heated, the temperature doesn't begin to rise until all the ice has melted.

The atomic model can give us some insight into what is taking place. The thermal energy of a solid is the kinetic energy of the vibrating atoms plus the potential energy of stretched and compressed molecular bonds. If you heat a solid, at some point the thermal energy gets so large that the molecular bonds begin to break, allowing the atoms to move around—the solid begins to melt. Continued heating will produce further melting, as the energy is used to break more bonds. The temperature will not rise until all of the bonds are broken. If you wish to return the liquid to a solid, this energy must be removed.

FIGURE 12.21 The temperature as a function of time as water is transformed from solid to liquid to gas.

(a)

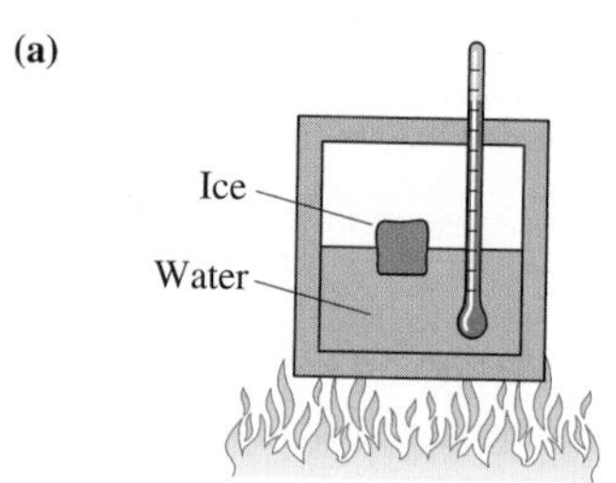

(b)

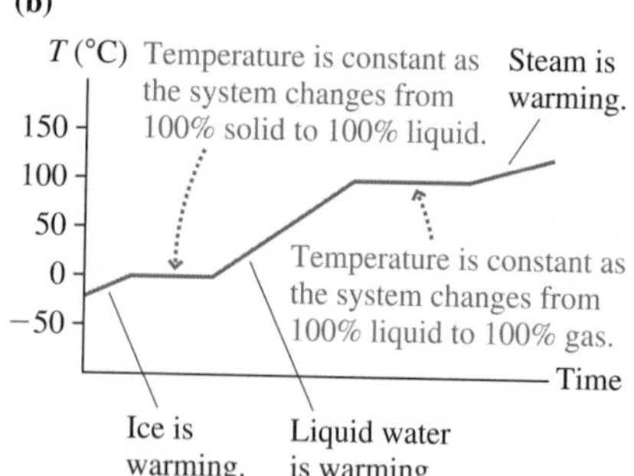

NOTE ► In everyday language, the three phases of water are called *ice, water,* and *steam.* The term "water" implies the liquid phase. Scientifically, these are the solid, liquid, and gas phases of the compound called *water.* When we are working with different phases of water, we'll use the term "water" in the scientific sense of a collection of H_2O molecules. We'll say either *liquid* or *liquid water* to denote the liquid phase. ◄

CONCEPTUAL EXAMPLE 12.13 Strategy for cooling a drink

If you have a warm soda that you wish to cool, is it more effective to add 25 g of liquid water at 0°C or 25 g of water ice at 0°C?

REASON If you add liquid water at 0°C, heat will be transferred from the soda to the water, raising the temperature of the water and lowering that of the soda. If you add water ice at 0°C, heat first will be transferred from the soda to the ice to melt it, transforming the 0°C ice to 0°C liquid water, then will be transferred to the liquid water to raise its temperature. Thus more thermal energy will be removed from the soda, giving it a lower final temperature, if ice is used rather than liquid water.

ASSESS This makes sense because you know that this is what you do in practice. To cool a drink, you drop in an ice cube.

The temperature at which a solid becomes a liquid or, if the thermal energy is reduced, a liquid becomes a solid is called the **melting point** or the **freezing point.** Melting and freezing are *phase changes.* A system at the melting point is in **phase equilibrium,** meaning that any amount of solid can coexist with any amount of liquid. Raise the temperature ever so slightly and the entire system soon becomes liquid. Lower it slightly and it all becomes solid.

You can see another region of phase equilibrium of water in Figure 12.21b at 100°C. This is a phase equilibrium between the liquid phase and the gas phase, and any amount of liquid can coexist with any amount of gas at this temperature. As heat is added to the system, the temperature stays the same. The added energy is used to break bonds between the liquid molecules, allowing them to move into the gas phase. The temperature at which a gas becomes a liquid or a liquid becomes a gas is called the **condensation point** or the **boiling point.**

NOTE ▶ Liquid water becomes solid ice at 0°C, but that doesn't mean the temperature of ice is always 0°C. Ice reaches the temperature of its surroundings. If the air temperature in a freezer is −20°C, then the ice temperature is −20°C, Likewise, steam can be heated to temperatures above 100°C. That doesn't happen when you boil water on the stove because the steam escapes, but steam can be heated far above 100°C in a sealed container. ◀

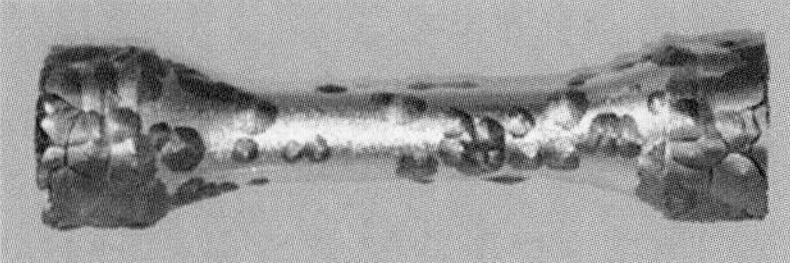

Tin pest Some substances have multiple solid phases in which the atoms are arranged in lattices with different structures. Such substances can undergo a phase change between the solid phases; the atoms rearrange themselves, and the properties of the substance can change. The element tin slowly changes from the usual "white" phase to a brittle "gray" phase at low temperatures. The tin sample in this photo was chilled for one year; the resulting "tin pest" is quite obvious. Some historians speculate that this phase change destroyed the tin buttons on the coats of Napoleon's troops during their winter invasion of Russia in 1812, contributing to their defeat as they froze in the bitter cold.

CONCEPTUAL EXAMPLE 12.14 Fast or slow boil?

You are cooking pasta on the stove; the water is at a slow boil. Will the pasta cook more quickly if you turn up the burner on the stove so that the water is at a fast boil?

REASON Water boils at 100°C; no matter how vigorously the water is boiling, the temperature is the same. It is the temperature of the water that determines how fast the cooking takes place. Adding heat at a faster rate will make the water boil away more rapidly but will not change the temperature—and will not alter the cooking time.

ASSESS This result may seem counterintuitive, but you can try the experiment next time you cook pasta!

Heat of Transformation

In Figure 12.21b, the phase changes appeared as horizontal line segments on the graph. During these segments, heat is being transferred to the system but the temperature isn't changing. The thermal energy continues to increase during a phase change, but, as noted, the additional energy goes into breaking molecular bonds rather than speeding up the molecules. **A phase change is characterized by a change in thermal energy without a change in temperature.**

The amount of heat energy that causes 1 kg of a substance to undergo a phase change is called the **heat of transformation** of that substance. For example, laboratory experiments show that 333,000 J of heat are needed to melt 1 kg of ice at 0°C. The symbol for heat of transformation is *L*. The heat required for the entire system of mass *M* to undergo a phase change is

$$Q = ML \tag{12.22}$$

"Heat of transformation" is a generic term that refers to any phase change. Two particular heats of transformation are the **heat of fusion** L_f, the heat of transformation between a solid and a liquid, and the **heat of vaporization** L_v, the heat of transformation between a liquid and a gas. The heat needed for these phase changes is

$$Q = \begin{cases} \pm ML_f & \text{Heat needed to melt/freeze mass } M \\ \pm ML_v & \text{Heat needed to boil/condense mass } M \end{cases} \tag{12.23}$$

where the ± indicates that heat must be *added* to the system during melting or boiling but *removed* from the system during freezing or condensing. **You must explicitly include the minus sign when it is needed.**

Frozen frogs BIO It seems impossible, but common wood frogs survive the winter with much of their bodies frozen. When you dissolve substances in water, the freezing point lowers. Although the liquid water *between* cells in the frogs' bodies freezes, the water *inside* their cells remains liquid because of high concentrations of dissolved glucose. This prevents the cell damage that accompanies the freezing and subsequent thawing of tissues. When spring arrives, the frogs thaw and appear no worse for their winter freeze.

Lava—molten rock—undergoes a phase change from liquid to solid when it contacts liquid water; the transfer of heat to the water causes the water to undergo a phase change from liquid to gas.

Table 12.5 lists some heats of transformation. Notice that the heat of vaporization is always much higher than the heat of fusion, which our atomic model can explain. Melting breaks just enough molecular bonds to allow the system to lose rigidity and flow. Even so, the molecules in a liquid remain close together and loosely bonded. Vaporization breaks all bonds completely and sends the molecules flying apart. This process requires a larger increase in the thermal energy and thus a larger quantity of heat.

TABLE 12.5 Melting and boiling temperatures and heats of transformation at standard atmospheric pressure

Substance	T_m (°C)	L_f (J/kg)	T_b (°C)	L_v (J/kg)
Nitrogen (N_2)	−210	0.26×10^5	−196	1.99×10^5
Ethyl alcohol	−114	1.09×10^5	78	8.79×10^5
Mercury	−39	0.11×10^5	357	2.96×10^5
Water	0	3.33×10^5	100	22.6×10^5
Lead	328	0.25×10^5	1750	8.58×10^5

EXAMPLE 12.15 Melting a popsicle

A girl eats a 45 g frozen popsicle that was taken out of a −10°C freezer. How much energy does her body use to bring the popsicle up to body temperature?

PREPARE We can assume that the popsicle is pure water. Normal body temperature is 37°C. The specific heats of ice and liquid water are given in Table 12.4; the heat of fusion of water is given in Table 12.5.

SOLVE There are three parts to the problem: The popsicle must be warmed to 0°C, then melted, and then the resulting water must be warmed to body temperature. The heat needed to warm the frozen water by $\Delta T = 10°\text{C} = 10\text{ K}$ to the melting point is

$$Q_1 = Mc_{\text{ice}}\,\Delta T = (0.045\text{ kg})(2090\text{ J/kg}\cdot\text{K})(10\text{ K}) = 940\text{ J}$$

Note that we use the specific heat of water ice, not liquid water, in this equation. Melting 45 g of ice requires heat

$$Q_2 = ML = (0.045\text{ kg})(3.33 \times 10^5\text{ J/kg}) = 15{,}000\text{ J}$$

The liquid water must now be warmed to body temperature; this requires heat

$$Q_3 = Mc_{\text{water}}\,\Delta T = (0.045\text{ kg})(4190\text{ J/kg}\cdot\text{K})(37\text{ K})$$
$$= 7000\text{ J}$$

The total energy is the sum of these three values: $Q_{\text{total}} =$ 23,000 J.

ASSESS More energy is needed to melt the ice than to warm the water, as we would expect. A commercial popsicle has 40 Calories, which is about 170 kJ. Roughly 15% of the chemical energy in this frozen treat is used to bring it up to body temperature!

Evaporation

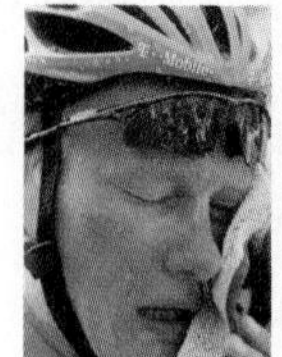

Keeping your cool BIO Humans (and cattle and horses) have sweat glands, so we can perspire to moisten our skin, allowing evaporation to cool our bodies. Animals that do not perspire can also use evaporation to keep cool. Dogs, goats, rabbits, and even birds pant, evaporating water from their respiratory passages. Elephants spray water on their skin; other animals may lick their fur.

Water boils at 100°C. But individual molecules of water can move from the liquid phase to the gas phase at lower temperatures. This process is known as **evaporation.** Water evaporates as sweat from your skin at a temperature well below 100°C. We can use our atomic model to explain this.

In gases, we have seen that there is a variation in particle speeds. The same principle applies to water and other liquids. At any temperature, some molecules will be moving fast enough to go into the gas phase. And they will do so, carrying away thermal energy as they go. The molecules that leave the liquid are the ones that have the highest kinetic energy, so evaporation reduces the average kinetic energy (and thus the temperature) of the liquid left behind.

The evaporation of water, both from sweat and in moisture you exhale, is one of the body's methods of exhausting the excess heat of metabolism to the environment, allowing you to maintain a steady body temperature. The heat to evaporate a mass M of water is $Q = ML_v$, and this amount of heat is removed from your body. However, the heat of vaporization L_v is a little larger than the value for boiling. At a skin temperature of 30°C, the heat of vaporization of water is $L_v = 24 \times 10^5$ J/kg, or 6% higher than the value at 100°C in Table 12.5.

EXAMPLE 12.16 Computing heat loss by perspiration

The human body can produce approximately 30 g of perspiration per minute. At what rate is it possible to exhaust heat by the evaporation of perspiration?

SOLVE The evaporation of 30 g of perspiration at normal body temperature requires heat energy

$$Q = ML_v = (0.030 \text{ kg})(24 \times 10^5 \text{ J/kg}) = 7.2 \times 10^4 \text{ J}$$

This is the heat lost per minute; the rate of heat loss is

$$\frac{Q}{\Delta t} = \frac{7.2 \times 10^4 \text{ J}}{60 \text{ s}} = 1200 \text{ W}$$

ASSESS Given the metabolic power required for different activities, as listed in Chapter 11, this rate of heat removal is sufficient to keep the body cool even when exercising in hot weather—as long as the person drinks enough water to keep up this rate of perspiration.

STOP TO THINK 12.5 1 kg of barely molten lead, at 328°C, is poured into a large beaker holding liquid water right at the boiling point, 100°C. What is the mass of the water that will be boiled away as the lead solidifies?

A. 0 kg B. <1 kg C. 1 kg D. >1 kg

12.6 Calorimetry

At one time or another you've probably put an ice cube into a hot drink to cool it quickly. You were engaged, in a trial-and-error way, in a practical aspect of heat transfer known as **calorimetry,** the quantitative measurement of the heat transferred between systems or evolved in reactions. You know that heat energy will be transferred from the hot drink into the cold ice cube, reducing the temperature of the drink, as shown in FIGURE 12.22.

FIGURE 12.22 Cooling hot coffee with ice.

Heat

Let's make this qualitative picture more precise. FIGURE 12.23 shows two systems that can exchange heat with each other but that are isolated from everything else. Suppose they start at different temperatures T_1 and T_2. As you know, heat energy will be transferred from the hotter to the colder system until they reach a common final temperature T_f. In the cooling coffee example of Figure 12.22, the coffee is system 1, the ice is system 2, and the insulating barrier is the mug.

The insulation prevents any heat energy from being transferred to or from the environment, so energy conservation tells us that any energy leaving the hotter system must enter the colder system. The concept is straightforward, but to state the idea mathematically we need to be careful with signs.

Let Q_1 be the energy transferred to system 1 as heat. Q_1 is positive if energy *enters* system 1, negative if energy *leaves* system 1. Similarly, Q_2 is the energy transferred to system 2. The fact that the systems are merely exchanging energy can be written as $|Q_1| = |Q_2|$. But Q_1 and Q_2 have opposite signs, so $Q_1 = -Q_2$. No energy is exchanged with the environment, so it makes more sense to write this relationship as

$$Q_{net} = Q_1 + Q_2 = 0 \tag{12.24}$$

This idea is easily extended. No matter how many systems interact thermally, Q_{net} must equal zero.

FIGURE 12.23 Two systems interact thermally. In this particular example, where heat is being transferred from system 1 to system 2, $Q_2 > 0$ and $Q_1 < 0$.

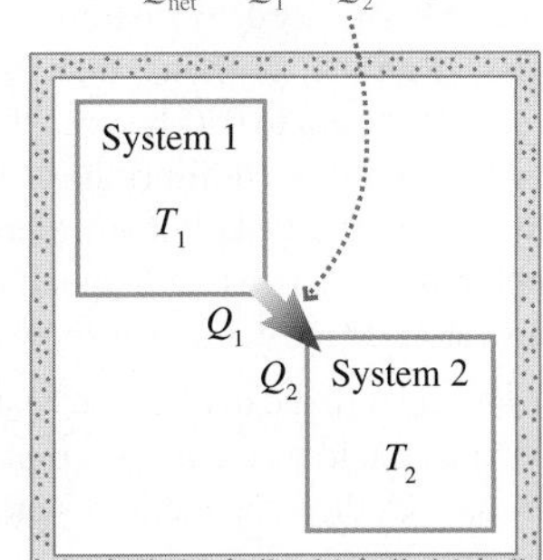
Heat energy is transferred from system 1 to system 2. Energy conservation requires
$|Q_1| = |Q_2|$
Opposite signs mean that
$Q_{net} = Q_1 + Q_2 = 0$
System 1
T_1
Q_1
Q_2
System 2
T_2

NOTE ► The signs are very important in calorimetry problems. ΔT is always $T_f - T_i$, so ΔT and Q are negative for any system whose temperature decreases. The proper sign of Q for any phase change must be supplied *by you,* depending on the direction of the phase change. ◄

PROBLEM-SOLVING STRATEGY 12.1 **Calorimetry problems**

PREPARE Identify the individual interacting systems. Assume that they are isolated from the larger environment. List known information and identify what you need to find. Convert all quantities to SI units.

SOLVE The statement of energy conservation is

$$Q_{net} = Q_1 + Q_2 + \cdots = 0$$

- For systems that undergo a temperature change, $Q_{\Delta T} = Mc(T_f - T_i)$. Be sure to have the temperatures T_i and T_f in the correct order.
- For systems that undergo a phase change, $Q_{phase} = \pm ML$. Supply the correct sign by observing whether energy enters or leaves the system during the transition.
- Some systems may undergo a temperature change *and* a phase change. Treat the changes separately. The heat energy is $Q = Q_{\Delta T} + Q_{phase}$.

ASSESS The final temperature should be in the middle of the initial temperatures. A T_f that is higher or lower than all initial temperatures is an indication that something is wrong, usually a sign error.

EXAMPLE 12.17 Using calorimetry to identify a metal

200 g of an unknown metal is heated to 90.0°C, then dropped into 50.0 g of water at 20.0°C in an insulated container. The water temperature rises within a few seconds to 27.7°C, then changes no further. Identify the metal.

PREPARE The metal and the water interact thermally; there are no phase changes. We know all the initial and final temperatures. We will label the temperatures as follows: The initial temperature of the metal is T_m; the initial temperature of the water is T_w. The common final temperature is T_f. For water, $c_w = 4190$ J/kg · K is known from Table 12.4. Only the specific heat c_m of the metal is unknown.

SOLVE Energy conservation requires $Q_w + Q_m = 0$. Using $Q = Mc(T_f - T_i)$ for each, we have

$$Q_w + Q_m = M_w c_w (T_f - T_w) + M_m c_m (T_f - T_m) = 0$$

This is solved for the unknown specific heat:

$$c_m = \frac{-M_w c_w (T_f - T_w)}{M_m (T_f - T_m)}$$

$$= \frac{-(0.0500 \text{ kg})(4190 \text{ J/kg} \cdot \text{K})(27.7°\text{C} - 20.0°\text{C})}{(0.200 \text{ kg})(27.7°\text{C} - 90.0°\text{C})}$$

$$= 129 \text{ J/kg} \cdot \text{K}$$

Referring to Table 12.4, we find we have either 200 g of gold or, if we made an ever-so-slight experimental error, 200 g of lead!

ASSESS The temperature of the unknown metal changed much more than the temperature of the water. This means that the specific heat of the metal must be much less than that of water, which is exactly what we found.

EXAMPLE 12.18 Calorimetry with a phase change

Your 500 mL diet soda, with a mass of 500g, is at 20°C, room temperature, so you add 100 g of ice from the −20°C freezer. Does all the ice melt? If so, what is the final temperature? If not, what fraction of the ice melts? Assume that you have a well-insulated cup.

PREPARE We need to distinguish between two possible outcomes. If all the ice melts, then $T_f > 0°$C. It's also possible that the soda will cool to 0°C before all the ice has melted, leaving the ice and liquid in equilibrium at 0°C. We need to distinguish between these before knowing how to proceed. All the initial temperatures, masses, and specific heats are known. The final temperature of the combined soda + ice system is unknown.

SOLVE Let's first calculate the heat needed to melt all the ice and leave it as liquid water at 0°C. To do so, we must warm the ice by 20 K to 0°C, then change it to water. The heat input for this two-stage process is

This is the energy to raise the temperature of the ice from −20°C to 0°C. $\Delta T = 20$ K.

This is the energy to melt the ice once it reaches 0°C.

$$Q_{melt} = M_{ice} c_{ice} (20 \text{ K}) + M_{ice} L_f = 37{,}000 \text{ J}$$

where L_f is the heat of fusion of water.

Q_{melt} is a *positive* quantity because we must *add* heat to melt the ice. Next, let's calculate how much heat energy will leave the 500 g soda if it cools all the way to 0°C:

$$Q_{cool} = M_{soda} c_{water} (-20 \text{ K}) = -42{,}000 \text{ J}$$

where $\Delta T = -20$ K because the temperature decreases. Because $|Q_{cool}| > Q_{melt}$, the soda has sufficient energy to melt all the ice. Hence the final state will be all liquid at $T_f > 0$. (Had we found

$|Q_{cool}| < Q_{melt}$, then the final state would have been an ice-liquid mixture at 0°C.)

Energy conservation requires $Q_{ice} + Q_{soda} = 0$. The heat Q_{ice} consists of three terms: warming the ice to 0°C, melting the ice to water at 0°C, then warming the 0°C water to T_f. The mass will still be M_{ice} in the last of these steps because it is the "ice system," but we need to use the specific heat of *liquid water*. Thus

$$
\begin{aligned}
Q_{ice} + Q_{soda} = [&M_{ice}c_{ice}(20\text{ K}) + M_{ice}L_f \\
&+ M_{ice}c_{water}(T_f - 0°\text{C})] \\
&+ M_{soda}c_{water}(T_f - 20°\text{C}) = 0
\end{aligned}
$$

We've already done part of the calculation, allowing us to write

$$37{,}000\text{ J} + M_{ice}c_{water}(T_f - 0°\text{C}) + M_{soda}c_{water}(T_f - 20°\text{C}) = 0$$

Solving for T_f gives

$$T_f = \frac{20M_{soda}c_{water} - 37{,}000}{M_{ice}c_{water} + M_{soda}c_{water}} = 1.9°\text{C}$$

ASSESS A good deal of ice has been put in the soda, so it ends up being cooled nearly to the freezing point, as we might expect.

STOP TO THINK 12.6 1 kg of lead at 100°C is dropped into a container holding 1 kg of water at 0°C. Once the lead and water reach thermal equilibrium, the final temperature is

A. <50°C B. 50°C C. >50°C

12.7 Thermal Properties of Gases

8.7

Now it's time to turn back to gases. What happens when they are heated? Can we assign specific heats to gases as we did for solids and liquids? As we'll see, gases are harder to characterize than solids or liquids because the heat required to cause a specified temperature change depends on the *process* by which the gas changes state.

Just as for solids and liquids, heating a gas changes its temperature. But by how much? **FIGURE 12.24** shows two isotherms on the pV diagram for a gas. Processes A and B, which start on the T_i isotherm and end on the T_f isotherm, have the *same* temperature change $\Delta T = T_f - T_i$, so we might expect them both to require the same amount of heat. But it turns out that process A, which takes place at constant volume, requires *less* heat than does process B, which occurs at constant pressure. The reason is that work is done in process B but not in process A.

FIGURE 12.24 Processes A and B have the same ΔT and the same ΔE_{th}, but they require different amounts of heat.

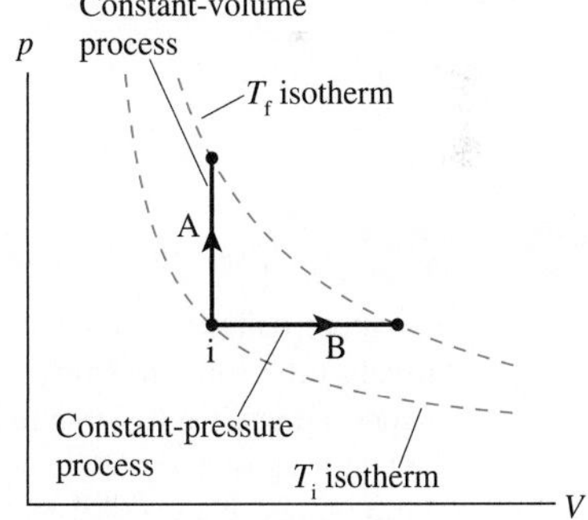

It is useful to define two different versions of the specific heat of gases: one for constant-volume processes and one for constant-pressure processes. We will define these as *molar* specific heats because we usually do gas calculations using moles instead of mass. The quantity of heat needed to change the temperature of n moles of gas by ΔT is, for a constant-volume process,

$$Q = nC_V\,\Delta T \tag{12.25}$$

and for a constant-pressure process,

$$Q = nC_P\,\Delta T \tag{12.26}$$

C_V is the **molar specific heat at constant volume,** and C_P is the **molar specific heat at constant pressure.** Table 12.6 on the next page gives the values of C_V and C_P for some common gases. The units are J/mol · K. The values for air are essentially equal to those for N_2.

It's interesting that all the monatomic gases in Table 12.6 have the same values for C_P and for C_V. Why should this be? Monatomic gases are really close to ideal, so let's go back to our atomic model of the ideal gas. We know that the thermal energy of an ideal gas of N atoms is $E_{th} = \frac{3}{2}Nk_BT = \frac{3}{2}nRT$. If the temperature of an ideal gas changes by ΔT, its thermal energy changes by

$$\Delta E_{th} = \frac{3}{2}nR\,\Delta T \tag{12.27}$$

TABLE 12.6 Molar specific heats of gases (J/mol · K) at 20°C

Gas	C_P	C_V
Monatomic Gases		
He	20.8	12.5
Ne	20.8	12.5
Ar	20.8	12.5
Diatomic Gases		
H_2	28.7	20.4
N_2	29.1	20.8
O_2	29.2	20.9
Triatomic Gas		
Water vapor	33.3	25.0

If we keep the volume of the gas constant, so that no work is done, this energy change can come only from heat, so

$$Q = \frac{3}{2} nR\,\Delta T \tag{12.28}$$

Comparing Equation 12.28 with the definition of molar specific heat in Equation 12.25, we see that the molar specific heat at constant volume must be

$$C_V\,(\text{monatomic gas}) = \frac{3}{2}R = 12.5\ \text{J/mol}\cdot\text{K} \tag{12.29}$$

This *predicted* value from the ideal-gas model is exactly the *measured* value of C_V for the monatomic gases in Table 12.6, a good check that the model we have been using is correct.

The molar specific heat at constant pressure is different. If you heat a gas in a sealed container so that there is no change in volume, then no work is done. But if you heat a sample of gas in a cylinder with a piston to keep it at constant pressure, the gas must expand, and it will do work as it expands. The expression for ΔE_{th} in Equation 12.27 is still valid, but now, according to the first law of thermodynamics, $Q = \Delta E_{th} + W_{gas}$. The work done by the gas in an isobaric process is $W_{gas} = p\,\Delta V$, so the heat required is

$$Q = \Delta E_{th} + W_{gas} = \frac{3}{2} nR\,\Delta T + p\,\Delta V \tag{12.30}$$

The ideal-gas law, $pV = nRT$, implies that if p is constant and only V and T change, then $p\,\Delta V = nR\,\Delta T$. Using this result in Equation 12.30, we find the heat needed to change the temperature by ΔT in a constant-pressure process is

$$Q = \frac{3}{2} nR\,\Delta T + nR\,\Delta T = \frac{5}{2} nR\,\Delta T$$

A comparison with the definition of molar specific heat shows that

$$C_P = \frac{5}{2}R = 20.8\ \text{J/mol}\cdot\text{K} \tag{12.31}$$

This is larger than C_V, as expected, and in perfect agreement for all the monatomic gases in Table 12.6.

EXAMPLE 12.19 Work done by an expanding gas

A typical weather balloon is made of a thin latex envelope that takes relatively little force to stretch, so the pressure inside the balloon is approximately equal to atmospheric pressure. The balloon is filled with a gas that is less dense than air, typically hydrogen or helium. Suppose a weather balloon filled with 180 mol of helium is waiting for launch on a cold morning at a high-altitude station. The balloon warms in the sun, which raises the temperature of the gas from 0°C to 30°C. As the balloon expands, how much work is done by the expanding gas?

PREPARE The work done is equal to $p\Delta V$, but we don't know the pressure (it's not sea level and we don't know the altitude) and we don't know the volume of the balloon. Instead, we'll use the first law of thermodynamics. We can rewrite Equation 12.17 as

$$W_{gas} = Q - \Delta E_{th}$$

The change in temperature of the gas is 30°C, so $\Delta T = 30$ K. We can compute how much heat energy is transferred to the balloon as it warms because this is a temperature change at constant pressure, and we can compute how much the thermal energy of the gas increases because we know ΔT.

SOLVE The heat required to increase the temperature of the gas is given by Equation 12.26:

$$Q = nC_P\,\Delta T = (180\,\text{mol})(20.8\ \text{J/mol}\cdot\text{K})(30\,\text{K}) = 110\ \text{kJ}$$

The change in thermal energy depends on the change in temperature according to Equation 12.18:

$$E_{th} = \frac{3}{2} nR\,\Delta T = \frac{3}{2}(180\ \text{mol})(8.31\ \text{J/mol}\cdot\text{K})(30\ \text{K})$$
$$= 67\ \text{kJ}$$

The work done by the expanding balloon is just the difference between these two values:

$$W_{gas} = Q - \Delta E_{th} = 110\ \text{kJ} - 67\ \text{kJ} = 43\ \text{kJ}$$

ASSESS The numbers are large—it's a lot of heat and a large change in thermal energy—but it's a big balloon with a lot of gas, so this seems reasonable.

Now let's turn to the diatomic gases in Table 12.6. The molar specific heats are higher than for monatomic gases, and our atomic model explains why. The thermal energy of a monatomic gas consists exclusively of the translational kinetic energy of the atoms; heating a monatomic gas simply means that the atoms move faster. The thermal energy of a diatomic gas is more than just the translational energy, as shown in **FIGURE 12.25**. Heating a diatomic gas makes the molecules move faster, but it also causes them to rotate more rapidly. Energy goes into the translational kinetic energy of the molecules (thus increasing the temperature), but some goes into rotational kinetic energy. Because some of the added heat goes into rotation and not into translation, the specific heat of a diatomic gas is higher than that of a monatomic gas, as we see in Table 12.6. The table shows that water vapor, a triatomic gas, has an even higher specific heat, as we might expect.

FIGURE 12.25 Thermal energy of a diatomic gas.

The thermal energy of a diatomic gas is the sum of the translational kinetic energy of the molecules . . .
. . . plus the rotational kinetic energy of the molecules.
$\vec{v}$

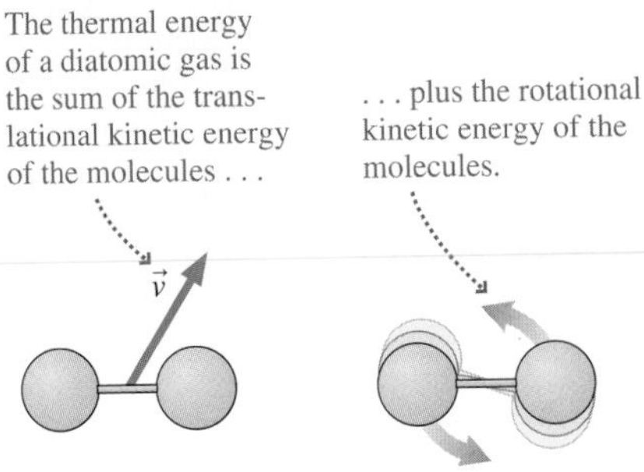

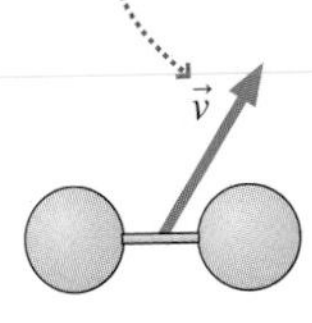

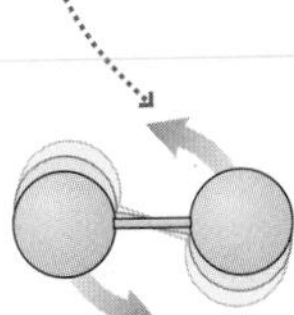

Heat Engines

Chapter 11 introduced the idea of a *heat engine,* a device that takes in heat energy from a hot reservoir, does useful work, and exhausts waste heat to a cold reservoir. Many real heat engines rely on gas processes, and we've now reached the point where we can examine heat engines more closely. A diagram of the (somewhat simplified) operation of a real heat engine, a two-stroke gasoline engine, is shown in **FIGURE 12.26**.

FIGURE 12.26 A simplified model of a gasoline engine.

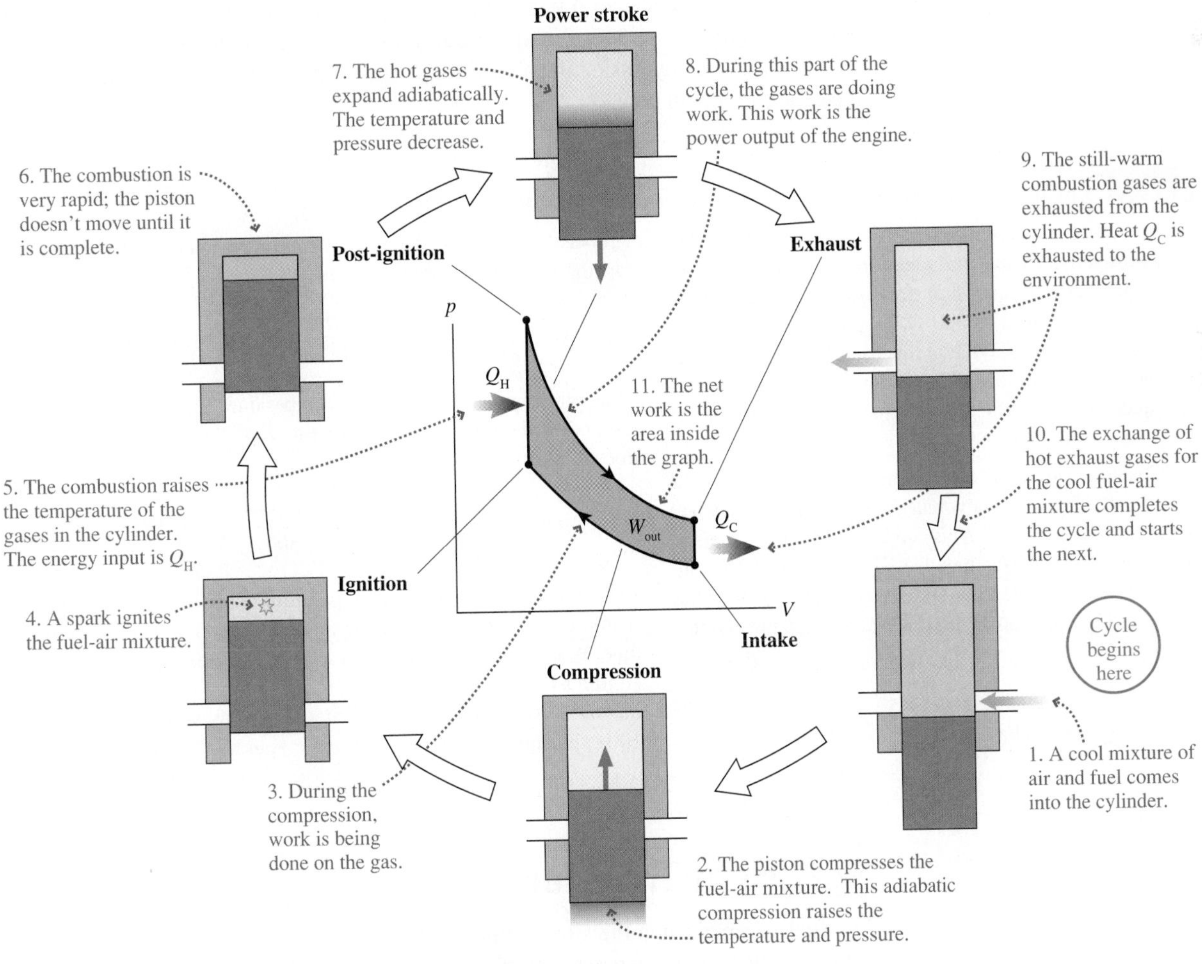

Power stroke
7. The hot gases expand adiabatically. The temperature and pressure decrease.
8. During this part of the cycle, the gases are doing work. This work is the power output of the engine.
6. The combustion is very rapid; the piston doesn't move until it is complete.
Post-ignition
Exhaust
9. The still-warm combustion gases are exhausted from the cylinder. Heat Q_C is exhausted to the environment.
p
Q_H
11. The net work is the area inside the graph.
10. The exchange of hot exhaust gases for the cool fuel-air mixture completes the cycle and starts the next.
5. The combustion raises the temperature of the gases in the cylinder. The energy input is Q_H.
W_{out}
Q_C
Ignition
V
4. A spark ignites the fuel-air mixture.
Intake
Cycle begins here
Compression
1. A cool mixture of air and fuel comes into the cylinder.
3. During the compression, work is being done on the gas.
2. The piston compresses the fuel-air mixture. This adiabatic compression raises the temperature and pressure.

NOTE ► It is common to use "gas" as shorthand for "gasoline." In this chapter, "gas" always refers to the state of matter. We will use "gasoline" or "fuel" to refer to the chemical energy source that you put in your tank. ◄

In this engine, the gases in the cylinder expand and contract in a repeating cycle as the piston moves back and forth. The cycle begins at the noted point with the introduction of gases (a mixture of air and fuel) into the cylinder. From this point, we can use our knowledge of gas processes to analyze the operation of the engine as it goes through its cycle. Figure 12.26 shows the connection between the operation of the engine and relevant points on the pV diagram of the gases in the engine's cylinder.

There are several important characteristics of this heat engine that apply generally to all heat engines:

- The sequence of processes forms a closed loop in the pV diagram. This must be true for the engine to go through its cycle over and over.
- Heat Q_H is taken in from a hot reservoir (in this case, from the burning fuel) at some point during the cycle. To bring the temperature back down at the end of the cycle, heat Q_C must be exhausted to a cold reservoir. This heat is carried out of the engine by the hot exhaust gases.
- The work done by the engine, W_{out}, is the area *inside* the closed loop on the pV diagram.
- Because the temperature returns to its starting point at the end of each cycle, there's no net change in thermal energy. Thus from energy conservation, $W_{out} = Q_H - Q_C$.

Work is the area inside the closed loop because of our earlier observation that the work done in a single process is the area under the graph. The work is negative for the compression, because ΔV is negative, and positive for the expansion. Thus the net work is the area under the upper curve *minus* the area under the lower curve, which is just the area inside the closed loop. This is true for all heat engines, not just the example shown here.

8.12, 8.13, 8.14 ActivPhysics ONLINE

EXAMPLE 12.20 Finding the pressure and temperature in a gasoline engine

At the start of a cycle in an internal combustion engine, 0.021 mol of air is introduced into the cylinder along with 0.018 g of gasoline. The adiabatic compression stroke raises the pressure to 17 atm and the temperature to 300°C. A spark then ignites the gasoline, which burns fully. What are the pressure and temperature immediately after ignition? (The combustion consumes oxygen but produces other gases; assume that the total number of moles of gas doesn't change.)

PREPARE Chapter 11 gives the energy content of gasoline as 44 kJ per gram, so we can figure out how much energy is liberated when the gas in the cylinder is burned. This combustion happens rapidly, so there is no time for the piston to move. The gas is heated at constant volume. The gas in the cylinder is mostly air, so we'll use $C_V = 20.8$ J/mol · K as the molar specific heat at constant volume.

SOLVE The energy released in the combustion of 0.018 g of gasoline is

$$Q = (44 \text{ kJ/g})(0.018 \text{ g}) = 790 \text{ J}$$

This heat is added to the air in the cylinder at constant volume, causing the temperature to increase by

$$\Delta T = \frac{Q}{nC_V} = \frac{790 \text{ J}}{(0.021 \text{ mol})(20.8 \text{ J/mol}\cdot\text{K})} = 1800°\text{C}$$

Thus igniting the fuel raises the temperature from $T_i = 300°$C at the end of the compression stroke to $T_f = 2100°$C. The volume doesn't change, so the new pressure is

$$p_f = p_i \frac{V_i}{V_f}\frac{T_f}{T_i} = (17 \text{ atm})(1)\frac{2100 + 273}{300 + 273} = 70 \text{ atm}$$

ASSESS The pressure and temperature are both very large, but that's to be expected following an explosion in a confined volume. When you drive your car, a series of such explosions in your engine's cylinders provide the necessary force to move your car down the road. In a typical car moving at highway speeds, over 100 explosions are taking place under your hood each second.

12.8 Heat Transfer

You feel warmer when the sun is shining on you, colder when you are sitting on a cold concrete bench or when a stiff wind is blowing. This is due to the transfer of heat. Although we've talked about heat and heat transfers a lot in these last two chapters, we haven't said much about *how* heat is transferred from a hotter object to a colder object. There are four basic mechanisms, described in the table on the next page, by which objects exchange heat with other objects or their surroundings. Evaporation was treated in an earlier section; in this section we will consider the other mechanisms.

Heat-transfer mechanisms

When two objects are in direct physical contact, such as the soldering iron and the circuit board, heat is transferred by *conduction.* **Energy is transferred by direct contact.**

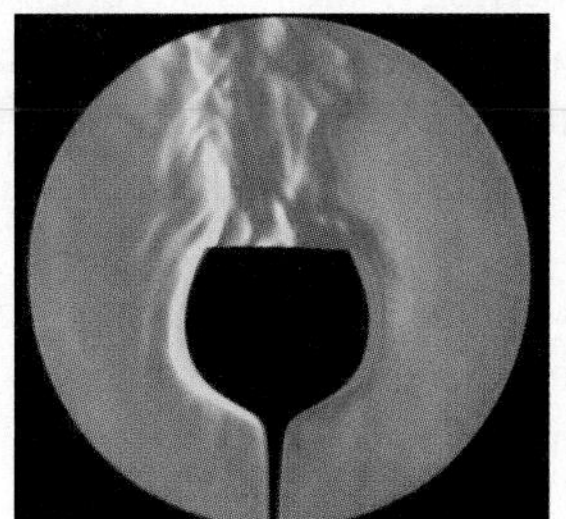

This special photograph shows air currents near a warm glass of water. Air near the glass is warmed and rises, taking thermal energy with it in a process known as *convection.* **Energy is transferred by the bulk motion of molecules with high thermal energy.**

The lamp shines on the lambs huddled below, warming them. The energy is transferred by infrared *radiation,* a form of electromagnetic waves. **Energy is transferred by electromagnetic waves.**

As we saw in an earlier section, the *evaporation* of liquid can carry away significant quantities of thermal energy. When you blow on a cup of cocoa, this increases the rate of evaporation, rapidly cooling it. **Energy is transferred by the removal of molecules with high thermal energy.**

Conduction

If you hold a metal spoon in a cup of hot coffee, the handle of the spoon soon gets warm. Thermal energy is transferred along the spoon from the coffee to your hand. The difference in temperature between the two ends drives this heat transfer by a process known as **conduction.** Conduction is the transfer of thermal energy directly through a physical material.

FIGURE 12.27 shows a copper rod placed between a hot reservoir (a fire) and a cold reservoir (a block of ice). We can use our atomic model to see how thermal energy is transferred along the rod by the interaction between atoms in the rod; fast-moving atoms at the hot end transfer energy to slower-moving atoms at the cold end.

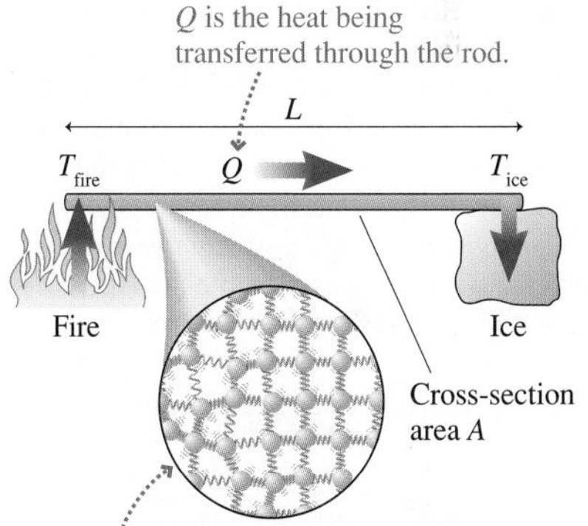
Q is the heat being transferred through the rod.
L
T_{fire}
Q
T_{ice}
Fire
Ice
Cross-section area A
The particles on the left side of the rod are vibrating more vigorously than the particles on the right. The particles on the left transfer energy to the particles on the right via the bonds connecting them.

FIGURE 12.27 Conduction of heat in a solid rod.

Suppose we set up a series of experiments to measure the heat Q transferred through various rods. We would find the following trends in our data:

- Q increases if the temperature difference ΔT between the hot end and the cold end is increased.
- Q increases if the cross-section area A of the rod is increased.
- Q decreases if the length L of the rod is increased.
- Some materials (such as metals) transfer heat quite readily. Other materials (such as wood) transfer very little heat.

The final observation is one that is familiar to you: If you are stirring a pot of hot soup on the stove, you generally use a wood or plastic spoon rather than a metal one.

These experimental observations about heat conduction can be summarized in a single formula. If heat Q is transferred in a time interval Δt, the *rate* of heat transfer (joules per second) is $Q/\Delta t$. For a material of cross-section area A and length L, spanning a temperature difference ΔT, the rate of heat transfer is

$$\frac{Q}{\Delta t} = \left(\frac{kA}{L}\right)\Delta T \tag{12.32}$$

Rate of conduction of heat across a temperature difference

TABLE 12.7 Thermal conductivity values (measured at 20°C)

Material	k (W/m · K)
Diamond	1000
Silver	420
Copper	400
Iron	72
Stainless steel	14
Ice	1.7
Concrete	0.8
Plate glass	0.75
Skin	0.50
Muscle	0.46
Fat	0.21
Wood	0.2
Carpet	0.04
Fur, feathers	0.02–0.06
Air (27°C, 100 kPa)	0.026

Carpet or tile? If you walk around your house barefoot, you will notice that tile floors feel much colder than carpeted floors. The tile floor isn't really colder—the temperatures of all the floors in your house are nearly the same. But tile has a much higher thermal conductivity than carpet, so more heat will flow from your feet into the tile than into carpet. This causes the tile to feel colder.

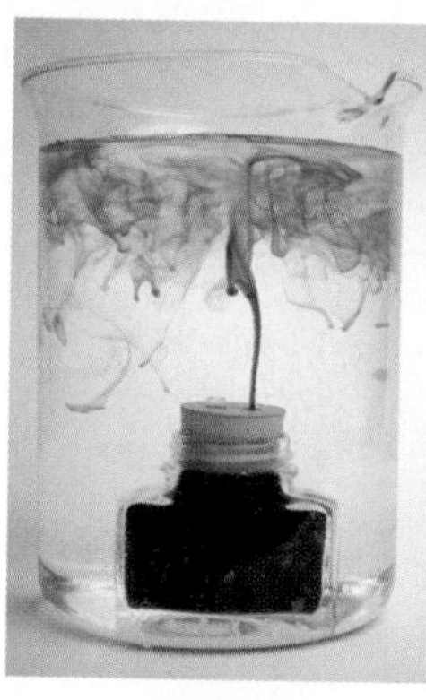

Warm water (colored) moves by convection.

The quantity k, which characterizes whether the material is a good conductor of heat or a poor conductor, is called the **thermal conductivity** of the material. Because the heat transfer rate J/s is a *power,* measured in watts, the units of k are W/m · K. Values of k for some common materials are listed in Table 12.7; a larger number for k means a material is a better conductor of heat.

Good conductors of electricity, such as silver and copper, are usually good conductors of heat. Air, like all gases, is a poor conductor of heat because there are no bonds between adjacent atoms.

The weak bonds between molecules make most biological materials poor conductors of heat. Muscle is a better conductor than fat, so sea mammals have thick layers of fat for insulation. Land mammals insulate their bodies with fur, birds with feathers. Both fur and feathers trap a good deal of air, so their conductivity is similar to that of air, as Table 12.7 shows.

EXAMPLE 12.21 Heat loss by conduction

At the start of the section we noted that you "feel cold" when you sit on a cold concrete bench. "Feeling cold" really means that your body is losing a significant amount of heat. How significant? Suppose you are sitting on a 10°C concrete bench. You are wearing thin clothing that provides negligible insulation. In this case, most of the insulation that protects your body's core (temperature 37°C) from the cold of the bench is provided by a 1.0-cm-thick layer of fat on the part of your body that touches the bench. (The thickness varies from person to person, but this is a reasonable average value.) A good estimate of the area of contact with the bench is 0.10 m^2. Given these details, what is the heat loss by conduction?

PREPARE Heat is lost to the bench by conduction through the fat layer, so we can compute the rate of heat loss by using Equation 12.32. The thickness of the conducting layer is 0.010 m, the area is 0.10 m^2, and the thermal conductivity of fat is given in Table 12.7. The temperature difference is the difference between your body's core temperature (37°C) and the temperature of the bench (10°C), a difference of 27°C, or 27 K.

SOLVE We have all of the data we need to use Equation 12.32 to compute the rate of heat loss:

$$\frac{Q}{\Delta t} = \left(\frac{(0.21\ \text{W/m}\cdot\text{K})(0.10\ \text{m}^2)}{0.010\ \text{m}}\right)(27\,\text{K}) = 57\ \text{W}$$

ASSESS 57 W is more than half your body's resting power, which we learned in Chapter 11 is approximately 100 W. That's a significant loss, so your body will feel cold, a result that seems reasonable if you've ever sat on a cold bench for any length of time.

Convection

We noted that air is a poor conductor of heat. In fact, the data in Table 12.7 show that air has a thermal conductivity comparable to that of feathers. So why, on a cold day, will you be more comfortable if you are wearing a down jacket?

In conduction, faster-moving atoms transfer thermal energy to adjacent atoms. But in fluids such as water or air, there is a more efficient means to move energy: by transferring the faster-moving atoms themselves. When you place a pan of cold water on a burner on the stove, it's heated on the bottom. This heated water expands and becomes less dense than the water above it, so it rises to the surface while cooler, denser water sinks to take its place. This transfer of thermal energy by the motion of a fluid is known as **convection.**

Convection is usually the main mechanism for heat transfer in fluid systems. On a small scale, convection mixes the pan of water that you heat on the stove; on a large scale, convection is responsible for making the wind blow and ocean currents circulate. Air is a very poor thermal conductor, but it is very effective at transferring energy by convection. To use air for thermal insulation, it is necessary to trap the air in small pockets to limit convection. And that's exactly what feathers, fur, double-paned windows, and fiberglass insulation do. Convection is much more rapid in water than in air, which is why people can die of hypothermia in 65°F water but can live quite happily in 65°F air.

A feather coat BIO Penguins are birds and thus have feathers, even though they do not fly. The feathers are more obvious on this penguin chick than on the adults. A penguin's short, dense feathers serve a different role than the flight feathers of other birds: They trap air to provide thermal insulation. The feathers are equipped with muscles at the base of the shafts to flatten them and eliminate these air pockets; otherwise, the penguins would be too buoyant to swim underwater.

Radiation

You *feel* the warmth from the glowing red coals in a fireplace. On a cool day, you prefer to sit in the sun rather than the shade so that the sunlight keeps you warm. In both cases heat energy is being transferred to your body in the form of **radiation.**

Radiation consists of electromagnetic waves—a topic we will explore further in later chapters—that transfer energy from the object that emits the radiation to the object that absorbs it. Visible light, the red-to-violet colors of the rainbow, consists of electromagnetic waves, and objects that are hot enough to glow emit part of their radiation as visible light. We say these objects are "red hot" or, at a high enough temperature, "white hot." The white light from an incandescent lightbulb is radiation emitted by a thin wire filament that has been heated to a very high temperature by an electric current.

Electromagnetic waves of somewhat lower frequency are called *infrared radiation;* we can't see these waves, but we can sometimes sense them as "heat" as they are absorbed on our skin. Glowing objects emit part of their radiation as visible light but much of it as infrared radiation—the warmth that you feel from the glowing coals. Objects near room temperature, including your body, also radiate, but they emit nearly all of their radiation as infrared radiation. Thermal images, like the one of the teapot in FIGURE 12.28, are made with cameras having special infrared-sensitive detectors that allow them to "see" radiation that our eyes don't respond to.

FIGURE 12.28 A thermal image of a teapot.

NOTE ► The word "radiation" comes from "radiate," which means "to beam." You have likely heard the word used to refer to x rays and radioactive materials. This is not the sense we will use in this chapter. Here, we will use "radiation" to mean electromagnetic waves that "beam" from an object. ◄

Radiation is a significant part of the energy balance that keeps your body at the proper temperature. Radiation from the sun or from warm, nearby objects is absorbed by your skin, increasing your body's thermal energy. At the same time, your body radiates energy away. You can change your body temperature by absorbing more radiation (sitting next to a fire) or by emitting more radiation (taking off your hat and scarf to expose more skin).

The energy radiated by an object shows a strong dependence on temperature. In Figure 12.28, the hot teapot glows much more brightly than the cooler hand. In the photo that opened the chapter, the person's warm hands emit quite a bit more energy than the cooler spider. We can quantify the dependence of radiated energy on temperature. If heat energy Q is radiated in a time interval Δt by an object with surface area A and absolute temperature T, the *rate* of heat transfer $Q/\Delta t$ (joules per second) is found to be

$$\frac{Q}{\Delta t} = e\sigma AT^4 \tag{12.33}$$

Rate of heat transfer by radiation at temperature T (Stefan's law)

Quantities in this equation are defined as follows:

- e is the **emissivity** of a surface, a measure of the effectiveness of radiation. The value of e ranges from 0 to 1. Human skin is a very effective radiator at body temperature, with $e = 0.97$, regardless of skin color.

Global heat transfer This satellite image shows radiation emitted by the waters of the ocean off the east coast of the United States. You can clearly see the warm waters of the Gulf Stream, a large-scale convection that transfers heat to northern latitudes. The satellite "sees" the radiation from the earth, so this radiation must readily pass through the atmosphere into space. In fact, the *only* means by which the earth as a whole can reduce its thermal energy is by radiation, because there is no conduction or convection in the vacuum of space. This has consequences for the energy balance of the earth, as we consider in a section on the greenhouse effect at the end of Part III.

- T is the absolute temperature in kelvin.
- A is the surface area in m^2.
- σ is a constant known as the Stefan-Boltzmann constant, with the value $\sigma = 5.67 \times 10^{-8}\ \text{W/m}^2 \cdot \text{K}^4$.

Notice the very strong fourth-power dependence on temperature. Doubling the absolute temperature of an object increases the radiant heat transfer by a factor of 16!

The amount of energy radiated by an animal can be surprisingly large. An adult human with bare skin in a room at a comfortable temperature has a skin temperature of approximately 33°C, or 306 K. A typical value for the skin's surface area is 1.8 m^2. With these values and the emissivity of skin noted above, we can calculate the rate of heat transfer via radiation from the skin:

$$\frac{Q}{\Delta t} = e\sigma A T^4 = (0.97)\left(5.67 \times 10^{-8}\ \frac{\text{W}}{\text{m}^2 \cdot \text{K}^4}\right)(1.8\ \text{m}^2)(306\ \text{K})^4 = 870\ \text{W}$$

As we learned in Chapter 11, the body at rest generates approximately 100 W of thermal energy. If the body radiated energy at 870 W, it would quickly cool. At this rate of emission, the body temperature would drop by 1°C every 7 minutes! Clearly, there must be some mechanism to balance this emitted radiation, and there is: the radiation *absorbed* by the body.

When you sit in the sun, your skin warms due to the radiation you absorb. Even if you are not in the sun, you are absorbing the radiation emitted by the objects surrounding you. Suppose an object at temperature T is surrounded by an environment at temperature T_0. The *net* rate at which the object radiates heat energy—that is, radiation emitted minus radiation absorbed—is

$$\frac{Q_{\text{net}}}{\Delta t} = e\sigma A(T^4 - T_0^4) \tag{12.34}$$

This makes sense. An object should have no net energy transfer by radiation if it's in thermal equilibrium ($T = T_0$) with its surroundings. Note that the emissivity e appears for absorption as well; objects that are good emitters are also good absorbers.

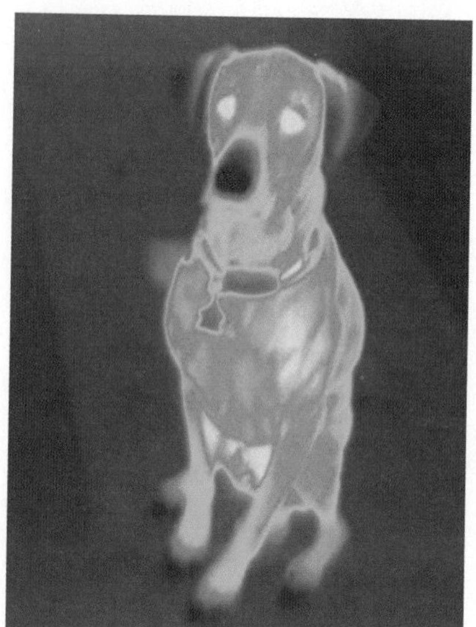

The infrared radiation emitted by a dog is captured in this image. His cool nose and paws radiate much less energy than the rest of his body.

EXAMPLE 12.22 Determining energy loss by radiation for the body

A person with a skin temperature of 33°C is in a room at 24°C. What is the net rate of heat transfer by radiation?

PREPARE Body temperature is $T = 33 + 273 = 306$ K; the temperature of the room is $T_0 = 24 + 273 = 297$ K.

SOLVE The net radiation rate, given by Equation 12.34, is

$$\frac{Q_{\text{net}}}{\Delta t} = e\sigma A(T^4 - T_0^4)$$

$$= (0.97)\left(5.67 \times 10^{-8}\frac{\text{W}}{\text{m}^2 \cdot \text{K}^4}\right)(1.8\ \text{m}^2)[(306\ \text{K})^4 - (297\ \text{K})^4] = 98\ \text{W}$$

ASSESS This is a reasonable value, roughly matching your resting metabolic power. When you are dressed (little convection) and sitting on wood or plastic (little conduction), radiation is your body's primary mechanism for dissipating the excess thermal energy of metabolism.

STOP TO THINK 12.7 Suppose you are an astronaut in the vacuum of space, hard at work in your sealed spacesuit. The only way that you can transfer heat to the environment is by

A. Conduction. B. Convection. C. Radiation. D. Evaporation.

INTEGRATED EXAMPLE 12.23 Breathing in cold air

On a cold day, breathing costs your body energy; as the cold air comes into contact with the warm tissues of your lungs, the air warms due to heat transferred from your body. The thermal image of the person inflating a balloon in **FIGURE 12.29** shows that exhaled air is quite a bit warmer than the surroundings.

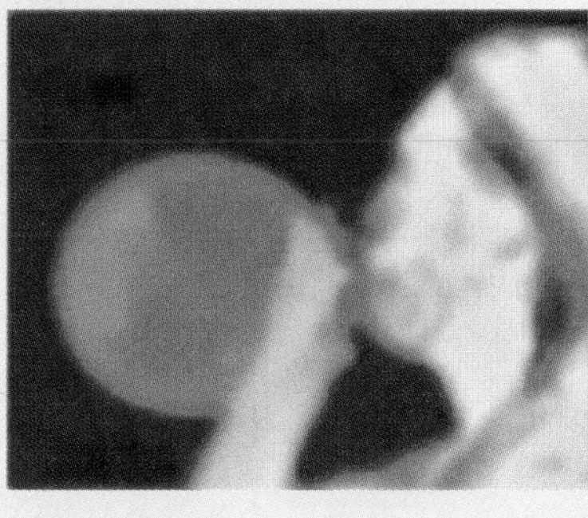

FIGURE 12.29 A thermal image of a person blowing up a balloon on a cold day.

FIGURE 12.30 shows this process for a frosty $-10°C$ (14°F) day. The inhaled air warms to nearly the temperature of the interior of your body, 37°C. When you exhale, some heat is retained by the body, but most is lost; the exhaled air is still about 30°C.

FIGURE 12.30 Breathing warms the air.

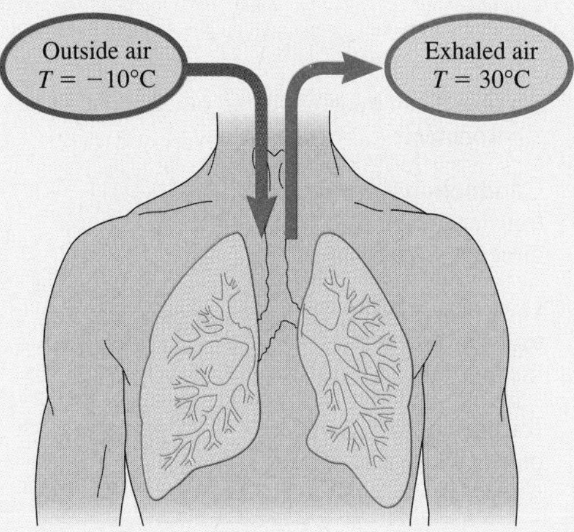

Your lungs hold several liters of air, but only a small part of the air is exchanged during each breath. A typical person takes 12 breaths each minute, with each breath drawing in 0.50 L of outside air. If the air warms up from $-10°C$ to 30°C, what is the volume of the air exhaled with each breath? What fraction of your body's resting power goes to warming the air? (Note that gases are exchanged as you breathe—oxygen to carbon dioxide—but to a good approximation the number of atoms, and thus the number of moles, stays the same.) Consider only the energy required to warm the air, not the energy lost to evaporation from the tissues of the lungs.

PREPARE There are two parts to the problem, which we'll solve in two different ways. First, we need to figure out the volume of exhaled air. When your body warms the air, its temperature increases, so its volume will increase as well. The initial and final states represented in Figure 12.30 are at atmospheric pressure, so we can treat this change as an isobaric (constant-pressure) process. We'll need the absolute temperatures of the initial and final states:

$$T_i = -10°C + 273 = 263\text{ K}$$

$$T_f = 30°C + 273 = 303\text{ K}$$

Next, we'll determine how much heat energy is required to raise the temperature of the air. Because the initial and final pressures are the same, we'll do this by computing the heat needed to raise the temperature of a gas at constant pressure. The change in temperature is $+40°C$, so we can use $\Delta T = 40\text{ K}$. Air is a mix of nitrogen and oxygen with a small amount of other gases. C_P for nitrogen and oxygen is the same to two significant figures, so we will assume the gas has $C_P = 29\text{ J/mol}\cdot\text{K}$.

SOLVE The change in volume of the air in your lungs is a constant-pressure process. 0.50 L of air is breathed in; this is V_i. When the temperature increases, so does the volume. The gas isn't in a sealed container, but we are considering the same "parcel" of gas before and after, so we can use the ideal-gas law to find the volume after the temperature increase:

$$V_f = V_i\frac{p_i T_f}{p_f T_i} = (0.50\text{ L}) \times 1 \times \frac{303\text{ K}}{263\text{ K}} = 0.58\text{ L}$$

The volume increases by a little over 10%, from 0.50 L to 0.58 L.

Now we can move on to the second part: determining the energy required. We'll need the number of moles of gas, which we can compute from the ideal-gas law:

$$n = \frac{pV}{RT} = \frac{(101.3 \times 10^3\text{ Pa})(0.50 \times 10^{-3}\text{ m}^3)}{(8.31\text{ J/mol}\cdot\text{K})(263\text{ K})} = 0.023\text{ mol}$$

In doing this calculation, we used $1\text{ m}^3 = 1000\text{ L}$ to convert the 0.50 L volume to m^3. Now we can compute the heat needed to warm one breath, using Equation 12.26:

$$Q(\text{one breath}) = nC_P\,\Delta T = (0.023\text{ mol})(29\text{ J/mol}\cdot\text{K})(40\text{ K}) = 27\text{ J}$$

If you take 12 breaths a minute, a single breath takes 1/12 of a minute, or 5.0 s. Thus the heat power your body supplies to warm the incoming air is

$$P = \frac{Q}{\Delta t} = \frac{27\text{ J}}{5.0\text{ s}} = 5.4\text{ W}$$

At rest, your body typically uses 100 W, so this is just over 5% of your body's resting power. You breathe in a good deal of air each minute and warm it by quite a bit, but the specific heat of air is small enough that the energy required is reasonably modest.

ASSESS The air expands slightly and a small amount of energy goes into heating it. If you've been outside on a cold day, you know that neither change is dramatic, so this final result seems reasonable. The energy loss is noticeable but reasonably modest; other forms of energy loss are more important when you are outside on a cold day.

SUMMARY

The goal of Chapter 12 has been to use the atomic model of matter to explain and explore many macroscopic phenomena associated with heat, temperature, and the properties of matter.

GENERAL PRINCIPLES

Atomic Model

We model matter as being made of simple basic particles. The relationship of these particles to each other defines the phase.

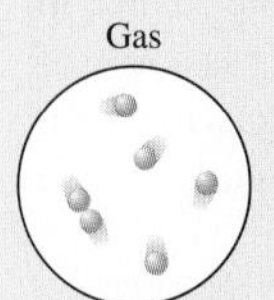

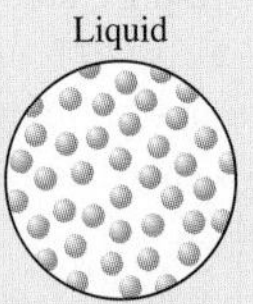

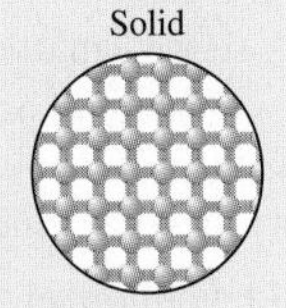

The atomic model explains thermal expansion, specific heat, and heat transfer.

Atomic Model of a Gas

Macroscopic properties of gases can be explained in terms of the atomic model of the gas. The speed of the particles is related to the temperature:

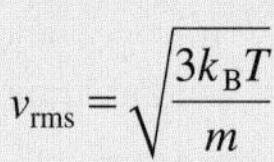

$$v_{rms} = \sqrt{\frac{3k_BT}{m}}$$

The collisions of particles with each other and with the walls of the container determine the pressure.

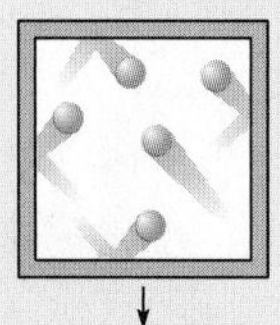

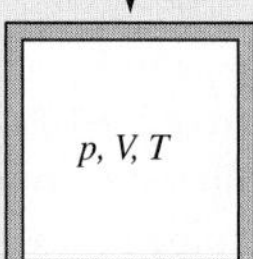

Ideal-Gas Law

The ideal gas law relates the pressure, volume, and temperature in a sample of gas. We can express the law in terms of the number of atoms or the number of moles in the sample:

$$pV = Nk_BT$$

$$pV = nRT$$

For a gas process in a sealed container,

$$\frac{p_iV_i}{T_i} = \frac{p_fV_f}{T_f}$$

IMPORTANT CONCEPTS

Effects of heat transfer

A system that is heated can either change temperature or change phase.

The **specific heat *c*** of a material is the heat required to raise 1 kg by 1 K. $Q = Mc\,\Delta T$

The **heat of transformation** is the energy necessary to change the phase of 1 kg of a substance. Heat is added to change a solid to a liquid or a liquid to a gas; heat is removed to reverse these changes.

$$Q = \begin{cases} \pm ML_f & \text{(melt/freeze)} \\ \pm ML_v & \text{(boil/condense)} \end{cases}$$

The **molar specific heat** of a gas depends on the process.

- For a constant-volume process: $Q = nC_V\,\Delta T$
- For a constant-pressure process: $Q = nC_P\,\Delta T$

Mechanisms of heat transfer

An object can transfer heat to other objects or to its environment:

Conduction is the transfer of heat by direct physical contact.

$$\frac{Q}{\Delta t} = \left(\frac{kA}{L}\right)\Delta T$$

Convection is the transfer of heat by the motion of a fluid.

Radiation is the transfer of heat by electromagnetic waves.

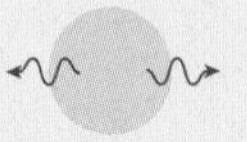

$$\frac{Q}{\Delta t} = e\sigma AT^4$$

A ***pV* (pressure-volume) diagram** is a useful means of looking at a process involving a gas.

A **constant-volume** process has no change in volume.

An **isobaric** process happens at a constant pressure.

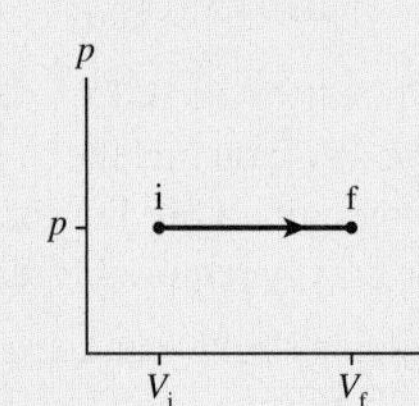

An **isothermal** process happens at a constant temperature.

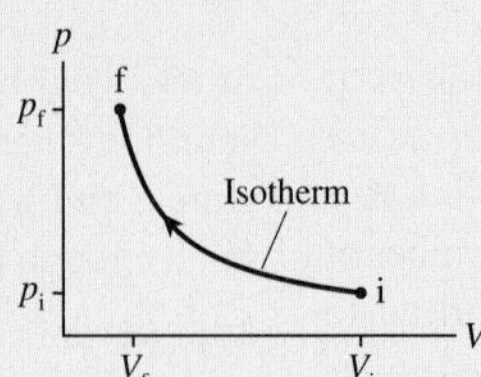

An **adiabatic** process involves no transfer of heat; the temperature changes.

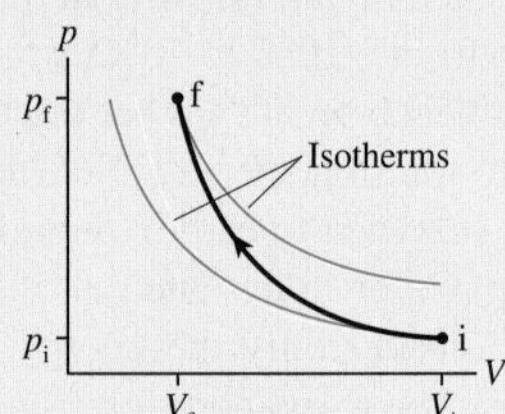

The work done by a gas is the area under the graph.

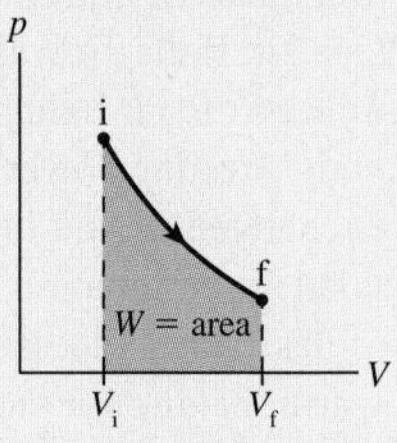

APPLICATIONS

Thermal expansion Objects experience an increase in volume and an increase in length when their temperature changes:

$$\Delta V = \beta V_i\,\Delta T \qquad \Delta L = \alpha L_i\,\Delta T$$

Calorimetry When two or more systems interact thermally, they come to a common final temperature determined by

$$Q_{net} = Q_1 + Q_2 + Q_3 + \cdots = 0$$

The number of **moles** is

$$n = \frac{M\text{ (in grams)}}{M_{mol}}$$

For homework assigned on MasteringPhysics, go to www.masteringphysics.com

Problem difficulty is labeled as | (straightforward) to ||||| (challenging).

Problems labeled INT integrate significant material from earlier chapters; BIO are of biological or medical interest.

QUESTIONS

Conceptual Questions

1. Which has more mass, a mole of Ne gas or a mole of N_2 gas?
2. If you launch a projectile upward with a high enough speed, its kinetic energy is sufficient to allow it to escape the earth's gravity—it will go up and not come back down. Given enough time, hydrogen and helium gas atoms in the earth's atmosphere will escape, so these elements are not present in our atmosphere. Explain why hydrogen and helium atoms have the necessary speed to escape but why other elements, such as oxygen and nitrogen, do not.
3. You may have noticed that latex helium balloons tend to shrink rather quickly; a balloon filled with air lasts a lot longer. Balloons shrink because gas diffuses out of them. The rate of diffusion is faster for smaller particles and for particles of higher speed. Diffusion is also faster when there is a large difference in concentration between two sides of a membrane. Given these facts, explain why an air-filled balloon lasts longer than a helium balloon.
4. If you buy a sealed bag of potato chips in Miami and drive with it to Denver, where the atmospheric pressure is lower, you will find that the bag gets very "puffy." Explain why.
5. If you double the typical speed of the molecules in a gas, by what factor does the pressure change? Give a simple explanation why the pressure changes by this factor.
6. Two gases have the same number of molecules per cubic meter (N/V) and the same rms speed. The molecules of gas 2 are more massive than the molecules of gas 1.
   a. Do the two gases have the same pressure? If not, which is larger?
   b. Do the two gases have the same temperature? If not, which is larger?
7. a. Which contains more particles, a mole of helium gas or a mole of oxygen gas? Explain.
   b. Which contains more particles, a gram of helium gas or a gram of oxygen gas? Explain.
8. You have 100 g of aluminum and 100 g of lead.
   a. Which contains a larger number of moles? Explain.
   b. Which contains more atoms? Explain.
9. Suppose you could suddenly increase the speed of every atom in a gas by a factor of 2.
   a. Does the thermal energy of the gas change? If so, by what factor? If not, why not?
   b. Does the molar specific heat change? If so, by what factor? If not, why not?
10. A gas cylinder contains 1.0 mol of helium at a temperature of 20°C. A second identical cylinder contains 1.0 mol of neon at 20°C. The helium atoms are moving with a higher average speed, but the gas pressure in the two containers is the same. Explain how this is possible.
11. A gas is in a sealed container. By what factor does the gas pressure change if
    a. The volume is doubled and the temperature is tripled?
    b. The volume is halved and the temperature is tripled?
12. A gas is in a sealed container. By what factor does the gas temperature change if
    a. The volume is doubled and the pressure is tripled?
    b. The volume is halved and the pressure is tripled?
13. What is the maximum amount of work that a gas can do during a constant-volume process?
14. You need to precisely measure the dimensions of a large wood panel for a construction project. Your metal tape measure was left outside for hours in the sun on a hot summer day, and now the tape is so hot it's painful to pick up. How will your measurements differ from those taken by your coworker, whose tape stayed in the shade? Explain.
15. Your car's radiator is made of steel and is filled with water. You fill the radiator to the very top with cold water, then drive off without remembering to replace the cap. As the water and the steel radiator heat up, will the level of water drop or will it rise and overflow? Explain.
16. Materials A and B have equal densities, but A has a larger specific heat than B. You have 100 g cubes of each material. Cube A, initially at 0°C, is placed in good thermal contact with cube B, initially at 200°C. The cubes are inside a well-insulated container where they don't interact with their surroundings. Is their final temperature greater than, less than, or equal to 100°C? Explain.
17. Two containers hold equal masses of nitrogen gas at equal temperatures. You supply 10 J of heat to container A while not allowing its volume to change, and you supply 10 J of heat to container B while not allowing its pressure to change. Afterward, is temperature T_A greater than, less than, or equal to T_B? Explain.
18. You need to raise the temperature of a gas by 10°C. To use the smallest amount of heat energy, should you heat the gas at constant pressure or at constant volume? Explain.
19. A sample of ideal gas is in a cylinder with a movable piston. 600 J of heat is added to the gas in an isothermal process. As the gas expands, pushing against the piston, how much work does it do?
20. A student is heating chocolate in a pan on the stove. He uses a cooking thermometer to measure the temperature of the chocolate and sees it varies as shown in Figure Q12.20. Describe what is happening to the chocolate in each of the three portions of the graph.

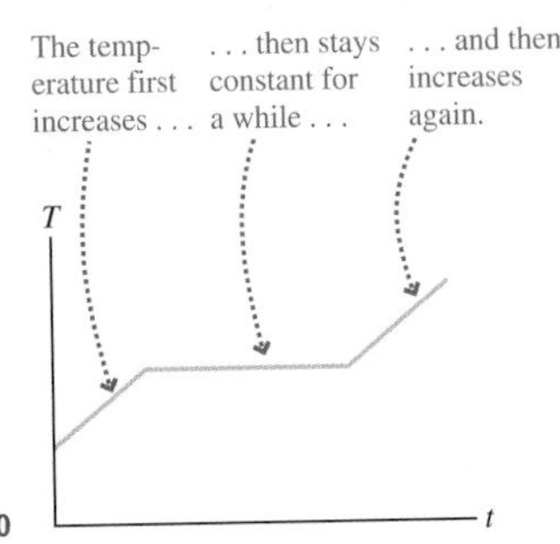

FIGURE Q12.20

21. If you bake a cake at high elevation, where atmospheric pressure is lower than at sea level, you will need to adjust the recipe. You will need to cook the cake for a longer time, and you will need to add less baking powder. (Baking powder is a leavening agent. As it heats, it releases gas bubbles that cause the cake to rise.) Explain why those adjustments are necessary.
22. The specific heat of aluminum is higher than that of iron. 1 kg blocks of iron and aluminum are heated to 100°C, and each is then dropped into its own 1 L beaker of 20°C water. Which beaker will end up with the warmer water? Explain.
23. A student is asked to sketch a pV diagram for a gas that goes through a cycle consisting of (a) an isobaric expansion, (b) a constant-volume reduction in temperature, and (c) an isothermal process that returns the gas to its initial state. The student draws the diagram shown in Figure Q12.23. What, if anything, is wrong with the student's diagram?

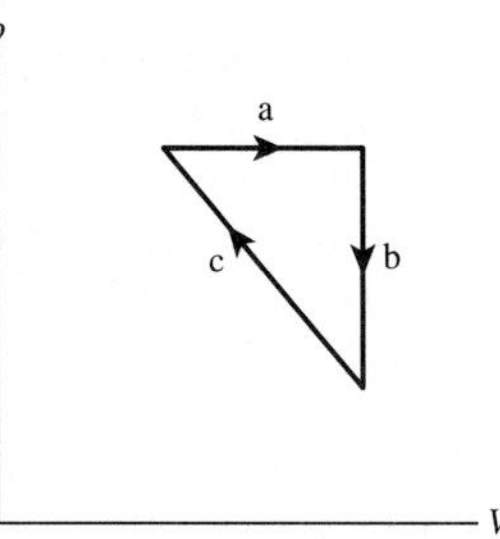

FIGURE Q12.23

24. If you have two spoons of the same size, one silver and one stainless steel, there is a quick test to tell which is which. Hold the end of a spoon in each hand, then lower them both into a cup of very hot water. One spoon will feel hot first. Is that the silver spoon or the stainless steel spoon? Explain.
25. If you live somewhere with cold, clear nights, you may have noticed some mornings when there was frost on open patches of ground but not under trees. This is because the ground under trees does not get as cold as open ground. Explain how tree cover keeps the ground under trees warmer.

Multiple-Choice Questions

26. | A tire is inflated to a gauge pressure of 35 psi. The absolute pressure in the tire is
    A. Less than 35 psi.
    B. Equal to 35 psi.
    C. Greater than 35 psi.
27. | The number of atoms in a container is increased by a factor of 2 while the temperature is held constant. The pressure
    A. Decreases by a factor of 4.
    B. Decreases by a factor of 2.
    C. Stays the same.
    D. Increases by a factor of 2.
    E. Increases by a factor of 4.
28. || A gas is compressed by an isothermal process that decreases its volume by a factor of 2. In this process, the pressure
    A. Does not change.
    B. Increases by a factor of less than 2.
    C. Increases by a factor of 2.
    D. Increases by a factor of more than 2.
29. ||| A gas is compressed by an adiabatic process that decreases its volume by a factor of 2. In this process, the pressure
    A. Does not change.
    B. Increases by a factor of less than 2.
    C. Increases by a factor of 2.
    D. Increases by a factor of more than 2.
30. | Suppose you do a calorimetry experiment to measure the specific heat of a penny. You take a number of pennies, measure their mass, heat them to a known temperature, and then drop them into a container of water at a known temperature. You then deduce the specific heat of a penny by measuring the temperature change of the water. Unfortunately, you didn't realize that you dropped one penny on the floor while transferring them to the water. This will
    A. Cause you to underestimate the specific heat.
    B. Cause you to overestimate the specific heat.
    C. Not affect your calculation of specific heat.
31. | A cup of water is heated with a heating coil that delivers 100 W of heat. In one minute, the temperature of the water rises by 20°C. What is the mass of the water?
    A. 72 g B. 140 g
    C. 720 g D. 1.4 kg
32. | Three identical beakers each hold 1000 g of water at 20°C. 100 g of liquid water at 0°C is added to the first beaker, 100 g of ice at 0°C is added to the second beaker, and the third beaker gets 100 g of aluminum at 0°C. The contents of which container end up at the lowest final temperature?
    A. The first beaker.
    B. The second beaker.
    C. The third beaker.
    D. All end up at the same temperature.
33. |||| 100 g of ice at 0°C and 100 g of steam at 100°C interact thermally in a well-insulated container. The final state of the system is
    A. An ice-water mixture at 0°C.
    B. Water at a temperature between 0°C and 50°C.
    C. Water at 50°C.
    D. Water at a temperature between 50°C and 100°C.
    E. A water-steam mixture at 100°C.
34. | Suppose the 600 W of radiation emitted in a microwave oven is absorbed by 250 g of water in a very lightweight cup. Approximately how long will it take to heat the water from 20°C to 80°C?
    A. 50 s B. 100 s
    C. 150 s D. 200 s
35. || 40,000 J of heat is added to 1.0 kg of ice at −10°C. How much ice melts?
    A. 0.012 kg B. 0.057 kg
    C. 0.12 kg D. 1.0 kg
36. || Steam at 100°C causes worse burns than liquid water at 100°C. This is because
    BIO
    A. The steam is hotter than the water.
    B. Heat is transferred to the skin as steam condenses.
    C. Steam has a higher specific heat than water.
    D. Evaporation of liquid water on the skin causes cooling.

PROBLEMS

Section 12.1 The Atomic Model of Matter

1. | Which contains the most moles: 10 g of hydrogen gas, 100 g of carbon, or 50 g of lead?
2. |||| How many grams of water (H_2O) have the same number of oxygen atoms as 1.0 mol of oxygen gas?
3. |||| How many atoms of hydrogen are in 100 g of hydrogen peroxide (H_2O_2)?
4. || How many cubic millimeters (mm^3) are in 1 L?
5. || A box is 200 cm wide, 40 cm deep, and 3.0 cm high. What is its volume in m^3?

Section 12.2 The Atomic Model of an Ideal Gas

6. ||| Dry ice is frozen carbon dioxide. If you have 1.0 kg of dry ice, what volume will it occupy if you heat it enough to turn it into a gas at a temperature of 20°C?
7. | What is the absolute pressure of the air in your car's tires, in psi, when your pressure gauge indicates they are inflated to 35.0 psi? Assume you are at sea level.
8. |||| Total lung capacity of a typical adult is approximately 5.0 L. BIO Approximately 20% of the air is oxygen. At sea level and at an average body temperature of 37°C, how many moles of oxygen do the lungs contain at the end of an inhalation?
9. |||| Many cultures around the world still use a simple weapon BIO called a blowgun, a tube with a dart that fits tightly inside. A sharp breath into the end of the tube launches the dart. When exhaling forcefully, a healthy person can supply air at a gauge pressure of 6.0 kPa. What force does this pressure exert on a dart in a 1.5-cm-diameter tube?
10. ||| When you stifle a sneeze, you can damage delicate tissues BIO because the pressure of the air that is not allowed to escape may rise by up to 45 kPa. If this extra pressure acts on the inside of your 8.4-mm-diameter eardrum, what is the outward force?
11. |||| 7.5 mol of helium are in a 15 L cylinder. The pressure gauge on the cylinder reads 65 psi. What are (a) the temperature of the gas in °C and (b) the average kinetic energy of a helium atom?
12. || Mars has an atmosphere composed almost entirely of carbon dioxide, with an average temperature of –63°C. What is the rms speed of a molecule in Mars's atmosphere?
13. || 3.0 mol of gas at a temperature of −120°C fills a 2.0 L container. What is the gas pressure?
14. | 265 m/s is a typical cruising speed for a jet airliner. At what temperature (in °C) do the molecules of nitrogen gas have an rms speed of 265 m/s?
15. |||| 10 g of liquid water is placed in a flexible bag, the air is excluded, and the bag is sealed. It is then placed in a microwave oven where the water is boiled to make steam at 100°C. What is the volume of the bag after all the water has boiled? Assume that the pressure inside the bag is equal to atmospheric pressure.

Section 12.3 Ideal-Gas Processes

16. || A cylinder contains 3.0 L of oxygen at 300 K and 2.4 atm. The gas is heated, causing a piston in the cylinder to move outward. The heating causes the temperature to rise to 600 K and the volume of the cylinder to increase to 9.0 L. What is the final gas pressure?
17. ||| A gas with initial conditions p_i, V_i, and T_i expands isothermally until $V_f = 2V_i$. What are (a) T_f and (b) p_f?
18. ||| 0.10 mol of argon gas is admitted to an evacuated 50 cm^3 container at 20°C. The gas then undergoes heating at constant volume to a temperature of 300°C.
    a. What is the final pressure of the gas?
    b. Show the process on a pV diagram. Include a proper scale on both axes.
19. | 0.10 mol of argon gas is admitted to an evacuated 50 cm^3 container at 20°C. The gas then undergoes an isobaric heating to a temperature of 300°C.
    a. What is the final volume of the gas?
    b. Show the process on a pV diagram. Include a proper scale on both axes.
20. ||| 0.10 mol of argon gas is admitted to an evacuated 50 cm^3 container at 20°C. The gas then undergoes an isothermal expansion to a volume of 200 cm^3.
    a. What is the final pressure of the gas?
    b. Show the process on a pV diagram. Include a proper scale on both axes.
21. || 0.0040 mol of gas undergoes the process shown in Figure P12.21.
    a. What type of process is this?
    b. What are the initial and final temperatures?

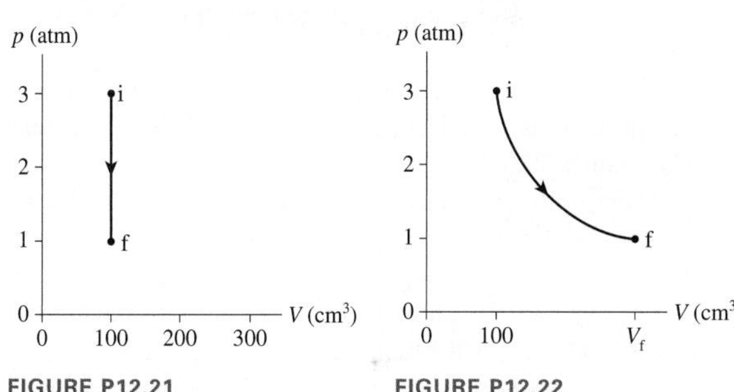

FIGURE P12.21 FIGURE P12.22

22. ||| 0.0040 mol of gas follows the hyperbolic trajectory shown in Figure P12.22.
    a. What type of process is this?
    b. What are the initial and final temperatures?
    c. What is the final volume V_f?
23. || A gas with an initial temperature of 900°C undergoes the process shown in Figure P12.23.
    a. What type of process is this?
    b. What is the final temperature?
    c. How many moles of gas are there?

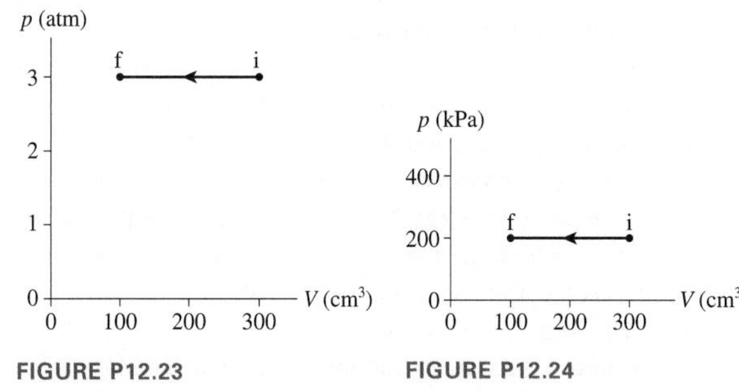

FIGURE P12.23 FIGURE P12.24

24. ||| How much work is done on the gas in the process shown in Figure P12.24?

25. || It is possible to make a thermometer by sealing gas in a rigid container and measuring the absolute pressure. Such a constant-volume gas thermometer is placed in an ice-water bath at 0.00°C. After reaching thermal equilibrium, the gas pressure is recorded as 55.9 kPa. The thermometer is then placed in contact with a sample of unknown temperature. After the thermometer reaches a new equilibrium, the gas pressure is 65.1 kPa. What is the temperature of this sample?
26. || A 1.0 cm^3 air bubble is released from the sandy bottom of a warm, shallow sea, where the gauge pressure is 1.5 atm. The bubble rises slowly enough that the air inside remains at the same constant temperature as the water.
    a. What is the volume of the bubble as it reaches the surface?
    b. As the bubble rises, is heat energy transferred from the water to the bubble or from the bubble to the water? Explain.
27. || A weather balloon rises through the atmosphere, its volume expanding from 4.0 m^3 to 12 m^3 as the temperature drops from 20°C to –10°C. If the initial gas pressure inside the balloon is 1.0 atm, what is the final pressure?

Section 12.4 Thermal Expansion

28. ||| A straight rod consists of a 1.2-cm-long piece of aluminum attached to a 2.0-cm-long piece of steel. By how much will the length of this rod change if its temperature is increased from 20°C to 40°C?
29. || The length of a steel beam increases by 0.73 mm when its temperature is raised from 22°C to 35°C. What is the length of the beam at 22°C?
30. || Older railroad tracks in the U.S. are made of 12-m-long pieces of steel. When the tracks are laid, gaps are left between the sections to prevent buckling when the steel thermally expands. If a track is laid at 16°C, how large should the gaps be if the track is not to buckle when the temperature is as high as 50°C?
31. |||| The temperature of an aluminum disk is increased by 120°C. By what percentage does its volume increase?

Section 12.5 Specific Heat and Heat of Transformation

32. ||| How much energy must be removed from a 200 g block of ice to cool it from 0°C to −30°C?
33. |||| How much heat is needed to change 20 g of mercury at 20°C into mercury vapor at the boiling point?
34. || a. 100 J of heat energy are transferred to 20 g of mercury. By how much does the temperature increase?
    b. How much heat is needed to raise the temperature of 20 g of water by the same amount?
35. || The maximum amount of water an adult in temperate BIO climates can perspire in one hour is typically 1.8 L. However, after several weeks in a tropical climate the body can adapt, increasing the maximum perspiration rate to 3.5 L/h. At what rate, in watts, is energy being removed when perspiring that rapidly? Assume all of the perspired water evaporates. At body temperature, the heat of vaporization of water is $L_v = 24 \times 10^5$ J/kg.
36. || Alligators and other reptiles don't use enough metabolic BIO energy to keep their body temperatures constant. They cool off at night and must warm up in the sun in the morning. Suppose a 300 kg alligator with an early-morning body temperature of 25°C is absorbing radiation from the sun at a rate of 1200 W. How long will the alligator need to warm up to a more favorable 30°C? (Assume that the specific heat of the reptilian body is the same as that of the mammalian body.)

37. || When air is inhaled, it quickly becomes saturated with water BIO vapor as it passes through the moist upper airways. When breathing dry air, about 25 mg of water are exhaled with each breath. At 12 breaths/min, what is the rate of energy loss due to evaporation? Express your answer in both watts and Calories per day. At body temperature, the heat of vaporization of water is $L_v = 24 \times 10^5$ J/kg.
38. ||| It is important for the body to have mechanisms to effectively BIO cool itself; if not, moderate exercise could easily increase body temperatures to dangerous levels. Suppose a 70 kg man runs on a treadmill for 30 min, using a metabolic power of 1000 W. Assume that all of this power goes to thermal energy in the body. If he couldn't perspire or otherwise cool his body, by how much would his body temperature rise during this exercise?
39. ||| What minimum heat is needed to bring 100 g of water at 20°C to the boiling point and completely boil it away?

Section 12.6 Calorimetry

40. ||| 30 g of copper pellets are removed from a 300°C oven and immediately dropped into 100 mL of water at 20°C in an insulated cup. What will the new water temperature be?
41. ||| A copper block is removed from a 300°C oven and dropped into 1.00 kg of water at 20.0°C. The water quickly reaches 25.5°C and then remains at that temperature. What is the mass of the copper block?
42. |||| A 750 g aluminum pan is removed from the stove and plunged into a sink filled with 10.0 kg of water at 20.0°C. The water temperature quickly rises to 24.0°C. What was the initial temperature of the pan?
43. || A 500 g metal sphere is heated to 300°C, then dropped into a beaker containing 4.08 kg of mercury at 20.0°C. A short time later the mercury temperature stabilizes at 99.0°C. Identify the metal.
44. || Brewed coffee is often too hot to drink right away. You can cool it with an ice cube, but this dilutes it. Or you can buy a device that will cool your coffee without dilution—a 200 g aluminum cylinder that you take from your freezer and place in a mug of hot coffee. If the cylinder is cooled to –20°C, a typical freezer temperature, and then dropped into a large cup of coffee (essentially water, with a mass of 500 g) at 85°C, what is the final temperature of the coffee?
45. ||||| Marianne really likes coffee, but on summer days she doesn't want to drink a hot beverage. If she is served 200 mL of coffee at 80°C in a well-insulated container, how much ice at 0°C should she add to obtain a final temperature of 30°C?
46. || If a person has a dangerously high fever, submerging her in BIO ice water is a bad idea, but an ice pack can help to quickly bring her body temperature down. How many grams of ice at 0°C will be melted in bringing down a 60 kg patient's fever from 40°C to 39°C?

Section 12.7 Thermal Properties of Gases

47. | A container holds 1.0 g of argon at a pressure of 8.0 atm.
    a. How much heat is required to increase the temperature by 100°C at constant volume?
    b. How much will the temperature increase if this amount of heat energy is transferred to the gas at constant pressure?
48. || A container holds 1.0 g of oxygen at a pressure of 8.0 atm.
    a. How much heat is required to increase the temperature by 100°C at constant pressure?
    b. How much will the temperature increase if this amount of heat energy is transferred to the gas at constant volume?
49. | What is the temperature change of 1.0 mol of a monatomic gas if its thermal energy is increased by 1.0 J?
50. || The temperature of 2.0 g of helium is increased at constant volume by ΔT. What mass of oxygen can have its temperature increased by the same amount at constant volume using the same amount of heat?
51. ||| How much work is done per cycle by a gas following the pV trajectory of Figure P12.51?

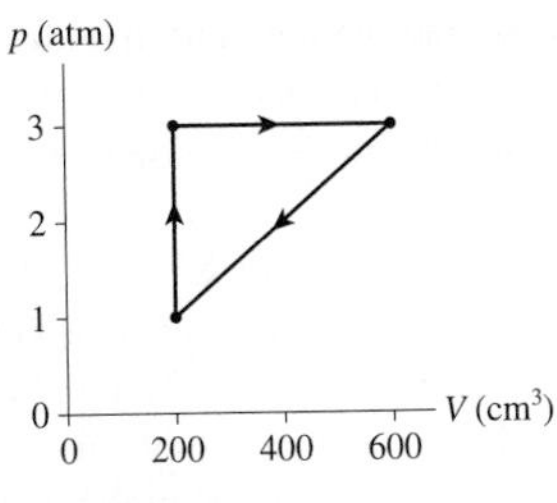

FIGURE P12.51

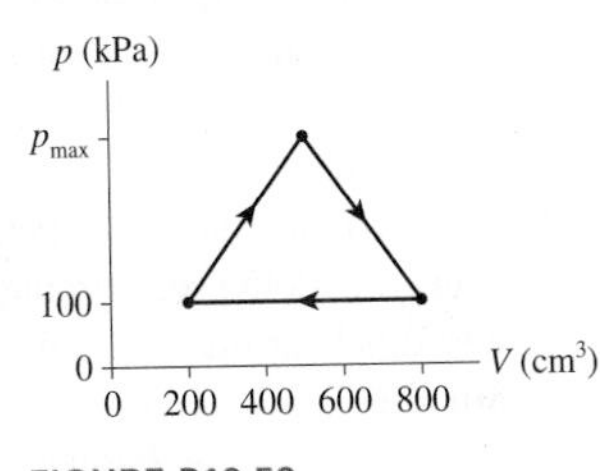

FIGURE P12.52

52. ||| A gas following the pV trajectory of Figure P12.52 does 60 J of work per cycle. What is p_{max}?

Section 12.8 Heat Transfer

53. ||| A 1.8-cm-thick wood floor covers a 4.0 m × 5.5 m room. The subfloor on which the flooring sits is at a temperature of 16.2°C, while the air in the room is at 19.6°C. What is the rate of heat conduction through the floor?
54. |||| A copper-bottomed kettle, its bottom 24 cm in diameter and 3.0 mm thick, sits on a burner. The kettle holds boiling water, and energy flows into the water from the kettle bottom at 800 W. What is the temperature of the bottom surface of the kettle?
55. |||| What is the greatest possible rate of energy transfer by radiation from a metal cube 2.0 cm on a side that is at 700°C? Its emissivity is 0.20.
56. ||| What is the greatest possible rate of energy transfer by radiation for a 5.0-cm-diameter sphere that is at 100°C?
57. ||| BIO Seals may cool themselves by using *thermal windows,* patches on their bodies with much higher than average surface temperature. Suppose a seal has a 0.030 m² thermal window at a temperature of 30°C. If the seal's surroundings are a frosty −10°C, what is the net rate of energy loss by radiation? Assume an emissivity equal to that of a human.

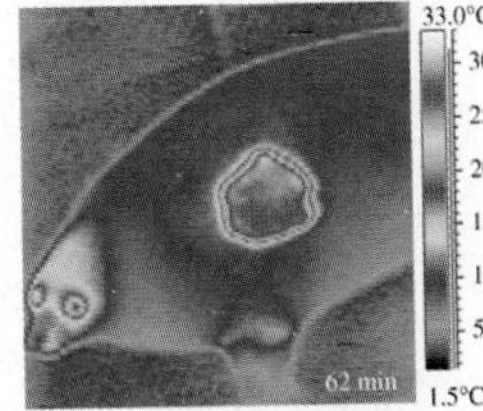

58. || Electronics and inhabitants of the International Space Station generate a significant amount of thermal energy that the station must get rid of. The only way that the station can exhaust thermal energy is by radiation, which it does using thin, 1.8-m-by-3.6-m panels that have a working temperature of about 6°C. How much power is radiated from each panel? Assume that the panels are in the shade so that the absorbed radiation will be negligible. Assume that the emissivity of the panels is 1.0. **Hint:** Don't forget that the panels have two sides!
59. || The glowing filament in a lamp is radiating energy at a rate of 60 W. At the filament's temperature of 1500°C, the emissivity is 0.23. What is the surface area of the filament?
60. ||| BIO If you lie on the ground at night with no cover, you get cold rather quickly. Much of this is due to energy loss by radiation. At night in a dry climate, the temperature of the sky can drop to −40°C. If you are lying on the ground with thin clothing that provides little insulation, the surface temperature of your skin and clothes will be about 30°C. Estimate the net rate at which your body loses energy by radiation to the night sky under these conditions. **Hint:** What area should you use?

General Problems

61. || A rigid container holds 2.0 mol of gas at a pressure of 1.0 atm and a temperature of 30°C.
    a. What is the container's volume?
    b. What is the pressure if the temperature is raised to 130°C?
62. || A 15-cm-diameter compressed-air tank is 50 cm tall. The pressure at 20°C is 150 atm.
    a. How many moles of air are in the tank?
    b. What volume would this air occupy at STP?
63. |||| A 10-cm-diameter cylinder of helium gas is 30 cm long and at 20°C. The pressure gauge reads 120 psi.
    a. How many helium atoms are in the cylinder?
    b. What is the mass of the helium?
64. || Party stores sell small tanks containing 30 g of helium gas. If you use such a tank to fill 0.010 m³ foil balloons (which don't stretch, and so have an internal pressure that is very close to atmospheric pressure), how many balloons can you expect to fill?
65. ||||| BIO Suppose you take and hold a deep breath on a chilly day, inhaling 3.0 L of air at 0°C. Assume that air pressure is 1.0 atm.
    a. How much heat must your body supply to warm the air to your internal body temperature of 37°C?
    b. How much does the air's volume increase as it is warmed?
66. || On average, each person in the industrialized world is responsible for the emission of 10,000 kg of carbon dioxide (CO_2) every year. This includes CO_2 that you generate directly, by burning fossil fuels to operate your car or your furnace, as well as CO_2 generated on your behalf by electric generating stations and manufacturing plants. CO_2 is a greenhouse gas that contributes to global warming. If you were to store your yearly CO_2 emissions in a cube at STP, how long would each edge of the cube be?
67. ||| On a cool morning, when the temperature is 15°C, you measure the pressure in your car tires to be 30 psi. After driving 20 mi on the freeway, the temperature of your tires is 45°C. What pressure will your tire gauge now show?
68. ||| Suppose you inflate your car tires to 35 psi on a 20°C day. Later, the temperature drops to 0°C. What is the pressure in your tires now?
69. | The volume in a constant-*pressure* gas thermometer is directly proportional to the absolute temperature. A constant-pressure thermometer is calibrated by adjusting its volume to 1000 mL while it is in contact with a reference cell at 0.01°C. The volume increases to 1638 mL when the thermometer is placed in contact with a sample. What is the sample's temperature in °C?

70. || A compressed-air cylinder is known to fail if the pressure exceeds 110 atm. A cylinder that was filled to 25 atm at 20°C is stored in a warehouse. Unfortunately, the warehouse catches fire and the temperature reaches 950°C. Does the cylinder explode?

71. || A rigid sphere has a valve that can be opened or closed. The sphere with the valve open is placed in boiling water in a room where the air pressure is 1.0 atm. After a long period of time has elapsed, the valve is closed. What will be the pressure inside the sphere if it is then placed in (a) a mixture of ice and water and (b) an insulated box filled with dry ice, which is at -78.5°C?

72. ||| 80 J of work are done on the gas in the process shown in Figure P12.72. What is V_f in cm^3?

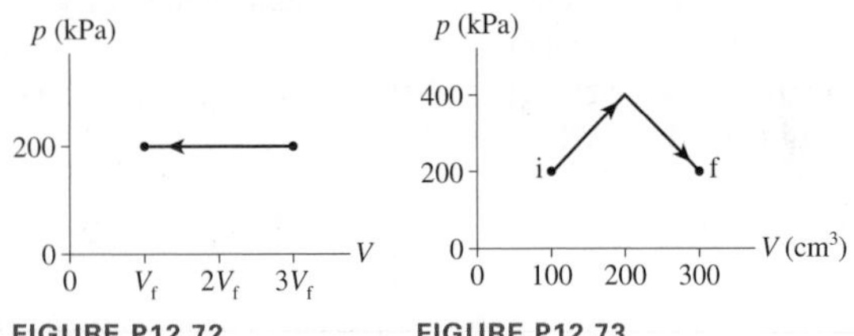

FIGURE P12.72 FIGURE P12.73

73. ||| How much work is done by the gas in the process shown in Figure P12.73?

74. | 0.10 mol of gas undergoes the process $1 \rightarrow 2$ shown in Figure P12.74.
a. What are temperatures T_1 and T_2?
b. What type of process is this?
c. The gas undergoes constant-volume heating from point 2 until the pressure is restored to the value it had at point 1. What is the final temperature of the gas?

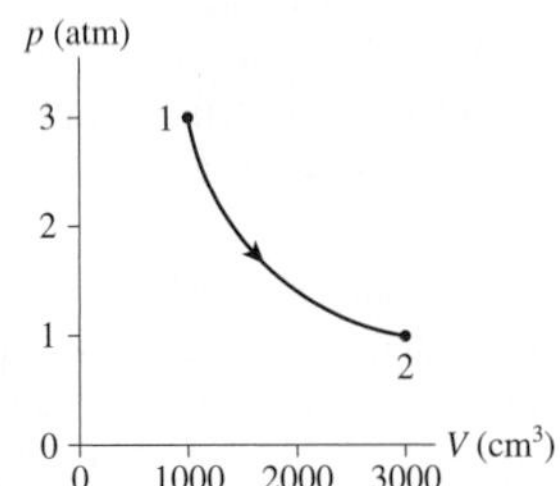

FIGURE P12.74

75. || 10 g of dry ice (solid CO_2) is placed in a 10,000 cm^3 container, then all the air is quickly pumped out and the container sealed. The container is warmed to 0°C, a temperature at which CO_2 is a gas.
a. What is the gas pressure? Give your answer in atm.

The gas then undergoes an isothermal compression until the pressure is 3.0 atm, immediately followed by an isobaric compression until the volume is 1000 cm^3.
b. What is the final temperature of the gas?
c. Show the process on a pV diagram.

76. |||| A large freshwater fish has a swim bladder with a volume of
BIO 5.0×10^{-4} m^3. The fish descends from a depth where the absolute pressure is 3.0 atm to deeper water where the swim bladder is compressed to 60% of its initial volume. As the fish descends, the gas pressure in the swim bladder is always equal to the water pressure, and the temperature of the gas remains at the internal temperature of the fish's body. To adapt to its new location, the fish must add gas to reinflate its swim bladder to the original volume. This takes energy to accomplish. What's the minimum amount of work required to expand the swim bladder back to its original volume?

77. ||| A 5.0-m-diameter garden pond holds 5.9×10^3 kg of water. Solar energy is incident on the pond at an average rate of 400 W/m^2. If the water absorbs all the solar energy and does not exchange energy with its surroundings, how many hours will it take to warm from 15°C to 25°C?

78. |||| 0.030 mol of an ideal monatomic gas undergoes an adiabatic compression that raises its temperature from 10°C to 50°C. How much work is done on the gas to compress it?

79. |||| 0.15 mol of an ideal monatomic gas undergoes an adiabatic expansion, cooling from 20°C to –10°C. How much work is done by the gas during the expansion?

80. |||| Susan, whose mass is 68 kg, climbs 59 m to the top of the
BIO Cape Hatteras lighthouse.
INT a. During the climb, by how much does her potential energy increase?
b. For a typical efficiency of 25%, what metabolic energy does she require to complete the climb?
c. When exercising, the body must perspire and use other mechanisms to cool itself to avoid potentially dangerous increases on body temperature. If we assume that Susan doesn't perspire or otherwise cool herself and that all of the "lost" energy goes into increasing her body temperature, by how much would her body temperature increase during this climb?

81. ||| A typical nuclear reactor generates 1000 MW of electric energy. In doing so, it produces "waste heat" at a rate of 2000 MW, and this heat must be removed from the reactor. Many reactors are sited next to large bodies of water so that they can use the water for cooling. Consider a reactor where the intake water is at 18°C. State regulations limit the temperature of the output water to 30°C so as not to harm aquatic organisms. How many kilograms of cooling water have to be pumped through the reactor each minute?

82. |||| A 68 kg woman cycles at a constant 15 km/h. All of the meta-
BIO bolic energy that does not go to forward propulsion is converted
INT to thermal energy in her body. If the only way her body has to keep cool is by evaporation, how many kilograms of water must she lose to perspiration each hour to keep her body temperature constant?

83. ||| A 1200 kg car traveling at 60 mph quickly brakes to a halt.
INT The kinetic energy of the car is converted to thermal energy of the disk brakes. The brake disks (one per wheel) are iron disks with a mass of 4.0 kg. Estimate the temperature rise in each disk as the car stops.

84. ||| A 5000 kg African elephant
BIO has a resting metabolic rate of
INT 2500 W. On a hot day, the elephant's environment is likely to be nearly the same temperature as the animal itself, so cooling by radiation is not effective. The only plausible way to keep cool is by evaporation, and elephants spray water on their body to accomplish this. If this is the only possible means of cooling, how many kilograms of water per hour must be evaporated from an elephant's skin to keep it at a constant temperature?

85. ||| Suppose you drop a water balloon from a height of 10 m. If
INT the balloon doesn't break on impact, its kinetic energy will be converted to thermal energy. Estimate the temperature rise of the water. Is this likely to be noticeable?

86. ||| What is the maximum mass of lead you could melt with 1000 J of heat, starting from 20°C?

87. || An experiment measures the temperature of a 200 g substance while steadily supplying heat to it. Figure P12.87 shows the results of the experiment. What are (a) the specific heat of the liquid phase and (b) the heat of vaporization?

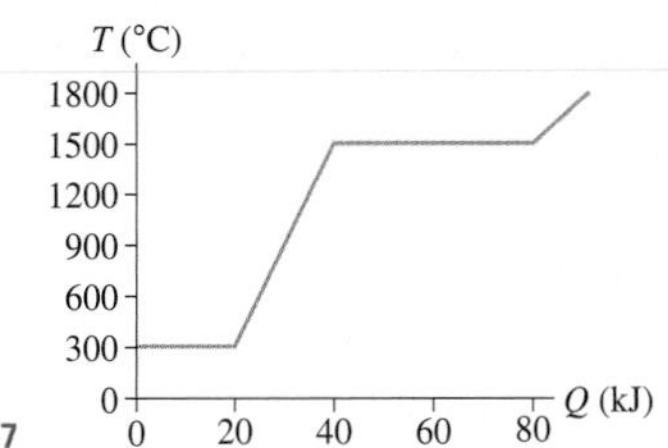

FIGURE P12.87

88. |||| 10 g of aluminum at 200°C and 20 g of copper are dropped into 50 cm^3 of ethyl alcohol at 15°C. The temperature quickly comes to 25°C. What was the initial temperature of the copper?

89. ||| A 100 g ice cube at −10°C is placed in an aluminum cup whose initial temperature is 70°C. The system comes to an equilibrium temperature of 20°C. What is the mass of the cup?

90. ||| A 50.0 g thermometer is used to measure the temperature of 200 g of water. The specific heat of the thermometer, which is mostly glass, is 750 J/kg · K, and it reads 20.0°C while lying on the table. After being completely immersed in the water, the thermometer's reading stabilizes at 71.2°C. What was the actual water temperature before it was measured?

91. |||| Your 300 mL cup of coffee is too hot to drink when served at 90°C. What is the mass of an ice cube, taken from a −20°C freezer, that will cool your coffee to a pleasant 60°C?

92. || A gas is compressed from 600 cm^3 to 200 cm^3 at a constant pressure of 400 kPa. At the same time, 100 J of heat energy is transferred out of the gas. What is the change in thermal energy of the gas during this process?

93. |||| An expandable cube, initially 20 cm on each side, contains 3.0 g of helium at 20°C. 1000 J of heat energy are transferred to this gas. What are (a) the final pressure if the process is at constant volume and (b) the final volume if the process is at constant pressure?

94. ||| 0.10 mol of a monatomic gas follows the process shown in Figure P12.94.

a. How much heat energy is transferred to or from the gas during process 1 → 2?
b. How much heat energy is transferred to or from the gas during process 2 → 3?
c. What is the total change in thermal energy of the gas?

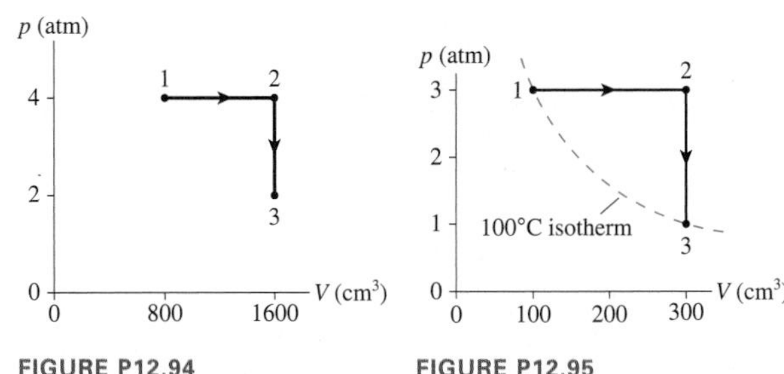

FIGURE P12.94

FIGURE P12.95

95. || A monatomic gas follows the process 1 → 2 → 3 shown in Figure P12.95. How much heat is needed for (a) process 1 → 2 and (b) process 2 → 3?

96. ||| What are (a) the heat Q_H extracted from the hot reservoir and
INT (b) the efficiency for a heat engine described by the pV diagram of Figure P12.96?

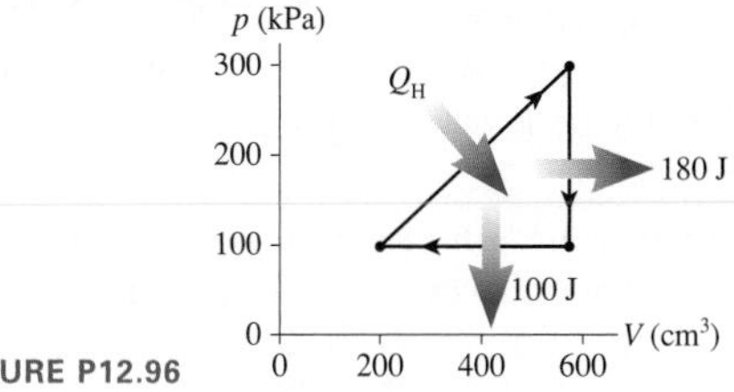

FIGURE P12.96

97. ||| The top layer of your goose down sleeping bag has a thickness of 5.0 cm and a surface area of 1.0 m^2. When the outside temperature is −20°C, you lose 25 Cal/h by heat conduction through the bag (which remains at a cozy 35°C inside). Assume that you're sleeping on an insulated pad that eliminates heat conduction to the ground beneath you. What is the thermal conductivity of the goose down?

98. |||| Suppose you go outside in your fiber-filled jacket on a windless but very cold day. The thickness of the jacket is 2.5 cm, and it covers 1.1 m^2 of your body. The purpose of fiber- or down-filled jackets is to trap a layer of air, and it's really the air layer that provides the insulation. If your skin temperature is 34°C while the air temperature is −20°C, at what rate is heat being conducted through the jacket and away from your body?

99. ||||| Two thin, square copper plates are radiating energy at the same rate. The edge length of plate 2 is four times that of plate 1. What is the ratio of absolute temperatures T_1/T_2 of the plates?

100. || The surface area of an adult human is about 1.8 m^2. Suppose
BIO a person with a skin temperature of 34°C is standing with bare skin in a room where the air is 25°C but the walls are 17°C.

a. There is a "dead-air" layer next to your skin that acts as insulation. If the dead-air layer is 5.0 mm thick, what is the person's rate of heat loss by conduction?
b. What is the person's net radiation loss to the walls? The emissivity of skin is 0.97.
c. Does conduction or radiation contribute more to the total rate of energy loss?
d. If the person is metabolizing food at a rate of 155 W, does the person feel comfortable, chilly, or too warm?

Passage Problems

Thermal Properties of the Oceans

Seasonal temperature changes in the ocean only affect the top layer of water, to a depth of 500 m or so. This "mixed" layer is thermally isolated from the cold, deep water below. The average temperature of this top layer of the world's oceans, which has area 3.6×10^8 km^2, is approximately 17°C.

In addition to seasonal temperature changes, the oceans have experienced an overall warming trend over the last century that is expected to continue as the earth's climate changes. A warmer ocean means a larger volume of water; the oceans will rise. Suppose the average temperature of the top layer of the world's oceans were to increase from a temperature T_i to a temperature T_f. The area of the oceans will not change, as this is fixed by the size of the ocean basin, so any thermal expansion of the water will cause the water level to rise, as shown in Figure P12.101. The original volume is the product of the original depth and the surface area, $V_i = Ad_i$. The change in volume is given by $\Delta V = A\ \Delta d$.

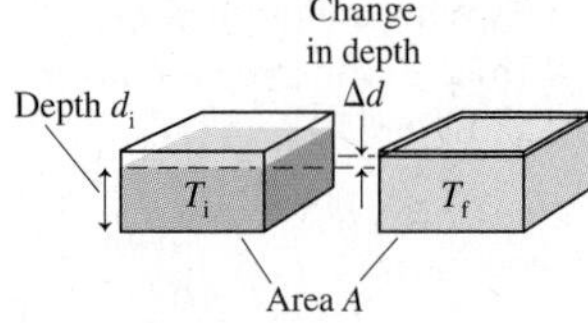

FIGURE P12.101

101. | If the top 500 m of ocean water increased in temperature from 17°C to 18°C, what would be the resulting rise in ocean height?
A. 0.11 m
B. 0.22 m
C. 0.44 m
D. 0.88 m

102. | Approximately how much energy would be required to raise the temperature of the top layer of the oceans by 1°C? (1 m^3 of water has a mass of 1000 kg.)
A. 1×10^{24} J
B. 1×10^{21} J
C. 1×10^{18} J
D. 1×10^{15} J

103. | Water's coefficient of expansion varies with temperature. For water at 2°C, an increase in temperature of 1°C would cause the volume to
A. Increase.
B. Stay the same.
C. Decrease.

104. | The ocean is mostly heated from the top, by light from the sun. The warmer surface water doesn't mix much with the colder deep ocean water. This lack of mixing can be ascribed to a lack of
A. Conduction.
B. Convection.
C. Radiation.
D. Evaporation.

STOP TO THINK ANSWERS

Stop to Think 12.1: A. Both helium and neon are monatomic gases, where the basic particles are atoms. 5 mol of helium contain 5 times as many atoms as 1 mol of neon, though both samples have the same mass.

Stop to Think 12.2: A, B. An increase in temperature means that the atoms have a larger average kinetic energy and will thus have a larger rms speed. Because the thermal energy of the gas is simply the total kinetic energy of the atoms, this must increase as well. The pressure *could* change, but it's not required; T and V could increase by the same factor, which would keep p constant. The container is sealed, so the number of molecules does not change.

Stop to Think 12.3: B. The product pV/T is constant. During the process, pV decreases by a factor of 2, so T must decrease by a factor of 2 as well.

Stop to Think 12.4: A. The thermal expansion coefficients of aluminum are greater than those of iron. Heating the rod and the ring will expand the outer diameter of the rod and the inner diameter of the ring, but the ring's expansion will be greater.

Stop to Think 12.5: B. To solidify the lead, heat must be removed; this heat boils the water. The heat of vaporization of water is 10 times the heat of fusion of lead, so much less than 1 kg of water vaporizes as 1 kg of lead solidifies.

Stop to Think 12.6: A. The lead cools and the water warms as heat is transferred from the lead to the water. The specific heat of water is much larger than that of lead, so the temperature change of the water is much less than that of the lead.

Stop to Think 12.7: C. With a sealed suit and no matter around you, there is no way to transfer heat to the environment except by radiation.

13 Fluids

The 20,000 pound boat floats while the 130 pound diver sinks. What determines whether an object floats or sinks?

LOOKING AHEAD

The goal of Chapter 13 is to understand the static and dynamic properties of fluids.

Pressure in Fluids

The pressure in a **fluid**—a gas or a liquid—exerts a force on the fluid's container as well as on all parts of the fluid itself.

The pressure in a liquid increases with depth. The high pressure at the base of this water tower allows water to be distributed throughout the city.

Looking Back
5.1 Equilibrium
12.3 Pressure in gases

Measuring Pressure

Pressure is measured by the force it exerts on a known area. An important kind of pressure gauge is the **manometer.**

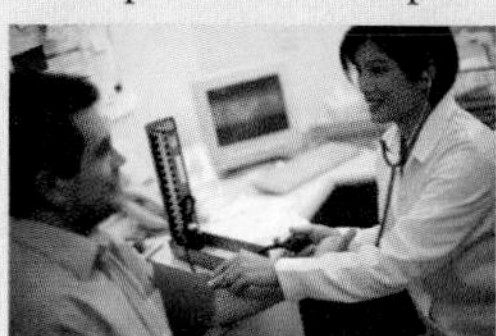

Blood pressure can be measured by the height of the mercury in this manometer.

Buoyancy

Why do some objects float while others sink? You'll learn that the **bouyant force**, the upward force of a fluid on an immersed object, is described by a simple but powerful concept called **Archimedes' principle.**

Teams from across the country compete in the annual National Concrete Canoe Competition. You'll learn how such a heavy object can stay afloat.

Fluid Dynamics

The volume of a moving fluid is conserved. This requires flowing fluids to speed up as they squeeze through a narrow section of a tube.

You'll learn why holding your thumb over the end of a hose makes the water come out faster.

Looking Back
4.6, 5.2 Newton's second law
10.2 Work
10.6 Conservation of energy

In studying moving fluids, we'll use many of the same ideas we used for studying the motion of particles: Newton's laws and conservation of energy.

Moving fluids can exert large forces on objects. **Bernoulli's equation**, a statement of conservation of energy applied to fluids, gives us a way to calculate pressures and forces due to moving fluids.

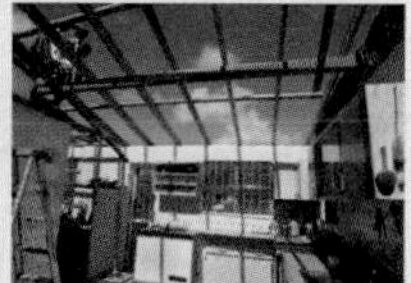

You'll learn how forces due to the motion of fluids can be enough to lift a massive airplane or lift the roof off a house during a hurricane.

Viscosity

You know from experience that some liquids, such as water, flow freely while others, like honey, are hard to pour. Honey has a much higher **viscosity** than water. You'll learn how viscosity affects the flow of fluids.

13.1 Fluids and Density

FIGURE 13.1 Simple atomic-level models of gases and liquids.

(a) A gas

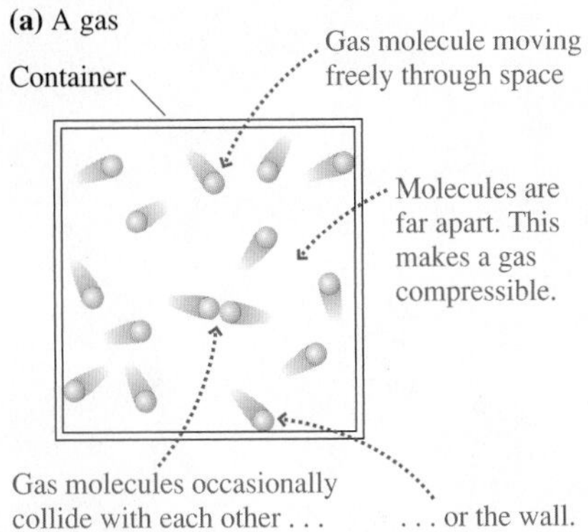

Container
Gas molecule moving freely through space
Molecules are far apart. This makes a gas compressible.
Gas molecules occasionally collide with each other . . .
. . . or the wall.

(b) A liquid

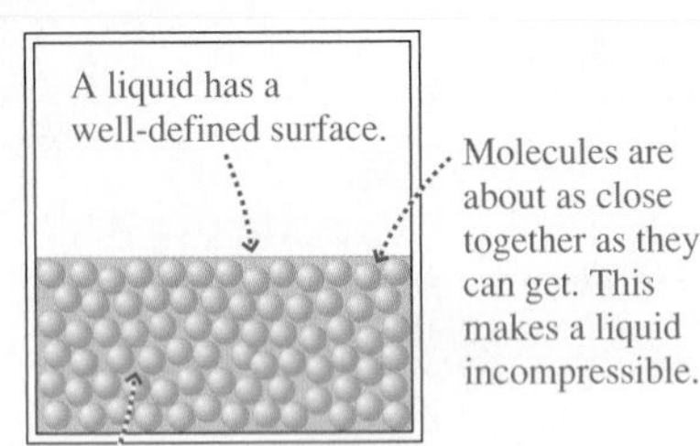

A liquid has a well-defined surface.
Molecules are about as close together as they can get. This makes a liquid incompressible.
Molecules make weak bonds with each other that keep them close together. But the molecules can slide around each other, allowing the liquid to flow and conform to the shape of its container.

A **fluid** is simply a substance that flows. Because they flow, fluids take the shape of their container rather than retaining a shape of their own. You may think that gases and liquids are quite different, but both are fluids, and their similarities are often more important than their differences.

As you learned in Chapter 12, a gas, as shown in FIGURE 13.1a, is a system in which each molecule moves freely through space until, on occasion, it collides with another molecule or with the wall of the container. The gas you are most familiar with is air, a mixture of mostly nitrogen and oxygen molecules. Gases are *compressible;* that is, the volume of a gas is easily increased or decreased, a consequence of the "empty space" between the molecules in a gas.

Liquids are more complicated than either gases or solids. Liquids, like solids, are essentially *incompressible.* This property tells us that the molecules in a liquid, as in a solid, are about as close together as they can get without coming into contact with each other. At the same time, a liquid flows and deforms to fit the shape of its container. The fluid nature of a liquid tells us that the molecules are free to move around.

Together, these observations suggest the model of a liquid shown in FIGURE 13.1b. Here you see a system in which the molecules are loosely held together by weak molecular bonds. The bonds are strong enough that the molecules never get far apart but not strong enough to prevent the molecules from sliding around each other.

Density

An important parameter that characterizes a macroscopic system is its *density.* Suppose you have several blocks of copper, each of different size. Each block has a different mass m and a different volume V. Nonetheless, all the blocks are copper, so there should be some quantity that has the *same* value for all the blocks, telling us, "This is copper, not some other material." The most important such parameter is the *ratio* of mass to volume, which we call the **mass density** ρ (lowercase Greek rho):

$$\rho = \frac{m}{V} \tag{13.1}$$

Mass density of an object of mass m and volume V

Conversely, an object of mass density ρ and volume V has mass

$$m = \rho V \tag{13.2}$$

The SI units of mass density are kg/m^3. Nonetheless, units of g/cm^3 are widely used. You need to convert these to SI units before doing most calculations. You must convert both the grams to kilograms and the cubic centimeters to cubic meters. The net result is the conversion factor

$$1\ g/cm^3 = 1000\ kg/m^3$$

The mass density is independent of the object's size. That is, mass and volume are parameters that characterize a *specific piece* of some substance—say, copper—whereas mass density characterizes the substance itself. All pieces of copper have the same mass density, which differs from the mass density of almost any other substance. Thus mass density allows us to talk about the properties of copper in general without having to refer to any specific piece of copper.

The mass density is usually called simply "the density" if there is no danger of confusion. However, we will meet other types of density as we go along, and sometimes it is important to be explicit about which density you are using. Table 13.1 provides a short list of the mass densities of various fluids. Notice the enormous difference between the densities of gases and liquids. Gases have lower densities because the molecules in gases are farther apart than in liquids. Also, the density of a

TABLE 13.1 Densities of fluids at 1 atm pressure

Substance	ρ (kg/m^3)
Helium gas (20°C)	0.166
Air (20°C)	1.20
Air (0°C)	1.28
Gasoline	680
Ethyl alcohol	790
Oil (typical)	900
Water	1000
Seawater	1030
Blood (whole)	1060
Glycerin	1260
Mercury	13,600

liquid varies only slightly with temperature because its molecules are always nearly in contact. The density of a gas, such as air, has a larger variation with temperature because it's easy to change the already large distance between the molecules.

What does it *mean* to say that the density of gasoline is 680 kg/m^3? Recall in Chapter 1 we discussed the meaning of the word "per." We found that it meant "for each," so that 2 miles per hour means you travel 2 miles *for each* hour that passes. In the same way, saying that the density of gasoline is 680 kg per cubic meter means that there are 680 kg of gasoline *for each* 1 cubic meter of the liquid. If we have 2 m^3 of gasoline, each will have a mass of 680 kg, so the total mass will be 2×680 kg $= 1360$ kg. The product ρV is the number of cubic meters times the mass of each cubic meter—that is, the total mass of the object.

EXAMPLE 13.1 Weighing the air in a living room

What is the mass of air in a living room with dimensions $4.0 \text{ m} \times 6.0 \text{ m} \times 2.5 \text{ m}$?

PREPARE Table 13.1 gives air density at a temperature of 20°C, which is about room temperature.

SOLVE The room's volume is

$$V = (4.0 \text{ m}) \times (6.0 \text{ m}) \times (2.5 \text{ m}) = 60 \text{ m}^3$$

The mass of the air is

$$m = \rho V = (1.20 \text{ kg/m}^3)(60 \text{ m}^3) = 72 \text{ kg}$$

ASSESS This is perhaps more mass—about that of an adult person—than you might have expected from a substance that hardly seems to be there. For comparison, a swimming pool this size would contain 60,000 kg of water.

STOP TO THINK 13.1 A piece of glass is broken into two pieces of different size. Rank in order, from largest to smallest, the mass densities of pieces 1, 2, and 3.

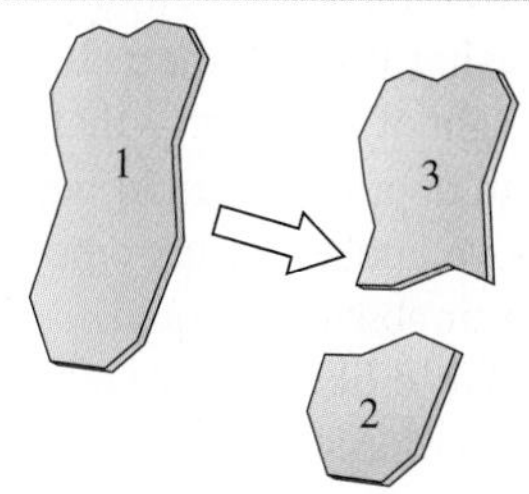

13.2 Pressure

In Chapter 12, you learned how a gas exerts a force on the walls of its containers. Liquids also exert forces on the walls of their containers, as shown in **FIGURE 13.2**, where a force $\vec{F}$ due to the liquid pushes against a small area A of the wall. Just as for a gas, we define the pressure at this point in the fluid to be the ratio of the force to the area on which the force is exerted:

$$p = \frac{F}{A} \tag{13.3}$$

This is the same as Equation 12.9 of Chapter 12. Recall also from Chapter 12 that the SI unit of pressure, the pascal, is defined as

$$1 \text{ pascal} = 1 \text{ Pa} = 1 \frac{\text{N}}{\text{m}^2}$$

It's important to realize that the force due to a fluid's pressure pushes not only on the walls of its container, but on *all* parts of the fluid itself. If you punch holes in a container of water, the water spurts out from the holes, as in **FIGURE 13.3**. It is the force due to the pressure of the water behind each hole that pushes the water forward through the holes.

FIGURE 13.2 The fluid presses against area A with force $\vec{F}$.

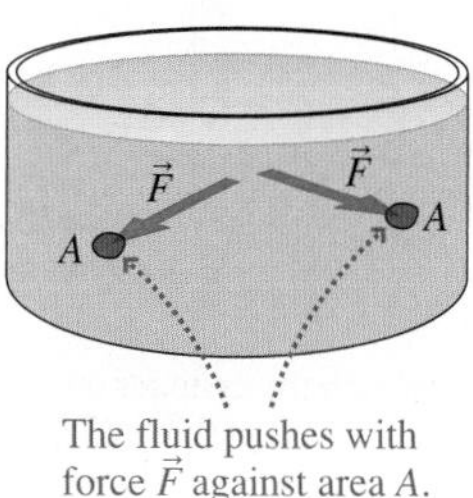

FIGURE 13.3 Pressure pushes the water *sideways,* out of the holes.

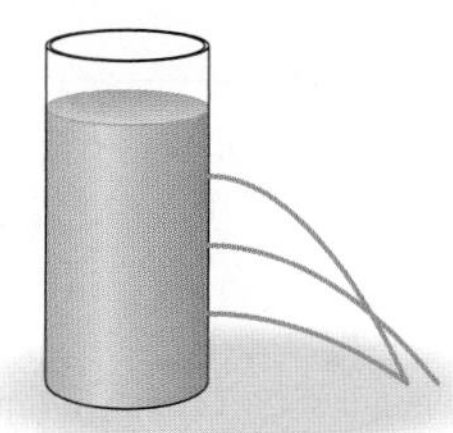

To measure the pressure at any point within a fluid we can use the simple pressure-measuring device shown in FIGURE 13.4a. Because the spring constant k and the area A are known, we can determine the pressure by measuring the compression of the spring. Once we've built such a device, we can place it in various liquids and gases to learn about pressure. FIGURE 13.4b shows what we can learn from simple experiments.

FIGURE 13.4 Learning about pressure.

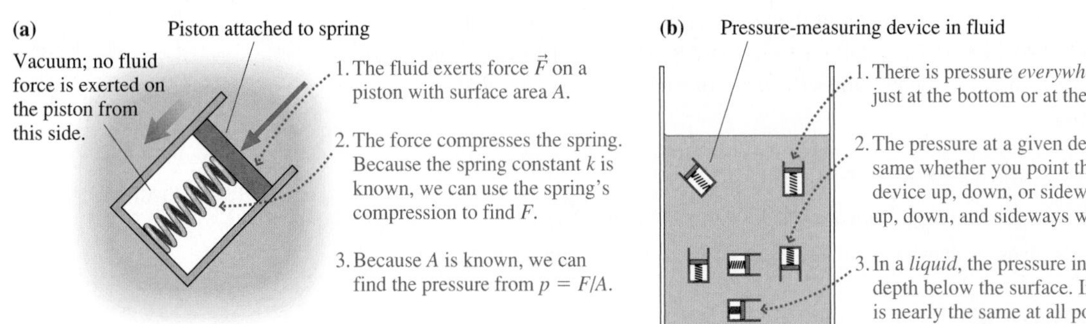

(a)
Piston attached to spring
Vacuum; no fluid force is exerted on the piston from this side.
1. The fluid exerts force $\vec{F}$ on a piston with surface area A.
2. The force compresses the spring. Because the spring constant k is known, we can use the spring's compression to find F.
3. Because A is known, we can find the pressure from $p = F/A$.
(b)
Pressure-measuring device in fluid
1. There is pressure *everywhere* in a fluid, not just at the bottom or at the walls of the container.
2. The pressure at a given depth in the fluid is the same whether you point the pressure-measuring device up, down, or sideways. The fluid pushes up, down, and sideways with equal strength.
3. In a *liquid*, the pressure increases rapidly with depth below the surface. In a *gas*, the pressure is nearly the same at all points (at least in laboratory-size containers).

The first statement in Figure 13.4b emphasizes again that pressure exists at *all* points within a fluid, not just at the walls of the container. You may recall that tension exists at *all* points in a string, not only at its ends where it is tied to an object. We understood tension as the different parts of the string *pulling* against each other. Pressure is an analogous idea, except that the different parts of a fluid are *pushing* against each other.

Pressure in Liquids

If you introduce a liquid into a container, the force of gravity pulls the liquid down, causing it to fill the bottom of the container. It is this force of gravity—that is, the weight of the liquid—that is responsible for the pressure in a liquid. Pressure increases with depth in a liquid because the liquid below is being squeezed by all the liquid above, including any other liquid floating on the first liquid, as well as the pressure of the air above the liquid.

We'd like to determine the pressure at a depth d below the surface of the liquid. We will assume that the liquid is at rest; flowing liquids will be considered later in this chapter. The darker shaded cylinder of liquid in FIGURE 13.5 extends from the surface to depth d. This cylinder, like the rest of the liquid, is in static equilibrium with $\vec{F}_{net} = \vec{0}$. Several forces act on this cylinder: its weight mg, a downward force p_0A due to the pressure p_0 at the surface of the liquid, an upward force pA due to the liquid beneath the cylinder pushing up on the bottom of the cylinder, and inward-directed forces due to the liquid pushing in on the sides of the cylinder. The forces due to the liquid pushing on the cylinder are a consequence of our earlier observation that different parts of a fluid push against each other. Pressure p, the pressure at the bottom of the cylinder, is what we're trying to find.

FIGURE 13.5 Measuring the pressure at depth d in a liquid.

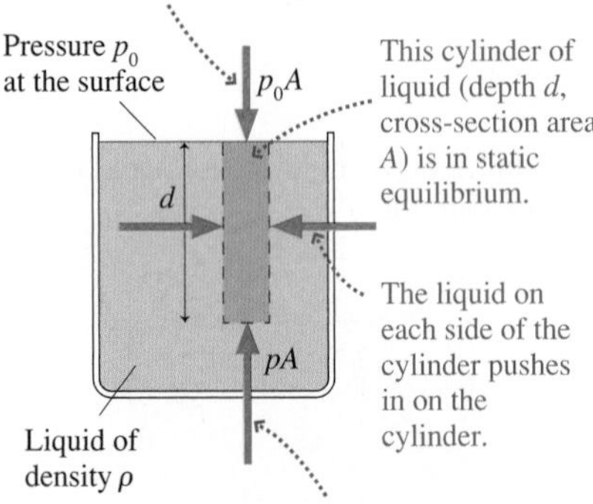

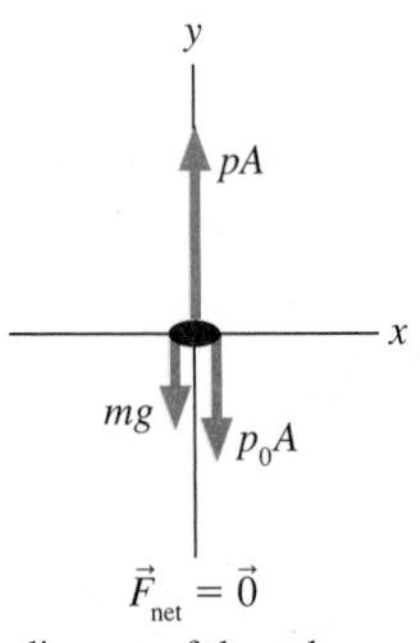

Free-body diagram of the column of liquid. The horizontal forces cancel and are not shown.

The horizontal forces cancel each other. The upward force balances the two downward forces, so

$$pA = p_0A + mg \tag{13.4}$$

The liquid is a cylinder of cross-section area A and height d. Its volume is $V = Ad$ and its mass is $m = \rho V = \rho Ad$. Substituting this expression for the mass of the liquid into Equation 13.4, we find that the area A cancels from all terms. The pressure at depth d in a liquid is then

$$p = p_0 + \rho g d \tag{13.5}$$

Pressure of a liquid with density ρ at depth d

Because of our assumption that the fluid is at rest, the pressure given by Equation 13.5 is called the **hydrostatic pressure.** The fact that g appears in Equation 13.5 reminds us that the origin of this pressure is the gravitational force on the fluid.

As expected, $p = p_0$ at the surface, where $d = 0$. Pressure p_0 is usually due to the air or other gas above the liquid. For a liquid that is open to the air at sea level, $p_0 = 1$ atm $= 101.3$ kPa, as we learned in Section 12.2. In other situations, p_0 might be the pressure due to a piston or a closed surface pushing down on the top of the liquid.

NOTE ► Equation 13.5 assumes that the fluid is *incompressible;* that is, its density ρ doesn't increase with depth. This is an excellent assumption for liquids, but not a good one for a gas. Equation 13.5 should not be used for calculating the pressure of a gas. Gas pressure is found with the ideal-gas law. ◄

EXAMPLE 13.2 **The pressure on a submarine**

A submarine cruises at a depth of 300 m. What is the pressure at this depth? Give the answer in both pascals and atmospheres.

SOLVE The density of seawater, from Table 13.1, is $\rho = 1030$ kg/m^3. At the surface, $p_0 = 1$ atm $= 101.3$ kPa. The pressure at depth $d = 300$ m is found from Equation 13.5 to be

$$p = p_0 + \rho g d$$
$$= (1.013 \times 10^5 \text{ Pa}) + (1030 \text{ kg/m}^3)(9.80 \text{ m/s}^2)(300 \text{ m})$$
$$= 3.13 \times 10^6 \text{ Pa}$$

Converting the answer to atmospheres gives

$$p = (3.13 \times 10^6 \text{ Pa}) \times \frac{1 \text{ atm}}{1.013 \times 10^5 \text{ Pa}} = 30.9 \text{ atm}$$

ASSESS The pressure deep in the ocean is very great. The research submarine *Alvin,* shown in the left photo, can safely dive as deep as 4500 m, where the pressure is over 450 atm! Its viewports are over 3.5 inches thick to withstand this pressure. As shown in the right photo, each viewport is tapered, with its larger face toward the sea. The water pressure then pushes the viewports firmly into their conical seats, helping to seal them tightly.

According to Equation 13.5, the hydrostatic pressure in a liquid depends on only the depth and the pressure at the surface. This observation has some important implications. **FIGURE 13.6a** shows two connected tubes. It's certainly true that the larger volume of liquid in the wide tube weighs more than the liquid in the narrow tube. You might think that this extra weight would push the liquid in the narrow tube higher than in the wide tube. But it doesn't. If d_1 were larger than d_2, then, according to the hydrostatic pressure equation, the pressure at the bottom of the narrow tube would be higher than the pressure at the bottom of the wide tube. This *pressure difference* would cause the liquid to *flow* from right to left until the heights were equal.

Thus a first conclusion: **A connected liquid in hydrostatic equilibrium rises to the same height in all open regions of the container.** This is the familiar observation that "water seeks its own level."

FIGURE 13.6b shows two connected tubes of different shape. The conical tube holds more liquid above the dotted line, so you might think that $p_1 > p_2$. But it isn't. Both points are at the same depth; thus $p_1 = p_2$. You can arrive at the same conclusion by thinking about the pressure at the bottoms of the tubes. If p_1 were greater than p_2, the pressure at the bottom of the left tube would be greater than the pressure at the bottom of the right tube. This would cause the liquid to flow until the pressures were equal. Thus a second conclusion: **In hydrostatic equilibrium, the pressure is the same at all points on a horizontal line through a connected liquid of a single kind.** (When the liquid is not the same kind at different points on the line, the pressure need not be the same.)

FIGURE 13.6 Some properties of a liquid in hydrostatic equilibrium are not what you might expect.

(a)

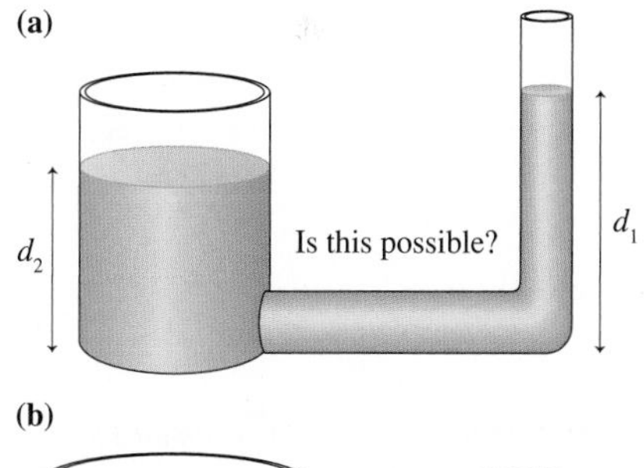

(b)

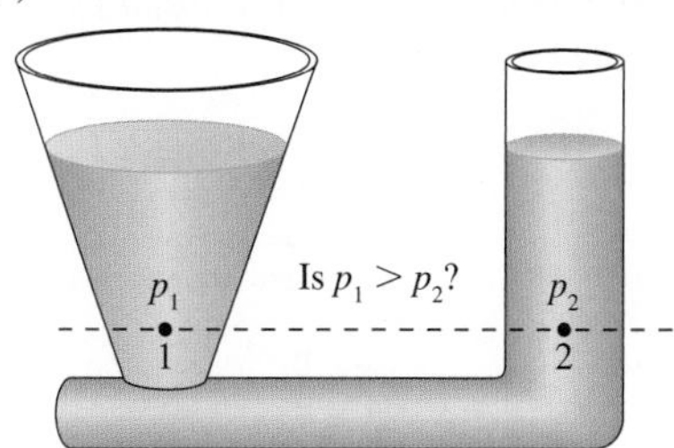

NOTE ► Both of these conclusions are restricted to liquids in hydrostatic equilibrium. The situation is entirely different for flowing fluids, as we'll see later on. ◄

EXAMPLE 13.3 **Pressure in a closed tube**

Water fills the tube shown in FIGURE 13.7. What is the pressure at the top of the closed tube?

FIGURE 13.7 A bent tube closed at one end.

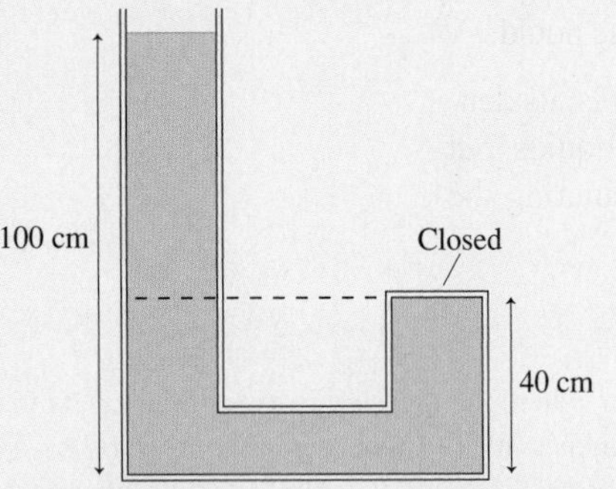

PREPARE This is a liquid in hydrostatic equilibrium. The closed tube is not an open region of the container, so the water cannot rise to an equal height. Nevertheless, the pressure is still the same at all points on a horizontal line. In particular, the pressure at the top of the closed tube equals the pressure in the open tube at the height of the dotted line. Assume $p_0 = 1$ atm.

SOLVE A point 40 cm above the bottom of the open tube is at a depth of 60 cm. The pressure at this depth is

$$p = p_0 + \rho g d$$
$$= (1.01 \times 10^5 \text{ Pa}) + (1000 \text{ kg/m}^3)(9.80 \text{ m/s}^2)(0.60 \text{ m})$$
$$= 1.07 \times 10^5 \text{ Pa} = 1.06 \text{ atm}$$

ASSESS The water in the open tube *pushes* the water in the closed tube up against the top of the tube. Consequently, in accordance with Newton's third law, the top of the tube *presses down on the liquid* with a force of magnitude $F = pA$. This explains why the pressure at the top of the closed tube is greater than atmospheric pressure.

We can draw one more conclusion from the hydrostatic pressure equation $p = p_0 + \rho g d$. If we change the pressure at the surface to $p_1 = p_0 + \Delta p$, so that Δp is the *change* in pressure, then the pressure at a point at a depth d becomes

$$p' = p_1 + \rho g d = (p_0 + \Delta p) + \rho g d = (p_0 + \rho g d) + \Delta p = p + \Delta p$$

That is, the pressure at depth d changes by the same amount as it did at the surface. This idea was first recognized by Blaise Pascal (the same Pascal for whom the pressure unit is named) and is called *Pascal's principle:*

Pascal's principle If the pressure at one point in an incompressible fluid is changed, the pressure at every other point in the fluid changes by the same amount.

For example, if we compress the air above the open tube in Example 13.3 to a pressure of 1.50 atm, an increase of 0.50 atm, the pressure at the top of the closed tube will increase to 1.56 atm.

STOP TO THINK 13.2 Water is slowly poured into the container until the water level has risen into tubes 1, 2, and 3. The water doesn't overflow from any of the tubes. How do the water depths in the three columns compare to each other?

A. $d_1 > d_2 > d_3$
B. $d_1 < d_2 < d_3$
C. $d_1 = d_2 = d_3$
D. $d_1 = d_2 > d_3$
E. $d_1 = d_2 < d_3$

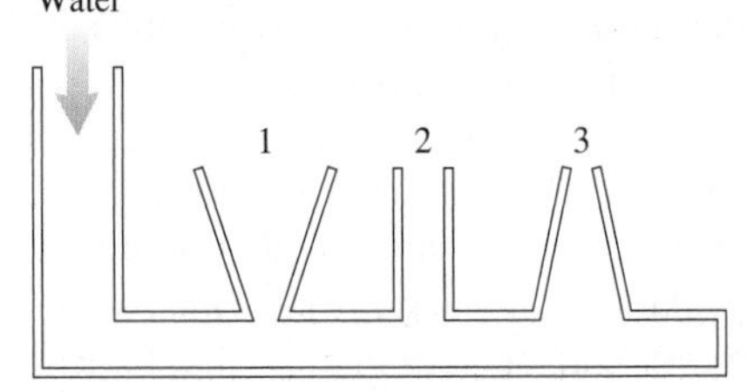

FIGURE 13.8 Atmospheric pressure and density.

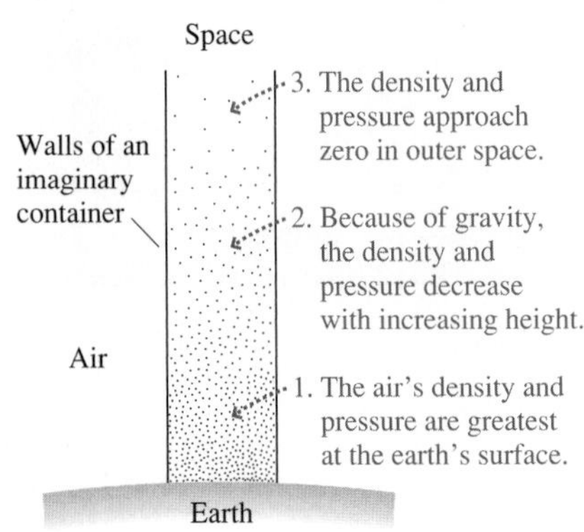

Atmospheric Pressure

We live at the bottom of a "sea" of air that extends up many kilometers. As FIGURE 13.8 shows, there is no well-defined top to the atmosphere; it just gets less and less dense with increasing height until reaching zero in the vacuum of space. Nonetheless, 99% of the air in the atmosphere is below about 30 km.

If we recall that a gas like air is quite compressible, we can see why the atmosphere becomes less dense with increasing altitude. In a liquid, pressure increases with depth because of the weight of the liquid above. The same holds true for the air in the atmosphere, but because the air is compressible, the weight of the air above compresses the air below, increasing its density. At high altitudes there is very little air above to push down, so the density is less.

We learned in Chapter 12 that the global average sea-level pressure, the *standard atmosphere,* is 1 atm = 101,300 Pa. The standard atmosphere, usually referred to simply as "atmospheres," is a commonly used unit of pressure. But it is not an SI unit, so you must convert atmospheres to pascals before doing most calculations with pressure.

NOTE ► Unless you happen to live right at sea level, the atmospheric pressure around you is not exactly 1 atm. A pressure gauge must be used to determine the actual atmospheric pressure. For simplicity, this textbook will always assume that the pressure of the air is $p_{\text{atmos}} = 1$ atm unless stated otherwise. ◄

Atmospheric pressure varies not only with altitude, but also with changes in the weather. Large-scale regions of low-pressure air are created at the equator, where hot air rises and flows to the north and south temperate zones. There the air falls, creating high-pressure zones. Local winds and weather are largely determined by presence and movement of air masses of differing pressure. You may have seen weather maps like the one shown in **FIGURE 13.9** on the evening news. The letters H and L denote regions of high and low atmospheric pressure.

FIGURE 13.9 High- and low-pressure zones on a weather map.

13.3 Measuring and Using Pressure

The pressure in a fluid is measured with a *pressure gauge,* which is often a device very similar to that shown in Figure 13.4. The fluid pushes against a spring, and the displacement of the spring is indicated on a scale. The familiar tire-pressure gauge shown in **FIGURE 13.10** works in the same way. The pressure p_{tire} exerts a force $p_{\text{tire}}A$ on the front area A of the piston, while atmospheric pressure exerts a force $p_{\text{atmos}}A$ on the back of the piston. Thus the *net* pressure force on the piston is $(p_{\text{tire}} - p_{\text{atmos}})A$. This force compresses the spring until equilibrium is reached. Thus the movement of the scale depends not on the absolute pressure in the tire, but on the *difference* between the tire pressure and atmospheric pressure. This type of gauge measures the *gauge pressure* $p_g = p_{\text{tire}} - p_{\text{atmos}}$, an idea introduced in Chapter 12.

FIGURE 13.10 A tire gauge measures the difference between the tire's pressure and atmospheric pressure.

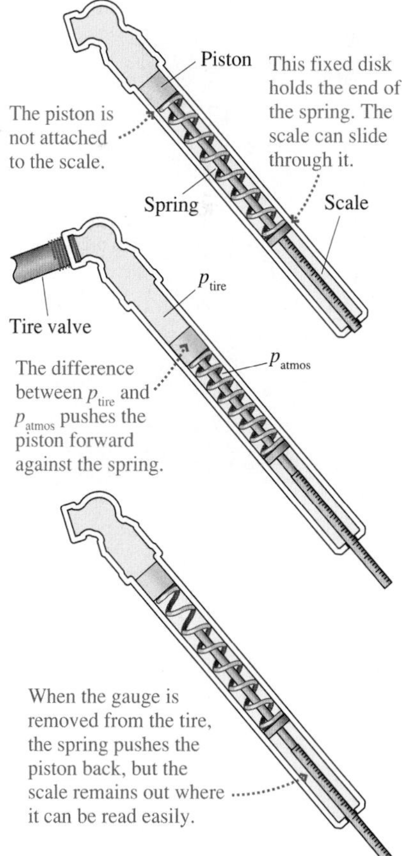
Piston
This fixed disk holds the end of the spring. The scale can slide through it.
The piston is not attached to the scale.
Spring
Scale
p_{tire}
Tire valve
p_{atmos}
The difference between p_{tire} and p_{atmos} pushes the piston forward against the spring.
When the gauge is removed from the tire, the spring pushes the piston back, but the scale remains out where it can be read easily.

Solving Hydrostatic Problems

We now have enough information to formulate a set of rules for working with hydrostatic problems.

TACTICS BOX 13.1 Hydrostatics

❶ **Draw a picture.** Show open surfaces, pistons, boundaries, and other features that affect pressure. Include height and area measurements and fluid densities. Identify the points at which you need to find the pressure.

❷ **Determine the pressure p_0 at surfaces.**
- **Surface open to the air:** $p_0 = p_{\text{atmos}}$, usually 1 atm.
- **Surface in contact with a gas:** $p_0 = p_{\text{gas}}$.
- **Closed surface:** $p_0 = F/A$, where F is the force that the surface, such as a piston, exerts on the fluid.

❸ **Use horizontal lines.** The pressure in a connected fluid (of one kind) is the same at any point along a horizontal line.

❹ **Allow for gauge pressure.** Pressure gauges read $p_g = p - 1$ atm.

❺ **Use the hydrostatic pressure equation:** $p = p_0 + \rho g d$.

Exercises 5–7

Manometers and Barometers

FIGURE 13.11 A manometer is used to measure gas pressure.

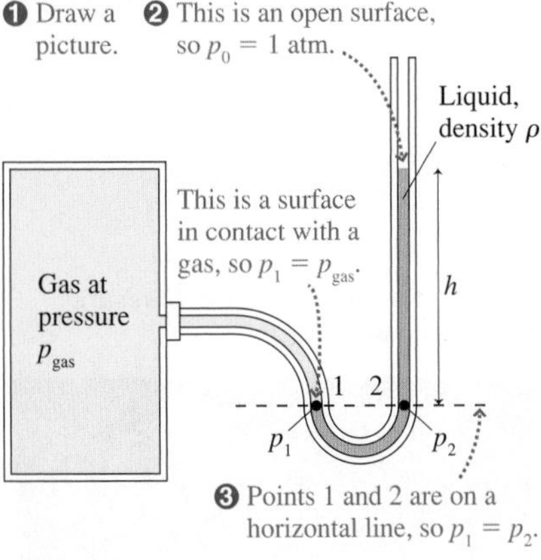

❶ Draw a picture.
❷ This is an open surface, so $p_0 = 1$ atm.
Liquid, density ρ
This is a surface in contact with a gas, so $p_1 = p_{gas}$.
Gas at pressure p_{gas}
h
1 2
p_1
p_2
❸ Points 1 and 2 are on a horizontal line, so $p_1 = p_2$.

Gas pressure is sometimes measured with a device called a *manometer.* A manometer, shown in FIGURE 13.11, is a U-shaped tube connected to the gas at one end and open to the air at the other end. The tube is filled with a liquid—often mercury—of density ρ. The liquid is in static equilibrium. A scale allows the user to measure the height h of the right side of the liquid above the left side.

Steps 1–3 from Tactics Box 13.1 lead to the conclusion that the pressures p_1 and p_2 must be equal. Pressure p_1, at the surface on the left, is simply the gas pressure: $p_1 = p_{gas}$. Pressure p_2 is the hydrostatic pressure at depth $d = h$ in the liquid on the right: $p_2 = 1 \text{ atm} + \rho g h$. Equating these two pressures gives

$$p_{gas} = 1 \text{ atm} + \rho g h \tag{13.6}$$

NOTE ▶ The height h has a *positive* value when the liquid is *higher* on the right than on the left ($p_{gas} > 1$ atm) and a *negative* value when the liquid is *lower* on the right than on the left ($p_{gas} < 1$ atm). ◀

FIGURE 13.12 A barometer.

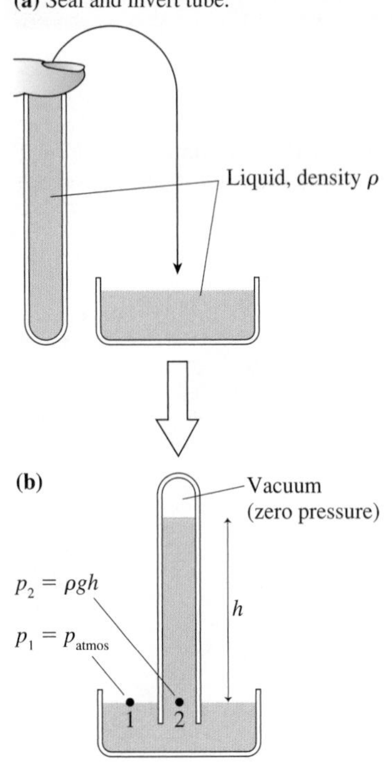

(a) Seal and invert tube.
Liquid, density ρ
(b)
Vacuum (zero pressure)
$p_2 = \rho g h$
$p_1 = p_{atmos}$
h
1 2

Another important pressure-measuring instrument is the *barometer,* which is used to measure atmospheric pressure p_{atmos}. FIGURE 13.12a shows a glass tube, sealed at the bottom, that has been completely filled with a liquid. If we temporarily seal the top end, we can invert the tube and place it in a beaker of the same liquid. When the temporary seal is removed, some, but not all, of the liquid runs out, leaving a liquid column in the tube that is a height h above the surface of the liquid in the beaker. This device, shown in FIGURE 13.12b, is a barometer. What does it measure? And why doesn't *all* the liquid in the tube run out?

We can analyze the barometer much as we did the manometer. Point 1 in Figure 13.12b is open to the atmosphere, so $p_1 = p_{atmos}$. The pressure at point 2 is the pressure due to the weight of the liquid in the tube plus the pressure of the gas above the liquid. But in this case there is no gas above the liquid! Because the tube had been completely full of liquid when it was inverted, the space left behind when the liquid ran out is essentially a vacuum, with $p_0 = 0$. Thus pressure p_2 is simply $\rho g h$.

Because points 1 and 2 are on a horizontal line, and the liquid is in hydrostatic equilibrium, the pressures at these two points must be equal. Equating these two pressures gives

$$p_{atmos} = \rho g h \tag{13.7}$$

Thus we can measure the atmosphere's pressure by measuring the height of the liquid column in a barometer.

Equation 13.7 shows that the liquid height is $h = p_{atmos}/\rho g$. If a barometer were made using water, with $\rho = 1000 \text{ kg/m}^3$, the liquid column would be more than 10 m high, which is impractical. Instead, mercury, with its high density of 13,600 kg/m^3, is usually used. The average air pressure at sea level causes a column of mercury in a mercury barometer to stand 760 mm above the surface. We can then use Equation 13.7 to find that the average atmospheric pressure is

$$\begin{aligned} p_{atmos} &= \rho_{Hg} g h = (13{,}600 \text{ kg/m}^3)(9.80 \text{ m/s}^2)(0.760 \text{ m}) \\ &= 1.013 \times 10^5 \text{ Pa} = 101.3 \text{ kPa} \end{aligned}$$

This is the value given earlier as "1 standard atmosphere."

Because of the importance of mercury-filled barometers in measuring pressure, the height of a column of mercury in millimeters is a common unit of pressure. From our discussion of barometers, 760 millimeters of mercury (abbreviated "mm Hg") corresponds to a pressure of 1 atm.

EXAMPLE 13.4 **Pressure in a tube with two liquids**

A U-shaped tube is closed at one end; the other end is open to the atmosphere. Water fills the side of the tube that includes the closed end, while oil, floating on the water, fills the side of the tube open to the atmosphere. The two liquids do not mix. The height of the oil above the point where the two liquids touch is 75 cm, while the height of the closed end of the tube above this point is 25 cm. What is the gauge pressure at the closed end?

FIGURE 13.13 A tube containing two different liquids.

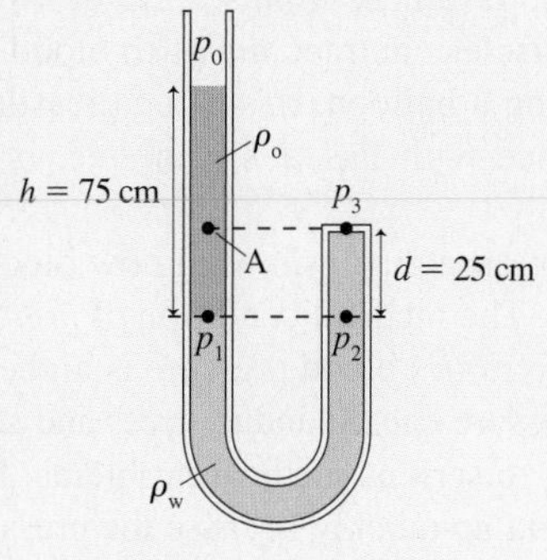

PREPARE Following the steps in Tactics Box 13.1, we start by drawing the picture shown in FIGURE 13.13. We know that the pressure at the open surface of the oil is $p_0 = 1$ atm. Pressures p_1 and p_2 are the same because they are on a horizontal line that connects two points in the *same* fluid. (The pressure at point A is *not* equal to p_3, even though point A and the closed end are on the same horizontal line, because the two points are in *different* fluids.)

We can apply the hydrostatic pressure equation twice: once to find the pressure p_1 by its known depth below the open end at pressure p_0, and again to find the pressure p_3 at the closed end once we know p_2 a distance d below it. We'll need the densities of water and oil, which are found in Table 13.1 to be $\rho_w = 1000$ kg/m^3 and $\rho_o = 900$ kg/m^3.

SOLVE The pressure at point 1, 75 cm below the open end, is

$$\begin{aligned} p_1 &= p_0 + \rho_o gh \\ &= 1 \text{ atm} + (900 \text{ kg/m}^3)(9.8 \text{ m/s}^2)(0.75 \text{ m}) \\ &= 1 \text{ atm} + 6600 \text{ Pa} \end{aligned}$$

(We will keep $p_0 = 1$ atm separate in this result because we'll eventually need to subtract exactly 1 atm to calculate the gauge pressure.) We can also use the hydrostatic pressure equation to find

$$\begin{aligned} p_2 &= p_3 + \rho_w gd \\ &= p_3 + (1000 \text{ kg/m}^3)(9.8 \text{ m/s}^2)(0.25 \text{ m}) \\ &= p_3 + 2500 \text{ Pa} \end{aligned}$$

But we know that $p_2 = p_1$, so

$$\begin{aligned} p_3 &= p_2 - 2500 \text{ Pa} = p_1 - 2500 \text{ Pa} \\ &= 1 \text{ atm} + 6600 \text{ Pa} - 2500 \text{ Pa} \\ &= 1 \text{ atm} + 4100 \text{ Pa} \end{aligned}$$

The gauge pressure at point 3, the closed end of the tube, is $p_3 - 1$ atm or 4100 Pa.

ASSESS The oil's open surface is 50 cm higher than the water's closed surface. Their densities are not too different, so we expect a pressure difference of roughly $\rho g(0.50 \text{ m}) = 5000$ Pa. This is not too far from our answer, giving us confidence that it's correct.

Pressure Units

In practice, pressure is measured in a number of different units. This plethora of units and abbreviations has arisen historically as scientists and engineers working on different subjects (liquids, high-pressure gases, low-pressure gases, weather, etc.) developed what seemed to them the most convenient units. These units continue in use through tradition, so it is necessary to become familiar with converting back and forth among them. Table 13.2 gives the basic conversions.

TABLE 13.2 Pressure units

Unit	Abbreviation	Conversion to 1 atm	Uses
pascal	Pa	101.3 kPa	SI unit: 1 Pa = 1 N/m^2 used in most calculations
atmosphere	atm	1 atm	general
millimeters of mercury	mm Hg	760 mm Hg	gases and barometric pressure
inches of mercury	in	29.92 in	barometric pressure in U.S. weather forecasting
pounds per square inch	psi	14.7 psi	U.S. engineering and industry

FIGURE 13.14 Blood pressure during one cycle of a heartbeat.

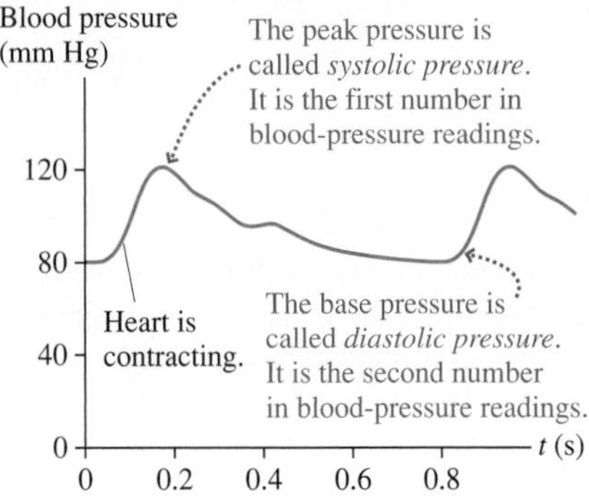

FIGURE 13.15 Measuring blood pressure with a manometer.

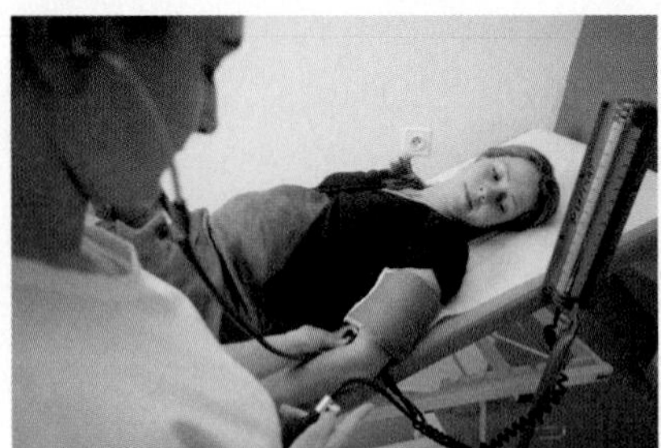

Blood Pressure

The last time you had a medical checkup, the doctor may have told you something like, "Your blood pressure is 120 over 80." What does that mean?

About every 0.8 s, assuming a pulse rate of 75 beats per minute, your heart "beats." The heart muscles contract and push blood out into your aorta. This contraction, like squeezing a balloon, raises the pressure in your heart. The pressure increase, in accordance with Pascal's principle, is transmitted through all your arteries.

FIGURE 13.14 is a pressure graph showing how blood pressure changes during one cycle of the heartbeat. The medical condition of *high blood pressure* usually means that your maximum (*systolic*) blood pressure is higher than necessary for blood circulation. The high pressure causes undue stress and strain on your entire circulatory system, often leading to serious medical problems. Low blood pressure can cause you to faint if you stand up quickly because the pressure isn't adequate to pump the blood up to your brain.

As shown in FIGURE 13.15, blood pressure is measured with a cuff that goes around your arm. The doctor or nurse pressurizes the cuff, places a stethoscope over the artery in your arm, then slowly releases the pressure while watching a pressure gauge. Initially, the cuff squeezes the artery shut and cuts off the blood flow. When the cuff pressure drops below the systolic pressure, the pressure pulse during each beat of your heart forces the artery open briefly and a squirt of blood goes through. The doctor or nurse records the pressure when he or she hears the blood start to flow. This is your systolic pressure.

This pulsing of the blood through your artery lasts until the cuff pressure reaches the diastolic pressure. Then the artery remains open continuously and the blood flows smoothly. This transition is easily heard in the stethoscope, and the doctor or nurse records your base or *diastolic* pressure.

Blood pressure is measured in millimeters of mercury. And it is a gauge pressure, the pressure in excess of 1 atm. A fairly typical blood pressure of a healthy young adult is 120/80, meaning that the systolic pressure is $p_g = 120$ mm Hg (absolute pressure $p = 880$ mm Hg) and the diastolic pressure is 80 mm Hg.

CONCEPTUAL EXAMPLE 13.5 Blood pressure and the height of the arm

In Figure 13.15, the patient's arm is held at about the same height as her heart. Why?

REASON The hydrostatic pressure of a fluid varies with height. Although flowing blood is not in hydrostatic equilibrium, it is still true that blood pressure increases with the distance below the heart and decreases above it. Because the upper arm when held beside the body is at the same height as the heart, the pressure here is the same as the pressure at the heart. If the patient held her arm straight up, the pressure cuff would be a distance $d \approx 25$ cm above her heart and the pressure would be *less* than the pressure at the heart by $\Delta p = \rho_{\text{blood}} g d \approx 20$ mm Hg.

ASSESS 20 mm Hg is a substantial fraction of the average blood pressure. Measuring pressure above or below heart level could lead to a misdiagnosis of the patient's condition.

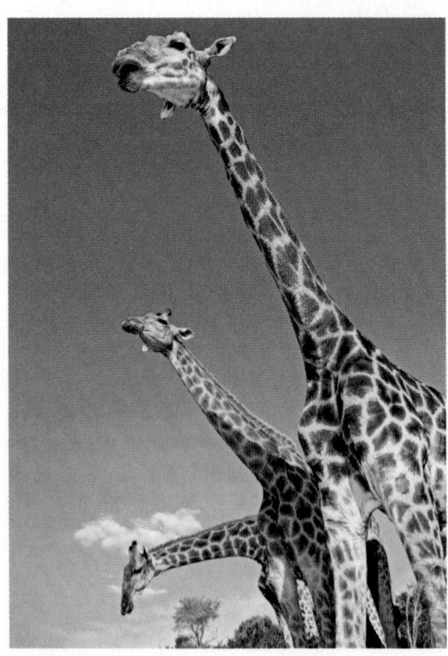

◀ **Pressure at the top** BIO A giraffe's head is some 2.5 m above its heart, compared to a distance of only about 30 cm for humans. To pump blood this extra height requires a blood pressure at the giraffe's heart that is some 170 mm Hg higher than a human's, making its blood pressure more than twice as high as a human's.

13.4 Buoyancy

A rock, as you know, sinks like a rock. Wood floats on the surface of a lake. A penny with a mass of a few grams sinks, but a massive steel aircraft carrier floats. How can we understand these diverse phenomena?

An air mattress floats effortlessly on the surface of a swimming pool. But if you've ever tried to push an air mattress underwater, you know it is nearly impossible. As you push down, the water pushes up. This upward force of a fluid is called the **buoyant force.**

The basic reason for the buoyant force is easy to understand. FIGURE 13.16 shows a cylinder submerged in a liquid. The pressure in the liquid increases with depth, so the pressure at the bottom of the cylinder is greater than at the top. Both cylinder ends have equal area, so force $\vec{F}_{up}$ is greater than force $\vec{F}_{down}$. (Remember that pressure forces push in *all* directions.) Consequently, the pressure in the liquid exerts a *net upward force* on the cylinder of magnitude $F_{net} = F_{up} - F_{down}$. This is the buoyant force.

FIGURE 13.16 The buoyant force arises because the fluid pressure at the bottom of the cylinder is greater than at the top.

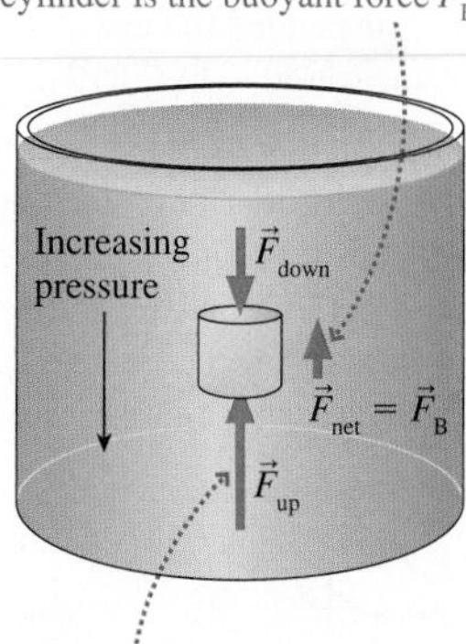
The net force of the fluid on the cylinder is the buoyant force $\vec{F}_B$.
Increasing pressure
$\vec{F}_{down}$
$\vec{F}_{net} = \vec{F}_B$
$\vec{F}_{up}$
$F_{up} > F_{down}$ because the pressure is greater at the bottom. Hence the fluid exerts a net upward force.

The submerged cylinder illustrates the idea in a simple way, but the result is not limited to cylinders or to liquids. Suppose we isolate a parcel of fluid of arbitrary shape and volume by drawing an imaginary boundary around it, as shown in FIGURE 13.17a. This parcel is in static equilibrium. Consequently, the parcel's weight force pulling it down must be balanced by an upward force. The upward force, which is exerted on this parcel of fluid by the surrounding fluid, is the buoyant force $\vec{F}_B$. The buoyant force matches the weight of the fluid: $F_B = w$.

Now imagine that we could somehow remove this parcel of fluid and instantaneously replace it with an object having exactly the same shape and size, as shown in FIGURE 13.17b. Because the buoyant force is exerted by the *surrounding* fluid, and the surrounding fluid hasn't changed, the buoyant force on this new object is *exactly the same* as the buoyant force on the parcel of fluid that we removed.

When an object (or a portion of an object) is immersed in a fluid, it *displaces* fluid that would otherwise fill that region of space. This fluid is called the **displaced fluid.** The displaced fluid's volume is exactly the volume of the portion of the object that is immersed in the fluid. Figure 13.17 leads us to conclude that the magnitude of the upward buoyant force matches the weight of this displaced fluid.

This idea was first recognized by the ancient Greek mathematician and scientist Archimedes, perhaps the greatest scientist of antiquity, and today we know it as *Archimedes' principle:*

FIGURE 13.17 The buoyant force on an object is the same as the buoyant force on the fluid it displaces.

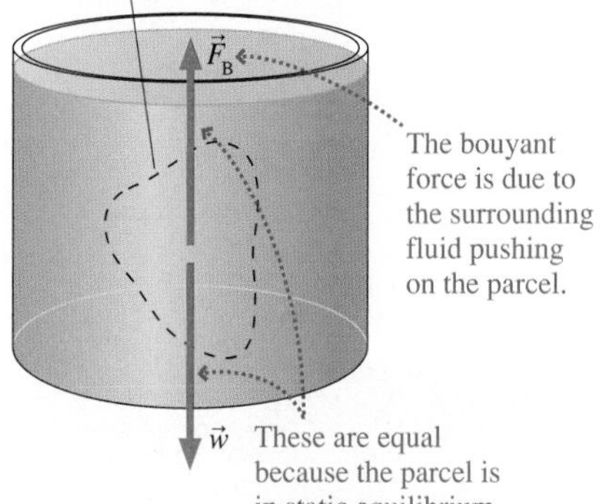
(a) Imaginary boundary around a parcel of fluid
$\vec{F}_B$
The bouyant force is due to the surrounding fluid pushing on the parcel.
$\vec{w}$
These are equal because the parcel is in static equilibrium.

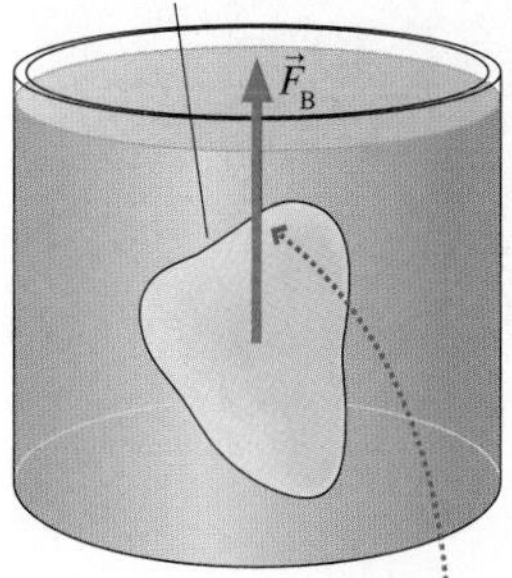
(b) Real object with same size and shape as the parcel of fluid
$\vec{F}_B$
The buoyant force on the object is the same as on the parcel of fluid because the *surrounding* fluid has not changed.

> **Archimedes' principle** A fluid exerts an upward buoyant force $\vec{F}_B$ on an object immersed in or floating on the fluid. The magnitude of the buoyant force equals the weight of the fluid displaced by the object.

Suppose the fluid has density ρ_f and the object displaces a volume V_f of fluid. The mass of the displaced fluid is then $m_f = \rho_f V_f$ and so its weight is $w_f = \rho_f V_f g$. Thus Archimedes' principle in equation form is

$$F_B = \rho_f V_f g \tag{13.8}$$

NOTE ► It is important to distinguish the density and volume of the displaced fluid from the density and volume of the object. To do so, we'll use subscript f for the fluid and o for the object. ◄

EXAMPLE 13.6 Is the crown gold?

Legend has it that Archimedes was asked by King Hiero of Syracuse to determine whether a crown was of pure gold or had been adulterated with a lesser metal by an unscrupulous goldsmith. It was this problem that led him to the principle that bears his name. In a modern version of his method, a crown weighing 8.30 N is suspended underwater from a string. The tension in the string is measured to be 7.81 N. Is the crown pure gold?

PREPARE To discover whether the crown is pure gold, we need to determine its density ρ_o and compare it to the known density of gold. **FIGURE 13.18** shows the forces acting on the crown. In addition to the familiar tension and weight forces, the water exerts an upward buoyant force on the crown. The size of the buoyant force is given by Archimedes' principle.

FIGURE 13.18 The forces acting on the submerged crown.

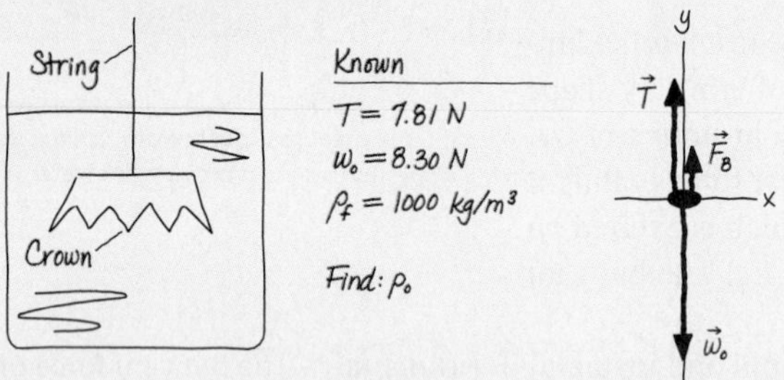

SOLVE Because the crown is in static equilibrium, its acceleration and the net force on it are zero. Newton's second law then reads

$$\sum F_y = F_B + T - w_o = 0$$

from which the buoyant force is

$$F_B = w_o - T = 8.30\text{ N} - 7.81\text{ N} = 0.49\text{ N}$$

According to Archimedes' principle, $F_B = \rho_f V_f g$, where V_f is the volume of the fluid displaced. Here, where the crown is completely submerged, the volume of the fluid displaced is equal to the volume V_o of the crown. Now the crown's weight is $w_o = m_o g = \rho_o V_o g$, so its volume is

$$V_o = \frac{w_o}{\rho_o g}$$

Inserting this volume into Archimedes' principle gives

$$F_B = \rho_f V_o g = \rho_f\left(\frac{w_o}{\rho_o g}\right)g = \frac{\rho_f}{\rho_o}w_o$$

or, solving for ρ_o,

$$\rho_o = \frac{\rho_f w_o}{F_B} = \frac{(1000\text{ kg/m}^3)(8.30\text{ N})}{0.49\text{ N}} = 17{,}000\text{ kg/m}^3$$

The crown's density is considerably lower than that of pure gold, which is 19,300 kg/m^3. The crown is not pure gold.

ASSESS For an object made of a dense material such as gold, the buoyant force is small compared to its weight.

Float or Sink?

If you *hold* an object underwater and then release it, it rises to the surface, sinks, or remains "hanging" in the water. How can we predict which it will do? Whether it heads for the surface or the bottom depends on whether the upward buoyant force F_B on the object is larger or smaller than the downward weight force w_o.

The magnitude of the buoyant force is $\rho_f V_f g$. The weight of a uniform object, such as a block of steel, is simply $\rho_o V_o g$. But a compound object, such as a scuba diver, may have pieces of varying density. If we define the **average density** to be $\rho_{avg} = m_o/V_o$, the weight of a compound object can be written as $w_o = \rho_{avg} V_o g$.

Comparing $\rho_f V_f g$ to $\rho_{avg} V_o g$, and noting that $V_f = V_o$ for an object that is fully submerged, we see that an object floats or sinks depending on whether the fluid density ρ_f is larger or smaller than the object's average density ρ_{avg}. If the densities are equal, the object is in static equilibrium and hangs motionless. This is called **neutral buoyancy.** These conditions are summarized in Tactics Box 13.2.

◀ **Submersible scales** BIO In Example 13.6, we saw how the density of an object could be determined by weighing it both underwater and in air. This idea is the basis of an accurate method for determining a person's percentage of body fat. Fat has a lower density than lean muscle or bone, so a lower overall body density implies a greater proportion of body fat. To determine a person's density, he is first weighed in air and then lowered completely into water and weighed again. Standard tables accurately relate body density to fat percentage.

TACTICS BOX 13.2 Finding whether an object floats or sinks

❶ Object sinks

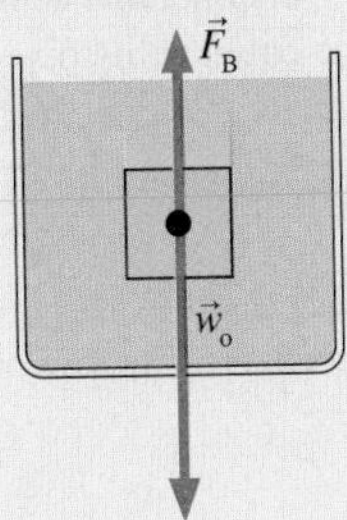

An object sinks if it weighs more than the fluid it displaces—that is, if its average density is greater than the density of the fluid:

$$\rho_{avg} > \rho_f$$

❷ Object floats

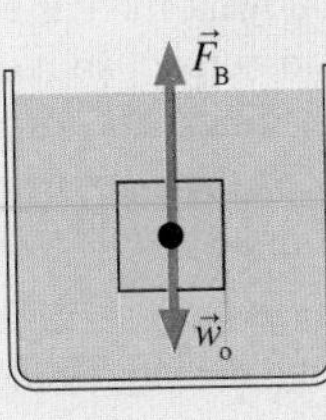

An object rises to the surface if it weighs less than the fluid it displaces—that is, if its average density is less than the density of the fluid:

$$\rho_{avg} < \rho_f$$

❸ Object has neutral buoyancy

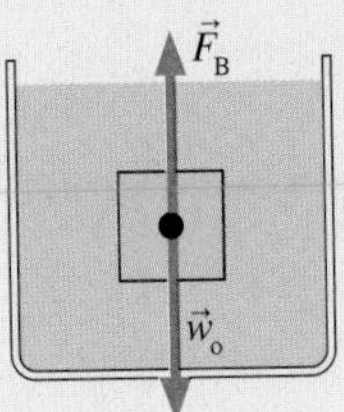

An object hangs motionless if it weighs exactly the same as the fluid it displaces—that is, if its average density equals the density of the fluid:

$$\rho_{avg} = \rho_f$$

Exercises 10–12

For example, steel is denser than water, so a chunk of steel sinks. Oil is less dense than water, so oil floats on water. Fish use *swim bladders* filled with air and scuba divers use weighted belts to adjust their average density to match the water. Both are examples of neutral buoyancy.

If you release a block of wood underwater, the net upward force causes the block to shoot to the surface. Then what? To understand floating, let's begin with a *uniform* object such as the block shown in **FIGURE 13.19**. This object contains nothing tricky, like indentations or voids. Because it's floating, it must be the case that $\rho_o < \rho_f$.

Now that the object is floating, it's in static equilibrium. Thus, the upward buoyant force, given by Archimedes' principle, exactly balances the downward weight of the object; that is,

$$F_B = \rho_f V_f g = w_o = \rho_o V_o g \tag{13.9}$$

For a floating object, the volume of the displaced fluid is *not* the same as the volume of the object. In fact, we can see from Equation 13.9 that the volume of fluid displaced by a floating object of uniform density is

$$V_f = \frac{\rho_o}{\rho_f} V_o \tag{13.10}$$

which is *less* than V_o because $\rho_o < \rho_f$.

NOTE ▶ Equation 13.10 applies only to *uniform* objects. It does not apply to boats, hollow spheres, or other objects of nonuniform composition. ◀

FIGURE 13.19 A floating object is in static equilibrium.

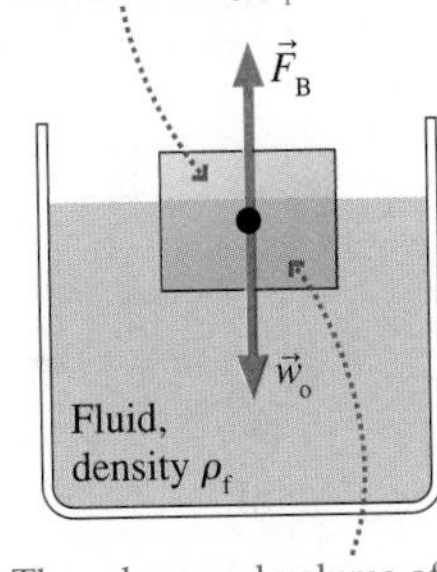

▶ **Hidden depths** You've probably heard it said that "90% of an iceberg is underwater." Equation 13.10 is the basis for that statement. Most icebergs break off glaciers and are fresh-water ice with a density of 917 kg/m^3. The density of seawater is 1030 kg/m^3. Thus

$$V_f = \frac{917\ \text{kg/m}^3}{1030\ \text{kg/m}^3} V_o = 0.89 V_o$$

V_f, the volume of the displaced water, is also the volume of the iceberg that is underwater. You can see that, indeed, 89% of the volume of an iceberg is underwater.

CONCEPTUAL EXAMPLE 13.7 Which has the greater buoyant force?

A block of iron sinks to the bottom of a vessel of water while a block of wood of the *same size* floats. On which is the buoyant force greater?

REASON The buoyant force is equal to the volume of water displaced. The iron block is completely submerged, so it displaces a volume of water equal to its own volume. The wood block floats, so it displaces only the fraction of its volume that is under water, which is *less* than its own volume. The buoyant force on the iron block is therefore greater than on the wood block.

ASSESS This result may seem counterintuitive, but remember that the iron block sinks because of its high density, while the wood block floats because of its low density. A smaller buoyant force is sufficient to keep it floating.

EXAMPLE 13.8 Measuring the density of an unknown liquid

You need to determine the density of an unknown liquid. You notice that a block floats in this liquid with 4.6 cm of the side of the block submerged. When the block is placed in water, it also floats but with 5.8 cm submerged. What is the density of the unknown liquid?

PREPARE Assume that the block is an object of uniform composition. FIGURE 13.20 shows the block as well as the cross-section area A and submerged lengths h_u in the unknown liquid and h_w in water.

FIGURE 13.20 A block floating in two liquids.

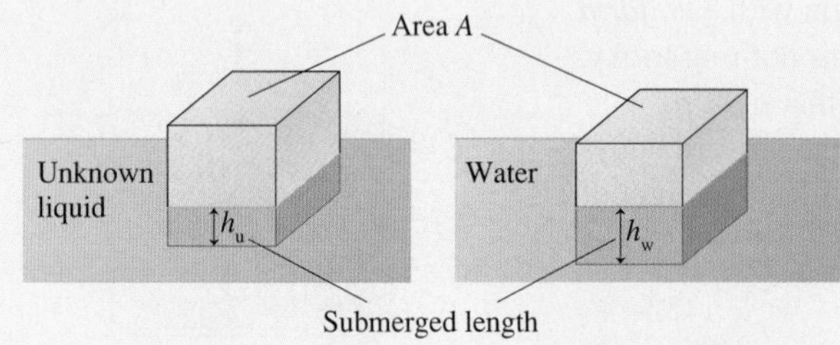

SOLVE The block is floating, so Equation 13.10 applies. The block displaces volume $V_u = Ah_u$ of the unknown liquid. Thus

$$V_u = Ah_u = \frac{\rho_o}{\rho_u} V_o$$

Similarly, the block displaces volume $V_w = Ah_w$ of the water, leading to

$$V_w = Ah_w = \frac{\rho_o}{\rho_w} V_o$$

Because there are two fluids, we've used subscripts w for water and u for the unknown in place of the fluid subscript f. The product $\rho_o V_o$ appears in both equations. In the first $\rho_o V_o = \rho_u A h_u$, and in the second $\rho_o V_o = \rho_w A h_w$. Equating the right-hand sides gives

$$\rho_u A h_u = \rho_w A h_w$$

The area A cancels, and the density of the unknown liquid is

$$\rho_u = \frac{h_w}{h_u}\rho_w = \frac{5.8\text{ cm}}{4.6\text{ cm}} 1000\text{ kg/m}^3 = 1300\text{ kg/m}^3$$

ASSESS Comparison with Table 13.1 shows that the unknown liquid is likely to be glycerin.

FIGURE 13.21 How a boat floats.

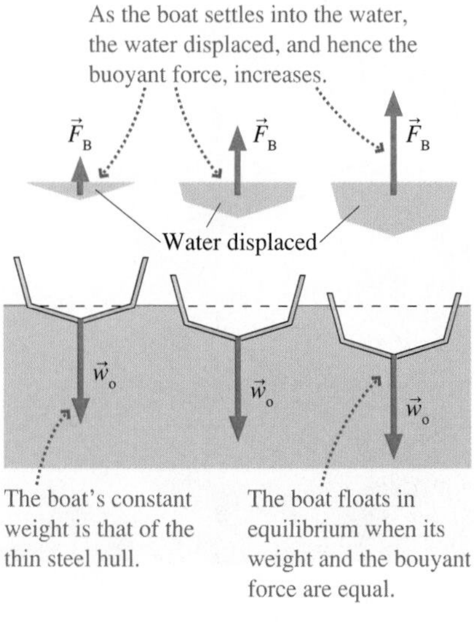

As the boat settles into the water, the water displaced, and hence the buoyant force, increases.

$\vec{F}_B$ $\vec{F}_B$ $\vec{F}_B$

Water displaced

$\vec{w}_o$ $\vec{w}_o$ $\vec{w}_o$

The boat's constant weight is that of the thin steel hull.

The boat floats in equilibrium when its weight and the bouyant force are equal.

Boats and Balloons

A chunk of steel sinks, so how does a steel-hulled boat float? As we've seen, an object floats if the upward buoyant force—the weight of the displaced water—balances the weight of the object. A boat is really a large hollow shell whose weight is determined by the volume of steel in the hull. As FIGURE 13.21 shows, the volume of water displaced by a shell is *much* larger than the volume of the hull itself. As a boat settles into the water, it sinks until the weight of the displaced water exactly matches the boat's weight. It is then in static equilibrium, so it floats at that level.

The concept of buoyancy and flotation applies to all fluids, not just liquids. An object immersed in a gas such as air feels a buoyant force as well. Because the density of air is so low, this buoyant force is generally negligible. Nonetheless, even though the buoyant force due to air is small, an object will float in air if it weighs less than the air that it displaces. This is why a floating balloon cannot be filled with regular air. If it were, then the weight of the air inside it would equal the weight of the air it displaced, so that it would have no net upward force on it. Adding in the weight of the balloon itself would then lead to a net downward force. For a balloon to float, it must be filled with a gas that has a *lower* density than that of air. The following example illustrates how this works.

EXAMPLE 13.9 How big does a balloon need to be?

What diameter must a helium-filled balloon have to float with neutral buoyancy? The mass of the empty balloon is 2.0 g.

PREPARE We'll model the balloon as a sphere. The balloon will float when its weight—the weight of the empty balloon plus the helium—equals the weight of the air it displaces. The densities of air and helium are given in Table 13.1.

SOLVE The volume of the balloon is V_{balloon}. Its weight is

$$w_{\text{balloon}} = m_{\text{balloon}}\, g + \rho_{\text{He}} V_{\text{balloon}}\, g$$

where m_{balloon} is the mass of the empty balloon. The weight of the displaced air is

$$w_{\text{air}} = \rho_{\text{air}} V_{\text{air}}\, g = \rho_{\text{air}} V_{\text{ballon}}\, g$$

where we noted that the volume of the displaced air is the volume of the balloon. The balloon will float when these two forces are equal, or when

$$\rho_{\text{air}} V_{\text{balloon}}\, g = m_{\text{balloon}}\, g + \rho_{\text{He}} V_{\text{balloon}}\, g$$

The g cancels, and we can solve for the volume of the balloon:

$$V_{\text{balloon}} = \frac{m_{\text{balloon}}}{\rho_{\text{air}} - \rho_{\text{He}}} = \frac{2.0 \times 10^{-3}\text{ kg}}{1.28\text{ kg/m}^3 - 0.17\text{ kg/m}^3} = 1.8 \times 10^{-3}\text{ m}^3$$

A sphere has volume $V = (4\pi/3)r^3$. Thus the balloon's radius is

$$r = \left(\frac{3V_{\text{balloon}}}{4\pi}\right)^{\frac{1}{3}} = \left(\frac{3 \times (1.8 \times 10^{-3}\text{ m}^3)}{4\pi}\right)^{\frac{1}{3}} = 0.075\text{ m}$$

The diameter of the balloon is twice this, or 15 cm.

ASSESS This example shows why a helium balloon will no longer float once its volume falls below a certain value.

Hot air rising A hot-air balloon is filled with a low-density gas: hot air! You learned in Chapter 12 that gases expand upon heating, thus lowering their density. The air at the top of a hot-air balloon is surprisingly hot—about 100°C, the temperature of boiling water. Using the ideal-gas law, you should be able to show that the density of 100°C air is only 79% that of room-temperature air at 20°C. A burst of flame lowers the density of the air and thus the balloon's weight. The balloon rises when its weight becomes less than the weight of the cooler air it has displaced.

STOP TO THINK 13.3 An ice cube is floating in a glass of water that is filled entirely to the brim. When the ice cube melts, the water level will

A. Fall. B. Stay the same. C. Rise, causing the water to spill.

13.5 Fluids in Motion

The wind blowing through your hair, a white-water river, and oil gushing from an oil well are examples of fluids in motion. We've focused thus far on fluid statics, but it's time to turn our attention to fluid *dynamics*.

Fluid flow is a complex subject. Many aspects of fluid flow, especially turbulence and the formation of eddies, are still not well understood and are areas of current research. We will avoid these difficulties by using a simplified *model* of an *ideal* fluid. Our model can be expressed in three assumptions about the fluid:

1. The fluid is *incompressible.* This is a very good assumption for liquids, but it also holds reasonably well for a moving gas, such as air. For instance, even when a 100 mph wind slams into a wall, its density changes by only about 1%.
2. The flow is *steady.* That is, the fluid velocity at each point in the fluid is constant; it does not fluctuate or change with time. Flow under these conditions is called **laminar flow,** and it is distinguished from *turbulent flow.*
3. The fluid is *nonviscous.* Water flows much more easily than cold pancake syrup because the syrup is a very viscous liquid. Viscosity is resistance to flow, and assuming a fluid is nonviscous is analogous to assuming the motion of a particle is frictionless. Gases have very low viscosity, and even many liquids are well approximated as being nonviscous.

Later, in Section 13.7, we'll relax assumption 3 and consider the effects of viscosity.

FIGURE 13.22 Rising smoke changes from laminar flow to turbulent flow.

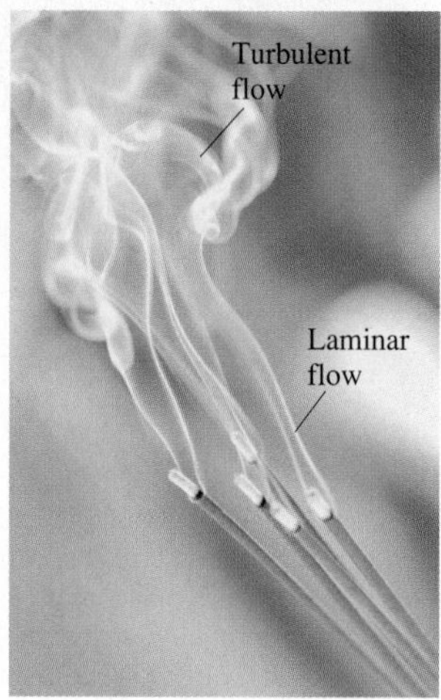
Turbulent flow
Laminar flow

The rising smoke in **FIGURE 13.22** begins as laminar flow, recognizable by the smooth contours, but at some point undergoes a transition to turbulent flow. A laminar-to-turbulent transition is not uncommon in fluid flow. Our model of fluids can be applied to the laminar flow, but not to the turbulent flow.

The Equation of Continuity

Consider a fluid flowing through a tube—oil through a pipe or blood through an artery. If the tube's diameter changes, as happens in **FIGURE 13.23**, what happens to the speed of the fluid?

When you squeeze a toothpaste tube, the volume of toothpaste that emerges matches the amount by which you reduce the volume of the tube. An *incompressible* fluid flowing through a rigid tube or pipe acts the same way. Fluid is neither created nor destroyed within the tube, and there's no place to store any extra fluid introduced into the tube. If volume V enters the tube during some interval of time Δt, then an equal volume of fluid must leave the tube.

To see the implications of this idea, suppose all the molecules of the fluid in Figure 13.23 are moving forward with speed v_1 at a point where the cross-section area is A_1. Farther along the tube, where the cross-section area is A_2, their speed is v_2. During an interval of time Δt, the molecules in the wider section move forward a distance $\Delta x_1 = v_1\,\Delta t$ and those in the narrower section move $\Delta x_2 = v_2\,\Delta t$. Because the fluid is incompressible, the volumes ΔV_1 and ΔV_2 must be equal; that is,

FIGURE 13.23 Flow speed changes through a tapered tube.

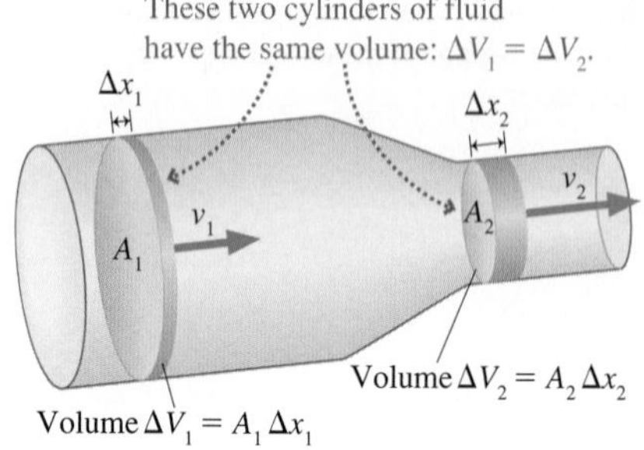
These two cylinders of fluid have the same volume: $\Delta V_1 = \Delta V_2$.
Δx_1
Δx_2
v_1
v_2
A_1
A_2
Volume $\Delta V_2 = A_2\,\Delta x_2$
Volume $\Delta V_1 = A_1\,\Delta x_1$

$$\Delta V_1 = A_1\,\Delta x_1 = A_1 v_1\,\Delta t = \Delta V_2 = A_2\,\Delta x_2 = A_2 v_2\,\Delta t \qquad (13.11)$$

Dividing both sides of the equation by Δt gives the **equation of continuity:**

$$v_1 A_1 = v_2 A_2 \qquad (13.12)$$

The equation of continuity relating the speed v of an incompressible fluid to the cross-section area A of the tube in which it flows

FIGURE 13.24 The speed of the water is inversely proportional to the diameter of the stream.

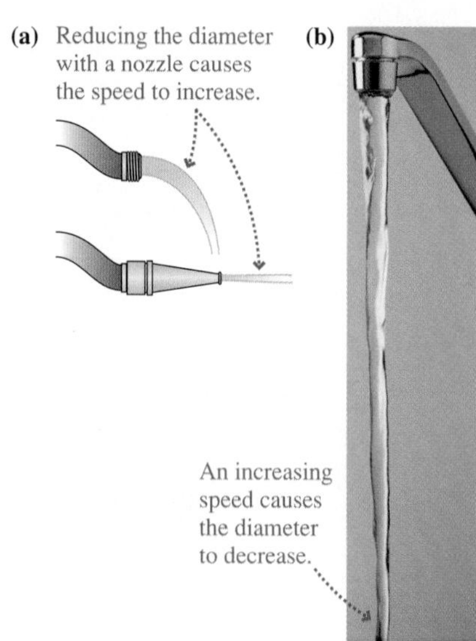
(a) Reducing the diameter with a nozzle causes the speed to increase.
(b)
An increasing speed causes the diameter to decrease.

Equations 13.11 and 13.12 say that **the volume of an incompressible fluid entering one part of a tube or pipe must be matched by an equal volume leaving downstream.**

An important consequence of the equation of continuity is that **flow is faster in narrower parts of a tube, slower in wider parts.** You're familiar with this conclusion from many everyday observations. The garden hose shown in **FIGURE 13.24a** squirts farther after you put a nozzle on it. This is because the narrower opening of the nozzle gives the water a higher exit speed. Water flowing from the faucet shown in **FIGURE 13.24b** picks up speed as it falls. As a result, the flow tube "necks down" to a smaller diameter.

The *rate* at which fluid flows through the tube—volume per second—is $\Delta V/\Delta t$. This is called the **volume flow rate** Q. We can see from Equation 13.11 that

$$Q = \frac{\Delta V}{\Delta t} = vA \qquad (13.13)$$

The SI units of Q are m^3/s, although in practice Q may be measured in cm^3/s, liters per minute, or, in the United States, gallons per minute and cubic feet per minute. Another way to express the meaning of the equation of continuity is to say that **the volume flow rate is constant at all points in a tube.**

EXAMPLE 13.10 Speed of water through a hose

A garden hose has an inside diameter of 16 mm. The hose can fill a 10 L bucket in 20 s.

a. What is the speed of the water out of the end of the hose?
b. What diameter nozzle would cause the water to exit with a speed 4 times greater than the speed inside the hose?

PREPARE Water is essentially incompressible, so the equation of continuity applies.

SOLVE

a. The volume flow rate is $Q = \Delta V/\Delta t = (10\text{ L})/(20\text{ s}) = 0.50\text{ L/s}$. To convert this to SI units, recall that $1\text{ L} = 1000\text{ mL} = 10^3\text{ cm}^3 = 10^{-3}\text{ m}^3$. Thus $Q = 5.0 \times 10^{-4}\text{ m}^3\text{/s}$. We can find the speed of the water from Equation 13.13:

$$v = \frac{Q}{A} = \frac{Q}{\pi r^2} = \frac{5.0 \times 10^{-4}\text{ m}^3\text{/s}}{\pi(0.0080\text{ m})^2} = 2.5\text{ m/s}$$

b. The quantity $Q = vA$ remains constant as the water flows through the hose and then the nozzle. To increase v by a factor of 4, A must be reduced by a factor of 4. The cross-section area depends on the square of the diameter, so the area is reduced by a factor of 4 if the diameter is reduced by a factor of 2. Thus the necessary nozzle diameter is 8 mm.

EXAMPLE 13.11 Blood flow in capillaries

The volume flow rate of blood leaving the heart to circulate throughout the body is about 5 L/min for a person at rest. All this blood eventually must pass through the smallest blood vessels, the capillaries. Microscope measurements show that a typical capillary is 6 μm in diameter and 1 mm long, and the blood flows through it at an average speed of 1 mm/s.

a. Estimate the total number of capillaries in the body.
b. Estimate the total surface area of all the capillaries.

The various lengths and areas are shown in **FIGURE 13.25**.

FIGURE 13.25 A capillary (not to scale).

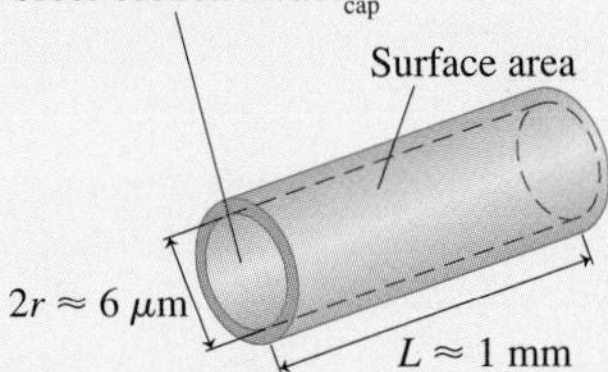

PREPARE We can use the equation of continuity: The 5 L of blood passing through the heart each minute must be the total flow through all the capillaries combined.

SOLVE

a. We'll start by converting Q to SI units:

$$Q = 5\frac{\text{L}}{\text{min}} \times \frac{1\text{ m}^3}{1000\text{ L}} \times \frac{1\text{ min}}{60\text{ s}} = 8 \times 10^{-5}\text{ m}^3\text{/s}$$

to one-significant-figure accuracy. Then the total cross-section area of all the capillaries together is

$$A_{\text{total}} = \frac{Q}{v} = \frac{8 \times 10^{-5}\text{ m}^3\text{/s}}{0.001\text{ m/s}} = 0.08\text{ m}^2$$

The cross-section area of each capillary is $A_{\text{cap}} = \pi r^2$, so the total number of capillaries is approximately

$$N = \frac{A_{\text{total}}}{A_{\text{cap}}} = \frac{0.08\text{ m}^2}{\pi(3 \times 10^{-6}\text{ m})^2} = 3 \times 10^9$$

b. The surface area of one capillary is

$$A = \text{circumference of capillary} \times \text{length}$$
$$= 2\pi r L = 2\pi(3 \times 10^{-6}\text{ m})(0.001\text{ m}) = 2 \times 10^{-8}\text{ m}^2$$

so the total surface area of all the capillaries is about

$$A_{\text{surface}} = NA = (3 \times 10^9)(2 \times 10^{-8}\text{ m}) = 60\text{ m}^2$$

ASSESS The number of capillaries is huge, and the total surface area is about the area of a two-car garage. As we saw in Chapter 12, oxygen and nutrients move from the blood into cells by the slow process of diffusion. Only by having this large surface area available for diffusion can the required rate of gas and nutrient exchange be attained.

Representing Fluid Flow: Streamlines and Fluid Elements

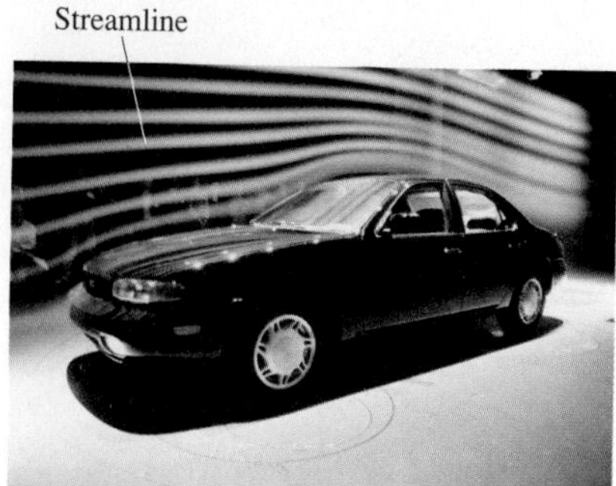

FIGURE 13.26 Smoke reveals the laminar air flow around a car in a wind tunnel.

Representing the flow of fluid is more complicated than representing the motion of a point particle because fluid flow is the collective motion of a vast number of particles. FIGURE 13.26 gives us an idea of one possible fluid-flow representation. Here smoke is being used to help engineers visualize the air flow around a car in a wind tunnel. The smoothness of the flow tells us this is laminar flow. But notice also how the individual smoke trails retain their identity. They don't cross or get mixed together. Each smoke trail represents a *streamline* in the fluid.

Imagine that we could inject a tiny colored drop of water into a stream of water undergoing laminar flow. Because the flow is steady and the water is incompressible, this colored drop would maintain its identity as it flowed along. The path or trajectory followed by this "particle of fluid" is called a **streamline.** Smoke particles mixed with the air allow you to see the streamlines in the wind-tunnel photograph of Figure 13.26. FIGURE 13.27 illustrates three important properties of streamlines.

FIGURE 13.27 Particles in a fluid move along streamlines.

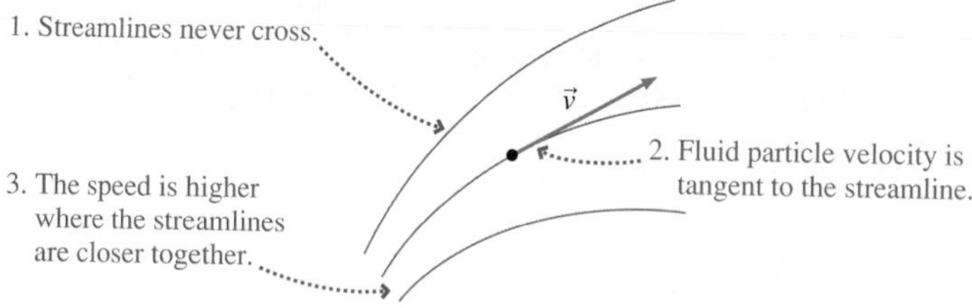

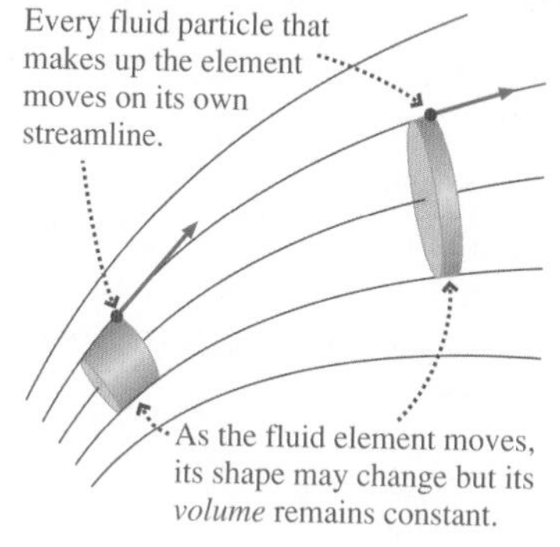

Every fluid particle that makes up the element moves on its own streamline.
As the fluid element moves, its shape may change but its *volume* remains constant.

FIGURE 13.28 Motion of a fluid element.

As we study the motion of a fluid, it is often also useful to consider a small *volume* of fluid, a volume containing many particles of fluid. Such a volume is called a **fluid element.** FIGURE 13.28 shows two important properties of a fluid element. Unlike a particle, a fluid element has an actual shape and volume. Although the shape of a fluid element may change as it moves, the equation of continuity requires that its volume remain constant. The progress of a fluid element as it moves along streamlines and changes shape is another very useful representation of fluid motion.

STOP TO THINK 13.4

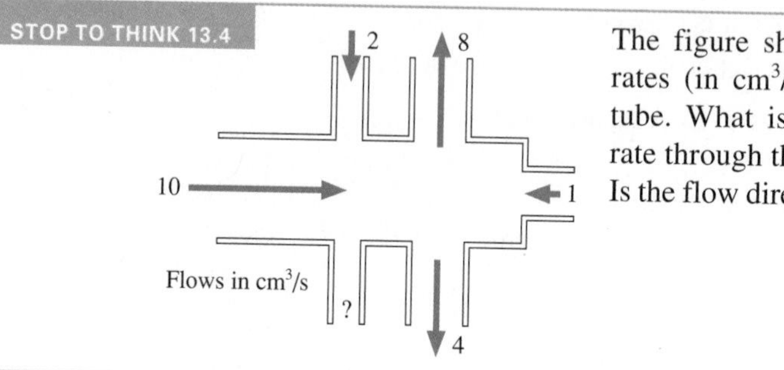

The figure shows volume flow rates (in cm^3/s) for all but one tube. What is the volume flow rate through the unmarked tube? Is the flow direction in or out?

13.6 Fluid Dynamics

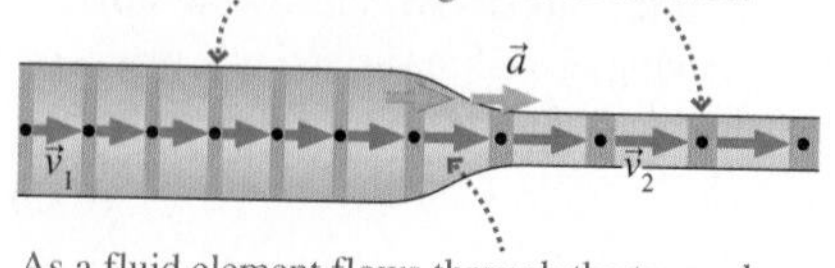

A fluid element coasts at steady speed through the constant-diameter segments of the tube.
$\vec{a}$
$\vec{v}_1$
$\vec{v}_2$
As a fluid element flows through the tapered section, it speeds up. Because it is accelerating, there must be a force acting on it.

FIGURE 13.29 A motion diagram of a fluid element moving through a narrowing tube.

The equation of continuity describes a moving fluid but doesn't tell us anything about *why* the fluid is in motion. To understand the dynamics, consider the ideal fluid moving from left to right through the tube shown in FIGURE 13.29. The fluid moves at a steady speed v_1 in the wider part of the tube. In accordance with the equation of continuity, its speed is a higher, but steady, v_2 in the narrower part of the tube. If we follow a fluid element through the tube, we see that it undergoes an *acceleration* from v_1 to v_2 in the tapered section of the tube.

In the absence of friction, Newton's first law tells us that a particle will coast forever at a steady speed. An ideal fluid is nonviscous and thus analogous to a particle moving without friction. This means a fluid element moving through the constant-diameter sections of the tube in Figure 13.29 requires no force to "coast" at steady speed. On the other hand, a fluid element moving through the tapered section of the tube accelerates, speeding up from v_1 to v_2. According to Newton's second law, there must be a net force acting on this fluid element to accelerate it.

What is the origin of this force? There are no external forces, and the horizontal motion rules out gravity. Instead, the fluid element is pushed from both ends by the *surrounding fluid*—that is, by *pressure forces.* The fluid element with cross-section area A in FIGURE 13.30 has a higher pressure on its left side than on its right. Thus the force $F_L = p_L A$ of the fluid pushing on its left side—a force pushing to the right—is greater than the force $F_R = p_R A$ of the fluid on its right side. The net force, which points from the higher-pressure side of the element to the lower-pressure side, is

$$F_{net} = F_L - F_R = (p_L - p_R)A = A\,\Delta p$$

In other words, there's a net force on the fluid element, causing it to change speed, if and only if there's a pressure *difference* Δp between the two faces.

FIGURE 13.30 The net force on a fluid element due to pressure points from high to low pressure.

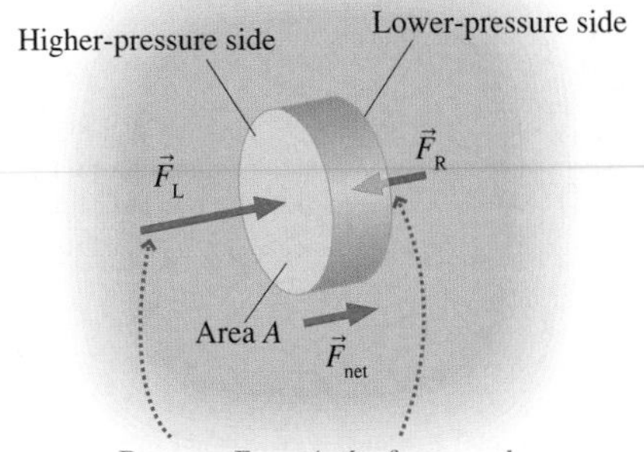

Thus, in order to accelerate the fluid elements in Figure 13.29 through the neck, the pressure p_1 in the wider section of the tube must be higher than the pressure p_2 in the narrower section. When the pressure is changing from one point in a fluid to another, we say that there is a **pressure gradient** in that region. We can also say that pressure forces are caused by pressure gradients, so **an ideal fluid accelerates wherever there is a pressure gradient.**

As a result, **the pressure is higher at a point along a streamline where the fluid is moving slower, lower where the fluid is moving faster.** This property of fluids was discovered in the 18th century by the Swiss scientist Daniel Bernoulli and is called the **Bernoulli effect.**

NOTE ► It is important to realize that it is the change in pressure from high to low that *causes* the fluid to speed up. A high fluid speed doesn't *cause* a low pressure any more than a fast-moving particle causes the force that accelerated it. ◄

This relationship between pressure and fluid speed can be used to measure the speed of a fluid with a device called a *Venturi tube.* FIGURE 13.31 shows a simple Venturi tube suitable for a flowing liquid. The high pressure at point 1, where the fluid is moving slowly, causes the fluid to rise in the vertical pipe to a total height d_1. Because there's no vertical motion of the fluid, we can use the hydrostatic pressure equation to find that the pressure at point 1 is $p_1 = p_0 + \rho g d_1$. At point 3, on the same streamline, the fluid is moving faster and the pressure is lower; thus the fluid in the vertical pipe rises to a lower height d_3. The pressure *difference* is $\Delta p = \rho g(d_1 - d_3)$, so the pressure difference across the neck of the pipe can be found by measuring the difference in heights of the fluid. We'll see later in this section how to relate this pressure difference to the increase of speed.

FIGURE 13.31 A Venturi tube measures flow speeds of a fluid.

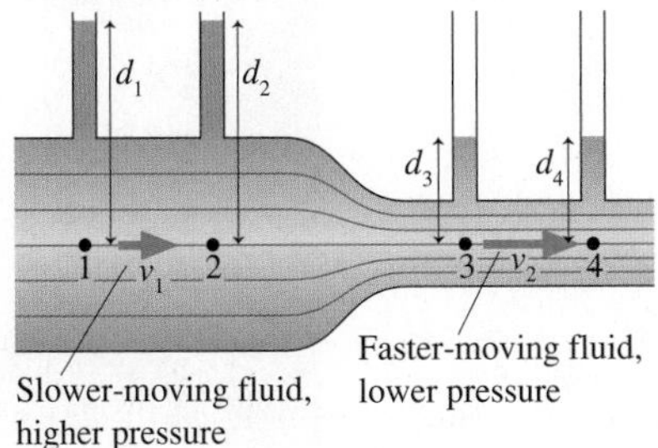

Note that the fluid height d_1 is the same as d_2, and d_3 is the same as d_4. For an ideal fluid, with no viscosity, no pressure difference is needed to keep the fluid moving at a constant speed, as it does in both the large- and small-diameter sections of the tube. In the next section we'll see how this result changes for a viscous fluid.

TRY IT YOURSELF

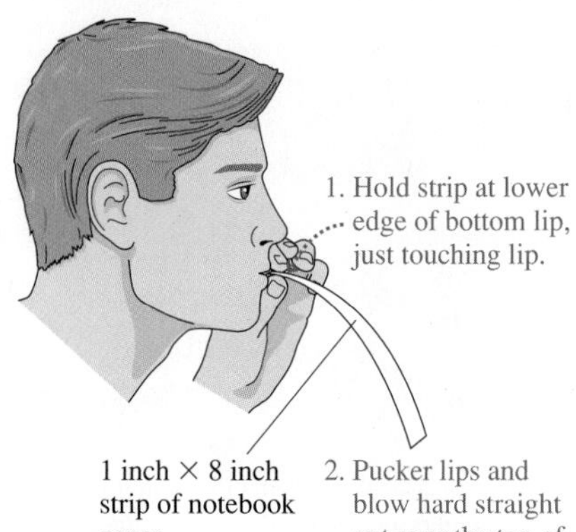

1. Hold strip at lower edge of bottom lip, just touching lip.
1 inch × 8 inch strip of notebook paper
2. Pucker lips and blow hard straight out over the top of the strip.

Blowing up Try the experiment in the figure. You might expect the strip to be pushed *down* by the force of your breath, but you'll find that the strip actually *rises*. Your breath moving over the curved strip is similar to wind blowing over a hill, and Bernoulli's effect likewise predicts a zone of lower pressure above the strip that causes it to rise.

Applications of the Bernoulli Effect

Many important applications of the Bernoulli effect can be understood by considering the flow of air over a hill, as shown in FIGURE 13.32. Far to the left, away from the hill, the wind blows at a constant speed, so its streamlines are equally spaced. But as the air moves over the hill, the hill forces the streamlines to bunch together so that the air speeds up. According to the Bernoulli effect, there exists a zone of *low pressure* at the crest of the hill, where the air is moving the fastest.

Using these ideas, we can understand *lift*, the upward force on the wing of a moving airplane that makes flight possible. FIGURE 13.33 shows an airplane wing, seen in cross section, for an airplane flying to the left. The figure is drawn in the reference frame of the airplane, so that the wing appears stationary and the air flows past it to the right. The shape of the wing is such that, just as for the hill in Figure 13.32, the streamlines of the air must squeeze together as they pass over the wing. This increased speed is, by the Bernoulli effect, accompanied by a region of low pressure above the wing. The air speed below the wing is actually slowed, so that there's a region of high pressure below. This high pressure pushes up on the wing, while the low-pressure air above it presses down, but less strongly. The result is a net upward force—the lift.

FIGURE 13.32 Wind speeds up as it crests a hill.

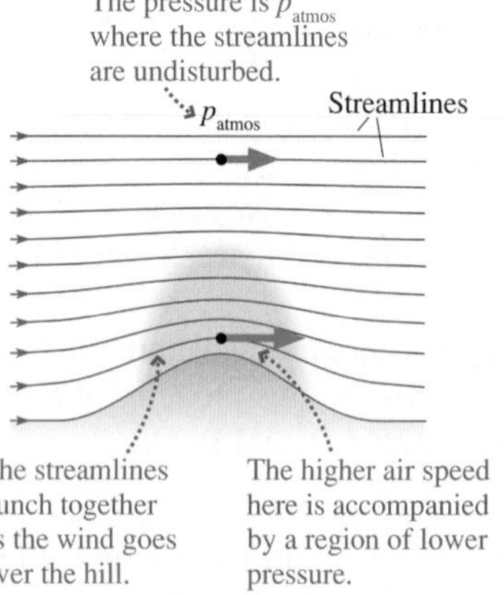

FIGURE 13.33 Air flow over a wing generates lift by creating unequal pressures above and below.

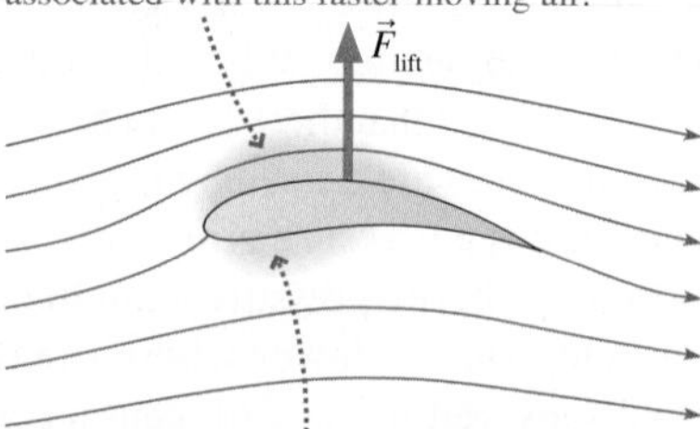

$\vec{F}_{lift}$

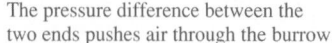

The Bernoulli effect also explains how hurricanes destroy the roofs of houses. The roofs are not "blown" off; they are *lifted* off by pressure differences caused by the Bernoulli effect. FIGURE 13.34 shows that the situation is similar to that for the hill and the airplane wing. The pressure difference causing the lift force is not very large—but the force is proportional to the *area* of the roof, which, being very large, can produce a lift force large enough to separate the roof from its supporting walls.

FIGURE 13.34 How high winds lift roofs off.

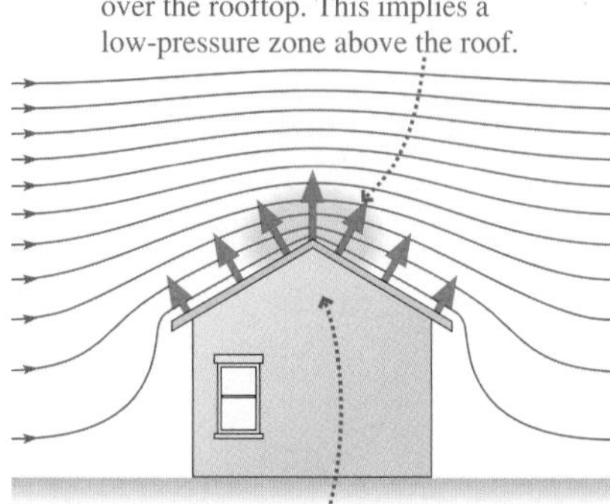

The wind speeds up as it squeezes over the rooftop. This implies a low-pressure zone above the roof.
The pressure inside the house is atmospheric pressure, which is higher. The result is a net upward force on the roof.

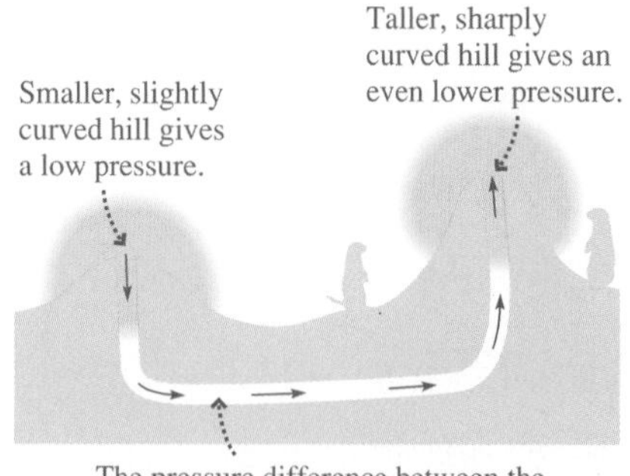

Smaller, slightly curved hill gives a low pressure.
Taller, sharply curved hill gives an even lower pressure.
The pressure difference between the two ends pushes air through the burrow.

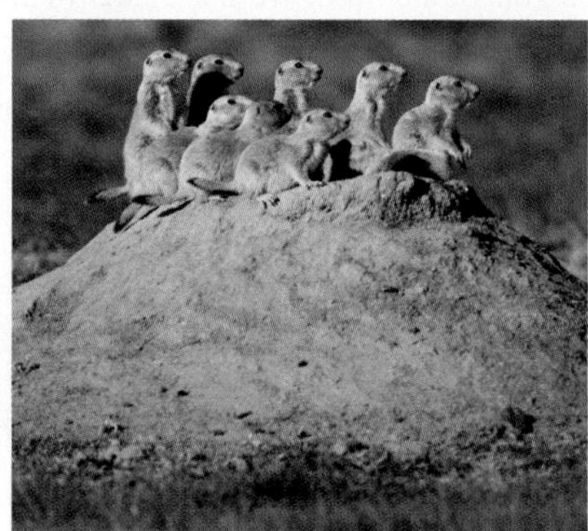

Nature's air conditioning Prairie dogs ventilate their underground burrows with the same aerodynamic forces and pressures that give airplanes lift. The two entrances to their burrows are surrounded by mounds, one higher than the other. When the wind blows across these mounds, the pressure is reduced at the top, just as for an airplane's wing. The taller mound, with its greater curvature, has the lower pressure of the two entrances. Air then is pushed through the burrow toward this lower-pressure side.

Bernoulli's Equation

We've seen that a pressure gradient causes a fluid to accelerate horizontally. Not surprisingly, gravity can also cause a fluid to speed up or slow down if the fluid changes elevation. These are the key ideas of fluid dynamics. Now we would like to make these ideas quantitative by finding a numerical relationship for pressure, height, and the speed of a fluid. We can do so by applying the statement of conservation of mechanical energy you learned in Chapter 10,

$$\Delta K + \Delta U = W$$

where U is the gravitational potential energy and W is the work done by other forces—in this case, pressure forces. Recall that we're still considering ideal fluids, with no friction or viscosity, so there's no dissipation of energy to thermal energy.

FIGURE 13.35a shows fluid flowing through a tube. The tube narrows from cross-section area A_1 to area A_2 as it bends uphill. Let's concentrate on the large volume of the fluid that is shaded in Figure 13.35a. **This moving segment of fluid will be our system for the purpose of applying conservation of energy.**

To use conservation of energy, we need to draw a before-and-after overview. The "before" situation is shown in FIGURE 13.35b. A short time Δt later, the system has moved along the tube a bit, as shown in the "after" drawing of FIGURE 13.35c. Because the tube is not of uniform diameter, the two ends of the fluid system do not move the same distance during Δt: The lower end moves distance Δx_1, while the upper end moves Δx_2. Thus the system moves *out of* a cylindrical volume $\Delta V_1 = A_1\,\Delta x_1$ at the lower end and *into* a volume $\Delta V_2 = A_2\,\Delta x_2$ at the upper end. The equation of continuity tells us that these two volumes must be the same, so $A_1\,\Delta x_1 = A_2\,\Delta x_2 = \Delta V$.

From the "before" situation to the "after" situation, the system *loses* the kinetic and potential energy it originally had in the volume ΔV_1 but *gains* the kinetic and potential energy in the volume ΔV_2 that it later occupies. (The energy it has in the region between these two small volumes is unchanged.) Let's find the kinetic energy in each of these small volumes. The mass of fluid in each cylinder is $m = \rho\,\Delta V$, where ρ is the density of the fluid. The kinetic energies of the two small volumes 1 and 2 are then

$$K_1 = \frac{1}{2}\underbrace{\rho\,\Delta V}_{m} v_1^2 \quad \text{and} \quad K_2 = \frac{1}{2}\underbrace{\rho\,\Delta V}_{m} v_2^2$$

Thus the net *change* in kinetic energy is

$$\Delta K = K_2 - K_1 = \frac{1}{2}\rho\,\Delta V v_2^2 - \frac{1}{2}\rho\,\Delta V v_1^2$$

Similarly, the net change in the gravitational potential energy of our fluid system is

$$\Delta U = U_2 - U_1 = \rho\,\Delta V g y_2 - \rho\,\Delta V g y_1$$

The final piece of our conservation of energy treatment is the work done on the system by the rest of the fluid. As the fluid system moves, positive work is done on it by the force $\vec{F}_1$ due to the pressure p_1 of the fluid to the left of the system, while negative work is done on the system by the force $\vec{F}_2$ due to the pressure p_2 of the fluid to the right of the system. The positive work is

$$W_1 = F_1\,\Delta x_1 = (p_1 A_1)\,\Delta x_1 = p_1(A_1\,\Delta x_1) = p_1\,\Delta V$$

Similarly the negative work is

$$W_2 = -F_2\,\Delta x_2 = -(p_2 A_2)\,\Delta x_2 = -p_2(A_2\,\Delta x_2) = -p_2\,\Delta V$$

FIGURE 13.35 An ideal fluid flowing through a tube.

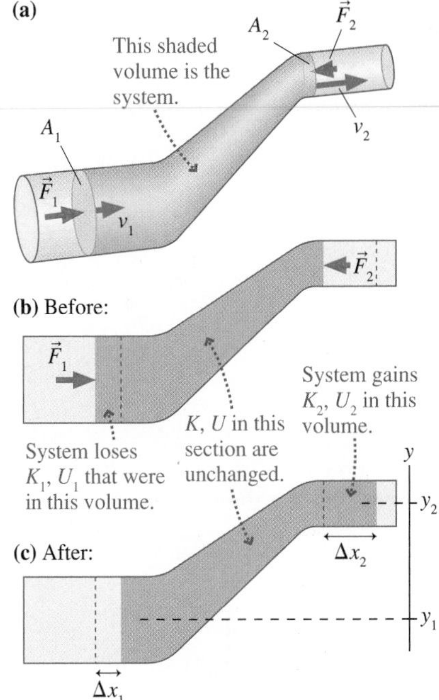
(a)
This shaded volume is the system.
A_1
A_2
$\vec{F}_1$
$\vec{F}_2$
v_1
v_2
(b) Before:
$\vec{F}_1$
$\vec{F}_2$
System loses K_1, U_1 that were in this volume.
K, U in this section are unchanged.
System gains K_2, U_2 in this volume.
(c) After:
y
y_2
y_1
Δx_2
Δx_1

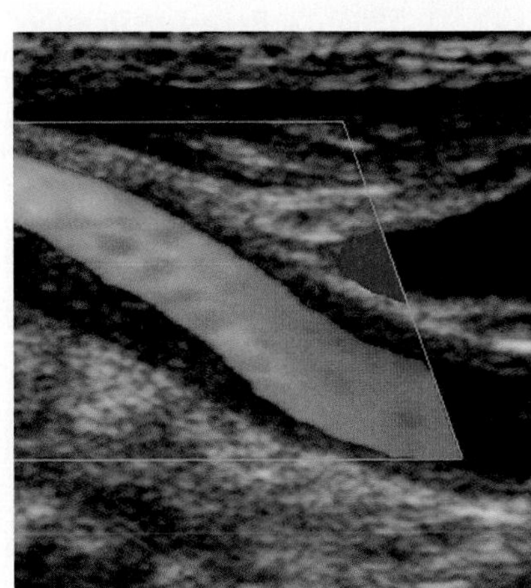

Living under pressure BIO When plaque builds up in major arteries, such as the carotid artery that supplies the head, dangerous drops in blood pressure can result. *Doppler ultrasound,* which you'll study in Chapter 15, uses sound waves to produce images of the interior of the body, and can detect the velocity of flowing blood. The image above shows the blood flow through a carotid artery with significant plaque buildup; yellow indicates a higher blood velocity than red. Once the velocities are known at two points along the flow, Bernoulli's equation can be used to deduce the corresponding pressure drop.

Thus the *net* work done on the system is

$$W = W_1 + W_2 = p_1\,\Delta V - p_2\,\Delta V = (p_1 - p_2)\,\Delta V$$

We can now use these expressions for ΔK, ΔU, and W to write the energy equation as

$$\underbrace{\frac{1}{2}\rho\,\Delta V v_2^2 - \frac{1}{2}\rho\,\Delta V v_1^2}_{\Delta K} + \underbrace{\rho\,\Delta V g y_2 - \rho\,\Delta V g y_1}_{\Delta U} = \underbrace{(p_1 - p_2)\,\Delta V}_{W}$$

The ΔV's cancel, and we can rearrange the remaining terms to get **Bernoulli's equation:**

$$p_2 + \frac{1}{2}\rho v_2^2 + \rho g y_2 = p_1 + \frac{1}{2}\rho v_1^2 + \rho g y_1 \qquad (13.14)$$

Bernoulli's equation relating pressure p, speed v, and height y at any two points along a streamline in an ideal fluid

Equation 13.14, a quantitative statement of the ideas we developed earlier in this section, is really nothing more than a statement about work and energy. Using Bernoulli's equation is very much like using the law of conservation of energy. Rather than identifying a "before" and "after," you want to identify two points on a streamline. As the following example shows, Bernoulli's equation is often used in conjunction with the equation of continuity.

EXAMPLE 13.12 Pressure in an irrigation system

Water flows through the pipes shown in FIGURE 13.36. The water's speed through the lower pipe is 5.0 m/s, and a pressure gauge reads 75 kPa. What is the reading of the pressure gauge on the upper pipe?

FIGURE 13.36 The water pipes of an irrigation system.

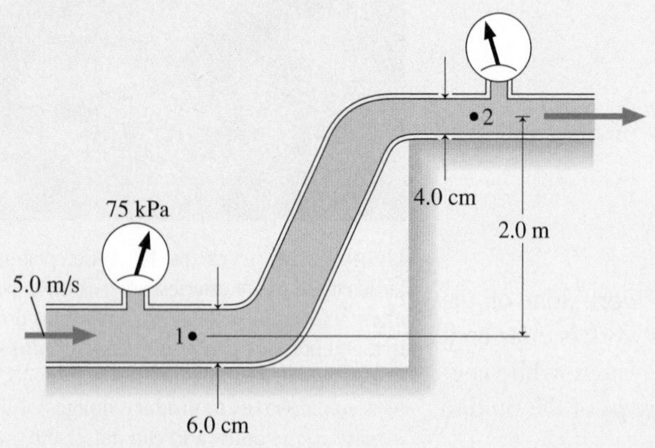

PREPARE Treat the water as an ideal fluid obeying Bernoulli's equation. Consider a streamline connecting point 1 in the lower pipe with point 2 in the upper pipe.

SOLVE Bernoulli's equation, Equation 13.14, relates the pressures, fluid speeds, and heights at points 1 and 2. It is easily solved for the pressure p_2 at point 2:

$$p_2 = p_1 + \frac{1}{2}\rho v_1^2 - \frac{1}{2}\rho v_2^2 + \rho g y_1 - \rho g y_2$$

$$= p_1 + \frac{1}{2}\rho(v_1^2 - v_2^2) + \rho g(y_1 - y_2)$$

All quantities on the right are known except v_2, and that is where the equation of continuity will be useful. The cross-section areas and water speeds at points 1 and 2 are related by

$$v_1 A_1 = v_2 A_2$$

from which we find

$$v_2 = \frac{A_1}{A_2}v_1 = \frac{r_1^2}{r_2^2}v_1 = \frac{(0.030\ \text{m})^2}{(0.020\ \text{m})^2}(5.0\ \text{m/s}) = 11.25\ \text{m/s}$$

The pressure at point 1 is $p_1 = 75\ \text{kPa} + 1\ \text{atm} = 176{,}300\ \text{Pa}$. We can now use the above expression for p_2 to calculate $p_2 = 105{,}900\ \text{Pa}$. This is the absolute pressure; the pressure gauge on the upper pipe will read

$$p_2 = 105{,}900\ \text{Pa} - 1\ \text{atm} = 4.6\ \text{kPa}$$

ASSESS Reducing the pipe size decreases the pressure because it makes $v_2 > v_1$. Gaining elevation also reduces the pressure.

STOP TO THINK 13.5 Rank in order, from highest to lowest, the liquid heights h_1 to h_4 in tubes 1 to 4. The air flow is from left to right.

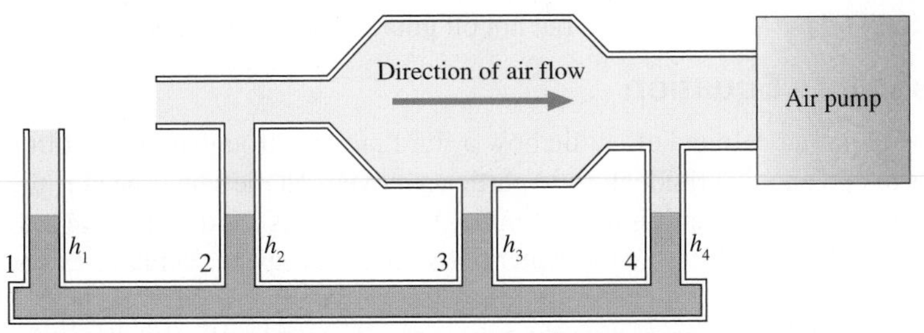
Direction of air flow
Air pump
1
h_1
2
h_2
3
h_3
4
h_4

13.7 Viscosity and Poiseuille's Equation

Thus far, the only thing we've needed to know about a fluid is its density. It is the density that determines how the pressure in a static fluid increases with depth, and density appears in Bernoulli's equation for the motion of a fluid. But you know from everyday experience that another property of fluids is often crucial in determining how fluids flow. The densities of honey and water are not too different, but there's a huge difference in the way honey and water pour. The honey is much "thicker." This property of a fluid, which measures its resistance to flow, is called **viscosity.** A very viscous fluid flows slowly when poured and is difficult to force through a pipe or tube. The viscous nature of fluids is of key importance in understanding a wide range of real-world applications, from blood flow to the flight of birds to throwing a curveball.

An ideal fluid, with no viscosity, will "coast" at constant speed through a constant-diameter tube with no change in pressure. That's why the fluid heights are the same in the first and second pressure-measuring columns on the Venturi tube of Figure 13.31. But as FIGURE 13.37 shows, any real fluid requires a *pressure difference* between the ends of a tube to keep the fluid moving at constant speed. The size of the pressure difference depends on the viscosity of the fluid. Think about how much harder you have to suck on a straw to drink a thick milkshake than to drink water or to pull air through the straw.

FIGURE 13.37 A viscous fluid needs a pressure difference to keep it moving. Compare this to Figure 13.31, which showed an ideal fluid.

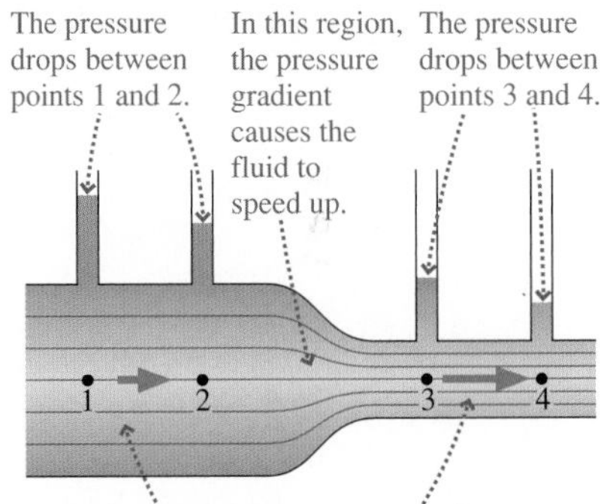
The pressure drops between points 1 and 2.
In this region, the pressure gradient causes the fluid to speed up.
The pressure drops between points 3 and 4.
1
2
3
4
In these regions, a pressure gradient is needed simply to keep the fluid moving with constant speed.

FIGURE 13.38 The pressure difference needed to keep the fluid flowing is proportional to the fluid's viscosity.

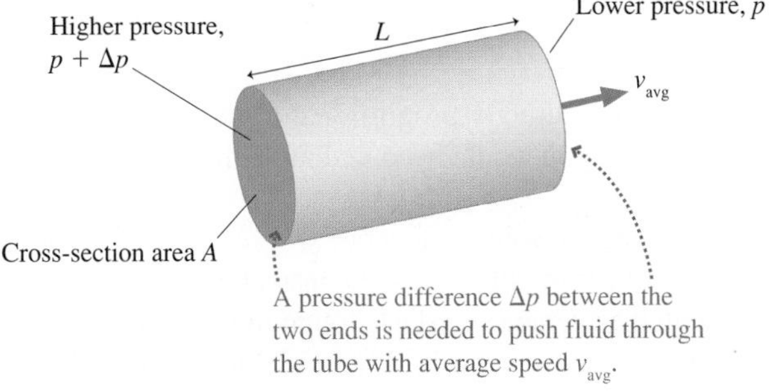
Higher pressure, $p + \Delta p$
L
Lower pressure, p
v_{avg}
Cross-section area A
A pressure difference Δp between the two ends is needed to push fluid through the tube with average speed v_{avg}.

FIGURE 13.38 shows a viscous fluid flowing with a constant average speed v_{avg} through a tube of length L and cross-section area A. Experiments show that the pressure difference needed to keep the fluid moving is proportional to v_{avg} and to L and inversely proportional to A. We can write

$$\Delta p = 8\pi\eta \frac{Lv_{avg}}{A} \tag{13.15}$$

where $8\pi\eta$ is the constant of proportionality and η (lowercase Greek eta) is called the **coefficient of viscosity** (or just the *viscosity*). (The 8π enters the equation from the technical definition of viscosity, which need not concern us.)

Equation 13.15 makes sense. A more viscous fluid needs a larger pressure difference to push it through the tube; an ideal fluid, with $\eta = 0$, will keep flowing without any

TABLE 13.3 Viscosities of fluids

Fluid	η (Pa · s)
Air (20°C)	1.8×10^{-5}
Water (20°C)	1.0×10^{-3}
Water (40°C)	0.7×10^{-3}
Water (60°C)	0.5×10^{-3}
Whole blood (37°C)	2.5×10^{-3}
Motor oil (−30°C)	3×10^{5}
Motor oil (40°C)	0.07
Motor oil (100°C)	0.01
Honey (15°C)	600
Honey (40°C)	20

pressure difference. We can also see from Equation 13.15 that the units of viscosity are N · s/m^2 or, equivalently, Pa · s. Table 13.3 gives values of η for some common fluids. Note that the viscosity of many liquids decreases *very* rapidly with temperature. Cold oil hardly flows at all, but hot oil pours almost like water.

Poiseuille's Equation

Viscosity has a profound effect on how a fluid moves through a tube. **FIGURE 13.39a** shows that in an ideal fluid, all fluid particles move with the same speed v, the speed that appears in the equation of continuity. For a viscous fluid, **FIGURE 13.39b** shows that the fluid moves fastest in the center of the tube. The speed decreases as you move away from the center of the tube until reaching zero on the walls of the tube. That is, the layer of fluid in contact with the tube doesn't move at all. Whether it be water through pipes or blood through arteries, the fact that the fluid at the outer edges "lingers" and barely moves allows deposits to build on the inside walls of a tube.

FIGURE 13.39 Viscosity alters the velocities of the fluid particles.

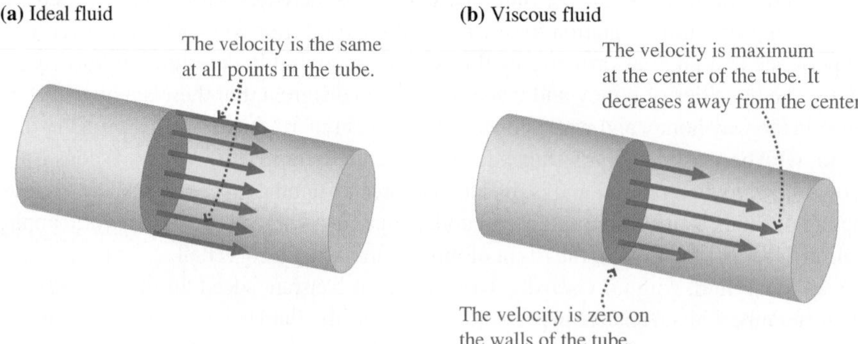

(a) Ideal fluid
The velocity is the same at all points in the tube.
(b) Viscous fluid
The velocity is maximum at the center of the tube. It decreases away from the center.
The velocity is zero on the walls of the tube.

Although we can't characterize the flow of a viscous liquid by a single flow speed v, we can still define the *average* flow speed. Suppose a fluid with viscosity η flows through a circular pipe with radius R and cross-section area $A = \pi R^2$. From Equation 13.15, a pressure difference Δp between the ends of the pipe causes the fluid to flow with average speed

$$v_{\text{avg}} = \frac{R^2}{8\eta L}\Delta p \tag{13.16}$$

The average flow speed is directly proportional to the pressure difference; for the fluid to flow twice as fast, you would need to double the pressure difference between the ends of the pipe.

Equation 13.13 defined the volume flow rate $Q = \Delta V/\Delta t$ and found that $Q = vA$ for an ideal fluid. For viscous flow, where v isn't constant throughout the fluid, we simply need to replace v with the average speed v_{avg} found in Equation 13.16. Using $A = \pi R^2$ for a circular tube, we see that a pressure difference Δp causes a volume flow rate

$$Q = v_{\text{avg}}A = \frac{\pi R^4 \Delta p}{8\eta L} \tag{13.17}$$

Poiseuille's equation for viscous flow through a tube of radius R and length L

This result is called **Poiseuille's equation** after the French scientist Jean Poiseuille (1797–1869) who first performed this calculation.

One surprising result of Poiseuille's equation is the very strong dependence of the flow on the tube's radius; the volume flow rate is proportional to the *fourth* power of R. If you double the radius of a tube, the flow rate will increase by a factor of $2^4 = 16$. This strong radius dependence comes from two factors. First, the flow rate depends on the area of the tube, which is proportional to R^2; larger tubes carry more fluid. Second, the average speed is faster in a larger tube because the center of the tube, where the flow is fastest, is farther from the "drag" exerted by the wall. As we've seen, the average speed is also proportional to R^2. These two terms combine to give the R^4 dependence.

CONCEPTUAL EXAMPLE 13.13 Blood pressure and cardiovascular disease

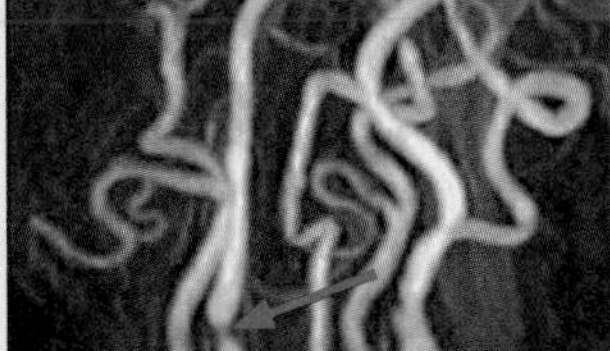
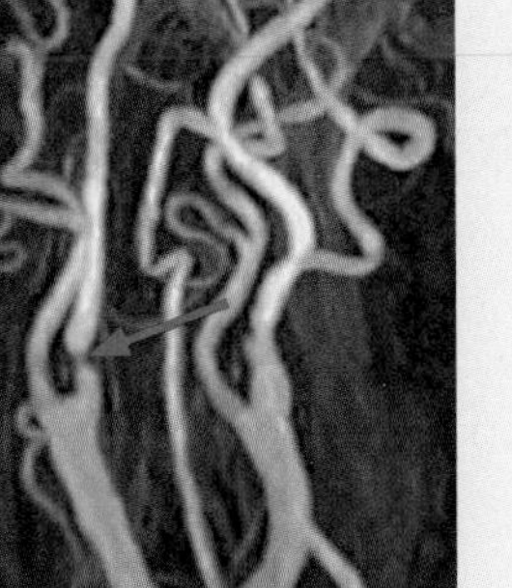

Cardiovascular disease is a narrowing of the arteries due to the buildup of plaque deposits on the interior walls. Magnetic resonance imaging, which you'll learn about in Chapter 24, can create exquisite three-dimensional images of the internal structure of the body. Shown are the carotid arteries that supply blood to the head, with a dangerous narrowing—a *stenosis*—indicated by the arrow.

If a section of an artery has narrowed by 8%, not nearly as much as the stenosis shown, by what percentage must the blood-pressure difference between the ends of the narrowed section increase to keep blood flowing at the same rate?

REASON According to Poiseuille's equation, the pressure difference Δp must increase to compensate for a decrease in the artery's radius R if the blood flow rate Q is to remain unchanged. If we write Poiseuille's equation as

$$R^4 \Delta p = \frac{8\eta L Q}{\pi}$$

we see that the product $R^4 \Delta p$ must remain unchanged if the artery is to deliver the same flow rate. Let the initial artery radius and pressure difference be R_i and Δp_i. Disease decreases the radius by 8%, meaning that $R_f = 0.92R_i$. The requirement

$$R_i^4 \Delta p_i = R_f^4 \Delta p_f$$

can be solved for the new pressure difference:

$$\Delta p_f = \frac{R_i^4}{R_f^4} \Delta p_i = \frac{R_i^4}{(0.92R_i)^4} \Delta p_i = 1.4\, \Delta p_i$$

The pressure difference must increase by 40% to maintain the flow.

ASSESS Because the flow rate depends on R^4, even a small change in radius requires a large change in Δp to compensate. Either the person's blood pressure must increase, which is dangerous, or he or she will suffer a significant reduction in blood flow. For the stenosis shown in the image, the reduction in radius is much greater than 8%, and the pressure difference will be large and very dangerous.

EXAMPLE 13.14 Pressure drop along a capillary

In Example 13.11 we examined blood flow through a capillary. Using the numbers from that example, calculate the pressure "drop" from one end of a capillary to the other.

PREPARE Example 13.11 gives enough information to determine the flow rate through a capillary. We can then use Poiseuille's equation to calculate the pressure difference between the ends.

SOLVE The measured volume flow rate leaving the heart was given as 5 L/min $= 8 \times 10^{-5}\ \text{m}^3/\text{s}$. This flow is divided among all the capillaries, which we found to number $N = 3 \times 10^9$. Thus the flow rate through each capillary is

$$Q_{\text{cap}} = \frac{Q_{\text{heart}}}{N} = \frac{8 \times 10^{-5}\ \text{m}^3/\text{s}}{3 \times 10^9} = 2.7 \times 10^{-14}\ \text{m}^3/\text{s}$$

Solving Poiseuille's equation for Δp, we get

$$\Delta p = \frac{8\eta L Q_{\text{cap}}}{\pi R^4} = \frac{8(2.5 \times 10^{-3}\ \text{Pa}\cdot\text{s})(0.001\ \text{m})(2.7 \times 10^{-14}\ \text{m}^3/\text{s})}{\pi(3 \times 10^{-6}\ \text{m})^4} = 2100\ \text{Pa}$$

Converting to mm of mercury, the units of blood pressure, the pressure drop across the capillary is $\Delta p = 16$ mm Hg.

ASSESS The average blood pressure provided by the heart (the average of the systolic and diastolic pressures) is about 100 mm Hg. A physiology textbook will tell you that the pressure has decreased to 35 mm by the time blood enters the capillaries, and it exits from capillaries into the veins at 17 mm. Thus the pressure drop across the capillaries is 18 mm Hg. Our calculation, based on the laws of fluid flow and some simple estimates of capillary size, is in almost perfect agreement with measured values.

STOP TO THINK 13.6 A viscous fluid flows through the pipe shown. The three marked segments are of equal length. Across which segment is the pressure difference the greatest?

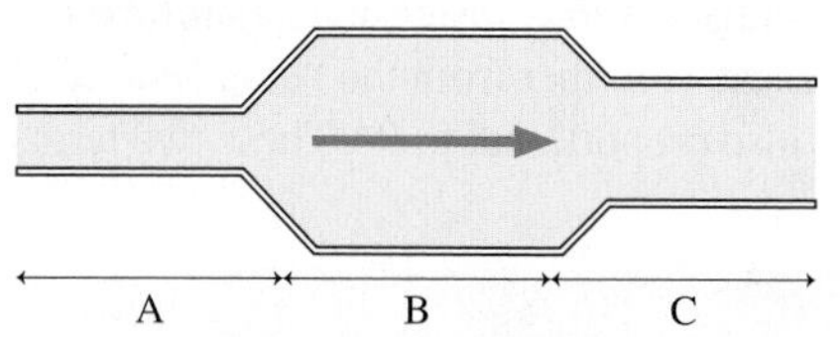

INTEGRATED EXAMPLE 13.15 An intravenous transfusion

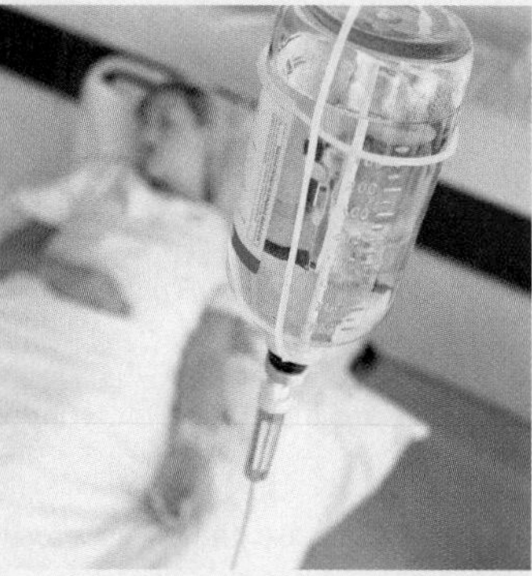

At the hospital, a patient often receives fluids via an intravenous (IV) infusion. A bag of the fluid is held at a fixed height above the patient's body. The fluid then travels down a large-diameter, flexible tube to a catheter—a short tube with a small diameter—inserted into the patient's vein.

1.0 L of saline solution, with a density of 1020 kg/m^3 and a viscosity of 1.1×10^{-3} Pa·s, is to be infused into a patient in 8.0 h. The catheter is 30 mm long and has an inner diameter of 0.30 mm. The pressure in the patient's vein is 20 mm Hg. How high above the patient should the bag be positioned to get the desired flow rate?

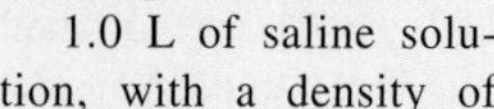

PREPARE FIGURE 13.40 shows a sketch of the situation, defines variables, and lists the known information. We're concerned with the flow of a viscous fluid. According to Poiseuille's equation, the flow rate depends inversely on the fourth power of a tube's radius. We are told that the tube from the elevated bag to the catheter has a large diameter, while the diameter of the catheter is small. Thus, we expect the flow rate to be determined entirely by the flow through the narrow catheter; the wide tube has a negligible effect on the rate.

FIGURE 13.40 Visual overview of an IV transfusion.

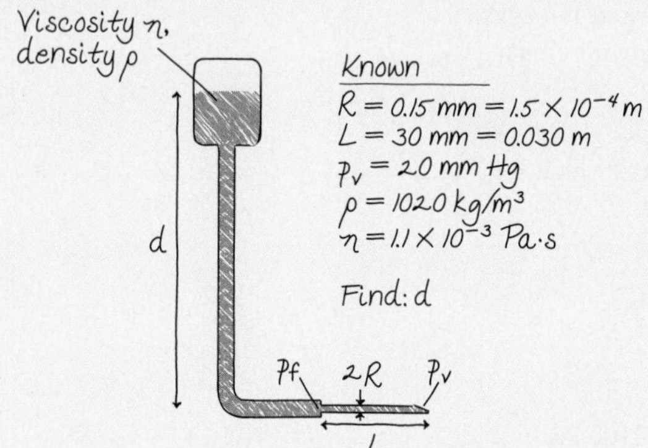

To use Poiseuille's equation for the catheter, we need the pressure difference Δp between the ends of the catheter. We know the pressure on the end of the catheter in the patient's vein is $p_v = 20$ mm Hg or, converting to SI units using Table 13.2,

$$p_v = 20 \text{ mm Hg} \times \frac{101 \times 10^3 \text{ Pa}}{760 \text{ mm Hg}} = 2700 \text{ Pa}$$

This is a gauge pressure, the pressure in excess of 1 atm. The pressure on the fluid side of the catheter is due to the hydrostatic pressure of the saline solution filling the bag and the flexible tube leading to the catheter. This pressure is given by the hydrostatic pressure equation $p = p_0 + \rho g d$, where d is the "depth" of the catheter below the bag. Thus we'll use Δp to find d.

SOLVE The desired volume flow rate is

$$Q = \frac{\Delta V}{\Delta t} = \frac{1.0 \text{ L}}{8.0 \text{ h}} = 0.125 \text{ L/h}$$

Converting to SI units using $1.0 \text{ L} = 1.0 \times 10^{-3} \text{ m}^3$, we have

$$Q = 0.125 \frac{\text{L}}{\text{h}} \times \frac{1.0 \times 10^{-3} \text{ m}^3}{\text{L}} \times \frac{1 \text{ h}}{3600 \text{ s}} = 3.5 \times 10^{-8} \text{ m}^3/\text{s}$$

Poiseuille's equation for viscous fluid flow is

$$Q = \frac{\pi R^4 \Delta p}{8 \eta L}$$

Thus the pressure difference needed between the ends of the tube to produce the desired flow rate Q is

$$\Delta p = \frac{8 \eta L Q}{\pi R^4}$$
$$= \frac{8(1.1 \times 10^{-3} \text{ Pa} \cdot \text{s})(0.030 \text{ m})(3.5 \times 10^{-8} \text{ m}^3/\text{s})}{\pi (1.5 \times 10^{-4} \text{ m})^4}$$
$$= 5800 \text{ Pa}$$

Now Δp is the difference between the fluid pressure p_f at one end of the catheter and the vein pressure p_v at the other end: $\Delta p = p_f - p_v$. We know p_v, so

$$p_f = p_v + \Delta p = 2700 \text{ Pa} + 5800 \text{ Pa} = 8500 \text{ Pa}$$

This pressure, like the vein pressure, is a gauge pressure. The true hydrostatic pressure at the catheter is $p = 1 \text{ atm} + 8500$ Pa. But the hydrostatic pressure in the fluid is

$$p = p_0 + \rho g d = 1 \text{ atm} + \rho g d$$

We see that $\rho g d = 8500$ Pa. Solving for d, we find the required elevation of the bag above the patient's arm:

$$d = \frac{p_f}{\rho g} = \frac{8500 \text{ Pa}}{(1020 \text{ kg/m}^3)(9.8 \text{ m/s}^2)} = 0.85 \text{ m}$$

ASSESS This height of about a meter seems reasonable for the height of an IV bag. In practice, the bag can be raised or lowered to adjust the fluid flow rate.

SUMMARY

The goal of Chapter 13 has been to understand the static and dynamic properties of fluids.

GENERAL PRINCIPLES

Fluid Statics

Gases

- Freely moving particles
- Compressible
- Pressure mainly due to particle collisions with walls

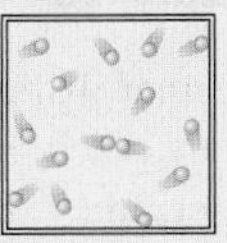

Liquids

- Loosely bound particles
- Incompressible
- Pressure due to the weight of the liquid
- Hydrostatic pressure at depth d is $p = p_0 + \rho g d$
- The pressure is the same at all points on a horizontal line through a liquid (of one kind) in hydrostatic equilibrium

Fluid Dynamics

Ideal-fluid model

- Incompressible
- Smooth, laminar flow
- Nonviscous

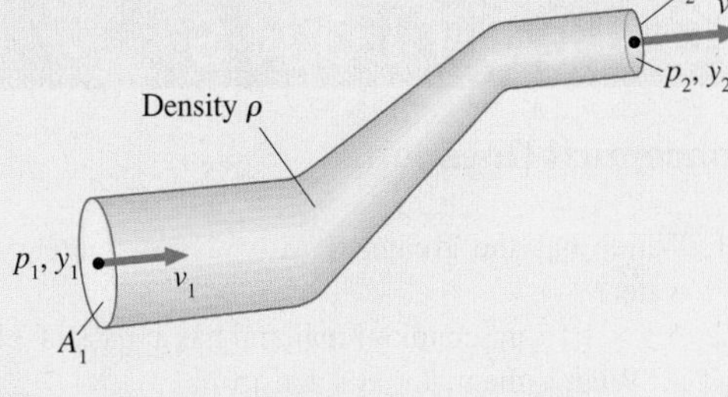

Equation of continuity

Volume flow rate $Q = \dfrac{\Delta V}{\Delta t} = v_1 A_1 = v_2 A_2$

Bernoulli's equation is a statement of energy conservation:

$$p_1 + \frac{1}{2}\rho v_1^2 + \rho g y_1 = p_2 + \frac{1}{2}\rho v_2^2 + \rho g y_2$$

Poiseuille's equation governs viscous flow through a tube:

$$Q = v_{avg} A = \frac{\pi R^4 \Delta p}{8 \eta L}$$

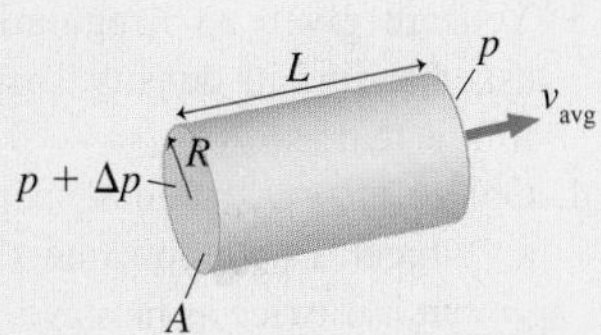

IMPORTANT CONCEPTS

Density $\rho = m/V$, where m is mass and V is volume.

Pressure $p = F/A$, where F is force magnitude and A is the area on which the force acts.

- Pressure exists at all points in a fluid.
- Pressure pushes equally in all directions.
- Gauge pressure $p_g = p - 1$ atm.

Viscosity η is the property of a fluid that makes it resist flowing.

Representing fluid flow

Streamlines are the paths of individual fluid particles.

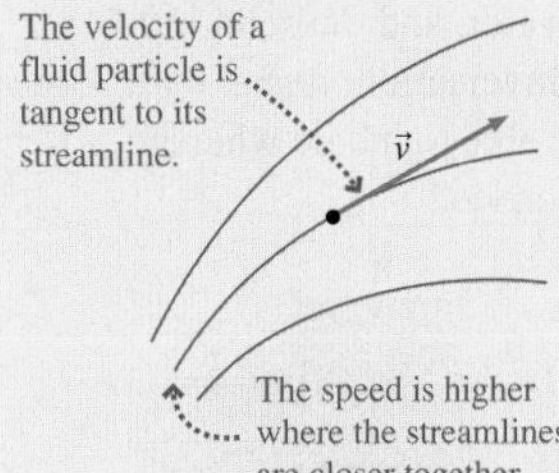

Fluid elements contain a fixed volume of fluid. Their shape may change as they move.

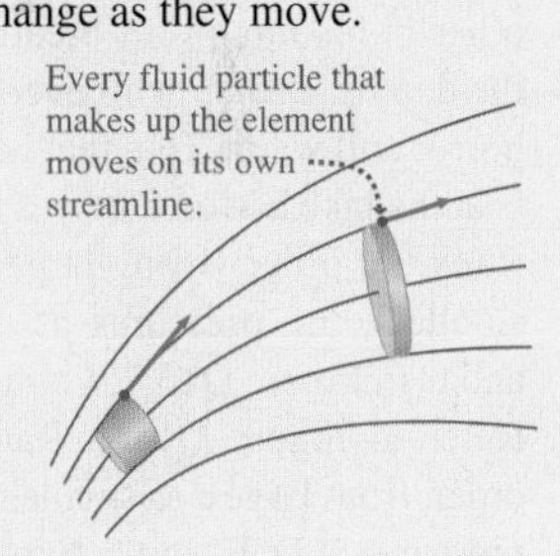

Every fluid particle that makes up the element moves on its own streamline.

APPLICATIONS

Buoyancy is the upward force of a fluid on an object immersed in the fluid.

Archimedes' principle: The magnitude of the buoyant force equals the weight of the fluid displaced by the object.

Sink: $\rho_{avg} > \rho_f \quad F_B < w_o$

Float: $\rho_{avg} < \rho_f \quad F_B > w_o$

Neutrally buoyant: $\rho_{avg} = \rho_f \quad F_B = w_o$

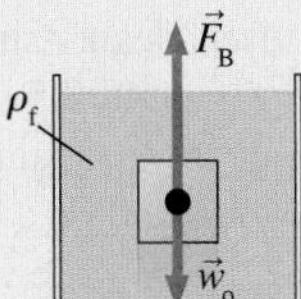

Barometers measure atmospheric pressure. Atmospheric pressure is related to the height of the liquid column by $p_{atmos} = \rho g h$.

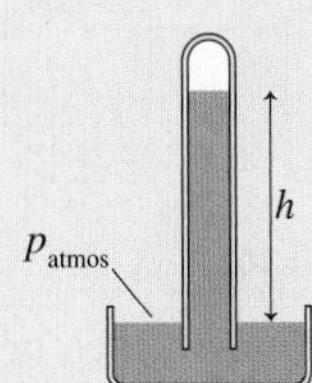

Manometers measure pressure. The pressure at the closed end of the tube is $p = 1 \text{ atm} + \rho g h$.

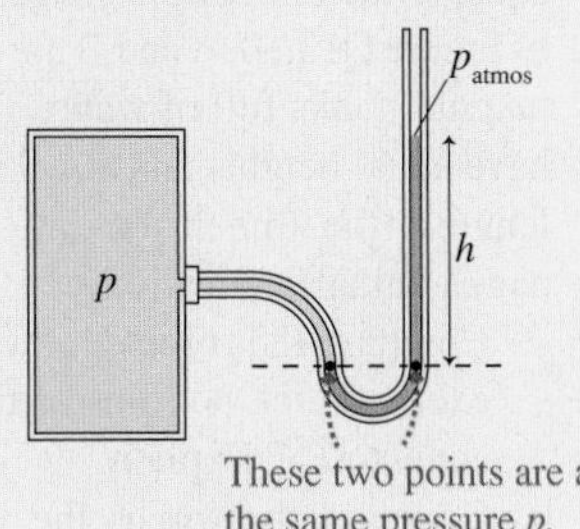

For homework assigned on MasteringPhysics, go to www.masteringphysics.com

Problem difficulty is labeled as | (straightforward) to ||||| (challenging).

Problems labeled INT integrate significant material from earlier chapters; BIO are of biological or medical interest.

QUESTIONS

Conceptual Questions

1. Which has the greater density, 1 g of mercury or 1000 g of water?
2. A $1 \times 10^{-3}\ \text{m}^3$ chunk of material has a mass of 3 kg.
   a. What is the material's density?
   b. Would a $2 \times 10^{-3}\ \text{m}^3$ chunk of the same material have the same mass? Explain.
   c. Would a $2 \times 10^{-3}\ \text{m}^3$ chunk of the same material have the same density? Explain.
3. You are given an irregularly shaped chunk of material and asked to find its density. List the *specific* steps that you would follow to do so.
4. Object 1 has an irregular shape. Its density is $4000\ \text{kg/m}^3$.
   a. Object 2 has the same shape and dimensions as object 1, but it is twice as massive. What is the density of object 2?
   b. Object 3 has the same mass and the same *shape* as object 1, but its size in all three dimensions is twice that of object 1. What is the density of object 3?
5. BIO When you get a blood transfusion the bag of blood is held above your body, but when you donate blood the collection bag is held below. Why is this?
6. BIO To explore the bottom of a 10-m-deep lake, your friend Tom proposes to get a long garden hose, put one end on land and the other in his mouth for breathing underwater, and descend into the depths. Susan, who overhears the conversation, reacts with horror and warns Tom that he will not be able to inhale when he is at the lake bottom. Why is Susan so worried?
7. Rank in order, from largest to smallest, the pressures at A, B, and C in Figure Q13.7. Explain.
8. Refer to Figure Q13.7. Rank in order, from largest to smallest, the pressures at D, E, and F. Explain.

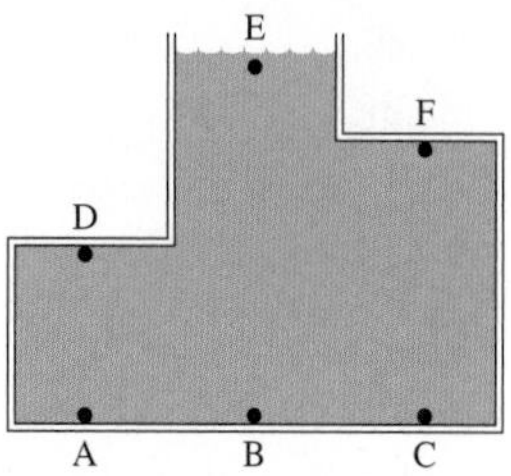

FIGURE Q13.7

9. Cylinders A and B contain liquids. The pressure p_A at the bottom of A is higher than the pressure p_B at the bottom of B. Is the ratio p_A/p_B of the absolute pressures larger, smaller, or equal to the ratio of the gauge pressures? Explain.
10. In Figure Q13.10, A and B are rectangular tanks full of water. They have equal heights and equal side lengths (the dimension into the page), but different widths.
    a. Compare the forces the water exerts on the bottoms of the tanks. Is F_A larger, smaller, or equal to F_B? Explain.
    b. Compare the forces the water exerts on the sides of the tanks. Is F_A larger, smaller, or equal to F_B? Explain.

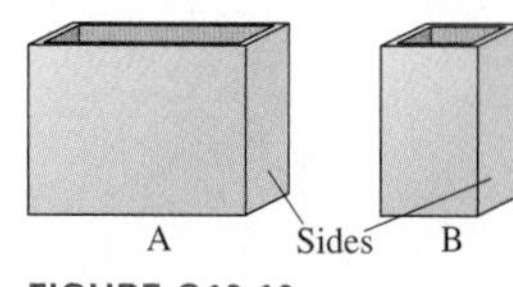

FIGURE Q13.10

11. Helium-filled weather balloons are spherical when they reach very high altitudes. However, they are only partially inflated with helium before they are released. Explain why this is done.

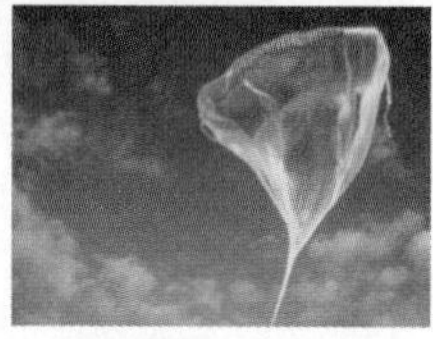

12. Water expands when heated. Suppose a beaker of water is heated from 10°C to 90°C. Does the pressure at the bottom of the beaker increase, decrease, or stay the same? Explain.
13. In Figure Q13.13, is p_A larger, smaller, or equal to p_B? Explain.
14. A beaker of water rests on a scale. A metal ball is then lowered into the beaker using a string tied to the ball. The ball doesn't touch the sides or bottom of the beaker, and no water spills from the beaker. Does the scale reading decrease, increase, or stay the same? Explain.

FIGURE Q13.13

15. Rank in order, from largest to smallest, the densities of objects A, B, and C in Figure Q13.15. Explain.

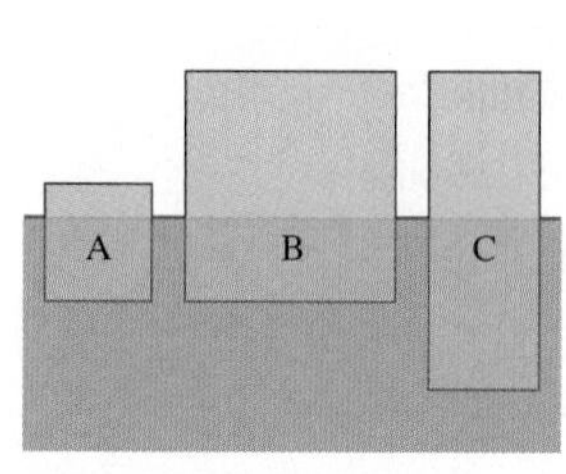

FIGURE Q13.15

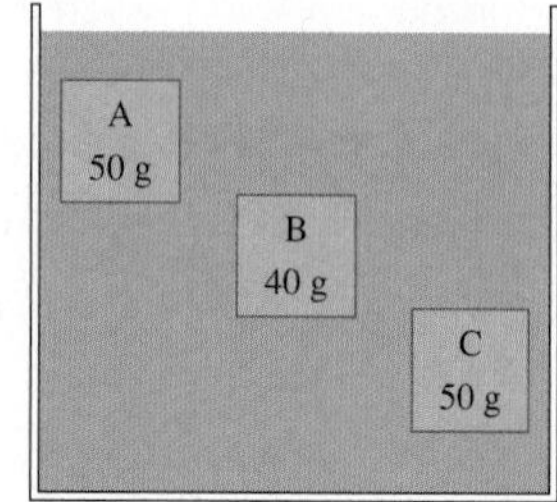

FIGURE Q13.16

16. Objects A, B, and C in Figure Q13.16 have the same volume. Rank in order, from largest to smallest, the sizes of the buoyant forces F_A, F_B, and F_C on A, B, and C. Explain.
17. Refer to Figure Q13.16. Now A, B, and C have the same density, but still have the masses given in the figure. Rank in order, from largest to smallest, the sizes of the buoyant forces on A, B, and C. Explain.
18. When you stand on a bathroom scale, it reads 700 N. Suppose a giant vacuum cleaner sucks half the air out of the room, reducing the pressure to 0.5 atm. Would the scale reading increase, decrease, or stay the same? Explain.
19. Suppose you stand on a bathroom scale that is on the bottom of a swimming pool. The water comes up to your waist. Does the scale read more, less, or the same as your true weight? Explain.
20. When you place an egg in water, it sinks. If you add salt to the water, after some time the egg floats. Explain.

21. Submerged submarines contain tanks filled with water. To rise to the surface, compressed air is used to force the water out of the tanks. Explain why this works.
22. Fish can adjust their buoyancy with an organ called the *swim* BIO *bladder.* The swim bladder is a flexible gas-filled sac; the fish can increase or decrease the amount of gas in the swim bladder so that it stays neutrally buoyant—neither sinking nor floating. Suppose the fish is neutrally buoyant at some depth and then goes deeper. What needs to happen to the volume of air in the swim bladder? Will the fish need to add or remove gas from the swim bladder to maintain its neutral buoyancy?
23. Figure Q13.23 shows two identical beakers filled to the same height with water. Beaker B has a plastic sphere floating in it. Which beaker, with all its contents, weighs more? Or are they equal? Explain.

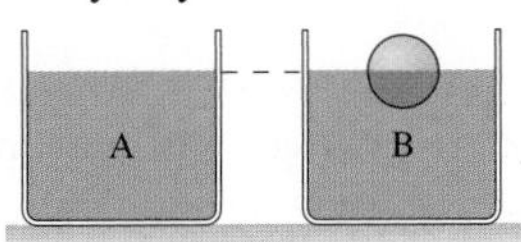

FIGURE Q13.23

24. A tub of water, filled to the brim, sits on a scale. Then a floating block of wood is placed in the tub, pushing some water over the rim. The water that overflows immediately runs off the scale. What happens to the reading of the scale?
25. Ships A and B have the same height and the same mass. Their cross-section profiles are shown in Figure Q13.25. Does one ship ride higher in the water (more height above the water line) than the other? If so, which one? Explain.

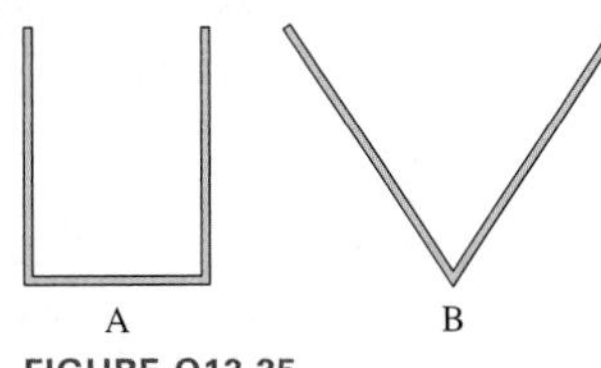

FIGURE Q13.25

26. Gas flows through a pipe, as shown in Figure Q13.26. The pipe's constant outer diameter is shown; you can't see into the pipe to know how the inner diameter changes. Rank in order, from largest to smallest, the gas speeds v_1 to v_3 at points 1, 2, and 3. Explain.

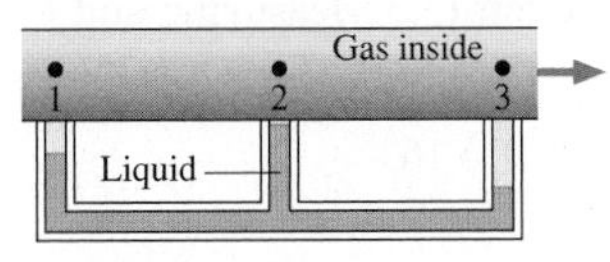

FIGURE Q13.26

27. Liquid flows through a pipe as shown in Figure Q13.27. The pipe's constant outer diameter is shown; you can't see into the pipe to know how the inner diameter changes. Rank in order, from largest to smallest, the flow speeds v_1 to v_3 at points 1, 2, and 3. Explain.

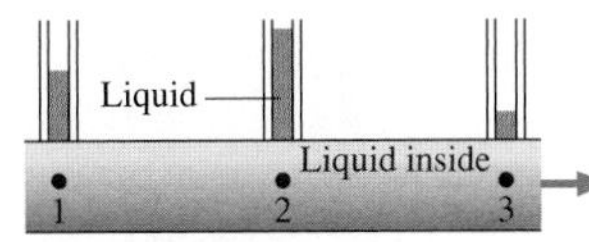

FIGURE Q13.27

28. A liquid with negligible viscosity flows through the pipe shown in Figure Q13.28. This is an overhead view.
    a. Rank in order, from largest to smallest, the flow speeds v_1 to v_4 at points 1 to 4. Explain.
    b. Rank in order, from largest to smallest, the pressures p_1 to p_4 at points 1 to 4. Explain.

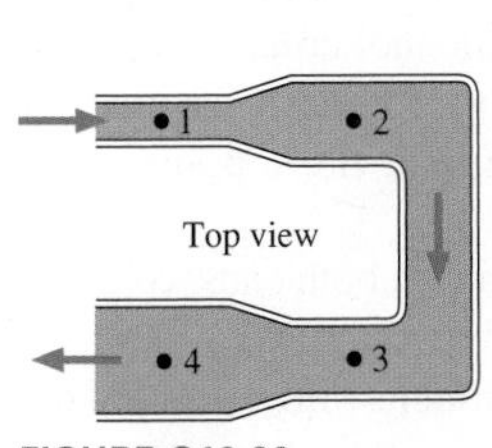

FIGURE Q13.28

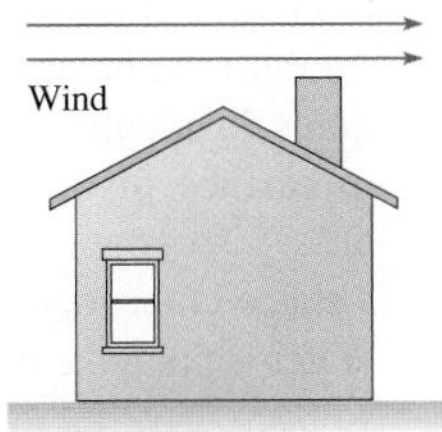

FIGURE Q13.29

29. Wind blows over the house shown in Figure Q13.29. A window on the ground floor is open. Is there an air flow through the house? If so, does the air flow in the window and out the chimney, or in the chimney and out the window? Explain.
30. Two pipes have the same inner cross-section area. One has a circular cross section and the other has a rectangular cross section with its height one-tenth its width. Through which pipe, if either, would it be easier to pump a viscous liquid? Explain.

Multiple-Choice Questions

31. | Figure Q13.31 shows a 100 g block of copper ($\rho = 8900\ \text{kg/m}^3$) and a 100 g block of aluminum ($\rho = 2700\ \text{kg/m}^3$) connected by a massless string that runs over two massless, frictionless pulleys. The two blocks exactly balance, since they have the same mass. Now suppose that the whole system is submerged in water. What will happen?

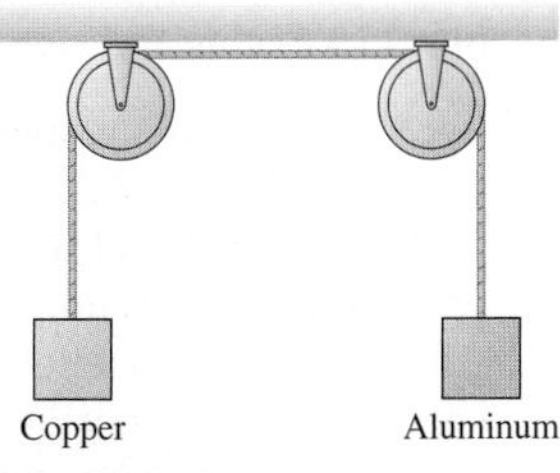

FIGURE Q13.31

    A. The copper block will fall, the aluminum block will rise.
    B. The aluminum block will fall, the copper block will rise.
    C. Nothing will change.
    D. Both blocks will rise.
32. | Masses A and B rest on very light pistons that enclose a fluid, as shown in Figure Q13.32. There is no friction between the pistons and the cylinders they fit inside. Which of the following is true?

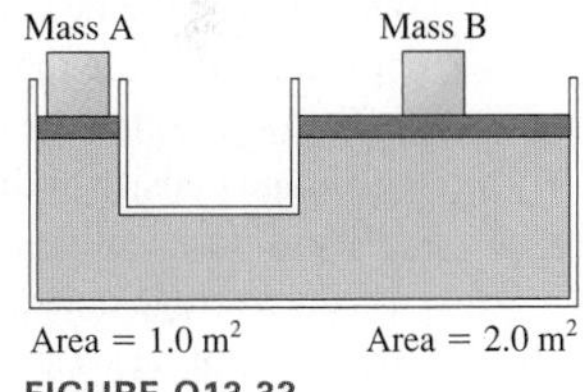

FIGURE Q13.32

    A. Mass A is greater. B. Mass B is greater.
    C. Mass A and mass B are the same.
33. | If you dive underwater, you notice an uncomfortable pres- BIO sure on your eardrums due to the increased pressure. The human eardrum has an area of about 70 mm² ($7 \times 10^{-5}\ \text{m}^2$), and it can sustain a force of about 7 N without rupturing. If your body had no means of balancing the extra pressure (which, in reality, it does), what would be the maximum depth you could dive without rupturing your eardrum?
    A. 0.3 m B. 1 m C. 3 m D. 10 m
34. || An 8.0 lb bowling ball has a diameter of 8.5 inches. When lowered into water, this ball will
    A. Float. B. Sink. C. Have neutral buoyancy.
35. | A basketball has a mass of 0.50 kg and a volume of $8.0 \times 10^{-3}\ \text{m}^3$. What is the magnitude of the net force on a basketball when it is fully submerged in water?
    A. 4.9 N B. 74 N C. 78 N D. 83 N
36. | An object floats in water, with 75% of its volume submerged. What is its approximate density?
    A. 250 kg/m³ B. 750 kg/m³
    C. 1000 kg/m³ D. 1250 kg/m³
37. | A syringe is being used to squirt water as shown in Figure Q13.37. The water is ejected from the nozzle at 10 m/s. At what speed is the plunger of the syringe being depressed?

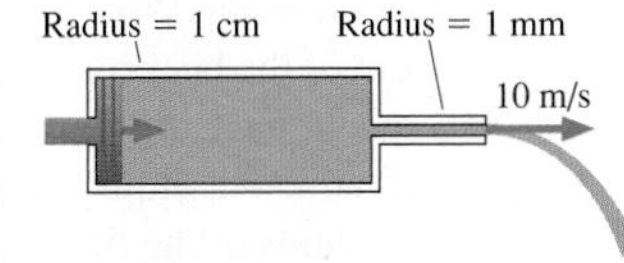

FIGURE Q13.37

    A. 0.01 m/s B. 0.1 m/s C. 1 m/s D. 10 m/s

38. || Water flows through a 4.0-cm-diameter horizontal pipe at a speed of 1.3 m/s. The pipe then narrows down to a diameter of 2.0 cm. Ignoring viscosity, what is the pressure difference between the wide and narrow sections of the pipe?
A. 850 Pa B. 3400 Pa C. 9300 Pa
D. 12,700 Pa E. 13,500 Pa

39. || A 15-m-long garden hose has an inner diameter of 2.5 cm. One end is connected to a spigot; 20°C water flows from the other end at a rate of 1.2 L/s. What is the gauge pressure at the spigot end of the hose?
A. 1900 Pa B. 2700 Pa C. 4200 Pa
D. 5800 Pa E. 7300 Pa

PROBLEMS

Section 13.1 Fluids and Density

1. || A 100 mL beaker holds 120 g of liquid. What is the liquid's density in SI units?
2. | Containers A and B have equal volumes. Container A holds helium gas at 1.0 atm pressure and 20°C. Container B is completely filled with a liquid whose mass is 7600 times the mass of helium gas in container A. Identify the liquid in B.
3. || Air enclosed in a sphere has density $\rho = 1.4\ \text{kg/m}^3$. What will the density be if the radius of the sphere is halved, compressing the air within?
4. || Air enclosed in a cylinder has density $\rho = 1.4\ \text{kg/m}^3$.
   a. What will be the density of the air if the length of the cylinder is doubled while the radius is unchanged?
   b. What will be the density of the air if the radius of the cylinder is halved while the length is unchanged?
5. || a. 50 g of gasoline are mixed with 50 g of water. What is the average density of the mixture?
   b. 50 cm^3 of gasoline are mixed with 50 cm^3 of water. What is the average density of the mixture?
6. ||| Ethyl alcohol has been added to 200 mL of water in a container that has a mass of 150 g when empty. The resulting container and liquid mixture has a mass of 512 g. What volume of alcohol was added to the water?
7. || The average density of the body of a fish is 1080 kg/m^3. To keep
BIO from sinking, the fish increases its volume by inflating an internal air bladder, known as a swim bladder, with air. By what percent must the fish increase its volume to be neutrally buoyant in fresh water? Use the Table 13.1 value for the density of air at 20°C.

Section 13.2 Pressure

8. || The deepest point in the ocean is 11 km below sea level, deeper than Mt. Everest is tall. What is the pressure in atmospheres at this depth?
9. || a. What volume of water has the same mass as 8.0 m^3 of ethyl alcohol?
   b. If this volume of water completely fills a cubic tank, what is the pressure at the bottom?
10. ||| A 1.0-m-diameter vat of liquid is 2.0 m deep. The pressure at the bottom of the vat is 1.3 atm. What is the mass of the liquid in the vat?
11. ||| A 35-cm-tall, 5.0-cm-diameter cylindrical beaker is filled to its brim with water. What is the downward force of the water on the bottom of the beaker?
12. || The gauge pressure at the bottom of a cylinder of liquid is $p_g = 0.40$ atm. The liquid is poured into another cylinder with twice the radius of the first cylinder. What is the gauge pressure at the bottom of the second cylinder?
13. |||| A research submarine has a 20-cm-diameter window 8.0 cm thick. The manufacturer says the window can withstand forces up to 1.0×10^6 N. What is the submarine's maximum safe depth in seawater? The pressure inside the submarine is maintained at 1.0 atm.
14. |||| The highest that George can suck water up a very long straw
BIO is 2.0 m. (This is a typical value.) What is the lowest pressure that he can maintain in his mouth?
15. ||| The two 60-cm-diameter cylinders in Figure P13.15, closed at one end, open at the other, are joined to form a single cylinder, then the air inside is removed.
    a. How much force does the atmosphere exert on the flat end of each cylinder?
    b. Suppose the upper cylinder is bolted to a sturdy ceiling. How many 100 kg football players would need to hang from the lower cylinder to pull the two cylinders apart?

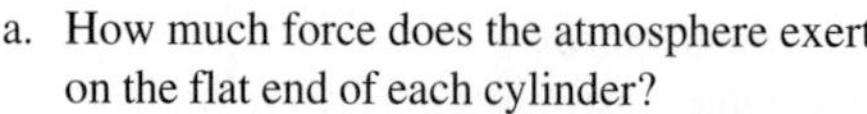

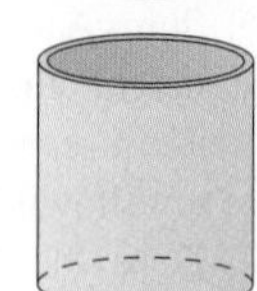

FIGURE P13.15

Section 13.3 Measuring and Using Pressure

16. |||| What is the gas pressure inside the box shown in Figure P13.16?

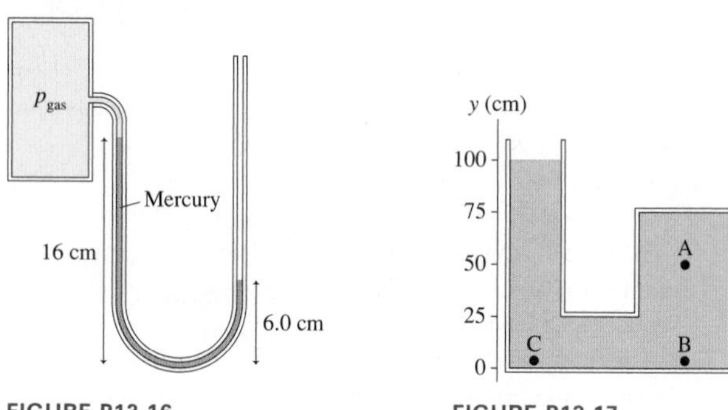

FIGURE P13.16 FIGURE P13.17

17. || The container shown in Figure P13.17 is filled with oil. It is open to the atmosphere on the left.
    a. What is the pressure at point A?
    b. What is the pressure difference between points A and B? Between points A and C?
18. ||| Glycerin is poured into an open U-shaped tube until the height in both sides is 20 cm. Ethyl alcohol is then poured into one arm until the height of the alcohol column is 20 cm. The two liquids do not mix. What is the difference in height between the top surface of the glycerin and the top surface of the alcohol?
19. ||| A U-shaped tube, open to the air on both ends, contains mercury. Water is poured into the left arm until the water column is 10.0 cm deep. How far upward from its initial position does the mercury in the right arm rise?

20. | What is the height of a water barometer at atmospheric pressure?

21. ||| Postural hypotension is the occurrence of low (systolic) BIO blood pressure when standing up too quickly from a reclined position, causing fainting or lightheadedness. For most people, a systolic pressure less than 90 mm Hg is considered low. If the blood pressure in your brain is 120 mm when you are lying down, what would it be when you stand up? Assume that your brain is 40 cm from your heart and that $\rho = 1060\ \text{kg/m}^3$ for your blood. Note: Normally, your blood vessels constrict and expand to keep your brain blood pressure stable when you change your posture.

Section 13.4 Buoyancy

22. || A 6.00-cm-diameter sphere with a mass of 89.3 g is neutrally buoyant in a liquid. Identify the liquid.

23. ||| A cargo barge is loaded in a saltwater harbor for a trip up a freshwater river. If the rectangular barge is 3.0 m by 20.0 m and sits 0.80 m deep in the harbor, how deep will it sit in the river?

24. || A 10 cm × 10 cm × 10 cm wood block with a density of 700 kg/m^3 floats in water.
   a. What is the distance from the top of the block to the water if the water is fresh?
   b. If it's seawater?

25. || What is the tension in the string in Figure P13.25?

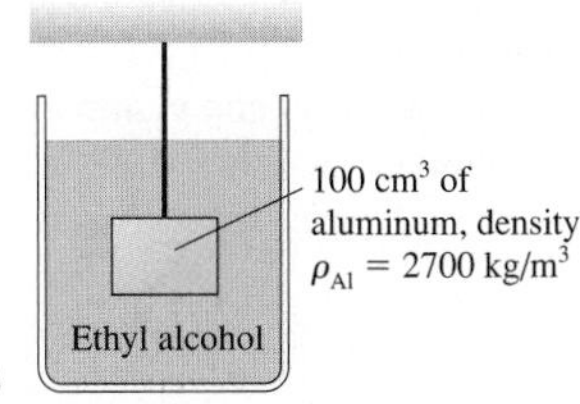

FIGURE P13.25

26. | A 10 cm × 10 cm × 10 cm block of steel ($\rho_{steel} = 7900\ \text{kg/m}^3$) is suspended from a spring scale. The scale is in newtons.
   a. What is the scale reading if the block is in air?
   b. What is the scale reading after the block has been lowered into a beaker of oil and is completely submerged?

27. ||| Styrofoam has a density of 300 kg/m^3. What is the maximum mass that can hang without sinking from a 50-cm-diameter Styrofoam sphere in water? Assume the volume of the mass is negligible compared to that of the sphere.

28. |||| Calculate the buoyant force due to the surrounding air on a man weighing 800 N. Assume his average density is the same as that of water.

Section 13.5 Fluids in Motion

29. ||| River Pascal with a volume flow rate of 5.0×10^5 L/s joins with River Archimedes, which carries 10.0×10^5 L/s, to form the Bernoulli River. The Bernoulli River is 150 m wide and 10 m deep. What is the speed of the water in the Bernoulli River?

30. || Water flowing through a 2.0-cm-diameter pipe can fill a 300 L bathtub in 5.0 min. What is the speed of the water in the pipe?

31. |||| A pump is used to empty a 6000 L wading pool. The water exits the 2.5-cm-diameter hose at a speed of 2.1 m/s. How long will it take to empty the pool?

32. || A 1.0-cm-diameter pipe widens to 2.0 cm, then narrows to 0.50 cm. Liquid flows through the first segment at a speed of 4.0 m/s.
   a. What are the speeds in the second and third segments?
   b. What is the volume flow rate through the pipe?

Section 13.6 Fluid Dynamics

33. || What does the top pressure gauge in Figure P13.33 read?

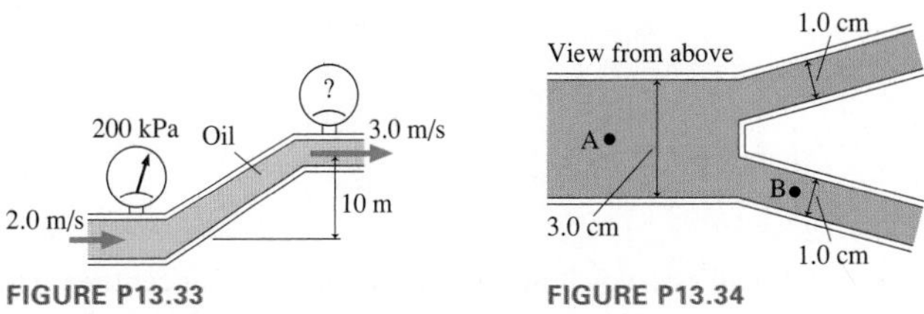

FIGURE P13.33

FIGURE P13.34

34. |||| The 3.0-cm-diameter water line in Figure P13.34 splits into two 1.0-cm-diameter pipes. All pipes are circular and at the same elevation. At point A, the water speed is 2.0 m/s and the gauge pressure is 50 kPa. What is the gauge pressure at point B?

35. ||| A rectangular trough, 2.0 m long, 0.60 m wide, and 0.45 m deep, is completely full of water. One end of the trough has a small drain plug right at the bottom edge. When you pull the plug, at what speed does water emerge from the hole?

Section 13.7 Viscosity and Poiseuille's Equation

36. |||| What pressure difference is required between the ends of a 2.0-m-long, 1.0-mm-diameter horizontal tube for 40°C water to flow through it at an average speed of 4.0 m/s?

37. |||| Water flows at 0.25 L/s through a 10-m-long garden hose 2.5 cm in diameter that is lying flat on the ground. The temperature of the water is 20°C. What is the gauge pressure of the water where it enters the hose?

38. || Figure P13.38 shows a water-filled syringe with a 4.0-cm-long needle. What is the gauge pressure of the water at the point P, where the needle meets the wider chamber of the syringe?

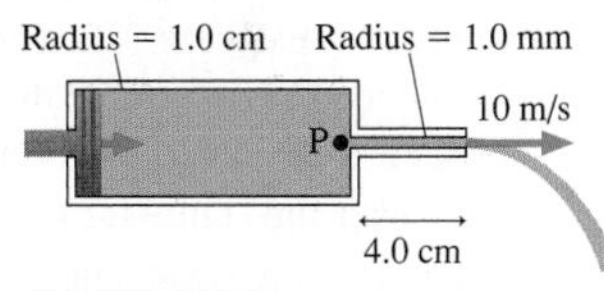

FIGURE P13.38

General Problems

39. || The density of gold is 19,300 kg/m^3. 197 g of gold is shaped into a cube. What is the length of each edge?

40. || The density of copper is 8920 kg/m^3. How many moles are in INT a 2.0 cm × 2.0 cm × 2.0 cm cube of copper?

41. || The density of aluminum is 2700 kg/m^3. How many atoms INT are in a 2.0 cm × 2.0 cm × 2.0 cm cube of aluminum?

42. || A 50-cm-thick layer of oil floats on a 120-cm-thick layer of water. What is the pressure at the bottom of the water layer?

43. |||| An oil layer floats on 85 cm of water in a tank. The absolute pressure at the bottom of the tank is 112.0 kPa. How thick is the oil?

44. || The little Dutch boy saved Holland by sticking his finger in the leaking dike. If the water level was 2.5 m above his finger, *estimate* the force of the water on his finger.

45. || a. In Figure P13.45, how much force does the fluid exert on the end of the cylinder at A?
b. How much force does the fluid exert on the end of the cylinder at B?

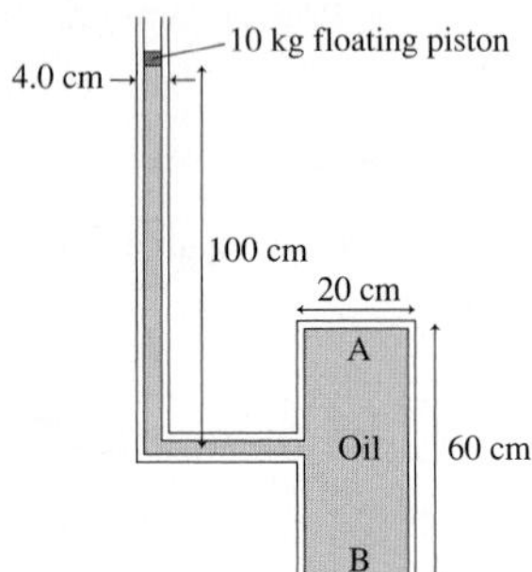

FIGURE P13.45

46. ||| A friend asks you how much pressure is in your car tires. You know that the tire manufacturer recommends 30 psi, but it's been a while since you've checked. You can't find a tire gauge in the car, but you do find the owner's manual and a ruler. Fortunately, you've just finished taking physics, so you tell your friend, "I don't know, but I can figure it out." From the owner's manual you find that the car's mass is 1500 kg. It seems reasonable to assume that each tire supports one-fourth of the weight. With the ruler you find that the tires are 15 cm wide and the flattened segment of the tire in contact with the road is 13 cm long. What answer will you give your friend?

47. ||| A diver 50 m deep in 10°C fresh water exhales a 1.0-cm-diameter bubble. What is the bubble's diameter just as it reaches the surface of the lake, where the water temperature is 20°C?
INT **Hint:** Assume that the air bubble is always in thermal equilibrium with the surrounding water.

48. || A 6.0-cm-tall cylinder floats in water with its axis perpendicular to the surface. The length of the cylinder above water is 2.0 cm. What is the cylinder's mass density?

49. || A sphere completely submerged in water is tethered to the bottom with a string. The tension in the string is one-third the weight of the sphere. What is the density of the sphere?

50. || You need to determine the density of a ceramic statue. If you suspend it from a spring scale, the scale reads 28.4 N. If you then lower the statue into a tub of water so that it is completely submerged, the scale reads 17.0 N. What is the density?

51. || A 5.0 kg rock whose density is 4800 kg/m^3 is suspended by a string such that half of the rock's volume is under water. What is the tension in the string?

52. || A flat slab of styrofoam, with a density of 32 kg/m^3, floats on a lake. What minimum volume must the slab have so that a 40 kg boy can sit on the slab without it sinking?

53. |||| A 2.0 mL syringe has an inner diameter of 6.0 mm, a needle
BIO inner diameter of 0.25 mm, and a plunger pad diameter (where you place your finger) of 1.2 cm. A nurse uses the syringe to inject medicine into a patient whose blood pressure is 140/100. Assume the liquid is an ideal fluid.
a. What is the minimum force the nurse needs to apply to the syringe?
b. The nurse empties the syringe in 2.0 s. What is the flow speed of the medicine through the needle?

54. ||| A child's water pistol shoots water through a 1.0-mm-diame-
INT ter hole. If the pistol is fired horizontally 70 cm above the ground, a squirt hits the ground 1.2 m away. What is the volume flow rate during the squirt? Ignore air resistance.

55. || The leaves of a tree lose water to the atmosphere via the
BIO process of *transpiration*. A particular tree loses water at the rate of 3×10^{-8} m^3/s; this water is replenished by the upward flow of sap through vessels in the trunk. This tree's trunk contains about 2000 vessels, each 100 μm in diameter. What is the speed of the sap flowing in the vessels?

56. || A hurricane wind blows across a 6.00 m $\times$ 15.0 m flat roof at a speed of 130 km/h.
a. Is the air pressure above the roof higher or lower than the pressure inside the house? Explain.
b. What is the pressure difference?
c. How much force is exerted on the roof? If the roof cannot withstand this much force, will it "blow in" or "blow out"?

57. |||| Water flows from the pipe shown in Figure P13.57 with a speed of 4.0 m/s.
a. What is the water pressure as it exits into the air?
b. What is the height h of the standing column of water?

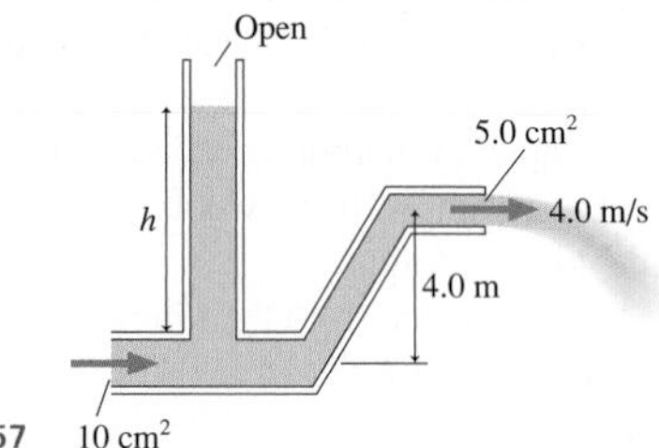

FIGURE P13.57

58. |||| Air flows through the tube shown in Figure P13.58. Assume that air is an ideal fluid.
a. What are the air speeds v_1 and v_2 at points 1 and 2?
b. What is the volume flow rate?

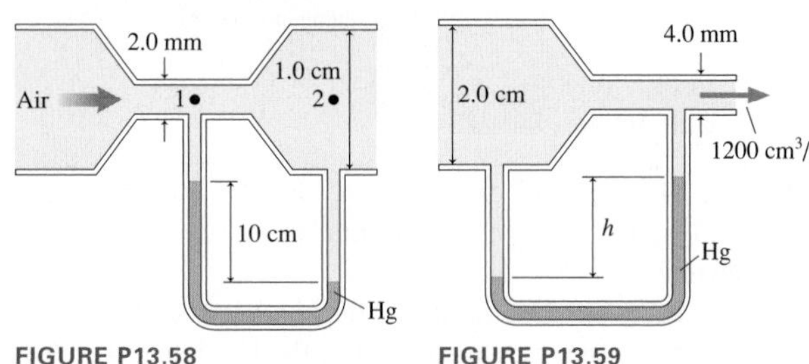

FIGURE P13.58 FIGURE P13.59

59. ||| Air flows through the tube shown in Figure P13.59 at a rate
INT of 1200 cm^3/s. Assume that air is an ideal fluid. What is the height h of mercury in the right side of the U-tube?

60. ||| Water flows at 5.0 L/s through a horizontal pipe that narrows smoothly from 10 cm diameter to 5.0 cm diameter. A pressure gauge in the narrow section reads 50 kPa. What is the reading of a pressure gauge in the wide section?

61. ||| The mercury manometer
INT shown in Figure P13.61 is attached to a gas cell. The mercury height h is 120 mm when the cell is placed in an ice-water mixture. The mercury height drops to 30 mm when the device is carried into an industrial freezer. What is the freezer temperature?
Hint: The right tube of the manometer is much narrower than the left tube. What reasonable assumption can you make about the gas volume?

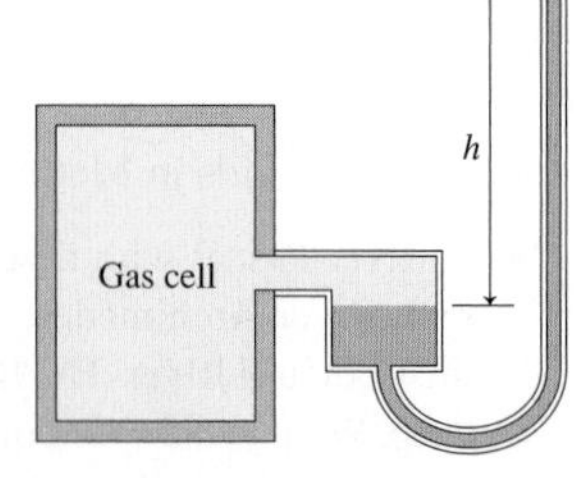

FIGURE P13.61

62. |||| Figure P13.62 shows a section of a long tube that narrows near its open end to a diameter of 1.0 mm. Water at 20°C flows out of the open end at 0.020 L/s. What is the gauge pressure at point P, where the diameter is 4.0 mm?

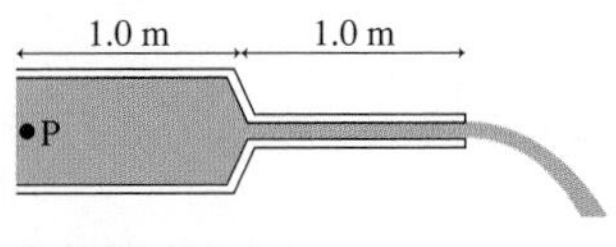

FIGURE P13.62

63. || Smoking tobacco is bad for your circulatory health. In an attempt to maintain the blood's capacity to deliver oxygen, the body increases its red blood cell production, and this increases the viscosity of the blood. In addition, nicotine from tobacco causes arteries to constrict. BIO

For a nonsmoker, with blood viscosity of 2.5×10^{-3} Pa · s, normal blood flow requires a pressure difference of 8.0 mm Hg between the two ends of an artery. If this person were to smoke regularly, his blood viscosity would increase to 2.7×10^{-3} Pa · s, and the arterial diameter would constrict to 90% of its normal value. What pressure difference would be needed to maintain the same blood flow?

64. ||| A stiff, 10-cm-long tube with an inner diameter of 3.0 mm is attached to a small hole in the side of a tall beaker. The tube sticks out horizontally. The beaker is filled with 20°C water to a level 45 cm above the hole, and it is continually topped off to maintain that level. What is the volume flow rate through the tube?

Passage Problems

Blood Pressure and Blood Flow BIO

The blood pressure at your heart is approximately 100 mm Hg. As blood is pumped from the left ventricle of your heart, it flows through the aorta, a single large blood vessel with a diameter of about 2.5 cm. The speed of blood flow in the aorta is about 60 cm/s. Any change in pressure as blood flows in the aorta is due to the change in height: the vessel is large enough that viscous drag is not a major factor. As the blood moves through the circulatory system, it flows into successively smaller and smaller blood vessels until it reaches the capillaries. Blood flows in the capillaries at the much lower speed of approximately 0.7 mm/s. The diameter of capillaries and other small blood vessels is so small that viscous drag is a major factor.

65. | There is a limit to how long your neck can be. If your neck were too long, no blood would reach your brain! What is the maximum height a person's brain could be above his heart, given the noted pressure and assuming that there are no valves or supplementary pumping mechanisms in the neck? The density of blood is 1060 kg/m^3.
A. 0.97 m B. 1.3 m C. 9.7 m D. 13 m

66. | Because the flow speed in your capillaries is much less than in the aorta, the total cross-section area of the capillaries considered together must be much larger than that of the aorta. Given the flow speeds noted, the total area of the capillaries considered together is equivalent to the cross-section area of a single vessel of approximately what diameter?
A. 25 cm B. 50 cm C. 75 cm D. 100 cm

67. | Suppose that in response to some stimulus a small blood vessel narrows to 90% its original diameter. If there is no change in the pressure across the vessel, what is the ratio of the new volume flow rate to the original flow rate?
A. 0.66 B. 0.73 C. 0.81 D. 0.90

68. | Sustained exercise can increase the blood flow rate of the heart by a factor of 5 with only a modest increase in blood pressure. This is a large change in flow. Although several factors come into play, which of the following physiological changes would most plausibly account for such a large increase in flow with a small change in pressure?
A. A decrease in the viscosity of the blood
B. Dilation of the smaller blood vessels to larger diameters
C. Dilation of the aorta to larger diameter
D. An increase in the oxygen carried by the blood

STOP TO THINK ANSWERS

Stop to Think 13.1: $\rho_1 = \rho_2 = \rho_3$. Density depends only on what the object is made of, not how big the pieces are.

Stop to Think 13.2: C. These are all open tubes, so the liquid rises to the same height in all three despite their different shapes.

Stop to Think 13.3: B. The weight of the displaced water equals the weight of the ice cube. When the ice cube melts and turns into water, that amount of water will exactly fill the volume that the ice cube is now displacing.

Stop to Think 13.4: 1 cm^3/s out. The fluid is incompressible, so the sum of what flows in must match the sum of what flows out. 13 cm^3/s is known to be flowing in while 12 cm^3/s flows out. An additional 1 cm^3/s must flow out to achieve balance.

Stop to Think 13.5: $h_2 > h_4 > h_3 > h_1$. The liquid level is higher where the pressure is lower. The pressure is lower where the flow speed is higher. The flow speed is highest in the narrowest tube, zero in the open air.

Stop to Think 13.6: A. All three segments have the same volume flow rate Q. According to Poiseuille's equation, the segment with the smallest radius R has the greatest pressure difference Δp.

PART III SUMMARY

Properties of Matter

The goal of Part III has been to understand macroscopic systems and their properties. We've introduced no new fundamental laws in these chapters. Instead, we've broadened and extended the scope of Newton's laws and the law of conservation of energy. In many ways, Part III has focused on *applications* of the general principles introduced in Parts I and II.

That matter exists in three phases—solids, liquids, gases—is perhaps the most basic fact about macroscopic systems. We looked at an atomic-level picture of the three phases of matter, and we also found a connection between *heat* and *phase changes.* Both liquids and gases are *fluids,* and we spent quite a bit of time investigating the properties of fluids, ranging from pressure and the ideal-gas law to buoyancy, fluid flow, and Bernoulli's equation.

Along the way, we introduced *mechanical properties* of matter, such as density and viscosity, and *thermal properties,* such as thermal-expansion coefficients and specific heats. You should now be able to use these properties to figure out what happens when you heat a system, whether an object will float, and how fast a fluid flows. These are all very practical, useful things to know, with many applications to the world around us.

Last, but not least, we discovered that many *macroscopic* properties are determined by the *microscopic* motions of atoms and molecules. For example, we found that the ideal-gas law is based on the idea that gas pressure is due to the collisions of atoms with the walls of the container. Similarly, regularities in the specific heats of gases were explained on the basis of how atoms and molecules share the thermal energy of a system. Atoms are important, even if we can't see them or follow their individual motions.

KNOWLEDGE STRUCTURE III Properties of Matter

BASIC GOAL How can we describe the macroscopic flows of energy and matter in heat transfer and fluid flow?

GENERAL PRINCIPLES **Phases of matter**

Gas: Particles are freely moving. Thermal energy is kinetic energy of motion of particles. Pressure is due to collisions of particles with walls of container.

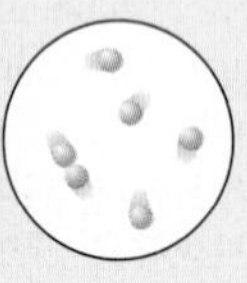

Liquid: Particles are loosely bound, but flow is possible. Heat is transferred by convection and conduction. Pressure is due to gravity.

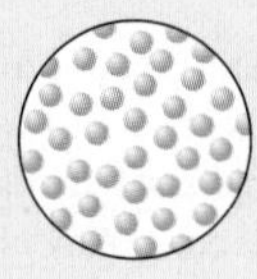

Solid: Particles are joined by spring-like bonds. Increasing temperature causes expansion. Heat is transferred by conduction only.

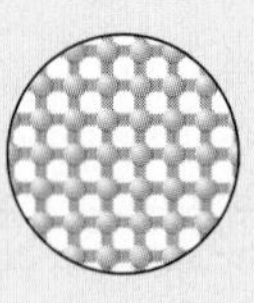

Heat must be added to change a solid to a liquid and a liquid to a gas. Heat must be removed to reverse these changes.

Ideal-gas processes

We can graph processes on a pV diagram. Work by the gas is the area under the graph.

For a gas process in a sealed container, the ideal-gas law relates values of pressure, volume, and temperature:

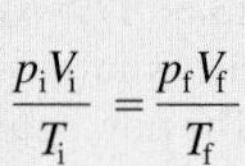

$$\frac{p_i V_i}{T_i} = \frac{p_f V_f}{T_f}$$

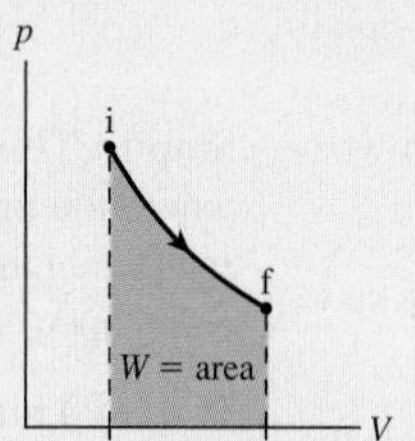

- Constant-volume process $V = \text{constant}$
- Isothermal process $T = \text{constant}$
- Isobaric process $p = \text{constant}$
- Adiabatic process $Q = 0$

Heat and heat transfer

The **specific heat** c of a material is the heat required to raise the temperature of 1 kg by 1 K:

$$Q = Mc\,\Delta T$$

When two or more systems interact thermally, they come to a common final temperature determined by

$$Q_{net} = Q_1 + Q_2 + Q_3 + \cdots = 0$$

Conduction transfers heat by direct physical contact.
Convection transfers heat by the motion of a fluid.
Radiation transfers heat by electromagnetic waves.

Fluid statics

Archimedes' principle: The magnitude of the buoyant force equals the weight of the fluid displaced by the object.

An object sinks if it is more dense than the fluid in which it is submerged; if it is less dense, it floats.

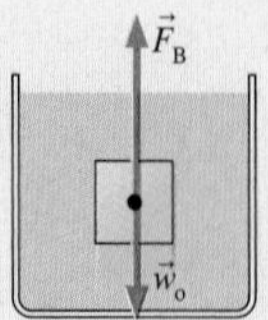

Pressure increases with depth in a liquid. The pressure at depth d is

$$p = p_0 + \rho g d$$

Fluid flow

Ideal fluid flow is laminar, incompressible, and nonviscous. The **equation of continuity** is

$$v_1 A_1 = v_2 A_2$$

Bernoulli's equation is

$$p_1 + \tfrac{1}{2}\rho v_1^2 + \rho g y_1 = p_2 + \tfrac{1}{2}\rho v_2^2 + \rho g y_2$$

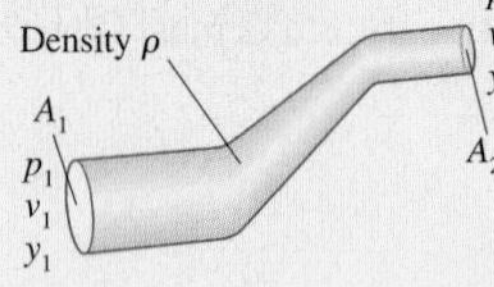

Size and Life

Physicists look for simple models and general principles that underlie and explain diverse physical phenomena. In the first 13 chapters of this book, you've seen that just a handful of general principles and laws can be used to solve a wide range of problems. Can this approach have any relevance to a subject like biology? It may seem surprising, but there *are* general "laws of biology" that apply, with quantitative accuracy, to organisms as diverse as elephants and mice.

Let's look at an example. An elephant uses more metabolic power than a mouse. This is not surprising, as an elephant is quite a bit bigger. But recasting the data shows an interesting trend. When we looked at the energy required to raise the temperature of different substances, we considered *specific heat*. The "specific" meant that we considered the heat required for 1 kilogram. For animals, rather than metabolic rate, we can look at the *specific metabolic rate,* the metabolic power used *per kilogram* of tissue. If we factor out the mass difference between a mouse and an elephant, are their specific metabolic powers the same?

In fact, the specific metabolic rate varies quite a bit among mammals, as the graph of specific metabolic rate versus mass shows. But there is an interesting trend: All of the data points lie on a single smooth curve. In other words, there really is a biological *law* we can use to predict a mammal's metabolic rate knowing only its mass M. In particular, the specific metabolic rate is proportional to $M^{-0.25}$. Because a 4000 kg elephant is 160,000 times more massive than a 25 g mouse, the mouse's specific metabolic power is $(160{,}000)^{0.25} = 20$ times that of the elephant. A law that shows how a property scales with the size of a system is called a *scaling law*.

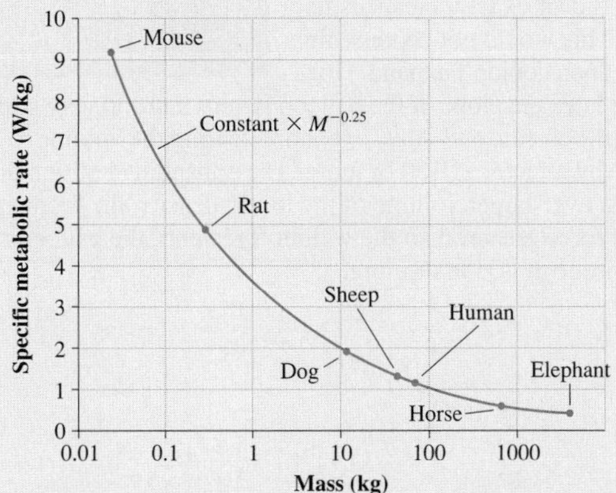

Specific metabolic rate as a function of body mass follows a simple scaling law.

A similar scaling law holds for birds, reptiles, and even bacteria. Why should a single simple relationship hold true for organisms that range in size from a single cell to a 100 ton blue whale? Interestingly, no one knows for sure. It is a matter of current research to find out just what this and other scaling laws tell us about the nature of life.

Perhaps the metabolic-power scaling law is a result of heat transfer. In Chapter 12, we noted that all metabolic energy used by an animal ends up as heat, which must be transferred to the environment. A 4000 kg elephant has 160,000 times the mass of a 25 g mouse, but it has only about 3,000 times the surface area. The heat transferred to the environment depends on the surface area; the more surface area, the greater the rate of heat transfer. An elephant with a mouse-sized metabolism simply wouldn't be able to dissipate heat fast enough—it would quickly overheat and die.

If heat dissipation were the only factor limiting metabolism, we can show that the specific metabolic rate should scale as $M^{-0.33}$, quite different from the $M^{-0.25}$ scaling observed. Clearly, another factor is at work. Exactly what underlies the $M^{-0.25}$ scaling is still a matter of debate, but some recent analysis suggests the scaling is due to limitations not of heat transfer but of fluid flow. Cells in mice, elephants, and all mammals receive nutrients and oxygen for metabolism from the bloodstream. Because the minimum size of a capillary is about the same for all mammals, the structure of the circulatory system must vary from animal to animal. The human aorta has a diameter of about 1 inch; in a mouse, the diameter is approximately 1/20th of this. Thus a mouse has fewer levels of branching to smaller and smaller blood vessels as we move from the aorta to the capillaries. The smaller blood vessels in mice mean that viscosity is more of a factor throughout the circulatory system. The circulatory system of a mouse is quite different from that of an elephant.

A model of specific metabolic rate based on blood-flow limitations predicts a $M^{-0.25}$ law, exactly as observed. The model also makes other testable predictions. For example, the model predicts that the smallest possible mammal should have a body mass of about 1 gram—exactly the size of the smallest shrew. Even smaller animals have different types of circulatory systems; in the smallest animals, nutrient transport is by diffusion alone. But the model can be extended to predict that the specific metabolic rate for these animals will follow a scaling law similar to that for mammals, exactly as observed. It is too soon to know if this model will ultimately prove to be correct, but it's indisputable that there are large-scale regularities in biology that follow mathematical relationships based on the laws of physics.

PART III PROBLEMS

The following questions are related to the passage "Size and Life" on the previous page. BIO

1. A typical timber wolf has a mass of 40 kg, a typical jackrabbit a mass of 2.5 kg. Given the scaling law presented in the passage, we'd expect the specific metabolic rate of the jackrabbit to be higher by a factor of
   A. 2 B. 4 C. 8 D. 16
2. A typical timber wolf has a mass of 40 kg, a typical jackrabbit a mass of 2.5 kg. Given the scaling law presented in the passage, we'd expect the wolf to use _______ times more energy than a jackrabbit in the course of a day.
   A. 2 B. 4 C. 8 D. 16
3. Given the data of the graph, approximately how much energy, in Calories, would a 200 g rat use during the course of a day?
   A. 10 B. 20 C. 100 D. 200
4. All other things being equal, species that inhabit cold climates tend to be larger than related species that inhabit hot climates. For instance, the Alaskan hare is the largest North American hare, with a typical mass of 5.0 kg, double that of a jackrabbit. A likely explanation is that
   A. Larger animals have more blood flow, allowing for better thermoregulation.
   B. Larger animals need less food to survive than smaller animals.
   C. Larger animals have larger blood volumes than smaller animals.
   D. Larger animals lose heat less quickly than smaller animals.
5. The passage proposes that there are quantitative "laws" of biology that have their basis in physical principles, using the scaling of specific metabolic rate with body mass as an example. Which of the following regularities among animals might also be an example of such a "law"?
   A. As a group, birds have better color vision than mammals.
   B. Reptiles have a much lower specific metabolic rate than mammals.
   C. Predators tend to have very good binocular vision; prey animals tend to be able to see over a very wide angle.
   D. Jump height varies very little among animals. Nearly all animals, ranging in size from a flea to a horse, have a maximum vertical leap that is quite similar.

The following passages and associated questions are based on the material of Part III.

Keeping Your Cool BIO

A 68 kg cyclist is pedaling down the road at 15 km/h, using a total metabolic power of 480 W. A certain fraction of this energy is used to move the bicycle forward, but the balance ends up as thermal energy in his body, which he must get rid of to keep cool. On a very warm day, conduction, convection, and radiation transfer little energy, and so he does this by perspiring, with the evaporation of water taking away the excess thermal energy.

6. If the cyclist reaches his 15 km/h cruising speed by rolling down a hill, what is the approximate height of the hill?
   A. 22 m B. 11 m C. 2 m D. 1 m
7. As he cycles at a constant speed on level ground, at what rate is chemical energy being converted to thermal energy in his body, assuming a typical efficiency of 25% for the conversion of chemical energy to the mechanical energy of motion?
   A. 480 W B. 360 W C. 240 W D. 120 W
8. To keep from overheating, the cyclist must get rid of the excess thermal energy generated in his body. If he cycles at this rate for 2 hours, how many liters of water must he perspire, to the nearest 0.1 liter?
   A. 0.4 L B. 0.9 L C. 1.1 L D. 1.4 L
9. Being able to exhaust this thermal energy is very important. If he isn't able to get rid of any of the excess heat, by how much will the temperature of his body increase in 10 minutes of riding, to the nearest 0.1°C?
   A. 0.3°C B. 0.6°C C. 0.9°C D. 1.2°C

Weather Balloons

The data used to generate weather forecasts are gathered by hundreds of weather balloons launched from sites throughout the world. A typical balloon is made of latex and filled with hydrogen. A packet of sensing instruments (called a *radiosonde*) transmits information back to earth as the balloon rises into the atmosphere.

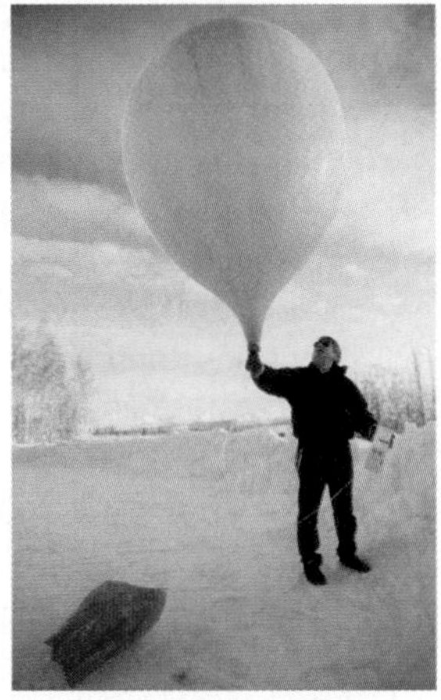

At the beginning of its flight, the average density of the weather balloon package (total mass of the balloon plus cargo divided by their volume) is less than the density of the surrounding air, so the balloon rises. As it does, the density of the surrounding air decreases, as shown in Figure III.1. The balloon will rise to the point at which the buoyant force of the air exactly balances its weight. This would not be very high if the balloon couldn't expand. However, the latex envelope of the balloon is very thin and very stretchy, so the balloon can, and does, expand, allowing the volume to increase by a factor of 100 or more. The expanding balloon displaces an ever-larger volume of the lower-density air, keeping the buoyant force greater than the weight force until the balloon rises to an altitude of 40 km or more.

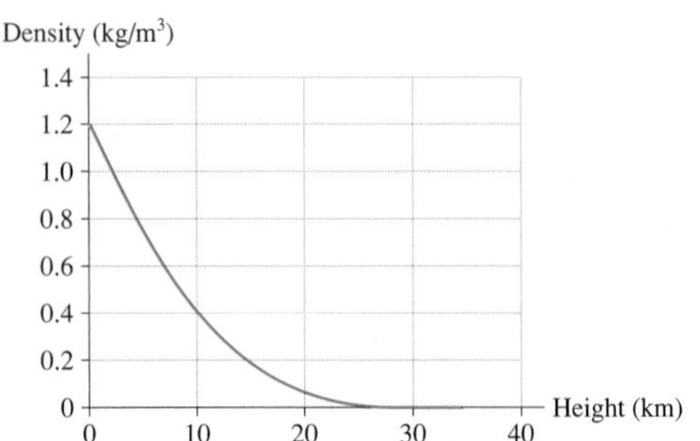

FIGURE III.1

10. A balloon launched from sea level has a volume of approximately 4 m^3. What is the approximate buoyant force on the balloon?
A. 50 N B. 40 N
C. 20 N D. 10 N
11. A balloon launched from sea level with a volume of 4 m^3 will have a volume of about 12 m^3 on reaching an altitude of 10 km. What is the approximate buoyant force now?
A. 50 N B. 40 N
C. 20 N D. 10 N
12. The balloon expands as it rises, keeping the pressures inside and outside the balloon approximately equal. If the balloon rises slowly, heat transfers will keep the temperature inside the same as the outside air temperature. A balloon with a volume of 4.0 m^3 is launched at sea level, where the atmospheric pressure is 100 kPa and the temperature is 15°C. It then rises slowly to a height of 5500 m, where the pressure is 50 kPa and the temperature is –20°C. What is the volume of the balloon at this altitude?
A. 5.0 m^3 B. 6.0 m^3
C. 7.0 m^3 D. 8.0 m^3
13. If the balloon rises quickly, so that no heat transfer is possible, the temperature inside the balloon will drop as the gas expands. If a 4.0 m^3 balloon is launched at a pressure of 100 kPa and rapidly rises to a point where the pressure is 50 kPa, the volume of the balloon will be
A. Greater than 8.0 m^3
B. 8.0 m^3
C. Less than 8.0 m^3
14. At the end of the flight, the radiosonde is dropped and falls to earth by parachute. Suppose the parachute achieves its terminal speed at a height of 30 km. As it descends into the atmosphere, how does the terminal speed change?
A. It increases.
B. It stays the same.
C. It decreases.

Passenger Balloons

Long-distance balloon flights are usually made using a hot-air-balloon/helium-balloon hybrid. The balloon has a sealed, flexible chamber of helium gas that expands or contracts to keep the helium pressure approximately equal to the air pressure outside. The helium chamber sits on top of an open (that is, air can enter or leave), constant-volume chamber of propane-heated air. Assume that the hot air and the helium are kept at a constant temperature by burning propane.

15. A balloon is launched at sea level, where the air pressure is 100 kPa. The helium has a volume of 1000 m^3 at this altitude. What is the volume of the helium when the balloon has risen to a height where the atmospheric pressure is 33 kPa?
A. 330 m^3 B. 500 m^3 C. 1000 m^3 D. 3000 m^3
16. A balloon is launched at sea level, where the air pressure is 100 kPa. The density in the hot-air chamber is 1.0 kg/m^3. What is the density of the air when the balloon has risen to a height where the atmospheric pressure is 33 kPa?
A. 3.0 kg/m^3
B. 1.0 kg/m^3
C. 0.66 kg/m^3
D. 0.33 kg/m^3
17. A balloon is at a height of 5.0 km and is descending at a constant rate. The buoyancy force is directed ________; the drag force is directed ________.
A. Up, up
B. Up, down
C. Down, up
D. Down, down

Additional Integrated Problems

18. When you exhale, all of the air in your lungs must exit through the trachea. If you exhale through your nose, this air subsequently leaves through your nostrils. The area of your nostrils is less than that of your trachea. How does the speed of the air in the trachea compare to that in the nostrils? BIO
19. Sneezing requires an increase in pressure of the air in the lungs; a typical sneeze might result in an extra pressure of 7.0 kPa. Estimate how much force this exerts on the diaphragm, the large muscle at the bottom of the ribcage. BIO
20. A 20 kg block of aluminum sits on the bottom of a tank of water. How much force does the block exert on the bottom of the tank?
21. We've seen that fish can control their buoyancy through the use of a swim bladder, a gas-filled organ inside the body. You can assume that the gas pressure inside the swim bladder is roughly equal to the external water pressure. A fish swimming at a particular depth adjusts the volume of its swim bladder to give it neutral buoyancy. If the fish swims upward or downward, the changing water pressure causes the bladder to expand or contract. Consequently, the fish must adjust the quantity of gas to restore the original volume and thus reestablish neutral buoyancy. Consider a large, 7.0 kg striped bass with a volume of 7.0 L. When neutrally buoyant, 7.0% of the fish's volume is taken up by the swim bladder. Assume a body temperature of 15°C. BIO
a. How many moles of air are in the swim bladder when the fish is at a depth of 80 ft?
b. What will the volume of the swim bladder be if the fish ascends to a 50 ft depth without changing the quantity of gas?
c. To return the swim bladder to its original size, how many moles of gas must be added?

PART IV

Oscillations and Waves

Wolves are social animals, and they howl to communicate over distances of several miles with other members of their pack. How are such sounds made? How do they travel through the air? And how are other wolves able to hear these sounds from such a great distance?

Motion That Repeats Again and Again

Up to this point in the book, we have generally considered processes that have a clear starting and ending point, such as a car accelerating from rest to a final speed, or a solid being heated from an initial to a final temperature. In Part IV, we begin to consider processes that are *periodic*—they repeat. A child on a swing, a boat bobbing on the water, and even the repetitive bass beat of a rock song are *oscillatory motions* that happen over and over without a starting or ending point. The *period*, the time for one cycle of the motion, will be a key parameter for us to consider as we look at oscillatory motion.

Our first goal will be to develop the language and tools needed to describe oscillations, ranging from the swinging of the bob of a pendulum clock to the bouncing of a car on its springs. Once we understand oscillations, we will extend our analysis to consider oscillations that travel—*waves*.

The Wave Model

We've had great success modeling the motion of complex objects as the motion of one or more particles. We were even able to explain the macroscopic properties of matter, such as pressure and temperature, in terms of the motion of the atomic particles that comprise all matter.

Now it's time to explore another way of looking at nature, the *wave model*. Familiar examples of waves include

- Ripples on a pond.
- The sound of thunder.
- The swaying ground of an earthquake.
- A vibrating guitar string.
- The colors of a rainbow.

Despite the great diversity of types and sources of waves, there is a single, elegant physical theory that is capable of describing them all. Our exploration of wave phenomena will call upon water waves, sound waves, and light waves for examples, but our goal will be to emphasize the unity and coherence of the ideas that are common to *all* types of waves. As was the case with the particle model, we will use the wave model to explain a wide range of phenomena.

When Waves Collide

The collision of two particles is a dramatic event. Energy and momentum are transferred as the two particles head off in different directions. Something much gentler happens when two waves come together—the two waves pass through each other unchanged. Where they overlap, we get a *superposition* of the two waves. We will finish our discussion of waves by analyzing the standing waves that result from the superposition of two waves traveling in opposite directions. The physics of standing waves will allow us to understand how your vocal tract can produce such a wide range of sounds, and how your ears are able to analyze them.

14 Oscillations

A gibbon swings below a branch, moving forward by pivoting about one handhold and then the next. How does an understanding of oscillatory motion allow us to analyze the swinging motion of a gibbon—and the walking motion of other animals?

LOOKING AHEAD ▸▸

The goal of Chapter 14 is to understand systems that oscillate with simple harmonic motion.

Simple Harmonic Motion

A cart attached to a spring oscillates back and forth. A graph of its motion is sinusoidal. This is a special kind of motion called **simple harmonic motion.**

Oscillation

An understanding of simple harmonic motion will build on many topics from past chapters.

Looking Back ◂◂

3.8 Period and frequency
6.2 Velocity and acceleration in circular motion
8.3 Springs and restoring forces
10.1 and10.6 Energy transformations and the conservation of energy

Spring Systems

Oscillations occur in any system having a restoring force that pushes the system back toward an equilibrium position.

A person bouncing up and down on elastic cords and the swaying of a tall building in the wind are both examples of simple harmonic motion.

An oscillation is characterized by the **period** (the time for one oscillation) and the **amplitude** (the size of the oscillation).

Pendulum Systems

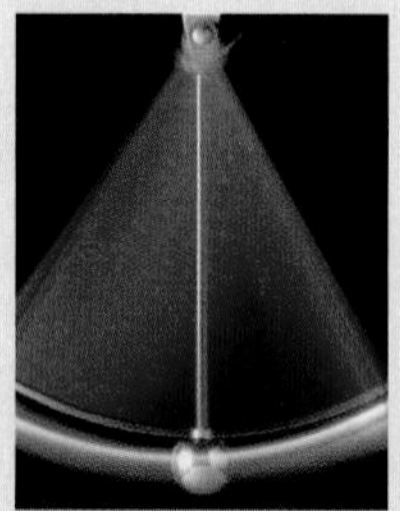

A mass swinging on the end of a rod or a cord is a **pendulum** —and another example of simple harmonic motion. Its motion is mathematically the same as that of a mass on a spring.

The period of a pendulum is determined by the pendulum length and the strength of gravity; the amplitude doesn't affect the period. This makes a pendulum the ideal basis for a clock.

Damping and Resonance

As time goes on, oscillating systems may lose energy. This **damping** results in a slow decay of the oscillation.

The pendulum swings back and forth, tracing a pattern in the sand. The drag from the sand makes each oscillation smaller than the last.

The tuning fork oscillates at a particular frequency. There is one particular spot on a membrane in the inner ear that also oscillates at this exact frequency.

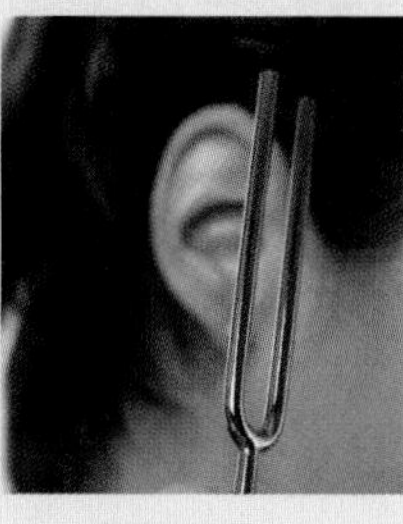

We'll see that the concept of **resonance** explains how your ear can distinguish different frequencies.

14.1 Equilibrium and Oscillation

Consider a marble that is free to roll inside a spherical bowl, as shown in FIGURE 14.1. The marble has an **equilibrium position** at the bottom of the bowl where it will rest with no net force on it. If you push the marble away from equilibrium, the marble's weight leads to a net force directed back toward the equilibrium position. We call this a **restoring force** because it acts to restore equilibrium. The magnitude of this restoring force increases if the marble is moved farther away from the equilibrium position.

If you pull the marble to the side and release it, it doesn't just roll back to the bottom of the bowl and stay put. It keeps on moving, rolling up and down each side of the bowl, repeatedly moving through its equilibrium position, as we see in FIGURE 14.2. We call such repetitive motion an **oscillation.** This oscillation is a result of an interplay between the restoring force and the marble's inertia, something we will see in all of the oscillations we consider.

We'll start our description by noting the most important fact about oscillatory motion: It repeats. Any oscillation is characterized by a *period,* the time for the motion to repeat. We met the concepts of period and frequency when we studied circular motion in Chapter 6. As a starting point then, let's review these ideas and see how they apply to oscillatory motion.

FIGURE 14.1 Equilibrium and restoring forces for a ball in a bowl.

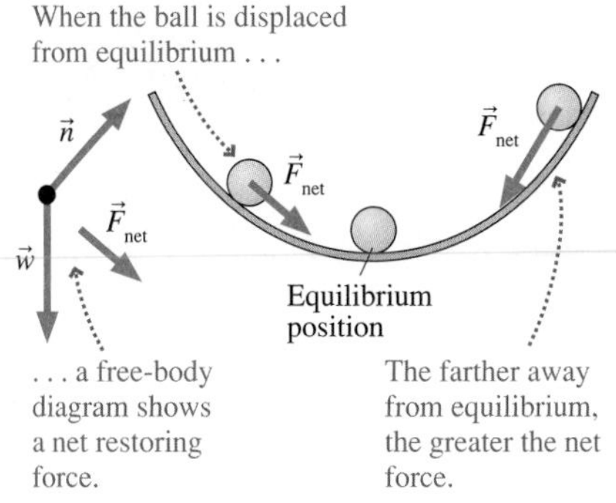

FIGURE 14.2 The motion of a ball rolling in a bowl.

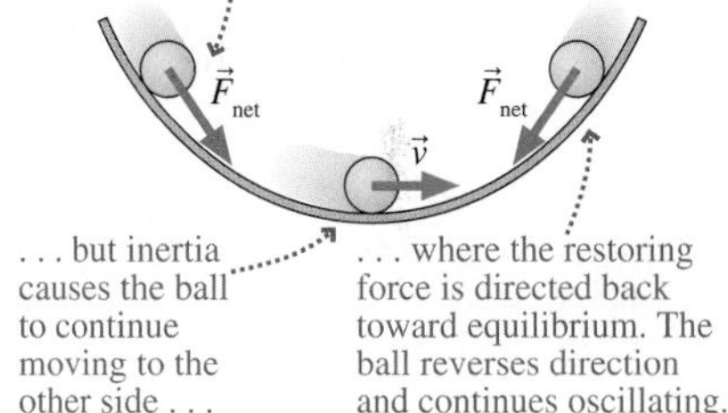

When the ball is moved from equilibrium and released, a restoring force pulls it back toward equilibrium . . .
$\vec{F}_{net}$
$\vec{F}_{net}$
$\vec{v}$
. . . but inertia causes the ball to continue moving to the other side . . .
. . . where the restoring force is directed back toward equilibrium. The ball reverses direction and continues oscillating.

Frequency and Period

An electrocardiogram (ECG), such as the one shown in FIGURE 14.3, is a record of the electrical signals of the heart as it beats. We will explore the ECG in some detail in Chapter 21. Although the shape of a typical ECG is rather complex, notice that it has a *repeating pattern.* For this or any oscillation, the time to complete one full cycle is called the **period** of the oscillation. Period is given the symbol T.

An equivalent piece of information is the number of cycles, or oscillations, completed per second. If the period is $\frac{1}{10}$ s, then the oscillator can complete 10 cycles in 1 second. Conversely, an oscillation period of 10 s allows only $\frac{1}{10}$ of a cycle to be completed per second. In general, T seconds per cycle implies that $1/T$ cycles will be completed each second. The number of cycles per second is called the **frequency** f of the oscillation. The relationship between frequency and period is therefore

$$f = \frac{1}{T} \quad \text{or} \quad T = \frac{1}{f} \tag{14.1}$$

The units of frequency are **hertz,** abbreviated Hz. By definition,

$$1 \text{ Hz} = 1 \text{ cycle per second} = 1 \text{ s}^{-1}$$

We will frequently deal with very rapid oscillations and make use of the units shown in Table 14.1.

NOTE ► Uppercase and lowercase letters *are* important. 1 MHz is 1 megahertz = 10^6 Hz, but 1 mHz is 1 millihertz = 10^{-3} Hz! ◄

FIGURE 14.3 An electrocardiogram has a well-defined period.

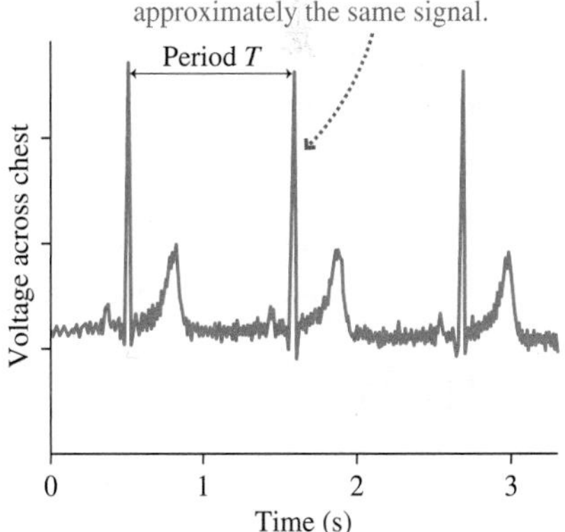

EXAMPLE 14.1 Frequency and period of a radio station

An FM radio station broadcasts an oscillating radio wave at a frequency of 100 MHz. What is the period of the oscillation?

SOLVE The frequency f of oscillations in the radio transmitter is 100 MHz = 1.0×10^8 Hz. The period is the inverse of the frequency; hence,

$$T = \frac{1}{f} = \frac{1}{1.0 \times 10^8 \text{ Hz}} = 1.0 \times 10^{-8} \text{ s} = 10 \text{ ns}$$

TABLE 14.1 Common units of frequency

Frequency	Period
10^3 Hz = 1 kilohertz = 1 kHz	1 ms
10^6 Hz = 1 megahertz = 1 MHz	1 μs
10^9 Hz = 1 gigahertz = 1 GHz	1 ns

Oscillatory Motion

Let's return to the marble in the bowl and describe its motion in more detail. We'll start by making a graph of the motion, with positions to the right of equilibrium positive and positions to the left of equilibrium negative. FIGURE 14.4 shows a series of "snapshots" of the motion and the corresponding points on the graph. This graph has the form of a *cosine function.* A graph or a function that has the form of a sine or cosine function is called **sinusoidal.** A sinusoidal oscillation is called **simple harmonic motion,** often abbreviated SHM.

FIGURE 14.4 Constructing a position-versus-time graph for a marble rolling in a bowl.

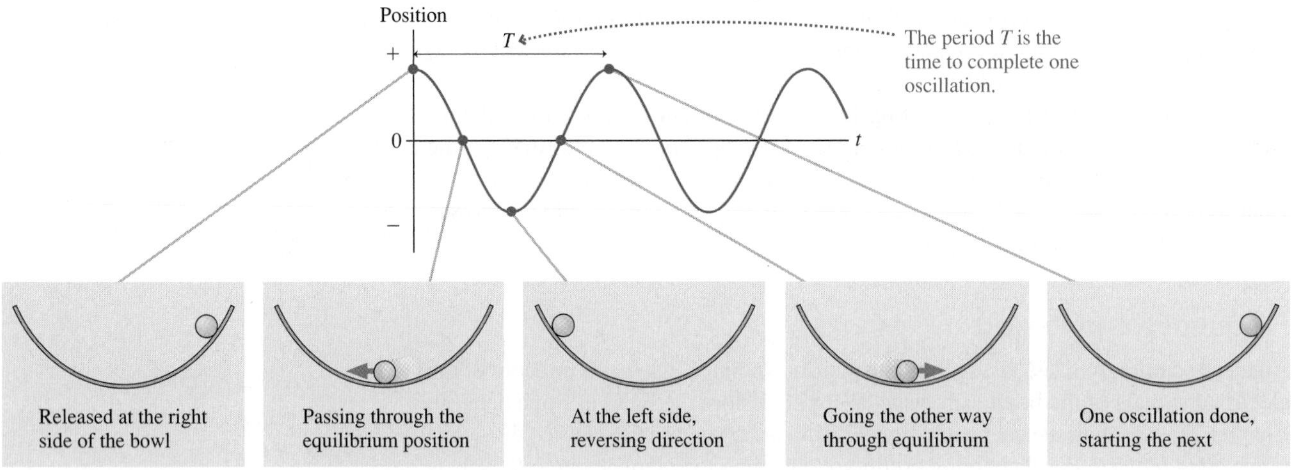

A marble rolling in the bottom of a bowl undergoes simple harmonic motion, as does a car bouncing up and down on its springs. SHM is very common, but we'll find that most cases of SHM can be modeled as one of two simple systems: a mass oscillating on a spring or a pendulum swinging back and forth. The following table shows two examples.

Examples of simple harmonic motion

Oscillating system		Related real-world example	
Mass on a spring 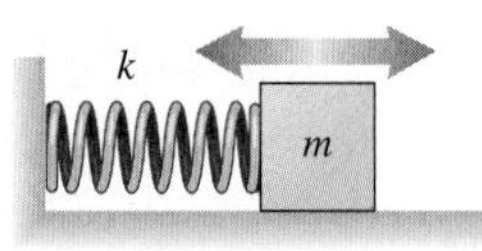	The mass oscillates back and forth due to the restoring force of the spring. The period depends on the mass and the stiffness of the spring.	Vibrations in the ear 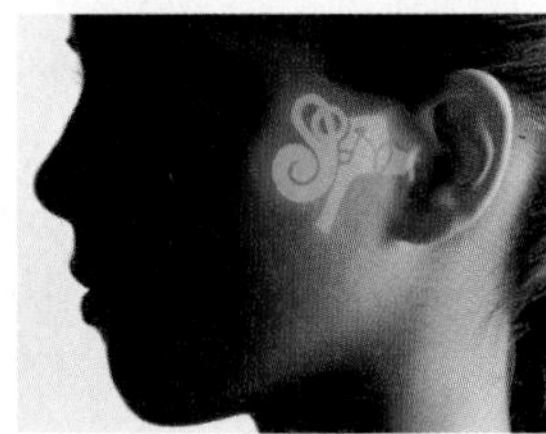	Sound waves entering the ear cause the oscillation of a membrane in the cochlea. The vibration can be modeled as a mass on a spring. The period of oscillation of a segment of the membrane depends on mass (the thickness of the membrane) and stiffness (the rigidity of the membrane).
Pendulum 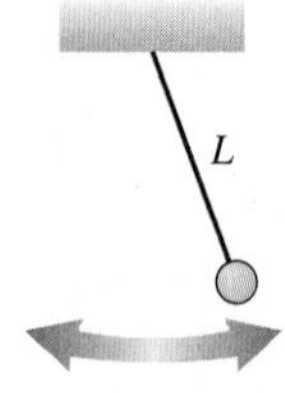	The mass oscillates back and forth due to the restoring gravitational force. The period depends on the length of the pendulum and the free-fall acceleration g.	Motion of legs while walking 	The motion of a walking animal's legs can be modeled as pendulum motion. The rate at which the legs swing depends on the length of the legs and the free-fall acceleration g.

STOP TO THINK 14.1 Two oscillating systems have periods T_1 and T_2, with $T_1 < T_2$. How are the frequencies of the two systems related?

A. $f_1 < f_2$ B. $f_1 = f_2$ C. $f_1 > f_2$

14.2 Linear Restoring Forces and Simple Harmonic Motion

Simple harmonic motion can occur when a system has a restoring force that pushes it back toward equilibrium. We will begin our study with a very simple system, analyzing it in detail before moving on to other oscillatory systems.

FIGURE 14.5 shows a glider that rides with very little friction on an air track. There is a spring connecting the glider to the end of the track. When the spring is neither stretched nor compressed, the net force on the glider is zero. The glider just sits there—this is the equilibrium position.

If the glider is now displaced from this equilibrium position by Δx, the spring exerts a force back toward equilibrium—a restoring force. In Section 8.3, we found that the spring force is given by Hooke's law: $(F_{sp})_x = -k\,\Delta x$, where k is the spring constant. (Recall that a "stiffer" spring has a larger value of k.) If we set the origin of our coordinate system at the equilibrium position, the displacement from equilibrium Δx is equal to x; thus the spring force can be written as $(F_{sp})_x = -kx$.

The net force on the glider is simply the spring force, so we can write

$$(F_{net})_x = -kx \tag{14.2}$$

$(F_{net})_x$ x p.37 PROPORTIONAL

The negative sign tells us that this is a restoring force because the force is in the direction opposite the displacement. If we pull the cart to the right (x is positive), the force is to the left (negative)—back toward equilibrium.

This is a **linear restoring force;** that is, **the net force is toward the equilibrium position and is proportional to the distance from equilibrium.**

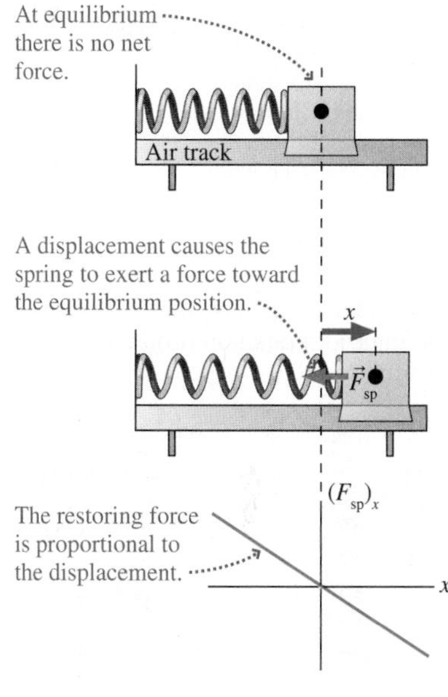

At equilibrium there is no net force.
Air track
A displacement causes the spring to exert a force toward the equilibrium position.
x
$\vec{F}_{sp}$
$(F_{sp})_x$
The restoring force is proportional to the displacement.
x

FIGURE 14.5 The restoring force on an air-track glider attached to a spring.

Motion of a Mass on a Spring

If we pull the air-track glider of Figure 14.5 a short distance to the right and release it, it will oscillate back and forth. **FIGURE 14.6** shows actual data from an experiment in which the position of a glider was measured 20 times every second. This is a position-versus-time graph that has been rotated 90° from its usual orientation in order for the x-axis to match the motion of the glider.

The object's maximum displacement from equilibrium is called the **amplitude** A of the motion. The object's position oscillates between $x = -A$ and $x = +A$.

NOTE ► When interpreting a graph, notice that the amplitude is the distance from equilibrium to the maximum, *not* the distance from the minimum to the maximum. ◄

The graph of the position is sinusoidal, so this is simple harmonic motion. **Oscillation about an equilibrium position with a linear restoring force is always simple harmonic motion.** There are many other circumstances in which a linear restoring force exists, and we will see simple harmonic motion in these cases as well.

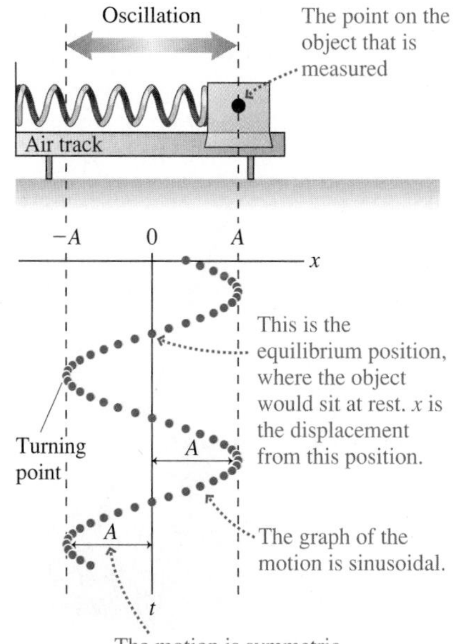

FIGURE 14.6 An experiment showing the oscillation of an air-track glider.

Vertical Mass on a Spring

FIGURE 14.7 on the following page shows a block of mass m hanging from a spring with spring constant k. Will this mass-spring system have the same simple harmonic

FIGURE 14.7 The equilibrium position of a mass on a vertical spring.

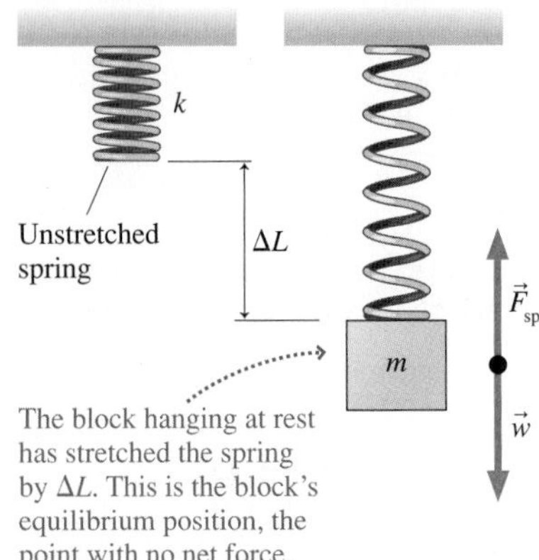

k
Unstretched spring
ΔL
m
F_sp
w
The block hanging at rest has stretched the spring by ΔL. This is the block's equilibrium position, the point with no net force.

motion as the horizontal system we just saw, or will gravity add an additional complication? An important fact to notice is that the equilibrium position of the block is *not* where the spring is at its unstretched length. At the equilibrium position of the block, where it hangs motionless, the spring has stretched by ΔL.

Finding ΔL is a static-equilibrium problem in which the upward spring force balances the downward weight force of the block. The y-component of the spring force is given by Hooke's law:

$$(F_{\text{sp}})_y = k\,\Delta L \tag{14.3}$$

Newton's first law for the block in equilibrium is

$$(F_{\text{net}})_y = (F_{\text{sp}})_y + w_y = k\,\Delta L - mg = 0 \tag{14.4}$$

from which we can find

$$\Delta L = \frac{mg}{k} \tag{14.5}$$

This is the distance the spring stretches when the block is attached to it.

FIGURE 14.8 Displacing the block from its equilibrium position produces a restoring force.

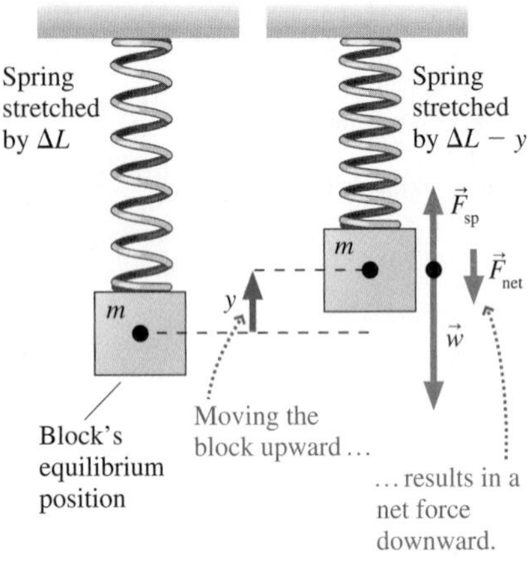

Suppose we now displace the block from this equilibrium position, as shown in FIGURE 14.8. We've placed the origin of the y-axis at the block's equilibrium position in order to be consistent with our analyses of oscillations throughout this chapter. If the block moves upward, as in the figure, the spring gets shorter compared to its equilibrium length, but the spring is still *stretched* compared to its unstretched length in Figure 14.7. When the block is at position y, the spring is stretched by an amount $\Delta L - y$ and hence exerts an *upward* spring force $F_{\text{sp}} = k(\Delta L - y)$. The net force on the block at this point is

$$(F_{\text{net}})_y = (F_{\text{sp}})_y + w_y = k(\Delta L - y) - mg = (k\,\Delta L - mg) - ky \tag{14.6}$$

But $k\,\Delta L - mg = 0$, from Equation 14.4, so the net force on the block is

$$(F_{\text{net}})_y = -ky \tag{14.7}$$

Equation 14.7 for a mass hung from a spring has the same form as Equation 14.2 for the horizontal spring, where we found $(F_{\text{net}})_x = -kx$. That is, the restoring force for vertical oscillations is identical to the restoring force for horizontal oscillations. **The role of gravity is to determine where the equilibrium position is, but it doesn't affect the restoring force for displacement from the equilibrium position.** Because it has a linear restoring force, **a mass on a vertical spring oscillates with simple harmonic motion.** The motion has the same form as that of the air-track glider.

The Pendulum

The chapter opened with a picture of a gibbon, whose body swings back and forth below a tree branch. The motion of the gibbon is essentially that of a **pendulum,** a mass suspended from a pivot point by a light string or rod. A pendulum oscillates about its equilibrium position, but is this simple harmonic motion? To answer this question, we need to examine the restoring force on the pendulum. If the restoring force is linear, the motion will be simple harmonic.

Pendulum motion at the playground.

FIGURE 14.9a shows a mass m attached to a string of length L and free to swing back and forth. The pendulum's position can be described by either the arc of length s or the angle θ, both of which are zero when the pendulum hangs straight down. Because angles are measured counterclockwise, s and θ are positive when the pendulum is to the right of center, negative when it is to the left.

Two forces are acting on the mass: the string tension $\vec{T}$ and the weight $\vec{w}$. The motion is along a circular arc. We choose a coordinate system on the mass with one axis along the radius of the circle and the other tangent to the circle. We divide the forces into tangential components, parallel to the motion (denoted with a subscript t), and components directed toward the center of the circle. These are shown on the free-body diagram of **FIGURE 14.9b**.

The mass must move along a circular arc, as noted. As Figure 14.9b shows, the net force in this direction is the tangential component of the weight:

$$(F_{\text{net}})_t = \sum F_t = w_t = -mg\sin\theta \qquad (14.8)$$

This is the restoring force pulling the mass back toward the equilibrium position.

This equation becomes much simpler if we restrict the pendulum's oscillations to *small angles* of about 10° (0.17 rad) or less. For such small angles, if the angle θ is in radians, it turns out that

$$\sin\theta \approx \theta$$

This result, which applies only when the angle is in radians, is called the **small-angle approximation.**

NOTE ► You can check this approximation on your calculator. Put your calculator in radian mode, enter some values that are less than 0.17 rad, take the sine, and see what you get. ◄

Recall from Chapter 6 that the angle is related to the arc length by $\theta = s/L$. Using this and the small-angle approximation, we can write the restoring force as

$$(F_{\text{net}})_t = -mg\sin\theta \approx -mg\theta = -mg\frac{s}{L} = -\left(\frac{mg}{L}\right)s \qquad (14.9)$$

The net force is directed toward the equilibrium position, and it is linearly proportional to the displacement s from equilibrium. In other words, **the force on a pendulum is a linear restoring force for small angles, so the pendulum will undergo simple harmonic motion.**

Whenever there is a linear restoring force, there can be simple harmonic motion. There are many examples beyond the few we have considered so far, such as sloshing water in a cup or the vibration of a bridge or a building swaying in the wind. We will see these and other examples in the coming sections.

FIGURE 14.9 Describing the motion of and force on a pendulum.

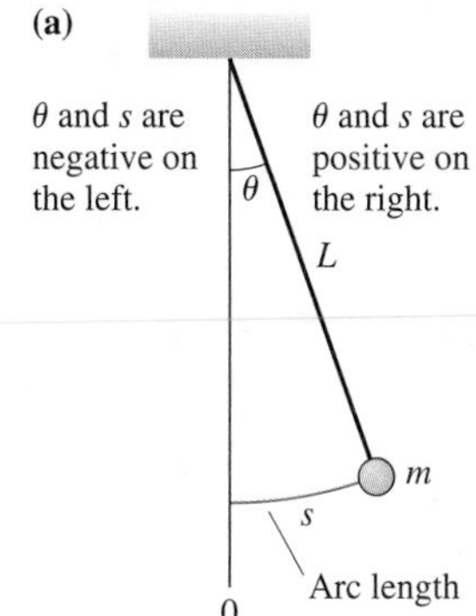
(a)
θ and s are negative on the left.
θ and s are positive on the right.
θ
L
m
s
Arc length
0

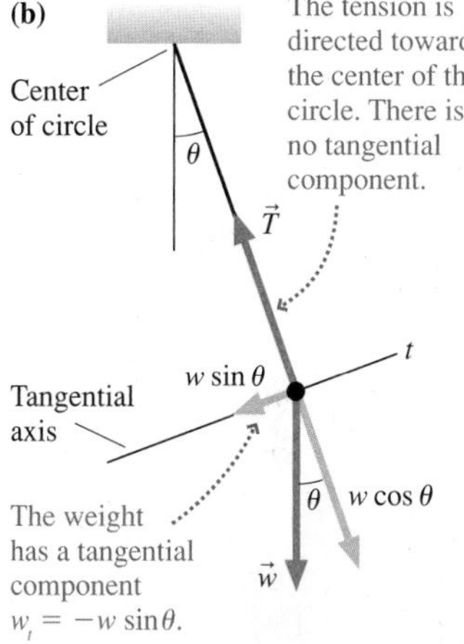
(b)
The tension is directed toward the center of the circle. There is no tangential component.
Center of circle
θ
T
t
w sin θ
Tangential axis
θ
w cos θ
The weight has a tangential component $w_t = -w\sin\theta$.
w

STOP TO THINK 14.2 A ball is hung from a rope, making a pendulum. When it is pulled 5° to the side, the restoring force is 1.0 N. What will be the magnitude of the restoring force if the ball is pulled 10° to the side?

A. 0.5 N B. 1.0 N C. 1.5 N D. 2.0 N

14.3 Describing Simple Harmonic Motion

Now that we know what *causes* simple harmonic motion, we can continue to develop graphical and mathematical descriptions. If we do this for one system, we can adapt the description to any system. The details will vary, but the basic form of the motion will stay the same.

Let's look in detail at one period of the oscillation of a mass on a vertical spring. As we saw in the preceding section, the mass has an equilibrium position at which the net force is zero. If the mass is displaced from equilibrium, a linear restoring force causes the mass to undergo simple harmonic motion. In the table on the next page we present snapshots of the motion together with graphs of the position, velocity, and acceleration.

Details of oscillatory motion

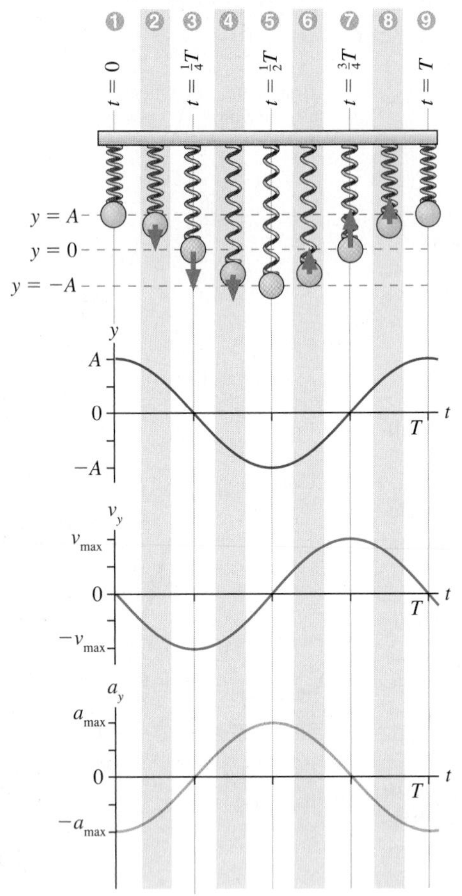

A mass is suspended from a vertical spring with an equilibrium position at $y = 0$. The mass is then lifted upward by a distance A and released. We measure position with respect to the equilibrium position, with positive positions above the equilibrium point, negative below.

1. The mass starts at its maximum positive displacement, $y = A$. The velocity is zero at the instant the mass is released, but the acceleration is negative because there is a net downward force.
2. The mass is now moving downward, so the velocity is negative and the distance from equilibrium is decreasing. As the mass nears equilibrium, the restoring force, and thus the magnitude of the acceleration, decreases.
3. At this time the mass is at the equilibrium position, so the net force—and thus the acceleration—is zero. The speed is at a maximum, but the velocity is negative because the motion is downward.
4. The velocity is still negative but its magnitude is decreasing, so the acceleration is positive.
5. At this time, the mass has reached the lowest point of its motion, with $y = -A$. This is a **turning point** of the motion. The velocity is zero. The spring is at its maximum extension, so there is a net upward force and the acceleration is positive.
6. The mass has begun moving upward; the velocity is positive, and the acceleration is positive.
7. The mass is passing through the equilibrium position again, but in the opposite direction. The acceleration is zero because there is no net force; the upward velocity is positive.
8. The mass continues moving upward. The velocity is positive but its magnitude is decreasing, so the acceleration is negative.
9. The mass is now back at its starting position. This is another turning point; the mass is at rest but will soon begin moving downward, and the whole cycle will repeat.

There are three general points to note about the description of simple harmonic motion in the table above:

- The graphs are for an oscillation in which the object just happened to be at $y = A$ at $t = 0$. You can certainly imagine a different set of *initial conditions,* with the object at $y = -A$ or somewhere in the middle of an oscillation. But even if an oscillation begins at a different position, it will certainly pass through the $y = A$ position and we can simply choose to set $t = 0$ at this instant.
- The position, velocity, and acceleration graphs are all sinusoidal functions. The graphs all have the same general shape. We'll return to this point later.
- If we did a series of experiments on this system, we would find that **the frequency does not depend on the amplitude of the motion;** small oscillations and large oscillations have the same period. This is a key feature of simple harmonic motion that we will explore in more detail later in the chapter. Keep in mind that all these features of the motion of a mass on a spring also apply to other examples of simple harmonic motion.

Now that we have a complete graphical description of the motion, let's develop a mathematical description.

The position-versus-time graph in the above table is a cosine curve. We can write the object's position as

$$x(t) = A\cos\left(\frac{2\pi t}{T}\right) \tag{14.10}$$

where the notation $x(t)$ indicates that the position x is a *function* of time t. Because $\cos(2\pi \text{ rad}) = \cos(0 \text{ rad})$, we see that the position at time $t = T$ is the same as the position at $t = 0$. In other words, this is a cosine function with period T. We can write Equation 14.10 in an alternative form. Because the oscillation frequency is $f = 1/T$, we can write

$$x(t) = A\cos(2\pi ft) \tag{14.11}$$

NOTE ► The argument $2\pi ft$ of the cosine function is in *radians*. That will be true throughout this chapter. Don't forget to set your calculator to radian mode before working oscillation problems. ◄

Just as the position graph was a cosine function, the velocity graph is an upside-down sine function with the same period T. The velocity v_x, which is a function of time, can be written as

$$v_x(t) = -v_{\max}\sin\left(\frac{2\pi t}{T}\right) = -v_{\max}\sin(2\pi ft) \tag{14.12}$$

NOTE ► $v_{\max}$ is the maximum *speed* and thus is inherently a *positive* number. The minus sign in Equation 14.12 is needed to turn the sine function upside down. ◄

Keeping the beat The metal rod in a metronome swings back and forth, making a loud click each time it passes through the center. This is simple harmonic motion, so the motion repeats, cycle after cycle, with the same period. Musicians use this steady click to help them keep a steady beat. The photo—a long exposure—shows another interesting aspect of the motion: You can see the rod most clearly at those points where it was momentarily at rest—at each end of its swing.

How about the acceleration? As we saw earlier in Equation 14.2, the restoring force that causes the mass to oscillate with simple harmonic motion is $(F_{\text{net}})_x = -kx$. Using Newton's second law, we see that this force causes an acceleration

$$a_x = \frac{(F_{\text{net}})_x}{m} = -\frac{k}{m}x \tag{14.13}$$

Acceleration is proportional to the position x, but with a minus sign. Consequently we expect the acceleration-versus-time graph to be an inverted form of the position-versus-time graph. This is, indeed, what we find; the acceleration-versus-time graph in the table on the previous page is clearly an upside-down cosine function with the same period T. We can write the acceleration as

$$a_x(t) = -a_{\max}\cos\left(\frac{2\pi t}{T}\right) = -a_{\max}\cos(2\pi ft) \tag{14.14}$$

In this and coming chapters, we will use sine and cosine functions extensively, so we will summarize some of the key aspects of mathematical relationships using these functions.

Sinusoidal relationships

A quantity that oscillates in time and can be written

$$x = A\sin\left(\frac{2\pi t}{T}\right)$$

or

$$x = A\cos\left(\frac{2\pi t}{T}\right)$$

is called a **sinusoidal function** with **period** T. The argument of the functions, $2\pi t/T$, is in radians.

The graphs of both functions have the same shape, but they have different initial values at $t = 0$ s.

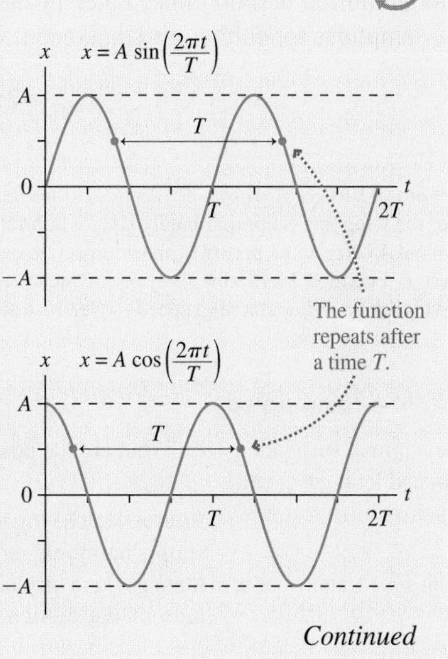

Continued

LIMITS If x is a sinusoidal function, then x is:

- *Bounded*—it can take only values between A and $-A$.
- *Periodic*—it repeats the same sequence of values over and over again. Whatever value x has at time t, it has the same value at $t + T$.

SPECIAL VALUES The function x has special values at certain times:

	$t=0$	$t=\frac{1}{4}T$	$t=\frac{1}{2}T$	$t=\frac{3}{4}T$	$t=T$
$x=A\sin(2\pi t/T)$	0	A	0	$-A$	0
$x=A\cos(2\pi t/T)$	A	0	$-A$	0	A

Exercise 6 

EXAMPLE 14.2 **Motion of a glider on a spring**

An air-track glider oscillates horizontally on a spring at a frequency of 0.50 Hz. Suppose the glider is pulled to the right of its equilibrium position by 12 cm and then released. Where will the glider be 1.0 s after its release? What is its velocity at this point?

PREPARE The glider undergoes simple harmonic motion with amplitude 12 cm. The frequency is 0.50 Hz, so the period is $T = 1/f = 2.0$ s. The glider is released at maximum extension from the equilibrium position, meaning that we can take this point to be $t = 0$.

SOLVE 1.0 s is exactly half the period. As the graph of the motion in **FIGURE 14.10** shows, half a cycle brings the glider to its left turning point, 12 cm to the left of the equilibrium position. The velocity at this point is zero.

FIGURE 14.10 Position-versus-time graph for the glider.

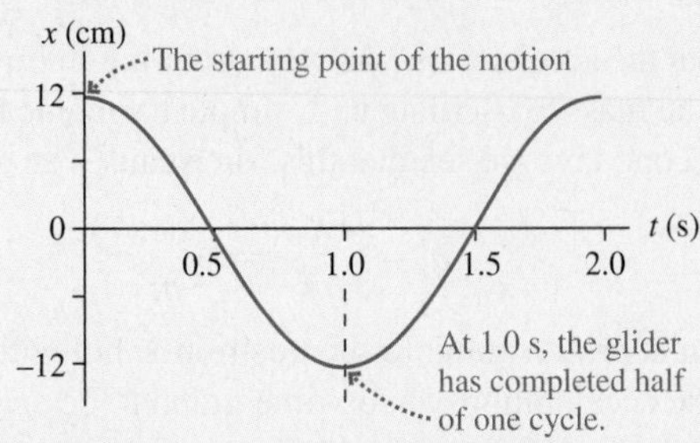

ASSESS Drawing a graph was an important step that helped us make sense of the motion.

We were able to determine the velocity in Example 14.2 because $t = 1.0$ s was a turning point where the instantaneous velocity was zero. To determine the velocity at other points, we need to know v_{max} and use Equation 14.12. How does v_{max} depend on other variables of the motion? Basic reasoning about the motion tells us a few things. A large amplitude implies a high speed because the glider moves a large distance. A small period implies a high speed as well because the glider must complete its motion in a short time. Later in the chapter we will show that both of these assumptions are correct, and that the maximum speed is

$$v_{max} = 2\pi f A = \frac{2\pi A}{T} \tag{14.15}$$

◀ **Small bird, fast wings** BIO As this rufous hummingbird flaps its wings, the motion of the wing tips is approximately simple harmonic. The frequency of the motion is about 45 Hz, so the period of the wingbeat is only a few hundredths of a second. As we see in Equation 14.15, this short period means that the tips of the hummingbird's tiny wings are moving at a high speed—over 15 m/s, or nearly 35 mph!

EXAMPLE 14.3 **Analyzing the motion of a hanging toy**

A classic children's toy consists of a wooden animal suspended from a spring. If you lift the toy up by 10 cm and let it go, it will gently bob up and down, completing 4 oscillations in 10 seconds.

a. What is the oscillation frequency?
b. When does the toy first reach its maximum speed, and what is this speed?
c. What are the position and velocity 4.0 s after you release the toy?

PREPARE The toy is a mass on a vertical spring, so it undergoes simple harmonic motion. The toy begins its motion at its maximum positive displacement of 10 cm, or 0.10 m; this is the amplitude of the motion. Because the toy is released at maximum positive displacement, we can set the moment of release as $t = 0$ s. The toy completes 4 oscillations in 10 seconds, so the period is 1/4 of 10 seconds, or $T = 2.5$ s. **FIGURE 14.11** shows the first two cycles of the oscillation.

FIGURE 14.11 A position graph for the spring toy.

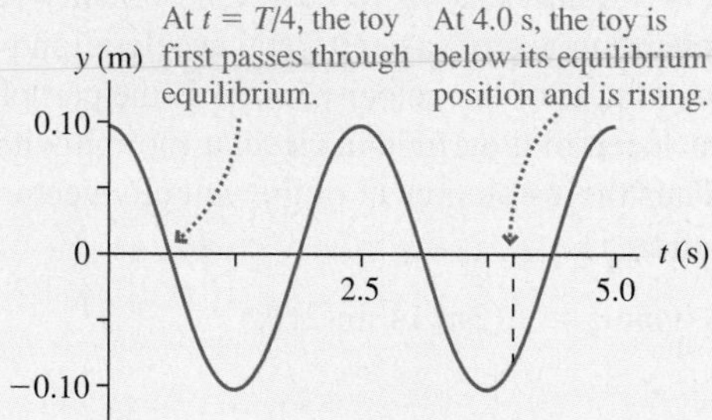

SOLVE a. The period is 2.5 s, so the frequency is

$$f = \frac{1}{T} = \frac{1}{2.5\text{ s}} = 0.40 \text{ oscillation/s} = 0.40 \text{ Hz}$$

b. Figure 14.11 shows that the toy first passes through the equilibrium position at $t = T/4 = (2.5\text{ s})/4 = 0.62$ s. This is a point of maximum speed, with the value given by Equation 14.15:

$$v_{max} = 2\pi f A = 2\pi(0.40\text{ Hz})(0.10\text{m}) = 0.25\text{ m/s}$$

c. 4.0 s after release is $t = 4.0$ s. Figure 14.11 shows that the toy is below the equilibrium position and rising at this time, so we expect that the position is negative and the velocity is positive. We can use Equations 14.11 and 14.12 to find the position and velocity at this time:

$$\begin{aligned} y(t = 4.0\text{ s}) &= A\cos(2\pi f t) \\ &= (0.10\text{ m})\cos[2\pi(0.40\text{ Hz})(4.0\text{ s})] \\ &= -0.081\text{ m} \end{aligned}$$

$$\begin{aligned} v_y(t = 4.0\text{ s}) &= -v_{max}\sin(2\pi f t) \\ &= -(0.25\text{ m/s})\sin[2\pi(0.40\text{ Hz})(4.0\text{ s})] \\ &= 0.15\text{ m/s} \end{aligned}$$

Note that your calculator *must* be in radian mode to do calculations like these.

ASSESS Our calculations for $t = 4.0$ s show that the toy is below the equilibrium position and moving upward, just as we expected from the graph in Figure 14.11. Drawing the graph provided a good check on our work.

Connection to Uniform Circular Motion

Both circular motion and simple harmonic motion are motions that repeat. Many of the concepts we have used for describing simple harmonic motion were introduced in our study of circular motion. We will use the close connection between the two to extend our knowledge of the kinematics of circular motion to simple harmonic motion.

We can demonstrate the relationship between circular and simple harmonic motion with a simple experiment. Suppose a turntable has a small ball glued to the edge. As **FIGURE 14.12a** shows, we can make a "shadow movie" of the ball by projecting a light past the ball and onto a screen. The ball's shadow oscillates back and forth as the turntable rotates.

If you place a real object on a real spring directly below the shadow, as shown in **FIGURE 14.12b**, and if you adjust the turntable to have the same period as the spring, you will find that the shadow's motion exactly matches the simple harmonic motion of the object on the spring. **Uniform circular motion projected onto one dimension is simple harmonic motion.**

FIGURE 14.12 A projection of the circular motion of a rotating ball matches the SHM of an object on a spring.

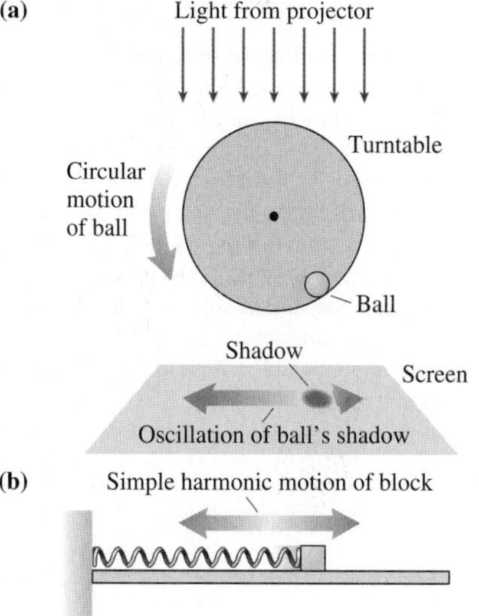

We can show why the projection of circular motion is simple harmonic motion by considering the particle in **FIGURE 14.13** on the next page. As in Chapter 6, we can locate the particle by the angle ϕ measured counterclockwise from the x-axis. Projecting the ball's shadow onto a screen in Figure 14.12a is equivalent to observing just the x-component of the particle's motion. Figure 14.13 shows that the x-component, when the particle is at angle ϕ, is

$$x = A\cos\phi$$

If the particle starts from $\phi_0 = 0$ at $t = 0$, its angle at a later time t is

$$\phi = \omega t$$

where ω is the particle's *angular velocity,* as defined in Chapter 6. Recall that the angular velocity is related to the frequency f by

$$\omega = 2\pi f$$

FIGURE 14.13 A particle in uniform circular motion with radius A and angular velocity ω.

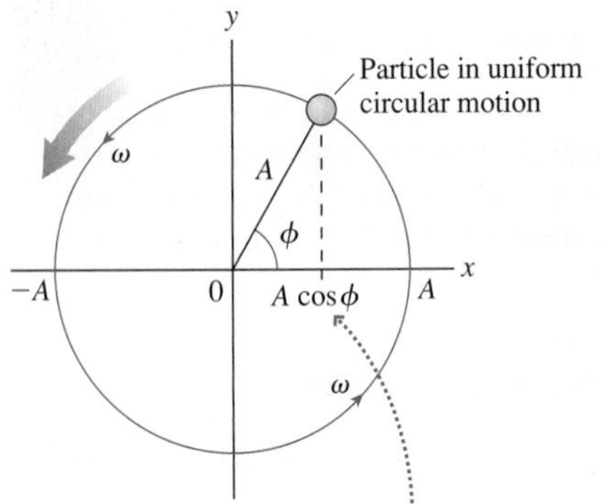

The x-component of the particle's position describes the position of the ball's shadow.

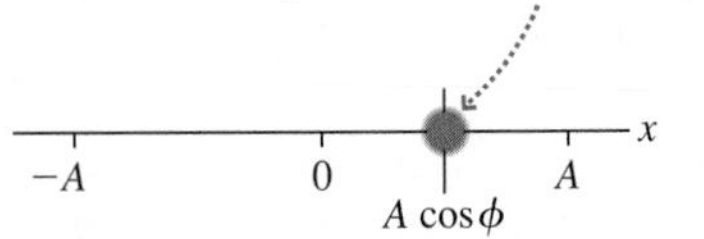

TRY IT YOURSELF

SHM in your microwave It's possible to do something like the turntable demonstration right at home. Place a tall (microwave safe) glass or cup filled with water on the outside edge of the turntable in your microwave oven. Start the oven. The turntable will rotate, moving the cup in a circle. Stand in front of the oven with your eyes level with the cup. Watch the cup, paying attention to the side-to-side motion. This motion is the horizontal component of the circular motion, and so it is simple harmonic motion.

The particle's x-component can therefore be expressed as

$$x(t) = A\cos(2\pi ft) \tag{14.16}$$

This is identical to Equation 14.11 for the position of a mass on a spring! **The x-component of a particle in uniform circular motion is simple harmonic motion.**

We can use this correspondence to deduce more details. **FIGURE 14.14** is the same motion we looked at in Figure 14.13, but here we've shown the velocity vector (tangent to the circle) and the acceleration vector (a centripetal acceleration pointing to the center of the circle). The magnitude of the velocity vector is the particle's speed. Recall from Chapter 6 that the speed of a particle in circular motion with radius A and frequency f is $v = 2\pi fA$. Thus the x-component of the velocity vector, which is pointing in the negative x-direction, is

$$v_x = -v\sin\phi = -(2\pi f)A\sin(2\pi ft)$$

FIGURE 14.14 Projection of the velocity and acceleration vectors.

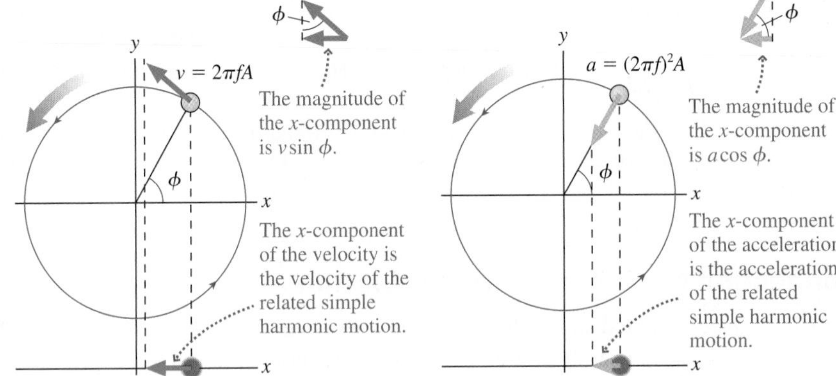

According to the correspondence between circular motion and simple harmonic motion, this is the velocity of an object in simple harmonic motion. This is, indeed, exactly Equation 14.12, which we deduced from the graph of the motion, if we define the maximum speed, as we did in Equation 14.15, to be

$$v_{max} = 2\pi fA = \frac{2\pi A}{T}$$

Similarly, the magnitude of the acceleration vector is the centripetal acceleration $a = v^2/A = (2\pi f)^2 A$. The x-component of the acceleration vector, which is the acceleration for simple harmonic motion, is

$$a_x = -a\cos\phi = -(2\pi f)^2 A\cos(2\pi ft)$$

The maximum acceleration is

$$a_{max} = (2\pi f)^2 A \tag{14.17}$$

We can now summarize our findings for the position, velocity, and acceleration of an object in simple harmonic motion:

$$\begin{aligned} x(t) &= A\cos(2\pi ft) \\ v_x(t) &= -(2\pi f)A\sin(2\pi ft) \\ a_x(t) &= -(2\pi f)^2 A\cos(2\pi ft) \end{aligned} \tag{14.18}$$

x, v_x, and a_x p. 453 t SINUSOIDAL

Position, velocity, and acceleration for an object in simple harmonic motion with frequency f and amplitude A

Any simple harmonic motion follows these equations. **If you know the amplitude and the frequency, the motion is completely specified.**

EXAMPLE 14.4 Measuring the sway of a tall building

The John Hancock Center in Chicago is 100 stories high. Strong winds can cause the building to sway, as is the case with all tall buildings. On particularly windy days, the top of the building is known to oscillate with an amplitude of 40 cm ($\approx$16 in) and a period of 7.7 s. What are the maximum speed and acceleration of the top of the building?

PREPARE We will assume that the oscillation of the building is simple harmonic motion with amplitude $A = 0.40$ m. The frequency can be computed from the period:

$$f = \frac{1}{T} = \frac{1}{7.7\text{ s}} = 0.13\text{ Hz}$$

SOLVE We can use Equations 14.15 and 14.17 for the maximum velocity and acceleration to compute:

$$v_{max} = 2\pi f A = 2\pi(0.13\text{ Hz})(0.40\text{ m}) = 0.33\text{ m/s}$$

$$a_{max} = (2\pi f)^2 A = [2\pi(0.13\text{ Hz})]^2(0.40\text{ m}) = 0.27\text{ m/s}^2$$

In terms of the free-fall acceleration, the maximum acceleration is $a_{max} = 0.027g$.

ASSESS The free-fall acceleration is quite small, as you would expect; if it were large, building occupants would certainly complain! Even if they don't notice the motion directly, office workers on high floors of high buildings may experience a bit of nausea when the oscillations are large because the acceleration affects the equilibrium organ in the inner ear.

STOP TO THINK 14.3 The figures show four identical oscillators at different points in their motion. Which is moving fastest at the time shown?

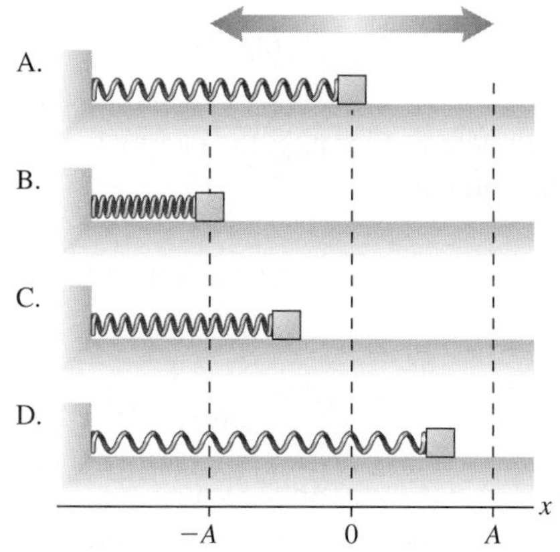

14.4 Energy in Simple Harmonic Motion

A bungee jumper falls, increasing in speed, until the elastic cords attached to his ankles start to stretch. The kinetic energy of his motion is transformed into the elastic potential energy of the cords. Once the cords reach their maximum stretch, his velocity reverses. He rises as the elastic potential energy of the cords is transformed back into kinetic energy. If he keeps bouncing (simple harmonic motion), this transformation happens again and again.

This interplay between kinetic and potential energy is very important to understanding simple harmonic motion. The five diagrams in FIGURE 14.15a on the next page show the position and velocity of a mass on a spring at successive points in time.

The object begins at rest, with the spring at a maximum extension; the kinetic energy is zero, and the potential energy is a maximum. As the spring contracts, the object speeds up until it reaches the center point of its oscillation, the equilibrium point. At this point, the potential energy is zero and the kinetic energy is a maximum. As the object continues to move, it slows down as it compresses the

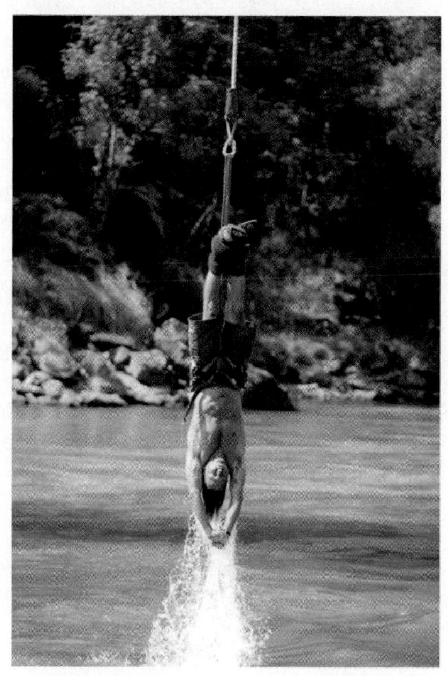

Elastic cords lead to the up-and-down motion of a bungee jump.

FIGURE 14.15 Energy transformations for a mass on a spring.

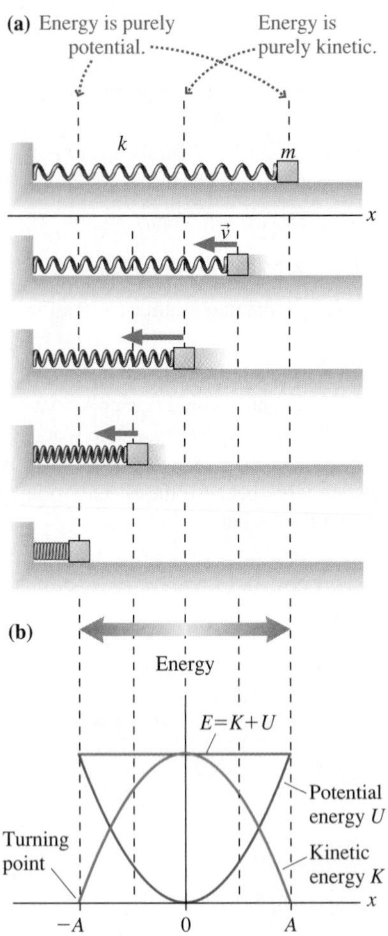

spring. Eventually, it reaches the turning point, where its instantaneous velocity is zero. At this point, the kinetic energy is zero and the potential energy is again a maximum.

Now we'll specify that the object has mass m, the spring has spring constant k, and the motion takes place on a frictionless surface. You learned in Chapter 10 that the elastic potential energy of a spring stretched by a distance x from its equilibrium position is

$$U = \frac{1}{2}kx^2 \tag{14.19}$$

The potential energy is zero at the equilibrium position and is a maximum when the spring is at its maximum extension or compression. There is no energy loss to thermal energy, so conservation of energy for this system can be written

$$E = K + U = \frac{1}{2}mv^2 + \frac{1}{2}kx^2 = \text{constant} \tag{14.20}$$

FIGURE 14.15b shows a graph of the potential energy, kinetic energy, and total energy for the object as it moves. You can see that, as the object goes through its motion, energy is transformed from potential to kinetic and then back to potential. At maximum displacement, with $x = \pm A$ and $v_x = 0$, the energy is purely potential, so the potential energy has its maximum value:

$$E(\text{at } x = \pm A) = U_{\text{max}} = \frac{1}{2}kA^2 \tag{14.21}$$

At $x = 0$, where $v_x = \pm v_{\text{max}}$, the energy is purely kinetic, so the kinetic energy has its maximum value:

$$E(\text{at } x = 0) = K_{\text{max}} = \frac{1}{2}m(v_{\text{max}})^2 \tag{14.22}$$

CONCEPTUAL EXAMPLE 14.5 Energy changes for a playground swing

You are at the park, undergoing simple harmonic motion on a swing. Describe the changes in energy that occur during one cycle of the motion, starting from when you are at the farthest forward point, motionless and just about to swing backward.

REASON The motion of the swing is that of a pendulum, with you playing the role of the pendulum bob. The energy at different points of the motion is illustrated in FIGURE 14.16. When you are motionless and at the farthest forward point, you are raised up; you have potential energy. As you swing back, your potential energy decreases and your kinetic energy increases, reaching a maximum when the swing is at the lowest point. The swing continues to move backward; you rise up, transforming kinetic energy into potential energy. The process then reverses as you move forward.

ASSESS This description matches the experience of anyone who has been on a swing. You know that you are momentarily motionless at the highest points, and that the motion is fastest at the lowest point.

FIGURE 14.16 Energy at different points of the motion of a swing.

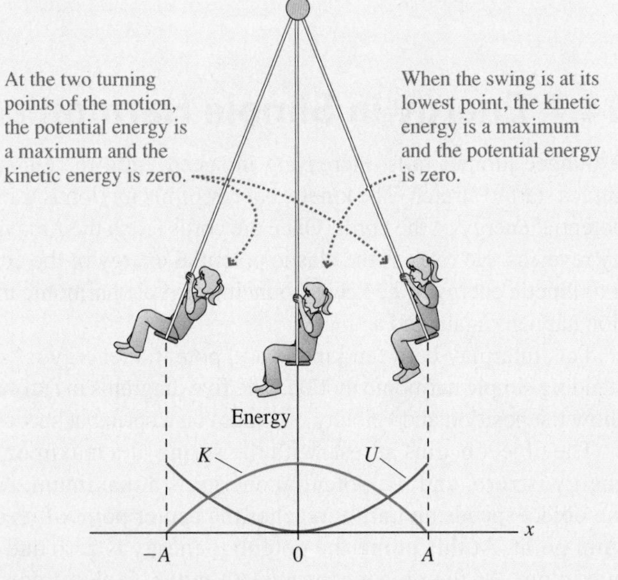

At the two turning points of the motion, the potential energy is a maximum and the kinetic energy is zero.
When the swing is at its lowest point, the kinetic energy is a maximum and the potential energy is zero.
Energy
K
U
x
−A
0
A

Finding the Frequency for Simple Harmonic Motion

Now that we have an energy description of simple harmonic motion, we can use what we know about energy to deduce other details of the motion. Let's return to the mass on a spring of Figure 14.15. The graph in Figure 14.15b shows energy being transformed back and forth between kinetic and potential energy. At the turning points, the energy is purely potential; at the equilibrium point, the energy is purely kinetic. Because the total energy doesn't change, the maximum kinetic energy given in Equation 14.22 must be equal to the maximum potential energy given in Equation 14.21:

$$\frac{1}{2}m(v_{\text{max}})^2 = \frac{1}{2}kA^2 \tag{14.23}$$

By solving Equation 14.23 for the maximum speed, we can see that it is related to the amplitude by

$$v_{\text{max}} = \sqrt{\frac{k}{m}}A \tag{14.24}$$

Earlier we found that

$$v_{\text{max}} = 2\pi fA \tag{14.25}$$

Comparing Equations 14.24 and 14.25, we see that the frequency, and thus the period, of an oscillating mass on a spring is determined by the spring constant k and the object's mass m:

$$f = \frac{1}{2\pi}\sqrt{\frac{k}{m}} \quad \text{and} \quad T = 2\pi\sqrt{\frac{m}{k}} \tag{14.26}$$

Frequency and period of SHM for mass m on a spring with spring constant k

$\sqrt{\frac{k}{m}}$ — f p.37 PROPORTIONAL

We can make two observations about these equations:

- **The frequency and period of simple harmonic motion are determined by the physical properties of the oscillator.** The frequency and period of a mass on a spring are determined by (1) the mass and (2) the stiffness of the spring, as shown in FIGURE 14.17. This dependence of frequency and period on a force term and an inertia term will also apply to other oscillators.
- **The frequency and period of simple harmonic motion do not depend on the amplitude A.** A small oscillation and a large oscillation have the same frequency and period.

FIGURE 14.17 Frequency dependence on mass and spring stiffness.

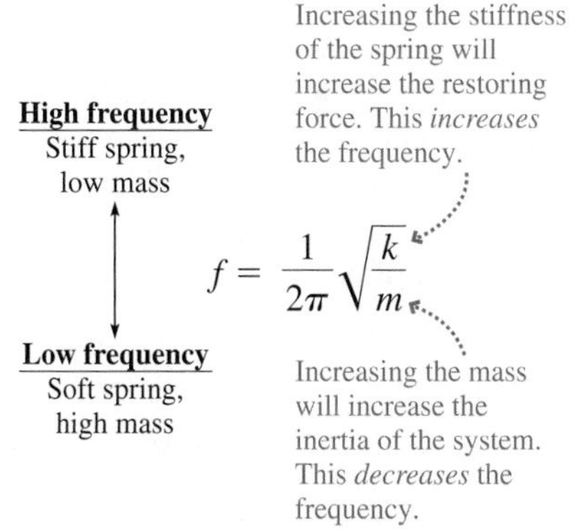

High frequency
Stiff spring, low mass
Low frequency
Soft spring, high mass
$f = \frac{1}{2\pi}\sqrt{\frac{k}{m}}$
Increasing the stiffness of the spring will increase the restoring force. This *increases* the frequency.
Increasing the mass will increase the inertia of the system. This *decreases* the frequency.

CONCEPTUAL EXAMPLE 14.6 Changing mass, changing period

An astronaut measures her mass each day using the Body Mass Measurement Device on the Space Shuttle, as described at right. During an 8-day flight, her mass steadily decreases. How does this change the frequency of her oscillatory motion on the device?

REASON The period and frequency of a mass-spring system depend on the mass of the object and the spring constant. The spring constant of the device won't change, so the only change that matters is the change in the astronaut's mass. Equation 14.26 shows that the frequency is proportional to $\sqrt{k/m}$, so a decrease in her mass will cause an increase in the frequency. The oscillation will be a bit more rapid.

ASSESS This makes sense. The force of the spring—which causes the oscillation—is the same, but the mass to be accelerated is less. We expect a higher frequency.

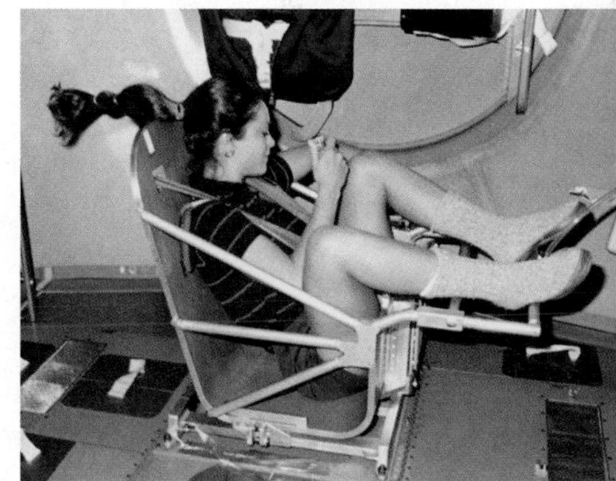

Measuring mass in space Astronauts on extended space flights monitor their mass to track the effects of weightlessness on their bodies. But because they are weightless, they can't just hop on a scale! Instead, they use an ingenious device in which an astronaut sitting on a platform oscillates back and forth due to the restoring force of a spring. The astronaut is the moving mass in a mass-spring system, so a measurement of the period of the motion allows a determination of an astronaut's mass.

Now that we have a complete description of simple harmonic motion in one system, we will summarize the details in a Tactics Box on the next page, showing how to use this information to solve oscillation problems.

9.3, 9.4, 9.6, 9.7, 9.8, 9.9 

TACTICS BOX 14.1 **Identifying and analyzing simple harmonic motion** 

1. If the net force acting on a particle is a linear restoring force, the motion is simple harmonic motion around the equilibrium position.
2. The position, velocity, and acceleration as a function of time are given in Equations 14.18. The equations are given here in terms of x, but they can be written in terms of y, θ, or some other variable if the situation calls for it.
3. The amplitude A is the maximum value of the displacement from equilibrium. The maximum speed and the maximum magnitude of the acceleration are $v_{max} = 2\pi f A$ and $a_{max} = (2\pi f)^2 A$.
4. The frequency f (and hence the period $T = 1/f$) depends on the physical properties of the particular oscillator, but f does *not* depend on A.

   For a mass on a spring, the frequency is given by $f = \dfrac{1}{2\pi}\sqrt{\dfrac{k}{m}}$.
5. The sum of potential energy plus kinetic energy is constant. As the oscillation proceeds, energy is transformed from kinetic into potential energy and then back again.

Exercise 11

EXAMPLE 14.7 **Finding the frequency of an oscillator**

A spring has an unstretched length of 10.0 cm. A 25 g mass is hung from the spring, stretching it to a length of 15.0 cm. If the mass is pulled down and released so that it oscillates, what will be the frequency of the oscillation?

PREPARE The spring provides a linear restoring force, so the motion will be simple harmonic, as noted in Tactics Box 14.1.

FIGURE 14.18 Visual overview of a mass suspended from a spring.

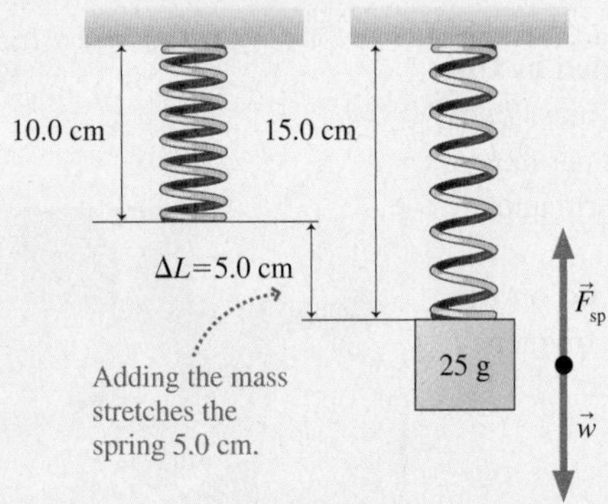

The oscillation frequency depends on the spring constant, which we can determine from the stretch of the spring. **FIGURE 14.18** gives a visual overview of the situation.

SOLVE When the mass hangs at rest, after stretching the spring to 15 cm, the net force on it must be zero. Thus the magnitude of the upward spring force equals the downward weight, giving $k\Delta L = mg$. The spring constant is thus

$$k = \frac{mg}{\Delta L} = \frac{(0.025\text{ kg})(9.8\text{ m/s}^2)}{0.050\text{ m}} = 4.9\text{ N/m}$$

Now that we know the spring constant, we can compute the oscillation frequency:

$$f = \frac{1}{2\pi}\sqrt{\frac{k}{m}} = \frac{1}{2\pi}\sqrt{\frac{4.9\text{ N/m}}{0.025\text{ kg}}} = 2.2\text{ Hz}$$

ASSESS 2.2 Hz is 2.2 oscillations per second. This seems like a reasonable frequency for a mass on a spring. A frequency in the kHz range (thousands of oscillations per second) would have been suspect!

EXAMPLE 14.8 **Weighing DNA molecules**

It has recently become possible to "weigh" individual DNA molecules by measuring the influence of their mass on a nanoscale oscillator. **FIGURE 14.19** shows a thin rectangular cantilever etched out of silicon. The cantilever has a mass of 3.7×10^{-16} kg. If pulled down and released, the end of the cantilever vibrates with simple harmonic motion, moving up and down like a diving board after a jump. When the end of the cantilever is bathed with DNA molecules whose ends have been modified to bind to a surface, one or more molecules may attach to the end of the cantilever. The addition of their mass causes a very slight—but measurable—decrease in the oscillation frequency.

FIGURE 14.19 A nanoscale cantilever.

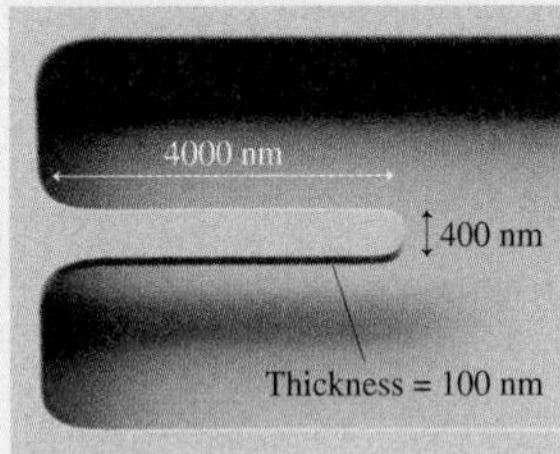

A vibrating cantilever of mass M can be modeled as a simple block of mass $\frac{1}{3}M$ attached to a spring. (The factor of $\frac{1}{3}$ arises from the moment of inertia of a bar pivoted at one end: $I = \frac{1}{3}ML^2$.) Neither the mass nor the spring constant can be determined very accurately—perhaps only to two significant figures—but the oscillation frequency can be measured with very high precision simply by counting the oscillations. In one experiment, the cantilever was initially vibrating at exactly 12 MHz. Attachment of a DNA molecule caused the frequency to decrease by 50 Hz. What was the mass of the DNA molecule?

PREPARE We will model the cantilever as a block of mass $m = \frac{1}{3}M = 1.2 \times 10^{-16}$ kg oscillating on a spring with spring constant k. When the mass increases to $m + m_{\text{DNA}}$, the oscillation frequency decreases from $f_0 = 12{,}000{,}000$ Hz to $f_1 = 11{,}999{,}950$ Hz.

SOLVE The oscillation frequency of a mass on a spring is given by Equation 14.26. Addition of mass doesn't change the spring constant, so solving this equation for k allows us to write

$$k = m(2\pi f_0)^2 = (m + m_{\text{DNA}})(2\pi f_1)^2$$

The 2π terms cancel, and we can rearrange this equation to give

$$\frac{m + m_{\text{DNA}}}{m} = 1 + \frac{m_{\text{DNA}}}{m} = \left(\frac{f_0}{f_1}\right)^2 = \left(\frac{12{,}000{,}000 \text{ Hz}}{11{,}999{,}950 \text{ Hz}}\right)^2 = 1.0000083$$

Subtracting 1 from both sides gives

$$\frac{m_{\text{DNA}}}{m} = 0.0000083$$

and thus

$$m_{\text{DNA}} = 0.0000083m = (0.0000083)(1.2 \times 10^{-16} \text{ kg}) = 1.0 \times 10^{-21} \text{ kg} = 1.0 \times 10^{-18} \text{ g}$$

ASSESS This is a reasonable mass for a DNA molecule. It's a remarkable technical achievement to be able to measure a mass this small. With a slight further improvement in sensitivity, scientists will be able to determine the number of base pairs in a strand of DNA simply by weighing it!

EXAMPLE 14.9 Slowing a mass with a spring collision

A 1.5 kg mass slides across a horizontal, frictionless surface at a speed of 2.0 m/s until it collides with and sticks to the free end of a spring with spring constant 50 N/m. The spring's other end is anchored to a wall. How far has the spring compressed when the mass is, at least for an instant, at rest? How much time does it take for the spring to compress to this point?

PREPARE This is a collision problem, but we can solve it using the tools of simple harmonic motion. **FIGURE 14.20** gives a visual overview of the problem. The motion is along the x-axis. We have set $x = 0$ at the uncompressed end of the spring, which is the point of collision. Once the mass hits and sticks, it will start oscillating with simple harmonic motion. The position-versus-time graph shows that the motion of the mass until it stops is $\frac{1}{4}$ of a cycle of simple harmonic motion—though the starting point is different from our usual choice. During this quarter cycle of motion, the kinetic energy of the mass is transformed into the potential energy of the compressed spring.

SOLVE Conservation of energy tells us that the potential energy of the spring when fully compressed is equal to the initial kinetic energy of the mass, so we write

$$\frac{1}{2}m(v_x)_i^2 = \frac{1}{2}kx_f^2$$

The final position of the mass is

$$x_f = (v_x)_i\sqrt{\frac{m}{k}} = (2.0 \text{ m/s})\sqrt{\frac{1.5 \text{ kg}}{50 \text{ N/m}}} = 0.35 \text{ m}$$

This is the compression of the spring.

Because the compression is $\frac{1}{4}$ of a cycle of simple harmonic motion, the time needed to stop the mass is $t_f = \frac{1}{4}T$. The period is computed using Equation 14.26, so we write

$$t_f = \frac{1}{4}T = \frac{1}{4}2\pi\sqrt{\frac{m}{k}} = \frac{\pi}{2}\sqrt{\frac{1.5 \text{ kg}}{50 \text{ N/m}}} = 0.27 \text{ s}$$

This is the time required for the spring to compress.

ASSESS Interestingly, the time needed to stop the mass does not depend on its initial velocity or on the distance the spring compresses. The motion is simple harmonic; thus the period depends only on the stiffness of the spring and the mass, not on the details of the motion.

FIGURE 14.20 Visual overview for the mass-spring collision.

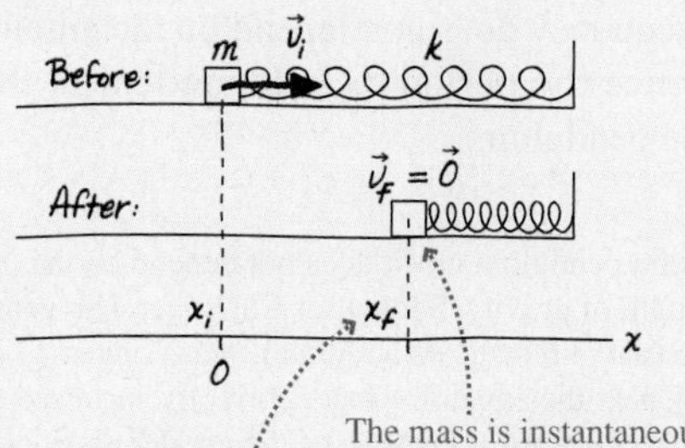

The spring compresses by x_f as the mass comes to rest.

The mass is instantaneously at rest when the spring reaches maximum compression.

Known
$m = 1.5$ kg
$k = 50$ N/m
$t_i = 0$ s
$x_i = 0$ m
$(v_x)_i = 2.0$ m/s
$(v_x)_f = 0$ m/s

Find
x_f, t_f

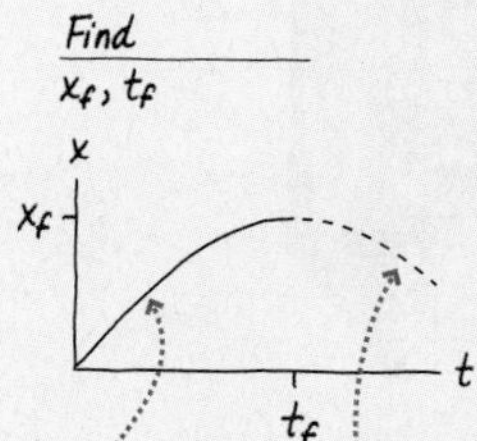

The graph is the start of a sine function, and the motion is $\frac{1}{4}$ of a cycle of SHM.

The mass would, after stopping, reverse direction and continue the cycle of SHM.

Automobile collision times When a car hits a stationary barrier, it takes approximately 0.1 s to come to rest, regardless of the initial speed, something we can understand with a simple model of the collision. The crumpling of the front of a car during a collision is quite complex, but for many cars the force is approximately proportional to the displacement during the compression of the front of the car. As long as we consider only this initial compression, we can model the body of the car as a mass and the front of the car as a spring. With this model, the collision is similar to that of Example 14.9, and, as in the example, the time for the car to come to rest does not depend on the initial speed.

Many real-world collisions (such as a pole vaulter landing on a foam pad) involve a moving object that is stopped by an elastic object. In other cases (such as a bouncing rubber ball) the object itself is elastic. The time of the collision is reasonably constant, independent of the speed of the collision, for the reasons noted in the above example.

STOP TO THINK 14.4 Four mass-spring systems have masses and spring constants shown here. Rank in order, from highest to lowest, the frequencies of the oscillations.

A. k, $4m$

B. $\frac{1}{2}k$, m

C. k, $2m$

D. $2k$, m

14.5 Pendulum Motion

9.10, 9.11, 9.12 ActivPhysics Online

As we've already seen, a simple pendulum—a mass at the end of a string or a rod that is free to pivot—is another system that exhibits simple harmonic motion. Everything we have learned about the mass on a spring can be applied to the pendulum as well.

In the first part of the chapter, we looked at the restoring force in the pendulum. For a pendulum of length L displaced by an arc length s, as in FIGURE 14.21, the tangential restoring force is

$$(F_{\text{net}})_t = -\frac{mg}{L}s \tag{14.27}$$

FIGURE 14.21 A simple pendulum.

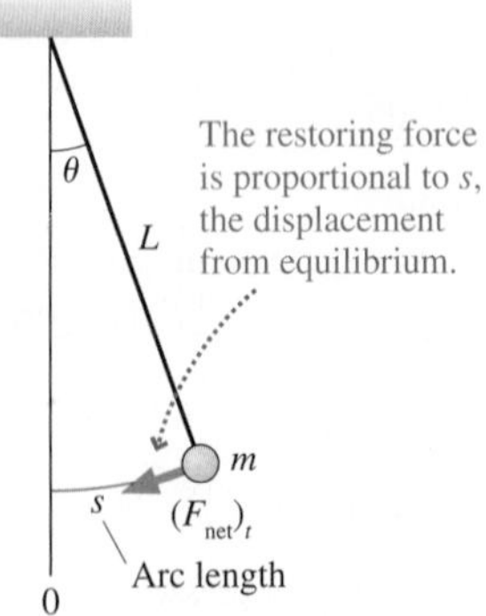

NOTE ► Recall that this equation holds only for small angles. ◄

This linear restoring force has exactly the same form as the net force in a mass-spring system, but with the constants mg/L in place of the constant k. Given this, we can quickly deduce the essential features of pendulum motion by replacing k, wherever it occurs in the oscillating spring equations, with mg/L.

- The oscillation of a pendulum is simple harmonic motion; the equations of motion can be written for the arc length or the angle:

$$s(t) = A\cos(2\pi ft) \qquad \text{or} \qquad \theta(t) = \theta_{\max}\cos(2\pi ft)$$

- The frequency can be obtained from the equation for the frequency of the mass on a spring by substituting mg/L in place of k:

$$f = \frac{1}{2\pi}\sqrt{\frac{g}{L}} \qquad \text{and} \qquad T = 2\pi\sqrt{\frac{L}{g}} \tag{14.28}$$

Frequency of a pendulum of length L with free-fall acceleration g

- As for a mass on a spring, the frequency does not depend on the amplitude. Note also that **the frequency, and hence the period, is independent of the mass.** It depends only on the length of the pendulum.

◄ **Pendulum prospecting** The period of a pendulum clock does not depend on the amplitude, but it does depend on the strength of gravity. Soon after Christiaan Huygens built an accurate pendulum clock in 1656 (the photo shows a replica), Jean Richer discovered that the clock ran more slowly near the equator. Richer correctly surmised that this was due to the weaker gravity near the equator because of the greater distance from the center of the earth. In later years, more accurate pendulums were built that could detect much smaller variations in gravity—small enough that they could sense the presence of dense mineral deposits or low-density strata containing petroleum.

Galileo was the first person to study the pendulum in detail. He realized that the pendulum's fixed frequency would serve as the basis of an accurate clock. Pendulum clocks were the most accurate timepieces available until well into the 20th century.

EXAMPLE 14.10 Designing a pendulum for a clock

A grandfather clock is designed so that one swing of the pendulum in either direction takes 1.00 s. What is the length of the pendulum?

PREPARE One period of the pendulum is two swings, so the period is $T = 2.00$ s.

SOLVE The period is independent of the mass and depends only on the length. From Equation 14.28,

$$T = \frac{1}{f} = 2\pi\sqrt{\frac{L}{g}}$$

Solving for L, we find

$$L = g\left(\frac{T}{2\pi}\right)^2 = (9.80 \text{ m/s}^2)\left(\frac{2.00 \text{ s}}{2\pi}\right)^2 = 0.993 \text{ m}$$

ASSESS A pendulum clock with a "tick" or "tock" each second requires a long pendulum of about 1 m—which is why these clocks were originally known as "tall case clocks."

Physical Pendulums and Locomotion

In Chapter 6, we computed maximum walking speed using the ideas of circular motion. We can also model the motion of your legs during walking as pendulum motion. When you walk, you push off with your rear leg and then let it swing forward for the next stride. At normal, comfortable walking speeds, you use very little force to bring your leg forward. Your leg swings forward under the influence of gravity—like a pendulum.

Try this: Stand on one leg, and gently swing your free leg back and forth. There is a certain frequency at which it will naturally swing. This is your leg's pendulum frequency. Now, try swinging your leg at twice this frequency. You can do it, but it is very difficult. The muscles that move your leg back and forth aren't very strong because under normal circumstances they don't need to apply much force.

A pendulum, like your leg, whose mass is distributed along its length is known as a **physical pendulum.** The motion of a physical pendulum is similar to that of a simple pendulum, but its frequency depends on the distribution of mass.

FIGURE 14.22 shows a simple pendulum and a physical pendulum of the same length. The position of the center of gravity of the physical pendulum is at a distance d from the pivot.

What will be the frequency of a physical pendulum? This is really a rotational motion problem similar to those we considered in Chapter 7. We would expect the frequency to depend on the moment of inertia and the distance to the center of gravity as follows:

- The moment of inertia I is a measure of an object's resistance to rotation. Increasing the moment of inertia while keeping other variables equal should cause the frequency to decrease. In an expression for the frequency of the physical pendulum, we would expect I to appear in the denominator.
- When the pendulum is pushed to the side, a gravitational torque pulls it back. The greater the distance d of the center of gravity from the pivot point, the greater the torque. Increasing this distance while keeping the other variables constant should cause the frequency to increase. In an expression for the frequency of the physical pendulum, we would expect d to appear in the numerator.

FIGURE 14.22 A simple pendulum and a physical pendulum of equal length.

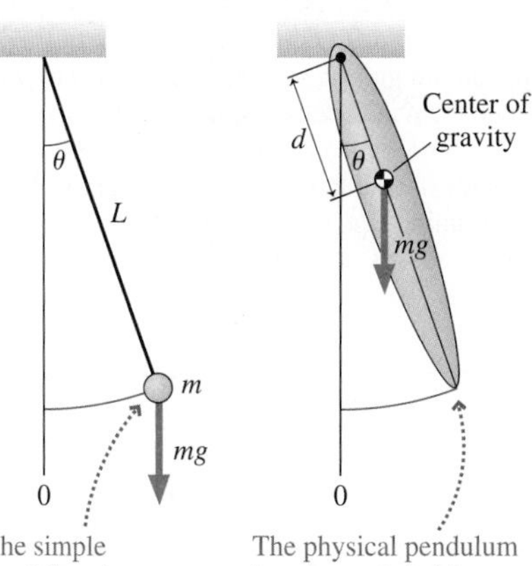
θ
L
m
mg
0
The simple pendulum is a small mass m at the end of a light rod of length L.
d
θ
Center of gravity
mg
0
The physical pendulum is an extended object with mass m, length L, and moment of inertia I.

A careful analysis of the motion of the physical pendulum produces a result for the frequency that matches these expectations:

$$f = \frac{1}{2\pi}\sqrt{\frac{mgd}{I}} \tag{14.29}$$

Frequency of a physical pendulum of mass m, moment of inertia I, with center of gravity distance d from the pivot

TRY IT YOURSELF

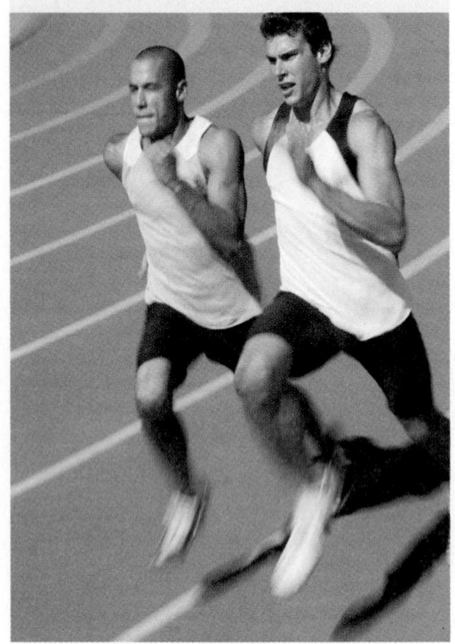

How do you hold your arms? You maintain your balance when walking or running by moving your arms back and forth opposite the motion of your legs. You hold your arms so that the natural period of their pendulum motion matches that of your legs. At a normal walking pace, your arms are extended and naturally swing at the same period as your legs. When you run, your gait is more rapid. To decrease the period of the pendulum motion of your arms to match, you bend them at the elbows, shortening their effective length and increasing the natural frequency of oscillation. To test this for yourself, try running fast with your arms fully extended. It's quite awkward!

EXAMPLE 14.11 Finding the frequency of a swinging leg

A student in a biomechanics lab measures the length of his leg, from hip to heel, to be 0.90 m. What is the frequency of the pendulum motion of the student's leg? What is the period?

PREPARE We can model a human leg reasonably well as a rod of uniform cross section, pivoted at one end (the hip). Recall from Chapter 7 that the moment of inertia of a rod pivoted about its end is $\frac{1}{3}mL^2$. The center of gravity of a uniform leg is at the midpoint, so $d = L/2$.

SOLVE The frequency of a physical pendulum is given by Equation 14.29. Before we put in numbers, we will use symbolic relationships and simplify:

$$f = \frac{1}{2\pi}\sqrt{\frac{mgd}{I}} = \frac{1}{2\pi}\sqrt{\frac{mg(L/2)}{\frac{1}{3}mL^2}} = \frac{1}{2\pi}\sqrt{\frac{3}{2}\frac{g}{L}}$$

The expression for the frequency is similar to that for the simple pendulum, but with an additional numerical factor of 3/2 inside the square root. The numerical value of the frequency is

$$f = \frac{1}{2\pi}\sqrt{\left(\frac{3}{2}\right)\left(\frac{9.8 \text{ m/s}^2}{0.90 \text{ m}}\right)} = 0.64 \text{ Hz}$$

The period is

$$T = \frac{1}{f} = 1.6 \text{ s}$$

ASSESS Notice that we didn't need to know the mass of the leg to find the period. The period of a physical pendulum does not depend on the mass, just as it doesn't for the simple pendulum. The period depends only on the *distribution* of mass. When you walk, swinging your free leg forward to take another stride corresponds to half a period of this pendulum motion. For a period of 1.6 s, this is 0.80 s. For a normal walking pace, one stride in just under one second sounds about right.

As you walk, your legs do swing as physical pendulums as you bring them forward. The frequency is fixed by the length of your legs and their distribution of mass; it doesn't depend on amplitude. Consequently, you don't increase your walking speed by taking more rapid steps—changing the frequency is quite difficult. You simply take longer strides, changing the amplitude but not the frequency.

Gibbons and other apes move through the forest canopy by a hand-over-hand swinging motion called *brachiation.* In this motion, the body swings under a pivot point where a hand grips a tree branch, a clear example of pendulum motion. A brachiating ape will increase its speed by taking bigger swings; it does this by "pumping" the swinging motion, much as you do when increasing your amplitude on a playground swing. But because the period of the pendulum motion is fixed, a brachiating gibbon can only go so fast. At some point a maximum speed is reached, and gibbons and other apes break into a different gait, launching themselves from branch to branch through the air.

STOP TO THINK 14.5 A pendulum clock is made with a metal rod. It keeps perfect time at a temperature of 20°C. At a higher temperature, the metal rod lengthens. How will this change the clock's timekeeping?

A. The clock will run fast; the dial will be ahead of the actual time.
B. The clock will keep perfect time.
C. The clock will run slow; the dial will be behind the actual time.

14.6 Damped Oscillations

A real pendulum clock must have some energy input; otherwise, the oscillation of the pendulum would slowly decrease in amplitude due to air resistance. If you strike a bell, the oscillation will soon die away as energy is lost to sound waves in the air and dissipative forces within the metal of the bell.

All real oscillators do run down—some very slowly but others quite quickly—as their mechanical energy is transformed into the thermal energy of the oscillator and its environment. An oscillation that runs down and stops is called a **damped oscillation.**

For a pendulum, the main energy loss is due to air resistance, which we called the *drag force* in Chapter 4. When we learned about the drag force, we noted that it depends on velocity: The faster the motion, the bigger the drag force. For this reason, the decrease in amplitude of an oscillating pendulum will be fastest at the start of the motion. For a pendulum or other oscillator with modest damping, we end up with a graph of motion like that in FIGURE 14.23a. The maximum displacement, x_{max}, decreases with time. As the oscillation decays, the *rate* of the decay decreases; the difference between successive peaks is less.

If we plot a smooth curve that connects the peaks of successive oscillations (we call such a curve an *envelope*), we get the dotted line shown in FIGURE 14.23b. It's possible, using calculus, to show that x_{max} decreases with time as

$$x_{max}(t) = Ae^{-t/\tau} \tag{14.30}$$

where $e \approx 2.718$ is the base of the natural logarithm and A is the *initial* amplitude. This steady decrease of x_{max} with time is called an **exponential decay.**

The constant τ (lowercase Greek tau) in Equation 14.30 is called the **time constant.** After one time constant has elapsed—that is, at $t = \tau$—the maximum displacement x_{max} has decreased to

$$x_{max}(\text{at } t = \tau) = Ae^{-1} = \frac{A}{e} \approx 0.37A$$

In other words, the oscillation has decreased after one time constant to about 37% of its initial value. The time constant τ measures the "characteristic time" during which damping causes the amplitude of the oscillation to decay away. An oscillation that decays quickly has a small time constant, whereas a "lightly damped" oscillator, which decays very slowly, has a large time constant.

Because we will see exponential decay again, we will look at it in more detail.

FIGURE 14.23 The motion of a damped oscillator.

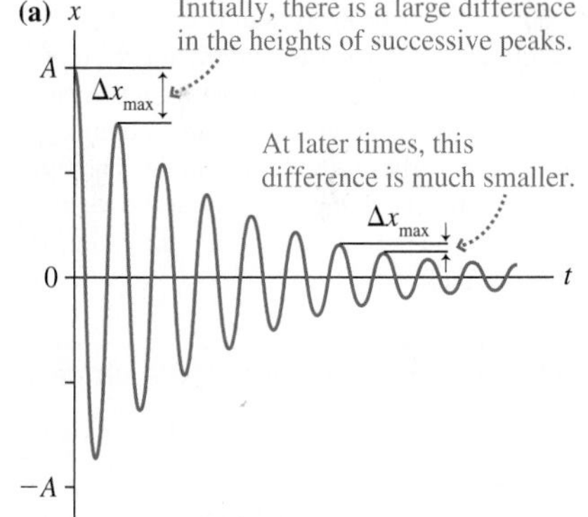
(a) x
Initially, there is a large difference in the heights of successive peaks.
A
Δx_max
At later times, this difference is much smaller.
Δx_max
0
t
−A

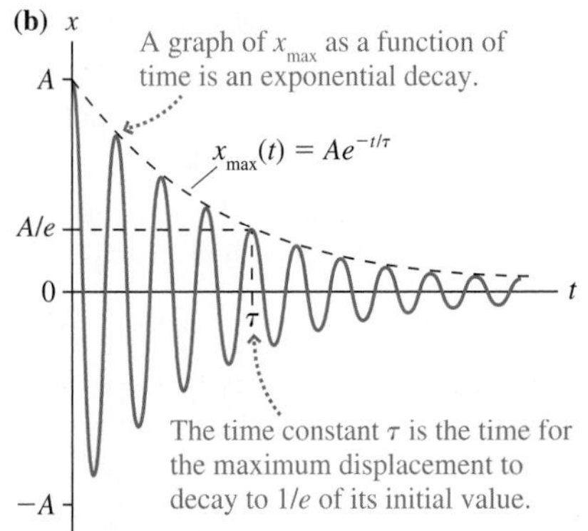
(b) x
A graph of x_max as a function of time is an exponential decay.
A
x_max(t) = Ae^{−t/τ}
A/e
0
τ
t
The time constant τ is the time for the maximum displacement to decay to 1/e of its initial value.
−A

Exponential decay

Exponential decay occurs when a quantity y is proportional to the number e taken to the power $-t/\tau$. The quantity τ is known as the **time constant.** We write this mathematically as

$$y = Ae^{-t/\tau}$$

y is proportional to $e^{-t/\tau}$

y starts with initial value A.

$y = Ae^{-t/\tau}$

y has decreased to 37% of its initial value.

A, A/e, A/e^2, 0, t, 0, τ, 2τ

Because t appears as the ratio t/τ, the important time intervals are τ, 2τ, and so on.

SCALING Whenever t increases by one time constant, y decreases by a factor of $1/e$. For instance:

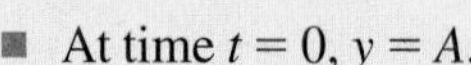

- At time $t = 0$, $y = A$.
- Increasing time to $t = \tau$ reduces y to A/e.
- A further increase to $t = 2\tau$ reduces y by another factor of $1/e$ to A/e^2.

Generally, we can say:

At $t = n\tau$, y has the value A/e^n.

LIMITS As t becomes large, y becomes very small and approaches zero.

Exercises 14–17

Damping smoothes the ride A car's wheels are attached to the car's body with springs so that the wheels can move up and down as the car moves over an uneven road. The car-spring system is a simple harmonic oscillator with a typical period of just under 1 second. You don't want the car to continue bouncing after hitting a bump, so a shock absorber provides damping. The time constant for the damping is about the same length as the period, so that the oscillation damps quickly.

Different Amounts of Damping

The damped oscillation shown in Figure 14.23 continues for a long time. The amplitude isn't zero after one time constant, or two, or three. . . . Mathematically, the oscillation never ceases, though the amplitude will eventually be so small as to be undetectable. For practical purposes, we can speak of the time constant τ as the *lifetime* of an oscillation—a measure of about how long it takes to decay. The best way to measure the relative size of the time constant is to compare it to the period. If $\tau \gg T$, the oscillation persists for many, many periods and the amplitude decrease from cycle to cycle is quite small. The oscillation of a bell after it is struck has $\tau \gg T$; the sound continues for a long time. Other oscillatory systems have very short time constants, as noted in the description of a car's suspension.

EXAMPLE 14.12 Finding a clock's decay time

The pendulum in a grandfather clock has a period of 1.00 s. If the clock's driving spring is allowed to run down, damping due to friction will cause the pendulum to slow to a stop. If the time constant for this decay is 300 s, how long will it take for the pendulum's swing to be reduced to half its initial amplitude?

PREPARE The time constant of 300 s is much greater than the 1.00 s period, so this is an example of modest damping as described by Equation 14.30.

SOLVE Equation 14.30 gives an expression for the decay of the maximum displacement of a damped harmonic oscillator:

$$x_{max}(t) = Ae^{-t/\tau}$$

As noted in Tactics Box 14.1, we can write this equation equally well in terms of the pendulum's angle in a straightforward manner:

$$\theta_{max}(t) = \theta_i e^{-t/\tau}$$

where θ_i is the initial angle of swing. At some time t, the time we wish to find, the amplitude has decayed to half its initial value. At this time,

$$\theta_{max}(t) = \theta_i e^{-t/\tau} = \frac{1}{2}\theta_i$$

The θ_i cancels, giving $e^{-t/\tau} = \frac{1}{2}$.

To solve this for t, we take the natural logarithm of both sides and use the logarithm property $\ln(e^a) = a$:

$$\ln(e^{-t/\tau}) = -\frac{t}{\tau} = \ln\left(\frac{1}{2}\right) = -\ln 2$$

In the last step we used the property $\ln(1/b) = -\ln b$. Now we can solve for t:

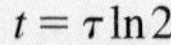

$$t = \tau \ln 2$$

The time constant was specified as $\tau = 300$ s, so

$$t = (300 \text{ s})(0.693) = 208 \text{ s}$$

It will take 208 s, or about 3.5 min, for the oscillations to decay by half after the spring has run down.

ASSESS The time is less than the time constant, which makes sense. The time constant is the time for the amplitude to decay to 37% of its initial value; we are looking for the time to decay to 50% of its initial value, which should be a shorter time. The time to decay to $\frac{1}{2}$ of the initial amplitude, $t = \tau \ln 2$, could be called the *half-life.* We will see this expression again when we work with radioactivity, another example of an exponential decay.

STOP TO THINK 14.6 Rank in order, from largest to smallest, the time constants τ_A to τ_D of the decays in the figures. The scales on all the graphs are the same.

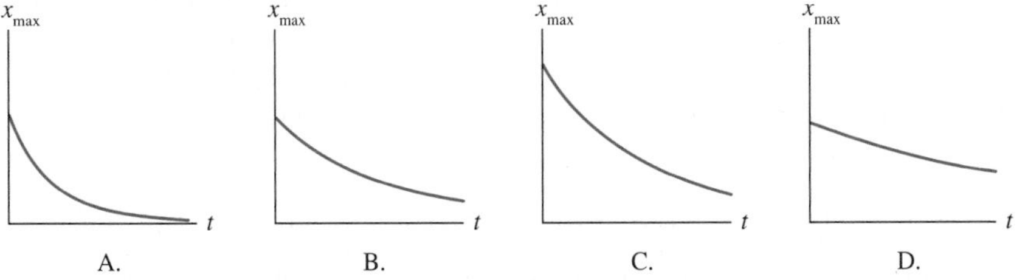

14.7 Driven Oscillations and Resonance

If you jiggle a cup of water, the water sloshes back and forth. This is an example of an oscillator (the water in the cup) subjected to a periodic external force (from your hand). This motion is called a **driven oscillation.**

We can give many examples of driven oscillations. The electromagnetic coil on the back of a loudspeaker cone provides a periodic magnetic force to drive the cone back and forth, causing it to send out sound waves. Earthquakes cause the surface of the earth to move back and forth; this motion causes buildings to oscillate, possibly producing damage or collapse.

Consider an oscillating system that, when left to itself, oscillates at a frequency f_0. We will call this the **natural frequency** of the oscillator. f_0 is simply the frequency of the system if it is displaced from equilibrium and released.

Suppose that this system is now subjected to a *periodic* external force of frequency f_{ext}. This frequency, which is called the **driving frequency,** is completely independent of the oscillator's natural frequency f_0. Somebody or something in the environment selects the frequency f_{ext} of the external force, causing the force to push on the system f_{ext} times every second. The external force causes the oscillation of the system, so it will oscillate at f_{ext}, the driving frequency, not at its natural frequency f_0.

Let's return to the example of the cup of water. If you nudge the cup, you will notice that the water sloshes back and forth at a particular frequency; this is the natural frequency f_0. Now shake the cup at some frequency; this is the driving frequency f_{ext}. As you shake the cup, the oscillation amplitude of the water depends very sensitively on the frequency f_{ext} of your hand. If the driving frequency is near the natural frequency of the system, the oscillation amplitude may become so large that water splashes out of the cup.

Any driven oscillator will show a similar dependence of amplitude on the driving frequency. Suppose a mass on a spring has a natural frequency $f_0 = 2$ Hz. We can use an external force to push and pull on the mass at frequency f_{ext}, measure the amplitude of the resulting oscillation, and then repeat this over and over for many different driving frequencies. A graph of amplitude versus driving frequency, such as the one in FIGURE 14.24, is called the oscillator's **response curve.**

Serious sloshing Water in a cup has a natural frequency at which it will slosh back and forth. The same is true of larger bodies of water. Water in Canada's Bay of Fundy would naturally move into or out of the bay with a period of 12 hours. This is nearly equal to the period of the tidal force of the moon; the two daily high tides are 12.5 hours apart. This *resonance,* a close match between the bay's natural frequency and the moon's driving frequency, produces a huge tidal amplitude. Low tide can be as much as 16 m below high tide, leaving boats high and dry.

FIGURE 14.24 The response curve shows the amplitude of a driven oscillator at frequencies near its natural frequency $f_0 = 2$ Hz.

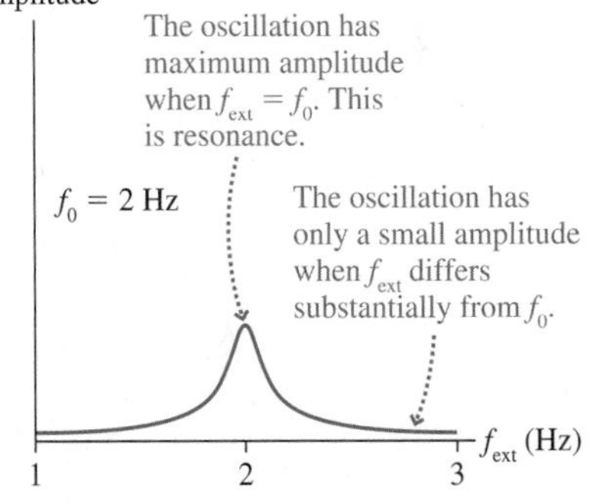

FIGURE 14.25 The response curve becomes taller and narrower as the damping is reduced.

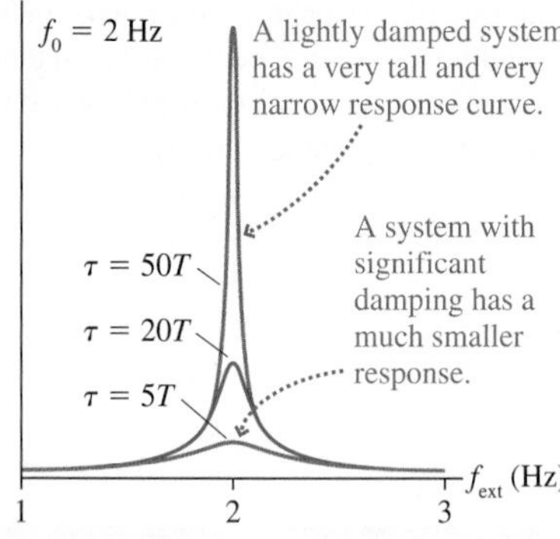
Amplitude
f_0 = 2 Hz
A lightly damped system has a very tall and very narrow response curve.
A system with significant damping has a much smaller response.
τ = 50T
τ = 20T
τ = 5T
f_{ext} (Hz)
1
2
3

At the right and left edges of Figure 14.24, the driving frequency is substantially different from the oscillator's natural frequency. The system oscillates, but its amplitude is very small. The system simply does not respond well to a driving frequency that differs much from f_0. As the driving frequency gets closer and closer to the natural frequency, the amplitude of the oscillation rises dramatically. After all, f_0 is the frequency at which the system "wants" to oscillate, so it is quite happy to respond to a driving frequency near f_0. Hence the amplitude reaches a maximum when the driving frequency matches the system's natural frequency: $f_{ext} = f_0$. This large-amplitude response to a driving force whose frequency matches the natural frequency of the system is a phenomenon called **resonance.** Within the context of driven oscillations, the natural frequency f_0 is often called the **resonance frequency.**

The amplitude can become exceedingly large when the frequencies match, especially if there is very little damping. FIGURE 14.25 shows the response curve of the oscillator of Figure 14.24 with different amounts of damping. Three different graphs are plotted, each with a different time constant for damping. The three graphs have damping that ranges from $\tau = 50T$ (very little damping) to $\tau = 5T$ (significant damping).

◀ **Simple harmonic music** A typical wine glass has a natural frequency of oscillation and a very small amount of damping. A tap on the rim of the glass causes it to "ring" like a bell. The time constant is hundreds of times longer than the period, so the sound will persist for several seconds. If you moisten your finger and slide it gently around the rim of the glass, it will stick and slip in quick succession. With some practice you can match the stick-slip to the frequency of oscillation of the glass. The resulting resonance creates a large amplitude and thus a very loud sound. You can tune the oscillation frequency by adding water, turning a set of glasses into an unusual musical instrument.

CONCEPTUAL EXAMPLE 14.13 Fixing an unwanted resonance

Railroad cars have a natural frequency at which they rock side to side. This can lead to problems on certain stretches of track that have bumps where the rails join. If the joints alternate sides, with a bump on the left rail and then on the right, a train car moving down the track is bumped one way and then the other. In some cases, bumps have caused rocking with amplitude large enough to derail the train. A train moving down the track at a certian speed is experiencing a large amplitude of oscillation due to alternating joints in the track. How can the driver correct this potentially dangerous situation?

REASON The large amplitude of oscillation is produced by a resonance, a match between the frequency at which the train car rocks back and forth and the frequency at which the car hits the bumps. To eliminate this resonance, the driver must either reduce the speed of the train—decreasing the driving frequency—or increase the speed of the train—thus increasing the driving frequency.

ASSESS It's perhaps surprising that increasing the speed of the train could produce a smoother ride. But increasing the frequency at which the train hits the bumps will eliminate the match with the natural rocking frequency just as surely as decreasing the speed.

Resonance and Hearing

Resonance in a system means that certain frequencies produce a large response and others do not. The phenomenon of resonance is responsible for the frequency discrimination of the ear.

As we will see in the next chapter, sound is a vibration in air. FIGURE 14.26 on the next page provides an overview of the structures by which sound waves that enter the ear produce vibrations in the cochlea, the coiled, fluid-filled, sound-sensing organ of the inner ear.

FIGURE 14.26 The structures of the ear.

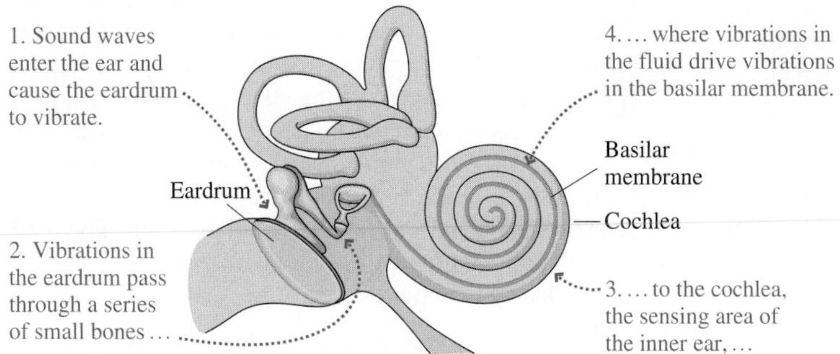

1. Sound waves enter the ear and cause the eardrum to vibrate.

FIGURE 14.27 shows a very simplified model of the cochlea. As a sound wave travels down the cochlea, it causes a large-amplitude vibration of the basilar membrane at the point where the membrane's natural oscillation frequency matches the sound frequency—a resonance. Lower-frequency sound causes a response farther from the stapes. Sensitive hair cells on the membrane sense the vibration and send nerve signals to your brain. The fact that different frequencies produce maximal response at different positions allows your brain to very accurately determine frequency because a small shift in frequency causes a detectable change in the position of the maximal response. People with no musical training can listen to two notes and easily determine which is at a higher pitch.

We now know a bit about how your ear responds to the vibration of a sound wave—but how does this vibration get from a source to your ear? This is a topic we will consider in the next chapter, when we look at *waves,* oscillations that travel.

FIGURE 14.27 Resonance plays a role in determining the frequencies of sounds we hear.

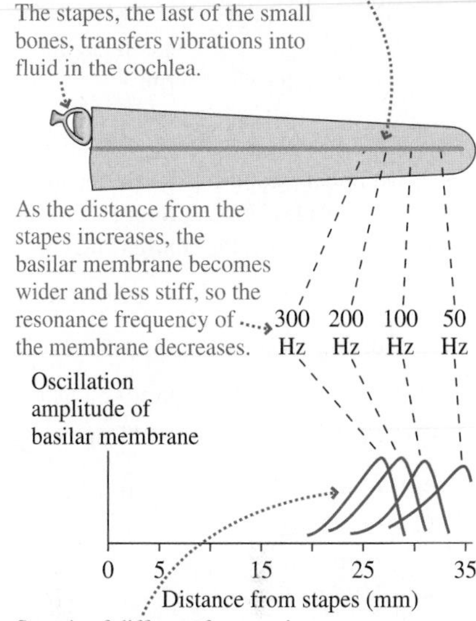

To analyze the cochlea, we imagine the spiral structure unrolled, with the basilar membrane separating two fluid-filled chambers.
The stapes, the last of the small bones, transfers vibrations into fluid in the cochlea.
As the distance from the stapes increases, the basilar membrane becomes wider and less stiff, so the resonance frequency of the membrane decreases.
300 Hz
200 Hz
100 Hz
50 Hz
Oscillation amplitude of basilar membrane
0
5
15
25
35
Distance from stapes (mm)
Sounds of different frequencies cause different responses in the basilar membrane.

INTEGRATED EXAMPLE 14.14 Springboard diving

Flexible diving boards designed for large deflections are called springboards. If a diver jumps up and lands on the end of the board, the resulting deflection of the diving board produces a linear restoring force that launches him into the air. But if the diver simply bobs up and down on the end of the board, we can effectively model his motion as that of a mass oscillating on a spring.

A light and flexible springboard deflects by 15 cm when a 65 kg diver stands on its end. He then jumps and lands on the end of the board, depressing it by a total of 25 cm, after which he moves up and down with the oscillations of the end of the board.

a. What is the frequency of the oscillation?
b. What is the maximum speed of his up-and-down motion?

Suppose the diver then drives the motion of the board with his legs, gradually increasing the amplitude of the oscillation. At some point the oscillation becomes large enough that his feet leave the board.

c. What is the amplitude of the oscillation when the diver just becomes airborne at one point of the cycle? What is the acceleration at this point?
d. A diver leaving a springboard can achieve a much greater height than a diver jumping from a fixed platform. Use energy concepts to explain how the spring of the board allows a greater vertical jump.

PREPARE We will model the diver on the board as a mass on a spring. As we've seen, the oscillation frequency is determined by the spring constant and the mass. The mass of the diver is given; we can determine the spring constant from the deflection of the springboard when the diver stands on the end.

FIGURE 14.28 is a sketch of the oscillation that will help us visualize the motion. The equilibrium position, with the diver standing motionless on the end of the board, corresponds to a deflection of 15 cm. When the diver jumps on the board, the total deflection is 25 cm, which means a deflection of an additional 10 cm beyond the equilibrium position, so 10 cm is the amplitude of the subsequent oscillation. When the board rises, it won't bend beyond its undeflected position, so the maximum possible amplitude is 15 cm.

FIGURE 14.28 Position-versus-time graph for the springboard.

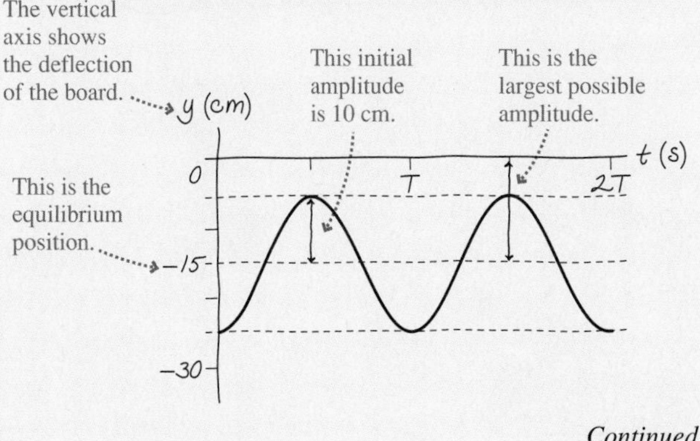

Continued

NOTE ▶ We've started the graph at the lowest point of the motion. Because we'll use only the equations for the maximum values of the speed and the acceleration, not the full equations that describe the motion, the exact starting point isn't critical. ◀

SOLVE a. **FIGURE 14.29** shows the forces on the diver as he stands motionless at the end of the board. The net force on him is zero, $\vec{F}_{net} = \vec{F}_{sp} + \vec{w} = \vec{0}$, so the two forces have equal magnitudes and we can write

$$F_{sp} = w$$

A linear restoring force means that the board obeys Hooke's law for a spring: $F_{sp} = k\Delta y$, where k is the spring constant. Thus the equilibrium equation is

$$k\Delta y = mg$$

Solving for the spring constant, we find

$$k = \frac{mg}{\Delta y} = \frac{(65\ \text{kg})(9.8\ \text{m/s}^2)}{0.15\ \text{m}} = 4.2 \times 10^3\ \text{N/m}$$

FIGURE 14.29 Forces on a springboard diver at rest at the end of the board.

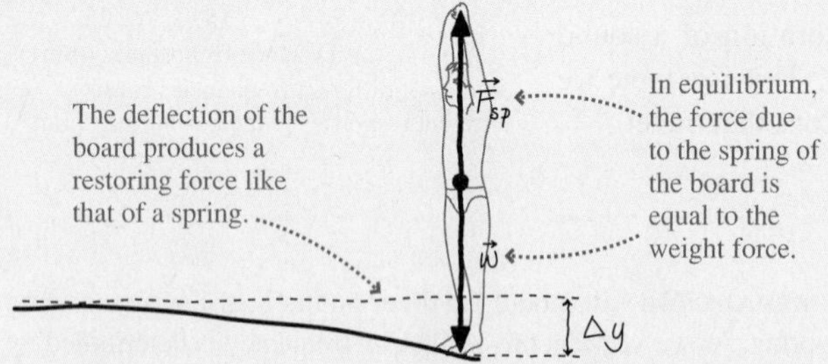

The deflection of the board produces a restoring force like that of a spring.
In equilibrium, the force due to the spring of the board is equal to the weight force.
$\vec{F}_{sp}$
$\vec{w}$
Δy

The frequency of the oscillation depends on the diver's mass and the spring constant of the board:

$$f = \frac{1}{2\pi}\sqrt{\frac{k}{m}} = \frac{1}{2\pi}\sqrt{\frac{4.2 \times 10^3\ \text{N/m}}{65\ \text{kg}}} = 1.3\ \text{Hz}$$

b. The maximum speed of the oscillation is given by Equation 14.15:

$$v_{max} = 2\pi f A = 2\pi(1.3\ \text{Hz})(0.10\ \text{m}) = 0.82\ \text{m/s}$$

c. We can see from Figure 14.28 that an amplitude of 15 cm returns the board to its undeflected position. At this point, the board exerts no upward force—no supporting normal force—on the diver, so the diver loses contact with the board. (His apparent weight becomes zero.) The acceleration at this point in the motion has the maximum possible magnitude but is negative because the acceleration graph is an upside-down version of the position graph. For a 15 cm oscillation amplitude, the acceleration at this point is computed using Equation 14.17:

$$a = -a_{max} = -(2\pi f)^2 A = [2\pi(1.3\ \text{Hz})]^2(0.15\ \text{m})$$
$$= -10\ \text{m/s}^2$$

d. The maximum jump height from a fixed platform is determined by the maximum speed at which a jumper leaves the ground. During a jump, chemical energy in the muscles is transformed into kinetic energy. This kinetic energy is transformed into potential energy as the jumper rises, back into kinetic energy as he falls, then into thermal energy as he hits the ground. The springboard recaptures and stores this kinetic energy rather than letting it degrade to thermal energy; a diver can jump up once and land on the board, storing the energy of his initial jump as elastic potential energy of the bending board. Then, as the board rebounds, turning the stored energy back into kinetic energy, the diver can push off from this moving platform, transforming even more chemical energy and thus further increasing his kinetic energy. This allows the diver to get "two jumps worth" of chemical energy in a single jump.

ASSESS The answer to part c, $-10\ \text{m/s}^2$ is very close to $a = -g$; only rounding errors in early steps kept our result from being exactly $a = -g$. This is to be expected. We learned in earlier chapters that an object loses contact with a surface—like a car coming off the track in a loop-the-loop—when its apparent weight becomes zero: $w_{app} = 0$. And in Chapter 5 we found that the apparent weight of an object in vertical motion becomes zero when $a = -g$—that is, when it enters free fall. This correspondence is a good check on our work; because part c has the answer we expect, we have confidence in our earlier steps.

SUMMARY

The goal of Chapter 14 has been to understand systems that oscillate with simple harmonic motion.

GENERAL PRINCIPLES

Restoring Forces

SHM occurs when a **linear restoring force** acts to return a system to an equilibrium position.

Mass on spring

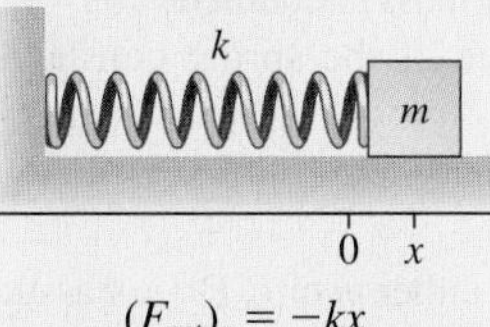

$(F_{net})_x = -kx$

The frequency of a mass on a spring depends on the mass and the spring constant:

$$f = \frac{1}{2\pi}\sqrt{\frac{k}{m}}$$

Pendulum

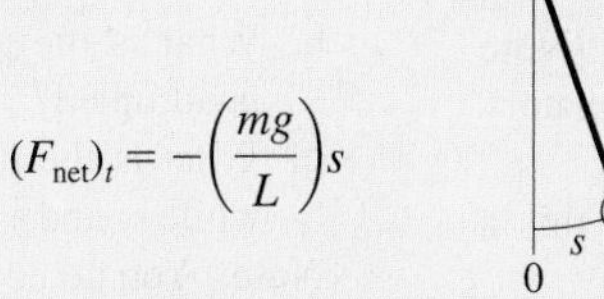

$$(F_{net})_t = -\left(\frac{mg}{L}\right)s$$

The frequency of a pendulum depends on the length and the free-fall acceleration:

$$f = \frac{1}{2\pi}\sqrt{\frac{g}{L}}$$

Energy

If there is no friction or dissipation, kinetic and potential energies are alternately transformed into each other in SHM, with the sum of the two conserved.

$$E = \frac{1}{2}mv_x^2 + \frac{1}{2}kx^2 = \frac{1}{2}m{v_{max}}^2 = \frac{1}{2}kA^2$$

All kinetic

−A 0 A x

All potential

IMPORTANT CONCEPTS

Oscillation

An **oscillation** is a repetitive motion about an equilibrium position. The **amplitude** A is the maximum displacement from equilibrium. The period T is the time for one cycle. We may also characterize an oscillation by its frequency f.

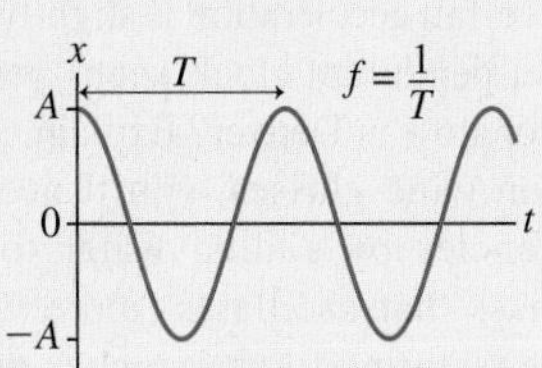

Simple Harmonic Motion (SHM)

SHM is an oscillation that is described by a sinusoidal function. All systems that undergo SHM can be described by the same functional forms.

Position-versus-time is a cosine function.

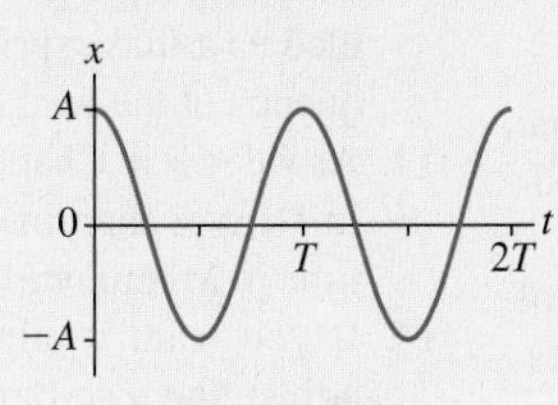

$x(t) = A\cos(2\pi ft)$

$x_{max} = A$

Velocity-versus-time is an inverted sine function.

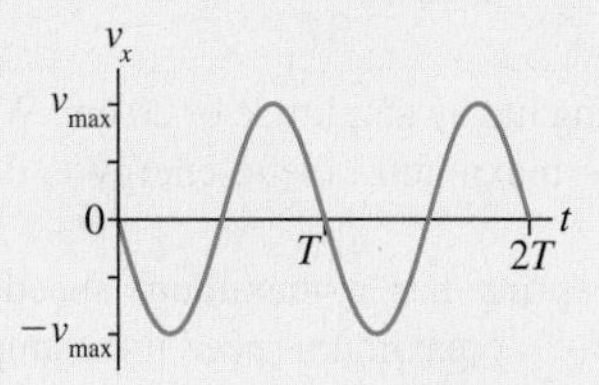

$v_x(t) = -v_{max}\sin(2\pi ft)$

$v_{max} = 2\pi fA$

Acceleration-versus-time is an inverted cosine function.

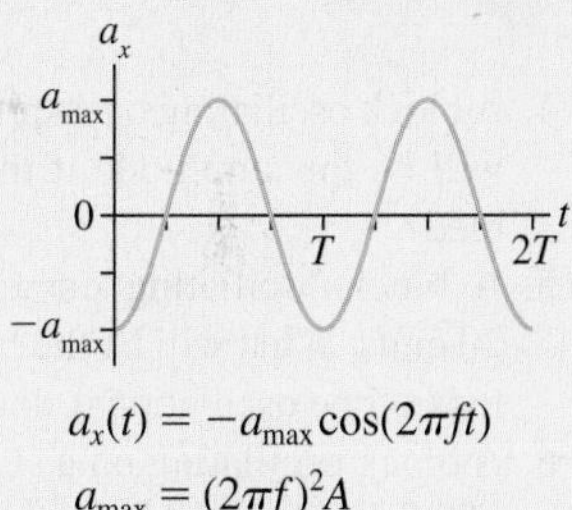

$a_x(t) = -a_{max}\cos(2\pi ft)$

$a_{max} = (2\pi f)^2 A$

APPLICATIONS

Damping

Simple harmonic motion with damping (due to drag) decreases in amplitude over time. The **time constant** τ determines how quickly the amplitude decays.

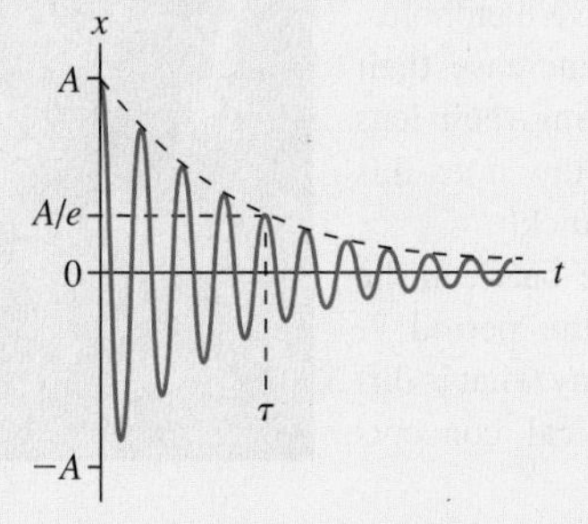

Resonance

A system that oscillates has a **natural frequency** of oscillation f_0. **Resonance** occurs if the system is driven with a frequency f_{ext} that matches this natural frequency. This may produce a large amplitude of oscillation.

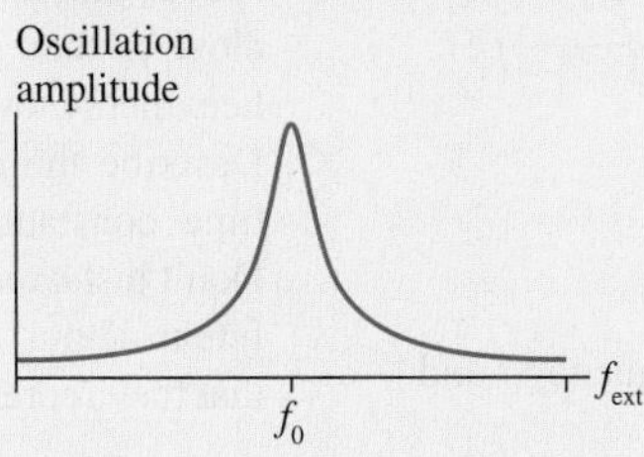

Physical pendulum

A **physical pendulum** is a pendulum with mass distributed along its length. The frequency depends on the position of the center of gravity and the moment of inertia.

The motion of legs during walking can be described using a physical pendulum model.

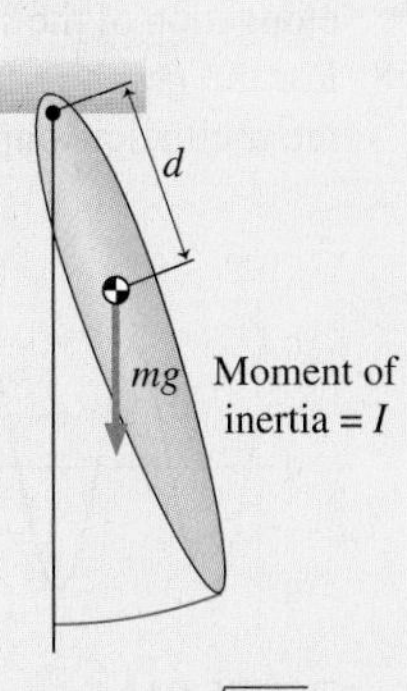

$$f = \frac{1}{2\pi}\sqrt{\frac{mgd}{I}}$$

For homework assigned on MasteringPhysics, go to www.masteringphysics.com

Problem difficulty is labeled as | (straightforward) to ||||| (challenging).

Problems labeled INT integrate significant material from earlier chapters; BIO are of biological or medical interest.

QUESTIONS

Conceptual Questions

1. Give three real-world examples of *oscillatory* motion. (Note that circular motion is similar to, but not the same as oscillatory motion.)
2. A person's heart rate is given in beats per minute. Is this a period or a frequency?
3. Figure Q14.3 shows the position-versus-time graph of a particle in SHM.
   a. At what time or times is the particle moving to the right at maximum speed?
   b. At what time or times is the particle moving to the left at maximum speed?
   c. At what time or times is the particle instantaneously at rest?

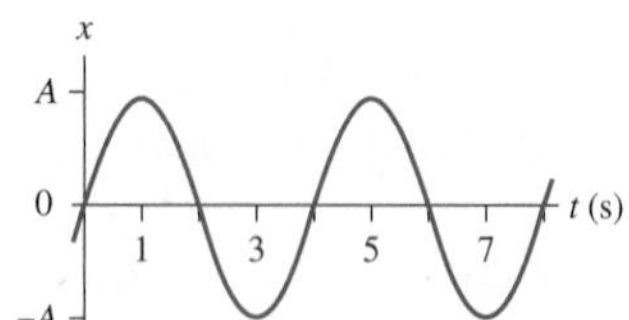

FIGURE Q14.3

4. A block oscillating on a spring has an amplitude of 20 cm. What will be the amplitude if the maximum kinetic energy is doubled?
5. A block oscillating on a spring has a maximum speed of 20 cm/s. What will be the block's maximum speed if the amplitude of the oscillation is doubled?
6. A block oscillating on a spring has a maximum kinetic energy of 2.0 J. What will be the maximum kinetic energy if the amplitude is doubled? Explain.
7. A block oscillating on a spring has a maximum speed of 30 cm/s. What will be the block's maximum speed if the initial elongation of the spring is doubled?
8. For the graph in Figure Q14.8, determine the frequency f and the oscillation amplitude A.

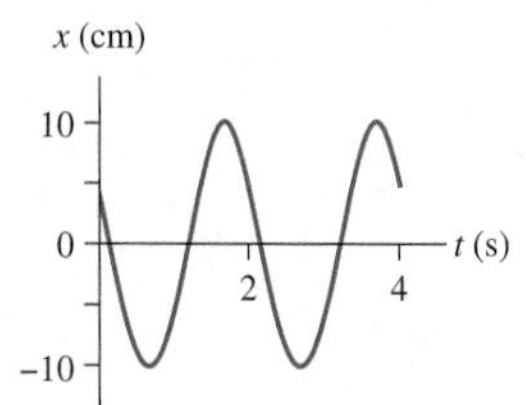

FIGURE Q14.8

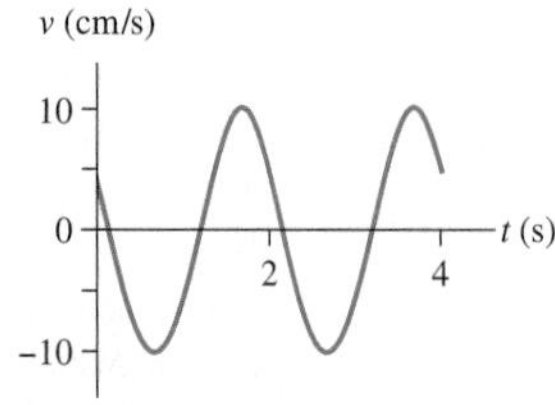

FIGURE Q14.9

9. For the graph in Figure Q14.9, determine the frequency f and the oscillation amplitude A.
10. A block oscillating on a spring has period $T = 2.0$ s.
    a. What is the period if the block's mass is doubled?
    b. What is the period if the value of the spring constant is quadrupled?
    c. What is the period if the oscillation amplitude is doubled while m and k are unchanged?

    Note: You do not know values for either m or k. Do *not* assume any particular values for them. The required analysis involves thinking about ratios.
11. A pendulum on Planet X, where the value of g is unknown, oscillates with a period of 2.0 s. What is the period of this pendulum if:
    a. Its mass is doubled?
    b. Its length is doubled?
    c. Its oscillation amplitude is doubled?

    Note: You do not know the values of m, L, or g, so do not assume any specific values.
12. BIO Flies flap their wings at frequencies much too high for pure muscle action. A hypothesis for how they achieve these high frequencies is that the flapping of their wings is the driven oscillation of a mass-spring system. One way to test this is to trim a fly's wings. If the oscillation of the wings can be modeled as a mass-spring system, how would this change the frequency of the wingbeats?
13. As we saw in Chapter 6, the free-fall acceleration is slightly less in Denver than in Miami. If a pendulum clock keeps perfect time in Miami, will it run fast or slow in Denver? Explain.
14. If you want to play a tune on wine glasses, you'll need to adjust the oscillation frequencies by adding water to the glasses. This changes the mass that oscillates (more water means more mass) but not the restoring force, which is determined by the stiffness of the glass itself. If you need to raise the frequency of a particular glass, should you add water or remove water?
15. BIO Sprinters push off from the ball of their foot, then bend their knee to bring their foot up close to the body as they swing their leg forward for the next stride. Why is this an effective strategy for running fast?
16. BIO Gibbons move through the trees by swinging from successive handholds, as we have seen. To increase their speed, gibbons may bring their legs close to their bodies. How does this help them move more quickly?

17. Describe the difference between the time constant τ and the period T. Don't just *name* them; say what is different about the physical concepts that they represent.

18. What is the difference between the driving frequency and the natural frequency of an oscillator?
19. Humans have a range of hearing of approximately 20 Hz to
BIO 20 kHz. Mice have auditory systems similar to humans, but all of the physical elements are smaller. Given this, would you expect mice to have a higher or lower frequency range than humans? Explain.
20. A person driving a truck on a "washboard" road, one with regularly spaced bumps, notices an interesting effect: When the truck travels at low speed, the amplitude of the vertical motion of the car is small. If the truck's speed is increased, the amplitude of the vertical motion also increases, until it becomes quite unpleasant. But if the speed is increased yet further, the amplitude decreases, and at high speeds the amplitude of the vertical motion is small again. Explain what is happening.
21. We've seen that stout tendons in the legs of hopping kangaroos
BIO store energy. When a kangaroo lands, much of the kinetic energy of motion is converted to elastic energy as the tendons stretch, returning to kinetic energy when the kangaroo again leaves the ground. If a hopping kangaroo increases its speed, it spends more time in the air with each bounce, but the contact time with the ground stays approximately the same. Explain why you would expect this to be the case.

Multiple-Choice Questions

22. | A spring has an unstretched length of 20 cm. A 100 g mass hanging from the spring stretches it to an equilibrium length of 30 cm.
    a. Suppose the mass is pulled down to where the spring's length is 40 cm. When it is released, it begins to oscillate. What is the amplitude of the oscillation?
    A. 5.0 cm B. 10 cm C. 20 cm D. 40 cm
    b. For the data given above, what is the frequency of the oscillation?
    A. 0.10 Hz B. 0.62 Hz C. 1.6 Hz D. 10 Hz
    c. Suppose this experiment were done on the moon, where the free-fall acceleration is approximately 1/6 of that on the earth. How would this change the frequency of the oscillation?
    A. The frequency would decrease.
    B. The frequency would increase.
    C. The frequency would stay the same.
23. | Figure Q14.23 represents the motion of a mass on a spring.
    a. What is the period of this oscillation?
    A. 12 s B. 24 s C. 36 s
    D. 48 s E. 50 s

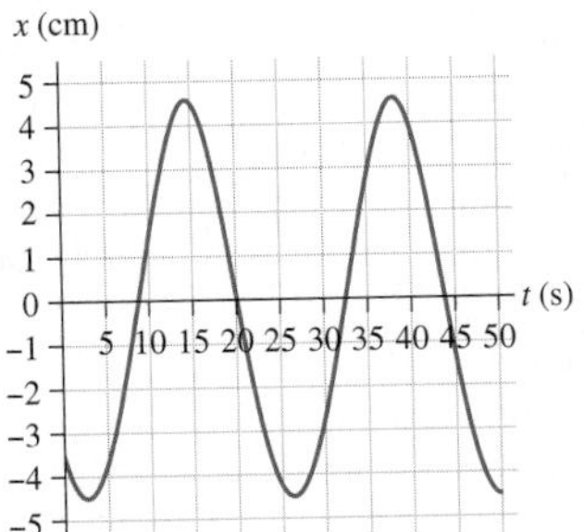

FIGURE Q14.23

    b. What is the amplitude of the oscillation?
    A. 1.0 cm B. 2.5 cm C. 4.5 cm
    D. 5.0 cm E. 9.0 cm
    c. What is the position of the mass at time $t = 30$ s?
    A. −4.5 cm B. −2.5 cm C. 0.0 cm
    D. 4.5 cm E. 30 cm
    d. When is the first time the velocity of the mass is zero?
    A. 0 s B. 2 s C. 8 s
    D. 10 s E. 13 s
    e. At which of these times does the kinetic energy have its maximum value?
    A. 0 s B. 8 s C. 13 s
    D. 26 s E. 30 s
24. | A ball of mass m oscillates on a spring with spring constant $k = 200$ N/m. The ball's position is $x = (0.350 \text{ m})\cos(15.0t)$, with t measured in seconds.
    a. What is the amplitude of the ball's motion?
    A. 0.175 m B. 0.350 m C. 0.700 m
    D. 7.50 m E. 15.0 m
    b. What is the frequency of the ball's motion?
    A. 0.35 Hz B. 2.39 Hz C. 5.44 Hz
    D. 6.28 Hz E. 15.0 Hz
    c. What is the value of the mass m?
    A. 0.45 kg B. 0.89 kg C. 1.54 kg
    D. 3.76 kg E. 6.33 kg
    d. What is the total mechanical energy of the oscillator?
    A. 1.65 J B. 3.28 J C. 6.73 J
    D. 10.1 J E. 12.2 J
    e. What is the ball's maximum speed?
    A. 0.35 m/s B. 1.76 m/s C. 2.60 m/s
    D. 3.88 m/s E. 5.25 m/s
25. || If you carry heavy weights in your hands, how will this affect the natural frequency at which your arms swing back and forth?
    A. The frequency will increase.
    B. The frequency will stay the same.
    C. The frequency will decrease.
26. | A heavy brass ball is used to make a pendulum with a period of 5.5 s. How long is the cable that connects the pendulum ball to the ceiling?
    A. 4.7 m B. 6.2 m
    C. 7.5 m D. 8.7 m
27. | Suppose you travel to the moon, and you take with you two timepieces: a pendulum clock and a wristwatch that runs with a wheel and a mainspring. (The wheel and spring work, essentially, like a mass on a spring, but the wheel rotates back and forth rather than moving up and down.) Which will keep good time on the moon?
    A. Only the pendulum clock
    B. Only the wristwatch
    C. Both timepieces
    D. Neither timepiece
28. | Very loud sounds can damage hearing by injuring the
BIO vibration-sensing hair cells on the basilar membrane. Suppose a person has injured hair cells on a segment of the basilar membrane close to the stapes. What type of sound is most likely to have produced this particular pattern of damage?
    A. Loud music with a mix of different frequencies
    B. A very loud, high-frequency sound
    C. A very loud, low-frequency sound

PROBLEMS

Section 14.1 Equilibrium and Oscillation

Section 14.2 Linear Restoring Forces and Simple Harmonic Motion

1. I When a guitar string plays the note "A," the string vibrates at 440 Hz. What is the period of the vibration?
2. I In the aftermath of an intense earthquake, the earth as a whole "rings" with a period of 54 minutes. What is the frequency (in Hz) of this oscillation?
3. I In taking your pulse, you count 75 heartbeats in 1 min. What are the period (in s) and frequency (in Hz) of your heart's oscillations? BIO
4. II Make a table with 3 columns and 8 rows. In row 1, label the columns θ (°), θ (rad), and $\sin\theta$. In the left column, starting in row 2, write 0, 2, 4, 6, 8, 10, and 12.
   a. Convert each of these angles, in degrees, to radians. Put the results in column 2. Show four decimal places.
   b. Calculate the sines. Put the results, showing four decimal places, in column 3.
   c. What is the first angle for which θ and $\sin\theta$ differ by more than 0.0010?
   d. Over what range of angles does the small-angle approximation appear to be valid?
5. I A heavy steel ball is hung from a cord to make a pendulum. The ball is pulled to the side so that the cord makes a 5° angle with the vertical. Holding the ball in place takes a force of 20 N. If the ball is pulled farther to the side so that the cord makes a 10° angle, what force is required to hold the ball?

Section 14.3 Describing Simple Harmonic Motion

6. II An air-track glider attached to a spring oscillates between the 10 cm mark and the 60 cm mark on the track. The glider completes 10 oscillations in 33 s. What are the (a) period, (b) frequency, (c) amplitude, and (d) maximum speed of the glider?
7. III An air-track glider is attached to a spring. The glider is pulled to the right and released from rest at $t = 0$ s. It then oscillates with a period of 2.0 s and a maximum speed of 40 cm/s.
   a. What is the amplitude of the oscillation?
   b. What is the glider's position at $t = 0.25$ s?
8. I What are the (a) amplitude and (b) frequency of the oscillation shown in Figure P14.8?

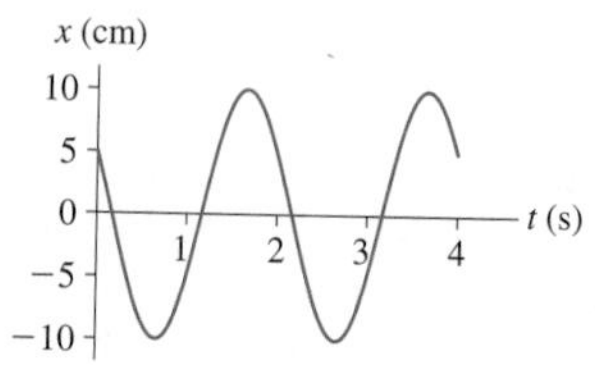

FIGURE P14.8

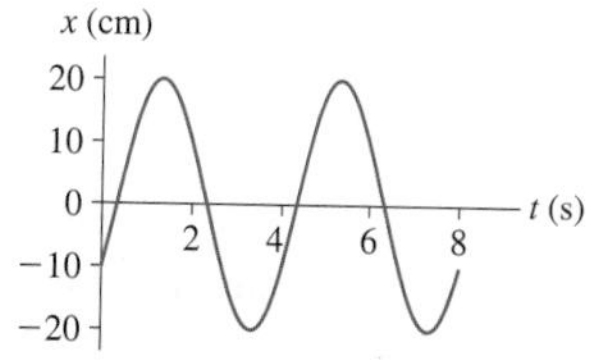

FIGURE P14.9

9. I What are the (a) amplitude and (b) frequency of the oscillation shown in Figure P14.9?
10. I An object in simple harmonic motion has an amplitude of 6.0 cm and a frequency of 0.50 Hz. Draw a position graph showing two cycles of the motion.
11. I During an earthquake, the top of a building oscillates with an amplitude of 30 cm at 1.2 Hz. What are the magnitudes of (a) the maximum displacement, (b) the maximum velocity, and (c) the maximum acceleration of the top of the building?
12. II Some passengers on an ocean cruise may suffer from motion sickness as the ship rocks back and forth on the waves. At one position on the ship, passengers experience a vertical motion of amplitude 1 m with a period of 15 s.
    a. To one significant figure, what is the maximum acceleration of the passengers during this motion?
    b. What fraction is this of g?
13. II A passenger car traveling down a rough road bounces up and down at 1.3 Hz with a maximum vertical acceleration of 0.20 m/s², both typical values. What are the (a) amplitude and (b) maximum speed of the oscillation?
14. III The New England Merchants Bank Building in Boston is 152 m high. On windy days it sways with a frequency of 0.17 Hz, and the acceleration of the top of the building can reach 2.0% of the free-fall acceleration, enough to cause discomfort for occupants. What is the total distance, side to side, that the top of the building moves during such an oscillation?

Section 14.4 Energy in Simple Harmonic Motion

15. III a. When the displacement of a mass on a spring is $\frac{1}{2}A$, what fraction of the mechanical energy is kinetic energy and what fraction is potential energy?
    b. At what displacement, as a fraction of A, is the energy half kinetic and half potential?
16. III A 1.0 kg block is attached to a spring with spring constant 16 N/m. While the block is sitting at rest, a student hits it with a hammer and almost instantaneously gives it a speed of 40 cm/s. What are
    a. The amplitude of the subsequent oscillations?
    b. The block's speed at the point where $x = \frac{1}{2}A$?
17. I A block attached to a spring with unknown spring constant oscillates with a period of 2.00 s. What is the period if
    a. The mass is doubled?
    b. The mass is halved?
    c. The amplitude is doubled?
    d. The spring constant is doubled?

    Parts a to d are independent questions, each referring to the initial situation.
18. II A 200 g air-track glider is attached to a spring. The glider is pushed 10.0 cm against the spring, then released. A student with a stopwatch finds that 10 oscillations take 12.0 s. What is the spring constant?
19. III The position of a 50 g oscillating mass is given by $x(t) = (2.0\text{ cm})\cos(10t)$, where t is in seconds. Determine:
    a. The amplitude.
    b. The period.
    c. The spring constant.
    d. The maximum speed.
    e. The total energy.
    f. The velocity at $t = 0.40$ s.

20. || A 200 g mass attached to a horizontal spring oscillates at a frequency of 2.0 Hz. At one instant, the mass is at $x = 5.0$ cm and has $v_x = -30$ cm/s. Determine:
    a. The period.
    b. The amplitude.
    c. The maximum speed.
    d. The total energy.
21. || A 507 g mass oscillates with an amplitude of 10.0 cm on a spring whose spring constant is 20.0 N/m. Determine:
    a. The period.
    b. The maximum speed.
    c. The total energy.
22. ||| A 300 g oscillator has a speed of 95.4 cm/s when its displacement is 3.00 cm and 71.4 cm/s when its displacement is 6.00 cm. What is the oscillator's maximum speed?

Section 14.5 Pendulum Motion

23. || A mass on a string of unknown length oscillates as a pendulum with a period of 4.00 s. What is the period if
    a. The mass is doubled?
    b. The string length is doubled?
    c. The string length is halved?
    d. The amplitude is halved?

    Parts a to d are independent questions, each referring to the initial situation.
24. ||| A 200 g ball is tied to a string. It is pulled to an angle of 8.00° and released to swing as a pendulum. A student with a stopwatch finds that 10 oscillations take 12.0 s. How long is the string?
25. || The angle of a pendulum is given by $\theta(t) = (0.10 \text{ rad})\cos(5t)$, where t is in seconds. Determine:
    a. The amplitude.
    b. The frequency.
    c. The length of the string.
    d. The angle at $t = 2.0$ s.
26. || It is said that Galileo discovered a basic principle of the pendulum—that the period is independent of the amplitude—by using his pulse to time the period of swinging lamps in the cathedral as they swayed in the breeze. Suppose that one oscillation of a swinging lamp takes 5.5 s. How long is the lamp chain?
27. || The free-fall acceleration on the moon is 1.62 m/s^2. What is the length of a pendulum whose period on the moon matches the period of a 2.00-m-long pendulum on the earth?
28. | Astronauts on the first trip to Mars take along a pendulum that has a period on earth of 1.50 s. The period on Mars turns out to be 2.45 s. What is the Martian free-fall acceleration?
29. ||| A building is being knocked down with a wrecking ball, which is a big metal sphere that swings on a 10-m-long cable. You are (unwisely!) standing directly beneath the point from which the wrecking ball is hung when you notice that the ball has just been released and is swinging directly toward you. How much time do you have to move out of the way?
30. ||| Interestingly, there have been several studies using cadavers to determine the moment of inertia of human body parts by letting them swing as a pendulum about a joint. In one study, the center of gravity of a 5.0 kg lower leg was found to be 18 cm from the knee. When pivoted at the knee and allowed to swing, the oscillation frequency was 1.6 Hz. What was the moment of inertia of the lower leg? BIO
31. ||| A pendulum clock keeps time by the swinging of a uniform solid rod pivoted at one end. The angular position of the rod is given by $\theta(t) = (0.175 \text{ rad})\sin(\pi t)$, where t is in seconds.
    a. What is the angular position of the rod at $t = 0.250$ s?
    b. What is the period of oscillation?
    c. How long is the rod?
32. ||| You and your friends find a rope that hangs down 15 m from a high tree branch right at the edge of a river. You find that you can run, grab the rope, and swing out over the river. You run at 2.0 m/s and grab the rope, launching yourself out over the river. How long must you hang on if you want to stay dry?
33. || A thin, circular hoop with a radius of 0.22 m is hanging from its rim on a nail. When pulled to the side and released, the hoop swings back and forth as a physical pendulum. The moment of inertia of a hoop for a rotational axis passing through its edge is $I = 2MR^2$. What is the period of oscillation of the hoop?
34. |||| An elephant's legs have a reasonably uniform cross section from top to bottom, and they are quite long, pivoting high on the animal's body. When an elephant moves at a walk, it uses very little energy to bring its legs forward, simply allowing them to swing like pendulums. For fluid walking motion, this time should be half the time for a complete stride; as soon as the right leg finishes swinging forward, the elephant plants the right foot and begins swinging the left leg forward. BIO

    a. An elephant has legs that stretch 2.3 m from its shoulders to the ground. How much time is required for one leg to swing forward after completing a stride?
    b. What would you predict for this elephant's stride frequency? That is, how many steps per minute will the elephant take?

Section 14.6 Damped Oscillations

35. | The amplitude of an oscillator decreases to 36.8% of its initial value in 10.0 s. What is the value of the time constant?
36. || Calculate and draw an accurate displacement graph from $t = 0$ s to $t = 10$ s of a damped oscillator having a frequency of 1.0 Hz and a time constant of 4.0 s.
37. || A small earthquake starts a lamppost vibrating back and forth. The amplitude of the vibration of the top of the lamppost is 6.5 cm at the moment the quake stops, and 8.0 s later it is 1.8 cm.
    a. What is the time constant for the damping of the oscillation?
    b. What was the amplitude of the oscillation 4.0 s after the quake stopped?
38. ||| When you drive your car over a bump, the springs connecting the wheels to the car compress. Your shock absorbers then damp the subsequent oscillation, keeping your car from bouncing up and down on the springs. Figure P14.38 shows real data for a car driven over a bump. Estimate the frequency and the time constant for this damped oscillation.

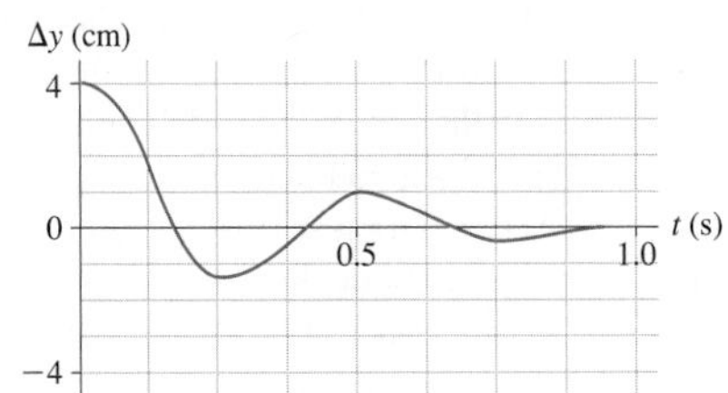

FIGURE P14.38

Section 14.7 Driven Oscillations and Resonance

39. || A 25 kg child sits on a 2.0-m-long rope swing. You are going to give the child a small, brief push at regular intervals. If you want to increase the amplitude of her motion as quickly as possible, how much time should you wait between pushes?

40. ||| Your car rides on springs, so it will have a natural frequency of oscillation. Figure P14.40 shows data for the amplitude of motion of a car driven at different frequencies. The car is driven at 20 mph over a washboard road with bumps spaced 10 feet apart; the resulting ride is quite bouncy. Should the driver speed up or slow down for a smoother ride?

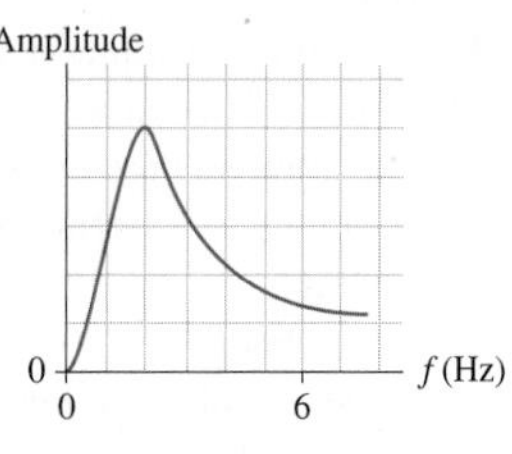

FIGURE P14.40

41. || Vision is blurred if the head is vibrated at 29 Hz because the vibrations are resonant with the natural frequency of the eyeball held by the musculature in its socket. If the mass of the eyeball is 7.5 g, a typical value, what is the effective spring constant of the musculature attached to the eyeball?
BIO

General Problems

42. || A spring has an unstretched length of 12 cm. When an 80 g ball is hung from it, the length increases by 4.0 cm. Then the ball is pulled down another 4.0 cm and released.
    a. What is the spring constant of the spring?
    b. What is the period of the oscillation?
    c. Draw a position-versus-time graph showing the motion of the ball for three cycles of the oscillation. Let the equilibrium position of the ball be $y = 0$. Be sure to include appropriate units on the axes so that the period and the amplitude of the motion can be determined from your graph.

43. || A 0.40 kg ball is suspended from a spring with spring constant 12 N/m. If the ball is pulled down 0.20 m from the equilibrium position and released, what is its maximum speed while it oscillates?

44. | A spring is hanging from the ceiling. Attaching a 500 g mass to the spring causes it to stretch 20.0 cm in order to come to equilibrium.
    a. What is the spring constant?
    b. From equilibrium, the mass is pulled down 10.0 cm and released. What is the period of oscillation?
    c. What is the maximum speed of the mass? At what position or positions does it have this speed?

45. || A spring with spring constant 15.0 N/m hangs from the ceiling. A ball is suspended from the spring and allowed to come to rest. It is then pulled down 6.00 cm and released. If the ball makes 30 oscillations in 20.0 s, what are its (a) mass and (b) maximum speed?

46. || A spring is hung from the ceiling. When a coffee mug is attached to its end, the spring stretches 2.0 cm before reaching its new equilibrium length. The mug is then pulled down slightly and released. What is the frequency of oscillation?

47. ||| On your first trip to Planet X you happen to take along a 200 g mass, a 40.0-cm-long spring, a meter stick, and a stopwatch. You're curious about the free-fall acceleration on Planet X, where ordinary tasks seem easier than on earth, but you can't find this information in your Visitor's Guide. One night you suspend the spring from the ceiling in your room and hang the mass from it. You find that the mass stretches the spring by 31.2 cm. You then pull the mass down 10.0 cm and release it. With the stopwatch you find that 10 oscillations take 14.5 s. Can you now satisfy your curiosity?

48. ||| An object oscillating on a spring has the velocity graph shown in Figure P14.48. Draw a velocity graph if the following changes are made.
    a. The amplitude is doubled and the frequency is halved.
    b. The amplitude and spring constant are kept the same, but the mass is quadrupled.

    Parts a and b are independent questions, each starting from the graph shown.

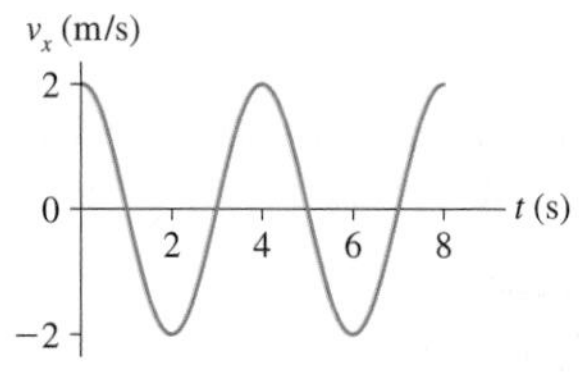

FIGURE P14.48

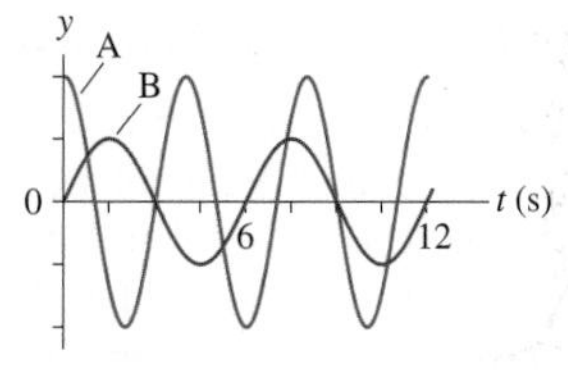

FIGURE P14.49

49. || The two graphs in Figure P14.49 are for two different vertical mass-spring systems.
    a. What is the frequency of system A? What is the first time at which the mass has maximum speed while traveling in the upward direction?
    b. What is the period of system B? What is the first time at which the mechanical energy is all potential?
    c. If both systems have the same mass, what is the ratio k_A/k_B of their spring constants?

50. ||| As we've seen, astronauts measure their mass by measuring the period of oscillation when sitting in a chair connected to a spring. The Body Mass Measurement Device on Skylab, a 1970s space station, had a spring constant of 606 N/m. The empty chair oscillated with a period of 0.901 s. What is the mass of an astronaut who oscillates with a period of 2.09 s when sitting in the chair?
BIO

51. |||| A 100 g ball attached to a spring with spring constant 2.50 N/m oscillates horizontally on a frictionless table. Its velocity is 20.0 cm/s when $x = -5.00$ cm.
    a. What is the amplitude of oscillation?
    b. What is the speed of the ball when $x = 3.00$ cm?

52. || The ultrasonic transducer used in a medical ultrasound imaging device is a very thin disk ($m = 0.10$ g) driven back and forth in SHM at 1.0 MHz by an electromagnetic coil.
BIO
    a. The maximum restoring force that can be applied to the disk without breaking it is 40,000 N. What is the maximum oscillation amplitude that won't rupture the disk?
    b. What is the disk's maximum speed at this amplitude?

53. || A compact car has a mass of 1200 kg. Assume that the car has one spring on each wheel, that the springs are identical, and that the mass is equally distributed over the four springs.
    a. What is the spring constant of each spring if the empty car bounces up and down 2.0 times each second?
    b. What will be the car's oscillation frequency while carrying four 70 kg passengers?

54. ||| Four people with a combined mass of 300 kg are riding in a 1100 kg car. When they drive down a washboard road with bumps spaced 5.0 m apart, they notice that the car bounces up and down with a maximum amplitude when the car is traveling at 6.0 m/s. The driver stops the car and everyone exits the vehicle. How much does the car rise up on its springs?

55. || A 500 g air-track glider attached to a spring with spring constant 10 N/m is sitting at rest on a frictionless air track. A 250 g glider is pushed toward it from the far end of the track at a speed of 120 cm/s. It collides with and sticks to the 500 g glider. What are the amplitude and period of the subsequent oscillations?
INT

56. ||||| A 1.00 kg block is attached to a horizontal spring with spring constant 2500 N/m. The block is at rest on a frictionless surface. A 10.0 g bullet is fired into the block, in the face opposite the spring, and sticks.
INT
   a. What was the bullet's speed if the subsequent oscillations have an amplitude of 10.0 cm?
   b. Could you determine the bullet's speed by measuring the oscillation frequency? If so, how? If not, why not?

57. |||| Figure P14.57 shows two springs, each with spring constant 20 N/m, connecting a 2.5 kg block to two walls. The block slides on a frictionless surface. If the block is displaced from equilibrium, it will undergo simple harmonic motion. What is the frequency of that motion?

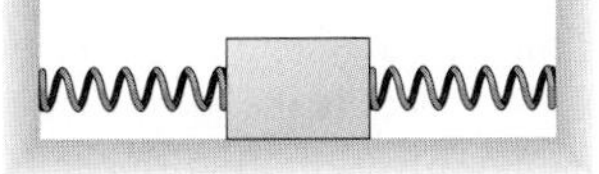

FIGURE P14.57

58. || Bungee Man is a superhero who does super deeds with the help of Super Bungee cords. The Super Bungee cords act like ideal springs no matter how much they are stretched. One day, Bungee Man stopped a school bus that had lost its brakes by hooking one end of a Super Bungee to the rear of the bus as it passed him, planting his feet, and holding on to the other end of the Bungee until the bus came to a halt. (Of course, he then had to quickly release the Bungee before the bus came flying back at him.) The mass of the bus, including passengers, was 12,000 kg, and its speed was 21.2 m/s. The bus came to a stop in 50.0 m.
   a. What was the spring constant of the Super Bungee?
   b. How much time after the Super Bungee was attached did it take the bus to stop?

59. |||| Two 50 g blocks are held 30 cm above a table. As shown in Figure P14.59, one of them is just touching a 30-cm-long spring. The blocks are released at the same time. The block on the left hits the table at exactly the same instant as the block on the right first comes to an instantaneous rest. What is the spring constant?
INT

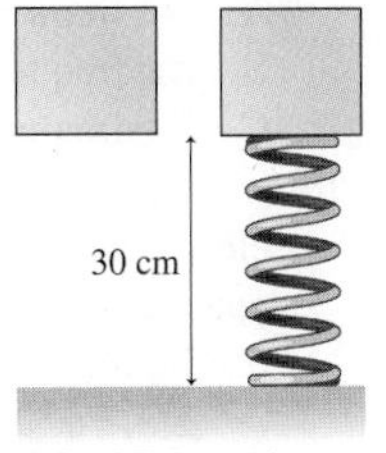

FIGURE P14.59

60. | The earth's free-fall acceleration varies from 9.78 m/s^2 at the equator to 9.83 m/s^2 at the poles. A pendulum whose length is precisely 1.000 m can be used to measure g. Such a device is called a *gravimeter.*
   a. How long do 100 oscillations take at the equator?
   b. How long do 100 oscillations take at the north pole?
   c. Suppose you take your gravimeter to the top of a high mountain peak near the equator. There you find that 100 oscillations take 201 s. What is g on the mountain top?

61. |||| A pendulum clock has a heavy bob supported on a very thin steel rod that is 1.00000 m long at 20°C.
INT
   a. To 6 significant figures, what is the clock's period? Assume that g is 9.80 m/s^2 exactly.
   b. To 6 significant figures, what is the period if the temperature increases by 10°C?
   c. The clock keeps perfect time at 20°C. At 30°C, after how many hours will the clock be off by 1.0 s?

62. ||| A pendulum consists of a massless, rigid rod with a mass at one end. The other end is pivoted on a frictionless pivot so that it can turn through a complete circle. The pendulum is inverted, so the mass is directly above the pivot point, then released. The speed of the mass as it passes through the lowest point is 5.0 m/s. If the pendulum later undergoes small-amplitude oscillations at the bottom of the arc, what will the frequency be?
INT

63. | Two side-by-side pendulum clocks have heavy bobs at the ends of rigid, very lightweight arms. One pendulum has a 38.8-cm-long rod, the other a 24.8-cm-long rod. Each clock makes one tick for each complete swing of its pendulum.
   a. Determine the frequencies and periods of the two clocks.
   b. Because the two pendulums have different frequencies, their ticks are usually "out of step." However, you notice that they do get back into step (tick at the same instant) at regular intervals. How much time elapses between such events?
   c. The getting-into-step phenomenon is, itself, periodic. What is the frequency of this phenomenon? Can you see a relationship between its frequency and the frequencies of the two clocks?

64. |||| Orangutans can move by brachiation, swinging like a pendulum beneath successive handholds. If an orangutan has arms that are 0.90 m long and repeatedly swings to a 20° angle, taking one swing immediately after another, estimate how fast it is moving in m/s.
BIO INT

65. ||| The 15 g head of a bobble-head doll oscillates in SHM at a frequency of 4.0 Hz.
   a. What is the spring constant of the spring on which the head is mounted?
   b. Suppose the head is pushed 2.0 cm against the spring, then released. What is the head's maximum speed as it oscillates?
   c. The amplitude of the head's oscillations decreases to 0.50 cm in 4.0 s. What is the head's time constant?

66. || An oscillator with a mass of 500 g and a period of 0.50 s has an amplitude that decreases by 2.0% during each complete oscillation. If the initial amplitude is 10 cm, what will be the amplitude after 25 oscillations?

67. |||| An infant's toy has a 120 g wooden animal hanging from a spring. If pulled down gently, the animal oscillates up and down with a period of 0.50 s. His older sister pulls the spring a bit more than intended. She pulls the animal 30 cm below its equilibrium position, then lets go. The animal flies upward and detaches from the spring right at the animal's equilibrium position. If the animal does not hit anything on the way up, how far above its equilibrium position will it go?
INT

68. |||| A jellyfish can propel itself with jets of water pushed out of its bell, a flexible structure on top of its body. The elastic bell and the water it contains function as a mass-spring system, greatly increasing efficiency. Normally, the jellyfish emits one jet right after the other, but we can get some insight into the jet system by looking at a single jet thrust. Figure P14.68 shows a graph of the motion of one point in the wall of the bell for such a single jet; this is the pattern of a damped oscillation. The spring constant for the bell can be estimated to be 1.2 N/m.
BIO
   a. What is the period for the oscillation?
   b. Estimate the effective mass participating in the oscillation. This is the mass of the bell itself plus the mass of the water.
   c. Consider the peaks of positive displacement in the graph. By what factor does the amplitude decrease over one period? Given this, what is the time constant for the damping?

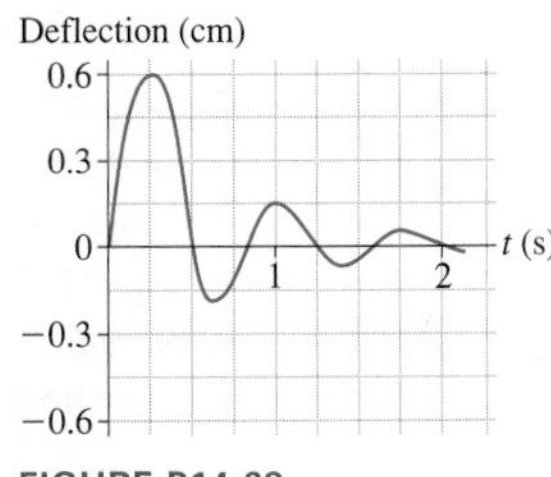

FIGURE P14.68

69. |||| A 200 g oscillator in a vacuum chamber has a frequency of 2.0 Hz. When air is admitted, the oscillation decreases to 60% of its initial amplitude in 50 s. How many oscillations will have been completed when the amplitude is 30% of its initial value?

70. ||| While seated on a tall bench, extend your lower leg a small BIO amount and then let it swing freely about your knee joint, with no muscular engagement. It will oscillate as a damped pendulum. Figure P14.70 is a graph of the lower leg angle versus time in such an experiment. Estimate (a) the period and (b) the time constant for this oscillation.

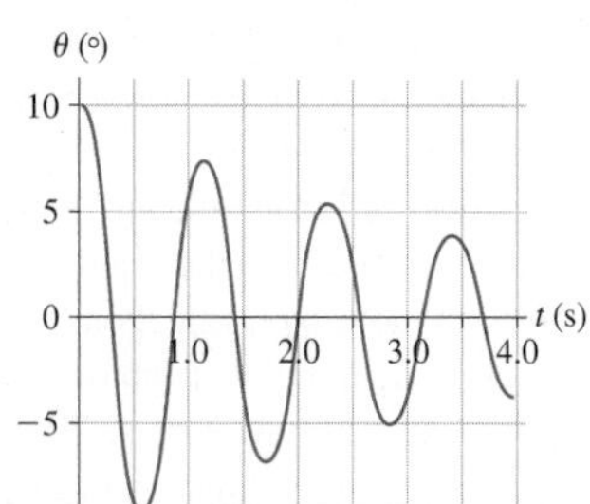

FIGURE P14.70

71. || A 2.0 kg block oscillates up and down on a spring with spring constant 220 N/m. Its initial amplitude is 15 cm. If the time constant for damping of the oscillation is 3.0 s, how much mechanical energy has been dissipated from the block-spring system after 6.0 s?

72. |||| In Chapter 10, we saw that the Achilles tendon will stretch BIO and then rebound, storing and returning energy during a step. We can model this motion as that of a mass on a spring. This is far from a perfect model, but it does give some insight. If a 60 kg person stands on a low wall with her full weight on the ball of one foot and the heel free to move, the stretch of the Achilles tendon will cause her center of gravity to lower by about 2.5 mm.
   a. What is the spring constant of her Achilles tendon?
   b. If she bounces a little, what is her oscillation period?
   c. When walking or running, the tendon spring begins to stretch as the ball of the foot takes the weight of a stride, transforming kinetic energy into elastic potential energy. Ideally, the cycle of the motion will have advanced so that potential energy has just finished being converted back to kinetic energy as the foot leaves the ground. What fraction of an oscillation period should the time between landing and liftoff correspond to? Given the period you calculated above, what is this time?
   d. Sprinters running a short race keep their foot in contact with the ground for about 0.10 s, some of which corresponds to the heel strike and subsequent rolling forward of the foot. Given this, does the answer to part c make sense?

Passage Problems

Web Spiders and Oscillations

All spiders have special organs that make them exquisitely sensitive to vibrations. Web spiders detect vibrations of their web to determine what has landed in their web, and where.

In fact, spiders carefully adjust the tension of strands to "tune" their web. Suppose an insect lands and is trapped in a web. The silk of the web serves as the spring in a spring-mass system while the body of the insect is the mass. The frequency of oscillation depends on the restoring force of the web and the mass of the insect. Spiders respond more quickly to larger—and therefore more valuable—prey, which they can distinguish by the web's oscillation frequency.

Suppose a 12 mg fly lands in the center of a horizontal spider's web, causing the web to sag by 3.0 mm.

73. | Assuming that the web acts like a spring, what is the spring constant of the web?
   A. 0.039 N/m B. 0.39 N/m
   C. 3.9 N/m D. 39 N/m

74. | Modeling the motion of the fly on the web as a mass on a spring, at what frequency will the web vibrate when the fly hits it?
   A. 0.91 Hz B. 2.9 Hz C. 9.1 Hz D. 29 Hz

75. | If the web were vertical rather than horizontal, how would the frequency of oscillation be affected?
   A. The frequency would be higher.
   B. The frequency would be lower.
   C. The frequency would be the same.

76. | Spiders are more sensitive to oscillations at higher frequencies. For example, a low-frequency oscillation at 1 Hz can be detected for amplitudes down to 0.1 mm, but a high-frequency oscillation at 1 kHz can be detected for amplitudes as small as 0.1 μm. For these low- and high-frequency oscillations, we can say that
   A. The maximum acceleration of the low-frequency oscillation is greater.
   B. The maximum acceleration of the high-frequency oscillation is greater.
   C. The maximum accelerations of the two oscillations are approximately equal.

STOP TO THINK ANSWERS

Stop to Think 14.1: C. The frequency is inversely proportional to the period, so a shorter period implies a higher frequency.

Stop to Think 14.2: D. The restoring force is proportional to the displacement. If the displacement from equilibrium is doubled, the force is doubled as well.

Stop to Think 14.3: A. The maximum speed occurs when the mass passes through its equilibrium position.

Stop to Think 14.4: $f_D > f_C = f_B > f_A$. The frequency is determined by the ratio of k to m.

Stop to Think 14.5: C. The increase in length will cause the frequency to decrease and thus the period will increase. The time between ticks will increase, so the clock will run slow.

Stop to Think 14.6: $\tau_D > \tau_B = \tau_C > \tau_A$. The time constant is the time to decay to 37% of the initial height. The time constant is independent of the initial height.

15 Traveling Waves and Sound

This bat's ears are much more prominent than its eyes. It would appear that hearing is a much more important sense than sight for bats. How does a bat use sound waves to locate prey?

LOOKING AHEAD ►►

The goal of Chapter 15 is to learn the basic properties of traveling waves.

The Wave Model

A **wave** is a disturbance traveling through a medium.

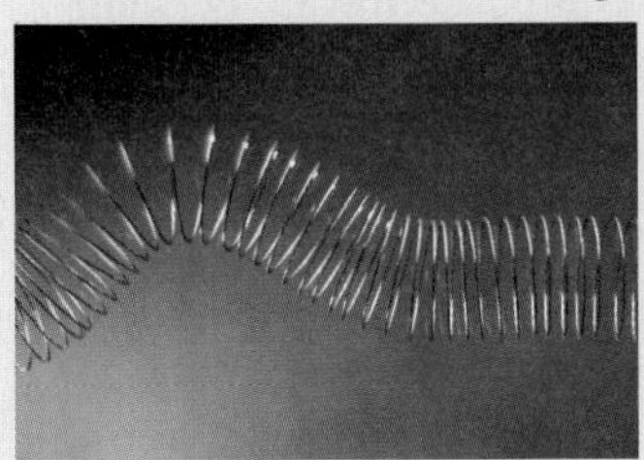

The wave propagates, but the particles of the medium don't. Here the coils of a stretched spring simply move up and back down as the wave passes.

Wave Properties

A few basic quantities can describe any type of wave.

This train of ocean waves is periodic. How fast do the waves move? That's the wave **speed.** What is the distance between successive wave crests? That's the **wavelength.** How many waves strike the beach each minute? That's the **frequency.**

The description of wave motion is closely related to that of simple harmonic motion.

Looking Back ◄◄
Section 14.3 Description of simple harmonic motion

Types of Waves

Our model can describe any type of wave, from ocean waves to vibrating strings, but two types of waves are especially important: **sound** and **light.**

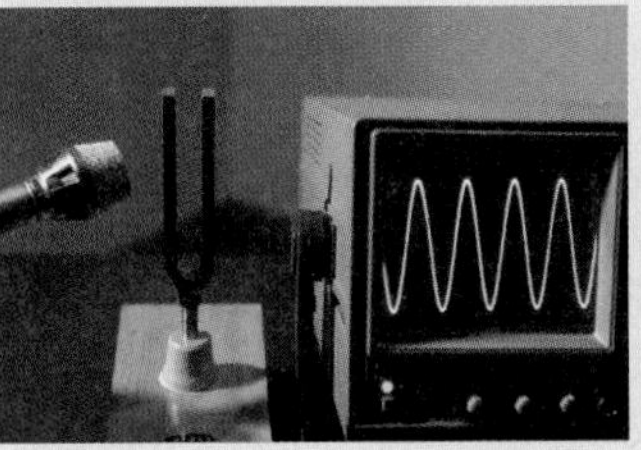

Displaying the sound waves from a tuning fork clearly shows their periodic nature.

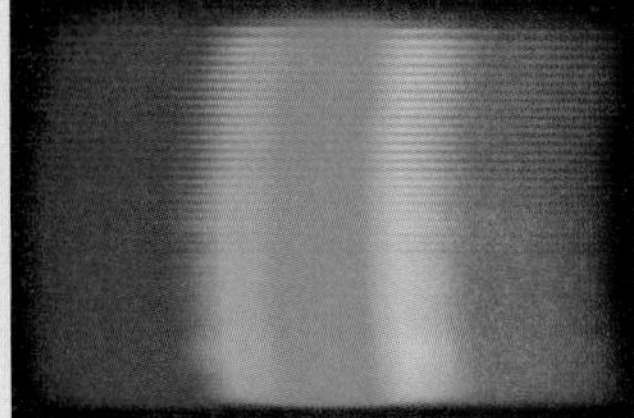

Visible light comes in a range of wavelengths corresponding to the colors of the rainbow.

Energy and Intensity

All waves carry energy. How much? That's a question of **intensity.** Your ears are sensitive to sounds over a remarkable range of intensities, so we use the logarithmic **decibel** scale for sound intensity level.

A lens focuses sunlight onto a small area, increasing the intensity.

Doppler Effect

The frequency and wavelength of waves are shifted when there is relative motion between the source and the observer of waves. We call this the **Doppler effect.**

Radar waves shift in frequency when they reflect from a moving vehicle; the higher the speed, the bigger the shift.

15.1 The Wave Model

The *particle model* that we have been using since Chapter 1 allowed us to simplify the treatment of motion of complex objects by considering them to be particles. Balls, cars, and rockets obviously differ from one another, but the general features of their motions are well described by treating them as particles. As we saw in Chapter 3, a ball or a rock or a car flying through the air will undergo the same motion. The particle model helps us understand this underlying simplicity.

In this chapter we will introduce the basic properties of waves with a **wave model** that emphasizes those aspects of wave behavior common to all waves. Although sound waves, water waves, and radio waves are clearly different, the wave model will allow us to understand many of the important features they share.

The wave model is built around the idea of a **traveling wave,** which is an organized disturbance that travels with a well-defined wave speed. This definition seems straightforward, but several new terms must be understood to gain a complete understanding of the concept of a traveling wave.

FIGURE 15.1 Ripples on a pond are a traveling wave.

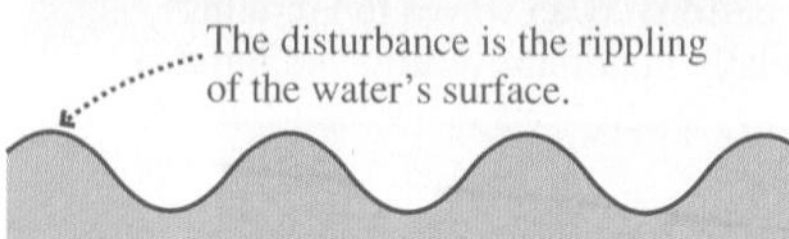
The disturbance is the rippling of the water's surface.
The water is the medium.

Mechanical Waves

Mechanical waves are waves that involve the motion of a substance through which they move, the **medium.** For example, the medium of a water wave is the water, the medium of a sound wave is the air, and the medium of a wave on a stretched string is the string.

As a wave passes through a medium, the atoms that make up the medium are displaced from equilibrium, much like pulling a spring away from its equilibrium position. This is a **disturbance** of the medium. The water ripples of FIGURE 15.1 are a disturbance of the water's surface.

A wave disturbance is created by a *source.* The source of a wave might be a rock thrown into water, your hand plucking a stretched string, or an oscillating loudspeaker cone pushing on the air. Once created, the disturbance travels outward through the medium at the **wave speed** v. This is the speed with which a ripple moves across the water or a pulse travels down a string.

The disturbance propagates through the medium, and a wave does transfer *energy,* but **the medium as a whole does not travel!** The ripples on the pond (the disturbance) move outward from the splash of the rock, but there is no outward flow of water. Likewise, the particles of a string oscillate up and down but do not move in the direction of a pulse traveling along the string. **A wave transfers energy, but it does not transfer any material or substance outward from the source.**

You may have been at a sporting event in which spectators do "The Wave." The wave moves around the stadium, but the spectators (the medium, in this case) stay right where they are. This is a clear example of the principle that a wave does not transfer any material.

Electromagnetic and Matter Waves

Mechanical waves require a medium, but there are waves that do not. Two important types of such waves are electromagnetic waves and matter waves.

Electromagnetic waves are waves of an *electromagnetic field.* Electromagnetic waves are very diverse, including visible light, radio waves, microwaves, and x rays. Electromagnetic waves require no material medium and can travel through a vacuum; light can travel through space, though sound cannot. At this point, we have not defined what an "electromagnetic field" is, so we won't worry about the precise nature of what is "waving" in electromagnetic waves. The wave model can describe many of the important aspects of these waves without a detailed description of their exact nature. We'll look more closely at electromagnetic waves in Chapter 25, once we have a full understanding of electric and magnetic fields.

One of the most significant discoveries of the 20th century was that material particles, such as electrons and atoms, have wave-like characteristics. We will learn how a full description of matter at an atomic scale requires an understanding of such **matter waves** in Chapter 28.

Transverse and Longitudinal Waves

Most waves fall into two general classes: *transverse* and *longitudinal.* For mechanical waves, these terms describe the relationship between the motion of the particles that carry the wave and the motion of the wave itself.

Two types of wave motion

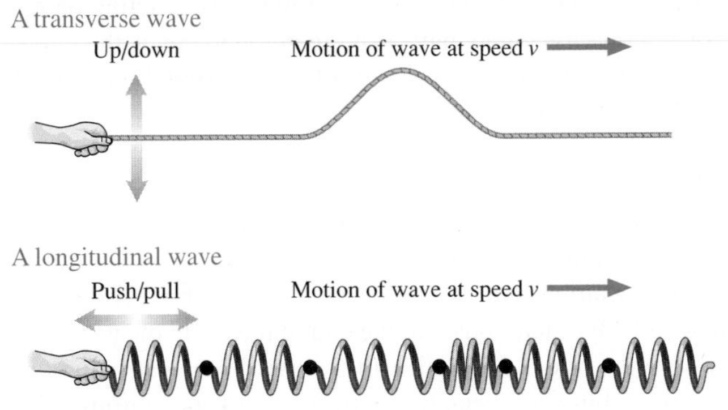

For mechanical waves, a **transverse wave** is a wave in which the particles in the medium move *perpendicular* to the direction in which the wave travels. Shaking the end of a stretched string up and down creates a wave that travels along the string in a horizontal direction while the particles that make up the string oscillate vertically.

In a **longitudinal wave,** the particles in the medium move *parallel* to the direction in which the wave travels. Here we see a chain of masses connected by springs. If you give the first mass in the chain a sharp push, a disturbance travels down the chain by compressing and expanding the springs.

The rapid motion of the earth's crust during an earthquake can produce a disturbance that travels through the earth. The two most important types of earthquake waves are S waves (which are transverse) and P waves (which are longitudinal), as shown in FIGURE 15.2. The longitudinal P waves are faster, but the transverse S waves are more destructive. Residents of a city a few hundred kilometers from an earthquake will feel the resulting P waves as much as a minute before the S waves, giving them a crucial early warning.

FIGURE 15.2 Different types of earthquake waves.

The passage of a P wave expands and compresses the ground. The motion is parallel to the direction of travel of the wave.

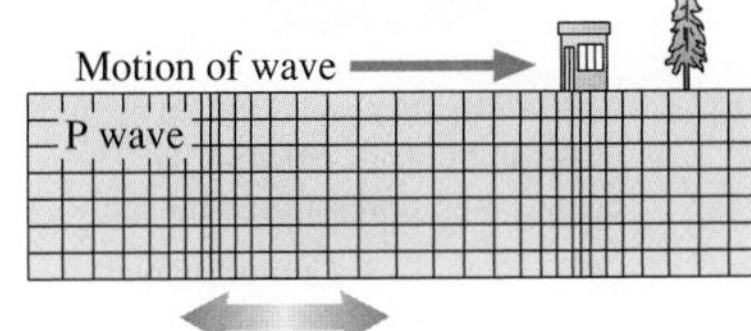

The passage of an S wave moves the ground up and down. The motion is perpendicular to the direction of travel of the wave.

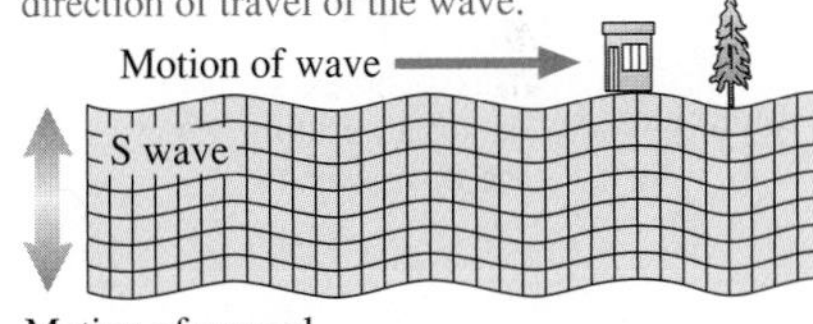

STOP TO THINK 15.1 Spectators at a sporting event do "The Wave," as shown in the photo on the preceding page. Is this a transverse or longitudinal wave?

15.2 Traveling Waves

When you drop a pebble in a pond, waves travel outward. But how does this happen? How does a mechanical wave travel through a medium? In answering this question, we must be careful to distinguish the motion of the wave from the motion of the particles that make up the medium. The wave itself is not a particle, so we cannot apply Newton's laws to the wave. However, we can use Newton's laws to examine how the medium responds to a disturbance.

Waves on a String

FIGURE 15.3 shows a transverse *wave pulse* traveling to the right along a stretched string. Imagine watching a little dot on the string as a wave pulse passes by. As the pulse approaches from the left, the string near the dot begins to curve. Once the string

FIGURE 15.3 The motion of a string as a wave passes.

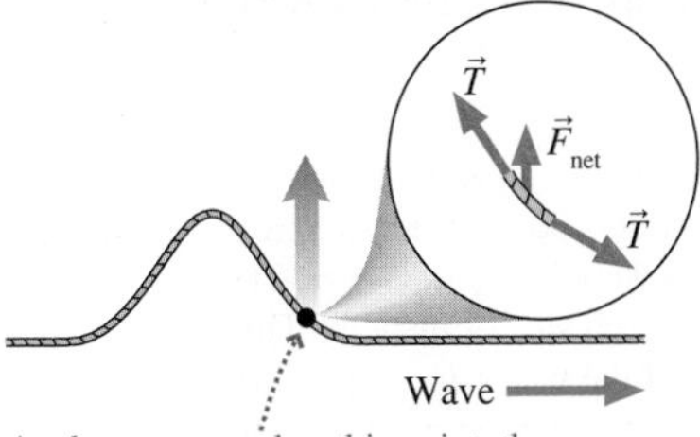

As the wave reaches this point, the curvature of the string leads to a net force that pulls the string upward.

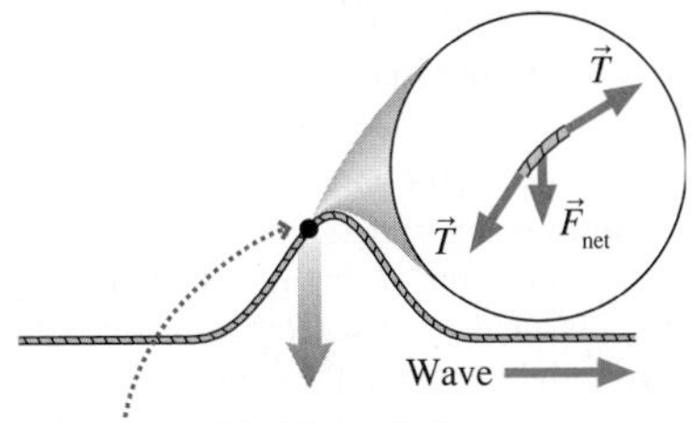

After the peak has passed, the curvature of the string leads to a net force that pulls the string downward.

curves, the tension forces pulling on a small segment of string no longer cancel each other. As the wave passes, the curvature of the string leads to a net force that first pulls each little piece of the string up and then, after the pulse passes, back down. Each point on the string moves perpendicular to the motion of the wave, so **a wave on a string is a transverse wave.**

No new physical principles are required to understand how this wave moves. The motion of a pulse along a string is a direct consequence of the tension acting on the segments of the string. An external force may have been required to create the pulse, but **once started, the pulse continues to move because of the internal dynamics of the medium.**

Sound Waves

FIGURE 15.4 A sound wave produced by a loudspeaker.

(a) The loudspeaker cone moves in and out in response to electrical signals.

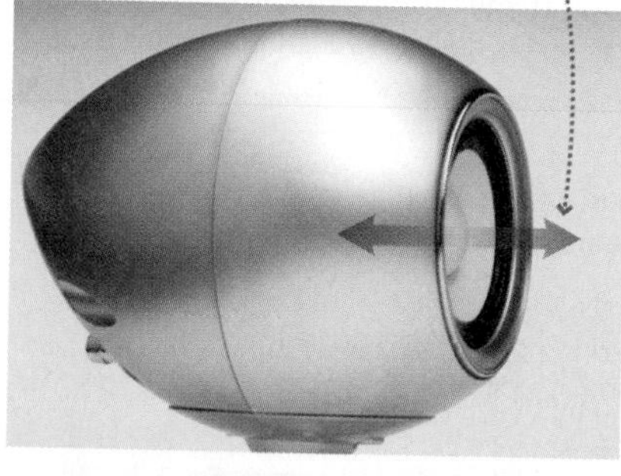

(b) The loudspeaker receives an electrical pulse and pushes out sharply. This creates a compression of the air.

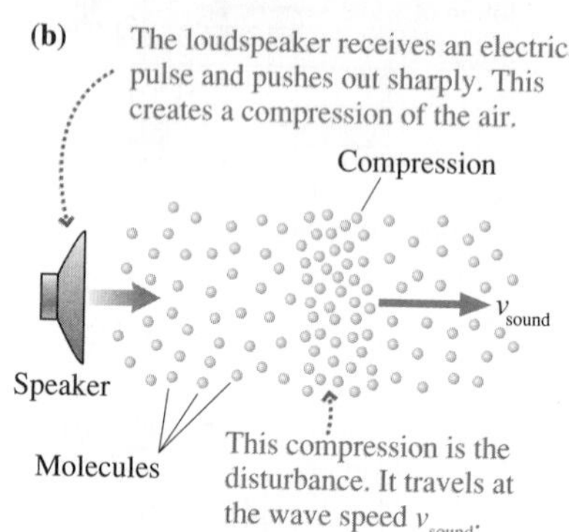
Compression
v_{sound}
Speaker
Molecules
This compression is the disturbance. It travels at the wave speed v_{sound}.

Next, let's see how a sound wave in air is created using a loudspeaker. When the loudspeaker cone in FIGURE 15.4a moves forward, it compresses the air in front of it, as shown in FIGURE 15.4b. The *compression* is the disturbance that travels forward through the air. This is much like the sharp push on the end of the chain of springs on the preceding page, so **a sound wave is a longitudinal wave.** We usually think of sound waves as traveling in air, but sound can travel through any gas, through liquids, and even through solids. A wave similar to that in Figure 15.4b is produced if you hit the end of a metal rod with a hammer.

The motion of a wave on a string is determined by the internal dynamics of the string. Similarly, the motion of a sound wave in air is determined by the physics of gases that we explored in Chapter 12. Once created, the wave in Figure 15.4b will propagate forward; its motion is entirely determined by the properties of the air.

Wave Speed Is a Property of the Medium

The above discussions of waves on a string and sound waves in air show that the motion of these waves depends on the properties of the medium. A full analysis of these waves would lead to the important and somewhat surprising conclusion that **the wave speed does not depend on the shape or size of the pulse, how the pulse was generated, or how far it has traveled.**

What properties of a string determine the speed of waves traveling along the string? The only likely candidates are the string's mass, length, and tension. Because a pulse doesn't travel faster on a longer and thus more massive string, neither the total mass m nor the total length L is important. Instead, the speed depends on the mass-to-length *ratio,* which is called the **linear density** μ of the string:

$$\mu = \frac{m}{L} \tag{15.1}$$

Linear density characterizes the *type* of string we are using. A fat string has a larger value of μ than a skinny string made of the same material. Similarly, a steel wire has a larger value of μ than a plastic string of the same diameter.

How does the speed of a wave on a string vary with the tension and the linear density? Using what we know about forces and motion, we can make some predictions:

- A string with a greater tension responds more rapidly, so the wave will move at a higher speed. **Wave speed increases with increasing tension.**
- A string with a greater linear density has more inertia. It will respond less rapidly, so the wave will move at a lower speed. **Wave speed decreases with increasing linear density.**

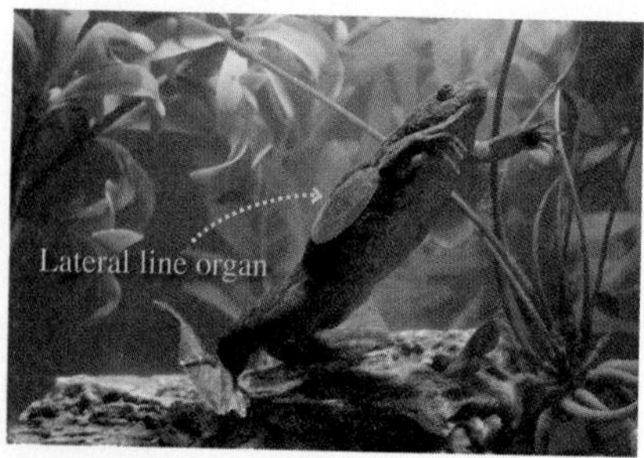
Lateral line organ

◄ **Sensing water waves** BIO The African clawed frog has a hunting strategy similar to that of many spiders: It sits and waits for prey to come to it. Like a spider, the frog detects prey animals by the vibrations they cause—not in a web, but in the water. The frog has an array of sensors called the lateral line organ on each side of its body. This organ, highlighted in the photo at left, detects oscillations of the water due to passing waves. The frog can determine where the waves come from and what type of animal made them, and thus whether a strike is called for.

A full analysis of the motion of the string leads to an expression for the speed of a wave that shows both of these trends:

$$v_{\text{string}} = \sqrt{\frac{T_s}{\mu}} \tag{15.2}$$

Wave speed on a stretched string with tension T_s and linear density μ

The subscript s on the symbol T_s for the string's tension will distinguish it from the symbol T for the *period* of oscillation.

Every point on a wave pulse travels with the speed given by Equation 15.2. You can increase the wave speed either by *increasing* the string's tension (make it tighter) or by *decreasing* the string's linear density (make it skinnier). We'll examine the implications for stringed musical instruments in Chapter 16.

10.2

EXAMPLE 15.1 When does the spider sense its lunch?

All spiders are very sensitive to vibrations. An orb spider will sit at the center of its large, circular web and monitor radial threads for vibrations created when an insect lands. Assume that these threads are made of silk with a linear density of 1.0×10^{-5} kg/m under a tension of 0.40 N. If an insect lands in the web 30 cm from the spider, how long will it take for the spider to find out?

PREPARE When the insect hits the web, a wave pulse will be transmitted along the silk fibers. The speed of the wave depends on the properties of the silk.

SOLVE First, we determine the speed of the wave:

$$v = \sqrt{\frac{T_s}{\mu}} = \sqrt{\frac{0.40\ \text{N}}{1.0 \times 10^{-5}\ \text{kg/m}}} = 200\ \text{m/s}$$

The time for the wave to travel a distance $d = 30$ cm to reach the spider is

$$\Delta t = \frac{d}{v} = \frac{0.30\ \text{m}}{200\ \text{m/s}} = 1.5\ \text{ms}$$

ASSESS Spider webs are made of very light strings under significant tension, so the wave speed is quite high and we expect a short travel time—important for the spider to quickly respond to prey caught in the web. Our answer makes sense.

What properties of a gas determine the speed of a sound wave traveling through the gas? It seems plausible that the speed of a sound pulse is related to the speed with which the molecules of the gas move—faster molecules should mean a faster sound wave. In Chapter 12, we found that the typical speed of an atom of mass m, the root-mean-square speed, is

$$v_{\text{rms}} = \sqrt{\frac{3k_B T}{m}}$$

where k_B is Boltzmann's constant and T the absolute temperature in kelvin. A thorough analysis finds that the sound speed is slightly less than this rms speed, but has the same dependence on the temperature and the molecular mass:

$$v_{\text{sound}} = \sqrt{\frac{\gamma k_B T}{m}} = \sqrt{\frac{\gamma R T}{M}} \tag{15.3}$$

v_{sound} / T — SQUARE ROOT p. 368

Sound speed in a gas at temperature T

10.3

In Equation 15.3 M is the molar mass (kg per mol) and γ is a constant that depends on the gas: $\gamma = 1.67$ for monatomic gases such as helium, $\gamma = 1.40$ for diatomic

TABLE 15.1 The speed of sound

Medium	Speed (m/s)
Air (0°C)	331
Air (20°C)	343
Helium (0°C)	970
Ethyl alcohol	1170
Water	1480
Human tissue (ultrasound)	1540
Lead	1200
Aluminum	5100
Granite	6000
Diamond	12,000

gases such as nitrogen and oxygen, and $\gamma \approx 1.3$ for a triatomic gas such as carbon dioxide or water vapor.

Certain trends in Equation 15.3 are worth mentioning:

- The speed of sound in air (and other gases) increases with temperature. **For calculations in this chapter, you can use the speed of sound in air at 20°C, 343 m/s, unless otherwise specified.**
- At a given temperature, the speed of sound increases as the molecular mass of the gas decreases. Thus the speed of sound in room-temperature helium is faster than that in room-temperature air.
- The speed of sound doesn't depend on the pressure or the density of the gas.

Table 15.1 lists the speeds of sound in various materials. The bonds between atoms in liquids and solids result in higher sound speeds in these phases of matter. Generally, sound waves travel faster in liquids than in gases, and faster in solids than in liquids. The speed of sound in a solid depends on its density and stiffness. Light, stiff solids (such as diamond) transmit sound at very high speeds. The sound speed is much lower in dense, soft solids such as lead.

EXAMPLE 15.2 The speed of sound on Mars

On a typical Martian morning, the very thin atmosphere (which is almost entirely carbon dioxide) is a frosty −100°C. What is the speed of sound? At what approximate temperature would the speed be double this value?

PREPARE Equation 15.3 gives the speed of sound in terms of the molar mass and the absolute temperature. If we assume that the atmosphere is composed of pure CO_2, the molar mass is the sum of the molar masses of the constituents: 12 g/mol for C and 16 g/mol for each O, giving $M = 12 + 16 + 16 = 44$ g/mol $= 0.044$ kg/mol. The absolute temperature is $T = 173$ K. Carbon dioxide is a triatomic gas, so $\gamma = 1.3$.

SOLVE The speed of sound with the noted conditions is

$$v_{\text{sound}} = \sqrt{\frac{\gamma RT}{M}} = \sqrt{\frac{(1.3)(8.31 \text{ J/mol})(173 \text{ K})}{0.044 \text{ kg/mol}}} = 210 \text{ m/s}$$

Rather than do a separate calculation to determine the temperature required for the higher speed, we can make an argument using proportionality. The speed is proportional to the square root of the temperature, so doubling the speed requires an increase in the temperature by a factor of 4 to about 690 K, or about 420°C.

ASSESS The speed of sound doesn't depend on pressure, so even though the atmosphere is "thin" we needn't adjust our calculation. The speed of sound is lower for heavier molecules and colder temperatures, so we expect that the speed of sound on Mars, with its cold, carbon-dioxide atmosphere, will be lower than that on earth, just as we found.

Electromagnetic waves, such as light, travel at a much higher speed than do mechanical waves. As we'll discuss in Chapter 25, all electromagnetic waves travel at the same speed in a vacuum. We call this speed the **speed of light,** which we represent with the symbol c. The value of the speed of light in a vacuum is

$$v_{\text{light}} = c = 3.00 \times 10^8 \text{ m/s} \tag{15.4}$$

This is almost one million times the speed of sound in air! At this speed, light could circle the earth 7.5 times in one second.

NOTE ▶ The speed of electromagnetic waves is lower when they travel through a material, but this value for the speed of light in a vacuum is also good for electromagnetic waves traveling through air. ◀

EXAMPLE 15.3 How far away was the lightning?

During a thunderstorm, you see a flash from a lightning strike. 8.0 seconds later, you hear the crack of the thunder. How far away did the lightning strike?

PREPARE Two different kinds of waves are involved, with very different wave speeds. The flash of the lightning generates light waves; these will travel from the point of the strike to your position essentially instantaneously. (It takes light about 5 μs to travel

1 mile—not something you will notice.) The strike also generates sound waves that you hear as thunder; these travel much more slowly. The delay between the flash and the thunder is the time it takes for the sound wave to travel.

SOLVE We will assume that the speed of sound has its room temperature $(20°C)$ value of 343 m/s. During the time between seeing the flash and hearing the thunder, the sound travels a distance

$$d = v\,\Delta t = (343 \text{ m/s})(8.0 \text{ s}) = 2.7 \times 10^3 \text{ m} = 2.7 \text{ km}$$

ASSESS This seems reasonable. As you know from casual observations of lightning storms, an 8-second delay between the flash of the lightning and the crack of the thunder means a strike that is close but not too close. A few km seems reasonable.

TRY IT YOURSELF

Distance to a lightning strike Sound travels approximately 1 km in 3 s, or 1 mi in 5s. When you see a lightning flash, start counting seconds. When you hear the thunder, stop counting. Divide the result by 3, and you will have the approximate distance to the lightning strike in kilometers; divide by 5 and you will have the approximate distance in miles.

STOP TO THINK 15.2 Suppose you shake the end of a stretched string to produce a wave. Which of the following actions would increase the speed of the wave down the string? There may be more than one correct answer; if so, give all that are correct.

A. Move your hand up and down more quickly as you generate the wave.
B. Move your hand up and down a greater distance as you generate the wave.
C. Use a heavier string of the same length, under the same tension.
D. Use a lighter string of the same length, under the same tension.
E. Stretch the string tighter to increase the tension.
F. Loosen the string to decrease the tension.

15.3 Graphical and Mathematical Descriptions of Waves

Now that we know a bit about waves and how they travel, it's time to develop our understanding further by describing waves with graphs and equations.

Describing waves and their motion takes a bit more thought than describing particles and their motion. Until now, we have been concerned with quantities, such as position and velocity, that depend only on time. We can write these, as we did in Chapter 14, as $x(t)$ or $v(t)$, indicating that x and v are *functions* of the time variable t. Functions of the single variable t are all right for a particle, because a particle is in only one place at a time, but a wave is not localized. It is spread out through space at each instant of time. We need a function that tells us what a wave is doing at an instant of time (when) at a particular point in space (where). We need a function that depends on *both* position and time.

NOTE ▶ The analysis that follows is for a wave on a string, which is easy to visualize, but the results apply to any traveling wave. ◀

Snapshot and History Graphs

When we considered the motion of particles, we developed a graphical description before the mathematical description. We'll follow a similar approach for waves. Consider the wave pulse shown moving along a stretched string in **FIGURE 15.5**. (We will consider somewhat artificial triangular and square-shaped wave pulses in this section to clearly show the edges of the pulse.) The graph shows the string's displacement y at a particular instant of time t_1 as a function of position x along the string. This is a "snapshot" of the wave, much like what you might make with a camera whose shutter is opened briefly at t_1. A graph that shows the wave's displacement as a function of position at a single instant of time is called a **snapshot graph.** For a wave on a string, a snapshot graph is literally a picture of the wave at this instant.

As the wave moves, we can take more snapshots. **FIGURE 15.6** shows a sequence of snapshot graphs as the wave of Figure 15.5 continues to move. These are like successive

FIGURE 15.5 A snapshot graph of a wave pulse on a string.

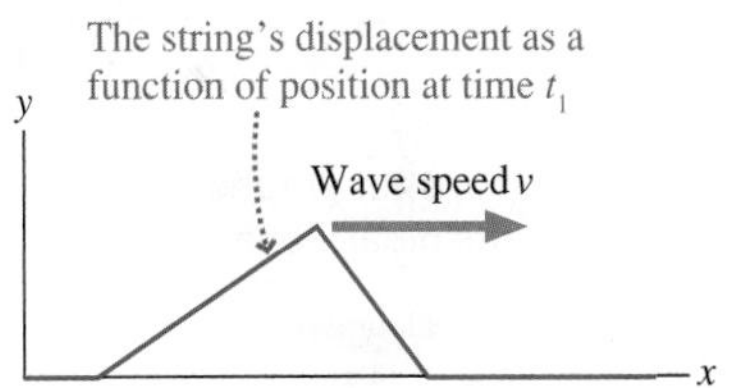

FIGURE 15.6 A sequence of snapshot graphs shows the wave in motion.

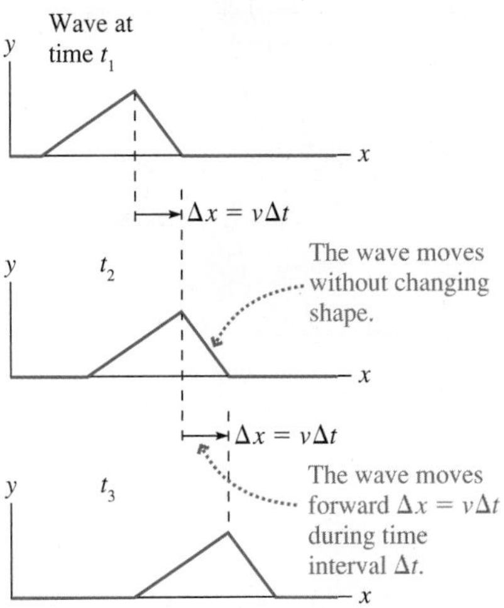

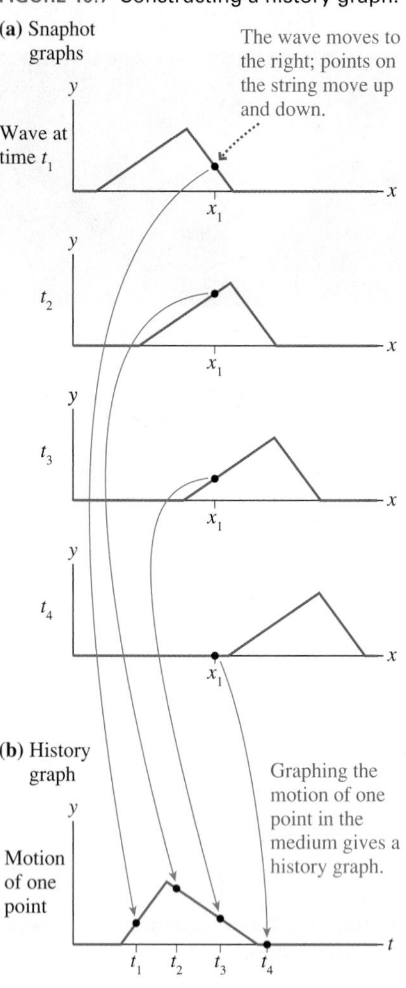

(a) Snapshot graphs
The wave moves to the right; points on the string move up and down.
Wave at time t_1
t_2
t_3
t_4
y
x
x_1

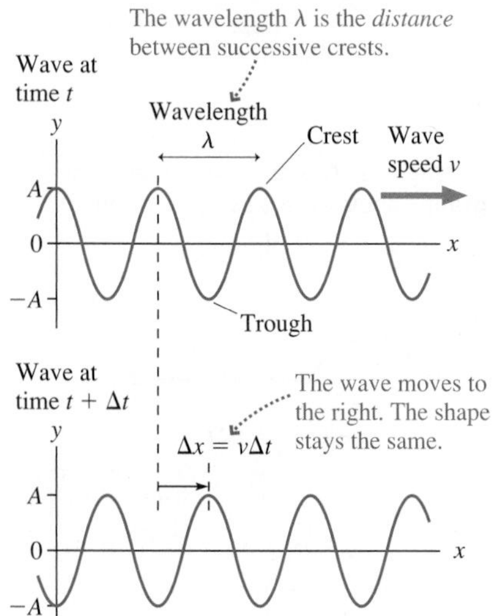

(b) History graph
Motion of one point
Graphing the motion of one point in the medium gives a history graph.
y
t
t_1 t_2 t_3 t_4

FIGURE 15.7 Constructing a history graph.

frames from a video, reminiscent of the sequences of pictures we saw in Chapter 1. The wave pulse moves forward a distance $\Delta x = v\,\Delta t$ during each time interval Δt; that is, the wave moves with constant speed.

A snapshot graph shows the motion of the *wave,* but that's only half the story. Now we want to consider the motion of the *medium*. For the wave of Figure 15.5 we can construct a different type of graph, as shown in FIGURE 15.7. Figure 15.7a shows four snapshot graphs of the wave as it travels. In each of the graphs we've placed a dot at a particular point on the string carrying the wave. As the wave travels horizontally, the dot moves vertically; this is a transverse wave. We can use information from these graphs to construct the graph in Figure 15.7b, which shows the motion of this one point on the string. We call this a **history graph** because it shows the history—the time evolution—of this particular point in the medium.

Take a close look at the history graph. The snapshot graphs show the steeper edge of the wave on the right; the history graph has the steeper edge on the left. Careful thought tells why. The history graph isn't a picture of the wave—it's a record of the motion of one point in the medium. As the wave moves toward the dot, the steep leading edge of the wave causes the dot to rise quickly. On the displacement-versus-time history graph, *earlier* times (smaller values of t) are to the *left* and *later* times (larger t) to the *right*. The rapid rise when the wave hits the dot is at an early time, and so appears on the left side of the Figure 15.7b history graph.

Sinusoidal Waves

Waves can come in many different shapes, but for the mathematical description of wave motion we will focus on a particular shape, the **sinusoidal wave.** This is the type of wave produced by a source that oscillates with simple harmonic motion. For example, a loudspeaker cone that oscillates in SHM radiates a sinusoidal sound wave. The sinusoidal electromagnetic waves broadcast by television and FM radio stations are generated by electrons oscillating back and forth in the antenna wire with SHM.

FIGURE 15.8 Snapshot graphs show the motion of a sinusoidal wave.

The wavelength λ is the *distance* between successive crests.
Wave at time t
y
Wavelength λ
Crest
Wave speed v
A
0
−A
x
Trough
Wave at time $t + \Delta t$
The wave moves to the right. The shape stays the same.
$\Delta x = v\Delta t$

The pair of snapshot graphs in FIGURE 15.8 show two successive views of a string carrying a sinusoidal wave, revealing the motion of the wave as it moves to the right. We define the **amplitude** A of the wave to be the maximum value of the displacement. The crests of the wave—the high points—have displacement $y_{crest} = A$, and the troughs—the low points—have displacement $y_{trough} = -A$. Because the wave is produced by a source undergoing SHM, which is periodic, the wave is periodic as well. As you move from left to right along the "frozen" wave in the top snapshot graph of Figure 15.8, the disturbance repeats itself over and over. The distance spanned by one cycle of the motion is called the **wavelength** of the wave. Wavelength is symbolized by λ (lowercase Greek lambda) and, because it is a length, it is measured in units of meters. The wavelength is shown in Figure 15.8 as the distance between two crests, but it could equally well be the distance between two troughs. As time passes, the wave moves to the right; comparing the two snapshot graphs in Figure 15.8 makes this motion apparent.

The snapshot graphs of Figure 15.8 show that the wave, at one instant in time, is a sinusoidal function of the distance x along the wave, with wavelength λ. At the time represented by the top graph of Figure 15.8, the displacement is given by

$$y(x) = A\cos\left(2\pi\frac{x}{\lambda}\right) \tag{15.5}$$

Next, let's look at the motion of a point in the medium as this wave passes. FIGURE 15.9 shows a history graph for a point in a string as the sinusoidal wave of Figure 15.8 passes by. This graph has exactly the same shape as the snapshot graphs of Figure 15.8—it's a sinusoidal function—but the meaning of the graph is different; it shows the motion of one point in the medium. This graph is identical to the graphs you worked with in Chapter 14 because **each point in the medium oscillates with simple harmonic motion as the wave passes.** The *period* T of the wave, shown on the graph, is the time interval to complete one cycle of the motion.

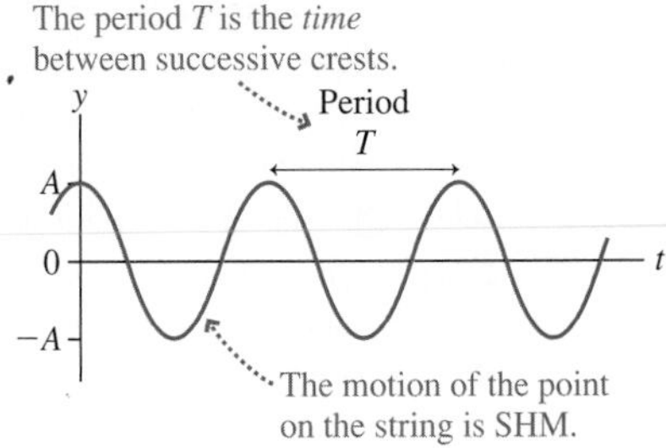

The period T is the time between successive crests.
y
Period
T
A
0
t
−A
The motion of the point on the string is SHM.

FIGURE 15.9 A history graph shows the motion of a point on a string carrying a sinusoidal wave.

NOTE ▶ Wavelength is the spatial analog of period. The period T is the *time* in which the disturbance at a single point in space repeats itself. The wavelength λ is the *distance* in which the disturbance at one instant of time repeats itself. ◀

The period is related to the wave *frequency* by $T = 1/f$, exactly as in SHM. Because each point on the string oscillates up and down in SHM with period T, we can describe the motion of a point on the string with the familiar expression

$$y(t) = A\cos\left(2\pi\frac{t}{T}\right) \tag{15.6}$$

NOTE ▶ As in Chapter 14, the argument of the cosine function is in radians. Make sure that your calculator is in radian mode before starting a calculation with the trigonometric equations in this chapter. ◀

Equation 15.5 gives the displacement as a function of position at one instant in time, and Equation 15.6 gives the displacement as a function of time at one point in space. How can we combine these two to form a single expression that is a complete description of the wave? As we'll verify, a wave traveling to the right is described by the equation

$$y(x, t) = A\cos\left(2\pi\left(\frac{x}{\lambda} - \frac{t}{T}\right)\right) \tag{15.7}$$

p. 453 SINUSOIDAL

Displacement of a traveling wave moving to the right with amplitude A, wavelength λ, and period T

The notation $y(x, t)$ indicates that the displacement y is a function of the *two* variables x and t. We must specify both where (x) and when (t) before we can calculate the displacement of the wave.

We can understand why this expression describes a wave traveling to the right by looking at **FIGURE 15.10**. In the figure, we have graphed Equation 15.7 at five instants of time, each separated by one-quarter of the period T, to make five snapshot graphs. The crest marked with the arrow represents one point on the wave. As the time t increases, so does the position x of this point—the wave moves to the right. One full period has elapsed between the first graph and the last. During this time, the wave has moved by one wavelength. And, as the wave has passed, each point in the medium has undergone one complete oscillation.

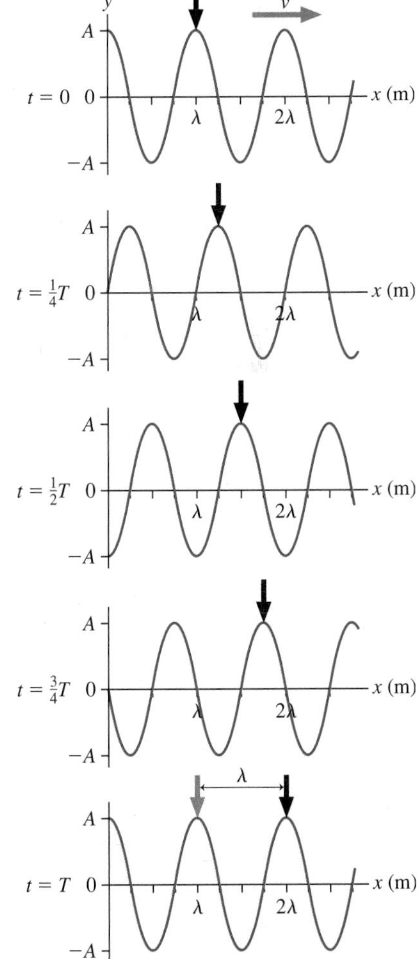

This crest is moving to the right.
y
v
A
t = 0 0
−A
λ
2λ
x (m)
t = ¼T 0
t = ½T 0
t = ¾T 0
t = T 0
During a time interval of exactly one period, the crest has moved forward exactly one wavelength.

FIGURE 15.10 Equation 15.7 graphed at intervals of one-quarter of the period. This is a traveling wave moving to the right.

For a wave traveling to the left, we have a slightly different form:

$$y(x, t) = A\cos\left(2\pi\left(\frac{x}{\lambda} + \frac{t}{T}\right)\right) \tag{15.8}$$

Displacement of a traveling wave moving to the left

NOTE ▶ A wave moving to the right (the $+x$-direction) has a $-$ in the expression, while a wave moving to the left (the $-x$-direction) has a $+$. Remember that the sign is *opposite* the direction of travel. ◀

EXAMPLE 15.4 **Determining the rise and fall of a boat**

A boat is moving to the right at 5.0 m/s with respect to the water. An ocean wave is moving to the left, opposite the motion of the boat. The waves have 2.0 m between the top of the crests and the bottom of the troughs. The period of the waves is 8.3 s, and their wavelength is 110 m. At one instant, the boat sits on a crest of the wave. 20 s later, what is the vertical displacement of the boat?

PREPARE We begin with the visual overview, as in **FIGURE 15.11**. Let $t = 0$ be the instant the boat is on the crest, and draw a snapshot graph of the traveling wave at that time. Because the wave is traveling to the left, we will use Equation 15.8 to represent the wave. The boat begins at a crest of the wave, so we see that the boat can start the problem at $x = 0$.

FIGURE 15.11 Visual overview for the boat.

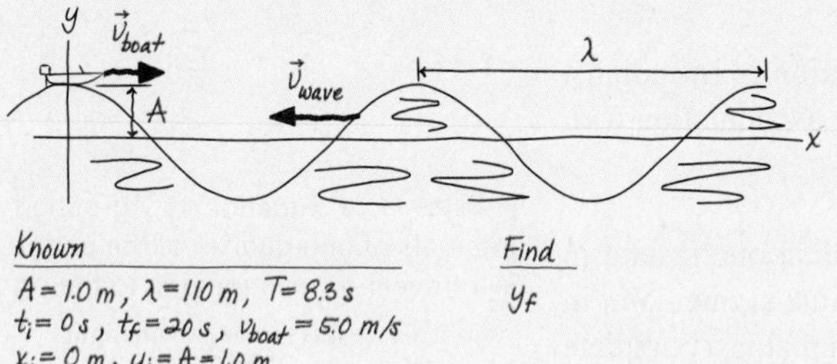

The distance between the high and low points of the wave is 2.0 m; the amplitude is half this, so $A = 1.0$ m. The wavelength and period are given in the problem.

SOLVE The boat is moving to the right at 5.0 m/s. At $t_f = 20$ s the boat is at position

$$x_f = (5.0 \text{ m/s})(20 \text{ s}) = 100 \text{ m}$$

We need to find the wave's displacement at this position and time. Substituting known values for amplitude, wavelength, and period into Equation 15.8, for a wave traveling to the left, we obtain the following equation for the wave:

$$y(x, t) = (1.0 \text{ m})\cos\left(2\pi\left(\frac{x}{110 \text{ m}} + \frac{t}{8.3 \text{ s}}\right)\right)$$

At $t_f = 20$ s and $x_f = 100$ m, the boat's displacement on the wave is

$$y_f = y(\text{at } 100 \text{ m}, 20 \text{ s}) = (1.0 \text{ m})\cos\left(2\pi\left(\frac{100 \text{ m}}{110 \text{ m}} + \frac{20 \text{ s}}{8.3 \text{ s}}\right)\right)$$

$$= -0.42 \text{ m}$$

ASSESS The final displacement is negative—meaning the boat is in a trough of a wave, not underwater. Don't forget that your calculator must be in radian mode when you make your final computation!

The Fundamental Relationship for Sinusoidal Waves

10.1 ActivPhysics ONLINE

In Figure 15.10, one critical observation is that the wave crest marked by the arrow has moved one full wavelength between the first graph and the last. That is, **during a time interval of exactly one period T, each crest of a sinusoidal wave travels forward a distance of exactly one wavelength λ.** Because speed is distance divided by time, the wave speed must be

$$v = \frac{\text{distance}}{\text{time}} = \frac{\lambda}{T} \tag{15.9}$$

Using $f = 1/T$, it is customary to write Equation 15.9 in the form

$$v = \lambda f \tag{15.10}$$

Relationship between velocity, wavelength, and frequency for sinusoidal waves

Although Equation 15.10 has no special name, it is *the* fundamental relationship for sinusoidal waves. When using it, keep in mind the *physical* meaning that a wave moves forward a distance of one wavelength during a time interval of one period.

We will frequently see problems in which we are given two of the variables in Equation 15.10 but we need to find the third, as in the following example.

EXAMPLE 15.5 **Writing the equation for a wave**

A sinusoidal wave with an amplitude of 1.5 cm and a frequency of 100 Hz travels at 200 m/s in the positive x-direction. Write the equation for the wave's displacement as it travels.

PREPARE The problem statement gives the following characteristics of the wave: $A = 1.5 \text{ cm} = 0.015$ m, $v = 200$ m/s, and $f = 100$ Hz.

SOLVE To write the equation for the wave, we need the amplitude, the wavelength, and the period. The amplitude was given in the problem statement. We can use the fundamental relationship for sinusoidal waves to find the wavelength:

$$\lambda = \frac{v}{f} = \frac{200 \text{ m/s}}{100 \text{ Hz}} = 2.0 \text{ m}$$

The period can be calculated from the frequency:

$$T = \frac{1}{f} = \frac{1}{100 \text{ Hz}} = 0.010 \text{ s}$$

With these values in hand, we can write the equation for the wave's displacement using Equation 15.7:

$$y(x, t) = (0.015 \text{ m})\cos\left(2\pi\left(\frac{x}{2.0 \text{ m}} - \frac{t}{0.010 \text{ s}}\right)\right)$$

ASSESS If the speed of a wave is known, you can use the fundamental relationship for sinusoidal waves to find the wavelength if you are given the frequency, or the frequency if you are given the wavelength. You'll often need to do this in the early stages of problems that you solve.

STOP TO THINK 15.3 Three waves travel to the right with the same speed. Which wave has the highest frequency?

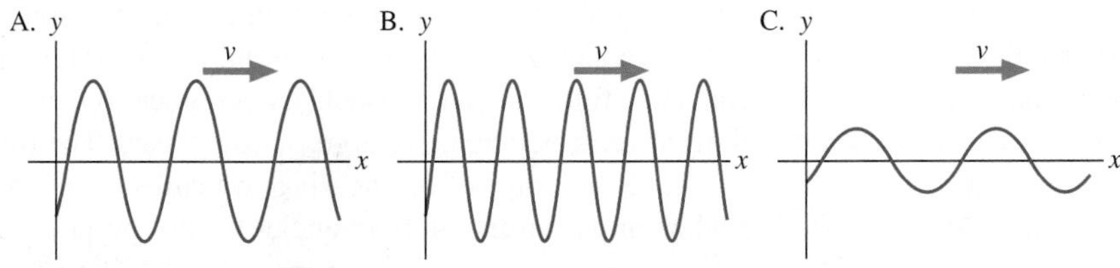

15.4 Sound and Light Waves

Think about how you are experiencing the world right now. Chances are, your senses of sight and sound are hard at work, detecting and interpreting light and sound waves from the world around you. Because these waves are so important to us, we'll spend some time exploring their properties.

Sound Waves

We saw in Figure 15.4 how a loudspeaker creates a sound wave. If the loudspeaker cone moves with simple harmonic motion, it will create a sinusoidal sound wave, as illustrated in **FIGURE 15.12**. Each time the cone moves forward, it moves the air molecules closer together, creating a region of higher pressure. A half cycle later, as the cone moves backward, the air has room to expand and the pressure decreases. These regions of higher and lower pressure are called **compressions** and **rarefactions,** respectively.

As Figure 15.12 suggests, it is often most convenient and informative to think of a sound wave as a pressure wave. As the graph of the pressure shows, the pressure oscillates sinusoidally around the atmospheric pressure p_{atmos}. This is a snapshot graph of the wave at one instant of time, and the distance between two adjacent crests (two points of maximum compression) is the wavelength λ.

When the wave reaches your ear, the oscillating pressure causes your eardrums to vibrate. This vibration is transferred through your inner ear to the cochlea, where it is sensed, as we learned in Chapter 14. Humans with normal hearing are able to detect sinusoidal sound waves with frequencies between about 20 Hz and 20,000 Hz, or 20 kHz. Low frequencies are perceived as a "low pitch" bass note, while high frequencies are heard as a "high pitch" treble note.

FIGURE 15.12 A sound wave is a pressure wave.

The loudspeaker cone moves back and forth, creating regions of higher and lower pressure—compressions and rarefactions.

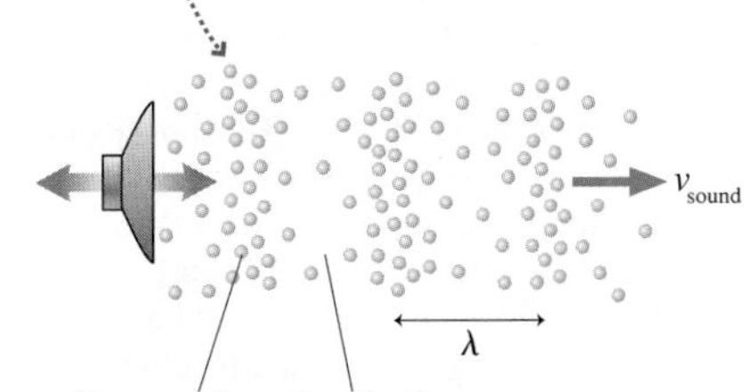

Compression Rarefaction

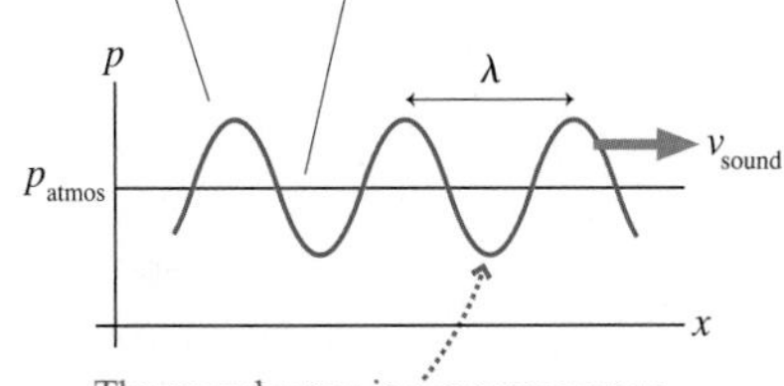

The sound wave is a pressure wave. Compressions are crests; rarefactions are troughs.

EXAMPLE 15.6 Range of wavelengths of sound

What are the wavelengths of sound waves at the limits of human hearing and at the midrange frequency of 500 Hz? Notes sung by human voices are near 500 Hz, as are notes played by striking keys near the center of a piano keyboard.

PREPARE We will do our calculation at room temperature, 20°C, so we will use $v = 343$ m/s for the speed of sound.

SOLVE We can solve for the wavelengths given the fundamental relationship $v = f\lambda$:

$$f = 20\text{ Hz:}\qquad \lambda = \frac{v}{f} = \frac{343\text{ m/s}}{20\text{ Hz}} = 17\text{ m}$$

$$f = 500\text{ Hz:}\qquad \lambda = \frac{v}{f} = \frac{343\text{ m/s}}{500\text{ Hz}} = 0.69\text{ m}$$

$$f = 20\text{ kHz:}\qquad \lambda = \frac{v}{f} = \frac{343\text{ m/s}}{20 \times 10^3\text{ Hz}} = 0.017\text{ m} = 1.7\text{ cm}$$

ASSESS The wavelength of a 20 kHz note is a small 1.7 cm. At the other extreme, a 20 Hz note has a huge wavelength of 17 m! A wave moves forward one wavelength during a time interval of one period, and a wave traveling at 343 m/s can move 17 m during the $\frac{1}{20}$ s period of a 20 Hz note.

TABLE 15.2 Range of hearing for animals

Animal	Range of hearing (Hz)
Elephant	<5–12,000
Owl	200–12,000
Human	20–20,000
Dog	30–45,000
Mouse	1000–90,000
Bat	2000–100,000
Porpoise	75–150,000

It is well known that dogs are sensitive to high frequencies that humans cannot hear. Other animals also have quite different ranges of hearing than humans; some examples are listed in Table 15.2.

Elephants communicate over great distances with vocalizations at frequencies far too low for us to hear. These low-frequency sounds are transmitted with less energy loss than high-frequency sounds, allowing elephants to hear each other from over 6 km away—important for these social animals that may forage over very large areas.

There are animals that use frequencies well above the range of our hearing as well. High-frequency sounds are useful for *echolocation*—emitting a pulse of sound and listening for its reflection. Bats, which generally feed at night, rely much more on their hearing than their sight. They find and catch insects by echolocation, emitting loud chirps whose reflections are detected by their large, sensitive ears. The frequencies that they use are well above the range of our hearing; we call such sound **ultrasound.** Why do bats and other animals use high frequencies for this purpose?

In Chapter 19, we will look at the *resolution* of optical instruments. The finest detail that your eye—or any optical instrument—can detect is limited by the wavelength of light. Shorter wavelengths allow for the imaging of smaller details.

The same limitations apply to the acoustic image of the world made by bats. In order to sense fine details of their surroundings, bats must use sound of very short wavelength (and thus high frequency). A 50 kHz chirp from a little brown bat has a wavelength of just over half a centimeter, allowing the bat to precisely locate an insect that reflects it. Other animals that use echolocation, such as porpoises, also produce and sense high-frequency sounds.

Sound travels very well though tissues in the body, and the reflections of sound waves from different tissues can be used to create an image of the body's interior. The fine details necessary for a clinical diagnosis require the short wavelengths of ultrasound. X rays create very good images of bones; ultrasound is used for creating images of soft tissues. It is also useful in cases where the radiation exposure of x rays should be avoided. You have certainly seen ultrasound images taken during pregnancy, where the use of x rays is clearly undesirable.

Short wavelengths mean sharper images Light doesn't travel very far in any but the clearest water. To navigate through and investigate their surroundings, vessels at sea can create images using sound instead of light, sending out sound waves and detecting and analyzing the reflections. This image of a shipwreck on the ocean bottom was made from the surface with 600 kHz ultrasound. This high frequency gave a very short wavelength of just under a quarter of a centimeter, making for very sharp outlines and fine details in the image.

EXAMPLE 15.7 Ultrasonic frequencies in medicine

To make a sufficiently detailed ultrasound image of a fetus in its mother's uterus, a physician has decided that a wavelength of 0.50 mm is needed. What frequency is required?

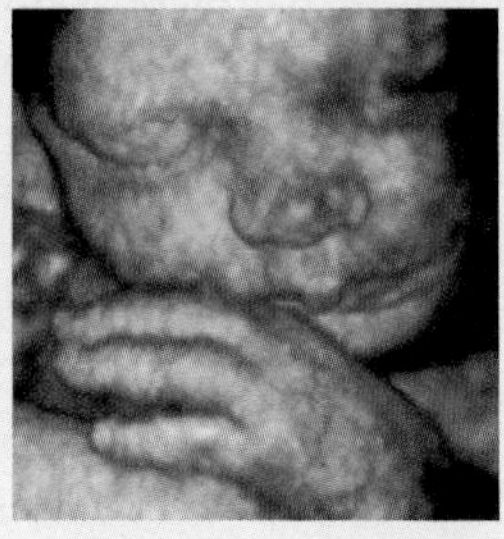

Computer processing of an ultrasound image shows fine detail.

PREPARE The speed of ultrasound in the body is given in Table 15.1 as 1540 m/s.

SOLVE We can use the fundamental relationship among speed, wavelength, and frequency to calculate

$$f = \frac{v}{\lambda} = \frac{1540 \text{ m/s}}{0.50 \times 10^{-3} \text{ m}} = 3.1 \times 10^6 \text{ Hz} = 3.1 \text{ MHz}$$

ASSESS This is a reasonable result. Clinical ultrasound uses frequencies in the range of 1–20 MHz. Lower frequencies have greater penetration; higher frequencies (and thus shorter wavelengths) show finer detail.

Light and Other Electromagnetic Waves

A light wave is an electromagnetic wave, an oscillation of the electromagnetic field. Other electromagnetic waves, such as radio waves, microwaves, and ultraviolet light, have the same physical characteristics as light waves even though we cannot sense them with our eyes. As we saw earlier, all electromagnetic waves, regardless of wavelength or frequency, travel through vacuum (or air) with the same speed: $c = 3.00 \times 10^8$ m/s.

The wavelengths of light are extremely short. Visible light is an electromagnetic wave with a wavelength (in air) of roughly 400 nm $(=400 \times 10^{-9}\text{ m})$ to 700 nm $(=700 \times 10^{-9}\text{ m})$. Each wavelength is perceived as a different color. Longer wavelengths, in the 600–700 nm range, are seen as orange or red light; shorter wavelengths, in the 400–500 nm range, are seen as blue or violet light.

FIGURE 15.13 shows that the visible spectrum is a small slice out of the much broader **electromagnetic spectrum.** We will have much more to say about light and the rest of the electromagnetic spectrum in future chapters.

FIGURE 15.13 The electromagnetic spectrum from 10^6 Hz to 10^{18} Hz.

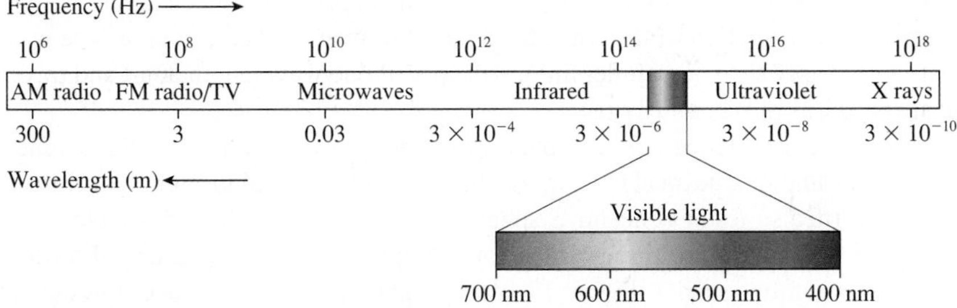

EXAMPLE 15.8 **Finding the frequency of microwaves**

The wavelength of microwaves in a microwave oven is 12 cm. What is the frequency of the waves?

PREPARE Microwaves are electromagnetic waves, so their speed is the speed of light, $c = 3.00 \times 10^8$ m/s.

SOLVE Using the fundamental relationship for sinusoidal waves, we find

$$f = \frac{v}{\lambda} = \frac{3.00 \times 10^8 \text{ m/s}}{0.12 \text{ m}} = 2.5 \times 10^9 \text{ Hz} = 2.5 \text{ GHz}$$

ASSESS This is a high frequency, but the speed of the waves is also very high, so our result seems reasonable.

We've just used the same equation to describe ultrasound and microwaves, which brings up a remarkable point: The wave model we've been developing applies equally well to all types of waves. Features such as wavelength, frequency, and speed are characteristics of waves in general. At this point we don't yet know any details about electromagnetic fields, yet we can use the wave model to say some significant things about the properties of electromagnetic waves.

It's also interesting to note that the 2.5 GHz frequency of microwaves in an oven we found in Example 15.8 is similar to the frequencies of other devices you use each day. Many cordless phones work at a frequency of 2.4 GHz, and cell phones at just under 2.0 GHz. Although these frequencies are not very different from those of the waves that heat your food in a microwave oven, the *intensity* is much less—which brings us to the topic of the next section.

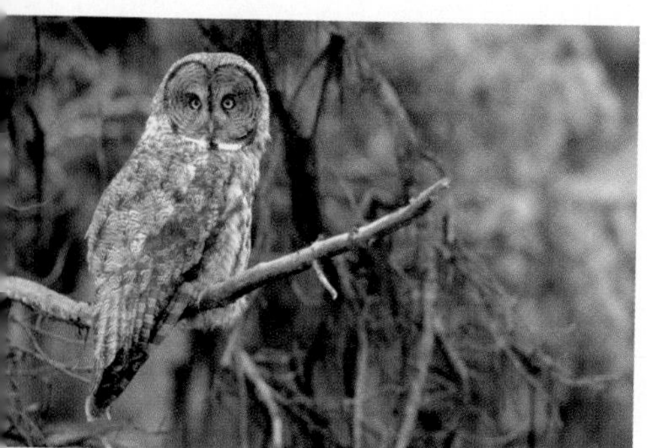

The better to hear you with BIO The great grey owl has its ears on the front of its face, hidden behind its facial feathers. Its round face works like a radar dish, collecting the energy of sound waves and "funneling" it into the ears. Collecting sound over a large area in this manner allows owls to sense very quiet sounds. Having ears on the front of the face allows them to precisely determine the source of sounds as well, an asset for a bird of prey. These owls can hear—and locate—mice moving underneath a thick blanket of snow.

15.5 Energy and Intensity

A traveling wave transfers energy from one point to another. The sound wave from a loudspeaker sets your eardrum into motion. Light waves from the sun warm the earth and, if focused with a lens, can start a fire. The *power* of a wave is the rate, in joules per second, at which the wave transfers energy. As you learned in Chapter 10, power is measured in watts. A person singing or shouting as loud as possible is emitting energy in the form of sound waves at a rate of about 1 W, or 1 J/s. A 25 W lightbulb emits about 1 W of visible light, with the other 24 W of power being emitted as heat, or infrared radiation, rather than as visible light. In this section, we will learn how to characterize the power of waves. A first step in doing so is to understand how waves change as they spread out.

Circular, Spherical, and Plane Waves

Suppose you were to take a photograph of ripples spreading on a pond. If you mark the location of the *crests* on the photo, your picture would look like **FIGURE 15.14a**. The lines that locate the crests are called **wave fronts,** and they are spaced precisely one wavelength apart. A wave like this is called a **circular wave.** It is a two-dimensional wave that spreads across a surface.

Although the wave fronts are circles, you would hardly notice the curvature if you observed a small section of the wave front very far away from the source. The wave fronts would appear to be parallel lines, still spaced one wavelength apart and traveling at speed v as in **FIGURE 15.14b**.

Many waves of interest, such as sound waves or light waves, move in three dimensions. For example, loudspeakers and lightbulbs emit **spherical waves.** The crests of the wave form a series of concentric, spherical shells separated by the wavelength λ. In essence, the waves are three-dimensional ripples. It will still be useful to draw wave-front diagrams such as Figure 15.14a, but now the circles are slices through the spherical shells locating the wave crests.

If you observe a spherical wave very far from its source, the small piece of the wave front that you can see is a little patch on the surface of a very large sphere. If the radius of the sphere is sufficiently large, you will not notice the curvature and this little patch of the wave front appears to be a plane. **FIGURE 15.15** illustrates the idea of a **plane wave.**

FIGURE 15.14 The wave fronts of a circular or spherical wave.

(a)

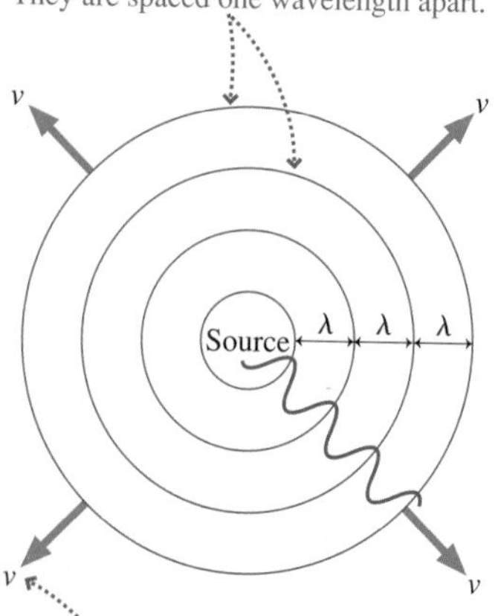

(b)

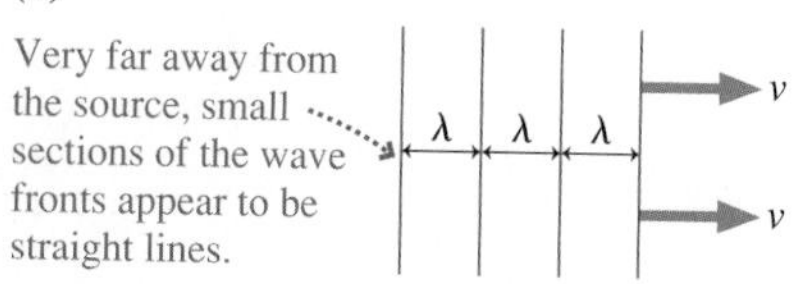

FIGURE 15.15 A plane wave.

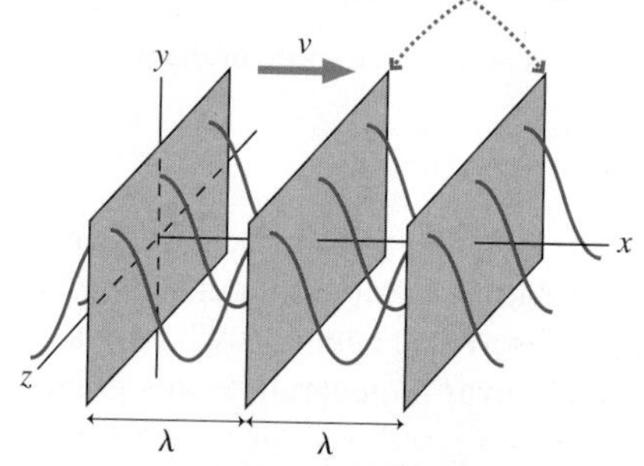

Very far from the source, small segments of spherical wave fronts appear to be planes. The wave is cresting at every point in these planes.
y
v
x
z
λ
λ

Power, Energy, and Intensity

Imagine doing two experiments with a lightbulb that emits 2 W of visible light. In the first, you hang the bulb in the center of a room and allow the light to illuminate the walls. In the second experiment, you use mirrors and lenses to "capture" the bulb's light and focus it onto a small spot on one wall. (This is what a projector does.) The energy emitted by the bulb is the same in both cases, but, as you know, the light is much brighter when focused onto a small area. We would say that the focused light is more *intense* than the diffuse light that goes in all directions. Similarly, a loudspeaker that beams its sound forward into a small area produces a louder sound in that area than a speaker of equal power that radiates the sound in all directions. Quantities such as brightness and loudness depend not only on the rate of energy transfer, or power, but also on the *area* that receives that power.

FIGURE 15.16 shows a wave impinging on a surface of area a. The surface is perpendicular to the direction in which the wave is traveling. This might be a real physical surface, such as your eardrum or a solar cell, but it could equally well be a mathematical surface in space that the wave passes right through. If the wave has power P, we define the **intensity** I of the wave as

$$I = \frac{P}{a} \tag{15.11}$$

The SI units of intensity are W/m^2. Because intensity is a power-to-area ratio, a wave focused onto a small area has a higher intensity than a wave of equal power that is spread out over a large area.

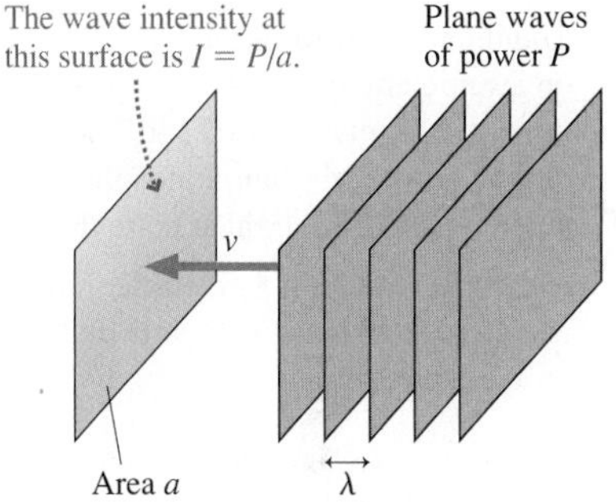

FIGURE 15.16 Plane waves of power P impinge on area a.

NOTE ► In this chapter we will use a for area to avoid confusion with amplitude, for which we use the symbol A. ◄

EXAMPLE 15.9 The intensity of a laser beam

A bright, tightly focused laser pointer emits 1.0 mW of light power into a beam that is 1.0 mm in diameter. What is the intensity of the laser beam?

SOLVE The light waves of the laser beam pass through a circle of diameter 1.0 mm. The intensity of the laser beam is

$$I = \frac{P}{a} = \frac{P}{\pi r^2} = \frac{0.0010\ \text{W}}{\pi(0.00050\ \text{m})^2} = 1300\ \text{W/m}^2$$

ASSESS This intensity is roughly equal to the intensity of sunlight at noon on a summer day. Such a high intensity for a low-power source may seem surprising, but the area is very small, so the energy is packed into a tiny spot. You know that the light from a laser pointer won't burn you but you don't want the beam to shine into your eye, so an intensity similar to that of sunlight seems reasonable.

Sound from a loudspeaker and light from a lightbulb become less intense as you get farther from the source. This is because spherical waves spread out to fill larger and larger volumes of space. To conserve energy, the wave's amplitude must decrease with increasing distance r.

If a source of spherical waves radiates uniformly in all directions, then, as FIGURE 15.17 shows, the power at distance r is spread uniformly over the surface of a sphere of radius r. The surface area of a sphere is $a = 4\pi r^2$, so the intensity of a uniform spherical wave is

$$I = \frac{P_{\text{source}}}{4\pi r^2} \tag{15.12}$$

Intensity at distance r of a spherical wave from a source of power P_{source}

INVERSE-SQUARE p. 186

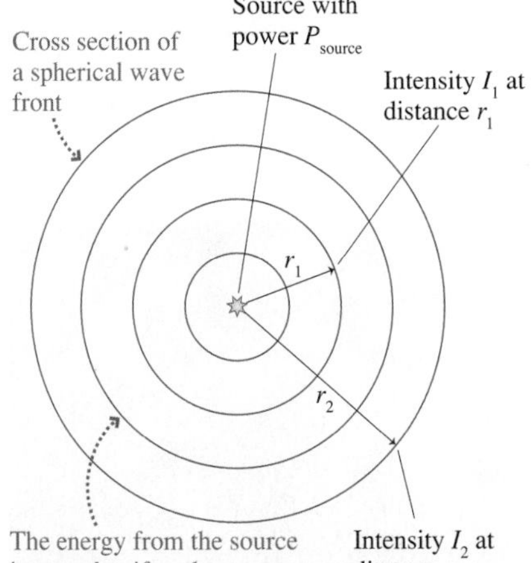

FIGURE 15.17 A source emitting uniform spherical waves.

The inverse-square dependence of r is really just a statement of energy conservation. The source emits energy at the rate of P joules per second. The energy is spread over a larger and larger area as the wave moves outward. Consequently, the *energy per area* must decrease in proportion to the surface area of a sphere.

If the intensity at distance r_1 is $I_1 = P_{\text{source}}/4\pi r_1^2$ and the intensity at r_2 is $I_2 = P_{\text{source}}/4\pi r_2^2$, then you can see that the intensity *ratio* is

$$\frac{I_1}{I_2} = \frac{r_2^2}{r_1^2} \tag{15.13}$$

You can use Equation 15.13 to compare the intensities at two distances from a source without needing to know the power of the source.

NOTE ► Wave intensities are strongly affected by reflections and absorption. Equations 15.12 and 15.13 apply to situations such as the light from a star and the sound from a firework exploding high in the air. Indoor sound does *not* obey a simple inverse-square law because of the many reflecting surfaces. ◄

EXAMPLE 15.10 Intensity of sunlight on Mars

The intensity of sunlight on earth is roughly 1300 W/m² at noon on a summer day. Mars orbits at a distance from the sun approximately 1.5 times that of earth. Assuming similar absorption of energy by the Martian atmosphere, what would you predict for the intensity of sunlight at noon during the Martian summer?

PREPARE We aren't given the power emitted by the sun or the distances from the sun to earth and to Mars, but we can solve this problem by using the ratios of distances and intensities.

SOLVE According to Equation 15.13,

$$I_{\text{Mars}} = I_{\text{earth}}\frac{r_{\text{earth}}^2}{r_{\text{Mars}}^2} = (1300\ \text{W/m}^2)\left(\frac{1}{1.5}\right)^2 = 580\ \text{W/m}^2$$

ASSESS The intensity of sunlight is quite a bit less on Mars than on earth, as we would expect, given the much greater distance of Mars from the sun.

STOP TO THINK 15.4 A plane wave, a circular wave, and a spherical wave all have the same intensity. Each of the waves travels the same distance. Afterward, which wave has the highest intensity?

A. The plane wave B. The circular wave C. The spherical wave

15.6 Loudness of Sound

Ten guitars playing in unison sound only about twice as loud as one guitar. 100 trumpets playing together seem to you only four times as loud as a soloist. Generally, **increasing the sound intensity by a factor of 10 results in an increase in perceived loudness by a factor of approximately 2.** Thus your ears are sensitive over a very wide range of intensities. The difference in intensity between the quietest sound you can detect and the loudest you can safely hear is a factor of 1,000,000,000,000! A normal conversation has 10,000 times the sound intensity of a whisper, but it sounds only about 16 times as loud.

The loudness of sound is measured by a quantity called the **sound intensity level.** Because of the wide range of intensities we can hear, and the fact that the difference in perceived loudness is much less than the actual difference in intensity, the sound intensity level is measured on a *logarithmic scale.* In this section we will explain what this means. The units of sound intensity level (i.e., of loudness) are *decibels,* a word you have likely heard.

◄ **The loudest animal in the world** BIO The blue whale is the largest animal in the world, up to 30 m (about 100 ft) long, weighing 150,000 kg or more. It is also the loudest. At close range in the water, the 10–30 second calls of the blue whale would be intense enough to damage tissues in your body. Blue whales produce sound with a larynx, much as humans do, but at much lower frequencies of 10–40 Hz that can travel great distances in water. Their loud, low-frequency calls can be heard by other whales hundreds of miles away.

The Decibel Scale

There is a lower limit to the intensity of sound that a human can hear. The exact value varies among individuals and varies with frequency, but an average value for the lowest-intensity sound that can be heard in an extremely quiet room is

$$I_0 = 1.0 \times 10^{-12}\ \text{W/m}^2$$

This intensity is called the *threshold of hearing*. A sound wave can have lower intensity than this, but you won't be able to perceive it.

It's convenient and logical to place the zero of our loudness scale at the threshold of hearing. All other sounds can then be referenced to this intensity. To create a loudness scale, we define the *sound intensity level,* expressed in **decibels** (dB), as

Quiet as a mouse 0 dB is the lower limit of human hearing, but other animals can hear quieter sounds. The harvest mouse has an especially well-developed sense of hearing and will detect and respond to sounds down to −10 dB, making it hard for predators to sneak up on it.

$$\beta = (10\ \text{dB})\log_{10}\left(\frac{I}{I_0}\right) \quad (15.14)$$

Sound intensity level in decibels for a sound of intensity I

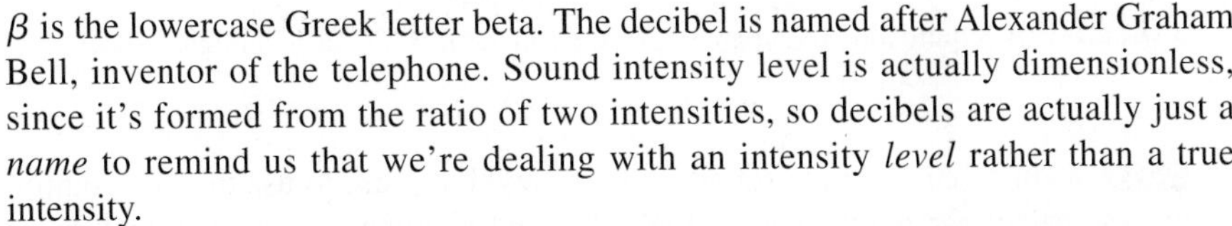

β is the lowercase Greek letter beta. The decibel is named after Alexander Graham Bell, inventor of the telephone. Sound intensity level is actually dimensionless, since it's formed from the ratio of two intensities, so decibels are actually just a *name* to remind us that we're dealing with an intensity *level* rather than a true intensity.

Equation 15.14 takes the base-10 logarithm of the intensity ratio I/I_0. As a reminder, logarithms work like this:

If you express a number as a power of 10 . . . . . . the logarithm is the exponent.

$$\log_{10}(1000) = \log_{10}(10^3) = 3$$

Right at the threshold of hearing, where $I = I_0$, the sound intensity level is

$$\beta = (10\ \text{dB})\log_{10}\left(\frac{I_0}{I_0}\right) = (10\ \text{dB})\log_{10}(1) = (10\ \text{dB})\log_{10}(10^0) = 0\ \text{dB}$$

The threshold of hearing corresponds to 0 dB, as we wanted.

Table 15.3 lists the intensities and sound intensity levels for a number of typical sounds. Notice that the sound intensity level increases by 10 dB each time the actual intensity increases by a factor of 10. For example, the sound intensity level increases from 70 dB to 80 dB when the sound intensity increases from $10^{-5}\ \text{W/m}^2$ to $10^{-4}\ \text{W/m}^2$. A 20 dB increase in the sound intensity level means a factor of 100 increase in intensity; 30 dB a factor of 1000. We found earlier that sound is perceived as "twice as loud" when the intensity increases by a factor of 10. In terms of decibels, we can say that the apparent loudness of a sound doubles with each 10 dB increase in the sound intensity level.

The range of sounds in Table 15.3 is very wide; the top of the scale, 130 dB, represents 10 trillion times the intensity of the quietest sound you can hear. Vibrations of this intensity will injure the delicate sensory apparatus of the ear and cause pain. Exposure to less intense sounds also is not without risk. A fairly short exposure to 120 dB can cause damage to the hair cells in the ear, but lengthy exposure to sound intensity levels of over 85 dB can produce damage as well.

TABLE 15.3 Intensity and sound intensity levels of common environmental sounds

Sound	β (dB)	I (W/m²)
Threshold of hearing	0	1.0×10^{-12}
Person breathing, at 3 m	10	1.0×10^{-11}
A whisper, at 1 m	20	1.0×10^{-10}
Classroom during test, no talking	30	1.0×10^{-9}
Residential street, no traffic	40	1.0×10^{-8}
Quiet restaurant	50	1.0×10^{-7}
Normal conversation, at 1 m	60	1.0×10^{-6}
Busy traffic	70	1.0×10^{-5}
Vacuum cleaner, for user	80	1.0×10^{-4}
Niagara Falls, at viewpoint	90	1.0×10^{-3}
Pneumatic hammer, at 2 m	100	0.010
Home stereo at max volume	110	0.10
Rock concert	120	1.0
Threshold of pain	130	10

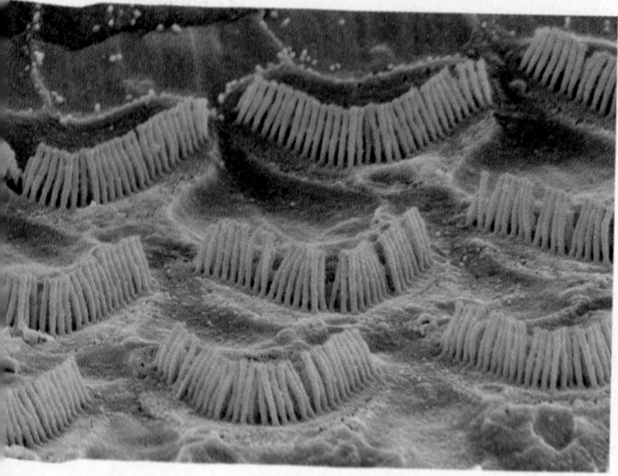

◀ **Hearing hairs** BIO This electron microscope picture shows the hair cells in the cochlea of the ear that are responsible for sensing sound. Each cell has a curved row of tiny hairs, which give your ears their remarkable sensitivity. Even very tiny vibrations transmitted into the fluid of the cochlea deflect the hairs, triggering a response in the cells. Motion of the basilar membrane by as little as 0.5 nm, about 5 atomic diameters, can produce an electrical response in the hair cells. With this remarkable sensitivity, it is no wonder that loud sounds can damage these structures.

EXAMPLE 15.11 **Finding the loudness of a shout**

A person shouting at the top of his lungs emits about 1.0 W of energy as sound waves. What is the sound intensity level 1.0 m from such a person?

PREPARE We will assume that the shouting person emits a spherical sound wave, in which case the intensity decreases according to Equation 15.12.

SOLVE At a distance of 1.0 m, the sound intensity is

$$I = \frac{P}{4\pi r^2} = \frac{1.0\text{ W}}{4\pi(1.0\text{ m})^2} = 0.080\text{ W/m}^2$$

Thus the sound intensity level is

$$\beta = (10\text{ dB})\log_{10}\left(\frac{0.080\text{ W/m}^2}{1.0\times10^{-12}\text{ W/m}^2}\right) = 110\text{ dB}$$

ASSESS This is quite loud (compare with values in Table 15.3), as you might expect.

NOTE ▶ In calculating the sound intensity level, be sure to use the $\log_{10}$ button on your calculator, not the natural logarithm button. On many calculators, $\log_{10}$ is labeled LOG and the natural logarithm is labeled LN. ◀

Equation 15.14 allows us to compute the sound intensity level β from the intensity I. We can do the reverse, finding an intensity from the sound intensity level, by taking the inverse of the $\log_{10}$ function. Recall, from the definition of the base-10 logarithm, that $10^{\log(x)} = x$. Applying this to Equation 15.14, we find

$$I = (1.0\times10^{-12}\text{ W/m}^2)10^{(\beta/10\text{ dB})} \tag{15.15}$$

EXAMPLE 15.12 **How far away can you hear a conversation?**

The sound intensity level 1.0 m from a person talking in a normal conversational voice is 60 dB. Suppose you are outside, 100 m from the person speaking. If it is a very quiet day with minimal background noise, will you be able to hear him or her?

PREPARE We know how sound intensity changes with distance, but not sound intensity level, so we need to break this problem into three steps. First, we will convert the sound intensity level at 1.0 m into intensity, using Equation 15.15. Second, we will compute the intensity at a distance of 100 m, using Equation 15.13. Finally, we will convert this result back to a sound intensity level, using Equation 15.14, so that we can judge the loudness.

SOLVE The sound intensity level of a normal conversation at 1.0 m is 60 dB. The intensity is

$$I(1\text{ m}) = (1.0\times10^{-12}\text{ W/m}^2)10^{(60\text{ dB}/10\text{ dB})}$$
$$= 1.0\times10^{-6}\text{ W/m}^2$$

The intensity at 100 m can be found using Equation 15.13:

$$\frac{I(100\text{ m})}{I(1\text{ m})} = \frac{(1\text{ m})^2}{(100\text{ m})^2}$$

$$I(100\text{ m}) = I(1\text{ m})(1.0\times10^{-4}) = 1.0\times10^{-10}\text{ W/m}^2$$

This intensity corresponds to a sound intensity level

$$\beta = (10\text{ dB})\log_{10}\left(\frac{1.0\times10^{-10}\text{ W/m}^2}{1.0\times10^{-12}\text{ W/m}^2}\right) = 20\text{ dB}$$

This is above the threshold for hearing—about the level of a whisper—so it could be heard on a very quiet day.

ASSESS The sound is well within what the ear can detect. However, normal background noise is rarely less than 40 dB, which would make the conversation difficult to decipher. Our result thus seems reasonable based on experience. The sound is at a level that can theoretically be detected, but it will be much quieter than the ambient level of sound, so it's not likely to be noticed.

15.7 The Doppler Effect and Shock Waves

In this final section, we will look at sounds from moving objects and for moving observers. You've likely noticed that the pitch of an ambulance's siren drops as it goes past you. A higher pitch suddenly becomes a lower pitch. This change in frequency, which is due to the motion of the ambulance, is called the *Doppler effect.* A more dramatic effect happens when an object moves faster than the speed of sound. The crack of a whip is a *shock wave* produced when the tip moves at a *supersonic* speed. These examples—and much of this section—concern sound waves, but the phenomena we will explore apply generally to all waves.

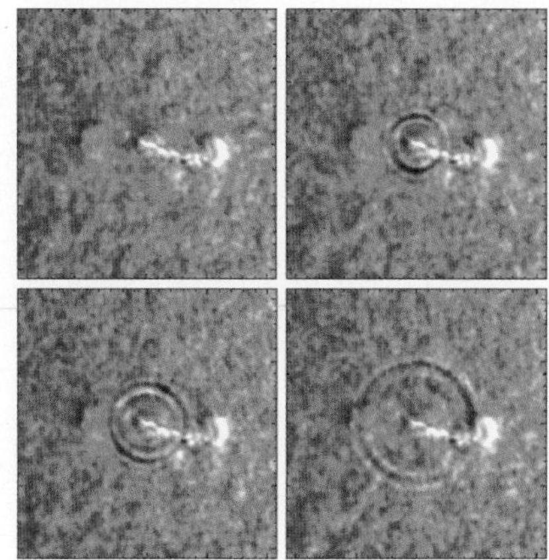

Catching a wave on the sun SOHO (the **S**olar and **H**eliospheric **O**bservatory) is a satellite studying the sun. The instrument on the satellite responsible for images like this measures the Doppler effect of light emitted from the surface of the sun. If the surface is rising toward the satellite, the light is shifted to higher frequencies; these positions are shown darker on the image. If the surface is falling, the light is shifted to lower frequencies and is shown lighter. This series of images of the sun's surface shows a wave produced by the disruption of a solar flare. This wave is similar to the wave from an earthquake on earth, but 20 times as fast and carrying 40,000 times as much energy as a typical earthquake wave.

Sound Waves from a Moving Source

FIGURE 15.18a shows a source of sound waves moving away from Pablo and toward Nancy at a steady speed v_s. The subscript s indicates that this is the speed of the source, not the speed of the waves. The source is emitting sound waves of frequency f_0 as it travels. Part a of the figure is a motion diagram showing the positions of the source at times $t = 0$, T, $2T$, and $3T$, where $T = 1/f_0$ is the period of the waves.

After a wave crest leaves the source, its motion is governed by the properties of the medium. The motion of the source cannot affect a wave that has already been emitted. Thus each circular wave front in FIGURE 15.18b is centered on the point from which it was emitted. You can see that the wave crests are bunched up in the direction in which the source is moving and are stretched out behind it. The distance between one crest and the next is one wavelength, so the wavelength λ_+ that Nancy measures is *less* than the wavelength $\lambda_0 = v/f_0$ that would be emitted if the source were at rest. Similarly, λ_- behind the source is larger than λ_0.

These crests move through the medium at the wave speed v. Consequently, the frequency $f_+ = v/\lambda_+$ detected by the observer whom the source is approaching is *higher* than the frequency f_0 emitted by the source. Similarly, $f_- = v/\lambda_-$ detected behind the source is *lower* than frequency f_0. This change of frequency when a source moves

FIGURE 15.18 A motion diagram showing the wave fronts emitted by a source as it moves to the right at speed v_s.

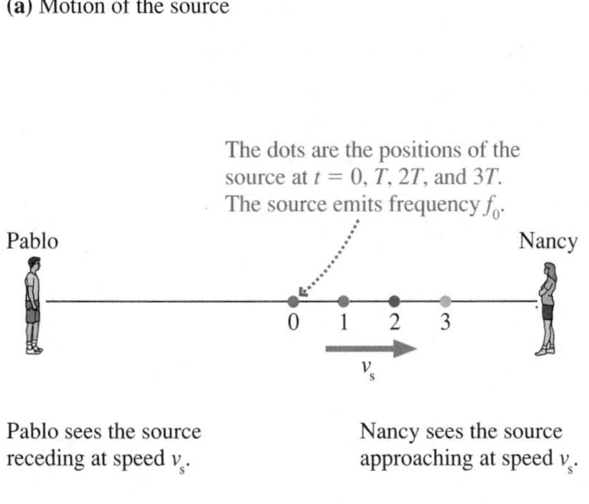
(a) Motion of the source
The dots are the positions of the source at t = 0, T, 2T, and 3T. The source emits frequency f0.
Pablo
Nancy
0 1 2 3
vs
Pablo sees the source receding at speed vs.
Nancy sees the source approaching at speed vs.

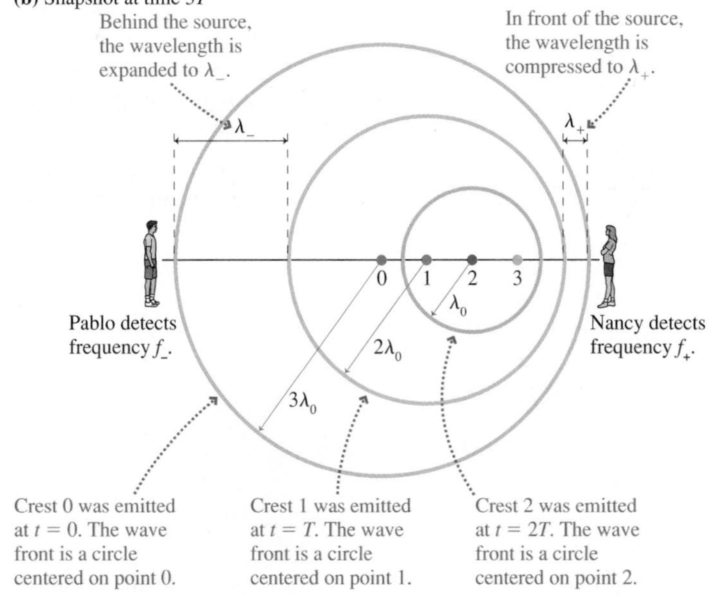
(b) Snapshot at time 3T
Behind the source, the wavelength is expanded to λ−.
In front of the source, the wavelength is compressed to λ+.
λ−
λ+
0 1 2 3
λ0
2λ0
3λ0
Pablo detects frequency f−.
Nancy detects frequency f+.
Crest 0 was emitted at t = 0. The wave front is a circle centered on point 0.
Crest 1 was emitted at t = T. The wave front is a circle centered on point 1.
Crest 2 was emitted at t = 2T. The wave front is a circle centered on point 2.

10.8, 10.9 ActivPhysics ONLINE

relative to an observer is called the **Doppler effect.** A quantitative analysis based on this information would show that the frequency heard by a stationary observer depends on whether the observer sees the source approaching or receding:

$$f_+ = \frac{f_0}{1 - v_s/v}$$

Observed frequency of a wave of speed v emitted from a source approaching at speed v_s (15.16)

$$f_- = \frac{f_0}{1 + v_s/v}$$

Observed frequency of a wave of speed v emitted from a source receding at speed v_s

As expected, $f_+ > f_0$ (the frequency is higher) for an approaching source because the denominator is less than 1, and $f_- < f_0$ (the frequency is lower) for a receding source.

EXAMPLE 15.13 How fast are the police driving?

A police siren has a frequency of 550 Hz as the police car approaches you, 450 Hz after it has passed you and is moving away. How fast are the police traveling?

PREPARE The siren's frequency is altered by the Doppler effect. The frequency is f_+ as the car approaches and f_- as it moves away. We can write two equations for these frequencies and solve for the speed of the police car, v_s.

SOLVE Because our goal is to find v_s, we rewrite Equations 15.16 as

$$f_0 = \left(1 + \frac{v_s}{v}\right) f_- \quad \text{and} \quad f_0 = \left(1 - \frac{v_s}{v}\right) f_+$$

Subtracting the second equation from the first, we get

$$0 = f_- - f_+ + \frac{v_s}{v}(f_- + f_+)$$

Now we can solve for the speed v_s:

$$v_s = \frac{f_+ - f_-}{f_+ + f_-} v = \frac{100 \text{ Hz}}{1000 \text{ Hz}} 343 \text{ m/s} = 34 \text{ m/s}$$

ASSESS This is pretty fast (about 75 mph) but reasonable for a police car speeding with the siren on.

A Stationary Source and a Moving Observer

Suppose the police car in Example 15.13 is at rest while you drive toward it at 34 m/s. You might think that this is equivalent to having the police car move toward you at 34 m/s, but there is an important difference. Mechanical waves move through a medium, and the Doppler effect depends not just on how the source and the observer move with respect to each other but also on how they move with respect to the medium. The frequency heard by an observer moving at speed v_o relative to a stationary source emitting frequency f_0 is given by

$$f_+ = \left(1 + \frac{v_o}{v}\right) f_0$$

Doppler effect for an observer approaching a source (15.17)

$$f_- = \left(1 - \frac{v_o}{v}\right) f_0$$

Doppler effect for an observer receding from a source

A quick calculation shows that the frequency of the police siren as you approach it at 34 m/s is 545 Hz, not the 550 Hz you heard as it approached you at 34 m/s.

The Doppler Effect for Light Waves

If a source of light waves is receding from you, the wavelength λ_- that you detect is longer than the wavelength λ_0 emitted by the source. Because the wavelength is shifted toward the red end of the visible spectrum, the longer wavelengths of light, this effect is called the **red shift.** Similarly, the light you detect from a source moving toward you is **blue shifted** to shorter wavelengths. For objects moving at normal speeds, this is a small effect; you won't see a Doppler shift of the flashing light of the police car in the above example! But for objects moving at very high speeds, the Doppler effect is significant.

In the 1920s, an analysis of the spectra of many distant galaxies showed that they *all* had a distinct red shift—all distant galaxies are moving away from us. How can we make sense of this observation? The astronomer Edwin Hubble concluded that the galaxies of the universe are *all* moving apart from each other. Extrapolating backward in time brings us to a point when all the matter of the universe—and even space itself, according to the theory of relativity—began rushing out of a primordial fireball. Many observations and measurements since have given support to the idea that the universe began in a *Big Bang* about 14 billion years ago.

The greater the distance to a galaxy, the faster it moves away from us. This photo shows galaxies at distances of up to 12 billion light years. The great distances imply large red shifts, making the light from the most distant galaxies appear distinctly red.

Frequency Shift on Reflection from a Moving Object

A wave striking a barrier or obstacle can *reflect* and travel back toward the source of the wave, a process we will examine more closely in the next chapter. For sound, the reflected wave is called an *echo.* A bat is able to determine the distance to a flying insect by measuring the time between the emission of an ultrasonic chirp and the detection of the echo from the insect. But if the bat is to catch the insect, it's just as important to know where the insect is going—its velocity. The bat can figure this out by noting the *frequency shift* in the reflected wave, another application of the Doppler effect.

Suppose a sound wave of frequency f_0 travels toward a moving object. The object would see the wave's frequency Doppler shifted to a higher frequency f_+, as given by Equation 15.17. The wave reflected back toward the sound source by the moving object is also Doppler shifted to a higher frequency because, to the source, the reflected wave is coming from a moving object. Thus the echo from a moving object is "double Doppler shifted." If the object's speed v_o is much lower than the wave speed v ($v_o \ll v$), the frequency *shift* of waves reflected from a moving object is

$$\Delta f = \pm 2 f_0 \frac{v_o}{v} \tag{15.18}$$

Frequency shift of waves reflected from an object moving at speed v_o

The + case is for objects moving toward the source of sound, the − for objects moving away. Notice that there is no shift (i.e., the reflected wave has the same frequency as the emitted wave) when the reflecting object is at rest.

The *Doppler blood flow meter* is an important application of the frequency shift of waves reflected from moving objects. If an ultrasound emitter is pressed against the skin, the sound waves reflect off tissues in the body. In most cases, the frequency of the reflected wave is the same as that of the emitted wave. However, some of the sound waves reflect from blood cells moving through arteries toward or away from the emitter. The moving blood cells produce a frequency shift Δf in the reflected wave. By measuring Δf, doctors can determine blood flow speeds in an entirely noninvasive manner.

Frequency shift on reflection is observed for all types of waves. Radar units emit pulses of radio waves and observe the reflected waves. The time between the emission of a pulse and its return gives an object's position. The change in frequency of the returned pulse gives the object's speed. This is the principle behind the radar guns used by traffic police, as well as the Doppler radar images you have seen in weather reports.

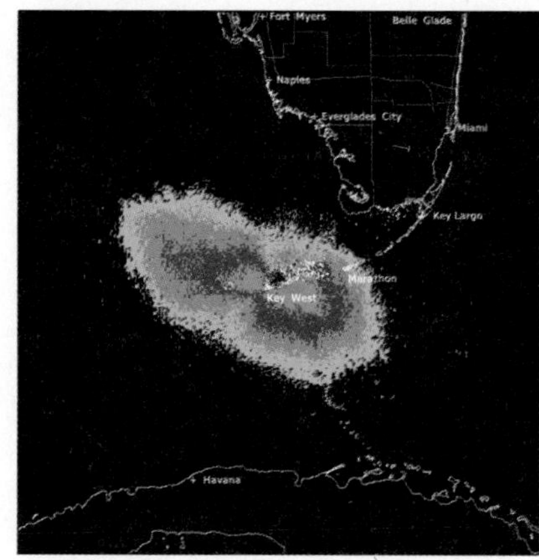

Keeping track of wildlife with radar Doppler radar is tuned to measure only those reflected radio waves that have a frequency shift, eliminating reflections from stationary objects and showing only objects in motion. Televised Doppler radar images of storms are made using radio waves reflected from moving water droplets. But this Doppler radar image of an area off the tip of Florida was made on a clear night with no rain. The blue and green patch isn't a moving storm, it's a moving flock of birds. Migratory birds frequently move at high altitudes at night, so Doppler radar is an excellent tool for analyzing their movements.

EXAMPLE 15.14 Ultrasound frequency to measure blood flow

A biomedical engineer is designing a Doppler blood flow meter to measure blood flow in an artery where a typical flow speed is known to be 0.60 m/s. What ultrasound frequency should she use to produce a frequency shift of 1500 Hz when this flow is detected?

SOLVE We can rewrite Equation 15.18 to calculate the frequency of the emitter:

$$f_0 = \left(\frac{\Delta f}{2}\right)\left(\frac{v}{v_{\text{blood}}}\right)$$

The values on the right side are all known. Thus the required ultrasound frequency is

$$f_0 = \left(\frac{1500\text{ Hz}}{2}\right)\left(\frac{1540\text{ m/s}}{0.60\text{ m/s}}\right) = 1.9\text{ MHz}$$

ASSESS Doppler units to measure blood flow in deep tissues actually work at about 2.0 MHz, so our answer is reasonable.

If we add a frequency-shift measurement to an ultrasound imaging unit, like the one in Example 15.7, we have a device—called *Doppler ultrasound*—that can show not only structure but also motion. *Doppler ultrasound* is a very valuable tool in cardiology because an image can reveal the motion of the heart muscle and the blood in the chambers of the heart in addition to the structure of the heart itself.

FIGURE 15.19 Waves emitted by a source traveling faster than the speed of the waves in a medium.

The source of waves is moving to the right at v_s. The positions at times $t = 0, t = T, t = 2T, \ldots$ are marked.

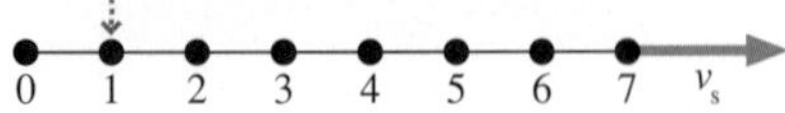
0 1 2 3 4 5 6 7 v_s

At each point, the source emits a wave that spreads out. A snapshot is taken at $t = 7T$.

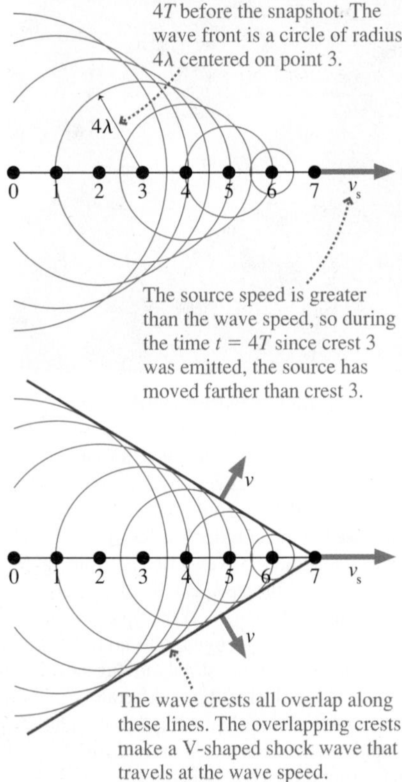
Crest 3 was emitted at $t = 3T$, $4T$ before the snapshot. The wave front is a circle of radius 4λ centered on point 3.
4λ
0 1 2 3 4 5 6 7 v_s
The source speed is greater than the wave speed, so during the time $t = 4T$ since crest 3 was emitted, the source has moved farther than crest 3.
v
0 1 2 3 4 5 6 7 v_s
v
The wave crests all overlap along these lines. The overlapping crests make a V-shaped shock wave that travels at the wave speed.

Shock Waves

When sound waves are emitted by a moving source, the frequency is shifted, as we have seen. Now, let's look at what happens when the source speed v_s exceeds the wave speed v and the source "outruns" the waves it produces.

Earlier in this section, in Figure 15.18, we looked at a motion diagram of waves emitted by a moving source. **FIGURE 15.19** is the same diagram, but in this case the source is moving faster than the waves. This motion causes the waves to overlap. (Compare Figure 15.19 to Figure 15.18.) The amplitudes of the overlapping waves add up to produce a very large amplitude wave—a **shock wave.**

Anything that moves faster than the speed of sound in air will create a shock wave. The speed of sound was broken on land in 1997 by a British team driving the Thrust SCC in Nevada's Black Rock Desert. **FIGURE 15.20a** shows the shock waves produced during a **supersonic** (faster than the speed of sound) run. You can see the distortion of the view of the landscape behind the car where the waves add to produce regions of high pressure. This shock wave travels along with the car. If the car went by you, the passing of the shock wave would produce a **sonic boom,** the distinctive loud sound you may have heard when a supersonic aircraft passes. The crack of a whip is a sonic boom as well, though on a much smaller scale.

FIGURE 15.20 Extreme and everyday examples of shock waves.

(a)

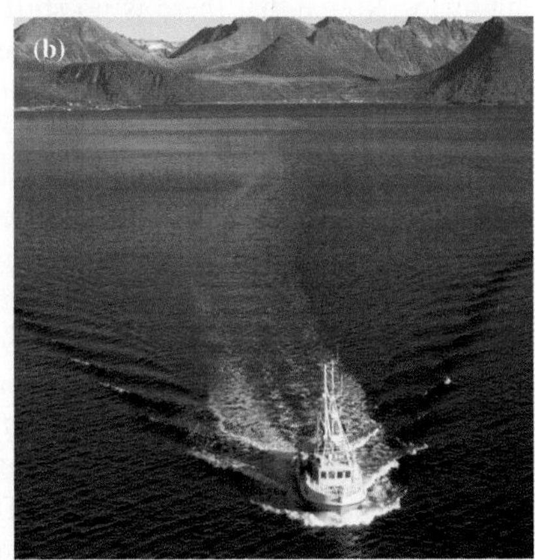
(b)

Shock waves are not necessarily an extreme phenomenon. A boat (or even a duck!) can easily travel faster than the speed of the water waves it creates; the resulting wake, with its characteristic "V" shape as shown in **FIGURE 15.20b**, is really a shock wave.

STOP TO THINK 15.5 Amy and Zack are both listening to the source of sound waves that is moving to the right. Compare the frequencies each hears.

A. $f_{Amy} > f_{Zack}$
B. $f_{Amy} = f_{Zack}$
C. $f_{Amy} < f_{Zack}$

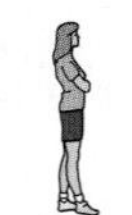

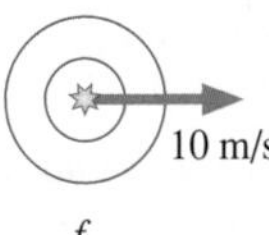

INTEGRATED EXAMPLE 15.15 Shaking the ground

Earthquakes are dramatic slips of the earth's crust. You may feel the waves generated by an earthquake even if you're some distance from the *epicenter,* the point where the slippage occurs. Earthquake waves are usually complicated, but some of the long-period waves shake the ground with motion that is approximately simple harmonic motion. **FIGURE 15.21** shows the vertical position of the ground recorded at a distant monitoring station following an earthquake that hit Japan in 2003. This particular wave traveled with a speed of 3500 m/s.

a. Was the wave transverse or longitudinal?
b. What were the wave's frequency and wavelength?
c. What were the maximum speed and the maximum acceleration of the ground during this earthquake wave?
d. Intense earthquake waves produce accelerations greater than free-fall acceleration of gravity. How does this wave compare?

FIGURE 15.21 The vertical motion of the ground during the passage of an earthquake wave.

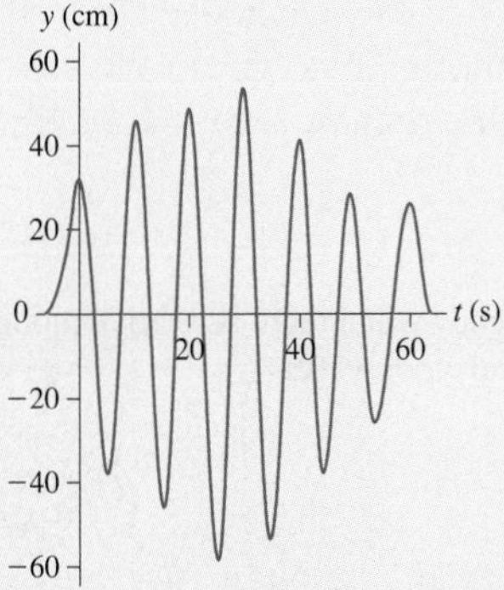

PREPARE There are two different but related parts to the solution: analyzing the motion of the ground, and analyzing the motion of the wave. The graph in Figure 15.21 is a history graph; it's a record of the motion of one point of the medium—the ground. The graph is approximately sinusoidal, so we can model the motion of the ground as simple harmonic motion. If we look at the middle portion of the wave, we can estimate the amplitude to be 0.50 m (it varies, but this is a reasonable average over a few cycles). Because the ground was in simple harmonic motion, the traveling wave was a sinusoidal wave. Consequently, we can use the fundamental relationships for sinusoidal waves to relate the wavelength, frequency, and speed.

SOLVE

a. The graph shows the *vertical* motion of the ground. This motion of the ground—the medium—was perpendicular to the horizontal motion of the wave traveling along the ground, so this was a transverse wave.
b. We can see from the graph that 6 cycles of the oscillation took 60 seconds, so the period was $T = 10$ s. The period of the wave was the same as that of the point on the ground, 10 s; thus the frequency was $f = 1/T = 1/10 \text{ s} = 0.10$ Hz. The speed of the wave was 3500 m/s, so the wavelength was

$$\lambda = \frac{v}{f} = \frac{3500 \text{ m/s}}{0.10 \text{ Hz}} = 35{,}000 \text{ m} = 35 \text{ km}$$

c. The motion of the ground was simple harmonic motion with frequency 0.10 Hz and amplitude 0.50 m. We can compute the maximum speed and acceleration using relationships from Chapter 14:

$$v_{max} = 2\pi fA = (2\pi)(0.10 \text{ Hz})(0.50 \text{ m}) = 0.31 \text{ m/s}$$

$$a_{max} = (2\pi f)^2 A = [2\pi(0.10 \text{ Hz})]^2(0.50 \text{ m}) = 0.20 \text{ m/s}^2$$

d. This was a reasonably gentle earthquake wave, with an acceleration of the ground quite small compared to the free-fall acceleration of gravity:

$$a_{max}(\text{in units of } g) = \frac{0.20 \text{ m/s}^2}{9.8 \text{ m/s}^2} = 0.020g$$

ASSESS The wavelength is quite long, as we might expect for such a fast wave with a long period. Given the relatively small amplitude and long period, it's no surprise that the maximum speed and acceleration are relatively modest. You'd certainly feel the passage of this wave, but it wouldn't knock buildings down. This value is less than the acceleration for the sway of the top of the building in Example 14.4 in Chapter 14!

SUMMARY

The goal of Chapter 15 has been to learn the basic properties of traveling waves.

GENERAL PRINCIPLES

The Wave Model

This model is based on the idea of a traveling wave, which is an organized disturbance traveling at a well-defined **wave speed** v.

- In transverse waves the particles of the medium move *perpendicular* to the direction in which the wave travels.

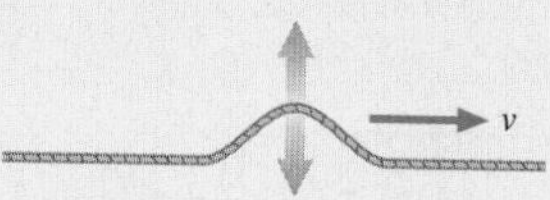

- In longitudinal waves the particles of the medium move *parallel* to the direction in which the wave travels.

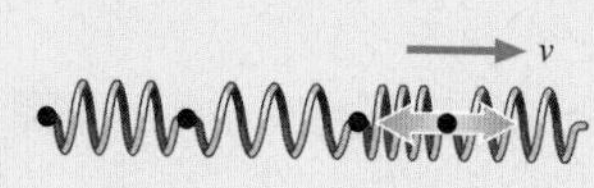

A wave transfers energy, but there is no material or substance transferred.

Mechanical waves require a material **medium.** The speed of the wave is a property of the medium, not the wave. The speed does not depend on the size or shape of the wave.

- For a **wave on a string,** the string is the medium.

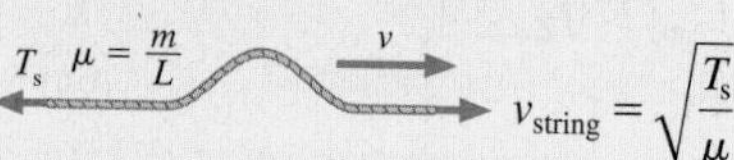

$$v_{string} = \sqrt{\frac{T_s}{\mu}}$$

- A **sound wave** is a wave of compressions and rarefactions of a medium such as air.

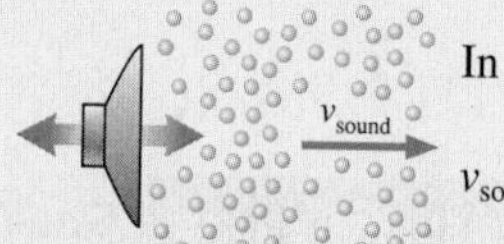

In a gas:

$$v_{sound} = \sqrt{\frac{\gamma RT}{M}}$$

Electromagnetic waves are waves of the electromagnetic field. They do not require a medium. All electromagnetic waves travel at the same speed in a vacuum, $c = 3.00 \times 10^8$ m/s.

IMPORTANT CONCEPTS

Graphical representation of waves

A snapshot graph is a picture of a wave at one instant in time. For a periodic wave, the **wavelength** λ is the distance between crests.

Fixed t:

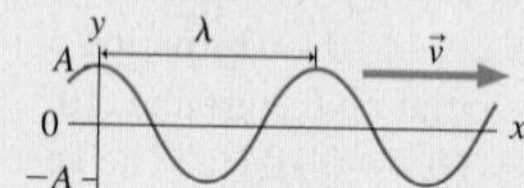

A history graph is a graph of the displacement of one point in a medium versus time. For a periodic wave, the **period** T is the time between crests.

Fixed x:

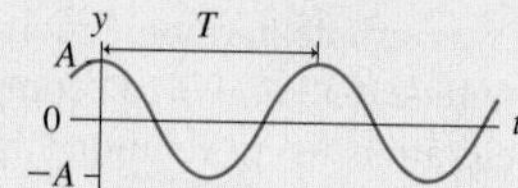

Mathematical representation of waves

Sinusoidal waves are produced by a source moving with simple harmonic motion. The equation for a sinusoidal wave is a function of position and time:

$$y(x, t) = A\cos\left(2\pi\left(\frac{x}{\lambda} \pm \frac{t}{T}\right)\right)$$

+: wave travels to left

−: wave travels to right

For sinusoidal and other periodic waves:

$$T = \frac{1}{f} \qquad v = f\lambda$$

The intensity of a wave is the ratio of the power to the area:

$$I = \frac{P}{A}$$

For a **spherical wave** the power decreases with the surface area of the spherical **wave fronts:**

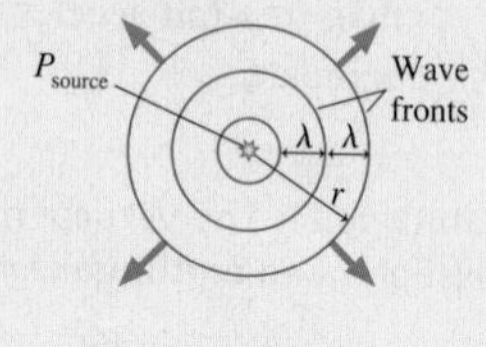

$$I = \frac{P_{source}}{4\pi r^2}$$

APPLICATIONS

The loudness of a sound is given by the sound intensity level. This is a logarithmic function of intensity and is in units of **decibels.**

- The usual **reference level** is the quietest sound that can be heard:

$$I_0 = 1.0 \times 10^{-12}\ \text{W/m}^2$$

- The sound intensity level in dB is computed relative to this value:

$$\beta = (10\ \text{dB})\log_{10}\left(\frac{I}{I_0}\right)$$

- A sound at the reference level corresponds to 0 dB.

The Doppler effect is a shift in frequency when there is relative motion of a wave source (frequency f_0, wave speed v) and an observer.

Moving source, stationary observer:

Receding source:

$$f_- = \frac{f_0}{1 + v_s/v}$$

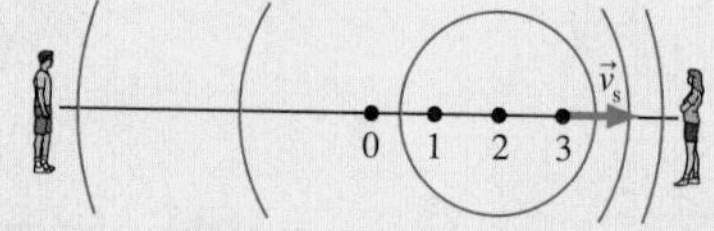

Approaching source:

$$f_+ = \frac{f_0}{1 - v_s/v}$$

Moving observer, stationary source:

Approaching the source:

$$f_+ = \left(1 + \frac{v_o}{v}\right)f_0$$

Moving away from the source:

$$f_- = \left(1 - \frac{v_o}{v}\right)f_0$$

Reflection from a moving object:

For $v_o \ll v$, $\Delta f = \pm 2f_0\frac{v_o}{v}$

When an object moves faster than the wave speed in a medium, a shock wave is formed.

For instructor-assigned homework, go to www.masteringphysics.com

Problems labeled INT integrate significant material from earlier chapters; BIO are of biological or medical interest.

Problem difficulty is labeled as | (straightforward) to ||||| (challenging).

QUESTIONS

Conceptual Questions

1. a. In your own words, define what a *transverse wave* is.
   b. Give an example of a wave that, from your own experience, you know is a transverse wave. What observations or evidence tells you this is a transverse wave?
2. a. In your own words, define what a *longitudinal wave* is.
   b. Give an example of a wave that, from your own experience, you know is a longitudinal wave. What observations or evidence tells you this is a longitudinal wave?
3. The wave pulses shown in Figure Q15.3 travel along the same string. Rank in order, from largest to smallest, their wave speeds v_1, v_2, and v_3. Explain.

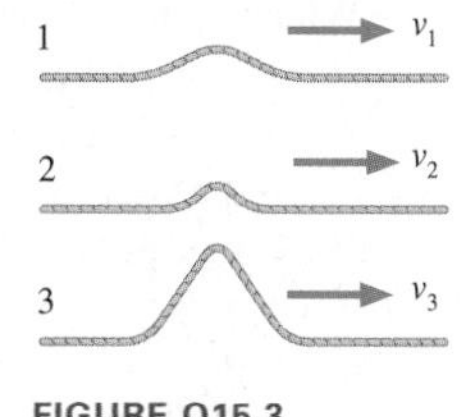

FIGURE Q15.3

4. Is it ever possible for one sound wave in air to overtake and pass another? Explain.
5. A wave pulse travels along a string at a speed of 200 cm/s. What will be the speed if:
   a. The string's tension is doubled?
   b. The string's mass is quadrupled (but its length is unchanged)?
   c. The string's length is quadrupled (but its mass is unchanged)?
   d. The string's mass and length are both quadrupled?
   Note that parts a–d are independent and refer to changes made to the original string.
6. An ultrasonic range finder sends out a pulse of ultrasound and measures the time between the emission of the pulse and the return of an echo from an object. This time is used to determine the distance to the object. To get good accuracy from the device, a user must enter the air temperature in the room. Why is this?
7. A thermostat on the wall of your house keeps track of the air temperature. This simple approach is of little use in the large volume of a covered sports stadium, but there are systems that determine an average temperature of the air in a stadium by measuring the time delay between the emission of a pulse of sound on one side of the stadium and its detection on the other. Explain how such a system works.
8. When water freezes, the density decreases and the bonds between molecules become stronger. Do you expect the speed of sound to be greater in liquid water or in water ice?
9. Figure Q15.9 shows a history graph of the motion of one point on a string as a wave traveling to the left passes by. Sketch a snapshot graph for this wave.

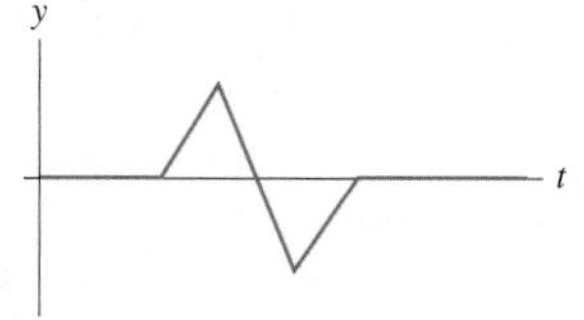

FIGURE Q15.9

10. Figure Q15.10 shows a history graph *and* a snapshot graph for a wave pulse on a string. They describe the same wave from two perspectives.
    a. In which direction is the wave traveling? Explain.
    b. What is the speed of this wave?

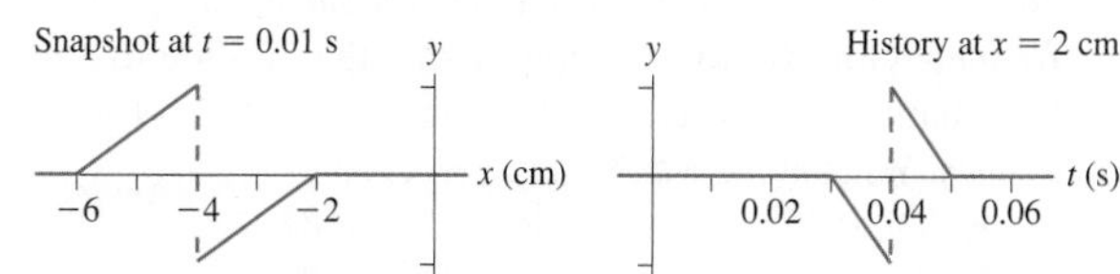

FIGURE Q15.10

11. Rank in order, from largest to smallest, the wavelengths λ_1 to λ_3 for sound waves having frequencies $f_1 = 100$ Hz, $f_2 = 1000$ Hz, and $f_3 = 10{,}000$ Hz. Explain.
12. Explain why there is a factor of 2π in Equation 15.5.
13. BIO Bottlenose dolphins use echolocation pulses with a frequency of about 100 kHz, higher than the frequencies used by most bats. Why might you expect these water-dwelling creatures to use higher echolocation frequencies than bats?

14. A laser beam has intensity I_0.
    a. What is the intensity, in terms of I_0, if a lens focuses the laser beam to 1/10 its initial diameter?
    b. What is the intensity, in terms of I_0, if a lens defocuses the laser beam to 10 times its initial diameter?
15. Sound wave A delivers 2 J of energy in 2 s. Sound wave B delivers 10 J of energy in 5 s. Sound wave C delivers 2 mJ of energy in 1 ms. Rank in order, from largest to smallest, the sound powers P_A, P_B, and P_C of these three waves. Explain.
16. When you want to "snap" a towel, the best way to wrap the towel is so that the end that you hold and shake is thick, and the far end is thin. When you shake the thick end, a wave travels down the towel. How does wrapping the towel in a tapered shape help make for a good snap?
    Hint: Think about the speed of the wave as it moves down the towel.
17. The volume control on your stereo is likely designed so that each time you turn it by one click, the loudness increases by a certain number of dB. Does each click increase the output power by a fixed amount as well?
18. A bullet can travel at a speed of over 1000 m/s. When a bullet is fired from a rifle, the actual firing makes a distinctive sound, but people at a distance may hear a second, different sound that is even louder. Explain the source of this sound.

19. You are standing at $x = 0$ m, listening to seven identical sound sources described by Figure Q15.19. At $t = 0$ s, all seven are at $x = 343$ m and moving as shown below. The sound from all seven will reach your ear at $t = 1$ s. Rank in order, from highest to lowest, the seven frequencies f_1 to f_7 that you hear at $t = 1$ s. Explain.

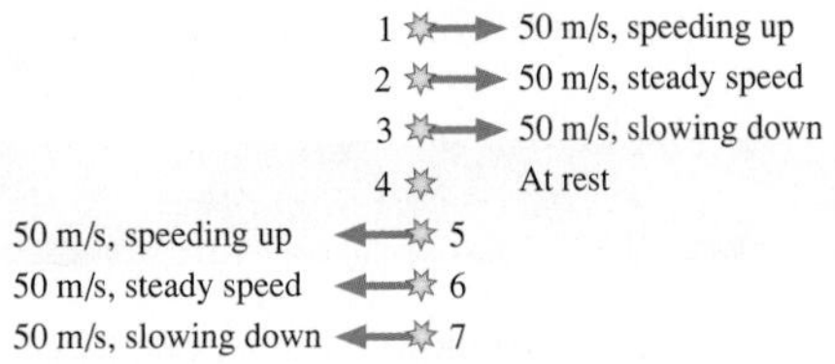

1 → 50 m/s, speeding up
2 → 50 m/s, steady speed
3 → 50 m/s, slowing down
4 At rest
50 m/s, speeding up ← 5
50 m/s, steady speed ← 6
50 m/s, slowing down ← 7

FIGURE Q15.19

Multiple-Choice Questions

20. | Denver, Colorado, has an oldies station that calls itself "KOOL 105." This means that they broadcast radio waves at a frequency of 105 MHz. Suppose that they decide to describe their station by its wavelength (in meters), instead of by its frequency. What name would they now use?
A. KOOL 0.35 B. KOOL 2.85
C. KOOL 3.5 D. KOOL 285

21. | What is the frequency of blue light with a wavelength of 400nm?
A. 1.33×10^3 Hz B. 7.50×10^{12} Hz
C. 1.33×10^{14} Hz D. 7.50×10^{14} Hz

22. | Ultrasound can be used to deliver energy to tissues for therapy. It can penetrate tissue to a depth approximately 200 times its wavelength. What is the approximate depth of penetration of ultrasound at a frequency of 5.0 MHz?
BIO
A. 0.29 mm B. 1.4 cm
C. 6.2 cm D. 17 cm

23. | A sinusoidal wave traveling on a string has a period of 0.20 s, a wavelength of 32 cm, and an amplitude of 3 cm. The speed of this wave is
A. 0.60 cm/s. B. 6.4 cm/s.
C. 15 cm/s. D. 160 cm/s.

24. || Two strings of different linear density are joined together and pulled taut. A sinusoidal wave on these strings is traveling to the right, as shown in Figure Q15.24. When the wave goes across the boundary from string 1 to string 2, the frequency is unchanged. What happens to the velocity?
A. The velocity increases.
B. The velocity stays the same.
C. The velocity decreases.

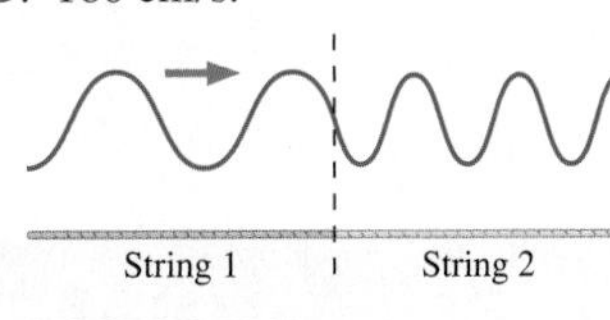

FIGURE Q15.24

25. || You stand at $x = 0$ m, listening to a sound that is emitted at frequency f_0. Figure Q15.25 shows the frequency you hear during a four-second interval. Which of the following describes the motion of the sound source?
A. It moves from left to right and passes you at $t = 2$ s.
B. It moves from right to left and passes you at $t = 2$ s.
C. It moves toward you but doesn't reach you. It then reverses direction at $t = 2$ s.
D. It moves away from you until $t = 2$ s. It then reverses direction and moves toward you but doesn't reach you.

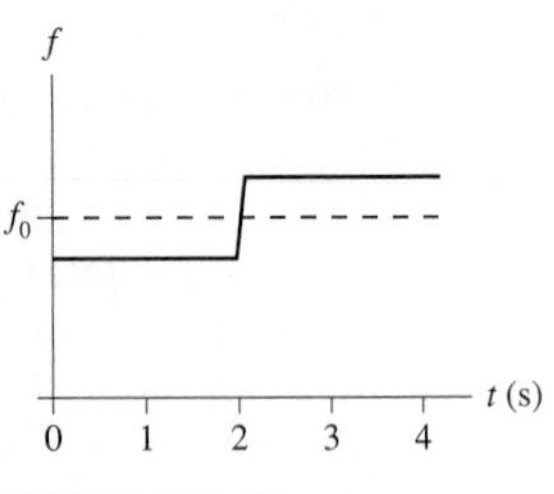

FIGURE Q15.25

PROBLEMS

Section 15.1 The Wave Model

Section 15.2 Traveling Waves

1. || The wave speed on a string under tension is 200 m/s. What is the speed if the tension is doubled?
2. || The wave speed on a string is 150 m/s when the tension is 75.0 N. What tension will give a speed of 180 m/s?
3. || A wave travels along a string at a speed of 280 m/s. What will be the speed if the string is replaced by one made of the same material and under the same tension but having twice the radius?
4. || The back wall of an auditorium is 26.0 m from the stage. If you are seated in the middle row, how much time elapses between a sound from the stage reaching your ear directly and the same sound reaching your ear after reflecting from the back wall?
5. |||| A hammer taps on the end of a 4.00-m-long metal bar at room temperature. A microphone at the other end of the bar picks up two pulses of sound, one that travels through the metal and one that travels through the air. The pulses are separated in time by 11.0 ms. What is the speed of sound in this metal?
6. || In an early test of sound propagation through the ocean, an underwater explosion of 1 pound of dynamite in the Bahamas was detected 3200 km away on the coast of Africa. How much time elapsed between the explosion and the detection?

Section 15.3 Graphical and Mathematical Descriptions of Waves

7. || Figure P15.7 is a snapshot graph of a wave at $t = 0$ s. Draw the history graph for this wave at $x = 6$ m, for $t = 0$ s to 6 s.

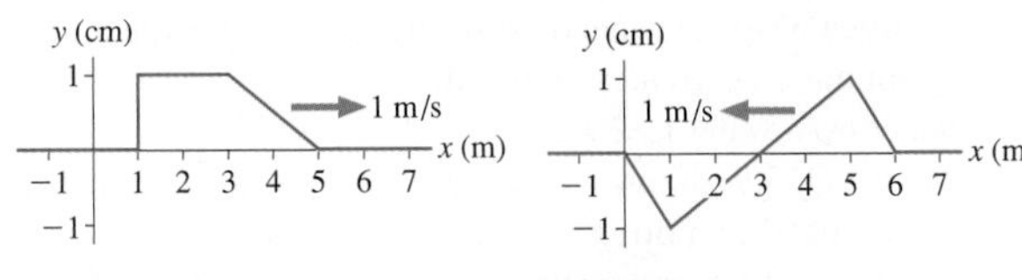

FIGURE P15.7 FIGURE P15.8

8. || Figure P15.8 is a snapshot graph of a wave at $t = 2$ s. Draw the history graph for this wave at $x = 0$ m, for $t = 0$ s to 8 s.

9. ||| Figure P15.9 is a history graph at $x = 0$ m of a wave moving to the right at 1 m/s. Draw a snapshot graph of this wave at $t = 1$ s.

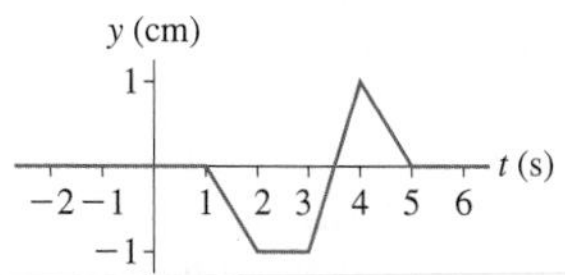

FIGURE P15.9

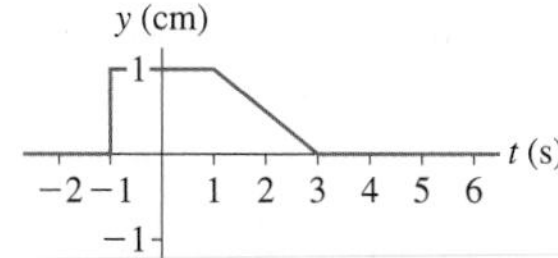

FIGURE P15.10

10. ||| Figure P15.10 is a history graph at $x = 2$ m of a wave moving to the left at 1 m/s. Draw the snapshot graph of this wave at $t = 0$ s.

11. | A sinusoidal wave has period 0.20 s and wavelength 2.0 m. What is the wave speed?

12. | A sinusoidal wave travels with speed 200 m/s. Its wavelength is 4.0 m. What is its frequency?

13. || The motion detector used in a physics lab sends and receives 40 kHz ultrasonic pulses. A pulse goes out, reflects off the object being measured, and returns to the detector. The lab temperature is 20°C.
   a. What is the wavelength of the waves emitted by the motion detector?
   b. How long does it take for a pulse that reflects off an object 2.5 m away to make a round trip?

14. | The displacement of a wave traveling in the positive x-direction is $y(x, t) = (3.5\text{ cm})\cos(2.7x - 92t)$, where x is in m and t is in s. What are the (a) frequency, (b) wavelength, and (c) speed of this wave?

15. || The displacement of a wave traveling in the negative x-direction is $y(x, t) = (5.2\text{ cm})\cos(5.5x + 72t)$, where x is in m and t is in s. What are the (a) frequency, (b) wavelength, and (c) speed of this wave?

16. ||| A traveling wave has displacement given by $y(x, t) = (2.0\text{cm}) \times \cos(2\pi x - 4\pi t)$, where x is measured in cm and t in s.
   a. Draw a snapshot graph of this wave at $t = 0$ s.
   b. On the same set of axes, use a dotted line to show the snapshot graph of the wave at $t = 1/8$ s.
   c. What is the speed of the wave?

17. | Figure P15.17 is a snapshot graph of a wave at $t = 0$ s. What are the amplitude, wavelength, and frequency of this wave?

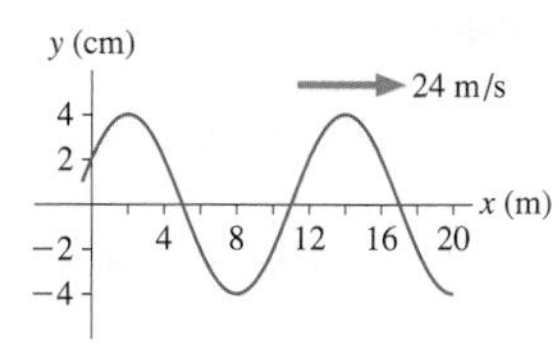

FIGURE P15.17

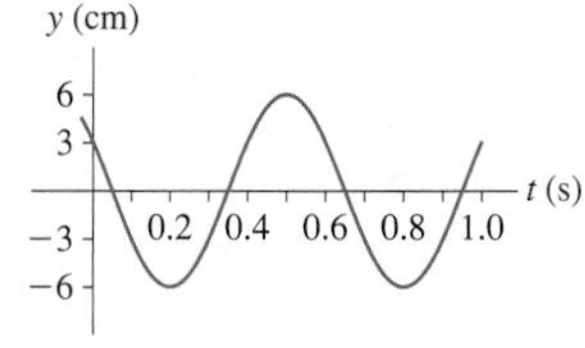

FIGURE P15.18

18. | Figure P15.18 is a history graph at $x = 0$ m of a wave moving to the right at 2 m/s. What are the amplitude, frequency, and wavelength of this wave?

19. |||| A boat is traveling at 4.0 m/s in the same direction as an ocean wave of wavelength 30 m and speed 6.8 m/s. If the boat is on the crest of a wave, how much time will elapse until the boat is next on a crest?

20. ||| In the deep ocean, a water wave with wavelength 95 m travels at 12 m/s. Suppose a small boat is at the crest of this wave, 1.2 m above the equilibrium position. What will be the vertical position of the boat 5.0 s later?

Section 15.4 Sound and Light Waves

21. | A dolphin emits ultrasound at 100 kHz and uses the timing of
BIO reflections to determine the position of objects in the water. What is the wavelength of this ultrasound?

22. || a. What is the wavelength of a 2.0 MHz ultrasound wave traveling through aluminum?
   b. What frequency of electromagnetic wave would have the same wavelength as the ultrasound wave of part a?

23. | a. At 20°C, what is the frequency of a sound wave in air with a wavelength of 20 cm?
   b. What is the frequency of an electromagnetic wave with a wavelength of 20 cm?
   c. What would be the wavelength of a sound wave in water that has the same frequency as the electromagnetic wave of part b?

24. || a. What is the frequency of blue light that has a wavelength of 450 nm?
   b. What is the frequency of red light that has a wavelength of 650 nm?

25. || a. Telephone signals are often transmitted over long distances by microwaves. What is the frequency of microwave radiation with a wavelength of 3.0 cm?
   b. Microwave signals are beamed between two mountaintops 50 km apart. How long does it take a signal to travel from one mountaintop to the other?

26. || a. An FM radio station broadcasts at a frequency of 101.3 MHz. What is the wavelength?
   b. What is the frequency of a sound source that produces the same wavelength in 20°C air?

Section 15.5 Energy and Intensity

27. || Sound is detected when a sound wave causes the eardrum
BIO to vibrate (see Figure 14.26). Typically, the diameter of the
INT eardrum is about 8.4 mm in humans. When someone speaks to you in a normal tone of voice, the sound intensity at your ear is approximately $1.0 \times 10^{-6}\text{ W/m}^2$. How much energy is delivered to your eardrum each second?

28. ||| At a rock concert, the sound intensity 1.0 m in front of the
BIO bank of loudspeakers is 0.10 W/m^2. A fan is 30 m from the
INT loudspeakers. Her eardrums have a diameter of 8.4 mm. How much sound energy is transferred to each eardrum in 1.0 second?

29. ||| The intensity of electromagnetic waves from the sun is 1.4 kW/m^2 just above the earth's atmosphere. Eighty percent of this reaches the surface at noon on a clear summer day. Suppose you model your back as a 30 cm $\times$ 50 cm rectangle. How many joules of solar energy fall on your back as you work on your tan for 1.0 hr?

30. || The sun emits electromagnetic waves with a power of 4.0×10^{26} W. Determine the intensity of electromagnetic waves from the sun just outside the atmospheres of (a) Venus, (b) Mars, and (c) Saturn. Refer to the table of astronomical data inside the back cover.

31. ||| A large solar panel on a spacecraft in Earth orbit produces 1.0 kW of power when the panel is turned toward the sun. What power would the solar cell produce if the spacecraft were in orbit around Saturn, 9.5 times as far from the sun?

32. ||| Solar cells convert the energy of incoming light to electric energy; a good quality cell operates at an efficiency of 15%. Each person in the United States uses energy (for lighting, heating, transportation, etc.) at an average rate of 11 kW. Although

sunlight varies with season and time of day, solar energy falls on the United States at an average intensity of 200 W/m^2. Assuming you live in an average location, what total solar-cell area would you need to provide all of your energy needs with energy from the sun?

33. | LASIK eye surgery uses pulses of laser light to shave off tissue from the cornea, reshaping it. A typical LASIK laser emits a 1.0-mm-diameter laser beam with a wavelength of 193 nm. Each laser pulse lasts 15 ns and contains 1.0 mJ of light energy.
BIO
    a. What is the power of one laser pulse?
    b. During the very brief time of the pulse, what is the intensity of the light wave?

Section 15.6 Loudness of Sound

34. | What is the sound intensity level of a sound with an intensity of 3.0×10^{-6} W/m^2?
35. ||| What is the sound intensity of a whisper at a distance of 2.0 m, in W/m^2? What is the corresponding sound intensity level in dB?
36. || You hear a sound at 65 dB. What is the sound intensity level if the intensity of the sound is doubled?
37. || The sound intensity from a jack hammer breaking concrete is 2.0 W/m^2 at a distance of 2.0 m from the point of impact. This is sufficiently loud to cause permanent hearing damage if the operator doesn't wear ear protection. What are (a) the sound intensity and (b) the sound intensity level for a person watching from 50 m away?
38. ||| A concert loudspeaker suspended high off the ground emits 35 W of sound power. A small microphone with a 1.0 cm^2 area is 50 m from the speaker. What are (a) the sound intensity and (b) the sound intensity level at the position of the microphone?
39. || A rock band playing an outdoor concert produces sound at 120 dB 5.0 m away from their single working loudspeaker. What is the sound intensity level 35 m from the speaker?
40. ||| Your ears are sensitive to differences in pitch, but they are not very sensitive to differences in intensity. You are not capable of detecting a difference in sound intensity level of less than 1 dB. By what factor does the sound intensity increase if the sound intensity level increases from 60 dB to 61 dB?
BIO

Section 15.7 The Doppler Effect and Shock Waves

41. | An opera singer in a convertible sings a note at 600 Hz while cruising down the highway at 90 km/hr. What is the frequency heard by
    a. A person standing beside the road in front of the car?
    b. A person standing beside the road behind the car?
42. || An osprey's call is a distinct whistle at 2200 Hz. An osprey calls while diving at you, to drive you away from her nest. You hear the call at 2300 Hz. How fast is the osprey approaching?
BIO
43. || A whistle you use to call your hunting dog has a frequency of 21 kHz, but your dog is ignoring it. You suspect the whistle may not be working, but you can't hear sounds above 20 kHz. To test it, you ask a friend to blow the whistle, then you hop on your bicycle. In which direction should you ride (toward or away from your friend) and at what minimum speed to know if the whistle is working?
44. | A friend of yours is loudly singing a single note at 400 Hz while driving toward you at 25.0 m/s on a day when the speed of sound is 340 m/s.
    a. What frequency do you hear?
    b. What frequency does your friend hear if you suddenly start singing at 400 Hz?
45. |||| While anchored in the middle of a lake, you count exactly three waves hitting your boat every 10 s. You raise anchor and start motoring slowly in the same direction the waves are going. When traveling at 1.5 m/s, you notice that exactly two waves are hitting the boat from behind every 10 s. What is the speed of the waves on the lake?
46. |||| A Doppler blood flow unit emits ultrasound at 5.0 MHz. What is the frequency shift of the ultrasound reflected from blood moving in an artery at a speed of 0.20 m/s?
BIO

General Problems

47. | You're watching a carpenter pound a nail. He hits the nail twice a second, but you hear the sound of the strike when his hammer is fully raised. What is the minimum distance from you to the carpenter? Assume the air temperature is 20°C.
48. || A 2.50 kHz sound wave is transmitted through an aluminum rod.
    a. What is its wavelength in the aluminum?
    b. What is the sound wave's frequency when it passes into the air?
    c. What is its wavelength in air?
49. ||| Oil explorers set off explosives to make loud sounds, then listen for the echoes from underground oil deposits. Geologists suspect that there is oil under 500-m-deep Lake Physics. It's known that Lake Physics is carved out of a granite basin. Explorers detect a weak echo 0.94 s after exploding dynamite at the lake surface. If it's really oil, how deep will they have to drill into the granite to reach it?
50. ||| A 2.0-m-long string is under 20 N of tension. A pulse travels the length of the string in 50 ms. What is the mass of the string?
51. ||| A stout cord is stretched between two fixed supports. You vigorously shake one end of the string and send a sinusoidal wave of wavelength 4.0 m along it at 16 m/s. The amplitude of the motion is 2.0 cm. What are the maximum speed and maximum acceleration of a point on the string as the wave passes?
INT
52. ||| A female orb spider has a mass of 0.50 g. She is suspended from a tree branch by a 1.1 m length of 0.0020-mm-diameter silk. Spider silk has a density of 1300 kg/m^3. If you tap the branch and send a vibration down the thread, how long does it take to reach the spider?
BIO INT
53. || Andy (mass 80 kg) uses a 3.0-m-long rope to pull Bob (mass 60 kg) across the floor ($\mu_k = 0.20$) at a constant speed of 1.0 m/s. Bob signals to Andy to stop by "plucking" the rope, sending a wave pulse forward along the rope. The pulse reaches Andy 150 ms later. What is the mass of the rope?
INT
54. || If a bungee cord is stretched horizontally to a length of 2.5 m, the tension in the cord is 2.1 N. A transverse pulse on the cord travels from one end to the other in 0.80 s. If the cord is stretched to a length of 3.5 m, the pulse takes the same time of 0.80 s to travel from one end to the other. What is the tension in the cord when it is stretched to this length?
55. || String 1 in Figure P15.55 has linear density 2.0 g/m and string 2 has linear density 4.0 g/m. A student sends pulses in both directions by quickly pulling up on the knot, then releasing it. What should the string lengths L_1 and L_2 be if the pulses are to reach the ends of the strings simultaneously?

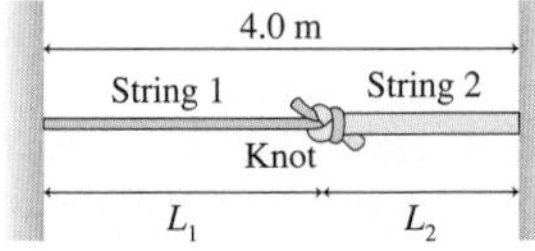

FIGURE P15.55

56. ||| In 2003, an earthquake in Japan generated 1.1 Hz waves that traveled outward at 7.0 km/s. 200 km to the west, seismic instruments recorded a maximum acceleration of 0.25g along the east-west axis.
INT

a. How much time elapsed between the earthquake and the first detection of the waves?
b. Was this a transverse or a longitudinal wave?
c. What was the wavelength?
d. What was the maximum horizontal displacement of the ground as the wave passed?

57. ||| A coyote can locate a sound source with good accuracy by
BIO comparing the arrival times of a sound wave at its two ears. Suppose a coyote is listening to a bird whistling at 1000 Hz. The bird is 3.0 m away, directly in front of the coyote's right ear. The coyote's ears are 15 cm apart.
a. What is the distance between the bird and the coyote's left ear?
b. What is the difference in the arrival time of the sound at the left ear and the right ear?
c. What is the ratio of this time difference to the period of the sound wave?

Hint: You are looking for the difference between two numbers that are nearly the same. What does this near equality imply about the necessary precision during intermediate stages of the calculation?

58. || An earthquake produces longitudinal P waves that travel outward at 8000 m/s and transverse S waves that move at 4500 m/s. A seismograph at some distance from the earthquake records the arrival of the S waves 2.0 min after the arrival of the P waves. How far away was the earthquake? You can assume that the waves travel in straight lines, although actual seismic waves follow more complex routes.

59. |||| One way to monitor global warming is to measure the average temperature of the ocean. Researchers are doing this by measuring the time it takes sound pulses to travel underwater over large distances. At a depth of 1000 m, where ocean temperatures hold steady near 4°C, the average sound speed is 1480 m/s. It's known from laboratory measurements that the sound speed increases 4.0 m/s for every 1.0°C increase in temperature. In one experiment, where sounds generated near California are detected in the South Pacific, the sound waves travel 8000 km. If the smallest time change that can be reliably detected is 1.0 s, what is the smallest change in average temperature that can be measured?

60. ||| Figure P15.60 shows two snapshot graphs taken 10 ms apart, with the blue curve being the first snapshot. What are the (a) wavelength, (b) speed, (c) frequency, and (d) amplitude of this wave?

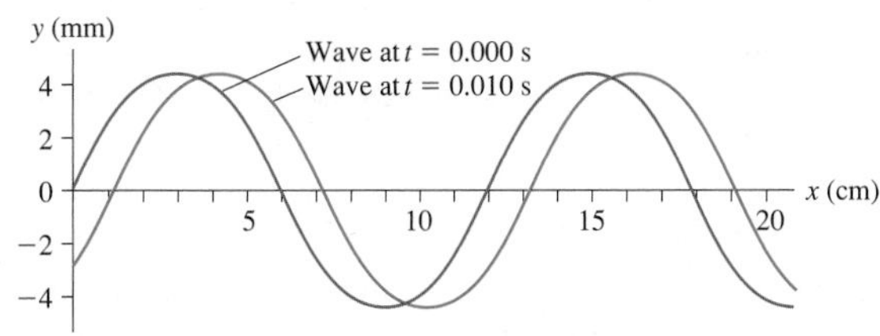

FIGURE P15.60

61. || Low-frequency vertical oscillations are one possible cause of
BIO motion sickness, with 0.3 Hz having the strongest effect. Your boat is bobbing in place at just the right frequency to cause you the maximum discomfort.
a. How much time elapses between two waves hitting the ship?
b. If the wave crests appear to be about 30 m apart, what would you estimate to be the speed of the waves?

62. ||| A wave on a string is described by $y(x, t) = (3.0 \text{ cm}) \times \cos[2\pi(x/(2.4 \text{ m}) + t/(0.20 \text{ s}))]$, where x is in m and t in s.
a. In what direction is this wave traveling?
b. What are the wave speed, frequency, and wavelength?
c. At $t = 0.50$ s, what is the displacement of the string at $x = 0.20$ m?

63. | Write the y-equation for a wave traveling in the negative x-direction with wavelength 50 cm, speed 4.0 m/s, and amplitude 5.0 cm.

64. | Write the y-equation for a wave traveling in the positive x-direction with frequency 200 Hz, speed 400 m/s, and amplitude 0.010 mm.

65. | A wave is described by the expression $y(x, t) = (3.0 \text{ cm}) \times \cos(1.5x - 50t)$, where x is in m and t is in s.
a. Draw an accurate snapshot graph of this wave.
b. What is the speed of the wave and in what direction is it traveling?

66. ||| A point on a string undergoes simple harmonic motion as a
INT sinusoidal wave passes. When a sinusoidal wave with speed 24 m/s, wavelength 30 cm, and amplitude of 1.0 cm passes, what is the maximum speed of a point on the string?

67. ||| A simple pendulum is
INT made by attaching a small cup of sand with a hole in the bottom to a 1.2-m-long string. The pendulum is mounted on the back of a small motorized car. As the car drives forward, the pendulum swings from side to side and leaves a trail of sand as shown in Figure P15.67. How fast was the car moving?

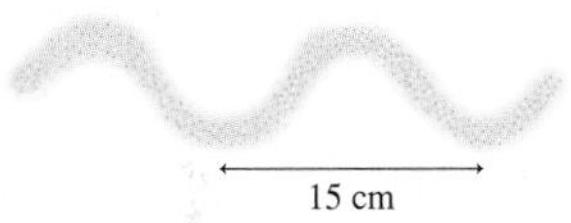

15 cm

FIGURE P15.67

68. || a. A typical 100 W lightbulb produces 4.0 W of visible light. (The other 96 W are dissipated as heat and infrared radiation.) What is the light intensity on a wall 2.0 m away from the lightbulb?
b. A krypton laser produces a cylindrical red laser beam 2.0 mm in diameter with 2.0 W of power. What is the light intensity on a wall 2.0 m away from the laser?

69. || An AM radio station broadcasts with a power of 25 kW at a frequency of 920 kHz. Estimate the intensity of the radio wave at a point 10 km from the broadcast antenna.

70. ||| The earth's average distance from the sun is 1.50×10^{11} m. At this distance, the intensity of radiation from the sun is 1.38 kW/m^2. The earth's radius is 6.37×10^6 m. What is the total solar power received by the earth? (For comparison, total human power consumption is roughly 10^{13} W.)

71. || Lasers can be used to drill or cut material. One such laser generates a series of high-intensity pulses rather than a continuous beam of light. Each pulse contains 500 mJ of energy and lasts 10 ns. The laser fires 10 such pulses per second.
a. What is the *peak power* of the laser light? The peak power is the power output during one of the 10 ns pulses.
b. What is the average power output of the laser? The average power is the total energy delivered per second.
c. A lens focuses the laser beam to a 10-μm-diameter circle on the target. During a pulse, what is the light intensity on the target?
d. The intensity of sunlight at the surface of the earth at midday is about 1100 W/m^2. What is the ratio of the laser intensity on the target to the intensity of the midday sun?

72. || The quietest sound you can hear is 0 dB. Estimate the diame-
BIO ter of your ear canal and compute an approximate area. At 0 dB, how much sound power is "captured" by one ear? (Ignore any focusing of energy by the pinna, the external folds of your ear.)

73. || The sound intensity 50 m from a wailing tornado siren is $0.10\ \text{W/m}^2$.
   a. What is the sound intensity level?
   b. In a noisy neighborhood, the weakest sound likely to be heard over background noise is 60 dB. Estimate the maximum distance at which the siren can be heard.
74. ||| A harvest mouse can detect sounds as quiet as -10 dB. Suppose you are sitting in a field on a very quiet day while a harvest mouse sits nearby at the entrance to its nest. A very gentle breeze causes a leaf 1.5 m from your head to rustle, generating a faint sound right at the limit of your ability to hear it. The sound of the rustling leaf is also right at the threshold of hearing of the harvest mouse. How far is the harvest mouse from the leaf?
BIO
75. |||| A speaker at an open-air concert emits 600 W of sound power, radiated equally in all directions.
   a. What is the intensity of the sound 5.0 m from the speaker?
   b. What sound intensity level would you experience there if you did not have any protection for your ears?
   c. Earplugs you can buy in the drugstore have a noise reduction rating of 23 decibels. If you are wearing those earplugs but your friend Phil is not, how far from the speaker should Phil stand to experience the same loudness as you?
76. || A bat locates insects by emitting ultrasonic "chirps" and then listening for echoes. The lowest-frequency chirp of a big brown bat is 26 kHz. How fast would the bat have to fly, and in what direction, for you to just barely be able to hear the chirp at 20 kHz?
BIO
77. | A physics professor demonstrates the Doppler effect by tying a 600 Hz sound generator to a 1.0-m-long rope and whirling it around her head in a horizontal circle at 100 rpm. What are the highest and lowest frequencies heard by a student in the classroom? Assume the room temperature is 20°C.
INT
78. ||| Ocean waves with wavelength 1.2 m and period 1.5 s are moving past a pier. A boy runs along the pier, in the direction opposite to the motion of the wave, at 3.5 m/s. How many wave crests pass the boy each second?
79. || A source of sound moves toward you at speed v_s and away from Jane, who is standing on the other side of it. You hear the sound at twice the frequency as Jane. What is the speed of the source? Assume that the speed of sound is 340 m/s.
80. ||||| When the heart pumps blood into the aorta, the *pressure gradient*—the difference between the blood pressure inside the heart and the blood pressure in the artery—is an important diagnostic measurement. A direct measurement of the pressure gradient is difficult, but an indirect determination can be made by measuring the Doppler shift of reflected ultrasound. Blood is essentially at rest in the heart; when it leaves and enters the aorta, it speeds up significantly and—according to Bernoulli's equation—the pressure must decrease. A doctor using ultrasound of 2.5 MHz measures a 6000 Hz frequency shift as the ultrasound reflects from blood ejected from the heart.
BIO INT
   a. What is the speed of the blood in the aorta?
   b. What is the difference in blood pressure between the inside of the heart and the aorta? Assume that the patient is lying down and that there is no difference in height as the blood moves from the heart into the aorta.

Passage Problems

Echolocation BIO

As discussed in the chapter, many species of bats find flying insects by emitting pulses of ultrasound and listening for the reflections. This technique is called **echolocation**. Bats possess several adaptations that allow them to echolocate very effectively.

81. | Although we can't hear them, the ultrasonic pulses are very loud. In order not to be deafened by the sound they emit, bats can temporarily turn off their hearing. Muscles in the ear cause the bones in their middle ear to separate slightly, so that they don't transmit vibrations to the inner ear. After an ultrasound pulse ends, a bat can hear an echo from an object a minimum of 1 m away. Approximately how much time after a pulse is emitted is the bat ready to hear its echo?
   A. 0.5 ms B. 1 ms C. 3 ms D. 6 ms
82. | Bats are sensitive to very small changes in frequency of the reflected waves. What information does this allow them to determine about their prey?
   A. Size B. Speed C. Distance D. Species
83. | Some bats have specially shaped noses that they use to focus the ultrasound pulses in the forward direction. Why is this useful?
   A. They are not distracted by echoes from several directions.
   B. The energy of the pulse is concentrated in a smaller area, so the intensity is larger.
   C. The pulse goes forward only, so it doesn't affect the bat's hearing.
84. | Some bats utilize a sound pulse with a rapidly decreasing frequency. A decreasing-frequency pulse has
   A. Decreasing wavelength.
   B. Decreasing speed.
   C. Increasing wavelength.
   D. Increasing speed.

STOP TO THINK ANSWERS

Stop to Think 15.1: Transverse. The wave moves horizontally through the crowd, but individual spectators move up and down, transverse to the motion of the wave.

Stop to Think 15.2: D, E. Shaking your hand faster or farther will change the shape of the wave, but this will not change the wave speed; the speed is a property of the medium. Changing the linear density of the string or its tension will change the wave speed. To increase the speed, you must decrease the linear density or increase the tension.

Stop to Think 15.3: B. All three waves have the same speed, so the frequency is highest for the wave that has the shortest wavelength. (Imagine the three waves moving to the right; the one with the crests closest together has the crests passing by most rapidly.)

Stop to Think 15.4: A. The plane wave does not spread out, so its intensity will be constant. The other two waves spread out, so their intensity will decrease.

Stop to Think 15.5: C. The source is moving toward Zack, so he observes a higher frequency. The source is moving away from Amy, so she observes a lower frequency.

16 Superposition and Standing Waves

The didgeridoo is a simple musical instrument played by aboriginal tribes in Australia. It consists of a hollow tube that is open at both ends. How can the player get a wide range of notes out of such a simple device?

LOOKING AHEAD ➡

The goal of Chapter 16 is to use the idea of superposition to understand the phenomena of interference and standing waves.

Superposition

Traveling waves can pass through each other. As they do, their displacements add together. This is the **principle of superposition.**

The surface of the water supports multiple waves. It looks like the waves simply stack on top of each other, which, in fact, they do.

Looking Back ◀◀
15.2–15.3 The fundamental properties of traveling waves

Standing Waves

Traveling waves bounce back and forth between the ends of a string that is clamped at both ends. The superposition of these reflected waves makes the string vibrate up and down. We call this a **standing wave.**

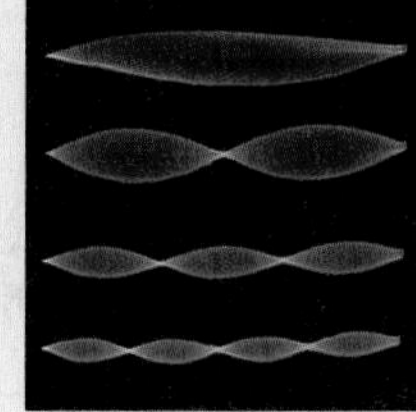

Standing waves occur only as well-defined patterns called **modes,** each with its own distinctive frequency. These are the **resonant modes** of the medium. Some points on the wave, called **nodes,** do not oscillate at all.

Looking Back ◀◀
14.7 The principle of resonance

Constructive and Destructive Interference

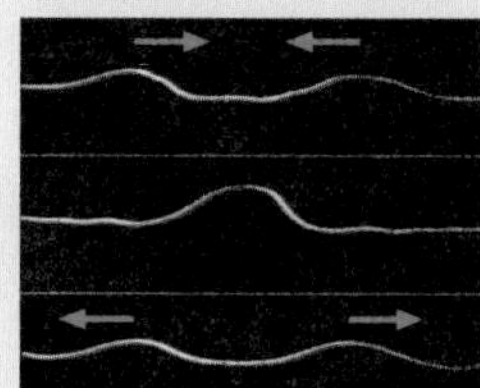

Two waves on a string each displace the string upward. Where the two waves overlap, the displacement is twice that of the individual waves. This is **constructive interference.**

Noise-canceling headphones create a sound wave that is inverted from the ambient sound. When the waves are added, they cancel, producing a much smaller wave. This is **destructive interference.**

Music and Speech

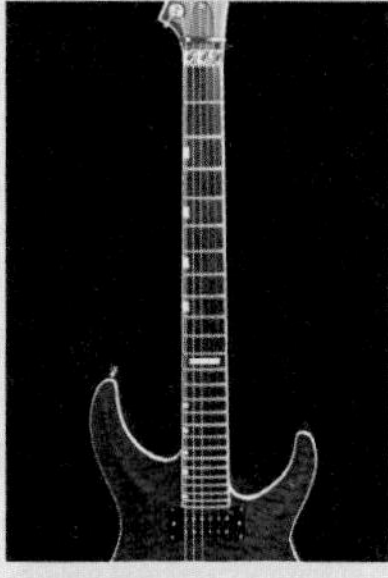

Standing waves on the strings of the guitar allow it to produce different musical notes. The frequency is determined by the length, mass, and tension of the string.

A tube can support a standing wave as well—a standing sound wave. We'll see how to calculate the possible standing waves and how these determine the notes a wind instrument can produce.

Your vocal system also depends on standing waves. A vibration of your vocal cords is amplified by the resonances of the tube of your vocal tract. We'll see how these elements work together to make speech.

Looking Back ◀◀
15.4 The nature of sound waves

16.1 The Principle of Superposition

FIGURE 16.1a shows two baseball players, Alan and Bill, at batting practice. Unfortunately, someone has turned the pitching machines so that pitching machine A throws baseballs toward Bill while machine B throws toward Alan. If two baseballs are launched at the same time and with the same speed, they collide at the crossing point and bounce away. Two baseballs cannot occupy the same point of space at the same time.

FIGURE 16.1 Two baseballs cannot pass through each other. Two waves can.

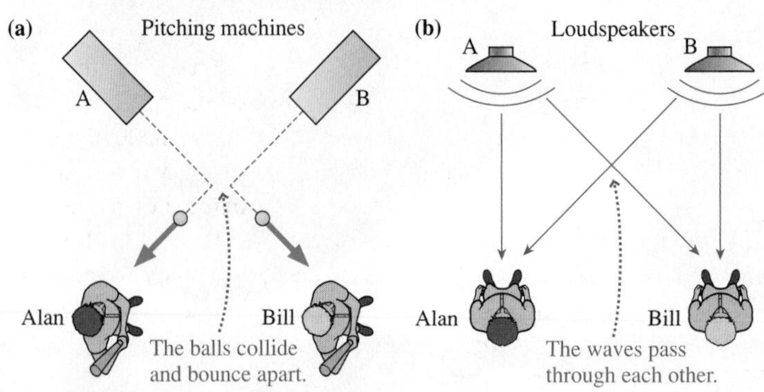

FIGURE 16.2 Two wave pulses on a stretched string pass through each other.

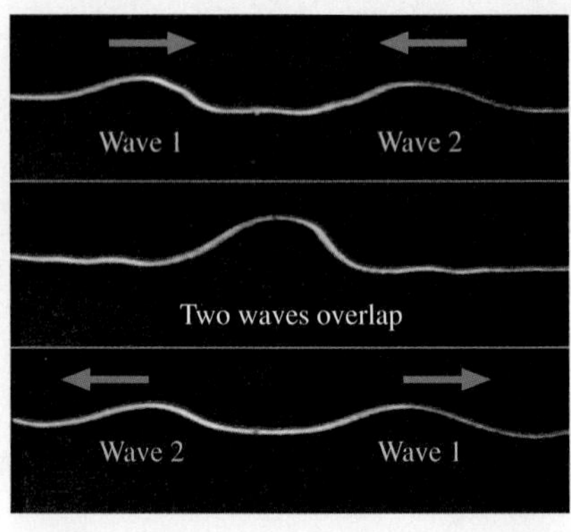

But unlike baseballs, sound waves *can* pass directly through each other. In FIGURE 16.1b, Alan and Bill are listening to the stereo system in the locker room after practice. Both hear the music quite well, without distortion or missing sound, so the sound wave that travels from speaker A toward Bill must pass through the wave traveling from speaker B toward Alan, with no effect on either. This is a basic property of waves.

What happens to the medium at a point where two waves are present simultaneously? What is the displacement of the medium at this point? FIGURE 16.2 shows a sequence of photos of two wave pulses traveling along a stretched string. In the first photo, the waves are approaching each other. In the second, the waves overlap, and the displacement of the string is larger than it was for either of the individual waves. A careful measurement would reveal that the displacement is the sum of the displacements of the two individual waves. In the third frame, the waves have passed through each other and continue on as if nothing had happened.

This result is not limited to stretched strings; the outcome is the same whenever two waves of any type pass through each other. This is known as the *principle of superposition:*

Principle of superposition When two or more waves are *simultaneously* present at a single point in space, the displacement of the medium at that point is the sum of the displacements due to each individual wave.

FIGURE 16.3 The superposition of two waves on a string as they pass through each other.

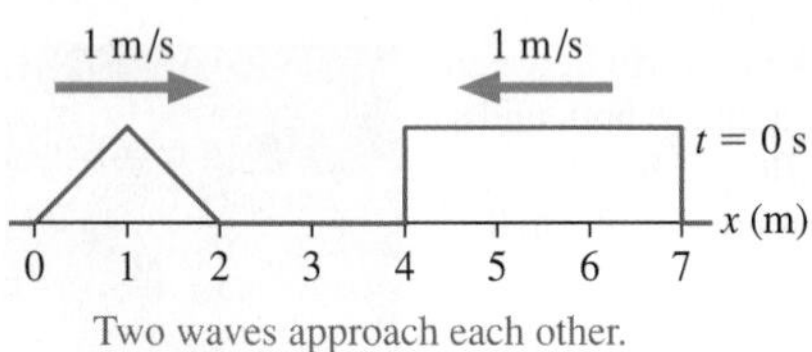

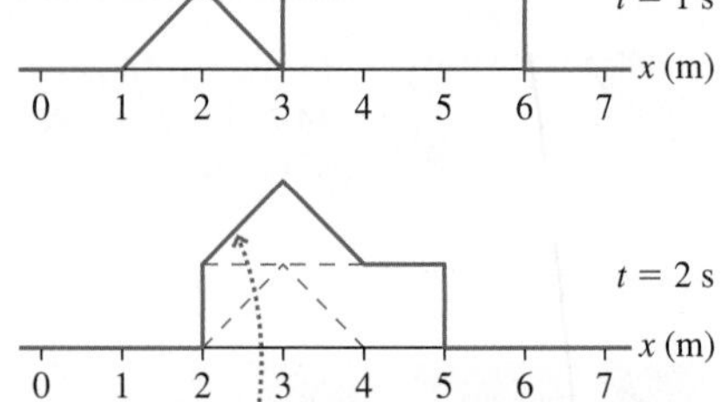

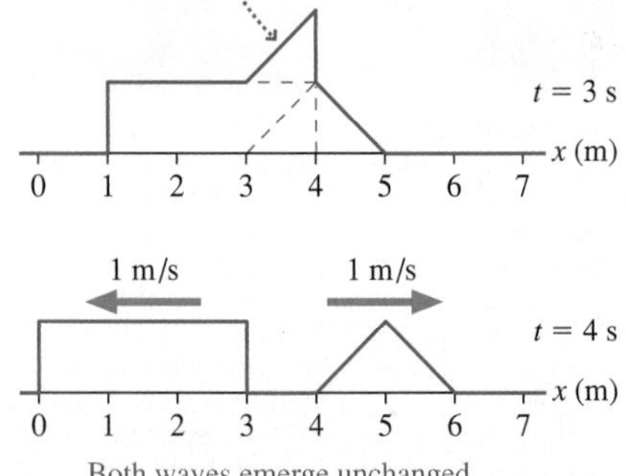

To use the principle of superposition you must know the displacement that each wave would cause if it traveled through the medium alone. Then you go through the medium *point by point* and add the displacements due to each wave *at that point* to find the net displacement at that point. The outcome will be different at each point in the medium because the displacements are different at each point.

Let's illustrate this principle with an idealized example. FIGURE 16.3 shows five snapshot graphs taken 1 s apart of two waves traveling at the same speed (1 m/s) in opposite directions along a string. The displacement of each wave is shown as a dotted line. The solid line is the sum *at each point* of the two displacements at that point. This is the displacement that you would actually observe as the two waves pass through each other.

Constructive and Destructive Interference

The superposition of two waves is often called **interference.** The displacements of the waves in Figure 16.3 are both positive, so the total displacement of the medium where they overlap is larger than it would be due to either of the waves separately. We call this **constructive interference.**

FIGURE 16.4 shows another series of snapshot graphs of two counterpropagating waves, but this time one has a negative displacement. The principle of superposition still applies, but now the displacements are opposite each other. The displacement of the medium where the waves overlap is *less* than it would be due to either of the waves separately. We call this **destructive interference.**

In the series of graphs in Figure 16.4 the displacement of the medium at $x = 3.5$ m is always zero. The positive displacement of the wave traveling to the right and the negative displacement of the wave traveling to the left always exactly cancel at this spot. The complete cancellation at one point of two waves traveling in opposite directions is something we will see again. When the displacements of the waves cancel, where does the energy of the wave go? We know that the waves continue on unchanged after their interaction, so no energy is dissipated. Consider the graph at $t = 1.5$ s. There is no net displacement at any point of the medium at this instant, but the *string is moving rapidly.* The energy of the waves hasn't vanished—it is in the form of the kinetic energy of the medium.

FIGURE 16.4 Two waves with opposite displacements produce destructive interference.

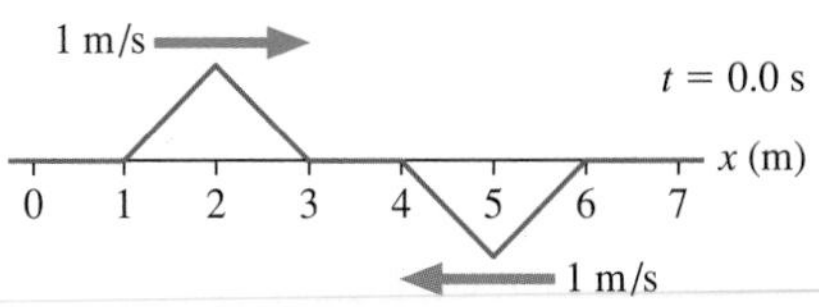

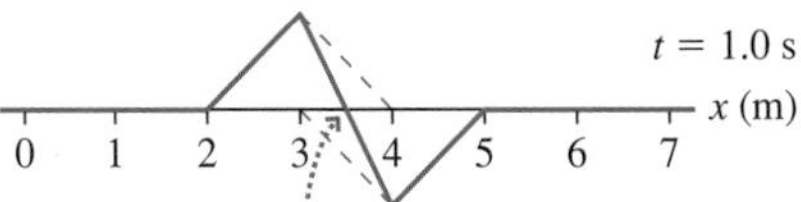

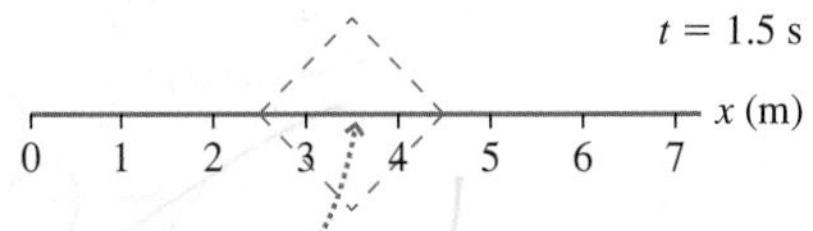

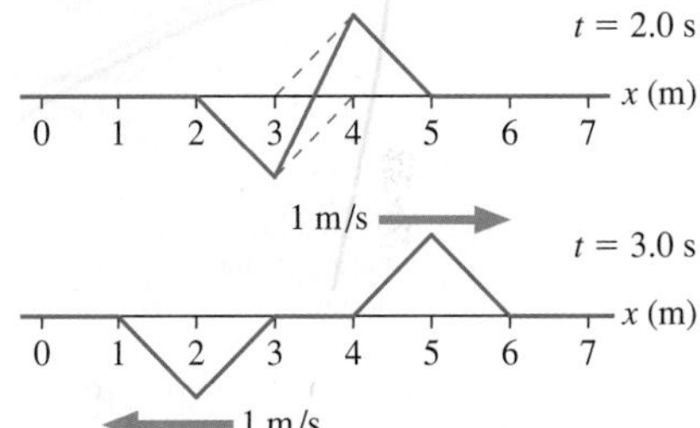

STOP TO THINK 16.1 Two pulses on a string approach each other at speeds of 1 m/s. What is the shape of the string at $t = 6$ s?

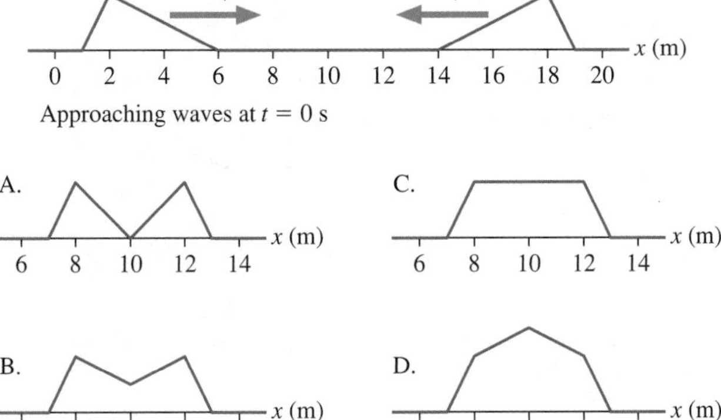

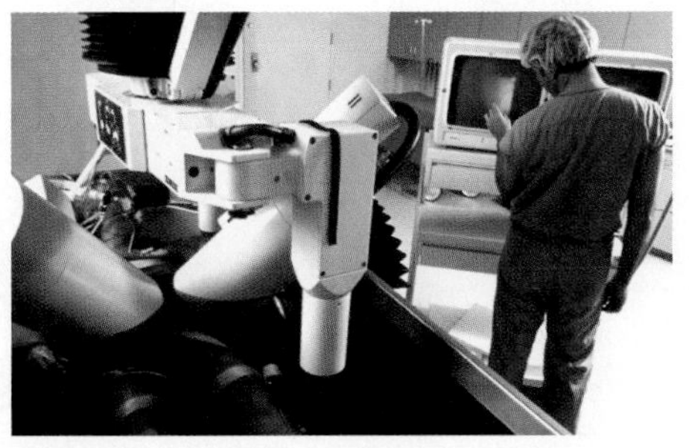

▶ **Breaking stones with sound** BIO As we saw in Chapter 15, waves carry energy. The energy of high-intensity ultrasonic waves can be used to break up kidney stones so that they can be cleared from the body, a technique known as *shock wave lithotripsy.* This machine uses two generators, each of which produces ultrasonic waves. The two waves, which enter the body at different points, are directed so that they overlap and produce constructive interference at the position of a stone. This allows the individual waves to have lower intensity, minimizing tissue damage as they pass through the body, while still providing a region of high intensity right where it is needed.

16.2 Standing Waves

When you pluck a guitar string or a rubber band stretched between your fingers, you create waves. But how is this possible? There isn't really anywhere for the waves to go, because the string or the rubber band is held between two fixed ends. **FIGURE 16.5** shows a strobe photograph of waves on a stretched elastic cord. This is a wave, though it may not look like one, because it doesn't "travel" either right or left. Waves that are "trapped" between two boundaries, like those in the photo or on a guitar

FIGURE 16.5 The motion of a standing wave on a string.

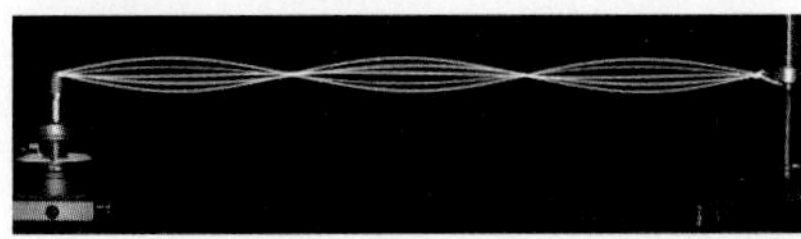

string, are what we call *standing waves.* **Individual points on the string oscillate up and down, but the wave itself does not travel.** It is called a **standing wave** because the crests and troughs "stand in place" as it oscillates. As we'll see, a standing wave isn't a totally new kind of wave; it is simply the superposition of two traveling waves moving in opposite directions.

Superposition Creates a Standing Wave

Suppose we have a string on which two sinusoidal waves of equal wavelength and amplitude travel in opposite directions, as in FIGURE 16.6a. When the waves meet, the displacement of the string will be a superposition of these two waves. FIGURE 16.6b shows nine snapshot graphs, at intervals of $\frac{1}{8}T$, of the two waves as they move through each other. The red and orange dots identify particular crests of each of the waves to help you see that the red wave is traveling to the right and the orange wave to the left. At *each point,* the net displacement of the medium is found by adding the red displacement and the orange displacement. The resulting blue wave is the superposition of the two traveling waves.

FIGURE 16.6 Two sinusoidal waves traveling in opposite directions.

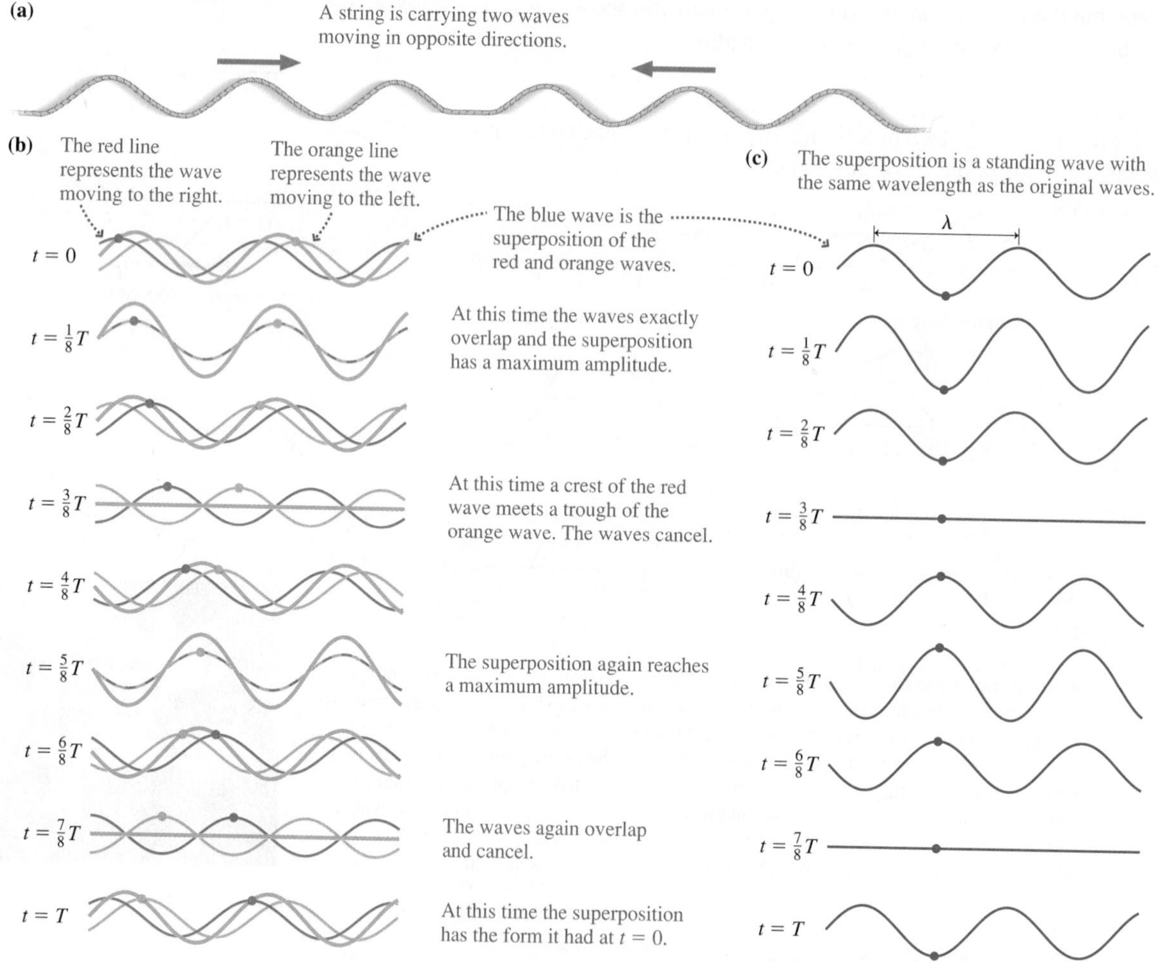

(a)
A string is carrying two waves moving in opposite directions.
(b)
The red line represents the wave moving to the right.
The orange line represents the wave moving to the left.
The blue wave is the superposition of the red and orange waves.
At this time the waves exactly overlap and the superposition has a maximum amplitude.
At this time a crest of the red wave meets a trough of the orange wave. The waves cancel.
The superposition again reaches a maximum amplitude.
The waves again overlap and cancel.
At this time the superposition has the form it had at $t = 0$.
$t = 0$
$t = \frac{1}{8}T$
$t = \frac{2}{8}T$
$t = \frac{3}{8}T$
$t = \frac{4}{8}T$
$t = \frac{5}{8}T$
$t = \frac{6}{8}T$
$t = \frac{7}{8}T$
$t = T$
(c)
The superposition is a standing wave with the same wavelength as the original waves.
λ

FIGURE 16.6c shows just the superposition of the two waves. This is the wave that you would actually observe in the medium. The blue dot shows that the wave in Figure 16.6c is moving neither right nor left. The superposition of the two counter-propagating traveling waves is a standing wave. Notice that the wavelength of the standing wave, the distance between two crests or two troughs, is the same as the wavelengths of the two traveling waves that combine to produce it.

Nodes and Antinodes

In FIGURE 16.7 we have superimposed the nine snapshot graphs of Figure 16.6c into a single graphical representation of this standing wave. The graphs at different times overlap, much as the photos of the string at different times in the strobe photograph of Figure 16.5. The motion of individual points on the standing wave is now clearly seen. A striking feature of a standing-wave pattern is points that *never move!* These points, which are spaced $\lambda/2$ apart, are called **nodes.** Halfway between the nodes are points where the particles in the medium oscillate with maximum displacement. These points of maximum amplitude are called **antinodes,** and you can see that they are also spaced $\lambda/2$ apart. This means that **the wavelength of a standing wave is *twice* the distance between successive nodes or successive antinodes.**

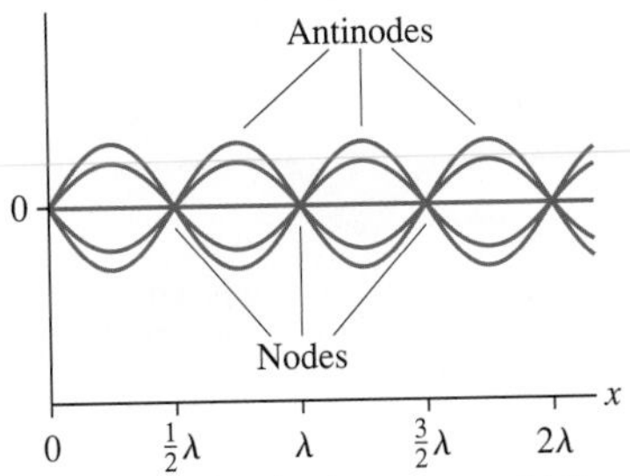

FIGURE 16.7 Superimposing multiple snapshot graphs of a standing wave clearly shows the nodes and antinodes.

It seems surprising and counterintuitive that some particles in the medium have no motion at all. This happens for the same reason we saw in Figure 16.4: The two waves exactly offset each other at that point. Look carefully at the two traveling waves in Figure 16.6b. You will see that the nodes occur at points where at *every instant* of time the displacements of the two traveling waves have equal magnitudes but *opposite signs.* Thus the superposition of the displacements at these points is always zero—they are points of destructive interference. The antinodes have large displacements. They correspond to points where the two displacements have equal magnitudes and the *same sign* at all times. Constructive interference at these points gives a displacement twice that of each individual wave.

The intensity of a wave is largest at points where it oscillates with maximum amplitude. FIGURE 16.8 shows that the points of maximum intensity along the standing wave occur at the antinodes; the intensity is zero at the nodes. For a standing sound wave, the loudness varies from zero (no sound) at the nodes to a maximum at the antinodes and then back to zero. The key idea is that **the intensity is maximum at points of constructive interference and zero at points of destructive interference.**

FIGURE 16.8 Intensity of a standing wave.

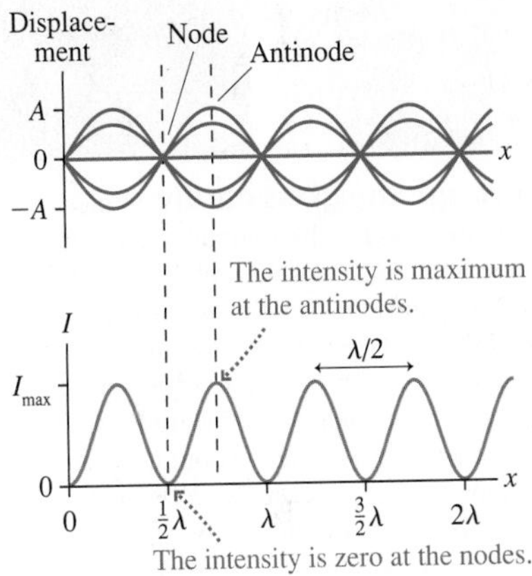

EXAMPLE 16.1 Setting up a standing wave

Two children hold an elastic cord at each end. Each child shakes her end of the cord 2.0 times per second, sending waves at 3.0 m/s toward the middle, where the two waves combine to create a standing wave. What is the distance between adjacent nodes?

SOLVE The distance between adjacent nodes is $\lambda/2$. The wavelength, frequency, and speed are related as $v = f\lambda$, as we saw in Chapter 15, so the wavelength is

$$\lambda = \frac{v}{f} = \frac{3.0 \text{ m/s}}{2.0 \text{ Hz}} = 1.5 \text{ m}$$

The distance between adjacent nodes is $\lambda/2$ and thus is 0.75 m.

16.3 Standing Waves on a String

10.4, 10.5, 10.6

The oscillation of a guitar string is a standing wave. A standing wave is naturally produced on a string when both ends are fixed (i.e., tied down), as in the case of a guitar string or the string in the photo of Figure 16.5. We also know that a standing wave is produced when there are two counterpropagating traveling waves. But you don't shake both ends of a guitar string to produce the standing wave! How do we actually get two traveling waves on a string with both ends fixed? Before we can answer this question, we need a brief explanation of what happens when a traveling wave encounters a boundary or a discontinuity.

Reflections

We know that light reflects from mirrors; it can also reflect from the surface of a pond or from a pane of glass. As we saw in Chapter 15, sound waves reflect as well;

FIGURE 16.9 A wave reflects when it encounters a boundary.

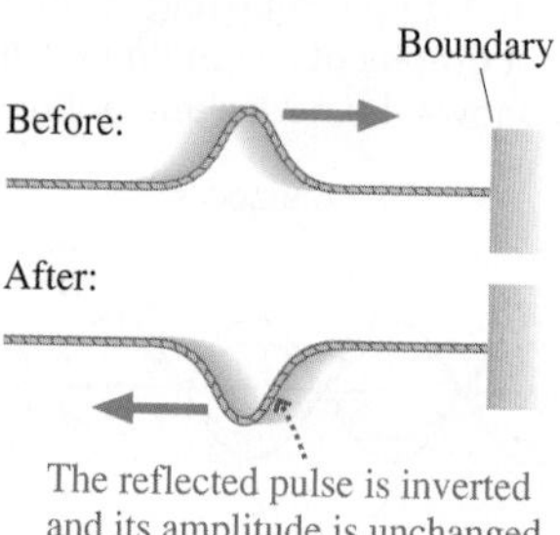

that's how an echo is produced. To understand reflections, we'll look at waves on a string, but the results will be completely general and can be applied to other waves as well.

Suppose we have a string that is attached to a wall or other fixed support, as in FIGURE 16.9. The wall is what we will call a *boundary*—it's the end of the medium. When the pulse reaches this boundary, it reflects, moving away from the wall. *All* the wave's energy is reflected; hence **the amplitude of a wave reflected from a boundary is unchanged.** Figure 16.9 shows that the amplitude doesn't change when the pulse reflects, but the pulse is inverted.

Waves also reflect from what we will call a *discontinuity,* a point where there is a change in the properties of the medium. FIGURE 16.10a shows a discontinuity where a string with a large linear density connects with a string with a small linear density. The tension is the same in both strings, so the wave speed is slower on the left, faster on the right. Whenever a wave encounters a discontinuity, some of the wave's energy is *transmitted* forward and some is reflected. Because energy must be conserved, both the transmitted and the reflected pulses have a smaller amplitude than the initial pulse in this case.

TRY IT YOURSELF

Through the glass darkly A piece of window glass is a discontinuity to a light wave, so it both transmits and reflects light. To verify this, look at the windows in a brightly lit room at night. The small percentage of the interior light that reflects from windows is more intense than the light coming in from outside, so reflection dominates and the windows show a mirror-like reflection of the room. Now turn out the lights. With no more reflected interior light you will be able to see the transmitted light from outside.

FIGURE 16.10 The reflection of a wave at a discontinuity.

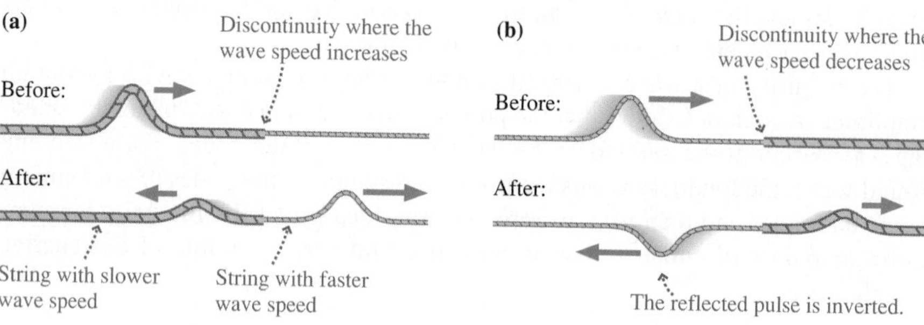

In FIGURE 16.10b, an incident wave encounters a discontinuity at which the wave speed decreases. Once again, some of the wave's energy is transmitted and some is reflected.

In Figure 16.10a, the reflected pulse is right-side up. The string on the right is light and provides little resistance, so the junction moves up and down as the pulse passes. This motion of the string is like the original "snap" of the string that started the pulse, so the reflected pulse has the same orientation as the original pulse. In Figure 16.10b, the string on the right is more massive, so it looks more like the fixed boundary in Figure 16.9 and the reflected pulse is again inverted.

Creating a Standing Wave

FIGURE 16.11 Reflections at the two boundaries cause a standing wave on the string.

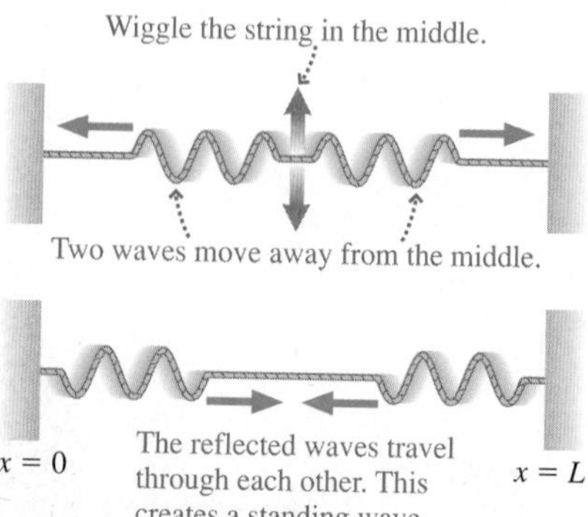

Now that we understand reflections, let's create a standing wave. FIGURE 16.11 shows a string of length L that is tied at $x = 0$ and $x = L$. This string has *two* boundaries where reflections can occur. If you wiggle the string in the middle, sinusoidal waves travel outward in both directions and soon reach the boundaries, where they reflect. The reflections at the ends of the string cause two waves of *equal amplitude and wavelength* to travel in opposite directions along the string. As we've just seen, these are the conditions that cause a standing wave!

What kind of standing waves might develop on the string? There are two conditions that must be met:

- Because the string is fixed at the ends, the displacements at $x = 0$ and $x = L$ must be zero at all times. Stated another way, we require nodes at both ends of the string.
- We know that standing waves have a spacing of $\lambda/2$ between nodes. This means that the nodes must be equally spaced.

FIGURE 16.12 shows the first three possible waves that meet these conditions. These are called the standing-wave **modes** of the string. To help quantify the possible waves, we can assign a **mode number** m to each. The first wave in Figure 16.12, with a node at each end, has mode number $m = 1$. The next wave is $m = 2$, and so on.

NOTE ▶ Figure 16.12 shows only the first three modes, for $m = 1$, $m = 2$, and $m = 3$. But there are many more modes, for all possible values of m. ◀

The distance between adjacent nodes is $\lambda/2$, so the different modes have different wavelengths. For the first mode in Figure 16.12, the distance between nodes is the length of the string, so we can write

$$\lambda_1 = 2L$$

The subscript identifies the mode number; in this case $m = 1$. For $m = 2$, the distance between nodes is $L/2$; this means that $\lambda_2 = L$. Generally, for any mode m the wavelength is given by the following equation:

$$\lambda_m = \frac{2L}{m} \qquad m = 1, 2, 3, 4, \ldots \tag{16.1}$$

Wavelengths of standing-wave modes of a string of length L

These are the only possible wavelengths for standing waves on the string. **A standing wave can exist on the string *only* if its wavelength is one of the values given by Equation 16.1.**

NOTE ▶ Other wavelengths, which would be perfectly acceptable wavelengths for a traveling wave, cannot exist as a *standing* wave of length L because they do not meet the constraint of having a node at each end of the string. ◀

If standing waves are possible only for certain wavelengths, then only specific oscillation frequencies are allowed. Because $\lambda f = v$ for a sinusoidal wave, the oscillation frequency corresponding to wavelength λ_m is

$$f_m = \frac{v}{\lambda_m} = \frac{v}{2L/m} = m\left(\frac{v}{2L}\right) \qquad m = 1, 2, 3, 4, \ldots \tag{16.2}$$

Frequencies of standing-wave modes of a string of length L

FIGURE 16.13 shows the first three modes with their wavelengths and frequencies. You can see that **the mode number m is equal to the number of antinodes of the standing wave.** You can therefore tell a string's mode of oscillation by counting the number of antinodes (*not* the number of nodes).

In Chapter 14, we looked at the concept of *resonance*. A mass on a spring has a certain frequency at which it "wants" to oscillate. If the system is driven at its resonance frequency, it will develop a large amplitude of oscillation. A stretched string will support standing waves, meaning it has a series of frequencies at which it "wants" to oscillate: the frequencies of the different standing-wave modes. We can call these **resonant modes,** or more simply, **resonances.** A small oscillation of a stretched string at a frequency near one of its resonant modes will cause it to develop a standing wave with a large amplitude. FIGURE 16.14 shows photographs of the first four standing-wave modes on a string, corresponding to four different driving frequencies.

NOTE ▶ When we draw standing-wave modes, as in Figure 16.13, we usually show only the *envelope* of the wave, the greatest extent of the motion of the string. The string's motion is actually continuous and goes through all intermediate positions as well, as we see from the time-exposure photographs of standing waves in Figure 16.14. ◀

FIGURE 16.12 The first three possible standing waves on a string of length L.

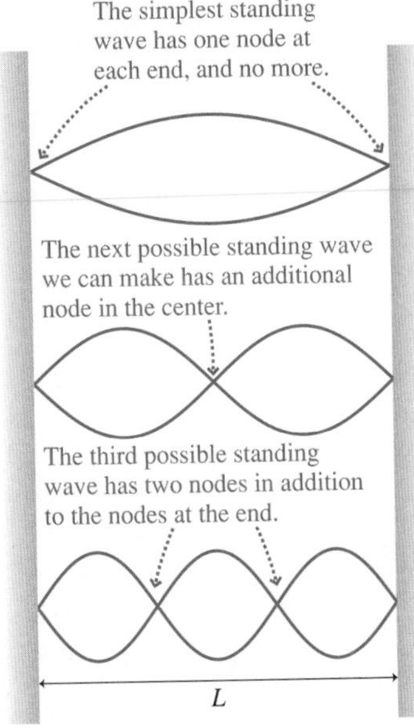

The simplest standing wave has one node at each end, and no more.
The next possible standing wave we can make has an additional node in the center.
The third possible standing wave has two nodes in addition to the nodes at the end.
L

FIGURE 16.13 Possible standing waves of a string fixed at both ends.

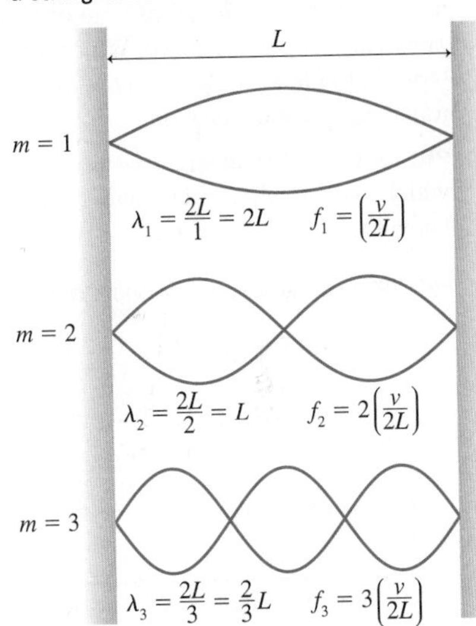

FIGURE 16.14 Resonant modes of a stretched string.

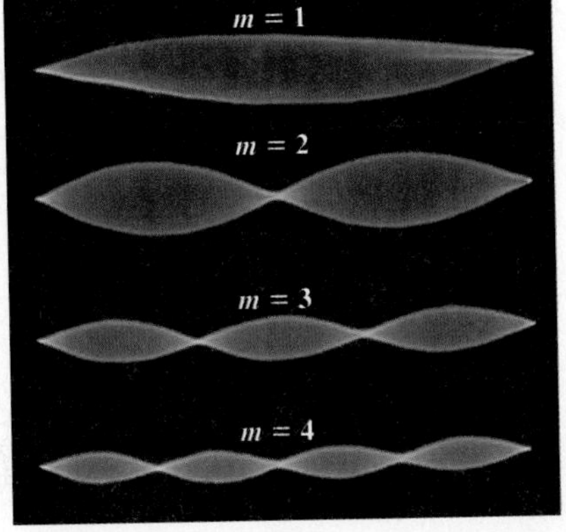

The Fundamental and the Higher Harmonics

The sequence of possible frequencies for a standing wave on a string has an interesting pattern that is worth exploring. The first mode has frequency

$$f_1 = \frac{v}{2L} \qquad (16.3)$$

PROPORTIONAL p.37 INVERSE p.114

We call this the **fundamental frequency** of the string. All of the other modes have frequencies that are multiples of this fundamental frequency. We can rewrite Equation 16.2 in terms of the fundamental frequency as

$$f_m = mf_1 \qquad m = 1, 2, 3, 4, \ldots \qquad (16.4)$$

The allowed standing-wave frequencies are all integer multiples of the fundamental frequency. This sequence of possible frequencies is called a set of **harmonics.** The fundamental frequency f_1 is also known as the *first harmonic,* the $m = 2$ wave at frequency f_2 is called the *second harmonic,* the $m = 3$ wave is called the *third harmonic,* and so on. The frequencies above the fundamental frequency, the harmonics with $m = 2, 3, 4, \ldots$, are referred to as the **higher harmonics.**

EXAMPLE 16.2 Identifying harmonics on a string

A 2.50-m-long string vibrates as a 100 Hz standing wave with nodes at 1.00 m and 1.50 m from one end of the string and at no points in between these two. Which harmonic is this? What is the string's fundamental frequency? And what is the speed of the traveling waves on the string?

PREPARE We begin with the visual overview in **FIGURE 16.15**, in which we sketch this particular standing wave and note the known and unknown quantities. We set up an x-axis with one end of the string at $x = 0$ m and the other end at $x = 2.50$ m. The ends of the string are nodes, and there are nodes at 1.00 m and 1.50 m as well, with no nodes in between. We know that standing-wave nodes are equally spaced, so there must be other nodes on the string, as shown in Figure 16.15a. Figure 16.15b is a sketch of the standing-wave mode with this node structure.

FIGURE 16.15 A visual overview of the string.

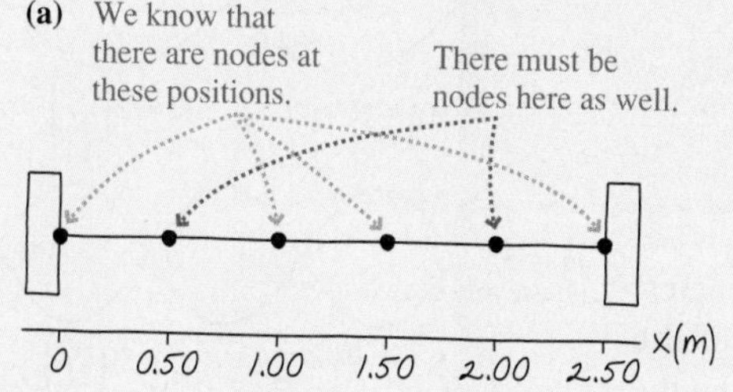

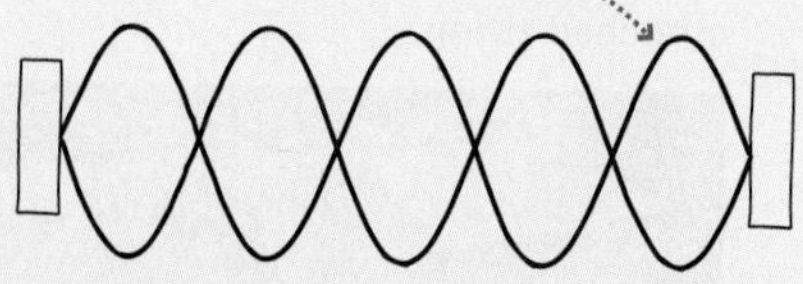

There must be nodes here as well.

(b) The mode with nodes at these positions looks like this.

Known
$L = 2.50$ m
$f_m = 100$ Hz
Find
m, f_1, v

SOLVE We count the number of antinodes of the standing wave to deduce the mode number; this is mode $m = 5$. This is the fifth harmonic. The frequencies of the harmonics are given by $f_m = mf_1$, so the fundamental frequency is

$$f_1 = \frac{f_5}{5} = \frac{100\text{ Hz}}{5} = 20\text{ Hz}$$

The wavelength of the fundamental mode is $\lambda_1 = 2L = 2(2.50\text{ m}) = 5.00$ m, so can find the wave speed using the fundamental relationship for sinusoidal waves:

$$v = f_1\lambda_1 = (20\text{ Hz})(5.00\text{ m}) = 100\text{ m/s}$$

ASSESS We can calculate the speed of the wave using any possible mode, which gives us a way to check our work. The distance between successive nodes is $\lambda/2$. Figure 16.15 shows that the nodes are spaced by 0.50 m, so the wavelength of the $m = 5$ mode is 1.00 m. The frequency of this mode is 100 Hz, so we calculate

$$v = f_5\lambda_5 = (100\text{ Hz})(1.00\text{ m}) = 100\text{ m/s}$$

This is the same speed that we calculated earlier, which gives us confidence in our results.

Stringed Musical Instruments

Think about stringed musical instruments, such as the guitar, the piano, and the violin. These instruments all have strings that are fixed at both ends and tightened to create tension. A disturbance is generated on the string by plucking, striking, or bowing. Regardless of how it is generated, the disturbance creates standing waves on the string. Understanding the sound of a stringed musical instrument means understanding standing waves.

In Chapter 15, we saw that the speed of a wave on a stretched string depended on T_s, the tension in the string, and μ, the linear density, as $v = \sqrt{T_s/\mu}$. Combining this with Equation 16.3, we find that the fundamental frequency of a stretched string is

$$f_1 = \frac{v}{2L} = \frac{1}{2L}\sqrt{\frac{T_s}{\mu}} \tag{16.5}$$

When you pluck or bow a string, you initially excite a wide range of frequencies. However, resonance sees to it that the only frequencies to persist are those of the possible standing waves. The string will support a wave of the fundamental frequency f_1 plus waves of all the higher harmonics f_2, f_3, f_4, and so on. Your brain interprets the sound as a musical note of frequency f_1; the higher harmonics determine the *tone quality*, a concept we will explore later in the chapter.

NOTE ► In Equation 16.5, v is the wave speed *on the string,* not the speed of sound in air. ◄

For instruments like the guitar or the violin, the strings are all the same length and under approximately the same tension. Were that not the case, the neck of the instrument would tend to twist toward the side of higher tension. The strings have different frequencies because they differ in linear density. The lower-pitched strings are "fat" while the higher-pitched strings are "skinny." This difference changes the frequency by changing the wave speed. Small adjustments are then made in the tension to bring each string to the exact desired frequency.

Standing waves on a bridge This photo shows the Tacoma Narrows suspension bridge on the day in 1940 when it experienced a catastrophic oscillation that led to its collapse. Aerodynamic forces caused the amplitude of a particular resonant mode of the bridge to increase dramatically until the bridge failed. In this photo, the red line shows the original line of the deck of the bridge. You can clearly see the large amplitude of the oscillation and the node at the center of the span.

CONCEPTUAL EXAMPLE 16.3 Tuning and playing a guitar

A guitar has strings of a fixed length. Plucking a string makes a particular musical note. A player can make other notes by pressing the string against frets, metal bars on the neck of the guitar, as shown in **FIGURE 16.16**. The fret becomes the new end of the string, making the effective length shorter.

FIGURE 16.16 Guitar frets.

a. A guitar player plucks a string to play a note. He then presses down on a fret to make the string shorter. Does the new note have a higher or lower frequency?

b. The frequency of one string is too low. (Musically, we say the note is "flat.") How must the tension be adjusted to bring the string to the right frequency?

REASON

a. The fundamental frequency, the note we hear, is $f_1 = (1/2L)\sqrt{T_s/\mu}$. Because f_1 is inversely proportional to L, decreasing the string length increases the frequency.

b. Because f is proportional to the square root of T_s, the player must increase the tension to increase the fundamental frequency.

ASSESS If you watch someone play a guitar, you can see that he or she plays higher notes by moving the fingers to shorten the strings.

EXAMPLE 16.4 Setting the tension in a guitar string

The fifth string on a guitar plays the musical note A, at a frequency of 110 Hz. On a typical guitar, this string is stretched between two fixed points 0.640 m apart, and this length of string has a mass of 2.86 g. What is the tension in the string?

PREPARE Strings sound at their fundamental frequency, so 110 Hz is f_1.

SOLVE The linear density of the string is

$$\mu = \frac{m}{L} = \frac{2.86 \times 10^{-3}\ \text{kg}}{0.640\ \text{m}} = 4.47 \times 10^{-3}\ \text{kg/m}$$

We can rearrange Equation 16.5 for the fundamental frequency to solve for the tension in terms of the other variables:

$$T_s = (2Lf_1)^2\mu = [2(0.640\ \text{m})(110\ \text{Hz})]^2\,(4.47 \times 10^{-3}\ \text{kg/m})$$
$$= 88.6\ \text{N}$$

ASSESS If you have ever strummed a guitar, you know that the tension is quite large, so this result seems reasonable. If each of the guitar's six strings has approximately the same tension, the total force on the neck of the guitar is a bit more than 500 N.

FIGURE 16.17 A laser contains a standing light wave between two parallel mirrors.

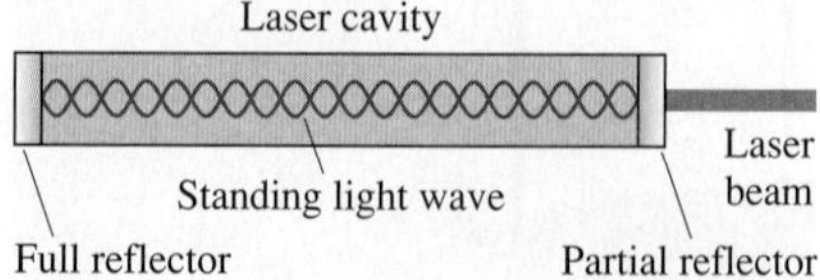

Microwave modes A microwave oven uses a type of electromagnetic wave—microwaves, with a wavelength of about 12 cm—to heat food. The inside walls of a microwave oven are reflective to microwaves, so we have the correct conditions to set up a standing wave. This isn't a good thing! A standing wave has high intensity at the antinodes and low intensity at the nodes, so your oven has hot spots and cold spots, as we see in this thermal image showing the interior of an oven with a thin layer of water that has been "cooked" for a short time. A turntable in a microwave oven keeps the food moving so that no part of your dinner remains at a node or an antinode.

Standing Electromagnetic Waves

The standing-wave descriptions we've found for a vibrating string are valid for any transverse wave, including an electromagnetic wave. For example, standing light waves can be established between two parallel mirrors that reflect the light back and forth. The mirrors are boundaries, analogous to the boundaries at the ends of a string. In fact, this is exactly how a laser works. The two facing mirrors in FIGURE 16.17 form what is called a *laser cavity.*

Because the mirrors act exactly like the points to which a string is tied, the light wave must have a node at the surface of each mirror. (To allow some of the light to escape the laser cavity and form the *laser beam,* one of the mirrors lets some of the light through. This doesn't affect the node.)

EXAMPLE 16.5 Finding the mode number for a laser

A helium-neon laser emits light of wavelength $\lambda = 633$ nm. A typical cavity for such a laser is 15.0 cm long. What is the mode number of the standing wave in this cavity?

PREPARE Because a light wave is a transverse wave, Equation 16.1 for λ_m applies to a laser as well as a vibrating string.

SOLVE The standing light wave in a laser cavity has a mode number m that is roughly

$$m = \frac{2L}{\lambda} = \frac{2 \times 0.150 \text{ m}}{633 \times 10^{-9} \text{ m}} = 474{,}000$$

ASSESS The wavelength of light is very short, so we'd expect the nodes to be closely spaced. A high mode number seems reasonable.

STOP TO THINK 16.2 A standing wave on a string is shown. Which of the modes shown below (on the same string) has twice the frequency of the original wave?

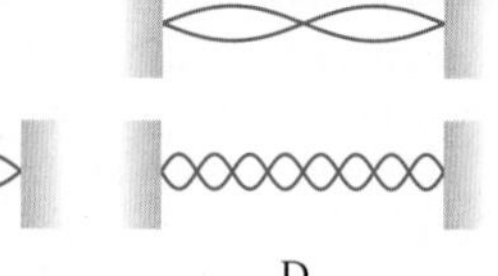

A. B. C. D.

16.4 Standing Sound Waves

Wind instruments like flutes, trumpets, and didgeridoos work very differently from stringed instruments. The player blows air into one end of a tube, producing standing waves of sound that make the notes we hear. In this section, we will look at the properties of such standing sound waves.

Recall that a sound wave is a longitudinal pressure wave. The air molecules oscillate back and forth parallel to the direction in which the wave is traveling, creating *compressions* (regions of higher pressure) and *rarefactions* (regions of lower pressure). Consider a sound wave confined to a long, narrow column of air, such as the air in a tube. A wave traveling down the tube eventually reaches the end, where it encounters the atmospheric pressure of the surrounding environment. This is a discontinuity, much like the small rope meeting the big rope in Figure 16.10b. Part of the wave's energy is transmitted out into the environment, allowing you to hear the sound, and part is reflected back into the tube. Reflections at both ends of the tube create waves traveling both directions inside the tube, and their superposition, like that of the reflecting waves on a string, is a standing wave.

We start by looking at a sound wave in a tube open at both ends. What kind of standing waves can exist in such a tube? Because the ends of the tube are open to the atmosphere, the pressure at the ends is fixed at atmospheric pressure and cannot vary. This is analogous to a stretched string that is fixed at the end. As a result, **the open end of a column of air must be a node of the pressure wave.**

FIGURE 16.18a shows a column of air open at both ends. We call this an *open-open tube.* Whereas the antinodes of a standing wave on a string are points where the string oscillates with maximum displacement, the antinodes of a standing sound wave are where the pressure has the largest variation, creating, alternately, the maximum compression and the maximum rarefaction. In Figure 16.18a, the air molecules squeeze together on the left side of the tube. Then, in FIGURE 16.18b, half a cycle later, the air molecules squeeze together on the right side. The varying density creates a variation in pressure across the tube, as the graphs show. FIGURE 16.18c combines the information of Figures 16.18a and 16.18b into a graph of the pressure of the standing sound wave in the tube.

As the standing wave oscillates, the air molecules "slosh" back and forth along the tube with the wave frequency, alternately pushing together (maximum pressure at the antinode) and pulling apart (minimum pressure at the antinode). This makes sense, because sound is a longitudinal wave in which the air molecules oscillate parallel to the tube.

NOTE ▶ The variation in pressure from atmospheric pressure in a real standing sound wave is much smaller than Figure 16.18 implies. When we display graphs of the pressure in sound waves, we won't generally graph the pressure p. Instead, we will graph Δp, the variation from atmospheric pressure. ◀

Many musical instruments, such as a flute, can be modeled as open-open tubes. The flutist blows across one end to create a standing wave inside the tube, and a note of this frequency is emitted from both ends of the flute. The possible standing waves in tubes, like standing waves on strings, are resonances of the system. A gentle puff of air across the mouthpiece of a flute can cause large standing waves at these resonant frequencies.

Other instruments work differently from the flute. A trumpet or a clarinet is a column of air open at the bell end but *closed* by the player's lips at the mouthpiece. To be complete in our treatment of sound waves in tubes, we need to consider tubes that are closed at one or both ends. At a closed end, the air molecules can alternately rush toward the wall, creating a compression, and then rush away from the wall, leaving a rarefaction. Thus **a closed end of an air column is an antinode of pressure.**

FIGURE 16.19 shows graphs of the first three standing-wave modes of a tube open at both ends (an *open-open tube*), a tube closed at both ends (a *closed-closed tube*), and a tube open at one end but closed at the other (an *open-closed tube*), all with the same length L. These are graphs of the pressure wave, with a node at open ends and an antinode at closed ends. The standing wave in the closed-closed tube looks like the wave in the open-open tube except that the positions of the nodes and antinodes are interchanged. In both cases there are m half-wavelength segments between the

FIGURE 16.18 The $m = 2$ standing sound wave inside an open-open column of air.

(a) At one instant

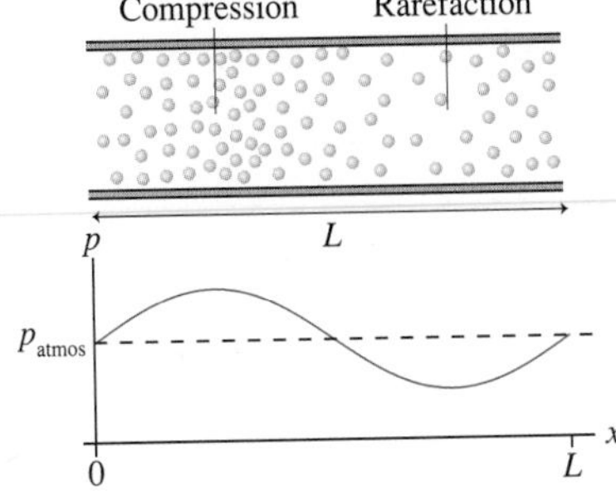

(b) Half a cycle later

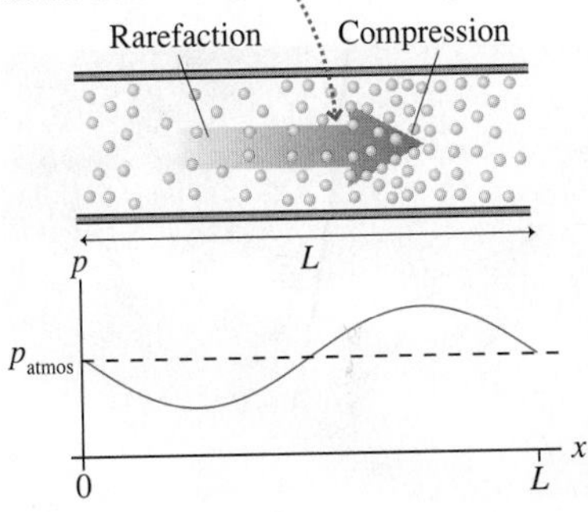

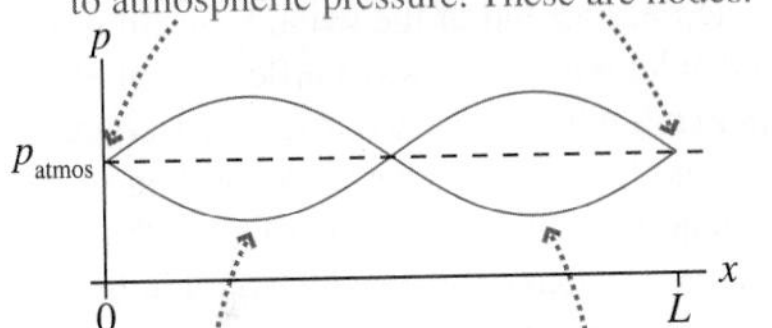

(c) At the ends of the tube, the pressure is equal to atmospheric pressure. These are nodes.
p
p_atmos
0
L
x
At the antinodes, each cycle sees a change from compression to rarefaction and back to compression.

FIGURE 16.19 The first three standing sound wave modes in columns of air with different ends. These graphs show the pressure variation in the tube.

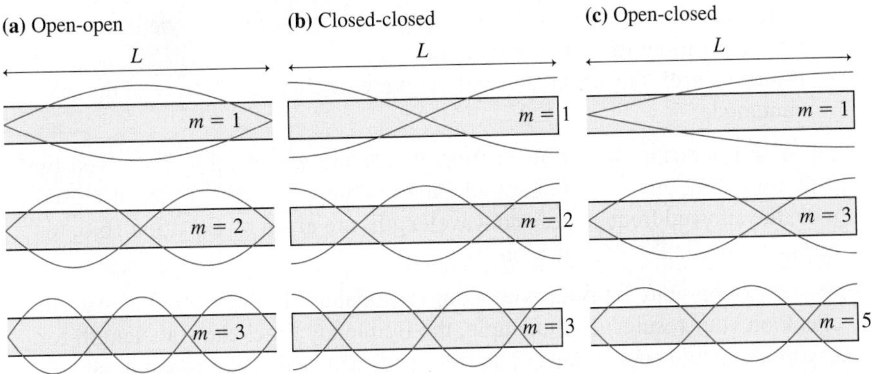

ends; thus the wavelengths and frequencies of an open-open tube and a closed-closed tube are the same as those of a string tied at both ends:

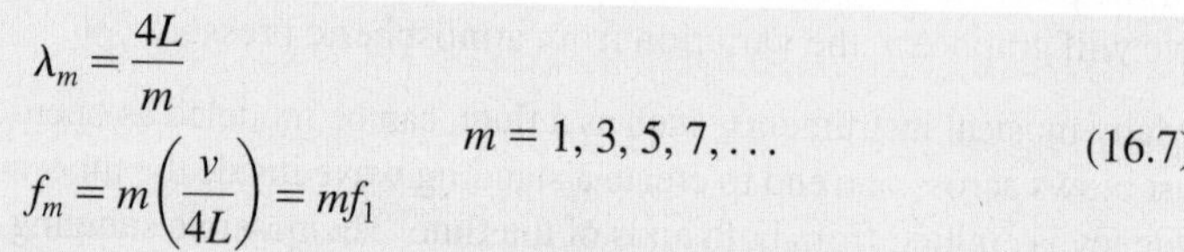

$$\lambda_m = \frac{2L}{m} \qquad m = 1, 2, 3, 4, \ldots \qquad (16.6)$$
$$f_m = m\left(\frac{v}{2L}\right) = mf_1$$

Wavelengths and frequencies of standing sound wave modes in an open-open or closed-closed tube

The open-closed tube is different, as we can see from Figure 16.19. The $m = 1$ mode has a node at one and an antinode at the other, and so has only one-quarter of a wavelength in a tube of length L. The $m = 1$ wavelength is $\lambda_1 = 4L$, twice the $m = 1$ wavelength of an open-open or a closed-closed tube. Consequently, **the fundamental frequency of an open-closed tube is half that of an open-open or a closed-closed tube of the same length.**

The wavelength of the next mode of the open-closed tube is $4L/3$. Because this is 1/3 of λ_1, we assign $m = 3$ to this mode. The wavelength of the subsequent mode is $4L/5$, so this is $m = 5$. In other words, an open-closed tube allows only odd-numbered modes. Consequently, the possible wavelengths and frequencies are

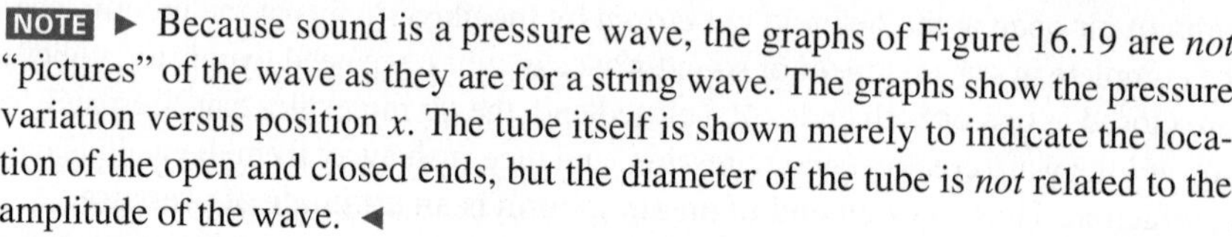

$$\lambda_m = \frac{4L}{m} \qquad m = 1, 3, 5, 7, \ldots \qquad (16.7)$$
$$f_m = m\left(\frac{v}{4L}\right) = mf_1$$

Wavelengths and frequencies of standing sound wave modes in an open-closed tube

Fiery interference In this apparatus, a speaker at one end of the metal tube emits a sinusoidal wave. The wave reflects from the other end, which is closed, to make a counter-propagating wave and set up a standing sound wave in the tube. The tube is filled with propane gas that exits through small holes on top. The burning propane allows us to easily discern the nodes and the antinodes of the standing sound wave. An exciting demonstration—but one you shouldn't try yourself!

NOTE ► Because sound is a pressure wave, the graphs of Figure 16.19 are *not* "pictures" of the wave as they are for a string wave. The graphs show the pressure variation versus position x. The tube itself is shown merely to indicate the location of the open and closed ends, but the diameter of the tube is *not* related to the amplitude of the wave. ◄

We are now in a position to suggest the following problem-solving strategy, not just for sound waves, but for any standing wave.

PROBLEM-SOLVING STRATEGY 16.1 Standing waves

PREPARE

- For sound waves, determine what sort of pipe or tube you have: open-open, closed-closed, or open-closed.
- For string or light waves, the ends will be fixed points.
- Determine known values: length of the tube or string, frequency, wavelength, positions of nodes or antinodes.
- It may be useful to sketch a visual overview, including a picture of the relevant mode.

SOLVE For a string, the allowed frequencies and wavelengths are given by Equations 16.1 and 16.2. For sound waves in an open-open or closed-closed tube, the allowed frequencies and wavelengths are given by Equation 16.6; for an open-closed tube, by Equation 16.7.

ASSESS Does your final answer seem reasonable? Is there another way to check on your results? For example, the frequency times the wavelength for any mode should equal the wave speed—you can check to see that it does.

EXAMPLE 16.6 **Resonances of the ear canal**

The eardrum, which transmits vibrations to the sensory organs of your ear, lies at the end of the ear canal. As **FIGURE 16.20** shows, the ear canal in adults is about 2.5 cm in length. What frequency standing waves can occur within the ear canal that are within the range of human hearing? The speed of sound in the warm air of the ear canal is 350 m/s.

FIGURE 16.20 The anatomy of the ear.

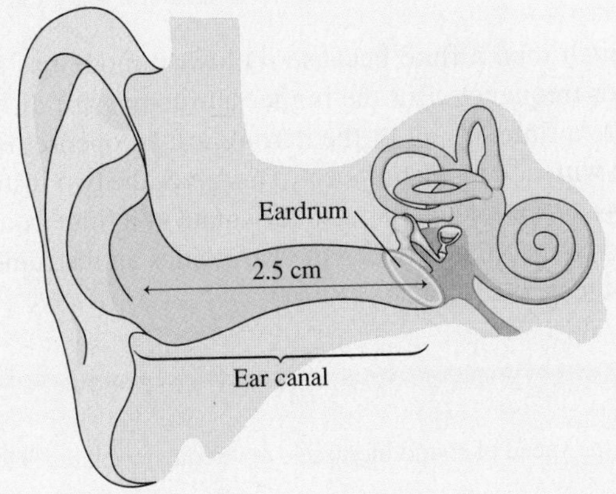

PREPARE We proceed according to the steps in Problem-Solving Strategy 16.1. We can treat the ear canal as an open-closed tube: open to the atmosphere at the external end, closed by the eardrum at the other end. The possible standing-wave modes appear as in Figure 16.19c. The length of the tube is 2.5 cm.

SOLVE Equation 16.7 gives the allowed frequencies in an open-closed tube. We are looking for the frequencies in the range 20 Hz–20,000 Hz. The fundamental frequency is

$$f_1 = \frac{v}{4L} = \frac{350 \text{ m/s}}{4(0.025 \text{ m})} = 3500 \text{ Hz}$$

The higher harmonics are odd multiples of this frequency:

$$f_3 = 3(3500 \text{ Hz}) = 10{,}500 \text{ Hz}$$

$$f_5 = 5(3500 \text{ Hz}) = 17{,}500 \text{ Hz}$$

These three modes lie within the range of human hearing; higher modes are greater than 20,000 Hz.

ASSESS The ear canal is short, so we expect the resonant frequencies to be high; our answers seem reasonable.

How important are the resonances of the ear canal calculated in Example 16.6? In Chapter 15, you learned about the decibel scale for measuring loudness of sound. In fact, your ears have varying sensitivity to sounds of different frequencies. **FIGURE 16.21** shows a curve of *equal perceived loudness,* the sound intensity level (in dB) required to give the *impression* of equal loudness for sinusoidal waves of the noted frequency. Lower values mean that your ear is more sensitive at that frequency. In general, the curve decreases to about 1000 Hz, the frequency at which your hearing is most acute, then slowly rises at higher frequencies. However, this general trend is punctuated by two dips in the curve, showing two frequencies at which a quieter sound produces the same perceived loudness. As you can see, these two dips correspond to the resonances f_1 and f_3 of the ear canal. Incoming sounds at these frequencies produce a larger oscillation, resulting in an increased sensitivity to these frequencies.

FIGURE 16.21 A curve of equal perceived loudness.

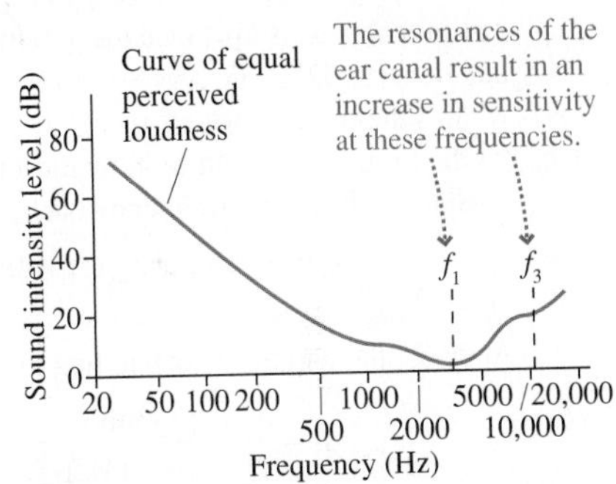

Wind Instruments

With a wind instrument, blowing into the mouthpiece creates a standing sound wave inside a tube of air. The player changes the notes by using her fingers to cover holes or open valves, changing the effective length of the tube. The first open hole becomes a node because the tube is open to the atmosphere at that point. The fact that the holes are on the side, rather than literally at the end, makes very little difference. The length of the tube determines the standing-wave resonances, and thus the musical note that the instrument produces.

Many wind instruments have a "buzzer" at one end of the tube, such as a vibrating reed on a saxophone or clarinet, or the musician's vibrating lips on a trumpet or trombone. Buzzers like these generate a continuous range of frequencies rather than single notes, which is why they sound like a "squawk" if you play on just the mouthpiece without the rest of the instrument. When the buzzer is connected to the body of the instrument, most of those frequencies cause little response. But the frequencies from the buzzer that match the resonant frequencies of the instrument cause the buildup of large amplitudes at these frequencies—standing-wave resonances. The combination of these frequencies makes the musical note that we hear.

Truly classical music The oldest known musical instruments are bone flutes from burial sites in central China. The flutes in the photo are up to 9000 years old and are made from naturally hollow bones from crowned cranes. The positions of the holes determine the frequencies that the flutes can produce. Soon after the first flutes were created, the design was standardized so that different flutes would play the same notes—including the notes in the modern Chinese musical scale.

CONCEPTUAL EXAMPLE 16.7 **Comparing the flute and the clarinet**

A flute and a clarinet have about the same length, but the lowest note that can be played on the clarinet is much lower than the lowest note that can be played on the flute. Why is this?

REASON A flute is an open-open tube; the frequency of the fundamental mode is $f_1 = v/2L$. A clarinet is open at one end, but the player's lips and the reed close it at the other end. The clarinet is thus an open-closed tube with a fundamental frequency $f_1 = v/4L$. This is about half the fundamental frequency of the flute, so the lowest note on the clarinet has a much lower pitch. In musical terms, it's about an octave lower than the flute.

ASSESS A quick glance at Figure 16.19 shows that the wavelength of the lowest mode of the open-closed tube is longer than that of the open-open tube, so we expect a lower frequency for the clarinet.

A clarinet has a lower pitch than a flute because its lowest mode has a longer wavelength and thus a lower frequency. But the higher harmonics are different as well. An open-open tube like a flute has all of the harmonics; an open-closed tube like a clarinet has only those with odd mode numbers. This gives the two instruments a very different tone quality—it's easy to distinguish the sound of a flute from that of a clarinet. We'll explore this connection between the harmonics an instrument produces and its tone quality in the next section.

EXAMPLE 16.8 **The importance of warming up**

Wind instruments have an adjustable joint to change the tube length. Players know that they may need to adjust this joint to stay in tune—that is, to stay at the correct frequency. To see why, suppose a "cold" flute plays the note A at 440 Hz when the air temperature is 20°C.

a. How long is the tube? At 20°C, the speed of sound in air is 343 m/s.
b. As the player blows air through the flute, the air inside the instrument warms up. Once the air temperature inside the flute has risen to 32°C, increasing the speed of sound to 350 m/s, what is the frequency?
c. At the higher temperature, how must the length of the tube be changed to bring the frequency back to 440 Hz?

SOLVE A flute is an open-open tube with fundamental frequency $f_1 = v/2L$.

a. At 20°C, the length corresponding to 440 Hz is

$$L = \frac{v}{2f} = \frac{343 \text{ m/s}}{2(440 \text{ Hz})} = 0.390 \text{ m}$$

b. As the speed of sound increases, the frequency changes to

$$f_1(\text{at } 32°\text{C}) = \frac{350 \text{ m/s}}{2(0.390 \text{ m})} = 449 \text{ Hz}$$

c. To bring the flute back into tune, the length must be increased to give a frequency of 440 Hz with a speed of 350 m/s. The new length is

$$L = \frac{v}{2f_1} = \frac{350 \text{ m/s}}{2(440 \text{ Hz})} = 0.398 \text{ m}$$

Thus the flute must be increased in length by 8 mm.

ASSESS A small change in the absolute temperature produces a correspondingly small change in the speed of sound. We expect that this will require a small change in length, so our answer makes sense.

STOP TO THINK 16.3 A tube that is open at both ends supports a standing wave with harmonics at 300 Hz and 400 Hz, with no harmonics between. What is the fundamental frequency of this tube?

A. 50 Hz B. 100 Hz C. 150 Hz D. 200 Hz E. 300 Hz

16.5 Speech and Hearing

When you hear a particular note played on a guitar, it sounds very different from the same note played on a trumpet. And you have perhaps been to a lecture in which the speaker talked at essentially the same pitch the entire time—but you could still understand what was being said. Clearly, there is more to your brain's perception of sound than pitch alone. How do you tell the difference between a guitar and a trumpet? How do you distinguish between an "oo" vowel sound and an "ee" vowel sound at the same pitch?

The Frequency Spectrum

To this point, we have pictured sound waves as sinusoidal waves, with a well-defined frequency. In fact, most of the sounds that you hear are not pure sinusoidal waves. Most sounds are a mix, or superposition, of different frequencies. For example, we have seen how certain standing-wave modes are possible on a stretched string. When you pluck a string on a guitar, you generally don't excite just one standing-wave mode—you simultaneously excite many different modes.

If you play the note "middle C" on a guitar, the fundamental frequency is 262 Hz. There will be a standing wave at this frequency, but there will also be standing waves at the frequencies 524 Hz, 786 Hz, 1048 Hz, . . . , all the higher harmonics predicted by Equation 16.4.

FIGURE 16.22a is a bar chart showing all the frequencies present in the sound of the vibrating guitar string. The height of each bar shows the relative intensity of that harmonic. The fundamental frequency has the highest intensity, but many other harmonics have significant intensities as well. A bar chart showing the relative intensities of the different frequencies is called the **frequency spectrum** of the sound.

When your brain interprets the mix of frequencies from the guitar in Figure 16.22a, it identifies the fundamental frequency as the *pitch.* 262 Hz corresponds to middle C, so you will identify the pitch as middle C, even though the sound consists of many different frequencies. Your brain uses the higher harmonics to determine the **tone quality,** which is also called the *timbre.* The tone quality—and therefore the higher harmonics—is what makes a middle C played on a guitar sound quite different from a middle C played on a trumpet. The frequency spectrum of a trumpet would show a very different pattern of the relative intensities of the higher harmonics, and this different mix of higher harmonics gives the trumpet a different sound.

The actual sound wave produced by a guitar playing middle C is shown in FIGURE 16.22b. The sound wave is periodic, with a period of 3.82 ms that corresponds to the 262 Hz fundamental frequency. But the wave doesn't have a simple sinusoidal shape; it is more complex. **The higher harmonics don't change the period of the sound wave; they change only its shape.** The sound wave of a trumpet playing middle C would also have a 3.82 ms period, but its shape would be entirely different.

FIGURE 16.22 The frequency spectrum and a graph of the sound wave of a guitar playing a note with fundamental frequency 262 Hz.

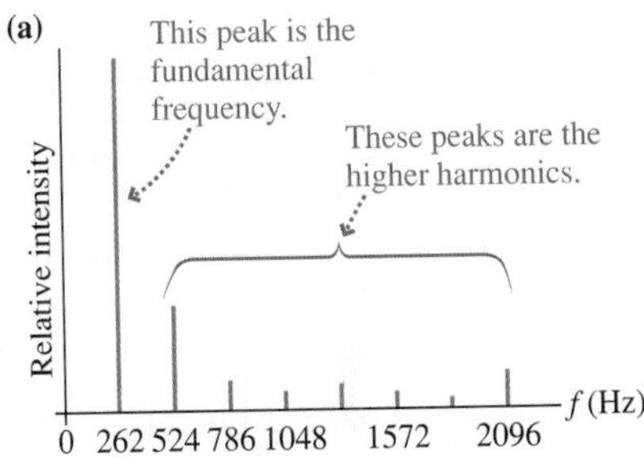
(a)
This peak is the fundamental frequency.
These peaks are the higher harmonics.
Relative intensity
f (Hz)
0 262 524 786 1048 1572 2096

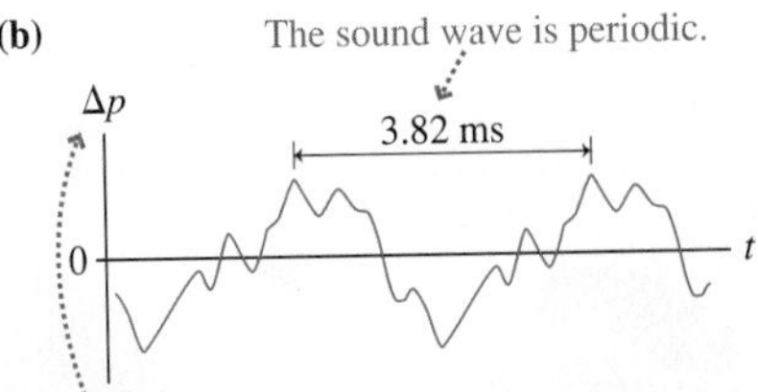
(b)
The sound wave is periodic.
Δp
3.82 ms
0
t
The vertical axis is the change in pressure from atmospheric pressure due to the sound wave.

CONCEPTUAL EXAMPLE 16.9 Playing the didgeridoo

The didgeridoo, a musical instrument developed by aboriginal Australians, is deceptively simple. It consists of a tube (a eucalyptus stem or branch hollowed out by termites) of $1\frac{1}{2}$m or more in length. The player presses his lips against the end and blows air through his lips as with a trumpet. He may also make sounds with his vocal cords. Skilled players can make a wide variety of sounds. How is this possible with such a simple instrument?

REASON Because the lips seal one end, a didgeridoo has the resonances of an open-closed tube, given by Equation 16.7. The instrument can therefore produce many different frequencies. Changing the vibration of the lips and the sounds from the vocal cords can change the mix of standing-wave modes that are produced, leading to very different sounds.

One instrument, many sounds A synthesizer can be adjusted to sound like a flute, a clarinet, a trumpet, a piano—or any other musical instrument. The keys on a synthesizer determine what fundamental frequency to produce. The other controls adjust the mix of higher harmonics to match the frequency spectrum of various musical instruments, effectively mimicking their sounds.

Vowels and Formants

Try this: Keep your voice at the same pitch and say the "ee" sound, as in "beet," then the "oo" sound, as in "boot." Pay attention to how you reshape your mouth as you

FIGURE 16.23 The frequency spectrum from the vocal cords, and after passing through the vocal tract.

(a) Frequencies from the vocal cords

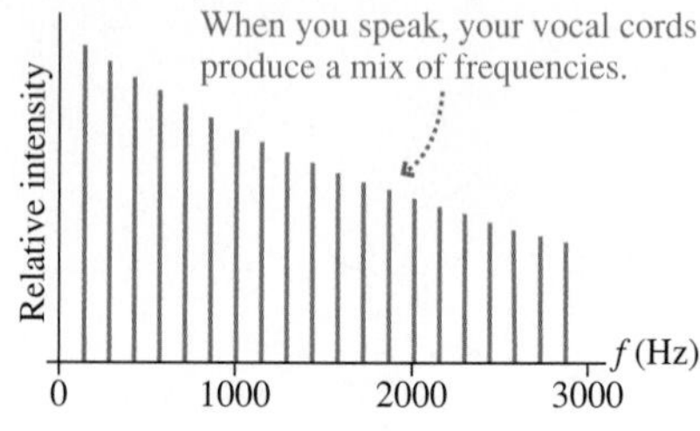

(b) Actual spoken frequencies (vowel sound "ee")

When you form your vocal tract to make a certain vowel sound, it increases the amplitudes of certain frequencies and suppresses others.

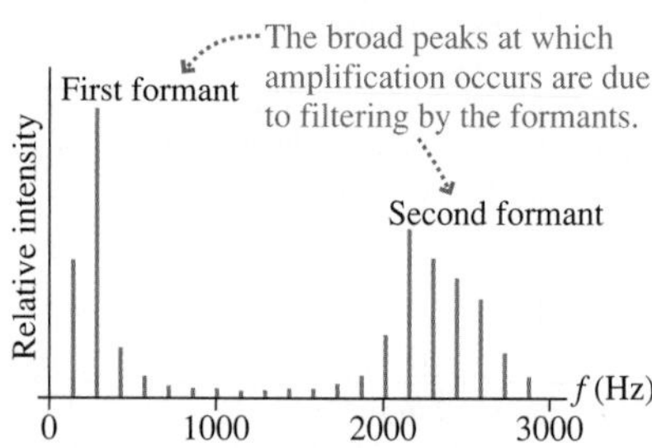

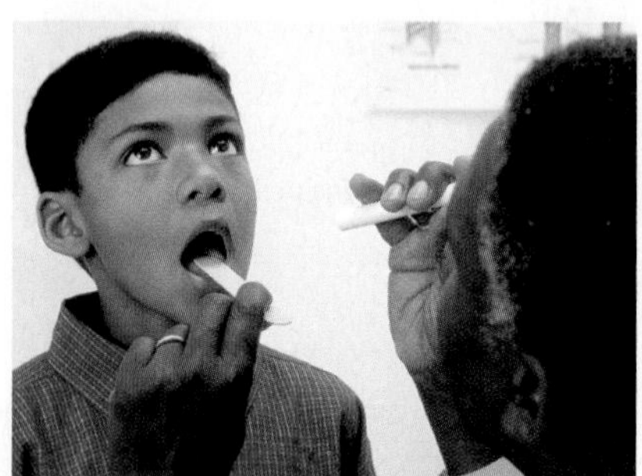

Saying "ah" BIO Why, during a throat exam, does a doctor ask you to say "ah"? This particular vowel sound is formed by opening the mouth and the back of the throat wide—giving a clear view of the tissues of the throat.

move back and forth between the two sounds. The two vowel sounds are at the same pitch, and thus have the same period, but they sound quite different. The difference in sound arises from the difference in the higher harmonics, just as for musical instruments. As you speak, you adjust the properties of your vocal tract to produce different mixes of harmonics that make the "ee," "oo," "ah," and other vowel sounds.

Speech begins with the vibration of your vocal cords, stretched bands of tissue in your throat. The vibration is similar to that of a wave on a stretched string. In ordinary speech, the average fundamental frequency for adult males and females is about 150 Hz and 250 Hz, respectively, but you can change the vibration frequency by changing the tension of your vocal cords. That's how you make your voice higher or lower as you sing.

Your vocal cords produce a mix of different frequencies as they vibrate—the fundamental frequency and a rich mixture of higher harmonics. If you put a microphone in your throat and measured the sound waves right at your vocal cords the frequency spectrum would appear as in FIGURE 16.23a.

There is more to the story though. Before reaching the opening of your mouth, sound from your vocal cords must pass through your vocal tract—a series of hollow cavities including your throat, mouth, and nose. The vocal tract acts like a series of tubes, and, as in any tube, certain frequencies will set up standing-wave resonances. The rather broad standing-wave resonances of the vocal tract are called **formants.** Harmonics of your vocal cords at or near the formant frequencies will be amplified; harmonics far from a formant will be suppressed. FIGURE 16.23b shows the formants of an adult male making an "ee" sound. The filtering of the vocal cord harmonics by the formants is clear.

You can change the shape and frequencies of the formants, and thus the sound you make, by changing the shape and length of your vocal tract. You do this by changing your mouth opening and the shape and position of your tongue. The first two formants for an "ee" sound are at roughly 270 Hz and 2300 Hz, but for an "oo" sound they are 300 Hz and 870 Hz. The much lower second formant of the "oo" emphasizes midrange frequencies, making a "calming" sound, while the more strident sound of "ee" comes from enhancing the higher frequencies.

CONCEPTUAL EXAMPLE 16.10 High-frequency hearing loss

As you age, your hearing sensitivity will decrease. This decrease is not uniform; for most people, the loss of sensitivity is greater for higher frequencies. The loss of sensitivity at high frequencies may make it difficult to understand what others are saying. Why is this?

REASON It is the high-frequency components of speech that allow us to distinguish different vowel sounds. A decrease in sensitivity to these higher frequencies makes it more difficult to make such distinctions.

ASSESS This result makes sense. In Figure 16.23b, the lowest frequency is less than 200 Hz, but the second formant is over 2000 Hz. There's a big difference between what we hear as the pitch of someone's voice and the frequencies we use to interpret speech.

STOP TO THINK 16.4 These sound waves represent notes played on different musical instruments. Which has the highest pitch?

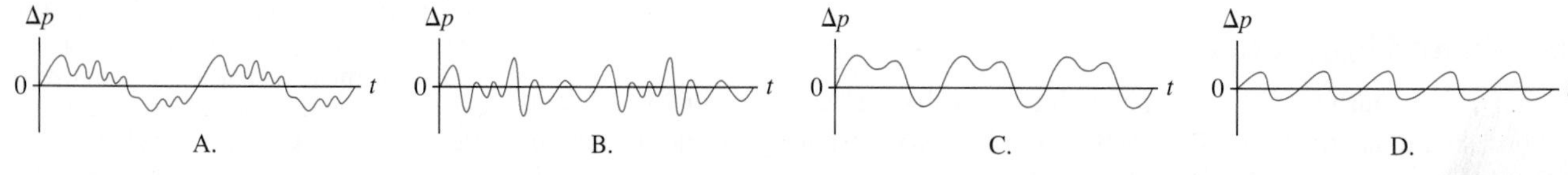

16.6 The Interference of Waves from Two Sources

Perhaps you have seen headphones that offer "active noise reduction." When you turn on the headphones, they produce sound that somehow *cancels* noise from the external environment. How does adding sound to a system make it quieter?

We began the chapter by noting that waves, unlike particles, can pass through each other. Where they do, the principle of superposition tells us that the displacement of the medium is the sum of the displacements due to each wave acting alone. Consider the two loudspeakers in FIGURE 16.24, both emitting sound waves with the same frequency. In Figure 16.24a, sound from loudspeaker 2 passes loudspeaker 1, then two overlapped sound waves travel to the right along the x-axis. What sound is heard at the point indicated with the dot? And what about at the dot in Figure 16.24b, where the speakers are side by side? These are two cases we will consider in this section. Although we'll use sound waves for our discussion, the results are general and apply to all waves. In Chapter 17, we will use these ideas to study the interference of light waves.

FIGURE 16.24 Interference of waves from two sources.

(a) Two sound waves overlapping along a line

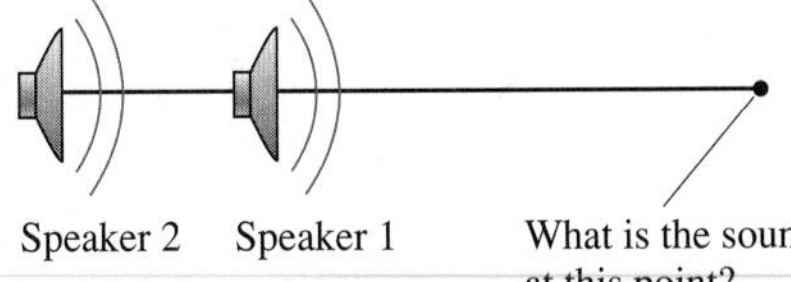

Speaker 2
Speaker 1
What is the sound at this point?

(b) Two overlapping spherical sound waves

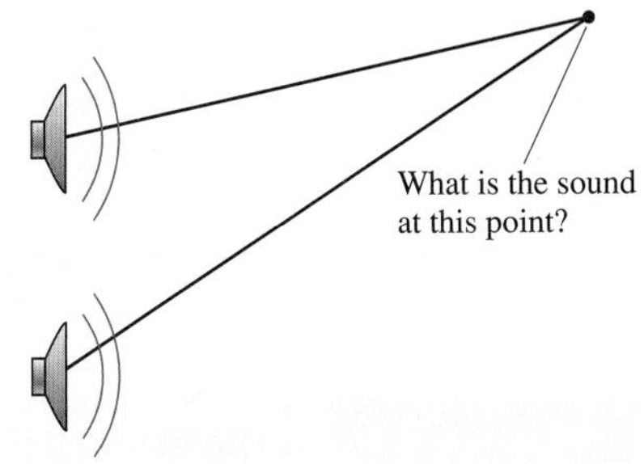

What is the sound at this point?

Interference Along a Line

FIGURE 16.25 shows traveling waves from two loudspeakers spaced exactly one wavelength apart. The graphs are slightly displaced from each other so that you can see what each wave is doing, but the *physical situation* is one in which the waves are traveling *on top of* each other. We assume that the two speakers emit sound waves of identical frequency f, wavelength λ, and amplitude A.

At every point along the line, the net sound pressure wave will be the sum of the pressures from the individual waves. That's the principle of superposition. Because the two speakers are separated by one wavelength, the two waves are aligned crest-to-crest and trough-to-trough. Waves aligned this way are said to be **in phase;** waves that are in phase march along "in step" with each other. The superposition is a traveling wave with wavelength λ and twice the amplitude of the individual waves. This is constructive interference.

NOTE ▶ Textbook pictures can be misleading because they're frozen in time. The net sound wave in Figure 16.25 is a *traveling* wave, moving to the right with the speed of sound v. It differs from the two individual waves only by having twice the amplitude. This is not a standing wave with nodes and antinodes that remain in one place. ◀

FIGURE 16.25 Constructive interference of two waves traveling along the x-axis.

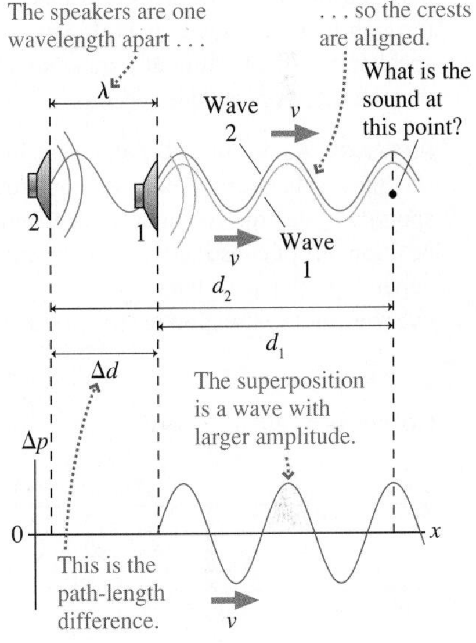

If d_1 and d_2 are the distances from loudspeakers 1 and 2 to a point at which we want to know the combined sound wave, their difference $\Delta d = d_2 - d_1$ is called the **path-length difference.** It is the *extra* distance traveled by wave 2 on the way to the point where the two waves are combined. In Figure 16.25, we see that constructive interference results from a path-length difference $\Delta d = \lambda$. But increasing Δd by an additional λ would produce exactly the same result, so we will also have constructive interference for $\Delta d = 2\lambda$, $\Delta d = 3\lambda$, and so on. In other words, **two waves will be in phase and will produce constructive interference any time their path-length difference is a whole number of wavelengths.**

In FIGURE 16.26, the two speakers are separated by half a wavelength. Now the crests of one wave align with the troughs of the other, and the waves march along "out of step" with each other. We say that the two waves are **out of phase.** When two waves are out of phase, they are equal and opposite at every point. Consequently, the sum of the two waves is zero *at every point.* This is destructive interference.

The destructive interference of Figure 16.26 results from a path-length difference $\Delta d = \frac{1}{2}\lambda$. Again, increasing Δd by an additional λ would produce a picture that looks exactly the same, so we will also have destructive interference for $\Delta d = 1\frac{1}{2}\lambda$, $\Delta d = 2\frac{1}{2}\lambda$, and so on. That is, **two waves will be out of phase and will produce destructive interference any time their path-length difference is a whole number of wavelengths plus half a wavelength.**

FIGURE 16.26 Destructive interference of two waves traveling along the x-axis.

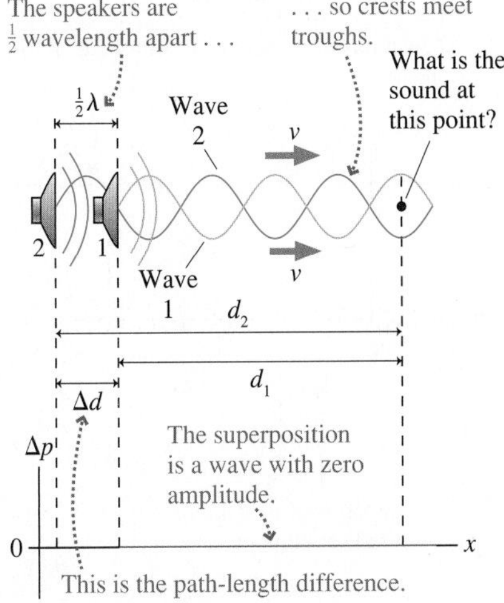

Summing up, for two identical sources of waves, constructive interference occurs when the path-length difference is

$$\Delta d = m\lambda \qquad m = 0, 1, 2, 3, \ldots \tag{16.8}$$

and destructive interference occurs when the path-length difference is

$$\Delta d = \left(m + \frac{1}{2}\right)\lambda \qquad m = 0, 1, 2, 3, \ldots \tag{16.9}$$

NOTE ► The path-length difference needed for constructive or destructive interference depends on the wavelength and hence the frequency. If one particular frequency interferes destructively, another may not. ◄

The path-length difference is not necessarily the distance between the speakers, as we see in the next example. It is simply the difference in the distances traveled by the two waves.

EXAMPLE 16.11 Is the sound loud or quiet? Part I

Two loudspeakers 42.0 m apart and facing each other emit identical 115 Hz sinusoidal sound waves. Susan is walking along a line between the speakers. As she walks, she finds herself moving through loud and quiet spots. If Susan stands 19.5 m from one speaker, is she standing at a quiet spot or a loud spot? Assume that the speed of sound is 345 m/s.

PREPARE As Susan walks along the line between the speakers, she moves between points of constructive interference (loud spots) and destructive interference (quiet spots). Is her current position one of constructive or destructive interference? This will depend on the path-length difference. We start with a visual overview of the situation in **FIGURE 16.27**.

FIGURE 16.27 Visual overview of loudspeakers.

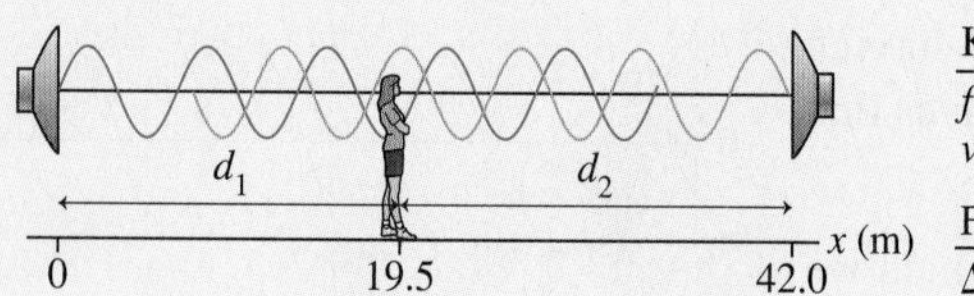

SOLVE At Susan's position, the distances the two waves travel to reach her are

$$d_1 = 19.5 \text{ m} \qquad d_2 = 42.0 \text{ m} - 19.5 \text{ m} = 22.5 \text{ m}$$

At the point where the two waves reach Susan and interfere, their path-length difference is

$$\Delta d = d_2 - d_1 = 3.0 \text{ m}$$

To know if we have constructive or destructive interference, we need to compare this with the wavelength:

$$\lambda = \frac{v}{f} = \frac{345 \text{ m/s}}{115 \text{ Hz}} = 3.0 \text{ m}$$

Because the path-length difference is exactly one wavelength, Susan is standing at a point of constructive interference; that is, she is standing at a loud spot.

ASSESS There's a nice way that we can check to see that our answer makes sense. As you'll recall from earlier in the chapter, two counterpropagating waves create a standing wave. In this case, the conditions for constructive and destructive interference are the same as the conditions for antinodes and nodes, respectively, that we found earlier. Susan is at an antinode of the standing wave. The ideas of interference give us a different perspective on standing waves.

FIGURE 16.28 Opposite waves cancel.

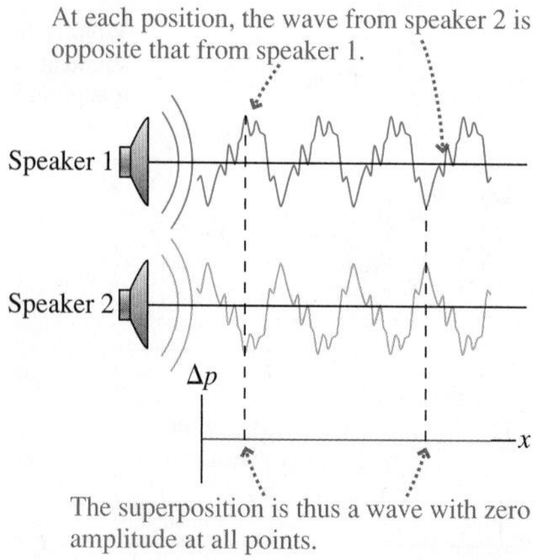

At each position, the wave from speaker 2 is opposite that from speaker 1.

Speaker 1

Speaker 2

Δp

x

The superposition is thus a wave with zero amplitude at all points.

The analysis above assumed that the two loudspeakers were emitting identical waves. Another interesting and important case of interference, illustrated in **FIGURE 16.28**, occurs when one loudspeaker emits a sound wave that is *the exact inverse* of the wave from the other speaker. If the speakers are side by side, so that $\Delta d = 0$, the superposition of these two waves will result in destructive interference; they will completely cancel. This destructive interference does not require the waves to have any particular frequency or any particular shape.

Headphones with *active noise reduction* use this technique. A microphone on the outside of the headphones measures ambient sound. A circuit inside the headphones produces an inverted version of the microphone signal and sends it to the headphone speakers. The ambient sound and the inverted version of the sound from the speakers arrive at the ears together and interfere destructively, reducing the sound intensity. In this case, *adding* sound results in a *lower* overall intensity inside the headphones!

Interference of Spherical Waves

Interference along a line illustrates the idea of interference, but it's not very realistic. In practice, sound waves from a loudspeaker or light waves from a lightbulb spread out as spherical waves. **FIGURE 16.29** shows a wave-front diagram for a spherical wave. Recall that the wave fronts represent the *crests* of the wave and are spaced by the wavelength λ. Halfway between two wave fronts is a trough of the wave. What happens when two spherical waves overlap? For example, imagine two loudspeakers emitting identical waves radiating sound in all directions. **FIGURE 16.30** shows the wave fronts of the two waves. This is a static picture, of course, so you have to imagine the wave fronts spreading out as new circular rings are born at the speakers. The waves overlap as they travel, and, as was the case in one dimension, this causes interference.

Consider a particular point like that marked by the red dot in Figure 16.30. The two waves each have a crest at this point, so there is constructive interference here. But at other points, such as that marked by the black dot, a crest overlaps a trough, so this is a point of destructive interference.

Notice—simply by counting the wave fronts—that the red dot is three wavelengths from speaker 2 ($r_2 = 3\lambda$) but only two wavelengths from speaker 1 ($r_1 = 2\lambda$). The path-length difference of the two waves arriving at the red dot is $\Delta r = r_2 - r_1 = \lambda$. That is, the wave from speaker 2 has to travel one full wavelength more than the wave from speaker 1, so the waves are in phase (crest aligned with crest) and interfere constructively. You should convince yourself that Δr is a *whole number of wavelengths* at every point where two wave fronts intersect.

Similarly, the path-length difference at the black dot, where the interference is destructive, is $\Delta r = \frac{1}{2}\lambda$. As with interference along a line, destructive interference results when the path-length difference is a whole number of wavelengths plus half a wavelength.

Thus the general rule for determining whether there is constructive or destructive interference at any point is the same for spherical waves as for waves traveling along a line. For identical sources, constructive interference occurs when the path-length difference is

$$\Delta r = m\lambda \qquad m = 0, 1, 2, 3, \ldots \tag{16.10}$$

Destructive interference occurs when the path-length difference is

$$\Delta r = \left(m + \frac{1}{2}\right)\lambda \qquad m = 0, 1, 2, 3, \ldots \tag{16.11}$$

The conditions for constructive and destructive interference are the same for spherical waves as for waves along a line. And the treatment we have seen for sound waves can be applied to any wave, as we have noted. For any two wave sources, the following Tactics Box sums up how to determine if the interference at a point is constructive or destructive.

FIGURE 16.29 A spherical wave.

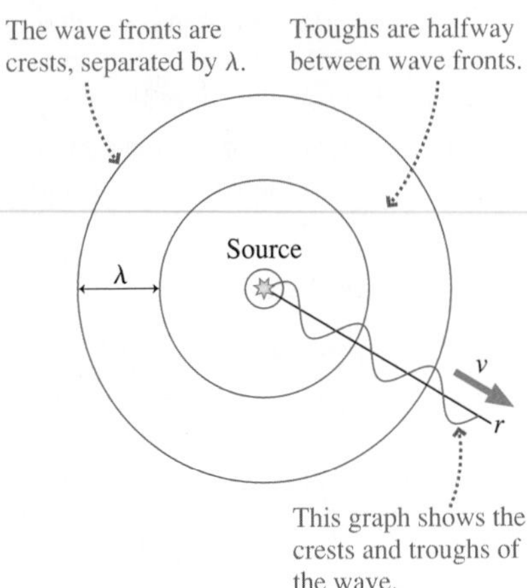
The wave fronts are crests, separated by λ.
Troughs are halfway between wave fronts.
Source
λ
v
r
This graph shows the crests and troughs of the wave.

FIGURE 16.30 The overlapping wave patterns of two sources.

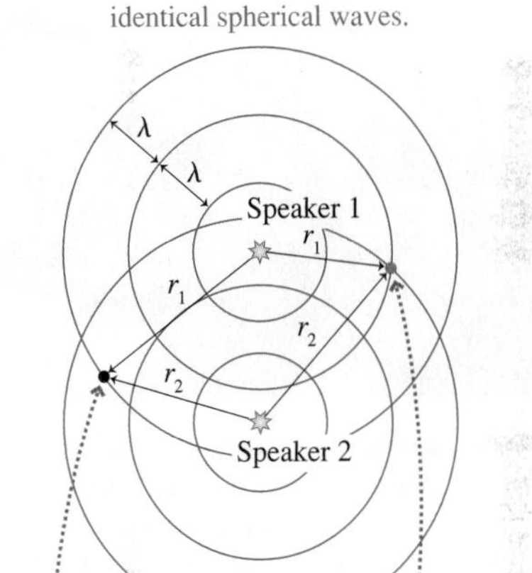
Two sources emit identical spherical waves.
λ
λ
Speaker 1
r1
r1
r2
r2
Speaker 2
Destructive interference occurs where a crest overlaps a trough.
Constructive interference occurs where two crests overlap.

TACTICS BOX 16.1 Identifying constructive and destructive interference (MP)™

1. Identify the path length from each source to the point of interest. Compute the path-length difference $\Delta r = |r_2 - r_1|$.
2. Find the wavelength, if it is not specified.
3. If the path-length difference is a whole number of wavelengths ($\lambda, 2\lambda, 3\lambda, \ldots$), crests are aligned with crests and there is constructive interference.
4. If the path-length difference is a whole number of wavelengths plus a half wavelength ($1\frac{1}{2}\lambda, 2\frac{1}{2}\lambda, 3\frac{1}{2}\lambda, \ldots$), crests are aligned with troughs and there is destructive interference.

Exercises 9,10

NOTE ► Keep in mind that interference is determined by Δr, the path-length *difference*, not by r_1 or r_2. ◄

EXAMPLE 16.12 Is the sound loud or quiet? Part II

Two speakers are 3.0 m apart and play identical tones of frequency 170 Hz. Sam stands directly in front of one speaker at a distance of 4.0 m. Is this a loud spot or a quiet spot? Assume that the speed of sound in air is 340 m/s.

PREPARE FIGURE 16.31 shows a visual overview of the situation, showing the positions of and path lengths from each speaker.

FIGURE 16.31 Visual overview of two speakers.

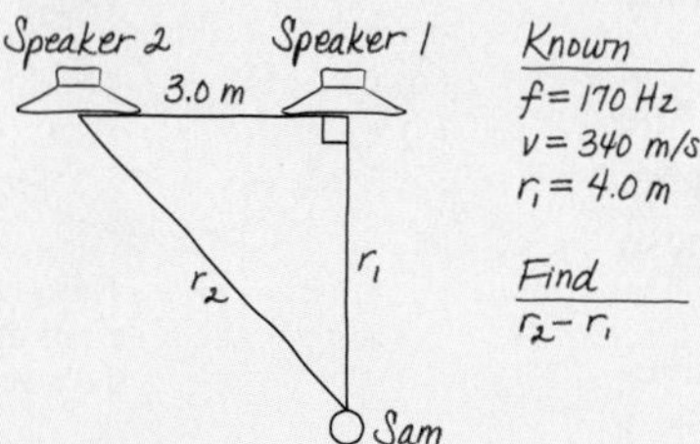

SOLVE Following the steps in Tactics Box 16.1, we first compute the path-length difference. r_1, r_2, and the distance between the speakers form a right triangle, so we can use the Pythagorean theorem to find

$$r_2 = \sqrt{(4.0\text{ m})^2 + (3.0\text{ m})^2} = 5.0\text{ m}$$

Thus the path-length difference is

$$\Delta r = r_2 - r_1 = 1.0\text{ m}$$

Next, we compute the wavelength:

$$\lambda = \frac{v}{f} = \frac{340\text{ m/s}}{170\text{ Hz}} = 2.0\text{ m}$$

The path-length difference is $\frac{1}{2}\lambda$, so this is a point of destructive interference. Sam is at a quiet spot.

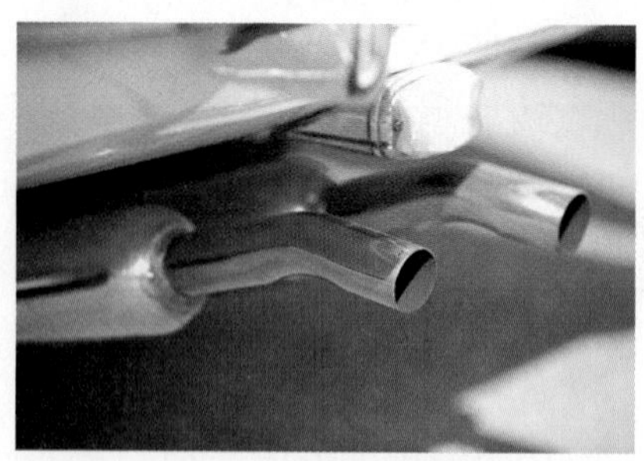

Taming and tuning exhaust noise It's possible to use destructive interference to cut automobile exhaust noise using a device called a *resonator*. The wide section of pipe at the end of the exhaust system is a tube with one closed end. A sound wave enters the tube, reflects from the end, and reenters the exhaust pipe. If the length of the tube is just right, the reflected wave will be out of phase with the sound wave in the pipe, producing destructive interference. The resonator is tuned to eliminate the loudest frequencies from the engine, but other frequencies will produce constructive interference, enhancing them and giving the exhaust a certain "note."

So far, we have looked at interference only at particular points. What can we say about the overall pattern of points at which we have constructive or destructive interference? For instance, the red dot in Figure 16.30 is only one point where $\Delta r = \lambda$; you should be able to locate several more. Taken together, all the points with $\Delta r = \lambda$ form a curved line along which constructive interference is occurring. Another curved line of constructive interference connects all the points at which $\Delta r = 2\lambda$, another connects the $\Delta r = 3\lambda$ points, and so on. These lines, shown as red lines in FIGURE 16.32, are called **antinodal lines.** They are analogous to the antinodes of a standing wave—hence the name. An antinode is a *point* of constructive interference; for spherical waves, oscillation at maximum amplitude occurs along a continuous *line.* To understand this idea better, imagine the static picture of Figure 16.32 evolving with time. As the wave fronts expand, the *intersection point* of two rings moves outward along one of the red lines.

FIGURE 16.32 The points of constructive and destructive interference fall along antinodal and nodal lines.

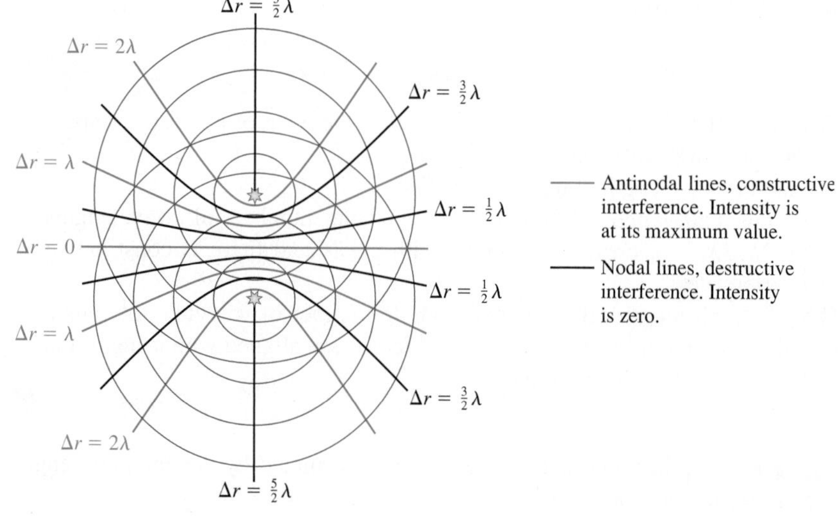

Similarly, we can connect together points where Δr is a multiple of λ plus $\frac{1}{2}\lambda$. The black dot in Figure 16.30 is just one of many points with $\Delta r = \frac{1}{2}\lambda$. Together, these points of destructive interference form a **nodal line** along which the displacement is always zero. The nodal lines in Figure 16.32 are shown in black.

The two water waves overlap, leading to patterns of constructive and destructive interference.

You are regularly exposed to sound from two separated sources: stereo speakers. When you walk across a room in which a stereo is playing, why don't you hear a pattern of loud and soft sounds as you cross antinodal and nodal lines? First, we don't listen to single frequencies. Music is a complex sound wave with many frequencies, but only one frequency at a time satisfies the condition for constructive or destructive interference. Most of the sound frequencies are not affected. Second, reflections of sound waves from walls and furniture make the situation much more complex than the idealized two-source picture in Figure 16.32. Sound wave interference can be heard, but it takes careful selection of a pure tone and a room with no hard, reflecting surfaces. Interference that's rather tricky to demonstrate with sound waves is easy to produce with light waves, as we'll see in the next chapter.

STOP TO THINK 16.5 These speakers emit identical sound waves with a wavelength of 1.0 m. At the point indicated, is the interference constructive, destructive, or something in between?

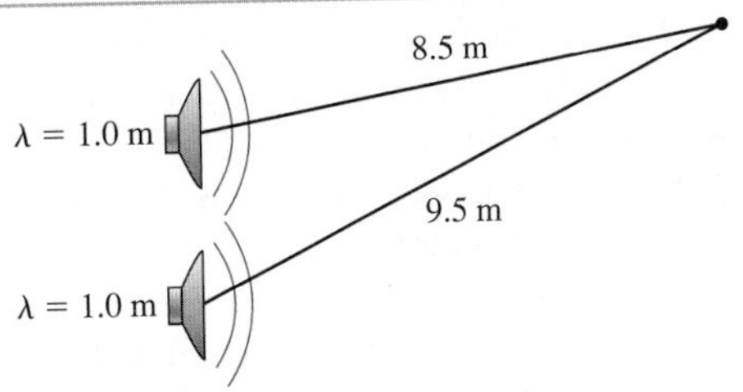

16.7 Beats

10.7

Thus far we have looked at the superposition of waves from sources having the same frequency. We can also use the principle of superposition to investigate a phenomenon that is easily demonstrated with two sources of slightly *different* frequencies.

Suppose two sinusoidal waves are traveling toward your ear, as shown in FIGURE 16.33. The two waves have the same amplitude but slightly different frequencies: The red wave has a slightly higher frequency (and thus a slightly shorter wavelength) than the orange wave. This slight difference causes the waves to combine in a manner that alternates between constructive and destructive interference. Their superposition, drawn in blue below the two waves, is a wave whose amplitude shows

FIGURE 16.33 The superposition of two sound waves with slightly different frequencies.

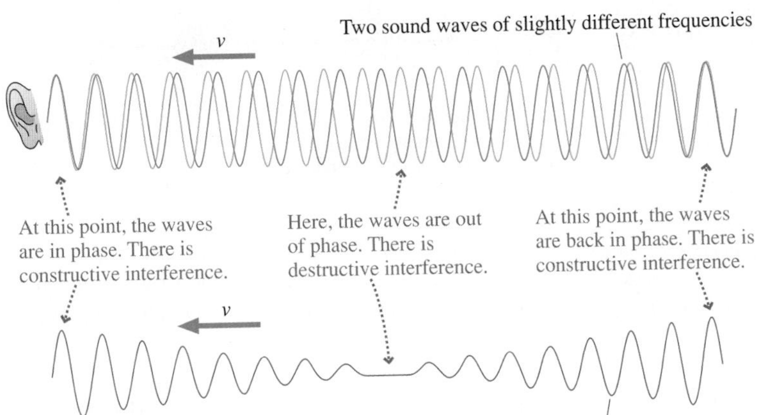

Two sound waves of slightly different frequencies
v
At this point, the waves are in phase. There is constructive interference.
Here, the waves are out of phase. There is destructive interference.
At this point, the waves are back in phase. There is constructive interference.
v
Superposition of the two sound waves

a periodic variation. As the waves reach your ear, you will hear a single tone whose intensity is *modulated.* That is, the sound goes up and down in volume, loud, soft, loud, soft, . . . , making a distinctive sound pattern called **beats.**

Suppose the two waves have frequencies f_1 and f_2 that differ only slightly, so that $f_1 \approx f_2$. A complete mathematical analysis would show that the air oscillates against your eardrum at frequency

$$f_{osc} = \frac{1}{2}(f_1 + f_2)$$

This is the *average* of f_1 and f_2, and it differs little from either since the two frequencies are nearly equal. Further, the intensity of the sound is modulated at a frequency called the *beat frequency:*

$$f_{beat} = |f_1 - f_2| \qquad (16.12)$$

The beat frequency is the *difference* between the two individual frequencies.

FIGURE 16.34 is a history graph of the wave at the position of your ear. You can see both the sound wave oscillation at frequency f_{osc} and the much slower intensity oscillation at frequency f_{beat}. Frequency f_{osc} determines the pitch you hear, while f_{beat} determines the frequency of the loud-soft-loud modulations of the sound intensity.

FIGURE 16.34 The modulated sound of beats.

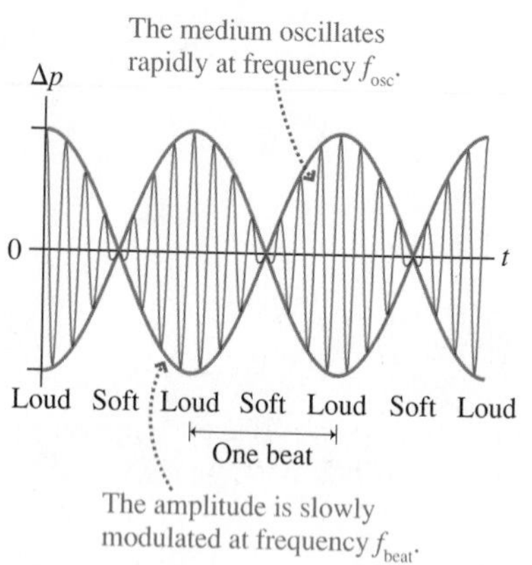
The medium oscillates rapidly at frequency f_{osc}.
Δp
0
t
Loud Soft Loud Soft Loud Soft Loud
One beat
The amplitude is slowly modulated at frequency f_{beat}.

Musicians can use beats to tune their instruments. If one flute is properly tuned at 440 Hz but another plays at 438 Hz, the flutists will hear two loud-soft-loud beats per second. The second flutist is "flat" and needs to shorten her flute slightly to bring the frequency up to 440 Hz.

Many measurement devices use beats to determine an unknown frequency by comparing it to a known frequency. For example, Chapter 15 described a Doppler blood flow meter that used the Doppler shift of ultrasound reflected from moving blood to determine its speed. The meter determines this very small frequency shift by combining the emitted wave and the reflected wave and measuring the resulting beat frequency. The beat frequency is equal to the shift in frequency on reflection, exactly what is needed to determine the blood speed. Another example of using beats to make a measurement is given in the following example.

EXAMPLE 16.13 Detecting bats using beats

The little brown bat is a common bat species in North America. It emits echolocation pulses at a frequency of 40 kHz, well above the range of human hearing. To allow observers to "hear" these bats, the bat detector shown in FIGURE 16.35 combines the bat's sound wave at frequency f_1 with a wave of frequency f_2 from a tunable oscillator. The resulting beat frequency is isolated with a filter, then amplified and sent to a loudspeaker. To what frequency should the tunable oscillator be set to produce an audible beat frequency of 3 kHz?

SOLVE The beat frequency is $f_{beat} = |f_1 - f_2|$, so the oscillator frequency and the bat frequency need to *differ* by 3 kHz. An oscillator frequency of either 37 kHz or 43 kHz will work nicely.

FIGURE 16.35 The operation of a bat detector.

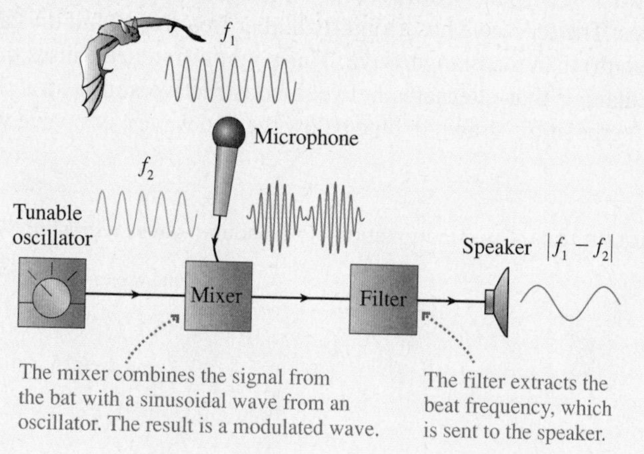
f_1
Microphone
f_2
Tunable oscillator
Speaker $|f_1 - f_2|$
Mixer
Filter
The mixer combines the signal from the bat with a sinusoidal wave from an oscillator. The result is a modulated wave.
The filter extracts the beat frequency, which is sent to the speaker.

STOP TO THINK 16.6 You hear three beats per second when two sound tones are generated. The frequency of one tone is known to be 610 Hz. The frequency of the other is

A. 604 Hz
B. 607 Hz
C. 613 Hz
D. 616 Hz
E. Either A or D
F. Either B or C

INTEGRATED EXAMPLE 16.14 The size of a dog determines the sound of its growl

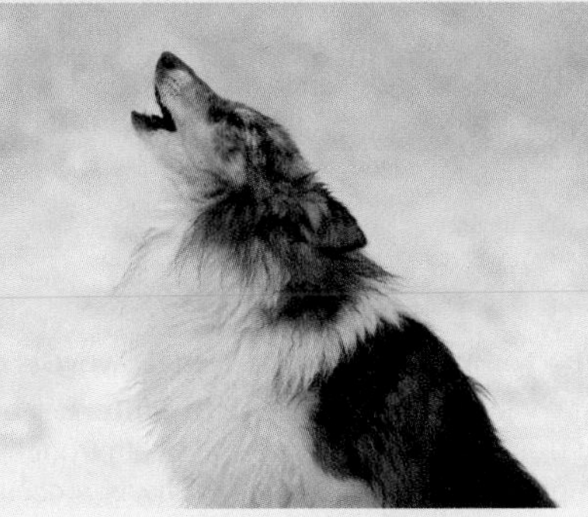

The sounds of the human vocal system result from the interplay of two different oscillations: the oscillation of the vocal cords and the standing-wave resonances of the vocal tract. A dog's vocalizations are based on similar principles, but the canine vocal tract is quite a bit simpler than that of a human. When a dog growls or howls, the shape of the vocal tract is essentially a tube closed at the larynx and open at the lips.

All dogs growl at a low pitch because the fundamental frequency of the vocal cords is quite low. But the growls of small dogs and big dogs differ because they have very different formants. The frequency of the formants is determined by the length of the vocal tract, and the vocal-tract length is a pretty good measure of the size of a dog. A larger dog has a longer vocal tract and a correspondingly lower-frequency formant, which sends an important auditory message to other dogs.

The masses of two different dogs and the frequencies of their formants are given in Table 16.1.

a. What is the approximate vocal-tract length of these two dogs? Assume that the speed of sound is 350 m/s at a dog's body temperature.
b. Growls aren't especially loud; at a distance of 1.0 m, a dog's growl is about 60 dB—the same as normal conversation. What is the acoustic power emitted by a 60 dB growl?
c. The lower-pitched growl of the Doberman will certainly sound more menacing, but which dog's 60 dB growl sounds louder to a human? **Hint:** Assume that most of the acoustic energy is at frequencies near the first formant; then look at the curve of equal perceived loudness in Figure 16.21.

TABLE 16.1 Mass and acoustic data for two growling dogs

Breed	Mass (kg)	First formant (Hz)	Second formant (Hz)
West Highland Terrier (Westie)	8.0	650	1950
Doberman	38	350	1050

PREPARE FIGURE 16.36 shows an idealized model of a dog's vocal tract as an open-closed tube. The formants will correspond to the standing-wave resonances of this system. The first two modes are shown in the figure. The first formant corresponds to $m = 1$; the second formant corresponds to $m = 3$ because an open-closed tube has only odd harmonics. The frequencies of the formants will allow us to determine the length of the tube that produces them.

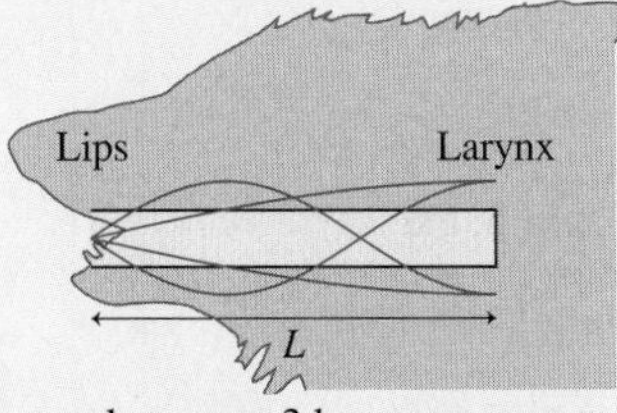

FIGURE 16.36 A model of the canine vocal tract.

For the question about loudness, we can determine the sound intensity from the sound intensity level. Knowing the distance, we can use this value to determine the emitted acoustic power.

SOLVE a. We can use the frequency of the first formant to find the length of the vocal tract. Recall that the standing-wave frequencies of an open-closed tube are given by

$$f_m = m\frac{v}{4L} \qquad m = 1, 3, 5, \ldots$$

Rearranging this to solve for L, with $m = 1$, we find

$$L(\text{Westie}) = m\frac{v}{4f_m} = (1)\frac{350\text{ m/s}}{4(650\text{ Hz})} = 0.13\text{ m}$$

$$L(\text{Doberman}) = m\frac{v}{4f_m} = (1)\frac{350\text{ m/s}}{4(350\text{ Hz})} = 0.25\text{ m}$$

The length is greater for the Doberman, as we would predict, given the relative masses of the dogs.

b. Equation 15.15 lets us compute the sound intensity for a given sound intensity level:

$$I = (1.0\times10^{-12}\text{ W/m}^2)10^{(\beta/10\text{dB})}$$
$$= (1.0\times10^{-12}\text{ W/m}^2)10^{(60\text{ dB}/10\text{dB})} = 1.0\times10^{-6}\text{ W/m}^2$$

This is the intensity at a distance of 1.0 m. The sound spreads out in all directions, so we can use Equation 15.12, $I = P_{\text{source}}/4\pi r^2$, to compute the power of the source:

$$P_{\text{source}} = I\cdot4\pi r^2 = (1.0\times10^{-6}\text{ W/m}^2)\cdot4\pi(1.0\text{ m})^2 = 13\ \mu\text{W}$$

A growl may sound menacing, but that's not because of its power!

c. Figure 16.21 is a curve of equal perceived loudness. The curve steadily decreases until reaching $\approx$3500 Hz, so the acoustic power needed to produce the same sensation of loudness decreases with frequency up to this point. That is, below 3500 Hz your ear is more sensitive to higher frequencies than to lower frequencies. Much of the acoustic power of a dog growl is concentrated at the frequencies of the first two formants. The Westie's growl has its energy at higher frequencies than that of the Doberman, so the Westie's growl will sound louder—though the lower formants of the Doberman's growl will make it sound more formidable.

ASSESS The lengths of the vocal tract that we calculated—13 cm (5 in) for a small terrier and 25 cm (10 in) for a Doberman—seem reasonable given the size of the dogs. We can also check our work by looking at the second formant, corresponding to $m = 3$; using this harmonic in the second expression in Equation 16.7, we find

$$L(\text{Westie}) = m\frac{v}{4f_m} = (3)\frac{350\text{ m/s}}{4(1950\text{ Hz})} = 0.13\text{ m}$$

$$L(\text{Doberman}) = m\frac{v}{4f_m} = (3)\frac{350\text{ m/s}}{4(1050\text{ Hz})} = 0.25\text{ m}$$

This exact match to our earlier calculations gives us confidence in our model and in our results.

SUMMARY

The goal of Chapter 16 has been to use the idea of superposition to understand the phenomena of interference and standing waves.

GENERAL PRINCIPLES

Principle of Superposition

The displacement of a medium when more than one wave is present is the sum of the displacements due to each individual wave.

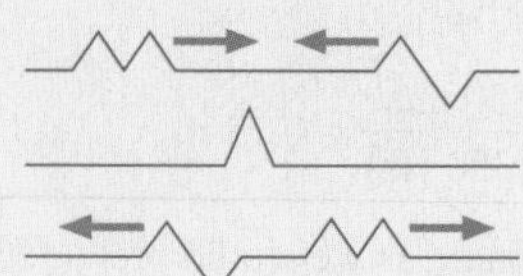

Interference

In general, the superposition of two or more waves into a single wave is called interference.

Constructive interference occurs when crests are aligned with crests and troughs with troughs. We say the waves are in phase. It occurs when the path-length difference Δd is a whole number of wavelengths.

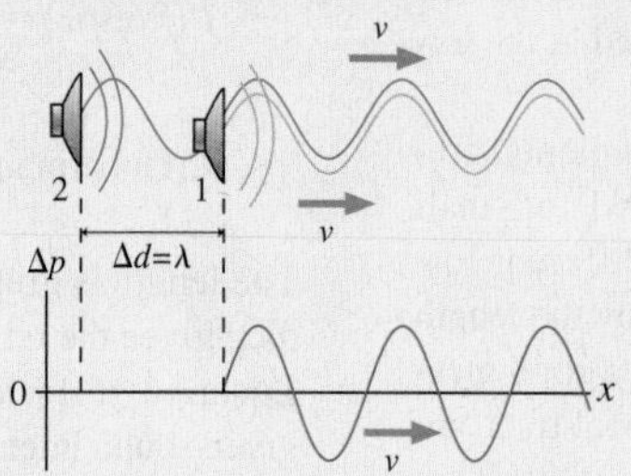

Destructive interference occurs when crests are aligned with troughs. We say the waves are out of phase. It occurs when the path-length difference Δd is a whole number of wavelengths plus half a wavelength.

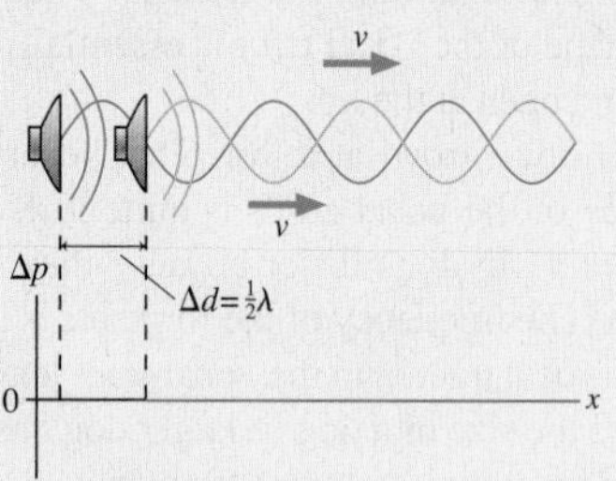

IMPORTANT CONCEPTS

Standing Waves

Two identical traveling waves moving in opposite directions create a standing wave.

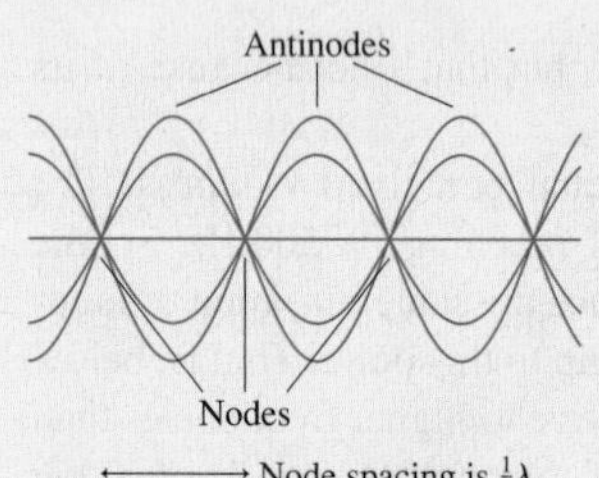

The boundary conditions determine which standing-wave frequencies and wavelengths are allowed. The allowed standing waves are **modes** of the system.

A **standing wave on a string** has a node at each end. Possible modes:

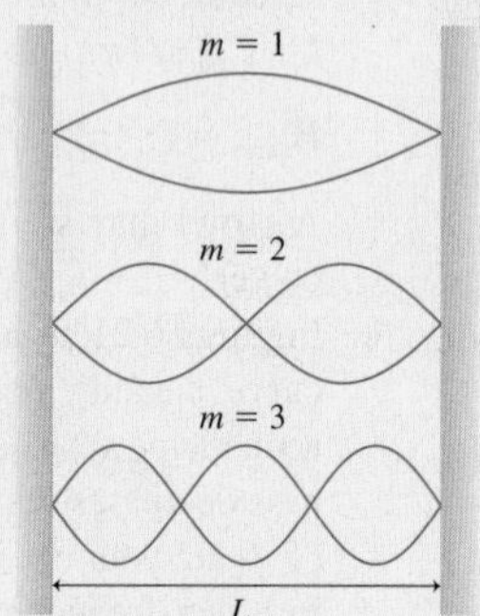

$$\lambda_m = \frac{2L}{m} \qquad f_m = m\left(\frac{v}{2L}\right) = mf_1$$

$$m = 1, 2, 3, \ldots$$

A **standing sound wave in a tube** can have different boundary conditions: open-open, closed-closed, or open-closed.

Open-open

$$f_m = m\left(\frac{v}{2L}\right)$$
$$m = 1, 2, 3, \ldots$$

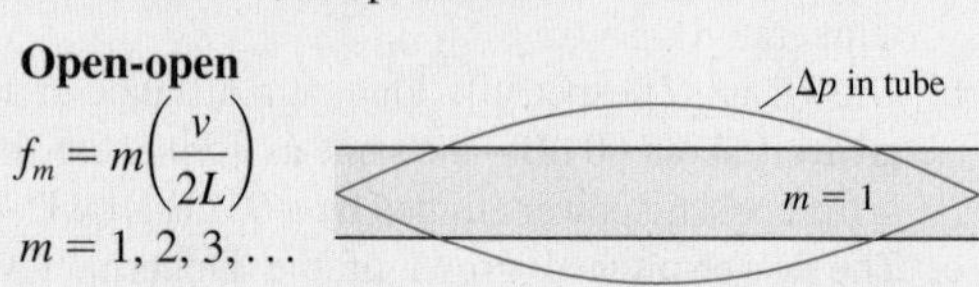

Closed-closed

$$f_m = m\left(\frac{v}{2L}\right)$$
$$m = 1, 2, 3, \ldots$$

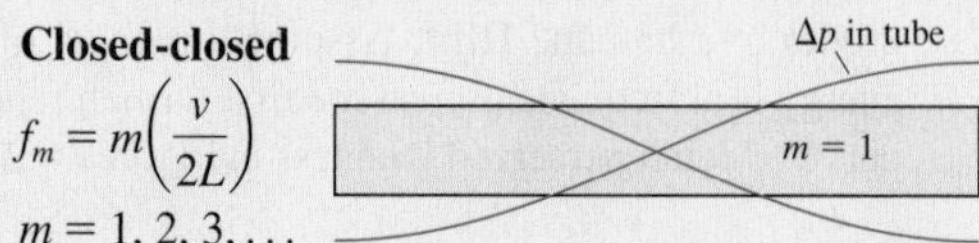

Open-closed

$$f_m = m\left(\frac{v}{4L}\right)$$
$$m = 1, 3, 5, \ldots$$

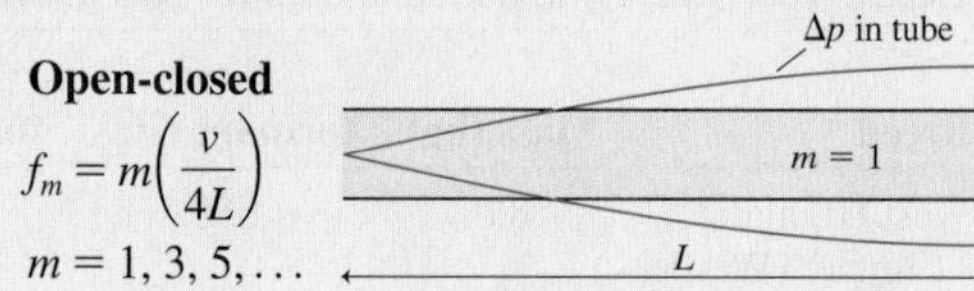

APPLICATIONS

Beats (loud-soft-loud-soft modulations of intensity) are produced when two waves of slightly different frequencies are superimposed.

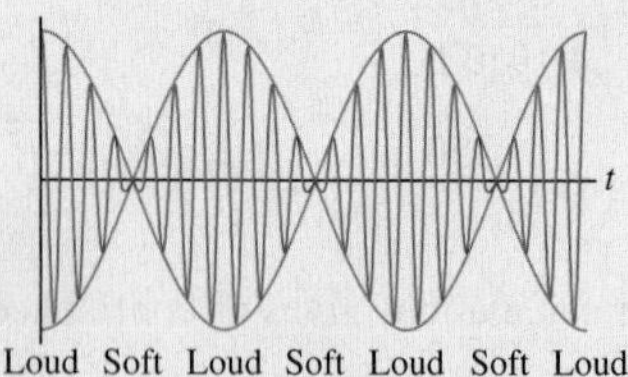

$$f_{\text{beat}} = |f_1 - f_2|$$

Standing waves are multiples of a **fundamental frequency,** the frequency of the lowest mode. The higher modes are the higher **harmonics.**

For sound, the fundamental frequency determines the perceived **pitch;** the higher harmonics determine the **tone quality.**

Our vocal cords create a range of harmonics. The mix of higher harmonics is changed by our vocal tract to create different vowel sounds.

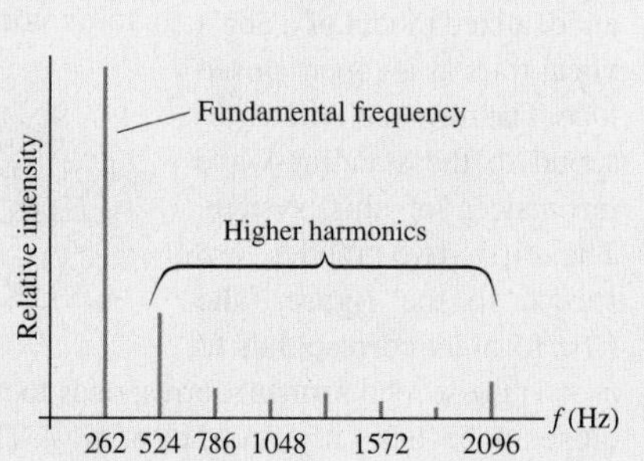

For homework assigned on MasteringPhysics, go to www.masteringphysics.com

Problems labeled INT integrate significant material from earlier chapters; BIO are of biological or medical interest.

Problem difficulty is labeled as | (straightforward) to ||||| (challenging).

QUESTIONS

Conceptual Questions

1. Light can pass easily through water and through air, but light will reflect from the surface of a lake. What does this tell you about the speed of light in air and in water?

2. Ocean waves are partially reflected from the entrance to a harbor, where the depth of the water is suddenly less. What does this tell you about the speed of waves in water of different depths?
3. A string has an abrupt change in linear density at its midpoint so that the speed of a pulse on the left side is 2/3 of that on the right side.
   a. On which side is the linear density greater? Explain.
   b. From which side would you start a pulse so that its reflection from the midpoint would not be inverted? Explain.
4. A guitarist finds that the frequency of one of her strings is too low by 1.4%. Should she increase or decrease the tension of the string? Explain.
5. BIO Certain illnesses inflame your vocal cords, causing them to swell. How does this affect the pitch of your voice? Explain.
6. Figure Q16.6 shows a standing wave on a string that is oscillating at frequency f_0. How many antinodes will there be if the frequency is doubled to $2f_0$? Explain.

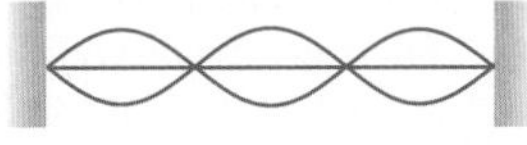

FIGURE Q16.6

7. Figure Q16.7 shows a standing sound wave in a tube of air that is open at both ends.
   a. Which mode (value of m) standing wave is this?
   b. Is the air vibrating horizontally or vertically?

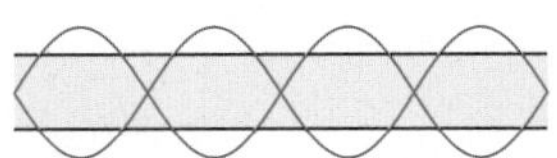

FIGURE Q16.7

8. A typical flute is about 66 cm long. A piccolo is a very similar instrument, though it is smaller, with a length of about 32 cm. How does the pitch of a piccolo compare to that of a flute?
9. Some pipes on a pipe organ are open at both ends, others are closed at one end. For pipes that play low-frequency notes, there is an advantage to using pipes that are closed at one end. What is the advantage?
10. A flute player tunes her instrument when the air (and the flute) is cold. As she plays, the flute and the air inside it warm up. Both the changing speed of sound in the air inside and the thermal expansion of the flute affect the frequency of the sound wave produced by the flute. After some time, she finds that her playing is "sharp"—the frequencies are too high. Which change produced this effect: the warming of the air or the warming of the body of the flute?
11. BIO A friend's voice sounds different over the telephone than it does in person. This is because telephones do not transmit frequencies over about 3000 Hz. 3000 Hz is well above the normal frequency of speech, so why does eliminating these high frequencies change the sound of a person's voice?
12. Suppose you were to play a trumpet after breathing helium, in which the speed of sound is much greater than in air. Would the pitch of the instrument be higher or lower than normal, or would it be unaffected by being played with helium inside the tube rather than air?
13. If you pour liquid in a tall, narrow glass, you may hear sound with a steadily rising pitch. What is the source of the sound, and why does the pitch rise as the glass fills?
14. BIO When you speak after breathing helium, in which the speed of sound is much greater than in air, your voice sounds quite different. The frequencies emitted by your vocal cords do not change since they are determined by the mass and tension of your vocal cords. So what *does* change when your vocal tract is filled with helium rather than air?
15. BIO Sopranos can sing notes at very high frequencies—over 1000 Hz. When they sing such high notes, it can be difficult to understand the words they are singing. Use the concepts of harmonics and formants to explain this.
16. When you hit a baseball with a bat, the bat flexes and then vibrates. We can model this vibration as a transverse standing wave. The modes of this standing wave are similar to the modes of a stretched string, but with one important difference: The ends of the bat are antinodes instead of nodes, because the ends of the bat are free to move. The modes thus look like the modes of a stretched string with antinodes replacing nodes and nodes replacing antinodes. If the ball hits the bat near an antinode of a standing-wave mode, the bat will start oscillating in this mode. The batter holds the bat at one end, which is also an antinode, so a large vibration of the bat causes an unpleasant vibration in the batter's hands. This can be avoided if the ball hits the bat at what players call the "sweet spot," which is a node of the standing-wave pattern. The first standing-wave mode of a vibrating bat is the $m = 2$ mode. Sketch the appearance of this vibrational mode of the bat, then estimate the approximate distance of the sweet spot (as a fraction of the bat's length) from the end of the bat.
17. BIO If a cold gives you a stuffed-up nose, it changes the way your voice sounds, even if your vocal cords are not affected. Explain why this is so.
18. BIO A small boy and a grown woman both speak at approximately the same pitch. Nonetheless, it's easy to tell which is which from listening to the sounds of their voices. How are you able to make this determination?

19. Figure Q16.19 shows wave fronts of a circular wave. Are the displacements at the following pairs of positions *in phase* or *out of phase?* Explain.
   a. A and B b. C and D c. E and F

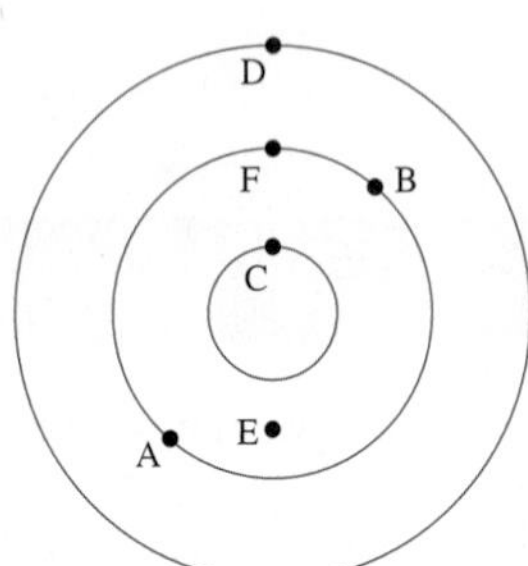

FIGURE Q16.19

Multiple-Choice Questions

Questions 20 through 22 refer to the snapshot graph Figure Q16.20.

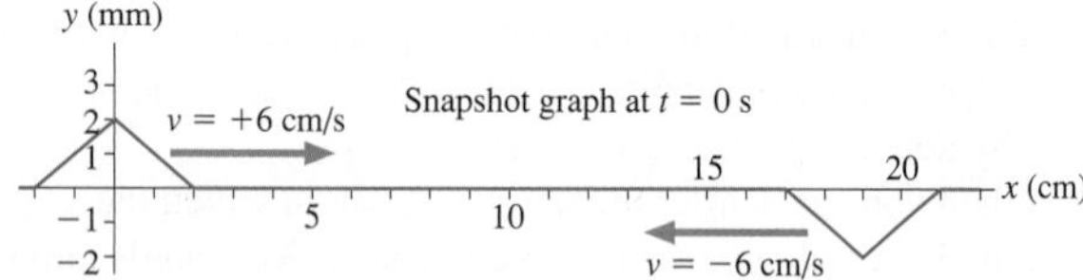

FIGURE Q16.20

20. | At $t = 1$ s, what is the displacement y of the string at $x = 7$ cm?
   A. −1.0 mm B. 0 mm C. 0.5 mm
   D. 1.0 mm E. 2.0 mm
21. | At $x = 3$ cm, what is the earliest time that y will equal 2 mm?
   A. 0.5 s B. 0.7 s C. 1.0 s
   D. 1.5 s E. 2.5 s
22. | At $t = 1.5$ s, what is the value of y at $x = 10$ cm?
   A. −2.0 mm B. −1.0 mm C. −0.5 mm
   D. 0 mm E. 1.0 mm
23. || Two sinusoidal waves with the same amplitude A and frequency f travel in opposite directions along a long string. You stand at one point and watch the string. The maximum displacement of the string at that point is
   A. A B. $2A$ C. 0
   D. There is not enough information to decide.
24. | A student in her physics lab measures the standing-wave modes of a tube. The lowest frequency that makes a resonance is 20 Hz. As the frequency is increased, the next resonance is at 60 Hz. What will be the next resonance after this?
   A. 80 Hz B. 100 Hz C. 120 Hz D. 180 Hz
25. | An organ pipe is tuned to exactly 384 Hz when the temperature in the room is 20°C. Later, when the air has warmed up to 25°C, the frequency is
   A. Greater than 384 Hz.
   B. 384 Hz.
   C. Less than 384 Hz.
26. || Two guitar strings made of the same type of wire have the same length. String 1 has a higher pitch than string 2. Which of the following is true?
   A. The wave speed of string 1 is greater than that of string 2.
   B. The tension in string 2 is greater than that in string 1.
   C. The wavelength of the lowest standing-wave mode on string 2 is longer than that on string 1.
   D. The wavelength of the lowest standing-wave mode on string 1 is longer than that on string 2.
27. | The frequency of the lowest standing-wave mode on a 1.0-m-long string is 20 Hz. What is the wave speed on the string?
   A. 10 m/s B. 20 m/s C. 30 m/s D. 40 m/s
28. | Suppose you pluck a string on a guitar and it produces the note A at a frequency of 440 Hz. Now you press your finger down on the string against one of the frets, making this point the new end of the string. The newly shortened string has 4/5 the length of the full string. When you pluck the string, its frequency will be
   A. 350 Hz B. 440 Hz C. 490 Hz D. 550 Hz

PROBLEMS

Section 16.1 The Principle of Superposition

1. | Figure P16.1 is a snapshot graph at $t = 0$ s of two waves on a taut string approaching each other at 1 m/s. Draw six snapshot graphs, stacked vertically, showing the string at 1 s intervals from $t = 1$ s to $t = 6$ s.

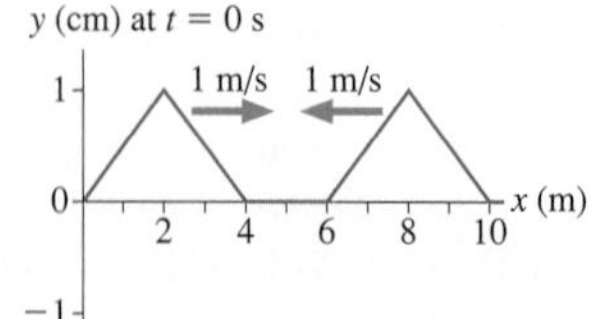

FIGURE P16.1

2. || Figure P16.2 is a snapshot graph at $t = 0$ s of two waves approaching each other at 1 m/s. Draw four snapshot graphs, stacked vertically, showing the string at $t = 2$, 4, 6, and 8 s.

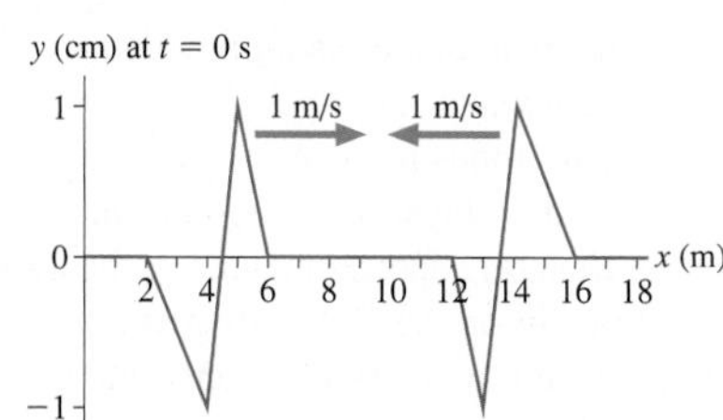

FIGURE P16.2

3. || Figure P16.3a is a snapshot graph at $t = 0$ s of two waves on a string approaching each other at At what time was the snapshot graph in Figure P16.3b taken?

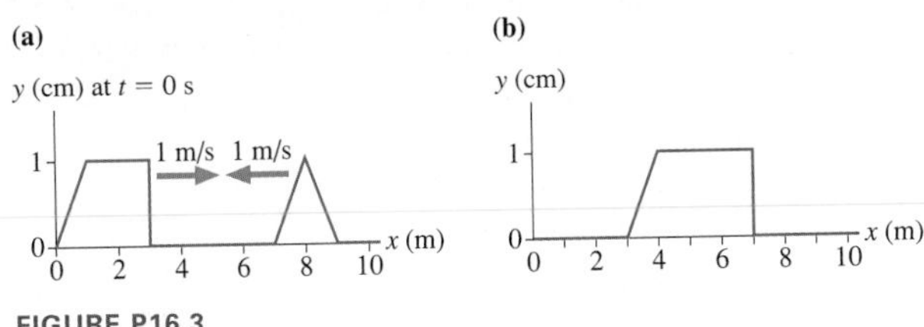

FIGURE P16.3

Section 16.2 Standing Waves

Section 16.3 Standing Waves on a String

4. ||| Figure P16.4 is a snapshot graph at $t = 0$ s of a pulse on a string moving to the right at 1 m/s. The string is fixed at $x = 5$ m. Draw a history graph spanning the time interval $t = 0$ s to $t = 10$ s for the location $x = 3$ m on the string.

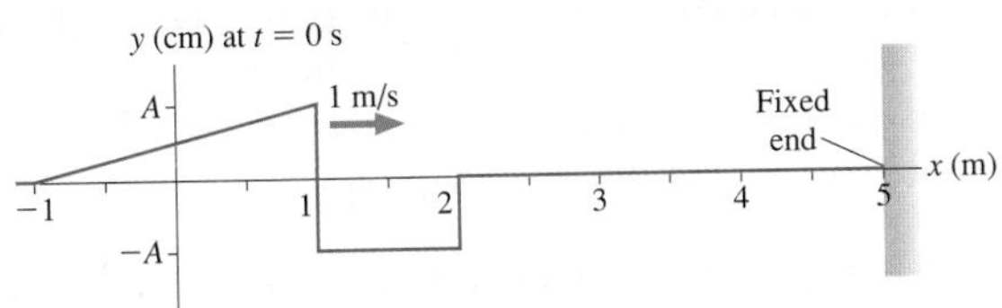

FIGURE P16.4

5. ||| At $t = 0$ s, a small "upward" (positive y) pulse centered at $x = 6.0$ m is moving to the right on a string with fixed ends at $x = 0.0$ m and $x = 10.0$ m. The wave speed on the string is 4.0 m/s. At what time will the string next have the same appearance that it did at $t = 0$ s?
6. ||| You are holding one end of an elastic cord that is fastened to a wall 3.0 m away. You begin shaking the end of the cord at 3.5 Hz, creating a continuous sinusoidal wave of wavelength 1.0 m. How much time will pass until a standing wave fills the entire length of the string?
7. ||| A 2.0-m-long string is fixed at both ends and tightened until the wave speed is 40 m/s. What is the frequency of the standing wave shown in Figure P16.7?

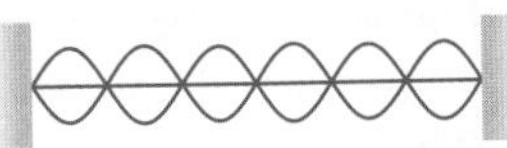

FIGURE P16.7

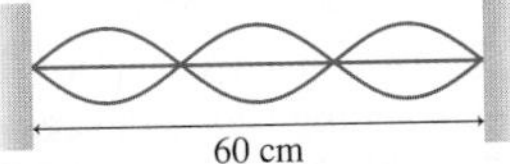

FIGURE P16.8

8. || Figure P16.8 shows a standing wave oscillating at 100 Hz on a string. What is the wave speed?
9. || A bass guitar string is 89 cm long with a fundamental frequency of 30 Hz. What is the wave speed on this string?
10. ||| The fundamental frequency of a guitar string is 384 Hz. What is the fundamental frequency if the tension in the string is halved?
11. || a. What are the three longest wavelengths for standing waves on a 240-cm-long string that is fixed at both ends?
    b. If the frequency of the second-longest wavelength is 50.0 Hz, what is the frequency of the third-longest wavelength?
12. || A 121-cm-long, 4.00 g string oscillates in its $m = 3$ mode with a frequency of 180 Hz and a maximum amplitude of 5.00 mm. What are (a) the wavelength and (b) the tension in the string?
13. || A guitar string with a linear density of 2.0 g/m is stretched between supports that are 60 cm apart. The string is observed to form a standing wave with three antinodes when driven at a frequency of 420 Hz. What are (a) the frequency of the fifth harmonic of this string and (b) the tension in the string?
14. || A violin string has a standard length of 32.8 cm. It sounds the musical note A (440 Hz) when played without fingering. How far from the end of the string should you place your finger to play the note C (523 Hz)?
15. || The lowest note on a grand piano has a frequency of 27.5 Hz. The entire string is 2.00 m long and has a mass of 400 g. The vibrating section of the string is 1.90 m long. What tension is needed to tune this string properly?

Section 16.4 Standing Sound Waves

16. | The lowest frequency in the audible range is 20 Hz. (a) What are the lengths of (a) the shortest open-open tube and (b) the shortest open-closed tube needed to produce this frequency?
17. | The contrabassoon is the wind instrument capable of sounding the lowest pitch in an orchestra. It is folded over several times to fit its impressive 18 ft length into a reasonable size instrument.
    a. If we model the instrument as an open-closed tube, what is its fundamental frequency? The sound speed inside is 350 m/s because the air is warmed by the player's breath.
    b. The actual fundamental frequency of the contrabassoon is 27.5 Hz, which should be different from your answer in part a. This means the model of the instrument as an open-closed tube is a bit too simple. But if you insist on using that model, what is the "effective length" of the instrument?
18. | Figure P16.18 shows a standing sound wave in an 80-cm-long tube. The tube is filled with an unknown gas. What is the speed of sound in this gas?

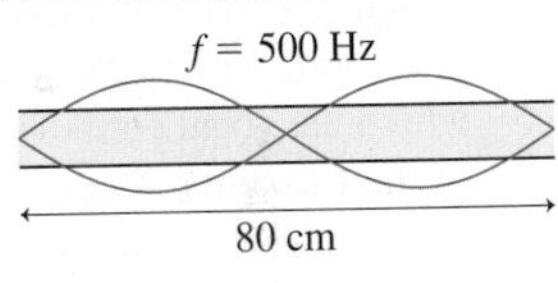

FIGURE P16.18

19. || What are the three longest wavelengths for standing sound waves in a 121-cm-long tube that is (a) open at both ends and (b) open at one end, closed at the other?
20. | The lowest pedal note on a large pipe organ has a fundamental frequency of 16 Hz. This extreme bass note is more felt as a rumble than heard with the ears. What is the length of the open-closed pipe that makes that note?
21. || The fundamental frequency of an open-open tube is 1500 Hz
INT when the tube is filled with 0°C helium. What is its frequency when filled with 0°C air?
22. | *Parasaurolophus* was a
BIO dinosaur whose distinguishing feature was a hollow crest on the head. The 1.5-m-long hollow tube in the crest had connections to the nose and throat, leading some investigators to hypothesize that the tube was a resonant chamber for vocalization. If you model the tube as an open-closed system, what are the first three resonant frequencies?

23. || A drainage pipe running under a freeway is 30.0 m long. Both ends of the pipe are open, and wind blowing across one end causes the air inside to vibrate.
   a. If the speed of sound on a particular day is 340 m/s, what will be the fundamental frequency of air vibration in this pipe?
   b. What is the frequency of the lowest harmonic that would be audible to the human ear?
   c. What will happen to the frequency in the later afternoon as the air begins to cool?

24. | Although the vocal tract is quite complicated, we can make a
BIO simple model of it as an open-closed tube extending from the opening of the mouth to the diaphragm, the large muscle separating the abdomen and the chest cavity. What is the length of this tube if its fundamental frequency equals a typical speech frequency of 200 Hz? Assume a sound speed of 350 m/s. Does this result for the tube length seem reasonable, based on observations on your own body?

25. |||| A child has an ear canal that is 1.3 cm long. At what sound
BIO frequencies in the audible range will the child have increased hearing sensitivity?

Section 16.5 Speech and Hearing

26. || The first formant of your vocal system can be modeled as the
BIO resonance of an open-closed tube, the closed end being your vocal cords and the open end your lips. Estimate the frequency of the first formant from the graph of Figure 16.23, and then estimate the length of the tube of which this is a resonance. Does your result seem reasonable?

27. ||| Deep-sea divers often breathe a mixture of helium and oxy-
BIO gen to avoid the complications of breathing high-pressure
INT nitrogen. At great depths the mix is almost entirely helium, which has the side effect of making the divers' voices sound very odd. Breathing helium doesn't affect the frequency at which the vocal cords vibrate, but it does affect the frequencies of the formants. The text gives the frequencies of the first two formants for an "ee" vowel sound as 270 and 2300 Hz. What will these frequencies be for a helium-oxygen mixture in which the speed of sound at body temperature is 750 m/s?

Section 16.6 The Interference of Waves from Two Sources

28. ||| Two loudspeakers in a 20°C room emit 686 Hz sound waves along the x-axis. What is the smallest distance between the speakers for which the interference of the sound waves is destructive?

29. ||| Two loudspeakers emit sound waves along the x-axis. The sound has maximum intensity when the speakers are 20 cm apart. The sound intensity decreases as the distance between the speakers is increased, reaching zero at a separation of 30 cm.
   a. What is the wavelength of the sound?
   b. If the distance between the speakers continues to increase, at what separation will the sound intensity again be a maximum?

30. |||| Two identical loudspeakers separated by distance d emit 170 Hz sound waves along the x-axis. As you walk along the axis, away from the speakers, you don't hear anything even though both speakers are on. What are three possible values for d? Assume a sound speed of 340 m/s.

31. | Figure P16.31 shows the circular wave fronts emitted by two sources. Make a table with rows labeled P, Q, and R and columns labeled r_1, r_2, Δr, and C/D. Fill in the table for points P, Q, and R, giving the distances as multiples of λ and indicating, with a C or a D, whether the interference at that point is constructive or destructive.

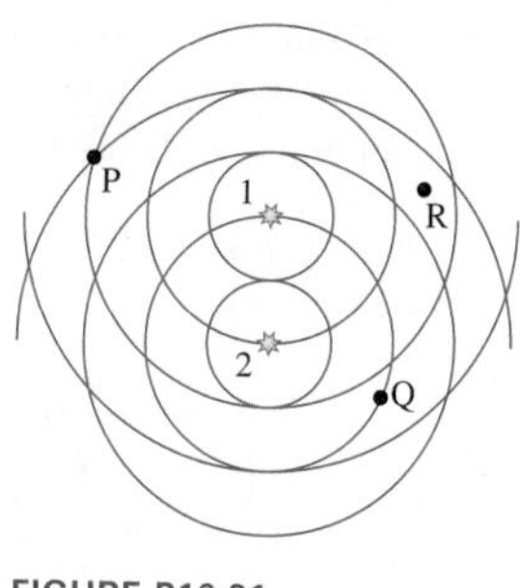

FIGURE P16.31

32. ||| Two identical loudspeakers 2.0 m apart are emitting 1800 Hz sound waves into a room where the speed of sound is 340 m/s. Is the point 4.0 m directly in front of one of the speakers, perpendicular to the plane of the speakers, a point of maximum constructive interference, perfect destructive interference, or something in between?

Section 16.7 Beats

33. | Two strings are adjusted to vibrate at exactly 200 Hz. Then the tension in one string is increased slightly. Afterward, three beats per second are heard when the strings vibrate at the same time. What is the new frequency of the string that was tightened?

34. || A flute player hears four beats per second when she compares her note to a 523 Hz tuning fork (the note C). She can match the frequency of the tuning fork by pulling out the "tuning joint" to lengthen her flute slightly. What was her initial frequency?

General Problems

35. | The fundamental frequency of a standing wave on a 1.0-m-long string is 440 Hz. What would be the wave speed of a pulse moving along this string?

36. ||| In addition to producing images, ultrasound can be used to
BIO heat tissues of the body for therapeutic purposes. When a sound
INT wave hits the boundary between soft tissue and air, or between soft tissue and bone, most of the energy is reflected; only 0.11% is transmitted. This means that standing waves can be set up in the body, creating excess thermal energy in the tissues at an antinode. Suppose 0.75 MHz ultrasound is directed through a layer of tissue with a bone 0.50 cm below the surface. Will standing waves be created? Explain.

37. ||| An 80-cm-long steel string with a linear density of 1.0 g/m is
INT under 200 N tension. It is plucked and vibrates at its fundamental frequency. What is the wavelength of the sound wave that reaches your ear in a 20°C room?

38. |||| Tendons are, essentially, elastic cords stretched between
BIO two fixed ends; as such, they can support standing waves.
INT These resonances can be undesirable. The Achilles tendon connects the heel with a muscle in the calf. A woman has a 20-cm-long tendon with a cross-section area of 110 mm^2. The density of tendon tissue is 1100 kg/m^3. For a reasonable tension of 500 N, what will be the resonant frequencies of her Achilles tendon?

39. | A string, stretched between two fixed posts, forms standing-wave resonances at 325 Hz and 390 Hz. What is the largest possible value of its fundamental frequency?

40. |||| Spiders may "tune" strands of their webs to give enhanced response at frequencies corresponding to the frequencies at which desirable prey might struggle. Orb web silk has a typical diameter of 0.0020 mm, and spider silk has a density of 1300 kg/m^3. To give a resonance at 100 Hz, to what tension must a spider adjust a 12-cm-long strand of silk?
BIO INT

41. ||| A violinist places her finger so that the vibrating section of a 1.0 g/m string has a length of 30 cm, then she draws her bow across it. A listener nearby in a 20°C room hears a note with a wavelength of 40 cm. What is the tension in the string?
INT

42. || A particularly beautiful note reaching your ear from a rare Stradivarius violin has a wavelength of 39.1 cm. The room is slightly warm, so the speed of sound is 344 m/s. If the string's linear density is 0.600 g/m and the tension is 150 N, how long is the vibrating section of the violin string?
INT

43. || A heavy piece of hanging sculpture is suspended by a 90-cm-long, 5.0 g steel wire. When the wind blows hard, the wire hums at its fundamental frequency of 80 Hz. What is the mass of the sculpture?
INT

44. | An experimenter finds that standing waves on a 0.80-m-long string, fixed at both ends, occur at 24 Hz and 32 Hz, but at no frequencies in between.
   a. What is the fundamental frequency?
   b. What is the wave speed on the string?
   c. Draw the standing-wave pattern for the string at 32 Hz.

45. ||| Astronauts visiting Planet X have a 2.5-m-long string whose mass is 5.0 g. They tie the string to a support, stretch it horizontally over a pulley 2.0 m away, and hang a 1.0 kg mass on the free end. Then the astronauts begin to excite standing waves on the string. Their data show that standing waves exist at frequencies of 64 Hz and 80 Hz, but at no frequencies in between. What is the value of g, the free-fall acceleration, on Planet X?
INT

46. |||| A 75 g bungee cord has an equilibrium length of 1.2 m. The cord is stretched to a length of 1.8 m, then vibrated at 20 Hz. This produces a standing wave with two antinodes. What is the spring constant of the bungee cord?
INT

47. ||||| A 2.5-cm-diameter steel cable (with density 7900 kg/m^3) that is part of the suspension system for a footbridge stretches 14 m between the tower and the ground. After walking over the bridge, a hiker finds that the cable is vibrating in its fundamental mode with a period of 0.40 s. What is the tension in the cable?

48. || Lake Erie is prone to remarkable *seiches*—standing waves that slosh water back and forth in the lake basin from the west end at Toledo to the east end at Buffalo. Figure P16.48 shows smoothed data for the displacement from normal water levels along the lake at the high point of one particular seiche. 3 hours later the water was at normal levels throughout the basin; 6 hours later the water was high in Toledo and low in Buffalo.
   a. What is the wavelength of this standing wave?
   b. What is the frequency?
   c. What is the wave speed?

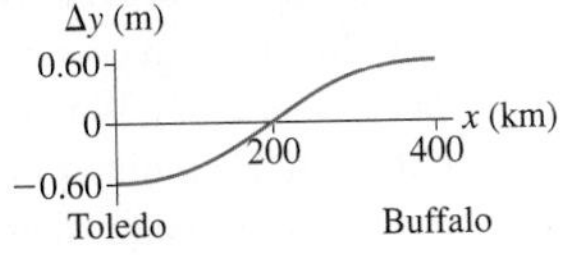

FIGURE P16.48

49. ||| A steel wire is used to stretch a spring, as shown in Figure P16.49. An oscillating magnetic field drives the steel wire back and forth. A standing wave with three antinodes is created when the spring is stretched 8.0 cm. What stretch of the spring produces a standing wave with two antinodes?
INT

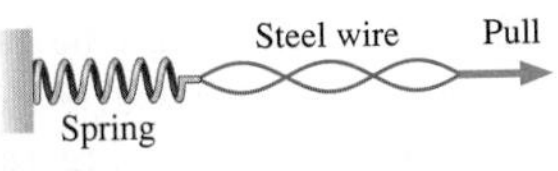

FIGURE P16.49

50. || Just as you are about to step into a nice hot bath, a small earthquake rattles your bathroom. Immediately afterward, you notice that the water in the tub is oscillating. The water in the center seems to be motionless while the water at the two ends alternately rises and falls, like a seesaw. You happen to know that your bathtub is 1.4 m long, and you count 10 complete oscillations of the water in 20 s.
   a. What is the wavelength of this standing wave?
   b. What is the speed of the waves that are reflecting back and forth inside the tub to create the standing wave?

51. ||| A microwave generator can produce microwaves at any frequency between 10 GHz and 20 GHz. As Figure P16.51 shows, the microwaves are aimed, through a small hole, into a "microwave cavity" that consists of a 10-cm-long cylinder with reflective ends.
   a. Which frequencies between 10 GHz and 20 GHz will create standing waves in the microwave cavity?
   b. For which of these frequencies is the cavity midpoint an antinode?

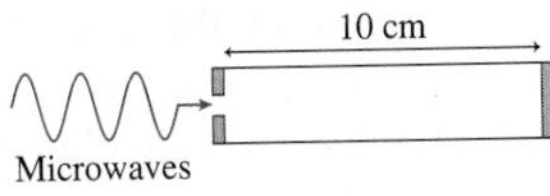

FIGURE P16.51

52. || An open-open organ pipe is 78.0 cm long. An open-closed pipe has a fundamental frequency equal to the third harmonic of the open-open pipe. How long is the open-closed pipe?

53. ||| A carbon-dioxide laser emits infrared light with a wavelength of 10.6 μm.
INT
   a. What is the length of a tube that will oscillate in the $m = 100{,}000$ mode?
   b. What is the frequency?
   c. Imagine a pulse of light bouncing back and forth between the ends of the tube. How many round trips will the pulse make in each second?

54. ||| In 1866, the German scientist Adolph Kundt developed a technique for accurately measuring the speed of sound in various gases. A long glass tube, known today as a Kundt's tube, has a vibrating piston at one end and is closed at the other. Very finely ground particles of cork are sprinkled in the bottom of the tube before the piston is inserted. As the vibrating piston is slowly moved forward, there are a few positions that cause the cork particles to collect in small, regularly spaced piles along the bottom. Figure P16.54 shows an experiment in which the tube is filled with pure oxygen and the piston is driven at 400 Hz.

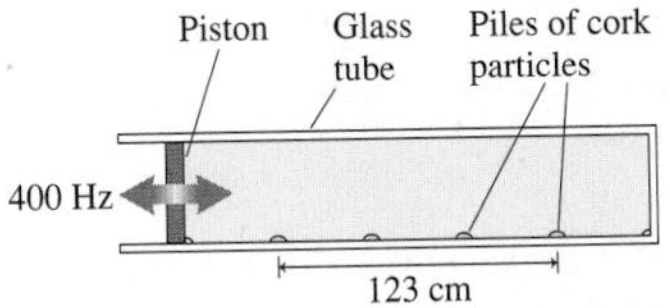

FIGURE P16.54

   a. Do the cork particles collect at standing-wave nodes or antinodes?
   Hint: Consider the appearance of the ends of the tube.
   b. What is the speed of sound in oxygen?

55. || A 40-cm-long tube has a 40-cm-long insert that can be pulled in and out, as shown in Figure P16.55. A vibrating tuning fork is held next to the tube. As the insert is slowly pulled out, the sound from the tuning fork creates standing waves in the tube when the total length L is 42.5 cm, 56.7 cm, and 70.9 cm. What is the frequency of the tuning fork? The air temperature is 20°C.

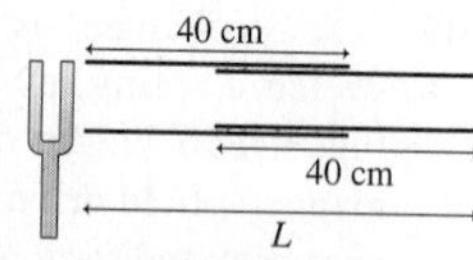

FIGURE P16.55

• 56. |||| A 1.0-m-tall vertical tube is filled with 20°C water. A tuning fork vibrating at 580 Hz is held just over the top of the tube as the water is slowly drained from the bottom. At what water heights, measured from the bottom of the tube, will there be a standing sound wave in the air at the top of the tube?

57. || A 50-cm-long wire with a mass of 1.0 g and a tension of 440 N passes across the open end of an open-closed tube of air. The wire, which is fixed at both ends, is bowed at the center so as to vibrate at its fundamental frequency and generate a sound wave. Then the tube length is adjusted until the fundamental frequency of the tube is heard. What is the length of the tube? Assume the speed of sound is 340 m/s.

58. ||| A 25-cm-long wire with a linear density of 20 g/m passes across the open end of an 85-cm-long open-closed tube of air. If the wire, which is fixed at both ends, vibrates at its fundamental frequency, the sound wave it generates excites the second vibrational mode of the tube of air. What is the tension in the wire? Assume the speed of sound is 340 m/s.

59. ||| Two loudspeakers located along the x-axis as shown in Figure P16.59 produce sounds of equal frequency. Speaker 1 is at the origin, while the location of speaker 2 can be varied by a remote control wielded by the listener. He notices maxima in the sound intensity when speaker 2 is located at $x = 0.75$ m and 1.00 m, but at no points in between. What is the frequency of the sound? Assume the speed of sound is 340 m/s.

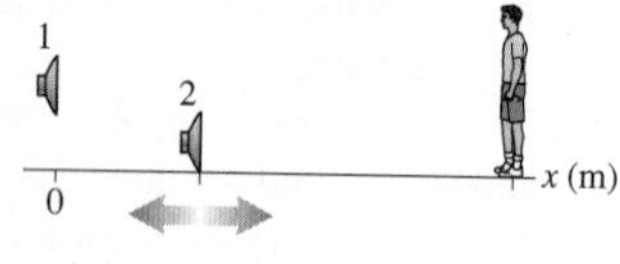

FIGURE P16.59

60. ||| You are standing 2.50 m directly in front of one of the two loudspeakers shown in Figure P16.60. They are 3.00 m apart and both are playing a 686 Hz tone in phase. As you begin to walk directly away from the speaker, at what distances from the speaker do you hear a *minimum* sound intensity? The room temperature is 20°C.

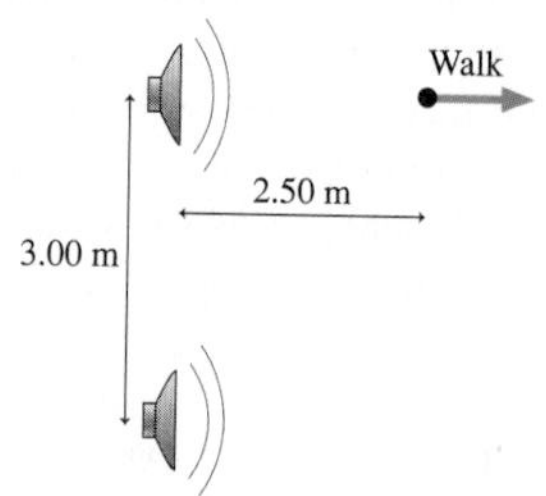

FIGURE P16.60

61. || FM station KCOM ("All commercials, all the time") transmits simultaneously, at a frequency of 99.9 MHz, from two broadcast towers placed precisely 31.5 m apart along a north-south line.
   a. What is the wavelength of KCOM's transmissions?
   b. Suppose you stand 90.0 m due east of the point halfway between the two towers with your portable FM radio. Will you receive a strong or weak signal at this position? Why?
   c. You then stand 90.0 m due north of the northern tower with your radio. Will you receive a strong or weak signal at this position? Why?

62. ||| Two loudspeakers, 4.0 m apart and facing each other, play identical sounds of the same frequency. You stand halfway between them, where there is a maximum of sound intensity. Moving from this point toward one of the speakers, you encounter a minimum of sound intensity when you have moved 0.25 m.
   a. What is the frequency of the sound?
   b. If the frequency is then increased while you remain 0.25 m from the center, what is the first frequency for which that location will be a maximum of sound intensity?

63. |||| Two radio antennas are separated by 2.0 m. Both broadcast identical 750 MHz waves. If you walk around the antennas in a circle of radius 10 m, how many maxima will you detect?

64. | Certain birds produce vocalizations consisting of two distinct
BIO frequencies that are not harmonically related—that is, the two frequencies are not harmonics of a common fundamental frequency. These two frequencies must be produced by two different vibrating structures in the bird's vocal tract.
   a. Wood ducks have been observed to make a call with approximately equal intensities at 850 Hz and 1200 Hz. The membranes that produce the vocalizations do not seem to vibrate at frequencies less than 500 Hz. Given this limitation, could these two frequencies be higher harmonics of a lower-frequency fundamental?
   b. If we model the duck's vocal tract as an open-closed tube, what length has a fundamental frequency equal to the lower of the two frequencies in part a?

65. || Piano tuners tune pianos by listening to the beats between the harmonics of two different strings. When properly tuned, the note A should have the frequency 440 Hz and the note E should be at 659 Hz. The tuner can determine this by listening to the beats between the third harmonic of the A and the second harmonic of the E.
   a. A tuner first tunes the A string very precisely by matching it to a 440 Hz tuning fork. She then strikes the A and E strings simultaneously and listens for beats between the harmonics. What beat frequency indicates that the E string is properly tuned?
   b. The tuner starts with the tension in the E string a little low, then tightens it. What is the frequency of the E string when she hears four beats per second?

66. ||| A flutist assembles her flute in a room where the speed of sound is 342 m/s. When she plays the note A, it is in perfect tune with a 440 Hz tuning fork. After a few minutes, the air inside her flute has warmed to where the speed of sound is 346 m/s.
   a. How many beats per second will she hear if she now plays the note A as the tuning fork is sounded?
   b. How far does she need to extend the "tuning joint" of her flute to be in tune with the tuning fork?

67. ||| A student waiting at a stoplight notices that her turn signal, which has a period of 0.85 s, makes one blink exactly in sync with the turn signal of the car in front of her. The blinker of the car ahead then starts to get ahead, but 17 s later the two are exactly in sync again. What is the period of the blinker of the other car?

68. ||| Musicians can use beats to tune their instruments. One flute is properly tuned and plays the musical note A at exactly 440 Hz. A second player sounds the same note and hears that her instrument is slightly "flat" (i.e., at too low a frequency). Playing at the same time as the first flute, she hears two loud-soft-loud beats per second. Must she shorten or lengthen her flute, and by how much, to bring it into tune? Assume a speed of sound of 350 m/s.

69. ||| Police radars determine speed by measuring the Doppler INT shift of radio waves reflected by a moving vehicle. They do so by determining the beat frequency between the reflected wave and the 10.5 GHz emitted wave. Some units can be calibrated by using a tuning fork; holding a vibrating fork in front of the unit causes the display to register a speed corresponding to the vibration frequency. A tuning fork is labeled "55 mph." What is the frequency of the tuning fork?

70. || A Doppler blood flow meter emits ultrasound at a frequency BIO INT of 5.0 MHz. What is the beat frequency between the emitted waves and the waves reflected from blood cells moving away from the emitter at 0.15 m/s?

71. | An ultrasound unit is being used to measure a patient's heart- BIO INT beat by combining the emitted 2.0 MHz signal with the sound waves reflected from the moving tissue of one point on the heart. The beat frequency between the two signals has a maximum value of 520 Hz. What is the maximum speed of the heart tissue?

Passage Problems

Harmonics and Harmony

You know that certain musical notes sound good together—harmonious—whereas others do not. This harmony is related to the various harmonics of the notes.

The musical notes C (262 Hz) and G (392 Hz) make a pleasant sound when played together; we call this consonance. As Figure P16.72 shows, the harmonics of the two notes are either far from each other or very close to each other (within a few Hz). This is the key to consonance: harmonics that are spaced either far apart or very close. The close harmonics have a beat frequency of a few Hz that is perceived as pleasant. If the harmonics of two notes are close but not too close, the rather high beat frequency between the two is quite unpleasant. This is what we hear as dissonance. Exactly how much a difference is maximally dissonant is a matter of opinion, but harmonic separations of 30 or 40 Hz seem to be quite unpleasant for most people.

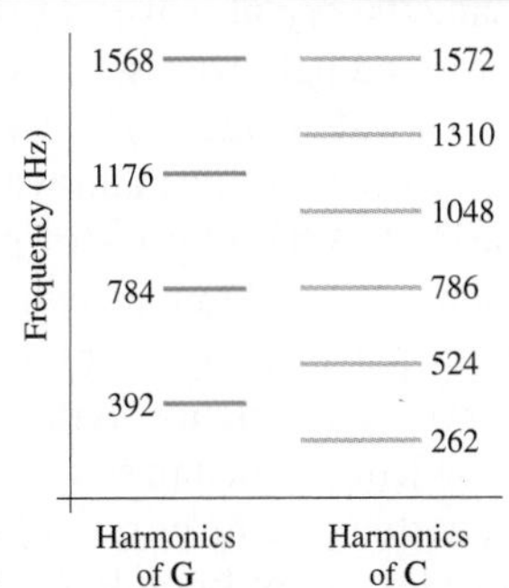

FIGURE P16.72

72. | What is the beat frequency between the second harmonic of G and the third harmonic of C?
A. 1 Hz B. 2 Hz C. 4 Hz D. 6 Hz

73. | Would a G-flat (frequency 370 Hz) and a C played together be consonant or dissonant?
A. Consonant
B. Dissonant

74. | An organ pipe open at both ends is tuned so that its fundamental frequency is a G. How long is the pipe?
A. 43 cm B. 87 cm C. 130 cm D. 173 cm

75. | If the C were played on an organ pipe that was open at one end and closed at the other, which of the harmonic frequencies in Figure 16.72 would be present?
A. All of the harmonics in the figure would be present.
B. 262, 786, and 1310 Hz
C. 524, 1048, and 1572 Hz
D. 262, 524, and 1048 Hz

STOP TO THINK ANSWERS

Stop to Think 16.1: C. The figure shows the two waves at $t = 6$ s and their superposition. The superposition is the *point-by-point* addition of the displacements of the two individual waves.

Stop to Think 16.2: C. Standing-wave frequencies are $f_m = mf_1$. The original wave has frequency $f_2 = 2f_1$ because it has two antinodes. The wave with frequency $2f_2 = 4f_1$ is the $m = 4$ mode with four antinodes.

Stop to Think 16.3: B. 300 Hz and 400 Hz are not f_1 and f_2 because 400 Hz $\neq 2 \times 300$ Hz. Instead, both are multiples of the fundamental frequency. Because the difference between them is 100 Hz, we see that $f_3 = 3 \times 100$ Hz and $f_4 = 4 \times 100$ Hz. Thus $f_1 = 100$ Hz.

Stop to Think 16.4: D. Highest pitch, or highest frequency, corresponds to the shortest period. For a complex wave, the period is the time required for the entire wave pattern to repeat.

Stop to Think 16.5: Constructive interference. The path-length difference is $\Delta r = 1.0 \text{ m} = \lambda$. Interference is constructive when the path-length difference is a whole number of wavelengths.

Stop to Think 16.6: F. The beat frequency is the difference between the two frequencies.

PART IV SUMMARY

Oscillations and Waves

As we have studied oscillations and waves, one point we have emphasized is the *unity* of the basic physics. The mathematics of oscillation describes the motion of a mass on a spring or the motion of a gibbon swinging from a branch. The same theory of waves works for string waves and sound waves and light waves. A few basic ideas enable us to understand a wide range of physical phenomena.

The physics of oscillations and waves is not quite as easily summarized as the physics of particles. Newton's laws and the conservation laws are two very general sets of principles about particles, principles that allowed us to develop the powerful problem-solving strategies of Parts I and II. The knowledge structure of oscillations and waves, shown below, rests more heavily on *phenomena* than on general principles. This knowledge structure doesn't contain a problem-solving strategy or a wide range of general principles, but is instead a logical grouping of the major topics you studied. This is a different way of structuring knowledge, but it still provides you with a mental framework for analyzing and thinking about wave problems.

The physics of oscillations and waves will be with us for the rest of the book. Part V is an exploration of optics, beginning with a detailed study of light as a wave. In Part VI, we will study the nature of electromagnetic waves in more detail. Finally, in Part VII, we will see that matter has a wave nature. There, the wave ideas we have developed in Part IV will lead us into the exciting world of quantum physics, finishing with some quite remarkable insights about the nature of light and matter.

KNOWLEDGE STRUCTURE IV Oscillations and Waves

BASIC GOALS
How can we describe oscillatory motion?
How does a wave travel through a medium?
What are the distinguishing features of waves?
What happens when two waves meet?

GENERAL PRINCIPLES
Simple harmonic motion occurs when a linear restoring force acts to return a system to equilibrium.
The period of simple harmonic motion depends on physical properties of the system but not the amplitude.
All sinusoidal waves, whether water waves, sound waves, or light waves, have the same functional form.
When two waves meet, they pass through each other, combining where they overlap by superposition.

Simple harmonic motion

$x(t) = A\cos(2\pi f t)$

$v_x(t) = -v_{max}\sin(2\pi f t)$

$v_{max} = 2\pi f A$

$a_x(t) = -a_{max}\cos(2\pi f t)$

$a_{max} = (2\pi f)^2 A$

The frequency of a mass on a spring depends on the spring constant and the mass:

$$f = \frac{1}{2\pi}\sqrt{\frac{k}{m}}$$

The frequency of a pendulum depends on the length and the free-fall acceleration:

$$f = \frac{1}{2\pi}\sqrt{\frac{g}{L}}$$

Traveling waves

All traveling waves are described by the same equation:

$$y = A\cos\left(2\pi\left(\frac{x}{\lambda} \pm \frac{t}{T}\right)\right)$$

+: wave travels to left
−: wave travels to right

The wave speed depends on the properties of the medium. $v_{string} = \sqrt{\dfrac{T_s}{\mu}}$

The speed, wavelength, period, and frequency of a wave are related. $T = \dfrac{1}{f}$ $\quad v = f\lambda$

Superposition and interference

The superposition of two waves is the sum of the displacements of the two individual waves.

Constructive interference occurs when crests are aligned with crests and troughs with troughs. Constructive interference occurs when the path-length difference is a whole number of wavelengths.

Destructive interference occurs when crests are aligned with troughs. Destructive interference occurs when the path-length difference is a whole number of wavelengths plus half a wavelength.

Standing waves

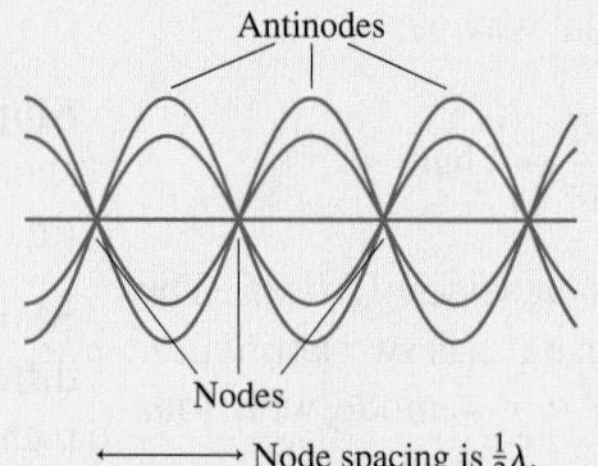

Standing waves are due to the superposition of two traveling waves moving in opposite directions.

The boundary conditions determine which standing-wave frequencies and wavelengths are allowed. The allowed standing waves are modes of the system.

For a string of length L, the modes have wavelength and frequency

$$\lambda_m = \frac{2L}{m} \qquad f_m = m\left(\frac{v_{string}}{2L}\right) = mf_1 \qquad m = 1, 2, 3, 4, \ldots$$

Waves in the Earth and the Ocean

In December 2004, a large earthquake off the coast of Indonesia produced a devastating water wave, called a *tsunami,* that caused tremendous destruction thousands of miles away from the earthquake's epicenter. The tsunami was a dramatic illustration of the energy carried by waves.

It was also a call to action. Many of the communities hardest hit by the tsunami were struck hours after the waves were generated, long after seismic waves from the earthquake that passed through the earth had been detected at distant recording stations, long after the possibility of a tsunami was first discussed. With better detection and more accurate models of how a tsunami is formed and how a tsunami propagates, the affected communities could have received advance warning. The study of physics may seem an abstract undertaking with few practical applications, but on this day a better scientific understanding of these waves could have averted tragedy.

Let's use our knowledge of waves to explore the properties of a tsunami. In Chapter 15, we saw that a vigorous shake of one end of a rope causes a pulse to travel along it, carrying energy as it goes. The earthquake that produced the Indian Ocean tsunami of 2004 caused a sudden upward displacement of the seafloor that produced a corresponding rise in the surface of the ocean. This was the *disturbance* that produced the tsunami, very much like a quick shake on the end of a rope. The resulting wave propagated through the ocean, as we see in the figure.

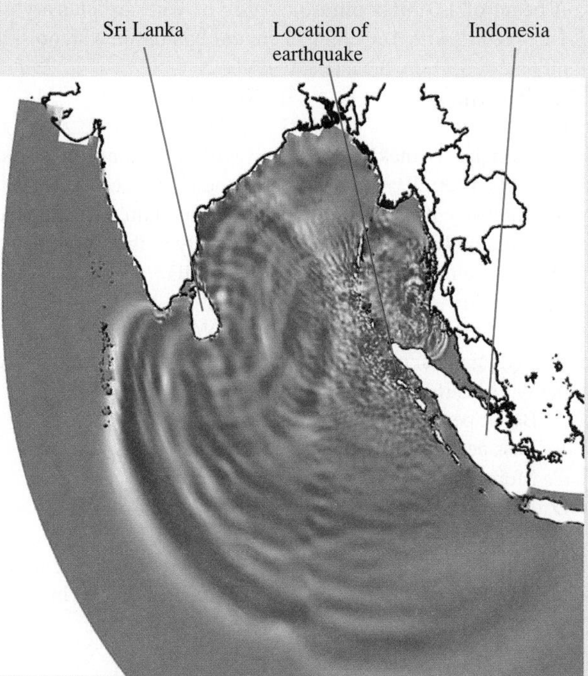

One frame from a computer simulation of the Indian Ocean tsunami three hours after the earthquake that produced it. The disturbance propagating outward from the earthquake is clearly seen, as are wave reflections from the island of Sri Lanka.

This simulation of the tsunami looks much like the ripples that spread when you drop a pebble into a pond. But there is a big difference—the scale. The fact that you can see the individual waves on this diagram that spans 5000 km is quite revealing. To show up so clearly, the individual wave pulses must be very wide—up to hundreds of kilometers from front to back.

A tsunami is actually a "shallow water wave," even in the deep ocean, because the depth of the ocean is much less than the width of the wave. Consequently, a tsunami travels differently than normal ocean waves. In Chapter 15 we learned that wave speeds are fixed by the properties of the medium. That is true for normal ocean waves, but the great width of the wave causes a tsunami to "feel the bottom." Its wave speed is determined by the depth of the ocean: The greater the depth, the greater the speed. In the deep ocean, a tsunami travels at hundreds of kilometers per hour, much faster than a typical ocean wave. Near shore, as the ocean depth decreases, so does the speed of the wave.

The height of the tsunami in the open ocean was about half a meter. Why should such a small wave—one that ships didn't even notice as it passed—be so fearsome? Again, it's the *width* of the wave that matters. Because a tsunami is the wave motion of a considerable mass of water, great energy is involved. As the front of a tsunami wave nears shore, its speed decreases, and the back of the wave moves faster than the front. Consequently, the width decreases. The water begins to pile up, and the wave dramatically increases in height.

The Indian Ocean tsunami had a height of up to 15 m when it reached shore, with a width of up to several kilometers. This tremendous mass of water was still moving at high speed, giving it a great deal of energy. A tsunami reaching the shore isn't like a typical wave that breaks and crashes. It is a kilometers-wide wall of water that moves onto the shore and just keeps on coming. In many places, the water reached 2 km inland.

The impact of the Indian Ocean tsunami was devastating, but it was the first tsunami for which scientists were able to use satellites and ocean sensors to make planet-wide measurements. An analysis of the data has helped us better understand the physics of these ocean waves. We won't be able to stop future tsunamis, but with a better knowledge of how they are formed and how they travel, we will be better able to warn people to get out of their way.

PART IV PROBLEMS

The following questions are related to the passage "Waves in the Earth and the Ocean" on the previous page.

1. Rank from fastest to slowest the following waves according to their speed of propagation:
   A. An earthquake wave B. A tsunami
   C. A sound wave in air D. A light wave
2. The increase in height as a tsunami approaches shore is due to
   A. The increase in frequency as the wave approaches shore.
   B. The increase in speed as the wave approaches shore.
   C. The decrease in speed as the wave approaches shore.
   D. The constructive interference with the wave reflected from shore.
3. In the middle of the Indian Ocean, the tsunami referred to in the passage was a train of pulses approximating a sinusoidal wave with speed 200 m/s and wavelength 150 km. What was the approximate period of these pulses?
   A. 1 min B. 3 min
   C. 5 min D. 15 min
4. If a train of pulses moves into shallower water as it approaches a shore,
   A. The wavelength increases.
   B. The wavelength stays the same.
   C. The wavelength decreases.
5. The tsunami described in the passage produced a very erratic pattern of damage, with some areas seeing very large waves and nearby areas seeing only small waves. Which of the following is a possible explanation?
   A. Certain areas saw the wave from the primary source, others only the reflected waves.
   B. The superposition of waves from the primary source and reflected waves produced regions of constructive and destructive interference.
   C. A tsunami is a standing wave, and certain locations were at nodal positions, others at antinodal positions.

The following passages and associated questions are based on the material of Part IV.

Deep-Water Waves

Water waves are called *deep-water waves* when the depth of the water is much greater than the wavelength of the wave. The speed of deep-water waves depends on the wavelength as follows:

$$v = \sqrt{\frac{g\lambda}{2\pi}}$$

Suppose you are on a ship at rest in the ocean, observing the crests of a passing sinusoidal wave. You estimate that the crests are 75 m apart.

6. Approximately how much time elapses between one crest reaching your ship and the next?
   A. 3 s
   B. 5 s
   C. 7 s
   D. 12 s
7. The captain starts the engines and sails directly opposite the motion of the waves at 4.5 m/s. Now how much time elapses between one crest reaching your ship and the next?
   A. 3 s
   B. 5 s
   C. 7 s
   D. 12 s
8. In the deep ocean, a longer-wavelength wave travels faster than a shorter-wavelength wave. Thus, a higher-frequency wave travels ______ a lower-frequency wave.
   A. Faster than
   B. At the same speed as
   C. Slower than

Attenuation of Ultrasound BIO

Ultrasound is absorbed in the body; this complicates the use of ultrasound to image tissues. The intensity of a beam of ultrasound decreases by a factor of 2 after traveling a distance of 40 wavelengths. Each additional travel of 40 wavelengths results in a decrease by another factor of 2.

9. A beam of 1.0 MHz ultrasound begins with an intensity of 1000 W/m^2. After traveling 12 cm through tissue with no significant reflection, the intensity is about
   A. 750 W/m^2 B. 500 W/m^2
   C. 250 W/m^2 D. 125 W/m^2
10. A physician is making an image with ultrasound of initial intensity 1000 W/m^2. When the frequency is set to 1.0 MHz, the intensity drops to 500 W/m^2 at a certain depth in the patient's body. What will be the intensity at this depth if the physician changes the frequency to 2.0 MHz?
    A. 750 W/m^2 B. 500 W/m^2
    C. 250 W/m^2 D. 125 W/m^2
11. A physician is using ultrasound to make an image of a patient's heart. Increasing the frequency will provide
    A. Better penetration and better resolution.
    B. Less penetration but better resolution.
    C. More penetration but worse resolution.
    D. Less penetration and worse resolution.
12. A physician is using Doppler ultrasound to measure the motion of a patient's heart. The device measures the beat frequency between the emitted and the reflected waves. Increasing the frequency of the ultrasound will
    A. Increase the beat frequency.
    B. Not affect the beat frequency.
    C. Decrease the beat frequency.

Measuring the Speed of Sound

A student investigator is measuring the speed of sound by looking at the time for a brief, sinusoidal pulse from a loudspeaker to travel down a tube, reflect from the closed end, and reach a microphone. The apparatus is shown in Figure IV.1a; typical data recorded by the microphone are graphed in Figure IV.1b. The first pulse is the sound directly from the loudspeaker; the second pulse is the reflection from the closed end. A portion of the returning wave reflects from the open end of the tube and makes another round trip before being detected by the microphone; this is the third pulse seen in the data.

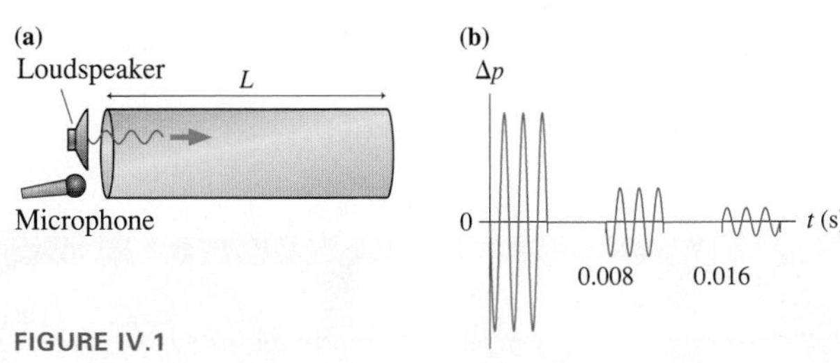

FIGURE IV.1

13. What was the approximate frequency of the sound wave used in this experiment?
    A. 250 Hz
    B. 500 Hz
    C. 750 Hz
    D. 1000 Hz
14. What can you say about the reflection of sound waves at the ends of a tube?
    A. Sound waves are inverted when reflected both from open and closed tube ends.
    B. Sound waves are inverted when reflected from a closed end, not inverted when reflected from an open end.
    C. Sound waves are inverted when reflected from an open end, not inverted when reflected from a closed end.
    D. Sound waves are not inverted when reflected either from open or closed tube ends.
15. What was the approximate length of the tube?
    A. 0.35 m
    B. 0.70 m
    C. 1.4 m
    D. 2.8 m
16. An alternative technique to determine sound speed is to measure the frequency of a standing wave in the tube. What is the wavelength of the lowest resonance of this tube?
    A. $L/2$
    B. L
    C. $2L$
    D. $4L$

In the Swing

A rope swing is hung from a tree right at the edge of a small creek. The rope is 5.0 m long; the creek is 3.0 m wide.

17. You sit on the swing, and your friend gives you a gentle push so that you swing out over the creek. How long will it be until you swing back to where you started?
    A. 4.5 s B. 3.4 s
    C. 2.2 s D. 1.1 s
18. Now you switch places with your friend, who has twice your mass. You give your friend a gentle push so that he swings out over the creek. How long will it be until he swings back to where he started?
    A. 4.5 s B. 3.4 s
    C. 2.2 s D. 1.1 s
19. Your friend now pushes you over and over, so that you swing higher and higher. At some point you are swinging all the way across the creek—at the top point of your arc you are right above the opposite side. How fast are you moving when you get back to the lowest point of your arc?
    A. 6.3 m/s B. 5.4 m/s
    C. 4.4 m/s D. 3.1 m/s

Additional Integrated Problems

20. BIO The jumping gait of the kangaroo is efficient because energy is stored in the stretch of stout tendons in the legs; the kangaroo literally bounces with each stride. We can model the bouncing of a kangaroo as the bouncing of a mass on a spring. A 70 kg kangaroo hits the ground, the tendons stretch to a maximum length, and the rebound causes the kangaroo to leave the ground approximately 0.10 s after its feet first touch.
    a. Modeling this as the motion of a mass on a spring, what is the period of the motion?
    b. Given the kangaroo mass and the period you've calculated, what is the spring constant?
    c. If the kangaroo speeds up, it must bounce higher and farther with each stride, and so must store more energy in each bounce. How does this affect the time and the amplitude of each bounce?
21. BIO A brand of earplugs reduces the sound intensity level by 27 dB. By what factor do these earplugs reduce the acoustic intensity?
22. BIO Sperm whales, just like bats, use echolocation to find prey. A sperm whale's vocal system creates a single sharp click, but the emitted sound consists of several equally spaced clicks of decreasing intensity. Researchers use the time interval between the clicks to estimate the size of the whale that created them. Explain how this might be done. **Hint:** The head of a sperm whale is complex, with air pockets at either end.

PART

V

Optics

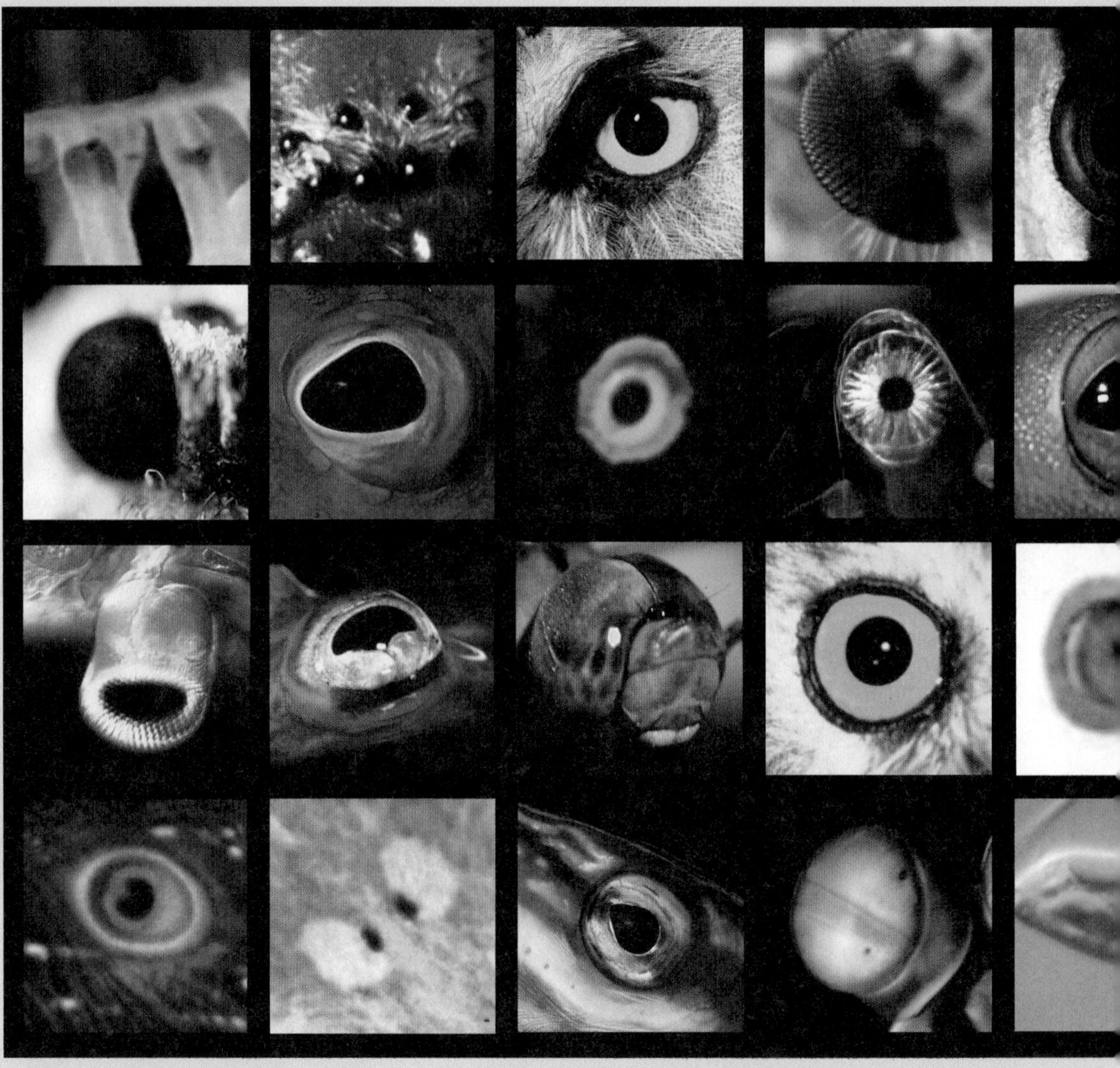

These images of animal eyes and light-sensing organs reveal the wide range of structures that can produce a visual sense. There is a certain similarity among eyes as well; did you spot the false eyes, patterns that are designed to mimic the appearance of an eye?

OVERVIEW

Light Is a Wave

Isaac Newton is best known for his studies of mechanics and the three laws that bear his name, but he also did important early work on optics. He was the first person to carefully study how a prism breaks white light into colors. Newton was a strong proponent of the "corpuscle" theory of light, arguing that light consists of a stream of particles.

In fact, Newton wasn't quite right. As you will see, the beautiful colors of a peacock's feathers and the shimmery rainbow of a soap bubble both depend on the fact that light is a wave, not a particle. In particular, light is an electromagnetic wave, although these chapters depend on nothing more than the "waviness" of light waves for your understanding. The wave theory we developed in Part IV will be put to good use in Part V as we begin our investigation of light and optics with an analysis of the *wave model* of light.

The Ray Model

Yet Newton was correct in his observation that light seems to travel in straight lines, something we wouldn't expect a wave to do. Consequently, our investigations of how light works will be greatly aided by another model of light, the *ray model,* in which light travels in straight lines, bounces from mirrors, and is bent by lenses.

The ray model will be an excellent tool for analyzing many of the practical applications of optics. When you look in a mirror, you see an *image* of yourself that appears to be behind the mirror. We will use the ray model of light to determine just how it is that mirrors and lenses form images. At the same time, we will need to reconcile the wave and ray models, learning how they are related to each other and when it is appropriate to use each.

Working with Light

The nature of light is quite subtle and elusive. In Parts VI and VII, we will turn to the question of just what light is. As we will see, light has both wave-like *and* particle-like aspects. For now, however, we will set this question aside and work with the wave and ray models to develop a practical understanding of light. This will lead us, in Chapter 19, to an analysis of some common optical instruments. We will explore how a camera captures images and how telescopes and microscopes work.

Ultimately, the fact that you are reading this book is due to the optics of the first optical instrument you ever used, your eye! We will investigate the optics of the eye, learn how the cornea and lens bend light to create an image on your retina, and see how glasses or contact lenses can be used to correct the image should it be out of focus.

17 Wave Optics

The vivid colors of this hummingbird's feathers have a sheen unlike that of ordinary pigments, and they change subtly depending on the angle at which they're viewed. How does light interact with the feathers to produce this bright display?

LOOKING AHEAD ►►

The goal of Chapter 17 is to understand and use the wave model of light.

What Is Light?

You'll learn that light has aspects of waves, rays, and particles. We'll develop models for each of these in this and coming chapters; this chapter will concentrate on the **wave model of light**.

The colors of soap bubbles can be understood using the **wave model** of light.

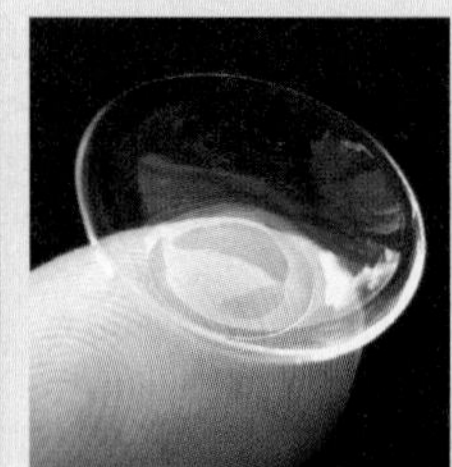

To understand the bending of light by a contact lens, the **ray model** (Chapter 18) is appropriate.

Solar cells generate electricity from sunlight. We'll use the **photon model** (Chapter 28) to understand how.

Looking Back ◄◄
15.4 Light and electromagnetic waves

Diffraction

One of the most basic aspects of waves is that they can bend, or *diffract,* around the edges of objects. The diffraction of light, although difficult to observe because of light's small wavelength, is an indication that light is a wave.

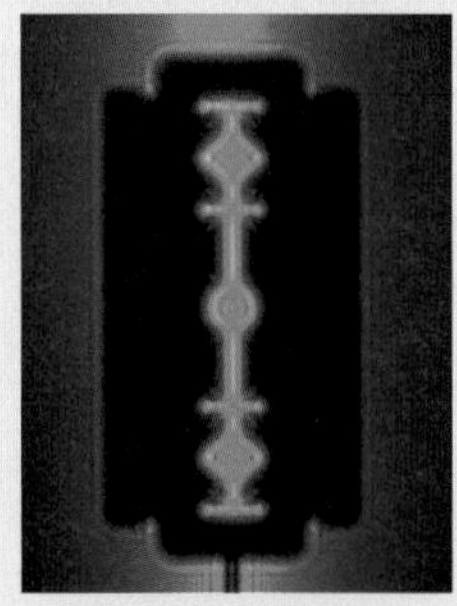

A careful examination of the shadow of this razor blade shows fringes due to diffraction of light.

Interference of Light

Like all waves, light waves of the same frequency can interfere constructively or destructively.

Double-slit interference

When light shines on two narrow, closely spaced slits, interference fringes are seen on a screen behind the slits. Interference is a clear indication of the wave nature of light.

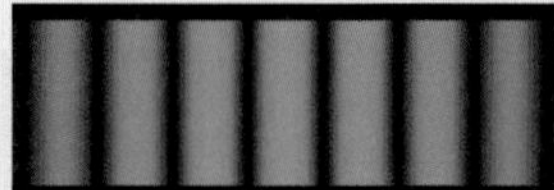

Fringes seen with green light.

The diffraction grating

Many closely spaced slits or grooves form a **diffraction grating,** capable of breaking white light into its component colors.

The microscopic grooves in a CD act as a diffraction grating, leading to its colorful appearance.

Thin-film interference

Interference is also possible between waves reflecting off the front and back surfaces of a thin transparent film.

Looking Back ◄◄
15.5 Circular and plane waves
16.3 Reflections
16.6 Interference

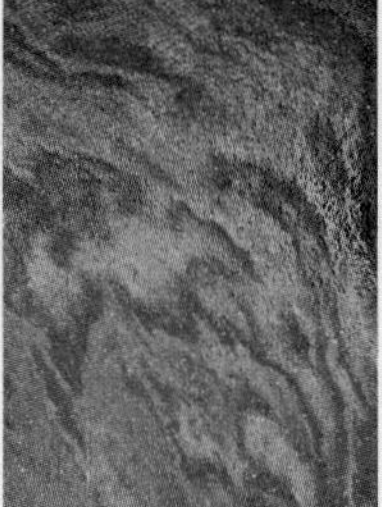

The colors in an oil slick result from thin-film interference.

17.1 What Is Light?

What is light? The first Greek scientists and philosophers did not make a distinction between light and vision. Light, to them, was not something that existed apart from seeing. But gradually there arose a view that light actually "exists," that light is some sort of physical entity that is present regardless of whether or not someone is looking. Newton, in addition to doing pioneering work in mathematics and mechanics in the 1660s, was also one of the early investigators of light. He believed that light consists of very small, light, fast particles, which he called *corpuscles,* traveling in straight lines. Newton was vigorously opposed by Robert Hooke (of Hooke's law) and the Dutch scientist Christiaan Huygens, who argued that light was some sort of wave. Although the debate was lively, and sometimes acrimonious, Newton eventually prevailed. The belief that light consists of corpuscles was not seriously questioned for more than a hundred years after Newton's death.

The situation changed dramatically in 1801, when the English scientist Thomas Young announced that he had produced *interference* between two waves of light. Young's experiment, which we will analyze in this chapter, was a painstakingly difficult experiment with the technology of his era. Nonetheless, Young's experiment appeared to settle the debate in favor of a wave theory of light because interference is a distinctly wave-like phenomenon. But if light is a wave, why does it sometimes seem to travel in straight lines? And just what kind of wave is it?

Models of Light

As Newton, Young, and others found, the nature of light is elusive. Light is the chameleon of the physical world. Under some circumstances, light acts like particles traveling in straight lines. But change the circumstances, and light shows the same kinds of wave-like behavior as sound waves or water waves. Change the circumstances yet again, and light exhibits behavior that is neither wave-like nor particle-like but has characteristics of both.

Rather than an all-encompassing "theory of light," it will be better to develop three **models of light.** Each model successfully explains the behavior of light within a certain domain—that is, within a certain range of physical situations. Our task will be twofold:

1. To develop clear and distinct models of light.
2. To learn the conditions and circumstances for which each model is valid.

The second task is especially important.

We'll begin with a brief summary of all three models, so that you will have a road map of where we're headed. Each of these models will be developed in the coming chapters.

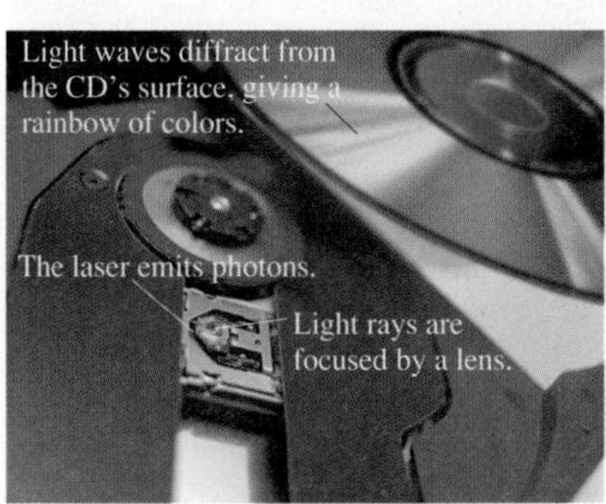

Light waves diffract from the CD's surface, giving a rainbow of colors.
The laser emits photons.
Light rays are focused by a lens.

All three models of light are needed to explain the operation of a CD player and the rainbow colors of the CD itself.

The wave model: The wave model of light is the most widely applicable model, responsible for the widely known "fact" that light is a wave. It is certainly true that, under many circumstances, light exhibits the same behavior as sound or water waves. Lasers and electro-optical devices, critical technologies of the 21st century, are best understood in terms of the wave model of light. Some aspects of the wave model of light were introduced in Chapters 15 and 16, and the wave model is the primary focus of this chapter. The study of light as a wave is called **wave optics.**

The ray model: An equally well-known "fact" is that light travels in a straight line. These straight-line paths are called *light rays.* In Newton's view, light rays are the trajectories of particle-like corpuscles of light. The properties of prisms, mirrors, lenses, and optical instruments such as telescopes and microscopes are best understood in terms of light rays. Unfortunately, it's difficult to reconcile the statement "light travels in a straight line" with the statement "light is a wave." For the most part, waves and rays are mutually exclusive models of light. One of our most important tasks will be to learn when each model is appropriate. The ray model of light, the basis of **ray optics,** is the subject of the next chapter.

FIGURE 17.1 Light passing through an opening makes a sharp-edged shadow.

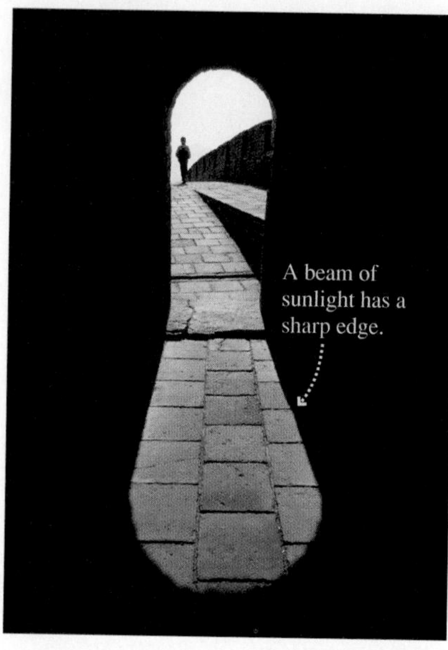

A beam of sunlight has a sharp edge.

The photon model: Modern technology is increasingly reliant on quantum physics. In the quantum world, light behaves like neither a wave nor a particle. Instead, light consists of *photons* that have both wave-like and particle-like properties. Photons are the *quanta* of light. Much of the quantum theory of light is beyond the scope of this textbook, but we will take a peek at the important ideas in Chapters 25 and 28 of this text.

The Propagation of Light Waves

FIGURE 17.1 shows an everyday observation about light. The sunlight passing through the door makes a sharp-edged shadow as it falls upon the floor. This behavior is exactly what you would expect if light consisted of little particles traveling outward from the source in straight lines. Indeed, it was the observation of sharp-edged shadows that led Newton to propose the idea of corpuscles of light.

Is this behavior, where light seems to travel in straight lines, consistent with the motion of a wave? Light waves oscillate so rapidly—at some 10^{14} Hz—that we cannot hope to directly observe the crests and troughs of its electric or magnetic fields. Instead, let's use ordinary water waves to illustrate some basic properties of wave motion common to all waves, including light. FIGURE 17.2 shows a water wave, as seen looking straight down on the water surface. The wave enters from the top of the picture, then passes though a window-like opening in a barrier. After passing through the opening, the wave *spreads out* to fill the space behind the opening. This spreading of a wave is the phenomenon called **diffraction.** Diffraction is a sure sign that whatever is passing through the opening is a wave.

FIGURE 17.2 A water wave passing through a narrow opening in a barrier.

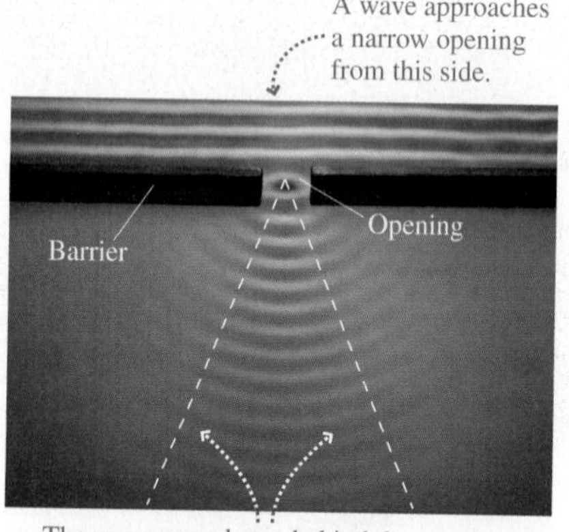

A wave approaches a narrow opening from this side.
Barrier
Opening

The wave spreads out behind the opening.

The straight-line travel of light appears to be incompatible with this spreading of a wave. Notice, however, that the width of the opening in Figure 17.2 is only slightly larger than the wavelength of the water wave. As FIGURE 17.3 shows, something quite different occurs when we make the opening much wider. Rather than spreading out, now the wave continues to move straight forward, with a well-defined boundary between where the wave is moving and its "shadow," where there is no wave. This is just the behavior observed for light in Figure 17.1.

Whether a wave spreads out or travels straight ahead, with sharp shadows on both sides, evidently depends on the size of the objects with which the wave interacts. The spreading of diffraction becomes noticeable only when an opening or object is "narrow," comparable in size to the wavelength of the wave. Thus we would expect a light wave to spread out in this way only when it passes objects that are comparable in size to the wavelength of light.

But the wavelength of light is *extremely* short. Later in this chapter we'll see how its wavelength can be measured, and we'll find that a typical wavelength of light is only about 0.5 μm. When light waves interact with everyday-sized objects, such as the opening in Figure 17.1, the situation is like that of the water wave in Figure 17.3. The wave travels straight ahead, and we'll be able to use the ray model of light. Only when the size of an object or an opening approaches the wavelength of light does the wave-like spreading become important. For example, FIGURE 17.4 on the next page shows the shadow of a 0.5-mm-diameter pin. For an object this small, the shadow is *not* sharp, but shows a series of bright and dark *fringes* as we move away from the center of the shadow. One of the goals of this chapter is to understand the origin of such fringes, a goal for which we'll need the wave model of light.

FIGURE 17.3 A water wave passing through an opening that's many wavelengths wide.

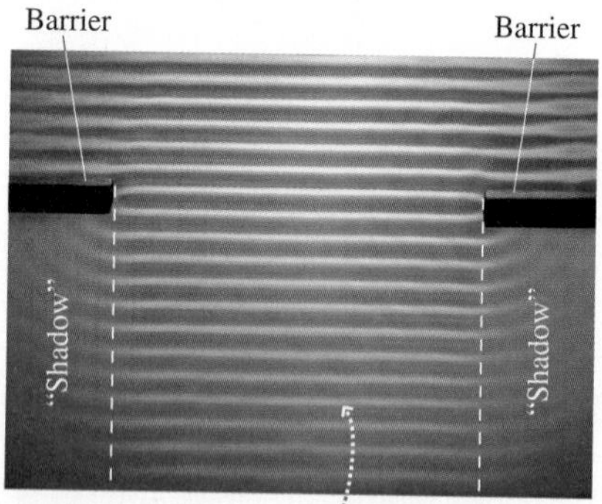

The wave moves straight forward.

Light Is an Electromagnetic Wave

If light is a wave, what is it that is waving? This was the question posed at the start of the 19th century by Young's demonstration of the interference of light. As we briefly noted in Chapter 15, it was ultimately established, through theoretical and experimental efforts by numerous scientists, that light is an *electromagnetic wave,* an oscillation of electric and magnetic fields. We will examine the nature of electromagnetic waves in more detail in Part VI after we introduce the ideas of electric and magnetic fields. For now we can say that light waves are a "self-sustaining oscillation of the electromagnetic field." Being self-sustaining means that electromagnetic waves require *no material medium* in order to travel; hence electromagnetic waves are not mechanical waves. Fortunately, we can learn about the wave properties of light without having to understand electromagnetic fields. In fact, the discovery that light propagates as a wave was made 60 years before it was realized that light is an electromagnetic wave. We, too, will be able to learn much about the wave nature of light without having to know just what it is that is waving.

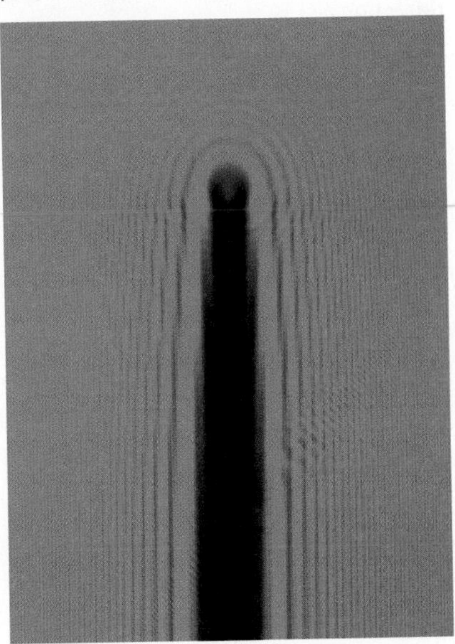

FIGURE 17.4 Diffraction of light by the point of a pin.

Recall that all electromagnetic waves, including light waves, travel in a vacuum at the same speed, called the *speed of light.* Its value is

$$v_{\text{light}} = c = 3.00 \times 10^8 \text{ m/s}$$

where the symbol c is used to designate the speed of light.

Recall also that the wavelengths of light are extremely small, ranging from about 400 nm for violet light to 700 nm for red light. Electromagnetic waves with wavelengths outside this range are not visible to the human eye. A prism is able to spread the different wavelengths apart, from which we learn that "white light" is all the colors, or wavelengths, combined. The spread of colors seen with a prism, or seen in a rainbow, is called the *visible spectrum.*

If the wavelengths of light are incredibly small, the oscillation frequencies are unbelievably high. The frequency for a 600 nm wavelength of light is

$$f = \frac{v}{\lambda} = \frac{3.00 \times 10^8 \text{ m/s}}{600 \times 10^{-9} \text{ m}} = 5.00 \times 10^{14} \text{ Hz}$$

The frequencies of light waves are roughly a factor of a trillion (10^{12}) higher than sound frequencies.

White light passing through a prism is spread out into a band of colors called the *visible spectrum.*

The Index of Refraction

Light waves travel with speed c in a vacuum, but they slow down as they pass through transparent materials such as water or glass or even, to a very slight extent, air. The slowdown is a consequence of interactions between the electromagnetic field of the wave and the electrons in the material. The speed of light in a material is characterized by the material's **index of refraction** n, defined as

$$n = \frac{\text{speed of light in a vacuum}}{\text{speed of light in the material}} = \frac{c}{v} \qquad (17.1)$$

where v is the speed of light in the material. The index of refraction of a material is always greater than 1 because v is always less than c. A vacuum has $n = 1$ exactly. Table 17.1 lists the indices of refraction for several materials. Liquids and solids have higher indices of refraction than gases, simply because they have a much higher density of atoms for the light to interact with.

NOTE ► An accurate value for the index of refraction of air is relevant only in very precise measurements. We will assume $n_{\text{air}} = 1.00$ in this text. ◄

TABLE 17.1 Typical indices of refraction

Material	Index of refraction
Vacuum	1 exactly
Air	1.0003
Water	1.33
Glass	1.50
Diamond	2.42

If the speed of a light wave changes as it enters into a transparent material, such as glass, what happens to the light's frequency and wavelength? Because $v = \lambda f$, either λ or f or both have to change when v changes.

As an analogy, think of a sound wave in the air as it impinges on the surface of a pool of water. As the air oscillates back and forth, it periodically pushes on the surface of the water. These pushes generate the compressions of the sound wave that continues on into the water. Because each push of the air causes one compression of the water, the frequency of the sound wave in the water must be *exactly the same* as the frequency of the sound wave in the air. In other words, **the frequency of a wave does not change as the wave moves from one medium to another.**

FIGURE 17.5 Light passing through a transparent material with index of refraction n.

A transparent material in which light travels slower, at speed $v = c/n$

Vacuum $n = 1$ Index n $n = 1$

λ_{vac} $\lambda = \lambda_{vac}/n$

The wavelength inside the material decreases, but the frequency doesn't change.

The same is true for electromagnetic waves, although the pushes are a bit more complex as the electric and magnetic fields of the wave interact with the atoms at the surface of the material. Nonetheless, the frequency does not change as the wave moves from one material to another.

FIGURE 17.5 shows a light wave passing through a transparent material with index of refraction n. As the wave travels through a vacuum it has wavelength λ_{vac} and frequency f_{vac} such that $\lambda_{vac} f_{vac} = c$. In the material, $\lambda_{mat} f_{mat} = v = c/n$. The frequency does not change as the wave enters ($f_{mat} = f_{vac}$), so the wavelength must change. The wavelength in the material is

$$\lambda_{mat} = \frac{v}{f_{mat}} = \frac{c}{nf_{mat}} = \frac{c}{nf_{vac}} = \frac{\lambda_{vac}}{n} \tag{17.2}$$

The wavelength in the transparent material is shorter than the wavelength in a vacuum. This makes sense. Suppose a marching band is marching at one step per second at a speed of 1 m/s. Suddenly they slow their speed to $\frac{1}{2}$ m/s but maintain their march at one step per second. The only way to go slower while marching at the same pace is to take *smaller steps*. When a light wave enters a material, the only way it can go slower while oscillating at the same frequency is to have a *shorter wavelength.*

EXAMPLE 17.1 Analyzing light traveling through glass

Orange light with a wavelength of 600 nm is incident on a 1.00-mm-thick glass microscope slide.

a. What is the light speed in the glass?
b. How many wavelengths of the light are inside the slide?

SOLVE

a. From Table 17.1 we see that the index of refraction of glass is $n_{glass} = 1.50$. Thus the speed of light in glass is

$$v_{glass} = \frac{c}{n_{glass}} = \frac{3.00 \times 10^8 \text{ m/s}}{1.50} = 2.00 \times 10^8 \text{ m/s}$$

b. Because $n_{air} = 1.00$, the wavelength of the light is the same in air and vacuum: $\lambda_{vac} = \lambda_{air} = 600$ nm. Thus the wavelength inside the glass is

$$\lambda_{glass} = \frac{\lambda_{vac}}{n_{glass}} = \frac{600 \text{ nm}}{1.50} = 400 \text{ nm} = 4.00 \times 10^{-7} \text{ m}$$

N wavelengths span a distance $d = N\lambda$, so the number of wavelengths in $d = 1.00$ mm is

$$N = \frac{d}{\lambda} = \frac{1.00 \times 10^{-3} \text{ m}}{4.00 \times 10^{-7} \text{ m}} = 2500$$

ASSESS The fact that 2500 wavelengths fit within 1 mm shows how small the wavelengths of light are.

STOP TO THINK 17.1 A light wave travels through three transparent materials of equal thickness. Rank in order, from the highest to lowest, the indices of refraction n_1, n_2, and n_3.

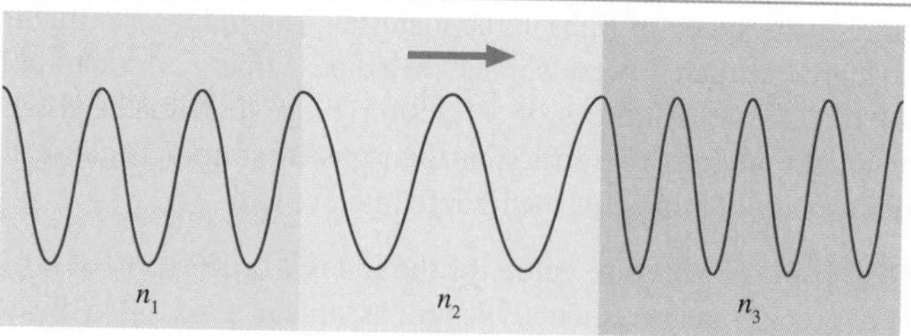

17.2 The Interference of Light

Consider the situation shown in **FIGURE 17.6**. Light passes through a narrow opening—a *slit*—that is only 0.1 mm wide, about twice the width of a human hair. This situation is similar to Figure 17.2, where water waves spread out after passing through a narrow opening. When light waves pass through a narrow slit, they too spread out behind the slit, just as the water wave did behind the opening in the barrier. The light is exhibiting *diffraction*, the sure sign of waviness. We will look at diffraction in more detail later in the chapter. For now, we merely need the *observation* that light does, indeed, spread out behind an opening that is sufficiently small.

FIGURE 17.6 Light, just like a water wave, spreads out behind a slit in a screen if the slit is sufficiently small.

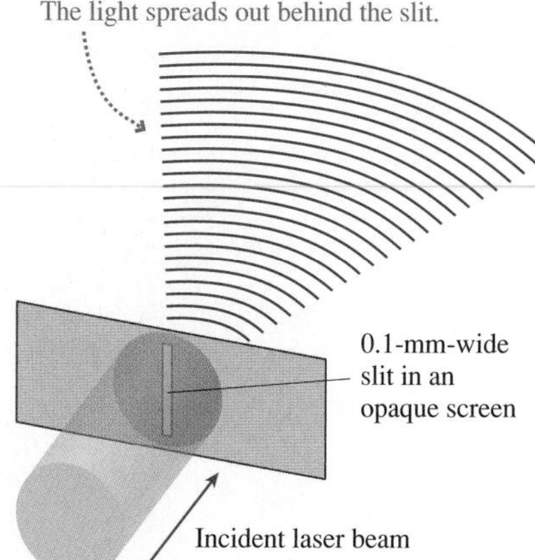

Young's Double-Slit Experiment

Rather than one small slit, suppose we use two. FIGURE 17.7a shows an experiment in which a laser beam is aimed at an opaque screen containing two long, narrow slits that are very close together. This pair of slits is called a **double slit,** and in a typical experiment they are ≈0.1 mm wide and spaced ≈0.5 mm apart. We will assume that the laser beam illuminates both slits equally, and any light passing through the slits impinges on a viewing screen. This is the essence of Young's experiment of 1801, although he used sunlight rather than a laser beam.

What should we expect to see on the screen? FIGURE 17.7b is a view from above the experiment, looking down on the top ends of the slits and the top edge of the viewing screen. Because the slits are very narrow, **light spreads out behind each slit** as it did in Figure 17.6, and these two spreading waves overlap in the region between the slits and the screen.

FIGURE 17.7 A double-slit interference experiment.

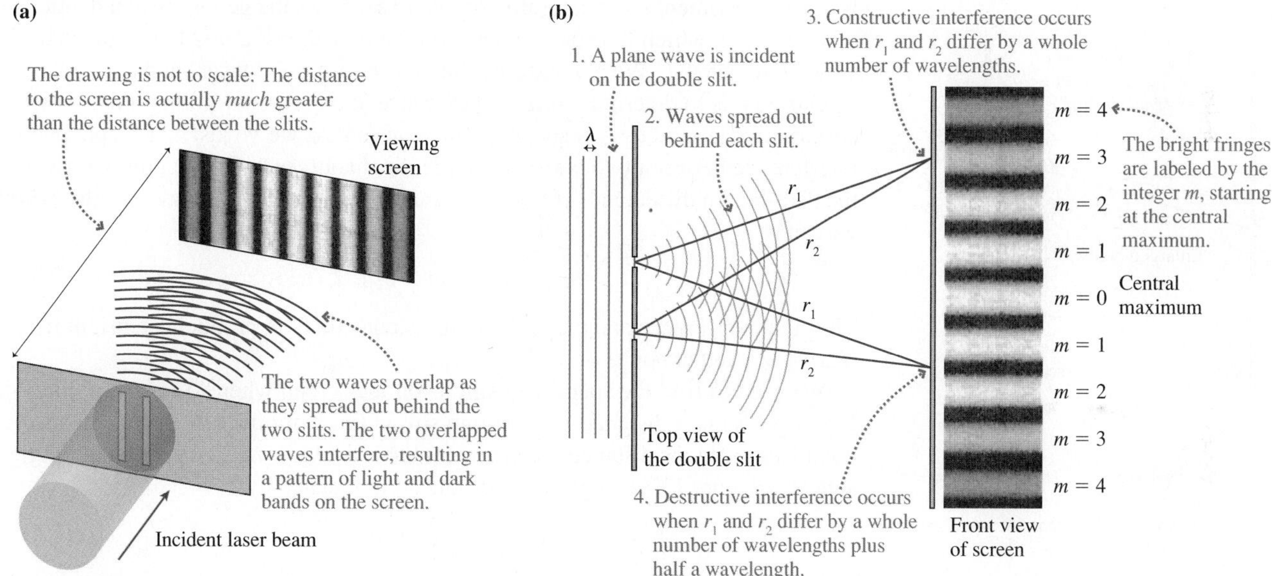

(a)
The drawing is not to scale: The distance to the screen is actually *much* greater than the distance between the slits.
Viewing screen
The two waves overlap as they spread out behind the two slits. The two overlapped waves interfere, resulting in a pattern of light and dark bands on the screen.
Incident laser beam
(b)
1. A plane wave is incident on the double slit.
λ
2. Waves spread out behind each slit.
3. Constructive interference occurs when r_1 and r_2 differ by a whole number of wavelengths.
r_1
r_2
r_1
r_2
Top view of the double slit
4. Destructive interference occurs when r_1 and r_2 differ by a whole number of wavelengths plus half a wavelength.
Front view of screen
$m = 4$
$m = 3$
$m = 2$
$m = 1$
$m = 0$
$m = 1$
$m = 2$
$m = 3$
$m = 4$
The bright fringes are labeled by the integer *m*, starting at the central maximum.
Central maximum

One primary conclusion of Chapter 16 was that two overlapped waves of equal wavelength produce interference. In fact, because both slits are illuminated by the *same* plane wave, they act as sources of *identical* waves. Thus Figure 17.7b is equivalent to the waves emitted by two identical loudspeakers, a situation we analyzed in Section 16.6. Nothing in that analysis depended on what type of wave it was, so the conclusions apply equally well to two overlapped light waves.

In Chapter 16 we found that *constructive* interference occurs at a given position when the distances r_1 and r_2 from speakers 1 and 2 to that position differ by a *whole number* of wavelengths. For light waves, we thus expect constructive interference to occur, and the intensity of light on the screen to be high, when the distances r_1 and r_2 from the two slits to a point on the screen differ by a whole number of wavelengths. And, just as we found for sound waves, destructive interference for light waves occurs at positions on the screen for which r_1 and r_2 differ by a whole number of wavelengths plus half a wavelength. At these positions the screen will be dark.

The photograph in Figure 17.7b shows how the screen looks. As we move along the screen, the difference $\Delta r = r_2 - r_1$ alternates between being a whole number of wavelengths and a whole number of wavelengths plus half a wavelength, leading to a series of alternating bright and dark bands of light called **interference fringes.** The fringes are numbered by an integer $m = 1, 2, 3, \ldots$, going outward from the center. The brightest fringe, at the midpoint of the viewing screen, has $m = 0$ and is called the **central maximum.**

STOP TO THINK 17.2 Suppose the viewing screen in Figure 17.7 is moved closer to the double slit. What happens to the interference fringes?

A. They get brighter but otherwise do not change.
B. They get brighter and closer together.
C. They get brighter and farther apart.
D. They get out of focus.
E. They fade out and disappear.

FIGURE 17.8 Geometry of the double-slit experiment.

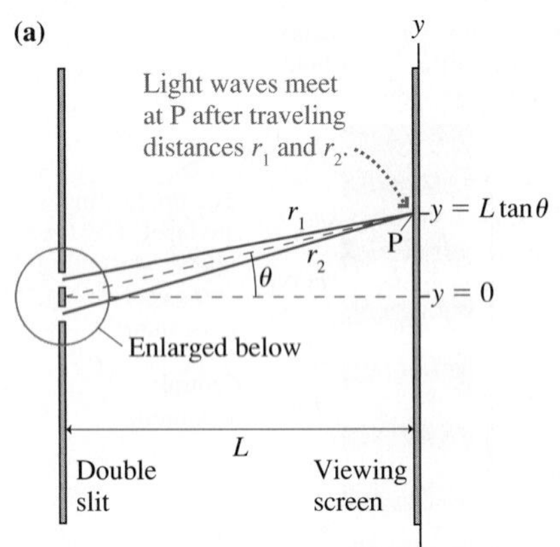

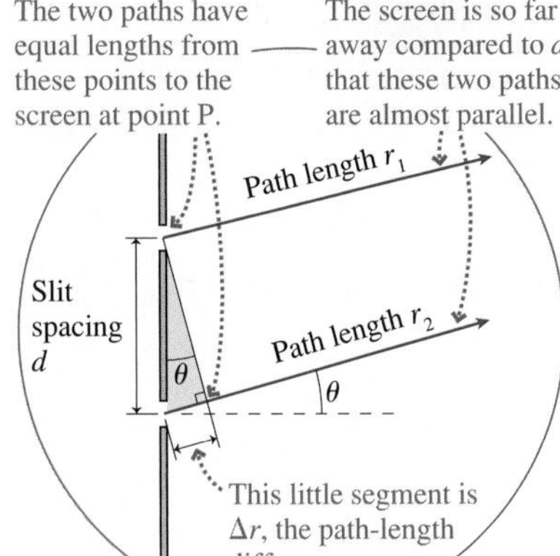

Analyzing Double-Slit Interference

Figure 17.7b showed qualitatively that interference is produced behind a double slit by the overlap of the light waves spreading out behind each opening. Now let's analyze the experiment more carefully. FIGURE 17.8a shows the geometry of a double-slit experiment in which the spacing between the two slits is d and the distance to the viewing screen is L. **We will assume that L is *very* much larger than d.**

Our goal is to determine if the interference at a particular point on the screen is constructive, destructive, or something in between. As we've just noted, constructive interference between two waves from identical sources occurs at points for which the path-length difference $\Delta r = r_2 - r_1$ is an integer number of wavelengths, which we can write as

$$\Delta r = m\lambda \qquad m = 0, 1, 2, 3, \ldots \tag{17.3}$$

Thus the interference at a particular point is constructive, producing a bright fringe, if $\Delta r = m\lambda$ at that point.

We need to find the specific positions on the screen where $\Delta r = m\lambda$. Point P in Figure 17.8a is a distance r_1 from one slit and r_2 from the other. We can specify point P either by its distance y from the center of the viewing screen or by the angle θ shown in Figure 17.8a; angle θ and distance y are related by

$$y = L\tan\theta \tag{17.4}$$

Because the screen is very far away compared to the spacing between the slits, the two paths to point P are virtually parallel, and thus the small triangle that is shaded green in the enlargement of FIGURE 17.8b is a right triangle whose angle is also θ. The path-length difference between the two waves is the short side of this triangle, so

$$\Delta r = d\sin\theta \tag{17.5}$$

Bright fringes due to constructive interference then occur at angles θ_m such that

$$\Delta r = d\sin\theta_m = m\lambda \qquad m = 0, 1, 2, 3, \ldots \tag{17.6}$$

We have added the subscript m to denote that θ_m is the angle of the mth bright fringe, starting with $m = 0$ at the center.

The center of the viewing screen at $y = 0$ is equally distant from both slits, so $\Delta r = 0$. This point of constructive interference, with $m = 0$, is the bright fringe identified as the central maximum in Figure 17.7b. The path-length difference increases as you move away from the center of the screen, and the $m = 1$ fringes occur at the positions where $\Delta r = 1\lambda$. That is, one wave has traveled exactly one wavelength farther than the other. In general, **the mth bright fringe occurs where one wave has traveled m wavelengths farther than the other and thus $\Delta r = m\lambda$.**

16.1–16.3 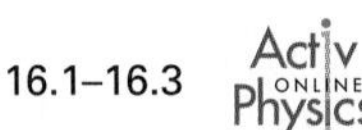

In practice, the angle θ in a double-slit experiment is almost always a very small angle ($<1°$). Recall, from Chapter 14, the *small-angle approximation* $\sin\theta \approx \theta$, where θ must be in *radians*. We can use this approximation to write Equation 17.6 as

$$\theta_m = m\frac{\lambda}{d} \qquad m = 0, 1, 2, 3, \ldots \qquad (17.7)$$

Angles (in radians) of bright fringes for double-slit interference with slit spacing d

This gives the angular positions *in radians* of the bright fringes in the interference pattern.

It is usually more convenient to measure the *position* of the mth bright fringe, as measured from the center of the viewing screen. Using the small-angle approximation once again, this time in the form $\tan\theta \approx \theta$, we can substitute θ_m from Equation 17.7 for $\tan\theta_m$ in Equation 17.4 to find that the mth bright fringe occurs at position

$$y_m = \frac{m\lambda L}{d} \qquad m = 0, 1, 2, 3, \ldots \qquad (17.8)$$

Positions of bright fringes for double-slit interference at screen distance L

The interference pattern is symmetrical, so there is an mth bright fringe at the same distance on both sides of the center. You can see this in Figure 17.7b.

NOTE ▶ Equations 17.7 and 17.8 do *not* apply to the interference of sound waves from two loudspeakers. The approximations we've used (small angles, $L \gg d$) are usually not valid for the much longer wavelengths of sound waves. ◀

TRY IT YOURSELF

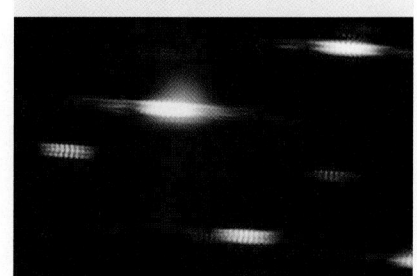

Observing interference It's actually not that hard to observe double-slit interference. Place a piece of aluminum foil on a hard surface and cut two parallel slits, about 1 mm apart, using a razor blade. Now hold the slits up to your eye and look at a distant small bright light, such as a streetlight at night. Because of diffraction, you'll see the light source spread out in a direction *perpendicular* to the long direction of the slits. Superimposed on the diffraction pattern is a fine pattern of interference maxima and minima, as seen in the photo above of Christmas tree lights taken through two slits in foil.

EXAMPLE 17.2 **How far do the waves travel?**

Light from a helium-neon laser ($\lambda = 633$ nm) illuminates two slits spaced 0.40 mm apart. A viewing screen is 2.0 m behind the slits. A bright fringe is observed at a point 9.5 mm from the center of the screen. What is the fringe number m, and how much farther does the wave from one slit travel to this point than the wave from the other slit?

PREPARE A bright fringe is observed when one wave has traveled an integer number of wavelengths farther than the other. Thus we know that Δr must be $m\lambda$, where m is an integer. We can find m from Equation 17.8.

SOLVE Solving Equation 17.8 for m gives

$$m = \frac{y_m d}{\lambda L} = \frac{(9.5 \times 10^{-3}\ \text{m})(0.40 \times 10^{-3}\ \text{m})}{(633 \times 10^{-9}\ \text{m})(2.0\ \text{m})} = 3$$

Then the extra distance traveled by one wave compared to the other is

$$\Delta r = m\lambda = 3(633 \times 10^{-9}\ \text{m}) = 1.9 \times 10^{-6}\ \text{m}$$

ASSESS The path-length differences in two-slit interference are generally very small, just a few wavelengths of light. Here, Δr is only about one part in a million of the 2 m distance traveled by the waves!

Equation 17.8 predicts that **the interference pattern is a series of equally spaced bright lines** on the screen, exactly as shown in Figure 17.7b. How do we know the fringes are equally spaced? The **fringe spacing** between fringe m and fringe $m + 1$ is

$$\Delta y = y_{m+1} - y_m = \frac{(m+1)\lambda L}{d} - \frac{m\lambda L}{d} = \frac{\lambda L}{d} \qquad (17.9)$$

Because Δy is independent of m, *any* two bright fringes have the same spacing.

The dark fringes in the photograph are bands of destructive interference. You learned in Chapter 16 that destructive interference occurs at positions where the

FIGURE 17.9 Symbols used to describe two-slit interference.

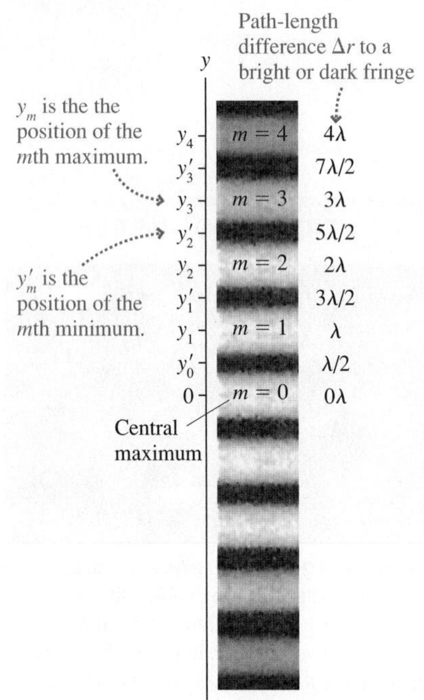

path-length difference of the waves is a whole number of wavelengths plus half a wavelength:

$$\Delta r = \left(m + \frac{1}{2}\right)\lambda \qquad m = 0, 1, 2, 3, \ldots \tag{17.10}$$

We can use Equation 17.6 for Δr and the small-angle approximation to find that the dark fringes are located at positions

$$y'_m = \left(m + \frac{1}{2}\right)\frac{\lambda L}{d} \qquad m = 0, 1, 2, 3, \ldots \tag{17.11}$$

Positions of dark fringes for double-slit interference

We have used y'_m, with a prime, to distinguish the location of the mth minimum from the mth maximum at y_m. You can see from Equation 17.11 that **the dark fringes are located exactly halfway between the bright fringes.** FIGURE 17.9 summarizes the symbols we use to describe two-slit interference.

FIGURE 17.10 is a graph of the double-slit intensity versus y. Notice the unusual orientation of the graph, with the intensity increasing toward the left so that the y-axis can match the experimental layout. You can see that the intensity oscillates between dark fringes, where the intensity is zero, and equally spaced bright fringes of maximum intensity. The maxima occur at positions where $y_m = m\lambda L/d$.

FIGURE 17.10 Intensity of the interference fringes in the double-slit experiment.

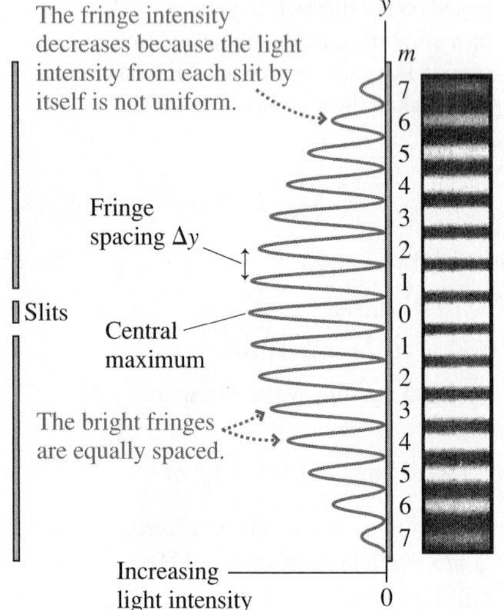

EXAMPLE 17.3 Measuring the wavelength of light

A double-slit interference pattern is observed on a screen 1.0 m behind two slits spaced 0.30 mm apart. From the center of one particular fringe to the center of the ninth bright fringe from this one is 1.6 cm. What is the wavelength of the light?

PREPARE It is not always obvious which fringe is the central maximum. Slight imperfections in the slits can make the interference fringe pattern less than ideal. However, you do not need to identify the $m = 0$ fringe because you can make use of the fact that the fringe spacing Δy is uniform. The interference pattern looks like the photograph of Figure 17.7b.

SOLVE The fringe spacing is

$$\Delta y = \frac{1.6 \text{ cm}}{9} = 1.8 \times 10^{-3} \text{ m}$$

Using this fringe spacing in Equation 17.9, we find that the wavelength is

$$\lambda = \frac{d}{L}\Delta y = \frac{3.0 \times 10^{-4} \text{ m}}{1.0 \text{ m}}(1.8 \times 10^{-3} \text{ m})$$
$$= 5.4 \times 10^{-7} \text{ m} = 540 \text{ nm}$$

It is customary to express the wavelengths of light in nanometers. Be sure to do this as you solve problems.

ASSESS You learned in Chapter 15 that visible light spans the wavelength range 400–700 nm, so finding a wavelength in this range is reasonable. In fact, it's because of experiments like the double-slit experiment that we're able to measure the wavelengths of light.

STOP TO THINK 17.3 Light of wavelength λ_1 illuminates a double slit, and interference fringes are observed on a screen behind the slits. When the wavelength is changed to λ_2, the fringes get closer together. Is λ_2 larger or smaller than λ_1?

17.3 The Diffraction Grating

Suppose we were to replace the double slit with an opaque screen that has N closely spaced slits. When illuminated from one side, each of these slits becomes the source of a light wave that diffracts, or spreads out, behind the slit. Such a multi-slit device is called a **diffraction grating.** The light intensity pattern on a screen behind a diffraction grating is due to the interference of N overlapped waves.

FIGURE 17.11 shows a diffraction grating in which N slits are equally spaced a distance d apart. This is a top view of the grating, as we look down on the experiment, and the slits extend above and below the page. Only 10 slits are shown here, but a practical grating will have hundreds or even thousands of slits. Suppose a plane wave of wavelength λ approaches from the left. The crest of a plane wave arrives *simultaneously* at each of the slits, causing the wave emerging from each slit to be *in phase* with the wave emerging from every other slit—that is, all the emerging waves crest and trough simultaneously. Each of these emerging waves spreads out, just like the light wave in Figure 17.6, and after a short distance they all overlap with each other and interfere.

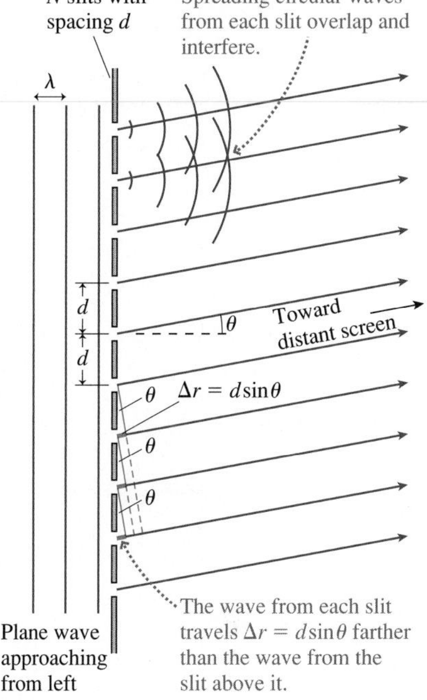

FIGURE 17.11 Top view of a diffraction grating with $N = 10$ slits.

We want to know how the interference pattern will appear on a screen behind the grating. The light wave at the screen is the superposition of N waves, from N slits, as they spread and overlap. As we did with the double slit, we'll assume that the distance L to the screen is very large in comparison with the slit spacing d; hence the path followed by the light from one slit to a point on the screen is *very nearly* parallel to the path followed by the light from neighboring slits. The paths cannot be perfectly parallel, of course, or they would never meet to interfere, but the slight deviation from perfect parallelism is too small to notice. You can see in Figure 17.11 that the wave from one slit travels distance $\Delta r = d\sin\theta$ farther than the wave from the slit above it and $\Delta r = d\sin\theta$ less than the wave below it. This is the same reasoning we used in Figure 17.8 to analyze the double- slit experiment.

FIGURE 17.12 Interference for a grating with five slits.

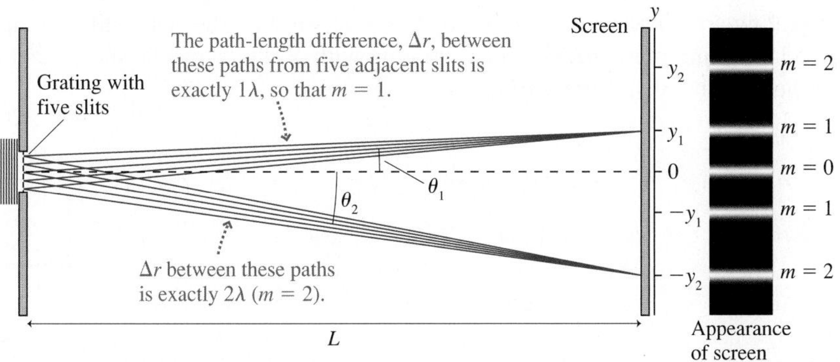

Figure 17.11 was a magnified view of the slits. FIGURE 17.12 steps back to where we can see the viewing screen, for a grating with five slits. If the angle θ is such that $\Delta r = d\sin\theta = m\lambda$, where m is an integer, then the light wave arriving at the screen from one slit will travel *exactly m* wavelengths more or less than light from the two slits next to it, so these waves will be *exactly in phase* with each other. But each of those waves is in phase with waves from the slits next to them, and so on until we reach the end of the grating. In other words, **N light waves, from N different slits, will *all* be in phase with each other when they arrive at a point on the screen at angle θ_m such that**

$$d\sin\theta_m = m\lambda \qquad m = 0, 1, 2, 3, \ldots \qquad (17.12)$$

Angles of bright fringes due to a diffraction grating with slits distance d apart

16.4, 16.5 ActivPhysics ONLINE

The screen will have bright constructive-interference fringes at the values of θ_m given by Equation 17.12. When this happens, we say that the light is "diffracted at angle θ_m." Because it's usually easier to measure distances rather than angles, the position y_m of the mth maximum is

$$y_m = L\tan\theta_m \tag{17.13}$$

Positions of bright fringes due to a diffraction grating distance L from screen

The integer m is called the **order** of the diffraction. Practical gratings, with very small values for d, display only a few orders. Because d is usually very small, it is customary to characterize a grating by the number of *lines per millimeter.* Here "line" is synonymous with "slit," so the number of lines per millimeter is simply the inverse of the slit spacing d in millimeters.

NOTE ► The condition for constructive interference in a grating of N slits is identical to Equation 17.6 for just two slits. Equation 17.12 is simply the requirement that the path-length difference between adjacent slits, be they two or N, is $m\lambda$. But unlike the angles in double-slit interference, the angles of constructive interference from a diffraction grating are generally *not* small angles. The reason is that the slit spacing d in a diffraction grating is usually so small that λ/d is not a small number. Thus you *cannot* use the small-angle approximation to simplify Equations 17.12 and 17.13. ◄

Although the condition for constructive interference is the same for a diffraction grating as it was for the double slit, the intensity of the fringes for a grating differs from that for a double slit in two important ways. First, the bright fringes of a diffraction grating are *narrower* than those of a double slit with the same slit spacing d. Second, the fringes are *brighter.* For a grating with N slits, it can be shown that the maximum intensity of a bright fringe is

$$I_{max} = N^2 I_1 \tag{17.14}$$

Maximum intensity of a bright fringe for a diffraction grating with N slits

I_{max} vs N, QUADRATIC, p.47

where I_1 is the intensity of the wave from a single slit. For a double slit, with $N = 2$, the intensity of the central maximum is 4 times that of each slit alone. For a practical diffraction grating, with hundreds of slits, the bright fringes are enormously brighter.

Not only do the fringes get brighter as N increases, they also get narrower. If you double the number of slits, for instance, twice as much light will reach the screen. But according to Equation 17.14, the maximum intensity of a bright fringe would increase by a factor of 4. Since there's only twice as much light to work with, the fringes must be narrower in order to concentrate the light in each fringe enough to make it four times brighter.

To demonstrate the significance of the number of slits, **FIGURE 17.13** on the next page shows the interference pattern due to three diffraction gratings with the same slit spacing d but increasing numbers of slits. The case with $N = 2$ is that of double-slit interference. As the number of slits is increased, the bright fringes become much narrower. For a realistic diffraction grating, with $N > 1000$, the interference pattern consists of a small number of *very* bright and *very* narrow fringes while most of the screen remains dark.

FIGURE 17.13 The intensity on the screen due to three diffraction gratings. Notice that the intensity axes have different scales.

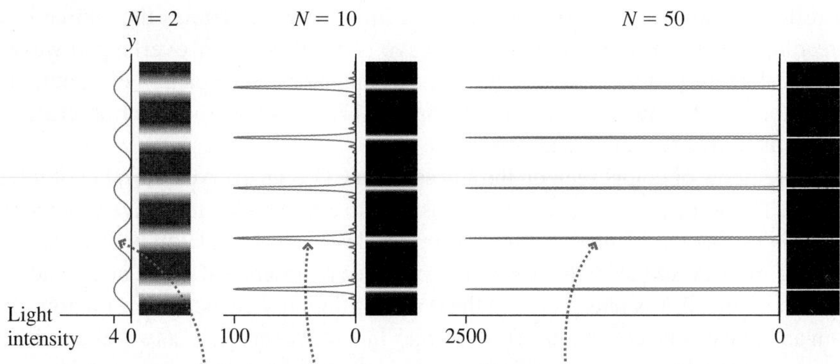

$N = 2$
$N = 10$
$N = 50$
y
Light intensity
4 0
100
0
2500
0
As the number of slits in the grating increases, the fringes get narrower and brighter.

FIGURE 17.14 A diffraction grating can be used to measure wavelengths of light.

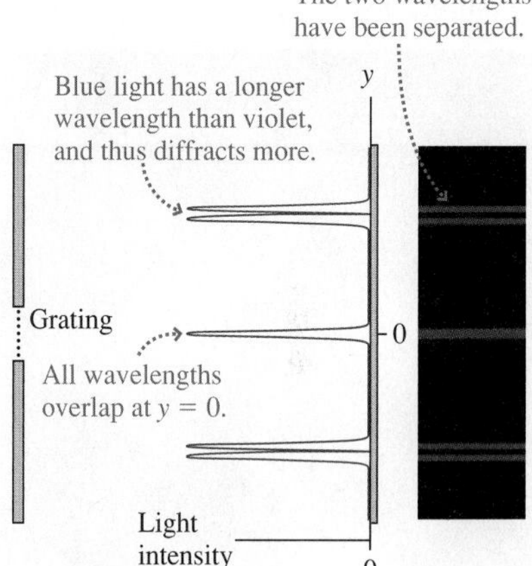

The two wavelengths have been separated.
Blue light has a longer wavelength than violet, and thus diffracts more.
y
Grating
0
All wavelengths overlap at y = 0.
Light intensity
0

Spectroscopy

As we'll see in Chapter 29, each atomic element in the periodic table, if appropriately excited by light, electricity, or collisions with other atoms, emits light at only certain well-defined wavelengths. By accurately measuring these wavelengths, we can deduce the various elements in a sample of unknown composition. Molecules also emit light that is characteristic of their composition. The science of measuring the wavelengths of atomic and molecular emissions is called **spectroscopy.**

Because their bright fringes are so distinct, diffraction gratings are an ideal tool for spectroscopy. Suppose the light incident on a grating consists of two slightly different wavelengths. According to Equation 17.12, each wavelength will diffract at a slightly different angle and, if N is sufficiently large, we'll see two distinct fringes on the screen. FIGURE 17.14 illustrates this idea. By contrast, the fringes in a double-slit experiment are so broad that it would not be possible to distinguish the fringes of one wavelength from those of the other.

EXAMPLE 17.4 Measuring wavelengths emitted by sodium atoms

Light from a sodium lamp passes through a diffraction grating having 1000 slits per millimeter. The interference pattern is viewed on a screen 1.000 m behind the grating. Two bright yellow fringes are visible 72.88 cm and 73.00 cm from the central maximum. What are the wavelengths of these two fringes?

PREPARE This situation is similar to that in Figure 17.14. The two fringes are very close together, so we expect the wavelengths to be only slightly different. No other yellow fringes are mentioned, so we will assume these two fringes are the first-order diffraction ($m = 1$).

SOLVE The distance y_m of a bright fringe from the central maximum is related to the diffraction angle by $y_m = L\tan\theta_m$. Thus the diffraction angles of these two fringes are

$$\theta_1 = \tan^{-1}\left(\frac{y_1}{L}\right) = \begin{cases} 36.08° & \text{fringe at 72.88 cm} \\ 36.13° & \text{fringe at 73.00 cm} \end{cases}$$

These angles must satisfy the interference condition $d\sin\theta_1 = \lambda$, so the wavelengths are

$$\lambda = d\sin\theta_1$$

What is d? If a 1 mm length of the grating has 1000 slits, then the spacing from one slit to the next must be 1/1000 mm, or $d = 1.00 \times 10^{-6}$ m. Thus the wavelengths creating the two bright fringes are

$$\lambda = d\sin\theta_1 = \begin{cases} 589.0 \text{ nm} & \text{fringe at 72.88 cm} \\ 589.6 \text{ nm} & \text{fringe at 73.00 cm} \end{cases}$$

ASSESS In Chapter 15 you learned that yellow light has a wavelength of about 600 nm, so our answer is reasonable.

Instruments that measure and analyze spectra, called *spectrophotometers,* are widely used in chemistry, biology, and medicine. Because each molecule has a distinct spectrum—a "fingerprint"—spectroscopy is used to identify specific biomolecules in tissue, drugs in urine, and chlorophyll in seawater.

Reflection Gratings

We have analyzed what is called a *transmission grating,* with many parallel slits. It's difficult to make such a grating with many closely spaced slits. In practice, most diffraction gratings are manufactured as *reflection gratings.* The simplest reflection

FIGURE 17.15 A reflection grating.

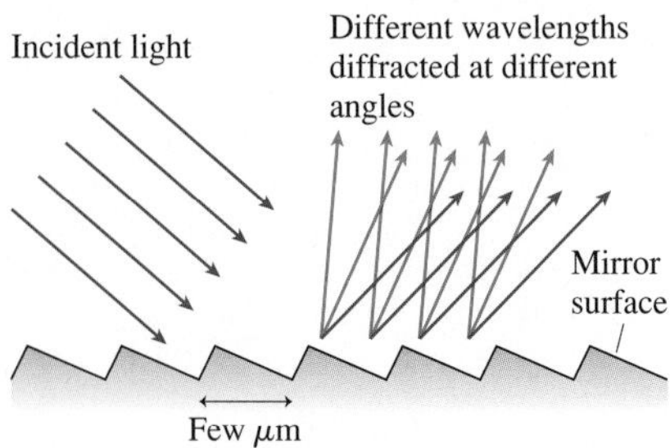

A reflection grating can be made by cutting parallel grooves in a mirror surface. These can be very precise, for scientific use, or mass produced in plastic.

grating, shown in FIGURE 17.15, is a mirror with hundreds or thousands of narrow, parallel grooves cut into the surface. The grooves divide the surface into many parallel reflective stripes, each of which, when illuminated, becomes the source of a spreading wave. Thus an incident light wave is divided into N overlapped waves. The interference pattern is exactly the same as the interference pattern of light transmitted through N parallel slits, and so **Equation 17.13 applies to reflection gratings as well as to transmission gratings.**

The rainbow of colors seen on the surface of a CD is an everyday display of interference. The surface of a CD is smooth plastic with a mirror-like reflective coating. As shown in FIGURE 17.16, billions of microscopic holes, each about 1 μm in diameter, are "burned" into the surface with a laser. The presence or absence of a hole at a particular location on the disk is interpreted as the 0 or 1 of digitally encoded information. But from an optical perspective, the array of holes in a shiny surface is a two-dimensional version of the reflection grating shown in Figure 17.15. Less precise plastic reflection gratings can be manufactured at very low cost simply by stamping holes or grooves into a reflective surface, and these are widely sold as toys and novelty items. Rainbows of color are seen as each wavelength of white light is diffracted at a unique angle.

FIGURE 17.16 A CD's colors are caused by diffraction.

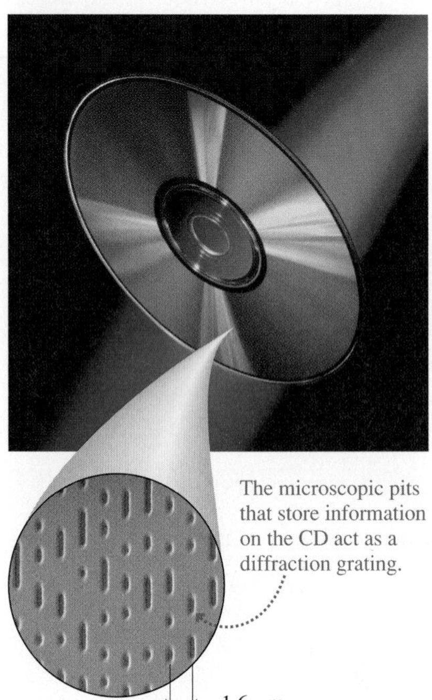

STOP TO THINK 17.4 White light passes through a diffraction grating and forms rainbow patterns on a screen behind the grating. For each rainbow,

A. The red side is on the right, the violet side on the left.
B. The red side is on the left, the violet side on the right.
C. The red side is closest to the center of the screen, the violet side is farthest from the center.
D. The red side is farthest from the center of the screen, the violet side is closest to the center.

17.4 Thin-Film Interference

In Chapter 16 you learned about the interference of sound waves in one dimension. Depending on whether they are in phase or out of phase, two sound waves of the same frequency, traveling in the same direction, can undergo constructive or destructive interference. Light waves can also interfere in this way. Equal-frequency light waves are produced when *partial reflection* at a boundary splits a light wave into a reflected wave and a transmitted wave. The interference of light waves reflected from the two boundaries of a thin film, such as the thin film of water that makes a soap bubble, is called **thin-film interference.**

Thin-film interference has important applications in the optics industry. Thin-film coatings, less than 1 μm (10^{-6} m) thick, are used for the antireflection coatings on the lenses in cameras, microscopes, and other optical equipment. The bright colors of oil slicks and soap bubbles are also due to thin-film interference.

FIGURE 17.17 Two reflections are visible in the window, one from each surface.

Interference of Reflected Light Waves

As you know, and as we discussed in Chapter 16, a light wave encountering a piece of glass is partially transmitted and partially reflected. In fact, a light wave is partially reflected from *any* boundary between two transparent media with different indices of refraction. Thus light is partially reflected not only from the front surface of a sheet of glass, but from the back surface as well, as it exits from the glass into the air. This leads to the *two* reflections seen in FIGURE 17.17.

Another important aspect of wave reflections was shown for strings in Figure 16.10b of the last chapter. If a wave moves from a string with a higher wave speed to a string with a lower wave speed, the reflected wave is *inverted* with respect to the incoming wave. It is not inverted if the wave moves from a string with a lower wave speed to a string with a higher wave speed.

The same thing happens for light waves. When a light wave moves from a medium with a higher light speed (lower index of refraction) to a medium with a lower light speed (higher index of refraction), the reflected wave is inverted. This inversion of the wave, called a *phase change,* is equivalent to adding an extra half-wavelength $\lambda/2$ to the distance the wave travels. You can see this in FIGURE 17.18, where a reflected wave with a phase change is compared to a reflection without a phase change. In summary, we can say that **a light wave undergoes a phase change if it reflects from a boundary at which the index of refraction increases.** There's no phase change at a boundary where the index of refraction decreases.

FIGURE 17.18 Reflected waves with and without a phase change.

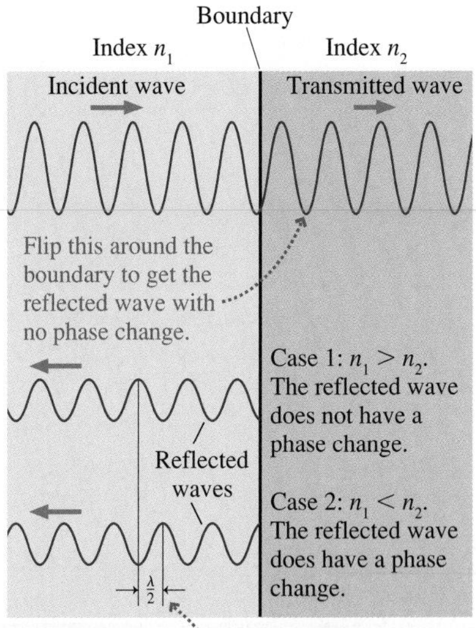

Consider a thin, transparent film with thickness t and index of refraction n coated onto a piece of glass. FIGURE 17.19 shows a light wave of wavelength λ approaching the film. Most of the light is transmitted into the film, but, as we've seen, a bit is reflected off the first (air-film) boundary. Further, a bit of the wave that continues into the film is reflected off the second (film-glass) boundary. The two reflected waves, which have exactly the same frequency, travel back out into the air where they overlap and interfere. As we learned in Chapter 16, the two reflected waves will interfere constructively to cause a *strong reflection* if they are *in phase* (i.e., if their crests overlap). If the two reflected waves are *out of phase,* with the crests of one wave overlapping the troughs of the other, they will interfere destructively to cause a *weak reflection* or, if their amplitudes are equal, *no reflection* at all.

We found the interference of two sound waves to be constructive if their path-length difference is $\Delta d = m\lambda$ and destructive if $\Delta d = \left(m + \frac{1}{2}\right)\lambda$, where m is an integer. The same idea holds true for reflected light waves, for which the path-length difference is the extra distance traveled by the wave that reflects from the second surface. Because this wave travels twice through a film of thickness t, the path-length difference is $\Delta d = 2t$.

We noted above that the phase change when a light wave reflects from a boundary with a higher index of refraction is equivalent to adding an extra half-wavelength to the distance traveled. This leads to two situations:

1. If *neither* or *both* waves have a phase change due to reflection, the net addition to the path-length difference is zero. The *effective path-length difference* is $\Delta d_{\text{eff}} = 2t$.
2. If only *one* wave has a phase change due to reflection, the effective path-length difference is increased by one half-wavelength to $\Delta d_{\text{eff}} = 2t + \frac{1}{2}\lambda$.

FIGURE 17.19 In thin-film interference, two reflections, one from the film and one from the glass, overlap and interfere.

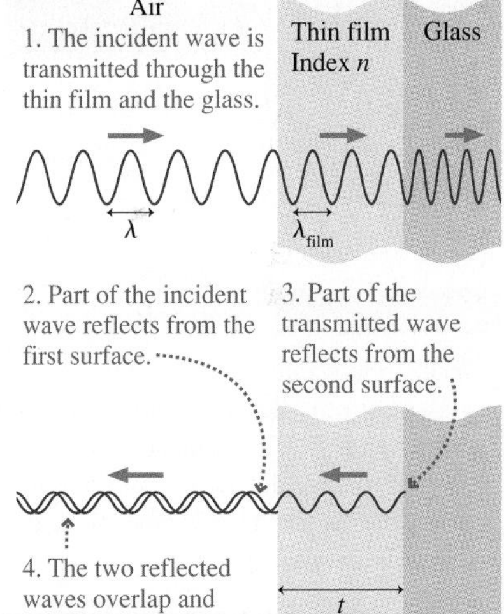

The interference of the two reflected waves is then constructive if $\Delta d_{\text{eff}} = m\lambda_{\text{film}}$ and destructive if $\Delta d_{\text{eff}} = \left(m + \frac{1}{2}\right)\lambda_{\text{film}}$. Why λ_{film}? Because the extra distance is traveled inside the film, so we need to compare $2t$ to the wavelength in the film. Further, the film's index of refraction is n, so the wavelength in the film is $\lambda_{\text{film}} = \lambda/n$, where λ is the wavelength of the light in vacuum or air.

With this information, we can write the conditions for constructive and destructive interference of the light waves reflected by a thin film:

$$2t = m\frac{\lambda}{n} \qquad m = 0, 1, 2, \ldots \tag{17.15}$$

Condition for constructive interference with either 0 or 2 reflective phase changes
Condition for destructive interference with only 1 reflective phase change

$$2t = \left(m + \frac{1}{2}\right)\frac{\lambda}{n} \qquad m = 0, 1, 2, \ldots \tag{17.16}$$

Condition for destructive interference with either 0 or 2 reflective phase changes
Condition for constructive interference with only 1 reflective phase change

NOTE ► Equations 17.15 and 17.16 give the film thicknesses that yield constructive or destructive interference. At other thicknesses, the waves will interfere neither fully constructively nor fully destructively, and the reflected intensity will fall somewhere between these two extremes. ◄

These conditions are the basis of a procedure to analyze thin-film interference.

TACTICS BOX 17.1 Analyzing thin-film interference

Follow the light wave as it passes through the film. The wave reflecting from the second boundary travels an extra distance $2t$.

1. Note the indices of refraction of the three media: the medium before the film, the film itself, and the medium beyond the film. The first and third may be the same. There's a reflective phase change at any boundary where the index of refraction increases.
2. If *neither* or *both* reflected waves undergo a phase change, the phase changes cancel and the effective path-length difference is $\Delta d = 2t$. Use Equation 17.15 for constructive interference and 17.16 for destructive interference.
3. If *only one* wave undergoes a phase change, the effective path-length difference is $\Delta d = 2t + \frac{1}{2}\lambda$. Use Equation 17.15 for destructive interference and 17.16 for constructive interference.

Exercises 12, 13 

×10,000

◄ **Iridescent feathers** BIO The gorgeous colors of the hummingbird shown at the beginning of this chapter are due not to pigments but to interference. This *iridescence,* present in some bird feathers and insect shells, arises from biological structures whose size is similar to the wavelength of light. The sheen of an insect, for instance, is due to thin-film interference from multiple thin layers in its shell. Peacock feathers are also a layered structure, but each layer itself consists of nearly parallel rods of melanin, as shown in the micrograph, that act as a diffraction grating. Thus a peacock feather combines thin-film interference and grating-like diffraction to produce its characteristic multi-colored iridescent hues.

EXAMPLE 17.5 Designing an antireflection coating

To keep unwanted light from reflecting from the surface of eyeglasses or other lenses, a thin film of a material with an index of refraction $n = 1.38$ is coated onto the plastic lens ($n = 1.55$). It is desired to have destructive interference for $\lambda = 550$ nm because that is the center of the visible spectrum. What is the thinnest film that will do this?

The glasses on the top have an antireflection coating on them. Those on the bottom do not.

PREPARE We follow the steps of Tactics Box 17.1. As the light traverses the film, it first reflects at the front surface of the coating. Here, the index of refraction increases from that of air ($n = 1.00$) to that of the film ($n = 1.38$), so there will be a reflective phase change. The light then reflects from the rear surface of the coating. The index again increases from that of the film ($n = 1.38$) to that of the plastic ($n = 1.55$). With two phase changes, Tactics Box 17.1 tells us that we should use Equation 17.16 for destructive interference.

SOLVE We can solve Equation 17.16 for the thickness t that causes destructive interference:

$$t = \frac{\lambda}{2n}\left(m + \frac{1}{2}\right)$$

The thinnest film is the one for which $m = 0$, giving

$$t = \frac{550\text{ nm}}{2(1.38)} \times \frac{1}{2} = 100\text{ nm}$$

ASSESS Interference effects occur when path-length differences are on the order of a wavelength, so our answer of 100 nm seems reasonable.

Thin Films of Air

A film need not be of a solid material. A thin layer of air sandwiched between two glass surfaces also exhibits thin-film interference due to the waves that reflect off both interior air-glass boundaries. FIGURE 17.20 shows two microscope slides being pressed together. The light and dark "fringes" occur because the slides are not exactly flat and they touch each other only at a few points. Everywhere else there is a thin layer of air between them. At some points, the air layer's thickness is such as to give constructive interference (light fringes), while at other places its thickness gives destructive interference (dark fringes). These fringes can be used to accurately measure the flatness of two glass plates, as Example 17.6 will show.

FIGURE 17.20 Light and dark fringes caused by thin-film interference due to the air layer between two microscope slides.

EXAMPLE 17.6 Finding the fringe spacing from a wedge-shaped film of air

Two 15-cm-long flat glass plates are separated by a 10-μm-thick spacer at one end, leaving a thin wedge of air between them, as shown in FIGURE 17.21. The plates are illuminated by light from a sodium lamp with wavelength $\lambda = 589$ nm. Alternating bright and dark fringes are observed. What is the spacing between two bright fringes?

FIGURE 17.21 Two glass plates with an air wedge between them.

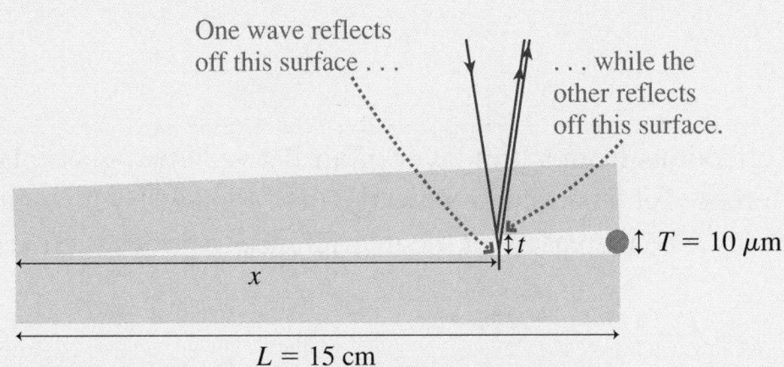

PREPARE The wave reflected from the lower plate has a reflective phase change, but the top reflection does not because the index of refraction decreases at the glass-air boundary. According to Tactics Box 17.1, we should use Equation 17.16 for constructive interference:

$$2t = \left(m + \frac{1}{2}\right)\frac{\lambda}{n}$$

This is a film of air, so here n is the index of refraction of air. Each integer value of m corresponds to a wedge thickness t for which there is constructive interference and thus a bright fringe.

SOLVE Let x be the distance from the left end to a bright fringe. From Figure 17.21, by similar triangles we have

$$\frac{t}{x} = \frac{T}{L}$$

or $t = xT/L$. From the condition for constructive interference, we then have

$$2\frac{xT}{L} = \left(m + \frac{1}{2}\right)\frac{\lambda}{n}$$

There will be a bright fringe for any integer value of m, and so the position of the mth fringe, as measured from the left end, is

$$x_m = \frac{\lambda L}{2nT}\left(m + \frac{1}{2}\right)$$

We want to know the spacing between two adjacent fringes, m and $m + 1$, which is

$$\Delta x = x_{m+1} - x_m = \frac{\lambda L}{2nT}\left(m + 1 + \frac{1}{2}\right) - \frac{\lambda L}{2nT}\left(m + \frac{1}{2}\right) = \frac{\lambda L}{2nT}$$

Evaluating, we find

$$\Delta x = \frac{\lambda L}{2nT} = \frac{(5.89 \times 10^{-7}\ \text{m})(0.15\ \text{m})}{2(1.00)(10 \times 10^{-6}\ \text{m})} = 4.4\ \text{mm}$$

ASSESS As the photo shows, if the two plates are very flat, the fringes will appear as straight lines perpendicular to the direction of increasing air thickness. However, if the plates are not quite flat, the fringes will appear curved. The amount of curvature indicates the departure of the plates from perfect flatness.

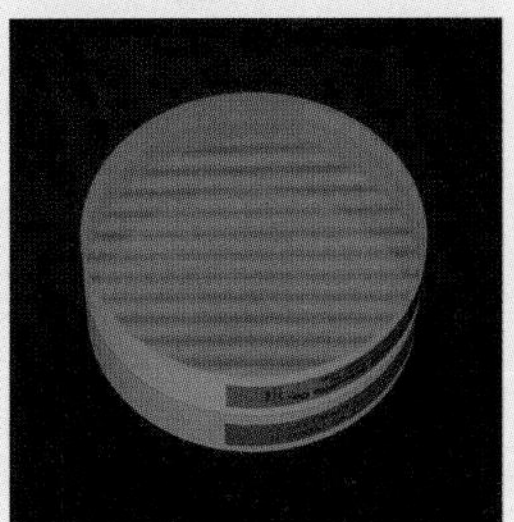

The Colors of Soap Bubbles and Oil Slicks

So far we have considered thin-film interference only for single wavelengths of light. The bright colors of soap bubbles and oil slicks on water are due to thin-film interference of white light, which, as we've seen, is a mixture of *all* wavelengths.

A soap bubble is a very thin, spherical film of soapy water ($n = 1.33$). Consider a soap film with thickness $t = 470$ nm. Light waves reflect from both surfaces of the film, and these reflected waves interfere. The light reflecting from the front (air-water) surface has a reflective phase change, but the back reflection does not. Thus Equation 17.16 describes constructive interference and Equation 17.15 is destructive interference. Table 17.2 shows wavelengths of constructive and destructive interference for three values of m.

TABLE 17.2 Wavelengths for constructive and destructive interference from a 470-nm-thick soap bubble. Visible wavelengths are shown in **bold.**

Equation	$m = 1$	$m = 2$	$m = 3$
$\lambda_{con} = \dfrac{2nt}{m + \frac{1}{2}}$	833 nm	**500 nm**	357 nm
$\lambda_{des} = \dfrac{2nt}{m}$	1250 nm	**625 nm**	**417 nm**

FIGURE 17.22 The colors of a soap bubble.

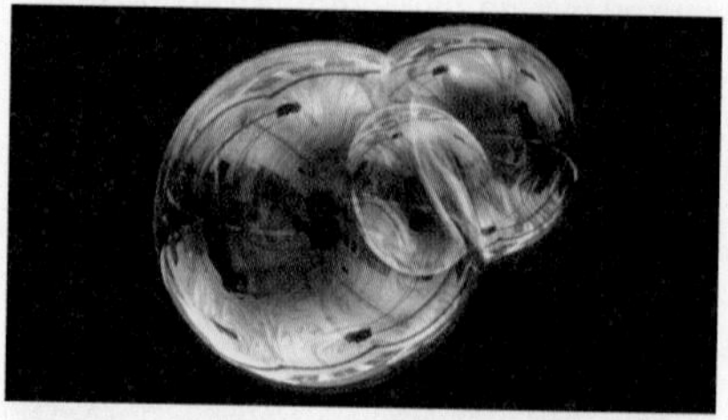

Light near the red end (625 nm) and violet end (417) of the spectrum undergoes destructive interference; these colors are *not* reflected by the film. At the same time, light near 500 nm (green) interferes constructively and so is strongly reflected. Consequently, a soap film of this thickness will appear green.

Real soap bubbles and oil slicks have thicknesses that vary from point to point. At some thicknesses, green light is strongly reflected, while at others red or violet light is. FIGURE 17.22 shows that the colors of a soap bubble are predominately greens and red/violets.

CONCEPTUAL EXAMPLE 17.7 Colors in a vertical soap film

FIGURE 17.23 shows a soap film in a metal ring. The ring is held vertically. Explain the colors seen in the film.

FIGURE 17.23 A soap film in a metal ring.

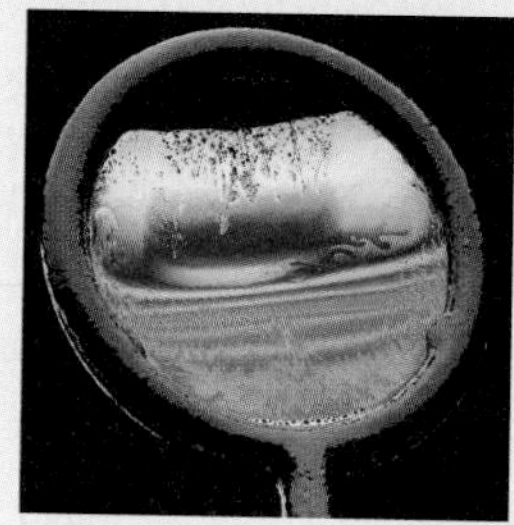

REASON Because of gravity, the film is thicker near the bottom and thinner at the top. It thus has a wedge shape, and the interference pattern consists of lines of alternating constructive and destructive interference, just as for the air wedge of Example 17.6. Because this soap film is illuminated by white light, colors form as just discussed for any soap film.

Notice that the very top of the film, which is extremely thin, appears black. This means that it is reflecting no light at all. When the film is very thin—much thinner than the wavelength of light—there is almost no path-length difference between the two waves reflected off the front and the back of the film. However, the wave reflected off the back undergoes a reflective phase change and is out of phase with the wave reflected off the front. The two waves thus *always* interfere destructively, no matter what their wavelength.

ASSESS This simple experiment shows directly that the two reflected waves have different reflective phase changes.

STOP TO THINK 17.5 Reflections from a thin layer of air between two glass plates cause constructive interference for a particular wavelength of light λ. By how much must the thickness of this layer be increased for the interference to be destructive?

A. $\lambda/8$ B. $\lambda/4$ C. $\lambda/2$ D. λ

17.5 Single-Slit Diffraction

We opened this chapter with a photograph of a water wave passing through a hole in a barrier, then spreading out on the other side. You then saw a photograph showing that light, after passing a narrow pin, also spreads out on the other side. This phenomenon is called *diffraction.* We're now ready to look at the details of diffraction.

FIGURE 17.24 again shows the experimental arrangement for observing the diffraction of light through a narrow slit of width a. Diffraction through a tall, narrow slit of width a is known as **single-slit diffraction.** A viewing screen is placed a distance L behind the slit, and we will assume that $L \gg a$. The light pattern on the viewing screen consists of a *central maximum* flanked by a series of weaker **secondary maxima** and dark fringes. Notice that the central maximum is significantly broader than the secondary maxima. It is also significantly brighter than the secondary maxima, although that is hard to tell here because this photograph has been overexposed to make the secondary maxima show up better.

FIGURE 17.24 A single-slit diffraction experiment.

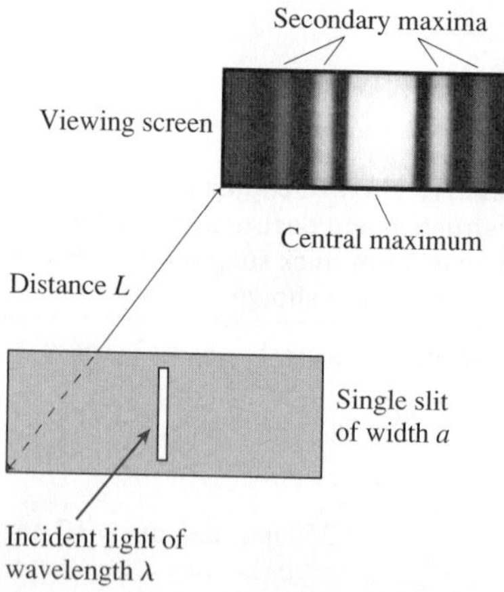

Huygens' Principle

Our analysis of the superposition of waves from distinct sources, such as two loudspeakers or the two slits in a double-slit experiment, has tacitly assumed that the sources are *point sources,* with no measurable extent. To understand diffraction, we need to think about the propagation of an *extended* wave front. This problem was first considered by the Dutch scientist Christiaan Huygens, a contemporary of Newton who argued that light is a wave.

Huygens lived before a mathematical theory of waves had been developed, so he developed a geometrical model of wave propagation. His idea, which we now call **Huygens' principle,** has two steps:

1. Each point on a wave front is the source of a spherical *wavelet* that spreads out at the wave speed.
2. At a later time, the shape of the wave front is the curve that is tangent to all the wavelets.

FIGURE 17.25 illustrates Huygens' principle for a plane wave and a spherical wave. As you can see, the curve tangent to the wavelets of a plane wave is a plane that has propagated to the right. The curve tangent to the wavelets of a spherical wave is a larger sphere.

Huygens' principle is a visual device, not a theory of waves. Nonetheless, the full mathematical theory of waves, as it developed in the 19th century, justifies Huygens' basic idea, although it is beyond the scope of this textbook to prove it.

FIGURE 17.25 Huygens' principle applied to the propagation of plane waves and spherical waves.

(a) Plane wave

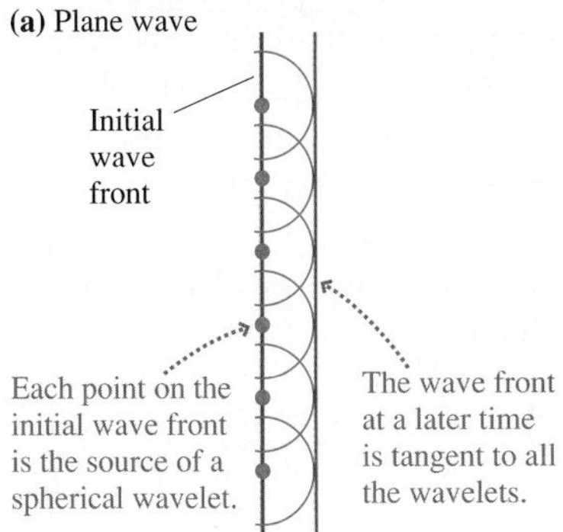
Initial wave front
Each point on the initial wave front is the source of a spherical wavelet.
The wave front at a later time is tangent to all the wavelets.

(b) Spherical wave

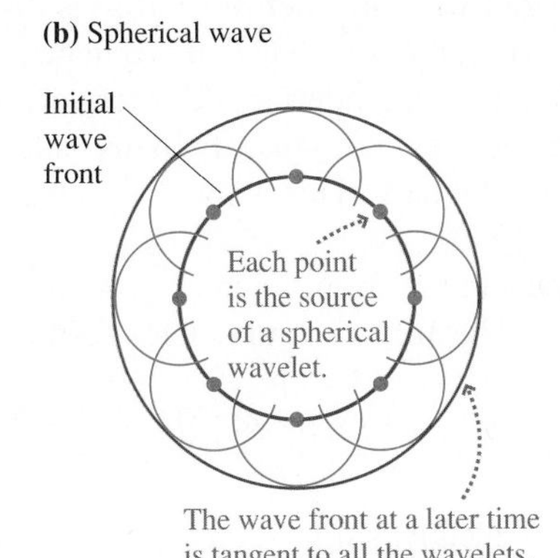
Initial wave front
Each point is the source of a spherical wavelet.
The wave front at a later time is tangent to all the wavelets.

Analyzing Single-Slit Diffraction

FIGURE 17.26a shows a wave front passing through a narrow slit of width a. According to Huygens' principle, each point on the wave front can be thought of as the source of a spherical wavelet. These wavelets overlap and interfere, producing the diffraction pattern seen on the viewing screen. The full mathematical analysis, using *every*

Activ Online Physics
16.6

FIGURE 17.26 Each point on the wave front is a source of spherical wavelets. The superposition of these wavelets produces the diffraction pattern on the screen.

(a) Greatly magnified view of slit

Initial wave front

Slit width a

The wavelets from each point on the initial wave front overlap and interfere, creating a diffraction pattern on the screen.

(b)

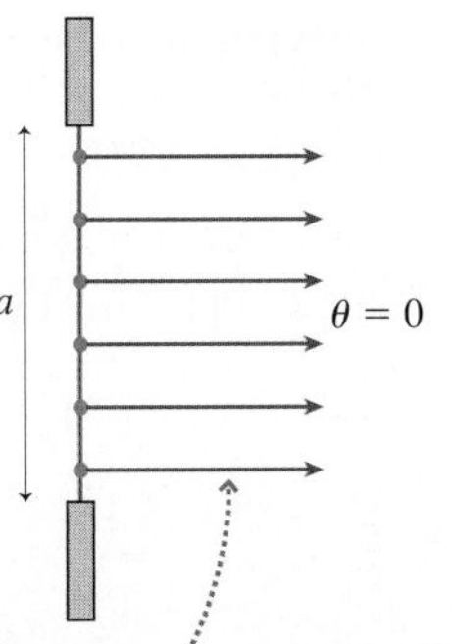
a
θ = 0

The wavelets going straight forward all travel the same distance to the screen. Thus they arrive in phase and interfere constructively to produce the central maximum.

(c)

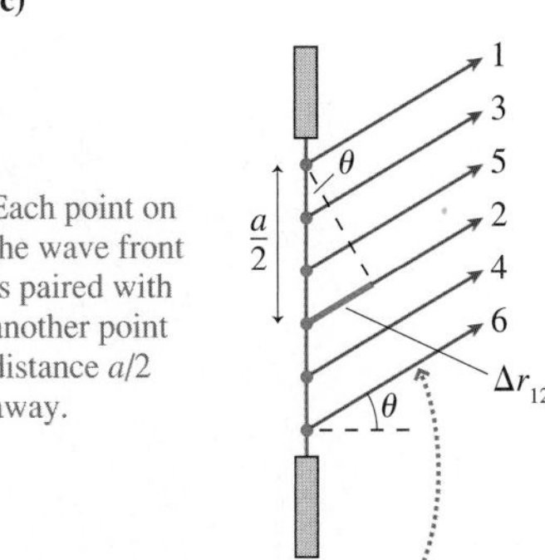
Each point on the wave front is paired with another point distance a/2 away.
a/2
θ
1
3
5
2
4
6
Δr12

These wavelets all meet on the screen at angle θ. Wavelet 2 travels distance $\Delta r_{12} = (a/2)\sin\theta$ farther than wavelet 1.

Water waves can be seen diffracting behind the "slit" between the two breakwaters. The wave pattern can be understood using Huygens' principle.

point on the wave front, is a fairly difficult problem in calculus. We'll be satisfied with a geometrical analysis based on just a few wavelets.

FIGURE 17.26b shows the paths of several wavelets as they travel straight ahead to the central point on the screen. (The screen is *very* far to the right in this magnified view of the slit.) The paths to the screen are very nearly parallel to each other; thus all the wavelets travel the same distance and arrive at the screen *in phase* with each other. The *constructive interference* between these wavelets produces the central maximum of the diffraction pattern at $\theta = 0$.

The situation is different at points away from the center of the screen. Wavelets 1 and 2 in FIGURE 17.26c start from points that are distance $a/2$ apart. Suppose that Δr_{12}, the extra distance traveled by wavelet 2, happens to be $\lambda/2$. In that case, wavelets 1 and 2 arrive out of phase and interfere destructively. But if Δr_{12} is $\lambda/2$, then the difference Δr_{34} between paths 3 and 4 and the difference Δr_{56} between paths 5 and 6 are also $\lambda/2$. Those pairs of wavelets also interfere destructively. The superposition of all the wavelets produces perfect destructive interference.

Figure 17.26c happens to show six wavelets, but our conclusion is valid for any number of wavelets. The key idea is that **every point on the wave front can be paired with another point that is distance *a/2* away.** If the path-length difference is $\lambda/2$, the wavelets that originate at these two points will arrive at the screen out of phase and interfere destructively. When we sum the displacements of all N wavelets, they will—pair by pair—add to zero. The viewing screen at this position will be dark. This is the main idea of the analysis, one worth thinking about carefully.

You can see from Figure 17.26c that $\Delta r_{12} = (a/2)\sin\theta$. This path-length difference will be $\lambda/2$, the condition for destructive interference, if

$$\Delta r_{12} = \frac{a}{2}\sin\theta_1 = \frac{\lambda}{2} \tag{17.17}$$

or, equivalently, $\sin\theta_1 = \lambda/a$.

NOTE ► Equation 17.17 cannot be satisfied if the slit width a is less than the wavelength λ. If a wave passes through an opening smaller than the wavelength, the central maximum of the diffraction pattern expands to completely fill the space behind the opening. There are no minima or dark spots at any angle. This situation is uncommon for light waves, because λ is so small, but quite common in the diffraction of sound and water waves. ◄

We can extend this idea to find other angles of perfect destructive interference. Suppose each wavelet is paired with another wavelet from a point $a/4$ away. If Δr between these wavelets is $\lambda/2$, then all N wavelets will again cancel in pairs to give complete destructive interference. The angle θ_2 at which this occurs is found by replacing $a/2$ in Equation 17.17 with $a/4$, leading to the condition $a\sin\theta_2 = 2\lambda$. This process can be continued, and we find that the general condition for complete destructive interference is

$$a\sin\theta_p = p\lambda \qquad p = 1, 2, 3, \ldots \tag{17.18}$$

When $\theta_p \ll 1$ rad, which is almost always true for light waves, we can use the small-angle approximation to write

$$\theta_p = p\frac{\lambda}{a} \qquad p = 1, 2, 3, \ldots \tag{17.19}$$

Angles (in radians) of *dark* fringes in single-slit diffraction with slit width a

Equation 17.19 gives the angles *in radians* to the dark minima in the diffraction pattern of a single slit. Notice that $p = 0$ is explicitly *excluded.* $p = 0$ corresponds to the straight-ahead position at $\theta = 0$, but you saw in Figures 17.6 and 17.26b that $\theta = 0$ is the central *maximum,* not a minimum.

NOTE ► Equations 17.18 and 17.19 are *mathematically* the same as the condition for the *m*th *maximum* of the double-slit interference pattern. But the physical meaning here is quite different. Equation 17.19 locates the *minima* (dark fringes) of the single-slit diffraction pattern. ◄

It is possible, although beyond the scope of this textbook, to calculate the entire light intensity pattern. The results of such a calculation are shown graphically in FIGURE 17.27. You can see the bright central maximum at $\theta = 0$, the weaker secondary maxima, and the dark points of destructive interference at the angles given by Equation 17.19. Compare this graph to the photograph of Figure 17.24 and make sure you see the agreement between the two.

FIGURE 17.27 A graph of the intensity of a single-slit diffraction pattern.

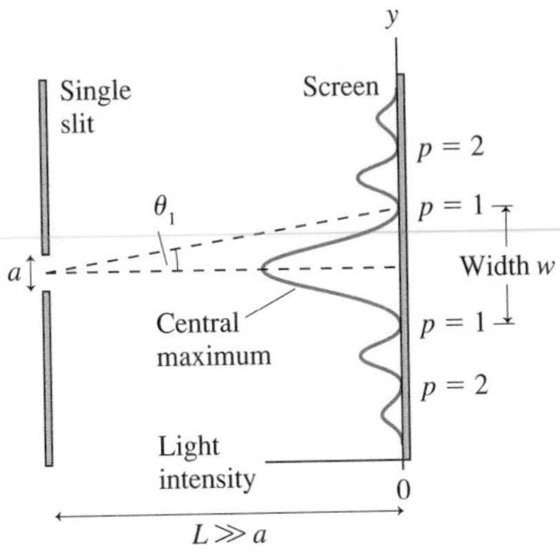

The Width of a Single-Slit Diffraction Pattern

We'll find it useful, as we did for the double slit, to measure positions on the screen rather than angles. The position of the *p*th dark fringe, at angle θ_p, is $y_p = L\tan\theta_p$, where L is the distance from the slit to the viewing screen. Using Equation 17.19 for θ_p and the small-angle approximation $\tan\theta_p \approx \theta_p$, we find that the dark fringes in the single-slit diffraction pattern are located at

$$y_p = \frac{p\lambda L}{a} \qquad p = 1, 2, 3, \ldots \qquad (17.20)$$

Positions of dark fringes for single-slit diffraction with screen distance L

Again, $p = 0$ is explicitly excluded because the midpoint on the viewing screen is the central maximum, not a dark fringe.

A diffraction pattern is dominated by the central maximum, which is much brighter than the secondary maxima. The width w of the central maximum, shown in Figure 17.27, is defined as the distance between the two $p = 1$ minima on either side of the central maximum. Because the pattern is symmetrical, the width is simply $w = 2y_1$. This is

$$w = \frac{2\lambda L}{a} \qquad (17.21)$$

Width of the central maximum for single-slit diffraction

The width of the central maximum is *twice* the spacing $\lambda L/a$ between the dark fringes on either side. The farther away the screen (larger L), the wider the pattern of light on it becomes. In other words, the light waves are *spreading out* behind the slit, and they fill a wider and wider region as they travel farther.

An important implication of Equation 17.21, one contrary to common sense, is that a narrower slit (smaller a) causes a *wider* diffraction pattern. **The smaller the opening a wave squeezes through, the *more* it spreads out on the other side.**

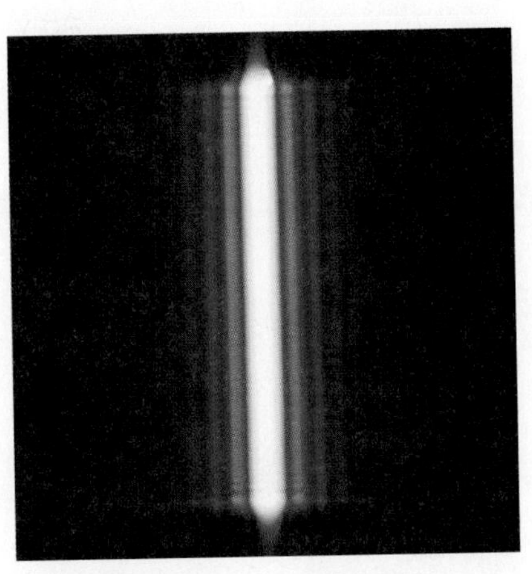

The central maximum of this single-slit diffraction pattern appears white because it is overexposed. The width of the central maximum is clear.

EXAMPLE 17.8 Finding the width of a slit

Light from a helium-neon laser ($\lambda = 633$ nm) passes through a narrow slit and is seen on a screen 2.0 m behind the slit. The first minimum in the diffraction pattern is 1.2 cm from the middle of the central maximum. How wide is the slit?

PREPARE The first minimum in a diffraction pattern corresponds to $p = 1$. The position of this minimum is given as $y_1 = 1.2$ cm. We can then use Equation 17.20 to find the slit width a.

SOLVE Equation 17.20 gives

$$a = \frac{p\lambda L}{y_p} = \frac{(1)(633 \times 10^{-9}\text{ m})(2.0\text{ m})}{0.012\text{ m}}$$
$$= 1.1 \times 10^{-4}\text{ m} = 0.11\text{ mm}$$

ASSESS This value is typical of the slit widths used to observe single-slit diffraction. You can see that the small-angle approximation is well satisfied.

STOP TO THINK 17.6 The figure shows two single-slit diffraction patterns. The distance between the slit and the viewing screen is the same in both cases. Which of the following could be true?

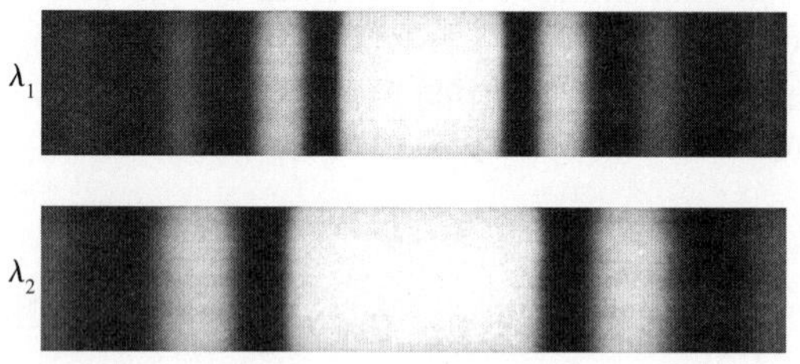

A. The slits are the same for both; $\lambda_1 > \lambda_2$.
B. The slits are the same for both; $\lambda_2 > \lambda_1$.
C. The wavelengths are the same for both; $a_1 > a_2$.
D. The wavelengths are the same for both; $a_2 > a_1$.
E. The slits and the wavelengths are the same for both; $p_1 > p_2$.
F. The slits and the wavelengths are the same for both; $p_2 > p_1$.

17.6 Circular-Aperture Diffraction

16.7 ActivPhysics Online

Diffraction occurs if a wave passes through an opening of any shape. Diffraction by a single slit establishes the basic ideas of diffraction, but a common situation of practical importance is diffraction of a wave by a **circular aperture.** Circular diffraction is mathematically more complex than diffraction from a slit, and we will present results without derivation.

Consider some examples. A loudspeaker cone generates sound by the rapid oscillation of a diaphragm, but the sound wave must pass through the circular aperture defined by the outer edge of the speaker cone before it travels into the room beyond. This is diffraction by a circular aperture. Telescopes and microscopes are the reverse. Light waves from outside need to enter the instrument. To do so, they must pass through a circular lens. In fact, the performance limit of optical instruments is determined by the diffraction of the circular openings through which the waves must pass. This is an issue we'll look at more closely in Chapter 19.

FIGURE 17.28 shows a circular aperture of diameter D. Light waves passing through this aperture spread out to generate a *circular* diffraction pattern. You should compare this to Figure 17.24 for a single slit to note the similarities and differences. The diffraction pattern still has a *central maximum,* now circular, and it is surrounded by a series of secondary bright fringes. Most of the intensity is contained within the central maximum.

FIGURE 17.28 The diffraction of light by a circular opening.

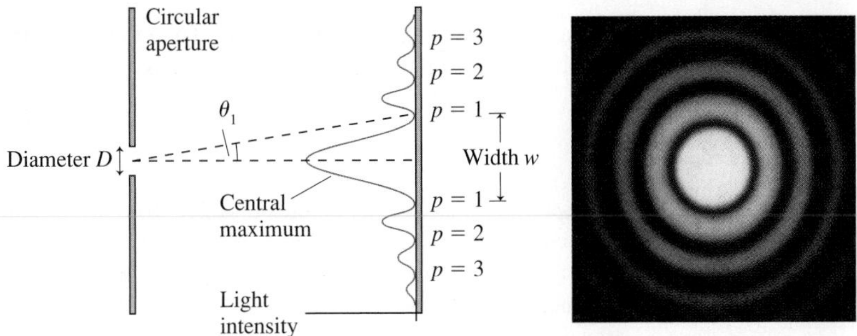

Angle θ_1 locates the first minimum in the intensity, where there is perfect destructive interference. A mathematical analysis of circular diffraction finds that

$$\theta_1 = \frac{1.22\lambda}{D} \tag{17.22}$$

where D is the *diameter* of the circular opening. This is very similar to the result for a single slit, but not quite the same. Equation 17.22 has assumed the small-angle approximation, which is almost always valid for the diffraction of light but usually is *not* valid for the diffraction of longer-wavelength sound waves.

Within the small-angle approximation, the width of the central maximum is

$$w = 2y_1 = 2L\tan\theta_1 \approx \frac{2.44\lambda L}{D} \tag{17.23}$$

Width of central maximum for diffraction from a circular aperture of diameter D

Again, this is similar to, but not quite identical with, the width of the central maximum in a single-slit diffraction pattern. The diameter of the diffraction pattern increases with distance L, showing that light spreads out behind a circular aperture, but it decreases if the size D of the aperture is increased.

EXAMPLE 17.9 Finding the right viewing distance

Light from a helium-neon laser ($\lambda = 633$ nm) passes through a 0.50-mm-diameter hole. How far away should a viewing screen be placed to observe a diffraction pattern whose central maximum is 3.0 mm in diameter?

SOLVE Equation 17.23 gives us the appropriate screen distance:

$$L = \frac{wD}{2.44\lambda} = \frac{(3.0 \times 10^{-3}\text{ m})(5.0 \times 10^{-4}\text{ m})}{2.44(633 \times 10^{-9}\text{ m})} = 0.97\text{ m}$$

As we've seen, we need to use the wave model of light to understand the passage of light through narrow apertures, where "narrow" means comparable in size to the wavelength of the light. In the next two chapters, we'll consider the interaction of light with objects much larger than the wavelength. There, the ray model of light will be more appropriate for describing how light reflects from mirrors and refracts from lenses. But the wave model will reappear in Chapter 19 when we study telescopes and microscopes. We'll find that the *resolution* of these instruments has a fundamental limit set by the wave nature of light.

TRY IT YOURSELF

Observing diffraction To observe diffraction from a circular aperture, use a pin to prick a *very* small hole in two pieces of aluminum foil. Tape one piece up to a window that faces the sun. Then, holding the other hole close to your eye, observe the sun through the pinhole on the window. You will see clear diffraction fringes around that pinhole. How do the fringes vary if you change the size of the pinhole near your eye? (Don't look directly at the sun except through the two pinholes!)

INTEGRATED EXAMPLE 17.10 Laser range finding

Scientists use *laser range finding* to measure the distance to the moon with great accuracy. A very brief (100 ps) laser pulse, with a wavelength of 532 nm, is fired at the moon, where it reflects off an array of 100 4.0-cm-diameter mirrors placed there by Apollo 15 astronauts in 1971. The reflected laser light returns to earth, where it is collected by a telescope and detected. The average earth-moon distance is 384,000 km.

The laser beam spreads out on its way to the moon because of diffraction, reaching the mirrors with an intensity of 300 W/m^2. The reflected beam spreads out even more on its way back because of diffraction due to the circular aperture of the mirrors.

a. What is the round-trip time for the laser pulse to travel to the moon and back?
b. If we want to measure the distance to the moon to an accuracy of 1.0 cm, how accurately must the arrival time of the returning pulse be measured?
c. Because of the spread of the beam due to diffraction, the light arriving at earth from one of the mirrors will be spread over a circular spot. Estimate the diameter of this spot.
d. What is the intensity of the laser beam when it arrives back at the earth?

PREPARE When the light reflects from one of the circular mirrors, diffraction causes it to spread out as shown in **FIGURE 17.29**. The width of the central maximum for circular-aperture diffraction is given by Equation 17.23.

FIGURE 17.29 The geometry of the returning laser beam.

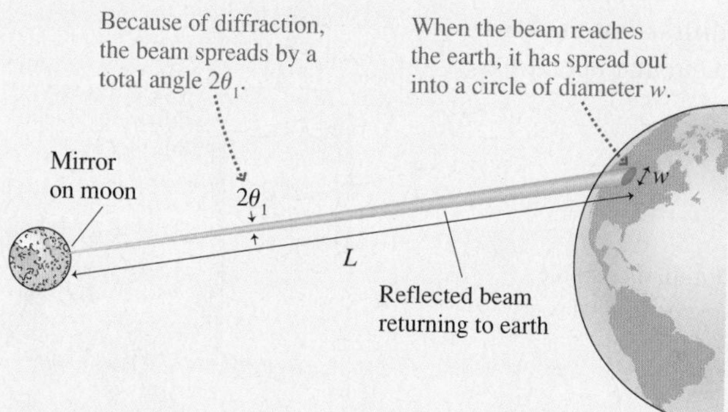
Because of diffraction, the beam spreads by a total angle $2\theta_1$.
When the beam reaches the earth, it has spread out into a circle of diameter w.
Mirror on moon
$2\theta_1$
w
L
Reflected beam returning to earth

Because we know the intensity of the laser beam as it strikes the mirrors, and we can easily find a mirror's area, we can use Equation 15.11 to calculate the power of the beam as it leaves a mirror. The intensity of the beam when it arrives back on earth will be much lower, and can also be found from Equation 15.11.

SOLVE a. The round-trip distance is $2L$. Thus the round-trip travel time for the pulse, traveling at speed c, is

$$\Delta t = \frac{2L}{c} = \frac{2(3.84 \times 10^8 \text{ m})}{3.00 \times 10^8 \text{ m/s}} = 2.56 \text{ s}$$

b. If we wish to measure the moon's distance from the earth to an accuracy of ± 1.0 cm, then, because the laser beam travels both to and from the moon, we need to know the round-trip distance to an accuracy of ± 2.0 cm. The time it takes light to travel $\Delta x = 2.0$ cm is

$$\Delta t = \frac{\Delta x}{c} = \frac{0.020 \text{ m}}{3.00 \times 10^8 \text{ m/s}} = 6.6 \times 10^{-11} = 66 \text{ ps}$$

Thus the arrival time of the pulses must be timed to an accuracy of about 70 ps.

c. The light arriving at the moon reflects from circular mirrors of diameter D. Diffraction by these circular apertures causes the returning light to spread out with angular width $2\theta_1$, where θ_1 is the angle of the first minimum on either side of the central maximum. Equation 17.23 found that the width of the central maximum—the diameter of the circular spot of light when it reaches the earth—is

$$w = \frac{2.44\lambda L}{D} = \frac{2.44(532 \times 10^{-9} \text{ m})(3.84 \times 10^8 \text{ m})}{0.040 \text{ m}}$$
$$= 12{,}000 \text{ m}$$

d. The light is reflected from circular mirrors of radius $r = D/2$ and area $a = \pi r^2 = \pi D^2/4$. The power reflected from one mirror is, according to Equation 15.11,

$$P = Ia = I\pi\frac{D^2}{4} = (300 \text{ W/m}^2)\pi\frac{(0.040 \text{ m})^2}{4} = 0.38 \text{ W}$$

When the pulse returns to earth, it is now spread over the large area $\pi w^2/4$. Thus the intensity at the earth's surface from one mirror reflection is

$$I_1 = \frac{P}{\pi\frac{w^2}{4}} = \frac{0.38 \text{ W}}{\pi\frac{(12{,}000 \text{ m})^2}{4}} = 3.3 \times 10^{-9} \text{ W/m}^2$$

There are 100 reflecting mirrors in the array, so the total intensity reaching earth is $I = 100I_1 = 3.3 \times 10^{-7}$ W/m^2. A large telescope is needed to detect this very small intensity.

ASSESS A 1.0-m-diameter telescope collects only a few *photons*, or particles of light, in each returning pulse. Despite the difficult challenge of detecting this very weak signal, the accuracy of these measurements is astounding: The latest experiments can measure the instantaneous distance to the moon to ± 1 mm, a precision of 3 parts in a trillion!

SUMMARY

The goal of Chapter 17 has been to understand and apply the wave model of light.

GENERAL PRINCIPLES

The Wave Model

The wave model considers light to be a wave propagating through space. Interference and diffraction are important. The wave model is appropriate when light interacts with objects whose size is comparable to the wavelength of light, or roughly less than about 0.1 mm.

Huygens' principle says that each point on a wave front is the source of a spherical wavelet. The wave front at a later time is tangent to all the wavelets.

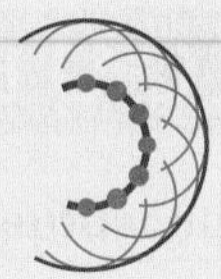

IMPORTANT CONCEPTS

The **index of refraction** of a material determines the speed of light in that material: $v = c/n$. The index of refraction of a material is always greater than 1, so that v is always less than c.

The wavelength λ in a material with index of refraction n is *shorter* than the wavelength λ_{vac} in a vacuum: $\lambda = \lambda_{vac}/n$.

The *frequency* of light does not change as it moves from one material to another.

Diffraction is the spreading of a wave after it passes through an opening.

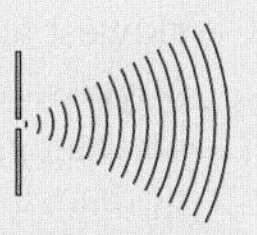

Constructive and destructive **interference** are due to the overlap of two or more waves as they spread behind openings.

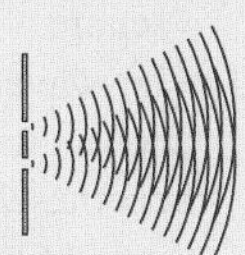

APPLICATIONS

Diffraction from a single slit

A single slit of width a has a bright **central maximum** of width

$$w = \frac{2\lambda L}{a}$$

that is flanked by weaker **secondary maxima.**

Dark fringes are located at angles such that

$$a\sin\theta_p = p\lambda \qquad p = 1, 2, 3, \ldots$$

If $\lambda/a \ll 1$, then from the small-angle approximation,

$$\theta_p = \frac{p\lambda}{a} \qquad y_p = \frac{p\lambda L}{a}$$

Interference from multiple slits

Waves overlap as they spread out behind slits. Bright fringes are seen on the viewing screen at positions where the path-length difference Δr between successive slits is equal to $m\lambda$, where m is an integer.

Double slit with separation d

Equally spaced bright fringes are located at

$$\theta_m = \frac{m\lambda}{d} \qquad y_m = \frac{m\lambda L}{d} \qquad m = 0, 1, 2, \ldots$$

The **fringe spacing** is $\Delta y = \dfrac{\lambda L}{d}$

Diffraction grating with slit spacing d

Very bright and narrow fringes are located at angles and positions

$$d\sin\theta_m = m\lambda \qquad y_m = L\tan\theta_m$$

Circular aperture of diameter D

A bright central maximum of diameter

$$w = \frac{2.44\lambda L}{D}$$

is surrounded by circular secondary maxima. The first dark fringe is located at

$$\theta_1 = \frac{1.22\lambda}{D} \qquad y_1 = \frac{1.22\lambda L}{D}$$

For an aperture of any shape, a smaller opening causes a greater spreading of the wave behind the opening.

Thin-film interference

Interference occurs between the waves reflected from the two surfaces of a thin film with index of refraction n. A wave that reflects from a surface at which the index of refraction increases has a phase change.

Interference	0 or 2 phase changes	1 phase change
Constructive	$2t = m\dfrac{\lambda}{n}$	$2t = \left(m + \dfrac{1}{2}\right)\dfrac{\lambda}{n}$
Destructive	$2t = \left(m + \dfrac{1}{2}\right)\dfrac{\lambda}{n}$	$2t = m\dfrac{\lambda}{n}$

For homework assigned on MasteringPhysics, go to www.masteringphysics.com

Problems labeled INT integrate significant material from earlier chapters; BIO are of biological or medical interest.

Problem difficulty is labeled as | (straightforward) to ||||| (challenging).

QUESTIONS

Conceptual Questions

1. The frequency of a light wave in air is 5.3×10^{14} Hz. Is the frequency of this wave higher, lower, or the same after the light enters a piece of glass?
2. Rank in order the following according to their speeds, from slowest to fastest: (i) 425-nm-wavelength light through a pane of glass, (ii) 500-nm-wavelength light through air, (iii) 540-nm-wavelength light through water, (iv) 670-nm-wavelength light through a diamond, and (v) 670-nm-wavelength light through a vacuum.
3. The wavelength of a light wave is 700 nm in air; this light appears red. If this wave enters a pool of water, its wavelength becomes $\lambda_{air}/n = 530$ nm. If you were swimming underwater, the light would still appear red. Given this, what property of a wave determines its color?
4. A double-slit interference experiment shows fringes on a screen. The entire experiment is then immersed in water. Do the fringes on the screen get closer together, farther apart, remain the same, or disappear entirely? Explain.
5. Figure Q17.5 shows the fringes observed in a double-slit interference experiment when the two slits are illuminated by white light. The central maximum is white, but as we move away from the central maximum, the fringes become less distinct and more colorful. What is special about the central maximum that makes it white? Explain the presence of colors in the outlying fringes.

FIGURE Q17.5

6. In a double-slit interference experiment, interference fringes are observed on a distant screen. The width of both slits is then doubled without changing the distance between their centers.
   a. What happens to the spacing of the fringes? Explain.
   b. What happens to the intensity of the bright fringes? Explain.
7. Figure Q17.7 shows the viewing screen in a double-slit experiment with monochromatic light. Fringe C is the central maximum.
   a. What will happen to the fringe spacing if the wavelength of the light is decreased?
   b. What will happen to the fringe spacing if the spacing between the slits is decreased?
   c. What will happen to the fringe spacing if the distance to the screen is decreased?
   d. Suppose the wavelength of the light is 500 nm. How much farther is it from the dot on the screen in the center of fringe E to the left slit than it is from the dot to the right slit?

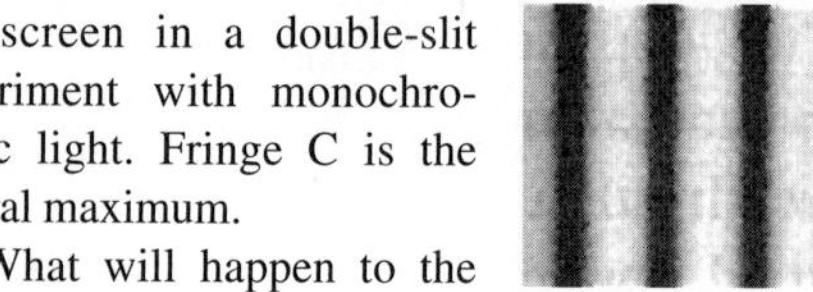

FIGURE Q17.7

8. Figure Q17.7 is the interference pattern seen on a viewing screen behind 2 slits. Suppose the 2 slits were replaced by 20 slits having the same spacing d between adjacent slits.
   a. Would the number of fringes on the screen increase, decrease, or stay the same?
   b. Would the fringe spacing increase, decrease, or stay the same?
   c. Would the width of each fringe increase, decrease, or stay the same?
   d. Would the brightness of each fringe increase, decrease, or stay the same?
9. Figure Q17.9 shows the light intensity on a viewing screen behind a single slit of width a. The light's wavelength is λ. Is $\lambda < a$, $\lambda = a$, $\lambda > a$, or is it not possible to tell? Explain.

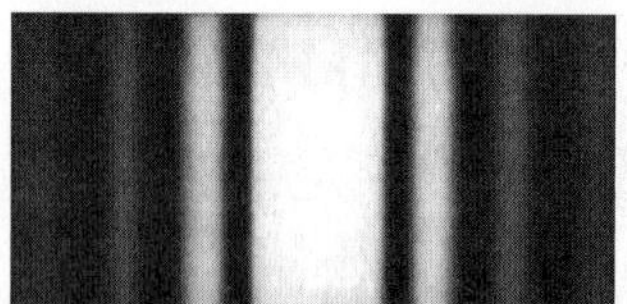

FIGURE Q17.9

10. Figure Q17.10 shows the light intensity on a viewing screen behind a circular aperture. What happens to the width of the central maximum if
    a. The wavelength is increased?
    b. The diameter of the aperture is increased?
    c. How will the screen appear if the aperture diameter is less than the light wavelength?

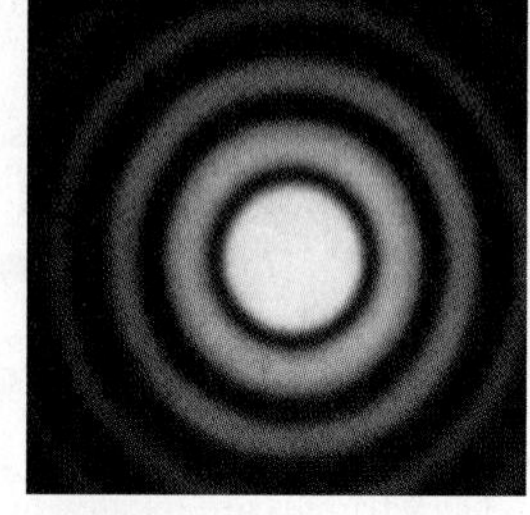

FIGURE Q17.10

11. BIO Why does light reflected from peacock feathers change color when you see the feathers at a different angle?
12. White light is incident on a diffraction grating. What color is the central maximum of the interference pattern?
13. A soap bubble usually pops because some part of it becomes too thin due to evaporation or drainage of fluid. The change in thickness also changes the color of light the bubble reflects. Why?
14. An oil film on top of water has one patch that is much thinner than the wavelength of visible light. The index of refraction of the oil is less than that of water. Will the reflection from that extremely thin part of the film be bright or dark? Explain.
15. Should the antireflection coating of a microscope objective lens designed for use with ultraviolet light be thinner, thicker, or the same thickness as the coating on a lens designed for visible light?
16. If the thin wedge of air between the two plates of glass in Figure 17.21 were replaced by water, would the distance between the fringes increase, decrease, or remain the same? Explain.

17. Example 17.5 showed that a thin film whose thickness is one-quarter of the wavelength of light in the film serves as an antireflection coating when coated on glass. In Example 17.5, $n_{film} < n_{glass}$. If a quarter-wave thickness film with $n_{film} > n_{glass}$ were used instead, would the film still serve as an antireflection coating? Explain.
18. You are standing against the wall near a corner of a large building. A friend is standing against the wall that is around the corner from you. You can't see your friend. How is it that you can hear her when she talks to you?

Multiple-Choice Questions

19. | Light of wavelength 500 nm in air enters a glass block with index of refraction $n = 1.5$. When the light enters the block, which of the following properties of the light will not change?
   A. The speed of the light
   B. The frequency of the light
   C. The wavelength of the light
20. | The frequency of a light wave in air is 4.6×10^{14} Hz. What is the wavelength of this wave after it enters a pool of water?
   A. 300 nm B. 490 nm C. 650 nm D. 870 nm
21. | Light passes through a diffraction grating with a slit spacing of 0.001 mm. A viewing screen is 100 cm behind the grating. If the light is blue, with a wavelength of 450 nm, at about what distance from the center of the interference pattern will the first-order maximum appear?
   A. 5 cm B. 25 cm C. 50 cm D. 100 cm
22. || Blue light of wavelength 450 nm passes through an interference grating with a slit spacing of 0.001 mm and makes an interference pattern on the wall. How many bright fringes will be seen?
   A. 1 B. 3 C. 5 D. 7
23. | Yellow light of wavelength 590 nm passes through a diffraction grating and makes an interference pattern on a screen 80 cm away. The first bright fringes are 1.9 cm from the center of the pattern. How many lines per mm does this grating have?
   A. 20 B. 40 C. 80 D. 200
24. | Light passes through a 10-μm-wide slit and is viewed on a screen 1 m behind the slit. If the width of the slit is narrowed, the band of light on the screen will
   A. Become narrower.
   B. Become wider.
   C. Stay about the same.
25. || Blue light of wavelength 450 nm passes through a 0.20-mm-wide slit and illuminates a screen 1.2 m away. How wide is the central maximum of the diffraction pattern?
   A. 1.2 mm B. 2.0 mm
   C. 2.7 mm D. 5.4 mm
26. || A green laser beam of wavelength 540 nm passes through a pinhole and illuminates a dartboard 3.0 m past the pinhole. The first minimum in the intensity coincides with the ring surrounding the bull's-eye, 12 mm in diameter. What is the diameter of the pinhole?
   A. 0.14 mm B. 0.33 mm
   C. 0.59 mm D. 1.2 mm

PROBLEMS

Section 17.1 What Is Light?

1. ||| a. How long does it take light to travel through a 3.0-mm-thick piece of window glass?
   b. Through what thickness of water could light travel in the same amount of time?
2. | a. How long (in ns) does it take light to travel 1.0 m in vacuum?
   b. What distance does light travel in water, glass, and diamond during the time that it travels 1.0 m in a vacuum?
3. |||| A 5.0-cm-thick layer of oil ($n = 1.46$) is sandwiched between a 1.0-cm-thick sheet of glass and a 2.0-cm-thick sheet of polystyrene plastic ($n = 1.59$). How long (in ns) does it take light incident perpendicular to the glass to pass through this 8.0-cm-thick sandwich?
4. || A light wave has a 670 nm wavelength in air. Its wavelength in a transparent solid is 420 nm.
   a. What is the speed of light in this solid?
   b. What is the light's frequency in the solid?
5. || How much time does it take a pulse of light to travel through 150 m of water?
6. || A helium-neon laser beam has a wavelength in air of 633 nm. It takes 1.38 ns for the light to travel through 30.0 cm of an unknown liquid. What is the wavelength of the laser beam in the liquid?

Section 17.2 The Interference of Light

7. || Two narrow slits 50 μm apart are illuminated with light of wavelength 500 nm. What is the angle of the $m = 2$ bright fringe in radians? In degrees?
8. ||| Light from a sodium lamp ($\lambda = 589$ nm) illuminates two narrow slits. The fringe spacing on a screen 150 cm behind the slits is 4.0 mm. What is the spacing (in mm) between the two slits?
9. || Two narrow slits are illuminated by light of wavelength λ. The slits are spaced 20 wavelengths apart. What is the angle, in radians, between the central maximum and the $m = 1$ bright fringe?
10. || A double-slit experiment is performed with light of wavelength 600 nm. The bright interference fringes are spaced 1.8 mm apart on the viewing screen. What will the fringe spacing be if the light is changed to a wavelength of 400 nm?
11. |||| Light from a helium-neon laser ($\lambda = 633$ nm) is used to illuminate two narrow slits. The interference pattern is observed on a screen 3.0 m behind the slits. Twelve bright fringes are seen, spanning a distance of 52 mm. What is the spacing (in mm) between the slits?
12. || Two narrow slits are 0.12 mm apart. Light of wavelength 550 nm illuminates the slits, causing an interference pattern on a screen 1.0 m away. Light from each slit travels to the $m = 1$ maximum on the right side of the central maximum. How much farther did the light from the left slit travel than the light from the right slit?

13. ||| Consider a point P on the viewing screen of a double-slit interference experiment. This point is 75% of the way from the center of the 3rd bright fringe to the center of the 4th bright fringe. If the wavelength of the light is 600 nm, what is the extra distance that the wave from one slit traveled compared to the wave from the other?

Section 17.3 The Diffraction Grating

14. ||| A diffraction grating with 750 slits/mm is illuminated by light that gives a first-order diffraction angle of 34.0°. What is the wavelength of the light?
15. |||| A 1.0-cm-wide diffraction grating has 1000 slits. It is illuminated by light of wavelength 550 nm. What are the angles of the first two diffraction orders?
16. ||| Light of wavelength 600 nm illuminates a diffraction grating. The second-order maximum is at angle 39.5°. How many lines per millimeter does this grating have?
17. ||| A lab technician uses laser light with a wavelength of 670 nm to test a diffraction grating. When the grating is 40.0 cm from the screen, the first-order maxima appear 6.00 cm from the center of the pattern. How many lines per millimeter does this grating have?
18. || The human eye can readily detect wavelengths from about 400 nm to 700 nm. If white light illuminates a diffraction grating having 750 lines/mm, over what range of angles does the visible $m = 1$ spectrum extend?
19. ||| A diffraction grating with 600 lines/mm is illuminated with light of wavelength 500 nm. A very wide viewing screen is 2.0 m behind the grating.
    a. What is the distance between the two $m = 1$ fringes?
    b. How many bright fringes can be seen on the screen?
20. || A 500 line/mm diffraction grating is illuminated by light of wavelength 510 nm. How many diffraction orders are seen, and what is the angle of each?

Section 17.4 Thin-Film Interference

21. || What is the thinnest film of MgF_2 ($n = 1.38$) on glass that produces a strong reflection for orange light with a wavelength of 600 nm?
22. |||| A very thin oil film ($n = 1.25$) floats on water ($n = 1.33$). What is the thinnest film that produces a strong reflection for green light with a wavelength of 500 nm?
23. || A film with $n = 1.60$ is deposited on glass. What is the thinnest film that will produce constructive interference in the reflection of light with a wavelength of 550 nm?
24. || Antireflection coatings can be used on the *inner* surfaces of BIO eyeglasses to reduce the reflection of stray light into the eye, thus reducing eyestrain.
    a. A 90-nm-thick coating is applied to the lens. What must be the coating's index of refraction to be most effective at 480 nm? Assume that the coating's index of refraction is less than that of the lens.
    b. If the index of refraction of the coating is 1.38, what thickness should the coating be so as to be most effective at 480 nm? The thinnest possible coating is best.
25. || Solar cells are given antireflection coatings to maximize their efficiency. Consider a silicon solar cell ($n = 3.50$) coated with a layer of silicon dioxide ($n = 1.45$). What is the minimum coating thickness that will minimize the reflection at the wavelength of 700 nm, where solar cells are most efficient?
26. ||| A thin film of MgF_2 ($n = 1.38$) coats a piece of glass. Constructive interference is observed for the reflection of light with wavelengths of 500 nm and 625 nm. What is the thinnest film for which this can occur?
27. | Looking straight downward into a rain puddle whose surface is covered with a thin film of gasoline, you notice a swirling pattern of colors caused by interference inside the gasoline film. The point directly beneath you is colored a beautiful iridescent green. You happen to remember that the index of refraction of gasoline is 1.38 and that the wavelength of green light is about 540 nm. What is the minimum possible thickness of the gasoline layer directly beneath you?

Section 17.5 Single-Slit Diffraction

28. || A helium-neon laser ($\lambda = 633$ nm) illuminates a single slit and is observed on a screen 1.50 m behind the slit. The distance between the first and second minima in the diffraction pattern is 4.75 mm. What is the width (in mm) of the slit?
29. ||| For a demonstration, a professor uses a razor blade to cut a thin slit in a piece of aluminum foil. When she shines a laser pointer ($\lambda = 680$ nm) through the slit onto a screen 5.5 m away, a diffraction pattern appears. The bright band in the center of the pattern is 8.0 cm wide. What is the width of the slit?
30. || A 0.50-mm-wide slit is illuminated by light of wavelength 500 nm. What is the width of the central maximum on a screen 2.0 m behind the slit?
31. ||| The second minimum in the diffraction pattern of a 0.10-mm-wide slit occurs at 0.70°. What is the wavelength of the light?
32. || What is the width of a slit for which the first minimum is at 45° when the slit is illuminated by a helium-neon laser ($\lambda = 633$ nm)?
    Hint: The small-angle approximation is not valid at 45°.

Section 17.6 Circular-Aperture Diffraction

33. ||| A 0.50-mm-diameter hole is illuminated by light of wavelength 500 nm. What is the width of the central maximum on a screen 2.0 m behind the slit?
34. || Light from a helium-neon laser ($\lambda = 633$ nm) passes through a circular aperture and is observed on a screen 4.0 m behind the aperture. The width of the central maximum is 2.5 cm. What is the diameter (in mm) of the hole?
35. ||| You want to photograph a circular diffraction pattern whose central maximum has a diameter of 1.0 cm. You have a helium-neon laser ($\lambda = 633$ nm) and a 0.12-mm-diameter pinhole. How far behind the pinhole should you place the viewing screen?
36. || Infrared light of wavelength 2.5 μm illuminates a 0.20-mm-diameter hole. What is the angle of the first dark fringe in radians? In degrees?

General Problems

37. ||| An advanced computer sends information to its various parts via infrared light pulses traveling through silicon fibers ($n = 3.50$). To acquire data from memory, the central processing unit sends a light-pulse request to the memory unit. The

memory unit processes the request, then sends a data pulse back to the central processing unit. The memory unit takes 0.50 ns to process a request. If the information has to be obtained from memory in 2.00 ns, what is the maximum distance the memory unit can be from the central processing unit?

38. ||| Figure P17.38 shows the light intensity on a screen behind a double slit. The slit spacing is 0.20 mm and the wavelength of the light is 600 nm. What is the distance from the slits to the screen?

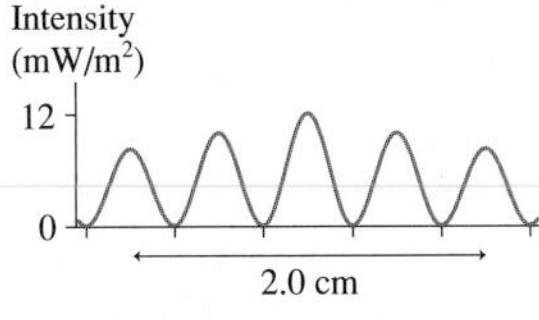

FIGURE P17.38

39. |||| Figure P17.38 shows the light intensity on a screen behind a double slit. The slit spacing is 0.20 mm and the screen is 2.0 m behind the slits. What is the wavelength of the light?

40. || Your friend has been given a laser for her birthday. Unfortunately, she did not receive a manual with it and so she doesn't know the wavelength that it emits. You help her by performing a double-slit experiment, with slits separated by 0.36 mm. You find that the two bright fringes are 5.5 mm apart on a screen 1.6 m from the slits. What is the wavelength the laser emits?

41. | A double slit is illuminated simultaneously with orange light of wavelength 600 nm and light of an unknown wavelength. The $m = 4$ bright fringe of the unknown wavelength overlaps the $m = 3$ bright orange fringe. What is the unknown wavelength?

42. |||| A laser beam, with a wavelength of 532 nm, is directed exactly perpendicular to a screen having two narrow slits spaced 0.15 mm apart. Interference fringes, including a central maximum, are observed on a screen 1.0 m away. The direction of the beam is then slowly rotated around an axis parallel to the slits to an angle of 1.0°. By what distance does the central maximum on the screen move?

43. | A laser beam of wavelength 670 nm shines through a diffraction grating that has 750 lines/mm. Sketch the pattern that appears on a screen 1.0 m behind the grating, noting distances on your drawing and explaining where these numbers come from.

44. || The two most prominent wavelengths in the light emitted by a hydrogen discharge lamp are 656 nm (red) and 486 nm (blue). Light from a hydrogen lamp illuminates a diffraction grating with 500 lines/mm, and the light is observed on a screen 1.50 m behind the grating. What is the distance between the first-order red and blue fringes?

45. ||| A triple-slit experiment illuminates three equally spaced, narrow slits with light of wavelength λ. The intensity of the wave from each slit is I_1. Consider a point on a distant screen at an angle such that the path-length difference between any two adjacent slits is $\lambda/2$. What is the intensity at this point? Give your answer as a multiple of I_1.

46. || A diffraction grating consists of 100 slits. If the number of slits is increased to 200, with the same spacing, by what factor does the maximum intensity of the bright fringes on the screen increase?

47. || A diffraction grating produces a first-order maximum at an angle of 20.0°. What is the angle of the second-order maximum?

48. | A diffraction grating is illuminated simultaneously with red light of wavelength 660 nm and light of an unknown wavelength. The fifth-order maximum of the unknown wavelength exactly overlaps the third-order maximum of the red light. What is the unknown wavelength?

49. ||| White light (400–700 nm) is incident on a 600 line/mm diffraction grating. What is the width of the first-order rainbow on a screen 2.0 m behind the grating?

50. ||| For your science fair project you need to design a diffraction grating that will disperse the visible spectrum (400–700 nm) over 30.0° in first order.
    a. How many lines per millimeter does your grating need?
    b. What is the first-order diffraction angle of light from a sodium lamp ($\lambda = 589$ nm)?

51. || Figure P17.51 shows the interference pattern on a screen 1.0 m behind an 800 line/mm diffraction grating. What is the wavelength of the light?

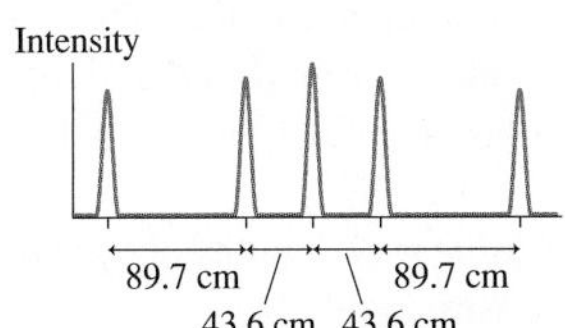

FIGURE P17.51

52. || Figure P17.51 shows the interference pattern on a screen 1.0 m behind a diffraction grating. The wavelength of the light is 600 nm. How many lines per millimeter does the grating have?

53. || Because sound is a wave, it is possible to make a diffraction
INT grating for sound from a large board with several parallel slots for the sound to go through. When 10 kHz sound waves pass through such a grating, listeners 10 m from the grating report "loud spots" 1.4 m on both sides of center. What is the spacing between the slots? Use 340 m/s for the speed of sound.

54. ||| The shiny surface of a CD is imprinted with millions of tiny pits, arranged in a pattern of thousands of essentially concentric circles that act like a reflection grating when light shines on them. You decide to determine the distance between those circles by aiming a laser pointer (with $\lambda = 680$ nm) perpendicular to the disk and measuring the diffraction pattern reflected onto a screen 1.5 m from the disk. The central bright spot you expected to see is blocked by the laser pointer itself. You do find two other bright spots separated by 1.4 m, one on either side of the missing central spot. The rest of the pattern is apparently diffracted at angles too great to show on your screen. What is the distance between the circles on the CD's surface?

55. || If sunlight shines straight onto a peacock feather, the feather
BIO appears bright blue when viewed from 15° on either side of the incident beam of sunlight. The blue color is due to diffraction from the melanin bands in the feather barbules, as was shown in the photograph on page 558. Blue light with a wavelength of 470 nm is diffracted at 15° by these bands (this is the first-order diffraction) while other wavelengths in the sunlight are diffracted at different angles. What is the spacing of the melanin bands in the feather?

56. || The wings of some beetles
BIO have closely spaced parallel lines of melanin, causing the wing to act as a reflection grating. Suppose sunlight shines straight onto a beetle wing. If the melanin lines on the wing are spaced 2.0 μm apart, what is the first-order diffraction angle for green light ($\lambda = 550$ nm)?

57. ||| A diffraction grating having 500 lines/mm diffracts visible light at 30°. What is the light's wavelength?

58. ||| Light emitted by Element X passes through a diffraction grating having 1200 lines/mm. The interference pattern is observed on a screen 75.0 cm behind the grating. Bright fringes are *seen* on the screen at distances of 56.2 cm, 65.9 cm, and 93.5 cm from the central maximum. No other fringes are seen.
   a. What is the value of m for each of these diffracted wavelengths? Explain why only one value is possible.
   b. What are the wavelengths of light emitted by Element X?

59. || Helium atoms emit light at several wavelengths. Light from a helium lamp illuminates a diffraction grating and is observed on a screen 50.00 cm behind the grating. The emission at wavelength 501.5 nm creates a first-order bright fringe 21.90 cm from the central maximum. What is the wavelength of the bright fringe that is 31.60 cm from the central maximum?

60. || A sheet of glass is coated with a 500-nm-thick layer of oil ($n = 1.42$).
   a. For what *visible* wavelengths of light do the reflected waves interfere constructively?
   b. For what *visible* wavelengths of light do the reflected waves interfere destructively?
   c. What is the color of reflected light? What is the color of transmitted light?

61. | A soap bubble is essentially a thin film of water surrounded by air. The colors you see in soap bubbles are produced by interference. What visible wavelengths of light are strongly reflected from a 390-nm-thick soap bubble? What color would such a soap bubble appear to be?

62. || In a single-slit experiment, the slit width is 200 times the wavelength of the light. What is the width of the central maximum on a screen 2.0 m behind the slit?

63. ||| You need to use your cell phone, which broadcasts an 830 MHz signal, but you're in an alley between two massive, radio-wave-absorbing buildings that have only a 15 m space between them. What is the angular width, in degrees, of the electromagnetic wave after it emerges from between the buildings?

64. ||| Light from a sodium lamp ($\lambda = 589$ nm) illuminates a narrow slit and is observed on a screen 75 cm behind the slit. The distance between the first and third dark fringes is 7.5 mm. What is the width (in mm) of the slit?

65. ||| INT The opening to a cave is a tall, 30-cm-wide crack. A bat that is preparing to leave the cave emits a 30 kHz ultrasonic chirp. How wide is the "sound beam" 100 m outside the cave opening? Use $v_{sound} = 340$ m/s.

66. | For what slit-width-to-wavelength ratio does the first minimum of a single-slit diffraction pattern appear at (a) 30°, (b) 60°, and (c) 90°?
   Hint: The small-angle approximation is not valid.

67. ||| Figure P17.67 shows the light intensity on a screen behind a single slit. The wavelength of the light is 500 nm and the screen is 1.0 m behind the slit. What is the width (in mm) of the slit?

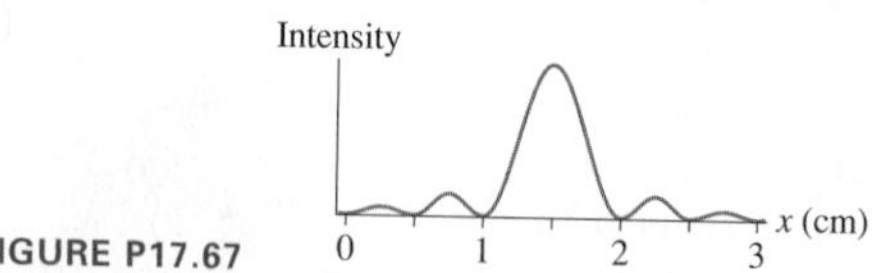

FIGURE P17.67

68. ||| Figure P17.67 shows the light intensity on a screen behind a single slit. The wavelength of the light is 600 nm and the slit width is 0.15 mm. What is the distance from the slit to the screen?

69. || Figure P17.69 shows the light intensity on a screen 2.5 m behind an aperture. The aperture is illuminated with light of wavelength 600 nm.
   a. Is the aperture a single slit or a double slit? Explain.
   b. If the aperture is a single slit, what is its width? If it is a double slit, what is the spacing between the slits?

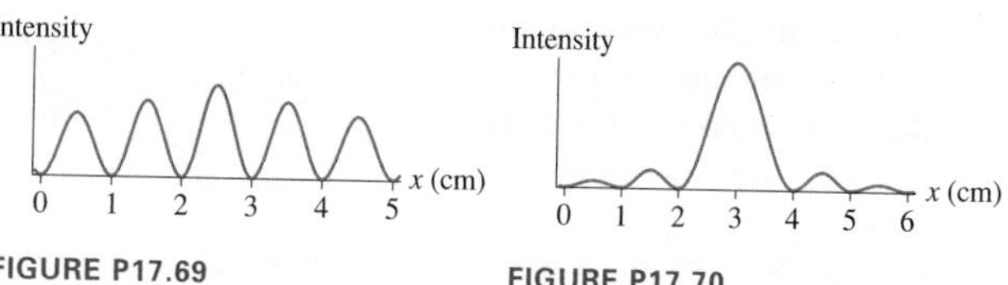

FIGURE P17.69 FIGURE P17.70

70. || Figure P17.70 shows the light intensity on a screen 2.5 m behind an aperture. The aperture is illuminated with light of wavelength 600 nm.
   a. Is the aperture a single slit or a double slit? Explain.
   b. If the aperture is a single slit, what is its width? If it is a double slit, what is the spacing between the slits?

71. ||| One day, after pulling down your window shade, you notice that sunlight is passing through a pinhole in the shade and making a small patch of light on the far wall. Having recently studied optics in your physics class, you're not too surprised to see that the patch of light seems to be a circular diffraction pattern. It appears that the central maximum is about 3 cm across, and you estimate that the distance from the window shade to the wall is about 3 m. Knowing that the average wavelength of sunlight is about 500 nm, estimate the diameter of the pinhole.

72. |||| INT A radar for tracking aircraft broadcasts a 12 GHz microwave beam from a 2.0-m-diameter circular radar antenna. From a wave perspective, the antenna is a circular aperture through which the microwaves diffract.
   a. What is the diameter of the radar beam at a distance of 30 km?
   b. If the antenna emits 100 kW of power, what is the average microwave intensity at 30 km?

73. ||||| A helium-neon laser ($\lambda = 633$ nm), shown in Figure P17.73, is built with a glass tube of inside diameter 1.0 mm. One mirror is partially transmitting to allow the laser beam out. An electrical discharge in the tube causes it to glow like a neon light. From an optical perspective, the laser beam is a light wave that diffracts out through a 1.0-mm-diameter circular opening.
   a. Explain why a laser beam can't be *perfectly* parallel, with no spreading.
   b. The angle θ_1 to the first minimum is called the *divergence angle* of a laser beam. What is the divergence angle of this laser beam?
   c. What is the diameter (in mm) of the laser beam after it travels 3.0 m?
   d. What is the diameter of the laser beam after it travels 1.0 km?

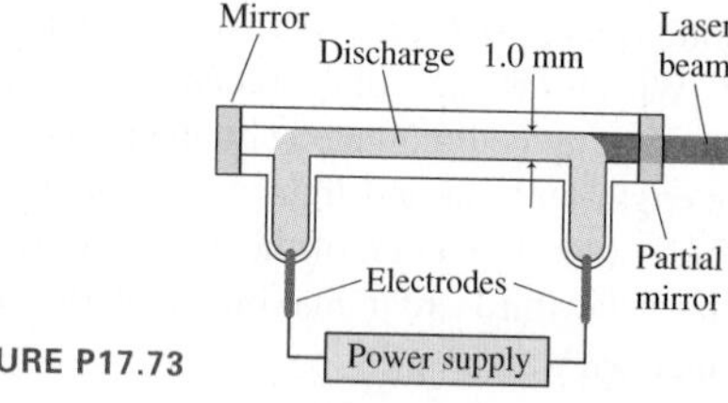

FIGURE P17.73

74. || In the laser range-finding experiments of Example 17.10, the laser beam fired toward the moon spreads out as it travels because it diffracts through a circular exit as it leaves the laser. In order for the reflected light to be bright enough to detect, the laser spot on the moon must be no more than 1 km in diameter. Staying within this diameter is accomplished by using a special large-diameter laser. If $\lambda = 532$ nm, what is the minimum diameter of the circular opening from which the laser beam emerges? The earth-moon distance is 384,000 km.

Passage Problems

The Blue Morpho Butterfly BIO

The brilliant blue color of a blue morpho butterfly is, like the colors of peacock feathers, due to interference. Figure P17.75a shows an easy way to demonstrate this: If a drop of the clear solvent acetone is placed on the wing of a blue morpho butterfly, the color changes from a brilliant blue to an equally brilliant green—returning to blue once the acetone evaporates. There would be no change if the color were due to pigment.

(a)

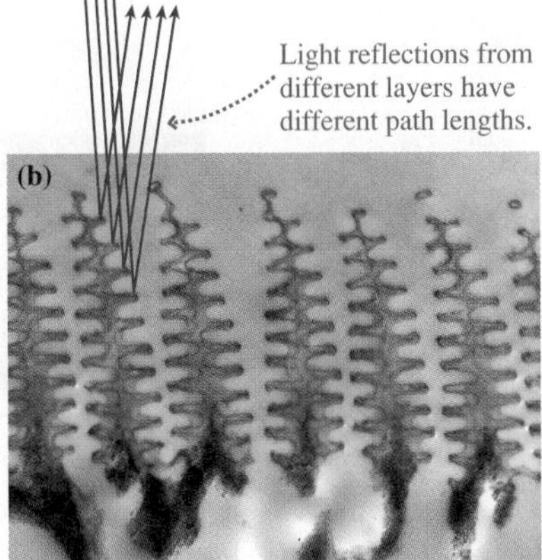
(b)
Light reflections from different layers have different path lengths.

FIGURE P17.75

A cross section of a scale from the wing of a blue morpho butterfly reveals the source of the butterfly's color. As Figure P17.75b shows, the scales are covered with structures that look like small Christmas trees. Light striking the wings reflects from different layers of these structures, and the differing path lengths cause the reflected light to interfere constructively or destructively, depending on the wavelength. For light at normal incidence, blue light experiences constructive interference while other colors undergo destructive interference and cancel. Acetone fills the spaces in the scales with a fluid of index of refraction $n = 1.38$; this changes the conditions for constructive interference and results in a change in color.

75. | The coloring of the blue morpho butterfly is protective. As the butterfly flaps its wings, the angle at which light strikes the wings changes. This causes the butterfly's color to change and makes it difficult for a predator to follow. This color change is because:
    A. A diffraction pattern appears only at certain angles.
    B. The index of refraction of the wing tissues changes as the wing flexes.
    C. The motion of the wings causes a Doppler shift in the reflected light.
    D. As the angle changes, the differences in paths among light reflected from different surfaces change, resulting in constructive interference for a different color.
76. | The change in color when acetone is placed on the wing is due to the difference between the indices of refraction of acetone and air. Consider light of some particular color. In acetone,
    A. The frequency of the light is less than in air.
    B. The frequency of the light is greater than in air.
    C. The wavelength of the light is less than in air.
    D. The wavelength of the light is greater than in air.
77. | The scales on the butterfly wings are actually made of a transparent material with index of refraction 1.56. Light reflects from the surface of the scales because
    A. The scales' index of refraction is different from that of air.
    B. The scales' index of refraction is similar to that of glass.
    C. The scales' density is different from that of air.
    D. Different colors of light have different wavelengths.

STOP TO THINK ANSWERS

Stop to Think 17.1: $n_3 > n_1 > n_2$. $\lambda = \lambda_{vac}/n$, so a shorter wavelength corresponds to a higher index of refraction.

Stop to Think 17.2: B. When the screen is closer, you don't have to move as far from the center to reach the point where the path-length difference is one wavelength.

Stop to Think 17.3: Smaller. The fringe spacing Δy is directly proportional to the wavelength λ.

Stop to Think 17.4: D. Longer wavelengths have larger diffraction angles. Red light has a longer wavelength than violet light, so red light is diffracted farther from the center.

Stop to Think 17.5: B. An extra path difference of $\lambda/2$ must be added to change from constructive to destructive interference. In thin-film interference, one wave passes *twice* through the film. To increase the path length by $\lambda/2$, the thickness needs to be increased by only one-half this, or $\lambda/4$.

Stop to Think 17.6: B or C. The width of the central maximum, which is proportional to λ/a, has increased. This could occur either because the wavelength has increased or because the slit width has decreased.

18 Ray Optics

These thin beams of light at this laser show are well described using ray optics. How do these light beams behave when they reflect from shiny surfaces or pass through transparent materials?

LOOKING AHEAD ▸▸

The goal of Chapter 18 is to understand and apply the ray model of light.

The Ray Model of Light

The ray model applies when light interacts with objects that are large compared to its wavelength. It describes how shadows are formed and how mirrors and lenses work. You'll learn that ...

Looking Back ◂◂
17.1 Models of light

... light rays travel in straight lines unless they are ...

... *reflected* by a surface or ...

... *refracted* at the surface of a transparent material.

Reflection

Light rays can bounce, or **reflect**, off of a surface. The ray model explains what we see in two important cases.

Light rays reflecting from a smooth surface, such as that of still water, can form an image of objects.

Light rays reflect in all directions when they strike a rough surface such as paper, allowing us to see the page.

Refraction

When light rays travel from one transparent material to another, they can change direction, or **refract**, at the interface.

Looking Back ◂◂
17.1 Index of refraction

A diamond sparkles as light reflects and refracts at its many facets.

Light refracting at the water's surface is responsible for the two images of the turtle.

Images Formed by Lenses and Mirrors

You'll learn how lenses form images, starting with a graphical method—**ray tracing**—to find image positions, sizes, and orientations. We'll then develop the **thin-lens equation** for more accurate results.

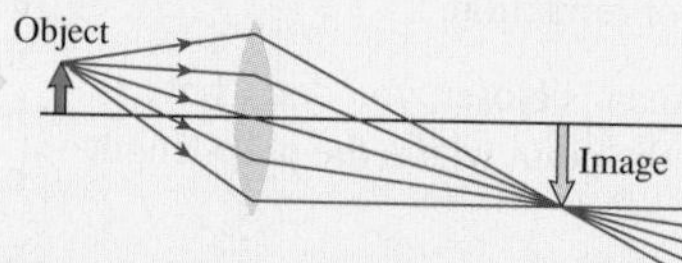

We can use ray tracing to show how this lens creates a **real image** on the opposite side of the lens from the object.

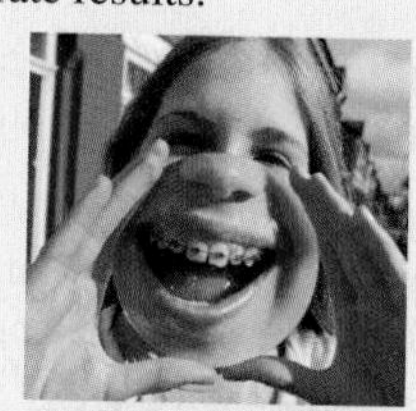

The **virtual image** seen through this lens is on the same side of the lens as the object.

Curved mirrors also can be used to create images. Again, we'll use both graphical and mathematical methods to understand image formation by mirrors.

Your passenger-side rearview mirror is curved, allowing you to see a wider field of view.

18.1 The Ray Model of Light

A flashlight makes a beam of light through the night's darkness. Sunbeams stream into a darkened room through a small hole in the shade, and laser beams are even more well defined. Our everyday experience that light travels in straight lines is the basis of the ray model of light.

The ray model is an oversimplification of reality, but nonetheless is very useful within its range of validity. As we saw in Chapter 17, diffraction and other wave aspects of light are important only for apertures and objects comparable in size to the wavelength of light. Because the wavelength is so small, typically 0.5 μm, the wave nature of light is not apparent when light interacts with ordinary-sized objects. The ray model of light, which ignores diffraction, is valid as long as any apertures through which the light passes (lenses, mirrors, holes, and the like) are larger than about 1 mm.

To begin, let us define a **light ray** as a line in the direction along which light energy is flowing. A light ray is an abstract idea, not a physical entity or a "thing." Any narrow beam of light, such as the laser beam in FIGURE 18.1, is actually a bundle of many parallel light rays. You can think of a single light ray as the limiting case of a laser beam whose diameter approaches zero. Laser beams are good approximations of light rays, certainly adequate for demonstrating ray behavior, but any real laser beam is a bundle of many parallel rays.

FIGURE 18.1 A laser beam is a bundle of parallel light rays.

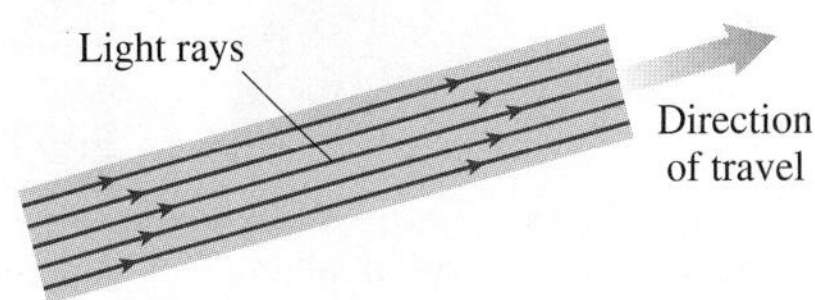

The following table outlines five basic ideas and assumptions of the ray model of light.

The ray model of light

Light rays travel in straight lines.

Light travels through a vacuum or a transparent material in straight lines called light rays. The speed of light in a material is $v = c/n$, where n is the index of refraction of the material.

Light rays can cross.

Light rays do not interact with each other. Two rays can cross without either being affected in any way.

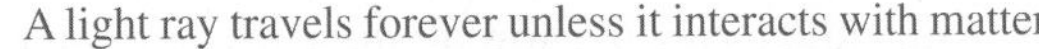

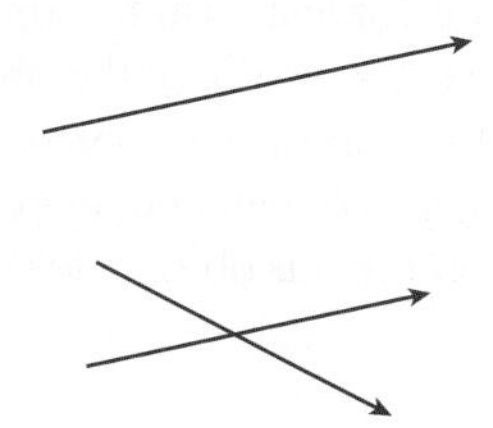

A light ray travels forever unless it interacts with matter.

A light ray continues forever unless it has an interaction with matter that causes the ray to change direction or to be absorbed. Light interacts with matter in four different ways:

- At an interface between two materials, light can be *reflected, refracted,* or both.
- Within a material, light can be either *scattered* or *absorbed.*

These interactions are discussed later in the chapter.

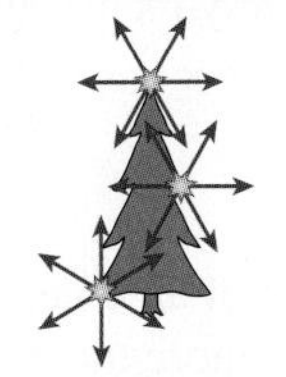

An object is a source of light rays.

An **object** is a source of light rays. Rays originate from *every* point on the object, and each point sends rays in *all* directions. Objects may be self-luminous—they create light rays—or they may be reflective objects that only reflect rays that originate elsewhere.

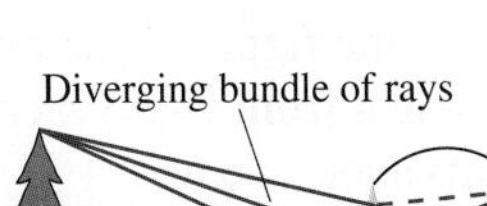

The eye sees by focusing a bundle of rays.

The eye sees an object when *diverging* bundles of rays from each point on the object enter the pupil and are focused to an image on the retina. Imaging is discussed later in the chapter, and the eye will be treated in much greater detail in Chapter 19.

Sources of Light Rays

In the ray model, there are two kinds of objects. **Self-luminous objects** (or *sources*) directly create light rays. Self-luminous objects include lightbulbs and the sun. Other objects, such as a piece of paper or a tree, are **reflective objects** that reflect rays originating in self-luminous objects. The table below shows four important kinds of self-luminous sources. Note that although ray and point sources are idealizations, they are useful in understanding the propagation of rays.

Self-luminous objects

A ray source	A point source	An extended source	A parallel-ray source
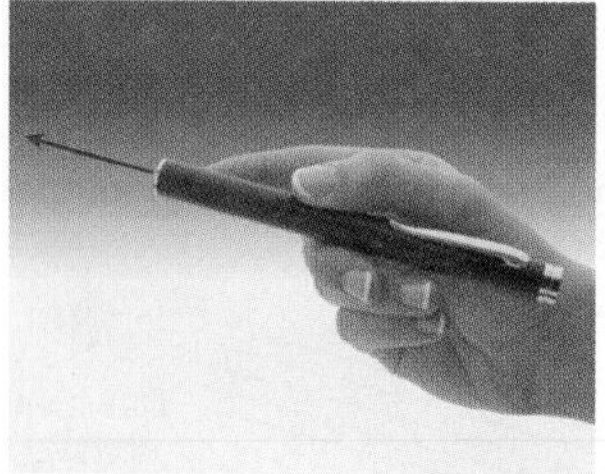			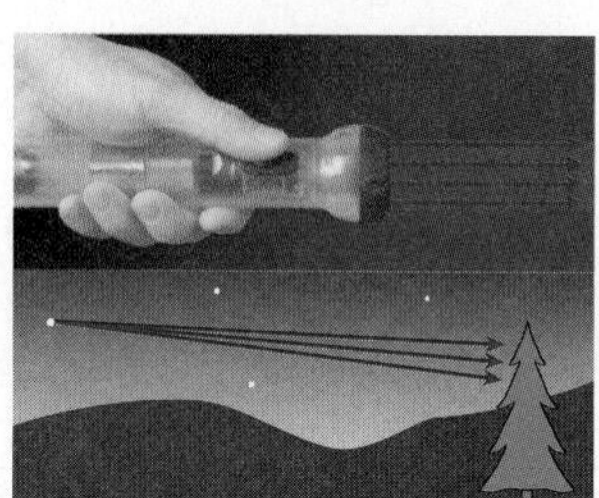
Since a light ray is an idealization, there are no true ray sources. Still, the thin beam of a laser is often a good approximation of a single ray.	A point source is also an idealized source of light. It is infinitely small and emits light rays in every direction. The tiny filaments of these bulbs approximate point sources.	This is the most common light source. The *entire surface* of an extended source is luminous, so that **every point of an extended source acts as a point source.** Lightbulbs, flames, and the sun are extended sources.	Certain sources, such as flashlights and movie projectors, produce a bundle of parallel rays. Rays from a very distant object, such as a star, are very nearly parallel.

Reflective objects, such as a newspaper, a face, or a mirror, can also be considered as sources of light rays. However, the origin of these rays is not in the object itself. Instead, light rays from a self-luminous object strike a reflective object and "bounce" off of it. These rays can then illuminate other objects, or enter our eyes and form images of the reflective object, just as rays from self-luminous objects do.

Ray Diagrams

FIGURE 18.2 A ray diagram simplifies the situation by showing only a few rays.

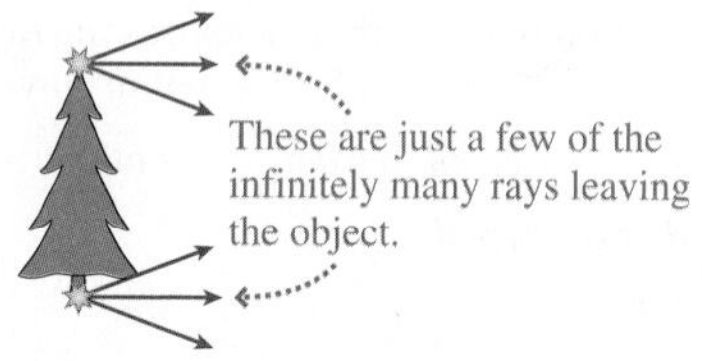

Rays originate from *every* point on an object and travel outward in *all* directions, but a diagram trying to show all these rays would be hopelessly messy and confusing. To simplify the picture, we usually use a **ray diagram** that shows only a few rays. For example, FIGURE 18.2 is a ray diagram showing only a few rays leaving the top and bottom points of the object and traveling to the right. These rays will be sufficient to show us how the object is imaged by lenses or mirrors.

NOTE ► Ray diagrams are the basis for a *visual overview* that we'll use throughout this chapter. Be careful not to think that a ray diagram shows all of the rays. The rays shown on the diagram are just a subset of the infinitely many rays leaving the object. ◄

FIGURE 18.3 A laser beam traveling through air is invisible.

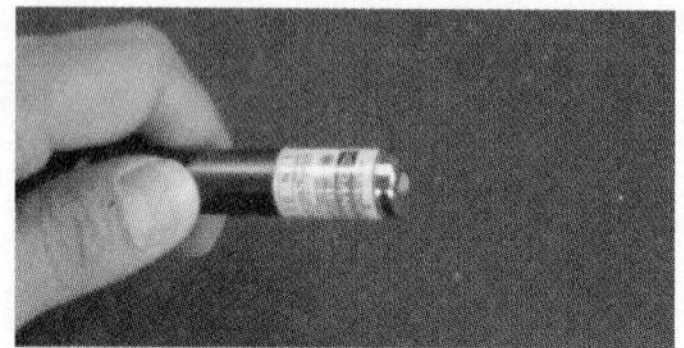

You can't see a laser beam crossing the room because no light ray enters your eye.

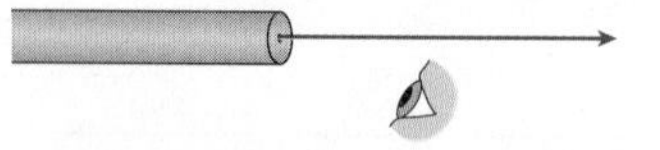

Seeing Objects

How do we *see* an object? The eye works by focusing an image of an object on the retina, a process we'll examine in Chapter 19. For now, we'll ignore the details of image formation and instead make use of a simpler fact from the figure in the ray model table above: **In order for our eye to see an object, rays from that object must enter the eye.** This idea helps explain some subtle points about seeing.

Consider a ray source such as a laser. Can you see a laser beam traveling across the room? Under ordinary circumstances, the answer is no. As we've seen, a laser beam is a good approximation of a single ray. This ray travels in a straight line from the laser to whatever it eventually strikes. As FIGURE 18.3 shows, no light ray enters the eye, so the beam is invisible. The same argument holds for a parallel-ray source.

A point source and an extended source behave differently, as shown in **FIGURE 18.4**. Because a point source emits rays in *every* direction, some of these rays will enter the eye no matter where it is located. Thus a point source is visible to everyone looking at it. And, since every point on the surface of an extended source is itself a point source, all parts of an extended source (not blocked by something else) can be viewed by all observers as well.

FIGURE 18.4 Point and extended sources can be seen by all observers.

A point source — Everyone can see a point source.

An extended source — All points of an extended source are visible.

We can also use our simple model of seeing to explain how we see non-luminous objects. As we've already mentioned, such objects *reflect* rays that strike them. Most ordinary objects—paper, skin, grass—reflect incident light in every direction, a process called **diffuse reflection.** **FIGURE 18.5** illustrates the idea. Single rays are broken into many weaker rays that leave in all directions, a process called **scattering.**

Scattered light is what allows you to read a book by lamplight. As shown in Figure 18.5, every point on the surface of the page is struck by a ray (or rays) from the lamp. Then, because of diffuse reflection, these rays scatter in every direction; some of the scattered rays reach your eye, allowing you to see the page.

FIGURE 18.5 Reading a book by scattered light.

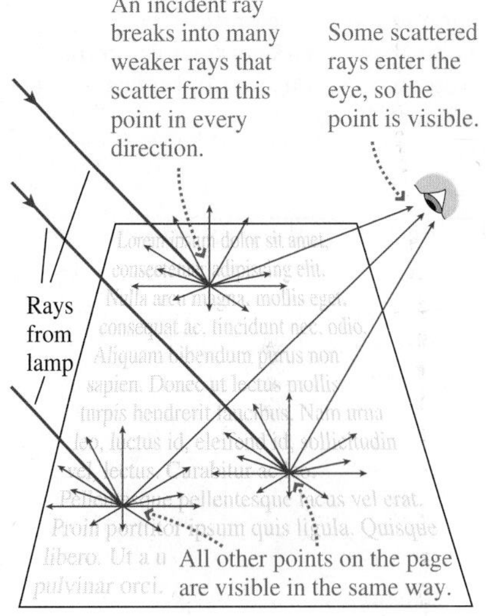

Rays from lamp
All other points on the page are visible in the same way.

It is possible to make a laser beam visible by scattering it from very small particles suspended in air. These particles can be dust, smoke, or water droplets such as fog. **FIGURE 18.6** shows that as the beam strikes such a particle, it scatters rays in every direction. Some of these rays enter the eye, making each particle in the path of the beam visible and outlining the beam's path across the room.

FIGURE 18.6 A laser beam is visible if it travels through smoke or dust.

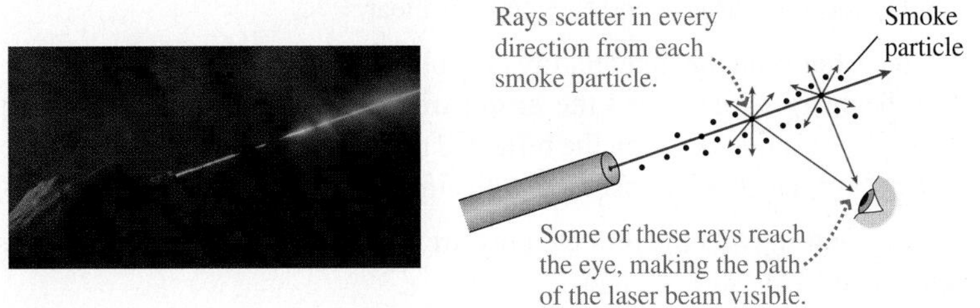

Shadows

Our ray model of light also explains the common phenomenon of *shadows.* Suppose an opaque object (such as a cardboard disk) is placed between a source of light and a screen. The object intercepts some of the rays, leaving a dark area behind it. Other rays travel on to a screen and illuminate it. The simplest shadows are those cast by a point source of light. This process is shown in **FIGURE 18.7a**. With a point source, the shadow is completely dark, and the edges of the shadow are sharp.

FIGURE 18.7 Shadows produced by point and extended sources of light.

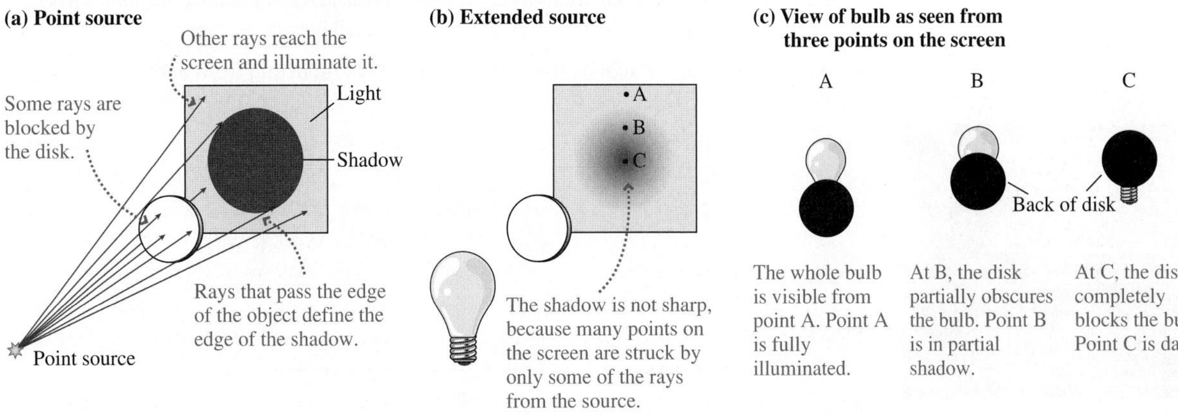

(a) Point source
Other rays reach the screen and illuminate it.
Some rays are blocked by the disk.
Light
Shadow
Rays that pass the edge of the object define the edge of the shadow.
Point source
(b) Extended source
A
B
C
The shadow is not sharp, because many points on the screen are struck by only some of the rays from the source.
(c) View of bulb as seen from three points on the screen
A
B
C
Back of disk
The whole bulb is visible from point A. Point A is fully illuminated.
At B, the disk partially obscures the bulb. Point B is in partial shadow.
At C, the disk completely blocks the bulb. Point C is dark.

During a solar eclipse, the sun—a small but extended source—casts a shadow of the moon on the earth. The moon's shadow has a dark center surrounded by a region of increasing brightness, just as in Figure 18.7b.

Shadows cast by extended sources are more complicated. An extended source is a collection of a large number of point sources, each of which casts its own shadow. However, as shown in **FIGURE 18.7b**, the patterns of shadow and light from each point overlap and thus the shadow region is no longer sharp. **FIGURE 18.7c** shows the view of the *bulb* as seen from three points on the screen. Depending on the size of the source, there is often a true shadow that no light reaches, surrounded by a fuzzy region of increasing brightness.

18.2 Reflection

Reflection of light is a familiar, everyday experience. You see your reflection in the bathroom mirror first thing every morning, reflections in your car's rearview mirror as you drive to school, and the sky reflected in puddles of standing water. Reflection from a smooth, shiny surface, such as a mirror or a piece of polished metal, is called **specular reflection.**

FIGURE 18.8a shows a bundle of parallel light rays reflecting from a mirror-like surface. You can see that the incident and reflected rays are both in a plane that is normal, or perpendicular, to the reflective surface. A three-dimensional perspective accurately shows the relation between the light rays and the surface, but figures such as this are hard to draw by hand. Instead, it is customary to represent reflection with the simpler visual overview of **FIGURE 18.8b**. In this figure,

- The incident and reflected rays are in the plane of the page. The reflective surface extends into and out of the page.
- A *single* light ray represents the entire bundle of parallel rays. This is oversimplified, but it keeps the figure and the analysis clear.

FIGURE 18.8 Specular reflection of light.

(a)

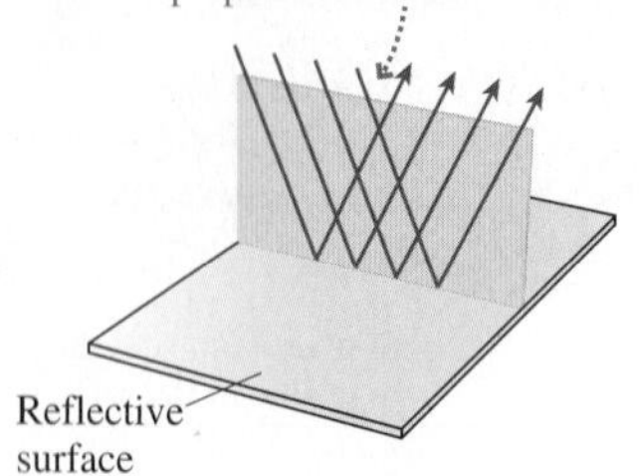

(b)

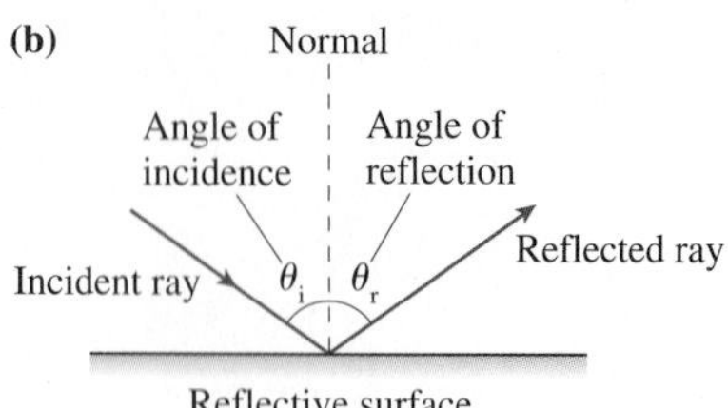
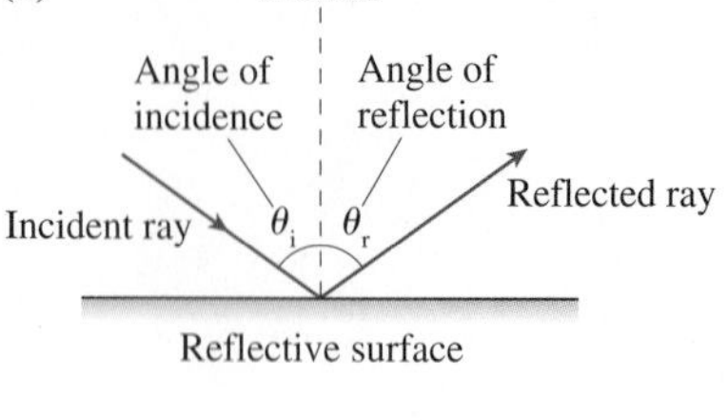

The angle θ_i between the incident ray and a line perpendicular to the surface—the *normal* to the surface—is called the **angle of incidence.** Similarly, the **angle of reflection** θ_r is the angle between the reflected ray and the normal to the surface. The **law of reflection,** easily demonstrated with simple experiments, states that

1. The incident ray and the reflected ray are both in the same plane, which is perpendicular to the surface, and
2. The angle of reflection equals the angle of incidence: $\theta_r = \theta_i$.

NOTE ▶ Optics calculations *always* use the angle measured from the normal, not the angle between the ray and the surface. ◀

EXAMPLE 18.1 Light reflecting from a mirror

A full-length mirror on a closet door is 2.0 m tall. The bottom touches the floor. A bare lightbulb hangs 1.0 m from the closet door, 0.5 m above the top of the mirror. How long is the streak of reflected light across the floor?

PREPARE Treat the lightbulb as a point source and use the ray model of light. **FIGURE 18.9** is a visual overview of the light rays. We need to consider only the two rays that strike the edges of the mirror. All other reflected rays will fall between these two.

FIGURE 18.9 Visual overview of the light rays reflecting from a mirror.

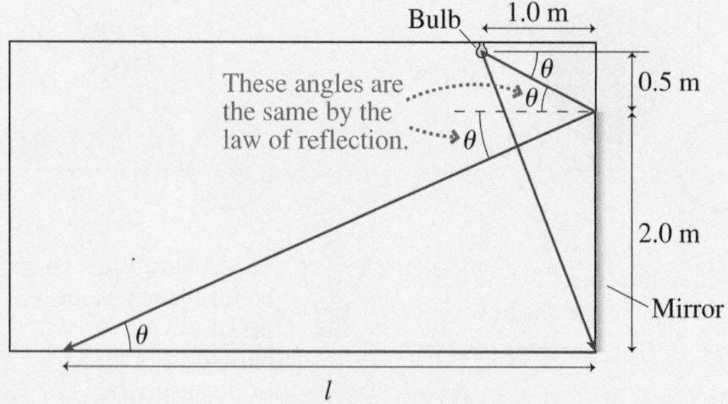

Reflection is an everyday experience.

SOLVE The ray that strikes the bottom of the mirror reflects from it and hits the floor just where the mirror meets the floor. For the top ray, Figure 18.9 has used the law of reflection to set the angle of reflection equal to the angle of incidence; we call both θ. By simple geometry, the other angles shown are also equal to θ. From the small triangle at the upper right,

$$\theta = \tan^{-1}\left(\frac{0.5\text{ m}}{1.0\text{ m}}\right) = 26.6°$$

But we also have $\tan\theta = (2.0\text{ m})/l$, or

$$l = \frac{2.0\text{ m}}{\tan\theta} = \frac{2.0\text{ m}}{\tan 26.6°} = 4.0\text{ m}$$

Since the lower ray struck right at the mirror's base, the total length of the reflected streak is 4.0 m.

Diffuse Reflection

We've already discussed diffuse reflection, the reflection of light rays off of a surface such as paper or cloth that is not shiny like a mirror. If you magnify the surface of a diffuse reflector, you'll find that on the microscopic scale it is quite rough. The law of reflection $\theta_r = \theta_i$ is still obeyed at each point, but the irregularities of the surface cause the reflected rays to leave in many random directions. This situation is shown in **FIGURE 18.10**. Diffuse reflection is actually much more common than the mirror-like specular reflection.

FIGURE 18.10 Diffuse reflection from an irregular surface.

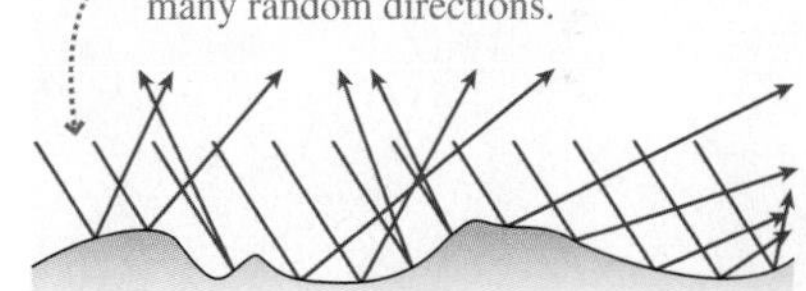

The Plane Mirror

One of the most commonplace observations is that you can see yourself in a mirror. How? **FIGURE 18.11a** shows rays from point source P reflecting from a flat mirror, called a **plane mirror.** Consider the particular ray shown in **FIGURE 18.11b**. The reflected ray travels along a line that passes through point P′ on the "back side" of the mirror. Because $\theta_r = \theta_i$, simple geometry dictates that P′ is the same distance behind the mirror as P is in front of the mirror. That is, the **image distance** s' is equal to the **object distance** s:

$$s' = s \qquad \text{(plane mirror)} \tag{18.1}$$

FIGURE 18.11 The light rays reflecting from a plane mirror.

(a)

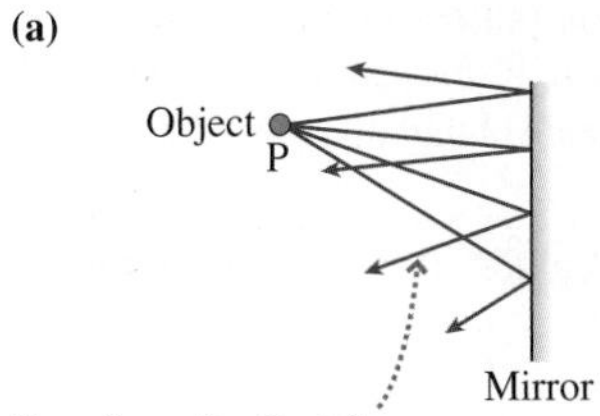

Rays from P reflect from the mirror. Each ray obeys the law of reflection.

(b)

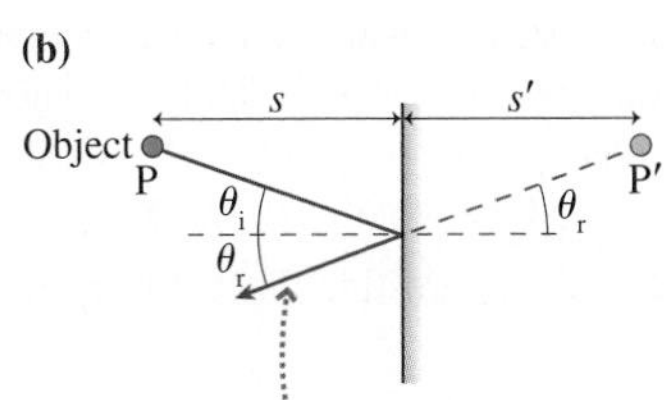

This reflected ray appears to have come from point P′.

(c)

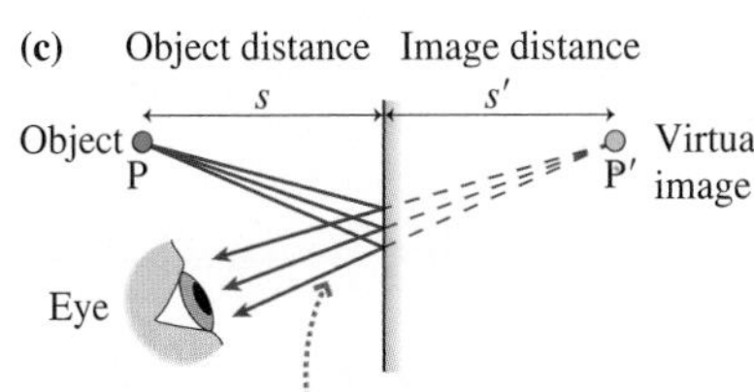

The reflected rays *all* diverge from P′, which appears to be the source of the reflected rays. Your eye collects the bundle of diverging rays and "sees" the light coming from P′.

The reflected ray in Figure 18.11b appears to have come from point P′. But because our argument applies to any incoming ray, all reflected rays appear to be coming from point P′, as **FIGURE 18.11c** shows. We call P′, the point from which the reflected rays diverge, the **virtual image** of P. The image is "virtual" in the sense that no rays actually leave P′, which is in darkness behind the mirror. But as far as your eye is concerned, the light rays act exactly *as if* the light really originated at P′. So while you may say "I see P in the mirror," what you are actually seeing is the virtual image of P.

FIGURE 18.12 Each point on the extended object has a corresponding image point an equal distance on the opposite side of the mirror.

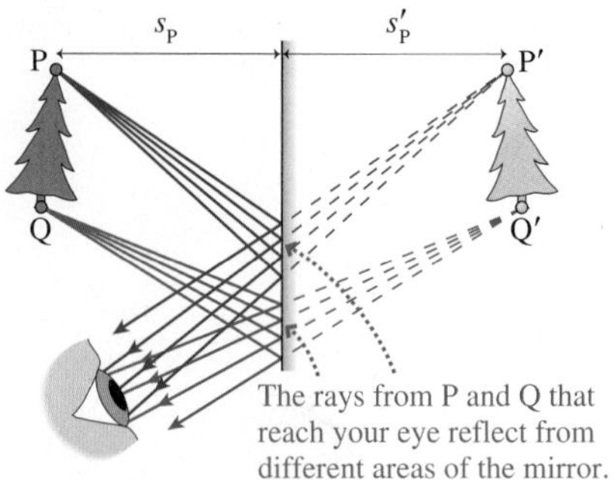

Your eye intercepts only a very small fraction of all the reflected rays.

A floating image The student stands on his leg that is hidden behind the mirror. Because every point of his body to the left of the mirror has its image to the right, when he raises his other leg he appears to float above the floor.

For an extended object, such as the one in **FIGURE 18.12**, each point on the object has a corresponding image point an equal distance on the opposite side of the mirror. The eye captures and focuses diverging bundles of rays from each point of the image in order to see the full image in the mirror. Two facts are worth noting:

1. Rays from each point on the object spread out in all directions and strike *every point* on the mirror. Only a very few of these rays enter your eye, but the other rays are very real and might be seen by other observers.
2. Rays from points P and Q enter your eye after reflecting from *different* areas of the mirror. This is why you can't always see the full image of an object in a very small mirror.

EXAMPLE 18.2 **How high is the mirror?**

If your height is h, what is the shortest mirror on the wall in which you can see your full image? Where must the top of the mirror be hung?

PREPARE Use the ray model of light. **FIGURE 18.13** is a visual overview of the light rays. We need to consider only the two rays that leave the top of your head and your feet and reflect into your eye.

FIGURE 18.13 Visual overview of light rays from your head and feet reflecting into your eye.

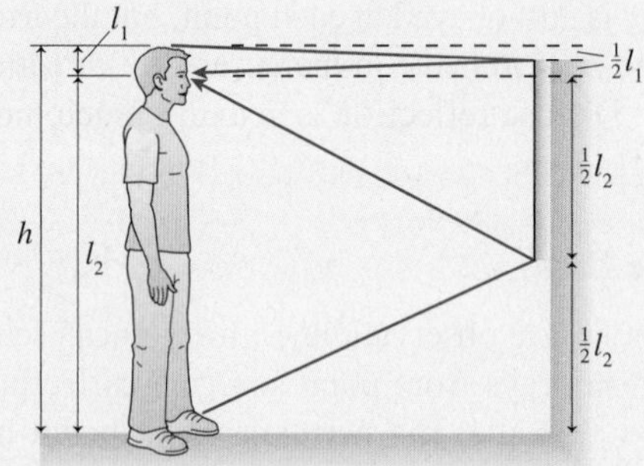

SOLVE Let the distance from your eyes to the top of your head be l_1 and the distance to your feet be l_2. Your height is $h = l_1 + l_2$. A light ray from the top of your head that reflects from the mirror at $\theta_r = \theta_i$ and enters your eye must, by congruent triangles, strike the mirror a distance $\frac{1}{2}l_1$ above your eyes. Similarly, a ray from your foot to your eye strikes the mirror a distance $\frac{1}{2}l_2$ below your eyes. The distance between these two points on the mirror is $\frac{1}{2}l_1 + \frac{1}{2}l_2 = \frac{1}{2}h$. A ray from anywhere else on your body will reach your eye if it strikes the mirror between these two points. Pieces of the mirror outside these two points are irrelevant, not because rays don't strike them but because the reflected rays don't reach your eye. Thus the shortest mirror in which you can see your full reflection is $\frac{1}{2}h$. But this will work only if the top of the mirror is hung midway between your eyes and the top of your head.

ASSESS It is interesting that the answer does not depend on how far you are from the mirror.

STOP TO THINK 18.1 Two plane mirrors form a right angle. How many images of the ball can you see in the mirrors?

A. 1
B. 2
C. 3
D. 4

Observer

18.3 Refraction

In FIGURE 18.14, two things happen when a light ray crosses the boundary between the air and the glass:

1. Part of the light *reflects* from the boundary, obeying the law of reflection. This is how you see reflections from pools of water or storefront windows, even though water and glass are transparent.
2. Part of the light continues into the second medium. It is *transmitted* rather than reflected, but the transmitted ray changes direction as it crosses the boundary. The transmission of light from one medium to another, but with a change in direction, is called **refraction.**

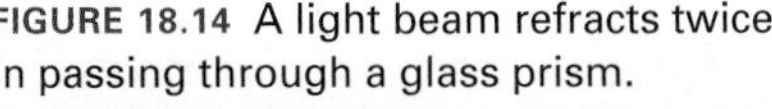
FIGURE 18.14 A light beam refracts twice in passing through a glass prism.

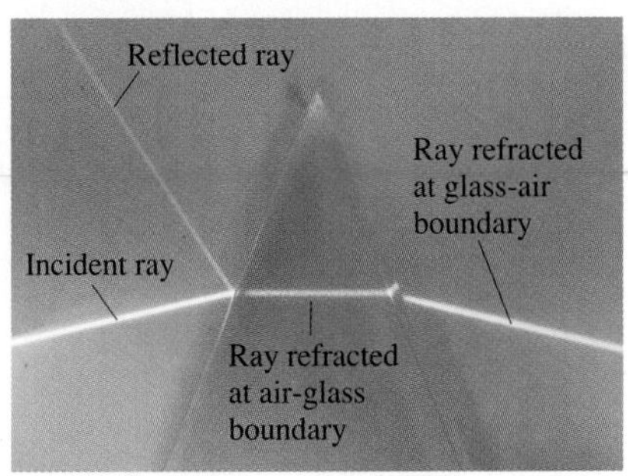

In Figure 18.14, notice the refraction of the light beam as it passes through the prism. Notice also that the ray direction changes as the light enters and leaves the glass. You can also see a weak reflection leaving the first surface of the prism.

Reflection from the boundary between transparent media is usually weak. Typically 95% of the light is transmitted and only 5% is reflected. Our goal in this section is to understand refraction, so we will usually ignore the weak reflection and focus on the transmitted light.

NOTE ▶ The transparent material through which light travels is called the *medium* (plural *media*). ◀

ActivPhysics OnLine 15.1, 15.2, 15.3

FIGURE 18.15a shows the refraction of light rays from a parallel beam of light, such as a laser beam, and rays from a point source. These pictures remind us that an infinite number of rays are incident on the boundary, but our analysis will be simplified if we focus on a single light ray. FIGURE 18.15b is a ray diagram showing the refraction of a single ray at a boundary between medium 1 and medium 2. Let the angle between the ray and the normal be θ_1 in medium 1 and θ_2 in medium 2. Just as for reflection, the angle between the incident ray and the normal is the *angle of incidence*. The angle on the transmitted side, *measured from the normal,* is called the **angle of refraction.** Notice that θ_1 is the angle of incidence in Figure 18.15b but is the angle of refraction in FIGURE 18.15c, where the ray is traveling in the opposite direction.

FIGURE 18.15 Refraction of light rays.

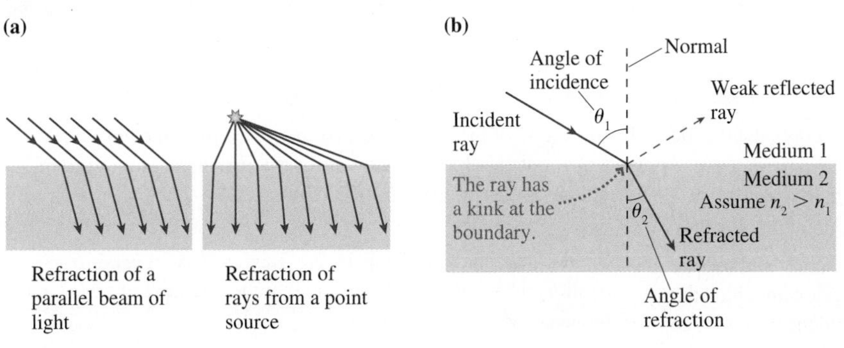

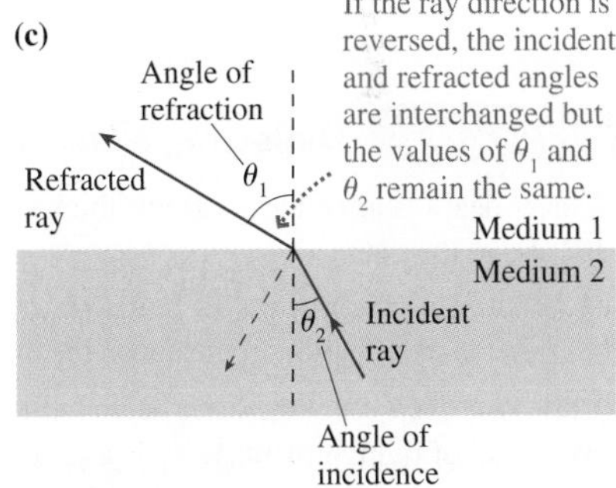

Refraction was first studied experimentally by the Arab scientist Ibn al-Haitham in about the year 1000. His work arrived in Europe six hundred years later, where it influenced the Dutch scientist Willebrord Snell. In 1621, Snell proposed a mathematical statement of the "law of refraction" or, as we know it today, Snell's law. If a ray refracts between medium 1 and medium 2, having indices of refraction n_1 and n_2, the ray angles θ_1 and θ_2 in the two media are related by

$$n_1 \sin\theta_1 = n_2 \sin\theta_2 \tag{18.2}$$

Snell's law for refraction between two media

Notice that Snell's law does not mention which is the incident angle and which the refracted angle.

TABLE 18.1 Indices of refraction

Medium	n
Vacuum	1 exactly
Air (actual)	1.0003
Air (accepted)*	1.00
Water	1.33
Ethyl alcohol	1.36
Oil	1.46
Glass (typical)	1.50
Polystyrene plastic	1.59
Cubic zirconia	2.18
Diamond	2.42
Silicon (infrared)	3.50

*Use this value in problems.

Table 18.1 lists the indices of refraction for several media. It is interesting to note that the n in the law of refraction, Equation 18.2, is the same index of refraction n we studied in Chapter 17. There we found that the index of refraction determines the speed of a light wave in a medium according to $v = c/n$. Here, it appears to play a different role, determining by how much a light ray is bent when crossing the boundary between two different media. Although we won't do so here, it is possible to use Huygens' principle to show that Snell's law is a *consequence* of the change in the speed of light as it moves across a boundary between media.

Examples of Refraction

Look back at Figure 18.15. As the ray in Figure 18.15b moves from medium 1 to medium 2, where $n_2 > n_1$, it bends closer to the normal. In Figure 18.15c, where the ray moves from medium 2 to medium 1, it bends away from the normal. This is a general conclusion that follows from Snell's law:

- When a ray is transmitted into a material with a higher index of refraction, it bends to make a smaller angle with the normal.
- When a ray is transmitted into a material with a lower index of refraction, it bends to make a larger angle with the normal.

This rule becomes a central idea in a procedure for analyzing refraction problems.

TACTICS BOX 18.1 Analyzing refraction

❶ **Draw a ray diagram.** Represent the light beam with one ray.
❷ **Draw a line normal (perpendicular) to the boundary.** Do this at each point where the ray intersects a boundary.
❸ **Show the ray bending in the correct direction.** The angle is larger on the side with the smaller index of refraction. This is the qualitative application of Snell's law.
❹ **Label angles of incidence and refraction.** Measure all angles from the normal.
❺ **Use Snell's law.** Calculate the unknown angle or unknown index of refraction.

Exercises 10–13

EXAMPLE 18.3 Deflecting a laser beam

A laser beam is aimed at a 1.0-cm-thick sheet of glass at an angle 30° above the glass.

a. What is the laser beam's direction of travel in the glass?
b. What is its direction in the air on the other side of the glass?

PREPARE Represent the laser beam with a single ray and use the ray model of light. **FIGURE 18.16** is a visual overview in which the first four steps of Tactics Box 18.1 have been identified. Notice that the angle of incidence must be measured from the normal, so $\theta_1 = 60°$, not the 30° value given in the problem. The index of refraction of glass was taken from Table 18.1.

FIGURE 18.16 The ray diagram of a laser beam passing through a sheet of glass.

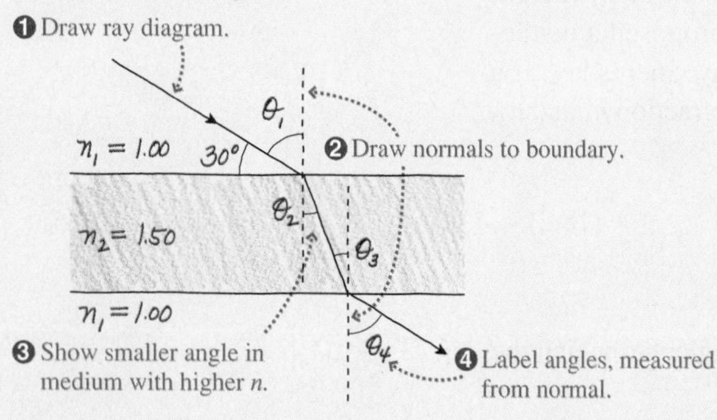

SOLVE

a. Snell's law, the final step in the Tactics Box, is $n_1 \sin\theta_1 = n_2 \sin\theta_2$. Using $\theta_1 = 60°$, we find that the direction of travel in the glass is

$$\theta_2 = \sin^{-1}\left(\frac{n_1 \sin\theta_1}{n_2}\right) = \sin^{-1}\left(\frac{\sin 60°}{1.5}\right)$$
$$= \sin^{-1}(0.577) = 35.3°$$

or 35° to two significant figures.

b. Snell's law at the second boundary is $n_2 \sin\theta_3 = n_1 \sin\theta_4$. You can see from Figure 18.16 that the interior angles are equal: $\theta_3 = \theta_2 = 35.3°$. Thus the ray emerges back into the air traveling at angle

$$\theta_4 = \sin^{-1}\left(\frac{n_2 \sin\theta_3}{n_1}\right) = \sin^{-1}(1.5 \sin 35.3°)$$
$$= \sin^{-1}(0.867) = 60°$$

This is the same as θ_1, the original angle of incidence.

ASSESS As expected, the laser beam bends toward the normal as it moves into the higher-index glass, and away from the normal as it moves back into air. The beam exits the glass still traveling in the same direction as it entered, but its path is *displaced.* This is a general result for light traveling through a medium with parallel sides. As the glass becomes thinner, the displacement becomes less; there is no displacement as the glass thickness becomes zero. This will be an important observation when we later study lenses.

EXAMPLE 18.4 Measuring the index of refraction

FIGURE 18.17 shows a laser beam deflected by a 30°-60°-90° prism. What is the prism's index of refraction?

FIGURE 18.17 A prism deflects a laser beam.

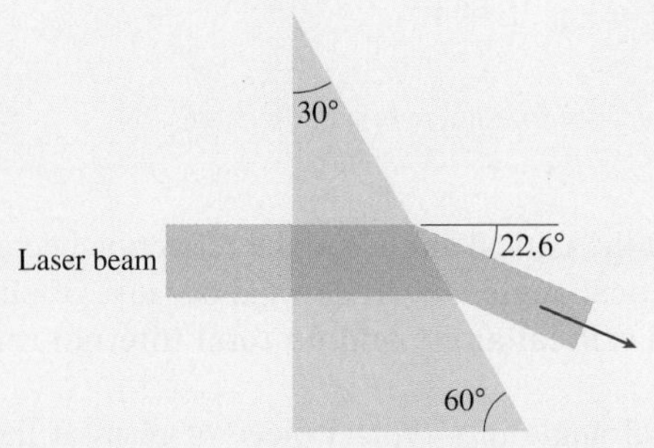

PREPARE Represent the laser beam with a single ray and use the ray model of light. **FIGURE 18.18** uses the steps of Tactics Box 18.1 to draw a ray diagram. The ray is incident perpendicular to the front face of the prism ($\theta_i = 0°$); thus it is transmitted through the first boundary without deflection. At the second boundary it is especially important to *draw the normal to the surface* at the point of incidence and to *measure angles from the normal.*

FIGURE 18.18 Visual overview of a laser beam passing through the prism.

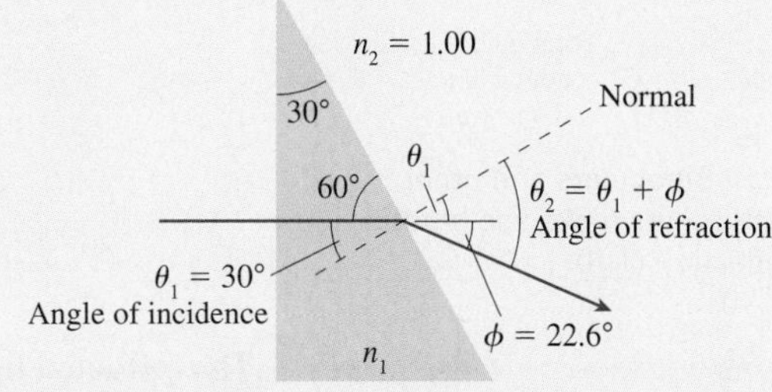

SOLVE From the geometry of the triangle you can find that the laser's angle of incidence on the hypotenuse of the prism is $\theta_1 = 30°$, the same as the apex angle of the prism. The ray exits the prism at angle θ_2 such that the deflection is $\phi = \theta_2 - \theta_1 = 22.6°$. Thus $\theta_2 = 52.6°$. Knowing both angles and $n_2 = 1.00$ for air, we can use Snell's law to find n_1:

$$n_1 = \frac{n_2 \sin\theta_2}{\sin\theta_1} = \frac{1.00 \sin 52.6°}{\sin 30°} = 1.59$$

ASSESS Referring to the indices of refraction in Table 18.1, we see that the prism is made of polystyrene plastic.

▶ **Optical image stabilization** When you make a video with a handheld video camera, the inevitable movement of your hands shows up as unwanted motion in the video itself. High-end video cameras largely eliminate this motion using *optical image stabilization.* The angle of an internal liquid-filled prism is automatically adjusted so that rays entering the prism are refracted at the second glass plate to always hit the camera's light sensor in the correct spot.

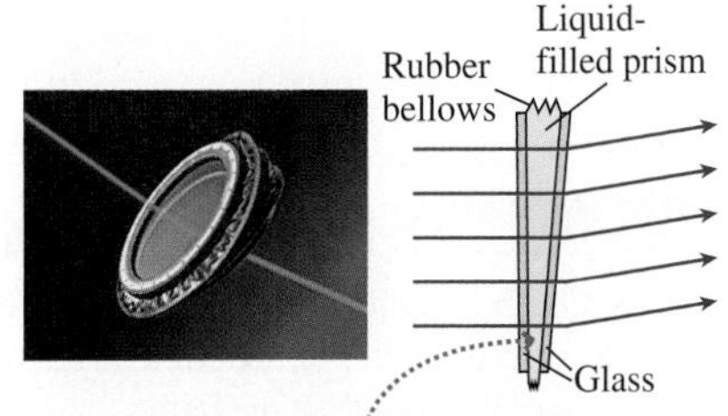

Total Internal Reflection

What would have happened in Example 18.4 if the prism angle had been 45° rather than 30°? The light rays would approach the rear surface of the prism at an angle of incidence $\theta_1 = 45°$. When we try to calculate the angle of refraction at which the ray emerges into the air, we find

$$\sin\theta_2 = \frac{n_1}{n_2}\sin\theta_1 = \frac{1.59}{1.00}\sin 45° = 1.12$$

$$\theta_2 = \sin^{-1}(1.12) = ???$$

Angle θ_2 cannot be computed because the sine of an angle can't be greater than 1. The ray is unable to refract through the boundary. Instead, 100% of the light *reflects* from the boundary back into the prism. This process is called **total internal reflection,** often abbreviated TIR. That it really happens is illustrated in **FIGURE 18.19**. Here, three light beams strike the surface of the water at increasing angles of incidence. The

FIGURE 18.19 One of the three beams of light undergoes total internal reflection.

two beams with the smallest angles of incidence refract out of the water, but the beam with the largest angle of incidence undergoes total internal reflection at the water's surface.

FIGURE 18.20 Refraction and reflection of rays as the angle of incidence increases.

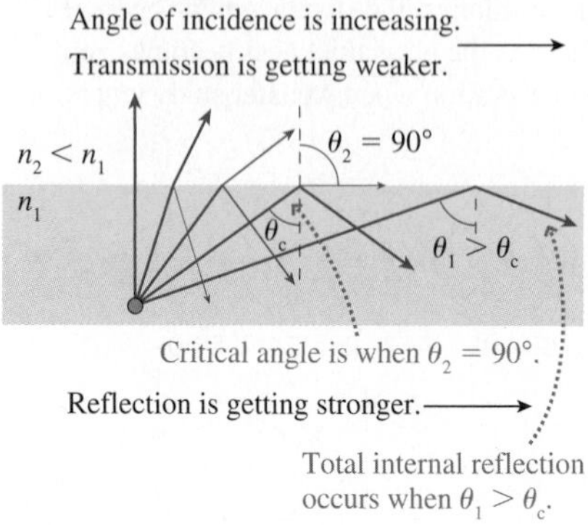

Angle of incidence is increasing.
Transmission is getting weaker.
$n_2 < n_1$
n_1
$\theta_2 = 90°$
θ_c
$\theta_1 > \theta_c$
Critical angle is when $\theta_2 = 90°$.
Reflection is getting stronger.
Total internal reflection occurs when $\theta_1 > \theta_c$.

FIGURE 18.20 shows several rays leaving a point source in a medium with index of refraction n_1. The medium on the other side of the boundary has $n_2 < n_1$. As we've seen, crossing a boundary into a material with a lower index of refraction causes the ray to bend away from the normal. Two things happen as angle θ_1 increases. First, the refraction angle θ_2 approaches 90°. Second, the fraction of the light energy that is transmitted decreases while the fraction reflected increases.

A **critical angle** θ_c is reached when $\theta_2 = 90°$. Snell's law becomes $n_1 \sin\theta_c = n_2 \sin 90°$, or

$$\theta_c = \sin^{-1}\left(\frac{n_2}{n_1}\right) \qquad (18.3)$$

Critical angle of incidence for total internal reflection

FIGURE 18.21 Binoculars and other optical instruments make use of total internal reflection (TIR).

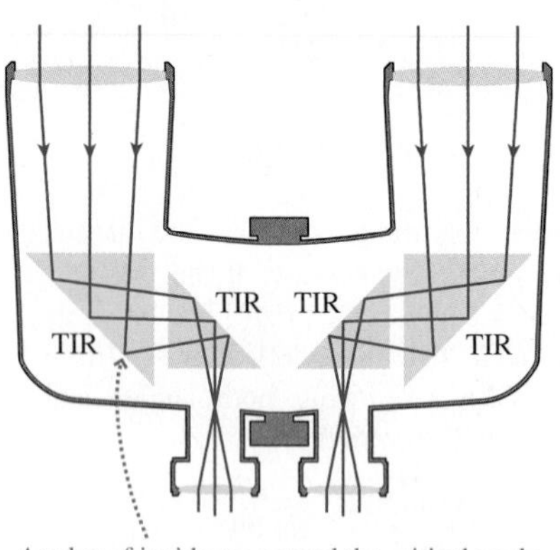
TIR
TIR
TIR
TIR
Angles of incidence exceed the critical angle.

The refracted light vanishes at the critical angle and the reflection becomes 100% for any angle $\theta_1 \geq \theta_c$. The critical angle is well defined because of our assumption that $n_2 < n_1$. **There is no critical angle and no total internal reflection if $n_2 > n_1$.**

We can compute the critical angle in a typical piece of glass at the glass-air boundary as

$$\theta_{c\ glass} = \sin^{-1}\left(\frac{1.00}{1.50}\right) = 42°$$

The fact that the critical angle is smaller than 45° has important applications. For example, **FIGURE 18.21** shows a pair of binoculars. The lenses are much farther apart than your eyes, so the light rays need to be brought together before exiting the eyepieces. Rather than using mirrors, which get dirty, are easily scratched, and require alignment, binoculars use a pair of prisms on each side. Thus the light undergoes two TIRs and emerges from the eyepiece. (The actual prism arrangement in binoculars is a bit more complex, but this illustrates the basic idea.)

EXAMPLE 18.5 Seeing a submerged light

A lightbulb is set in the bottom of a 3.0-m-deep swimming pool. What is the diameter of the circle inside which a duck swimming on the surface could see the bulb?

PREPARE Represent the lightbulb as a point source and use the ray model of light. **FIGURE 18.22** is a visual overview of the light rays. The lightbulb emits rays at all angles, but only some of the rays refract into the air where they can be seen from above. Rays striking the surface at greater than the critical angle undergo TIR back down into the water. The diameter of the circle of light is the distance D between the two points at which rays strike the surface at the critical angle.

FIGURE 18.22 Visual overview of the rays leaving a lightbulb at the bottom of a swimming pool.

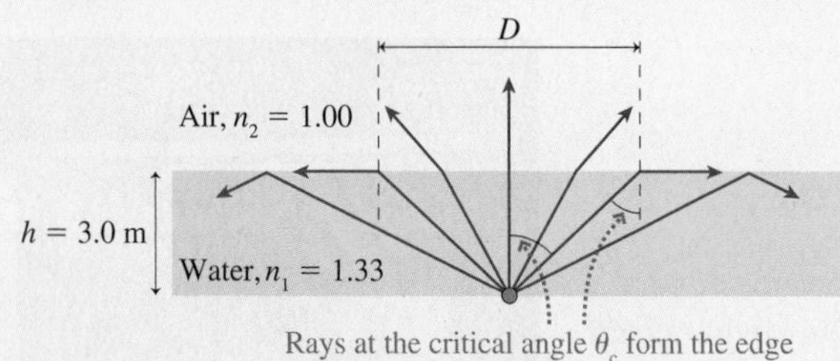
D
Air, $n_2 = 1.00$
$h = 3.0$ m
Water, $n_1 = 1.33$
Rays at the critical angle θ_c form the edge of the circle of light seen from above.

SOLVE From trigonometry, the circle diameter is $D = 2h\tan\theta_c$, where h is the depth of the water. The critical angle for a water-air boundary is $\theta_c = \sin^{-1}(1.00/1.33) = 48.7°$. Thus

$$D = 2(3.0\text{ m})\tan 48.7° = 6.8\text{ m}$$

ASSESS Light rays emerging at the edge of the circle actually skim the surface of the water. By reversing the direction of the rays, we can understand what a diver sees when she's underwater. This idea is explored further in the discussion on the next page.

▶ **Snell's window** We can understand what a diver sees when she's underwater by reversing the direction of all the rays in Figure 18.22 in the preceding example. The drawing shows that she can see the sun overhead, and clouds at larger angles. She can even see objects sitting at the waterline—but they appear at the edge of a circle as she looks up. Anything outside of this circle is a reflection of something in the water. The photo shows what she sees: a bright circle from the sky above—*Snell's window*—surrounded by the dark reflection of the water below.

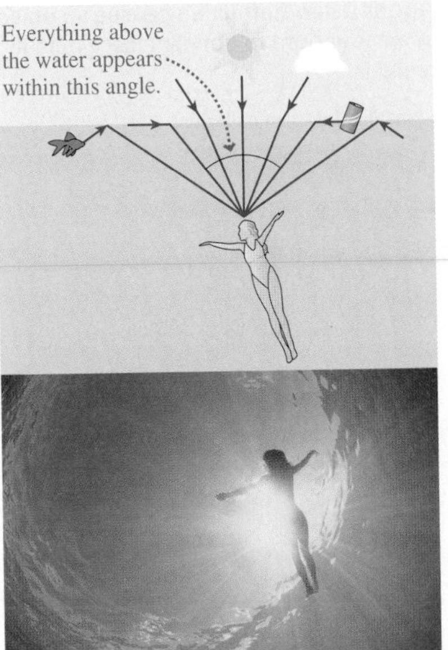

Everything above the water appears within this angle.

Fiber Optics

The most important modern application of total internal reflection is the transmission of light through optical fibers. FIGURE 18.23a shows a laser beam shining into the end of a long, narrow-diameter glass fiber. The light rays pass easily from the air into the glass, but they then strike the inside wall of the fiber at an angle of incidence θ_1 approaching 90°. This is much larger than the critical angle, so the laser beam undergoes TIR and remains inside the glass. The laser beam continues to "bounce" its way down the fiber as if the light were inside a pipe. Indeed, optical fibers are sometimes called "light pipes." The rays have an angle of incidence *smaller* than the critical angle ($\theta_1 \approx 0$) when they finally reach the flat end of the fiber; thus they refract out without difficulty and can be detected.

FIGURE 18.23 Light rays are confined within an optical fiber by total internal reflection.

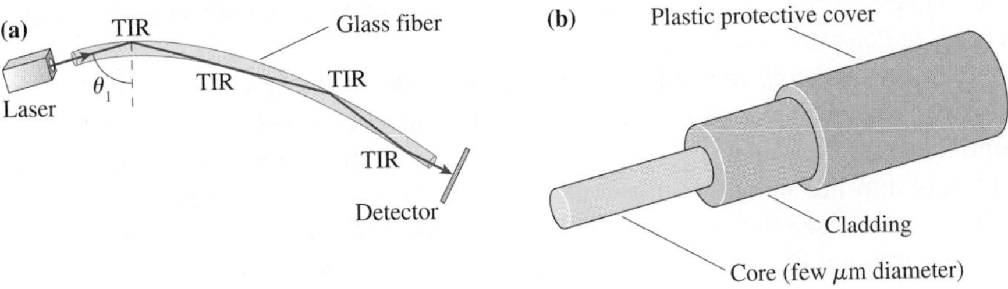

While a simple glass fiber can transmit light, reliance on a glass-air boundary is not sufficiently reliable for commercial use. Any small scratch on the side of the fiber alters the rays' angle of incidence and allows leakage of light. FIGURE 18.23b shows the construction of a practical optical fiber. A small-diameter glass *core* is surrounded by a layer of glass *cladding*. The glasses used for the core and the cladding have $n_{core} > n_{cladding}$. Thus, light undergoes TIR at the core-cladding boundary and remains confined within the core. This boundary is not exposed to the environment and hence retains its integrity even under adverse conditions.

Optical fibers have found important applications in medical diagnosis and treatment. Thousands of small fibers can be fused together to make a flexible bundle capable of transmitting high-resolution images along its length. Such *endoscopes* are used for minimally invasive inspection of body cavities, joints, and internal organs. One end of the endoscope bundle has a lens that allows an image of what's in front of it to be sent up the bundle where it can be observed on a television monitor by the physician. Optical fibers are also used to send light *down* the bundle to provide illumination, and small instruments such as forceps and clamps can be operated along its length to retrieve samples for biopsy.

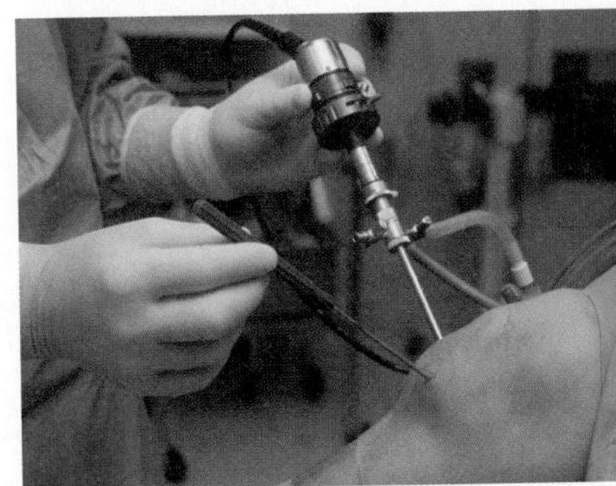

Arthroscopic surgery BIO Operations on injured joints can often be performed using an endoscope inserted through a small incision. The endoscope allows the surgeon to observe the procedure, which is performed with instruments inserted through another incision. The recovery time for such surgery is usually much shorter than for conventional operations requiring a full incision to expose the interior of the joint.

STOP TO THINK 18.2 A light ray travels from medium 1 to medium 3 as shown. For these media,

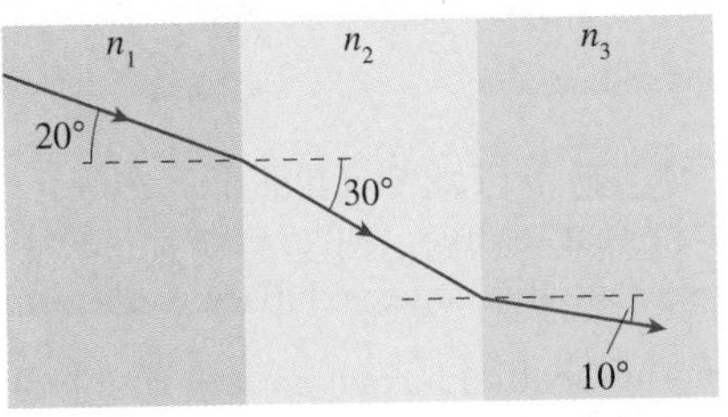

A. $n_3 > n_1$
B. $n_3 = n_1$
C. $n_3 < n_1$
D. We can't compare n_1 to n_3 without knowing n_2.

FIGURE 18.24 Refraction causes an object in an aquarium to appear closer than it really is.

(a) A ruler in an aquarium

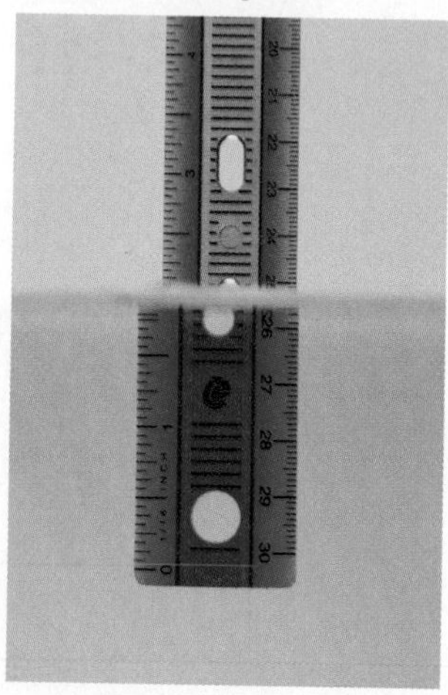

(b) Finding the image of the ruler

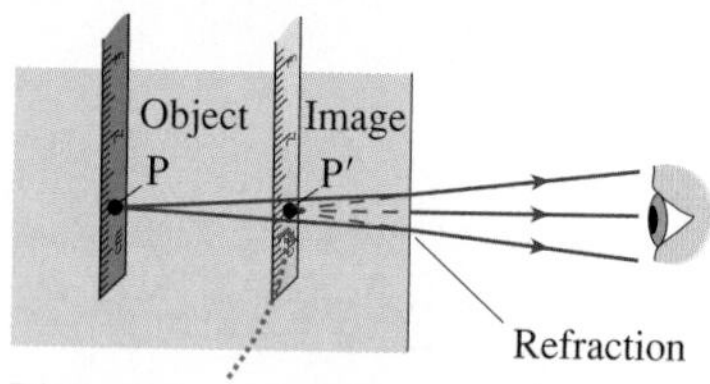

18.4 Image Formation by Refraction

FIGURE 18.24a shows a photograph of a ruler as seen through the front of an aquarium tank. The part of the ruler below the waterline appears *closer* than the part that is above water. FIGURE 18.24b shows why this is so. Rays that leave point P on the ruler refract away from the normal at the water-air boundary. (The thin glass wall of the aquarium has little effect on the refraction of the rays and can be ignored.) To your eye, outside the aquarium, these rays appear to diverge not from the object at point P, but instead from point P′ that is *closer* to the boundary. The same argument holds for every point on the ruler, so that the **ruler appears closer than it really is because of refraction of light at the boundary.**

We found that the rays reflected from a mirror diverge from a point that is not the object point. We called that point a *virtual image.* Similarly, if rays from an object point P refract at a boundary between two media such that the rays then diverge from a point P′ and *appear* to come from P′, we call P′ a virtual image of point P. The virtual image of the ruler is what you see.

Let's examine this image formation a bit more carefully. FIGURE 18.25 shows a boundary between two transparent media having indices of refraction n_1 and n_2. Point P, a source of light rays, is the object. Point P′, from which the rays *appear* to diverge, is the virtual image of P. The figure assumes $n_1 > n_2$, but this assumption isn't necessary. Distance s, measured from the boundary, is the object distance. Our goal is to determine the image distance s'.

The line through the object and perpendicular to the boundary is called the **optical axis.** Consider a ray that leaves the object at angle θ_1 with respect to the optical axis. θ_1 is also the angle of incidence at the boundary, where the ray refracts into the second medium at angle θ_2. By tracing the refracted ray backward, you can see that θ_2 is also the angle between the refracted ray and the optical axis at point P′.

The distance l is common to both the incident and the refracted rays, and you can see that $l = s\tan\theta_1 = s'\tan\theta_2$. Thus

$$s' = \frac{\tan\theta_1}{\tan\theta_2}s \tag{18.4}$$

Snell's law relates the sines of angles θ_1 and θ_2; that is,

$$\frac{\sin\theta_1}{\sin\theta_2} = \frac{n_2}{n_1} \tag{18.5}$$

FIGURE 18.25 Finding the virtual image P′ of an object at P.

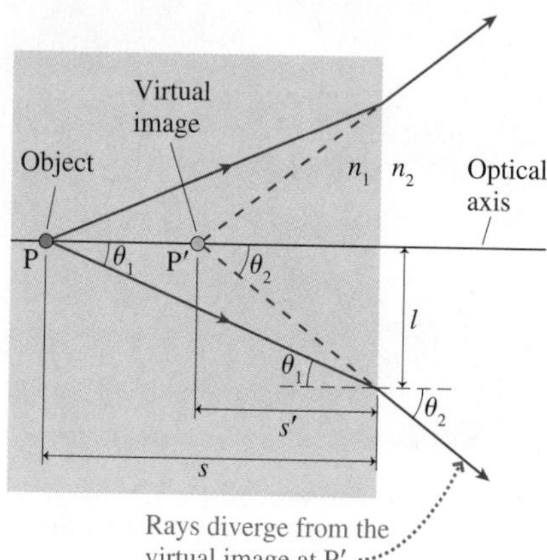

In practice, the angle between any of these rays and the optical axis is very small because the pupil of your eye is very much smaller than the distance between the object and your eye. (The angles in the figure have been greatly exaggerated.) The small-angle approximation $\sin\theta \approx \tan\theta \approx \theta$, where θ is in radians, is therefore applicable. Consequently,

$$\frac{\tan\theta_1}{\tan\theta_2} \approx \frac{\sin\theta_1}{\sin\theta_2} = \frac{n_2}{n_1} \tag{18.6}$$

Using this result in Equation 18.4, we find that the image distance is

$$s' = \frac{n_2}{n_1}s \tag{18.7}$$

NOTE ► The fact that the result for s' is independent of θ_1 implies that *all* rays appear to diverge from the same point P′. This property of the diverging rays is essential in order to have a well-defined image. ◄

This section has given us a first look at image formation via refraction. We will extend this idea to image formation with lenses in the next section.

EXAMPLE 18.6 An air bubble in a window

A fish and a sailor look at each other through a 5.0-cm-thick glass porthole in a submarine. There happens to be a small air bubble right in the center of the glass. How far behind the surface of the glass does the air bubble appear to the fish? To the sailor?

PREPARE Represent the air bubble as a point source and use the ray model of light. Light rays from the bubble refract into the air on one side and into the water on the other. The ray diagram looks like Figure 18.25.

SOLVE The index of refraction of the glass is $n_1 = 1.50$. The bubble is in the center of the window, so the object distance from either side of the window is $s = 2.5$ cm. From the water side, the fish sees the bubble at an image distance

$$s' = \frac{n_2}{n_1}s = \frac{1.33}{1.50}(2.5 \text{ cm}) = 2.2 \text{ cm}$$

This is the apparent depth of the bubble. The sailor, in air, sees the bubble at an image distance

$$s' = \frac{n_2}{n_1}s = \frac{1.00}{1.50}(2.5 \text{ cm}) = 1.7 \text{ cm}$$

ASSESS The image distance is *shorter* for the sailor because of the *larger* difference between the two indices of refraction.

18.5 Thin Lenses: Ray Tracing

A **lens** is a transparent material that uses refraction of light rays at *curved* surfaces to form an image. In this section we want to establish a pictorial method of understanding image formation. This method is called **ray tracing.** We will defer a mathematical analysis of the image formation by lenses until the next section.

FIGURE 18.26 Converging and diverging lenses.

(a) Converging lenses, which are thicker in the center than at the edges, refract parallel rays toward the optical axis.

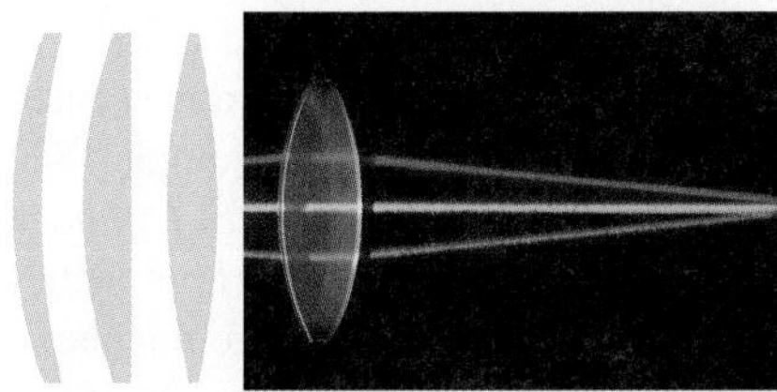

(b) Diverging lenses, which are thinner in the center than at the edges, refract parallel rays away from the optical axis.

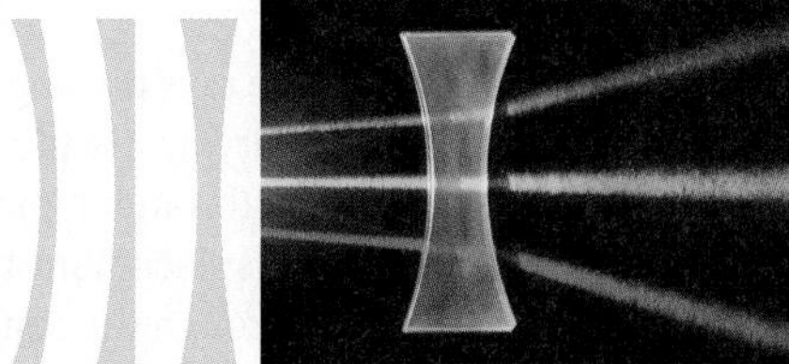

FIGURE 18.26 shows parallel light rays entering two different lenses. The lens in Figure 18.26a, called a **converging lens,** causes the rays to refract *toward* the optical axis. FIGURE 18.27 shows how, for a converging lens, a ray refracts toward the optical axis at both the first, air-to-glass boundary and the second, glass-to-air boundary. The common point through which initially parallel rays pass is called the **focal point** of the lens. The distance of the focal point from the lens is called the **focal length** f of the lens. The lens in Figure 18.26b, called a **diverging lens,** refracts parallel rays *away from* the optical axis. This lens also has a focal point, but it is not as obvious in the figure.

FIGURE 18.28 on the next page clarifies the situation. In the case of a diverging lens, a backward projection of the diverging rays shows that they all *appear* to have started from the same point. This is the focal point of a diverging lens, and its distance from the lens is the focal length of the lens. For both types of lenses, **the focal length is the distance from the lens to the point at which rays parallel to the optical axis converge or from which they appear to diverge.**

FIGURE 18.27 Both surfaces of a converging lens bend an incident ray toward the optical axis.

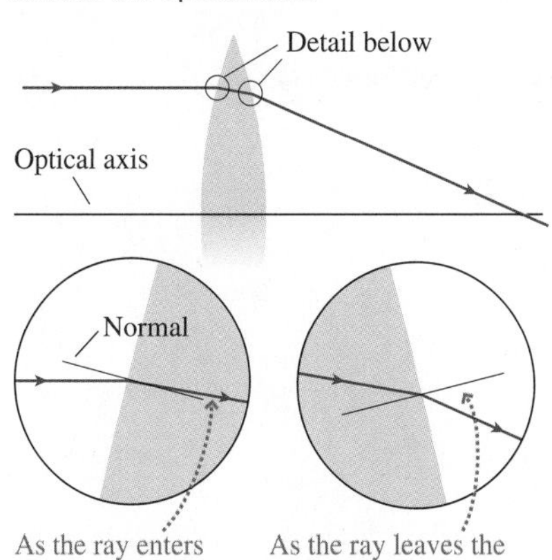
Detail below
Optical axis
Normal
As the ray enters the lens, it bends toward the normal and the optical axis.
As the ray leaves the lens, it bends away from the normal, but still toward the optical axis.

FIGURE 18.28 The focal point and focal length of converging and diverging lenses.

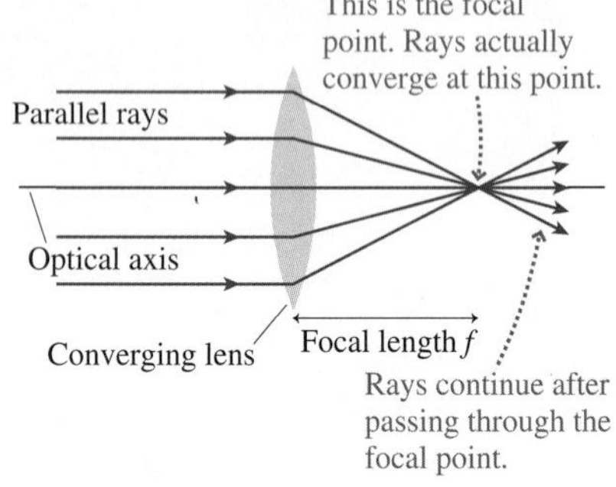
This is the focal point. Rays actually converge at this point.
Parallel rays
Optical axis
Converging lens
Focal length f
Rays continue after passing through the focal point.

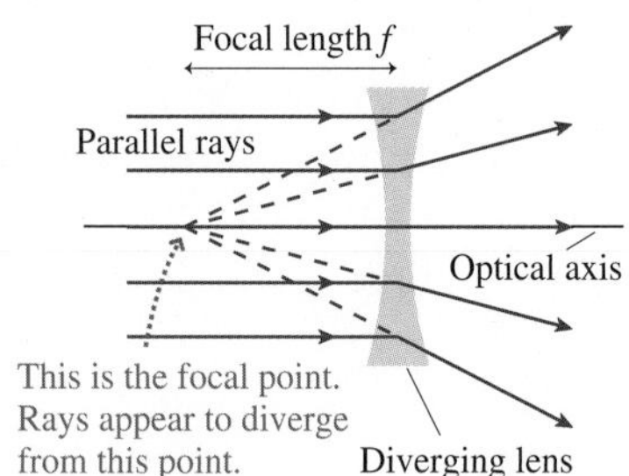
Focal length f
Parallel rays
Optical axis
This is the focal point. Rays appear to diverge from this point.
Diverging lens

NOTE ► The focal length f is a property *of the lens,* independent of how the lens is used. The focal length characterizes a lens in much the same way that a mass m characterizes an object or a spring constant k characterizes a spring. ◄

Converging Lenses

These basic observations about lenses are enough to understand image formation by a **thin lens,** an idealized lens whose thickness is zero and that lies entirely in a plane called the **lens plane.** Within this *thin-lens approximation,* **all refraction occurs as the rays cross the lens plane, and all distances are measured from the lens plane.** Fortunately, the thin-lens approximation is quite good for most practical applications of lenses.

NOTE ► We'll *draw* lenses as if they have a thickness, because that is how we expect lenses to look, but our analysis will not depend on the shape or thickness of a lens. ◄

FIGURE 18.29 shows three important situations of light rays passing through a thin, converging lens. Part (a) is familiar from Figure 18.28. If the direction of each of the rays in Figure 18.29a is reversed, Snell's law tells us that each ray will exactly retrace its path and emerge from the lens parallel to the optical axis. This leads to Figure 18.29b, which is the "mirror image" of part (a). Notice that the lens actually has *two* focal points, located at distances f on either side of the lens.

FIGURE 18.29 Three important sets of rays passing through a thin, converging lens.

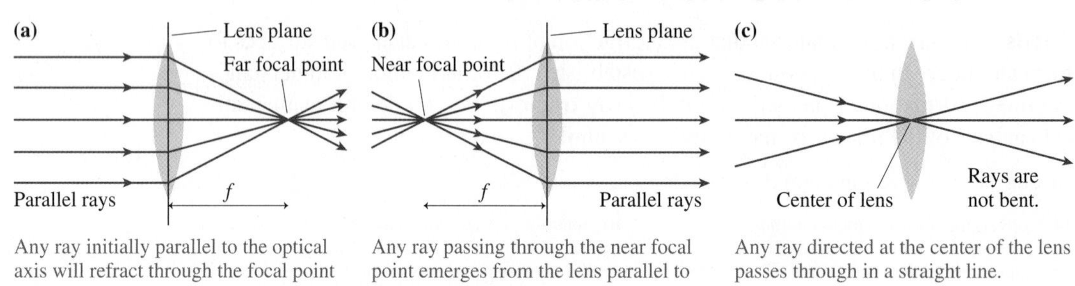
(a)
Lens plane
Far focal point
Parallel rays
f
Any ray initially parallel to the optical axis will refract through the focal point on the far side of the lens.
(b)
Lens plane
Near focal point
f
Parallel rays
Any ray passing through the near focal point emerges from the lens parallel to the optical axis.
(c)
Center of lens
Rays are not bent.
Any ray directed at the center of the lens passes through in a straight line.

Figure 18.29c shows three rays passing through the *center* of the lens. At the center, the two sides of a lens are very nearly parallel to each other. Earlier, in Example 18.3, we found that a ray passing through a piece of glass with parallel sides is *displaced* but *not bent* and that the displacement becomes zero as the thickness approaches zero. Consequently, a ray through the center of a thin lens, which has zero thickness, is neither bent nor displaced but travels in a straight line.

These three situations form the basis for ray tracing.

Real Images

FIGURE 18.30 on the next page shows a lens and an object whose distance s from the lens is larger than the focal length. Rays from point P on the object are refracted by the lens so as to converge at point P′ on the opposite side of the lens, at a distance s' from the lens. If rays diverge from an object point P and interact with a lens such that the refracted rays *converge* at point P′, actually meeting at P′, then we call P′ a **real image** of point P. Contrast this with our prior definition of a *virtual image* as a point from which rays appear to *diverge.*

All points on the object that are in the same plane, the **object plane,** converge to image points in the **image plane.** Points Q and R in the object plane of Figure 18.30 have image points Q′ and R′ in the same plane as point P′. Once we locate *one* point in the image plane, such as point P′, we know that the full image lies in the same plane.

There are two important observations to make about Figure 18.30. First, as also seen in FIGURE 18.31, the image is upside down with respect to the object. This is

FIGURE 18.30 Rays from an object point P are refracted by the lens and converge to a real image at point P′.

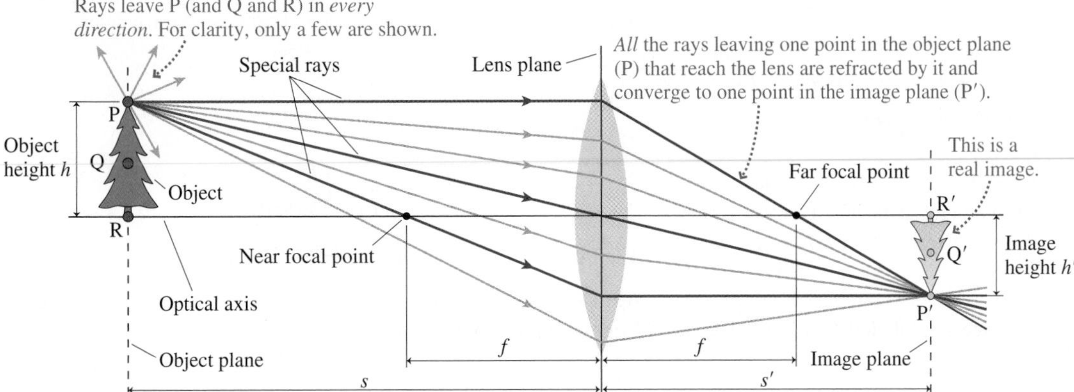

called an **inverted image,** and it is a standard characteristic of real-image formation with a converging lens. Second, rays from point P *fill* the entire lens surface, so that all portions of the lens contribute to the image. A larger lens will "collect" more rays and thus make a brighter image.

FIGURE 18.32 is a close-up view of the rays and images very near the image plane. The rays don't stop at P′ unless we place a screen in the image plane. When we do so, we see a sharp, well-focused image on the screen. If a screen is placed other than in the image plane, an image is produced on the screen, but it's blurry and out of focus.

NOTE ► Our ability to see a real image on a screen sets real images apart from *virtual* images. But keep in mind that we need not *see* a real image in order to *have* an image. A real image exists at a point in space where the rays converge even if there's no viewing screen in the image plane. ◄

Figure 18.30 highlights the three "special rays" that are based on the three situations of Figure 18.29. Notice that these three rays alone are sufficient to locate the image point P′. That is, we don't need to draw all the rays shown in Figure 18.30. The procedure known as *ray tracing* consists of locating the image by the use of just these three rays.

FIGURE 18.31 The lamp's image is upside down.

FIGURE 18.32 A close-up look at the rays and images near the image plane.

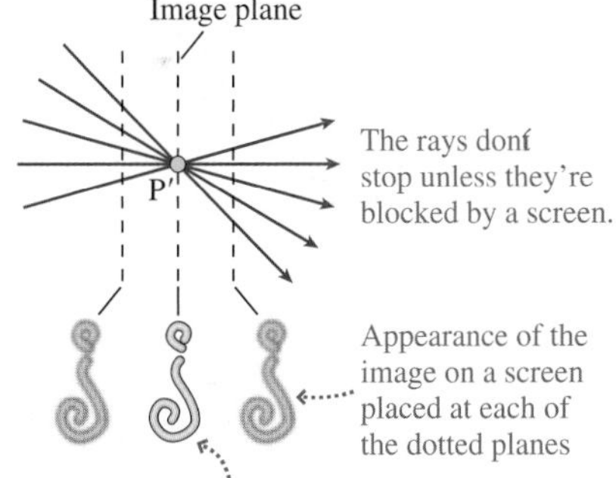

TACTICS BOX 18.2 Ray tracing for a converging lens MP™

❶ **Draw an optical axis.** Use graph paper or a ruler! Establish an appropriate scale.

❷ **Center the lens on the axis.** Mark and label the focal points at distance f on either side.

❸ **Represent the object with an upright arrow at distance s.** It's usually best to place the base of the arrow on the axis and to draw the arrow about half the radius of the lens.

❹ **Draw the three "special rays" from the tip of the arrow.** Use a straightedge or a ruler.
   a. A ray initially parallel to the axis refracts through the far focal point.
   b. A ray that enters the lens along a line through the near focal point emerges parallel to the axis.
   c. A ray through the center of the lens does not bend.

❺ **Extend the rays until they converge.** The rays converge at the image point. Draw the rest of the image in the image plane. If the base of the object is on the axis, then the base of the image will also be on the axis.

❻ **Measure the image distance s'.** Also, if needed, measure the image height relative to the object height. The magnification can be found from Equation 18.8.

Exercise 19

EXAMPLE 18.7 Finding the image of a flower

A 4.0-cm-diameter flower is 200 cm from the 50-cm-focal-length lens of a camera. How far should the plane of the camera's light detector be placed behind the lens to record a well-focused image? What is the diameter of the image on the detector?

PREPARE The flower is in the object plane. Use ray tracing to locate the image.

SOLVE **FIGURE 18.33** shows the ray-tracing diagram and the steps of Tactics Box 18.2. The image has been drawn in the plane where the three special rays converge. You can see *from the drawing* that the image distance is $s' \approx 65$ cm. This is where the detector needs to be placed to record a focused image. The heights of the object and image are labeled h and h'. The ray through the center of the lens is a straight line; thus the object and image both subtend the same angle θ. Using similar triangles,

$$\frac{h'}{s'} = \frac{h}{s}$$

Solving for h' gives

$$h' = h\frac{s'}{s} = (4.0 \text{ cm})\frac{65 \text{ cm}}{200 \text{ cm}} = 1.3 \text{ cm}$$

The flower's image has a diameter of 1.3 cm.

ASSESS We've been able to learn a great deal about the image from a simple geometric procedure.

FIGURE 18.33 Ray-tracing diagram for the image of a flower.

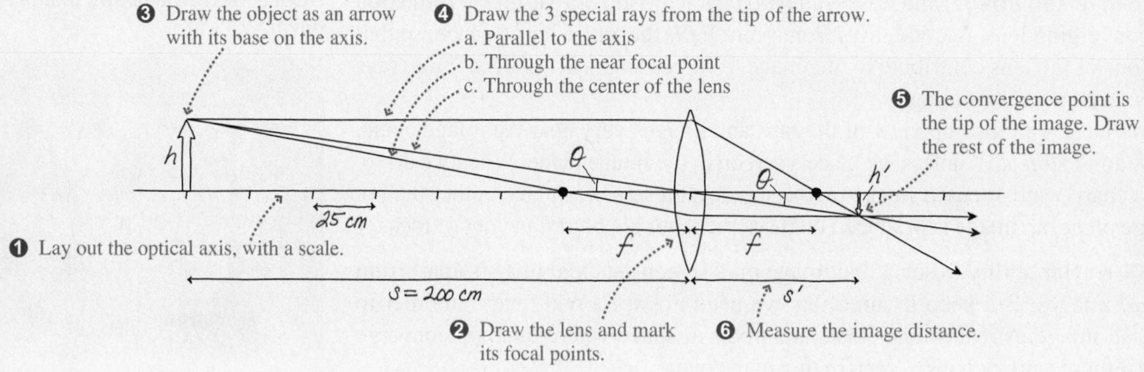
③ Draw the object as an arrow with its base on the axis.
④ Draw the 3 special rays from the tip of the arrow.
a. Parallel to the axis
b. Through the near focal point
c. Through the center of the lens
⑤ The convergence point is the tip of the image. Draw the rest of the image.
h
θ
θ
h'
25 cm
① Lay out the optical axis, with a scale.
f
f
s = 200 cm
s'
② Draw the lens and mark its focal points.
⑥ Measure the image distance.

Magnification

The image can be either larger or smaller than the object, depending on the location and focal length of the lens. Because the image height scales with that of the object, we're usually interested in the *ratio* h'/h of the image height to the object height. This ratio is greater than 1 when the image is taller than the object, and less than 1 when the image is shorter than the object.

But there's more to a description of the image than just its size. We also want to know its *orientation* relative to the object; that is, is the image upright or inverted? It is customary to combine image size and orientation information in a single number, the **magnification** m, defined as

$$m = -\frac{s'}{s} \tag{18.8}$$

Magnification of a lens or mirror

You saw in Example 18.7 that $s'/s = h'/h$. Consequently, we interpret the magnification m as follows:

1. The absolute value of m gives the ratio of image height to object height: $h'/h = |m|$.
2. A positive value of m indicates that the image is upright relative to the object. A negative value of m indicates that the image is inverted relative to the object.

The magnification in Example 18.7 would be

$$m = -\frac{s'}{s} = -\frac{65 \text{ cm}}{200 \text{ cm}} = -0.33$$

indicating that the image is 33% the size of the object and, because of the minus sign, is inverted.

NOTE ► Equation 18.8 applies to real or virtual images produced by both lenses and mirrors. Although s and s' are both positive in Example 18.7, leading to a negative magnification, we'll see that this is *not* the case for virtual images. ◄

STOP TO THINK 18.3 A lens produces a sharply focused, inverted image on a screen. What will you see on the screen if the lens is removed?

A. The image will be inverted and blurry.
B. The image will be upright and sharp.
C. The image will be upright and blurry.
D. The image will be much dimmer but otherwise unchanged.
E. There will be no image at all.

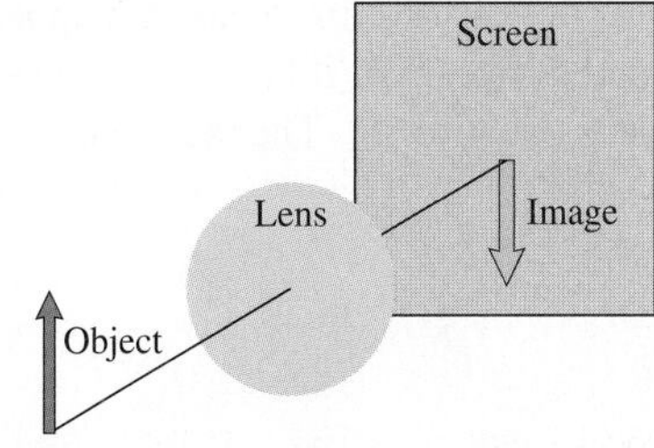
Screen
Lens
Image
Object

Virtual Images

The preceding section considered a converging lens with the object at distance $s > f$; that is, the object was outside the focal point. What if the object is inside the focal point, at distance $s < f$? **FIGURE 18.34** shows just this situation.

The special rays initially parallel to the axis and through the center of the lens present no difficulties. However, a ray through the near focal point would travel toward the left and would never reach the lens! Referring back to Figure 18.29b, you can see that the rays emerging parallel to the axis entered the lens *along a line* passing through the near focal point. It's the angle of incidence on the lens that is important, not whether the light ray actually passes through the focal point. This was the basis for the wording of step 4b in Tactics Box 18.2 and is the third special ray shown in Figure 18.34.

The three refracted rays don't converge. Instead, all three rays appear to *diverge* from point P′. This is the situation we found for rays reflecting from a mirror and for the rays refracting out of an aquarium. Point P′ is a *virtual image* of the object point P. Furthermore, it is an **upright image,** having the same orientation as the object.

The refracted rays, which are all to the right of the lens, *appear* to come from P′, but none of the rays were ever at that point. No image would appear on a screen placed in the image plane at P′. So what good is a virtual image?

Your eye collects and focuses bundles of diverging rays. Thus, as **FIGURE 18.35a** shows, you can "see" a virtual image by looking *through* the lens. This is exactly what you do with a magnifying glass, producing a scene like the one in **FIGURE 18.35b**. In fact, you view a virtual image any time you look *through* the eyepiece of an optical instrument such as a microscope or binoculars.

FIGURE 18.34 Rays from an object at distance $s < f$ are refracted by the lens and diverge to form a virtual image at point P′.

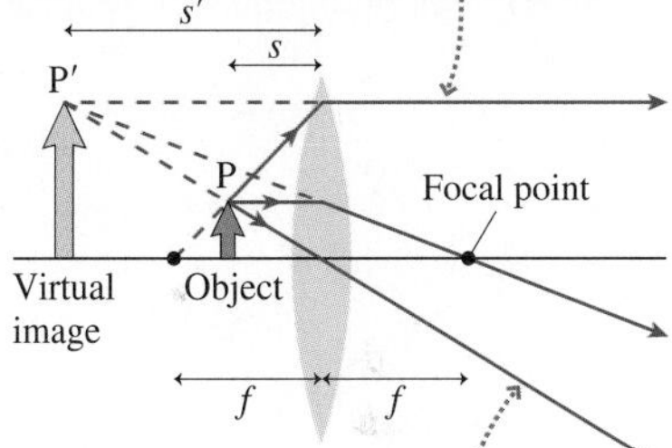
A ray *along a line* through the near focal point refracts parallel to the optical axis.
s′
s
P′
P
Focal point
Virtual image
Object
f
f
The refracted rays are diverging. They appear to come from point P′.

FIGURE 18.35 A converging lens is a magnifying glass when the object distance is $< f$.

(a)

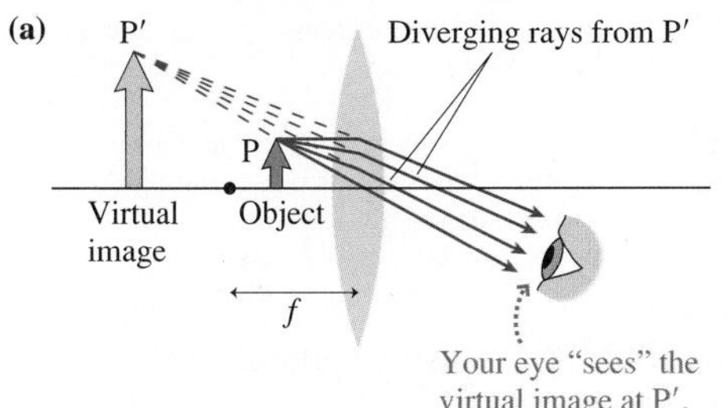
P′
Diverging rays from P′
P
Virtual image
Object
f
Your eye "sees" the virtual image at P′.

(b)

Insert
Home
Page Up
Delete
End
Page Down

NOTE ► Recall that a lens thicker in the middle than at the edges is classified as a converging lens. The light rays from an object *can* converge to form a real image after passing through such a lens, but only if the object distance is greater than the focal length of the lens: $s > f$. If $s < f$, the rays leaving a converging lens diverge to produce a virtual image. ◄

Because a virtual image is upright, the magnification $m = -s'/s$ is positive. This means that the ratio s'/s must be *negative*. We can ensure this if **we define the image distance s' to be negative for a virtual image,** indicating that the image is on the *same* side of the lens as the object. This is our first example of a **sign convention** for the various distances that appear in understanding image formation from lenses and mirrors. We'll have more to say about sign conventions when we study the thin-lens equation in a later section.

EXAMPLE 18.8 Magnifying a flower

To see a flower better, you hold a 6.0-cm-focal-length magnifying glass 4.0 cm from the flower. What is the magnification?

PREPARE The flower is in the object plane. Use ray tracing to locate the image. Once the image distance is known, Equation 18.8 can be used to find the magnification.

SOLVE FIGURE 18.36 shows the ray-tracing diagram. The three special rays diverge from the lens, but we can use a straightedge to extend the rays backward to the point from which they diverge. This point, the image point, is seen to be 12 cm to the left of the lens. Because this is a virtual image, the image distance is $s' = -12$ cm. From Equation 18.8 the magnification is

$$m = -\frac{s'}{s} = -\frac{-12\text{ cm}}{4.0\text{ cm}} = 3.0$$

FIGURE 18.36 Ray-tracing diagram for a magnifying glass.

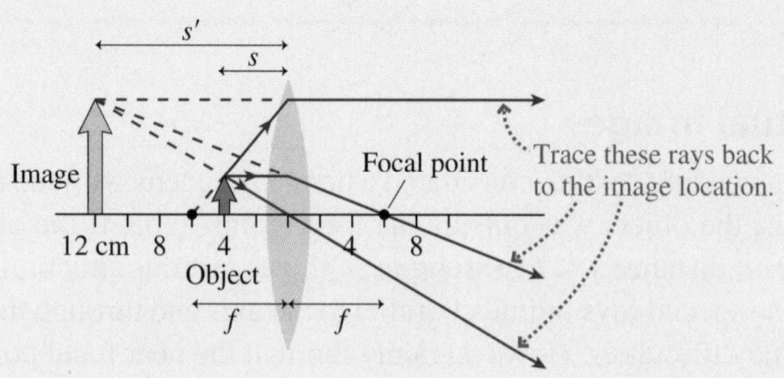

ASSESS The image is three times as large as the object and, as we see from the ray-tracing diagram and the fact that $m > 0$, upright.

Diverging Lenses

As Figure 18.26b showed, a *diverging lens* is one that is thinner at its center than at its edge. FIGURE 18.37 shows three important sets of rays passing through a diverging lens. These are based on Figures 18.26 and 18.28, where you saw that rays initially parallel to the axis diverge after passing through a diverging lens.

FIGURE 18.37 Three important sets of rays passing through a thin, diverging lens.

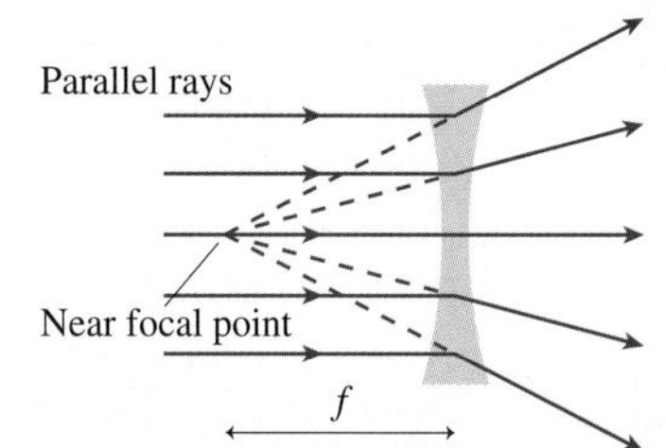

Any ray initially parallel to the optical axis diverges along a line through the near focal point.

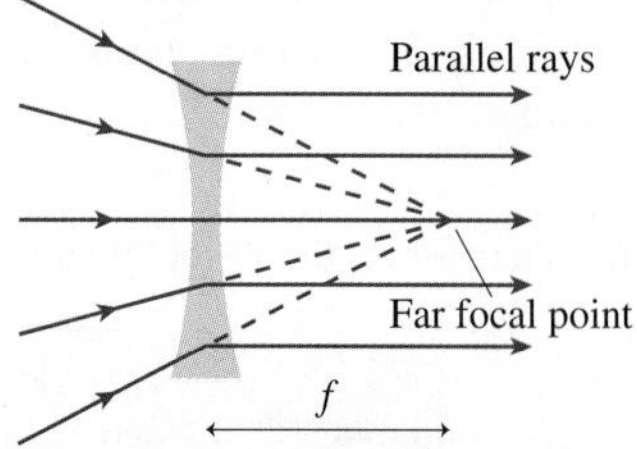

Any ray directed along a line toward the far focal point emerges from the lens parallel to the optical axis.

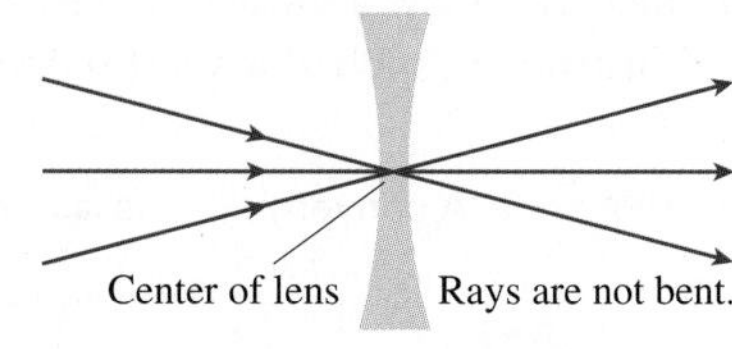

Any ray directed at the center of the lens passes through in a straight line.

15.9 

Ray tracing follows the steps of Tactics Box 18.2 for a converging lens *except* that two of the three special rays in step 4 are different.

TACTICS BOX 18.3 **Ray tracing for a diverging lens**

❶–❸ **Follow steps 1 through 3 of Tactics Box 18.2.**

❹ **Draw the three "special rays" from the tip of the arrow.** Use a straightedge or a ruler.

a. A ray parallel to the axis diverges along a line through the near focal point.
b. A ray along a line toward the far focal point emerges parallel to the axis.
c. A ray through the center of the lens does not bend.

❺ **Trace the diverging rays backward.** The point from which they are diverging is the image point, which is always a virtual image.

❻ **Measure the image distance** s', which, because the image is virtual, we will take as a negative number. Also, if needed, measure the image height relative to the object height. The magnification can be found from Equation 18.8.

EXAMPLE 18.9 **Demagnifying a flower**

A diverging lens with a focal length of 50 cm is placed 100 cm from a flower. Where is the image? What is its magnification?

PREPARE The flower is in the object plane. Use ray tracing to locate the image. Then Equation 18.8 can be used to find the magnification.

SOLVE FIGURE 18.38 shows the ray-tracing diagram. The three special rays (labeled a, b, and c to match the Tactics Box) do not converge. However, they can be traced backward to an intersection ≈33 cm to the left of the lens. Because the rays appear to diverge from the image, this is a virtual image and s' is < 0. The magnification is

$$m = -\frac{s'}{s} = -\frac{-33\text{ cm}}{100\text{ cm}} = 0.33$$

FIGURE 18.38 Ray-tracing diagram for demagnifying.

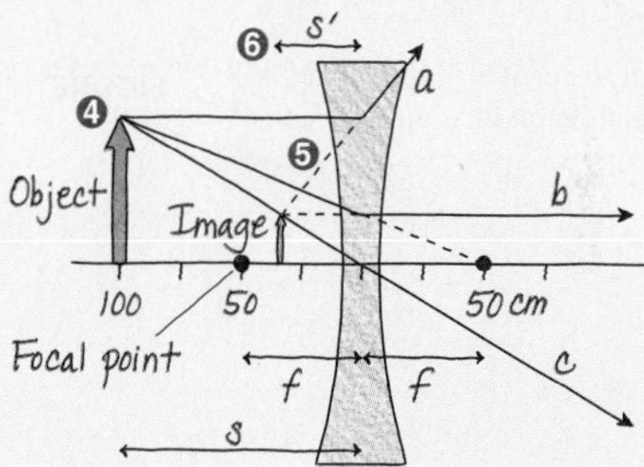

The image, which can be seen by looking *through* the lens, is one-third the size of the object and upright.

ASSESS Ray tracing with a diverging lens is somewhat trickier than with a converging lens, so this example is worth careful study.

Diverging lenses *always* make virtual images and, for this reason, are rarely used alone. However, they have important applications when used in combination with other lenses. Cameras, eyepieces, and eyeglasses often incorporate diverging lenses.

18.6 Image Formation with Spherical Mirrors

Curved mirrors can also be used to form images. Such mirrors are commonly used in telescopes, security and rearview mirrors, and searchlights. Their images can be analyzed with ray diagrams similar to those used with lenses. We'll consider only the important case of **spherical mirrors,** whose surface is a section of a sphere.

FIGURE 18.39 shows parallel light rays approaching two spherical mirrors. The upper mirror, where the edges curve toward the light source, is called a **concave mirror.** Parallel rays reflect off the shiny front surface of the mirror and pass through a single point on the optical axis. This is the focal point of the mirror. The lower mirror, where the edges curve away from the light source, is called a **convex mirror.** Parallel rays that reflect off its surface appear to have come from a point behind the mirror. This is the focal point for a convex mirror. For both mirrors, the focal length is the distance from the mirror surface to the focal point.

FIGURE 18.39 The focal point and focal length of concave and convex mirrors.

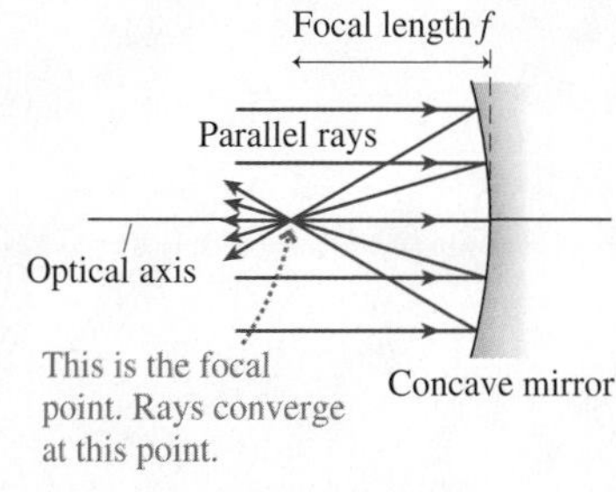

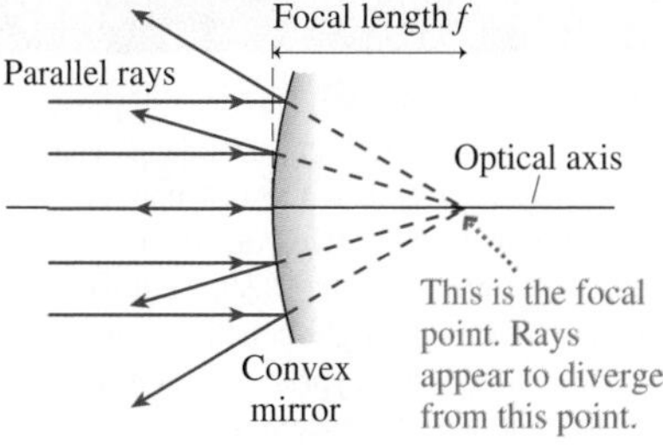

FIGURE 18.40 Three special rays for a concave mirror.

(a)

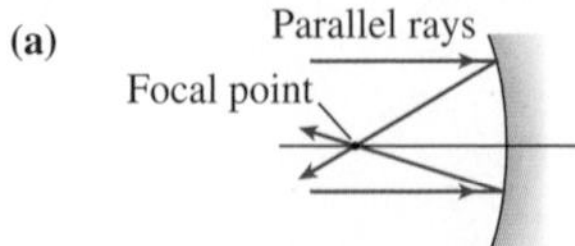

Any ray initially parallel to the optical axis will reflect through the focal point.

(b)

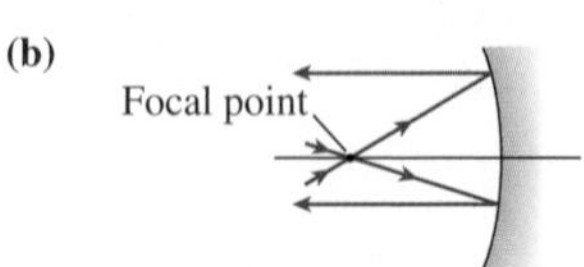

Any ray passing through the focal point will, after reflection, emerge parallel to the optical axis.

(c)

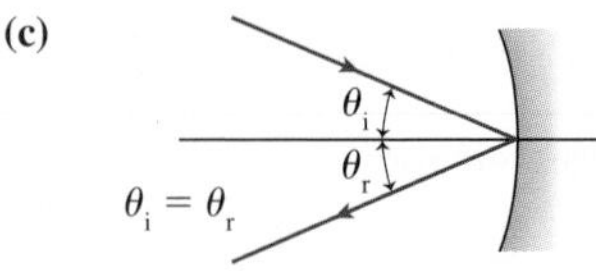

Any ray directed at the center of the mirror will reflect at an equal angle on the opposite side of the optical axis.

Concave Mirrors

To understand image formation by a concave mirror, consider the three special rays shown in FIGURE 18.40. These rays are closely related to those used for ray tracing with lenses. Figure 18.40a shows two incoming rays parallel to the optical axis. As Figure 18.39 showed, these rays reflect off the mirror and pass through the focal point.

Figure 18.40b is the same as Figure 18.40a, but with the directions of the rays reversed. Here we see that rays passing through the focal point emerge parallel to the axis. Finally, Figure 18.40c shows what happens to a ray that is directed toward the center of the mirror. Right at its center, the surface of the mirror is perpendicular to the optical axis. The law of reflection then tells us that the incoming ray will reflect at the same angle, but on the opposite side of the optical axis.

Let's begin by considering the case where the object's distance s from the mirror is greater than the focal length ($s > f$), as shown in FIGURE 18.41. The three special rays just discussed are enough to locate the position and size of the image. Recall that when ray tracing a thin lens, although we drew the lens as having an actual thickness, the rays refracted at an imaginary plane centered on the lens. Similarly, when ray tracing mirrors, the incoming rays reflect off the **mirror plane** as shown in Figure 18.41, not off the curved surface of the mirror. We see that the image is *real* because rays converge at the image point P′. Further, the image is *inverted.*

FIGURE 18.41 A real image formed by a concave mirror.

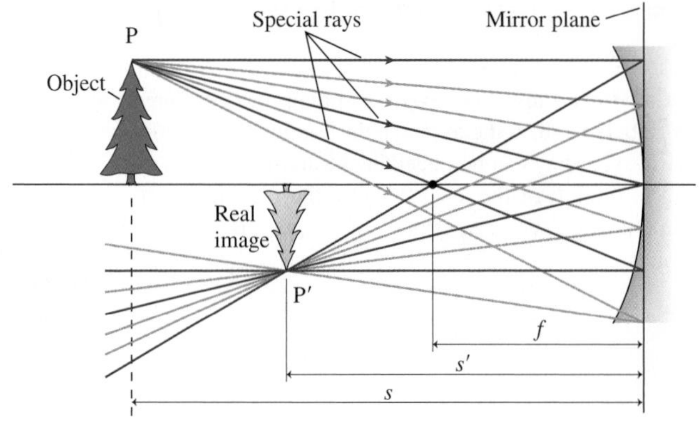

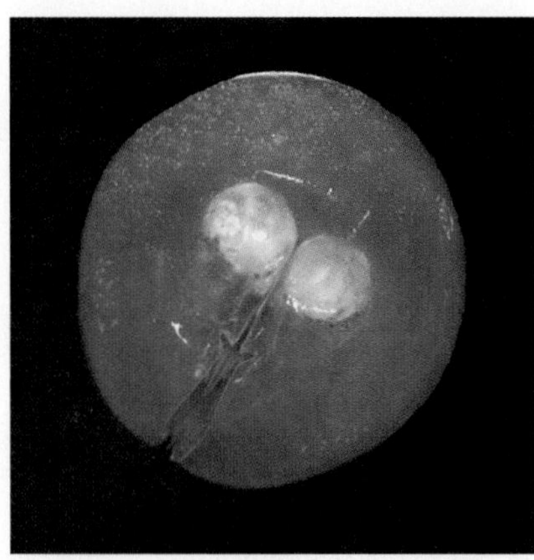

Look into my eyes BIO The eyes of most animals use lenses to focus an image. The *gigantocypris,* a deep-sea crustacean, is unusual in that it uses two concave mirrors to focus light onto its retina. Because it lives at depths where no sunlight penetrates, it is believed that gigantocypris uses its mirror eyes to hunt bioluminescent animals.

Figure 18.41 suggests the following Tactics Box for using ray tracing with a concave mirror:

TACTICS BOX 18.4 Ray tracing for a concave mirror

❶ **Draw an optical axis.** Use graph paper or a ruler! Establish an appropriate scale.

❷ **Center the mirror on the axis.** Mark and label the focal point at distance f from the mirror's surface.

❸ **Represent the object with an upright arrow at distance s.** It's usually best to place the base of the arrow on the axis and to draw the arrow about half the radius of the mirror.

❹ **Draw the three "special rays" from the tip of the arrow.** Use a straightedge or a ruler. The rays should reflect off the mirror plane.

a. A ray parallel to the axis reflects through the focal point.
b. An incoming ray that passes through the focal point emerges parallel to the axis.
c. A ray that strikes the center of the mirror reflects at an equal angle on the opposite side of the optical axis.

Continued

❺ **Extend the rays until they converge.** The rays converge at the image point. Draw the rest of the image in the image plane. If the base of the object is on the axis, then the base of the image will also be on the axis.

❻ **Measure the image distance s'.** Also, if needed, measure the image height relative to the object height. The magnification can be found from Equation 18.8.

Exercises 21a, 22 

EXAMPLE 18.10 Analyzing a concave mirror

A 3.0-cm-high object is located 60 cm from a concave mirror. The mirror's focal length is 40 cm. Use ray tracing to find the position, height, and magnification of the image.

PREPARE FIGURE 18.42 shows the ray-tracing diagram and the steps of Tactics Box 18.4.

SOLVE After preparing a careful drawing, we can use a ruler to find that the image position is $s' \approx 120$ cm. The magnification is thus

$$m = -\frac{s'}{s} \approx -\frac{120 \text{ cm}}{60 \text{ cm}} = -2.0$$

The negative sign indicates that the image is inverted. The image height is thus twice the object height, or $h' \approx 6$ cm.

ASSESS The image is a *real* image because light rays converge at the image point.

FIGURE 18.42 Ray-tracing diagram for a concave mirror.

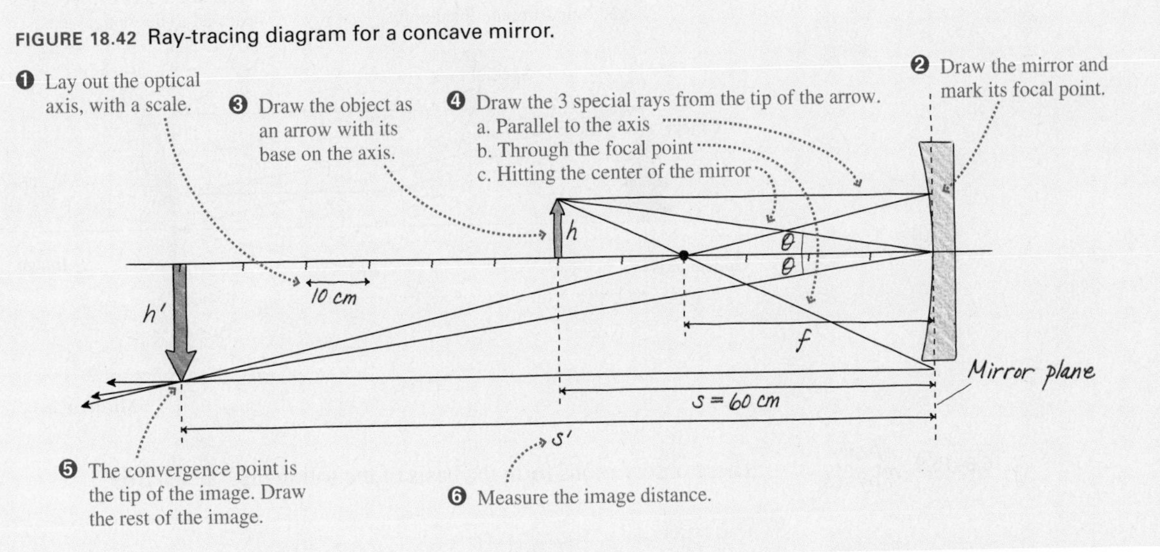

❶ Lay out the optical axis, with a scale.
❷ Draw the mirror and mark its focal point.
❸ Draw the object as an arrow with its base on the axis.
❹ Draw the 3 special rays from the tip of the arrow.
a. Parallel to the axis
b. Through the focal point
c. Hitting the center of the mirror
h
θ
θ
10 cm
h'
f
Mirror plane
s = 60 cm
s'
❺ The convergence point is the tip of the image. Draw the rest of the image.
❻ Measure the image distance.

If the object is inside the focal point ($s < f$), ray tracing can be used to show that the image is a virtual image. This situation is analogous to the formation of a virtual image by a lens when the object is inside the focal point.

Convex Mirrors

A common example of a convex mirror is a silvered ball, such as a tree ornament. You may have noticed that if you look at your reflection in such a ball, your image appears right side up but is quite small. FIGURE 18.43 shows a self-portrait of the Dutch artist M. C. Escher that illustrates these observations. Let's use ray tracing to see why the image appears this way.

Once more, there are three special rays we can use to find the location of the image. These rays are shown in FIGURE 18.44 on the next page; they are similar to the special rays we've already studied in other situations.

FIGURE 18.43 Self-portrait of M. C. Escher.

FIGURE 18.44 Three special rays for a convex mirror.

(a)

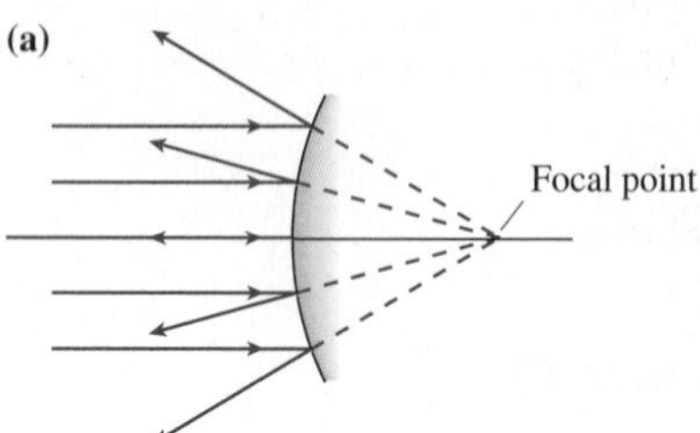

Any ray initially parallel to the optical axis will reflect as though it came from the focal point.

(b)

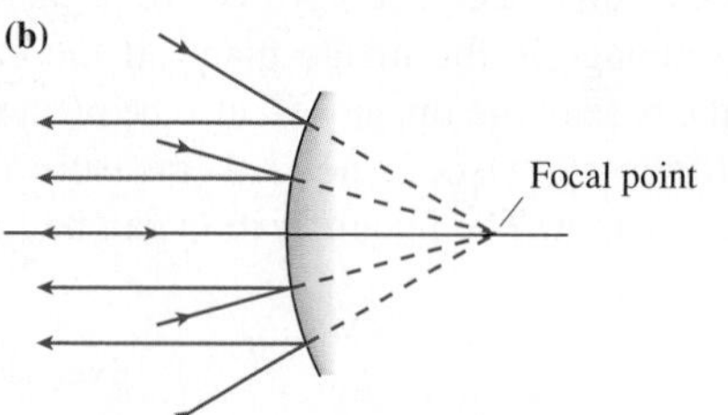

Any ray initially directed toward the focal point will reflect parallel to the optical axis.

(c)

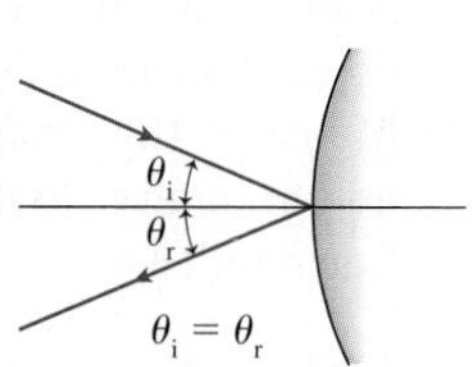

Any ray directed at the center of the mirror will reflect at an equal angle on the opposite side of the optical axis.

We can use these three special rays to find the image of an object, as shown in FIGURE 18.45. We see that the image is virtual—no actual rays converge at the image point P′. Instead, diverging rays *appear* to have come from this point. The image is also upright and much smaller than the object, in accord with our experience and the drawing of Figure 18.43.

FIGURE 18.45 Rays from point P reflect from the mirror and appear to have come from point P′.

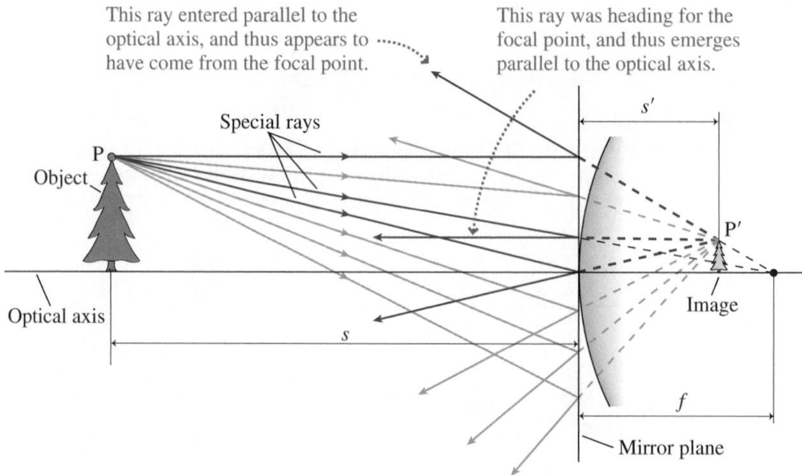

15.5–15.8 ActivPhysics

These observations form the basis of the following Tactics Box.

TACTICS BOX 18.5 Ray tracing for a convex mirror

❶-❸ **Follow steps 1 through 3 of Tactics Box 18.4.**

❹ **Draw the three "special rays" from the tip of the arrow.** Use a straight-edge or a ruler.

a. A ray parallel to the axis reflects as though it came from the focal point.
b. A ray initially directed toward the focal point reflects parallel to the axis.
c. A ray that strikes the center of the mirror reflects at an equal angle on the opposite side of the optical axis.

❺ **Extend the emerging rays *behind the mirror* until they converge.** The point of convergence is the image point. Draw the rest of the image in the image plane. If the base of the object is on the axis, then the base of the image will also be on the axis.

❻ **Measure the image distance s'.** Also, if needed, measure the image height relative to the object height. The magnification can be found from Equation 18.8.

Exercises 21b, 23

The small image in a convex mirror allows a wide-angle view of the store to be visible.

Convex mirrors are used for a variety of safety and monitoring applications, such as passenger-side rearview mirrors and the round mirrors used in stores to keep an eye on the customers. The idea behind such mirrors can be understood from Figure 18.45. When an object is reflected in a convex mirror, the image appears smaller. Because the image is, in a sense, a miniature version of the object, you can *see much more of it* within the edges of the mirror than you could with an equal-sized flat mirror. This wide-angle view is clearly useful for checking traffic behind you, or for checking up on your store.

CONCEPTUAL EXAMPLE 18.11 Driver and passenger mirrors

The rearview mirror on the driver's side of a car is a plane (flat) mirror, while the mirror on the passenger's side is convex. Why is this?

REASON It is important for the driver to have a wide field of view from either mirror. He sits close to the driver-side mirror, so it appears large and can reflect a fairly wide view of what's behind. The passenger-side mirror is quite far from the driver, so it appears relatively small. If it were flat, it would offer only a narrow view of what's behind. Making it convex, like the security mirror discussed above, provides a wider field of view, but the trade-off is a smaller image. That's why the passenger-side mirror usually contains a warning: Objects in mirror are closer than they appear!

STOP TO THINK 18.4 A concave mirror of focal length f forms an image of the moon. Where is the image located?

A. At the mirror's surface
B. Almost exactly a distance f behind the mirror
C. Almost exactly a distance f in front of the mirror
D. At a distance behind the mirror equal to the distance of the moon in front of the mirror

18.7 The Thin-Lens Equation

Ray tracing is an important tool for quickly grasping the overall positions and sizes of an object and its image. For more precise work, however, we would like a mathematical expression that relates the three fundamental quantities of an optical system: the focal length f of the lens or mirror, the object distance s, and the image distance s'. We can find such an expression by considering the converging lens in **FIGURE 18.46**. Two of the special rays are shown: one initially parallel to the optical axis that then passes through the far focal point, and the other traveling undeviated through the center of the lens.

FIGURE 18.46 Deriving the thin-lens equation.

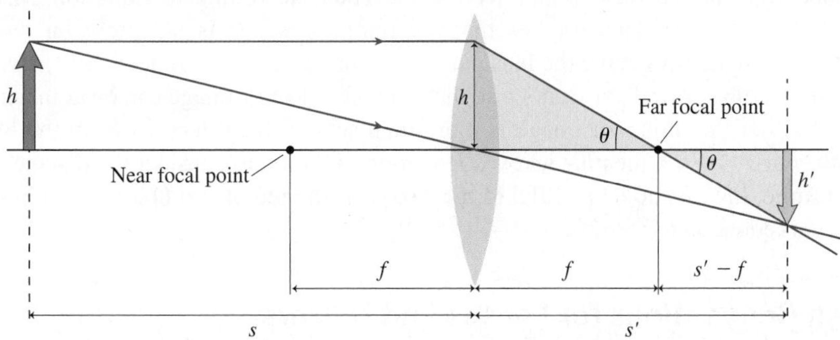

Consider the two right triangles highlighted in green and pink. Because they both have one 90° angle and a second angle θ that is the same for both, the two triangles are *similar.* This means that they have the same shape, although their sizes may be different. For similar triangles, the ratios of any two similar sides are the same. Thus we have

$$\frac{h'}{h} = \frac{s' - f}{f} \tag{18.9}$$

Further, we found in Example 18.7 that

$$\frac{h'}{h} = \frac{s'}{s} \tag{18.10}$$

Combining Equations 18.9 and 18.10 gives

$$\frac{s'}{s} = \frac{s' - f}{f}$$

Dividing both sides by s' gives

$$\frac{1}{s} = \frac{s' - f}{s'f} = \frac{1}{f} - \frac{1}{s'}$$

which we can write as

$$\frac{1}{s} + \frac{1}{s'} = \frac{1}{f} \tag{18.11}$$

Thin-lens equation (also works for mirrors) relating object and image distances to focal length

This equation, the **thin-lens equation,** relates the three important quantities f, s, and s'. In particular, if we know the focal length of a lens and the object's distance from the lens, we can use the thin-lens equation to find the position of the image.

NOTE ► Although we derived the thin-lens equation for a converging lens that produced a real image, it works equally well for *any* image—real or virtual—produced by both converging and diverging lenses. And, in spite of its name, the thin-lens equation also describes the images formed by *mirrors.* ◄

It's worth checking that the thin-lens equation describes what we already know about lenses. In Figure 18.29a, we saw that rays initially parallel to the optical axis are focused at the focal point of a converging lens. Initially parallel rays come from an object extremely far away, with $s \to \infty$. Because $1/\infty = 0$, the thin-lens equation tells us that the image distance is $s' = f$, as we expected. Or suppose an object is located right at the focal point, with $s = f$. Then, according to Equation 18.11, $1/s' = 1/f - 1/s = 0$. This implies that the image distance is infinitely far away ($s' = \infty$), so the rays leave the lens parallel to the optical axis. Indeed, this is what Figure 18.29b showed. Now, it's true that no real object or image can be at infinity. But if either the object or image is more than several focal lengths from the lens ($s \gg f$ or $s' \gg f$), then it's an excellent approximation to consider the distance to be infinite, the rays to be parallel to the axis, and the reciprocal ($1/s$ or $1/s'$) in the thin-lens equation to be zero.

Sign Conventions for Lenses and Mirrors

We've already noted that the image distance s' is positive for real images and negative for virtual images. In the thin-lens equation, the sign of the focal length can also be either positive or negative, depending on the type of lens or mirror. Tactics Box 18.6 summarizes the sign conventions that we will use in this text.

TACTICS BOX 18.6 Using sign conventions for lenses and mirrors

Quantity	Positive when		Negative when	
Object distance s	Always		We won't treat this case, which can occur for two lenses or mirrors in combination.	
Image distance s'	**Real image**		**Virtual image**	
	Image is on the opposite side of the lens from the object.	Image is in front of the mirror.	Image is on the same side of the lens as the object.	Image is behind the mirror.
Focal length f	Converging lens or concave mirror		Diverging lens or convex mirror	
Magnification m	Image is upright.		Image is inverted.	

NOTE ► When using the thin-lens equation, the focal length must be taken as positive or negative according to Tactics Box 18.6. In the pictorial method of ray tracing, however, focal lengths are just *distances* and are therefore always positive. ◄

EXAMPLE 18.12 Analyzing a magnifying lens

A stamp collector uses a magnifying lens that sits 2.0 cm above the stamp. The magnification is 4. What is the focal length of the lens?

FIGURE 18.47 Ray-tracing diagram of a magnifying lens.

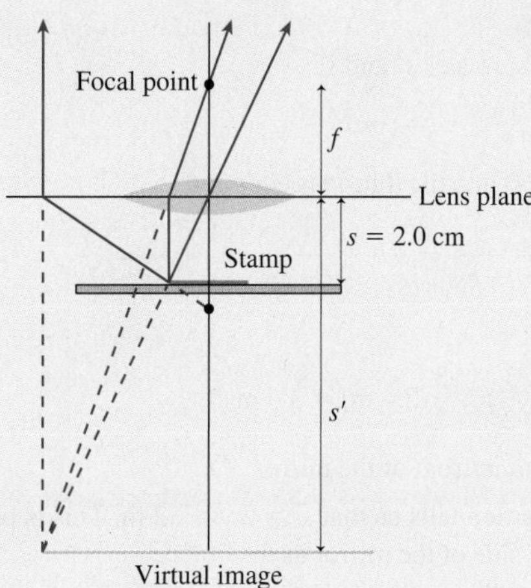

PREPARE A magnifying lens is a converging lens with the object distance less than the focal length ($s < f$). Assume it is a thin lens. The user looks *through* the lens and sees a virtual image. **FIGURE 18.47** shows the lens and a ray-tracing diagram.

SOLVE A virtual image is upright, so $m = +4$. The magnification is $m = -s'/s$; thus

$$s' = -4s = -4(2.0\text{ cm}) = -8.0\text{ cm}$$

We can use s and s' in the thin-lens equation to find the focal length:

$$\frac{1}{f} = \frac{1}{s} + \frac{1}{s'} = \frac{1}{2.0\text{ cm}} + \frac{1}{-8.0\text{ cm}} = 0.375\text{ cm}^{-1}$$

Thus

$$f = \frac{1}{0.375\text{ cm}^{-1}} = 2.7\text{ cm}$$

ASSESS $f > 2$ cm, as expected because the object has to be inside the focal point.

EXAMPLE 18.13 What is the focal length?

An object is 38.0 cm to the left of a lens. Its image is found to be 22.0 cm from the lens on the same side as the object. What is the focal length of the lens? Draw a ray diagram for the lens and object.

PREPARE We know the object distance is $s = 38.0$ cm. According to Tactics Box 18.6, the image distance is a negative number because the image is on the same side of the lens as the object, so $s' = -22.0$ cm.

Continued

SOLVE The thin-lens equation, Equation 18.11, is

$$\frac{1}{f} = \frac{1}{s} + \frac{1}{s'} = \frac{1}{38.0\text{ cm}} + \frac{1}{-22.0\text{ cm}} = -0.0191\text{ cm}^{-1}$$

from which we find

$$f = \frac{1}{-0.0191\text{ cm}^{-1}} = -52.3\text{ cm}$$

Because f is negative, Tactics Box 18.6 indicates that the lens is a diverging lens. We can use the techniques learned in Section 18.5 to draw the ray diagram shown in FIGURE 18.48. Notice that the negative value of s' corresponds to a virtual image.

FIGURE 18.48 Ray-tracing diagram of a diverging lens.

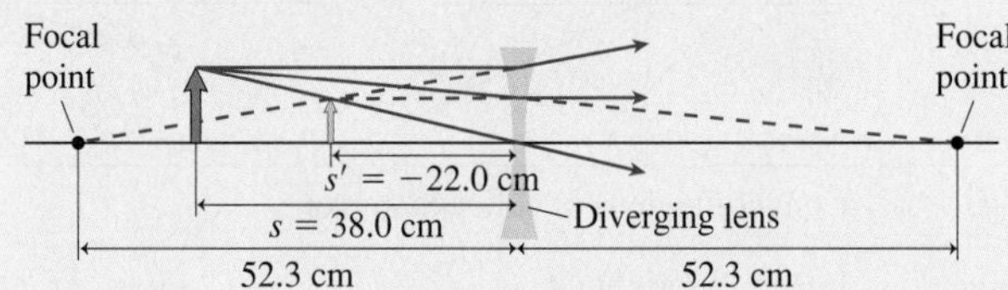

ASSESS This problem would be difficult to solve graphically because you need to locate the focal points in order to draw a ray diagram. Using the thin-lens equation gives rapid and accurate results. Nonetheless, sketching a ray diagram once f is known is very helpful in fully understanding the situation.

EXAMPLE 18.14 Finding the mirror image of a candle

A concave mirror has a focal length of 2.4 m. Where should a candle be placed so that its image is inverted and twice as large as the object?

PREPARE Even though this is a mirror, we can use the thin-lens equation. We have $f = +2.4$ m because, according to Tactics Box 18.6, a concave mirror has a positive focal length. We also know that the magnification is $m = -2$ because an inverted image implies a negative magnification.

SOLVE Equation 18.8, $m = -s'/s$, relates s' and s:

$$s' = -ms$$

We can insert this expression for s' into the thin-lens equation:

$$\frac{1}{f} = \frac{1}{s} + \frac{1}{s'} = \frac{1}{s} + \frac{1}{-ms} = \frac{1}{s} + \frac{1}{-(-2)s} = \frac{1}{s} + \frac{1}{2s} = \frac{3}{2s}$$

from which we get

$$s = \frac{3}{2}f = \frac{3}{2}(2.4\text{ m}) = 3.6\text{ m}$$

The candle should be placed 3.6 m in front of the mirror.

ASSESS The magnification equation tells us that $s' = 2s = 7.2$ m. This is positive, so the image is real and on the same side of the mirror as the object.

STOP TO THINK 18.5 A candle is placed in front of a converging lens. A well-focused image of the flame is seen on a screen on the opposite side of the lens. If the candle is moved farther away from the lens, how must the screen be adjusted to keep showing a well-focused image?

A. The screen must be moved closer to the lens.
B. The screen must be moved farther away from the lens.
C. The screen does not need to be moved.

INTEGRATED EXAMPLE 18.15 Optical fiber imaging

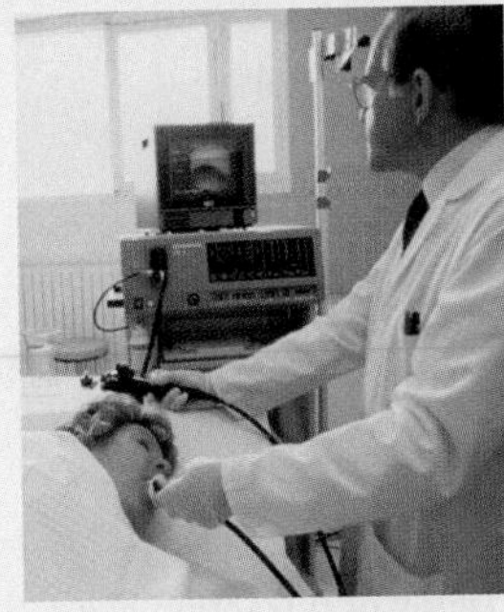

An *endoscope* is a narrow bundle of optical fibers that can be inserted through a bodily opening or a small incision to view the interior of the body. As **FIGURE 18.49** shows, an *objective* lens focuses a real image onto the entrance face of the fiber bundle. Individual fibers—using total internal reflection—transport the light to the exit face, where it emerges. The doctor observes a magnified image of the exit face by viewing it through an *eyepiece* lens.

FIGURE 18.49 An endoscope.

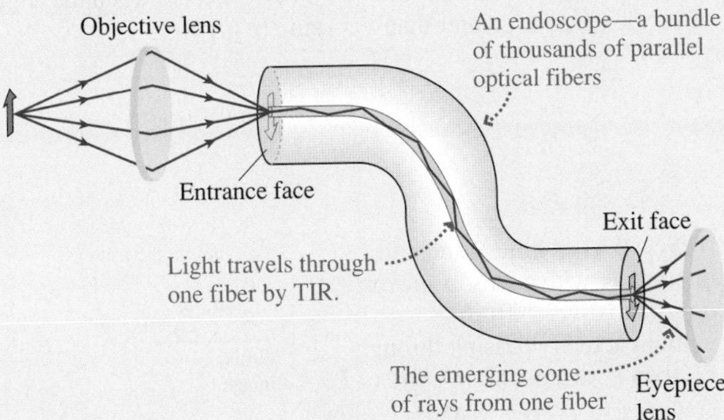

A single optical fiber, as discussed in Section 18.3 and shown in **FIGURE 18.50**, consists of a lower-index glass core surrounded by a higher-index cladding layer. To remain in the fiber, light rays propagating through the core must strike the cladding boundary at angles of incidence greater than the critical angle θ_c. As Figure 18.50 shows, this means that rays that enter the core at angles of incidence smaller than θ_{max} are totally internally reflected down the fiber; those that enter at angles larger than θ_{max} are not totally internally reflected, and escape from the fiber.

FIGURE 18.50 Cross section of an optical fiber (greatly magnified).

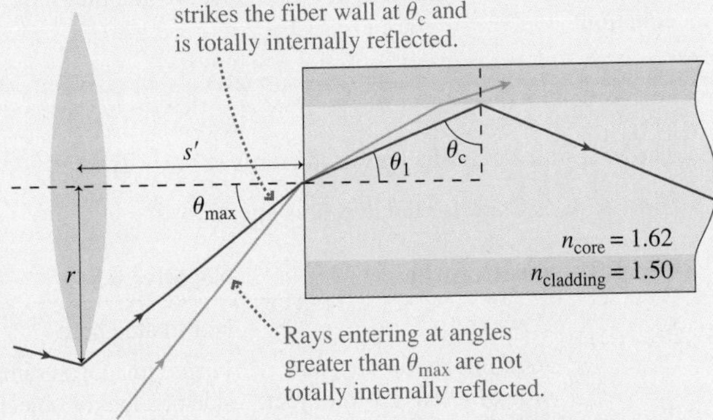

The objective lens of an endoscope must be carefully matched to the fiber. Ideally, the lens diameter is such that rays from the edge of the lens enter the fiber at θ_{max}, as shown in Figure 18.50. A lens larger than this is not useful because rays from its outer regions will enter the fiber at an angle larger than θ_{max} and will thus escape from the fiber. A lens smaller than this will suffer from a reduced light-gathering power.

a. What is θ_{max} for the fiber shown in Figure 18.50?
b. A typical objective lens is 3.0 mm in diameter and can focus on an object 3.0 mm in front of it. What focal length should the lens have so that rays from its edge just enter the fiber at angle θ_{max}?
c. What is the magnification of this lens?

PREPARE From Figure 18.50 we can find the critical angle θ_c and then use geometry to find θ_1. Snell's law can then be used to find θ_{max}. Once θ_{max} is known, we can find the image distance s', and, because we know that the object distance is $s = 3$ mm, we can use the thin-lens equation to solve for f.

SOLVE

a. The critical angle for total internal reflection is given by Equation 18.3:

$$\theta_c = \sin^{-1}\left(\frac{n_2}{n_1}\right) = \sin^{-1}\left(\frac{n_{cladding}}{n_{core}}\right) = \sin^{-1}\left(\frac{1.50}{1.62}\right) = 68°$$

Because of the right triangle, this ray's angle of refraction from the fiber's face is $\theta_1 = 90° - 68° = 22°$. Then, by Snell's law, this ray enters from the air at angle θ_{max} such that

$$n_{air}\sin\theta_{max} = 1.00\sin\theta_{max} = n_{core}\sin\theta_1 = 1.62(\sin 22°) = 0.61$$

Thus

$$\theta_{max} = \sin^{-1}(0.61) = 38°$$

b. As Figure 18.50 shows, the distance of the lens from the fiber's face—the image distance s'—is related to the lens radius $r = 1.5$ mm by

$$\frac{r}{s'} = \tan\theta_{max}$$

Thus the image distance is

$$s' = \frac{r}{\tan\theta_{max}} = \frac{1.5\text{ mm}}{\tan 38°} = 1.9\text{ mm}$$

Then the thin-lens equation gives

$$\frac{1}{f} = \frac{1}{s} + \frac{1}{s'} = \frac{1}{3.0\text{ mm}} + \frac{1}{1.9\text{ mm}} = 0.85\text{ mm}^{-1}$$

so that

$$f = \frac{1}{0.85\text{ mm}^{-1}} = 1.2\text{ mm}$$

c. The magnification is

$$m = -\frac{s'}{s} = -\frac{1.9\text{ mm}}{3.0\text{ mm}} = -0.63$$

ASSESS The object distance of 3.0 mm is greater than the 1.2 mm focal length we calculated, as must be the case when a converging lens produces a real image.

SUMMARY

The goal of Chapter 18 has been to understand and apply the ray model of light.

GENERAL PRINCIPLES

Reflection

Law of reflection: $\theta_r = \theta_i$

Reflection can be **specular** (mirror-like) or **diffuse** (from rough surfaces).

Plane mirrors: A virtual image is formed at P′ with $s' = s$, where s is the **object distance** and s' is the **image distance.**

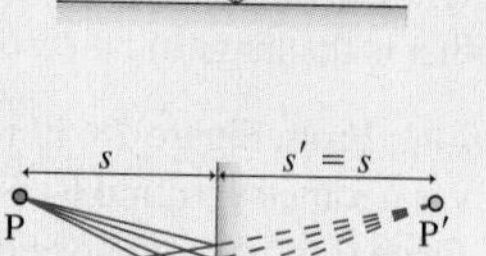

Refraction

Snell's law of refraction:

$$n_1 \sin\theta_1 = n_2 \sin\theta_2$$

Index of refraction is $n = c/v$. The ray is closer to the normal on the side with the larger index of refraction.

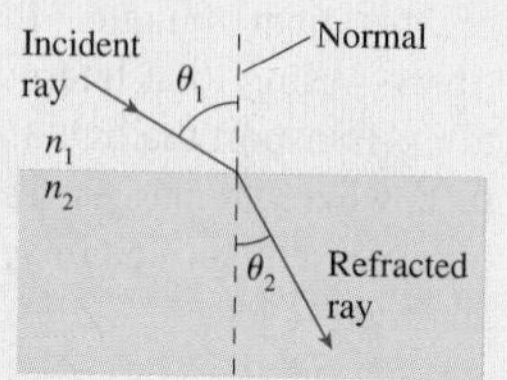

If $n_2 < n_1$, **total internal reflection** (TIR) occurs when the angle of incidence θ_1 is greater than $\theta_c = \sin^{-1}(n_2/n_1)$.

IMPORTANT CONCEPTS

The ray model of light

Light travels along straight lines, called **light rays,** at speed $v = c/n$.

A light ray continues forever unless an interaction with matter causes it to reflect, refract, scatter, or be absorbed.

Light rays come from self-luminous or reflective **objects.** Each point on the object sends rays in all directions.

Ray diagrams represent all the rays emitted by an object by only a few select rays.

In order for the eye to see an object (or image), rays from the object or image must enter the eye.

Image formation

If rays diverge from P and, after interacting with a lens or mirror, *appear* to diverge from P′ without actually passing through P′, then P′ is a **virtual image** of P.

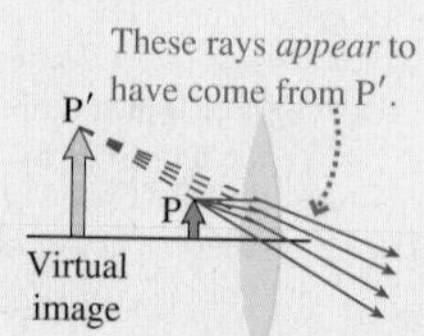

If rays diverge from P and interact with a lens or mirror so that the refracted rays *converge* at P′, then P′ is a **real image** of P. Rays actually pass through a real image.

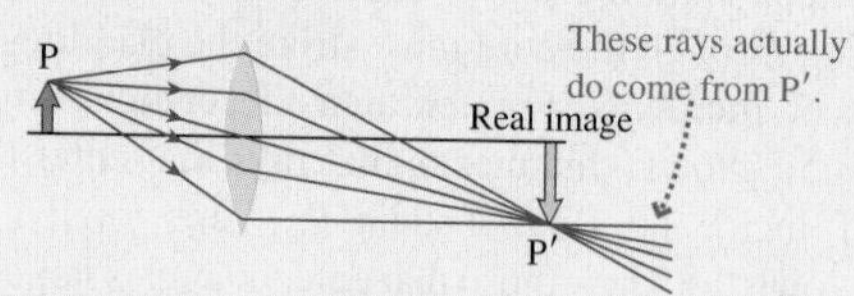

APPLICATIONS

Ray tracing for lenses

Three special rays in three basic situations:

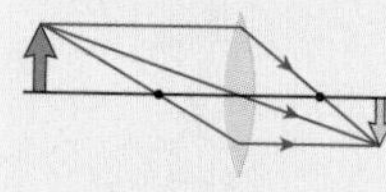

Converging lens
Real image

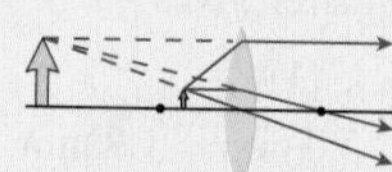

Converging lens
Virtual image

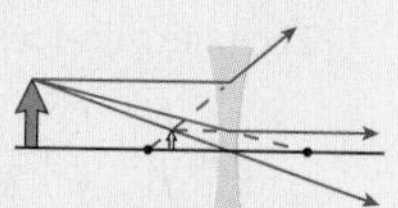

Diverging lens
Virtual image

Ray tracing for mirrors

Three special rays in three basic situations:

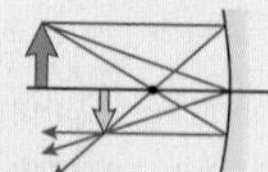

Concave mirror
Real image

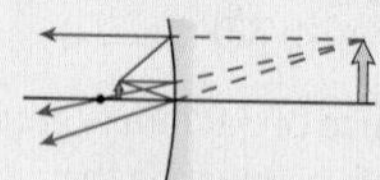

Concave mirror
Virtual image

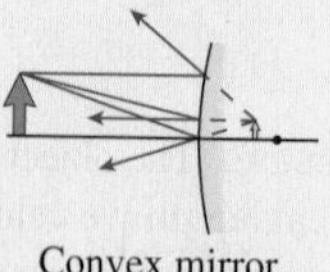

Convex mirror
Virtual image

The thin-lens equation

For a lens or curved mirror, the object distance s, the image distance s', and the focal length f are related by the thin-lens equation:

$$\frac{1}{s} + \frac{1}{s'} = \frac{1}{f}$$

The **magnification** of a lens or mirror is $m = -s'/s$.

Sign conventions for the thin-lens equation:

Quantity	Positive when	Negative when
s	Always	Not treated here
s'	*Real* image; on opposite side of a lens from object, or in front of a mirror	*Virtual* image; on same side of a lens as object, or behind a mirror
f	Converging lens or concave mirror	Diverging lens or convex mirror
m	Image is upright.	Image is inverted.

For homework assigned on MasteringPhysics, go to www.masteringphysics.com

Problem difficulty is labeled as | (straightforward) to ||||| (challenging).

Problems labeled INT integrate significant material from earlier chapters; BIO are of biological or medical interest.

QUESTIONS

Conceptual Questions

1. The idea of light rays goes back to the ancient Greeks. However, they believed that "visual rays" were *emitted* by eyes. If you were transported back in time, what arguments would you present to those early scientists to convince them that vision has something to do with rays going into, rather than out of, eyes?
2. Is there any property that distinguishes a light ray emitted by a light bulb and one that has been diffusely reflected by the page of a book? Explain.
3. If you turn on your car headlights during the day, the road ahead of you doesn't appear to get brighter. Why not?
4. Can you see the rays from the sun on a clear day? Why or why not? How about when they stream through a forest on a foggy morning? Why or why not?
5. If you take a walk on a summer night along a dark, unpaved road in the woods, with a flashlight pointing at the ground several yards ahead to guide your steps, any water-filled potholes are noticeable because they appear much darker than the surrounding dry road. Explain why.
6. You are looking at the image of a pencil in a mirror, as shown in Figure Q18.6.
   a. What happens to the image you see if the top half of the mirror, down to the midpoint, is covered with a piece of cardboard? Explain.
   b. What happens to the image you see if the bottom half of the mirror is covered with a piece of cardboard?

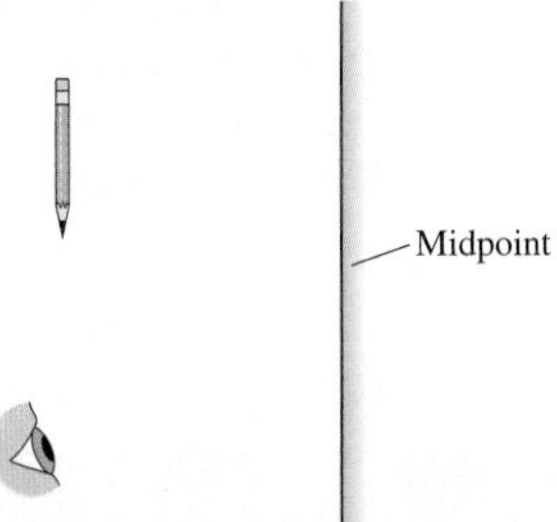

FIGURE Q18.6

7. In *The Toilet of Venus* by Velázquez (see Figure Q18.7), we can see the face of Venus in the mirror. Can she see her own face in the mirror, when the mirror is held as shown in the picture? If yes, explain why; if not, what does she see instead?

FIGURE Q18.7

Diego de Silva Velazquez (1599-1660), "Venus and Cupid," 1650. Oil on canvas. National Gallery, London. Erich Lessing/Art Resource, N.Y.

8. In Manet's *A Bar at the Folies-Bergère* (see Figure Q18.8) the reflection of the barmaid is visible in the mirror behind her. Is this the reflection you would expect if the mirror's surface is parallel to the bar? Where is the man seen facing her in the mirror actually standing?

FIGURE Q18.8

Edouard Manet 1832-1883, "Bar at the Folies-Bergere". 1881/82. Oil on Canvas. 37 13/16″ × 51″ (90 × 130 cm). Courtauld Institute Galleries, London. AKG-Images.

9. Explain why ambulances have the word "AMBULANCE" written backward on the front of them.
10. a. Consider *one* point on an object near a lens. What is the minimum number of rays needed to locate its image point?
    b. For each point on the object, how many rays from this point actually strike the lens and refract to the image point?
11. When you look at your reflection in the bowl of a spoon, it is upside down. Why is this?
12. A concave mirror brings the sun's rays to a focus at a distance of 30 cm from the mirror. If the mirror were submerged in a swimming pool, would the sun's rays be focused nearer to, further from, or at the same distance from the mirror?
13. A student draws the ray diagram shown in Figure Q18.13 but forgets to label the object, the image, or the type of lens used. Using the diagram, explain whether the lens is converging or diverging, which arrow represents the object, and which represents the image.

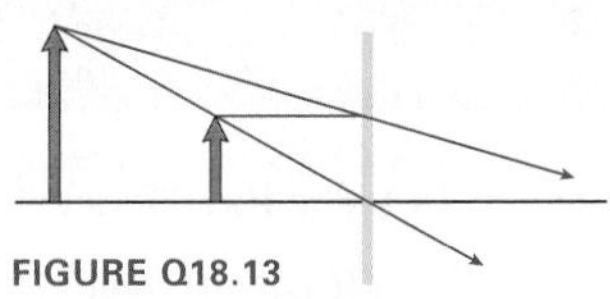

FIGURE Q18.13

14. An object at distance s from a concave mirror of focal length f produces a real image at distance s' from the mirror. Suppose the mirror is replaced by a new mirror, at the same location, with focal length $\frac{1}{2}f$. Will the new image be real or virtual? Will its distance from the mirror be more or less than s'? Explain.
15. A lens can be used to start a fire by focusing an image of the sun onto a piece of flammable material. All other things being equal, would a lens with a short focal length or a long focal length be better as a fire starter? Explain.

Multiple-Choice Questions

Questions 16 through 18 are concerned with the situation sketched in Figure Q18.16, in which a beam of light in the air encounters a transparent block with index of refraction $n = 1.53$. Some of the light is reflected and some is refracted.

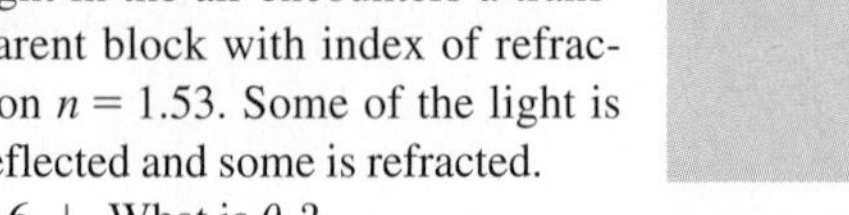

FIGURE Q18.16

16. | What is θ_1?
   A. 40° B. 45°
   C. 50° D. 90°
17. | What is θ_2?
   A. 20° B. 30°
   C. 50° D. 60°
18. | Is there an angle of incidence between 0° and 90° such that all of the light will be reflected?
   A. Yes, at an angle greater than 50°
   B. Yes, at an angle less than 50° C. No
19. | A 2.0-m-tall man is 5.0 m from the converging lens of a camera. His image appears on a detector that is 50 mm behind the lens. How tall is his image on the detector?
   A. 10 mm B. 20 mm
   C. 25 mm D. 50 mm
20. || You are 2.4 m from a plane mirror, and you would like to take a picture of yourself in the mirror. You need to manually adjust the focus of the camera by dialing in the distance to what you are photographing. What distance do you dial in?
   A. 1.2 m B. 2.4 m C. 3.6 m D. 4.8 m
21. || Figure Q18.21 shows an object and lens positioned to form a well-focused, inverted image on a viewing screen. Then a piece of cardboard is lowered just in front of the lens to cover the *top half* of the lens. What happens to the image on the screen?

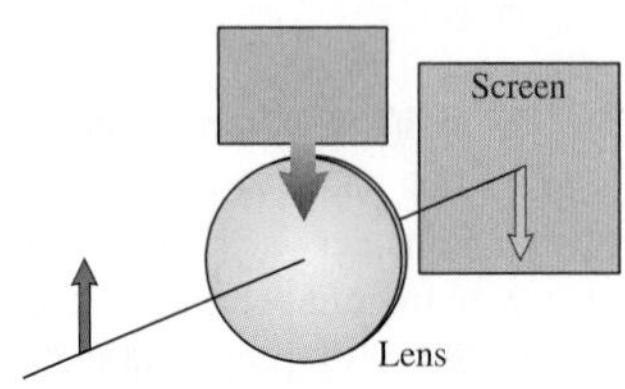

FIGURE Q18.21

   A. Nothing.
   B. The upper half of the image will vanish.
   C. The lower half of the image will vanish.
   D. The image will become fuzzy and out of focus.
   E. The image will become dimmer.
22. || A real image of an object can be formed by
   A. A converging lens.
   B. A plane mirror.
   C. A convex mirror.
   D. Any of the above.
23. | An object is 40 cm from a converging lens with a focal length of 30 cm. A real image is formed on the other side of the lens, 120 cm from the lens. What is the magnification?
   A. 2.0 B. 3.0 C. 4.0
   D. 1.33 E. 0.33
24. | The lens in Figure Q18.24 is used to produce a real image of a candle flame. What is the focal length of the lens?
   A. 9.0 cm
   B. 12 cm
   C. 24 cm
   D. 36 cm
   E. 48 cm

FIGURE Q18.24

25. | A converging lens of focal length 20 cm is used to form a real image 1.0 m away from the lens. How far from the lens is the object?
   A. 20 cm B. 25 cm C. 50 cm D. 100 cm
26. | You look at yourself in a convex mirror. Your image is
   A. Erect. B. Inverted.
   C. It's impossible to tell without knowing how far you are from the mirror and its focal length.
27. || An object is 50 cm from a diverging lens with a focal length of −20 cm. How far from the lens is the image, and on which side of the lens is it?
   A. 14 cm, on the same side as the object
   B. 14 cm, on the opposite side from the object
   C. 30 cm, on the same side as the object
   D. 33 cm, on the same side as the object
   E. 33 cm, on the opposite side from the object

PROBLEMS

Section 18.1 The Ray Model of Light

1. || A 5.0-ft-tall girl stands on level ground. The sun is 25° above the horizon. How long is her shadow?
2. |||| A 10-cm-diameter disk emits light uniformly from its surface. 20 cm from this disk, along its axis, is an 8.0-cm-diameter opaque black disk; the faces of the two disks are parallel. 20 cm beyond the black disk is a white viewing screen. The lighted disk illuminates the screen, but there's a shadow in the center due to the black disk. What is the diameter of the *completely dark* part of this shadow?
3. ||||| A point source of light illuminates an aperture 2.00 m away. A 12.0-cm-wide bright patch of light appears on a screen 1.00 m behind the aperture. How wide is the aperture?

Section 18.2 Reflection

4. | The mirror in Figure P18.4 deflects a horizontal laser beam by 60°. What is the angle ϕ?

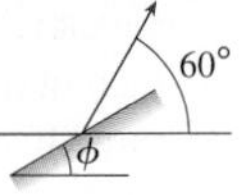

FIGURE P18.4

5. | Figure P18.5 shows an object O in front of a plane mirror. Use ray tracing to determine from which locations A–D the object's image is visible.

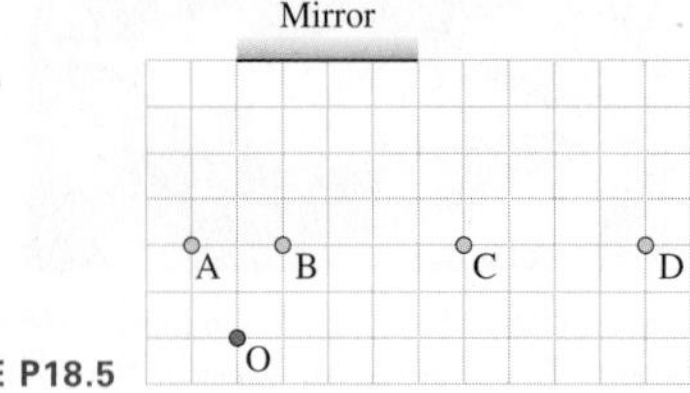

FIGURE P18.5

6. ||| A light ray leaves point A in Figure P18.6, reflects from the mirror, and reaches point B. How far below the top edge does the ray strike the mirror?

FIGURE P18.6

7. |||| It is 165 cm from your eyes to your toes. You're standing 200 cm in front of a tall mirror. How far is it from your eyes to the image of your toes?

8. ||| A ray of light impinges on a mirror as shown in Figure P18.8. A second mirror is fastened at 90° to the first.
   a. After striking both mirrors, at what angle relative to the incoming ray does the outgoing ray emerge?
   b. What is the answer if the incoming angle is 30°?

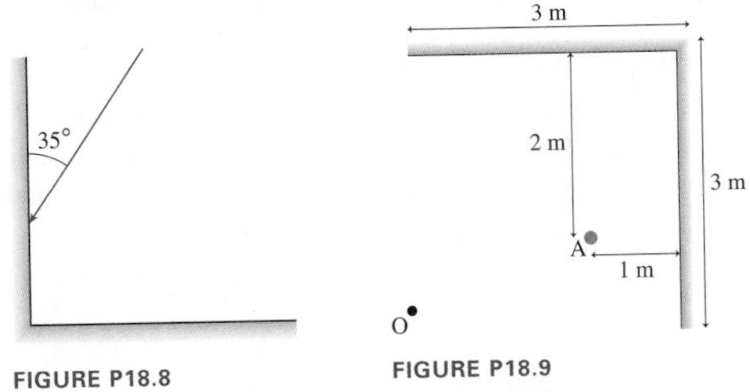

FIGURE P18.8

FIGURE P18.9

9. ||| A red ball is placed at point A in Figure P18.9.
   a. How many images are seen by an observer at point O?
   b. Where is each image located?
   c. Draw a ray diagram showing the formation of each image.

Section 18.3 Refraction

10. ||| An underwater diver sees the sun 50° above horizontal. How high is the sun above the horizon to a fisherman in a boat above the diver?

11. || A laser beam in air is incident on a liquid at an angle of 37° with respect to the normal. The laser beam's angle in the liquid is 26°. What is the liquid's index of refraction?

12. ||| A 1.0-cm-thick layer of water stands on a horizontal slab of glass. A light ray in the air is incident on the water 60° from the normal. After entering the glass, what is the ray's angle from the normal?

13. ||| A 4.0-m-wide swimming pool is filled to the top. The bottom of the pool becomes completely shaded in the afternoon when the sun is 20° above the horizon. How deep is the pool?

14. || A diamond is underwater. A light ray enters one face of the diamond, then travels at an angle of 30° with respect to the normal. What was the ray's angle of incidence on the diamond?

15. || A thin glass rod is submerged in oil. What is the critical angle for light traveling inside the rod?

Section 18.4 Image Formation by Refraction

16. ||| A biologist keeps a specimen of his favorite beetle embedded in a cube of polystyrene plastic. The hapless bug appears to be 2.0 cm within the plastic. What is the beetle's actual distance beneath the surface?

17. || A fish in a flat-sided aquarium sees a can of fish food on the counter. To the fish's eye, the can looks to be 30 cm outside the aquarium. What is the actual distance between the can and the aquarium? (You can ignore the thin glass wall of the aquarium.)

18. | A swim mask has a pocket of air between your eyes and the flat glass front.
   a. If you look at a fish while swimming underwater with a swim mask on, does the fish appear closer or farther than it really is? Draw a ray diagram to explain.
   b. Does the fish see your face closer or farther than it really is? Draw a ray diagram to explain.

Section 18.5 Thin Lenses: Ray Tracing

19. | An object is 30 cm in front of a converging lens with a focal length of 10 cm. Use ray tracing to determine the location of the image. Is the image upright or inverted? Is it real or virtual?

20. | An object is 6.0 cm in front of a converging lens with a focal length of 10 cm. Use ray tracing to determine the location of the image. Is the image upright or inverted? Is it real or virtual?

21. || An object is 20 cm in front of a diverging lens with a focal length of 10 cm. Use ray tracing to determine the location of the image. Is the image upright or inverted? Is it real or virtual?

22. | An object is 15 cm in front of a diverging lens with a focal length of 10 cm. Use ray tracing to determine the location of the image. Is the image upright or inverted? Is it real or virtual?

Section 18.6 Image Formation with Spherical Mirrors

23. | A concave cosmetic mirror has a focal length of 40 cm. A 5-cm-long mascara brush is held upright 20 cm from the mirror. Use ray tracing to determine the location and height of its image. Is the image upright or inverted? Is it real or virtual?

24. | A light bulb is 60 cm from a concave mirror with a focal length of 20 cm. Use ray tracing to determine the location of its image. Is the image upright or inverted? Is it real or virtual?

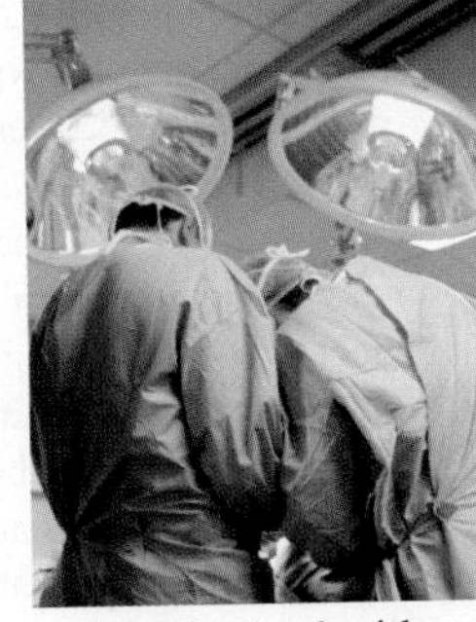

25. | The illumination lights in an
BIO operating room use a concave mirror to focus an image of a bright lamp onto the surgical site. One such light has a mirror with a focal length of 15.0 cm. Use ray tracing to find the position of its lamp when the patient is positioned 1.0 m from the mirror (you'll need a careful drawing to get a good answer).

26. || A dentist uses a curved mirror to view the back side of
BIO teeth on the upper jaw. Suppose she wants an erect image with a magnification of 2.0 when the mirror is 1.2 cm from a tooth. (Treat this problem as though the object and image lie along a straight line.) Use ray tracing to decide whether a concave or convex mirror is needed, and to estimate its focal length.

27. | A convex mirror, like the passenger-side rearview mirror on a car, has a focal length of 2.0 m. An object is 4.0 m from the mirror. Use ray tracing to determine the location of its image. Is the image upright or inverted? Is it real or virtual?

28. | An object is 6 cm in front of a convex mirror with a focal length of 10 cm. Use ray tracing to determine the location of the image. Is the image upright or inverted? Is it real or virtual?

Section 18.7 The Thin-Lens Equation

For Problems 29 through 38, calculate the image position and height.

29. | A 2.0-cm-tall object is 40 cm in front of a converging lens that has a 20 cm focal length.
30. || A 1.0-cm-tall object is 10 cm in front of a converging lens that has a 30 cm focal length.
31. || A 2.0-cm-tall object is 15 cm in front of a converging lens that has a 20 cm focal length.
32. | A 1.0-cm-tall object is 75 cm in front of a converging lens that has a 30 cm focal length.
33. | A 2.0-cm-tall object is 15 cm in front of a diverging lens that has a −20 cm focal length.
34. | A 1.0-cm-tall object is 60 cm in front of a diverging lens that has a −30 cm focal length.
35. || A 3.0-cm-tall object is 15 cm in front of a convex mirror that has a −25 cm focal length.
36. || A 3.0-cm-tall object is 45 cm in front of a convex mirror that has a −25 cm focal length.
37. | A 3.0-cm-tall object is 15 cm in front of a concave mirror that has a 25 cm focal length.
38. | A 3.0-cm-tall object is 45 cm in front of a concave mirror that has a 25 cm focal length.

General Problems

39. ||| Starting 3.5 m from a department store mirror, Suzanne walks toward the mirror at 1.5 m/s for 2.0 s. How far is Suzanne from her image in the mirror after 2.0 s?
INT
40. || You slowly back away from a plane mirror at a speed of 0.10 m/s. With what speed does your image appear to be moving away from you?
INT
41. || At what angle ϕ should the laser beam in Figure P18.41 be aimed at the mirrored ceiling in order to hit the midpoint of the far wall?

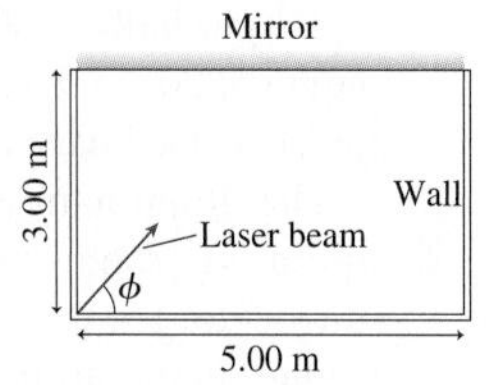

FIGURE P18.41

42. |||| You're helping with an experiment in which a vertical cylinder will rotate about its axis by a very small angle. You need to devise a way to measure this angle. You decide to use what is called an *optical lever.* You begin by mounting a small mirror on top of the cylinder. A laser 5.0 m away shoots a laser beam at the mirror. Before the experiment starts, the mirror is adjusted to reflect the laser beam directly back to the laser. Later, you measure that the reflected laser beam, when it returns to the laser, has been deflected sideways by 2.0 mm. How many degrees has the cylinder rotated?
43. || Figure P18.43 shows a light ray incident on a polished metal cylinder. At what angle θ will the ray be reflected?

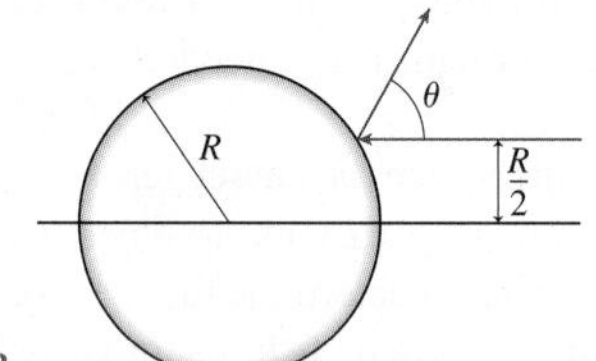

FIGURE P18.43

44. ||| The place you get your hair cut has two nearly parallel mirrors 5.0 m apart. As you sit in the chair, your head is 2.0 m from the nearer mirror. Looking toward this mirror, you first see your face and then, farther away, the back of your head. (The mirrors need to be slightly nonparallel for you to be able to see the back of your head, but you can treat them as parallel in this problem.) How far away does the back of your head appear to be? Neglect the thickness of your head.
45. | You shine your laser pointer through the flat glass side of a rectangular aquarium at an angle of incidence of 45°. The index of refraction of this type of glass is 1.55.
    a. At what angle from the normal does the beam from the laser pointer enter the water inside the aquarium?
    b. Does your answer to part a depend on the index of refraction of the glass?
46. || A ray of light traveling through air encounters a 1.2-cm-thick sheet of glass at a 35° angle of incidence. How far does the light ray travel in the glass before emerging on the far side?
47. || What is the angle of incidence in air of a light ray whose angle of refraction in glass is half the angle of incidence?
48. |||| Figure P18.48 shows a light ray incident on a glass cylinder. What is the angle α of the ray after it has entered the cylinder?

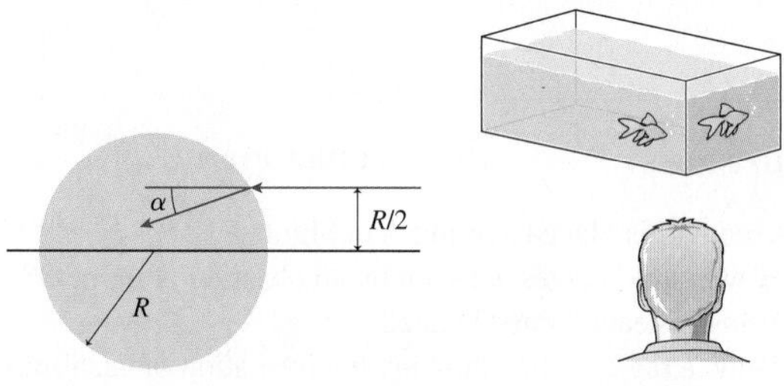

FIGURE P18.48 FIGURE P18.49

49. | If you look at a fish through the corner of a rectangular aquarium you sometimes see two fish, one on each side of the corner, as shown in Figure P18.49. Sketch some of the light rays that reach your eye from the fish to show how this can happen.
50. ||||| It's nighttime, and you've dropped your goggles into a swimming pool that is 3.0 m deep. If you hold a laser pointer 1.0 m directly above the edge of the pool, you can illuminate the goggles if the laser beam enters the water 2.0 m from the edge. How far are the goggles from the edge of the pool?
51. |||| One of the contests at the school carnival is to throw a spear at an underwater target lying flat on the bottom of a pool. The water is 1.0 m deep. You're standing on a small stool that places your eyes 3.0 m above the bottom of the pool. As you look at the target, your gaze is 30° below horizontal. At what angle below horizontal should you throw the spear in order to hit the target? Your raised arm brings the spear point to the level of your eyes as you throw it, and over this short distance you can assume that the spear travels in a straight line rather than a parabolic trajectory.
52. ||| Figure P18.52 shows a meter stick lying on the bottom of a 100-cm-long tank with its zero mark against the left edge. You look into the tank at a 30° angle, with your line of sight just grazing the upper left edge of the tank. What mark do you see on the meter stick if the tank is (a) empty, (b) half full of water, and (c) completely full of water?

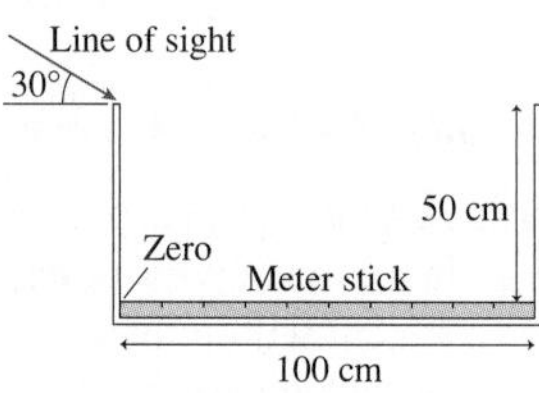

FIGURE P18.52

53. ||| There is just one angle of incidence β onto a prism for which the light inside an isosceles prism travels parallel to the base and emerges at that same angle β, as shown in Figure P18.53.
    a. Find an expression for β in terms of the prism's apex angle α and index of refraction n.
    b. A laboratory measurement finds that $\beta = 52.2°$ for a prism that is shaped as an equilateral triangle. What is the prism's index of refraction?

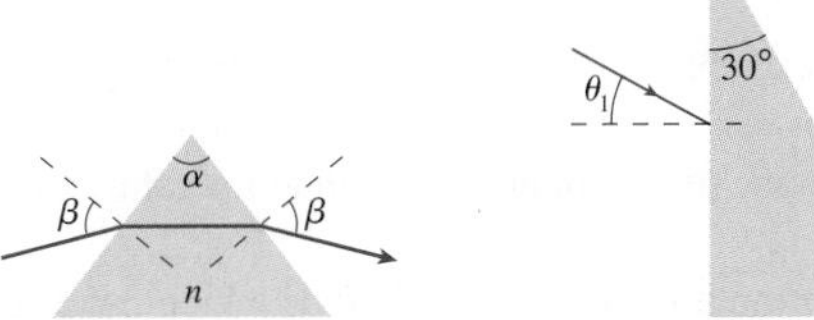

FIGURE P18.53 FIGURE P18.54

54. |||| What is the smallest angle θ_1 for which a laser beam will undergo total internal reflection on the hypotenuse of the glass prism in Figure P18.54?
55. ||| A 1.0-cm-thick layer of water stands on a horizontal slab of glass. Light from within the glass is incident on the glass-water boundary. What is the maximum angle of incidence for which a light ray can emerge into the air above the water?
56. |||| The glass core of an optical fiber has index of refraction 1.60. The index of refraction of the cladding is 1.48. What is the maximum angle between a light ray and the wall of the core if the ray is to remain inside the core?
57. || A swimmer looks upward from the bottom of a 3.0-m-deep swimming pool. The end of the diving board is directly above him, 2.0 m above the water's surface. How far from the swimmer does the board appear to be?
58. |||| A 150-cm-tall diver is standing completely submerged on the bottom of a swimming pool full of water. You are sitting on the end of the diving board, almost directly over her. How tall does the diver appear to be?
59. || To a fish, the 4.00-mm-thick aquarium walls appear only 3.50 mm thick. What is the index of refraction of the walls?
60. || A microscope is focused on an amoeba. When a 0.15-mm-thick cover glass ($n = 1.50$) is placed over the amoeba, by how far must the microscope objective be moved to bring the organism back into focus? Must it be raised or lowered?
BIO
61. || A ray diagram can be used to find the location of an object if you are given the location of its image and the focal length of the mirror. Draw a ray diagram to find the height and position of an object that makes a 2.0-cm-high upright virtual image that appears 8.0 cm behind a convex mirror of focal length 20 cm.
62. | A 2.0-cm-tall object is located 8.0 cm in front of a converging lens with a focal length of 10 cm. Use ray tracing to determine the location and height of the image. Is the image upright or inverted? Is it real or virtual?
63. || The image produced by a converging lens is typically a different size from the object itself. However, for a lens with focal length f there is one object distance that will yield an image the same size as the object. What is that object distance?
64. ||| A near-sighted person might correct his vision by wearing diverging lenses with focal length $f = -50$ cm. When wearing his glasses, he looks not at actual objects but at the virtual images of those objects formed by his glasses. Suppose he looks at a 12-cm-long pencil held vertically 2.0 m from his glasses. Use ray tracing to determine the location and height of the image.
BIO
65. | A 1.0-cm-tall object is 20 cm in front of a converging lens that has a 10 cm focal length. Use ray tracing to find the position and height of the image. To do this accurately, use a ruler or paper with a grid. Determine the image distance and image height by making measurements on your diagram.
66. || A 2.0-cm-tall object is 20 cm in front of a converging lens that has a 60 cm focal length. Use ray tracing to find the position and height of the image. To do this accurately, use a ruler or paper with a grid. Determine the image distance and image height by making measurements on your diagram.
67. || A 1.0-cm-tall object is 7.5 cm in front of a diverging lens that has a 10 cm focal length. Use ray tracing to find the position and height of the image. To do this accurately, use a ruler or paper with a grid. Determine the image distance and image height by making measurements on your diagram.
68. || A 1.5-cm-tall object is 90 cm in front of a diverging lens that has a 45 cm focal length. Use ray tracing to find the position and height of the image. To do this accurately, use a ruler or paper with a grid. Determine the image distance and image height by making measurements on your diagram.
69. | A 1.6-m-tall woman stands 2.0 m in front of a convex funhouse mirror with a focal length of 2/3 m. Use ray tracing to determine the location and height of her image.
70. ||| A 2.0-cm-tall candle flame is 2.0 m from a wall. You happen to have a lens with a focal length of 32 cm. How many places can you put the lens to form a well-focused image of the candle flame on the wall? For each location, what are the height and orientation of the image?
71. || A 2.0-cm-diameter spider is 2.0 m from a wall. Determine the focal length and position (measured from the wall) of a lens that will make a half-size image of the spider on the wall.
72. ||| Figure P18.72 shows a meter stick held lengthwise along the optical axis of a concave mirror. How long is the image of the meter stick?

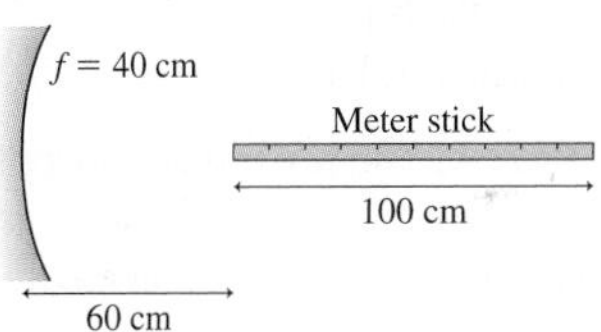

FIGURE P18.72

73. || A slide projector needs to create a 98-cm-high image of a 2.0-cm-tall slide. The screen is 300 cm from the slide.
    a. What focal length does the lens need? Assume that it is a thin lens.
    b. How far should you place the lens from the slide?
74. ||| The writing on the passenger-side mirror of your car says "Warning! Objects in mirror are closer than they appear." There is no such warning on the driver's mirror. Consider a typical convex passenger-side mirror with a focal length of -80 cm. A 1.5-m-tall cyclist on a bicycle is 25 m from the mirror. You are 1.0 m from the mirror, and suppose, for simplicity, that the mirror, you, and the cyclist all lie along a line.

    a. How far are you from the image of the cyclist?
    b. How far would you have been from the image if the mirror were flat?
    c. What is the image height?
    d. What would the image height have been if the mirror were flat?
    e. Why is there a label on the passenger-side mirror?

Passage Problems

Mirages

There is an interesting optical effect you have likely noticed while driving along a flat stretch of road on a sunny day. A small, distant dip in the road appears to be filled with water. You may even see the reflection of an oncoming car. But, as you get closer, you find no puddle of water after all; the shimmering surface vanishes, and you see nothing but empty road. It was only a *mirage,* the name for this phenomenon.

The mirage is due to the different index of refraction of hot and cool air. The actual bending of the light rays that produces the mirage is subtle, but we can make a simple model as follows. When air is heated, its density decreases and so does its index of refraction. Consequently, a pocket of hot air in a dip in a road has a lower index of refraction than the cooler air above it. Incident light rays with large angles of incidence (that is, nearly parallel to the road, as shown in Figure P18.75) experience total internal reflection. The mirage that you see is due to this reflection. As you get nearer, the angle goes below the critical angle and there is no more total internal reflection; the "water" disappears!

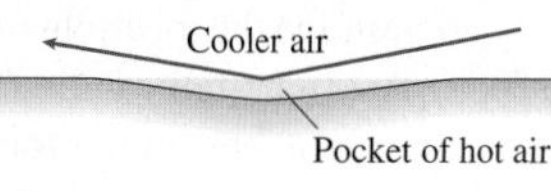

FIGURE P18.75

75. | The pocket of hot air appears to be a pool of water because
   A. Light reflects at the boundary between the hot and cool air.
   B. Its density is close to that of water.
   C. Light refracts at the boundary between the hot and cool air.
   D. The hot air emits blue light that is the same color as the daytime sky.
76. | Which of these changes would allow you to get closer to the mirage before it vanishes?
   A. Making the pocket of hot air nearer in temperature to the air above it
   B. Looking for the mirage on a windy day, which mixes the air layers
   C. Increasing the difference in temperature between the pocket of hot air and the air above it
   D. Looking at it from a greater height above the ground
77. | If you could clearly see the image of an object that was reflected by a mirage, the image would appear
   A. Magnified.
   B. With up and down reversed.
   C. Farther away than the object.
   D. With right and left reversed.

STOP TO THINK ANSWERS

Stop to Think 18.1: C. There's one image behind the vertical mirror and a second behind the horizontal mirror. A third image in the corner arises from rays that reflect twice, once off each mirror.

Stop to Think 18.2: A. The ray travels closer to the normal in both media 1 and 3 than in medium 2, so n_1 and n_3 are both larger than n_2. The angle is smaller in medium 3 than in medium 1, so $n_3 > n_1$.

Stop to Think 18.3: E. The rays from the object are diverging. Without a lens, the rays cannot converge to form any kind of image on the screen.

Stop to Think 18.4: C. For a converging mirror, the focal length f is the distance from the mirror at which incoming parallel rays meet. The moon is so distant that rays from any point on the moon are very nearly parallel. Thus the image of the moon would be very nearly at a distance f in front of the mirror.

Stop to Think 18.5: A. The thin-lens equation is $1/s + 1/s' = 1/f$. The focal length of the lens is fixed. Because $1/s$ gets smaller as s is increased, $1/s'$ must get larger to compensate. Thus s' must decrease.

19 Optical Instruments

This *anableps* is called the "four-eyed fish." How must the top half of its eye differ from the lower half so that it has clear vision both above and below the waterline at the same time?

LOOKING AHEAD ▶▶

The goal of Chapter 19 is to understand how common optical instruments work.

Cameras

A **camera** uses a lens to project a real image onto a light-sensitive detector.

Although a modern digital camera is an extremely complex device, at its heart it is just a light-tight box with a lens to focus an image.

Looking Back ◀◀
18.5 Lenses, images, and magnification

Optical Systems That Magnify

Lenses and mirrors can be used to magnify nearby objects (magnifiers and microscopes) or distant ones (telescopes).

Looking Back ◀◀
18.6 Image formation with curved mirrors

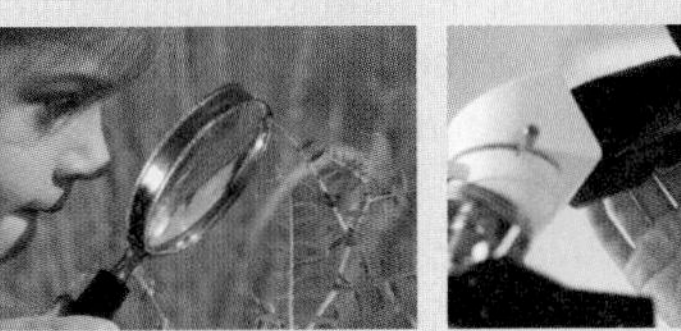

A **magnifier** uses a convex lens to achieve magnifications up to about 20.

Microscopes use two sets of lenses in combination to magnify objects as much as 1000 times.

Large **telescopes** use a curved mirror as their main optical element. The mirror of this telescope is 8 m in diameter!

The Human Eye

The eye functions much like a camera: The cornea and lens focus a real image onto the light-sensitive surface of the retina.

Lens
Retina
Cornea

You'll learn how the lenses in eyeglasses and contacts can correct for near- and farsightedness.

Color and Dispersion

White light entering a prism is **dispersed** into a rainbow of colors.

A prism separates white light into its constituent colors.

The dispersion of sunlight by raindrops causes the colors seen in a rainbow.

Resolution of Optical Instruments

There are limits to the ability of a microscope or telescope to make out the fine details of an object being viewed.

Seen at high power, the diffraction patterns of these two nearby stars almost merge, making it hard to tell them apart.

This simple lens suffers from distortions and chromatic (color) aberrations.

Looking Back ◀◀
17.6 Circular-aperture diffraction

A pinhole eye BIO The chambered nautilus is the only animal with a true pinhole "camera" as an eye. Light rays passing through the small opening form a crude image on the back surface of the eye, where the rays strike light-sensitive cells. The image may be poor, but it's sufficient to allow the nautilus to catch prey and escape predators.

19.1 The Camera

This chapter will investigate a number of optical instruments in which a combination of lenses and mirrors performs a useful function. We'll start with an instrument familiar to everyone: the camera. A **camera** is a device that projects a real image onto a plane surface, where the image can be recorded onto film or, in today's digital cameras, an electronic detector. Although modern cameras are marvels of optical engineering, it is possible to produce decent images using only a light-proof box with a small hole punched in it. Such a **pinhole camera** is shown in FIGURE 19.1a. FIGURE 19.1b uses the ray model of light passing through a small hole to illustrate how the pinhole camera works. Each point on an object emits light rays in all directions, but, ideally, only one of these rays passes through the hole and reaches the film. Each point on the object thus illuminates just one point on the film, forming the image. As the figure illustrates, the geometry of the rays causes the image to be upside down.

FIGURE 19.1 A pinhole camera.

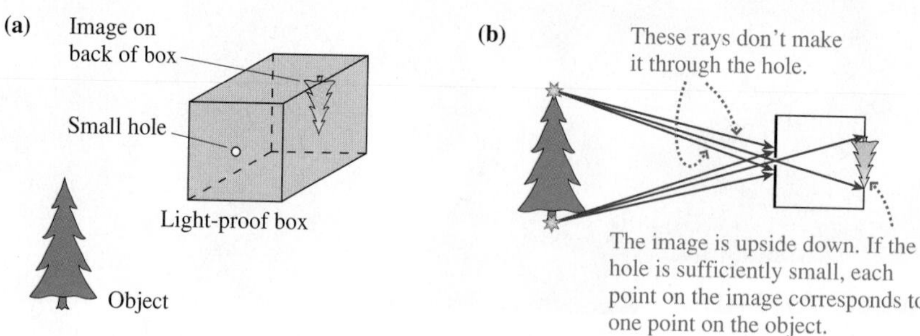

Actually, as you may have realized, each *point* on the object illuminates a small but finite *patch* on the film. This is because the finite size of the hole allows several rays from each point on the object to pass through at slightly different angles. As a result, the image is slightly blurred. Maximum sharpness is achieved by making the hole smaller and smaller, which makes the image dimmer and dimmer. (Diffraction also becomes an issue if the hole gets too small.) A real pinhole camera has to accept a small amount of blurring as the trade-off for having an image bright enough to be practical.

FIGURE 19.2 A camera.

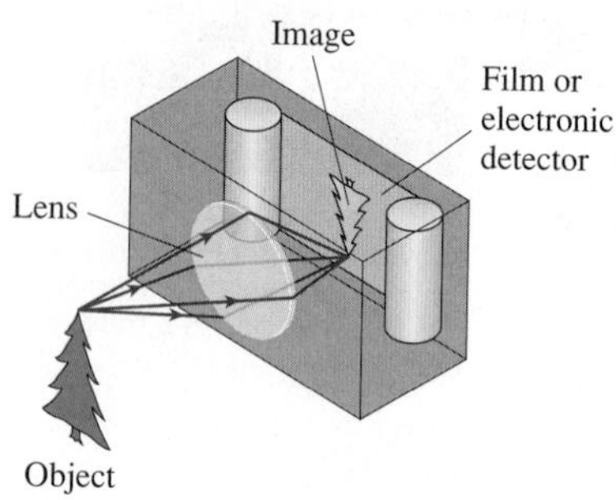

In a standard camera, a dramatic improvement is made possible by using a *lens* in place of a pinhole. FIGURE 19.2 shows how a camera's converging lens projects an inverted real image onto the film, just as a pinhole camera does. Unlike a pinhole, however, a lens can be large, letting in plenty of light while still giving a sharply focused image. A shutter (not shown), an opaque barrier, is briefly moved out of the way in order for light to pass through the lens.

FIGURE 19.3 shows light rays from an object passing through the lens and converging at the image plane B. Here, a single point P on the object focuses to a single point P′

FIGURE 19.3 Focusing a camera.

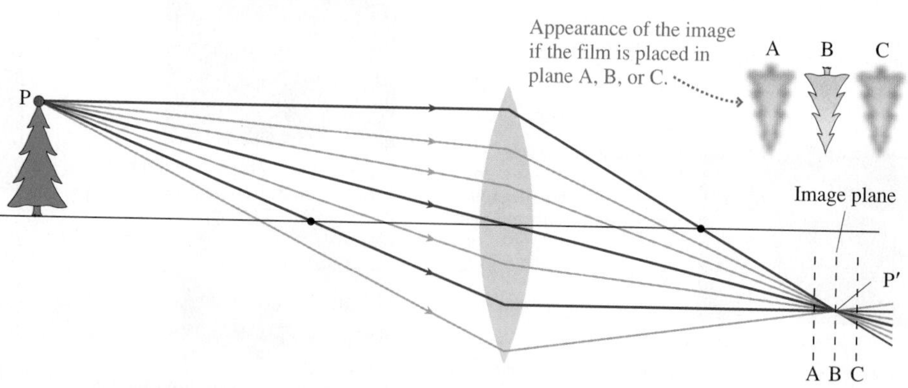

in the image plane. If the film or electronic detector is located in this plane, a sharp image will form on it. If, however, the film had been located a bit in *front* of the image plane, at position A, rays from point P would not yet have completely converged and would form a small blurry *circle* on the film instead of a sharp point. Thus the image would appear blurred, as shown. Similarly, if the film is placed at C, *behind* the image plane, the rays will be diverging from their perfect focus and again form a blurry image.

Thus to get a sharp image, the film must be accurately located in the image plane. Figure 19.3 shows that one way to do this is to move the film until it coincides with the image plane. More commonly, however, a camera is **focused** by moving the *lens* either toward or away from the film plane until the image is sharp. In either case, the lens-film distance is varied.

EXAMPLE 19.1 **Focusing a camera**

A digital camera whose lens has a focal length of 8.0 mm is used to take a picture of an object 30 cm away. What must be the distance from the lens to the light-sensitive detector in order for the image to be in focus?

PREPARE As shown in Figure 19.3, the image will be in focus when the detector is in the image plane. Thus we need to find the image distance, knowing the object distance $s = 30$ cm and the lens's focal length $f = 8.0$ mm.

SOLVE We can rearrange the thin-lens equation, Equation 18.11, to solve for the image distance s':

$$\frac{1}{s'} = \frac{1}{f} - \frac{1}{s} = \frac{1}{0.0080 \text{ m}} - \frac{1}{0.30 \text{ m}} = 122 \text{ m}^{-1}$$

Thus $s' = 1/122 \text{ m}^{-1} = 0.0082 \text{ m} = 8.2$ mm. The lens-detector distance has to be 8.2 mm.

ASSESS When the object is infinitely far away, the image, by definition, is at the focal length: $s' = f = 8.0$ mm. If the object is brought to 30 cm, the lens has to move forward a distance of only 8.2 mm − 8.0 mm = 0.2 mm to bring the object into focus. In general, camera lenses don't need to move far.

If a digital picture is magnified enough, you can see the individual pixels that make it up.

For traditional cameras, the light-sensitive surface is **film.** Light striking microscopic silver halide crystals creates small particles of silver metal on the crystals. The development process converts the entire crystal to dark metallic silver. Thus the film appears *dark* where light has hit it, forming a *negative* image. To get a positive image—your print—the negative image is projected onto paper with its own photosensitive coating. This results in a second negative process, leading to a positive print. Today's *digital cameras* use an electronic light-sensitive detector called a *charge-coupled device* or **CCD.** A CCD consists of a rectangular array of many millions of small detectors called *pixels.* When light hits one of these pixels, it generates an electric charge proportional to the light intensity. Thus an image is recorded on the CCD in terms of little packets of charge. After the CCD has been exposed, the charges are read out, the signal levels are digitized, and the picture is stored in the digital memory of the camera.

FIGURE 19.4 shows a CCD "chip" and, schematically, the magnified appearance of the pixels on its surface. To record color information, different pixels are covered by red, green, or blue filters; a pixel covered by a green filter, for instance, records only the intensity of the green light hitting it. Later, the camera's microprocessor interpolates nearby colors to give each pixel an overall true color. The structure of the retina of the eye is remarkably similar, as we'll see in Chapter 25.

FIGURE 19.4 A CCD chip used in a digital camera.

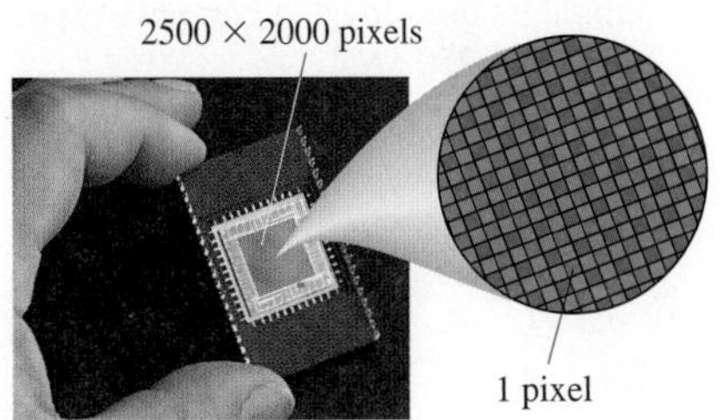

Controlling the Exposure

The camera also must control the amount of light reaching the detector. As FIGURE 19.5 shows, too little light results in photos that are *underexposed;* too much light gives *overexposed* pictures. Both the shutter and the lens diameter help control the exposure.

The *shutter* is "opened" for a selected amount of time as the image is recorded. Older cameras used a spring-loaded mechanical shutter that literally opened and closed; digital cameras electronically control the amount of time the detector is active. Either way, the exposure—the amount of light captured by the detector—is directly proportional to the time the shutter is open. Typical exposure times range from 1/1000 s or less for a sunny scene to 1/30 s or more for dimly lit or indoor scenes. The exposure time is generally referred to as the *shutter speed;* a very short exposure, such as 1/1000 s is called a "fast shutter speed," while a much longer exposure is a "slow shutter speed."

FIGURE 19.5 Photos with increasing amounts of light reaching the film.

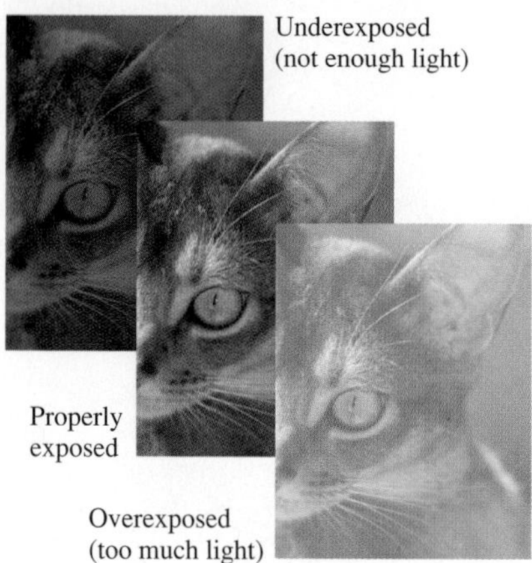

A second means of controlling the exposure is to effectively change the diameter d of the lens. A small lens will let in less light than a large one. Of course, the actual diameter of the lens cannot be changed. Instead, an *iris diaphragm,* placed behind the lens, is adjusted to change the amount of light entering the camera, as shown in FIGURE 19.6.

An important measure of the light-gathering ability of a lens is its ***f*-number,** defined as

$$f\text{-number} = \frac{\text{lens focal length}}{\text{lens diameter}} = \frac{f}{d}$$

FIGURE 19.6 A camera's iris can change the effective diameter of the lens.

We use the notation $f/11$ ("f 11"), for instance, to indicate an f-number of 11. The f-number of a lens directly determines the *brightness* of the image on the film. Somewhat contrary to common sense, the *smaller* the f-number, the *brighter* the image. Here's why: For a given focal length, more light enters a lens with a larger diameter than one with a smaller diameter. But since the f-number is inversely proportional to d, the larger-diameter, brighter lens has the lower f-number. Further, a lens of a given diameter will give a brighter image when the focal length f is *small.* To see why, consider a lens imaging a distant object, so that the image is focused at the far focal point. If f is large, the image is large and its light is spread out and dim. If f is small, the image is small and the light is concentrated and bright.

A lens with its diaphragm fully open will let in the most possible light. But under bright light conditions we may need to close, or *stop down,* the diaphragm to control the exposure. How much must the diaphragm be closed in order to let in half as much light? The amount of light entering the lens is proportional to the *area* of the lens, or to d^2. To let in half as much light, we must reduce d^2 by a factor of 2 or, equivalently, reduce the diameter of the lens by a factor of $1/\sqrt{2}$. This means that the f-number, which is inversely proportional to the diameter, will *increase* by a factor of $\sqrt{2} \approx 1.4$. On most camera lenses, the ***f*-stops** increase in the series 1.4, 2, 2.8, 4, 5.6, 8, 11, 16. Each f-number is $\sqrt{2}$ times larger than the preceding one, so an increase of one f-stop cuts the light intensity in half.

CONCEPTUAL EXAMPLE 19.2 Adjusting a camera lens

A camera takes a perfectly exposed picture when the shutter speed is 1/60 s and the lens diaphragm is set to $f/11$. What f-stop should be used to get a correctly exposed picture with a shutter speed of 1/250 s?

REASON We need the same amount of light hitting the film or CCD in both cases. For the second picture the shutter is open only about one-fourth as long, since

$$\frac{1/250\text{ s}}{1/60\text{ s}} = \frac{60}{250} = \frac{1}{4.17} \approx \frac{1}{4}$$

To compensate for the reduced time, we need to increase the light intensity by a factor of 4. Because each stop lets in twice as much light as the preceding stop, the light intensity will be 4 times as great if we open up the diaphragm by two stops. From the sequence given above, opening the iris by two stops from $f/11$ is $f/5.6$.

ASSESS Camera film and CCD detectors yield a good image over a rather wide range of exposures. Thus the fact that shutter speeds and f-numbers change the exposure by factors of 2 still allows us to get the exposure close enough for a good picture.

STOP TO THINK 19.1 A camera takes a correctly exposed picture with a certain lens. The lens is then replaced by one with the same diameter but twice the focal length. To get the correct exposure when the focal length is doubled, the shutter speed should

A. Be increased. B. Be decreased. C. Remain the same.

19.2 The Human Eye

The human eye functions much like a camera. Like the camera, it has three main functional groups: an optical system to focus the incoming light, a diaphragm to adjust the amount of light entering the eye, and a light-sensitive surface to detect the resulting image. The parts of the eye making up these three groups are shown in **FIGURE 19.7**. The *cornea,* the *aqueous humor,* and the *lens* are together responsible for refracting incoming light rays and producing an image. The adjustable *iris* determines how much light enters the eye, in much the same way as does the diaphragm of a camera. And the *retina* is the light-sensitive surface on which the image is formed. The retina is the biological equivalent of the CCD in a digital camera.

FIGURE 19.7 The human eye.

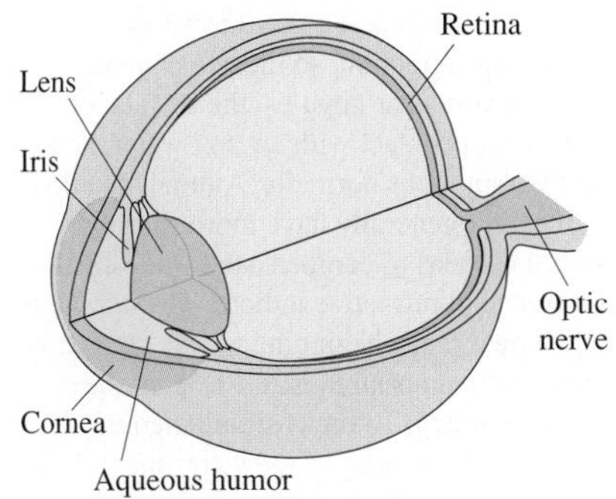

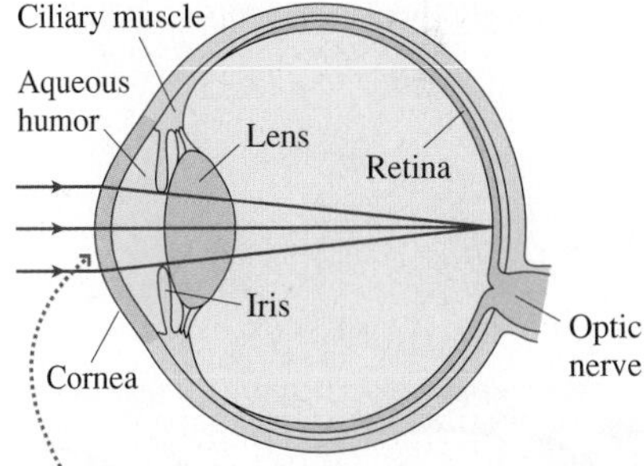

Most of the refraction occurs at the cornea's surface, where Δn is the largest.

Focusing and Accommodation

Like a camera, the eye works by focusing incoming rays onto a light-sensitive surface, here the retina. To do so, light is refracted by, in turn, the cornea, the aqueous humor, and the lens, as shown in Figure 19.7. The indices of refraction of these parts of the eye vary somewhat, but average around 1.4. Perhaps surprisingly, most of the eye's refraction occurs not in the lens but at the surface of the cornea. This is due both to the strong curvature of the cornea and to the large difference between the indices of refraction on either side of the surface. Light refracts less as it passes through the lens because the lens's index of refraction doesn't differ much from that of the fluid in which it is embedded. If the lens is surgically removed, which it often is for people with cataracts (a clouding of the lens), the cornea alone still provides a marginal level of vision.

The eye must constantly refocus as it views distant objects, then closer ones. It does this so automatically that we're not normally aware of the process. A camera focuses by changing the distance between the lens and the film, but your eye focuses in a different way: by changing the focal length of the lens itself. As shown in **FIGURE 19.8**, it does so by using the ciliary muscles to *change the shape* of the lens. This process of changing the lens shape as the eye focuses at different distances is called **accommodation.**

FIGURE 19.8 Accommodation by the eye.

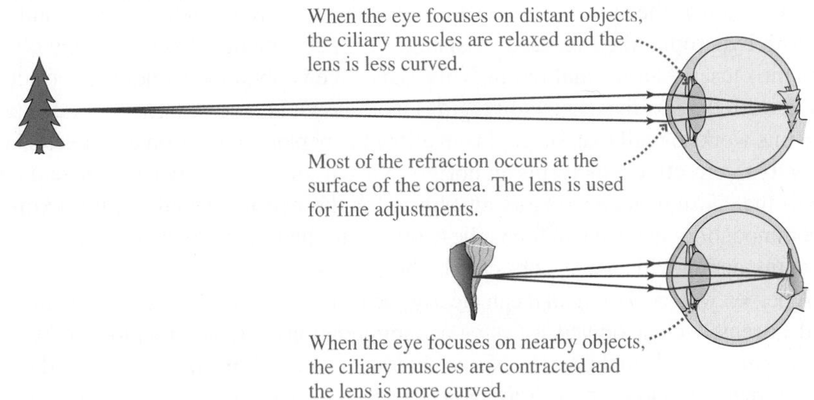

When the eye focuses on distant objects, the ciliary muscles are relaxed and the lens is less curved.
Most of the refraction occurs at the surface of the cornea. The lens is used for fine adjustments.
When the eye focuses on nearby objects, the ciliary muscles are contracted and the lens is more curved.

TRY IT YOURSELF

Inverted vision Just like a camera, the lens of the eye produces an *inverted* image on the retina. The brain is wired to "flip" this inverted image and interpret it as being upright. To show this directly, try this simple experiment. Poke a small hole in a card using a pin and, holding the card a few inches away, look through the hole at a lightbulb. While doing so, move the head of the pin between the hole and your eye; you'll see an *upside-down* pinhead. The hole acts as a point source that casts an *erect* shadow of the pin on your retina. The brain then inverts this erect shadow, making it appear inverted.

Seeing underwater BIO When you swim underwater, the difference in refractive indices between the cornea ($n = 1.38$) and water ($n = 1.33$) is too small to allow significant refraction, so the eye cannot focus. If you wear goggles, the surface of the cornea is in contact with air, not water, and the eye can focus normally. Animals that live underwater generally have more sharply curved corneas to compensate for the small difference in refractive indices. The *anableps* fish shown at the beginning of this chapter is particularly unusual in that it lives at the water's surface. To focus simultaneously on objects on both sides of the waterline, it has evolved a very asymmetrical cornea that is more strongly curved below the waterline.

The most distant point on which the completely relaxed eye can focus is called the eye's **far point** (FP). For normal vision, the far point is at infinity. The closest point on which the eye can focus, with the ciliary muscles fully contracted, is called the **near point** (NP). Objects closer than the near point cannot be brought into sharp focus.

Vision Defects and Their Correction

The near point of normal vision is considered to be 25 cm, but the near point of an individual changes with age. The near point of young children can be as little as 10 cm. The "normal" 25 cm near point is characteristic of young adults, but the near point of most individuals begins to move outward by age 40 or 45 and can reach 200 cm by age 60. This loss of accommodation, which arises because the lens loses flexibility, is called **presbyopia.** Even if their vision is otherwise normal, individuals with presbyopia need reading glasses to bring their near point back to 25 or 30 cm, a comfortable distance for reading.

Presbyopia is known as a *refractive error* of the eye. Two other common refractive errors are *hyperopia* and *myopia*. All three can be corrected with lenses—either eye-glasses or contact lenses—that assist the eye's focusing. Corrective lenses are prescribed not by their focal length but by their **refractive power.** The refractive power of a lens is the inverse of its focal length:

$$P = \frac{1}{f} \qquad (19.1)$$

Refractive power of a lens with focal length f

INVERSE p.114

A lens with higher refractive power (shorter focal length) causes light rays to refract through a larger angle. The SI unit of lens refractive power is the **diopter,** abbreviated D, defined as $1\text{ D} = 1\text{ m}^{-1}$. Thus a lens with $f = 50\text{ cm} = 0.50\text{ m}$ has refractive power $P = 2.0\text{ D}$.

When writing prescriptions, optometrists don't write the D because the lens maker already knows that prescriptions are in diopters. If you look at your eyeglass prescription next time you visit the optometrist, it will look something like +2.5/+2.7. This says that your right eye needs a corrective lens with $P = +2.5$ D, the + indicating a converging lens with a positive focal length. Your left eye needs a lens with $P = +2.7$ D. Most people's eyes are not the same, so each eye usually gets a slightly different lens. Prescriptions with negative numbers indicate diverging lenses with negative focal lengths.

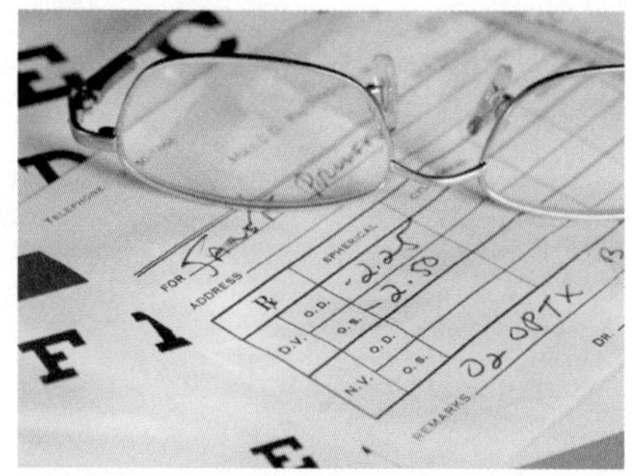

The optometrist's prescription is −2.25 D for the right eye (top) and −2.50 D for the left (bottom).

A person who is *farsighted* can see faraway objects (but even then must use some accommodation rather than a relaxed eye), but his near point is larger than 25 cm, often much larger, so he cannot focus on nearby objects. The cause of farsightedness—called **hyperopia**—is an eyeball that is too short for the refractive power of the cornea and lens. As FIGURES 19.9a and b on the next page show, no amount of accommodation allows the eye to focus on an object 25 cm away, the normal near point.

With hyperopia, the eye needs assistance to focus the rays from a nearby object onto the closer-than-normal retina. This assistance is obtained by adding refractive power with the positive (i.e., converging) lens shown in FIGURE 19.9c. To understand why this works, recall that the goal is to allow the person to focus on an object 25 cm away. If a corrective lens forms an upright, virtual image at the person's actual near point, that virtual image acts as an object for the eye itself and, with maximum accommodation, the eye can focus these rays onto the retina. Presbyopia, the loss of accommodation with age, is corrected in the same way.

A person who is *nearsighted* can clearly see nearby objects when the eye is relaxed (and extremely close objects by using accommodation), but no amount of relaxation allows her to see distant objects. Nearsightedness—called **myopia**—is caused by an eyeball that is too long. As FIGURE 19.10a on the next page shows, rays from a distant

FIGURE 19.9 Hyperopia.

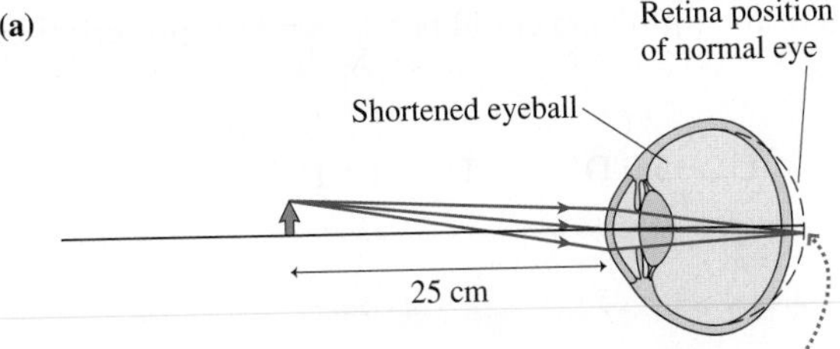
(a)
Retina position of normal eye
Shortened eyeball
25 cm
Even with maximum accommodation, the image is focused behind the retina. Thus the image is blurry.

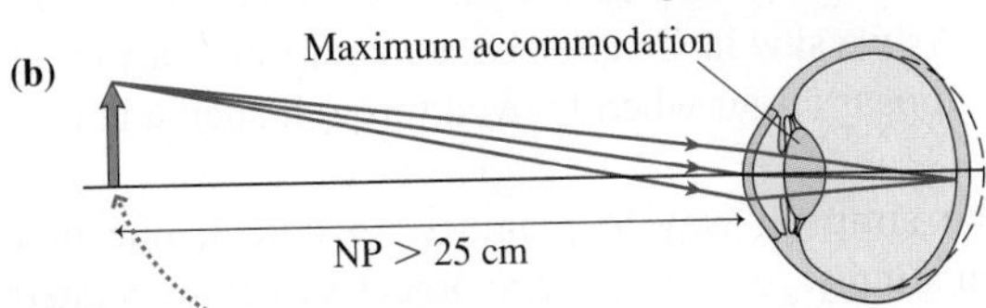
(b)
Maximum accommodation
NP > 25 cm
This is the closest point at which the eye can focus.

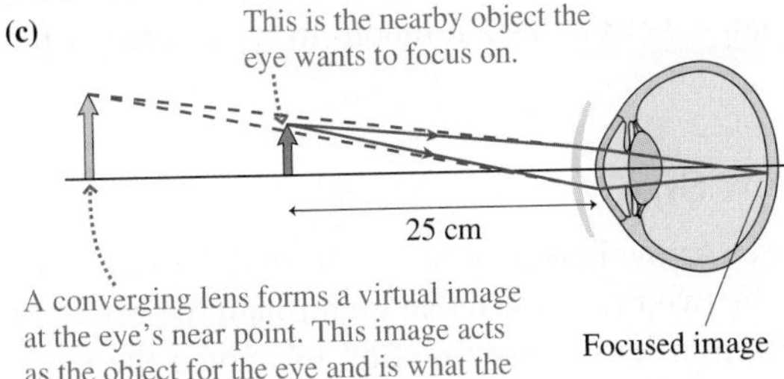
(c)
This is the nearby object the eye wants to focus on.
25 cm
A converging lens forms a virtual image at the eye's near point. This image acts as the object for the eye and is what the eye actually focuses on.
Focused image

FIGURE 19.10 Myopia.

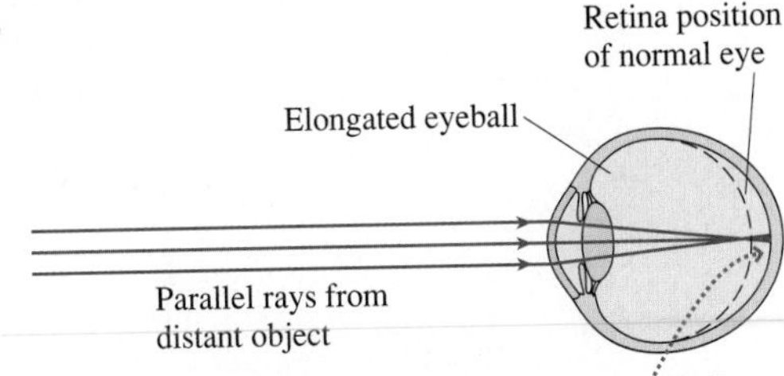
(a)
Retina position of normal eye
Elongated eyeball
Parallel rays from distant object
A fully relaxed eye focuses the image in front of the actual retina. The image is blurry.

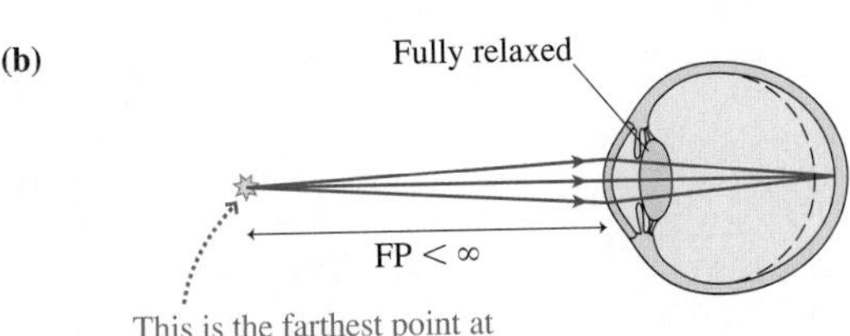
(b)
Fully relaxed
FP < ∞
This is the farthest point at which the eye can focus.

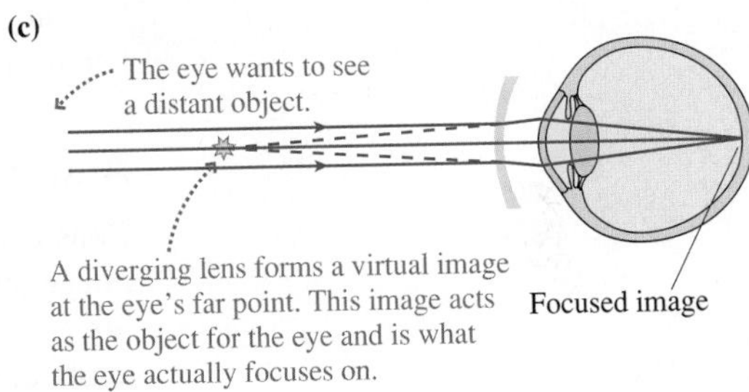
(c)
The eye wants to see a distant object.
A diverging lens forms a virtual image at the eye's far point. This image acts as the object for the eye and is what the eye actually focuses on.
Focused image

object come to a focus in front of the retina and have begun to diverge by the time they reach the retina. The eye's far point, shown in FIGURE 19.10b, is less than infinity.

To correct myopia, we needed a diverging lens, as shown in FIGURE 19.10c, to slightly defocus the rays and move the image point back to the retina. To focus on a very distant object, the person needs a corrective lens that forms an upright, virtual image at her actual far point. That virtual image acts as an object for the eye itself and, when fully relaxed, the eye can focus these rays onto the retina.

EXAMPLE 19.3 Correcting hyperopia

Sanjay has hyperopia, The near point of his left eye is 150 cm. What prescription lens will restore normal vision?

PREPARE Normal vision will allow Sanjay to focus on an object 25 cm away. In measuring distances, we'll ignore the small space between the lens and his eye.

SOLVE Because Sanjay can see objects at 150 cm, using maximum accommodation, we want a lens that creates a virtual image at position $s' = -150$ cm (negative because it's a virtual image) of an object at $s = 25$ cm. From the thin-lens equation,

$$\frac{1}{f} = \frac{1}{s} + \frac{1}{s'} = \frac{1}{0.25\text{ m}} + \frac{1}{-1.50\text{ m}} = 3.3\text{ m}^{-1}$$

$1/f$ is the lens power, and m^{-1} are diopters. Thus the prescription is for a lens with power $P = 3.3$ D.

ASSESS Hyperopia is always corrected with a converging lens.

EXAMPLE 19.4 Correcting myopia

Martina has myopia. The far point of her left eye is 200 cm. What prescription lens will restore normal vision?

PREPARE Normal vision will allow Martina to focus on a very distant object. In measuring distances, we'll ignore the small space between the lens and her eye.

SOLVE Because Martina can see objects at 200 cm with a fully relaxed eye, we want a lens that will create a virtual image at position $s' = -200$ cm (negative because it's a virtual image) of an object at $s = \infty$ cm. From the thin-lens equation,

$$\frac{1}{f} = \frac{1}{s} + \frac{1}{s'} = \frac{1}{\infty\text{ m}} + \frac{1}{-2.0\text{ m}} = -0.5\text{ m}^{-1}$$

Thus the prescription is for a lens with power $P = -0.5$ D.

ASSESS Myopia is always corrected with a diverging lens.

STOP TO THINK 19.2 With her right eye, Maria can focus on a vase 0.5 m away, but not on a tree 10 m away. Which of the following could be the eyeglass prescription for her right eye?

A. +3.0 D B. +10 D C. −5.0 D D. −1.5 D

19.3 The Magnifier

You've no doubt used a magnifier, or magnifying glass, to get a better look at a small object such as an insect or a coin. As we saw in Chapter 18, a magnifier is a simple converging lens, but why objects appear larger when viewed through such a lens is actually rather subtle.

Let's begin by considering the simplest way to magnify an object, one that requires no extra optics at all. You simply get closer to the object you're interested in. The closer you get, the bigger the object appears. Obviously the actual size of the object is unchanged as you approach it, so what exactly is getting "bigger"? A penny, held at arm's length, will more than cover the distant moon. In what sense is the penny "larger" than the moon?

Angular Size and Apparent Size

Consider an object such as the green arrow in FIGURE 19.11a. To find the size of its image on the retina, we can trace the two rays shown that go through the center of the eye lens. (Here, we'll use the thin-lens approximation to reduce the eye's entire optical system to one thin lens.) As we've learned, such rays are undeviated as they pass through the lens, so we can use them to locate the image, shown in green, on the retina. If the arrow is then brought closer to the eye, as shown in red, ray tracing reveals that the size of the arrow's image is *larger.* Our brain interprets a larger image on the retina as representing a larger-appearing *object.* As the object is moved closer, its size doesn't change, but its **apparent size** gets larger.

Angles θ_1 and θ_2 in Figure 19.11a are the angles *subtended* by the green and red arrows. The angle subtended by an object is called its **angular size.** As you

FIGURE 19.11 How the apparent size of an object is determined.

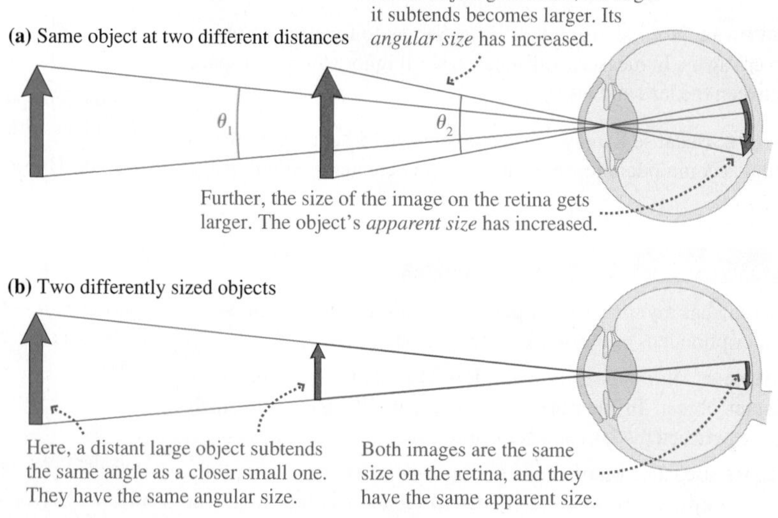
(a) Same object at two different distances
As the object gets closer, the angle it subtends becomes larger. Its angular size has increased.
θ_1
θ_2
Further, the size of the image on the retina gets larger. The object's apparent size has increased.
(b) Two differently sized objects
Here, a distant large object subtends the same angle as a closer small one. They have the same angular size.
Both images are the same size on the retina, and they have the same apparent size.

can see from the figure, **objects that subtend a larger angle appear larger to the eye.** It's also possible, as FIGURE 19.11b shows, for objects with different actual sizes to have the same angular size and thus the same apparent size.

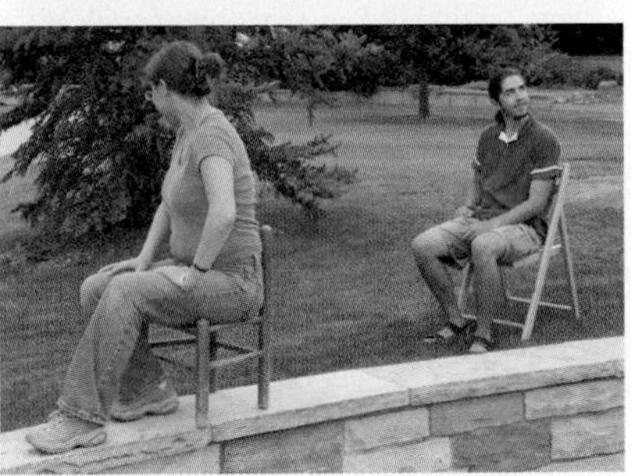

Movie magic? Even if the more distant of two equally sized objects *appears* smaller, we don't usually believe it actually *is* smaller because there are abundant visual clues that tell our brain that it's farther away. If those clues are removed, however, the brain readily accepts the illusion that the farther object is smaller. The technique of *forced perspective* is a special effect used in movies to give this illusion. Here, Camelia, who is actually closer to the camera, looks like a giant compared to Kevin. The lower photo shows how the trick was done.

Using a Magnifier

There are many ways to use a magnifier. You can hold the lens close to or far from the object, and you can bring your eye right up to the lens or hold it somewhat farther away. However, there is a simple case that reflects the way we usually use a magnifier. The lens is held such that the object is at or just inside the lens's focal point. As shown graphically in FIGURE 19.12, this produces a virtual image that is quite far from the lens. Your eye, looking through the lens, "sees" the virtual image. This is a convenient image location, because your eye's muscles are fully relaxed when looking at a distant image. Thus you can use the magnifier in this way for a long time without eye strain.

FIGURE 19.12 The magnifier.

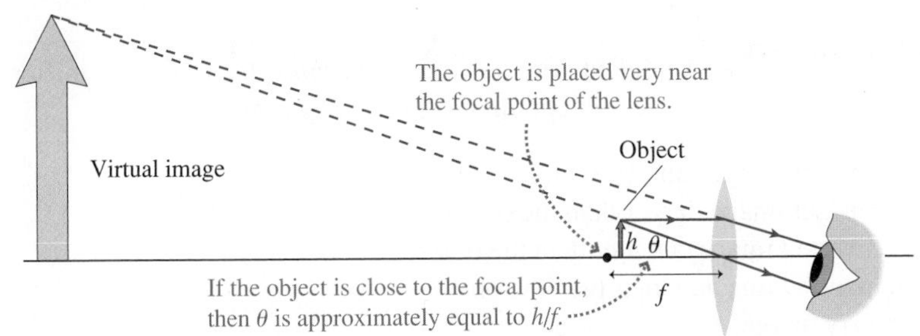
The object is placed very near the focal point of the lens.
Object
Virtual image
h
θ
f
If the object is close to the focal point, then θ is approximately equal to h/f.

We can analyze the situation using Figure 19.12. Suppose that the object is almost exactly at the focal point, a distance f from the lens. Then, tracing the ray that goes through the lens's center, we can see that the angular size θ of the image is such that $\tan\theta = h/f$. If this is a fairly small angle, which it usually is, we can use the small-angle approximation $\tan\theta \approx \theta$ to write

$$\theta \approx \frac{h}{f} \qquad (19.2)$$

Is this an improvement over using no magnifier? With no magnifier, the object has its largest angular size when we bring it as close as possible to the eye. The closest position at which we can focus is the near point of the eye, as shown in FIGURE 19.13. The object at the near point would have angular size

FIGURE 19.13 Angular size without a magnifier.

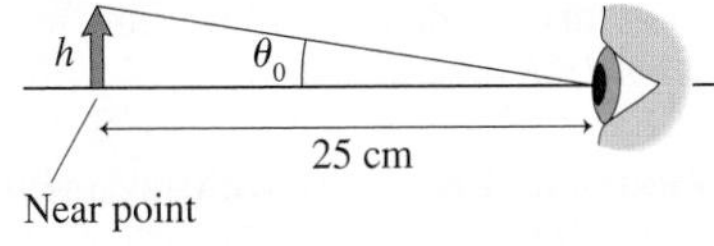
h
θ0
25 cm
Near point

$$\theta_0 \approx \frac{h}{25\text{ cm}}$$

where we have taken the conventional value of 25 cm as the near-point distance. Thus the angular size θ when using the magnifier is larger than that without the magnifier by a factor of

$$M = \frac{\theta}{\theta_0} = \frac{h/f}{h/25\text{ cm}} = \frac{25\text{ cm}}{f} \qquad (19.3)$$

where, for this calculation, the focal length f must be in cm. M is called the **angular magnification** of the magnifier. With a lens of short focal length it is possible to get magnifications as high as about 20.

CONCEPTUAL EXAMPLE 19.5 **The angular size of a magnified image**

An object is placed right at the focal point of a magnifier. How does the apparent size of the image depend on where the *eye* is placed relative to the lens?

REASON When the object is precisely at the focal point of the lens, Equation 19.2 holds exactly. The angular size is equal to h/f *independent* of the position of the eye. Thus the object's apparent size is independent of the eye's position as well. FIGURE 19.14 shows a calculator at the focal point of a magnifier. The apparent size of the COS button is the same whether the camera taking the picture is close to or far from the lens.

ASSESS When the object is at the magnifier's focus, we've seen that the image is at infinity. The situation is similar to observing any "infinitely" distant object, such as the moon. If you walk closer to or farther from the moon, its apparent size doesn't change at all. The same holds for a virtual *image* at infinity: Its apparent size is independent of the point from which you observe it.

FIGURE 19.14 Viewing a magnifier with the object at its focus.

STOP TO THINK 19.3 A student tries to use a *diverging* lens as a magnifier. She observes a coin placed at the focal point of the lens. She sees

A. An upright image, smaller than the object.
B. An upright image, larger than the object.
C. An inverted image, smaller than the object.
D. An inverted image, larger than the object.
E. A blurry image.

19.4 The Microscope

To get higher magnifications than are possible using a simple magnifier, a *combination* of lenses must be used. This is how microscopes and telescopes are constructed. A simple rule governs how two lenses work in combination: **The image from the first lens acts as the object for the second lens.** The following example illustrates this rule.

EXAMPLE 19.6 **Finding the image for two lenses in combination**

A 5.0-cm-focal-length converging lens is 16.0 cm in front of a 10.0-cm-focal-length diverging lens. A 4.0-cm-tall object is placed 11.0 cm in front of the converging lens. What are the position and size of the final image?

PREPARE Let's start with ray tracing. A ray-tracing diagram helps us understand the situation and tells us what to expect for an answer. A diagram can often alert you to a calculation error. FIGURE 19.15a first uses the three special rays of the converging lens to locate its image. We see that the image of the first lens is a real image falling between the two lenses. According to the rule, we then use this image as the object for the second lens. This is done in FIGURE 19.15b, where we see that the final image is inverted, virtual, and roughly 4 cm to the left of the diverging lens.

Mathematically, we can use the thin-lens equation to find the image location and size due to the first lens, then use this as the object for the second lens in a second use of the lens equation.

FIGURE 19.15 Two lenses in combination.

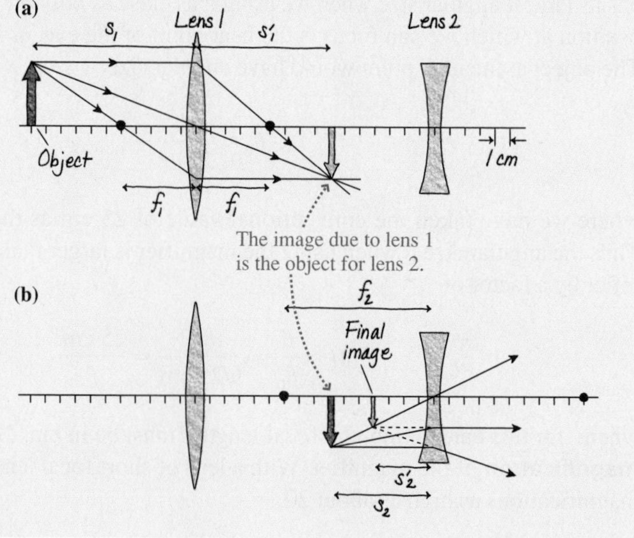

SOLVE We first solve for the image due to lens 1. We have

$$\frac{1}{s_1'} = \frac{1}{f_1} - \frac{1}{s_1} = \frac{1}{5.0 \text{ cm}} - \frac{1}{11.0 \text{ cm}}$$

from which $s_1' = 9.2$ cm. Because this is a positive image distance, the image is real and located to the right of the first lens. The magnification of the first lens is $m_1 = -s_1'/s_1 = -(9.2 \text{ cm})/(11.0 \text{ cm}) = -0.84$.

This image is the object for the second lens. Because it is 9.2 cm to the right of the first lens, and the lenses are 16.0 cm apart, it is 16.0 cm − 9.2 cm = 6.8 cm in front of the second lens. Thus $s_2 = 6.8$ cm.

Applying the thin-lens equation again, we have

$$\frac{1}{s_2'} = \frac{1}{f_2} - \frac{1}{s_2} = \frac{1}{-10.0 \text{ cm}} - \frac{1}{6.8 \text{ cm}}$$

from which we find $s_2' = -4.0$ cm. Because this image distance is negative, the image is virtual and located to the left of lens 2, as shown in Figure 19.15b. The magnification of the second lens is $m_2 = -s_2'/s_2 = -(-4.0 \text{ cm})/(6.8 \text{ cm}) = 0.59$.

Thus the final image size is

$$h_2' = m_2 h_2 = m_2 h_1' = m_2(m_1 h_1) = m_1 m_2 h_1$$
$$= (-0.84)(0.59)(4.0 \text{ cm}) = -2.0 \text{ cm}$$

Here we used the fact that the object height h_2 of the second lens is equal to the image height h_1' of the first lens.

ASSESS When calculating the final image size, we found that $h_2' = m_1 m_2 h_1$. This shows the important fact that **the total magnification of a combination of lenses is the *product* of the magnifications for each lens alone.**

A microscope, whose major parts are shown in **FIGURE 19.16**, attains a magnification of up to 1000 by using two lenses in combination. A specimen to be observed is placed on the *stage* of the microscope, directly beneath the **objective lens** (or simply the **objective**), a converging lens with a relatively short focal length. The objective creates a magnified real image that is further enlarged by the **eyepiece,** a lens used as an ordinary magnifier. In most modern microscopes a prism bends the path of the rays from the object so that the eyepiece can be held at a comfortable angle. However, we'll consider a simplified version of a microscope without a prism. The light then travels along a straight tube.

FIGURE 19.16 A microscope.

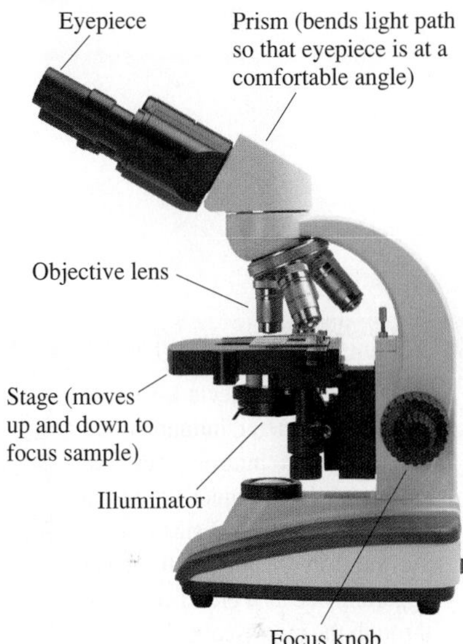

Let's examine the magnification process in more detail. In **FIGURE 19.17** we draw a microscope tilted horizontally. The object distance is just slightly greater than the focal length f_o of the objective lens, so the objective forms a highly magnified real image of the object at a distance $s' = L$. This distance, known as the **tube length,** has been standardized for most biological microscopes at $L = 160$ mm. Most microscopes, such as the one shown in Figure 19.16, are focused by moving the sample stage up and down, using the focusing knob, until the object distance is correct for placing the image at L.

FIGURE 19.17 A horizontal view of the optics in a microscope.

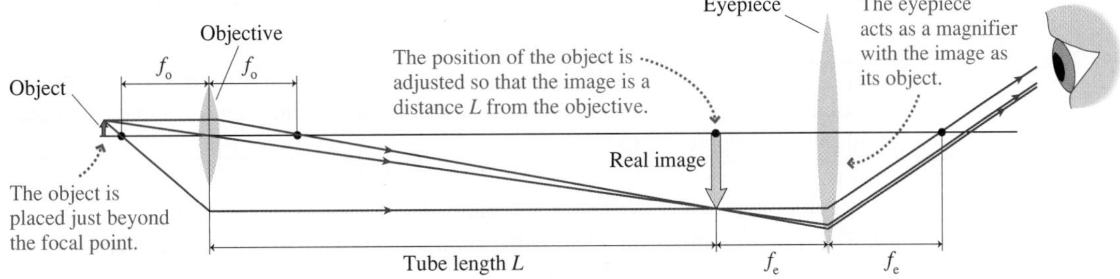

From the magnification equation, Equation 18.8, the magnification of the objective lens is

$$m_o = -\frac{s'}{s} \approx -\frac{L}{f_o} \tag{19.4}$$

Here we used the fact that the image distance s' is equal to the tube length L and the object distance s is very close to the focal length f_o of the objective. The minus sign tells us that the image is inverted with respect to the object.

The image of the objective acts as the object for the eyepiece, which functions as a simple magnifier. The angular magnification of the eyepiece is given by Equation 19.3: $M_e = (25\text{ cm})/f_e$. Together, the objective and eyepiece produce a total angular magnification

$$M = m_o M_e = -\frac{L}{f_o}\frac{25\text{ cm}}{f_e} \tag{19.5}$$

The minus sign shows that the image seen in a microscope is inverted.

EXAMPLE 19.7 Finding the focal length of a microscope objective

A biological microscope objective is labeled "20×." What is its focal length?

PREPARE The "20×" means that the objective has a magnification m_o of −20. We can use Equation 19.4 with L as 160 mm, which we've seen is the standard length for a biological microscope.

SOLVE From Equation 19.4 we have

$$f_o = -\frac{L}{m_o} = -\frac{160\text{ mm}}{-20} = 8.0\text{ mm}$$

ASSESS Microscope objectives are specified by their magnification, or "power," not by their focal length. Equation 19.4 relates these two important specifications.

Many microscopes have a set of objectives that can be pivoted into place to change the overall magnification. A complete set of objectives might include 5×, 10×, 20×, 40×, and 100×. Eyepieces are also specified by their magnification and are available with magnifications in the range of 10× to 20×. With these lenses, the lowest magnification available would be 5 × 10 = 50×, while the highest magnification would be 100 × 20 = 2000×.

EXAMPLE 19.8 Viewing blood cells

A pathologist inspects a sample of 7-μm-diameter human blood cells under a microscope. She selects a 40× objective and a 10× eyepiece. What size object, viewed from 25 cm, has the same apparent size as a blood cell seen through the microscope?

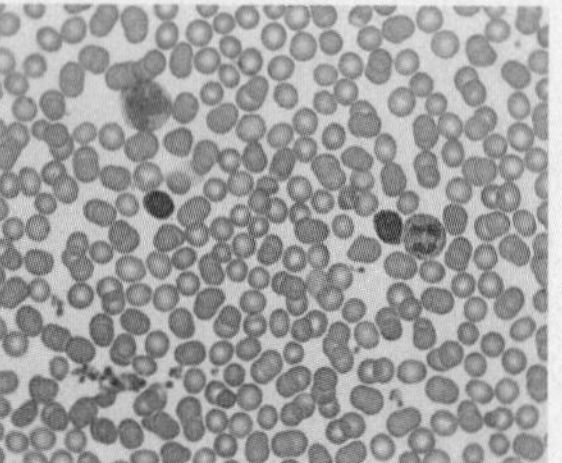

PREPARE Angular magnification compares the magnified angular size to the angular size seen at the near-point distance of 25 cm.

SOLVE The microscope's angular magnification is $M = -(40) \times (10) = -400$. The magnified cells will have the same apparent size as an object $400 \times 7\ \mu\text{m} \approx 3$ mm in diameter seen from a distance of 25 cm.

ASSESS 3 mm is about the size of a capital O in this textbook, so a blood cell seen through the microscope will have about the same apparent size as an O seen from a comfortable reading distance.

STOP TO THINK 19.4 A biologist rotates the turret of a microscope to replace the 20× objective with a 10× objective. To keep the magnification the same, the focal length of the eyepiece must

A. Be doubled. B. Be halved. C. Remain the same.
D. The magnification cannot stay the same if the objective power is changed.

19.5 The Telescope

The microscope magnifies small objects that can be placed near its objective lens. A *telescope* is used to magnify distant objects. The two-lens arrangement, shown in FIGURE 19.18 on the next page, is similar to that of the microscope, but the objective lens has a long focal length instead of the very short focal length of a microscope objective. Because the object is very far away ($s \approx \infty$), the converging objective lens forms a real image of the distant object at the lens's focal point. A second lens,

FIGURE 19.18 The telescope.

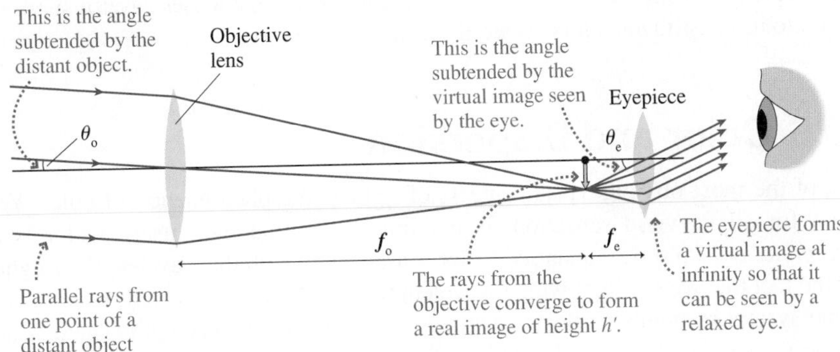

the eyepiece, is then used as a simple magnifier to enlarge this real image for final viewing by the eye.

We can use Figure 19.18 to find the magnification of a telescope. The original object subtends an angle θ_o. Because the object is distant, its image is formed in the focal plane of the objective lens, a distance f_o from the objective. From the geometry of Figure 19.18, the height of this image (negative because it's inverted) is

$$h' \approx -f_o\theta_o$$

where we have used the small-angle approximation $\tan\theta_o \approx \theta_o$. This image is now the object for the eyepiece lens, which functions as a magnifier. The height of the "object" it views is h', so, from Equation 19.2, the angular size θ_e of the virtual image formed by the eyepiece is

$$\theta_e = \frac{h'}{f_e} = \frac{-f_o\theta_o}{f_e}$$

where f_e is the focal length of the eyepiece. The telescope's angular magnification is the ratio of the angular size seen when looking through the telescope to that seen without the telescope, so we have

$$M = \frac{\theta_e}{\theta_o} = -\frac{f_o}{f_e} \tag{19.6}$$

The minus sign indicates that you see an upside-down image when looking through a simple telescope. This is not a problem when looking at astronomical objects, but it could be disconcerting to a bird watcher. More sophisticated telescope designs produce an upright image.

To get a high magnification with a telescope, the focal length of the objective should be large and that of the eyepiece small. Contrast this with the magnification of a microscope, Equation 19.5, which is high when the focal lengths of both objective and eyepiece are small.

The telescope shown in Figure 19.18 uses a lens as its objective and hence is known as a *refracting telescope*. It is also possible to make a *reflecting telescope* with a concave mirror instead of a lens, as shown in FIGURE 19.19. One problem with this arrangement, for small telescopes, is that the image is formed in front of the mirror where it's hard to magnify with an eyepiece without getting one's head in the way. Newton, who built the first such telescope, used a small angled plane mirror, called a *secondary mirror,* to deflect the image to an eyepiece on the side of the telescope.

For large telescopes, such as those used in astronomy, mirrors have two important advantages over lenses. First, objectives of astronomical telescopes must be quite large in order to gather as much light from faint objects as possible. The Subaru Telescope in Hawaii is the world's largest single-mirror telescope; its mirror has a diameter of 8.3 m (27 ft)! A giant lens of this diameter would sag under its own weight.

A clearer view The performance of telescopes on the earth is limited by the atmosphere. Even at night, the atmosphere glows faintly, interfering with the long exposures needed to photograph faint astronomical objects. Further, atmospheric turbulence—visible to the naked eye as the twinkling of stars—obscures the finest details of the object being observed. Because of this, the Hubble Space Telescope orbits the earth high above the atmosphere. With its 2.4-m-diameter mirror, it has produced some of the most spectacular images of astronomical objects, such as this gas cloud surrounding the star V838 Monocerotis.

FIGURE 19.19 A reflecting telescope.

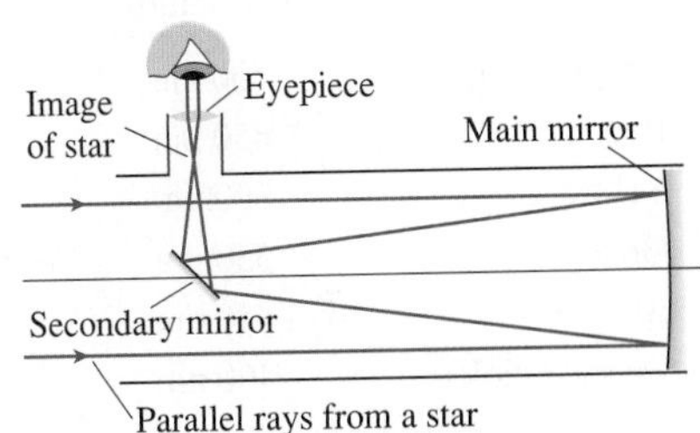

A mirror, on the other hand, can be supported along its entire back surface. Second, mirrors are free from chromatic aberration, the tendency that lenses have of splitting light into its constituent colors, as we'll see in a later section.

19.6 Color and Dispersion

I procured me a triangular glass prism to try therewith the celebrated phenomena of colors.

Isaac Newton

One of the most obvious visual aspects of light is the phenomenon of color. Yet color, for all its vivid sensation, is not inherent in the light itself. Color is a *perception,* not a physical quantity. Color is associated with the wavelength of light, but the fact that we see light with a wavelength of 650 nm as "red" tells us how our visual system responds to electromagnetic waves of this wavelength. There is no "redness" associated with the light wave itself.

Most of the results of optics do not depend on color. We generally don't need to know the color of light—or, to be more precise, its wavelength—to use the laws of reflection and refraction. Nonetheless, color is an interesting subject, one worthy of a short digression.

Color

FIGURE 19.20 Newton used prisms to study color.

(a)

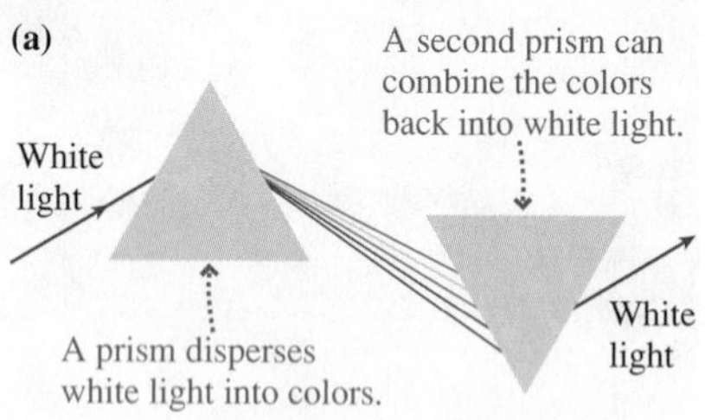

A second prism can combine the colors back into white light.
White light
White light
A prism disperses white light into colors.

(b)

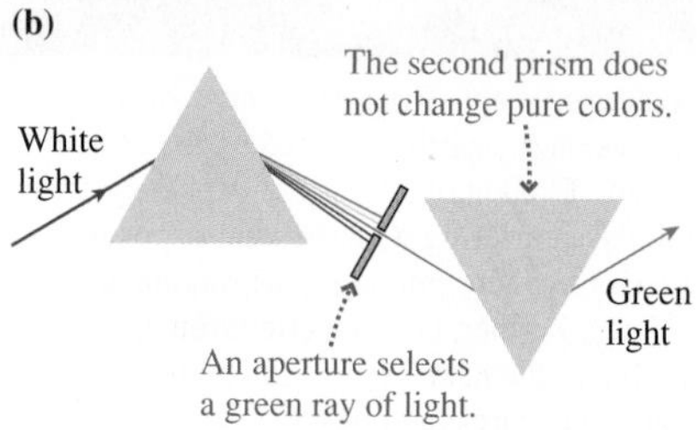

The second prism does not change pure colors.
White light
Green light
An aperture selects a green ray of light.

It has been known since antiquity that irregularly shaped glass and crystals cause sunlight to be broken into various colors. A common idea was that the glass or crystal somehow altered the properties of the light by *adding* color to the light. Newton suggested a different explanation. He first passed a sunbeam through a prism, producing the familiar rainbow of light. We say that the prism *disperses* the light. Newton's novel idea, shown in FIGURE 19.20a, was to use a second prism, inverted with respect to the first, to "reassemble" the colors. He found that the light emerging from the second prism was a beam of pure white light.

But the emerging light beam is white only if *all* the rays are allowed to move between the two prisms. Blocking some of the rays with small obstacles, as in FIGURE 19.20b, causes the emerging light beam to have color. This suggests that color is associated with the light itself, not with anything that the prism is "doing" to the light. Newton tested this idea by inserting a small aperture between the prisms to pass only the rays of a particular color, such as green. If the prism alters the properties of light, then the second prism should change the green light to other colors. Instead, the light emerging from the second prism is unchanged from the green light entering the prism.

These and similar experiments show that:

1. What we perceive as white light is a mixture of all colors. White light can be dispersed into its various colors and, equally important, mixing all the colors produces white light.
2. The index of refraction of a transparent material differs slightly for different colors of light. Glass has a slightly higher index of refraction for violet light than for green light or red light. Consequently, different colors of light refract at slightly different angles. A prism does not alter the light or add anything to the light; it simply causes the different colors that are inherent in white light to follow slightly different trajectories.

TABLE 19.1 A brief summary of the visible spectrum of light

Color	Approximate wavelength
Deepest red	700 nm
Red	650 nm
Yellow	600 nm
Green	550 nm
Blue	450 nm
Deepest violet	400 nm

Dispersion

It was Thomas Young, with his two-slit interference experiment, who showed that different colors are associated with light of different wavelengths. The longest wavelengths are perceived as red light and the shortest wavelengths are perceived as violet light. Table 19.1 is a brief summary of the *visible spectrum* of light. Visible-light wavelengths are used so frequently that it is well worth committing this short table to memory.

The slight variation of index of refraction with wavelength is known as **dispersion.** FIGURE 19.21 shows the *dispersion curves* of two common glasses. Notice that ***n* is *higher* when the wavelength is *shorter;*** thus violet light refracts more than red light.

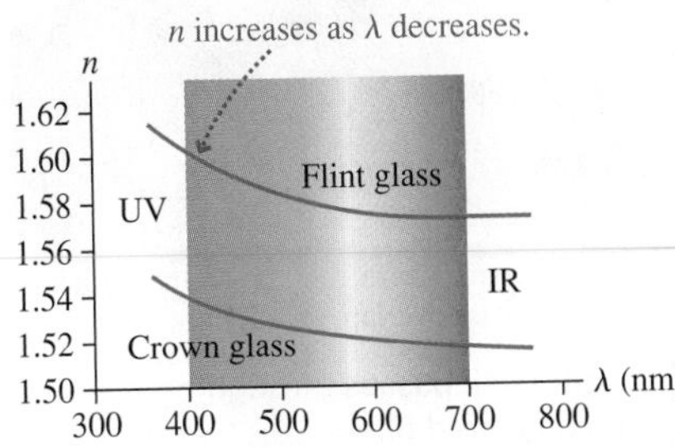

n increases as λ decreases.
n
1.62
1.60
1.58
1.56
1.54
1.52
1.50
UV
Flint glass
IR
Crown glass
λ (nm)
300 400 500 600 700 800

FIGURE 19.21 Dispersion curves show how the index of refraction varies with wavelength.

EXAMPLE 19.9 **Dispersing light with a prism**

Example 18.4 in Chapter 18 found that a ray incident on a 30° prism is deflected by 22.6° if the prism's index of refraction is 1.59. Suppose this is the index of refraction of deep violet light, and that deep red light has an index of refraction of 1.54.

a. What is the deflection angle for deep red light?
b. If a beam of white light is dispersed by this prism, how wide is the rainbow spectrum on a screen 2.0 m away?

PREPARE Figure 18.18 in Example 18.4 showed the geometry. A ray is incident on the hypotenuse of the prism at $\theta_1 = 30°$.

SOLVE a. If $n_1 = 1.54$ for deep red light, the refraction angle is

$$\theta_2 = \sin^{-1}\left(\frac{n_1 \sin\theta_1}{n_2}\right) = \sin^{-1}\left(\frac{1.54 \sin 30°}{1.00}\right) = 50.4°$$

Example 18.4 showed that the deflection angle is $\phi = \theta_2 - \theta_1$, so deep red light is deflected by $\phi_{red} = 20.4°$. This angle is slightly smaller than the previously observed $\phi_{violet} = 22.6°$.

b. The entire spectrum is spread between $\phi_{red} = 20.4°$ and $\phi_{violet} = 22.6°$. The angular spread is

$$\delta = \phi_{violet} - \phi_{red} = 2.2° = 0.038 \text{ rad}$$

At distance r, the spectrum spans an arc length

$$s = r\delta = (2.0 \text{ m})(0.038 \text{ rad}) = 0.076 \text{ m} = 7.6 \text{ cm}$$

ASSESS Notice that we needed three significant figures for ϕ_{red} and ϕ_{violet} in order to determine δ, the *difference* between the two angles, to two significant figures. The angle is so small that there's no appreciable difference between arc length and a straight line. The spectrum will be 7.6 cm wide at a distance of 2.0 m.

Rainbows

One of the most interesting sources of color in nature is the rainbow. The details get somewhat complicated, but FIGURE 19.22a shows that the basic cause of the rainbow is a combination of refraction, reflection, and dispersion.

FIGURE 19.22 Light seen in a rainbow has undergone refraction + reflection + refraction in a raindrop.

(a)

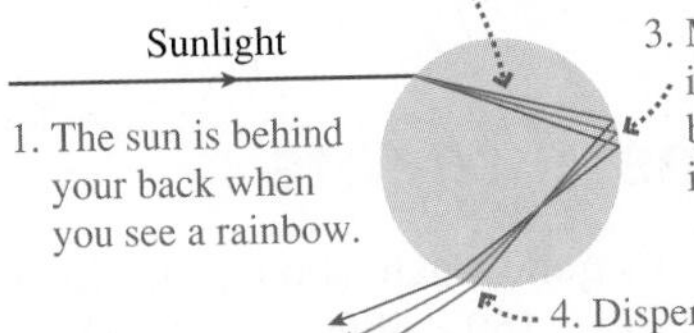

2. Dispersion causes different colors to refract at different angles.
Sunlight
3. Most of the light refracts into the air at this point, but a little reflects back into the drop.
1. The sun is behind your back when you see a rainbow.
4. Dispersion separates the colors even more as the rays refract back into the air.

(b)

Sunlight

42.5°

40.8°

Eye

Red light is refracted predominantly at 42.5°. The red light reaching your eye comes from drops higher in the sky.

Violet light is refracted predominantly at 40.8°. The violet light reaching your eye comes from drops lower in the sky.

You see a rainbow with red on the top, violet on the bottom.

Figure 19.22a might lead you to think that the top edge of a rainbow is violet. In fact, the top edge is red, and violet is on the bottom. The rays leaving the drop in Figure 19.22a are spreading apart, so they can't all reach your eye. As FIGURE 19.22b shows, a ray of red light reaching your eye comes from a drop *higher* in the sky than a ray of violet light. In other words, the colors you see in a rainbow refract toward

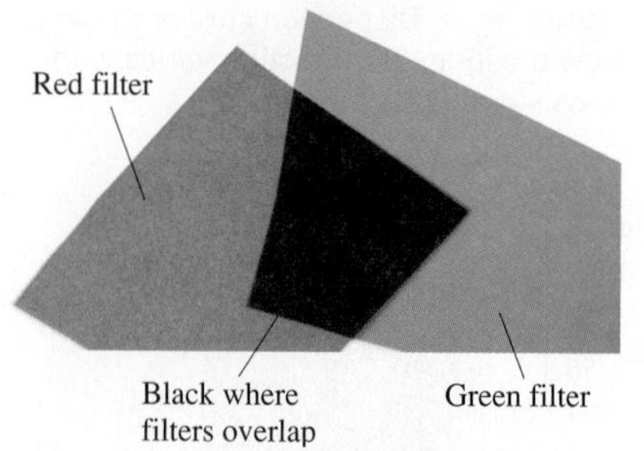
Red filter
Black where filters overlap
Green filter

No light at all passes through both a green and a red filter.

your eye from different raindrops, not from the same drop. You have to look higher in the sky to see the red light than to see the violet light.

Colored Filters and Colored Objects

White light passing through a piece of green glass emerges as green light. A possible explanation would be that the green glass *adds* "greenness" to the white light, but Newton found otherwise. Green glass is green because it *removes* any light that is "not green." More precisely, a piece of colored glass *absorbs* all wavelengths except those of one color, and that color is transmitted through the glass without hindrance. We can think of a piece of colored glass or plastic as a *filter* that removes all wavelengths except a chosen few.

CONCEPTUAL EXAMPLE 19.10 Filtering light

White light passes through a green filter and is observed on a screen. Describe how the screen will look if a second green filter is placed between the first filter and the screen. Describe how the screen will look if a red filter is placed between the green filter and the screen.

REASON The first filter removes all light except for wavelengths near 550 nm that we perceive as green light. A second green filter doesn't have anything to do. The nongreen wavelengths have already been removed, and the green light emerging from the first filter will pass through the second filter without difficulty. The screen will continue to be green and its intensity will not change.

A red filter, by contrast, absorbs all wavelengths except those near 650 nm. The red filter will absorb the green light, and *no* light will reach the screen. The screen will be dark.

Opaque objects appear colored by virtue of *pigments* that absorb light of some wavelengths but *reflect* light of other wavelengths. For example, red paint contains pigments that reflect light of wavelengths near 650 nm while absorbing all other wavelengths. Pigments in paints, inks, and natural objects are responsible for most of the color we observe in the world, from the red of lipstick to the blue of a bluebird's feathers.

As an example, FIGURE 19.23 shows the absorption curve of *chlorophyll.* Chlorophyll is essential for photosynthesis in green plants. The chemical reactions of photosynthesis are able to use red light and blue/violet light; thus chlorophyll has evolved to absorb red light and blue/violet light from sunlight and put it to use. But green and yellow light are not absorbed. Instead, these wavelengths are mostly *reflected* to give the object a greenish-yellow color. When you look at the green leaves on a tree, you're seeing the light that was reflected because it *wasn't* needed for photosynthesis.

FIGURE 19.23 The absorption curve of chlorophyll.

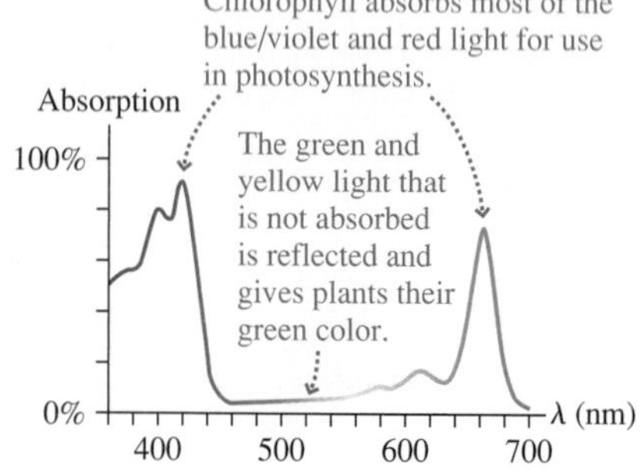
Chlorophyll absorbs most of the blue/violet and red light for use in photosynthesis.
Absorption
100%
The green and yellow light that is not absorbed is reflected and gives plants their green color.
0%
400
500
600
700
λ (nm)

STOP TO THINK 19.5 A red apple is viewed through a green filter. The apple appears

A. Red. B. Green. C. Yellow. D. Black.

19.7 Resolution of Optical Instruments

16.8 

Suppose you wanted to study the *E. coli* bacterium. It's quite small, about 2 μm long and 0.5 μm wide. You might imagine that you could pair a 150× objective (the highest magnification available) with a 25× eyepiece to get a total magnification of 3750! At that magnification, the *E. coli* would appear about 8 mm across—about the size of Lincoln's head on a penny—with much fine detail revealed. But if you tried this, you'd be disappointed. Although you would see the general shape of a bacterium, you wouldn't be able to make out any real details. All real optical instruments are limited in the details they can observe. Some limits are practical: Lenses are never perfect, suffering from **aberrations.** But even a perfect lens would have a fundamental limit to the smallest details that could be seen. As we'll see, this limit is set by the diffraction of light, and so is intimately related to the wave nature of light itself. Together, lens aberrations and diffraction set a limit on an optical system's **resolution**—its ability to make out the fine details of an object.

Aberrations

Consider the simple lens shown in FIGURE 19.24 imaging an object located at infinity, so that the incoming rays are parallel. An ideal lens would focus all the rays to a single point. However, for a real lens with spherical surfaces, the rays that pass near the lens's center come to a focus a bit farther from the lens than those that pass near its edge. There is no single focal point; even at the "best" focus the image is a bit blurred. This inability of a real lens to focus perfectly is called **spherical aberration.**

FIGURE 19.24 Spherical aberration.

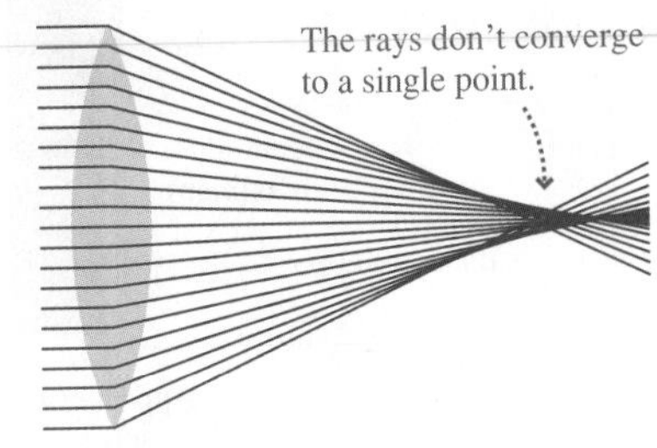
The rays don't converge to a single point.

A careful examination of Figure 19.24 shows that the outer rays are most responsible for the poor focus. Consequently, the effects of spherical aberration can be minimized by using an iris diaphragm to pass only rays near the optical axis. This "stopping down" of a lens improves its imaging characteristics at the expense of its light-gathering capabilities. Part of the function of the iris of the human eye is to improve vision in this way. Our vision is poorer at night with the iris wide open—but our ancestors probably avoided many a predator with this poor but sensitive night vision.

As we learned in the previous section, glass has *dispersion;* that is, the index of refraction of glass varies slightly with wavelength. The higher a lens's index of refraction, the more it bends incoming light rays. Because the index of refraction for violet light is higher than that for red light, a lens's focal length is slightly shorter for violet light than for red light. Consequently, different colors of light come to a focus at slightly different distances from the lens. If red light is sharply focused on a viewing screen, then blue and violet wavelengths are not well focused. This imaging error, illustrated in FIGURE 19.25, is called **chromatic aberration.**

FIGURE 19.25 Chromatic aberration.

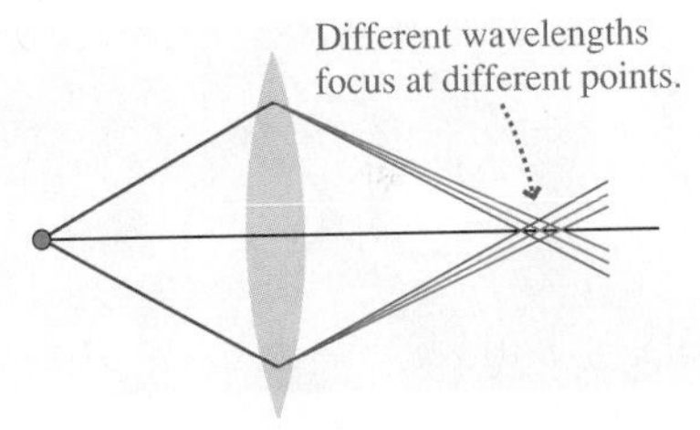
Different wavelengths focus at different points.

Correcting Aberrations

Single lenses always have aberrations of some kind. For high-quality optics, such as those used in microscopes or telescopes, the aberrations are minimized by using a careful *combination* of lenses. An important example is the **achromatic doublet** (achromatic = "without color"), two lenses used in combination to greatly reduce chromatic aberration. FIGURE 19.26 shows how this works. A converging lens is paired with a weaker diverging lens; the combination has an overall positive refractive power and so is converging. However, the glasses are chosen so that the weaker diverging lens has a greater dispersion than the stronger converging lens. In this way the colors of white light, separated at first by the converging lens, are brought back together by the diverging lens. Achromatic doublets also minimize spherical aberration. Real microscope objectives are even more complex, but are based on the same principle as the achromatic doublet.

FIGURE 19.26 An achromatic lens.

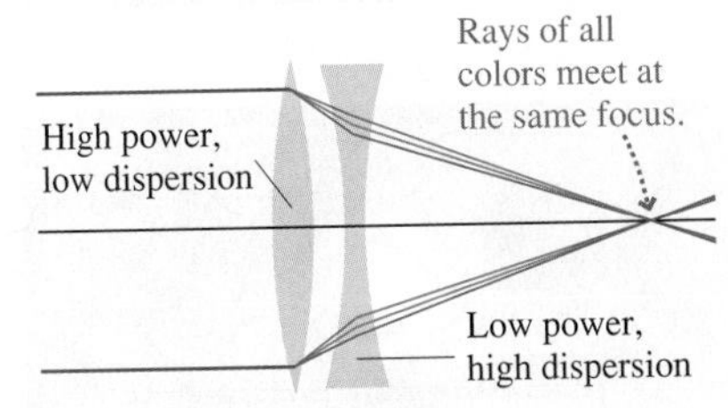
Rays of all colors meet at the same focus.
High power, low dispersion
Low power, high dispersion

Resolution and the Wave Nature of Light

Modern lenses can be well corrected for aberrations, so they might be expected to focus perfectly. According to the ray model of light, a perfect lens should focus parallel rays to a single point in the focal plane. However, we've already hinted that there's a more fundamental limit to the performance of an optical instrument, a limit set by the wave nature of light.

► **Eyeglasses in space** After launch, it was discovered that the mirror of the Hubble Space Telescope had been ground to the wrong shape, giving it severe spherical aberration. In a later service mission, corrective optics—in essence, very high-tech glasses—were put in place to correct for this spherical aberration. The photos to the right show an image of a galaxy before and after the corrective optics were added.

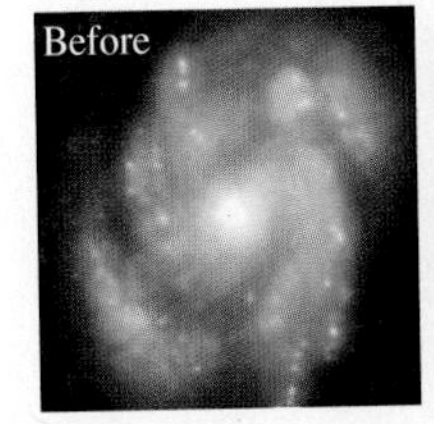
Before

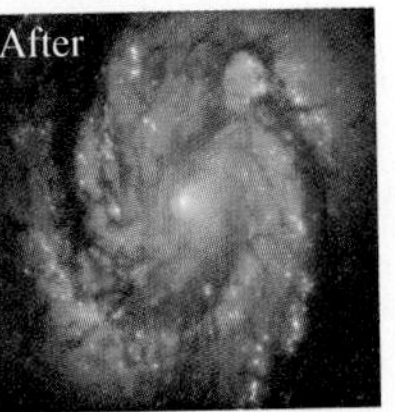
After

FIGURE 19.27 The image of a distant point source is a circular diffraction pattern.

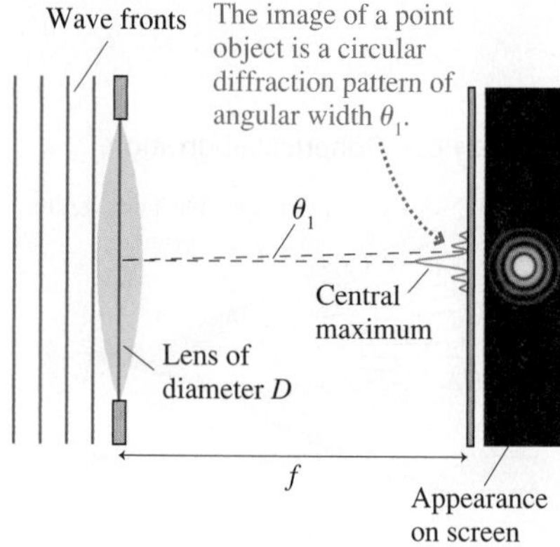

FIGURE 19.27 shows a plane wave from a distant point source such as a star being focused by a lens of diameter D. Only those waves passing *through* the lens can be focused, so the lens acts like a circular aperture of diameter D in an opaque barrier. In other words, the lens both focuses *and diffracts* light waves.

You learned in Chapter 17 that a circular aperture produces a diffraction pattern with a bright central maximum surrounded by dimmer circular fringes. Consequently, as Figure 19.27 shows, light from a distant point source focuses not to a perfect point but, instead, to a small circular diffraction pattern. Equation 17.22 in Chapter 17 gave the angle θ_1 of the outer edge of the central maximum as

$$\theta_1 = \frac{1.22\lambda}{D} \tag{19.7}$$

Because the wavelength of light λ is so much shorter than the lens diameter D of an ordinary lens, the angular size of the central maximum is very small—but it is not zero.

The fact that light is focused to a small spot, not a perfect point, has important consequences for how well a telescope can resolve two stars separated by only a small angle in the sky. FIGURE 19.28a shows how two nearby stars would appear in a telescope. Instead of perfect points, they appear as two diffraction images. Nonetheless, because they're clearly two separate stars, we say they are *resolved.* FIGURE 19.28b shows two stars that are closer together. Here the two diffraction patterns overlap, and it is becoming difficult to see them as two independent stars: They are barely resolved. The two very nearby stars in Figure 19.28c are so close together that we can't resolve them at all.

FIGURE 19.28 The resolution of a telescope.

(a) Stars resolved

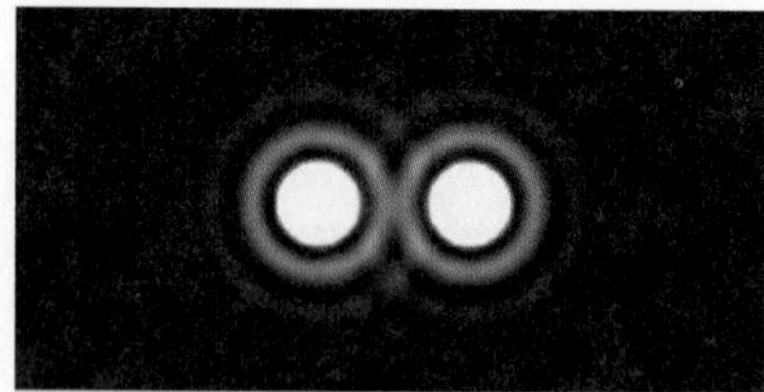

(b) Stars just resolved

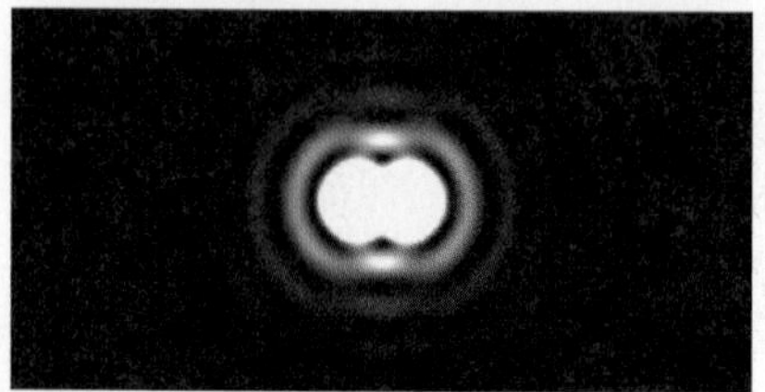

(c) Stars not resolved

How close can the two diffraction patterns be before you can no longer resolve them? One of the major scientists of the 19th century, Lord Rayleigh, studied this problem and suggested a reasonable rule that today is called **Rayleigh's criterion.** In Figure 19.28b, where the two stars are just resolved, *the central maximum of the diffraction pattern of one star lies on top of the first dark fringe of the diffraction pattern of the other star.* Because the angle between the central maximum and the first dark fringe is θ_1, this means that the centers of the two stars are separated by angle $\theta_1 = 1.22\lambda/D$. Thus Rayleigh's criterion is:

Two objects are resolvable if they are separated by an angle θ that is greater than $\theta_1 = 1.22\lambda/D$. If their angular separation is less than θ_1, then they are not resolvable. If their separation is equal to θ_1, then they are just barely resolvable.

For telescopes, the angle $\theta_1 = 1.22\lambda/D$ is called the *angular resolution* of the telescope. The angular resolution depends only on the lens diameter and the wavelength; the magnification is not a factor. Two overlapped, unresolved images will remain overlapped and unresolved no matter what the magnification. For visible light, where λ is pretty much fixed, the only parameter over which the astronomer has any control is the diameter of the lens or mirror of the telescope. The urge to build ever-larger telescopes is motivated, in part, by a desire to improve the angular resolution. (Another important motivation is to increase the light-gathering power so as to see objects farther away.)

The Resolution of a Microscope

A microscope differs from a telescope in that it magnifies objects that are very close to the lens, not far away. Nonetheless, the wave nature of light still sets a limit on the ultimate resolution of a microscope. FIGURE 19.29 shows the objective lens of a microscope that is observing two small objects. An analysis based on Rayleigh's criterion finds that the smallest resolvable separation between the two objects is

$$d_{min} = \frac{0.61\lambda}{n \sin \phi_0} \tag{19.8}$$

Here, ϕ_0, defined in Figure 19.29, is the angular size of the objective lens and n is the index of refraction of the medium between the objective lens and the specimen being observed. Usually this medium is air, so that $n = 1$, but biologists often use an *oil-immersion microscope* in which this space is filled with oil having $n \approx 1.5$. From Equation 19.8, you can see that this higher value of n reduces d_{min}, allowing objects that are closer together to be resolved.

The quantity $n \sin \phi_0$ is called the **numerical aperture** NA of the objective when immersed in a fluid of index n. For a microscope whose objective has numerical aperture NA, the minimum resolvable distance, also called the **resolving power** RP, is

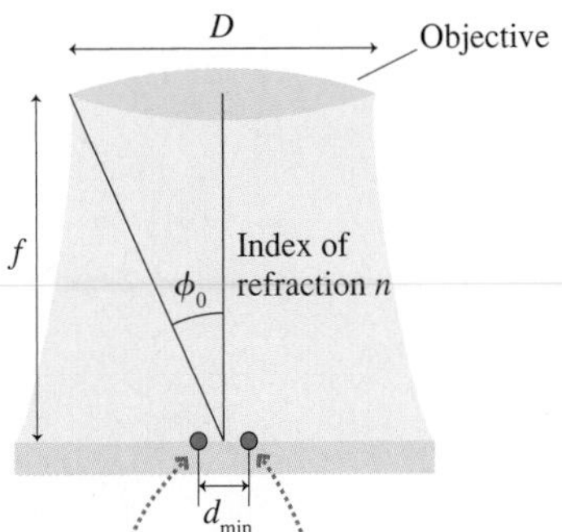

FIGURE 19.29 The resolution of a microscope.

$$\text{RP} = d_{min} = \frac{0.61\lambda_0}{\text{NA}} \tag{19.9}$$

Resolving power of a microscope with numerical aperture NA

The lower the resolving power, the *better* the objective is at seeing small details.

In principle, it would appear from Equation 19.9 that the resolving power of a microscope could be made as low as desired simply by increasing the numerical aperture. But there are rather severe practical limits on how high the numerical aperture can be made. The highest possible numerical aperture for a high-magnification 100× objective used in air is about 0.95. For an oil-immersion objective, the numerical aperture might be as high as 1.3. With such an objective, the resolving power would be

$$\text{RP} \approx 0.5\lambda_0$$

This illustrates the fundamental fact that **the minimum resolving power of a microscope, and thus the size of the smallest detail observable, is about half the wavelength of light.** This is a *fundamental limit* set by the wave nature of light. For $\lambda \approx 400$ nm, at the short-wavelength edge of the visible spectrum, the maximum possible resolving power is RP ≈ 200 nm.

FIGURE 19.30 shows an actual micrograph of the bacillus *E. coli.* The width is about equal to the wavelength of light in the middle of the spectrum, about 500 nm, and the smallest resolved features are about half this. A higher magnification would not reveal any more detail because this micrograph is at the resolution limit set by the diffraction of light.

In contrast, the **electron microscope** micrograph of *E. coli* shows a wealth of detail unobservable in the optical picture. In Chapter 28 we'll find out why the resolving power of an electron microscope is so much lower than that of an optical microscope.

FIGURE 19.30 Optical and electron micrographs of *E. coli.*

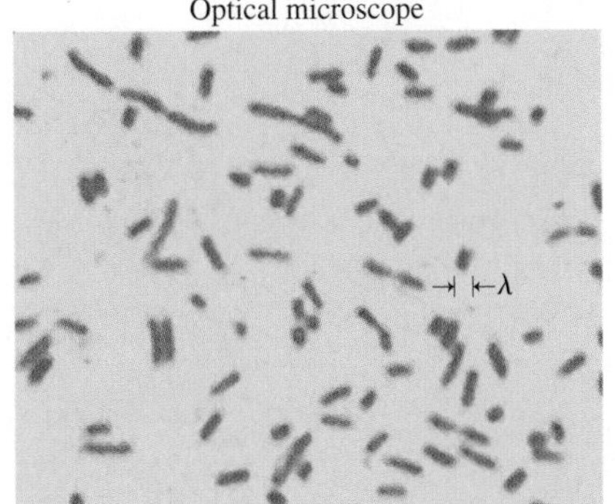

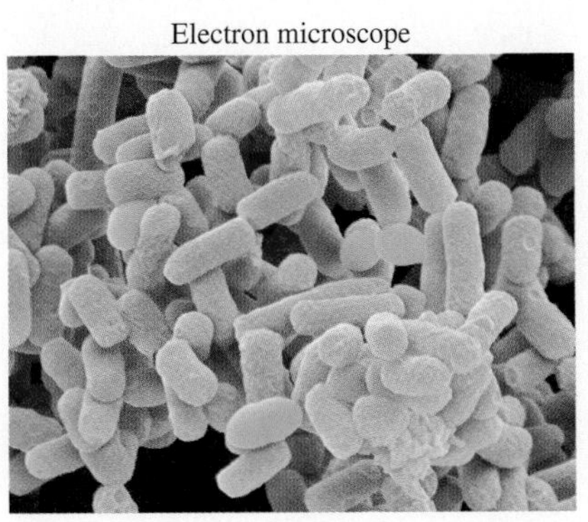

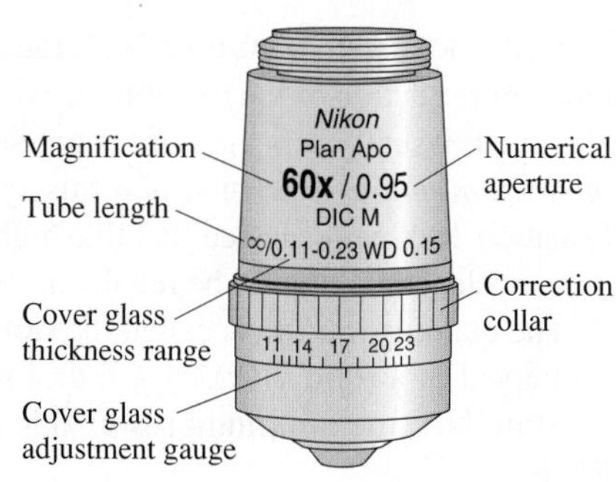

▶ The anatomy of a microscope objective The two most important specifications of a microscope objective, its magnification and its numerical aperture, are prominently displayed on its barrel. Other important information is shown as well. This objective is designed to project its real image at a tube length of infinity (∞), instead of at the standard distance of 160 mm. Extra lenses in the microscope move this image to just in front of the eyepiece. Many biological studies are conducted through a cover glass. This cover glass can introduce spherical aberration, blurring the image. By turning the correction collar, you can adjust this objective to correct for the exact thickness of the cover glass used.

EXAMPLE 19.11 Finding the resolving power of a microscope

A microscope objective lens has a diameter of 6.8 mm and a focal length of 4.0 mm. For a sample viewed in air, what is the resolving power of this objective in red light? In blue light?

PREPARE We can use Equation 19.9 to find the resolving power. We'll need the numerical aperture of the objective, given as $\text{NA} = n\sin\phi_0$.

SOLVE From the geometry of Figure 19.29,

$$\tan\phi_0 = \frac{D/2}{f} = \frac{3.4\text{ mm}}{4.0\text{ mm}} = 0.85$$

from which $\phi_0 = \tan^{-1}0.85 = 40.4°$ and $\sin\phi_0 = \sin 40.4° = 0.65$. Hence the numerical aperture is (since $n = 1$ in air)

$$\text{NA} = n\sin\phi_0 = 1 \times 0.65 = 0.65$$

Then, from Equation 19.9, the resolving power is

$$\text{RP} = \frac{0.61\lambda}{0.65} = 0.94\lambda_0$$

Wavelengths of different colors of light were listed in Table 19.1. For red light, with $\lambda_0 = 650$ nm, RP = 610 nm, while blue light, with $\lambda_0 = 450$ nm, has RP = 420 nm.

ASSESS We see that shorter-wavelength light yields a higher resolution (lower RP). Unfortunately, wavelengths much shorter than 400 nm are invisible, and glass lenses are opaque to light of very short wavelength.

STOP TO THINK 19.6 Four lenses are used as microscope objectives, all for light with the same wavelength λ. Rank in order, from highest to lowest, the resolving powers RP_1 to RP_4 of the lenses.

$f = 10$ mm — lens 1, 2 mm
$f = 5$ mm — lens 2, 2 mm
$f = 10$ mm — lens 3, 4 mm
$f = 24$ mm — lens 4, 8 mm

INTEGRATED EXAMPLE 19.12 The visual acuity of a kestrel

Like most birds of prey, the American kestrel has excellent eyesight. The smallest angular separation between two objects that an eye can resolve is called its *visual acuity;* a smaller visual acuity means better eyesight because objects closer together can be resolved. The eye of a particular kestrel has a pupil diameter of 3.0 mm. The fixed distance from its lens to the retina is 9.0 mm, and the space within the eye is filled with a clear liquid whose index of refraction is 1.31. Assume that the kestrel's optical system can be adequately modeled as a thin lens and a detector.

a. As the bird focuses on an insect sitting 0.80 m away, what is the focal length of its lens?
b. Laboratory measurements indicate that the kestrel can just resolve two small objects that have an angular separation of only 0.013°. How does this result compare with the visual acuity predicted by Rayleigh's criterion? Take the wavelength of light in air to be 550 nm.
c. What is the distance on the retina between the images of two small objects that can just be resolved? How does this distance compare to the 2.0 μm distance between two photoreceptors, the light-sensitive cells on the retina? Does this comparison make sense from the standpoint of vision?

PREPARE a. Recall that an eye focuses by changing the focal length of its lens. The image distance s' from the lens to the image plane (at the retina) is unchanged as the bird focuses from a distant object to a nearby one, so, as **FIGURE 19.31** shows, $s' = 9.0$ mm. We can then use the thin-lens equation to find f.

FIGURE 19.31 The kestrel's eye observing two closely spaced objects.

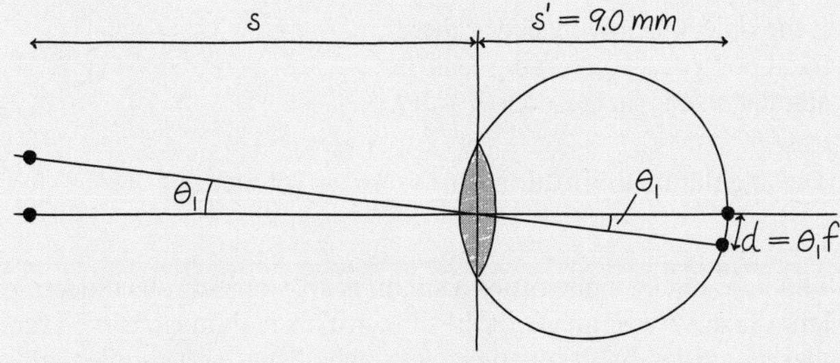

b. By Rayleigh's criterion, visual acuity—the smallest resolvable angle of an ideal lens—is proportional to the wavelength of light used. Inside the eye, however, the wavelength of light is shorter than its wavelength λ_0 in air because of the index of refraction of the liquid within the eye. Thus we must use $\lambda = \lambda_0/n = (550 \text{ nm})/(1.31) = 420$ nm in Rayleigh's criterion.

c. Figure 19.31 shows that, within the small-angle approximation, the distance d on the retina between the images of two small objects subtending an angle θ_1 is simply $d = \theta_1 f$, where θ_1 is in radians.

SOLVE a. From the thin-lens equation, Equation 18.11, we have

$$\frac{1}{f} = \frac{1}{s} + \frac{1}{s'} = \frac{1}{800 \text{ mm}} + \frac{1}{9.0 \text{ mm}} = 0.112 \text{ mm}^{-1}$$

Thus $f = 1/0.112 \text{ mm}^{-1} = 8.9$ mm.

b. For a perfect lens, with no aberrations, the smallest resolvable angular separation between two objects is given by Rayleigh's criterion as

$$\theta_1 = \frac{1.22\lambda}{D} = \frac{1.22(420 \times 10^{-9} \text{ m})}{3.0 \times 10^{-3} \text{ m}} = 1.7 \times 10^{-4} \text{ rad}$$

Recalling that there are 360° in 2π rad, this angle is

$$\theta_1 = (1.7 \times 10^{-4} \text{ rad}) \times \frac{360°}{2\pi \text{ rad}} = 0.0098°$$

The observed visual acuity of 0.013° is about 30% greater than this theoretical value. Presumably this is due to aberrations in the optical system of the eye.

c. The angle of 0.013° corresponds to 2.3×10^{-4} rad. Thus the distance between the images of the two small objects is

$$d = \theta_1 f = (2.3 \times 10^{-4} \text{ rad})(9.0 \text{ mm})$$
$$= 2.1 \times 10^{-3} \text{ mm} = 2.1\ \mu\text{m}$$

This is just about the same as the photoreceptor distance. This makes sense. If d were significantly greater than the photoreceptor distance, then many receptors would be wasted. If d were smaller, the eye's resolution would be determined by the photoreceptor spacing and would not reach the visual acuity predicted by Rayleigh's criterion.

ASSESS An object far from a converging lens gives an image distance close to the lens's focal length. Thus our focal length of 8.9 mm—close to the image distance of 9.0 mm—is reasonable.

Although the visual acuity of the kestrel is impressive—equivalent to being able to resolve a mouse at a distance of 600 feet—it turns out not to be significantly greater than that for the human eye. Raptors are also aided in hunting their prey by a highly evolved cerebral function that allows them to pick out small movements that humans would miss.

SUMMARY

The goal of Chapter 19 has been to understand how common optical instruments work.

IMPORTANT CONCEPTS

Color and dispersion

The eye perceives light of different wavelengths as having different colors.

Dispersion is the dependence of the index of refraction n of a transparent medium on the wavelength of light: Long wavelengths have the lowest n, short wavelengths the highest n.

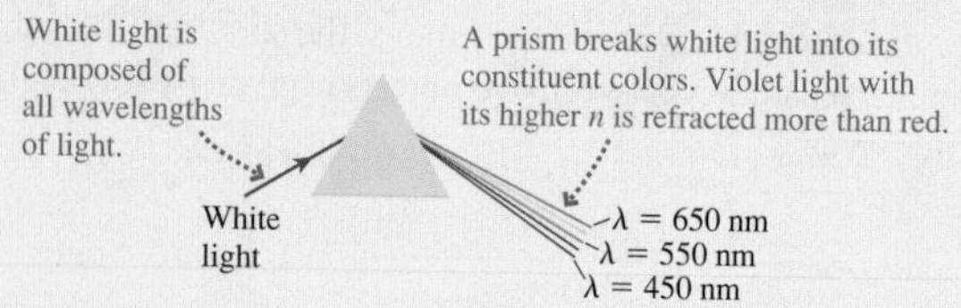

White light is composed of all wavelengths of light.
A prism breaks white light into its constituent colors. Violet light with its higher n is refracted more than red.
White light
λ = 650 nm
λ = 550 nm
λ = 450 nm

Resolution of optical instruments

The **resolution** of a telescope or microscope is limited by imperfections, or **aberrations,** in the optical elements, and by the more fundamental limits imposed by diffraction.

For a *microscope,* the minimum resolvable distance between two objects is

$$d_{min} = \frac{0.61\lambda}{NA}$$

For a *telescope,* the minimum resolvable angular separation between two objects is

$$\theta_1 = \frac{1.22\lambda}{D}$$

Lenses in combination

When two lenses are used in combination, the image from the first lens serves as the object for the second.

The **refractive power** P of a lens is the inverse of its focal length: $P = 1/f$. Refractive power is measured in diopters:

$$1\text{ D} = 1\text{ m}^{-1}$$

Angular and apparent size

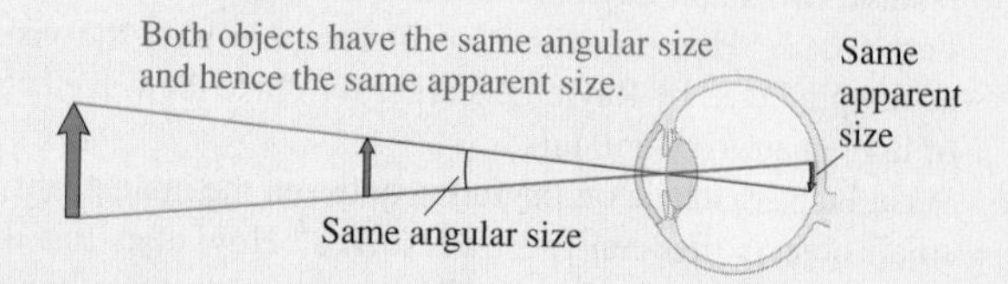

APPLICATIONS

The camera and the eye

Both the camera and the eye work by focusing an image on a light-sensitive surface.

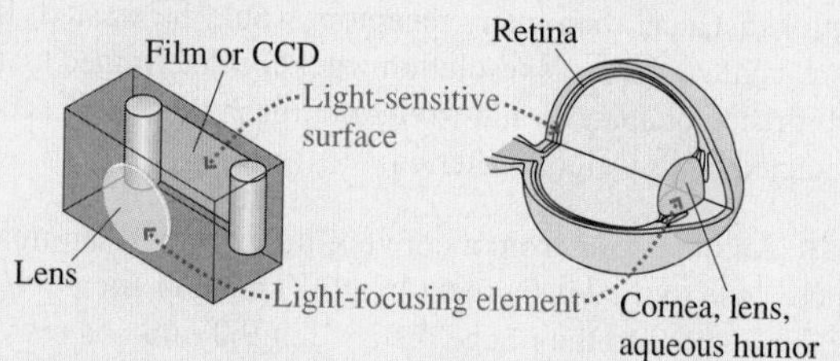

The camera focuses by changing the lens-film distance, while the eye focuses by changing the focal length of its lens.

The magnifier

Without a lens, an object cannot be viewed closer than the eye's near point of ≈25 cm. Its angular size θ_0 is $h/25$ cm.

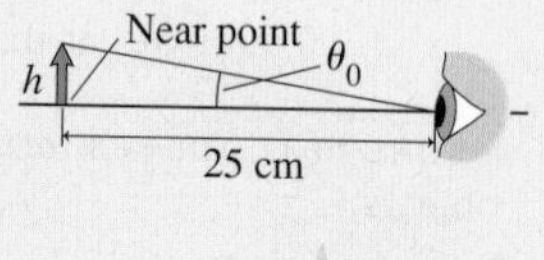

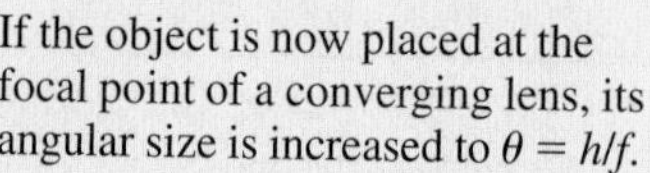

If the object is now placed at the focal point of a converging lens, its angular size is increased to $\theta = h/f$.

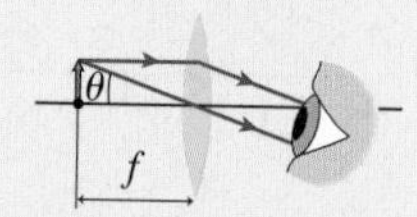

The angular magnification is $M = \theta/\theta_0 = 25\text{ cm}/f$.

The telescope magnifies distant objects. The objective lens creates a real image of the distant object. This real image is then magnified by the eyepiece lens, which acts as a simple magnifier. The angular magnification is $M = -f_o/f_e$.

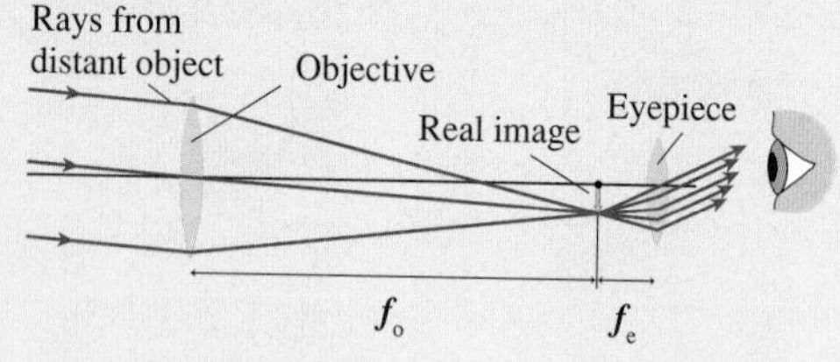

The microscope magnifies a small, nearby object. The objective lens creates a real image of the object. This real image is then further magnified by the eyepiece lens, which acts as a simple magnifier. The angular magnification is

$$M = -\frac{L \times 25\text{ cm}}{f_o f_e}$$

For homework assigned on MasteringPhysics, go to www.masteringphysics.com

Problems labeled INT integrate significant material from earlier chapters; BIO are of biological or medical interest.

Problem difficulty is labeled as | (straightforward) to ||||| (challenging).

QUESTIONS

Conceptual Questions

1. On a sunny summer day, with the sun overhead, you can stand under a tree and look on the ground at the pattern of light that has passed through gaps between the leaves. You may see illuminated circles of varying brightness. Why are there circles, when the gaps between the leaves have irregular shapes?
2. Suppose you have two pinhole cameras. The first has a small round hole in the front of the camera. The second is identical in every regard, except that it has a square hole of the same area as the round hole in the first camera. Would the pictures taken by these two cameras, under the same conditions, be different in any obvious way? Explain.
3. A photographer focuses his camera on his subject. The subject then moves closer to the camera. To refocus, should the lens be moved closer to or farther from the film? Explain.
4. Many cameras have a *zoom lens.* This is a lens whose focal length and distance from the film can be varied. If the camera's exposure is correct when the lens has a focal length of 8.0 mm, will it be overexposed, underexposed, or still correct when the focal length is increased to 16.0 mm (assume the lens diameter remains constant)? Explain.
5. A nature photographer taking a close-up shot of an insect replaces the standard lens on his camera with a lens that has a shorter focal length and is positioned farther from the film. Explain why he does this.
6. The CCD detector in a certain camera has a width of 8 mm. The photographer realizes that with the lens she is currently using, she can't fit the entire landscape she is trying to photograph into her picture. Should she switch to a lens with a longer or shorter focal length? Explain.
7. BIO All humans have what is known as a *blind spot,* where the optic nerve exits the eye and no light-sensitive cells exist. To locate your blind spot, look at the figure of the cross. Close your left eye and place your index finger on the cross. Slowly move your finger to the left while following it with your right eye. At a certain point the cross will disappear. Is your right eye's blind spot on the right or left side of your retina? Explain. ⊗
8. Suppose you wanted special glasses designed to wear underwater, without a face mask. Should the glasses use a converging or diverging lens in order for you to be able to focus under water? Explain.
9. You have lenses with the following focal lengths: $f = 25$ mm, 50 mm, 100 mm, and 200 mm. Which lens or pair of lenses would you use, and in what arrangement, to get the highest-power magnifier, microscope, and telescope? Explain.
10. BIO An 8-year-old child and a 75-year-old man both use the same magnifier to observe a bug. For whom does the magnifier more likely have the higher magnification? Explain.
11. A friend lends you the eyepiece of his microscope to use on your own microscope. He claims that since his eyepiece has the same diameter as yours but twice the focal length, the resolving power of your microscope will be doubled. Is his claim valid? Explain.
12. An astronomer is using a telescope to observe two distant stars. The stars are marginally resolved when she looks at them through a filter that passes green light near 550 nm. Which of the following actions would improve the resolution? Assume that the resolution is not limited by the atmosphere.
    a. Changing the filter to a different wavelength? If so, should she use a shorter or a longer wavelength?
    b. Using a telescope with an objective lens of the same diameter but a different focal length? If so, should she select a shorter or a longer focal length?
    c. Using a telescope with an objective lens of the same focal length but a different diameter? If so, should she select a larger or a smaller diameter?
    d. Using an eyepiece with a different magnification? If so, should she select an eyepiece with more or less magnification?
13. A pair of binoculars has a magnification of 7×. What would be their magnification if you were to look through them the wrong way, that is, through one of their objective lenses instead of the eyepieces?
14. BIO Is the wearer of the glasses in Figure Q19.14 nearsighted or farsighted? How can you tell?
15. A red card is illuminated by red light. What color does it appear to be? What if it's illuminated by blue light?

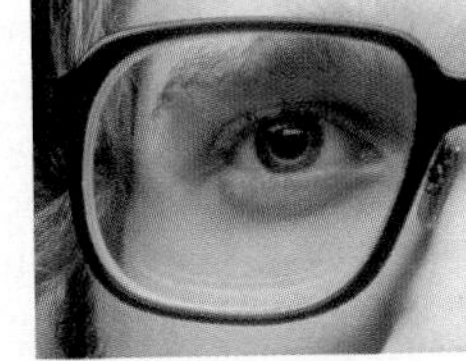
FIGURE Q19.14

Multiple-Choice Questions

16. | A photographer takes a perfectly exposed picture at an *f*-number of *f*/4.0 and a shutter speed of 1/125 s. Now he wishes to use a shutter speed of 1/250 s. What *f*-number should he choose to get a correctly exposed picture?
    A. *f*/2.0 B. *f*/2.8 C. *f*/5.6 D. *f*/8.0
17. | A microscope has a tube length of 20 cm. What combination of objective and eyepiece focal lengths will give an overall magnification of 100?
    A. 1.5 cm, 3 cm B. 2 cm, 2 cm
    C. 1 cm, 5 cm D. 3 cm, 8 cm

18. | | The distance between the objective and eyepiece of a telescope is 55 cm. The focal length of the eyepiece is 5.0 cm. What is the angular magnification of this telescope?
A. −10 B. −11 C. −50 D. −275

19. | A nearsighted person has a near point of 20 cm and a far point of 40 cm. When he is wearing glasses to correct his distant vision, what is his near point?
BIO
A. 10 cm B. 20 cm C. 40 cm D. 1.0 m

20. | A nearsighted person has a near point of 20 cm and a far point of 40 cm. What power lens is necessary to correct this person's vision to allow her to see distant objects?
BIO
A. −5.0 D B. −2.5 D C. +2.5 D D. +5.0 D

21. | A 60-year-old man has a near point of 100 cm, making it impossible to read. What power reading glasses would he need to focus on a newspaper held at a comfortable distance of 40 cm?
BIO
A. −2.5 D B. −1.5 D C. +1.5 D D. +2.5 D

22. | A person looking through a −10 D lens sees an image that appears 8.0 cm from the lens. How far from the lens is the object?
A. 10 cm B. 20 cm C. 25 cm D. 40 cm

23. | In a darkened room, red light shines on a red cup, a white card, and a blue toy. The cup, card, and toy will appear, respectively,
A. Red, red, blue.
B. Red, white, blue.
C. Red, red, black.
D. Red, black, blue.

24. | | An amateur astronomer looks at the moon through a telescope with a 15-cm-diameter objective. What is the minimum separation between two objects on the moon that she can resolve with this telescope? Assume her eye is most sensitive to light with a wavelength of 550 nm.
A. 120 m B. 1.7 km C. 26 km D. 520 km

PROBLEMS

Section 19.1 The Camera

1. | The human eye has a lot in common with a pinhole camera, being essentially a small box with a hole in the front (the pupil) and "film" at the back (the retina). The distance from the pupil to the retina is approximately 24 mm.
BIO
   a. Suppose you look at a 180-cm-tall friend who is standing 7.4 m in front of you. Assuming your eye functions like a pinhole camera, what will be the height, in mm, of your friend's image on your retina?
   b. Suppose your friend's image begins to get bigger. How does your brain interpret this information?

2. | A student has built a 20-cm-long pinhole camera for a science fair project. She wants to photograph the Washington Monument, which is 167 m (550 ft) tall, and to have the image on the film be 5.0 cm high. How far should she stand from the Washington Monument?

3. | | A pinhole camera is made from an 80-cm-long box with a small hole in one end. If the hole is 5.0 m from a 1.8-m-tall person, how tall will the image of the person on the film be?

4. | | A photographer uses his camera, whose lens has a 50 mm focal length, to focus on an object 2.0 m away. He then wants to take a picture of an object that is 40 cm away. How far, and in which direction, must the lens move to focus on this second object?

5. | | | A camera takes a perfectly exposed picture when the lens diaphragm is set to *f*/4 and the shutter speed is 1/250 s. If the diaphragm is changed to *f*/11, what should the new shutter speed be so that the exposure is still correct? (Standard camera shutter speeds include 1/250 s, 1/125 s, 1/60 s, 1/30 s, and 1/15 s.)

6. | | | In Figure P19.6 the camera lens has a 50 mm focal length. How high is the man's well-focused image on the film?

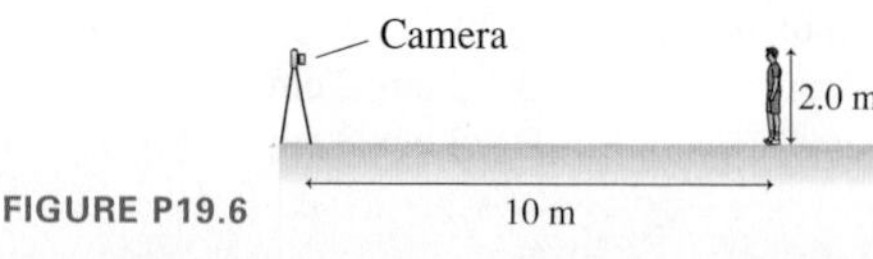

FIGURE P19.6

7. | A telephoto lens with focal length of 135 mm has *f*-numbers ranging from *f*/2.8 to *f*/22. What is the diameter of the lens aperture at these two *f*-numbers?

Section 19.2 The Human Eye

8. | a. Estimate the diameter of your eyeball.
BIO
   b. Bring this page up to the closest distance at which the text is sharp—not the closest at which you can still read it, but the closest at which the letters remain sharp. If you wear glasses or contact lenses, leave them on. This distance is called the *near point* of your (possibly corrected) eye. Record it.
   c. Estimate the effective focal length of your eye. The effective focal length includes the focusing due to the lens, the curvature of the cornea, and any corrections you wear. Ignore the effects of the fluid in your eye.

9. | | A farsighted person has a near point of 50 cm rather than the normal 25 cm. What strength lens, in diopters, should be prescribed to correct this vision problem?
BIO

10. | A nearsighted woman has a far point of 300 cm. What kind of lens, converging or diverging, should be prescribed for her to see distant objects more clearly? What power should the lens have?
BIO

11. | The relaxed human eye is about 2 cm from front to back. If the iris of the human eye can be opened to 7 mm at its widest, what is the *f*-number of the human eye?
BIO

12. | The near point for your myopic uncle is 10 cm. Your own vision is normal; that is, your near point is 25 cm. Suppose you and your uncle hold dimes (which are 1.7 cm in diameter) at your respective near points.
   a. What is the dime's angular size, in radians, according to you?
   b. What is the dime's angular size, in radians, according to your uncle?
   c. Do these calculations suggest any benefit to near-sightedness?

13. || For a patient, a doctor prescribes glasses with a converging
BIO lens having a power of 4.0 D.
   a. Is the patient nearsighted or farsighted?
   b. If the patient is nearsighted, what is the location of her eye's far point? If she is farsighted, what is the location of her near point?

14. ||| Rank the following people from the most nearsighted to the
BIO most farsighted, indicating any ties:
   A. Bernie has a prescription of +2.0 D.
   B. Carol needs diverging lenses with a focal length of −0.35 m.
   C. Maria Elena wears converging lenses with a focal length of 0.50 m.
   D. Janet has a prescription of +2.5 D.
   E. Warren's prescription is −3.2 D.

Section 19.3 The Magnifier

15. | The diameter of a penny is 19 mm. How far from your eye must it be held so that it has the same apparent size as the moon? (Use the astronomical data inside the back cover.)

16. | What is the angular size of the moon? (Use the astronomical data inside the back cover.)

17. | A magnifier has a magnification of 5×. How far from the lens should an object be placed so that its (virtual) image is at the near-point distance of 25 cm?

18. ||| A farsighted man has a near point of 40 cm. What power lens should he use as a magnifier to see clearly at a distance of 10 cm without wearing his glasses?

Section 19.4 The Microscope

19. || An inexpensive microscope has a tube length of 12.0 cm, and its objective lens is labeled with a magnification of 10×.
   a. Calculate the focal length of the objective lens.
   b. What focal length eyepiece lens should the microscope have to give an overall magnification of 150×?

20. ||| A standard biological microscope is required to have a magnification of 200×.
   a. When paired with a 10× eyepiece, what power objective is needed to get this magnification?
   b. What is the focal length of the objective?

21. |||| A forensic scientist is using a standard biological microscope with a 15× objective and a 5× eyepiece to examine a hair from a crime scene. How far from the objective is the hair?

22. ||| A microscope with an 8.0-mm-focal-length objective has a tube length of 16.0 cm. For the microscope to be in focus, how far should the objective lens be from the specimen?

23. ||| The distance between the objective and eyepiece lenses in a microscope is 20 cm. The objective lens has a focal length of 5.0 mm. What eyepiece focal length will give the microscope an overall angular magnification of 350?

Section 19.5 The Telescope

24. || For the combination of two identical lenses shown in Figure P19.24, find the position, size, and orientation of the final image of the 2.0-cm-tall object.

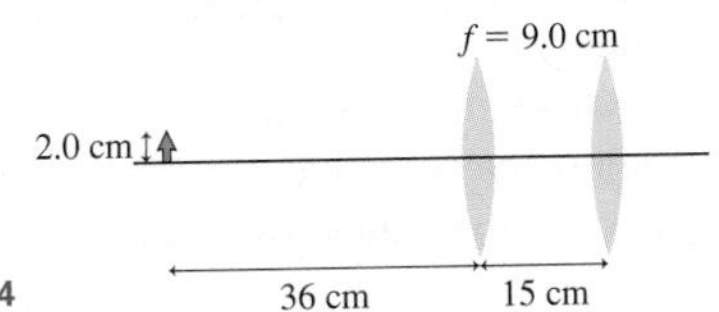

FIGURE P19.24

25. || For the combination of two lenses shown in Figure P19.25, find the position, size, and orientation of the final image of the 1.0-cm-tall object.

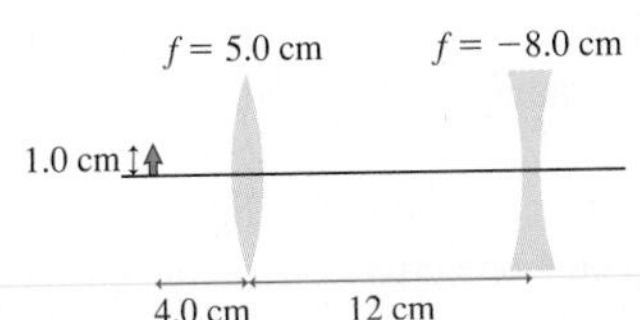

FIGURE P19.25

26. || A researcher is trying to shoot a tranquilizer dart at a 2.0-m-tall rhino that is 150 m away. Its angular size as seen through the rifle telescope is 9.1°. What is the magnification of the telescope?

27. || The objective lens of the refracting telescope at the Lick Observatory in California has a focal length of 57 ft.
   a. What is the refractive power of this lens?
   b. What focal length (mm) eyepiece would give a magnification of 1000× for this telescope?

28. ||| You use your 8× binoculars to focus on a yellow-rumped warbler (length 14 cm) in a tree 18 m away from you. What angle (in degrees) does the image of the warbler subtend on your retina?

29. || Your telescope has a 700-mm-focal-length objective and a 26-mm-focal-length eyepiece. One evening you decide to look at the full moon, which has an angular size of 0.52° when viewed with the naked eye.
   a. What angle (in degrees) does the image of the moon subtend when you look at it through your telescope?
   b. Suppose you decide to take a photograph of the moon using your telescope. You position film so that it captures the image produced by the objective lens. What is the diameter of that image?

Section 19.6 Color and Dispersion

30. |||| A narrow beam of light with wavelengths from 450 nm to 700 nm is incident perpendicular to one face of a 40.00° prism made of crown glass, for which the index of refraction ranges from $n = 1.533$ to $n = 1.517$ for those wavelengths. What is the angular spread of the beam after passing through the prism?

31. ||| A ray of white light strikes the surface of a 4.0-cm-thick slab of flint glass as shown in Figure P19.31. As the ray enters the glass, it is dispersed into its constituent colors. Estimate how far apart the rays of deepest red and deepest violet light are as they exit the bottom surface. Which exiting ray is closer to point P?

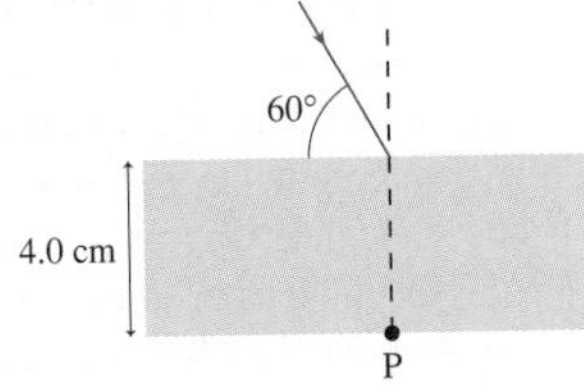

FIGURE P19.31

32. |||| A ray of red light, for which $n = 1.54$, and a ray of violet light, for which $n = 1.59$, travel through a piece of glass. They meet right at the boundary between the glass and the air, and emerge into the air as one ray with an angle of refraction of 22.5°. What is the angle between the two rays in the glass?

Section 19.7 Resolution of Optical Instruments

33. ||| Two lightbulbs are 1.0 m apart. From what distance can these light bulbs be marginally resolved by a small telescope with a 4.0-cm-diameter objective lens? Assume that the lens is limited only by diffraction and $\lambda = 600$ nm.
34. | A 1.0-cm-diameter microscope objective has a focal length of 2.8 mm. It is used in visible light with a wavelength of 550 nm.
    a. What is the objective's resolving power if used in air?
    b. What is the resolving power of the objective if it is used in an oil-immersion microscope with $n_{oil} = 1.45$?
35. || A microscope with an objective of focal length 1.6 mm is used to inspect the tiny features of a computer chip. It is desired to resolve two objects only 400 nm apart. What diameter objective is needed if the microscope is used in air with light of wavelength 550 nm?

General Problems

36. | Suppose you point a pinhole camera at a 15-m-tall tree that is 75 m away.
    a. If the film is 22 cm behind the pinhole, what will be the size of the tree's image on the film?
    b. If you would like the image to be larger, should you get closer to the tree or farther from the tree? Explain.
    c. If you had time, you could make the image larger by rebuilding the camera, changing the length or the pinhole size. What one change would give a larger image?
37. |||| "Jason uses a lens with focal length of 10.0 cm as a magnifier by holding it right up to his eye. He is observing an object that is 8.0 cm from the lens. What is the angular magnification of the lens used this way if Jason's near-point distance is 25 cm?
38. | A magnifier is labeled "5×." What would its magnification be if used by a person with a near-point distance of 50 cm?
39. || A 20× microscope objective is designed for use in a microscope with a 16 cm tube length. The objective is marked NA = 0.40. What is the diameter of the objective lens?
40. |||| Two converging lenses with focal lengths of 40 cm and 20 cm are 10 cm apart. A 2.0-cm-tall object is 15 cm in front of the 40-cm-focal-length lens. INT
    a. Use ray tracing to find the position and height of the image. To do this accurately use a ruler or paper with a grid. Determine the image distance and image height by making measurements on your diagram.
    b. Calculate the image height and image position relative to the second lens. Compare with your ray-tracing answers in part a.
41. ||| A converging lens with a focal length of 40 cm and a diverging lens with a focal length of −40 cm are 160 cm apart. A 2.0-cm-tall object is 60 cm in front of the converging lens. INT
    a. Use ray tracing to find the position and height of the image. To do this, accurately use a ruler or paper with a grid. Determine the image distance and image height by making measurements on your diagram.
    b. Calculate the image height and image position relative to the second lens. Compare with your ray-tracing answers in part a.
42. || A lens with a focal length of 25 cm is placed 40 cm in front of a lens with a focal length of 5.0 cm. How far from the second lens is the final image of an object infinitely far from the first lens? Is this image in front of or behind the second lens?
43. |||| A microscope with a 5× objective lens images a 1.0-mm-diameter specimen. What is the diameter of the real image of this specimen formed by the objective lens?
44. || Your task in physics lab is to make a microscope from two lenses. One lens has a focal length of 10 cm, the other a focal length of 3.0 cm. You plan to use the more powerful lens as the objective, and you want its image to be 16 cm from the lens, as in a standard biological microscope.
    a. How far should the objective lens be from the object to produce a real image 16 cm from the objective?
    b. What will be the magnification of your microscope?
45. || A 20× objective and 10× eyepiece give an angular magnification of 200× when used in a microscope with a 160 mm tube length. What magnification would this objective and eyepiece give if used in a microscope with a 200 mm tube length?
46. || The objective lens and the eyepiece lens of a telescope are 1.0 m apart. The telescope has an angular magnification of 50. Find the focal lengths of the eyepiece and the objective.
47. ||| Your telescope has an objective lens with a focal length of 1.0 m. You point the telescope at the moon, only to realize that the eyepiece is missing. Even so, you can still see the real image of the moon formed by the objective lens if you place your eye a little past the image so as to view the rays diverging from the image plane, just as rays would diverge from an object at that location. What is the angular magnification of the moon if you view its real image from 25 cm away, your near-point distance?
48. ||| The 200-inch-diameter objective mirror of the reflecting telescope at the Mt. Palomar Observatory has a focal length of 17 m.
    a. The *f*-number of a mirror is defined exactly the same as the *f*-number of a lens. What is the *f*-number of this mirror?
    b. The *f*-number of the 200-inch telescope is well within the range of *f*-numbers of a cheap camera. So why not just use the camera to take pictures of distant galaxies, instead of constructing this very expensive telescope?
49. | Marooned on a desert island and with a lot of time on your hands, you decide to disassemble your glasses to make a crude telescope with which you can scan the horizon for rescuers. Luckily you're farsighted, and as for most people your two eyes have somewhat different lens prescriptions. Your left eye uses a lens of power +4.5 D and your right eye's lens is +3.0 D.
    a. Which lens should you use for the objective and which for the eyepiece? Explain.
    b. What will be the magnification of your telescope?
    c. Approximately how far apart should the two lenses be when you focus on distant objects?
50. ||| A spy satellite uses a telescope with a 2.0-m-diameter mirror. It orbits the earth at a height of 220 km. What minimum spacing must there be between two objects on the earth's surface if they are to be resolved as distinct objects by this telescope? Assume the telescope's resolution is limited only by diffraction and that it is recording light with a wavelength of 500 nm.
51. ||||| Two stars have an angular separation of 3.3×10^{-6} rad. What diameter telescope objective is necessary to just resolve these two stars, using light with a wavelength of 650 nm?

52. ||| The planet Neptune is 4.5×10^{12} m from the earth. Its diameter is 4.9×10^7 m. What diameter telescope objective would be necessary to just barely see Neptune as a disk rather than as a point of light? Assume a wavelength of 550 nm.

53. |||| What is the angular resolution of the Hubble Space Telescope's 2.4-m-diameter mirror when viewing light with a wavelength of 550 nm? The resolution of a reflecting telescope is calculated exactly the same as for a refracting telescope.

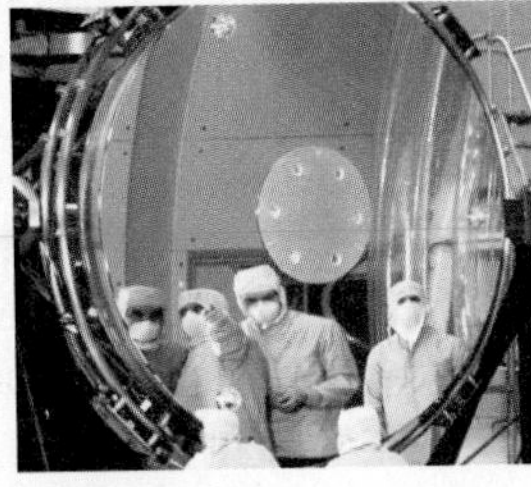

54. ||| The Hubble Space Telescope has a mirror diameter of 2.4 m. Suppose the telescope is used to photograph stars near the center of our galaxy, 30,000 light years away, using red light with a wavelength of 650 nm.
   a. What is the distance (in km) between two stars that are marginally resolved? The resolution of a reflecting telescope is calculated exactly the same as for a refracting telescope.
   b. For comparison, what is this distance as a multiple of the distance of Jupiter from the sun?

55. ||| Once dark adapted, the pupil of your eye is approximately BIO 7 mm in diameter. The headlights of an oncoming car are 120 cm apart. If the lens of your eye is limited only by diffraction, at what distance are the two headlights marginally resolved? Assume the light's wavelength in air is 600 nm and the index of refraction inside the eye is 1.33. (Your eye is not really good enough to resolve headlights at this distance, due both to aberrations in the lens and to the size of the receptors in your retina, but it comes reasonably close.)

56. |||| The normal human eye has maximum visual acuity with a BIO pupil size of about 3 mm. For larger pupils, acuity decreases due to increasing aberrations; for smaller pupils, acuity decreases due to increasing effects of diffraction. If your pupil diameter is 2.0 mm, as it would be in fairly bright light, what is the smallest diameter circle that you can barely see as a circle, rather than just a dot, if the circle is at your near point, 25 cm from your eye? Assume the light's wavelength in air is 600 nm and the index of refraction inside the eye is 1.33.

57. ||| Microtubules are filamentous structures in cells that maintain BIO cell shape and facilitate the movement of molecules within the cell. They are long, hollow cylinders with a diameter of about 25 nm. It is possible to incorporate fluorescent molecules into microtubules; when illuminated by an ultraviolet light, the fluorescent molecules emit visible light that can be imaged by the optical system of a microscope. If the emitted light has a wavelength of 500 nm and the NA of the microscope objective is 1.4, can a biologist looking through the microscope tell whether she is looking at a single microtubule or at two microtubules lying side by side?

Passage Problems

Surgical Vision Correction BIO

Light that enters your eyes is focused to form an image on your retina. The optics of your visual system have a total power of about +60 D—about +20 D from the lens in your eye and +40 D from the curved shape of your cornea. Surgical procedures to correct vision generally do not work on the lens; they work to reshape the cornea. In the most common procedure, a laser is used to remove tissue from the center of the cornea, reducing its curvature. This change in shape can correct certain kinds of vision problems.

58. | Flattening the cornea would be a good solution for someone who was
   A. Nearsighted. B. Farsighted.
   C. Either nearsighted or farsighted.

59. | Suppose a woman has a far point of 50 cm. How much should the focusing power of her cornea be changed to correct her vision?
   A. −2.0 D B. −1.0 D C. +1.0 D D. +2.0 D

60. | A *cataract* is a clouding or opacity that develops in the eye's lens, often in older people. In extreme cases, the lens of the eye may need to be removed. This would have the effect of leaving a person
   A. Nearsighted. B. Farsighted.
   C. Neither nearsighted nor farsighted.

61. | The length of your eye decreases slightly as you age, making the lens a bit closer to the retina. Suppose a man had his vision surgically corrected at age 30. At age 70, once his eyes had decreased slightly in length, he would be
   A. Nearsighted. B. Farsighted.
   C. Neither nearsighted nor farsighted.

STOP TO THINK ANSWERS

Stop to Think 19.1: B. The diameter d of the lens is constant, so increasing the focal length increases the f-number f/d of the lens. A lens with a higher f-number needs *more* light for a correct exposure, requiring a *slower* shutter speed.

Stop to Think 19.2: D. Because Maria can focus on an object 0.5 m away, but not on one 10 m away, her far point must lie between these two distances. Following Example 19.4, we see that the prescription for her lens must then lie between $1/(-10\text{ m}) = -0.1$ D and $1/(-0.5\text{ m}) = -2$ D. Only the −1.5 D prescription falls in this range.

Stop to Think 19.3: A. Ray tracing shows why.

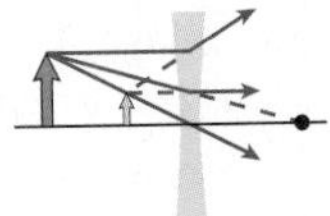

Stop to Think 19.4: B. The total magnification is the product of the objective magnification m_o and the eyepiece angular magnification M_e. If m_o is halved, from 20× to 10×, M_e must be doubled. Because M_e is inversely proportional to the eyepiece focal length, the focal length of the eyepiece must be halved.

Stop to Think 19.5: D. A green filter lets through only green light, so it blocks the red light from the apple. No light from the apple can pass through the filter, so it appears black.

Stop to Think 19.6: $RP_1 > RP_4 > RP_2 = RP_3$. The resolving power is $RP = 0.61\lambda/\sin\phi_0$ for objectives used in air ($n = 1$), so the resolving power is higher (worse resolution) when the angle ϕ_0 is smaller. From Figure 19.29 you can see that ϕ_0 is smaller when the ratio D/f is smaller. These ratios are $(D/f)_1 = 1/5$, $(D/f)_2 = 2/5$, $(D/f)_3 = 2/5$, and $(D/f)_4 = 1/3$.

PART V SUMMARY

Optics

Light is an elusive entity. It is everywhere around us, but exactly what *is* it? One of the more curious aspects of light is that its basic properties depend on the circumstances under which it's studied. Thus it's difficult to develop a single theory of light that applies under all circumstances. Because of this, we have developed two *models* of light in Part V, the wave model and the ray model. We found that each model has its particular realm of applicability.

Many experiments show that light has distinct wave-like properties. Light waves exhibit interference and diffraction, just as water and sound waves do. However, we're usually not aware of the wave aspects of light because the wavelengths of visible light are so short. Wave phenomena become apparent only when light interacts with objects or holes whose size is less than about 0.1 mm.

We can usually ignore the wave nature of light when we consider the propagation of light on larger length scales. In this case, we model light as traveling outward in straight lines, or *rays,* from its source. Light rays change direction at an interface between two media with different indices of refraction (different light speeds). At this interface the rays both reflect, heading back into their original medium, and refract, moving into the new medium but in a new direction. These processes are governed by the laws of reflection and refraction.

Despite light's subtle nature, the practical applications of optics are crucial to many of today's technologies. Cameras, telescopes, and microscopes all employ basic ideas of image formation with lenses and mirrors. We found that the ultimate resolution of an optical instrument is set by the wave nature of light, bringing our study of optics full circle.

KNOWLEDGE STRUCTURE V Optics

BASIC GOALS What are the consequences of the wave nature of light?
In the ray model, how do light rays refract and reflect to form images?

GENERAL PRINCIPLES Light is understood using two models, the **wave model,** in which light exhibits wave properties such as interference and diffraction, and the **ray model,** in which light travels in straight lines until it reflects or refracts.

Wave model

- Light spreads out when passing through a narrow opening. This is **diffraction.**

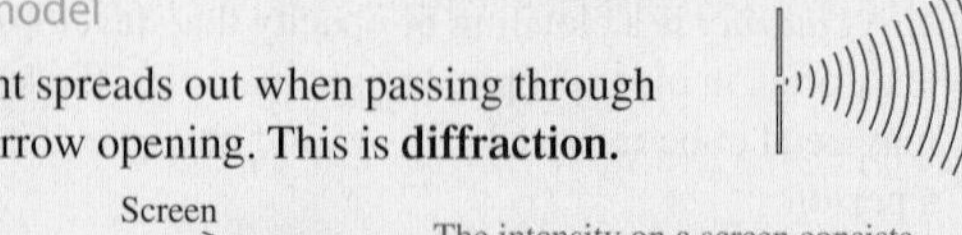

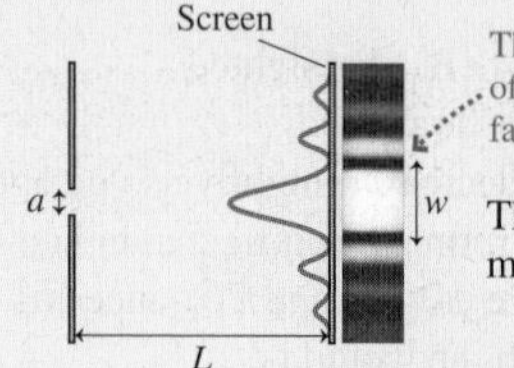

The intensity on a screen consists of a bright central maximum and fainter secondary maxima.

The width of the central maximum is

$$w = \frac{2\lambda L}{a}$$

- Light waves from multiple slits in a screen **interfere** where they overlap. The light intensity is large where the interfering waves are in phase, and small where they are out of phase.

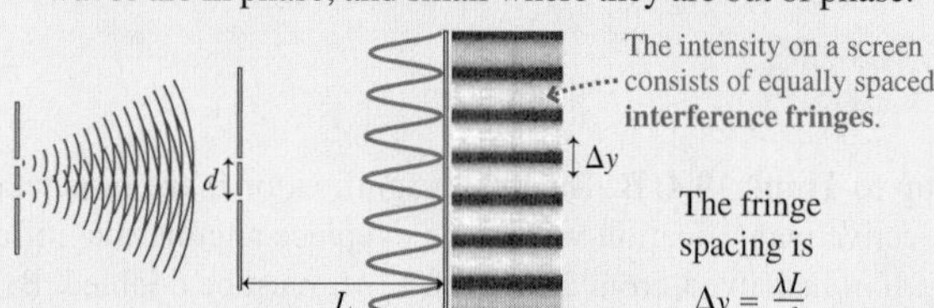

The intensity on a screen consists of equally spaced interference fringes.

The fringe spacing is

$$\Delta y = \frac{\lambda L}{d}$$

- Light waves reflected from the two surfaces of a thin transparent film also interfere.

The resolution of optical instruments

Diffraction limits how close together two point objects can be and still be resolved.

For a *telescope,* the minimum resolvable angular separation between two objects is

$$\theta_1 = \frac{1.22\lambda}{D}$$

For a *microscope,* the minimum resolvable distance between two objects is

$$d_{min} = \frac{0.61\lambda}{NA}$$

where the *numerical aperture* NA is a characteristic of the microscope objective.

Ray model

- Light travels out from its source in straight lines, called **rays.**
- Rays reflect off a surface between two media, obeying the **law of reflection,** $\theta_i = \theta_r$.

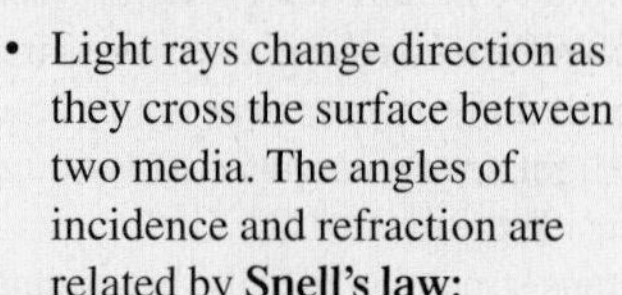

- Light rays change direction as they cross the surface between two media. The angles of incidence and refraction are related by **Snell's law:**

$$n_1 \sin\theta_1 = n_2 \sin\theta_2$$

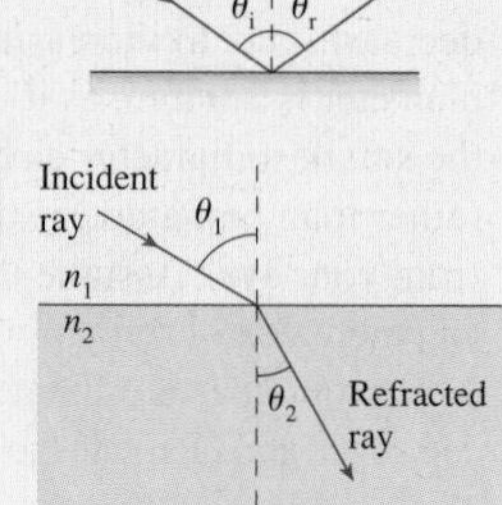

where n is the **index of refraction.** The speed of light in a transparent material is $v = c/n$.

Image formation by lenses and mirrors

A lens or mirror has a characteristic **focal length** f. Rays parallel to the optical axis come to focus a distance f from the lens or mirror.

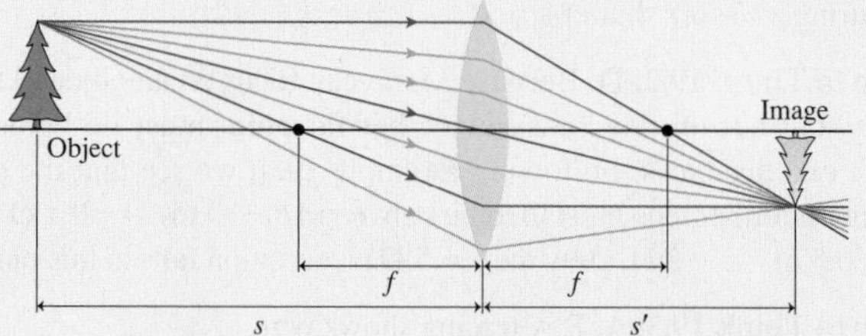

The **object distance** s, the **image distance** s', and the focal length are related by the *thin-lens equation,* which also works for mirrors:

$$\frac{1}{s} + \frac{1}{s'} = \frac{1}{f}$$

Scanning Confocal Microscopy

Although modern microscopes are marvels of optical engineering, their basic design is not too different from the 1665 compound microscope of Robert Hooke. Recently, advances in optics, lasers, and computer technology have made practical a new kind of optical microscope, the *scanning confocal microscope.* This microscope is capable of taking images of breathtaking clarity.

The figure shows the microscope's basic principle of operation. The left part of the figure shows how the translucent specimen is illuminated by light from a laser. The laser beam is converted to a diverging bundle of rays by suitable optics, reflected off a mirror, then directed through a microscope objective lens to a focus within the sample. The microscope objective focuses the laser beam to a very small ($\approx 0.5\ \mu m$) spot. Note that light from the laser passes through other regions of the specimen but, because the rays are not focused in those regions, they are not as intensely illuminated as is the point at the focus. This is the first important aspect of the design: Very intensely illuminate one very small volume of the sample while leaving other regions only weakly illuminated.

As shown in the right half of the figure, light is reflected from all illuminated points in the sample and passes back through the objective lens. The mirror that had reflected the laser light downward is actually a *partially transparent mirror* that reflects 50% of the light and transmits 50%. Thus half of the light reflected upward from the sample passes through the mirror and is focused on a screen containing a small hole. Because of the hole, only light rays that emanate from the brightly illuminated volume in the sample can completely pass through the hole and reach the light detector behind it. Rays from other points in the sample either miss the hole completely or are out of focus when they reach the screen, so that only a small fraction of them pass through the hole. This second key design aspect limits the detected light to only those rays that are emitted from the point in the sample at which the laser light was originally focused.

So we see that (a) the point in the sample that is at the focus of the objective is much more intensely illuminated than any other point, so it reflects more rays than any other point, and (b) the hole serves to further limit the detected rays to only those that emanate from the focus. Taken together, these design aspects ensure the detected light comes from a very small, very well-defined volume in the sample.

The microscope as shown would only be useful for examining one small point in the sample. To make an actual *image,* the objective is *scanned* across the sample while the intensity is recorded by a computer. This procedure builds up an image of the sample one *scan line* at a time. The final result is a picture of the sample in the very narrow plane in which the laser beam is focused. Different planes within the sample can be imaged by moving the objective up or down before scanning. It is actually possible to make three-dimensional images of a specimen in this way.

The improvement in contrast and resolution over conventional microscopy can be striking. The images show a section of a human medulla taken using conventional and confocal microscopy. Because light reflected from all parts of the specimen reach the camera in a conventional microscope, that image appears blurred and has low contrast. The confocal microscope image represents a single plane or slice of the sample, and many details become apparent that are invisible in the conventional image.

Light detector
The screen blocks out light reflected from points other than the focus of the laser beam.
Hole in screen
Partially transparent mirror reflects 50%, transmits 50%.
Laser source
Objective lens
The intensity at the focus of the laser beam is very high.
$\approx 0.5\ \mu m$
Specimen
The intensity at other points is much lower.
Bright light reflected from point at focus
Faint light reflected from some other point

A confocal microscope.

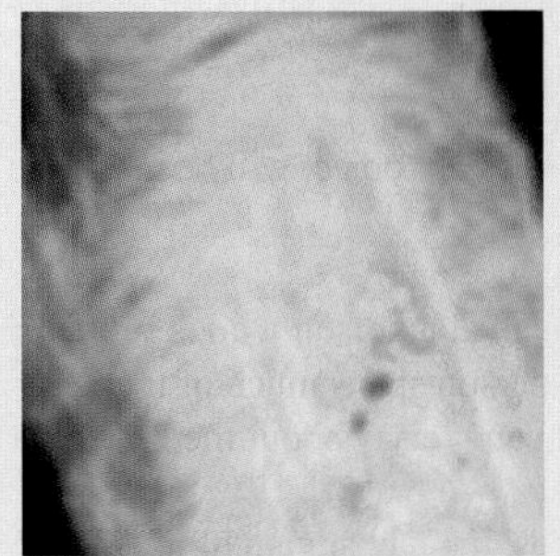

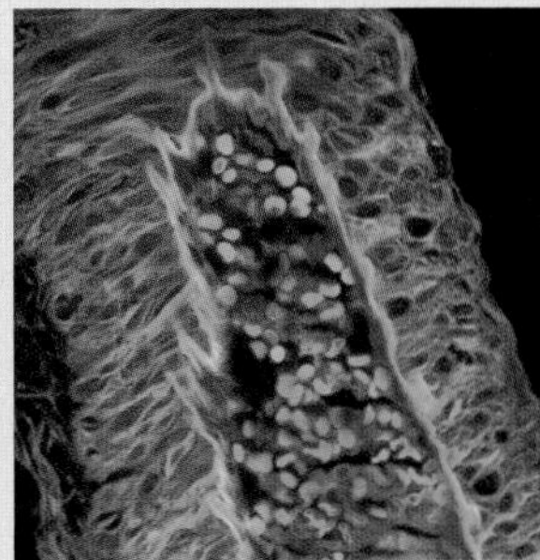

A thick section of fluorescently stained human medulla imaged using standard optical microscopy (left) and scanning confocal microscopy (right).

PART V PROBLEMS

The following questions are related to the passage "Scanning Confocal Microscopy" on the previous page.

1. A laser beam consists of parallel rays of light. To convert this light to the diverging rays required for a scanning confocal microscope requires
   A. A converging lens. B. A diverging lens.
   C. Either a converging or a diverging lens.
2. If, because of a poor-quality objective, the light from the laser illuminating the sample in a scanning confocal microscope is focused to a larger spot,
   A. The image would be dimmer because the light illuminating the point imaged would be dimmer.
   B. The image would be blurry because light from more than one point would reach the detector.
   C. The image would be dimmer and blurry—both of the above problems would exist.
3. The resolution of a scanning confocal microscope is limited by diffraction, just as for a regular microscope. In principle, switching to a laser with a shorter wavelength would provide
   A. Greater resolution.
   B. Lesser resolution.
   C. The same resolution.
4. In the optical system shown in the passage, the distance from the source of the diverging light rays to the sample is ________ the distance from the sample to the screen.
   A. greater than
   B. the same as
   C. less than

The following passages and associated questions are based on the material of Part V.

Horse Sense BIO

The ciliary muscles in a horse's eye can make only small changes to the shape of the lens, so a horse can't change the shape of the lens to focus on objects at different distances as humans do. Instead, a horse relies on the fact that its eyes aren't spherical. As Figure V.1 shows, different points at the back of the eye are at somewhat different distances from the front of the eye. We say that the eye has a "ramped retina"; images that form on the top of the retina are farther from the cornea and lens than those that form at lower positions. The horse uses this ramped retina to focus on objects at different distances, tipping its head so that light from an object forms an image at a vertical location on the retina that is at the correct distance for sharp focus.

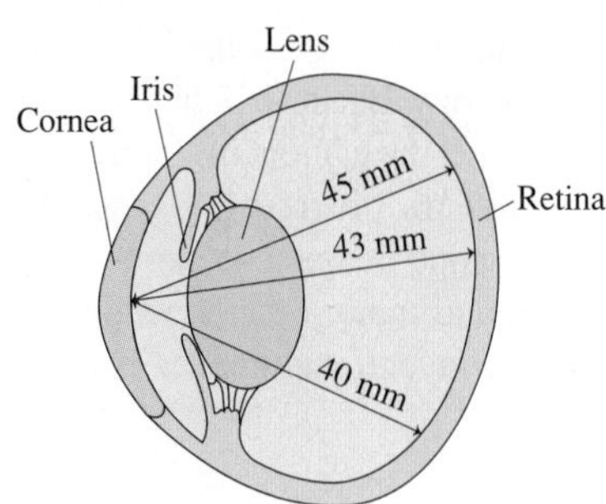

FIGURE V.1

5. In a horse's eye, the image of a close object will be in focus
   A. At the top of the retina.
   B. At the bottom of the retina.
6. In a horse's eye, the image of a distant object will be in focus
   A. At the top of the retina.
   B. At the bottom of the retina.
7. A horse is looking straight ahead at a person who is standing quite close. The image of the person spans much of the vertical extent of the retina. What can we say about the image on the retina?
   A. The person's head is in focus; the feet are out of focus.
   B. The person's feet are in focus; the head is out of focus.
   C. The person's head and feet are both in focus.
   D. The person's head and feet are both out of focus.
8. Certain medical conditions can change the shape of a horse's eyeball; these changes can affect vision. If the lens and cornea are not changed but all of the distances in the Figure V.1 are increased slightly, then the horse will be
   A. Nearsighted.
   B. Farsighted.
   C. Unable to focus clearly at any distance.

The Fire in the Eye BIO

You have certainly seen the reflected light from the eyes of a cat or a dog at night. This "eye shine" is the reflection of light from a layer at the back of the eye called the *tapetum lucidum* (Latin for "bright carpet"). The tapetum is a common structure in the eyes of animals that must see in low light. Light that passes through the retina is reflected by the tapetum back through the cells of the retina, giving them a second chance to detect the light.

Sharks and related fish have a very well-developed tapetum. Figure V.2a shows a camera flash reflected from a shark's eye back toward the camera. This reflected light is much brighter than the diffuse reflection from the body of the shark. How is this bright reflection created?

Figure V.2b shows a typical tapetum structure for a fish. (The tapetum in land animals such as cats, dogs, and deer uses similar principles but has a different structure.) The reflection comes from the interfaces between two layers of nearly transparent cells (whose index of refraction is essentially that of water) and a stack of guanine crystals sandwiched between. Light is reflected from the interface at both sides of the stack of crystals. For certain wavelengths, constructive interference leads to an especially strong reflection.

Bright light from a distant source is focused by the lens of a shark's eye to a point on the retina, as shown in Figure V.2c. The tapetum reflects these rays back through the lens, where refraction bends them into parallel rays traveling back toward the source of the light. Because the reflected light from the tapetum is directional, it is much brighter than the diffuse reflection from the shark's body. But the bright reflection is seen by an observer—or a camera—only at or near the source of the flash that produced the reflection.

(a) "Eye shine" in a flash photo of a shark

(b) Structure of the tapetum

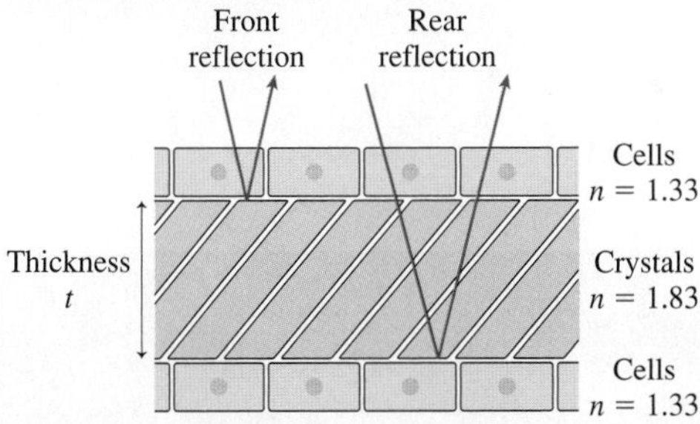

(c) Reflection of light rays

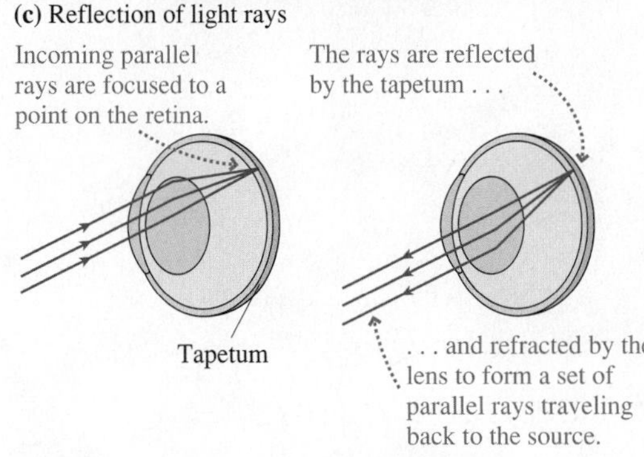

FIGURE V.2

9. Light of wavelength 600 nm in air passes into the layer of guanine crystals. What is the wavelength of the light in this layer?
   A. 1100 nm B. 600 nm
   C. 450 nm D. 330 nm
10. Figure V.2b shows rays that reflect from the two interfaces in the tapetum. Given the indices of refraction of the cells and the crystals, there will be a phase shift on reflection for
    A. The front reflection and the rear reflection.
    B. The front reflection only.
    C. The rear reflection only.
    D. Neither the front nor the rear reflection.
11. What is the (approximate) smallest thickness of the crystal layer that would lead to constructive interference between the front reflection and the rear reflection for light of wavelength 600 nm?
    A. 80 nm B. 160 nm
    C. 240 nm D. 320 nm
12. In human vision, the curvature of the cornea provides much of the power of the visual system. This is not the case in fish; in Figure V.2c, the light rays are bent by the lens but are not bent when they enter the cornea. This is because
    A. Fish eyes work in water, and the index of refraction of the fluids in the eye is similar to that of water.
    B. Fish eyes have a much smaller curvature of the cornea.
    C. Most fish have eyes that are more sensitive to light than the eyes of typical land animals.
    D. The reflection of the tapetum interferes with the refraction of the cornea.
13. Flash photographs of cats will generally show the tapetum reflection unless you are careful to avoid it. If you want to take a flash photograph of your cat while minimizing the "eye shine," which of the following strategies will *not* work?
    A. Take the photographs in dim light so that the irises of your cat's eyes are wide open.
    B. Use a flash on a stand at some distance from the camera.
    C. Use a diffuser so that the light from the flash is spread over a wide area.
    D. Use multiple flashes at different positions around the room.
14. Figure V.2c shows the lens of the eye bringing parallel rays together right at the retina. The retina is located
    A. In front of the focal point of the lens.
    B. At the focal point of the lens.
    C. Behind the focal point of the lens.

Additional Integrated Problems

15. The pupil of your eye is smaller in bright light than in dim BIO light. Explain how this makes images seen in bright light appear sharper than images seen in dim light.
16. People with good vision can make out an 8.8-mm-tall letter on BIO an eye chart at a distance of 6.1 m. Approximately how large is the image of the letter on the retina? Assume that the distance from the lens to the retina is 24 mm.
17. A photographer uses a lens with $f = 50$ mm to form an image of a distant object on the CCD detector in a digital camera. The image is 1.2 mm high, and the intensity of light on the detector is 2.5 W/m^2. She then switches to a lens with $f = 300$ mm that is the same diameter as the first lens. What are the height of the image and the intensity now?
18. Sound and other waves undergo diffraction just as light does. Suppose a loudspeaker in a 20°C room is emitting a steady tone of 1200 Hz. A 1.0-m-wide doorway in front of the speaker diffracts the sound wave. A person on the other side walks parallel to the wall in which the door is set, staying 12 m from the wall. When he is directly in front of the doorway, he can hear the sound clearly and loudly. As he continues walking, the sound intensity decreases. How far must he walk from the point where he was directly in front of the door until he reaches the first quiet spot?

PART

VI

Electricity and Magnetism

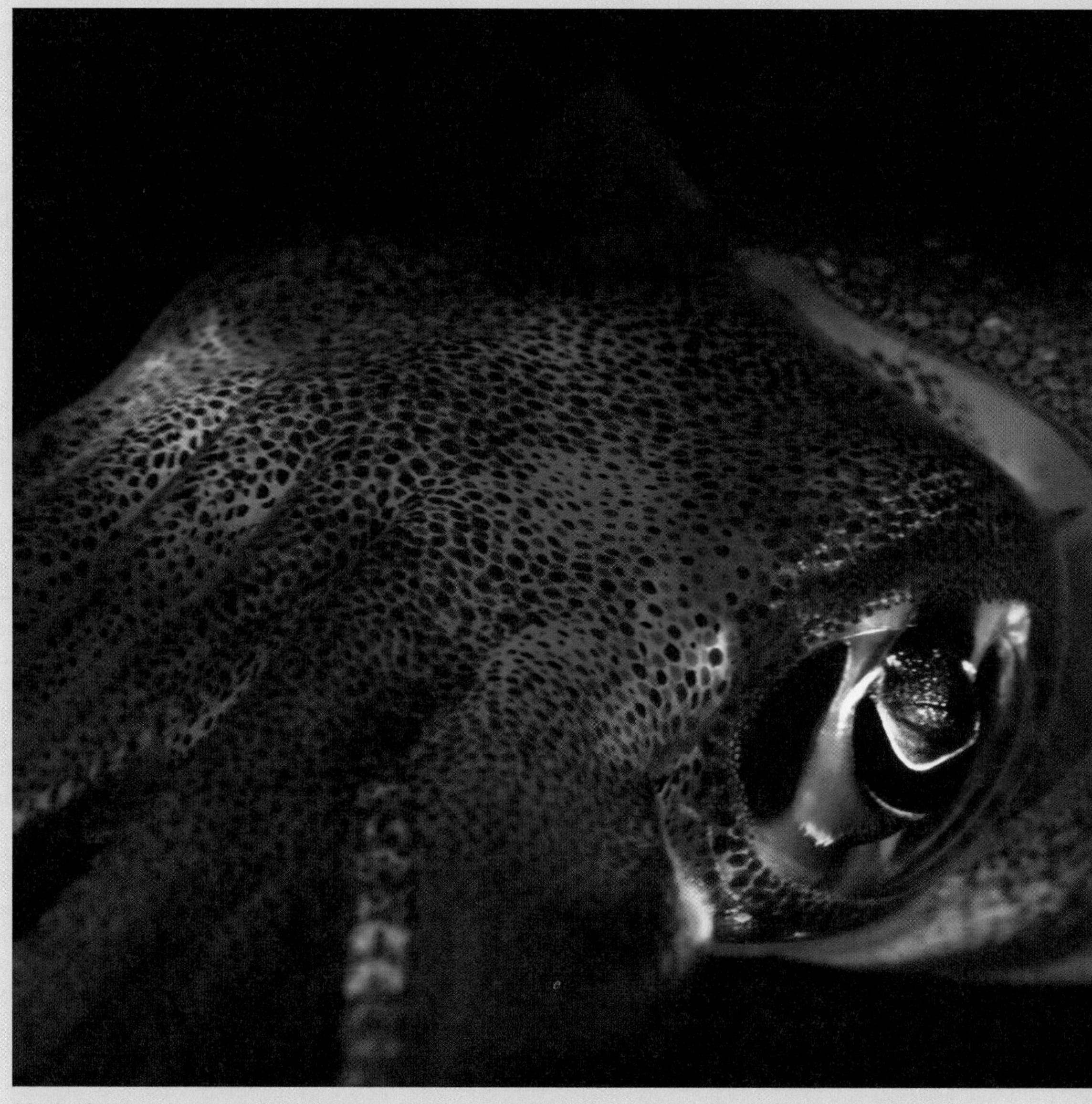

Much of what is known about your nervous system comes from the study of an animal that seems quite different from humans—the squid. Nerve fibers conduct electrical signals along their length, allowing the brain to direct the actions of the body. How is an electrical signal generated and transmitted in the nervous system of a human or a squid?

Charges, Currents, and Fields

The early Greeks discovered that a piece of amber that has been rubbed briskly can attract feathers or small pieces of straw. They also found that certain stones from the region they called *Magnesia* can pick up pieces of iron. These first experiences with the forces of electricity and magnetism began a chain of investigations that has led to today's high-speed computers, lasers, fiber-optic communications, and magnetic resonance imaging, as well as mundane modern-day miracles such as the lightbulb.

The development of a successful electromagnetic theory, which occupied the leading physicists of Europe for most of the nineteenth century, led to sweeping revolutions in both science and technology. The complete formulation of the theory of the electromagnetic field has been called by no less than Einstein "the most important event in physics since Newton's time."

The basic phenomena of electricity and magnetism are not as familiar to most people as those of mechanics. We will deal with this lack of experience by placing a large emphasis on these basic phenomena. We will begin where the Greeks did, by looking at the forces between objects that have been briskly rubbed, exploring the concept of *electric charge.* It is easy to make systematic observations of how charges behave, and we will be led to consider the forces between charges and how charges behave in different materials. *Electric current,* whether it be for lighting a lightbulb or changing the state of a computer memory element, is simply a controlled motion of charges through conducting materials. One of our goals will be to understand how charges move through electric circuits.

When we turn to magnetic behavior, we will again start where the Greeks did, noting how magnets stick to some metals. Magnets also affect compass needles. And, as we will see, an electric current can affect a compass needle in exactly the same way as a magnet. This observation shows the close connection between electricity and magnetism, which leads us to the phenomenon of *electromagnetic waves.*

Our theory of electricity and magnetism will introduce the entirely new concept of a *field.* Electricity and magnetism are about the long-range interactions of charges, both static charges and moving charges, and the field concept will help us understand how these interactions take place.

Microscopic Models

The field theory provides a macroscopic perspective on the phenomena of electricity and magnetism, but we can also take a microscopic view. At the microscopic level, we want to know what charges are, how they are related to atoms and molecules, and how they move through various kinds of materials. Electromagnetic waves are composed of electric and magnetic fields. The interaction of electromagnetic waves with matter can be analyzed in terms of the interactions of these fields with the charges in matter. When you heat food in a microwave oven, you are using the interactions of electric and magnetic fields with charges in a very fundamental way.

20 Electric Fields and Forces

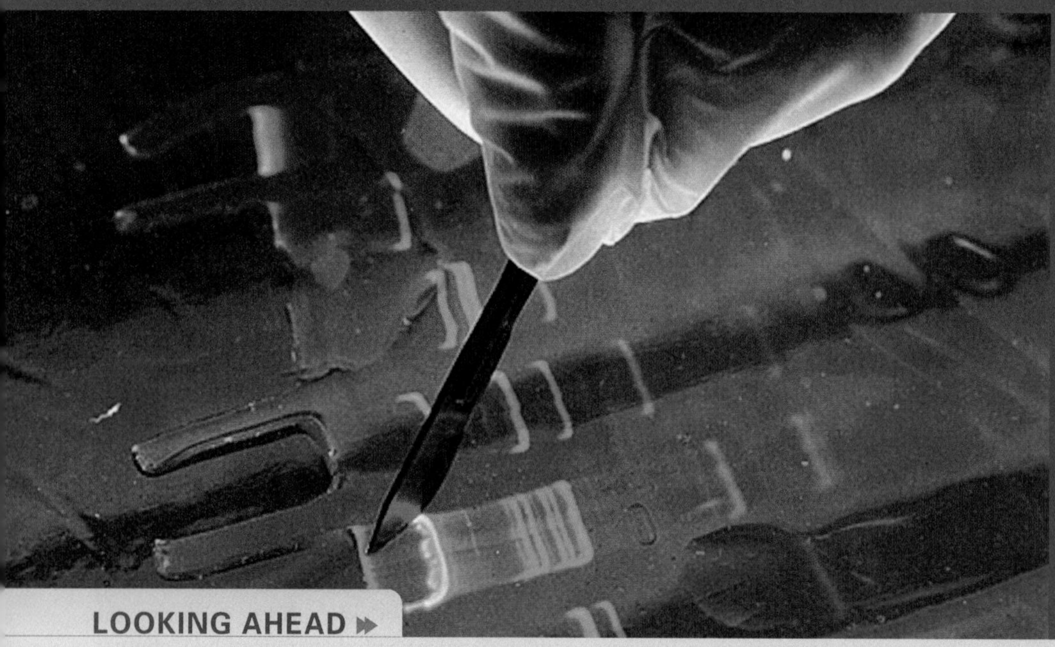

DNA analysis is often done using gel electrophoresis. A solution of DNA segments is placed in a well at one end of a plate of gel. Different segments migrate through the gel at different rates, leading to the lines in the photo. What force causes the DNA segments to move through the gel?

LOOKING AHEAD ▸▸

The goal of Chapter 20 is to develop a basic understanding of electric phenomena in terms of charges, forces, and fields.

Charges and Forces

We'll find that basic electric phenomena can be understood in terms of a **charge model** of electricity:

- There are two kinds of charge, called "positive" and "negative" charge.
- There is an attractive force between charges of opposite kind, a repulsive force between charges of the same kind.

When you brush your hair, why do the strands fly away from each other?

Why will a pencil rubbed through your hair pick up small pieces of paper?

Charges, Atoms, and Molecules

To understand charging processes, we'll need to review our contemporary understanding of the atomic nature of matter.

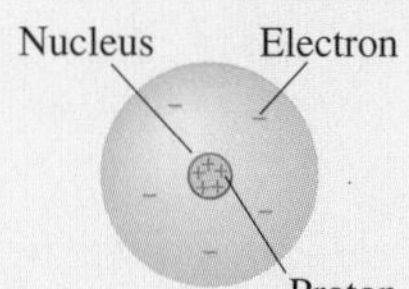

You'll learn that electrons and protons are the basic charges of ordinary matter.

The process of charging an object by rubbing can be understood as a *transfer* of electrons from one material to another.

Coulomb's Law

The attractive and repulsive forces between two charged particles can be calculated from **Coulomb's law**.

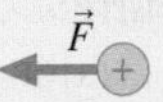
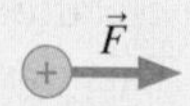
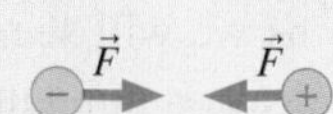

Coulomb's law tells us how the force between charges depends on their charge and the distance between them.

Looking Back ◂◂
3.1–3.3 Vectors and components
6.6 Newton's law of gravity

The Electric Field

Every charge alters the space around it, creating an **electric field** at every point. This electric field then exerts a force on other charges.

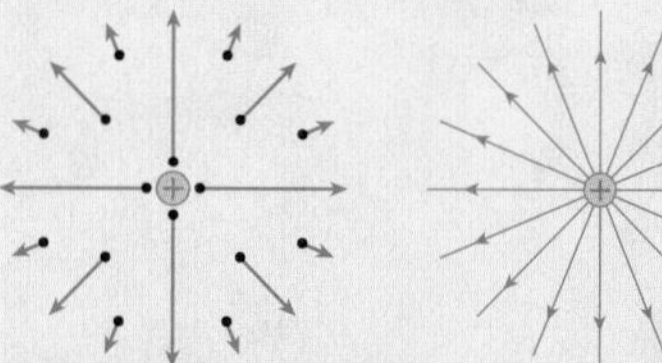

You'll learn how to represent the electric field using **field diagrams** and **field lines**.

Forces and Torques in Electric Fields

Charges in an electric field experience forces and torques due to the field.

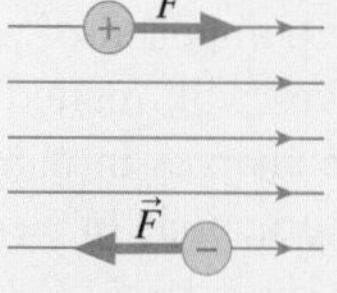

The force on a positive charge is in the same direction as the field; the force on a negative charge is opposite the field.

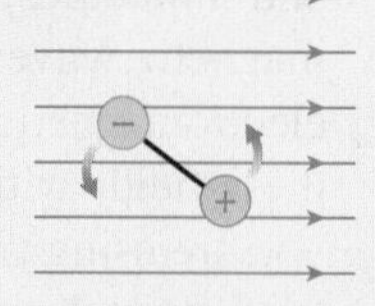

Two equal but opposite charges form an **electric dipole.** A dipole in an electric field experiences a *torque.*

Strong electric fields are used to remove charged particulates from this power plant's emissions.

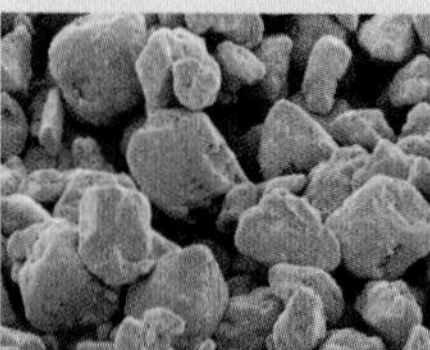

In a laser printer, toner particles like these (shown magnified 500×) stick to the paper by electric forces.

Looking Back ◂◂
7.2 Torque

20.1 Charges and Forces

You can receive a mildly unpleasant shock and produce a little spark if you touch a metal doorknob after scuffing your shoes across a carpet. A plastic comb that you've run through your hair will pick up bits of paper and other small objects. In both of these cases, two objects are *rubbed* together. Why should rubbing an object cause forces and sparks? What kind of forces are these? These are the questions with which we begin our study of electricity.

Our first goal is to develop a model for understanding electric phenomena in terms of *charges* and *forces.* We will later use our contemporary knowledge of atoms to understand electricity on a microscopic level, but the basic concepts of electricity make *no* reference to atoms or electrons. The theory of electricity was well established long before the electron was discovered.

The ancient Greeks first noted the electrical nature of matter by observing amber, a form of fossilized tree resin. When rubbed with fur, amber buttons would attract bits of feather, hair, or straw. The Greek word for amber, "elektron," is the source of our words "electric," "electricity," and—of course—"electron."

Experimenting with Charges

Let us enter a laboratory where we can make observations of electric phenomena. This is a modest laboratory, much like one you would have found in the year 1800. The major tools in the lab are:

- A variety of plastic, glass, and wood rods, each several inches long. These can be held in your hand or suspended by threads from a support.
- A few metal rods with wood handles.
- Pieces of wool and silk.
- Small metal spheres, an inch or two in diameter, on wood stands.

We will manipulate and use these tools with the goal of developing a theory to explain the phenomena we see. The experiments and observations described below are very much like those of early investigators of electric phenomena.

Discovering electricity I

Experiment 1

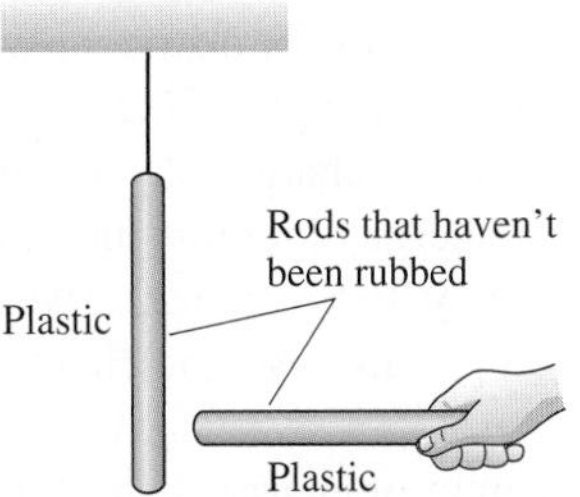

Take a plastic rod that has been undisturbed for a long period of time and hang it by a thread. Pick up another undisturbed plastic rod and bring it close to the hanging rod. Nothing happens to either rod.

Interpretation: There are no special electrical properties to these undisturbed rods. We say that they are **neutral.**

Experiment 2

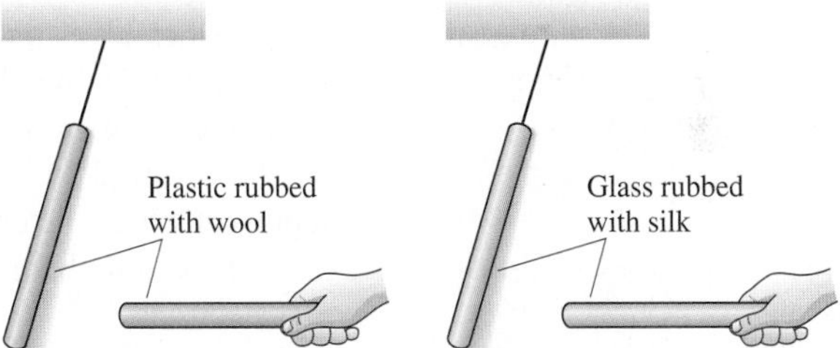

Vigorously rub both the hanging plastic rod and the handheld plastic rod with wool. Now the hanging rod *moves away* from the handheld rod when you bring the two close together. Rubbing two glass rods with silk produces the same result: The two rods repel each other.

Interpretation: Rubbing a rod somehow changes its properties so that forces now act between two such rods. We call this process of rubbing **charging** and say that the rubbed rod is *charged*, or that it has *acquired a* ***charge.***

Experiment 2 shows that there is a *long-range repulsive force* (i.e., a force requiring no contact) between two identical objects that have been charged in the *same* way, such as two plastic rods both rubbed with wool or two glass rods rubbed with silk. The force between charged objects is called the **electric force.** We have seen a long-range force before, gravity, but the gravitational force is always attractive. This is the first time we've observed a repulsive long-range force. However, the electric force is not always repulsive, as the next experiment shows.

Discovering electricity II

Experiment 3

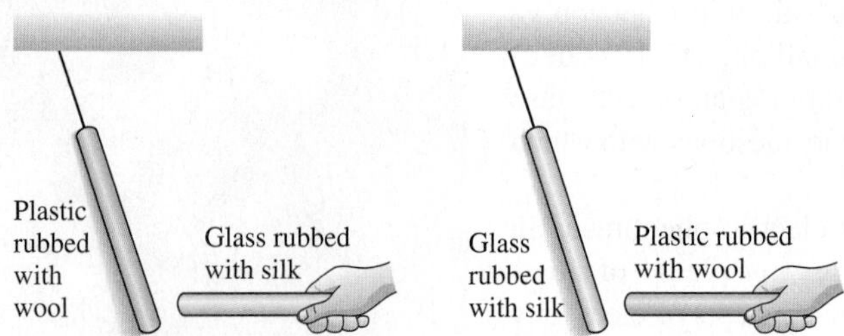

Bring a glass rod that has been rubbed with silk close to a hanging plastic rod that has been rubbed with wool. These two rods *attract* each other.

Interpretation: We can explain this experiment as well as Experiment 2 by assuming that there are two *different* kinds of charge that a material can acquire. We *define* the kind of charge acquired by a glass rod as *positive* charge, and that acquired by a plastic rod as *negative* charge. Then these two experiments can be summarized as **like charges** (positive/positive or negative/negative) exert repulsive forces on each other, while **opposite charges** (positive/negative) exert attractive forces on each other.

Experiment 4

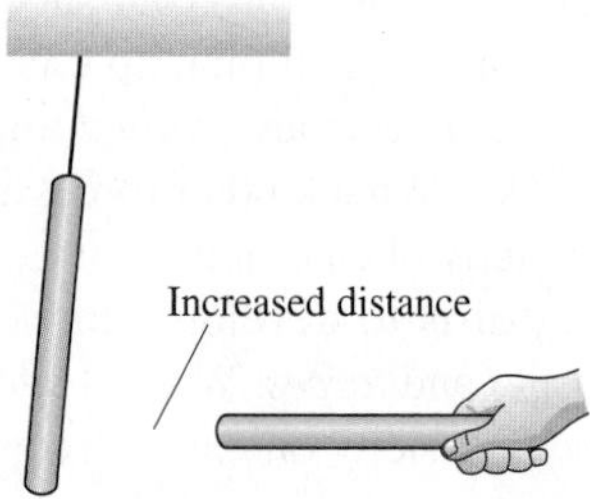

- If the two rods are held farther from each other, the force between them decreases.
- The strength of the force is greater for rods that have been rubbed more vigorously.

Interpretation: Like the gravitational force, the electric force decreases with the distance between the charged objects. And, the greater the charge on the two objects, the greater the force between them.

Although we showed experimental results only for plastic rods rubbed with wool and glass rods rubbed with silk, further experiments show that there are *only* two kinds of charge, positive and negative. For instance, when you rub a balloon on your hair the balloon becomes negatively charged, while nylon rubbed with a polyester cloth becomes positively charged.

Visualizing Charge

FIGURE 20.1 Visualizing charge.

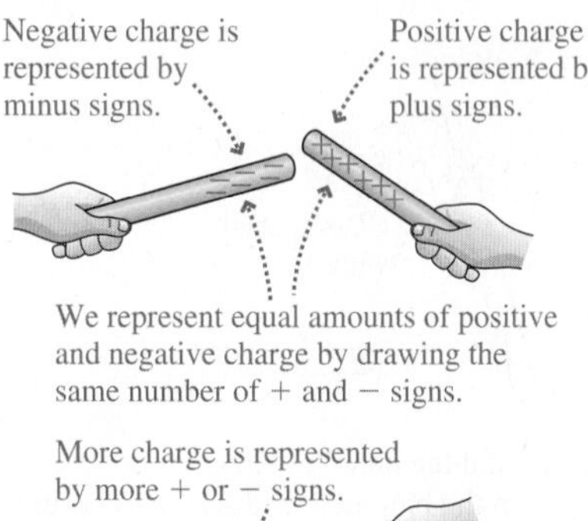

Negative charge is represented by minus signs.
Positive charge is represented by plus signs.
We represent equal amounts of positive and negative charge by drawing the same number of + and − signs.
More charge is represented by more + or − signs.

Diagrams are going to be an important tool for understanding and explaining charges and the forces between charged objects. FIGURE 20.1 shows how to draw a *charge diagram,* which gives a schematic picture of the distribution of charge on an object. It's important to realize that the + and − signs drawn in Figure 20.1 do not represent "individual" charges. At this point, we are thinking of charge only as something that can be acquired by an object by rubbing, so in charge diagrams the + and − signs represent where the charge is only in a general way. In Section 20.2 we'll look at an atomic view of charging and learn about the single microscopic charges of protons and electrons.

We can gain an important insight into the nature of charge by investigating what happens when we bring together a plastic rod and the wool used to charge it, as the following experiment shows.

Discovering electricity III

Experiment 5

Start with a neutral, uncharged hanging plastic rod and a piece of wool. Rub the plastic rod with the wool, then hold the wool close to the rod. The rod is *attracted* to the wool.

Interpretation: From Experiment 3 we know that the plastic rod has a negative charge. Because the wool attracts the rod, the wool must have a *positive* charge.

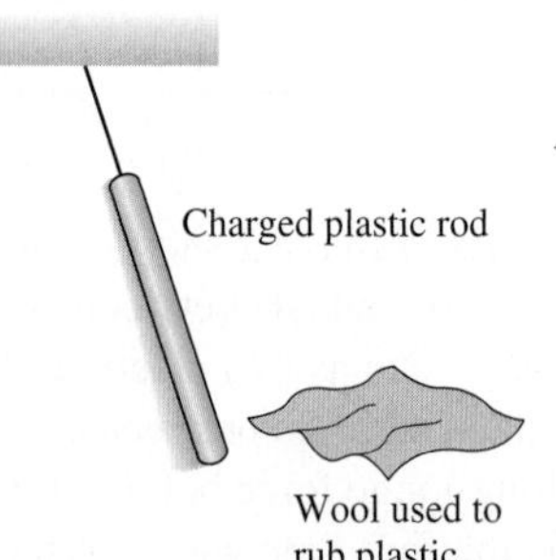

Experiment 5 shows that when a plastic rod is rubbed by wool, not only does the plastic rod acquire a negative charge, but also the wool used to rub it acquires a positive charge. This observation can be explained if we postulate that a neutral object is not one that has no charge at all; rather, **a neutral object contains *equal amounts* of positive and negative charge.** Just as in ordinary addition, where $2 + (-2) = 0$, equal amounts of opposite charge "cancel," leaving no overall or *net* charge.

In this model, an object becomes positively charged if the amount of positive charge on it exceeds the amount of negative charge; the mathematical analogy of this is $2 + (-1) = +1$. Similarly, an object is negatively charged when the amount of negative charge on it is greater than the amount of positive charge, analogous to $2 + (-3) = -1$.

As **FIGURE 20.2** shows, the rubbing process works by *transferring* charge from one object to the other. When the two are rubbed together, negative charge is transferred from the wool to the rod. This clearly leaves the rod with an excess of negative charge. But the wool, having lost some of its negative charge to the rod, now has an *excess* of positive charge, leaving it positively charged. (We'll see in Section 20.2 why it is usually negative charge that moves.)

FIGURE 20.2 How a plastic rod and wool acquire charge during the rubbing process.

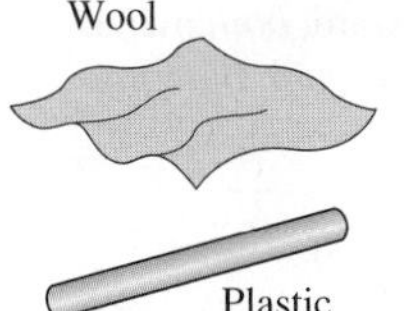

1. The rod and wool are initially neutral.

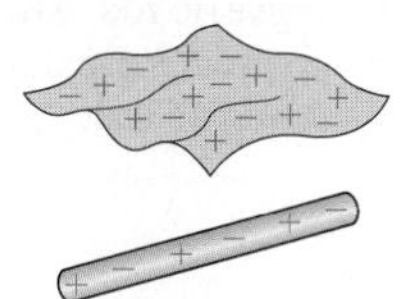

2. "Neutral" actually means that they each have equal amounts of positive and negative charge.

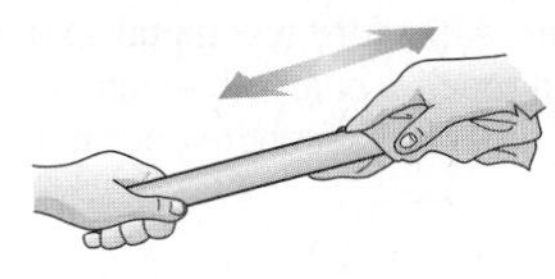

3. Now we rub the plastic and wool together. As we do so, negative charge moves from the wool to the rod.

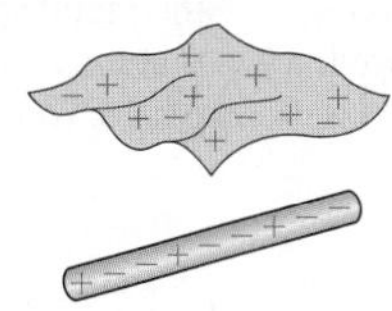

4. This leaves the rod with extra negative charge. The wool is left with more positive charge than negative.

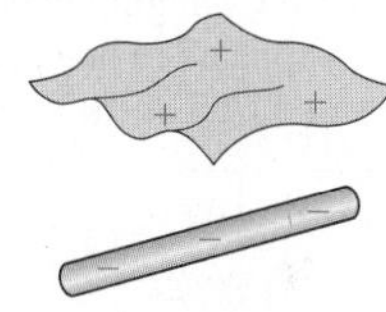

5. In a charge diagram, we draw only the *excess* charge. This shows how the rod and wool acquire their charge.

There is another crucial fact about charge implicit in Figure 20.2: Nowhere in the rubbing process was charge either created or destroyed. Charge was merely transferred from one place to another. It turns out that this fact is a fundamental law of nature, the **law of conservation of charge.** If a certain amount of positive charge appears somewhere, an equal amount of negative charge must appear elsewhere so that the net charge doesn't change.

The results of our experiments, and our interpretation of them in terms of positive and negative charge, can be summarized in the following **charge model.**

Charge model, part I The basic postulates of our model are:

1. Frictional forces, such as rubbing, add something called *charge* to an object or remove it from the object. The process itself is called *charging*. More vigorous rubbing produces a larger quantity of charge.
2. There are two kinds of charge, positive and negative.
3. Two objects with *like charge* (positive/positive or negative/negative) exert repulsive forces on each other. Two objects with *opposite charge* (positive/negative) exert attractive forces on each other. We call these *electric forces*.
4. The force between two charged objects is a long-range force. The magnitude of the force increases as the quantity of charge increases and decreases as the distance between the charges increases.
5. *Neutral* objects have an *equal mixture* of positive and negative charge.
6. The rubbing process charges the objects by *transferring* charge (usually negative) from one to the other. The objects acquire equal but opposite charges.
7. Charge is conserved: It cannot be created or destroyed.

TRY IT YOURSELF

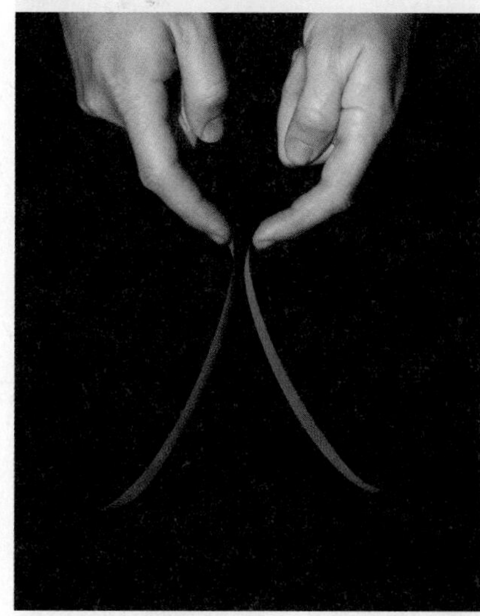

Charges on tape Pull a piece of transparent tape about 6″ long off a roll. Now, pull a second piece of tape off the roll, and hold the two pieces near each other. They show a strong repulsive force. What does this tell you about the charges on the two pieces of tape? How could you prepare the two strips of tape so that they attract?

Insulators and Conductors

Experiments 2, 3, and 5 involved a transfer of charge from one object to another. Let's do some more experiments with charge to look at how charge *moves* on different materials.

Discovering electricity IV

Experiment 6

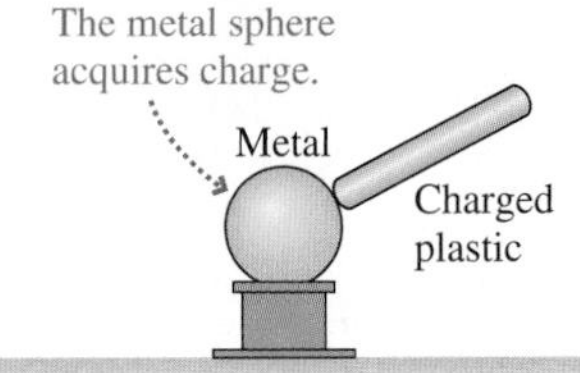

Charge a plastic rod by rubbing it with wool. Touch a neutral metal sphere with the rubbed area of the rod. The metal sphere then repels a charged, hanging plastic rod. The metal sphere appears to have acquired a charge of the same sign as the plastic rod.

Experiment 7

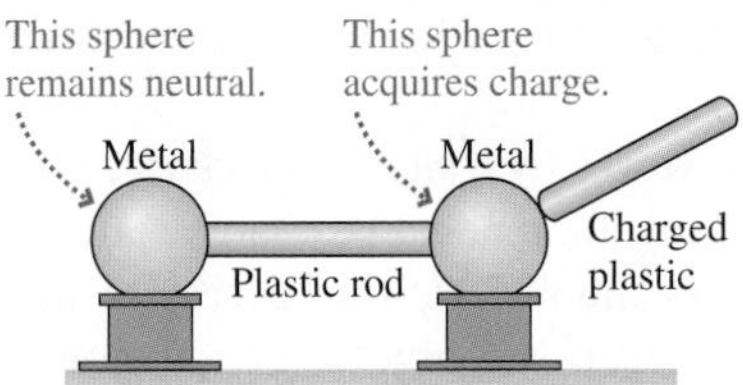

Place two metal spheres close together with a plastic rod connecting them. Charge a second plastic rod, by rubbing, and touch it to one of the metal spheres. Afterward, the metal sphere that was touched repels a charged, hanging plastic rod. The other metal sphere does not.

Experiment 8

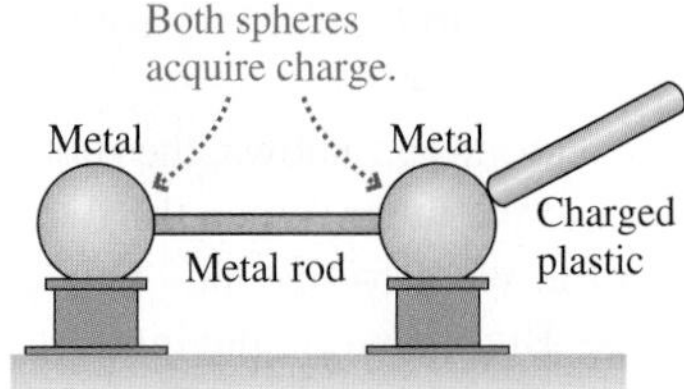

Repeat Experiment 7 with a metal rod connecting the two metal spheres. Touch one metal sphere with a charged plastic rod. Afterward, *both* metal spheres repel a charged, hanging plastic rod.

Our final set of experiments has shown that charge can be transferred from one object to another only when the objects *touch*. Contact is required. Removing charge from an object, which you can do by touching it, is called **discharging.**

In Experiments 7 and 8, charge is transferred from the charged rod to the metal sphere as the two are touched together. In Experiment 7, the other sphere remains neutral, indicating that no charge moved along the plastic rod connecting the two spheres. In Experiment 8, by contrast, the other sphere is found to be charged; evidently charge has moved along the metal rod connecting the spheres, transferring some charge from the first sphere to the second. We define **conductors** as those materials through or along which charge easily moves and **insulators** as those materials on or in which charges remain immobile. Glass and plastic are insulators; metal is a conductor.

This new information allows us to add more postulates to our charge model:

Charge model, part II

8. There are two types of materials. Conductors are materials through or along which charge easily moves. Insulators are materials on or in which charges remain fixed in place.
9. Charge can be transferred from one object to another by contact.

NOTE ► Both insulators and conductors can be charged. They differ in the ability of charge to *move*. ◄

◄A dry day, a plastic slide, and a child with clothes of the right fabric lead to a startling demonstration of electric charges and forces. The rubbing of the child's clothes on the slide has made her build up charge. The body is a good conductor, so the charges spread across her body and her hair. The resulting repulsion produces a dramatic result!

CONCEPTUAL EXAMPLE 20.1 **Transferring charge**

In Experiment 8, touching a metal sphere with a charged plastic rod caused a second metal sphere, connected by a metal rod to the first, to become charged with the same type of charge as the rod. Use the postulates of the charge model to construct a charge diagram for the process.

REASON We need the following ideas from the charge model:

1. **Charge is transferred upon contact.** The plastic rod was charged by rubbing with wool, giving it a negative charge. The charge doesn't move around on the rod, an insulator, but some of the charge is transferred to the metal upon contact.
2. **Metal is a conductor.** Once in the metal, which is a conductor, the charges are free to move around.
3. **Like charges repel.** Because like charges repel, these negative charges quickly move as far apart as they possibly can. Some move through the connecting metal rod to the second sphere. Consequently, the second sphere acquires a net negative charge. The repulsive forces drive the negative charges as far apart as they can possibly get, causing them to end up on the *surfaces* of the conductors.

The charge diagram in **FIGURE 20.3** illustrates these three steps.

FIGURE 20.3 A charge diagram for Experiment 8.

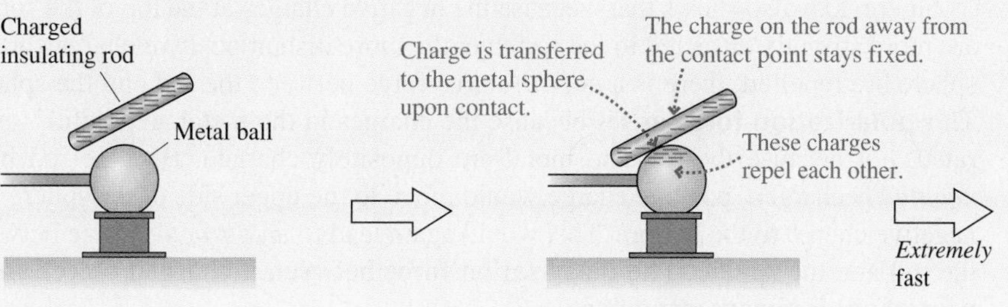

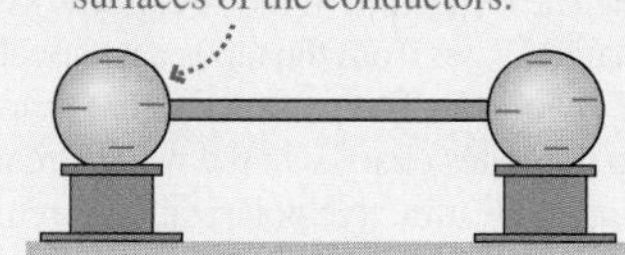

In Conceptual Example 20.1, once the charge is placed on the conductor it rapidly distributes itself over the conductor's surface. This movement of charge is *extremely* fast. Other than this very brief interval during which the charges are adjusting, the charges on an isolated conductor are in static equilibrium with the charges at rest. This condition is called **electrostatic equilibrium.**

CONCEPTUAL EXAMPLE 20.2 **Drawing a charge diagram for an electroscope**

Many electricity demonstrations are carried out with the help of an *electroscope* like the one shown in **FIGURE 20.4**. Touching the sphere at the top of an electroscope with a charged plastic rod causes the leaves to fly apart and remain hanging at an angle. Use charge diagrams to explain why.

REASON We will use the charge model and our understanding of insulators and conductors to make a series of charge diagrams in **FIGURE 20.5** that shows the charging of the electroscope.

ASSESS The charges move around, but, because charge is conserved, the total number of negative charges doesn't change from picture to picture.

FIGURE 20.4 A charged electroscope.

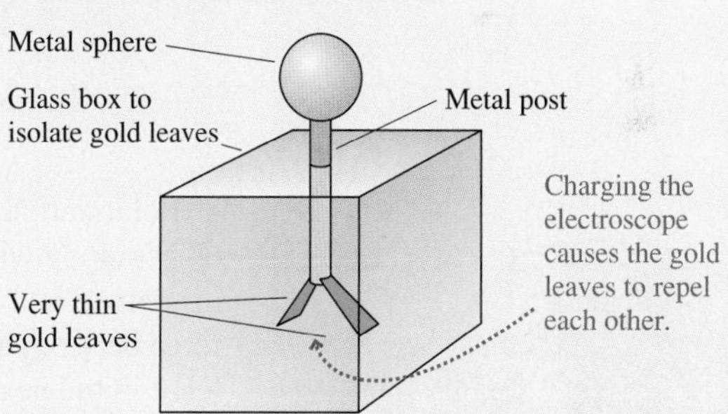

FIGURE 20.5 Charging an electroscope.

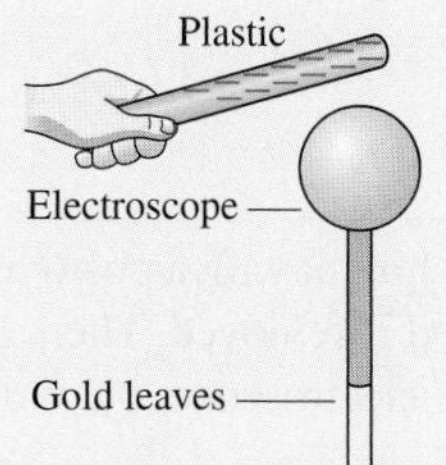

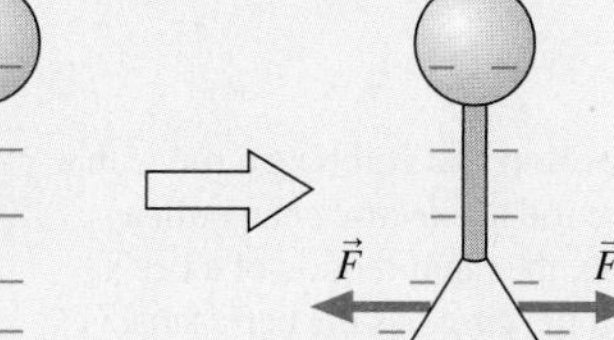

1. Negative charges are transferred from the rod to the metal sphere upon contact.

2. Metal is a conductor. Therefore charge spreads (very rapidly) throughout the entire electroscope. The leaves become negatively charged.

3. Like charges repel. The negatively charged leaves exert repulsive forces on each other, causing them to spread apart.

TRY IT YOURSELF

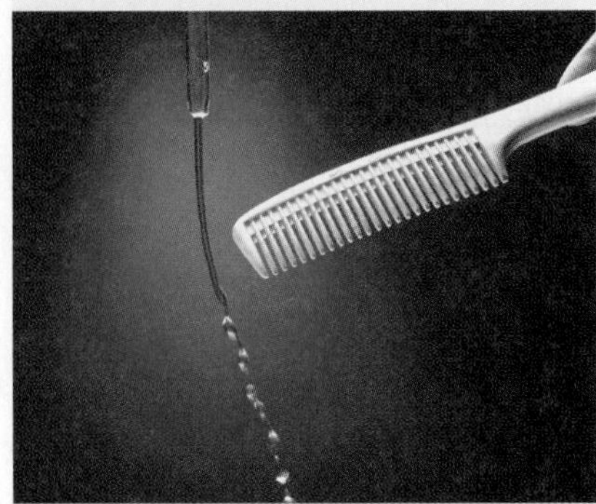

Pulling water Turn on your tap so that a thin stream of water flows. Next, run a comb briskly through your hair, and bring the comb close to—but not touching—the stream of water. The deflection of the stream can be quite dramatic! Water from the tap is a reasonably good conductor. The presence of the charged comb separates charges in the water stream, leading to an attractive polarization force.

Polarization

At the beginning of this chapter we showed a picture of a small feather being picked up by a piece of amber that had been rubbed with fur. The amber was charged by rubbing, but the feather had not been rubbed—it was *neutral.* How can our charge model explain the attraction of a neutral object toward a charged one?

Although a feather is an insulator, it's easiest to understand this phenomenon by first considering how a neutral *conductor* is attracted to a charged object. FIGURE 20.6 shows how this works. Because the charged rod doesn't touch the sphere, no charge is added to or removed from the sphere. Instead, the rod attracts some of the sphere's negative charge to the side of the sphere near the rod. This leaves a deficit of negative charge on the opposite side of the sphere, so that side is now positively charged. This slight *separation* of the positive and negative charge in a neutral object when a charged object is brought near is called **charge polarization.**

Figure 20.6 also shows that, because the negative charges at the top of the sphere are more strongly attracted to the rod than the more distant positive charges on the sphere are repelled, there is a net *attractive* force between the rod and the sphere. This **polarization force** arises because the charges in the metal are slightly separated, *not* because the rod and metal are oppositely charged. Had the rod been *negatively* charged, positive charge would move to the upper side of the sphere and negative charge to the bottom. This would again lead to an *attractive* force between the rod and the sphere. **The polarization force between a charged object and a neutral one is always attractive.**

FIGURE 20.6 Why a neutral metal object is attracted to a charged object.

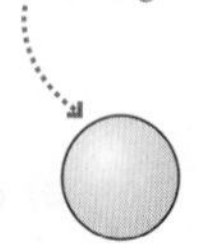

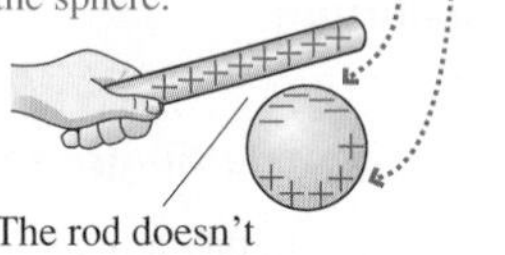

Negative charge is attracted to the positive rod. This leaves behind positive charge on the other side of the sphere.
The rod doesn't touch the sphere

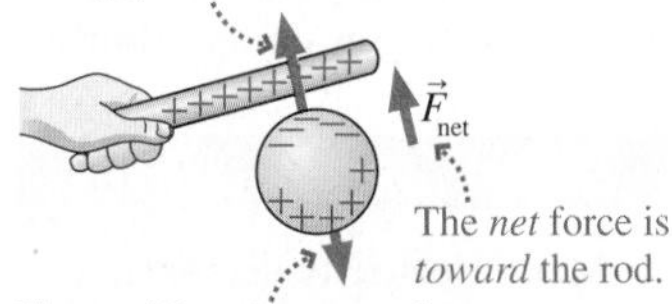

Picking up pollen BIO Rubbing a rod with a cloth gives the rod an electric charge. In a similar fashion, the rapid motion of a bee's wings through the air gives the bee a small positive electric charge. As small pieces of paper are attracted to a charged rod, so are tiny grains of pollen attracted to the charged bee, helping it collect and hold the pollen.

Polarization explains why forces arise between a charged object and a metal object along which charge can freely move. But the feathers attracted to amber are insulators, and charge can't move through an insulator. Nevertheless, the attractive force between a charged object and an insulator is also a polarization force. As we'll learn in the next section, the charge in each *atom* that makes up an insulator can be slightly polarized. Although the charge separation in one atom is exceedingly small, the net effect over all the countless atoms in an insulator is to shift a perceptible amount of charge from one side of the insulator to the other. This is just what's needed to allow a polarization force to arise.

STOP TO THINK 20.1 An electroscope is charged by touching it with a positive glass rod. The electroscope leaves spread apart and the glass rod is removed. Then a negatively charged plastic rod is brought close to the top of the electroscope, but it doesn't touch. What happens to the leaves?

A. The leaves get closer together.
B. The leaves spread farther apart.
C. The leaves do not change their position.

20.2 Charges, Atoms, and Molecules

We have been speaking about giving objects positive or negative charge without explaining what is happening at an atomic level. You already know that the basic constituents of atoms—the nucleus and the electrons surrounding it—are charged. In this section we will connect our observations of the previous section with our understanding of the atomic nature of matter.

Our current model of the atom is that it is made up of a very small and dense positively charged *nucleus,* containing positively charged *protons* as well as neutral particles called *neutrons,* surrounded by much-less-massive orbiting negatively charged *electrons* that form an **electron cloud** surrounding the nucleus, as illustrated in FIGURE 20.7. The atom is held together by the attractive electric force between the positive nucleus and the negative electrons.

FIGURE 20.7 Our modern view of the atom.

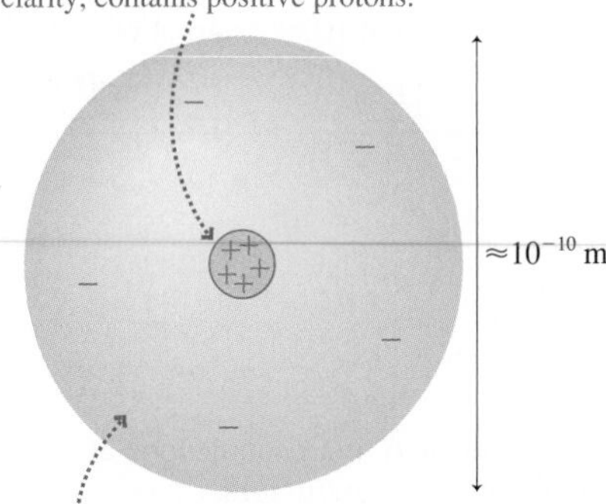

The nucleus, exaggerated in size for clarity, contains positive protons.
≈10^-10 m
The electron cloud is negatively charged.

Experiments show that **charge, like mass, is an inherent property of electrons and protons.** It's no more possible to have an electron without charge than it is to have an electron without mass.

An Atomic View of Charging

Electrons and protons are the basic charges in ordinary matter. **There are no other sources of charge.** Consequently, the various observations we made in Section 20.1 need to be explained in terms of electrons and protons.

Experimentally, it's found that electrons and protons have charges of opposite sign but *exactly* equal magnitude. Thus, because charge is due to electrons and protons, **an object is charged if it has an unequal number of electrons and protons.** An object with a negative charge has more electrons than protons; an object with a positive charge has more protons than electrons. Most macroscopic objects have an *equal number* of protons and electrons. Such an object has no *net* charge; we say it is *electrically neutral.*

In practice, objects acquire a positive charge not by gaining protons but by losing electrons. Protons are *extremely* tightly bound within the nucleus and cannot be added to or removed from atoms. Electrons, on the other hand, are bound much more loosely than the protons and can be removed with little effort.

The process of removing an electron from the electron cloud of an atom is called **ionization.** An atom that is missing an electron is called a *positive ion.* Some atoms can accommodate an *extra* electron and thus become a *negative ion.* FIGURE 20.8 shows positive and negative ions.

FIGURE 20.8 Positive and negative ions.

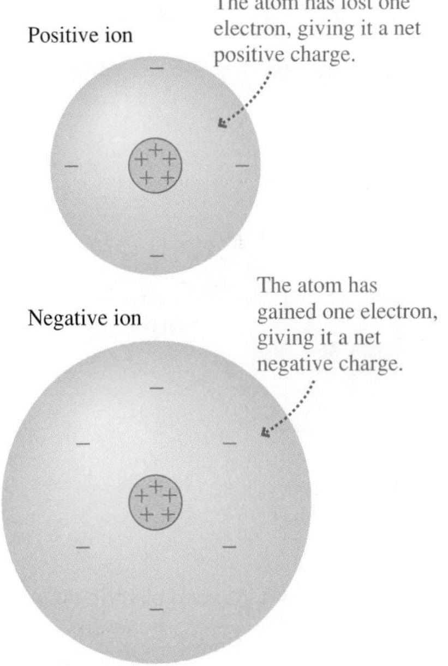

Positive ion
The atom has lost one electron, giving it a net positive charge.
Negative ion
The atom has gained one electron, giving it a net negative charge.

The charging processes we observed in Section 20.1 involved rubbing and friction. The forces of friction often cause molecular bonds at the surface to break as two materials slide past each other. Molecules are electrically neutral, but FIGURE 20.9 shows that *molecular ions* can be created when one of the bonds in a large molecule is broken. If the positive molecular ions remain on one material and the negative ions on the other, one of the objects being rubbed ends up with a net positive charge and the other with a net negative charge. This is the way in which a plastic rod is charged by rubbing with wool or a comb is charged by passing through your hair.

FIGURE 20.9 Charging by friction may result from molecular ions produced as bonds are broken.

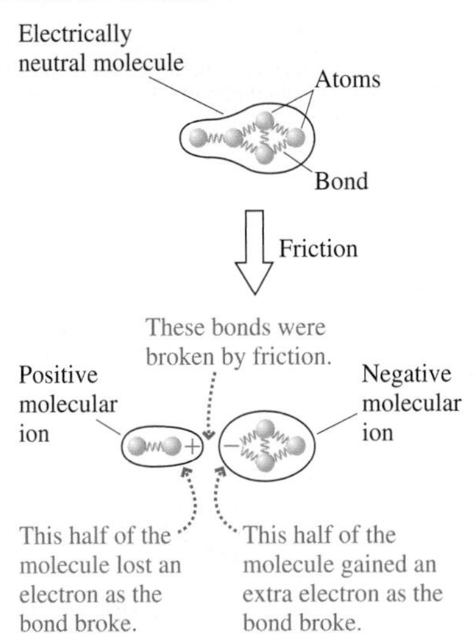

These bonds were broken by friction.

This half of the molecule lost an electron as the bond broke.
This half of the molecule gained an extra electron as the bond broke.

Charge Conservation

Charge is represented by the symbol q (or sometimes Q). The SI unit of charge is the **coulomb** (C), named for French scientist Charles Coulomb, one of many scientists investigating electricity in the late 18th century.

Protons and electrons, the charged particles in ordinary matter, have the same amount of charge, but of opposite signs. We use the symbol e for the **fundamental charge,** the magnitude of the charge of an electron or a proton. The fundamental charge e has been measured to have the value

$$e = 1.60 \times 10^{-19}\ \text{C}$$

TABLE 20.1 Protons and electrons

Particle	Mass (kg)	Charge (C)
Proton	1.67×10^{-27}	$+e = 1.60 \times 10^{-19}$
Electron	9.11×10^{-31}	$-e = -1.60 \times 10^{-19}$

Table 20.1 lists the masses and charges of protons and electrons.

NOTE ► The amount of charge produced by rubbing plastic or glass rods is typically in the range 1 nC (10^{-9} C) to 100 nC (10^{-7} C). This corresponds to an excess or deficit of 10^{10} to 10^{12} electrons. But, because of the enormous number of atoms in a macroscopic object, this represents an excess or deficit of only perhaps 1 electron in 10^{13}. ◄

That charge is associated with electrons and protons explains why charge is conserved. Because electrons and protons are neither created nor destroyed in ordinary processes, their associated charge is conserved as well.

FIGURE 20.10 A microscopic look at insulators and conductors.

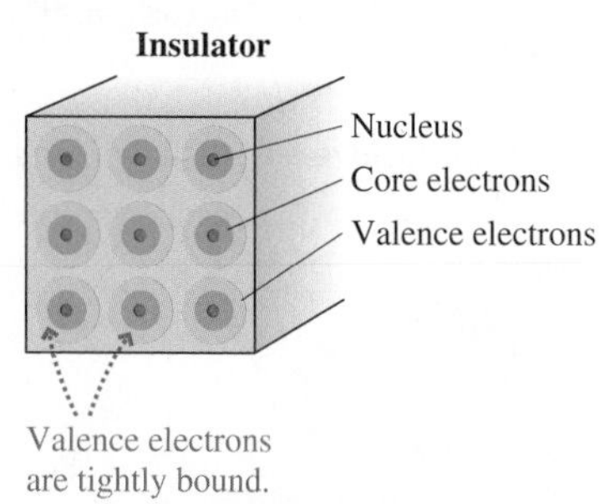
Insulator
Nucleus
Core electrons
Valence electrons
Valence electrons
are tightly bound.

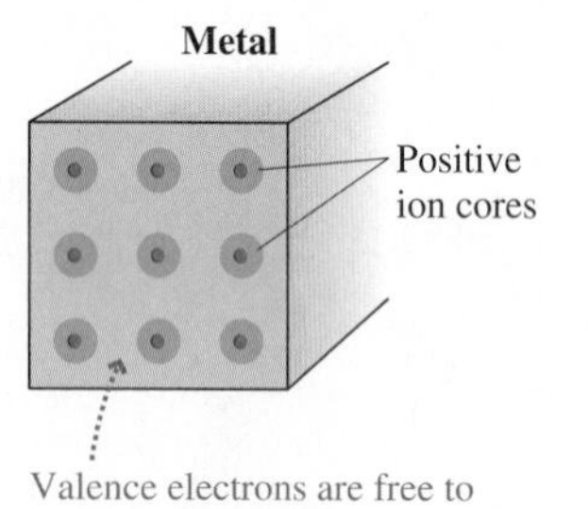
Metal
Positive
ion cores
Valence electrons are free to
move throughout the metal.

Insulators and Conductors

FIGURE 20.10 looks inside an insulator and a metallic conductor. The electrons in the insulator are all tightly bound to the positive nuclei and not free to move around. Charging an insulator by friction leaves patches of molecular ions on the surface, but these patches are immobile.

In metals, the outer atomic electrons (called the *valence electrons* in chemistry) are only weakly bound to the nuclei. As the atoms come together to form a solid, these outer electrons become detached from their parent nuclei and are free to wander about through the entire solid. The solid *as a whole* remains electrically neutral, because we have not added or removed any electrons, but the electrons are now rather like a negatively charged gas or liquid—what physicists like to call a **sea of electrons**—permeating an array of positively charged **ion cores.** However, although the electrons are highly mobile *within* the metal, they are still weakly bound to the ion cores and will not leave the metal.

Electric Dipoles

In the last section we noted that an insulator, such as paper, becomes polarized when brought near a charged object. We can use an atomic description of matter to see why.

Consider what happens if we bring a positive charge near a neutral atom. As **FIGURE 20.11** shows, the charge polarizes the atom by attracting the electron cloud while repelling the nucleus. The polarization of just one atom is a very small effect, but there are an enormous number of atoms in an insulator. Added together, their net polarization—and the resulting polarization force—can be quite significant. This is how the rubbed amber picks up a feather, exerting an upward polarization force on it larger than the downward force of gravity.

FIGURE 20.11 An induced electric dipole.

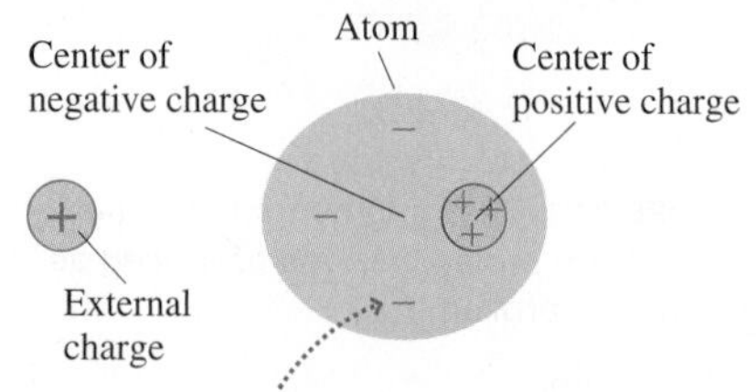
Atom
Center of
negative charge
Center of
positive charge
External
charge
Forces from the external charge cause
the atom's negative charge and the
positive charge to be slightly offset.

Two equal but opposite charges with a separation between them are called an **electric dipole.** In this case, where the polarization is caused by the external charge, the atom has become an *induced electric dipole.* Because the negative end of the dipole is slightly closer to the positive charge, the attractive force on the negative end slightly exceeds the repulsive force on the positive end, and there is a net force toward the external charge. If a charged rod causes all the atoms in a piece of paper to become induced electric dipoles, the net force is enough to lift the paper to the rod.

Hydrogen Bonding

Some molecules have an asymmetry in their charge distribution that makes them *permanent electric dipoles.* An important example is the water molecule. Bonding between the hydrogen and oxygen atoms results in an unequal sharing of charge that, as shown in **FIGURE 20.12**, leaves the hydrogen atoms with a small positive charge and the oxygen atom with a small negative charge.

When two water molecules are close, the attractive electric force between the positive hydrogen atom of one molecule and the negative oxygen atom of the second molecule can form a weak bond, called a **hydrogen bond,** as illustrated in **FIGURE 20.13**. These weak bonds result in a certain "stickiness" between water molecules that is

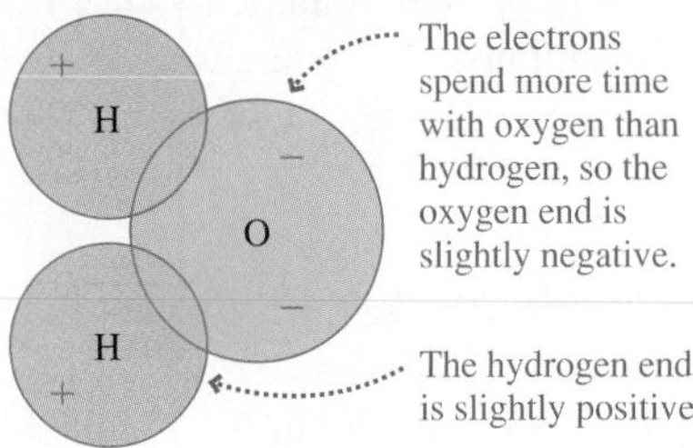

H
O
H
+
–
–
+
The electrons spend more time with oxygen than hydrogen, so the oxygen end is slightly negative.
The hydrogen end is slightly positive.

FIGURE 20.12 A water molecule is a permanent electric dipole.

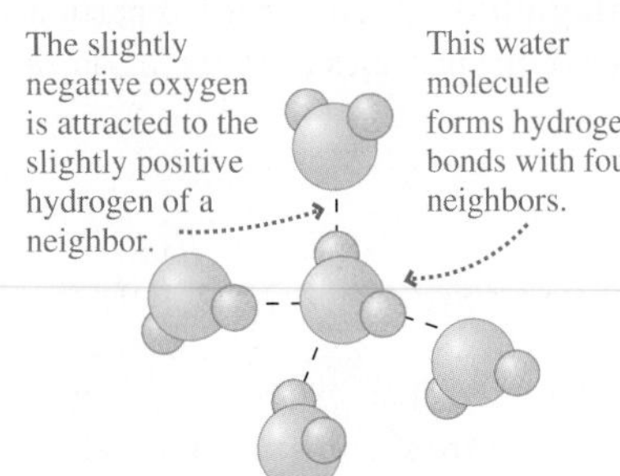

The slightly negative oxygen is attracted to the slightly positive hydrogen of a neighbor.
This water molecule forms hydrogen bonds with four neighbors.

FIGURE 20.13 Hydrogen bonds between water molecules.

responsible for many of water's special properties, including its expansion on freezing, the wide range of temperatures over which it is liquid, and its high heat of vaporization.

Hydrogen bonds are extremely important in biological systems. As you know, the DNA molecule has the structure of a double helix. Information in DNA is coded in the *nucleotides,* the four molecules guanine, thymine, adenine, and cytosine. The nucleotides on one strand of the DNA helix form hydrogen bonds with the nucleotides on the opposite strand.

The nucleotides bond only in certain pairs: Cytosine always forms a bond with guanine, adenine with thymine. This preferential bonding is crucial to DNA replication. When the two strands of DNA are taken apart, each separate strand of the DNA forms a template on which another complementary strand can form, creating two identical copies of the original DNA molecule.

The preferential bonding of nucleotide base pairs in DNA is explained by hydrogen bonding. In each of the nucleotides, the hydrogen atoms have a small positive charge, oxygen and nitrogen a small negative charge. The positive hydrogen atoms on one nucleotide attract the negative oxygen or nitrogen atoms on another. As the detail in FIGURE 20.14 shows, the geometry of the nucleotides allows cytosine to form a hydrogen bond only with guanine, adenine only with thymine.

FIGURE 20.14 Hydrogen bonds in DNA base pairs.

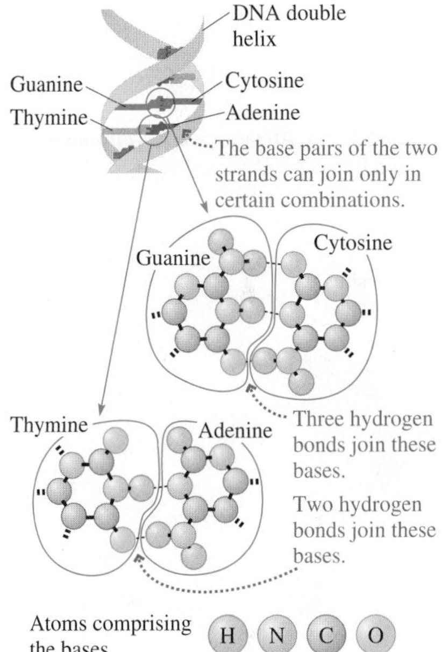

STOP TO THINK 20.2 Rank in order, from most positive to most negative, the charges q_A to q_E of these five systems.

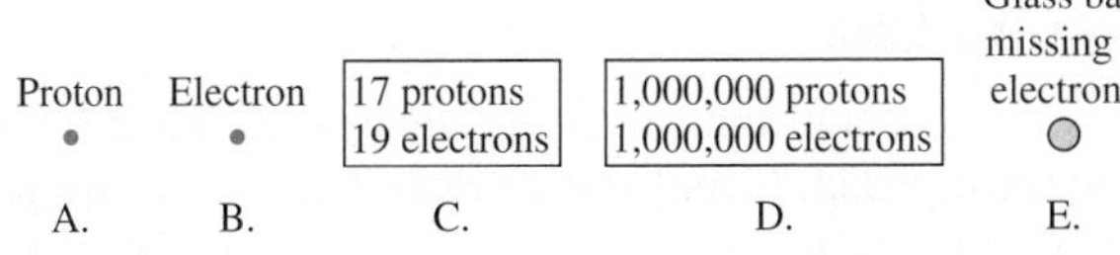

20.3 Coulomb's Law

11.1, 11.2, 11.3

The last two sections established a *model* of charges and electric forces. This model is very good at explaining electric phenomena and providing a general understanding of electricity. Now we need to become quantitative. Experiment 4 in Section 20.1 found that the electric force increases for objects with more charge and decreases as charged objects are moved farther apart. The force law that describes this behavior is known as *Coulomb's law.*

In the mathematical formulation of Coulomb's law, we will use the magnitude of the charge only, not the sign. We show this by using the absolute value notation we used earlier in the book. $|q|$ therefore represents the magnitude of the charge. It is always a positive number, whether the charge is positive or negative.

Coulomb's law

Magnitude: If two charged particles having charges q_1 and q_2 are a distance r apart, the particles exert forces on each other of magnitude

$$F_{1\,\text{on}\,2} = F_{2\,\text{on}\,1} = \frac{K|q_1||q_2|}{r^2} \tag{20.1}$$

INVERSE-SQUARE p. 186

where the charges are in coulombs (C), and $K = 8.99 \times 10^9\ \text{N} \cdot \text{m}^2/\text{C}^2$ is called the **electrostatic constant.** These forces are an action/reaction pair, equal in magnitude and opposite in direction. It is customary to round K to $9.0 \times 10^9\ \text{N} \cdot \text{m}^2/\text{C}^2$ for all but extremely precise calculations, and we will do so.

Direction: The forces are directed along the line joining the two particles. The forces are *repulsive* for two like charges and *attractive* for two opposite charges.

FIGURE 20.15 Attractive and repulsive forces between charges.

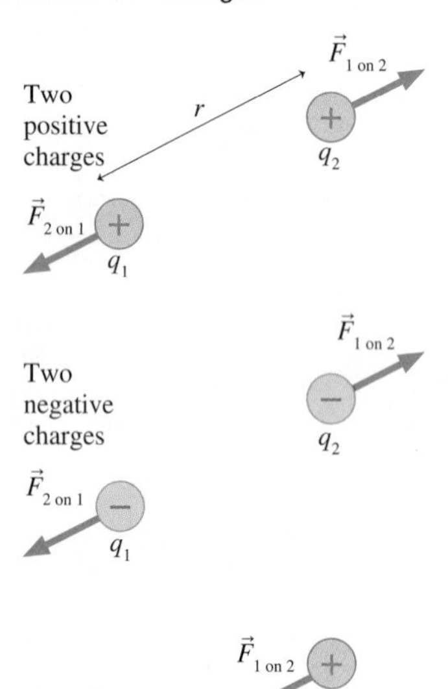

We sometimes speak of the "force between charge q_1 and charge q_2," but keep in mind that we are really dealing with charged *objects* that also have a mass, a size, and other properties. Charge is not some disembodied entity that exists apart from matter. Coulomb's law describes the force between charged *particles.*

NOTE ► Coulomb's law applies only to *point charges.* A point charge is an idealized material object with charge and mass but with no size or extension. For practical purposes, two charged objects can be modeled as point charges if they are much smaller than the separation between them. ◄

Coulomb's law looks much like Newton's law of gravity, but there is a key difference: The charge q can be either positive or negative, so the forces can be attractive or repulsive. Consequently, the absolute value signs in Equation 20.1 are especially important. The first part of Coulomb's law gives only the *magnitude* of the force, which is always positive. The direction must be determined from the second part of the law. FIGURE 20.15 shows the forces between different combinations of positive and negative charges.

Using Coulomb's Law

Coulomb's law is a force law, and forces are vectors. **Electric forces, like other forces, can be superimposed.** If multiple charges 1, 2, 3, . . . are present, the *net* electric force on charge j due to all other charges is therefore the sum of all the individual forces due to each charge; that is,

$$\vec{F}_{\text{net}} = \vec{F}_{1\,\text{on}\,j} + \vec{F}_{2\,\text{on}\,j} + \vec{F}_{3\,\text{on}\,j} + \cdots \tag{20.2}$$

where each of the forces $\vec{F}_{i\,\text{on}\,j}$ is given by Equation 20.1. These conditions are the basis of a strategy for using Coulomb's law to solve electric force problems.

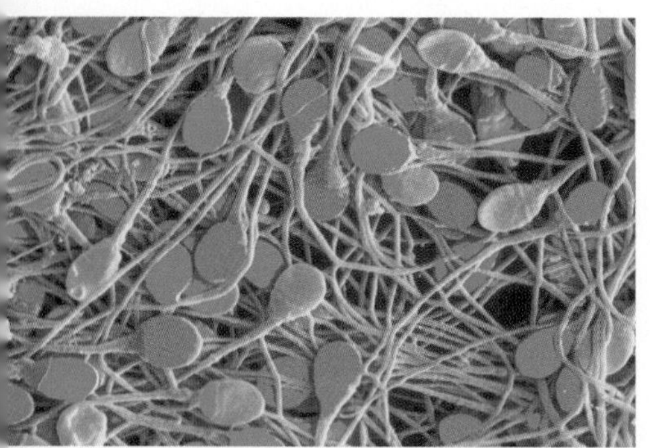

◄ **Separating the girls from the boys** BIO Sperm cells can be sorted according to whether they contain an X or a Y chromosome. The cells are put into solution, and the solution is forced through a nozzle, which breaks the solution into droplets. Suppose a droplet contains a sperm cell. An optical test measures which type of chromosome, X or Y, the cell has. The droplet is then given a positive charge if it has an X sperm cell, negative if it contains a Y. The droplets fall between two oppositely charged plates where they are pushed left or right depending on their charge—separating the X from the Y.

PROBLEM-SOLVING STRATEGY 20.1 **Electric forces and Coulomb's law** 

PREPARE Identify point charges or objects that can be modeled as point charges. Create a visual overview in which you establish a coordinate system, show the positions of the charges, show the force vectors on the charges, define distances and angles, and identify what the problem is trying to find.

SOLVE The magnitude of the force between point charges is given by Coulomb's law:

$$F_{1 \text{ on } 2} = F_{2 \text{ on } 1} = \frac{K|q_1||q_2|}{r^2}$$

Use your visual overview as a guide to the use of this law:

- Show the directions of the forces—repulsive for like charges, attractive for opposite charges—on the visual overview.
- When possible, do graphical vector addition on the visual overview. While not exact, it tells you the type of answer you should expect.
- Write each force vector in terms of its x- and y-components, then add the components to find the net force. Use the visual overview to determine which components are positive and which are negative.

ASSESS Check that your result has the correct units, is reasonable, and answers the question.

Exercise 21 

EXAMPLE 20.3 Adding electric forces in one dimension

Two +10 nC charged particles are 2.0 cm apart on the x-axis. What is the net force on a +1.0 nC charge midway between them? What is the net force if the charged particle on the right is replaced by a −10 nC charge?

PREPARE We proceed using the steps of Problem-Solving Strategy 20.1. We model the charged particles as point charges. The visual overview of **FIGURE 20.16** establishes a coordinate system and shows the forces $\vec{F}_{1 \text{ on } 3}$ and $\vec{F}_{2 \text{ on } 3}$. Figure 20.16a shows a +10 nC charge on the right; Figure 20.16b shows a −10 nC charge.

FIGURE 20.16 A visual overview of the forces for the two cases.

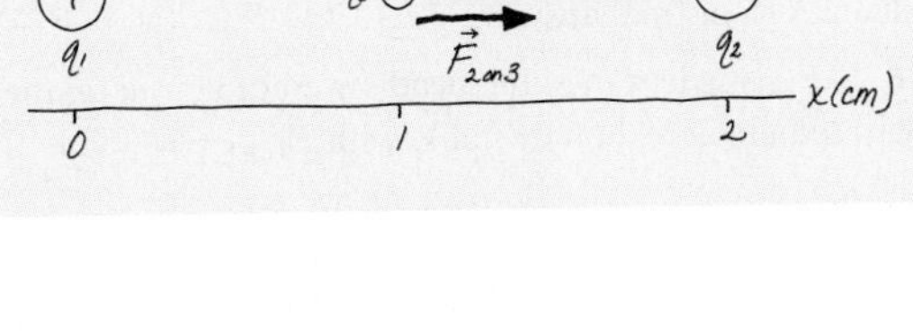

SOLVE Electric forces are vectors, and the net force on q_3 is the *vector* sum $\vec{F}_{\text{net}} = \vec{F}_{1 \text{ on } 3} + \vec{F}_{2 \text{ on } 3}$. Charges q_1 and q_2 each exert a repulsive force on q_3, but these forces are equal in magnitude and opposite in direction. Consequently, $\vec{F}_{\text{net}} = \vec{0}$. The situation changes if q_2 is negative, as in Figure 20.16b. In this case, the two forces are equal in magnitude but in the *same* direction, so $\vec{F}_{\text{net}} = 2\vec{F}_{1 \text{ on } 3}$. The magnitude of the force is given by Coulomb's law. The force due to q_1 is

$$F_{1 \text{ on } 3} = \frac{K|q_1||q_3|}{r_{13}^2}$$

$$= \frac{(9.0 \times 10^9 \text{ N} \cdot \text{m}^2/\text{C}^2)(10 \times 10^{-9} \text{ C})(1.0 \times 10^{-9} \text{ C})}{(0.010 \text{ m})^2}$$

$$= 9.0 \times 10^{-4} \text{ N}$$

There is an equal force due to q_2, so the net force on the 1.0 nC charge is $\vec{F}_{\text{net}} = (1.8 \times 10^{-3} \text{ N, to the right})$.

ASSESS This example illustrates the important idea that electric forces are *vectors*. An important part of assessing our answer is to see if it is "reasonable." In the second case, the net force on the charge is approximately 1 mN. Generally, charges of a few nC separated by a few cm experience forces in the range from a fraction of a mN to several mN. With this guideline, the answer appears to be reasonable.

EXAMPLE 20.4 Adding electric forces in two dimensions

Three charged particles with $q_1 = -50$ nC, $q_2 = +50$ nC, and $q_3 = +30$ nC are placed as shown in **FIGURE 20.17**. What is the net force on charge q_3 due to the other two charges?

FIGURE 20.17 The arrangement of the charges.

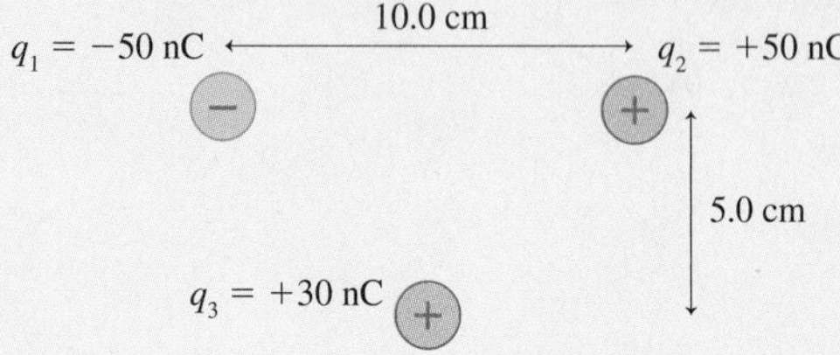

PREPARE We solve for the net force using the steps of Problem-Solving Strategy 20.1, beginning with the visual overview shown in **FIGURE 20.18a**. We have defined a coordinate system, with charge q_3 at the origin. We have drawn the forces on charge q_3, with directions determined by the signs of the charges. We can see from the geometry that the forces $\vec{F}_{1\text{on}3}$ and $\vec{F}_{2\text{on}3}$ are at the angles noted in the figure. The vector addition in **FIGURE 20.18b** shows the anticipated direction of the net force; this will be a good check on our final result. The distance between charges q_1 and q_3 is the same as that between charges q_2 and q_3; this distance r is $= \sqrt{(5.0 \text{ cm})^2 + (5.0 \text{ cm})^2} = 7.07$ cm.

FIGURE 20.18 A visual overview of the charges and forces.

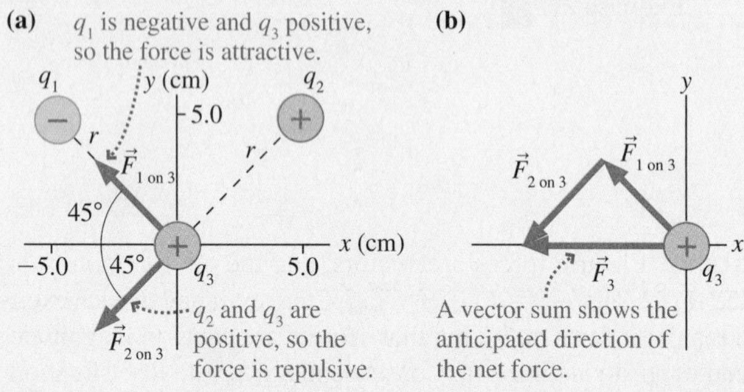

SOLVE We are interested in the net force on charge q_3. Let's start by using Coulomb's law to compute the magnitudes of the two forces on charge q_3:

$$F_{1\text{on}3} = \frac{K|q_1||q_3|}{r^2}$$

$$= \frac{(9.0 \times 10^9 \text{ N}\cdot\text{m}^2/\text{C}^2)(50 \times 10^{-9} \text{ C})(30 \times 10^{-9} \text{ C})}{(0.0707 \text{ m})^2}$$

$$= 2.7 \times 10^{-3} \text{ N}$$

The magnitudes of the charges and the distance are the same for $F_{2\text{on}3}$, so

$$F_{2\text{on}3} = \frac{K|q_2||q_3|}{r^2} = 2.7 \times 10^{-3} \text{ N}$$

The components of these forces are illustrated in **FIGURE 20.19a**.

FIGURE 20.19 The net force on q_3 is to the left.

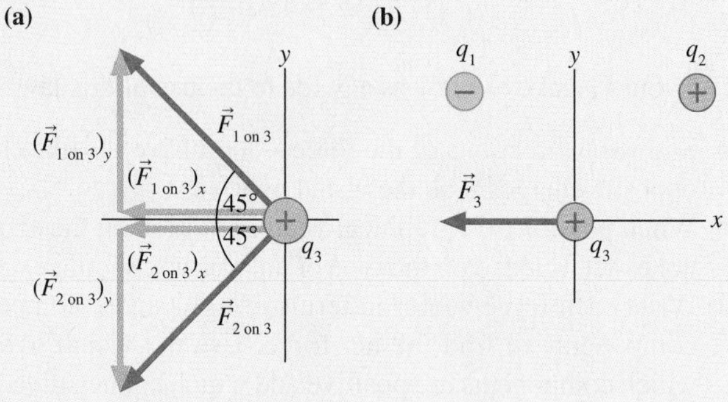

Computing values for the components, we find

$$(F_{1\text{on}3})_x = -(2.7 \times 10^{-3} \text{ N})\cos 45° = -1.9 \times 10^{-3} \text{ N}$$

$$(F_{1\text{on}3})_y = (2.7 \times 10^{-3} \text{ N})\sin 45° = 1.9 \times 10^{-3} \text{ N}$$

$$(F_{2\text{on}3})_x = -(2.7 \times 10^{-3} \text{ N})\cos 45° = -1.9 \times 10^{-3} \text{ N}$$

$$(F_{2\text{on}3})_y = -(2.7 \times 10^{-3} \text{ N})\sin 45° = -1.9 \times 10^{-3} \text{ N}$$

Next, we add components of the net force:

$$F_{3x} = (F_{1\text{on}3})_x + (F_{2\text{on}3})_x = -1.9 \times 10^{-3} \text{ N} - 1.9 \times 10^{-3} \text{ N}$$
$$= -3.8 \times 10^{-3} \text{ N}$$

$$F_{3y} = (F_{1\text{on}3})_y + (F_{2\text{on}3})_y$$
$$= +1.9 \times 10^{-3} \text{ N} - 1.9 \times 10^{-3} \text{ N} = 0$$

Thus the net force, as shown in **FIGURE 20.19b**, is

$$\vec{F}_3 = (3.8 \times 10^{-3} \text{ N}, -x\text{-direction})$$

ASSESS The net force is directed to the left, as we anticipated. The magnitude of the net force, a few mN, seems reasonable as well.

EXAMPLE 20.5 Comparing electric and gravitational forces

A small plastic sphere is charged to −10 nC. It is held 1.0 cm above a small glass bead at rest on a table. The bead has a mass of 15 mg and a charge of +10 nC. Will the glass bead "leap up" to the plastic sphere?

PREPARE We model the plastic sphere and glass bead as point charges. **FIGURE 20.20** establishes a y-axis, identifies the plastic sphere as q_1 and the glass bead as q_2, and shows a free-body diagram. We don't yet know the relative magnitudes of the gravitational and electric forces. In our diagram, we assume that $F_{1\text{on}2} < w$, so there is an additional upward normal force. This choice allows us to complete the diagram, but it does not affect our calculations or final answer.

SOLVE If $F_{1\text{on}2}$ is less than the bead's weight $w = m_2 g$, then the bead will remain at rest on the table with $\vec{F}_{1\text{on}2} + \vec{w} + \vec{n} = \vec{0}$. But

FIGURE 20.20 A visual overview showing the charges and forces.

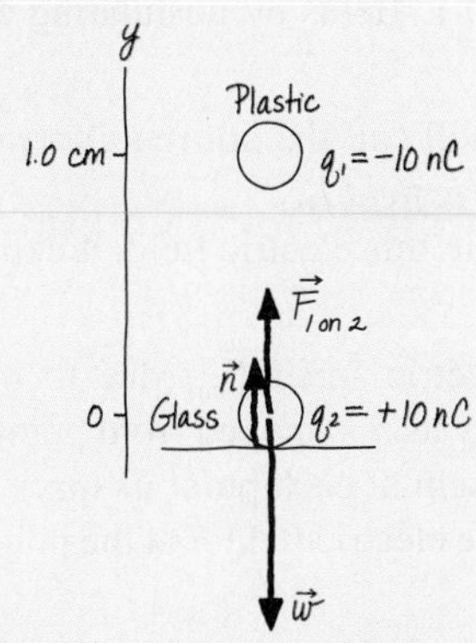

if $F_{1 \text{ on } 2}$ is greater than the bead's weight, the glass bead will accelerate upward from the table. Using the values provided, we have

$$F_{1 \text{ on } 2} = \frac{K|q_1||q_2|}{r^2} = 9.0 \times 10^{-3} \text{ N}$$

$$w = m_2 g = 1.5 \times 10^{-4} \text{ N}$$

$F_{1 \text{ on } 2}$ exceeds the bead's weight by a factor of 60, so the glass bead will leap upward.

ASSESS The answer is different from what we assumed in the diagram, but this assumption did not affect the final result. The values used in this example are realistic for spheres ≈2 mm in diameter. In general, as in this example, electric forces are *significantly* larger than weight forces. Consequently, we can neglect weight forces when working electric-force problems unless the particles are fairly massive.

STOP TO THINK 20.3 Charges 1 and 2 exert repulsive forces on each other. $q_1 = 4q_2$. Which statement is true?

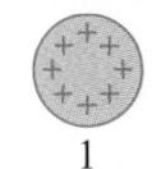

A. $F_{1 \text{ on } 2} > F_{2 \text{ on } 1}$ B. $F_{1 \text{ on } 2} = F_{2 \text{ on } 1}$ C. $F_{1 \text{ on } 2} < F_{2 \text{ on } 1}$.

20.4 The Concept of the Electric Field

Coulomb's law is the basic law of electrostatics. We can use Coulomb's law to calculate the force a positive charge exerts on a nearby negative charge. But there is an unanswered question: How does the negative charge "know" that the positive charge is there? Coulomb's law tells us how to calculate the magnitude and direction of the force, but it doesn't tell us how the force is transmitted through empty space from one charge to the other. To answer this question, we will introduce the *field model,* first suggested in the early 19th century by Michael Faraday, a British investigator of electricity and magnetism.

FIGURE 20.21 shows a photograph of the surface of a shallow pan of oil with tiny grass seeds floating on it. When charged spheres, one positive and one negative, touch the surface of the oil, the grass seeds line up to form a regular pattern. The pattern suggests that some kind of electric influence from the charges *fills the space* around the charges. Perhaps the grass seeds are reacting to this influence, creating the pattern that we see. This alteration of the space around the charges could be the *mechanism* by which the long-range Coulomb's law force is exerted.

FIGURE 20.21 Visualizing the electric field.

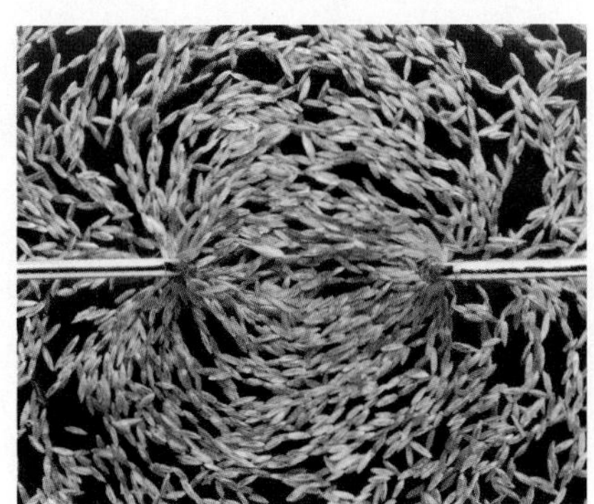

This is the essence of the field model. Consider the attractive force between a positive charge A and a negative charge B. **FIGURE 20.22** shows the difference between the force model, which we have been using, and the field model. **In the field model, it is the alteration of space around charge A that is the *agent* that exerts a force on charge B.** This alteration of space is what we call a **field.** The charge makes an alteration *everywhere* in space. Other charges then respond to the alteration at their position.

The field model applies to many branches of physics. The space around a charge is altered to create the **electric field.** The alteration of the space around a mass is called the **gravitational field.** The alteration of the space around a magnet is called the **magnetic field,** which we will consider in Chapter 24.

FIGURE 20.22 The force and field models for the interaction between two charges.

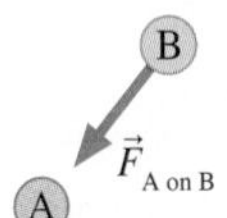

In the force model, A exerts a force directly on B.

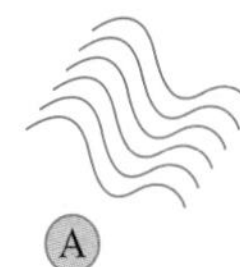

In the field model, A alters the space around it. (The wavy lines are poetic license. We'll soon learn a better representation.)

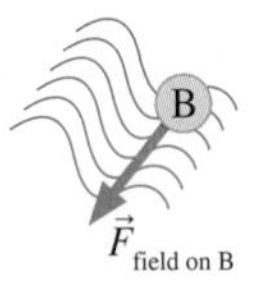

Particle B then responds to the altered space. The altered space is the agent that exerts the force on B.

The Field Model

The field model might seem arbitrary and abstract. We can already calculate forces between charges; why introduce another way of looking at things? We will find that the field model is a very useful tool for visualizing and calculating forces for complex arrangements of charges. However, there's a more fundamental reason for introducing the electric field. When we begin to study fields that change with time, we'll find phenomena that can be understood *only* in terms of fields.

We begin our investigation of electric fields by postulating a **field model** that describes how charges interact:

1. A group of charges, which we will call the **source charges,** alter the space around them by creating an *electric field* $\vec{E}$.
2. If another charge is then placed in this electric field, it experiences a force $\vec{F}$ exerted *by the field.*

Suppose charge q experiences an electric force $\vec{F}_{\text{on}\,q}$ due to other charges. The strength and direction of this force vary as q is moved from point to point in space. This suggests that "something" is present at each point in space to cause the force that charge q experiences. We define the electric field $\vec{E}$ at the point (x, y, z) as

$$\vec{E} \text{ at } (x, y, z) = \frac{\vec{F}_{\text{on}\,q} \text{ at } (x, y, z)}{q} \tag{20.3}$$

Electric field at a point defined by the force on charge q

We're *defining* the electric field as a force-to-charge ratio; hence the units of the electric field are newtons per coulomb, or N/C. The magnitude E of the electric field is called the **electric field strength.** Typical electric field strengths are given in Table 20.2.

You can think of using charge q as a *probe* to determine if an electric field is present at a point in space. If charge q experiences an electric force at a point in space, as FIGURE 20.23a shows, then there is an electric field at that point causing the force. Further, we *define* the electric field at that point to be the vector given by Equation 20.3. FIGURE 20.23b shows the electric field at two points, but you can imagine "mapping out" the electric field by moving the charge q all through space.

FIGURE 20.23 Charge q is a probe of the electric field.

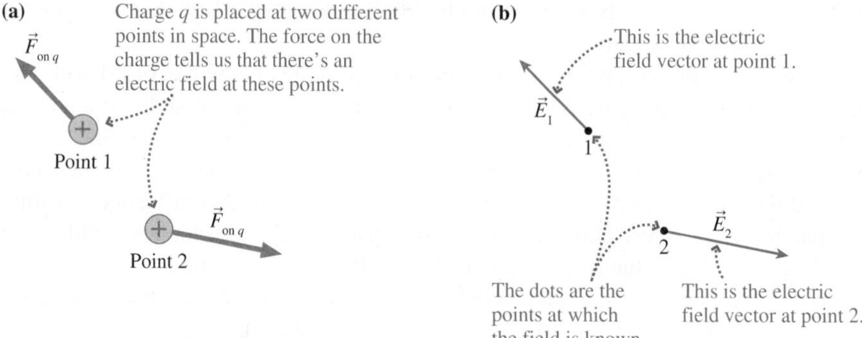

TABLE 20.2 Typical electric field strengths

Field	Field strength (N/C)
Inside a current-carrying wire	10^{-2}
Earth's field, near the earth's surface	10^{2}
Near objects charged by rubbing	10^{3} to 10^{6}
Needed to cause a spark in air	10^{6}
Inside a cell membrane	10^{7}
Inside an atom	10^{11}

The basic idea of the field model is that **the field is the agent that exerts an electric force on charge q.** Notice three important things about the field:

1. The electric field, a vector, exists at every point in space. Electric field diagrams will show a sample of the vectors, but there is an electric field vector at every point whether one is shown or not.
2. If the probe charge q is positive, the electric field vector points in the same direction as the force on the charge; if negative, the electric field vector points opposite the force.
3. Because q appears in Equation 20.3, it may seem that the electric field depends on the magnitude of the charge used to probe the field. It doesn't. We know from Coulomb's law that the force $\vec{F}_{\text{on}\,q}$ is proportional to q. Thus the electric field defined in Equation 20.3 is *independent* of the charge q that probes the field. The electric field depends only on the source charges that create the field.

The Electric Field of a Point Charge

FIGURE 20.24a shows a point source charge q that creates an electric field at all points in space. We can use a second charge, shown as q' in **FIGURE 20.24b**, to serve as a probe of the electric field created by charge q.

For the moment, assume both charges are positive. The force on q', which is repulsive and points directly away from q, is given by Coulomb's law:

$$\vec{F}_{\text{on } q'} = \left(\frac{Kqq'}{r^2}, \text{ away from } q\right) \tag{20.4}$$

Equation 20.3 defines the electric field in terms of the force on the probe charge as $\vec{E} = \vec{F}_{\text{on } q'}/q'$, so for a positive charge q,

$$\vec{E} = \left(\frac{Kq}{r^2}, \text{ away from } q\right) \tag{20.5}$$

The electric field is shown in **FIGURE 20.24c**.

If q is negative, the magnitude of the force on the probe charge is the same as in Equation 20.5, but the direction is toward q, so the general expression for the field is

$$\vec{E} = \left(\frac{K|q|}{r^2}, \begin{bmatrix} \text{away from } q \text{ if } q > 0 \\ \text{toward } q \text{ if } q < 0 \end{bmatrix}\right) \tag{20.6}$$

Electric field of point charge q at a distance r from the charge

E, r, p. 186, INVERSE-SQUARE

NOTE ▶ The expression for the electric field is similar to Coulomb's law. To distinguish the two, remember that Coulomb's law has the product of two charges in the numerator. It describes the force between *two* charges. The electric field has a single charge in the numerator. It is the field of a *single* charge. ◀

FIGURE 20.24 Charge q' is used to probe the electric field of point charge q.

(a) What is the electric field of q at this point?

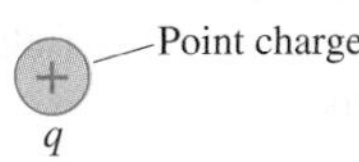

(b) 1. Place q' at the point to probe the field.
$\vec{F}_{\text{on } q'}$
q'
r
q
2. Measure the force on q'.

(c)

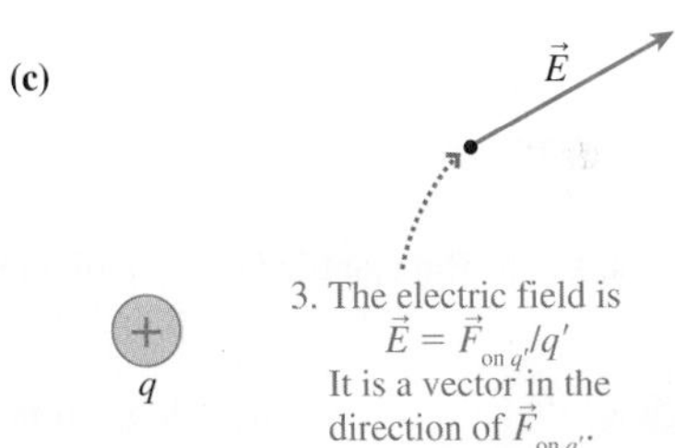

$\vec{E}$
q
3. The electric field is $\vec{E} = \vec{F}_{\text{on } q'}/q'$
It is a vector in the direction of $\vec{F}_{\text{on } q'}$.

EXAMPLE 20.6 Finding the electric field of a proton

The electron in a hydrogen atom orbits the proton at a radius of 0.053 nm. What is the electric field due to the proton at the position of the electron?

SOLVE The proton's charge is $q = e$. At the distance of the electron, the magnitude of the field is

$$E = \frac{Ke}{r^2} = \frac{(9.0 \times 10^9 \text{ N} \cdot \text{m}^2/\text{C}^2)(1.60 \times 10^{-19} \text{ C})}{(5.3 \times 10^{-11} \text{ m})^2}$$

$$= 5.1 \times 10^{11} \text{ N/C}$$

Because the proton is positive, the electric field is directed away from the proton:

$$\vec{E} = (5.1 \times 10^{11} \text{ N/C, outward from the proton})$$

ASSESS This is a large field, but Table 20.2 shows that this is the correct magnitude for the field within an atom.

FIGURE 20.25 The electric field near a point charge.

(a) The electric field diagram of a positive point charge

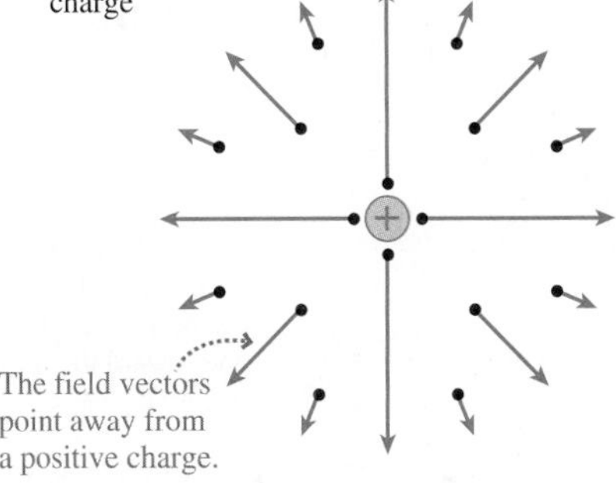

(b) The electric field diagram of a negative point charge

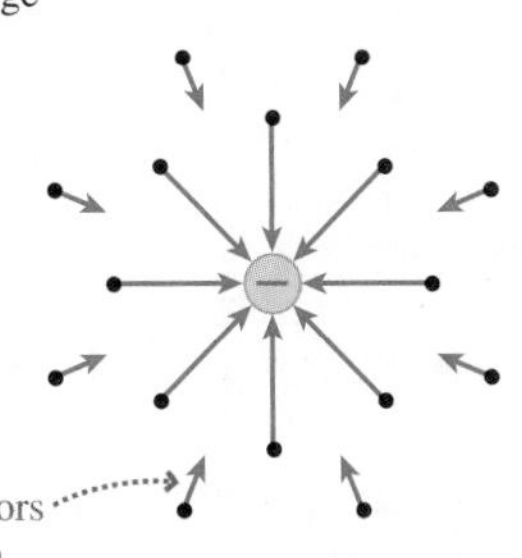

By drawing electric field vectors at a number of points around a positive point charge, we can construct an **electric field diagram** such as the one shown in **FIGURE 20.25a**. Notice that the field vectors all point straight away from charge q. We can draw a field diagram for a negative point charge in a similar fashion, as in **FIGURE 20.25b**. In this case, the field vectors point toward the charge, as this would be the direction of the force on a positive probe charge.

In the coming sections, as we use electric field diagrams, keep these points in mind:

1. The diagram is just a representative sample of electric field vectors. The field exists at all the other points. A well-drawn diagram gives a good indication of what the field would be like at a neighboring point.

2. The arrow indicates the direction and the strength of the electric field *at the point to which it is attached*—at the point where the *tail* of the vector is placed. The length of any vector is significant only relative to the lengths of other vectors.
3. Although we have to draw a vector across the page, from one point to another, an electric field vector does not "stretch" from one point to another. Each vector represents the electric field at *one point* in space.

STOP TO THINK 20.4 Rank in order, from largest to smallest, the electric field strengths E_A to E_D at points A to D.

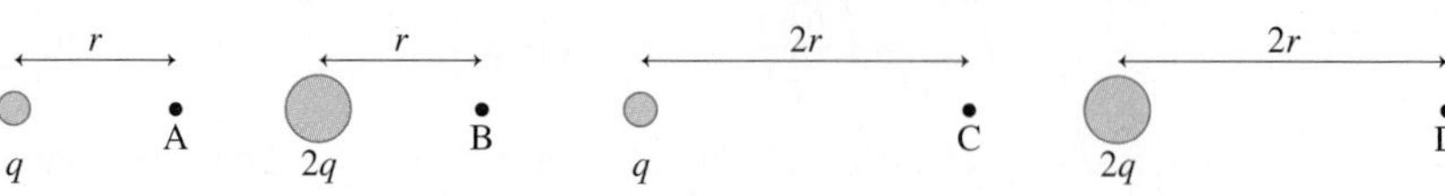

20.5 Applications of the Electric Field

Suppose we want to find the electric field due to more than one source charge. No matter what the number of source charges, the electric field at a point in space can be found by looking at the force on a probe charge. Because the net force on the probe charge is the vector sum of the forces due to all of the individual charges, **the electric field due to multiple charges is the vector sum of the electric field due to each of the charges.**

EXAMPLE 20.7 Finding the field near a dipole

A dipole consists of a positive and negative charge separated by 1.2 cm, as shown in FIGURE 20.26. What is the electric field strength along the line connecting the charges at a point 1.2 cm to the right of the positive charge?

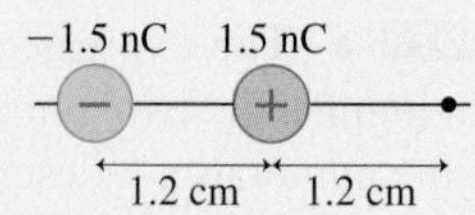

FIGURE 20.26 Charges and distances for a dipole.

PREPARE We define the x-axis to be along the line connecting the two charges, as in FIGURE 20.27. The dipole has no net charge, but it does have a net electric field. The point at which we calculate the field is 1.2 cm from the positive charge and 2.4 cm from the negative charge. Thus the electric field of the positive charge will be larger, as shown in Figure 20.27. The net electric field of the dipole is the vector sum of these two fields, so the electric field of the dipole at this point is in the positive x-direction.

FIGURE 20.27 Visual overview for finding the electric field.

The dipole electric field at this point is in the positive x-direction.

−1.5 nC 1.5 nC $\vec{E}_{dipole}$ x $\vec{E}_-$ $\vec{E}_+$

$E_- < E_+$ because the + charge is closer.

SOLVE The magnitudes of the fields of the two charges are given by Equation 20.6, so the magnitude of the dipole field is

$$E_{dipole} = E_+ - E_-$$

$$= \frac{\left(9.0 \times 10^9 \, \frac{\text{N} \cdot \text{m}^2}{\text{C}^2}\right)(1.5 \times 10^{-9} \text{ C})}{(0.012 \text{ m})^2} - \frac{\left(9.0 \times 10^9 \, \frac{\text{N} \cdot \text{m}^2}{\text{C}^2}\right)(1.5 \times 10^{-9} \text{ C})}{(0.024 \text{ m})^2}$$

$$= 7.0 \times 10^4 \text{ N/C}$$

ASSESS Table 20.2 lists the fields due to objects charged by rubbing as typically 10^3 to 10^6 N/C, and we've already seen that charges caused by rubbing are in the range of 1–10 nC. Our answer is in this range and thus is reasonable.

FIGURE 20.28 The electric field of a dipole.

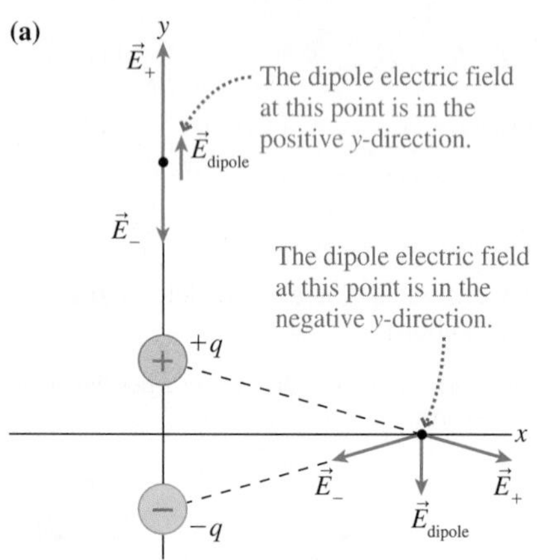

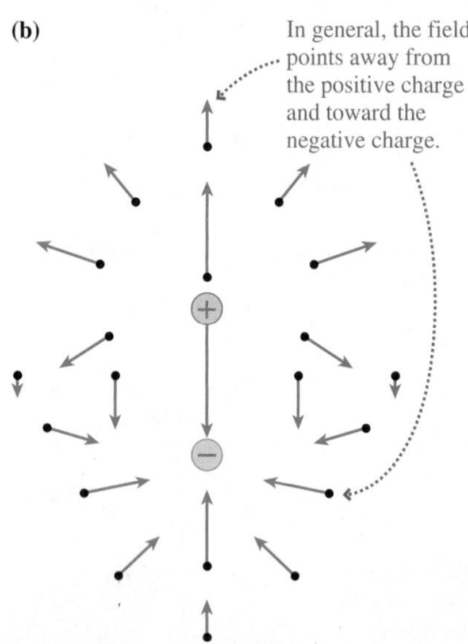

(b)
In general, the field points away from the positive charge and toward the negative charge.

The electric dipole is an important charge distribution that we will see many times, so it's worth exploring the full field diagram. FIGURE 20.28 shows a dipole oriented along the y-axis. We can determine the field at any point by a vector addition of the fields of the two charges, as shown in Figure 20.28a. If we repeat this process at many points, we end up with the field diagram of Figure 20.28b. This is more complex than the field of a single charge, but it accurately shows how two charges alter the space around them.

Uniform Electric Fields

FIGURE 20.29a shows another important practical situation, one we'll meet many times. Two conducting plates, called **electrodes,** are face-to-face with a narrow gap between them. One electrode has total charge $+Q$ and the other has total charge $-Q$. This arrangement of two electrodes, closely spaced and charged equally but oppositely, is called a **parallel-plate capacitor.** What is the electric field between the two plates? To keep things simple, we will focus on the field in the central region, far from the edges. **FIGURE 20.29b** shows a blown-up cross-section view of a region near the center of the plates.

At any point, the electric field is the vector sum of the fields from all of the positive charges and all of the negative charges on the plates. However, the field of a point charge decreases inversely with the square of its distance, so in practice only the nearby charges contribute to the field. As **FIGURE 20.30a** shows, the horizontal components of the individual fields cancel, while the vertical components add to give an electric field vector pointing from the positive plate toward the negative plate. The exact position of the point we've chosen is not crucial; moving either right or left would produce a similar result.

By mapping the electric field at many points, we find that the field inside a parallel-plate capacitor is the same—in both strength and direction—at every point. This is called a **uniform electric field. FIGURE 20.30b** shows that a uniform electric field is represented with parallel electric field vectors of equal length. A more detailed analysis finds that the electric field inside a parallel-plate capacitor is

$$\vec{E}_{\text{capacitor}} = \left(\frac{Q}{\epsilon_0 A}, \text{ from positive to negative}\right) \tag{20.7}$$

Electric field in a parallel-plate capacitor with plate area A and charge Q

E p.37 Q PROPORTIONAL

Equation 20.7 introduces a new constant ϵ_0, pronounced "epsilon zero" or "epsilon naught," called the **permittivity constant.** Its value is related to the electrostatic constant as

$$\epsilon_0 = \frac{1}{4\pi K} = 8.85 \times 10^{-12}\ \text{C}^2/\text{N}\cdot\text{m}^2$$

There are a few things to note about the field in a parallel-plate capacitor:

- The field depends on the charge-to-area ratio Q/A, which is often called the *charge density.* If the charges are packed more closely, the field will be larger.
- Our analysis requires that the separation of the plates be small compared to their size. If this is true, the spacing between the plates does not affect the electric field, *and this spacing does not appear in Equation 20.7.*
- Although Figure 20.29 shows circular electrodes, the shape of the electrodes—circular or square or any other shape—is not relevant as long as the electrodes are very close together.

NOTE ► The charges on the plates are equal and opposite, $+Q$ and $-Q$, so the net charge is zero. The symbol Q in Equation 20.7 is the *magnitude* of the charge on each plate. ◄

FIGURE 20.29 A parallel-plate capacitor.

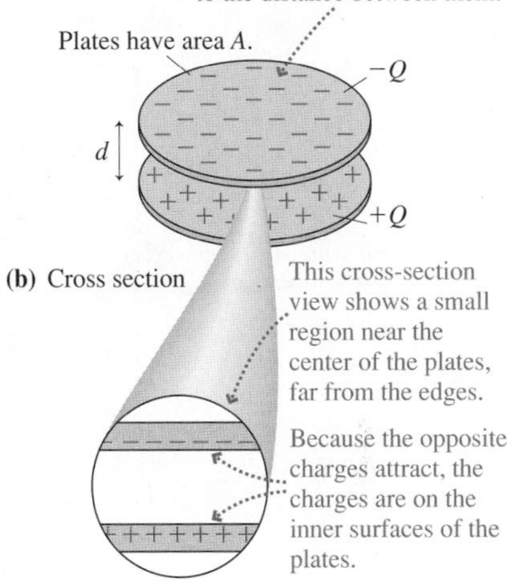

FIGURE 20.30 The electric field inside a parallel-plate capacitor.

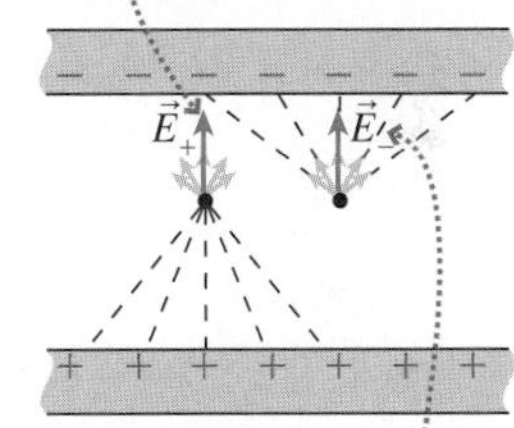

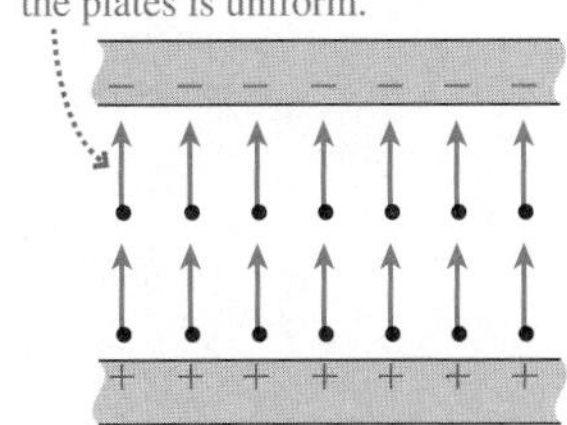

(b) The electric field between the plates is uniform.

EXAMPLE 20.8 **Finding the field in an air cleaner**

Long highway tunnels must have air cleaners to remove dust and soot coming from passing cars and trucks. In one type, known as an *electrostatic precipitator,* air passes between two oppositely charged metal plates, as in **FIGURE 20.31**. The large electric field between the plates ionizes dust and soot particles, which then feel a force due to the field. This force causes the charged particles to move toward and stick to one or the other plate, removing them from the air. A typical unit has dimensions and charges as shown in Figure 20.31. What is the electric field between the plates?

FIGURE 20.31 An electrostatic precipitator.

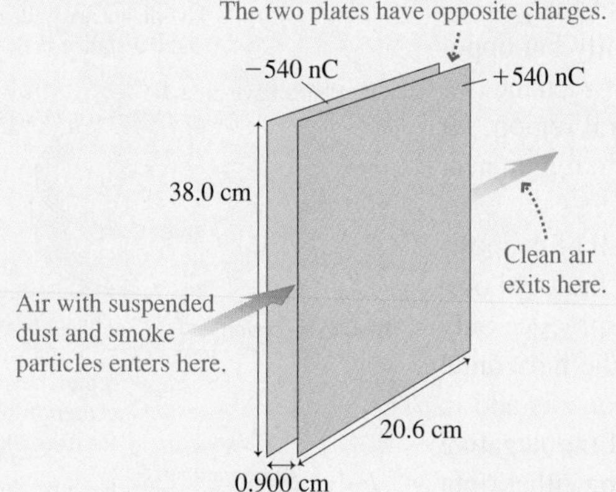
The two plates have opposite charges.
−540 nC
+540 nC
38.0 cm
Clean air exits here.
Air with suspended dust and smoke particles enters here.
20.6 cm
0.900 cm

PREPARE Because the spacing between the plates is much smaller than their size, this is a parallel-plate capacitor with a uniform electric field between the plates.

SOLVE We find the field using Equation 20.7. The direction is from the positive to the negative plate, which is to the left. The area of the plates is $A = (0.206\text{ m})(0.380\text{ m}) = 0.0783\text{ m}^2$, so the field strength between the plates is

$$E = \frac{Q}{\epsilon_0 A} = \frac{540 \times 10^{-9}\text{ C}}{(8.85 \times 10^{-12}\text{ C}^2/\text{N}\cdot\text{m}^2)(0.0783\text{ m}^2)}$$
$$= 7.79 \times 10^5\text{ N/C}$$

The question asked for the electric field, a vector, not just for the field strength. The electric field between the plates is

$$\vec{E} = (7.79 \times 10^5\text{ N/C, to the left})$$

ASSESS Table 20.2 shows that a field of 10^6 N/C will create a spark in air. The field we calculated between the plates is just a bit less than this, which makes sense. The field should be large, but not large enough to make a spark jump between the plates!

Electric Field Lines

11.4, 11.5, 11.6 ActivPhysics ONLINE

We can't see the electric field, so we use pictorial tools like electric field diagrams to help us visualize the electric field in a region of space. Another way to picture the field is to draw **electric field lines.** These are imaginary lines drawn through a region of space so that

- The tangent to a field line at any point is in the direction of the electric field $\vec{E}$ at that point, and
- The field lines are closer together where the electric field strength is greater.

FIGURE 20.32a shows the relationship between electric field lines and electric field vectors in one region of space. If we know what the field vectors look like, we can extrapolate to the field lines, as in **FIGURES 20.32b** and **20.32c** for the electric field lines near a positive charge and between the plates of a capacitor.

FIGURE 20.32 Field vectors and field lines.

(a) Relationship between field vectors and field lines

The electric field vector is tangent to the electric field line.

Field vector

Field line

The electric field is stronger where the electric field vectors are longer and where the electric field lines are closer together.

(b) Field lines of a positive point charge

The field is directed away from the positive charge, so the field lines are directed radially outward.

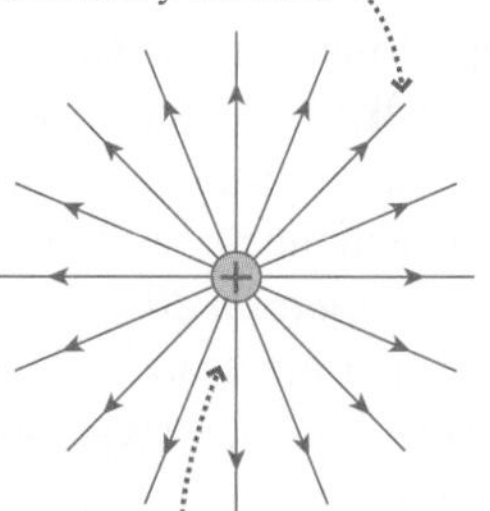

The field lines are closest together near the charge, where the field strength is greatest.

(c) Field lines in an ideal capacitor

The field vectors are directed from the positive to the negative plate, so the field lines are as well.

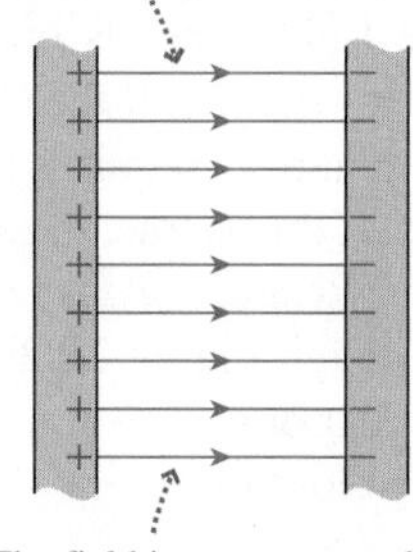

The field is constant, so the field lines are evenly spaced.

If we have an arrangement of charges, we can draw field lines as a guide to what the field looks like. As you generate a field line picture, there are two rules to keep in mind:

- Field lines cannot cross. The tangent to the field line is the electric field vector, which indicates the direction of the force on a positive charge. The force must be in a unique, well-defined direction, so two field lines cannot cross.
- The electric field is created by charges. Field lines start on a positive charge and end on a negative charge.

You can use the above information as the basis of a technique for sketching a field-line picture for an arrangement of charges. Draw field lines starting on positive charges and moving toward negative charges. Draw the lines tangent to the field vector at each point. Make the lines close together where the field is strong, far apart where the field is weak. For example, FIGURE 20.33 pictures the electric field of a dipole using electric field lines. You should compare this to Figure 20.28b, which illustrated the field with field vectors.

FIGURE 20.33 Electric field lines for a dipole.

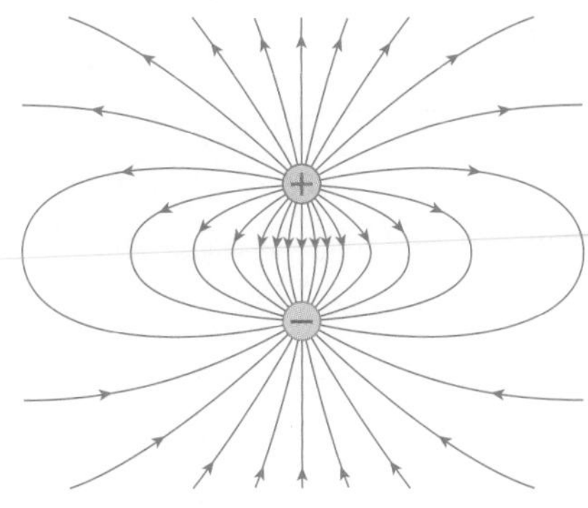
+
−

The Electric Field of the Heart

Nerve and muscle cells have a prominent electrical nature. As we will see in detail in Chapter 23, a cell membrane is an insulator that encloses a conducting fluid and is surrounded by conducting fluid. While resting, the membrane is *polarized* with positive charges on the outside of the cell, negative charges on the inside. When a nerve or a muscle cell is stimulated, the polarity of the membrane switches; we say that the cell *depolarizes*. Later, when the charge balance is restored, we say that the cell *repolarizes*.

All nerve and muscle cells generate an electrical signal when depolarization occurs, but the largest electrical signal in the body comes from the heart. The rhythmic beating of the heart is produced by a highly coordinated wave of depolarization that sweeps across the tissue of the heart. As FIGURE 20.34a shows, the surface of the heart is positive on one side of the boundary between tissue that is depolarized and tissue that is not yet depolarized, negative on the other. In other words, the heart is a large electric dipole. The orientation and strength of the dipole change during each beat of the heart as the depolarization wave sweeps across it.

The electric dipole of the heart generates a dipole electric field that extends throughout the torso, as shown in FIGURE 20.34b. As we will see in Chapter 21, an *electrocardiogram* measures the changing electric field of the heart as it beats. Measurement of the heart's electric field can be used to diagnose the operation of the heart.

FIGURE 20.34 The beating heart generates a dipole electric field.

(a) The electric dipole of the heart

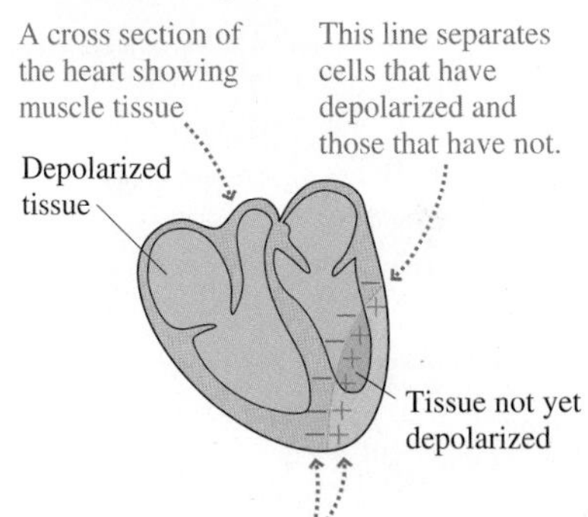
A cross section of the heart showing muscle tissue
This line separates cells that have depolarized and those that have not.
Depolarized tissue
Tissue not yet depolarized
The charge separation at the line between the two regions creates an electric dipole.

(b) The field of the heart in the body

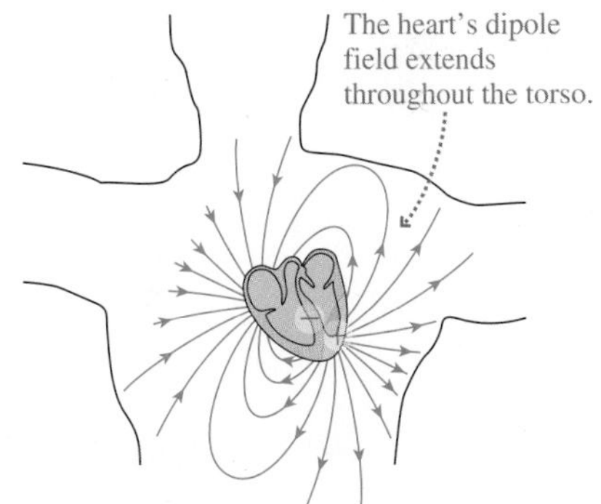
The heart's dipole field extends throughout the torso.

STOP TO THINK 20.5 Which of the following is the correct representation of the electric field created by two positive charges?

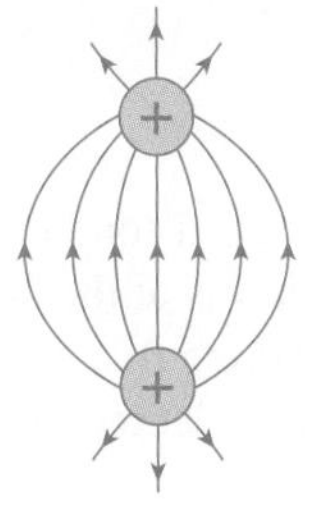
A.

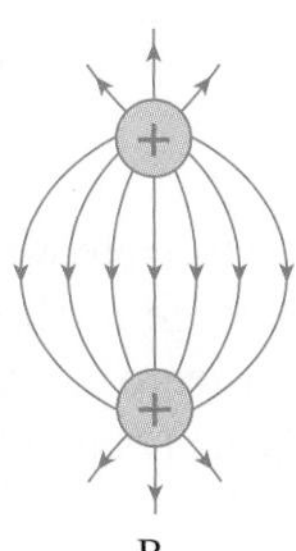
B.

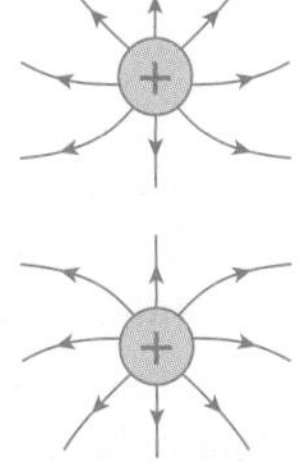
C.

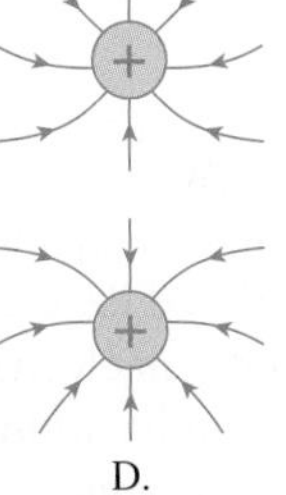
D.

FIGURE 20.35 The electric field inside and outside a charged conductor.

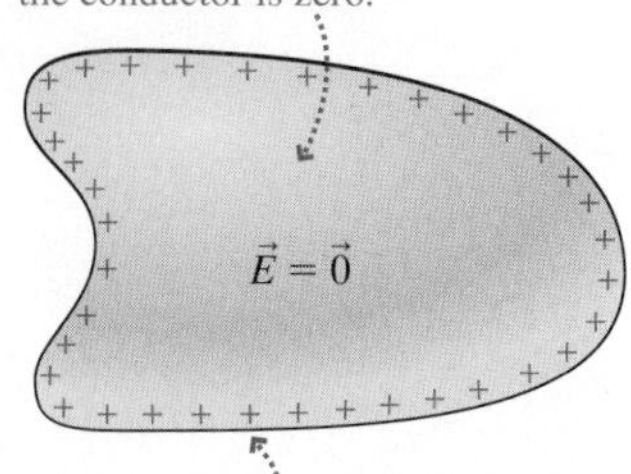

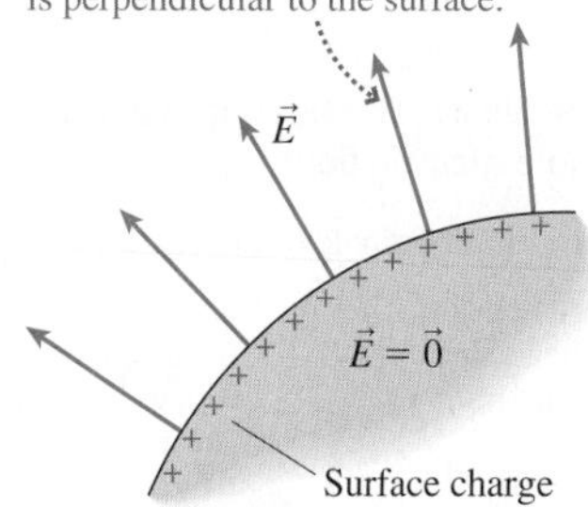

20.6 Conductors and Electric Fields

Consider a conductor in electrostatic equilibrium (recall that this means that none of the charges are moving). Suppose there were an electric field inside the conductor. Electric fields exert forces on charges, so an internal electric field would exert forces on the charges in the conductor. Because charges in a conductor are free to move, these forces would cause the charges to move. But that would violate the assumption that all the charges are at rest. Thus we're forced to conclude that **the electric field is zero at all points inside a conductor in electrostatic equilibrium.**

Because the electric field inside a conductor in electrostatic equilibrium is zero, any *excess* charge on the conductor must lie at its surface, as shown in FIGURE 20.35a. Any charge in the interior of the conductor would create an electric field there, in violation of our conclusion that the field inside is zero. Physically, excess charge ends up on the surface because the repulsive forces between like charges cause them to move as far apart as possible without leaving the conductor.

FIGURE 20.35b shows that **the electric field right at the surface of a charged conductor is perpendicular to the surface.** To see that this is so, suppose $\vec{E}$ had a component tangent to the surface. This component of $\vec{E}$ would exert a force on charges at the surface and cause them to move along the surface, thus violating the assumption that all charges are at rest. The only exterior electric field consistent with electrostatic equilibrium is one that is perpendicular to the surface.

CONCEPTUAL EXAMPLE 20.9 Drawing electric field lines for a charged sphere and a plate

FIGURE 20.36 shows a positively charged metal sphere above a conducting plate with a negative charge. Sketch the electric field lines.

FIGURE 20.36 The charged sphere and plate.

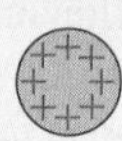

REASON Field lines start on positive charges and end on negative charges. Thus we draw the field lines from the positive sphere to the negative plate, perpendicular to both surfaces, as shown in FIGURE 20.37. The single field line that goes upward tells us that there is a field above the sphere, but that it is weak.

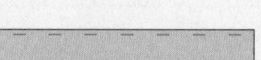

FIGURE 20.37 Drawing field lines from sphere to plate.

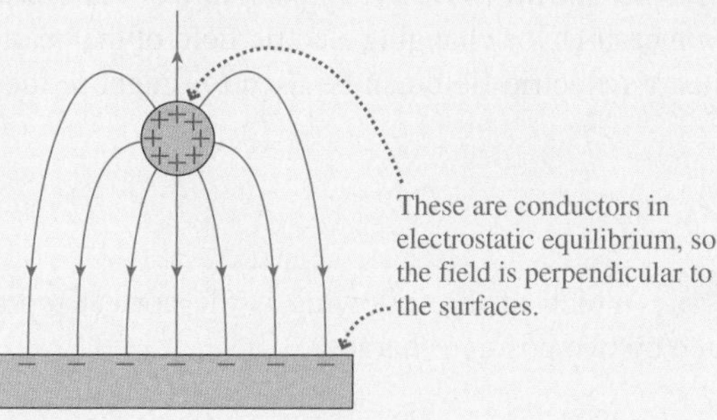

These are conductors in electrostatic equilibrium, so the field is perpendicular to the surfaces.

FIGURE 20.38 A region of space enclosed by conducting walls is screened from electric fields.

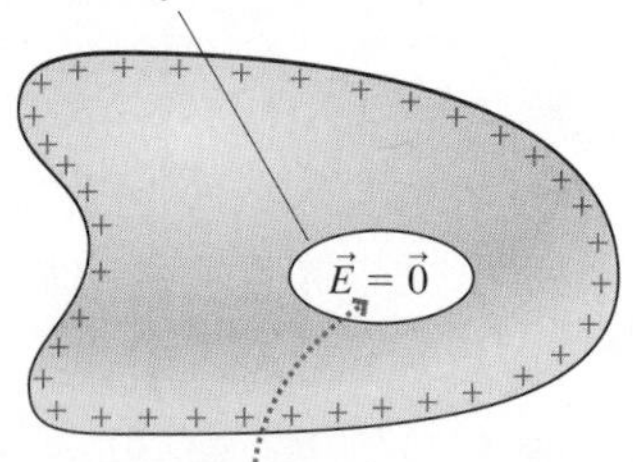

FIGURE 20.38 shows a practical use of these ideas. Here we see a charged conductor with a completely enclosed void. The excess charge on the conductor is at the surface and the electric field within the conductor is zero, so there's nothing that could create an electric field within the enclosure. We can conclude that **the electric field within a conducting enclosure is zero.**

A conducting box can be used to exclude electric fields from a region of space; this is called **screening.** Solid metal walls are ideal, but in practice wire screen or wire mesh provides sufficient screening for all but the most sensitive applications.

CONCEPTUAL EXAMPLE 20.10 **Analyzing static protection**

Computer chips and other electronic components are very sensitive to electric charges and fields. Even a small static charge or field may damage them. Such components are shipped and stored in conducting bags. How do these bags protect the components stored inside?

REASON Such a bag, when sealed, forms a conducting shell around its interior. All excess charge is on the surface of the bag, and the electric field inside is zero. A chip or component inside the bag is protected from damaging charges and fields.

Although any excess charge on a conductor will be found on the surface, it may not be uniformly distributed. **FIGURE 20.39** shows a charged conductor that is more pointed at one end than the other. It turns out that the density of charge is highest—and thus the electric field is strongest—at the pointed end.

FIGURE 20.39 The electric field is strongest at the pointed end.

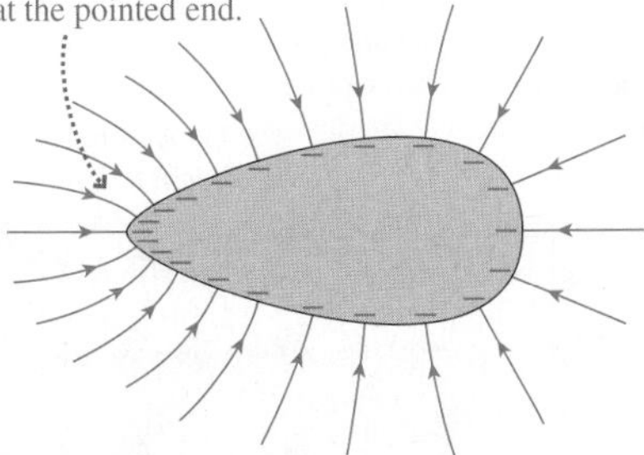
The charges are closer together and the electric field is strongest at the pointed end.

The sharper the point, the stronger the field. The electric field near very sharp points may be strong enough to ionize the air around it. Lightning rods on buildings have such a point at the top. If charge begins to accumulate on the building, meaning a lightning strike might be imminent, a large field develops at the tip of the rod. Once the field ionizes the air, excess charge from the building can dissipate into the air, reducing the electric field and thus reducing the probability of a lightning strike. A lightning rod is intended to *prevent* a lightning strike.

▶ **Electrolocation** BIO Many fish have stacks of specially adapted cells called *electrocytes* that develop electric charges across them. The electrocytes in the tail of this fish (called an elephant nose) form an electric dipole that produces an electric field in the water around it. The elephant nose has sensors along its body that can detect very small changes in this electric field. A nearby conductor—such as another fish—will alter this field. The elephant nose uses these changes in the field to "see" around it. These fish live in very murky water where vision is of little use, so you can see the advantage of having such an alternative form of perception.

20.7 Forces and Torques in Electric Fields

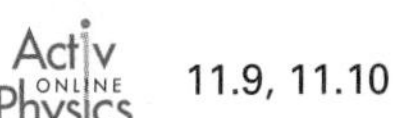
Activ ONLINE Physics
11.9, 11.10

The electric field was defined in terms of the force on a charge. In practice, we often want to turn the definition around to find the force exerted on a charge in a known electric field. If a charge q is placed at a point in space where the electric field is $\vec{E}$, then according to Equation 20.3 the charge experiences an electric force

$$\vec{F}_{\text{on } q} = q\vec{E} \tag{20.8}$$

Force on a charge due to an electric field

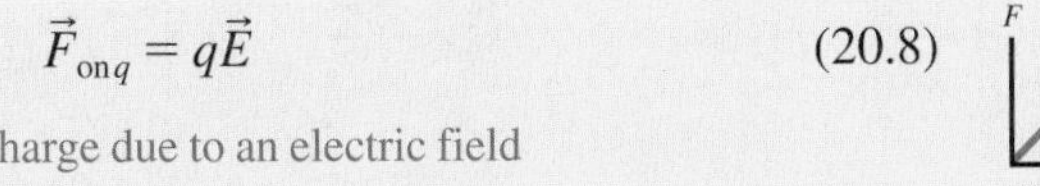
F
p.37
E
PROPORTIONAL

If q is positive, the force on charge q is in the direction of $\vec{E}$. The force on a negative charge is *opposite* the direction of $\vec{E}$.

EXAMPLE 20.11 **Finding the force on an electron in the atmosphere**

Under normal circumstances, the earth's electric field outdoors near ground level is uniform, about 100 N/C, directed down. What is the electric force on a free electron in the atmosphere? What acceleration does this force cause?

PREPARE The electric field is uniform, as shown in the field diagram of **FIGURE 20.40**. Whatever the position of the electron, it experiences the same field. Because the electron is negative, the force on it is opposite the field—upward.

FIGURE 20.40 An electron in the earth's electric field.

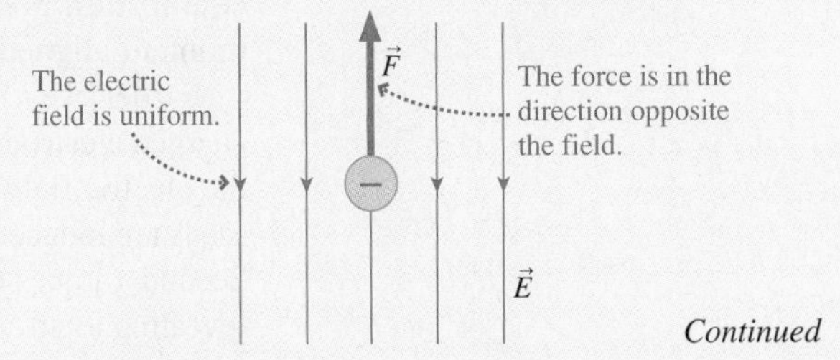
The electric field is uniform.
$\vec{F}$
The force is in the direction opposite the field.
$\vec{E}$

Continued

SOLVE The magnitude of the force is given by Equation 20.8:

$$F = eE = (1.6 \times 10^{-19}\ \text{C})(100\ \text{N/C}) = 1.6 \times 10^{-17}\ \text{N}$$

Thus the force on the electron is

$$\vec{F} = (1.6 \times 10^{-17}\ \text{N, upward})$$

The electron will accelerate upward, in the direction of the force. The magnitude of the acceleration is

$$a = \frac{F}{m} = \frac{1.6 \times 10^{-17}\ \text{N}}{9.1 \times 10^{-31}\ \text{kg}} = 1.8 \times 10^{13}\ \text{m/s}^2$$

ASSESS This everyday field produces an extremely large acceleration on a free electron. Forces and accelerations at the atomic scale are quite different from what we are used to for macroscopic objects.

FIGURE 20.41 Electrophoresis of DNA samples.

The electric field between the electrodes exerts a force on the negatively charged DNA fragments.

DNA samples begin here.

Different fragments have different sizes and migrate at different rates.

Gel

The photo at the start of the chapter showed the colored lines produced by gel electrophoresis of a sample of DNA. The first step of the analysis is to put the sample of DNA into solution. The DNA is then cut into fragments by enzymes. In solution, these fragments have a negative charge. Drops of solution containing the charged DNA fragments are placed in wells at one end of a container of gel. Electrodes at opposite ends of the gel create an electric field that exerts an electric force on the DNA fragments in the solution, as illustrated in **FIGURE 20.41**. The electric force makes the fragments move through the gel, but drag forces cause fragments of different sizes to migrate at different rates, with smaller fragments migrating faster than larger ones. After some time, the fragments sort themselves into distinct lines, creating a "genetic fingerprint." Two identical samples of DNA will produce the same set of fragments and thus the same pattern in the gel, but the odds are extremely small that two unrelated DNA samples would produce the same pattern.

If an electric dipole is placed in a uniform electric field, as shown in **FIGURE 20.42a**, the electric force on its negative charge is equal in magnitude but opposite in direction to the force on its positive charge. Thus an electric dipole in a uniform electric field experiences *no net force*. However, as you can see from **FIGURE 20.42b**, there is a net *torque* on the dipole that causes it to *rotate*.

FIGURE 20.42 Forces and torques on an electric dipole.

(a) Because the forces on the positive and negative charges are equal in magnitude but oppositely directed, there is no net force on the dipole.

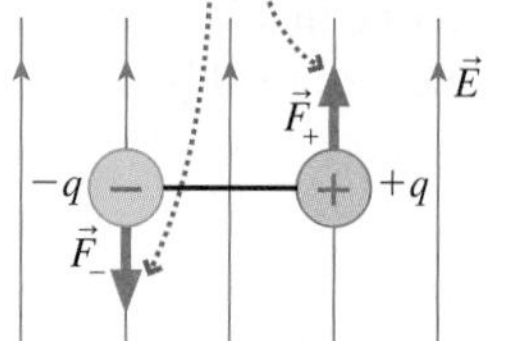

(b) However, there is a net *torque* on the dipole that causes it to *rotate*.

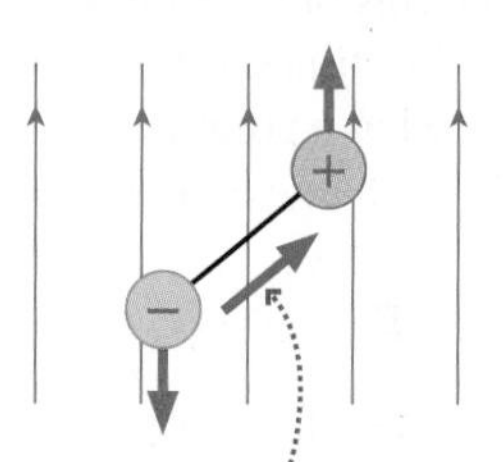

The *dipole moment* is a vector that points from the negative to the positive charge.

(c) When the dipole lines up with the field, the net torque is zero. The dipole is in static equilibrium.

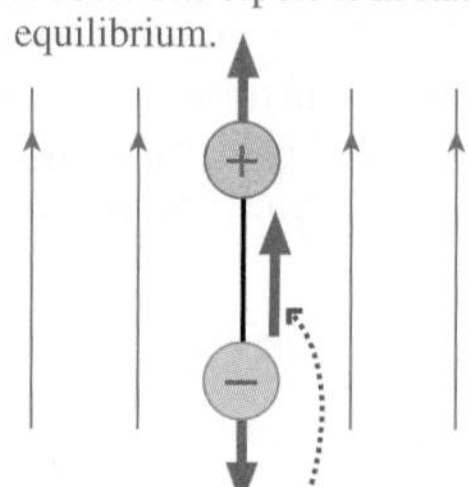

We can say that the dipole moment tries to align itself with the field.

It is useful to define the **electric dipole moment,** a vector pointing from the negative to the positive charge of a dipole. As Figures 20.42b and **20.42c** show, an electric dipole in a uniform electric field experiences a torque that causes it to rotate. **The equilibrium position of a dipole in an electric field is with the electric dipole moment aligned with the field.**

Earlier, we saw a photo of grass seeds lined up with the electric field from two charged electrodes. Now we can understand why the seeds line up as they do. First, the electric field polarizes the seeds, inducing opposite charges on their ends. The seeds are induced electric dipoles, with dipole moments along the axis of each seed. Second, torques on the dipole moments cause them to line up with the electric field, revealing its structure.

STOP TO THINK 20.6 Rank in order, from largest to smallest, the forces F_A to F_E a proton would experience if placed at points A to E in this parallel-plate capacitor.

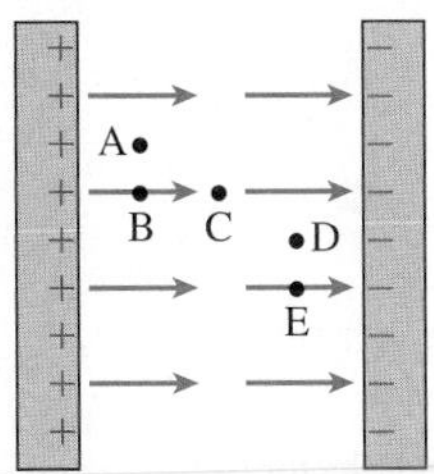

INTEGRATED EXAMPLE 20.12 A cathode-ray tube

Some televisions, older computer monitors, and other electronic equipment use a *cathode-ray tube*, or CRT, to create an image on a screen. In a CRT, electrons are accelerated by an electric field inside an electron "gun," creating a beam of electrons all moving along in a straight line at the same high speed. A second electric field then steers these electrons to a particular point on a phosphor-coated glass screen, causing the phosphor to glow brightly at that point. By rapidly varying the steering electric field and the intensity of the electron beam, the spot of electrons can be swept over the entire screen, resulting in the familiar glowing picture of a television.

FIGURE 20.43 shows a simplified model of the internal structure of a CRT. Electrons—emitted from a hot filament—start with zero speed at the negative plate of a parallel-plate capacitor. The electric field inside this capacitor accelerates the electrons toward the positive plate, where they exit the capacitor with speed v_1 through a small hole. They then coast along at this speed until they enter the steering electric field of the deflector. This field causes them to follow a curved trajectory, exiting at an angle θ with respect to their original direction.

a. The CRT designer has specified that the electrons must leave the 4.0-cm-wide electron gun with a speed of 6.0×10^7 m/s. What electric field strength is needed inside the electron-gun capacitor?

b. The steering electric field has a constant strength of 1.5×10^5 N/C over the 5.0 cm length of the deflector. By what angle θ are the electrons deflected?

PREPARE We'll use a coordinate system in which the x-axis is horizontal and the y-axis vertical.

a. We can use constant-acceleration kinematics to find the electron's acceleration inside the electron gun. Newton's second law then gives the force on the electron, which we can relate to the electric field using Equation 20.3: $\vec{E} = \vec{F}_{\text{on } q}/q$.

b. Because the electric field is vertically down, the force on a negative electron is vertically up. An electron will accelerate vertically, but not horizontally, so the x-component of its velocity remains unchanged and equal to v_1 as it passes through the deflector. This is exactly analogous to the motion of a projectile, and the electrons follow a projectile-like parabolic trajectory. Just as with projectile motion, we'll use the horizontal motion to find the time interval, then use the time interval to find the final velocity in the y-direction. As **FIGURE 20.44** shows, the ratio of the electron's y- and x-components of velocity can be used to find θ.

FIGURE 20.44 The exit velocity of the electron.

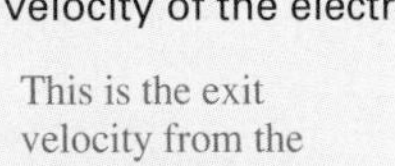

This is the exit velocity from the electron deflector

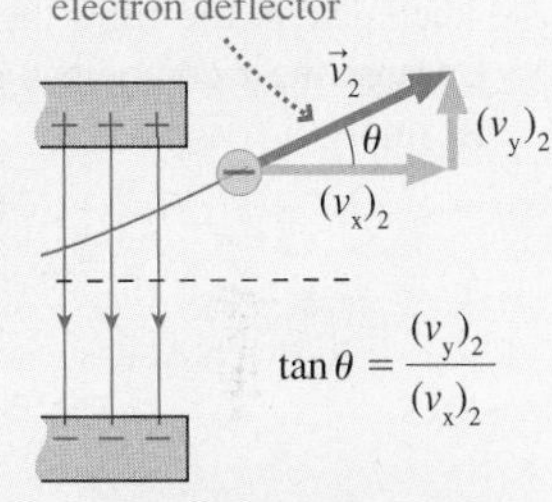

SOLVE a. One of the constant-acceleration kinematic equations from Chapter 2 was $(v_x)_1^2 = (v_x)_0^2 + 2a_x\Delta x$. Using $(v_x)_0 = 0$ and $\Delta x = d = 4.0$ cm, we find that an electron's acceleration inside the electron gun is

$$a_x = \frac{(v_x)_1^2}{2d} = \frac{(6.0 \times 10^7 \text{ m/s})^2}{2(0.040 \text{ m})} = 4.5 \times 10^{16} \text{ m/s}^2$$

Newton's second law tells us that the force causing this acceleration is

$$F_x = ma_x = (9.1 \times 10^{-31} \text{ kg})(4.5 \times 10^{16} \text{ m/s}^2) = 4.1 \times 10^{-14} \text{ N}$$

FIGURE 20.43 The electron gun and electron deflector of a CRT.

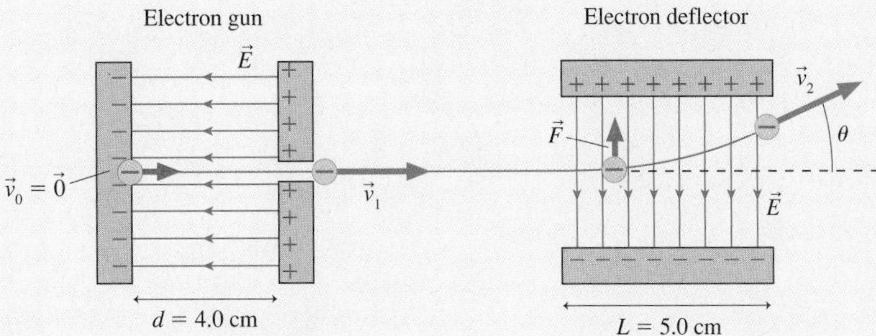

Continued

Then, by Equation 20.3, the electric field is

$$E_x = \frac{F_x}{q} = \frac{4.1 \times 10^{-14}\ \text{N}}{-1.6 \times 10^{-19}\ \text{C}} = -2.6 \times 10^5\ \text{N/C}$$

This is a field in the negative x-direction, as we can see in Figure 20.43, with strength 2.6×10^5 N/C.

b. The y-component of the electron's acceleration in the deflector is

$$a_y = \frac{F_y}{m} = \frac{eE_y}{m} = \frac{(-1.6 \times 10^{-19}\ \text{C})(-1.5 \times 10^5\ \text{N/C})}{9.1 \times 10^{-31}\ \text{kg}}$$

$$= 2.6 \times 10^{16}\ \text{m/s}^2$$

The negative electrons have an upward (positive) acceleration. This acceleration causes an electron to leave the deflector with a y-component of velocity

$$(v_y)_2 = a_y \Delta t$$

where Δt is the time the electron spends in the deflector. Because the x-component of the velocity is constant, this time is simply

$$\Delta t = \frac{L}{(v_x)_1} = \frac{0.050\ \text{m}}{6.0 \times 10^7\ \text{m/s}} = 8.3 \times 10^{-10}\ \text{s}$$

Thus

$$(v_y)_2 = a_y \Delta t = (2.6 \times 10^{16}\ \text{m/s}^2)(8.3 \times 10^{-10}\ \text{s})$$

$$= 2.2 \times 10^7\ \text{m/s}$$

Referring to Figure 20.44, and using $(v_x)_2 = (v_x)_1$ because there's no horizontal acceleration, we see that

$$\tan\theta = \frac{(v_y)_2}{(v_x)_2} = \frac{2.2 \times 10^7\ \text{m/s}}{6.0 \times 10^7\ \text{m/s}} = 0.37$$

so that

$$\theta = \tan^{-1}(0.37) = 20°$$

ASSESS A strong field accelerates the electrons in the x-direction, while a weaker one accelerates them in the y-direction. Thus it is reasonable that the ratio of the y- to the x-component of velocity is significantly less than 1.

The CRT shown in Figure 20.43 deflects electrons only vertically. A real CRT has a second electron deflector, rotated 90°, to provide a horizontal deflection. The two deflectors working together can scan the electron beam over all points on the television screen.

SUMMARY

The goal of Chapter 20 has been to develop a basic understanding of electric phenomena in terms of charges, forces, and fields.

GENERAL PRINCIPLES

Charge

There are two kinds of charges, called **positive** and **negative.**

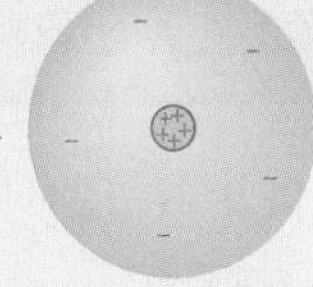

- Atoms consist of a nucleus containing positively charged protons surrounded by a cloud of negatively charged electrons.
- The **fundamental charge** e is the magnitude of the charge on an electron or proton: $e = 1.60 \times 10^{-19}$ C.
- Matter with equal amounts of positive and negative charge is **neutral**.
- Charge is conserved; it can't be created or destroyed.

Coulomb's Law

The forces between two charged particles q_1 and q_2 separated by distance r are

$$F_{1\,\text{on}\,2} = F_{2\,\text{on}\,1} = \frac{K|q_1||q_2|}{r^2}$$

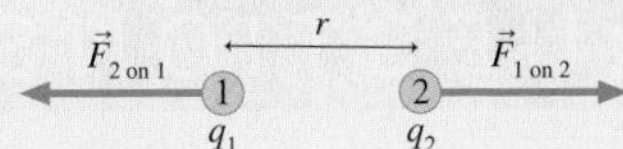

where $K = 8.99 \times 10^9\ \text{N} \cdot \text{m}^2/\text{C}^2$ is the **electrostatic constant.** These forces are an action/reaction pair directed along the line joining the particles.

- The forces are repulsive for two like charges, attractive for two opposite charges.
- The net force on a charge is the vector sum of the forces from all other charges.
- The unit of charge is the coulomb (C).

IMPORTANT CONCEPTS

The Electric Field

Charges interact with each other via the electric field $\vec{E}$.

- Charge A alters the space around it by creating an electric field.

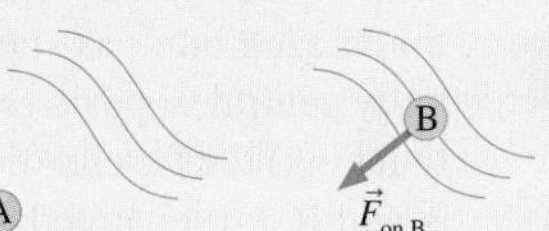

- The field is the agent that exerts a force on charge B. $\vec{F}_{\text{on B}} = q_B\vec{E}$
- An electric field is identified and measured in terms of the force on a probe charge q. The unit of the electric field is N/C. $\vec{E} = \dfrac{\vec{F}_{\text{on}\,q}}{q}$
- The electric field is a vector. The electric field from multiple charges is the vector sum of the fields from the individual charges. $\vec{E}_{\text{total}} = \vec{E}_1 + \vec{E}_2 + \cdots$

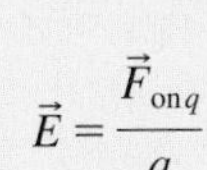

Visualizing the electric field

The electric field exists at all points in space.

- An electric field vector shows the field only at one point, the point at the tail of the vector.
- A **field diagram** shows field vectors at several points.

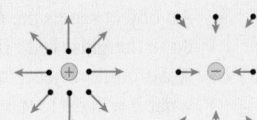

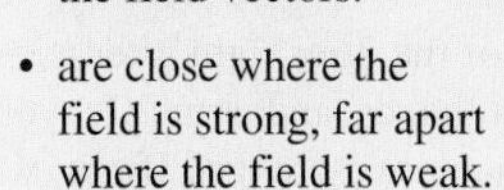

- **Electric field lines:**
  - are always parallel to the field vectors.
  - are close where the field is strong, far apart where the field is weak.
  - go from positive to negative charges.

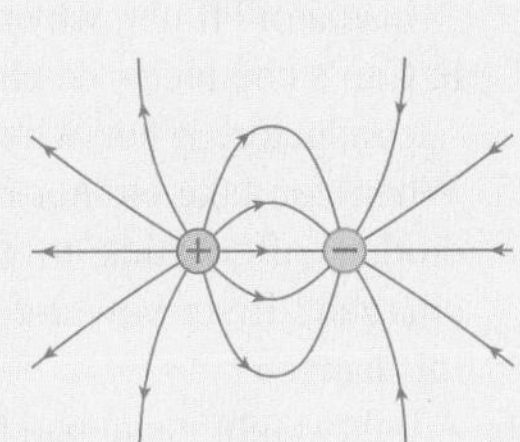

APPLICATIONS

There are two types of material, **insulators** and **conductors.**

- Charge remains fixed on an insulator.
- Charge moves easily through conductors.
- Charge is transferred by contact between objects.

A **dipole** has no net charge, but has a field because the two charges are separated.

A dipole will rotate to align with an electric field.

Electric fields: important cases

The electric field of a **point charge** is

$$\vec{E} = \left(\frac{K|q|}{r^2}, \begin{bmatrix} \text{away from } q \text{ if } q > 0 \\ \text{toward } q \text{ if } q < 0 \end{bmatrix}\right)$$

The electric field inside a **parallel-plate capacitor** is uniform:

$$\vec{E} = \left(\frac{Q}{\epsilon_0 A}, \text{from positive to negative}\right)$$

where $\epsilon_0 = 8.85 \times 10^{-12}\ \text{C}^2/\text{N} \cdot \text{m}^2$ is the **permittivity constant.**

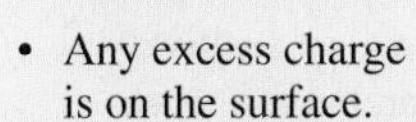

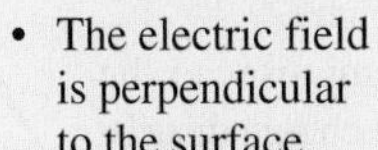

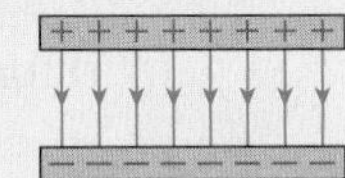

Conductors in electric fields

- The electric field inside a conductor in **electrostatic equilibrium** is zero.
- Any excess charge is on the surface.

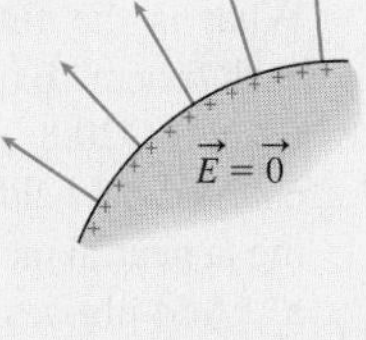

- The electric field is perpendicular to the surface.
- The density of charge and the electric field are highest near a pointed end.

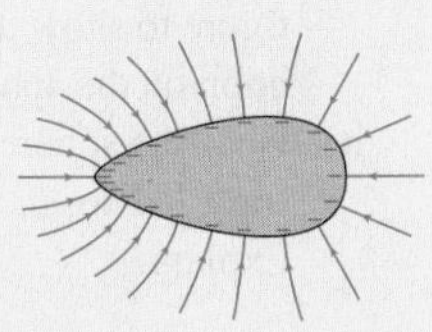

For homework assigned on MasteringPhysics, go to www.masteringphysics.com

Problems labeled INT integrate significant material from earlier chapters; BIO are of biological or medical interest.

Problem difficulty is labeled as | (straightforward) to ||||| (challenging).

QUESTIONS

Conceptual Questions

1. What is alike about charges when we say "two like charges"? Do they look, smell, or taste the same?
2. Four lightweight balls A, B, C, and D are suspended by threads. Ball A has been touched by a plastic rod that was rubbed with wool. When the balls are brought close together, without touching, the following observations are made:
   - Balls B, C, and D are attracted to ball A.
   - Balls B and D have no effect on each other.
   - Ball B is attracted to ball C.

   What are the charge states (positive, negative, or neutral) of balls A, B, C, and D? Explain.
3. Plastic and glass rods that have been charged by rubbing with wool and silk, respectively, hang by threads.
   a. An object repels the plastic rod. Can you predict what it will do to the glass rod? If so, what? If not, why not? Explain.
   b. A different object attracts the plastic rod. Can you predict what it will do to the glass rod? If so, what? If not, why not? Explain.
4. a. Can an insulator be charged? If so, how would you charge an insulator? If not, why not?
   b. Can a conductor be charged? If so, how would you charge a conductor? If not, why not?
5. When you take clothes out of the drier right after it stops, the clothes often stick to your hands and arms. Is your body charged? If so, how did it acquire a charge? If not, why does this happen?
6. A lightweight metal ball hangs by a thread. When a charged rod is held near, the ball moves toward the rod, touches the rod, then quickly "flies away" from the rod. Explain this behavior.
7. As shown in Figure Q20.7, metal sphere A has 4 units of negative charge and metal sphere B has 2 units of positive charge. The two spheres are brought into contact. What is the final charge state of each sphere? Explain.

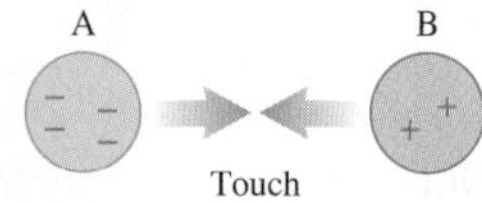

FIGURE Q20.7

8. Figure Q20.8 shows a positively charged rod held near, but not touching, a neutral metal sphere.
   a. Add plusses and minuses to the figure to show the charge distribution on the sphere.
   b. Does the sphere experience a net force? If so, in which direction? Explain.

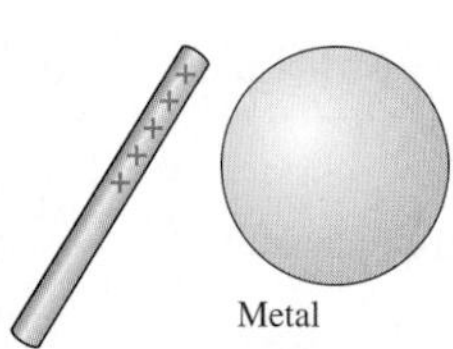

FIGURE Q20.8

9. A plastic balloon that has been rubbed with wool will stick to a wall.
   a. Can you conclude that the wall is charged? If not, why not? If so, where does the charge come from?
   b. Draw a charge diagram showing how the balloon is held to the wall.
10. You are given two metal spheres on portable insulating stands, a glass rod, and a piece of silk. Explain how to give the spheres *exactly* equal but opposite charges.
11. BIO A honeybee acquires a positive electric charge as it flies through the air. This charge causes pollen grains to be attracted to the bee. Explain, using words and diagrams, how a neutral, conducting pollen grain will be attracted to a positively charged bee.
12. A metal rod A and a metal sphere B, on insulating stands, touch each other as shown in Figure Q20.12. They are originally neutral. A positively charged rod is brought near (but not touching) the far end of A. While the charged rod is still close, A and B are separated. The charged rod is then withdrawn. Is the sphere then positively charged, negatively charged, or neutral? Explain.

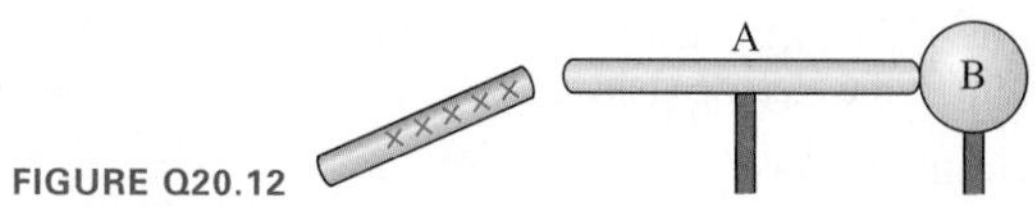

FIGURE Q20.12

13. Each part of Figure Q20.13 shows two points near two charges. Compare the electric field strengths E_1 and E_2 at these two points. Is $E_1 > E_2$, $E_1 = E_2$, or $E_1 < E_2$?

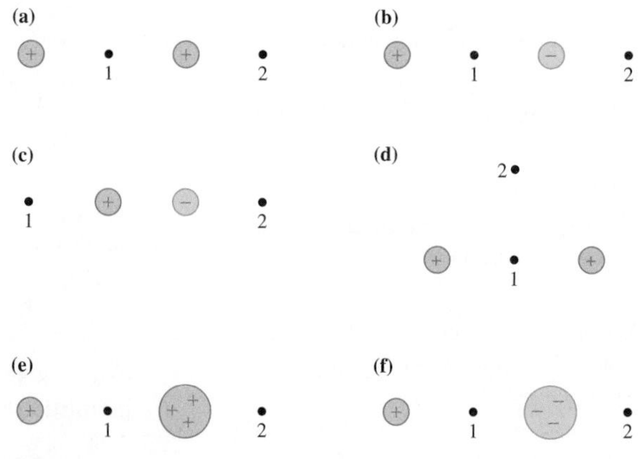

FIGURE Q20.13

14. Iontophoresis is a noninvasive BIO process that transports drugs through the skin without needles. In the photo, the red electrode is positive and the black electrode is negative. The electric field between the electrodes will drive the negatively charged molecules of an anesthetic through the skin. Should the drug be placed at the red or the black electrode? Explain.

15. Rank in order, from largest to smallest, noting any ties, the electric field strengths E_1 to E_4 at points 1 to 4 in Figure Q20.15.

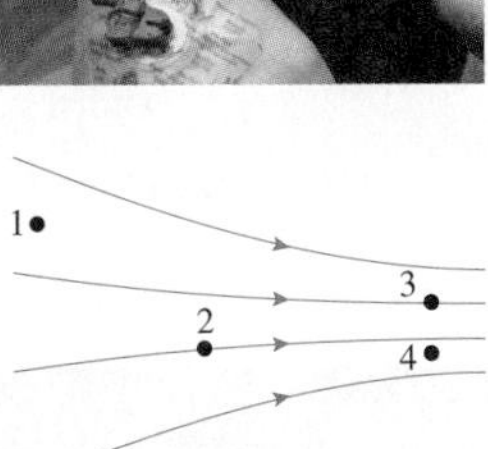

FIGURE Q20.15

16. A 10 nC charge sits at a point in space where the magnitude of the electric field is 1200 N/C. What will the magnitude of the field be if the 10 nC charge is replaced by a 20 nC charge?

17. When DNA breaks into fragments in a cell, electrostatic forces BIO may actually inhibit repair. Explain why this happens.

18. A hollow soda straw is uniformly charged, as shown in Figure Q20.18. What is the electric field at the center (inside) of the straw? Explain.

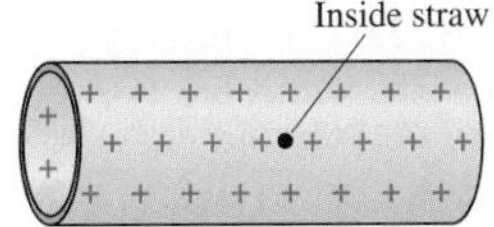

FIGURE Q20.18

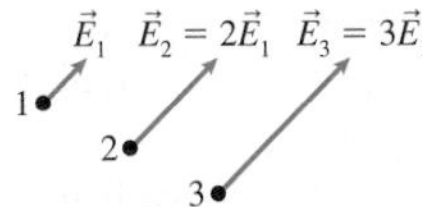

FIGURE Q20.19

19. A small positive charge q experiences a force of magnitude F_1 when placed at point 1 in Figure Q20.19. In terms of F_1:
   a. What is the magnitude of the force on charge q at point 3?
   b. What is the magnitude of the force on a charge $3q$ at point 1?
   c. What is the magnitude of the force on a charge $2q$ at point 2?
   d. What is the magnitude of the force on a charge $-2q$ at point 2?

20. A typical commercial airplane is struck by lightning about once per year. When this happens, the external metal skin of the airplane might be burned, but the people and equipment inside the aircraft experience no ill effects. Explain why this is so.

21. Microbes such as bacteria have small positive charges when in BIO solution. Public health agencies are exploring a new way to measure the presence of small numbers of microbes in drinking water by using electric forces to concentrate the microbes. Water is sent between the two oppositely charged electrodes of a parallel-plate capacitor. Any microbes in the water will collect on one of the electrodes.
   a. On which electrode will the microbes collect?
   b. How could the microbes be easily removed from the electrodes for analysis?

22. a. Is there a point between a 10 nC charge and a 20 nC charge at which the electric field is zero? If so, which charge is this point closer to? If not, why not?
   b. Repeat part a for the case of a 10 nC charge and a −20 nC charge.

Multiple-Choice Questions

23. | Two lightweight, electrically neutral conducting balls hang from threads. Choose the diagram in Figure Q20.23 that shows how the balls hang after:
   a. Both are touched by a negatively charged rod.
   b. Ball 1 is touched by a negatively charged rod and ball 2 is touched by a positively charged rod.
   c. Both are touched by a negatively charged rod but ball 2 picks up more charge than ball 1.
   d. Only ball 1 is touched by a negatively charged rod.

   Note that parts a through d are independent; these are not actions taken in sequence.

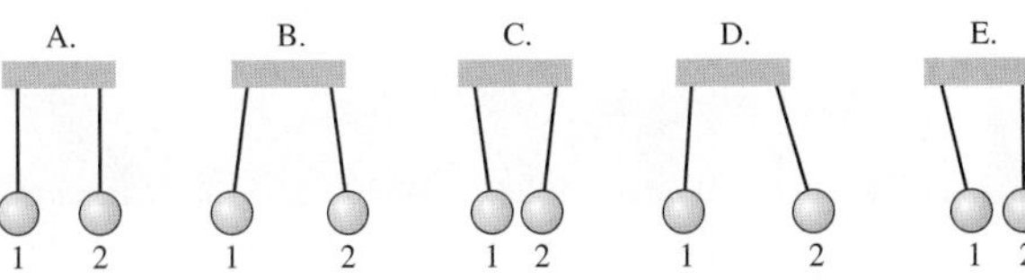

FIGURE Q20.23

24. | All the charges in Figure Q20.24 have the same magnitude. In which case does the electric field at the dot have the largest magnitude?

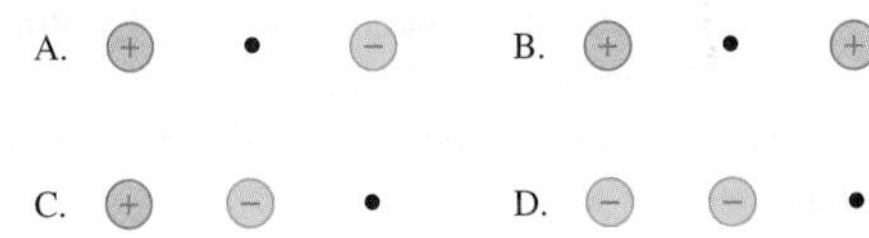

FIGURE Q20.24

25. | All the charges in Figure Q20.25 have the same magnitude. In which case does the electric field at the dot have the largest magnitude?

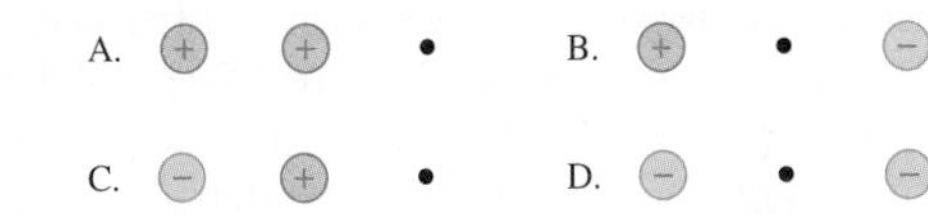

FIGURE Q20.25

26. | All the charges in Figure Q20.26 have the same magnitude. In which case does the electric field at the dot have the largest magnitude?

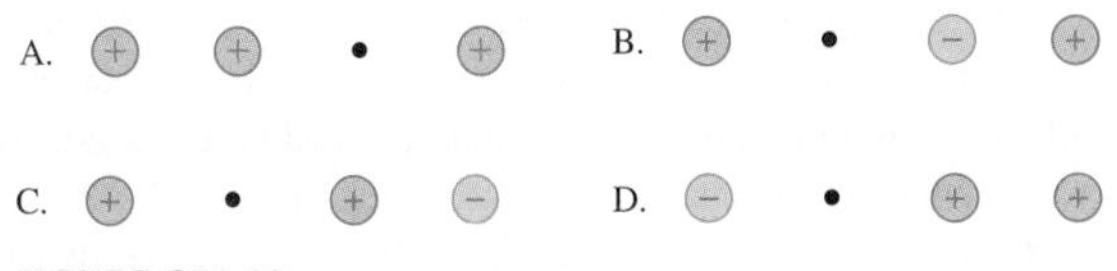

FIGURE Q20.26

27. | A glass bead charged to +3.5 nC exerts an 8.0×10^{-4} N repulsive electric force on a plastic bead 2.9 cm away. What is the charge on the plastic bead?
   A. +2.1 nC
   B. +7.4 nC
   C. +21 nC
   D. +740 nC

28. | A +7.5 nC point charge and a −2.0 nC point charge are 3.0 cm apart. What is the electric field strength at the midpoint between the two charges?
   A. 3.3×10^3 N/C
   B. 5.7×10^3 N/C
   C. 2.2×10^5 N/C
   D. 3.8×10^5 N/C

29. || Three point charges are arranged as shown in Figure Q20.29. Which arrow best represents the direction of the electric field vector at the position of the dot?

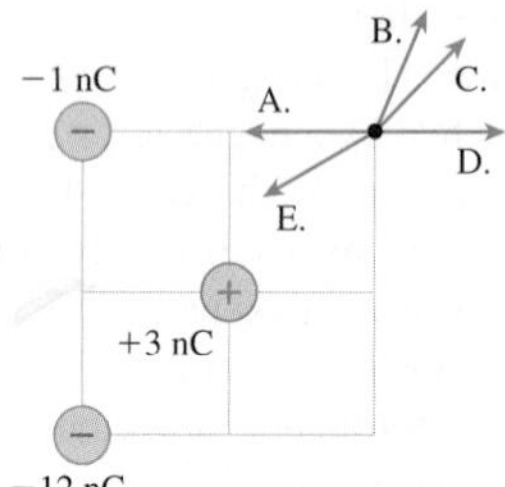

FIGURE Q20.29

30. || A rod has positive charge $+q$ at one end and negative charge $-q$ at the other, forming a dipole. The dipole is placed in a nonuniform electric field represented by the field lines in Figure Q20.30. Which arrow best indicates the direction of the net electric force on the dipole?

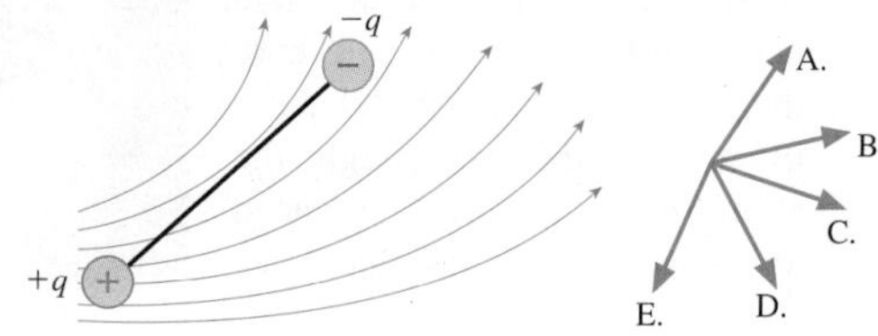

FIGURE Q20.30

PROBLEMS

Section 20.1 Charges and Forces

Section 20.2 Charges, Atoms, and Molecules

1. ||| A glass rod is charged to +5.0 nC by rubbing.
   a. Have electrons been removed from the rod or protons added? Explain.
   b. How many electrons have been removed or protons added?
2. ||| A plastic rod is charged to −20 nC by rubbing.
   a. Have electrons been added to the rod or protons removed? Explain.
   b. How many electrons have been added or protons removed?
3. ||| Suppose you have 1.0 mol of O_2 gas. How many coulombs of positive charge are contained in the atomic nuclei of this gas? INT
4. || A plastic rod that has been charged to −15.0 nC touches a metal sphere. Afterward, the rod's charge is −10.0 nC.
   a. What kind of charged particle was transferred between the rod and the sphere, and in which direction? That is, did it move from the rod to the sphere or from the sphere to the rod?
   b. How many charged particles were transferred?
5. || A glass rod that has been charged to +12.0 nC touches a metal sphere. Afterward, the rod's charge is +8.0 nC.
   a. What kind of charged particle was transferred between the rod and the sphere, and in which direction? That is, did it move from the rod to the sphere or from the sphere to the rod?
   b. How many charged particles were transferred?
6. ||| Two identical metal spheres A and B are connected by a metal rod. Both are initially neutral. 1.0×10^{12} electrons are added to sphere A, then the connecting rod is removed. Afterward, what are the charge of A and the charge of B?
7. || Two identical metal spheres A and B are connected by a plastic rod. Both are initially neutral. 1.0×10^{12} electrons are added to sphere A, then the connecting rod is removed. Afterward, what are the charge of A and the charge of B?
8. || If two identical conducting spheres are in contact, any excess charge will be evenly distributed between the two. Three identical metal spheres are labeled A, B, and C. Initially, A has charge q, B has charge $-q/2$, and C is uncharged.
   a. What is the final charge on each sphere if C is touched to B, removed, and then touched to A?
   b. Starting again from the initial conditions, what is the charge on each sphere if C is touched to A, removed, and then touched to B?

Section 20.3 Coulomb's Law

9. || Two 1.0 kg masses are 1.0 m apart on a frictionless table. Each has +1.0 μC of charge. INT
   a. What is the magnitude of the electric force on one of the masses?
   b. What is the initial acceleration of each mass if they are released and allowed to move?
10. ||| Two small plastic spheres each have a mass of 2.0 g and a charge of −50.0 nC. They are placed 2.0 cm apart.
    a. What is the magnitude of the electric force between the spheres?
    b. By what factor is the electric force on a sphere larger than its weight?
11. ||| A small plastic sphere with a charge of −5.0 nC is near another small plastic sphere with a charge of −12 nC. If the spheres repel one another with a force of magnitude 8.2×10^{-4} N, what is the distance between the spheres?
12. ||| A small metal bead, labeled A, has a charge of 25 nC. It is touched to metal bead B, initially neutral, so that the two beads share the 25 nC charge, but not necessarily equally. When the two beads are then placed 5.0 cm apart, the force between them is 5.4×10^{-4} N. What are the charges q_A and q_B on the beads?
13. |||| A small glass bead has been charged to +20 nC. A tiny ball bearing 1.0 cm above the bead feels a 0.018 N downward electric force. What is the charge on the ball bearing?
14. | What are the magnitude and direction of the electric force on charge A in Figure P20.14?

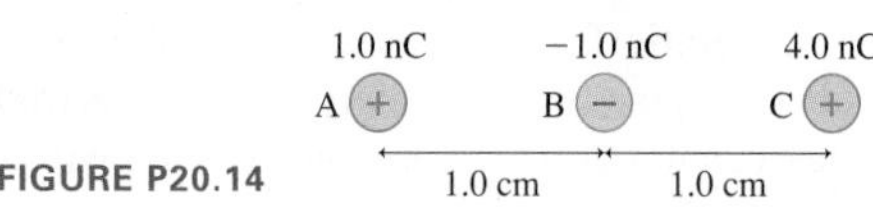

FIGURE P20.14

15. ||| In Figure P20.15, charge q_2 experiences no net electric force. What is q_1?

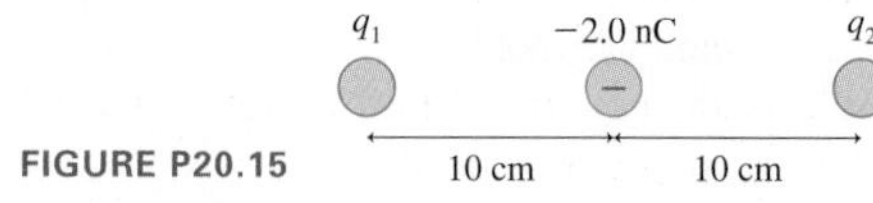

FIGURE P20.15

16. | Object A, which has been charged to $+10$ nC, is at the origin. Object B, which has been charged to -20 nC, is at $(x, y) = (0.0 \text{ cm}, 2.0 \text{ cm})$. What are the magnitude and direction of the electric force on each object?

17. | A small glass bead has been charged to $+20$ nC. What are the magnitude and direction of the acceleration of (a) a proton and (b) an electron that is 1.0 cm from the center of the bead?
INT

Section 20.4 The Concept of the Electric Field

18. || What magnitude charge creates a 1.0 N/C electric field at a point 1.0 m away?

19. || What are the strength and direction of the electric field 2.0 cm from a small glass bead that has been charged to $+6.0$ nC?

20. || A 30 nC charged particle and a 50 nC charged particle are near each other. There are no other charges nearby. The electric force on the 30 nC particle is 0.035 N. The 50 nC particle is then moved very far away. Afterward, what is the magnitude of the electric field at its original position?

21. | What are the strength and direction of the electric field 1.0 mm from (a) a proton and (b) an electron?

22. | A $+10$ nC charge is located at the origin.
   a. What are the strengths of the electric fields at the positions $(x, y) = (5.0 \text{ cm}, 0.0 \text{ cm})$, $(-5.0 \text{ cm}, 5.0 \text{ cm})$, and $(-5.0 \text{ cm}, -5.0 \text{ cm})$?
   b. Draw a field diagram showing the electric field vectors at these points.

23. | A -10 nC charge is located at the origin.
   a. What are the strengths of the electric fields at the positions $(x, y) = (0.0 \text{ cm}, 5.0 \text{ cm})$, $(-5.0 \text{ cm}, -5.0 \text{ cm})$, and $(-5.0 \text{ cm}, 5.0 \text{ cm})$?
   b. Draw a field diagram showing the electric field vectors at these points.

24. ||| What are the strength and direction of the electric field at the position indicated by the dot in Figure P20.24? Specify the direction as an angle above or below horizontal.

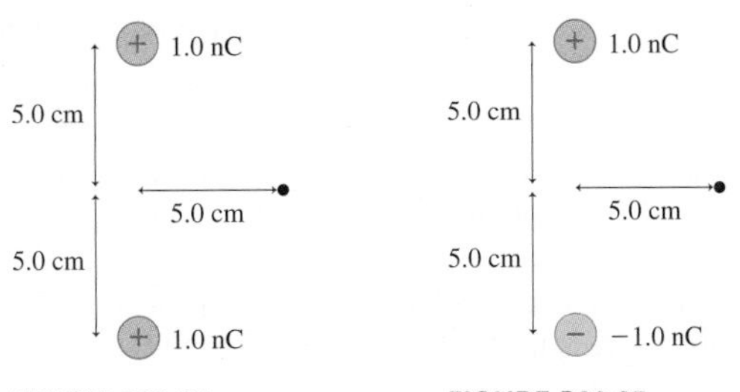

FIGURE P20.24 FIGURE P20.25

25. || What are the strength and direction of the electric field at the position indicated by the dot in Figure P20.25? Specify the direction as an angle above or below horizontal.

Section 20.5 Applications of the Electric Field

26. || What are the strength and direction of an electric field that will balance the weight of a 1.0 g plastic sphere that has been charged to -3.0 nC?
INT

27. | What are the strength and direction of an electric field that will balance the weight of (a) a proton and (b) an electron?
INT

28. || A 0.10 g plastic bead is charged by the addition of 1.0×10^{10} excess electrons. What electric field $\vec{E}$ (strength and direction) will cause the bead to hang suspended in the air?
INT

29. || A parallel-plate capacitor is constructed of two square plates, size $L \times L$, separated by distance d. The plates are given charge $\pm Q$. What is the ratio E_f/E_i of the final electric field strength E_f to the initial electric field strength E_i if:
   a. Q is doubled?
   b. L is doubled?
   c. d is doubled?

30. || A parallel-plate capacitor is formed from two 4.0 cm $\times$ 4.0 cm electrodes spaced 2.0 mm apart. The electric field strength inside the capacitor is 1.0×10^6 N/C. What is the charge (in nC) on each electrode?

31. | Two identical closely spaced circular disks form a parallel-plate capacitor. Transferring 1.5×10^9 electrons from one disk to the other causes the electric field strength between them to be 1.0×10^5 N/C. What are the diameters of the disks?

Section 20.6 Conductors and Electric Fields

32. ||| Storm clouds may build up large negative charges near their bottom edges. The earth is a good conductor, so the charge on the cloud attracts an equal and opposite charge on the earth under the cloud. The electric field strength near the earth depends on the shape of the earth's surface, as we can explain with a simple model. The top metal plate in Figure 20.32 has uniformly distributed negative charge. The bottom metal plate, which has a high point, has an equal and opposite charge that is free to move.
   a. Sketch the two plates and the region between them, showing the distribution of positive charge on the bottom plate.
   b. Complete your diagram by sketching electric field lines between the two plates. Be sure to note the direction of the field. Where is the field strongest?
   c. Explain why it is more dangerous to be on top of a hill or mountain during a lightning storm than on level ground.

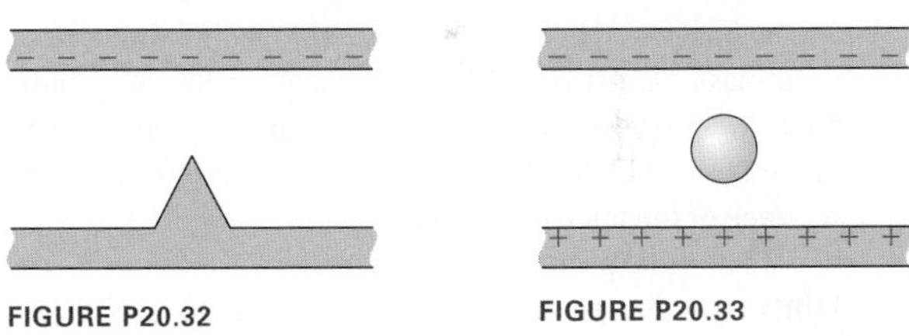

FIGURE P20.32 FIGURE P20.33

33. ||| A neutral conducting sphere is between two parallel charged plates, as shown in Figure P20.33. Sketch the electric field lines in the region between the plates. Be sure to include the effect of the conducting sphere.

Section 20.7 Forces and Torques in Electric Fields

34. ||| Two small plastic spheres, one charged to 17 nC and the other to -17 nC, are connected by a 25-mm-long insulating rod. Suppose this dipole is placed in a uniform electric field with strength 7.4×10^5 N/C. What is the maximum possible torque on the dipole?
INT

35. || A protein molecule in an electrophoresis gel has a negative charge. The exact charge depends on the pH of the solution, but 30 excess electrons is typical. What is the magnitude of the electric force on a protein with this charge in a 1500 N/C electric field?
BIO

36. || Large electric fields in cell membranes cause ions to move through the cell wall, as we will explore in Chapter 23. The field strength in a typical membrane is 1.0×10^7 N/C. What is the magnitude of the force on a calcium ion with charge $+e$?
BIO

37. ||| Molecules of carbon monoxide are permanent electric dipoles due to unequal sharing of electrons between the carbon and oxygen atoms. Figure P20.37 shows the distance and charges. Suppose a carbon monoxide molecule with a horizontal axis is in a vertical electric field of strength 15,000 N/C.
INT

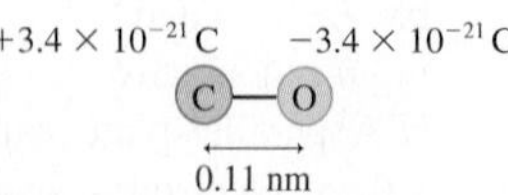

FIGURE P20.37

a. What is the magnitude of the net force on the molecule?
b. What is the magnitude of the torque on the molecule?

General Problems

38. |||| A 2.0-mm-diameter copper ball is charged to +50 nC. What fraction of its electrons have been removed? The density of copper is 8900 kg/m^3.
INT

39. | Pennies today are copper-covered zinc, but older pennies are 3.1 g of solid copper. What are the total positive charge and total negative charge in a solid copper penny that is electrically neutral? The density of copper is 8900 kg/m^3.
INT

40. || Two protons are 2.0 fm apart. (1 fm = 1 femtometer = 1×10^{-15} m.)
INT
a. What is the magnitude of the electric force on one proton due to the other proton?
b. What is the magnitude of the gravitational force on one proton due to the other proton?
c. What is the ratio of the electric force to the gravitational force?

41. |||| The nucleus of a ^{125}Xe atom (an isotope of the element xenon with mass 125 u) is 6.0 fm in diameter. It has 54 protons and charge $q = +54e$. (1 fm = 1 femtometer = 1×10^{-15} m.)
INT
a. What is the electric force on a proton 2.0 fm from the surface of the nucleus?
b. What is the proton's acceleration?
Hint: Treat the spherical nucleus as a point charge.

42. ||| Two equally charged, 1.00 g spheres are placed with 2.00 cm between their centers. When released, each begins to accelerate at 225 m/s^2. What is the magnitude of the charge on each sphere?
INT

43. || Objects A and B are both positively charged. Both have a mass of 100 g, but A has twice the charge of B. When A and B are placed with 10 cm between their centers, B experiences an electric force of 0.45 N.
a. How large is the force on A?
b. What are the charges q_A and q_B?

44. |||| An electric dipole is formed from ±1.0 nC point charges spaced 2.0 mm apart. The dipole is centered at the origin, oriented along the y-axis. What is the electric field strength at the points (a) $(x, y) = (10$ mm, 0 mm$)$ and (b) $(x, y) =$ (0 mm, 10 mm)?

45. ||| What are the strength and direction of the electric field at the position indicated by the dot in Figure P20.45? Specify the direction as an angle above or below horizontal.

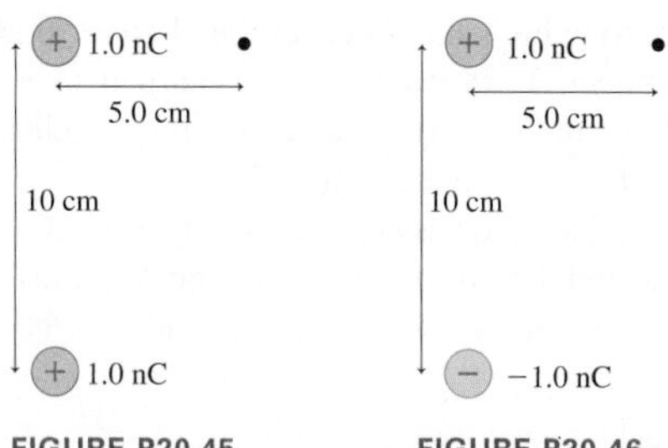

FIGURE P20.45 FIGURE P20.46

46. |||| What are the strength and direction of the electric field at the position indicated by the dot in Figure P20.46? Specify the direction as an angle above or below horizontal.

47. || What is the force on the 1.0 nC charge in Figure P20.47? Give your answer as a magnitude and a direction.

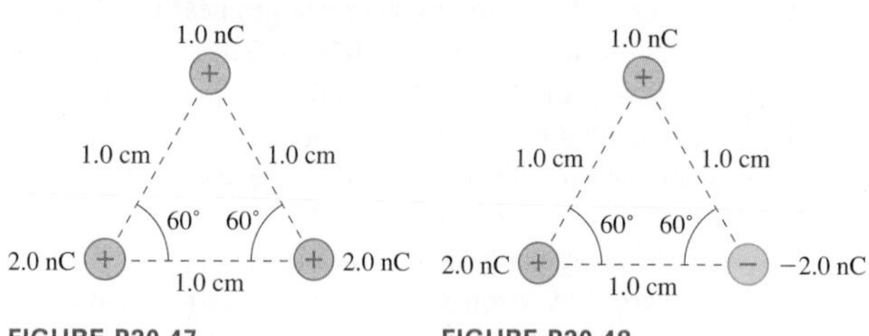

FIGURE P20.47 FIGURE P20.48

48. || What is the force on the 1.0 nC charge in Figure P20.48? Give your answer as a magnitude and a direction.

49. | What is the magnitude of the force on the 1.0 nC charge in the middle of Figure P20.49 due to the four other charges?

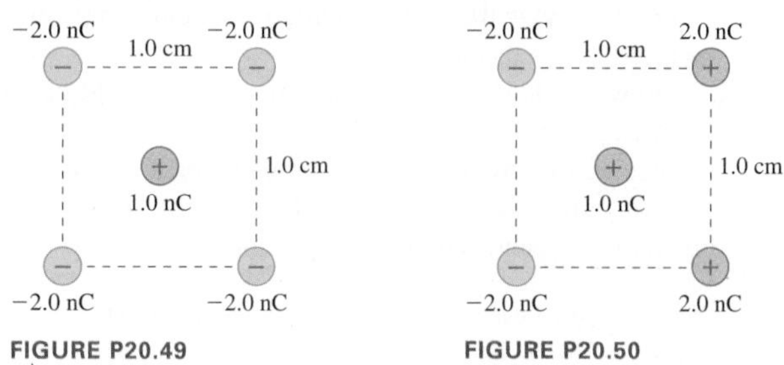

FIGURE P20.49 FIGURE P20.50

50. ||| What are the magnitude and direction of the force on the 1.0 nC charge in the middle of Figure P20.50 due to the four other charges?

51. || What are the magnitude and direction of the force on the 1.0 nC charge at the bottom of Figure P20.51?

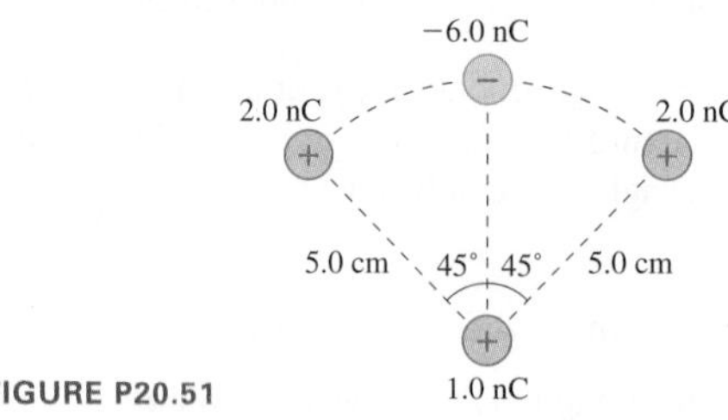

FIGURE P20.51

52. |||| A 5.0 nC point charge sits at $x = 0$. At the same time, a 4500 N/C uniform electric field (created by distant source charges) points in the positive x-direction. At what point along the x-axis, if any, would (a) a proton and (b) an electron experience no net force?

53. || The net force on the 1.0 nC charge in Figure P20.53 is zero. What is q?

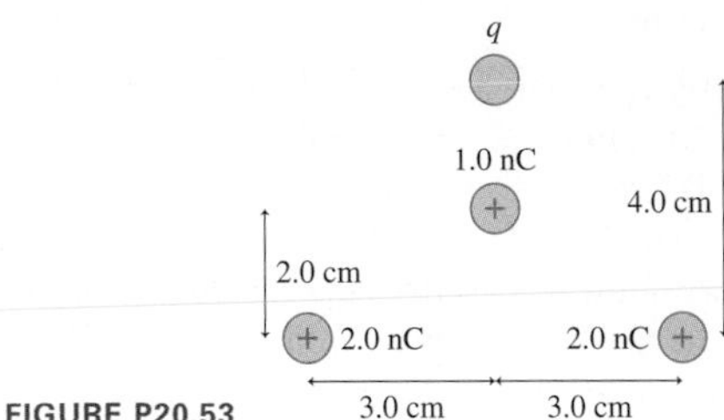

FIGURE P20.53

54. || Two particles have positive charges q and Q. A third charged particle is placed halfway between them. What must this particle's charge be so that the net force on charge Q is zero?

55. ||||| Figure P20.55 shows four charges at the corners of a square of side L. Assume q and Q are positive.
   a. Draw a diagram showing the three forces on charge q due to the other charges. Give your vectors the correct relative lengths.
   b. Find an expression for the magnitude of the net force on q.

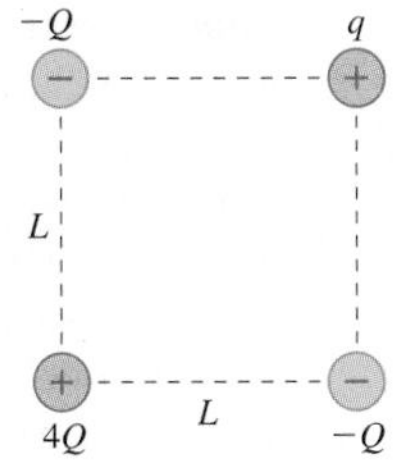

$-Q$ q L $4Q$ L $-Q$

FIGURE P20.55

56. | Suppose the magnitude of the proton charge differs from that of the electron charge by a mere 1 part in 10^9.
   a. What would be the force between two 2.0-mm-diameter copper spheres 1.0 cm apart? Assume that each copper atom has an equal number of electrons and protons. The density of copper is 8900 kg/m^3.
   b. Would this amount of force be detectable? What can you conclude from the fact that no such forces are observed?

57. ||| In a simple model of the hydrogen atom, the electron moves in a circular orbit of radius 0.053 nm around a stationary proton. How many revolutions per second does the electron make?
   Hint: What must be true for a force that causes circular motion?

58. ||| A 0.10 g honeybee acquires a charge of +23 pC while flying.
   a. The electric field near the surface of the earth is typically 100 N/C, directed downward. What is the ratio of the electric force on the bee to the bee's weight?
   b. What electric field strength and direction would allow the bee to hang suspended in the air?

59. ||||| A +10 nC charge is located at $(x, y) = (0 \text{ cm}, 10 \text{ cm})$ and a −5.0 nC charge is located $(x, y) = (5.0 \text{ cm}, 0 \text{ cm})$. Where would a −10 nC charge need to be located in order that the electric field at the origin be zero?

60. ||| Two 2.0-cm-diameter disks face each other, 1.0 mm apart. They are charged to ±10 nC.
   a. What is the electric field strength between the disks?
   b. A proton is shot from the negative disk toward the positive disk. What launch speed must the proton have to just barely reach the positive disk?

61. |||| The electron gun in a television tube uses a uniform electric field to accelerate electrons from rest to 5.0×10^7 m/s in a distance of 1.2 cm. What is the electric field strength?

62. |||| A 0.020 g plastic bead hangs from a lightweight thread. Another bead is fixed in position beneath the point where the thread is tied. If both beads have charge q, the moveable bead swings out to the position shown in Figure P20.62. What is q?

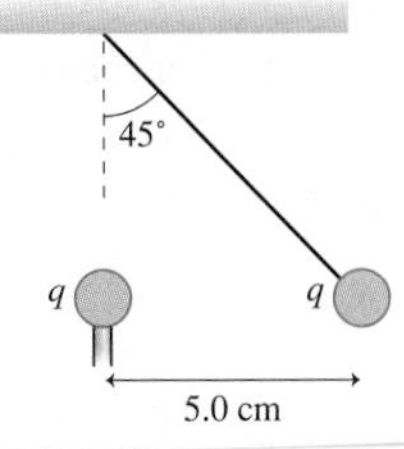

FIGURE P20.62

63. |||| You have a lightweight spring whose unstretched length is 4.0 cm. You're curious to see if you can use this spring to measure charge. First, you attach one end of the spring to the ceiling and hang a 1.0 g mass from it. This stretches the spring to a length of 5.0 cm. You then attach two small plastic beads to the opposite ends of the spring, lay the spring on a frictionless table, and give each plastic bead the same charge. This stretches the spring to a length of 4.5 cm. What is the magnitude of the charge (in nC) on each bead?

64. |||| Two 3.0 g spheres on 1.0-m-long threads repel each other after being equally charged, as shown in Figure P20.64. What is the charge q?

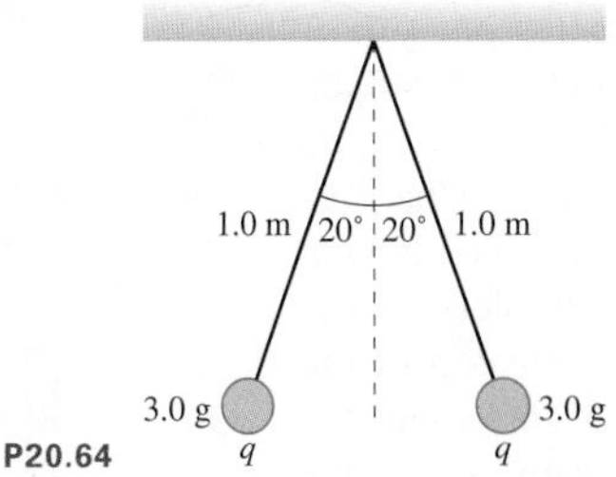

FIGURE P20.64

65. |||| An electric field $\vec{E} = (100{,}000 \text{ N/C, right})$ causes the 5.0 g ball in Figure P20.65 to hang at a 20° angle. What is the charge on the ball?

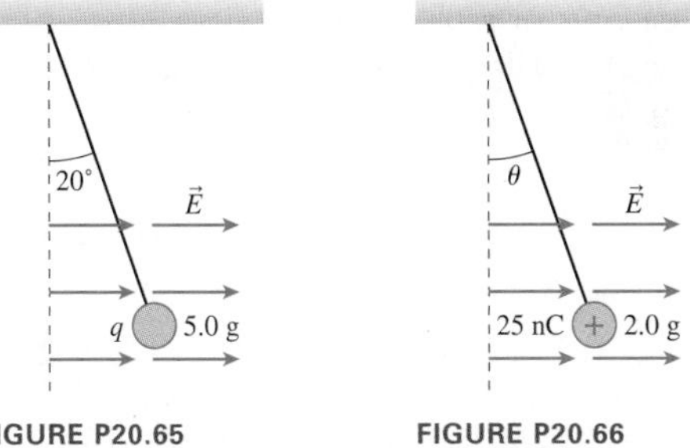

FIGURE P20.65

FIGURE P20.66

66. |||| An electric field $\vec{E} = (200{,}000 \text{ N/C, right})$ causes the 2.0 g ball in Figure P20.66 to hang at an angle. What is θ?

67. || A small charged bead has a mass of 1.0 g. It is held in a uniform electric field $\vec{E} = (200{,}000 \text{ N/C, up})$. When the bead is released, it accelerates upward with an acceleration of 20 m/s^2. What is the charge on the bead?

68. || A bead with a mass of 0.050 g and a charge of 15 nC is free to slide on a vertical rod. At the base of the rod is a fixed 10 nC charge. In equilibrium, at what height above the fixed charge does the bead rest?

69. || A small bead with a positive charge q is free to slide on a horizontal wire of length 4.0 cm. At the left end of the wire is a fixed charge q, and at the right end is a fixed charge $4q$. How far from the left end of the wire does the bead come to rest?

In Problems 70 and 71 you are given the equation used to solve a problem. For each of these,

a. Write a realistic problem for which this is the correct equation.
b. Finish the solution of the problem.

70. ||| $\dfrac{(9.0 \times 10^9 \text{ N} \cdot \text{m}^2/\text{C}^2) \times N \times (1.60 \times 10^{-19} \text{ C})}{(1.0 \times 10^{-6} \text{ m})^2}$
$= 1.5 \times 10^6 \text{ N/C}$

71. ||| $\dfrac{(9.0 \times 10^9 \text{ N} \cdot \text{m}^2/\text{C}^2)q^2}{(0.015 \text{ m})^2} = 0.020 \text{ N}$

Passage Problems

Flow Cytometry BIO

Flow cytometry, illustrated in Figure P20.72, is a technique used to sort cells by type. The cells are placed in a conducting saline solution which is then forced from a nozzle. The stream breaks up into small droplets, each containing one cell. A metal collar surrounds the stream right at the point where the droplets separate from the stream. Charging the collar polarizes the conducting liquid, causing the droplets to become charged as they break off from the stream. A laser beam probes the solution just upstream from the charging collar, looking for the presence of certain types of cells. All droplets containing one particular type of cell are given the same charge by the charging collar. Droplets with other desired types of cells receive a different charge, and droplets with no desired cell receive no charge. The charged droplets then pass between two parallel charged electrodes where they receive a horizontal force that directs them into different collection tubes, depending on their charge.

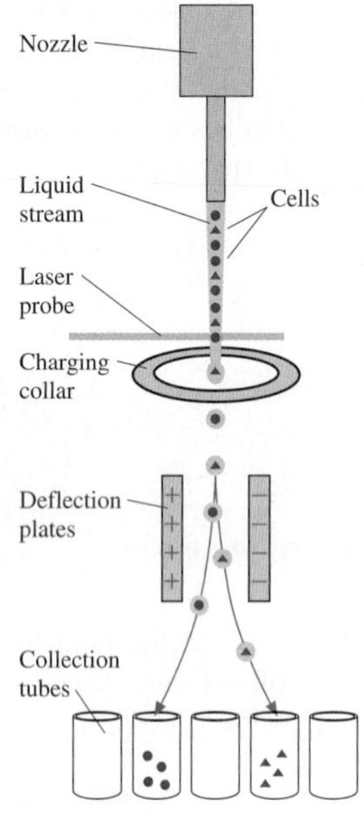

FIGURE P20.72

72. | If the charging collar has a positive charge, the net charge on a droplet separating from the stream will be
A. Positive. B. Negative. C. Neutral.
D. The charge will depend on the type of cell.

73. | Which of the following describes the charges on the droplets that end up in the five tubes, moving from left to right?
A. $+2q, +q, 0, -q, -2q$
B. $+q, +2q, 0, -2q, -q$
C. $-q, -2q, 0, +2q, +q$
D. $-2q, -q, 0, +q, +2q$

74. | Because the droplets are conductors, a droplet's positive and negative charges will separate while the droplet is in the region between the deflection plates. Suppose a neutral droplet passes between the plates. The droplet's dipole moment will point
A. Up. B. Down. C. Left. D. Right.

75. | Another way to sort the droplets would be to give each droplet the same charge, then vary the electric field between the deflection plates. For the apparatus as sketched, this technique will not work because
A. Several droplets are between the plates at one time, and they would all feel the same force.
B. The cells in the solution have net charges that would affect the droplet charge.
C. A droplet with a net charge would always experience a net force between the plates.
D. The droplets would all repel each other, and this force would dominate the deflecting force.

STOP TO THINK ANSWERS

Stop to Think 20.1: A. The electroscope is originally given a positive charge. The charge spreads out, and the leaves repel each other. When a rod with a negative charge is brought near, some of the positive charge is attracted to the top of the electroscope, away from the leaves. There is less charge on the leaves, and so they move closer together.

Stop to Think 20.2: $q_E(+3e) > q_A(+1e) > q_D(0) > q_B(-1e) > q_C(-2e)$.

Stop to Think 20.3: B. The two forces are an action/reaction pair, opposite in direction but *equal* in magnitude.

Stop to Think 20.4: $E_B > E_A > E_D > E_C$. The field is proportional to the charge, and inversely proportional to the square of the distance.

Stop to Think 20.5: C. Electric field lines *start* on positive charges. Very near to each of the positive charges, the field lines should look like the field lines of a single positive charge.

Stop to Think 20.6: $F_A = F_B = F_C = F_D = F_E$. The field inside a capacitor is the same at all points. Because the field is uniform, the force on the proton will be the same at all points. The electric field exists at all points whether or not a vector is shown at that point.

21 Electric Potential

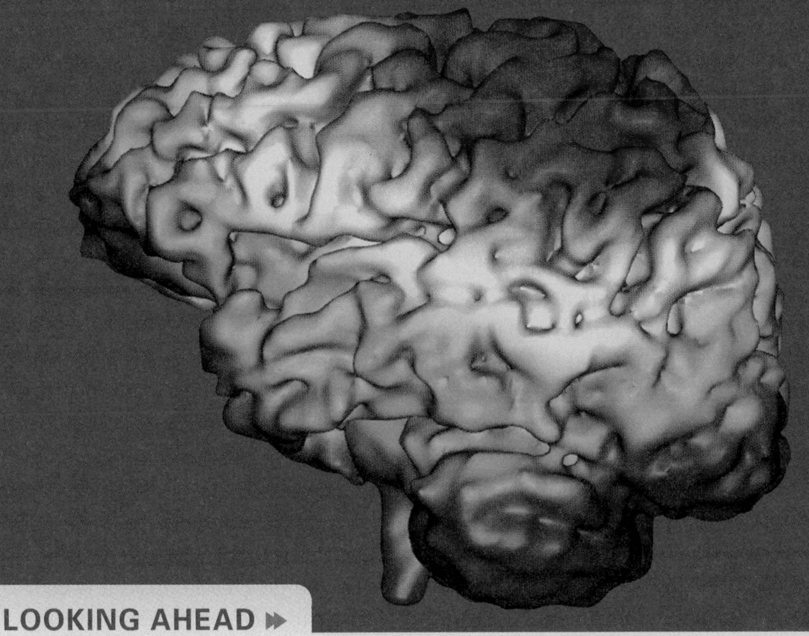

The colors on this patient's brain are a map of the electric potential on the brain's surface after the patient was given a particular sensory stimulus. How does visualizing the electric potential help us understand the electric properties of the brain?

LOOKING AHEAD ▸▸

The goal of Chapter 21 is to calculate and use the electric potential and electric potential energy.

Electric Potential Energy

A charged particle acquires **electric potential energy** when it is brought near other charges.

Lightning is a dramatic example of conversion of electric potential energy to light and thermal energy.

Looking Back ◂◂
10.2–10.4 Work, kinetic energy, and potential energy

Electric Potential

The **electric potential** at a given location determines the electric potential energy that a charge would have if placed there. Electric potential differences are caused by the *separation of charge.*

Batteries create a charge separation by chemical means, leading to a potential difference between their terminals.

Using the Electric Potential

Depending on the sign of its charge, a charged particle speeds up or slows down as it moves through a potential difference.

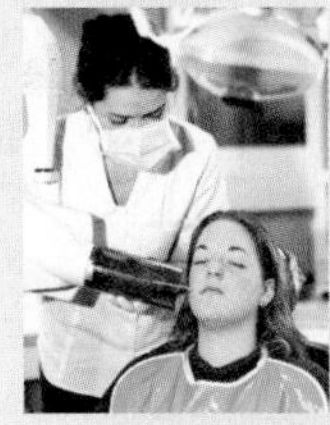

X rays are produced when electrons, accelerated through a large potential difference, collide with a metal target.

Looking Back ◂◂
20.3 Conservation of energy

Capacitors

Capacitors store charge and electric potential energy. They're used in devices ranging from computers to defibrillators.

Each cylindrical element in this circuit is a capacitor, capable of rapidly storing and releasing charge and energy.

Looking Back ◂◂
20.5 Parallel-plate capacitors

Connecting Potential and Field

The electric field and electric potential are intimately connected. You'll learn how to move from the field representation to the potential representation and back again.

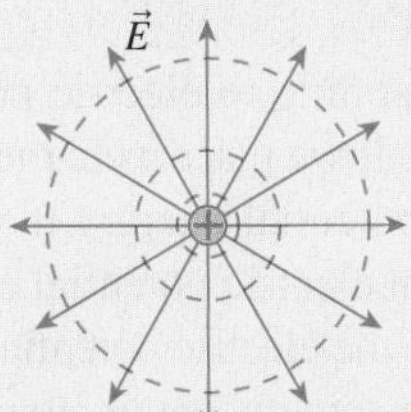

The electric field and the electric potential can be related to each other graphically.

Looking Back ◂◂
20.4 The electric field

Calculating the Electric Potential

You'll learn to calculate the electric potential for several important charge distributions.

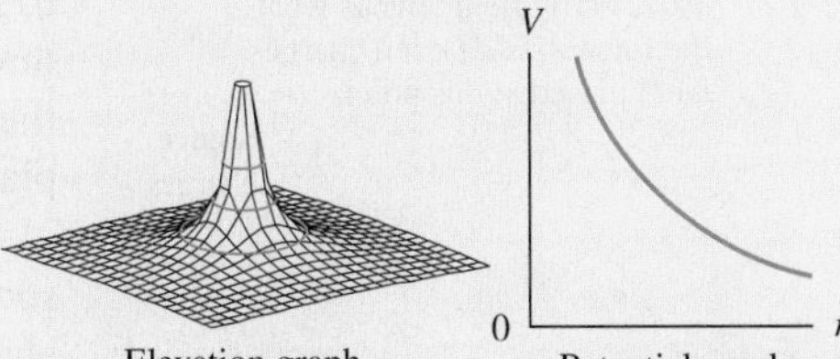

Because the electric potential is an abstract idea, we'll develop several different ways of *visualizing* the electric potential, as shown here for the important case of the potential due to a point charge.

FIGURE 21.1 The elastic potential energy of a spring and the electric potential energy of a system of charges.

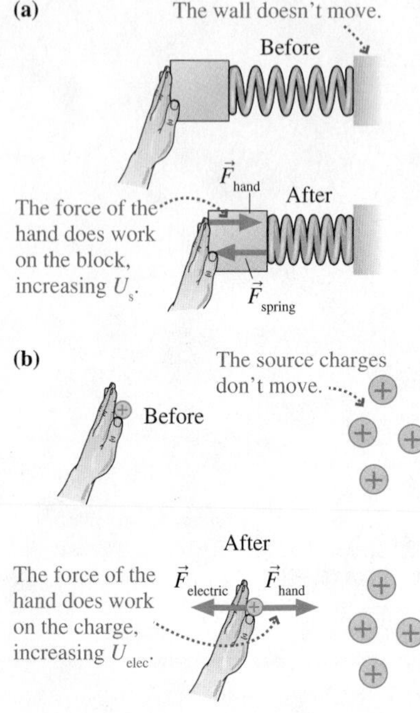
(a)
The wall doesn't move.
Before
$\vec{F}_{hand}$
After
The force of the hand does work on the block, increasing U_s.
$\vec{F}_{spring}$
(b)
The source charges don't move.
Before
After
The force of the hand does work on the charge, increasing U_{elec}.
$\vec{F}_{electric}$
$\vec{F}_{hand}$

21.1 Electric Potential Energy and Electric Potential

Conservation of energy was a powerful tool for understanding the motion of mechanical systems. In Chapter 10 you learned that the total energy of an isolated system remains constant. However, a system's energy can be changed by doing *work* on it. You will recall that work is the *transfer* of energy to or from a system by external forces that act on it as it undergoes a displacement. Depending on how the work is done, the energy transferred to the system can appear as kinetic energy K, potential energy U, or, as we'll investigate in this chapter, forms of energy associated with *electric* forces acting on charges.

To remind ourselves of how conservation of energy works for a mechanical system, FIGURE 21.1a shows a hand pushing a block against a spring in such a way that the block moves at a constant speed. (The spring also exerts a force on the block, directed opposite the hand force, so that the net force is zero and the block's speed is constant.) The force of the hand is a force *external* to the block + spring system, so this force does work on the system, increasing its energy. In this case, the energy transferred to the system by the hand appears as the elastic potential energy U_s stored in the spring. If the hand is removed, this stored energy will shoot the block away as the elastic potential energy is transformed into kinetic energy.

Let's apply these same ideas to the system of charged particles in FIGURE 21.1b. Several charges have been identified as source charges, and these—like the wall in Figure 21.1a—don't move. Suppose the hand pushes charge q at a constant speed toward the source charges. (The charge q is also subject to the electric force due to the source charges.) The force of the hand does work as it pushes the charge through a displacement, increasing the system's energy. Just as in the block + spring system, this energy appears as increased potential energy—in this case, as **electric potential energy** U_{elec}. If the hand is removed, this stored energy will shoot the charge q back out, in exact analogy with how the block was shot out by the spring.

In both cases, the energy transfer is $\Delta U = W$; that is, the work done increases the system's potential energy. This means that **we can determine electric potential energy by computing how much work must be done to assemble a set of charged particles.**

NOTE ► The electric potential energy in Figure 21.1b is an energy of the *system*—of charge q *and* the fixed source charges. However, because we're usually focused only on the moving charge, we often speak of the electric potential energy *of the charge*. ◄

Electric Potential

We introduced the concept of the *electric field* in Chapter 20. In the field model, the electric field is the agent by which charges exert a long-range force on another charge q. As shown in FIGURE 21.2, the source charges first alter the space around them by creating an electric field $\vec{E}$ at every point in space. It is then this electric field—not the source charges themselves—that exerts a force $\vec{F}_{elec} = q\vec{E}$ on charge q. An important idea was that the electric field of the source charges is present throughout space whether or not charge q is present to experience it. In other words, the electric field tells us what the force on the charge *would be* if the charge were placed there.

FIGURE 21.2 What is the electric potential energy near some source charges?

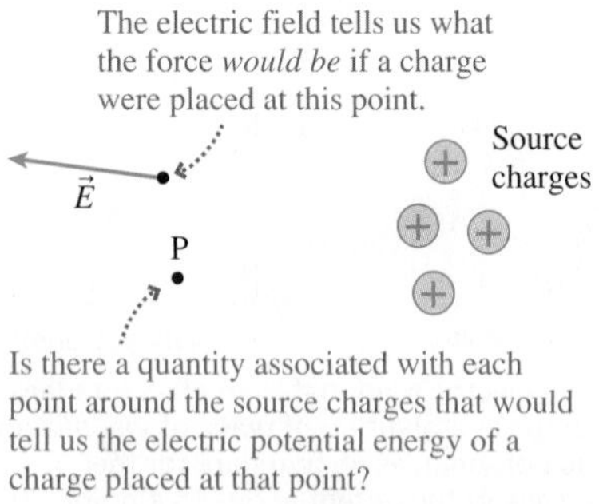

The electric field tells us what the force *would be* if a charge were placed at this point.

$\vec{E}$

Source charges

P

Is there a quantity associated with each point around the source charges that would tell us the electric potential energy of a charge placed at that point?

Can we apply similar reasoning to electric potential energy? Consider again the source charges in Figure 21.2. If we place a charge q at point P near the source charges, charge q will have electric potential energy. If we place a different charge q' at P, charge q' will have a different electric potential energy. Is there a quantity associated with P that would tell us what the electric potential energy of q, q', or any other charge *would have* at point P, without actually having to place the charge there?

To find out if this is possible, we'll have to understand in a bit more detail how to find the electric potential energy of a charge q. Suppose, to be specific, we take $q = 10$ nC in **FIGURE 21.3a** and, for convenience, we let $(U_{\text{elec}})_A = 0$ J when the charge is at point A.

As we've already seen, to find charge q's electric potential energy at any other point, such as point B or C, we need to find the amount of work it takes to move the charge from A to B, or A to C. In **FIGURE 21.3b** it takes the hand 4 μJ of work to move the charge from A to B; thus its electric potential energy at B is $(U_{\text{elec}})_B = 4\ \mu\text{J}$. (Recall that the SI unit of energy is the joule, J.) Similarly, q's electric potential energy at point C is $(U_{\text{elec}})_C = 6\ \mu\text{J}$ because it took 6 μJ of work to move it to point C.

What if we were to repeat this experiment with a different charge—say, $q = 20$ nC? According to Coulomb's law, the electric force on this charge due to the source charges will be twice that on the 10 nC charge. Consequently, the hand will have to push with twice as much force and thus do twice as much work in moving this charged particle from A to B. As a result, the 20 nC particle has $(U_{\text{elec}})_B = 8\ \mu\text{J}$ at B. A 5 nC particle would have $(U_{\text{elec}})_B = 2\ \mu\text{J}$ at B because the hand would have to work only half as hard to move it to B as it did to move the 10 nC particle to B. In general, **a charged particle's potential energy is proportional to its charge.**

When two quantities are proportional to each other, their *ratio* is constant. We can see this directly for the electric potential energy of a charged particle at point B by calculating the ratio

$$\frac{U_{(\text{elec})B}}{q} = \frac{2\ \mu\text{J}}{5\ \text{nC}} = \frac{4\ \mu\text{J}}{10\ \text{nC}} = \frac{8\ \mu\text{J}}{20\ \text{nC}} = 400\ \frac{\text{J}}{\text{C}}$$

U/q for $q = 5$ nC; U/q for $q = 10$ nC; U/q for $q = 20$ nC; All three ratios are the *same*.

Thus we can write a simple expression for the electric potential energy that *any* charge q *would have* if placed at point B:

$$(U_{\text{elec}})_B = \left(400\ \frac{\text{J}}{\text{C}}\right)q$$

This number is associated with point B. This part depends on the charge we place at B.

The number 400 J/C, which is associated with point B, tells us the *potential* for creating potential energy (there's a mouthful!) if a charge q is placed at point B. Thus this value is called the **electric potential,** and it is given the symbol V. At B, then, the electric potential is 400 J/C, as shown in **FIGURE 21.3c**. By similar reasoning, the electric potential at point C is 600 J/C.

This idea can be generalized. Any source charges create an electric potential at every point in the space around them. At a point where the potential is V, the electric potential energy of a charged particle q is

$$U_{\text{elec}} = qV \qquad (21.1)$$

Relationship between electric potential and electric potential energy

U_{elec} — V — p.37 — PROPORTIONAL

Notice the similarity to $\vec{F}_{\text{elec}} = q\vec{E}$. The electric potential, like the electric field, is created by the source charges and is present at all points in space. The electric potential is there whether or not charge q is present to experience it. While the electric field tells us how the source charges would exert a *force* on q, the electric potential tells us how the source charges would provide q with *potential energy*. Although we used the work done on a positive charge to justify Equation 21.1, it is also valid if q is negative.

FIGURE 21.3 Finding the electric potential energy and the electric potential.

(a) The electric potential energy of a 10 nC charge at A is zero. What is its potential energy at point B or C?

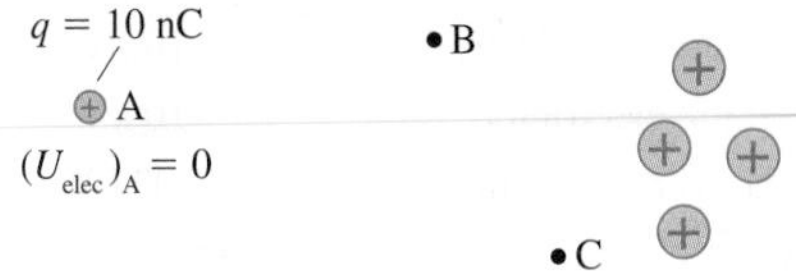

(b) The charge's electric potential energy at any point is equal to the amount of work done in moving it there from point A.

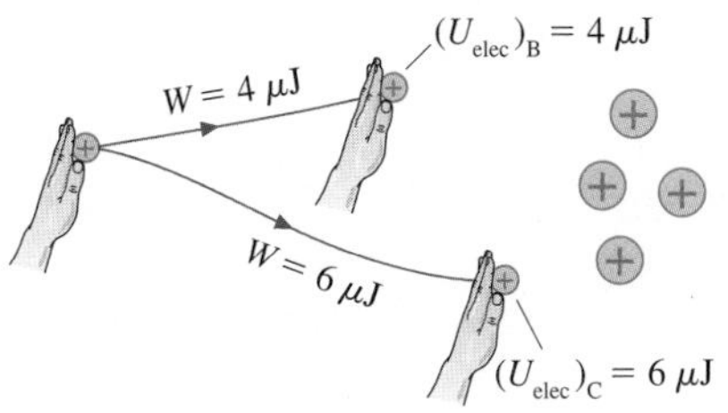

(c) The electric potential is created by the source charges. It exists at *every* point in space, not only at A, B, and C.

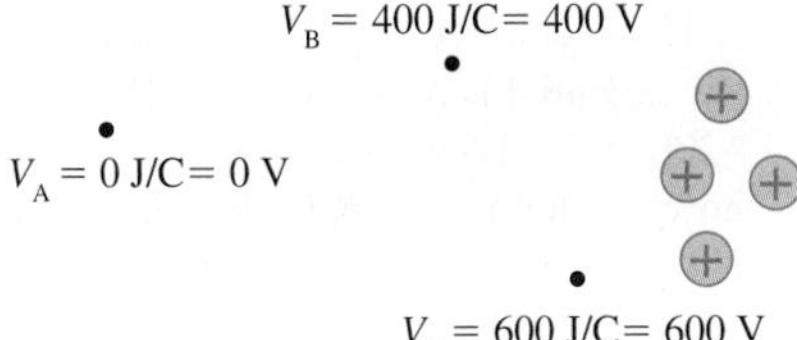

TABLE 21.1 Typical electric potentials

Source of potential	Approximate potential
Cells in human body	100 mV
Battery	1–10 V
Household electricity	100 V
Static electricity	10 kV
Transmission lines	500 kV

NOTE ► For potential energy, we found that we could choose a particular configuration of the system to have $U = 0$, the zero of potential energy. The same idea holds for electric potential. We can choose any point in space, wherever is convenient, to be $V = 0$. It will turn out that only *changes* in the electric potential are important, so the choice of a point to be $V = 0$ has no physical consequences. ◄

The unit of electric potential is the joule per coulomb, called the **volt** V:

$$1 \text{ volt} = 1 \text{ V} = 1 \text{ J/C}$$

This unit is named for Alessandro Volta, who invented the battery in 1800. Microvolts (μV), millivolts (mV), and kilovolts (kV) are commonly used units. Table 21.1 lists some typical electric potentials. We can now recognize that the electric potential in Figure 21.2—a potential due to the source charges—is 0 V at A and 400 V at B. This is shown in Figure 21.3c.

NOTE ► The symbol *V* is widely used to represent the *volume* of an object, and now we're introducing the same symbol to mean *potential.* To make matters more confusing, V is the abbreviation for *volts.* In printed text, *V* for potential is italicized while V for volts is not, but you can't make such a distinction in handwritten work. This is not a pleasant state of affairs, but these are the commonly accepted symbols. You must be especially alert to the *context* in which a symbol is used. ◄

EXAMPLE 21.1 Finding the change in a charge's electric potential energy

A 15 nC charged particle moves from point A, where the electric potential is 300 V, to point B, where the electric potential is −200 V. By how much does the electric potential change? By how much does the particle's electric potential energy change? How would your answers differ if the particle's charge were −15 nC?

PREPARE The change in the electric potential ΔV is the potential at the final point B minus the potential at the initial point A. From Equation 21.1, we can find the change in the electric potential energy by noting that $\Delta U_{elec} = (U_{elec})_B - (U_{elec})_A = q(V_B - V_A) = q\Delta V$.

SOLVE We have

$$\Delta V = V_B - V_A = (-200 \text{ V}) - (300 \text{ V}) = -500 \text{ V}$$

This change is *independent* of the charge q because the electric potential is created by source charges.

The change in the particle's electric potential energy is

$$\Delta U_{elec} = q\Delta V = (15 \times 10^{-9} \text{ C})(-500 \text{ V}) = -7.5 \ \mu\text{J}$$

A −15 nC charge would have ΔU_{elec} + 7.5 μJ because q changes sign while ΔV remains unchanged.

ASSESS Because the electric potential at B is lower than that at A, the positive (+15 nC) charge will lose electric potential energy, while the negative (−15 nC) charge will gain energy.

STOP TO THINK 21.1 A positively charged particle moves from point 1 to point 2. As it does, its electric potential energy

A. Increases.
B. Decreases.
C. Stays the same.

21.2 Sources of Electric Potential

How is an electric potential created in the first place? Consider the uncharged capacitor shown in **FIGURE 21.4a**. There's no force on charge q, so no work is required to move it from A to B. Consequently, charge q's electric potential energy remains *unchanged* as it is moved from A to B, so that $(U_{elec})_B = (U_{elec})_A$. Then, because $U_{elec} = qV$, it must be the case that $V_B = V_A$. We say that the **potential difference** $\Delta V = V_B - V_A$ is *zero.*

FIGURE 21.4 Potential differences are created by charge separation.

(a) The force on charge q is zero. No work is needed to move it from A to B, so there is no potential difference between A and B.

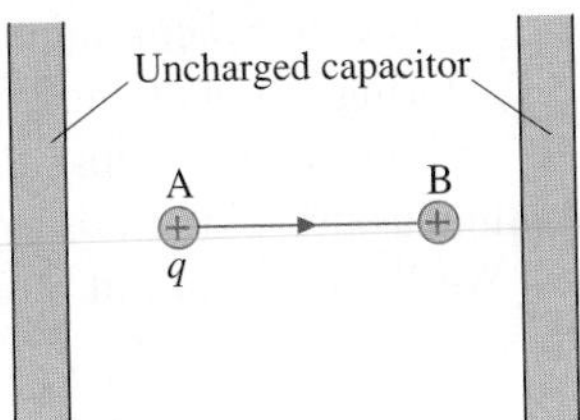

(b) The capacitor still has no net charge, but charge has been *separated* to give the plates charges $+Q$ and $-Q$.

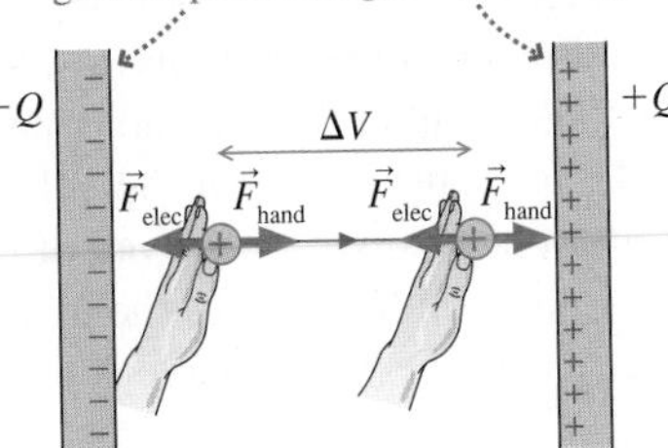

Now, because the separated charge exerts an electric force, the hand must do work on q to push it from A to B. The charge's electric potential energy increases, so there must be an *electric potential difference* ΔV between A and B.

Now consider what happens if electrons are transferred from the right side of the capacitor to the left, giving the left electrode charge $-Q$ and the right electrode charge Q. The capacitor still has no net charge, but the charge has been *separated*. These separated charges exert a force $\vec{F}_{elec}$ on q, as FIGURE 21.4b shows, so that the hand must now do work on q to move it from A to B, increasing its electric potential energy so that $(U_{elec})_B > (U_{elec})_A$. This means that the potential difference ΔV between A and B is no longer zero. What we've shown here is quite general: **A potential difference is created by *separating* positive charge from negative charge.**

Perhaps the most straightforward way to create a separation of charge is by the frictional transfer of charge discussed in Chapter 20. As you shuffle your feet across a carpet, the friction between your feet and the carpet transfers charge to your body, causing a potential difference between your body and, say, a nearby doorknob. The potential difference between you and a doorknob can be many tens of thousands of volts—enough to create a spark as the excess charge on your body moves from higher to lower potential.

Lightning is the result of large potential differences built up by charge separation within clouds.

Lightning is the result of a charge separation that occurs in clouds. As small ice particles in the clouds collide, they become charged by frictional rubbing. The details are still not well understood, but heavier particles, which fall to the bottom of the cloud, gain a negative charge, while lighter particles, which are lifted to the cloud's top, become positive. Thus the top of the cloud becomes positively charged and the bottom negatively charged. This natural charge separation creates a huge potential difference—as much as 100 millions volts—between the top and bottom of the cloud. The negative charge in the bottom of the cloud causes positive charge to accumulate in the ground below. A lightning strike occurs when the potential difference between the cloud and the ground becomes too large for the air to sustain.

Charge separation, and hence potential differences, can also be created by chemical processes. A common and important means of creating a fixed potential difference is the **battery.** We'll study batteries in Chapter 22, but all batteries use chemical reactions to create an internal charge separation. This separation proceeds until a characteristic potential difference—about 1.5 V for a standard alkaline battery—appears between the two terminals of the battery. Different kinds of batteries maintain different potential differences between their terminals.

NOTE ▶ The potential difference between two points is often called the **voltage.** Thus we say that a battery's voltage is 1.5 V or 12 V, and we speak of the potential difference between a battery's terminals as the voltage "across" the battery. ◀

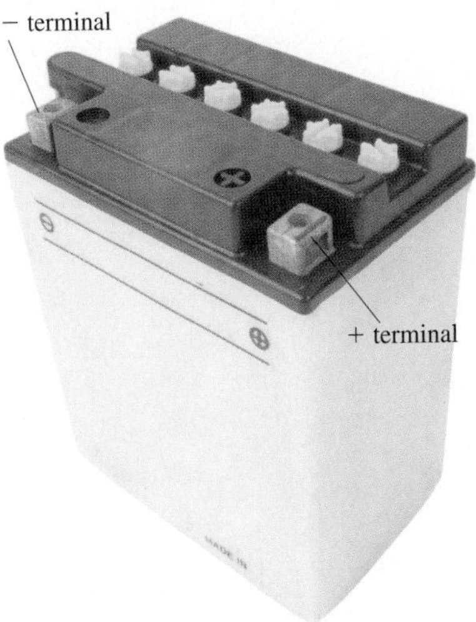

A car battery maintains a fixed potential difference of 12 V between its + and – terminals.

Chemical means of producing potential differences are also crucial in biological systems. For example, there's a potential difference of about 70 mV between the inside and outside of a cell, with the inside of the cell more negative than the outside.

FIGURE 21.5 The membrane potential of a cell is due to a charge separation.

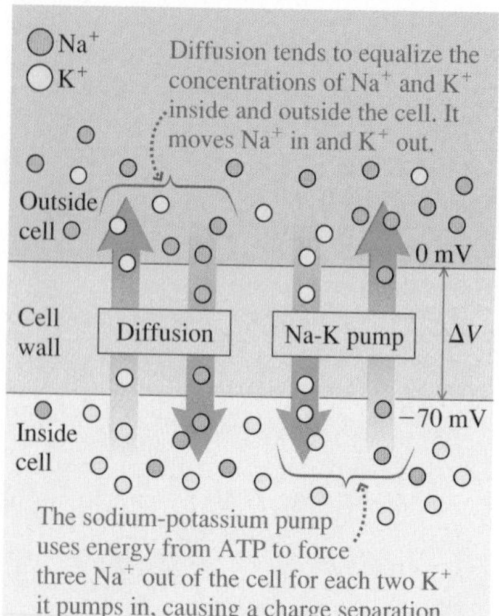

As illustrated in FIGURE 21.5, this *membrane potential* is caused by an imbalance of potassium (K^+) and sodium (Na^+) ions. The molar concentration of K^+ is higher inside the cell than outside, while the molar concentration of Na^+ is higher outside than inside. To keep the charge separated in the face of diffusion, which tends to equalize the ion concentrations, a *sodium-potassium exchange pump* continuously pumps sodium out of the cell and potassium into the cell. During one pumping cycle, three Na^+ are pushed out of the cell but only two K^+ are pushed in, giving a net transfer of one positive charge out of the cell. This continuous pumping leads to the charge separation that causes the membrane potential. We'll have a careful look at the electrical properties of nerve cells in Chapter 23.

Measuring Electric Potential

Measurements of the electric potential play an important role in a broad range of applications. An electrocardiogram measures the potential difference between several locations on the body to diagnose possible heart problems; temperature is often measured using a *thermocouple,* a device that develops a potential difference proportional to temperature; your digital camera can sense when its battery is low by measuring the potential difference between its terminals.

Note that in all these applications, it's the potential *difference* between two points that's measured. The actual value of the potential at a given point depends on where we choose V to be zero, but the difference in potential between two points is independent of this choice. Because of this, a **voltmeter,** the basic instrument for measuring potential differences, always has *two* inputs. Probes are connected from these inputs to the two points between which the potential difference is to be measured. We'll learn more in Chapter 23 about how voltmeters work.

A voltmeter always uses two probes to measure a potential difference. Here, we see that the potential difference of a fresh 9 V battery is closer to 9.7 V.

As small as cells are, the membrane potential difference between the inside and outside of a cell can be measured by a (very small) probe connected to a voltmeter. FIGURE 21.6 is a micrograph of a nerve cell whose membrane potential is being measured. A very small glass pipette, filled with conductive fluid, is actually inserted through the cell's membrane. This pipette is one of the probes. The second probe need not be so small; it is simply immersed in the conducting fluid that surrounds the cell and can be quite far from the cell.

FIGURE 21.6 Measuring the membrane potential.

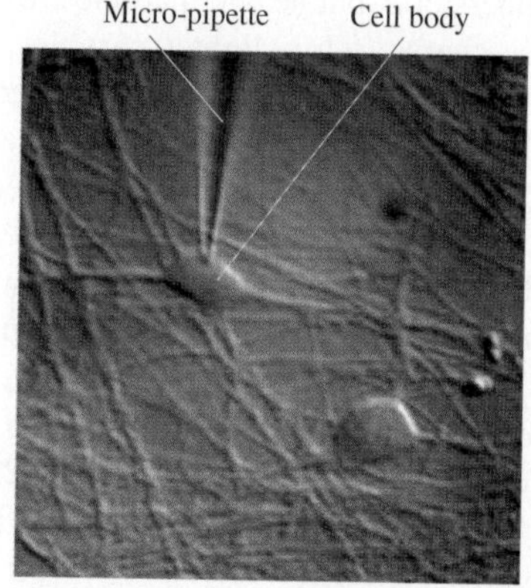

21.3 Electric Potential and Conservation of Energy

The potential energy of a charged particle is determined by the electric potential: $U_{\text{elec}} = qV$. Although potential and potential energy are related and have similar names, they are not the same thing. Table 21.2 will help you distinguish between electric potential and electric potential energy.

Because energy is conserved, a particle of charge q speeds up or slows down as it moves through a region of changing potential. The conservation of energy equation is

$$K_f + (U_{elec})_f = K_i + (U_{elec})_i$$

which we can write in terms of the electric potential V as

$$K_f + qV_f = K_i + qV_i \tag{21.2}$$

where, as usual, the subscripts i and f stand for the initial and final situations.

FIGURE 21.7 shows two positive charges moving through a region of changing electric potential. This potential has been created by source charges that aren't shown; our concern is only with the *effect* of this potential on the moving charge. Notice that we've used the before-and-after visual overview introduced earlier when we studied conservation of momentum and energy.

TABLE 21.2 Distinguishing electric potential and potential energy

The *electric potential* is a property of the source charges. The electric potential is present whether or not a charged particle is there to experience it. Potential is measured in J/C, or V.

The *electric potential energy* is the interaction energy of a charged particle with the source charges. Potential energy is measured in J.

FIGURE 21.7 A charged particle speeds up or slows down as it moves through a potential difference.

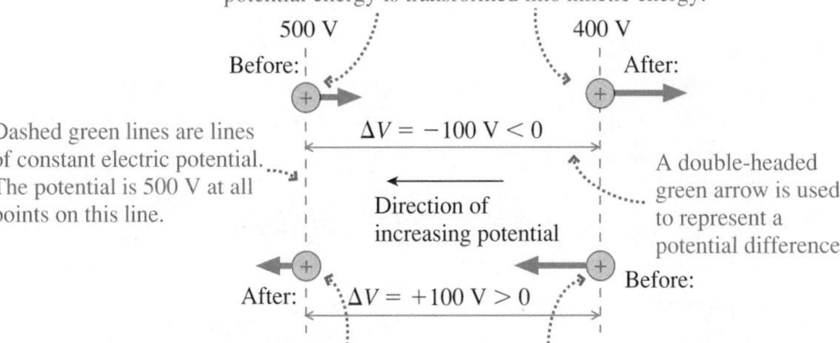
A positive charge speeds up ($\Delta K > 0$) as it moves from higher to lower potential ($\Delta V < 0$). Electric potential energy is transformed into kinetic energy.
500 V
400 V
Before:
After:
Dashed green lines are lines of constant electric potential. The potential is 500 V at all points on this line.
$\Delta V = -100\text{ V} < 0$
Direction of increasing potential
A double-headed green arrow is used to represent a potential difference.
Before:
After:
$\Delta V = +100\text{ V} > 0$
A positive charge slows down ($\Delta K < 0$) as it moves from lower to higher potential ($\Delta V > 0$). Kinetic energy is transformed into electric potential energy.

We can understand the motion of the charges if we rewrite Equation 21.2 as $K_f - K_i = -q(V_f - V_i)$, or

$$\Delta K = -q\Delta V \tag{21.3}$$

For the upper charge in Figure 21.7, the change in potential—that is, the potential difference—as it moves from left to right is

$$\Delta V = V_f - V_i = 400\text{ V} - 500\text{ V} = -100\text{ V}$$

which is negative. Equation 21.3 then shows that ΔK is *positive,* indicating that the particle *speeds up* as it moves from higher to lower potential. Conversely, for the lower charge in Figure 21.7 the potential difference is +100 V; Equation 21.3 then shows that ΔK is negative, so the particle *slows down* in moving from lower to higher potential.

NOTE ► The situation is reversed for a negative charge. If $q < 0$, Equation 21.3 requires K to increase as V increases. A negative charge speeds up if it moves into a region of higher potential. ◄

Conservation of energy is the basis of a powerful problem-solving strategy.

PROBLEM-SOLVING STRATEGY 21.1 **Conservation of energy in charge interactions**

PREPARE Draw a before-and-after visual overview. Define symbols that will be used in the problem, list known values, and identify what you're trying to find.

SOLVE The mathematical representation is based on the law of conservation of mechanical energy:

$$K_f + qV_f = K_i + qV_i$$

- Find the electric potential at both the initial and final positions. You may need to calculate it from a known expression for the potential, such as that of a point charge.
- K_i and K_f are the total kinetic energies of all moving particles.
- Some problems may need additional conservation laws, such as conservation of charge or conservation of momentum.

ASSESS Check that your result has the correct units, is reasonable, and answers the question.

Exercise 18

EXAMPLE 21.2 A speeding proton

A proton moves through an electric potential created by a number of source charges. Its speed is 2.5×10^5 m/s at a point where the potential is 1500 V. What will be the proton's speed a short time later when it reaches a point where the potential is –500 V?

PREPARE The mass of a proton is $m = 1.7 \times 10^{-27}$ kg. The positively charged proton moves from a region of higher potential to one of lower potential, so the change in potential is negative and the proton loses electric potential energy. Conservation of energy then requires its kinetic energy to increase. We can use Equation 21.2 to find the proton's final speed.

SOLVE Conservation of energy gives

$$K_f + qV_f = K_i + qV_i$$

or

$$\tfrac{1}{2}mv_f^2 + qV_f = \tfrac{1}{2}mv_i^2 + qV_i$$

Solving for v_f^2 gives

$$v_f^2 = v_i^2 + \frac{2}{m}(qV_i - qV_f) = v_i^2 + \frac{2q}{m}(V_i - V_f)$$

or

$$v_f^2 = (2.5 \times 10^5 \text{ m/s})^2 + \frac{2(1.6 \times 10^{-19}\text{ C})}{1.7 \times 10^{-27}\text{ kg}}[1500 \text{ V} - (-500 \text{ V})]$$

$$= 4.4 \times 10^{11} \text{ (m/s)}^2$$

Solving for the final speed gives

$$v_f = 6.6 \times 10^5 \text{ m/s}$$

ASSESS This problem is very similar to the situation in the upper part of Figure 21.7. A positively charged particle speeds up as it moves from higher to lower potential, analogous to a particle speeding up as it slides down a hill from higher gravitational potential energy to lower gravitational potential energy.

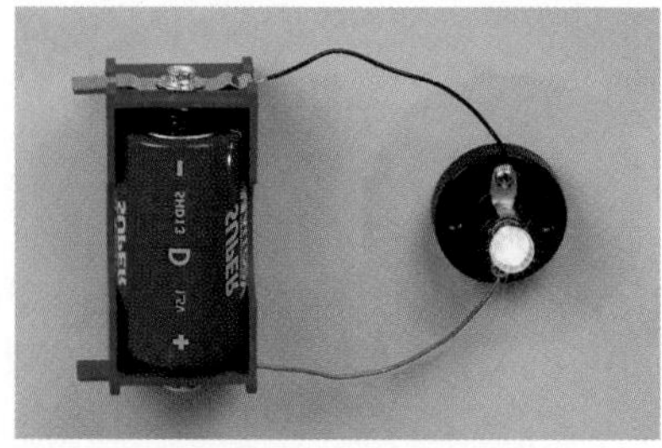

FIGURE 21.8 A transformation of electric potential energy into thermal energy.

So far we've considered only the transformation of electric potential energy into kinetic energy as a charged particle moves from higher to lower electric potential. That is, we've studied the energy transformation $\Delta K = -q\Delta V$. But it's worth noting that electric potential energy can also be transformed into other kinds of energy; this is the basis of many applications of electricity. In FIGURE 21.8, for example, charges move in the wires from the high-potential terminal of the battery, through the lightbulb, and back to the low-potential terminal. In the bulb, their electric potential energy is transformed into thermal energy E_{th}, making the bulb hot enough to glow brightly. This energy transformation is $\Delta E_{th} = -q\Delta V$. Or, as charges move from the high to the low potential terminals of an elevator motor, their electric potential energy is transformed into gravitational potential energy as the elevator and its passengers are lifted, so in this case $\Delta U_g = -q\Delta V$.

We'll have much more to say about these and other transformations of electric energy in chapters to come!

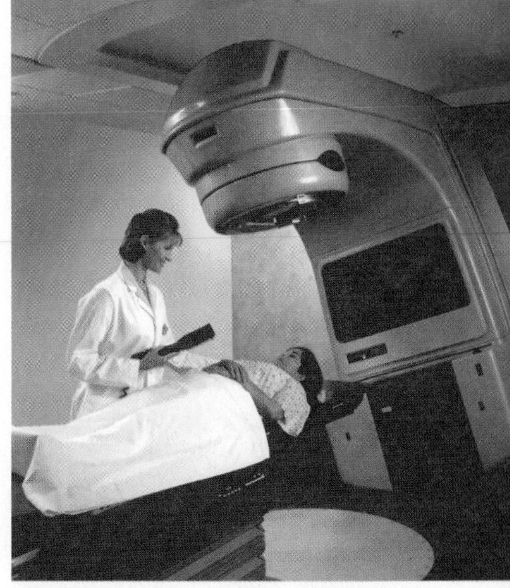

Cancer-fighting electrons BIO Tightly focused beams of x rays can be used in radiation therapy for cancer patients. The x rays are generated by directing a high-energy beam of electrons at a metal target. The electrons gain their energy by being accelerated in a *linear accelerator* through a potential difference of 20 MV, so their final kinetic energy is 20 MeV.

The Electron Volt

The joule is a unit of appropriate size in mechanics and thermodynamics, where we deal with macroscopic objects, but it will be very useful to have an energy unit appropriate to atomic and nuclear events.

Suppose an electron accelerates through a potential difference $\Delta V = 1$ V. The electron might be accelerating from 0 V to 1 V, or from 1000 V to 1001 V. Regardless of the actual voltages, an electron, being negative, *speeds up* when it moves toward a higher potential, so that, according to Equation 21.3, the 1 V potential difference causes the electron, with $q = -e$, to gain kinetic energy

$$\Delta K = -q\Delta V = e\Delta V = (1.60 \times 10^{-19}\ \text{C})(1\ \text{V}) = 1.60 \times 10^{-19}\ \text{J}$$

Let us define a new unit of energy, called the **electron volt,** as

$$1\ \text{electron volt} = 1\ \text{eV} = 1.60 \times 10^{-19}\ \text{J}$$

With this definition, the kinetic energy gained by the electron in our example is

$$\Delta K = 1\ \text{eV}$$

In other words, **1 electron volt is the kinetic energy gained by an electron (or proton) if it accelerates through a potential difference of 1 volt.**

NOTE ► The abbreviation eV uses a lowercase e but an uppercase V. Units of keV (10^3 eV), MeV (10^6 eV), and GeV (10^9 eV) are common. ◄

The electron volt can be a troublesome unit. One difficulty is its unusual name, which looks less like a unit than, say, "meter" or "second." A more significant difficulty is that the name suggests a relationship to volts. But *volts* are units of electric potential, whereas this new unit is a unit of energy! It is crucial to distinguish between the *potential V*, measured in volts, and an *energy* that can be measured either in joules or in electron volts. You can now use electron volts anywhere that you would previously have used joules. Doing so is no different from converting back and forth between pressure units of pascals and atmospheres.

NOTE ► The joule remains the SI unit of energy. It will be useful to express energies in eV, but you *must* convert this energy to joules before doing most calculations. ◄

EXAMPLE 21.3 The speed of a proton

Atomic particles are often characterized by their kinetic energy in MeV. What is the speed of an 8.7 MeV proton?

SOLVE The kinetic energy of this particle is 8.7×10^6 eV. First, we convert the energy to joules:

$$K = 8.7 \times 10^6\ \text{eV} \times \frac{1.60 \times 10^{-19}\ \text{J}}{1.0\ \text{eV}} = 1.4 \times 10^{-12}\ \text{J}$$

Now we can find the speed from

$$K = \frac{1}{2}mv^2$$

which gives

$$v = \sqrt{\frac{2K}{m}} = \sqrt{\frac{2(1.4 \times 10^{-12}\ \text{J})}{1.7 \times 10^{-27}\ \text{kg}}} = 4.1 \times 10^7\ \text{m/s}$$

Because the proton's charge and the electron's charge have the same magnitude, a general rule is that a proton or electron that accelerates (decelerates) through a potential difference of V volts gains (loses) V eV of kinetic energy. In Example 21.2, with a 2000 V potential difference, the proton gained 2000 eV of kinetic energy. In Example 21.3, the proton had to accelerate through a 8.7×10^6 V = 8.7 MV potential difference to acquire 8.7 MeV of kinetic energy.

STOP TO THINK 21.2 A proton is released from rest at point Q, where the potential is 0 V. Afterward, the proton

A. Remains at rest at Q.
B. Moves toward P with a steady speed.
C. Moves toward P with an increasing speed.
D. Moves toward R with a steady speed.
E. Moves toward R with an increasing speed.

−100 V 0 V +100 V
P• Q• R•

21.4 Calculating the Electric Potential

11.11 ActivPhysics ONLINE

Let's put these ideas to work by calculating the electric potential for some important cases. We'll do so using Equation 21.1, the relationship between the potential energy of a charge q at a point in space and the electric potential at that point. Rewriting Equation 21.1 slightly, we have

$$V = \frac{U_{\text{elec}}}{q} \tag{21.4}$$

Our prescription for finding the potential at a certain point in space, then, is to first calculate the electric potential *energy* of a charge q placed at that point. Then we can use Equation 21.4 to find the electric potential.

The Electric Potential Inside a Parallel-Plate Capacitor

FIGURE 21.9 Finding the potential of a parallel-plate capacitor.

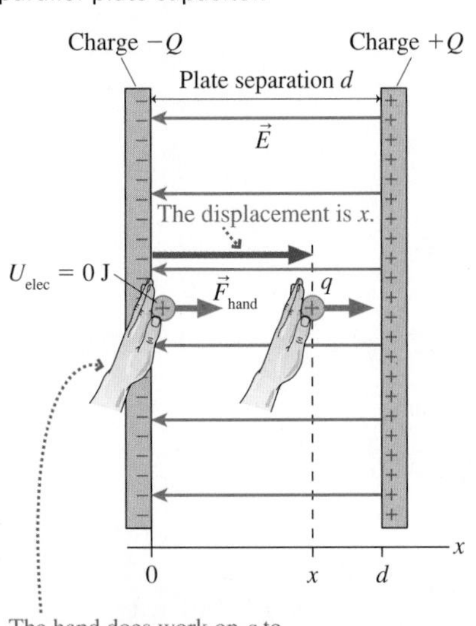
Charge −Q
Charge +Q
Plate separation d
E
The displacement is x.
U_elec = 0 J
F_hand
q
0
x
d
x
The hand does work on q to move it "uphill" against the field, thus giving the charge electric potential energy.

In Chapter 20 we learned that a *uniform* electric field can be created by placing equal but opposite charges on two parallel conducting plates—the **parallel-plate capacitor.** Thus finding the electric potential inside a parallel-plate capacitor is equivalent to finding the potential for the very important case of a uniform electric field.

FIGURE 21.9 shows a cross-section view of a charged parallel-plate capacitor with separation d between the plates. The charges $\pm Q$ on the plates are the source charges that create both the electric field and an electric potential in the space between the plates. As we found in Chapter 20, the electric field is $\vec{E} = (Q/\epsilon_0 A$, from positive to negative). We'll choose a coordinate system with $x = 0$ at the negative plate and $x = d$ at the positive plate.

We're free to choose the point of zero potential energy anywhere that's convenient, so let $U_{\text{elec}} = 0$ when a mobile charge q is at the negative plate. The charge's potential energy at any other position x is then the amount of work an external force must do to move the charge at steady speed from the negative plate to that position. We'll represent the external force by a hand, although that's not really how charges get moved around.

The electric field in Figure 21.9 points to the left, so the force $\vec{F}_{\text{elec}} = q\vec{E}$ of the field on the charge is also to the left. To move the charge to the right at constant speed ($\vec{F}_{\text{net}} = \vec{0}$), the external force $\vec{F}_{\text{hand}}$ must push to the right with a force of the same magnitude: $F_{\text{hand}} = qE$. This force does work on the system as the charge is moved, changing the potential energy of the system.

Because the external force is constant and is parallel to the displacement, the work to move the charge to position x is

$$W = \text{force} \times \text{displacement} = F_{\text{hand}}x = qEx$$

Consequently, the electric potential energy when charge q is at position x is

$$U_{\text{elec}} = W = qEx$$

As the final step, we can use Equation 21.4 to find that the electric potential of the parallel-plate capacitor at position x, measured from the positive plate, is

$$V = \frac{U_{\text{elec}}}{q} = Ex = \frac{Q}{\epsilon_0 A}x \qquad (21.5)$$

where, in the last step, we wrote the electric field strength in terms of the amount of charge on the capacitor plates. Keep in mind that this is the potential due to the source charges on the plates. We used the movable charge q to find the potential, but q does not cause or contribute to the potential.

A first point to notice is that the electric potential increases linearly from the negative plate at $x = 0$, where $V = V_- = 0$, to the positive plate at $x = d$, where $V = V_+ = Ed$. Let's define the potential difference ΔV_C between the two capacitor plates to be

$$\Delta V_C = V_+ - V_- = Ed \qquad (21.6)$$

People who work with circuits would call ΔV_C "the voltage across the capacitor" or simply "the voltage."

In many cases, the capacitor voltage is fixed at some value ΔV_C by connecting its plates to a battery with a known voltage. In this case, the electric field strength inside the capacitor is determined from Equation 21.6 as

$$E = \frac{\Delta V_C}{d} \qquad (21.7)$$

This means that we can establish an electric field of known strength by applying a voltage across a capacitor whose plate spacing is known.

Equation 21.7 implies that the units of electric field are volts per meter, or V/m. We have been using electric field units of newtons per coulomb. In fact, as you can show as a homework problem, these units are equivalent to each other; that is,

$$1\ \text{N/C} = 1\ \text{V/m}$$

NOTE ► Volts per meter are the electric field units used by scientists and engineers in practice. We will now adopt them as our standard electric field units. ◄

Returning to the electric potential, we can substitute Equation 21.7 for E into Equation 21.5 for V. In terms of the capacitor voltage ΔV_C, the electric potential at position x inside the capacitor is

$$V = \frac{x}{d}\Delta V_C \qquad (21.8)$$

You can see that the potential increases linearly from $V = 0$ at $x = 0$ (the negative plate) to $V = \Delta V_C$ at $x = d$ (the positive plate).

Let's explore the electric potential inside the capacitor by looking at several different, but related, ways that the potential can be represented. In this example, a battery has established a 1.5 V potential difference across a parallel-plate capacitor with a 3 mm plate spacing.

Graphical representations of the electric potential inside a capacitor

A graph of potential versus x. You can see the potential increasing from 0 V at the negative plate to 1.5 V at the positive plate.

A three-dimensional view showing **equipotential surfaces.** These are mathematical surfaces, not physical surfaces, that have the same value of V at every point. The equipotential surfaces of a capacitor are planes parallel to the capacitor plates. The capacitor plates are also equipotential surfaces.

A two-dimensional **equipotential map.** The green dashed lines represent slices through the equipotential surfaces, so V has the same value everywhere along such a line. We call these lines of constant potential **equipotential lines** or simply **equipotentials.**

A three-dimensional elevation graph. The potential is graphed vertically versus the x- and y-coordinates on the other axes. Viewing the front face of the elevation graph gives you the potential graph.

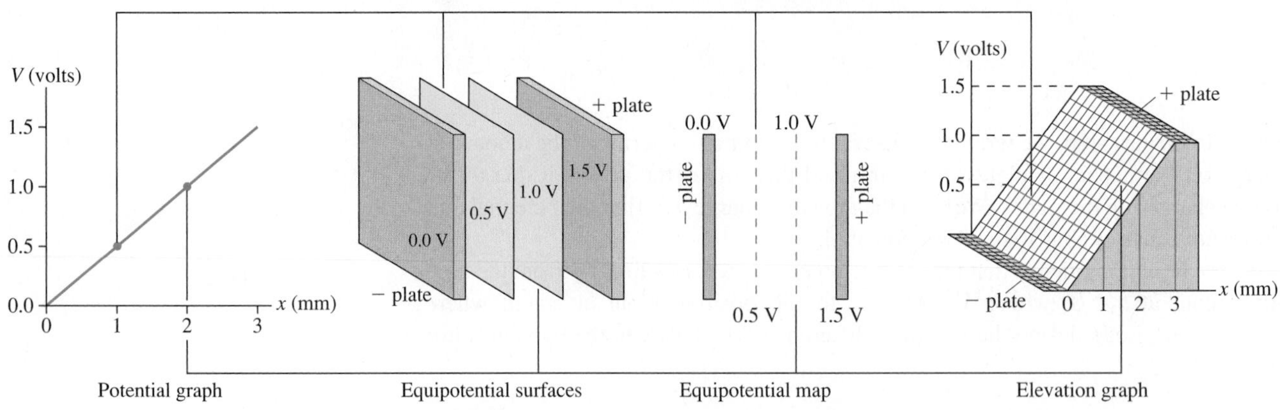

NOTE ▶ Equipotential lines are just the intersections of equipotential surfaces with the two-dimensional plane of the paper, and so are really just another way of representing equipotential surfaces. Because of this, we'll often use the terms "equipotential lines," "equipotential surfaces," and "equipotentials" interchangeably. ◀

EXAMPLE 21.4 A proton in a capacitor

A parallel-plate capacitor is constructed of two disks spaced 2.00 mm apart. It is charged to a potential difference of 500 V. A proton is shot through a small hole in the negative plate with a speed of 2.0×10^5 m/s. Does it reach the other side? If not, what is the farthest distance from the negative plate that the proton reaches?

PREPARE Energy is conserved. The proton's potential energy inside the capacitor can be found from the capacitor's electric potential. **FIGURE 21.10** is a before-and-after visual overview of the proton in the capacitor.

FIGURE 21.10 A before-and-after visual overview of a proton moving in a capacitor.

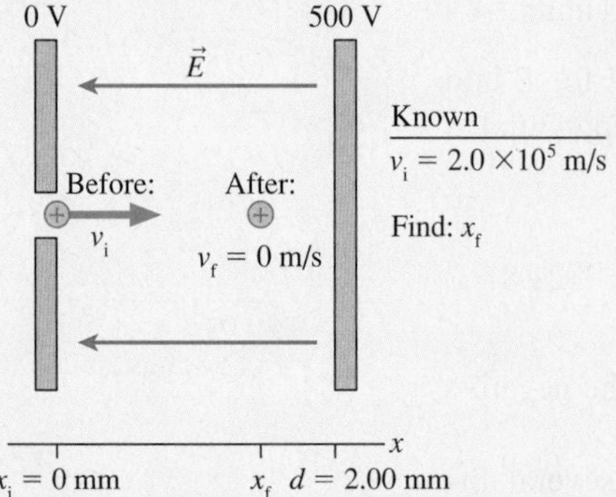

SOLVE The proton has charge $q = e$, and its potential energy at a point where the capacitor's potential is V is $U_{elec} = eV$. It will gain potential energy $\Delta U_{elec} = e\,\Delta V_C$ if it moves all the way across the capacitor. The increase in potential energy comes at the expense of kinetic energy, so the proton has sufficient kinetic energy to make it all the way across only if

$$K_i \geq e\,\Delta V_C$$

We can calculate that $K_i = \frac{1}{2}mv_i^2 = 3.34 \times 10^{-17}$ J and that $e\,\Delta V_C = 8.00 \times 10^{-17}$ J. The proton does *not* have sufficient kinetic energy to be able to gain 8.00×10^{-17} J of potential energy, so it will not make it across. Instead, the proton will reach a turning point and reverse direction.

The proton starts at the negative plate, where $x_i = 0$ mm. Let the turning point be at x_f. The potential inside the capacitor is given by $V = \Delta V_C x/d$ with $d = 0.0020$ m and $\Delta V_C = 500$ V. Conservation of energy requires $K_f + eV_f = K_i + eV_i$. This is

$$0 + e\,\Delta V_C \frac{x_f}{d} = \frac{1}{2}mv_i^2 + 0$$

where we used $V_i = 0$ V at the negative plate ($x_i = 0$) and $K_f = 0$ at the turning point. The solution for the turning point is

$$x_f = \frac{mdv_i^2}{2e\,\Delta V_C} = 0.84 \text{ mm}$$

The proton travels 0.84 mm, less than halfway across, before being turned back.

ASSESS We were able to use the electric potential inside the capacitor to determine the proton's potential energy.

The Potential of a Point Charge

The simplest possible source charge is a single fixed point charge q. To find the electric potential, we'll again start by first finding the electric potential energy when a second charge, which we'll call q', is distance r from charge q. As usual, we'll do this by calculating the work needed to bring q' from a point where $U_{elec} = 0$ to distance r from q. We're free to choose $U_{elec} = 0$ at any point that's convenient. Because the influence of a point charge extends infinitely far, our result for the electric potential will have an especially simple form if we choose $U_{elec} = 0$ (and hence $V = 0$) at a point that is infinitely distant from q.

We can't use the simple expression $W = Fd$ to find the work done in moving q'; this expression is valid only for a *constant* force F and, as we know from Coulomb's law, the force on q' gets larger and larger as it approaches q. To do this calculation properly requires the methods of calculus. However, we can understand *qualitatively* how the potential energy depends on the distance r between the two charges.

FIGURE 21.11 shows q' at two different distances r from the fixed charge q. When q' is relatively far from q, the electric force on q' is small. Not much external force is needed to push q' closer to q by a small displacement d, so the work done on q' is small and the *change* ΔU_{elec} in the electric potential energy is small as well. This implies, as Figure 21.11 shows, that the graph of U_{elec} versus r is fairly flat when q' is far from q.

On the other hand, the force on q' is quite large when it gets near q, so the work required to move it through the same small displacement is much greater than before. The change in U_{elec} is large, so the graph of U_{elec} versus r is steeper when q' is near q.

The general shape of the graph of U_{elec} must be as shown in Figure 21.11. When q' is far from q, the potential energy is small (U_{elec} must go to zero as r goes to infinity). As q' approaches q, the potential energy gets larger and larger. An exact calculation finds the potential energy of two point charges to be

FIGURE 21.11 The electric potential energy of two point charges.

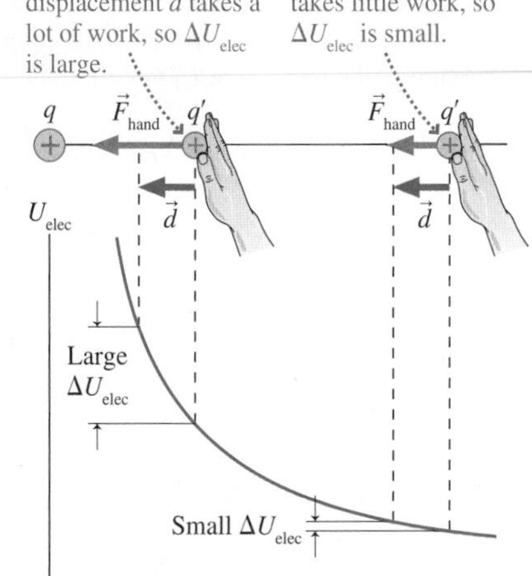
Near q the force is large. To move q' by displacement d takes a lot of work, so ΔU_{elec} is large.
Far from q the force is small. To move q' by d takes little work, so ΔU_{elec} is small.
q
$\vec{F}_{hand}$
q'
U_{elec}
$\vec{d}$
Large ΔU_{elec}
Small ΔU_{elec}
r

$$U_{elec} = K\frac{qq'}{r} = \frac{1}{4\pi\epsilon_0}\frac{qq'}{r} \quad (21.9)$$

U p.114 INVERSE

Electric potential energy of two charges q and q' separated by distance r

NOTE ▶ This expression is very similar to Coulomb's law. The difference is that the electric potential energy depends on the *inverse* of r—that is, on $1/r$—instead of the inverse-square dependence of Coulomb's law. Make sure you remember which is which! ◀

Figure 21.11 was a graph of the potential energy for two like charges, where the product qq' is positive. But Equation 21.9 is equally valid for *opposite* charges. In this case, the potential energy of the charges is *negative*. As **FIGURE 21.12** shows, the potential energy of the two charges *decreases* as they get closer together. A particle speeds up as its potential energy decreases ($U \rightarrow K$), so charge q' accelerates toward the fixed charge q. The graph lets us think about the attractive force between opposite charges from the perspective of energy.

FIGURE 21.12 Potential-energy diagram for two opposite charges.

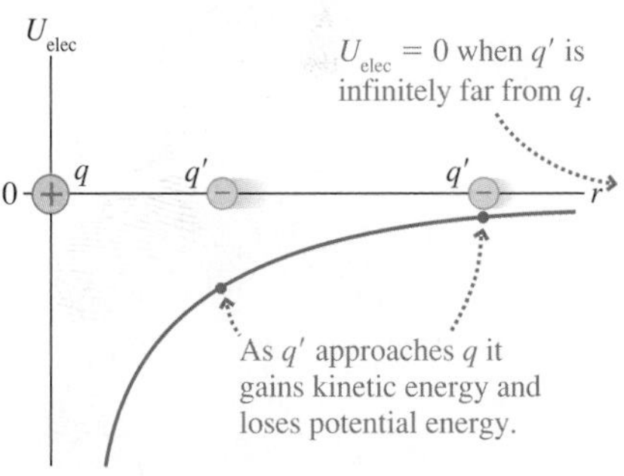
U_{elec}
$U_{elec} = 0$ when q' is infinitely far from q.
0
q
q'
r
As q' approaches q it gains kinetic energy and loses potential energy.

EXAMPLE 21.5 Finding the escape velocity

An interaction between two elementary particles causes an electron and a positron (a positively charged electron) to be shot out back-to-back with equal speeds. What minimum speed must each particle have when they are 100 fm apart in order to end up far from each other? (Recall that 1 fm = 10^{-15} m.)

PREPARE Energy is conserved. The particles end up "far from each other," which we interpret as sufficiently far to make $(U_{elec})_f \approx 0$ J. **FIGURE 21.13** shows the before-and-after visual

FIGURE 21.13 The before-and-after visual overview of an electron and a positron flying apart.

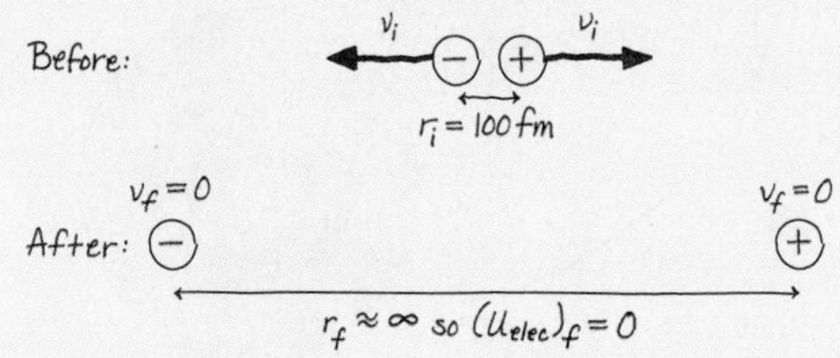
Before:
v_i
v_i
$r_i = 100$ fm
After:
$v_f = 0$
$v_f = 0$
$r_f \approx \infty$ so $(U_{elec})_f = 0$

Continued

overview. The minimum speed to escape is the speed that allows the particles to reach $r_f = \infty$ with $v_f = 0$.

SOLVE Here it is essential to interpret U_{elec} as the potential energy of the electron + positron system. Similarly, K is the *total* kinetic energy of the system. The electron and the positron, with equal masses and equal speeds, have equal kinetic energies. Conservation of energy $K_f + U_f = K_i + U_i$ is

$$0 + 0 = \left(\frac{1}{2}mv_i^2 + \frac{1}{2}mv_i^2\right) + K\frac{q_e q_p}{r_i} = mv_i^2 - \frac{Ke^2}{r_i}$$

Using $r_i = 100\text{ fm} = 1.0 \times 10^{-13}$ m, we can calculate the minimum initial speed to be

$$v_i = \sqrt{\frac{Ke^2}{mr_i}} = 5.0 \times 10^7 \text{ m/s}$$

ASSESS v_i is a little more than 10% the speed of light, just about the limit of what a "classical" calculation can predict. We would need to use the theory of relativity if v_i were much larger.

Equation 21.9 gives the potential energy of a charge q' when it is a distance r from a point charge q. We know that the electric *potential* is related to the potential energy by $V = U_{elec}/q'$. Thus the electric potential of charge q is

$$V = K\frac{q}{r} = \frac{1}{4\pi\epsilon_0}\frac{q}{r} \qquad (21.10)$$

Electric potential at distance r from a point charge q

INVERSE p.114

Notice that only the *source* charge q appears in this expression. This is in line with our picture that a source charge *creates* the electric potential around it.

EXAMPLE 21.6 Calculating the potential of a point charge

What is the electric potential 1.0 cm from a 1.0 nC charge? What is the potential difference between a point 1.0 cm away and a second point 3.0 cm away?

PREPARE We can use Equation 21.10 to find the potential at the two distances from the charge.

SOLVE The potential at $r = 1.0$ cm is

$$V_{1\text{cm}} = K\frac{q}{r} = (9.0 \times 10^9 \text{ N} \cdot \text{m}^2/\text{C}^2)\left(\frac{1.0 \times 10^{-9}\text{ C}}{0.010\text{ m}}\right)$$
$$= 900 \text{ V}$$

We can similarly calculate $V_{3\text{cm}} = 300$ V. Thus the potential difference between these two points is $\Delta V = V_{1\text{cm}} - V_{3\text{cm}} = 600$ V.

ASSESS 1 nC is typical of the electrostatic charge produced by rubbing, and you can see that such a charge creates a fairly large potential nearby. Why aren't we shocked and injured when working with the "high voltages" of such charges? As we'll learn in Chapter 26, the sensation of being shocked is a result of current, not potential. Some high-potential sources simply do not have the ability to generate much current.

FIGURE 21.14 shows four graphical representations of the electric potential of a point charge. These match the four representations of the electric potential inside a capacitor,

FIGURE 21.14 Four graphical representations of the electric potential of a point charge.

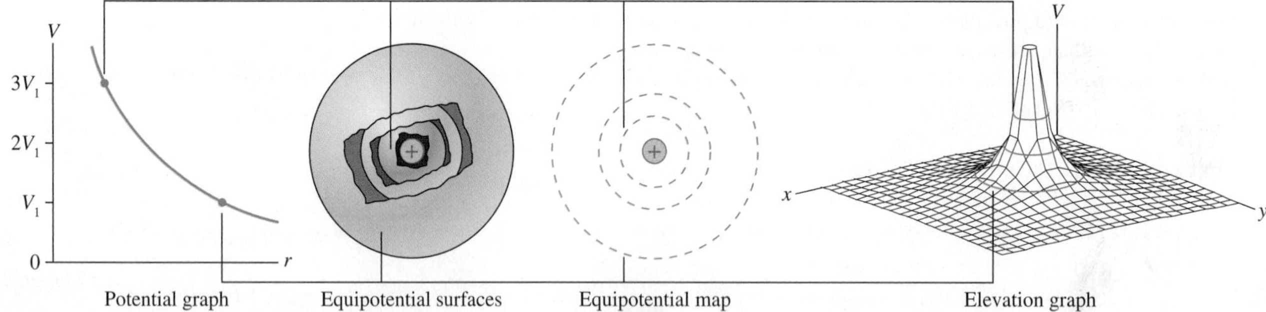

and a comparison of the two is worthwhile. This figure assumes that q is positive; you may want to think about how the representations would change if q were negative.

The Electric Potential of a Charged Sphere

Equation 21.10 gives the electric potential of a point charge. It can be shown that the electric potential outside a charged sphere is the *same* as that of a point charge; that is,

$$V = K\frac{Q}{r} = \frac{1}{4\pi\epsilon_0}\frac{Q}{r} \qquad (21.11)$$

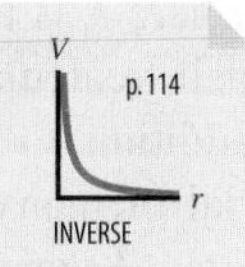

Electric potential at a distance $r > R$ from the center of a sphere of radius R and with charge Q

We can cast this result in a more useful form. It is common to charge a metal object, such as a sphere, "to" a certain potential; for instance, this can be done by connecting the sphere to, say, a 30 volt battery. This potential, which we will call V_0, is the potential right on the surface of the sphere. We can see from Equation 21.11 that

$$V_0 = V(\text{at } r = R) = \frac{Q}{4\pi\epsilon_0 R} \qquad (21.12)$$

Consequently, a sphere of radius R that is charged to potential V_0 has total charge

$$Q = 4\pi\epsilon_0 RV_0 \qquad (21.13)$$

If we substitute this expression for Q into Equation 21.11, we can write the potential outside a sphere that is charged to potential V_0 as

$$V = \frac{R}{r}V_0 \qquad (21.14)$$

Equation 21.14 tells us that the potential of a sphere is V_0 on the surface and decreases inversely with the distance. Thus the potential at $r = 3R$ is $\frac{1}{3}V_0$.

EXAMPLE 21.7 Releasing a proton from a charged sphere

A proton is released from rest at the surface of a 1.0-cm-diameter sphere that has been charged to +1000 V.

a. What is the charge of the sphere?
b. What is the proton's speed when it is 1.0 cm from the sphere?
c. When the proton is 1.0 cm from the sphere, what is its kinetic energy in eV?

PREPARE Energy is conserved. The potential outside the charged sphere is the same as the potential of a point charge at the center. **FIGURE 21.15** is a before-and-after visual overview.

FIGURE 21.15 A before-and-after visual overview of a sphere and a proton.

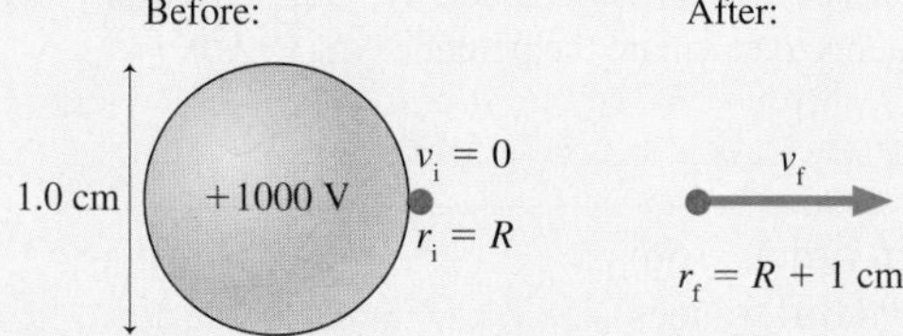

SOLVE

a. We can use the sphere's potential in Equation 21.13 to find that the charge of the sphere is

$$Q = 4\pi\epsilon_0 RV_0 = 0.56 \times 10^{-9}\ \text{C} = 0.56\ \text{nC}$$

b. A sphere charged to $V_0 = +1000$ V is positively charged. The proton will be repelled by this charge and move away from the sphere. The conservation of energy equation $K_f + eV_f = K_i + eV_i$, with Equation 21.14 for the potential of a sphere, is

$$\frac{1}{2}mv_f^2 + \frac{eR}{r_f}V_0 = \frac{1}{2}mv_i^2 + \frac{eR}{r_i}V_0$$

The proton starts from the surface of the sphere, $r_i = R$, with $v_i = 0$. When the proton is 1.0 cm from the *surface* of the sphere, it has $r_f = 1.0\ \text{cm} + R = 1.5$ cm. Using these values, we can solve for v_f:

$$v_f = \sqrt{\frac{2eV_0}{m}\left(1 - \frac{R}{r_f}\right)} = 3.6 \times 10^5\ \text{m/s}$$

c. The kinetic energy is

$$K_f = \tfrac{1}{2}mv_f^2 = \tfrac{1}{2}(1.7 \times 10^{-27}\ \text{kg})(3.6 \times 10^5\ \text{m/s})^2$$
$$= 1.1 \times 10^{-16}\ \text{J}$$

Converting to electron volts, we find

$$1.1 \times 10^{-16}\ \text{J} \times \frac{1\ \text{eV}}{1.6 \times 10^{-19}\ \text{J}} = 690\ \text{eV}$$

ASSESS A proton at the surface of the sphere, where $V = 1000$ V would, by definition, have 1000 eV of electric potential energy. It is reasonable that after it's moved some distance away, it has transformed 690 eV of this to kinetic energy.

The Electric Potential of Many Charges

Suppose there are many source charges $q_1, q_2, \ldots$. The electric potential V at a point in space is the *sum* of the potentials due to each charge:

$$V = \sum_i K\frac{q_i}{r_i} = \sum_i \frac{1}{4\pi\epsilon_0}\frac{q_i}{r_i} \tag{21.15}$$

where r_i is the distance from charge q_i to the point in space where the potential is being calculated. Unlike the electric field of multiple charges, which required vector addition, the electric potential is a simple scalar sum. This makes finding the potential of many charges considerably easier than finding the corresponding electric field. As an example, the equipotential map and elevation graph in **FIGURE 21.16** show that the potential of an electric dipole is the sum of the potentials of the positive and negative charges. We'll see later that the electric potential of the heart has the form of an electric dipole.

FIGURE 21.16 The electric potential of an electric dipole.

Equipotential map

Elevation graph

Equipotential lines

EXAMPLE 21.8 Finding the potential of two charges

What is the electric potential at the point indicated in **FIGURE 21.17**?

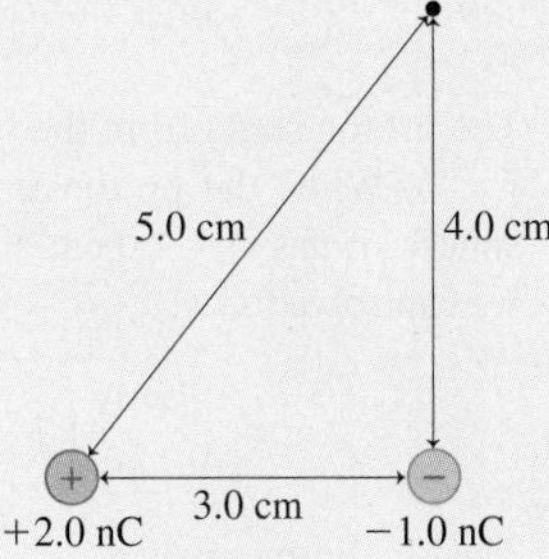

FIGURE 21.17 Finding the potential of two charges.

PREPARE The potential is the sum of the potentials due to each charge.

SOLVE The potential at the indicated point is

$$V = \frac{Kq_1}{r_1} + \frac{Kq_2}{r_2}$$

$$= (9.0 \times 10^9\ \text{N} \cdot \text{m}^2/\text{C}^2)\left(\frac{2.0 \times 10^{-9}\ \text{C}}{0.050\ \text{m}} + \frac{-1.0 \times 10^{-9}\ \text{C}}{0.040\ \text{m}}\right)$$

$$= 140\ \text{V}$$

ASSESS As noted, the potential is a *scalar,* so we found the net potential by adding two scalars. We don't need any angles or components to calculate the potential.

STOP TO THINK 21.3 Rank in order, from largest to smallest, the potential differences ΔV_{12}, ΔV_{13}, and ΔV_{23} between points 1 and 2, points 1 and 3, and points 2 and 3.

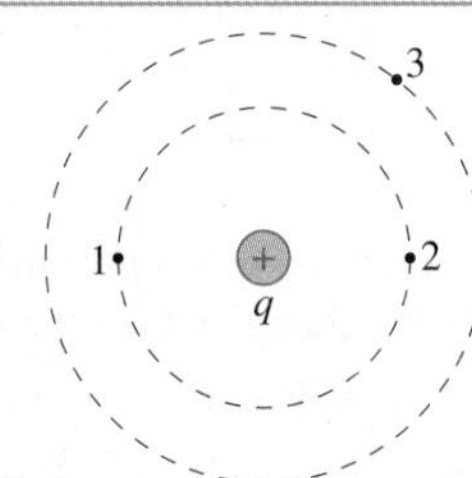

21.5 Connecting Potential and Field

In Chapter 20, we learned how source charges create an electric field around them; in this chapter, we found that source charges also create an electric potential everywhere in the space around them. But these two concepts of field and potential are clearly linked. For instance, we can calculate potential differences by considering the work done on a charge as it is pushed against the electric force due to the field. In this section, we'll build on this idea and will find that **the electric potential and electric field are not two distinct entities but, instead, two different perspectives or two different mathematical representations of how source charges alter the space around them.**

To make the connection between potential and field, FIGURE 21.18 shows an equipotential map of the electric potential due to some source charges (which aren't shown here). Suppose a charge q moves a short distance along one of the equipotential surfaces. Because it moves along an equipotential, its potential and hence its potential energy are the *same* at the beginning and end of its displacement. This means that no work is done in moving the charge. As Figure 21.18 shows, the only way that no work can be done is if the electric field is *perpendicular* to the equipotential. (You should recall, from Chapter 10, that no work is done by a force perpendicular to a particle's displacement.) This, then, is our first discovery about the connection between field and potential: **The electric field at a point is perpendicular to the equipotential surface at that point.**

FIGURE 21.18 The electric field is always perpendicular to an equipotential surface.

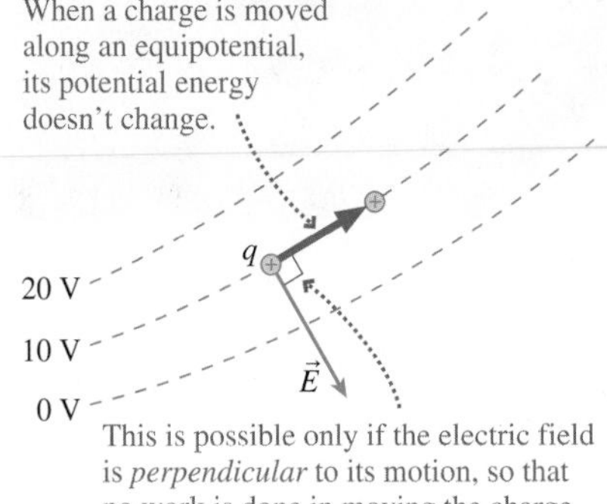

When a charge is moved along an equipotential, its potential energy doesn't change.
q
20 V
10 V
0 V
E
This is possible only if the electric field is *perpendicular* to its motion, so that no work is done in moving the charge.

In Figure 21.18, there are actually two directions that are perpendicular to the equipotential surface: the one shown by $\vec{E}$ in the figure, and the other pointing opposite $\vec{E}$. Which direction is correct?

FIGURE 21.19 shows a positive charge released from rest starting at the 10 V equipotential. You learned in Section 21.3 that a positive charge speeds up as it moves from higher to lower potential, so this charge will speed up as it moves toward the lower 0 V equipotential. Because it is the electric field $\vec{E}$ in Figure 21.19 that pushes on the charge, causing it to speed up, $\vec{E}$ must point as shown, from higher potential (10 V) to lower potential (0 V). This is our second discovery about the connection between field and potential: **The electric field points in the direction of *decreasing* potential.**

FIGURE 21.19 The electric field points "downhill."

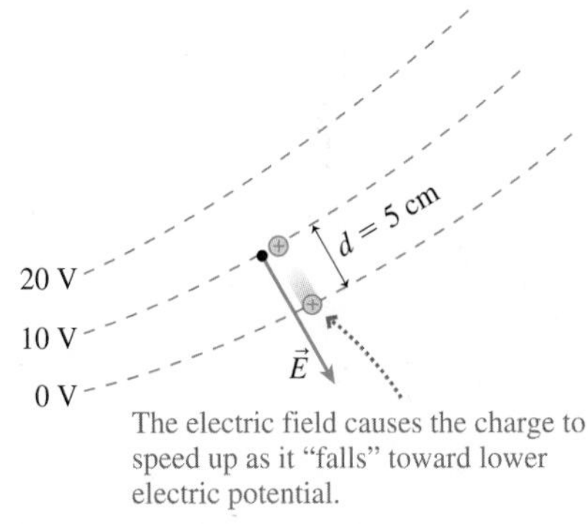

Finally, we can find an expression for the *magnitude* of $\vec{E}$ by considering the work required to move the charge, at constant speed, through a displacement that is directed *opposite* $\vec{E}$, as shown in FIGURE 21.20. The displacement is small enough that the electric field in this region can be considered as nearly constant. Conservation of energy requires that

$$W = \Delta U_{\text{elec}} = q\Delta V \tag{21.16}$$

FIGURE 21.20 Finding the strength of the electric field.

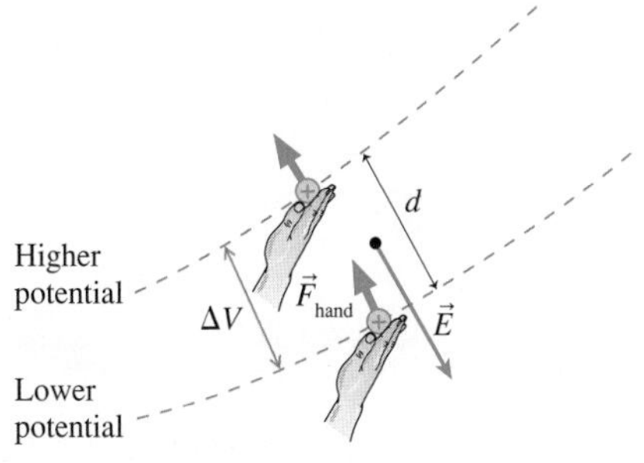

Because the charge moves at a constant speed, the magnitude of the force of the hand $\vec{F}_{\text{hand}}$ is exactly equal to that of the electric force $q\vec{E}$. Thus, the work done on the charge in moving it through displacement d is

$$W = F_{\text{hand}}d = qEd$$

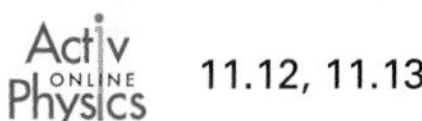

11.12, 11.13

Comparing this result with Equation 21.16 shows that the strength of the electric field is

$$E = \frac{\Delta V}{d} \tag{21.17}$$

Electric field strength in terms of the potential difference ΔV between two equipotential surfaces a distance d apart

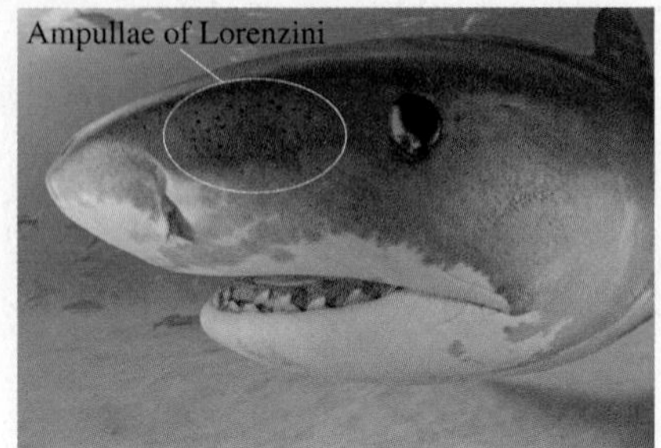
Ampullae of Lorenzini

◀ **Electroreception in sharks** BIO Sharks have a "sixth sense" that aids them in detecting prey: They are able to detect the weak electric potentials created by other fish. A shark has an array of sensor cells, the *ampullae of Lorenzini,* distributed around its snout. These appear as small black spots on the blue shark shown. These cells are highly sensitive to potential differences and can measure voltages as small as a few billionths of a volt! As we saw in Chapter 20, there are electric fields and potentials associated with every beat of the heart; muscle contractions also create potential differences. The shark can detect these small potential differences, even using them to sense animals buried in the sand.

For example, in Figure 21.19 the magnitude of the potential difference is $\Delta V = (10\text{ V}) - (0\text{ V}) = 10\text{ V}$, while $d = 5\text{ cm} = 0.05\text{ m}$. Thus the electric field strength between these two equipotential surfaces is

$$E = \frac{\Delta V}{d} = \frac{10\text{ V}}{0.05\text{ m}} = 200\text{ V/m}$$

NOTE ▶ This expression for the electric field strength is similar to the result $E = \Delta V_C/d$ for the electric field strength inside a parallel-plate capacitor. A capacitor has a uniform field, the same at all points, so we can calculate the field using the full potential difference ΔV_C and the full spacing d. In contrast, Equation 21.17 applies only *locally,* at a point where the spacing between two adjacent equipotential lines is d. As the spacing varies, so too does the field strength. ◀

FIGURE 21.21 summarizes what we've learned about the connection between potential and field.

FIGURE 21.21 The geometry of potential and field.

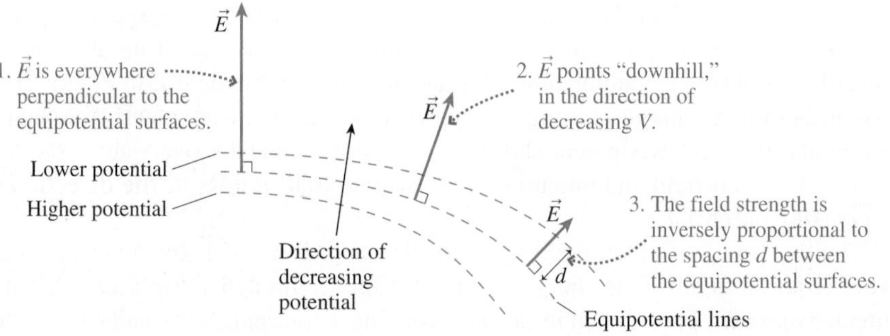
$\vec{E}$
1. $\vec{E}$ is everywhere perpendicular to the equipotential surfaces.
2. $\vec{E}$ points "downhill," in the direction of decreasing V.
$\vec{E}$
Lower potential
Higher potential
Direction of decreasing potential
$\vec{E}$
d
3. The field strength is inversely proportional to the spacing d between the equipotential surfaces.
Equipotential lines

FIGURE 21.22 shows three important arrangements of charges that we've studied in this chapter and Chapter 20. Both electric field lines and equipotentials are shown, and you can see how the connections between field and potential summarized in Figure 21.21 apply in each case.

FIGURE 21.22 Electric field lines and equipotentials for three important cases.

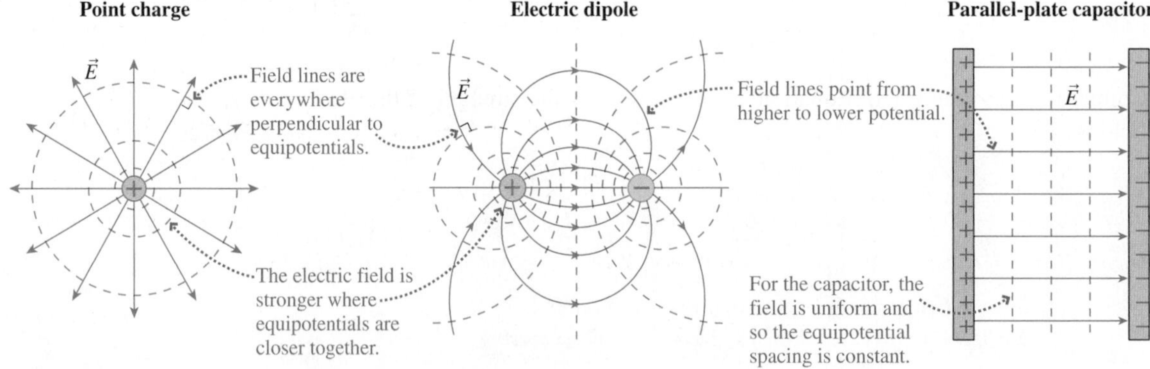
Point charge
Electric dipole
Parallel-plate capacitor
$\vec{E}$
Field lines are everywhere perpendicular to equipotentials.
$\vec{E}$
Field lines point from higher to lower potential.
$\vec{E}$
The electric field is stronger where equipotentials are closer together.
For the capacitor, the field is uniform and so the equipotential spacing is constant.

EXAMPLE 21.9 Finding the electric field from equipotential lines

In FIGURE 21.23 a 1 cm × 1 cm grid is superimposed on an equipotential map of the potential. Estimate the strength and direction of the electric field at points 1, 2, and 3. Show your results graphically by drawing the electric field vectors on the equipotential map.

FIGURE 21.23 Equipotential lines.

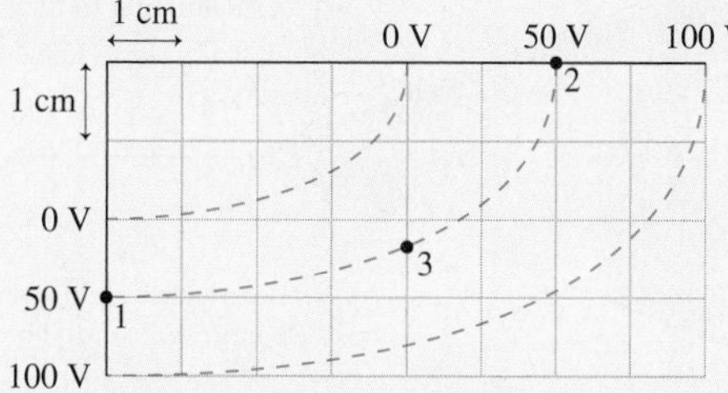

PREPARE The electric field is perpendicular to the equipotential lines, points "downhill," and depends on the spacing between the equipotential lines. The potential is highest on the bottom and the right. An elevation graph of the potential would look like the lower-right quarter of a bowl or a football stadium.

SOLVE Some distant but unseen source charges have created an electric field and potential. We do not need to see the source charges to relate the field to the potential. Because $E = \Delta V/d$, the electric field is stronger where the equipotential lines are closer together and weaker where they are farther apart.

FIGURE 21.24 shows how measurements of d from the grid are combined with values of ΔV to determine $\vec{E}$. Point 3 requires an estimate of the spacing between the 0 V and the 100 V lines. Notice that we're using the 0 V and 100 V equipotential lines to determine $\vec{E}$ at a point on the 50 V equipotential.

FIGURE 21.24 The electric field at points 1, 2, and 3.

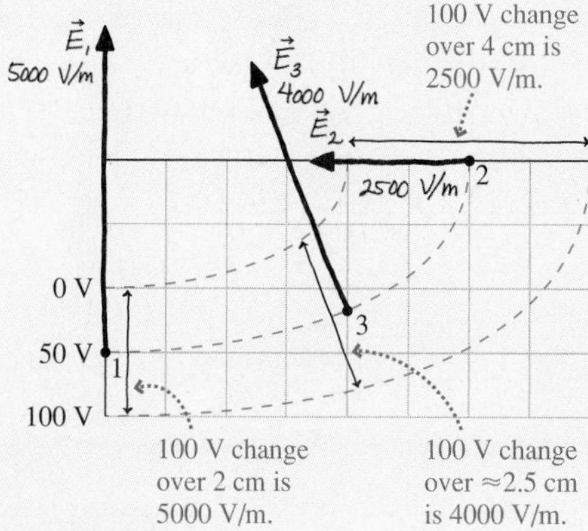

ASSESS The *directions* of $\vec{E}$ are found by drawing downhill vectors perpendicular to the equipotentials. The distances between the equipotential lines are needed to determine the field strengths.

A Conductor in Electrostatic Equilibrium

In Chapter 20, you learned four important properties about conductors in electrostatic equilibrium:

1. Any excess charge is on the surface.
2. The electric field inside is zero.
3. The exterior electric field is perpendicular to the surface.
4. The field strength is largest at sharp corners.

Now we can add a fifth important property:

5. The entire conductor is at the same potential, and thus the surface is an equipotential surface.

At pointed metal tips, the electric field can be strong enough to ionize air molecules. The field accelerates these ions, which collide, releasing energy in the bluish glow seen here.

To see why this is so, FIGURE 21.25 shows two points inside a conductor connected by a line that remains entirely inside the conductor. We can find the potential difference $\Delta V = V_2 - V_1$ between these points by using an external force to push a charge along the line from 1 to 2 and calculating the work done. But because $\vec{E} = \vec{0}$, there is no force on the charge. The work is zero, and so $\Delta V = 0$. In other words, **any two points inside a conductor in electrostatic equilibrium are at the same potential.**

When a conductor is in electrostatic equilibrium, the *entire conductor* is at the same potential. If we charge a metal electrode, the entire electrode is at a single potential. The facts that the surface is an equipotential surface and that the exterior electric field is perpendicular to the surface can now be seen as a special case of our conclusion from the preceding section that electric fields are always perpendicular to equipotentials.

Conductor
$\vec{E} = \vec{0}$
$\Delta V = 0$
1
2

▶ FIGURE 21.25 All points inside a conductor in electrostatic equilibrium are at the same potential.

FIGURE 21.26 summarizes what we know about conductors in electrostatic equilibrium. These are important and practical conclusions because conductors are the primary components of electrical devices.

FIGURE 21.26 Electrical properties of a conductor in electrostatic equilibrium.

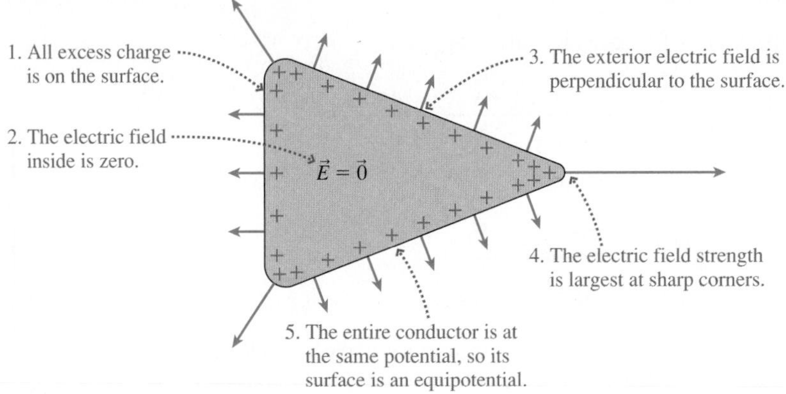
1. All excess charge is on the surface.
2. The electric field inside is zero.
$\vec{E} = \vec{0}$
3. The exterior electric field is perpendicular to the surface.
4. The electric field strength is largest at sharp corners.
5. The entire conductor is at the same potential, so its surface is an equipotential.

STOP TO THINK 21.4 Which set of equipotential surfaces matches this electric field?

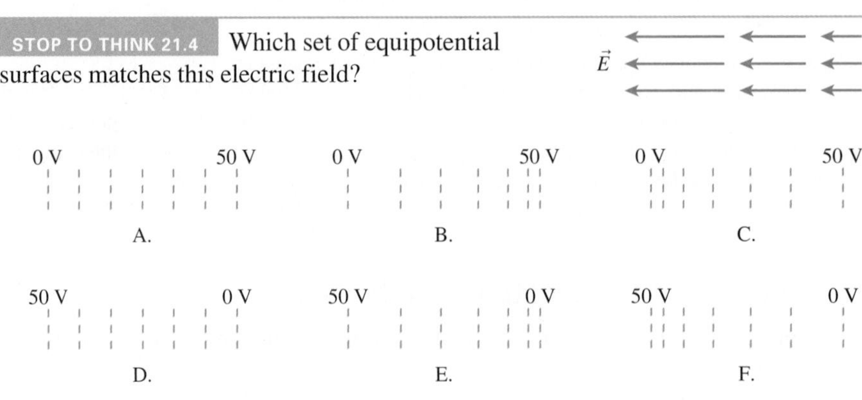
$\vec{E}$
0 V
50 V
A.
B.
C.
50 V
0 V
D.
E.
F.

21.6 The Electrocardiogram

As we saw in Chapter 20, the electrical activity of cardiac muscle cells makes the beating heart an electric dipole. A resting nerve or muscle cell is *polarized,* meaning that the outside is positive and the inside negative. Figure 21.5 showed this situation. Initially, all muscle cells in the heart are polarized. When triggered by an electrical impulse from the heart's sino-auricular node in the right atrium, heart cells begin to *depolarize,* moving ions through the cell wall until the outside becomes negative. This causes the muscle to contract. The depolarization of one cell triggers depolarization in an adjacent cell, causing a "wave" of depolarization to spread across the tissues of the heart.

At any instant during this process, a boundary divides the negative charges of depolarized cells from the positive charges of cells that have not depolarized. As FIGURE 21.27a shows, this separation of charges creates an electric dipole and produces a dipole electric field and potential. FIGURE 21.27b shows the equipotential surfaces of the heart's dipole at one instant of time. These equipotential surfaces match those shown earlier in Figure 21.16.

NOTE ► The convention is to draw diagrams of the heart as if you were facing the person whose heart is being drawn. The left side of the heart is thus on the right side of the diagram. ◄

FIGURE 21.27 A contracting heart is an electric dipole.

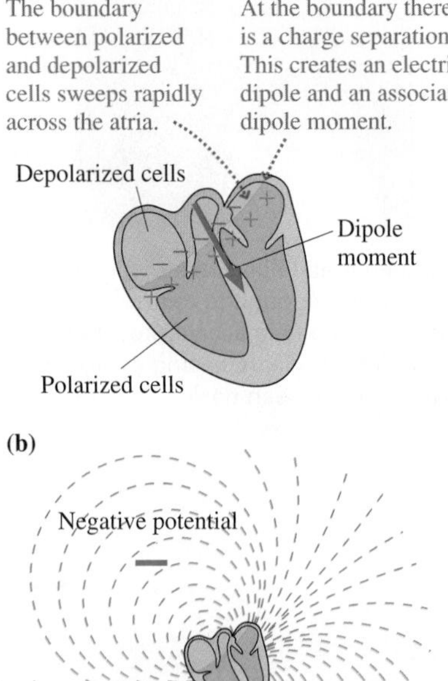
(a)
The boundary between polarized and depolarized cells sweeps rapidly across the atria.
At the boundary there is a charge separation. This creates an electric dipole and an associated dipole moment.
Depolarized cells
Dipole moment
Polarized cells
(b)
Negative potential
Positive potential

A measurement of the electric potential of the heart is an invaluable diagnostic tool. Recall, however, that only potential *differences* are meaningful, so we need to measure the potential difference between two points on the torso. In practice, as FIGURE 21.28 shows, the potential difference is measured between several pairs of *electrodes* (often called *leads*). A chart of the potential differences is known as an **electrocardiogram,** abbreviated either ECG or, from its European origin, EKG. A common method of performing an EKG uses 12 leads and records 12 pairs of potential differences.

FIGURE 21.29 shows a simplified model of electrocardiogram measurement using only two electrodes, one on each arm. As the wave of depolarization moves across the heart muscle during each heart beat, the dipole moment vector of the heart changes its magnitude and direction. As Figure 21.29 shows, both of these affect the potential difference between the electrodes, so each point on the EKG graph corresponds to a particular magnitude and orientation of the dipole moment.

FIGURE 21.28 Measuring an EKG.

Many electrodes are attached to the torso.

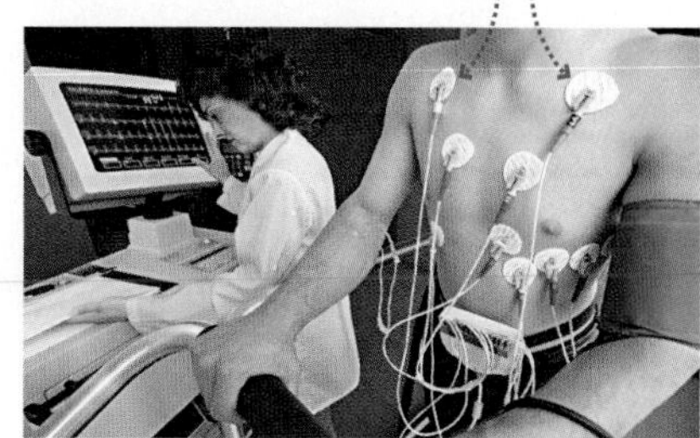

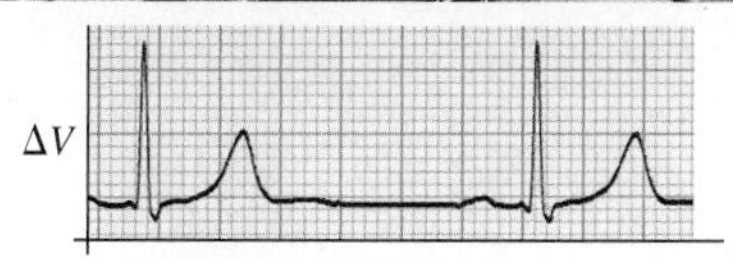
ΔV
t

Records of the potential difference between various pairs of electrodes allow the doctor to analyze the heart's condition.

FIGURE 21.29 The potential difference between the electrodes changes as the heart beats.

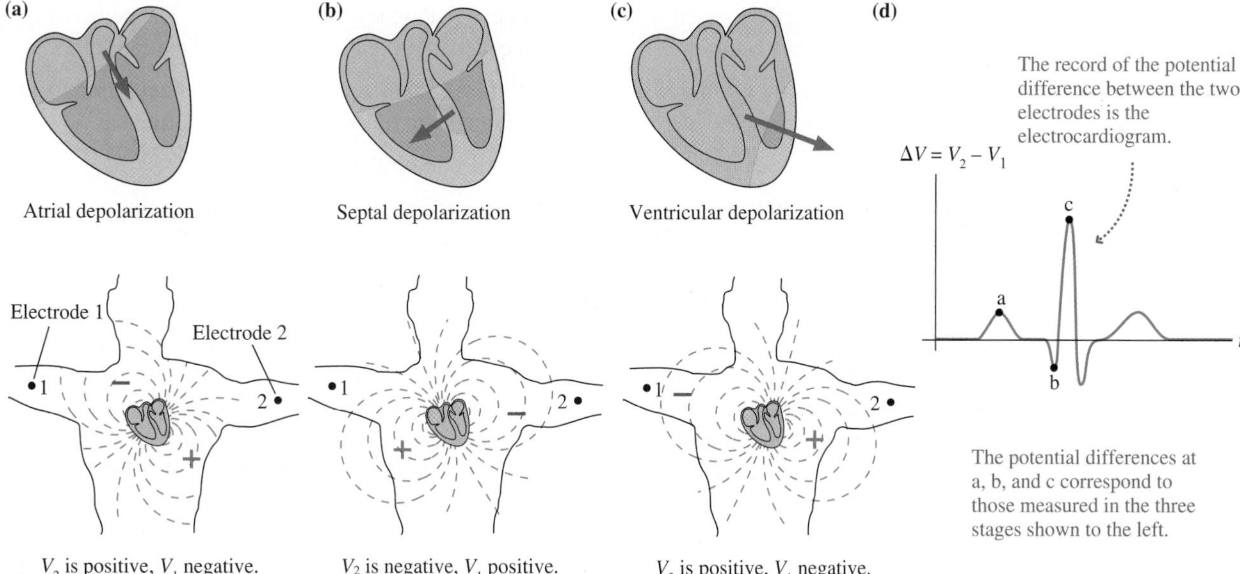
(a)
(b)
(c)
(d)
Atrial depolarization
Septal depolarization
Ventricular depolarization
Electrode 1
Electrode 2
1
2
V_2 is positive, V_1 negative.
V_2 is negative, V_1 positive.
V_2 is positive, V_1 negative.
The record of the potential difference between the two electrodes is the electrocardiogram.
$\Delta V = V_2 - V_1$
a
b
c
t
The potential differences at a, b, and c correspond to those measured in the three stages shown to the left.

21.7 Capacitance and Capacitors

In Section 21.2 we found that potential differences are caused by the separation of charge. One common method of creating a charge separation, shown in FIGURE 21.30 on the next page, is to move charge Q from one initially uncharged conductor to a second initially uncharged conductor. This results in charge $+Q$ on one conductor and $-Q$ on the other. Two conductors with equal but opposite charge form a **capacitor.** The two conductors that make up a capacitor are its *electrodes* or *plates*. We've already looked at some of the properties of parallel-plate capacitors; now we want to study capacitors that might have any shape. Capacitors can be used to store charge, making them invaluable in all kinds of electronic circuits.

As Figure 21.30 on the next page shows, the electric field strength E and the potential difference ΔV_C increase as the charge on each electrode increases. If we double the amount of charge on each electrode, the work required to move a charge from one electrode to the other doubles. This implies a doubling of the potential difference between the electrodes. Thus **the potential difference between the electrodes is directly proportional to their charge.**

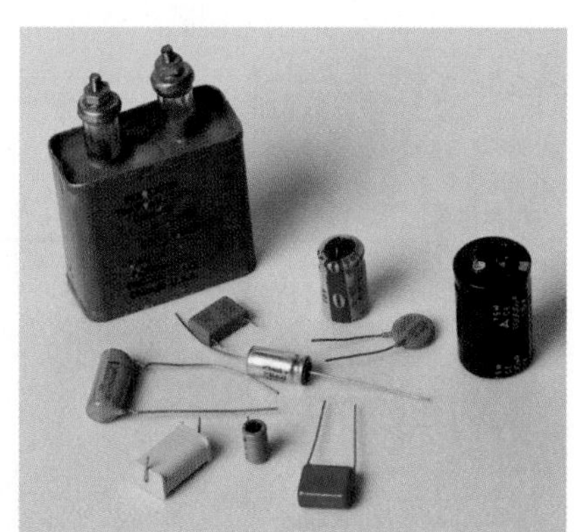

Capacitors are important elements in electric circuits. They come in a wide variety of sizes and shapes.

FIGURE 21.30 Charging a capacitor.

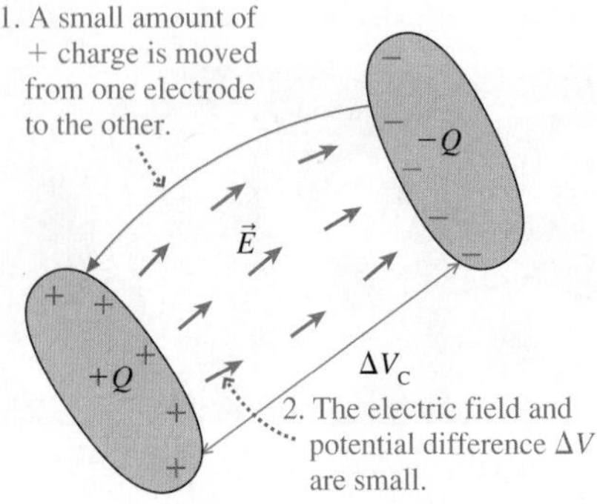

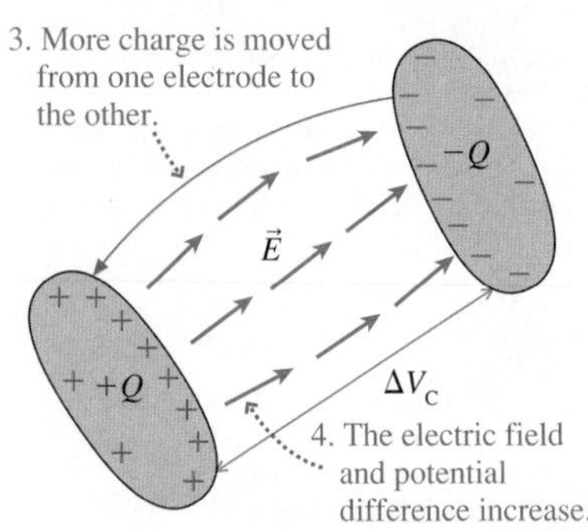

Stated another way, **the charge of a capacitor is directly proportional to the potential difference between its electrodes.** As we'll see, this is actually the most common way that a capacitor is used: A source of potential difference, such as a battery, is connected between the electrodes, causing a charge proportional to the potential difference to be moved from one electrode to the other. We can write the relationship between charge and potential difference as

$$Q = C\,\Delta V_C \tag{21.18}$$

Charge on a capacitor with potential difference ΔV_C

PROPORTIONAL p.37

The constant of proportionality C between Q and ΔV_C is called the **capacitance** of the capacitor. Capacitance depends on the shape, size, and spacing of the two electrodes. A capacitor with a large capacitance holds more charge for a given potential difference than one with a small capacitance.

NOTE ► We will consider only situations where the charge on the electrodes is equal in magnitude but opposite in sign. When we say "A capacitor has charge Q," we mean that one electrode has charge $+Q$ and the other charge $-Q$. The potential difference between the electrodes is called the potential difference *of* the capacitor. ◄

The SI unit of capacitance is the **farad,** named in honor of Michael Faraday, the originator of the idea of electric and magnetic fields. One farad is defined as

$$1 \text{ farad} = 1 \text{ F} = 1 \text{ coulomb/volt} = 1 \text{ C/V}$$

12.6 ActivPhysics ONLINE

One farad turns out to be a very large capacitance. Practical capacitors are usually measured in units of microfarads (μF) or picofarads (1 pF $= 10^{-12}$ F).

Charging a Capacitor

To "charge" a capacitor, we need to move charge from one electrode to the other. The simplest way to do this is to use a source of potential difference such as a battery, as shown in FIGURE 21.31. We learned earlier that a battery uses its internal chemistry to maintain a fixed potential difference between its terminals. If we connect a capacitor to a battery, charge flows from the negative electrode of the capacitor, through the battery, and onto the positive electrode. This flow of charge continues until the potential difference between the capacitor's electrodes is the same as the fixed potential difference of the battery. If the battery is then removed, the capacitor remains charged with a potential equal to that of the battery that charged it because there's no conducting path for charge on the positive electrode to move back to the negative electrode. Thus **a capacitor can be used to store charge.**

FIGURE 21.31 Charging a capacitor using a battery.

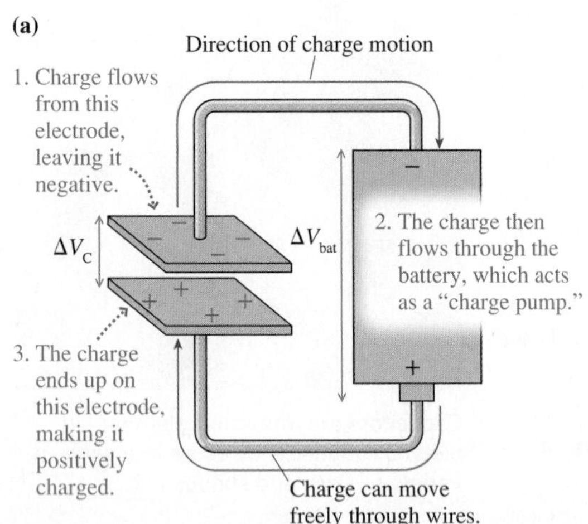

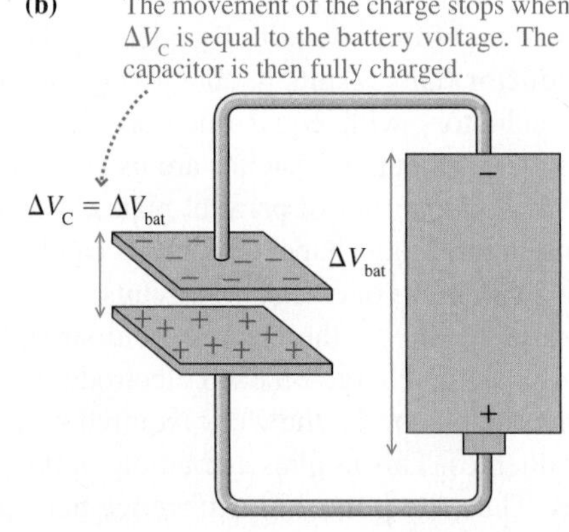

(b)
The movement of the charge stops when ΔV_C is equal to the battery voltage. The capacitor is then fully charged.
$\Delta V_C = \Delta V_{bat}$
ΔV_{bat}
3. The charge ends up on this electrode, making it positively charged.
Charge can move freely through wires.

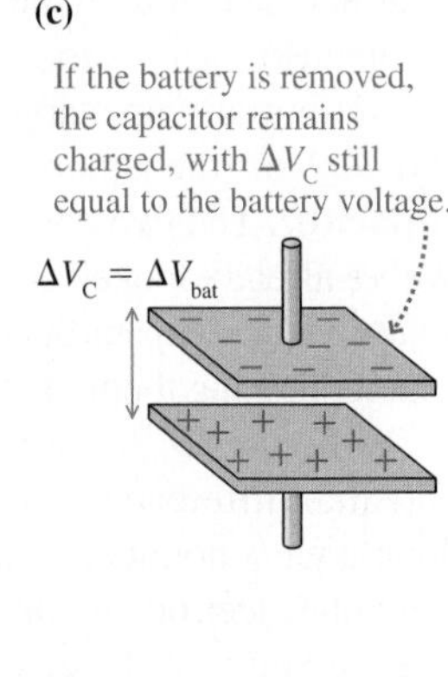

(c)
If the battery is removed, the capacitor remains charged, with ΔV_C still equal to the battery voltage.
$\Delta V_C = \Delta V_{bat}$

EXAMPLE 21.10 Charging a capacitor

A 1.3 μF capacitor is connected to a 1.5 V battery. What is the charge on the capacitor?

PREPARE Charge flows through the battery from one capacitor electrode to the other until the potential difference ΔV_C between the electrodes equals that of the battery, or 1.5 V.

SOLVE The charge on the capacitor is given by Equation 21.18:

$$Q = C\,\Delta V_C = (1.3 \times 10^{-6}\ \text{F})(1.5\ \text{V}) = 2.0 \times 10^{-6}\ \text{C}$$

ASSESS This is the charge on the positive electrode; the other electrode has a charge of -2.0×10^{-6} C.

The Parallel-Plate Capacitor

As we've seen, the *parallel-plate capacitor* is important because it creates a uniform electric field between its flat electrodes. In Chapter 20, we found that the electric field of a parallel-plate capacitor is

$$\vec{E} = \left(\frac{Q}{\epsilon_0 A},\ \text{from positive to negative}\right)$$

where A is the surface area of the electrodes and Q is the charge on the capacitor. We can use this result to find the capacitance of a parallel-plate capacitor.

Earlier in this chapter, we found that the electric field strength of a parallel-plate capacitor is related to the potential difference ΔV_C and the plate spacing d by

$$E = \frac{\Delta V_C}{d}$$

Combining these two results, we see that

$$\frac{Q}{\epsilon_0 A} = \frac{\Delta V_C}{d}$$

or, equivalently,

$$Q = \frac{\epsilon_0 A}{d}\Delta V_C \qquad (21.19)$$

If we compare Equation 21.19 to Equation 21.18, the definition of capacitance, we see that the capacitance of the parallel-plate capacitor is

$$C = \frac{\epsilon_0 A}{d} \qquad (21.20)$$

Capacitance of a parallel-plate capacitor with plate area A and separation d

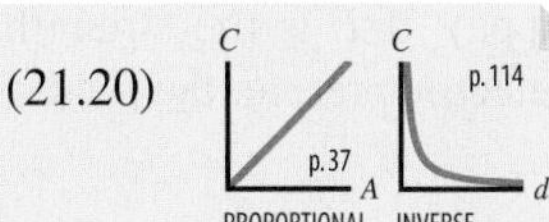

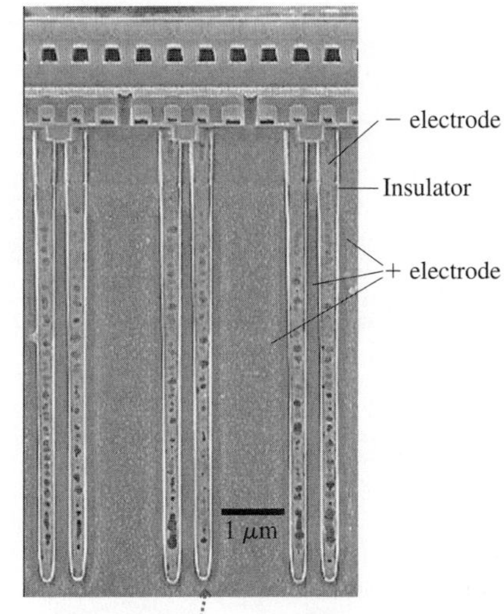

Each long structure is one capacitor.

A capacity for memory Your computer's random-access memory, or RAM, uses tiny capacitors to store the digital ones and zeroes that make up your data. A charged capacitor represents a one and an uncharged capacitor a zero. For a billion or more capacitors to fit on a single chip they must be very small. The micrograph is a cross section through the silicon wafer that makes up the memory chip. Each capacitor consists of a very long electrode separated by a thin insulating layer from the common electrode shared by all capacitors. Each capacitor's capacitance is only about 30×10^{-15} F!

NOTE ► From Equation 21.20 you can see that the units of ϵ_0 can be written as F/m. These units are useful when working with capacitors. ◄

EXAMPLE 21.11 Charging a parallel-plate capacitor

The spacing between the plates of a 1.0 μF parallel-plate capacitor is 0.070 mm.

a. What is the surface area of the plates?
b. How much charge is on the plates if this capacitor is attached to a 1.5 V battery?

SOLVE

a. From the definition of capacitance,

$$A = \frac{dC}{\epsilon_0} = \frac{(0.070 \times 10^{-3}\ \text{m})(1.0 \times 10^{-6}\ \text{F})}{8.85 \times 10^{-12}\ \text{F/m}} = 7.9\ \text{m}^2$$

b. The charge is $Q = C\,\Delta V_C = (1.0 \times 10^{-6}\ \text{F})(1.5\ \text{V}) = 1.5 \times 10^{-6}\ \text{C} = 1.5\ \mu\text{C}$.

ASSESS The surface area needed to construct a 1.0 μF capacitor (a fairly typical value) is enormous and hardly practical. We'll see in the next section that real capacitors can be reduced to a more manageable size by placing an insulator between the capacitor plates.

STOP TO THINK 21.5 If the potential difference across a capacitor is doubled, its capacitance

A. Doubles. B. Halves. C. Remains the same.

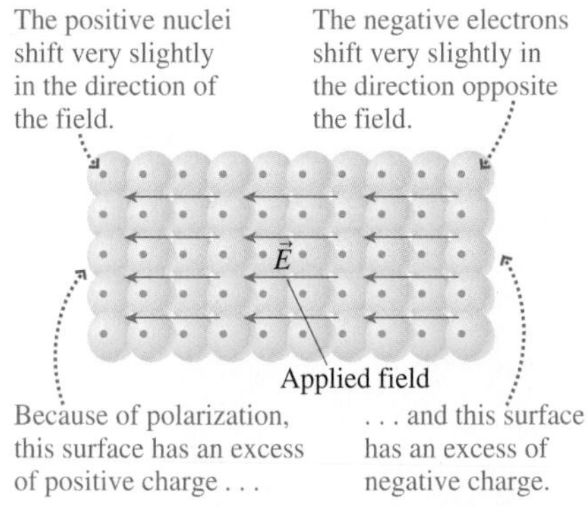
The positive nuclei shift very slightly in the direction of the field.
The negative electrons shift very slightly in the direction opposite the field.
$\vec{E}$
Applied field
Because of polarization, this surface has an excess of positive charge . . .
. . . and this surface has an excess of negative charge.

FIGURE 21.32 An insulator in an electric field becomes polarized.

21.8 Dielectrics and Capacitors

An insulator consists of vast numbers of atoms. When an insulator is placed in an electric field, each of its atoms polarizes. Recall, from Chapter 20, that *polarization* occurs when an atom's negative electron cloud and positive nucleus shift very slightly in opposite directions in response to an applied electric field. The net effect of all these tiny atomic polarizations is shown in FIGURE 21.32: An *induced* positive charge builds up on one surface of the insulator, and an induced negative charge on the other surface. The insulator has no net charge, but it has become polarized.

An insulator placed between the plates of a capacitor is called a **dielectric.** FIGURE 21.33a shows a dielectric in the uniform field of a parallel-plate capacitor. The capacitor's electric field polarizes the dielectric, and positive and negative charges build up on opposite surfaces. Notice that this distribution of charge—two equal but opposite layers—is identical to that of a parallel-plate capacitor, so this induced charge will create an electric field identical to that of a parallel-plate capacitor.

However, as Figure 21.33a shows, the field due to the induced charge on the surface of the dielectric is *opposite* the applied electric field due to the charge on the capacitor plates that established the polarization in the first place. The *net* electric field inside the dielectric is the vector sum of these two contributions: the applied field of the capacitor plates plus the field due to the polarization of the dielectric. As FIGURE 21.33b shows, these two fields add to give a net field in the same direction as the applied field, but smaller. Thus **the electric field inside a dielectric is smaller than the applied field.**

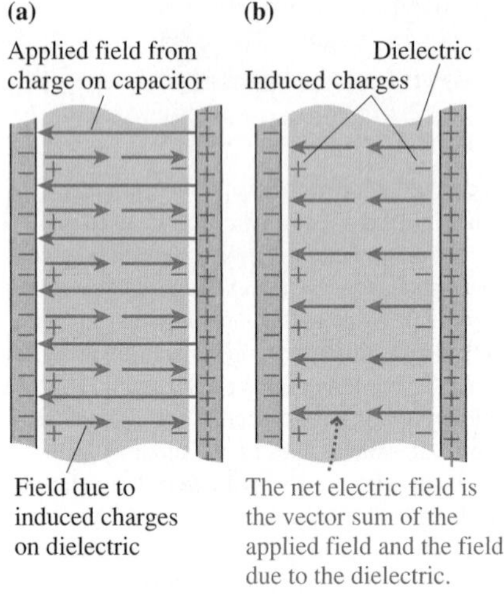
(a)
(b)
Applied field from charge on capacitor
Dielectric
Induced charges
Field due to induced charges on dielectric
The net electric field is the vector sum of the applied field and the field due to the dielectric.

FIGURE 21.33 The electric field inside a dielectric.

Because some atoms are more easily polarized than others, the factor by which the electric field is reduced depends on the dielectric material. This factor is called the **dielectric constant** of the material, and it is given the symbol κ (Greek letter kappa). If E is the strength of the electric field inside the capacitor without the dielectric present, the field strength E' with the dielectric present is

$$E' = \frac{E}{\kappa} \tag{21.21}$$

Table 21.3 lists the dielectric constants for a number of common substances. Note the high dielectric constant of water, which is of great importance in regulating the chemistry of biological processes. As we saw in Section 20.2, a molecule of water is a *permanent* electric dipole because the oxygen atom has a slight negative charge while the two hydrogen atoms have a slight positive charge. Because the charge in a water molecule is *already* separated, the molecules' dipoles easily turn to line up with an applied electric field, leading to water's very high dielectric constant.

TABLE 21.3 Dielectric constants of some materials at 20°C

Material	Dielectric constant κ
Vacuum	1 (exactly)
Air	1.00054*
Teflon	2.0
Paper	3.0
Pyrex glass	4.8
Cell membrane	9.0
Ethanol	24
Water	80
Strontium titanate	300

*Use 1.00 in all calculations.

If we insert a dielectric between the plates of a capacitor, the electric field between the plates will be reduced by a factor of κ. The electric potential difference between the plates, given by $\Delta V_C = Ed$, is reduced by the same factor, and thus the capacitance $C = Q/\Delta V_C$ is *increased* by a factor of κ. Therefore the capacitance of a dielectric-containing parallel-plate capacitor is

$$C = \frac{\kappa \epsilon_0 A}{d} \tag{21.22}$$

Capacitance of a parallel-plate capacitor with a dielectric of dielectric constant κ

PROPORTIONAL p.37 INVERSE p.114

EXAMPLE 21.12 Finding the dielectric constant

A parallel-plate capacitor is charged using a 100 V battery; then the battery is removed. If a dielectric slab is slid between the plates, filling the space inside, the capacitor voltage drops to 30 V. What is the dielectric constant of the dielectric?

PREPARE The capacitor voltage remains $(\Delta V_C)_1 = 100$ V when it is disconnected from the battery. Placing the dielectric between the plates reduces the voltage to $(\Delta V_C)_2 = 30$ V.

SOLVE The electric field strength between the capacitor plates is $E = \Delta V_C/d$. The plate separation d doesn't change, so

$$\frac{E_1}{E_2} = \frac{(\Delta V_C)_1/d}{(\Delta V_C)_2/d} = \frac{(\Delta V_C)_1}{(\Delta V_C)_2} = \frac{100\text{ V}}{30\text{ V}} = 3.3$$

But $E_2 = E_1/\kappa$, from Equation 21.21, so

$$\kappa = \frac{E_1}{E_2} = 3.3$$

ASSESS The amount of charge on the capacitor didn't change. The capacitor voltage decreased because the dielectric reduced the electric field strength inside the capacitor.

In Example 21.11, we found that impractically large electrodes would be needed to create a parallel-plate capacitor with $C = 1\ \mu$F. Practical capacitors take advantage of dielectrics with $\kappa > 100$. As shown in **FIGURE 21.34**, a typical capacitor is a "sandwich" of two modest-sized pieces of aluminum foil separated by a very thin dielectric that increases the capacitance by a large factor. This sandwich is then folded, rolled up, and sealed in a small plastic cylinder. The two wires extending from a capacitor connect to the two electrodes and allow the capacitor to be charged. Even though the electrodes are no longer planes, the capacitance is reasonably well predicted from the parallel-plate-capacitor equation if A is the area of the foils before being folded and rolled.

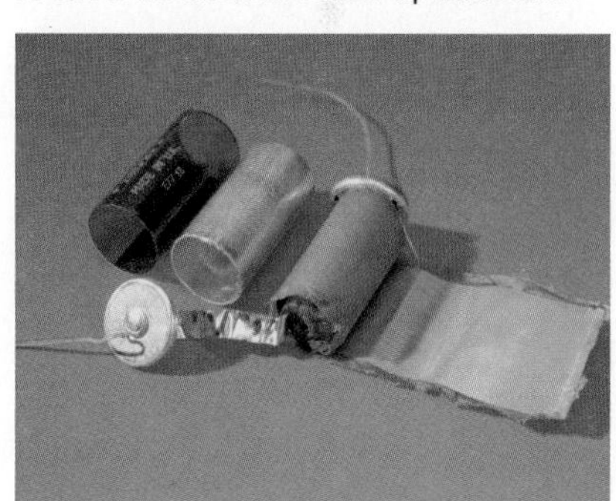

FIGURE 21.34 A capacitor disassembled to show its internal rolled-up structure.

STOP TO THINK 21.6 A 100 V battery is connected across the plates of a parallel-plate capacitor. If a Teflon slab is slid between the plates, without disconnecting the battery, the electric field between the plates

A. Increases. B. Decreases. C. Remains the same.

21.9 Energy and Capacitors

We've seen that a practical way of charging a capacitor is to attach its plates to a battery. Charge then flows from one plate, through the battery, and onto the other plate, leaving one plate with charge $-Q$ and the other with charge $+Q$. The battery must do *work* to transfer the charge; this work increases the electric potential energy of the charge on the capacitor. **A charged capacitor stores energy as electric potential energy.**

To find out how much energy is stored in a charged capacitor, recall that a charge q moved through a potential difference ΔV gains potential energy $U = q\Delta V$. When a capacitor is charged, total charge Q is moved from the negative plate to the positive plate. At first, when the capacitor is uncharged, the potential difference is $\Delta V = 0$. The last little bit of charge to be moved, as the capacitor reaches full charge, has to move through a potential difference $\Delta V \approx \Delta V_C$. *On average,* the potential difference across a capacitor while it's being charged is $\Delta V_{\text{average}} = \frac{1}{2}\Delta V_C$. Thus it seems plausible

Taking a picture in a flash When you take a flash picture, the flash is fired using electric potential energy stored in a capacitor. Batteries are unable to deliver the required energy rapidly enough, but capacitors can discharge all their energy in only microseconds. A battery is used to slowly charge up the capacitor, which then rapidly discharges through the flashlamp. This slow recharging process is why you must wait some time between taking flash pictures.

(and can be proved in a more advanced treatment) that the potential energy U_C stored in a charged capacitor is

$$U_C = Q\Delta V_{\text{average}} = \frac{1}{2}Q\Delta V_C$$

We can use $Q = C\Delta V_C$ to write this in two different ways:

$$U_C = \frac{1}{2}\frac{Q^2}{C} = \frac{1}{2}C(\Delta V_C)^2 \qquad (21.23)$$

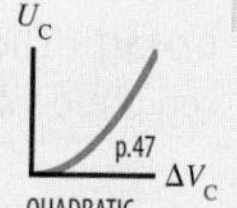

Electric potential energy of a capacitor with charge Q and potential difference ΔV_C

The potential energy stored in a capacitor depends on the *square* of the potential difference across it. This result is reminiscent of the potential energy $U_s = \frac{1}{2}k(\Delta x)^2$ stored in a spring, and a charged capacitor really is analogous to a stretched spring. A stretched spring holds energy until we release it; then that potential energy is transformed into kinetic energy. Likewise, a charged capacitor holds energy until we discharge it.

EXAMPLE 21.13 Energy in a camera flash

How much energy is stored in a 220 μF camera-flash capacitor that has been charged to 330 V? What is the average power delivered to the flash lamp if this capacitor is discharged in 1.0 ms?

SOLVE The energy stored in the capacitor is

$$U_C = \frac{1}{2}C(\Delta V_C)^2 = \frac{1}{2}(220 \times 10^{-6}\text{ F})(330\text{ V})^2 = 12\text{ J}$$

If this energy is released in 1.0 ms, the average power is

$$P = \frac{\Delta E}{\Delta t} = \frac{12\text{ J}}{1.0 \times 10^{-3}\text{ s}} = 12{,}000\text{ W}$$

ASSESS The stored energy is equivalent to raising a 1 kg mass by 1.2 m. This is a rather large amount of energy; imagine the damage a 1 kg object could do after falling 1.2 m. When this energy is released very quickly, as is possible in an electronic circuit, the power is very high.

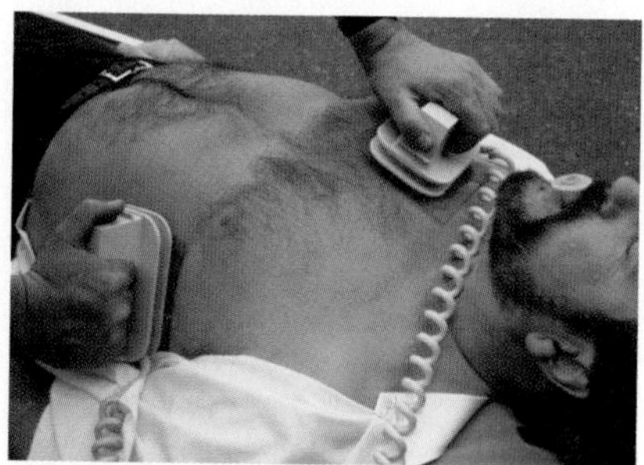

A defibrillator, which can restore a heartbeat, discharges a capacitor through the patient's chest.

The usefulness of a capacitor stems from the fact that it can be charged very slowly, over many seconds, and then can release the energy very quickly. A mechanical analogy would be using a crank to slowly stretch the spring of a catapult, then quickly releasing the energy to launch a massive rock.

An important medical application of the ability of capacitors to rapidly deliver energy is the *defibrillator.* A heart attack or a serious injury can cause the heart to enter a state known as *fibrillation* in which the heart muscles twitch randomly and cannot pump blood. A strong electric shock through the chest can sometimes stop the fibrillation and allow a normal heart rhythm to be restored. A defibrillator has a large capacitor that can store up to 360 J of energy. This energy is released in about 2 ms through two "paddles" pressed against the patient's chest. It takes several seconds to charge the capacitor, which is why, on television medical shows, you hear an emergency room doctor or nurse shout, "Charging!"

The Energy in the Electric Field

We can "see" the potential energy of a stretched spring in the tension of the coils. If a charged capacitor is analogous to a stretched spring, where is the stored energy? It's in the electric field!

FIGURE 21.35 shows a parallel-plate capacitor in which the plates have area A and are separated by distance d. The potential difference across the capacitor is related to the electric field inside the capacitor by $\Delta V_C = Ed$. The capacitance, which we found in Equation 21.22, is $C = \kappa\epsilon_0 A/d$. Substituting these into Equation 21.23, we find that the energy stored in the capacitor is

$$U_C = \frac{1}{2}C(\Delta V_C)^2 = \frac{1}{2}\frac{\kappa\epsilon_0 A}{d}(Ed)^2 = \frac{\kappa\epsilon_0}{2}(Ad)E^2 \qquad (21.24)$$

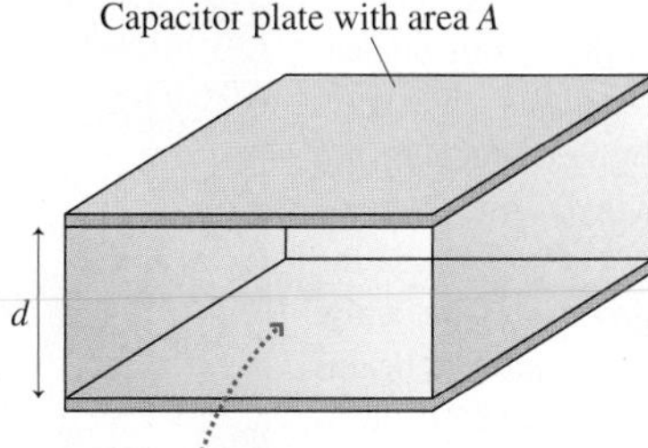
Capacitor plate with area A
d
The capacitor's energy is stored in the electric field in volume Ad between the plates.

FIGURE 21.35 A capacitor's energy is stored in the electric field.

The quantity Ad is the volume *inside* the capacitor, the region in which the capacitor's electric field exists. (Recall that an ideal capacitor has $\vec{E} = \vec{0}$ everywhere except between the plates.) Although we talk about "the energy stored in the capacitor," Equation 21.24 suggests that, strictly speaking, **the energy is stored in the capacitor's electric field.**

Because Ad is the volume in which the energy is stored, we can define an **energy density** u_E of the electric field:

$$u_E = \frac{\text{energy stored}}{\text{volume in which it is stored}} = \frac{U_C}{Ad} = \frac{1}{2}\kappa\epsilon_0 E^2 \qquad (21.25)$$

The energy density has units J/m^3. We've derived Equation 21.25 for a parallel-plate capacitor, but it turns out to be the correct expression for any electric field.

From this perspective, charging a capacitor stores energy in the capacitor's electric field as the field grows in strength. Later, when the capacitor is discharged, the energy is released as the field collapses.

We first introduced the electric field as a way to visualize how a long-range force operates. But if the field can store energy, the field must be real, not merely a pictorial device. We'll explore this idea further in Chapter 25, where we'll find that the energy transported by a light wave—the very real energy of warm sunshine—is the energy of electric and magnetic fields.

EXAMPLE 21.14 Finding the energy density for a defibrillator

A defibrillator unit contains a 150 μF capacitor that is charged to 2000 V. The capacitor plates are separated by a 0.010-mm-thick dielectric with $\kappa = 300$.

a. What is the total area of the capacitor plates?
b. What is the energy density stored in the electric field when the capacitor is charged?

PREPARE Assume the capacitor can be modeled as a parallel-plate capacitor with a dielectric.

SOLVE a. Equation 21.22 for the capacitance of a dielectric-filled parallel-plate capacitor gives the surface area of the electrodes:

$$A = \frac{dC}{\kappa\epsilon_0} = \frac{(1.0\times10^{-5}\text{ m})(150\times10^{-6}\text{ F})}{(300)(8.85\times10^{-12}\text{ F/m})} = 0.56\text{ m}^2$$

b. The electric field strength is

$$E = \frac{\Delta V_C}{d} = \frac{2000\text{ V}}{1.0\times10^{-5}\text{ m}} = 2.0\times10^8\text{ V/m}$$

Consequently, the energy density in the electric field is

$$\begin{aligned} u_E &= \frac{1}{2}\kappa\epsilon_0 E^2 \\ &= \frac{1}{2}(300)(8.85\times10^{-12}\text{ F/m})(2.0\times10^8\text{ V/m})^2 \\ &= 5.3\times10^7\text{ J/m}^3 \end{aligned}$$

ASSESS For comparison, the energy density of gasoline is about 3×10^9 J/m^3, about 60 times higher than this capacitor. Capacitors store less energy than some other devices, but they can deliver this energy *very* rapidly.

STOP TO THINK 21.7 The plates of a parallel-plate capacitor are connected to a battery. If the distance between the plates is halved, the energy of the capacitor

A. Increases by a factor of 4.
B. Doubles.
C. Remains the same.
D. Is halved.
E. Decreases by a factor of 4.

INTEGRATED EXAMPLE 21.15 Proton fusion in the sun

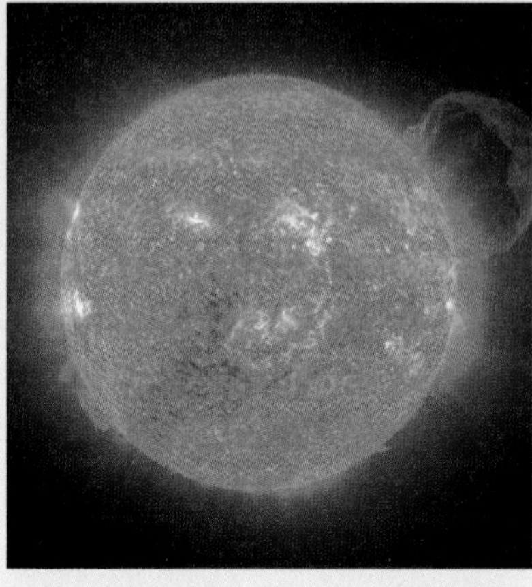

The sun's energy comes from nuclear reactions that fuse lighter nuclei into heavier ones, releasing energy in the process. The solar fusion process begins when two protons (the nuclei of hydrogen atoms) merge to produce a *deuterium* nucleus. Deuterium is the "heavy" isotope of hydrogen, with a nucleus consisting of a proton *and* a neutron. To become deuterium, one of the protons that fused has to turn into a neutron. The nuclear-physics process by which this occurs—and which releases the energy—will be studied in Chapter 30. Our interest for now lies not with the nuclear physics but with the conditions that allow fusion to occur.

Before two protons can fuse, they must come into contact. However, the energy required to bring two protons into contact is considerable because the electric potential energy of the two protons increases rapidly as they approach each other. Fusion occurs in the core of the sun because the ultra-high temperature there gives the protons the kinetic energy they need to come together.

a. A proton can be modeled as a charged sphere of diameter $d_p = 1.6 \times 10^{-15}$ m with total charge e. When two protons are in contact, what is the electric potential of one proton at the center of the other?
b. Two protons are approaching each other head-on, each with the same speed v_0. What value of v_0 is required for the protons to just come into contact with each other?
c. What does the temperature of the sun's core need to be so that the rms speed v_{rms} of protons is equal to v_0?

PREPARE Energy is conserved, so Problem-Solving Strategy 21.1 is the basis of our solution. **FIGURE 21.36** shows a before-and-after visual overview. Both protons are initially moving with speeds $v_i = v_0$, so both contribute to the initial kinetic energy. We will assume that they start out so far apart that $U_i \approx 0$. To "just touch" means that they've instantaneously come to rest ($K_f = 0$) at the point where the distance between their centers is equal to the diameter of a proton. We can use the potential of a charged sphere and the energy-conservation equation to find the speed v_0 required to achieve contact. Then we can use the results of Chapter 12 to find the temperature at which v_0 is the rms speed of the protons.

FIGURE 21.36 Visual overview of two protons coming into contact.

Before: $v_i = v_0$ (+) → ← (+) $v_i = v_0$, $r_i \approx \infty$

After: $v_f = 0$ (+)(+) $r_f = d_p$

Known: $d_p = 1.6 \times 10^{-15}$ m

Find: v_0

SOLVE a. The electric potential at distance r from a charged sphere was found to be $V = KQ/r$. When the protons are in contact, the distance between their centers is $r_f = d_p = 1.6 \times 10^{-15}$ m. Thus the potential of one proton, with $Q = e$, at the center of the other is

$$V = \frac{Ke}{r_f} = \frac{Ke}{d_p} = \frac{(9.0 \times 10^9 \text{ N} \cdot \text{m}^2/\text{C}^2)(1.6 \times 10^{-19} \text{ C})}{1.6 \times 10^{-15} \text{ m}}$$

$$= 9.0 \times 10^5 \text{ V}$$

b. The conservation of energy equation $K_f + qV_f = K_i + qV_i$ is

$$(0 + 0) + eV_f = \left(\frac{1}{2}mv_0^2 + \frac{1}{2}mv_0^2\right) + 0$$

where, as noted above, both protons contribute to the initial kinetic energy, both end up at rest as the protons touch, and they started far enough apart that the initial potential energy (and potential) is effectively zero. When the protons meet, their potential energy is the charge of one proton (e) multiplied by the electric potential of the other—namely, the potential found in part a: $V_f = 9.0 \times 10^5$ V. Solving the energy equation for v_0, we get

$$v_0 = \sqrt{\frac{eV_f}{m}} = \sqrt{\frac{(1.6 \times 10^{-19} \text{ C})(9.0 \times 10^5 \text{ V})}{1.7 \times 10^{-27} \text{ kg}}}$$

$$= 9.2 \times 10^6 \text{ m/s}$$

c. In Chapter 12, we found that the temperature of a gas is related to the average kinetic energy of the particles and thus to the rms speed of the particles by the equation

$$T = \frac{mv_{rms}^2}{3k_B}$$

It may seem strange to think of protons as a gas, but in the center of the sun, where all the atoms are ionized into nuclei and electrons, the protons are zooming around and do, indeed, act like a gas. For v_{rms} of the protons to be equal to v_0 that we calculated in part b, the temperature would have to be

$$T = \frac{mv_0^2}{3k_B} = \frac{(1.7 \times 10^{-27} \text{ kg})(9.2 \times 10^6 \text{ m/s})^2}{3(1.4 \times 10^{-23} \text{ J/K})} = 3.4 \times 10^9 \text{ K}$$

ASSESS An extraordinarily high temperature—over 3 billion kelvin—is required to give an average solar proton a speed of 9.2×10^6 m/s. In fact, the core temperature of the sun is "only" about 14 million kelvin, a factor of $\approx$200 less than we calculated. Protons can fuse at this lower temperature both because there are always a few protons moving much faster than average and because protons can reach each other even if their speeds are too low by the quantum-mechanical process of *tunneling*, which you'll learn about in Chapter 28. Still, because of the core's relatively "low" temperature, most protons bounce around in the sun for several billion years before fusing!

SUMMARY

The goal of Chapter 21 has been to calculate and use the electric potential and electric potential energy.

GENERAL PRINCIPLES

Electric Potential and Potential Energy

The electric potential V is created by charges and exists at every point surrounding those charges.

These charges create the electric potential.
$V = -100$ V
$V = 300$ V $V = 200$ V

When a charge q is brought near these charges, it acquires an electric potential energy

$$U_{elec} = qV$$

at a point where the other charges have created an electric potential V.

Energy is conserved for a charged particle in an electric potential:

$$K_f + qV_f = K_i + qV_i$$

or

$$\Delta K = -q\Delta V$$

Sources of Potential

Potential differences ΔV are created by a *separation of charge*. Two important sources of potential difference are

- A *battery*, which uses chemical means to separate charge and produce a potential difference.
- The opposite charges on the plates of a *capacitor*, which create a potential difference between the plates.

The electric potential of a point charge q is $V = K\dfrac{q}{r}$

Connecting potential and field

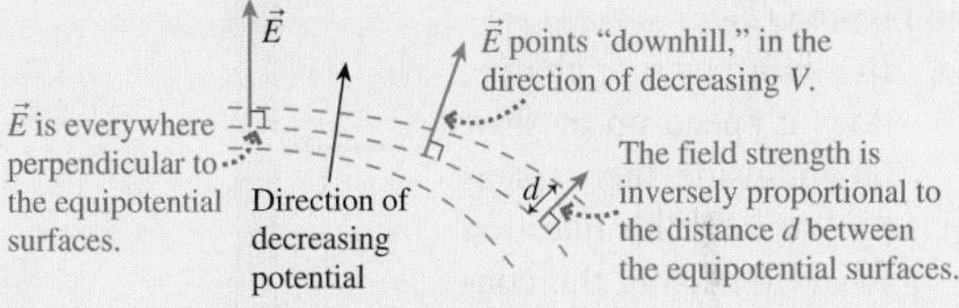
$\vec{E}$
$\vec{E}$ points "downhill," in the direction of decreasing V.
$\vec{E}$ is everywhere perpendicular to the equipotential surfaces.
Direction of decreasing potential
d
The field strength is inversely proportional to the distance d between the equipotential surfaces.

IMPORTANT CONCEPTS

For a **conductor in electrostatic equilibrium**

- Any excess charge is on the surface.
- The electric field inside is zero.
- The exterior electric field is perpendicular to the surface.
- The field strength is largest at sharp corners.
- The entire conductor is at the same potential and so the surface is an equipotential.

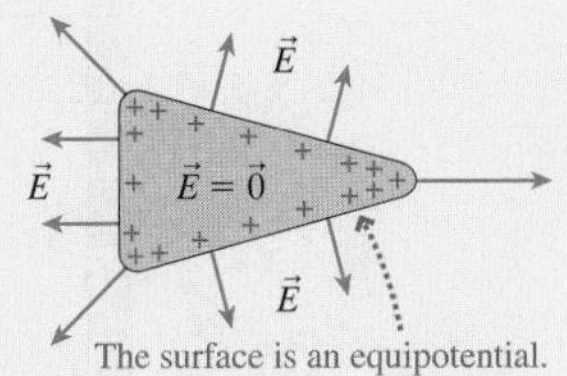
$\vec{E}$
$\vec{E} = \vec{0}$
The surface is an equipotential.

Graphical representations of the potential

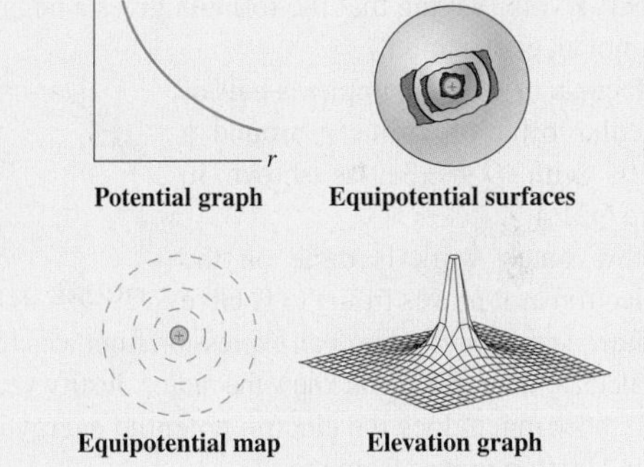
V
r
Potential graph
Equipotential surfaces
Equipotential map
Elevation graph

APPLICATIONS

Capacitors and dielectrics

The potential difference ΔV_C between two conductors charged to $\pm Q$ is proportional to the charge:

$-Q$
$+Q$
ΔV_C

$$\Delta V_C = Q/C$$

where C is the **capacitance** of the two conductors.

A **parallel-plate capacitor** with plates of area A and separation d has a capacitance

$$C = \kappa\epsilon_0 A/d$$

When a **dielectric** is inserted between the plates of a capacitor, its capacitance is increased by a factor κ, the **dielectric constant** of the material.

The **energy stored in a capacitor** is $U_C = \frac{1}{2}C(\Delta V_C)^2$.

This energy is stored in the electric field, which has energy density

$$u_E = \tfrac{1}{2}\kappa\epsilon_0 E^2$$

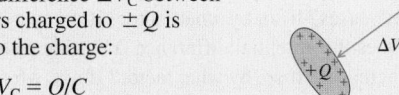

Parallel-plate capacitor

For a capacitor charged to ΔV_C the potential at distance x from the negative plate is

$$V = \frac{x}{d}\Delta V_C$$

The electric field inside is

$$E = \Delta V_C/d$$

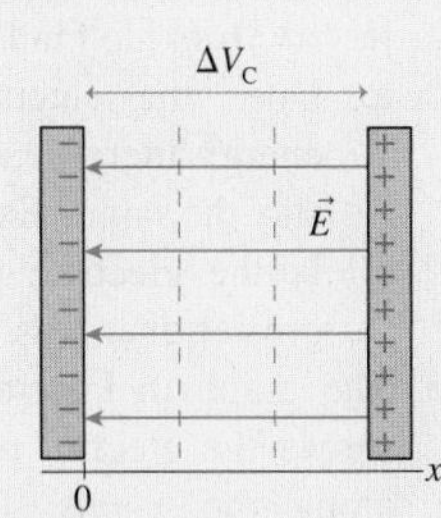
ΔV_C
$\vec{E}$
0
x

Units

- Electric potential: 1 V = 1 J/C
- Electric field: 1 V/m = 1 N/C
- Energy: 1 electron volt = 1 eV = 1.60×10^{-19} J is the kinetic energy gained by an electron upon accelerating through a potential difference of 1 V.

For homework assigned on MasteringPhysics, go to www.masteringphysics.com

Problems labeled INT integrate significant material from earlier chapters; BIO are of biological or medical interest.

Problem difficulty is labeled as | (straightforward) to ||||| (challenging).

QUESTIONS

Conceptual Questions

1. By moving a 10 nC charge from point A to point B, you determine that the electric potential at B is 150 V. What would be the potential at B if a 20 nC charge were moved from A to B?
2. Charge q is fired through a small hole in the positive plate of a capacitor, as shown in Figure Q21.2.
   a. If q is a positive charge, does it speed up or slow down inside the capacitor? Answer this question twice: (i) Using the concept of force. (ii) Using the concept of energy.
   b. Repeat part a if q is a negative charge.

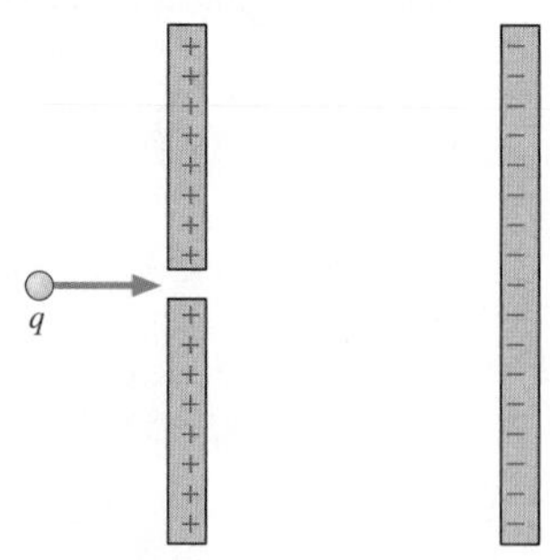

FIGURE Q21.2

3. *Why* is the potential energy of two opposite charges a negative number? (Note: Saying that the formula gives a negative number is not an explanation.)
4. An electron ($q = -e$) completes half of a circular orbit of radius r around a nucleus with $Q = +3e$, as shown in Figure Q21.4.
   a. How much work is done on the electron as it moves from i to f? Give either a numerical value or an expression from which you could calculate the value if you knew the radius. Justify your answer.
   b. By how much does the electric potential energy change as the electron moves from i to f?
   c. Is the electron's speed at f greater than, less than, or equal to its speed at i?
   d. Are your answers to parts a and c consistent with each other?

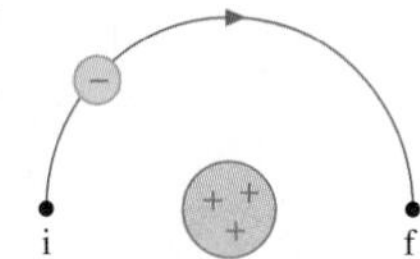

FIGURE Q21.4

5. An electron moves along the trajectory from i to f in Figure Q21.5.
   a. Does the electric potential energy increase, decrease, or stay the same? Explain.
   b. Is the electron's speed at f greater than, less than, or equal to its speed at i? Explain.

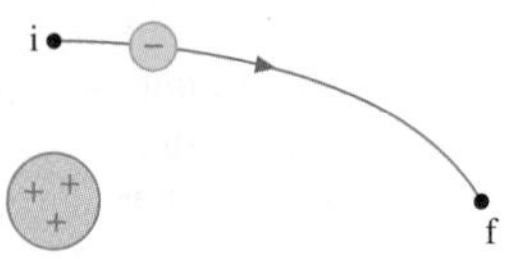

FIGURE Q21.5

6. The graph in Figure Q21.6 shows the electric potential along the x-axis. Draw a graph of the potential energy of a 0.10 C charged particle in this region of space, providing a numerical scale on the energy axis.

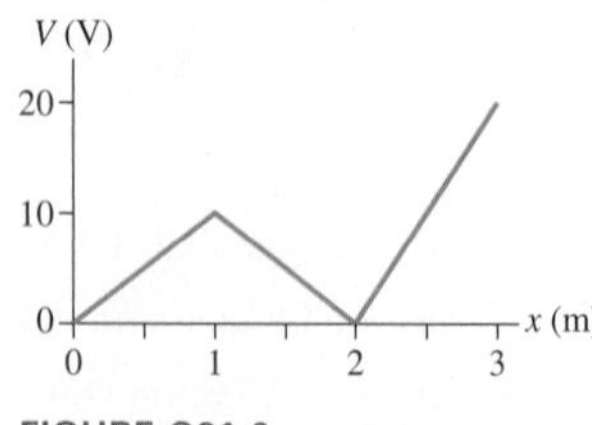

FIGURE Q21.6

7. As shown in Figure Q21.7, two protons are launched with the same speed from point 1 inside a parallel-plate capacitor. One proton moves along the path from 1 to 2, the other from 1 to 3. Points 2 and 3 are the same distance from the positive plate.
   a. Is $\Delta U_{1\to 2}$, the change in potential energy along the path $1 \to 2$, larger than, smaller than, or equal to $\Delta U_{1\to 3}$? Explain.
   b. Is the proton's speed v_2 at point 2 larger than, smaller than, or equal to the proton's speed v_3 at point 3? Explain.

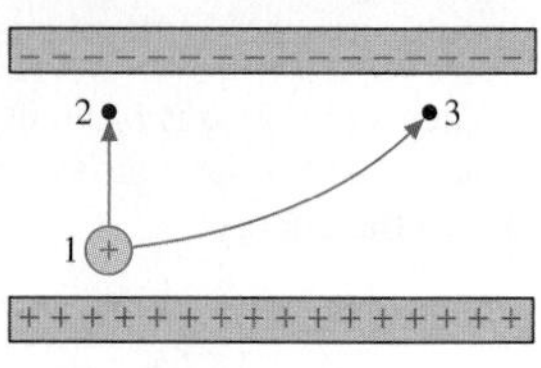

FIGURE Q21.7

8. Figure Q21.8 shows two points inside a capacitor. Let $V = 0$ V at the negative plate.
   a. What is the ratio V_2/V_1 of the electric potential at these two points? Explain.
   b. What is the ratio E_2/E_1 of the electric field strength at these two points? Explain.

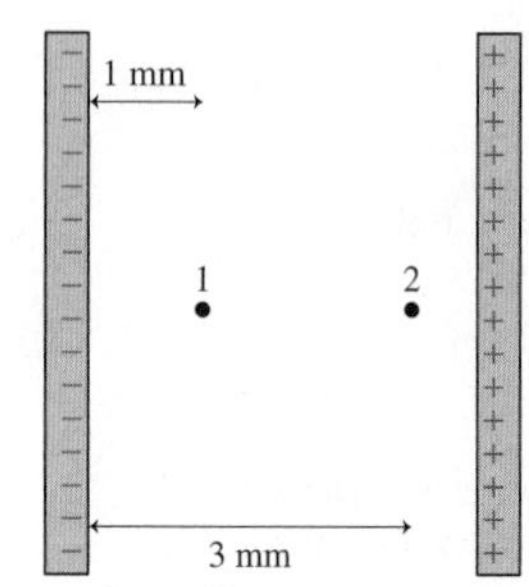

FIGURE Q21.8

9. A capacitor with plates separated by distance d is charged to a potential difference ΔV_C. All wires and batteries are disconnected, then the two plates are pulled apart (with insulated handles) to a new separation of distance $2d$.
   a. Does the capacitor charge Q change as the separation increases? If so, by what factor? If not, why not?
   b. Does the electric field strength E change as the separation increases? If so, by what factor? If not, why not?
   c. Does the potential difference ΔV_C change as the separation increases? If so, by what factor? If not, why not?
10. Rank in order, from most positive to most negative, the electric potentials V_1 to V_5 at points 1 to 5 in Figure Q21.10. Explain.

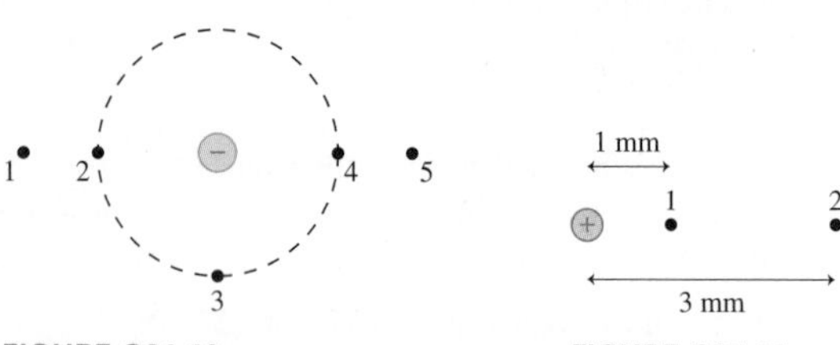

FIGURE Q21.10

FIGURE Q21.11

11. Figure Q21.11 shows two points near a positive point charge.
    a. What is the ratio V_1/V_2 of the electric potentials at these two points? Explain.
    b. What is the ratio E_1/E_2 of the electric field strengths at these two points? Explain.
12. Each part of Figure Q21.12 shows three points in the vicinity of two point charges. The charges have equal magnitudes. Rank in order, from largest to smallest, the potentials V_1, V_2, and V_3.

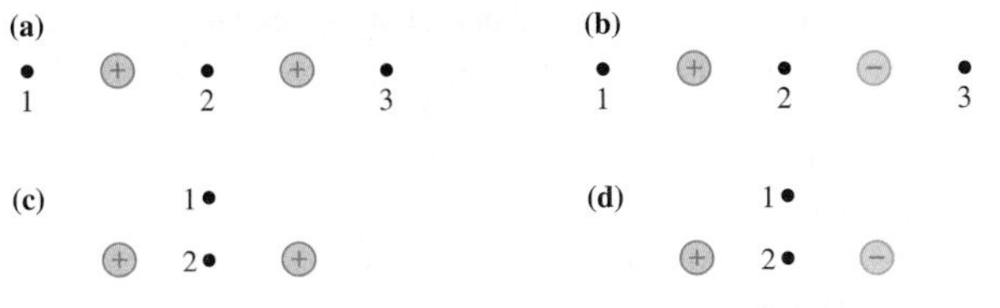

FIGURE Q21.12

13. a. Suppose that $\vec{E} = \vec{0}$ throughout some region of space. Can you conclude that $V = 0$ V in this region? Explain.
    b. Suppose that $V = 0$ V throughout some region of space. Can you conclude that $\vec{E} = \vec{0}$ in this region? Explain.
14. Rank in order, from largest to smallest, the electric field strengths E_1, E_2, E_3, and E_4 at the four labeled points in Figure Q21.14. Explain.

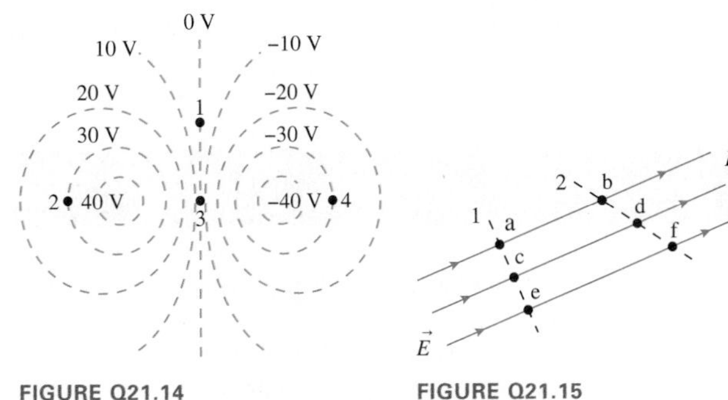

FIGURE Q21.14 FIGURE Q21.15

15. Figure Q21.15 shows an electric field diagram. Dotted lines 1 and 2 are two surfaces in space, not physical objects.
    a. Is the electric potential at point a higher than, lower than, or equal to the electric potential at point b? Explain.
    b. Rank in order, from largest to smallest, the potential differences ΔV_{ab}, ΔV_{cd}, and ΔV_{ef}. Explain.
    c. Is surface 1 an equipotential surface? What about surface 2? Explain why or why not.
16. Figure Q21.16 shows a negatively charged electroscope. The gold leaf stands away from the rigid metal post. Is the electric potential of the leaf higher than, less than, or equal to the potential of the post? Explain.

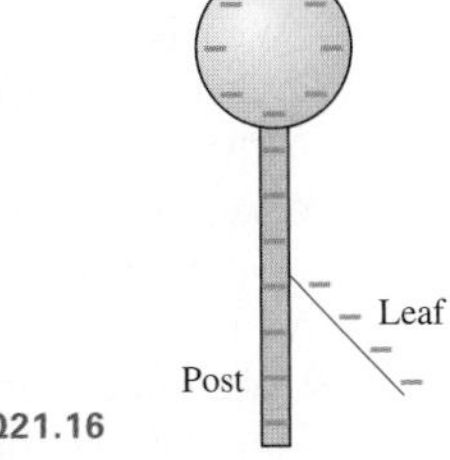

FIGURE Q21.16

17. Rank in order, from largest to smallest, the energies $(U_C)_1$ to $(U_C)_4$ stored in each of the capacitors in Figure Q21.17. Explain.

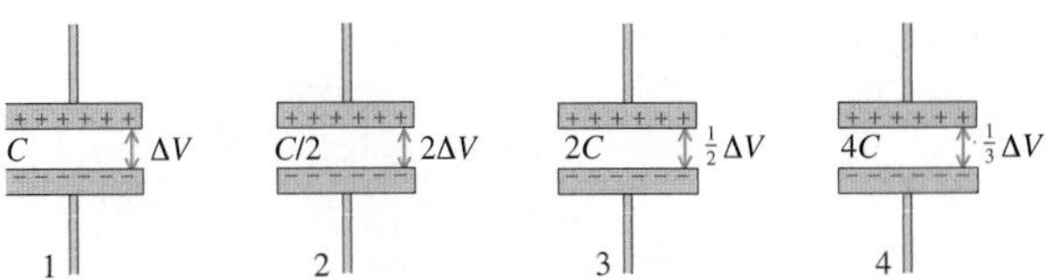

FIGURE Q21.17

18. A parallel-plate capacitor with plate separation d is connected to a battery that has potential difference ΔV_{bat}. Without breaking any of the connections, insulating handles are used to increase the plate separation to $2d$.
    a. Does the potential difference ΔV_C change as the separation increases? If so, by what factor? If not, why not?
    b. Does the capacitance change? If so, by what factor? If not, why not?
    c. Does the capacitor charge Q change? If so, by what factor? If not, why not?
19. The gap between the capacitor plates shown in Figure Q21.19 is *partially* filled with a dielectric. The capacitor was charged by a 9 V battery, then disconnected from the battery. Rank in order, from smallest to largest, the electric field strengths E_1, E_2, and E_3 at the points labeled in the figure, as well as the field strength E_4 between the plates if the dielectric is removed. Explain.

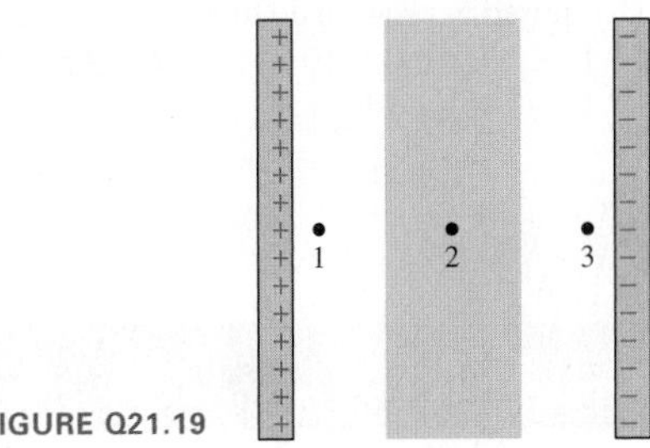

FIGURE Q21.19

Multiple-Choice Questions

20. | A 1.0 nC positive point charge is located at point A in Figure Q21.20. The electric potential at point B is
    A. 9.0 V
    B. $9.0\sin 30°$ V
    C. $9.0\cos 30°$ V
    D. $9.0\tan 30°$ V

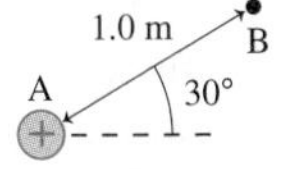

FIGURE Q21.20

21. | For the capacitor shown in Figure Q21.21, the potential difference ΔV_{ab} between points a and b is
    A. 6 V
    B. $6\sin 30°$ V
    C. $6\cos 30°$ V
    D. $6/\sin 30°$ V
    E. $6/\cos 30°$ V

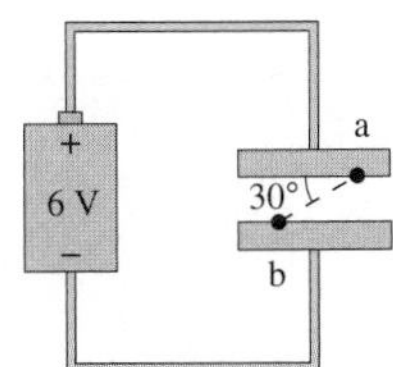

FIGURE Q21.21

22. | The electric potential is 300 V at $x = 0$ cm, is −100 V at $x = 5$ cm, and varies linearly with x. If a positive charge is released from rest at $x = 2.5$ cm, and is subject only to electric forces, the charge will
    A. Move to the right.
    B. Move to the left.
    C. Stay at $x = 2.5$ cm.
    D. Not enough information to tell.

Questions 23 through 27 refer to Figure Q21.23, which shows equipotential lines in a region of space. The equipotential lines are spaced by the same difference in potential, and several of the potentials are given.

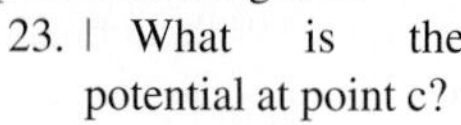

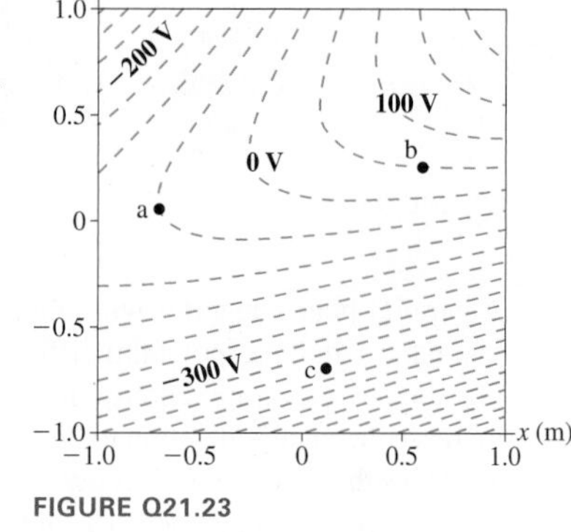

FIGURE Q21.23

23. | What is the potential at point c?
    A. −400 V
    B. −350 V
    C. −100 V
    D. 350 V
    E. 400 V
24. | At which point, a, b, or c, is the magnitude of the electric field the greatest?
25. | What is the approximate magnitude of the electric field at point c?
    A. 100 V/m B. 300 V/m C. 800 V/m
    D. 1500 V/m E. 3000 V/m
26. | The direction of the electric field at point b is closest to which direction?
    A. Right B. Up C. Left D. Down
27. || A +10 nC charge is moved from point c to point a. How much work is required in order to do this?
    A. 3.5×10^{-6} J B. 4.0×10^{-6} J C. 3.5×10^{-3} J
    D. 4.0×10^{-3} J E. 3.5 J
28. | A bug zapper consists of two metal plates connected to a high-voltage power supply. The voltage between the plates is set to give an electric field slightly less than 1×10^6 V/m. When a bug flies between the two plates, it increases the field enough to initiate a spark that incinerates the bug. If a bug zapper has a 4000 V power supply, what is the approximate separation between the plates?
    A. 0.05 cm B. 0.5 cm
    C. 5 cm D. 50 cm
29. | An atom of helium and one of argon are singly ionized—one electron is removed from each. The two ions are then accelerated from rest by the electric field between two plates with a potential difference of 150 V. After accelerating from one plate to the other,
    A. The helium ion has more kinetic energy.
    B. The argon ion has more kinetic energy.
    C. Both ions have the same kinetic energy.
    D. There is not enough information to say which ion has more kinetic energy.
30. || The dipole moment of the heart is shown
BIO at a particular instant in Figure Q21.30. Which of the following potential differences will have the largest positive value?
    A. $V_1 - V_2$
    B. $V_1 - V_3$
    C. $V_2 - V_1$
    D. $V_3 - V_1$

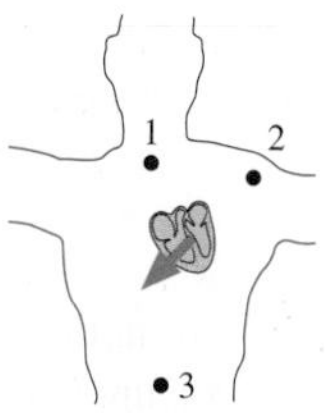

FIGURE Q21.30

PROBLEMS

Section 21.1 Electric Potential Energy and Electric Potential

Section 21.2 Sources of Electric Potential

1. ||| Moving a charge from point A, where the potential is 300 V, to point B, where the potential is 150 V, takes 4.5×10^{-4} J of work. What is the value of the charge?
2. |||| The graph in Figure P21.2 shows the electric potential energy as a function of separation for two point charges. If one charge is +0.44 nC, what is the other charge?

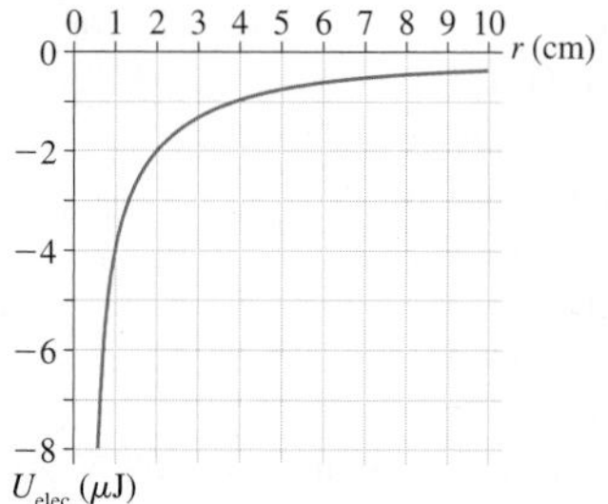

FIGURE P21.2

3. ||| It takes 3.0 μJ of work to move a 15 nC charge from point A to B. It takes −5.0 μJ of work to move the charge from C to B. What is the potential difference $V_C - V_A$?
4. | A 20 nC charge is moved from a point where $V = 150$ V to a point where $V = -50$ V. How much work is done by the force that moves the charge?
5. | At one point in space, the electric potential energy of a 15 nC charge is 45 μJ.
    a. What is the electric potential at this point?
    b. If a 25 nC charge were placed at this point, what would its electric potential energy be?

Section 21.3 Electric Potential and Conservation of Energy

6. | An electron has been accelerated from rest through a potential difference of 1000 V.
    a. What is its kinetic energy, in electron volts?
    b. What is its kinetic energy, in joules?
    c. What is its speed?
7. | A proton has been accelerated from rest through a potential difference of −1000 V.
    a. What is its kinetic energy, in electron volts?
    b. What is its kinetic energy, in joules?
    c. What is its speed?
8. ||||| What potential difference is needed to accelerate a He^+ ion (charge $+e$, mass 4 u) from rest to a speed of 1.0×10^6 m/s?
9. || An electron with an initial speed of 500,000 m/s is brought to rest by an electric field.
    a. Did the electron move into a region of higher potential or lower potential?
    b. What was the potential difference that stopped the electron?
    c. What was the initial kinetic energy of the electron, in electron volts?

10. ||| A proton with an initial speed of 800,000 m/s is brought to rest by an electric field.
   a. Did the proton move into a region of higher potential or lower potential?
   b. What was the potential difference that stopped the proton?
   c. What was the initial kinetic energy of the proton, in electron volts?

19. || a. What is the potential difference between the terminals of an ordinary AA or AAA battery? (If you're not sure, find one and look at the label.)
   b. An AA battery is connected to a parallel-plate capacitor having 4.0-cm-diameter plates spaced 2 mm apart. How much charge does the battery move from one plate to the other?

Section 21.4 Calculating the Electric Potential

11. || The electric potential at a point that is halfway between two identical charged particles is 300 V. What is the potential at a point that is 25% of the way from one particle to the other?
12. || A 2.0 cm × 2.0 cm parallel-plate capacitor has a 2.0 mm spacing. The electric field strength inside the capacitor is 1.0×10^5 V/m.
   a. What is the potential difference across the capacitor?
   b. How much charge is on each plate?
13. ||| Two 2.00 cm × 2.00 cm plates that form a parallel-plate capacitor are charged to ± 0.708 nC. What are the electric field strength inside and the potential difference across the capacitor if the spacing between the plates is (a) 1.00 mm and (b) 2.00 mm?
14. | a. In Figure P21.14, which capacitor plate, left or right, is the positive plate?
   b. What is the electric field strength inside the capacitor?
   c. What is the potential energy of a proton at the midpoint of the capacitor?

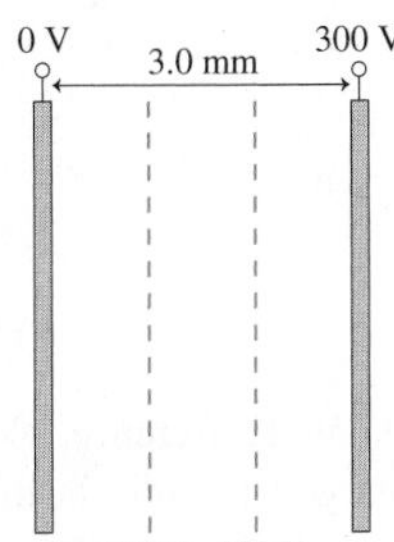

FIGURE P21.14

15. | A +25 nC charge is at the origin.
   a. What are the radii of the 1000 V, 2000 V, 3000 V, and 4000 V equipotential surfaces?
   b. Draw an equipotential map in the *xy*-plane showing the charge and these four surfaces.
16. || a. What is the electric potential at points A, B, and C in Figure P21.16?
   b. What is the potential energy of an electron at each of these points?
   c. What are the potential differences ΔV_{AB} and ΔV_{BC}?

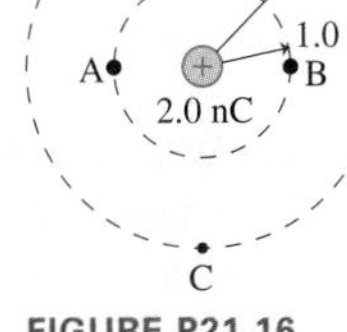

FIGURE P21.16

17. ||| A 1.0-mm-diameter ball bearing has 2.0×10^9 excess electrons. What is the ball bearing's potential?
18. || What is the electric potential at the point indicated with the dot in Figure P21.18?

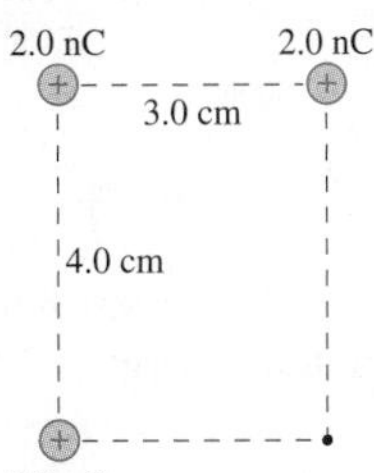

FIGURE P21.18

Section 21.5 Connecting Potential and Field

20. || a. In Figure P21.20, which point, A or B, has a higher electric potential?
   b. What is the potential difference between A and B?

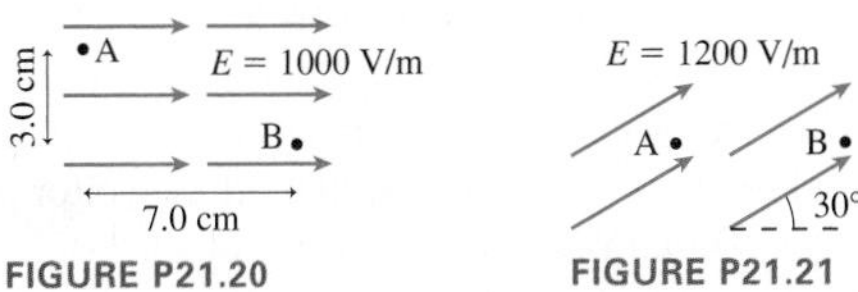

FIGURE P21.20 FIGURE P21.21

21. ||| In Figure P21.21, the electric potential at point A is −300 V. What is the potential at point B, which is 5.0 cm to the right of A?
22. || What is the potential difference between $x_i = 10$ cm and $x_f = 30$ cm in the uniform electric field $E_x = 1000$ V/m?
23. | What are the magnitude and direction of the electric field at the dot in Figure P21.23?

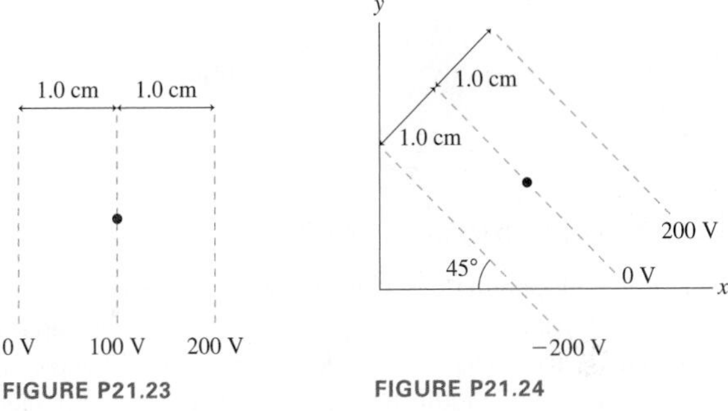

FIGURE P21.23 FIGURE P21.24

24. | What are the magnitude and direction of the electric field at the dot in Figure P21.24?

Section 21.6 The Electrocardiogram

25. | BIO One standard location for a pair of electrodes during an EKG is shown in Figure P21.25. The potential difference $\Delta V_{31} = V_3 - V_1$ is recorded. For each of the three instants a, b, and c during the heart's cycle shown in Figure 21.29, will ΔV_{31} be positive or negative? Explain.

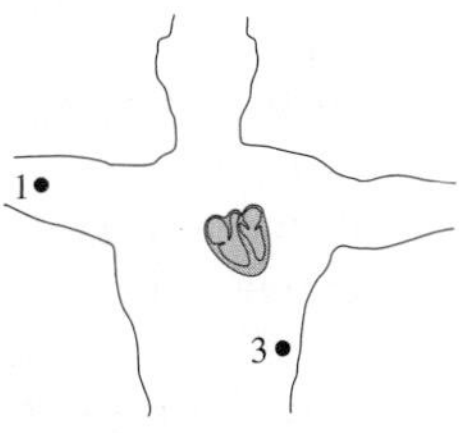

FIGURE P21.25

26. | BIO Three electrodes, 1–3, are attached to a patient as shown in Figure P21.26. During ventricular depolarization (see Figure 21.29), across which pair of electrodes is the magnitude of the potential difference likely to be the smallest? Explain.

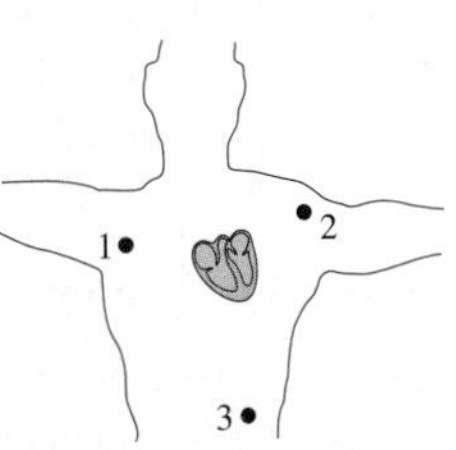

FIGURE P21.26

Section 21.7 Capacitance and Capacitors

27. ||| Two 2.0 cm × 2.0 cm square aluminum electrodes, spaced 0.50 mm apart, are connected to a 100 V battery.
    a. What is the capacitance?
    b. What is the charge on the positive electrode?
28. ||| An uncharged capacitor is connected to the terminals of a 3.0 V battery, and 6.0 μC flows to the positive plate. The 3.0 V battery is then disconnected and replaced with a 5.0 V battery, with the positive and negative terminals connected in the same manner as before. How much additional charge flows to the positive plate?
29. ||| You need to construct a 100 pF capacitor for a science project. You plan to cut two $L \times L$ metal squares and place spacers between them. The thinnest spacers you have are 0.20 mm thick. What is the proper value of L?
30. | A switch that connects a battery to a 10 μF capacitor is closed. Several seconds later you find that the capacitor plates are charged to ± 30 μC. What is the battery voltage?
31. | What is the voltage of a battery that will charge a 2.0 μF capacitor to ± 48 μC?
32. | Two electrodes connected to a 9.0 V battery are charged to ± 45 nC. What is the capacitance of the electrodes?
33. | Initially, the switch in Figure P21.33 is open and the capacitor is uncharged. How much charge flows through the switch after the switch is closed?

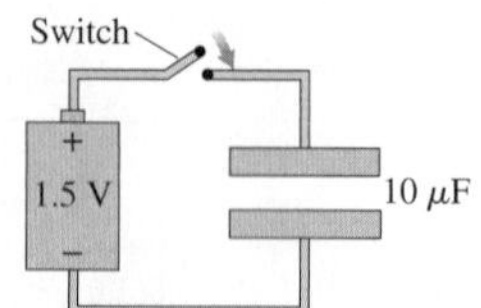

FIGURE P21.33

Section 21.8 Dielectrics and Capacitors

34. || A 1.2 nF parallel-plate capacitor has an air gap between its plates. Its capacitance increases by 3.0 nF when the gap is filled by a dielectric. What is the dielectric constant of that dielectric?
35. ||| A science-fair radio uses a homemade capacitor made of two 35 cm × 35 cm sheets of aluminum foil separated by a 0.25-mm-thick sheet of paper. What is its capacitance?
36. ||| A 25 pF parallel-plate capacitor with an air gap between the plates is connected to a 100 V battery. A Teflon slab is then inserted between the plates and completely fills the gap. What is the change in the charge on the positive plate when the Teflon is inserted?
37. || Two 2.0-cm-diameter electrodes with a 0.10-mm-thick sheet of Teflon between them are attached to a 9.0 V battery. Without disconnecting the battery, the Teflon is removed. What are the charge, potential difference, and electric field (a) before and (b) after the Teflon is removed?
38. |||| A capacitor with its plates separated by paper stores 4.4 nC of charge when it is connected to a particular battery. An otherwise identical capacitor, but with its plates separated by Pyrex glass, is connected to the same battery. How much charge does that capacitor store?

Section 21.9 Energy and Capacitors

39. ||| To what potential should you charge a 1.0 μF capacitor to store 1.0 J of energy?
40. || A pair of 10 μF capacitors in a high-power laser are charged to 1.7 kV.
    a. What charge is stored in each capacitor?
    b. How much energy is stored in each capacitor?
41. | Capacitor 2 has half the capacitance and twice the potential difference as capacitor 1. What is the ratio $(U_C)_1/(U_C)_2$?
42. |||| Two uncharged metal spheres, spaced 15.0 cm apart, have a capacitance of 24.0 pF. How much work would it take to move 12.0 nC of charge from one sphere to the other?
43. |||| 50 pJ of energy is stored in a 2.0 cm × 2.0 cm × 2.0 cm region of uniform electric field. What is the electric field strength?

General Problems

44. || A 2.0-cm-diameter parallel-plate capacitor with a spacing of 0.50 mm is charged to 200 V. What are (a) the total energy stored in the electric field and (b) the energy density?
45. ||| What is the change in electric potential energy of a 3.0 nC point charge when it is moved from point A to point B in Figure P21.45?

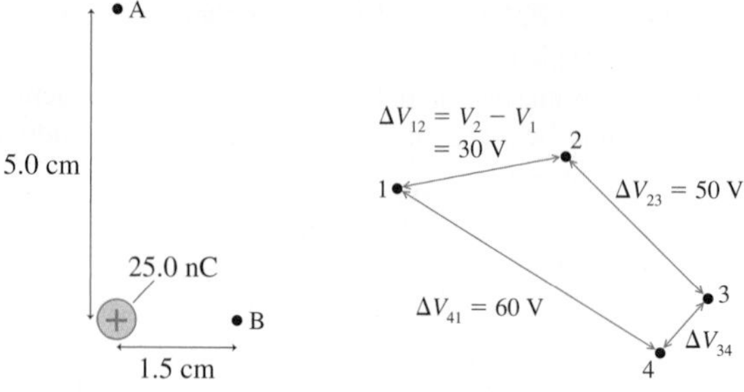

FIGURE P21.45 FIGURE P21.46

46. || What is the potential difference ΔV_{34} in Figure P21.46?
47. || INT A −50 nC charged particle is in a uniform electric field $\vec{E} = (10\text{ V/m, east})$. An external force moves the particle 1.0 m north, then 5.0 m east, then 2.0 m south, and finally 3.0 m west. The particle begins and ends its motion with zero velocity.
    a. How much work is done on it by the external force?
    b. What is the potential difference between the particle's final and initial positions?
48. || INT At a distance r from a point charge, the electric potential is 3000 V and the magnitude of the electric field is 2.0×10^5 V/m.
    a. What is the distance r?
    b. What are the electric potential and the magnitude of the electric field at distance $r/2$ from the charge?
49. || What is the electric potential energy of the electron in Figure P21.49? The protons are fixed and can't move.

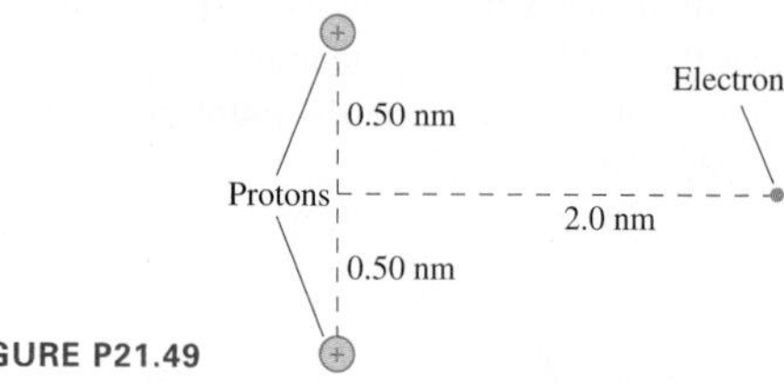

FIGURE P21.49

50. |||| Two point charges 2.0 cm apart have an electric potential energy −180 μJ. The total charge is 30 nC. What are the two charges?

51. || Two positive point charges are 5.0 cm apart. If the electric potential energy is 72 μJ, what is the magnitude of the force between the two charges?
INT

52. ||||| A $+3.0$ nC charge is at $x = 0$ cm and a -1.0 nC charge is at $x = 4$ cm. At what point or points on the x-axis is the electric potential zero?

53. ||||| A -3.0 nC charge is on the x-axis at $x = -9$ cm and a $+4.0$ nC charge is on the x-axis at $x = 16$ cm. At what point or points on the y-axis is the electric potential zero?

54. || A -2.0 nC charge and a $+2.0$ nC charge are located on the x-axis at $x = -1.0$ cm and $x = +1.0$ cm, respectively.
INT
   a. At what position or positions on the x-axis is the electric field zero?
   b. At what position or positions on the x-axis is the electric potential zero?
   c. Draw graphs of the electric field strength and the electric potential along the x-axis.

55. ||| A -10.0 nC point charge and a $+20.0$ nC point charge are 15.0 cm apart on the x-axis.
INT
   a. What is the electric potential at the point on the x-axis where the electric field is zero?
   b. What are the magnitude and direction of the electric field at the point on the x-axis, between the charges, where the electric potential is zero?

56. ||||| A 2.0-mm-diameter glass bead is positively charged. The potential difference between a point 2.0 mm from the bead and a point 4.0 mm from the bead is 500 V. What is the charge on the bead?

57. | In a semiclassical model of the hydrogen atom, the electron orbits the proton at a distance of 0.053 nm.
   a. What is the electric potential of the proton at the position of the electron?
   b. What is the electron's potential energy?

58. | What is the electric potential at the point indicated with the dot in Figure P21.58?

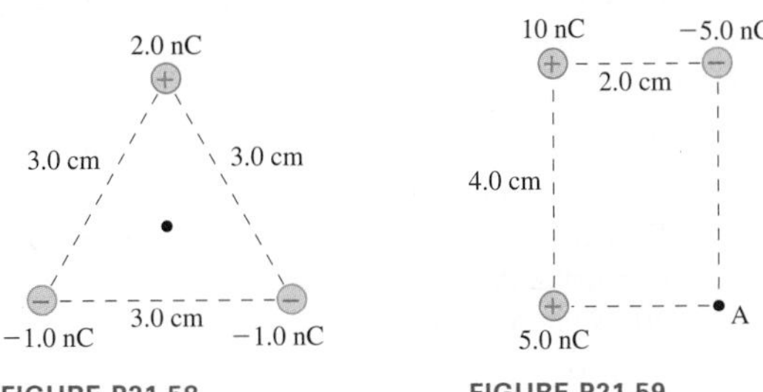

FIGURE P21.58

FIGURE P21.59

59. | a. What is the electric potential at point A in Figure P21.59?
   b. What is the potential energy of a proton at point A?

60. ||| A proton's speed as it passes point A is 50,000 m/s. It follows the trajectory shown in Figure P21.60. What is the proton's speed at point B?

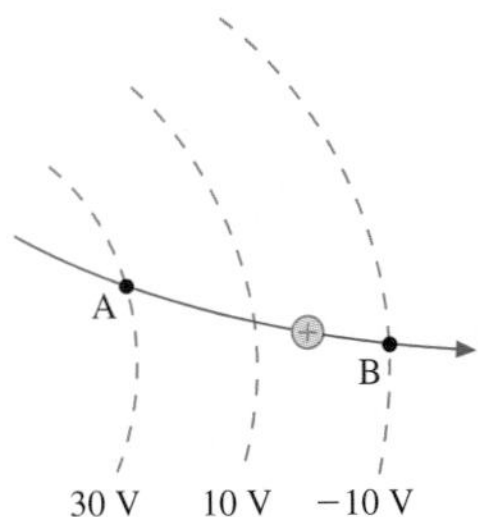

FIGURE P21.60

61. || Electric outlets have a voltage of approximately 120 V between the two parallel slots. Estimate the electric field strength between these two slots.

62. | Estimate the magnitude of the electric field in a cell membrane with a thickness of 8 nm.
BIO

63. || A Na^+ ion moves from inside a cell, where the electric potential is -70 mV, to outside the cell, where the potential is 0 V. What is the change in the ion's electric potential energy as it moves from inside to outside the cell? Does its energy increase or decrease?
BIO

64. ||| Suppose that a molecular ion with charge $-10e$ is embedded within the 5.0-nm-thick cell membrane of a cell with membrane potential -70 mV. What is the electric force on the molecule?
BIO

65. ||| The electric field strength is 50,000 V/m inside a parallel-plate capacitor with a 2.0 mm spacing. A proton is released from rest at the positive plate. What is the proton's speed when it reaches the negative plate?

66. ||||| An alpha particle (the nucleus of a helium atom, with charge $+2e$ and a mass four times that of a proton) and an antiproton (which has the same mass as a proton but charge $-e$) are released from rest a great distance apart. They are oppositely charged, so each accelerates toward the other. What are the speeds of the two particles when they are 2.5 nm apart?
INT
Hint: You'll need to use *two* conservation laws. And what does "a great distance" suggest about the initial value of r?

67. ||| A proton is released from rest at the positive plate of a parallel-plate capacitor. It crosses the capacitor and reaches the negative plate with a speed of 50,000 m/s. What will be the proton's final speed if the experiment is repeated with double the amount of charge on each capacitor plate?

68. || The electric field strength is 20,000 V/m inside a parallel-plate capacitor with a 1.0 mm spacing. An electron is released from rest at the negative plate. What is the electron's speed when it reaches the positive plate?

69. || In the early 1900s, Robert Millikan used small charged droplets of oil, suspended in an electric field, to make the first quantitative measurements of the electron's charge. A 0.70-μm-diameter droplet of oil, having a charge of $+e$, is suspended in midair between two horizontal plates of a parallel-plate capacitor. The upward electric force on the droplet is exactly balanced by the downward force of gravity. The oil has a density of 860 kg/m^3, and the capacitor plates are 5.0 mm apart. What must the potential difference between the plates be to hold the droplet in equilibrium?
INT

70. |||| Two 2.0-cm-diameter disks spaced 2.0 mm apart form a parallel-plate capacitor. The electric field between the disks is 5.0×10^5 V/m.
   a. What is the voltage across the capacitor?
   b. How much charge is on each disk?
   c. An electron is launched from the negative plate. It strikes the positive plate at a speed of 2.0×10^7 m/s. What was the electron's speed as it left the negative plate?

71. || In *proton-beam therapy,* a high-energy beam of protons is fired at a tumor. The protons come to rest in the tumor, depositing their kinetic energy and breaking apart the tumor's DNA, thus killing its cells. For one patient, it is desired that 0.10 J of proton energy be deposited in a tumor. To create the proton beam, the protons are accelerated from rest through a 10 MV potential difference. What is the total charge of the protons that must be fired at the tumor to deposit the required energy?
BIO

72. |||| A 2.5-mm-diameter sphere is charged to −4.5 nC. An electron fired directly at the sphere from far away comes to within 0.30 mm of the surface of the target before being reflected.
 a. What was the electron's initial speed?
 b. At what distance from the surface of the sphere is the electron's speed half of its initial value?
 c. What is the acceleration of the electron at its turning point?

73. || A proton is fired from far away toward the nucleus of an iron atom. Iron is element number 26, and the diameter of the nucleus is 9.0 fm. (1 fm $= 10^{-15}$ m.) What initial speed does the proton need to just reach the surface of the nucleus? Assume the nucleus remains at rest.

74. || Two 10.0-cm-diameter electrodes 0.50 cm apart form a parallel-plate capacitor. The electrodes are attached by metal wires to the terminals of a 15 V battery. After a long time, the capacitor is disconnected from the battery but is not discharged. What are the charge on each electrode, the electric field strength inside the capacitor, and the potential difference between the electrodes
 a. Right after the battery is disconnected?
 b. After insulating handles are used to pull the electrodes away from each other until they are 1.0 cm apart?

75. || Two 10.0-cm-diameter electrodes 0.50 cm apart form a parallel-plate capacitor. The electrodes are attached by metal wires to the terminals of a 15 V battery. What are the charge on each electrode, the electric field strength inside the capacitor, and the potential difference between the electrodes
 a. While the capacitor is attached to the battery?
 b. After insulating handles are used to pull the electrodes away from each other until they are 1.0 cm apart? The electrodes remain connected to the battery during this process.

76. ||| Determine the magnitude and direction of the electric field at points 1 and 2 in Figure P21.76.

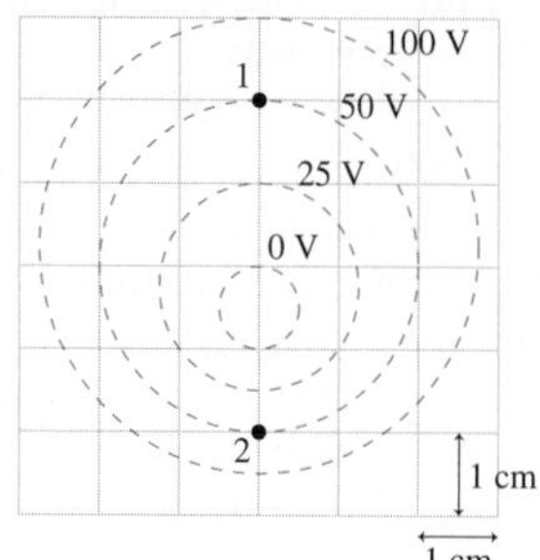

FIGURE P21.76

77. | Figure P21.77 shows a series of equipotential curves.
 a. Is the electric field strength at point A larger than, smaller than, or equal to the field strength at point B? Explain.
 b. Is the electric field strength at point C larger than, smaller than, or equal to the field strength at point D? Explain.
 c. Determine the electric field $\vec{E}$ at point D. Express your answer as a magnitude and direction.

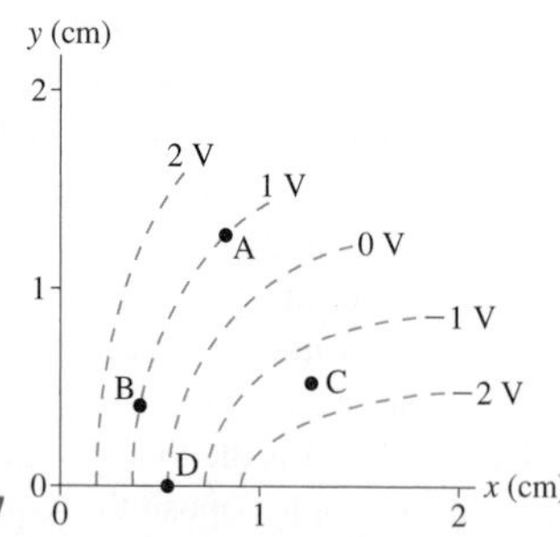

FIGURE P21.77

78. || Figure P21.78 shows the electric potential on a grid whose squares are 5.0 cm on a side.
 a. Reproduce this figure on your paper, then draw the 50 V, 75 V, and 100 V equipotential surfaces.
 b. Estimate the electric field (strength and direction) at points A, B, C, and D.
 c. Draw the electric field vectors at points A, B, C, and D on your diagram.

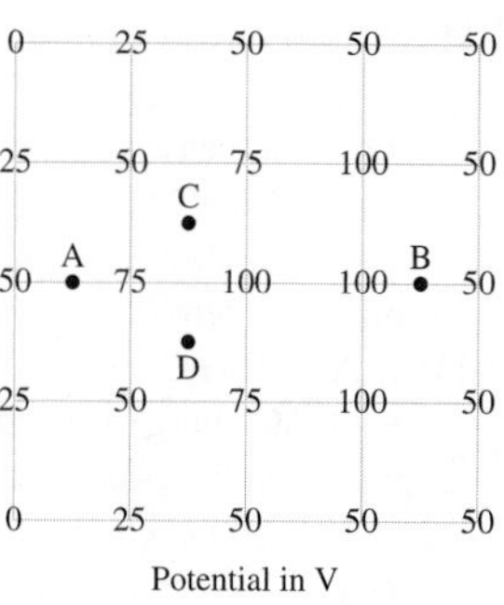

FIGURE P21.78

79. || The plates of a 3.0 nF parallel-plate capacitor are each 0.27 m^2 in area.
 a. How far apart are the plates if there's air between them?
 b. If the plates are separated by a Teflon sheet, how thick is the sheet?

80. |||| The dielectric in a capacitor serves two purposes. It increases the capacitance, compared to an otherwise identical capacitor with an air gap, and it increases the maximum potential difference the capacitor can support. If the electric field in a material is sufficiently strong, the material will suddenly become able to conduct, creating a spark. The critical field strength, at which breakdown occurs, is 3.0 MV/m for air, but 60 MV/m for Teflon.
 a. A parallel-plate capacitor consists of two square plates, 15 cm on a side, spaced 0.50 mm apart with only air between them. What is the maximum energy that can be stored by the capacitor?
 b. What is the maximum energy that can be stored if the plates are separated by a 0.50-mm-thick Teflon sheet?

81. ||||| The flash unit in a camera uses a special circuit to "step up" INT the 3.0 V from the batteries to 300 V, which charges a capacitor. The capacitor is then discharged through a flashlamp. The discharge takes 10 μs, and the average power dissipated in the flashlamp is 10^5 W. What is the capacitance of the capacitor?

In Problems 82 through 85 you are given the equation(s) used to solve a problem. For each of these,
 a. Write a realistic problem for which this is the correct equation(s).
 b. Finish the solution of the problem.

82. | $$\frac{(9.0\times10^9\ \text{N}\cdot\text{m}^2/\text{C}^2)q_1q_2}{0.030\ \text{m}} = 90\times10^{-6}\ \text{J}$$
$$q_1 + q_2 = 40\ \text{nC}$$

83. || $$\tfrac{1}{2}(1.67\times10^{-27}\ \text{kg})(2.5\times10^6\ \text{m/s})^2 + 0 = \tfrac{1}{2}(1.67\times10^{-27}\ \text{kg})v_i^2 + \frac{(9.0\times10^9\ \text{N}\cdot\text{m}^2/\text{C}^2)(2.0\times10^{-9}\ \text{C})(1.60\times10^{-19}\ \text{C})}{0.0010\ \text{m}}$$

84. || $$\frac{(9.0\times10^9\ \text{N}\cdot\text{m}^2/\text{C}^2)(3.0\times10^{-9}\ \text{C})}{0.030\ \text{m}} + \frac{(9.0\times10^9\ \text{N}\cdot\text{m}^2/\text{C}^2)(3.0\times10^{-9}\ \text{C})}{(0.030\ \text{m}) + d} = 1200\ \text{V}$$

85. || $$400\ \text{nC} = (100\ \text{V})\,C$$
$$C = \frac{(8.85\times10^{-12}\ \text{F/m})(0.10\ \text{m}\times0.10\ \text{m})}{d}$$

Passage Problems

A Lightning Strike

Storm clouds build up large negative charges, as described in the chapter. The charges dwell in *charge centers,* regions of concentrated charge. Suppose a cloud has -25 C in a 1.0-km-diameter spherical charge center located 10 km above the ground, as sketched in Figure P21.88. The negative charge center attracts a similar amount of positive charge that is spread on the ground below the cloud.

The charge center and the ground function as a charged capacitor, with a potential difference of approximately 4×10^8 V. The large electric field between these two "electrodes" may ionize the air, leading to a conducting path between the cloud and the ground. Charges will flow along this conducting path, causing a discharge of the capacitor—a lightning strike.

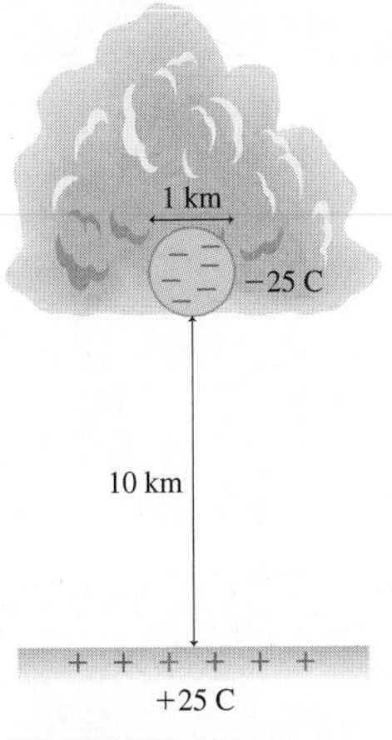

FIGURE P21.88

86. | What is the approximate magnitude of the electric field between the charge center and the ground?
A. 4×10^4 V/m B. 4×10^5 V/m
C. 4×10^6 V/m D. 4×10^7 V/m

87. | Which of the curves sketched in Figure P21.87 best approximates the shape of an equipotential drawn halfway between the charge center and the ground?

A. B. C. D.

FIGURE P21.87

88. | What is the approximate capacitance of the charge center + ground system?
A. 6×10^{-8} F B. 2×10^7 F
C. 4×10^6 F D. 8×10^6 F

89. | If 12.5 C of charge is transferred from the cloud to the ground in a lightning strike, what fraction of the stored energy is dissipated?
A. 12% B. 25% C. 50% D. 75%

90. | If the cloud transfers all of its charge to the ground via several rapid lightning flashes lasting a total of 1 s, what is the average power?
A. 1 GW B. 2 GW C. 5 GW D. 10 GW

STOP TO THINK ANSWERS

Stop to Think 21.1: B If the charge were moved from 1 to 2 at a constant speed by a hand, the force exerted by the hand would need to be to the left, to oppose the rightward-directed electric force on the charge due to the source charges. Because the force due to the hand would be opposite the displacement, the hand would do *negative* work on the charge, decreasing its electric potential energy.

Stop to Think 21.2: C. The proton gains speed by losing potential energy. It loses potential energy by moving in the direction of decreasing electric potential.

Stop to Think 21.3: $\Delta V_{13} = \Delta V_{23} > \Delta V_{12}$. The potential depends only on the *distance* from the charge, not the direction. $\Delta V_{12} = 0$ because these points are at the same distance.

Stop to Think 21.4: C. $\vec{E}$ points "downhill," so V must decrease from right to left. E is larger on the left than on the right, so the equipotential lines must be closer together on the left.

Stop to Think 21.5: C. Capacitance is a property of the shape and position of the electrodes. It does not depend on the potential difference or charge.

Stop to Think 21.6: C. The electric field is $\Delta V_C/d$. With ΔV_C fixed by the battery, introducing the dielectric does not change E. More charge flows from the battery to compensate for the dielectric.

Stop to Think 21.7: B. The energy is $\frac{1}{2}C(\Delta V_C)^2$. ΔV_C is constant, but C doubles when the distance is halved.

22 Current and Resistance

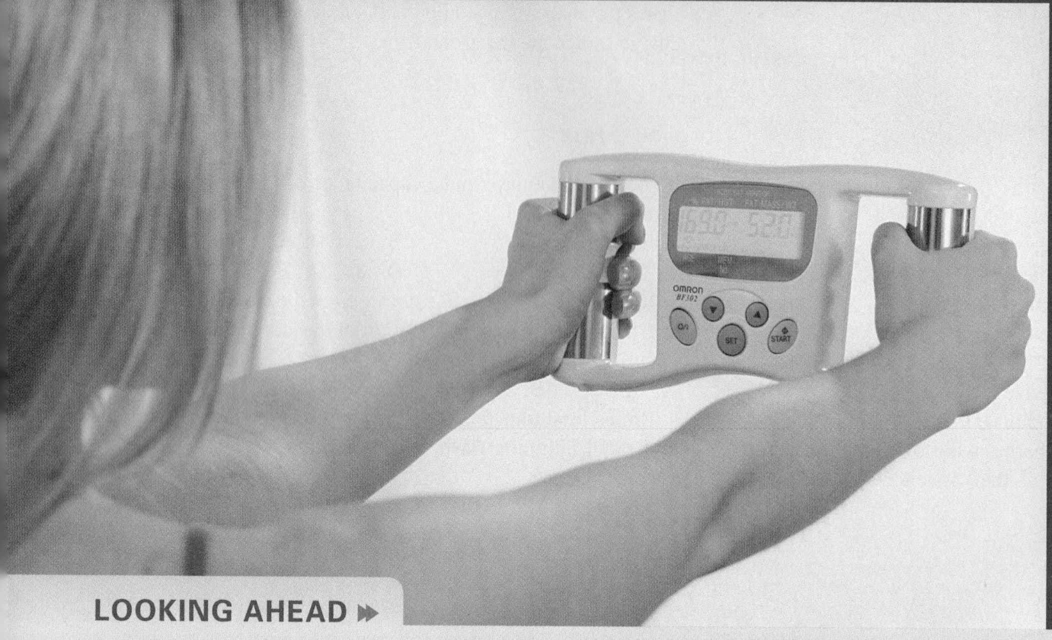

This woman is measuring her percentage body fat by gripping a device that passes a small electric current through her body. How does a measurement of the current reveal such details of the body's structure?

LOOKING AHEAD ▸▸

The goal of Chapter 22 is to learn how and why charge moves through a conductor as what we call a current.

A Basic Circuit

Connecting a bulb to a battery makes the bulb glow. How is the chemical energy of the battery transferred to the bulb? In this chapter, you'll learn to explain this process in terms of electric current.

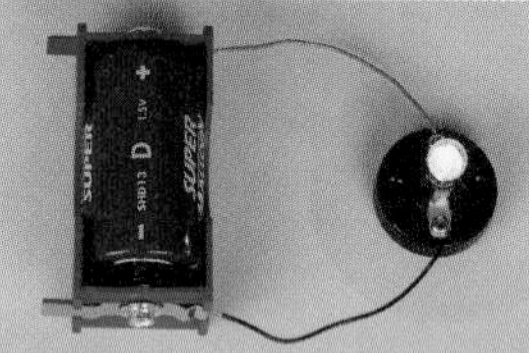

Looking Back ◂◂

20.1–20.2 Charges and conductors
20.7 Forces on charges in an electric field
21.3–21.5 Electric potential

In this simple circuit, the battery's chemical energy lifts charges "uphill" as if they were on a charge escalator. Charges then move "downhill" through the wire and bulb. This flow of charge—a **current**—warms the bulb until it glows.

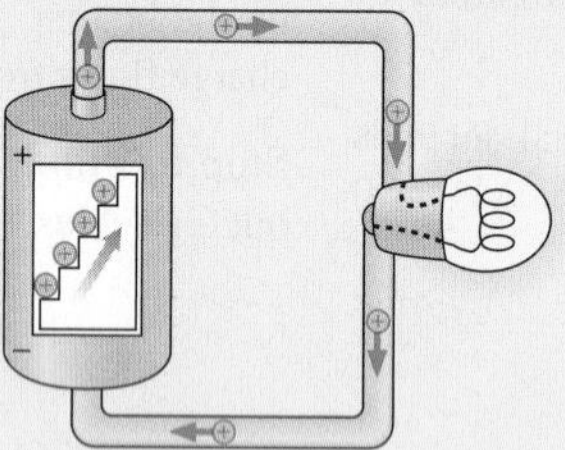

Developing the model of current in a circuit will draw on many concepts from past chapters.

Current

Current is the flow of charge, and **charge is conserved:** it isn't used up in a circuit. These bulbs are connected one after the other. All of the current that goes through one bulb goes through the next and the next, so all the bulbs are equally bright.

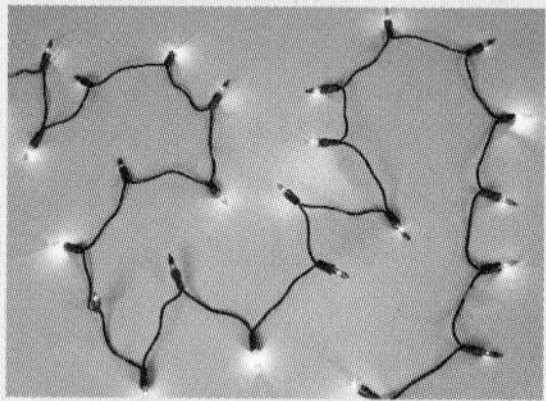

You'll use this basic property of current to analyze simple circuits.

Energy and Power

Passing a current through a **resistor** converts electric energy to thermal energy, increasing the temperature of the resistor. Many practical devices are based on this principle.

How much current does it take to warm the hot air from this hair dryer? You'll find out how to do this calculation.

Ohm's Law

The current in a circuit depends not only on the resistance but also on the voltage. **Ohm's law** is a simple expression that relates current, voltage, and resistance.

Turning up the heat on the stove means turning up the voltage across the resistance of the burner.

Resistance

The wire filament in a lightbulb has a **resistance** to the flow of charge. Decreasing the resistance increases the current through the filament, leading to a brighter bulb.

How long must the wire inside the bulb be to have the correct resistance? You'll learn how to figure this out.

22.1 A Model of Current

FIGURE 22.1 A capacitor is discharged by a metal wire.

(a)

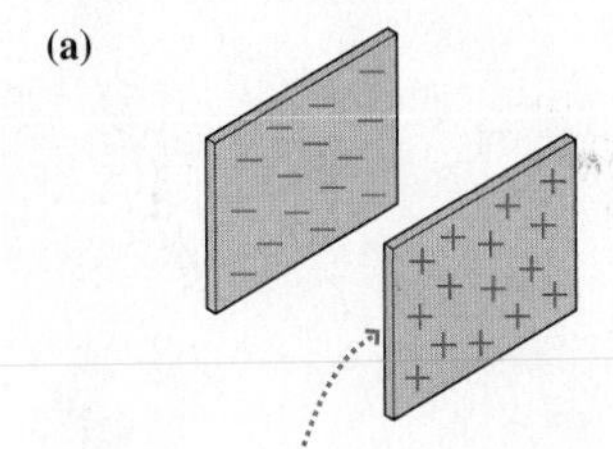

A charged parallel-plate capacitor

(b)

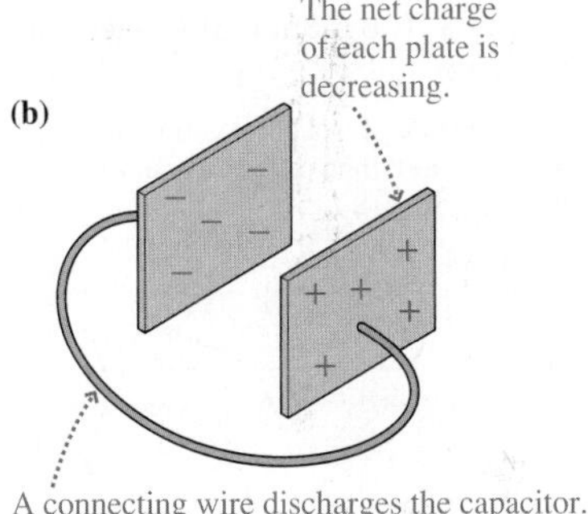
The net charge of each plate is decreasing.
A connecting wire discharges the capacitor.

Let's start our exploration of current with a very simple experiment. FIGURE 22.1a shows a charged parallel-plate capacitor. If we connect the two capacitor plates to each other with a metal wire, as shown in FIGURE 22.1b, the plates quickly become neutral. We say that the capacitor has been *discharged.*

The wire is a conductor, a material through which charge easily moves. In Chapter 20, we defined a *current* as the motion of charges through a material. Later in this chapter we will develop a quantitative expression for current, but for now this definition will suffice. Apparently the excess charge on one capacitor plate is able to move through the wire to the other plate, neutralizing both plates. **The capacitor is discharged by a current in the connecting wire.**

NOTE ► Current is defined as the motion of charges, so we don't say that "current flows." It is *charges* that flow, not current. Current *is* the flow. ◄

If we observe the capacitor discharge, we see other effects. As FIGURE 22.2 shows, the connecting wire gets warmer. If the wire is very thin in places, such as the thin filament in a lightbulb, the wire gets hot enough to glow. The current-carrying wire also deflects a compass needle. We will explore the connection between currents and magnetism in Chapter 24. For now, we will use "makes the wire warmer" and "deflects a compass needle" as *indicators* that a current is present in a wire. We can use the brightness of a lightbulb to tell us the magnitude of the current; **more current means a brighter bulb.**

FIGURE 22.2 Properties of a current.

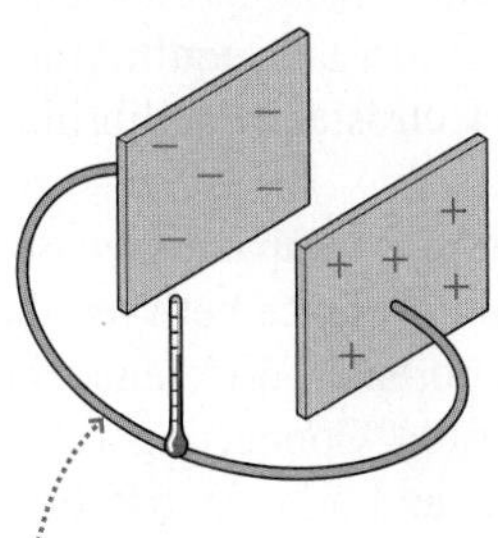

The connecting wire gets warm.

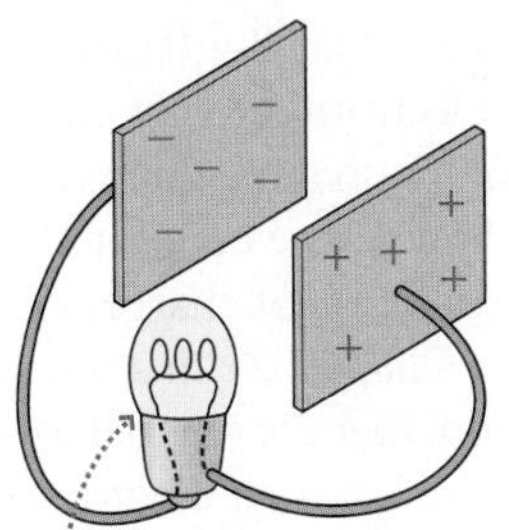

A lightbulb glows. The lightbulb filament is part of the connecting wire.

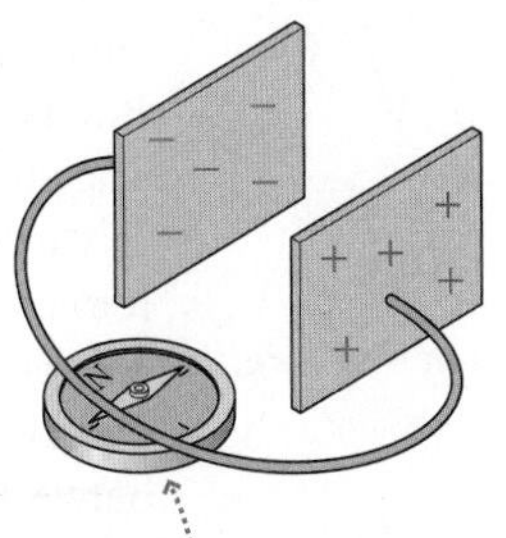

A compass needle is deflected.

Charge Carriers

Opposite charges attract, but the oppositely charged plates of a capacitor don't spontaneously discharge because the charges can't leap from one plate to the other. A connecting wire discharges the capacitor by providing a pathway for charge to move from one side of the capacitor to the other. But does positive charge move toward the negative plate, or does negative charge move toward the positive plate?

The charges that move in a current are called the *charge carriers.* The first experiments that could distinguish between positive and negative charge carriers didn't take place until the early 20th century, but the experimental evidence is now clear: **The charge carriers in metals are electrons.** As FIGURE 22.3 shows, it is the motion of the *conduction electrons,* which are free to move around, that forms a current—a flow of charge—in the metal. An *insulator* does not have such free charges and cannot carry a current. A *semiconductor* is an intermediate case, with relatively few charge carriers, which can be either positive or negative. It will carry a current, but not as easily as a conductor.

NOTE ► Electrons are the charge carriers in *metals.* Other materials, such as semiconductors, may have different charge carriers. In ionic solutions, such as seawater, blood, and intercellular fluids, the charge carriers are ions, both positive and negative. ◄

FIGURE 22.3 Conduction electrons in a metal.

Ions (the metal atoms minus one electron) occupy fixed positions.

The conduction electrons (one per atom) are bound to the solid as a whole, not any particular atom. They are free to move around.

The metal as a whole is electrically neutral.

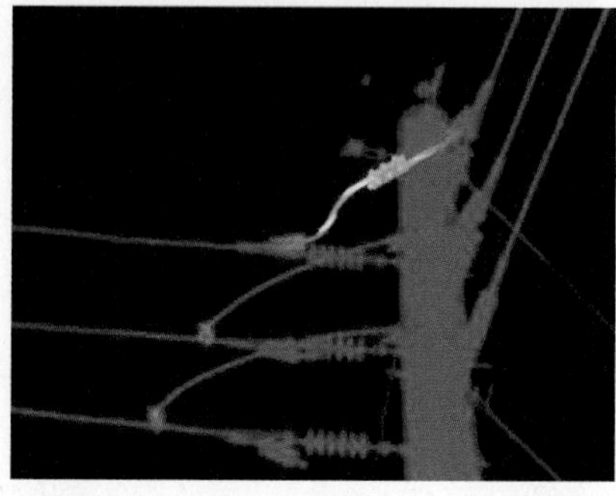

◀ **Hot wire** This thermal camera image of power lines shows that the lines are warm—as we'd expect, given the large currents that they carry. Spots where corrosion has thinned the wires get especially warm, making such images helpful for monitoring the condition of power lines.

Creating a Current

Suppose you want to slide a book across the table to your friend. You give it a quick push to start it moving, but it begins slowing down because of friction as soon as you take your hand off of it. The book's kinetic energy is transformed into thermal energy, leaving the book and the table slightly warmer. The only way to keep the book moving at a *constant* speed is to continue pushing it.

Something similar happens in a conductor. As we saw in Chapter 20, we can use an electric field to push on the electrons in a conductor. Suppose we take a piece of metal and apply an electric field, as in FIGURE 22.4. The field exerts a force on the electrons, and they begin to move. But the electrons aren't moving in a vacuum. Collisions between the electrons and the atoms of the metal slow them down, transforming the electrons' kinetic energy into the thermal energy of the metal, making the metal warmer. (Recall that "makes the wire warmer" is one of our indicators of a current.) The motion of the electrons will cease *unless you continue pushing*. To keep the electrons moving, we must maintain an electric field. In a constant field, an electron's average motion will be opposite the field. We call this motion the electron's *drift velocity*. If the field goes to zero, so does the drift velocity.

FIGURE 22.4 The motion of an electron in a conductor.

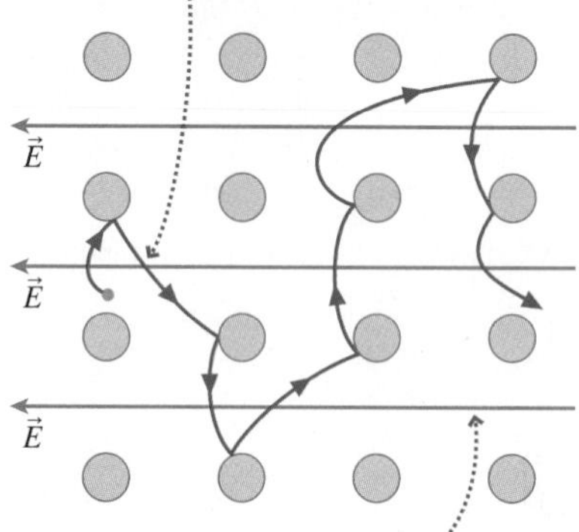

How can you have an electric field inside a conductor? One of the important conclusions of Chapter 20 was that $\vec{E} = \vec{0}$ inside a conductor in electrostatic equilibrium. But a conductor with electrons moving through it is *not* in electrostatic equilibrium. The charges are in motion, so the field need not be zero.

FIGURE 22.5 shows how a wire connected between the plates of a capacitor causes it to discharge. The separation of charges creates a potential difference between the two plates. We saw in Chapter 21 that whenever there's a potential difference, an electric field points from higher potential toward lower potential. Connecting a wire between the plates establishes an electric field in the wire, and this electric field causes electrons to flow from the negative plate (which has an excess of electrons) toward the positive plate. **The potential difference creates the electric field that drives the current in the wire.**

FIGURE 22.5 Creating a current in a wire.

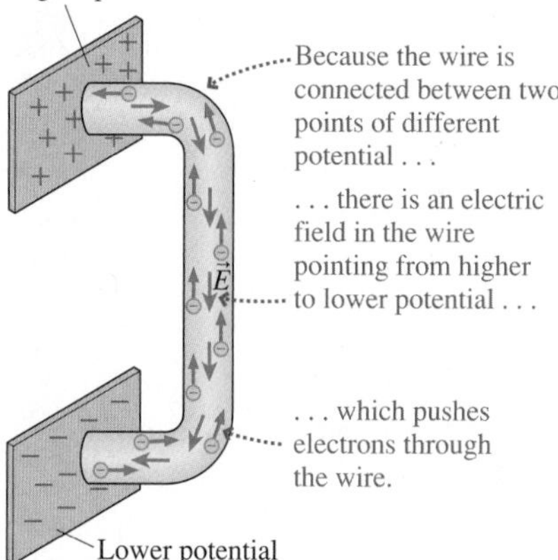

Higher potential
Because the wire is connected between two points of different potential . . .
. . . there is an electric field in the wire pointing from higher to lower potential . . .
$\vec{E}$
. . . which pushes electrons through the wire.
Lower potential

As the current continues, and the charges flow, the plates discharge and the potential difference decreases. At some point, the plates will be completely discharged, meaning no more potential difference, no more field—and no more current. Finally, $\vec{E} = \vec{0}$ inside the conducting wire, and we have equilibrium.

Conservation of Current

In FIGURE 22.6 a lightbulb has been added to the wire connecting two capacitor plates. The bulb glows while the current is discharging the capacitor. How does the current at point A compare to the current at point B? Are the currents at these points the same? Or is one larger than the other?

FIGURE 22.6 How does the current at A compare to the current at B?

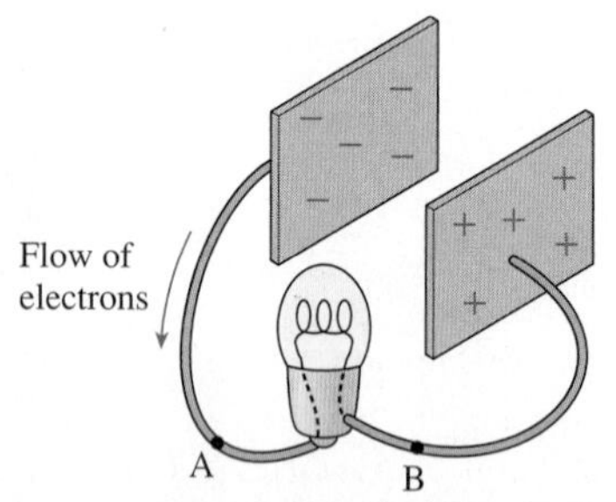

You might have predicted that the current at B is less than the current at A because the bulb, in order to glow, must use up some of the current. It's easy to test this prediction; for instance, we could compare the currents at A and B by comparing how far two compass needles at these positions are deflected. Any such test gives the same result: The current at point B is *exactly equal* to the current at point A. **The current leaving a lightbulb is exactly the same as the current entering the lightbulb.**

This is an important observation, one that demands an explanation. After all, "something" makes the bulb glow, so why don't we observe a decrease in the current? Electrons are charged particles. The lightbulb can't destroy electrons without

Conservation of Current at a Junction

FIGURE 22.11 shows a wire splitting into two and two wires merging into one. A point where a wire branches is called a **junction.** The presence of a junction doesn't change the fact that current is conserved. We cannot create or destroy charges in the wire, and neither can we store them in the junction. The rate at which electrons flow into one *or many* wires must be exactly balanced by the rate at which they flow out of others. For a *junction,* the law of conservation of charge requires that

$$\sum I_{in} = \sum I_{out} \tag{22.3}$$

where, as usual, the $\sum$ symbol means "the sum of."

This basic conservation statement—that the sum of the currents into a junction equals the sum of the currents leaving—is called **Kirchhoff's junction law.** The junction law isn't a new law of physics; it is a consequence of the conservation of charge.

FIGURE 22.11 The sum of the currents into a junction must equal the sum of the currents leaving the junction.

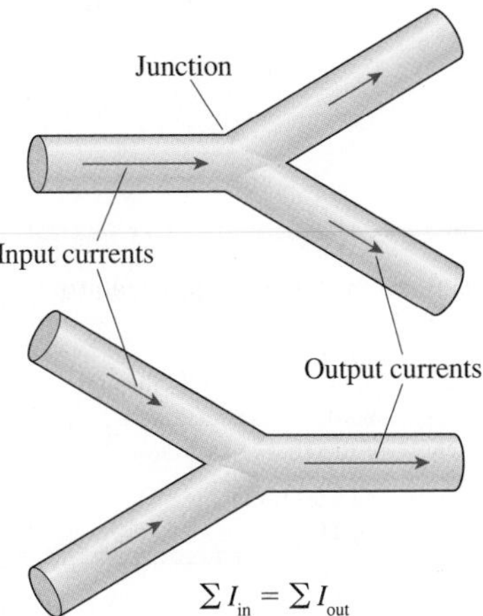

EXAMPLE 22.3 Currents in a junction

Four wires have currents as noted in FIGURE 22.12. What are the direction and the magnitude of the current in the fifth wire?

FIGURE 22.12 The junction of five wires.

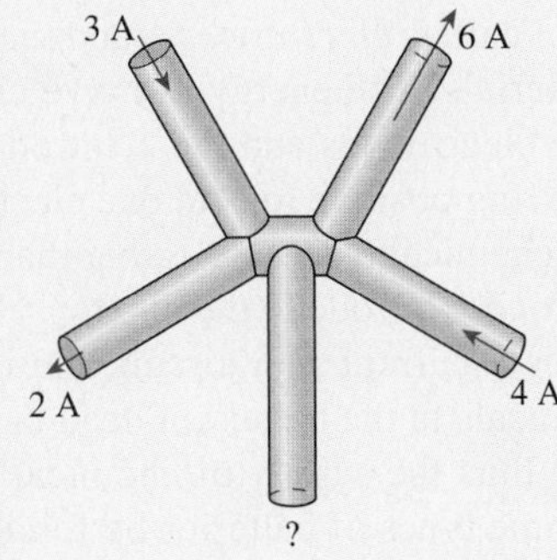

PREPARE This is a conservation of current problem. We compute the sum of the currents coming into the junction and the sum of the currents going out of the junction, and then compare these two sums. The unknown current is whatever is required to make the currents into and out of the junction "balance."

SOLVE Two of the wires have currents into the junction:

$$\sum I_{in} = 3\text{ A} + 4\text{ A} = 7\text{ A}$$

Two of the wires have currents out of the junction:

$$\sum I_{out} = 6\text{ A} + 2\text{ A} = 8\text{ A}$$

To conserve current, the fifth wire must carry a current of 1 A into the junction.

ASSESS If the unknown current is 1 A into the junction, a total of 8 A flows in—exactly what is needed to balance the current going out.

STOP TO THINK 22.1 The discharge of a capacitor lights three bulbs. Comparing the current in bulbs 1 and 2, we can say that

A. The current in bulb 1 is greater than the current in bulb 2.
B. The current in bulb 1 is less than the current in bulb 2.
C. The current in bulb 1 is equal to the current in bulb 2.

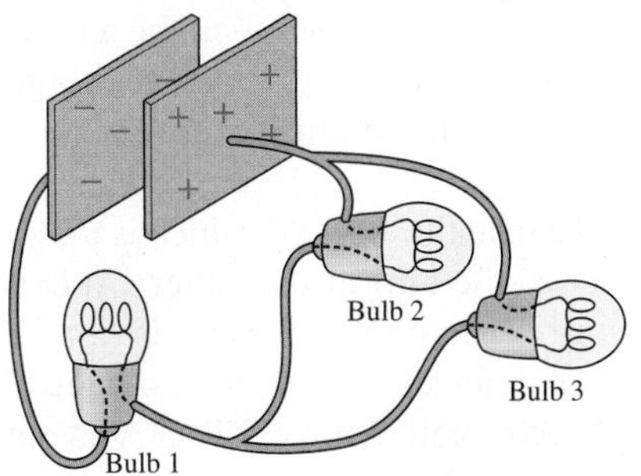

22.3 Batteries and emf

There are practical devices, such as a camera flash, that use the charge on a capacitor to drive a current. But a camera flash gives a single, bright flash of light; the capacitor discharges and the current ceases. If you want a light to illuminate your way along a dark path, you need a *continuous* source of light like a flashlight. Continuous light requires the current to be continuous as well.

FIGURE 22.13 shows a wire connecting the two terminals of a battery, much like the wire that connected the capacitor plates in Figure 22.1. Just like that wire, the wire

FIGURE 22.13 There is a current in a wire connecting the terminals of a battery.

Battery

A lightbulb lights and a compass needle deflects, just as they do for a wire that discharges a capacitor.

connecting the battery terminals gets warm, deflects a compass needle, and makes a lightbulb inserted into it glow brightly. These indicators tell us that charges flow through the wire from one terminal to the other. The current in the wire is the same whether it is supplied by a capacitor or a battery. Everything you've learned so far about current applies equally well to the current supplied by a battery, with one important difference—the duration of the current.

The wire connecting the battery terminals *continues* to deflect the compass needle and *continues* to light the lightbulb. The capacitor quickly runs out of excess charge, but the battery can keep the charges in motion.

FIGURE 22.14 The charge escalator model of a battery.

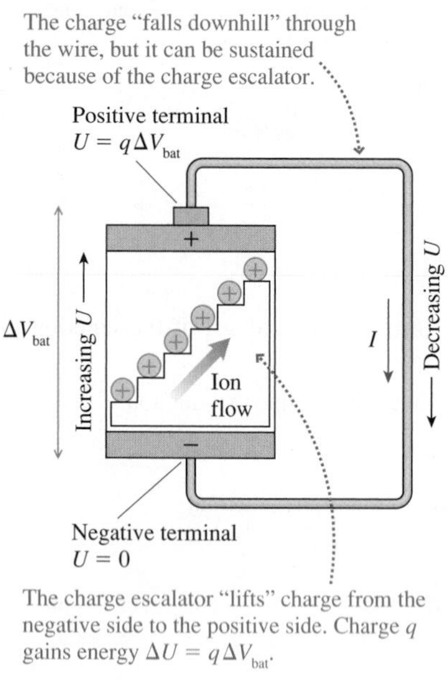

The charge "falls downhill" through the wire, but it can be sustained because of the charge escalator.

Positive terminal $U = q\Delta V_{bat}$

ΔV_{bat} Increasing U Ion flow I Decreasing U

Negative terminal $U = 0$

The charge escalator "lifts" charge from the negative side to the positive side. Charge q gains energy $\Delta U = q\Delta V_{bat}$.

How does a battery produce this sustained motion of charge? A real battery involves a series of chemical reactions, but FIGURE 22.14 shows a simple model of a battery that illustrates the motion of charges. The inner workings of a battery act like a *charge escalator* between the two terminals. Charges are removed from the negative terminal and "lifted" to the positive terminal. It is the charge escalator that sustains the current in the wire by providing a continuously renewed supply of charges at the positive terminal.

Once a charge reaches the positive terminal, it is able to flow downhill through the wire as a current until it reaches the negative terminal. The charge escalator then lifts the charge back to the positive terminal where it can start the loop all over again. This flow of charge in a continuous loop is what we call a **complete circuit.**

The charge escalator in the battery must be powered by some external source of energy. It is lifting the electrons "uphill" against an electric field. A battery consists of chemicals, called *electrolytes,* sandwiched between two electrodes made of different materials. The energy to move charges comes from chemical reactions between the electrolytes and the electrodes. These chemical reactions separate charge by moving positive ions to one electrode and negative ions to the other. In other words, chemical reactions, rather than a mechanical conveyor belt, transport charge from one electrode to the other.

As a battery creates a current in a circuit, the reactions that run the charge escalator deplete chemicals in the battery. A dead battery is one in which the supply of chemicals, and thus the supply of chemical energy, has been exhausted. You can "recharge" some types of batteries by forcing a current into the positive terminal, reversing the chemical reactions that move the charges, thus replenishing the chemicals and storing energy as chemical energy.

By separating charge, the charge escalator establishes the potential difference ΔV_{bat} that is shown between the terminals of the battery in Figure 22.14. Chemical reactions do work W_{chem} to move charge q from the negative to the positive terminal. If there are no internal energy losses, the charge gains electric potential energy $\Delta U = W_{chem}$.

The quantity W_{chem}/q, which is the work done *per charge* by the charge escalator, is called the **emf** of the battery. It is pronounced as the sequence of three letters "e-m-f." The symbol for emf is $\mathcal{E}$, a script E, and its units are those of the electric potential: joules per coulomb, or volts. The term emf was originally an abbreviation of "electromotive force." That is an outdated term, so today we just call it emf and it's not an abbreviation of anything.

The *rating* of a battery, such as 1.5 V, is the battery's emf. It is determined by the specific chemical reactions employed by the battery. An alkaline battery has an emf of 1.5 V; a rechargeable NiCd battery has an emf of 1.2 V. Larger emfs are created by using several smaller "cells" in a row, much like going from the first to the fourth floor by taking three separate escalators.

By definition, the electric potential is related to the electric potential energy of charge q by $\Delta V = \Delta U/q$. But $\Delta U = W_{chem}$ for the charges in the battery; hence the potential difference between the terminals is

$$\Delta V_{bat} = \frac{W_{chem}}{q} = \mathcal{E} \qquad (22.4)$$

TRY IT YOURSELF

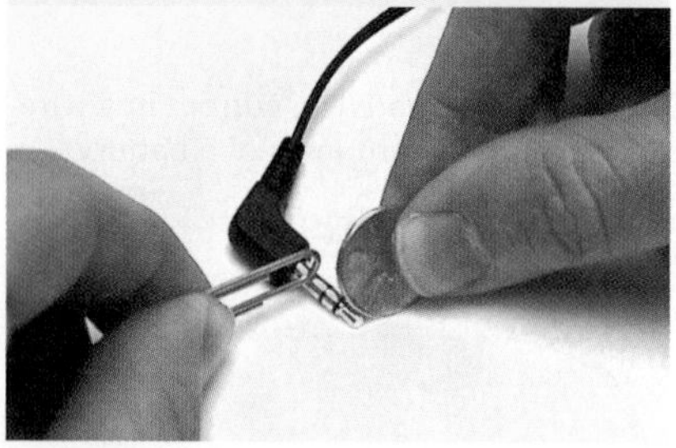

Listen to your potential Put on a set of earphones from a portable music player and place the plug on the table. Moisten your fingertips and hold a penny in one hand and a paper clip in the other. This makes a very weak battery; the penny and the clip are the electrodes and your moist skin the electrolyte. Touch the paper clip to the innermost contact on the headphone plug and the penny to the outermost. You will hear a *very* soft click as the potential difference causes a small current in the headphones.

The potential difference between the terminals of a battery, often called the **terminal voltage,** is ideally the battery's emf. In practice, inevitable energy losses within the battery cause the terminal voltage of a real battery to be slightly less than the emf. We'll overlook this small difference and assume $\Delta V_{bat} = \mathcal{E}$.

Electric generators, photocells, and power supplies use different means to separate charge, but otherwise they function much like a battery. The common feature of all such devices is that **they use some source of energy to separate charge and, thus, to create a potential difference.** The emf of the device is the work done per charge to separate the charge. In contrast, a capacitor stores separated charges, but a capacitor has no means to *do* the separation. Hence a capacitor has a potential difference, but not an emf.

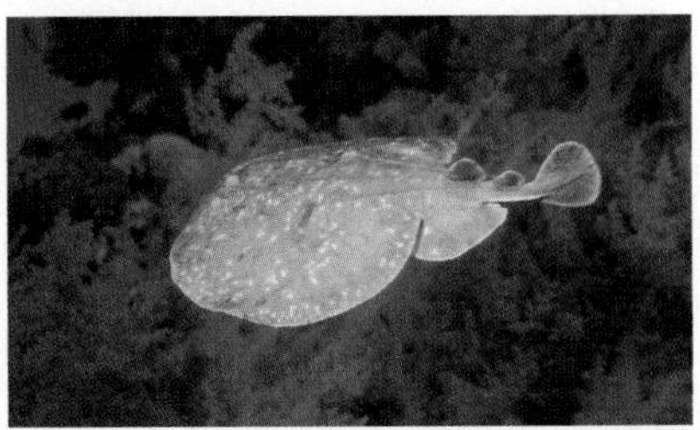

A shocking predator? BIO The torpedo ray captures and eats fish by paralyzing them with electricity. As we will see in Chapter 23, cells in the body use chemical energy to separate charge, just as in a battery. Special cells in the body of the ray called *electrocytes* produce an emf of a bit more than 0.10 V for a short time when stimulated. Such a small emf will not produce a large effect, but the torpedo ray has organs that contain clusters of hundreds of these electrocytes connected in a row. The total emf can be 50 V or more, enough to immobilize nearby prey.

NOTE ► The term *emf,* often capitalized as EMF, is widely used in popular science articles in newspapers and magazines to mean "electromagnetic field." This is *not* how we will use the term *emf.* ◄

CONCEPTUAL EXAMPLE 22.4 Potential difference for batteries in series

Three batteries are connected one after the other as shown in FIGURE 22.15; we say they are connected in *series.* What's the total potential difference?

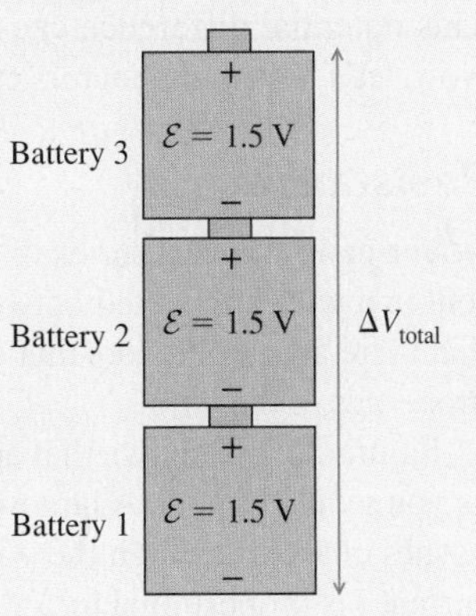

FIGURE 22.15 Three batteries in series.

REASON We can think of this as three charge escalators, one after the other. Each one lifts charges to a higher potential. Because each battery raises the potential by 1.5 V, the total potential difference of the three batteries in series is 4.5 V.

ASSESS Common AA and AAA batteries are 1.5 V batteries. Many consumer electronics, such as digital cameras, use two or four of these batteries. Wires inside the device connect the batteries in series to produce a total 3.0 V or 6.0 V potential difference.

Electricity generators, such as coal-burning power plants, burn fuel (a source of chemical energy), transform the resulting thermal energy into the mechanical energy of a spinning turbine, and then use a generator, which we'll discuss in Chapter 24, to transform the mechanical energy into electricity. There are unavoidable thermodynamic inefficiencies associated with this process, as we learned in Chapter 11.

A more elegant solution to the generation of electricity is the *fuel cell.* A particular fuel cell, one that combines hydrogen fuel with oxygen, is illustrated in FIGURE 22.16. Rather than burning the fuel, as in a power plant, with the resulting thermal losses, the fuel cell's specially designed electrodes allow an electrochemical reaction that transforms the chemical energy of the hydrogen directly into electric energy. A fuel cell thus works like a battery with the chemicals supplied externally. As long as fuel and oxygen are coming in, a fuel cell can produce electricity.

FIGURE 22.16 The operation of a fuel cell.

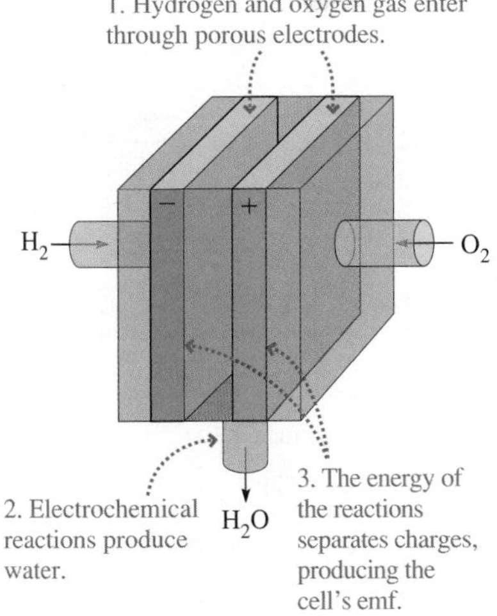

STOP TO THINK 22.2 A battery produces a current in a wire. As the current continues, which of the following quantities (perhaps more than one) decreases?

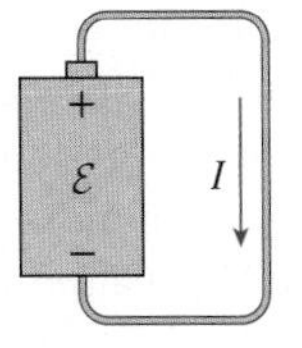

A. The positive charge in the battery
B. The emf of the battery
C. The chemical energy in the battery

FIGURE 22.17 The electric field and the current inside the wire.

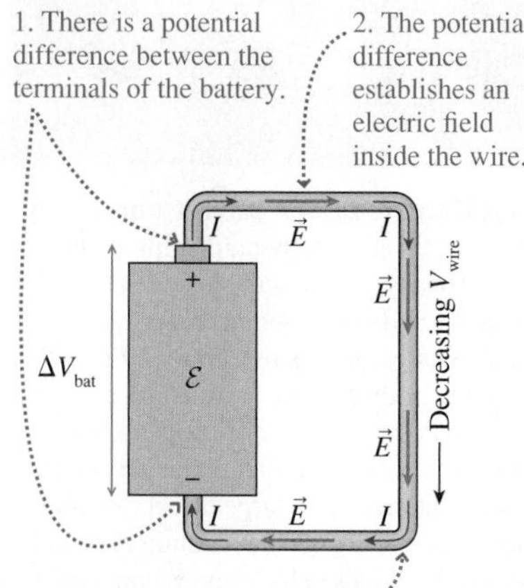

22.4 Connecting Potential and Current

An important conclusion of the charge escalator model is that **a battery is a source of potential difference.** When charges flow through a wire that connects the battery terminals, this current is a *consequence* of the battery's potential difference. You can think of the battery's emf as being the *cause.* Current, heat, light, sound, and so on are all *effects* that happen when the battery is used in certain ways.

Figure 22.14 showed how the charge escalator model explained the motion of charges in a wire connected between the terminals of a battery: The charge escalator in the battery raises the charges "uphill," and then they flow "downhill" through the wire. Let's extend this analysis and look at the connection between the potential difference and the current.

In FIGURE 22.17, a battery produces a potential difference that causes a current in a wire connecting the terminals. You learned in Chapter 21 that the potential difference between any two points is independent of the path between them. Consequently, the potential difference between the two ends of the wire, along a path through the wire, is equal to the potential difference between the two terminals of the battery:

$$\Delta V_{wire} = \Delta V_{bat} \tag{22.5}$$

This potential difference causes a current in the direction of decreasing potential. Now, let's look at the factors that determine the magnitude of this current.

Resistance

FIGURE 22.18 Factors affecting the current in a wire.

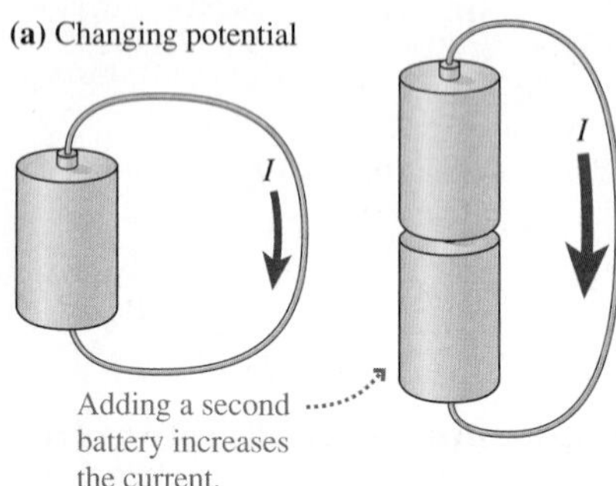

(a) Changing potential
I
Adding a second battery increases the current.

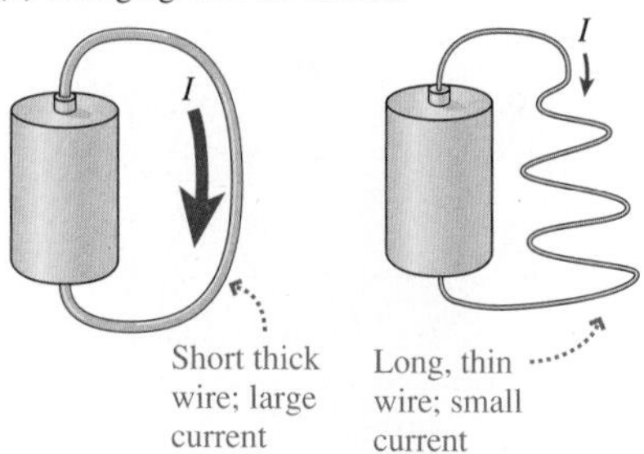

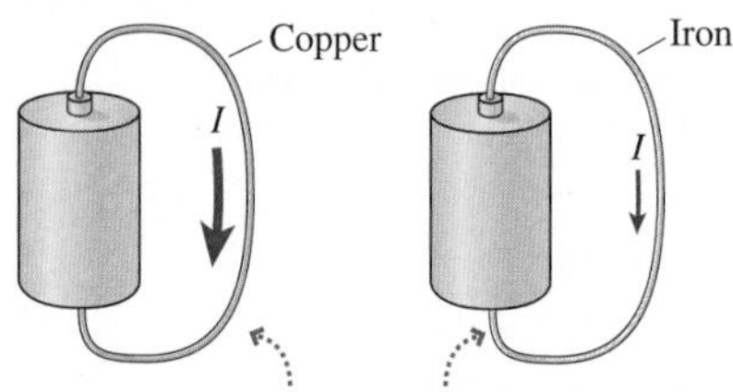

FIGURE 22.18 shows a series of experiments to determine what factors affect the current in a wire connected between the terminals of a battery. The experiments show that there are two factors that determine the current: the potential difference and the properties of the wire.

Figure 22.18a shows that adding a second battery in series increases the current, as you would expect. A larger potential difference creates a larger electric field that pushes charges through the wire faster. Careful measurements would show that the current I is proportional to ΔV_{wire}.

Figures 22.18b and 22.18c illustrate the two properties of the wire that affect the current: the dimensions and the material of which the wire is made. Figure 22.18b shows that increasing the length of the wire decreases the current, while increasing the thickness of the wire increases the current. This seems reasonable because it should be harder to push charges through a long wire than a short one, and an electric field should be able to push more charges through a fat wire than a skinny one. Figure 22.18c shows that wires of different materials will carry different currents—some materials are better conductors than others.

For any particular wire, we can define a quantity called the **resistance** that is a measure of how hard it is to push charges through the wire. We use the symbol R for resistance. A large resistance implies that it is hard to move the charges through the wire; in a wire with small resistance, the charges move much more easily. The current in the wire depends on the potential difference ΔV_{wire} between the ends of the wire and the wire's resistance R:

$$I = \frac{\Delta V_{wire}}{R} \tag{22.6}$$

Establishing a potential difference ΔV_{wire} between the ends of a wire of resistance R creates an electric field that, in turn, causes a current $I = \Delta V_{wire}/R$ in the wire. As we would expect, the smaller the resistance, the larger the current.

We can think of Equation 22.6 as the definition of resistance. If a potential difference V_{wire} causes current I in a wire, the wire's resistance is

$$R = \frac{\Delta V_{wire}}{I} \tag{22.7}$$

The SI unit of resistance is the **ohm,** defined as

$$1 \text{ ohm} = 1\ \Omega = 1 \text{ V/A}$$

where Ω is an uppercase Greek omega. The unit takes its name from the German physicist Georg Ohm. The ohm is the basic unit of resistance, although kilohms ($1\ \text{k}\Omega = 10^3\ \Omega$) and megohms ($1\ \text{M}\Omega = 10^6\ \Omega$) are widely used.

EXAMPLE 22.5 **Resistance of a lightbulb**

The glowing element in an incandescent lightbulb is the *filament,* a long, thin piece of tungsten wire that is heated by the electric current through it. When connected to the 120 V of an electric outlet, a 60 W bulb carries a current of 0.50 A. What is the resistance of the filament in the lamp?

SOLVE We can use Equation 22.7 to compute the resistance:

$$R = \frac{\Delta V_{\text{wire}}}{I} = \frac{120 \text{ V}}{0.50 \text{ A}} = 240\ \Omega$$

ASSESS As we will see below, the resistance of the filament varies with temperature. This value holds for the lightbulb only when the bulb is glowing and the filament is hot.

Resistivity

Figure 22.18c showed that the resistance of a wire depends on what it is made of. We define a quantity called **resistivity,** for which we use the symbol ρ (lowercase Greek rho), to characterize the electrical properties of materials. Materials that are good conductors have low resistivity; materials that are poor conductors (and thus that are good insulators) have high resistivity. The resistivity ρ has units of $\Omega \cdot \text{m}$. The resistivities of some common materials are listed in Table 22.1.

Metals are generally good conductors (and so have very low resistivity), but metals such as copper are much better conductors than metals such as nichrome, an alloy of nickel and chromium that is used to make heating wires. Water is a poor conductor, but the dissolved salts in seawater produce ions that can carry charge, so seawater is a good conductor, with a resistivity one million times less than that of pure water. Glass is an excellent insulator with a resistivity in excess of $10^{14}\ \Omega \cdot \text{m}$, 10^{22} times that of copper.

The resistivity of a material depends on the temperature, as you can see from the two values for tungsten listed in Table 22.1. As the temperature increases, so do the thermal vibrations of the atoms. This makes them "bigger targets" for the moving electrons, causing collisions to be more frequent. Thus the resistivity of a metal increases with increasing temperature.

The resistance of a wire depends both on the resistivity of its material and on the dimensions of the wire. A wire made of a material of resistivity ρ, with length L and cross-section area A, has resistance

TABLE 22.1 Resistivity of materials

Material	Resistivity ($\Omega \cdot \text{m}$)
Copper	1.7×10^{-8}
Tungsten (20°C)	5.6×10^{-8}
Tungsten (1500°C)	5.0×10^{-7}
Iron	9.7×10^{-8}
Nichrome	1.5×10^{-6}
Seawater	0.22
Blood (average)	1.6
Muscle	13
Fat	25
Pure water	2.4×10^{5}
Cell membrane	3.6×10^{7}

$$R = \frac{\rho L}{A} \qquad (22.8)$$

PROPORTIONAL p. 37 INVERSE p. 114

Resistance of a wire in terms of resistivity and dimensions

Resistance is a property of a *specific* wire or conductor because it depends on the conductor's length and diameter as well as on the resistivity of the material from which it is made.

NOTE ► It is important to distinguish between resistivity and resistance. *Resistivity* is a property of the *material,* not any particular piece of it. All copper wires (at the same temperature) have the same resistivity. *Resistance* characterizes a specific piece of the conductor having a specific geometry. A short, thick copper wire has a smaller resistance than a long, thin copper wire. The relationship between resistivity and resistance is analogous to that between density and mass. ◄

EXAMPLE 22.6 The length of a lightbulb filament

We calculated in Example 22.5 that a 60 W lightbulb has a resistance of 240 Ω. At the operating temperature of the tungsten filament, the resistivity is approximately $5.0 \times 10^{-7}\ \Omega \cdot \text{m}$. If the wire used to make the filament is 0.040 mm in diameter (a typical value), how long must the filament be?

PREPARE The resistance of a wire depends on its length, its cross-section area, and the material of which it is made.

SOLVE The cross-section area of the wire is $A = \pi r^2 = \pi(2.0 \times 10^{-5}\ \text{m})^2 = 1.26 \times 10^{-9}\ \text{m}^2$. Rearranging Equation 22.8 shows us that the filament must be of length

$$L = \frac{AR}{\rho} = \frac{(1.26 \times 10^{-9}\ \text{m}^2)(240\ \Omega)}{5.0 \times 10^{-7}\ \Omega \cdot \text{m}} = 0.60\ \text{m}$$

ASSESS This is quite long—nearly two feet. This result may seem surprising, but some reflection shows that it makes sense. The resistivity of tungsten is low, so the filament must be quite thin and long.

◄**Coils of coils** A close view of a typical lightbulb's filament shows that it is made of very thin wire that is coiled and then coiled again. The double-coil structure is necessary to fit the great length of the filament into the small space of the bulb's globe.

EXAMPLE 22.7 Making a heater

An amateur astronomer uses a heater to warm her telescope eyepiece so moisture does not collect on it. The heater is a 20-cm-long, 0.50-mm-diameter nichrome wire that wraps around the eyepiece. When the wire is connected to a 1.5 V battery, what is the current in the wire?

PREPARE The current in the wire depends on the emf of the battery and the resistance of the wire. The resistance of the wire depends on the resistivity of nichrome, given in Table 22.1, and the dimensions of the wire. Converted to meters, the relevant dimensions of the wire are $L = 0.20$ m and $r = 2.5 \times 10^{-4}$ m.

SOLVE The wire's resistance is

$$R = \frac{\rho L}{A} = \frac{\rho L}{\pi r^2} = \frac{(1.5 \times 10^{-6}\ \Omega \cdot \text{m})(0.20\ \text{m})}{\pi(2.5 \times 10^{-4}\ \text{m})^2} = 1.5\ \Omega$$

The wire is connected to the battery, so $\Delta V_{\text{wire}} = \Delta V_{\text{bat}} = 1.5$ V. The current in the wire is

$$I = \frac{\Delta V_{\text{wire}}}{R} = \frac{1.5\ \text{V}}{1.5\ \Omega} = 1.0\ \text{A}$$

ASSESS The emf of the battery is small, but so is the resistance of the wire, so this is a reasonable current, enough to warm the wire and the eyepiece.

Electrical Measurements of Physical Properties

Measuring resistance is quite straightforward. Because resistance depends sensitively on the properties of materials, a measurement of resistance can be a simple but effective probe of other quantities of interest. For example, the resistivity of water is strongly dependent on dissolved substances in the water, so it is easy to make a quick test of water purity by making a measurement of resistivity.

CONCEPTUAL EXAMPLE 22.8 **Testing drinking water**

A house gets its drinking water from a well that has an intermittent problem with salinity. Before the water is pumped into the house, it passes between two electrodes in the circuit shown in **FIGURE 22.19**. The current passing through the water is measured with a meter. Which corresponds to increased salinity—an increased current or a decreased current?

REASON Increased salinity causes the water's resistivity to decrease. This decrease causes a decrease in resistance between the electrodes. Current is inversely proportional to resistance, so this leads to an increase in current.

ASSESS Increasing salinity means more ions in solution and thus more charge carriers, so an increase in current is expected. Electrical systems similar to this can therefore provide a quick check of water purity.

FIGURE 22.19 A water-testing circuit.

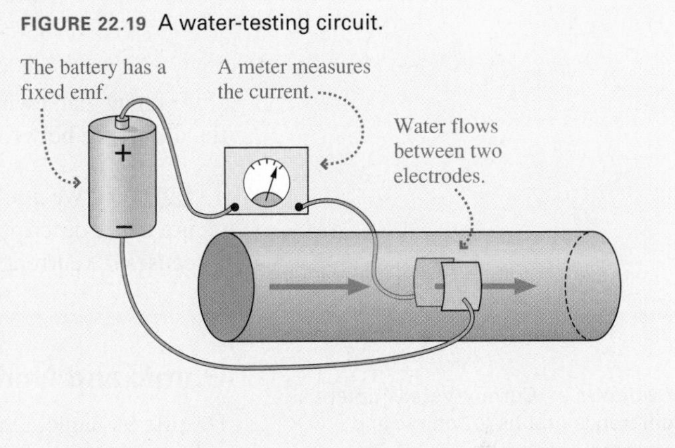
The battery has a fixed emf.
A meter measures the current.
Water flows between two electrodes.
+
−

Different tissues in the body have different resistivities, as we see in Table 22.1. For example, fat has a higher resistivity than muscle. Consequently, a routine test to estimate the percentage of fat in a person's body is based on a measurement of the body's resistance, as illustrated in the photo at the start of the chapter. A higher resistance of the body means a higher proportion of fat.

More careful measurements of resistance can provide more detailed diagnostic information. Passing a small, safe current between pairs of electrodes on opposite sides of a person's torso permits a measurement of the resistance of the intervening tissue, a technique known as *electrical impedance tomography*. (Impedance is similar to resistance, but it also applies to AC circuits, which we will explore in Chapter 26.) **FIGURE 22.20** shows an image of a patient's torso generated from measurements of resistance between many pairs of electrodes. The image shows the change in resistance between two subsequent measurements; decreasing resistance shows in red, increasing resistance in blue. This image was created during the resting phase of the heart, when blood was leaving the lungs and entering the heart. Blood is a better conductor than the tissues of the heart and lungs, so the motion of blood decreased the resistance of the heart and increased that of the lungs. This patient was healthy, but in a patient with circulatory problems any deviation from normal blood flow would lead to abnormal patterns of resistance that would be revealed in such an image.

FIGURE 22.20 An electrical impedance map showing the cross section of a healthy patient's torso.

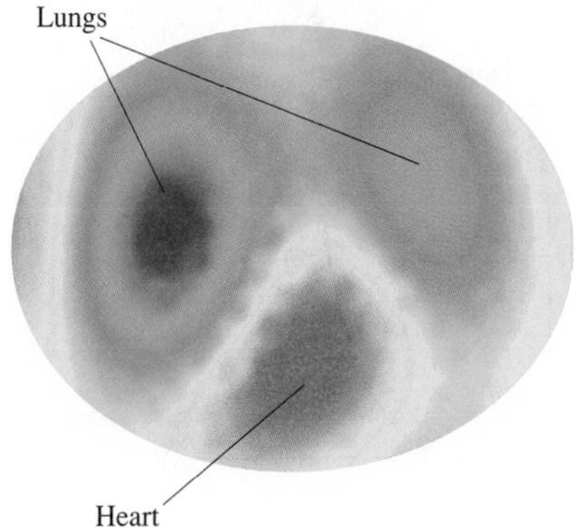
Lungs
Heart

STOP TO THINK 22.3 A wire connected between the terminals of a battery carries a current. The wire is removed and stretched, decreasing its cross-section area and increasing its length. When the wire is reconnected to the battery, the new current is

A. Larger than the original current.
B. The same as the original current.
C. Smaller than the original current.

22.5 Ohm's Law and Resistor Circuits

The relationship between the potential difference across a conductor and the current passing through it that we saw in the preceding section was first deduced by Georg Ohm and is known as **Ohm's law:**

$$I = \frac{\Delta V}{R} \qquad (22.9)$$

Ohm's law for a conductor of resistance R

PROPORTIONAL p. 37 INVERSE p. 114

If we know that a wire of resistance R carries a current I, we can compute the potential difference between the ends of the wire as $\Delta V = IR$.

NOTE ► We could write the equation for Ohm's law as $\Delta V = IR$, but $I = \Delta V/R$ is a better description of cause and effect because it is the potential difference that causes the current. ◄

Ohmic and Nonohmic Materials

FIGURE 22.21 Current-versus-potential-difference graphs for ohmic and nonohmic materials.

(a) Ohmic material: copper wire

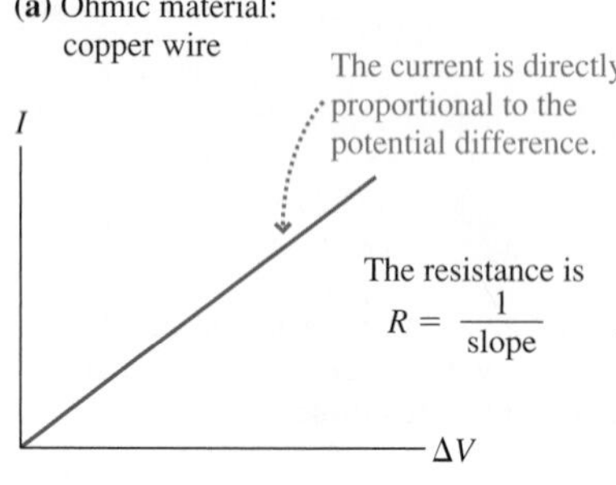
I
The current is directly proportional to the potential difference.
The resistance is $R = \frac{1}{\text{slope}}$
ΔV

(b) Nonohmic material: diode (semiconductor device)

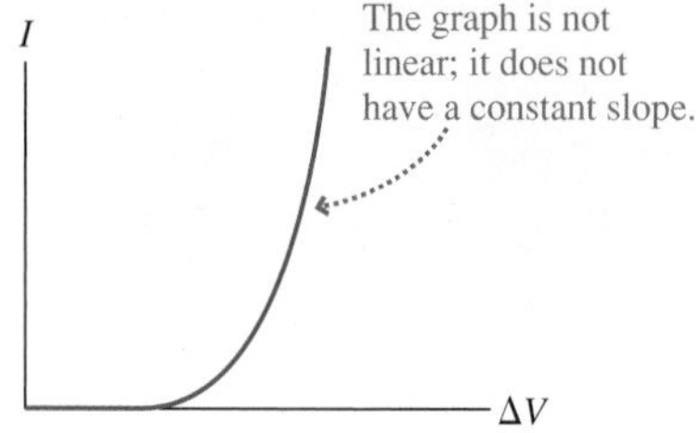
I
The graph is not linear; it does not have a constant slope.
ΔV

Despite its name, Ohm's law is *not* a law of nature. It is limited to those materials whose resistance R remains constant—or very nearly so—during use. Materials to which Ohm's law applies are called **ohmic.** FIGURE 22.21a shows that the current through an ohmic material is directly proportional to the potential difference; doubling the potential difference results in a doubling of the current. This is a linear relationship, and the resistance R can be determined from the slope of the graph.

Many materials are ohmic over a reasonable range of operating conditions. The resistance of metals varies slightly with temperature, but a metal wire is ohmic as long as the temperature is reasonably constant, and we can give it a fixed resistance value.

Other materials and devices are **nonohmic,** meaning that the current through the device is *not* directly proportional to the potential difference. An example is a semiconductor device known as a *diode.* The graph in FIGURE 22.21b shows that a diode does not have a well-defined resistance. Three important examples of nonohmic devices are:

1. Batteries, where $\Delta V = \mathcal{E}$ is determined by chemical reactions, independent of I
2. Semiconductor devices, where the I-versus-ΔV curve is far from linear
3. Capacitors, where, as you'll learn in Chapter 26, the relationship between I and ΔV is very different from that of a resistor

The main point to remember is that Ohm's law does *not* apply to these nonohmic devices.

Resistors

The word "resistance" may have negative connotations—who needs something that slows charges and robs energy? In some cases resistance *is* undesirable. But in many other cases, circuit elements are designed to have a certain resistance for very practical reasons. We call these circuit elements **resistors.** There are a few basic types that will be very important as we start to look at electric circuits in detail.

Types of resistors

Power resistors

We've seen that passing a current through a wire increases the temperature. This transformation of electric energy into thermal energy is the basis for many practical devices. The nichrome wire in this toaster is a resistor that gets hot enough to glow. Hotter wires produce the visible light of incandescent bulbs; cooler wires defrost the rear windows of cars.

Fixed resistors

If you open up an electronic device and look inside, you'll see a circuit board with many cylinders with colored bands. These cylinders are resistors. Each has a specified value of resistance that is revealed by the colors of the bands. We'll see examples of the use of resistors in timing circuits and other applications in the next chapter.

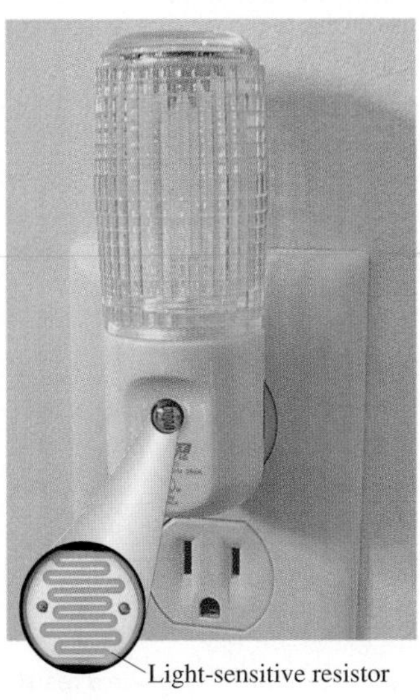

Variable resistors

Not all resistors have a fixed value of resistance; in many circuits and devices a variable resistor allows for changing circumstances. A volume control on a stereo is a variable resistor; so is the sensor on this nightlight. The resistance decreases when light shines on it. During the day, the resistance is low; at night, the resistance rises. A circuit monitors the resistance and switches on the light when the resistance is above a certain value.

CONCEPTUAL EXAMPLE 22.9 The changing current in a toaster

When you press the lever on a toaster, a switch connects the heating wires to 120 V. The wires are initially cool, but the current in the wires raises the temperature until they are hot enough to glow. As the wire heats up, how does the current in the toaster change?

REASON As the wire heats up, its resistivity increases, as noted above, so the resistance of the wires increases. Because the potential difference stays the same, an increasing resistance causes the current to decrease. The current through a toaster is largest when the toaster is first turned on.

ASSESS This result makes sense. As the wire's temperature increases, the current decreases. This makes the system stable. If, instead, the current increased as the temperature increased, higher temperature could lead to more current, leading to even higher temperatures, and the toaster could overheat.

Analyzing a Simple Circuit

FIGURE 22.22a shows the anatomy of a lightbulb. The important point is that a lightbulb, like a wire, has two "ends" and current passes *through* the bulb. Connections to the filament in the bulb are made at the tip and along the side of the metal cylinder. It is often useful to think of a lightbulb as a resistor that happens to give off light when a current is present. Now, let's look at a circuit using a battery, a lightbulb, and wires to make connections, as in FIGURE 22.22b. This is the basic circuit in a flashlight.

A typical flashlight bulb has a resistance of $\approx 3\ \Omega$, while a wire that one would use to connect such a bulb to a battery has a resistance of $\approx 0.01\ \Omega$. The resistance of the wires is so much less than that of the bulb that we can, with very little error,

FIGURE 22.22 The basic circuit of a battery and a bulb.

(a)

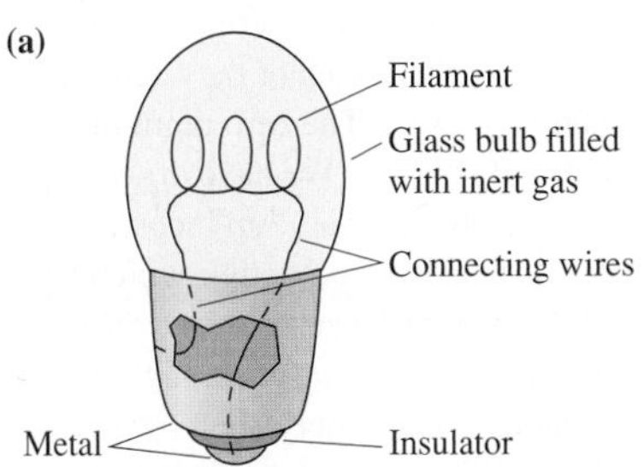

(b)

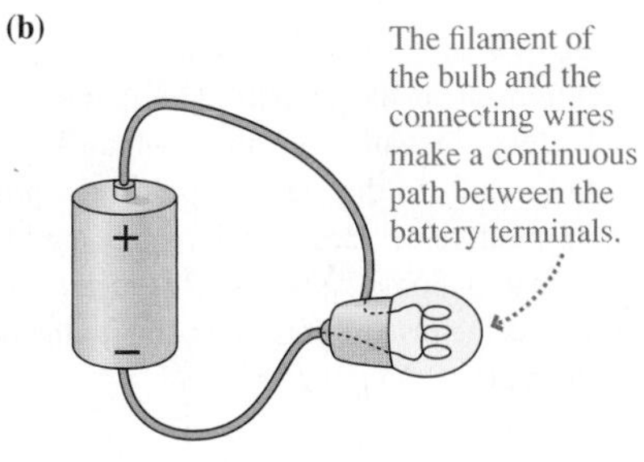

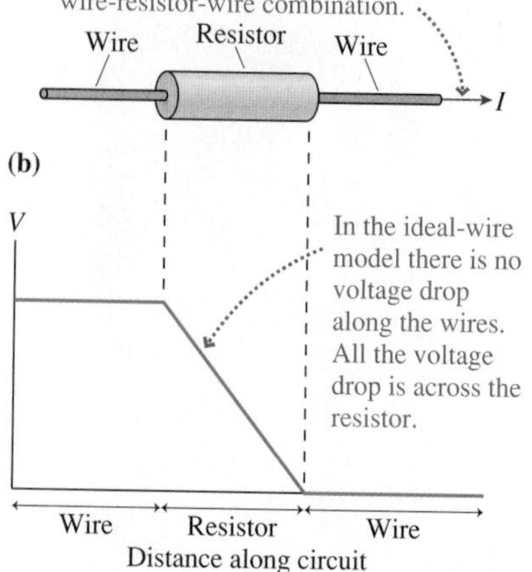

FIGURE 22.23 The potential along a wire-resistor-wire combination.

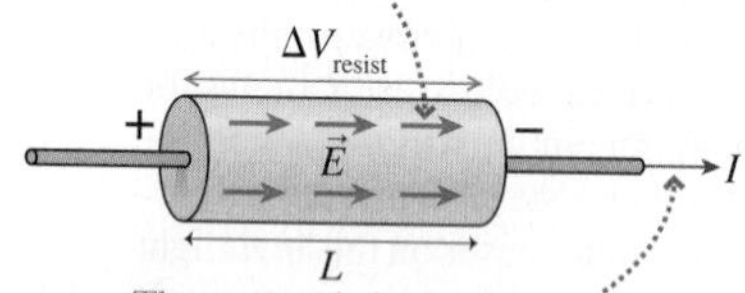

FIGURE 22.24 Electric field inside a resistor.

adopt the *ideal-wire model* and assume that any connecting wires in a circuit are ideal. An **ideal wire** has $R = 0\ \Omega$; hence the potential difference between the ends of an ideal wire is $\Delta V = 0$ V *even if there is a current in it.*

NOTE ▶ We know that, physically, the potential difference can't be zero. There must be an electric field in the wire for the charges to move, so it must have a potential difference. But in practice this potential difference is so small that we can assume it to be zero with little error. ◀

FIGURE 22.23 shows how the ideal-wire model is used in the analysis of circuits. The resistor in Figure 22.23a is connected at each end to a wire, and current I flows through all three. The current requires a potential difference $\Delta V_{\text{resist}} = IR_{\text{resist}}$ across the resistor, but there's no potential difference ($\Delta V_{\text{wire}} = 0$) for the ideal wires. Figure 22.23b shows this idea graphically by displaying the potential along the wire-resistor-wire combination. Current moves in the direction of decreasing potential, so there is a large *voltage drop*—a decrease in potential—across the resistor as we go from left to right, the direction of the current. The segments of the graph corresponding to the wires are horizontal because there's no voltage change along an ideal wire.

The linear variation in the potential across the resistor is similar to the linear variation in potential between the plates of a parallel-plate capacitor. In a capacitor, this linear variation in potential corresponds to a uniform electric field; the same will be true here. As we see in **FIGURE 22.24**, the electric field in a resistor carrying a current in a circuit is uniform; the strength of the electric field is

$$E = \frac{\Delta V}{L}$$

in analogy to Equation 21.6 for a parallel-plate capacitor. A larger potential difference corresponds to a larger field, as we would expect.

EXAMPLE 22.10 Analyzing a single-resistor circuit

A 15 Ω resistor is connected to the terminals of a 1.5 V battery.

a. Draw a graph showing the potential as a function of distance traveled through the circuit, starting from $V = 0$ V at the negative terminal of the battery.
b. What is the current in the circuit?

PREPARE To help us visualize the change in potential as charges move through the circuit, we begin with the sketch of the circuit in **FIGURE 22.25**. The zero point of potential is noted. We have drawn our sketch so that "up" corresponds to higher potential, which will help us make sense of the circuit. Charges are raised to higher potential in the battery, then travel "downhill" from the positive terminal through the resistor and back to the negative terminal. We assume ideal wires.

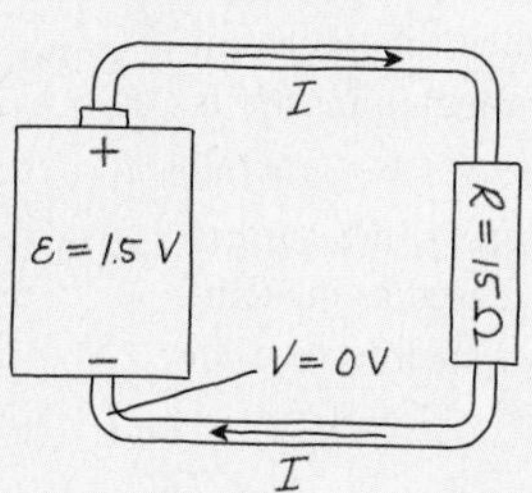

FIGURE 22.25 A single-resistor circuit.

SOLVE a. **FIGURE 22.26** is a graphical representation of the potential in the circuit. The distance s is measured from the battery's negative terminal, where $V = 0$ V. As we move around the circuit to the starting point, the potential must return to its original value. Because the wires are ideal, there is no change in potential along the wires. This means that the potential difference across the resistor must be equal to the potential difference across the battery: $\Delta V_R = \mathcal{E} = 1.5$ V.

FIGURE 22.26 Potential-versus-position graph.

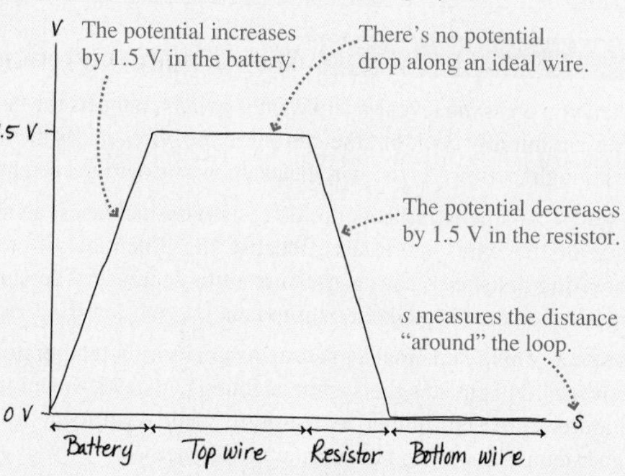

V
The potential increases by 1.5 V in the battery.
There's no potential drop along an ideal wire.
1.5 V
The potential decreases by 1.5 V in the resistor.
s measures the distance "around" the loop.
0 V
s
Battery
Top wire
Resistor
Bottom wire

b. Now that we know the potential difference of the resistor, we can compute the current in the resistor by using Ohm's law:

$$I = \frac{\Delta V_R}{R} = \frac{1.5\ \text{V}}{15\ \Omega} = 0.10\ \text{A}$$

Because current is conserved, this is the current at any point in the circuit. In other words, the battery's charge escalator lifts charge at the rate 0.10 C/s, and charge flows through the wires and the resistor at the rate 0.10 C/s.

ASSESS This is a reasonable value of the current in a battery-powered circuit.

As we noted, there are many devices whose resistance varies as a function of a physical variable that we might like to measure, such as light intensity, temperature, or sound intensity. As the following example shows, we can use resistance measurements to monitor a physical variable.

EXAMPLE 22.11 Using a thermistor

A thermistor is a device whose resistance varies with temperature in a well-defined way. A certain thermistor has a resistance of 2.8 kΩ at 20°C and 0.39 kΩ at 70°C. This thermistor is used in a water bath in a lab to monitor the temperature. The thermistor is connected in a circuit with a 1.5 V battery, and the current measured. What is the change in current in the circuit as the temperature rises from 20°C to 70°C?

SOLVE We can use Ohm's law to find the current in each case:

$$I(20°\text{C}) = \frac{\Delta V}{R} = \frac{1.5\ \text{V}}{2.8 \times 10^3\ \Omega} = 0.54\ \text{mA}$$

$$I(70°\text{C}) = \frac{\Delta V}{R} = \frac{1.5\ \text{V}}{0.39 \times 10^3\ \Omega} = 3.8\ \text{mA}$$

The change in current is thus 3.3 mA.

ASSESS A modest change in temperature leads to a large change in current, which is reasonable—this is a device intended to provide a sensitive indication of a temperature change.

STOP TO THINK 22.4 Two identical batteries are connected in series in a circuit with a single resistor. $V = 0$ V at the negative terminal of the lower battery. Rank in order, from highest to lowest, the potentials V_A to V_E at the labeled points, noting any ties. Assume the wires are ideal.

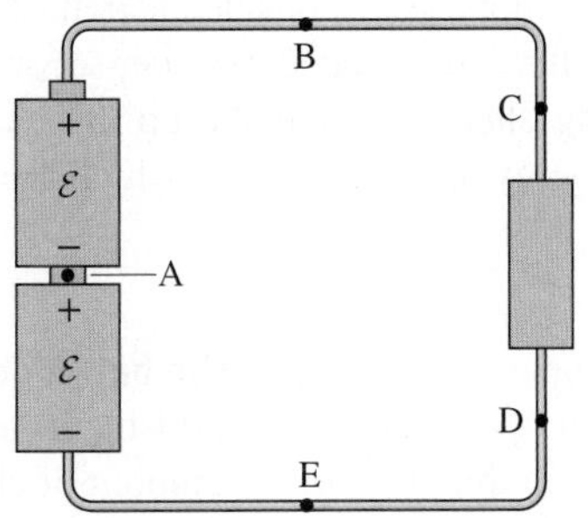

22.6 Energy and Power

When you flip the switch on a flashlight, a battery is connected to a lightbulb, which then begins to glow. The bulb is radiating energy. Where does this energy come from?

A battery not only supplies a potential difference but also supplies energy, as shown in the battery and bulb circuit of **FIGURE 22.27**. The charge escalator is an energy-transfer process, transferring the chemical energy E_{chem} stored in the battery to the electric potential energy U of the charges. That energy is then dissipated as the charges move through the lightbulb, keeping the filament warm and glowing.

Recall that charge q gains potential energy $\Delta U = q\,\Delta V$ as it moves through a potential difference ΔV. The potential difference of a battery is $\Delta V_{\text{bat}} = \mathcal{E}$, so the battery supplies energy $\Delta U = q\mathcal{E}$ to charge q as it lifts the charge up the charge escalator from the negative to the positive terminal.

It's more useful to know the *rate* at which the battery supplies energy. You learned in Chapter 10 that the rate at which energy is transformed is *power,* measured in joules per second or *watts.* Suppose an amount of charge Δq moves through the battery in a time Δt. The charge Δq will increase its potential energy by $\Delta U = (\Delta q)\mathcal{E}$. The *rate* at which energy is transferred from the battery to the moving charges is

$$P_{\text{bat}} = \text{rate of energy transfer} = \frac{\Delta U}{\Delta t} = \frac{\Delta q}{\Delta t}\mathcal{E} \tag{22.10}$$

FIGURE 22.27 Energy transformations in a circuit with a battery and a lightbulb.

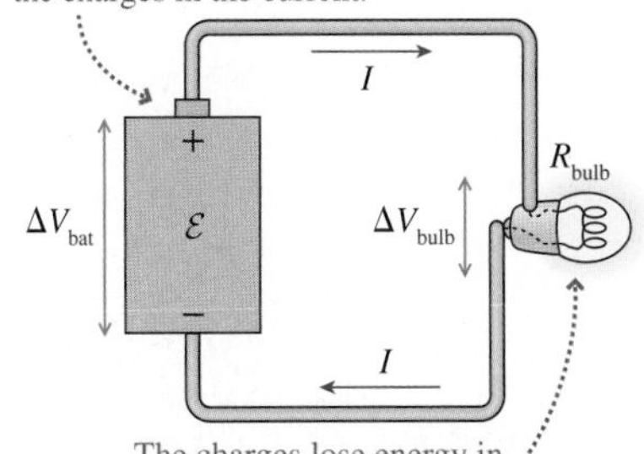

Chemical energy in the battery is transferred to potential energy of the charges in the current.
I
R_{bulb}
ΔV_{bat}
+
ℰ
−
ΔV_{bulb}
I
The charges lose energy in collisions as they pass through the filament of the bulb. This energy is transformed into the thermal energy of the glowing filament.

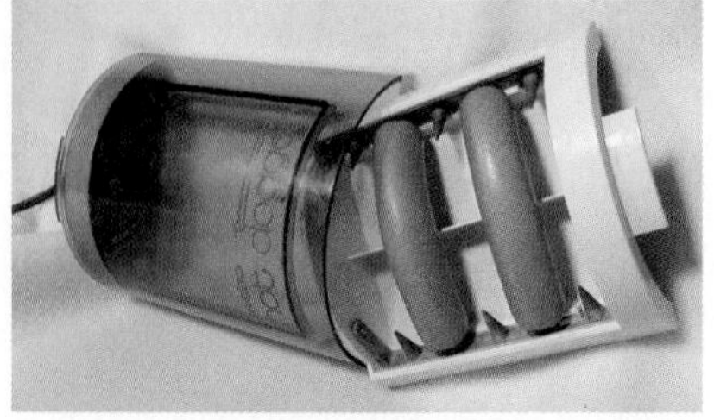

Hot dog resistors Before microwave ovens were common, there were devices that used a decidedly lower-tech approach to the rapid cooking of hot dogs. Prongs connected the hot dog to the 120 V of household electricity, making it the resistor in a circuit. The current through the hot dog dissipated energy as thermal energy, cooking the hot dog in about 2 minutes.

But $\Delta q/\Delta t$, the rate at which charge moves through the battery, is the current I. Hence the power supplied by a battery or any emf is

$$P_{\text{emf}} = I\mathcal{E} \tag{22.11}$$

Power delivered by an emf

$I\mathcal{E}$ has units of J/s, or W.

EXAMPLE 22.12 Power delivered by a car battery

A car battery has $\mathcal{E} = 12$ V. When the car's starter motor is running, the battery current is 320 A. What power does the battery supply?

SOLVE The power is the product of the emf of the battery and the current:

$$P_{\text{bat}} = I\mathcal{E} = (320 \text{ A})(12 \text{ V}) = 3.8 \text{ kW}$$

ASSESS This is a lot of power (about 5 hp), but this amount makes sense because turning over a car's engine is hard work. Car batteries are designed to reliably provide such intense bursts of power for starting the engine.

FIGURE 22.28 The power from the battery is dissipated in the resistor.

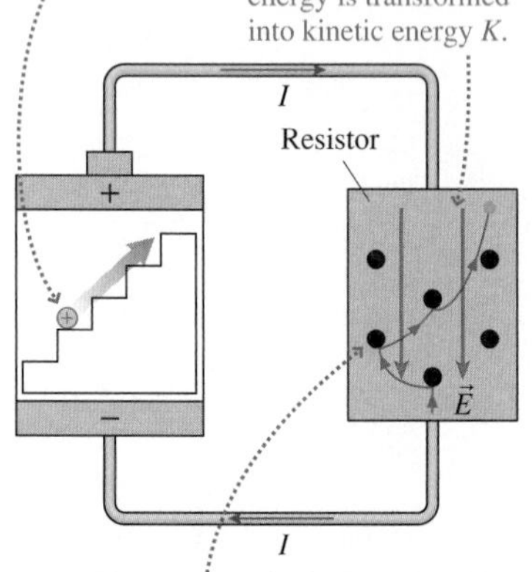

1. Charges gain potential energy U in the battery.
2. As charges accelerate in the electric field in the resistor, potential energy is transformed into kinetic energy K.
I
Resistor
+
−
$\vec{E}$
I
3. Collisions with atoms in the resistor transform the kinetic energy of the charges into thermal energy E_{th} of the resistor.

Suppose we consider a circuit consisting of a battery and a single resistor. $P_{\text{bat}} = I\mathcal{E}$ is the energy transferred per second from the battery's store of chemicals to the moving charges that make up the current. FIGURE 22.28 shows the entire sequence of energy transformations, which looks like

$$E_{\text{chem}} \rightarrow U \rightarrow K \rightarrow E_{\text{th}}$$

The net result is that **the battery's chemical energy is transferred to the thermal energy of the resistor,** raising its temperature.

In the resistor, the amount of charge Δq loses potential energy $\Delta U = (\Delta q)(\Delta V_{\text{R}})$ as this energy is transformed into kinetic energy and then into the resistor's thermal energy. Thus the rate at which energy is transferred from the current to the resistor is

$$P_{\text{R}} = \frac{\Delta U}{\Delta t} = \frac{\Delta q}{\Delta t}\Delta V_{\text{R}} = I\,\Delta V_{\text{R}} \tag{22.12}$$

We say that this power—so many joules per second—is *dissipated* by the resistor as charge flows through it.

Our analysis of the single-resistor circuit in Example 22.10 found that $\Delta V_{\text{R}} = \mathcal{E}$. That is, the potential difference across the resistor is exactly the emf supplied by the battery. Because the current is the same in the battery and the resistor, a comparison of Equations 22.11 and 22.12 shows that

$$P_{\text{R}} = P_{\text{bat}} \tag{22.13}$$

The power dissipated in the resistor is exactly equal to the power supplied by the battery. The *rate* at which the battery supplies energy is exactly equal to the *rate* at which the resistor dissipates energy. This is, of course, exactly what we would have expected from energy conservation.

Most household appliances, such as a 100 W lightbulb or a 1500 W hair dryer, have a power rating. These appliances are intended for use at a standard household voltage of 120 V, and their rating is the power they will dissipate if operated with a potential difference of 120 V. Their power consumption will differ from the rating if they are operated at any other potential difference—for instance, if you use a lightbulb with a dimmer switch.

EXAMPLE 22.13 Finding the current in a lightbulb

How much current is "drawn" by a 75 W lightbulb connected to a 120 V outlet?

PREPARE We can model the lightbulb as a resistor.

SOLVE Because the lightbulb is operating as intended, it will dissipate 75 W of power. We can rearrange Equation 22.12 to find

$$I = \frac{P_R}{\Delta V_R} = \frac{75\ \text{W}}{120\ \text{V}} = 0.63\ \text{A}$$

ASSESS We've said that we expect currents on the order of 1 A for lightbulbs and other household items, so our result seems reasonable.

A resistor obeys Ohm's law: $I = \Delta V_R / R$. This gives us two alternative ways of writing the power dissipated by a resistor. We can either substitute IR for ΔV_R or substitute $\Delta V_R / R$ for I. Thus

$$P_R = I\,\Delta V_R = I^2 R = \frac{(\Delta V_R)^2}{R} \qquad (22.14)$$

Power dissipated by resistance R with current I and potential difference ΔV_R

P_R vs. I: QUADRATIC p.47; P_R vs. ΔV: QUADRATIC p.47

It is worth writing the different forms of this equation to illustrate that the power varies as the square of both the current and the potential difference.

EXAMPLE 22.14 Finding the power of a dim bulb

How much power is dissipated by a 60 W (120 V) lightbulb when operated, using a dimmer switch, at 100 V?

PREPARE The 60 W rating is for operation at 120 V. We will assume that the resistance doesn't change if the bulb is run at a lower power—not quite right, but a reasonable approximation for this case in which the voltage is only slightly different from the rated value. We can compute the resistance for this case and then compute the power with the dimmer switch.

SOLVE The lightbulb dissipates 60 W at $\Delta V_R = 120$ V. Thus the filament's resistance is

$$R = \frac{(\Delta V_R)^2}{P_R} = \frac{(120\ \text{V})^2}{60\ \text{W}} = 240\ \Omega$$

The power dissipation when operated at $\Delta V_R = 100$ V is

$$P_R = \frac{(\Delta V_R)^2}{R} = \frac{(100\ \text{V})^2}{240\ \Omega} = 42\ \text{W}$$

ASSESS Reducing the voltage by 17% leads to a 30% reduction of the power. This makes sense; the power is proportional to the square of the voltage, so we expect a proportionally larger change in power.

EXAMPLE 22.15 Determining the voltage of a stereo

Most stereo speakers are designed to have a resistance of 8.0 Ω. If an 8.0 Ω speaker is connected to a stereo amplifier with a rating of 100 W, what is the maximum possible potential difference the amplifier can apply to the speakers?

PREPARE The rating of an amplifier is the *maximum* power it can deliver. Most of the time it delivers far less, but the maximum might be needed for brief, intense sounds. The maximum potential difference will occur when the amplifier is providing the maximum power, so we will make our computation with this figure. We can model the speaker as a resistor.

SOLVE The maximum potential difference occurs when the power is a maximum. At the maximum power of 100 W,

$$P_R = 100\text{ W} = \frac{(\Delta V_R)^2}{R} = \frac{(\Delta V_R)^2}{8.0\ \Omega}$$

$$\Delta V_R = \sqrt{(8.0\ \Omega)(100\text{ W})} = 28\text{ V}$$

This is the maximum potential difference the amplifier might provide.

ASSESS As a check on our result, we note that the resistance of the speaker is less than that of a lightbulb, so a smaller potential difference can provide 100 W of power.

STOP TO THINK 22.5 Rank in order, from largest to smallest, the powers P_A to P_D dissipated in resistors A to D.

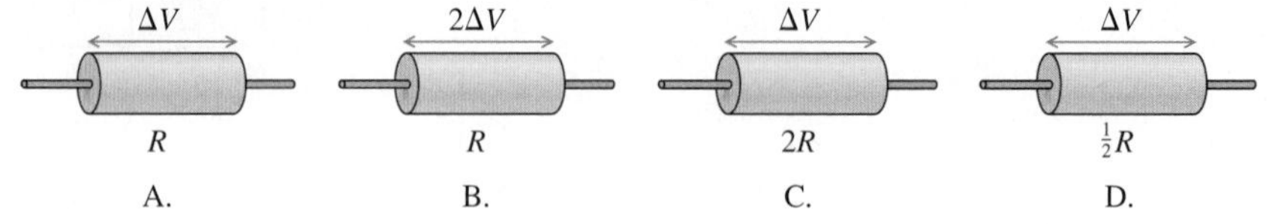

INTEGRATED EXAMPLE 22.16 Electrical measurements of body composition

The woman in the photo at the start of the chapter is gripping a device that passes a small current through her body. How does this permit a determination of body fat?

The exact details of how the device works are beyond the scope of this chapter, but the basic principle is quite straightforward: The device applies a small potential difference and measures the resulting current. Comparing multiple measurements allows the device to determine the resistance of one part of the body, the upper arm. The resistance of the upper arm depends sensitively on the percentage of body fat in the upper arm, and the percentage of body fat in the upper arm is a good predictor of the percentage of fat in the body overall. Let's make a simple model of the upper arm to show how the resistance of the upper arm varies with percentage body fat.

The model of a person's upper arm in **FIGURE 22.29** ignores the nonconductive elements (such as the skin and the mineralized portion of the bone) and groups the conductive elements into two distinct sections—muscle and fat—that form two parallel segments. The resistivity of each tissue type is shown. This simple model isn't a good description of the actual structure of the arm, but it predicts the electrical character quite well.

FIGURE 22.29 A simple model of the resistance of the upper arm.

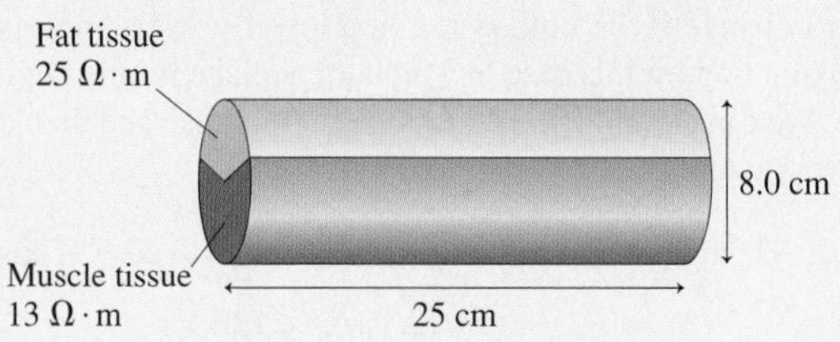

a. An experimental subject's upper arm, with the dimensions shown in the figure, is 40% fat and 60% muscle. A potential difference of 0.60 V is applied between the elbow and the shoulder. What current is measured?
b. A 0.60 V potential difference applied to the upper arm of a second subject with an arm of similar dimensions gives a current of 0.87 mA. What are the percentages of muscle and fat in this person's upper arm?

PREPARE FIGURE 22.30 shows how we can model the upper arm as two resistors that are connected together at the ends. We use the ideal-wire model in which there's no "loss" of potential along the wires. Consequently, the potential difference across each of the two segments is the full 0.60 V of the battery. The current "splits" at the junction between the two resistors, and conservation of current tells us that

$$I_{total} = I_{muscle} + I_{fat}$$

The resistance of each segment depends on its resistivity (given in Figure 22.29) and on its length and cross-section area. The cross-section area of the whole arm is $A = \pi r^2 = \pi(0.040\ \text{m})^2 = 0.0050\ \text{m}^2$; the area of each segment is this number multiplied by the appropriate fraction.

SOLVE a. An object's resistance is related to its geometry and the resistivity of the material by $R = \rho L/A$. Thus the resistances of the muscle (60% of the area) and fat (40% of the area) segments are

FIGURE 22.30 Current through the tissues of the upper arm.

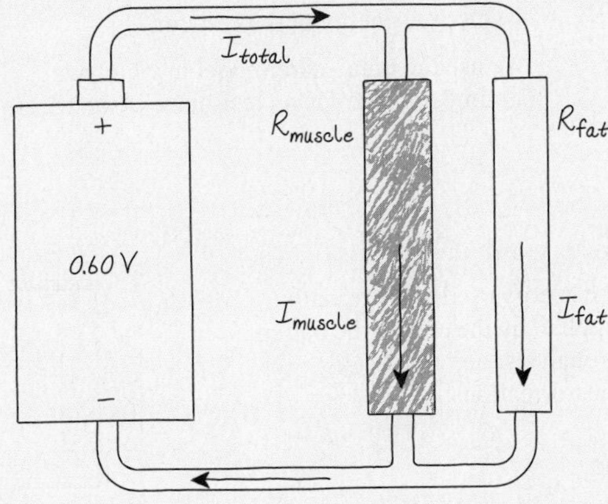

$$R_{muscle} = \frac{\rho_{muscle} L}{A_{muscle}} = \frac{(13\ \Omega \cdot \text{m})(0.25\ \text{m})}{(0.60)(0.0050\ \text{m}^2)} = 1100\ \Omega$$

$$R_{fat} = \frac{\rho_{fat} L}{A_{fat}} = \frac{(25\ \Omega \cdot \text{m})(0.25\ \text{m})}{(0.40)(0.0050\ \text{m}^2)} = 3100\ \Omega$$

The potential difference across each segment is 0.60 V. We can then use Ohm's law, $I = \Delta V/R$, to find that the current in each segment is

$$I_{muscle} = \frac{0.60\ \text{V}}{1100\ \Omega} = 0.55\ \text{mA}$$

$$I_{fat} = \frac{0.60\ \text{V}}{3100\ \Omega} = 0.19\ \text{mA}$$

The conservation of current equation then gives the total current as the sum of these two values:

$$I_{total} = 0.55\ \text{mA} + 0.19\ \text{mA} = 0.74\ \text{mA}$$

b. If we know the current, we can determine the amount of muscle and fat. Let the fraction of muscle tissue be x; the fraction of fat tissue is then $(1 - x)$. We repeat the steps of the above calculation with these expressions in place:

$$R_{muscle} = \frac{\rho_{muscle} L}{A} = \frac{(13\ \Omega \cdot \text{m})(0.25\ \text{m})}{(x)(0.0050\ \text{m}^2)} = \frac{650\ \Omega}{x}$$

$$R_{fat} = \frac{\rho_{fat} L}{A} = \frac{(25\ \Omega \cdot \text{m})(0.25\ \text{m})}{(1 - x)(0.0050\ \text{m}^2)} = \frac{1250\ \Omega}{1 - x}$$

In terms of these values, the current in each segment is:

$$I_{muscle} = \frac{0.60\ \text{V}}{650\ \Omega}x = 0.92(x)\ \text{mA}$$

$$I_{fat} = \frac{0.60\ \text{V}}{1250\ \Omega}(1 - x) = 0.48(1 - x)\ \text{mA}$$

The sum of these currents is the total current:

$$I_{total} = 0.87\ \text{mA} = 0.92(x)\ \text{mA} + 0.48(1 - x)\ \text{mA}$$

Rearranging the terms on the right side gives

$$0.87\ \text{mA} = (0.48 + 0.44x)\ \text{mA}$$

Finally, we can solve for x:

$$x = 0.89$$

This person therefore has 89% muscle and 11% fatty tissue in the upper arm.

ASSESS A good check on our work is that the total current we find in part a is small—important for safety—and reasonably close to the value given in part b; the arms are the same size, and the variation in body fat between individuals isn't all that large, so we expect the numbers to be similar. The current given in part b is greater than we found in part a. Muscle has a lower resistance than fat, so the subject of part b must have a higher percentage of muscle—exactly what we found.

SUMMARY

The goal of Chapter 22 has been to learn how and why charge moves through a conductor as a current.

GENERAL PRINCIPLES

Current

The **current** is defined to be the motion of positive charges

$$I = \frac{\Delta q}{\Delta t}$$

Conservation of current dictates that the current is the same at all points in the circuit.

The battery does work to raise the potential of the charges. The potential difference of the battery is its **emf** $\mathcal{E}$.

The battery creates an electric field in the circuit that causes charges to move. Positive charges move in the direction of the electric field, which is the direction of decreasing potential.

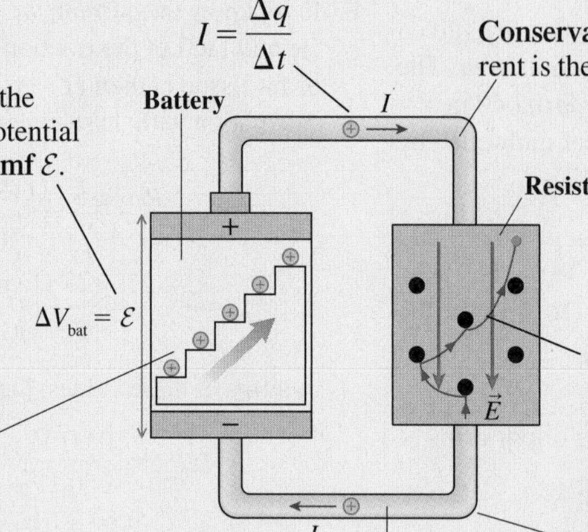

A **battery** is a source of potential difference. Chemical processes in the battery separate charges. We use a **charge escalator** model to show the lifting of charges to higher potential.

The actual charge carriers are electrons. Their random collisions with atoms impede the flow of charge and are the source of **resistance.** The collisions increase the thermal energy of the **resistor.**

We use the **ideal-wire model** in which we assume that there is no resistance in the wires.

IMPORTANT CONCEPTS

Resistance, resistivity, and Ohm's law

The **resistivity** ρ is a property of a material, a measure of how good a conductor the material is.

- Good conductors have low resistivity.
- Poor conductors have high resistivity.

The **resistance** is a property of a particular wire or conductor. The resistance of a wire depends on its resistivity and dimensions.

$$R = \frac{\rho L}{A}$$

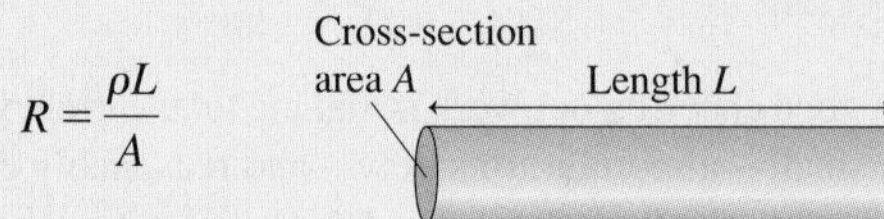

Ohm's law describes the relationship between potential difference and current in a resistor:

$$I = \frac{\Delta V}{R}$$

ΔV

R

I

Energy and power

The energy used by a circuit is supplied by the emf of the battery through a series of energy transformations:

I

+

$\mathcal{E}$

−

R

ΔV

I

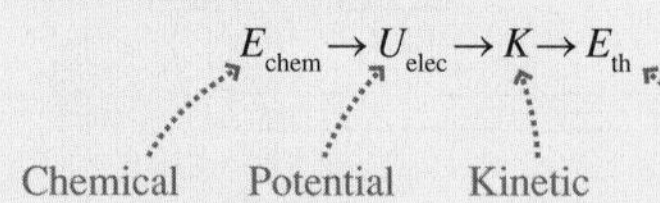

The battery *supplies* power at the rate

$$P_{emf} = I\mathcal{E}$$

The resistor *dissipates* power at the rate

$$P_R = I\,\Delta V_R = I^2 R = \frac{(\Delta V_R)^2}{R}$$

APPLICATIONS

Conducting materials

When a potential difference is applied to a wire, if the relationship between potential difference and current is linear, the material is **ohmic.**

The resistance is

$$R = \frac{1}{\text{slope}}$$

I

ΔV

Resistors are made of ohmic materials and have a well-defined value of resistance:

$$R = \frac{\Delta V}{I}$$

If the variation is not linear, the material is **nonohmic.**

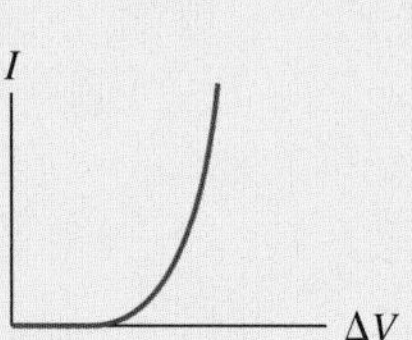

For homework assigned on MasteringPhysics, go to www.masteringphysics.com

Problem difficulty is labeled as | (straightforward) to ||||| (challenging).

Problems labeled INT integrate significant material from earlier chapters; BIO are of biological or medical interest.

QUESTIONS

Conceptual Questions

1. Two wires connect a lightbulb to a battery, completing a circuit and causing the bulb to glow. Do the simple observations and measurements that you can make on this circuit prove that something is *flowing* through the wires? If so, state the observations and/or measurements that are relevant and the steps by which you can then infer that something must be flowing. If not, can you offer an alternative hypothesis about why the bulb glows that is at least plausible and that could be tested?
2. Two wires connect a lightbulb to a battery, completing a circuit and causing the bulb to glow. Are the simple observations and measurements you can make on this circuit able to distinguish a current composed of positive charge carriers from a current composed of negative charge carriers? If so, describe how you can tell which it is. If not, why not?
3. What *causes* electrons to move through a wire as a current?
4. A lightbulb is connected to a battery by two copper wires of equal lengths but different thicknesses. A thick wire connects one side of the lightbulb to the positive terminal of the battery and a thin wire connects the other side of the bulb to the negative terminal.
   a. Which wire carries a greater current? Or is the current the same in both? Explain.
   b. If the two wires are switched, will the bulb get brighter, dimmer, or stay the same? Explain.
5. All wires in Figure Q22.5 are made of the same material and have the same diameter. Rank in order, from largest to smallest, the currents I_1 to I_4. Explain.

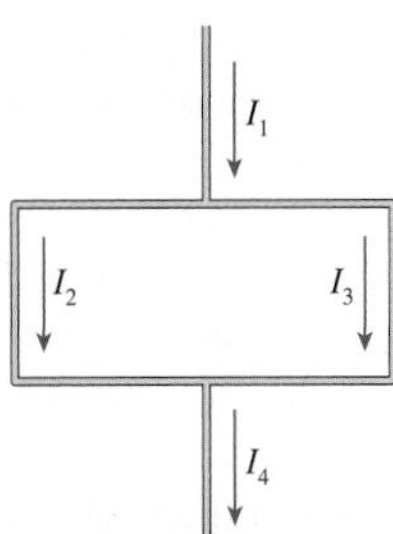

FIGURE Q22.5

6. A wire carries a 4 A current. What is the current in a second wire that delivers twice as much charge in half the time?
7. Metal 1 and metal 2 are each formed into 1-mm-diameter wires. The electric field needed to cause a 1 A current in metal 1 is larger than the electric field needed to cause a 1 A current in metal 2. Which metal has the larger resistivity? Explain.
8. BIO Cells in the nervous system have a potential difference of 70 mV across the cell membrane separating the interior of the cell from the extracellular fluid. This potential difference is maintained by ion pumps that move charged ions across the membrane. Is this an emf?
9. a. Which direction—clockwise or counterclockwise—does an electron travel through the wire in Figure Q22.9? Explain.
   b. Does an electron's electric potential energy increase, decrease, or stay the same as it moves through the wire? Explain.
   c. If you answered "decrease" in part b, where does the energy go? If you answered "increase" in part b, where does the energy come from?
   d. Which way—up or down—does an electron move through the *battery?* Explain.
   e. Does an electron's electric potential energy increase, decrease, or stay the same as it moves through the battery? Explain.
   f. If you answered "decrease" in part e, where does the energy go? If you answered "increase" in part e, where does the energy come from?

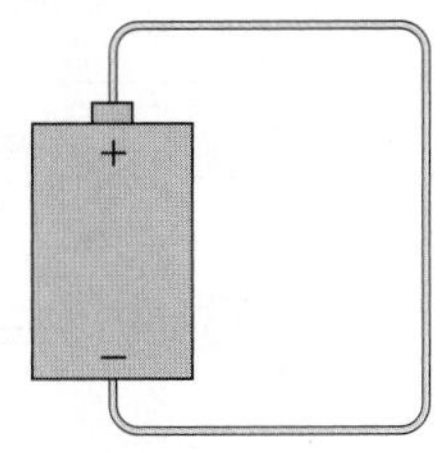

FIGURE Q22.9

10. If you change the temperature of a segment of metal wire, the dimensions change and the resistivity changes. How does each of these changes affect the resistance of the wire?
11. The wires in Figure Q22.11 are all made of the same material; the length and radius of each wire is noted. Rank in order, from largest to smallest, the resistances R_1 to R_5 of these wires. Explain.

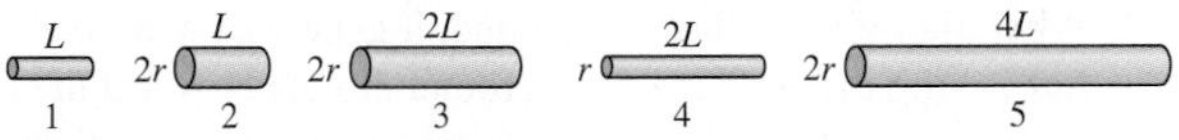

L, r, 1 | L, 2r, 2 | 2L, 2r, 3 | 2L, r, 4 | 4L, 2r, 5

FIGURE Q22.11

12. The two circuits in Figure Q22.12 use identical batteries and wires of equal diameters. Rank in order, from largest to smallest, the currents I_1, I_2, I_3, and I_4 at points 1 to 4.

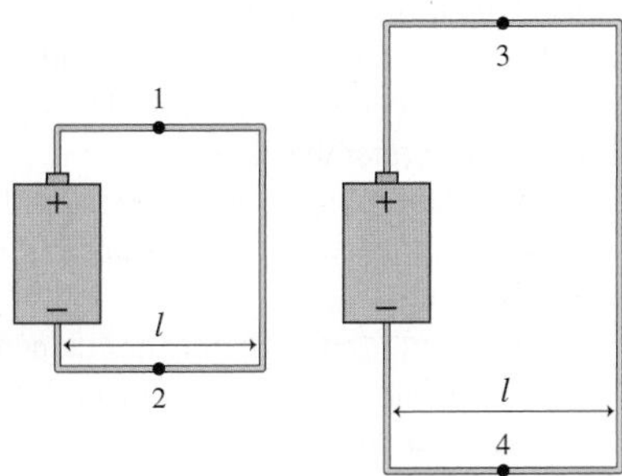

FIGURE Q22.12

13. The two circuits in Figure Q22.13 use identical batteries and wires of equal diameters. Rank in order, from largest to smallest, the currents I_1 to I_7 at points 1 to 7. Explain.

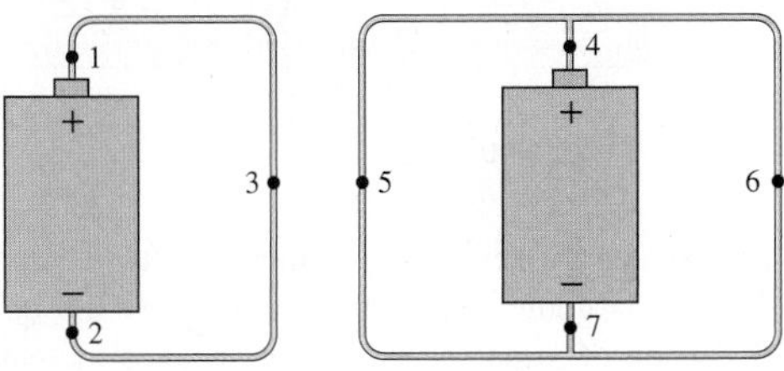

FIGURE Q22.13

14. Which, if any, of these statements are true? (More than one may be true.) Explain your choice or choices.
    a. A battery supplies energy to a circuit.
    b. A battery is a source of potential difference. The potential difference between the terminals of the battery is always the same.
    c. A battery is a source of current. The current leaving the battery is always the same.
15. Rank in order, from largest to smallest, the currents I_1 to I_4 through the four resistors in Figure Q22.15. Explain.

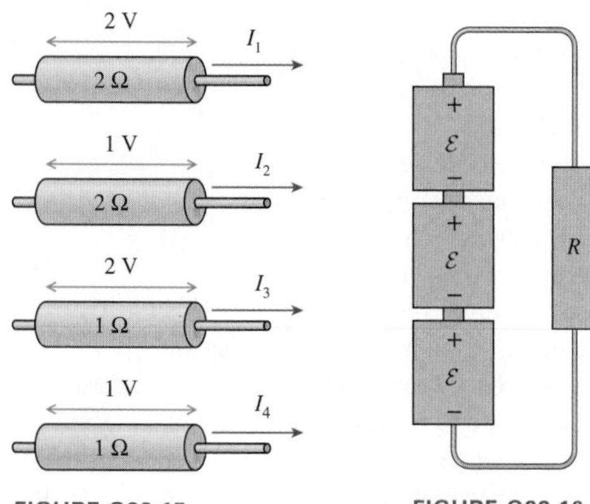

FIGURE Q22.15

FIGURE Q22.16

16. The circuit in Figure Q22.16 has three batteries of emf $\mathcal{E}$ in series. Assuming the wires are ideal, sketch a graph of the potential as a function of distance traveled around the circuit, starting from $V = 0$ V at the negative terminal of the bottom battery. Note all important points on your graph.
17. When lightning strikes the ground, it generates a large electric field along the surface of the ground directed toward the point of the strike. People near a lightning strike are often injured not by the lightning itself but by a large current that flows up one leg and down the other due to this electric field. To minimize this possibility, you are advised to stand with your feet close together if you are trapped outside during a lightning storm. Explain why this is beneficial.
    Hint: The current path through your body, up one leg and down the other, has a certain resistance. The larger the current along this path, the greater the damage.
18. One way to find out if a wire has corroded is to measure its resistance. Explain why the resistance of a wire increases if it becomes corroded.
19. Over time, atoms "boil off" the hot filament in an incandescent bulb and the filament becomes thinner. How does this affect the brightness of the lightbulb?
20. Rank in order, from largest to smallest, the powers P_1 to P_4 dissipated by the four resistors in Figure Q22.20.

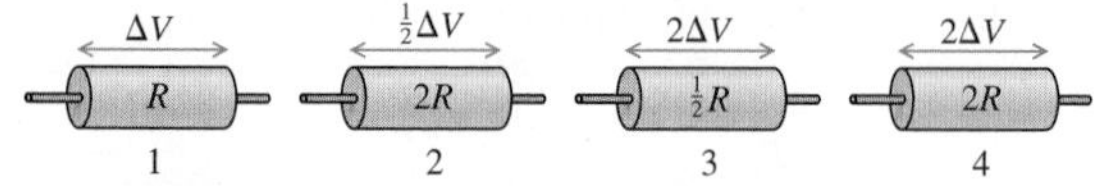

FIGURE Q22.20

21. We can model the rear window defroster in a car as a resistor that is connected to the car's 12 V battery. The defroster is made of a material whose resistance increases rapidly as the temperature increases. When the defroster is cold, its resistance is low; when the defroster is warm, its resistance is high. Why is it better to make a defroster with a material like this than with a material whose resistance is independent of temperature? Think about how the resistance, the current, and the power will change as the window warms.

Multiple-Choice Questions

22. | Lightbulbs are typically rated by their power dissipation when operated at a given voltage. Which of the following lightbulbs has the largest current through it when operated at the voltage for which it's rated?
    A. 0.8 W, 1.5 V B. 6 W, 3 V
    C. 4 W, 4.5 V D. 8 W, 6 V
23. || Lightbulbs are typically rated by their power dissipation when operated at a given voltage. Which of the following lightbulbs has the largest resistance when operated at the voltage for which it's rated?
    A. 0.8 W, 1.5 V B. 6 W, 3 V
    C. 4 W, 4.5 V D. 8 W, 6 V
24. | A copper wire is stretched so that its length increases and its diameter decreases. As a result,
    A. The wire's resistance decreases, but its resistivity stays the same.
    B. The wire's resistivity decreases, but its resistance stays the same.
    C. The wire's resistance increases, but its resistivity stays the same.
    D. The wire's resistivity increases, but its resistance stays the same.
25. | The potential difference across a length of wire is increased. Which of the following does *not* increase as well?
    A. The electric field in the wire
    B. The power dissipated in the wire
    C. The resistance of the wire
    D. The current in the wire
26. ||| A stereo amplifier creates a 5.0 V potential difference across a speaker. To double the power output of the speaker, the amplifier's potential difference must be increased to
    A. 7.1 V B. 10 V C. 14 V D. 25 V
27. | If a 1.5 V battery stores 5.0 kJ of energy (a reasonable value for an inexpensive C cell), for how many minutes could it sustain a current of 1.2 A?
    A. 2.7 B. 6.9 C. 9.0 D. 46
28. | Figure Q22.28 shows a side view of a wire of varying circular cross section. Rank in order the currents flowing in the three sections.

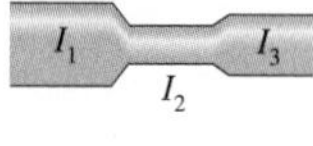

FIGURE Q22.28

    A. $I_1 > I_2 > I_3$
    B. $I_2 > I_3 > I_1$
    C. $I_1 = I_2 = I_3$
    D. $I_1 > I_3 > I_2$
29. ||| A person gains weight by adding fat—and therefore adding girth—to his body and his limbs, with the amount of muscle remaining constant. How will this affect the electrical resistance of his limbs?
BIO
    A. The resistance will increase.
    B. The resistance will stay the same.
    C. The resistance will decrease.

PROBLEMS

Section 22.1 A Model of Current

Section 22.2 Defining and Describing Current

1. ||| The current in an electric hair dryer is 10 A. How much charge and how many electrons flow through the hair dryer in 5.0 min?
2. || 2.0×10^{13} electrons flow through a transistor in 1.0 ms. What is the current through the transistor?
3. | A wire carries a 1.0 A current for 30 s. How many electrons move past a point in the wire?
4. | When a nerve cell depolarizes, charge is transferred across BIO the cell membrane, changing the potential difference. For a typical nerve cell, 9.0 pC of charge flows in a time of 0.50 ms. What is the average current?
5. | A wire carries a 15 μA current. How many electrons pass a given point on the wire in 1.0 s?
6. || In a typical lightning strike, 2.5 C flows from cloud to ground in 0.20 ms. What is the current during the strike?
7. || A capacitor is charged to 6.0×10^{-4} C, then discharged by connecting a wire between the two plates. 40 μs after the discharge begins, the capacitor still holds 13% of its original charge. What was the average current during the first 40 μs of the discharge?
8. || In an ionic solution, 5.0×10^{15} positive ions with charge $+2e$ pass to the right each second while 6.0×10^{15} negative ions with charge $-e$ pass to the left. What are the magnitude and direction of current in the solution?
9. || The starter motor of a car engine draws a current of 150 A from the battery. The copper wire to the motor is 5.0 mm in diameter and 1.2 m long. The starter motor runs for 0.80 s until the car engine starts. How much charge passes through the starter motor?
10. | A car battery is rated at 90 A·hr, meaning that it can supply a 90 A current for 1 hr before being completely discharged. If you leave your headlights on until the battery is completely dead, how much charge leaves the positive terminal of the battery?
11. || What are the values of currents I_B and I_C in Figure P22.11? The directions of the currents are as noted.

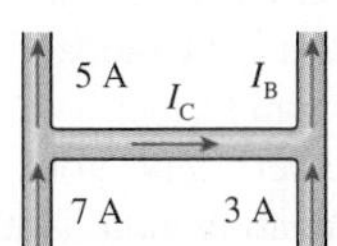

FIGURE P22.11

12. | The currents through several segments of a wire object are shown in Figure P22.12. What are the magnitudes and directions of the currents I_B and I_C in segments B and C?

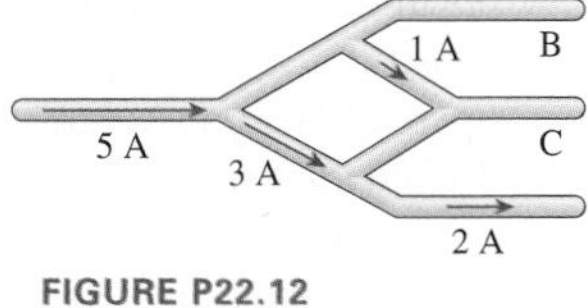

FIGURE P22.12

Section 22.3 Batteries and emf

13. || A battery supplies a steady 1.5 A current to a circuit. If the charges moving in the battery are positive ions with charge e, how many ions per second are transported from the negative terminal to the positive terminal?
14. | How much work is done to move 1.0 μC of charge from the negative terminal to the positive terminal of a 1.5 V battery?
15. | What is the emf of a battery that does 0.60 J of work to transfer 0.050 C of charge from the negative to the positive terminal?
16. || A 9.0 V battery supplies a 2.5 mA current to a circuit for 5.0 hr.
    a. How much charge has been transferred from the negative to the positive terminal?
    b. How much work has been done on the charges that passed through the battery?
17. | An individual hydrogen-oxygen fuel cell has an output of 0.75 V. How many cells must be connected in series to drive a 24.0 V motor?
18. | An electric catfish can generate BIO a significant potential difference using stacks of special cells called *electrocytes*. Each electrocyte develops a potential difference of 110 mV. How many cells must be connected in series to give the 350 V a large catfish can produce?

Section 22.4 Connecting Potential and Current

19. | A wire with resistance R is connected to the terminals of a 6.0 V battery. What is the potential difference ΔV_{ends} between the ends of the wire and the current I through it if the wire has the following resistances? (a) 1.0 Ω (b) 2.0 Ω (c) 3.0 Ω.
20. | Wires 1 and 2 are made of the same metal. Wire 2 has twice the length and twice the diameter of wire 1. What are the ratios (a) ρ_2 / ρ_1 of the resistivities and (b) R_2 / R_1 of the resistances of the two wires?
21. ||| A wire has a resistance of 0.010 Ω. What will the wire's resistance be if it is stretched to twice its original length without changing the volume of the wire?
22. ||| Resistivity measurements on the leaves of corn plants are a BIO good way to assess stress and overall health. The leaf of a corn plant has a resistance of 2.0 MΩ measured between two electrodes placed 20 cm apart along the leaf. The leaf has a width of 2.5 cm and is 0.20 mm thick. What is the resistivity of the leaf tissue? Is this greater than or less than the resistivity of muscle tissue in the human body?
23. || What is the resistance of
    a. A 1.0-m-long copper wire that is 0.50 mm in diameter?
    b. A 10-cm-long piece of iron with a 1.0 mm $\times$ 1.0 mm square cross section?
24. ||| A motorcyclist is making an electric vest that, when connected to the motorcycle's 12 V battery, will warm her on cold rides. She is using 0.25-mm-diameter copper wire, and she wants a current of 4.0 A in the wire. What length wire must she use?
25. || The femoral artery is the large artery that carries blood to the leg. BIO A person's femoral artery has an inner diameter of 1.0 cm. What is the resistance of a 20-cm-long column of blood in this artery?
26. || A 3.0 V potential difference is applied between the ends of a 0.80-mm-diameter, 50-cm-long nichrome wire. What is the current in the wire?
27. ||| A 1.0-mm-diameter, 20-cm-long copper wire carries a 3.0 A current. What is the potential difference between the ends of the wire?

28. |||| The relatively high resistivity of dry skin, about BIO $1 \times 10^6\ \Omega \cdot \text{m}$, can safely limit the flow of current into deeper tissues of the body. Suppose an electrical worker places his palm on an instrument whose metal case is accidently connected to a high voltage. The skin of the palm is about 1.5 mm thick. Estimate the area of skin on the worker's palm that would contact a flat panel, then calculate the approximate resistance of the skin of the palm.
29. | a. How long must a 0.60-mm-diameter copper wire be to carry a 0.50 A current when connected to the terminals of a 1.5 V flashlight battery?
    b. What is the current if the wire is half this length?

Section 22.5 Ohm's Law and Resistor Circuits

30. | Figure P22.30 shows the current-versus-potential-difference graph for a resistor.
    a. What is the resistance of this resistor?
    b. Suppose the length of the resistor is doubled while keeping its cross section the same. (This requires doubling the amount of material the resistor is made of.) Copy the figure and add to it the current-versus-potential-difference graph for the longer resistor.

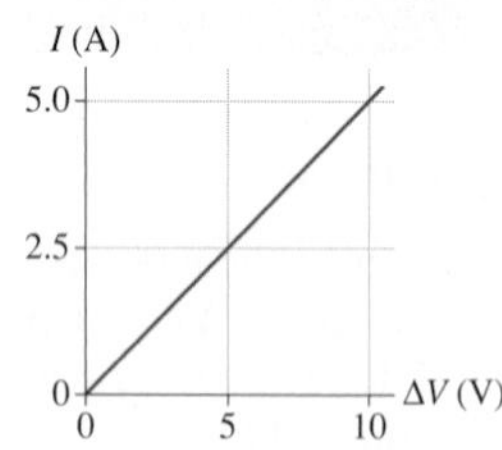

FIGURE P22.30

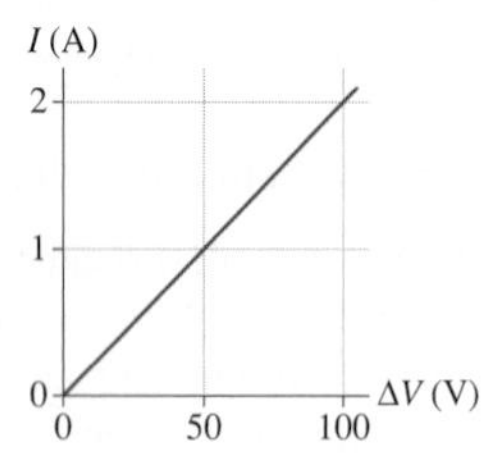

FIGURE P22.31

31. | Figure P22.31 is a current-versus-potential-difference graph for a cylinder. What is the cylinder's resistance?
32. | In Example 22.6 the length of a 60 W, 240 Ω lightbulb filament was calculated to be 60 cm.
    a. If the potential difference across the filament is 120 V, what is the strength of the electric field inside the filament?
    b. Suppose the length of the bulb's filament were doubled without changing its diameter or the potential difference across it. What would the electric field strength be in this case?
    c. Remembering that the current in the filament is proportional to the electric field, what is the current in the filament following the doubling of its length?
    d. What is the resistance of the filament following the doubling of its length?
33. | The electric field inside a 30-cm-long copper wire is 0.010 V/m. What is the potential difference between the ends of the wire?
34. || A small electric lap blanket contains a 40-foot-long wire wrapped back and forth inside. An 18 V supply creates a current in this wire, warming it and thus the blanket. What is the electric field strength inside this wire?
35. ||| Two identical lightbulbs are connected in series to a single 9.0 V battery.
    a. Sketch the circuit.
    b. Sketch a graph showing the potential as a function of distance through the circuit, starting with $V = 0\ \text{V}$ at the negative terminal of the battery.

Section 22.6 Energy and Power

36. | a. What is the resistance of a 1500 W (120 V) hair dryer?
    b. What is the current in the hair dryer when it is used?
37. | You've brought your 1000 W (120 V) hair dryer on vacation to Europe, where the standard outlet voltages are 230 V. Assuming the hair dryer can operate safely at the higher voltage, can you actually use it if the outlet can provide at most 15 A, or will it draw more current than this?
38. || A 70 W electric blanket runs at 18 V.
    a. What is the resistance of the wire in the blanket?
    b. How much current does the wire carry?
39. | A 60-cm-long heating wire is connected to a 120 V outlet. If the wire dissipates 45 W, what are (a) the current in and (b) the resistance of the wire?
40. ||| An electric eel develops a potential difference of 450 V, driving BIO a current of 0.80 A for a 1.0 ms pulse. For this pulse, find (a) the power, (b) the total energy, and (c) the total charge that flows.
41. |||| The total charge a household battery can supply is given in units of mA · hr. For example, a 9.0 V alkaline battery is rated 450 mA · hr, meaning that such a battery could supply a 1 mA current for 450 hr, a 2 mA current for 225 hr, etc. How much energy, in joules, is this battery capable of supplying?

General Problems

42. | A 3.0 V battery powers a flashlight bulb that has a resistance of 6.0 Ω. How much charge moves through the battery in 10 min?
43. |||| A sculptor has asked you to help electroplate gold onto a INT brass statue. You know that the charge carriers in the ionic solution are monovalent (charge e) gold ions, and you've calculated that you must deposit 0.50 g of gold to reach the necessary thickness. How much current do you need, in mA, to plate the statue in 3.0 hr?
44. |||| Older freezers developed a coating of ice inside that had to be INT melted periodically; an electric heater could speed this defrosting process. Suppose you're melting ice from your freezer using a heating wire that carries a current of 5.0 A when connected to 120 V.
    a. What is the resistance of the wire?
    b. How long will it take the heater to melt 720 g of accumulated ice at –10°C? Assume that all of the heat goes into warming and melting the ice, and that the melt water runs out and doesn't warm further.
45. |||| For a science experiment you need to electroplate a INT 100-nm-thick zinc coating onto both sides of a very thin, 2.0 cm × 2.0 cm copper sheet. You know that the charge carriers in the ionic solution are divalent (charge $2e$) zinc ions. The density of zinc is 7140 kg/m^3. If the electroplating apparatus operates at 1.0 mA, how long will it take the zinc to reach the desired thickness?
46. ||| The hot dog cooker described in the chapter heats hot dogs INT by connecting them to 120 V household electricity. A typical hot dog has a mass of 60 g and a resistance of 150 Ω. How long will it take for the cooker to raise the temperature of the hot dog from 20°C to 80°C? The specific heat of a hot dog is approximately 2500 J/kg · K.
47. |||| Air isn't a perfect electric insulator, but it has a very high INT resistivity. Dry air has a resistivity of approximately $3 \times 10^{13}\ \Omega \cdot \text{m}$. A capacitor has square plates 10 cm on a side

separated by 1.2 mm of dry air. If the capacitor is charged to 250 V, what fraction of the charge will flow across the air gap in 1 minute? Make the approximation that the potential difference doesn't change as the charge flows.

48. || The biochemistry that takes place inside cells depends on BIO various elements, such as sodium, potassium, and calcium, that are dissolved in water as ions. These ions enter cells through narrow pores in the cell membrane known as *ion channels*. Each ion channel, which is formed from a specialized protein molecule, is selective for one type of ion. Measurements with microelectrodes have shown that a 0.30-nm-diameter potassium ion (K^+) channel carries a current of 1.8 pA. How many potassium ions pass through if the ion channel opens for 1.0 ms?

49. || High-resolution measurements have shown that an ion chan- BIO nel (see Problem 48) is a 0.30-nm-diameter cylinder with length of 5.0 nm. The intracellular fluid filling the ion channel has resistivity 0.60 $\Omega \cdot$ m. What is the resistance of the ion channel?

50. | When an ion channel opens in a cell wall (see Problem 48), BIO monovalent (charge e) ions flow through the channel at a rate of 1.0×10^7 ions/s.
   a. What is the current through the channel?
   b. The potential difference across the ion channel is 70 mV. What is the power dissipation in the channel?

51. ||| The total charge a battery can supply is rated in mA $\cdot$ hr, the product of the current (in mA) and the time (in hr) that the battery can provide this current. A battery rated at 1000 mA $\cdot$ hr can supply a current of 1000 mA for 1.0 hr, 500 mA current for 2.0 hr, and so on. A typical AA rechargeable battery has a voltage of 1.2 V and a rating of 1800 mA $\cdot$ hr. For how long could this battery drive current through a long, thin wire of resistance 22 Ω?

52. || The heating element of a simple heater consists of a 2.0-m-long, 0.60-mm-diameter nichrome wire. When plugged into a 120 V outlet, the heater draws 8.0 A of current when hot.
   a. What is the wire's resistance when it is hot?
   b. Use your answer to part a to calculate the resistivity of nichrome in this situation. Why is it not the same as the value of ρ given for nichrome in Table 22.1?

53. ||| Variations in the resistivity of blood can give valuable clues BIO to changes in the blood's viscosity and other properties. The resistivity is measured by applying a small potential difference and measuring the current. Suppose a medical device attaches electrodes into a 1.5-mm-diameter vein at two points 5.0 cm apart. What is the blood resistivity if a 9.0 V potential difference causes a 230 μA current through the blood in the vein?

54. |||| A 40 W (120 V) lightbulb has a tungsten filament of thickness 0.040 mm. The filament's operating temperature is 1500°C.
   a. How long is the filament?
   b. What is the resistance of the filament at 20°C?

55. || Wires aren't really ideal. The voltage drop across a current-carrying wire can be significant unless the resistance of the wire is quite low. Suppose a 50 ft extension cord is being used to provide power to an electric lawn mower. The cord carries a 10 A current. The copper wire in a typical extension cord has a 1.3 mm diameter. What is the voltage drop across a 50 ft length of wire at this current?

56. ||| When the starter motor on a car is engaged, there is a 300 A current in the wires between the battery and the motor. Suppose the wires are made of copper and have a total length of 1.0 m. What minimum diameter can the wires have if the voltage drop along the wires is to be less than 0.50 V?

57. ||| The electron beam inside a television picture tube is 0.40 mm INT in diameter and carries a current of 50 μA. This electron beam impinges on the inside of the picture tube screen.
   a. How many electrons strike the screen each second?
   b. The electrons move with a velocity of 4.0×10^7 m/s. What electric field strength is needed to accelerate electrons from rest to this velocity in a distance of 5.0 mm?
   c. Each electron transfers its kinetic energy to the picture tube screen upon impact. What is the *power* delivered to the screen by the electron beam?

   Hint: What potential difference produced the field that accelerated electrons? This is an emf.

58. | The two segments of the wire in Figure P22.58 have equal diameters and equal lengths but different resistivities ρ_1 and ρ_2. Current I passes through this wire. If the resistivities have the ratio $\rho_2/\rho_1 = 2$, what is the ratio $\Delta V_1/\Delta V_2$ of the potential differences across the two segments of the wire?

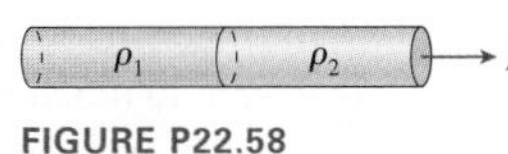

FIGURE P22.58

59. || A 15-cm-long nichrome wire is connected between the terminals of a 1.5 V battery. If the current in the wire is 2.0 A, what is the wire's diameter?

60. || A wire is 2.3 m long and has a diameter of 0.38 mm. When connected to a 1.2 V battery, there is a current of 0.61 A. What material is the wire likely made of?

61. | The filament of a 100 W (120 V) lightbulb is a tungsten wire 0.035 mm in diameter. At the filament's operating temperature, the resistivity is 5.0×10^{-7} $\Omega \cdot$ m. How long is the filament?

62. ||| You've made the finals of the Science Olympics! As one of INT your tasks, you're given 1.0 g of copper and asked to make a wire, using all the metal, with a resistance of 1.0 Ω. Copper has a density of 8900 kg/m^3. What length and diameter will you choose for your wire?

63. || Not too long ago houses were protected from excessive currents by fuses rather than circuit breakers. Sometimes a fuse blew out and a replacement wasn't at hand. Because a copper penny happens to have almost the same diameter as a fuse, some people replaced the fuse with a penny. Unfortunately, a penny never blows out, no matter how large the current, and the use of pennies in fuse boxes caused many house fires. Make the appropriate measurements on a penny, then calculate the resistance between the two faces of a solid-copper penny. (Modern pennies have the same dimensions, but are made of zinc with a copper coating.)

64. || An immersion heater used to boil water for a single cup of tea INT plugs into a 120 V outlet and is rated at 300 W.
   a. What is the resistance of the heater?
   b. Suppose your super-size, super-insulated tea mug contains 400 g of water at a temperature of 18°C. How long will this heater take to bring the water to a boil? You can ignore the energy needed to raise the temperature of the mug and the heater itself.

65. |||| The graph in Figure P22.65 shows the current through a 1.0 Ω resistor as a function of time.
   a. How much charge flowed through the resistor during the 10 s interval shown?
   b. What was the total energy dissipated by the resistor during this time?

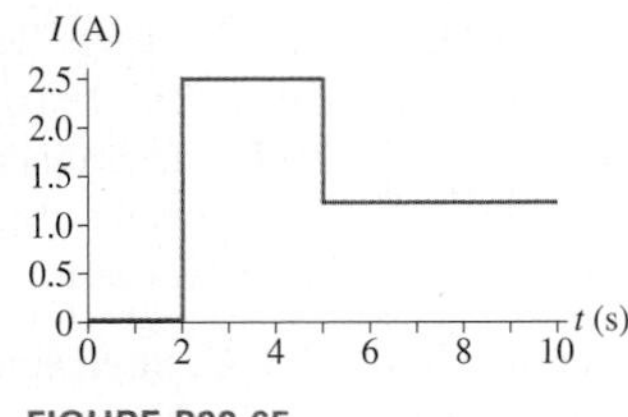

FIGURE P22.65

66. | | | It's possible to estimate the percentage of fat in the body by measuring the resistance of the upper leg rather than the upper arm; the calculation is similar. A person's leg measures 40 cm between the knee and the hip, with an average leg diameter (ignoring bone and other poorly conducting tissue) of 12 cm. A potential difference of 0.75 V causes a current of 1.6 mA. What are the fractions of (a) muscle and (b) fat in the leg?
BIO

67. | If you touch the two terminals of a power supply with your two fingertips on opposite hands, the potential difference will produce a current through your torso. The maximum safe current is approximately 5 mA.
BIO
    a. If your hands are completely dry, the resistance of your body from fingertip to fingertip is approximately 500 kΩ. If you accidentally touch both terminals of your 120 V household electricity supply with dry fingers, will you receive a dangerous shock?
    b. If your hands are moist, your resistance drops to approximately 1 kΩ. If you accidentally touch both terminals of your 120 V household supply with moist fingers, will you receive a dangerous shock?

68. | The average resistivity of the human body (apart from surface resistance of the skin) is about 5.0 Ω · m. The conducting path between the right and left hands can be approximated as a cylinder 1.6 m long and 0.10 m in diameter. The skin resistance can be made negligible by soaking the hands in salt water.
BIO
    a. What is the resistance between the hands if the skin resistance is negligible?
    b. If skin resistance is negligible, what potential difference between the hands is needed for a lethal shock current of 100 mA? Your result shows that even small potential differences can produce dangerous currents when skin is damp.

Passage Problems

Lightbulb Failure

You've probably observed that the most common time for an incandescent lightbulb to fail is the moment when it is turned on. Let's look at the properties of the bulb's filament to see why this happens.

The current in the tungsten filament of a lightbulb heats the filament until it glows. The filament is so hot that some of the atoms on its surface fly off and end up sticking on a cooler part of the bulb. Thus the filament gets progressively thinner as the bulb ages. There will certainly be one spot on the filament that is a bit thinner than elsewhere. This thin segment will have a higher resistance than the surrounding filament. More power will be dissipated at this spot, so it won't only be a thin spot, it also will be a hot spot.

Now, let's look at the resistance of the filament. The graph in Figure P22.69 shows data for the current in a lightbulb as a function of the potential difference across it. The graph is not linear, so the filament is not an ohmic material with a constant resistance. However, we can define the resistance at any particular potential difference ΔV to be $R = \Delta V/I$. This ratio, and hence the resistance, increases with ΔV and thus with temperature.

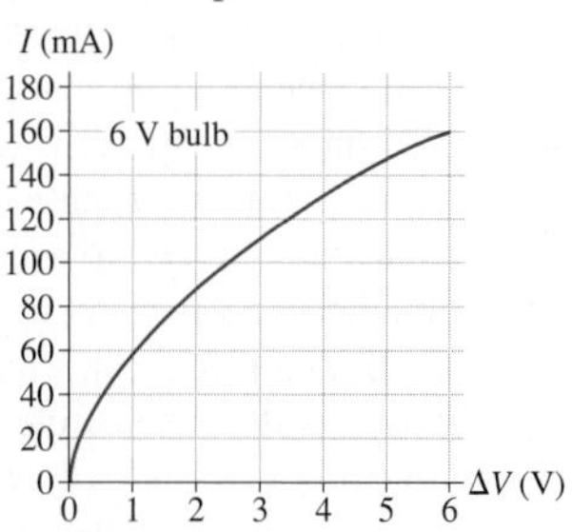

FIGURE P22.69

When the bulb is turned on, the filament is cold and its resistance is much lower than during normal, high-temperature operation. The low resistance causes a surge of higher-than-normal current lasting a fraction of a second until the filament heats up. Because power dissipation is I^2R, the power dissipated during this first fraction of a second is much larger than the bulb's rated power. This current surge concentrates the power dissipation at the high-resistance thin spot, perhaps melting it and breaking the filament.

69. | For the bulb in Figure P22.69, what is the approximate resistance of the bulb at a potential difference of 6.0 V?
    A. 7.0 Ω B. 17 Ω C. 27 Ω D. 37 Ω

70. | As the bulb ages, the resistance of the filament
    A. Increases. B. Decreases.
    C. Stays the same.

71. | Which of the curves in Figure P22.71 best represents the expected variation in current as a function of time in the short time interval immediately after the bulb is turned on?

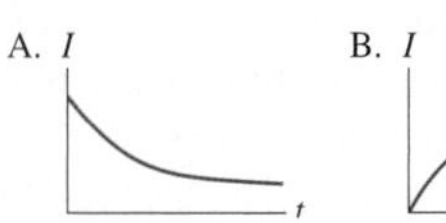

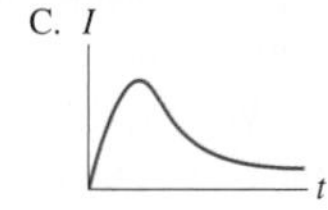

FIGURE P22.71

72. | There are devices to put in a light socket that control the current through a lightbulb, thereby increasing its lifetime. Which of the following strategies would increase the lifetime of a bulb without making it dimmer?
    A. Reducing the average current through the bulb
    B. Limiting the maximum current through the bulb
    C. Increasing the average current through the bulb
    D. Limiting the minimum current through the bulb

STOP TO THINK ANSWERS

Stop to Think 22.1: A. From Kirchhoff's junction law the current through bulb 1 is the sum of the currents through bulbs 2 and 3. Bulb 1 carries a larger current than bulb 2, so it will be brighter.

Stop to Think 22.2: C. Charge flows out of one terminal of the battery but back into the other; the amount of charge in the battery does not change. The emf is determined by the chemical reactions in the battery and is constant. But the chemical energy in the battery steadily decreases as the battery converts it to the potential energy of charges.

Stop to Think 22.3: C. Stretching the wire decreases the area and increases the length. Both of these changes increase the resistance of the wire. When the wire is reconnected to the battery, the resistance is greater but the potential difference is the same as in the original case, so the current will be smaller.

Stop to Think 22.4: $V_B = V_C > V_A > V_D = V_E$. There's no potential difference along ideal wires, so $V_B = V_C$ and $V_D = V_E$. Potential increases in going from the − to the + terminal of a battery, so $V_A > V_E$ and $V_B > V_A$. These imply $V_C > V_D$, which was expected because potential decreases as current passes through a resistor.

Stop to Think 22.5: $P_B > P_D > P_A > P_C$. The power dissipated by a resistor is $P_R = (\Delta V_R)^2/R$. Increasing R decreases P_R; increasing ΔV_R increases P_R. But changing the potential has a larger effect because P_R depends on the square of ΔV_R.

23 Circuits

The electric eel isn't really an eel; it's a fish. But it is electric, producing pulses of up to 600 V that it uses to stun prey. How does the fish produce such a large potential difference?

LOOKING AHEAD ▸▸

The goal of Chapter 23 is to understand the fundamental physical principles that govern electric circuits.

Analyzing Circuits

We've seen different elements that you can use to create a circuit—batteries, resistors, capacitors. In this chapter, we'll explore how to analyze circuits built with these parts.

Most practical circuits consist of many different elements connected together. We'll see how to take complex circuits and break them down into manageable pieces, allowing us to analyze their operation.

Circuit analysis will draw on material of the two preceding chapters:

Looking Back ◂◂

21.5 Relating field and potential
21.7–21.9 Capacitors, dielectrics, and energy in capacitors
22.2 Current
22.4–22.6 Resistors and circuits, energy and power

Series and Parallel Circuits

As we look for ways to simplify circuits, we'll find patterns in how circuit elements are connected. The most basic connections come in two types: **series circuits** and **parallel circuits.**

In a series circuit, the elements are connected one after the other. The simple wiring of a series connection is a good choice for these inexpensive minilights. But there's a cost: If you remove one light, the string goes dark.

In a parallel circuit, each element "sees" the full voltage of the battery. If one is lost, the others stay on. Car headlights are wired as a parallel circuit, for safety.

Capacitor Circuits

Capacitors store energy, as we've seen. In this chapter we'll explore different uses of capacitors in circuits.

Your camera flash is powered by a charged capacitor. After the flash fires, it takes a certain amount of time for the battery to recharge the capacitor for another flash. Capacitor circuits often have such a characteristic time.

Electricity in the Body

We will use our knowledge of electric fields, potentials, resistance, and capacitance—and an understanding of basic circuits—to explain how electrical signals propagate in your nervous system.

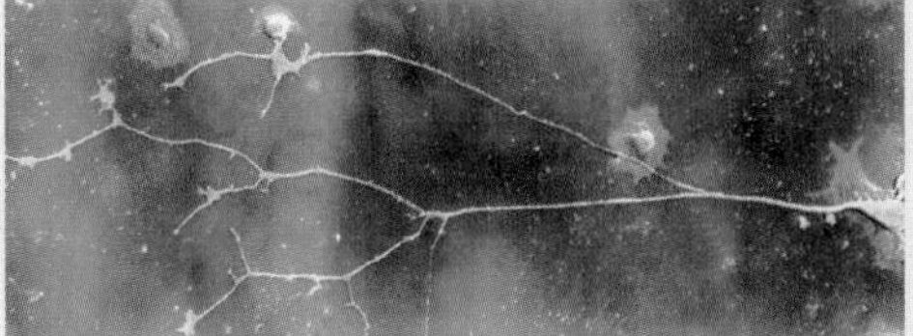

The long, yellow fibers are called axons. They transmit electrical signals from cell to cell. We'll make a simple model to understand how this transmission works.

23.1 Circuit Elements and Diagrams

In Chapter 22 we analyzed a very simple circuit, a resistor connected to a battery. In this chapter, we will explore more complex circuits involving more and different elements. As was the case with other topics in the book, we will learn a good deal by making appropriate drawings. To do so, we need a system for representing circuits symbolically in a manner that highlights their essential features.

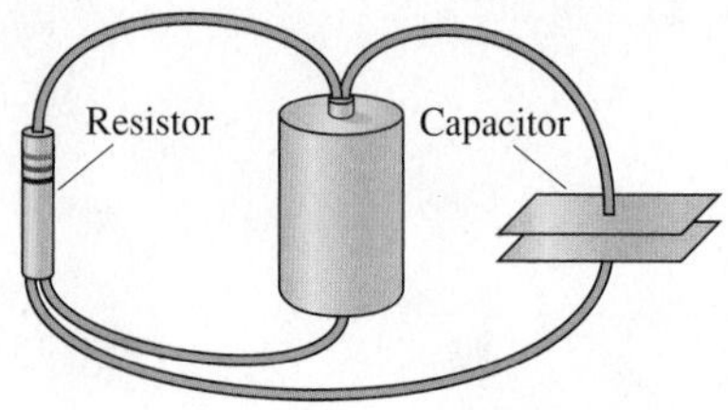

FIGURE 23.1 An electric circuit.

FIGURE 23.1 shows an electric circuit in which a resistor and a capacitor are connected by wires to a battery. To understand the operation of this circuit, we do not need to know whether the wires are bent or straight, or whether the battery is to the right or to the left of the resistor. The literal picture of Figure 23.1 provides many irrelevant details. It is customary when describing or analyzing circuits to use a more abstract picture called a **circuit diagram.** A circuit diagram is a *logical* picture of what is connected to what. The actual circuit, once it is built, may *look* quite different from the circuit diagram, but it will have the same logic and connections.

In a circuit diagram we replace pictures of the circuit elements with symbols. FIGURE 23.2 shows the basic symbols that we will need.

FIGURE 23.2 A library of basic symbols used for electric circuit drawings.

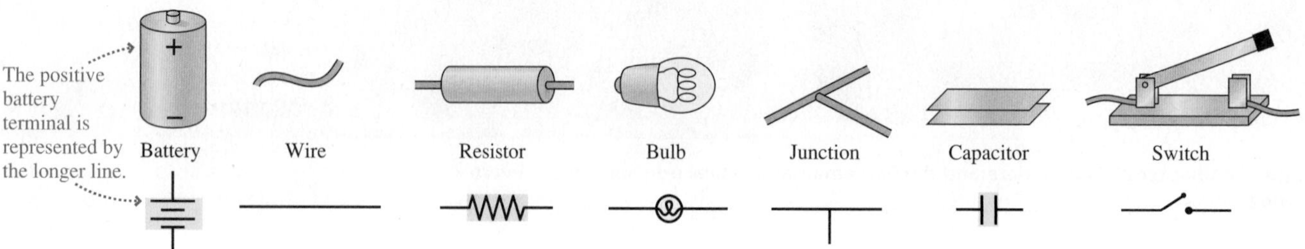

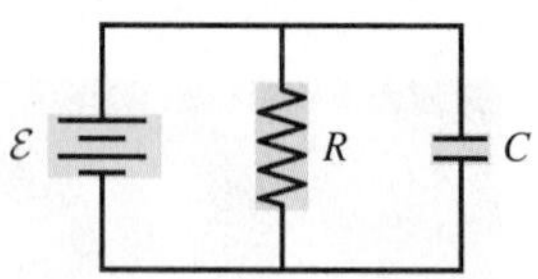

FIGURE 23.3 A circuit diagram for the circuit of Figure 23.1.

FIGURE 23.3 is a circuit diagram of the circuit shown in Figure 23.1. Notice how circuit elements are labeled. The battery's emf $\mathcal{E}$ is shown beside the battery, and the resistance R of the resistor and capacitance C of the capacitor are written beside them. We would use numerical values for $\mathcal{E}$, R, and C if we knew them. The wires, which in practice may bend and curve, are shown as straight-line connections between the circuit elements. The positive potential of the battery is at the top of the diagram; in general, we try to put higher potentials toward the top. You should get into the habit of drawing your own circuit diagrams in a similar fashion.

STOP TO THINK 23.1 Which of these diagrams represent the same circuit?

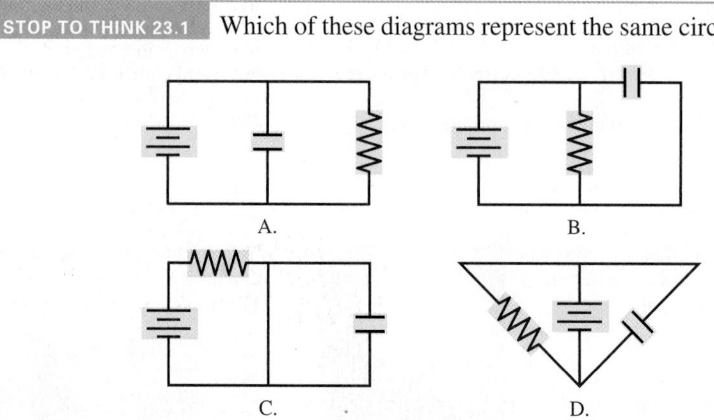

23.2 Kirchhoff's Laws

Once we have a diagram for a circuit, we can analyze it. Our tools and techniques for analyzing circuits will be based on the physical principles of potential differences and currents.

You learned in Chapter 22 that, as a result of charge and current conservation, the total current into a junction must equal the total current leaving the junction, as in FIGURE 23.4. This result was called *Kirchhoff's junction law,* which we wrote as

$$\sum I_{in} = \sum I_{out} \quad (23.1)$$

Kirchhoff's junction law

FIGURE 23.4 Kirchhoff's junction law.

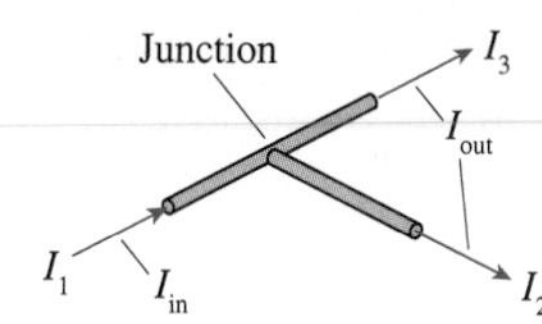

Junction law: $I_1 = I_2 + I_3$

Kirchhoff's junction law isn't a new law of nature. It's an application of a law we already know: the conservation of charge. We can also apply the law of conservation of energy to circuits. When we learned about gravitational potential energy in Chapter 10, we saw that the gravitational potential energy of an object depends on its position, not on the path it took to get to that position. The same is true of electric potential energy, as you learned in Chapter 21 and as we discussed in Chapter 22. If a charged particle moves around a closed loop and returns to its starting point, there is no net change in its electric potential energy: $\Delta U_{elec} = 0$. Because $V = U_{elec}/q$, **the net change in the electric potential around any loop or closed path must be zero** as well.

FIGURE 23.5a shows a circuit consisting of a battery and two resistors. If we start at the lower left corner, at the negative terminal of the battery, and plot the potential around the loop, we get the graph shown in the figure. The potential increases as we move "uphill" through the battery, then decreases in two "downhill" steps, one for each resistor. Ultimately, the potential ends up where it started, as it must. This is a general principle that we can apply to any circuit, as shown in FIGURE 23.5b. If we add all of the potential differences around the loop formed by the circuit, the sum must be zero. This result is known as **Kirchhoff's loop law:**

$$\Delta V_{loop} = \sum_i \Delta V_i = 0 \quad (23.2)$$

Kirchhoff's loop law

FIGURE 23.5 Kirchhoff's loop law.

(a)

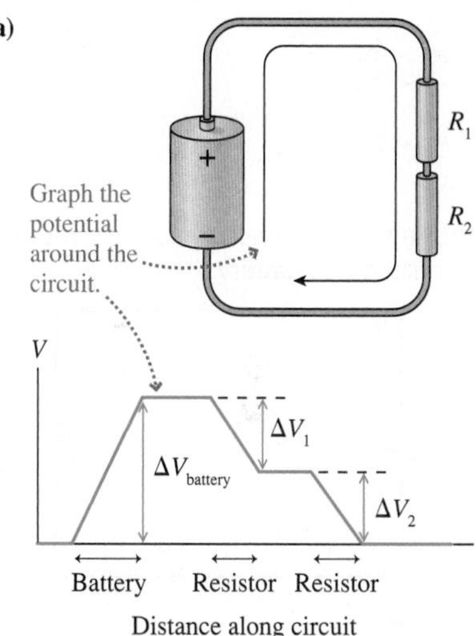

(b)

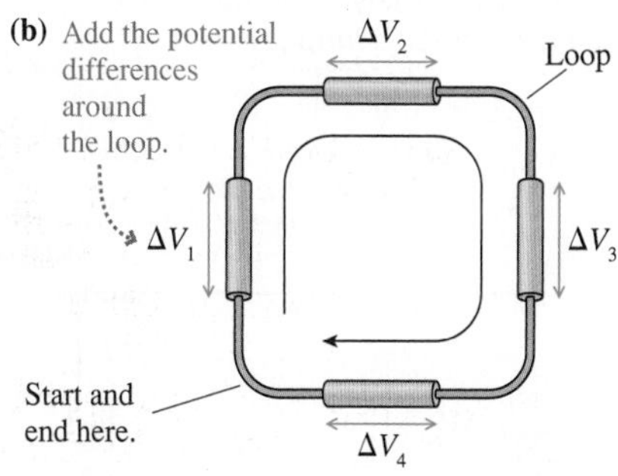

Loop law: $\Delta V_1 + \Delta V_2 + \Delta V_3 + \Delta V_4 = 0$

In Equation 23.2, ΔV_i is the potential difference of the *i*th component in the loop.

Kirchhoff's loop law can be true only if at least one of the potential differences ΔV_i is negative. To apply the loop law, we need to explicitly identify which potential differences are positive and which are negative.

TACTICS BOX 23.1 Using Kirchhoff's loop law (MP)™

1. **Draw a circuit diagram.** Label all known and unknown quantities.
2. **Assign a direction to the current.** Draw and label a current arrow I to show your choice. Choose the direction of the current based on how the batteries or sources of emf "want" the current to go. If you choose the current direction opposite the actual direction, the final value for the current that you calculate will have the correct magnitude but will be negative, letting you know that the direction is opposite the direction you chose.

Continued

❸ **"Travel" around the loop.** Start at any point in the circuit: then go all the way around the loop in the direction you assigned to the current in step 2. As you go through each circuit element, ΔV is interpreted to mean $\Delta V = V_{\text{downstream}} - V_{\text{upstream}}$.

- For a battery with current in the negative-to-positive direction:

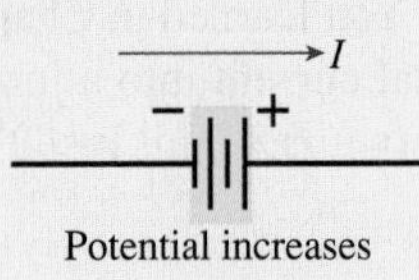
I
−
+
Potential increases

$$\Delta V_{\text{bat}} = +\mathcal{E}$$

- For a battery in the positive-to-negative direction (i.e., the current is going into the positive terminal of the battery):

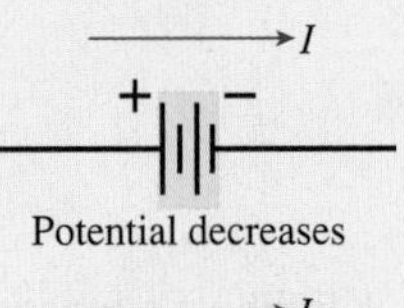
I
+
−
Potential decreases

$$\Delta V_{\text{bat}} = -\mathcal{E}$$

- For a resistor: $\Delta V_R = -IR$

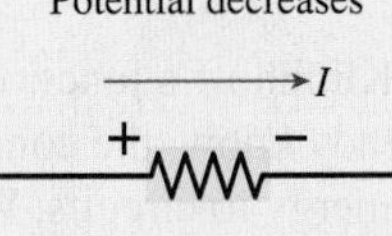
I
+
−
Potential decreases

❹ **Apply the loop law:** $\sum \Delta V_i = 0$

Exercises 7, 8

ΔV_{bat} can be positive or negative for a battery, but ΔV_R for a resistor is always negative because the potential in a resistor *decreases* along the direction of the current—charge flows "downhill," as we saw in Chapter 22. Because the potential across a resistor always decreases, we often speak of the *voltage drop* across the resistor.

NOTE ► The equation for ΔV_R in Tactics Box 23.1 seems to be the opposite of Ohm's law, but Ohm's law was concerned only with the *magnitude* of the potential difference. Kirchhoff's law requires us to recognize that the electric potential inside a resistor *decreases* in the direction of the current. ◄

FIGURE 23.6 The basic circuit of a resistor connected to a battery.

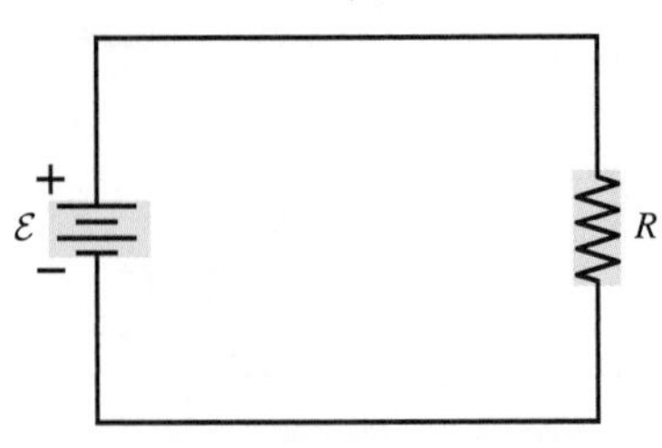
+
$\mathcal{E}$
−
R

The most basic electric circuit is a single resistor connected to the two terminals of a battery, as in FIGURE 23.6. We considered this circuit in Chapter 22, but let's now apply Kirchhoff's laws to its analysis.

This circuit of Figure 23.6 has no junctions, so the current is the same in all parts of the circuit. Kirchhoff's junction law is not needed. Kirchhoff's loop law is the tool we need to analyze this circuit, and FIGURE 23.7 shows the first three steps of Tactics Box 23.1. Notice that we're assuming the ideal-wire model in which there are no potential differences along the connecting wire. The fourth step is to apply Kirchhoff's loop law, $\sum \Delta V_i = 0$:

FIGURE 23.7 Analysis of the basic circuit using Kirchhoff's loop law.

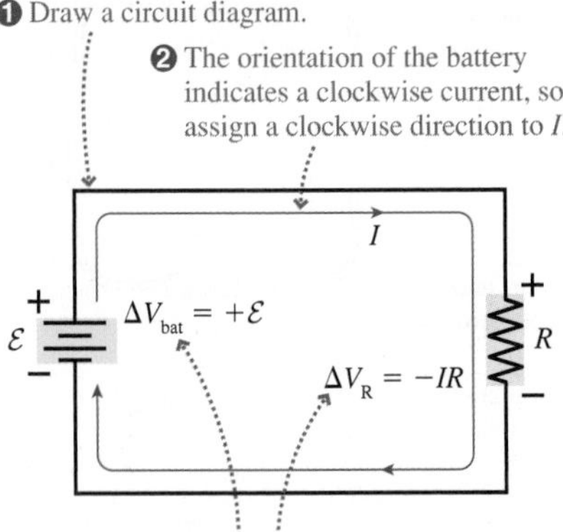
❶ Draw a circuit diagram.
❷ The orientation of the battery indicates a clockwise current, so assign a clockwise direction to I.
I
+
$\mathcal{E}$
−
$\Delta V_{\text{bat}} = +\mathcal{E}$
$\Delta V_R = -IR$
+
R
−
❸ Determine ΔV for each circuit element.

$$\Delta V_{\text{loop}} = \sum_i \Delta V_i = \Delta V_{\text{bat}} + \Delta V_R = 0 \qquad (23.3)$$

Let's look at each of the two terms in Equation 23.3:

1. The potential *increases* as we travel through the battery on our clockwise journey around the loop, as we see in the conventions in Tactics Box 23.1. We enter the negative terminal and, farther downstream, exit the positive terminal after having gained potential $\mathcal{E}$. Thus
$$\Delta V_{\text{bat}} = +\mathcal{E}$$
2. The *magnitude* of the potential difference across the resistor is $\Delta V = IR$, but Ohm's law does not tell us whether this should be positive or negative—and the difference is crucial. The potential of a resistor *decreases* in the direction of the current, which we've indicated with the + and − signs in Figure 23.7. Thus
$$\Delta V_R = -IR$$

With this information about ΔV_{bat} and ΔV_R, the loop equation becomes

$$\mathcal{E} - IR = 0 \tag{23.4}$$

We can solve the loop equation to find that the current in the circuit is

$$I = \frac{\mathcal{E}}{R} \tag{23.5}$$

This is exactly the result we saw in Chapter 22. Notice again that the current in the circuit depends on the size of the resistance. The emf of a battery is a fixed quantity; the current that the battery delivers depends jointly on the emf and the resistance.

EXAMPLE 23.1 Analyzing a circuit with two batteries

What is the current in the circuit of FIGURE 23.8? What is the potential difference across each resistor?

FIGURE 23.8 The circuit with two batteries.

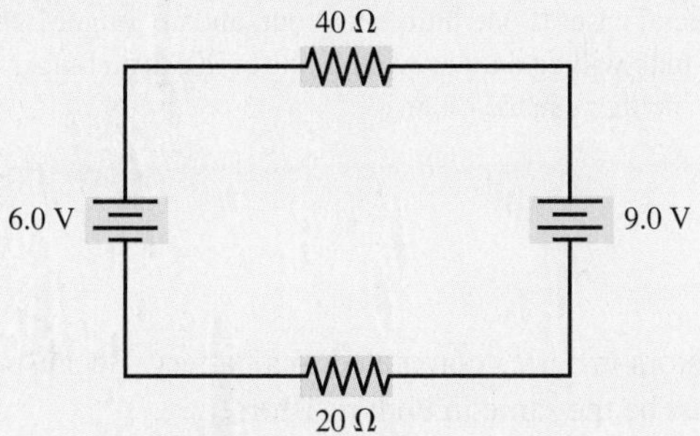

PREPARE We will solve this circuit using Kirchhoff's loop law, as outlined in Tactics Box 23.1. But how do we deal with *two* batteries? What happens when charge flows "backward" through a battery, from positive to negative? Consider the charge escalator analogy. Left to itself, a charge escalator lifts charge from lower to higher potential. But it *is* possible to run down an up escalator, as many of you have probably done. If two escalators are placed "head to head," whichever is "stronger" will, indeed, force the charge to run down the up escalator of the other battery. The current in a battery *can* be from positive to negative if driven in that direction by a larger emf from a second battery. Indeed, this is how batteries are "recharged." In this circuit, the current goes in the direction that the larger emf—the 9.0 V battery—"wants" it to go. We have redrawn the circuit in FIGURE 23.9, showing the direction of the current and the direction of the potential difference for each circuit element.

FIGURE 23.9 Analyzing the circuit.

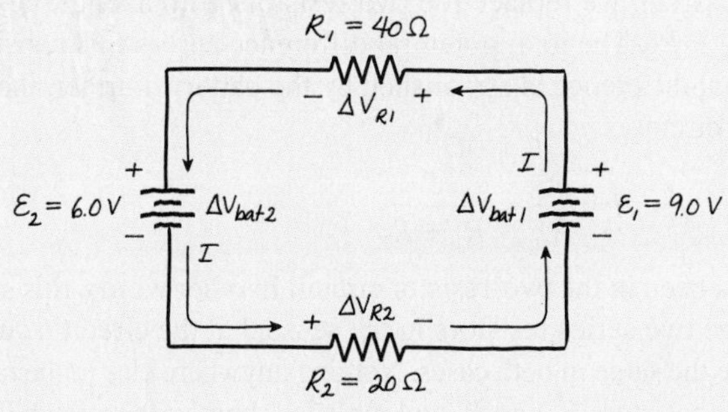

SOLVE Kirchhoff's loop law requires us to add the potential differences as we travel around the circuit in the direction of the current. Let's do this starting at the negative terminal of the 9.0 V battery:

$$\sum_i \Delta V_i = +9.0\text{ V} - I(40\ \Omega) - 6.0\text{ V} - I(20\ \Omega) = 0$$

The 6.0 V battery has $\Delta V_{bat} = -\mathcal{E}$, in accord with Tactics Box 23.1, because the potential decreases as we travel through this battery in the positive-to-negative direction. We can solve this equation for the current:

$$I = \frac{3.0\text{ V}}{60\ \Omega} = 0.050\text{ A} = 50\text{ mA}$$

Now that the current is known, we can use Ohm's law, $\Delta V = IR$, to find the potential difference across each resistor. For the 40 Ω resistor,

$$\Delta V_1 = (0.050\text{ A})(40\ \Omega) = 2.0\text{ V}$$

and for the 20 Ω resistor,

$$\Delta V_2 = (0.050\text{ A})(20\ \Omega) = 1.0\text{ V}$$

ASSESS The Assess step will be very important in circuit problems. There are generally other ways that you can analyze a circuit to check your work. In this case, you can do a final application of the loop law. If we start at the lower right-hand corner of the circuit and travel clockwise around the loop, the potential increases by 9.0 V in the first battery, then decreases by 2.0 V in the first resistor, decreases by 6.0 V in the second battery, and decreases by 1.0 V in the second resistor. The total decrease is 9.0 V, so the charge returns to its starting potential, a good check on our calculations.

23.3 Series and Parallel Circuits

Example 23.1 involved a circuit with multiple elements—two batteries and two resistors. As we introduce more circuit elements, we have possibilities for different types of connections. Suppose you use a single battery to light two lightbulbs. There are two possible ways that you can connect the circuit, as shown in FIGURE 23.10. These *series* and *parallel* circuits have very different properties. We will consider these two cases in turn.

FIGURE 23.10 Series and parallel circuits.

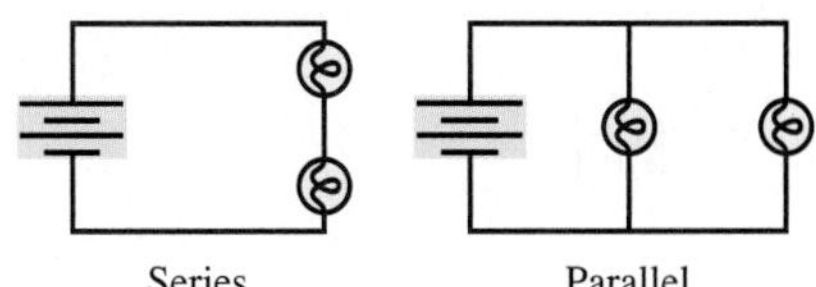

We say two bulbs are connected in **series** if they are connected directly to each other with no junction in between. All series circuits share certain characteristics.

CONCEPTUAL EXAMPLE 23.2 **Brightness of bulbs in series**

FIGURE 23.11 shows two identical lightbulbs connected in series. Which bulb is brighter: A or B? Or are they equally bright?

FIGURE 23.11 Two bulbs in series.

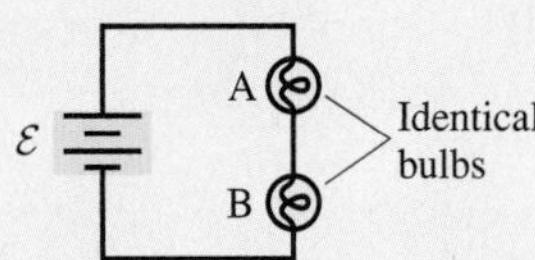

REASON Current is conserved, and there are no junctions in the circuit. Thus the current is the same at all points, as we see in FIGURE 23.12.

FIGURE 23.12 The current in the series circuit.

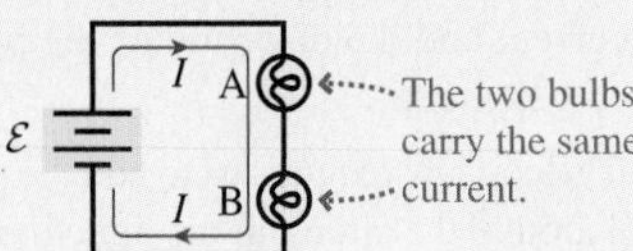

We learned in Chapter 22 that the power dissipated by a resistor is $P = I^2R$. If the two bulbs are identical (i.e., the same resistance) and have the same current through them, the power dissipated by each bulb is the same. This means that the brightness of the bulbs must be the same. The voltage across each of the bulbs will be the same as well because $\Delta V = IR$.

ASSESS It's perhaps tempting to think that bulb A will be brighter than bulb B, thinking that something is "used up" before the current gets to bulb B. It is true that *energy* is being transformed in each bulb, but current must be conserved and so both bulbs dissipate energy at the same rate. We can extend this logic to a special case: If one bulb burns out, and no longer lights, the second bulb will go dark as well. If one bulb can no longer carry a current, neither can the other.

Series Resistors

FIGURE 23.13 Replacing two series resistors with an equivalent resistor.

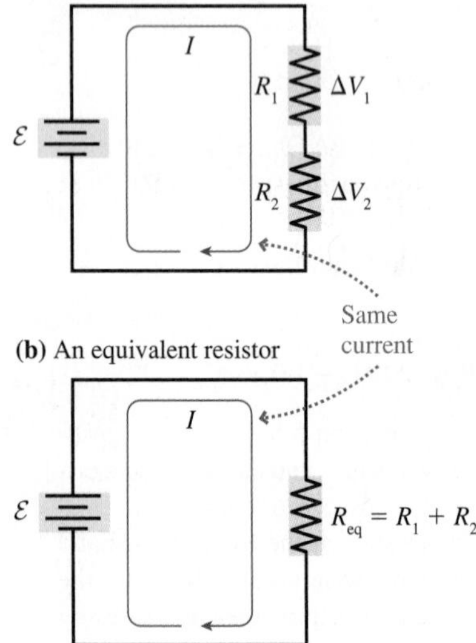

FIGURE 23.13a shows two resistors in series connected to a battery. Because there are no junctions, the current I must be the same in both resistors.

We can use Kirchhoff's loop law to look at the potential differences. Starting at the battery's negative terminal and following the current clockwise around the circuit, we find

$$\sum_i \Delta V_i = \mathcal{E} + \Delta V_1 + \Delta V_2 = 0 \tag{23.6}$$

The voltage drops across the two resistors, in the direction of the current, are $\Delta V_1 = -IR_1$ and $\Delta V_2 = -IR_2$, so we can use Equation 23.6 to find the current in the circuit:

$$\mathcal{E} = -\Delta V_1 - \Delta V_2 = IR_1 + IR_2$$

$$I = \frac{\mathcal{E}}{R_1 + R_2} \tag{23.7}$$

Suppose, as in FIGURE 23.13b, we replace the two resistors with a single resistor having the value $R_{eq} = R_1 + R_2$. The total potential difference across this resistor is still $\mathcal{E}$ because the potential difference is established by the battery. Further, the current in this single-resistor circuit is

$$I = \frac{\mathcal{E}}{R_{eq}} = \frac{\mathcal{E}}{R_1 + R_2}$$

which is the same as it had been in the two-resistor circuit. In other words, this single resistor is *equivalent* to the two series resistors in the sense that the circuit's current and potential difference are the same in both cases. Nothing anywhere else in the circuit would differ if we took out resistors R_1 and R_2 and replaced them with resistor R_{eq}.

We can extend this analysis to a case with more resistors. If we have N resistors in series, their **equivalent resistance** is the sum of the N individual resistances:

$$R_{eq} = R_1 + R_2 + \cdots + R_N \tag{23.8}$$

Equivalent resistance of N series resistors

The behavior of the circuit will be unchanged if the N series resistors are replaced by the single resistor R_{eq}.

EXAMPLE 23.3 Potential difference of Christmas-tree minilights

A string of Christmas-tree minilights consists of 50 bulbs wired in series. What is the potential difference across each bulb when the string is plugged into a 120 V outlet?

PREPARE FIGURE 23.14 shows the minilight circuit, which has 50 bulbs in series. The current in each of the bulbs is the same because they are in series.

FIGURE 23.14 50 bulbs connected in series.

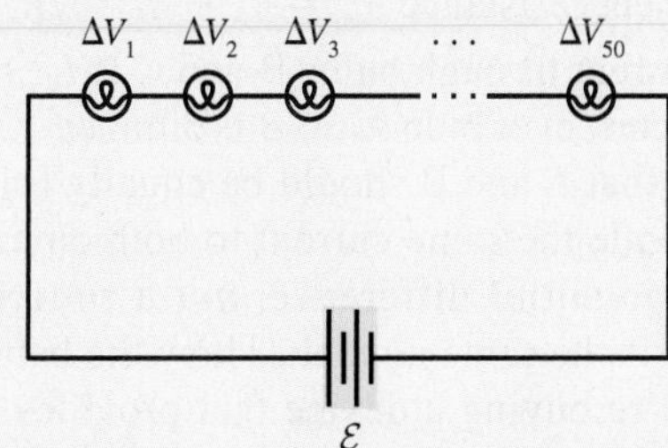

SOLVE Applying Kirchhoff's loop law around the circuit, we find

$$\mathcal{E} = \Delta V_1 + \Delta V_2 + \cdots + \Delta V_{50}$$

The bulbs are all identical and, because the current in the bulbs is the same, all of the bulbs have the same potential difference. The potential difference across a single bulb is thus

$$\Delta V_1 = \frac{\mathcal{E}}{50} = \frac{120\text{ V}}{50} = 2.4\text{ V}$$

ASSESS This result seems reasonable. The potential difference is "shared" by the bulbs in the circuit. Since the potential difference is shared among 50 bulbs, the potential difference across each bulb will be quite small.

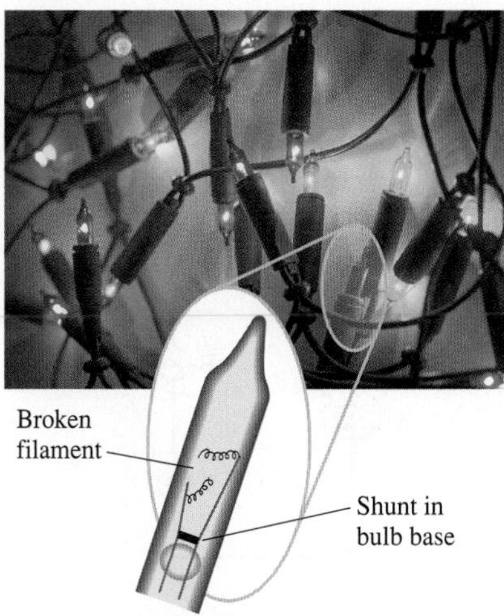

A seasonal series circuit puzzle
Christmas-tree minilights are connected in series. This is easy to verify: When you remove one bulb from a string of lights, the circuit is not complete, and the entire string of lights goes out. But when one bulb *burns* out, meaning its filament has broken, the string of lights stays lit. How is this possible? The secret is a *shunt* in the base of the bulb. Initially, the shunt is a good insulator. But if the filament breaks, the shunt is activated, and its resistance drops. The shunt can now carry the current, so the other bulbs will stay lit.

Minilights are wired in series because the bulbs can be inexpensive low-voltage bulbs. But there is a drawback that is true of all series circuits: If one bulb is removed, there is no longer a complete circuit, and there will be no current. Indeed, if you remove a bulb from a string of minilights, the entire string will go dark.

EXAMPLE 23.4 Analyzing a series resistor circuit

What is the current in the circuit of FIGURE 23.15?

PREPARE The three resistors are in series, so we can replace them with a single equivalent resistor as shown in FIGURE 23.16.

SOLVE The equivalent resistance is calculated using Equation 23.8:

$$R_{eq} = 25\ \Omega + 31\ \Omega + 19\ \Omega = 75\ \Omega$$

The current in the equivalent circuit of Figure 23.16 is

$$I = \frac{\mathcal{E}}{R_{eq}} = \frac{9.0\text{ V}}{75\ \Omega} = 0.12\text{ A}$$

This is also the current in the original circuit.

ASSESS The current in the circuit is the same whether there are three resistors or a single equivalent resistor. The equivalent resistance is the sum of the individual resistance values, and so it is always greater than any of the individual values. This is a good check on your work.

FIGURE 23.15 A series resistor circuit.

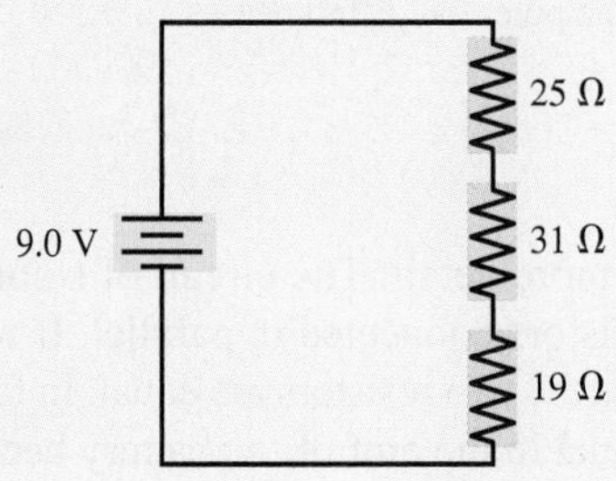

FIGURE 23.16 Analyzing a circuit with series resistors.

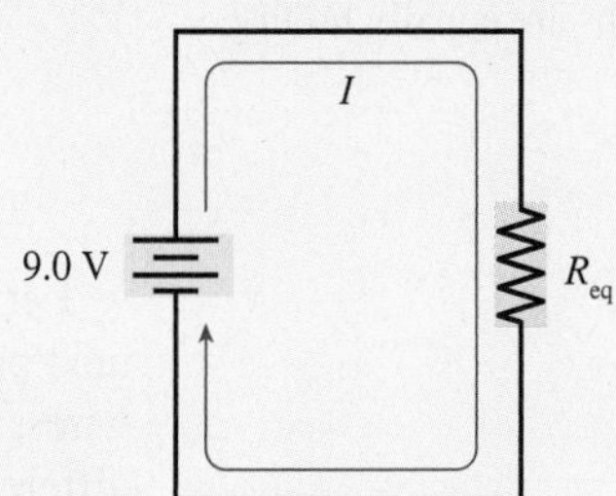

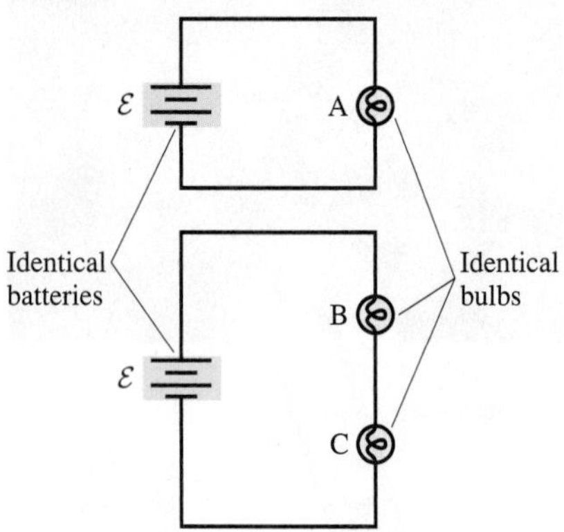

$\mathcal{E}$
A
Identical batteries
Identical bulbs
B
$\mathcal{E}$
C

FIGURE 23.17 How does the brightness of bulb B compare to that of bulb A?

Let's use our knowledge of series circuits to look at another lightbulb puzzle. FIGURE 23.17 shows two different circuits, one with one battery and one lightbulb and a second with one battery and two lightbulbs. All of the batteries and bulbs are identical. You now know that B and C, which are connected in series, are equally bright, but how does the brightness of B compare to that of A?

Suppose the resistance of each identical lightbulb is R. In the first circuit, the battery drives current $I_A = \mathcal{E}/R$ through bulb A. In the second circuit, bulbs B and C are in series, with an equivalent resistance $R_{eq} = R_A + R_B = 2R$, but the battery has the same emf $\mathcal{E}$. Thus the current through bulbs B and C is $I_{B+C} = \mathcal{E}/R_{eq} = \mathcal{E}/2R = \frac{1}{2}I_A$. Bulb B has only half the current of bulb A, so B is dimmer.

Many people predict that A and B should be equally bright. It's the same battery, so shouldn't it provide the same current to both circuits? No—recall that **a battery is a source of potential difference, *not* a source of current.** In other words, the battery's emf is the same no matter how the battery is used. When you buy a 1.5 V battery you're buying a device that provides a specified amount of potential difference, not a specified amount of current. The battery does provide the current to the circuit, but the *amount* of current depends on the resistance. Your 1.5 V battery causes 1 A to pass through a 1.5 Ω resistor but only 0.1 A to pass through a 15 Ω resistor.

This is a critical idea for understanding circuits. A battery provides a fixed emf (potential difference). It does *not* provide a fixed and unvarying current. **The amount of current depends jointly on the battery's emf *and* the resistance of the circuit attached to the battery.**

Parallel Resistors

In the next example, we consider the second way of connecting two bulbs in a circuit. The two bulbs in Figure 23.18 are connected at *both* ends. We say that they are connected in **parallel.**

CONCEPTUAL EXAMPLE 23.5 Brightness of bulbs in parallel

Which lightbulb in the circuit of FIGURE 23.18 is brighter: A or B? Or are they equally bright?

FIGURE 23.18 Two bulbs in parallel.

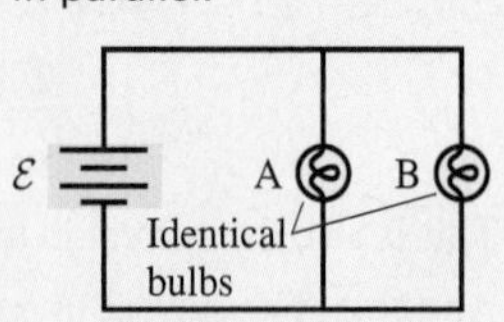

REASON Both ends of the two lightbulbs are connected together by wires. Because there's no potential difference along ideal wires, the potential at the top of bulb A must be the same as the potential at the top of bulb B. Similarly, the potentials at the bottoms of the bulbs must be the same. This means that the potential *difference* ΔV across the two bulbs must be the same, as we see in FIGURE 23.19. Because the bulbs are identical (i.e., equal resistances), the currents $I = \Delta V/R$ through the two bulbs are equal and thus the bulbs are equally bright.

FIGURE 23.19 The potential differences of the bulbs.

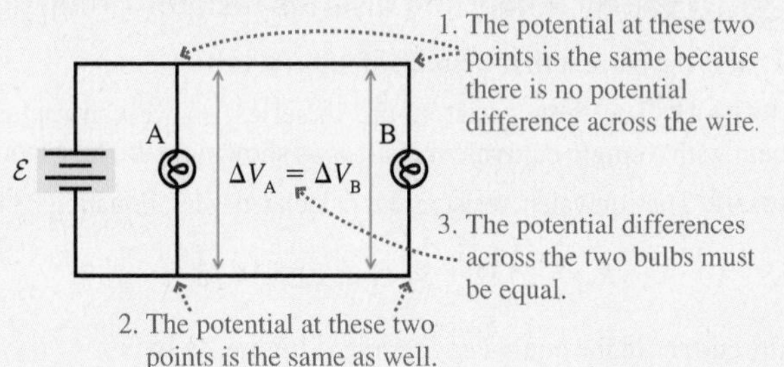

1. The potential at these two points is the same because there is no potential difference across the wire.
A
B
$\mathcal{E}$
$\Delta V_A = \Delta V_B$
3. The potential differences across the two bulbs must be equal.
2. The potential at these two points is the same as well.

ASSESS One might think that A would be brighter than B because current takes the "shortest route." But current is determined by potential difference, and two bulbs connected in parallel have the same potential difference.

Let's look at parallel circuits in more detail. The circuit of FIGURE 23.20a on the next page has a battery and two resistors connected in parallel. If we assume ideal wires, the potential differences across the two resistors are equal. In fact, the potential difference across each resistor is equal to the emf of the battery because both resistors are connected directly to the battery with ideal wires; that is, $\Delta V_1 = \Delta V_2 = \mathcal{E}$.

Now we apply Kirchhoff's junction law. The current I_{bat} from the battery splits into currents I_1 and I_2 at the top junction noted in **FIGURE 23.20b**. According to the junction law,

$$I_{bat} = I_1 + I_2 \tag{23.9}$$

We can apply Ohm's law to each resistor to find that the battery current is

$$I_{bat} = \frac{\Delta V_1}{R_1} + \frac{\Delta V_2}{R_2} = \frac{\mathcal{E}}{R_1} + \frac{\mathcal{E}}{R_2} = \mathcal{E}\left(\frac{1}{R_1} + \frac{1}{R_2}\right) \tag{23.10}$$

Can we replace a group of parallel resistors with a single equivalent resistor as we did for series resistors? To be equivalent, the potential difference across the equivalent resistor must be $\Delta V = \mathcal{E}$, the same as for the two resistors it replaces. Further, so that the battery can't know there's been any change, the current through the equivalent resistor must be $I = I_{bat}$. A resistor with this current and potential difference must have resistance

$$R_{eq} = \frac{\Delta V}{I} = \frac{\mathcal{E}}{I_{bat}} = \left(\frac{1}{R_1} + \frac{1}{R_2}\right)^{-1} \tag{23.11}$$

where we used Equation 23.10 for I_{bat}. This is the *equivalent resistance,* so a single resistor R_{eq} acts exactly the same as the two resistors R_1 and R_2 as shown in **FIGURE 23.20c**.

We can extend this analysis to the case of N resistors in parallel. For this circuit, the equivalent resistance is the inverse of the sum of the inverses of the N individual resistances:

$$R_{eq} = \left(\frac{1}{R_1} + \frac{1}{R_2} + \cdots + \frac{1}{R_N}\right)^{-1} \tag{23.12}$$

Equivalent resistance of N parallel resistors

FIGURE 23.20 Replacing two parallel resistors with an equivalent resistor.

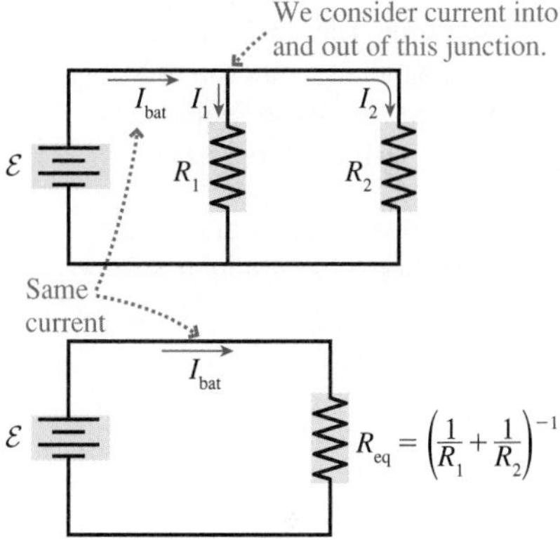

Activ Physics ONLINE 12.2, 12.5

The behavior of the circuit will be unchanged if the N parallel resistors are replaced by the single resistor R_{eq}.

NOTE ► When you use Equation 23.12, don't forget to take the inverse of the sum that you compute. ◄

In Figure 23.20 each of the resistors "sees" the full potential difference of the battery. If one resistor were removed, the conditions of the second resistor would not change. This is an important property of parallel circuits.

► **Parallel circuits for safety** You have certainly seen cars with only one headlight lit. This tells us that automobile headlights are connected in parallel: The currents in the two bulbs are independent, so the loss of one bulb doesn't affect the other. The parallel wiring is very important so that the failure of one headlight will not leave the car without illumination.

Now, let's look at another lightbulb puzzle. **FIGURE 23.21** shows two different circuits: one with one battery and one lightbulb and a second with one battery and two lightbulbs. As before, the batteries and the bulbs are identical. You know that B and C, which are connected in parallel, are equally bright, but how does the brightness of B compare to that of A?

Each of the bulbs A, B, and C is connected to the same potential difference, that of the battery, so they each have the *same* brightness. Though all of the bulbs have the same brightness, there is a difference between the circuits. In the second circuit, the battery must power two lightbulbs, and so it must provide twice as much current. Recall that the battery is a source of fixed potential difference; the current depends on the circuit that is connected to the battery. Adding a second lightbulb doesn't change the potential difference, but it does increase the current from the battery.

FIGURE 23.21 How does the brightness of bulb B compare to that of bulb A?

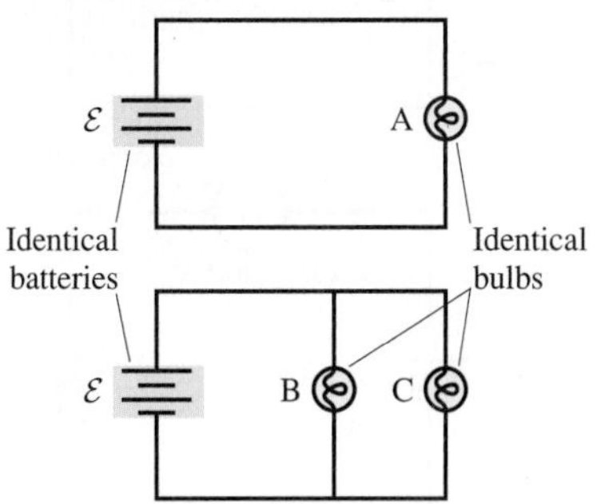

EXAMPLE 23.6 Current in a parallel resistor circuit

The three resistors of **FIGURE 23.22** are connected to a 12 V battery. What current is provided by the battery?

FIGURE 23.22 A parallel resistor circuit.

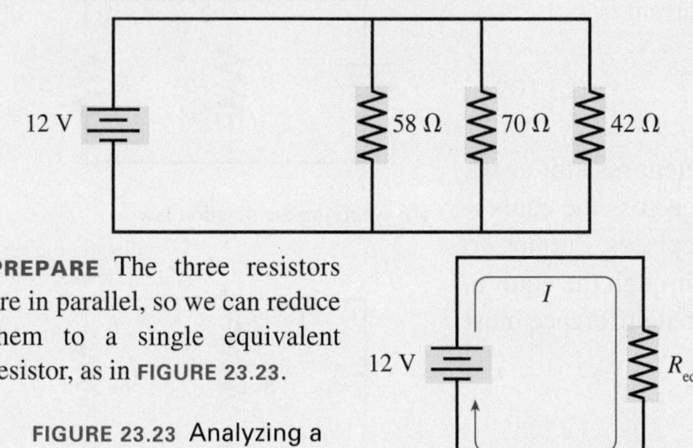

PREPARE The three resistors are in parallel, so we can reduce them to a single equivalent resistor, as in **FIGURE 23.23**.

FIGURE 23.23 Analyzing a circuit with parallel resistors.

SOLVE We can use Equation 23.12 to calculate the equivalent resistance:

$$R_{eq} = \left(\frac{1}{58\ \Omega} + \frac{1}{70\ \Omega} + \frac{1}{42\ \Omega}\right)^{-1} = 18\ \Omega$$

Once we know the equivalent resistance, we can use Ohm's law to calculate the current leaving the battery:

$$I = \frac{\mathcal{E}}{R_{eq}} = \frac{12\ \text{V}}{18\ \Omega} = 0.67\ \text{A}$$

Because the battery can't tell the difference between the original three resistors and this single equivalent resistor, the battery in Figure 23.22 provides a current of 0.67 A to the circuit.

ASSESS As we'll see, the equivalent resistance of a group of parallel resistors is less than the resistance of any of the resistors in the group. 18 Ω is less than any of the individual values, a good check on our work.

The value of the total resistance in this example may seem surprising. The equivalent of a parallel combination of 58 Ω, 70 Ω, and 42 Ω is 18 Ω. Shouldn't more resistors imply more resistance? The answer is yes for resistors in series, but not for resistors in parallel. Even though a resistor is an obstacle to the flow of charge, parallel resistors provide more pathways for charge to get through. Consequently, **the equivalent of several resistors in parallel is always *less* than any single resistor in the group.** As an analogy, think about driving in heavy traffic. If there is an alternate route or an extra lane for cars to travel, more cars will be able to "flow."

STOP TO THINK 23.2 Rank in order, from brightest to dimmest, the identical bulbs A to D.

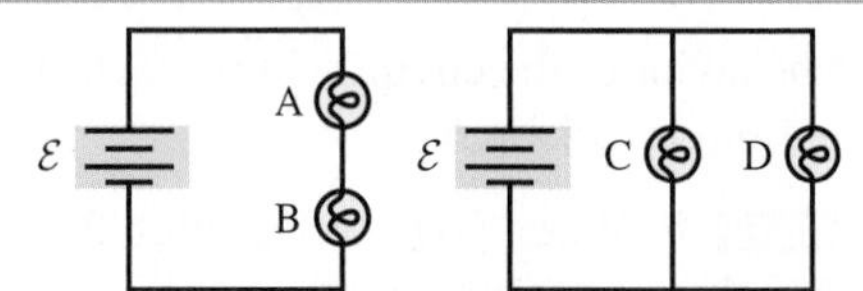

23.4 Measuring Voltage and Current

12.1, 12.4 

When you use a meter to measure the voltage or the current in a circuit, how do you connect the meter? The connection depends on the quantity you wish to measure.

A device that measures the current in a circuit element is called an **ammeter.** Because charge flows *through* circuit elements, an ammeter must be placed *in series* with the circuit element whose current is to be measured.

FIGURE 23.24 An ammeter measures the current in a circuit.

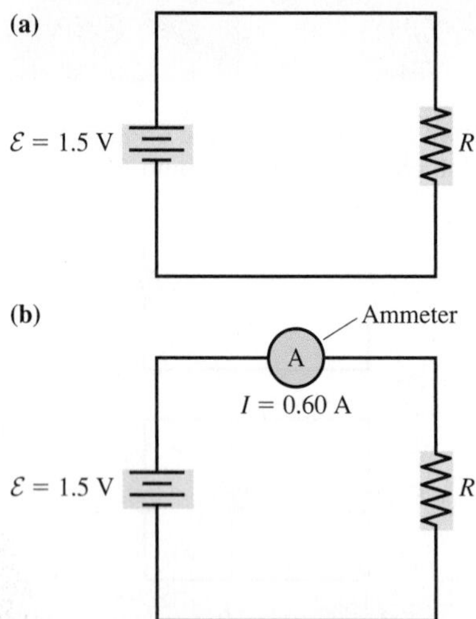

FIGURE 23.24a shows a simple one-resistor circuit with a fixed emf $\mathcal{E} = 1.5$ V and an unknown resistance R. To determine the resistance, we must know the current in the circuit, which we measure using an ammeter. We insert the ammeter in the circuit as shown in **FIGURE 23.24b**. We have to *break the connection* between the battery and the resistor in order to insert the ammeter. The resistor and the ammeter now have the same current because they are in series, so the reading of the ammeter is the current through the resistor.

Because the ammeter is in series with resistor R, the total resistance seen by the battery is $R_{eq} = R + R_{meter}$. In order to *measure* the current without *changing* the current, the ammeter's resistance must be much less than R. Thus **the resistance of an ideal ammeter is zero.** Real ammeters come quite close to this ideal.

The ammeter in Figure 23.24b reads 0.60 A, meaning that the current in the ammeter—and in the resistor—is $I = 0.60$ A. If the ammeter is ideal, which we will assume, then there is no potential difference across the ammeter ($\Delta V = IR = 0$ if $R = 0\ \Omega$) and thus the potential difference across the resistor is $\Delta V = \mathcal{E}$. The resistance can then be calculated as

$$R = \frac{\mathcal{E}}{I} = \frac{1.5\text{ V}}{0.60\text{ A}} = 2.5\ \Omega$$

As we saw in Chapter 21, we can use a **voltmeter** to measure potential differences in a circuit. Because a potential difference is measured *across* a circuit element, from one side to the other, a voltmeter is placed in *parallel* with the circuit element whose potential difference is to be measured. We want to *measure* the voltage without *changing* the voltage—without affecting the circuit. Because the voltmeter is in parallel with the resistor, the voltmeter's resistance must be very large so that it draws very little current. **An ideal voltmeter has infinite resistance.** Real voltmeters come quite close to this ideal.

FIGURE 23.25a shows a simple circuit in which a 24 Ω resistor is connected in series with an unknown resistance, with the pair of resistors connected to a 9.0 V battery. To determine the unknown resistance, we first characterize the circuit by measuring the potential difference across the known resistor with a voltmeter as shown in FIGURE 23.25b. The voltmeter is connected in parallel with the resistor; using a voltmeter does *not* require that we break the connections. The resistor and the voltmeter have the same potential difference because they are in parallel, so the reading of the voltmeter is the voltage across the resistor.

The voltmeter in Figure 23.25b tells us that the potential difference across the 24 Ω resistor is 6.0 V, so the current through the resistor is

$$I = \frac{\Delta V}{R} = \frac{6.0\text{ V}}{24\ \Omega} = 0.25\text{ A} \tag{23.13}$$

The two resistors are in series, so this is also the current in unknown resistor R. We can use Kirchhoff's loop law and the voltmeter reading to find the potential difference across the unknown resistor:

$$\sum_i \Delta V_i = 9.0\text{ V} + \Delta V_R - 6.0\text{ V} = 0 \tag{23.14}$$

from which we find $\Delta V_R = -3.0$ V. We can now use $\Delta V_R = -IR$ to calculate

$$R = \frac{-\Delta V_R}{I} = -\frac{(-3.0\text{ V})}{0.25\text{ A}} = 12\ \Omega \tag{23.15}$$

FIGURE 23.25 A voltmeter measures the potential difference across a circuit element.

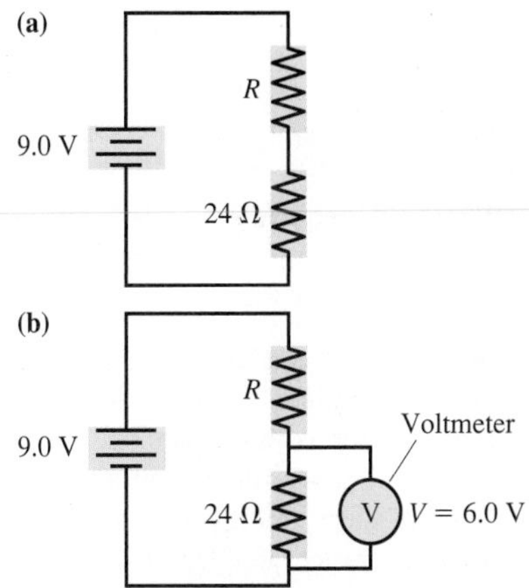

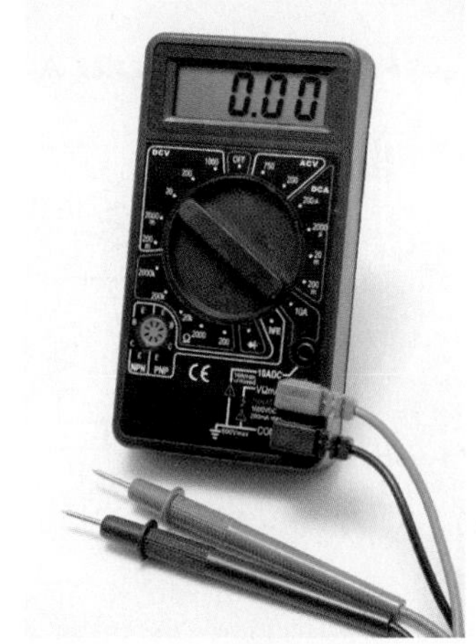

▲ In this text, we speak of ammeters and voltmeters, but in practice we generally make measurements with a *multimeter.* A dial on the front sets the meter to measure voltage, current, or other electrical quantities. When set to measure current, it works as an ammeter and must be connected in series. When set to measure voltage, it works as a voltmeter and must be connected in parallel. It can also work as an *ohmmeter,* directly measuring the resistance of a resistor not in a circuit.

STOP TO THINK 23.3 Which is the right way to connect the meters to measure the potential difference across and the current through the resistor?

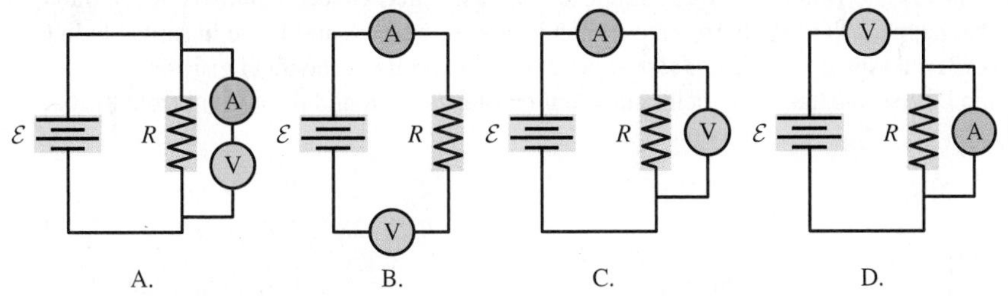

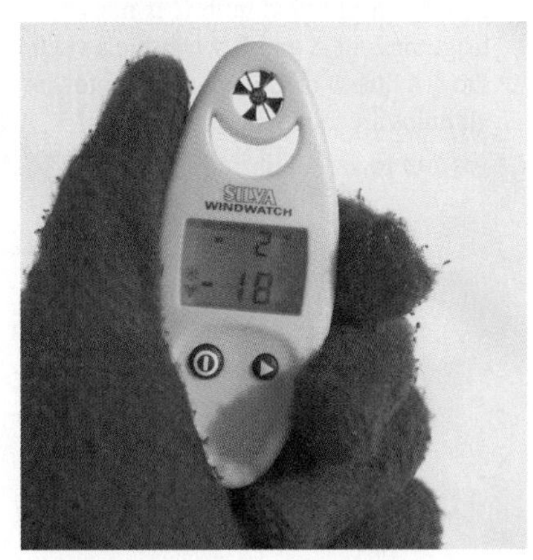

► **A circuit for all seasons** This device displays wind speed and temperature, but these are computed from basic measurements of voltage and current. The wind turns a propeller attached to a generator; a rapid spin means a high voltage. A circuit in the device contains a *thermistor,* whose resistance varies with temperature; low temperatures mean high resistance and thus a small current.

23.5 More Complex Circuits

In this section, we will consider circuits that involve both series and parallel resistors. Combinations of resistors can often be reduced to a single equivalent resistance through a step-by-step application of the series and parallel rules.

EXAMPLE 23.7 Combining resistors

What is the equivalent resistance of the group of resistors shown in FIGURE 23.26?

PREPARE We can analyze this circuit by reducing combinations of series and parallel resistors. We will do this in a series of steps, redrawing the circuit after each step.

FIGURE 23.26 A resistor circuit.

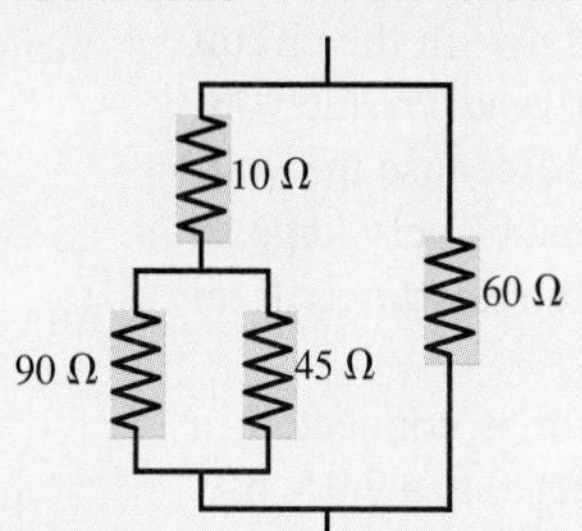

SOLVE The process of simplifying the circuit is shown in FIGURE 23.27. Note that the 10 Ω and 60 Ω resistors are *not* in parallel. They are connected at their top ends but not at their bottom ends. Resistors must be connected at *both* ends to be in parallel. Similarly, the 10 Ω and 45 Ω resistors are *not* in series because of the junction between them.

ASSESS The last step in the process is to reduce a combination of parallel resistors. The resistance of parallel resistors is always less than the smallest of the individual resistance values, so our final result must be less than 40 Ω. This is a good check on the result.

FIGURE 23.27 A combination of resistors is reduced to a single equivalent resistor.

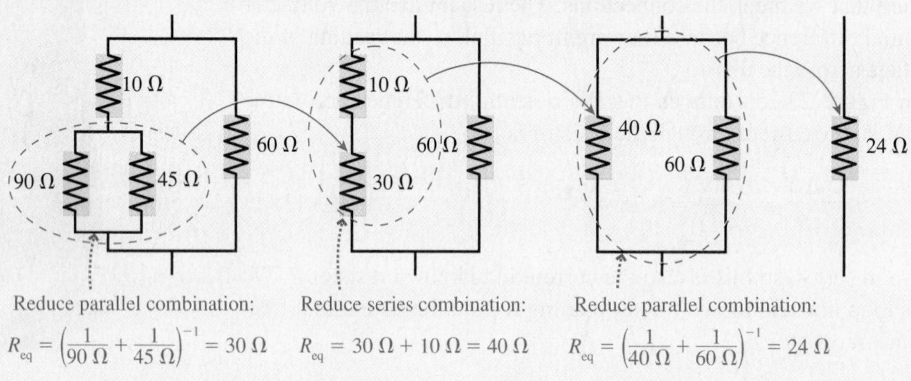

Reduce parallel combination: $R_{eq} = \left(\frac{1}{90\ \Omega} + \frac{1}{45\ \Omega}\right)^{-1} = 30\ \Omega$

Reduce series combination: $R_{eq} = 30\ \Omega + 10\ \Omega = 40\ \Omega$

Reduce parallel combination: $R_{eq} = \left(\frac{1}{40\ \Omega} + \frac{1}{60\ \Omega}\right)^{-1} = 24\ \Omega$

Two special cases (worth remembering for reducing circuits) are the equivalent resistances of two identical resistors $R_1 = R_2 = R$ in series and in parallel:

Two identical resistors in series: $R_{eq} = 2R$

Two identical resistors in parallel: $R_{eq} = \frac{R}{2}$

EXAMPLE 23.8 How does the brightness change?

Initially the switch in FIGURE 23.28 is open. Bulbs A and B are equally bright, and bulb C is not glowing. What happens to the brightness of A and B when the switch is closed? And how does the brightness of C then compare to that of A and B? Assume that all bulbs are identical.

FIGURE 23.28 A lightbulb circuit.

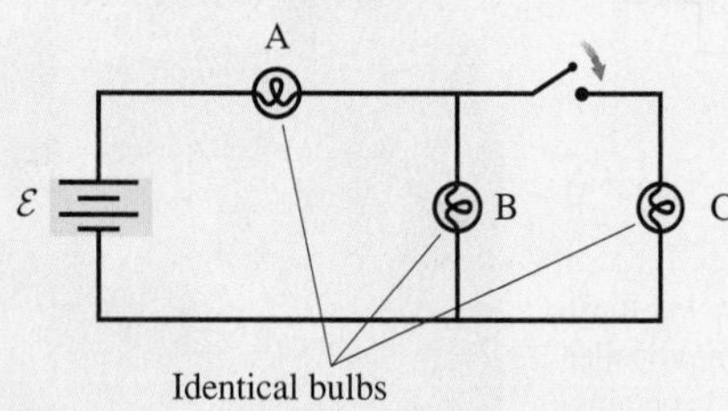

SOLVE Suppose the resistance of each bulb is R. Initially, before the switch is closed, bulbs A and B are in series; bulb C is not part of the circuit. A and B are identical resistors in series, so their equivalent resistance is $2R$ and the current from the battery is

$$I_{before} = \frac{\mathcal{E}}{R_{eq}} = \frac{\mathcal{E}}{2R} = \frac{1}{2}\frac{\mathcal{E}}{R}$$

This is the initial current in bulbs A and B, so they are equally bright.

Closing the switch places bulbs B and C in parallel with each other. The equivalent resistance of the two identical resistors in parallel is $R_{B+C} = R/2$. This equivalent resistance of B and C is in series with bulb A; hence the total resistance of the circuit is $R_{eq} = R + \frac{1}{2}R = \frac{3}{2}R$, and the current leaving the battery is

$$I_{after} = \frac{\mathcal{E}}{R_{eq}} = \frac{\mathcal{E}}{3R/2} = \frac{2}{3}\frac{\mathcal{E}}{R} > I_{before}$$

Closing the switch *decreases* the total circuit resistance and thus *increases* the current leaving the battery.

All the current from the battery passes through bulb A, so A *increases* in brightness when the switch is closed. The current I_{after} then splits at the junction. Bulbs B and C have equal resistance, so the current divides equally. The current in B is $\frac{1}{3}(\mathcal{E}/R)$, which is *less* than I_{before}. Thus B *decreases* in brightness when the switch is closed. With the switch closed, bulbs B and C are in parallel, so bulb C has the same brightness as bulb B.

ASSESS Our final results make sense. Initially, bulbs A and B are in series, and all of the current that goes through bulb A goes through bulb B. But when we add bulb C, the current has another option—it can go through bulb C. This will increase the total current, and all that current must go through bulb A, so we expect a brighter bulb A. But now the current through bulb A can go through bulb B or C. The current splits, so we'd expect that bulb B will be dimmer than before.

Analyzing Complex Circuits

We can use the information in this chapter to analyze more complex but more realistic circuits. This will give us a chance to bring together the many ideas of this chapter and to see how they are used in practice. The techniques that we use for this analysis are general, so we present them as a Problem-Solving Strategy. You can use these steps to analyze any resistor circuit, as we show in the next example.

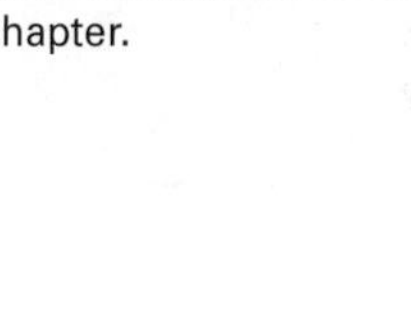

This x-ray image of a cell phone shows the complex circuitry inside. Though there are thousands of components, the analysis of such a circuit starts with the same basic rules we are studying in this chapter.

PROBLEM-SOLVING STRATEGY 23.1 **Resistor circuits**

PREPARE Draw a circuit diagram. Label all known and unknown quantities.

SOLVE Base your mathematical analysis on Kirchhoff's laws and on the rules for series and parallel resistors:

- Step by step, reduce the circuit to the smallest possible number of equivalent resistors.
- Determine the current through and potential difference across the equivalent resistors.
- Rebuild the circuit, using the facts that the current is the same through all resistors in series and the potential difference is the same across all parallel resistors.

ASSESS Use two important checks as you rebuild the circuit.

- Verify that the sum of the potential differences across series resistors matches ΔV for the equivalent resistor.
- Verify that the sum of the currents through parallel resistors matches I for the equivalent resistor.

Exercise 22

12.3

EXAMPLE 23.9 Analyzing a complex circuit

Find the current through and the potential difference across each of the four resistors in the circuit shown in **FIGURE 23.29**.

FIGURE 23.29 A multiple-resistor circuit.

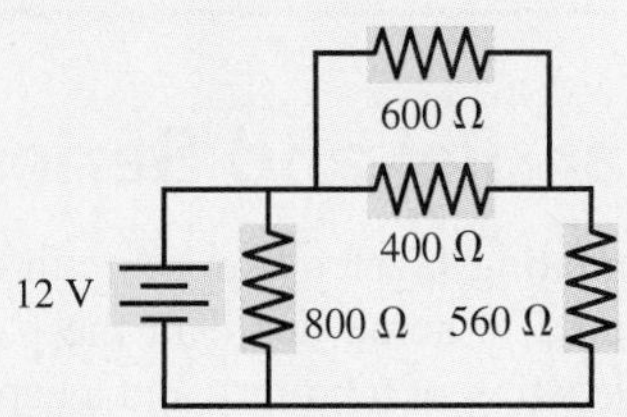

PREPARE FIGURE 23.30 shows the circuit diagram. We'll keep redrawing the diagram as we analyze the circuit.

SOLVE First, we break down the circuit, step-by-step, into one with a single resistor. Figure 23.30a does this in three steps, using the rules for series and parallel resistors. The final battery-and-resistor circuit is one we know well how to analyze. The potential difference across the 400 Ω equivalent resistor is $\Delta V_{400} = \Delta V_{bat} = \mathcal{E} = 12$ V. The current is

$$I = \frac{\mathcal{E}}{R} = \frac{12\text{ V}}{400\ \Omega} = 0.030\text{ A} = 30\text{ mA}$$

Continued

Second, we rebuild the circuit, step-by-step, finding the currents and potential differences at each step. Figure 23.30b repeats the steps of Figure 23.30a exactly, but in reverse order. The 400 Ω resistor came from two 800 Ω resistors in parallel. Because $\Delta V_{400} = 12$ V, it must be true that each $\Delta V_{800} = 12$ V. The current through each 800 Ω is then $I = \Delta V/R = 15$ mA. A check on our work is to note that 15 mA + 15 mA = 30 mA.

The right 800 Ω resistor was formed by combining 240 Ω and 560 Ω in series. Because $I_{800} = 15$ mA, it must be true that $I_{240} = I_{560} = 15$ mA. The potential difference across each is $\Delta V = IR$, so $\Delta V_{240} = 3.6$ V and $\Delta V_{560} = 8.4$ V. Here the check on our work is to note that 3.6 V + 8.4 V = 12 V = ΔV_{800}, so the potential differences add as they should.

Finally, the 240 Ω resistor came from 600 Ω and 400 Ω in parallel, so they each have the same 3.6 V potential difference as their 240 Ω equivalent. The currents are $I_{600} = 6.0$ mA and $I_{400} = 9.0$ mA. Note that 6.0 mA + 9.0 mA = 15 mA, which is a third check on our work. We now know all currents and potential differences.

ASSESS We *checked our work* at each step of the rebuilding process by verifying that currents summed properly at junctions and that potential differences summed properly along a series of resistances. This "check as you go" procedure is extremely important. It provides you, the problem solver, with a built-in error finder that will immediately inform you if a mistake has been made.

FIGURE 23.30 The step-by-step circuit analysis.

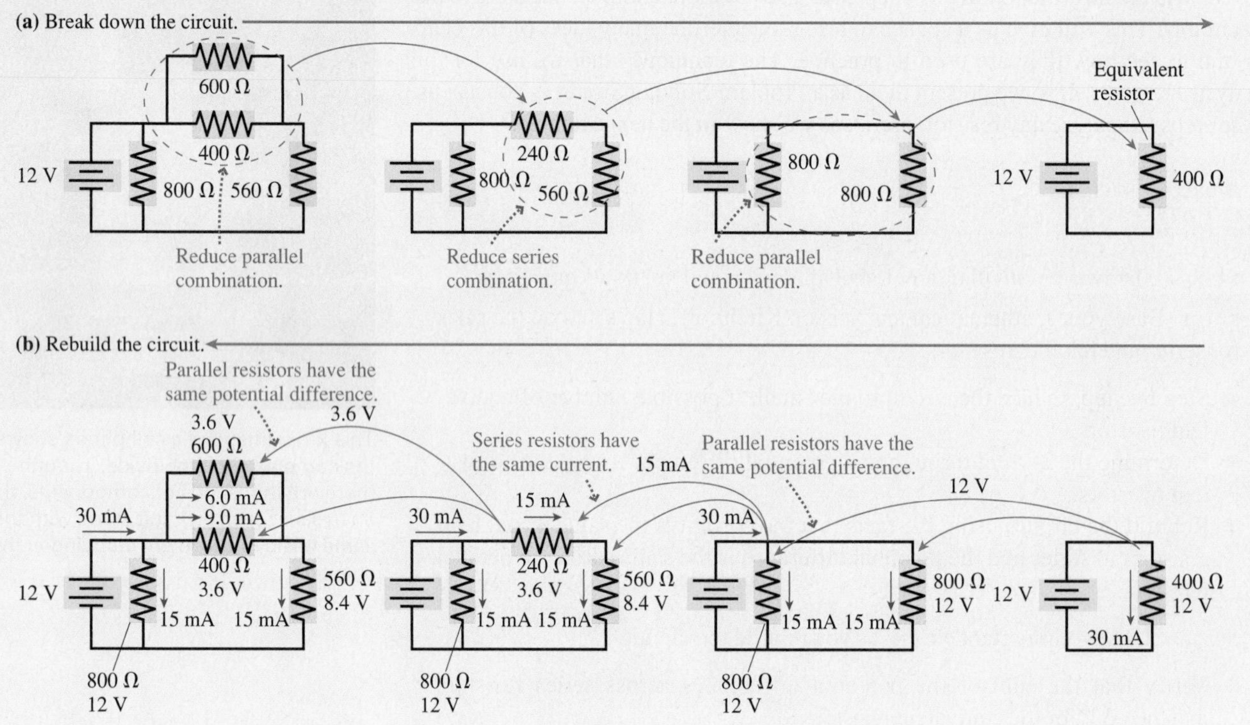

STOP TO THINK 23.4 Rank in order, from brightest to dimmest, the identical bulbs A to D.

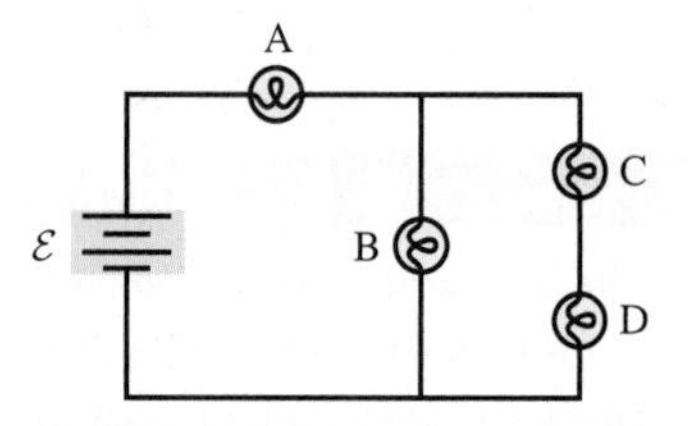

23.6 Capacitors in Parallel and Series

FIGURE 23.31 Simple capacitor circuit.

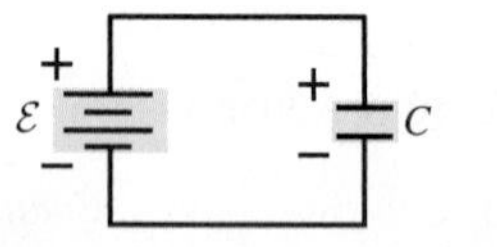

Two conductors separated by an insulating layer make a circuit element called a *capacitor,* a device we have considered in some detail in the past few chapters. FIGURE 23.31 shows a basic circuit consisting of a battery and a capacitor. When we connect the capacitor to the battery, charge will flow to the capacitor, increasing its potential difference until $\Delta V_C = \mathcal{E}$. Once the capacitor is fully charged, there will be no further current. We saw in Chapter 21 that the magnitude of the charge on each plate of the capacitor at this point will be $Q = C\,\Delta V_C = C\mathcal{E}$.

In resistor circuits, we often combine multiple resistors; we can do the same with capacitors. FIGURE 23.32 illustrates two basic combinations: parallel capacitors and series capacitors.

NOTE ► The terms "parallel capacitors" and "parallel-plate capacitor" do not describe the same thing. The former term describes how two or more capacitors are connected to each other: the latter describes how a particular capacitor is constructed. ◄

FIGURE 23.32 Parallel and series capacitors.

(a) Parallel capacitors

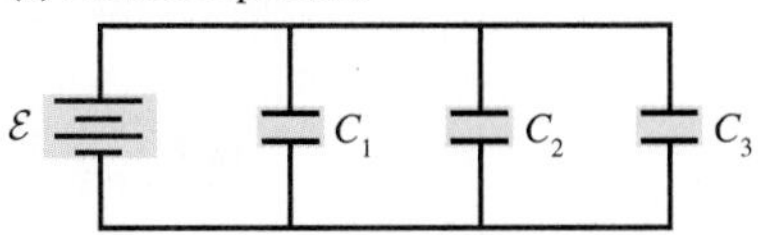

(b) Series capacitors

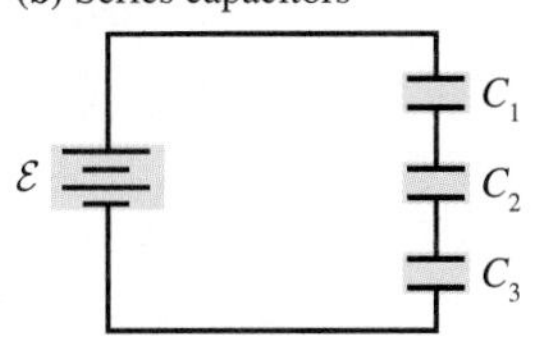

Parallel or series capacitors can be represented by a single **equivalent capacitance,** though the rules for the combinations are different from those for resistors. Let's start our analysis with the two parallel capacitors C_1 and C_2 of FIGURE 23.33a.

Suppose we start with a circuit consisting of only the battery and capacitor C_1. Adding capacitor C_2 in parallel won't change the potential difference across or the charge on capacitor C_1, but there will be a change in the circuit. C_2 sees the same potential difference as C_1, so the battery must charge up the second capacitor to ΔV_C. As a result, the combination of two capacitors stores more charge than the single capacitor, at the same voltage. This means an increased overall capacitance.

The total charge is the sum of the charges on the two capacitors:

$$Q = Q_1 + Q_2 = C_1\Delta V_C + C_2\Delta V_C = (C_1 + C_2)\Delta V_C$$

We can replace the two capacitors by a single equivalent capacitance C_{eq}, as shown in FIGURE 23.33b. The equivalent capacitance is the sum of the individual capacitance values:

$$C_{eq} = \frac{Q}{\Delta V_C} = \frac{(C_1 + C_2)\Delta V_C}{\Delta V_C} = C_1 + C_2 \tag{23.16}$$

We could easily extend this analysis to more than two capacitors. If N capacitors are in parallel, their equivalent capacitance is the sum of the individual capacitances:

$$C_{eq} = C_1 + C_2 + C_3 + \cdots + C_N \tag{23.17}$$

Equivalent capacitance of N parallel capacitors

Neither the battery nor any other part of a circuit can tell if the parallel capacitors are replaced by a single capacitor having capacitance C_{eq}.

FIGURE 23.33 Replacing two parallel capacitors with an equivalent capacitor.

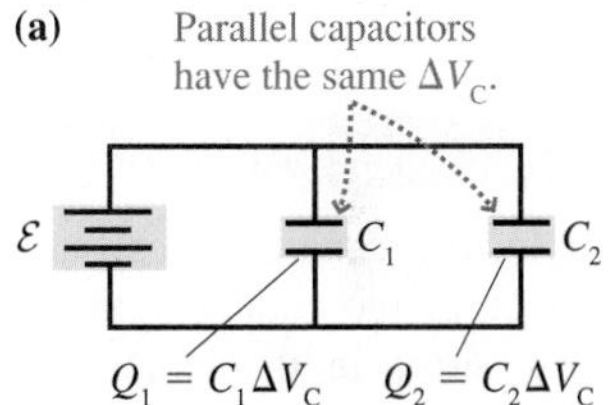

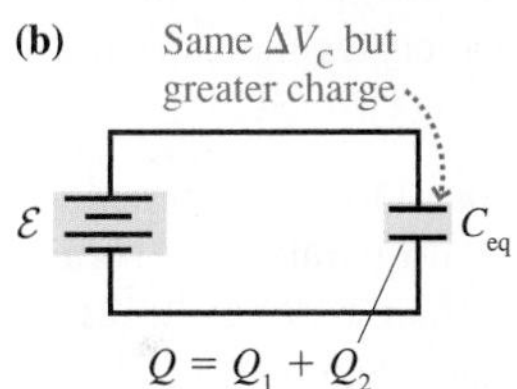

NOTE ► Adding another capacitor in parallel adds more capacitance. The formula for *parallel* capacitors is thus similar to the formula for *series* resistors. ◄

Now let's look at two capacitors connected in series. As we'll see, in this case the capacitors have the same charge, but each sees less than the full voltage of the battery.

Take a look at the circuit consisting of two series capacitors in FIGURE 23.34a. The center section, consisting of the bottom plate of C_1, the top plate of C_2, and the connecting wire, is electrically isolated. The battery cannot remove charge from or add charge to this section. If it starts out with no net charge, it must end up with no net charge. As a consequence, the two capacitors in series have equal charges $\pm Q$. The battery transfers Q from the bottom of C_2 to the top of C_1. This transfer polarizes the center section, but it still has $Q_{net} = 0$.

The potential differences across the two capacitors are $\Delta V_1 = Q/C_1$ and $\Delta V_2 = Q/C_2$. The total potential difference across both capacitors is $\Delta V_C = \Delta V_1 + \Delta V_2$. Suppose, as in FIGURE 23.34b, we replaced the two capacitors with a single capacitor having charge Q and potential difference $\Delta V_C = \Delta V_1 + \Delta V_2$. This capacitor is equivalent to the original two because the battery has to establish the same potential difference and move the same amount of charge in either case.

FIGURE 23.34 Replacing two series capacitors with an equivalent capacitor.

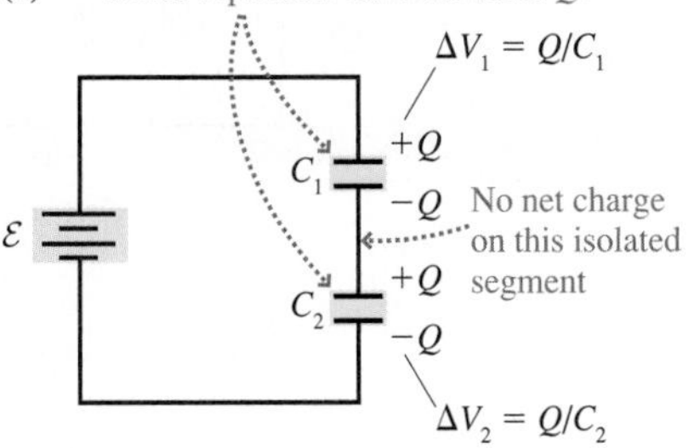

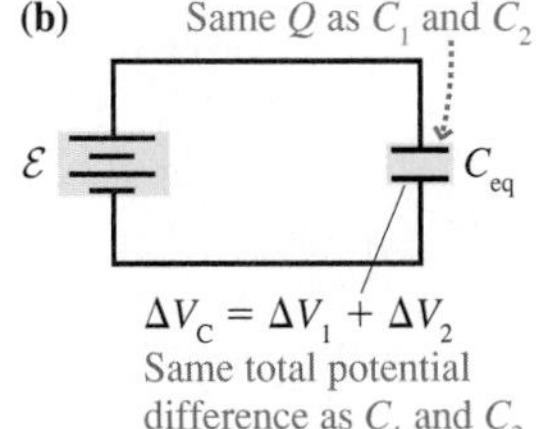

The inverse of the capacitance of this equivalent capacitor is

$$\frac{1}{C_{eq}} = \frac{\Delta V_C}{Q} = \frac{\Delta V_1 + \Delta V_2}{Q} = \frac{\Delta V_1}{Q} + \frac{\Delta V_2}{Q} = \frac{1}{C_1} + \frac{1}{C_2} \quad (23.18)$$

12.7 ActivPhysics ONLINE

This analysis hinges on the fact that **series capacitors each have the same charge Q.**

We could easily extend this analysis to more than two capacitors. If N capacitors are in series, their equivalent capacitance is the inverse of the sum of the inverses of the individual capacitances:

$$C_{eq} = \left(\frac{1}{C_1} + \frac{1}{C_2} + \frac{1}{C_3} + \cdots + \frac{1}{C_N}\right)^{-1} \quad (23.19)$$

Equivalent capacitance of N series capacitors

For series capacitors, the equivalent capacitance is less than that of the individual capacitors.

NOTE ▶ The total charge on the capacitors is the charge on each individual capacitor, but each capacitor sees only a fraction of the voltage. Adding capacitors in *series* reduces the total capacitance, just like adding resistors in *parallel*. ◀

EXAMPLE 23.10 Analyzing a capacitor circuit

a. Find the equivalent capacitance of the combination of capacitors in the circuit of FIGURE 23.35.

b. What charge flows through the battery as the capacitors are being charged?

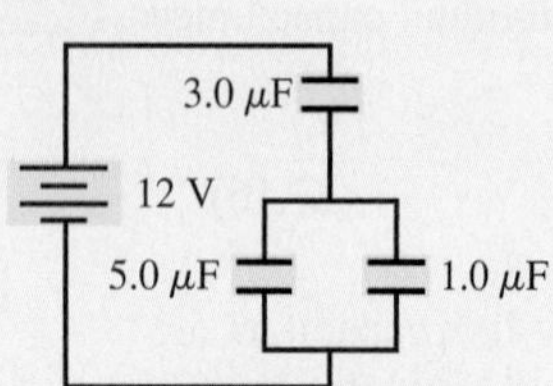

FIGURE 23.35 A capacitor circuit.

PREPARE We can use the relationships for parallel and series capacitors to reduce the capacitors to a single equivalent capacitance, much as we did for resistor circuits. We can then compute the charge through the battery using this value of capacitance.

SOLVE

a. FIGURE 23.36 shows how we find the equivalent capacitance by reducing parallel and series combinations.

b. The battery sees a capacitance of 2.0 μF. To establish a potential difference of 12 V, the charge that must flow is

$$Q = C_{eq}\,\Delta V_C = (2.0 \times 10^{-6}\ \text{F})(12\ \text{V}) = 2.4 \times 10^{-5}\ \text{C}$$

ASSESS We solve capacitor circuit problems in a manner very similar to what we followed for resistor circuits.

FIGURE 23.36 Analyzing a capacitor circuit.

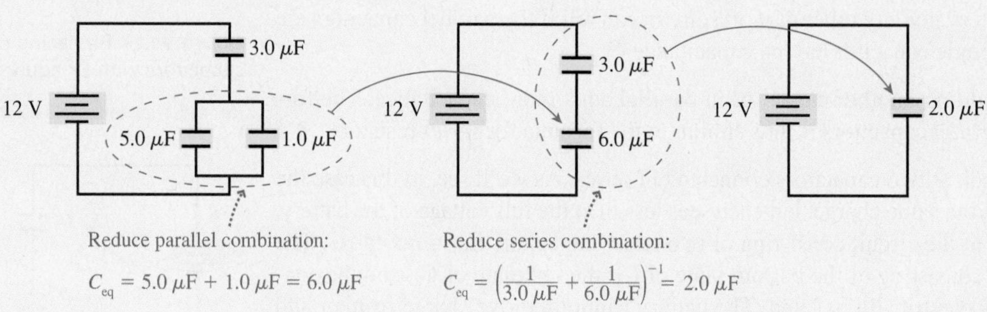

STOP TO THINK 23.5 Rank in order, from largest to smallest, the equivalent capacitance $(C_{eq})_A$ to $(C_{eq})_C$ of circuits A to C.

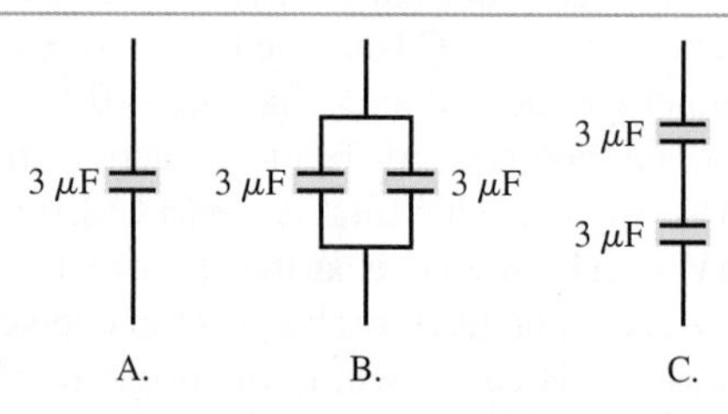

23.7 *RC* Circuits

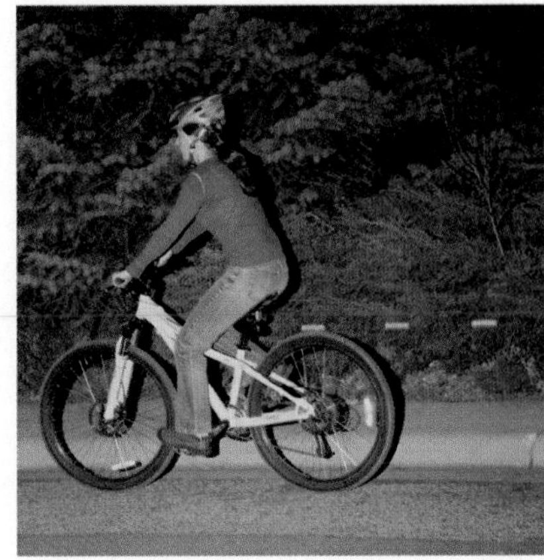
The rear flasher on a bike blinks on and off. The timing is controlled by an *RC* circuit.

The resistor circuits we have seen have a steady current. If we add a capacitor to a resistor circuit, we can make a circuit in which the current varies with time. Circuits containing resistors and capacitors are known as ***RC* circuits.** They are very common in electronic equipment. A simple example of an *RC* circuit is the flashing bike light in the photograph. As we will see, the values of the resistance and capacitance in an *RC* circuit determine the *time* it takes the capacitor to charge or discharge. In the case of the bike light, this time determines the time between flashes. A large capacitance causes a slow cycle of on-off-on; a smaller capacitance means a more rapid flicker.

FIGURE 23.37a shows an *RC* circuit consisting of a charged capacitor, an open switch, and a resistor. The capacitor has initial charge Q_0 and potential difference $(\Delta V_C)_0 = Q_0/C$. There is no current, so the potential difference across the resistor is zero. Then, at $t = 0$, the switch closes and the capacitor begins to discharge through the resistor, as in FIGURE 23.37b.

FIGURE 23.37 Discharging an *RC* circuit.

(a) Before the switch closes

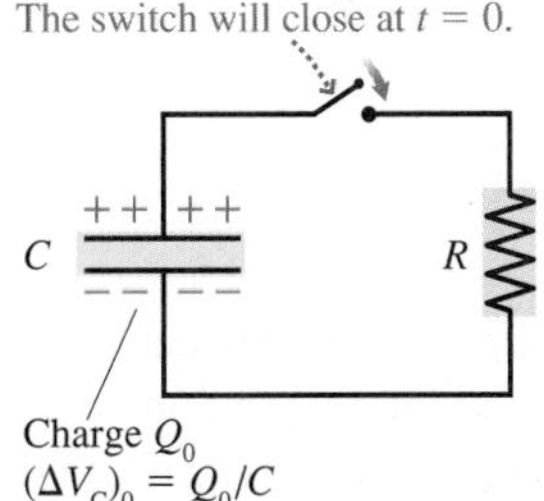

(b) Immediately after the switch closes

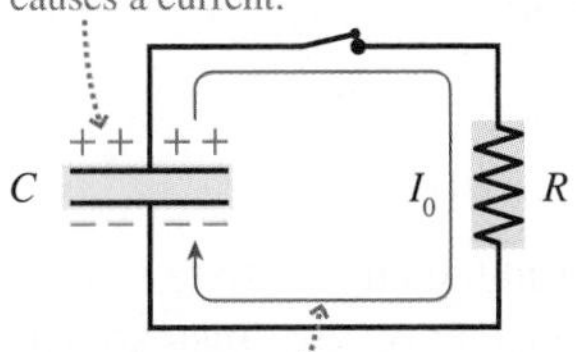

(c) At a later time

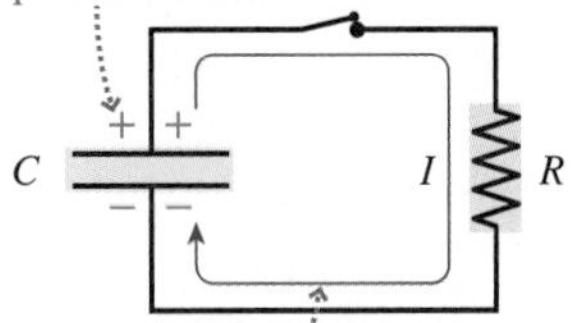

The current has reduced the charge on the capacitor. This reduces the potential difference.
C
I
R
The reduced potential difference leads to a reduced current.

How long does the capacitor take to discharge? How does the current through the resistor vary as a function of time? Let's look at the behavior of the circuit after the switch is closed.

Figure 23.37b shows the circuit *immediately* after the switch closes. The capacitor voltage is still $(\Delta V_C)_0$ because the capacitor hasn't yet had time to lose any charge, but now there's a current I_0 in the circuit that's starting to discharge the capacitor. Applying Kirchhoff's loop law, going around the loop clockwise, we find

$$\sum_i \Delta V_i = \Delta V_C + \Delta V_R = (\Delta V_C)_0 - I_0 R = 0$$

Thus the *initial* current—the initial rate at which the capacitor begins to discharge—is

$$I_0 = \frac{(\Delta V_C)_0}{R} \tag{23.20}$$

As time goes by, the current continues and the charge on the capacitor decreases. FIGURE 23.37c shows the circuit some time after the switch is closed; as we can see, both the charge on the capacitor (and thus the potential difference) and the current in the circuit have decreased. When the capacitor voltage has decreased to ΔV_C, the current has decreased to

$$I = \frac{\Delta V_C}{R} \tag{23.21}$$

The current discharges the capacitor, which causes ΔV_C to decrease. The capacitor voltage ΔV_C drives the current, so the current I decreases as well. The current I and the voltage ΔV_C both decrease until the capacitor is fully discharged and the current is zero.

If we use a voltmeter and an ammeter to measure the capacitor voltage and the current in the circuit of Figure 23.37 as a function of time, we find the variation

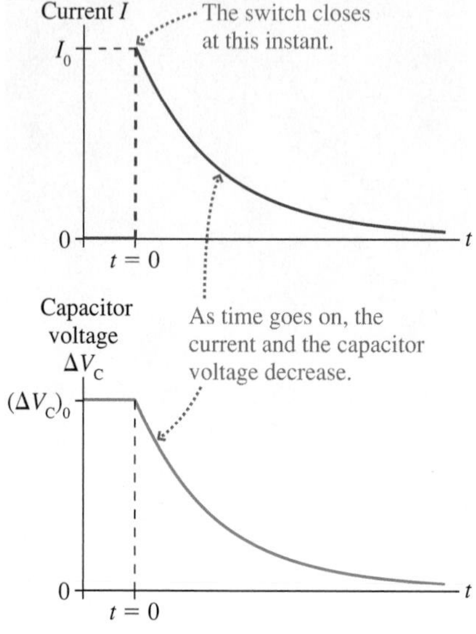

FIGURE 23.38 Current and capacitor voltage in an *RC* discharge circuit.

shown in the graphs of **FIGURE 23.38**. At $t = 0$, when the switch closes, the potential difference across the capacitor is $(\Delta V_C)_0$ and the current suddenly jumps to I_0. The current and the capacitor voltage then "decay" to zero, but *not* linearly.

The graphs in Figure 23.38 have the same shape as that for the decay of the amplitude of a damped simple harmonic oscillator we saw in Chapter 14. Both the voltage and the current are *exponential decays* given by the equations

$$I = I_0 e^{-t/RC}$$
$$\Delta V_C = (\Delta V_C)_0 e^{-t/RC} \qquad (23.22)$$

Current and voltage during a capacitor discharge

p. 464 EXPONENTIAL

In Chapter 14, we saw that we could characterize exponential decay by a **time constant** τ. The time constant is really a *characteristic time* for the circuit. A long time constant implies a slow decay; a short time constant, a rapid decay. The time constant for the decay of current and voltage in an *RC* circuit is

$$\tau = RC \qquad (23.23)$$

If you work with the units, you can show that the product of ohms and farads is seconds, so the quantity *RC* really is a time. In terms of this time constant, the current and voltage equations are

$$I = I_0 e^{-t/\tau}$$
$$\Delta V_C = (\Delta V_C)_0 e^{-t/\tau} \qquad (23.24)$$

12.8 ActivPhysics Online

The current and voltage in the circuit do not drop to zero after one time constant; that's not what the time constant means. Instead, each increase in time by one time constant causes the voltage and current to decrease by a factor of $e^{-1} = 0.37$, as we see in **FIGURE 23.39**.

FIGURE 23.39 The meaning of the time constant in an *RC* circuit.

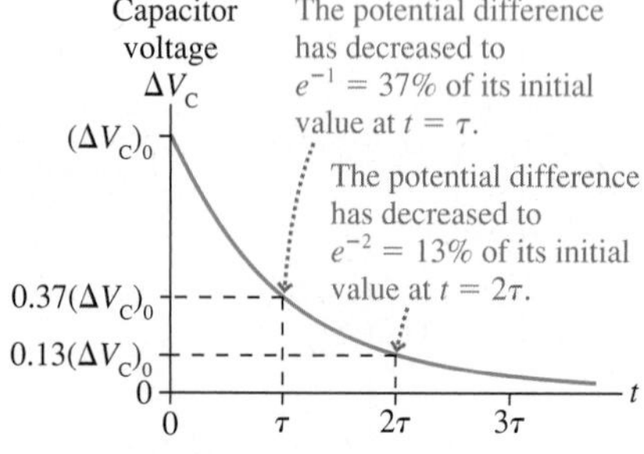

We can understand why the time constant has the form $\tau = RC$. A large value of resistance opposes the flow of charge, so increasing R increases the decay time. A larger capacitance stores more charge, so increasing C also increases the decay time.

After one time constant, the current and voltage in a capacitor circuit have decreased to 37% of their initial values. When is the capacitor fully discharged? There is no exact time that we can specify because ΔV_C approaches zero gradually. But after 5τ the voltage and current have decayed to less than 1% of their initial values. For most purposes, we can say that the capacitor is discharged at this time.

EXAMPLE 23.11 Finding the current in an *RC* circuit

The switch in the circuit of **FIGURE 23.40** has been in position a for a long time, so the capacitor is fully charged. The switch is changed to position b at $t = 0$. What is the current in the circuit immediately after the switch is closed? What is the current in the circuit 25 μs later?

FIGURE 23.40 The *RC* circuit.

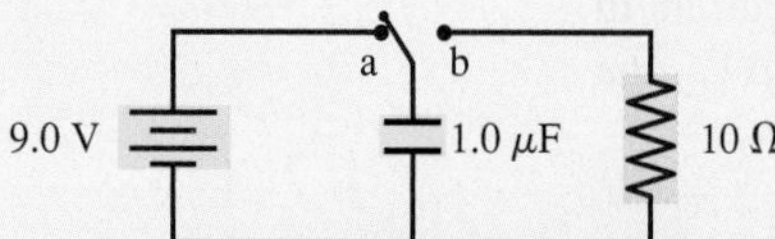

SOLVE The capacitor is connected across the battery terminals, so initially it is charged to $(\Delta V_C)_0 = 9.0$ V. When the switch is closed, the initial current is given by Equation 23.20:

$$I_0 = \frac{(\Delta V_C)_0}{R} = \frac{9.0\ \text{V}}{10\ \Omega} = 0.90\ \text{A}$$

As charge flows, the capacitor discharges. The time constant for the decay is given by Equation 23.23:

$$\tau = (10\ \Omega)(1.0 \times 10^{-6}\ \text{F}) = 1.0 \times 10^{-5}\ \text{s} = 10\ \mu\text{s}$$

The current in the circuit as a function of time is given by Equation 23.22. 25 μs after the switch is closed, the current is

$$I = I_0 e^{-t/\tau} = (0.90\ \text{A})e^{-(25\ \mu\text{s})/(10\ \mu\text{s})} = 0.074\ \text{A}$$

ASSESS This result makes sense. 25 μs after the switch has closed is 2.5 time constants, so we expect the current to decrease to a small fraction of the initial current. Notice that we left times in units of μs; this is one of the rare cases where we needn't convert to SI units. Because the exponent is $-t/\tau$, which involves a ratio of two times, we need only be certain that both t and τ are in the same units.

Charging a Capacitor

FIGURE 23.41a shows a circuit that charges a capacitor. After the switch is closed, the potential difference of the battery causes a current in the circuit, and the capacitor begins to charge. As the capacitor charges, it develops a potential difference that opposes the current, so the current decreases. As the current decreases, so does the rate of charging of the capacitor. The capacitor charges until $\Delta V_C = \mathcal{E}$, when the charging current ceases.

If we measure the current in the circuit and the potential difference across the capacitor as a function of time, we find that they vary according to the graphs in **FIGURE 23.41b**. The characteristic time for this charging circuit is the same as for the discharge, the time constant $\tau = RC$.

When the switch is first closed, the potential difference across the uncharged capacitor is zero, so the initial current is

$$I_0 = \frac{\mathcal{E}}{R}$$

The equations that describe the capacitor voltage and the current as a function of time are

$$\Delta V_C = \mathcal{E}(1 - e^{-t/\tau})$$
$$I = I_0 e^{-t/\tau} \tag{23.25}$$

Current and voltage while charging a capacitor

FIGURE 23.41 A circuit for charging a capacitor.

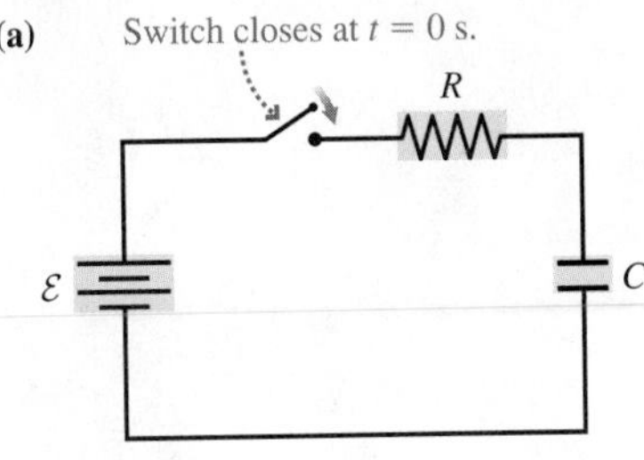

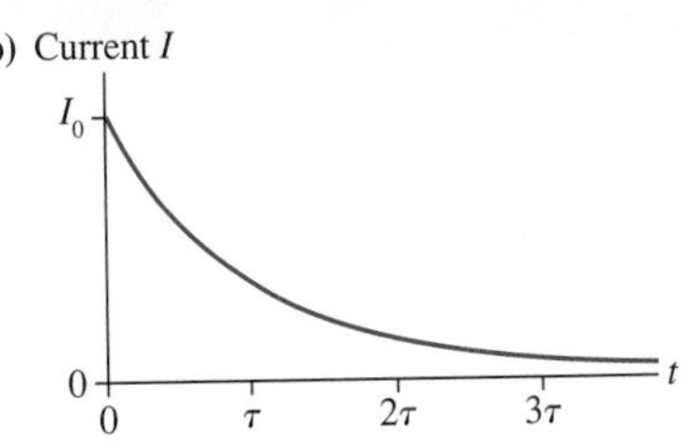

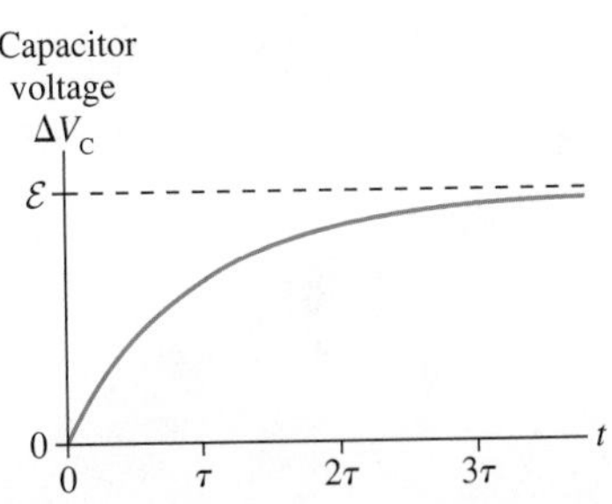

The time constant τ in an RC circuit can be used to control the behavior of a circuit. For example, a bike flasher uses an RC circuit that alternately charges and discharges, over and over, as a switch opens and closes. A separate circuit turns the light on when the capacitor voltage exceeds some threshold voltage and turns the light off when the capacitor voltage goes below this threshold. The time constant of the RC circuit determines how long the capacitor voltage stays above the threshold and thus sets the length of the flashes. More complex RC circuits provide timing in computers and other digital electronics. As we will see in the next section, we can also use RC circuits to model the transmission of nerve impulses, and the time constant will be a key factor in determining the speed at which signals can be propagated in the nervous system.

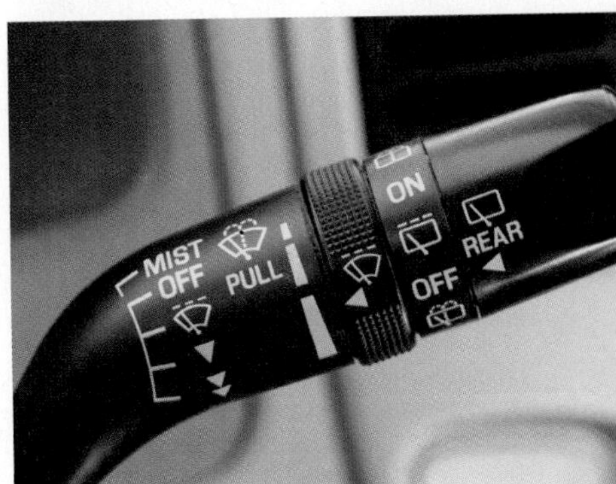

A rainy-day *RC* circuit When you adjust the dial to control the delay of the intermittent windshield wipers in your car, you are adjusting a variable resistor in an RC circuit that triggers the wipers. Increasing the resistance increases the time constant and thus produces a longer delay between swipes of the blades. A light mist calls for a long time constant and thus a large resistance.

STOP TO THINK 23.6 The time constant for the discharge of this capacitor is

A. 5 s.
B. 4 s.
C. 2 s.
D. 1 s.

2 Ω 2 Ω 1 F

23.8 Electricity in the Nervous System

In the late 1700s, the Italian scientist Galvani discovered that animal tissue has an electrical nature. He found that a frog's leg would twitch when stimulated with electricity, even when no longer attached to the frog. Further investigations by Galvani and others revealed that electrical signals can animate muscle cells, and that a small

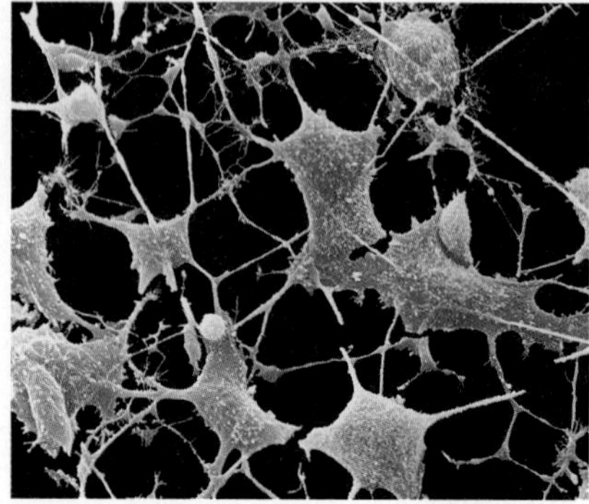

The long fibers connecting these nerve cells are *axons,* the transmission lines for electrical signals between cells.

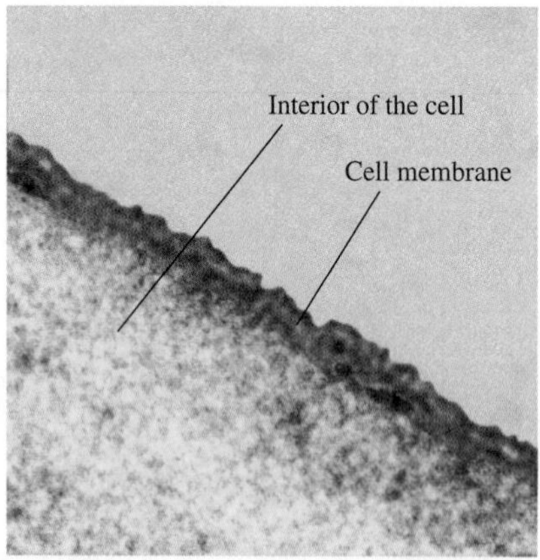

Interior of the cell
Cell membrane

A close-up view of the cell membrane, the insulating layer that divides the interior of a cell from the conducting fluid outside.

potential applied to the *axon* of a nerve cell can produce a signal that propagates down its length.

Our goal in this section will be to understand the nature of electrical signals in the nervous system. When your brain orders your hand to move, how does the signal get from your brain to your hand? Answering this question will use our knowledge of fields, potential, resistance, capacitance, and circuits, all of the knowledge and techniques that we have learned so far in Part VI.

The Electrical Nature of Nerve Cells

We start our analysis with a very simple *model* of a nerve cell that allows us to describe its electrical properties. The model begins with a *cell membrane,* an insulating layer of lipids approximately 7 nm thick that separates regions of conducting fluid inside and outside the cell.

As we saw in Chapter 21, the cell membrane is not a passive structure. It has channels and pumps that transport ions between the inside and the outside of the cell. In our simple model we will consider the transport of only two positive ions, sodium (Na^+) and potassium (K^+), though other ions are also important to cell function. Ions, rather than electrons, are the charge carriers of the cell. These ions can slowly diffuse across the cell membrane. In addition, sodium and potassium ions are transported via the following structures:

- *Sodium-potassium exchange pumps.* These pump Na^+ ions out of the cell and K^+ ions in. In the cell's resting state, the concentration of sodium ions outside the cell is about ten times the concentration on the inside. Potassium ions are more concentrated on the inside.
- *Sodium and potassium channels.* These channels in the cell membrane are usually closed. When they are open, ions move in the direction of lower concentration. Thus Na^+ ions flow into the cell and K^+ ions flow out.

FIGURE 23.42 A simple model of a nerve cell.

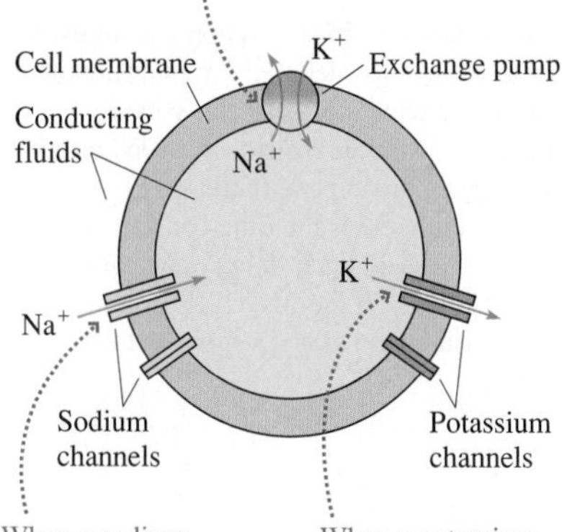

The pump moves sodium out of the cell and potassium in, so the sodium concentration is higher outside the cell, the potassium concentration is higher inside.
Cell membrane
K+
Exchange pump
Conducting fluids
Na+
K+
Na+
Sodium channels
Potassium channels
When a sodium channel is open, the higher sodium concentration outside the cell causes ions to flow into the cell.
When a potassium channel is open, the higher potassium concentration inside the cell causes ions to flow out of the cell.

Our simple model, illustrated in FIGURE 23.42, neglects many of the features of real cells, but it allows us to accurately describe the reaction of nerve cells to a stimulus and the conduction of electrical signals.

The ion exchange pumps act much like the charge escalator of a battery, using chemical energy to separate charge by transporting ions. The transport and subsequent diffusion of charged ions lead to a separation in charge across the cell membrane. Consequently, **a living cell generates an emf.** This emf takes energy to create and maintain. The ion pumps that produce the emf of neural cells account for 25–40% of the energy usage of the brain.

The charge separation produces an electric field inside the cell membrane and results in a potential difference between the inside and the outside of the cell, as shown in FIGURE 23.43 on the next page. The potential inside a nerve cell is typically 70 mV less than that outside the cell. This is called the cell's *resting potential.* Because this potential difference is produced by a charge separation across the membrane, we say that the membrane is *polarized.* Because the potential difference is entirely across the membrane, we may call this potential difference the *membrane potential.*

EXAMPLE 23.12 Electric field in a cell membrane

The thickness of a typical nerve cell membrane is 7.0 nm. What is the electric field inside the membrane of a resting nerve cell?

PREPARE The potential difference across the membrane of a resting nerve cell is -70 mV. The inner and outer surfaces of the membrane are equipotentials. We learned in Chapter 21 that the electric field is perpendicular to the equipotentials and is related to the potential difference by $E = \Delta V/d$.

SOLVE The magnitude of the potential difference between the inside and the outside of the cell is 70 mV. The field strength is thus

$$E = \frac{\Delta V}{d} = \frac{70 \times 10^{-3}\ \text{V}}{7.0 \times 10^{-9}\ \text{m}} = 1.0 \times 10^{7}\ \text{V/m}$$

The field points from positive to negative, so the electric field is

$$\vec{E} = (1.0 \times 10^{7}\ \text{V/m, inward})$$

ASSESS This is a very large electric field; in air it would be large enough to cause a spark! But we expect the fields to be large to explain the cell's strong electrical character.

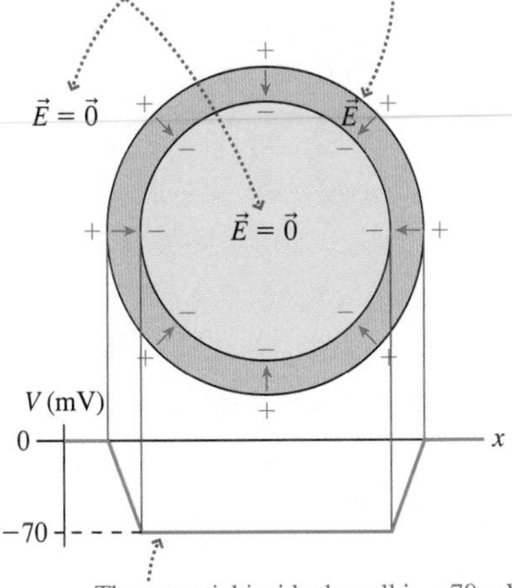

Charges on the inside and outside surfaces of the insulating membrane create a field inside it.
The conducting fluids inside and outside the cell have zero field.
$\vec{E} = \vec{0}$
$\vec{E} = \vec{0}$
$\vec{E}$
V (mV)
0
−70
x
The potential inside the cell is −70 mV.

FIGURE 23.43 The resting potential of a nerve cell.

EXAMPLE 23.13 Finding the resistance of a cell membrane

Charges can move across the cell membrane, so it is not a perfect insulator; the cell membrane will have a certain resistance. The resistivity of the cell membrane was given in Chapter 22 as $36 \times 10^{6}\ \Omega \cdot \text{m}$. What is the resistance of the 7.0-nm-thick membrane of a spherical cell with diameter 0.050 mm?

PREPARE The membrane potential will cause charges to move *through* the membrane. As we learned in Chapter 22, an object's resistance depends on its resistivity, length, and cross-section area. What this means for a cell membrane is noted in **FIGURE 23.44**.

FIGURE 23.44 The cell membrane can be modeled as a resistor.

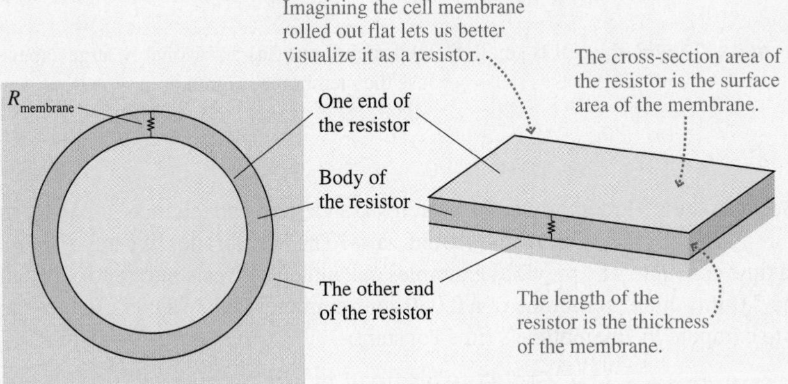

Imagining the cell membrane rolled out flat lets us better visualize it as a resistor.
The cross-section area of the resistor is the surface area of the membrane.
R_{membrane}
One end of the resistor
Body of the resistor
The other end of the resistor
The length of the resistor is the thickness of the membrane.

SOLVE The area of the membrane is the surface area of a sphere, $4\pi r^2$. We can calculate the resistance using the equation for the resistance of a conductor of length L and cross-section area A from Chapter 22:

$$R_{\text{membrane}} = \frac{\rho L}{A} = \frac{(3.6 \times 10^{7}\ \Omega \cdot \text{m})(7.0 \times 10^{-9}\ \text{m})}{4\pi(2.5 \times 10^{-5}\ \text{m})^2}$$

$$= 3.2 \times 10^{7}\ \Omega = 32\ \text{M}\Omega$$

ASSESS The resistance is quite high; the membrane is a good insulator, as we noted.

We can associate a resistance with the cell membrane, but we can associate other electrical quantities as well. The fluids inside and outside of the membrane are good conductors; they are separated by the membrane, which is not. Charges therefore accumulate on the inside and outside surfaces of the membrane. A cell thus looks like two charged conductors separated by an insulator—a capacitor.

EXAMPLE 23.14 Finding the capacitance of a cell membrane

What is the capacitance of the membrane of the spherical cell specified in Example 23.13? The dielectric constant of a cell membrane is approximately 9.0.

PREPARE If we imagine opening up a cell membrane and flattening it out, we would get something that looks like a parallel-plate capacitor with the plates separated by a dielectric as illustrated in FIGURE 23.45. The relevant dimensions are the same as those in Example 23.13.

FIGURE 23.45 The cell membrane can also be modeled as a capacitor.

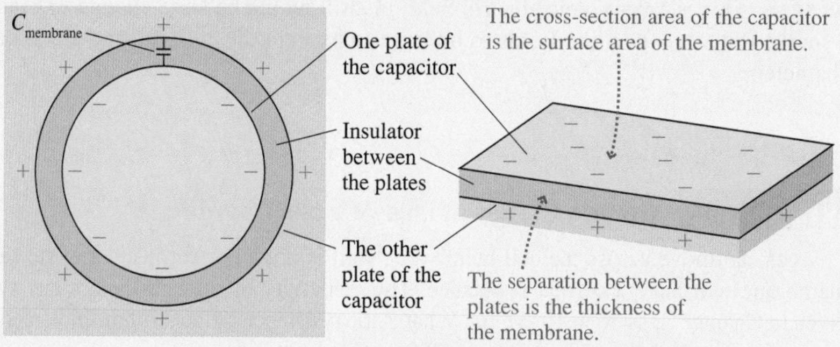

$C_{membrane}$
One plate of the capacitor
Insulator between the plates
The other plate of the capacitor
The cross-section area of the capacitor is the surface area of the membrane.
The separation between the plates is the thickness of the membrane.

SOLVE The capacitance of the membrane is that of a parallel-plate capacitor filled with a dielectric, which was given in Equation 21.22. Inserting the dimensions from Example 23.13, we find

$$C_{membrane} = \frac{\kappa\epsilon_0 A}{d} = \frac{9.0(8.85 \times 10^{-12}\ \text{C}^2/\text{N}\cdot\text{m}^2)4\pi(2.5 \times 10^{-5}\ \text{m})^2}{7.0 \times 10^{-9}\ \text{m}}$$

$$= 8.9 \times 10^{-11}\ \text{F}$$

ASSESS Though the cell is small, the cell membrane has a reasonably large capacitance of $\approx$90 pF. This makes sense because the membrane is quite thin.

FIGURE 23.46 The cell membrane can be modeled as an *RC* circuit.

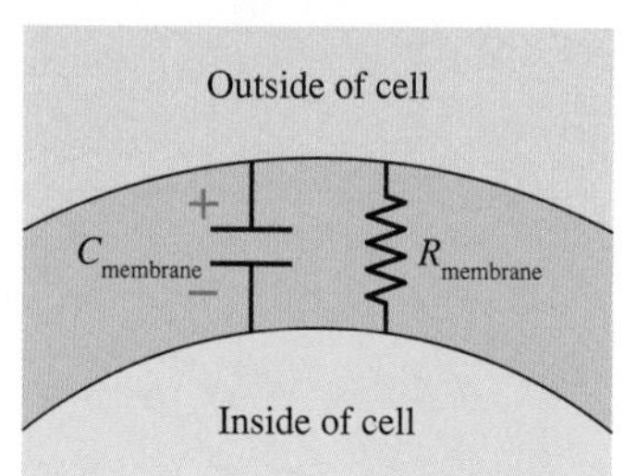

Outside of cell
$C_{membrane}$
$R_{membrane}$
Inside of cell

Because the cell membrane has both resistance and capacitance, it can be modeled as an *RC* circuit, as shown in FIGURE 23.46. The membrane, like any *RC* circuit, has a time constant. The previous examples calculated the resistance and capacitance of the 7.0-nm-thick membrane of a 0.050-mm-diameter cell. We can use these numbers to compute the membrane's time constant:

$$\tau = RC = (3.2 \times 10^7\ \Omega)(8.9 \times 10^{-11}\ \text{F}) = 2.8 \times 10^{-3}\ \text{s} \approx 3\ \text{ms}$$

Indeed, if we raise the membrane potential of a real nerve cell by 10 mV (large enough to easily measure but not enough to trigger a response in the cell), the potential will decay back to its resting value with a time constant of a few ms.

But the real action happens when some stimulus *is* large enough to trigger a response in the cell. In this case, ion channels open and the potential changes in much less time than the cell's time constant, as we will see next.

The Action Potential

Suppose a nerve cell is sitting quietly at its resting potential. The membrane potential is approximately −70 mV. However, this potential can change drastically in response to a *stimulus*. *Neurons*—nerve cells—can be stimulated by neurotransmitter chemicals released at synapse junctions. A neuron can also be electrically stimulated by a changing potential, which is why Galvani saw the frog's leg jump. Whatever the stimulus, the result is a rapid change called an *action potential;* this is the "firing" of a nerve cell. There are three phases in the action potential, as outlined below.

The action potential

Depolarization

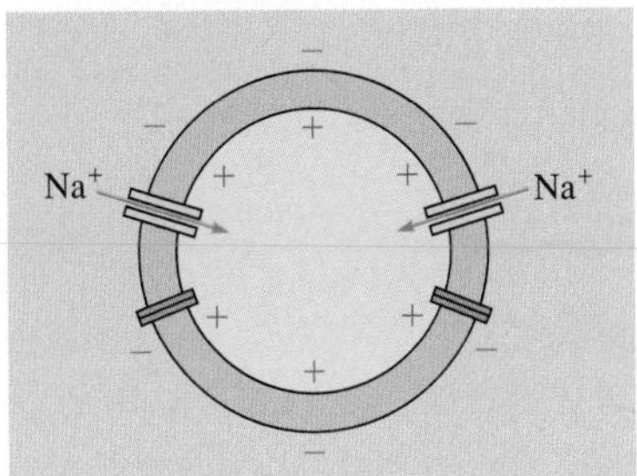

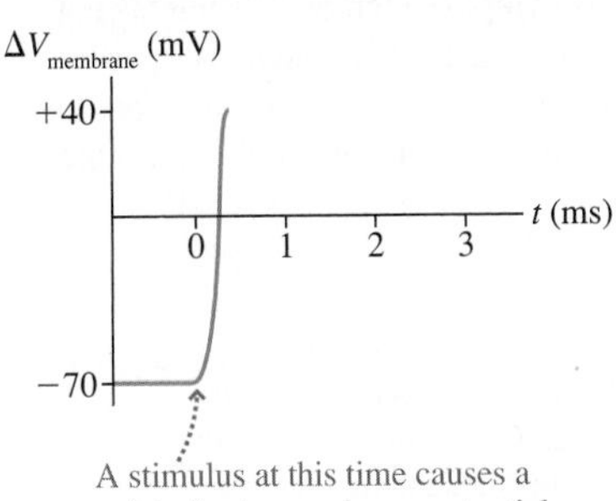
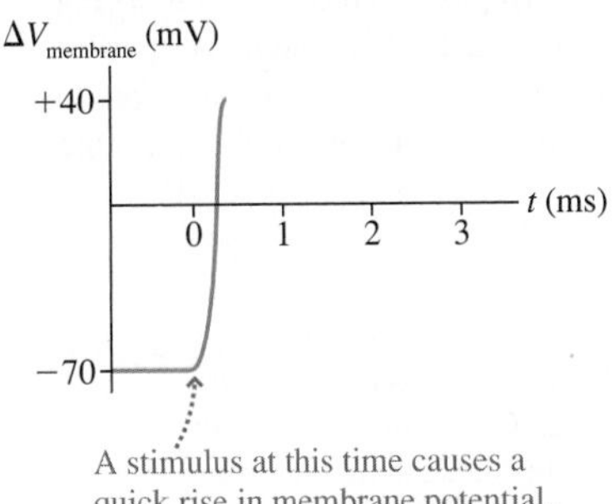

A stimulus at this time causes a quick rise in membrane potential.

A stimulus causes the cell to "fire"; the first step is the opening of the sodium channels. The concentration of sodium ions is much higher outside the cell, so sodium ions flow rapidly into the cell. In less than 1 ms, this influx of positive ions raises the membrane potential from −70 mV to +40 mV, at which point the sodium channels close. This phase of the action potential is called *depolarization*.

Repolarization

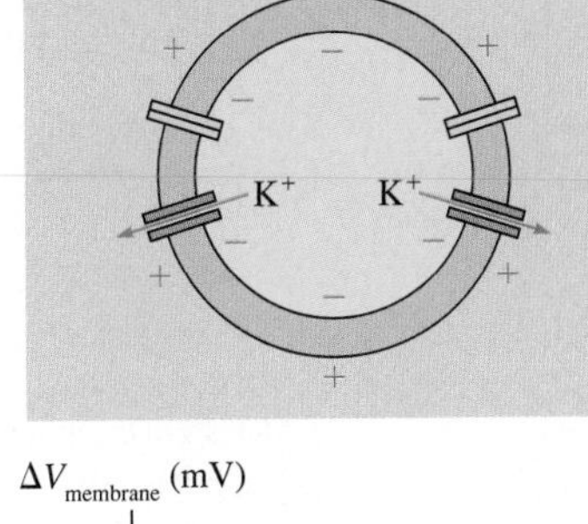

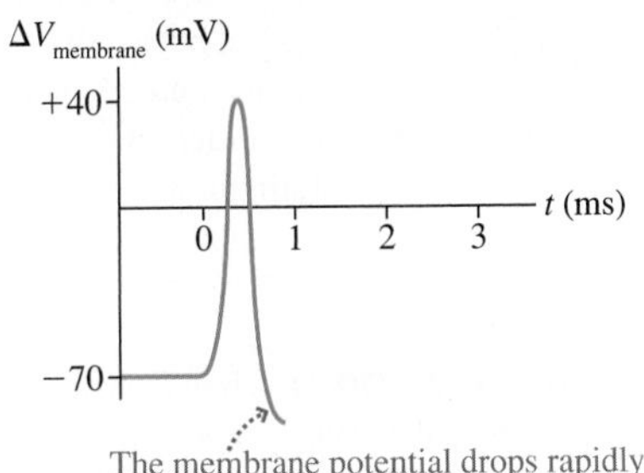

The membrane potential drops rapidly, overshooting its initial value.

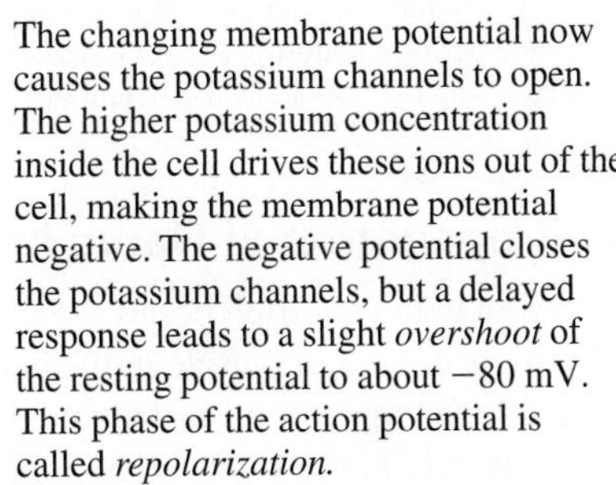

The changing membrane potential now causes the potassium channels to open. The higher potassium concentration inside the cell drives these ions out of the cell, making the membrane potential negative. The negative potential closes the potassium channels, but a delayed response leads to a slight *overshoot* of the resting potential to about −80 mV. This phase of the action potential is called *repolarization*.

Reestablishing resting potential

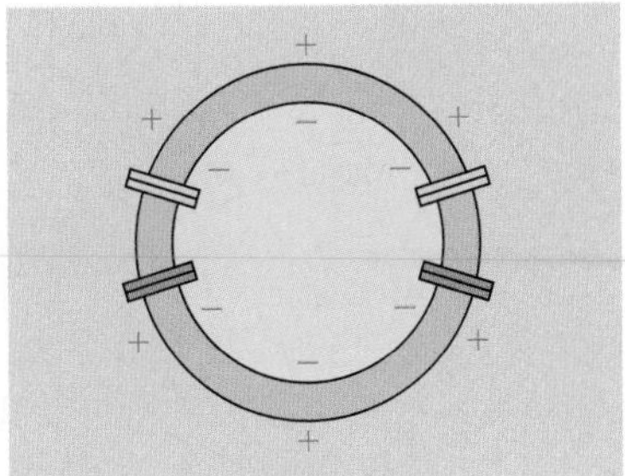

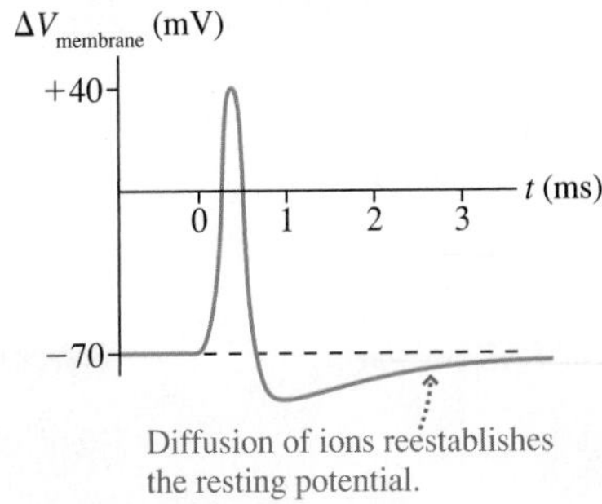

Diffusion of ions reëstablishes the resting potential.

The reestablishment of the resting potential after the sodium and potassium channels close is a relatively slow process controlled by the motion of ions across the membrane.

After the action potential is complete, there is a brief resting period, after which the cell is ready to be triggered again. The action potential is driven by ionic conduction through sodium and potassium channels, so the potential changes are quite rapid. The time for the potential to rise and then to fall is much less than the 3 ms time constant of the membrane.

The above discussion concerned nerve cells, but muscle cells undergo a similar cycle of depolarization and repolarization. The resulting potential changes are responsible for the signal that is measured by an electrocardiogram, which we learned about in Chapter 21. The potential differences in the human body are small because the changes in potential are small. But some fish have electric organs in which the action potentials of thousands of specially adapted cells are added in series, leading to very large potential differences—hundreds of volts in the case of the electric eel.

EXAMPLE 23.15 Counting ions through a channel

Investigators can measure the ion flow through a single ion channel with the *patch clamp* technique, as illustrated in FIGURE 23.47. A micropipette, a glass tube $\approx 1\ \mu\text{m}$ in diameter, makes a seal on a patch of cell membrane that includes one sodium channel. This tube is filled with a conducting saltwater solution, and a very sensitive ammeter measures the current as sodium ions flow into the cell. A sodium channel passes an average current of 4.0 pA during the 0.40 ms that the channel is open during an action potential. How many sodium ions pass through the channel?

Continued

FIGURE 23.47 Measuring the current in a single sodium channel.

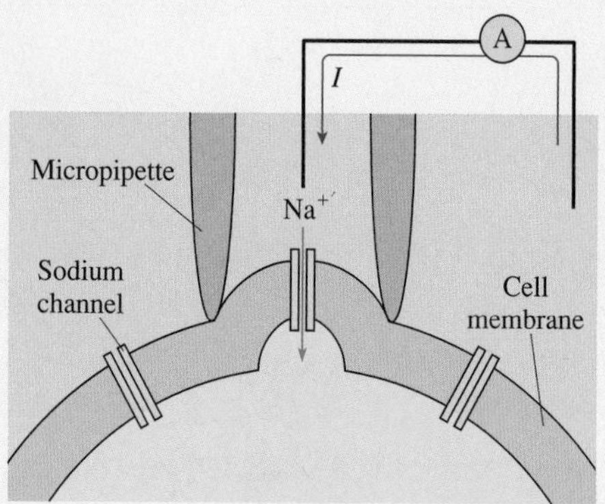
A
I
Micropipette
Na+
Sodium channel
Cell membrane

PREPARE Current is the rate of flow of charge. Each ion has charge $q = e$.

SOLVE In Chapter 22, we saw that the charge delivered by a steady current in time Δt is $Q = I\Delta t$. The amount of charge flowing through the channel in $\Delta t = 4.0 \times 10^{-4}$ s is

$$Q = I\Delta t = (4.0 \times 10^{-12}\text{ A})(4.0 \times 10^{-4}\text{ s}) = 1.6 \times 10^{-15}\text{ C}$$

This charge is due to N ions, each with $q = e$, so the number of ions is

$$N = \frac{Q}{e} = \frac{1.6 \times 10^{-15}\text{ C}}{1.6 \times 10^{-19}\text{ C}} = 10{,}000$$

ASSESS The number of ions flowing through one channel is not large, but a cell has a great many channels. The patch clamp technique and other similar procedures have allowed investigators to elucidate the details of the response of the cell membrane to a stimulus.

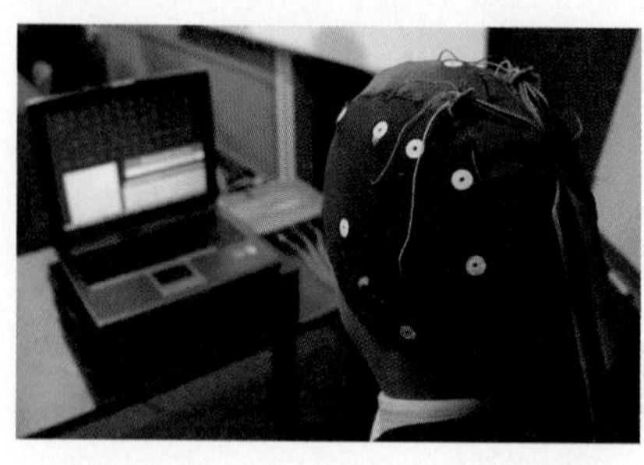

◀**Touchless typing** Different thought processes lead to different patterns of action potentials among the many neurons of the brain. The electrical activity of the cells and the motion of ions through the conducting fluid surrounding them lead to measurable differences in potential between points on the scalp. You can't use these potential differences to "read someone's mind," but it is possible to program a computer to recognize patterns and perform actions when they are detected. This man is using his thoughts—and the resulting pattern of electric potentials—to select and enter letters.

The Propagation of Nerve Impulses

Let's return to the question posed at the start of the section: How is a signal transmitted from the brain to a muscle in the hand? The primary cells of the nervous system responsible for signal transmission are known as *neurons*. The transmission of a signal to a muscle is the function of a *motor neuron*, whose structure is sketched in FIGURE 23.48. The transmission of signals takes place along the *axon* of the neuron, a long fiber—up to 1 m in length—that connects the cell body to a muscle fiber. This particular neuron has a myelin sheath around the axon, though not all neurons do.

How is a signal transmitted along an axon? The axon is long enough that different points on its membrane may have different potentials. When one point on

FIGURE 23.48 A motor neuron.

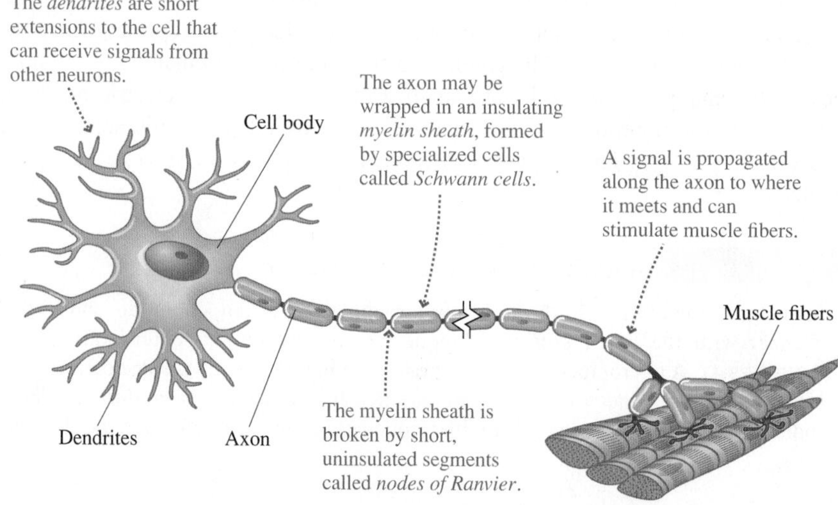
The *dendrites* are short extensions to the cell that can receive signals from other neurons.
Cell body
The axon may be wrapped in an insulating *myelin sheath*, formed by specialized cells called *Schwann cells*.
A signal is propagated along the axon to where it meets and can stimulate muscle fibers.
Muscle fibers
The myelin sheath is broken by short, uninsulated segments called *nodes of Ranvier*.
Dendrites
Axon

the axon's membrane is stimulated, the membrane will depolarize at this point. The resulting action potential may trigger depolarization in adjacent parts of the membrane. Stimulating the axon's membrane at one point can trigger a *wave* of action potential—a nerve impulse—that travels along the axon. When this signal reaches a muscle cell, the muscle cell depolarizes and produces a mechanical response.

Let's look at this process in more detail. We will start with a simple model of an axon with no myelin sheath in FIGURE 23.49a. The sodium channels are normally closed, but if the potential at some point is raised by ≈ 15 mV, from the resting potential of -70 mV to ≈ -55 mV, the sodium channels suddenly open, sodium ions rush into the cell, and an action potential is triggered. This is the key idea: **A small increase in the potential difference across the membrane causes the sodium channels to open, triggering a large action-potential response.**

This process begins at the cell body, in response to signals the neuron receives at its dendrites. If the cell body potential goes up by ≈ 15 mV, an action potential is initiated in the cell body. As the cell body potential quickly rises to a peak of $+40$ mV, it causes the potential on the nearest section of the axon—where the axon attaches to the cell body—to rise by 15 mV. This triggers an action potential in this first section of the axon. The action potential in the first section of the axon triggers an action potential in the next section of the axon, which triggers an action potential in the next section, and so on down the axon until reaching the end.

As FIGURE 23.49b shows, this causes a wave of action potential to propagate down the axon. The signal moves relatively slowly. At each point on the membrane, channels must open and ions must diffuse through, which takes time. On a typical axon with no myelin sheath, the action potential propagates at a speed of about 1 m/s. If all nerve signals traveled at this speed, a signal telling your hand to move would take about 1 s to travel from your brain to your hand. Clearly, at least some neurons in the nervous system must transmit signals at a higher speed than this!

One way to make the signals travel more quickly is to increase an axon's diameter. The giant axon in the squid triggers a rapid escape response when the squid is threatened. This axon may have a diameter of 1 mm, a thousand times that of a typical axon, providing for the necessary rapid signal transmission. But your nervous system consists of 300 billion neurons, and they can't all be 1 mm in diameter—there simply isn't enough space in your body. In your nervous system, higher neuron signal speed is achieved in a totally different manner.

FIGURE 23.49 Propagation of a nerve impulse.

(a) A model of a neuron

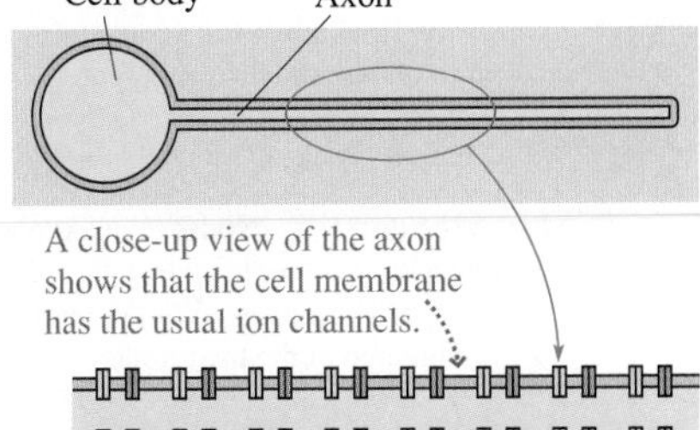

Cell body
Axon
A close-up view of the axon shows that the cell membrane has the usual ion channels.

(b) Signal propagation in the axon

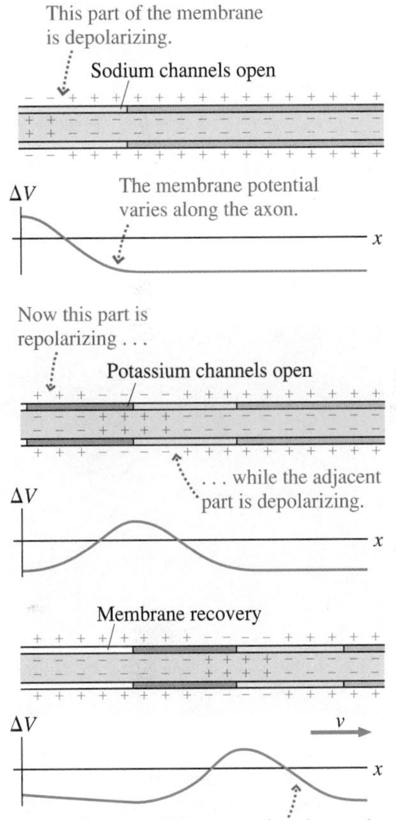

This part of the membrane is depolarizing.
Sodium channels open
ΔV
The membrane potential varies along the axon.
x
Now this part is repolarizing . . .
Potassium channels open
. . . while the adjacent part is depolarizing.
ΔV
x
Membrane recovery
ΔV
v
x
A changing potential at one point triggers the membrane to the right, leading to a wave of action potential that moves along the axon.

Increasing Speed by Insulation

The axons of motor neurons and most other neurons in your body can transmit signals at very high speeds because they are insulated with a myelin sheath. Look back at the diagram of a motor neuron in Figure 23.48. Schwann cells wrap the axon with myelin, insulating it electrically and chemically, with breaks at the nodes of Ranvier. The ion channels are concentrated in these nodes because this is the only place where the extracellular fluid is in contact with the cell membrane. In an insulated axon, a signal propagates by jumping from one node to the next. This process is called *saltatory conduction,* from the Latin *saltare,* "to leap."

FIGURE 23.50 on the next page shows a model for saltatory conduction. The membrane is triggered to depolarize at the left node. An action potential is produced here, but it can't travel down the axon as before because the axon is insulated. Instead, the

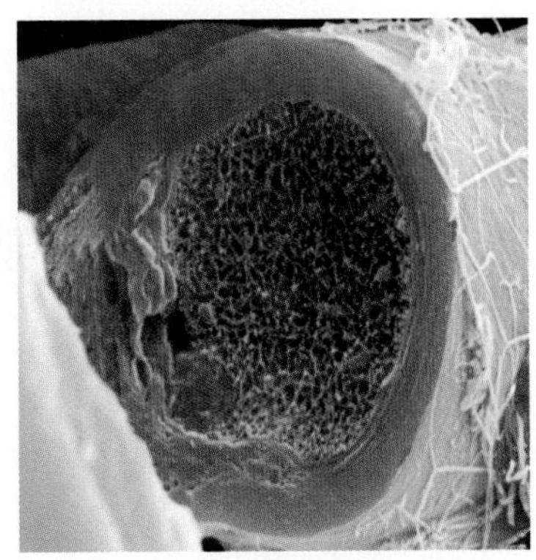

▶ The interior of this axon is insulated from the surrounding intercellular fluids by a thick myelin sheath, clearly visible in the cross-section view.

FIGURE 23.50 Nerve propagation along a myelinated axon.

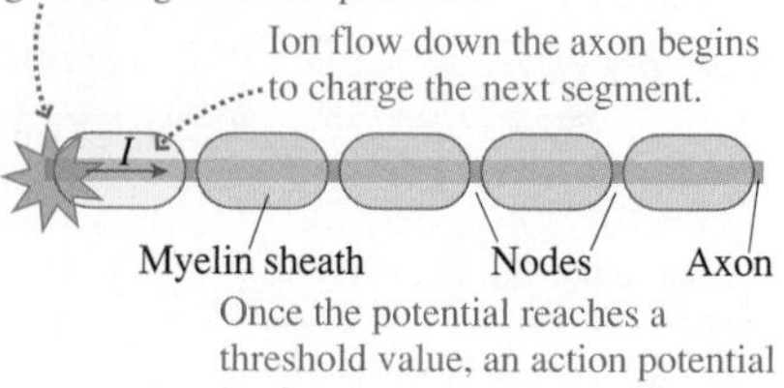

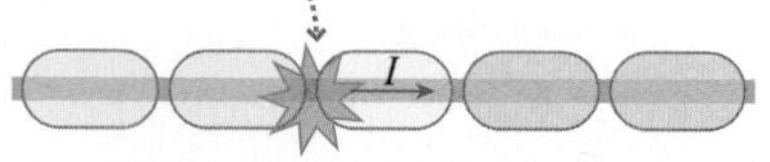

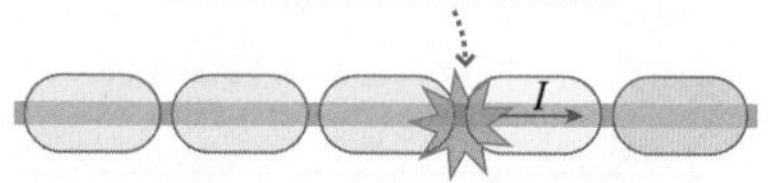

potential difference between this node and the next causes ions to flow in the body of the axon. As the charges flow down the axon, the potential at the next node rises. When the potential has risen by ≈ 15 mV, the next node is triggered and an action potential is produced at this node. This process continues, with the depolarization "jumping" from node to node down the axon. This dramatically increases the speed of propagation, as we can show.

How rapidly does a pulse move down a myelinated axon? FIGURE 23.51 provides a model of the process based on RC circuits. The critical time for propagation is the time constant $\tau = RC$ for charging the capacitance of the segments of the axon.

The resistance of an axon between one node and the next is $\approx 25\ \text{M}\Omega$. The myelin insulation increases the separation between the inner conducting fluid and the outer conducting fluid. Because the capacitance of a capacitor depends inversely on the electrode spacing d, the myelin reduces the capacitance of the membrane from the ≈ 90 pF we calculated earlier to ≈ 1.6 pF per segment. With these values, the time constant for charging the capacitor in one segment is

$$\tau = R_{\text{axon}} C_{\text{membrane}} = (25 \times 10^{6}\ \Omega)(1.6 \times 10^{-12}\ \text{F}) = 40\ \mu\text{s}$$

We've modeled the axon as a series of such segments, and the time constant is a good estimate of how much time it takes for a signal to jump from one node to the next. Because the nodes of Ranvier are spaced about 1 mm apart, the speed at which the nerve impulse travels down the axon is approximately

$$v = \frac{L_{\text{node}}}{\tau} = \frac{1.0 \times 10^{-3}\ \text{m}}{40 \times 10^{-6}\ \text{s}} = 25\ \text{m/s}$$

Although our model of nerve-impulse propagation is very simple, this predicted speed is just about right for saltatory conduction of signals in myelinated axons. This speed is 25 times faster than that in unmyelinated axons; at this speed, your brain can send a signal to your hand in $\approx \frac{1}{25}$ s.

Your electrical nature might not be as apparent as that of the electric eel, but the operation of your nervous system is inherently electrical. When you decide to move your hand, the signal from your brain travels to your hand in a process that is governed by the electrical nature of the cells in your body.

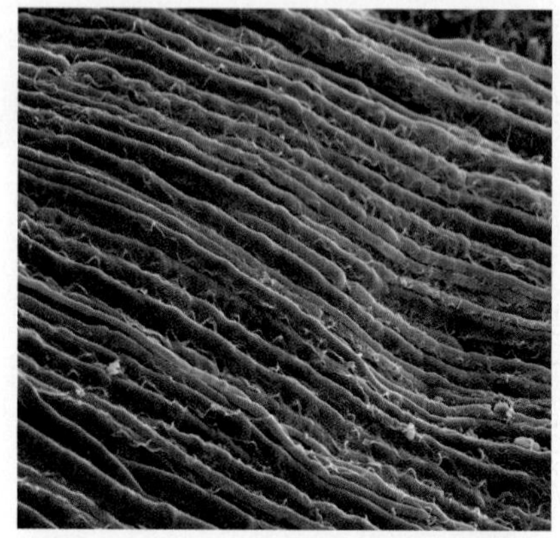

Myelinated axons in the spinal cord carry electrical signals between the brain and body. You can see the nodes of Ranvier on some of the axons.

FIGURE 23.51 A circuit model of nerve-impulse propagation along myelinated axons.

(a) A model of a myelinated axon

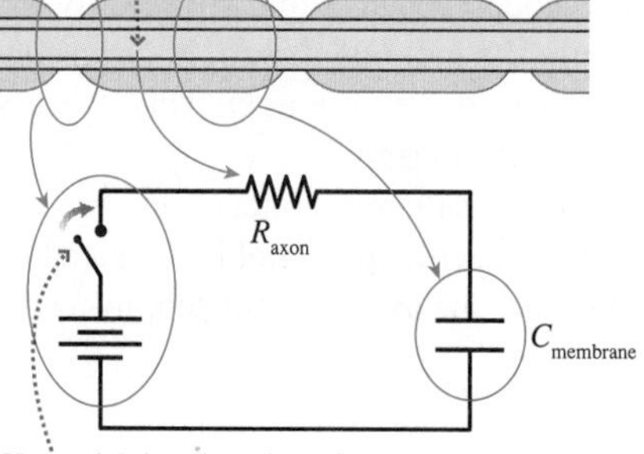

The fluid in the axon has a certain resistivity. The axon is thin, so the resistance is large.

We model the triggering of an action potential as closing a switch connected to a battery.

(b) Signal propagation in the myelinated axon

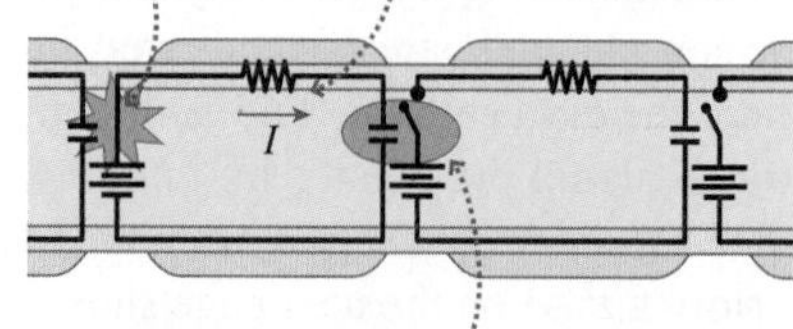

1. An action potential is triggered at this node; we close the switch.
2. Once the switch is closed, the action potential emf drives a current down the axon and charges the capacitance of the membrane.
I
3. When the voltage on the capacitor exceeds a threshold, it triggers an action potential at this node—the next switch is closed.

INTEGRATED EXAMPLE 23.16 Soil moisture measurement

The moisture content of soil is given in terms of its *volumetric fraction,* the ratio of the volume of water in the soil to the volume of the soil itself. Making this measurement directly is quite time-consuming, so soil scientists are eager to find other means to reliably measure soil moisture. Water has a very large dielectric constant, so the dielectric constant of soil increases as its moisture content increases. **FIGURE 23.52** shows data for the dielectric constant of soil versus the volumetric fraction; the increase in dielectric constant with soil moisture is quite clear. This strong dependence of dielectric constant on soil moisture allows a very sensitive—and simple—electrical test of soil moisture.

FIGURE 23.52 Variation of the dielectric constant with soil moisture.

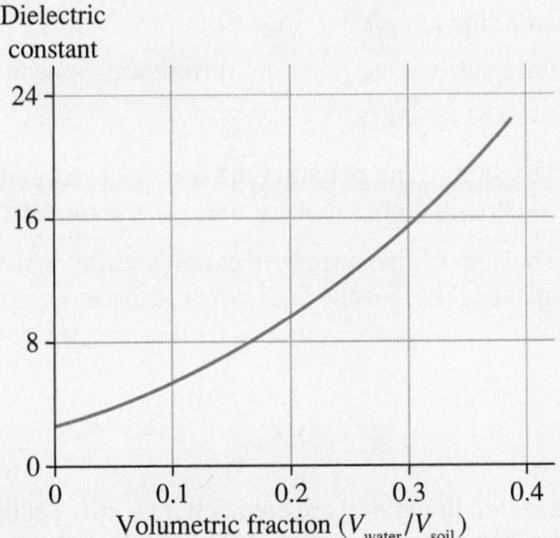

A soil moisture meter has a probe with two separated electrodes. When the probe is inserted into the soil, the electrodes form a capacitor whose capacitance depends on the dielectric constant of the soil between them. A circuit charges the capacitor probe to 3.0 V, then discharges it through a resistor. The decay time depends on the capacitance—and thus the soil moisture—so a measurement of the time for the capacitor to discharge allows a determination of the amount of moisture in the soil.

In air, the probe's capacitance takes 15 μs to discharge from 3.0 V to 1.0 V. In one particular test, when the probe was inserted into the ground, this discharge required 150 μs. What was the approximate volumetric fraction of water for the soil in this test?

PREPARE **FIGURE 23.53** is a sketch of the measurement circuit of the soil moisture meter. The capacitor is first charged, then connected across a resistor to form an *RC* circuit. The decay of the capacitor voltage is governed by the time constant for the circuit. The time constant depends on the resistance and on the electrode capacitance, which depends on the dielectric constant of the soil between the electrodes.

The capacitance of the probe in air is that of a parallel-plate capacitor: $C_{air} = \epsilon_0 A/d$. The capacitance of the probe in soil differs only by the additional factor of the dielectric constant of the medium between the plates—in this case, the soil:

$$C_{soil} = \kappa_{soil}\left(\frac{\epsilon_0 A}{d}\right) = \kappa_{soil} C_{air}$$

FIGURE 23.53 The measurement circuit of the soil moisture meter.

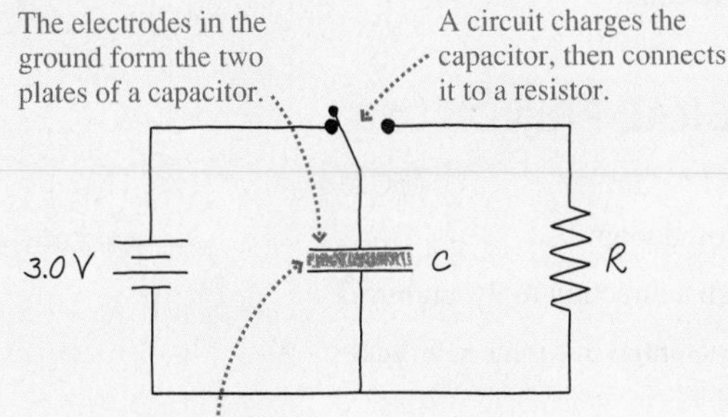

The electrodes in the ground form the two plates of a capacitor.
A circuit charges the capacitor, then connects it to a resistor.
3.0 V
C
R
The soil is the dielectric material between the plates of the capacitor.

The ratio of the capacitance values gives the dielectric constant:

$$\kappa_{soil} = \frac{C_{soil}}{C_{air}}$$

Once we know the dielectric constant, we can determine the volumetric fraction of water from the graph.

SOLVE We are given the times for the decay in air and in soil, but not the capacitance or the resistance of the probe. That's not a problem, though; we don't actually need the capacitance, only the *ratio* of the capacitances in air and in soil. Equation 23.22 gives the voltage decay of an *RC* circuit: $\Delta V_C = (\Delta V_C)_0 e^{-t/RC}$. In air, the decay is

$$1.0\text{ V} = (3.0\text{ V})e^{-(15\times10^{-6}\text{ s})/RC_{air}}$$

In soil, the decay is

$$1.0\text{ V} = (3.0\text{ V})e^{-(150\times10^{-6}\text{ s})/RC_{soil}}$$

Because the starting and ending points for the decay are the same, the exponents of the two expressions must be equal:

$$\frac{15\times10^{-6}\text{ s}}{RC_{air}} = \frac{150\times10^{-6}\text{ s}}{RC_{soil}}$$

We can solve this for the ratio of the capacitances in soil and air:

$$\frac{C_{soil}}{C_{air}} = \frac{150\times10^{-6}\text{ s}}{15\times10^{-6}\text{ s}} = 10$$

We saw above that this ratio is the dielectric constant, so $\kappa_{soil} = 10$. We then use the graph of Figure 23.52 to determine that this dielectric constant corresponds to a volumetric water fraction of approximately 0.20.

ASSESS The decay times in air and in soil differ by a factor of 10, so the capacitance in the soil is much larger than that in air. This implies a large dielectric constant, meaning that there is a lot of water in the soil. A volumetric fraction of 0.20 means that 20% of the soil's volume is water (that is, 1.0 cm^3 of soil contains 0.20 cm^3 of water)—which is quite a bit, so our result seems reasonable.

SUMMARY

The goal of Chapter 23 has been to understand the fundamental physical principles that govern electric circuits.

GENERAL PRINCIPLES

Kirchhoff's loop law

For a closed loop:

- Assign a direction to the current.
- Add potential differences around the loop:

$$\sum_i \Delta V_i = 0$$

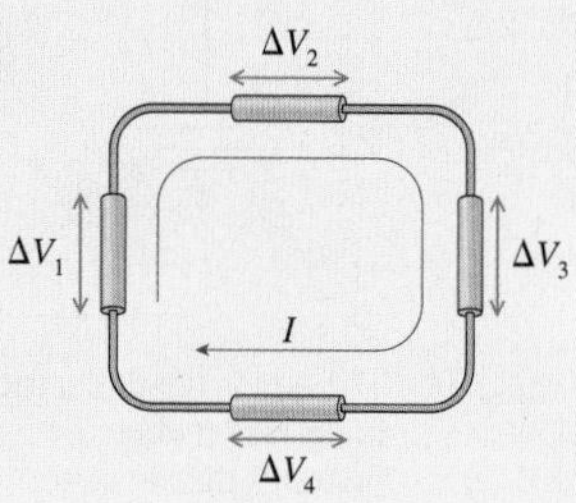

Kirchhoff's junction law

For a junction:

$$\sum I_{\text{in}} = \sum I_{\text{out}}$$

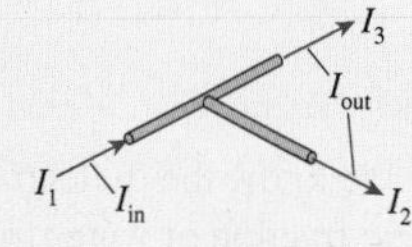

Analyzing Circuits

PREPARE Draw a circuit diagram.

SOLVE *Break the circuit down:*

- Reduce the circuit to the smallest possible number of equivalent resistors.
- Find the current and potential difference.

Rebuild the circuit:

- Find current and potential difference for each resistor.

ASSESS Verify that

- The sum of the potential differences across series resistors matches that for the equivalent resistor.
- The sum of the currents through parallel resistors matches that for the equivalent resistor.

IMPORTANT CONCEPTS

Series elements

A series connection has no junction.
The current in each element is the same.

Resistors in series can be reduced to an equivalent resistance:

$$R_{\text{eq}} = R_1 + R_2 + R_3 + \cdots$$

Capacitors in series can be reduced to an equivalent capacitance:

$$C_{\text{eq}} = \left(\frac{1}{C_1} + \frac{1}{C_2} + \frac{1}{C_3} + \cdots\right)^{-1}$$

Parallel elements

Elements connected in parallel are connected by wires at both ends.
The potential difference across each element is the same.

Resistors in parallel can be reduced to an equivalent resistance:

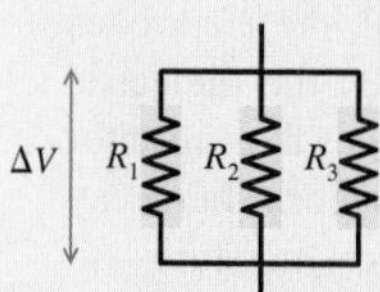

$$R_{\text{eq}} = \left(\frac{1}{R_1} + \frac{1}{R_2} + \frac{1}{R_3} + \cdots\right)^{-1}$$

Capacitors in parallel can be reduced to an equivalent capacitance:

$$C_{\text{eq}} = C_1 + C_2 + C_3 + \cdots$$

APPLICATIONS

RC circuits

The discharge of a capacitor through a resistor is an exponential decay:

$$\Delta V_C = (\Delta V_C)_0 e^{-t/\tau}$$

The **time constant** for the decay is

$$\tau = RC$$

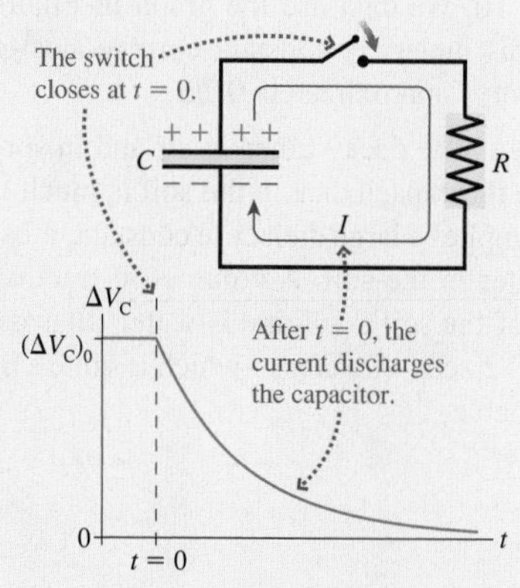

Electricity in the nervous system

Cells in the nervous system maintain a negative potential inside the cell membrane. When triggered, the membrane depolarizes and generates an *action potential.*

An action potential travels as a wave along the axon of a neuron. More rapid saltatory conduction can be achieved by insulating the axon with myelin, causing the action potential to jump from node to node.

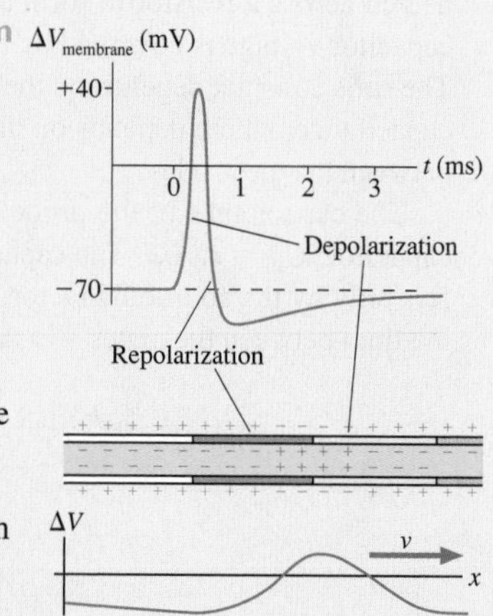

For homework assigned on MasteringPhysics, go to www.masteringphysics.com

Problems labeled INT integrate significant material from earlier chapters; BIO are of biological or medical interest.

Problem difficulty is labeled as | (straightforward) to ||||| (challenging).

QUESTIONS

Conceptual Questions

1. The tip of a flashlight bulb is touching the top of a 3 V battery as shown in Figure Q23.1. Does the bulb light? Why or why not?

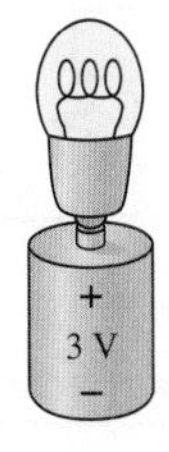

FIGURE Q23.1

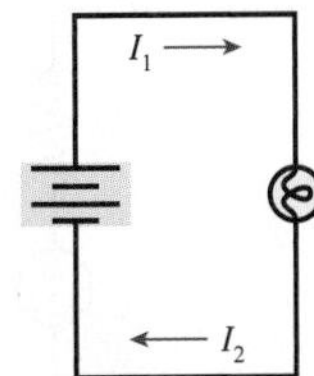

FIGURE Q23.2

2. A flashlight bulb is connected to a battery and is glowing; the circuit is shown in Figure Q23.2. Is current I_2 greater than, less than, or equal to current I_1? Explain.
3. Current I_{in} flows into three resistors connected together one after the other as shown in Figure Q23.3. The accompanying graph shows the value of the potential as a function of position.
   a. Is I_{out} greater than, less than, or equal to I_{in}? Explain.
   b. Rank in order, from largest to smallest, the three resistances R_1, R_2, and R_3. Explain.

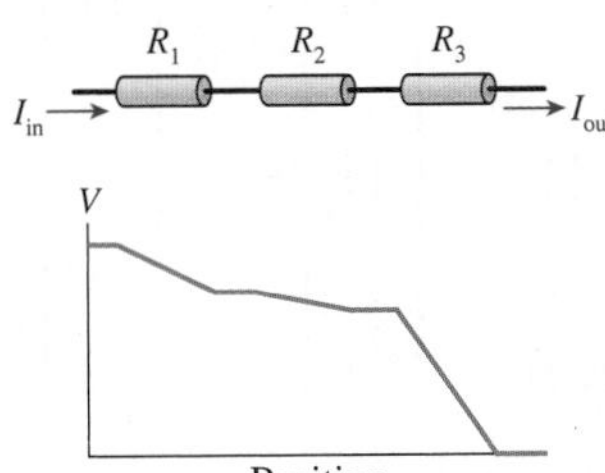

FIGURE Q23.3

4. The circuit in Figure Q23.4 has two resistors, with $R_1 > R_2$. Which resistor dissipates the larger amount of power? Explain.

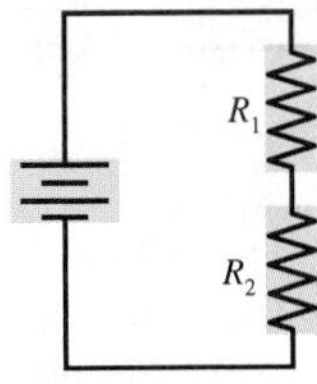

FIGURE Q23.4

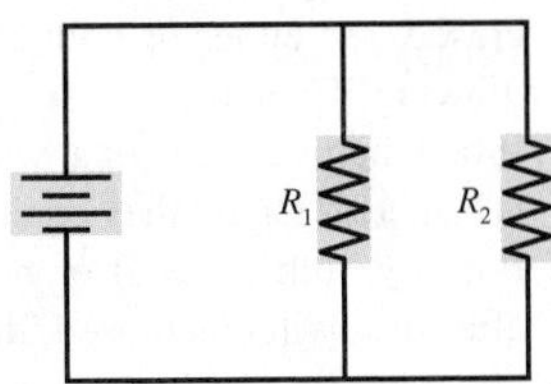

FIGURE Q23.5

5. The circuit in Figure Q23.5 has a battery and two resistors, with $R_1 > R_2$. Which resistor dissipates the larger amount of power? Explain.
6. In the circuit shown in Figure Q23.6, bulbs A and B are glowing. Then the switch is closed. What happens to each bulb? Does it get brighter, stay the same, get dimmer, or go out? Explain.

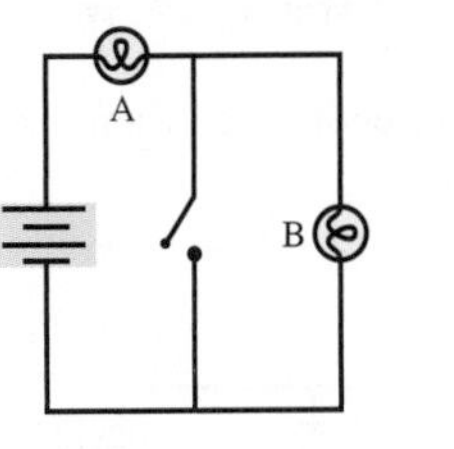

FIGURE Q23.6

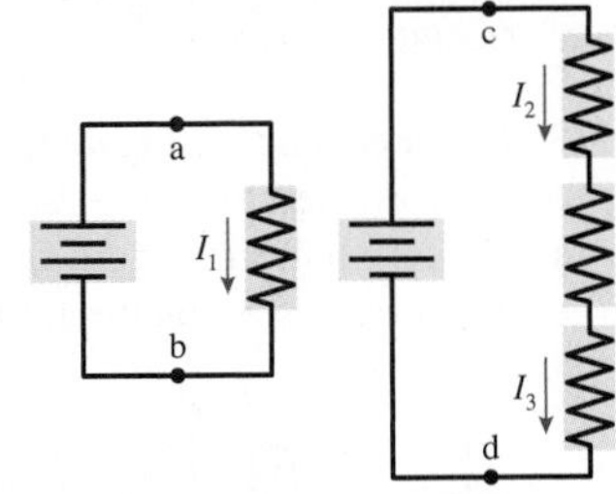

FIGURE Q23.7

7. Figure Q23.7 shows two circuits. The two batteries are identical and the four resistors all have exactly the same resistance.
   a. Is ΔV_{ab} larger than, smaller than, or equal to ΔV_{cd}? Explain.
   b. Rank in order, from largest to smallest, the currents I_1, I_2, and I_3. Explain.
8. Figure Q23.8 shows two circuits. The two batteries are identical and the four resistors all have exactly the same resistance.
   a. Compare ΔV_{ab}, ΔV_{cd}, and ΔV_{ef}. Are they all the same? If not, rank them in order from largest to smallest. Explain.
   b. Rank in order, from largest to smallest, the five currents I_1 to I_5. Explain.

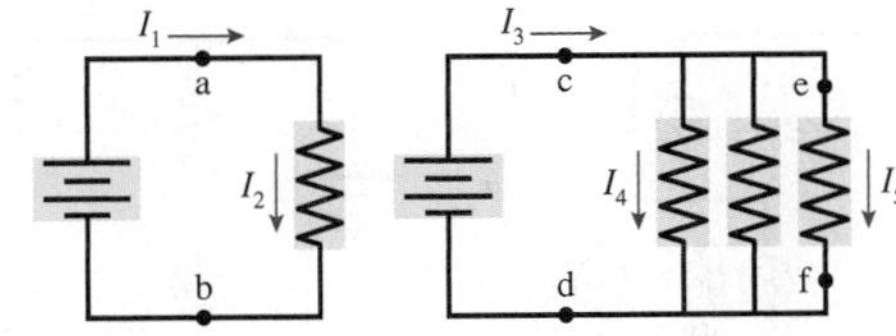

FIGURE Q23.8

9. a. In Figure Q23.9, what fraction of current I goes through the 3 Ω resistor?
   b. If the 9 Ω resistor is replaced with a larger resistor, will the fraction of current going through the 3 Ω resistor increase, decrease, or stay the same?

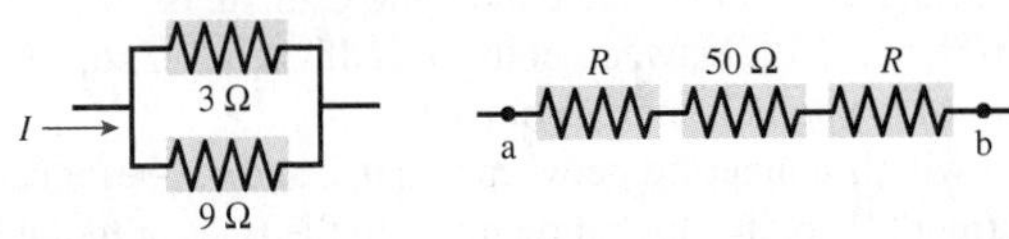

FIGURE Q23.9

FIGURE Q23.10

10. Two of the three resistors in Figure Q23.10 are unknown but equal. Is the total resistance between points a and b less than, greater than, or equal to 50 Ω? Explain.
11. Two of the three resistors in Figure Q23.11 are unknown but equal. Is the total resistance between points a and b less than, greater than, or equal to 200 Ω? Explain.

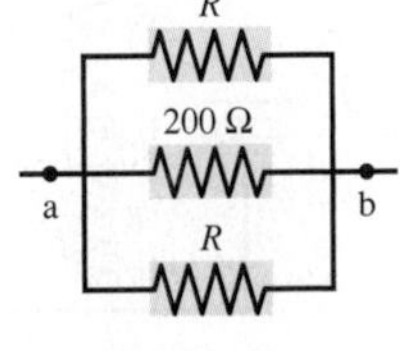

FIGURE Q23.11

12. Rank in order, from largest to smallest, the currents I_1, I_2, and I_3 in the circuit diagram in Figure Q23.12.

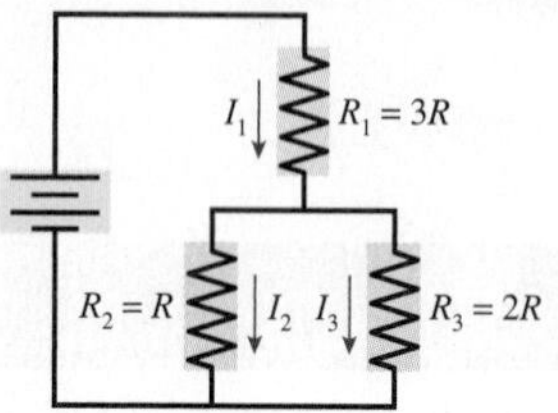

FIGURE Q23.12

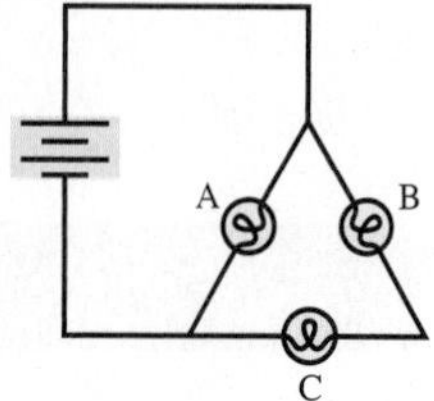

FIGURE Q23.13

13. The three bulbs in Figure Q23.13 are identical. Rank the bulbs from brightest to dimmest. Explain.
14. The four bulbs in Figure Q23.14 are identical. Rank the bulbs from brightest to dimmest. Explain.

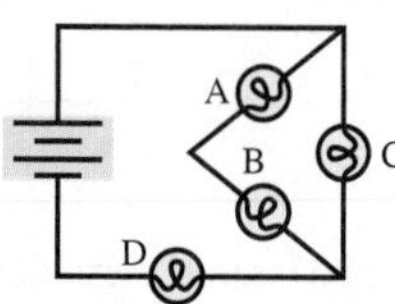

FIGURE Q23.14

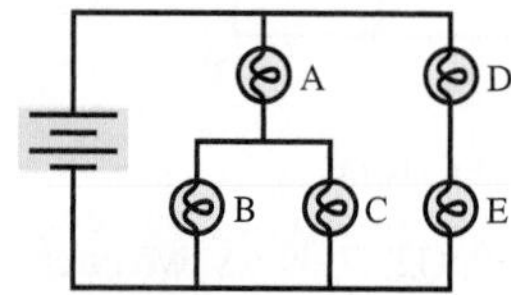

FIGURE Q23.15

15. Figure Q23.15 shows five identical bulbs connected to a battery. All the bulbs are glowing. Rank the bulbs from brightest to dimmest. Explain.
16. a. The three bulbs in Figure Q23.16 are identical. Rank the bulbs from brightest to dimmest. Explain.
    b. Suppose a wire is connected between points 1 and 2. What happens to each bulb? Does it get brighter, stay the same, get dimmer, or go out? Explain.

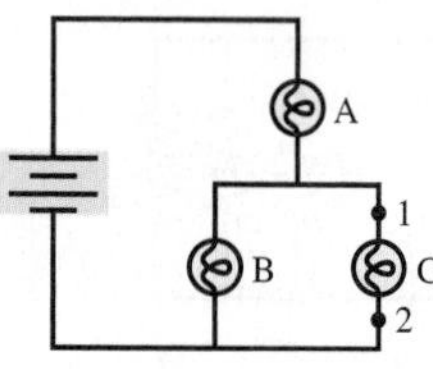

FIGURE Q23.16

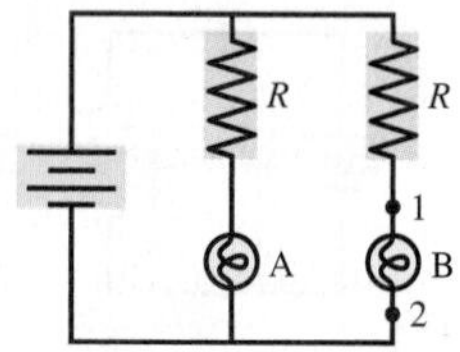

FIGURE Q23.17

17. Initially, bulbs A and B in Figure Q23.17 are both glowing. Bulb B is then removed from its socket. Does removing bulb B cause the potential difference ΔV_{12} between points 1 and 2 to increase, decrease, stay the same, or become zero? Explain.
18. a. Consider the points a and b in Figure Q23.18. Is the potential difference ΔV_{ab} between points a and b zero? If so, why? If not, which point is more positive?
    b. If a wire is connected between points a and b, does it carry a current? If so, in which direction—to the right or to the left? Explain.

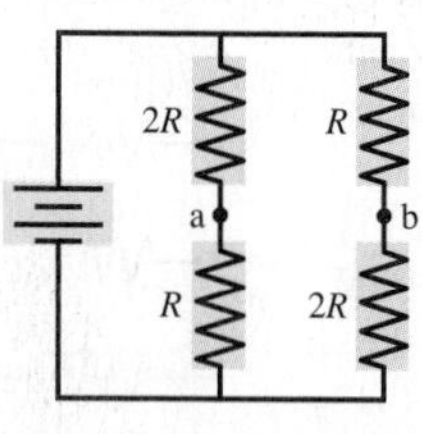

FIGURE Q23.18

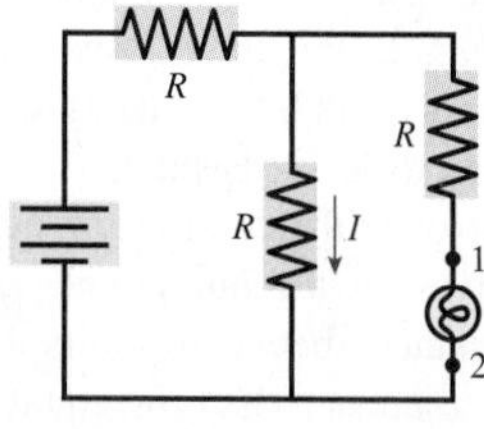

FIGURE Q23.19

19. Initially the lightbulb in Figure Q23.19 is glowing. It is then removed from its socket.
    a. What happens to the current I when the bulb is removed? Does it increase, stay the same, or decrease? Explain.
    b. What happens to the potential difference ΔV_{12} between points 1 and 2? Does it increase, stay the same, decrease, or become zero? Explain.
20. A voltmeter is (incorrectly) inserted into a circuit as shown in Figure Q23.20.
    a. What is the current in the circuit?
    b. What does the voltmeter read?
    c. How would you change the circuit to correctly connect the voltmeter to measure the potential difference across the resistor?

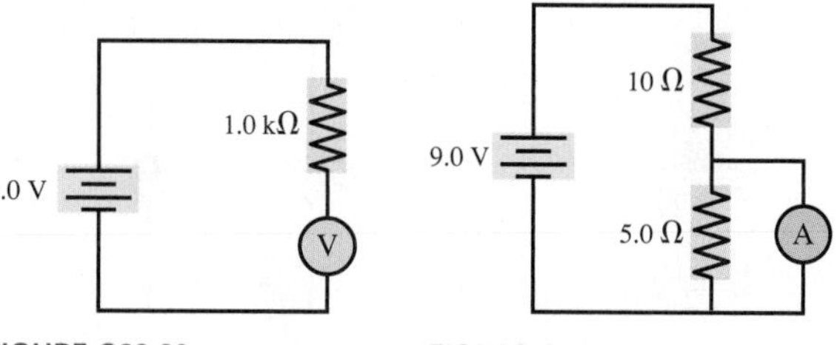

FIGURE Q23.20 FIGURE Q23.21

21. An ammeter is (incorrectly) inserted into a circuit as shown in Figure Q23.21.
    a. What is the current through the 5.0 Ω resistor?
    b. How would you change the circuit to correctly connect the ammeter to measure the current through the 5.0 Ω resistor?
22. Rank in order, from largest to smallest, the equivalent capacitances $(C_{eq})_1$ to $(C_{eq})_4$ of the four groups of capacitors shown in Figure Q23.22.

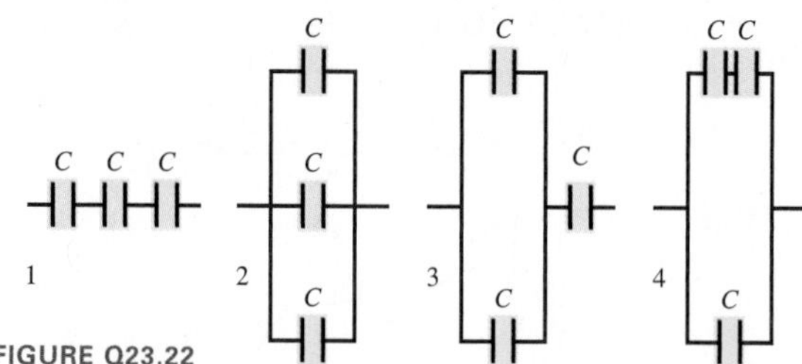

FIGURE Q23.22

23. Figure Q23.23 shows a circuit consisting of a battery, a switch, two identical lightbulbs, and a capacitor that is initially uncharged.
    a. *Immediately* after the switch is closed, are either or both bulbs glowing? Explain.
    b. If both bulbs are glowing, which is brighter? Or are they equally bright? Explain.
    c. For any bulb (A or B or both) that lights up immediately after the switch is closed, does its brightness increase with time, decrease with time, or remain unchanged? Explain.

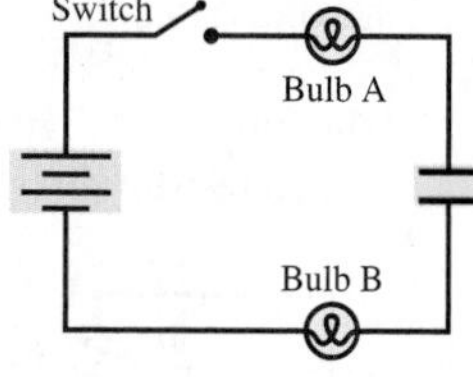

FIGURE Q23.23

24. Figure Q23.24 shows the voltage as a function of time across a capacitor as it is discharged (separately) through three different resistors. Rank in order, from largest to smallest, the values of the resistances R_1 to R_3.

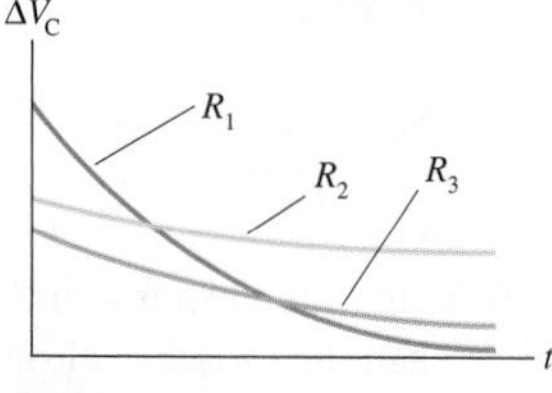

FIGURE Q23.24

25. A charged capacitor could be connected to two identical resistors in either of the two ways shown in Figure Q23.25. Which configuration will discharge of the capacitor in the shortest time once the switch is closed? Explain.

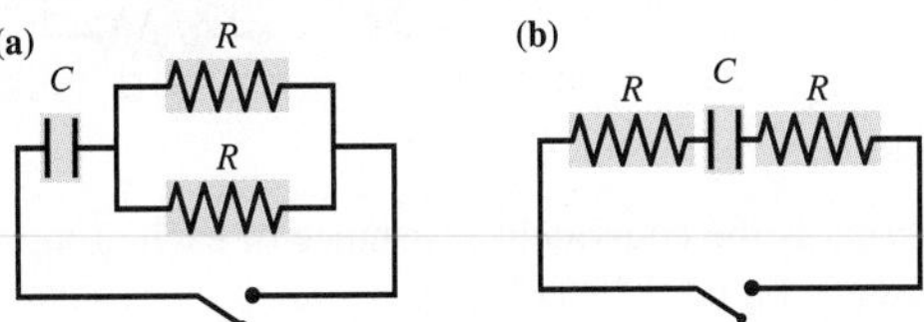

FIGURE Q23.25

26. A flashing light is controlled by the charging and discharging of an *RC* circuit. If the light is flashing too rapidly, describe two changes that you could make to the circuit to reduce the flash rate.

27. A device to make an electrical measurement of skin moisture
BIO has electrodes that form two plates of a capacitor; the skin is the dielectric between the plates. Adding moisture to the skin means adding water, which has a large dielectric constant. If a circuit repeatedly charges and discharges the capacitor to determine the capacitance, how will an increase in skin moisture affect the charging and discharging time? Explain.

28. Consider the model of nerve conduction in myelinated axons
BIO presented in the chapter. Suppose the distance between the nodes of Ranvier was halved for a particular axon.
 a. How would this affect the resistance and the capacitance of one segment of the axon?
 b. How would this affect the time constant for the charging of one segment?
 c. How would this affect the signal propagation speed for the axon?

29. Adding a myelin sheath to an axon results in faster signal prop-
BIO agation. It also means that less energy is required for a signal to propagate down the axon. Explain why this is so.

Multiple-Choice Questions

30. | What is the current in the circuit of Figure Q23.30?
 A. 1.0 A B. 1.7 A
 C. 2.5 A D. 4.2 A

31. | Which resistor in Figure Q23.30 dissipates the most power?
 A. The 4.0 Ω resistor.
 B. The 6.0 Ω resistor.
 C. Both dissipate the same power.

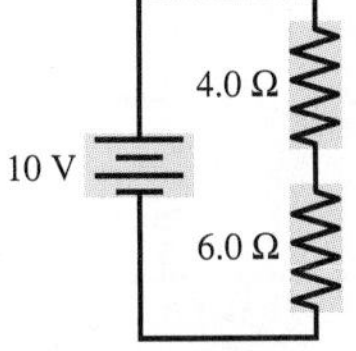

FIGURE Q23.30

32. || Normally, household lightbulbs are connected in parallel to a power supply. Suppose a 40 W and a 60 W lightbulb are, instead, connected in series, as shown in Figure Q23.32. Which bulb is brighter?
 A. The 60 W bulb.
 B. The 40 W bulb.
 C. The bulbs are equally bright.

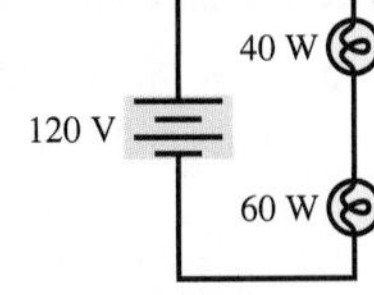

FIGURE Q23.32

33. ||| A metal wire of resistance R is cut into two pieces of equal length. The two pieces are connected together side by side. What is the resistance of the two connected wires?
 A. $R/4$ B. $R/2$ C. R
 D. $2R$ E. $4R$

34. | What is the value of resistor R in Figure Q23.34?
 A. 4.0 Ω
 B. 12 Ω
 C. 36 Ω
 D. 72 Ω
 E. 96 Ω

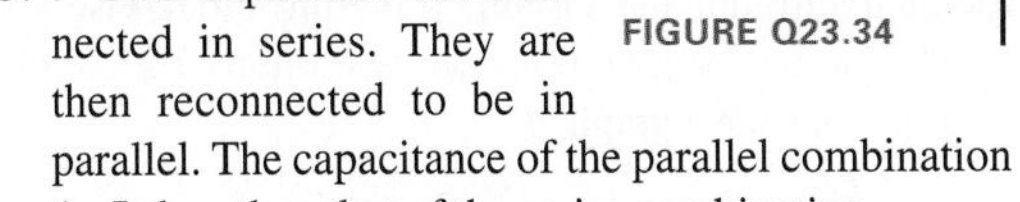

FIGURE Q23.34

35. | Two capacitors are connected in series. They are then reconnected to be in parallel. The capacitance of the parallel combination
 A. Is less than that of the series combination.
 B. Is more than that of the series combination.
 C. Is the same as that of the series combination.
 D. Could be more or less than that of the series combination depending on the values of the capacitances.

36. | If a cell's membrane thickness doubles but the cell stays the same size, how do the resistance and the capacitance of the cell
BIO membrane change?
 A. The resistance and the capacitance would increase.
 B. The resistance would increase, the capacitance would decrease.
 C. The resistance would decrease, the capacitance would increase.
 D. The resistance and the capacitance would decrease.

37. ||| If a cell's diameter is reduced by 50% without changing the membrane thickness, how do the resistance and capacitance of
BIO the cell membrane change?
 A. The resistance and the capacitance would increase.
 B. The resistance would increase, the capacitance would decrease.
 C. The resistance would decrease, the capacitance would increase.
 D. The resistance and the capacitance would decrease.

PROBLEMS

Section 23.1 Circuit Elements and Diagrams

1. || Draw a circuit diagram for the circuit of Figure P23.1.

2. || Draw a circuit diagram for the circuit of Figure P23.2.

3. || Draw a circuit diagram for the circuit of Figure P23.3.

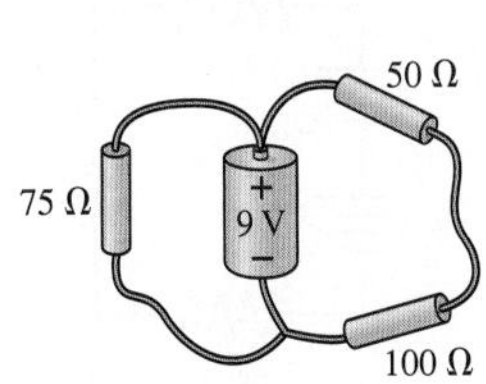

FIGURE P23.1

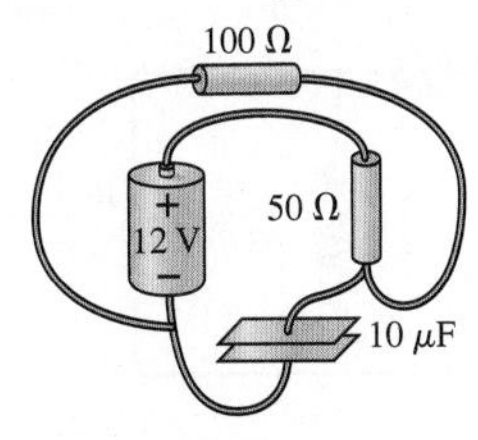

FIGURE P23.2

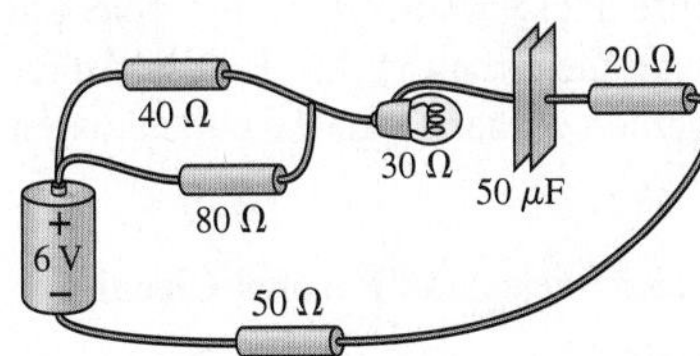

FIGURE P23.3

Section 23.2 Kirchhoff's Laws

4. || In Figure P23.4, what is the current in the wire above the junction? Does charge flow toward or away from the junction?

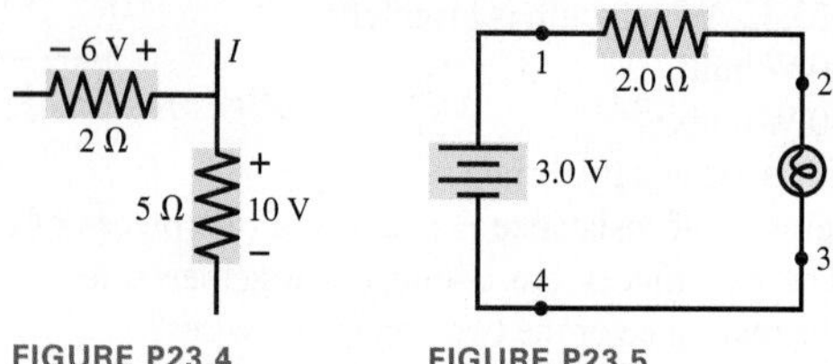

FIGURE P23.4 FIGURE P23.5

5. || The lightbulb in the circuit diagram of Figure P23.5 has a resistance of 1.0 Ω. Consider the potential difference between pairs of points in the figure.
   a. What are the values of ΔV_{12}, ΔV_{23}, and ΔV_{34}?
   b. What are the values if the bulb is removed?
6. | a. What are the magnitude and direction of the current in the 30 Ω resistor in Figure P23.6?
   b. Draw a graph of the potential as a function of the distance traveled through the circuit, traveling clockwise from $V = 0$ V at the lower left corner. See Figure P23.9 for an example of such a graph.

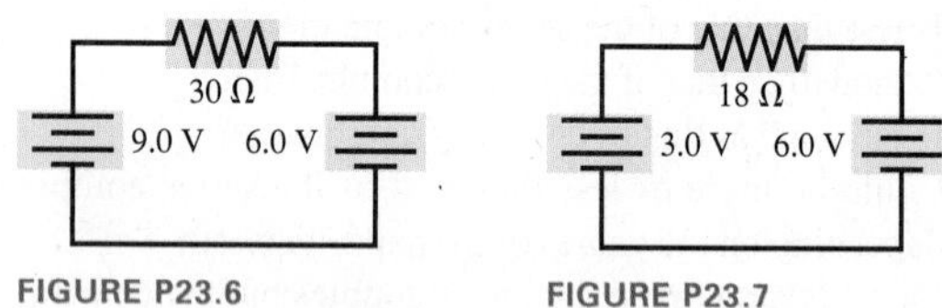

FIGURE P23.6 FIGURE P23.7

7. || a. What are the magnitude and direction of the current in the 18 Ω resistor in Figure P23.7?
   b. Draw a graph of the potential as a function of the distance traveled through the circuit, traveling clockwise from $V = 0$ V at the lower left corner. See Figure P23.9 for an example of such a graph.
8. | a. What is the potential difference across each resistor in Figure P23.8?
   b. Draw a graph of the potential as a function of the distance traveled through the circuit, traveling clockwise from $V = 0$ V at the lower left corner. See Figure P23.9 for an example of such a graph.

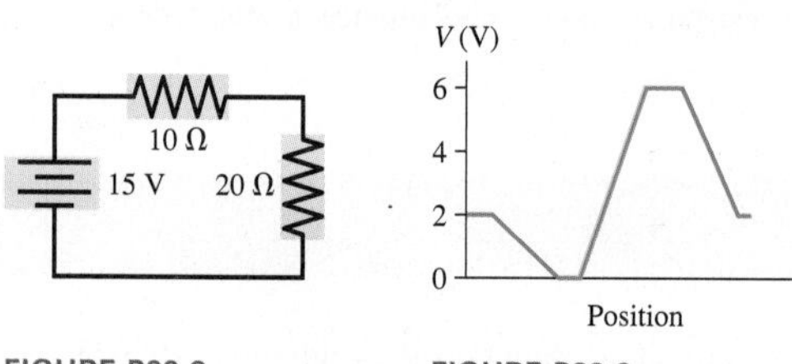

FIGURE P23.8 FIGURE P23.9

9. | The current in a circuit with only one battery is 2.0 A. Figure P23.9 shows how the potential changes when going around the circuit in the clockwise direction, starting from the lower left corner. Draw the circuit diagram.

Section 23.3 Series and Parallel Circuits

10. | What is the equivalent resistance of each group of resistors shown in Figure P23.10?

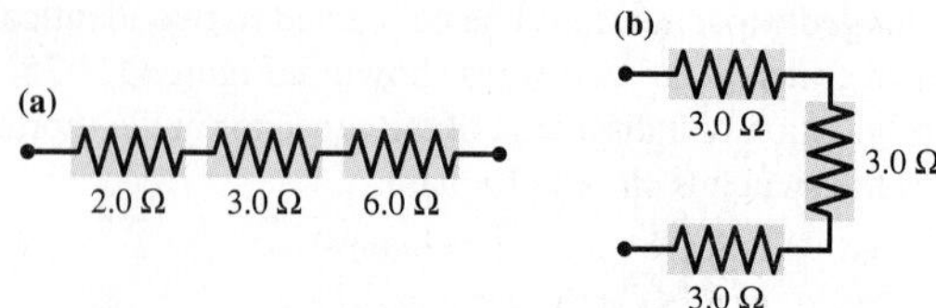

FIGURE P23.10

11. | What is the equivalent resistance of each group of resistors shown in Figure P23.11?

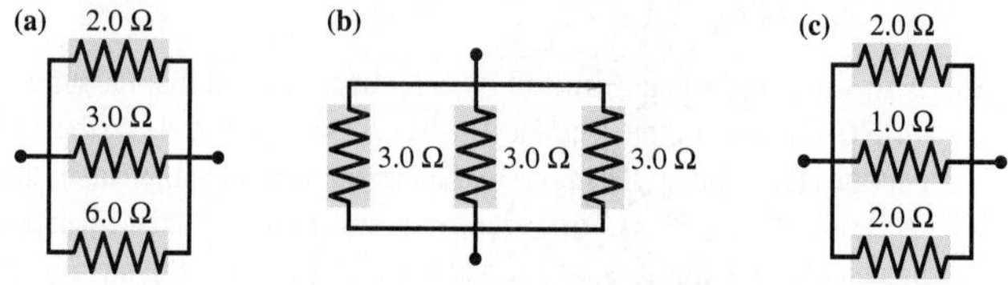

FIGURE P23.11

12. |||| An 80-cm-long wire is made by welding a 1.0-mm-diameter,
INT 20-cm-long copper wire to a 1.0-mm-diameter, 60-cm-long iron wire. What is the resistance of the composite wire?
13. | You have a collection of 1.0 kΩ resistors. How can you connect four of them to produce an equivalent resistance of 0.25 kΩ?
14. | You have a collection of six 1.0 kΩ resistors. What is the smallest resistance you can make by combining them?
15. | You have three 6.0 Ω resistors and one 3.0 Ω resistor. How can you connect them to produce an equivalent resistance of 5.0 Ω?
16. || You have six 1.0 kΩ resistors. How can you connect them to produce a total equivalent resistance of 1.5 kΩ?

Section 23.4 Measuring Voltage and Current

Section 23.5 More Complex Circuits

17. || What is the equivalent resistance between points a and b in Figure P23.17?

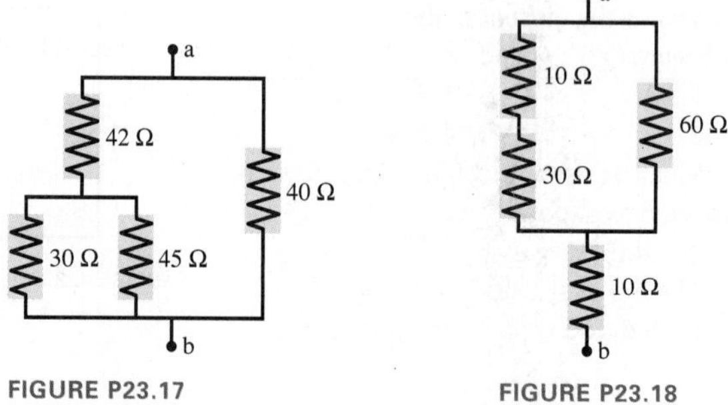

FIGURE P23.17 FIGURE P23.18

18. | What is the equivalent resistance between points a and b in Figure P23.18?
19. || The currents in two resistors in a circuit are shown in Figure P23.19. What is the value of resistor R?

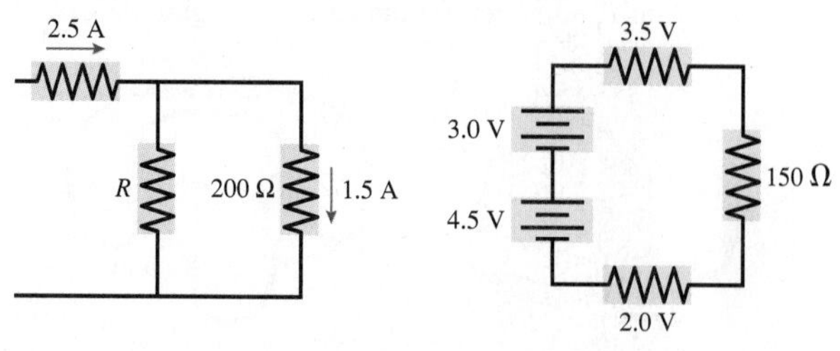

FIGURE P23.19 FIGURE P23.20

20. || Two batteries supply current to the circuit in Figure P23.20. The figure shows the potential difference across two of the resistors and the value of the third resistor. What current is supplied by the batteries?

21. | Part of a circuit is shown in Figure P23.21.
a. What is the current through the 3.0 Ω resistor?
b. What is the value of the current I?

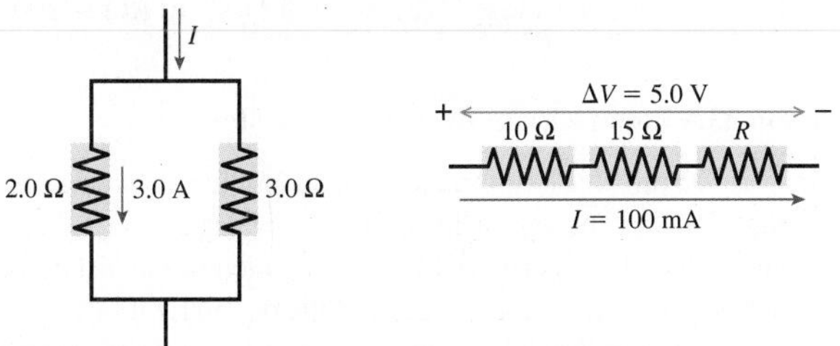

FIGURE P23.21 FIGURE P23.22

22. | What is the value of resistor R in Figure P23.22?

23. || What are the resistance R and the emf of the battery in Figure P23.23?

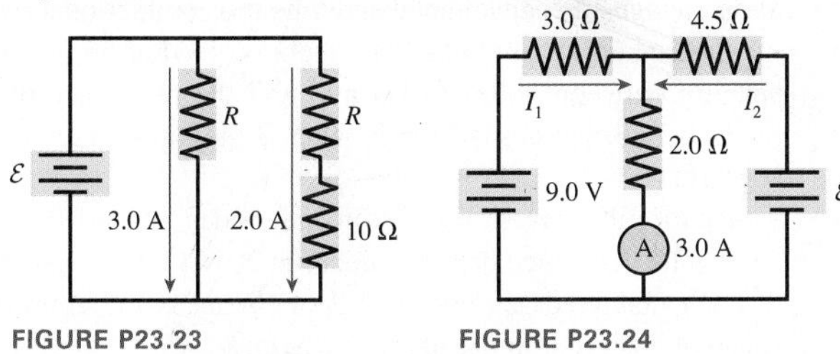

FIGURE P23.23 FIGURE P23.24

24. || The ammeter in Figure P23.24 reads 3.0 A. Find I_1, I_2, and $\mathcal{E}$.

25. ||| Find the current through and the potential difference across each resistor in Figure P23.25.

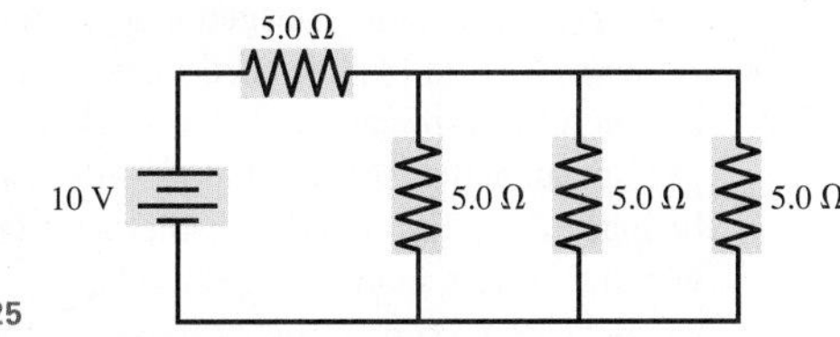

FIGURE P23.25

26. ||| Find the current through and the potential difference across each resistor in Figure P23.26.

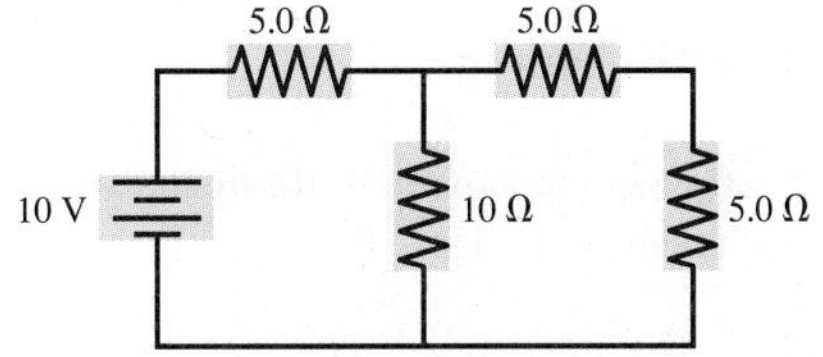

FIGURE P23.26

27. || For the circuit shown in Figure P23.27, find the current through and the potential difference across each resistor. Place your results in a table for ease of reading.

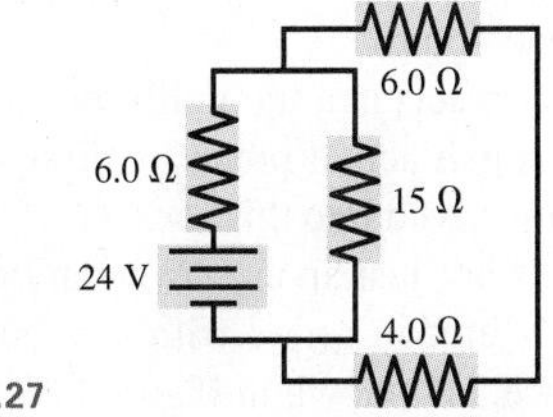

FIGURE P23.27

28. |||| In the circuit of Figure P23.28, what are the values of ΔV_{14}, ΔV_{24}, and ΔV_{34}?

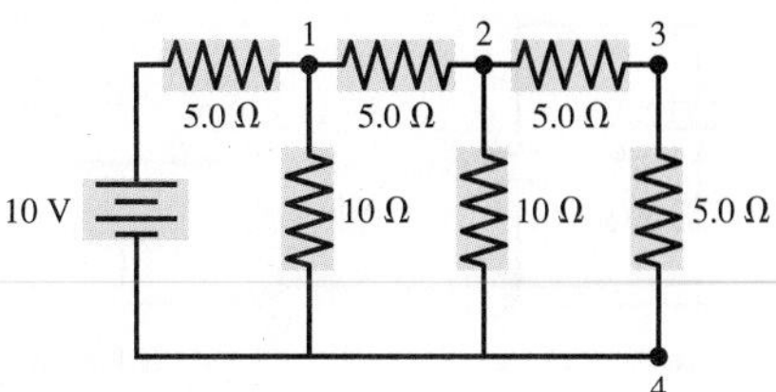

FIGURE P23.28

29. || For the circuit shown in Figure P23.29, find the current through and the potential difference across each resistor. Place your results in a table for ease of reading.

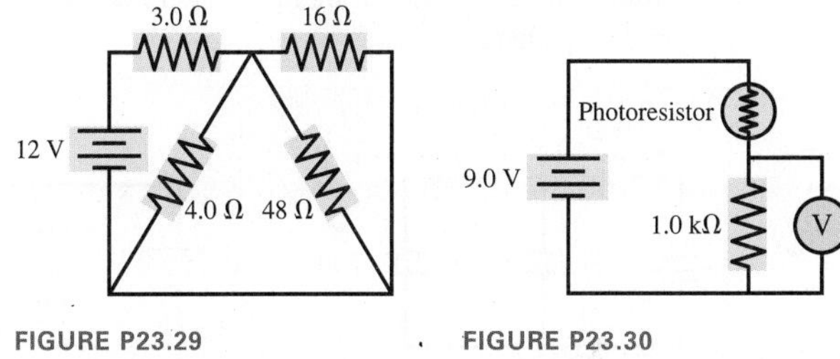

FIGURE P23.29 FIGURE P23.30

30. || A photoresistor, whose resistance decreases with light intensity, is connected in the circuit of Figure P23.30. On a sunny day, the photoresistor has a resistance of 0.56 kΩ. On a cloudy day, the resistance rises to 4.0 kΩ. At night, the resistance is 20 kΩ.
a. What does the voltmeter read for each of these conditions?
b. Does the voltmeter reading increase or decrease as the light intensity increases?

31. || A photoresistor, whose resistance decreases with light intensity, is connected in the circuit of Figure P23.31.
a. Draw a circuit diagram to illustrate how you would use a voltmeter and an ammeter to determine the resistance of the photoresistor in this circuit.
b. What do the two meters read when the resistance of the photoresistor is 2.5 kΩ?

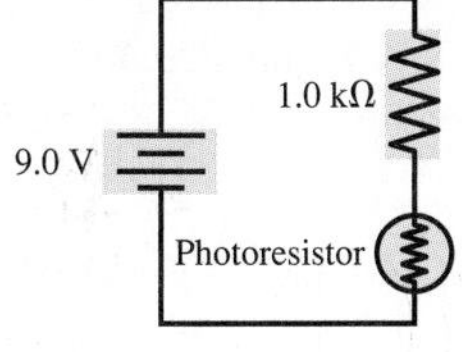

FIGURE P23.31

Section 23.6 Capacitors in Parallel and Series

32. | A 6.0 μF capacitor, a 10 μF capacitor, and a 16 μF capacitor are connected in parallel. What is their equivalent capacitance?

33. | A 6.0 μF capacitor, a 10 μF capacitor, and a 16 μF capacitor are connected in series. What is their equivalent capacitance?

34. | You need a capacitance of 50 μF, but you don't happen to have a 50 μF capacitor. You do have a 30 μF capacitor. What additional capacitor do you need to produce a total capacitance of 50 μF? Should you join the two capacitors in parallel or in series?

35. | You need a capacitance of 50 μF, but you don't happen to have a 50 μF capacitor. You do have a 75 μF capacitor. What additional capacitor do you need to produce a total capacitance of 50 μF? Should you join the two capacitors in parallel or in series?

36. || What is the equivalent capacitance of the three capacitors in Figure P23.36?

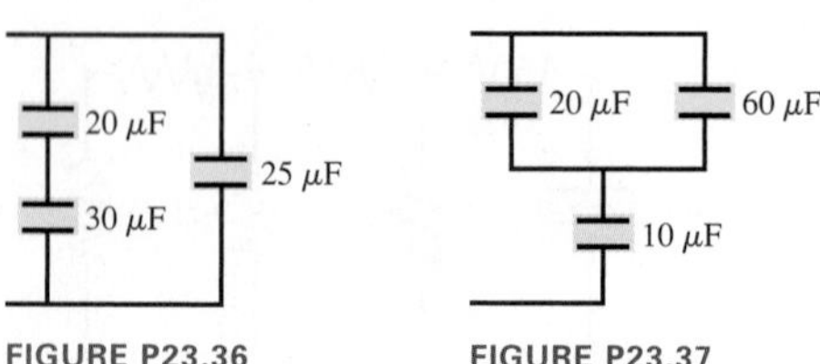

FIGURE P23.36

FIGURE P23.37

37. | What is the equivalent capacitance of the three capacitors in Figure P23.37?

38. ||| For the circuit of Figure P23.38,
   a. What is the equivalent capacitance?
   b. How much charge flows through the battery as the capacitors are being charged?

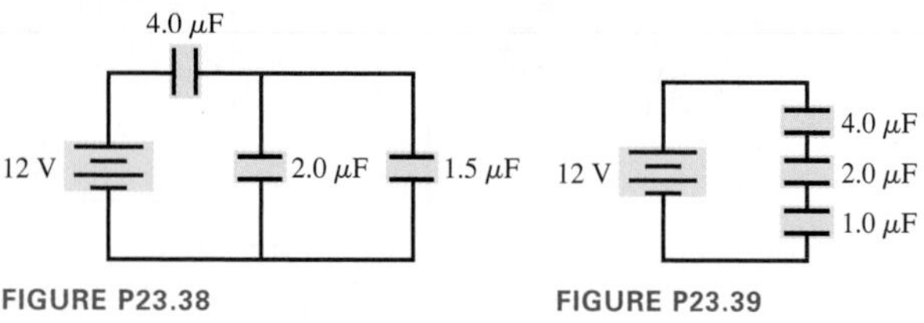

FIGURE P23.38

FIGURE P23.39

39. ||| For the circuit of Figure P23.39,
   a. What is the equivalent capacitance?
   b. What is the charge of each of the capacitors?

Section 23.7 *RC* Circuits

40. || What is the time constant for the discharge of the capacitor in Figure P23.40?

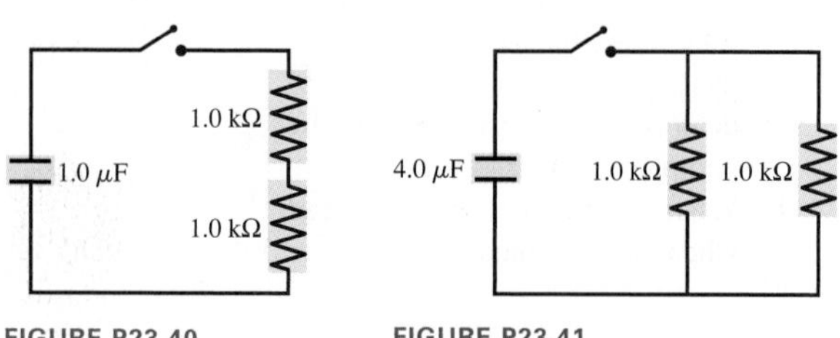

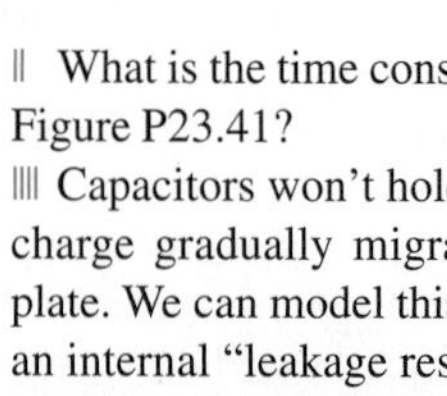

FIGURE P23.40

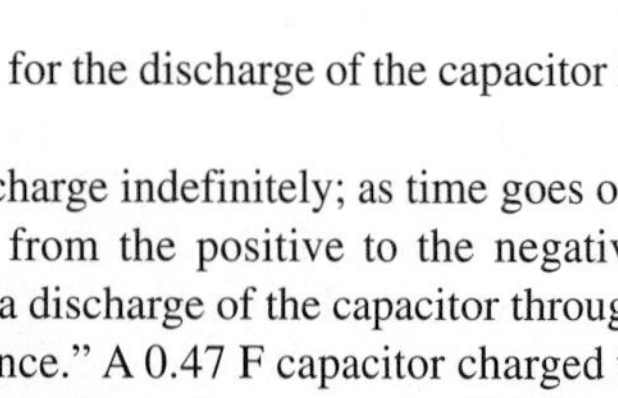

FIGURE P23.41

41. || What is the time constant for the discharge of the capacitor in Figure P23.41?

42. |||| Capacitors won't hold a charge indefinitely; as time goes on, charge gradually migrates from the positive to the negative plate. We can model this as a discharge of the capacitor through an internal "leakage resistance." A 0.47 F capacitor charged to 2.5 V will initially discharge with a leakage current of 0.25 mA.
   a. What is the leakage resistance?
   b. How long will it take for the capacitor voltage to drop to 1.0 V?

43. ||| A 10 μF capacitor initially charged to 20 μC is discharged through a 1.0 kΩ resistor. How long does it take to reduce the capacitor's charge to 10 μC?

44. || The switch in Figure P23.44 has been in position a for a long time. It is changed to position b at $t = 0$ s. What are the charge Q on the capacitor and the current I through the resistor (a) immediately after the switch is closed? (b) At $t = 50\ \mu$s? (c) At $t = 200\ \mu$s?

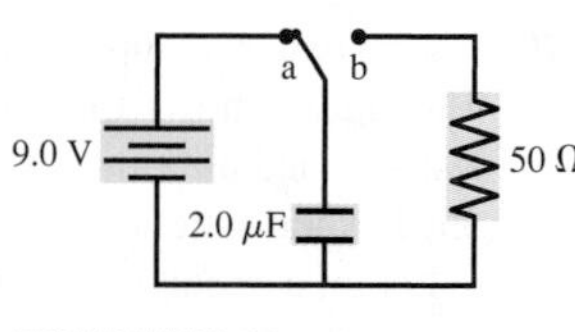

FIGURE P23.44

Section 23.8 Electricity in the Nervous System

45. | A 9.0-nm-thick cell membrane undergoes an action potential
BIO that follows the curve in the table on page 761. What is the strength of the electric field inside the membrane just before the action potential and at the peak of the depolarization?

46. ||| A cell membrane has a resistance and a capacitance and thus
BIO a characteristic time constant. What is the time constant of a 9.0-nm-thick membrane surrounding a 0.040-mm-diameter spherical cell?

47. | Changing the thickness of the myelin sheath surrounding an
BIO axon changes its capacitance and thus the conduction speed. A myelinated nerve fiber has a conduction speed of 55 m/s. If the spacing between nodes is 1.0 mm and the resistance of segments between nodes is 25 MΩ, what is the capacitance of each segment?

48. ||| A particular myelinated axon has nodes spaced 0.80 mm
BIO apart. The resistance between nodes is 20 MΩ; the capacitance of each insulated segment is 1.2 pF. What is the conduction speed of a nerve impulse along this axon?

49. | To measure signal propagation in a nerve in the arm, the
BIO nerve is triggered near the armpit. The peak of the action potential is measured at the elbow and then, 4.0 ms later, 24 cm away from the elbow at the wrist.
   a. What is the speed of propagation along this nerve?
   b. A determination of the speed made by measuring the time between the application of a stimulus at the armpit and the peak of an action potential at the elbow or the wrist would be inaccurate. Explain the problem with this approach, and why the noted technique is preferable.

50. || A myelinated axon conducts nerve impulses at a speed of
BIO 40 m/s. What is the signal speed if the thickness of the myelin sheath is halved but no other changes are made to the axon?

General Problems

51. || How much power is dissipated by
INT each resistor in Figure P23.51?

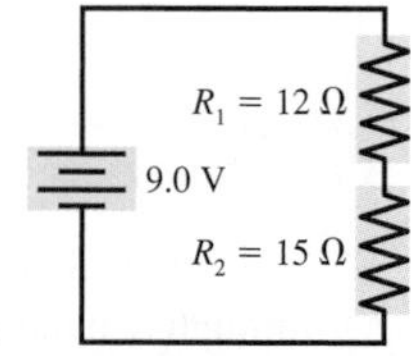

FIGURE P23.51

52. |||| Two 75 W (120 V) lightbulbs are wired in series, then the
INT combination is connected to a 120 V supply. How much power is dissipated by each bulb?

53. |||| The corroded contacts in a lightbulb socket have 5.0 Ω total
INT resistance. How much actual power is dissipated by a 100 W (120V) lightbulb screwed into this socket?

54. |||| A real battery is not just an emf. We can model a real 1.5 V
INT battery as a 1.5 V emf in series with a resistor known as the "internal resistance," as shown in Figure P23.54. A typical bat-

tery has 1.0 Ω internal resistance due to imperfections that limit current through the battery. When there's no current through the battery, and thus no voltage drop across the internal resistance, the potential difference between its terminals is 1.5 V, the value of the emf. Suppose the terminals of this battery are connected to a 2.0 Ω resistor.

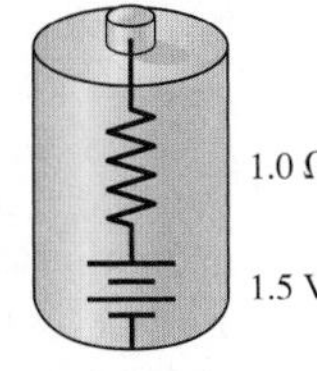

FIGURE P23.54

a. What is the potential difference between the terminals of the battery?
b. What fraction of the battery's power is dissipated by the internal resistance?

55. ||| For the real battery shown in Figure P23.54, calculate the power dissipated by a resistor R connected to the battery when (a) $R = 0.25\ \Omega$, (b) $R = 0.50\ \Omega$, (c) $R = 1.0\ \Omega$, (d) $R = 2.0\ \Omega$, and (e) $R = 4.0\ \Omega$. (Your results should suggest that maximum power dissipation is achieved when the external resistance R equals the internal resistance. This is true in general.)

56. ||| INT Batteries are recharged by connecting them to a power supply (i.e., another battery) of greater emf in such a way that the current flows *into* the positive terminal of the battery being recharged, as was shown in Example 23.1. This reverse current through the battery replenishes its chemicals. The current is kept fairly low so as not to overheat the battery being recharged by dissipating energy in its internal resistance.
a. Suppose the real battery of Figure P23.54 is rechargeable. What emf power supply should be used for a 0.75 A recharging current?
b. If this power supply charges the battery for 10 minutes, how much energy goes into the battery? How much is dissipated as thermal energy in the internal resistance?

57. ||| When two resistors are connected in parallel across a battery of unknown voltage, one resistor carries a current of 3.2 A while the second carries a current of 1.8 A. What current will be supplied by the same battery if these two resistors are connected to it in series?

58. || INT The 10 Ω resistor in Figure P23.58 is dissipating 40 W of power. How much power are the other two resistors dissipating?

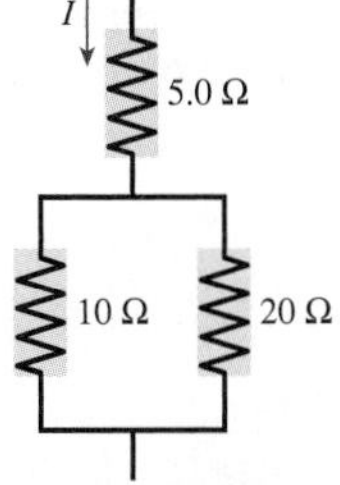

FIGURE P23.58

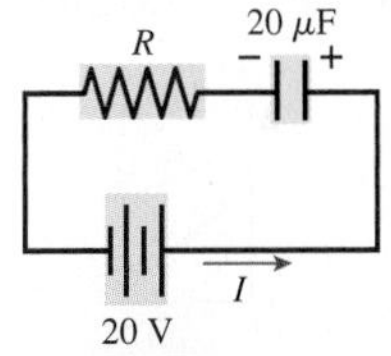

FIGURE P23.59

59. ||| INT At this instant, the current in the circuit of Figure P23.59 is 20 mA in the direction shown and the capacitor charge is 200 μC. What is the resistance R?

60. | What is the equivalent resistance between points a and b in Figure P23.60?

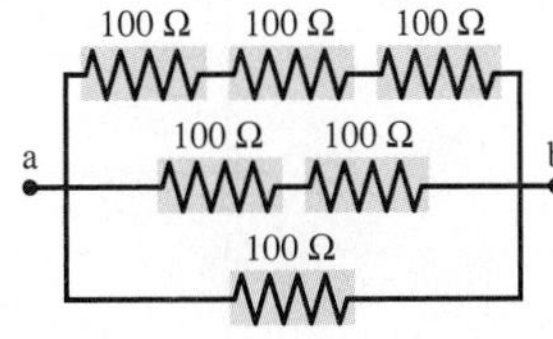

FIGURE P23.60

61. || You have three 12 Ω resistors. Draw diagrams showing how you could arrange all three so that their equivalent resistance is (a) 4.0 Ω, (b) 8.0 Ω, (c) 18 Ω, and (d) 36 Ω.

62. ||| INT A 9.0 V battery is connected to a wire made of three segments of different metals connected one after another: 10 cm of copper wire, then 12 cm of iron wire, then 18 cm of tungsten wire. All of the wires are 0.26 mm in diameter. Find the potential difference across each piece of wire.

63. ||| You have a device that needs a voltage reference of 3.0 V, but you have only a 9.0 V battery. Fortunately, you also have several 10 kΩ resistors. Show how you can use the resistors and the battery to make a circuit that provides a potential difference of 3.0 V.

64. | INT There is a current of 0.25 A in the circuit of Figure P23.64.
a. What is the direction of the current? Explain.
b. What is the value of the resistance R?
c. What is the power dissipated by R?
d. Make a graph of potential versus position, starting from $V = 0$ V in the lower left corner and proceeding clockwise. See Figure P23.9 for an example.

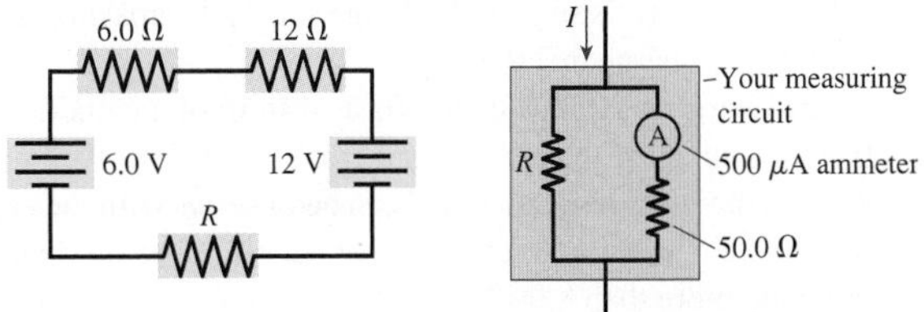

FIGURE P23.64

FIGURE P23.65

65. || A circuit you're building needs an ammeter that goes from 0 mA to a full-scale reading of 50.0 mA. Unfortunately, the only ammeter in the storeroom goes from 0 μA to a full-scale reading of only 500 μA. Fortunately, you can make this ammeter work by putting it in a measuring circuit, as shown in Figure P23.65 This lets a certain fraction of the current pass through the meter; knowing this value, you can deduce the total current. Assume that the ammeter is ideal.
a. What value of R must you use so that the meter will go to full scale when the current I is 50.0 mA?

Hint: When $I = 50.0$ mA, the ammeter should be reading its maximum value.

b. What is the equivalent resistance of your measuring circuit?

66. || A circuit you're building needs a voltmeter that goes from 0 V to a full-scale reading of 5.0 V. Unfortunately, the only meter in the storeroom is an *ammeter* that goes from 0 μA to a full-scale reading of 500 μA. It is possible to use this meter to measure voltages by putting in a measuring circuit as shown in Figure P23.66. What value of R must you use so that the meter will go to full scale when the potential difference ΔV is 5.0 V? Assume that the ammeter is ideal.

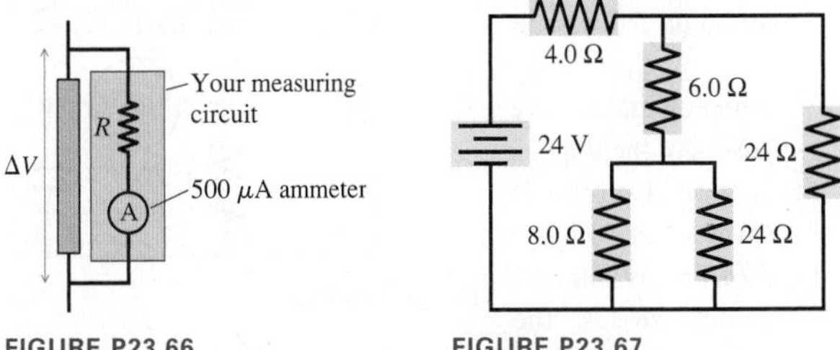

FIGURE P23.66

FIGURE P23.67

67. || For the circuit shown in Figure P23.67, find the current through and the potential difference across each resistor. Place your results in a table for ease of reading.

68. || You have three capacitors. Draw diagrams showing how you could arrange all three so that their equivalent capacitance is (a) 4.0 μF, (b) 8.0 μF, (c) 18 μF, and (d) 36 μF.

69. ‖ Initially, the switch in Figure P23.69 is in position a and capacitors C_2 and C_3 are uncharged. Then the switch is flipped to position b. Afterward, what are the charge on and the potential difference across each capacitor?

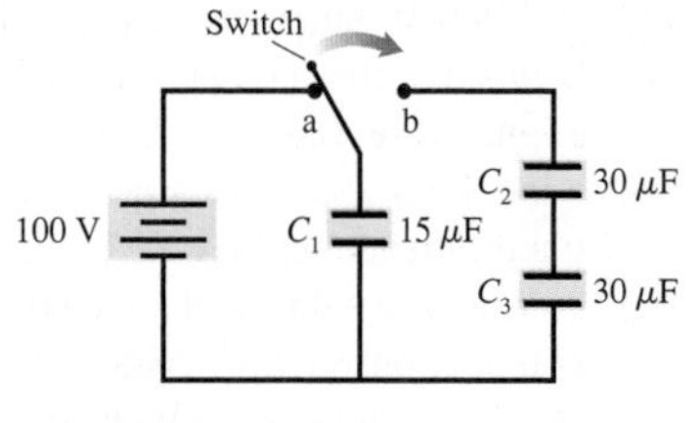

FIGURE P23.69

70. ‖ The capacitor in an RC circuit with a time constant of 15 ms INT is charged to 10 V. The capacitor begins to discharge at $t = 0$ s.
   a. At what time will the charge on the capacitor be reduced to half its initial value?
   b. At what time will the energy stored in the capacitor be reduced to half its initial value?

71. ‖ What value resistor will discharge a 1.0 μF capacitor to 10% of its initial charge in 2.0 ms?

72. ‖‖ The charging circuit for the flash system of a camera uses a 100 μF capacitor that is charged from a 250 V power supply. What is the most resistance that can be in series with the capacitor if the capacitor is to charge to at least 87% of its final voltage in no more than 8.0 s?

73. ‖ A capacitor is discharged through a 100 Ω resistor. The discharge current decreases to 25% of its initial value in 2.5 ms. What is the value of the capacitor?

74. ‖‖ A 50 μF capacitor that had been charged to 30 V is discharged through a resistor. Figure P23.74 shows the capacitor voltage as a function of time. What is the value of the resistance?

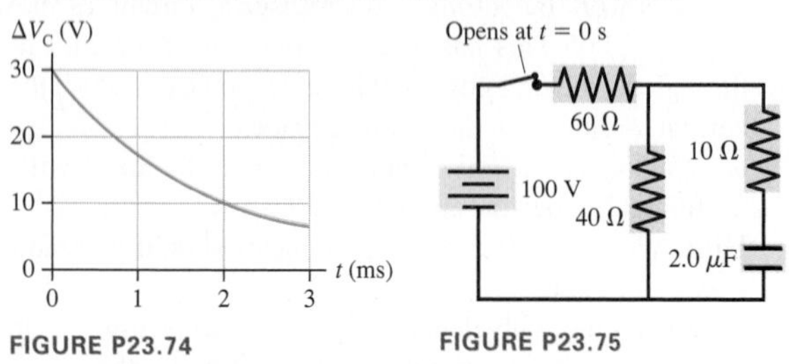

FIGURE P23.74 FIGURE P23.75

75. ‖‖‖ The switch in Figure P23.75 has been closed for a very long time.
   a. What is the charge on the capacitor?
   b. The switch is opened at $t = 0$ s. At what time has the charge on the capacitor decreased to 10% of its initial value?

76. ‖‖‖‖ Intermittent windshield wipers use a variable resistor in an RC circuit to set the delay between successive passes of the wipers. A typical circuit is shown in Figure P23.76. When the switch closes, the capacitor (initially uncharged) begins to charge and the potential at point b begins to increase. A sensor measures the potential difference between points a and b, triggering a pass of the wipers when $V_b = V_a$. (Another part of the circuit, not shown, discharges the capacitor at this time so that the cycle can start again.)

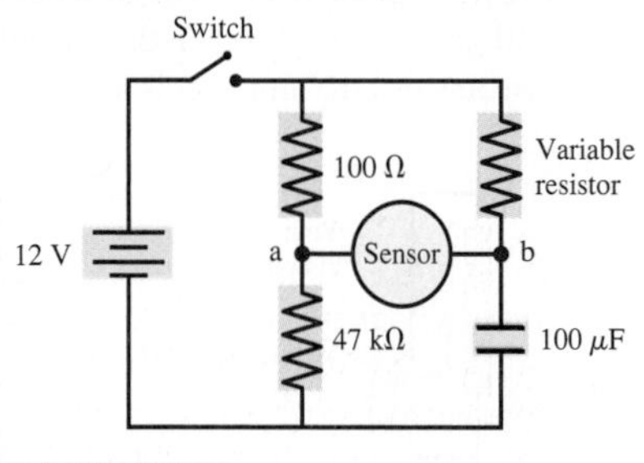

FIGURE P23.76

   a. What value of the variable resistor will give 12 seconds from the start of a cycle to a pass of the wipers?
   b. To decrease the time, should the variable resistance be increased or decreased?

77. ‖‖ In Example 23.14 we estimated the capacitance of the cell BIO membrane to be 89 pF, and in Example 23.15 we found that approximately 10,000 Na^+ ions flow through an ion channel when it opens. Based on this information and what you learned in this chapter about the action potential, estimate the total number of sodium ion channels in the membrane of a nerve cell.

78. ‖‖‖ The giant axon of a squid is 0.5 mm in diameter, 10 cm long, BIO INT and not myelinated. Unmyelinated cell membranes behave as capacitors with 1 μF of capacitance per square centimeter of membrane area. When the axon is charged to the -70 mV resting potential, what is the energy stored in this capacitance?

79. ‖ A cell has a 7.0-nm-thick membrane with a total membrane BIO area of 6.0×10^{-9} m^2.
   a. We can model the cell as a capacitor, as we have seen. What is the magnitude of the charge on each "plate" when the membrane is at its resting potential of -70 mV?
   b. How many sodium ions does this charge correspond to?

Passage Problems

The Defibrillator BIO

A defibrillator is designed to pass a large current through a patient's torso in order to stop dangerous heart rhythms. Its key part is a capacitor that is charged to a high voltage. The patient's torso plays the role of a resistor in an RC circuit. When a switch is closed, the capacitor discharges through the patient's torso. A jolt from a defibrillator is intended to be intense and rapid; the maximum current is very large, so the capacitor discharges quickly. This rapid pulse depolarizes the heart, stopping all electrical activity. This allows the heart's internal nerve circuitry to reestablish a healthy rhythm.

A typical defibrillator has a 32 μF capacitor charged to 5000 V. The electrodes connected to the patient are coated with a conducting gel that reduces the resistance of the skin to where the effective resistance of the patient's torso is 100 Ω.

80. | Which pair of graphs in Figure P23.80 best represents the capacitor voltage and the current through the torso as a function of time after the switch is closed?

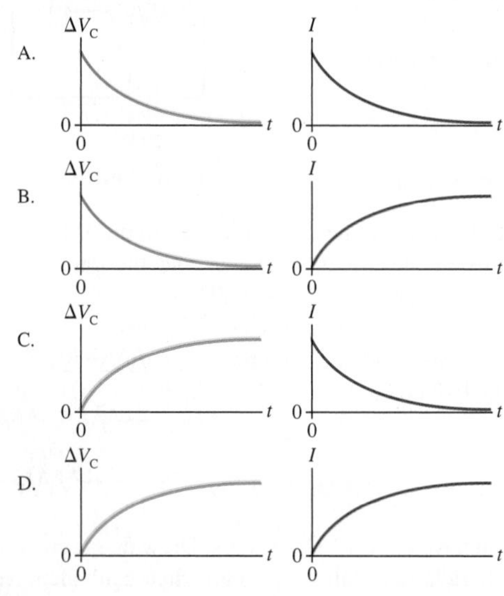

FIGURE P23.80

81. | For the values noted in the passage above, what is the time constant for the discharge of the capacitor?
A. 3.2 μs B. 160 μs C. 3.2 ms D. 160 ms

82. | If a patient receives a series of jolts, the resistance of the torso may increase. How does such a change affect the initial current and the time constant of subsequent jolts?
A. The initial current and the time constant both increase.
B. The initial current decreases, the time constant increases.
C. The initial current increases, the time constant decreases.
D. The initial current and the time constant both decrease.

83. | In some cases, the defibrillator may be charged to a lower voltage. How will this affect the time constant of the discharge?
A. The time constant will increase.
B. The time constant will not change.
C. The time constant will decrease.

Electric Fish BIO INT

The voltage produced by a single nerve or muscle cell is quite small, but there are many species of fish that use multiple action potentials in series to produce significant voltages. The electric organs in these fish are composed of specialized disk-shaped cells called *electrocytes*. The cell at rest has the usual potential difference between the inside and the outside, but the net potential difference *across* the cell is zero. An electrocyte is connected to nerve fibers that initially trigger a depolarization in one side of the cell but not the other. For the very short time of this depolarization, there is a net potential difference across the cell, as shown in Figure P23.84. Stacks of these cells connected in series can produce a large total voltage. Each stack can produce a small current; for more total current, more stacks are needed, connected in parallel.

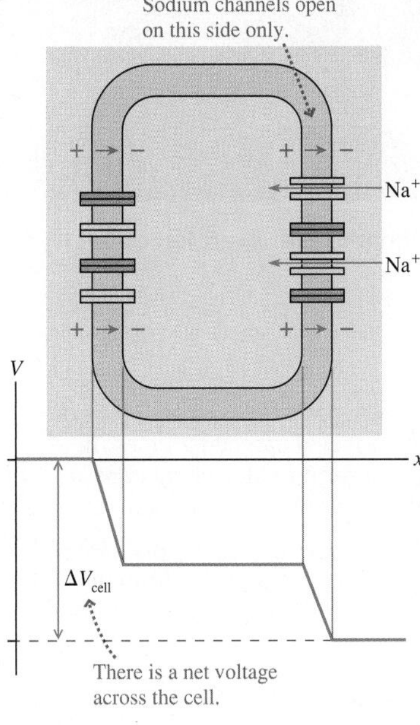
Electrolyte during depolarization of one side of cell
Sodium channels open on this side only.
+
−
Na^+
Na^+
V
x
ΔV_{cell}
There is a net voltage across the cell.

FIGURE P23.84

84. | In an electric eel, each electrocyte can develop a voltage of 150 mV for a short time. For a total voltage of 450 V, how many electrocytes must be connected in series?
A. 300
B. 450
C. 1500
D. 3000

85. | An electric eel produces a pulse of current of 0.80 A at a voltage of 500 V. For the short time of the pulse, what is the instantaneous power?
A. 400 W
B. 500 W
C. 625 W
D. 800 W

86. | Electric eels live in fresh water. The torpedo ray is an electric fish that lives in salt water. The electrocytes in the ray are grouped differently than in the eel; each stack of electrocytes has fewer cells, but there are more stacks in parallel. Which of the following best explains the ray's electrocyte arrangement?
A. The lower resistivity of salt water requires more current but lower voltage.
B. The lower resistivity of salt water requires more voltage but lower current.
C. The higher resistivity of salt water requires more current but lower voltage.
D. The higher resistivity of salt water requires more voltage but lower current.

87. | The electric catfish is another electric fish that produces a voltage pulse by means of stacks of electrocytes. As the fish grows in length, the magnitude of the voltage pulse the fish produces grows as well. The best explanation for this change is that, as the fish grows,
A. The voltage produced by each electrocyte increases.
B. More electrocytes are added.to each stack.
C. More stacks of electrocytes are added in parallel to the existing stacks.
D. The thickness of the electrocytes increases.

STOP TO THINK ANSWERS

Stop to Think 23.1: A, B, and D. These three are the same circuit because the logic of the connections is the same. In each case, there is a junction that connects one side of each circuit element and a second junction that connects the other side. In C, the functioning of the circuit is changed by the extra wire connecting the two sides of the capacitor.

Stop to Think 23.2: C = D > A = B. The two bulbs in series are of equal brightness, as are the two bulbs in parallel. But the two bulbs in series have a larger resistance than a single bulb, so there will be less current through the bulbs in series than the bulbs in parallel.

Stop to Think 23.3: C. The voltmeter must be connected in parallel with the resistor, and the ammeter in series.

Stop to Think 23.4: A > B > C = D. All the current from the battery goes through A, so it is brightest. The current divides at the junction, but not equally. Because B is in parallel with C + D, but has half the resistance of the two bulbs together, twice as much current travels through B as through C + D. So B is dimmer than A but brighter than C and D. C and D are equally bright because of conservation of current.

Stop to Think 23.5: $(C_{eq})_B > (C_{eq})_A > (C_{eq})_C$. Two capacitors in parallel have a larger capacitance than either alone; two capacitors in series have a smaller capacitance than either alone.

Stop to Think 23.6: B. The two 2 Ω resistors are in series and equivalent to a 4 Ω resistor. Thus $\tau = RC = 4$ s.

24 Magnetic Fields and Forces

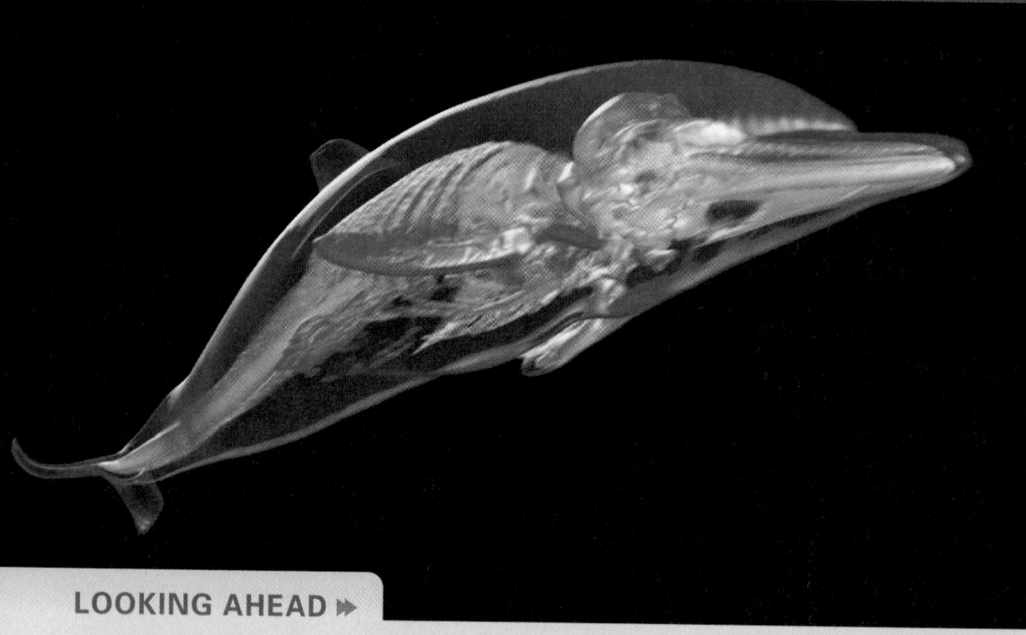

This detailed image of the skeletal system of a dolphin wasn't made with x rays; it was made with magnetism. How is this done?

LOOKING AHEAD ▸▸

The goal of Chapter 24 is to learn about magnetic fields and how magnetic fields exert forces on currents and moving charges.

Magnetic Fields

We've seen how to describe electric forces by using electric fields; now we'll look at **magnetic forces** and **magnetic fields.**

It's the magnetic force that causes a compass to line up with the earth's magnetic field.

Looking Back ◂◂
20.4 The electric field

Forces on Moving Charges

Magnetic fields exert forces on moving charged particles. In a uniform field, a charged particle moves in a circular path.

The aurora is due to the motion of charged particles from the sun in the magnetic field of the earth.

Looking Back ◂◂
6.3 Dynamics of circular motion

A current is simply the motion of charges, so magnetic fields exert forces on currents.

A loudspeaker works by the magnetic force acting on a current in a coil of wire at the base of the loudspeaker's cone.

Magnetic Field Sources

Iron filings work like little compasses to show magnetic field patterns. We'll see that magnetic fields are created by permanent magnets and by electric currents.

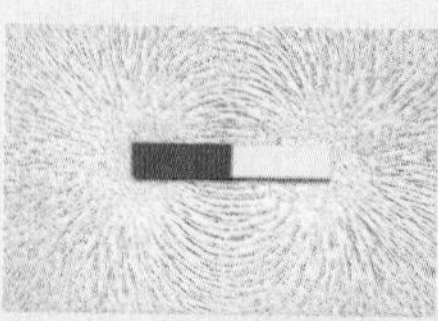

The simplest magnet is a bar magnet. It has two poles, north and south, and so creates a dipole field.

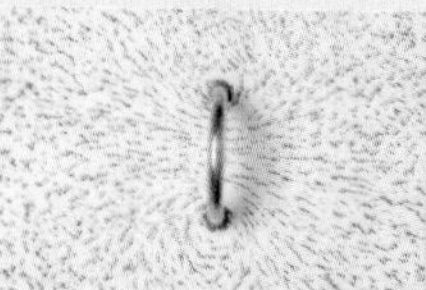

A loop of current creates a dipole field as well. You will learn how to compute magnetic fields resulting from currents in wires, loops, and coils.

Looking Back ◂◂
20.5 Electric dipoles

Magnetic Materials

Iron and a few other elements can exhibit **permanent magnetism.** The permanent alignment of electron dipoles leads to a large, fixed magnetic field in these materials.

You will see how the atomic behavior of electrons in atoms leads to the familiar observation that magnets stick to a refrigerator.

Dipoles and Torques

A compass, a loop of wire, and electrons and protons all are **magnetic dipoles.** All dipoles experience a torque in a magnetic field that rotates them to line up with the field.

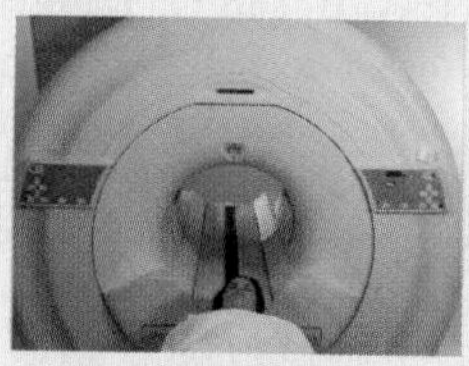

We'll explore how the alignment of atomic dipoles by the large magnetic field of an MRI solenoid can be used to create an image.

Magnetic torque on these coils causes this computer fan motor to turn.

Looking Back ◂◂
7.2 Calculating torque

24.1 Magnetism

We began our investigation of electricity in Chapter 20 by looking at the results of simple experiments with charged rods. Let's try a similar approach with magnetism.

Exploring magnetism

Experiment 1

If a bar magnet is taped to a piece of cork and allowed to float in a dish of water, it turns to align itself in an approximate north-south direction. The end of a magnet that points north is called the *north-seeking pole,* or simply the **north pole.** The other end is the **south pole.**

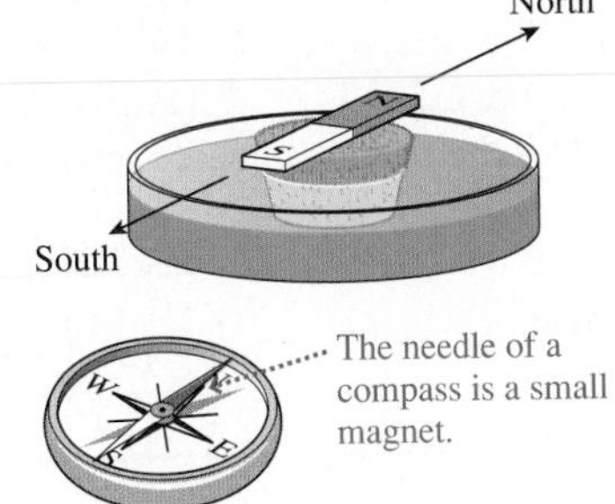

A magnet that is free to pivot like this is called a **compass.**

Experiment 2

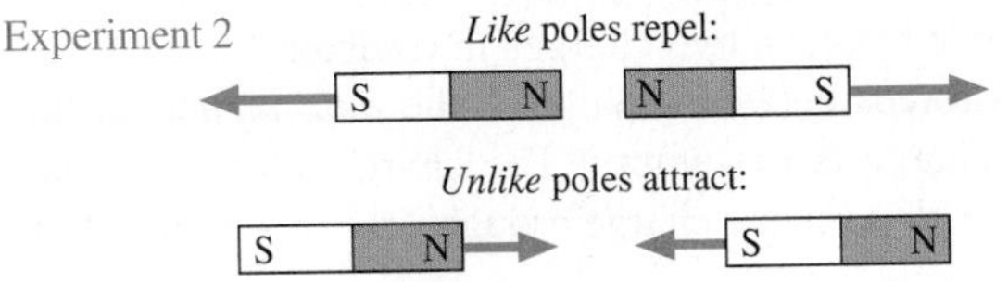

If the north pole of one magnet is brought near the north pole of another magnet, they repel each other. Two south poles also repel each other, but the north pole of one magnet exerts an attractive force on the south pole of another magnet.

Experiment 3

Since a compass needle is itself a little bar magnet, the north pole of a bar magnet attracts the south pole of a compass needle and repels the north pole.

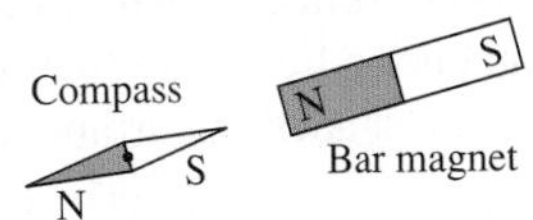

Experiment 4

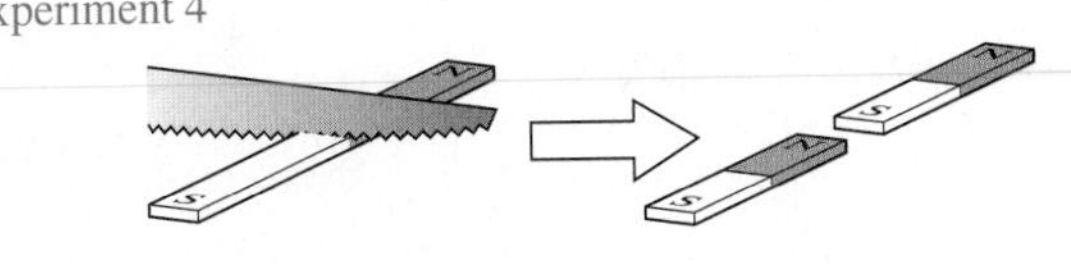

Cutting a bar magnet in half produces two weaker but still complete magnets, each with a north pole and a south pole. No matter how small the magnets are cut, even down to microscopic sizes, each piece remains a complete magnet with two poles.

Experiment 5

Magnets can pick up some objects, such as paper clips, but not all. If an object is attracted to one pole of a magnet, it is also attracted to the other pole. Most materials, including copper, aluminum, glass, and plastic, experience no force from a magnet.

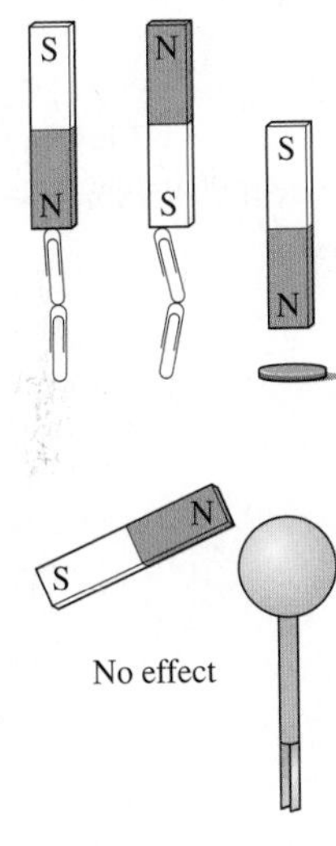

Experiment 6

When a magnet is brought near an electroscope, the leaves of the electroscope remain undeflected. If a charged rod is brought near a magnet, there is a weak *attractive* force on *both* ends of a magnet. However, the force is the same as the force on a metal bar that isn't a magnet, so it is simply a polarization force like the ones we studied in Chapter 21. Other than polarization forces, charged objects have *no effects* on magnets.

No effect

What do these experiments tell us?

1. Experiment 6 reveals that **magnetism is not the same as electricity.** Magnetic poles and electric charges share some similar behavior, but they are not the same.
2. Experiment 5 reveals that **magnetism is a long-range force.**
3. **Magnets have two types of poles,** called north and south poles. Two like poles exert repulsive forces on each other; two unlike poles exert attractive forces on each other. The behavior is *analogous* to that of electric charges, but, as noted, magnetic poles and electric charges are *not* the same. (In this text we will indicate the north pole of a magnet with reddish shading and the south pole with white.)
4. The poles of a bar magnet can be identified by using it as a compass. Other magnets aren't so easily made into a compass, but their poles can be identified by testing them against a bar magnet. **A pole that repels a known south pole and attracts a known north pole must be a south magnetic pole.**
5. Materials that are attracted to a magnet or that a magnet sticks to are called **magnetic materials.** The most common magnetic material is iron. Others include nickel and cobalt. Magnetic materials are attracted to *both* poles of a magnet. This attraction is analogous to how neutral objects are attracted to both positively and negatively charged rods by the polarization force. The difference is that *all* neutral objects are attracted to a charged rod, whereas only a few materials are attracted to a magnet.

6. Experiment 4 shows that cutting a magnet in half yields two weaker but still complete magnets, each with a north pole and a south pole. The basic unit of magnetism is thus a **magnetic dipole.** A magnetic dipole is analogous to an electric dipole, but the two charges in an electric dipole can be separated and used individually. This is *not* true for a magnetic dipole. The needle of a compass is a small, straight magnet, and so a compass needle is an especially simple magnetic dipole.

STOP TO THINK 24.1 Does the compass needle rotate?

A. Yes, clockwise.
B. Yes, counterclockwise.
C. No, not at all.

24.2 The Magnetic Field

When we studied the *electric* force between two charges in Chapter 20, we developed a new way to think about forces between charges—the *field model.* In this viewpoint, the space around a charge is not empty: The charge alters the space around it by creating an *electric field.* A second charge brought into this electric field then feels a force due to the *field.*

In Experiment 3 above, we learned that if the north pole of a bar magnet is brought near a compass, the compass needle will turn so that its south pole is toward the magnet. The concept of a field can also be used to describe the force that turns the compass: **Every magnet sets up a *magnetic* field in the space around it.** If another magnet—such as a compass needle—is then brought into this field, the second magnet will feel a force from the *field* of the first magnet. In this section, we'll see how to define the magnetic field, and then we'll study what the magnetic field looks like for some common shapes and arrangements of magnets.

FIGURE 24.1 Dipoles in electric and magnetic fields.

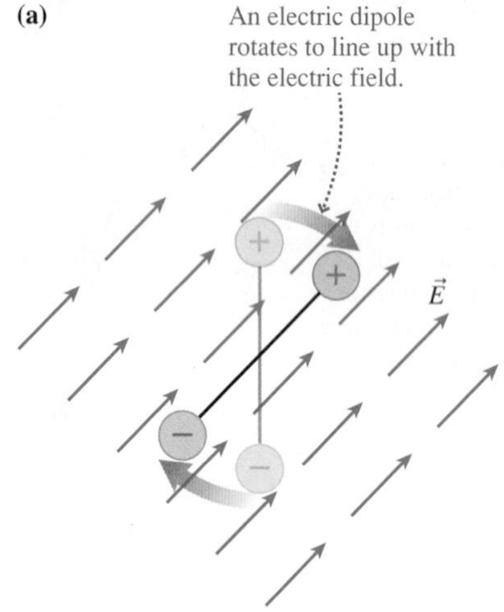

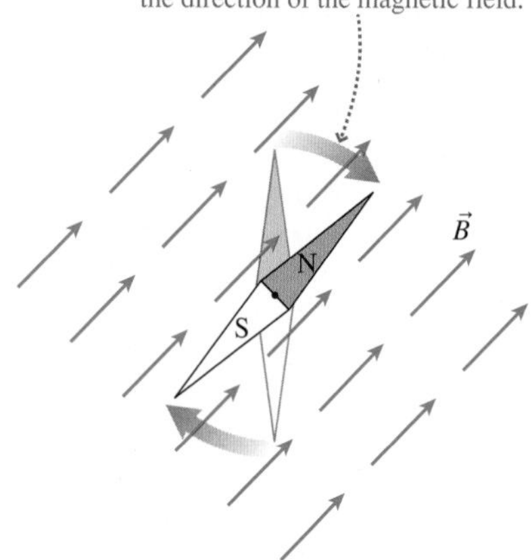

Measuring the Magnetic Field

What does the direction in which a compass needle points tell us about the magnetic field at the position of the compass? We can gain some insight by recalling how an *electric* dipole behaves when placed in an electric field, as shown in **FIGURE 24.1a.** In Chapter 20 we learned that an electric dipole experiences a *torque* when placed in an electric field, a torque that tends to align the axis of the dipole with the field. This means that the *direction* of the electric field is the same as the direction of the dipole's axis. Furthermore, the torque on the dipole is greater when the electric field is stronger; hence, the *magnitude* of the field, which we will often call the *strength* of the field, is proportional to the torque on the dipole.

The magnetic dipole of a compass needle behaves very similarly when it is in a magnetic field. The magnetic field exerts a torque on the compass needle, causing the needle to point in the field direction, as shown in **FIGURE 24.1b.**

Because the magnetic field has both a direction and a magnitude, we need to represent it using a *vector.* We will use the symbol $\vec{B}$ to represent the magnetic field and B to represent the magnitude of the field. We can use compasses to determine the magnitude and direction of the magnetic field, as shown below.

- The *direction* of the magnetic field is the direction that the north pole of a compass needle points.

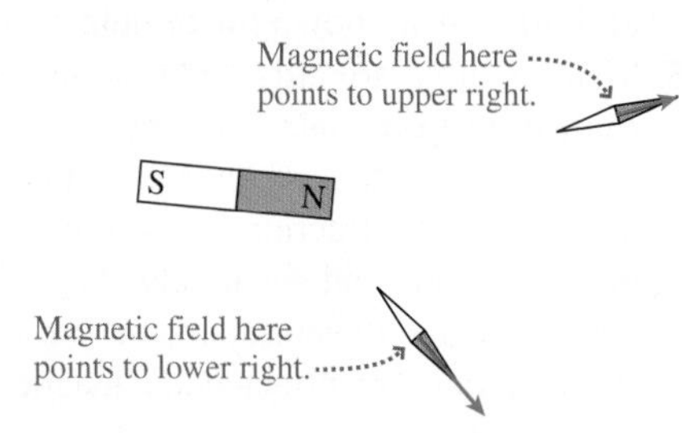

Magnetic field here points to upper right.
S
N
Magnetic field here points to lower right.

■ The *strength* of the magnetic field is proportional to the torque felt by the compass needle if it is turned slightly away from the field direction.

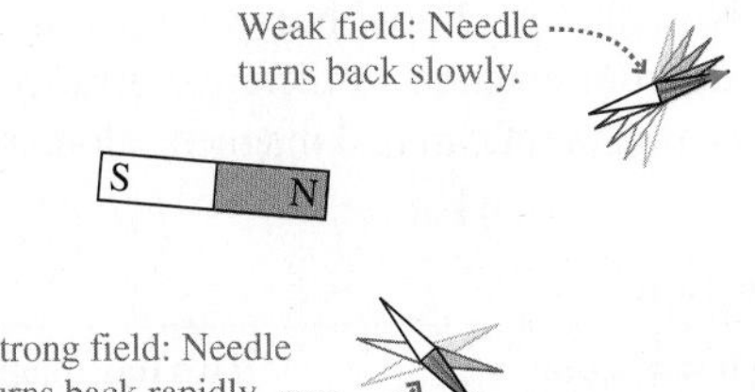

Weak field: Needle turns back slowly.

Strong field: Needle turns back rapidly.

We can produce a "picture" of the magnetic field by using *iron filings*—very small elongated grains of iron. If there are enough grains, iron filings can give a very detailed representation of the magnetic field, as shown in FIGURE 24.2. The compasses that we use to determine field direction show us that **the magnetic field of a magnet points *away* from the north pole and *toward* the south pole.**

FIGURE 24.2 Revealing the field of a bar magnet using iron filings.

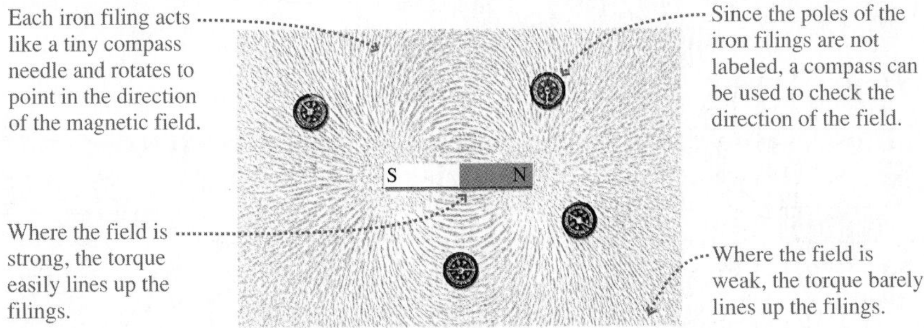

Each iron filing acts like a tiny compass needle and rotates to point in the direction of the magnetic field.
Since the poles of the iron filings are not labeled, a compass can be used to check the direction of the field.
S
N
Where the field is strong, the torque easily lines up the filings.
Where the field is weak, the torque barely lines up the filings.

Magnetic Field Vectors and Field Lines

We can draw the field of a magnet such as the one shown in Figure 24.2 in either of two ways. When we want to represent the magnetic field at one particular point, the **magnetic field vector** representation is especially useful. But if we want an overall representation of the field, **magnetic field lines** are often simpler to use. These two representations are similar to the *electric field* vectors and lines used in Chapter 20, and we'll use similar rules to draw them.

Let's start by drawing some magnetic field vectors that represent the magnetic field of our bar magnet. As shown in FIGURE 24.3, we can imagine placing a number of compasses near the magnet to measure the direction and magnitude of the magnetic field. To represent the field at the location of one of the compasses, we then draw a vector with its *tail* at that location. Figure 24.3 shows how to choose the direction and magnitude of this vector. Although we've drawn magnetic field vectors at only a few points around the magnet, it's important to remember that the magnetic field exists at *every* point around the magnet.

We can also represent the magnetic field using magnetic field lines. The rules for drawing these lines are similar to those for drawing the electric field lines of Chapter 20 and are shown for magnetic fields in FIGURE 24.4.

FIGURE 24.3 Mapping out the field of a bar magnet using compasses.

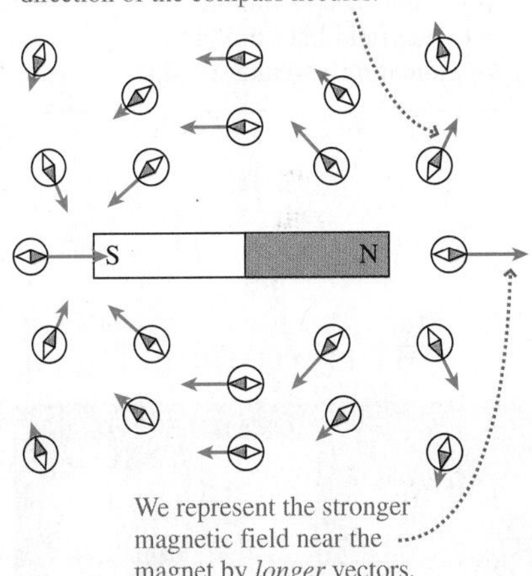

The magnetic field vectors point in the direction of the compass needles.
S
N
We represent the stronger magnetic field near the magnet by *longer* vectors.

FIGURE 24.4 Drawing the magnetic field lines of a bar magnet.

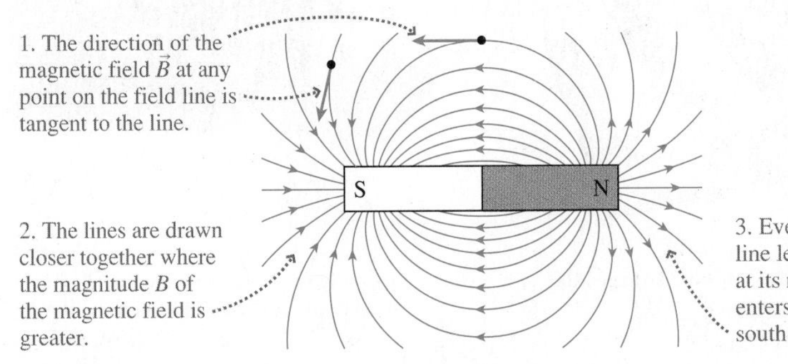

Now that we know how to think about magnetic fields, let's look at magnetic fields from magnets of different arrangements. We'll use the iron filing method to show the lines from real magnets, along with a drawing of the field lines.

An atlas of magnetic fields produced by magnets

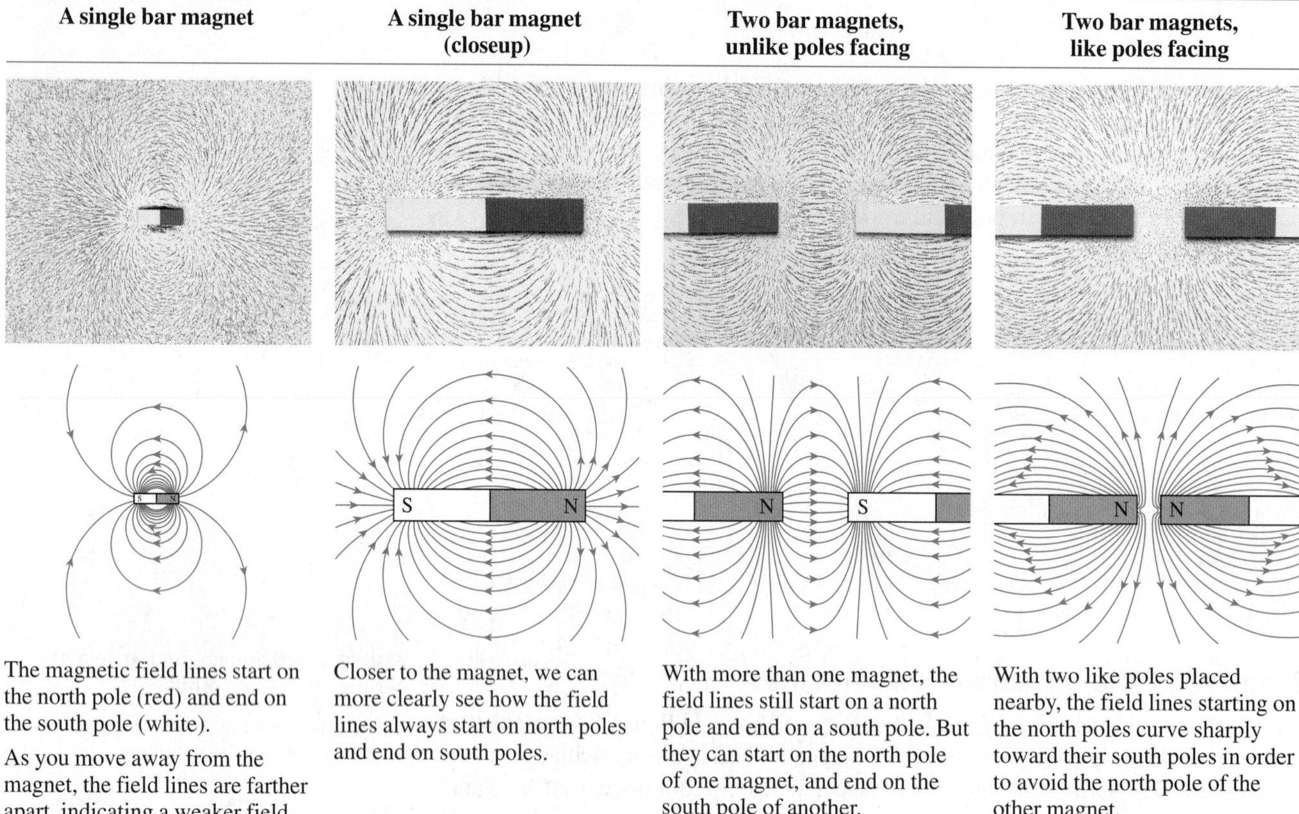

A single bar magnet
A single bar magnet (closeup)
Two bar magnets, unlike poles facing
Two bar magnets, like poles facing
S
N
S
N
N
S
N
N
The magnetic field lines start on the north pole (red) and end on the south pole (white).
As you move away from the magnet, the field lines are farther apart, indicating a weaker field.
Closer to the magnet, we can more clearly see how the field lines always start on north poles and end on south poles.
With more than one magnet, the field lines still start on a north pole and end on a south pole. But they can start on the north pole of one magnet, and end on the south pole of another.
With two like poles placed nearby, the field lines starting on the north poles curve sharply toward their south poles in order to avoid the north pole of the other magnet.

Magnetic Fields Around Us

We live in a magnetic field that is created in the earth's core, and magnetic fields are used in a wide range of technological applications. An everyday example is the flexible refrigerator magnet. As seen in FIGURE 24.5, these magnets have an unusual arrangement of long, striped poles. This arrangement forces most of the field to exit the brown side of the magnet, which is why this side sticks better to your refrigerator than the label side (see Conceptual Example 24.14).

FIGURE 24.5 The magnetic field of a refrigerator magnet.

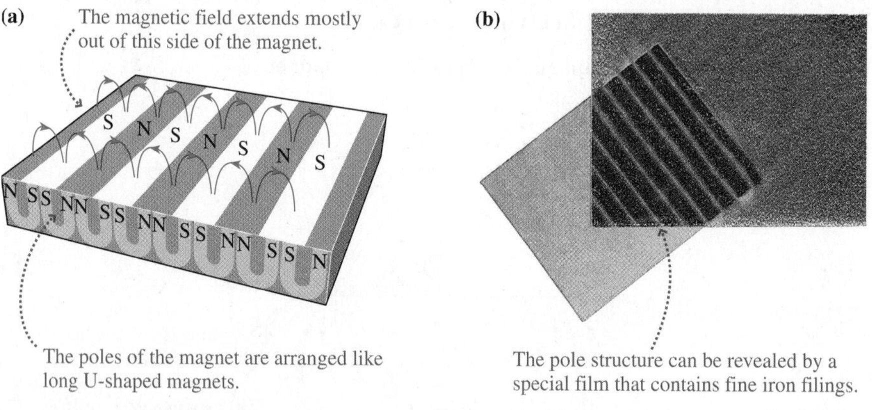

The poles of the magnet are arranged like long U-shaped magnets.

The pole structure can be revealed by a special film that contains fine iron filings.

TRY IT YOURSELF

Buzzing magnets You can use two *identical* flexible refrigerator magnets for a nice demonstration of their alternating pole structure. Place the two magnets together, back to back, then quickly pull them across each other, noting the alternating attraction and repulsion from the alternating poles. If you pull them quickly enough, you will hear a buzz as the magnets are rapidly pushed apart and then pulled together.

Another application is the use of magnetic materials to store information on computer hard disk drives. As shown in FIGURE 24.6, a hard drive consists of a rapidly rotating disk with a thin magnetic coating on its surface. Information for computers is stored as digital zeros and ones, and on the disk these are stored as tiny magnetic dipoles. Each dipole is less than 100 nm long! The direction of the dipoles can be changed by the *write head*—a tiny switchable magnet that skims over the surface of the disk. The information can then be retrieved by the *read head*—a small probe that is sensitive to the magnetic fields of the tiny dipoles.

FIGURE 24.6 Computer hard disks store information using magnetic fields.

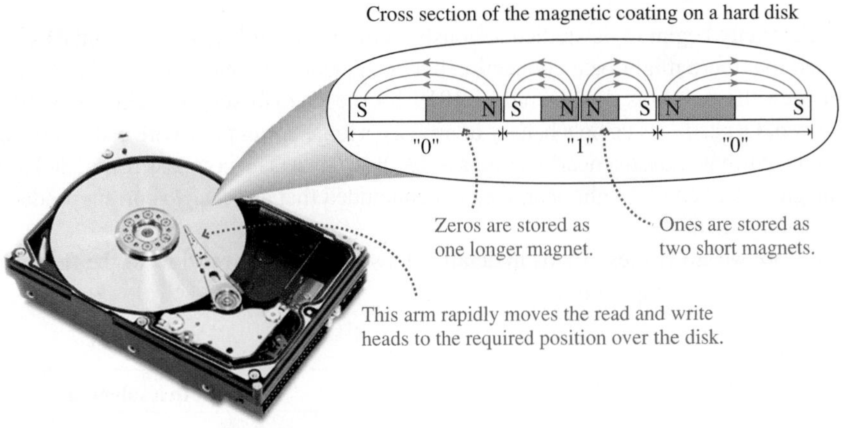

As we've seen, a bar magnet that is free to pivot—a compass—always swings so that its north pole points geographically north. But we've also seen that if a magnet is brought near a compass, the compass swings so that its north pole faces the south pole of the magnet. These observations can be reconciled if the earth itself is a large magnet, as shown in FIGURE 24.7a, where **the south pole of the earth's magnet is located near—but not exactly coincident with—the north geographic pole of the earth.** You can see from the figure that the north pole of a compass needle placed, for example, at the equator, will point toward the south pole of the earth's magnet—that is, to the north!

FIGURE 24.7b shows that the earth's magnetic field has components both parallel to the ground (horizontal) and perpendicular to the ground (vertical). An ordinary north-pointing compass responds only to the horizontal component, but a compass free to pivot vertically tilts downward at an angle called the **dip angle.** The dip angle varies with latitude on the earth's surface, and measuring the dip angle is one way to determine your latitude.

FIGURE 24.7 The earth's magnetic field.

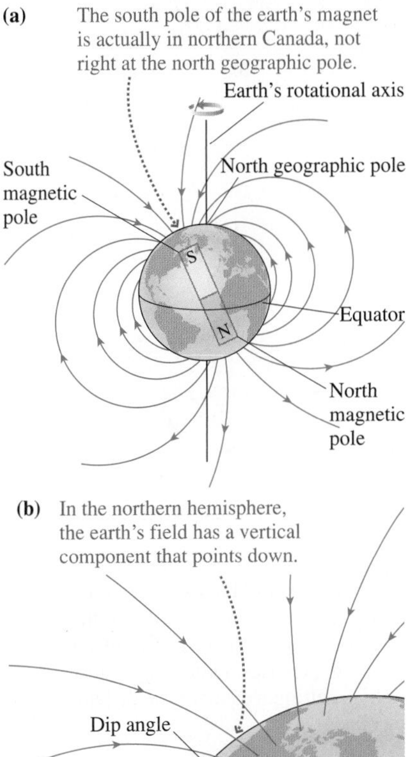

CONCEPTUAL EXAMPLE 24.1 Balancing a compass

Compasses made for use in northern latitudes are weighted so that the south pole of their needle is slightly heavier than the north pole. Explain why this is done.

REASON Figure 24.7b shows that, at northern latitudes, the magnetic field of the earth has a large vertical component. A compass needle that pivots to line up with the field has its north pole pointing north, but the north pole also tips down to follow the field. To keep the compass balanced, there must be an extra force on the south end of the compass. A small weight on the south pole provides a force that keeps the needle balanced.

ASSESS This strategy makes sense. Keeping the needle horizontal when the field is not horizontal requires some extra force.

STOP TO THINK 24.2 A compass is placed next to a bar magnet as shown. Which figure shows the correct alignment of the compass?

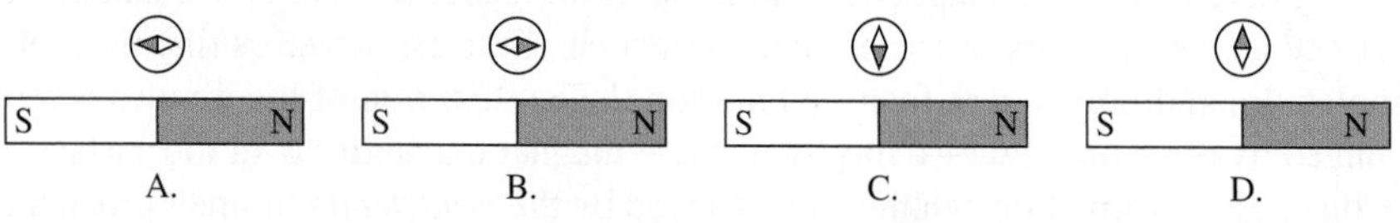

24.3 Electric Currents Also Create Magnetic Fields

As electricity began to be studied seriously in the 18th century, some scientists speculated that there might be a connection between electricity and magnetism. The link between the two was discovered in 1819 by the Danish scientist Hans Christian Oersted. Oersted was using a battery to produce a large current in a wire and he noticed that the current caused a nearby compass needle to turn. The compass responded as if a magnet had been brought near. Oersted concluded that an *electric* current produces a *magnetic* field.

Before we study these fields in detail, let's start with an overview of the fields we will see in the coming sections.

An atlas of magnetic fields produced by currents

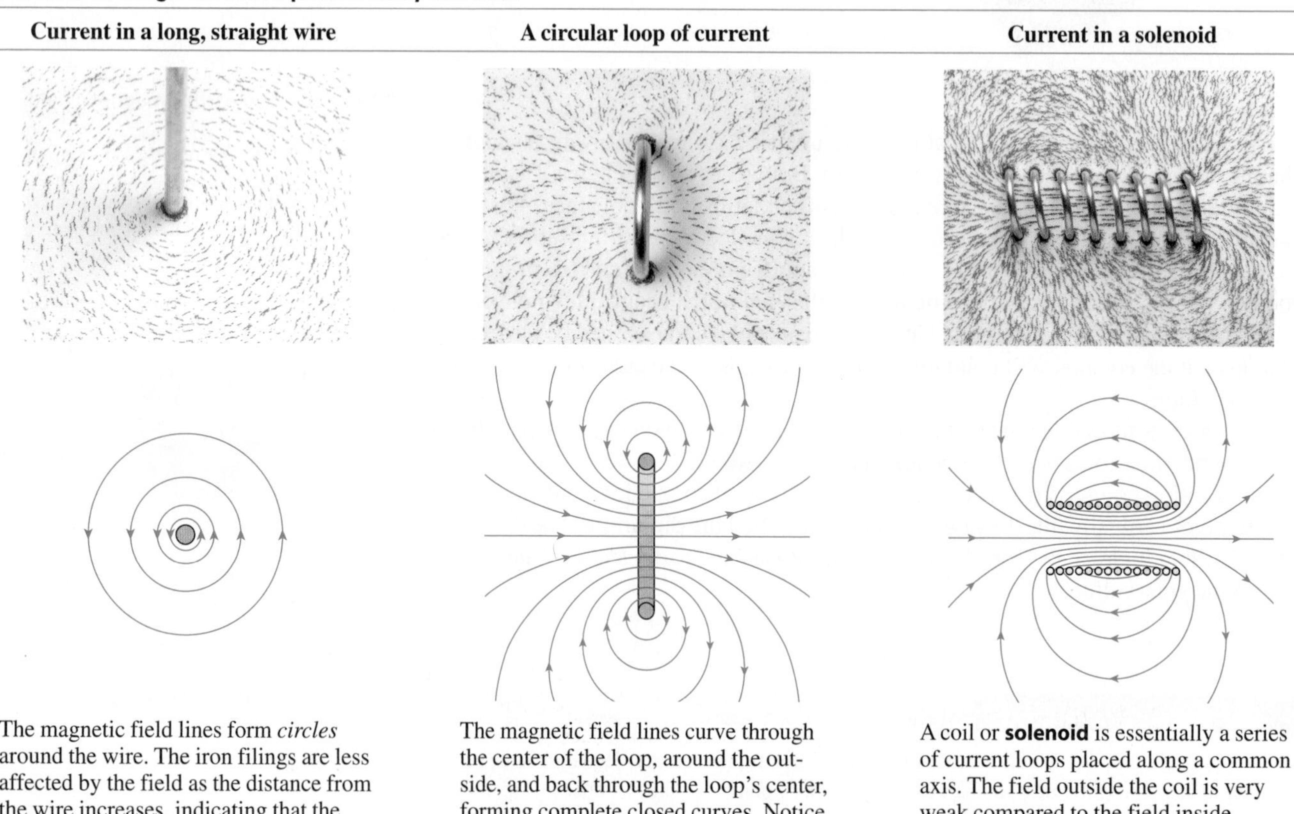

The magnetic field lines form *circles* around the wire. The iron filings are less affected by the field as the distance from the wire increases, indicating that the field is getting weaker as the distance increases.

The magnetic field lines curve through the center of the loop, around the outside, and back through the loop's center, forming complete closed curves. Notice that field lines far from the loop look like the field lines far from a bar magnet.

A coil or **solenoid** is essentially a series of current loops placed along a common axis. The field outside the coil is very weak compared to the field inside. Inside the solenoid, the magnetic field lines are reasonably evenly spaced. The magnetic field inside a solenoid is nearly uniform.

There are striking similarities between the iron filing patterns created by currents in wires and those created by magnets. As we develop our understanding of magnetic fields, we'll look at the similarities and differences between fields from these two different sources.

Earlier we noted that the field lines of magnets start and end on magnetic poles. However, the field lines due to currents have no start or end: They form complete closed curves. If we consider the field lines continuing *inside* a magnet, however, we find that these lines also form complete closed curves, as shown in FIGURE 24.8. In fact, *all* magnetic field lines form complete curves.

In this and the next section, we'll explore in some detail the magnetic fields created by currents; later, we'll look again at fields due to magnets. Ordinary magnets are often called **permanent magnets** to distinguish their unchanging magnetism from that caused by currents that can be switched on and off. We look at magnetism in this order because magnetism from currents is easier to understand, but keep in mind that currents and magnets are both equally important sources of magnetic fields.

FIGURE 24.8 Field lines form closed curves for magnets, too.

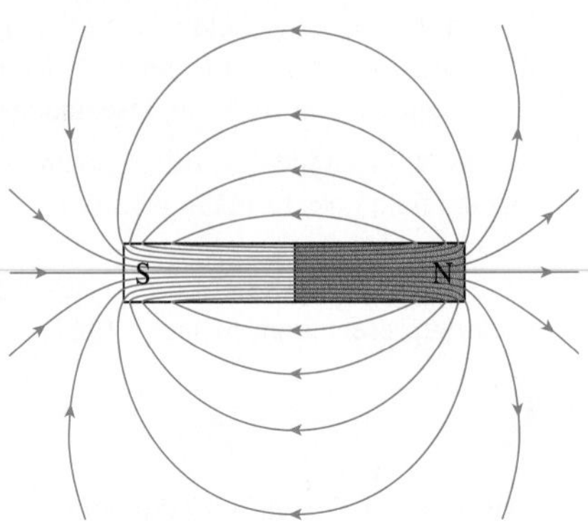

The Magnetic Field of a Straight, Current-Carrying Wire

From the previous atlas picture, we see that the iron filings line up in *circles* around a straight, current-carrying wire. As FIGURE 24.9 shows, we also can use our basic instrument, the compass, to determine the direction of the magnetic field.

FIGURE 24.9 How compasses respond to a current-carrying wire.

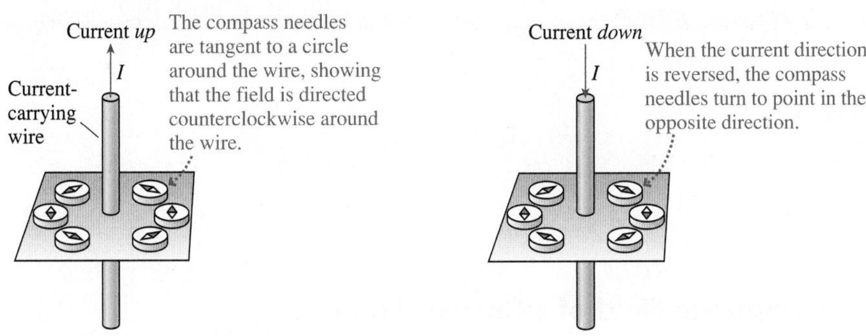

To help remember in which direction compasses will point, we use the *right-hand rule* shown in Tactics Box 24.1. We'll use this same rule later to find the direction of the magnetic field due to several other shapes of current-carrying wire, so we'll call this rule the **right-hand rule for fields.**

TACTICS BOX 24.1 Right-hand rule for fields

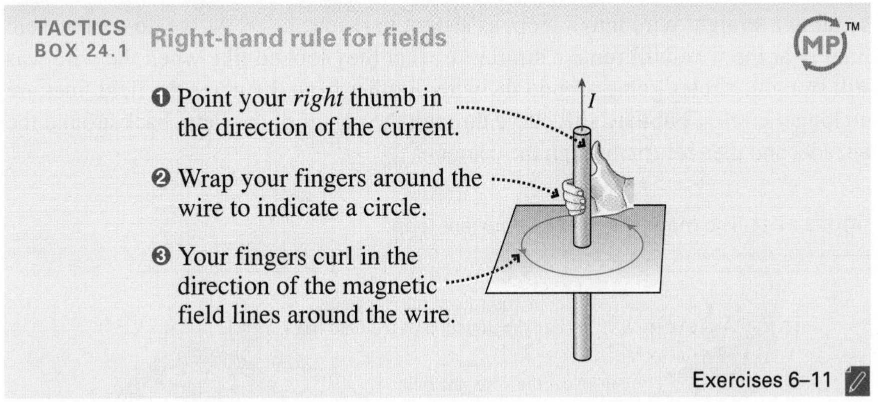

MP
❶ Point your *right* thumb in the direction of the current.
❷ Wrap your fingers around the wire to indicate a circle.
❸ Your fingers curl in the direction of the magnetic field lines around the wire.
I

Exercises 6–11

FIGURE 24.10 The notation for vectors and currents that are perpendicular to the page.

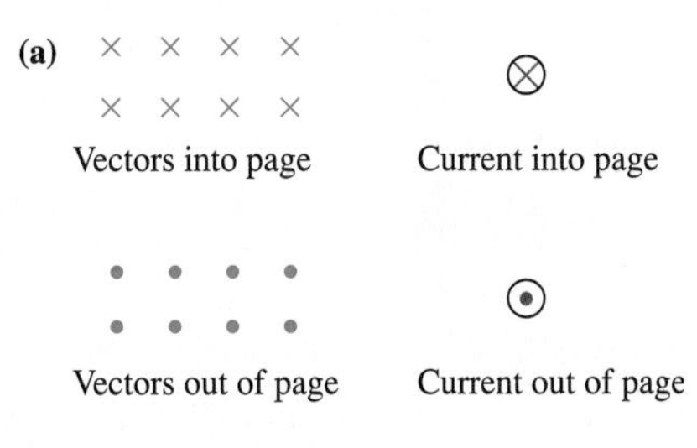

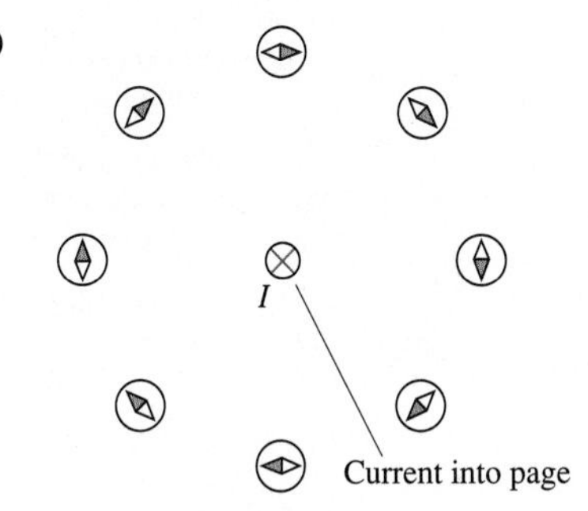

Magnetism is more demanding than electricity in often requiring a three-dimensional perspective of the sort shown in Tactics Box 24.1. But since two-dimensional figures are easier to draw, we will make as much use of them as we can. Consequently, we will often need to indicate field vectors or currents that are perpendicular to the page. FIGURE 24.10a shows the notation we will use. Then FIGURE 24.10b demonstrates this notation by showing the compasses around a current that is directed into the page. To use the right-hand rule with this drawing, point your right thumb into the page. Your fingers will curl clockwise, and that is the direction in which the north poles of the compass needles point.

CONCEPTUAL EXAMPLE 24.2 **Drawing the magnetic field of a current-carrying wire**

Sketch the magnetic field of a long, current-carrying wire, with the current going into the paper. Draw both magnetic field line and magnetic field vector representations.

REASON From the iron filing picture in the atlas, we have seen that the field lines form circles around the wire, and the magnetic field becomes weaker as the distance from the wire is increased. **FIGURE 24.11** shows how we construct both field line and field vector representations of such a field.

ASSESS Figure 24.11 shows the key features of the field, but it's important that you understand its limitations. Although we've drawn only a few circles, the magnetic field actually exists at *all* points around the wire, out to arbitrarily great distances—although it gets quite weak as we move far from the wire. Figure 24.11 shows the field lines only in a single plane perpendicular to the wire. A more complex 3-D drawing, such as **FIGURE 24.12**, is necessary to convey the idea that the field lines also exist in *every* plane along the length of the wire.

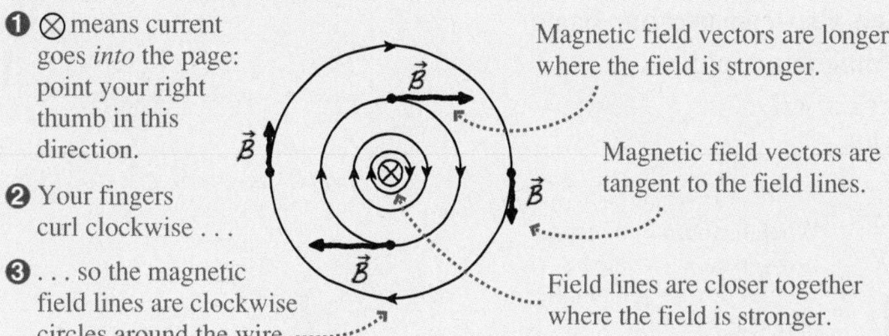

❶ ⊗ means current goes *into* the page: point your right thumb in this direction.
❷ Your fingers curl clockwise . . .
❸ . . . so the magnetic field lines are clockwise circles around the wire.
Magnetic field vectors are longer where the field is stronger.
Magnetic field vectors are tangent to the field lines.
Field lines are closer together where the field is stronger.
$\vec{B}$

FIGURE 24.11 Drawing the magnetic field of a long, straight, current-carrying wire.

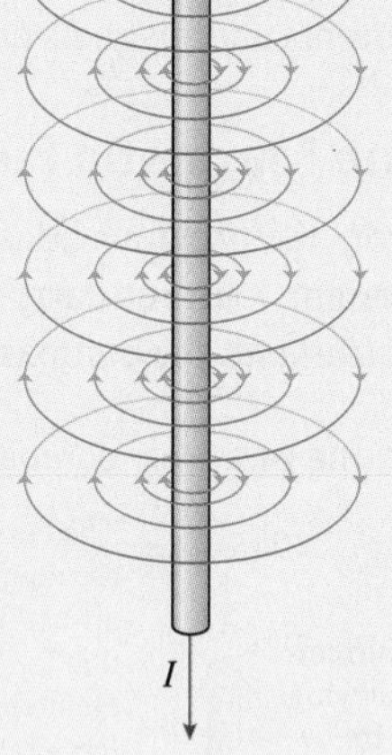

FIGURE 24.12 Field lines exist everywhere along the wire.

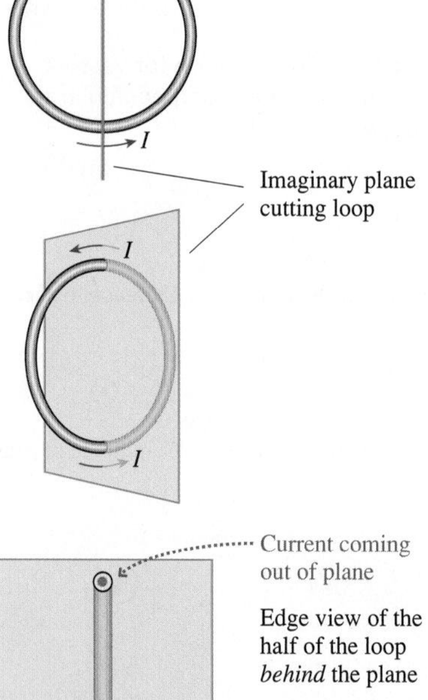

FIGURE 24.13 Three views of a current loop.

The Magnetic Field of a Current Loop

We can extend our understanding of the field from a long, straight, current-carrying wire to the fields due to other shapes of current-carrying wires. Let's look at the other two configurations described in the atlas: a simple circular loop of wire (a *current loop*) and a tightly wound coil (a *solenoid*).

Let's start with the simple circular current-carrying loop, shown in three views in **FIGURE 24.13**. To see what the field due to a current loop looks like, we can imagine bending a straight wire into a loop, as shown in **FIGURE 24.14**. As we do so, the field lines near the wire will remain similar to what they looked like when the wire was still straight: circles going around the wire. Farther from the wires the field lines are no longer circles, but they still curve through the center of the loop, back around the outside, and then return through the center.

FIGURE 24.14 The magnetic field of a current loop.

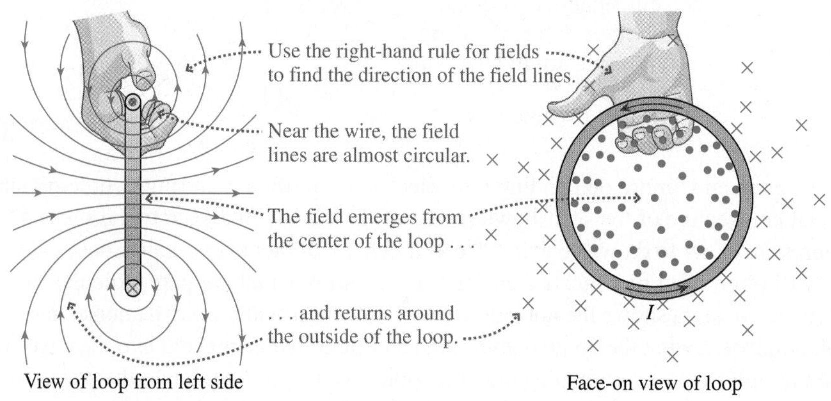

If we reverse the direction of the current in the loop, all the field lines reverse direction as well. Because a current loop is essentially a straight wire bent into a circle, the same right-hand rule of Tactics Box 24.1, used to find the field direction for a long, straight wire, can also be used to find the field direction for a current loop. As shown in Figure 24.14, you again point your thumb in the direction of the current in the loop and let your fingers curl through the center of the loop. Your fingers are then pointing in the direction in which $\vec{B}$ passes through the *center* of the loop.

FIGURE 24.15 A solenoid.

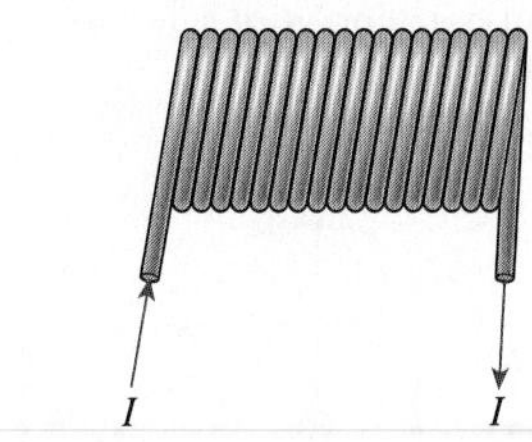

The Magnetic Field of a Solenoid

There are many applications of magnetism, such as the MRI system used to make the image at the beginning of this chapter, for which we would like to generate a **uniform magnetic field,** a field that has the same magnitude and the same direction at every point within some region of space. As we've seen, a reasonably uniform magnetic field can be generated with a **solenoid.** A solenoid, as shown in FIGURE 24.15, is a coil of wire with the same current I passing through each loop in the coil. Solenoids may have hundreds or thousands of coils, often called *turns,* sometimes wrapped in several layers.

The iron filing picture in the atlas on page 782 shows us that **the field within the solenoid is strong, mainly parallel to the axis, and reasonably uniform, whereas the field outside the solenoid is very weak.** FIGURE 24.16 reviews these points and shows why the field inside is much stronger than the field outside. The field direction inside the solenoid can be determined by using the right-hand rule for any of the loops that form it.

FIGURE 24.16 The field of a solenoid.

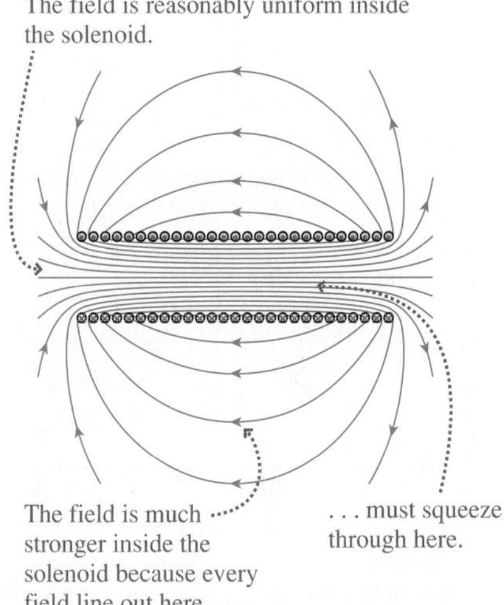

The field is reasonably uniform inside the solenoid.
The field is much stronger inside the solenoid because every field line out here . . .
. . . must squeeze through here.

STOP TO THINK 24.3 A compass is placed a few centimeters above a very long wire with no current. Because of the earth's field, the needle points north. When a large current is turned on in the direction shown, in which direction will the compass point?

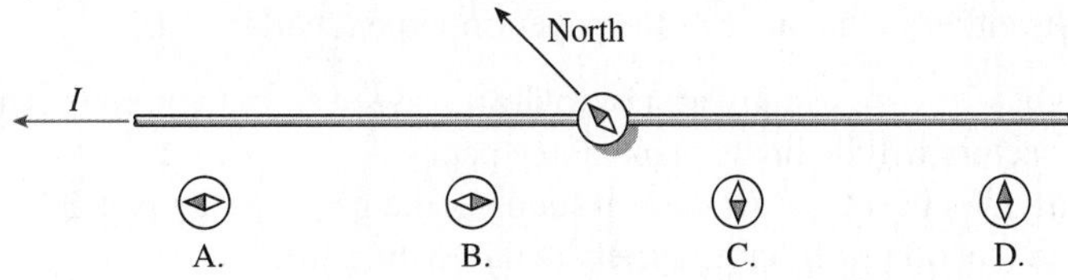

24.4 Calculating the Magnetic Field Due to a Current

13.1

The previous section showed qualitatively how the magnetic field looks for several shapes of current-carrying wires. In this section, we'll give quantitative expressions for these magnetic fields.

Let's start with the simplest case—that of a long, straight, current-carrying wire. We've seen already that the magnetic field lines form circles around the wire and that the field gets weaker as the distance from the wire increases. Not surprisingly, the magnitude of the field also depends on the *current* through the wire, increasing in proportion to the current. The magnitude of the magnetic field a distance r from the wire carrying current I is given by the expression

$$B = \frac{\mu_0 I}{2\pi r} \tag{24.1}$$

Magnetic field due to a long, straight, current-carrying wire

INVERSE p.114

TABLE 24.1 Typical magnetic field strengths

Field location	Field strength (T)
Surface of the earth	5×10^{-5}
Refrigerator magnet	5×10^{-3}
Laboratory magnet	0.1 to 1
Superconducting magnet	10

Although this equation is exact only for an infinitely long wire, it is quite accurate whenever the length of the wire is significantly greater than the distance r from the wire. FIGURE 24.17 shows how the field varies with distance from the wire.

The SI unit of the magnetic field is the **tesla,** abbreviated as T. One tesla is quite a large field. Table 24.1 on the previous page shows some typical magnetic field strengths. Most magnetic fields are a small fraction of a tesla. We sometimes express magnetic fields in μT (10^{-6} T) or mT (10^{-3} T). The strength of the earth's field varies from place to place, but a good average strength, which you can use for solving problems, is 50 μT.

The constant μ_0, which relates the strength of the magnetic field to the currents that produce it, in Equation 24.1 is called the **permeability constant.** Its role in magnetic field expressions is similar to the role of the permittivity constant ϵ_0 in electric field expressions. Its value is

$$\mu_0 = 4\pi \times 10^{-7}\ \text{T} \cdot \text{m/A} = 1.257 \times 10^{-6}\ \text{T} \cdot \text{m/A}$$

FIGURE 24.17 The magnetic field due to a long, current-carrying wire.

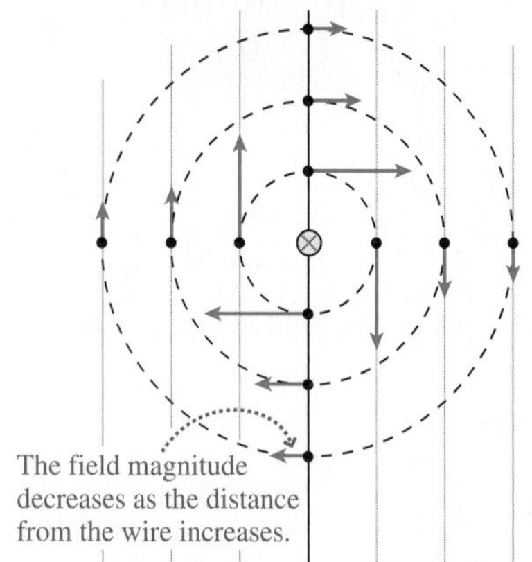

The field magnitude decreases as the distance from the wire increases.

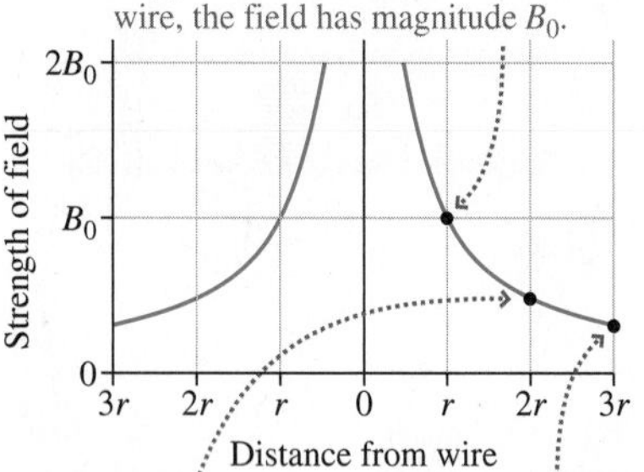

Twice as far away, the field is half as strong.
Three times as far away, the field is one-third as strong.

Magnetic Fields from More Than One Source

When two or more sources of magnetic fields are brought near each other, how do we find the total magnetic field at any particular point in space? For electric fields, we used the principle of superposition: The total electric field at any point is the *vector* sum of the individual fields at that point. The same principle holds for magnetic fields as well. FIGURE 24.18 illustrates the principle of superposition applied to *magnetic* fields.

We've now learned enough to suggest the following general problem-solving strategy for finding the magnetic field at a point due to known sources of magnetic field.

FIGURE 24.18 Adding fields due to more than one source.

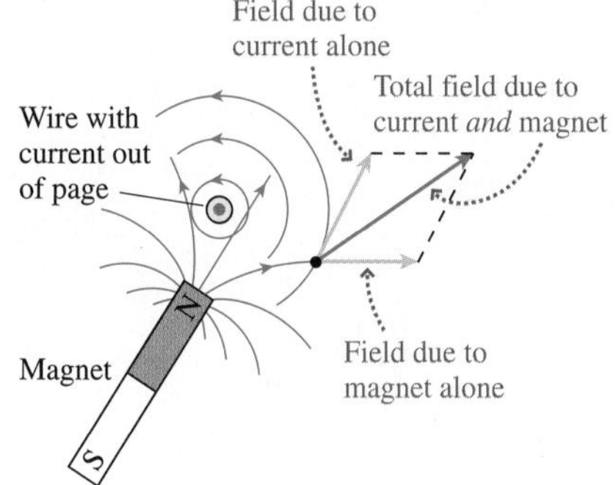

Field due to current alone

Wire with current out of page

Field due to magnet alone

PROBLEM-SOLVING STRATEGY 24.1 **Magnetic field problems**

PREPARE Because current-carrying wires do not lie in the same plane as the fields they produce, you'll need to prepare an especially careful drawing. Generally, you should choose the plane of your drawing so that the magnetic field vectors lie either in the plane of the paper or perpendicular to it.

- Straight wires are usually easiest to draw as seen from their ends. Then the field vectors will lie in the plane of the paper.
- Usually, it's best to draw current loops in the plane of the paper. Then the field in the loop's center is perpendicular to the paper.
- Solenoids can be drawn either end-on (field perpendicular to the plane of the paper) or as seen from the side (field in the plane of the paper).
- If the problem has more than one source of magnetic field, it's usually best that your drawing shows the field vectors in the plane of the paper. Then they can be added as vectors, using coordinate systems and components if needed.

SOLVE Find the directions of the fields due to each source (wire, loop, solenoid) by using the right-hand rule for fields. Then find the magnitude of each field using the expressions given in this section for a wire, a loop, or a solenoid. Finally, add these fields together using the rules for vector addition.

ASSESS Check if the magnitude and direction of the total field seem reasonable.

Exercise 17 

EXAMPLE 24.3 Finding the magnetic field of two parallel wires

Two long, straight wires lie parallel to each other, as shown in FIGURE 24.19. They each carry a current of 5.0 A, but in opposite directions. What is the magnetic field at point P?

▶ FIGURE 24.19 Two parallel current-carrying wires.

PREPARE Redraw the wires as seen from their right ends, as in FIGURE 24.20, and add a coordinate system. In this view, the right-hand rule for fields tells us that the magnetic field at P from wire 1 points to the right and that from wire 2 to the left. Because wire 2 is twice as far from P as wire 1, we've drawn its field $\vec{B}_2$ half as long as the field $\vec{B}_1$ from wire 1.

FIGURE 24.20 View of the wires from their right ends.

SOLVE From Figure 24.20, we can see that the total field $\vec{B}$—the vector sum of the two fields $\vec{B}_1$ and $\vec{B}_2$—points to the right. To find the magnitude of $\vec{B}$, we'll need the magnitudes of $\vec{B}_1$ and $\vec{B}_2$

We can use Equation 24.1 to find these magnitudes. From Figure 24.20, r for wire 1 is 8.0 cm (or 0.080 m), while r for wire 2 is 16 cm.

We then have

$$B_1 = \frac{\mu_0 I}{2\pi r} = \frac{(4\pi \times 10^{-7}\ \text{T}\cdot\text{m/A})(5.0\ \text{A})}{2\pi(0.080\ \text{m})} = 1.25 \times 10^{-5}\ \text{T}$$

and

$$B_2 = \frac{\mu_0 I}{2\pi r} = \frac{(4\pi \times 10^{-7}\ \text{T}\cdot\text{m/A})(5.0\ \text{A})}{2\pi(0.16\ \text{m})} = 0.63 \times 10^{-5}\ \text{T}$$

Because $\vec{B}_2$ points to the left, its x-component is negative. Thus, the x-component of the total field $\vec{B}$ is

$$\begin{aligned} B_x &= (B_1)_x + (B_2)_x \\ &= 1.25 \times 10^{-5}\ \text{T} - 0.63 \times 10^{-5}\ \text{T} = 0.62 \times 10^{-5}\ \text{T} \\ &= 6.2\ \mu\text{T} \end{aligned}$$

In terms of the original view of the problem in Figure 24.19, we can write

$$\vec{B} = (6.2\ \mu\text{T, into the page})$$

ASSESS That B_x is positive tells us that the total field points to the right, in the same direction as the field due to wire 1. This makes sense, because wire 1 is closer to P than is wire 2.

Current Loops

The magnetic field due to a current loop is more complex than that of a straight wire, as we can see from FIGURE 24.21, but there is a simple expression for the field at an important place: the *center* of the loop. Because the loop can be thought of as a wire bent into a circle (of radius R), the expression for the field at the center is very similar to that of a wire. The field at the center is given by

$$B = \frac{\mu_0 I}{2R} \qquad (24.2)$$

Magnetic field at the center of a current loop of radius R

13.2

FIGURE 24.21 The magnetic field at the center of a current loop.

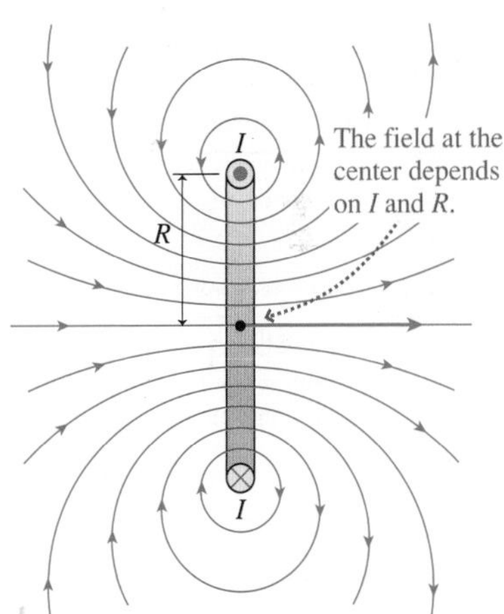

where I is the current in the loop. We have already seen how to find the *direction* of the field in the center, using our right-hand rule for fields.

If N loops of wire carrying the same current I are all tightly wound into a single thin coil, then the field at the center is just N times bigger (since we're superimposing N individual current loops):

$$B = \frac{\mu_0 N I}{2R} \qquad (24.3)$$

Magnetic field at the center of a thin coil with N turns

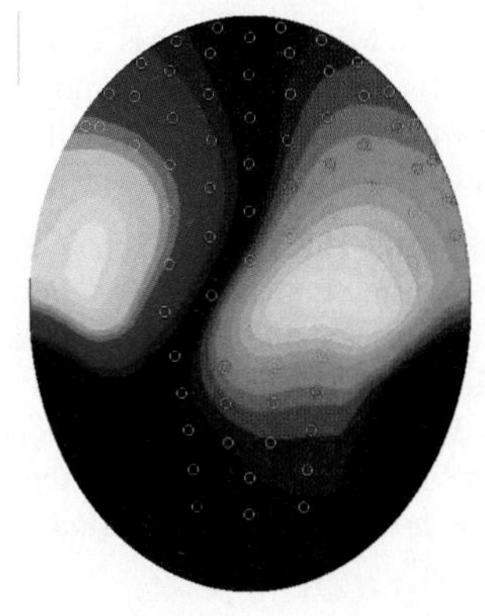

▶ **The magnetocardiogram** BIO When the heart muscle contracts, action potentials create a dipole electric field that can be measured to create an electrocardiogram. These action potentials also cause charges to circulate around the heart, creating a current loop. This current loop creates a small, but measurable, magnetic field. A record of the heart's magnetic field, a *magnetocardiogram,* can provide useful information about the heart in cases where an electrocardiogram is not possible. This image shows the magnetic field of a fetal heartbeat measured at the surface of the mother's abdomen. Here, blue represents a field pointing into the body and red a field out of the body. This is just the field expected from a current loop whose plane lies along the black line between the two colored areas.

EXAMPLE 24.4 Canceling the earth's field

Green turtles are thought to navigate by using the dip angle of the earth's magnetic field. To test this hypothesis, green turtle hatchlings were placed in a 72-cm-diameter tank with a 50-turn coil of wire wrapped around the outside. A current in the coil created a magnetic field at the center of the tank that exactly canceled the vertical component of earth's 50 μT field. At the location of the test, the earth's field was directed 60° below the horizontal. What was the current in the coil?

PREPARE FIGURE 24.22 shows the earth's field passing downward through the coil. To cancel the vertical component of this field, the current in the coil must generate an upward field of equal magnitude. We can use the right-hand rule (see Figure 24.14) to find that the current must circulate around the coil as shown. Viewed from above, the current will be counterclockwise.

FIGURE 24.22 The coil field needed to cancel the earth's field.

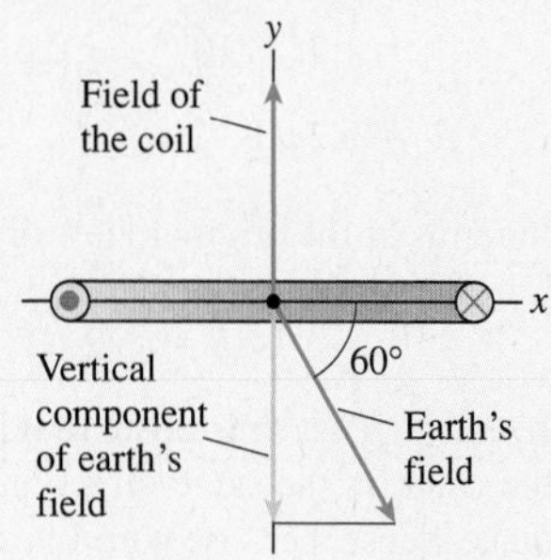
y
Field of the coil
x
60°
Vertical component of earth's field
Earth's field

SOLVE The vertical component of the earth's field is

$$(B_{\text{earth}})_y = -(50 \times 10^{-6}\ \text{T})\sin(60°) = -43 \times 10^{-6}\ \text{T}$$

The field of the coil, given by Equation 24.3, must have the same magnitude at the center. The $2R$ in the equation is just the diameter of the coil, 72 cm or 0.72 m. Thus

$$B_{\text{coil}} = \frac{\mu_0 NI}{2R} = 43 \times 10^{-6}\ \text{T}$$

$$I = \frac{(43 \times 10^{-6}\ \text{T})(2R)}{(\mu_0 N)}$$

$$= \frac{(43 \times 10^{-6}\ \text{T})(0.72\ \text{m})}{(4\pi \times 10^{-7}\ \text{T/m} \cdot \text{A})(50)} = 0.49\ \text{A}$$

As noted, this current is counterclockwise as viewed from above.

ASSESS This seems like a reasonable current. The earth's field isn't very strong, so the coil need not carry a large current to cancel it.

Solenoids

13.3 ActivPhysics ONLINE

As we've seen, the field inside a solenoid is fairly uniform, while the field outside is quite small; the greater a solenoid's length in comparison to its diameter, the better these statements hold. Measurements that need a uniform magnetic field are often conducted inside a solenoid, which can be built quite large. The cylinder that surrounds a patient undergoing magnetic resonance imaging (MRI), such as the one shown on the next page, contains a large solenoid made of superconducting wire, allowing it to carry the very large currents needed to generate a strong uniform magnetic field. Consider a solenoid of length L having N turns of wire, as in FIGURE 24.23. We expect that the more turns we can pack into a solenoid of a given length—that is, the greater the ratio N/L—the stronger the field inside will be. We further expect that the strength of the field will be proportional to the current I in the turns. Somewhat surprisingly, the field inside a solenoid does *not* depend on its radius. For this reason, the radius R doesn't appear in the equation for the field inside a solenoid:

FIGURE 24.23 The magnetic field inside a solenoid.

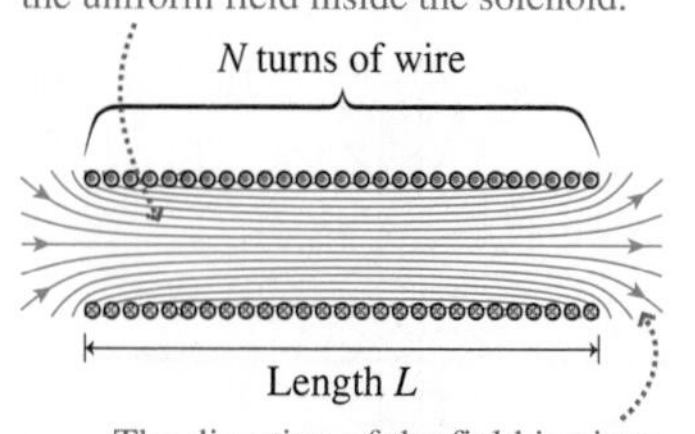
Equation 24.4 gives the magnitude of the uniform field inside the solenoid.
N turns of wire
Length L
The direction of the field is given by the right-hand rule for fields.

$$B = \mu_0 I \frac{N}{L} \tag{24.4}$$

Magnetic field inside a solenoid of length L with N turns

EXAMPLE 24.5 Generating an MRI magnetic field

A typical MRI solenoid has a length of about 1 m and a diameter of about 1 m. A typical field inside such a solenoid is about 1 T. How many turns of wire must the solenoid have to produce this field if the largest current the wire can carry is 100 A?

PREPARE This solenoid is not very long compared to its diameter, so using Equation 24.4 will give only an approximate result. This is acceptable, since we have only rough estimates of the field B and the length L.

Equation 24.4 gives the magnetic field B of a solenoid in terms of the current I, the number of turns N, and the length L. Here, however, we want to find the number of turns in terms of the other variables. We'll need $B = 1$ T, $I = 100$ A, and $L = 1$ m.

SOLVE We can solve Equation 24.4 for N to get

$$N = \frac{LB}{\mu_0 I} = \frac{(1\ \text{m})(1\ \text{T})}{(4\pi \times 10^{-7}\ \text{T} \cdot \text{m/A})(100\ \text{A})} = 8000\ \text{turns}$$

to one significant figure.

ASSESS The number of turns required is quite large, but the field is quite large, so this makes sense.

The amount of wire in an MRI solenoid is surprisingly large. In the above example, the diameter of the solenoid was about 1 m. The length of wire in each turn is thus $\pi \times 1$ m, or about 3 m. The total length of wire is then about 8000 turns $\times$ 3 m/turn = 24,000 m = 24 km $\approx$ 15 miles! If this magnet used ordinary copper wire large enough to carry 100 A, the total resistance R would be about 35 Ω. We learned in Chapter 23 that the power dissipated by a resistor is equal to I^2R, so the total power would be about $(100\text{ A})^2(35\ \Omega) = 350{,}000$ W, a huge and impractical value. MRI magnets must use *superconducting* wire, which when cooled near absolute zero has *zero* resistance. This allows 24 km of wire to carry 100 A with no power dissipation at all.

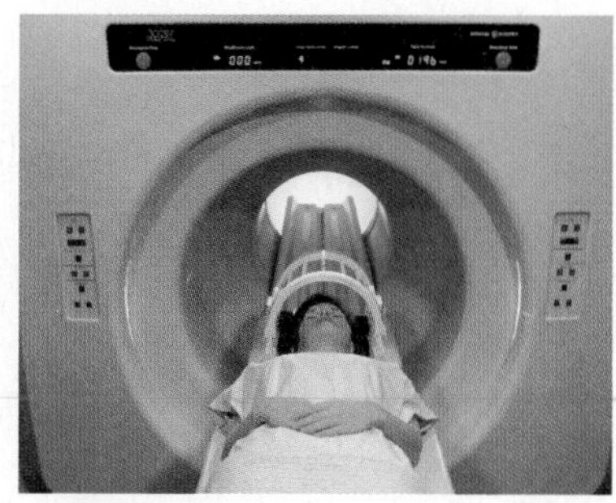

A patient's head in the solenoid of an MRI scanner.

24.5 Magnetic Fields Exert Forces on Moving Charges

It's time to switch our attention from what magnetic fields look like and how they are generated to what they actually *do:* exert forces and torques on moving charges and currents. As we saw earlier in the chapter, Oersted discovered that a current passing through a wire deflects a nearby compass. Upon hearing of Oersted's discovery, the French scientist André Marie Ampère reasoned that the current was acting like a magnet and that two current-carrying wires should exert magnetic forces on each other, just as two magnets do.

Ampère set up two parallel wires that could carry large currents in either the same direction or in opposite directions. FIGURE 24.24 shows the outcome of his experiment: For currents, "likes" attract and "opposites" repel. This is the opposite of what would have happened had the wires been charged and thus exerting electric forces on each other.

Ampère's experiment showed that **a magnetic field exerts a force on a current.** Since currents consist of moving charges, Ampère's experiment therefore also implies that **a magnetic field exerts a force on a moving charged particle.** Let's explore the nature of this force.

FIGURE 24.24 The forces between parallel current-carrying wires.

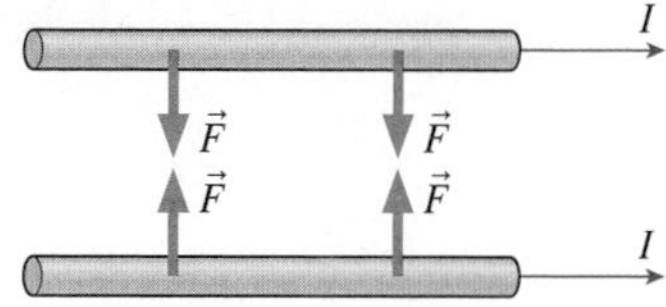

"Like" currents attract.

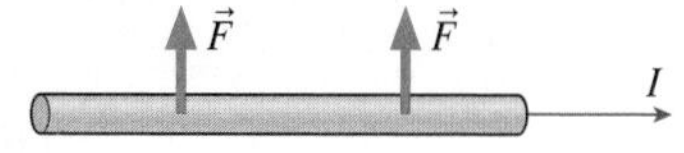

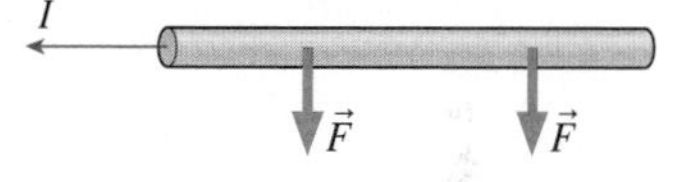

"Opposite" currents repel.

The Magnetic Force on a Moving Charged Particle

The following series of experiments illustrates the nature of the magnetic force on a moving charged particle.

The force on a charged particle moving in a magnetic field

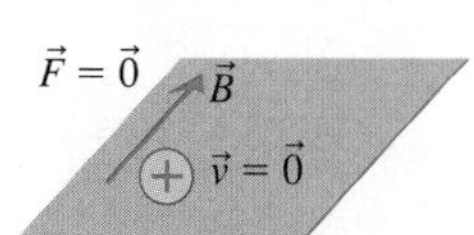

Only a *moving* charged particle experiences a magnetic force. There is no magnetic force on a charge at rest ($v = 0$) in a magnetic field.

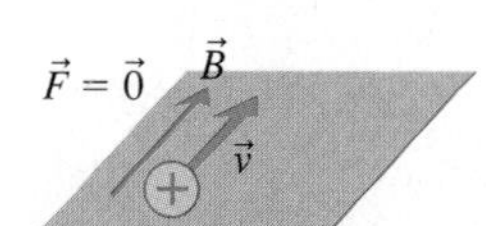

There is no force on a charged particle moving *parallel* to a magnetic field. There is a force only if the charged particle is moving at an angle to the field.

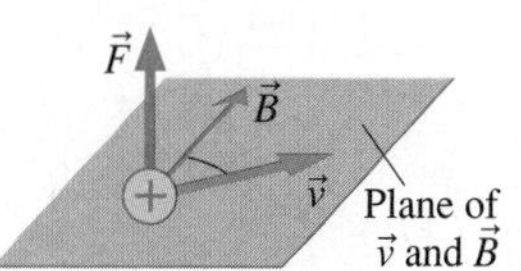

A charged particle moving at an angle to the field *does* experience a force. This force is perpendicular to *both* $\vec{v}$ and $\vec{B}$. That is, the force is at right angles to the plane containing $\vec{v}$ and $\vec{B}$.

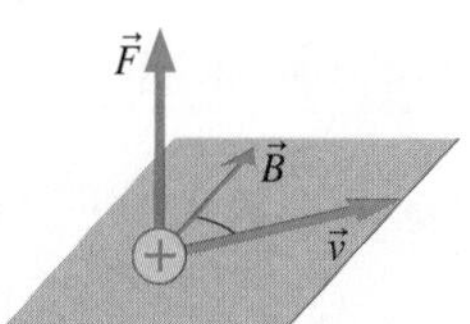

The force is greater if the particle moves faster or if the magnitude of the charge is increased.

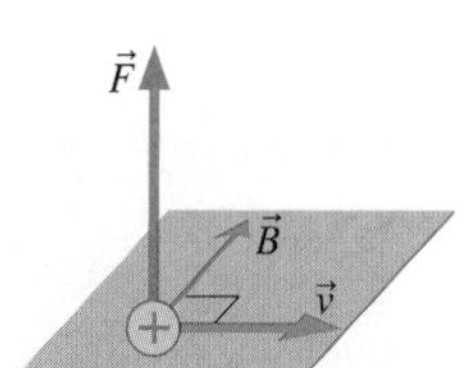

The magnitude of the force depends on the angle between $\vec{v}$ and $\vec{B}$. For given values of v and B, the force is greatest when the angle between $\vec{v}$ and $\vec{B}$ is 90°.

The magnetic force is quite different from the electric force: A charged particle in an electric field feels a force that is *parallel* to the field, but a charged particle moving in a magnetic field feels a force that is *perpendicular* to the field.

The magnetic force on a moving charged particle is perpendicular to the field and the velocity, but this is not sufficient to uniquely determine the direction of the force. We also need the **right-hand rule for forces,** as shown in FIGURE 24.25.

NOTE ► The right-hand rule for forces is different from the right-hand rule for fields. ◄

FIGURE 24.25 The right-hand rule for forces.

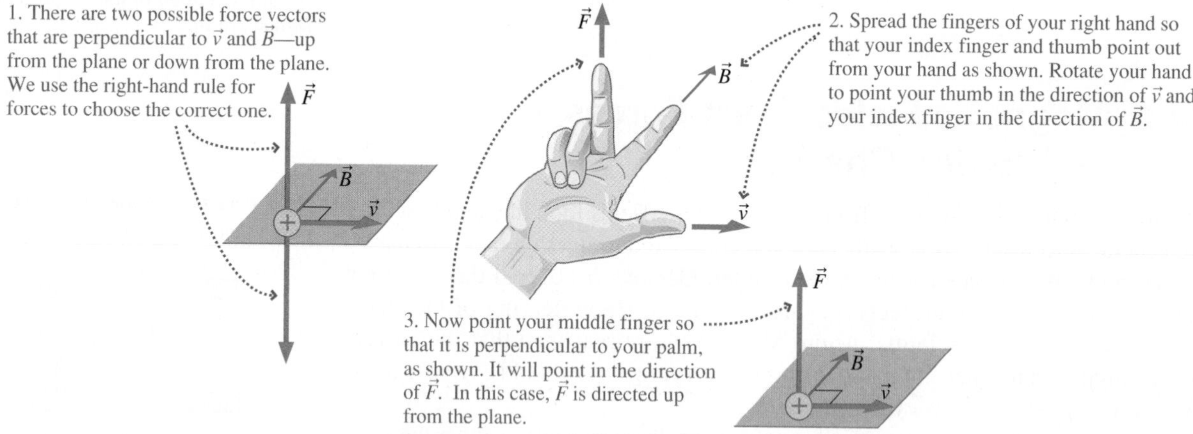

1. There are two possible force vectors that are perpendicular to $\vec{v}$ and $\vec{B}$—up from the plane or down from the plane. We use the right-hand rule for forces to choose the correct one.
$\vec{F}$
$\vec{B}$
$\vec{v}$
2. Spread the fingers of your right hand so that your index finger and thumb point out from your hand as shown. Rotate your hand to point your thumb in the direction of $\vec{v}$ and your index finger in the direction of $\vec{B}$.
3. Now point your middle finger so that it is perpendicular to your palm, as shown. It will point in the direction of $\vec{F}$. In this case, $\vec{F}$ is directed up from the plane.

NOTE ► The right-hand rule for forces gives the direction of the force on a *positive* charge. For a negative charge, the force is in the opposite direction. ◄

13.4 ActivPhysics ONLINE

We can organize all of the experimental information about the magnetic force on a moving charged particle into a single equation. If a particle of charge q moves with a velocity $\vec{v}$ at an angle α to a magnetic field $\vec{B}$, the force is

$$\vec{F} = (|q|vB\sin\alpha, \text{ direction given by the right-hand rule for forces}) \tag{24.5}$$

Force on a charged particle moving in a magnetic field

The velocity and the magnetic field are perpendicular in many practical situations. In this case α is 90°, and we can simplify Equation 24.5 to

$$\vec{F} = (|q|vB, \text{ direction given by the right-hand rule for forces}) \tag{24.6}$$

The following Tactics Box summarizes and shows how to use the above information.

TACTICS BOX 24.2 Determining the magnetic force on a moving charged particle

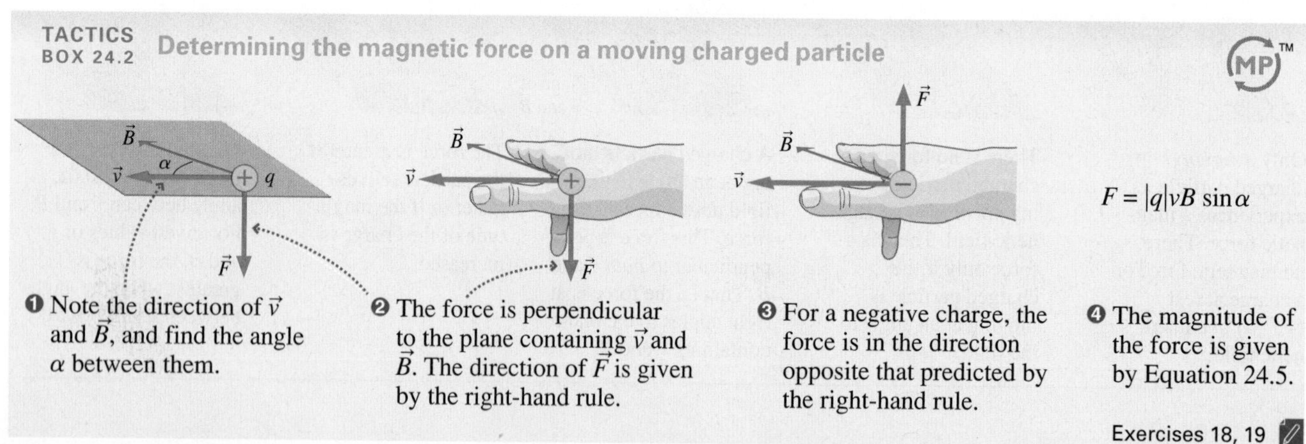

❶ Note the direction of $\vec{v}$ and $\vec{B}$, and find the angle α between them.

❷ The force is perpendicular to the plane containing $\vec{v}$ and $\vec{B}$. The direction of $\vec{F}$ is given by the right-hand rule.

❸ For a negative charge, the force is in the direction opposite that predicted by the right-hand rule.

❹ The magnitude of the force is given by Equation 24.5.

Exercises 18, 19

CONCEPTUAL EXAMPLE 24.6 **Determining the force on a moving electron**

An electron is moving to the right in a magnetic field that points upward, as in FIGURE 24.26. What is the direction of the magnetic force?

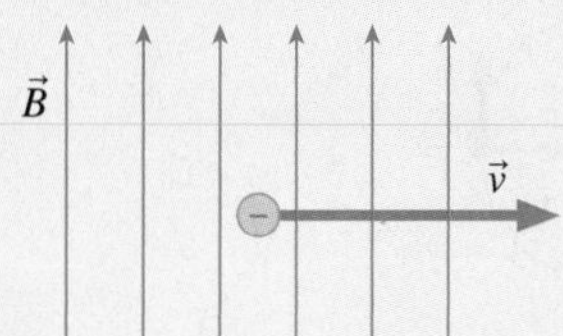

FIGURE 24.26 An electron moving in a magnetic field.

REASON FIGURE 24.27 shows how the right-hand rule for forces is applied to this situation:

- Point your right thumb in the direction of the electron's velocity and your index finger in the direction of the magnetic field.
- Bend your middle finger to be perpendicular to your index finger. Your middle finger, which now points out of the page, is the direction of the force on a positive charge. But the electron is negative, so the force on the electron is *into* the page.

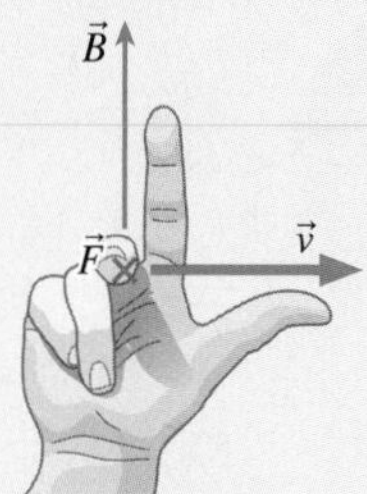

FIGURE 24.27 Using the right-hand rule.

ASSESS The force is perpendicular to both the velocity and the magnetic field, as it must be. The force on an electron is into the page; the force on a proton would be out of the page.

CONCEPTUAL EXAMPLE 24.7 **Determining the force on a charged particle moving near a current-carrying wire**

A proton is moving to the right above a horizontial wire that carries a current to the right. What is the direction of the magnetic force on the proton?

REASON The current in the wire creates a magnetic field; this magnetic field exerts a force on the moving proton. We follow three steps to solve the problem:

1. Sketch the situation, as shown in FIGURE 24.28a.
2. Determine the direction of the field at the position of the proton due to the current in the wire (FIGURE 24.28b).
3. Determine the direction of the force that this field exerts on the proton (FIGURE 24.28c). In this case, the force is down.

ASSESS At the start of this section, where we looked at the forces between currents, we saw that "like" currents attract each other. So we'd expect that a proton moving to the right, which is essentially a small current in the same direction as the current in the wire, will feel a force toward the wire.

FIGURE 24.28 Determining the direction of the force.

(a)

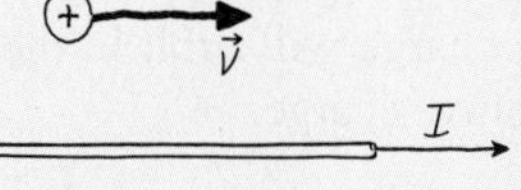

1. The proton moves above the wire. The proton velocity and the current in the wire are shown.

(b)

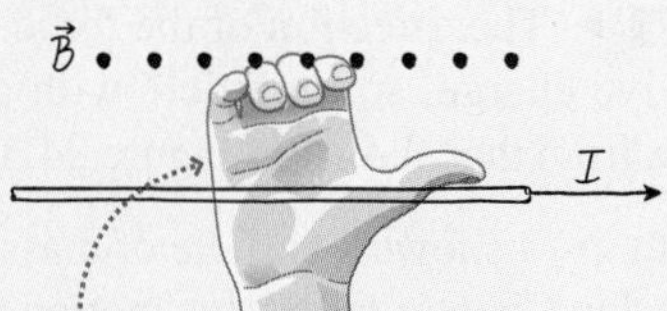

2. The right-hand rule for fields shows that, at the position of the proton, above the wire, the field points out of the page.

(c)

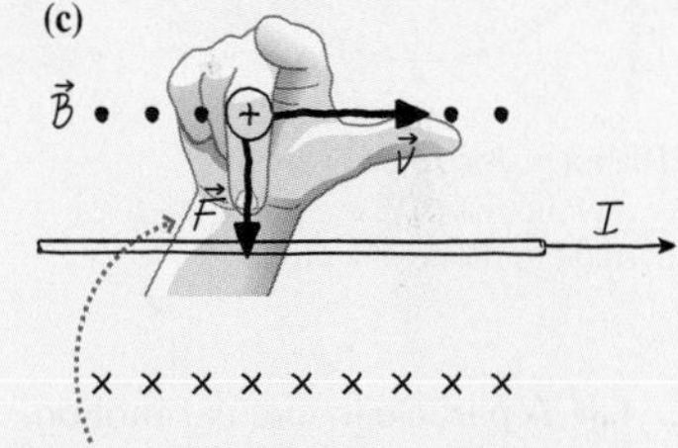

3. The right-hand rule for forces shows that the field of the wire exerts a force on the moving proton that points down, toward the wire.

EXAMPLE 24.8 **Force on a charged particle in the earth's field**

The sun emits streams of charged particles (in what is called the *solar wind*) that move toward the earth at very high speeds. A proton is moving toward the equator of the earth at a speed of 500 km/s. At this point, the earth's field is 5.0×10^{-5} T directed parallel to the earth's surface. What are the direction and the magnitude of the force on the proton?

PREPARE Our first step is, as usual, to draw a picture. As we saw in Figure 24.7a, the field lines of the earth go from the earth's south pole to the earth's north pole. FIGURE 24.29 shows the proton entering the field, which is directed north.

We also need to convert the proton's velocity to m/s:

$$500 \text{ km/s} = 5.0 \times 10^2 \text{ km/s} \times \frac{1 \times 10^3 \text{ m}}{1 \text{ km}} = 5.0 \times 10^5 \text{ m/s}$$

SOLVE We use the steps of Tactics Box 24.2 to determine the force.

❶ $\vec{v}$ and $\vec{B}$ are perpendicular, so $\alpha = 90°$.

❷ The right-hand rule for forces tells us that the force will be into the page in Figure 24.29. That is, the force is toward the east.

❹ We compute the magnitude of the force using Equation 24.6:

$$F = |q|vB = (1.6 \times 10^{-19} \text{ C})(5.0 \times 10^5 \text{ m/s})(5.0 \times 10^{-5} \text{ T})$$
$$= 4.0 \times 10^{-18} \text{ N}$$

ASSESS This is a small force, but the proton has an extremely small mass of 1.67×10^{-27} kg. Consequently, this force produces a very large acceleration: 200 million times the acceleration due to gravity!

Continued

FIGURE 24.29 A proton in the field of the earth.

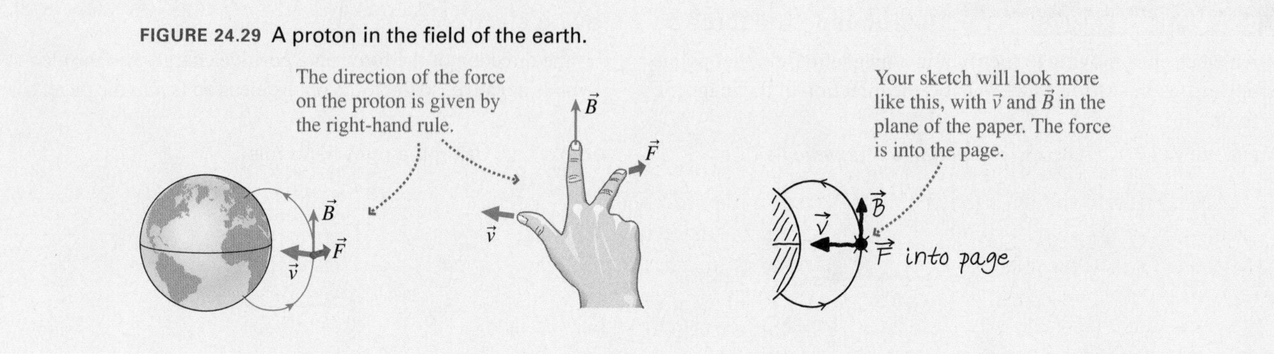
The direction of the force on the proton is given by the right-hand rule.
Your sketch will look more like this, with $\vec{v}$ and $\vec{B}$ in the plane of the paper. The force is into the page.
$\vec{B}$
$\vec{F}$
$\vec{v}$
$\vec{F}$ into page

Example 24.8 looked at the force on a proton from the sun as it reaches the magnetic field of the earth. How does this force affect the motion of the proton? Let's consider the general question of how charged particles move in magnetic fields.

Paths of Charged Particles in Magnetic Fields

FIGURE 24.30 A charged particle moving perpendicular to a uniform magnetic field.

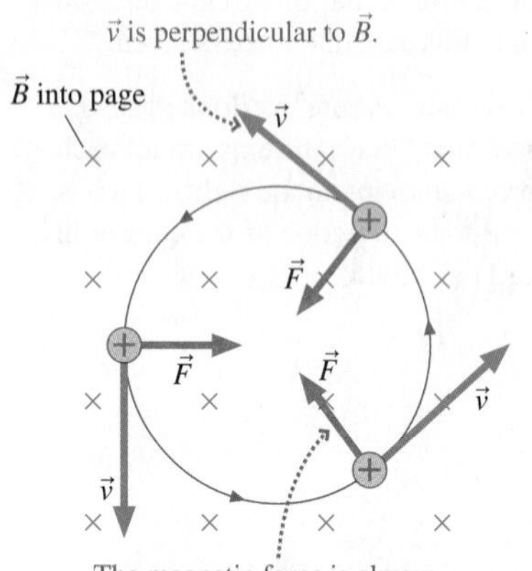
$\vec{v}$ is perpendicular to $\vec{B}$.
$\vec{B}$ into page
$\vec{v}$
$\vec{F}$
The magnetic force is always perpendicular to $\vec{v}$, causing the particle to move in a circle.

We know that the magnetic force on a moving charged particle is always perpendicular to its velocity. This means that the force changes the *direction* of the velocity but not the *magnitude*. The particle's path will bend, but the particle will not speed up or slow down. Suppose a positively charged particle is moving perpendicular to a uniform magnetic field $\vec{B}$, as shown in FIGURE 24.30. In Chapter 6, we looked at the motion of objects subject to a force that was always perpendicular to the velocity. The net result was *circular motion at a constant speed.* For a mass moving in a circle at the end of a string, the tension force is always perpendicular to $\vec{v}$. For a satellite moving in a circular orbit, the gravitational force is always perpendicular to $\vec{v}$. Now, for a charged particle moving in a magnetic field, it is the magnetic force that is always perpendicular to $\vec{v}$ and so causes the particle to move in a circle. Thus, **a particle moving perpendicular to a uniform magnetic field undergoes uniform circular motion at constant speed.**

NOTE ► The direction of the force on a negative charge is opposite to that on a positive charge, so a particle with a negative charge will orbit in the opposite sense from that shown in Figure 24.30 for a positive charge. ◄

FIGURE 24.31 A particle in circular motion.

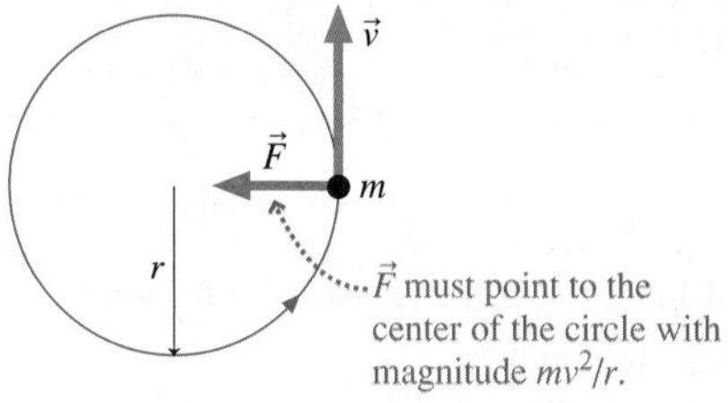
$\vec{v}$
$\vec{F}$
m
r
$\vec{F}$ must point to the center of the circle with magnitude mv^2/r.

FIGURE 24.31 shows a particle of mass m moving at a speed v in a circle of radius r. We found in Chapter 6 that this motion requires a force directed toward the center of the circle with magnitude

$$F = \frac{mv^2}{r} \tag{24.7}$$

For a charged particle moving in a magnetic field, this force is provided by the magnetic force. In Figure 24.30 we assumed that the velocity was perpendicular to the magnetic field, so the magnitude of the force on the charged particle due to the magnetic field is given by Equation 24.6. This is the force that produces the circular motion, so we can equate it to the force in Equation 24.7:

$$F = |q|vB = \frac{mv^2}{r}$$

Solving for the radius of the orbit, we get

$$r = \frac{mv}{|q|B} \tag{24.8}$$

The radius depends on the ratio of the mass of the particle to its charge, a fact we will use later. The radius also depends on the particle's speed and the magnetic field strength: Increasing the speed will increase the radius of the circular motion, while increasing the field will decrease the radius.

A particle moving *perpendicular* to a magnetic field moves in a circle. In the table at the start of this section, we saw that a particle moving *parallel* to a magnetic field experiences no magnetic force, and so continues in a straight line. A more general situation in which a charged particle's velocity $\vec{v}$ is neither parallel to nor perpendicular to the field $\vec{B}$ is shown in FIGURE 24.32. The net result is a circular motion due to the perpendicular component of the velocity coupled with a constant velocity parallel to the field: The charged particle spirals around the magnetic field lines in a helical trajectory.

High-energy particles stream out from the sun in the solar wind. Some of the charged particles of the solar wind become trapped in the earth's magnetic field. As FIGURE 24.33 shows, the particles spiral in helical trajectories along the earth's magnetic field lines. Some of these particles enter the atmosphere near the north and south poles, ionizing gas and creating the ghostly glow of the **aurora**.

FIGURE 24.32 A charged particle in a magnetic field follows a helical trajectory.

(a) The velocity can be broken into components parallel and perpendicular to the field. The parallel component will continue without change.

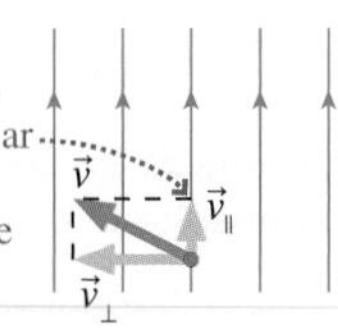

(b) A top view shows that the perpendicular component will change, leading to circular motion.

$\vec{B}$

$\vec{v}_{\perp}$

(c) The net result is a helical path that spirals around the field lines.

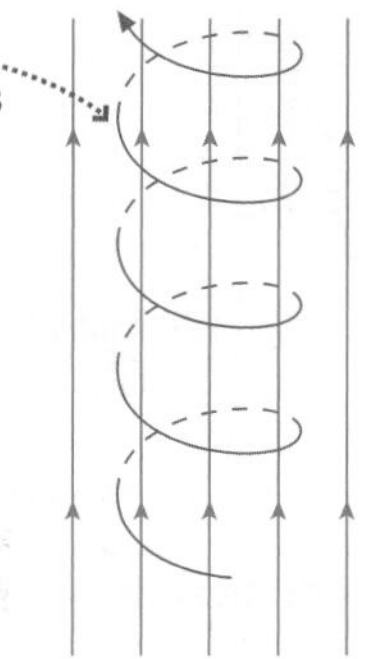

FIGURE 24.33 Charged particles in the earth's magnetic field create the aurora.

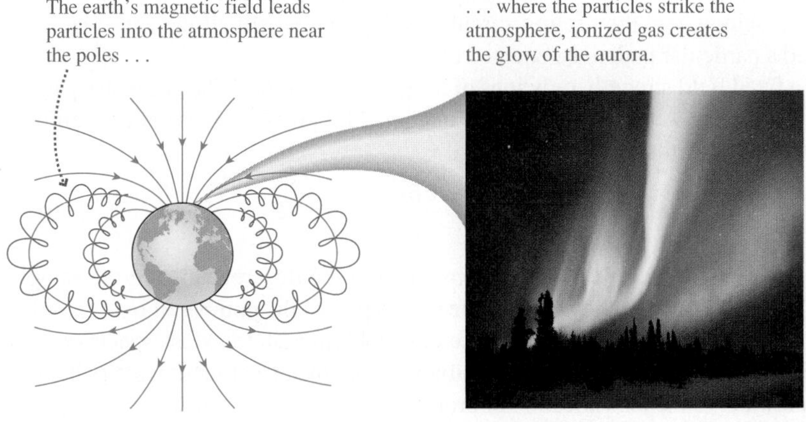

The earth's magnetic field leads particles into the atmosphere near the poles . . .
. . . where the particles strike the atmosphere, ionized gas creates the glow of the aurora.

EXAMPLE 24.9 Force on a charged particle in the earth's field, revisited

In Example 24.8, we considered a proton in the solar wind moving toward the equator of the earth, where the earth's field is 5.0×10^{-5} T, at a speed of 500 km/s (5.0×10^5 m/s). We now know that the proton will move in a circular orbit around the earth's field lines. What are the radius and the period of the orbit?

PREPARE We begin with a sketch of the situation, noting the proton's orbit, as shown in FIGURE 24.34.

FIGURE 24.34 A proton orbits the earth's field lines.

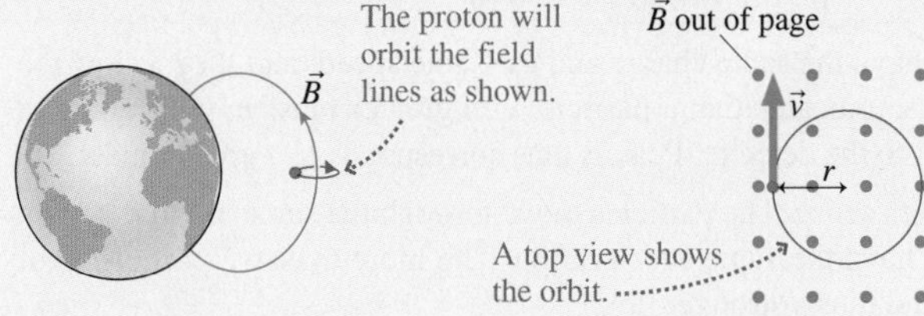

SOLVE Before we use any numbers, we will do some work with symbols. The radius r of the orbit of the proton is given by Equation 24.8. The period T for one orbit is just the distance of one orbit (the circumference $2\pi r$) divided by the speed:

$$T = \frac{2\pi r}{v}$$

We substitute Equation 24.8 for the radius of the orbit of a charged particle moving in a magnetic field to get

$$T = \frac{2\pi}{v} r = \frac{2\pi}{\cancel{v}}\left(\frac{m\cancel{v}}{qB}\right) = \frac{2\pi m}{qB}$$

The speed cancels, and doesn't appear in the final expression. All protons in the earth's field orbit with the same period, regardless of their speed. A higher speed just means a larger circle, completed in the same time. Using values for mass, charge, and field, we compute the radius and the period of the orbit:

$$r = \frac{(1.67 \times 10^{-27}\ \text{kg})(5.0 \times 10^5\ \text{m/s})}{(1.60 \times 10^{-19}\ \text{C})(5.0 \times 10^{-5}\ \text{T})} = 100\ \text{m}$$

$$T = \frac{2\pi(1.67 \times 10^{-27}\ \text{kg})}{(1.60 \times 10^{-19}\ \text{C})(5.0 \times 10^{-5}\ \text{T})} = 0.0013\ \text{s}$$

ASSESS We can do a quick check on our math. We've found the radius of the orbit and the period, so we can compute the speed:

$$v = \frac{2\pi r}{T} = \frac{2\pi(100\ \text{m})}{0.0013\ \text{s}} = 5 \times 10^5\ \text{m/s}$$

This is the speed that we were given in the problem statement, which is a good check on our work.

13.7 ActivPhysics Online

The Mass Spectrometer

In 1995, the Galileo spacecraft dropped a probe into Jupiter's atmosphere. One instrument on board, called a **mass spectrometer,** was used to determine the composition of the atmosphere. The data, shown in FIGURE 24.35, are called a **mass spectrum,** a record of the masses of atoms and molecules that were encountered.

FIGURE 24.35 Mass spectrum of the atmosphere of Jupiter.

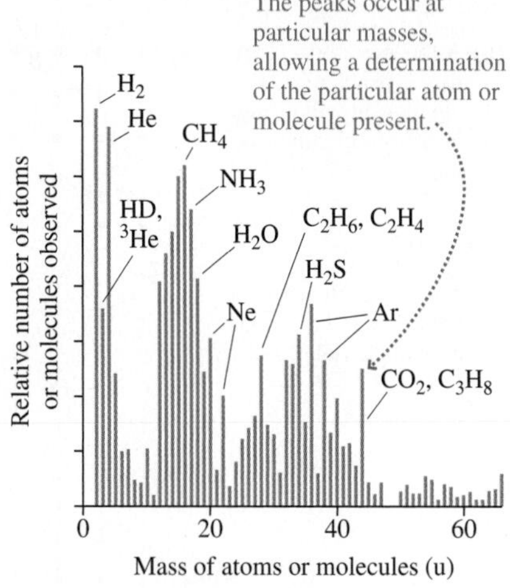
The peaks occur at particular masses, allowing a determination of the particular atom or molecule present.
H_2
He
CH_4
NH_3
HD, ^{3}He
H_2O
C_2H_6, C_2H_4
H_2S
Ne
Ar
CO_2, C_3H_8
Relative number of atoms or molecules observed
0
20
40
60
Mass of atoms or molecules (u)

It's possible to determine the mass of a charged atom or molecule by observing its circular path in a uniform magnetic field. As we've seen, the radius of the orbit depends on the magnetic field, the velocity of the particle, the charge of the particle, and, most important, the mass of the particle.

The operation of such a mass spectrometer is illustrated in FIGURE 24.36. A sample to be analyzed is collected and, if necessary, vaporized to form a gas. The atoms and molecules in this sample are then ionized: An electron is removed from each atom or molecule, leaving a positive ion. The ions are accelerated through an electric field, and ions of a particular velocity selected. These ions travel into a region of a uniform magnetic field, where they follow circular paths. An exit slit allows ions that have followed a particular path to be counted by a detector.

FIGURE 24.36 A mass spectrometer.

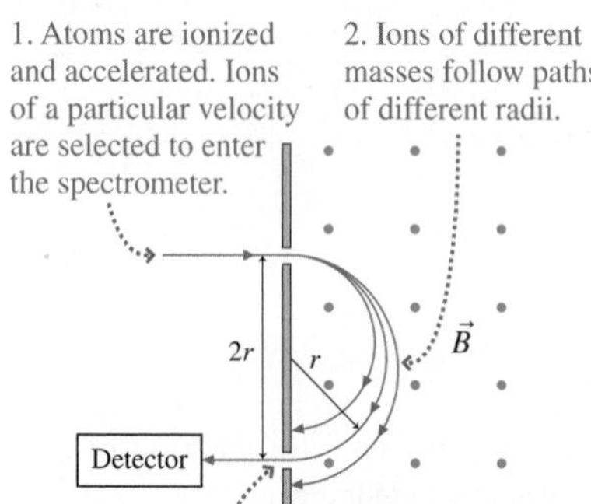
1. Atoms are ionized and accelerated. Ions of a particular velocity are selected to enter the spectrometer.
2. Ions of different masses follow paths of different radii.
$2r$
r
$\vec{B}$
Detector
3. Only ions of a particular mass reach the exit slit and continue to the detector.

For a fixed field strength, only ions of a certain mass follow the necessary path to reach the detector. A rearrangement of Equation 24.8 gives the mass of the particles that reach the detector as

$$m = \frac{qBr}{v} \tag{24.9}$$

Varying the magnetic field scans, one by one, all the different ions in the sample across the exit slit where they can be detected. A plot of the number of ions recorded versus mass is a mass spectrum like the one in Figure 24.35. Mass spectrometers find wide application in chemistry and biology, from the detection of trace pollutants in groundwater to the identification of proteins in biological systems.

CONCEPTUAL EXAMPLE 24.10 Using a mass spectrometer

A mass spectrometer is measuring singly ionized atoms that enter the detector at the same speed. The detector is at a fixed position, and the field is varied to measure different ions. There are two clear peaks in the spectrum, as shown in FIGURE 24.37. Which one corresponds to a greater mass?

FIGURE 24.37 Mass spectrum for singly ionized atoms.

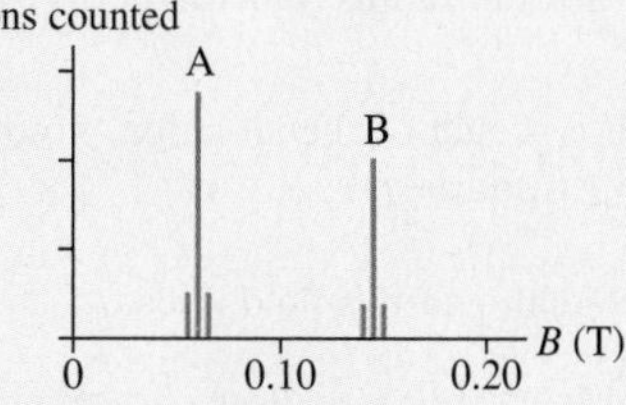
Ions counted
A
B
0
0.10
0.20
B (T)

REASON The atoms all have the same charge and the same speed, and they follow the same path. Equation 24.9 indicates that a particle with greater mass m will require a stronger field B to send it to the detector. Peak B thus corresponds to a greater mass.

ASSESS Our result makes sense. The particles move through the same circular path at the same speed, so they have the same acceleration. The more massive particles will require a larger force—and thus a stronger field.

Electromagnetic Flowmeters

Blood contains many kinds of ions, such as Na^+ and Cl^-. When blood flows through a vessel, these ions move with the blood. An applied magnetic field will produce a force on these moving charges. We can use this principle to make a completely noninvasive device for measuring the blood flow in an artery: an *electromagnetic flowmeter.*

A flowmeter probe clamped to an artery has two active elements: magnets that apply a strong field across the artery and electrodes that contact the artery on opposite sides, as shown in FIGURE 24.38. The blood flowing in an artery carries a mix of positive and negative ions. Because these ions are in motion, the magnetic field exerts a force on them that produces a measurable voltage. We know from Equation 24.5 that the faster the blood's ions are moving, the greater the forces separating the positive and negative ions. The greater the forces, the greater the degree of separation and the larger the voltage. The measured voltage is therefore directly proportional to the velocity of the blood.

Go with the flow Many scientists and resource managers rely on accurate measurements of stream flows. The easiest way to get a quick measurement of the speed of a river or creek is to use an electromagnetic flowmeter similar to the one used for measuring flow in blood vessels. Water flows between the poles of a strong magnet. Two electrodes measure the resulting potential difference, which is proportional to the flow speed.

FIGURE 24.38 The operation of an electromagnetic flowmeter.

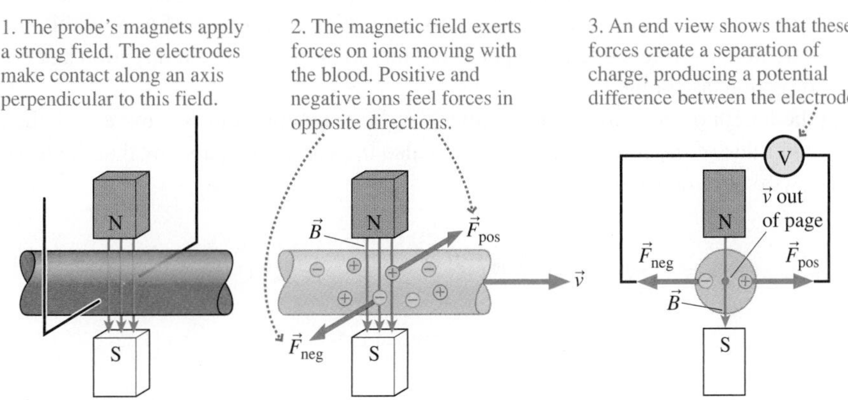

STOP TO THINK 24.4 These charged particles are traveling in circular orbits with velocities and field directions as noted. Which particles have a negative charge?

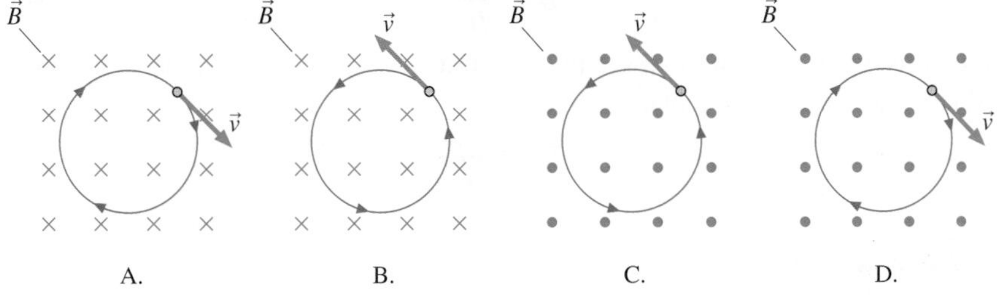

▶ **Magnets and television screens** The image on a cathode-ray tube television screen is drawn by an electron beam that is steered by magnetic fields from coils of wire. Other magnetic fields can also exert forces on the moving electrons. If you place a strong magnet near the TV screen, the electrons will be forced along altered trajectories and will strike different places on the screen than they are supposed to, producing an array of bright colors. *(The magnet can magnetize internal components and permanently alter the image, so do not do this to your television!)*

DON'T TRY IT YOURSELF

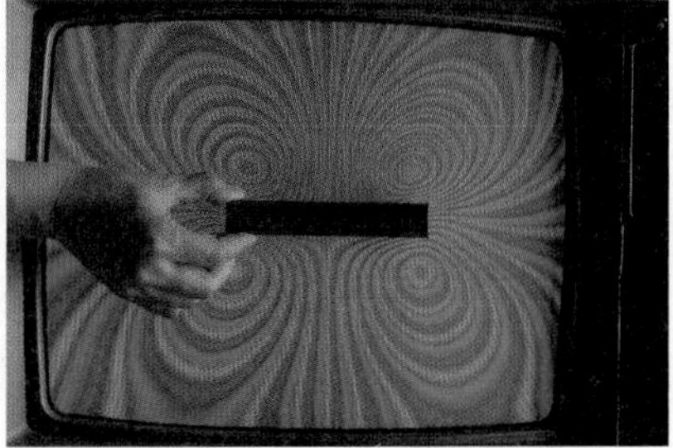

24.6 Magnetic Fields Exert Forces on Currents

We have seen that a magnetic field exerts a force on a current. This force is responsible for the operation of loudspeakers, electric motors, and many other devices.

The Form of the Magnetic Force on a Current

In the table at the start of Section 24.5, we saw that a magnetic field exerts no force on a charged particle moving parallel to a magnetic field. If a current-carrying wire is *parallel* to a magnetic field, we also find that the force on it is zero.

FIGURE 24.39 Magnetic force on a current-carrying wire.

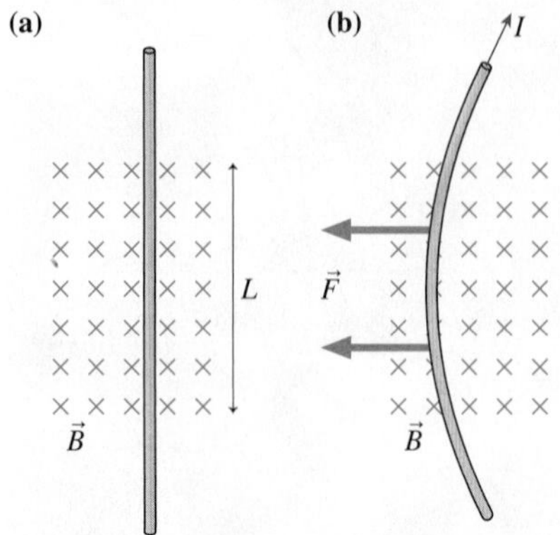
(a)
(b)
I
L
$\vec{F}$
$\vec{B}$
$\vec{B}$

A wire is perpendicular to an externally created magnetic field.

If the wire carries a current, the magnetic field will exert a force on the moving charges, causing a deflection of the wire.

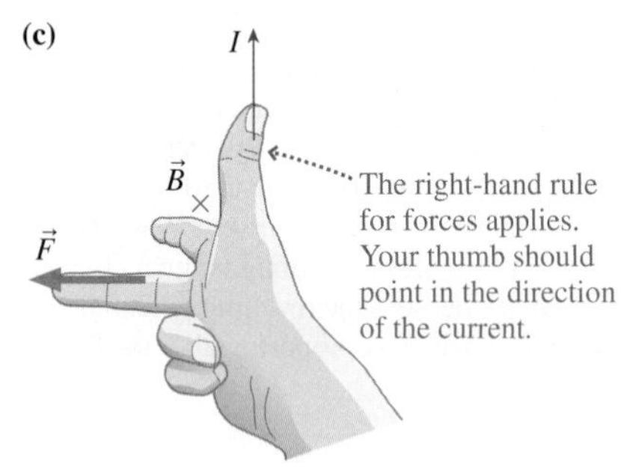
(c)
I
$\vec{B}$
$\vec{F}$
The right-hand rule for forces applies. Your thumb should point in the direction of the current.

However, there is a force on a current-carrying wire that is *perpendicular* to a magnetic field, as shown in FIGURE 24.39.

NOTE ► The magnetic field is an external field, created by a permanent magnet or by other currents; it is *not* the field of the current I in the wire. ◄

The direction of the force on the current is found by considering the force on each charge in the current. We model current as the flow of positive charge, so **the right-hand rule for forces applies to currents in the same way it does for moving charges.** With your fingers aligned as usual, point your right thumb in the direction of the current (the direction of the motion of positive charges) and your index finger in the direction of $\vec{B}$. Your middle finger is then pointing in the direction of the force $\vec{F}$ on the wire, as in FIGURE 24.39c. Consequently, the entire length of wire within the magnetic field experiences a force perpendicular to both the current direction and the field direction, as shown in FIGURE 24.39b.

If the length of the wire L, the current I, or the magnetic field B is increased, then the magnitude of the force on the wire will also increase. We can show that the force on the wire is given by

$$F_{\text{wire}} = ILB \tag{24.10}$$

Magnitude of the force on a current-carrying wire of length L perpendicular to a magnetic field B

If the wire is at an angle α to the field, the force will depend on this angle:

$$F_{\text{wire}} = ILB \sin \alpha \tag{24.11}$$

It is sometimes useful to rewrite Equation 24.10 as

$$B = \frac{F_{\text{wire}}}{IL} \tag{24.12}$$

Given this expression, we can see that the unit for magnetic field, the tesla, can be defined in terms of other units:

$$1\text{ T} = 1\,\frac{\text{N}}{\text{A}\cdot\text{m}}$$

EXAMPLE 24.11 Magnetic force on a power line

A DC power line near the equator runs east-west. At this location, the earth's magnetic field is parallel to the ground, points north, and has magnitude 50 μT. A 400 m length of the heavy cable that spans the distance between two towers has a mass of 1000 kg. What direction and magnitude of current would be necessary to offset the force of gravity and "levitate" the wire? (The power line will actually carry a current that is much less than this; 850 A is a typical value.)

PREPARE First, we sketch a top view of the situation, as in FIGURE 24.40. The magnetic force on the wire must be opposite that of gravity. An application of the right-hand rule for forces shows that a current to the east will result in an upward force—out of the page.

FIGURE 24.40 Top view of a power line near the equator.

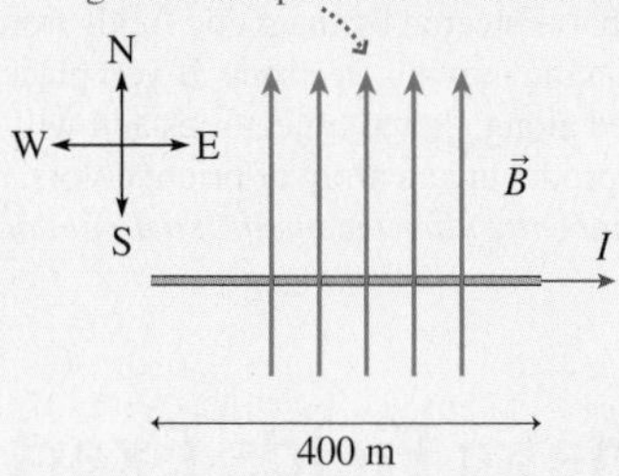
The field of the earth near the equator is parallel to the ground and points to the north.
N
W
E
S
$\vec{B}$
I
400 m

SOLVE The magnetic field is perpendicular to the current, so the magnitude of the magnetic force is given by Equation 24.10. To levitate the wire, this force must be opposite to the weight force but equal in magnitude, so we can write

$$mg = ILB$$

where m and L are the mass and length of the wire and B is the magnitude of the earth's field. Solving for the current, we find

$$I = \frac{mg}{LB} = \frac{(1000\text{ kg})(9.8\text{ m/s}^2)}{(400\text{ m})(50\times10^{-6}\text{ T})} = 4.9\times10^5\text{ A}$$

directed to the east.

ASSESS The current is much larger than a typical current, as we expected.

Forces Between Currents

Because a current produces a magnetic field, and a magnetic field exerts a force on a current, it follows that two current-carrying wires will exert forces on each other, as Ampère discovered. It will be a good check on our results to this point to show that the experimental results we saw earlier are consistent with our rules for determining magnetic fields from currents and determining forces on currents due to magnetic fields.

Activ ONLINE Physics 13.5

Suppose we have two parallel wires of length L a distance d apart, each carrying a current. **FIGURE 24.41a** shows the currents I_1 and I_2 in the same direction and **FIGURE 24.41b** in opposite directions. We will assume that the wires are sufficiently long to allow us to use the earlier result, Equation 24.1, for the magnetic field of a long, straight wire: $B = \mu_0 I/2\pi r$.

FIGURE 24.41 Forces between currents.

(a) Currents in the same direction attract.

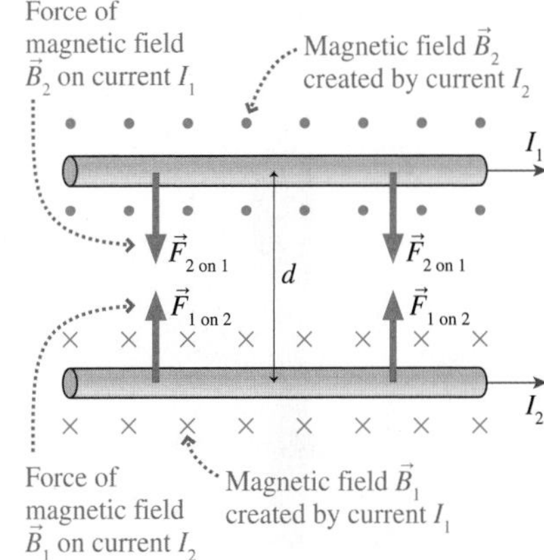

(b) Currents in opposite directions repel.

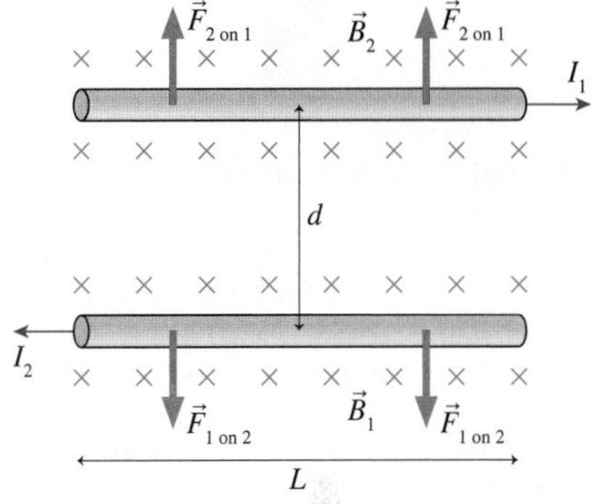

Let's look at the situation of Figure 24.41a. There are three steps in our analysis:

1. The current I_2 in the lower wire creates a magnetic field $\vec{B}_2$ at the position of the upper wire. This field $\vec{B}_2$ points out of the page, perpendicular to the current I_1. **It is this field $\vec{B}_2$, due to the lower wire, that exerts a magnetic force on the upper wire.** At the position of the upper wire, which is a constant distance $r = d$ from the lower wire, the field has the same value at all points along the wire. The field is

$$\vec{B}_2 = \left(\frac{\mu_0 I_2}{2\pi d}, \text{ out of the page}\right) \tag{24.13}$$

2. For the upper wire, the current is to the right and the field $\vec{B}_2$ from the lower wire points out of the page. Using the right-hand rule for forces, you can see that the force on the upper wire is downward, attracting it toward the lower wire.
3. The magnitude of the force on the upper wire is given by Equation 24.10. Using the field from Equation 24.13, we compute

$$F_{\text{parallel wires}} = I_1 L B_2 = I_1 L \frac{\mu_0 I_2}{2\pi d}$$

$$F_{\text{parallel wires}} = \frac{\mu_0 L I_1 I_2}{2\pi d} \tag{24.14}$$

Magnetic force between two parallel current-carrying wires

The current in the upper wire exerts an upward-directed magnetic force on the lower wire with exactly the same magnitude. (You know that this must be the case: The two forces form a Newton's third law pair.) You should convince yourself, using the right-hand rule, that the forces are repulsive and tend to push the wires apart if the two currents are in opposite directions, as shown in Figure 24.41b. Our rules predict exactly what the experimental results showed: Parallel wires carrying currents in the same direction attract each other; parallel wires carrying currents in opposite directions repel each other.

EXAMPLE 24.12 Finding the force between wires in jumper cables

You may have used a set of jumper cables connected to a running vehicle to start a car with a dead battery. Jumper cables are a matched pair of wires, red and black, joined together along their length. Suppose we have a set of jumper cables in which the two wires are separated by 1.2 cm along their 3.7 m (12 ft) length. While starting a car, the wires each carry a current of 150 A, in opposite directions. What is the force between the two wires?

FIGURE 24.42 Jumper cables carrying opposite currents.

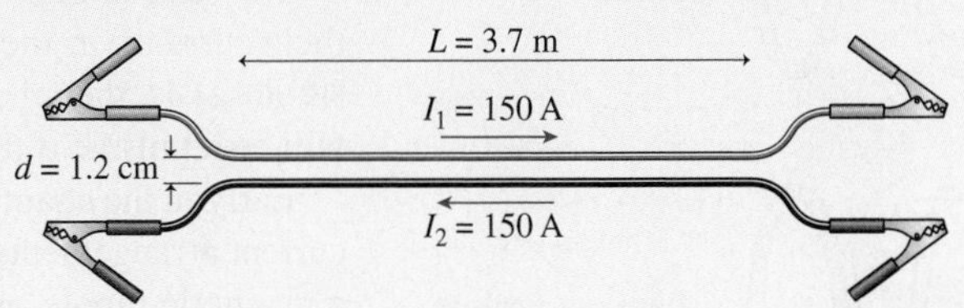

PREPARE Our first step is to sketch the situation, noting distances and currents, as shown in **FIGURE 24.42**. Because the currents in the two wires are in opposite directions, the force between the two wires is repulsive.

Continued

SOLVE The currents are parallel all along the length of the wires, so we can use Equation 24.14 to compute the force between the wires. The current in each is 150 A, so

$$F = \frac{\mu_0 L I_1 I_2}{2\pi d}$$

$$= \frac{(1.26 \times 10^{-6}\ \text{T}\cdot\text{m/A})(3.7\ \text{m})(150\ \text{A})(150\ \text{A})}{2(\pi)(0.012\ \text{m})}$$

$$= 1.4\ \text{N}$$

ASSESS These wires are long, close together, and carry very large currents. But the force between them is quite small—much less than the weight of the wires. In practice, the forces between currents are not an important consideration unless there are many coils of wire, leading to a large total force. This is the case in an MRI solenoid, as we will discuss on page 801.

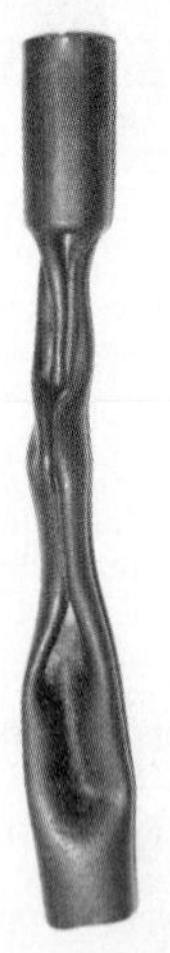

Crushed by currents The forces between currents are quite small for ordinary currents, even the large currents in the jumper cables of the previous example. But for a current of tens of thousands of amps, it's a different story. This lightning rod is hollow. When struck by lightning, it carried an enormous current for a very short time. The currents in all parts of the rod were parallel, so they attracted each other. The tremendous size of the currents led to attractive forces strong enough to crush the rod.

We're now able to summarize the steps in finding the magnetic force on a charge or current due to known magnetic fields:

PROBLEM-SOLVING STRATEGY 24.2 **Magnetic force problems**

PREPARE There are two key factors to identify in magnetic force problems:

- The source of the magnetic field.
- The charges or currents that feel a force due to this magnetic field.

SOLVE First, determine the magnitude and direction of the magnetic field at the position of the charges or currents that are of interest.

- In some problems the magnetic field is given.
- If the field is due to a current, use the right-hand rule for fields to determine the field direction and the appropriate equation to determine its magnitude (see Problem-Solving Strategy 24.1).

Next, determine the force this field produces. Working with the charges or currents you identified previously,

- Use the right-hand rule for forces to determine the direction of the force on any moving charge or current.
- Use the appropriate equation to determine the magnitude of the force on any moving charge or current.

ASSESS Are the forces you determine perpendicular to velocities of moving charges and to currents? Are the forces perpendicular to the fields? Do the magnitudes of the forces seem reasonable?

Exercise 34

FIGURE 24.43 Forces between parallel current loops.

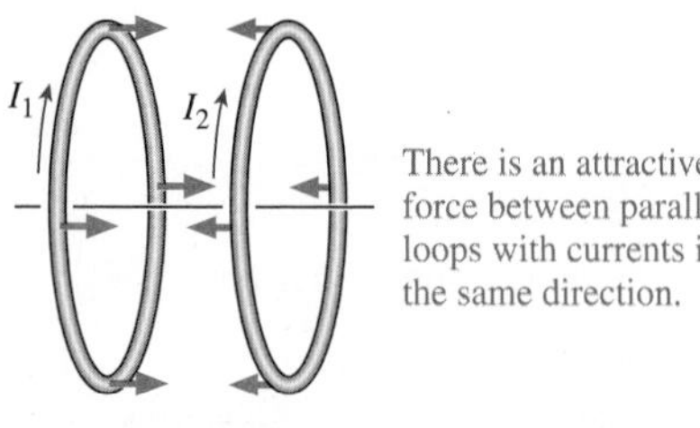

There is an attractive force between parallel loops with currents in the same direction.

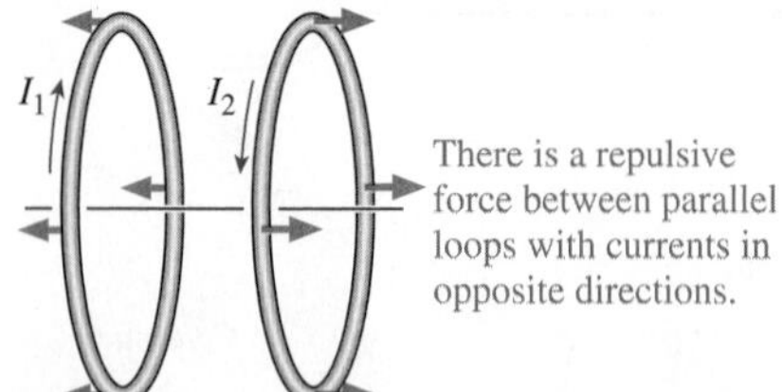

There is a repulsive force between parallel loops with currents in opposite directions.

Forces Between Current Loops

We will now consider the forces between two current loops. Doing so will allow us to begin to make connections with some of the basic phenomena of magnetism that we saw earlier in the chapter.

We've seen that there is an attractive force between two parallel wires that have their currents in the same direction. If these two wires are bent into loops, as in FIGURE 24.43, then the force between the two loops will also be attractive. The forces will be repulsive if the currents are in opposite directions.

Early in the chapter, we examined the fields from various permanent magnets and current arrangements. FIGURE 24.44a on the next page reminds us that a bar magnet is a magnetic dipole, with a north and a south pole. Field lines come out of its north pole, loop back around, and go into the south pole. FIGURE 24.44b shows that the field of a current loop is very similar to that of a bar magnet. This leads us to the conclusion that **a current loop, like a bar magnet, is a magnetic dipole,** with a north and a south pole, as indicated in Figure 24.44b. We can use this conclusion to understand the forces between current loops. In FIGURE 24.45, we show how the poles of each current loop can be represented by a magnet. This model helps us understand the forces between current loops; in the next section, we'll use this model to understand torques on current loops as well.

FIGURE 24.45 Forces between current loops can be understood in terms of their magnetic poles.

Because currents loops have north and south poles, we can picture a current loop as a small bar magnet.

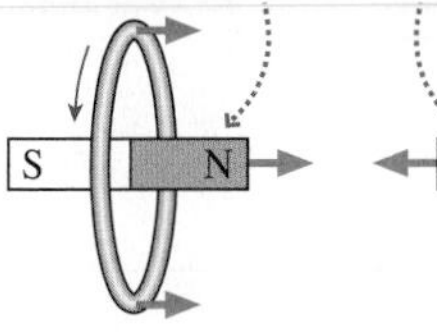

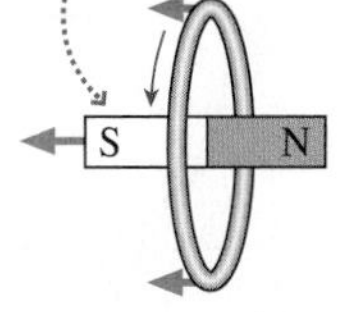

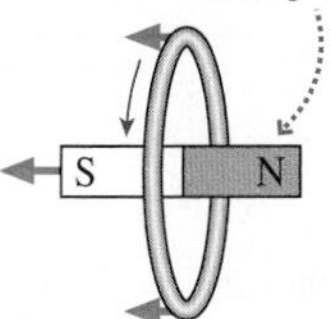

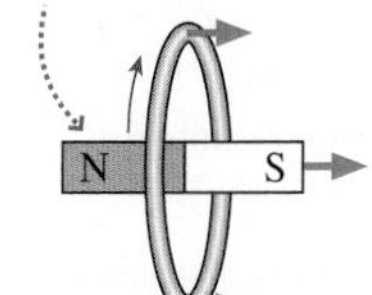

FIGURE 24.44 We can picture the loop as a small bar magnet.

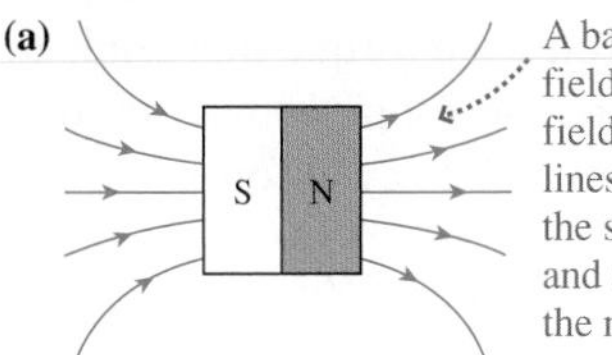

(a)
S
N
A bar magnet's field is a dipole field. Field lines go into the south pole and come out the north pole.

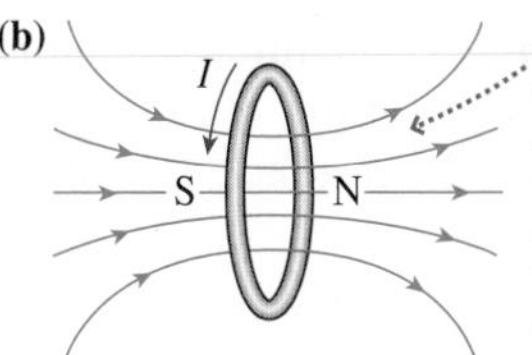

STOP TO THINK 24.5 Four wires carry currents in the directions shown. A uniform magnetic field is directed into the paper as shown. Which wire experiences a force to the left?

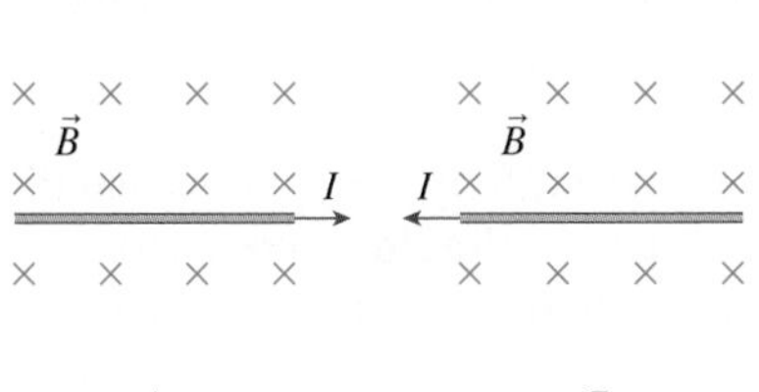

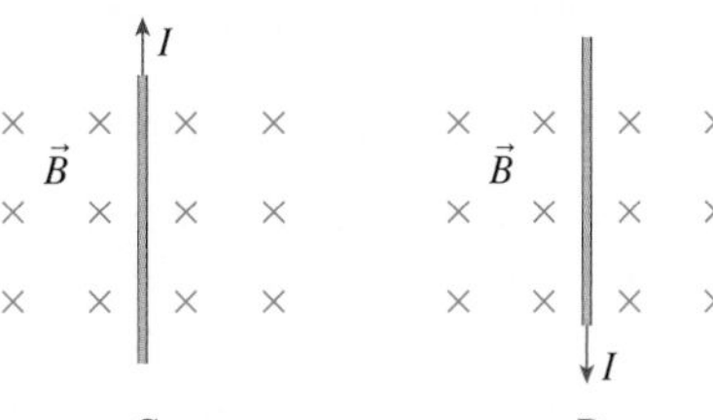

24.7 Magnetic Fields Exert Torques on Dipoles

One of our first observations at the beginning of the chapter was that a compass needle is a small magnet. The fact that it pivots to line up with an external magnetic field means that it experiences a *torque*. In this section, we'll use what we've learned about magnetic forces to understand the torque on a magnetic dipole. We will consider only the case of a current loop (which you'll recall is a magnetic dipole), but the results will be equally applicable to permanent magnets and magnetic dipoles such as compass needles.

A Current Loop in a Uniform Field

FIGURE 24.46 shows a current loop—a magnetic dipole—in a *uniform* magnetic field. To make our analysis more straightforward, we consider a square current loop, but any shape would do. The current in each of the four sides of the loop experiences a magnetic force due to the external field $\vec{B}$. Because the field is uniform, the forces on opposite sides of the loop are of equal magnitude. The direction of each force is determined by the right-hand rule for forces. The forces $\vec{F}_{front}$ and $\vec{F}_{back}$ produce no net force or torque. The forces $\vec{F}_{top}$ and $\vec{F}_{bottom}$ also add to give no net force, but because $\vec{F}_{top}$ and $\vec{F}_{bottom}$ don't act along the same line, they will rotate the loop by exerting a torque on it.

Although we've shown a current loop, the conclusion is true for any magnetic dipole: **In a uniform field, a dipole experiences a torque but no net force.** For example, a compass needle is not attracted to the earth's poles; it merely feels a torque that lines it up with the earth's field.

FIGURE 24.46 A loop in a uniform magnetic field experiences a torque.

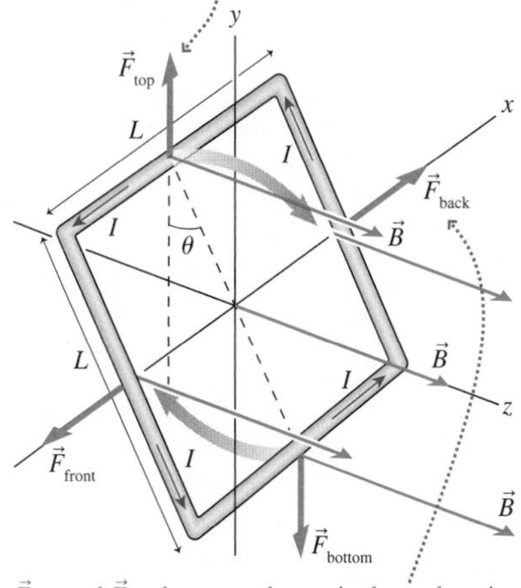

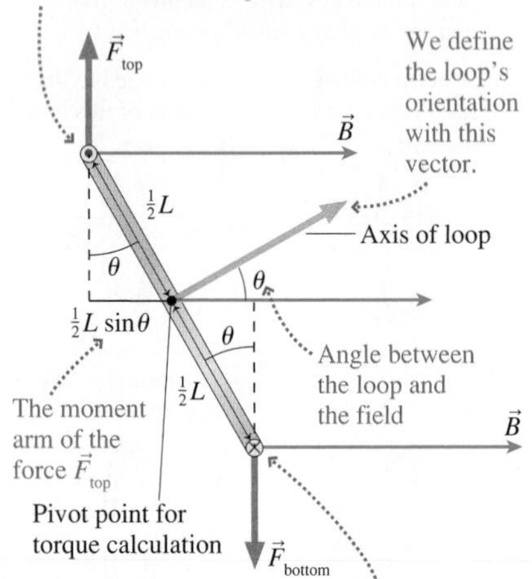

FIGURE 24.47 Calculating the torque on a current loop in a uniform magnetic field.

We can calculate the torque by looking at a side view of the current loop of Figure 24.46. This is shown in FIGURE 24.47. The angle between the loop and the field will be important. We've drawn a vector from the center of the loop that we'll use to define this angle. We also use this angle to compute the torques on segments of the loop. In Chapter 7, we calculated the torque by multiplying the force by the moment arm. The loop will rotate about a pivot point through the center, and we will compute our moment arms from this point. As you can see, the moment arm of $\vec{F}_{top}$ is $\frac{1}{2}L\sin\theta$. $\vec{F}_{top}$ and $\vec{F}_{bottom}$ each produce a torque that tends to rotate the loop. The torque τ_{top} on the top segment of the loop rotates the loop clockwise, as does the torque τ_{bottom} on the bottom segment. The net torque is

$$\begin{aligned}\tau = \tau_{top} + \tau_{bottom} &= F_{top}(\tfrac{1}{2}L\sin\theta) + F_{bottom}(\tfrac{1}{2}L\sin\theta)\\ &= (ILB)(\tfrac{1}{2}L\sin\theta) + (ILB)(\tfrac{1}{2}L\sin\theta)\\ &= (IL^2)B\sin\theta\end{aligned}$$

L^2 is the area A of the square loop. Using this, we can generalize the result to any loop of area A:

$$\tau = (IA)B\sin\theta \qquad (24.15)$$

There are two things to note about this torque:

1. **The torque depends on properties of the current loop:** its area (A) and the current (I). The quantity IA, known as the **magnetic dipole moment** (or simply *magnetic moment*), is a measure of how much torque a dipole will feel in a magnetic field. A permanent magnet can be assigned a magnetic dipole moment as well. We'll represent the magnetic moment as an arrow pointing in the direction of the dipole's magnetic field, as in FIGURE 24.48. This figure shows the magnetic dipole moment for a current loop and a bar magnet. You can now see that the vector defining the axis of the loop in Figure 24.47 is simply the magnetic dipole moment.
2. **The torque depends on the angle between the magnetic dipole moment and the magnetic field.** The torque is maximum when θ is 90°, when the magnetic moment is perpendicular to the field. The torque is zero when θ is 0° (or 180°), when the magnetic moment is parallel to the field. As we see in FIGURE 24.49, a magnetic dipole free to rotate in a field will do so until θ is zero, at which point it will be stable. **A magnetic dipole will rotate to line up with a magnetic field** just as an electric dipole will rotate to line up with an electric field.

FIGURE 24.48 The dipole moment vector.

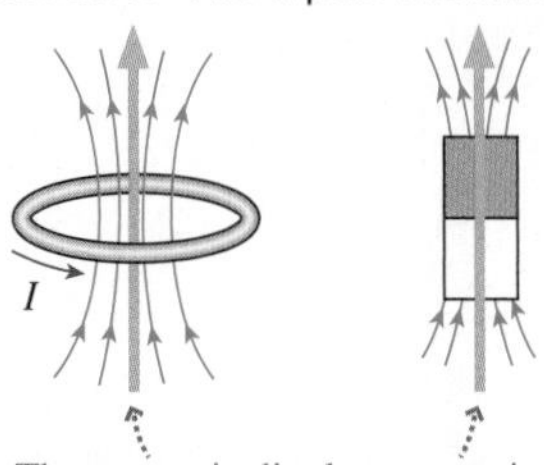

I
The magnetic dipole moment is represented as a vector that points in the direction of the dipole's field. A longer vector means a stronger field.

FIGURE 24.49 Torque on a dipole in an externally created magnetic field.

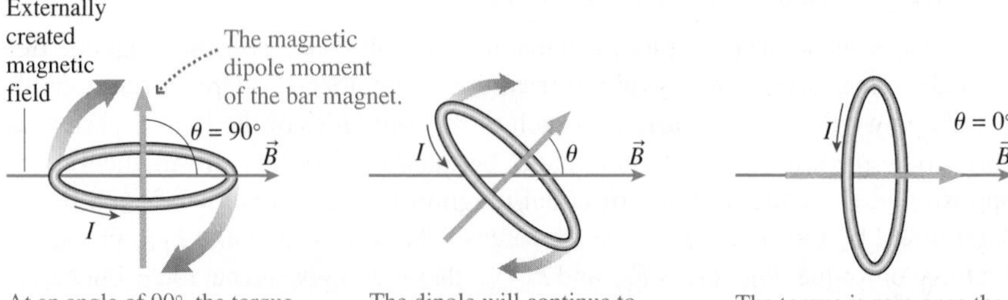

Externally created magnetic field
The magnetic dipole moment of the bar magnet.
θ = 90°
θ = 0°
At an angle of 90°, the torque is maximum. A dipole free to rotate will do so.
The dipole will continue to rotate; as the angle θ decreases, the torque decreases.
The torque is zero once the dipole is lined up so that the angle θ is zero.

A compass needle, which is a dipole, also rotates until its north pole is in the direction of the magnetic field, as we noted at the very start of the chapter.

CONCEPTUAL EXAMPLE 24.13 Does the loop rotate?

Two nearby current loops are oriented as shown in FIGURE 24.50. Loop 1 is fixed in place; loop 2 is free to rotate. Will it do so?

FIGURE 24.50 Will loop 2 rotate?

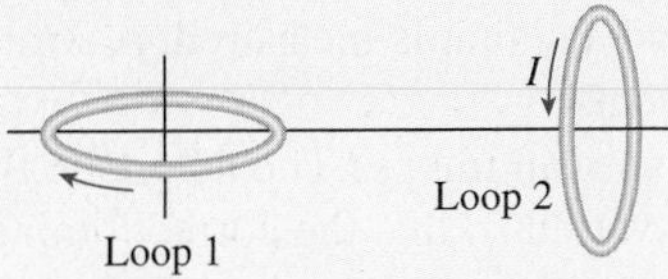

REASON The current in loop 1 generates a magnetic field. As FIGURE 24.51 shows, the field of loop 1 is upward as it passes loop 2. Because the field is perpendicular to the magnetic moment of loop 2, the field exerts a torque on loop 2 and causes loop 2 to rotate until its magnetic moment lines up with the field from loop 1.

ASSESS These two loops align so that their magnetic moments point in opposite directions. We can think of this in terms of their poles: When aligned this way, the north pole of loop 1 is closest to the south pole of loop 2. This makes sense, because these opposite poles attract each other.

FIGURE 24.51 How the field of loop 1 affects loop 2.

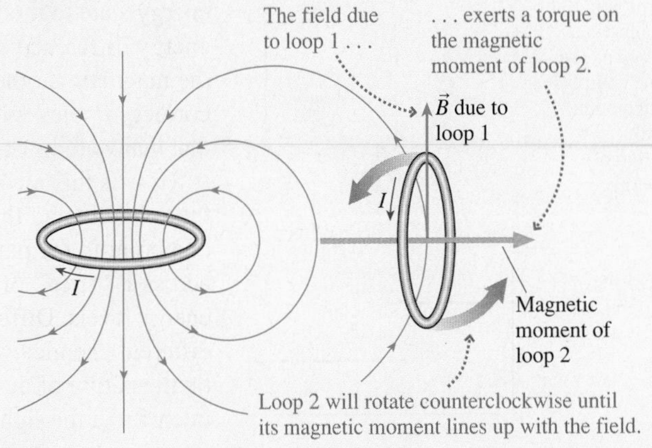

For any dipole in a field, there are actually two angles for which the torque is zero, $\theta = 0°$ and $\theta = 180°$, but there is a difference between these two cases. The $\theta = 0°$ case is *stable:* Once the dipole is in this configuration, it will stay there. The $\theta = 180°$ case is *unstable.* There is no torque if the alignment is perfect, but the slightest rotation will result in a torque that rotates the dipole until it reaches $\theta = 0°$.

We can make a gravitational analogy with this situation in FIGURE 24.52. For an upside-down pendulum, there will be no torque if the mass is directly above the pivot point. But, as we saw in Chapter 8, this is a position of unstable equilibrium. If displaced even slightly, the mass will rotate until it is below the pivot point. This is the point of lowest potential energy.

We can see, by analogy with the upside-down pendulum, that the unstable alignment of a magnetic dipole has a higher energy. Given a chance, the magnet will rotate "downhill" to the position of lower energy and stable equilibrium. This difference in energy is the key to understanding how the magnetic properties of atoms can be used to image tissues in the body in MRI.

FIGURE 24.52 Going from unstable to stable equilibrium.

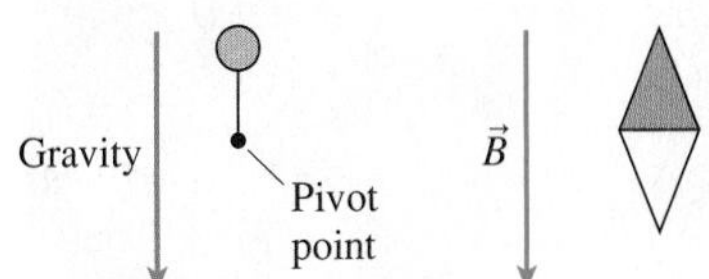

The pendulum balancing upside down and the magnet aligned opposite the field are in unstable equilibrium. A small nudge . . .

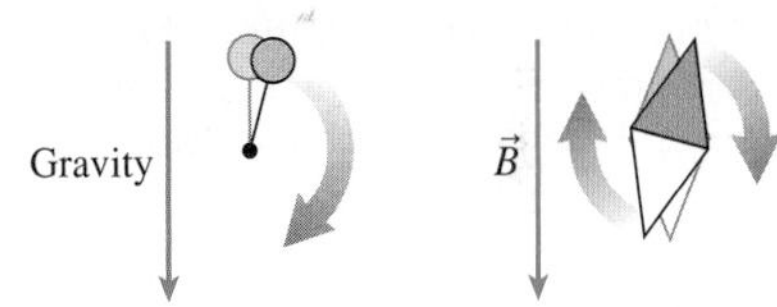

. . . will lead to a torque that will cause a rotation that will continue until . . .

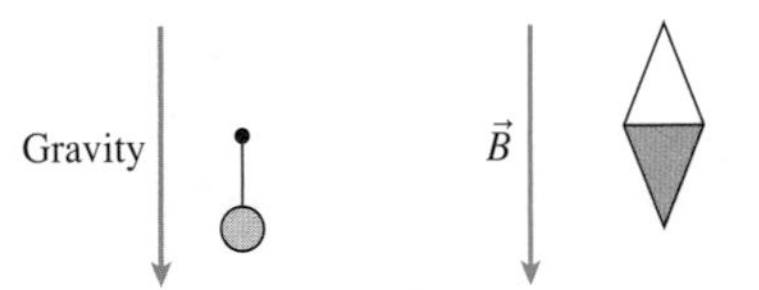

. . . the condition of stable equilibrium is reached.

Magnetic Resonance Imaging (MRI)

Magnetic resonance imaging is a modern diagnostic tool that provides detailed images of tissues and structures in the body with no radiation exposure. The key to this imaging technique is the magnetic nature of atoms. **The nuclei of individual atoms have magnetic moments and behave like magnetic dipoles.** Atoms of different elements have different magnetic moments; therefore, a magnetic field exerts different torques on different kinds of atoms.

A person receiving an MRI scan is placed in a large solenoid that has a strong magnetic field along its length. Think about one hydrogen atom in this person's body. The nucleus of the hydrogen atom is a single proton. The proton has a magnetic moment, but the proton is a bit different from a simple bar magnet: It is subject to the rules of quantum mechanics, which we will learn about in Chapter 28. A bar magnet can have any angle with the field, but the proton can only line up either *with the field* (the low-energy state) or *opposed to the field* (the high-energy state), as we see in FIGURE 24.53 on the next page.

The energy difference ΔE between these two orientations of the proton depends on two key parameters: the magnetic moment of the proton and the strength of the magnetic field.

FIGURE 24.53 The energy difference between the two possible orientations of a proton's magnetic moment during MRI.

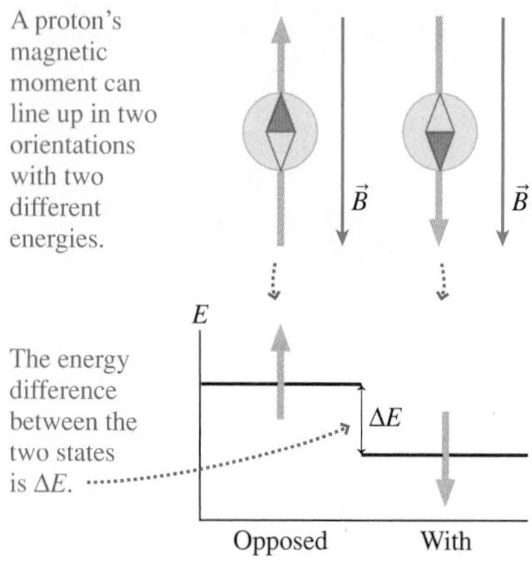
A proton's magnetic moment can line up in two orientations with two different energies.
$\vec{B}$
$\vec{B}$
E
The energy difference between the two states is ΔE.
ΔE
Opposed
With

A tissue sample will have many hydrogen atoms with many protons. Some of the protons will be in the high-energy state and some in the low-energy state. A second magnetic field, called a *probe field,* can be applied to "flip" the dipoles from the low-energy state to the high-energy state. The probe field must be precisely tuned to the energy difference ΔE for this to occur. The probe field is selected to correspond to the magnetic moment of a particular nucleus, in this case hydrogen. If the tuning is correct, dipoles will change state and a signal is measured. A strong signal means that many atoms of this kind are present.

How is this measurement turned into an image? The magnetic field strengths of the solenoid and the probe field are varied so that the correct tuning occurs at only one point in the patient's body. The position of the point of correct tuning is swept across a "slice" of tissue. Combining atoms into molecules slightly changes the energy levels. Different tissues in the body have different concentrations of atoms in different chemical states, so the strength of the signal at a point will vary depending on the nature of the tissue at that point. As the point of correct tuning is moved, the intensity of the signal is measured at each point; a record of the intensity versus position gives an image of the structure of the interior of the body, as in FIGURE 24.54.

FIGURE 24.54 Cross-sectional image from an MRI scan.

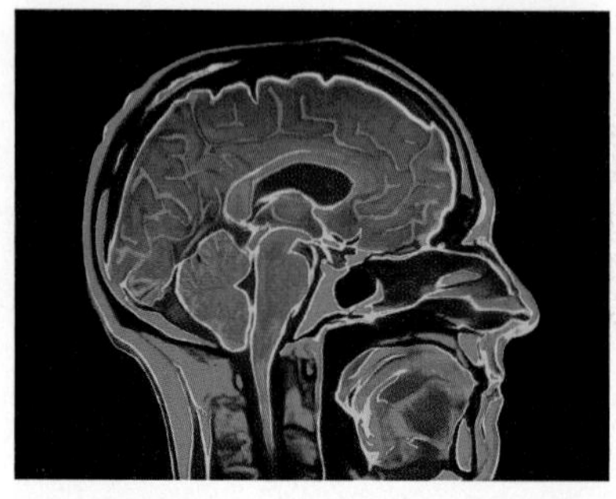

Electric Motors

The torque on a current loop in a magnetic field is the basis for how an electric motor works. The *armature* of a motor is a loop of wire wound on an axle that is free to rotate. This loop is in a strong magnetic field. A current in the loop causes the wires of the loop to feel forces due to this field, as FIGURE 24.55 shows. The loop becomes a magnetic dipole that will rotate to align itself with the external field. If the current were steady, the armature would simply rotate until its magnetic moment was in its stable position. To keep the motor turning, a device called a *commutator* reverses the current direction in the loop every 180°. As the loop reaches its stable configuration, the direction of current in the loop switches, putting the loop back into the unstable configuration, so it will keep rotating to line up in the other direction. This process continues: Each time the loop nears a stable point, the current switches. The loop will keep rotating as long as the current continues.

▶ FIGURE 24.55 The operation of a simple motor depends on the torque on a current loop in a magnetic field.

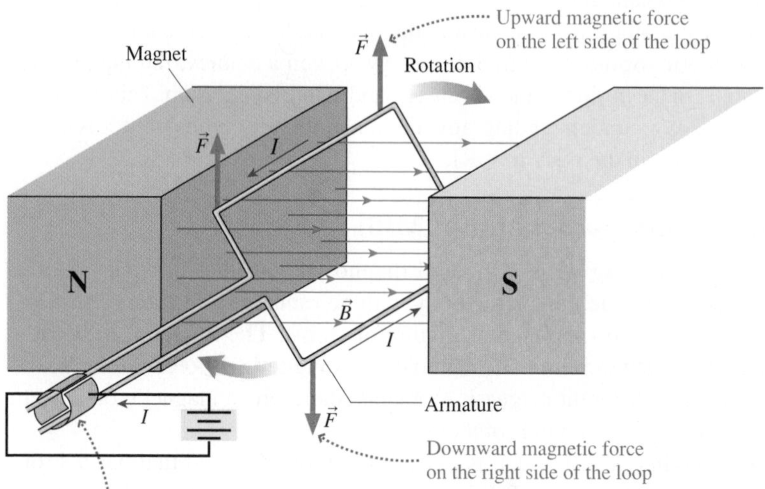
Upward magnetic force on the left side of the loop
Magnet
$\vec{F}$
Rotation
$\vec{F}$
I
N
S
$\vec{B}$
I
I
Armature
$\vec{F}$
Downward magnetic force on the right side of the loop
The commutator reverses the current in the loop every half cycle so that the force is always upward on the left side of the loop.

STOP TO THINK 24.6 Which way will this current loop rotate?

A. Clockwise.
B. Counterclockwise.
C. The loop will not rotate.

24.8 Magnets and Magnetic Materials

We started the chapter by looking at permanent magnets. We know that permanent magnets produce a magnetic field, but what is the source of this field? There are no electric currents in these magnets. Why can you make a magnet out of certain materials but not others? Why does a magnet stick to the refrigerator? The goal of this section is to answer these questions by developing an atomic-level view of the magnetic properties of matter.

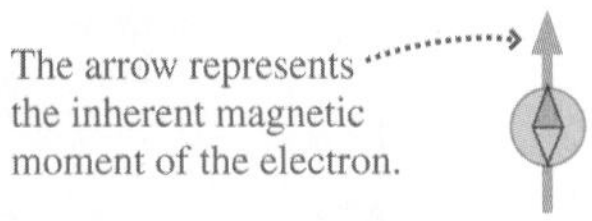

Ferromagnetism

Iron, nickel, and cobalt are elements that have very strong magnetic behavior: A chunk of iron (or steel, which is mostly iron) will stick to a magnet, and the chunk can be magnetized so that it is itself a magnet. Other metals—such as aluminum and copper—do not exhibit this property. We call materials that are strongly attracted to magnets and that can be magnetized **ferromagnetic** (from the Latin for iron, *ferrum*).

The key to understanding magnetism at the atomic level was the 1922 discovery that electrons, just like protons and nuclei, have an *inherent magnetic moment*, as we see in FIGURE 24.56. Magnetism, at an atomic level, is due to the inherent magnetic moment of electrons.

FIGURE 24.56 Magnetic moment of the electron.

The arrow represents the inherent magnetic moment of the electron.

If the magnetic moments of all the electrons in an atom pointed in the same direction, the atom would have a very strong magnetic moment. But this doesn't happen. In atoms with many electrons, the electrons usually occur in pairs with magnetic moments in opposite directions, as we'll see in Chapter 29. Only the electrons that are unpaired are able to give the atom a net magnetic moment.

Even so, atoms with magnetic moments don't necessarily form a solid with magnetic properties. For most elements whose atoms have magnetic moments, the magnetic moments of the atoms are randomly arranged when the atoms join together to form a solid. As FIGURE 24.57 shows, this random arrangement produces a solid whose net magnetic moment is very close to zero. Ferromagnetic materials have atoms with net magnetic moments that tend to line up and reinforce each other as in FIGURE 24.58. This alignment of moments occurs in only a few elements and alloys, and this is why a small piece of iron has such a strong overall magnetic moment. Such a piece has a north and a south magnetic pole, generates a magnetic field, and aligns parallel to an external magnetic field. In other words, it is a magnet—and a very strong one at that.

FIGURE 24.57 The random magnetic moments of the atoms in a typical solid.

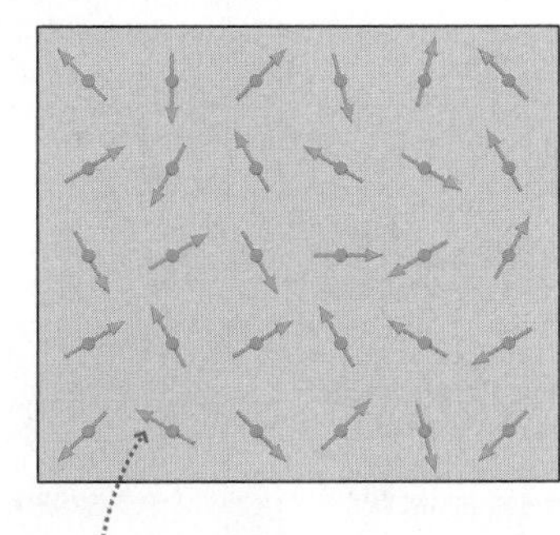

The atomic magnetic moments due to unpaired electrons point in random directions. The sample has no net magnetic moment.

FIGURE 24.58 In a ferromagnetic solid, the atomic magnetic moments align.

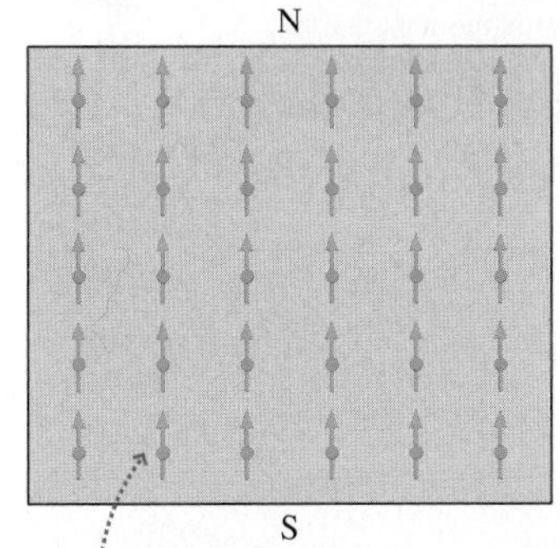

The atomic magnetic moments are aligned. The sample has north and south magnetic poles.

In a large sample of iron, the magnetic moments will be lined up in local regions called **domains,** each looking like the ordered situation of Figure 24.58, but there will be no long-range ordering. Inside a domain, the atomic magnetic moments will be aligned, but the magnetic moments of individual domains will be randomly oriented. The individual domains are quite small—on the order of 0.1 mm or so—so a piece of iron the size of a nail has thousands of domains. The random orientation of the magnetic moments of the domains, as shown in FIGURE 24.59, means that there is no overall magnetic moment. So, how can you magnetize a nail?

FIGURE 24.59 Magnetic domains in a ferromagnetic material.

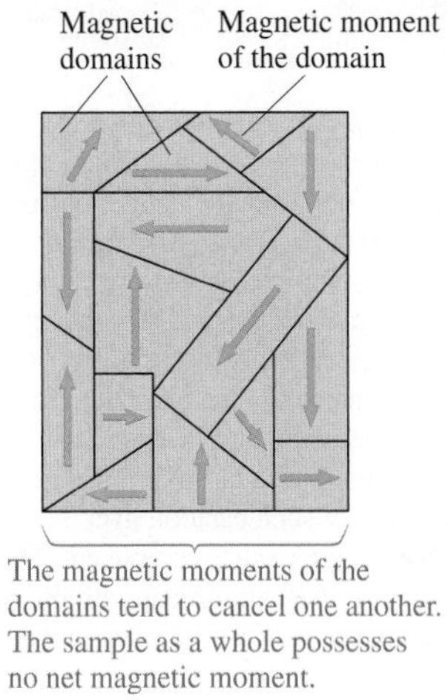

Magnetic domains

Magnetic moment of the domain

The magnetic moments of the domains tend to cancel one another. The sample as a whole possesses no net magnetic moment.

Induced Magnetic Moments

When you bring a magnet near a piece of iron, as in FIGURE 24.60 on the next page, the magnetic field of the magnet penetrates the iron and creates torques on the atomic magnetic moments. The atoms will stay organized in domains, but certain domains will now have more favorable alignments. Domain boundaries move: Domains aligned with the external field become larger at the expense of domains opposed to the field. After this shift in domain boundaries, the magnetic moments of the domains no longer cancel out. The iron will have a net magnetic moment that is aligned with the external field. The iron will have developed an *induced magnetic moment*.

FIGURE 24.60 Inducing a magnetic moment in a piece of iron.

Unmagnetized piece of iron

1. Initially, the magnetic moments of the domains cancel each other; there is no net magnetic moment.

2. Adding the magnetic field from a magnet causes favorably oriented domains (shown in green) to grow at the expense of other domains.

3. The resulting domain structure has a net magnetic moment that is attracted to the magnet. The piece of iron has been magnetized.

N

S

N

S

NOTE ► Inducing a magnetic moment with a magnetic field is analogous to inducing an electric dipole with an electric field, which we saw in Chapter 21. ◄

Looking at the pole structure of the induced magnetic moment in the iron, we can see that the iron will now be attracted to the magnet. The fact that a magnet attracts and picks up ferromagnetic objects was one of the basic observations about magnetism with which we started the chapter. Now we have an explanation of how this works, based on three facts:

1. Electrons are microscopic magnets due to their inherent magnetic moment.
2. A ferromagnetic material in which the magnetic moments are aligned is organized into magnetic domains.
3. The individual domains shift in response to an external magnetic field to produce an induced magnetic moment for the entire object. This induced magnetic moment will be attracted by a magnet that produced the orientation.

When a piece of iron is near a magnet, the iron becomes a magnet as well. But when the applied field is taken away, the domain structure will (generally) return to where it began: The induced magnetic moment will disappear. In the presence of a *very* strong field, however, a piece of iron can undergo more significant changes to its domain structure, and some domains may permanently change orientation. When the field is removed, the iron may retain some of this magnetic character: The iron will have become permanently magnetized. But pure iron is a rather poor permanent magnetic material; it is very easy to disrupt the ordering of the domains that has created the magnetic moment of the iron. For instance, if you heat (or even just drop!) a piece of magnetized iron, the resulting random atomic motions tend to destroy the alignment of the domains, destroying the magnetic character in the process.

Alloys of ferromagnetic materials often possess more robust magnetic characters. Alloys of iron and other ferromagnetic elements with rare earth elements can make permanent magnets of incredible strength.

CONCEPTUAL EXAMPLE 24.14 Sticking things to the refrigerator

Everyone has used a magnet to stick papers to the fridge. Why does the magnet stick to the fridge through a layer of paper?

REASON When you bring a magnet near the steel door of the refrigerator, the magnetic field of the magnet induces a magnetic moment in the steel. The direction of this induced moment will be such that it is attracted to the magnet—thus the magnet will stick to the fridge. Magnetic fields go through nonmagnetic materials such as paper, so the magnet will hold the paper to the fridge.

ASSESS This result helps makes sense of another observation you've no doubt made: You can't stick a thick stack of papers to the fridge. This is because the magnetic field of the magnet decreases rapidly with distance. Because the field is weaker, the induced magnetic moment is smaller.

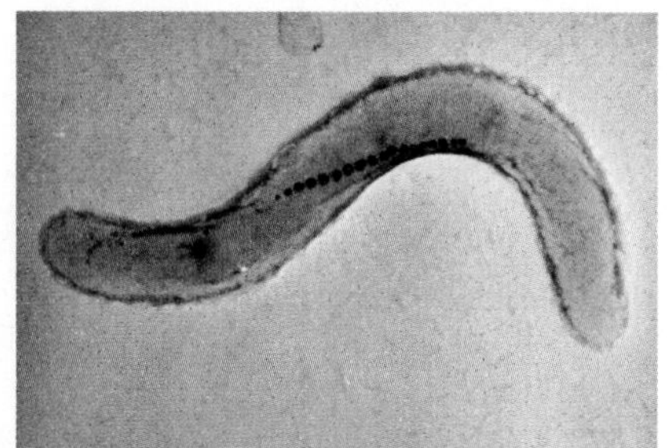

Magnetotactic bacteria BIO Several organisms use the earth's magnetic field to navigate. The clearest example of this is *magnetotactic bacteria*. The dark dots in this image are small pieces of iron; each piece is a single domain and hence a very strong magnet. Such a bacterium possesses a very strong magnetic moment: The bacterium itself acts like a bar magnet, and lines up with the earth's magnetic field. In the temperate regions where such bacteria live, the earth's field has a large vertical component. The bacteria use their alignment with this vertical field component to navigate up and down.

Electromagnets

The magnetic domains in a ferromagnetic material have a strong tendency to line up with an applied magnetic field. This means that it is possible to use a piece of iron or other ferromagnetic material to increase the strength of the field from a current-carrying wire. For example, suppose a solenoid is wound around a piece of iron. When current is passed through the wire, the solenoid's magnetic field lines up the domains in the iron, thus magnetizing it. The resulting **electromagnet** may produce a field that is hundreds of times stronger than the field due to the solenoid itself.

In the past few sections, we've begun to see examples of the deep connection between electricity and magnetism. This connection was one of the most important scientific discoveries of the 1800s, and is something we will explore in more detail in the next chapter.

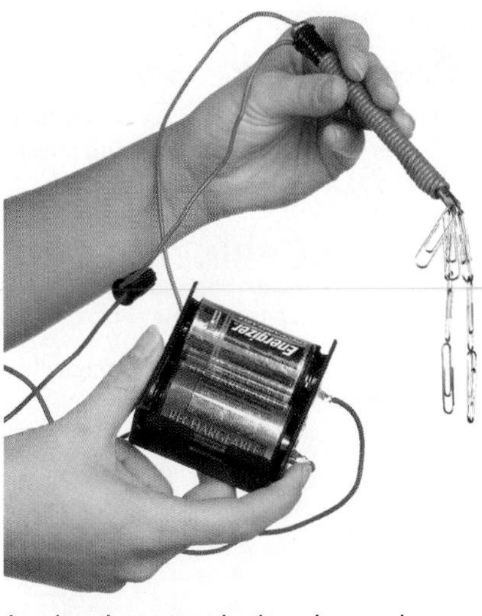
An electric current in the wire produces a magnetic field that magnetizes the nail around which the wire is wound.

STOP TO THINK 24.7 A chain of paper clips is hung from a permanent magnet. Which diagram shows the correct induced pole structure of the paper clips?

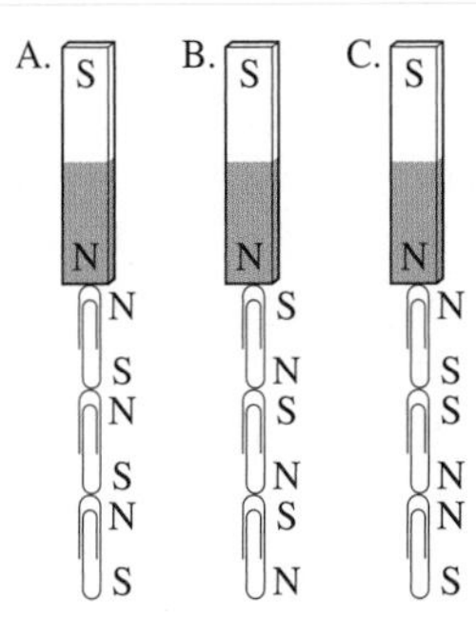

INTEGRATED EXAMPLE 24.15 **Making music with magnetism**

A loudspeaker creates sound by pushing air back and forth with a paper cone that is driven by a magnetic force on a wire coil at the base of the cone. **FIGURE 24.61** shows the details. The bottom of the cone is wrapped with several turns of fine wire. This coil of wire sits in the gap between the poles of a circular magnet, the black disk in the photo. The magnetic field exerts a force on a current in the wire, pushing the cone and thus pushing the air.

FIGURE 24.61 The arrangement of the coil and magnet poles in a loudspeaker.

(a)

(b)
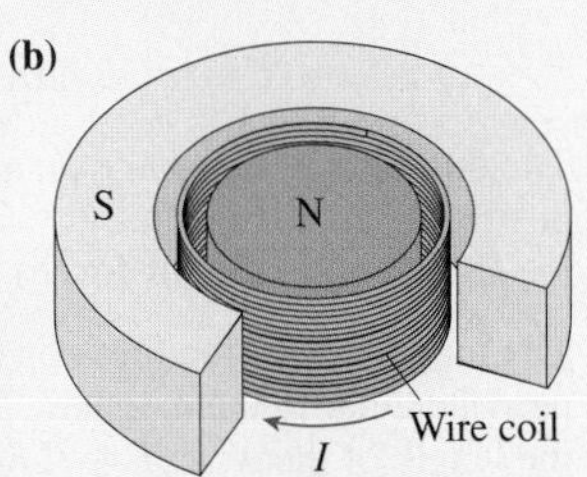

There is a 0.18 T field in the gap between the poles. The coil of wire that sits in this gap has a diameter of 5.0 cm, contains 20 turns of wire, and has a resistance of 8.0 Ω. The speaker is connected to an amplifier whose instantaneous output voltage of 6.0 V creates a clockwise current in the coil as seen from above. What is the magnetic force on the coil at this instant?

PREPARE The current in the coil experiences a force due to the magnetic field between the poles. Let's start with a sketch of the field to determine the direction of this force. Magnetic field lines go from the north pole to the south pole of a magnet, so the field lines for the loudspeaker magnet appear as in **FIGURE 24.62**. The field is at all points perpendicular to the current, and the right-hand rule shows us that, for a clockwise current, the force at each point of the wire is out of the page.

FIGURE 24.62 The magnetic field in the gap and the current in the coil.

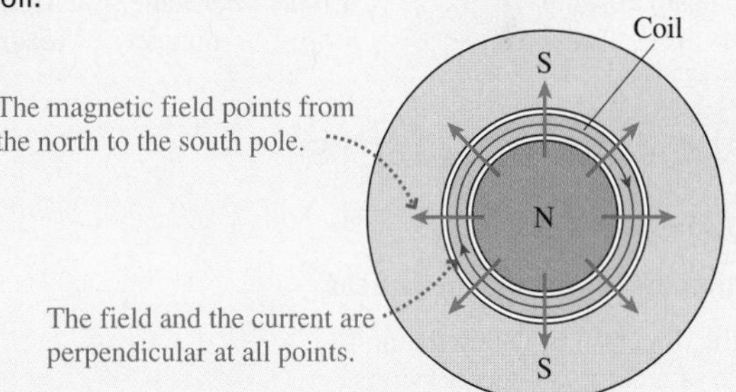

SOLVE The current in the wire is produced by the amplifier. The current is related to the potential difference and the resistance of the wire by Ohm's law:

$$I = \frac{\Delta V}{R} = \frac{6.0 \text{ V}}{8.0 \ \Omega} = 0.75 \text{ A}$$

Because the current is perpendicular to the field, we can use Equation 24.10 to determine the force on this current. We know the field and the current, but we need to know the length of the wire in the field region. The coil has diameter 5.0 cm and thus circumference $\pi(0.050 \text{ m})$. The coil has 20 turns, so the total length of the wire in the field is

$$L = 20\pi(0.050 \text{ m}) = 3.1 \text{ m}$$

The magnitude of the force is then given by Equation 24.10 as

$$F = ILB = (0.75 \text{ A})(3.1 \text{ m})(0.18 \text{ T}) = 0.42 \text{ N}$$

This force is directed out of the page, as already noted.

ASSESS The force is small, but this is reasonable. A loudspeaker cone is quite light, so only a small force is needed for a large acceleration. The force for a clockwise current is out of the page, but when the current switches direction to counterclockwise, the force will switch directions as well. A current that alternates direction will cause the cone to oscillate in and out—just what is needed for making music.

SUMMARY

The goal of Chapter 24 has been to learn about magnetic fields and how magnetic fields exert forces on currents and moving charges.

GENERAL PRINCIPLES

Sources of Magnetism

At its most fundamental level, magnetism is an interaction between moving charges. Magnetic fields can be created by either:

- Electric currents or
- Permanent magnets

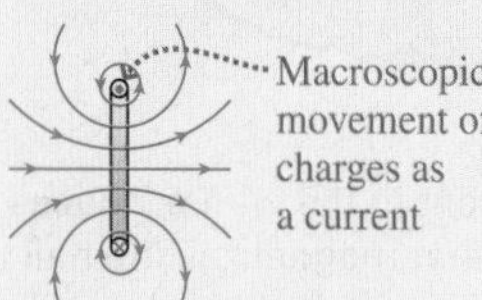

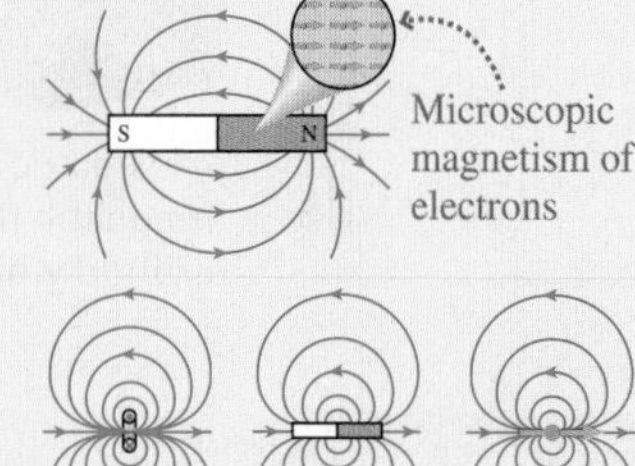

The most basic unit of magnetism is the **magnetic dipole,** which consists of a north and a south pole.

Three basic kinds of dipoles are: Current loop, Permanent magnet, Atomic magnet

Consequences of Magnetism

Magnetic fields exert long-range forces on magnetic materials and on moving charges (or currents).

- Unlike poles of magnets attract each other; like poles repel each other.

- Parallel wires with currents in the same direction attract each other; when the currents are in opposite directions, the wires repel each other.

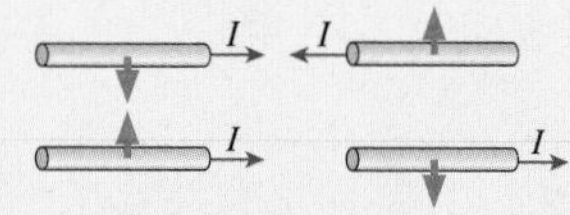

Magnetic fields exert torques on magnetic dipoles, lining them up with the field.

If two or more sources of magnetic field are present, the **principle of superposition** applies.

IMPORTANT CONCEPTS

Magnetic Fields

The **direction of the magnetic field**

- is the direction in which the north pole of a compass needle points.
- due to a current can be found from the **right-hand rule for fields.**

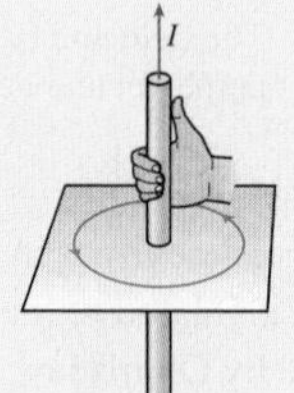

The **strength of the magnetic field** is

- proportional to the torque on a compass needle when turned slightly from the field direction.
- measured in tesla (T)

Magnetic Forces and Torques

The magnitude of the magnetic force on a *moving* charge depends on its charge q, its speed v, and the angle α between the velocity and the field:

$$F = |q|vB\sin\alpha$$

The direction of this force on a positive charge is given by the **right-hand rule for forces.**

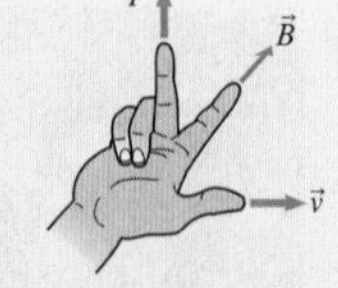

The magnitude of the force on a *current-carrying wire* perpendicular to the magnetic field depends on the current and the length of the wire: $F = ILB$.

The torque on a *current loop* in a magnetic field depends on the current, the loop's area, and how the loop is oriented in the field: $\tau = (IA)B\sin\theta$.

APPLICATIONS

Fields due to common currents

Long straight wire

$$B = \frac{\mu_0 I}{2\pi r}$$

Current loop

$$B = \frac{\mu_0 I}{2R}$$

Solenoid

$$B = \mu_0 I \frac{N}{L}$$

Charged-particle motion

No force if $\vec{v}$ is parallel to $\vec{B}$.

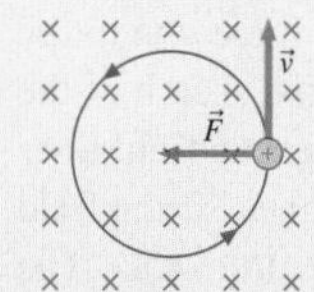

If $\vec{v}$ is perpendicular to $\vec{B}$, the particle undergoes uniform circular motion with radius $r = mv/|q|B$.

Stability of magnetic dipoles

A magnetic dipole is stable (in a lower energy state) when aligned with the external magnetic field. It is unstable (in a higher energy state) when aligned opposite to the field.

The probe field of an MRI scanner measures the flipping of magnetic dipoles between these two orientations.

For homework assigned on MasteringPhysics, go to www.masteringphysics.com

Problem difficulty is labeled as | (straightforward) to ||||| (challenging).

Problems labeled INT integrate significant material from earlier chapters; BIO are of biological or medical interest.

QUESTIONS

Conceptual Questions

1. The north pole of a bar magnet is brought near the *center* of another bar magnet, as shown in Figure Q24.1. Will the force between the magnets be attractive, repulsive, or zero? Why?

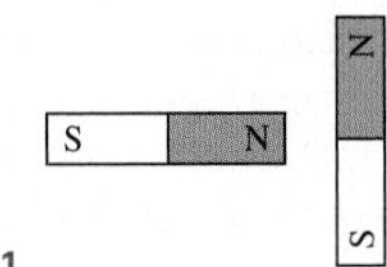

FIGURE Q24.1

2. You have a bar magnet whose poles are not marked. How can you find which pole is north and which is south by using only a piece of string?
3. When you are in the southern hemisphere, does a compass point north or south?
4. BIO Green turtles use the earth's magnetic field to navigate. They seem to use the field to tell them their latitude—how far north or south of the equator they are. Explain how knowing the direction of the earth's field could give this information.
5. A *horseshoe* magnet consists of a bar magnet bent into a U-shape, as shown in Figure Q24.5. Sketch the magnetic field lines for a horseshoe magnet.

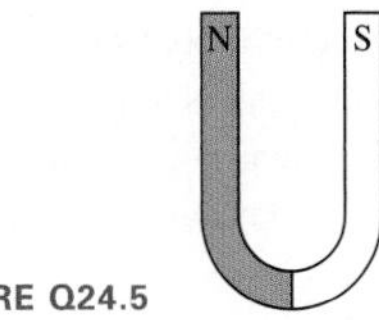

FIGURE Q24.5

6. What is the current direction in the wire of Figure Q24.6? Explain.

FIGURE Q24.6 FIGURE Q24.7

7. What is the current direction in the wire of Figure Q24.7?
8. Since the wires in the walls of your house carry current, you might expect that you could use a compass to detect the positions of the wires. In fact, a compass will experience no deflection when brought near a current-carrying wire because the current is AC (meaning "alternating current"—the current switches direction 120 times each second). Explain why a compass doesn't react to an AC current.
9. Two wires carry currents in opposite directions, as in Figure Q24.9. The field is 2.0 mT at a point below the lower wire. What are the strength and direction of the field at point 1 (midway between the two wires) and at point 2 (the same distance above the upper wire as the 2.0 mT point is below the lower wire)?

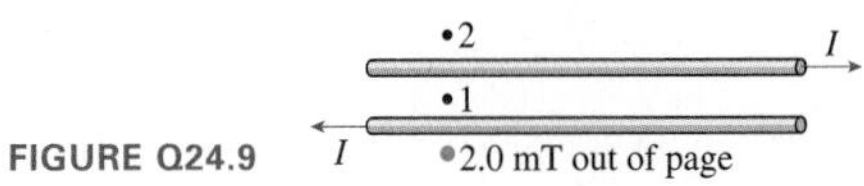

FIGURE Q24.9

10. As shown in Figure Q24.10, a uniform magnetic field points upward, in the plane of the paper. A long wire perpendicular to the paper initially carries no current. When a current is turned on in the wire in the direction shown, the magnetic field at point 1 is found to be zero. Draw the magnetic field vector at point 2 when the current is on.

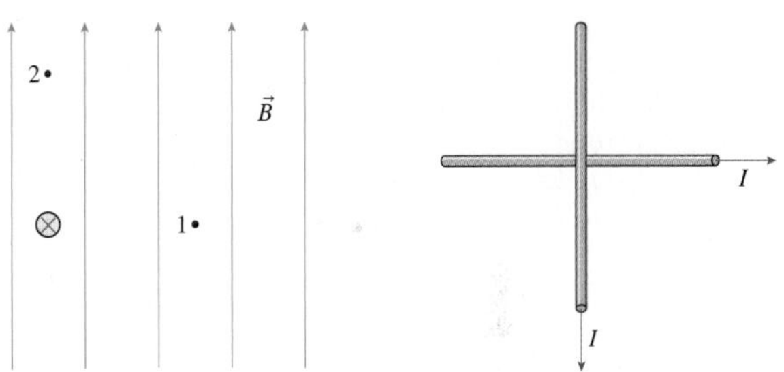

FIGURE Q24.10 FIGURE Q24.11

11. Two long wires carry currents in the directions shown in Figure Q24.11. One wire is 10 cm above the other. In which direction is the magnetic field at a point halfway between them?
12. If an electron is not moving, is it possible to set it in motion using a magnetic field? Explain.
13. Figure Q24.13 shows a solenoid as seen in cross section. Compasses are placed at points 1 and 2. In which direction will each compass point when there is a large current in the direction shown? Explain.

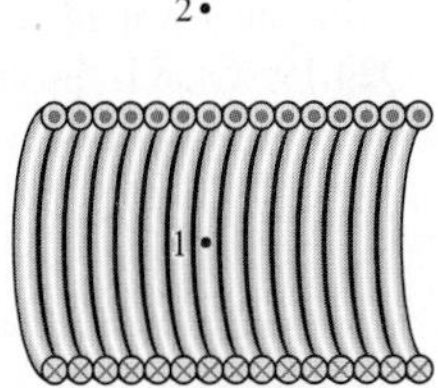

FIGURE Q24.13

14. One long solenoid is placed inside another solenoid with twice the diameter but the same length. Each solenoid carries the same current but in opposite directions, as shown in Figure Q24.14. If they also have the same number of turns, in which direction does the magnetic field in the center point? Explain.

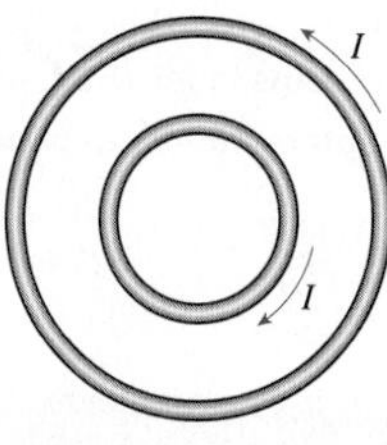

FIGURE Q24.14

15. What is the *initial* direction of deflection for the charged particles entering the magnetic fields shown in Figure Q24.15?

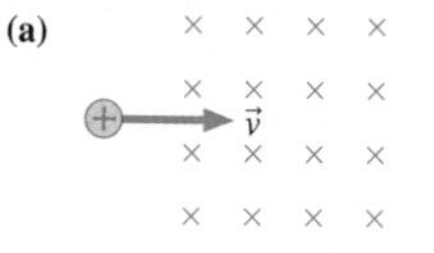

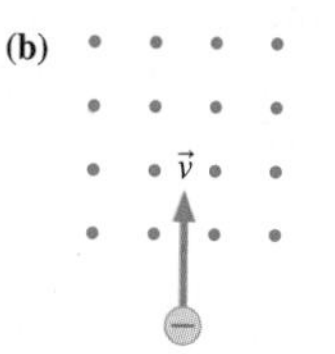

FIGURE Q24.15

16. What is the *initial* direction of deflection for the charged particles entering the magnetic fields shown in Figure Q24.16?

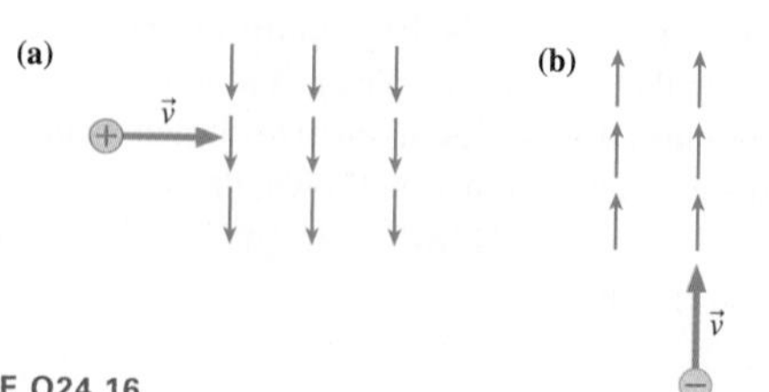

FIGURE Q24.16

17. Determine the magnetic field direction that causes the charged particles shown in Figure Q24.17 to experience the indicated magnetic forces.

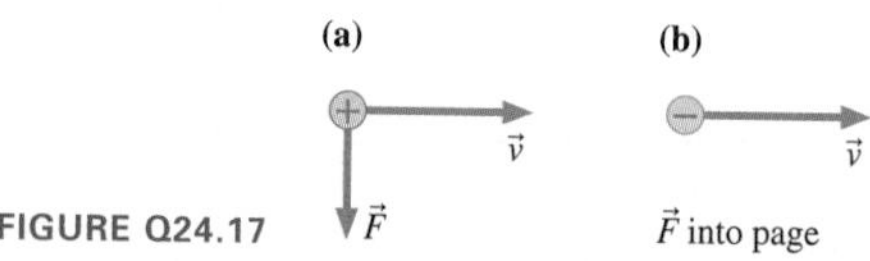

FIGURE Q24.17

18. Determine the magnetic field direction that causes the charged particles shown in Figure Q24.18 to experience the indicated magnetic forces.

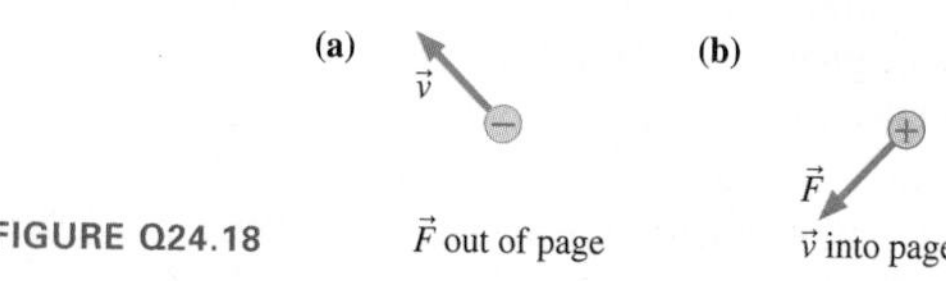

FIGURE Q24.18

19. An electron is moving near a long, current-carrying wire, as shown in Figure Q24.19. What is the direction of the magnetic force on the electron?

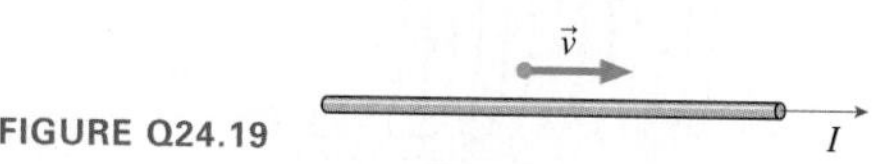

FIGURE Q24.19

20. Two positive charges are moving in a uniform magnetic field with velocities, as shown in Figure Q24.20. The magnetic force on each charge is also shown. In which direction does the magnetic field point?

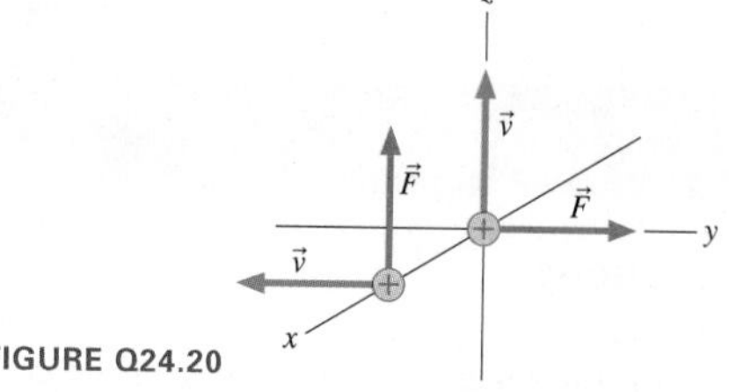
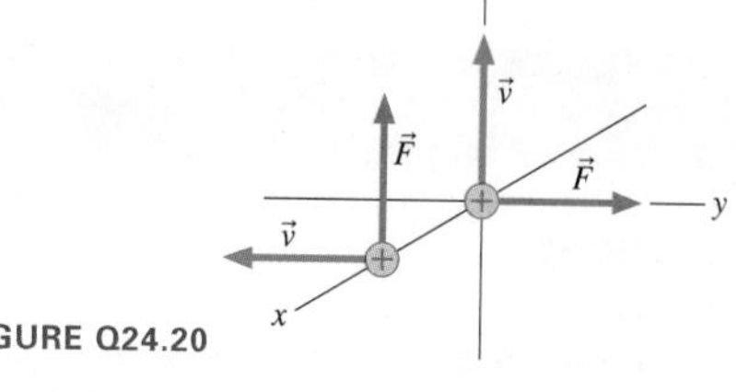

FIGURE Q24.20

21. An electron is moving in a circular orbit in the earth's magnetic field directly above the north magnetic pole. Viewed from above, is the rotation clockwise or counterclockwise?

22. A proton moves in a region of uniform magnetic field, as shown in Figure Q24.22. The velocity at one instant is shown. Will the subsequent motion be a clockwise or counterclockwise orbit?

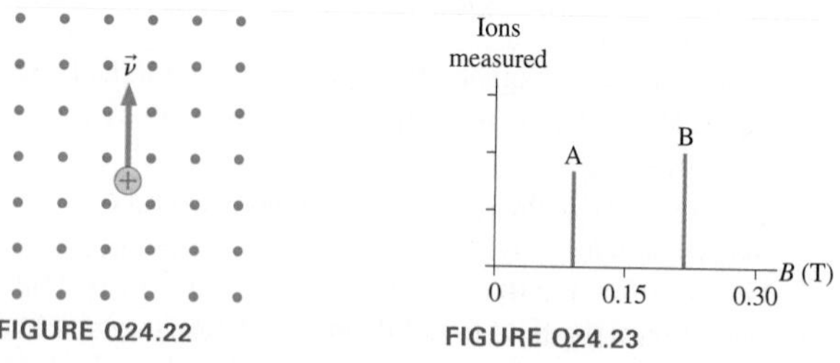

FIGURE Q24.22

FIGURE Q24.23

23. The detector in a mass spectrometer records the number of ions measured at a fixed position as the field is varied. For a sample consisting of a single atomic species, two peaks were found where one was expected, as shown in Figure Q24.23. The most likely explanation is that the atoms received different charges when ionized. If the two peaks correspond to ions with charges $+e$ and $+2e$, which peak is which? Explain.

24. A proton is moving near a long, current-carrying wire. When the proton is at the point shown in Figure Q24.24, in which direction is the force on it?

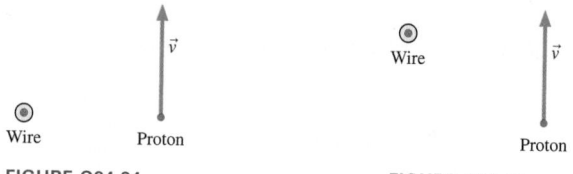

FIGURE Q24.24

FIGURE Q24.25

25. A proton is moving near a long, current-carrying wire. When the proton is at the point shown in Figure Q24.25, in which direction is the force on it?

26. A long wire and a square loop lie in the plane of the paper. Both carry a current in the direction shown in Figure Q24.26. In which direction is the net force on the loop? Explain.

FIGURE Q24.26

27. The computers that control MRI machines cannot have CRT monitors. Explain why this is so.

28. A Slinky is a child's toy that is a long coil spring. Suppose you take a Slinky and let it hang down and stretch out so that the coils do not touch each other, as in Figure Q24.28. Now you connect the Slinky to a power supply and pass a large DC current through it. Think about the current in the coils. Will the Slinky expand or contract?

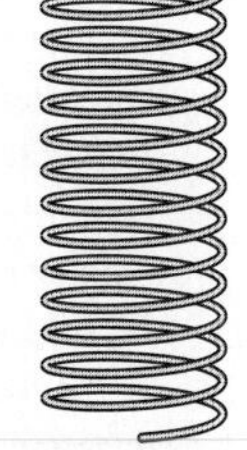

FIGURE Q24.28

29. A solenoid carries a current that produces a field inside it. A wire carrying a current lies inside the solenoid, at the center, carrying a current along the solenoid's axis. Is there a force on this wire due to the field of the solenoid? Explain.
30. You want to make an electromagnet by wrapping wire around a nail. Should you use bare copper wire or wire coated with insulating plastic? Explain.
31. The moon does not have a molten iron core like the earth, but the moon does have a small magnetic field. What might be the source of this field?
32. Archaeologists can use instruments that measure small variations in magnetic field to locate buried walls made of fired brick, as shown in Figure Q24.32. When fired, the magnetic moments in the clay become randomly aligned; as the clay cools, the magnetic moments line up with the earth's field and retain this alignment even if the bricks are subsequently moved. Explain how this leads to a measurable magnetic field variation over a buried wall.

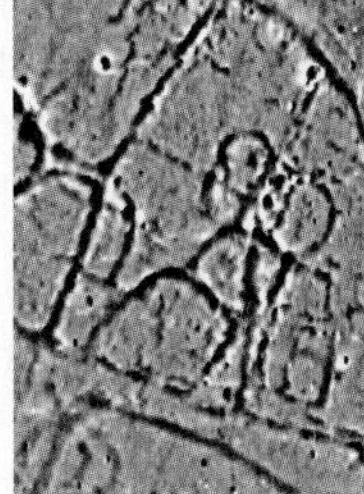

FIGURE Q24.32

Multiple-Choice Questions

33. || An unmagnetized metal sphere hangs by a thread. When the north pole of a bar magnet is brought near, the sphere is strongly attracted to the magnet, as shown in Figure Q24.33. Then the magnet is reversed and its south pole is brought near the sphere. How does the sphere respond?

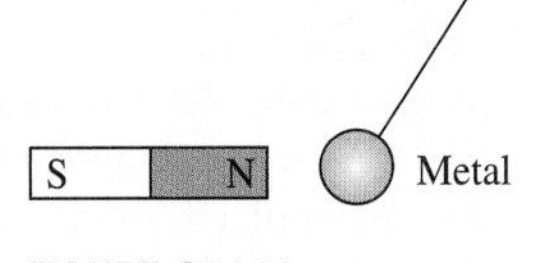

FIGURE Q24.33

A. It is strongly attracted to the magnet.
B. It is weakly attracted to the magnet.
C. It does not respond.
D. It is weakly repelled by the magnet.
E. It is strongly repelled by the magnet.

34. || If a compass is placed above a current-carrying wire, as in Figure Q24.34, the needle will line up with the field of the wire. Which of the views shows the correct orientation of the needle for the noted current direction?

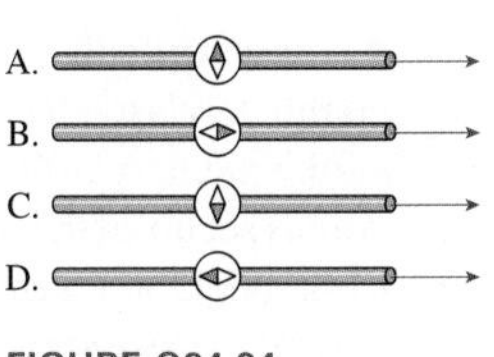

FIGURE Q24.34

35. | Two wires carry equal and opposite currents, as shown in Figure Q24.35. At a point directly between the two wires, the field is

FIGURE Q24.35

A. Directed up, toward the top of the page.
B. Directed down, toward the bottom of the page.
C. Directed to the left.
D. Directed to the right.
E. Zero.

36. | Figure Q24.36 shows four particles moving to the right as they enter a region of uniform magnetic field, directed into the paper as noted. All particles move at the same speed and have the same charge. Which particle has the largest mass?

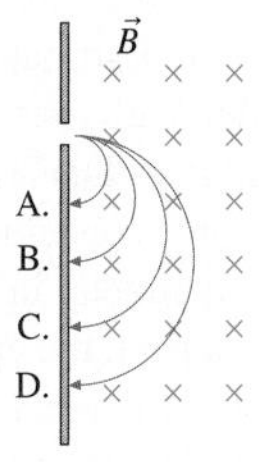

FIGURE Q24.36

37. | Four particles of identical charge and mass enter a region of uniform magnetic field and follow the trajectories shown in Figure Q24.37. Which particle has the highest velocity?

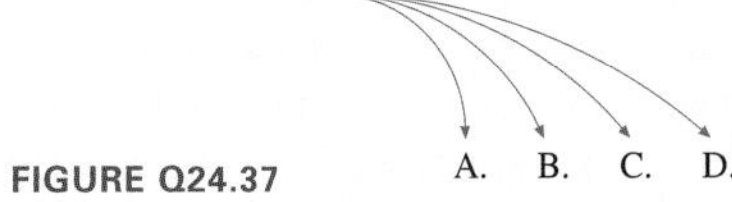

FIGURE Q24.37

38. | If all of the particles shown in Figure Q24.37 are electrons, what is the direction of the magnetic field that produced the indicated deflection?
A. Up (toward the top of the page).
B. Down (toward the bottom of the page).
C. Out of the plane of the paper.
D. Into the plane of the paper.
39. | If two compasses are brought near enough to each other, the magnetic fields of the compasses themselves will be larger than the field of the earth, and the needles will line up with each other. Which of the arrangements of two compasses shown in Figure Q24.39 is a possible stable arrangement?

A.
B.
C.
D.

FIGURE Q24.39

PROBLEMS

Section 24.1 Magnetism

Section 24.2 The Magnetic Field

Section 24.3 Electric Currents Also Create Magnetic Fields

Section 24.4 Calculating the Magnetic Field Due to a Current

1. | What currents are needed to generate the magnetic field strengths of Table 24.1 at a point 1.0 cm from a long, straight wire?
2. | At what distances from a very thin, straight wire carrying a 10 A current would the magnetic field strengths of Table 24.1 be generated?
3. || The magnetic field at the center of a 1.0-cm-diameter loop is 2.5 mT.
   a. What is the current in the loop?
   b. A long, straight wire carries the same current you found in part a. At what distance from the wire is the magnetic field 2.5 mT?

4. ||| For a particular scientific experiment, it is important to be completely isolated from any magnetic field, including the earth's field. A 1.00-m-diameter current loop with 200 turns of wire is set up so that the field at the center is exactly equal to the earth's field in magnitude but opposite in direction. What is the current in the current loop?

5. | What are the magnetic field strength and direction at points 1 to 3 in Figure P24.5?

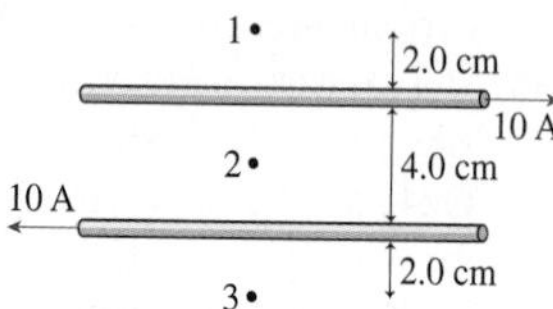

FIGURE P24.5

6. | Although the evidence is weak, there has been concern in BIO recent years over possible health effects from the magnetic fields generated by transmission lines. A typical high-voltage transmission line is 20 m off the ground and carries a current of 200 A. Estimate the magnetic field strength on the ground underneath such a line. What percentage of the earth's magnetic field does this represent?

7. | Some consumer groups urge pregnant women not to use elec- BIO tric blankets, in case there is a health risk from the magnetic fields from the approximately 1 A current in the heater wires.
   a. Estimate, stating any assumptions you make, the magnetic field strength a fetus might experience. What percentage of the earth's magnetic field is this?
   b. It is becoming standard practice to make electric blankets with minimal external magnetic field. Each wire is paired with another wire that carries current in the opposite direction. How does this reduce the external magnetic field?

8. ||| A long wire carrying a 5.0 A current perpendicular to the xy-plane intersects the x-axis at $x = -2.0$ cm. A second, parallel wire carrying a 3.0 A current intersects the x-axis at $x = +2.0$ cm. At what point or points on the x-axis is the magnetic field zero if (a) the two currents are in the same direction and (b) the two currents are in opposite directions?

9. || The element niobium, which is a metal, is a superconductor (i.e., no electrical resistance) at temperatures below 9 K. However, the superconductivity is destroyed if the magnetic field at the surface of the wire of the metal reaches or exceeds 0.10 T. What is the maximum current in a straight, 3.0-mm-diameter superconducting niobium wire?
Hint: You can assume that all the current flows in the center of the wire.

10. | The small currents in axons corresponding to nerve impulses BIO produce measurable magnetic fields. A typical axon carries a peak current of 0.040 μA. What is the strength of the field at a distance of 1.0 mm?

11. || A solenoid used to produce magnetic fields for research purposes is 2.0 m long, with an inner radius of 30 cm and 1000 turns of wire. When running, the solenoid produces a field of 1.0 T in the center. Given this, how large a current does it carry?

12. | Two concentric current loops lie in the same plane. The smaller loop has a radius of 3.0 cm and a current of 12 A. The bigger loop has a current of 20 A. The magnetic field at the center of the loops is found to be zero. What is the radius of the bigger loop?

13. | The magnetic field of the brain has been measured to be BIO approximately 3.0×10^{-12} T. Although the currents that cause this field are quite complicated, we can get a rough estimate of their size by modeling them as a single circular current loop 16 cm (the width of a typical head) in diameter. What current is needed to produce such a field at the center of the loop?

14. || A researcher would like to perform an experiment in zero magnetic field, which means that the field of the earth must be cancelled. Suppose the experiment is done inside a solenoid of diameter 1.0 m, length 4.0 m, with a total of 5000 turns of wire. The solenoid is oriented to produce a field that opposes and exactly cancels the field of the earth. What current is needed in the solenoid's wire?

15. ||| What is the magnetic field at the center of the loop in Figure P24.15?

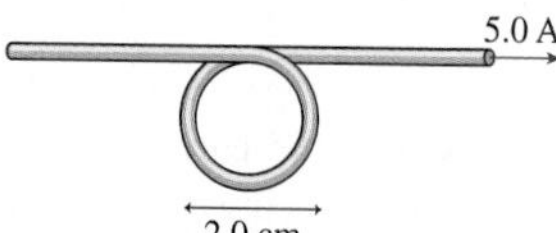

FIGURE P24.15

16. | Experimental tests have shown that hammerhead sharks can BIO detect magnetic fields. In one such test, 100 turns of wire were wrapped around a 7.0-m-diameter cylindrical shark tank. A magnetic field was created inside the tank when this coil of wire carried a current of 1.5 A. Sharks trained by getting a food reward when the field was present would later unambiguously respond when the field was turned on.
   a. What was the magnetic field strength in the center of the tank due to the current in the coil?
   b. Is the strength of the coil's field at the center of the tank larger or smaller than that of the earth?

17. | We have seen that the heart produces a magnetic field that BIO can be used to diagnose problems with the heart. The magnetic field of the heart is a dipole field produced by a loop current in the outer layers of the heart. Suppose the field at the center of the heart is 90 pT (a pT is 10^{-12} T) and that the heart has a diameter of approximately 12 cm. What current circulates around the heart to produce this field?

18. |||| You have a 1.0-m-long copper wire. You want to make an N-turn current loop that generates a 1.0 mT magnetic field at the center when the current is 1.0 A. You must use the entire wire. What will be the diameter of your coil?

19. ||||| In the Bohr model of the hydrogen atom, the electron moves in a circular orbit of radius 5.3×10^{-11} m with speed 2.2×10^6 m/s. According to this model, what is the magnetic field at the center of a hydrogen atom due to the motion of the electron?
Hint: Determine the *average* current of the orbiting electron.

Section 24.5 Magnetic Fields Exert Forces on Moving Charges

20. | A proton moves with a speed of 1.0×10^7 m/s in the directions shown in Figure P24.20. A 0.50 T magnetic field points in the positive x-direction. For each, what is magnetic force on the proton? Give your answers as a magnitude and a direction.

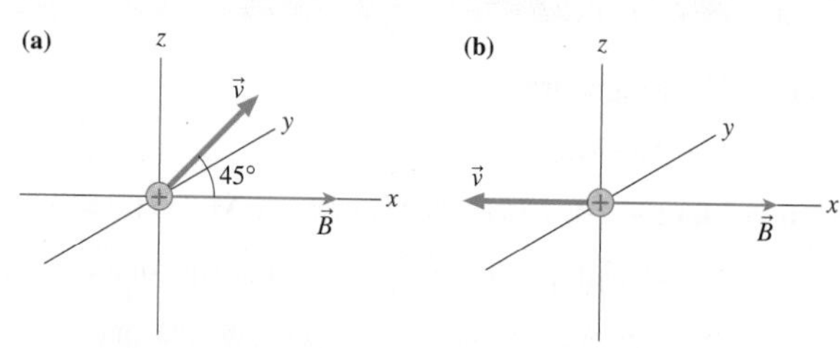

FIGURE P24.20

21. ‖ An electron moves with a speed of $1..0 \times 10^7$ m/s in the directions shown in Figure P24.21. A 0.50 T magnetic field points in the positive x-direction. For each, what is magnetic force $\vec{F}$ on the electron? Give your answers as a magnitude and a direction.

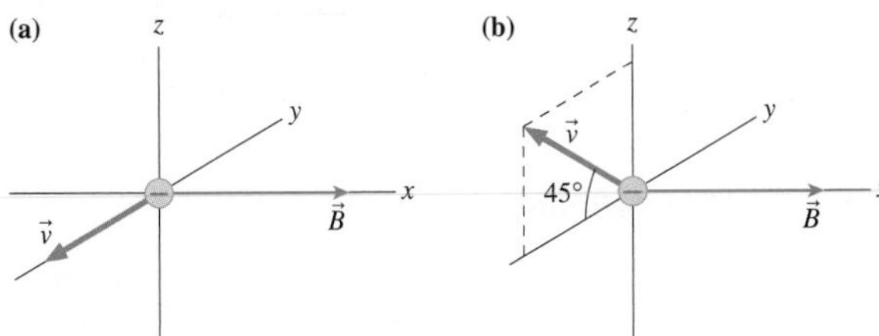

FIGURE P24.21

22. | An electromagnetic flowmeter applies a magnetic field of 0.20 T
BIO to blood flowing through a coronary artery at a speed of 15 cm/s. What force is felt by a chlorine ion with a single negative charge?

23. | The aurora is caused when electrons and protons, moving in the earth's magnetic field of $\approx 5.0 \times 10^{-5}$ T, collide with molecules of the atmosphere and cause them to glow. What is the radius of the circular orbit for
   a. An electron with speed 1.0×10^6 m/s?
   b. A proton with speed 5.0×10^4 m/s?

24. ‖| Problem 24.23 describes two particles that orbit the earth's magnetic field lines. What is the *frequency* of the circular orbit for
   a. An electron with speed 1.0×10^6 m/s?
   b. A proton with speed 5.0×10^4 m/s?

25. ‖‖ The microwaves in a microwave oven are produced in a special tube called a *magnetron.* The electrons orbit in a magnetic field at a frequency of 2.4 GHz, and as they do so they emit 2.4 GHz electromagnetic waves. What is the strength of the magnetic field?

26. ‖| A mass spectrometer similar to the one in Figure 24.36 is designed to separate protein fragments. The fragments are ionized by the removal of a single electron, then they enter a 0.80 T uniform magnetic field at a speed of 2.3×10^5 m/s. If a fragment has a mass 85 times the mass of the proton, what will be the distance between the points where the ion enters and exits the magnetic field?

27. | In a certain mass spectrometer, particles with a charge of $+e$ are sent into the spectrometer with a velocity of 2.5×10^5 m/s. They are found to move in a circular path with a radius of 0.21 m. If the magnetic field of the spectrometer is 0.050 T, what kind of particles are these likely to be?

28. ‖| At $t = 0$ s, a proton is moving with a speed of 5.5×10^5 m/s at an angle of 30° from the x-axis, as shown in Figure P24.28. A uniform magnetic field of magnitude 1.50 T is pointing in the positive y-direction. What will be the y-coordinate of the proton 10 μs later?

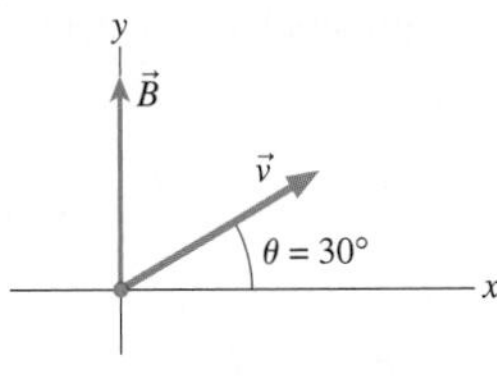

FIGURE P24.28

29. ‖| Early black-and-white television sets used an electron beam
INT to draw a picture on the screen. The electrons in the beam were accelerated by a voltage of 3.0 kV; the beam was then steered to different points on the screen by coils of wire that produced a magnetic field of up to 0.65 T.
   a. What is the speed of electrons in the beam?
   b. What acceleration do they experience due to the magnetic field, assuming that it is perpendicular to their path? What is this acceleration in units of g?
   c. If the electrons were to complete a full circular orbit, what would be the radius?
   d. A magnetic field can be used to redirect the beam, but the electrons are brought to high speed by an electric field. Why can't we use a magnetic field for this task?

Section 24.6 Magnetic Fields Exert Forces on Currents

30. | What magnetic field strength and direction will levitate the 2.0 g wire in Figure P24.30?

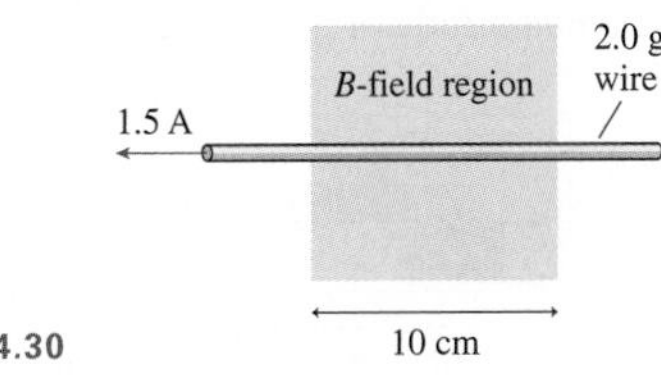

FIGURE P24.30

31. | What is the net force (magnitude and direction) on each wire in Figure P24.31?

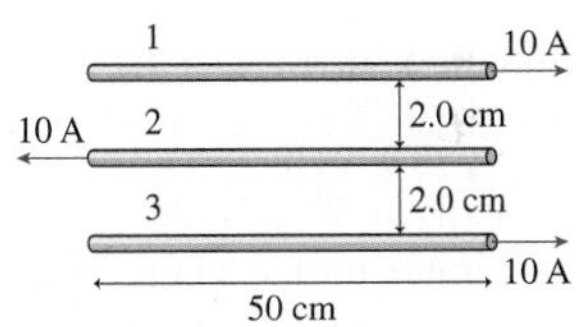

FIGURE P24.31

32. | The unit of current, the ampere, is defined in terms of the force between currents. If two 1.0-meter-long sections of very long wires a distance 1.0 m apart each carry a current of 1.0 A, what is the force between them? (If the force between two actual wires has this value, the current is defined to be exactly 1 A.)

33. | A uniform 2.5 T magnetic field points to the right. A 3.0-m-long wire, carrying 15 A, is placed at an angle of 30° to the field, as shown in Figure 24.33. What is the force (magnitude and direction) on the wire?

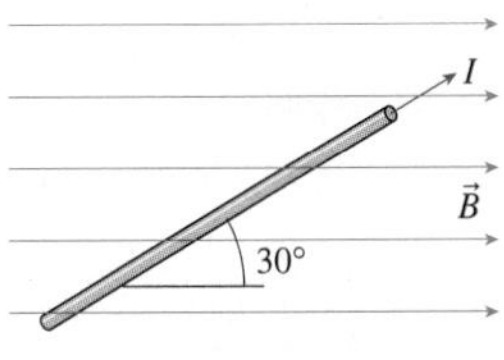

FIGURE P24.33

34. | The four wires in Figure P24.34 are tilted at 20° with respect to a uniform 0.35 T field. If each carries 4.5 A and is 0.35 m long, what is the force (direction and magnitude) on each?

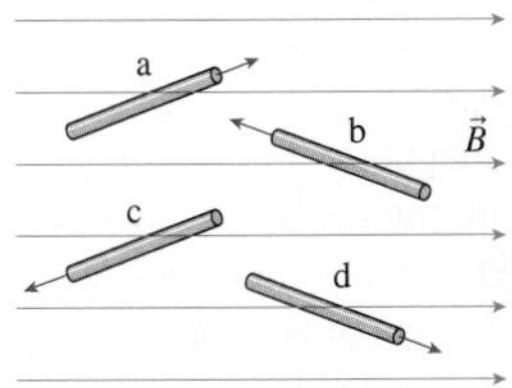

FIGURE P24.34

35. II Magnetic information on hard drives is accessed by a read INT head that must move rapidly back and forth across the disk. The force to move the head is generally created with a *voice coil actuator*, a flat coil of fine wire that moves between the poles of a strong magnet, as in Figure P24.35. Assume that the coil is a square 1.0 cm on a side made of 200 turns of fine wire with total resistance 1.5 Ω. The field between the poles of the magnet is 0.30 T; assume that the field does not extend beyond the edge of the magnet. The coil and the mount that it rides on have a total mass of 12 g.
  a. If a voltage of 5.0 V is applied to the coil, what is the current?
  b. If the current is clockwise viewed from above, what are the magnitude and direction of the net force on the coil?
  c. What is the magnitude of the acceleration of the coil?

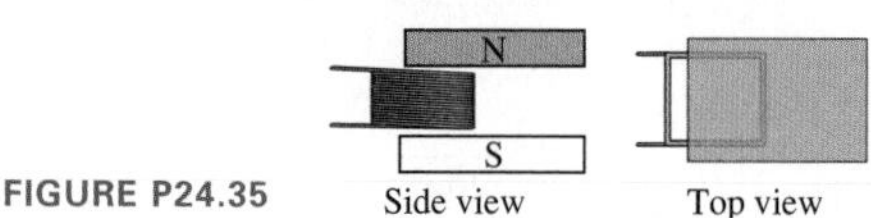

FIGURE P24.35

Section 24.7 Magnetic Fields Exert Torques on Dipoles

36. II A current loop in a motor has an area of 0.85 cm^2. It carries a 240 mA current in a uniform field of 0.62 T. What is the magnitude of the maximum torque on the current loop?
37. II A square current loop 5.0 cm on each side carries a 500 mA current. The loop is in a 1.2 T uniform magnetic field. The axis of the loop, perpendicular to the plane of the loop, is 30° away from the field direction. What is the magnitude of the torque on the current loop?
38. I Figure P24.38 shows two square current loops. The loops are far apart and do not interact with each other.
  a. Use force diagrams to show that both loops are in equilibrium, having a net force of zero and no torque.
  b. One of the loop positions is stable. That is, the forces will return it to equilibrium if it is rotated slightly. The other position is unstable, like an upside-down pendulum: If rotated slightly, it will not return to the position shown. Which is which? Explain.

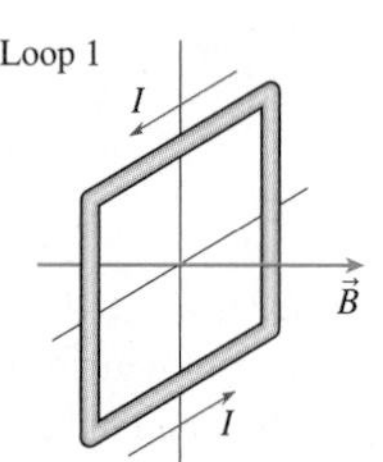

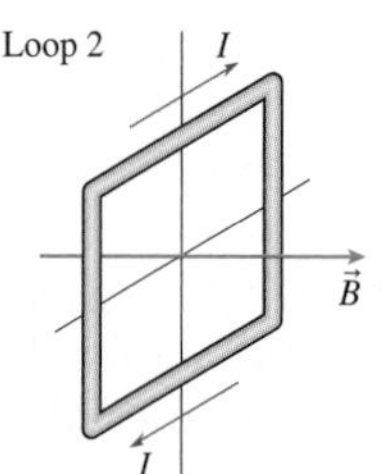

FIGURE P24.38

39. III The earth's magnetic dipole moment of 8.0×10^{22} A · m^2 is generated by currents within the molten iron of the earth's outer core. (The inner core is solid iron.) As a simple model, consider the outer core to be a 3000-km-diameter current loop. What is the current in the current loop?

40. II a. What is the magnitude of the torque on the circular current loop in Figure P24.40?
  b. What is the loop's equilibrium position?

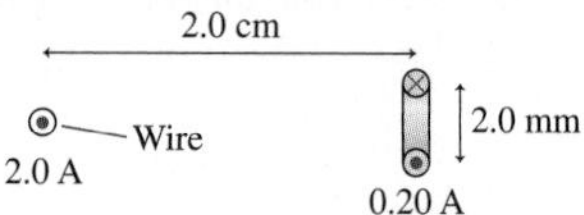

FIGURE P24.40

Section 24.8 Magnets and Magnetic Materials

41. II A computer diskette is a plastic disk with a ferromagnetic coating. A single magnetic domain can have its magnetic moment oriented to point either up or down, and these two orientations can be interpreted as a binary 0 (up) or 1 (down). Each 0 or 1 is called a *bit* of information. A diskette stores roughly 500,000 *bytes* of data on one side, and each byte contains eight bits. Estimate the width of a magnetic domain, and compare your answer to the typical domain size given in the text. List any assumptions you use in your estimate.
42. I All ferromagnetic materials have a *Curie temperature,* a temperature above which they will cease to be magnetic. Explain in some detail why you might expect this to be so.

General Problems

43. II In Figure P24.43, a compass sits 1.0 cm above a wire in a cir- INT cuit containing a 1.0 F capacitor charged to 5.0 V, a 1.0 Ω resistor, and an open switch. The compass is lined up with the earth's magnetic field. The switch is then closed, so there is a current in the circuit, and the switch remains closed until the capacitor has completely discharged.
  a. At the position of the compass, what is the magnitude of the magnetic field due to the current in the wire right after the switch is closed? How does this compare with the magnitude of the field of the earth?
  b. Describe how the compass orientation changes right after the switch is closed, and how the compass orientation changes as time goes on.

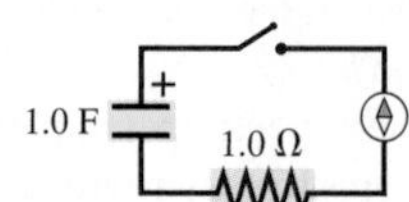

FIGURE P24.43

44. I The right edge of the circuit in Figure P24.44 extends into a INT 50 mT uniform magnetic field. What are the magnitude and direction of the net force on the circuit?

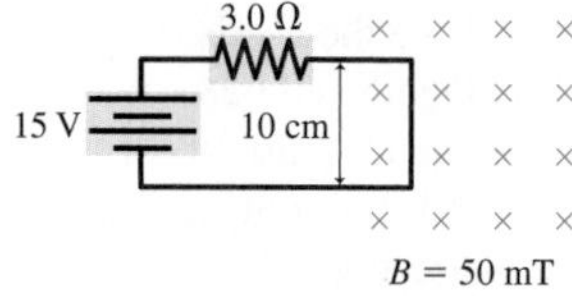

FIGURE P24.44

45. II The two 10-cm-long parallel wires in Figure P24.45 are sepa- INT rated by 5.0 mm. For what value of the resistor R will the force between the two wires be 5.4×10^{-5} N?

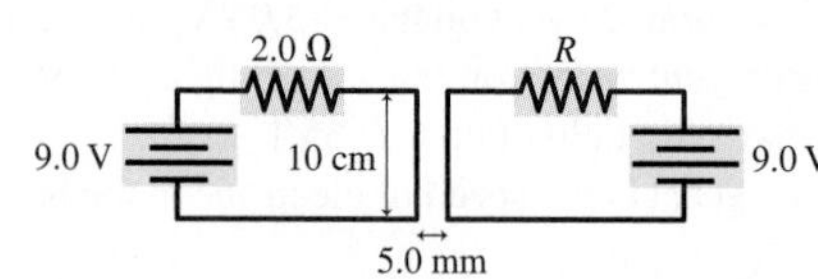

FIGURE P24.45

46. ||| The capacitor in Figure P24.46 is charged to 50 V. The switch INT closes at $t = 0$ s. Draw a graph showing the magnetic field strength as a function of time at the position of the dot. On your graph indicate the maximum field strength and provide an appropriate numerical scale on the horizontal axis.

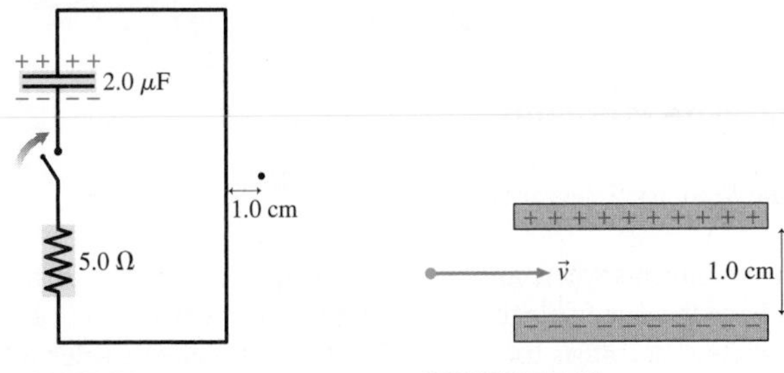

FIGURE P24.46

FIGURE P24.47

47. || An electron travels with speed 1.0×10^7 m/s between the INT two parallel charged plates shown in Figure P24.47. The plates are separated by 1.0 cm and are charged by a 200 V battery. What magnetic field strength and direction will allow the electron to pass between the plates without being deflected?

48. || The two springs in Figure P24.48 each have a spring constant INT of 10 N/m. They are stretched by 1.0 cm when a current passes through the wire. How big is the current?

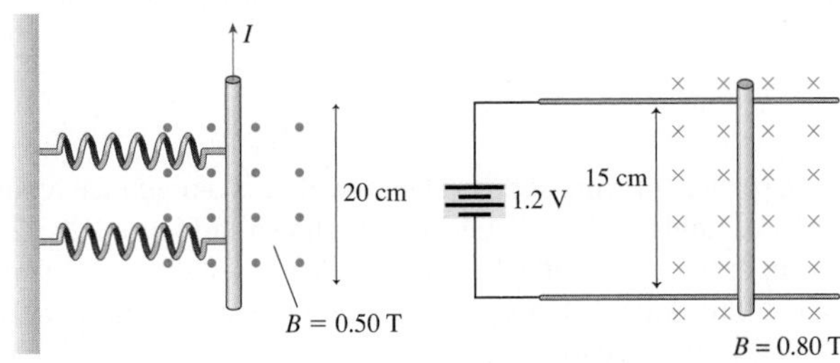

FIGURE P24.48

FIGURE P24.49

49. || A device called a *railgun* uses the magnetic force on currents INT to launch projectiles at very high speeds. An idealized model of a railgun is illustrated in Figure 24.49. A 1.2 V power supply is connected to two conducting rails. A segment of copper wire, in a region of uniform magnetic field, slides freely on the rails. The wire has a 0.85 mΩ resistance and a mass of 5.0 g. Ignore the resistance of the rails. When the power supply is switched on,
   a. What is the current?
   b. What are the magnitude and direction of the force on the wire?
   c. What will be the wire's speed after it has slid a distance of 6.0 cm?

50. || An antiproton (which has INT the same properties as a proton except that its charge is $-e$) is moving in the combined electric and magnetic fields of Figure P24.50.
   a. What are the magnitude and direction of the antiproton's acceleration at this instant?
   b. What would be the magnitude and direction of the acceleration if $\vec{v}$ were reversed?

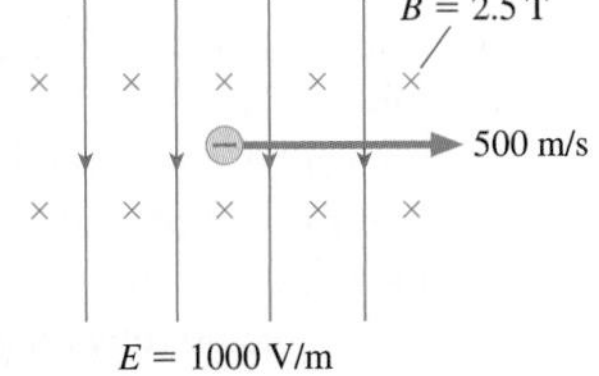

FIGURE P24.50

51. | Typical blood velocities in the coronary arteries range from INT BIO 10 to 30 cm/s. An electromagnetic flowmeter applies a magnetic field of 0.25 T to a coronary artery with a blood velocity of 15 cm/s. As we saw in Figure 24.38, this field exerts a force on ions in the blood, which will separate. The ions will separate until they make an electric field that exactly balances the magnetic force. This electric field produces a voltage that can be measured.
   a. What force is felt by a singly ionized (positive) sodium ion?
   b. Charges in the blood will separate until they produce an electric field that cancels this magnetic force. What will be the resulting electric field?
   c. What voltage will this electric field produce across an artery with a diameter of 3.0 mm?

52. | A power line consists of two wires, each carrying a current of 400 A in the same direction. The lines are perpendicular to the earth's magnetic field, and are separated by a distance of 5.0 m. Which is larger: the force of the earth's magnetic field on each wire or the magnetic force between the wires?

53. ||| Bats are capable of navigating using the earth's field—a plus BIO for an animal that may fly great distances from its roost at night. If, while sleeping during the day, bats are exposed to a field of a similar magnitude but different direction than the earth's field, they are more likely to lose their way during their next lengthy night flight. Suppose you are a researcher doing such an experiment in a location where the earth's field is 50 μT at a 60° angle below horizontal. You make a 50-cm-diameter, 100-turn coil around a roosting box; the sleeping bats are at the center of the coil. You wish to pass a current through the coil to produce a field that, when combined with the earth's field, creates a net field with the same strength and dip angle (60° below horizontal) as the earth's field but with a horizontal component that points south rather than north. What are the proper orientation of the coil and the necessary current?

54. ||| At the equator, the earth's field is essentially horizontal; near BIO the north pole, it is nearly vertical. In between, the angle varies. As you move farther north, the dip angle, the angle of the earth's field below horizontal, steadily increases. Green turtles seem to use this dip angle to determine their latitude. Suppose you are a researcher wanting to test this idea. You have gathered green turtle hatchlings from a beach where the magnetic field strength is 50 μT and the dip angle is 56°. You then put the turtles in a 1.2-m-diameter circular tank and monitor the direction in which they swim as you vary the magnetic field in the tank. You change the field by passing a current through a 100-turn horizontal coil wrapped around the tank. This creates a field that adds to that of the earth. What current should you pass through the coil, and in what direction, to produce a net field in the center of the tank that has a dip angle of 62°?

55. || Internal components of cathode-ray-tube televisions and computer monitors can become magnetized; the resulting magnetic field can deflect the electron beam and distort the colors on the screen. Demagnetization can be accomplished with a coil of wire whose current switches direction rapidly and gradually decreases in amplitude. Explain what effect this will have on the magnetic moments of the magnetic materials in the device, and how this might eliminate any magnetic ordering.

56. ||| A 1.0-m-long, 1.0-mm-diameter copper wire carries a current INT of 50.0 A to the east. Suppose we create a magnetic field that produces an upward force on the wire exactly equal in magnitude to the wire's weight, causing the wire to "levitate." What are the field's direction and magnitude?

57. | An insulated copper wire is wrapped around an iron nail. The resulting coil of wire consists of 240 turns of wire that cover 1.8 cm of the nail, as shown in Figure P24.57. A current of 0.60 A passes through the wire. If the ferromagnetic properties of the nail increase the field by a factor of 100, what is the magnetic field strength inside the nail?

240 turns

1.8 cm

FIGURE P24.57

58. |||| Figure P24.58 is a cross section through three long wires with linear mass density 50 g/m. They each carry equal currents in the directions shown. The lower two wires are 4.0 cm apart and are attached to a table. What current I will allow the upper wire to "float" so as to form an equilateral triangle with the lower wires?
INT

4.0 cm

FIGURE P24.58

59. ||| A long, straight wire with a linear mass density of 50 g/m is suspended by threads, as shown in Figure P24.59. There is a uniform magnetic field pointing vertically downward. A 10 A current in the wire experiences a horizontal magnetic force that deflects it to an equilibrium angle of 10°. What is the strength of the magnetic field $\vec{B}$?
INT

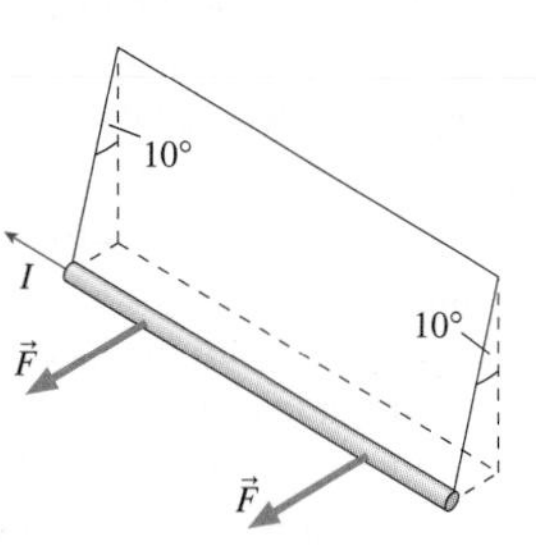

FIGURE P24.59

60. || A mass spectrometer is designed to separate atoms of carbon to determine the fraction of different isotopes. (Isotopes of an element, as we will see in Chapter 30, have the same atomic number but different atomic mass, due to different numbers of neutrons.) There are three main isotopes of carbon, with the following atomic masses:

Atomic masses	
^{12}C	1.99×10^{-26} kg
^{13}C	2.16×10^{-26} kg
^{14}C	2.33×10^{-26} kg

The atoms of carbon are singly ionized and enter a mass spectrometer with magnetic field strength $B = 0.200$ T at a speed of 1.50×10^5 m/s. The ions move along a semicircular path and exit through an exit slit. How far from the entrance will the beams of the different isotope ions end up?

61. | A solenoid is near a piece of iron, as shown in Figure P24.61. When a current is present in the solenoid, a magnetic field is created. This magnetic field will magnetize the iron, and there will be a net force between the solenoid and the iron.

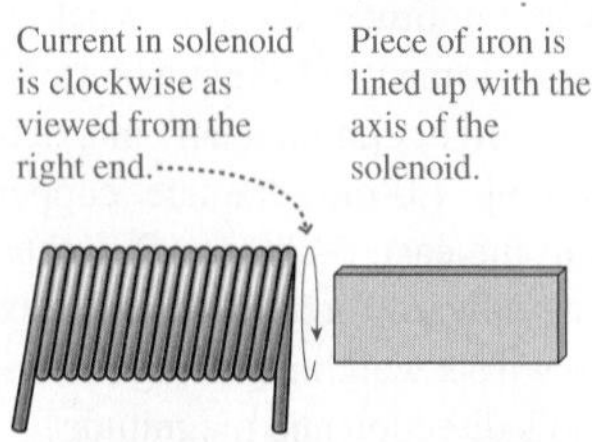

Current in solenoid is clockwise as viewed from the right end.
Piece of iron is lined up with the axis of the solenoid.

FIGURE P24.61

a. Make a sketch showing the direction of the magnetic field from the solenoid. On your sketch, label the induced north magnetic pole and the induced south magnetic pole in the iron.
b. Will the force on the iron be attractive or repulsive?
c. Suppose this force moves the iron. Which way will the iron move?

Passage Problems

The Velocity Selector INT ActivPhysics Online 13.8

In experiments where all the charged particles in a beam are required to have the same velocity (for example, when entering a mass spectrometer), scientists use a *velocity selector.* A velocity selector has a region of uniform electric and magnetic fields that are perpendicular to each other and perpendicular to the motion of the charged particles. Both the electric and magnetic fields exert a force on the charged particles. If a particle has precisely the right velocity, the two forces exactly cancel and the particle is not deflected. Equating the forces due to the electric field and the magnetic field gives the following equation:

$$qE = qvB$$

Solving for the velocity, we get:

$$v = \frac{E}{B}$$

A particle moving at this velocity will pass through the region of uniform fields with no deflection, as shown in Figure P24.62. For higher or lower velocities than this, the particles will feel a net force and will be deflected. A slit at the end of the region allows only the particles with the correct velocity to pass.

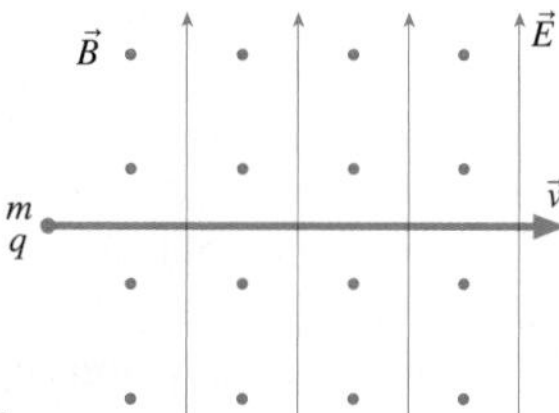

FIGURE P24.62

62. | Assuming the particle in Figure P24.62 is positively charged, what are the directions of the forces due to the electric field and to the magnetic field?
A. The force due to the electric field is directed up (toward the top of the page); the force due to the magnetic field is directed down (toward the bottom of the page).
B. The force due to the electric field is directed down (toward the bottom of the page); the force due to the magnetic field is directed up (toward the top of the page).
C. The force due to the electric field is directed out of the plane of the paper; the force due to the magnetic field is directed into the plane of the paper.
D. The force due to the electric field is directed into the plane of the paper; the force due to the magnetic field is directed out of the plane of the paper.

63. | How does the kinetic energy of the particle in Figure P24.62 change as it traverses the velocity selector?
   A. The kinetic energy increases.
   B. The kinetic energy does not change.
   C. The kinetic energy decreases.
64. | Suppose a particle with twice the velocity of the particle in Figure P24.62 enters the velocity selector. The path of this particle will curve
   A. Upward (toward the top of the page).
   B. Downward (toward the bottom of the page).
   C. Out of the plane of the paper.
   D. Into the plane of the paper.
65. | Next, a particle with the same mass and velocity as the particle in Figure P24.62 enters the velocity selector. This particle has a charge of $2q$—twice the charge of the particle in Figure P24.62. In this case, we can say that
   A. The force of the electric field on the particle is greater than the force of the magnetic field.
   B. The force of the magnetic field on the particle is greater than the force of the electric field.
   C. The forces of the electric and magnetic fields on the particle are still equal.

Ocean Potentials INT

The ocean is salty because it contains many dissolved ions. As these charged particles move with the water in strong ocean currents, they feel a force from the earth's magnetic field. Positive and negative charges are separated until an electric field develops that balances this magnetic force. This field produces measurable potential differences that can be monitored by ocean researchers.

The Gulf Stream moves northward off the east coast of the United States at a speed of up to 3.5 m/s. Assume that the current flows at this maximum speed and that the earth's field is 50 μT tipped 60° below horizontal.

66. | What is the direction of the magnetic force on a singly ionized negative chlorine ion moving in this ocean current?
   A. East
   B. West
   C. Up
   D. Down
67. | What is the magnitude of the force on this ion?
   A. 2.8×10^{-23} N
   B. 2.4×10^{-23} N
   C. 1.6×10^{-23} N
   D. 1.4×10^{-23} N
68. | What magnitude electric field is necessary to exactly balance this magnetic force?
   A. 1.8×10^{-4} N/C
   B. 1.5×10^{-4} N/C
   C. 1.0×10^{-4} N/C
   D. 0.9×10^{-4} N/C
69. | The electric field produces a potential difference. If you place one electrode 10 m below the surface of the water, you will measure the greatest potential difference if you place the second electrode
   A. At the surface.
   B. At a depth of 20 m.
   C. At the same depth 10 m to the north.
   D. At the same depth 10 m to the east.

STOP TO THINK ANSWERS

Stop to Think 24.1: C. The compass needle will not rotate since there is no force between the stationary charges on the rod and the magnetic poles of the compass needle.

Stop to Think 24.2: A. The compass needle will rotate to line up with the field of the magnet, which goes from the north to the south pole.

Stop to Think 24.3: D. The compass needle will rotate to line up with the field circling the wire. The right-hand rule for fields shows this to be toward the top of the paper in the figure.

Stop to Think 24.4: A, C. The force to produce these circular orbits is directed toward the center of the circle. Using the right-hand rule for forces, we see that this will be true for the situations in A and C if the particles are negatively charged.

Stop to Think 24.5: C. The right-hand rule for forces gives the direction of the force. With the field into the paper, the force is to the left if the current is toward the top of the paper.

Stop to Think 24.6: B. Looking at the forces on the top and the bottom of the loop, we can see that the loop will rotate counterclockwise. Alternatively, we can look at the dipole structure of the loop: With a north pole on the left and a south pole on the right, the loop will rotate counterclockwise.

Stop to Think 24.7: B. All of the induced dipoles will be aligned with the field of the bar magnet.

25 EM Induction and EM Waves

The photo of the flower on the left shows how it appears to our eyes, in visible light. But there's more to the story! The false-color view of the flower on the right shows its appearance in the ultraviolet, beyond the range of human vision, revealing pigments we can't see. Whose eyes are these pigments intended for?

LOOKING AHEAD

The goal of Chapter 25 is to understand the nature of electromagnetic induction and electromagnetic waves.

Connecting Electric and Magnetic Phenomena

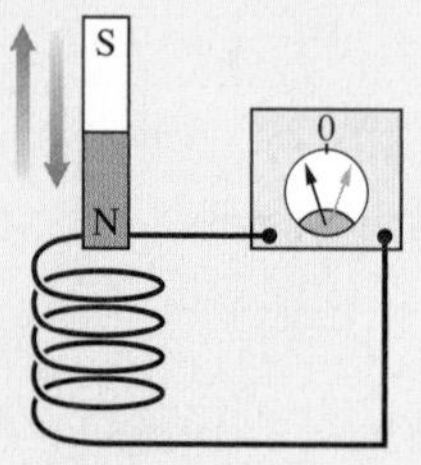

Moving a magnet into or out of a coil of wire creates a momentary current in the wire. The changing *magnetic* field creates an *electric* current in the coil, an example of the close connection between electric and magnetic phenomena we explore in this chapter.

This chapter will draw on all that we have learned about electricity and magnetism. You should especially review these sections:

Looking Back
20.4 The electric field
21.4 The electric potential
24.2–24.6 Magnetic fields and forces

Induced emf and Induced Currents

A *changing* magnetic field creates an emf—which can create a current in a conductor. An emf produced this way is called an **induced emf**; the resulting current is called an **induced current.**

Shaking a magnet back and forth through a coil induces a current that runs the flashlight.

Turning blades rotate a coil of wire in a magnetic field. This change induces enough current to power many homes.

A rapidly changing magnetic field from the base unit induces a current in a coil inside the phone, charging it with no wire connection.

Electromagnetic Waves

The connection between electricity and magnetism is seen most clearly in the existence of **electromagnetic waves,** traveling waves of electric and magnetic fields.

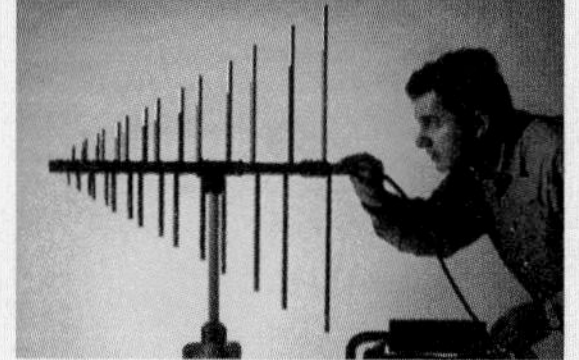

The metal bars on this antenna detect the vertical electric field of an electromagnetic wave.

Everything you've learned about waves applies to electromagnetic waves.

Looking Back
15.3–15.5 Basic wave concepts, light waves, plane waves, intensity

The Electromagnetic Spectrum

You are already familiar with electromagnetic waves: Radio waves, microwaves, light waves, and x rays are all electromagnetic waves with different wavelengths. We'll explore the **spectrum** of possible waves.

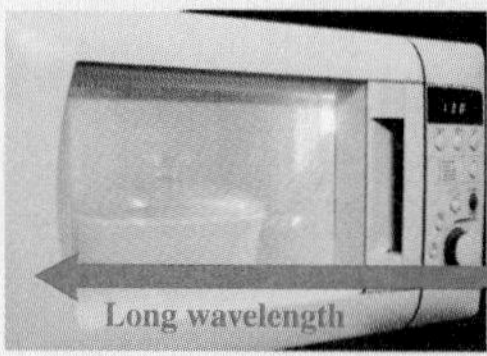

Long-wavelength microwaves rotate water molecules in food.

The vibrations of atoms in the filament emit visible light.

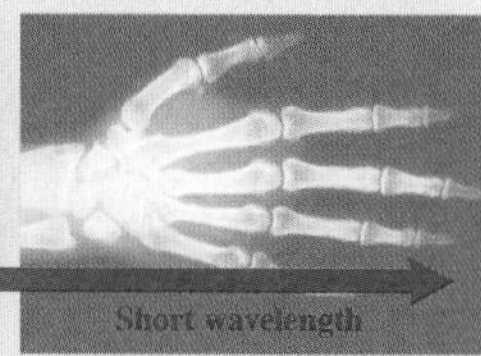

Very-short-wavelength x rays act rather like particles that we'll call **photons.**

25.1 Induced Currents

In Chapter 24, we learned that a current can create a magnetic field. As soon as this discovery was widely known, investigators began considering a related question: Can a magnetic field create a current?

One of the early investigators was Michael Faraday, who was experimenting with two coils of wire wrapped around an iron ring, as shown in FIGURE 25.1, when he made a remarkable discovery. He had hoped that the magnetic field generated by a current in the coil on the left would create a magnetic field in the iron, and that the magnetic field in the iron might then somehow produce a current in the circuit on the right.

This technique failed to generate a steady current, but Faraday noticed that the needle of the current meter jumped ever so slightly at the instant when he closed the switch in the circuit on the left. After the switch was closed, the needle immediately returned to zero. Faraday's observation suggested to him that a current was generated only if the magnetic field was *changing* as it passed through the coil. Faraday set out to test this hypothesis through a series of experiments.

FIGURE 25.1 Faraday's discovery of electromagnetic induction.

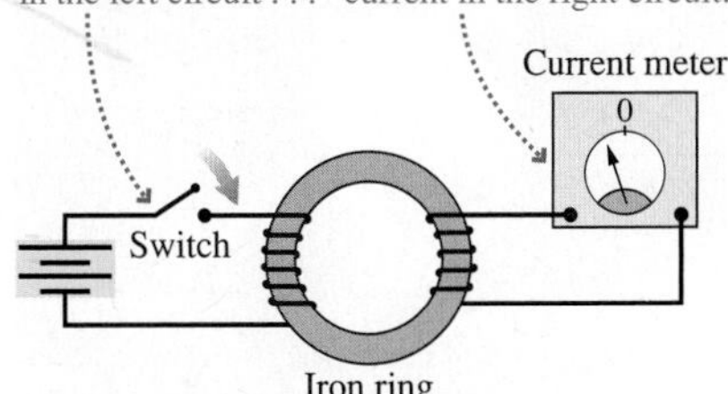
Closing the switch in the left circuit . . .
. . . causes a *momentary* current in the right circuit.
Current meter
0
Switch
Iron ring

Faraday investigates electromagnetic induction

Faraday placed one coil directly above the other, without the iron ring. There was no current in the lower circuit while the switch was in the closed position, but a momentary current appeared whenever the switch was opened or closed.

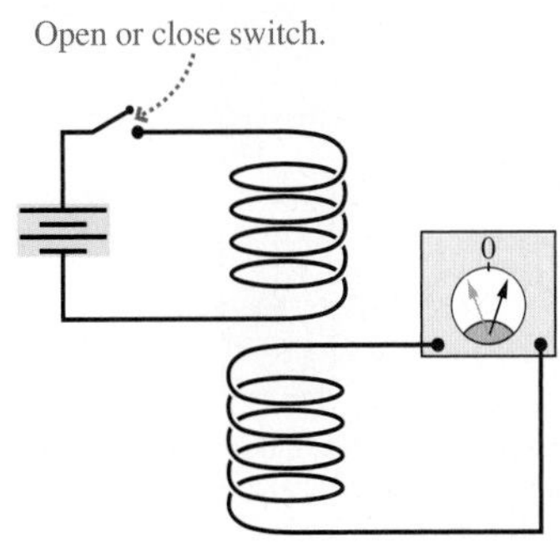
Open or close switch.
0

Opening or closing the switch creates a momentary current.

Faraday pushed a bar magnet into a coil of wire. This action caused a momentary deflection of the needle in the current meter, although *holding* the magnet inside the coil had no effect. A quick withdrawal of the magnet deflected the needle in the other direction.

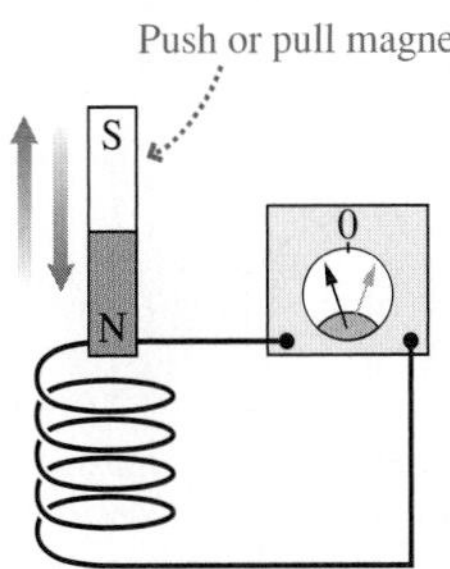
Push or pull magnet.
S
N
0

Pushing the magnet into the coil or pulling it out creates a momentary current.

Must the magnet move? Faraday created a momentary current by rapidly pulling a coil of wire out of a magnetic field, although there was no current if the coil was stationary in the magnetic field. Pushing the coil *into* the magnet caused the needle to deflect in the opposite direction.

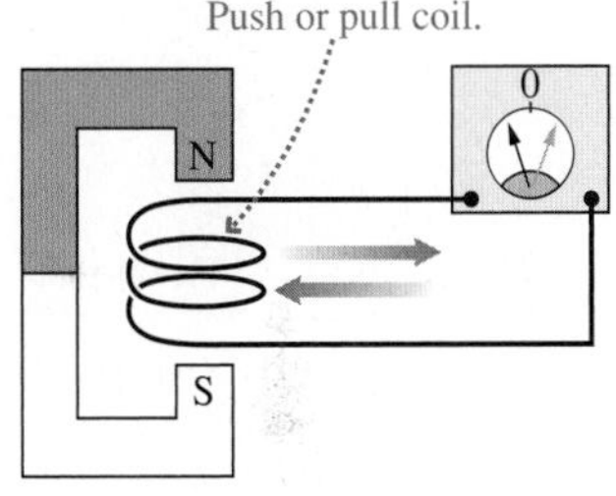
Push or pull coil.
N
S
0

Pushing the coil into the magnet or pulling it out creates a momentary current.

All of these experiments served to bolster Faraday's hypothesis: **Faraday found that there is a current in a coil of wire if and only if the magnetic field passing through the coil is *changing*.** It makes no difference what causes the magnetic field to change: current stopping or starting in a nearby circuit, moving a magnet through the coil, or moving the coil into and out of a magnet. The effect is the same in all cases. There is no current if the field through the coil is not changing, so it's not the magnetic field itself that is responsible for the current but, instead, it is the *changing of the magnetic field.*

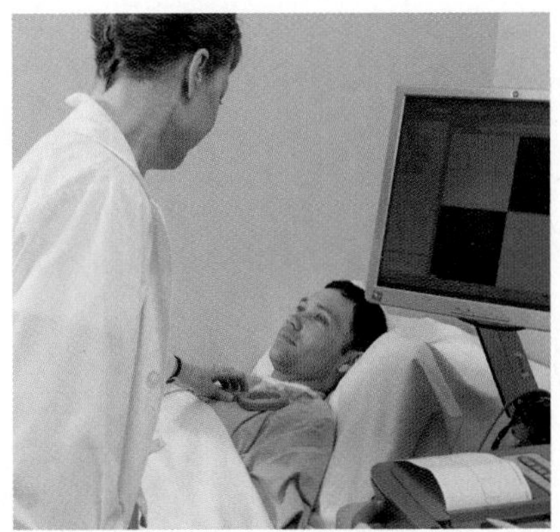

▶ **External pacemaker programming** BIO A circuit that creates a changing magnetic field can induce a current in a second circuit with no direct electrical connection to the first. In this photo a coil carrying an alternating current creates a changing magnetic field that induces currents in a sensing circuit in a cardiac pacemaker inside the patient's chest. These currents adjust settings for the pacemaker. The magnetic fields can penetrate the body so the pacemaker can be programmed with no need for surgery.

The current in a circuit due to a changing magnetic field is called an **induced current.** Opening the switch or moving the magnet *induces* a current in a nearby circuit. An induced current is not caused by a battery; it is a completely new way to generate a current. The creation of an electric current by a changing magnetic field is our first example of **electromagnetic induction.**

A satellite tethered to the Space Shuttle.

25.2 Motional emf

In 1996, astronauts on the Space Shuttle deployed a satellite at the end of a 20-km-long conducting tether. A potential difference of up to 3500 V developed between the shuttle and the satellite as the wire between the two swept through the earth's magnetic field. Why would the motion of a wire in a magnetic field produce such a large voltage? Let's explore the mechanism behind this *motional emf.*

To begin, consider a conductor of length l that moves with velocity $\vec{v}$ through a uniform magnetic field $\vec{B}$, as shown in FIGURE 25.2. The charge carriers inside the conductor—assumed to be positive, as in our definition of current—also move with velocity $\vec{v}$, so they each experience a magnetic force. For simplicity, we will assume that $\vec{v}$ is perpendicular to $\vec{B}$, in which case the magnitude of the force is $F_B = qvB$. This force causes the charge carriers to move. For the geometry of Figure 25.2, the right-hand rule tells us that the positive charges move toward the top of the moving conductor, leaving an excess of negative charge at the bottom.

FIGURE 25.2 The magnetic force on the charge carriers in a moving conductor creates an electric field inside the conductor.

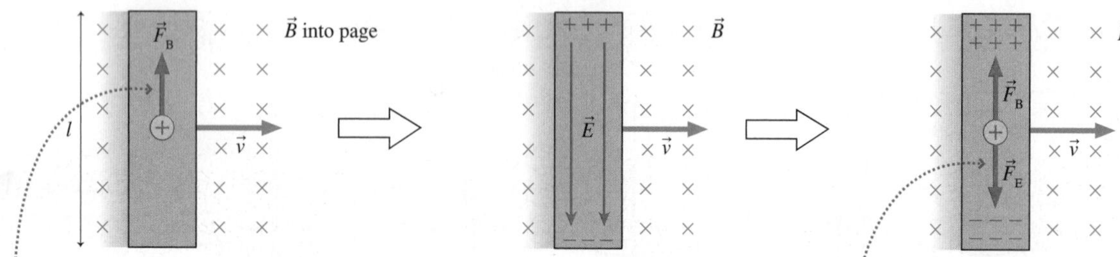
$\vec{B}$ into page
$\vec{F}_B$
l
$\vec{v}$
Charge carriers in the conductor experience a force of magnitude $F_B = qvB$. Positive charges are free to move and drift upward.
$\vec{B}$
$\vec{E}$
$\vec{v}$
The resulting charge separation creates an electric field in the conductor. $\vec{E}$ increases as more charge flows.
$\vec{B}$
$\vec{F}_B$
$\vec{F}_E$
$\vec{v}$
The charge flow continues until the electric and magnetic forces balance. For a positive charge carrier, the upward magnetic force $\vec{F}_B$ is equal to the downward electric force $\vec{F}_E$.

This motion of the charge carriers cannot continue forever. The separation of the charge carriers creates an electric field. The resulting electric force *opposes* the separation of charge, so the charge separation continues only until the electric force has grown to exactly balance the magnetic force:

$$F_E = qE = F_B = qvB$$

When this balance occurs, the charge carriers experience no net force and thus undergo no further motion. The electric field strength at equilibrium is

$$E = vB \tag{25.1}$$

Thus, **the magnetic force on the charge carriers in a moving conductor creates an electric field $E = vB$ inside the conductor.**

The electric field, in turn, creates an electric potential difference between the two ends of the moving conductor. We found in Chapter 21 that the potential difference between two points separated by distance l parallel to an electric field E is $\Delta V = El$. Thus the motion of the wire through a magnetic field *induces* a potential difference

$$\Delta V = vlB \tag{25.2}$$

between the ends of the conductor. The potential difference depends on the strength of the magnetic field and on the wire's speed through the field. This is similar to the action of the electromagnetic flowmeter that we saw in the preceding chapter.

There's an important analogy between this potential difference and the potential difference of a battery. FIGURE 25.3a reminds you that a battery uses a nonelectric force—which we called the charge escalator—to separate positive and negative charges. We refer to a battery, where the charges are separated by chemical reactions, as a source of *chemical emf.* The moving conductor of FIGURE 25.3b develops a potential difference because of the work done to separate the charges. The emf of the conductor is due to its motion, rather than to chemical reactions inside, so we can define the **motional emf** of a conductor of length l moving with velocity $\vec{v}$ perpendicular to a magnetic field $\vec{B}$ to be

$$\mathcal{E} = vlB \quad (25.3)$$

FIGURE 25.3 Two different ways to generate an emf.

(a) Chemical reactions separate the charges and cause a potential difference between the ends. This is a chemical emf.

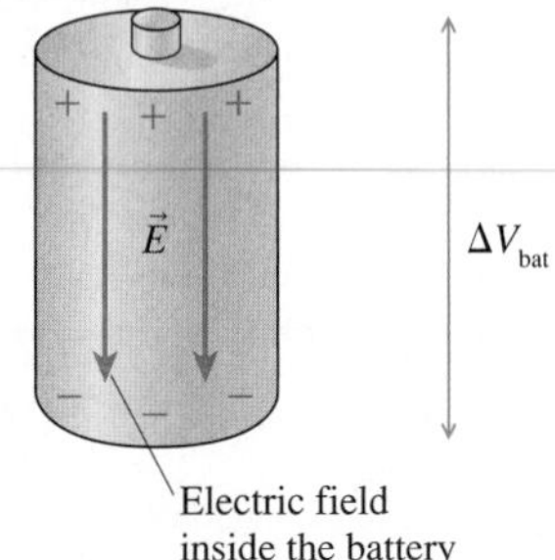

(b) Magnetic forces separate the charges and cause a potential difference between the ends. This is a motional emf.

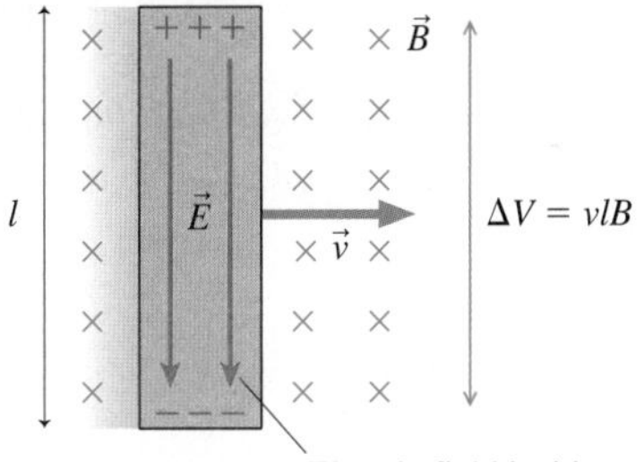

EXAMPLE 25.1 Finding the motional emf for an airplane

A Boeing 747 aircraft with a wingspan of 65 m is cruising at 260 m/s over northern Canada, where the magnetic field of the earth (magnitude 5.0×10^{-5} T) is directed straight down. What is the potential difference between the tips of the wings?

PREPARE The wing is a conductor moving through a magnetic field, so there will be a motional emf. We can visualize a top view of this situation exactly as in Figure 25.3b, with the wing as the moving conductor.

SOLVE The magnetic field is perpendicular to the velocity, so we can compute the potential difference using Equation 25.3:

$$\Delta V = vlB = (260 \text{ m/s})(65 \text{ m})(5.0 \times 10^{-5} \text{ T}) = 0.85 \text{ V}$$

ASSESS The earth's magnetic field is small, so the motional emf will be small as well unless the speed and the length are quite large. The tethered satellite generated a much higher voltage due to its much greater speed and the great length of the tether, the moving conductor.

▶ **A head for magnetism?** BIO Observations of hammerhead sharks imply that they navigate using the earth's magnetic field, and controlled laboratory experiments verify that they can indeed reliably detect such modest fields. How do they do it? One possibility is that they use their keen electric sense, described in Chapter 21. Sharks can detect small potential differences; perhaps they detect magnetic fields by sensing the motional emf as they move through the water. If so, the width of their oddly shaped heads would be an asset because the magnitude of the potential difference is proportional to the length of the moving conductor.

Induced Current in a Circuit

The moving conductor of Figure 25.3 had an emf, but it couldn't sustain a current because the charges had nowhere to go. We can change this by including the moving conductor in a circuit.

FIGURE 25.4 shows a length of wire with resistance R sliding with speed v along a fixed U-shaped conducting rail. The wire and the rail together form a closed conducting loop—a circuit.

Suppose a magnetic field $\vec{B}$ is perpendicular to the plane of the circuit. Charges in the moving wire will be pushed to the ends of the wire by the magnetic force, just as they were in Figure 25.3, but now the charges can continue to flow around the circuit. The moving wire acts like the battery in a circuit.

The current in the circuit is an induced current, due to magnetic forces on moving charges. In this example, the induced current is counterclockwise. The total

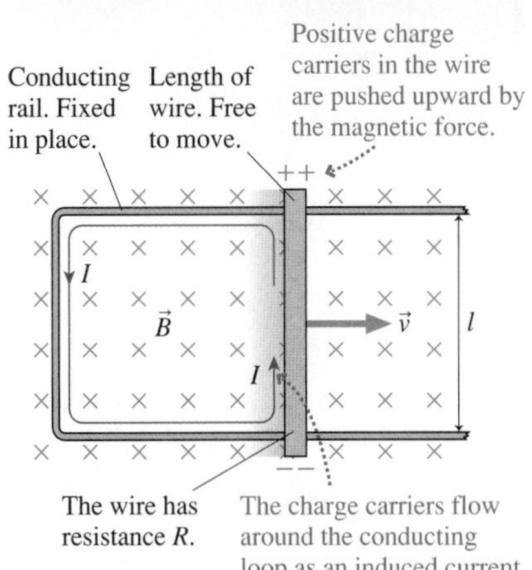

FIGURE 25.4 A current is induced in the circuit as the wire moves through a magnetic field.

FIGURE 25.5 A pulling force is needed to move the wire to the right.

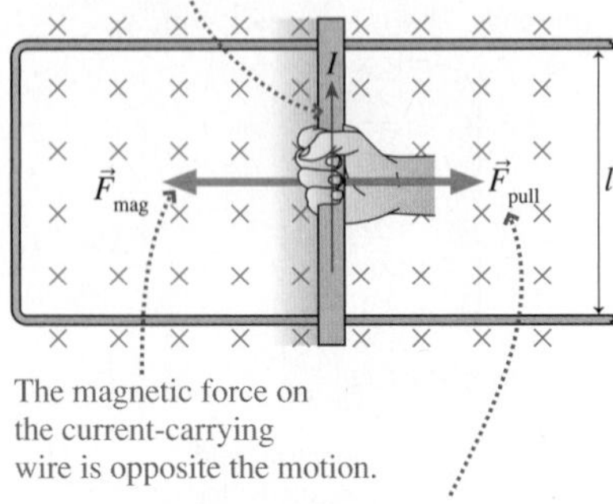
The induced current flows through the moving wire.
I
$\vec{F}_{mag}$
$\vec{F}_{pull}$
l
The magnetic force on the current-carrying wire is opposite the motion.
A pulling force to the right must balance the magnetic force to keep the wire moving at constant speed.

resistance of the circuit is just the resistance R of the moving wire, so the induced current is given by Ohm's law:

$$I = \frac{\mathcal{E}}{R} = \frac{vlB}{R} \tag{25.4}$$

We've assumed that the wire is moving along the rail at constant speed. But we must apply a continuous pulling force $\vec{F}_{pull}$ to make this happen; FIGURE 25.5 shows why. The moving wire, which now carries induced current I, is in a magnetic field. You learned in Chapter 24 that a magnetic field exerts a force on a current-carrying wire. According to the right-hand rule, the magnetic force $\vec{F}_{mag}$ on the moving wire points to the left. This "magnetic drag" will cause the wire to slow down and stop *unless* we exert an equal but opposite pulling force $\vec{F}_{pull}$ to keep the wire moving.

NOTE ▶ Think about this carefully. As the wire moves to the right, the magnetic force $\vec{F}_B$ pushes the charge carriers *parallel* to the wire. Their motion, as they continue around the circuit, is the induced current I. Now, because we have a current, a second magnetic force $\vec{F}_{mag}$ enters the picture. This force on the current is *perpendicular* to the wire and acts to slow the wire's motion. ◀

The magnitude of the magnetic force on a current-carrying wire was found in Chapter 24 to be $F_{mag} = IlB$. Using that result, along with Equation 25.4 for the induced current, we find that the force required to pull the wire with a constant speed v is

$$F_{pull} = F_{mag} = IlB = \left(\frac{vlB}{R}\right)lB = \frac{vl^2B^2}{R} \tag{25.5}$$

Energy Considerations

FIGURE 25.6 Power into and out of an induced-current circuit.

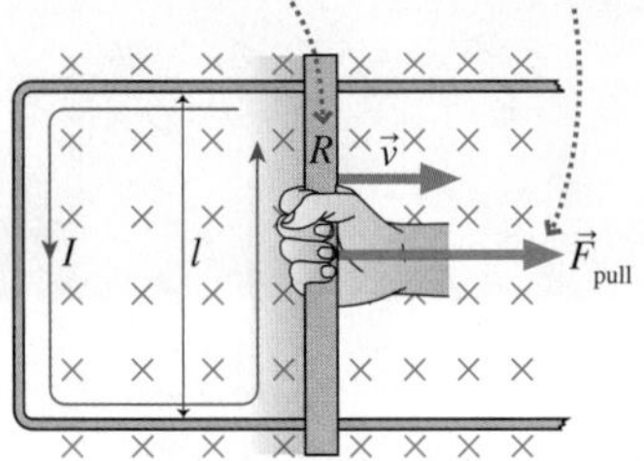
Because there is a current, power is dissipated in the resistance of the rail.
Pulling to the right takes work. This is a power input to the system.
R
$\vec{v}$
I
l
$\vec{F}_{pull}$

FIGURE 25.6 is another look at the wire moving on a conducting rail. Because a force is needed to pull the wire through the magnetic field at a constant speed, we must do work to keep the wire moving. You learned in Chapter 10 that the power exerted by a force pushing or pulling an object with velocity v is $P = Fv$, so the power provided to the circuit by the force pulling on the wire is

$$P_{input} = F_{pull}v = \frac{v^2l^2B^2}{R} \tag{25.6}$$

This is the rate at which energy is added to the circuit by the pulling force.

But the circuit dissipates energy in the resistance of the circuit. You learned in Chapter 22 that the power dissipated by current I as it passes through resistance R is $P = I^2R$. Equation 25.4 for the induced current I gives us the power dissipated by the circuit of Figure 25.6:

$$P_{dissipated} = I^2R = \frac{v^2l^2B^2}{R} \tag{25.7}$$

Equations 25.6 and 25.7 have identical results. This makes sense: The rate at which work is done on the circuit is exactly balanced by the rate at which energy is dissipated. The fact that our final result is consistent with energy conservation is a good check on our work.

Generators

A device that converts mechanical energy to electric energy is called a **generator.** The example of Figure 25.6 is a simple generator, but it is not very practical. Rather than move a straight wire, it's more practical to rotate a coil of wire, as in FIGURE 25.7.

As the coil rotates, the left edge always moves upward through the magnetic field while the right edge always moves downward. The motion of the wires through the magnetic field induces a current to flow as noted in the figure. The induced current is removed from the rotating loop by *brushes* that press up against rotating *slip rings*. The circuit is completed as shown in the figure.

FIGURE 25.7 A generator using a rotating loop of wire.

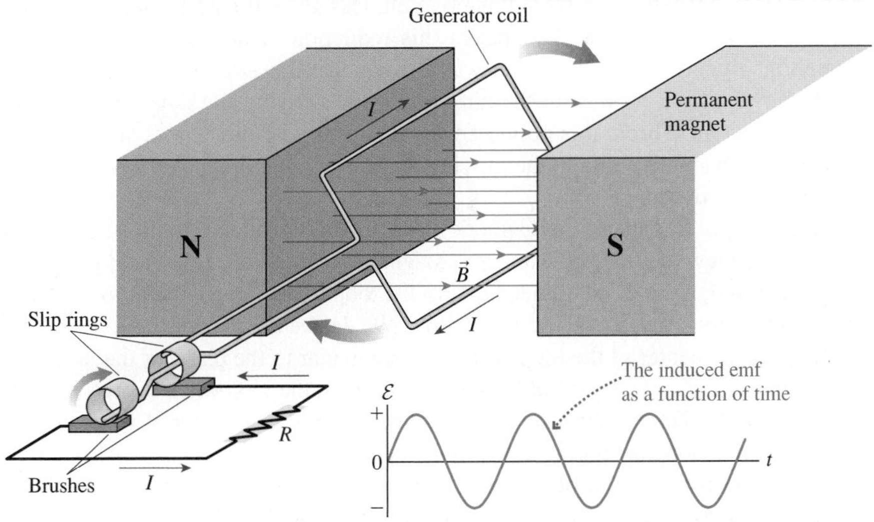

As the coil in the generator of Figure 25.7 rotates, the sense of the emf changes, giving a sinusoidal variation of emf as a function of time. Electricity is produced using generators of this sort; the electricity in your house has a varying voltage. The alternating sign of the voltage produces an *alternating current,* so we call such electricity AC. We'll have more to say about this type of electricity in Chapter 26.

TRY IT YOURSELF

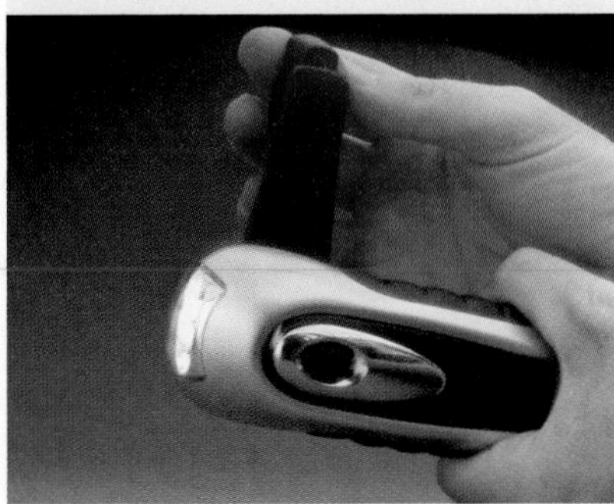

No work, no light Turning the crank on a generator flashlight rotates a coil of wire in the magnetic field of a permanent magnet. With the switch off, there is no current and no drag force; it's easy to turn the crank. Closing the switch allows an induced current to flow through the coil, so the bulb lights. But current in the wire experiences a drag force in the magnetic field, so you must do work to keep the crank turning. This is the source of the output power of the circuit, the light of the bulb. Commercial generators use water flowing through a dam, wind turning a propeller, or turbines spun by expanding steam to rotate much larger coils, but the principle is the same. When you flip a switch, something, somewhere has to do a bit more work to turn a crank.

STOP TO THINK 25.1 A square conductor moves through a uniform magnetic field. Which of the figures shows the correct charge distribution on the conductor?

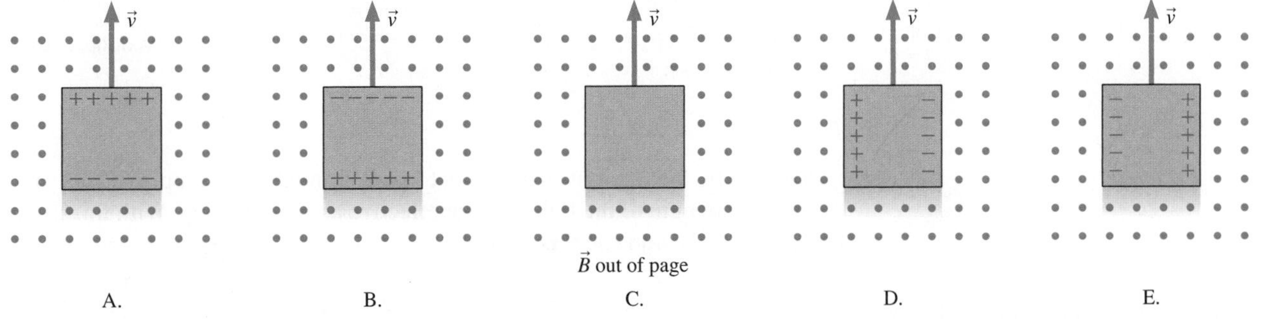

25.3 Magnetic Flux

We've begun our exploration of electromagnetic induction by analyzing a circuit in which one wire moves through a magnetic field. You might be wondering what this has to do with Faraday's discovery. Faraday found that a current is induced when the amount of magnetic field passing through a coil or a loop of wire

FIGURE 25.8 The amount of air flowing through a loop depends on the angle.

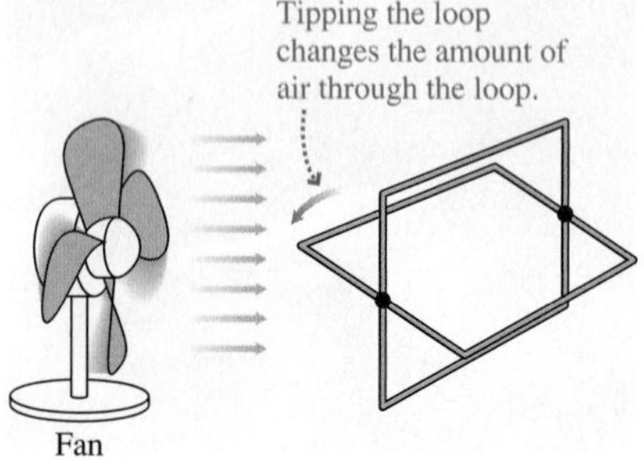
(a) A fan blows air through a loop
Tipping the loop changes the amount of air through the loop.
Fan

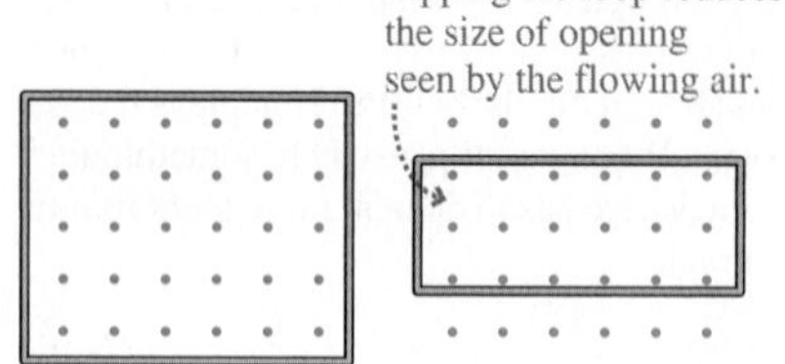
(b) Side view of the air through the loop
Tipping the loop reduces the amount of air that flows through.

(c) Front view of the air through the loop
Tipping the loop reduces the size of opening seen by the flowing air.

changes. But that's exactly what happens as the slide wire moves down the rail in Figure 25.4! As the circuit expands, more magnetic field passes through the larger loop. It's time to define more clearly what we mean by "the amount of field passing through a loop."

Imagine holding a rectangular loop of wire in front of a fan, as shown in FIGURE 25.8a. The arrows represent the flow of the air. If you want to get the most air through the loop, you know that you should hold the loop perpendicular to the direction of the flow. If you tip the loop from this position, less air will pass through the loop. FIGURE 25.8b is a side view that makes this reduction clear—fewer arrows pass through the tipped loop. Yet another way to visualize this situation is FIGURE 25.8c, which shows a front view with the air coming toward you; the dots represent the front of the arrows. From this point of view it's clear why the flow is smaller: The loop presents a smaller area to the moving air. We say that the *effective area* of the loop has been reduced.

We can apply this idea to a magnetic field passing through a loop. FIGURE 25.9a shows a side view of a loop in a uniform magnetic field. To have the most field vectors going through the loop, we need to turn the loop to be perpendicular to the magnetic field vectors, just as we did for airflow. We define the *axis* of the loop to be a line through the center of the loop that is perpendicular to the plane of the loop. We see that the largest number of field vectors go through the loop when its axis is lined up with the field. Tipping the loop by an angle θ reduces the number of vectors passing through the loop, just for the air from the fan.

FIGURE 25.9b is front view of the loop with the dimensions noted. When the loop is tipped by an angle θ, fewer field vectors pass through the loop because the effective area is smaller. We can define the effective area as

$$A_{\text{eff}} = ab\cos\theta = A\cos\theta \tag{25.8}$$

FIGURE 25.9 The amount of magnetic field passing through a loop depends on the angle.

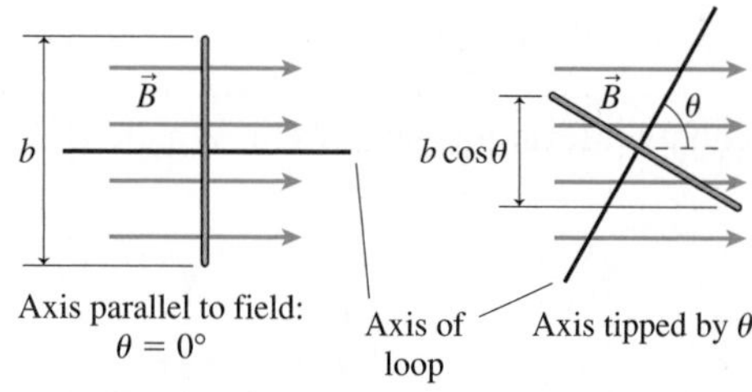
(a) Loop seen from the side
$\vec{B}$
b
Axis parallel to field: $\theta = 0°$
Axis of loop
$b\cos\theta$
$\vec{B}$
θ
Axis tipped by θ

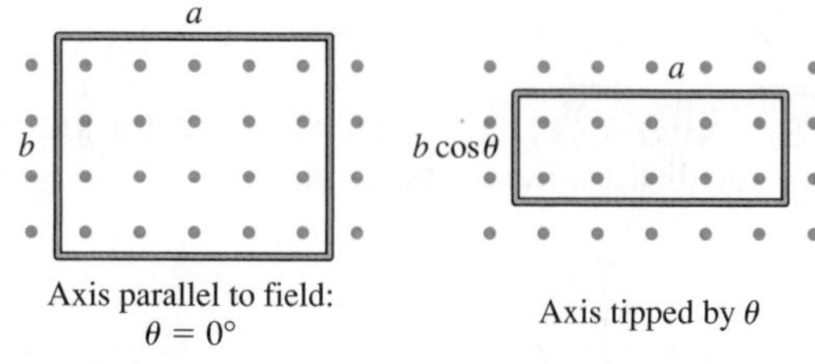
(b) Loop seen looking toward the magnetic field
a
b
Axis parallel to field: $\theta = 0°$
a
$b\cos\theta$
Axis tipped by θ

Ultimately, the amount of field that "goes through" the loop depends on two things: the strength of the field and the effective area of the loop. With this in mind, let's define the **magnetic flux** Φ as

$$\Phi = A_{\text{eff}}B = AB\cos\theta \tag{25.9}$$

Magnetic flux through area A at angle θ to field B

FIGURE 25.10 Definition of magnetic flux.

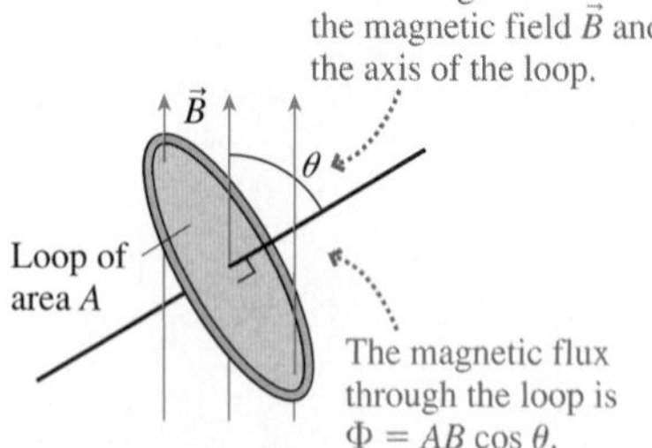
θ is the angle between the magnetic field $\vec{B}$ and the axis of the loop.
$\vec{B}$
θ
Loop of area A
The magnetic flux through the loop is $\Phi = AB\cos\theta$.

The magnetic flux measures the amount of magnetic field passing through a loop of area A if the loop is tilted at angle θ from the field. The SI unit of magnetic flux is the **weber.** From Equation 25.9 you can see that

$$1 \text{ weber} = 1 \text{ Wb} = 1 \text{ T} \cdot \text{m}^2$$

The relationship of Equation 25.9 is illustrated in FIGURE 25.10.

EXAMPLE 25.2 Finding the flux of the earth's field through a vertical loop

At a particular location, the earth's magnetic field is 50 μT tipped at an angle of 60° below horizontal. A 10-cm-diameter circular loop of wire sits flat on a table. What is the magnetic flux through the loop?

PREPARE FIGURE 25.11 shows the loop and the field of the earth. The field is tipped by 60°, so the angle of the field with respect to the axis of the loop is $\theta = 30°$. The radius of the loop is 5.0 cm, so the area of the loop is $A = \pi r^2 = \pi(0.050\text{ m})^2 = 0.0079\text{ m}^2$.

FIGURE 25.11 Finding the flux of the earth's field through a loop.

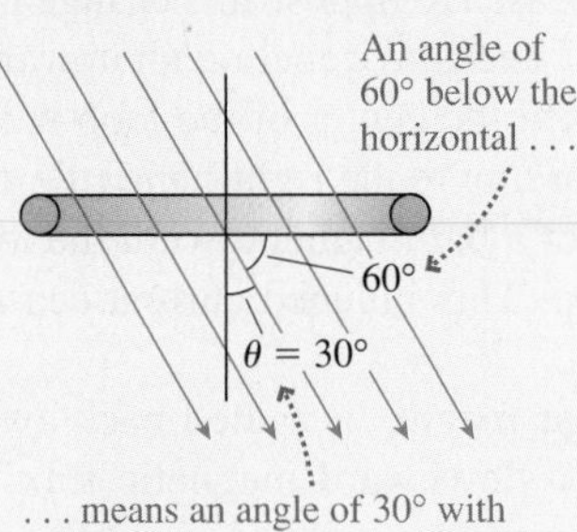
An angle of 60° below the horizontal . . .
60°
$\theta = 30°$
. . . means an angle of 30° with respect to the axis of the loop.

SOLVE The flux through the loop is given by Equation 25.9, with the angle and area as above:

$$\Phi = AB\cos\theta = (0.0079\text{ m}^2)(50 \times 10^{-6}\text{ T})\cos(30°)$$
$$= 3.4 \times 10^{-7}\text{ Wb}$$

ASSESS It's a small loop and a small field, so a very small flux seems reasonable.

Lenz's Law

Some of the induction experiments from earlier in the chapter could be explained in terms of motional emf, but others had no motion. What they all have in common, though, is that one way or another the magnetic flux through the coil or loop *changes*. We can summarize all of the discoveries as follows: **Current is induced in a loop of wire when the magnetic flux through the loop changes.**

For example, a momentary current is induced in the loop of FIGURE 25.12 as the bar magnet is pushed toward the loop because the flux through the loop increases. Pulling the magnet away from the loop, which decreases the flux, causes the current meter to deflect in the opposite direction. How can we predict the *direction* of the current in the loop?

FIGURE 25.12 Pushing a bar magnet toward the loop induces a current in the loop.

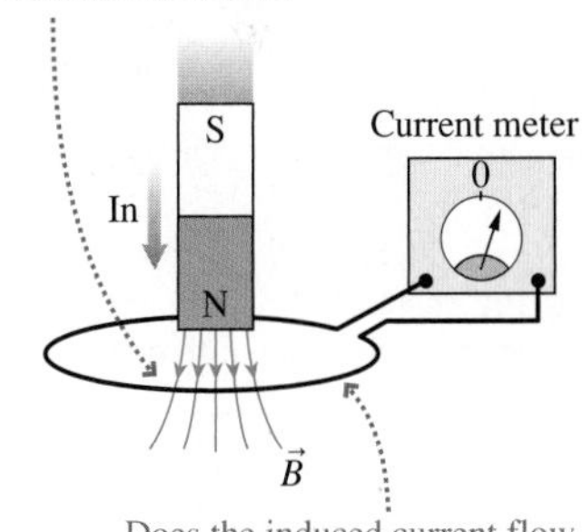
Pushing a bar magnet toward a loop increases the flux through the loop and induces a current to flow.
S
N
In
Current meter
0
$\vec{B}$
Does the induced current flow clockwise or counterclockwise?

The German physicist Heinrich Lenz began to study electromagnetic induction after learning of Faraday's discovery. Lenz developed a rule for determining the direction of the induced current. We now call his rule **Lenz's law,** and it can be stated as follows:

> **Lenz's law** There is an induced current in a closed, conducting loop if and only if the magnetic flux through the loop is changing. The direction of the induced current is such that the induced magnetic field opposes the *change* in the flux.

Lenz's law is rather subtle, and it takes some practice to see how to apply it.

NOTE ▶ One difficulty with Lenz's law is the term "flux," from a Latin root meaning "flow." In everyday language, the word "flux" may imply that something is changing. Think of the phrase "The situation is in flux." In physics, "flux" simply means "passes through." A steady magnetic field through a loop creates a steady, *un*changing magnetic flux. ◀

Lenz's law tells us to look for situations where the flux is *changing*. This can happen in three ways:

1. The magnetic field through the loop changes (increases or decreases).
2. The loop changes in area or angle.
3. The loop moves into or out of a magnetic field.

We can understand Lenz's law this way: If the flux through a loop changes, a current is induced in a loop. That current generates *its own* magnetic field $\vec{B}_{\text{induced}}$. **It is this induced field that opposes the flux change.** Let's look at an example to clarify what we mean by this statement.

FIGURE 25.13 The induced current is counterclockwise.

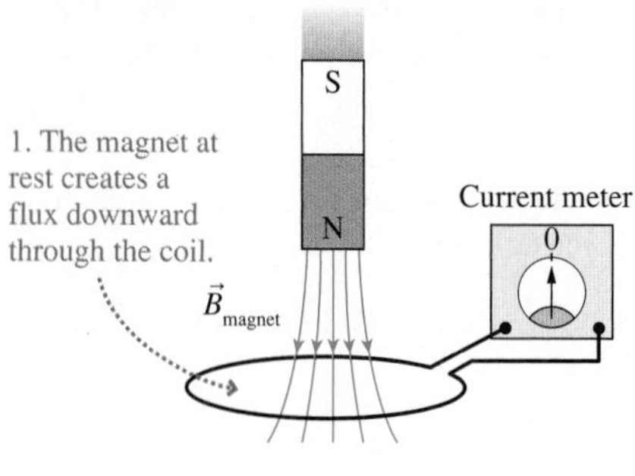
S
N
1. The magnet at rest creates a flux downward through the coil.
Current meter
0
$\vec{B}_{magnet}$

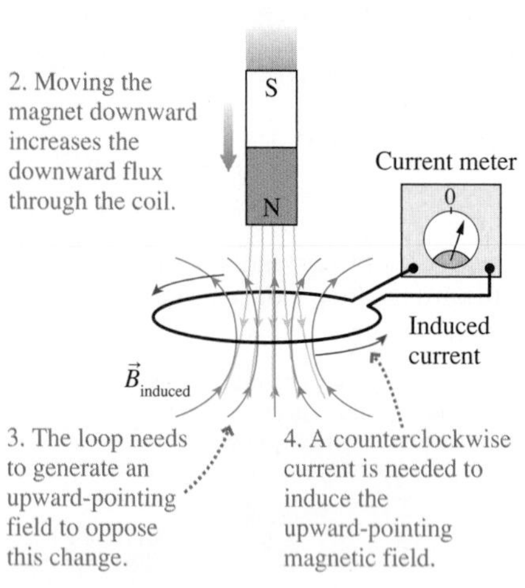
S
N
2. Moving the magnet downward increases the downward flux through the coil.
Current meter
0
Induced current
$\vec{B}_{induced}$
3. The loop needs to generate an upward-pointing field to oppose this change.
4. A counterclockwise current is needed to induce the upward-pointing magnetic field.

FIGURE 25.14 Pulling the magnet away induces a clockwise current.

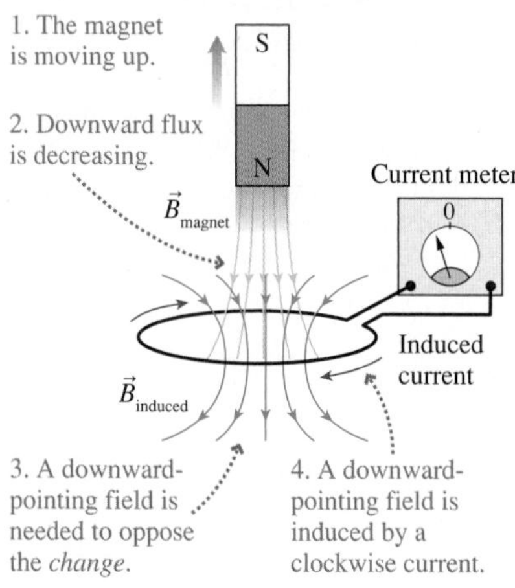
1. The magnet is moving up.
S
N
2. Downward flux is decreasing.
$\vec{B}_{magnet}$
Current meter
0
Induced current
$\vec{B}_{induced}$
3. A downward-pointing field is needed to oppose the *change*.
4. A downward-pointing field is induced by a clockwise current.

The top part of **FIGURE 25.13** shows a magnet at rest above a coil of wire. The field of the magnet creates a downward flux through the loop. In the lower part of the figure, the magnet is moving toward the loop. This causes the downward magnetic flux through the loop to increase. To oppose this change in flux, which is what Lenz's law requires, the loop itself needs to generate an upward-pointing magnetic field. The induced magnetic field at the center of the loop will point upward if the current is counterclockwise, according to the right-hand rule you learned in Chapter 24. Thus pushing the north end of a bar magnet toward the loop induces a counterclockwise current around the loop. This induced current ceases as soon as the magnet stops moving.

Now suppose the bar magnet is pulled back away from the loop, as shown in **FIGURE 25.14**. There is a downward magnetic flux through the loop, but the flux *decreases* as the magnet moves away. According to Lenz's law, the induced magnetic field of the loop will *oppose this decrease.* To do so, the induced field needs to point in the *downward* direction, as shown in Figure 25.14. Thus as the magnet is withdrawn, the induced current is clockwise, opposite the induced current of Figure 25.13.

NOTE ► Notice that the magnetic field of the bar magnet is pointing downward in both Figures 25.13 and 25.14. It is not the *flux* due to the magnet that the induced current opposes, but the *change* in the flux. This is a subtle but critical distinction. When the field of the magnet points down and is increasing, the induced current opposes the increase by generating an upward field. When the field of the magnet points down but is decreasing, the induced current opposes the decrease by generating a downward field. ◄

TACTICS BOX 25.1 Using Lenz's law

1. Determine the direction of the applied magnetic field. The field must pass through the loop.
2. Determine how the flux is changing. Is it increasing, decreasing, or staying the same?
3. Determine the direction of an induced magnetic field that will oppose the *change* in the flux:
   - Increasing flux: The induced magnetic field points opposite the applied magnetic field.
   - Decreasing flux: The induced magnetic field points in the same direction as the applied magnetic field.
   - Steady flux: There is no induced magnetic field.
4. Determine the direction of the induced current. Use the right-hand rule to determine the current direction in the loop that generates the induced magnetic field you found in step 3.

Exercises 9–12 

EXAMPLE 25.3 Applying Lenz's law 1

The switch in the top circuit of **FIGURE 25.15** on the next page has been closed for a long time. What happens in the lower loop when the switch is opened?

PREPARE The current in the upper loop creates a magnetic field. This magnetic field produces a flux through the lower loop. When you open a switch, the current doesn't immediately drop to zero; it falls off over a short time. As the current changes in the upper loop, the flux in the lower loop changes.

SOLVE **FIGURE 25.16** on the next page shows the four steps of using Lenz's law to find the current in the lower loop. Opening the switch induces a counterclockwise current in the lower loop. This is a momentary current, lasting only until the magnetic field of the upper loop drops to zero.

ASSESS The induced current is in the same direction as the original current. This makes sense, because the induced current is opposing the change, a decrease in the current.

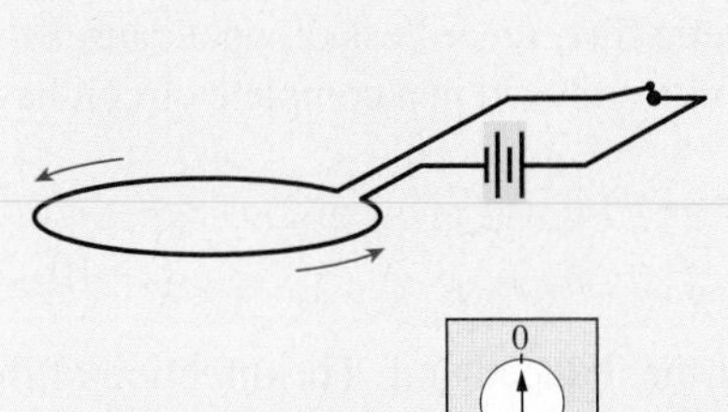

FIGURE 25.15 Circuits for Example 25.3.

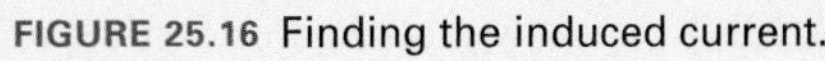

FIGURE 25.16 Finding the induced current.

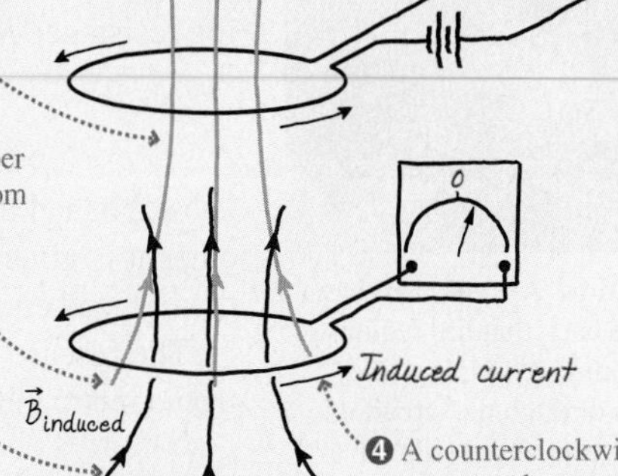

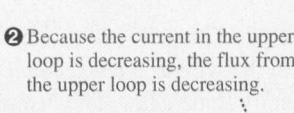

EXAMPLE 25.4 Applying Lenz's law 2

A loop is moved toward a current-carrying wire as shown in FIGURE 25.17. As the wire is moving, is there a clockwise current around the loop, a counterclockwise current, or no current?

FIGURE 25.17 The moving loop.

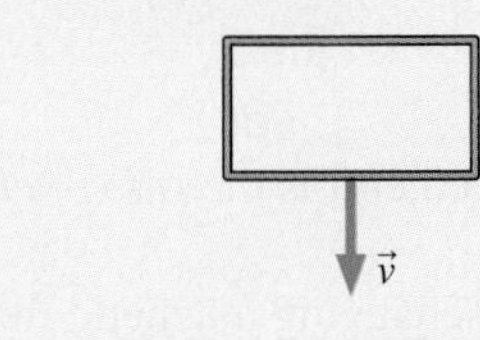

PREPARE FIGURE 25.18 shows that the magnetic field above the wire points into the page. We learned in Chapter 24 that the magnetic field of a straight, current-carrying wire is proportional to $1/r$, where r is the distance away from the wire, so the field is stronger closer to the wire.

SOLVE As the loop moves toward the wire, the flux through the loop increases. To oppose the *change* in the flux—the increase into the page—the magnetic field of the induced current must point out of the page. Thus, according to the right-hand rule, a counterclockwise current is induced, as shown in Figure 25.18.

FIGURE 25.18 The motion of the loop changes the flux through the loop and induces a current.

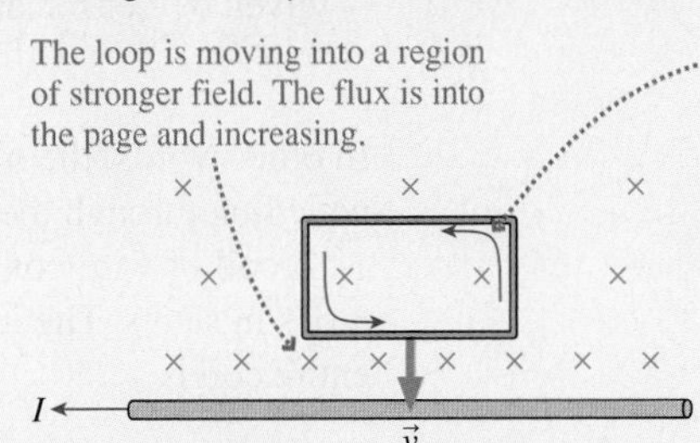

The loop is moving into a region of stronger field. The flux is into the page and increasing.
The induced current must create a magnetic field out of the page to oppose the change, so the right-hand rule tells us that the induced current is counterclockwise.
I
$\vec{v}$

ASSESS We could have solved this problem using the concept of motional emf, but treating it as a flux-change problem is more straightforward. The loop moves into a region of stronger field. To oppose the increasing flux, the induced field should be opposite the existing field, so our answer makes sense.

STOP TO THINK 25.2 As a coil moves to the right at constant speed, it passes over the north pole of a magnet and then moves beyond it. Which graph best represents the current in the loop for the time of the motion? A counterclockwise current as viewed from above the loop is a positive current, clockwise is a negative current.

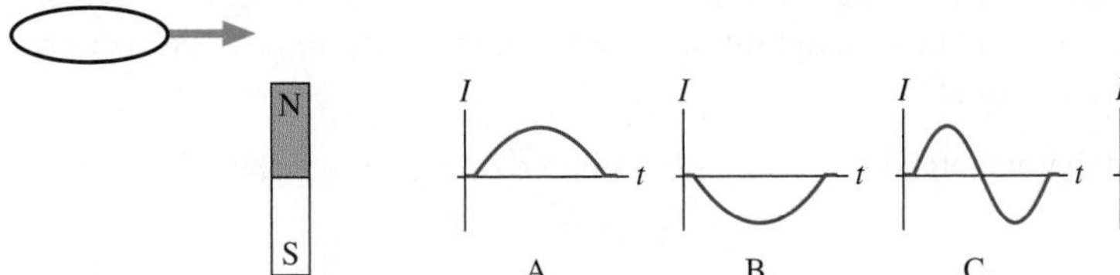

25.4 Faraday's Law

Faraday discovered that a current is induced when the magnetic flux through a conducting loop changes. Lenz's law allows us to find the direction of the induced current. To put electromagnetic induction to practical use, we also need to know the *size* of the induced current.

13.9–13.10

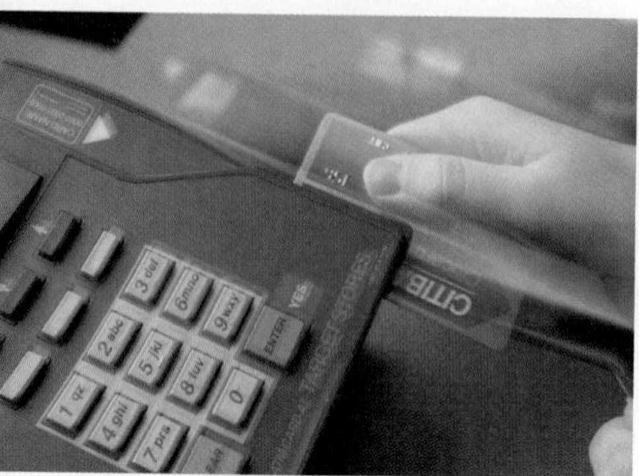

Keep the (flux) change A credit card has a magnetic strip on the back that has regions of alternating magnetization. "Swiping" the card moves this strip through the circuit of the reader. Different parts of the strip have different fields, so the motion of the strip causes a series of flux changes that induce currents. The pattern of magnetization of the card determines the pattern of the induced currents—and so the card is "read."

In the preceding examples, a change in flux caused a current to flow in a loop of wire. But we know that charges don't start moving spontaneously. A current requires an emf to provide the energy. There *must* be an emf in these circuits, even though the mechanism for this emf is not yet clear.

The emf associated with a changing magnetic flux, regardless of what causes the change, is called an **induced emf** $\mathcal{E}$. If this emf is induced in a complete circuit having resistance R, a current

$$I_{\text{induced}} = \frac{\mathcal{E}}{R} \tag{25.10}$$

is established in the wire as a *consequence* of the induced emf. The direction of the current is given by Lenz's law. The last piece of information we need is the size of the induced emf $\mathcal{E}$.

The research of Faraday and others led to the discovery of the basic law of electromagnetic induction, which we now call **Faraday's law.**

Faraday's law An emf $\mathcal{E}$ is induced in a conducting loop if the magnetic flux through the loop changes. If the flux changes by $\Delta\Phi$ during time interval Δt, the magnitude of the emf is

$$\mathcal{E} = \left|\frac{\Delta\Phi}{\Delta t}\right| \tag{25.11}$$

and the direction of the emf is such as to drive an induced current in the direction given by Lenz's law.

In other words, the magnitude of the induced emf is the *rate of change* of the magnetic flux through the loop.

A coil of wire consisting of N turns in a changing magnetic field acts like N batteries in series. The induced emf of each of the coils adds, so the induced emf of the entire coil is

$$\mathcal{E}_{\text{coil}} = N\left|\frac{\Delta\Phi_{\text{per coil}}}{\Delta t}\right| \tag{25.12}$$

Faraday's law allows us to find the *magnitude* of induced emfs and currents; Lenz's law allows us to determine the *direction.*

Using Faraday's Law

Most electromagnetic induction problems can be solved with a three-step strategy.

PROBLEM-SOLVING STRATEGY 25.1 Electromagnetic induction

PREPARE Make simplifying assumptions about wires and magnetic fields. Draw a picture or a circuit diagram. Use Lenz's law to determine the direction of the induced current.

SOLVE The mathematical representation is based on Faraday's law

$$\mathcal{E} = \left|\frac{\Delta\Phi}{\Delta t}\right|$$

For an N-turn coil, multiply by N. The size of the induced current is $I = \mathcal{E}/R$.

ASSESS Check that your result has the correct units, is reasonable, and answers the question.

Exercise 15

Let's return to the situation of Figure 25.4, where a wire moves through a magnetic field by sliding on a U-shaped conducting rail. We looked at this problem as an example of motional emf; now, let's look at it using Faraday's law.

EXAMPLE 25.5 Finding the emf using Faraday's law

FIGURE 25.19 shows a wire of resistance R sliding on a U-shaped conducting rail. Assume that the conducting rail is an ideal wire. Use Faraday's law and the steps of Problem-Solving Strategy 25.1 to derive an expression for the current in the wire.

FIGURE 25.19 A wire sliding on a rail.

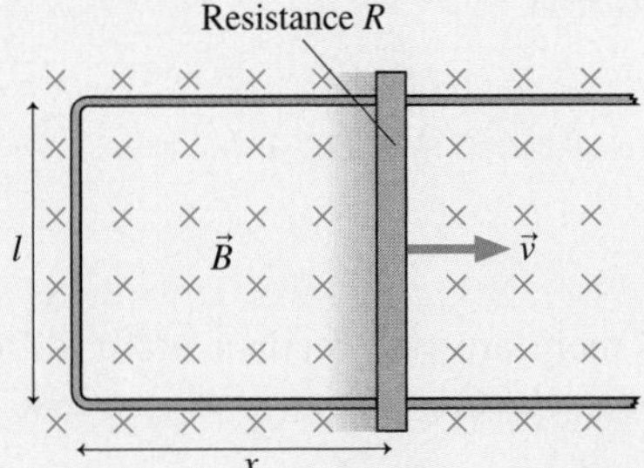

PREPARE **FIGURE 25.20** shows the current loop formed by the wire and the rail. Even though the magnetic field is constant, the flux is changing because the loop is increasing in area. The flux is into the loop and increasing. According to Lenz's law, the induced current must be counterclockwise so as to oppose the change, because the induced magnetic field must be out of the loop.

FIGURE 25.20 Induced current in the sliding wire.

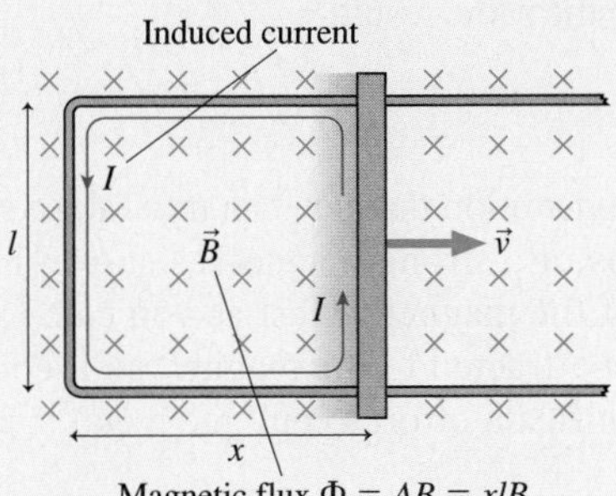

SOLVE The magnetic field $\vec{B}$ is perpendicular to the plane of the loop, so $\theta = 0°$ and the magnetic flux is $\Phi = AB$, where A is the area of the loop. If the sliding wire is distance x from the end, as in Figure 25.20, the area of the loop is $A = xl$ and the flux at that instant of time is

$$\Phi = AB = xlB$$

The flux through the loop increases as the wire moves and x increases. This flux change induces an emf, according to Faraday's law, so we write

$$\mathcal{E} = \left|\frac{\Delta\Phi}{\Delta t}\right| = \left|\frac{\Delta(AB)}{\Delta t}\right| = \left|\frac{\Delta(xlB)}{\Delta t}\right|$$

The only quantity in the final ratio that is changing is the position x, so we can write

$$\mathcal{E} = \left|\frac{\Delta(xlB)}{\Delta t}\right| = lB\left|\frac{\Delta x}{\Delta t}\right|$$

But $|\Delta x/\Delta t|$ is the wire's speed v, so the induced emf is

$$\mathcal{E} = vlB$$

The wire and the loop have a total resistance R, thus the magnitude of the induced current is

$$I = \frac{\mathcal{E}}{R} = \frac{vlB}{R}$$

ASSESS This is exactly the same result we found in Section 25.2, where we analyzed this situation by considering the force on moving charge carriers. This is a good check on our work, and a nice connection between the ideas of motional emf and Faraday's law.

EXAMPLE 25.6 Finding the induced current in a circular loop

A patient having an MRI scan has neglected to remove a copper bracelet. The bracelet is 6.0 cm in diameter and has a resistance of 0.010 Ω. The magnetic field in the MRI solenoid is directed along the person's body from head to foot; her bracelet is perpendicular to $\vec{B}$. As a scan is taken, the magnetic field in the solenoid decreases from 1.00 T to 0.40 T in 1.2 s. What are the magnitude and direction of the current induced in the bracelet?

PREPARE We follow the steps in Problem-Solving Strategy 25.1, beginning with a sketch of the situation. **FIGURE 25.21** shows the bracelet and the applied field looking down along the patient's body. The field is directed down through the loop; as the applied field decreases, the flux into the loop decreases. To oppose the decreasing flux, as required by Lenz's law, the field from the induced current must be in the direction of the applied field. Thus, from the right-hand rule, the induced current in the bracelet must be clockwise.

FIGURE 25.21 A circular conducting loop in a decreasing magnetic field.

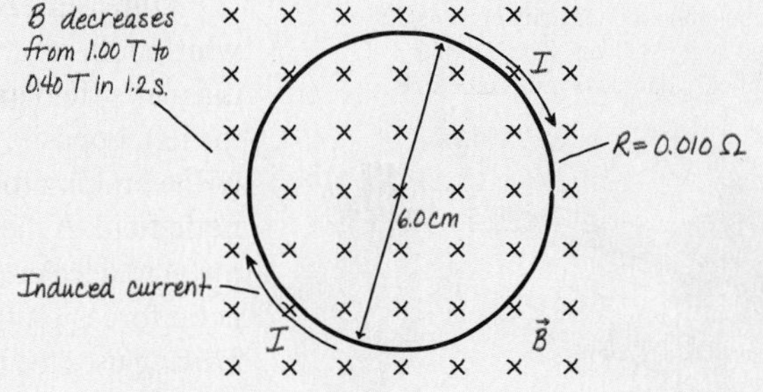

Continued

SOLVE The magnetic field is perpendicular to the plane of the loop; hence $\theta = 0°$ and the magnetic flux is $\Phi = AB = \pi r^2 B$. The area of the loop doesn't change with time, but B does, so $\Delta\Phi = \Delta(AB) = A\Delta B = \pi r^2 \Delta B$. The change in the magnetic field during $\Delta t = 1.2$ s is $\Delta B = 0.40\text{ T} - 1.00\text{ T} = -0.60\text{ T}$. According to Faraday's law, the magnitude of the induced emf is

$$\mathcal{E} = \left|\frac{\Delta\Phi}{\Delta t}\right| = \pi r^2 \left|\frac{\Delta B}{\Delta t}\right| = \pi(0.030\text{ m})^2 \left|\frac{-0.60\text{ T}}{1.2\text{ s}}\right|$$

$$= \pi(0.030\text{ m})^2(0.50\text{ T/s}) = 0.0014\text{ V}$$

The current induced by this emf is

$$I = \frac{\mathcal{E}}{R} = \frac{0.0014\text{ V}}{0.010\ \Omega} = 0.14\text{ A}$$

The decreasing magnetic field causes a 0.14 A clockwise current during the 1.2 s that the field is decreasing.

ASSESS The emf is quite small, but, because the resistance of the metal bracelet is also very small, the current is respectable. We know that electromagnetic induction produces currents large enough for practical applications, so this result seems plausible. The induced current could easily distort the readings of the MRI machine, and a larger current could cause enough heating to be potentially dangerous. For these and other reasons, operators are careful to have patients remove all metal before an MRI scan.

FIGURE 25.22 Eddy currents.

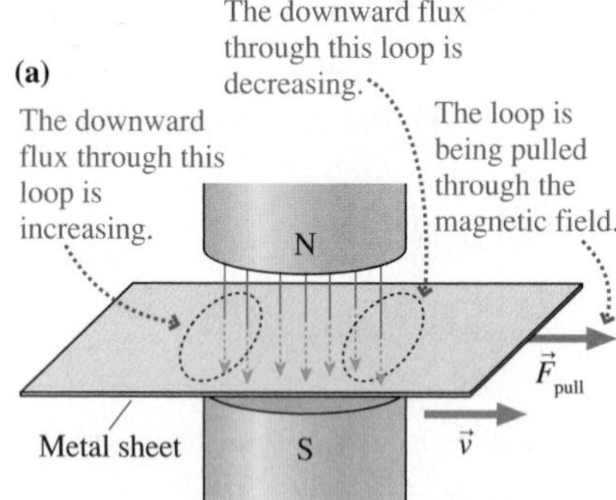

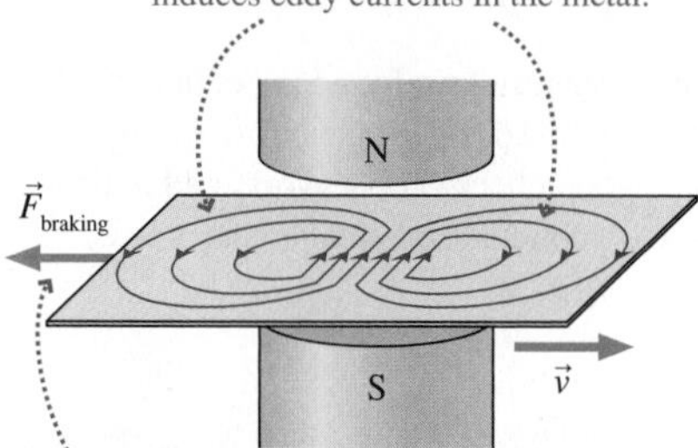

FIGURE 25.23 Eddy-current damping in a balance.

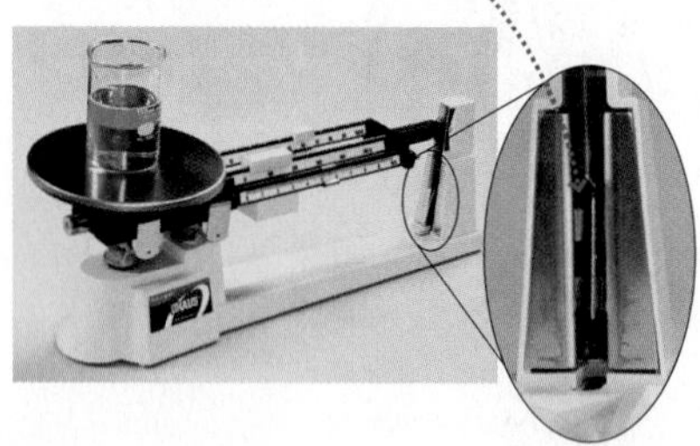

Strong magnets on either side of an aluminum vane induce eddy currents in the vane as it oscillates. The braking force quickly dampens out oscillations.

As these two examples show, there are two fundamentally different ways to change the magnetic flux through a conducting loop:

1. The loop can move or expand or rotate, creating a motional emf.
2. The magnetic field can change.

Faraday's law tells us that the induced emf is simply the rate of change of the magnetic flux through the loop, *regardless* of what causes the flux to change. Motional emf is included within Faraday's law as one way of changing the flux, but any other way of changing the flux will have the same result.

Eddy Currents

Here is a remarkable physics demonstration that you can try: Take a sheet of copper and place it between the pole tips of a strong magnet, as shown in FIGURE 25.22a. Now, pull the copper sheet out of the magnet as fast as you can. Copper is not a magnetic material and thus is not attracted to the magnet, so it comes as quite a surprise to find that it takes a significant effort to pull the metal through the magnetic field.

Let's analyze this situation to discover the origin of the force. Figure 25.22a shows two "loops" lying entirely inside the metal sheet. The loop on the right is leaving the magnetic field, and the flux through it is decreasing. According to Faraday's law, the flux change will induce a current to flow around this loop, just as in a loop of wire, even though this current does not have a wire to define its path. As a consequence, a clockwise—as given by Lenz's law— "whirlpool" of current begins to circulate in the metal, as shown in FIGURE 25.22b. Similarly, the loop on the left is entering the field, and the flux through it is increasing. Lenz's law requires this whirlpool of current to circulate the opposite way. These spread-out whirlpools of induced current in a solid conductor are called **eddy currents.**

Figure 25.22b shows the direction of the eddy currents. Notice that both whirlpools are moving in the same direction as they pass through the magnet. The magnet's field exerts a force on this current. By the right-hand rule, this force is to the left, opposite the direction of the pull, and thus it acts as a *braking* force. Because of the braking force, **an external force is required to pull a metal through a magnetic field.** If the pulling force ceases, the magnetic braking force quickly causes the metal to decelerate until it stops. No matter which way the metal is moved, the magnetic forces on the eddy currents act to oppose the motion of the metal. *Magnetic braking* uses the braking force associated with eddy currents to slow trains and transit-system vehicles. It can also provide valuable vibration damping on a much smaller scale, as we see in FIGURE 25.23.

Eddy currents can also be induced by changing fields; this has practical applications as well. In a technique known as *transcranial magnetic stimulation* (TMS), a large oscillating magnetic field is applied to the head via a current-carrying coil. FIGURE 25.24 illustrates how this field produces small eddy currents that stimulate neurons in the tissue of the brain. This produces a short-term inhibitory effect on the neurons in the stimulated region that can produce long-term clinical effects. The technique is also useful in research. By inhibiting the action of specific regions of the brain, researchers can determine the importance of these regions to certain perceptions or tasks.

FIGURE 25.24 Transcranial magnetic stimulation.

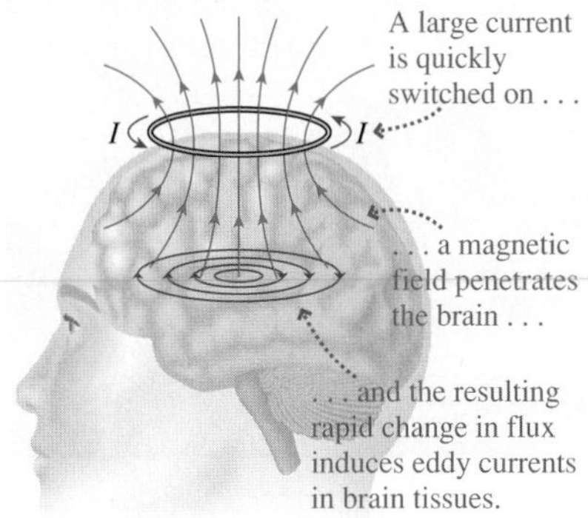

A large current is quickly switched on . . .
I
I
. . . a magnetic field penetrates the brain . . .
. . . and the resulting rapid change in flux induces eddy currents in brain tissues.

25.5 Induced Fields and Electromagnetic Waves

We will start this section with the puzzle shown in FIGURE 25.25. A long, tightly wound solenoid of radius r_1 passes through the center of a conducting loop having a larger radius r_2. The solenoid carries a current and generates a magnetic field. What happens to the loop if the solenoid current changes?

You learned in Chapter 24 that the magnetic field is strong inside a long solenoid but essentially zero outside. Even so, changing the field inside the solenoid causes the flux through the loop to change, so our theory predicts an induced current in the loop. Indeed, you would find an induced current if you did this experiment.

But the loop is completely outside the solenoid, where the magnetic field is zero. How can the charge carriers in the conducting loop possibly "know" that the magnetic field inside the solenoid is changing?

FIGURE 25.25 A changing current in the solenoid induces a current in the loop.

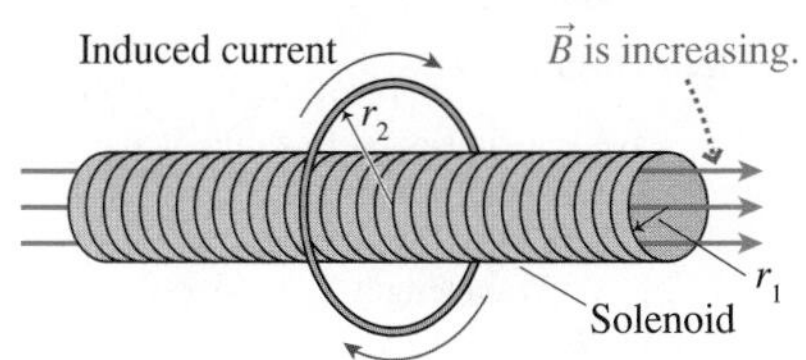

Induced Electric Fields

In order to answer this question, we will first consider another related question: When a changing flux through a loop induces a current, what actually *causes* the current? What *force* pushes the charges around the loop against the resistive forces of the metal? When we considered currents in Chapter 22, it was an *electric field* that moved charges through a conductor. Somehow, changing a magnetic field must create an electric field.

In fact, a changing magnetic field *does* cause what we call an **induced electric field.** FIGURE 25.26a shows a conducting loop in an increasing magnetic field. According to Lenz's law, there is an induced current in the counterclockwise direction. Something has to act on the charge carriers to make them move, so we can infer that the current is produced by an induced electric field tangent to the loop at all points. The induced electric field is the *mechanism* we were seeking that creates a current when there is a changing magnetic field inside a stationary loop.

But the induced electric field exists whether there is a conducting loop or not. The space in which the magnetic field is changing is filled with the pinwheel pattern of induced electric fields shown in FIGURE 25.26b. Charges will move if a conducting path is present, but the induced electric field is there as a direct consequence of the changing magnetic field, whether a conducting path is present or not.

Making this connection between electric and magnetic fields has brought together all of the pieces we have explored so far in this chapter. But there's more to the story of induced fields. Faraday's discovery was the source of inspiration for further investigations that had very far-reaching and practical implications.

FIGURE 25.26 An induced electric field creates a current in the loop.

(a)

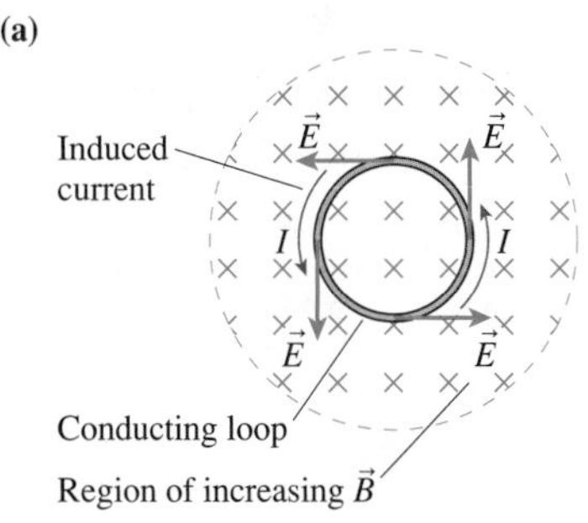

(b)

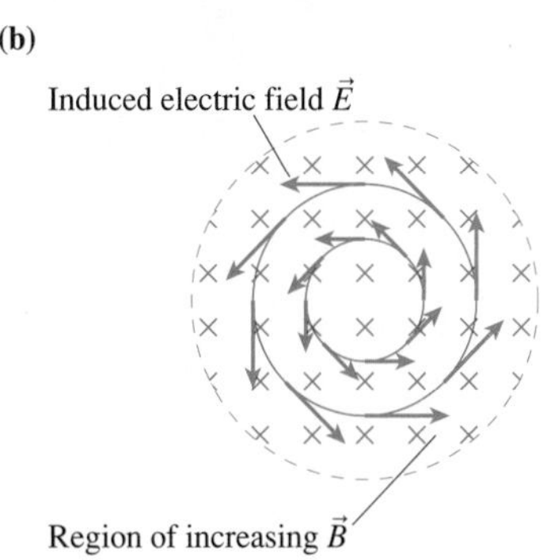

Maxwell's Theory

The development of a theory of electricity and magnetism continued with work by the Scottish physicist James Clerk Maxwell, who asked the following question: What about a changing *electric* field? Could this induce a *magnetic* field? Maxwell thought the symmetry was compelling. He proposed that a changing electric field creates an **induced magnetic field.**

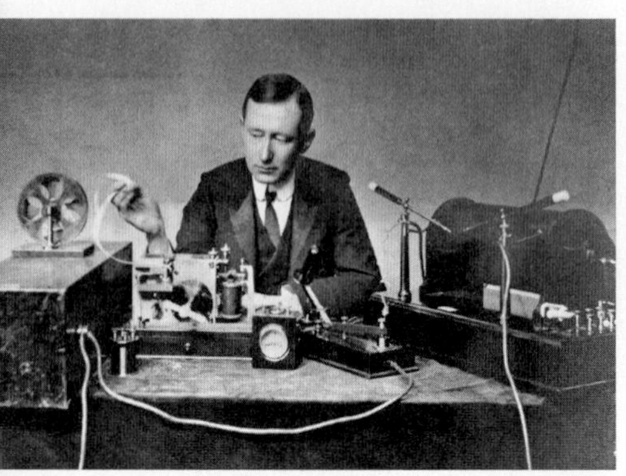

◀ **The grandfather of the cell phone** In this photo, Guglielmo Marconi, who is credited with the development of radio communication, or "wireless telegraphy," sits surrounded by high-voltage equipment used in his experiments in the late 1800s. The German scientist Heinrich Hertz discovered that a spark in one circuit emitted a disturbance that affected a second circuit at some distance. Soon after, Marconi developed the technology for producing and detecting these *electromagnetic waves,* transmitting signals, with no wires, over long distances—something we have come to take for granted.

If a changing magnetic field can induce an electric field in the absence of any charges, and a changing electric field can induce a magnetic field in the absence of any currents, Maxwell soon realized that it would be possible to establish self-sustaining electric and magnetic fields independent of any charges or currents. That is, a changing electric field $\vec{E}$ creates a magnetic field $\vec{B}$, which then changes in just the right way to recreate the electric field, which then changes in just the right way to again recreate the magnetic field, with the fields continuously recreated through electromagnetic induction.

Maxwell was able to show that these electric and magnetic fields would be able to sustain themselves, free from charges and currents, if they took the form of an **electromagnetic wave.** The wave would have to have a very specific geometry, shown in FIGURE 25.27a, in which $\vec{E}$ and $\vec{B}$ are perpendicular to each other as well as perpendicular to the direction of travel. FIGURE 25.27b shows a right-hand rule for determining the relative orientations of the fields and the velocity, reminiscent of the right-hand rule for forces from Chapter 24.

FIGURE 25.27 A sinusoidal electromagnetic wave.

(a) Electromagnetic wave

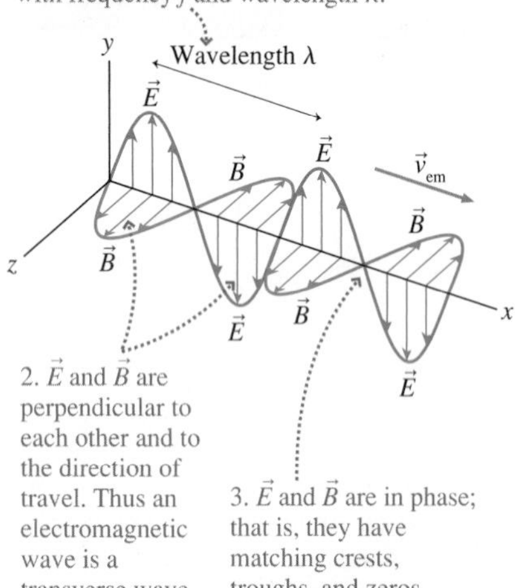
1. The wave is a sinusoidal traveling wave, with frequency f and wavelength λ.
Wavelength λ
y
z
x
$\vec{E}$
$\vec{B}$
$\vec{v}_{em}$
2. $\vec{E}$ and $\vec{B}$ are perpendicular to each other and to the direction of travel. Thus an electromagnetic wave is a transverse wave.
3. $\vec{E}$ and $\vec{B}$ are in phase; that is, they have matching crests, troughs, and zeros.

(b) Right-hand rule for electromagnetic waves

Spread the fingers of your right hand so that your index finger, thumb, and middle finger point out from your hand as shown. Your thumb, index, and middle fingers give the directions of $\vec{v}$, $\vec{E}$, and $\vec{B}$, as shown.

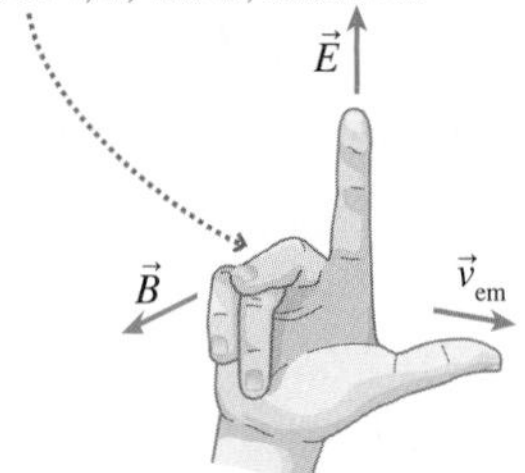
$\vec{E}$
$\vec{B}$
$\vec{v}_{em}$

Maxwell's theory predicted that an electromagnetic wave would travel with speed

$$v_{em} = \frac{1}{\sqrt{\epsilon_0 \mu_0}} \tag{25.13}$$

where ϵ_0 and μ_0 are the permittivity and permeability constants from our expressions for electric and magnetic fields. Maxwell computed that an electromagnetic wave would travel with speed $v_{em} = 3.00 \times 10^8$ m/s.

This is a value you have seen before—it is the speed of light. Making a bold leap of imagination, Maxwell concluded that **light is an electromagnetic wave.** We studied the wave properties of light in Part V, but at that time we were not able to determine just what is "waving." Now we know—light is a wave of electric and magnetic fields.

Maxwell was able to establish the following properties of electromagnetic waves:

- An electromagnetic wave is a *transverse* wave, as shown in Figure 25.27a.
- Electromagnetic waves can exist at any frequency, not just the frequencies of visible light.
- In a vacuum, all electromagnetic waves travel with the same speed, a speed that we call the *speed of light,* for which we use the symbol c.
- At any point on the wave, the electric and magnetic field strengths are related by $E = cB$.

Figure 25.27a shows the values of the electric and magnetic fields at points along a single line, the x-axis.

NOTE ▶ An $\vec{E}$ vector pointing in the y-direction says that *at that point* on the x-axis, where the vector's tail is, the electric field points in the y-direction and has a certain strength. Nothing is "reaching" to a point in space above the x-axis. ◀

Suppose this electromagnetic wave is a *plane wave* traveling in the x-direction. Recall from Chapter 15 that the displacement of a plane wave is the same at *all points* in any plane perpendicular to the direction of motion. In this case, the fields are the same in any yz-plane. If you were standing on the x-axis as the wave moves toward you, the electric and magnetic fields would vary as in the series of pictures of FIGURE 25.28 on the next page. The $\vec{E}$ and $\vec{B}$ fields at each point in the yz-plane oscillate in time, but they are always synchronized with all the other points in the plane. As the plane wave passed you, you would see a uniform oscillation of the $\vec{E}$ and $\vec{B}$ fields of the wave.

FIGURE 25.28 The fields of an electromagnetic plane wave moving toward you, shown every one-eighth period for half a cycle.

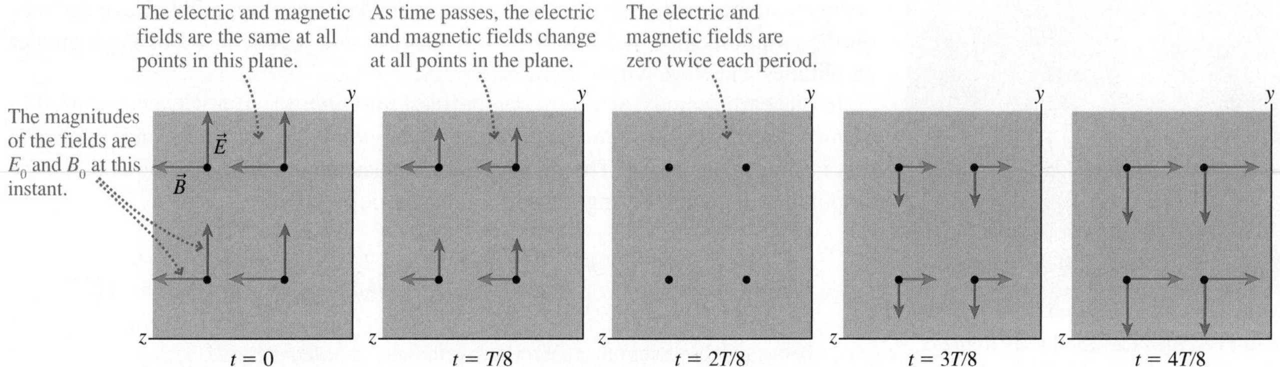

We can adapt our equation for traveling waves from Chapter 15 to electromagnetic waves. If a plane electromagnetic wave moves in the x-direction with the electric field along the y-axis, then the magnetic field is along the z-axis. The equations for the electric and magnetic fields of a wave with wavelength λ and period T are

$$E_y = E_0 \sin\left(2\pi\left(\frac{x}{\lambda} - \frac{t}{T}\right)\right) \qquad B_z = B_0 \sin\left(2\pi\left(\frac{x}{\lambda} - \frac{t}{T}\right)\right) \tag{25.14}$$

E_0 and B_0 are the amplitudes of the oscillating fields. The amplitudes of the fields must have a particular relationship:

$$E_0 = cB_0 \tag{25.15}$$

Relationship between field amplitudes for an electromagnetic wave

STOP TO THINK 25.3 An electromagnetic wave is traveling in the direction shown. What is the direction of the magnetic field at this instant?

A. Toward the top of the page
B. Toward the bottom of the page
C. To the left
D. To the right
E. Into the page
F. Out of the page

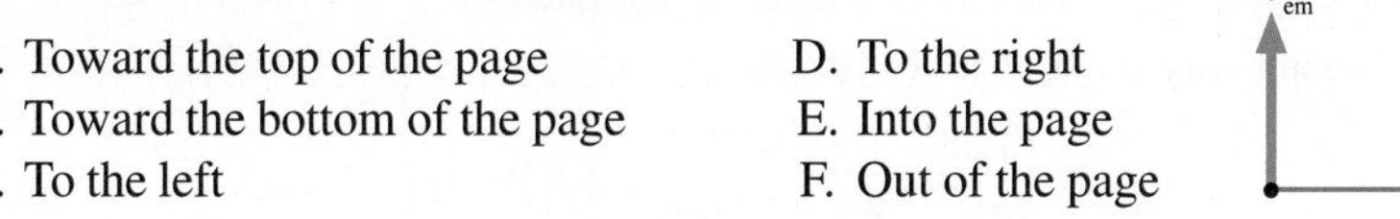

25.6 Properties of Electromagnetic Waves

Electromagnetic waves are oscillations of the electric and magnetic field, but they are still waves, and all the general principles we have learned about waves apply.

In Chapter 15, we learned that we could characterize sinusoidal waves by their speed, wavelength, and frequency, with these related by the fundamental relationship $v = \lambda f$. All electromagnetic waves move at the speed of light, $v_{em} = c$, so this relationship becomes

$$c = \lambda f \tag{25.16}$$

The *spectrum* of electromagnetic waves ranges from waves of long wavelength and (relatively) low frequency (radio waves and microwaves) to waves of short wavelength and high frequency (visible light, ultraviolet, and x rays), as we saw in Chapter 15. Later in the chapter we will explore the properties of different parts of the electromagnetic spectrum. For now, we will concentrate on physical properties that all electromagnetic waves share.

Energy of Electromagnetic Waves

All waves transfer energy. Ocean waves erode beaches, sound waves set your eardrum to vibrating, and light from the sun warms the earth. In all of these cases, the waves carry energy from the point where they are emitted to another point where

A field-fired furnace The energy from the sun is carried through space by electromagnetic waves—that is, it is carried by electric and magnetic fields. The mirrors of this solar furnace in the Pyrenees in southern France concentrate the electromagnetic waves from the sun to an intensity 1000 times that of normal sunlight, allowing researchers to test the high-temperature properties of materials at up to 3800°C.

their energy is transferred to an object. In water waves, the wave energy is the kinetic and gravitational potential energy of water; for electromagnetic waves, the wave energy is in the form of electric and magnetic fields. The energy of the wave depends on the amplitudes of these fields. If the electric and magnetic fields have greater amplitudes, the wave will carry more energy.

In our earlier study of waves, we defined the *intensity* of a wave (measured in W/m^2) to be $I = P/A$, where P is the power (energy transferred per second) of a wave that impinges on area A. The intensity of an electromagnetic wave depends on the amplitudes of the oscillating electric and magnetic fields:

$$I = \frac{P}{A} = \frac{1}{2}c\epsilon_0 E_0^2 = \frac{1}{2}\frac{c}{\mu_0}B_0^2 \tag{25.17}$$

Intensity of an electromagnetic wave with field amplitudes E_0 and B_0

The intensity of a plane wave, such as that of a laser beam, does not change with distance. As we saw in Chapter 15, the intensity of a spherical wave, spreading out from a point, must decrease with the square of the distance to conserve energy. If a source with power P_{source} emits waves *uniformly* in all directions, the wave intensity at distance r is

$$I = \frac{P_{\text{source}}}{4\pi r^2} \tag{25.18}$$

The intensities of the electromagnetic waves from antennas, cell phones, and other "point sources" are reasonably well described by Equation 25.18.

EXAMPLE 25.7 Electric and magnetic fields of a cell phone

A digital cell phone emits 0.60 W of 1.9 GHz radio waves. What are the amplitudes of the electric and magnetic fields at a distance of 10 cm?

PREPARE We can approximate the cell phone as a point source of waves, so we can use Equation 25.18 to determine the intensity of the waves at the noted distance. Once we know the intensity, we can compute the field amplitudes.

SOLVE The intensity at a distance of 10 cm is

$$I = \frac{P_{\text{source}}}{4\pi r^2} = \frac{0.60\ \text{W}}{4\pi(0.10\ \text{m})^2} = 4.8\ \text{W/m}^2$$

We can rearrange Equation 25.17 to solve for the amplitude of the electric field:

$$E_0 = \sqrt{\frac{2I}{c\varepsilon_0}} = \sqrt{\frac{2(4.8\ \text{W/m}^2)}{(3.0\times 10^8\ \text{m/s})(8.85\times 10^{-12}\ \text{C}^2/\text{N}\cdot\text{m}^2)}} = 60\ \text{V/m}$$

We can then use Equation 25.15, which relates the amplitudes of the electric and magnetic fields for an electromagnetic wave, to find the amplitude of the magnetic field:

$$B_0 = \frac{E_0}{c} = 2.0\times 10^{-7}\ \text{T}$$

ASSESS The electric field amplitude is reasonably small. For comparison, the typical electric field due to atmospheric electricity is 100 V/m; the field near a charged Van de Graaf generator can be 1000 times larger than this. This seems reasonable; we know that the electric fields near a cell phone's antenna aren't large enough to produce significant effects. The magnetic field is smaller yet, only 1/250th of the earth's field, which, as you know, is quite weak. The interaction of electromagnetic waves with matter is mostly due to the electric field; the magnetic field is generally small enough that we can ignore its effects.

FIGURE 25.29 The polarization of an electromagnetic wave is determined by the plane in which the electric field oscillates.

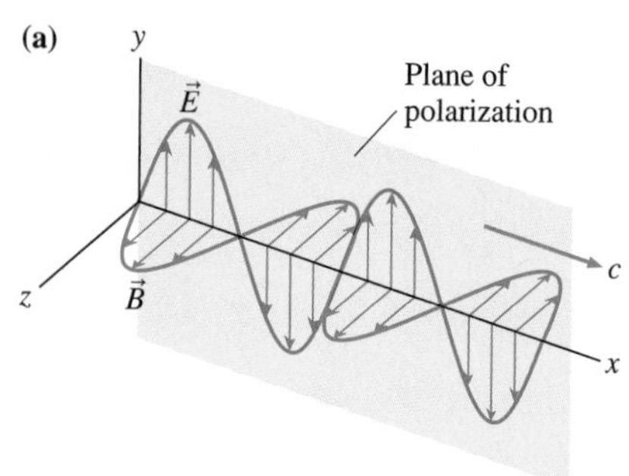

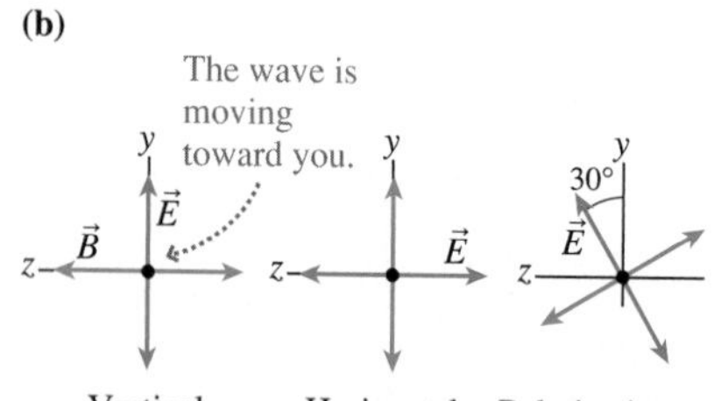

Polarization

The electric field vectors of an electromagnetic wave lie in a plane perpendicular to the direction of propagation. The plane containing the electric field vectors is called the **plane of polarization.** In FIGURE 25.29a, the wave is traveling along the x-axis, and the plane of polarization is the xy-plane. If the wave were moving toward you, it would appear as in the first diagram in FIGURE 25.29b. This particular wave is *vertically polarized* ($\vec{E}$ oscillating along the y-axis). The second diagram of Figure 25.29b shows a wave that is *horizontally polarized* ($\vec{E}$ oscillating along the z-axis). The plane of polarization needn't be horizontal or vertical; it can have any orientation, as in the third diagram.

ActivPhysics ONLINE 16.9

NOTE ► This use of the term "polarization" is completely independent of the idea of *charge polarization* that you learned about in Chapter 20. ◄

Most natural sources of electromagnetic radiation are *unpolarized.* Each atom in the sun's hot atmosphere emits light independently of all the other atoms, as does each tiny piece of metal in the incandescent filament of a lightbulb. An electromagnetic wave that you see or measure is a superposition of waves from each of these tiny emitters. Although the wave from each individual emitter is polarized, it is polarized in a random direction with respect to the waves from all its neighbors. The resulting wave, a superposition of waves with electric fields in all possible directions, is *unpolarized.*

We can create polarized light by sending unpolarized light through a *polarizing filter.* A typical polarizing filter is a plastic sheet containing long organic molecules called polymers, as shown in FIGURE 25.30. The molecules are aligned to form a grid, like the metal bars in a barbecue grill, then treated so they conduct electrons along their length.

As a light wave travels through a polarizing filter, the component of the electric field oscillating parallel to the polymer grid drives the electrons up and down the molecules. The electrons absorb energy from the light wave, so the parallel component of $\vec{E}$ is absorbed in the filter. But the conduction electrons can't oscillate perpendicular to the molecules, so the component of $\vec{E}$ perpendicular to the polymer grid passes through without absorption. Thus the light wave emerging from a polarizing filter is polarized perpendicular to the polymer grid. We call the direction of the transmitted polarization the axis of the polarizer.

FIGURE 25.30 A polarizing filter.

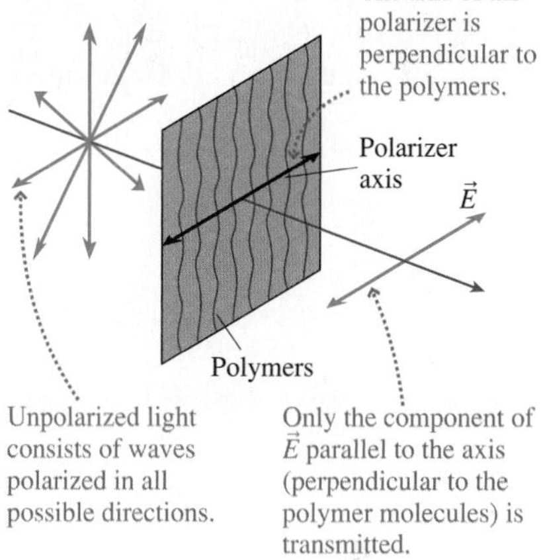
The axis of the polarizer is perpendicular to the polymers.
Polarizer axis
$\vec{E}$
Polymers
Unpolarized light consists of waves polarized in all possible directions.
Only the component of $\vec{E}$ parallel to the axis (perpendicular to the polymer molecules) is transmitted.

Suppose a *polarized* light wave with electric field amplitude E_{incident} approaches a polarizing filter with a vertical axis (that is, the filter transmits only vertically polarized light). What is the intensity of the light that passes through the filter? FIGURE 25.31 shows that the oscillating electric field of the polarized light can be decomposed into horizontal and vertical components. The vertical component will pass; the horizontal component will be blocked. As we see in Figure 25.31, the magnitude of electric field of the light transmitted by the filter is

$$E_{\text{transmitted}} = E_{\text{incident}} \cos\theta \tag{25.19}$$

Because the intensity depends on the square of the electric field amplitude, the transmitted intensity is related to the incident intensity by what is known as **Malus's law:**

$$I_{\text{transmitted}} = I_{\text{incident}}(\cos\theta)^2 \tag{25.20}$$

Malus's law for transmission of polarized light by a polarizing filter

FIGURE 25.31 An incident electric field can be decomposed into components parallel and perpendicular to a polarizer's axis.

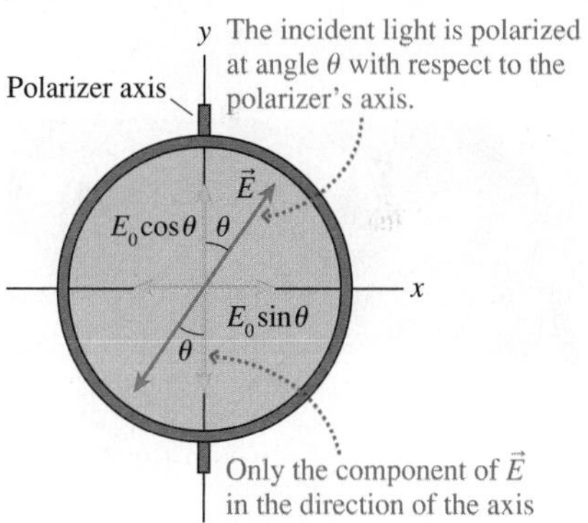
y
The incident light is polarized at angle θ with respect to the polarizer's axis.
Polarizer axis
$\vec{E}$
$E_0\cos\theta$
θ
x
$E_0\sin\theta$
θ
Only the component of $\vec{E}$ in the direction of the axis is transmitted.

FIGURE 25.32 shows how Malus's law can be demonstrated with two polarizing filters. The first, called the *polarizer,* is used to produce polarized light of intensity I_0.

FIGURE 25.32 The intensity of the transmitted light depends on the angle between the polarizing filters.

(a)

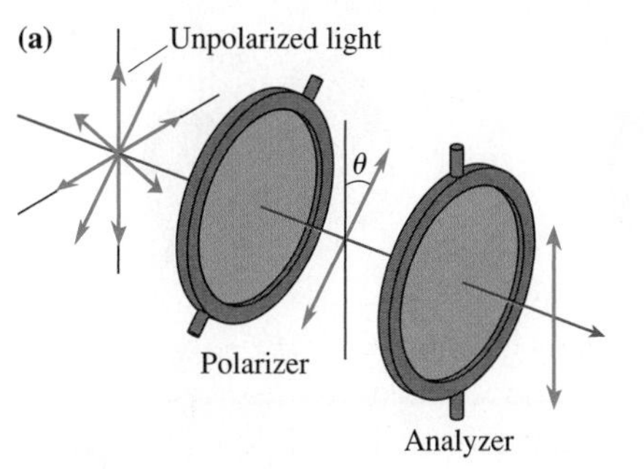
Unpolarized light
θ
Polarizer
Analyzer

(b)

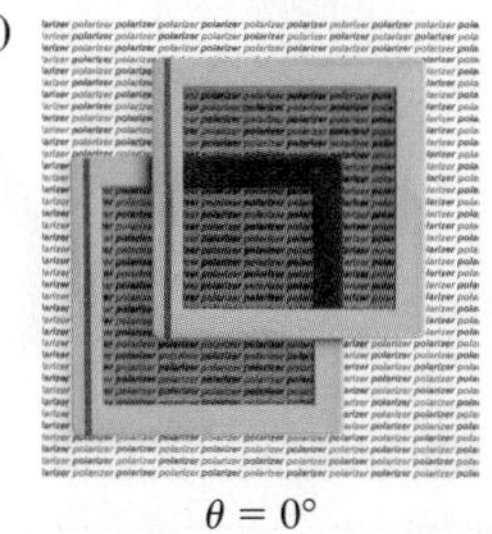
$\theta = 0°$

$\theta = 45°$

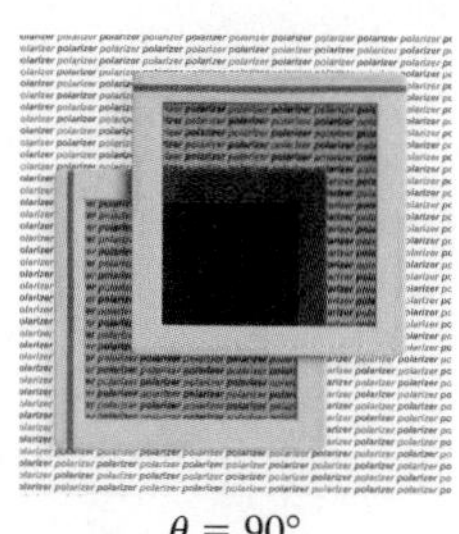
$\theta = 90°$

The red lines show the axes of the polarizers.

FIGURE 25.33 Polarized light micrograph of a thin section of molar teeth.

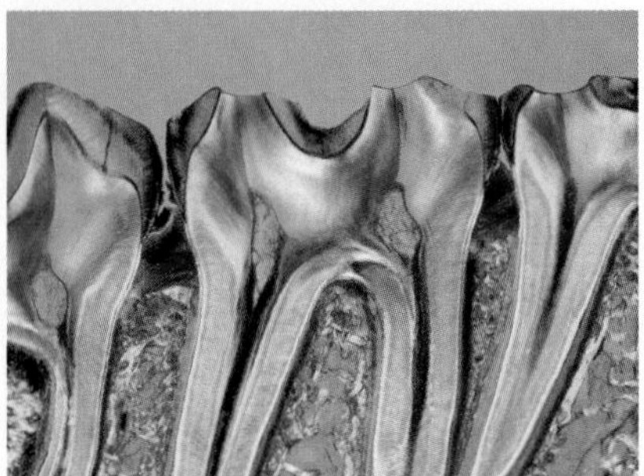

The second, called the *analyzer*, is rotated by angle θ relative to the polarizer. As the photographs of Figure 25.32b show, the transmission of the analyzer is (ideally) 100% when $\theta = 0°$ and steadily decreases to zero when $\theta = 90°$. Two polarizing filters with perpendicular axes, called *crossed polarizers,* block all the light.

Suppose you place an object between two crossed polarizers. Normally, no light would make it through the analyzer, and the object would appear black. But if the object is able to *change* the polarization of the light, some of the light emerging from the object will be able to pass through the analyzer. This can be a valuable analytical technique. Many minerals, crystals, and biological molecules do, indeed, change the polarization of light. As an example, FIGURE 25.33 shows a micrograph of a very thin section of molar teeth as it appears when viewed between crossed polarizers. Different minerals and different materials in the teeth affect the polarization of the light in different ways. The result is an image that clearly highlights the different tissues in the teeth.

In polarizing sunglasses, the polarization axis is vertical (when the glasses are in the normal orientation) so that the glasses transmit only vertically polarized light. *Glare*—the reflection of the sun and the skylight from lakes and other horizontal surfaces—has a strong horizontal polarization. This light is almost completely blocked, so the sunglasses "cut glare" without affecting the main scene you wish to see.

The molecules in a polarizing filter absorb light if its electric field is oriented along the axis of the molecules. Light-sensing *photopigments* in the eye are also long-chain molecules with similar properties; they are much more likely to absorb and respond to light that is polarized along the axis of the molecules. In human eyes, the orientation of the photopigments is random, so we do not have a polarization sense. But honeybees and other insects have light-sensing cells in their eyes in which the axes of the photopigment molecules are aligned. The response of these cells varies with the polarization of the light. Bees and other insects use this polarization sense to navigate, as shown in the photo below.

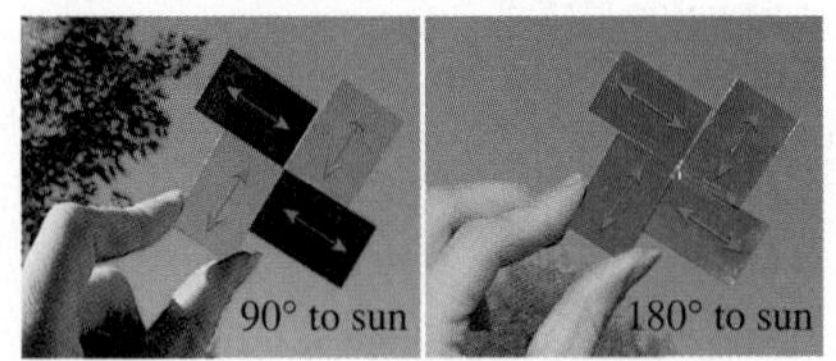

◀ **Making a beeline** BIO At an angle of 90° to the sun, skylight is partially polarized because of scattering from air molecules. In the left photo, this polarization causes different transmission through polarizers with different axes. Opposite the sun in the sky, the skylight is unpolarized, causing equal transmission for all of the polarizers in the right photo. We can't sense this sky polarization without external filters, but honeybees can. Light-sensing cells in honeybee eyes have different polarization axes, so they sense the polarization of the sky in a similar fashion to the photos. A bee can determine the sun's position from a tiny patch of clear sky, so bees can reliably navigate even in dense forest cover.

STOP TO THINK 25.4 Unpolarized light of equal intensity is incident on four pairs of polarizing filters. Rank in order, from largest to smallest, the intensities I_A to I_D transmitted through the second polarizer of each pair.

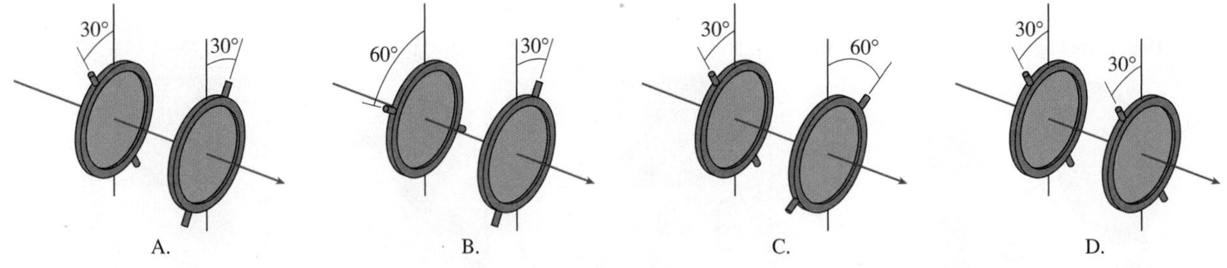

25.7 The Photon Model of Electromagnetic Waves

FIGURE 25.34 shows three photographs made with a camera in which the film has been replaced by a special high-sensitivity detector. A correct exposure, at the bottom, shows a perfectly normal photograph of a woman. But with very faint illumination (top), the picture is *not* just a dim version of the properly exposed photo. Instead, it is a collection of dots. A few points on the detector have registered the presence of light, but most have not. As the illumination increases, the density of these dots increases until the dots form a full picture.

This is not what we might expect. If light is a wave, reducing its intensity should cause the picture to grow dimmer and dimmer until it disappears, but the entire picture would remain present. It should be like turning down the volume on your stereo until you can no longer hear the sound. Instead, the top photograph in Figure 25.34 looks as if someone randomly threw "pieces" of light at the detector, causing full exposure at some points but no exposure at others.

If we did not know that light is a wave, we would interpret the results of this experiment as evidence that light is a stream of some type of particle-like object. If these particles arrive frequently enough, they overwhelm the detector and it senses a steady "river" instead of the individual particles in the stream. Only at very low intensities do we become aware of the individual particles.

As we will see in Chapter 28, many experiments convincingly lead to the surprising result that **electromagnetic waves, although they are waves, have a particle-like nature.** These particle-like components of electromagnetic waves are called **photons.**

The **photon model** of electromagnetic waves consists of three basic postulates:

1. Electromagnetic waves consist of discrete, massless units called photons. A photon travels in vacuum at the speed of light, 3.00×10^8 m/s.
2. Each photon has energy

$$E_{\text{photon}} = hf \tag{25.21}$$

where f is the frequency of the wave and h is a *universal constant* called **Planck's constant.** The value of Planck's constant is

$$h = 6.63 \times 10^{-34}\ \text{J} \cdot \text{s}$$

In other words, the electromagnetic waves come in discrete "chunks" of energy hf. The higher the frequency, the more energetic the chunks.

3. The superposition of a sufficiently large number of photons has the characteristics of a continuous electromagnetic wave.

FIGURE 25.34 Photographs made with an increasing level of light intensity.

The photo at very low light levels shows individual points, as if particles are arriving at the detector.

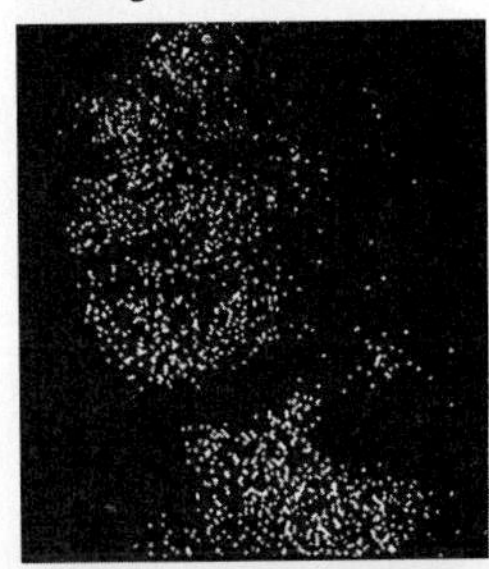

Increasing light intensity

The particle-like behavior is not noticeable at higher light levels.

EXAMPLE 25.8 Finding the energy of a photon of visible light

550 nm is the approximate average wavelength of visible light.

a. What is the energy of a photon with a wavelength of 550 nm?
b. A 40 W incandescent lightbulb emits about 1 J of visible light energy every second. Estimate the number of visible light photons emitted per second.

SOLVE a. The frequency of the photon is

$$f = \frac{c}{\lambda} = \frac{3.00 \times 10^8\ \text{m/s}}{550 \times 10^{-9}\ \text{m}} = 5.4 \times 10^{14}\ \text{Hz}$$

Equation 25.21 gives us the energy of this photon:

$$E_{\text{photon}} = hf = (6.63 \times 10^{-34}\ \text{J} \cdot \text{s})(5.4 \times 10^{14}\ \text{Hz}) = 3.6 \times 10^{-19}\ \text{J}$$

Continued

This is an extremely small energy! In fact, photon energies are so small that they are usually measured in electron volts (eV) rather than joules. Recall that $1 \text{ eV} = 1.60 \times 10^{-19}$ J. With this, we find that the photon energy is

$$E_{\text{photon}} = 3.6 \times 10^{-19}\text{ J} \times \frac{1\text{ eV}}{1.60 \times 10^{-19}\text{ J}} = 2.3\text{ eV}$$

b. The photons emitted by a lightbulb span a range of energies, because the light spans a range of wavelengths, but the *average* photon energy corresponds to a wavelength near 550 nm. Thus we can estimate the number of photons in 1 J of light as

$$N \approx \frac{1\text{ J}}{3.6 \times 10^{-19}\text{ J/photon}} \approx 3 \times 10^{18}\text{ photons}$$

A typical lightbulb emits about 3×10^{18} photons every second.

ASSESS The number of photons emitted per second is staggeringly large. It's not surprising that in our everyday life we sense only the river and not the individual particles within the flow.

TABLE 25.1 Energies of some atomic and molecular processes

Process	Energy
Breaking a hydrogen bond between two water molecules	0.24 eV
Energy released in metabolizing one molecule of ATP	0.32 eV
Breaking the bond between atoms in a water molecule	4.7 eV
Ionizing a hydrogen atom	13.6 eV

As we saw, a single photon of light at a wavelength of 550 nm has an energy of 2.3 eV. It is worthwhile to see just what 2.3 eV "buys" in interactions with atoms and molecules. Table 25.1 shows some energies required for typical atomic and molecular processes. These values show that 2.3 eV is a significant amount of energy on an atomic scale. It is certainly enough to cause a molecular transformation (as it does in the sensory system of your eye), and photons with a bit more energy (shorter wavelength) can break a covalent bond. The photon model of light will be essential as we explore the interaction of electromagnetic waves with matter.

STOP TO THINK 25.5 Two FM radio stations emit radio waves at frequencies of 90.5 MHz and 107.9 MHz. Each station emits the same total power. If you think of the radio waves as photons, which station emits the larger number of photons per second?

A. The 90.5 MHz station.
B. The 107.9 MHz station.
C. Both stations emit the same number of photons per second.

FIGURE 25.35 The electromagnetic spectrum.

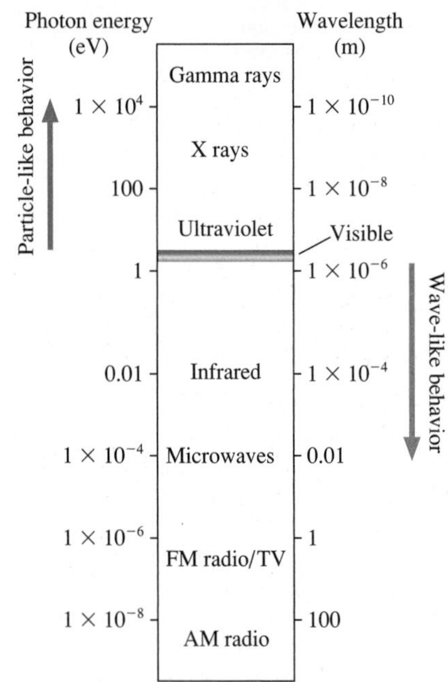

25.8 The Electromagnetic Spectrum

We have now seen two very different ways to look at electromagnetic waves: as oscillating waves of the electric and magnetic fields, and as particle-like units of the electromagnetic field called photons. This dual nature of electromagnetic waves is something we will discuss at length in Chapter 28. For now, we will note that each view is appropriate in certain circumstances. For example, we speak of radio *waves* but of x *rays*. The "ray" terminology tells us that x rays are generally better described as photons than as waves.

FIGURE 25.35 shows the *electromagnetic spectrum* with photon energy (in eV) and wavelength (in m) scales. As you can see, electromagnetic waves span an extraordinarily wide range of wavelengths and energies. Radio waves have wavelengths of many meters but very low photon energies—only a few billionths of an eV. Because the photon energies are so small, radio waves are well described by Maxwell's theory of electromagnetic waves. At the other end of the spectrum, x rays and gamma rays have very short wavelengths and very high photon energies—large enough to ionize atoms and break molecular bonds. Consequently, x rays and gamma rays, although they do have wave-like characteristics, are best described as photons. Visible light is in the middle. As we will see in Chapter 28, we must consider *both* views to fully understand the nature of visible light.

Radio Waves and Microwaves

An electromagnetic wave is self-sustaining, independent of charges or currents. However, charges and currents are needed at the *source* of an electromagnetic wave. Radio waves and microwaves are generally produced by the motion of charged particles in an antenna.

FIGURE 25.36 reminds you what the electric field of an electric dipole looks like. If the dipole is vertical, the electric field $\vec{E}$ at points along the horizontal axis in the figure is also vertical. Reversing the dipole, by switching the charges, reverses $\vec{E}$. If the charges were to *oscillate* back and forth, switching position at frequency f, then $\vec{E}$ would oscillate in a vertical plane. The changing $\vec{E}$ would then create an induced magnetic field $\vec{B}$, which could then create an $\vec{E}$, which could then create a $\vec{B}$, . . . , and a vertically polarized electromagnetic wave at frequency f would radiate out into space.

This is exactly what an **antenna** does. **FIGURE 25.37** shows two metal wires attached to the terminals of an oscillating voltage source. The figure shows an instant when the top wire is negative and the bottom is positive, but these will reverse in half a cycle. The wire is basically an oscillating dipole, and it creates an oscillating electric field. The oscillating $\vec{E}$ induces an oscillating $\vec{B}$, and they take off as an electromagnetic wave at speed $v_{em} = c$. The wave does need oscillating charges as a *wave source,* but once created it is self-sustaining and independent of the source.

Radio waves are *detected* by antennas as well. The electric field of a vertically polarized radio wave drives a current up and down a vertical conductor, producing a potential difference that can be amplified. For best reception, the antenna length should be about $\frac{1}{4}$ of a wavelength. A typical cell phone works at 1.9 GHz, with wavelength $\lambda = c/f = 16$ cm. Thus a cell phone antenna should be about 4 cm long, or about $1\frac{1}{2}$ inches. The antenna on your cell phone may seem quite short, but it is the right length to do its job.

AM radio has a lower frequency and thus a longer wavelength—typically 300 m. Having an antenna that is $\frac{1}{4}$ of a wavelength—75 m long!—is simply not practical. Instead, the antenna in an AM radio consists of a coil of wire wrapped around a core of magnetic material. This antenna detects the *magnetic* field of the radio wave. The changing flux of the wave's magnetic field induces an emf in the coil that is detected and amplified by the receiver.

FIGURE 25.36 The electric field of an oscillating dipole.

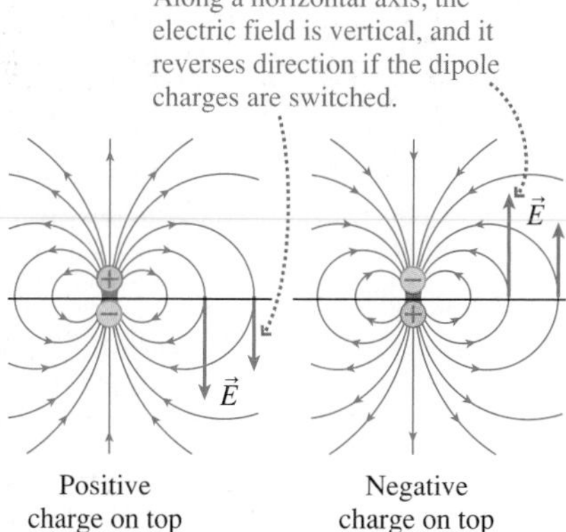
Along a horizontal axis, the electric field is vertical, and it reverses direction if the dipole charges are switched.
$\vec{E}$
$\vec{E}$
Positive charge on top
Negative charge on top

FIGURE 25.37 An antenna generates a self-sustaining electromagnetic wave.

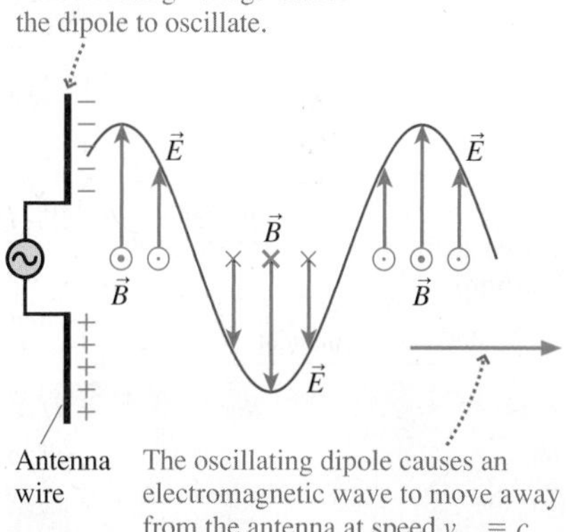
An oscillating voltage causes the dipole to oscillate.
$\vec{E}$
$\vec{E}$
$\vec{B}$
$\vec{B}$
$\vec{B}$
$\vec{E}$
Antenna wire
The oscillating dipole causes an electromagnetic wave to move away from the antenna at speed $v_{em} = c$.

CONCEPTUAL EXAMPLE 25.9 **Orienting a coil antenna**

A vertically polarized AM radio wave is traveling to the right. How should you orient a coil antenna to detect the oscillating magnetic field component of the wave?

REASON You want the oscillating magnetic field of the wave to produce the maximum possible induced emf in the coil, which requires the maximum changing flux. The flux is maximum when the coil is perpendicular to the magnetic field of the electromagnetic wave, as in **FIGURE 25.38**. Thus the plane of the coil should match the wave's plane of polarization.

FIGURE 25.38 A coil antenna.

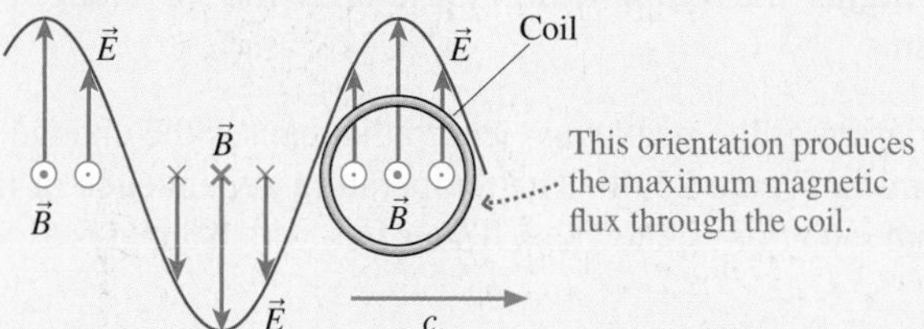
$\vec{E}$
$\vec{E}$
Coil
$\vec{B}$
$\vec{B}$
$\vec{B}$
This orientation produces the maximum magnetic flux through the coil.
$\vec{E}$
c

ASSESS Coil antennas are highly directional. If you turn an AM radio—and thus the antenna—in certain directions, you will no longer have the correct orientation of the magnetic field and the coil, and reception will be poor.

TRY IT YOURSELF

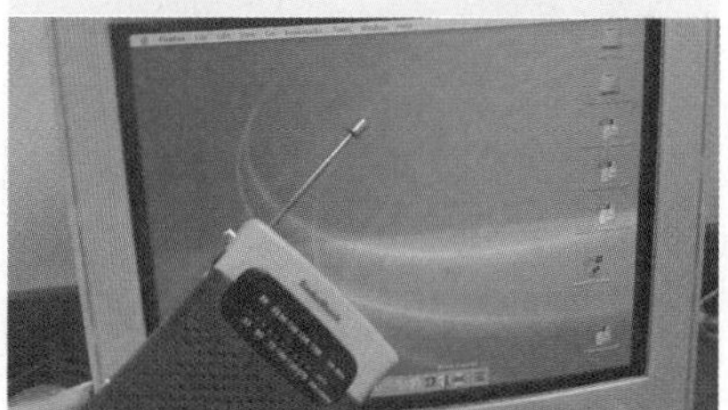

Unwanted transmissions Airplane passengers are asked to turn off all portable electronic devices during takeoff and landing. To see why, hold an AM radio near your computer and adjust the tuning as the computer performs basic operations, such as opening files. You will pick up intense static because the rapid switching of voltages in circuits causes computers—and other electronic devices—to emit radio waves, whether they're designed for communications or not. These electromagnetic waves could interfere with the airplane's electronics.

FIGURE 25.39 A radio wave interacts with matter.

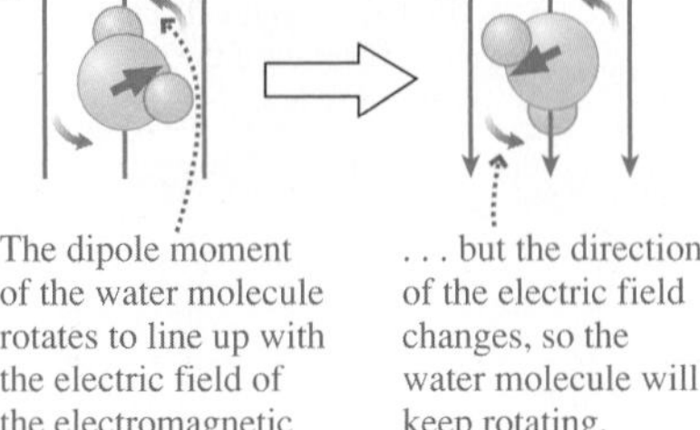

In materials with no free charges, the electric fields of radio waves and microwaves can interact with matter by exerting a torque on molecules, such as water, that have a permanent electric dipole moment, as shown in FIGURE 25.39. The molecules acquire kinetic energy from the wave; then their collisions with other molecules transform that energy into thermal energy, increasing the temperature.

This is how a microwave oven heats food. Water molecules, with their large dipole moment, rotate in response to the electric field of the microwaves, then transfer this energy to the food via molecular collisions.

Infrared, Visible Light, and Ultraviolet

Radio waves can be produced by oscillating charges in an antenna. At the higher frequencies of infrared, visible light, and ultraviolet, the "antennas" are individual atoms. This portion of the electromagnetic spectrum is *atomic radiation.*

Nearly all the atomic radiation in our environment is *thermal radiation* due to the thermal motion of the atoms in an object. As we saw in Chapter 12, thermal radiation—a form of heat transfer—is described by Stefan's law: If heat energy Q is radiated in a time interval Δt by an object with surface area A and absolute temperature T, the *rate* of heat transfer $Q/\Delta t$ (joules per second) is

$$\frac{Q}{\Delta t} = e\sigma AT^4 \tag{25.22}$$

The constant e in this equation is the object's emissivity, a measure of its effectiveness at emitting electromagnetic waves, and σ is the Stefan-Boltzmann constant, $\sigma = 5.67 \times 10^{-8}\ \text{W/(m}^2 \cdot \text{K}^4)$.

FIGURE 25.40 The brightness of the bulb varies with the temperature of the filament.

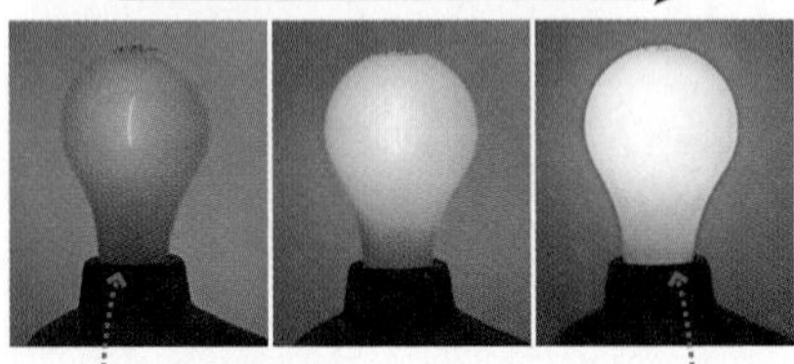

Increasing filament temperature
At lower filament temperatures, the bulb is dim and the light is noticeably reddish.
When the filament is hotter, the bulb is brighter and the light is whiter.

In Chapter 12 we considered the amount of energy radiated and its dependence on temperature. The filament of an incandescent bulb glows simply because it is hot. If you increase the current through a lightbulb filament, the filament temperature increases and so does the total energy emitted by the bulb, in accordance with Stefan's law. The three pictures in FIGURE 25.40 show a glowing lightbulb with the filament at successively higher temperatures. We can clearly see an increase in brightness in the sequence of three photographs.

But it's not just the brightness that varies. The *color* of the emitted radiation changes as well. At low temperatures, the light from the bulb is quite red. (A dim bulb doesn't look this red to your eye because your brain, knowing that the light "should" be white, compensates. But the camera doesn't lie.) Looking at the change in color as the temperature of the bulb rises in Figure 25.40, we see that **the spectrum of thermal radiation changes with temperature.**

FIGURE 25.41 A thermal emission spectrum depends on the temperature.

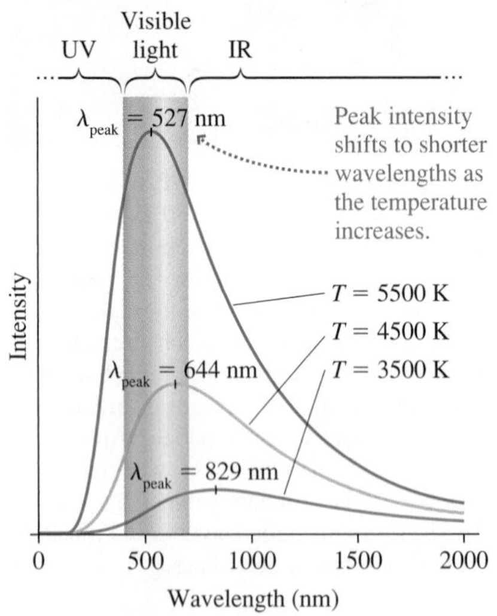

If we measure the intensity of thermal radiation as a function of wavelength for an object at three temperatures, 3500 K, 4500 K, and 5500 K, the data appear as in FIGURE 25.41. Notice two important features:

- Increasing the temperature increases the intensity at all wavelengths. **Making the object hotter causes it to emit more radiation across the entire spectrum.**
- Increasing the temperature causes the peak intensity to shift to a shorter wavelength. **The higher the temperature, the shorter the wavelength of the peak of the spectrum.**

It is this variation of the peak wavelength that causes the change in color of the glowing filament in Figure 25.40. The temperature dependence of the peak wavelength of thermal radiation is known as *Wien's law,* which appears as follows:

$$\lambda_{\text{peak}}(\text{in nm}) = \frac{2.9 \times 10^6\ \text{nm} \cdot \text{K}}{T(\text{in K})} \tag{25.23}$$

Wien's law for the peak wavelength of a thermal emission spectrum

λ_{peak} p. 114 T INVERSE

EXAMPLE 25.10 Finding peak wavelengths

What are the wavelengths of peak intensity and the corresponding spectral regions for radiating objects at (a) normal human body temperature of 37°C, (b) the temperature of the filament in an incandescent lamp, 1500°C, and (c) the temperature of the surface of the sun, 5800 K?

PREPARE All of the objects emit thermal radiation, so the peak wavelengths are given by Equation 25.23.

SOLVE First, we convert temperatures to kelvin. The temperature of the human body is $T = 37 + 273 = 310$ K, and the filament temperature is $T = 1500 + 273 = 1773$ K. Equation 25.23 then gives the wavelengths of peak intensity as

a. $\lambda_{\text{peak}}(\text{body}) = \dfrac{2.9 \times 10^6\ \text{nm}\cdot\text{K}}{310\ \text{K}} = 9.4 \times 10^3\ \text{nm} = 9.4\ \mu\text{m}$

b. $\lambda_{\text{peak}}(\text{filament}) = \dfrac{2.9 \times 10^6\ \text{nm}\cdot\text{K}}{1773\ \text{K}} = 1600\ \text{nm}$

c. $\lambda_{\text{peak}}(\text{sun}) = \dfrac{2.9 \times 10^6\ \text{nm}\cdot\text{K}}{5800\ \text{K}} = 500\ \text{nm}$

ASSESS The peak of the emission curve at body temperature is far into the infrared region of the spectrum, well below the range of sensitivity of human vision. You don't see someone "glow," although people do indeed emit significant energy in the form of electromagnetic waves, as we saw in Chapter 12. The sun's emission peaks right in the middle of the visible spectrum, which seems reasonable. Interestingly, most of the energy radiated by an incandescent bulb is *not* visible light. The tail of the emission curve extends into the visible region, but the peak of the emission curve—and most of the emitted energy—is in the infrared region of the spectrum. A 100 W bulb emits only a few watts of visible light.

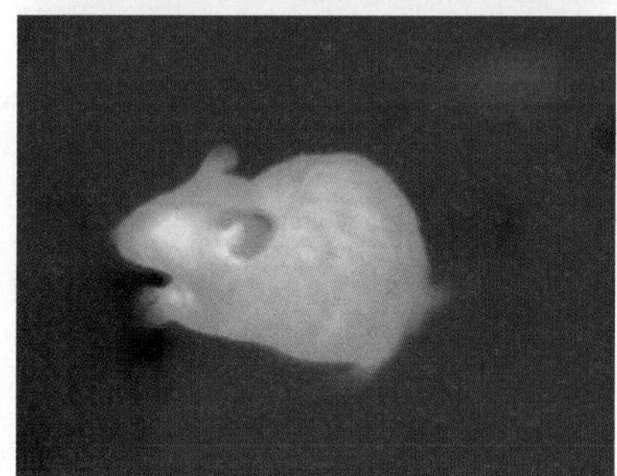

▶ **It's the pits ...** BIO Certain snakes—including rattlesnakes and other *pit vipers*—can hunt in total darkness. Prey animals are warm, and warm objects emit thermal radiation. In the top photo, notice the pits in front of the viper's eyes. These pits are actually a second set of vision organs; they have sensitive tissue at the bottom that allow them to sense this thermal radiation. The pits are sensitive to infrared wavelengths of $\approx 10\ \mu\text{m}$, near the wavelength of peak emission at mammalian body temperatures. Pit vipers sense the electromagnetic waves *emitted* by warm-blooded animals, such as the thermal radiation emitted by the mouse, shown in the lower image. They need no light to "see" you. You emit a "glow" they can detect.

Infrared radiation, with its relatively long wavelength and low photon energy, produces effects in tissue similar to those of microwaves—heating—but the penetration is much less than for microwaves. Infrared is absorbed mostly by the top layer of your skin and simply warms you up, as you know from sitting in the sun or under a heat lamp. The wave picture is generally most appropriate for infrared.

In contrast, ultraviolet photons have enough energy to interact with molecules in entirely different ways, ionizing molecules and breaking molecular bonds. The cells in skin are altered by ultraviolet radiation, causing sun tanning and sun burning. DNA molecules can be permanently damaged by ultraviolet radiation. There is a sharp threshold for such damage at 290 nm (corresponding to 4.3 eV photon energy). At longer wavelengths, damage to cells is slight; at shorter wavelengths, it is extensive. The interactions of ultraviolet radiation with matter are best understood from the photon perspective, with the absorption of each photon being associated with a particular molecular event.

Visible light is at a transition point in the electromagnetic spectrum. Your studies of wave optics in Chapter 17 showed you that light has a wave nature. At the same time, the energy of photons of visible light is large enough to cause molecular transitions—which is how your eye detects light. The bending of light by the lens of the eye requires us to think of light as a wave, but the detection of light by the cells in the retina requires us to think of light as photons. When we work with visible light, we will often move back and forth between the wave and photon models.

EXAMPLE 25.11 Finding the photon energy for ultraviolet light

Ultraviolet radiation with a wavelength of 254 nm is used in germicidal lamps. What is the photon energy in eV for such a lamp?

SOLVE The photon energy is $E = hf$:

$$E = hf = \frac{hc}{\lambda} = \frac{(6.63 \times 10^{-34}\ \text{J}\cdot\text{s})(3.00 \times 10^8\ \text{m/s})}{254 \times 10^{-9}\ \text{m}}$$

$$= 7.83 \times 10^{-19}\ \text{J}$$

In eV, this is

$$E = 7.83 \times 10^{-19}\ \text{J} \times \frac{1\ \text{eV}}{1.60 \times 10^{-19}\ \text{J}} = 4.89\ \text{eV}$$

ASSESS Table 25.1 shows that this energy is sufficient to break the bonds in a water molecule. It will be enough energy to break other bonds as well, leading to damage on a cellular level.

Color Vision

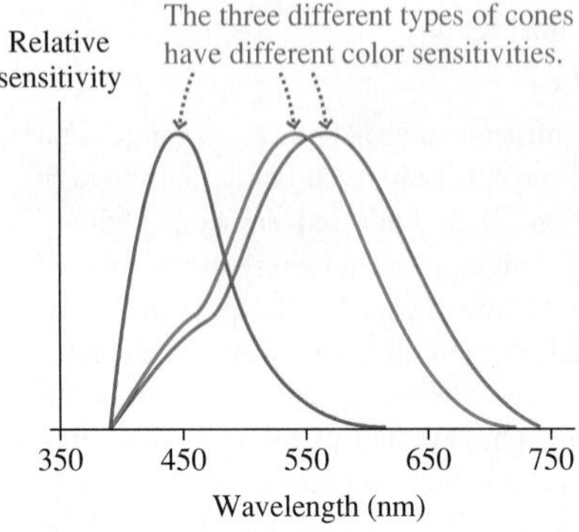

FIGURE 25.42 The sensitivity of different cones in the human eye.

The cones, the color-sensitive cells in the retina of the eye, each contain one of three slightly different forms of a light-sensitive photopigment. A single photon of light can trigger a reaction in a photopigment molecule, which ultimately leads to a signal being produced by a cell in the retina. The energy of the photon must be matched to the energy of a molecular transition for absorption of the photon energy to take place. Each photopigment has a range of photon energies to which it is sensitive. Our color vision is a result of the differential response of three types of cones containing three slightly different pigments, shown in FIGURE 25.42.

Humans have three types of cone in the eye, mice have two, and chickens four—giving a chicken keener color vision than a human. The three color photopigments that bees possess give them excellent color vision, but a bee's color sense is different from a human's. The peak sensitivities of a bee's photopigments are in the yellow, blue, and ultraviolet regions of the spectrum. A bee can't see the red of a rose, but it is quite sensitive to ultraviolet wavelengths well beyond the range of human vision. The flower in the right-hand photo at the start of the chapter looks pretty to us, but its coloration is really intended for other eyes. The ring of ultraviolet-absorbing pigments near the center of the flower, which is invisible to humans, helps bees zero in on the pollen.

X Rays and Gamma Rays

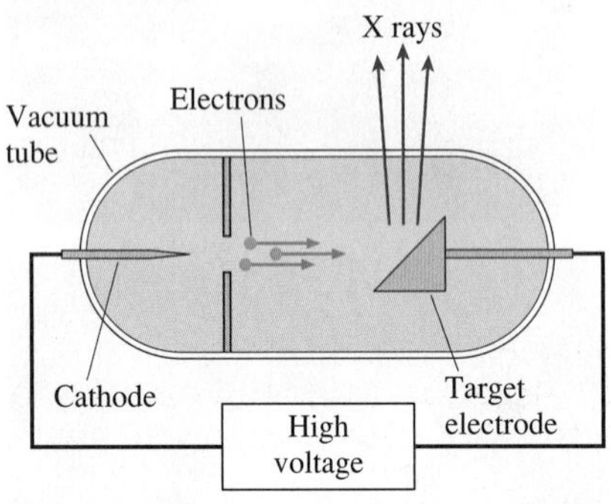

FIGURE 25.43 A simple x-ray tube.

At the highest energies of the electromagnetic spectrum we find x rays and gamma rays. There is no sharp dividing line between these two regions of the spectrum; the difference is the source of radiation. High-energy photons emitted by electrons are called x rays. If the source is a nuclear process, we call them gamma rays.

We will look at the emission of x rays in atomic processes and gamma rays in nuclear processes in Part VII. For now, we will focus on the "artificial" production of x rays in an x-ray tube, such as the one shown in FIGURE 25.43. Electrons are emitted from a cathode and accelerated to a kinetic energy of several thousand eV by the electric field between two electrodes connected to a high-voltage power supply. The electrons make a sudden stop when they hit a metal target electrode. The rapid deceleration of an electron can cause the emission of a single photon with a significant fraction of the electron's kinetic energy. These photons, with energies well in excess of 1000 eV, are x rays. The x rays pass through a window in the tube and then may be used to produce an image or to treat a disease.

EXAMPLE 25.12 Determining x-ray energies

An x-ray tube used for medical work has an accelerating voltage of 30 kV. What is the maximum energy of an x-ray photon that can be produced in this tube? What is the wavelength of this x ray?

SOLVE An electron accelerated through a potential difference of 30 kV acquires a kinetic energy of 30 keV. When this electron hits the metal target and stops, energy may be converted to an x ray. The maximum energy that could be converted is 30 keV, so this is the maximum possible energy of an x-ray photon from the tube. In joules, this energy is

$$E = 30 \times 10^3 \text{ eV} \times \frac{1.60 \times 10^{-19} \text{ J}}{1 \text{ eV}} = 4.8 \times 10^{-15} \text{ J}$$

For electromagnetic waves, $c = f\lambda$, so we can calculate

$$\lambda = \frac{c}{f} = \frac{c}{E/h} = \frac{hc}{E} = \frac{(6.63 \times 10^{-34} \text{ J}\cdot\text{s})(3.00 \times 10^8 \text{ m/s})}{4.8 \times 10^{-15} \text{ J}}$$

$$= 4.1 \times 10^{-11} \text{ m} = 0.041 \text{ nm}$$

ASSESS This is a very short wavelength, comparable to the spacing between atoms in a solid.

X rays and gamma rays (and the short-wavelength part of the ultraviolet spectrum) are **ionizing radiation;** the individual photons have sufficient energy to ionize atoms. When such radiation strikes tissue, the resulting ionization can produce cellular damage. When people speak of "radiation" they often mean "ionizing radiation."

X rays and gamma rays are very penetrating, but the absorption of these high-energy photons is greater in materials made of atoms with more electrons. This is why x rays are used in medical and dental imaging. The calcium in bones has many more electrons and thus is much more absorbing than the hydrogen, carbon, and oxygen that make up most of our soft tissue, so we can use x rays to image bones and teeth.

At several points in this chapter we have hinted at places where a full understanding of the phenomena requires some new physics. We have used the photon model of electromagnetic waves, and we have mentioned that nuclear processes can give rise to gamma rays. There are other questions that we did not raise, such as why the electromagnetic spectrum of a hot object has the shape that it does. These puzzles began to arise in the late 1800s and early 1900s, and it soon became clear that the physics of Newton and Maxwell was not sufficient to fully describe the nature of matter and energy. Some new rules, some new models, were needed. We will return to these puzzles as we begin to explore the exciting notions of quantum physics in Part VII.

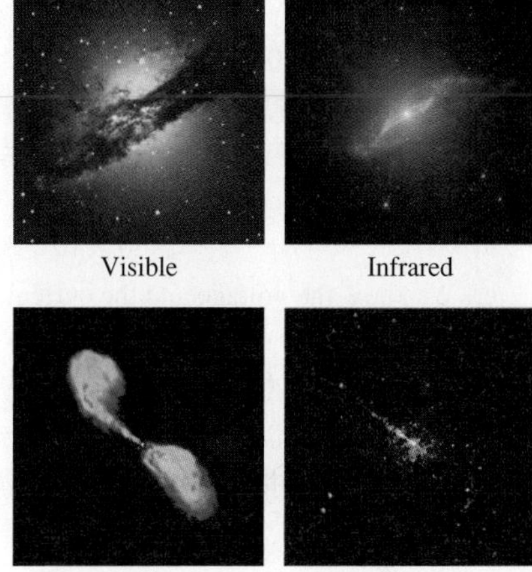

▶ **Seeing the universe in a different light** These four images of the Centaurus A galaxy have the same magnification and orientation, but they are records of different types of electromagnetic waves. (All but the visible-light image are false-color images.) The visible-light image shows a dark dust lane cutting across the galaxy. In the infrared, this dust lane glows quite brightly—telling us that the dust particles are hot. The radio and x-ray images show jets of matter streaming out of the galaxy's center, hinting at the presence of a massive black hole. Views of the cosmos beyond the visible range are important tools of modern astronomy.

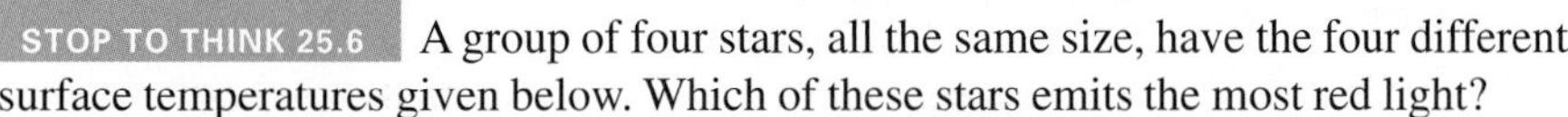

STOP TO THINK 25.6 A group of four stars, all the same size, have the four different surface temperatures given below. Which of these stars emits the most red light?

A. 3000 K B. 4000 K C. 5000 K D. 6000 K

INTEGRATED EXAMPLE 25.13 Space circuits

The very upper part of the atmosphere, where the Space Shuttle orbits, is called the *ionosphere*. The few atoms and molecules that remain at this altitude are mostly ionized by intense ultraviolet radiation from the sun. The thin gas of the ionosphere thus consists largely of positive ions and negative electrons, so it can carry an electric current. This was crucial to the operation of the tethered satellite system that was tested on the Space Shuttle in the 1990s. As described in the chapter, a probe was deployed on a conducting wire that tethered the probe at a great distance from the craft. As the Space Shuttle orbited, the wire moved through the earth's magnetic field, creating a potential difference between the ends of the wire. Charge flowing through the ionosphere back to the shuttle created a complete circuit.

In the final test of the tethered satellite system, a potential difference of 3500 V was generated across 20 km of cable as the shuttle orbited at 7800 m/s.

a. To produce the noted potential difference, what was the component of the magnetic field perpendicular to the wire?
b. The 3500 V potential created a current of 480 mA in the ionosphere. What was the total resistance of the circuit thus formed?
c. How much power was dissipated in this circuit?
d. What is the drag force on the wire due to its motion in the earth's field?
e. The ionization of the upper atmosphere is due to solar radiation at wavelengths of 95 nm and shorter. In what part of the spectrum is this radiation? What is the lowest-energy photon, in eV, that contributes to the ionization?

PREPARE The motion of the wire connecting the tethered satellite to the Space Shuttle leads to a motional emf that drives a current through this wire and back through the ionosphere. FIGURE 25.44 shows how we can model this process as an electric circuit. We know the voltage and the current in this circuit, so we can find the resistance and the power. Because the wire carries a current, the earth's field will exert a force on it, which is the drag force we are asked to find.

FIGURE 25.44 The tethered satellite circuit.

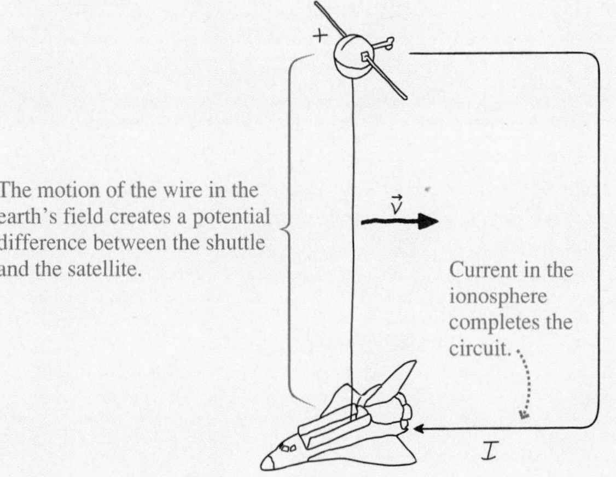

+
$\vec{v}$
The motion of the wire in the earth's field creates a potential difference between the shuttle and the satellite.
Current in the ionosphere completes the circuit.
I

Continued

SOLVE a. We know the magnitude of the velocity and the length of the wire, so we can use the equation for the motional emf, Equation 25.3, to find the magnitude of the component of the field perpendicular to the wire:

$$B = \frac{\mathcal{E}}{vl} = \frac{3500\text{ V}}{(7800\text{ m/s})(20 \times 10^3\text{ m})} = 2.2 \times 10^{-5}\text{ T} = 22\ \mu\text{T}$$

b. The 3500 V potential difference produced a current of 480 mA. From Ohm's law, the resistance of the circuit was thus

$$R = \frac{\Delta V}{I} = \frac{3500\text{ V}}{480 \times 10^{-3}\text{ A}} = 7300\ \Omega$$

c. We know the voltage and the current, so we can compute the power:

$$P = I\,\Delta V = (0.48\text{ A})(3500\text{ V}) = 1700\text{ W}$$

d. The component of the magnetic field perpendicular to the current-carrying wire exerts a drag force on the wire. We learned in Chapter 24 that the force on a current-carrying wire is $F = IlB$. Thus

$$F = IlB = (0.48\text{ A})(20 \times 10^3\text{ m})(2.2 \times 10^{-5}\text{ T}) = 0.21\text{ N}$$

e. Radiation with wavelengths of 95 nm and shorter is in the ultraviolet region of the spectrum. The lowest-energy photon in this region has the lowest frequency and thus the longest wavelength—namely, 95 nm. The frequency is

$$f = \frac{c}{\lambda} = \frac{3.0 \times 10^8\text{ m/s}}{95 \times 10^{-9}\text{ m}} = 3.2 \times 10^{15}\text{ Hz}$$

The photon energy is then given by Equation 25.21:

$$E_{\text{photon}} = hf = (6.63 \times 10^{-34}\text{ J} \cdot \text{s})(3.1 \times 10^{15}\text{ Hz})$$
$$= 2.1 \times 10^{-18}\text{ J}$$

Converting to eV, we find

$$E_{\text{photon}} = 2.1 \times 10^{-18}\text{ J} \times \frac{1\text{ eV}}{1.6 \times 10^{-19}\text{ J}} = 13\text{ eV}$$

ASSESS We have many good chances to check our work to verify that it makes sense. First, the field component that we calculate is about half the value we typically use for the earth's field, which seems reasonable—we'd be suspicious if the field we calculated was more than the earth's field.

The product of the drag force and the speed is the power dissipated by the drag force:

$$P = Fv = (0.22\text{ N})(7800\text{ m/s}) = 1700\text{ W}$$

This is exactly what we found for the electric power dissipated in the circuit, a good check on our work. The two values must be equal, as they are.

A final check on our work is the value we calculate for the photon energy. Table 25.1 shows that it takes about 13 eV to ionize a hydrogen atom. Photons with wavelengths shorter than 95 nm are able to ionize hydrogen atoms, so it seems likely they would also ionize the nitrogen and oxygen molecules of the upper atmosphere.

SUMMARY

The goal of Chapter 25 has been to understand the nature of electromagnetic induction and electromagnetic waves.

GENERAL PRINCIPLES

Electromagnetic Induction

The **magnetic flux** measures the amount of magnetic field passing through a surface:

$$\Phi = AB\cos\theta$$

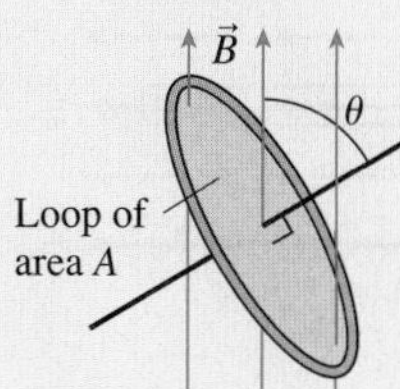

Lenz's law specifies that there is an induced current in a closed conducting loop if the magnetic flux through the loop is changing. The direction of the induced current is such that the induced magnetic field opposes the *change* in flux.

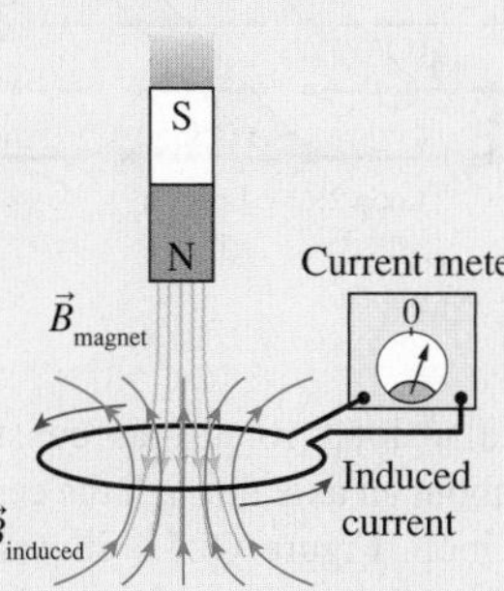

Faraday's law specifies the magnitude of the induced emf in a closed loop:

$$\mathcal{E} = \left|\frac{\Delta\Phi}{\Delta t}\right|$$

Multiply by N for an N-turn coil.

The size of the induced current is

$$I = \frac{\mathcal{E}}{R}$$

Electromagnetic Waves

An electromagnetic wave is a self-sustaining oscillation of electric and magnetic fields.

- The wave is a transverse wave with $\vec{E}$, $\vec{B}$, and $\vec{v}$ mutually perpendicular.
- The wave propagates with speed
$$v_{em} = c = \frac{1}{\sqrt{\epsilon_0\mu_0}} = 3.00 \times 10^8 \text{ m/s}$$
- The wavelength, frequency, and speed are related by
$$c = f\lambda$$
- The amplitudes of the fields are related by
$$E_0 = cB_0$$

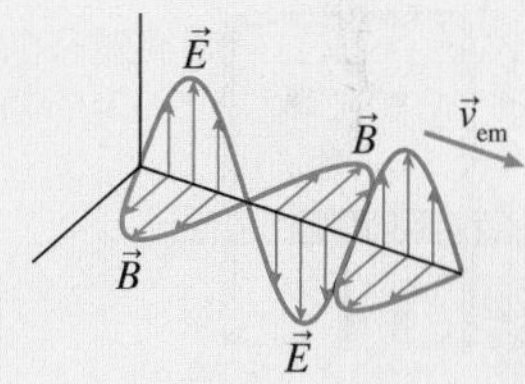

IMPORTANT CONCEPTS

Motional emf

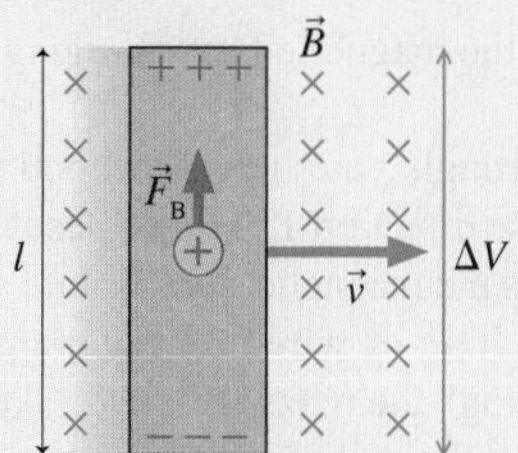

The motion of a conductor through a magnetic field produces a force on the charges. The separation of charges leads to an emf:

$$\mathcal{E} = vlB$$

The photon model

Electromagnetic waves appear to be made of discrete units called photons. The energy of a photon of frequency f is

$$E = hf$$

This photon view becomes increasingly important as the photon energy increases.

The electromagnetic spectrum

Electromagnetic waves come in a wide range of wavelengths and photon energies.

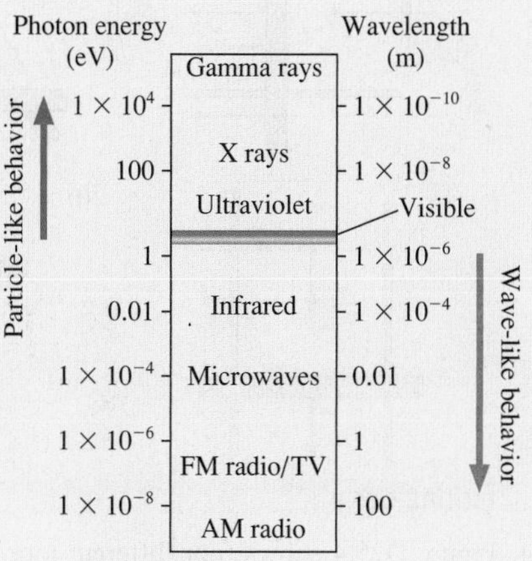

APPLICATIONS

A changing flux in a solid conductor creates **eddy currents**.

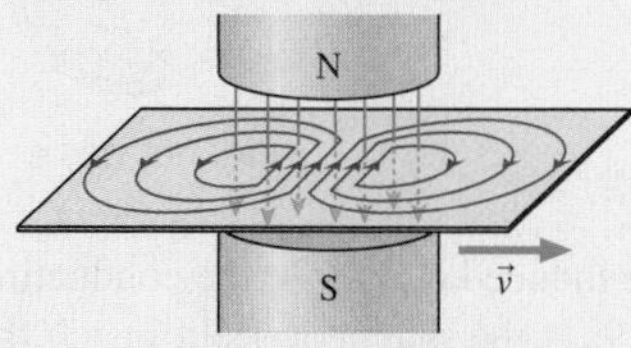

The plane of the electric field of an electromagnetic wave defines its **polarization.** The intensity of polarized light transmitted through a polarizing filter is given by **Malus's law:**

$$I = I_0\cos^2\theta$$

where θ is the angle between the electric field and the polarizer axis.

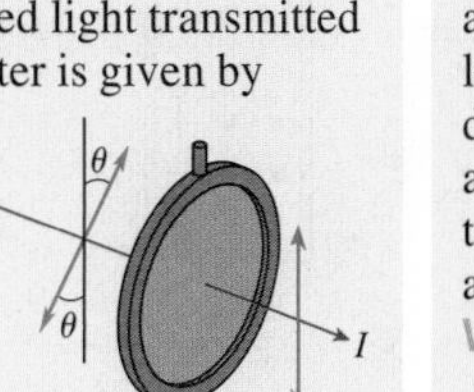

Thermal radiation has a peak wavelength that depends on an object's temperature according to **Wien's law:**

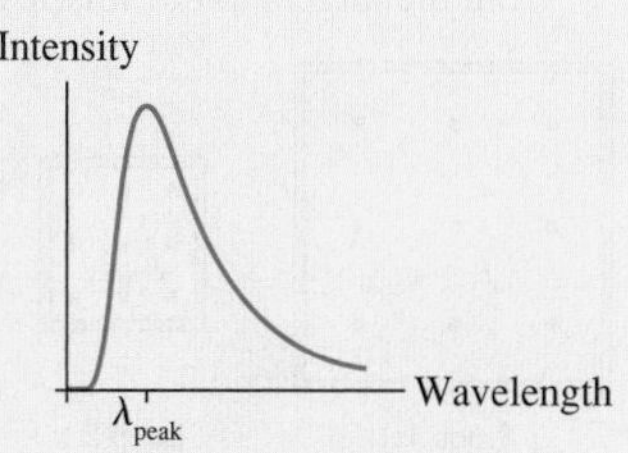

$$\lambda_{peak}(\text{in nm}) = \frac{2.9 \times 10^6 \text{ nm} \cdot \text{K}}{T}$$

For homework assigned on MasteringPhysics, go to www.masteringphysics.com

Problem difficulty is labeled as | (straightforward) to ||||| (challenging).

Problems labeled INT integrate significant material from earlier chapters; BIO are of biological or medical interest.

QUESTIONS

Conceptual Questions

1. The world's strongest magnet can produce a steady field of 45 T. If a circular wire loop of radius 10 cm were held in this magnetic field, what current would be induced in the loop?
2. BIO The rapid vibration accompanying the swimming motions of mayflies has been measured by gluing a small magnet to a swimming mayfly and recording the emf in a small coil of wire placed nearby. Explain how this technique works.
3. Parts a through f of Figure Q25.3 show one or more metal wires sliding on fixed metal rails in a magnetic field. For each, determine if the induced current is clockwise, counterclockwise, or is zero.

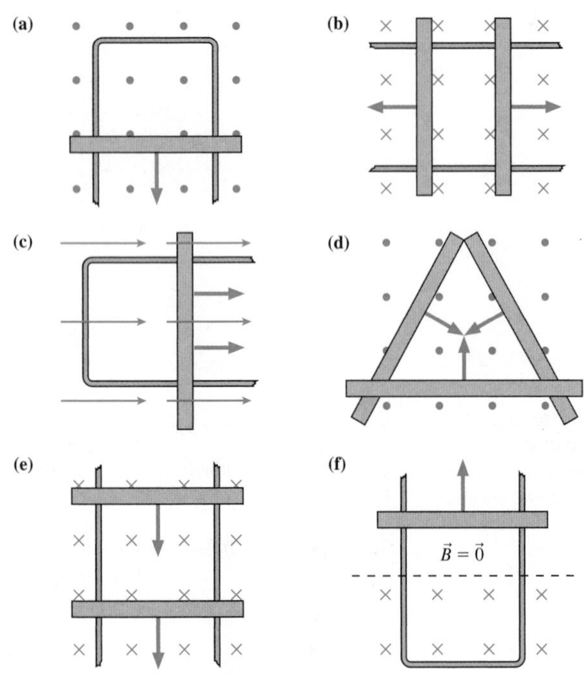

FIGURE Q25.3

4. Figure Q25.4 shows four different loops in a magnetic field. The numbers indicate the lengths of the sides and the strength of the field. Rank in order the magnetic fluxes Φ_1 through Φ_4, from the largest to the smallest. Some may be equal. Explain.

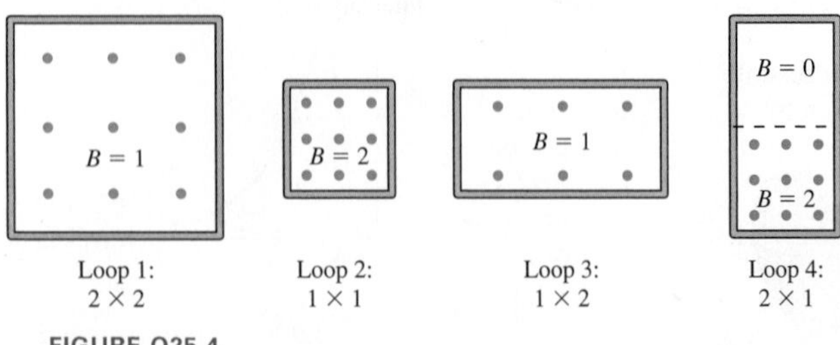

FIGURE Q25.4

5. Figure Q25.5 shows four different circular loops that are perpendicular to the page. The radius of loops 3 and 4 is twice that of loops 1 and 2. The magnetic field is the same for each. Rank in order the magnetic fluxes Φ_1 through Φ_4, from the largest to the smallest. Some may be equal. Explain.

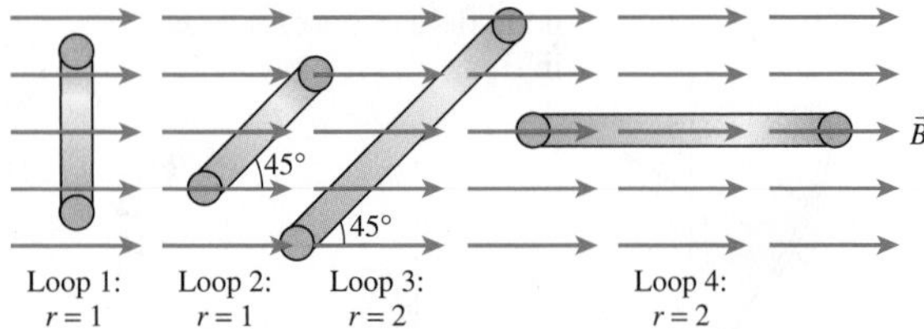

FIGURE Q25.5

6. A circular loop rotates at constant speed about an axle through the center of the loop. Figure Q25.6 shows an edge view and defines the angle ϕ, which increases from 0° to 360° as the loop rotates.
   a. At what angle or angles is the magnetic flux a maximum?
   b. At what angle or angles is the magnetic flux a minimum?
   c. At what angle or angles is the magnetic flux *changing* most rapidly?

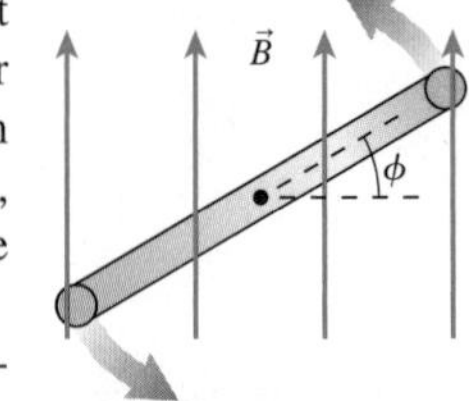

FIGURE Q25.6

7. The power lines that run through your neighborhood carry *alternating currents* that reverse direction 120 times per second. As the current changes, so does the magnetic field around a line. Suppose you wanted to put a loop of wire up near the power line to extract power by "tapping" the magnetic field. Sketch a picture of how you would orient the coil of wire next to a power line to develop the maximum emf in the coil. (Note that this is dangerous and illegal, and not something you should try.)
8. The magnetic flux passing through a coil of wire varies as shown in Figure Q25.8. During which time interval(s) will an induced current be present in the coil? Explain.

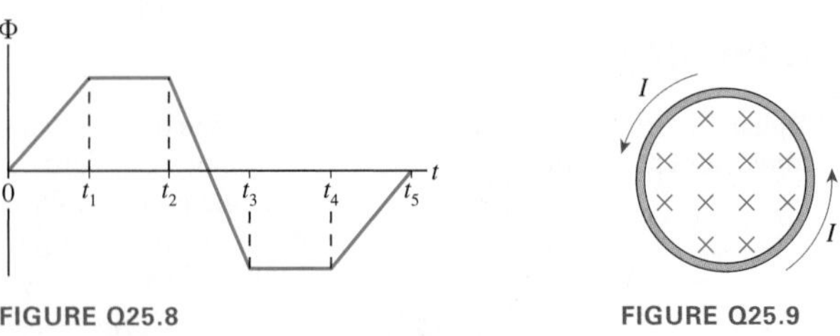

FIGURE Q25.8

FIGURE Q25.9

9. There is a counterclockwise induced current in the conducting loop shown in Figure Q25.9. Is the magnetic field inside the loop increasing in strength, decreasing in strength, or steady?

10. Cars on the "Tower of Doom" carnival ride are dropped from a great height, giving riders a few seconds of free fall. To stop the cars, strong permanent magnets attached to the bottoms of the cars pass very close to aluminum vanes sticking up from the bottom of the track. Explain how this braking system works.
11. A magnet dropped through a clear plastic tube accelerates as expected in free fall. If dropped through an aluminum tube of exactly the same length and diameter, the magnet falls much more slowly. Explain the behavior of the second magnet.
12. The conducting loop in Figure Q25.12 is moving into the region between the magnetic poles shown.
    a. Is the induced current (viewed from above) clockwise or counterclockwise?
    b. Is there an attractive magnetic force that tends to pull the loop in, like a magnet pulls on a paper clip? Or do you need to push the loop in against a repulsive force?

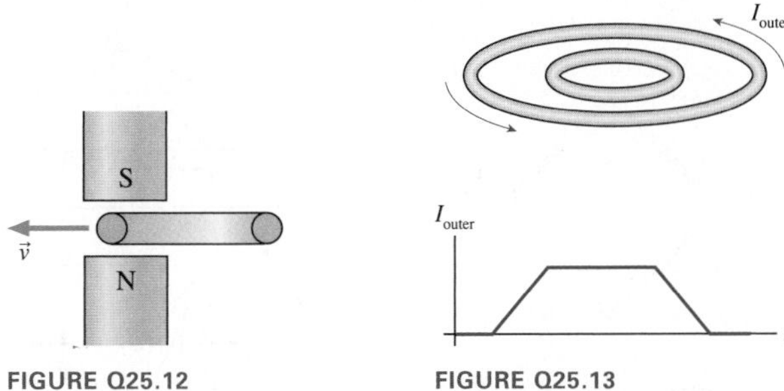

FIGURE Q25.12

FIGURE Q25.13

13. Figure Q25.13 shows two concentric, conducting loops. We will define a counterclockwise current (viewed from above) to be positive, a clockwise current to be negative. The graph shows the current in the outer loop as a function of time. Sketch a graph that shows the induced current in the inner loop. Explain.
14. Two loops of wire are stacked vertically, one above the other, as shown in Figure Q25.14. Does the upper loop have a clockwise current, a counterclockwise current, or no current at the following times? Explain your reasoning.

FIGURE Q25.14

    a. Before the switch is closed
    b. Immediately after the switch is closed
    c. Long after the switch is closed
    d. Immediately after the switch is reopened
15. A loop of wire is horizontal. A bar magnet is pushed toward the loop from below, along the axis of the loop, as shown in Figure Q25.15.
    a. In what direction is the current in the loop? Explain.
    b. Is there a magnetic force on the loop? If so, in which direction? Explain.

    Hint: Recall that a current loop is a magnetic dipole.

    c. Is there a magnetic force on the magnet? If so, in which direction?

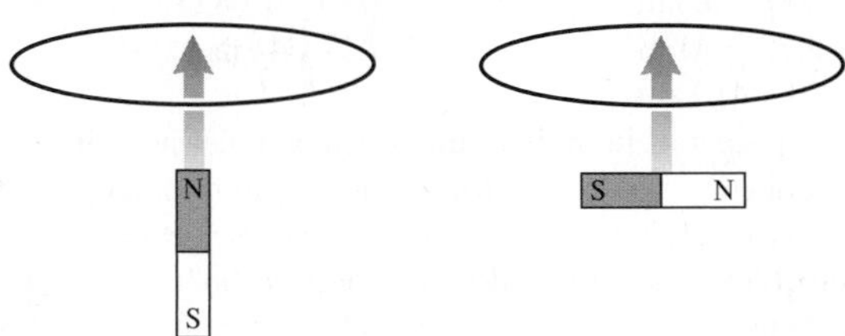

FIGURE Q25.15

FIGURE Q25.16

16. A bar magnet is pushed toward a loop of wire, as shown in Figure Q25.16. Is there a current in the loop? If so, in which direction? If not, why not?
17. A conducting loop around a region of strong magnetic field contains two light bulbs, as shown in Figure Q25.17. The wires connecting the bulbs are ideal. The magnetic field is increasing rapidly.
    a. Do the bulbs glow? Why or why not?
    b. If they glow, which bulb is brighter? Or are they equally bright? Explain.

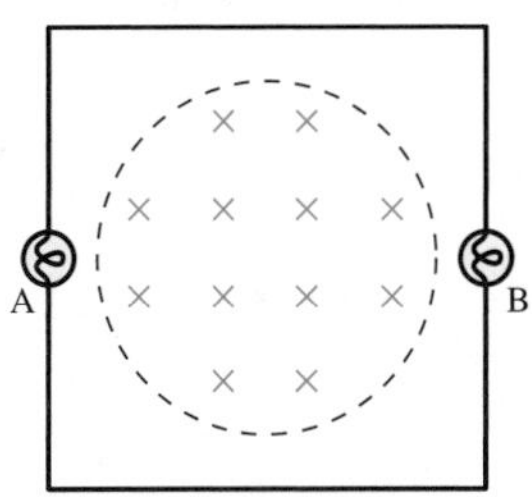

FIGURE Q25.17

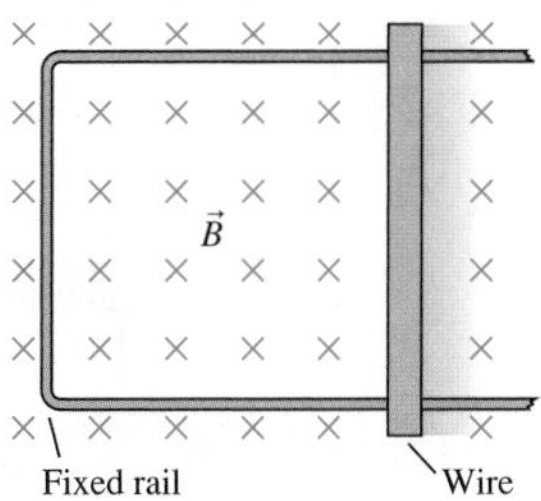

FIGURE Q25.18

18. A metal wire is resting on a U-shaped conducting rail, as shown in Figure Q25.18. The rail is fixed in position, but the wire is free to move.
    a. If the magnetic field is increasing in strength, what does the wire do? Does it remain in place? Or does it move to the right or left, or up or down, or out of the plane of the page (breaking contact with the rail)? Does it rotate clockwise or counterclockwise? Does it both move and rotate? Explain.
    b. If the magnetic field is decreasing in strength, what does the wire do?
19. Though sunlight is unpolarized, the light that reflects from smooth surfaces may be partially polarized in the direction parallel to the plane of the reflecting surface. How should the long axis of the polarizing molecules in polarized sunglasses be oriented—vertically or horizontally—to reduce the glare from a horizontal surface such as a road or a lake?
20. Two polarizers are oriented with axes at 90°, so no light passes through the pair. A piece of plastic is placed between the two. If the plastic is stressed, by being squeezed, light that passes through the first polarizer and the plastic now passes through the second polarizer. The dark and light lines allow the pattern of stress to be determined. What effect does the stressed plastic have on the polarization of light passing through it?

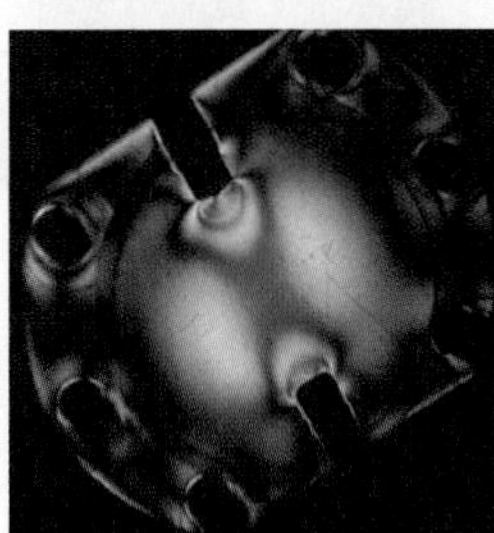

With polarizing filters

Without polarizing filters

21. Old-fashioned roof-mounted television antennas were designed to pick up signals across a broad frequency range. Explain why these antennas had metal bars of many different lengths.

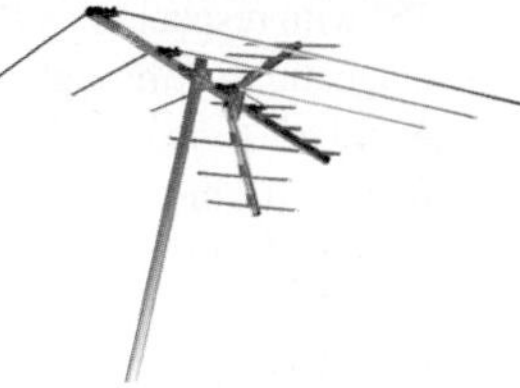

22. An AM radio detects the oscillating magnetic field of the radio wave with an antenna consisting of a coil of wire wrapped around a ferrite bar, as shown in Figure Q25.22. Ferrite is a magnetic material that "amplifies" the magnetic field of the wave.
    a. Explain how the radio antenna detects the magnetic field of the radio wave.
    b. If a radio station is located due north of you, how must the ferrite bar be oriented for best reception? Assume that the station broadcasts with a vertical antenna like the one shown in Figure 25.37.

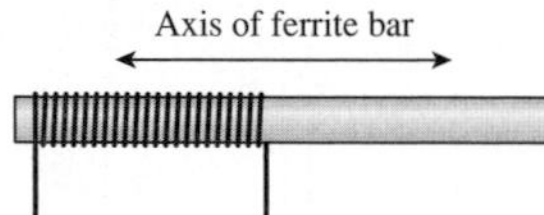

FIGURE Q25.22

23. Three laser beams have wavelengths $\lambda_1 = 400$ nm, $\lambda_2 = 600$ nm, and $\lambda_3 = 800$ nm. The power of each laser beam is 1 W.
    a. Rank in order, from largest to smallest, the photon energies E_1, E_2, and E_3 in these three laser beams. Explain.
    b. Rank in order, from largest to smallest, the number of photons per second N_1, N_2, and N_3 delivered by the three laser beams. Explain.
24. The intensity of a beam of light is increased but the light's frequency is unchanged. As a result, which of the following (perhaps more than one) are true? Explain.
    A. The photons travel faster.
    B. Each photon has more energy.
    C. The photons are larger.
    D. There are more photons per second.
25. The frequency of a beam of light is increased but the light's intensity is unchanged. As a result, which of the following (perhaps more than one) are true? Explain.
    A. The photons travel faster.
    B. Each photon has more energy.
    C. There are fewer photons per second.
    D. There are more photons per second.
26. Arc welding uses electric current to make an extremely hot electric arc that can melt metal. The arc emits ultraviolet light that can cause sunburn and eye damage if a welder is not wearing protective gear. Why does the arc give off ultraviolet light?

Multiple-Choice Questions

27. | A circular loop of wire has an area of 0.30 m^2. It is tilted by 45° with respect to a uniform 0.40 T magnetic field. What is the magnetic flux through the loop?
    A. 0.085 T·m^2
    B. 0.12 T·m^2
    C. 0.38 T·m^2
    D. 0.75 T·m^2
    E. 1.3 T·m^2
28. | In Figure Q25.28, a square loop is rotating in the plane of the page around an axis through its center. A uniform magnetic field is directed into the page. What is the direction of the induced current in the loop?
    A. Clockwise.
    B. Counterclockwise.
    C. There is no induced current.

FIGURE Q25.28

29. | A diamond-shaped loop of wire is pulled at a constant velocity through a region where the magnetic field is directed into the paper in the left half and is zero in the right half, as shown in Figure Q25.29. As the loop moves from left to right, which graph best represents the induced current in the loop as a function of time? Let a clockwise current be positive and a counterclockwise current be negative.

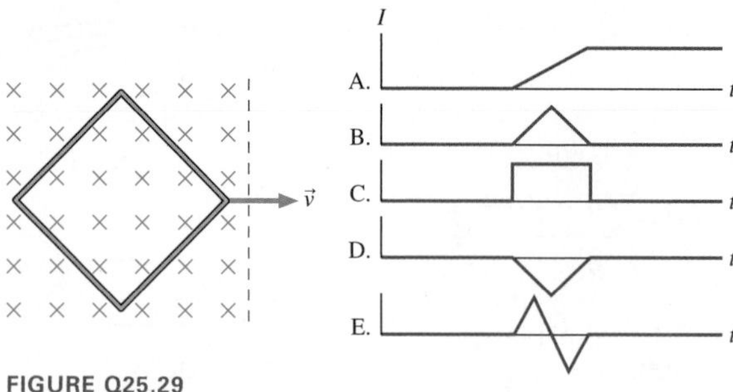

FIGURE Q25.29

30. || Figure Q25.30 shows a triangular loop of wire in a uniform magnetic field. If the field strength changes from 0.30 to 0.10 T in 50 ms, what is the induced emf in the loop?
    A. 0.08 V
    B. 0.12 V
    C. 0.16 V
    D. 0.24 V
    E. 0.36 V

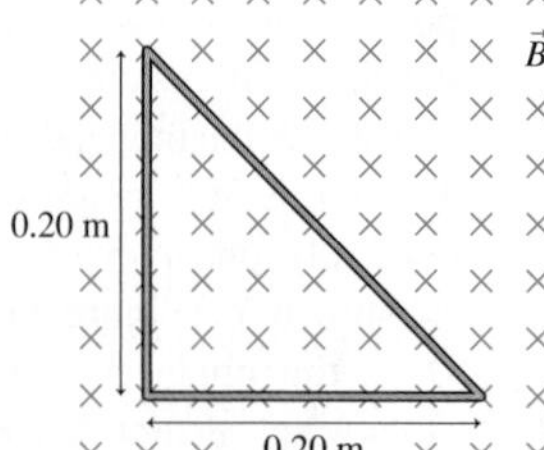

FIGURE Q25.30

31. || A device called a *flip coil* can be used to measure the earth's magnetic field. The coil has 100 turns and an area of 0.010 m^2. It is oriented with its plane perpendicular to the earth's magnetic field, then flipped 180° so the field goes through the coil in the opposite direction. The earth's magnetic field is 0.050 mT, and the coil flips over in 0.50 s. What is the average emf induced in the coil during the flip?
    A. 0.050 mV B. 0.10 mV
    C. 0.20 mV D. 1.0 mV
32. | The electromagnetic waves that carry FM radio range in frequency from 87.9 MHz to 107.9 MHz. What is the range of wavelengths of these radio waves?
    A. 500–750 nm B. 0.87–91.08 m
    C. 2.78–3.41 m D. 278–341 m
    E. 234–410 km
33. | A spacecraft in orbit around the moon measures its altitude by reflecting a pulsed 10 MHz radio signal from the surface. If the spacecraft is 10 km high, what is the time between the emission of the pulse and the detection of the echo?
    A. 33 ns B. 67 ns
    C. 33 μs D. 67 μs

34. | A 6.0 mW vertically polarized laser beam passes through a polarizing filter whose axis is 75° from vertical. What is the laser-beam power after passing through the filter?
A. 0.40 mW B. 1.0 mW
C. 1.6 mW D. 5.6 mW

35. | Communication with submerged submarines via radio waves is difficult because seawater is conductive and absorbs electromagnetic waves. Penetration into the ocean is greater at longer wavelengths, so the United States has radio installations that transmit at 76 Hz for submarine communications. What is the approximate wavelength of those extremely low-frequency waves?
A. 500 km B. 1000 km
C. 2000 km D. 4000 km

36. || How many photons are emitted during 5.0 s of operation of a red laser pointer? The device outputs 2.8 mW at a 635 nm wavelength.
A. 4.5×10^{10} B. 4.5×10^{11}
C. 4.5×10^{15} D. 4.5×10^{16}

PROBLEMS

Section 25.1 Induced Currents

Section 25.2 Motional emf

1. | A potential difference of 0.050 V is developed across the 10-cm-long wire in Figure P25.1 as it moves through a magnetic field at 5.0 m/s. The magnetic field is perpendicular to the axis of the wire. What are the direction and strength of the magnetic field?

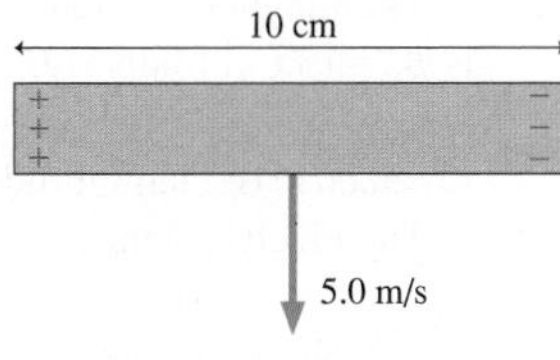

FIGURE P25.1

2. || A scalloped hammerhead shark swims at a steady speed of
BIO 1.5 m/s with its 85-cm-wide head perpendicular to the earth's 50 μT magnetic field. What is the magnitude of the emf induced between the two sides of the shark's head?

3. || A 10-cm-long wire is pulled along a U-shaped conducting rail in a perpendicular magnetic field. The total resistance of the wire and rail is 0.20 Ω. Pulling the wire with a force of 1.0 N causes 4.0 W of power to be dissipated in the circuit.
   a. What is the speed of the wire when pulled with a force of 1.0 N?
   b. What is the strength of the magnetic field?

4. | Figure P25.4 shows a 15-cm-long metal rod pulled along two frictionless, conducting rails at a constant speed of 3.5 m/s. The rails have negligible resistance, but the rod has a resistance of 0.65 Ω.
   a. What is the current induced in the rod?
   b. What force is required to keep the rod moving at a constant speed?

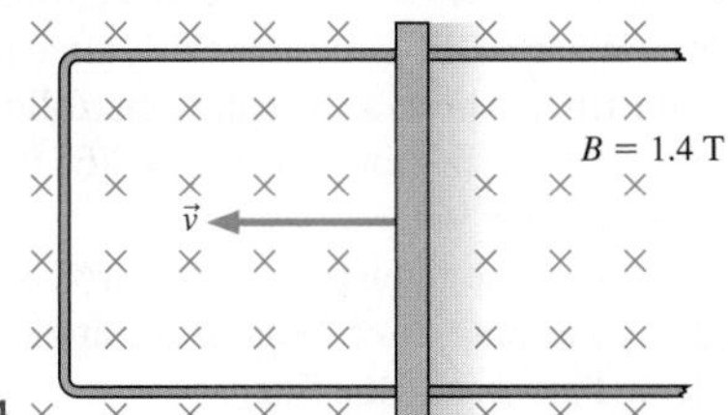

FIGURE P25.4

5. ||| A 50 g horizontal metal bar, 12 cm long, is free to slide up and down between two tall, vertical metal rods that are 12 cm apart. A 0.060 T magnetic field is directed perpendicular to the plane of the rods. The bar is raised to near the top of the rods, and a 1.0 Ω resistor is connected across the two rods at the top. Then the bar is dropped. What is the terminal speed at which the bar falls? Assume the bar remains horizontal and in contact with the rods at all times.

6. || A delivery truck with 2.8-m-high aluminum sides is driving west at 75 km/hr in a region where the earth's magnetic field is $\vec{B} = (5.0 \times 10^{-5}$ T, north).
   a. What is the potential difference between the top and the bottom of the truck's side panels?
   b. Will the tops of the panels be positive or negative relative to the bottoms?

Section 25.3 Magnetic Flux

7. | Figure P25.7 is an edge-on view of a 10-cm-diameter circular loop rotating in a uniform 0.050 T magnetic field. What is the magnetic flux through the loop when θ is 0°, 30°, 60°, and 90°?

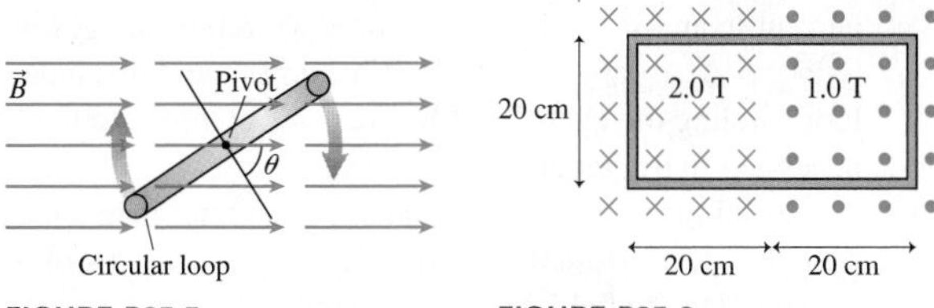

FIGURE P25.7 FIGURE P25.8

8. || What is the magnetic flux through the loop shown in Figure P25.8?

9. ||| The 2.0-cm-diameter solenoid in Figure P25.9 passes through the center of a 6.0-cm-diameter loop. The magnetic field inside the solenoid is 0.20 T. What is the magnetic flux through the loop (a) when it is perpendicular to the solenoid and (b) when it is tilted at a 60° angle?

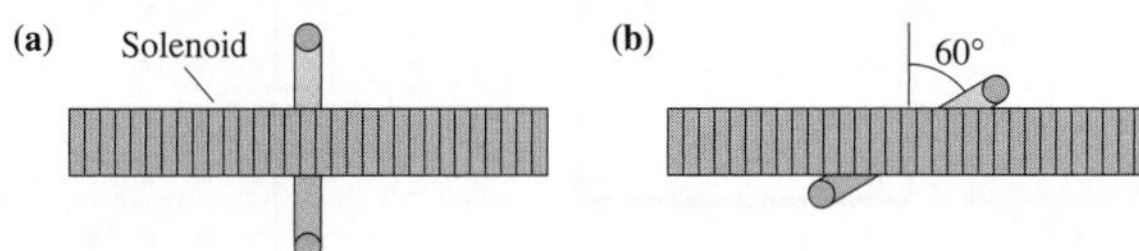

FIGURE P25.9

10. || At a typical location in the United States, the earth's magnetic field has a magnitude of 5.0×10^{-5} T and is at a 65° angle from the horizontal. What is the flux through the 22 cm × 28 cm front cover of your textbook if it is flat on your desk?

11. | The metal equilateral triangle in Figure P25.11, 20 cm on each side, is halfway into a 0.10 T magnetic field.
   a. What is the magnetic flux through the triangle?
   b. If the magnetic field strength decreases, what is the direction of the induced current in the triangle?

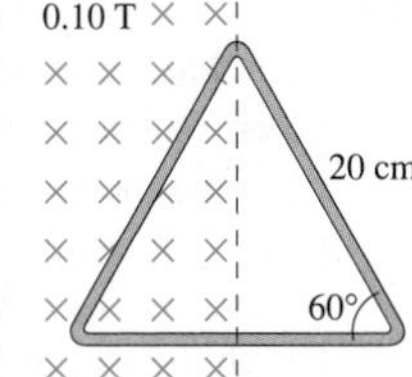

FIGURE P25.11

Section 25.4 Faraday's Law

12. I Figure P25.12 shows a 10-cm-diameter loop in three different magnetic fields. The loop's resistance is 0.10 Ω. For each case, determine the induced emf, the induced current, and the direction of the current.

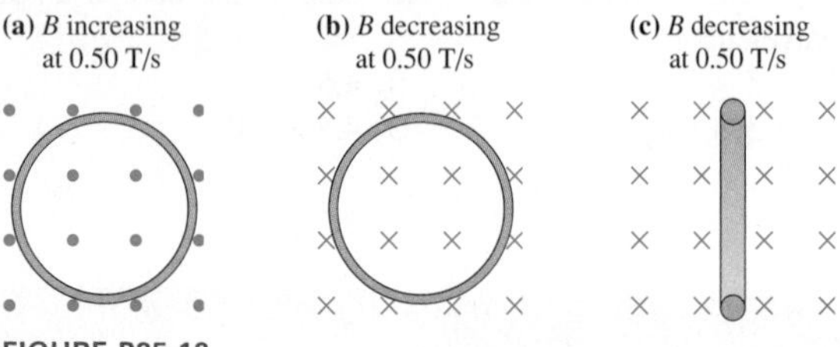

FIGURE P25.12

13. II A loop of wire is perpendicular to a magnetic field. Rank, from greatest to least, the magnitudes of the loop's induced emf for the following situations:
    A. The magnetic field strength increases from 0 to 1 T in 6 s.
    B. The magnetic field strength increases from 1 T to 4 T in 2 s.
    C. The magnetic field strength remains at 4 T for 1 min.
    D. The magnetic field strength decreases from 4 T to 3 T in 4 s.
    E. The magnetic field strength decreases from 3 T to 0 T in 1 s.
14. II Patients undergoing an MRI occasionally report seeing
BIO flashes of light. Some practitioners assume that this results from electric stimulation of the eye by the emf induced by the rapidly changing fields of an MRI solenoid. We can do a quick calculation to see if this is a reasonable assumption. The human eyeball has a diameter of approximately 25 mm. Rapid changes in current in an MRI solenoid can produce rapid changes in field, with $\Delta B/\Delta t$ as large as 50 T/s. What emf would this induce in a loop circling the eyeball? How does this compare to the 15 mV necessary to trigger an action potential?
15. III A 1000-turn coil of wire 2.0 cm in diameter is in a magnetic field that drops from 0.10 T to 0 T in 10 ms. The axis of the coil is parallel to the field. What is the emf of the coil?
16. II The loop in Figure P25.16 has an induced current as shown. The loop has a resistance of 0.10 Ω. Is the magnetic field strength increasing or decreasing? What is the rate of change of the field, $\Delta B/\Delta t$?

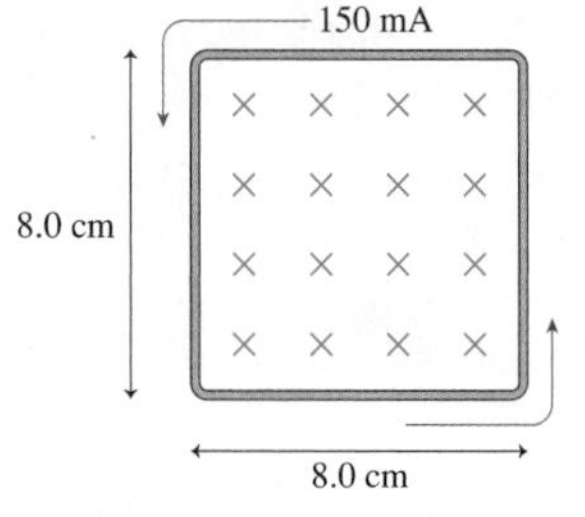

FIGURE P25.16

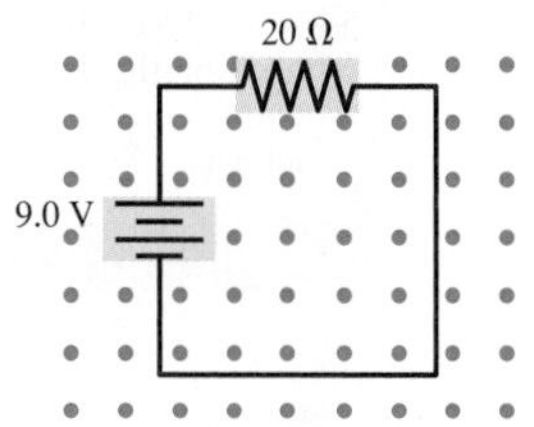

FIGURE P25.17

17. II The circuit of Figure P25.17 is a square 5.0 cm on a side. The magnetic field increases steadily from 0 T to 0.50 T in 10 ms. What is the current in the resistor during this time?
18. III A 5.0-cm-diameter loop of wire has resistance 1.2 Ω. A nearby solenoid generates a uniform magnetic field perpendicular to the loop that varies with time as shown in Figure P25.18. Graph the magnitude of the current in the loop over the same time interval.

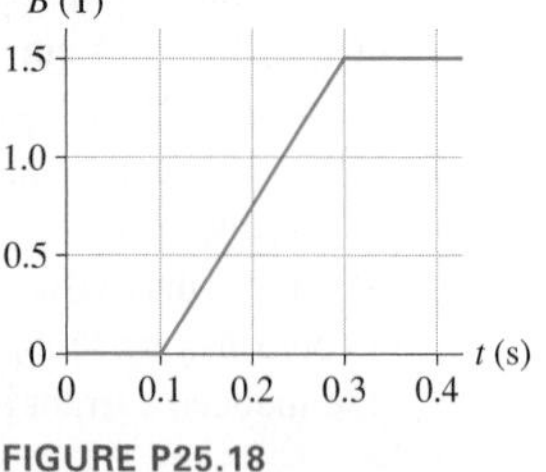

FIGURE P25.18

Section 25.5 Induced Fields and Electromagnetic Waves

Section 25.6 Properties of Electromagnetic Waves

19. I What is the electric field amplitude of an electromagnetic wave whose magnetic field amplitude is 2.0 mT?
20. I What is the magnetic field amplitude of an electromagnetic wave whose electric field amplitude is 10 V/m?
21. II A microwave oven operates at 2.4 GHz with an intensity inside the oven of 2500 W/m^2. What are the amplitudes of the oscillating electric and magnetic fields?
22. II The maximum allowed leakage of microwave radiation from a microwave oven is 5.0 mW/cm^2. If microwave radiation outside an oven has the maximum value, what is the amplitude of the oscillating electric field?
23. III A typical helium-neon laser found in supermarket checkout scanners emits 633-nm-wavelength light in a 1.0-mm-diameter beam with a power of 1.0 mW. What are the amplitudes of the oscillating electric and magnetic fields in the laser beam?
24. I The magnetic field of an electromagnetic wave in a vacuum is $B_z = (3.0\ \mu\text{T})\sin((1.0 \times 10^7)x - 2\pi ft)$, where x is in m and t is in s. What are the wave's (a) wavelength, (b) frequency, and (c) electric field amplitude?
25. II The electric field of an electromagnetic wave in a vacuum is $E_y = (20\ \text{V/m})\sin((6.28 \times 10^8)x - 2\pi ft)$, where x is in m and t is in s. What are the wave's (a) wavelength, (b) frequency, and (c) magnetic field amplitude?
26. II A radio receiver can detect signals with electric field amplitudes as small as 300 μV/m. What is the intensity of the smallest detectable signal?
27. II A 200 MW laser pulse is focused with a lens to a diameter of
INT 2.0 μm.
    a. What is the laser beam's electric field amplitude at the focal point?
    b. What is the ratio of the laser beam's electric field to the electric field that keeps the electron bound to the proton of a hydrogen atom? The radius of the electron's orbit is 0.053 nm.
28. III A radio antenna broadcasts a 1.0 MHz radio wave with 25 kW of power. Assume that the radiation is emitted uniformly in all directions.
    a. What is the wave's intensity 30 km from the antenna?
    b. What is the electric field amplitude at this distance?
29. II At what distance from a 10 W point source of electromagnetic waves is the electric field amplitude (a) 100 V/m and (b) 0.010 V/m?
30. I The intensity of a polarized electromagnetic wave is 10 W/m^2. What will be the intensity after passing through a polarizing filter whose axis makes the following angles with the plane of polarization? (a) $\theta = 0°$ (b) $\theta = 30°$ (c) $\theta = 45°$ (d) $\theta = 60°$ (e) $\theta = 90°$.
31. II Only 25% of the intensity of a polarized light wave passes through a polarizing filter. What is the angle between the electric field and the axis of the filter?
32. II A 200 mW horizontally polarized laser beam passes through a polarizing filter whose axis is 25° from vertical. What is the power of the laser beam as it emerges from the filter?
33. II The polarization of a helium-neon laser can change with time. The light from a 1.5 mW laser is initially horizontally polarized; as the laser warms up, the light changes to be vertically polarized. Suppose the laser beam passes through a polarizer whose axis is 30° from horizontal. By what percent does the light intensity transmitted through the polarizer decrease as the laser warms up?

Section 25.7 The Photon Model of Electromagnetic Waves

34. | What is the energy (in eV) of a photon of visible light that has a wavelength of 500 nm?
35. | What is the energy (in eV) of an x-ray photon that has a wavelength of 1.0 nm?
36. | What is the wavelength of a photon whose energy is twice that of a photon with a 600 nm wavelength?
37. || One recent study has shown that x rays with a wavelength of
BIO 0.0050 nm can produce mutations in human cells.
    a. Calculate the energy in eV of a photon of radiation with this wavelength.
    b. Assuming that the bond energy holding together a water molecule is typical, use Table 25.1 to estimate how many molecular bonds could be broken with this energy.
38. | Rod cells in the retina of the eye detect light using a pho-
BIO topigment called rhodopsin. 1.8 eV is the lowest photon energy that can trigger a response in rhodopsin. What is the maximum wavelength of electromagnetic radiation that can cause a transition? In what part of the spectrum is this?
39. ||| What is the energy of 1 mol of photons that have a wavelength of 1.0 μm?
40. || The thermal emission of the human body has maximum
BIO intensity at a wavelength of approximately 9.5 μm. What photon energy corresponds to this wavelength?
41. || The intensity of electromagnetic radiation from the sun reaching the earth's upper atmosphere is 1.37 kW/m^2. Assuming an average wavelength of 680 nm for this radiation, find the number of photons per second that strike a 1.00 m^2 solar panel directly facing the sun on an orbiting satellite.
42. ||| The human eye can barely detect a star whose intensity at the
BIO earth's surface is 1.6×10^{-11} W/m^2. If the dark-adapted eye has a pupil diameter of 7.0 mm, how many photons per second enter the eye from the star? Assume the starlight has a wavelength of 550 nm.

Section 25.8 The Electromagnetic Spectrum

43. || The spectrum of a glowing filament has its peak at a wavelength of 1200 nm. What is the temperature of the filament, in °C?
44. ||| While using a dimmer switch to investigate a new type of incandescent light bulb, you notice that the light changes both its spectral characteristics and its brightness as the voltage is increased.
    a. If the wavelength of maximum intensity decreases from 1800 nm to 1600 nm as the bulb's voltage is increased, by how many °C does the filament temperature increase?
    b. By what factor does the total radiation from the filament increase due to this temperature change?
45. | The photon energies used in different types of medical x-ray
BIO imaging vary widely, depending upon the application. Single dental x rays use photons with energies of about 25 keV. The photon energy used for x-ray microtomography, a process that allows repeated imaging in single planes at varying depths within the sample, is 2.5 times greater. What are the wavelengths of the x rays used for these two purposes?

General Problems

46. || A 10 cm × 10 cm square is bent at a 90° angle as shown in Figure P25.46. A uniform 0.050 T magnetic field points downward at a 45° angle. What is the magnetic flux through the loop?

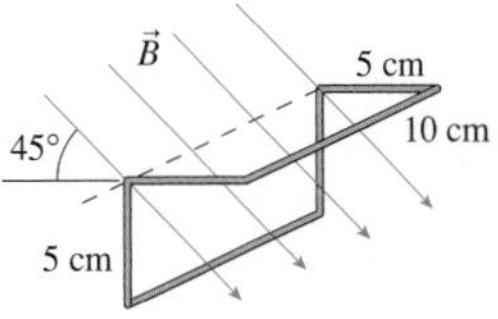

FIGURE P25.46

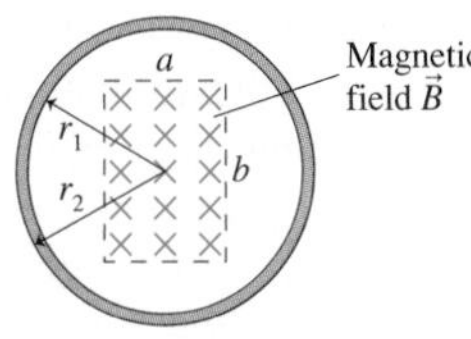

FIGURE P25.47

47. | What is the magnetic flux through the loop shown in Figure P25.47?
48. || a. A circular loop antenna has a diameter of 20 cm. If the plane of the loop is perpendicular to the earth's 50 μT magnetic field, what is the flux through the loop?
    b. What is the flux if the loop is rotated by 30°?
49. || An 1.1-m-diameter MRI solenoid with a length of 2.4 m has a magnetic field of 1.5 T along its axis. If the current is turned off in a time of 1.2 s, what is the induced emf in one turn of the solenoid's windings?
50. || A magnet and a coil are oriented as shown in Figure P25.50. The magnet is moved rapidly into the coil, held stationary in the coil for a short time, and then rapidly pulled back out of the coil. Sketch a graph showing the reading of the ammeter as a function of time. The ammeter registers a positive value when current goes into the "+" terminal.

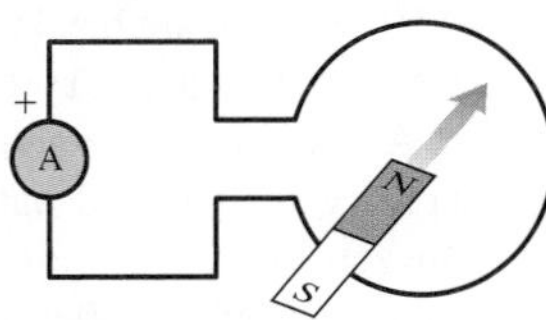

FIGURE P25.50

51. ||| A wire loop with an area of 0.020 m^2 is in a magnetic field of 0.30 T directed at a 30° angle to the plane of the loop. If the field drops to zero in 45 ms, what is the average induced emf in the loop?
52. ||| A 100-turn, 2.0-cm-diameter coil is at rest in a horizontal plane. A uniform magnetic field 60° away from vertical increases from 0.50 T to 1.50 T in 0.60 s. What is the induced emf in the coil?
53. |||| A 25-turn, 10.0-cm-diameter coil is oriented in a vertical plane with its axis aligned east-west. A magnetic field pointing to the northeast decreases from 0.80 T to 0.20 T in 2.0 s. What is the emf induced in the coil?
54. |||| People immersed in strong unchanging magnetic fields occa-
BIO sionally report sensing a metallic taste. Some investigators suspect that motion in the constant field could produce a changing flux and a resulting emf that could stimulate nerves in the tongue. We can make a simple model to see if this is reasonable by imagining a somewhat extreme case. Suppose a patient having an MRI is immersed in a 3.0 T field along the axis of his body. He then quickly tips his head to the side, toward his right shoulder, tipping his head by 30° in the rather short time of 0.15 s. Estimate the area of the tongue; then calculate the emf that could be induced in a loop around the outside of the tongue by this motion of the head. How does this emf compare to the approximately 15 mV necessary to trigger an action potential? Does it seem reasonable to suppose that an induced emf is responsible for the noted effect?
55. |||| A 20 cm length of 0.32-mm-diameter nichrome wire is
INT welded into a circular loop. The loop is placed between the poles of an electromagnet, and a field of 0.55 T is switched on in a time of 15 ms. What is the induced current in the loop?
56. |||| Currents induced by rapid field changes in an MRI solenoid
BIO can, in some cases, heat tissues in the body, but under normal
INT circumstances the heating is small. We can do a quick estimate to show this. Consider the "loop" of muscle tissue shown in Figure P25.56. This might be muscle circling the bone of

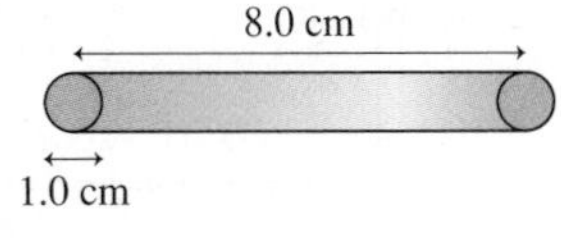

FIGURE P25.56

your arm or leg. Muscle tissue is not a great conductor, but current will pass through muscle and so we can consider this a conducting loop with a rather high resistance. Suppose the magnetic field along the axis of the loop drops from 1.6 T to 0 T in 0.30 s, as it might in an MRI solenoid.

a. How much energy is dissipated in the loop?

b. By how much will the temperature of the tissue increase? Assume that muscle tissue has resistivity 13 $\Omega \cdot$ m, density 1.1×10^3 kg/m^3, and specific heat 3600 J/kg $\cdot$ K.

57. ||| A 100-turn, 8.0-cm-diameter coil is made of 0.50-mm-diameter copper wire. A magnetic field is perpendicular to the coil. At what rate must B increase to induce a 2.0 A current in the coil?
INT

58. || The loop in Figure P25.58 is being pushed into the 0.20 T magnetic field at 50 m/s. The resistance of the loop is 0.10 Ω. What are the direction and magnitude of the current in the loop?

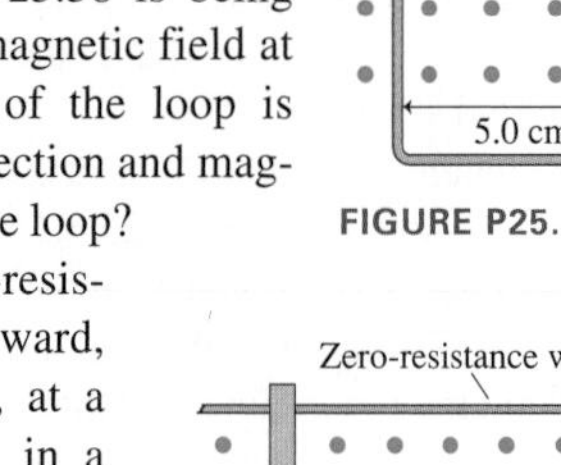

FIGURE P25.58

59. ||| A 20-cm-long, zero-resistance wire is pulled outward, on zero-resistance rails, at a steady speed of 10 m/s in a 0.10 T magnetic field. (See Figure P25.59.) On the opposite side, a 1.0 Ω carbon resistor completes the circuit by connecting the two rails. The mass of the resistor is 50 mg.
INT

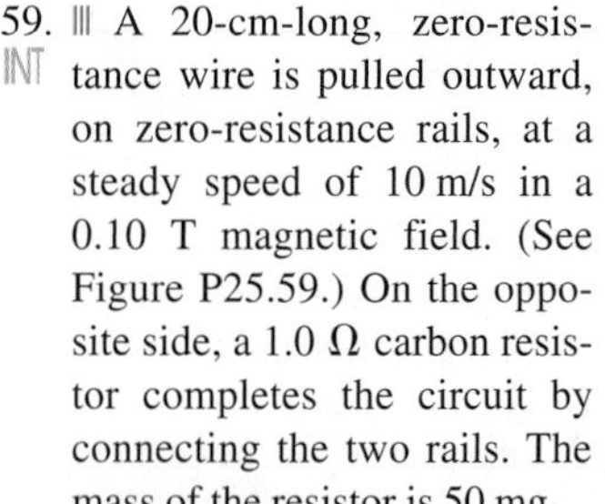

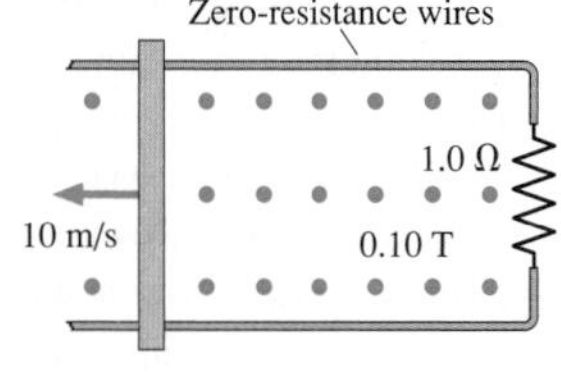

FIGURE P25.59

a. What is the induced current in the circuit?

b. How much force is needed to pull the wire at this speed?

c. How much does the temperature of the carbon increase if the wire is pulled for 10 s? The specific heat of carbon is 710 J/kg $\cdot$ K. Neglect thermal energy transfer out of the resistor.

60. |||| A TMS (transcranial magnetic stimulation) device creates very rapidly changing magnetic fields. The field near a typical pulsed-field machine rises from 0 T to 2.5 T in 200 μs. Suppose a technician holds his hand near the device so that the axis of his 2.0-cm-diameter wedding band is parallel to the field.
BIO INT

a. What emf is induced in the ring as the field changes?

b. If the band is gold with a cross-section area of 4.0 mm^2, what is the induced current?

Can you see why TMS technicians are advised to remove all jewelry?

61. || The 10-cm-wide, zero-resistance wire shown in Figure P25.61 is pushed toward the 2.0 Ω resistor at a steady speed of 0.50 m/s. The magnetic field strength is 0.50 T.

a. How big is the pushing force?

b. How much power does the pushing force supply to the wire?

c. What are the direction and magnitude of the induced current?

d. How much power is dissipated in the resistor?

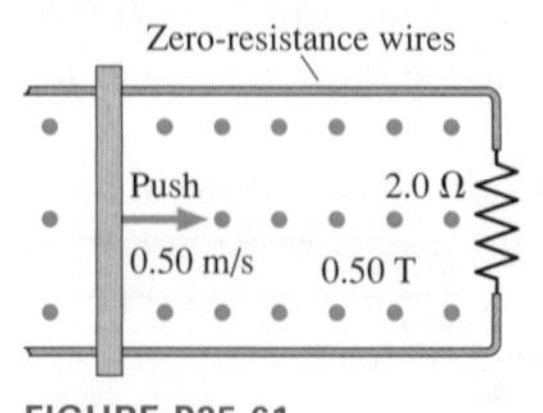

FIGURE P25.61

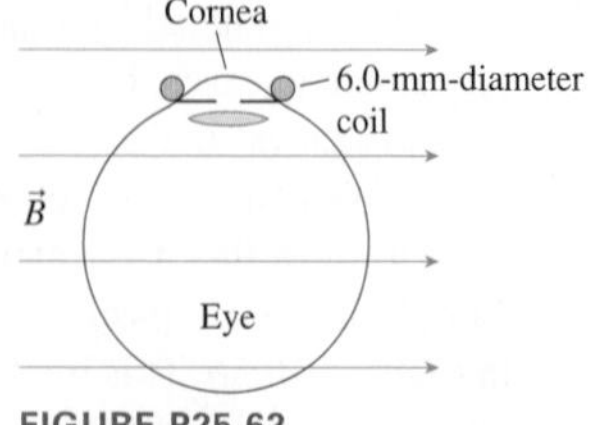

FIGURE P25.62

62. |||| Experiments to study vision often need to track the movements of a subject's eye. One way of doing so is to have the subject sit in a magnetic field while wearing special contact lenses that have a coil of very fine wire circling the edge. A current is induced in the coil each time the subject rotates his eye. Consider an experiment in which a 20-turn, 6.0-mm-diameter coil of wire circles the subject's cornea while a 1.0 T magnetic field is directed as shown in Figure P25.62. The subject begins by looking straight ahead. What emf is induced in the coil if the subject shifts his gaze by 5.0° in 0.20 s?
BIO

63. ||| The filament in the center of a 100 W incandescent bulb emits approximately 4.0 W of visible light. If you assume that all of this light is emitted at a single wavelength, estimate the electric and magnetic field strength at the surface of the bulb.

64. || A LASIK vision correction system uses a laser that emits 10-ns-long pulses of light, each with 2.5 mJ of energy. The laser is focused to a 0.85-mm-diameter circle. (a) What is the average power of each laser pulse? (b) What is the electric field strength of the laser light at the focus point?
BIO INT

65. | When the Voyager 2 spacecraft passed Neptune in 1989, it was 4.5×10^9 km from the earth. Its radio transmitter, with which it sent back data and images, broadcast with a mere 21 W of power. Assuming that the transmitter broadcast equally in all directions,

a. What signal intensity was received on the earth?

b. What electric field amplitude was detected?

(The received signal was slightly stronger than your result because the spacecraft used a directional antenna.)

66. || A new cordless phone emits 4.0 mW at 5.8 GHz. The manufacturer claims that the phone has a range of 100 feet. If we assume that the wave spreads out evenly with no obstructions, what is the electric field strength at the base unit 100 feet from the phone?

67. ||| 633-nm-wavelength light from a helium-neon laser is vertically polarized. Suppose a laser beam passes through a polarizer with its axis 45° from the vertical. What is the ratio of the laser beam's electric field strengths before and after the polarizer?

68. || In reading the instruction manual that came with your garage-door opener, you see that the transmitter unit in your car produces a 250 mW signal and that the receiver unit is supposed to respond to a radio wave of the correct frequency if the electric field amplitude exceeds 0.10 V/m. You wonder if this is really true. To find out, you put fresh batteries in the transmitter and start walking away from your garage while opening and closing the door. Your garage door finally fails to respond when you're 42 m away. Are the manufacturer's claims true?

69. ||| Unpolarized light passes through a vertical polarizing filter, emerging with an intensity I_0. The light then passes through a horizontal filter, which blocks all of the light; the intensity transmitted through the pair of filters is zero. Suppose a third polarizer with axis 45° from vertical is inserted between the first two. What is the transmitted intensity now?

70. | a. What is the wavelength of a gamma-ray photon with energy 1.0×10^{-13} J?

b. How many visible-light photons with a wavelength of 500 nm would you need to match the energy of this one gamma-ray photon?

71. | Gamma rays with the very high energy of 2.0×10^{13} eV are occasionally observed from distant astrophysical sources. What are the wavelength and frequency corresponding to this photon energy?

72. | A 1000 kHz AM radio station broadcasts with a power of 20 kW. How many photons does the transmitting antenna emit each second?

73. || Fireflies emit flashes of light to communicate, directly converting chemical energy into the energy of visible-light photons. A firefly emits a 0.15-s-long pulse of light with an average wavelength of 590 nm. During the flash, the emitted power is 40 μW.
BIO
   a. What is the energy of one light photon having the average wavelength?
   b. Assume that all photons have the average wavelength. How many photons does the firefly emit during each flash?

74. |||| The human body has a surface area of approximately 1.8 m^2, a surface temperature of approximately 30°C, and a typical emissivity at infrared wavelengths of $e = 0.97$. If we make the approximation that all photons are emitted at the wavelength of peak intensity, how many photons per second does the body emit?
BIO

75. ||| For radio and microwaves, the depth of penetration into the human body is approximately proportional to $\sqrt{\lambda}$. If 27 MHz radio waves penetrate to a depth of 14 cm, estimate the depth of penetration of 2.4 GHz microwaves.
BIO

Passage Problems

The Metal Detector

Metal detectors use induced currents to sense the presence of any metal—not just magnetic materials such as iron. A metal detector, shown in Figure P25.76, consists of two coils: a transmitter coil and a receiver coil. A high-frequency oscillating current in the transmitter coil generates an oscillating magnetic field along the axis and a changing flux through the receiver coil. Consequently, there is an oscillating induced current in the receiver coil.

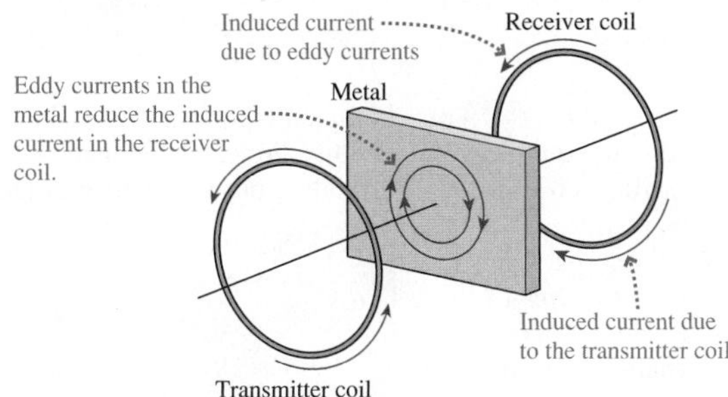

FIGURE P25.76

If a piece of metal is placed between the transmitter and the receiver, the oscillating magnetic field in the metal induces eddy currents in a plane parallel to the transmitter and receiver coils. The receiver coil then responds to the superposition of the transmitter's magnetic field and the magnetic field of the eddy currents. Because the eddy currents attempt to prevent the flux from changing, in accordance with Lenz's law, the net field at the receiver decreases when a piece of metal is inserted between the coils. Electronic circuits detect the current decrease in the receiver coil and set off an alarm.

76. | The metal detector will not detect insulators because
   A. Insulators block magnetic fields.
   B. No eddy current can be produced in an insulator.
   C. No emf can be produced in an insulator.
   D. An insulator will increase the field at the receiver.

77. | A metal detector can detect the presence of metal screws used to repair a broken bone inside the body. This tells us that
   A. The screws are made of magnetic materials.
   B. The tissues of the body are conducting.
   C. The magnetic fields of the device can penetrate the tissues of the body.
   D. The screws must be perfectly aligned with the axis of the device.

78. | Suppose the magnetic field from the transmitter coil in Figure P25.76 points toward the receiver coil and is increasing with time. As viewed along this axis, the induced currents are
   A. Clockwise in the metal, clockwise in the receiver coil
   B. Clockwise in the metal, counterclockwise in the receiver coil
   C. Counterclockwise in the metal, clockwise in the receiver coil
   D. Counterclockwise in the metal, counterclockwise in the receiver coil

79. | Which of the following changes would *not* produce a larger eddy current in the metal?
   A. Increasing the frequency of the oscillating current in the transmitter coil
   B. Increasing the magnitude of the oscillating current in the transmitter coil
   C. Increasing the resistivity of the metal
   D. Decreasing the distance between the metal and the transmitter

STOP TO THINK ANSWERS

Stop to Think 25.1: E. According to the right-hand rule, the magnetic force on a positive charge carrier is to the right.

Stop to Think 25.2: D. The field of the bar magnet emerges from the north pole and points upward. As the coil moves toward the pole, the flux through it is upward and increasing. To oppose the increase, the induced field must point downward. This requires a clockwise (negative) current. As the coil moves away from the pole, the upward flux is decreasing. To oppose the decrease, the induced field must point upward. This requires a counterclockwise (positive) current.

Stop to Think 25.3: E. The right-hand rule requires $\vec{B}$ to point into the page if the wave is to propagate upward.

Stop to Think 25.4: $I_D > I_A > I_B = I_C$. The intensity depends upon $\cos^2\theta$, where θ is the angle *between* the axes of the two filters. The filters in D have $\theta = 0°$, so all light is transmitted. The two filters in both B and C are crossed ($\theta = 90°$) and transmit no light at all.

Stop to Think 25.5: A. The photon energy is proportional to the frequency. The photons of the 90.5 MHz station each have lower energy, so more photons must be emitted per second.

Stop to Think 25.6: D. A hotter object emits more radiation across the *entire* spectrum than a cooler object. The 6000 K star has its maximum intensity in the blue region of the spectrum, but it still emits more red radiation than the somewhat cooler stars.

26 AC Electricity

Transmission lines carry alternating current at voltages that can exceed 500,000 V. Why are high voltages used? And why can birds perch safely on high-voltage wires?

LOOKING AHEAD ▸▸

The goal of Chapter 26 is to understand and apply basic principles of AC electricity, electricity transmission, and household electricity.

Transformers

Two coils of wire wrapped around an iron core makes a **transformer,** a device that can change the voltage of AC electricity.

You likely have several portable devices that charge up or run off power packs. These power packs use transformers, converting the 120 V of the wall outlet to the lower voltage needed by electronic devices.

Looking Back ◂◂
25.3–25.4 Lenz's law and Faraday's law

Electricity Transmission

The wires that transmit electricity across large distances—the grid—do so at very high voltage. We'll see why this is so.

Transformers in your neighborhood change the locally transmitted high-voltage electricity to a more modest voltage that is safe to use in your home.

AC Electricity

We've said that normal household electricity is 120 V, but that's not quite true. The actual voltage oscillates 60 times a second with a peak of approximately 170 V. We'll see that 120 V is what's called the *root-mean-square voltage*. The oscillating voltage causes the current to alternate directions, so we know this as AC, or **alternating current,** electricity.

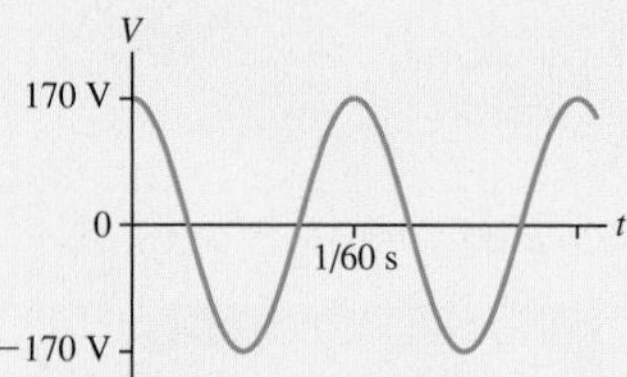

Household Electricity and Electrical Safety

The electric nature of your nervous system means that modest electric shocks can be dangerous. You'll learn about household electricity and the safety features built in.

GFI circuits, which detect potentially dangerous electric leakages, provide additional protection at bathroom and kitchen outlets.

AC Circuits

Many common devices make use of AC circuits. We can understand AC circuits by extending our knowledge of DC circuits.

The trackpad on a laptop computer uses a network of AC capacitor circuits to determine the position of your fingertip.

Turning your radio tuning dial adjusts the resonant frequency of an AC oscillation circuit.

Looking Back ◂◂
14.1 and 14.6–14.7 Simple harmonic motion, damped oscillations, and resonance
21.7 Capacitors
23.2 Fundamentals of circuit analysis

26.1 Alternating Current

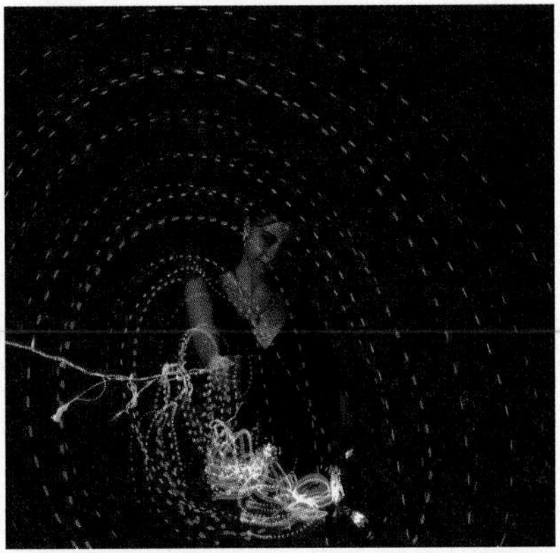

LED minilights flash on and off 60 times a second.

A battery creates a constant emf. In a battery-powered flashlight, the bulb carries a constant current and glows with a steady light. The electricity distributed to homes in your neighborhood is different. The picture on the right is a long-exposure photo of a string of LED minilights swung through the air. Each bulb appears as a series of dashes because each bulb in the string flashes on and off 60 times each second. This isn't a special property of the bulbs, but of the electricity that runs them. Household electricity does not have a constant emf; it has a sinusoidal variation that causes the light output of the bulbs to vary as well, although the resulting flicker is too rapid to notice under normal circumstances.

In Chapter 25 we saw that an electrical generator—whether powered by steam, water, or wind—works by rotating a coil of wire in a magnetic field. The steady rotation of the coil causes the emf and the induced current in the coil to oscillate sinusoidally, alternately positive and then negative. This oscillation forces the charges to flow first in one direction and then, a half cycle later, in the other—an **alternating current,** abbreviated as AC. (If the emf is constant and the current is always in the same direction, we call the electricity *direct current,* abbreviated as DC.) The electricity from power outlets in your house is *AC electricity,* with an emf oscillating at a frequency of 60 Hz. Audio, radio, television, computer, and telecommunication equipment also make extensive use of AC circuits, with frequencies ranging from approximately 10^2 Hz in audio circuits to approximately 10^9 Hz in cell phones.

The instantaneous emf of an AC voltage source, shown graphically in FIGURE 26.1, can be written as

$$\mathcal{E} = \mathcal{E}_0 \cos(2\pi f t) = \mathcal{E}_0 \cos\left(\frac{2\pi t}{T}\right) \qquad (26.1)$$

Emf of an AC voltage source

SINUSOIDAL p. 451

FIGURE 26.1 The emf of an AC voltage source.

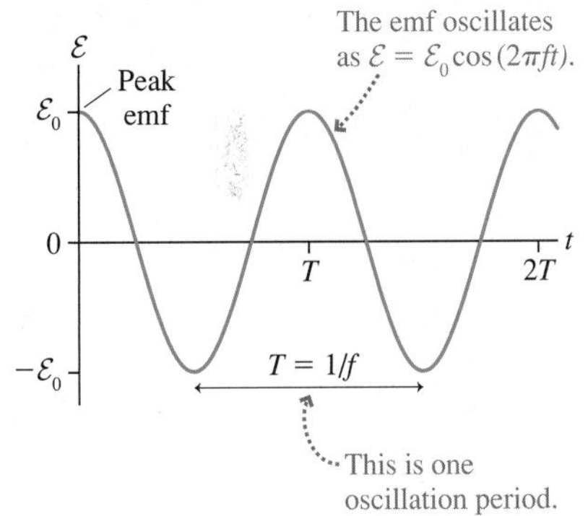

where $\mathcal{E}_0$ is the peak or maximum emf (recall that the units of emf are volts), T is the period of oscillation (in s), and $f = 1/T$ is the oscillation frequency (in cycles per second, or Hz).

Resistor Circuits

In Chapter 23 you learned to analyze a circuit in terms of the current I and potential difference ΔV. Now, because the current and voltage are oscillating, we will use a lowercase i to represent the *instantaneous* current through a circuit element, the value of the current at a particular instant of time. Similarly, we will use a lowercase v for the circuit element's instantaneous voltage.

FIGURE 26.2 shows the instantaneous current i_R through a resistor R. The potential difference across the resistor, which we call the *resistor voltage* v_R, is given by Ohm's law:

$$v_R = i_R R \qquad (26.2)$$

FIGURE 26.2 The instantaneous current through a resistor.

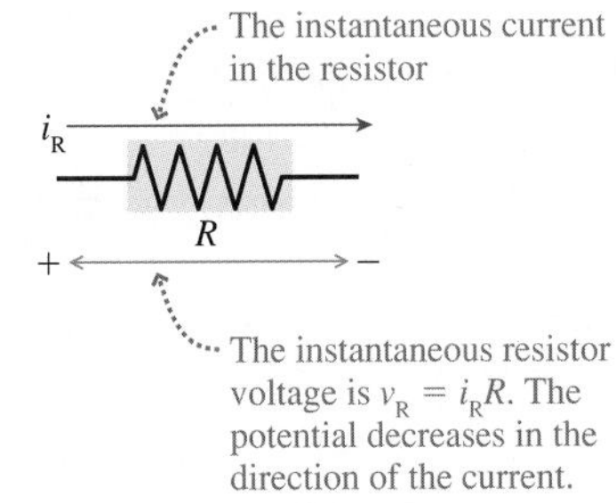

FIGURE 26.3 shows a resistor R connected across an AC emf $\mathcal{E}$. The circuit symbol for an AC generator is -(~)-. We can analyze this circuit in exactly the same way we analyzed a DC resistor circuit. Kirchhoff's loop law says that the sum of all the potential differences around a closed path is zero so we can write:

$$\sum \Delta V = \Delta V_{source} + \Delta V_R = \mathcal{E} - v_R = 0 \qquad (26.3)$$

The minus sign appears, just as it did in the equation for a DC circuit, because the potential *decreases* when we travel through a resistor in the direction of the current. Thus we find from the loop law that $v_R = \mathcal{E} = \mathcal{E}_0 \cos(2\pi f t)$. This isn't surprising because the resistor is connected directly across the terminals of the emf.

FIGURE 26.3 An AC resistor circuit.

This is the current direction when $\mathcal{E} > 0$. A half cycle later it will be in the opposite direction.

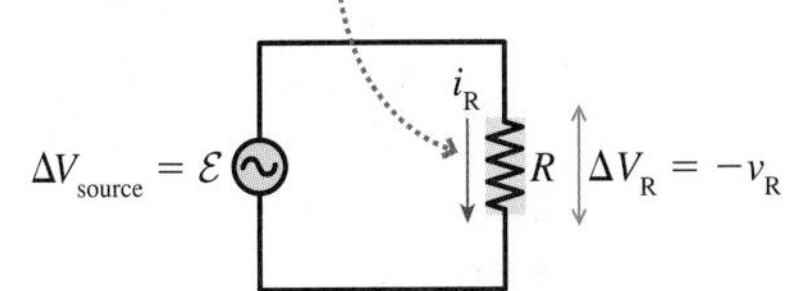

Because the resistor voltage is a sinusoidal voltage at frequency f, it is useful to write

$$v_R = V_R \cos(2\pi f t) \tag{26.4}$$

In this equation V_R is the peak or maximum voltage, the amplitude of the sinusoidally varying voltage. You can see that $V_R = \mathcal{E}_0$ in the single-resistor circuit of Figure 26.3. Thus the current through the resistor is

$$i_R = \frac{v_R}{R} = \frac{V_R \cos(2\pi f t)}{R} = I_R \cos(2\pi f t) \tag{26.5}$$

where $I_R = V_R/R$ is the peak current.

NOTE ▶ It is important to understand the distinction between instantaneous and peak quantities. The instantaneous current i_R, for example, is a quantity that is changing with time according to Equation 26.5. The peak current I_R is the maximum value that the instantaneous current reaches. The instantaneous current oscillates between $+I_R$ and $-I_R$. ◀

FIGURE 26.4 Graph of the current through and voltage across a resistor.

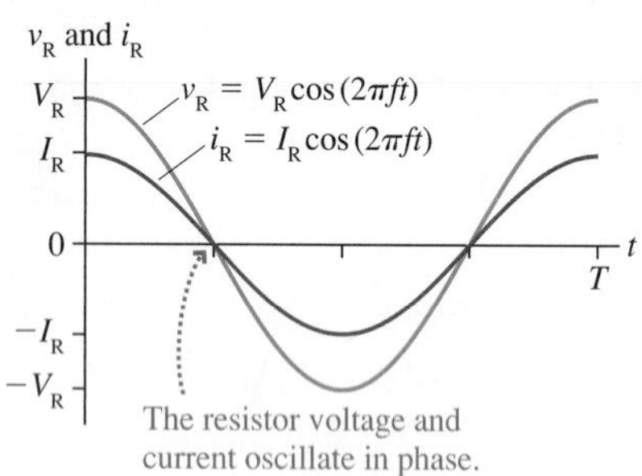

The resistor's instantaneous current and voltage are *in phase,* both oscillating as $\cos(2\pi f t)$. FIGURE 26.4 shows the voltage and the current simultaneously on a graph. The fact that the peak current I_R is drawn as being less than V_R has no significance. Current and voltage are measured in different units, so in a graph like this you can't compare the value of one to the value of the other. Showing the two different quantities on a single graph—a tactic that can be misleading if you're not careful—simply illustrates that they oscillate *in phase:* **The current is at its maximum value when the voltage is at its maximum, and the current is at its minimum value when the voltage is at its minimum.**

AC Power in Resistors

In Chapter 23 you learned that the power dissipated by a resistor is $P = I\Delta V_R = I^2R$. In an AC circuit, the resistor current i_R and voltage v_R are constantly changing, as we saw in Figure 26.4, so the instantaneous power loss $p = i_R v_R = i_R^2 R$ (note the lowercase p) is constantly changing as well. We can use Equations 26.4 and 26.5 to write this instantaneous power as

$$p = i_R^2 R = [I_R \cos(2\pi f t)]^2 R = I_R^2 R[\cos(2\pi f t)]^2 \tag{26.6}$$

FIGURE 26.5 The instantaneous power loss in a resistor.

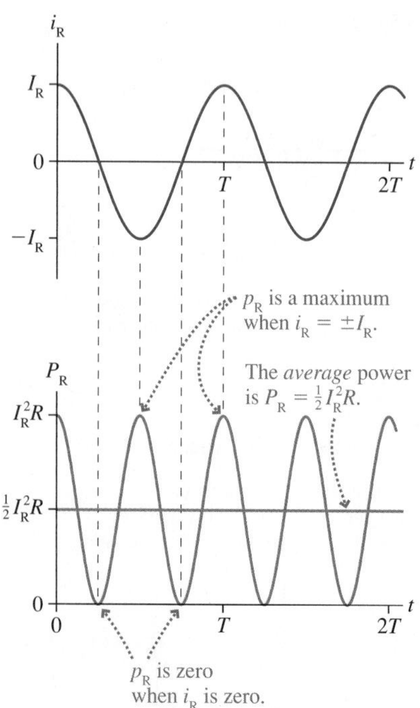

FIGURE 26.5 shows the instantaneous power graphically. You can see that, because the cosine is squared, the power oscillates *twice* during every cycle of the emf: The energy dissipation peaks both when $i_R = I_R$ and when $i_R = -I_R$. The energy dissipation doesn't depend on the current's direction through the resistor.

The current in a household lightbulb reverses direction 120 times per second (twice per cycle), so the power reaches a maximum 120 times a second. But the hot filament of the bulb glows steadily, so it makes more sense to pay attention to the *average power* than the instantaneous power. Figure 26.5 shows the **average power** P_R is related to the peak power as follows:

$$P_R = \frac{1}{2} I_R^2 R \tag{26.7}$$

In a DC circuit, the power is $P_R = I^2R$. It's conventional to write Equation 26.7 in a form that resembles this DC expression. We do so by writing

$$P_R = \left(\frac{I_R}{\sqrt{2}}\right)^2 R = (I_{rms})^2 R \tag{26.8}$$

where the quantity

$$I_{rms} = \frac{I_R}{\sqrt{2}} \tag{26.9}$$

is called the **root-mean-square current,** or rms current, I_{rms}. We saw this idea before in Chapter 12, where we introduced the rms speed of molecules in a gas.

Using the rms current allows us to compare Equation 26.8 directly to the energy dissipated by a resistor in a DC circuit: $P = I^2R$. The average power loss of a resistor in an AC circuit with $I_{rms} = 1$ A is the same as in a DC circuit with $I = 1$ A. **As far as average power is concerned, an rms current is equivalent to a DC current.**

Similarly, we can define the root-mean-square voltage:

$$V_{rms} = \frac{V_R}{\sqrt{2}} \tag{26.10}$$

The resistor's average power loss can be written in terms of the rms quantities as

$$P_R = (I_{rms})^2 R = \frac{(V_{rms})^2}{R} = I_{rms}V_{rms} \tag{26.11}$$

Average power loss in a resistor

The "120 V" on this lightbulb is its operating rms voltage. The "100 W" is its average power dissipation at this voltage.

NOTE ▶ As long as we work with rms voltages and currents, all the expressions you learned for DC power carry over to AC power. ◀

AC voltages and currents are usually given as rms values. For instance, we've noted that household lamps and appliances in the United States operate at the 120 V present at wall outlets. This voltage is the rms value $\mathcal{E}_{rms}$; the peak voltage is higher by a factor of $\sqrt{2}$, so $\mathcal{E}_0 = 170$ V.

EXAMPLE 26.1 The resistance and current of a toaster

The hot wire in a toaster dissipates 580 W when plugged into a 120 V outlet.

a. What is the wire's resistance?
b. What are the rms and peak currents through the wire?

PREPARE The filament has resistance R. It dissipates 580 W when there's an rms voltage of 120 V across it. We can solve Equation 26.11 for R and then use Equations 26.11 and 26.9 to find the rms current and the peak current.

SOLVE

a. We can rearrange Equation 26.11 to find the resistance from the rms voltage and the average power:

$$R = \frac{(V_{rms})^2}{P_R} = \frac{(120\text{ V})^2}{580\text{ W}} = 25\ \Omega$$

b. A second rearrangement of Equation 26.11 allows us to find the current in terms of the power and the resistance, both of which are known:

$$I_{rms} = \sqrt{\frac{P_R}{R}} = \sqrt{\frac{580\text{ W}}{25\ \Omega}} = 4.8\text{ A}$$

From Equation 26.9, the peak current is

$$I_R = \sqrt{2}I_{rms} = \sqrt{2}(4.8\text{ A}) = 6.8\text{ A}$$

ASSESS We can do a quick check on our work by calculating the power for the rated voltage and computed current:

$$P_R = I_{rms}V_{rms} = (4.8\text{ A})(120\text{ V}) = 580\text{ W}$$

This agrees with the value given in the problem statement, giving us confidence in our solution.

STOP TO THINK 26.1 An AC current with a peak value of 1.0 A passes through bulb A. A DC current of 1.0 A passes through an identical bulb B. Which bulb is brighter?

A. Bulb A. B. Bulb B. C. Both bulbs are equally bright.

26.2 AC Electricity and Transformers

Your cell phone runs at about 3.5 V. To charge it up from a wall outlet at $V_{rms} = 120$ V requires the use of a **transformer,** a device that takes an *alternating* voltage as an input and produces either a higher or lower voltage as its output. As

Your cell phone charger incorporates a transformer that provides the necessary voltage reduction.

we'll see, the operation of a transformer is based on the emf produced by changing magnetic fields, so the input must be AC electricity.

Transformer Operation

FIGURE 26.6 shows a simplified version of a transformer, consisting of two coils of wire wrapped on a single iron core. The left coil is called the **primary coil** (or simply the *primary*). It has N_1 turns of wire connected to an AC voltage of amplitude V_1. This AC voltage creates an alternating current. The current in the coil creates a magnetic field that magnetizes the iron of the core to produce a much stronger net field—and a large flux through the primary coil. The field lines tend to follow the iron core, as shown in the figure, so nearly all of the flux from the left primary coil also goes through the right coil of wire, which has N_2 turns and is called the **secondary coil** (or the *secondary*). The current in the primary coil is an alternating current, so it creates an oscillating magnetic field in the iron core; the changing magnetic field means that there is a changing flux in the secondary coil. This changing flux induces an emf, an AC voltage of amplitude V_2, in the secondary coil. To complete the picture, we connect this emf to a resistor R, which we call the *load*. Current in this resistor will dissipate power.

FIGURE 26.6 A transformer.

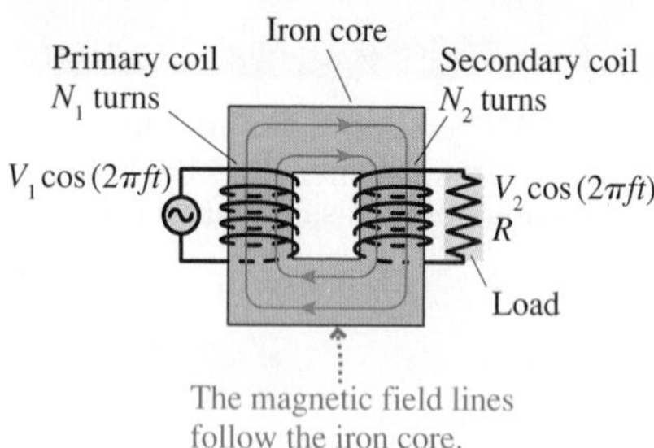

The point of a transformer is to change the voltage, so we need to define how the voltage V_2 at the secondary is related to the voltage V_1 at the primary. Suppose at some instant of time the instantaneous current through the primary is i_1. This current creates a magnetic flux Φ through the primary coil. This flux is changing, because i_1 is changing, and the change induces an emf across the coil. According to Faraday's law (Equation 25.12), the instantaneous voltage v_1 across the N_1 turns of the primary coil is

$$v_1 = \mathcal{E}_1 = N_1\frac{\Delta\Phi}{\Delta t} \tag{26.12}$$

In an **ideal transformer,** all of the flux is "guided" by the iron core through the secondary coil. Consequently, the rate at which the flux changes through the secondary coil is also $\Delta\Phi/\Delta t$. This changing flux induces an emf across the secondary coil given by

$$v_2 = \mathcal{E}_2 = N_2\frac{\Delta\Phi}{\Delta t} \tag{26.13}$$

Because $\Delta\Phi/\Delta t$ is the same in Equations 26.12 and 26.13, we can write

$$\frac{v_1}{v_2} = \frac{N_1}{N_2} \tag{26.14}$$

Equation 26.14 gives the ratio of the instantaneous voltages. This equation also applies at the instant at which these voltages are at their peak values V_1 and V_2, and the rms voltages are related to these peak values by a factor of $1/\sqrt{2}$. We can thus relate the peak and rms voltages of the primary and secondary coils:

$$V_2 = \frac{N_2}{N_1}V_1 \quad \text{and} \quad (V_2)_{\text{rms}} = \frac{N_2}{N_1}(V_1)_{\text{rms}} \tag{26.15}$$

Transformer voltages for primary and secondary coils with N_1 and N_2 turns

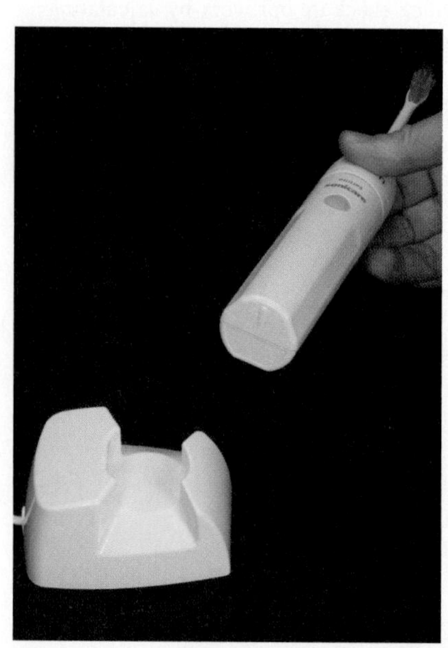

◀ **Getting a charge** There is no direct electrical contact between the primary and secondary coils in a transformer; the energy is carried from one coil to the other by the magnetic field. This makes it possible to charge devices that are completely sealed, with no external electrical contacts, such as the electric toothbrush shown. A primary coil in the base creates an alternating magnetic field that induces an alternating emf in a secondary coil in the brush's handle. This emf, after conversion to DC, is used to charge the toothbrush's battery.

Depending on the ratio N_2/N_1, the voltage V_2 across the load can be transformed to a higher or a lower voltage than V_1. A *step-up transformer,* with $N_2 > N_1$, increases the voltage, while a *step-down transformer,* with $N_2 < N_1$, lowers the voltage.

Equation 26.15 relates the voltage at a transformer's secondary to the voltage at its primary. We can also relate the *currents* in the secondary and primary by considering energy conservation. If a transformer is connected to a load that draws an rms current $(I_2)_{rms}$ from the secondary, the average power supplied to the load by the transformer, given by Equation 26.11, is

$$P_2 = (V_2)_{rms}\,(I_2)_{rms}$$

The primary coil draws a current $(I_1)_{rms}$ from the voltage source to which it's connected. This source provides power

$$P_1 = (V_1)_{rms}\,(I_1)_{rms}$$

to the transformer.

We'll assume that our ideal transformer has no loss of electric energy, so $P_1 = P_2$, or $(V_1)_{rms}(I_1)_{rms} = (V_2)_{rms}(I_2)_{rms}$. Thus we have

$$(I_2)_{rms} = \frac{(V_1)_{rms}}{(V_2)_{rms}}(I_1)_{rms} = \frac{N_1}{N_2}(I_1)_{rms} \quad (26.16)$$

Transformer currents for primary and secondary coils with N_1 and N_2 turns

Here we used Equation 26.15 to relate the ratio of voltages to the ratio of turns. Comparing Equations 26.15 and 26.16, you can see that a step-up transformer *raises* voltage but *lowers* current. This must be the case in order to conserve energy. Similarly, a step-down transformer *lowers* the voltage but *raises* the current. We've made some assumptions about the ideal transformer in completing this derivation. Real transformers come quite close to this ideal, so you can use the above equations for computations on real transformers.

EXAMPLE 26.2 Analyzing a step-up transformer

A book light has a 1.4 W, 4.8 V bulb that is powered by a transformer connected to a 120 V electric outlet. The secondary coil of the transformer has 20 turns of wire. How many turns does the primary coil have? What is the current in the primary coil?

PREPARE The circuit is the basic transformer circuit of Figure 26.6; the load is the bulb. We know the voltages of the primary and the secondary, so we can compute the turns in the primary coil using Equation 26.15. We know the voltage and the power of the bulb, so we can find the current in the bulb—the current in the secondary—which we can then use to compute the current in the primary using Equation 26.16.

SOLVE The bulb is rated at 4.8 V; this is the rms voltage at the secondary, so $(V_2)_{rms} = 4.8$ V. The power outlet has the usual $(V_1)_{rms} = 120$ V, so we can rearrange Equation 26.15 to find

$$N_1 = N_2\frac{(V_1)_{rms}}{(V_2)_{rms}} = (20\text{ turns})\left(\frac{120\text{ V}}{4.8\text{ V}}\right) = 500\text{ turns}$$

The bulb connected to the secondary dissipates 1.4 W at 4.8 V; this is an rms voltage, so the rms current in the secondary is

$$(I_2)_{rms} = \frac{P_2}{(V_2)_{rms}} = \frac{1.4\text{ W}}{4.8\text{ V}} = 0.29\text{ A}$$

We can then rearrange Equation 26.16 to find the rms current in the primary:

$$(I_1)_{rms} = (I_2)_{rms}\frac{N_2}{N_1} = (0.29\text{ A})\frac{20}{500} = 0.012\text{ A}$$

ASSESS We can check our results by looking at the power supplied by the wall outlet. This is $P_1 = (120\text{ V})(0.012\text{ A}) = 1.4$ W, the same as the power dissipated by the bulb, as must be the case because we've assumed the transformer is ideal.

Power Transmission

Long-distance electrical transmission lines run at very high voltages—up to 1,000,000 volts! This is a higher voltage than the output of an electrical generator, and a much higher voltage than you use in your home. Why is power transmitted at such high voltages? In a word: efficiency.

To understand why power transmission is more efficient at high voltages, let's look at the power transmitted and the power lost for the simple model of a power plant, transmission line, and city shown in FIGURE 26.7. To provide power to the city, the power plant generates an emf $\mathcal{E}_{rms}$ and delivers a current I_{rms} through the wires. If the power plant is 50 km from the city, a fairly typical distance, the 100 km of wire used in the transmission line (to the city and back) has a resistance of about 7 Ω.

FIGURE 26.7 How electric power is delivered from the power plant to a city.

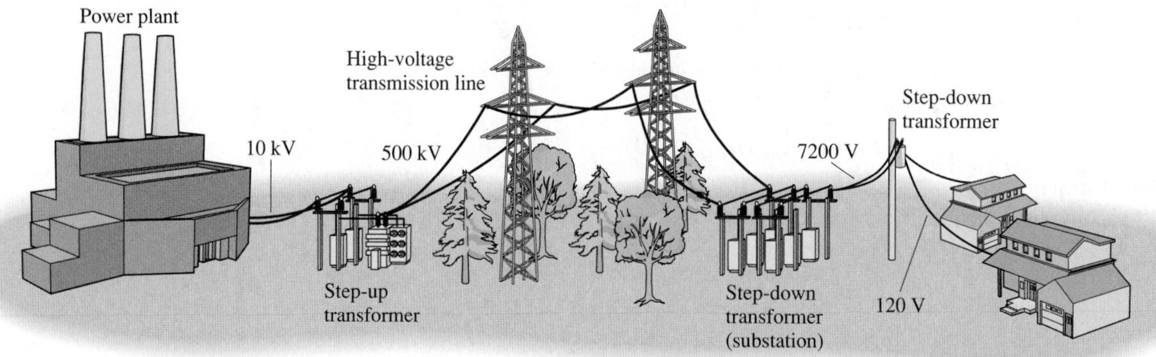

A modest-sized city of 100,000 inhabitants uses approximately 120 MW of electric power—this is the power that must be transmitted. To transmit this power at the outlet voltage of 120 V, we can use Equation 26.11 to find that the current in the transmission line would have to be

$$I_{rms} = \frac{P_{city}}{V_{rms}} = \frac{120 \times 10^6 \text{ W}}{120 \text{ V}} = 10^6 \text{ A}$$

This is an extraordinarily large current, which would incur correspondingly large losses. We know the resistance of the wire; passing this 10^6 A through a transmission line with 7 Ω of resistance would transform a good deal of energy to thermal energy in the resistance of the wire. The power "lost" would be

$$P_{lost} = (I_{rms})^2 R = (10^6 \text{ A})^2 (7 \text{ }\Omega) = 7 \times 10^{12} \text{ W}$$

In other words, 60,000 times more power would be lost in the transmission line than is used by the city! This is clearly impractical and unrealistic.

But suppose we use a step-up transformer at the power plant to boost the voltage to 500,000 V. To transmit 120 MW of power at 500 kV requires a current of only

$$I_{rms} = \frac{P_{city}}{V_{rms}} = \frac{120 \times 10^6 \text{ W}}{500{,}000 \text{ V}} = 240 \text{ A}$$

This is still a large current, but one that can be handled by typical aluminum transmission line cables with a diameter of roughly 1 inch. At this current, the power loss in the 7 Ω resistance of the transmission line is $P_{lost} = (I_{rms})^2 R = 400$ kW. This is less than 1% of the power used by the city, an acceptable loss. Because power loss in the transmission lines depends on the *square* of the current, higher voltages and thus smaller currents provide a huge improvement in efficiency.

Using electricity at 500,000 V at the outlet is clearly dangerous and impractical. Step-down transformers at the city lower the voltage to the usual 120 V. Only AC electricity can use transformers to change voltages and currents for long-distance power transmission. This is the primary reason we use AC electricity in our homes and not DC.

26.3 Household Electricity

The electricity in your home can be understood using the techniques of circuit analysis we've developed, but we need to add one more concept. So far we've only dealt with potential differences. Although we are free to choose the zero point of potential anywhere that is convenient, our analysis of circuits has not suggested any need to establish a zero point. Potential differences are all we have needed.

Difficulties can begin to arise, however, if you want to connect two different circuits together. Perhaps you would like to connect your CD player to your amplifier or connect your computer monitor to the computer itself. Incompatibilities can arise unless all the circuits to be connected have a common reference point for the potential. This is the reason for having an electric *ground*.

Getting Grounded

You learned previously that the earth itself is a conductor. Suppose we have two circuits. If we connect one point of each circuit to the earth by an ideal wire, then both circuits have a common reference point. A circuit connected to the earth in this way is said to be **grounded.** In practice, we also agree to call the potential of the earth $V_{\text{earth}} = 0\text{ V}$. FIGURE 26.8 shows a circuit with a ground connection. Under normal circumstances, the ground connection does not carry any current because it is not part of a complete circuit. In this case, it does not alter the behavior of the circuit.

Grounding serves two functions. First, it provides a common reference potential so that different circuits or instruments can be correctly interconnected. Second, it is an important safety feature. As we will see in the next section, a current to ground can quickly open a circuit breaker if an electric appliance malfunctions.

FIGURE 26.8 A grounded circuit.

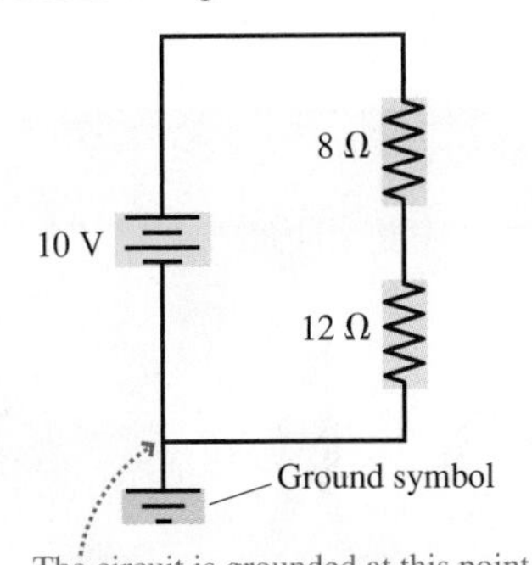

Electric Outlets Are Grounded Parallel Circuits

FIGURE 26.9 shows a circuit diagram for the outlets in your house. The 120 V electric supply is provided by the power company. It is transmitted to outlets throughout your house by wires in the walls. One terminal of the electric supply is grounded; we call this the **neutral** side. The other side is at a varying potential; we call this the **hot** side. Each electric outlet has two slots, one connected to the hot side and one connected to the neutral side. When you insert a plug into an electric outlet, the prongs of the plug connect to the two terminals of the electric supply. The device you've plugged in completes a circuit between the two terminals, the potential difference across the device leads to a current, and the device turns on.

FIGURE 26.9 Multiple outlets on one circuit.

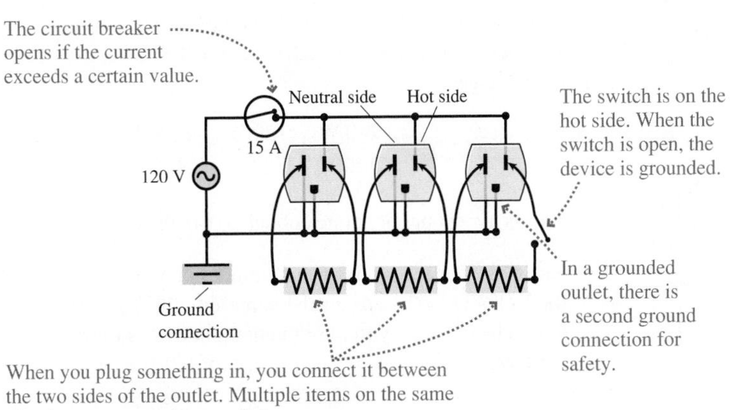

The multiple outlets in a room or area of your house are connected in parallel so that each works when the others are not being used. Because the outlets on a single circuit are in parallel, when you plug in another device, the total current in the circuit increases. This can create a problem; although the wires in your walls are good

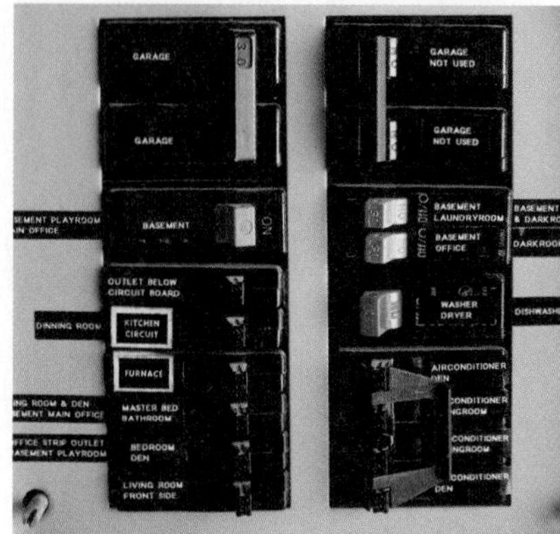

A typical circuit breaker panel. Each breaker corresponds to a different circuit in the house.

conductors, they aren't ideal. The wires have a small resistance and heat up when carrying a current. If there is too much current, the wires could get hot enough to cause a fire.

The circuits in your house are protected with circuit breakers to limit the current in each circuit. A circuit breaker consists of a switch and an ammeter that measures the current in the circuit. If the ammeter measures too much current (typically $I_{rms} \geq 15$ A), it sends a signal to open the switch to disconnect the circuit. You have probably had the experience of having a circuit breaker "trip" if you have too many things plugged in. To keep the problem from recurring, you must reduce the current in the circuit. Some things need to be turned off, unplugged, or moved to a different circuit.

Grounding of household circuits provides an important reference potential, as noted above, but the main reason for grounding is safety. The two slots in a standard outlet are different sizes; the neutral slot is a bit larger. Most electric devices are fitted with plugs that can be inserted into an outlet in only one orientation. A lamp, for instance, will almost certainly have this sort of plug. This is an important safety feature; when you turn the lamp off, the switch disconnects the hot wire, not the neutral wire. The lamp is then grounded, and thus safe, when it is switched off.

The round hole in a standard electric outlet is a second ground connection that serves a second safety function. If a device has a metal case, whether it's a microwave oven or an electric drill, the case will likely be connected to the ground. If a wire comes loose inside the device and contacts the metal case, a person touching the case could get a shock. But if the case is grounded, its potential is always 0 V and it is always safe to touch. In addition, a hot wire touching the grounded case would be a short circuit, causing a sudden very large current that would trip the circuit breaker, disconnecting the hot wire and preventing any danger. If you plug something in and the circuit breaker trips, the device likely has an electrical fault and should be repaired or discarded.

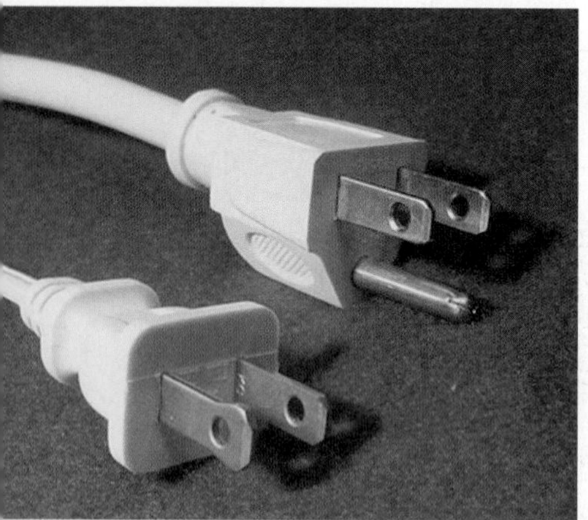
◀ A two-prong "polarized" plug has one large prong and one small one. When plugged into a standard outlet, the large prong is grounded. A three-prong plug has a round pin that makes a second ground connection.

EXAMPLE 26.3 Will the circuit breaker open?

A circuit in a student's room has a 15 A circuit breaker. One evening, she plugs in a computer (240 W), a lamp (with two 60 W bulbs), and a space heater (1200 W). Will this be enough to trip the circuit breaker?

PREPARE We start by sketching the circuit, as in **FIGURE 26.10**. Because the three devices are in the same circuit, they are connected in parallel. We can model each of them as a resistor.

FIGURE 26.10 The circuit with the circuit breaker.

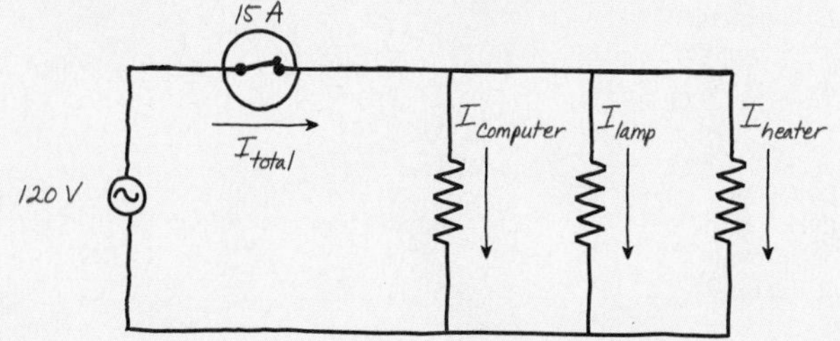

SOLVE The current in the circuit is the sum of the currents in the individual devices:

$$(I_{total})_{rms} = (I_{computer})_{rms} + (I_{lamp})_{rms} + (I_{heater})_{rms}$$

Equation 26.11 gives the power as the rms current times the rms voltage, so the current in each device is the power divided by the rms voltage:

$$(I_{total})_{rms} = \frac{240\text{ W}}{120\text{ V}} + \frac{120\text{ W}}{120\text{ V}} + \frac{1200\text{ W}}{120\text{ V}} = 13\text{ A}$$

This is almost but not quite enough to trip the circuit breaker.

ASSESS Generally all of the outlets in one room (and perhaps the lights as well) are on the same circuit. You can see that it would be quite easy to plug in enough devices to trip the circuit breaker.

Kilowatt Hours

The product of watts and seconds is joules, the SI unit of energy. However, your local electric company prefers to use a different unit, called *kilowatt hours,* to measure the energy you use each month.

A device in your home that consumes P kW of electricity for Δt hours has used $P\Delta t$ kilowatt hours of energy, abbreviated kWh. For example, suppose you run a 1500 W electric water heater for 10 hours. The energy used in kWh is $(1.5 \text{ kW})(10 \text{ hr}) = 15 \text{ kWh}$.

Despite the rather unusual name, a kilowatt hour is a unit of energy, as it is a power multiplied by a time. The conversion between kWh and J is

$$1.00 \text{ kWh} = (1.00 \times 10^3 \text{ W})(3600 \text{ s}) = 3.60 \times 10^6 \text{ J}$$

Your monthly electric bill specifies the number of kilowatt hours you used last month. This is the amount of energy that the electric company delivered to you, via an electric current, and that you transformed into light and thermal energy inside your home.

KILOWATTHOURS
0 0 1 0 3
The electric meter on the side of your house or apartment records the kilowatt hours of electricity you use each month.

EXAMPLE 26.4 Computing the cost of electric energy

A typical electric space heater draws an rms current of 12.5 A on its highest setting. If electricity costs 10¢ per kilowatt hour (an approximate national average), how much does it cost to run the heater for 2 hours?

SOLVE The power dissipated by the heater is

$$P = V_{\text{rms}} I_{\text{rms}} = (120 \text{ V})(12.5 \text{ A}) = 1500 \text{ W} = 1.5 \text{ kW}$$

In 2 hours, the energy used is

$$\mathcal{E} = (1.5 \text{ kW})(2.0 \text{ hr}) = 3.0 \text{ kWh}$$

At 10¢ per kWh, the cost is 30¢.

FIGURE 26.11 High voltages are not necessarily dangerous.

26.4 Biological Effects and Electrical Safety

You can handle an ordinary 9 V battery without the slightest danger, but the 120 volts from an electric outlet can lead to a nasty shock. Yet the girl in FIGURE 26.11 is safely touching a Van de Graaff generator at a potential of 400,000 V. What makes electricity either safe or dangerous?

The relative safety of electric sources isn't governed by the voltage but by the current. Current—the flow of charges through the body—is what produces physiological effects and damage because it mimics nerve impulses and causes muscles to involuntarily contract.

Higher voltages are generally more dangerous than lower voltages because they tend to produce larger currents, but the amount of current also depends on resistance and on the ability of the voltage source to deliver current. The Van de Graaff generator in Figure 26.11 is at a high potential with respect to the ground, but the girl is standing on an insulating platform. The high resistance of the platform means that very little current is passing through her to the ground; she won't feel a thing. Even if she touches a grounded object, the current will be modest—the total charge on the generator is quite small, so a dangerous current simply isn't possible. You can get a much worse shock from a 120 V household circuit because it is capable of providing a much larger current.

Table 26.1 lists approximate values of current that produce different physiological effects. Currents through the chest cavity are particularly dangerous because they can interfere with respiration and the proper rhythm of the heart. An AC current larger than 100 mA can induce fibrillation of the heart, in which it beats in a rapid, chaotic, uncontrolled fashion.

To calculate likely currents through the body, we model the body as several connected resistors, as shown in FIGURE 26.12. (These are averages; there is significant individual variation around these values.) Because of its high saltwater content, the resistance of the interior of the body is fairly low. But current must pass through the skin before getting inside the body, and the skin generally has a fairly high

TABLE 26.1 Physiological effects of currents passing through the body

Physiological effect	AC current (rms) (mA)	DC current (mA)
Threshold of sensation	1	3
Paralysis of respiratory muscles	15	60
Heart fibrillation, likely fatal	> 100	> 500

FIGURE 26.12 Resistance model of the body.

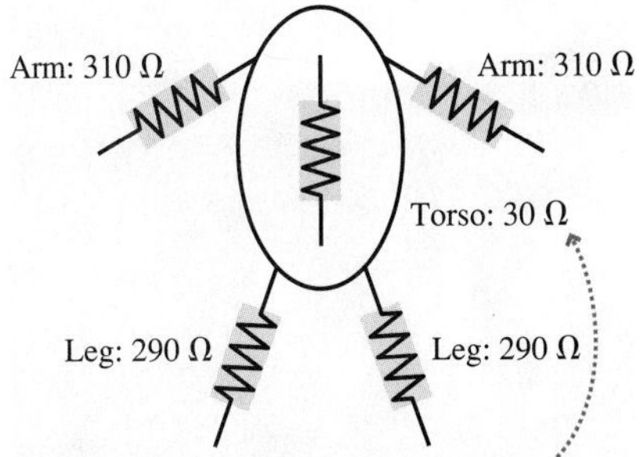
Arm: 310 Ω
Arm: 310 Ω
Torso: 30 Ω
Leg: 290 Ω
Leg: 290 Ω
When current traverses the torso between any two points (arm to arm, arm to leg, leg to leg) this adds a resistance of 30 Ω.

resistance. If you touch a wire with the dry skin of a finger, the skin's resistance might be greater than 1 MΩ. Moist skin and larger contact areas can reduce the skin's resistance to less than 10 kΩ. **The skin's resistance is in series with the resistances shown in Figure 26.12.**

EXAMPLE 26.5 Is the worker in danger?

A worker in a plant grabs a bare wire that he does not know is connected to a 480 V AC supply. His other hand is holding a grounded metal railing. The skin resistance of each of his hands, in full contact with a conductor, is 2200 Ω. He will receive a shock. Will it be large enough to be dangerous?

PREPARE We can draw a circuit model for this situation as in FIGURE 26.13a; the worker's body completes a circuit between two points at a potential difference of 480 V. The current will depend on this potential difference and the resistance of his body, including the resistance of the skin.

SOLVE Following the model of Figure 26.12, the current path goes through the skin of one hand, up one arm, across the torso, down the other arm, and through the skin of the other hand, as in FIGURE 26.13b. The equivalent resistance of the series combination is 5050 Ω, so the AC current through his body is

$$I = \frac{\Delta V}{R_{eq}} = \frac{480\ \text{V}}{5050\ \Omega} = 95\ \text{mA}$$

FIGURE 26.13 A circuit model for the worker.

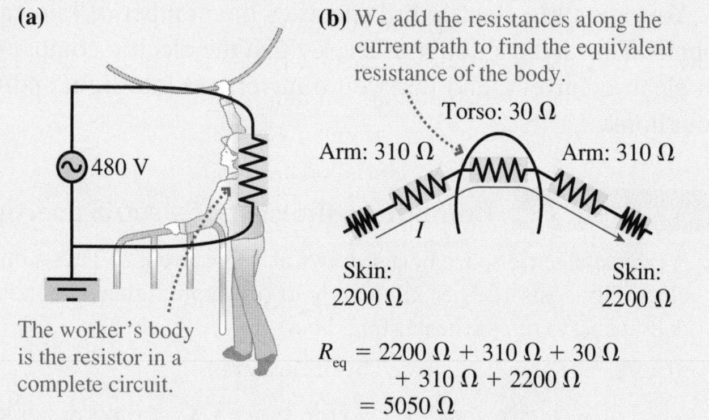

From Table 26.1 we see that this is a very dangerous, possibly fatal, current.

ASSESS The voltage is high and the resistance relatively low, so it's no surprise to find a dangerous level of current.

FIGURE 26.14 Wearing electrically insulating boots increases resistance.

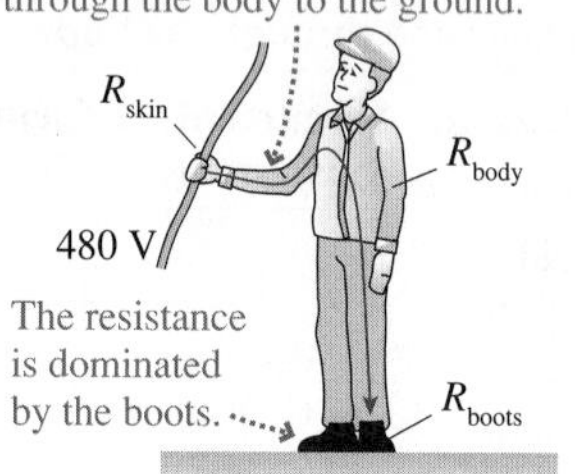

There is a current path from the wire through the body to the ground.
R_skin
R_body
480 V
The resistance is dominated by the boots.
R_boots

Because of the danger of electrocution, workers who might accidentally contact electric lines wear protective clothing. The soles of boots designed to protect against electric shock have a resistance of 10 MΩ or more. Suppose a worker wearing such boots touches a live wire at 480 V, as shown in FIGURE 26.14. The resistance of the current path through the body is now dominated by the resistance of the boots, which is much larger than the resistance of his body and skin. Assuming 10 MΩ for the resistance of the soles of the boots, the current passing through the worker's body is only

$$I = \frac{\Delta V}{R_{eq}} \approx \frac{\Delta V}{R_{boots}} = \frac{480\ \text{V}}{10\ \text{M}\Omega} = 48\ \mu\text{A}$$

In this case, the worker will be fine. Because the current is much less than the threshold for sensation, he won't even feel a shock!

The current in these examples was due to a potential difference between the hands or between a hand and the feet. How about the birds sitting on the wire at the start of the chapter? How are they able to perch on the high-voltage wire? The wire is at an elevated potential with respect to the ground, but there is no complete circuit—the birds have both feet on the same wire. Each foot is at a high potential, but there is only a very small potential *difference* between the feet due to the potential decrease along the wire, too small to produce any noticeable effect. If a bird touched the wire and a grounded pole, or two neighboring wires, the result would be very different.

◀ **The lightning crouch** You don't need to be hit by lightning to be hurt by lightning. When lightning strikes the ground, a tremendous amount of charge flows outward, implying a large electric field along the ground. If you are standing with your feet separated by a large distance d, a large potential difference $\Delta V = Ed$ will develop between your feet, causing a potentially dangerous current to flow up one leg and down the other. If you're caught outdoors in a lightning storm and can't get to safety, experts recommend that you assume the crouched position shown in the photo. By placing your feet as close together as possible and lifting your heels off the ground, you can minimize the potential difference and thus the current due to a nearby lightning strike.

GFI Circuits

If you are standing in good electrical contact with the ground, you are grounded. If you then accidentally touch a hot wire with your hand, a dangerous current could pass through you to ground. In kitchens and bathrooms, where grounding on damp floors is a good possibility, building codes require *ground fault interrupter* outlets, abbreviated GFI. Some devices, such as hair dryers, are generally constructed with a GFI in the power cord.

GFI outlets have a built-in sensing circuit that compares the currents in the hot and neutral wires of the outlet. In normal operation, all the current coming in through the hot wire passes through the device and then back out through the neutral wire, so the currents in the hot and neutral wires should always be equal. If the current in the hot wire does *not* equal the current in the neutral wire, some current from the hot wire is finding an alternative path to ground—perhaps through a person. This is a *ground fault,* and the GFI disconnects the circuit. GFIs are set to trip at current differences of about 5 mA—large enough to feel, but not large enough to be dangerous.

TRY IT YOURSELF

Testing GFI circuits You can test a GFI circuit by pressing the black "Test" button. This creates an electrical connection between the hot wire and the ground connection, so the currents in the hot and neutral wires will not be equal. You should hear a click as the circuit disconnects. You can then reset the outlet by pressing the red button. If an outlet does not respond like this, it should be replaced.

26.5 Capacitor Circuits

In Chapter 23 we analyzed the one-time charging or discharging of a capacitor in an *RC* circuit. In this chapter we will look at capacitors in circuits with an AC source of emf that repeatedly charges and discharges the capacitor.

FIGURE 26.15a shows a current i_C charging a capacitor with capacitance C. The instantaneous capacitor voltage is $v_C = q/C$, where $\pm q$ is the charge on the two capacitor plates at this instant of time. **FIGURE 26.15b**, where capacitance C is connected across an AC source of emf $\mathcal{E}$, is the most basic capacitor circuit. The capacitor is in parallel with the source, so the capacitor voltage equals the emf: $v_C = \mathcal{E} = \mathcal{E}_0\cos(2\pi ft)$. It is useful to write

$$v_C = V_C \cos(2\pi ft) \tag{26.17}$$

where V_C is the peak or maximum voltage across the capacitor. $V_C = \mathcal{E}_0$ in this single-capacitor circuit.

Charge flows to and from the capacitor plates but not through the gap between the plates. But the charges $\pm q$ on the opposite plates are always of equal magnitude, so the currents into and out of the capacitor must be equal. We can find the current i_C by considering how the charge q on the capacitor varies with time. In **FIGURE 26.16a** on the next page we have plotted the oscillating voltage v_C across the capacitor. Because the charge and the capacitor voltage are directly proportional, with $q = Cv_C$, the graph of q, shown in **FIGURE 26.16b**, looks like the graph of v_C. We say that the charge is *in phase* with the voltage.

In Chapter 22, we defined current to be $\Delta q/\Delta t$. The capacitor current, i_C, will thus be related to the charge on the capacitor by

$$i_C = \frac{\Delta q}{\Delta t}$$

The main factor determining i_C is Δq, the *change* of charge during a time interval Δt, not the amount of charge q. As the figure shows, when the current is large, the charge on the capacitor is changing rapidly, which makes sense. But these times of large current and rapid change occur when the voltage on the capacitor is small. The maximum current doesn't occur at the same time as the maximum voltage.

A capacitor's voltage and current are *not* in phase, as they were for a resistor. You can see from Figure 26.16 that the current peaks at $\frac{3}{4}$ T, one-quarter period *before* the voltage peaks. We say that **the AC current through a capacitor *leads* the capacitor voltage.**

FIGURE 26.15 An AC capacitor circuit.

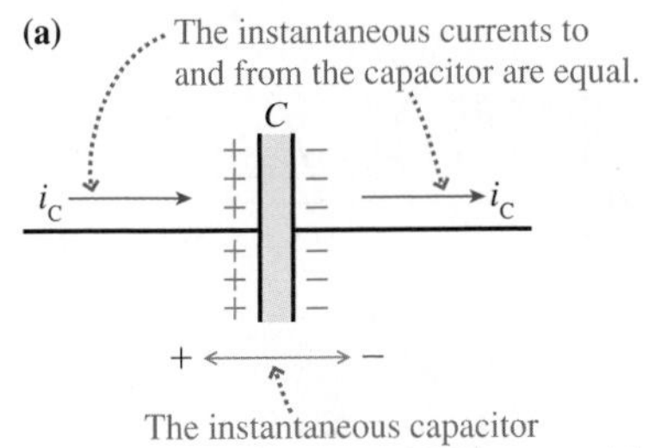
(a)
The instantaneous currents to and from the capacitor are equal.
C
i_C
i_C
+
−
The instantaneous capacitor voltage is $v_C = q/C$. The potential decreases from + to −.

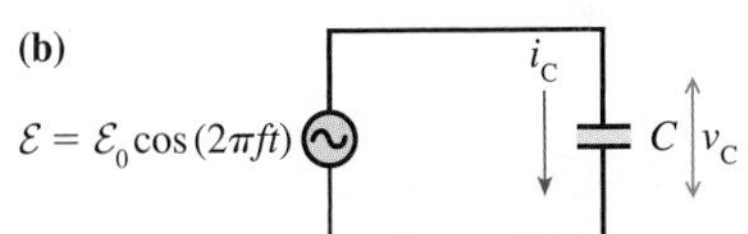
(b)
$\mathcal{E} = \mathcal{E}_0\cos(2\pi ft)$
i_C
C
v_C

FIGURE 26.16 Voltage, charge, and current graphs for a capacitor in an AC circuit.

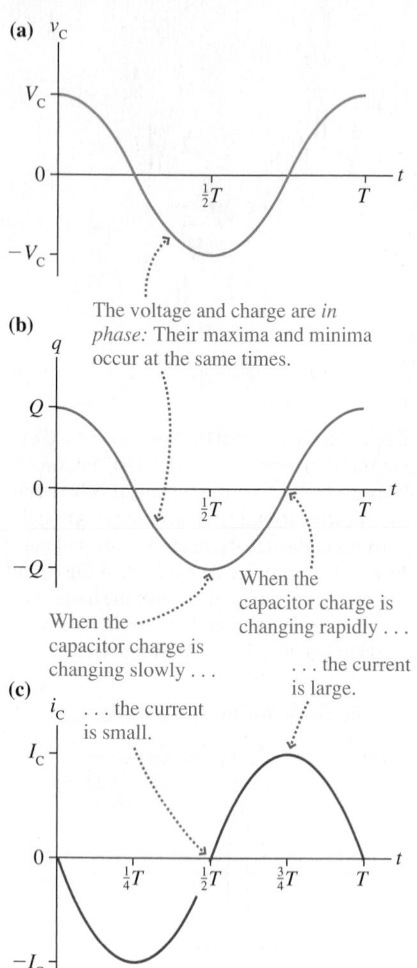

(a) v_C
V_C
0
$\frac{1}{2}T$
T
t
$-V_C$
The voltage and charge are *in phase:* Their maxima and minima occur at the same times.
(b) q
Q
0
$\frac{1}{2}T$
T
t
$-Q$
When the capacitor charge is changing slowly . . .
When the capacitor charge is changing rapidly . . .
. . . the current is large.
(c) i_C
. . . the current is small.
I_C
0
$\frac{1}{4}T$
$\frac{1}{2}T$
$\frac{3}{4}T$
T
t
$-I_C$

Replacing a mouse Under the surface of a laptop's trackpad is an array of tiny capacitors. When you touch the pad, your finger's high dielectric constant (it's largely water) changes the capacitance of nearby capacitors. This alters the current in these capacitor circuits, telling the computer the location of your finger. Try touching a trackpad with the eraser end of a pencil. The dielectric constant of the eraser will likely be too small to make an effect.

Capacitive Reactance

The phase of the current is not the whole story. We would like to know its peak value, which we can also determine from $i_C = \Delta q/\Delta t$. Because $q = Cv_C$, this can also be written as

$$i_C = C\frac{\Delta v_C}{\Delta t} \tag{26.18}$$

Let's now reason by analogy. In Chapter 14 the position of a simple harmonic oscillator was given by $v = A\cos(2\pi ft)$. Further, the oscillator's velocity $v = \Delta x/\Delta t$ was found to be $x = -v_{max}\sin(2\pi ft)$, with a maximum or peak value $v_{max} = 2\pi fA$. Here we have an oscillating voltage $v_C = V_C\cos(2\pi ft)$, analogous to position, and we need to find the quantity $\Delta v_C/\Delta t$, analogous to velocity. Thus $\Delta v_C/\Delta t$ must have a maximum or peak value $(\Delta v_C/\Delta t)_{max} = 2\pi fV_C$.

The peak capacitor current I_C occurs when $\Delta v_C/\Delta t$ is maximum, so the peak current is

$$I_C = C(2\pi fV_C) = (2\pi fC)V_C \tag{26.19}$$

For a resistor, the peak current and voltage are related through Ohm's law:

$$I_R = \frac{V_R}{R}$$

We can write Equation 26.19 in a form similar to Ohm's law if we define the **capacitive reactance** X_C to be

X_C p.114 f INVERSE

$$X_C = \frac{1}{2\pi fC} \tag{26.20}$$

With this definition of capacitive reactance, Equation 26.19 becomes

$$I_C = \frac{V_C}{X_C} \quad \text{or} \quad V_C = I_CX_C \tag{26.21}$$

Peak current through or voltage across a capacitor

The units of reactance, like those of resistance, are ohms.

NOTE ► Reactance relates the *peak* voltage V_C and current I_C. It does *not* relate the *instantaneous* capacitor voltage and current because they are out of phase; that is, $v_C \neq i_CX_C$. ◄

A resistor's resistance R is independent of the frequency of the emf. In contrast, as **FIGURE 26.17** shows, a capacitor's reactance X_C depends inversely on the frequency. The reactance becomes very large at low frequencies (i.e., the capacitor is a large impediment to current). The reactance decreases as the frequency increases until, at very high frequencies, $X_C \approx 0$ and the capacitor begins to act like an ideal wire.

FIGURE 26.17 The capacitive reactance as a function of frequency.

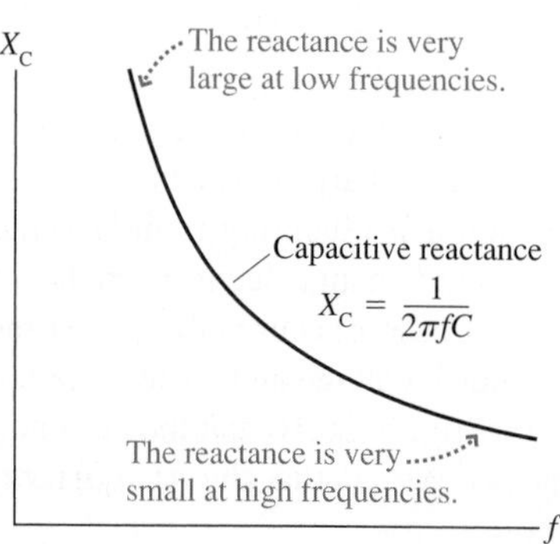

X_C
The reactance is very large at low frequencies.
Capacitive reactance
$X_C = \frac{1}{2\pi fC}$
The reactance is very small at high frequencies.
f

EXAMPLE 26.6 Finding the capacitive reactance

What is the capacitive reactance of a 0.100 μF capacitor at a 100 Hz audio frequency and at a 100 MHz FM-radio frequency?

SOLVE At 100 Hz,

$$X_C \text{(at 100 Hz)} = \frac{1}{2\pi f C} = \frac{1}{2\pi(100\text{ Hz})(1.00 \times 10^{-7}\text{ F})} = 15{,}900\ \Omega$$

Increasing the frequency by a factor of 10^6 decreases X_C by a factor of 10^6, giving

$$X_C \text{(at 100 MHz)} = 0.0159\ \Omega$$

ASSESS A capacitor with a substantial reactance at audio frequencies has virtually no reactance at FM radio frequencies.

EXAMPLE 26.7 Finding a capacitor's current

A 10 μF capacitor is connected to a 1000 Hz oscillator with a peak emf of 5.0 V. What is the peak current through the capacitor?

PREPARE The circuit diagram is as in Figure 26.15b. This is a simple one-capacitor circuit.

SOLVE The capacitive reactance at $f = 1000$ Hz is

$$X_C = \frac{1}{2\pi f C} = \frac{1}{2\pi(1000\text{ Hz})(10 \times 10^{-6}\text{ F})} = 16\ \Omega$$

The peak voltage across the capacitor is $V_C = \mathcal{E}_0 = 5.0$ V; hence the peak current is

$$I_C = \frac{V_C}{X_C} = \frac{5.0\text{ V}}{16\ \Omega} = 0.31\text{ A}$$

ASSESS Using reactance and Equation 26.21 is just like using resistance and Ohm's law, but don't forget that it applies to only the *peak* current and voltage, not the instantaneous values. Further, reactance, unlike resistance, depends on the frequency of the signal.

STOP TO THINK 26.2 A capacitor is attached to an AC voltage source. Which change will result in a doubling of the current?

A. Halving the voltage and doubling the frequency
B. Doubling the frequency
C. Halving the frequency
D. Doubling the voltage and halving the frequency

26.6 Inductors and Inductor Circuits

FIGURE 26.18 shows a length of wire formed into a coil, making a solenoid. In Chapter 24 you learned that current in a solenoid creates a magnetic field inside the solenoid. If this current is *increasing,* as shown in Figure 26.18a, then the magnetic field—and thus the flux—inside the coil increases as well. According to Faraday's law, this changing flux causes an emf—a potential difference—to develop across the coil. The *direction* of the emf can be inferred from Lenz's law: Its direction must be such to *oppose* the increase in the flux; that is, it must oppose the increase in the current. The emf will have the opposite sign if the current through the coil is decreasing, as shown in Figure 26.18b.

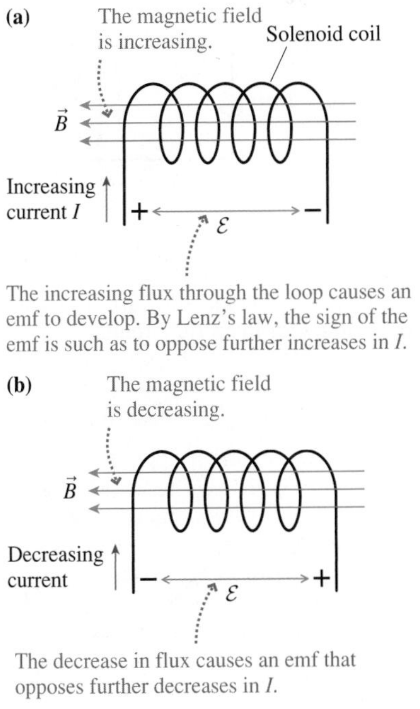

Coils of this kind, called **inductors,** are widely used in AC circuits. The circuit symbol for an inductor is —0000—. There are two primary things to remember about an inductor. First, **an inductor develops a potential difference across it if the current through it is *changing.*** Second, because the direction of this potential difference opposes the change in the current, **an inductor resists changes in the current through it.**

FIGURE 26.18 A changing current through a solenoid induces an emf across the solenoid.

(a)
The magnetic field is increasing.
Solenoid coil
$\vec{B}$
Increasing current I
+
$\mathcal{E}$
−
The increasing flux through the loop causes an emf to develop. By Lenz's law, the sign of the emf is such as to oppose further increases in I.
(b)
The magnetic field is decreasing.
$\vec{B}$
Decreasing current
−
$\mathcal{E}$
+
The decrease in flux causes an emf that opposes further decreases in I.

Is anybody up there? Detectors are often installed beneath the pavement to sense the presence of cars waiting at intersections and on stretches of road before traffic lights. A slot is cut in the pavement. A wire is then sealed into the slot, forming an inductor consisting of a single loop. The steel in a car over the loop acts just like the iron core of an ordinary inductor, greatly increasing the inductance of the loop. This change in inductance signals that a car is present.

Inductance

By Faraday's law, the voltage developed across an inductor is proportional to the rate at which the flux through the coil changes. And, because the flux is proportional to the coil's current, the instantaneous inductor voltage v_L must be proportional to $\Delta i_L/\Delta t$, the rate at which the current through the inductor changes. Thus we can write

$$v_L = L\frac{\Delta i_L}{\Delta t} \tag{26.22}$$

The constant of proportionality L is called the **inductance** of the inductor. The inductance is determined by the shape and size of the coil. A coil with many turns has a higher inductance than a similarly sized coil with fewer turns. Inductors often have an iron core inside their windings to increase their inductance. The magnetic field from the current magnetizes the iron core, which greatly increases the overall field through the windings. This gives a larger change in flux through the windings and hence a larger induced emf. Equation 26.22 shows that this implies a larger value of L.

From Equation 26.22 we see that inductance has units of V · s/A. It's convenient to define an SI unit of inductance called the **henry,** in honor of Joseph Henry, an early investigator of magnetism. We have

$$1 \text{ henry} = 1 \text{ H} = 1 \text{ V} \cdot \text{s/A}$$

Practical inductances are usually in the range of millihenries (mH) or microhenries (μH).

FIGURE 26.19 An AC inductor circuit.

(a)

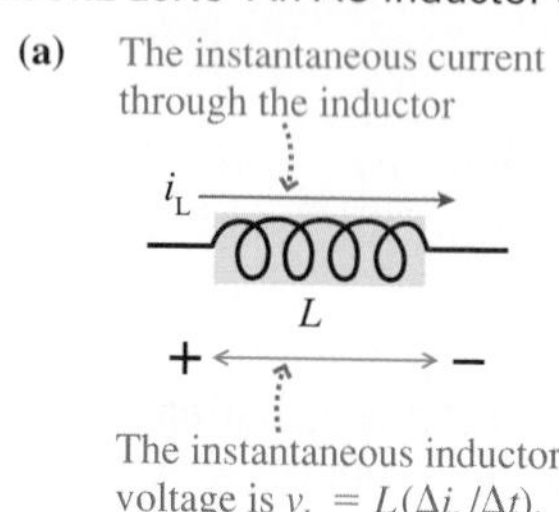

(b)

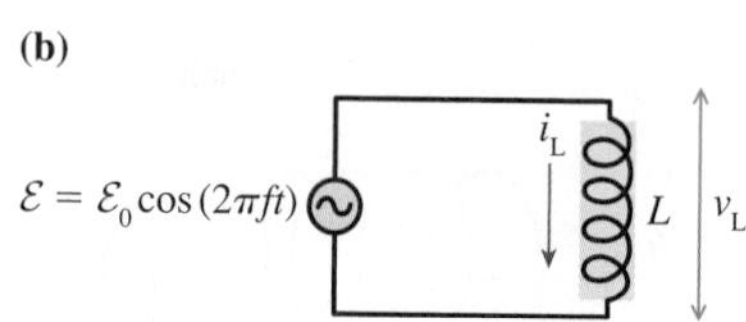

Inductor Circuits

FIGURE 26.19a shows the instantaneous current i_L through an inductor. If the current is changing, the instantaneous inductor voltage is given by Equation 26.22.

FIGURE 26.19b, where inductance L is connected across an AC source of emf $\mathcal{E}$, is the simplest inductor circuit. The inductor is in parallel with the source, so the inductor voltage equals the emf: $v_L = \mathcal{E} = \mathcal{E}_0\cos(2\pi ft)$. We can write

$$v_L = V_L\cos(2\pi ft) \tag{26.23}$$

where V_L is the peak or maximum voltage across the inductor. You can see that $V_L = \mathcal{E}_0$ in this single-inductor circuit.

We can find the inductor current i_L by considering again Equation 26.22, which tells us that the inductor voltage is high when the current is changing rapidly (i.e., $\Delta i_L/\Delta t$ is large) and is low when the current is changing slowly (i.e., $\Delta i_L/\Delta t$ is small). From this, we can graphically find the current, as shown in FIGURE 26.20.

Just as for a capacitor, the current and voltage are not in phase. There is again a phase difference of one-quarter cycle, but for an inductor the current peaks one-quarter period *after* the voltage peaks. **The AC current through an inductor *lags* the inductor voltage.**

FIGURE 26.20 Voltage and current graphs for an inductor in an AC circuit.

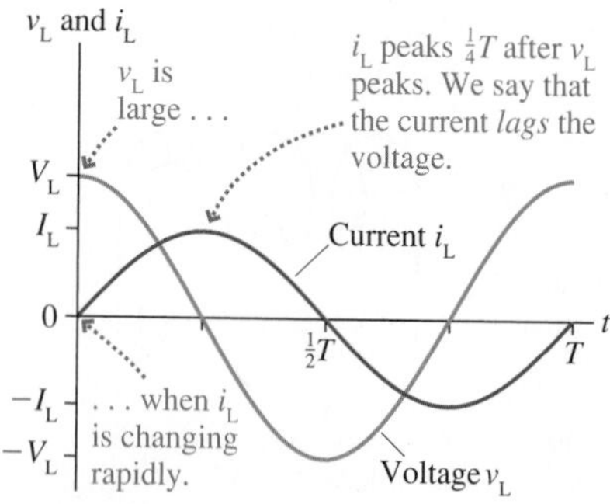

Inductive Reactance

To find the relationship between the peak values of the inductor's current and voltage, we can use Equation 26.22 and, as we did for the capacitor, the analogy with simple harmonic motion. For the oscillating capacitor voltage, with peak value $(v_C)_{max} = V_C$, the maximum value of $\Delta v_C/\Delta t$ was $(\Delta v_C/\Delta t)_{max} = 2\pi f V_C$. If an oscillating inductor current has peak value $(i_L)_{max} = I_L$, then, by exactly the same reasoning, the maximum value of $\Delta i_L/\Delta t$ is $(\Delta i_L/\Delta t)_{max} = 2\pi f I_L$. If we use this result in Equation 26.22, we see that the maximum or peak value of the inductor voltage is

$$V_L = L(2\pi f I_L) = (2\pi f L)I_L \tag{26.24}$$

We can write Equation 26.24 in a form reminiscent of Ohm's law:

$$I_L = \frac{V_L}{X_L} \quad \text{or} \quad V_L = I_L X_L \tag{26.25}$$

Peak current through or voltage across an inductor

where the **inductive reactance,** analogous to the capacitive reactance, is defined as

$$X_L = 2\pi f L \tag{26.26}$$

X_L p.37 f PROPORTIONAL

FIGURE 26.21 shows that the inductive reactance increases linearly as the frequency increases. This makes sense. Faraday's law tells us that the induced voltage across a coil increases as the rate of change of B increases, and B is directly proportional to the inductor current. For a given peak current I_L, B changes more rapidly at higher frequencies than at lower frequencies, and thus V_L is larger at higher frequencies than at lower frequencies.

FIGURE 26.21 The inductive reactance as a function of frequency.

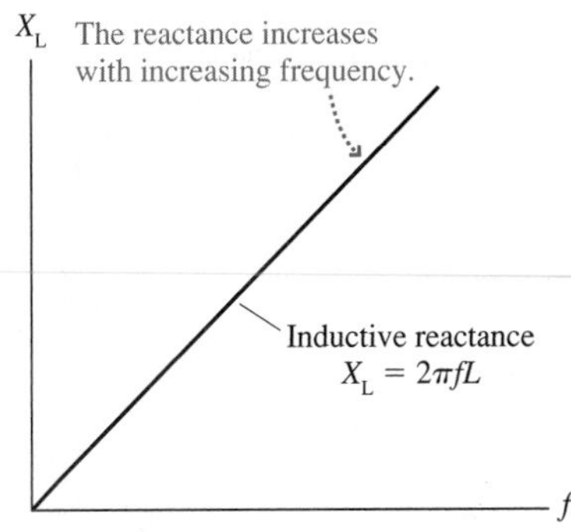
X_L
The reactance increases with increasing frequency.
Inductive reactance $X_L = 2\pi f L$
f

Clean power The digital circuits inside computers generate AC signals with frequencies of a few MHz to a few GHz. High-frequency AC currents can "leak" through the computer's power supply and propagate through your household electricity. To help prevent the transmission of such currents into or out of the computer, there will be one or more inductors on the board that connects your household electricity to the internal circuitry. At high frequencies, the inductive reactance of this inductor is very high and the high-frequency current entering or leaving is dramatically reduced.

EXAMPLE 26.8 Finding the current and voltage of a radio's inductor

A 0.25 μH inductor is used in an FM radio circuit that oscillates at 100 MHz. The current through the inductor reaches a peak value of 2.0 mA at $t = 5.0$ ns. What is the peak inductor voltage, and when, closest to $t = 5.0$ ns, does it occur?

PREPARE The inductor current lags the voltage; Figure 26.20 shows that the voltage reaches its peak value one-quarter period *before* the current peaks. The circuit looks like that in Figure 26.19b.

SOLVE The inductive reactance at $f = 100$ MHz $= 1.0 \times 10^8$ Hz is

$$X_L = 2\pi f L = 2\pi(1.0 \times 10^8 \text{ Hz})(0.25 \times 10^{-6} \text{ H}) = 160\ \Omega$$

Thus the peak voltage is $V_L = I_L X_L = (0.0020 \text{ A})(160\ \Omega) = 0.32$ V. The voltage peak occurs one-quarter period before the current peaks, and we know that the current peaks at $t = 5.0$ ns. The period of a 100 MHz oscillation is 10 ns, so the voltage peaks at

$$t = 5.0 \text{ ns} - \frac{10 \text{ ns}}{4} = 2.5 \text{ ns}$$

STOP TO THINK 26.3 An inductor is attached to an AC voltage source. Which change will result in a halving of the current?

A. Halving the voltage and doubling the frequency
B. Doubling the frequency
C. Halving the frequency
D. Doubling the voltage and halving the frequency

26.7 Oscillation Circuits

All of the radio stations in your city are broadcasting all the time, but you can tune a radio to pick up one station and no other. This is done using an *oscillation circuit*, a circuit that is designed to have a particular frequency at which it "wants" to oscillate. Tuning your radio means adjusting the frequency of the oscillation circuit to equal that of the station you want to listen to. Oscillation circuits are ubiquitous in modern communications devices, but they have applications in many other areas as well.

FIGURE 26.22 An LC circuit.

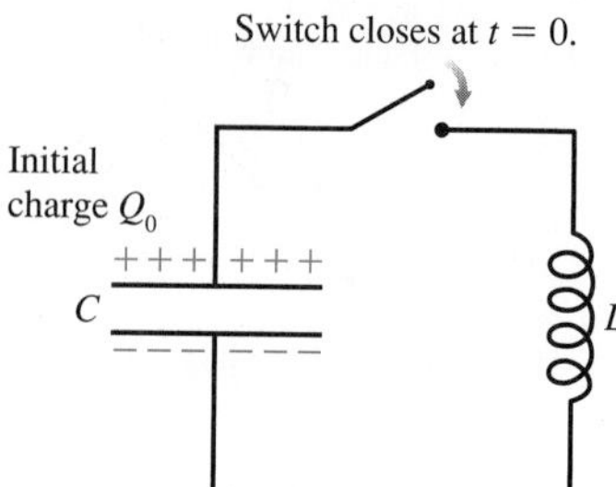

LC Circuits

You learned in Chapter 23 that the voltage across a charged capacitor decays exponentially if the capacitor is connected to a resistor to form an RC circuit. Something very different occurs if the resistor is replaced with an *inductor.* Instead of decaying to zero, the capacitor voltage now undergoes sinusoidal *oscillations.*

To understand how this occurs, let's start with the capacitor and inductor shown in the ***LC* circuit** of FIGURE 26.22. Initially, the capacitor has charge Q_0 and there is no current in the inductor. Then, at $t = 0$, the switch is closed. How does the circuit respond?

As FIGURE 26.23 shows, the inductor provides a conducting path for discharging the capacitor. However, the discharge current has to pass through the inductor, and, as we've seen, an inductor resists changes in current. Consequently, the current doesn't stop when the capacitor charge reaches zero.

FIGURE 26.23 The capacitor charge oscillates much like a block attached to a spring.

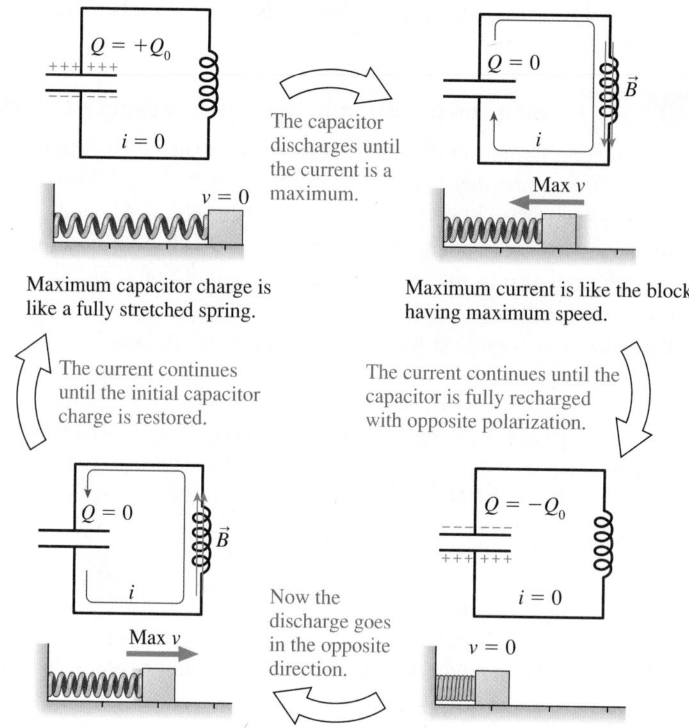

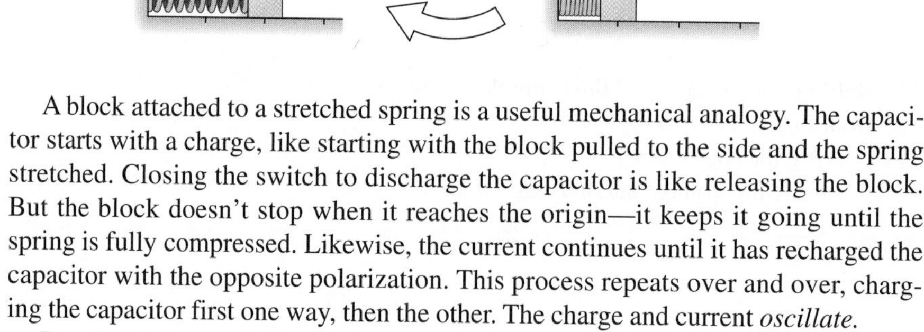

A *Tesla coil* uses the driven oscillation of an LC circuit to produce very high voltages. Careful tuning of the circuit is necessary to produce the dramatic discharges shown.

A block attached to a stretched spring is a useful mechanical analogy. The capacitor starts with a charge, like starting with the block pulled to the side and the spring stretched. Closing the switch to discharge the capacitor is like releasing the block. But the block doesn't stop when it reaches the origin—it keeps it going until the spring is fully compressed. Likewise, the current continues until it has recharged the capacitor with the opposite polarization. This process repeats over and over, charging the capacitor first one way, then the other. The charge and current *oscillate.*

Recall that the oscillation frequency of a mass m on a spring with spring constant k is

$$f = \frac{1}{2\pi}\sqrt{\frac{k}{m}}$$

The oscillation frequency thus depends only on the two basic parameters k and m of the system and not on the amplitude of the oscillation. Similarly, the frequency of an LC oscillator is determined solely by the values of its inductance and capacitance. We would expect larger values of L and C to cause an oscillation with a lower frequency, because a larger inductance means the current changes more slowly and a larger capacitance takes longer to discharge. A detailed analysis shows that the frequency has a form reminiscent of that for a mass and spring:

$$f = \frac{1}{2\pi\sqrt{LC}} \tag{26.27}$$

Frequency of an LC oscillator

EXAMPLE 26.9 **The frequency of an *LC* oscillator**

An LC circuit consists of a 10 μH inductor and a 500 pF capacitor. What is the oscillator's frequency?

SOLVE From Equation 26.27 we have

$$f = \frac{1}{2\pi\sqrt{LC}} = \frac{1}{2\pi\sqrt{(10\times 10^{-6}\text{ H})(500\times 10^{-12}\text{ F})}} = 2.3\times 10^{6}\text{ Hz} = 2.3\text{ MHz}$$

ASSESS The frequencies of LC oscillators are generally *much* higher than those of mechanical oscillators. Frequencies in the MHz or even GHz (10^9 Hz) range are typical.

FIGURE 26.24 An *RLC* circuit.

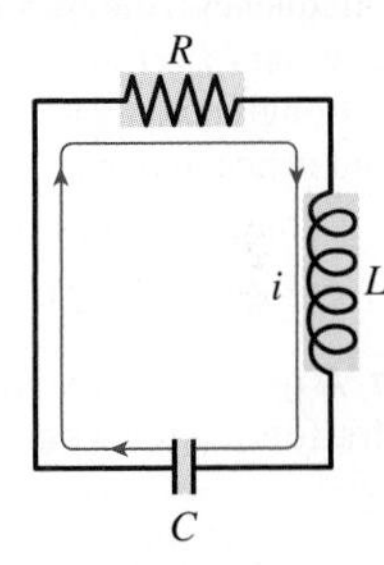

RLC Circuits

Once the oscillations of the ideal LC circuit of Figure 26.23 are started, they will continue forever. All real circuits, however, have some resistance, either from the small resistance of the wires that make up the circuit or from actual circuit resistors. These resistors dissipate energy, causing the amplitudes of the voltages and currents to decay.

We can model this situation by adding a series resistor R to our LC circuit, as shown in FIGURE 26.24. The circuit still oscillates, but the peak values of the voltage and current decrease with time as the current through the resistor transforms the electric and magnetic energy of the circuit into thermal energy. Because the power loss is proportional to R, a larger resistance causes the oscillations to decay more rapidly, as shown in FIGURE 26.25. This behavior is completely analogous to the damped harmonic oscillator discussed in Chapter 14.

FIGURE 26.25 An *RLC* circuit exhibits damped oscillations.

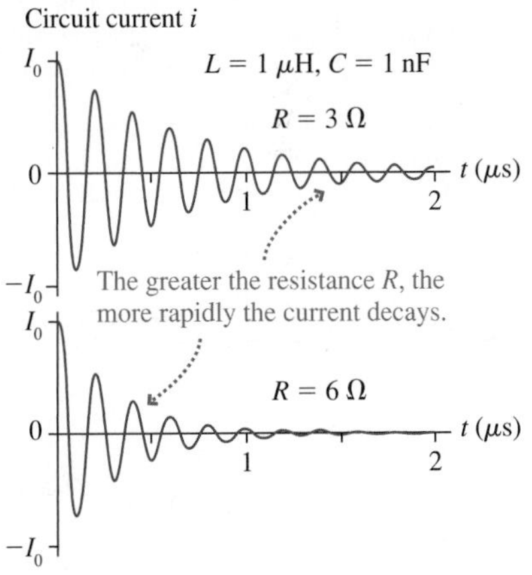

Circuit current i

$L = 1\ \mu$H, $C = 1$ nF

$R = 3\ \Omega$

The greater the resistance R, the more rapidly the current decays.

$R = 6\ \Omega$

t (μs)

To keep the circuit oscillating, we need to add an AC source to the circuit. This makes the **driven *RLC* circuit** of FIGURE 26.26. Because the reactances of the capacitor and inductor vary with the frequency of the AC source, the current in this circuit varies with frequency as well.

Recall that the reactance of a capacitor or inductor plays the same role for the peak quantities I and V as does the resistance of a resistor. When the reactance is large, the current through the capacitor or inductor is small. Because this is a series circuit, with the same current throughout, any circuit element with a large resistance or reactance can block the current.

If the AC source frequency f is very small, the capacitor's reactance $X_C = 1/(2\pi fC)$ is extremely large. If the source frequency f becomes very large, the inductor's reactance $X_L = 2\pi fL$ becomes extremely large. This has two consequences. First, the current in the driven RLC circuit will approach zero at very low and very high frequencies. Second, there must be some intermediate frequency, where neither X_C nor X_L is too large, at which the circuit current I is a maximum. The frequency f_0 at which the current is at its maximum value is called the **resonance frequency.**

FIGURE 26.26 A driven *RLC* circuit.

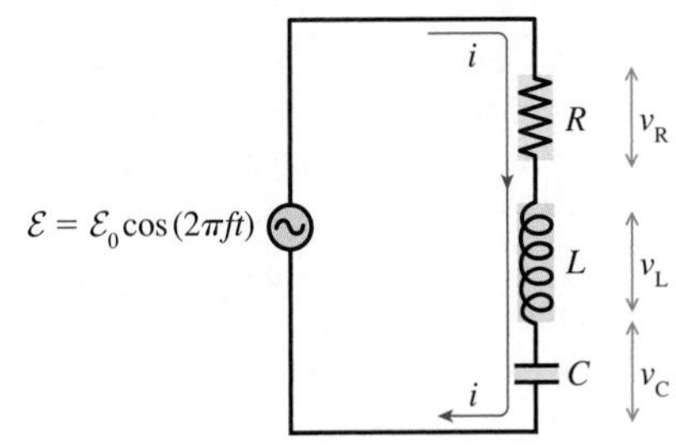

Resonance occurs at the frequency at which the capacitive reactance equals the inductive reactance. If the reactances are equal, the capacitor voltage and the inductor voltage are the same. But these two voltages are out of phase with the current, with the current *leading* the capacitor voltage and *lagging* the inductor voltage. At

Nuclear magnetic resonance, or *nmr,* is an important analytic technique in chemistry and biology. In a large magnetic field, the magnetic moments of atomic nuclei rotate at tens to hundreds of MHz. This motion generates a changing magnetic field that induces an AC emf in a coil placed around the sample. Each kind of nucleus rotates at a characteristic frequency. Adding a capacitor to the coil creates an *RLC* resonance circuit that responds strongly to only one frequency—and hence to one kind of nucleus.

resonance, the capacitor and inductor voltages have equal magnitudes but are exactly out of phase with each other. When we add all the voltages around the loop in Kirchhoff's loop law, the capacitor and inductor voltages cancel and the current is then limited only by the resistance R. Thus the condition for resonance is $X_C = X_L$, or

$$\frac{1}{2\pi f_0 C} = 2\pi f_0 L$$

which gives

$$f_0 = \frac{1}{2\pi\sqrt{LC}} \tag{26.28}$$

This is the same frequency at which the circuit would oscillate if it had no resistor. At this frequency maximum current is determined by the magnitude of the emf and the resistance:

$$I_{max} = \frac{\mathcal{E}_0}{R} \tag{26.29}$$

The driven *RLC* circuit is directly analogous to the driven, damped oscillator that you studied in Chapter 14. A mechanical oscillator exhibits resonance by having a large-amplitude response when the driving frequency matches the system's natural frequency. Equation 26.28 is the natural frequency of the driven *RLC* circuit, the frequency at which the current would like to oscillate. The circuit has a large current response when the oscillating emf matches this frequency.

FIGURE 26.27 shows the peak current I of a driven *RLC* circuit as the emf frequency f is varied. Notice how the current increases until reaching a maximum at frequency f_0, then decreases. This is the hallmark of resonance.

As R decreases, causing the damping to decrease, the maximum current becomes larger and the curve in Figure 26.27 becomes narrower. You saw exactly the same behavior for a driven mechanical oscillator. The emf frequency must be very close to f_0 in order for a lightly damped system to respond, but the response at resonance is very large—exactly what is needed for a tuning circuit.

It is possible to derive an expression for the current graphs shown in Figure 26.27. The peak current in the *RLC* circuit is

FIGURE 26.27 A graph of the current I versus emf frequency for a series *RLC* circuit.

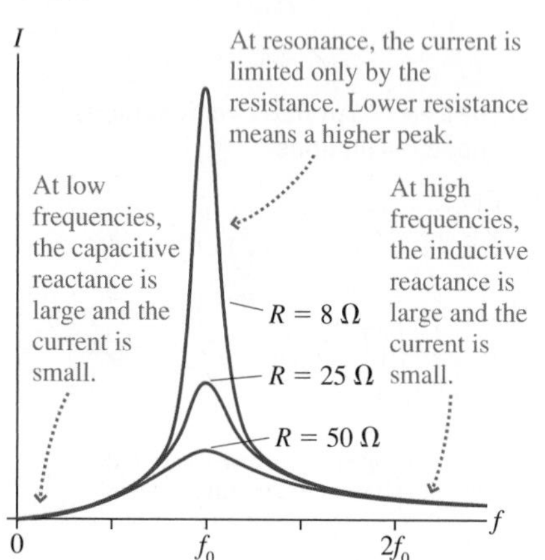

$$I = \frac{\mathcal{E}_0}{\sqrt{R^2 + (X_L - X_C)^2}} = \frac{\mathcal{E}_0}{\sqrt{R^2 + (2\pi f L - 1/2\pi f C)^2}} \tag{26.30}$$

Peak current in an *RLC* circuit

Note that the denominator is smallest, and hence the current is largest, when $X_L = X_C$; that is, at resonance. The three peak voltages, if you need them, are then found from $V_R = IR$, $V_L = IX_L$, and $V_C = IX_C$.

14.2, 14.3 ActivPhysics Online

EXAMPLE 26.10 Designing a radio receiver

An AM radio antenna picks up a 1000 kHz signal with a peak voltage of 5.0 mV. The tuning circuit consists of a 60 μH inductor in series with a variable capacitor. The inductor coil has a resistance of 0.25 Ω, and the resistance of the rest of the circuit is negligible.

a. To what value should the capacitor be tuned to listen to this radio station?
b. What is the peak current through the circuit at resonance?
c. A stronger station at 1050 kHz produces a 10 mV antenna signal. What is the current at this frequency when the radio is tuned to 1000 MHz?

PREPARE The inductor's 0.25 Ω resistance can be modeled as a resistance in series with the inductance; hence we have a series *RLC* circuit. The antenna signal at $f = 1000$ kHz $= 10^6$ Hz is the emf. The circuit looks like that in Figure 26.26.

SOLVE

a. The capacitor needs to be tuned so that the resonant frequency of the circuit is $f_0 = 1000$ kHz. Because $f_0 = 1/2\pi\sqrt{LC}$, the appropriate capacitance is

$$C = \frac{1}{L(2\pi f_0)^2} = \frac{1}{(60 \times 10^{-6}\ \text{H})(2\pi \times 10^6\ \text{Hz})^2}$$
$$= 4.2 \times 10^{-10}\ \text{F} = 420\ \text{pF}$$

b. $X_L = X_C$ at resonance, so the maximum current is

$$I_{max} = \frac{\mathcal{E}_0}{R} = \frac{5.0 \times 10^{-3}\ \text{V}}{0.25\ \Omega} = 0.020\ \text{A} = 20\ \text{mA}$$

c. The 1050 kHz signal is "off resonance," so the reactances X_L and X_C are not equal: $X_L = 2\pi fL = 396\ \Omega$ and $X_C = 1/2\pi fC = 361\ \Omega$ at $f = 1050$ kHz. The peak voltage of this signal is $\mathcal{E}_0 = 10$ mV. With these values, Equation 26.30 for the peak current is

$$I = \frac{\mathcal{E}_0}{\sqrt{R^2 + (X_L - X_C)^2}} = 0.29\ \text{mA}$$

ASSESS These are realistic values for the input stage of an AM radio. You can see that the signal from the 1050 kHz station is strongly suppressed when the radio is tuned to 1000 kHz. The resonant circuit has a large response to the selected station, but not nearby stations.

STOP TO THINK 26.4 A driven *RLC* circuit has $V_C = 5.0$ V, $V_R = 7.0$ V, and $V_L = 9.0$ V. Is the frequency higher than, lower than, or equal to the resonance frequency?

INTEGRATED EXAMPLE 26.11 The ground fault interrupter

As we've seen, a GFI disconnects a household circuit when the currents in the hot and neutral wires are unequal. Let's look inside a GFI outlet to see how this is done.

When you plug something in, the load completes the circuit between the hot and neutral wires. The current varies in the AC circuit thus produced, but at any instant the currents in the hot and neutral wires are equal and opposite, as **FIGURE 26.28** shows.

If someone accidentally touches a hot wire—a potentially dangerous situation—some of the current in the hot wire is diverted to a different path. The currents in the hot and neutral wires are no longer equal. This difference is sensed by the GFI, and the outlet is switched off.

FIGURE 26.28 Currents in the wires in an electric outlet.

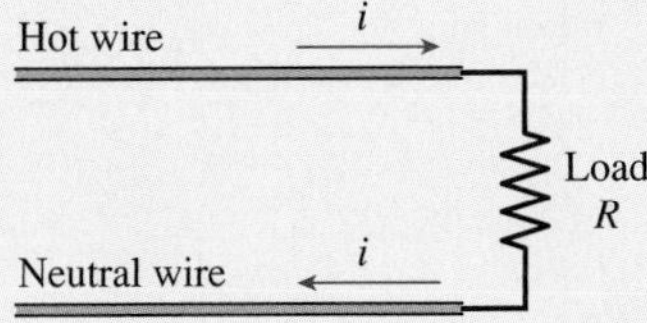

This determination is made inside the GFI outlet as shown in **FIGURE 26.29**. The hot and neutral wires thread through an iron ring. A coil of fine wire wraps around the ring; this coil is connected to a circuit that detects any current in the coil.

If the currents in the hot and neutral wires are equal and opposite, the magnetic fields of the two wires are equal and opposite. There is no net magnetic field in the iron ring, and thus no flux through the sensing coil. But if the two currents aren't equal, there is a net field. The current is AC, so this field isn't constant; it is changing. This field magnetizes the iron ring, increasing the overall field strength and producing a significant (and changing) flux through the sensing coil. According to Faraday's law, this changing flux induces a current in the coil.

FIGURE 26.29 The working elements of a GFI outlet.

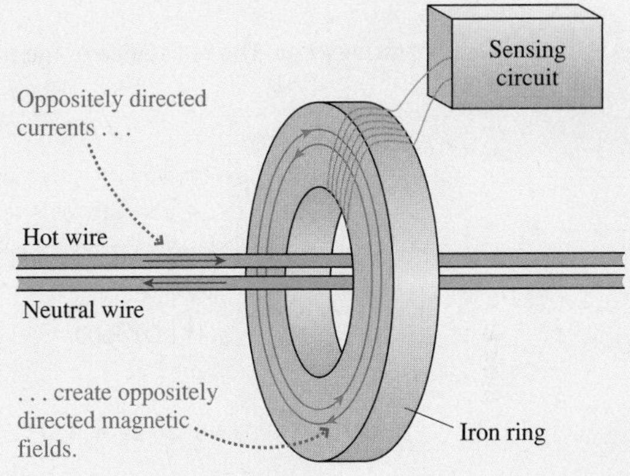

The sensing circuit detects this current and opens a small circuit breaker to turn the outlet off within a few milliseconds, in time to prevent any injury.

a. A GFI will break the circuit if the difference in current between the hot and neutral wires is 5.0 mA. Suppose that, at some instant, there is an excess current of 5.0 mA to the right in the hot wire of Figure 26.29. What are the direction and the magnitude of the magnetic field due to the wire in the iron ring? The ring has an average diameter of 0.80 cm.

b. Consider a worst-case scenario: A person immersed in the bathtub reaches out of the tub and accidentally touches a hot wire with a wet, soapy hand that has minimal skin resistance. What would be the peak current through the person's body? Would this be dangerous? Would a GFI disconnect the circuit?

c. If you connect a capacitor to an outlet, the current and voltage will be out of phase, as we've seen. Would this trigger a GFI?

Continued

PREPARE Part a is about the magnetic fields of the wires. The currents in the hot and neutral wires create a field around the wires, as we saw in Chapter 24. If there is an excess current in the hot wire to the right, the right-hand rule tells us that the field will be clockwise through the iron ring in the view of Figure 26.29. We can use Equation 24.1 to find the magnitude of the magnetic field. Part b is an electrical safety question; we will need to find the resistance of the current path through the body to find the current.

SOLVE a. The iron ring has an average diameter of 0.80 cm, so we need to find the field at a distance of 0.40 cm from the wire. Equal currents produce no net field, so we need only consider the field due to the excess current in the hot wire. We learned in Chapter 24 that the field of a long, straight, current-carrying wire is $B = \mu_0 I/2\pi r$. Thus

$$B = \frac{\mu_0 I}{2\pi r} = \frac{(4\pi \times 10^{-7}\,\text{T}\cdot\text{m/A})(5.0 \times 10^{-3}\,\text{A})}{2\pi(0.0040\,\text{m})}$$

$$= 2.5 \times 10^{-7}\,\text{T} = 0.25\,\mu T$$

We noted above that the direction of this field is clockwise.

b. The person in the tub is sitting in conducting water that is well grounded. If this person touches a hot wire, there is a current path from the hot wire to ground, as FIGURE 26.30a shows. Resistance values for elements of the body are as noted in Figure 26.12. Because of the negligible skin resistance of a wet, soapy hand, the equivalent resistance of the body is the sum of the resistance of the arm and the torso: $R_{eq} = 340\,\Omega$.

Given that the neutral wire of the electricity supply is also grounded, there is a complete circuit through the person, as shown in FIGURE 26.30b. The peak current in this circuit, occurring at the peak of the sinusoidal household AC voltage ($\mathcal{E}_0 = 170\,\text{V}$), is

$$I_R = \frac{\mathcal{E}_0}{R_{eq}} = \frac{170\,\text{V}}{340\,\Omega} = 0.50\,\text{A}$$

Table 26.1 shows that this is a very dangerous level of current, likely to be fatal. Fortunately, it's also well above the threshold for detection by the GFI, which will quickly disconnect the circuit.

c. A difference in phase between voltage and current won't affect the GFI. The GFI detects a difference in current between two wires. As we've seen, the currents into and out of a capacitor are always equal, with no difference between them, so the presence of a capacitor in the circuit will not affect the operation of the GFI.

ASSESS The worst-case scenario is one that you could imagine happening if a person in a bathtub touches a faulty radio or picks up a faulty hair dryer. It's a situation you would expect to be dangerous, in which we'd expect a GFI to come into play.

FIGURE 26.30 The resistance of the path through the body and the resulting circuit.

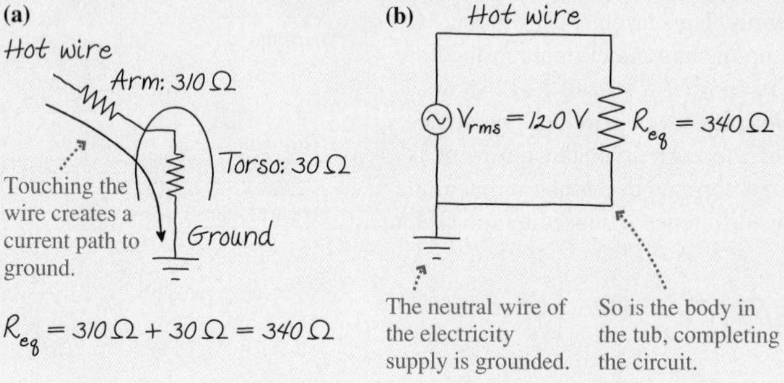

SUMMARY

The goal of Chapter 26 has been to understand and apply basic principles of AC electricity, electricity transmission, and household electricity.

IMPORTANT CONCEPTS

AC circuits

AC circuits are driven by an emf that oscillates with frequency f.

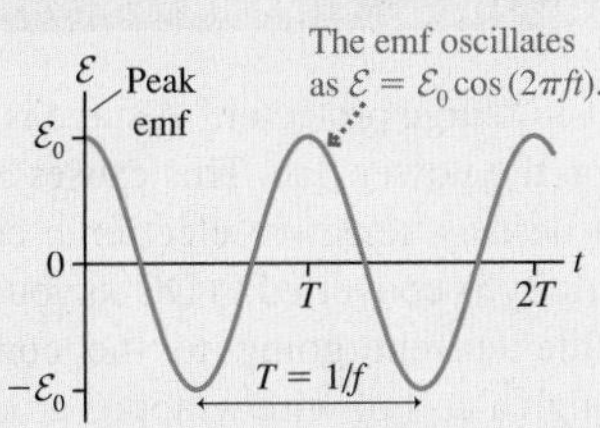

- Peak values of voltages and currents are denoted by capital letters: I, V.
- Instantaneous values of voltages and currents are denoted by lowercase letters: i, v.

Circuit elements used in AC circuits

	Resistor	Capacitor	Inductor
Symbol			
Reactance	Resistance R is constant	$X_C = 1/(2\pi fC)$	$X_L = 2\pi fL$
I and V	$V_R = I_R R$	$V_C = I_C X_C$	$V_L = I_L X_L$
Graph	v_R and i_R: i_R is in phase with v.	v_C and i_C: i_C leads v_C	v_L and i_L: i_L lags v_C

Power in AC resistor circuits

The average power dissipated by a resistor is

$$P_R = (I_{rms})^2 R = \frac{(V_{rms})^2}{R} = I_{rms} V_{rms}$$

where $I_{rms} = I_R/\sqrt{2}$ and $V_{rms} = V_R/\sqrt{2}$ are the root-mean-square (rms) voltage and current.

Electrical safety and biological effects

Currents passing through the body can produce dangerous effects. Currents larger than 15 mA (AC) and 60 mA (DC) are potentially fatal.

The body can be modeled as a network of resistors. If the body forms a circuit between two different voltages, the current is given by Ohm's law: $I = \Delta V/R$.

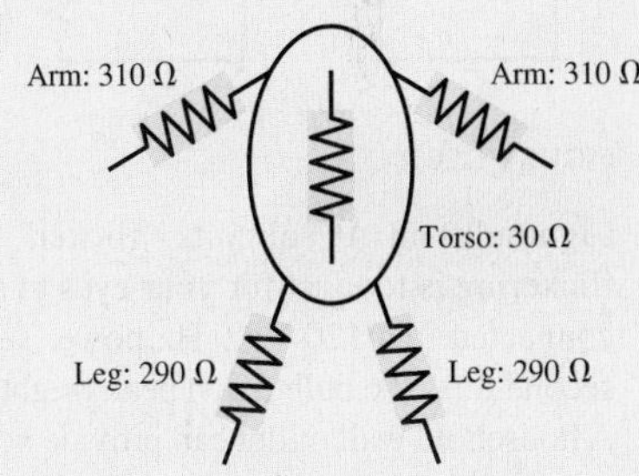

APPLICATIONS

Transformers

Transformers are used to increase or decrease an AC voltage. The rms voltage at the secondary is related to the rms voltage at the primary by

$$(V_2)_{rms} = \frac{N_2}{N_1}(V_1)_{rms}$$

Primary coil, N_1 turns

Secondary coil, N_2 turns

and the currents are related by

$$(I_2)_{rms} = \frac{N_1}{N_2}(I_1)_{rms}$$

LC and RLC circuits

In an LC circuit, the current and voltages oscillate with frequency

$$f = \frac{1}{2\pi\sqrt{LC}}$$

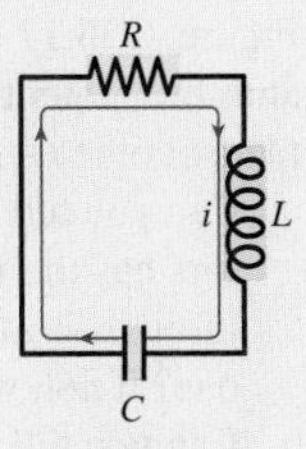

In the RLC circuit, the oscillations decay as energy is dissipated in the resistor.

The driven RLC circuit

If an AC source of amplitude $\mathcal{E}_0$ is placed in an RLC circuit, then the voltages and currents oscillate continuously.

$\mathcal{E} = \mathcal{E}_0 \cos(2\pi ft)$

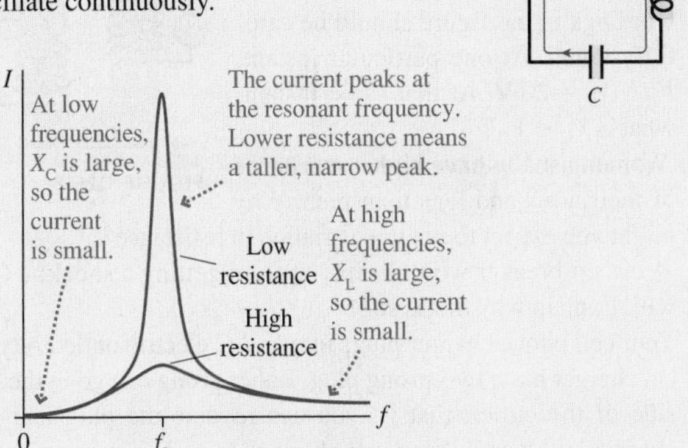

- The resonance frequency is $f_0 = \dfrac{1}{2\pi\sqrt{LC}}$.
- The maximum value of the current is $I_{max} = \mathcal{E}_0/R$.
- The peak current at any frequency f is given by

$$I = \frac{\mathcal{E}_0}{\sqrt{R^2 + (X_L - X_C)^2}}$$

For homework assigned on MasteringPhysics, go to www.masteringphysics.com

Problem difficulty is labeled as | (straightforward) to ||||| (challenging).

Problems labeled INT integrate significant material from earlier chapters; BIO are of biological or medical interest.

QUESTIONS

Conceptual Questions

1. Identical resistors are connected to separate 12 V AC sources. One source operates at 60 Hz, the other at 120 Hz. In which circuit, if either, does the resistor dissipate the greater average power?
2. Consider the three circuits in Figure Q26.2. Rank in order, from largest to smallest, the average total powers P_1 to P_3 dissipated by all the resistors in each circuit. Explain.

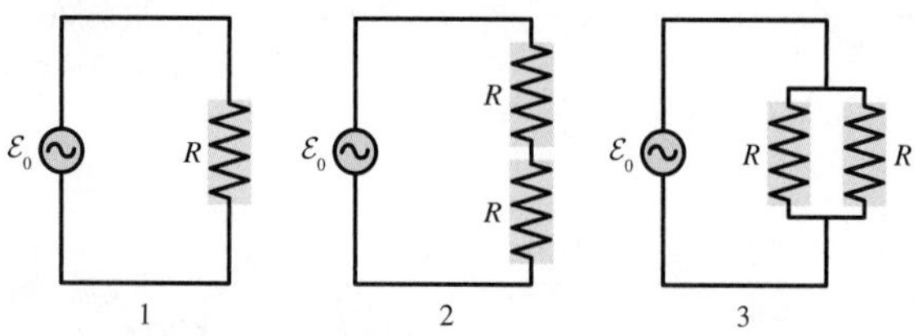

FIGURE Q26.2

3. Lightbulbs in AC circuits "flicker" in intensity, though the flickering is too fast for your eyes to sense it. If a light bulb is connected to a 120 V/60 Hz power supply, how many times a second does the bulb reach peak brightness?
4. A household wall outlet can provide no more than 15 A, but you can use this outlet to power an electric soldering iron that drives 150 A through the metal tip. How is this possible?
5. A 12 V DC power supply is connected to the primary coil of a transformer. The primary coil has 100 turns and the secondary coil has 200. What is the rms voltage across the secondary?
6. Figure Q26.6 shows three wires wrapped around an iron core. The number of turns and the directions of the windings in the figure should be carefully noted. At one particular instant, $V_A - V_B = 20$ V. At that same instant, what is $V_C - V_D$?

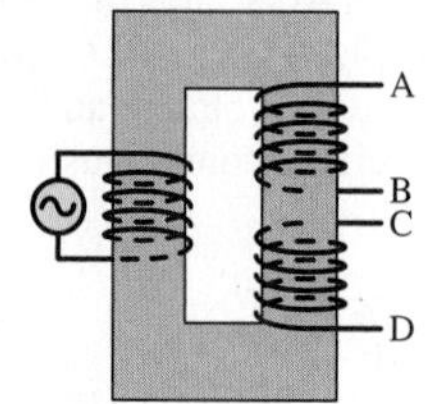

FIGURE Q26.6

7. BIO Women usually have higher resistance of their arms and legs than men. Why might you expect to see this variation in resistance?
8. BIO A circuit breaker won't keep you from getting a shock; a GFI will. Explain why this is so.
9. Your cell phone charger plugs into an AC electric outlet. A typical charger has a two-prong plug. Either prong can go in the hot side of the outlet; that is, you can reverse the plug and the charger still works. Your cell phone runs on DC current from a battery. If you reverse the battery, the phone won't operate. Explain why you can reverse the polarity of an AC source like an electric outlet, but not a DC source like a battery.
10. BIO New homes are required to have GFI-protected outlets in bathrooms, kitchens, and any outdoor locations. Why is GFI protection required in these locations but not, say, in bedrooms?
11. Your computer has a built-in "surge protector." A surge is a sudden increase in voltage on the power line. This causes a rapid "current spike" that can destroy sensitive electronic components. After the AC electricity is converted to DC in your computer's power supply, the current going to the computer processor must pass through a coil of wire wrapped around an iron core. Explain how this coil works as a surge suppressor.
12. The peak current through a resistor is 2.0 A. What is the peak current if
    a. The resistance R is doubled?
    b. The peak emf $\mathcal{E}_0$ is doubled?
    c. The frequency f is doubled?
13. The average power dissipated by a resistor is 4.0 W. What is the average power if
    a. The resistance R is doubled?
    b. The peak emf $\mathcal{E}_0$ is doubled?
    c. Both are doubled simultaneously?
14. The peak current through a capacitor is 2.0 A. What is the peak current if
    a. The peak emf $\mathcal{E}_0$ is doubled?
    b. The capacitance C is doubled?
    c. The frequency f is doubled?
15. Consider the four circuits in Figure Q26.15. Rank in order, from largest to smallest, the capacitive reactances $(X_C)_1$ to $(X_C)_4$. Explain.

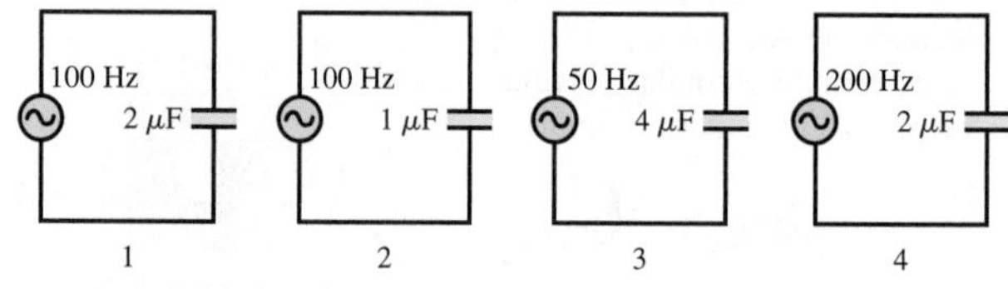

FIGURE Q26.15

16. An inductor is plugged into a 120 V/60 Hz wall outlet in the U.S. Would the peak current be larger, smaller, or unchanged if this inductor were plugged into a wall outlet in a country where the voltage is 120 V at 50 Hz? Explain.
17. Figure Q26.17 shows two inductors and the potential difference across them at time $t = 0$ s.
    a. Can you tell which of these inductors has the larger current flowing through it at $t = 0$ s? If so, which one? If not, why not?
    b. Can you tell through which inductor the current is changing more rapidly at $t = 0$ s? If so, which one? If not, why not?

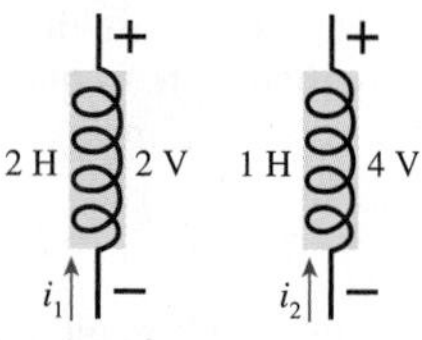

FIGURE Q26.17

18. The peak current passing through an inductor is 2.0 A. What is the peak current if:
 a. The peak emf $\mathcal{E}_0$ is doubled?
 b. The inductance L is doubled?
 c. The frequency f is doubled?
19. Consider the four circuits in Figure Q26.19. Rank in order, from largest to smallest, the inductive reactances $(X_L)_1$ to $(X_L)_4$. Explain.

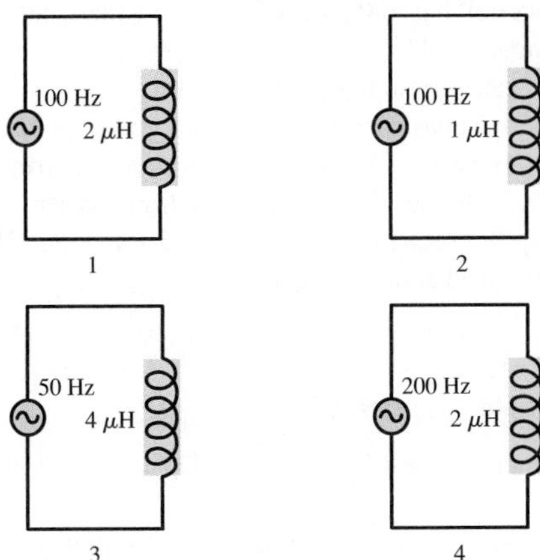

FIGURE Q26.19

20. The tuning circuit in a radio uses an *RLC* circuit. Will adjusting the resistance have any effect on the resonance frequency?
21. The resonance frequency of a driven *RLC* circuit is 1000 Hz. What is the resonance frequency if
 a. The resistance R is doubled?
 b. The inductance L is doubled?
 c. The capacitance C is doubled?
 d. The peak emf $\mathcal{E}_0$ is doubled?
 e. The emf frequency f is doubled?
22. Consider the four circuits in Figure Q26.22. They all have the same resonance frequency f_0 and are driven by the same emf. Rank in order, from largest to smallest, the maximum currents $(I_{max})_1$ to $(I_{max})_4$. Explain.

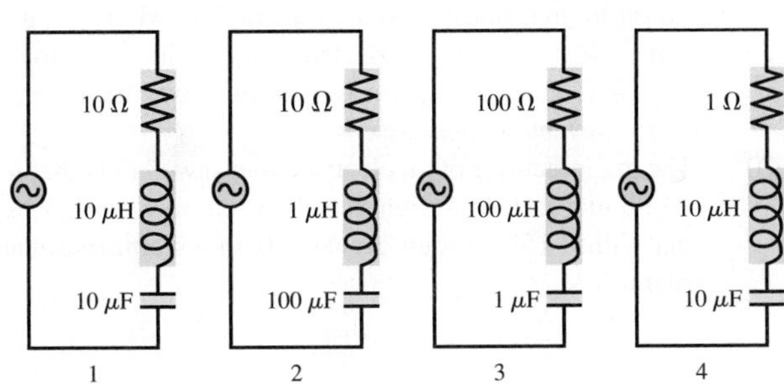

FIGURE Q26.22

Multiple-Choice Questions

23. | A transformer has 1000 turns in the primary coil and 100 turns in the secondary coil. If the primary coil is connected to a 120 V outlet and draws 0.050 A, what are the voltage and current of the secondary coil?
 A. 1200 V, 0.0050 A
 B. 1200 V, 0.50 A
 C. 12 V, 0.0050 A
 D. 12 V, 0.50 A
24. | An inductor is connected to an AC generator. As the generator's frequency is increased, the current in the inductor
 A. Increases.
 B. Decreases.
 C. Does not change.
25. | A capacitor is connected to an AC generator. As the generator's frequency is increased, the current in the capacitor
 A. Increases.
 B. Decreases.
 C. Does not change.
26. | An AC source is connected to a series combination of a lightbulb and a variable inductor. If the inductance is increased, the bulb's brightness
 A. Increases.
 B. Decreases.
 C. Does not change.
27. || An AC source is connected to a series combination of a lightbulb and a variable capacitor. If the capacitance is increased, the bulb's brightness
 A. Increases.
 B. Decreases.
 C. Does not change.
28. | The circuit shown in Figure Q26.28 has a resonance frequency of 15 kHz. What is the value of L?
 A. 1.6 μH
 B. 2.4 μH
 C. 5.2 μH
 D. 18 μH
 E. 59 μH

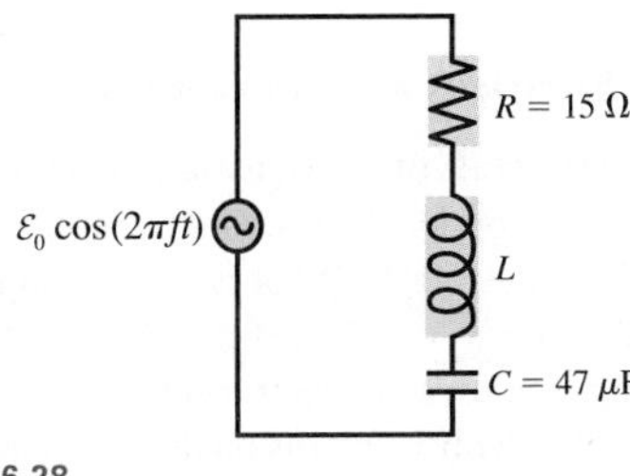

FIGURE Q26.28

PROBLEMS

Section 26.1 Alternating Current

1. | A 200 Ω resistor is connected to an AC source with $\mathcal{E}_0 = 10$ V. What is the peak current through the resistor if the emf frequency is (a) 100 Hz? (b) 100 kHz?
2. | Figure P26.2 shows voltage and current graphs for a resistor.
   a. What is the value of the resistance R?
   b. What is the emf frequency f?

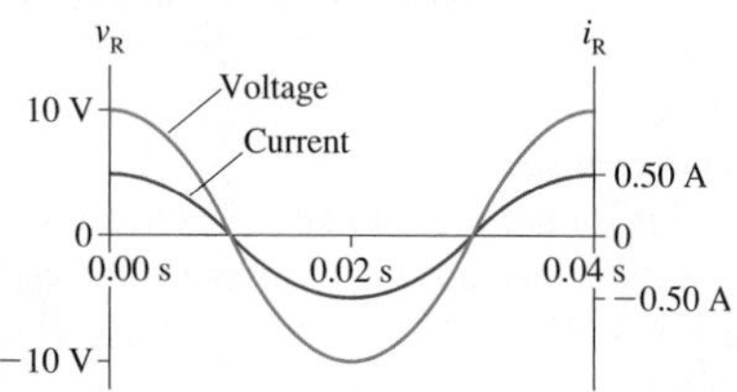

FIGURE P26.2

3. || A resistor dissipates 2.00 W when the rms voltage of the emf is 10.0 V. At what rms voltage will the resistor dissipate 10.0 W?
4. | The heating element of a hair dryer dissipates 1500 W when connected to a 120 V/60 Hz power line. What is its resistance?
5. || A toaster oven is rated at 1600 W for operation at 120 V/60 Hz.
   a. What is the resistance of the oven heater element?
   b. What is the peak current through it?
   c. What is the peak power dissipated by the oven?
6. || A 100 Ω resistor is connected to a 120 V/60 Hz power line. What is its average power loss?
7. || A generator produces 40 MW of power and sends it to town at an rms voltage of 75 kV. What is the rms current in the transmission lines?
8. | Soles of boots that are designed to protect workers from electric shock are rated to pass a maximum rms current of 1.0 mA when connected across an 18,000 V AC source. What is the minimum allowed resistance of the sole? BIO

Section 26.2 AC Electricity and Transformers

9. | The primary coil of a transformer is connected to a 120 V/60 Hz wall outlet. The secondary coil is connected to a lamp that dissipates 60 W. What is the rms current in the primary coil?
10. || A soldering iron uses an electric current in a wire to heat the tip. A transformer with 100 turns on the secondary coil provides 50 W at an rms voltage of 24 V.
    a. What is the resistance of the wire in the iron?
    b. How many turns are in the primary coil?
    c. What is the current in the primary coil?
11. || A power pack charging a cell phone battery has an output of 0.40 A at 5.2 V (both rms). What is the rms current at the 120 V/60 Hz wall outlet where the power pack is plugged in?
12. || A neon sign transformer has a 450 W AC output with an rms voltage of 15 kV when connected to a normal household outlet. There are 500 turns of wire in the primary coil.
    a. How many turns of wire does the secondary coil have?
    b. When the transformer is running at full power, what is the current in the secondary coil? The current in the primary coil?
13. || The "power cube" transformer for a portable CD player has an output of 4.5 V and 600 mA (both rms) when plugged into a 120 V/60 Hz outlet.
    a. If the primary coil has 400 turns of wire, how many turns are on the secondary coil?
    b. What is the peak current in the primary coil?
14. | A science hobbyist has purchased a surplus power-pole transformer that converts 7.2 kV from neighborhood distribution lines into 120 V for homes. He connects the transformer "backward," plugging the secondary coil into a 120 V outlet. What rms voltage is induced at the primary?

Section 26.3 Household Electricity

15. || A typical American family uses 1000 kWh of electricity a month.
    a. What is the average rms current in the 120 V power line to the house?
    b. On average, what is the resistance of a household?
16. || The wiring in the wall of your house to and from an outlet has a total resistance of typically 0.10 Ω. Suppose a device plugged into a 120 V outlet draws 10.0 A of current. INT
    a. What is the voltage drop along the wire?
    b. How much power is dissipated in the wire?
    c. What is the voltage drop across the device?
    d. At what rate does the device use electric energy?
17. | The following appliances are connected to a single 120 V, 15 A circuit in a kitchen: a 330 W blender, a 1000 W coffeepot, a 150 W coffee grinder, and a 750 W microwave oven. If these are all turned on at the same time, will they trip the circuit breaker? INT
18. || A 60 W (120 V) night light is turned on for an average 12 hr a day year round. What is the annual cost of electricity at a billing rate of \$0.10/kWh?
19. || Suppose you leave a 110 W television and two 100 W lightbulbs on in your house to scare off burglars while you go out dancing. If the cost of electric energy in your town is \$0.12/kWh and you stay out for 4.0 hr, how much does this robbery-prevention measure cost?
20. ||| The manufacturer of an electric table saw claims that it has a 3.0 hp motor. It is designed to be used on a normal 120 V outlet with a 15 A circuit breaker. Is this claim reasonable? Explain. INT

Section 26.4 Biological Effects and Electrical Safety

21. || If you touch the terminal of a battery, the small area of contact means that the skin resistance will be relatively large; 50 kΩ is a reasonable value. What current will pass through your body if you touch the two terminals of a 9.0 V battery with your two hands? Will you feel it? Will it be dangerous? BIO

22. || A person standing barefoot on the ground 20 m from the BIO point of a lightning strike experiences an instantaneous potential difference of 300 V between his feet. If we assume a skin resistance of 1.0 kΩ, how much current goes up one leg and back down the other?

23. || Electrodes used to make electrical measurements of the body BIO (such as those used when recording an electrocardiogram) use a conductive paste to reduce the skin resistance to very low values. If you have such an electrode on one wrist and one ankle, what are the lowest AC and DC voltages that will give you a perceptible shock? A potentially fatal shock?

24. || A fisherman has netted a torpedo ray. As he picks it up, this BIO electric fish creates a short-duration 50 V potential difference between his hands. His hands are wet with salt water, and so his skin resistance is a very low 100 Ω. What current passes through his body? Will he feel this DC pulse?

Problems 25 and 26 concern a high-voltage transmission line. Such lines are made of bare wire; they are not insulated. Assume that the wire is 100 km long, has a resistance of 7.0 Ω, and carries 200 A.

25. ||| A bird is perched on the wire with its feet 2.0 cm apart. What INT BIO is the potential difference between its feet?

26. ||| Would it be possible for a person to safely hang from this BIO wire? Assume that the hands are 15 cm apart, and assume a skin INT resistance of 2200 Ω.

Section 26.5 Capacitor Circuits

27. || A 0.30 μF capacitor is connected across an AC generator that produces a peak voltage of 10.0 V. What is the peak current through the capacitor if the emf frequency is (a) 100 Hz? (b) 100 kHz?

28. || A 20 μF capacitor is connected across an AC generator that produces a peak voltage of 6.0 V. The peak current is 0.20 A. What is the oscillation frequency in Hz?

29. | The peak current through a capacitor is 10.0 mA. What is the current if
   a. The emf frequency is doubled?
   b. The emf peak voltage is doubled (at the original frequency)?
   c. The frequency is halved and, at the same time, the emf is doubled?

30. || A 20 nF capacitor is connected across an AC generator that produces a peak voltage of 5.0 V.
   a. At what frequency f is the peak current 50 mA?
   b. What is the instantaneous value of the emf at the instant when $i_C = I_C$?

31. ||| A capacitor is connected to a 15 kHz oscillator that produces an rms voltage of 6.0 V. The peak current is 65 mA. What is the value of the capacitance C?

32. | The peak current through a capacitor is 8.0 mA when connected to an AC source with a peak voltage of 1.0 V. What is the capacitive reactance of the capacitor?

Section 26.6 Inductors and Inductor Circuits

33. ||| What is the potential difference across a 10 mH inductor if the current through the inductor drops from 150 mA to 50 mA in 10 μs?

34. || A 20 mH inductor is connected across an AC generator that produces a peak voltage of 10.0 V. What is the peak current through the inductor if the emf frequency is (a) 100 Hz? (b) 100 kHz?

35. | The peak current through an inductor is 10.0 mA. What is the current if
   a. The emf frequency is doubled?
   b. The emf peak voltage is doubled (at the original frequency)?
   c. The frequency is halved and, at the same time, the emf is doubled?

36. | A 500 μH inductor is connected across an AC generator that produces a peak voltage of 5.0 V.
   a. At what frequency f is the peak current 50 mA?
   b. What is the instantaneous value of the emf at the instant when $i_L = I_L$?

37. ||| An inductor is connected to a 15 kHz oscillator that produces an rms voltage of 6.0 V. The peak current is 65 mA. What is the value of the inductance L?

38. | The peak current through an inductor is 12.5 mA when connected to an AC source with a peak voltage of 1.0 V. What is the inductive reactance of the inductor?

39. | The superconducting magnet in a magnetic resonance imag- BIO ing system consists of a solenoid with 5.6 H of inductance. The solenoid is energized by connecting it to a 1.2 V DC power supply. Starting from zero, how long does it take for the solenoid to reach its operating current of 100 A?

Section 26.7 Oscillation Circuits

40. ||| In a nuclear magnetic resonance spectrometer, a 15 mH BIO coil is designed to detect an emf oscillating at 400 MHz. The coil is part of an *RLC* circuit. What value of the capacitance C should be used so that the coil circuit resonates at 400 MHz?

41. || A 2.0 mH inductor is connected in parallel with a variable capacitor. The capacitor can be varied from 100 pF to 200 pF. What is the range of oscillation frequencies for this circuit?

42. || An FM radio station broadcasts at a frequency of 100 MHz. What inductance should be paired with a 10 pF capacitor to build a receiver circuit for this station?

43. || The inductor in the *RLC* tuning circuit of an AM radio has a value of 350 mH. What should be the value of the variable capacitor in the circuit to tune the radio to 740 kHz?

44. || At what frequency f do a 1.0 μF capacitor and a 1.0 μH inductor have the same reactance? What is the value of the reactance at this frequency?

45. || Two *RLC* circuits have identical capacitors but different inductors. Circuit 1 has a resonance frequency f_0 while circuit 2 has a resonance at $2f_0$. What is the ratio L_2/L_1 of the inductance in circuit 2 to the inductance in circuit 1?

46. || What capacitor in series with a 100 Ω resistor and a 20 mH inductor will give a resonance frequency of 1000 Hz?

47. || What inductor in series with a 100 Ω resistor and a 2.5 μF capacitor will give a resonance frequency of 1000 Hz?

48. || A series *RLC* circuit has a 200 kHz resonance frequency. What is the resonance frequency if
   a. The resistor value is doubled?
   b. The capacitor value is doubled?

49. | A series RLC circuit has a 200 kHz resonance frequency. What is the resonance frequency if
   a. The resistor value is doubled?
   b. The capacitor value is doubled and, at the same time, the inductor value is halved?
50. ||| An RLC circuit with a 10 μF capacitor is connected to a variable-frequency power supply with an rms output voltage of 6.0 V. The current in the circuit as a function of the driving frequency appears as in Figure P26.50. What are the values of the resistor and the inductor?

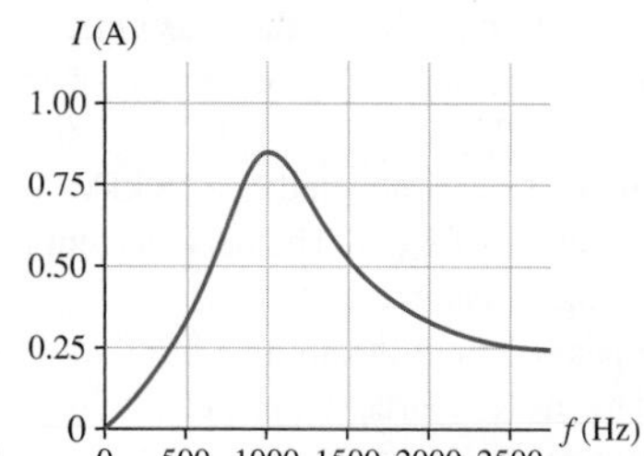

FIGURE P26.50

51. |||| A series RLC circuit consists of a 280 Ω resistor, a 25 μH inductor, and a 18 μF capacitor. What is the rms current if the emf is supplied by a standard 120 V/60 Hz wall outlet?

General Problems

52. ||| A step-down transformer converts 120 V to 24 V, which is connected to a load of resistance 8.0 Ω. What is the resistance "seen" by the power supply connected to the primary coil of the transformer?
   Hint: Resistance is defined as the ratio of two circuit quantities.
53. |||| A 15-km-long, 230 kV aluminum transmission line delivers
INT 34 MW to a city. If we assume a solid cylindrical cable, what minimum diameter is needed if the voltage decrease along this run is to be no more than 1.0% of the transmission voltage? The resistivity of aluminum is $2.7 \times 10^{-8}\ \Omega \cdot \text{m}$.
54. ||||| The outer membranes of cells are quite irregular at the sub-
BIO micron level, so the surface area of an apparently spherical cell differs from the value one would calculate from geometry. To determine the actual surface area of a cell membrane, electrophysiologists use intracellular electrodes to measure the membrane's capacitive reactance. They take advantage of the fact that the capacitance per unit area of all biological membranes is about 1.0 $\mu\text{F/cm}^2$.
   a. What is the surface area of a 20-μm-diameter sphere? Give your answer in square centimeters.
   b. The capacitive reactance of the outer membrane of a cell is measured to be $6.4 \times 10^6\ \Omega$ at 1.0 kHz. What is the surface area of the membrane, in cm^2? By what factor is the area greater than your answer to part a?

   An important biological process is the release of neurotransmitters that are involved in chemical communication between nerve cells and muscles. This release occurs when intracellular vesicles containing the neurotransmitter chemical fuse with the outer cell membrane and release their contents. The vesicle's membrane becomes part of the cell's outer membrane, increasing the latter's area. As a result, the fusion of vesicles with the outer membrane can be detected by monitoring the outer membrane's capacitive reactance. Indeed, the fusion of a single vesicle can be resolved.
   c. What is the percentage increase in the capacitive reactance of the cell considered in part b when a single vesicle's membrane, a smooth sphere 0.10 μm in diameter, fuses with the outer membrane of the cell?
55. || The voltage across a 60 μF capacitor is described by the equation $v_C = (18\text{ V})\cos(200t)$, where t is in seconds.
   a. What is the voltage across the capacitor at $t = 0.010$ s?
   b. What is the capacitive reactance?
   c. What is the peak current?
56. || The voltage across a 75 μH inductor is described by the equation $v_L = (25\text{ V})\cos(60t)$, where t is in seconds.
   a. What is the voltage across the inductor at $t = 0.10$ s?
   b. What is the inductive reactance?
   c. What is the peak current?
57. || An LC circuit is built with a 20 mH inductor and an 8.0 μF capacitor. The current has its maximum value of 0.50 A at $t = 0$ s. How long does it take until the capacitor is fully charged?
58. ||| An electronics hobbyist is building a radio set to receive the AM band, with frequencies from 520 to 1700 kHz. What range variable capacitor will she need to go with a 230 μH inductor?
59. || For the circuit of Figure P26.59,
   a. What is the resonance frequency?
   b. At resonance, what is the peak current through the circuit?

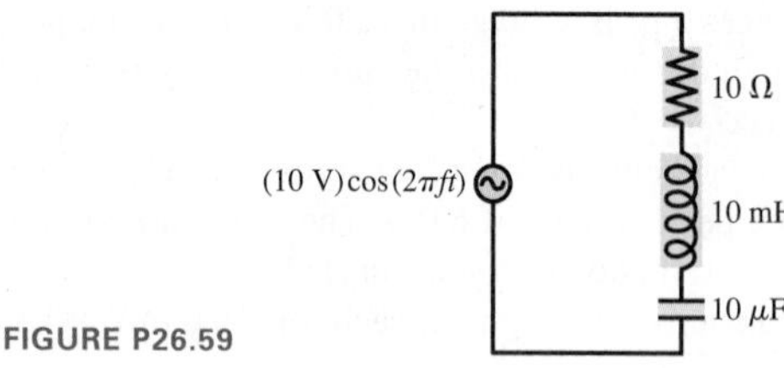

FIGURE P26.59

60. | For the circuit of Figure P26.60,
   a. What is the resonance frequency?
   b. At resonance, what is the peak current through the circuit?

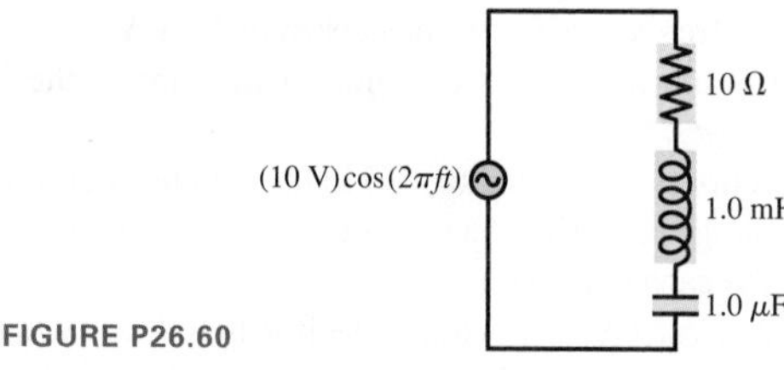

FIGURE P26.60

61. || An RLC circuit consists of a 48 Ω resistor, a 200 μF capacitor, and an inductor. The current is 2.5 A rms when the circuit is connected to a 120 V/60 Hz outlet.
   a. What is the inductance?
   b. What would be the current if this circuit were used in France, where the outlets are 220 V/50 Hz?

Passage Problems

Halogen Bulbs

Halogen bulbs have some differences from standard incandescent lightbulbs. They are generally smaller, the filament runs at a higher temperature, and they have a quartz (rather than glass) envelope. They may also operate at lower voltage. Consider a 12 V, 50 W halogen bulb for use in a desk lamp. The lamp plugs into a 120 V/60 Hz outlet, and it has a transformer in its base.

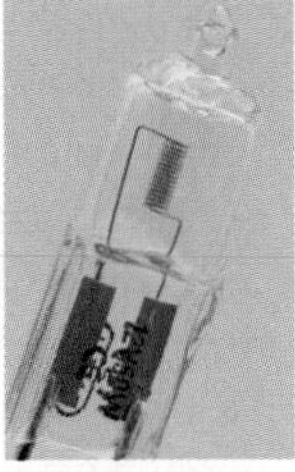

62. | The 12 V rating of the bulb refers to the rms voltage. What is the peak voltage across the bulb?
    A. 8.5 V
    B. 12 V
    C. 17 V
    D. 24 V

63. | Suppose the transformer in the base of the lamp has 500 turns of wire on its primary coil. How many turns are on the secondary coil?
    A. 50
    B. 160
    C. 500
    D. 5000

64. | How much current is drawn by the lamp at the outlet? That is, what is the rms current in the primary?
    A. 0.42 A
    B. 1.3 A
    C. 4.2 A
    D. 13 A

65. | What will be the voltage across the bulb if the lamp's power cord is accidentally plugged into a 240 V/60 Hz outlet?
    A. 12 V
    B. 24 V
    C. 36 V
    D. 48 V

STOP TO THINK ANSWERS

Stop to Think 26.1: B. The power in the AC circuit is proportional to the square of the rms current, or to $(I_{rms})^2 = (I_R/\sqrt{2})^2 = \frac{1}{2}I_R^2 = \frac{1}{2}(1\text{ A})^2$. The power in the DC circuit is proportional to the square of the DC current, or to $I_R^2 = (1\text{ A})^2$. Thus the power in the DC circuit is twice that in the AC circuit, so bulb B is brighter.

Stop to Think 26.2: B. The current is $I_C = V_C/X_C = 2\pi f C V_C$. Thus I_C is proportional to both f and V_C. Doubling the frequency while keeping V_C constant will double the current.

Stop to Think 26.3: B. The current is $I_L = V_L/X_L = V_L/2\pi f L$. Thus I_L is *inversely* proportional to f. Doubling the frequency while keeping V_L constant will halve the current.

Stop to Think 26.4: Higher than. $V_L > V_C$ tells us that $X_L > X_C$. This is the condition above resonance, where X_L is increasing with f while X_C is decreasing.

PART VI SUMMARY

Electricity and Magnetism

Mass and charge are the two most fundamental properties of matter. The first four parts of this text were about properties and interactions of masses. Part VI has been a study of charge—what charge is and how charges interact.

Electric and magnetic fields were introduced to enable us to understand the long-range forces of electricity and magnetism. One charge—the source charge—alters the space around it by creating an electric field and, if the charge is moving, a magnetic field. Other charges experience forces exerted *by the fields.* Electric and magnetic fields are the agents by which charges interact.

In addition to the electric field, we often describe electric interactions in terms of the electric potential. This is a particularly fruitful concept for dealing with electric circuits in which charges flow through wires, resistors, etc.

Electric and magnetic fields are *real* and can exist independently of charges. The clearest evidence for their independent existence is electromagnetic waves—the quintessential electromagnetic phenomenon. All electromagnetic waves, including light, are similar in structure, but they span a wide range of wavelengths and frequencies. In Part VI, we got our first hints that electromagnetic waves might have a particle-like nature, a concept we will explore further in Part VII.

Part VI has introduced many new phenomena, concepts, and laws. The knowledge structure below draws together the major ideas. These ideas build on what we've learned in the first five parts of this text. All the pieces are now in place to support the development of the ideas of modern physics in Part VII.

KNOWLEDGE STRUCTURE VI Electricity and Magnetism

BASIC GOALS How do charged particles interact? How do electric circuits work? What are the properties and characteristics of electric and magnetic fields?

GENERAL PRINCIPLES

Forces between charges: **Coulomb's law**	$F_{1\text{ on }2} = F_{2\text{ on }1} = \dfrac{K\lvert q_1\rvert\lvert q_2\rvert}{r^2}$	The force is along the line connecting the charges. For like charges, the force is repulsive; for opposite charges, attractive.
Electric force on a charge:	$\vec{F} = q\vec{E}$	The force is in the direction of the field for a positive charge; opposite the field for a negative charge.
Magnetic force on a moving charge:	$F = \lvert q\rvert vB\sin\alpha$	The force is perpendicular to the velocity and the field, with direction as specified by the right-hand rule for forces.
Induced emf: **Faraday's law**	$\mathcal{E} = \left\lvert \dfrac{\Delta\Phi}{\Delta t}\right\rvert$	The induced current $I = \mathcal{E}/R$ is such that the induced magnetic field opposes the *change* in the magnetic flux. This is **Lenz's law**.

Electric and magnetic fields

Charges and changing magnetic fields create electric fields.

- Electric fields exert forces on charges and torques on dipoles.
- The electric field is perpendicular to equipotential surfaces and points in the direction of decreasing potential.
- The electric field causes charges to move in conductors but not insulators.

Currents and permanent magnets create magnetic fields.

- Magnetic fields exert forces on currents (and moving charged particles) and torques on magnetic dipoles.
- A compass needle or other magnetic dipole will line up with a magnetic field.

Electric potential

The interaction of charged particles can also be described in terms of an electric potential V.

- Only potential differences ΔV are important.
- If the potential of a particle of charge q changes by ΔV, its potential energy changes by $\Delta U = q\Delta V$.
- Where two equipotential surfaces with potential difference ΔV are separated by distance d, the electric field strength is $E = \Delta V/d$.

Current and circuits

Potential differences ΔV drive current in circuits. Though electrons are the charge carriers in metals, the **current** I is defined to be the motion of positive charges.

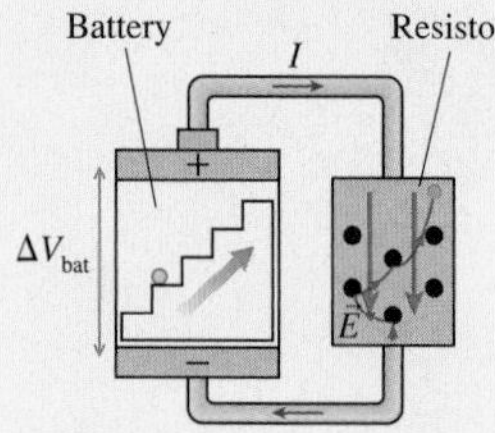

- Circuits obey Kirchhoff's loop law (conservation of energy) and Kirchhoff's junction law (conservation of charge).
- The current through a resistor is $I = \Delta V_R/R$. This is **Ohm's law.**

Electromagnetic waves

An electromagnetic wave is a self-sustaining oscillation of electric and magnetic fields.

- $\vec{E}$ and $\vec{B}$ are perpendicular to each other and to the direction of travel.
- All electromagnetic waves travel at the same speed, c.
- The **electromagnetic spectrum** is the spread of wavelengths and frequencies of electromagnetic waves, from radio waves through visible light to gamma rays.

The Greenhouse Effect and Global Warming

Electromagnetic waves are real, and we depend on them for our very existence; energy carried by electromagnetic waves from the sun provides the basis for all life on earth. Because of the sun's high surface temperature, it emits most of its thermal radiation in the visible portion of the electromagnetic spectrum. As the figure below shows, the earth's atmosphere is transparent to the visible and near-infrared radiation, so most of this energy travels through the atmosphere and warms the earth's surface.

Although seasons come and go, *on average* the earth's climate is very steady. To maintain this stability, the earth must radiate thermal energy—electromagnetic waves—back into space at exactly the same average rate that it receives energy from the sun. Because the earth is much cooler than the sun, its thermal radiation is long-wavelength infrared radiation that we cannot see. A straightforward calculation using Stefan's law finds that the average temperature of the earth should be $-18°C$, or $0°F$, for the incoming and outgoing radiation to be in balance.

This result is clearly not correct; at this temperature, the entire earth would be covered in snow and ice. The measured global average temperature is actually a balmier $15°C$, or $59°F$. The straightforward calculation fails because it neglects to consider the earth's atmosphere. At visible wavelengths, as the figure shows, the atmosphere has a wide "window" of transparency, but this is not true at the infrared wavelengths of the earth's thermal radiation. The atmosphere lets in the visible radiation from the sun, but the outgoing thermal radiation from the earth sees a much smaller "window." Most of this radiation is absorbed in the atmosphere.

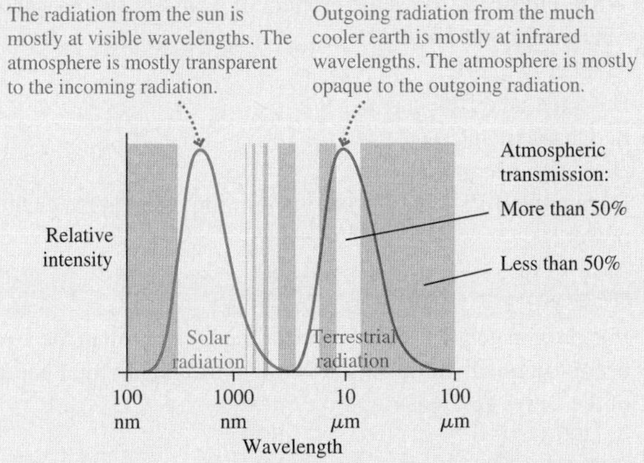

The radiation from the sun is mostly at visible wavelengths. The atmosphere is mostly transparent to the incoming radiation.

Outgoing radiation from the much cooler earth is mostly at infrared wavelengths. The atmosphere is mostly opaque to the outgoing radiation.

Atmospheric transmission:
More than 50%
Less than 50%
Relative intensity
Solar radiation
Terrestrial radiation
100 nm
1000 nm
10 μm
100 μm
Wavelength

Thermal radiation curves for the sun and the earth. The white and gray bars show regions for which the atmosphere is transparent (white) or opaque (tan) to electromagnetic radiation.

Because it's easier for visible radiant energy to get in than for infrared to get out, the earth is warmer than it would be without the atmosphere. The additional warming of the earth's surface because of the atmosphere is called the **greenhouse effect.** The greenhouse effect is a natural part of the earth's physics; it has nothing to do with human activities, although it's doubtful any advanced life forms would have evolved without it.

The atmospheric gases most responsible for the greenhouse effect are carbon dioxide and water vapor, both strong absorbers of infrared radiation. These **greenhouse gases** are of concern today because humans, through the burning of fossil fuels (oil, coal, and natural gas), are rapidly increasing the amount of carbon dioxide in the atmosphere. Preserved air samples show that carbon dioxide made up 0.027% of the atmosphere before the industrial revolution. In the last 150 years, human activities have increased the amount of carbon dioxide to 0.038%, a 40% increase. By 2050, the carbon dioxide concentration will likely increase to 0.054%, double the pre-industrial value, unless the use of fossil fuels is substantially reduced.

Carbon dioxide is a powerful absorber of infrared radiation. Adding more carbon dioxide makes it even harder for emitted thermal radiation to escape, increasing the average surface temperature of the earth. The net result is **global warming.**

There is strong evidence that the earth has warmed nearly $1°C$ in the last 100 years because of increased greenhouse gases. What happens next? Climate scientists, using sophisticated models of the earth's atmosphere and oceans, calculate that a doubling of the carbon dioxide concentration will likely increase the earth's average temperature by an additional $2°C$ ($\approx 3°F$) to $6°C$ ($\approx 9°F$). There is some uncertainty in these calculations; the earth is a large and complex system. Perhaps the earth will get cloudier as the temperature increases, moderating the increase. Or perhaps the arctic ice cap will melt, making the earth less reflective and leading to an even more dramatic temperature increase.

But the basic physics that leads to the greenhouse effect, and to global warming, is quite straightforward. Carbon dioxide in the atmosphere keeps the earth warm; more carbon dioxide will make it warmer. How much warmer? That's an important question, one that many scientists around the world are attempting to answer with ongoing research. But large or small, change *is* coming. Global warming is one of the most serious challenges facing scientists, engineers, and all citizens in the 21st century.

PART VI PROBLEMS

The following questions are related to the passage "The Greenhouse Effect and Global Warming" on the previous page.

1. The intensity of sunlight at the top of the earth's atmosphere is approximately 1400 W/m^2. Mars is about 1.5 times as far from the sun as the earth. What is the approximate intensity of sunlight at the top of Mars's atmosphere?
   A. 930 W/m^2
   B. 620 W/m^2
   C. 410 W/m^2
   D. 280 W/m^2
2. Averaged over day, night, seasons, and weather conditions, a square meter of the earth's surface receives an average of 240 W of radiant energy from the sun. The average power radiated back to space is
   A. Less than 240 W.
   B. More than 240 W.
   C. Approximately 240 W.
3. The thermal radiation from the earth's surface peaks at a wavelength of approximately 10 μm. If the surface of the earth warms, this peak will
   A. Shift to a longer wavelength.
   B. Stay the same.
   C. Shift to a shorter wavelength.
4. The thermal radiation from the earth's surface peaks at a wavelength of approximately 10 μm. What is the energy of a photon at this wavelength?
   A. 2.4 eV B. 1.2 eV
   C. 0.24 eV D. 0.12 eV
5. Electromagnetic waves in certain wavelength ranges interact with water molecules because the molecules have a large electric dipole moment. The electric field of the wave
   A. Exerts a net force on the water molecules.
   B. Exerts a net torque on the water molecules.
   C. Exerts a net force and a net torque on the water molecules.

The following passages and associated questions are based on the material of Part VI.

Taking an X Ray

X rays are a very penetrating form of electromagnetic radiation. X rays pass through the soft tissue of the body but are largely stopped by bones and other more dense tissues. This makes x rays very useful for medical and dental purposes, as you know.

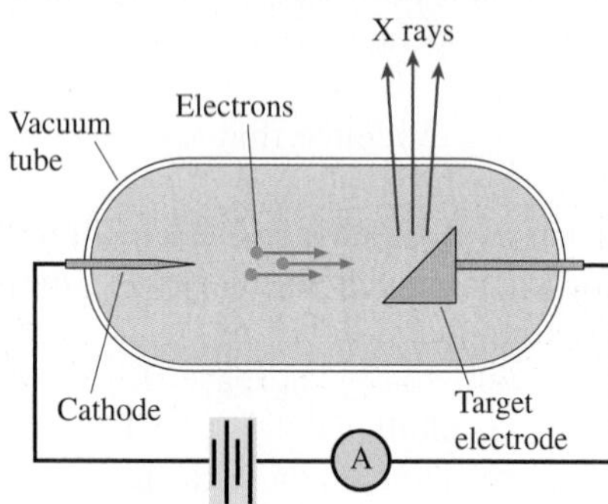

FIGURE VI.1

A schematic view of an x-ray tube and a driver circuit is given in Figure VI.1. A filament warms the cathode, freeing electrons. These electrons are accelerated by the electric field established by a high-voltage power supply connected between the cathode and a metal target. The electrons accelerate in the direction of the target. The rapid deceleration when they strike the target generates x rays. Each electron will emit one or more x rays as it comes to rest.

An x-ray image is essentially a shadow; x rays darken the film where they pass, but the film stays unexposed, and thus light, where bones or dense tissues block x rays. An x-ray technician adjusts the quality of an image by adjusting the energy and the intensity of the x-ray beam. This is done by adjusting two parameters: the accelerating voltage and the current through the tube. The accelerating voltage determines the energy of the x-ray photons, which can't be greater than the energy of the electrons. The current through the tube determines the number of electrons per second and thus the number of photons emitted. In clinical practice, the exposure is characterized by two values: "kVp" and "mAs." kVp is the peak voltage in kV. The value mAs is the product of the current (in mA) and the time (in s) to give a reading in mA · s. This is a measure of the total number of electrons that hit the target and thus the number of x rays emitted.

Typical values for a dental x ray are a kVp of 70 (meaning a peak voltage of 70 kV) and mAs of 7.5 (which comes from a current of 10 mA for 0.75 s, for a total of 7.5 mAs). Assume these values in all of the problems that follow.

6. In Figure VI.1, what is the direction of the electric field in the region between the cathode and the target electrode?
   A. To the left
   B. To the right
   C. Toward the top of the page
   D. Toward the bottom of the page
7. If the distance between the cathode and the target electrode is approximately 1.0 cm, what will be the maximum acceleration of the free electrons? Assume that the electric field is uniform.
   A. 1.2×10^{18} m/s^2 B. 1.2×10^{16} m/s^2
   C. 1.2×10^{15} m/s^2 D. 1.2×10^{12} m/s^2
8. What, physically, does the product of a current (in mA) and a time (in s) represent?
   A. Energy in mJ
   B. Potential difference in mV
   C. Charge in mC
   D. Resistance in mΩ
9. During the 0.75 s that the tube is running, what is the electric power?
   A. 7.0 kW B. 700 W
   C. 70 W D. 7.0 W
10. If approximately 1% of the electric energy ends up in the x-ray beam (a typical value), what is the approximate total energy of the x rays emitted?
    A. 500 J B. 50 J
    C. 5 J D. 0.5 J
11. What is the maximum energy of the emitted x-ray photons?
    A. 70×10^3 J B. 1.1×10^{-11} J
    C. 1.1×10^{-14} J D. 1.6×10^{-18} J

Electric Cars

In recent years, practical hybrid cars have hit the road—cars in which the gasoline engine runs a generator that charges batteries that run an electric motor. These cars offer increased efficiency, but significantly greater efficiency could be provided by a purely electric car run by batteries that you charge by plugging into an electric outlet in your house.

But there's a practical problem with such vehicles: the time necessary to recharge the batteries. If you refuel your car with gas at the pump, you add 130 MJ of energy per gallon. If you add 20 gallons, you add a total of 2.6 GJ in about 5 minutes. That's a lot of energy in a short time; the electric system of your house simply can't provide power at this rate.

There's another snag as well. Suppose there were electric filling stations that could provide very high currents to recharge your electric car. Conventional batteries can't recharge very quickly; it would still take longer for a recharge than to refill with gas.

One possible solution is to use capacitors instead of batteries to store energy. Capacitors can be charged much more quickly, and as an added benefit, they can provide energy at a much greater rate—allowing for peppier acceleration. Today's capacitors can't store enough energy to be practical, but future generations will.

12. A typical home's electric system can provide 100 A at a voltage of 220 V. If you had a charger that ran at this full power, approximately how long would it take to charge a battery with the equivalent of the energy in one gallon of gas?
A. 100 min B. 50 min
C. 20 min D. 5 min
13. The Tesla Roadster, a production electric car, has a 375 V battery system that can provide a power of 200 kW. At this peak power, what is the current supplied by the batteries?
A. 75 kA B. 1900 A
C. 530 A D. 75 A
14. To charge the batteries in a Tesla Roadster, a transformer is used to step up the voltage of the household supply. If you step a 220 V, 100 A system up to 400 V, what is the maximum current you can draw at this voltage?
A. 180 A B. 100 A
C. 55 A D. 45 A
15. One design challenge for a capacitor-powered electric car is that the voltage would change with time as the capacitors discharge. If the capacitors in a car were discharged to half their initial voltage, what fraction of energy would still be left?
A. 75% B. 67%
C. 50% D. 25%

Wireless Power Transmission

Your laptop has wireless communications connectivity, and you might even have a wireless keyboard or mouse. But there's one wire you haven't been able to get rid of yet—the power cord.

Researchers are working on ways to circumvent the need for a direct electrical connection for power, and they are experiencing some success. Recently, investigators were able to use current flowing through a primary coil to power a 60 W lightbulb connected to a secondary coil 2.0 m away, with approximately 15% efficiency. The coils were large and the efficiency low, but it's a start.

FIGURE VI.2

The wireless power transfer system is outlined in Figure VI.2. An AC supply generates a current through the primary coil, creating a varying magnetic field. This field induces a current in the secondary coil, which is connected to a resistance (the lightbulb) and a capacitor that sets the resonance frequency of the secondary circuit to match the frequency of the primary circuit.

16. At a particular moment, the current in the primary coil is clockwise, as viewed from the secondary coil. At the center of the secondary coil, the field from the primary coil is
A. To the right.
B. To the left.
C. Zero.
17. At a particular moment, the magnetic field from the primary coil points to the right and is increasing in strength. The field due to the induced current in the secondary coil is
A. To the right.
B. To the left.
C. Zero.
18. The power supply drives the primary coil at 9.9 MHz. If this frequency is doubled, how must the capacitor in the secondary circuit be changed?
A. Increase by a factor of 2
B. Increase by a factor of $\sqrt{2}$
C. Decrease by a factor of 2
D. Decrease by a factor of 4
19. What are the rms and peak currents for a 60 W bulb? (The rms voltage is the usual 120 V.)
A. 0.71 A, 0.71 A B. 0.71 A, 0.50 A
C. 0.50 A, 0.71 A D. 0.50 A, 0.50 A

Additional Integrated Problems

20. A 20 Ω resistor is connected across a 120 V source. The resistor is then lowered into an insulated beaker, containing 1.0 L of water at 20° C, for 60 s. What is the final temperature of the water?
21. As shown in Figure VI.3, a square loop of wire, with a mass of 200 g, is free to pivot about a horizontal axis through one of its sides. A 0.50 T horizontal magnetic field is directed as shown. What current I in the loop, and in what direction, is needed to hold the loop steady in a horizontal plane?

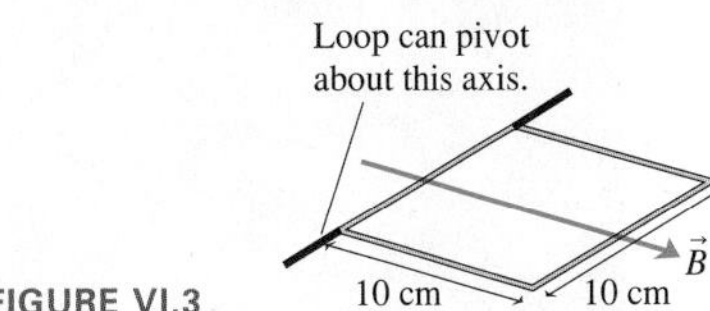

FIGURE VI.3

PART

VII

Modern Physics

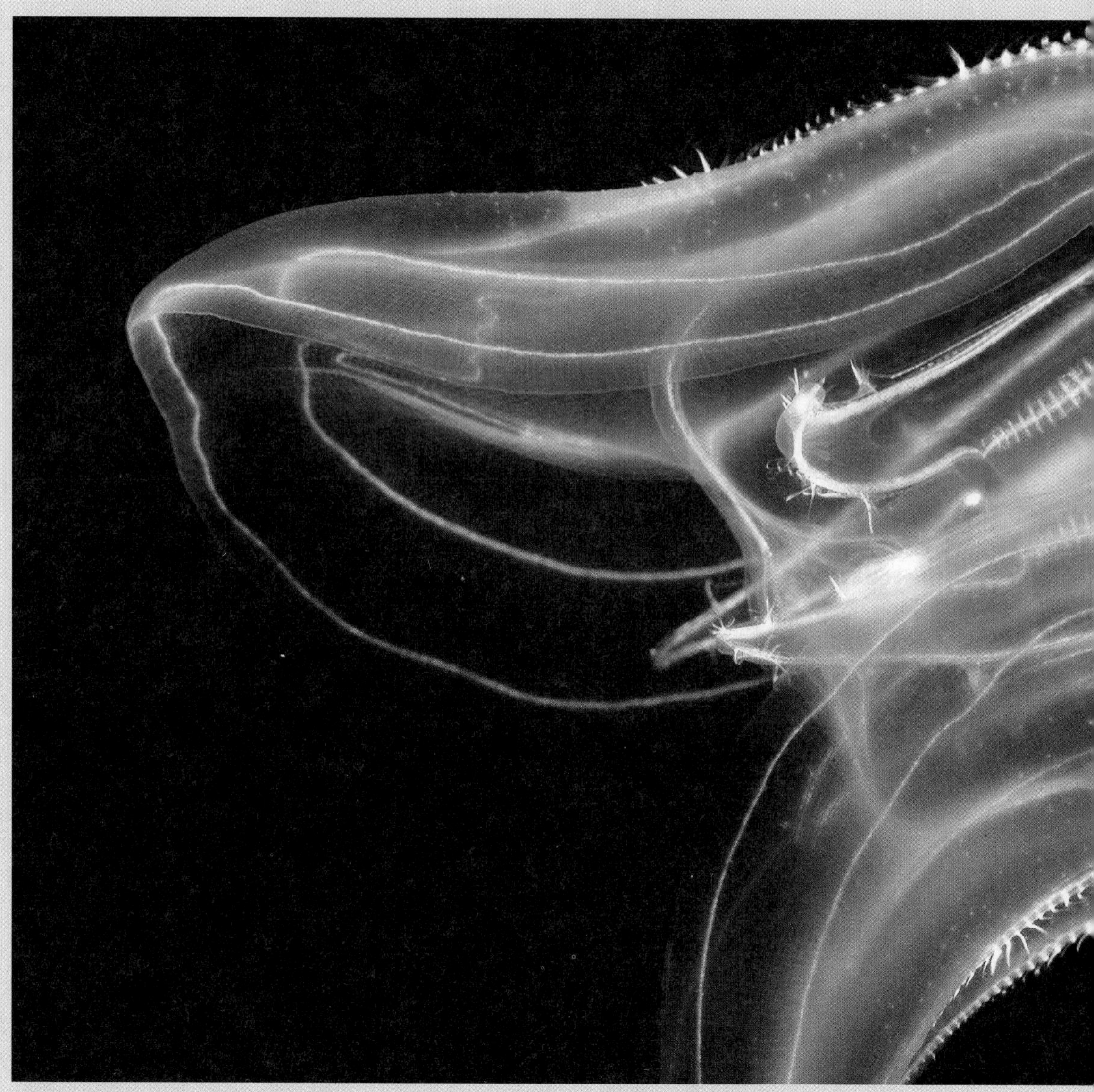

The eerie glow of this comb jelly is due to *bioluminescence.* Energy released in chemical reactions in special cells is turned directly into a cool blue light. How is this light different from the light from a hot, incandescent filament?

New Ways of Looking at the World

Newton's mechanics and Maxwell's electromagnetism are remarkable theories that explain a wide range of physical phenomena, as we have seen in the past 26 chapters—but our story doesn't stop there. In the early 20th century, a series of discoveries profoundly altered our understanding of the universe at the most fundamental level, forcing scientists to reconsider the very nature of space and time and to develop new models of light and matter.

Relativity

The idea of measuring distance with a meter stick and time with a clock or stopwatch seems self-evident. But Albert Einstein, as a young, unknown scientist, realized that Maxwell's theory of electromagnetism could be consistent only if an additional rather odd assumption was made: that the speed of light is the same for all observers, no matter how they might be moving with respect to each other or to the source of the light. This assumption changes the way that we think about space and time. When we study Einstein's theory of *relativity,* you will see how different observers can disagree about lengths and time intervals. We need to go beyond stopwatches and meter sticks. Time can pass at different rates for different observers; time is, as you will see, relative. Our exploration will end with the most famous equation in physics, Einstein's $E = mc^2$. Matter can be converted to energy, and energy to matter.

Quantum Physics

We've seen that light, a wave, sometimes acts like a particle, a photon. We'll now find that particles such as electrons or atoms sometimes behave like waves. All of the characteristic of waves, such as diffraction and interference, will also apply to particles. This odd notion—that there's no clear distinction between particles and waves—is the core of a new model of light and matter called *quantum physics.* The wave nature of particles will lead to the *quantization* of energy. A particle confined in a box—or an electron in an atom—can have only certain energies. This idea will be a fruitful one for us, allowing us to understand the spectra of gases and phenomena such as bioluminescence.

Atoms, Nuclei, and Particles

We have frequently used the atomic model, explaining the properties of matter by considering the behavior of the atoms that comprise it. But when you get right down to it, what *is* an atom? And what's inside an atom? As you know, an atom has a tiny core called a *nucleus.* We'll look at what goes on inside the nucleus. One remarkable discovery will be that the nuclei of certain atoms spontaneously decay, turning the atom from one element to another. This phenomenon of *radioactivity* will give us a window into the nature of atoms and the particles that comprise them.

Once we know that the nucleus of an atom is composed of protons and neutrons, there's a natural next question to ask: What's inside a proton or neutron? There is an answer to this question, an answer we will learn in the final chapter of this book.

27 Relativity

For a wildlife conservation study, this turtle has been fitted with a collar containing a global positioning system (GPS) receiver, allowing the turtle to be tracked over thousands of miles with an accuracy of ±15 m. How does the theory of relativity affect the accuracy of a GPS receiver?

LOOKING AHEAD ▸▸

The goal of Chapter 27 is to understand how Einstein's theory of relativity changes our concepts of space and time.

What's Relativity All About?

Einstein's theory of relativity is based on a very simple-sounding principle: All the laws of physics are the same in all **inertial reference frames**—frames that move at a constant velocity.

One important consequence of this principle is that light travels at the same speed c in all inertial reference frames. This seemingly innocuous result will lead us to completely rethink our ideas of space and time.

The effects of relativity are large only for objects moving at *relativistic* speeds, close to the speed of light.

Looking Back ◂◂
3.5 Relative motion

For everyday motion, velocities add: The speed of the javelin equals the speed at which the athlete throws it *plus* her running speed.

Light and objects with speeds near c do not obey this simple rule. The speed of the light from this plane's landing lights is the *same* whether the plane is moving or not.

Simultaneity

We'll find that two events that occur at the same time in one reference frame occur at *different* times in another frame moving with respect to the first.

To you, the lightning strikes hit the ground simultaneously. The strikes are not simultaneous to someone moving relative to you.

Time Dilation

Relativity shows that *a moving clock runs slow* compared to a stationary clock.

Your GPS receiver receives signals from some of 24 highly accurate clocks in satellites orbiting the earth at high speeds. These clocks must be corrected for relativistic effects in order for the GPS system to work.

Length Contraction

The length of a moving object is *shortened* compared to a stationary object.

In the reference frame of an electron moving along this 2-mile-long particle accelerator at nearly the speed of light, the accelerator appears to be only 3.3 cm long!

Mass and Energy

Einstein showed that mass and energy are equivalent, according to his famous equation $E = mc^2$.

The sun's energy comes from the conversion of 4 billion kilograms of matter into energy every second. Even so, the sun has enough mass to keep burning for another 4 billion years.

Looking Back ◂◂
10.3 Kinetic energy

27.1 Relativity: What's It All About?

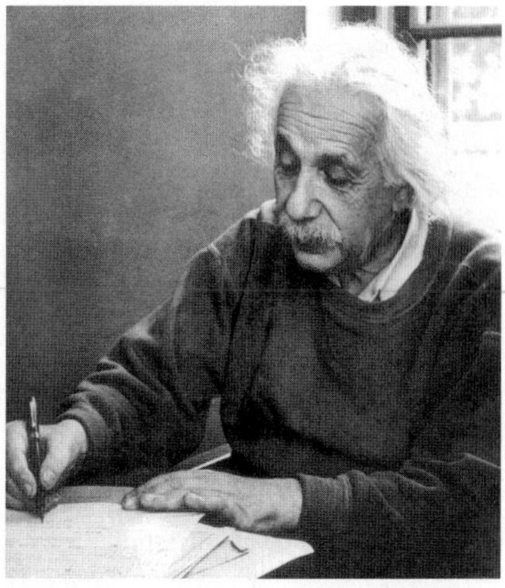

Albert Einstein (1879–1955) was one of the most influential thinkers in history.

What do you think of when you hear the phrase "theory of relativity"? A white-haired Einstein? $E = mc^2$? Black holes? Time travel? Perhaps you've heard that the theory of relativity is so complicated and abstract that only a handful of people in the whole world really understand it.

There is, without doubt, a certain mystique associated with relativity, an aura of the strange and exotic. The good news is that understanding the ideas of relativity is well within your grasp. Einstein's *special theory of relativity,* the portion of relativity we'll study, is not mathematically difficult at all. The challenge is conceptual because relativity questions deeply held assumptions about the nature of space and time. In fact, that's what relativity is all about—space and time.

In one sense, relativity is not a new idea at all. Certain ideas about relativity are part of Newtonian mechanics: Newton's laws appear to hold just as well on a fast-moving train as they do in a stationary laboratory. Einstein, however, thought that relativity should apply to *all* the laws of physics, not just mechanics. The difficulty, as you'll see, is that some aspects of relativity appear to be incompatible with the laws of electromagnetism, particularly the laws governing the propagation of light waves.

Lesser scientists might have concluded that relativity simply doesn't apply to electromagnetism. Einstein's genius was to see that the incompatibility arises from *assumptions* about space and time, assumptions no one had ever questioned because they seem so obviously true. Rather than abandon the ideas of relativity, Einstein changed our understanding of space and time.

Fortunately, you need not be a genius to follow a path that someone else has blazed. However, we will have to exercise the utmost care with regard to logic and precision. We will need to state very precisely just how it is that we know things about the physical world and then ruthlessly follow the logical consequences. The challenge is to stay on this path, not to let our prior assumptions—assumptions that are deeply ingrained in all of us—lead us astray.

What's Special About Special Relativity?

Einstein's first paper on relativity, in 1905, dealt exclusively with inertial reference frames, reference frames that move relative to each other with constant velocity. Ten years later, Einstein published a more encompassing theory of relativity that considers accelerated motion and its connection to gravity. The second theory, because it's more general in scope, is called *general relativity.* General relativity is the theory that describes black holes, curved spacetime, and the evolution of the universe. It is a fascinating theory but, alas, very mathematical and outside the scope of this textbook.

Motion at constant velocity is a "special case" of motion—namely, motion for which the acceleration is zero. Hence Einstein's first theory of relativity has come to be known as *special relativity.* It is special in the sense of being a restricted, special case of his more general theory, not special in the everyday sense of meaning distinctive or exceptional. Special relativity, with its conclusions about time dilation and length contraction, is what we will study.

27.2 Galilean Relativity

Galileo was the first to understand how the laws of physics depended on the relative motion between different observers. A firm grasp of *Galilean relativity* is necessary if we are to appreciate and understand what is new in Einstein's theory. Thus we begin with the ideas of relativity that are embodied in Newtonian mechanics.

You may have had the experience of sitting on a train and looking up to see another train moving slowly past. It can be hard to tell if the other train is moving past your stationary train, or if you're moving in the opposite direction past a stationary train. Only the *relative* velocity between the trains has meaning.

Reference Frames

Suppose you're passing me as we both drive in the same direction along a freeway. My car's speedometer reads 55 mph while your speedometer shows 60 mph. Is 60 mph your "true" speed? That is certainly your speed relative to someone standing beside the road, but your speed relative to me is only 5 mph. Your speed is 120 mph relative to a driver approaching from the other direction at 60 mph.

An object does not have a "true" speed or velocity. The very definition of velocity, $v = \Delta x/\Delta t$, assumes the existence of a coordinate system in which, during some time interval Δt, the displacement Δx is measured. The best we can manage is to specify an object's velocity relative to, or with respect to, the coordinate system in which it is measured.

Let's define a **reference frame** to be a coordinate system in which experimenters equipped with meter sticks, stopwatches, and any other needed equipment make position and time measurements on moving objects. Three ideas are implicit in our definition of a reference frame:

- A reference frame extends infinitely far in all directions.
- The experimenters are at rest in the reference frame.
- The number of experimenters and the quality of their equipment are sufficient to measure positions and velocities to any level of accuracy needed.

The first two points are especially important. It is often convenient to say "the laboratory reference frame" or "the reference frame of the rocket." These are shorthand expressions for "a reference frame, infinite in all directions, in which the laboratory (or the rocket) and a set of experimenters happen to be at rest."

NOTE ► A reference frame is not the same thing as a "point of view." That is, each person or each experimenter does not have his or her own private reference frame. **All experimenters at rest relative to each other share the same reference frame.** ◄

FIGURE 27.1 The standard reference frames S and S′.

1. The axes of S and S′ have the same orientation.

2. Frame S′ moves with velocity v relative to frame S. The relative motion is entirely along the x- and x'-axes.

y y' v S x S′ x'

3. The origins of S and S′ coincide at $t = 0$. This is our definition of $t = 0$.

FIGURE 27.1 shows how we represent two reference frames, S and S′, that are in relative motion. The coordinate axes in S are x, y, z and those in S′ are x', y', z'. Reference frame S′ moves with velocity v relative to S or, equivalently, S moves with velocity $-v$ relative to S′. There's no implication that either reference frame is "at rest." Notice that the zero of time, when experimenters start their stopwatches, is the instant when the origins of S and S′ coincide.

Inertial Reference Frames

FIGURE 27.2 Two reference frames.

(a)

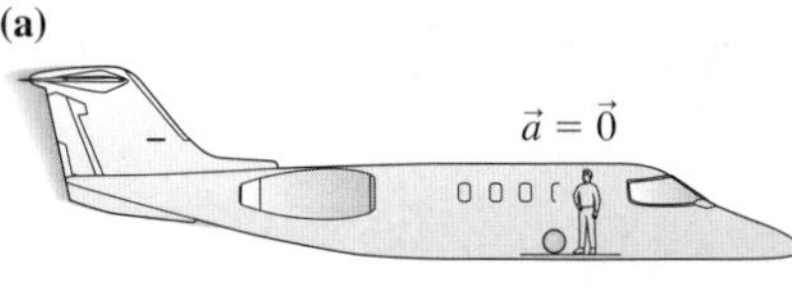

The ball stays in place.

A ball with no horizontal forces stays at rest in an airplane cruising at constant velocity. The airplane is an inertial reference frame.

(b)

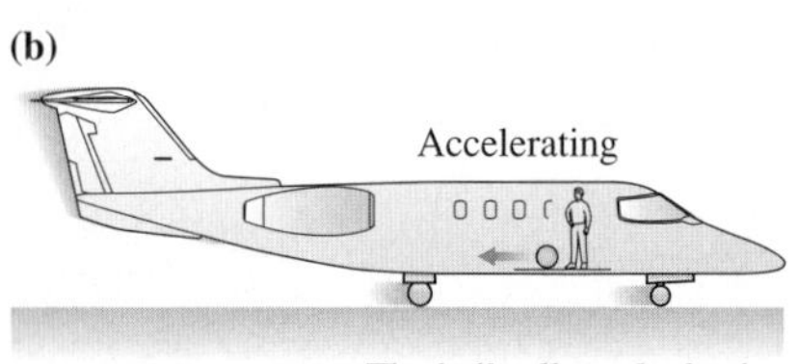

The ball rolls to the back of the plane during takeoff. An accelerating plane is not an inertial reference frame.

Certain reference frames are especially simple. FIGURE 27.2a shows a physics student cruising at constant velocity in an airplane. If the student places a ball on the floor, it stays there. There are no horizontal forces, and the ball remains at rest relative to the airplane. That is, $\vec{a} = \vec{0}$ in the airplane's coordinate system when $\vec{F}_{net} = \vec{0}$, so Newton's first law is satisfied. Similarly, if the student drops the ball, it falls straight down—relative to the student—with an acceleration of magnitude g, satisfying Newton's second law.

We define an **inertial reference frame** as one in which Newton's first law is valid. That is, an inertial reference frame is one in which an isolated particle, one on which there are no forces, either remains at rest or moves in a straight line at constant speed, as measured by experimenters at rest in that frame.

Not all reference frames are inertial. The physics student in FIGURE 27.2b conducts the same experiment during takeoff. He carefully places the ball on the floor just as the airplane starts to accelerate down the runway. You can imagine what happens. The ball rolls to the back of the plane as the passengers are being pressed back into their seats. If the student measures the ball's motion using a meter stick attached to the plane, he will find that the ball accelerates *in the plane's reference frame*. Yet he would be unable to identify any force on the ball that would act to accelerate it toward the back of the plane. This violates Newton's first law, so the plane is *not* an inertial reference frame during takeoff. **In general, accelerating reference frames are not inertial reference frames.**

NOTE ► An inertial reference frame is an idealization. A true inertial reference frame would need to be floating in deep space, far from any gravitational influence. In practice, an earthbound laboratory is a good approximation of an inertial reference frame because the accelerations associated with the earth's rotation and motion around the sun are too small to influence most experiments. ◄

These ideas are in accord with your everyday experience. If you're in a jet flying smoothly at 600 mph—an inertial reference frame—Newton's laws are valid: You can toss and catch a ball, or pour a cup of coffee, exactly as you would on the ground. But if the plane were diving, or shaking from turbulence, simple "experiments" like these would fail. A ball thrown straight up would land far from your hand, and the stream of coffee would bend and turn on its way to missing the cup. These apparently simple observations can be stated as the *Galilean principle of relativity:*

This flight attendant pours wine on a smoothly flying airplane moving at 600 mph just as easily as she does at the terminal. These ideas were first discussed by Galileo in 1632 in the context of pouring water while on a moving ship.

Galilean principle of relativity Newton's laws of motion are valid in all inertial reference frames.

In our study of relativity, we will restrict our attention to inertial reference frames. This implies that the relative velocity v between two reference frames is constant. Any reference frame that moves at constant velocity with respect to an inertial reference frame is itself an inertial reference frame. Conversely, a reference frame that accelerates with respect to an inertial reference frame is not an inertial reference frame. Although special relativity can be used for accelerating reference frames, we will confine ourselves here to the simple case of inertial reference frames moving with respect to each other at constant velocity.

STOP TO THINK 27.1 Which of these is an inertial reference frame (or a very good approximation)?

A. Your bedroom
B. A car rolling down a steep hill
C. A train coasting along a level track
D. A rocket being launched
E. A roller coaster going over the top of a hill
F. A sky diver falling at terminal speed

The Galilean Velocity Transformation

Special relativity is largely concerned with how physical quantities such as position and time are measured by experimenters in different reference frames. Let's begin by studying how, within Galilean relativity, the *velocity* of an object is measured in different reference frames. We have already touched on these ideas back in Section 3.5.

Suppose Sue is standing beside a highway as Jim drives by at 50 mph, as shown in FIGURE 27.3. Let S be Sue's reference frame—a reference frame attached to the ground—and let S′ be the reference frame moving with Jim, attached to his car. We see that the velocity of reference frame S′ relative to S is $v = 50$ mph.

Now suppose a motorcyclist blasts down the highway, traveling in the same direction as Jim. Sue measures the motorcycle's velocity to be $u = 75$ mph. What is

FIGURE 27.3 A motorcycle's velocity as seen by Sue and by Jim.

v is the relative velocity between Jim's reference frame and Sue's.
y
y′
Sue
u = 75 mph
v = 50 mph
Jim
S
x
S′
x′
u is the velocity of the motorcycle as measured in Sue's frame.

the cycle's velocity u' measured relative to Jim? We can answer this on the basis of common sense. If you're driving at 50 mph, and someone passes you going 75 mph, then his speed *relative to you* is 25 mph. This is the *difference* between his speed relative to the ground and your speed relative to the ground. Thus Jim measures the motorcycle's velocity to be $u' = 25$ mph.

NOTE ► In this chapter, we will use v to represent the velocity of one reference frame relative to another. We will use u and u' to represent the velocities of objects with respect to reference frames S and S′. In addition, we will assume that all motion is parallel to the x-axis. ◄

We can state this idea as a general rule. If u is the velocity of an object as measured in reference frame S, and u' its velocity as measured in S′, then the two velocities are related by

$$u' = u - v \quad \text{and} \quad u = u' + v \tag{27.1}$$

Galilean transformation of velocity, where v is the velocity of S′ relative to S

Equations 27.1 are the *Galilean transformation of velocity.* If you know the velocity of a particle as measured by the experimenters in one inertial reference frame, you can use Equations 27.1 to find the velocity that would be measured by experimenters in any other inertial reference frame.

EXAMPLE 27.1 Finding the speed of sound

An airplane is flying at speed 200 m/s with respect to the ground. Sound wave 1 is approaching the plane from the front, while sound wave 2 is catching up from behind. Both waves travel at 340 m/s relative to the ground. What is the velocity of each wave relative to the plane?

PREPARE Assume that the earth (frame S) and the airplane (frame S′) are inertial reference frames. Frame S′, in which the airplane is at rest, moves with velocity $v = 200$ m/s relative to frame S. FIGURE 27.4 shows the airplane and the sound waves.

FIGURE 27.4 Experimenters in the plane measure different speeds for the sound waves than do experimenters on the ground.

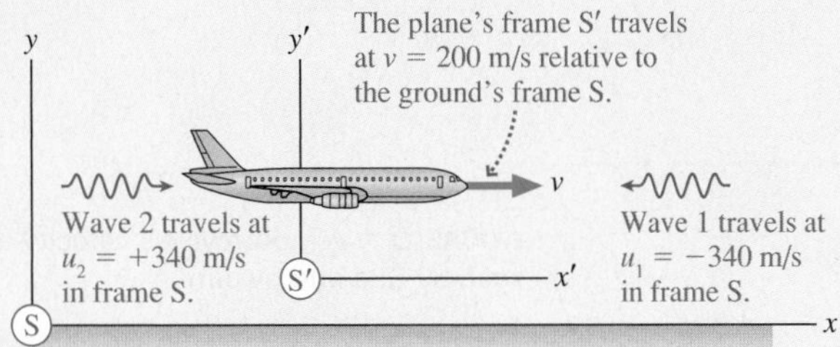
y
y′
The plane's frame S′ travels at v = 200 m/s relative to the ground's frame S.
v
Wave 2 travels at u_2 = +340 m/s in frame S.
Wave 1 travels at u_1 = −340 m/s in frame S.
S′
x′
S
x

SOLVE The speed of a mechanical wave, such as a sound wave or a wave on a string, is its speed *relative to its medium.* Thus the *speed of sound* is the speed of a sound wave through a reference frame in which the air is at rest. This is reference frame S, where wave 1 travels with velocity $u_1 = -340$ m/s and wave 2 travels with velocity $u_2 = +340$ m/s. Notice that the Galilean transformations use *velocities,* with appropriate signs, not just speeds.

The airplane travels to the right with reference frame S′ at velocity v. We can use the Galilean transformations of velocity to find the velocities of the two sound waves in frame S′:

$$u_1' = u_1 - v = -340 \text{ m/s} - 200 \text{ m/s} = -540 \text{ m/s}$$
$$u_2' = u_2 - v = 340 \text{ m/s} - 200 \text{ m/s} = 140 \text{ m/s}$$

Thus wave 1 approaches the plane with a *speed* of 540 m/s, while wave 2 approaches with a speed of 140 m/s.

ASSESS This isn't surprising. If you're driving at 50 mph, a car coming the other way at 55 mph is approaching you at 105 mph. A car coming up behind you at 55 mph seems to be gaining on you at the rate of only 5 mph. Wave speeds behave the same. Notice that a mechanical wave would appear to be stationary to a person moving at the wave speed. To a surfer, the crest of the ocean wave remains at rest under his or her feet.

STOP TO THINK 27.2 Ocean waves are approaching the beach at 10 m/s. A boat heading out to sea travels at 6 m/s. How fast are the waves moving in the boat's reference frame?

A. 16 m/s B. 10 m/s C. 6 m/s D. 4 m/s

27.3 Einstein's Principle of Relativity

The 19th century was an era of optics and electromagnetism. Thomas Young demonstrated in 1801 that light is a wave, and by midcentury scientists had devised techniques for measuring the speed of light. Faraday discovered electromagnetic induction in 1831, setting in motion a train of events leading to Maxwell's conclusion, in 1864, that light is an electromagnetic wave.

If light is a wave, what is the medium in which it travels? This was perhaps the most important scientific question in the second half of the 19th century. The medium in which light waves were assumed to travel was called the *ether.* Experiments to measure the speed of light were assumed to be measuring its speed through the ether. But just what is the ether? What are its properties? Can we collect a jar full of ether to study? Despite the significance of these questions, experimental efforts to detect the ether or measure its properties kept coming up empty-handed.

Maxwell's theory of electromagnetism didn't help the situation. The crowning success of Maxwell's theory was his prediction that light waves travel with speed

$$c = \frac{1}{\sqrt{\epsilon_0 \mu_0}} = 3.00 \times 10^8 \text{ m/s}$$

This is a very specific prediction with no wiggle room. The difficulty with such a specific prediction was the implication that Maxwell's laws of electromagnetism are valid *only* in the reference frame of the ether. After all, as FIGURE 27.5 shows, the light speed should certainly be faster or slower than c in a reference frame moving through the ether, just as the sound speed is different to someone moving through the air.

FIGURE 27.5 It seems as if the speed of light should differ from c in a reference frame moving through the ether.

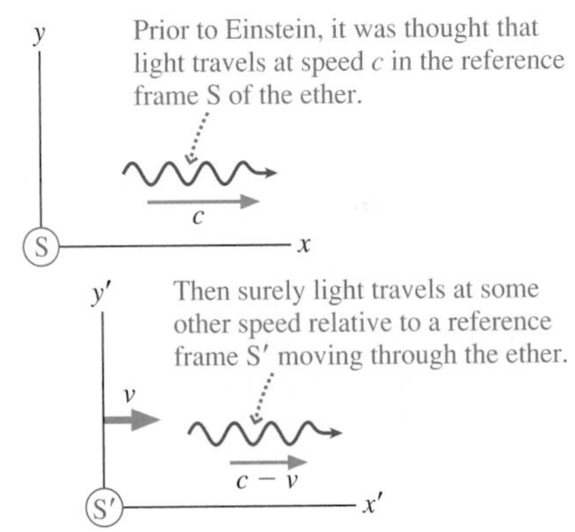

y
Prior to Einstein, it was thought that light travels at speed c in the reference frame S of the ether.
c
S
x
y′
Then surely light travels at some other speed relative to a reference frame S′ moving through the ether.
v
c − v
S′
x′

As the 19th century closed, it appeared that Maxwell's theory did not obey the classical principle of relativity. There was just one reference frame, the reference frame of the ether, in which the laws of electromagnetism seemed to be true. And to make matters worse, the fact that no one had been able to detect the ether meant that no one could identify the one reference frame in which Maxwell's equations "worked."

It was in this muddled state of affairs that a young Albert Einstein made his mark on the world. Even as a teenager, Einstein had wondered how a light wave would look to someone "surfing" the wave, traveling alongside the wave at the wave speed. You can do that with a water wave or a sound wave, but light waves seemed to present a logical difficulty. An electromagnetic wave sustains itself by virtue of the fact that a changing magnetic field induces an electric field and a changing electric field induces a magnetic field. But to someone moving with the wave, *the fields would not change.* How could there be an electromagnetic wave under these circumstances?

Several years of thinking about the connection between electromagnetism and reference frames led Einstein to the conclusion that *all* the laws of physics, not just the laws of mechanics, should obey the principle of relativity. In other words, the principle of relativity is a fundamental statement about the nature of the physical universe. The Galilean principle of relativity stated only that Newton's laws hold in any inertial reference frame. Einstein was able to state a much more general principle:

In certain rivers, tides send waves upriver that can be surfed for miles. From the reference frame of these surfers, the waves are standing still. If you could move along with a light wave, would the electric and magnetic fields appear motionless?

Principle of relativity All the laws of physics are the same in all inertial reference frames.

All of the results of Einstein's theory of relativity flow from this one simple statement.

The Constancy of the Speed of Light

If Maxwell's equations of electromagnetism are laws of physics, and there's every reason to think they are, then, according to the principle of relativity, Maxwell's equations must be true in *every* inertial reference frame. On the surface this seems to be an innocuous statement, equivalent to saying that the law of conservation of momentum is true in every inertial reference frame. But follow the logic:

1. Maxwell's equations are true in all inertial reference frames.
2. Maxwell's equations predict that electromagnetic waves, including light, travel at speed $c = 3.00 \times 10^8$ m/s.
3. Therefore, **light travels at speed c in all inertial reference frames.**

FIGURE 27.6 Light travels at speed c in all inertial reference frames, regardless of how the reference frames are moving with respect to the light source.

This light wave leaves Amy at speed c relative to Amy. It approaches Cathy at speed c relative to Cathy.

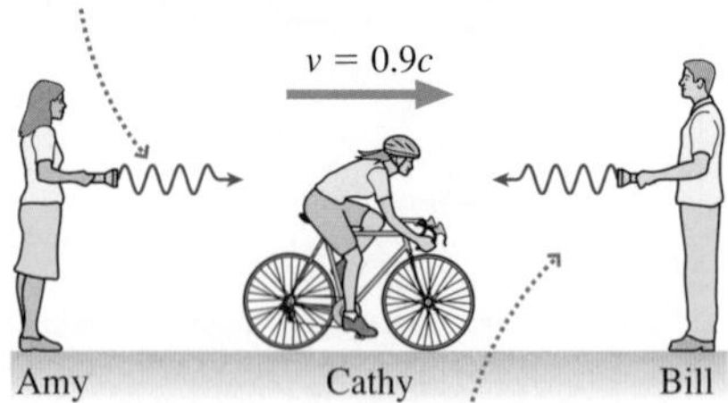

This light wave leaves Bill at speed c relative to Bill. It approaches Cathy at speed c relative to Cathy.

FIGURE 27.6 shows the implications of this conclusion. *All* experimenters, regardless of how they move with respect to each other, find that *all* light waves, regardless of the source, travel in their reference frame with the *same* speed c. If Cathy's velocity toward Bill and away from Amy is $v = 0.9c$, Cathy finds, by making measurements in her reference frame, that the light from Bill approaches her at speed c, not at $c + v = 1.9c$. And the light from Amy, which left Amy at speed c, catches up from behind at speed *c relative to Cathy,* not the $c - v = 0.1c$ you would have expected.

Although this prediction goes against all shreds of common sense, the experimental evidence for it is strong. Laboratory experiments are difficult because even the highest laboratory speed is insignificant in comparison to c. In the 1930s, however, the physicists R. J. Kennedy and E. M. Thorndike realized that they could use the earth itself as a laboratory. The earth's speed as it circles the sun is about 30,000 m/s. The velocity of the earth in January differs by 60,000 m/s from its velocity in July, when the earth is moving in the opposite direction. Kennedy and Thorndike were able to use a very sensitive and stable interferometer to show that the numerical values of the speed of light in January and July differ by less than 2 m/s.

More recent experiments have used unstable elementary particles, called π mesons, that decay into high-energy photons, or particles of light. The π mesons, created in a particle accelerator, move through the laboratory at 99.975% the speed of light, or $v = 0.99975c$, as they emit photons at speed c in the π meson's reference frame. As FIGURE 27.7 shows, you would expect the photons to travel through the laboratory with speed $c + v = 1.99975c$. Instead, the measured speed of the photons in the laboratory was, within experimental error, 3.00×10^8 m/s.

FIGURE 27.7 Experiments find that the photons travel through the laboratory with speed c, not the speed $1.99975c$ that you might expect.

A photon is emitted at speed c relative to the π meson. Measurements find that the photon's speed in the laboratory reference frame is also c.

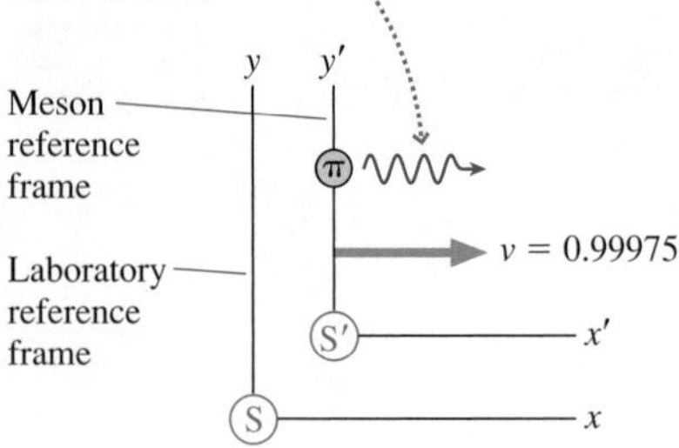

In summary, *every* experiment designed to compare the speed of light in different reference frames has found that light travels at 3.00×10^8 m/s in every inertial reference frame, regardless of how the reference frames are moving with respect to each other.

How Can This Be?

You're in good company if you find this impossible to believe. Suppose I shot a ball forward at 50 m/s while driving past you at 30 m/s. You would certainly see the ball traveling at 80 m/s relative to you and the ground. What we're saying with regard to light is equivalent to saying that the ball travels at 50 m/s relative to my car and *at the same time* travels at 50 m/s relative to the ground, even though the car is moving across the ground at 30 m/s. It seems logically impossible.

You might think that this is merely a matter of semantics. If we can just get our definitions and use of words straight, then the mystery and confusion will disappear. Or perhaps the difficulty is a confusion between what we "see" versus what "really happens." In other words, a better analysis, one that focuses on what really happens, would find that light "really" travels at different speeds in different reference frames.

Alas, what "really happens" is that light travels at 3.00×10^8 m/s in every inertial reference frame, regardless of how the reference frames are moving with respect to each other. It's not a trick. There remains only one way to escape the logical contradictions.

The definition of velocity is $u = \Delta x/\Delta t$, the ratio of a distance traveled to the time interval in which the travel occurs. Suppose you and I both make measurements on an object as it moves, but you happen to be moving relative to me. Perhaps I'm standing on the corner, you're jogging to the right, and we're both trying to measure the velocity of a bicycle moving to the right as it passes both of us. Further, suppose we have agreed in advance to measure the bicycle as it moves from the tree to the lamppost in FIGURE 27.8. Your $\Delta x'$ differs from my Δx because of your motion relative to me, causing you to calculate a bicycle velocity u' in your reference frame that differs from its velocity u in my reference frame. This is just the Galilean transformations showing up again.

FIGURE 27.8 Measuring the velocity of an object by appealing to the basic definition $u = \Delta x/\Delta t$.

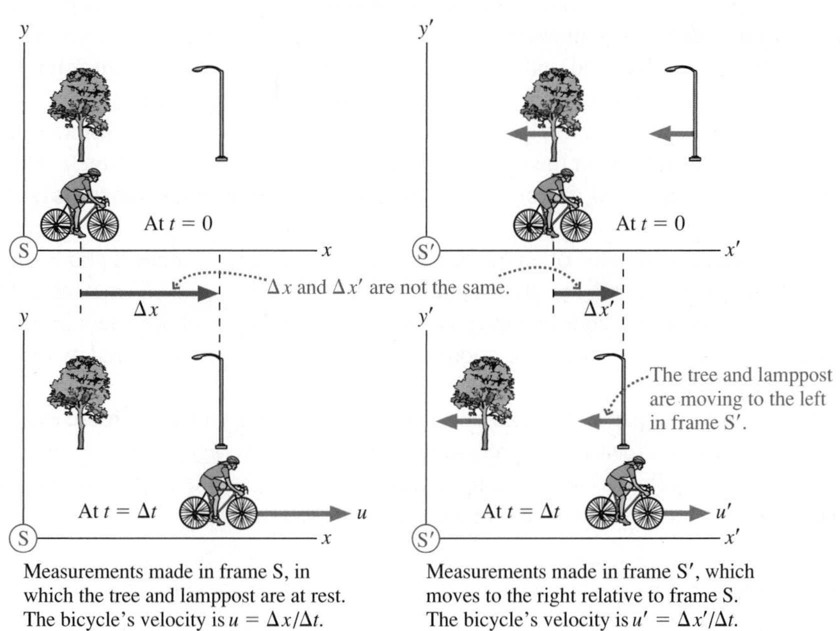
y
y′
At t = 0
At t = 0
S
x
S′
x′
Δx and Δx′ are not the same.
Δx
Δx′
y
y′
The tree and lamppost are moving to the left in frame S′.
At t = Δt
u
At t = Δt
u′
S
x
S′
x′
Measurements made in frame S, in which the tree and lamppost are at rest. The bicycle's velocity is u = Δx/Δt.
Measurements made in frame S′, which moves to the right relative to frame S. The bicycle's velocity is u′ = Δx′/Δt.

Now let's repeat the measurements, but this time let's measure the velocity of a light wave as it travels from the tree to the lamppost. Once again, your $\Delta x'$ differs from my Δx, although the difference will be pretty small unless you happen to be jogging at a speed that is an appreciable fraction of the speed of light. The obvious conclusion is that your light speed u' differs from my light speed u. But it doesn't. The experiments show that, for a light wave, we'll get the *same* values: $u' = u = c$.

The only way this can be true is if your Δt is not the same as my Δt. If the time it takes the light to move from the tree to the lamppost in your reference frame, a time we'll now call $\Delta t'$, differs from the time Δt it takes the light to move from the tree to the lamppost in my reference frame, then we might find that $\Delta x'/\Delta t' = \Delta x/\Delta t$. That is, $u' = u$ even though you are moving with respect to me.

We've assumed, since the beginning of this textbook, that time is simply time. It flows along like a river, and all experimenters in all reference frames simply use it. For example, suppose the tree and the lamppost both have big clocks that we both can see. Shouldn't we be able to agree on the time interval Δt the light needs to move from the tree to the lamppost?

Perhaps not. It's demonstrably true that $\Delta x' \neq \Delta x$. It's experimentally verified that $u' = u$ for light waves. Something must be wrong with *assumptions* that we've made about the nature of time. The principle of relativity has painted us into a corner, and our only way out is to reexamine our understanding of time.

27.4 Events and Measurements

To question some of our most basic assumptions about space and time requires extreme care. We need to be certain that no assumptions slip into our analysis unnoticed. Our goal is to describe the motion of a particle in a clear and precise way, making the barest minimum of assumptions.

Events

FIGURE 27.9 The location and time of an event are described by its spacetime coordinates.

An event has spacetime coordinates (x, y, z, t) in frame S and different spacetime coordinates (x', y', z', t') in frame S′.

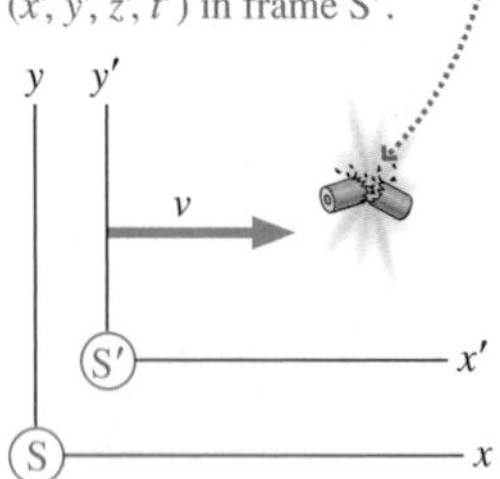

The fundamental entity of relativity is called an **event.** An event is a physical activity that takes place at a definite point in space and at a definite instant of time. A firecracker exploding is an event. A collision between two particles is an event. A light wave hitting a detector is an event.

Events can be observed and measured by experimenters in different reference frames. An exploding firecracker is as clear to you as you drive by in your car as it is to me standing on the street corner. We can quantify where and when an event occurs with four numbers: the coordinates (x, y, z) and the instant of time t. These four numbers, illustrated in FIGURE 27.9, are called the **spacetime coordinates** of the event.

The spatial coordinates of an event measured in reference frames S and S′ may differ. But it now appears that the instant of time recorded in S and S′ may also differ. Thus the spacetime coordinates of an event measured by experimenters in frame S are (x, y, z, t), and the spacetime coordinates of the *same event* measured by experimenters in frame S′ are (x', y', z', t').

The motion of a particle can be described as a sequence of two or more events. We introduced this idea in the preceding section when we agreed to measure the velocity of a bicycle and then of a light wave by comparing the object passing the tree (first event) to the object passing the lamppost (second event).

FIGURE 27.10 The spacetime coordinates of an event are measured by a lattice of meter sticks and clocks.

The spacetime coordinates of this event are measured by the nearest meter stick intersection and the nearest clock.

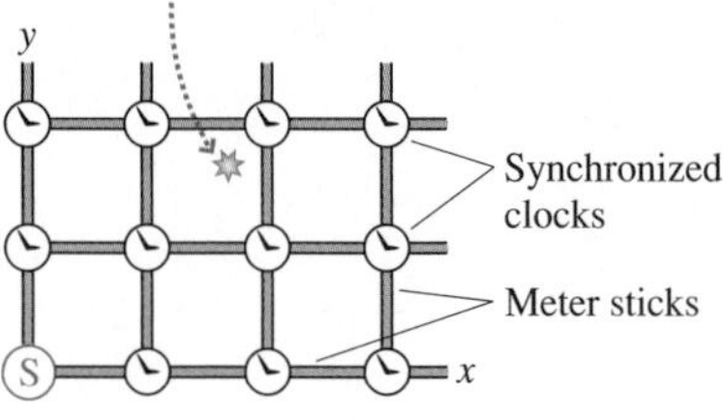

Reference frame S′ has its own meter sticks and its own clocks.

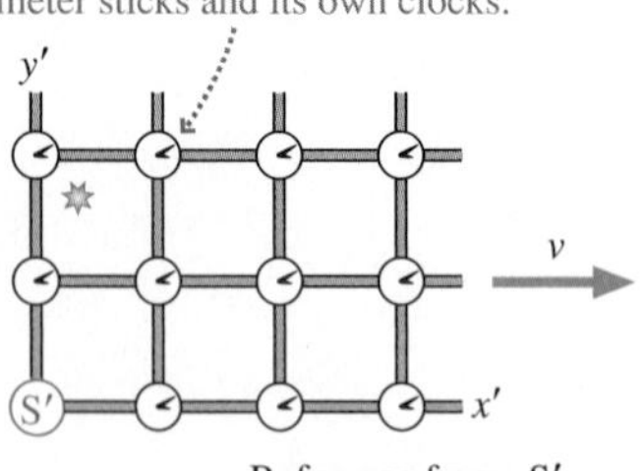

Measurements

Events are what "really happen," but how do we learn about an event? That is, how do the experimenters in a reference frame determine the spacetime coordinates of an event? This is a problem of *measurement.*

We defined a reference frame to be a coordinate system in which experimenters can make position and time measurements. That's a good start, but now we need to be more precise as to *how* the measurements are made. Imagine that a reference frame is filled with a cubic lattice of meter sticks, as shown in FIGURE 27.10. At every intersection is a clock, and all the clocks in a reference frame are *synchronized.* We'll return in a moment to consider how to synchronize the clocks, but assume for the moment it can be done.

Now, with our meter sticks and clocks in place, we can use a two-part measurement scheme:

- The (x, y, z) coordinates of an event are determined by the intersection of meter sticks closest to the event.
- The event's time t is the time displayed on the clock nearest the event.

You can imagine, if you wish, that each event is accompanied by a flash of light to illuminate the face of the nearest clock and make its reading known.

Several important issues need to be noted:

1. The clocks and meter sticks in each reference frame are imaginary, so they have no difficulty passing through each other.
2. Measurements of position and time made in one reference frame must use only the clocks and meter sticks in that reference frame.
3. There's nothing special about the sticks being 1 m long and the clocks 1 m apart. The lattice spacing can be altered to achieve whatever level of measurement accuracy is desired.
4. We'll assume that the experimenters in each reference frame have assistants sitting beside every clock to record the position and time of nearby events.
5. Perhaps most important, t is the time at which the event *actually happens,* not the time at which an experimenter sees the event or at which information about the event reaches an experimenter.
6. All experimenters in one reference frame agree on the spacetime coordinates of an event. In other words, **an event has a unique set of spacetime coordinates in each reference frame.**

STOP TO THINK 27.3 A carpenter is working on a house two blocks away. You notice a slight delay between seeing the carpenter's hammer hit the nail and hearing the blow. At what time does the event "hammer hits nail" occur?

A. At the instant you hear the blow
B. At the instant you see the hammer hit
C. Very slightly before you see the hammer hit
D. Very slightly after you see the hammer hit

Clock Synchronization

It's important that all the clocks in a reference frame be **synchronized,** meaning that all clocks in the reference frame have the same reading at any one instant of time. We would not be able to use a sequence of events to track the motion of a particle if the clocks differed in their readings. Thus we need a method of synchronization. One idea that comes to mind is to designate the clock at the origin as the *master clock.* We could then carry this clock around to every clock in the lattice, adjust that clock to match the master clock, and finally return the master clock to the origin.

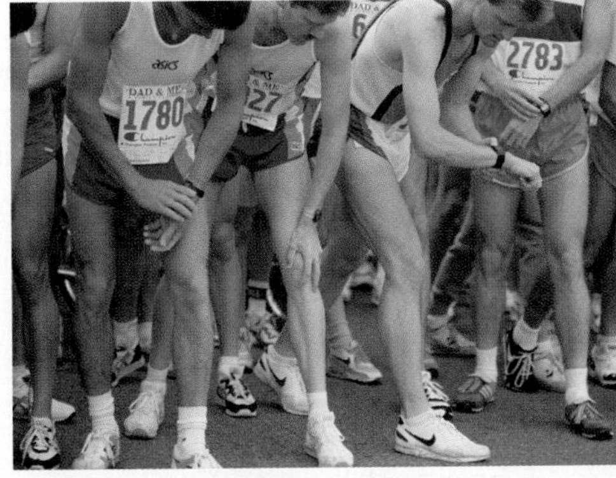

It's easy to synchronize clocks that are all in one place, but synchronizing distant clocks takes some care.

This would be a perfectly good method of clock synchronization in Newtonian mechanics, where time flows along smoothly, the same for everyone. But we've been driven to reexamine the nature of time by the possibility that time is different in reference frames moving relative to each other. Because the master clock would *move,* we cannot assume that the master clock keeps time in the same way as the stationary clocks.

We need a synchronization method that does not require moving the clocks. Fortunately, such a method is easy to devise. Each clock is resting at the intersection of meter sticks, so by looking at the meter sticks, the assistant knows, or can calculate, exactly how far each clock is from the origin. Once the distance is known, the assistant can calculate exactly how long a light wave will take to travel from the origin to each clock. For example, light will take 1.00 μs to travel to a clock 300 m from the origin.

NOTE ► It's handy for many relativity problems to know that the speed of light is $c = 300\ \text{m}/\mu\text{s}$. ◄

To synchronize the clocks, the assistants begin by setting each clock to display the light travel time from the origin, but they don't start the clocks. Next, as

FIGURE 27.11 Synchronizing the clocks.

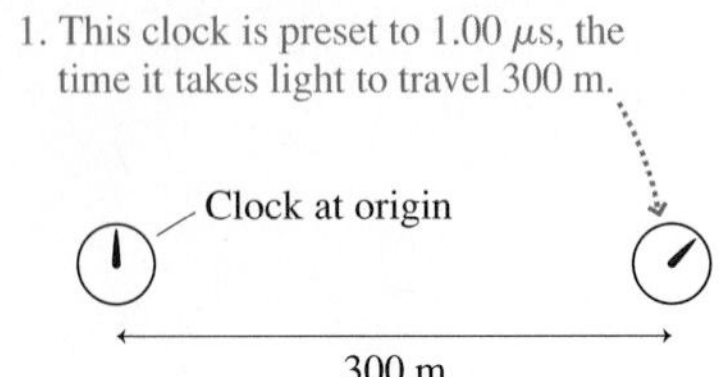
1. This clock is preset to 1.00 μs, the time it takes light to travel 300 m.
Clock at origin
300 m

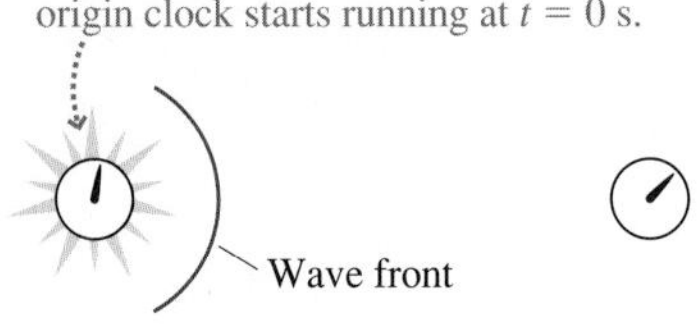
2. A light flashes at the origin and the origin clock starts running at t = 0 s.
Wave front

3. The clock starts when the light wave reaches it. It is now synchronized with the origin clock.

FIGURE 27.11 shows, a light flashes at the origin and, simultaneously, the clock at the origin starts running from $t = 0$ s. The light wave spreads out in all directions at speed c. A photodetector on each clock recognizes the arrival of the light wave and, without delay, starts the clock. The clock had been preset with the light travel time, so each clock as it starts reads exactly the same as the clock at the origin. Thus all the clocks will be synchronized after the light wave has passed by.

Events and Observations

We noted above that t is the time the event *actually happens.* This is an important point, one that bears further discussion. Light waves take time to travel. Messages, whether they're transmitted by light pulses, telephone, or courier on horseback, take time to be delivered. An experimenter *observes* an event, such as an exploding firecracker, only *at a later time* when light waves reach his or her eyes. But our interest is in the event itself, not the experimenter's observation of the event. The time at which the experimenter sees the event or receives information about the event is not when the event actually occurred.

Suppose at $t = 0$ s a firecracker explodes at $x = 300$ m. The flash of light from the firecracker will reach an experimenter at the origin at $t_1 = 1.0\ \mu$s. The sound of the explosion will reach the experimenter at $t_2 = 0.88$ s. Neither of these is the time t_{event} of the explosion, although the experimenter can work backward from these times, using known wave speeds, to determine t_{event}. In this example, the spacetime coordinates of the event—the explosion—are (300 m, 0 m, 0 m, 0 s).

EXAMPLE 27.2 Finding the time of an event

Experimenter A in reference frame S stands at the origin looking in the positive x-direction. Experimenter B stands at $x = 900$ m looking in the negative x-direction. A firecracker explodes somewhere between them. Experimenter B sees the light flash at $t = 3.00\ \mu$s. Experimenter A sees the light flash at $t = 4.00\ \mu$s. What are the spacetime coordinates of the explosion?

PREPARE Experimenters A and B are in the same reference frame and have synchronized clocks. **FIGURE 27.12** shows the two experimenters and the explosion at unknown position x.

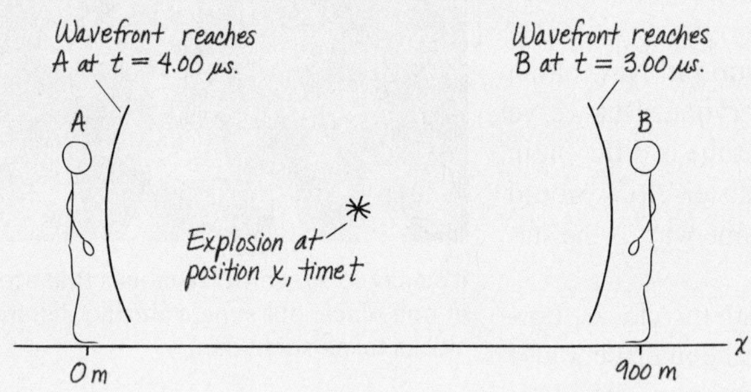
Wavefront reaches A at t = 4.00 μs.
Wavefront reaches B at t = 3.00 μs.
A
B
Explosion at position x, time t
0 m
900 m
x

SOLVE The two experimenters observe light flashes at two different instants, but there's only one event. Light travels at 300 m/μs, so the additional 1.00 μs needed for the light to reach experimenter A implies that distance ($x - 0$ m) from x to A is 300 m longer than distance (900 m $- x$) from B to x; that is,

$$(x - 0\text{ m}) = (900\text{ m} - x) + 300\text{ m}$$

This is easily solved to give $x = 600$ m as the position coordinate of the explosion. The light takes 1.00 μs to travel 300 m to experimenter B and 2.00 μs to travel 600 m to experimenter A. The light is received at 3.00 μs and 4.00 μs, respectively; hence it was emitted by the explosion at $t = 2.00\ \mu$s. The spacetime coordinates of the explosion are (600 m, 0 m, 0 m, 2.00 μs).

ASSESS Although the experimenters *see* the explosion at different times, they agree that the explosion actually *happened* at $t = 2.00\ \mu$s.

◀ **FIGURE 27.12** The light wave reaches the experimenters at different times. Neither of these is the time at which the event actually happened.

Simultaneity

Two events 1 and 2 that take place at different positions x_1 and x_2 but at the *same time* $t_1 = t_2$, as measured in some reference frame, are said to be **simultaneous** in that reference frame. Simultaneity is determined by when the events actually happen, not when they are seen or observed. In general, simultaneous events are not *seen* at the same time because of the difference in light travel times from the events to an experimenter.

EXAMPLE 27.3 Are the explosions simultaneous?

An experimenter in reference frame S stands at the origin looking in the positive x-direction. At $t = 3.0\ \mu s$ she sees firecracker 1 explode at $x = 600$ m. A short time later, at $t = 5.0\ \mu s$, she sees firecracker 2 explode at $x = 1200$ m. Are the two explosions simultaneous? If not, which firecracker exploded first?

PREPARE Light from both explosions travels toward the experimenter at 300 m/μs.

SOLVE The experimenter *sees* two different explosions, but perceptions of the events are not the events themselves. When did the explosions *actually* occur? Using the fact that light travels 300 m/μs, it's easy to see that firecracker 1 exploded at $t_1 = 1.0\ \mu s$ and firecracker 2 also exploded at $t_2 = 1.0\ \mu s$. The events *are* simultaneous.

STOP TO THINK 27.4 A tree and a pole are 3000 m apart. Each is suddenly hit by a bolt of lightning. Mark, who is standing at rest midway between the two, sees the two lightning bolts at the same instant of time. Nancy is at rest under the tree. Define event 1 to be "lightning strikes tree" and event 2 to be "lightning strikes pole." For Nancy, does event 1 occur before, after, or at the same time as event 2?

27.5 The Relativity of Simultaneity

We've now established a means for measuring the time of an event in a reference frame, so let's begin to investigate the nature of time. The following "thought experiment" is very similar to one suggested by Einstein.

FIGURE 27.13 shows a long railroad car traveling to the right with a velocity v that may be an appreciable fraction of the speed of light. A firecracker is tied to each end of the car, right above the ground. Each firecracker is powerful enough that, when it explodes, it will make a burn mark on the ground at the position of the explosion.

FIGURE 27.13 A railroad car traveling to the right with velocity v.

The firecrackers will make burn marks on the ground at the positions where they explode.

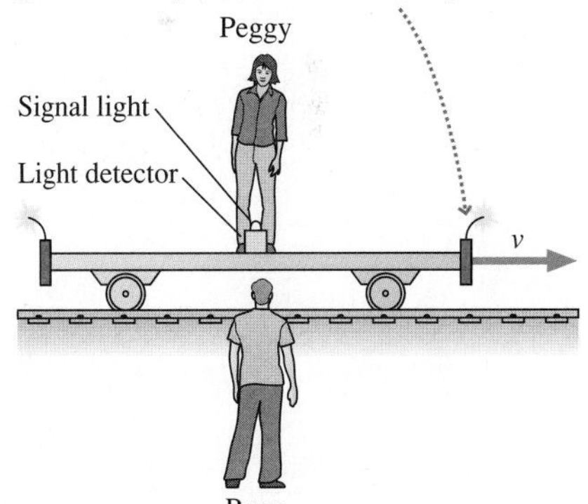

Ryan is standing on the ground, watching the railroad car go by. Peggy is standing in the exact center of the car with a special box at her feet. This box has two light detectors, one facing each way, and a signal light on top. The box works as follows:

1. If a flash of light is received at the right detector before a flash is received at the left detector, then the light on top of the box will turn green.
2. If a flash of light is received at the left detector before a flash is received at the right detector, or if two flashes arrive simultaneously, the light on top will turn red.

The firecrackers explode as the railroad car passes Ryan, and he sees the two light flashes from the explosions simultaneously. He then measures the distances to the two burn marks and finds that he was standing exactly halfway between the marks. Because light travels equal distances in equal times, Ryan concludes that the two explosions were simultaneous in his reference frame, the reference frame of the ground. Further, because he was midway between the two ends of the car, he was directly opposite Peggy when the explosions occurred.

FIGURE 27.14a on the next page shows the sequence of events in Ryan's reference frame. Light travels at speed c in all inertial reference frames, so, although the firecrackers were moving, the light waves are spheres centered on the burn marks. Ryan determines that the light wave coming from the right reaches Peggy and the box before the light wave coming from the left. Thus, according to Ryan, the signal light on top of the box turns green.

FIGURE 27.14 Exploding firecrackers seen in two different reference frames.

(a) The events in Ryan's frame

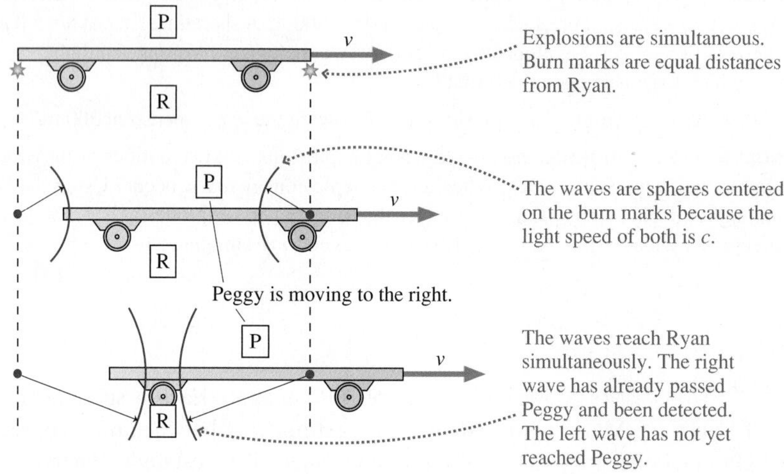

P
v
R
Explosions are simultaneous. Burn marks are equal distances from Ryan.
The waves are spheres centered on the burn marks because the light speed of both is c.
Peggy is moving to the right.
The waves reach Ryan simultaneously. The right wave has already passed Peggy and been detected. The left wave has not yet reached Peggy.

(b) The events in Peggy's frame

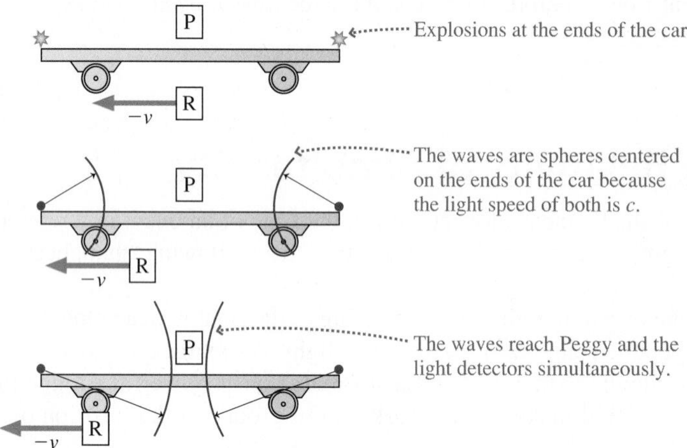

How do things look in Peggy's reference frame, a reference frame moving to the right at velocity v relative to the ground? As FIGURE 27.14b shows, Peggy sees Ryan moving to the left with speed v. Light travels at speed c in all inertial reference frames, so the light waves are spheres centered on the ends of the car. If the explosions are simultaneous, as Ryan has determined, the two light waves reach her and the box simultaneously. Thus, according to Peggy, the signal light on top of the box turns red!

Now the light on top must be either green or red. *It can't be both!* Later, after the railroad car has stopped, Ryan and Peggy can place the box in front of them. It has either a red light or a green light. Ryan can't see one color while Peggy sees the other. Hence we have a paradox. It's impossible for Peggy and Ryan both to be right. But who is wrong, and why?

What do we know with absolute certainty?

1. Ryan detected the flashes simultaneously.
2. Ryan was halfway between the firecrackers when they exploded.
3. The light from the two explosions traveled toward Ryan at equal speeds.

The conclusion that the explosions were simultaneous in Ryan's reference frame is unassailable. The light is green.

Peggy, however, made an assumption. It's a perfectly ordinary assumption, one that seems sufficiently obvious that you probably didn't notice, but an assumption nonetheless. Peggy assumed that the explosions were simultaneous.

Didn't Ryan find them to be simultaneous? Indeed, he did. Suppose we call Ryan's reference frame S, the explosion on the right event R, and the explosion on the left event L. Ryan found that $t_R = t_L$. But Peggy has to use a different set of clocks, the clocks in her reference frame S′, to measure the times t'_R and t'_L at which the explosions occurred. The fact that $t_R = t_L$ in frame S does *not* allow us to conclude that $t'_R = t'_L$ in frame S′.

In fact, the right firecracker must explode *before* the left firecracker in frame S′. Figure 27.14b, with its assumption about simultaneity, was incorrect. FIGURE 27.15 shows the situation in Peggy's reference frame with the right firecracker exploding first. Now the wave from the right reaches Peggy and the box first, as Ryan had concluded, and the light on top turns green.

One of the most disconcerting conclusions of relativity is that **two events occurring simultaneously in reference frame S are *not* simultaneous in any reference frame S′ that is moving relative to S.** This is called the **relativity of simultaneity.**

The two firecrackers *really* explode at the same instant of time in Ryan's reference frame. And the right firecracker *really* explodes first in Peggy's reference frame. It's not a matter of when they see the flashes. Our conclusion refers to the times at which the explosions actually occur.

The paradox of Peggy and Ryan contains the essence of relativity, and it's worth careful thought. First, review the logic until you're certain that there *is* a paradox, a logical impossibility. Then convince yourself that the only way to resolve the paradox is to abandon the assumption that the explosions are simultaneous in Peggy's reference frame. If you understand the paradox and its resolution, you've made a big step toward understanding what relativity is all about.

FIGURE 27.15 The real sequence of events in Peggy's reference frame.

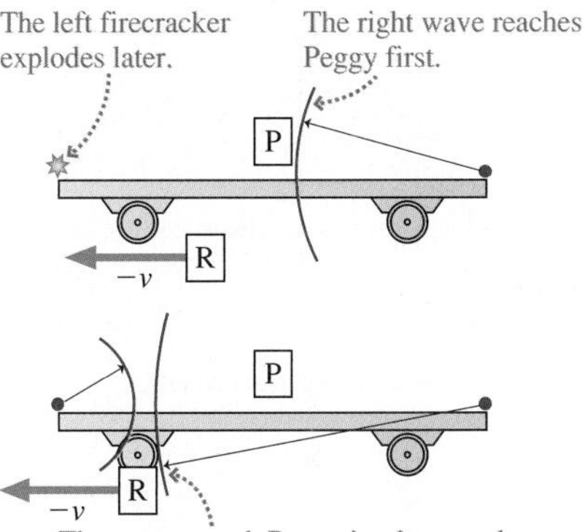

The right firecracker explodes first.
P
R
−v
The left firecracker explodes later.
The right wave reaches Peggy first.
P
R
−v
P
R
−v
The waves reach Ryan simultaneously.
The left wave has not reached Peggy.

STOP TO THINK 27.5 A tree and a pole are 3000 m apart. Each is suddenly hit by a bolt of lightning. Mark, who is standing at rest midway between the two, sees the two lightning bolts at the same instant of time. Nancy is flying her rocket at $v = 0.5c$ in the direction from the tree toward the pole. The lightning hits the tree just as she passes by it. Define event 1 to be "lightning strikes tree" and event 2 to be "lightning strikes pole." For Nancy, does event 1 occur before, after, or at the same time as event 2?

27.6 Time Dilation

The principle of relativity has driven us to the logical conclusion that the time at which an event occurs may not be the same for two reference frames moving relative to each other. Our analysis thus far has been mostly qualitative. It's time to start developing some quantitative tools that will allow us to compare measurements in one reference frame to measurements in another reference frame.

FIGURE 27.16a shows a special clock called a **light clock.** The light clock is a box of height h with a light source at the bottom and a mirror at the top. The light source emits a very short pulse of light that travels to the mirror and reflects back to a light detector beside the source. The clock advances one "tick" each time the detector receives a light pulse, and it immediately, with no delay, causes the light source to emit the next light pulse.

Our goal is to compare two measurements of the interval between two ticks of the clock: one taken by an experimenter standing next to the clock and the other by an experimenter moving with respect to the clock. To be specific, FIGURE 27.16b shows the clock at rest in reference frame S′. We call this the **rest frame** of the clock. Reference frame S′ moves to the right with velocity v relative to reference frame S.

Relativity requires us to measure *events,* so let's define event 1 to be the emission of a light pulse and event 2 to be the detection of that light pulse. Experimenters in

FIGURE 27.16 The ticking of a light clock can be measured by experimenters in two different reference frames.

(a) A light clock

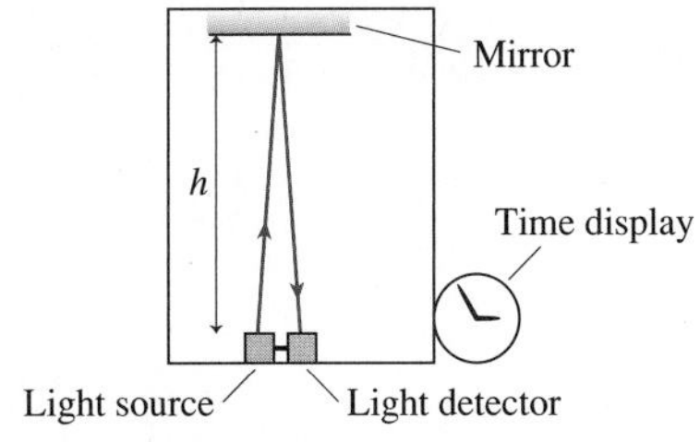

(b) The clock is at rest in frame S′.

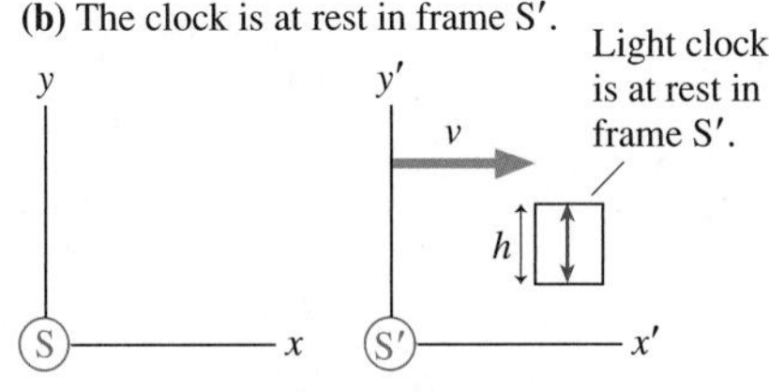

both reference frames are able to measure where and when these events occur *in their frame*. In frame S, the time interval $\Delta t = t_2 - t_1$ is one tick of the clock. Similarly, one tick in frame S′ is $\Delta t' = t_2' - t_1'$.

It's simple to calculate the tick interval $\Delta t'$ observed in frame S′, the rest frame of the clock, because the light simply goes straight up and back down. The total distance traveled by the light is $2h$, so the time of one tick is

$$\Delta t' = \frac{2h}{c} \tag{27.2}$$

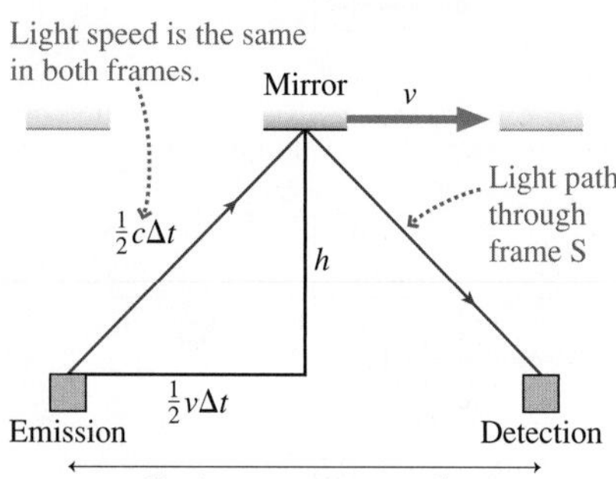

FIGURE 27.17 A light clock analysis in which the speed of light is the same in all reference frames.

FIGURE 27.17 shows the light clock as seen in frame S. As seen in S, the clock is moving to the right at speed v. Thus the mirror has moved a distance $\frac{1}{2}v(\Delta t)$ during the time $\frac{1}{2}(\Delta t)$ in which the light pulse moves from the source to the mirror. To move from the source to the mirror, as seen from frame S the light must move along the *diagonal path* shown. That is, the light must travel *farther* from source to mirror than it did in the rest frame of the clock.

The length of this diagonal is easy to calculate because, according to special relativity, the speed of light is equal to c in all inertial frames. Thus the diagonal length is simply

$$\text{distance} = \text{speed} \times \text{time} = c\left(\frac{1}{2}\,\Delta t\right) = \frac{1}{2}c\,\Delta t$$

We can then apply the Pythagorean theorem to the right triangle in Figure 27.17 to find that

$$h^2 + \left(\frac{1}{2}v\,\Delta t\right)^2 = \left(\frac{1}{2}c\,\Delta t\right)^2 \tag{27.3}$$

We can solve for Δt by first rewriting Equation 27.3 as

$$h^2 = \left(\frac{1}{2}c\,\Delta t\right)^2 - \left(\frac{1}{2}v\,\Delta t\right)^2 = \left[\left(\frac{1}{2}c\right)^2 - \left(\frac{1}{2}v\right)^2\right]\Delta t^2 = \left(\frac{1}{2}\right)^2(c^2 - v^2)\,\Delta t^2$$

so that

$$\Delta t^2 = \frac{h^2}{\left(\frac{1}{2}\right)^2(c^2 - v^2)} = \frac{(2h)^2}{c^2 - v^2} = \frac{(2h/c)^2}{1 - v^2/c^2}$$

from which we find

$$\Delta t = \frac{2h/c}{\sqrt{1 - v^2/c^2}} = \frac{\Delta t'}{\sqrt{1 - v^2/c^2}} \tag{27.4}$$

where we have used Equation 27.2 to write $2h/c$ as $\Delta t'$. The time interval between two ticks in frame S is *not* the same as in frame S′.

It's useful to define $\beta = v/c$, the speed as a fraction of the speed of light. For example, a reference frame moving with $v = 2.4 \times 10^8$ m/s has $\beta = 0.80$. In terms of β, Equation 27.4 is

$$\Delta t = \frac{\Delta t'}{\sqrt{1 - \beta^2}} \tag{27.5}$$

If reference frame S′ is at rest relative to frame S, then $\beta = 0$ and $\Delta t = \Delta t'$. In other words, experimenters in both reference frames measure time to be the same. But two experimenters moving relative to each other will measure *different* time intervals between the same two events. We're unaware of these differences in our everyday lives because our typical speeds are so small compared to c. But the differences are easily measured in a laboratory, and they do affect the precise timekeeping needed to make accurate location measurements with a GPS receiver.

Proper Time

Frame S′ has one important distinction. It is the *one and only* inertial reference frame in which the clock is at rest. Consequently, it is the one and only inertial reference frame in which the times of both events—the emission of the light and the detection of the light—are measured at the *same* position. You can see that the light pulse in Figure 27.16, the rest frame of the clock, starts and ends at the same position, while in Figure 27.17 the emission and detection take place at *different* positions in frame S.

17.1

The time interval between two events that occur at the *same position* is called the **proper time** $\Delta\tau$. Only one inertial reference frame measures the proper time, and it can do so with a single clock that is present at both events. Experimenters in an inertial reference frame moving with speed $v = \beta c$ relative to the proper-time frame must use two clocks to measure the time interval because the two events occur at different positions. We can rewrite Equation 27.5, where the time interval $\Delta t'$ is the proper time $\Delta\tau$, to find that the time interval in any other reference frame is

$$\Delta t = \frac{\Delta\tau}{\sqrt{1-\beta^2}} \geq \Delta\tau \qquad (27.6)$$

Time dilation in terms of proper time $\Delta\tau$ (where $\beta = v/c$)

Because $\beta = v/c$, and v is always less than c, β is always less than 1. This means that the factor $1/\sqrt{1-\beta^2}$ appearing in Equation 27.6 is always *equal to* (when $v = 0$) *or greater than 1*. Thus $\Delta t \geq \Delta\tau$. Recalling that Δt is the time between clock ticks in a frame such as S in which the clock is moving, while the proper time $\Delta\tau$ is the time between ticks in a frame at which the clock is at rest, we can interpret Equation 27.6 as saying that **the time interval between two ticks is the shortest in the reference frame in which the clock is at rest.** The time interval between two ticks is longer when it is measured in any reference frame in which the clock is moving. Because a longer tick interval implies a clock that runs more slowly, we can also say that a **moving clock runs slowly compared to an identical clock at rest.** This "stretching out" of the time interval implied by Equation 27.6 is called **time dilation.**

NOTE ► If $v > c$, so that $\beta > 1$, the factor $1 - \beta^2$ appearing in Equation 27.6 would be *negative*. In this case Δt would contain the square root of a negative number; that is, it would be an *imaginary* number. Time intervals certainly have to be real numbers, suggesting that $v > c$ is not physically possible. One of the predictions of relativity, as you've undoubtedly heard, is that nothing can travel faster than the speed of light. Now you can begin to see why. We'll examine this issue more closely later in the chapter. ◄

Equation 27.6 was derived using a light clock because the operation of a light clock is clear and easy to analyze. But the conclusion is really about time itself. Any clock, regardless of how it operates, behaves the same. For example, suppose you and a light clock are traveling in a very fast spaceship. The light clock happens to tick at the same rate as your heart beats—say, 60 times a minute, or once a second. Because the light clock is at rest in your frame, it measures the proper time between two successive beats of your heart; that is, $\Delta\tau = 1$ s. But to an experimenter stationed on the ground, watching you pass by at an enormous speed, the time interval $\Delta t = \Delta\tau/\sqrt{1-\beta^2}$ between two ticks of the clock—and hence two beats of your heart—would be *longer.* If, for instance, $\Delta t = 2$ s, the ground-based experimenter would conclude that your heart is beating only 30 times per minute. To the experimenter, *all* processes on your spaceship, including all your biological processes, would appear to run slowly.

EXAMPLE 27.4 **Journey time from the sun to Saturn**

Saturn is 1.43×10^{12} m from the sun. A rocket travels along a line from the sun to Saturn at a constant speed of exactly $0.9c$ relative to the solar system. How long does the journey take as measured by an experimenter on earth? As measured by an astronaut on the rocket?

PREPARE Let the solar system be in reference frame S and the rocket be in reference frame S′ that travels with velocity $v = 0.9c$ relative to S. Relativity problems must be stated in terms of *events.* Let event 1 be "the rocket and the sun coincide" (the experimenter on earth says that the rocket passes the sun; the astronaut on the rocket says that the sun passes the rocket) and event 2 be "the rocket and Saturn coincide."

FIGURE 27.18 shows the two events as seen from the two reference frames. Notice that the two events occur at the *same position* in S′, the position of the rocket.

FIGURE 27.18 Visual overview of the trip as seen in frames S and S′.

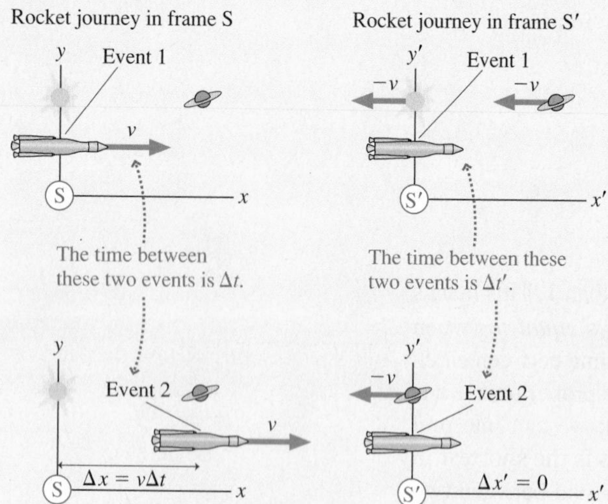

SOLVE The time interval measured in the solar system reference frame, which includes the earth, is simply

$$\Delta t = \frac{\Delta x}{v} = \frac{1.43 \times 10^{12}\ \text{m}}{0.9 \times (3.00 \times 10^{8}\ \text{m/s})} = 5300\ \text{s}$$

Relativity hasn't abandoned the basic definition $v = \Delta x/\Delta t$, although we do have to be sure that Δx and Δt are measured in just one reference frame and refer to the same two events.

How are things in the rocket's reference frame? The two events occur at the *same position* in S′. Thus the time measured by the astronauts is the *proper time* $\Delta\tau$ between the two events. We can then use Equation 27.6 with $\beta = 0.9$ to find

$$\Delta\tau = \sqrt{1-\beta^2}\,\Delta t = \sqrt{1-0.9^2}(5300\ \text{s}) = 2310\ \text{s}$$

ASSESS The time interval measured between these two events by the astronauts is less than half the time interval measured by experimenters on earth. The difference has nothing to do with when earthbound astronomers *see* the rocket pass the sun and Saturn. Δt is the time interval from when the rocket actually passes the sun, as measured by a clock at the sun, until it actually passes Saturn, as measured by a synchronized clock at Saturn. The interval between *seeing* the events from earth, which would have to allow for light travel times, would be something other than 5300 s. Δt and $\Delta\tau$ are different because *time is different* in two reference frames moving relative to each other.

STOP TO THINK 27.6 Molly flies her rocket past Nick at constant velocity v. Molly and Nick both measure the time it takes the rocket, from nose to tail, to pass Nick. Which of the following is true?

A. Both Molly and Nick measure the same amount of time.
B. Molly measures a shorter time interval than Nick.
C. Nick measures a shorter time interval than Molly.

The global positioning system (GPS) If you've ever used a GPS receiver, you know it can pinpoint your location anywhere in the world. The system uses a set of orbiting satellites whose positions are very accurately known. The satellites orbit at a speed of about 14,000 km/h, enough to make the moving satellite's clocks run slow by about 7 μs a day. This may not seem like much, but it would introduce an error of 2000 m into your position! To function properly, the clocks are carefully corrected for effects due to relativity (including effects due to general relativity).

Experimental Evidence

Is there any evidence for the crazy idea that clocks moving relative to each other tell time differently? Indeed, there's plenty. An experiment in 1971 sent an atomic clock around the world on a jet plane while an identical clock remained in the laboratory. This was a difficult experiment because the traveling clock's speed was so small compared to c, but measuring the small differences between the time intervals was just barely within the capabilities of atomic clocks. It was also a more complex experiment than we've analyzed because the clock accelerated as it moved around a circle. Nonetheless, the traveling clock, upon its return, was 200 ns behind the clock that stayed at home, which was exactly as predicted by relativity.

Very detailed studies have been done on unstable particles called *muons* that are created at the top of the atmosphere, at a height of about 60 km, when high-energy cosmic rays collide with air molecules. It is well known, from laboratory studies, that stationary muons decay with a *half-life* of 1.5 μs. That is, half the muons decay within 1.5 μs, half of those remaining decay in the next 1.5 μs, and so on. The decays can be used as a clock.

The muons travel down through the atmosphere at very nearly the speed of light. The time needed to reach the ground, assuming $v \approx c$, is $\Delta t \approx (60{,}000\ \text{m})/(3 \times 10^8\ \text{m/s}) = 200\ \mu\text{s}$. This is 133 half-lives, so the fraction of muons reaching the ground should be $\approx(1/2)^{133} = 10^{-40}$. That is, only 1 out of every 10^{40} muons should reach the ground. In fact, experiments find that about 1 in 10 muons reach the ground, an experimental result that differs by a factor of 10^{39} from our prediction!

The discrepancy is due to time dilation. In **FIGURE 27.19**, the two events "muon is created" and "muon hits ground" take place at two different places in the earth's reference frame. However, these two events occur at the *same position* in the muon's reference frame. (The muon is like the rocket in Example 27.4.) Thus the muon's internal clock measures the proper time. The time-dilated interval $\Delta t = 200\ \mu\text{s}$ in the earth's reference frame corresponds to a time of only $\Delta t' \approx 5\ \mu\text{s}$ in the muon's reference frame. That is, in the muon's reference frame it takes only 5 μs from its creation at the top of the atmosphere until the ground runs into it. This is 3.3 half-lives, so the fraction of muons reaching the ground is $(1/2)^{3.3} = 0.1$, or 1 out of 10. We wouldn't detect muons at the ground at all if not for time dilation.

FIGURE 27.19 We wouldn't detect muons at the ground if not for time dilation.

A muon travels $\approx$450 m in 1.5 μs. We would not detect muons at ground level if the half-life of a moving muon were 1.5 μs.

Muon is created.

Because of time dilation, the half-life of a high-speed muon is long enough in the earth's reference frame for about 1 in 10 muons to reach the ground.

Muon hits ground.

The details are beyond the scope of this textbook, but dozens of high-energy particle accelerators around the world that study quarks and other elementary particles have been designed and built on the basis of Einstein's theory of relativity. The fact that they work exactly as planned is strong testimony to the reality of time dilation.

The Twin Paradox

The most well-known relativity paradox is the twin paradox. George and Helen are twins. On their 25th birthday, Helen departs on a starship voyage to a distant star. Let's imagine, to be specific, that her starship accelerates almost instantly to a speed of $0.95c$ and that she travels to a star that is 9.5 light years (9.5 ly) from earth. Upon arriving, she discovers that the planets circling the star are inhabited by fierce aliens, so she immediately turns around and heads home at $0.95c$.

A **light year,** abbreviated ly, is the distance that light travels in one year. A light year is vastly larger than the diameter of the solar system. The distance between two neighboring stars is typically a few light years. For our purpose, we can write the speed of light as $c = 1$ ly/year. That is, light travels 1 light year per year.

This value for c allows us to determine how long, according to George and his fellow earthlings, it takes Helen to travel out and back. Her total distance is 19 ly and, due to her rapid acceleration and rapid turn around, she travels essentially the entire distance at speed $v = 0.95c = 0.95$ ly/year. Thus the time she's away, as measured by George, is

$$\Delta t_G = \frac{19\ \text{ly}}{0.95\ \text{ly/year}} = 20\ \text{years} \qquad (27.7)$$

George will be 45 years old when his sister Helen returns with tales of adventure.

While she's away, George takes a physics class and studies Einstein's theory of relativity. He realizes that time dilation will make Helen's clocks run more slowly than his clocks, which are at rest relative to him. Her heart—a clock—will beat fewer times and the minute hand on her watch will go around fewer times. In other words, she's aging more slowly than he is. Although she is his twin, she will be younger than he is when she returns.

Alpha Centauri (arrow) is one of the closest stars to the sun, at a distance of 4.3 ly. If you traveled there and back at $0.99c$, your earthbound friends would be 8.6 years older, while you would have aged only 1.2 years.

Calculating Helen's age is not hard. We simply have to identify Helen's clock, because it's always with Helen as she travels, as the clock that measures proper time $\Delta\tau$. From Equation 27.6,

$$\Delta t_H = \Delta\tau = \sqrt{1-\beta^2}\,\Delta t_G = \sqrt{1-0.95^2}\,(20 \text{ years}) = 6.25 \text{ years} \qquad (27.8)$$

George will have just celebrated his 45th birthday as he welcomes home his 31-year-and-3-month-old twin sister.

This may be unsettling, because it violates our commonsense notion of time, but it's not a paradox. There's no logical inconsistency in this outcome. So why is it called "the twin paradox"? Read on.

FIGURE 27.20 The twin paradox.

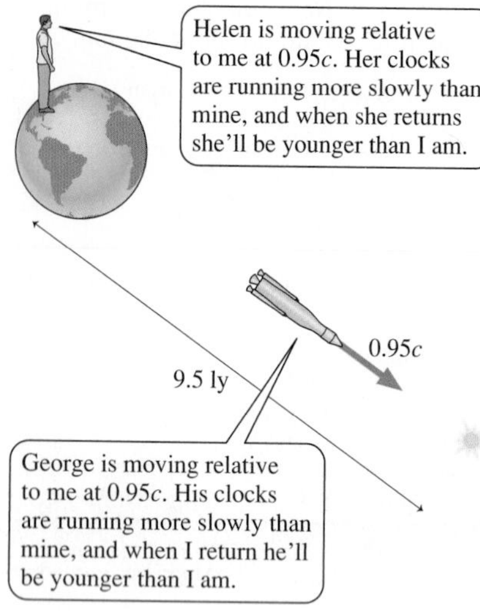

Helen is moving relative to me at 0.95c. Her clocks are running more slowly than mine, and when she returns she'll be younger than I am.
0.95c
9.5 ly
George is moving relative to me at 0.95c. His clocks are running more slowly than mine, and when I return he'll be younger than I am.

Helen, knowing that she had quite of bit of time to kill on her journey, brought along several physics books to read. As she learns about relativity, she begins to think about George and her friends back on earth. Relative to her, they are all moving away at $0.95c$. Later they'll come rushing toward her at $0.95c$. Time dilation will cause their clocks to run more slowly than her clocks, which are at rest relative to her. In other words, as FIGURE 27.20 shows, Helen concludes that people on earth are aging more slowly than she is. Alas, she will be much older than they when she returns.

Finally, the big day arrives. Helen lands back on earth and steps out of the starship. George is expecting Helen to be younger than he is. Helen is expecting George to be younger than she is.

Here's the paradox! It's logically impossible for each to be younger than the other at the time when they are reunited. Where, then, is the flaw in our reasoning? It seems to be a symmetrical situation—Helen moves relative to George and George moves relative to Helen—but symmetrical reasoning has led to a conundrum.

But are the situations really symmetrical? George goes about his business day after day without noticing anything unusual. Helen, on the other hand, experiences three distinct periods during which the starship engines fire, she's crushed into her seat, and free dust particles that had been floating inside the starship are no longer, in the starship's reference frame, at rest or traveling in a straight line at constant speed. In other words, George spends the entire time in an inertial reference frame, *but Helen does not.* The situation is *not* symmetrical.

The principle of relativity applies *only* to inertial reference frames. Our discussion of time dilation was for inertial reference frames. Thus George's analysis and calculations are correct. Helen's analysis and calculations are *not* correct because she was trying to apply an inertial reference frame result to a noninertial reference frame.

Helen is younger than George when she returns. This is strange, but not a paradox. It is a consequence of the fact that time flows differently in two reference frames moving relative to each other.

FIGURE 27.21 The length of a train car as measured by Ryan and by Peggy.

(a) In Ryan's frame S, Peggy moves to the right at speed v.

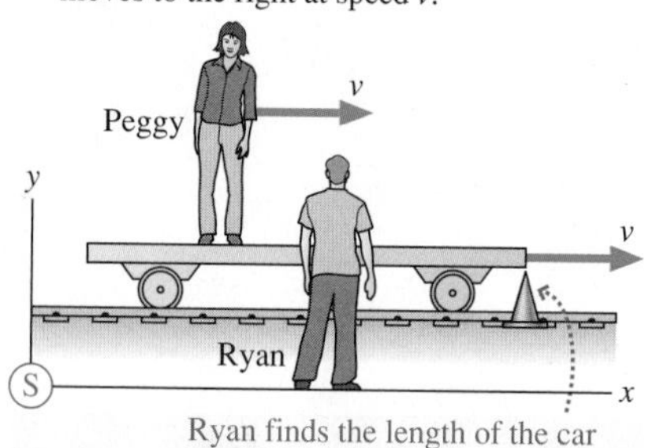

Ryan finds the length of the car by measuring the time Δt it takes to pass the cone. Then $L = v\Delta t$.

(b) In Peggy's frame S′, Ryan moves to the left at speed v.

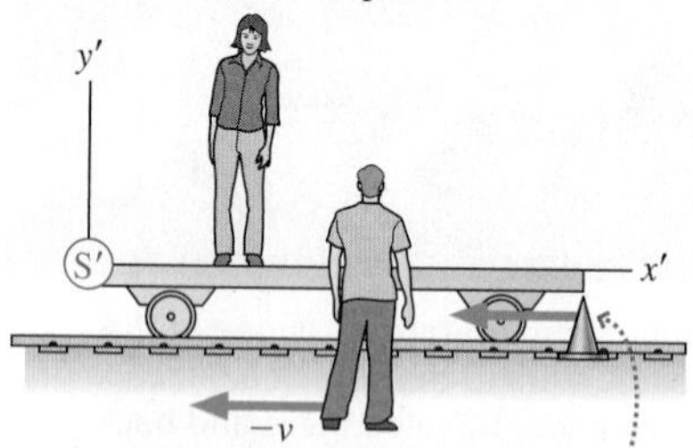

Peggy finds the length of the car by measuring the time $\Delta t'$ it takes the cone, moving at speed v, to pass her car. Then $L' = v\Delta t'$.

27.7 Length Contraction

We've seen that relativity requires us to rethink our idea of time. Now let's turn our attention to the concepts of space and distance. Consider again Peggy on her train car, which is reference frame S′, moving past Ryan, who is at rest in frame S, at relative speed v. Ryan wants to measure the length L of Peggy's car as it moves past him. As shown in FIGURE 27.21a, he can do so by measuring the time Δt that it takes the car to pass the fixed cone; he then calculates that $L = v\,\Delta t$.

FIGURE 27.21b shows the situation in Peggy's reference frame S′, where the car is at rest. Peggy wants to measure the length L' of her car; we'll soon see that L' need not be the same as L. Peggy can measure L' by finding the time $\Delta t'$ that the cone, moving at speed v, takes to move from one end of the car to the other. In this way, she finds that $L' = v\,\Delta t'$.

Speed v is the relative speed between S and S′ and is the same for both Ryan and Peggy. From Ryan's and Peggy's measurements of the car's length we can then write

ActivPhysics ONLINE 17.2

$$v = \frac{L}{\Delta t} = \frac{L'}{\Delta t'} \tag{27.9}$$

The time interval Δt measured in Ryan's frame S is the proper time $\Delta\tau$ because the two events that define the time intervals—the front end and back end of the car passing the cone—occur at the same position (the cone) in Ryan's frame. We can use the time-dilation result, Equation 27.6, to relate $\Delta\tau$ measured by Ryan to $\Delta t'$ measured by Peggy. Equation 27.9 then becomes

$$\frac{L}{\Delta\tau} = \frac{L'}{\Delta t'} = \frac{L'}{\Delta\tau/\sqrt{1-\beta^2}} \tag{27.10}$$

The $\Delta\tau$ cancels, so the car's length L in Ryan's frame is related to its length L' in Peggy's frame by

$$L = \sqrt{1-\beta^2}\,L' \tag{27.11}$$

Surprisingly, we find that **the length of the car in Ryan's frame is different from its length in Peggy's frame.**

Peggy's frame S′, in which the car's length is L', has one important distinction. It is the *one and only* inertial reference frame in which the car is at rest. Experimenters in frame S′ can take all the time they need to measure L' because the car isn't moving. The length of an object measured in the reference frame in which the object is at rest is called the **proper length** ℓ. When you measure the length of an everyday object, such as a curtain rod or tabletop, it is usually at rest in your reference frame, so everyday length measurements are of proper length.

We can use the proper length ℓ to write Equation 27.11 as

$$L = \sqrt{1-\beta^2}\,\ell \le \ell \tag{27.12}$$

Length contraction in terms of proper length ℓ

Two perspectives on a relativistic trip The Stanford Linear Accelerator is a 3.2-km-long electron accelerator that accelerates electrons to a speed of $0.99999999995c$. From our perspective, the electrons take a time $\Delta t = (3200\text{ m})/c = 11\ \mu\text{s}$ to make the trip. However, we see their "clocks" run slowly, ticking off only 110 ps during the trip. What do the electrons see? In their reference frame the end of the accelerator is coming toward them at $0.99999999995c$, but its length is contracted to only 3.3 cm. Thus it arrives in a time $(3.3\text{ cm})/c = 110$ ps. The same result, but from a different perspective.

Because $\beta \ge 0$, the factor $\sqrt{1-\beta^2}$ is less than or equal to 1. This means that $L \le \ell$. Because ℓ, the proper length, is measured in a reference frame in which the object is at rest while L is measured in a frame in which the object is moving, we see that **the length of an object is greatest in the reference frame in which the object is at rest.** This "shrinking" of the length of an object or the distance between two objects, as measured by an experimenter moving with respect to the object(s), is called **length contraction.**

NOTE ► A moving object's length is contracted only in the direction in which it's moving (its length along the x-axis in Figure 27.21). The object's length in the y- and z-directions doesn't change. ◄

EXAMPLE 27.5 Length contraction of a ladder

Dan holds a 5.0-m-long ladder parallel to the ground. He then gets up to a good sprint, eventually reaching 98% of the speed of light. How long is the ladder according to Dan, once he is running, and according to Carmen, who is standing on the ground as Dan goes by?

PREPARE Let reference frame S′ be attached to Dan. The ladder is at rest in this reference frame, so Dan measures the proper length of the ladder: $\ell = 5.0$ m. Dan's frame S′ moves relative to Carmen's frame S with velocity $v = 0.98c$.

SOLVE We can find the length of the ladder in Carmen's frame from Equation 27.12. We have

$$L = \sqrt{1-\beta^2}\,\ell = \sqrt{1-0.98^2}\,(5.0\text{ m}) = 1.0\text{ m}$$

ASSESS The length of the moving ladder as measured by Carmen is only one-fifth its length as measured by Dan. These lengths are different because *space is different* in two reference frames moving relative to each other.

The conclusion that space is different in reference frames moving relative to each other is a direct consequence of the fact that time is different. Experimenters in both reference frames agree on the relative velocity v, leading to Equation 27.9: $v = L/\Delta t = L'/\Delta t'$. Because of time dilation, Ryan (who measures proper time) finds that $\Delta t < \Delta t'$. Thus L *has* to be less than L'. That is the only way Ryan and Peggy can reconcile their measurements.

The Binomial Approximation

A useful mathematical tool is the **binomial approximation.** Suppose we need to evaluate the quantity $(1 + x)^n$. If x is much less than 1, it turns out that an excellent approximation is

$$(1 + x)^n \approx 1 + nx \qquad \text{if} \qquad x \ll 1 \tag{27.13}$$

You can try this on your calculator. Suppose you need to calculate 1.01^2. Comparing this expression with Equation 27.13, we see that $x = 0.01$ and $n = 2$. Equation 27.13 then tells us that

$$1.01^2 \approx 1 + 2 \times 0.01 = 1.02$$

The exact result, using a calculator, is 1.0201. The approximate answer is good to about 99.99%! The smaller the value of x, the better the approximation.

The binomial approximation is very useful when we need to calculate a relativistic expression for a speed much less than c, so that $v \ll c$. Because $\beta = v/c$, a reference frame moving with $v^2/c^2 \ll 1$ has $\beta^2 \ll 1$. In these cases, we can write

$$\begin{aligned} \sqrt{1-\beta^2} &= (1 - v^2/c^2)^{1/2} \approx 1 - \frac{1}{2}\frac{v^2}{c^2} \\ \frac{1}{\sqrt{1-\beta^2}} &= (1 - v^2/c^2)^{-1/2} \approx 1 + \frac{1}{2}\frac{v^2}{c^2} \end{aligned} \tag{27.14}$$

The following example illustrates the use of the binomial approximation.

EXAMPLE 27.6 The shrinking school bus

An 8.0-m-long school bus drives past at 30 m/s. By how much is its length contracted?

PREPARE The school bus is at rest in an inertial reference frame S′ moving at velocity $v = 30$ m/s relative to the ground frame S. The given length, 8.0 m, is the proper length ℓ in frame S′.

SOLVE In frame S, the school bus is length contracted to

$$L = \sqrt{1-\beta^2}\,\ell$$

The bus's speed v is much less than c, so we can use the binomial approximation to write

$$L \approx \left(1 - \frac{1}{2}\frac{v^2}{c^2}\right)\ell = \ell - \frac{1}{2}\frac{v^2}{c^2}\ell$$

The *amount* of the length contraction is

$$\begin{aligned} \ell - L &= \frac{1}{2}\frac{v^2}{c^2}\ell = \left(\frac{30 \text{ m/s}}{3.0 \times 10^8 \text{ m/s}}\right)^2 (4.0 \text{ m}) \\ &= 4.0 \times 10^{-14} \text{ m} = 40 \text{ fm} \end{aligned}$$

where 1 fm = 1 femtometer = 10^{-15} m.

ASSESS The amount the bus "shrinks" is only slightly larger than the diameter of the nucleus of an atom. It's no wonder that we're not aware of length contraction in our everyday lives. If you had tried to calculate this number exactly, your calculator would have shown $\ell - L = 0$. The difficulty is that the difference between ℓ and L shows up only in the 14th decimal place. A scientific calculator determines numbers to 10 or 12 decimal places, but that isn't sufficient to show the difference. The binomial approximation provides an invaluable tool for finding the very tiny difference between two numbers that are nearly identical.

27.8 Velocities of Objects in Special Relativity

In Section 27.2 we discussed Galilean relativity, which is applicable to objects that are moving at speeds much less than the speed of light. We found that if the velocity of an object is u in reference frame S, then its velocity measured in frame S′, moving at velocity v relative to frame S, is $u' = u - v$.

But we soon learned that this expression is invalid for objects moving at an appreciable fraction of the speed of light. In particular, light itself moves at speed c as measured by *all* observers, independent of their relative velocities. The Galilean transformation of velocity needs to be modified for objects moving at relativistic speeds.

Although a proof is beyond the scope of this text, Einstein's relativity includes a velocity-addition expression valid for *any* velocities. If u is the velocity of an object as measured in reference frame S, and u' its velocity as measured in S′, moving at velocity v relative to frame S, then the two velocities are related by

$$u' = \frac{u - v}{1 - uv/c^2} \quad \text{or} \quad u = \frac{u' + v}{1 + u'v/c^2} \qquad (27.15)$$

Lorentz transformation of velocity

These equations were discovered by Dutch physicist H. A. Lorentz a few years before Einstein published his theory of relativity, but Lorentz didn't completely understand their implications for space and time. Notice that the denominator is ≈ 1 if either u or v is much less than c. In other words, these equations agree with the Galilean velocity transformation when velocities are nonrelativistic (i.e., $\ll c$), but they differ as velocities approach the speed of light.

NOTE ► It is important to distinguish carefully between v, which is the relative velocity of the reference frames in which measurements are carried out, and u and u', which are the velocities of an *object* as measured in two different reference frames. ◄

EXAMPLE 27.7 A speeding bullet

A rocket flies past the earth at precisely $0.9c$. As it goes by, the rocket fires a bullet in the forward direction at precisely $0.95c$ with respect to the rocket. What is the bullet's speed with respect to the earth?

PREPARE The rocket and the earth are inertial reference frames. Let the earth be frame S and the rocket be frame S′. The velocity of frame S′ relative to frame S is $v = 0.9c$. The bullet's velocity in frame S′ is $u' = 0.95c$.

SOLVE We can use the Lorentz velocity transformation to find

$$u = \frac{u' + v}{1 + u'v/c^2} = \frac{0.95c + 0.90c}{1 + (0.95c)(0.90c)/c^2} = 0.997c$$

The bullet's speed with respect to the earth is 99.7% of the speed of light.

NOTE ► Many relativistic calculations are much easier when velocities are specified as a fraction of c. ◄

ASSESS The Galilean transformation of velocity would give $u = 1.85c$. Now, despite the very high speed of the rocket and of the bullet with respect to the rocket, the bullet's speed with respect to the earth remains less than c. This is yet more evidence that objects cannot exceed the speed of light.

Suppose the rocket in Example 27.7 fired a laser beam in the forward direction as it traveled past the earth at velocity v. The laser beam would travel away from the rocket at speed $u' = c$ in the rocket's reference frame S′. What is the laser beam's speed in the earth's frame S? According to the Lorentz velocity transformation, it must be

$$u = \frac{u' + v}{1 + u'v/c^2} = \frac{c + v}{1 + cv/c^2} = \frac{c + v}{1 + v/c} = \frac{c + v}{(c + v)/c} = c$$

Light travels at speed c in both frame S and frame S′. This important consequence of the principle of relativity is "built into" the Lorentz velocity transformation.

In this photograph, the track (shown in red) of a high-energy proton enters from the lower right. The proton then collides with other protons, sending them in all directions, where further collisions occur. Momentum is conserved, but at these high speeds the relativistic expression for momentum must be used.

27.9 Relativistic Momentum

In Newtonian mechanics, the total momentum of a system is a conserved quantity. Further, the law of conservation of momentum, $P_f = P_i$, is true in all inertial reference frames *if* the particle velocities in different reference frames are related by the Galilean velocity transformation.

The difficulty, of course, is that the Galilean transformation is not consistent with the principle of relativity. It is a reasonable approximation when all velocities are much less than c, but the Galilean transformation fails dramatically as velocities approach c. It's not hard to show that $P'_f \neq P'_i$ if the particle velocities in frame S′ are related to the particle velocities in frame S by the Lorentz velocity transformation.

There are two possibilities:

1. The so-called law of conservation of momentum is not really a law of physics. It is approximately true at low velocities but fails as velocities approach the speed of light.
2. The law of conservation of momentum really is a law of physics, but the expression $p = mu$ is not the correct way to calculate momentum when the particle velocity u becomes a significant fraction of c.

Momentum conservation is such a central and important feature of mechanics that it seems unlikely to fail in relativity. How else might the momentum of a particle be defined?

The classical momentum, for one-dimensional motion, is $p = mu = m(\Delta x/\Delta t)$. Δt is the time needed to move a displacement Δx. That seemed clear enough within a Newtonian framework, but now we've learned that experimenters in different reference frames disagree about the amount of time needed. So whose Δt should we use?

One possibility is to use the time measured *by the particle*. This is the proper time $\Delta\tau$ because the particle is at rest in its own reference frame. With this in mind, let's redefine the momentum of a particle of mass m moving with velocity $u = \Delta x/\Delta t$ to be

$$p = m\frac{\Delta x}{\Delta \tau} \tag{27.16}$$

We can relate this new expression for p to the familiar Newtonian expression by using the time-dilation result $\Delta\tau = (1 - u^2/c^2)^{1/2}\,\Delta t$ to relate the proper time interval measured by the particle to the more practical time interval Δt measured by experimenters in frame S. With this substitution, Equation 27.16 becomes

$$p = m\frac{\Delta x}{\Delta \tau} = m\frac{\Delta x}{\sqrt{1 - u^2/c^2}\,\Delta t} = \frac{mu}{\sqrt{1 - u^2/c^2}} \tag{27.17}$$

You can see that Equation 27.17 reduces to the classical expression $p = mu$ when the particle's speed $u \ll c$. That is an important requirement, but whether this is the "correct" expression for p depends on whether the total momentum P is conserved when the velocities of a system of particles are transformed with the Lorentz velocity transformation equations. The proof is rather long and tedious, so we will assert, without actual proof, that the momentum defined in Equation 27.17 is, indeed, conserved. **The law of conservation of momentum is still valid in all inertial reference frames *if* the momentum of each particle is calculated with Equation 27.17.**

To simplify our notation, let's define the quantity

$$\gamma = \frac{1}{\sqrt{1 - u^2/c^2}} \tag{27.18}$$

With this definition of γ, the momentum of a particle is

$$p = \gamma m u \tag{27.19}$$

Relativistic momentum for a particle with mass m and speed u

EXAMPLE 27.8 Momentum of a subatomic particle

Electrons in a particle accelerator reach a speed of $0.999c$ relative to the laboratory. One collision of an electron with a target produces a muon that moves forward with a speed of $0.950c$ relative to the laboratory. The muon mass is 1.90×10^{-28} kg. What is the muon's momentum in the laboratory frame and in the frame of the electron beam?

PREPARE Let the laboratory be reference frame S. The reference frame S′ of the electron beam (i.e., a reference frame in which the electrons are at rest) moves in the direction of the electrons at $v = 0.999c$. The muon velocity in frame S is $u = 0.95c$.

SOLVE γ for the muon in the laboratory reference frame is

$$\gamma = \frac{1}{\sqrt{1 - u^2/c^2}} = \frac{1}{\sqrt{1 - 0.95^2}} = 3.20$$

Thus the muon's momentum in the laboratory is

$$\begin{aligned} p &= \gamma m u \\ &= (3.20)(1.90 \times 10^{-28}\ \text{kg})(0.95 \times 3.00 \times 10^8\ \text{m/s}) \\ &= 1.73 \times 10^{-19}\ \text{kg} \cdot \text{m/s} \end{aligned}$$

The momentum is a factor of 3.2 larger than the Newtonian momentum mu. To find the momentum in the electron-beam reference frame, we must first use the velocity transformation equation to find the muon's velocity in frame S′:

$$u' = \frac{u - v}{1 - uv/c^2} = \frac{0.95c - 0.999c}{1 - (0.95c)(0.999c)/c^2} = -0.962c$$

In the laboratory frame, the faster electrons are overtaking the slower muon. Hence the muon's velocity in the electron-beam frame is negative. γ' for the muon in frame S′ is

$$\gamma' = \frac{1}{\sqrt{1 - u'^2/c^2}} = \frac{1}{\sqrt{1 - 0.962^2}} = 3.66$$

The muon's momentum in the electron-beam reference frame is

$$\begin{aligned} p' &= \gamma' m u' \\ &= (3.66)(1.90 \times 10^{-28}\ \text{kg})(-0.962 \times 3.00 \times 10^8\ \text{m/s}) \\ &= -2.01 \times 10^{-19}\ \text{kg} \cdot \text{m/s} \end{aligned}$$

ASSESS From the laboratory perspective, the muon moves only slightly slower than the electron beam. But it turns out that the muon moves faster with respect to the electrons, although in the opposite direction, than it does with respect to the laboratory.

The Cosmic Speed Limit

FIGURE 27.22a on the next page is a graph of momentum versus velocity. For a Newtonian particle, with $p = mu$, the momentum is directly proportional to the velocity. The relativistic expression for momentum agrees with the Newtonian value if $u \ll c$, but p approaches ∞ as $u \rightarrow c$.

The implications of this graph become clear when we relate momentum to force. Consider a particle subjected to a constant force, such as a rocket that never runs out of fuel. From the impulse-momentum theorem we have $\Delta p = F\Delta t$, or $p = mu = Ft$ if the rocket starts from rest at $t = 0$. If Newtonian physics were correct, a particle would go faster and faster as its velocity $u = p/m = (F/m)t$ increased without limit. But the relativistic result, shown in **FIGURE 27.22b**, is that the particle's velocity approaches the speed of light ($u \rightarrow c$) as p approaches ∞. Relativity gives a very different outcome than Newtonian mechanics.

FIGURE 27.22 The speed of a particle cannot reach the speed of light.

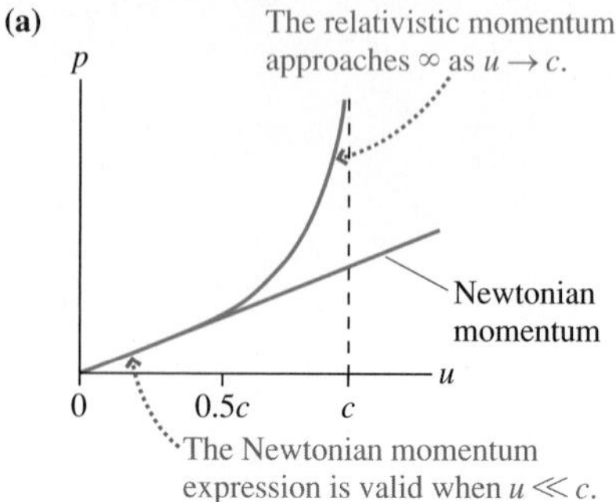
(a)
p
The relativistic momentum approaches ∞ as u → c.
Newtonian momentum
u
0
0.5c
c
The Newtonian momentum expression is valid when u ≪ c.

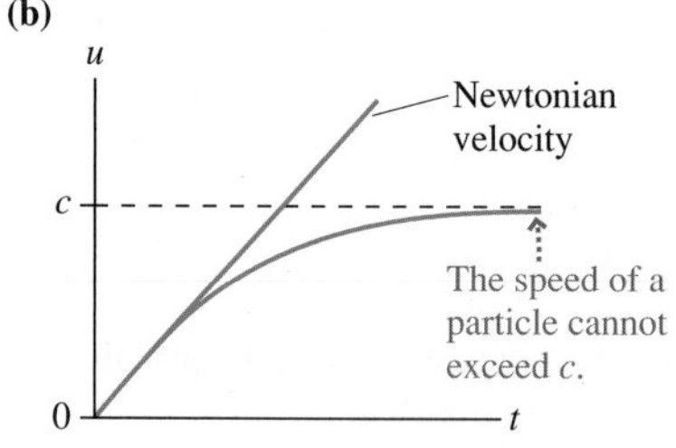
(b)
u
Newtonian velocity
c
The speed of a particle cannot exceed c.
0
t

The speed c is a "cosmic speed limit" for material particles. A force cannot accelerate a particle to a speed higher than c because the particle's momentum becomes infinitely large as the speed approaches c. The amount of effort required for each additional increment of velocity becomes larger and larger until no amount of effort can raise the velocity any higher.

Actually, at a more fundamental level, c is a speed limit for *any* kind of **causal influence.** If you throw a rock and break a window, your throw is the *cause* of the breaking window and the rock is the causal influence. A causal influence can be any kind of particle, wave, or information that travels from A to B and allows A to be the cause of B.

For two unrelated events—a firecracker explodes in Tokyo and a balloon bursts in Paris—the relativity of simultaneity tells us that in one reference frame the firecracker may explode before the balloon bursts, but in some other reference frame the balloon may burst first.

However, for two causally related events—A causes B—it would be nonsense for an experimenter in any reference frame to find that B occurs before A. No experimenter in any reference frame, no matter how it is moving, will find that you are born before your mother is born.

But according to relativity, a causal influence traveling faster than light could result in B causing A, a logical absurdity. Thus **no causal influence of any kind—a particle, wave, or other influence—can travel faster than c.**

The existence of a cosmic speed limit is one of the most interesting consequences of the theory of relativity. "Hyperdrive," in which a spaceship suddenly leaps to faster-than-light velocities, is simply incompatible with the theory of relativity. Rapid travel to the stars must remain in the realm of science fiction.

27.10 Relativistic Energy

Energy is our final topic in this chapter on relativity. Space, time, velocity, and momentum are changed by relativity, so it seems inevitable that we'll need a new view of energy. Indeed, one of the most profound results of relativity, and perhaps the one with the most far-reaching consequences, was Einstein's discovery of the fundamental relationship between energy and mass.

Consider an object of mass m moving at speed u. Einstein found that the **total energy** of such an object is

$$E = \frac{mc^2}{\sqrt{1 - u^2/c^2}} = \gamma mc^2 \qquad (27.20)$$

Total energy of an object of mass m moving at speed u

where $\gamma = 1/(1 - u^2/c^2)^{1/2}$ was defined in Equation 27.18.

To understand this expression, let's start by examining its behavior for objects traveling at speeds much less than the speed of light. In this case we can use the binomial approximation you learned in Section 27.7 to write

$$\gamma = \frac{1}{\sqrt{1 - u^2/c^2}} \approx 1 + \frac{1}{2}\frac{u^2}{c^2}$$

For low speeds u, then, the object's total energy is

$$E \approx mc^2 + \frac{1}{2}mu^2 \qquad (27.21)$$

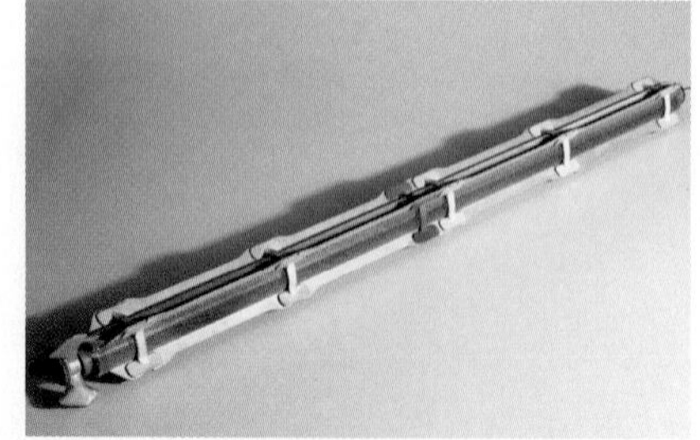

◀This fuel rod for a nuclear power reactor contains about 5 kg of uranium. Its usable energy content, which comes from the conversion of a small fraction of the uranium's mass to energy, is equivalent to that of about 10 million kg of coal.

The second term in this expression is the familiar Newtonian kinetic energy $K = \frac{1}{2}mu^2$, written here in terms of velocity u rather than v. But there is an additional term in the total energy, the **rest energy** given by

$$E_0 = mc^2 \tag{27.22}$$

When a particle is at rest, with $u = 0$, it still has energy E_0. Indeed, because c is so large, the rest energy can be enormous. Equation 27.22 is, of course, Einstein's famous $E = mc^2$, perhaps the most famous equation in all of physics. It tells us that there is a fundamental equivalence between mass and energy, an idea we'll explore later in this section.

EXAMPLE 27.9 The rest energy of an apple

What is the rest energy of a 200 g apple?

SOLVE From Equation 27.22 we have

$$E_0 = mc^2 = (0.20\text{ kg})(3.0 \times 10^8\text{ m/s})^2 = 1.8 \times 10^{16}\text{ J}$$

ASSESS This is an enormous energy, enough to power a medium-sized city for about a year.

Equation 27.21 suggests that the total energy of an object is the sum of a rest energy, which is a new idea, and the familiar kinetic energy. But Equation 27.21 is valid only when the object's speed is low compared to c. For higher speeds, we need to use the full energy expression, Equation 27.20. We can use Equation 27.20 to find a relativistic expression for the kinetic energy K by subtracting the rest energy E_0 from the total energy. Doing so gives

$$K = E - E_0 = \gamma mc^2 - mc^2 = (\gamma - 1)mc^2 = (\gamma - 1)E_0 \tag{27.23}$$

Thus we can write the total energy of an object of mass m as

$$E = \underbrace{mc^2}_{\text{Rest energy } E_0} + \underbrace{(\gamma - 1)mc^2}_{\text{Kinetic energy } K} \tag{27.24}$$

EXAMPLE 27.10 Comparing energies of a ball and an electron

Calculate the rest energy and the kinetic energy of (a) a 100 g ball moving with a speed of 100 m/s and (b) an electron with a speed of $0.999c$.

PREPARE The ball, with $u \ll c$, is a classical particle. We don't need to use the relativistic expression for its kinetic energy. The electron is highly relativistic.

SOLVE

a. For the ball, with $m = 0.100$ kg,

$$E_0 = mc^2 = 9.00 \times 10^{15}\text{ J}$$

$$K = \frac{1}{2}mu^2 = 500\text{ J}$$

b. For the electron, we start by calculating

$$\gamma = \frac{1}{\sqrt{1 - u^2/c^2}} = 22.4$$

Then, using $m_e = 9.11 \times 10^{-31}$ kg,

$$E_0 = mc^2 = 8.20 \times 10^{-14}\text{ J}$$

$$K = (\gamma - 1)E_0 = 175 \times 10^{-14}\text{ J}$$

ASSESS The ball's kinetic energy is a typical kinetic energy. Its rest energy, by contrast, is a staggeringly large number. For a relativistic electron, on the other hand, the kinetic energy is more important than the rest energy.

STOP TO THINK 27.7 An electron moves through the lab at a speed such that $\gamma = 1.5$. The electron's kinetic energy is

A. Greater than its rest energy.
B. Equal to its rest energy.
C. Less than its rest energy.

The Equivalence of Mass and Energy

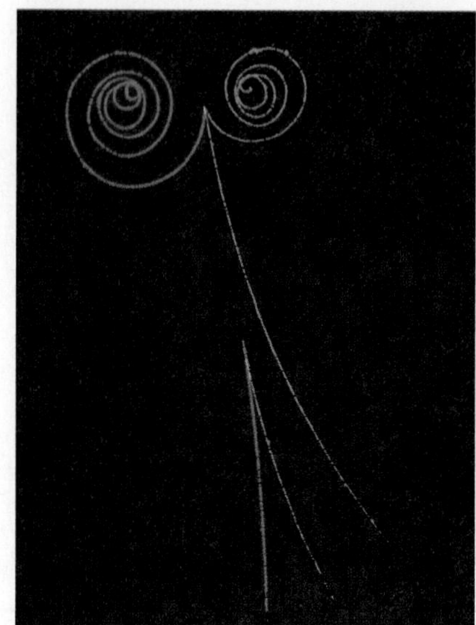

The tracks of elementary particles in a bubble chamber show the creation of an electron-positron pair. The negative electron and positive positron spiral in opposite directions in the magnetic field.

Now we're ready to explore the significance of Einstein's famous equation $E = mc^2$. FIGURE 27.23 shows an experiment that has been done countless times in the last 50 years at particle accelerators around the world. An electron that has been accelerated to $u \approx c$ is aimed at a target material. When a high-energy electron collides with an atom in the target, it can easily knock one of the electrons out of the atom. Thus we would expect to see two electrons leaving the target: the incident electron and the ejected electron. Instead, *four* particles emerge from the target: three electrons and a positron. A *positron,* or positive electron, is the antimatter version of an electron, identical to an electron in all respects other than having charge $q = +e$. In particular, a positron has the same mass m_e as an electron.

FIGURE 27.23 An inelastic collision between electrons can create an electron-positron pair.

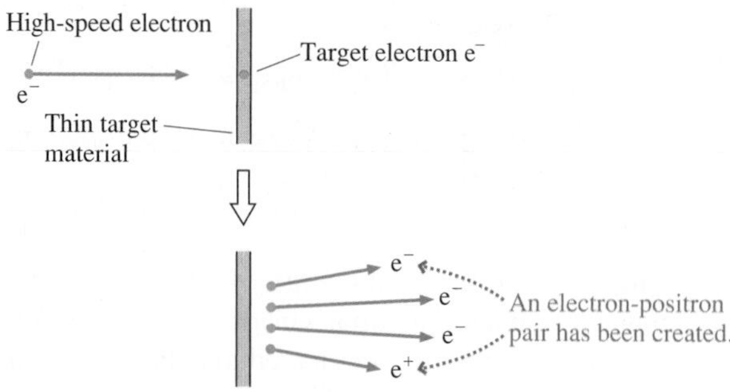
High-speed electron
Target electron e⁻
e⁻
Thin target material
e⁻
e⁻
e⁻
e⁺
An electron-positron pair has been created.

In chemical-reaction notation, the collision is

$$e^-(\text{fast}) + e^-(\text{at rest}) \rightarrow e^- + e^- + e^- + e^+$$

An electron and a positron have been created, apparently out of nothing. Mass $2m_e$ before the collision has become mass $4m_e$ after the collision. (Notice that charge has been conserved in this collision.)

Although the mass has increased, it wasn't created "out of nothing." If you measured the energies before and after the collision, you would find that the kinetic energy before the collision was *greater* than the kinetic energy after. In fact, the decrease in kinetic energy is exactly equal to the rest energy of the two particles that have been created: $\Delta K = 2m_e c^2$. The new particles have been created *out of energy!*

FIGURE 27.24 The annihilation of an electron-positron pair.

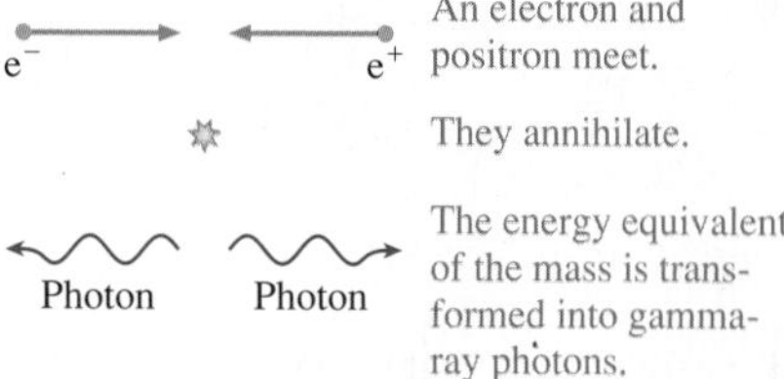
e⁻
e⁺
An electron and positron meet.
They annihilate.
The energy equivalent of the mass is transformed into gamma-ray photons.
Photon
Photon

Not only can particles be created from energy, particles can return to energy. FIGURE 27.24 shows an electron colliding with a positron, its antimatter partner. When a particle and its antiparticle meet, they *annihilate* each other. The mass disappears, and the energy equivalent of the mass is transformed into two high-energy photons. Photons have no mass and represent pure energy. Positron-electron annihilation is also the basis of the medical procedure known as a positron-emission tomography, or PET scans. We'll study this important diagnostic tool in detail in Chapter 30.

Conservation of Energy

The creation and annihilation of particles with mass, processes strictly forbidden in Newtonian mechanics, are vivid proof that neither mass nor the Newtonian definition of energy is conserved. Even so, the *total* energy—the kinetic energy *and* the energy equivalent of mass—remains a conserved quantity.

Law of conservation of total energy The energy $E = \Sigma E_i$ of an isolated system is conserved, where $E_i = \gamma_i m_i c^2$ is the total energy of particle i.

Mass and energy are not the same thing, but, as the last few examples have shown, they are *equivalent* in the sense that mass can be transformed into energy and energy can be transformed into mass as long as the total energy is conserved.

Probably the most well-known application of the conservation of total energy is nuclear fission. The uranium isotope ^{236}U, containing 236 protons and neutrons, does not exist in nature. It can be created when a ^{235}U nucleus absorbs a neutron, increasing its atomic mass from 235 to 236. The ^{236}U nucleus quickly fragments into two smaller nuclei and several extra neutrons, a process known as **nuclear fission.** The nucleus can fragment in several ways, but one is

$$n + {}^{235}\text{U} \rightarrow {}^{236}\text{U} \rightarrow {}^{144}\text{Ba} + {}^{89}\text{Kr} + 3n \qquad (27.25)$$

Ba and Kr are the atomic symbols for barium and krypton.

This reaction seems like an ordinary chemical reaction—until you check the masses. The masses of atomic isotopes are known with great precision from many decades of measurement in instruments called mass spectrometers. As shown in Table 27.1, if you add up the masses on both sides, you find that the mass of the products is 0.186 u less than the mass of the initial neutron and ^{235}U, where, you will recall, $1 \text{ u} = 1.66 \times 10^{-27}$ kg is the atomic mass unit. Converting to kilograms gives us the mass loss of 3.07×10^{-28} kg.

Mass has been lost, but the energy equivalent of the mass has not. As **FIGURE 27.25** shows, the mass has been converted to kinetic energy, causing the two product nuclei and three neutrons to be ejected at very high speeds. The kinetic energy is easily calculated: $\Delta K = m_{\text{lost}}c^2 = 2.8 \times 10^{-11}$ J.

This is a very tiny amount of energy, but it is the energy released from *one* fission. The number of nuclei in a macroscopic sample of uranium is on the order of N_A, Avogadro's number. Hence the energy available if *all* the nuclei fission is enormous. This energy, of course, is the basis for both nuclear power reactors and nuclear weapons.

We started this chapter with an expectation that relativity would challenge our basic notions of space and time. We end by finding that relativity changes our understanding of mass and energy. Most remarkable of all is that each and every one of these new ideas flows from one simple statement: The laws of physics are the same in all inertial reference frames.

TABLE 27.1 Mass before and after fission of ^{235}U

Initial nucleus	Initial mass (u)	Final nucleus	Final mass (u)
^{235}U	235.0439	^{144}Ba	143.9229
n	1.0087	^{89}Kr	88.9176
		3n	3.0260
Total	236.0526		235.8665

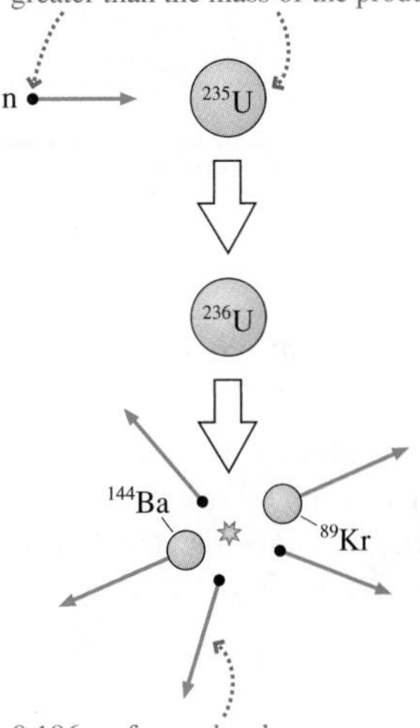
The mass of the reactants is 0.186 u greater than the mass of the products.
n
^{235}U
^{236}U
^{144}Ba
^{89}Kr
0.186 u of mass has been converted into kinetic energy.

FIGURE 27.25 In nuclear fission, the energy equivalent of lost mass is converted into kinetic energy.

INTEGRATED EXAMPLE 27.11 The global positioning system

The turtle in the photo that opens this chapter is being tracked using the global positioning system (GPS), a system of 24 satellites in circular orbits high above the earth. Each satellite, with an orbital speed of 3900 m/s, carries an atomic clock whose time is accurate to ±1 ns per day. Every 30 s, each satellite sends out a radio signal giving its precise location in space and the exact time the signal was sent.

The turtle's GPS receiver records the time at which the signal is received. Because the signal specifies the time at which it was sent, the receiver can easily calculate how long it took the signal to reach the turtle. Then, because radio waves are electromagnetic waves traveling at the speed of light, the precise distance to the satellite can be calculated.

The signal from one satellite actually locates the turtle's position only along a large sphere centered on the satellite. To pinpoint the location exactly requires the signals from four or more satellites. In this problem, we'll ignore these complicating effects.

a. As the satellite passes overhead, suppose two lights flash: one at the front of the satellite (the side toward which the satellite is moving) and one at the rear (the side opposite the direction of motion). In the reference frame of the satellite, the two flashes are simultaneous. Are they simultaneous to an experimenter on the earth? If not, which flash occurs first?
b. In one day, how much time does the clock running on the satellite gain or lose compared to an identical clock on earth?
c. If the clock error in part b were not properly taken into account, by how much would the turtle's position on earth be in error after one day?

PREPARE Consider an astronaut standing on the satellite halfway between the lights. Because the two flashes are simultaneous in the reference frame of the satellite, the light from these flashes will reach the astronaut at the same time. An experimenter on the earth will also see these flashes reaching the astronaut at the same time, but will not agree that the flashes occurred simultaneously. We can use these observations to decide which flash occurred first to an earthbound experimenter.

In part b, because of time dilation the clock on the moving satellite runs slow compared to one on the earth, so this clock will *lose* time.

Continued

SOLVE a. **FIGURE 27.26** shows the satellite (here represented by a rod) as seen by an experimenter on earth. When the light waves from the two flashes meet at the satellite's center, the one from the rear of the satellite will have traveled farther than the one from the front. This is because the flash coming from the rear needs to "catch up" with the center of the moving satellite. Since both waves travel at speed c, but the wave from the rear travels farther, the rear light must have flashed *earlier* according to an earthbound experimenter.

FIGURE 27.26 The satellite as seen by an observer on earth.

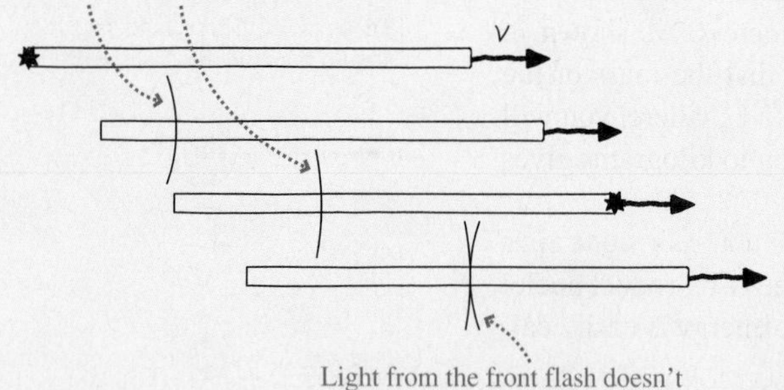

Because the satellite is moving, light from the rear flash — which is playing catch-up — has to travel farther to reach the satellite's center. This means it has to flash earlier.
v
Light from the front flash doesn't need to go as far. It flashes later.

b. The clock on the satellite measures proper time $\Delta\tau$ because this clock is at rest in the reference frame of the satellite. We want to know how much time the satellite's clock measures during an interval $\Delta t = 24\text{ h} = 86{,}400\text{ s}$ measured on the earth. We can rearrange Equation 27.6 as

$$\Delta\tau = \Delta t\sqrt{1-\beta^2}$$

Even for a fast-moving satellite, $\beta = v/c$ is so small that the term $\sqrt{1-\beta^2}$ will be exactly 1 on most calculators. We must therefore use the binomial expansion, Equation 27.14, to write

$$\Delta\tau = \Delta t\sqrt{1-\beta^2} \approx \Delta t\left(1 - \frac{1}{2}\frac{v^2}{c^2}\right) = \Delta t - \frac{1}{2}\frac{v^2}{c^2}\Delta t$$

The *difference* between the satellite and earth clocks is

$$\Delta t - \Delta\tau = \frac{1}{2}\frac{v^2}{c^2}\Delta t = \frac{1}{2}\left(\frac{3900\text{ m/s}}{3.0\times10^8\text{ m/s}}\right)^2(86{,}400\text{ s}) = 7.3\ \mu\text{s}$$

Because this result is positive, $\Delta\tau$ is less than Δt: The moving clock *loses* 7.3 μs per day compared to a clock on earth.

c. The radio signal from the satellite travels at speed c. In 7.3 μs, the signal travels a distance

$$\Delta x = c\Delta t = (3.0\times10^8\text{ m/s})(7.3\times10^{-6}\text{ s}) = 2200\text{ m}$$

If the satellite clocks were not corrected for this relativistic effect, all GPS receivers on earth would miscalculate their positions by 2.2 km after just one day.

ASSESS Even for a very fast-moving object like a satellite, the corrections due to relativity are small. But these small corrections can have large effects on high-precision measurements. Interestingly, the relativistic correction is implemented *before* the satellites are launched by setting the clocks to run slightly fast, by a factor of 1.000000000447. Once in orbit, the clocks slow down to match an earthbound clock exactly.

SUMMARY

The goal of Chapter 27 has been to understand how Einstein's theory of relativity changes our concepts of space and time.

GENERAL PRINCIPLES

Principle of Relativity

All the laws of physics are the same in all inertial reference frames.

- The speed of light c is the same in all inertial reference frames.
- No particle or causal influence can travel at a speed greater than c.

IMPORTANT CONCEPTS

Time

Time measurements depend on the motion of the experimenter relative to the events.

Proper time $\Delta\tau$ is the time interval between two events measured in a reference frame in which the events occur at the same position. The time interval Δt between the same two events, in a frame moving with relative velocity v, is

$$\Delta t = \Delta\tau/\sqrt{1-\beta^2} \geq \Delta\tau$$

where $\beta = v/c$. This is called **time dilation.**

Space

Spatial measurements depend on the motion of the experimenter relative to the events.

Proper length ℓ is the length of an object measured in a reference frame in which the object is at rest. The length L in a frame in which the object moves with velocity v is

$$L = \sqrt{1-\beta^2}\,\ell \leq \ell$$

This is called **length contraction.**

Momentum

The law of conservation of momentum is valid in all inertial reference frames if the momentum of a particle with velocity u is $p = \gamma m u$, where

$$\gamma = 1/\sqrt{1-u^2/c^2}$$

The momentum approaches ∞ as $u \to c$.

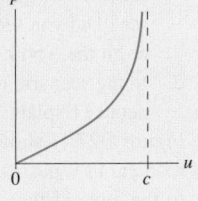

Energy

The **total energy** of a particle is $E = \gamma mc^2$. This can be written as

$$E = \underbrace{mc^2}_{\text{Rest energy } E_0} + \underbrace{(\gamma-1)mc^2}_{\text{Kinetic energy } K}$$

K approaches ∞ as $u \to c$.

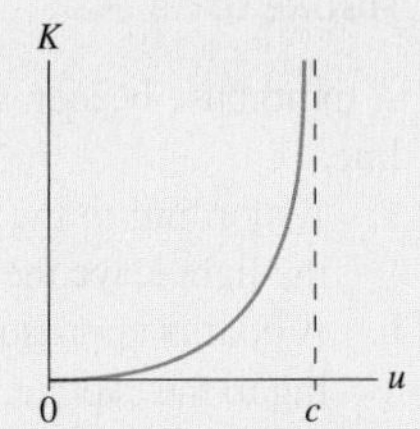

The total energy of an isolated system is conserved.

Simultaneity

Events that are simultaneous in reference frame S are not simultaneous in frame S′ moving relative to S.

Mass-energy equivalence

Mass m can be transformed into energy $E = mc^2$.

Energy can be transformed into mass $m = E/c^2$.

APPLICATIONS

An **event** happens at a specific place in space and time. Spacetime coordinates are (x, t) in frame S and (x', t') in frame S′.

A **reference frame** is a coordinate system with meter sticks and clocks for measuring events. Experimenters at rest relative to each other share the same reference frame.

If an object has velocity u in frame S and u' in frame S′, the two velocities are related by the **Lorentz velocity transformation:**

$$u' = \frac{u-v}{1-uv/c^2} \qquad u = \frac{u'+v}{1+u'v/c^2}$$

where v is the relative velocity between the two frames.

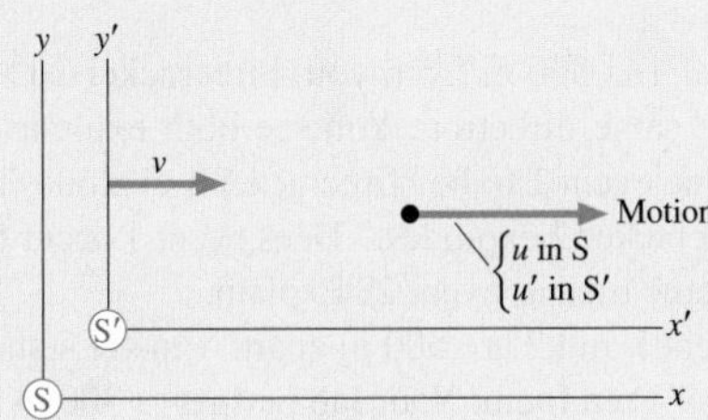

For homework assigned on MasteringPhysics, go to www.masteringphysics.com

Problems labeled INT integrate significant material from earlier chapters; BIO are of biological or medical interest.

Problem difficulty is labeled as | (straightforward) to ||||| (challenging).

QUESTIONS

Conceptual Questions

1. You are in an airplane cruising smoothly at 600 mph. What experiment, if any, could you do that would demonstrate that you are moving, while those on the ground are at rest?
2. Frame S′ moves relative to frame S as shown in Figure Q27.2.
   a. A ball is at rest in frame S′. What are the speed and direction of the ball in frame S?
   b. A ball is at rest in frame S. What are the speed and direction of the ball in frame S′?

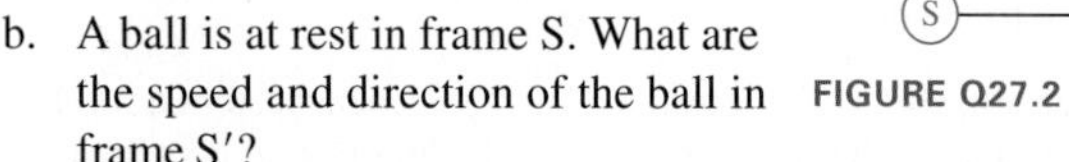

FIGURE Q27.2

3. a. Two balls move as shown in Figure Q27.3. What are the speed and direction of each ball in a reference frame that moves with ball 1?
   b. What are the speed and direction of each ball in a reference frame that moves with ball 2?

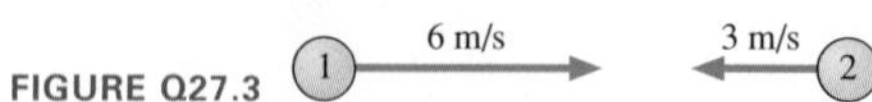

FIGURE Q27.3

4. A lighthouse beacon alerts ships to the danger of a rocky coastline.
   a. According to the lighthouse keeper, with what speed does the light leave the lighthouse?
   b. A boat is approaching the coastline at speed $0.5c$. According to the captain, with what speed is the light from the beacon approaching her boat?
5. As a rocket passes the earth at $0.75c$, it fires a laser perpendicular to its direction of travel as shown in Figure Q27.5.
   a. What is the speed of the laser beam relative to the rocket?
   b. What is the speed of the laser beam relative to the earth?

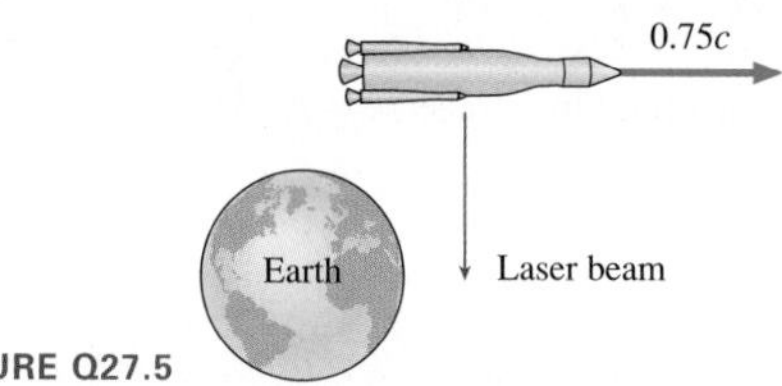

FIGURE Q27.5

6. At the instant that a clock standing next to you reads $t = 1.0\ \mu s$, you look at a second clock, 300 m away, and see that it reads $t = 0\ \mu s$. Are the two clocks synchronized? If not, which one is ahead?
7. Firecracker 1 is 300 m from you. Firecracker 2 is 600 m from you in the same direction. You see both explode at the same time. Define event 1 to be "firecracker 1 explodes" and event 2 to be "firecracker 2 explodes." Does event 1 occur before, after, or at the same time as event 2? Explain.
8. Firecrackers 1 and 2 are 600 m apart. You are standing exactly halfway between them. Your lab partner is 300 m on the other side of firecracker 1. You see two flashes of light, from the two explosions, at exactly the same instant of time. Define event 1 to be "firecracker 1 explodes" and event 2 to be "firecracker 2 explodes." According to your lab partner, based on measurements he or she makes, does event 1 occur before, after, or at the same time as event 2? Explain.
9. Your clocks and calendars are synchronized with the clocks and calendars in a star system exactly 10 ly from earth that is at rest relative to the earth. You receive a TV transmission from the star system that shows a date and time display. The date it shows is June 17, 2050. When you glance over at your own wall calendar, what date does it show?
10. Two trees are 600 m apart. You are standing exactly halfway between them and your lab partner is at the base of tree 1. Lightning strikes both trees.
    a. Your lab partner, based on measurements he makes, determines that the two lightning strikes were simultaneous. What did you see? Did you see the lightning hit tree 1 first, hit tree 2 first, or hit them both at the same instant of time? Explain.
    b. Lightning strikes again. This time your lab partner sees both flashes of light at the same instant of time. What did you see? Did you see the lightning hit tree 1 first, hit tree 2 first, or hit them both at the same instant of time? Explain.
    c. In the scenario of part b, were the lightning strikes simultaneous? Explain.
11. Figure Q27.11 shows Peggy standing at the center of her railroad car as it passes Ryan on the ground. Firecrackers attached to the ends of the car explode. A short time later, the flashes from the two explosions arrive at Peggy at the same time.
    a. Were the explosions simultaneous in Peggy's reference frame? If not, which exploded first? Explain.
    b. Were the explosions simultaneous in Ryan's reference frame. If not, which exploded first? Explain.

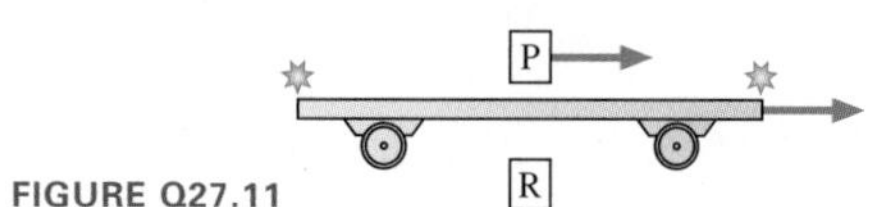

FIGURE Q27.11

12. In Figure Q27.12, clocks C_1 and C_2 in frame S are synchronized. Clock C′ moves at speed v relative to frame S. Clocks C′ and C_1 read exactly the same as C′ goes past. As C′ passes C_2, is the time shown on C′ earlier, later, or the same as the time shown on C_2? Explain.

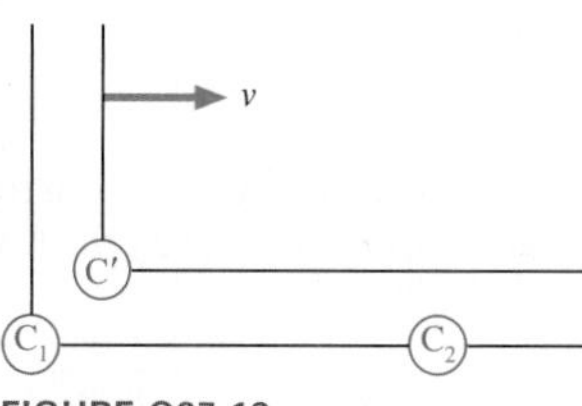

FIGURE Q27.12

13. A meter stick passes you at a speed of $0.5c$. Explain clearly how you would measure the length of this fast-moving object.

14. You're passing a car on the highway. You want to know how much time is required to completely pass the car, from no overlap between the cars to no overlap between the cars. Call your car A, the car you are passing B.
   a. Specify two events that can be given spacetime coordinates. In describing the events, refer to cars A and B and to their front bumpers and rear bumpers.
   b. In either reference frame, is there *one* clock that is present at both events?
   c. Who, if anyone, measures the proper time between the events?
15. Your friend flies from Los Angeles to New York. He determines the distance using the tried-and-true $d = vt$. You and your assistants on the ground also measure the distance using meter sticks and surveying equipment.
   a. Who, if anyone, measures the proper length?
   b. Who, if anyone, measures the shorter distance?
16. A 100-m-long train is heading for an 80-m-long tunnel. If the train moves sufficiently fast, is it possible, according to experimenters on the ground, for the entire train to be inside the tunnel at one instant of time? Explain.
17. Dan picks up a 15-m-long pole and begins running very fast, holding the pole horizontally and pointing in the direction he's running. He heads toward a barn that is 12 m long and has open doors at each end. Dan runs so fast that, to Farmer Brown standing by his barn, the ladder is only 5 m long. As soon as the pole is completely inside the barn, Farmer Brown closes both doors so that Dan and the pole are inside with both doors shut. Then, just before Dan reaches the far door, Farmer Brown opens both doors and Dan emerges, still moving at high speed. According to Dan, however, the barn is contracted to only 4 m and the pole has its full 15 m length. Farmer Brown sees the pole completely inside the barn with both doors closed. What does Dan see happening?
18. The rocket speeds shown in Figure Q27.18 are relative to the earth. Is the speed of A relative to B greater than, less than, or equal to 0.8c?

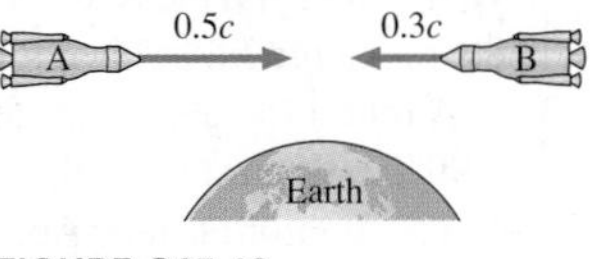

FIGURE Q27.18

19. Can a particle of mass m have total energy less than mc^2? Explain.
20. In your chemistry classes, you have probably learned that, in a chemical reaction, the mass of the products is equal to the mass of the reactants. That is, the mass of the substances produced in a chemical reaction is equal to the mass of the substances consumed in the reaction. Is this absolutely true, or is there actually a small difference? Explain.

Multiple-Choice Questions

21. | Lee and Leigh are twins. At their first birthday party, Lee is placed on a spaceship that travels away from the earth and back at a steady 0.866c. The spaceship eventually returns, landing in the swimming pool at Leigh's eleventh birthday party. When Lee emerges from the ship, it is discovered that
   A. He is still only 1 year old. B. He is 6 years old.
   C. He is also 11 years old. D. He is 21 years old.
22. || A space cowboy wants to eject from his spacecraft 100,000 km after passing a space buoy, as seen by spectators at rest with respect to the buoy. To do this, the cowboy sets a timer on his craft that will start as he passes the buoy. He plans to cruise by the buoy at 0.300c. How much time should he allow between passing the buoy and ejecting?
   A. 1.01 s B. 1.06 s C. 1.11 s
   D. 1.33 s E. 1.58 s
23. || Event 1 occurs at (0 m, 0 s) in reference frame S. The following events occur, also measured in reference frame S. Which of them could have been caused by event 1?
   A. (1 m, 2 ns) B. (−1 m, −2 ns) C. (−1 m, 2 ns)
   D. All of them
   E. None of them
24. | Energy in the sun is produced by the fusion of four protons into a helium nucleus. The process involves several steps, but the net reaction is simply 4p $\rightarrow$ ^{4}He + energy. Given this, you can say that
   A. One helium atom has more mass than four hydrogen atoms.
   B. One helium atom has less mass than four hydrogen atoms.
   C. One helium atom has the same mass as four hydrogen atoms.
25. || A particle moving at speed 0.40c has momentum p_0. The speed of the particle is increased to 0.80c. Its momentum is now
   A. Less than $2p_0$ B. Exactly $2p_0$ C. Greater than $2p_0$
26. | A particle moving at speed 0.40c has kinetic energy K_0. The speed of the particle is increased to 0.80c. The kinetic energy is now
   A. Less than $4K_0$ B. Exactly $4K_0$ C. Greater than $4K_0$

PROBLEMS

Section 27.2 Galilean Relativity

1. ||| A sprinter crosses the finish line of a race. The roar of the crowd in front approaches her at a speed of 360 m/s. The roar from the crowd behind her approaches at 330 m/s. What are the speed of sound and the speed of the sprinter?
2. | A baseball pitcher can throw a ball with a speed of 40 m/s. He is in the back of a pickup truck that is driving away from you. He throws the ball in your direction, and it floats toward you at a lazy 10 m/s. What is the speed of the truck?
3. | A boy on a skateboard coasts along at 5 m/s. He has a ball that he can throw at a speed of 10 m/s. What is the ball's speed relative to the ground if he throws the ball (a) forward or (b) backward?
4. || A boat takes 3.0 hours to travel 30 km down a river, then 5.0 hours to return. How fast is the river flowing?
5. || When the moving sidewalk at the airport is broken, as it often seems to be, it takes you 50 s to walk from your gate to baggage claim. When it is working and you stand on the moving sidewalk the entire way, without walking, it takes 75 s to travel the same distance. How long will it take you to travel from the gate to baggage claim if you walk while riding on the moving sidewalk?

6. || An assembly line has a staple gun that rolls to the left at 1.0 m/s while parts to be stapled roll past it to the right at 3.0 m/s. The staple gun fires 10 staples per second. How far apart are the staples in the finished part?

Section 27.3 Einstein's Principle of Relativity

7. || An out-of-control alien spacecraft is diving into a star at a speed of 1.0×10^8 m/s. At what speed, relative to the spacecraft, is the starlight approaching?
8. || A starship blasts past the earth at 2.0×10^8 m/s. Just after passing the earth, the starship fires a laser beam out its back. With what speed does the laser beam approach the earth?
9. | A positron moving in the positive x-direction at 2.0×10^8 m/s collides with an electron at rest. The positron and electron annihilate, producing two gamma-ray photons. Photon 1 travels in the positive x-direction and photon 2 travels in the negative x-direction. What is the speed of each photon?

Section 27.4 Events and Measurements

Section 27.5 The Relativity of Simultaneity

10. || Your job is to synchronize the clocks in a reference frame. You are going to do so by flashing a light at the origin at $t = 0$ s. To what time should the clock at $(x, y, z) = (30\text{ m}, 40\text{ m}, 0\text{ m})$ be preset?
11. || Bjorn is standing at $x = 600$ m. Firecracker 1 explodes at the origin and firecracker 2 explodes at $x = 900$ m. The flashes from both explosions reach Bjorn's eye at $t = 3.0\ \mu$s. At what time did each firecracker explode?
12. |||| Bianca is standing at $x = 600$ m. Firecracker 1, at the origin, and firecracker 2, at $x = 900$ m, explode simultaneously. The flash from firecracker 1 reaches Bianca's eye at $t = 3.0\ \mu$s. At what time does she see the flash from firecracker 2?
13. || You are standing at $x = 9.0$ km. Lightning bolt 1 strikes at $x = 0$ km and lightning bolt 2 strikes at $x = 12.0$ km. Both flashes reach your eye at the same time. Your assistant is standing at $x = 3.0$ km. Does your assistant see the flashes at the same time? If not, which does she see first and what is the time difference between the two?

14. || A light flashes at position $x = 0$ m. One microsecond later, a light flashes at position $x = 1000$ m. In a second reference frame, moving along the x-axis at speed v, the two flashes are simultaneous. Is this second frame moving to the right or to the left relative to the original frame?
15. ||| Jose is looking to the east. Lightning bolt 1 strikes a tree 300 m from him. Lightning bolt 2 strikes a barn 900 m from him in the same direction. Jose sees the tree strike $1.0\ \mu$s before he sees the barn strike. According to Jose, were the lightning strikes simultaneous? If not, which occurred first and what was the time difference between the two?
16. || You are flying your personal rocketcraft at $0.90c$ from Star A toward Star B. The distance between the stars, in the stars' reference frame, is 1.0 ly. Both stars happen to explode simultaneously in your reference frame at the instant you are exactly halfway between them. Do you see the flashes simultaneously? If not, which do you see first and what is the time difference between the two?

Section 27.6 Time Dilation

17. ||| A cosmic ray travels 60 km through the earth's atmosphere in 400 μs, as measured by experimenters on the ground. How long does the journey take according to the cosmic ray?
18. || At what speed relative to a laboratory does a clock tick at half the rate of an identical clock at rest in the laboratory? Give your answer as a fraction of c.
19. || An astronaut travels to a star system 4.5 ly away at a speed of $0.90c$. Assume that the time needed to accelerate and decelerate is negligible.
   a. How long does the journey take according to Mission Control on earth?
   b. How long does the journey take according to the astronaut?
   c. How much time elapses between the launch and the arrival of the first radio message from the astronaut saying that she has arrived?
20. ||| A starship voyages to a distant planet 10 ly away. The explorers stay 1 yr, return at the same speed, and arrive back on earth 26 yr after they left. Assume that the time needed to accelerate and decelerate is negligible.
   a. What is the speed of the starship?
   b. How much time has elapsed on the astronauts' chronometers?

Section 27.7 Length Contraction

21. | At what speed, as a fraction of c, will a moving rod have a length 60% that of an identical rod at rest?
22. | Jill claims that her new rocket is 100 m long. As she flies past your house, you measure the rocket's length and find that it is only 80 m. Should Jill be cited for exceeding the $0.5c$ speed limit?
23. || A muon travels 60 km through the atmosphere at a speed of $0.9997c$. According to the muon, how thick is the atmosphere?
24. || The Stanford Linear Accelerator (SLAC) accelerates electrons to $v = 0.99999997c$ in a 3.2-km-long tube. If they travel the length of the tube at full speed (they don't, because they are accelerating), how long is the tube in the electrons' reference frame?
25. | Our Milky Way galaxy is 100,000 ly in diameter. A spaceship crossing the galaxy measures the galaxy's diameter to be a mere 1.0 ly.
   a. What is the speed of the spaceship relative to the galaxy?
   b. How long is the crossing time as measured in the galaxy's reference frame?
26. ||| An optical interferometer can detect a displacement of about 50 nm. At what speed would a meter stick "shrink" by 50 nm? **Hint:** Use the binomial approximation.

Section 27.8 Velocities of Objects in Special Relativity

27. |||| A rocket cruising past earth at $0.800c$ shoots a bullet out the back door, opposite the rocket's motion, at $0.900c$ relative to the rocket. What is the bullet's speed relative to the earth?
28. ||| A base on Planet X fires a missile toward an oncoming space fighter. The missile's speed according to the base is $0.85c$. The space fighter measures the missile's speed as $0.96c$. How fast is the space fighter traveling relative to Planet X?

29. ||| A solar flare blowing out from the sun at $0.90c$ is overtaking a rocket as it flies away from the sun at $0.80c$. According to the crew on board, with what speed is the flare gaining on the rocket?

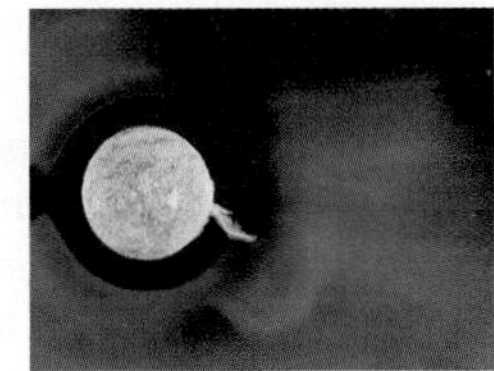

Section 27.9 Relativistic Momentum

30. | A proton is accelerated to $0.999c$.
    a. What is the proton's momentum?
    b. By what factor does the proton's momentum exceed its Newtonian momentum?
31. ||| A 1.0 g particle has momentum 400,000 kg · m/s. What is the particle's speed?
32. | At what speed is a particle's momentum twice its Newtonian value?
33. || What is the speed of a particle whose momentum is mc?

Section 27.10 Relativistic Energy

34. || What are the kinetic energy, the rest energy, and the total energy of a 1.0 g particle with a speed of $0.80c$?
35. | A quarter-pound hamburger with all the fixings has a mass of 200 g. The food energy of the hamburger (480 food calories) is 2 MJ.
    a. What is the energy equivalent of the mass of the hamburger?
    b. By what factor does the energy equivalent exceed the food energy?
36. || How fast must an electron move so that its total energy is 10% more than its rest mass energy?
37. | At what speed is a particle's kinetic energy twice its rest energy?

General Problems

38. ||| A firecracker explodes at $x = 0$ m, $t = 0$ μs. A second explodes at $x = 300$ m, $t = 2.0$ μs. What is the proper time between these events?
39. ||| You're standing on an asteroid when you see your best friend rocketing by in her new spaceship. As she goes by, you notice that the front and rear of her ship coincide exactly with the 400-m-diameter of another nearby asteroid that is stationary with respect to you. However, you happen to know that your friend's spaceship measured 500 m long in the showroom. What is your friend's speed relative to you?
40. || A subatomic particle moves through the laboratory at $0.90c$. Laboratory experimenters measure its lifetime, from creation to annihilation, to be 2.3 ps (1 ps = 1 picosecond = 10^{-12} s). According to the particle, how long did it live?
41. || You and Maria each own identical spaceships. As you fly past Maria, you measure her ship to be 90 m long and your own ship to be 100 m long.
    a. How long does Maria measure your ship to be?
    b. How fast is Maria moving relative to you?
42. ||| A very fast-moving train car passes you, moving to the right at $0.50c$. You measure its length to be 12 m. Your friend David flies past you to the right at a speed relative to you of $0.80c$. How long does David measure the train car to be?
43. || Two events in reference frame S occur 10 μs apart at the same point in space. Frame S′ travels at speed $v = 0.90c$ relative to frame S.
    a. What is the time interval between the events in reference frame S′?
    b. What is the distance between the events in frame S′?
44. ||| A 30-m-long rocket train car is traveling from Los Angeles to New York at $0.50c$ when a light at the center of the car flashes. When the light reaches the front of the car, it immediately rings a bell. Light reaching the back of the car immediately sounds a siren.
    a. Are the bell and siren simultaneous events for a passenger seated in the car? If not, which occurs first and by how much time?
    b. Are the bell and siren simultaneous events for a bicyclist waiting to cross the tracks? If not, which occurs first and by how much time?
45. ||| INT Because of the earth's rotation, a person living on top of a mountain moves at a faster speed than someone at sea level. The mountain dweller's clocks thus run slowly compared to those at sea level. If the average life span of a hermit is 80 years, on average how much longer would a hermit dwelling on the top of a 3000-m-high mountain live compared to a sea-level hermit?
46. ||| Two clocks are synchronized. One is placed in a race car that drives around a 2.0-km-diameter track at 100 m/s for 24 h. Afterward, by how much do the two clocks differ?
    Hint: Use the binomial approximation.
47. ||| You fly 5000 km across the United States on an airliner at 250 m/s. You return two days later at the same speed.
    a. Have you aged more or less than your friends at home?
    b. By how much?
    Hint: Use the binomial approximation.
48. || INT A cube has a density of 2000 kg/m^3 while at rest in the laboratory. What is the cube's density as measured by an experimenter in the laboratory as the cube moves through the laboratory at 90% of the speed of light in a direction perpendicular to one of its faces?
49. ||| In Section 27.6 we explained that muons can reach the ground because of time dilation. But how do things appear in the muon's reference frame, where the muon's half-life is only 1.5 μs? How can a muon travel the 60 km to reach the earth's surface before decaying? Resolve this apparent paradox. Be as quantitative as you can in your answer.
50. || A spaceship flies past an experimenter who measures its length to be one-half the length he had measured when the spaceship was at rest. An astronaut aboard the spaceship notes that his clock ticks at 1-second intervals. What is the time between ticks as measured by the experimenter?
51. ||| Marissa's spaceship approaches Joseph's at a speed of $0.99c$. As Marissa passes Joseph, they synchronize their clocks to both read $t = 0$ s. When Marissa's clock reads 100 s, she sends a light signal back to Joseph. According to his clock, when does he receive this signal?
52. ||| At a speed of $0.90c$, a spaceship travels to a star that is 9.0 ly distant.
    a. According to a scientist on earth, how long does the trip take?
    b. According to a scientist on the spaceship, how long does the trip take?
    c. According to the scientist on the spaceship, what is the distance traveled during the trip?
    d. At what speed do observers on the spaceship see the star approaching them?

53. || In an attempt to reduce the extraordinarily long travel times INT for voyaging to distant stars, some people have suggested traveling at close to the speed of light. Suppose you wish to visit the red giant star Betelgeuse, which is 430 ly away, and that you want your 20,000 kg rocket to move so fast that you age only 20 years during the round trip.
   a. How fast must the rocket travel relative to earth?
   b. How much energy is needed to accelerate the rocket to this speed?
   c. How many times larger is this energy than the total energy used by the United States in the year 2000, which was roughly 1.0×10^{20} J?

54. |||| A rocket traveling at $0.500c$ sets out for the nearest star, Alpha Centauri, which is 4.25 ly away from earth. It will return to earth immediately after reaching Alpha Centauri. What distance will the rocket travel and how long will the journey last according to (a) stay-at-home earthlings and (b) the rocket crew? (c) Which answers are the correct ones, those in part a or those in part b?

55. || A distant quasar is found to be moving away from the earth at $0.80c$. A galaxy closer to the earth and along the same line of sight is moving away from us at $0.20c$. What is the recessional speed of the quasar as measured by astronomers in the other galaxy?

56. || Two rockets approach each other. Each is traveling at $0.75c$ in the earth's reference frame. What is the speed of one rocket relative to the other?

57. |||| A military jet traveling at 1500 m/s has engine trouble and the pilot must bail out. Her ejection seat shoots her forward at 300 m/s relative to the jet. According to the Lorentz velocity transformation, by how much is her velocity relative to the ground less than the 1800 m/s predicted by Galilean relativity? **Hint:** Use the binomial approximation.

58. ||| James, Daniella, and Tara all possess identical clocks. As Daniella passes James in her rocket, James observes that her clock runs at 80% the rate of his clock. As Tara passes in her rocket, in the same direction as Daniella, James observes that her clock runs at 70% the rate of his clock. At what rate, relative to her clock, does Daniella observe Tara's clock to run?

59. ||| Two rockets approach earth from opposite directions at equal speeds relative to the earth. A scientist on earth notes that it takes 1 h and 10 min, according to her watch, for the clocks on the rockets to advance by 1 h.
   a. How fast are the rockets moving with respect to the earth?
   b. According to an astronaut on one rocket, how long does it take the clock on the other rocket to advance by 1 h?

60. ||| Two rockets, A and B, approach the earth from opposite directions at speed $0.800c$. The length of each rocket measured in its rest frame is 100 m. What is the length of rocket A as measured by the crew of rocket B?

61. ||| The highest-energy cosmic ray ever detected had an energy of about 3.0×10^{20} eV. Assume that this cosmic ray was a proton.
   a. What was the proton's speed as a fraction of c?
   b. If this proton started at the same time and place as a photon traveling at the speed of light, how far behind the photon would it be after traveling for 1 ly?

62. ||| What is the speed of an electron after being accelerated from INT rest through a 20×10^6 V potential difference?

63. ||| What is the speed of a proton after being accelerated from INT rest through a 50×10^6 V potential difference?

64. || The half-life of a muon at rest is 1.5 μs. Muons that have been accelerated to a very high speed and are then held in a circular storage ring have a half-life of 7.5 μs.
   a. What is the speed of the muons in the storage ring?
   b. What is the total energy of a muon in the storage ring? The mass of a muon is 207 times the mass of an electron.

65. || What is the momentum of a particle with speed $0.95c$ and total energy 2.0×10^{-10} J?

66. ||| What is the momentum of a particle whose total energy is four times its rest energy? Give your answer as a multiple of mc.

67. || What is the total energy, in MeV, of
INT a. A proton traveling at 99.0% of the speed of light?
   b. An electron traveling at 99.0% of the speed of light?

68. | What is the velocity, as a fraction of c, of
INT a. A proton with 500 GeV total energy?
   b. An electron with 2.0 GeV total energy?

69. |||| At what speed is the kinetic energy of a particle twice its Newtonian value?

70. | What is the speed of an electron whose total energy equals the rest energy of a proton?

71. || The factor γ appears in many relativistic expressions. A value $\gamma = 1.01$ implies that relativity changes the Newtonian values by approximately 1% and that relativistic effects can no longer be ignored. At what kinetic energy, in MeV, is $\gamma = 1.01$ for (a) an electron, and (b) a proton?

72. || The chemical energy of gasoline is 46 MJ/kg. If gasoline's mass could be completely converted into energy, what mass of gasoline would be needed to equal the chemical energy content of 1.0 kg of gasoline?

73. | The sun radiates energy at the rate 3.8×10^{26} W. The source INT of this energy is fusion, a nuclear reaction in which mass is transformed into energy. The mass of the sun is 2.0×10^{30} kg.
   a. How much mass does the sun lose each year?
   b. What percentage is this of the sun's total mass?
   c. Estimate the lifetime of the sun.

74. || The radioactive element radium (Ra) decays by a process known as *alpha decay,* in which the nucleus emits a helium nucleus. (These high-speed helium nuclei were named alpha particles when radioactivity was first discovered, long before the identity of the particles was established.) The reaction is $^{226}\text{Ra} \rightarrow {}^{222}\text{Rn} + {}^{4}\text{He}$, where Rn is the element radon. The accurately measured atomic masses of the three atoms are 226.025, 222.017, and 4.003. How much energy is released in each decay? (The energy released in radioactive decay is what makes nuclear waste "hot.")

75. | The nuclear reaction that powers the sun is the fusion of four protons into a helium nucleus. The process involves several steps, but the net reaction is simply $4\text{p} \rightarrow {}^{4}\text{He} + \text{energy}$. The mass of a helium nucleus is known to be 6.64×10^{-27} kg.
   a. How much energy is released in each fusion?
   b. What fraction of the initial rest mass energy is this energy?

76. || When antimatter (which we'll learn more about in Chapter 30) INT interacts with an equal mass of ordinary matter, both matter and antimatter are converted completely into energy in the form of photons. In an antimatter-fueled spaceship, a staple of science fiction, the newly created photons are shot from the back of the ship, propelling it forward. Suppose such a ship has a mass of 2.0×10^6 kg, and carries a mass of fuel equal to 1% of its mass, or 1.0×10^4 kg of matter and an equal mass of antimatter.

a. What is the final speed of the ship, assuming it starts from rest, if all energy released in the matter-antimatter annihilation is transformed into the kinetic energy of the ship?

b. Not only do photons have energy, as you learned in Chapter 25, they also have momentum. Explain why, when energy and momentum conservation are both considered, the final speed of the ship will be less than you calculated in part a.

Passage Problems

Pion Therapy BIO

Subatomic particles called *pions* are created when protons, accelerated to speeds very near c in a particle accelerator, smash into the nucleus of a target atom. Charged pions are unstable particles that decay into muons with a half-life of 1.8×10^{-8} s. Pions have been investigated for use in cancer treatment because they pass through tissue doing minimal damage until they decay, releasing significant energy at that point. The speed of the pions can be adjusted so that the most likely place for the decay is in a tumor.

Suppose pions are created in an accelerator, then directed into a medical bay 30 m away. The pions travel at the very high speed of $0.99995c$. Without time dilation, half of the pions would have decayed after traveling only 5.4 m, not far enough to make it to the medical bay. Time dilation allows them to survive long enough to reach the medical bay, enter tissue, slow down, and then decay where they are needed, in a tumor.

77. | What is the half-life of a pion in the reference frame of the patient undergoing pion therapy?
A. 1.8×10^{-10} s B. 1.8×10^{-8} s
C. 1.8×10^{-7} s D. 1.8×10^{-6} s

78. | According to the pion, what is the distance it travels from the accelerator to the medical bay?
A. 0.30 m B. 3.0 m C. 30 m D. 3000 m

79. | The proton collision that creates the pion also creates a gamma-ray photon traveling in the same direction as the pion. The photon will get to the medical bay first because it is moving faster. What is the speed of the photon in the pion's reference frame?
A. $0.00005c$ B. $0.5c$
C. $0.99995c$ D. c

80. | If the pion slows down to $0.99990c$, about what percentage of its kinetic energy is lost?
A. 0.03% B. 0.3% C. 3% D. 30%

STOP TO THINK ANSWERS

Stop to Think 27.1: A, C, and F. These move at constant velocity, or very nearly so. The others are accelerating.

Stop to Think 27.2: A. $u' = u - v = -10$ m/s $- 6$ m/s $= -16$ m/s. The *speed* is 16 m/s.

Stop to Think 27.3: C. Even the light has a slight travel time. The event is the hammer hitting the nail, not your seeing the hammer hit the nail.

Stop to Think 27.4: At the same time. Mark is halfway between the tree and the pole, so the fact that he *sees* the lightning bolts at the same time means they *happened* at the same time. It's true that Nancy *sees* event 1 before event 2, but the events actually occurred before she sees them. Mark and Nancy share a reference frame, because they are at rest relative to each other, and all experimenters in a reference frame, after correcting for any signal delays, *agree* on the spacetime coordinates of an event.

Stop to Think 27.5: After. This is the same as the case of Peggy and Ryan. In Mark's reference frame, as in Ryan's, the events are simultaneous. Nancy *sees* event 1 first, but the time when an event is seen is not when the event actually happens. Because all experimenters in a reference frame agree on the spacetime coordinates of an event, Nancy's position in her reference frame cannot affect the order of the events. If Nancy had been passing Mark at the instant the lightning strikes occur in Mark's frame, then Nancy would be equivalent to Peggy. Event 2, like the firecracker at the front of Peggy's railroad car, occurs first in Nancy's reference frame.

Stop to Think 27.6: C. Nick measures proper time because Nick's clock is present at both the "nose passes Nick" event and the "tail passes Nick" event. Proper time is the smallest measured time interval between two events.

Stop to Think 27.7: C. The kinetic energy is $(\gamma - 1)mc^2 = 0.5mc^2$, which is less than the rest energy mc^2.

28 Quantum Physics

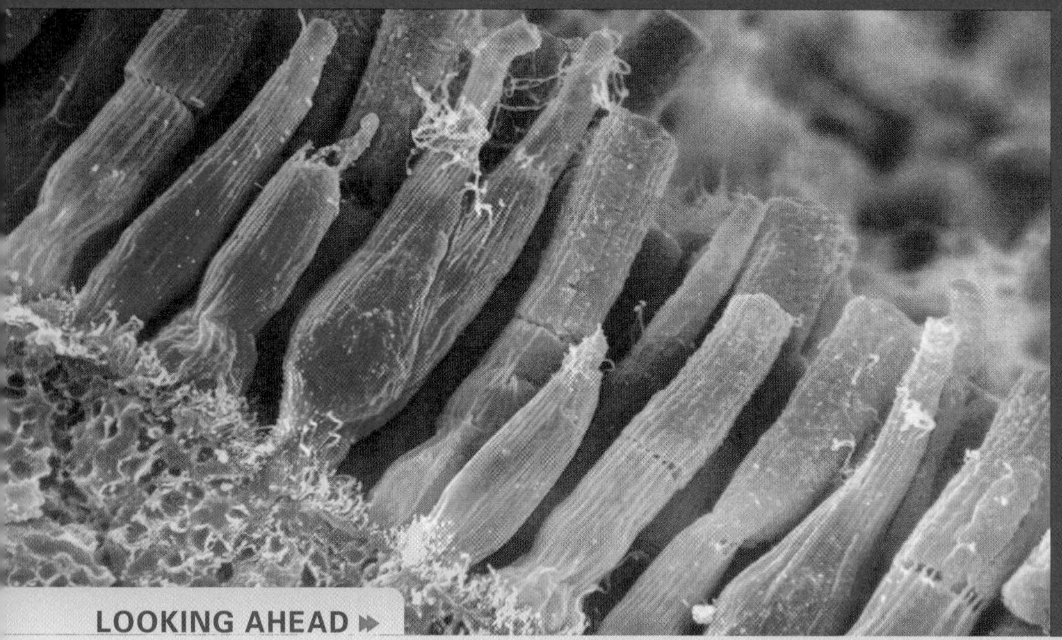

This false-color image showing individual rod cells (green) and cone cells (blue) on the human retina was made with an electron microscope. Such exquisite detail would not be possible in an image created with a light microscope. Why is greater resolution possible in an image made with a beam of electrons?

LOOKING AHEAD ▸▸

The goal of Chapter 28 is to understand the quantization of energy for light and matter.

Waves and Particles

Electromagnetic waves like light are waves of electric and magnetic fields. Light has the properties of a wave.

As we saw in Chapter 17, understanding diffraction and interference of light requires us to think of light as a wave.

But the picture is more complicated. Light and other electromagnetic waves also have a particle nature.

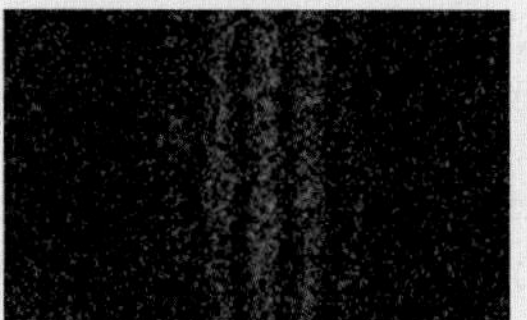

An interference pattern made with very low-intensity light clearly shows that the light hits the screen in "chunks." Sometimes, light looks like a particle.

As we've seen in earlier chapters, an electron has the properties of a particle.

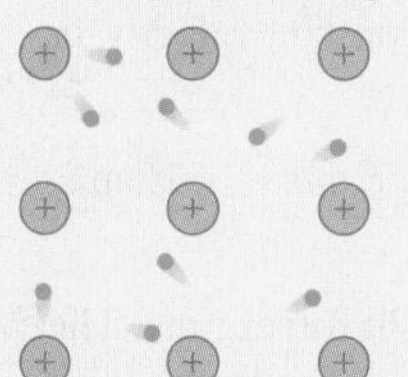

Our model of conduction in metals was based on the motion of particle-like electrons moving among fixed ions.

But the picture is more complicated. Electrons and other particles have a wave nature as well.

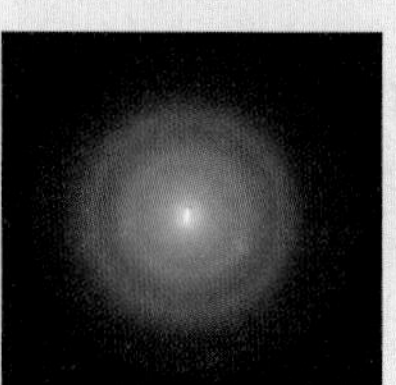

Shooting a beam of electrons through a crystal produces a diffraction pattern, something we'd expect for a wave.

Looking Back ◂◂
17.2–17.5 Diffraction and interference of light

Photons and the Photoelectric Effect

The particle nature of light requires us to think in terms of photons. As we've seen, the energy of a photon is proportional to the frequency of the light.

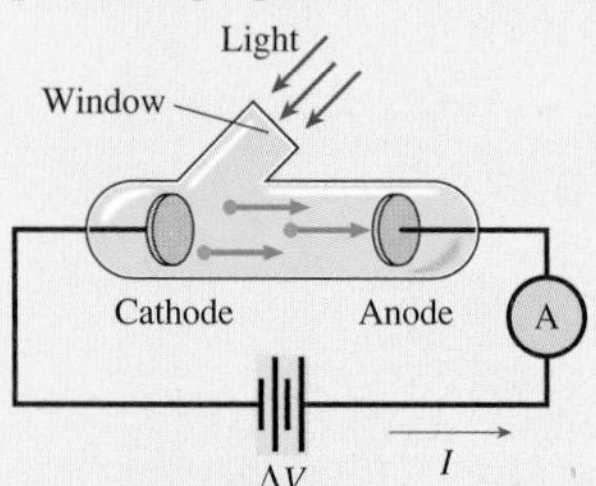

Light shining on a metal surface will eject electrons if the photons have sufficient energy.

Looking Back ◂◂
25.7 The photon model of EM waves
25.8 X rays

Matter Waves and Quantization

The wave nature of electrons has far-reaching consequences.

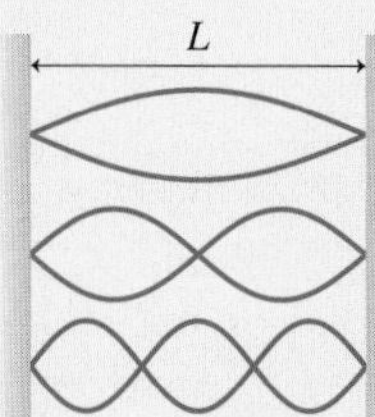

Just as a standing wave on a string stretched between fixed ends has only certain allowed modes ...

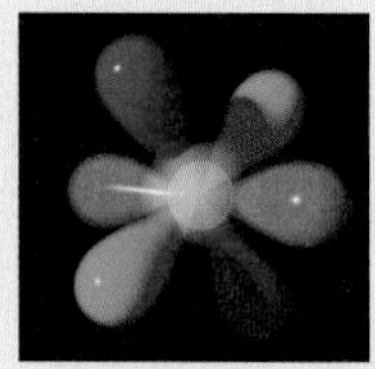

... an electron's wave nature means that electrons in atoms exist in only certain standing-wave modes.

Energy Levels

Atomic particles can have only certain allowed energies; their energy is **quantized.**

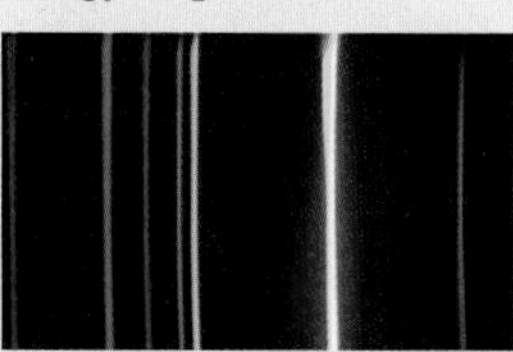

Electrons emit photons as they make quantum jumps between allowed energy levels. Only certain photon energies are possible, so we see a spectrum of discrete wavelengths.

Except for relativity, everything we have studied until this point in the book was known by 1900. Newtonian mechanics, thermodynamics, and the theory of electromagnetism form what we call *classical physics*. It is an impressive body of knowledge with immense explanatory power and a vast number of applications.

But a spate of discoveries right around 1900 showed that classical physics, though remarkable, was incomplete. Investigations into the nature of light and matter led to many astonishing discoveries that classical physics simply could not explain. Sometimes, as you will see, light refuses to act like a wave and seems more like a collection of particles. Other experiments found that electrons sometimes behave like waves. These discoveries eventually led to a radical new theory of light and matter called *quantum physics*.

This chapter will introduce you to this strange but wonderful quantum world. We will take a more historical approach than in previous chapters. As we introduce new ideas, we will describe in some detail the key experiments and the evolution of theories to explain them.

28.1 X Rays and X-Ray Diffraction

The rules of quantum physics apply at the scale of atoms and electrons. Experiments to elucidate the nature of the atom and the physics of atomic particles produced results that defied explanation with classical theories. Investigators saw things no one had ever seen before, phenomena that needed new principles and theories to explain them.

In 1895, the German physicist Wilhelm Röntgen was studying how electrons (called cathode rays at the time) travel through a vacuum. He sealed an electron-producing cathode and a metal target electrode into a vacuum tube. A high voltage pulled electrons from the cathode and accelerated them to very high speed before they struck the target electrode. One day, by chance, Röntgen left a sealed envelope containing film near the vacuum tube. He was later surprised to discover that the film had been exposed even though it had never been removed from the envelope. Some sort of penetrating radiation from the tube had exposed the film.

Röntgen had no idea what was coming from the tube, so he called them x rays, using the algebraic symbol x meaning "unknown." X rays were unlike anything, particle or wave, ever discovered before. Röntgen was not successful at reflecting the rays or at focusing them with a lens. He showed that they travel in straight lines, like particles, but they also pass right through most solid materials with very little absorption, something no known particle could do. The experiments of Röntgen and others led scientists to conclude that these mysterious rays were electromagnetic waves with very short wavelengths, as we learned in Chapter 25. These short-wavelength waves were produced in Röntgen's apparatus by the collision of fast electrons with a metal target. X rays are still produced this way, as shown in the illustration of the operation of a modern x-ray tube in FIGURE 28.1.

FIGURE 28.1 The operation of a modern x-ray tube.

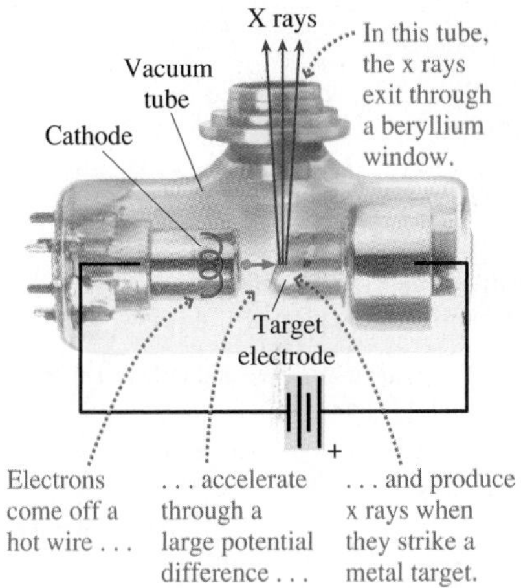

In this tube, the x rays exit through a beryllium window.

Electrons come off a hot wire . . .

. . . and produce x rays when they strike a metal target.

X-Ray Images

X rays are penetrating, and Röntgen immediately realized that x rays could be used to create an image of the interior of the body. One of Röntgen's first images showed the bones in his wife's hand, dramatically demonstrating the medical potential of these newly discovered rays. Substances with high atomic numbers, such as lead or the minerals in bone, are effective at stopping them; materials with low atomic numbers, such as the beryllium window of the x-ray tube in Figure 28.1 or the water and organic compounds of soft tissues in the body, diminish them only slightly. As illustrated in FIGURE 28.2, an x-ray image is essentially a shadow of the bones and dense components of the body; where these tissues stop the x rays, the film is not exposed. The basic procedure for producing an x-ray image on film is little changed from Röntgen's day.

FIGURE 28.2 Creating an x-ray image.

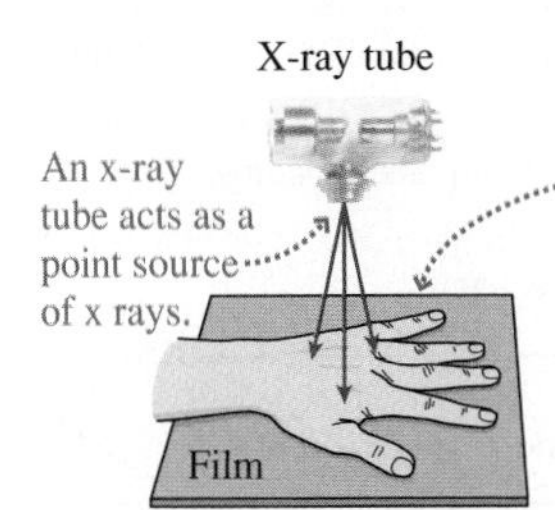

X-ray tube
An x-ray tube acts as a point source of x rays.
The part of the body to be imaged is on top of a piece of film. Dense tissues pass few x rays; the film is not exposed below these tissues.
Film

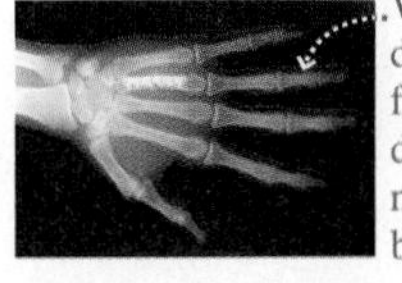

When the film is developed, the film is light where dense tissues or metal have blocked the x rays.

This use of x rays was of tremendous practical importance, but more important to the development of our story is the use of x rays to probe the structure of matter at an atomic scale.

X-Ray Diffraction

At about the same time scientists were first concluding that x rays were very-short-wavelength electromagnetic waves, researchers were also deducing that the size of an atom is ≈ 0.1 nm, and it was suggested that solids might consist of atoms arranged in a regular crystalline *lattice*. In 1912, the German scientist Max von Laue noted that x rays passing through a crystal ought to undergo diffraction from the "three-dimensional grating" of the crystal in much the same way that visible light diffracts from a diffraction grating. Such x-ray diffraction by crystals was soon confirmed experimentally, and measurements confirmed that x rays are indeed electromagnetic waves with wavelengths in the range 0.01 nm to 10 nm—a much shorter wavelength than visible light.

FIGURE 28.3 X rays incident on a simple cubic lattice crystal.

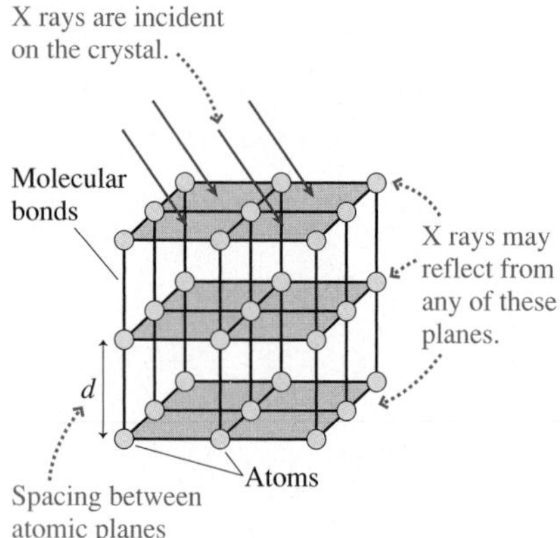
X rays are incident on the crystal.
Molecular bonds
X rays may reflect from any of these planes.
d
Atoms
Spacing between atomic planes

To understand x-ray diffraction, we begin by looking at the arrangement of atoms in a solid. FIGURE 28.3 shows x rays striking a crystal with a *simple cubic lattice*. This is a very straightforward arrangement, with the atoms in planes with spacing d between them.

FIGURE 28.4a shows a side view of the x rays striking the crystal, with the x rays incident at angle θ. Most of the x rays are transmitted through the plane, but a small fraction of the wave is reflected, much like the weak reflection of light from a sheet of glass. The reflected wave obeys the law of reflection—the angle of reflection equals the angle of incidence—and the figure has been drawn accordingly.

As we saw in Figure 28.3, a solid has not one single plane of atoms but many parallel planes. As x rays pass through a solid, a small fraction of the wave reflects from each of the parallel planes of atoms shown in FIGURE 28.4b. The *net* reflection from the solid is the *superposition* of the waves reflected by each atomic plane. For most angles of incidence, the reflected waves are out of phase and their superposition is very nearly zero. However, as in the thin-film interference we studied in Chapter 17, there are a few specific angles of incidence for which the reflected waves are in phase. For these angles of incidence, the reflected waves interfere constructively to produce a strong reflection. This strong x-ray reflection at a few specific angles of incidence is called **x-ray diffraction**.

FIGURE 28.4 X-ray reflections from parallel atomic planes.

(a) X rays are transmitted and reflected at one plane of atoms.

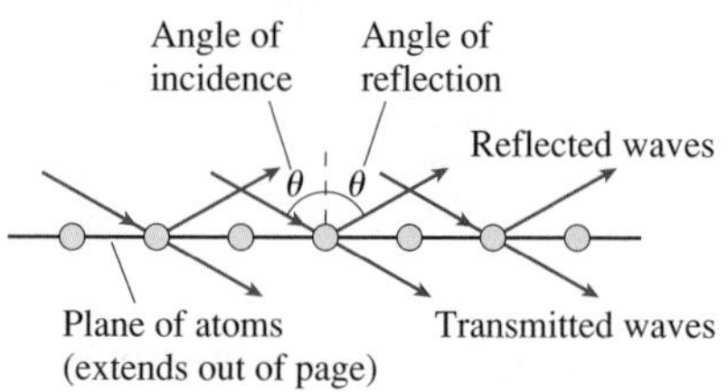
Angle of incidence
Angle of reflection
Reflected waves
θ
θ
Plane of atoms (extends out of page)
Transmitted waves

(b) The reflections from parallel planes interfere.

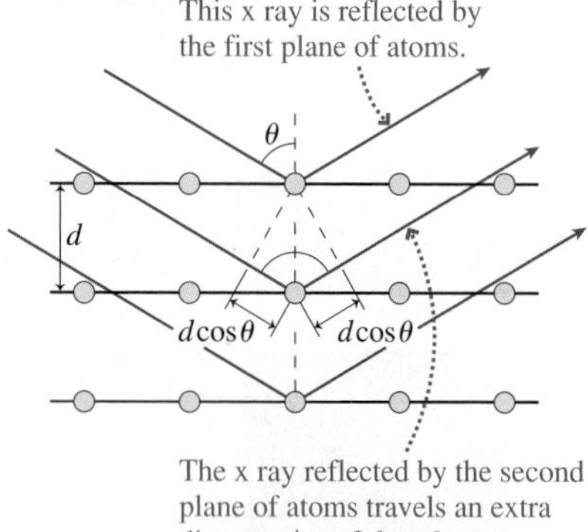
This x ray is reflected by the first plane of atoms.
θ
d
dcosθ
dcosθ
The x ray reflected by the second plane of atoms travels an extra distance Δr = 2dcosθ.

You can see from Figure 28.4b that the wave reflecting from any particular plane travels an extra distance $\Delta r = 2d\cos\theta$ before combining with the reflection from the plane immediately above it, where d is the spacing between the atomic planes. If Δr is a whole number of wavelengths, then these two waves will be in phase when they recombine. But if the reflections from two neighboring planes are in phase, then *all* the reflections from *all* the planes are in phase and will interfere constructively to produce a strong reflection. Consequently, x rays will reflect from the crystal when the angle of incidence θ_m satisfies the **Bragg condition:**

$$\Delta r = 2d\cos\theta_m = m\lambda \qquad m = 1, 2, 3, \ldots \tag{28.1}$$

The Bragg condition for constructive interference of x rays reflected from a solid

NOTE ► This formula is similar to that for constructive interference for light passed through a grating that we saw in Chapter 17. In both cases, we get constructive interference at only a few well-defined angles. ◄

EXAMPLE 28.1 Analyzing x-ray diffraction

X rays with a wavelength of 0.105 nm are diffracted by a crystal with a simple cubic lattice. Diffraction maxima are observed at angles 31.6° and 55.4° and at no angles between these two. What is the spacing between the atomic planes causing this diffraction?

PREPARE The angles must satisfy the Bragg condition. We don't know the values of m, but we know that they are two consecutive integers. In Equation 28.1 θ_m *decreases* as m increases, so 31.6° corresponds to the larger value of m. We will assume that 55.4° corresponds to m and 31.6° to $m + 1$.

SOLVE The values of d and λ are the same for both diffractions, so we can use the Bragg condition to find

$$\frac{m+1}{m} = \frac{\cos 31.6°}{\cos 55.4°} = 1.50 = \frac{3}{2}$$

Thus 55.4° is the second-order diffraction and 31.6° is the third-order diffraction. With this information we can use the Bragg condition again to find

$$d = \frac{2\lambda}{2\cos\theta_2} = \frac{0.105\text{ nm}}{\cos 55.4°} = 0.185\text{ nm}$$

ASSESS We learned above that the size of atoms is $\approx$0.1 nm, so this is a reasonable value for the atomic spacing in a crystal.

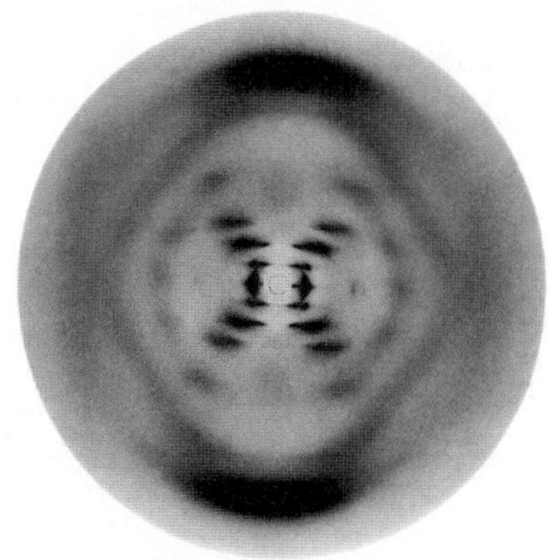

X marks the spot BIO Rosalind Franklin obtained this x-ray diffraction pattern for DNA in 1953. The cross of dark bands in the center of the diffraction pattern reveals something about the arrangement of atoms in the DNA molecule—that the molecule has the structure of a helix. This x-ray diffraction image was a key piece of information in the effort to unravel the structure of the DNA molecule.

Example 28.1 shows that an x-ray diffraction pattern reveals details of the crystal that produced it. The structure of the crystal was quite simple, so the example was straightforward. More complex crystals produce correspondingly complex patterns that can help reveal the structure of the crystals that produced them. As investigators developed theories of atoms and atomic structure, x rays were an invaluable tool—as they still are. X-ray diffraction is still widely used to decipher the three-dimensional structure of biological molecules such as proteins.

STOP TO THINK 28.1 The first-order diffraction of x rays from two crystals with simple cubic structure is measured. The first-order diffraction from crystal A occurs at an angle of 20°. The first-order diffraction of the same x rays from crystal B occurs at 30°. Which crystal has the larger atomic spacing?

28.2 The Photoelectric Effect

In Chapter 25, we introduced the idea that light can be thought of as *photons,* packets of energy of a particular size. This is an idea that you have likely heard before, but when it was first introduced, it was truly revolutionary. For such an odd idea to find broad acceptance, compelling experimental evidence was needed. This evidence was provided by studies of the *photoelectric effect,* which we will explore in detail in this section to recognize the rationale for and the impact of this startling new concept.

The first hints about the photon nature of light came in the late 1800s with the discovery that a negatively charged electroscope could be discharged by shining ultraviolet light on it. The English physicist J. J. Thomson found that the ultraviolet light was causing the electroscope to emit electrons, as illustrated in FIGURE 28.5. The emission of electrons from a substance due to light striking its surface came to be called the **photoelectric effect.** This seemingly minor discovery became a pivotal event that opened the door to the new ideas we discuss in this chapter.

FIGURE 28.5 Ultraviolet light discharges a negatively charged electroscope.

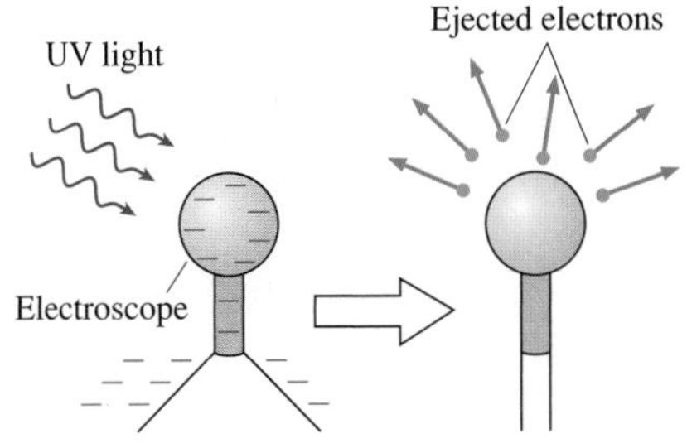

Ultraviolet light discharges a negatively charged electroscope by causing it to emit electrons.

17.3 Activ ONLINE Physics

Characteristics of the Photoelectric Effect

FIGURE 28.6 shows an evacuated glass tube with two facing electrodes and a window. When ultraviolet light shines on the cathode, a steady counterclockwise current (clockwise flow of electrons) passes through the ammeter. There are no junctions in this circuit, so the current must be the same all the way around the loop. The current in the space between the cathode and the anode consists of electrons moving freely through space (i.e., not inside a wire) at the *same rate* as the current in the wire. There is no current if the electrodes are in the dark, so electrons don't spontaneously leap off the cathode. Instead, the light causes electrons to be ejected from the cathode at a steady rate.

FIGURE 28.6 An experimental device to study the photoelectric effect.

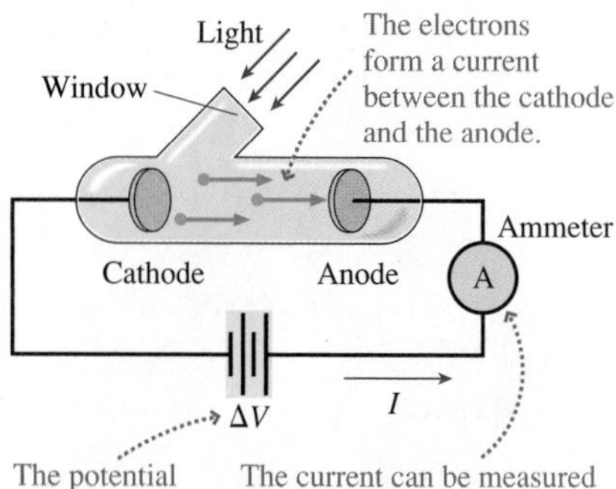
Light
Window
The electrons form a current between the cathode and the anode.
Ammeter
Cathode
Anode
A
ΔV
I
The potential difference can be changed or reversed.
The current can be measured as the potential difference, the light frequency, and the light intensity are varied.

The battery in Figure 28.6 establishes an adjustable potential difference ΔV between the two electrodes. With it, we can study how the current I varies as the potential difference and the light's wavelength and intensity are changed. Doing so reveals the following characteristics of the photoelectric effect:

1. The current I is directly proportional to the light intensity. If the light intensity is doubled, the current also doubles.
2. The current appears without delay when the light is applied.
3. Electrons are emitted *only* if the light frequency f exceeds a **threshold frequency** f_0. This is shown in the graph of FIGURE 28.7a.
4. The value of the threshold frequency f_0 depends on the type of metal from which the cathode is made.
5. If the potential difference ΔV is positive (anode positive with respect to the cathode), the current changes very little as ΔV is increased. If ΔV is made negative (anode negative with respect to the cathode), by reversing the battery, the current decreases until at some voltage $\Delta V = -V_{stop}$ the current reaches zero. The value of V_{stop} is called the **stopping potential.** This behavior is shown in FIGURE 28.7b.
6. The value of V_{stop} is the same for both weak light and intense light. A more intense light causes a larger current, but in both cases the current ceases when $\Delta V = -V_{stop}$.

FIGURE 28.7 The photoelectric current dependence on the light frequency f and the battery potential difference ΔV.

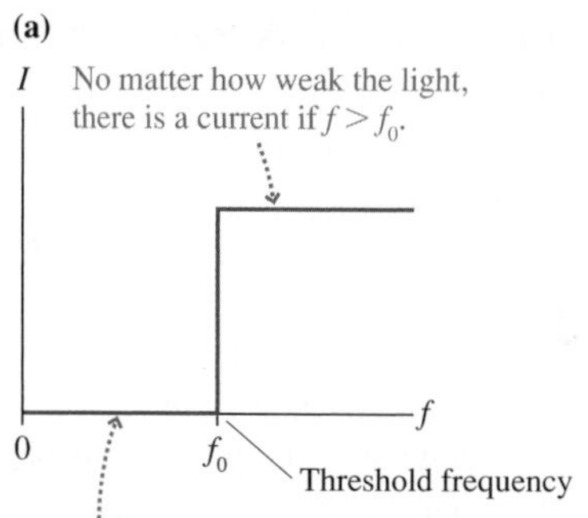
(a)
I
No matter how weak the light, there is a current if $f > f_0$.
f
0
f_0
Threshold frequency
No matter how intense the light, there is no current if $f < f_0$.

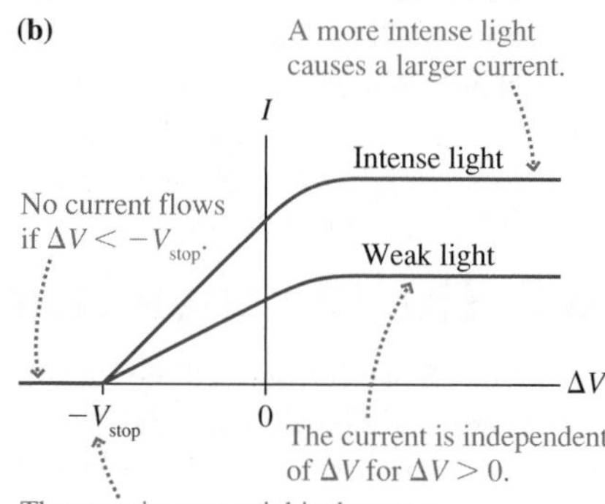
(b)
A more intense light causes a larger current.
I
Intense light
No current flows if $\Delta V < -V_{stop}$.
Weak light
ΔV
$-V_{stop}$
0
The current is independent of ΔV for $\Delta V > 0$.
The stopping potential is the same for intense light and weak light.

NOTE ► We're defining V_{stop} to be a *positive* number. The potential difference that stops the electrons is $\Delta V = -V_{stop}$, with an explicit minus sign. ◄

Understanding the Photoelectric Effect

You learned in Chapter 22 that electrons are the charge carriers in a metal and move around freely inside like a sea of negatively charged particles. The electrons are bound inside the metal and do not spontaneously spill out of an electrode at room temperature.

A useful analogy, shown in **FIGURE 28.8**, is the water in a swimming pool. Water molecules do not spontaneously leap out of the pool if the water is calm. To remove a water molecule, you must do *work* on it to lift it upward, against the force of gravity, to the edge of the pool. A minimum energy is needed to extract a water molecule—namely, the energy needed to lift a molecule that is right at the surface. Removing a water molecule that is deeper requires more than the minimum energy.

Similarly, a *minimum* energy is needed to free an electron from a metal. To extract an electron, you need to exert a force on it (i.e., do *work* on it) until its speed is fast enough to escape. The minimum energy E_0 needed to free an electron is called the **work function** of the metal. Some electrons, like deeper water molecules, may require more energy than E_0 to escape, but all will require *at least* E_0. Table 28.1 lists the work functions in eV of some elements. (Recall that the conversion to joules is $1 \text{ eV} = 1.60 \times 10^{-19}$ J.)

FIGURE 28.8 A swimming pool analogy of electrons in a metal.

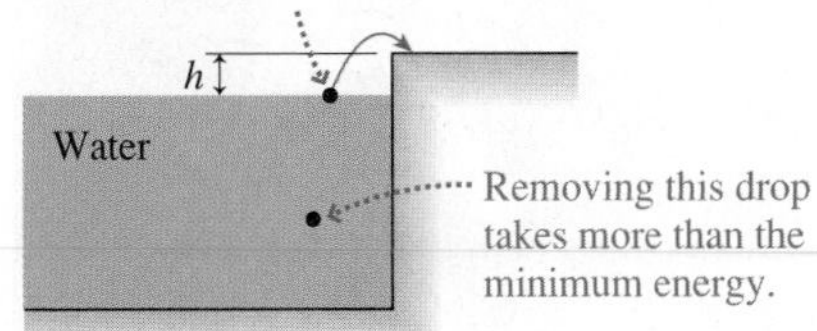
The *minimum* energy to remove a drop of water from the pool is *mgh*.
h
Water
Removing this drop takes more than the minimum energy.

TABLE 28.1 The work functions for some metals

Element	E_0 (eV)
Potassium	2.30
Sodium	2.75
Aluminum	4.28
Tungsten	4.55
Copper	4.65
Iron	4.70
Gold	5.10

Now, let's return to the photoelectric effect experiment of Figure 28.6. When ultraviolet light shines on the cathode, electrons leave with some kinetic energy. An electron with energy E_{elec} inside the metal loses energy ΔE as it escapes, so it emerges as an electron with kinetic energy $K = E_{elec} - \Delta E$. The work function energy E_0 is the *minimum* energy needed to remove an electron, so the *maximum* possible kinetic energy of an ejected electron is

$$K_{max} = E_{elec} - E_0$$

The electrons, after leaving the cathode, move out in all directions, as shown in **FIGURE 28.9**. If the potential difference between the cathode and the anode is $\Delta V = 0$, there will be no electric field between the plates. Some electrons will reach the anode, creating a measurable current, but many do not. The panels in the figure also show:

- If the anode is positive, it attracts *all* of the electrons to the anode. A further increase in ΔV does not cause any more electrons to reach the anode and thus does not cause a further increase in the current I. This is why the curves in Figure 28.7b become horizontal for positive ΔV.
- If the anode is negative, it repels the electrons. However, an electron leaving the cathode with sufficient kinetic energy can still reach the anode, just as a ball hits the ceiling if you toss it upward with sufficient kinetic energy. A slightly negative anode voltage turns back only the slowest electrons. The current steadily decreases as the anode voltage becomes increasingly negative until, as the left side of Figure 28.7b shows, at the stopping potential, *all* electrons are turned back and the current ceases.

We can use conservation of energy to relate the maximum kinetic energy to the stopping potential. When ΔV is negative, as in the bottom panel of Figure 28.9, electrons are "going uphill," converting kinetic energy to potential energy as they slow down. That is, $\Delta U = -e\Delta V = -\Delta K$, where we've used $q = -e$ for electrons and ΔK is negative because the electrons are losing kinetic energy. When $\Delta V = -V_{stop}$, where the current ceases, the very fastest electrons, with K_{max}, are being turned back *just* as they reach the anode. They're converting 100% of their kinetic energy into potential energy, so $\Delta K = -K_{max}$. Thus $e\Delta V_{stop} = K_{max}$, or

$$V_{stop} = \frac{K_{max}}{e} \tag{28.2}$$

In other words, **measuring the stopping potential tells us the maximum kinetic energy of the electrons.**

FIGURE 28.9 The effect of different voltages between the anode and cathode.

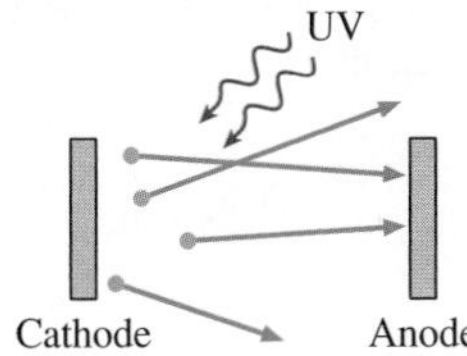
UV
Cathode
Anode

$\Delta V = 0$: The electrons leave the cathode in all directions. Only some reach the anode.

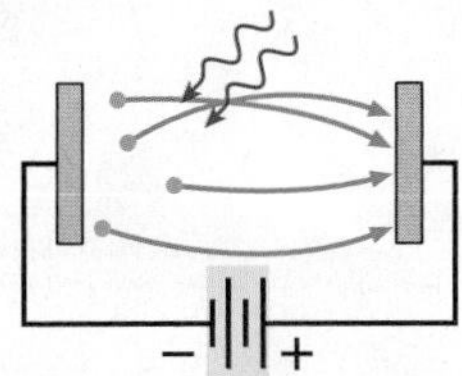
−
+

$\Delta V > 0$: Biasing the anode positive creates an electric field that pushes all the electrons to the anode.

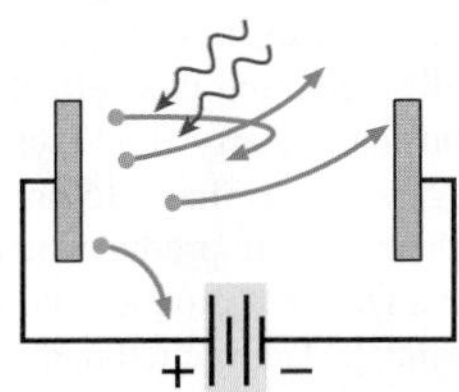
+
−

$\Delta V < 0$: Biasing the anode negative repels the electrons. Only the very fastest make it to the anode.

Einstein's Explanation

When light shines on the cathode in a photoelectric effect experiment, why do electrons leave the metal at all? Early investigators suggested explanations based on classical physics. A heated electrode spontaneously emits electrons, so it was natural to suggest that the light falling on the cathode simply heated it, causing it to emit

Einstein's "Miracle Year" Albert Einstein was a little-known young man of 26 in 1905. This photograph from the time bears little resemblance to the familiar picture of a white-haired older Einstein. In 1905, within the span of a single year, Einstein published three papers on three different topics, each of which would revolutionize physics. One was his initial paper on the theory of relativity, a subject you learned about in Chapter 27. Though relativity is the subject with which Einstein is most associated in the public mind, this paper received less attention at the time than the other two. A second paper used statistical mechanics to explain a phenomenon called *Brownian motion,* the random motion of small particles suspended in water. It is Einstein's third paper of 1905, on the nature of light, in which we are most interested in this chapter.

electrons. This would explain the photoelectric effect in terms of physics that was well accepted and understood.

But this simple explanation can't be correct. One way to see this is to consider the threshold frequency. If a weak intensity at a frequency just slightly above the threshold can generate a current, then certainly a strong intensity at a frequency just slightly below the threshold should be able to do so—it will heat the metal even more. There is no reason that a slight change in frequency should matter. Yet the experimental evidence shows a sharp frequency threshold, as we've seen.

A new physical theory was needed to fully explain the photoelectric-effect data. The currently accepted solution came in a 1905 paper by Albert Einstein in which he offered an exceedingly simple but amazingly bold idea that explained all of the noted features of the data.

Einstein's paper extended the work of the German physicist Max Planck, who had found that he could explain the form of the spectrum of a glowing, incandescent object that we saw in Chapter 25 only if he assumed that the oscillating atoms inside the heated solid vibrated in a particular way. The energy of an atom vibrating with frequency f had to be one of the specific energies $E = 0, hf, 2hf, 3hf, \ldots$, where h is a constant. That is, the vibration energies are **quantized.** The constant h, now called **Planck's constant,** is

$$h = 6.63 \times 10^{-34}\ \mathrm{J \cdot s} = 4.14 \times 10^{-15}\ \mathrm{eV \cdot s}$$

The first value, with SI units, is the proper one for most calculations, but you will find the second to be useful when energies are expressed in eV.

Einstein was the first to take Planck's idea seriously. Einstein went even further and suggested that **electromagnetic radiation itself is quantized!** That is, light is not really a continuous wave but, instead, arrives in small packets or bundles of energy. Einstein called each packet of energy a **light quantum,** and he postulated that the energy of one light quantum is directly proportional to the frequency of the light. That is, each quantum of light, which is now known as a **photon,** has energy

$$E = hf \qquad (28.3)$$

The energy of a photon, a quantum of light, of frequency f

where h is Planck's constant. Higher-frequency light is composed of higher-energy photons—it is composed of bundles of greater energy. This seemingly simple assumption allowed Einstein to explain all of the properties of the photoelectric effect. As we've seen, it can also explain many other observations about electromagnetic waves of different frequencies.

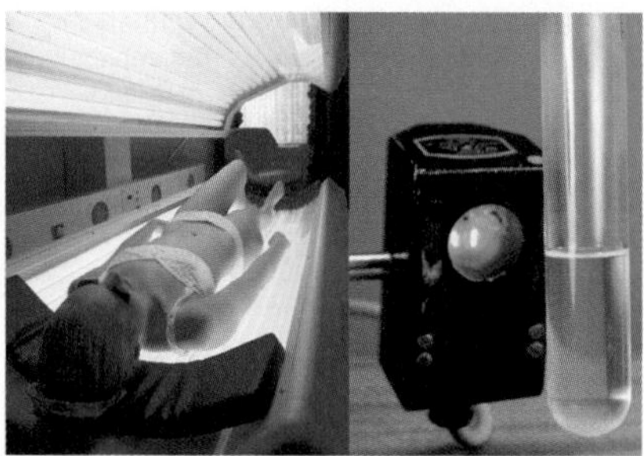

Not all ultraviolet is created equal BIO The sharp threshold for ultraviolet damage to tissue means that ultraviolet light sources with small differences in wavelength can have very different biological effects. Tanning beds emit nearly all of their energy at wavelengths greater than 315 nm. This light stimulates cells to produce melanin—resulting in a tan—but produces little short-term cell damage. Germicidal lamps use ultraviolet peaked at 254 nm, which will damage and even kill cells. Exposure to such a source will result in very painful sunburn.

EXAMPLE 28.2 Finding the energy of ultraviolet photons

Ultraviolet light at 290 nm does 250 times as much cellular damage as an equal intensity of ultraviolet at 310 nm; there is a clear threshold for damage at about 300 nm. What is the energy, in eV, of photons with a wavelength of 300 nm?

PREPARE The energy of a photon is related to its frequency by $E = hf$.

SOLVE The frequency at wavelength 300 nm is

$$f = \frac{c}{\lambda} = \frac{3.00 \times 10^{8}\ \mathrm{m/s}}{300 \times 10^{-9}\ \mathrm{m}} = 1.00 \times 10^{15}\ \mathrm{Hz}$$

We can now use Equation 28.3 to calculate the energy, using the value of h in $\mathrm{eV \cdot s}$:

$$E = hf = (4.14 \times 10^{-15}\ \mathrm{eV \cdot s})(1.00 \times 10^{15}\ \mathrm{Hz}) = 4.14\ \mathrm{eV}$$

ASSESS This number seems reasonable. We saw in Chapter 25 that splitting a bond in a water molecule requires an energy of 4.7 eV. We'd expect photons with energies in this range to be able to damage the complex organic molecules in a cell. As the problem notes, there is a sharp threshold for this damage. For energies larger than about 4.1 eV, photons can disrupt the genetic material of cells. Lower energies have little effect.

Einstein's Postulates and the Photoelectric Effect

The idea that light is quantized is now widely understood and accepted. But at the time of Einstein's paper, it was a truly revolutionary idea. Though we have used the photon model before, it is worthwhile to look at the theoretical underpinnings in more detail. In his 1905 paper, Einstein framed three postulates about light quanta and their interaction with matter:

1. Light of frequency f consists of discrete quanta, each of energy $E = hf$. Each photon travels at the speed of light c.
2. Light quanta are emitted or absorbed on an all-or-nothing basis. A substance can emit 1 or 2 or 3 quanta, but not 1.5. Similarly, an electron in a metal cannot absorb half a quantum but, instead, only an integer number.
3. A light quantum, when absorbed by a metal, delivers its entire energy to *one* electron.

NOTE ▶ These three postulates—that light comes in chunks, that the chunks cannot be divided, and that the energy of one chunk is delivered to one electron—are crucial for understanding the new ideas that will lead to quantum physics. ◀

Let's look at how Einstein's postulates apply to the photoelectric effect. We now think of the light shining on the metal as a torrent of photons, each of energy hf. Each photon is absorbed by *one* electron, giving that electron an energy $E_{elec} = hf$. This leads us to several interesting conclusions:

1. An electron that has just absorbed a quantum of light energy has $E_{elec} = hf$. FIGURE 28.10 shows that this electron can escape from the metal if its energy exceeds the work function E_0, or if

$$E_{elec} = hf \geq E_0 \qquad (28.4)$$

In other words, there is a *threshold frequency*

$$f_0 = \frac{E_0}{h} \qquad (28.5)$$

for the ejection of electrons. If f is less than f_0, even by just a small amount, none of the electrons will have sufficient energy to escape no matter how intense the light. But even very weak light with $f \geq f_0$ will give a few electrons sufficient energy to escape **because each photon delivers all of its energy to one electron.** This threshold behavior is exactly what the data show.

2. A more intense light delivers a larger number of photons to the surface. These eject a larger number of electrons and cause a larger current, exactly as observed.
3. There is a distribution of kinetic energies, because different electrons require different amounts of energy to escape, but the *maximum* kinetic energy is

$$K_{max} = E_{elec} - E_0 = hf - E_0 \qquad (28.6)$$

As we noted in Equation 28.2, the stopping potential V_{stop} is a measure of K_{max}. Einstein's theory predicts that the stopping potential is related to the light frequency by

$$V_{stop} = \frac{K_{max}}{e} = \frac{hf - E_0}{e} \qquad (28.7)$$

According to Equation 28.7, the stopping potential does *not* depend on the intensity of the light. Both weak light and intense light will have the same stopping potential. This agrees with the data.

4. If each photon transfers its energy hf to just one electron, that electron immediately has enough energy to escape. The current should begin instantly, with no delay, exactly as experiments had found.

FIGURE 28.10 The ejection of an electron.

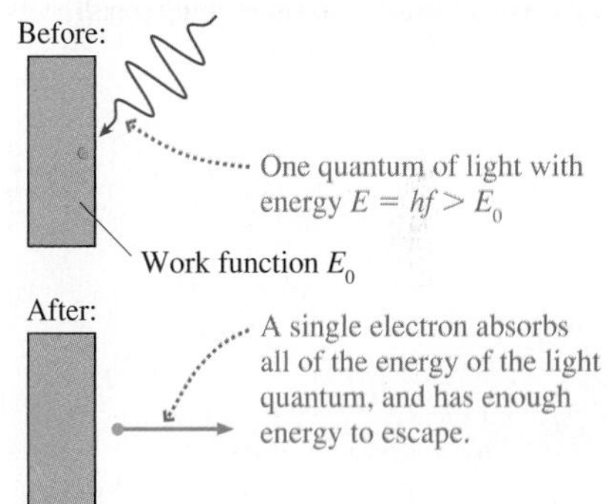
Before:
One quantum of light with energy $E = hf > E_0$
Work function E_0
After:
A single electron absorbs all of the energy of the light quantum, and has enough energy to escape.

Seeing the world in a different light BIO Many processes that are triggered by light have threshold frequencies. Plants use photosynthesis to convert the energy of light to chemical energy. Photons of visible light have sufficient energy to trigger the necessary molecular transitions, but photons of infrared do not. The leaves of trees absorb most of the visible light that falls on them, so trees will appear quite dark in a normal black-and-white landscape photo. But this photo wasn't made by visible light—the film was exposed by infrared. Infrared photons do not have enough energy to cause photosynthesis, so they are reflected, not absorbed, making the trees appear a ghostly white.

FIGURE 28.11 A pebble transfers energy to the water.

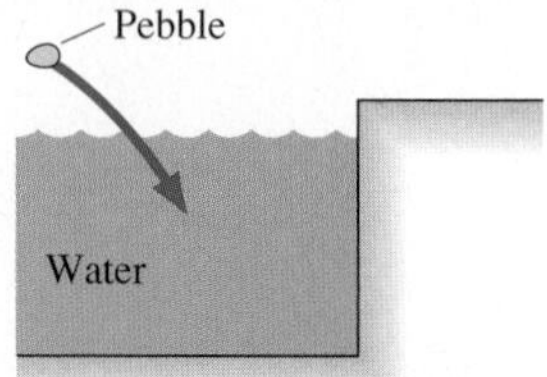

Classically, the energy of the pebble is shared by all the water molecules. One pebble causes only very small waves.

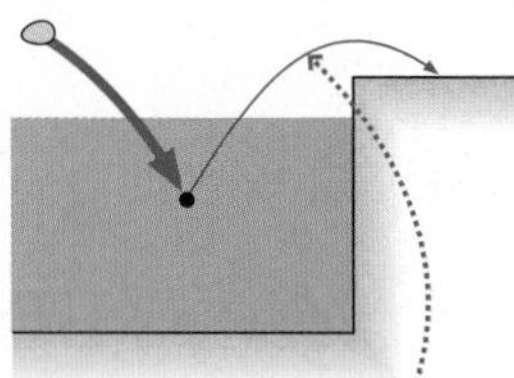

If the pebble could give *all* its energy to one drop, that drop could easily splash out of the pool.

Ultimately, Einstein's postulates are able to explain all of the observed features of the data for the photoelectric effect, though they require us to think of light in a very different way.

Let's use the swimming pool analogy again to help us visualize the photon model. FIGURE 28.11 shows a pebble being thrown into the pool. The pebble increases the energy of the water, but the increase is shared among all the molecules in the pool. The increase in the water's energy is barely enough to make ripples, not nearly enough to splash water out of the pool. But suppose *all* the pebble's energy could go to *one drop* of water that didn't have to share it. That one drop of water would easily have enough energy to leap out of the pool. Einstein's hypothesis that a light quantum transfers all its energy to one electron is equivalent to the pebble transferring all its energy to one drop of water.

Einstein was awarded the Nobel Prize in 1921 not for his theory of relativity, as many would suppose, but for his explanation of the photoelectric effect. Einstein showed convincingly that energy is quantized and that light, even though it exhibits wave-like interference, comes in the particle-like packets of energy we now call photons. This was the first big step in the development of the theory of quantum physics.

EXAMPLE 28.3 Finding the photoelectric threshold frequency

What are the threshold frequencies and wavelengths for electron emission from sodium and from aluminum?

PREPARE Table 28.1 gives the work function for sodium as $E_0 = 2.75$ eV and that for aluminum as $E_0 = 4.28$ eV.

SOLVE We can use Equation 28.5, with h in units of eV · s, to calculate

$$f_0 = \frac{E_0}{h} = \begin{cases} 6.64 \times 10^{14}\ \text{Hz} & \text{sodium} \\ 10.34 \times 10^{14}\ \text{Hz} & \text{aluminum} \end{cases}$$

These frequencies are converted to wavelengths with $\lambda = c/f$, giving

$$\lambda = \begin{cases} 452\ \text{nm} & \text{sodium} \\ 290\ \text{nm} & \text{aluminum} \end{cases}$$

ASSESS The photoelectric effect can be observed with sodium for $\lambda < 452$ nm. This includes blue and violet visible light but not red, orange, yellow, or green. Aluminum, with a larger work function, needs ultraviolet wavelengths, $\lambda < 290$ nm.

EXAMPLE 28.4 Determining the maximum electron speed

What are the maximum electron speed and the stopping potential if sodium is illuminated with light of wavelength 300 nm?

PREPARE The kinetic energy of the emitted electrons—and the potential difference necessary to stop them—depends on the energy of the incoming photons, $E = hf$, and the work function of the metal from which they are emitted, $E_0 = 2.75$ eV.

SOLVE The light frequency is $f = c/\lambda = 1.00 \times 10^{15}$ Hz, so each light quantum has energy $hf = 4.14$ eV. The maximum kinetic energy of an electron is

$$K_{max} = hf - E_0 = 4.14\ \text{eV} - 2.75\ \text{eV} = 1.39\ \text{eV}$$
$$= 2.22 \times 10^{-19}\ \text{J}$$

Because $K = \frac{1}{2}mv^2$, where m is the electron's mass, not the mass of the sodium atom, the maximum speed of an electron leaving the cathode is

$$v_{max} = \sqrt{\frac{2K_{max}}{m}} = 6.99 \times 10^5\ \text{m/s}$$

Note that K_{max} must be in J, the SI unit of energy, in order to calculate a speed in m/s.

Now that we know the maximum kinetic energy of the electrons, we can use Equation 28.7 to calculate the stopping potential:

$$V_{stop} = \frac{K_{max}}{e} = 1.39\ \text{V}$$

An anode voltage of -1.39 V will be just sufficient to stop the fastest electrons and thus reduce the current to zero.

ASSESS The stopping potential has the *same numerical value* as K_{max} expressed in eV, which makes sense. An electron with a kinetic energy of 1.39 eV can go "uphill" against a potential difference of 1.39 V, but no more.

STOP TO THINK 28.2 The work functions of metals A, B, and C are 3.0 eV, 4.0 eV, and 5.0 eV, respectively. Ultraviolet light shines on all three metals, causing electrons to be emitted. Rank in order, from largest to smallest, the stopping voltages for A, B, and C.

28.3 Photons

We've now seen compelling evidence for the photon nature of light, but this leaves an important question: Just what *are* photons? To begin our explanation, let's return to the experiment that showed most dramatically the wave nature of light—Young's double-slit interference experiment. We will make a change, though: We will dramatically lower the light intensity by inserting filters between the light source and the slits. The fringes will be too dim to see with the naked eye, so we will use a detector that can build up an image over time. (This is the same sort of detector we imagined using for the extremely low-light photograph in Chapter 25.)

FIGURE 28.12 shows the outcome of such an experiment at four different times. At early times, very little light has reached the detector, and it does not show bands at all. Instead, it shows dots; the detector is registering the arrival of particle-like objects.

As the detector builds up the image for a longer time, we see that the positions of the dots are not entirely random. They are grouped into bands at *exactly* the positions where we expect to see bright constructive-interference fringes. As the detector continues to gather light, the light and dark fringes become quite distinct. After a long time, the individual dots overlap and the image looks exactly like those we saw in Chapter 17.

The dots of light on the screen, which we'll attribute to the arrival of individual photons, are particle-like, but the overall picture clearly does not mesh with the classical idea of a particle. A classical particle, when faced with Young's double-slit apparatus, would go through one slit or the other. If light consisted of classical particles, we would see two bright areas on the screen, corresponding to light that has gone through one or the other slit. Instead, we see particle-like dots forming wave-like interference fringes.

This experiment was performed with a light level so low that only one photon at a time passed through the apparatus. If particle-like photons arrive at the detector in a banded pattern as a consequence of wave-like interference, as Figure 28.12 shows, but if only one photon at a time is passing through the experiment, what is it interfering with? The only possible answer is that the photon is somehow interfering *with itself.* Nothing else is present. But if each photon interferes with itself, rather than with other photons, then each photon, despite the fact that it is a particle-like object, must somehow go through *both* slits! This is something only a wave could do.

This all seems pretty crazy. But crazy or not, this is the way light behaves in real experiments. **Sometimes it exhibits particle-like behavior and sometimes it exhibits wave-like behavior.** The thing we call *light* is stranger and more complex than it first appeared, and there is no way to reconcile these seemingly contradictory behaviors. We have to accept nature as it is, rather than hoping that nature will conform to our expectations. Furthermore, as we will see, this half-wave/half-particle behavior is not restricted to light.

FIGURE 28.12 A double-slit experiment performed with light of very low intensity.

(a) Image after a very short time

(b) Image after a slightly longer time

(c) Continuing to build up the image

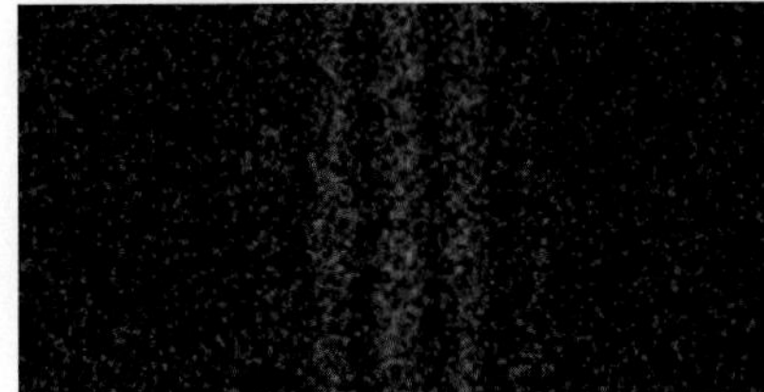

(d) Image after a very long time

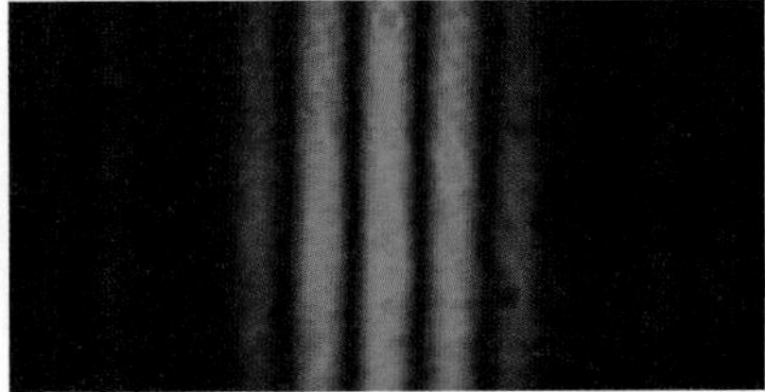

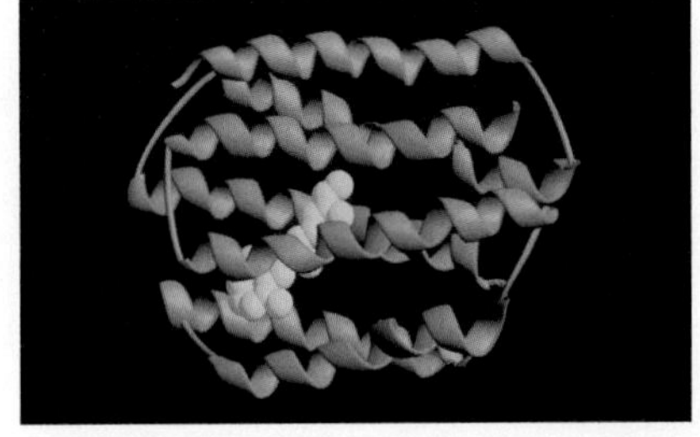

▶ **Seeing photons** The basis of vision is the detection of single photons by specially adapted molecules in the rod and cone cells of the eye. This image shows a molecule of *rhodopsin* (blue) with a molecule called *retinal* (yellow) nested inside. A single photon of the right energy triggers a transition of the retinal molecule, changing its shape so that it no longer fits inside the rhodopsin "cage". The rhodopsin then changes shape to eject the retinal, and this motion leads to an electrical signal in a nerve fiber. Slightly different versions of these molecules are "tuned" to different photon energies and thus different colors of light.

The Photon Rate

The photon nature of light isn't apparent in most cases. Most light sources with which you are familiar emit such vast numbers of photons that you are aware of only their wave-like superposition, just as you notice only the roar of a heavy rain on your roof and not the individual raindrops. Only at extremely low intensities does the light begin to appear as a stream of individual photons, like the random patter of raindrops when it is barely sprinkling.

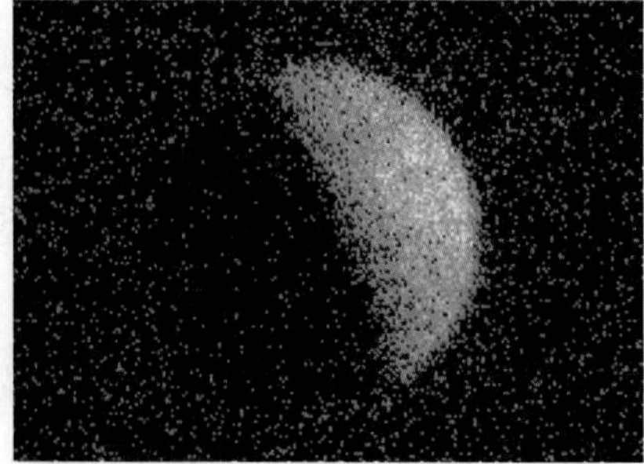

High-energy moonlight The sun emits vast quantities of visible-light photons. It also emits high-energy photons well beyond the range of the visible spectrum, but in much smaller numbers. This image of the moon was made with an orbiting telescope that detects x rays. Each dot in the image shows where one x-ray photon hit a detector in an orbiting telescope. The sunlit half of the moon "glows" with the reflection of x rays from the sun. The random dots seen everywhere are individual x-ray photons from our Milky Way galaxy, the *x-ray radiation background* of the cosmos.

EXAMPLE 28.5 How many photons per second does a laser emit?

The 1.0 mW light beam from a laser pointer ($\lambda = 670$ nm) shines on a screen. How many photons strike the screen each second?

PREPARE The power of the beam is 1.0 mW, or 1.0×10^{-3} J/s. Each second, 1.0×10^{-3} J of energy reaches the screen. It arrives as individual photons of energy given by Equation 28.3.

SOLVE The frequency of the photons is $f = c/\lambda = 4.5 \times 10^{14}$ Hz, so the energy of an individual photon is $E = hf = (6.6 \times 10^{-34}\ \text{J}\cdot\text{s})(4.5 \times 10^{14}\ \text{Hz}) = 3.0 \times 10^{-19}$ J. The number of photons reaching the screen each second is the total energy reaching the screen each second divided by the energy of an individual photon:

$$\frac{1.0 \times 10^{-3}\ \text{J/s}}{3.0 \times 10^{-19}\ \text{J/photon}} = 3.3 \times 10^{15}\ \text{photons per second}$$

ASSESS Each photon carries a small amount of energy, so there must be a huge number of photons per second to produce even this modest power.

CONCEPTUAL EXAMPLE 28.6 Comparing photon rates

A red laser pointer and a green laser pointer have the same power. Which one emits a larger number of photons per second?

REASON Red light has a longer wavelength and thus a lower frequency than green light, so the energy of a photon of red light is less than the energy of a photon of green light. The two pointers emit the same amount of light energy per second. Because the red laser emits light in smaller "chunks" of energy, it must emit more chunks per second to have the same power. The red laser emits more photons each second.

ASSESS This result can seem counterintuitive if you haven't thought hard about the implications of the photon model. Light of different wavelengths is made of photons of different energies, so these two lasers with different wavelengths—though they have the same power—must emit photons at different rates.

FIGURE 28.13 The operation of a solar cell.

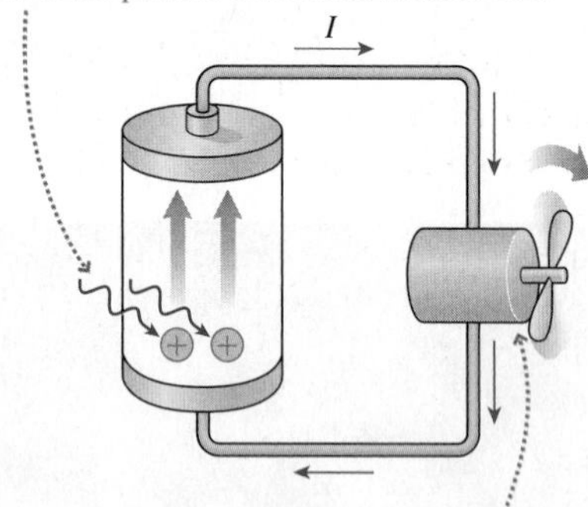

Photons with energy greater than the threshold give their energy to charge carriers, increasing their potential energy and lifting them to the positive terminal of the solar cell.
I
Charge carriers move "downhill" through the circuit. Their energy can be used to run useful devices.

Detecting Photons

Early light detectors, which used the photoelectric effect directly, consisted of a polished metal plate in a vacuum tube. When light fell on the plate, an electron current was generated that could trigger an action, such as sounding an alarm, or could provide a measurement of the light intensity.

Modern devices work on similar principles. In a *solar cell*, incoming photons give their energy to charge carriers, lifting them into higher-energy states. Recall the charge escalator model of a battery in Chapter 22. The solar cell works much like a battery, but the energy to lift charges to a higher potential comes from photons, not chemical reactions, as shown in FIGURE 28.13. The photon energy must exceed some minimum value to cause this transition, so solar cells have a threshold frequency, just like a device that uses the photoelectric effect directly. For a silicon-based solar cell, the most common type, the energy threshold is about 1.1 eV, corresponding to a wavelength of about 1200 nm, just beyond the range of the visible light spectrum, in the infrared.

EXAMPLE 28.7 Finding the current from a solar cell

1.0 W of monochromatic light of wavelength 550 nm illuminates a silicon solar cell, driving a current in a circuit. What is the maximum possible current this light could produce?

PREPARE The wavelength is shorter than the 1200 nm threshold wavelength noted for a silicon solar cell, so the photons will have sufficient energy to cause charge carriers to flow. Each photon of the incident light will give its energy to a single charge carrier. The maximum number of charge carriers that can possibly flow in each second is thus equal to the number of photons that arrive each second.

SOLVE The power of the light is $P = 1.0\text{ W} = 1.0\text{ J/s}$. The frequency of the light is $f = c/\lambda = 5.5 \times 10^{14}\text{ Hz}$, so the energy of individual photons is $E = hf = 3.6 \times 10^{-19}\text{ J}$. The number of photons arriving per second is $(1.0\text{ J/s})/(3.6 \times 10^{-19}\text{ J/photon}) = 2.8 \times 10^{18}$. Each photon can set at most one charge carrier into motion, so the maximum current is 2.8×10^{18} electrons/s. The current in amps—coulombs per second—is the electron flow rate multiplied by the charge per electron:

$$I_{\text{max}} = (2.8 \times 10^{18}\text{ electrons/s})(1.6 \times 10^{-19}\text{ C})$$
$$= 0.45\text{ C/s} = 0.45\text{ A}$$

ASSESS The key concept underlying the solution is that one photon gives its energy to a single charge carrier. We've calculated the current if all photons give their energy to charge carriers. The current in a real solar cell will be less than this because some photons will be reflected or otherwise "lost" and will not transfer their energy to charge carriers.

The *charge-coupled device* (CCD) or *complementary metal oxide semiconductor* (CMOS) detector in a digital camera consists of millions of *pixels,* each a microscopic silicon-based photodetector. Each photon hitting a pixel (if its frequency exceeds the threshold frequency) liberates one electron. These electrons are stored inside the pixel, and the total accumulated charge is directly proportional to the light intensity—the number of photons—hitting the pixel. After the exposure, the charge in each pixel is read and the value stored in memory; then the pixel is reset to be ready for the next picture.

TRY IT YOURSELF

Photographing photons
Photodetectors based on silicon can be triggered by photons with energy as low as 1.1. eV, corresponding to a wavelength in the infrared. The light-sensing chip in your digital camera can detect the infrared signal given off by a remote control. Press a button on your remote control, aim it at your digital camera, and snap a picture. The picture will clearly show the infrared emitted by the remote, though this signal is invisible to your eye. (Some cameras have infrared filters that may block most or nearly all of the signal.)

STOP TO THINK 28.3 The intensity of a beam of light is increased but the light's frequency is unchanged. Which one (or perhaps more than one) of the following is true?

A. The photons travel faster.
B. Each photon has more energy.
C. There are more photons per second.

28.4 Matter Waves

Prince Louis-Victor de Broglie was a French graduate student in 1924. It had been 19 years since Einstein had shaken the world of physics by introducing photons and thus blurring the distinction between a particle and a wave. As de Broglie thought about these issues, it seemed that nature should have some kind of symmetry. If light waves could have a particle-like nature, why shouldn't material particles have some kind of wave-like nature? In other words, could **matter waves** exist?

With no experimental evidence to go on, de Broglie reasoned by analogy with Einstein's equation $E = hf$ for the photon and with some of the ideas of his theory of relativity. De Broglie determined that *if* a material particle of momentum $p = mv$ has a wave-like nature, its wavelength must be given by

$$\lambda = \frac{h}{p} = \frac{h}{mv} \tag{28.8}$$

De Broglie wavelength for a moving particle

where h is Planck's constant. This wavelength is called the **de Broglie wavelength.**

EXAMPLE 28.8 Calculating the de Broglie wavelength of an electron

What is the de Broglie wavelength of an electron with a kinetic energy of 1.0 eV?

SOLVE An electron with kinetic energy $K = \frac{1}{2}mv^2 = 1.0\ \text{eV} = 1.6 \times 10^{-19}\ \text{J}$ has speed

$$v = \sqrt{\frac{2K}{m}} = 5.9 \times 10^5\ \text{m/s}$$

Although fast by macroscopic standards, the electron gains this speed by accelerating through a potential difference of a mere 1 V. The de Broglie wavelength is

$$\lambda = \frac{h}{mv} = 1.2 \times 10^{-9}\ \text{m} = 1.2\ \text{nm}$$

ASSESS The electron's wavelength is small, but it is larger than the wavelengths of x rays and larger than the approximately 0.1 nm spacing of atoms in a crystal. We can observe x-ray diffraction, so if an electron has a wave nature, it should be easily observable.

FIGURE 28.14 A double-slit interference pattern created with electrons.

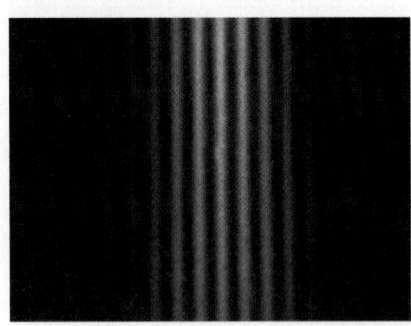

What would it mean for matter—an electron or a proton or a baseball—to have a wavelength? Would it obey the principle of superposition? Would it exhibit diffraction and interference? Surprisingly, **matter exhibits all of the properties that we associate with waves.** For example, FIGURE 28.14 shows the intensity pattern recorded after 50 keV electrons passed through two narrow slits separated by 1.0 μm. The pattern is clearly a double-slit interference pattern, and the spacing of the fringes is exactly as the theory of Chapter 17 would predict for a wavelength given by de Broglie's formula. **The electrons are behaving like waves!**

But if matter waves are real, why don't we see baseballs and other macroscopic objects exhibiting wave-like behavior? The key is the wavelength. We found in Chapter 17 that diffraction, interference, and other wave-like phenomena are observed when the wavelength is comparable to or larger than the size of an opening a wave must pass through. As Example 28.8 just showed, a typical electron wavelength is somewhat larger than the spacing between atoms in a crystal, so we expect to see wave-like behavior as electrons pass through matter or through microscopic slits. But the de Broglie wavelength is inversely proportional to an object's mass, so the wavelengths of macroscopic objects are millions or billions of times smaller than the wavelengths of electrons—vastly smaller than the size of any openings these objects might pass through. The wave nature of macroscopic objects is unimportant and undetectable because their wavelengths are so incredibly small, as the following example shows.

EXAMPLE 28.9 Calculating the de Broglie wavelength of a smoke particle

One of the smallest macroscopic particles we could imagine using for an experiment would be a very small smoke or soot particle. These are $\approx 1\ \mu$m in diameter, too small to see with the naked eye and just barely at the limits of resolution of a microscope. A particle this size has mass $m \approx 10^{-18}$ kg. Estimate the de Broglie wavelength for a 1-μm-diameter particle moving at the very slow speed of 1 mm/s.

SOLVE The particle's momentum is $p = mv \approx 10^{-21}\ \text{kg} \cdot \text{m/s}$. The de Broglie wavelength of a particle with this momentum is

$$\lambda = \frac{h}{p} \approx 7 \times 10^{-13}\ \text{m}$$

ASSESS The wavelength is much, much smaller than the particle itself—much smaller than an individual atom! We don't expect to see this particle exhibiting wave-like behavior.

The preceding example shows that a very small particle moving at a very slow speed has a wavelength that is too small to be of consequence. For larger objects moving at higher speeds, the wavelength is even smaller. A pitched baseball will have a wavelength of about 10^{-34} m, so a batter cannot use the wave nature of the ball as an excuse for not getting a hit. With such unimaginably small wavelengths, it is little wonder that we do not see macroscopic objects exhibiting wave-like behavior.

The Interference and Diffraction of Matter

17.5

Though de Broglie made his hypothesis in the absence of experimental data, experimental evidence was soon forthcoming. **FIGURES 28.15a** and **b** show diffraction patterns produced by x rays and electrons passing through an aluminum-foil target. The primary observation to make from Figure 28.15 is that **electrons diffract and interfere exactly like x rays.**

FIGURE 28.15 The diffraction patterns produced by x rays, electrons, and neutrons passing through an aluminum-foil target.

(a) X-ray diffraction pattern

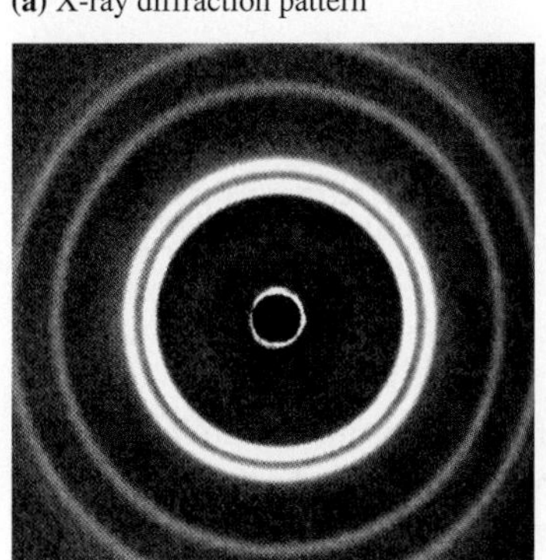

(b) Electron diffraction pattern

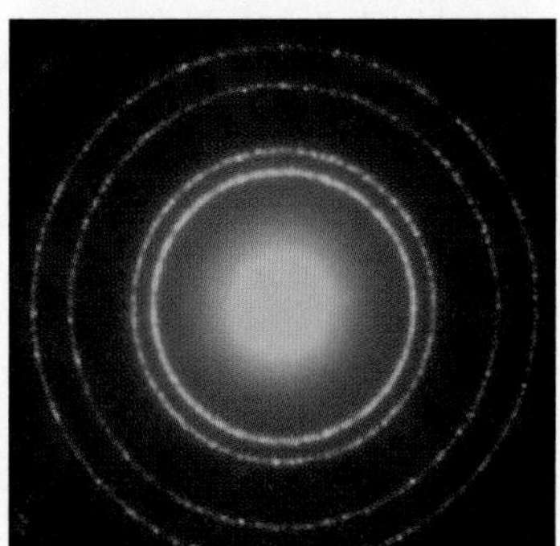

(c) Neutron diffraction pattern

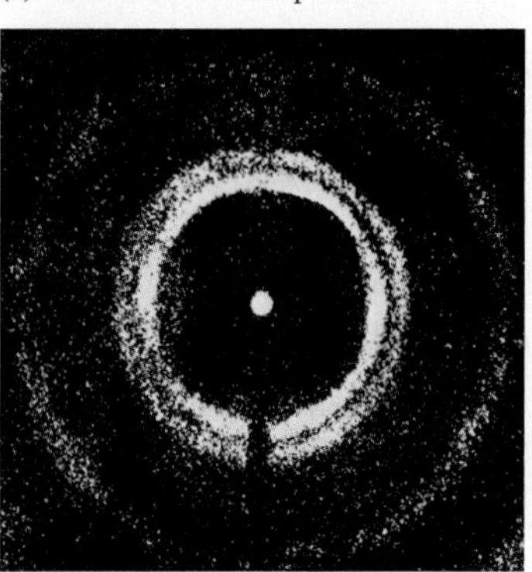

Later experiments demonstrated that de Broglie's hypothesis applies to other material particles as well. Neutrons have a much larger mass than electrons, which tends to decrease their de Broglie wavelength, but it is possible to generate very slow neutrons. The much smaller speed compensates for the heavier mass, so neutron wavelengths can be made comparable to electron wavelengths. **FIGURE 28.15c** shows a neutron diffraction pattern. It is similar to the x-ray and electron diffraction patterns, although of lower quality because neutrons are harder to detect. A neutron, too, is a matter wave. In recent years it has become possible to observe the interference and diffraction of atoms and even large molecules!

The Electron Microscope

Ray optics is based on the idea that light travels in straight lines—light rays—except when it crosses the boundary between two transparent media. Refraction at the boundary bends the rays, and we can use this idea to design lenses that bring parallel rays to a focus at a single point. If light really followed the ray model, carefully designed lenses would allow us to build a microscope with unlimited resolution and magnification. However, real microscopes are limited by the fact that light has wave-like properties. We learned in Chapter 19 that diffraction, a wave behavior, limits the resolving power of a microscope to, at best, about half the wavelength. For visible light, the smallest feature that can be resolved, even with perfect lenses, is about 200 or 250 nm. But the picture of the retina at the start of the chapter can show details much finer than this because it wasn't made with light—it was made with a beam of electrons.

The electron microscope, invented in the 1930s, works much like a light microscope. In the absence of electric or magnetic fields, electrons travel through a vacuum

FIGURE 28.16 The electron microscope.

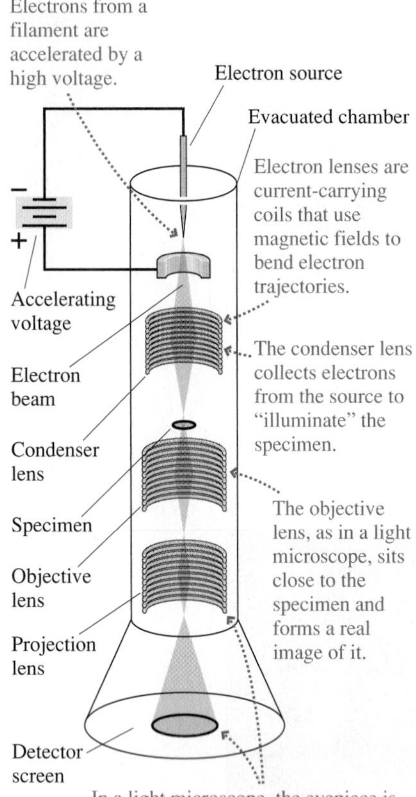

Electrons from a filament are accelerated by a high voltage.
Electron source
Evacuated chamber
Electron lenses are current-carrying coils that use magnetic fields to bend electron trajectories.
Accelerating voltage
Electron beam
The condenser lens collects electrons from the source to "illuminate" the specimen.
Condenser lens
Specimen
The objective lens, as in a light microscope, sits close to the specimen and forms a real image of it.
Objective lens
Projection lens
Detector screen
In a light microscope, the eyepiece is used to view the real image from the objective lens. An electron microscope has a projection lens that projects a magnified real image onto a detector.

FIGURE 28.17 TEM image of a pigment molecule from a crustacean shell.

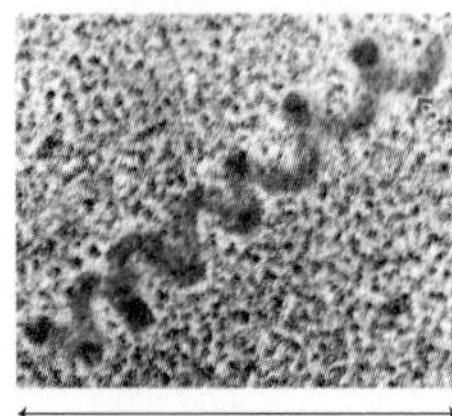

The great resolving power of the microscope allows the imaging of incredibly fine detail, in this case the actual structure of a molecule.
20 nm
The total image size is 1/10 the size of the smallest feature that can be resolved by a light microscope. A light microscope could not detect this individual molecule, let alone show its structure.

in straight lines much like light rays. Electron trajectories can be bent with electric or magnetic fields. A coil of wire carrying a current can produce a magnetic field that bends parallel electron trajectories so that they all cross at a single point; we call this an *electron lens*. An electron lens focuses electrons in the same way a glass lens bends and focuses light rays.

FIGURE 28.16 shows how a *transmission electron microscope (TEM)* works. This is purely classical physics; the electrons experience electric and magnetic forces, and they follow trajectories given by Newton's second law. Our ability to control electron trajectories allows electron microscopes to have magnifications far exceeding those of light microscopes. But just as in a light microscope, the resolution is ultimately limited by wave effects. Electrons are not classical point particles; they have wave-like properties and a de Broglie wavelength $\lambda = h/p$.

CONCEPTUAL EXAMPLE 28.10 Which wavelength is shorter?

An electron is accelerated through a potential difference ΔV. A second electron is accelerated through a potential difference that is twice as large. Which electron has a shorter de Broglie wavelength?

REASON The wavelength is inversely proportional to the speed. The electron that is accelerated through the larger potential difference will be moving faster and so will have a shorter de Broglie wavelength.

ASSESS Creating an electron micrograph requires high-speed electrons. Higher accelerating voltages mean higher speeds and shorter wavelengths, which would—in principle—allow for better resolution.

The reasoning used in Chapter 19 to determine maximum resolution applies equally well to electrons. Thus the resolving power is, at best, about half the electrons' de Broglie wavelength. For a 100 kV accelerating voltage, which is fairly typical, the de Broglie wavelength is $\lambda \approx 0.004$ nm (the electrons are moving fast enough that the momentum has to be calculated using relativity) and thus the theoretical resolving power of an electron microscope is about 0.002 nm.

In practice, the resolving powers of the best electron microscopes are limited by imperfections in the electron lenses to about 0.2 nm, not quite sufficient to resolve individual atoms with diameters of about 0.1 nm. This resolving power is about 1000 times smaller than can be achieved with light microscopes, as noted in FIGURE 28.17. Good light microscopes function at their theoretical limit, but there's still room to improve electron microscopes if a clever scientist or engineer can make a better electron lens.

STOP TO THINK 28.4 A beam of electrons, a beam of protons, and a beam of oxygen atoms each pass at the same speed through a 1-μm-wide slit. Which will produce the widest central maximum on a detector behind the slit?

A. The beam of electrons.
B. The beam of protons.
C. The beam of oxygen atoms.
D. All three patterns will be the same.
E. None of the beams will produce a diffraction pattern.

28.5 Energy Is Quantized

De Broglie hypothesized that material particles have wave-like properties, and you've now seen experimental evidence that this must be true. Not only is this bizarre, the implications are profound.

You learned in Chapter 16 that the waves on a string fixed at both ends form standing waves. Wave reflections from both ends create waves traveling in both directions, and the superposition of two oppositely directed waves produces a standing wave. Could we do something like this with particles? Is there such a thing as a "standing matter wave"? In fact, you are probably already familiar with standing matter waves—the atomic electron orbitals that you learned about in chemistry.

We'll have more to say about these orbitals in Chapter 29. For now, we'll start our discussion of standing matter waves with a simpler physical system called a "particle in a box." For simplicity, we'll consider one-dimensional motion, a particle that moves back and forth along the x-axis. The "box" is defined by two fixed ends, and the particle bounces back and forth between these boundaries as in **FIGURE 28.18**. We'll assume that collisions with the ends of the box are perfectly elastic, with no loss of kinetic energy.

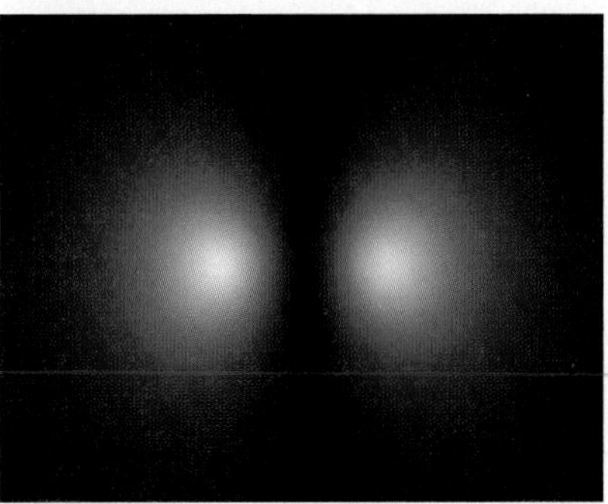

This computer simulation shows the p orbital of an atom. This orbital is an electron standing wave with a clear node at the center.

Figure 28.18a shows a classical particle, such as a ball or a dust particle, in the box. This particle simply bounces back and forth at constant speed. But if particles have wave-like properties, perhaps we should consider a *wave* reflecting back and forth from the ends of the box. The reflections will create the standing wave shown in Figure 28.18b. This standing wave is analogous to the standing wave on a string that is tied at both ends.

FIGURE 28.18 A particle of mass m confined in a box of length L.

(a) A classical particle of mass m bounces back and forth between two boundaries.

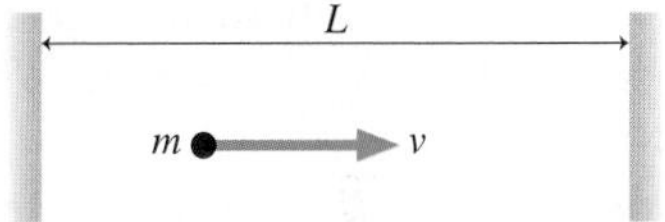

L
m
v

(b) Matter waves moving in opposite directions create standing waves.

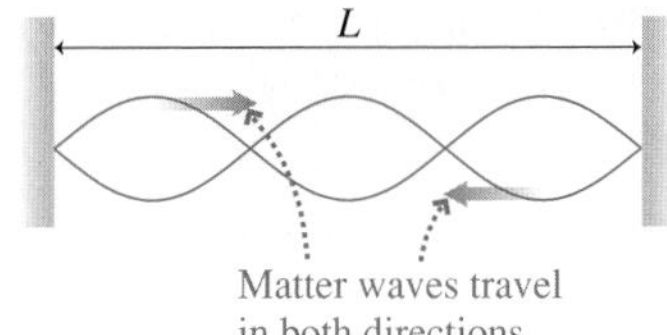

L
Matter waves travel in both directions.

What can we say about the properties of this standing matter wave? We can use what we know about matter waves and standing waves to make some deductions.

For waves on a string, we saw that there were only certain possible modes. The same will be true for the particle in a box; only certain states are possible. In Chapter 16, we found that the wavelength of a standing wave is related to the length L of the string by

$$\lambda_n = \frac{2L}{n} \qquad n = 1, 2, 3, 4, \ldots \tag{28.9}$$

The wavelength of the particle in a box will follow the same formula, but the wave describing the particle must also satisfy the de Broglie condition $\lambda = h/p$. Equating these two expressions for the wavelength gives

$$\frac{h}{p} = \frac{2L}{n} \tag{28.10}$$

Solving Equation 28.10 for the particle's momentum p, we find

$$p_n = n\left(\frac{h}{2L}\right) \qquad n = 1, 2, 3, 4, \ldots \tag{28.11}$$

This is a remarkable result; it is telling us that the momentum of the particle can have only certain values, the ones given by the equation. Other values simply aren't possible. The energy of the particle is related to its momentum by

$$E = \frac{1}{2}mv^2 = \frac{p^2}{2m} \tag{28.12}$$

If we use Equation 28.11 for the momentum, we find that the particle's energy is also restricted to a specific set of values:

$$E_n = \frac{1}{2m}\left(\frac{hn}{2L}\right)^2 = \frac{h^2}{8mL^2}n^2 \qquad n = 1, 2, 3, 4, \ldots \tag{28.13}$$

Allowed energies of a particle in a box

This conclusion is one of the most profound discoveries of physics. Because of the wave nature of matter, **a confined particle can have only certain energies.** This result—that a confined particle can have only discrete values of energy—is called the **quantization** of energy. More informally, we say that energy is *quantized.* The number n is called the **quantum number,** and each value of n characterizes one **energy level** of the particle in the box.

The lowest possible energy the particle in the box can have is

$$E_1 = \frac{h^2}{8mL^2} \tag{28.14}$$

We saw that, for a standing wave, the only possible frequencies were multiples of a lowest, fundamental frequency, $f_n = nf_1$. Similarly, for the particle in a box, the only possible energies are multiples of the lowest possible energy given by Equation 28.14; the only possible energies are

$$E_n = n^2 E_1 \tag{28.15}$$

This quantization is in stark contrast to the behavior of classical objects. It would be as if a baseball pitcher could throw a baseball only at 10 m/s, or 20 m/s, or 30 m/s, and so on, but at no speed in between. Baseball speeds aren't quantized, but the energy levels of a confined electron are—a result that has far-reaching implications.

EXAMPLE 28.11 Finding the allowed energies of a confined electron

An electron is confined to a region of space of length 0.19 nm—comparable in size to an atom. What are the first three allowed energies of the electron?

PREPARE We'll model this system as a particle in a box, with a box of length 0.19 m. The possible energies are given by Equation 28.13.

SOLVE The mass of an electron is $m = 9.11 \times 10^{-31}$ kg. Thus the first allowed energy is

$$E_1 = \frac{h^2}{8mL^2} = 1.6 \times 10^{-18}\ \text{J} = 10\ \text{eV}$$

This is the lowest allowed energy. The next two allowed energies are

$$E_2 = 2^2 E_1 = 40\ \text{eV}$$

$$E_3 = 3^2 E_1 = 90\ \text{eV}$$

ASSESS These energies are significant; E_1 is larger than the work function of any metal in Table 28.1. Confining an electron to a region the size of an atom limits its energy to states separated by significant differences in energy. Clearly, our treatment of electrons in atoms must be a quantum treatment.

The energies allowed by Equation 28.13 are inversely proportional to both m and L^2. Both m and L have to be exceedingly small before energy quantization has any significance. Classical physics still works for baseballs! It is only at the atomic scale that quantization effects become important, as the following example shows.

EXAMPLE 28.12 Determining the minimum energy of a smoke particle

What is the first allowed energy of the very small 1-μm-diameter particle of Example 28.9 if it is confined to a very small box 10 μm in length?

PREPARE As in Example 28.11, we'll model the system as a particle in a box, with the energy levels given by Equation 28.13.

SOLVE The particle's mass is given in Example 28.9 as $m \approx 10^{-18}$ kg; the length of the box is given in the problem statement as $L = 1.0 \times 10^{-6}$ m. The first allowed energy, $n = 1$, is

$$E_1 = \frac{h^2}{8mL^2} \approx 5 \times 10^{-40}\ \text{J}$$

ASSESS This is an unimaginably small amount of energy. By comparison, the kinetic energy of a 1-μm-diameter particle moving at a barely perceptible speed of 1 mm/s is $K = 5 \times 10^{-25}$ J, a factor of 10^{15} larger. There is no way we could ever observe or measure discrete energies this small, so it is not surprising that we are unaware of energy quantization for macroscopic objects.

An atom is certainly more complicated than a simple one-dimensional box, but an electron is "confined" within an atom. Thus the electron orbits must, in some sense, be standing waves, and **the energy of the electrons in an atom must be quantized.** This has important implications for the physics of atomic systems, as we'll see in the next section.

28.6 Energy Levels and Quantum Jumps

Einstein and de Broglie introduced revolutionary new ideas—a blurring of the distinction between waves and particles, and the quantization of energy—but the first to develop a full-blown theory of quantum physics, in 1925, was the Austrian physicist Erwin Schrödinger. Schrödinger's theory is now called *quantum mechanics*. It describes how to calculate the quantized energy levels of systems from the particle in a box to electrons in atoms. Quantum mechanics also describes another important piece of the puzzle: How does a quantized system gain or lose energy?

Erwin Schrödinger, one of the early architects of quantum mechanics.

Energy-Level Diagrams

We used the idea of a standing de Broglie wave to find the allowed energies of a particle in a one-dimensional box. The full theory of quantum mechanics is needed to predict the allowed energy of more realistic physical systems, such as atoms or semiconductors, but the final results share a key property: The energy is quantized. Only certain energies are allowed while all other energies are forbidden.

An **energy-level diagram** is a useful visual representation of the quantized energies. As an example, FIGURE 28.19 is the energy-level diagram for an electron in a 0.19-nm-long box. We computed these energies in Example 28.11. An energy-level diagram is less a graph than it is a picture. The vertical axis represents energy, but the horizontal axis is not a scale. Think of this as a ladder in which the energies are the rungs of the ladder. The lowest rung, with energy E_1, is called the **ground state.** Higher rungs, called **excited states,** are labeled by their quantum numbers, $n = 2, 3, 4, \ldots$. Whether it is a particle in a box, an atom, or the nucleus of an atom, quantum physics requires the system to be on one of the rungs of the ladder.

FIGURE 28.19 The energy-level diagram of an electron in a 0.19-nm-long box.

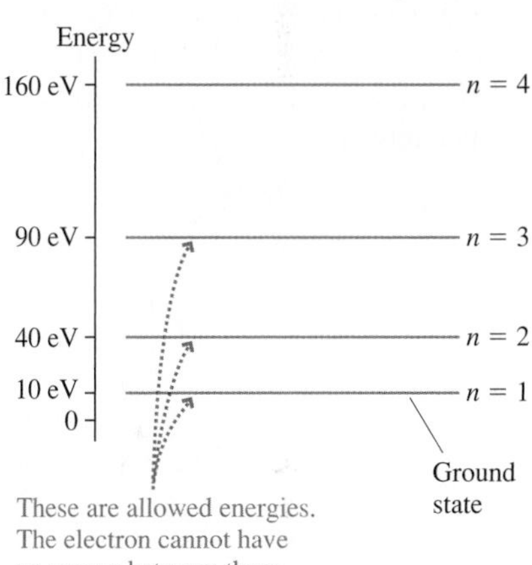
Energy
160 eV
90 eV
40 eV
10 eV
0
$n = 4$
$n = 3$
$n = 2$
$n = 1$
Ground state
These are allowed energies. The electron cannot have an energy between these.

If a quantum system changes from one state to another, there is a change in energy. One thing that has not changed in quantum physics is the conservation of energy—energy is still conserved in the quantum world. If a system drops from a higher energy level to a lower, the excess energy ΔE_{system} must go somewhere. In the systems we will consider, this energy generally ends up in the form of an emitted photon. A quantum system in energy level E_i that "jumps down" to energy level E_f loses an energy $\Delta E_{\text{system}} = |E_f - E_i|$. This jump must correspond to the emission of a photon of frequency

$$f_{\text{photon}} = \frac{\Delta E_{\text{system}}}{h} \tag{28.16}$$

Conversely, if the system absorbs a photon, it can "jump up" to a higher energy level. In this case, the frequency of the absorbed photon must follow Equation 28.16 as well. Such jumps are called **transitions** or **quantum jumps.** FIGURE 28.20 shows two transitions for the particle in a box system of Figure 28.19.

Notice that Equation 28.16 links Schrödinger's quantum theory to Einstein's earlier idea about the quantization of light energy. According to Einstein, a photon of frequency f has energy $E_{\text{photon}} = hf$. If a particle jumps from an initial state with energy E_i to a final state with *lower* energy E_f, energy will be conserved if the system emits a photon with $E_{\text{photon}} = \Delta E_{\text{system}}$. The photon must have exactly the frequency given by Equation 28.16 if it is to carry away exactly the right amount of energy. As we'll see in the next chapter, these photons form the *emission spectrum* of the quantum system.

Similarly, a particle can conserve energy while jumping to a higher-energy state, for which additional energy is needed, by absorbing a photon of frequency $f_{\text{photon}} = \Delta E_{\text{system}}/h$. The photon will not be absorbed unless it has exactly this frequency. The frequencies absorbed in these upward transitions form the system's *absorption spectrum.*

FIGURE 28.20 A particle can jump between energy levels by emitting or absorbing a photon.

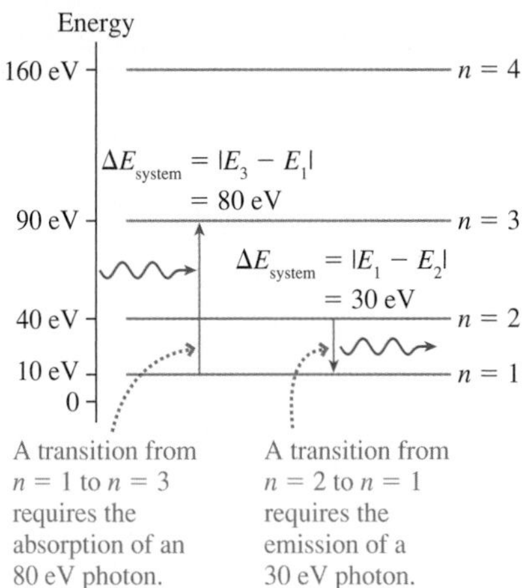
Energy
160 eV
90 eV
40 eV
10 eV
0
$n = 4$
$n = 3$
$n = 2$
$n = 1$
$\Delta E_{\text{system}} = |E_3 - E_1|$
$= 80$ eV
$\Delta E_{\text{system}} = |E_1 - E_2|$
$= 30$ eV
A transition from $n = 1$ to $n = 3$ requires the absorption of an 80 eV photon.
A transition from $n = 2$ to $n = 1$ requires the emission of a 30 eV photon.

The spectrum of helium gas shows a discrete set of wavelengths, corresponding to the energies of possible transitions.

Let's summarize what quantum physics has to say about the properties of atomic-level systems:

1. **The energies are quantized.** Only certain energies are allowed, all others are forbidden. This is a consequence of the wave-like properties of matter.
2. **The ground state is stable.** Quantum systems seek the lowest possible energy state. A particle in an excited state, if left alone, will jump to lower and lower energy states until it reaches the ground state. Once in its ground state, there are no states of any lower energy to which a particle can jump.
3. **Quantum systems emit and absorb a *discrete spectrum* of light.** Only those photons whose frequencies match the energy *intervals* between the allowed energy levels can be emitted or absorbed. Photons of other frequencies cannot be emitted or absorbed without violating energy conservation.

We'll use these ideas in the next two chapters to understand the properties of atoms and nuclei.

EXAMPLE 28.13 Determining an emission spectrum from quantum states

An electron in a quantum system has allowed energies $E_1 = 1.0$ eV, $E_2 = 4.0$ eV, and $E_3 = 6.0$ eV. What wavelengths are observed in the emission spectrum of this system?

PREPARE FIGURE 28.21 shows the energy-level diagram for this system. Photons are emitted when the system undergoes a quantum jump from a higher energy level to a lower energy level. There are three possible transitions.

FIGURE 28.21 The system's energy-level diagram and quantum jumps.

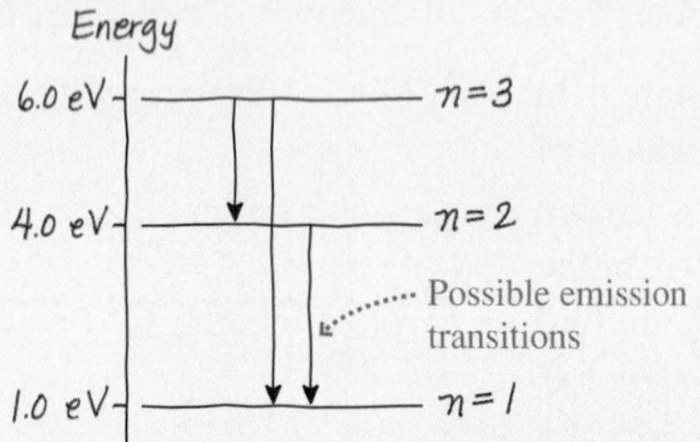

SOLVE This system will emit photons on the $3 \rightarrow 1$, $2 \rightarrow 1$, and $3 \rightarrow 2$ transitions, with $\Delta E_{3\rightarrow 1} = 5.0$ eV, $\Delta E_{2\rightarrow 1} = 3.0$ eV, and $\Delta E_{3\rightarrow 2} = 2.0$ eV. From $f_{\text{photon}} = \Delta E_{\text{system}}/h$ and $\lambda = c/f$, we find that the wavelengths in the emission spectrum are

$3 \rightarrow 1$ $\quad f = 5.0 \text{ eV}/h = 1.2 \times 10^{15}$ Hz

$\lambda = 250$ nm (ultraviolet)

$2 \rightarrow 1$ $\quad f = 3.0 \text{ eV}/h = 7.2 \times 10^{14}$ Hz

$\lambda = 420$ nm (blue)

$3 \rightarrow 2$ $\quad f = 2.0 \text{ eV}/h = 4.8 \times 10^{14}$ Hz

$\lambda = 620$ nm (orange)

ASSESS Transitions with a small energy difference, like $3 \rightarrow 2$, correspond to lower photon energies and thus longer wavelengths than transitions with a large energy difference like $3 \rightarrow 1$, as we would expect.

STOP TO THINK 28.5 A photon with a wavelength of 420 nm has energy $E_{\text{photon}} = 3.0$ eV. Do you expect to see a spectral line with $\lambda = 420$ nm in the emission spectrum of the system represented by this energy-level diagram?

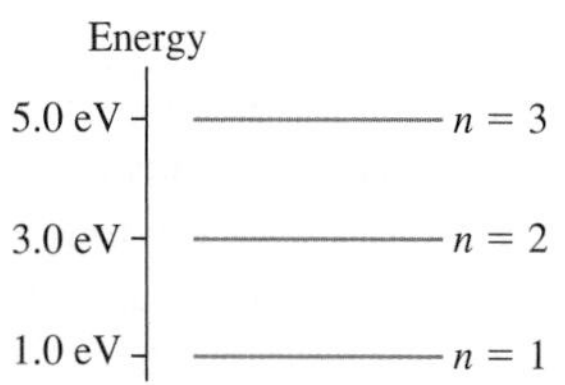

28.7 The Uncertainty Principle

One of the strangest aspects of the quantum view of the world is an inherent limitation on our knowledge: **For a particle such as an electron, if you know where it is, you can't know exactly where it is going.** This very counterintuitive notion is a result of the wave nature of matter and is worth a bit of explanation.

FIGURE 28.22 on the next page shows an experiment in which electrons moving along the y-axis pass through a slit of width a. We know the result: Because of the wave nature of electrons, the slit causes them to spread out and produce a diffraction pattern.

But we can think of the experiment in a different way—as making a measurement of the position of the electrons. As an electron goes through the slit, we know something about its horizontal position. Our knowledge isn't perfect; we just know it is somewhere within the slit. We can establish an *uncertainty*, a limit on our knowledge. The uncertainty in the horizontal position is $\Delta x = a$, the width of the slit.

But, after passing through the slit, the electrons spread out, as they must to produce a diffraction pattern. The electrons, which had been moving along the y-axis before reaching the slit, now have a component of velocity along the x-axis. Because the electrons strike the screen over a range of positions, the value of v_x must vary from electron to electron. By sending the electrons through a slit—by trying to pin down their horizontal position—we've created an uncertainty in their horizontal velocity. Gaining knowledge of the *position* of the electrons has introduced uncertainty into our knowledge of the *velocity* of the electrons.

FIGURE 28.22 An experiment to illustrate the uncertainty principle.

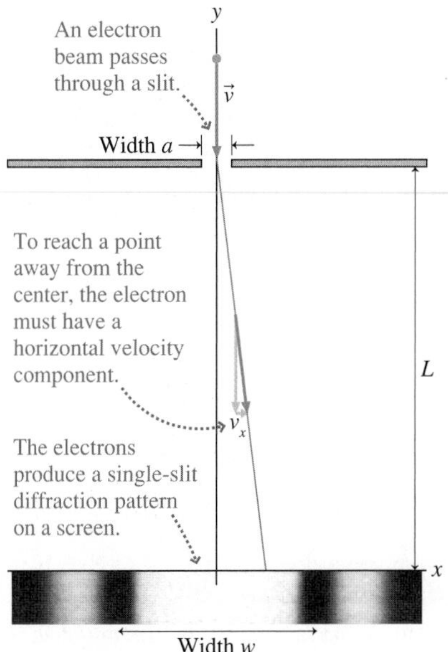
y
An electron beam passes through a slit.
$\vec{v}$
Width a
To reach a point away from the center, the electron must have a horizontal velocity component.
v_x
L
The electrons produce a single-slit diffraction pattern on a screen.
x
Width w
We can't predict with certainty where an electron will hit, but most land within the central maximum of the single-slit pattern.

CONCEPTUAL EXAMPLE 28.14 Changing the uncertainty

Suppose we narrow the slit in the above experiment, allowing us to determine the electron's horizontal position more precisely. How does this affect the diffraction pattern? How does this change in the diffraction pattern affect the uncertainty in the velocity?

REASON We learned in Chapter 17 that the width of the central maximum of the single-slit diffraction pattern is $w = 2\lambda L/a$. Making the slit narrower—decreasing the value of a—increases the value of w, making the central fringe wider. If the fringe is wider, the spread of horizontal velocities must be greater, so there is a greater uncertainty in the horizontal velocity.

ASSESS Improving our knowledge of the position decreases our knowledge of the velocity. This is the hallmark of the *uncertainty principle*.

We've made this argument by considering a particular experiment, but it is an example of a general principle. In 1927, the German physicist Werner Heisenberg proved that, for any particle, the product of the uncertainty Δx in its position and the uncertainty Δp_x in its x-momentum has a lower limit fixed by the expression

Activ ONLINE Physics
17.6

$$\Delta x \, \Delta p_x \geq \frac{h}{2\pi} \tag{28.17}$$

Heisenberg uncertainty principle for position and momentum

A decreased uncertainty in position—knowing more precisely where a particle is—comes at the expense of an increased uncertainty in velocity and thus in momentum. But the relationship also goes the other way: Knowing a particle's velocity or momentum more precisely requires an increase in the uncertainty about its position.

NOTE ► In statements of the uncertainty principle, the right side is sometimes $h/2\pi$, as we have it, but other formulations have $h/4\pi$ or $h/2$ because of different conventions for the definition of Δx and Δp_x. Don't worry about these small differences. The important idea is that the product of Δx and Δp_x for a particle cannot be significantly less than Planck's constant h. ◄

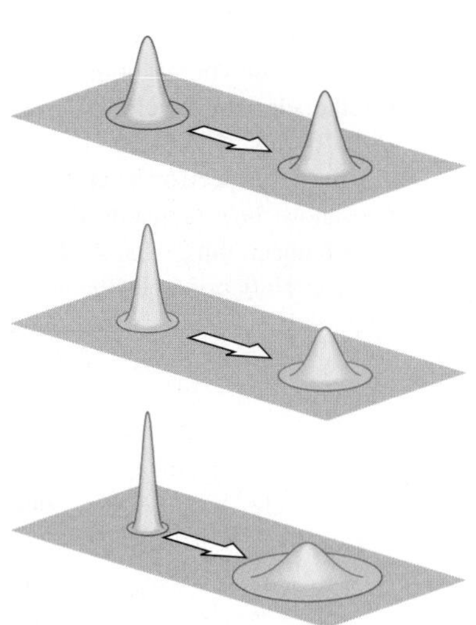

► **If I know where you are, I don't know where you're going** In quantum physics, we represent particles by a *wave function* that describes their wave nature. This series of diagrams shows simulations of the evolution of three traveling wave functions. The top diagram shows the broad wave function of a particle whose position is not precisely defined. The uncertainty in momentum (and thus velocity) is small, so the wave function doesn't spread out much as it travels. The lower graphs show particles with more sharply peaked wave functions, implying less uncertainty in their initial positions. A reduced uncertainty in position means a larger uncertainty in velocity, so the wave functions spread out more quickly.

Uncertainties are associated with all experimental measurements, but better procedures and techniques can reduce those uncertainties. Classical physics places no limits on how small the uncertainties can be. A classical particle at any instant of time has an exact position x and an exact momentum p_x, and with sufficient care we can measure both x and p_x with such precision that we can make the product $\Delta x\, \Delta p_x$ as small as we like. There are no inherent limits to our knowledge.

In the quantum world, it's not so simple. No matter how clever you are, and no matter how good your experiment, you *cannot* measure both x and p_x simultaneously with arbitrarily good precision. Any measurements you make are limited by the condition that $\Delta x\, \Delta p_x \geq h/2\pi$. **The position and the momentum of a particle are *inherently* uncertain.**

Why? Because of the wave-like nature of matter! The "particle" is spread out in space, so there simply is not a precise value of its position x. Our belief that position and momentum have precise values is tied to our classical concept of a particle. As we revise our ideas of what atomic particles are like, we must also revise our ideas about position and momentum.

Let's revisit particles in a one-dimensional "box," now looking at uncertainties.

EXAMPLE 28.15 Determining uncertainties

a. What range of velocities might an electron have if confined to a 0.10-nm-wide region, about the size of an atom?

b. A 1.0-μm-diameter dust particle ($m \approx 10^{-15}$ kg) is confined within a 10-μm-long box. Can we know with certainty if the particle is at rest? If not, within what range is its velocity likely to be found?

PREPARE Localizing a particle means specifying its position with some accuracy—so there must be an uncertainty in the velocity. We can estimate the uncertainty by using Heisenberg's uncertainty principle.

SOLVE a. We aren't given the exact position of the particle, only that it is within a 0.10-nm-wide region. This means that we have specified the electron's position within a range $\Delta x = 1.0 \times 10^{-10}$ m. The uncertainty principle thus specifies that the least possible uncertainty in the momentum is

$$\Delta p_x = \frac{h}{2\pi\,\Delta x}$$

The uncertainty in the velocity is thus approximately

$$\Delta v_x = \frac{\Delta p_x}{m} \approx \frac{h}{2\pi m\,\Delta x} \approx 1 \times 10^6 \text{ m/s}$$

Because the *average* velocity is zero, (the particle is equally likely to be moving right or left) the best we can do is to say that the electron's velocity is somewhere in the interval $-5 \times 10^5 \text{ m/s} \leq v_x \leq 5 \times 10^5$ m/s. **It is simply not possible to specify the electron's velocity more precisely than this.**

b. If we know *for sure* that the dust particle is at rest, then $p_x = 0$ with no uncertainty. That is, $\Delta p_x = 0$. But then, according to the uncertainty principle, the uncertainty in our knowledge of the particle's position would have to be $\Delta x = \infty$. In other words, we would have no knowledge at all about the particle's position—it could be anywhere! But that is not the case. We know the particle is *somewhere* in the box, so the uncertainty in our knowledge of its position is at most $\Delta x = L = 10\ \mu$m. With a finite Δx, the uncertainty Δp_x *cannot* be zero. **We cannot know with certainty if the particle is at rest inside the box.** No matter how hard we try to bring the particle to rest, the uncertainty in our knowledge of the particle's momentum will be approximately $\Delta p_x \approx h/(2\pi\,\Delta x) = h/(2\pi L)$. Consequently, the range of possible velocities is

$$\Delta v_x = \frac{\Delta p_x}{m} \approx \frac{h}{2\pi m L} \approx 1.0 \times 10^{-14} \text{ m/s}$$

This range of possible velocities will be centered on $v_x = 0$ m/s if we have done our best to have the particle be at rest. Therefore all we can know with certainty is that the particle's velocity is somewhere within the interval $-5 \times 10^{-15} \text{ m/s} \leq v_x \leq 5 \times 10^{-15}$ m/s.

ASSESS Our uncertainty about the electron's velocity is enormous—nearly 1% of the speed of light. For an electron confined to a region of this size, the best we can do is to state that its speed is less than one million miles per hour! The uncertainty principle clearly sets real, practical limits on our ability to describe electrons. The situation for the dust particle is different. We can't say for certain that the particle is absolutely at rest. But knowing that its speed is less than 5×10^{-15} m/s means that the particle is at rest for all practical purposes. At this speed, the dust particle would require nearly 6 hours to travel the width of one atom! Again we see that the quantum view has profound implications at the atomic scale but need not affect the way we think of macroscopic objects.

STOP TO THINK 28.6 The speeds of an electron and a proton have been measured to the same uncertainty. Which one has a larger uncertainty in position?

A. The proton, because it's more massive. B. The electron, because it's less massive.
C. The uncertainty in position is the same, because the uncertainty in velocity is the same.

28.8 Applications and Implications of Quantum Theory

Quantum theory seems bizarre to those of us living at a scale where the rather different rules of classical physics apply. In this section we consider some of the implications of quantum theory and some applications that confirm these unusual notions.

Tunneling and the Scanning Tunneling Microscope

The fact that particles have a wave nature allows for imaging at remarkably small scales—the scale of single atoms! The *scanning tunneling microscope* doesn't work like other microscopes you have seen, but instead builds an image of a solid surface by scanning a probe near the surface.

FIGURE 28.23 shows the tip of a very, very thin metal needle, called the *probe tip,* positioned above the surface of a solid sample. The space between the tip and the surface is about 0.5 nm, only a few atomic diameters. Electrons in the sample are attracted to the positive probe tip, but no current should flow, according to classical physics, because the electrons cannot cross the gap between the sample and the probe; it is an incomplete circuit. As we found with the photoelectric effect, the electrons are bound inside the sample and not free to leave.

However, electrons are not classical particles. The electron has a wave nature, and waves don't have sharp edges. One startling prediction of Schrödinger's quantum mechanics is that the electrons' wave functions extend very slightly beyond the edge of the sample. When the probe tip comes close enough to the surface, close enough to poke into an electron's wave function, an electron that had been in the sample might suddenly find itself in the probe tip. In other words, a quantum electron *can* cross the gap between the sample and the probe tip, thus causing a current to flow in the circuit. This process is called **tunneling** because it is rather like tunneling through an uncrossable mountain barrier to get to the other side. Tunneling is completely forbidden by the laws of classical physics, so the fact that it occurs is a testament to the reality of quantum ideas.

The probability that an electron will tunnel across the gap is very sensitive to the size of the gap, which makes the **scanning tunneling microscope,** or STM, possible. When the probe tip passes over an atom or over an atomic-level bump on the surface, the gap narrows and the current increases as more electrons are able to tunnel across. Similarly, the tunneling probability decreases when the probe tip passes across an atomic-size valley, and the current falls. The current-versus-position data are used to construct an image of the surface.

The STM was the first technology that allowed imaging of individual atoms, and it was one of a handful of inventions in the 1980s that jump-started the current interest in nanotechnology. STM images offer a remarkable view of the world at an atomic scale. The STM image of FIGURE 28.24a clearly shows the hexagonal arrangement of the individual atoms on the surface of pyrolytic graphite. The image of a DNA molecule in FIGURE 28.24b shows the actual twists of the double-helix structure. Current research efforts aim to develop methods for sequencing DNA with scanning tunneling microscopes and other nanoprobes—to directly "read" a single strand of DNA!

FIGURE 28.23 The scanning tunneling microscope.

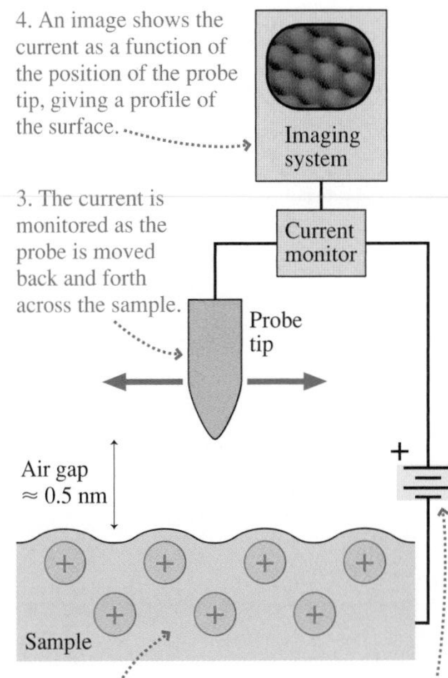

Imaging system
3. The current is monitored as the probe is moved back and forth across the sample.
Current monitor

1. The sample can be modeled as positive ion cores in an electron "sea."

FIGURE 28.24 STM images.

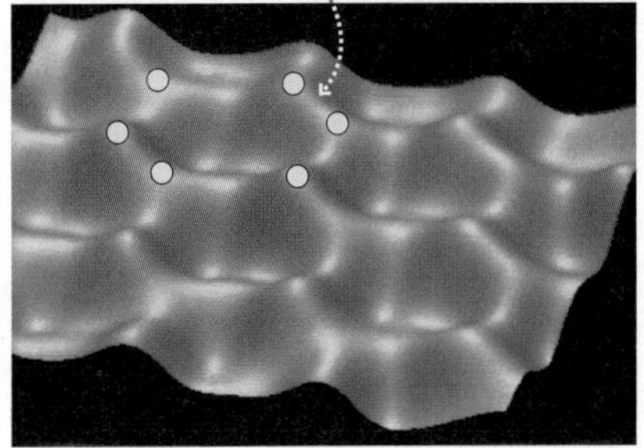

(a) Surface of graphite

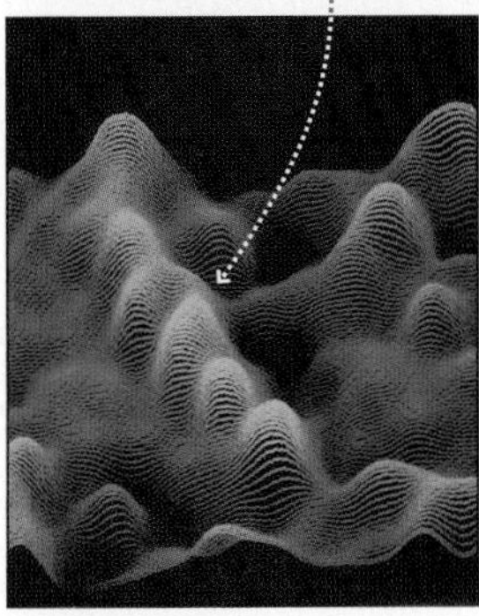

The light-colored peaks show the right-handed spiral of a DNA molecule.

(b) DNA molecule

Wave–Particle Duality

One common theme that has run through this chapter is the idea that, in quantum theory, things we think of as being waves have a particle nature, while things we think of as being particles have a wave nature. What is the true nature of light, or an electron? Are they particles or waves?

The various objects of classical physics are *either* particles *or* waves. There's no middle ground. Planets and baseballs are particles or collections of particles, while sound and light are clearly waves. Particles follow trajectories given by Newton's laws; waves obey the principle of superposition and exhibit interference. This wave–particle dichotomy seemed obvious until physicists encountered irrefutable evidence that light sometimes acts like a particle and, even stranger, that matter sometimes acts like a wave.

You might at first think that light and matter are *both* a wave *and* a particle, but that idea doesn't quite work. The basic definitions of particleness and waviness are mutually exclusive. Two sound waves can pass through each other and can overlap to produce a larger-amplitude sound wave; two baseballs can't. It is more profitable to conclude that light and matter are *neither* a wave *nor* a particle. At the microscopic scale of atoms and their constituents—a physical scale not directly accessible to our five senses—the classical concepts of particles and waves turn out to be simply too limited to explain the subtleties of nature.

Although matter and light have both wave-like aspects and particle-like aspects, they show us only one face at a time. If we arrange an experiment to measure a wave-like property, such as interference, we find photons and electrons acting like waves, not particles. An experiment to look for particles will find photons and electrons acting like particles, not waves. These two aspects of light and matter are *complementary* to each other, like a two-piece jigsaw puzzle. Neither the wave nor the particle model alone provides an adequate picture of light or matter, but taken together they provide us with a basis for understanding these elusive but most fundamental constituents of nature. This two-sided point of view is called *wave–particle duality.*

For over two hundred years, scientists and nonscientists alike felt that the clockwork universe of Newtonian physics was a fundamental description of reality. But wave–particle duality, along with Einstein's relativity, undermines the basic assumptions of the Newtonian worldview. The certainty and predictability of classical physics have given way to a new understanding of the universe in which chance and uncertainty play key roles—the universe of quantum physics.

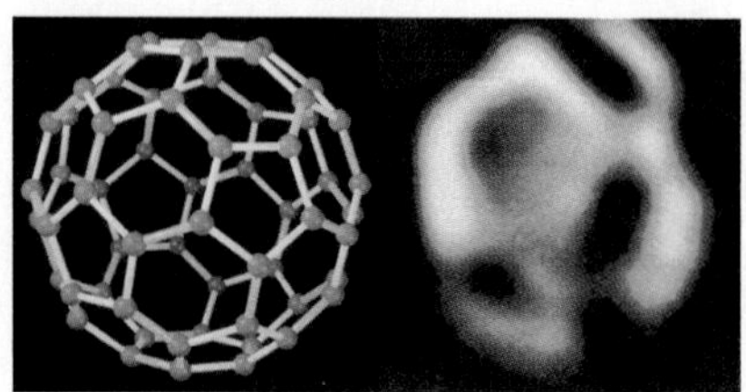

◀ **The dual nature of a buckyball** Treating atomic-level structures involves frequent shifts between particle and wave views. 60 carbon atoms can create the molecule diagrammed at left, known as C_{60}, or *buckminsterfullerene.* The scanning electron microscope image of a C_{60} molecule shown on the right is a particle-like view of the molecule with individual carbon atoms clearly visible. The C_{60} molecule, though we can make a picture of it—showing the atoms that make it up—also has a wave nature. A beam of C_{60} sent through a grating will produce a diffraction pattern!

INTEGRATED EXAMPLE 28.16 Magnetic resonance imaging

In Chapter 24, we learned that the magnetism of permanent magnets arises because the inherent magnetic moment of electrons causes them to act like little compass needles. Protons also have an inherent magnetic moment, and this is the basis for magnetic resonance imaging (MRI) in medicine.

Although a compass needle would prefer to align with a magnetic field, the needle can point in *any* direction. This isn't the case for the magnetic moment of a proton. Quantum physics tells us that the proton's energy must be quantized. There are only two possible energy levels—and thus two possible orientations—for protons in a magnetic field:

$E_1 = -\mu B$ magnetic moment aligned with the field

$E_2 = +\mu B$ magnetic moment aligned opposite the field

where $\mu = 1.41 \times 10^{-26}$ J/T is the known value of the proton's magnetic moment. **FIGURE 28.25** shows the two possible energy states. The magnetic moment, like a compass needle, "wants" to align with the field, so that is the lower-energy state.

FIGURE 28.25 Energy levels for a proton in a magnetic field.

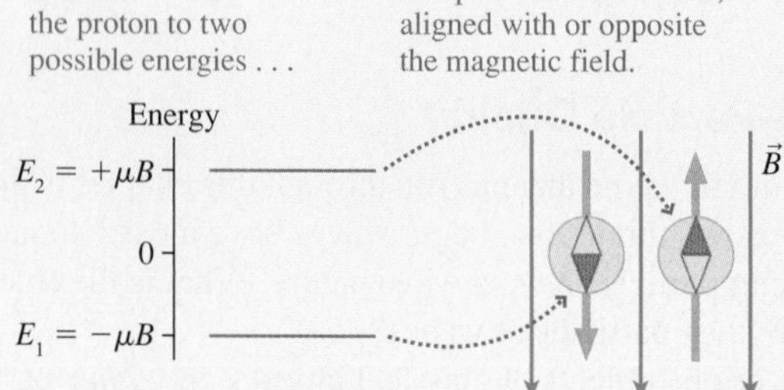
Quantum mechanics limits the proton to two possible energies . . .
. . . which correspond to two possible orientations, aligned with or opposite the magnetic field.
Energy
$E_2 = +\mu B$
0
$E_1 = -\mu B$
$\vec{B}$

Human tissue is mostly water. Each water molecule has two hydrogen atoms whose nuclei are single protons. In a magnetic field, the protons go into one or the other quantum state. A photon of just the right energy can "flip" the orientation of a proton's magnetic moment by causing a quantum jump from one state to the other. The energy difference between the states is small, so the relatively low-frequency photons are in the radio portion of the electromagnetic spectrum. These photons are provided by a *probe coil* that emits radio waves. When the probe is tuned to just the right frequency, the waves are *in resonance* with the energy levels of the protons, thus giving us the name magnetic *resonance* imaging.

The rate of absorption of these low-energy photons is proportional to the density of hydrogen atoms. Hydrogen density varies with tissue type, so an MRI image—showing different tissues—is formed by measuring the variation across the body of the rate at which photons cause quantum jumps between the two proton energy levels. A figure showing the absorption rate versus position in the body is an image of a "slice" through a patient's body, as in **FIGURE 28.26**.

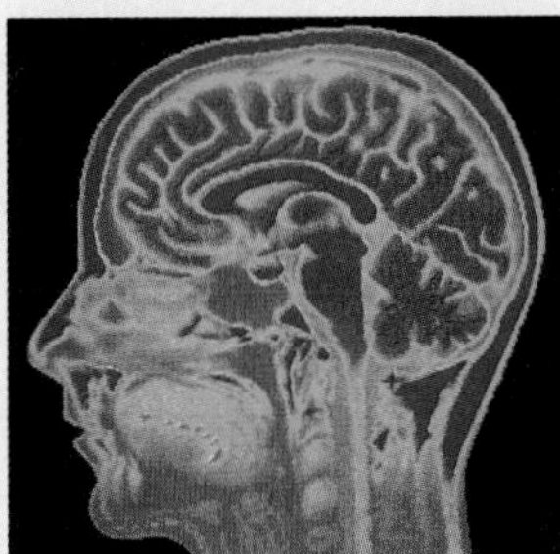

FIGURE 28.26 An MRI image shows the cross section of a patient's head.

a. An MRI patient is placed inside a solenoid that creates a strong magnetic field. If the field strength is 2.00 T, to what frequency must the probe coil be set? What is the wavelength of the photons produced?
b. In a uniform magnetic field, all protons in the body would absorb photons of the same frequency. To form an image of the body, the magnetic field is designed to vary from point to point in a known way. Because the field is different at each point in the body, each point has a unique frequency of photons that will be absorbed. The actual procedure is complex, but consider a simple model in which the field strength varies only along the axis of the patient's body, which we will call the x-axis. In particular, suppose that the magnetic field strength in tesla is given by $B = 2.00 + 1.60x$, where x, measured from a known reference point, is in meters. The probe coil is first tuned to the resonance frequency at the reference point. As the frequency is increased, a strong signal is observed at a frequency 4.7 MHz above the starting frequency. What is the location in the body, relative to the reference point, of the tissue creating this strong signal?

PREPARE If a photon has energy equal to the energy difference between the high- and low-energy states, it will be able to cause a quantum jump to the higher state—it will be absorbed. The photon energy E_{photon} must be equal to the energy difference between the two states: $\Delta E_{\text{system}} = 2\mu B$.

SOLVE a. At 2.00 T, the energy difference between the two proton states is

$$E_{\text{system}} = E_2 - E_1 = 2\mu B = 2(1.41 \times 10^{-26}\ \text{J/T})(2.00\ \text{T})$$
$$= 5.64 \times 10^{-26}\ \text{J}$$

This is a very low energy—only 3.5×10^{-7} eV. A photon will be absorbed if $E_{\text{photon}} = hf = E_{\text{system}}$. Thus the photon frequency must be

$$f = \frac{\Delta E_{\text{system}}}{h} = \frac{5.64 \times 10^{-26}\ \text{J}}{6.63 \times 10^{-34}\ \text{J}\cdot\text{s}}$$
$$= 85.1 \times 10^{6}\ \text{Hz} = 85.1\ \text{MHz}$$

This corresponds to a wavelength of $\lambda = c/f = 3.52$ m.

b. The magnetic field at the reference point ($x = 0$ m) is 2.00 T, so the probe frequency at this point is the 85.1 MHz we found in part a. The strong signal is 4.7 MHz above this, or 89.8 Hz. We can solve $E_{\text{photon}} = hf = E_{\text{system}} = 2\mu B$ to find the magnetic field at the point creating this strong signal:

$$B = \frac{hf}{2\mu} = 2.11\ \text{T}$$

We can then use the field-versus-distance formula given in the problem to find the position of this signal:

$$B = 2.11\ \text{T} = 2.00 + 1.60x$$
$$x = 0.069\ \text{m}$$

Thus there is a high density of protons 6.9 cm from the reference point.

ASSESS The frequency of the probe coil is in the radio portion of the electromagnetic spectrum, as we expected. The strong signal of part b is at a higher frequency, so this corresponds to a higher field and a positive value of x, as we found. The frequency is only slightly different from the original frequency, so we expect the point to be close to the reference position, as we found.

This is a simplified model of MRI, but the key features are present: A magnetic field that varies with position creates different energy levels for protons at different positions in the body, then tuned radio-wave photons measure the proton density at these different positions by causing and detecting quantum jumps between the two proton energy levels.

SUMMARY

The goal of Chapter 28 has been to understand the quantization of energy for light and matter.

GENERAL PRINCIPLES

Light has particle-like properties

- The energy of a light wave comes in discrete packets (light quanta) we call **photons.**
- For light of frequency f, the energy of each photon is $E = hf$, where h is **Planck's constant**.
- When light strikes a metal surface, all of the energy of a single photon is given to a single electron.

Matter has wave-like properties

- The **de Broglie wavelength** of a particle of mass m is $\lambda = h/mv$.
- The wave-like nature of matter is seen in the interference patterns of electrons, protons, and other particles.

Quantization of energy

When a particle is confined, it sets up a de Broglie standing wave.

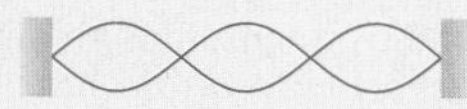

The fact that standing waves can have only certain allowed wavelengths leads to the conclusion that a confined particle can have only certain allowed energies.

Wave–particle duality

- Experiments designed to measure wave properties will show the wave nature of light and matter.
- Experiments designed to measure particle properties will show the particle nature of light and matter.

Heisenberg uncertainty principle

A particle with wave-like characteristics does not have a precise value of position x or a precise value of momentum p_x. Both are uncertain. The position uncertainty Δx and momentum uncertainty Δp_x are related by

$$\Delta x\, \Delta p_x \geq \frac{h}{2\pi}$$

The more you pin down the value of one, the less precisely the other can be known.

IMPORTANT CONCEPTS

Photoelectric effect

Light with frequency f can eject electrons from a metal only if $f \geq f_0 = E_0/h$, where E_0 is the metal's **work function.** Electrons will be ejected even if the intensity of the light is very small.

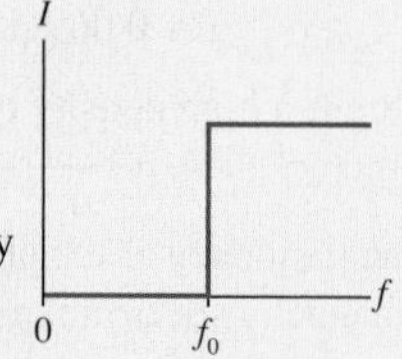

The **stopping potential** that stops even the fastest electrons is

$$V_{stop} = \frac{K_{max}}{e} = \frac{hf - E_0}{e}$$

I

ΔV

$-V_{stop}$ 0

The details of the photoelectric effect could not be explained with classical physics. New models were needed.

X-ray diffraction

X rays with wavelength λ undergo strong reflections from atomic planes spaced by d when the angle of incidence satisfies the **Bragg condition:**

$$2d\cos\theta = m\lambda$$

$$m = 1, 2, 3, \ldots$$

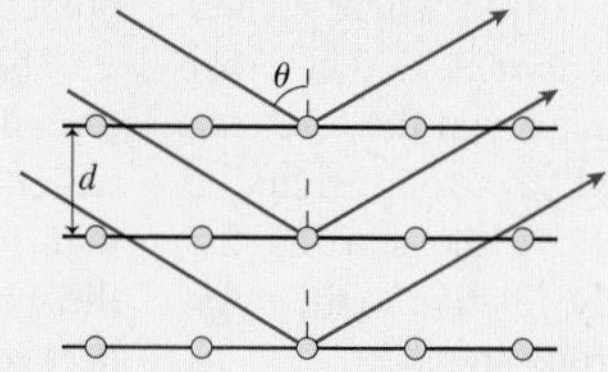

Energy levels and quantum jumps

The localization of electrons leads to quantized energy levels. An electron can exist only in certain energy states. An electron can jump to a higher level if a photon is absorbed, or to a lower level if a photon is emitted. The energy difference between the levels equals the photon energy.

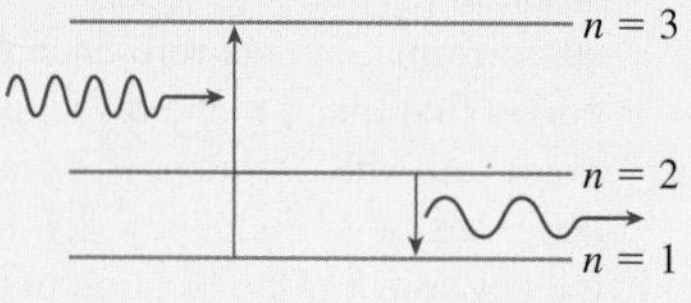

APPLICATIONS

The wave nature of light limits the resolution of a light microscope. A more detailed image may be made with an **electron microscope** because of the very small de Broglie wavelength of fast electrons.

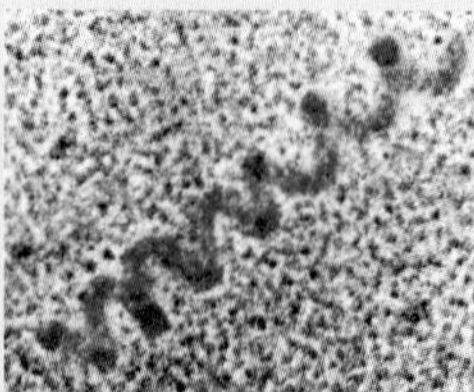

The wave nature of electrons allows them to **tunnel** across an insulating layer of air to the tip of a **scanning tunneling microscope,** revealing details of the atoms on a surface.

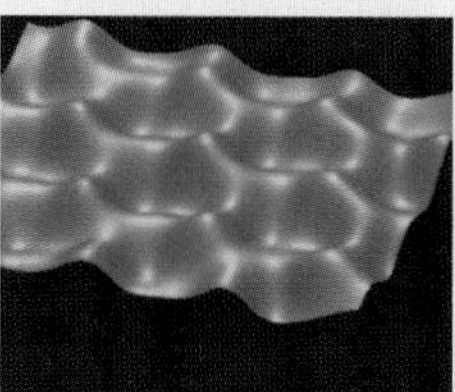

MP™ For homework assigned on MasteringPhysics, go to www.masteringphysics.com

Problem difficulty is labeled as | (straightforward) to ||||| (challenging).

Problems labeled INT integrate significant material from earlier chapters; BIO are of biological or medical interest.

QUESTIONS

Conceptual Questions

1. The first-order x-ray diffraction of monochromatic x rays from a crystal occurs at angle θ_1. The crystal is then compressed, causing a slight reduction in its volume. Does θ_1 increase, decrease, or stay the same? Explain.
2. Explain the reasoning by which we claim that the stopping potential V_{stop} measures the maximum kinetic energy of the electrons in a photoelectric-effect experiment.
3. How does Einstein's explanation account for each of these characteristics of the photoelectric effect?
   A. The photoelectric current is zero for frequencies below some threshold.
   B. The photoelectric current increases with increasing light intensity.
   C. The photoelectric current is independent of ΔV for $\Delta V > 0$.
   D. The photoelectric current decreases slowly as ΔV becomes more negative.
   E. The stopping potential is independent of the light intensity.

   Which of these *cannot* be explained by classical physics? Explain.
4. How would the graph of Figure 28.7a look if the emission of electrons from the cathode was due to the heating of the metal by light falling on it? Draw the graph and explain your reasoning. Assume that the light intensity remains constant as its frequency and wavelength are varied.
5. Figure Q28.5 shows the typical photoelectric behavior of a metal as the anode-cathode potential difference ΔV is varied.
   a. Why do the curves become horizontal for $\Delta V > 0$ V? Shouldn't the current increase as the potential difference increases? Explain.
   b. Why doesn't the current immediately drop to zero for $\Delta V < 0$ V? Shouldn't $\Delta V < 0$ V prevent the electrons from reaching the anode? Explain.
   c. The current is zero for $\Delta V < -2.0$ V. Where do the electrons go? Are no electrons emitted if $\Delta V < -2.0$ V? Or if they are, why is there no current? Explain.

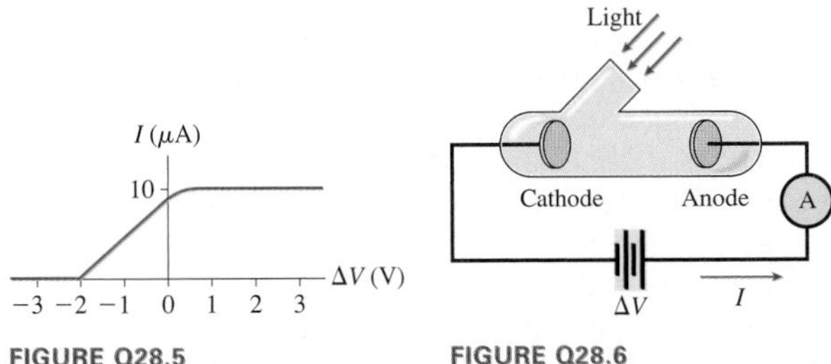

FIGURE Q28.5

FIGURE Q28.6

6. In the photoelectric effect experiment, as illustrated by Figure Q28.6, a current is measured while light is shining on the cathode. But this does not appear to be a complete circuit, so how can there be a current? Explain.
7. Metal surfaces on spacecraft in bright sunlight develop a net electric charge. Do they develop a negative or a positive charge? Explain.
8. Metal 1 has a larger work function than metal 2. Both are illuminated with the same short-wavelength ultraviolet light. Do electrons from metal 1 have a higher speed, a lower speed, or the same speed as electrons from metal 2? Explain.
9. A gold cathode is illuminated with light of wavelength 250 nm. It is found that the current is zero when $\Delta V = 1.0$ V. Would the current change if
   a. The light intensity is doubled?
   b. The anode-cathode potential difference is increased to $\Delta V = 5.5$ V?
10. Three laser beams have wavelengths $\lambda_1 = 400$ nm, $\lambda_2 = 600$ nm, and $\lambda_3 = 800$ nm. The power of each laser beam is 1 W.
    a. Rank in order, from largest to smallest, the photon energies E_1, E_2, and E_3 in these three laser beams. Explain.
    b. Rank in order, from largest to smallest, the number of photons per second N_1, N_2, and N_3 delivered by the three laser beams. Explain.
11. When we say that a photon is a "quantum of light," what does that mean? What is quantized?
12. An investigator is measuring the current in a photoelectric effect experiment. The cathode is illuminated by light of a single wavelength. What happens to the current if the intensity of the light is doubled while the wavelength is held constant?
13. An investigator is measuring the current in a photoelectric effect experiment. The cathode is illuminated by light of a single wavelength. What happens to the current if the wavelength of the light is reduced by a factor of two while keeping the intensity constant?
14. To have the best resolution, should an electron microscope use very fast electrons or very slow electrons? Explain.
15. An electron and a proton are accelerated from rest through potential differences of the same magnitude. Afterward, which particle has the larger de Broglie wavelength? Explain.
16. A neutron is shot straight up with an initial speed of 100 m/s. As it rises, does its de Broglie wavelength increase, decrease, or not change? Explain.
17. Double-slit interference of electrons occurs because:
    A. The electrons passing through the two slits repel each other.
    B. Electrons collide with each other behind the slits.
    C. Electrons collide with the edges of the slits.
    D. Each electron goes through both slits.
    E. The energy of the electrons is quantized.
    F. Only certain wavelengths of the electrons fit through the slits.

    Which of these (perhaps none, perhaps more than one) are correct? Explain.
18. Can an electron with a de Broglie wavelength of 2 μm pass through a slit that is 1 μm wide? Explain.

19. a. For the allowed energies of a particle in a box to be large, should the box be very big or very small? Explain.
    b. Which is likely to have larger values for the allowed energies: an atom in a molecule, an electron in an atom, or a proton in a nucleus? Explain.
20. Figure Q28.20 shows the standing de Broglie wave of a particle in a box.
    a. What is the quantum number?
    b. Can you determine from this picture whether the "classical" particle is moving to the right or to the left? If so, which is it? If not, why not?

FIGURE Q28.20 

21. A particle in a box of length L_a has $E_1 = 2$ eV. The same particle in a box of length L_b has $E_2 = 50$ eV. What is the ratio L_a/L_b?
22. Imagine that the horizontal box of Figure 28.18 is instead oriented vertically. Also imagine the box to be on a neutron star where the gravitational field is so strong that the particle in the box slows significantly, nearly stopping, before it hits the top of the box. Make a *qualitative* sketch of the $n = 3$ de Broglie standing wave of a particle in this box.
    Hint: The nodes are *not* uniformly spaced.
23. Figure Q28.23 shows a standing de Broglie wave.
    a. Does this standing wave represent a particle that travels back and forth between the boundaries with a constant speed or a changing speed? Explain.
    b. If the speed is changing, at which end is the particle moving faster and at which end is it moving slower?

FIGURE Q28.23 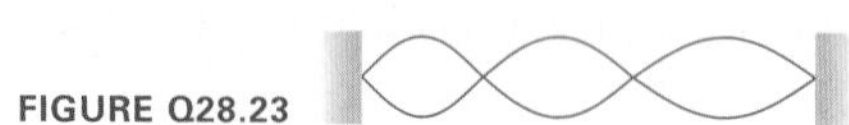

24. BIO The molecules in the rods and cones in the eye are tuned to absorb photons of particular energies. The retinal molecule, like many molecules, is a long chain. Electrons can freely move along one stretch of the chain but are reflected at the ends, thus behaving like a particle in a one-dimensional box. The absorption of a photon lifts an electron from the ground state into the first excited state. Do the molecules in a red cone (which are tuned to absorb red light) or the molecules in a blue cone (tuned to absorb blue light) have a longer "box"?
25. Science fiction movies often use devices that transport people and objects rapidly from one position to another. To "beam" people in this fashion means taking them apart atom by atom, carefully measuring each position, and then sending the atoms in a beam to the desired final location where they reassemble. How do the principles of quantum mechanics pose problems for this futuristic means of transportation?

Multiple-Choice Questions

26. | A light sensor is based on a photodiode that requires a minimum photon energy of 1.7 eV to create mobile electrons. What is the longest wavelength of electromagnetic radiation that the sensor can detect?
    A. 500 nm
    B. 730 nm
    C. 1200 nm
    D. 2000 nm
27. | In a photoelectric effect experiment, the frequency of the light is increased while the intensity is held constant. As a result,
    A. There are more electrons. B. The electrons are faster.
    C. Both A and B. D. Neither A nor B.
28. | In a photoelectric effect experiment, the intensity of the light is increased while the frequency is held constant. As a result,
    A. There are more electrons.
    B. The electrons are faster.
    C. Both A and B.
    D. Neither A nor B.
29. | In the photoelectric effect, electrons are never emitted from a metal if the frequency of the incoming light is below a certain threshold value. This is because
    A. Photons of lower-frequency light don't have enough energy to eject an electron.
    B. The electric field of low-frequency light does not vibrate the electrons rapidly enough to eject them.
    C. The number of photons in low-frequency light is too small to eject electrons.
    D. Low-frequency light does not penetrate far enough into the metal to eject electrons.
30. || Visible light has a wavelength of about 500 nm. A typical radio wave has a wavelength of about 1.0 m. How many photons of the radio wave are needed to equal the energy of one photon of visible light?
    A. 2,000 B. 20,000
    C. 200,000 D. 2,000,000
31. | Two radio stations have the same power output from their antennas. One broadcasts AM at a frequency of 1000 kHz and one broadcasts FM at a frequency of 100 MHz. Which statement is true?
    A. The FM station emits more photons per second.
    B. The AM station emits more photons per second.
    C. The two stations emit the same number of photons per second.
32. | An electron is accelerated through a 5000 V potential difference, strikes a metal target, and causes an x ray to be emitted. What is the (approximate) minimum wavelength of the emitted x ray?
    A. 0.25 nm B. 1.0 nm
    C. 2.5 nm D. 4.0 nm
33. || How many photons does a 5.0 mW helium-neon laser ($\lambda = 633$ nm) emit in 1 second?
    A. 1.2×10^{19} B. 4.0×10^{18}
    C. 8.0×10^{16} D. 1.6×10^{16}
34. | You shoot a beam of electrons through a double slit to make an interference pattern. After noting the properties of the pattern, you then double the speed of the electrons. What effect would this have?
    A. The fringes would get closer together.
    B. The fringes would get farther apart.
    C. The positions of the fringes would not change.
35. | Photon P in Figure Q28.35 moves an electron from energy level $n = 1$ to energy level $n = 3$. The electron jumps down to $n = 2$, emitting photon Q, and then jumps down to $n = 1$, emitting photon R. The spacing be-tween energy levels is drawn to scale. What is the correct relationship among the wavelengths of the photons?

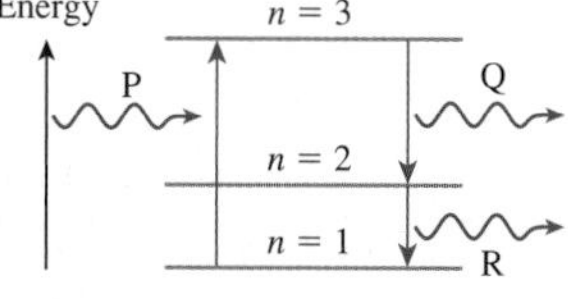

FIGURE Q28.35

    A. $\lambda_P < \lambda_Q < \lambda_R$ B. $\lambda_R < \lambda_P < \lambda_Q$
    C. $\lambda_Q < \lambda_P < \lambda_R$ D. $\lambda_R < \lambda_Q < \lambda_P$

PROBLEMS

Section 28.1 X Rays and X-Ray Diffraction

1. | X rays with a wavelength of 0.12 nm undergo first-order diffraction from a crystal at a 68° angle of incidence. What is the angle of second-order diffraction?
2. | X rays with a wavelength of 0.20 nm undergo first-order diffraction from a crystal at a 54° angle of incidence. At what angle does first-order diffraction occur for x rays with a wavelength of 0.15 nm?
3. | X rays diffract from a crystal in which the spacing between atomic planes is 0.175 nm. The second-order diffraction occurs at 45.0°. What is the angle of the first-order diffraction?
4. || The spacing between atomic planes in a crystal is 0.110 nm. If 12.0 keV x rays are diffracted by this crystal, what are the angles of (a) first-order and (b) second-order diffraction?
5. || X rays with a wavelength of 0.085 nm diffract from a crystal in which the spacing between atomic planes is 0.18 nm. How many diffraction orders are observed?

Section 28.2 The Photoelectric Effect

6. | Which metals in Table 28.1 exhibit the photoelectric effect for (a) light with $\lambda = 400$ nm and (b) light with $\lambda = 250$ nm?
7. | Electrons are emitted when a metal is illuminated by light with a wavelength less than 388 nm but for no greater wavelength. What is the metal's work function?
8. || Electrons in a photoelectric-effect experiment emerge from a copper surface with a maximum kinetic energy of 1.10 eV. What is the wavelength of the light?
9. || You need to design a photodetector that can respond to the entire range of visible light. What is the maximum possible work function of the cathode?
10. | A photoelectric-effect experiment finds a stopping potential of 1.93 V when light of 200 nm wavelength is used to illuminate the cathode.
    a. From what metal is the cathode made?
    b. What is the stopping potential if the intensity of the light is doubled?
11. || Zinc has a work function of 4.3 eV.
    a. What is the longest wavelength of light that will release an electron from a zinc surface?
    b. A 4.7 eV photon strikes the surface and an electron is emitted. What is the maximum possible speed of the electron?
12. || Image intensifiers used in night-vision devices create a bright image from dim light by letting the light first fall on a *photocathode*. Electrons emitted by the photoelectric effect are accelerated and then strike a phosphorescent screen, causing it to glow more brightly than the original scene. Recent devices are sensitive to wavelengths as long as 900 nm, in the infrared:

    a. If the threshold wavelength is 900 nm, what is the work function of the photocathode?
    b. If light of wavelength 700 nm strikes such a photocathode, what will be the maximum kinetic energy, in eV, of the emitted electrons?
13. || Light with a wavelength of 350 nm shines on a metal surface, which emits electrons. The stopping potential is measured to be 1.25 V.
    a. What is the maximum speed of emitted electrons?
    b. Calculate the work function and identify the metal.

Section 28.3 Photons

14. | BIO When an ultraviolet photon is absorbed by a molecule of DNA, the photon's energy can be converted into vibrational energy of the molecular bonds. Excessive vibration damages the molecule by causing the bonds to break. Ultraviolet light of wavelength less than 290 nm causes significant damage to DNA; ultraviolet light of longer wavelength causes minimal damage. What is the threshold photon energy, in eV, for DNA damage?
15. | The spacing between atoms in graphite is approximately 0.25 nm. What is the energy of an x-ray photon with this wavelength?
16. || BIO A firefly glows by the direct conversion of chemical energy to light. The light emitted by a firefly has peak intensity at a wavelength of 550 nm.

    a. What is the minimum chemical energy, in eV, required to generate each photon?
    b. One molecule of ATP provides 0.30 eV of energy when it is metabolized in a cell. What is the minimum number of ATP molecules that must be consumed in the reactions that lead to the emission of one photon of 550 nm light?
17. | BIO Your eyes have three different types of cones with maximum absorption at 437 nm, 533 nm, and 564 nm. What photon energies correspond to these wavelengths?
18. | What is the wavelength, in nm, of a photon with energy (a) 0.30 eV, (b) 3.0 eV, and (c) 30 eV? For each, is this wavelength visible light, ultraviolet, or infrared?
19. | What is the ratio of the energy of a photon of light at the far red end of the visible spectrum (700 nm) to that of a photon at the far blue end of the visible spectrum (400 nm)?
20. || BIO INT The wavelengths of light emitted by a firefly span the visible spectrum but have maximum intensity near 550 nm. A typical flash lasts for 100 ms and has a power of 1.2 mW. If we assume that all of the light is emitted at the peak-intensity wavelength of 550 nm, how many photons are emitted in one flash?
21. || Station KAIM in Hawaii broadcasts on the AM dial at 870 kHz, with a maximum power of 50,000 W. At maximun power, how many photons does the transmitting antenna emit each second?
22. ||| BIO At 510 nm, the wavelength of maximum sensitivity of the human eye, the dark-adapted eye can sense a 100-ms-long flash of light of total energy 4.0×10^{-17} J. (Weaker flashes of light may be detected, but not reliably.) If 60% of the incident light is lost to reflection and absorption by tissues of the eye, how many photons reach the retina from this flash?

23. | 550 nm is the average wavelength of visible light.
    a. What is the energy of a photon with a wavelength of 550 nm?
    b. A typical incandescent lightbulb emits about 1 J of visible light energy every second. Estimate the number of visible photons emitted per second.

24. || *Dinoflagellates* are single-cell creatures that float in the world's oceans; many types are bioluminescent. When disturbed by motion in the water, a typical bioluminescent dinoflagellate emits 100,000,000 photons in a 0.10-s-long flash of light of wavelength 460 nm. What is the power of the flash in watts?
BIO

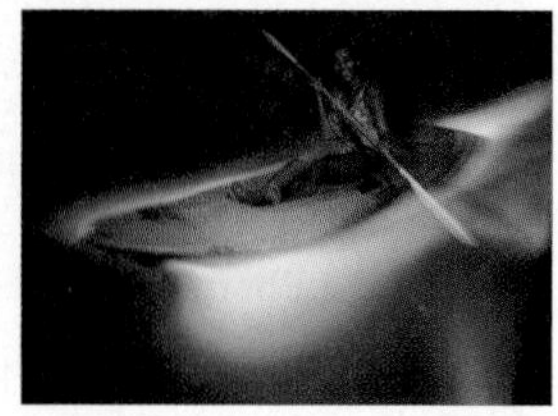

25. || A circuit employs a silicon solar cell to detect flashes of light lasting 0.25 s. The smallest current the circuit can detect reliably is 0.42 μA. Assuming that all photons reaching the solar cell give their energy to a charge carrier, what is the minimum power of a flash of light of wavelength 550 nm that can be detected?

Section 28.4 Matter Waves

26. | Estimate your de Broglie wavelength while walking at a speed of 1 m/s.

27. | a. What is the de Broglie wavelength of a 200 g baseball with a speed of 30 m/s?
    b. What is the speed of a 200 g baseball with a de Broglie wavelength of 0.20 nm?

28. | a. What is the speed of an electron with a de Broglie wavelength of 0.20 nm?
    b. What is the speed of a proton with a de Broglie wavelength of 0.20 nm?

29. || What is the kinetic energy, in eV, of an electron with a de Broglie wavelength of 1.0 nm?

30. || A paramecium is covered with motile hairs called cilia that propel it at a speed of 1 mm/s. If the paramecium has a volume of 2×10^{-13} m^3 and a density equal that of water, what is its de Broglie wavelength when in motion? What fraction of the paramecium's 150 μm length does this wavelength represent?
BIO INT

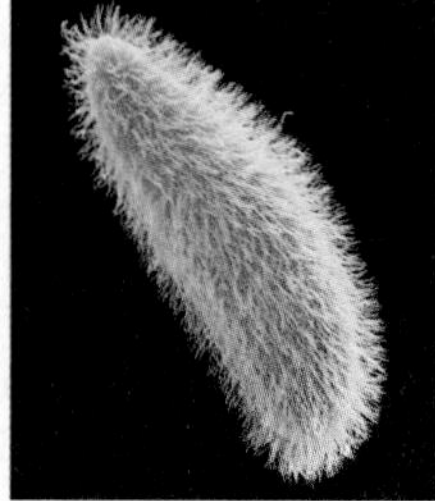

31. || The diameter of an atomic nucleus is about 10 fm (1 fm $= 10^{-15}$ m). What is the kinetic energy, in MeV, of a proton with a de Broglie wavelength of 10 fm?

32. || Rubidium atoms are cooled to 0.10 μK in an atom trap. What is their de Broglie wavelength? How many times larger is this than the 0.25 nm diameter of the atoms?
INT

33. || Through what potential difference must an electron be accelerated from rest to have a de Broglie wavelength of 500 nm?

Section 28.5 Energy Is Quantized

34. || What is the length of a box in which the minimum energy of an electron is 1.5×10^{-18} J?

35. ||| What is the length of a one-dimensional box in which an electron in the $n = 1$ state has the same energy as a photon with a wavelength of 600 nm?

36. ||| An electron confined in a one-dimensional box is observed, at different times, to have energies of 12 eV, 27 eV, and 48 eV. What is the length of the box?

37. | The nucleus of a typical atom is 5.0 fm (1 fm $= 10^{-15}$ m) in diameter. A very simple model of the nucleus is a one-dimensional box in which protons are confined. Estimate the energy of a proton in the nucleus by finding the first three allowed energies of a proton in a 5.0-fm-long box.

Section 28.6 Energy Levels and Quantum Jumps

38. || The allowed energies of a quantum system are 1.0 eV, 2.0 eV, 4.0 eV, and 7.0 eV. What wavelengths appear in the system's emission spectrum?

39. || Figure P28.39 is an energy-level diagram for a quantum system. What wavelengths appear in the system's emission spectrum?

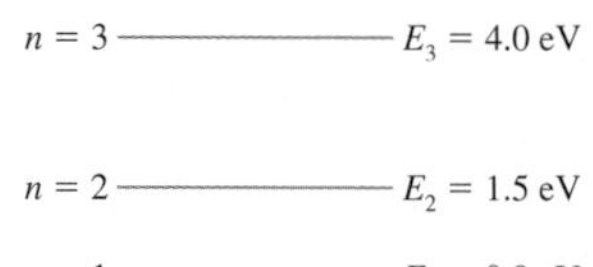

FIGURE P28.39

40. ||| The allowed energies of a quantum system are 0.0 eV, 4.0 eV, and 6.0 eV.
    a. Draw the system's energy-level diagram. Label each level with the energy and the quantum number.
    b. What wavelengths appear in the system's emission spectrum?

41. ||| The allowed energies of a quantum system are 0.0 eV, 1.5 eV, 3.0 eV, and 6.0 eV. How many different wavelengths appear in the emission spectrum?

Section 28.7 The Uncertainty Principle

42. || The speed of an electron is known to be between 3.0×10^6 m/s and 3.2×10^6 m/s. Estimate the uncertainty in its position.

43. || What is the smallest box in which you can confine an electron if you want to know for certain that the electron's speed is no more than 10 m/s?

44. ||| A spherical virus has a diameter of 50 nm. It is contained inside a long, narrow cell of length 1×10^{-4} m. What uncertainty does this imply for the velocity of the virus along the length of the cell? Assume the virus has a density equal to that of water.
BIO INT

45. ||| A thin solid barrier in the xy-plane has a 10-μm-diameter circular hole. An electron traveling in the z-direction with $v_x = 0$ m/s passes through the hole. Afterward, is v_x still zero? If not, within what range is v_x likely to be?

46. ||| A proton is confined within an atomic nucleus of diameter 4 fm (1 fm $= 10^{-15}$ m). Estimate the smallest range of speeds you might find for a proton in the nucleus.

General Problems

47. || X rays with a wavelength of 0.0700 nm diffract from a crystal. Two adjacent angles of x-ray diffraction are 45.6° and 21.0°. What is the distance in nm between the atomic planes responsible for the diffraction?

48. || Potassium and gold cathodes are used in a photoelectric-effect experiment. For each cathode, find:
    a. The threshold frequency
    b. The threshold wavelength
    c. The maximum electron ejection speed if the light has a wavelength of 220 nm
    d. The stopping potential if the wavelength is 220 nm

49. ||||| In a photoelectric-effect experiment, the maximum kinetic energy of electrons is 2.8 eV. When the wavelength of the light is increased by 50%, the maximum energy decreases to 1.1 eV. What are (a) the work function of the cathode and (b) the initial wavelength?

50. ||||| In a photoelectric-effect experiment, the stopping potential at a wavelength of 400 nm is 25.7% of the stopping potential at a wavelength of 300 nm. Of what metal is the cathode made?

51. || Light of constant intensity but varying wavelength was used to illuminate the cathode in a photoelectric-effect experiment. The graph of Figure P28.51 shows how the stopping potential depended on the frequency of the light. What is the work function, in eV, of the cathode?

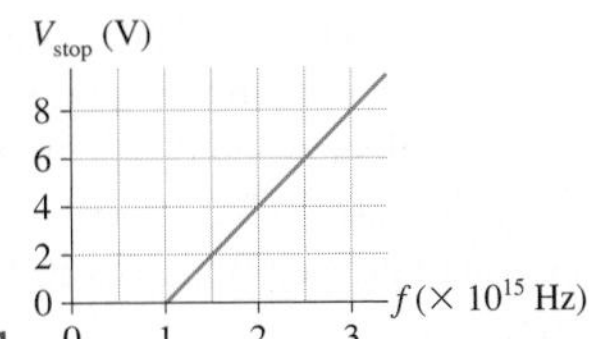

FIGURE P28.51

52. || What is the de Broglie wavelength of a red blood cell BIO with a mass of 1.00×10^{-11} g that is moving with a speed of 0.400 cm/s? Do we need to be concerned with the wave nature of the blood cells when we describe the flow of blood in the body?

53. || Suppose you need to image the structure of a virus with a BIO diameter of 50 nm. For a sharp image, the wavelength of the probing wave must be 5.0 nm or less. We have seen that, for imaging such small objects, this short wavelength is obtained by using an electron beam in an electron microscope. Why don't we simply use short-wavelength electromagnetic waves? There's a problem with this approach: As the wavelength gets shorter, the energy of a photon of light gets greater and could damage or destroy the object being studied. Let's compare the energy of a photon and an electron that can provide the same resolution.
   a. For light of wavelength 5.0 nm, what is the energy (in eV) of a single photon? In what part of the electromagnetic spectrum is this?
   b. For an electron with a de Broglie wavelength of 5.0 nm, what is the kinetic energy (in eV).

54. | Gamma rays are photons with very high energy.
   a. What is the wavelength of a gamma-ray photon with energy 625 keV?
   b. How many visible-light photons with a wavelength of 500 nm would you need to match the energy of this one gamma-ray photon?

55. || A red laser with a wavelength of 650 nm and a blue laser with a wavelength of 450 nm emit laser beams with the same light power. What is the ratio of the red laser's photon emission rate (photons per second) to the blue laser's photon emission rate?

56. || A typical incandescent lightbulb emits approximately BIO INT 3×10^{18} visible-light photons per second. Your eye, when it is fully dark adapted, can barely see the light from an incandescent lightbulb 10 km away. How many photons per second are incident at the image point on your retina? The diameter of a dark-adapted pupil is 6 mm.

57. || The intensity of sunlight hitting the surface of the earth on BIO a cloudy day is about 0.50 kW/m^2. Assuming your pupil can close down to a diameter of 2.0 mm and that the average wavelength of visible light is 550 nm, how many photons per second of visible light enter your eye if you look up at the sky on a cloudy day?

58. ||| A red LED (light emitting diode) is connected to a battery; it INT carries a current. As electrons move through the diode, they jump between states, emitting photons in the process. Assume that each electron that travels through the diode causes the emission of a single 630 nm photon. What current is necessary to produce 5.0 mW of emitted light?

59. || A ruby laser emits an intense pulse of light that lasts a mere 10 ns. The light has a wavelength of 690 nm, and each pulse has an energy of 500 mJ.
   a. How many photons are emitted in each pulse?
   b. What is the *rate* of photon emission, in photons per second, during the 10 ns that the laser is "on"?

60. ||| The human body emits thermal electromagnetic radiation, as BIO INT we've seen. Assuming that all radiation is emitted at the wavelength of peak intensity, for a skin temperature of 33°C and a surface area of 1.8 m^2, how many photons per second does the body emit?

61. ||| The wavelength of the radiation in a microwave oven is 12 cm. INT How many photons are absorbed by 200 g of water as it's heated from 20°C to 90°C?

62. ||| Exposure to a sufficient quantity of ultraviolet will redden BIO the skin, producing *erythema*—a sunburn. The amount of exposure necessary to produce this reddening depends on the wavelength. For a 1.0 cm^2 patch of skin, 3.7 mJ of ultraviolet light at a wavelength of 254 nm will produce reddening; at 300 nm wavelength, 13 mJ are required.
   a. What is the photon energy corresponding to each of these wavelengths?
   b. How many total photons does each of these exposures correspond to?
   c. Explain why there is a difference in the number of photons needed to provoke a response in the two cases.

63. ||| A silicon solar cell looks like a battery with a 0.50 V terminal INT voltage. Suppose that 1.0 W of light of wavelength 600 nm falls on a solar cell and that 50% of the photons give their energy to charge carriers, creating a current. What is the solar cell's efficiency—that is, what percentage of the energy incident on the cell is converted to electric energy?

64. ||| Electrons with a speed of 2.0×10^6 m/s pass through a double-INT slit apparatus. Interference fringes are detected with a fringe spacing of 1.5 mm.
   a. What will the fringe spacing be if the electrons are replaced by neutrons with the same speed?
   b. What speed must neutrons have to produce interference fringes with a fringe spacing of 1.5 mm?

65. | Electrons pass through a 1.0-μm-wide slit with a speed of INT 1.5×10^6 m/s. How wide is the electron diffraction pattern on a detector 1.0 m behind the slit?

66. | The electron interference pattern of Figure 28.14 was made INT by shooting electrons with 50 keV of kinetic energy through two slits spaced 1.0 μm apart. The fringes were recorded on a detector 1.0 m behind the slits.
   a. What was the speed of the electrons? (The speed is large enough to justify using relativity, but for simplicity do this as a nonrelativistic calculation.)
   b. Figure 28.14 is greatly magnified. What was the actual spacing on the detector between adjacent bright fringes?

67. |||| It is stated in the text that special relativity must be used to INT calculate the de Broglie wavelength of electrons in an electron microscope. Let us discover how much of an effect relativity has. Consider an electron accelerated through a potential difference of 1.00×10^5 V.
   a. Using the Newtonian (nonrelativistic) expressions for kinetic energy and momentum, what is the electron's de Broglie wavelength?
   b. The de Broglie wavelength is $\lambda = h/p$, but the momentum of a relativistic particle is not mv. Using the relativistic expressions for kinetic energy and momentum, what is the electron's de Broglie wavelength?
68. || An electron confined to a one-dimensional box of length 0.70 nm jumps from the $n = 2$ level to the ground state. What is the wavelength (in nm) of the emitted photon?
69. | a. What is the minimum energy of a 2.7 g Ping-Pong ball in a 10-cm-long box?
   b. What speed corresponds to this kinetic energy?
70. ||| The color of dyes results from the preferential absorption of certain wavelengths of light. Certain dye molecules consist of symmetric pairs of rings joined at the center by a chain of carbon atoms, as shown in Figure P28.70. Electrons of the bonds along the chain of carbon atoms are shared among the atoms in the chain, but are repelled by the nitrogen-containing rings at the end of the chain. These electrons are thus free to move along the chain but not beyond its ends. They look very much like a particle in a one-dimensional box. For the molecule shown, the effective length of the "box" is 0.85 nm. Assuming that the electrons start in the lowest energy state, what are the three longest wavelengths this molecule will absorb?

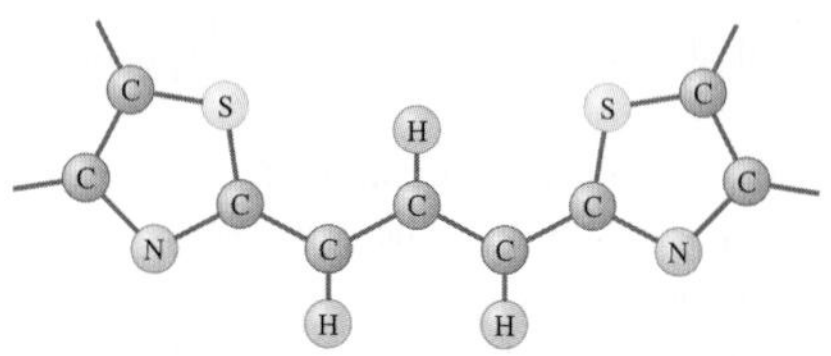

FIGURE P28.70

71. || What is the length of a box in which the difference between an electron's first and second allowed energies is 1.0×10^{-19} J?
72. || Two adjacent allowed energies of an electron in a one-dimensional box are 2.0 eV and 4.5 eV. What is the length of the box?
73. ||| An electron confined to a box has an energy of 1.28 eV. Another electron confined to an identical box has an energy of 2.88 eV. What is the smallest possible length for those boxes?
74. || Consider a small virus having a diameter of 10 nm. The BIO INT atoms of the intracellular fluid are confined within this "box." Suppose we model the virus as a one-dimensional box of length 10 nm. What is the ground-state energy (in eV) of a sodium ion confined in such a box?
75. ||| It can be shown that the allowed energies of a particle of mass m in a two-dimensional square box of side L are

$$E_{nm} = \frac{h^2}{8mL^2}(n^2 + l^2)$$

The energy depends on two quantum numbers, n and l, both of which must have an integer value 1, 2, 3, ...
   a. What is the minimum energy for a particle in a two-dimensional square box of side L?
   b. What are the five lowest allowed energies? Give your values as multiples of E_{min}.
76. ||| An electron confined in a one-dimensional box emits a 200 nm photon in a quantum jump from $n = 2$ to $n = 1$. What is the length of the box?
77. ||| A proton confined in a one-dimensional box emits a 2.0 MeV gamma-ray photon in a quantum jump from $n = 2$ to $n = 1$. What is the length of the box?
78. ||| As an electron in a one-dimensional box of length 0.600 nm jumps between two energy levels, a photon of energy 8.36 eV is emitted. What are the quantum numbers of the two levels?
79. || Magnetic resonance is used in imaging; it is also a useful tool for analyzing chemical samples. Magnets for magnetic resonance experiments are often characterized by the proton resonance frequency they create. What is the field strength of an 800 MHz magnet?
80. ||| The electron has a magnetic moment, so you can do magnetic INT resonance measurements on substances with unpaired electron spins. The electron has a magnetic moment $\mu = 9.3 \times 10^{-24}$ J/T. A sample is placed in a solenoid of length 15 cm with 1200 turns of wire carrying a current of 3.5 A. A probe coil provides radio waves to "flip" the spins. What is the necessary frequency for the probe coil?

Passage Problems

Compton Scattering

Further support for the photon model of electromagnetic waves comes from *Compton scattering,* in which x rays scatter from electrons, changing direction and frequency in the process. Classical electromagnetic wave theory cannot explain the change in frequency of the x rays on scattering, but the photon model can.

Suppose an x-ray photon is moving to the right. It has a collision with a slow-moving electron, as in Figure P28.81. The photon transfers energy and momentum to the electron, which recoils at a high speed. The x-ray photon loses energy, and the photon energy formula $E = hf$ tells us that its frequency must decrease. The collision looks very much like the collision between two particles.

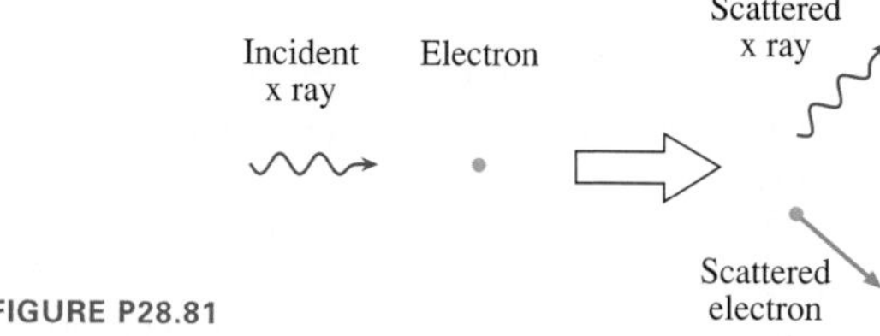

FIGURE P28.81

81. | When the x-ray photon scatters from the electron,
   A. Its speed increases.
   B. Its speed decreases.
   C. Its speed stays the same.
82. | When the x-ray photon scatters from the electron,
   A. Its wavelength increases.
   B. Its wavelength decreases.
   C. Its wavelength stays the same.

83. | When the electron is struck by the x-ray photon,
   A. Its de Broglie wavelength increases.
   B. Its de Broglie wavelength decreases.
   C. Its de Broglie wavelength stays the same.

84. | X-ray diffraction can also change the direction of a beam of x rays. Which statement offers the best comparison between Compton scattering and x-ray diffraction?
   A. X-ray diffraction changes the wavelength of x rays; Compton scattering does not.
   B. Compton scattering changes the speed of x rays; x-ray diffraction does not.
   C. X-ray diffraction relies on the particle nature of the x rays; Compton scattering relies on the wave nature.
   D. X-ray diffraction relies on the wave nature of the x rays; Compton scattering relies on the particle nature.

STOP TO THINK ANSWERS

Stop to Think 28.1: A. The Bragg condition $2d\sin\theta_1 = \lambda$ tells us that larger values of d go with smaller values of θ_1.

Stop to Think 28.2: $V_A > V_B > V_C$. For a given wavelength of light, electrons are ejected faster from metals with smaller work functions because it takes less energy to remove an electron. Faster electrons need a larger negative voltage to stop them.

Stop to Think 28.3: C. Photons always travel at c, and a photon's energy depends only on the light's frequency, not its intensity. Greater intensity means more energy each second, which means more photons.

Stop to Think 28.4: A. The widest diffraction pattern occurs for the largest wavelength. The de Broglie wavelength is inversely proportional to the particle's mass, and so will be largest for the least massive particle.

Stop to Think 28.5: No. The energy of an emitted photon is the energy *difference* between two allowed energies. The three possible quantum jumps have energy differences of 2.0 eV, 2.0 eV, and 4.0 eV.

Stop to Think 28.6: B. Because $\Delta p_x = m\Delta v_x$, the uncertainty in position is $\Delta x = \dfrac{h}{\Delta p_x} = \dfrac{h}{m\Delta v_x}$. A more massive particle has a smaller position uncertainty.

29 Atoms and Molecules

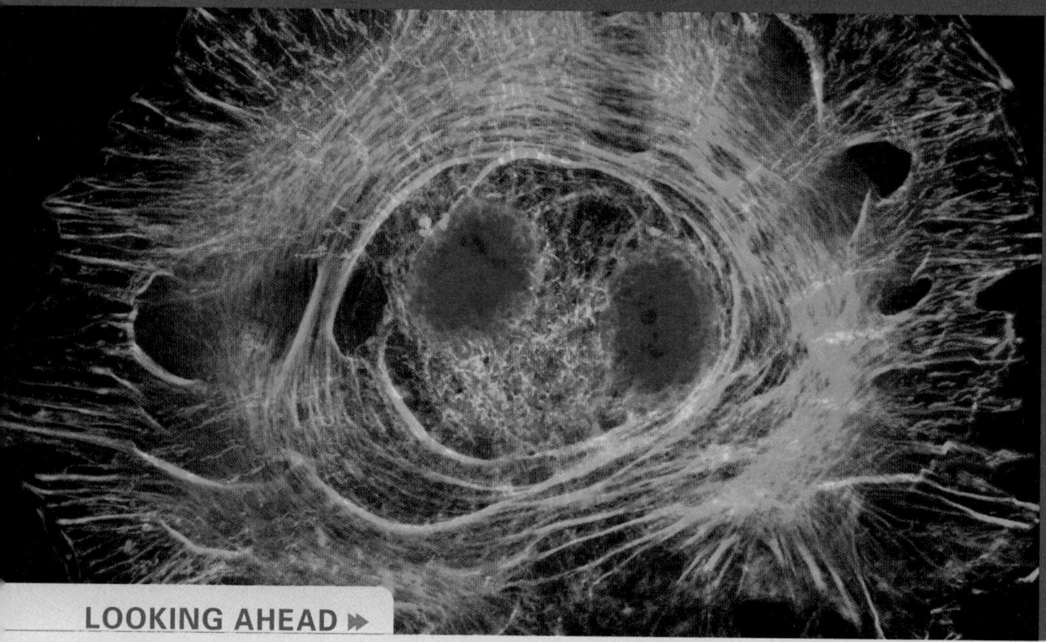

This microscopic image is a kidney cell from an African green monkey. The blue and green show, respectively, nuclear DNA and filamentous actin. What makes the different parts of this cell so brightly colored?

LOOKING AHEAD ▸▸

The goal of Chapter 29 is to use quantum physics to understand the properties of atoms, molecules, and their spectra.

Spectroscopy

You'll learn how each element can be identified by its distinct spectrum.

The discrete bright lines given off by these eight different elements (plus water) are important clues into the nature of the atom.

Looking Back ◂◂
17.3 The diffraction grating
25.8 Thermal radiation

Bohr's Atomic Model

Niels Bohr proposed a model of the atom in which quantization plays a central role.

Energy-level diagram — **Stationary states**

E_3, E_2, E_1

Atoms can exist in only certain *stationary states*, each with its own discrete quantized energy.

Photon emission

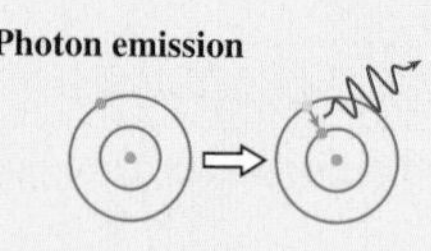

Photon absorption

As atoms "jump" from one energy level to another, they absorb and emit light as photons with discrete energies.

Looking Back ◂◂
28.5–28.6 Quantization and energy levels

Lasers

When atoms give off light, they can **stimulate** other atoms to emit light of exactly the same wavelength and phase. This leads to a powerful amplification of the light beam—the **laser**.

The intense beams of light at this laser light show result from the stimulated emission of photons by atoms.

Molecules

The absorption and emission of light by molecules are also described by quantum mechanics.

This false-color map uses the fluorescence of chlorophyll to show the distribution of algae near the coast of Florida.

Quantum-Mechanical Descriptions of the Atom

To understand the structure of atoms we need to use the principles of quantum mechanics. You'll learn that electrons have spin and that no two electrons can be in the same state.

The familiar structure of the periodic table of the elements comes directly from the quantum-mechanical description of atoms.

29.1 Spectroscopy

The interference and diffraction of light were well understood by the end of the 19th century, and the knowledge was used to design practical tools for measuring wavelengths with great accuracy. The primary instrument for measuring the wavelengths of light is a **spectrometer,** such as the one shown in FIGURE 29.1. The heart of a spectrometer is a diffraction grating that causes different wavelengths of light to diffract at different angles.

Each wavelength is focused to a different position on the photographic plate or, more likely today, a CCD detector like the one in your digital camera. The distinctive pattern of wavelengths emitted by a source of light and recorded on the detector is called the **spectrum** of the light. Spectroscopists discovered very early that there are two types of spectra: continuous spectra and discrete spectra.

- As you learned in Section 25.8, hot, self-luminous objects, such as the sun or an incandescent lightbulb, emit a **continuous spectrum** in which a rainbow is formed by light being emitted at every possible wavelength.
- In contrast, the light emitted by a gas discharge tube (such as those used to make neon signs) contains only certain discrete, individual wavelengths. Such a spectrum is called a **discrete spectrum.**

FIGURE 29.2 shows examples of spectra as they would appear on the photographic plate of a spectrometer. Each bright line in a discrete spectrum, called a **spectral line,** represents *one* specific wavelength present in the light emitted by the source. A discrete spectrum is sometimes called a *line spectrum* because of its appearance on the plate. You can see that a neon light has its familiar reddish-orange color because nearly all of the wavelengths emitted by neon atoms fall within the wavelength range 600–700 nm that we perceive as orange and red.

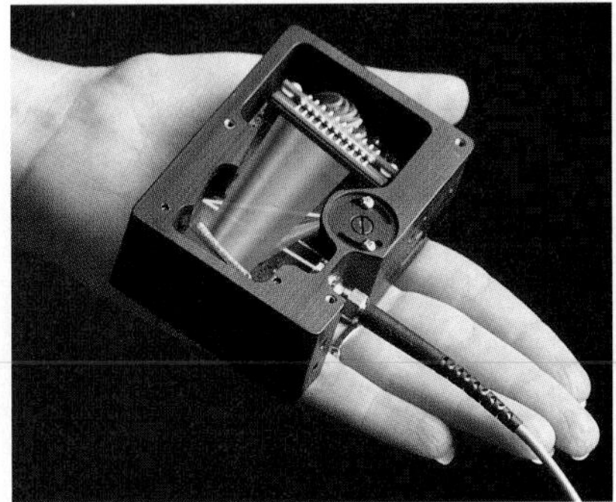

Some modern spectrometers are small enough to hold in your hand. (The rainbow has been added to show the paths that different colors take.)

FIGURE 29.1 A diffraction spectrometer.

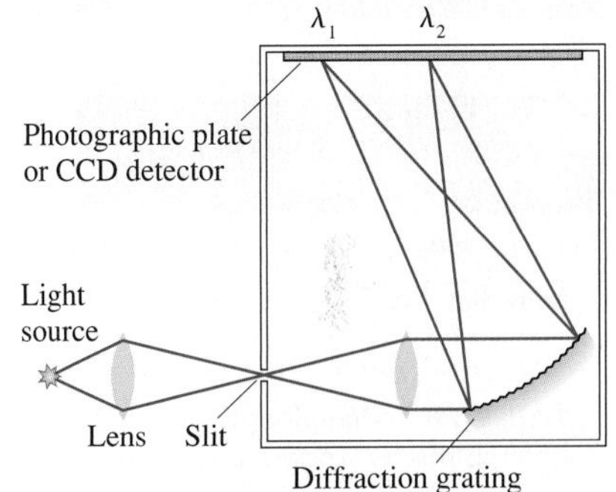

FIGURE 29.2 Examples of spectra in the visible wavelength range 400–700 nm. (a) is a continuous spectrum, while (b)–(d) are discrete spectra.

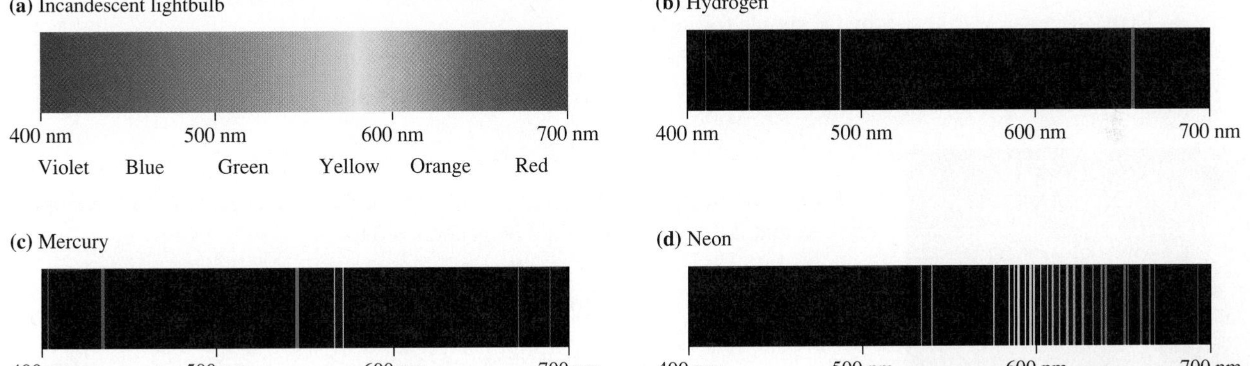

Discrete Spectra of the Elements

If a high voltage is applied to two electrodes sealed in a glass tube filled with a low-pressure gas, the gas begins to glow and emits light with a color that is unique to that gas. *Gas discharge tubes* were developed in the late 19th century, and scientists soon realized that the light was being emitted by the atoms in the gas. These *atomic spectra* were a new way to study the properties of the elements.

Two important conclusions had been established by the end of the 19th century:

1. The light emitted by atoms in a gas discharge tube has a discrete spectrum. Figure 29.2b–d showed some examples.
2. Every element in the periodic table has its own unique spectrum.

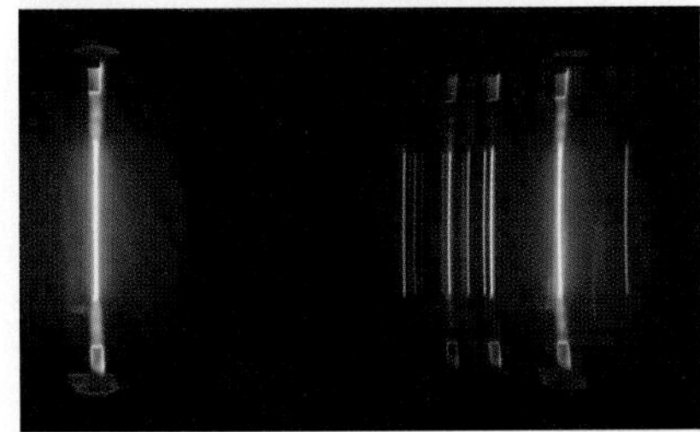

When the hydrogen discharge tube at the left is viewed through a diffraction grating, its unique discrete spectrum at the right is observed.

FIGURE 29.3 Measuring an absorption spectrum.

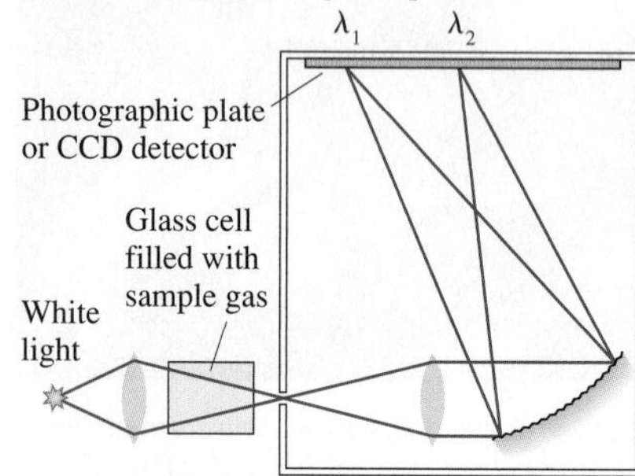

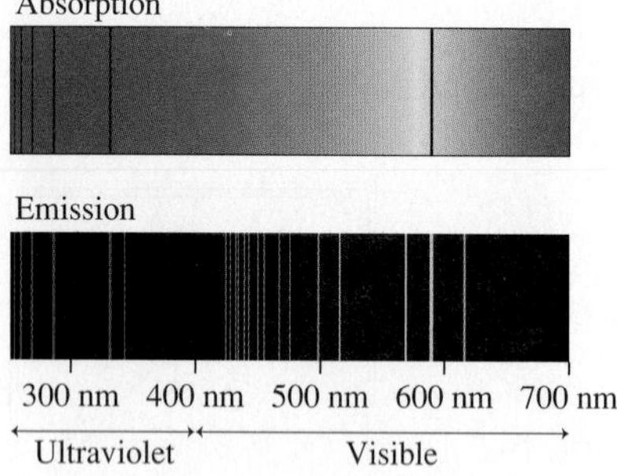

The fact that each element emits a unique spectrum means that atomic spectra can be used as "fingerprints" to identify elements. Consequently, atomic spectroscopy is the basis of many contemporary technologies for analyzing the composition of unknown materials, monitoring air pollutants, and studying the atmospheres of the earth and other planets.

Gases can also *absorb* light. In an absorption experiment, as shown in FIGURE 29.3a, a white-light source emits a continuous spectrum that, in the absence of a gas, exposes the film completely and uniformly. When a sample of gas is placed in the light's path, any wavelengths absorbed by the gas are missing and the film is dark at that wavelength.

Gases not only emit discrete wavelengths, but also absorb discrete wavelengths. But there is an important difference between the emission spectrum and the absorption spectrum of a gas: **Every wavelength that is absorbed by the gas is also emitted, but *not* every emitted wavelength is absorbed.** The wavelengths in the absorption spectrum appear as a subset of the wavelengths in the emission spectrum. As an example, FIGURE 29.3b shows both the emission and the absorption spectra of sodium atoms. All of the absorption wavelengths are prominent in the emission spectrum, but there are many emission lines for which no absorption occurs.

What causes atoms to emit or absorb light? Why a discrete spectrum? Why are some wavelengths emitted but not absorbed? Why is each element different? Nineteenth-century physicists struggled with these questions but could not answer them. Ultimately, classical physics was simply incapable of providing an understanding of atoms.

The only encouraging sign came from an unlikely source. While the spectra of other atoms have dozens or even hundreds of wavelengths, the visible spectrum of hydrogen, between 400 nm and 700 nm, consists of a mere four spectral lines (see Figure 29.2b and Table 29.1). If any spectrum could be understood, it should be that of the first element in the periodic table. The breakthrough came in 1885, not by an established and recognized scientist but by a Swiss school teacher, Johann Balmer. Balmer showed that the wavelengths in the hydrogen spectrum could be represented by the simple formula

TABLE 29.1 Wavelengths of visible lines in the hydrogen spectrum

656 nm
486 nm
434 nm
410 nm

$$\lambda = \frac{91.1\text{ nm}}{\left(\dfrac{1}{2^2} - \dfrac{1}{n^2}\right)} \qquad n = 3, 4, 5, \ldots \tag{29.1}$$

Later experimental evidence, as ultraviolet and infrared spectroscopy developed, showed that Balmer's result could be generalized to

$$\lambda = \frac{91.1\text{ nm}}{\left(\dfrac{1}{m^2} - \dfrac{1}{n^2}\right)} \tag{29.2}$$

m can be 1, 2, 3, . . .
If $m = 1$, n can be 2, 3, 4, . . .
If $m = 2$, n can be 3, 4, 5, . . .
If $m = 3$, n can be 4, 5, 6, . . .
and so on.

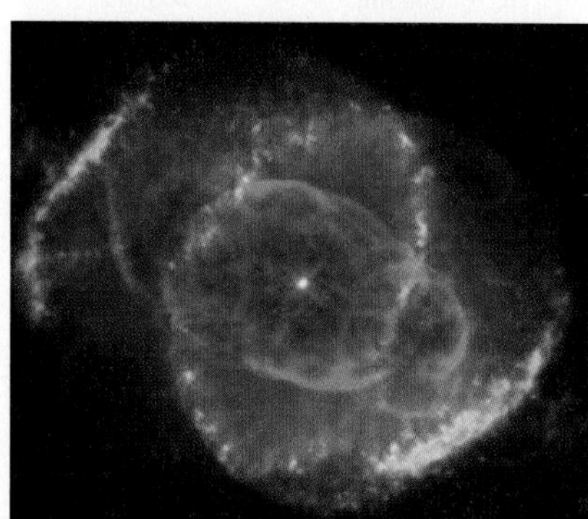

Astronomical colors The red color of this nebula is due to the emission of light from hydrogen atoms. The atoms are excited by intense ultraviolet light from the star in the center. They then emit red light, with $\lambda = 656$ nm, as predicted by the Balmer formula with $m = 2$ and $n = 3$. Spectroscopy of glowing gases from astronomical objects is an invaluable tool for astronomers, because it allows the gases present in objects many light years distant to be identified from earth.

We now refer to Equation 29.2 as the **Balmer formula,** although Balmer himself suggested only the original version in which $m = 2$. Other than at the very highest levels of resolution, where new details appear that need not concern us in this text, the Balmer formula accurately describes *every* wavelength in the emission spectrum of hydrogen.

The Balmer formula is what we call *empirical knowledge.* It is an accurate mathematical representation found through experimental evidence, but it does not rest on any physical principles or physical laws. Balmer's formula was useful, but no one was able to *derive* Balmer's formula from Newtonian mechanics or the theory of electromagnetism. Yet the formula was so simple that it must, everyone agreed, have a simple explanation. It would take 30 years to find it.

STOP TO THINK 29.1 The black lines show the emission or absorption lines observed in two spectra of the same element. Which one is an emission spectrum and which is an absorption spectrum?

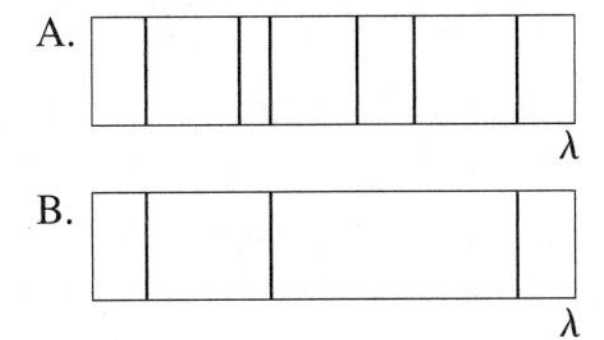
A.
λ
B.
λ

29.2 Atoms

It was the ancient Greeks who first had the idea of atoms as indivisible units of matter, but experimental evidence for atoms didn't appear until Dalton, Avogadro, and others began to formulate the laws of chemistry in the early 19th century. The existence of atoms with diameters of approximately 10^{-10} m was widely accepted by 1890, but it was still unknown if atoms were indivisible little spheres or if they had some kind of internal structure.

The 1897 discovery of the electron by J. J. Thomson had two important implications:

1. Atoms are not indivisible; they are built of smaller pieces. The electron was the first *subatomic* particle to be discovered.
2. The constituents of the atom are *charged particles.* Hence it seems plausible that the atom must be held together by electric forces.

Within a few years, measurements of the electron's charge

$$e = 1.60 \times 10^{-19}\ \text{C}$$

and mass

$$m_e = 9.11 \times 10^{-31}\ \text{kg}$$

revealed that the electron is much less massive than even the smallest atom.

Because the electrons are very small and light compared to the whole atom, it seemed reasonable to think that the positively charged part (protons were not yet known) had most of the mass and would take up most of the space. Thomson suggested that the atom consists of a spherical "cloud" of positive charge, roughly 10^{-10} m in diameter, in which the smaller negative electrons are embedded. The positive charge exactly balances the negative, so the atom as a whole has no net charge. This model of the atom has been called the "plum-pudding model" or the "raisin-cake model" for reasons that should be clear from the picture of **FIGURE 29.4**. However, Thomson's model of the atom did not stand the tests of time.

FIGURE 29.4 Thomson's raisin-cake model of the atom.

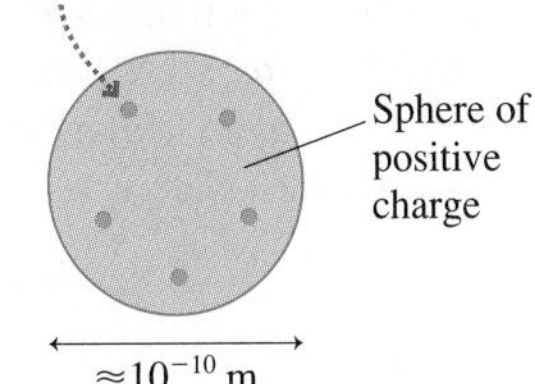
Thomson proposed that small negative electrons are embedded in a sphere of positive charge.
Sphere of positive charge
≈10^{-10} m

Almost simultaneously with Thomson's discovery of the electron, and just one year after Röntgen's discovery of x rays, the French physicist Antoine Henri Becquerel announced his discovery that some new form of "rays" were emitted by crystals of uranium. These rays, like x rays, could expose film and pass through objects, but they were emitted continuously from the uranium without having to "do" anything to it. This was the discovery of **radioactivity,** a topic we'll study in Chapter 30.

One of Thomson's former students, Ernest Rutherford, began a study of these new rays and quickly discovered that a uranium crystal actually emits two *different* rays. The first, which he called **alpha rays,** were easily absorbed by a piece of paper. The second, **beta rays,** could penetrate through at least 0.1 inch of metal and through much greater thicknesses of soft materials. The beta rays turned out to be high-speed electrons emitted by the uranium crystal. Rutherford soon showed that alpha rays are *positively* charged particles, and by 1906 he established that alpha rays (or alpha particles, as we now call them) consist of doubly ionized helium atoms (bare helium nuclei with mass $m = 6.64 \times 10^{-27}$ kg) emitted at high speed ($\approx 3 \times 10^7$ m/s) from the sample.

It had been a shock to discover that atoms are not indivisible—they have an inner structure. Now, with the discovery of radioactivity, it appeared that some atoms were not even stable but could spit out various kinds of charged particles!

EXAMPLE 29.1 The speed of an alpha particle

Alpha particles are usually characterized by their kinetic energy in MeV. What is the speed of an 8.3 MeV alpha particle?

SOLVE Recall that 1 eV is the energy acquired by an electron accelerating through a 1 V potential difference, with the conversion $1.00\text{ eV} = 1.60 \times 10^{-19}\text{ J}$. First, we convert the energy to joules:

$$K = 8.3 \times 10^6\text{ eV} \times \frac{1.60 \times 10^{-19}\text{ J}}{1.00\text{ eV}} = 1.3 \times 10^{-12}\text{ J}$$

Now, using the alpha-particle mass $m = 6.64 \times 10^{-27}$ kg given above, we can find the speed:

$$K = \frac{1}{2}mv^2 = 1.3 \times 10^{-12}\text{ J}$$

$$v = \sqrt{\frac{2K}{m}} = 2.0 \times 10^7\text{ m/s}$$

ASSESS This is quite fast, about 7% of the speed of light.

FIGURE 29.5 Rutherford's experiment to shoot high-speed alpha particles through a thin gold foil.

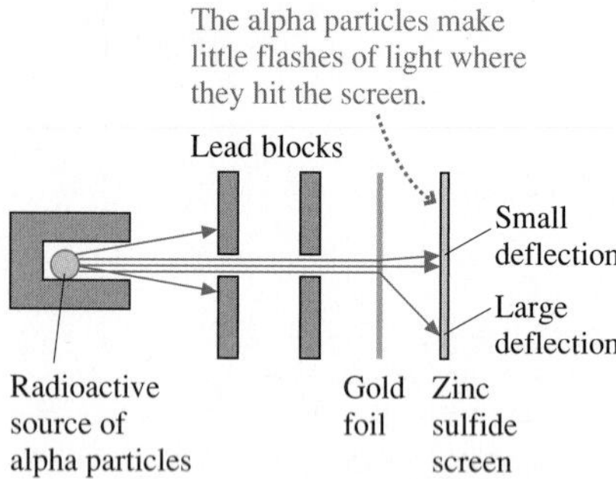

I remember two or three days later Geiger coming to me in great excitement and saying, "We have been able to get some of the alpha particles coming backward." It was quite the most incredible event that has ever happened to me in my life. It was almost as if you fired a 15-inch shell at a piece of tissue paper and it came back and hit you. . . . It was then that I had the idea of an atom with a minute massive center, carrying a charge.

Ernest Rutherford

The First Nuclear Physics Experiment

Rutherford soon realized that he could use these high-speed alpha particles as projectiles to probe inside other atoms. In 1909, Rutherford and his students Hans Geiger and Ernest Marsden set up the experiment shown in **FIGURE 29.5** to shoot alpha particles through very thin metal foils. The alpha particle is charged, and it experiences electric forces from the positive and negative charges of the atoms as it passes through the foil. According to Thomson's raisin-cake model of the atom, the forces exerted on the alpha particle by the positive atomic charges should roughly cancel the forces from the negative electrons, causing the alpha particles to experience only slight deflections. Indeed, this was the experimenters' initial observation.

At Rutherford's suggestion, Geiger and Marsden set up the apparatus to see if any alpha particles were deflected at *large* angles. It took only a few days to find the answer. Not only were alpha particles deflected at large angles, but a very few were reflected almost straight backward toward the source!

How can we understand this result? **FIGURE 29.6a** shows that an alpha particle passing through a Thomson atom would experience only a small deflection. But if an atom has a small positive core, such as the one in **FIGURE 29.6b**, a few of the alpha particles can come very close to the core. Because the electric force varies with the inverse square of the distance, the very large force of this very close approach can cause a large-angle scattering or even a backward deflection of the alpha particle. This is what Geiger and Marsden were observing.

FIGURE 29.6 Alpha particles interact differently with a concentrated positive nucleus than they would with the spread-out charges in Thomson's model.

Thomson model

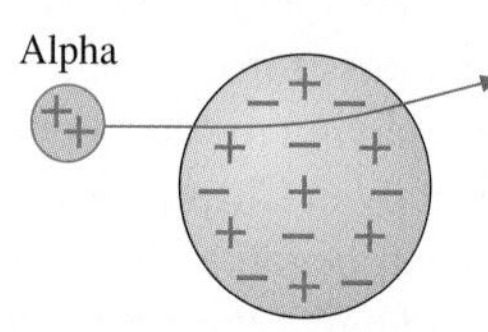

The alpha particle is only slightly deflected by a Thomson atom because forces from the spread-out positive and negative charges nearly cancel.

Nuclear model

Alpha

If the atom has a concentrated positive nucleus, some alpha particles will be able to come very close to the nucleus and thus feel a very strong repulsive force.

19.1 

The discovery of large-angle scattering of alpha particles quickly led Rutherford to envision an atom in which negative electrons orbit an unbelievably small, massive, positive **nucleus,** rather like a miniature solar system. This is the **nuclear model of the atom.** Further experiments showed that the diameter of the atomic nucleus is $\approx 1 \times 10^{-14}\text{ m} = 10\text{ fm}$ ($1\text{ fm} = 1\text{ femtometer} = 10^{-15}\text{ m}$), a mere 0.01% the diameter of the atom itself. Thus nearly all of the atom is merely empty space—the void!

EXAMPLE 29.2 Going for the gold!

An 8.3 MeV alpha particle is shot directly toward the nucleus of a gold atom (atomic number 79). What is the distance of closest approach of the alpha particle to the nucleus?

PREPARE Energy is conserved in electric interactions. Assume that the gold nucleus, which is much more massive than the alpha particle, does not move. Also recall that the electric field and potential of a sphere of charge can be found by treating the total charge as a point charge. **FIGURE 29.7** is a before-and-after visual overview; the "before" situation is when the alpha is very far from the gold nucleus, and the "after" situation is when the alpha is at its distance of closest approach to the nucleus. The motion is in and out along a straight line.

FIGURE 29.7 A before-and-after visual overview of an alpha particle colliding with a nucleus.

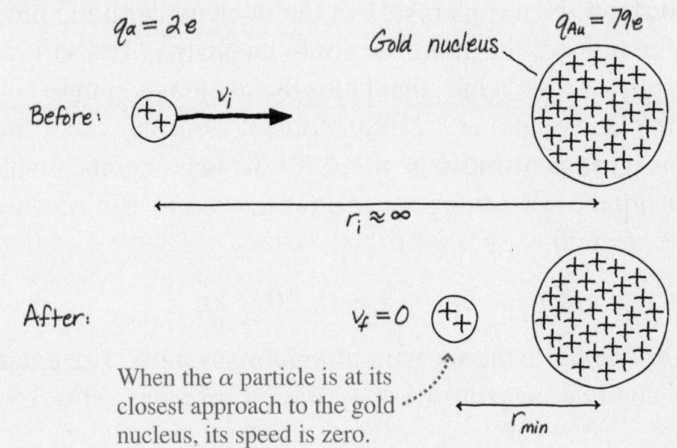

SOLVE Electric potential energy decreases rapidly with increasing separation, so initially, when the alpha particle is very far away, the system has only the initial kinetic energy of the alpha particle. At the moment of closest approach the alpha particle has just come to rest and so the system has only potential energy. The conservation of energy statement $K_f + U_f = K_i + U_i$ is

$$0 + \frac{1}{4\pi\epsilon_0}\frac{q_\alpha q_{Au}}{r_{min}} = \frac{1}{2}mv_i^2 + 0$$

where q_α is the alpha-particle charge and we've treated the gold nucleus as a point charge q_{Au}. The solution for r_{min} is

$$r_{min} = \frac{1}{4\pi\epsilon_0}\frac{2q_\alpha q_{Au}}{mv_i^2}$$

The mass of the alpha particle is $m = 6.64 \times 10^{-27}$ kg and its charge is $q_\alpha = 2e = 3.20 \times 10^{-19}$ C. Gold has atomic number 79, so $q_{Au} = 79e = 1.26 \times 10^{-17}$ C. In Example 29.1 we found that an 8.3 MeV alpha particle has speed $v = 2.0 \times 10^7$ m/s. With this information, we can calculate

$$r_{min} = 2.7 \times 10^{-14}\text{ m}$$

This is only about 1/10,000 the size of the atom itself!

ASSESS We ignored the gold atom's electrons in this example. In fact, they make almost no contribution to the alpha particle's trajectory. The alpha particle is exceedingly massive compared to the electrons, and the electrons are spread out over a distance very large compared to the size of the nucleus. Hence the alpha particle easily pushes them aside without any noticeable change in its velocity.

Using the Nuclear Model

The nuclear model of the atom makes it easy to picture atoms and understand such processes as ionization. For example, the **atomic number** of an element, its position in the periodic table of the elements, is the number of orbiting electrons (of a neutral atom) and the number of units of positive charge in the nucleus. The atomic number is represented by Z. Hydrogen, with $Z = 1$, has one electron orbiting a nucleus with charge $+1e$. Helium, with $Z = 2$, has two orbiting electrons and a nucleus with charge $+2e$. Because the orbiting electrons are very light, an x-ray photon or a rapidly moving particle, such as another electron, can knock one of the electrons away, creating a positive *ion*. Removing one electron makes a singly charged ion, with $q_{ion} = +e$. Removing two electrons creates a doubly charged ion, with $q_{ion} = +2e$. This is shown for lithium ($Z = 3$) in **FIGURE 29.8**.

Experiments soon led to the recognition that the positive charge of the nucleus is associated with a positive subatomic particle called the **proton.** The proton's charge is $+e$, equal in magnitude but opposite in sign to the electron's charge. Further, because nearly all the atomic mass is associated with the nucleus, the proton is about

FIGURE 29.8 Different ionization stages of the lithium atom ($Z = 3$).

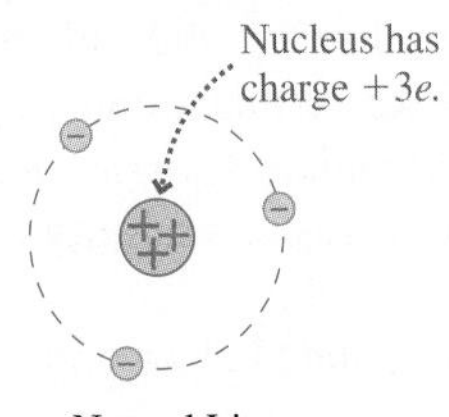

Neutral Li

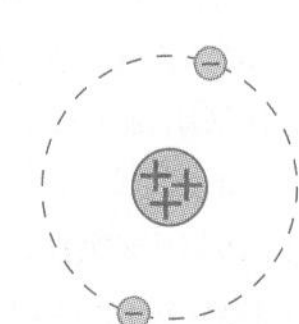
Singly charged Li^+

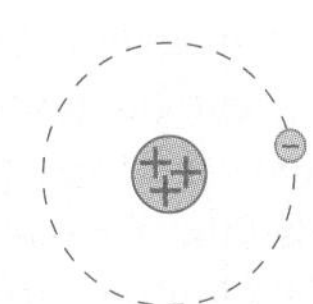
Doubly charged Li^{++}

1800 times more massive than the electron: $m_p = 1.67 \times 10^{-27}$ kg. Atoms with atomic number Z have Z protons in the nucleus, giving the nucleus charge $+Ze$.

This nuclear model of the atom allows us to understand why, during chemical reactions and when an object is charged by rubbing, electrons are easily transferred but protons are not. The protons are tightly bound in the nucleus, shielded by all the electrons, but outer electrons are easily stripped away. Rutherford's nuclear model has explanatory power that was lacking in Thomson's model.

But there was a problem. Helium, with atomic number 2, has twice as many electrons and protons as hydrogen. Lithium, $Z = 3$, has three electrons and protons. If a nucleus contains Z protons to balance the Z orbiting electrons, and if nearly all the atomic mass is contained in the nucleus, then helium should be twice as massive as hydrogen and lithium three times as massive. But it was known from chemistry measurements that helium is *four times* as massive as hydrogen and lithium is *seven times* as massive.

FIGURE 29.9 The nucleus of an atom contains protons and neutrons.

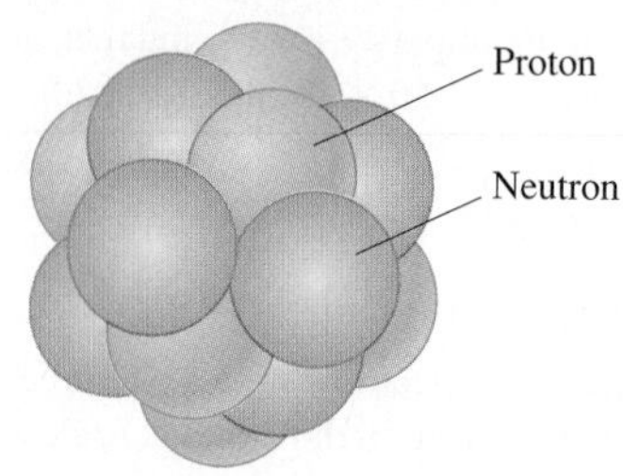

This difficulty was not resolved until the discovery, in 1932, of a third subatomic particle. This particle has essentially the same mass as a proton but *no* electric charge. It is called the **neutron**. Neutrons reside in the nucleus, with the protons, where they contribute to the mass of the atom but not to its charge. As you'll see in Chapter 30, neutrons help provide the "glue" that holds the nucleus together.

We now know that a nucleus contains Z protons plus N neutrons, as shown in FIGURE 29.9, giving the atom a **mass number** $A = Z + N$. The mass number, which is a dimensionless integer, is *not* the same thing as the atomic mass m. But because the proton and neutron masses are both ≈ 1 u, where

$$1 \text{ u} = 1 \text{ atomic mass unit} = 1.66 \times 10^{-27} \text{ kg}$$

the mass number A is *approximately* the mass in atomic mass units. For example, helium, with two protons and two neutrons ($A = 4$), has atomic mass $m = 6.646 \times 10^{-27}$ kg $= 4.003$ u ≈ 4 u.

FIGURE 29.10 The two isotopes of helium. ^{3}He is only 0.0001% abundant.

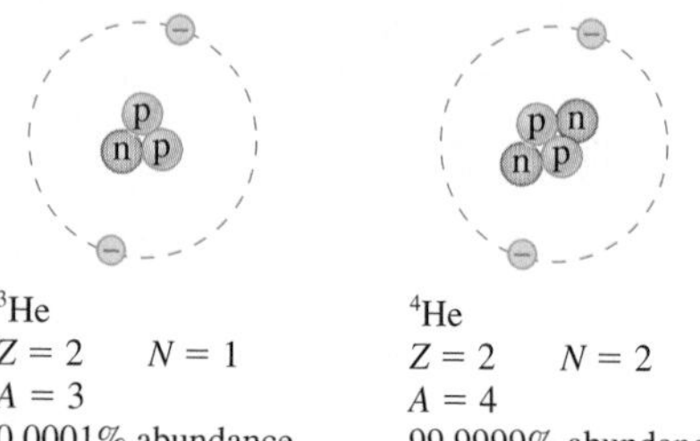

There are a *range* of neutron numbers that happily form a nucleus with Z protons, creating a series of nuclei with the same Z-value (i.e., they are all the same chemical element) but different masses. Such a series of nuclei are called **isotopes.** The notation used to label isotopes is ^AZ, where the mass number A is given as a *leading* superscript. The proton number Z is not specified by an actual number but, equivalently, by the chemical symbol for that element. The most common isotope of neon has $Z = 10$ protons and $N = 10$ neutrons. Thus it has mass number $A = 20$ and is labeled ^{20}Ne. The neon isotope ^{22}Ne has $Z = 10$ protons (that's what makes it neon) and $N = 12$ neutrons. Helium has the two isotopes shown in FIGURE 29.10. The rare ^{3}He is only 0.0001% abundant, but it can be isolated and has important uses in scientific research.

STOP TO THINK 29.2 Carbon is the sixth element in the periodic table. How many protons and how many neutrons are there in a nucleus of the isotope ^{14}C?

29.3 Bohr's Model of Atomic Quantization

18.1 ActivPhysics ONLINE

Rutherford's nuclear model was an important step toward understanding atoms, but it had two serious shortcomings. First, electrons orbiting the nucleus in a Rutherford atom are oscillating charged particles. According to Maxwell's theory of electricity and magnetism, these orbiting electrons should act as small antennas and radiate electromagnetic waves. That sounds encouraging, because we know that atoms can emit light, but it was easy to show that a Rutherford atom would radiate a *continuous* rainbow-like spectrum. Thus one failure of Rutherford's model was an inability to predict the discrete nature of emission and absorption spectra.

In addition, atoms would continuously lose energy as they radiated electromagnetic waves. As FIGURE 29.11 shows, this would cause the electrons to spiral into the nucleus! Calculations showed that a Rutherford atom can last no more than about a microsecond. In other words, classical Newtonian mechanics and electromagnetism predict that an atom with electrons orbiting a nucleus would be highly unstable and would immediately self-destruct. This clearly does not happen.

The experimental efforts of the late 19th and early 20th centuries had been impressive, and there could be no doubt about the existence of electrons, about the small positive nucleus, and about the unique discrete spectrum emitted by each atom. But the theoretical framework for understanding such observations had lagged behind. As the new century dawned, physicists could not explain the structure of atoms, could not explain the stability of matter, could not explain discrete spectra or why an element's absorption spectrum differs from its emission spectrum, and could not explain the origin of x rays or radioactivity.

A missing piece of the puzzle, although not recognized as such for a few years, was Einstein's 1905 introduction of light quanta. If light comes in discrete packets of energy, which we now call photons, and if atoms emit and absorb light, what does that imply about the structure of the atoms? This was the question posed by the Danish physicist Niels Bohr.

Bohr wanted to understand how a solar-system-like atom could be stable and not radiate away all its energy. He soon recognized that Einstein's light quanta had profound implications about the structure of atoms, and in 1913 Bohr proposed a radically new model of the atom in which he added quantization to Rutherford's nuclear atom. The basic assumptions of the **Bohr model of the atom** are as follows:

FIGURE 29.11 The fate of a Rutherford atom.

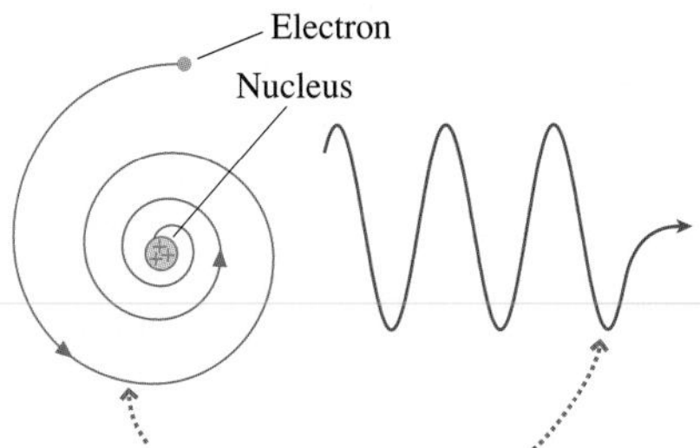

According to classical physics, an electron would spiral into the nucleus while radiating energy as an electromagnetic wave.

Understanding Bohr's model

Electrons can exist in only certain allowed orbits.

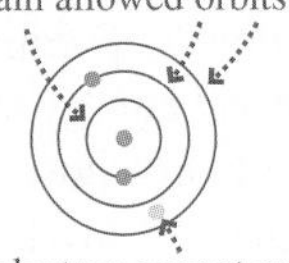

An electron cannot exist here, where there is no allowed orbit.

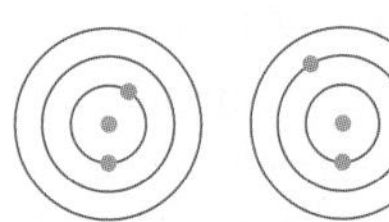

This is one stationary state. This is another stationary state.

1. The electrons in an atom can exist in only certain *allowed orbits*. A particular arrangement of electrons in these orbits is called a **stationary state.**

Energy-level diagram **Stationary states**

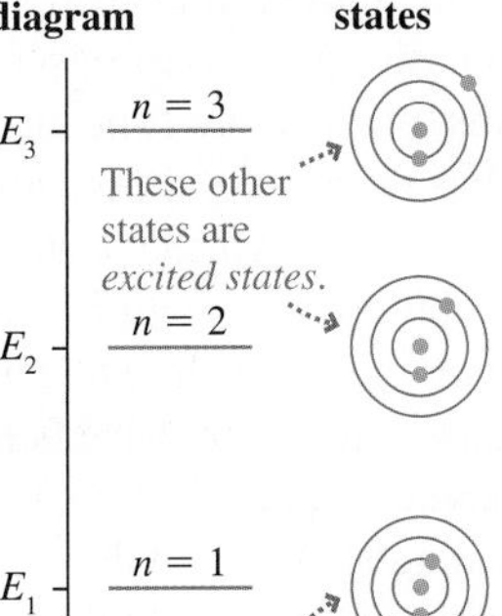

This state, with the lowest energy E_1, is the *ground state*. It is stable and can persist indefinitely.

2. Each stationary state has a discrete, well-defined energy E_n, as the energy-level diagram shows. That is, atomic energies are *quantized*. The stationary states are labeled by the *quantum number n* in order of increasing energy: $E_1 < E_2 < E_3 < \cdots$.

Photon emission

Excited-state electron

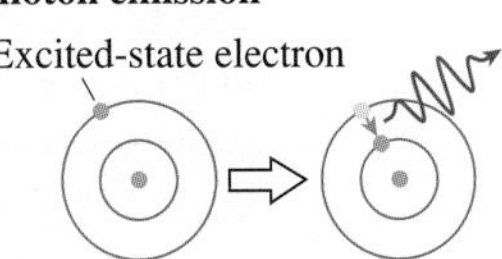

The electron jumps to a lower-energy stationary state and emits a photon.

Photon absorption

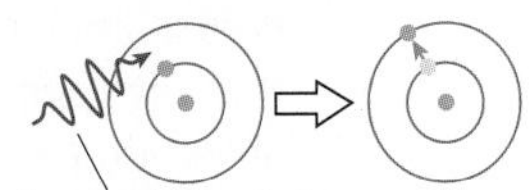

Approaching photon

The electron absorbs the photon and jumps to a higher-energy stationary state.

3. An atom can undergo a *transition* or *quantum jump* from one stationary state to another by emitting or absorbing a photon whose energy is exactly equal to the energy difference between the two stationary states.

Collisional excitation

Approaching particle

Particle loses energy.

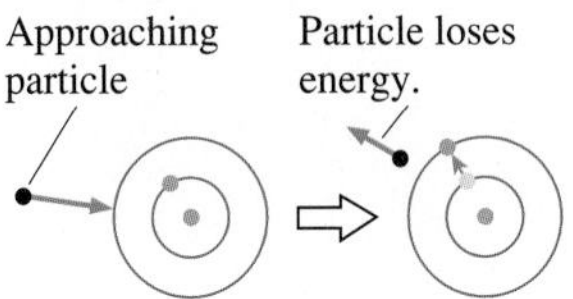

The particle transfers energy to the atom in the collision and excites the atom.

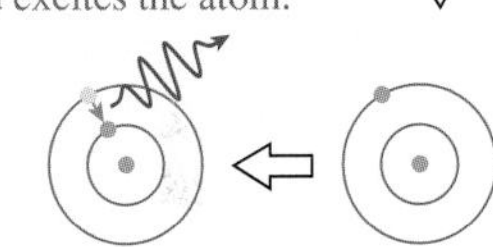

An atom in an excited state jumps to lower states, emitting a photon at each jump.

4. Atoms can also move from a lower-energy state to a higher-energy state by absorbing energy in a collision with an electron or other atom in a process called **collisional excitation.**

The excited atoms soon jump down to lower states, eventually ending in the stable ground state.

Assumption 3 of Bohr's model merges Einstein's idea of light quanta with the law of conservation of energy. Thus **an atom can jump from one stationary state to another by emitting or absorbing a photon of just the right frequency to conserve energy.** If an atom jumps from an initial state with energy E_i to a final state with *lower* energy E_f, energy will be conserved if the atom emits a photon with $E_{photon} = \Delta E_{atom} = |E_f - E_i|$. Because $E_{photon} = hf$, this photon must have frequency

$$f_{photon} = \frac{\Delta E_{atom}}{h} \qquad (29.3)$$

if it is to carry away exactly the right amount of energy. Similarly, an atom can jump to a higher-energy state, for which additional energy is needed, by absorbing a photon of frequency $f_{photon} = \Delta E_{atom}/h$. The total energy of the atom-plus-light system is conserved.

The Bohr model also introduces *collisions* as another important mechanism by which a quantum system can gain energy. For example, a fluorescent lightbulb contains mercury vapor at a very low pressure. When you turn on the light, electrons flow back and forth inside the tube. As they do, they occasionally collide with the mercury atoms. These are inelastic collisions in which some of the electron's energy is transferred to the mercury atom, kicking the atom up to a higher energy level. The atom then jumps back to a lower energy level, emitting an ultraviolet photon. The ultraviolet light causes the white paint on the inside surface of the tube to *fluoresce,* a process we'll study later in the chapter.

The implications of Bohr's model are profound. In particular:

1. **Matter is stable.** Once an atom is in its ground state, there are no states of any lower energy to which it can jump. It can remain in the ground state forever.
2. **Atoms emit and absorb a *discrete spectrum*.** Only those photons whose frequencies match the energy *intervals* between the stationary states can be emitted or absorbed. Photons of other frequencies cannot be emitted or absorbed without violating energy conservation.
3. **Emission spectra can be produced by collisions.** Energy from collisions can kick an atom up to an excited state. It then emits photons in a discrete emission spectrum as it jumps down to lower-energy states.
4. **Absorption wavelengths are a subset of the wavelengths in the emission spectrum.** Recall that all the lines seen in an absorption spectrum are also seen in emission, but many emission lines are *not* seen in absorption. According to Bohr's model, **most atoms, most of the time, are in their lowest energy state,** the $n = 1$ ground state. Thus the absorption spectrum consists of *only* those transitions such as $1 \rightarrow 2$, $1 \rightarrow 3, \ldots$ in which the atom jumps from $n = 1$ to a higher value of n by absorbing a photon. Transitions such as $2 \rightarrow 3$ are not observed because there are essentially no atoms in $n = 2$ at any instant of time to do the absorbing. On the other hand, atoms that have been excited to the $n = 3$ state by collisions can emit photons corresponding to transitions $3 \rightarrow 1$ *and* $3 \rightarrow 2$. Thus the wavelength corresponding to $\Delta E_{atom} = E_3 - E_1$ is seen in both emission and absorption, but photons with $\Delta E_{atom} = E_3 - E_2$ occur in emission only.
5. **Each element in the periodic table has a unique spectrum.** The energies of the stationary states are just the energies of the orbiting electrons. The atom has no other form of energy. Different elements, with different numbers of electrons, have different stable orbits and thus different stationary states. States with different energies will emit and absorb photons of different wavelengths.

EXAMPLE 29.3 Wavelengths in emission and absorption spectra

An atom has stationary states $E_1 = 0.0$ eV, $E_2 = 2.0$ eV, and $E_3 = 5.0$ eV. What wavelengths are observed in the absorption spectrum and in the emission spectrum of this atom?

PREPARE FIGURE 29.12 shows an energy-level diagram for the atom. Photons are emitted when an atom undergoes a quantum jump from a higher energy level to a lower energy level. Photons are absorbed in a quantum jump from a lower energy level to a higher energy level. However, most of the atoms are in the $n = 1$ ground state, so the only quantum jumps seen in the absorption spectrum start from the $n = 1$ state.

FIGURE 29.12 The atom's energy-level diagram.

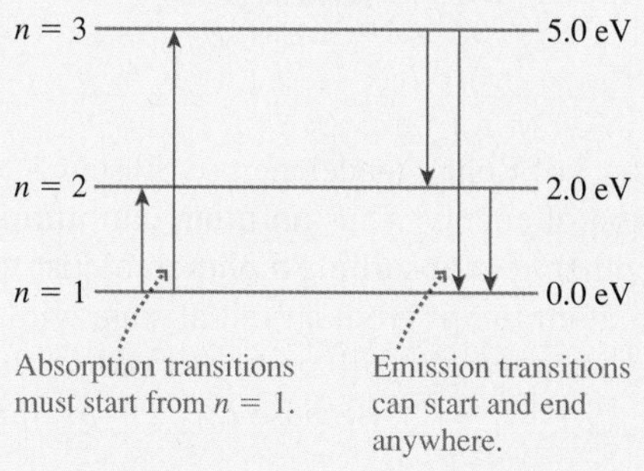
n = 3
5.0 eV
n = 2
2.0 eV
n = 1
0.0 eV
Absorption transitions must start from n = 1.
Emission transitions can start and end anywhere.

SOLVE This atom absorbs photons on the $1 \rightarrow 2$ and $1 \rightarrow 3$ transitions, with $\Delta E_{1\rightarrow 2} = 2.0$ eV and $\Delta E_{1\rightarrow 3} = 5.0$ eV. From $f_{photon} = \Delta E_{atom}/h$ and $\lambda = c/f$, we find that the wavelengths in the absorption spectrum are

$1 \rightarrow 3$ $f_{photon} = 5.0 \text{ eV}/h = 1.2 \times 10^{15}$ Hz

$\lambda = 250$ nm (ultraviolet)

$1 \rightarrow 2$ $f_{photon} = 2.0 \text{ eV}/h = 4.8 \times 10^{14}$ Hz

$\lambda = 620$ nm (orange)

The emission spectrum also has the 620 nm and 250 nm wavelengths due to the $2 \rightarrow 1$ and $3 \rightarrow 1$ quantum jumps. In addition, the emission spectrum contains the $3 \rightarrow 2$ quantum jump with $\Delta E_{3\rightarrow 2} = 3.0$ eV that is *not* seen in absorption because there are too few atoms in the $n = 2$ state to absorb. A similar calculation finds $f_{photon} = 7.3 \times 10^{14}$ Hz and $\lambda = c/f = 410$ nm. Thus the emission wavelengths are

$2 \rightarrow 1$ $\lambda = 620$ nm (orange)

$3 \rightarrow 2$ $\lambda = 410$ nm (blue)

$3 \rightarrow 1$ $\lambda = 250$ nm (ultraviolet)

STOP TO THINK 29.3 A photon with a wavelength of 410 nm has energy $E_{photon} = 3.0$ eV. Do you expect to see a spectral line with $\lambda = 410$ nm in the emission spectrum of the atom represented by this energy-level diagram? If so, what transition or transitions will emit it? Do you expect to see a spectral line with $\lambda = 410$ nm in the absorption spectrum? If so, what transition or transitions will absorb it?

$n = 4$	6.0 eV
$n = 3$	5.0 eV
$n = 2$	2.0 eV
$n = 1$	0.0 eV

29.4 The Bohr Hydrogen Atom

Bohr's hypothesis was a bold new idea, yet there was still one enormous stumbling block: What *are* the stationary states of an atom? Everything in Bohr's model hinges on the existence of these stationary states, of there being only certain electron orbits that are allowed. But nothing in classical physics provides any basis for such orbits. And Bohr's model describes only the *consequences* of having stationary states, not how to find them. If such states really exist, we will have to go beyond classical physics to find them.

To address this problem, Bohr did an explicit analysis of the hydrogen atom. The hydrogen atom, with only a single electron, was known to be the simplest atom. Furthermore, as we discussed in Section 29.1, Balmer had discovered a fairly simple formula that characterizes the wavelengths in the hydrogen emission spectrum. Anyone with a successful model of an atom was going to have to *derive* Balmer's formula for the hydrogen atom.

Bohr's paper followed a rather circuitous line of reasoning. That is not surprising, because he had little to go on at the time. But our goal is a clear explanation of the ideas, not a historical study of Bohr's methods, so we are going to follow a different analysis using de Broglie's matter waves. De Broglie did not propose matter waves until 1924, 11 years after Bohr, but with the clarity of hindsight we can see that treating the electron as a wave provides a more straightforward analysis of the hydrogen atom. Although our route will be different from Bohr's, we will arrive at the same point, and, in addition, we will be in a much better position to understand the work that came after Bohr.

NOTE ► Bohr's analysis of the hydrogen atom is sometimes called the *Bohr atom.* It's important not to confuse this analysis, which applies only to hydrogen, with the more general postulates of the *Bohr model of the atom.* Those postulates, which we looked at in the previous section, apply to any atom. To make the distinction clear, we'll call Bohr's analysis of hydrogen the *Bohr hydrogen atom.* ◄

The Stationary States of the Hydrogen Atom

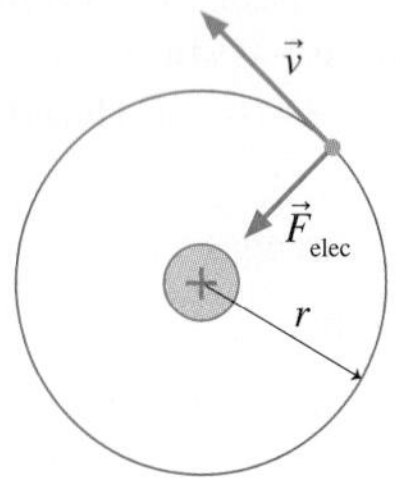

FIGURE 29.13 A Rutherford hydrogen atom. The size of the nucleus is greatly exaggerated.

FIGURE 29.13 shows a Rutherford hydrogen atom, with a single electron orbiting a proton. We will assume a circular orbit of radius r and speed v. We will also assume the proton remains stationary while the electron revolves around it, a reasonable assumption because the proton is roughly 1800 times as massive as the electron. With these assumptions, the atom's energy is the kinetic energy of the electron plus the potential energy of the electron-proton interaction. This is

$$E = K + U = \frac{1}{2}mv^2 + \frac{1}{4\pi\epsilon_0}\frac{q_{elec}\,q_{proton}}{r} = \frac{1}{2}mv^2 - \frac{e^2}{4\pi\epsilon_0 r} \tag{29.4}$$

where we used $q_{elec} = -e$ and $q_{proton} = +e$.

NOTE ► m is the mass of the electron, *not* the mass of the entire atom. ◄

Now the electron, as we are coming to understand it, has both particle-like and wave-like properties. First, let us treat the electron as a charged particle. The proton exerts a Coulomb electric force on the electron:

$$\vec{F}_{elec} = \left(\frac{1}{4\pi\epsilon_0}\frac{e^2}{r^2}, \text{ toward center}\right) \tag{29.5}$$

This force gives the electron an acceleration $\vec{a}_{elec} = \vec{F}_{elec}/m$ that also points to the center. This is a centripetal acceleration, causing the particle to move in its circular orbit. The centripetal acceleration of a particle moving in a circle of radius r at speed v is v^2/r, so that

$$a_{elec} = \frac{F_{elec}}{m} = \frac{e^2}{4\pi\epsilon_0 mr^2} = \frac{v^2}{r}$$

Rearranging, we find

$$v^2 = \frac{e^2}{4\pi\epsilon_0 mr} \tag{29.6}$$

Equation 29.6 is a *constraint* on the motion. The speed v and radius r must obey Equation 29.6 if the electron is to move in a circular orbit. This constraint is not unique to atoms; we earlier found a similar relationship between v and r for orbiting satellites.

Now let's treat the electron as a de Broglie wave. In Section 28.5 we found that a particle confined to a one-dimensional box sets up a standing wave as it reflects back and forth. A standing wave, you will recall, consists of two traveling waves moving in opposite directions. When the round-trip distance in the box is equal to an integer number of wavelengths ($2L = n\lambda$), the two oppositely traveling waves interfere constructively to set up the standing wave.

Suppose that, instead of traveling back and forth along a line, our wave-like particle travels around the circumference of a circle. The particle will set up a standing wave, just like the particle in the box, if there are waves traveling in both directions and if the round-trip distance is an integer number of wavelengths. This is the idea we want to carry over from the particle in a box. As an example, FIGURE 29.14 shows a standing wave around a circle with $n = 10$ wavelengths.

FIGURE 29.14 An $n = 10$ electron standing wave around the orbit's circumference.

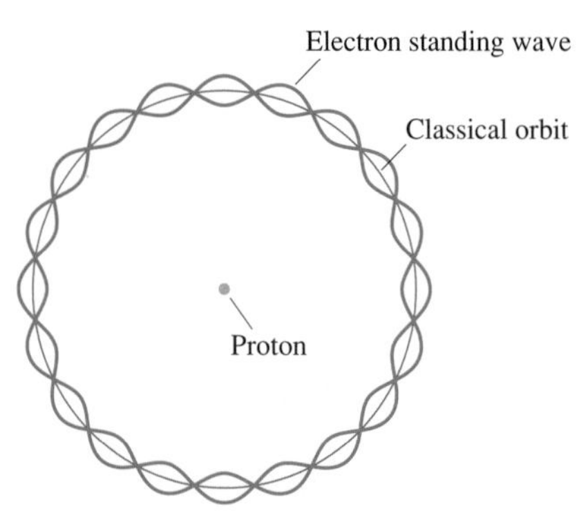

The mathematical condition for a circular standing wave is found by replacing the round-trip distance $2L$ in a box with the round-trip distance $2\pi r$ on a circle. Thus a circular standing wave will occur when

$$2\pi r = n\lambda \qquad n = 1, 2, 3, \ldots \tag{29.7}$$

But the de Broglie wavelength for a particle *has* to be $\lambda = h/p = h/mv$. Thus the standing-wave condition for a de Broglie wave is

$$2\pi r = n\frac{h}{mv}$$

This condition is true only if the electron's speed is

$$v = \frac{nh}{2\pi mr} \qquad n = 1, 2, 3, \ldots \tag{29.8}$$

In other words, the electron cannot have just any speed, only the discrete values given by Equation 29.8.

The quantity $h/2\pi$ occurs so often in quantum physics that it is customary to give it a special name. We define the quantity $\hbar$, pronounced "h bar," as

$$\hbar = \frac{h}{2\pi} = 1.05 \times 10^{-34}\ \text{J} \cdot \text{s} = 6.58 \times 10^{-16}\ \text{eV} \cdot \text{s}$$

With this definition, we can write Equation 29.8 as

$$v = \frac{n\hbar}{mr} \qquad n = 1, 2, 3, \ldots \tag{29.9}$$

This, like Equation 29.6, is another relationship between v and r. This is the constraint that arises from treating the electron as a wave.

Now if the electron can act as both a particle *and* a wave, then both the Equation 29.6 *and* Equation 29.9 constraints have to be obeyed. That is, v^2 as given by the Equation 29.6 particle constraint has to equal v^2 of the Equation 29.9 wave constraint. Equating these gives

$$v^2 = \frac{e^2}{4\pi\epsilon_0 mr} = \frac{n^2\hbar^2}{m^2r^2}$$

We can solve this equation to find that the radius r is

$$r_n = n^2\frac{4\pi\epsilon_0\hbar^2}{me^2} \qquad n = 1, 2, 3, \ldots \tag{29.10}$$

where we have added a subscript n to the radius r to indicate that it depends on the integer n.

The right-hand side of Equation 29.10, except for the n^2, is just a collection of constants. Let's group them all together and define the **Bohr radius** a_B to be

$$a_B = \text{Bohr radius} = \frac{4\pi\epsilon_0\hbar^2}{me^2} = 5.29 \times 10^{-11}\ \text{m} = 0.0529\ \text{nm}$$

With this definition, Equation 29.10 for the radius of the electron's orbit becomes

$$r_n = n^2 a_B \qquad n = 1, 2, 3, \ldots \tag{29.11}$$

Allowed radii of the Bohr hydrogen atom

The first few allowed values of r are

$$r_n = \begin{cases} 0.053\ \text{nm} & n = 1 \\ 0.212\ \text{nm} & n = 2 \\ 0.476\ \text{nm} & n = 3 \\ \vdots & \vdots \end{cases}$$

We have discovered stationary states! That is, **a hydrogen atom can exist *only* if the radius of the electron's orbit is one of the values given by Equation 29.11.** Intermediate values of the radius, such as $r = 0.100$ nm, cannot exist because the electron cannot set up a standing wave around the circumference. The possible orbits are *quantized,* and integer n is the quantum number.

Hydrogen Atom Energy Levels

Now we can make progress quickly. Knowing the possible radii, we can return to Equation 29.9 and find the possible electron speeds:

$$v_n = \frac{n\hbar}{mr_n} = \frac{1}{n}\frac{\hbar}{ma_B} = \frac{v_1}{n} \qquad n = 1, 2, 3, \ldots \tag{29.12}$$

where $v_1 = \hbar/ma_B = 2.19 \times 10^6$ m/s is the electron's speed in the $n = 1$ orbit. The speed decreases as n increases.

Finally, we can determine the energies of the stationary states by using Equations 29.11 and 29.12 for r and v in Equation 29.4 for the energy. The algebra is rather messy, but the result simplifies to

$$E_n = \frac{1}{2}mv_n^2 - \frac{e^2}{4\pi\epsilon_0 r_n} = -\frac{1}{n^2}\left(\frac{1}{4\pi\epsilon_0}\frac{e^2}{2a_B}\right) \tag{29.13}$$

Let's define

$$E_1 = \frac{1}{4\pi\epsilon_0}\frac{e^2}{2a_B} = 13.60 \text{ eV}$$

We can then write the energy levels of the stationary states of the hydrogen atom as

$$E_n = -\frac{E_1}{n^2} = -\frac{13.60 \text{ eV}}{n^2} \qquad n = 1, 2, 3, \ldots \tag{29.14}$$

Allowed energies of the Bohr hydrogen atom

This has been a lot of math, so we need to see where we are and what we have learned. Table 29.2 shows values of r_n, v_n, and E_n for quantum numbers $n = 1$ to 4. We do indeed seem to have discovered stationary states of the hydrogen atom. Each state, characterized by its quantum number n, has a unique radius, speed, and energy. These are displayed graphically in **FIGURE 29.15**, in which the orbits are drawn to scale. Notice how the atom's diameter increases very rapidly as n increases. At the same time, the electron's speed decreases.

TABLE 29.2 Radii, speeds, and energies for the first four states of the Bohr hydrogen atom

n	r_n (nm)	v_n (m/s)	E_n (eV)
1	0.053	2.19×10^6	−13.60
2	0.212	1.09×10^6	−3.40
3	0.476	0.73×10^6	−1.51
4	0.847	0.55×10^6	−0.85

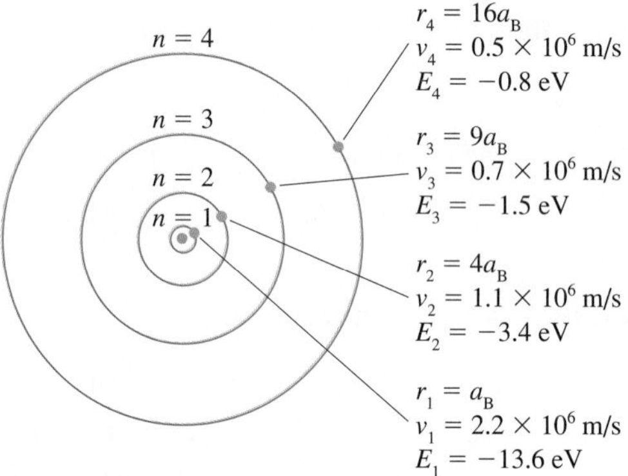

FIGURE 29.15 The first four stationary states, or allowed orbits, of the Bohr hydrogen atom drawn to scale.

EXAMPLE 29.4 Possible electron speeds in a hydrogen atom

Can an electron in a hydrogen atom have a speed of 3.60×10^5 m/s? If so, what are its energy and the radius of its orbit? What about a speed of 3.65×10^5 m/s?

PREPARE To be in a stationary state, the electron must have speed $v_n = v_1/n$, with $n = v_1/v$ an integer. Only if v_1/v is an integer is v an allowed electron speed.

SOLVE A speed of 3.60×10^5 m/s would require quantum number

$$n = \frac{v_1}{v} = \frac{2.19 \times 10^6 \text{ m/s}}{3.60 \times 10^5 \text{ m/s}} = 6.08$$

This is not an integer, so the electron can *not* have this speed. But if $v = 3.65 \times 10^5$ m/s, then

$$n = \frac{2.19 \times 10^6 \text{ m/s}}{3.65 \times 10^5 \text{ m/s}} = 6$$

This is the speed of an electron in the $n = 6$ excited state. An electron in this state has energy

$$E_6 = -\frac{13.60 \text{ eV}}{6^2} = -0.378 \text{ eV}$$

and the radius of its orbit is

$$r_6 = 6^2 a_B = 6^2(0.0529 \text{ nm}) = 1.90 \text{ nm}$$

It is important to understand why the energies of the stationary states are negative. An electron and a proton bound into an atom have *less* energy than they do when they're separated. We know this because we would have to do work (i.e., *add* energy) to pull the electron and proton apart.

When the electron and proton are completely separated ($r \to \infty$) and at rest ($v = 0$), their potential energy $U = q_1 q_2 / 4\pi\epsilon_0 r$ and kinetic energy $K = mv^2/2$ are zero. As the electron moves closer to the proton to form a hydrogen atom, its potential energy *decreases,* becoming negative, while the kinetic energy of the orbiting electron increases. Equation 29.13, however, shows that the potential energy decreases faster than the kinetic energy increases, leading to an overall negative energy for the atom. In general, negative energies are characteristic of *bound systems.*

Quantization of Angular Momentum

The angular momentum of a particle in circular motion, whether it is a planet or an electron, is

$$L = mvr$$

You will recall that angular momentum is conserved in orbital motion because a force directed toward a central point exerts no torque on the particle. Bohr used conservation of energy explicitly in his analysis of the hydrogen atom, but what role does conservation of angular momentum play?

The condition that a de Broglie wave for the electron set up a standing wave around the circumference was given, in Equation 29.7, as

$$2\pi r = n\lambda = n\frac{h}{mv}$$

We can rewrite this equation as

$$mvr = n\frac{h}{2\pi} = n\hbar \qquad (29.15)$$

But mvr is the angular momentum L for a particle in a circular orbit. It appears that the angular momentum of an orbiting electron cannot have just any value. Instead, it must satisfy

$$L = n\hbar \qquad n = 1, 2, 3, \ldots \qquad (29.16)$$

Quantization of angular momentum for the Bohr hydrogen atom

Thus angular momentum is also quantized! The atom's angular momentum must be an integer multiple of Planck's constant $\hbar$.

The quantization of angular momentum is a direct consequence of the wave-like nature of the electron. We will find that the quantization of angular momentum plays a major role in the behavior of more complex atoms, leading to the idea of electron shells that you likely have studied in chemistry.

STOP TO THINK 29.4 What is the quantum number of this hydrogen atom?

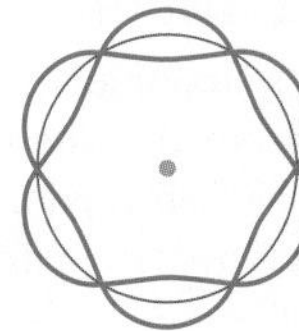

The Hydrogen Spectrum

FIGURE 29.16 The energy-level diagram of the hydrogen atom.

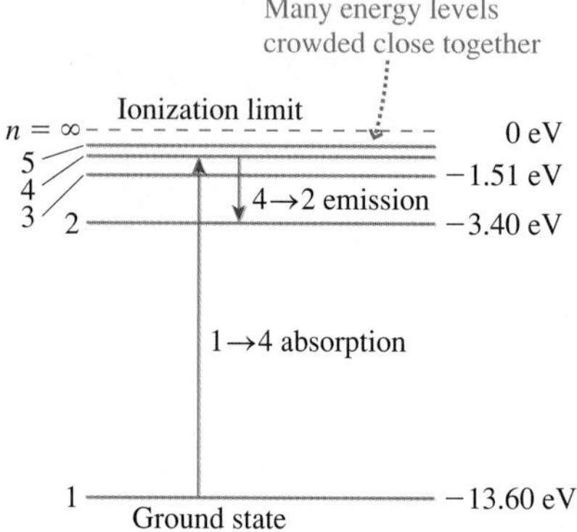

The most important experimental evidence that we have about the hydrogen atom is its spectrum, so the primary test of the Bohr hydrogen atom is whether it correctly predicts the spectrum. FIGURE 29.16 is an energy-level diagram for the hydrogen atom. As we noted in Chapter 28, the energies are like the rungs of a ladder. The lowest rung is the ground state, with $E_1 = -13.60$ eV. The top rung, with $E = 0$ eV, corresponds to a hydrogen ion in the limit $n \rightarrow \infty$. This top rung is called the **ionization limit.** In principle there are an infinite number of rungs, but only the lowest few are shown. The energy levels with higher values of n are all crowded together just below the ionization limit at $n = \infty$.

The figure shows a $1 \rightarrow 4$ transition in which a photon is absorbed and a $4 \rightarrow 2$ transition in which a photon is emitted. For two quantum states m and n, where $n > m$ and E_n is the higher-energy state, an atom can *emit* a photon in an $n \rightarrow m$ transition or *absorb* a photon in an $m \rightarrow n$ transition.

According to the fifth assumption of Bohr's model of atomic quantization, the frequency of the photon emitted in an $n \rightarrow m$ transition is

$$f_{\text{photon}} = \frac{\Delta E_{\text{atom}}}{h} = \frac{E_n - E_m}{h} \tag{29.17}$$

We can use Equation 29.14 for the energies E_n and E_m to predict that the emitted photon has frequency

$$f_{\text{photon}} = \frac{1}{h}\left(\frac{-13.60 \text{ eV}}{n^2} - \frac{-13.60 \text{ eV}}{m^2}\right) = \frac{13.60 \text{ eV}}{h}\left(\frac{1}{m^2} - \frac{1}{n^2}\right)$$

The frequency is a positive number because $m < n$ and thus $1/m^2 > 1/n^2$.

We are more interested in wavelength than frequency, because wavelengths are the quantity measured by experiment. The wavelength of the photon emitted in an $n \rightarrow m$ quantum jump is

$$\lambda_{n\rightarrow m} = \frac{c}{f_{\text{photon}}} = \frac{\lambda_0}{\left(\dfrac{1}{m^2} - \dfrac{1}{n^2}\right)} \qquad \begin{array}{l} m = 1, 2, 3, \ldots \\ n = m+1, m+2, \ldots \end{array} \tag{29.18}$$

with $\lambda_0 = 91.1$ nm. This should look familiar. It is the Balmer formula, Equation 29.2.

It works! Unlike previous atomic models, **the Bohr hydrogen atom correctly predicts the discrete spectrum of the hydrogen atom.** FIGURE 29.17 shows two series of transitions that give rise to wavelengths in the spectrum. The *Balmer series,* consisting of transitions ending on the $m = 2$ state, gives visible wavelengths, and this is the series that Balmer initially analyzed. The *Lyman series,* ending on the $m = 1$ ground state, is in the ultraviolet region of the spectrum and was not measured until later. These series, as well as others in the infrared, are observed in a discharge tube where collisions with electrons excite the atoms upward from the ground state to state n. They then decay downward by emitting photons. Only the Lyman series is observed in the absorption spectrum because, as noted previously, essentially all the atoms in a quiescent gas are in the ground state.

FIGURE 29.17 Transitions producing the Lyman series and the Balmer series of lines in the hydrogen spectrum.

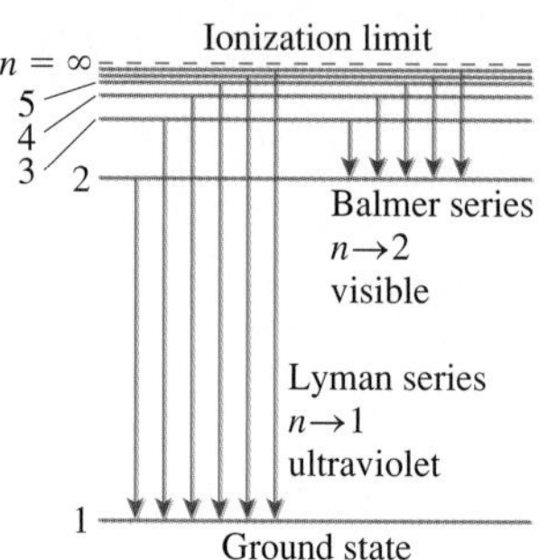

EXAMPLE 29.5 Wavelengths in galactic hydrogen absorption

Whenever astronomers look at distant galaxies, they find that the light has been strongly absorbed at the wavelength of the $1 \rightarrow 2$ transition in the Lyman series of hydrogen. This absorption tells us that interstellar space is filled with vast clouds of hydrogen left over from the Big Bang. What is the wavelength of the $1 \rightarrow 2$ absorption in hydrogen?

SOLVE Equation 29.18 predicts the *absorption* spectrum of hydrogen if we let $m = 1$. The absorption seen by astronomers is from the ground state of hydrogen ($m = 1$) to its first excited state ($n = 2$). The wavelength is

$$\lambda_{1\to 2} = \frac{91.1 \text{ nm}}{\left(\dfrac{1}{1^2} - \dfrac{1}{2^2}\right)} = 122 \text{ nm}$$

ASSESS This wavelength is far into the ultraviolet. Ground-based astronomy cannot observe this region of the spectrum because the wavelengths are strongly absorbed by the atmosphere, but with space-based telescopes, first widely used in the 1970s, astronomers see 122 nm absorption in nearly every direction they look.

29.5 The Quantum-Mechanical Hydrogen Atom

Bohr's analysis of the hydrogen atom seemed to be a resounding success. By introducing stationary states, together with Einstein's ideas about light quanta, Bohr was able to provide the first solid understanding of discrete spectra and, in particular, to predict the Balmer formula for the wavelengths in the hydrogen spectrum. And the Bohr hydrogen atom, unlike Rutherford's model, was stable. There was clearly some validity to the idea of stationary states.

But Bohr was completely unsuccessful at explaining the spectra of any other atom. His method did not work even for helium, the second element in the periodic table with a mere two electrons. Although Bohr was clearly on the right track, his inability to extend the Bohr hydrogen atom to more complex atoms made it equally clear that the complete and correct theory remained to be discovered. Bohr's theory was what we now call "semiclassical," a hybrid of classical Newtonian mechanics with the new ideas of quanta. Still missing was a complete theory of motion and dynamics in a quantized universe—a *quantum* mechanics.

In 1925, Erwin Schrödinger introduced this general theory of quantum mechanics, a theory capable of calculating the allowed energy levels (i.e., the stationary states) of any system. The calculations are mathematically difficult, even for a system as simple as the hydrogen atom, and we will present results without proof.

The Bohr hydrogen atom was characterized by a single quantum number n. In contrast, Schrödinger's quantum-mechanical analysis of the hydrogen atom found that it must be described by *four* quantum numbers.

1. Schrödinger found that the *energy* of the hydrogen atom is given by the same expression found by Bohr, or

$$E_n = -\frac{13.60 \text{ eV}}{n^2} \qquad n = 1, 2, 3, \ldots \qquad (29.19)$$

The integer n is called the **principal quantum number.**

2. The angular momentum L of the electron's orbit must be one of the values

$$L = \sqrt{l(l+1)}\,\hbar \qquad l = 0, 1, 2, 3, \ldots, n-1 \qquad (29.20)$$

The integer l is called the **orbital quantum number.**

3. The plane of the electron's orbit can be tilted, but only at certain discrete angles. Each allowed angle is characterized by a quantum number m, which must be one of the values

$$m = -l, -l+1, \ldots, 0, \ldots, l-1, l \qquad (29.21)$$

The integer m is called the **magnetic quantum number** because it becomes important when the atom is placed in a magnetic field.

4. The electron's *spin*—discussed below—can point only up or down. These two orientations are described by the **spin quantum number** m_s, which must be one of the values

$$m_s = -\frac{1}{2} \text{ or } +\frac{1}{2} \tag{29.22}$$

In other words, each stationary state of the hydrogen atom is identified by a quartet of quantum numbers (n, l, m, m_s), and each quantum number is associated with a physical property of the atom.

NOTE ▶ The energy of the stationary state depends on only the principal quantum number n, not on l, m, or m_s. ◀

EXAMPLE 29.6 Listing quantum numbers

List all possible states of a hydrogen atom that have energy $E = -3.40$ eV.

SOLVE Energy depends on only the principal quantum number n. From Equation 29.19, states with $E = -3.40$ eV have

$$n = \sqrt{\frac{-13.60 \text{ eV}}{-3.40 \text{ eV}}} = 2$$

An atom with principal quantum number $n = 2$ could have either $l = 0$ or $l = 1$, but $l \geq 2$ is ruled out. If $l = 0$, the only possible value for the magnetic quantum number m is $m = 0$. If $l = 1$, then the atom could have $m = -1$, $m = 0$, or $m = +1$. For each of these, the spin quantum number could be $m_s = +\frac{1}{2}$ or $m_s = -\frac{1}{2}$. Thus the possible sets of quantum numbers are

n	l	m	m_s	n	l	m	m_s
2	0	0	$+\frac{1}{2}$	2	0	0	$-\frac{1}{2}$
2	1	1	$+\frac{1}{2}$	2	1	1	$-\frac{1}{2}$
2	1	0	$+\frac{1}{2}$	2	1	0	$-\frac{1}{2}$
2	1	-1	$+\frac{1}{2}$	2	1	-1	$-\frac{1}{2}$

These eight states all have the same energy.

TABLE 29.3 Symbols used to represent quantum number l

l	Symbol
0	*s*
1	*p*
2	*d*
3	*f*

Hydrogen turns out to be unique. For all other elements, the allowed energies depend on both *n and l* (but not m or m_s). Consequently, it is useful to label the stationary states by their values of n and l. The lowercase letters shown in Table 29.3 are customarily used to represent the various values of quantum number l. These symbols come from spectroscopic notation used in prequantum-mechanics days, when some spectral lines were classified as sharp, others as principal, and so on.

Using these symbols, we call the ground state of the hydrogen atom, with $n = 1$ and $l = 0$, the 1*s* state. The 3*d* state has $n = 3$, $l = 2$. In Example 29.6, we found two 2*s* states (with $l = 0$) and six 2*p* states (with $l = 1$), all with the same energy.

Energy and Angular Momentum Are Quantized

The energy of the hydrogen atom is quantized. The allowed energies of hydrogen, given by Equation 29.19, depend on only the principal quantum number n, but for other atoms the energies will depend on both n and l. In anticipation of using both quantum numbers, **FIGURE 29.18** is an energy-level diagram for the hydrogen atom in which the rows are labeled by n and the columns by l. The left column contains all of the $l = 0$ *s* states, the next column is the $l = 1$ *p* states, and so on.

FIGURE 29.18 The energy-level diagram for the hydrogen atom.

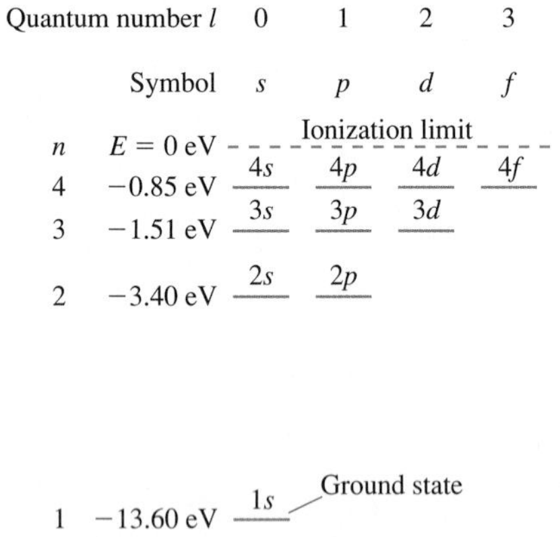

Because the quantum condition of Equation 29.20 requires $n > l$, the *s* states begin with $n = 1$, the *p* states begin with $n = 2$, and the *d* states with $n = 3$. That is, the lowest-energy *d* state is 3*d* because states with $n = 1$ or $n = 2$ cannot have $l = 2$. For hydrogen, where the energy levels do not depend on l, the energy-level diagram shows that the 3*s*, 3*p*, and 3*d* states have equal energy. Figure 29.18 shows only the first few energy levels for each value of l, but there really are an infinite number of levels, as $n \rightarrow \infty$, crowding together beneath $E = 0$. The dotted line at $E = 0$ is the atom's *ionization limit,* the energy of a hydrogen atom in which the electron has been moved infinitely far away to form an H^+ ion.

You should compare this energy-level diagram to Figure 29.16 for the Bohr atom. Because the energy levels of the quantum-mechanical hydrogen atom are exactly the same as for the Bohr atom, the quantum-mechanical solution also correctly predicts the discrete spectrum of hydrogen.

Classically, the angular momentum L of an orbiting electron can have any value. Not so in quantum mechanics. Equation 29.20 tells us that **the electron's orbital angular momentum is quantized.** The magnitude of the orbital angular momentum must be one of the discrete values

$$L = \sqrt{l(l+1)}\hbar = 0, \sqrt{2}\hbar, \sqrt{6}\hbar, \sqrt{12}\hbar, \ldots$$

where l is an integer. The Bohr atom also predicted quantized angular momentum, but the precise values of that prediction turned out to be in error. The quantum-mechanical prediction for L is more complex, but it agrees with experimental observations.

A particularly interesting prediction is that the ground state of hydrogen, with $l = 0$, has *no* angular momentum. A classical particle cannot orbit unless it has angular momentum, but apparently a quantum particle does not have this requirement.

STOP TO THINK 29.5 What are the quantum numbers n and l for a hydrogen atom with $E = -(13.60/9)$ eV and $L = \sqrt{2}\hbar$?

The Electron Spin

You learned in Chapter 24 that a current loop creates a *magnetic dipole moment,* acting like a bar magnet with a north and south pole. In the 1920s, physicists studying the magnetic properties of atoms discovered that the electron has a magnetic dipole moment. The electron has a mass, which allows it to experience gravitational forces, and an electric charge, which allows it to experience electric forces, so perhaps it's not surprising that the electron comes with a magnetic dipole moment, allowing it to experience magnetic forces.

It was first thought that the electron was a very tiny ball of negative charge spinning on its axis. This would give the electron both a magnetic dipole moment and *spin angular momentum.* However, a spinning ball of charge would violate the laws of relativity and other physical laws. As far as we know today, the electron is truly a point particle that happens to have an intrinsic magnetic dipole moment and angular momentum *as if* it were spinning. This inherent angular momentum is called the *electron spin,* but it is a convenient figure of speech, not a factual statement. **The electron has a spin, but it is not a spinning electron.**

The two possible spin quantum numbers $m_s = \pm\frac{1}{2}$ mean that the electron's intrinsic magnetic dipole points in the $+z$-direction or the $-z$-direction. These two orientations are called *spin up* and *spin down.* It is convenient to picture a little vector that can be drawn $\uparrow$ for a spin-up state and $\downarrow$ for a spin-down state. We will use this notation in the next section.

29.6 Multielectron Atoms

The quantum-mechanical solution for the hydrogen atom matches the experimental evidence, but so did the Bohr hydrogen atom. One of the first big successes of Schrödinger's quantum mechanics was an ability to calculate the stationary states and energy levels of *multielectron atoms*, atoms in which Z electrons orbit a nucleus with Z protons. A major difference between multielectron atoms and the simple one-electron hydrogen is that the energy of an electron in a multielectron atom depends on both quantum numbers *n and l*. Whereas the 2*s* and 2*p* states in hydrogen have the same energy, their energies are different in a multielectron atom. The difference arises from the electron-electron interactions that do not exist in a single-electron hydrogen atom.

FIGURE 29.19 An energy-level diagram for electrons in a multielectron atom.

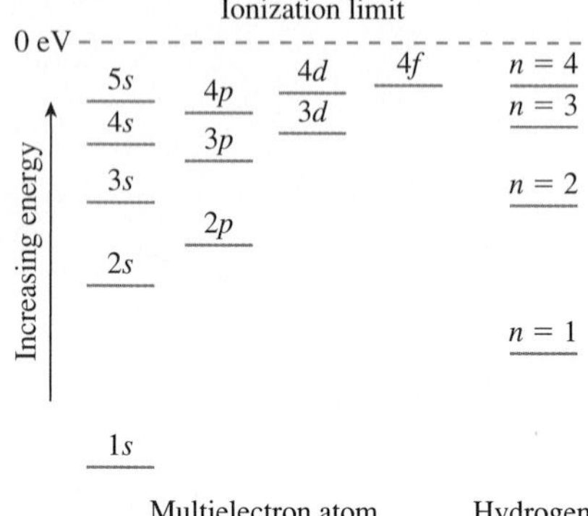

FIGURE 29.19 shows an energy-level diagram for the electrons in a multielectron atom. (Compare this to the hydrogen energy-level diagram in Figure 29.18.) For comparison, the hydrogen-atom energies are shown on the right edge of the figure. Two features of this diagram are of particular interest:

1. For each n, the energy increases as l increases until the maximum-l state has an energy very nearly that of the same n in hydrogen. States with small values of l are significantly lower in energy than the corresponding state in hydrogen.
2. As the energy increases, states with different n begin to alternate in energy. For example, the $3s$ and $3p$ states have lower energy than a $4s$ state, but the energy of an electron in a $3d$ state is slightly higher. This will have important implications for the structure of the periodic table of the elements.

The Pauli Exclusion Principle

By definition, the ground state of a quantum system is the state of lowest energy. What is the ground state of an atom having Z electrons and Z protons? Because the $1s$ state is the lowest energy state, it seems that the ground state should be one in which all Z electrons are in the $1s$ state. However, this idea is not consistent with the experimental evidence.

FIGURE 29.20 The ground state and first excited state of helium.

(a) He ground state

The horizontal lines are the allowed energies.

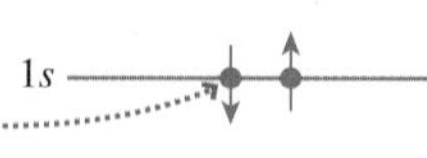

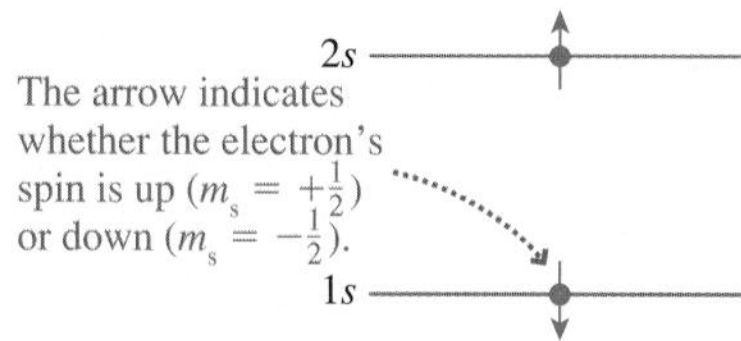

(b) He excited state

In 1925, the Austrian physicist Wolfgang Pauli hypothesized that no two electrons in a quantum system can be in the same quantum state. That is, **no two electrons can have exactly the same set of quantum numbers (n, l, m, m_s).** If one electron is present in a state, it *excludes* all others. This statement is called the **Pauli exclusion principle.** It turns out to be an extremely profound statement about the nature of matter.

The exclusion principle is not applicable to hydrogen, where there is only a single electron, but in helium, with $Z = 2$ electrons, we must make sure that the two electrons are in different quantum states. This is not difficult. For a $1s$ state, with $l = 0$, the only possible value of the magnetic quantum number is $m = 0$. But there are *two* possible values of m_s—namely, $-\frac{1}{2}$ and $+\frac{1}{2}$. If a first electron is in the spin-down $1s$ state $\left(1, 0, 0, -\frac{1}{2}\right)$, a second $1s$ electron can still be added to the atom as long as it is in the spin-up state $\left(1, 0, 0, +\frac{1}{2}\right)$. This is shown schematically in FIGURE 29.20a, where the dots represent electrons on the rungs of the "energy ladder" and the arrows represent spin down or spin up.

The Pauli exclusion principle does not prevent both electrons of helium from being in the $1s$ state as long as they have opposite values of m_s, so we predict this to be the ground state. A list of an atom's occupied energy levels is called its **electron configuration.** The electron configuration of the helium ground state is written $1s^2$, where the superscript 2 indicates two electrons in the $1s$ energy level. An excited state of the helium atom might be the electron configuration $1s2s$. This state is shown in FIGURE 29.20b. Here, because the two electrons have different values of n, there is no restriction on their values of m_s.

FIGURE 29.21 The ground state and first excited state of lithium.

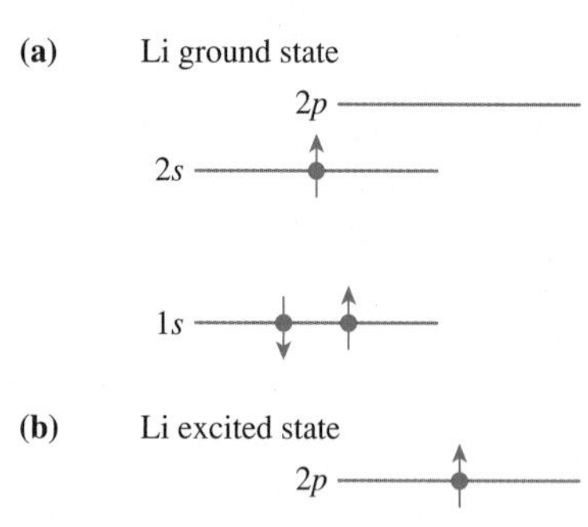

The states $\left(1, 0, 0, -\frac{1}{2}\right)$ and $\left(1, 0, 0, +\frac{1}{2}\right)$ are the only two states with $n = 1$. The ground state of helium has one electron in each of these states, so all the possible $n = 1$ states are filled. Consequently, the electron configuration $1s^2$ is called a **closed shell.**

The next element, lithium, has $Z = 3$ electrons. The first two electrons can go into $1s$ states, with opposite values of m_s, but what about the third electron? The $1s^2$ shell is closed, and there are no additional quantum states having $n = 1$. The only option for the third electron is the next energy state, $n = 2$. The $2s$ and $2p$ states had equal energies in the hydrogen atom, but they do *not* in a multielectron atom. As Figure 29.19 showed, a lower-l state has lower energy than a higher-l state with the same n. The $2s$ state of lithium is lower in energy than $2p$, so lithium's third ground-state electron will be $2s$. This requires $l = 0$ and $m = 0$ for the third electron, but the value of m_s is not relevant because there is only a single electron in $2s$. FIGURE 29.21a shows the electron configuration with the $2s$ electron being spin up, but it could equally well be spin down. The electron configuration for the lithium ground state is written $1s^2 2s$. This indicates two $1s$ electrons and a single $2s$ electron. One possible excited state of lithium is shown in FIGURE 29.21b.

The Periodic Table of the Elements

The 19th century was a time when chemists were discovering new elements and studying their chemical properties. The century opened with the atomic model still not completely validated, and with no one having any idea how many elements there might be. But chemistry developed quickly, and by mid-century it was clear that there were dozens of elements, but not hundreds.

The Russian chemist Dmitri Mendeléev was the first to propose, in 1867, a *periodic* arrangement of the elements based on the regular recurrence of chemical properties. He did so by explicitly pointing out "gaps" where, according to his hypothesis, undiscovered elements should exist. He could then predict the expected properties of the missing elements. The subsequent discovery of these elements verified Mendeléev's organizational scheme, which came to be known as the *periodic table of the elements.*

One of the great triumphs of the quantum-mechanical theory of multielectron atoms is that it explains the structure of the periodic table. We can understand this structure by looking at the energy-level diagram of **FIGURE 29.22**, which is an expanded version of the energy-level diagram of Figure 29.19. Just as for helium and lithium, atoms with larger values of Z are constructed by placing Z electrons into the lowest-energy levels that are consistent with the Pauli exclusion principle.

The s states of helium and lithium can each hold two electrons—one spin up and the other spin down—but the higher-angular-momentum states that will become filled for higher-Z atoms can hold more than two electrons. For each value l of the orbital quantum number, there are $2l + 1$ possible values of the magnetic quantum number m and, for each of these, two possible values of the spin quantum number m_s. Consequently, each energy *level* in Figure 29.22 is actually $2(2l + 1)$ different *states* that, taken together, are called a *subshell.* Table 29.4 lists the number of states in each subshell. Each state in a subshell is represented in Figure 29.22 by a colored dot. The dots' colors correspond to the periodic table in **FIGURE 29.23**, which is color coded to show which subshells are being filled as Z increases.

We can now use Figure 29.22 to construct the periodic table in Figure 29.23. We've already seen that lithium has two electrons in the $1s$ state and one electron in

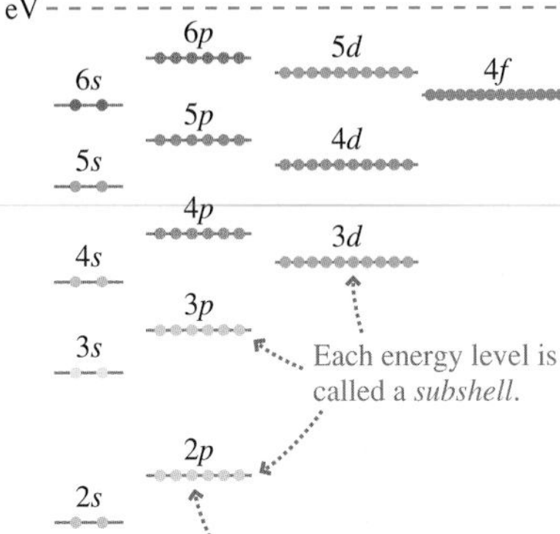
0 eV
6p
5d
4f
6s
5p
4d
5s
4p
3d
4s
3p
3s
Each energy level is called a subshell.
2p
2s
The number of dots indicates the number of states in a subshell. A p subshell has six states.
1s

FIGURE 29.22 An energy-level diagram showing how many electrons can occupy each subshell.

TABLE 29.4 Number of states in each subshell of an atom

Subshell	l	Number of states
s	0	2
p	1	6
d	2	10
f	3	14

FIGURE 29.23 The periodic table of the elements. The elements are color coded to the states in the energy-level diagram of Figure 29.22.

																		Subshells
1 H																	2 He	1s
3 Li	4 Be											5 B	6 C	7 N	8 O	9 F	10 Ne	2s, 2p
11 Na	12 Mg	Transition elements										13 Al	14 Si	15 P	16 S	17 Cl	18 Ar	3s, 3p
19 K	20 Ca	21 Sc	22 Ti	23 V	24 Cr	25 Mn	26 Fe	27 Co	28 Ni	29 Cu	30 Zn	31 Ga	32 Ge	33 As	34 Se	35 Br	36 Kr	4s, 3d, 4p
37 Rb	38 Sr	39 Y	40 Zr	41 Nb	42 Mo	43 Tc	44 Ru	45 Rh	46 Pd	47 Ag	48 Cd	49 In	50 Sn	51 Sb	52 Te	53 I	54 Xe	5s, 4d, 5p
55 Cs	56 Ba	71 Lu	72 Hf	73 Ta	74 W	75 Re	76 Os	77 Ir	78 Pt	79 Au	80 Hg	81 Tl	82 Pb	83 Bi	84 Po	85 At	86 Rn	6s, 5d, 6p
87 Fr	88 Ra	103 Lr	104 Rf	105 Db	106 Sg	107 Bh	108 Hs	109 Mt	110 Ds	111 Rg								7s, 6d

															Subshell
Lanthanides	57 La	58 Ce	59 Pr	60 Nd	61 Pm	62 Sm	63 Eu	64 Gd	65 Tb	66 Dy	67 Ho	68 Er	69 Tm	70 Yb	4f
Actinides	89 Ac	90 Th	91 Pa	92 U	93 Np	94 Pu	95 Am	96 Cm	97 Bk	98 Cf	99 Es	100 Fm	101 Md	102 No	5f

Inner transition elements

the 2s state. Four-electron beryllium ($Z = 4$) comes next. We see that there is still an empty state in the 2s subshell for this fourth electron to occupy, so beryllium closes the 2s subshell and has electron configuration $1s^2 2s^2$.

As Z increases further, the next six electrons can each occupy states in the 2p subshell. These are the elements boron (B) through neon (Ne), completing the second row of the periodic table. Neon, which completes the 2p subshell, has ground-state configuration $1s^2 2s^2 2p^6$.

The 3s subshell is the next to be filled, leading to the elements sodium and magnesium. Filling the 3p subshell gives aluminum through argon, completing the third row of the table.

Starting the fourth row of the periodic table are the two 4s states of potassium and calcium. Here the table begins to get more complicated. Following the pattern of rows two and three, you might expect the 4p subshell to start filling at this point. But, looking carefully at Figure 29.22, you will see that the 3d subshell has a slightly lower energy than the 4p subshell. Because the ground state of an atom is the *lowest energy state* consistent with the Pauli exclusion principle, the next element—scandium—finds it more favorable to add a 3d electron. The ten *transition elements* from scandium (Sc) through zinc (Zn) then fill the 10 states of the 3d subshell.

The same pattern applies to the fifth row, where the 5s, 4d, and 5p subshells fill in succession. In the sixth row, however, after the initial 6s states are filled, the 4f subshell has the lowest energy, so it begins to fill *before* the 5d states. The elements corresponding to the 4f subshell, lanthanum through ytterbium, are known as the lanthanides, and they are traditionally drawn as a row separated from the rest of the table. The seventh row follows this same pattern.

Thus the entire periodic table can be built up using our knowledge of the energy-level diagram of a multielectron atom along with the Pauli exclusion principle.

EXAMPLE 29.7 The ground state of arsenic

Predict the ground-state electron configuration of arsenic.

SOLVE The periodic table shows that arsenic (As) has $Z = 33$, so we must identify the states of 33 electrons. Arsenic is in the fourth row, following the first group of transition elements. Argon ($Z = 18$) filled the 3p subshell, then calcium ($Z = 20$) filled the 4s subshell. The next 10 elements, through zinc ($Z = 30$), filled the 3d subshell. The 4p subshell starts filling with gallium ($Z = 31$), and arsenic is the third element in this group, so it will have three 4p electrons. Thus the ground-state configuration of arsenic is

$$1s^2 2s^2 2p^6 3s^2 3p^6 4s^2 3d^{10} 4p^3$$

STOP TO THINK 29.6 Is the electron configuration $1s^2 2s^2 2p^4 3s$ a ground-state configuration or an excited-state configuration?

A. Ground-state.
B. Excited-state.
C. It's not possible to tell without knowing which element it is.

29.7 Excited States and Spectra

18.2 ActivPhysics ONLINE

The periodic table organizes information about the *ground states* of the elements. These states are chemically most important because most atoms spend most of the time in their ground states. All the chemical ideas of valence, bonding, reactivity, and so on are consequences of these ground-state atomic structures. But the periodic table does not tell us anything about the excited states of atoms. It is the excited states that hold the key to understanding atomic spectra, and that is the topic to which we turn next.

Sodium ($Z = 11$) is a multielectron atom that we will use to illustrate excited states. The ground-state electron configuration of sodium is $1s^2 2s^2 2p^6 3s$. The first 10 electrons completely fill the $1s$, $2s$, and $2p$ subshells, creating a *neon core* whose electrons are tightly bound together. The $3s$ electron, however, is a *valence electron* that can be easily excited to higher energy levels.

The excited states of sodium are produced by raising the valence electron to a higher energy level. The electrons in the neon core are unchanged. **FIGURE 29.24** is an energy-level diagram showing the ground state and some of the excited states of sodium. The $1s$, $2s$, and $2p$ states of the neon core are not shown on the diagram. These states are filled and unchanging, so only the states available to the valence electron are shown. Notice that the zero of energy has been shifted to the ground state. As we have discovered before, the zero of energy can be located where it is most convenient. With this choice, the excited-state energies tell us how far each state is above the ground state. The ionization limit now occurs at the value of the atom's ionization energy, which is 5.14 eV for sodium.

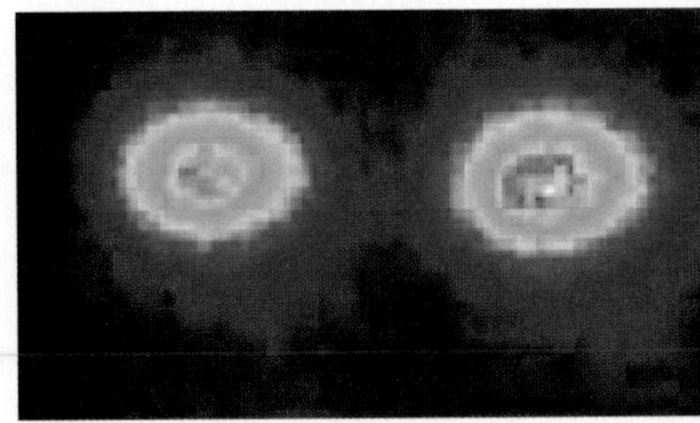

The dots of light are being emitted by two beryllium ions held in a device called an ion trap. Each ion, which is excited by an invisible ultraviolet laser, emits about 10^6 visible-light photons per second.

FIGURE 29.24 The $3s$ ground state of the sodium atom and some of the excited states.

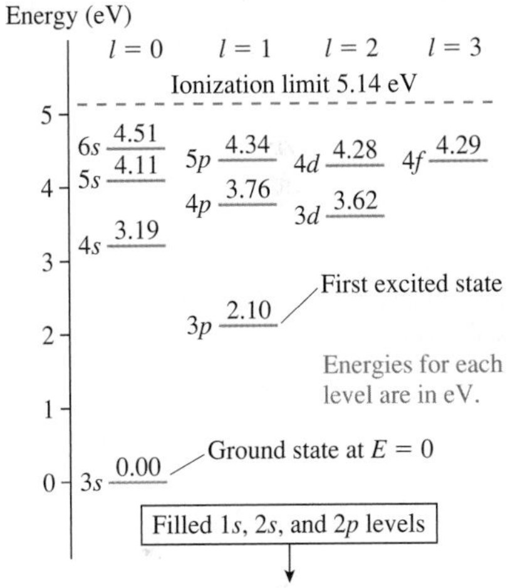

Excitation by Absorption

Left to itself, an atom will be in its lowest-energy ground state. How does an atom get into an excited state? The process of getting it there is called **excitation,** and there are two basic mechanisms: absorption and collision. We'll begin by looking at excitation by absorption.

One of the postulates of the basic Bohr model is that an atom can jump from one stationary state, of energy E_1, to a higher-energy state E_2 by absorbing a photon of frequency $f_{\text{photon}} = \Delta E_{\text{atom}}/h$. This process is shown in **FIGURE 29.25**. Because we are interested in spectra, it is more useful to write this in terms of the wavelength:

$$\lambda = \frac{c}{f_{\text{photon}}} = \frac{hc}{\Delta E_{\text{atom}}} = \frac{1240 \text{ eV} \cdot \text{nm}}{\Delta E_{\text{atom}}} \qquad (29.23)$$

The final expression, which uses the value $hc = 1240$ eV $\cdot$ nm, gives the wavelength in nanometers if ΔE_{atom} is in electron volts.

Not every quantum jump allowed by Equation 29.33 can actually occur in an atom. A quantum-mechanical analysis of how the electrons in an atom interact with a light wave shows that transitions must also satisfy the following **selection rule:** Transitions (either absorption or emission) from a state with orbital quantum number l can occur to only another state whose orbital quantum number differs from the original state by ± 1, or

$$\Delta l = l_2 - l_1 = \pm 1 \qquad (29.24)$$

Selection rule for emission and absorption

FIGURE 29.25 Excitation by photon absorption.

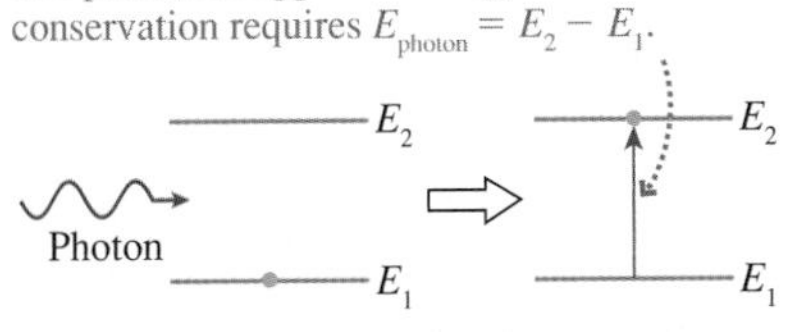

The photon disappears. Energy conservation requires $E_{\text{photon}} = E_2 - E_1$.

EXAMPLE 29.8 Analyzing absorption in sodium

What are the two longest wavelengths in the absorption spectrum of sodium? What are the transitions?

SOLVE As Figure 29.24 shows, the sodium ground state is $3s$. Starting from an s state, the selection rule permits quantum jumps only to p states. The lowest excited state is the $3p$ state and $3s \rightarrow 3p$ is an allowed transition ($\Delta l = 1$), so this will be the longest wavelength. You can see from the data in Figure 29.24 that $\Delta E_{\text{atom}} = 2.10 \text{ eV} - 0.00 \text{ eV} = 2.10$ eV for this transition.

The corresponding wavelength is

$$\lambda = \frac{1240 \text{ eV} \cdot \text{nm}}{2.10 \text{ eV}} = 590 \text{ nm}$$

(Because of rounding, the above calculation gives $\lambda = 590$ nm. The experimental value is actually 589 nm.)

Continued

The next excited state is $4s$, but a $3s \rightarrow 4s$ transition is not allowed by the selection rule. The next allowed transition is $3s \rightarrow 4p$, with $\Delta E_{atom} = 3.76$ eV. The wavelength of this transition is

$$\lambda = \frac{1240 \text{ eV} \cdot \text{nm}}{3.76 \text{ eV}} = 330 \text{ nm}$$

ASSESS If you look at the sodium spectrum shown in **FIGURE 29.26** (which is the same as that shown earlier in Figure 29.3), you will see that 589 nm and 330 nm are, indeed, the two longest wavelengths in the absorption spectrum.

FIGURE 29.26 The absorption spectrum of sodium.

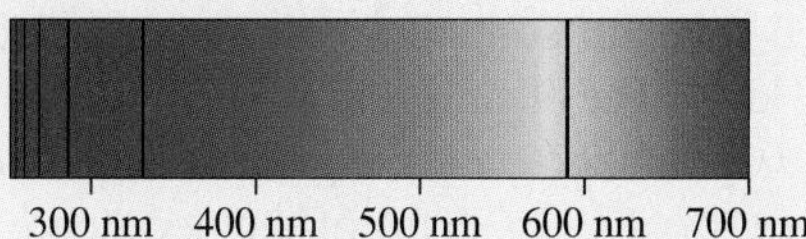

Collisional Excitation

An electron traveling with a speed of 1.0×10^6 m/s has a kinetic energy of 2.85 eV. If this electron collides with a ground-state sodium atom, a portion of its energy can be used to excite the atom to a higher-energy state, such as its $3p$ state. This process is called **collisional excitation** of the atom.

Collisional excitation differs from excitation by absorption in one very fundamental way. In absorption, the photon disappears. Consequently, *all* of the photon's energy must be transferred to the atom. Conservation of energy then requires $E_{photon} = \Delta E_{atom}$. In contrast, the electron is still present after collisional excitation and can still have some kinetic energy. That is, the electron does *not* have to transfer its entire energy to the atom. If the electron has an incident kinetic energy of 2.85 eV, it could transfer 2.10 eV to the sodium atom, thereby exciting it to the $3p$ state, and still depart the collision with a speed of 5.1×10^5 m/s and 0.75 eV of kinetic energy.

FIGURE 29.27 Excitation by electron collision.

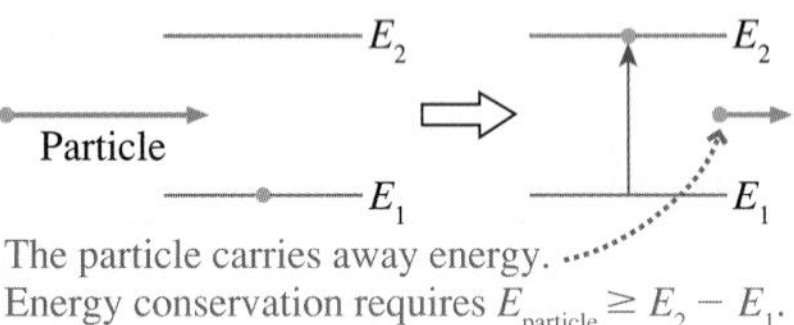

To excite the atom, the incident energy of the electron (or any other matter particle) merely has to *exceed* ΔE_{atom}; that is $E_{particle} \geq \Delta E_{atom}$. There's a threshold energy for exciting the atom, but no upper limit. It is all a matter of energy conservation. **FIGURE 29.27** shows the idea graphically.

Collisional excitation by electrons is the predominant method of excitation in electrical discharges such as fluorescent lights, street lights, and neon signs. A gas is sealed in a tube at reduced pressure ($\approx$1 mm Hg), then a fairly high voltage ($\approx$1000 V) between electrodes at the ends of the tube causes the gas to ionize, creating a current in which both ions and electrons are charge carriers. The electrons accelerate in the electric field, gaining several eV of kinetic energy, then transfer some of this energy to the gas atoms upon collision. The process does not work at atmospheric pressure because the average distance between collisions is too short for the electrons to gain enough kinetic energy to excite the atoms.

NOTE ▶ In contrast to photon absorption, there are no selection rules for collisional excitation. Any state can be excited if the colliding particle has sufficient energy. ◀

CONCEPTUAL EXAMPLE 29.9 **Possible excitation of hydrogen?**

Can an electron with a kinetic energy of 11.4 eV cause a hydrogen atom to emit the prominent red spectral line (λ = 656 nm, $E_{photon} = 1.89$ eV) in the Balmer series?

REASON The electron must have sufficient energy to excite the upper state of the transition. The electron's energy of 11.4 eV is significantly greater than the 1.89 eV energy of a photon with wavelength 656 nm, but don't confuse the energy of the photon with the energy of the excitation. The red spectral line in the Balmer series is emitted in an $n = 3 \rightarrow 2$ quantum jump with $\Delta E_{atom} = 1.89$ eV, but to cause this emission, the electron must excite an atom from its *ground state*, with $n = 1$, up to the $n = 3$ level. From Figure 29.16, the necessary excitation energy is

$$\Delta E_{atom} = E_3 - E_1 = (-1.51 \text{ eV}) - (-13.60 \text{ eV}) = 12.09 \text{ eV}$$

The electron does *not* have sufficient energy to excite the atom to the state from which the emission would occur.

ASSESS As our discussion of absorption spectra showed, almost all excitations of atoms begin from the ground state. Quantum jumps down in energy, however, can begin and end at any two states allowed by selection rules.

Emission Spectra

The absorption of light is an important process, but it is the emission of light that really gets our attention. The overwhelming bulk of sensory information that we perceive comes to us in the form of light. With the small exception of cosmic rays, all of our knowledge about the cosmos comes to us in the form of light and other electromagnetic waves emitted in various processes. And emission spectra are more than just scientific curiosities. Many of today's artificial light sources, from fluorescent lights to lasers, are applications of emission spectra.

Understanding emission hinges upon the three ideas shown in FIGURE 29.28. Once we have determined the energy levels of an atom, from quantum mechanics, we can immediately predict its emission spectrum. We might also ask *how long* an atom remains in an excited state before undergoing a quantum jump to a lower-energy state and emitting a photon. Just as the uncertainty principle prevents us from knowing exactly where an electron is, we also can't determine exactly how long any particular atom spends in the excited state. However, we can determine, both theoretically and experimentally, the *average* time, and the average time an atom spends in the excited state before emitting a photon is called the **lifetime** of that state. A typical excited-state lifetime is just a few nanoseconds.

As an example, FIGURE 29.29a shows some of the transitions and wavelengths observed in the emission spectrum of sodium. This diagram makes the point that each wavelength represents a quantum jump between two well-defined energy levels. Notice that the selection rule $\Delta l = \pm 1$ is obeyed in the sodium spectrum. The $5p$ levels can undergo quantum jumps to $3s$, $4s$, or $3d$ but *not* to $3p$ or $4p$.

FIGURE 29.29b shows the emission spectrum of sodium as it would be recorded in a spectrometer. (Many of the lines seen in this spectrum start from higher excited states that are not seen in the rather limited energy-level diagram of Figure 29.29a.) By comparing the spectrum to the energy-level diagram, you can recognize that the spectral lines at 589 nm, 330 nm, 286 nm, and 268 nm form a *series* of lines due to all the possible $np \rightarrow 3s$ transitions. They are the dominant features in the sodium spectrum.

The most obvious visual feature of sodium emission is its bright yellow color, produced by the 589 nm photons emitted in the $3p \rightarrow 3s$ transition. (The lifetime of the $3p$ state happens to be 17 ns.) This is the basis of the *flame test* used in chemistry to test for sodium: A sample is held in a Bunsen burner, and a bright yellow glow indicates the presence of sodium. The 589 nm emission is also prominent in the pinkish-yellow glow of the common sodium-vapor street lights. These operate by creating an electrical discharge in sodium vapor. Most sodium-vapor lights use high-pressure lamps to increase their light output. The high pressure, however, causes the formation of Na_2 molecules, and these molecules emit the pinkish portion of the light.

FIGURE 29.28 Generation of an emission spectrum.

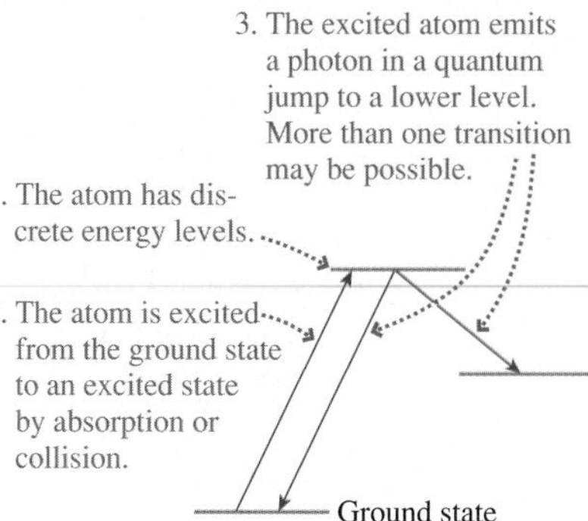

3. The excited atom emits a photon in a quantum jump to a lower level. More than one transition may be possible.
1. The atom has discrete energy levels.
2. The atom is excited from the ground state to an excited state by absorption or collision.
Ground state

FIGURE 29.29 The emission spectrum of sodium.

(a)

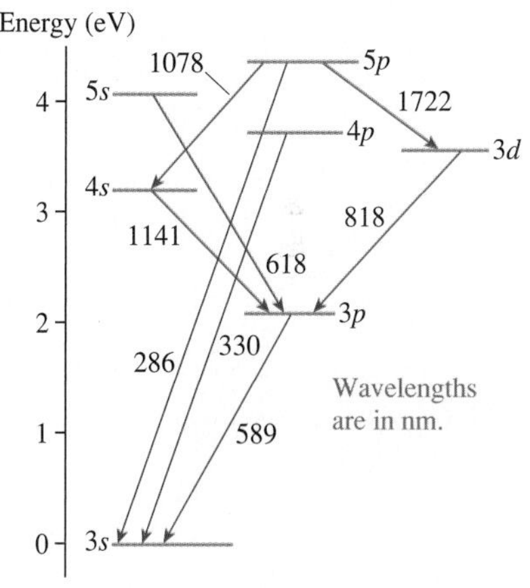

(b)

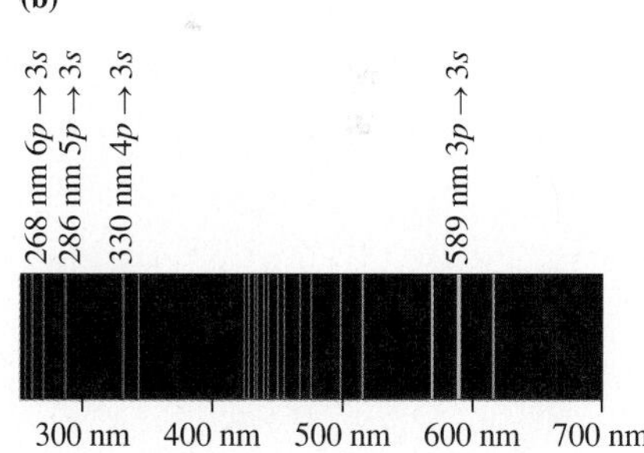

▲ **Seeing the light** Some cities close to astronomical observatories use low-pressure sodium lights, and these emit the distinctively yellow 589 nm light of sodium. The glow of city lights is a severe problem for astronomers, but the very specific 589 nm emission from sodium is easily removed with a *sodium filter*, an interference filter that lets all colors pass except the 589 nm yellow sodium light. The photos show the sky near a sodium streetlight without (left) and with (right) a sodium filter. The constellation of Orion, nearly obscured in the left photo, is clearly visible in the right photo.

FIGURE 29.30 The generation of x rays from copper atoms.

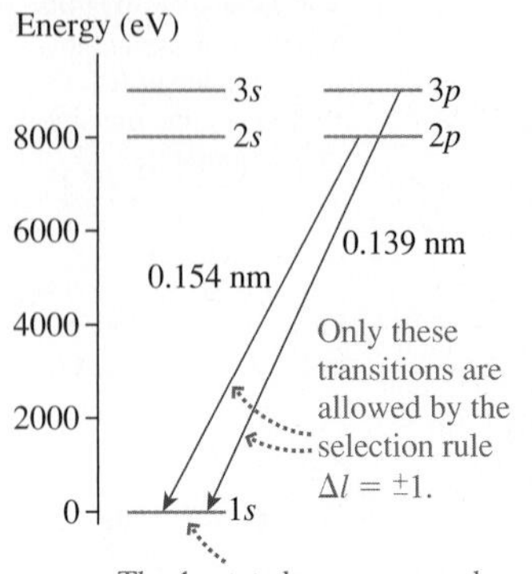

X Rays

Chapter 28 noted that x rays are produced by causing very-high-speed electrons, accelerated with potential differences of many thousands of volts, to crash into metal targets. Rather than exciting the atom's valence electrons, such as happens in a gas discharge tube, these high-speed projectiles are capable of knocking inner-shell electrons out of the target atoms, producing an *inner-shell vacancy.* As FIGURE 29.30 shows for copper atoms, this vacancy is filled when an electron from a higher shell undergoes a quantum jump into the vacancy, emitting a photon in the process.

In heavy elements, such as copper or iron, the energy difference between the inner and outer shells is very large—typically 10 keV. Consequently, the photon has energy $E_{photon} \approx 10$ keV and wavelength $\lambda \approx 0.1$ nm. These high-energy photons are the x rays discovered by Röntgen. X-ray photons are about 10,000 times more energetic than visible-light photons, and the wavelengths are about 10,000 times smaller. Even so, the underlying physics is the same: A photon is emitted when an electron in an atom undergoes a quantum jump.

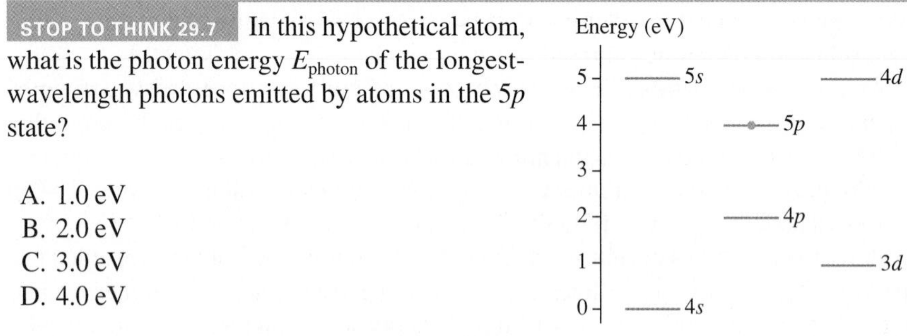

STOP TO THINK 29.7 In this hypothetical atom, what is the photon energy E_{photon} of the longest-wavelength photons emitted by atoms in the $5p$ state?

A. 1.0 eV
B. 2.0 eV
C. 3.0 eV
D. 4.0 eV

29.8 Molecules

Quantum mechanics applies to molecules just as it does to atoms, but molecules are more complex because they have internal modes of storing energy. In particular, molecules can *rotate* about their center of mass, and the atoms can *vibrate* back and forth as if the molecular bonds holding them together were little springs. For the most part, we'll overlook this internal motion and focus our attention on the electrons and the electron energy levels.

A quantum-mechanical analysis of molecules finds the following general results:

- The energy is quantized. Electrons can exist in only certain allowed energy levels.
- The number of energy levels is so extraordinarily high, and the energy levels are jammed so close together, that for all practical purposes the energy levels group into *bands* of allowed energy.
- In thermal equilibrium, nearly all molecules are in the very lowest energy levels.

FIGURE 29.31 The molecular energy-level diagram for a medium-size molecule.

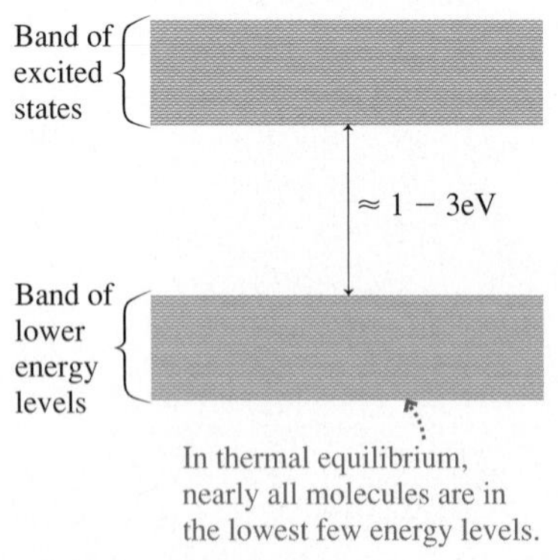

FIGURE 29.31 shows a generic molecular energy-level diagram for a medium-size molecule. Whereas an atom has a well-defined ground state, a molecule has a broad band of lower energy levels. Similarly, a single excited state, such as the $2s$ state of the hydrogen atom, has been replaced by a band of excited energy levels. Despite the vast number of allowed energy levels, nearly all molecules spend nearly all their time in the very lowest energy levels.

FIGURE 29.32 uses the energy-level diagram to explain two important phenomena of molecular spectroscopy: absorption and fluorescence. Whereas the absorption spectrum of an atom consists of discrete spectral lines, a molecule has a continuous *absorption band.* The absorption of light at a higher frequency (shorter wavelength) followed by the emission of light at a lower frequency (longer wavelength) is called **fluorescence.** Fluorescence occurs in molecules, but not atoms, because molecules can transform some of the absorbed energy into the vibrational energy of the atoms and thus increase the thermal energy of the molecules.

FIGURE 29.32 Molecular absorption and fluorescence.

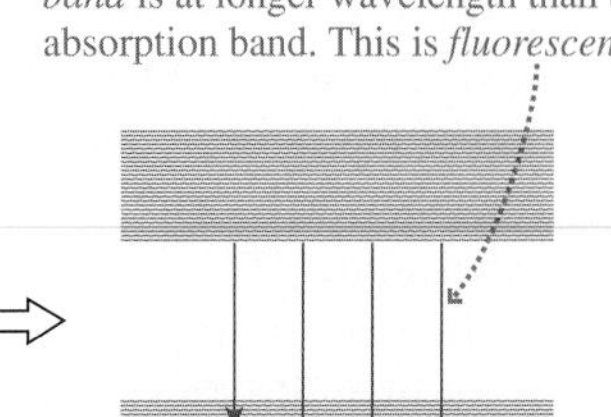

The molecules can absorb light over a range of wavelengths. This makes an *absorption band.*
The molecules rapidly transform some of the absorbed energy into molecular vibrations, causing the molecules to fall to the bottom edge of the excited band.
Quantum jumps back to the lower band have less energy than the original jumps up. Thus the *emission band* is at longer wavelength than the absorption band. This is *fluorescence.*
Absorption band
Emission band
Relative intensity
0
300 400 500 600
Wavelength (nm)

Fluorescence has many important applications. In biology, fluorescent dyes are used to stain tissue samples for microscopy. In the brilliantly colored photo that opened this chapter, the cells were stained with three different dyes, one of which (blue) was preferentially absorbed in the nucleus and another (green) by actin. The red dye was absorbed by other cell structures. When illuminated with ultraviolet light, which we can't see, each dye fluoresces with its own characteristic color.

▶ **Whiter than white** Laundry detergents often contain fluorescent dyes that absorb invisible ultraviolet light and then fluoresce in a broad band of visible wavelengths; that is, they give off white light. Sunlight contains lots of UV, so outdoors your white clothes not only reflect the white visible sunlight, but fluoresce even more white light—"whiter than white." In the photo, you can see how brightly the girl's shirt glows when illuminated with UV light (and, evidently, hair and teeth fluoresce this way as well).

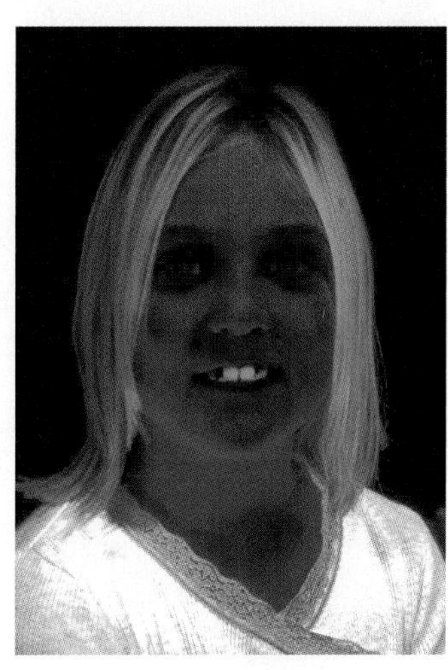

Many biological molecules are fluorescent. A good example is chlorophyll, the green pigment in plants that allows photosynthesis to happen. When illuminated by blue or ultraviolet light, chlorophyll fluoresces an unexpected dark red color. This property of chlorophyll is used by marine biologists to measure the concentration of phytoplankton—microscopic plants—in seawater.

Quite recently, scientists isolated a fluorescent protein from a species of jellyfish. This protein fluoresces green when illuminated with ultraviolet light, so it's been dubbed GFP, for *green fluorescent protein.* GFP has become an important tool in genetics because it can be used to identify when particular genes are being expressed in living cells. This is done by fusing the jellyfish GFP gene to the gene being studied. When the gene is active, the cell manufactures GFP in addition to the usual protein coded by the gene. If the cells are observed under ultraviolet light, a bright green glow from the GFP indicates that the gene is turned on. The cells are dark where the gene is not active. FIGURE 29.33 shows an example.

The color of GFP may remind you of *bioluminescence*—the summer flashes of fireflies or the green glow of various deep-sea fish—and, indeed, there is a connection. Bioluminescence is actually a form of *chemiluminescence,* the production of light in chemical reactions. Some chemical reactions create reactant molecules in an excited state. These molecules then emit light as they jump to lower energy levels, just as if they had first absorbed shorter-wavelength light. The emitted light has exactly the same spectrum as fluorescence, but the method by which the molecules are excited is different. Light sticks, which come in many colors, are an example of chemiluminescence.

FIGURE 29.33 Green fluorescent protein shows the locations in mosquito larvae at which a particular gene is being expressed.

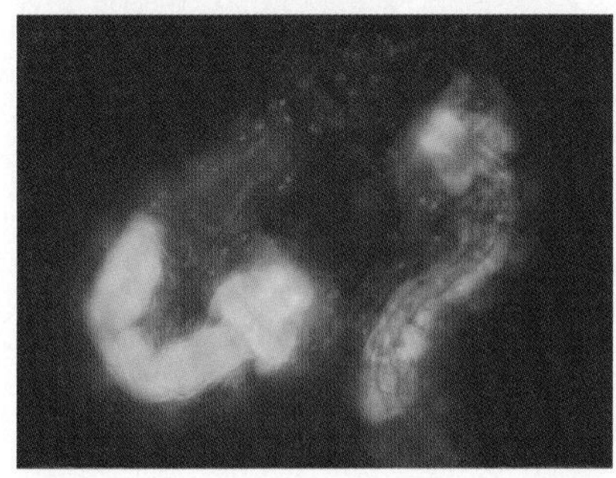

Bioluminescence is just chemiluminescence in a biological organism. Some biochemical reaction—often a catalyzed reaction in an organism—produces a molecule in an excited state, and light is emitted as the molecule jumps to a lower state. In fireflies, the reaction involves an enzyme with the intriguing name *luciferase.* In the jellyfish *Aequoria victoria,* the source of GFP, bioluminescent reactions actually create blue light. However, the blue light is absorbed by the green fluorescent protein—just as predicted by the absorption curve in Figure 29.32—and re-emitted as green fluorescence, giving the jellyfish an eerie green glow when seen in the dark ocean.

18.3 Activ Physics ONLINE

29.9 Stimulated Emission and Lasers

FIGURE 29.34 Three types of radiative transitions.

(a) Absorption

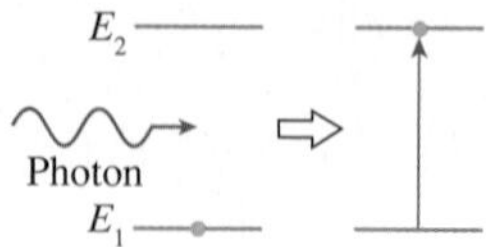

(b) Spontaneous emission

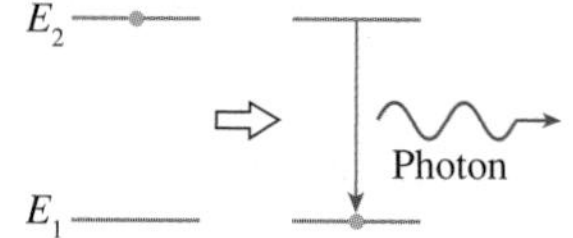

(c) Stimulated emission

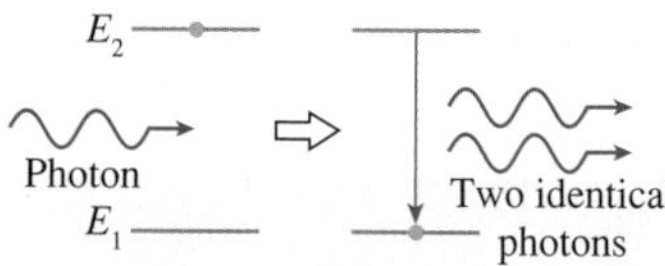

We have seen that an atom can jump from a lower-energy level E_1 to a higher-energy level E_2 by absorbing a photon. FIGURE 29.34a illustrates the basic absorption process, with a photon of frequency $f = \Delta E_{atom}/h$ disappearing as the atom jumps from level 1 to level 2. Once in level 2, as shown in FIGURE 29.34b, the atom can emit a photon of the same frequency as it jumps back to level 1. Because this transition occurs spontaneously, without the introduction of outside energy, it is called **spontaneous emission.**

In 1917, four years after Bohr's proposal of stationary states in atoms but still prior to de Broglie and Schrödinger, Einstein was puzzled by how quantized atoms reach thermodynamic equilibrium in the presence of electromagnetic radiation. Einstein found that the processes of absorption and spontaneous emission were not sufficient to allow a collection of atoms to reach thermodynamic equilibrium. To resolve this difficulty, Einstein proposed a third mechanism for the interaction of atoms with light.

The left half of FIGURE 29.34c shows a photon with frequency $f = \Delta E_{atom}/h$ approaching an *excited* atom. If a photon can induce the $1 \rightarrow 2$ transition of absorption, then Einstein proposed that it should also be able to induce a $2 \rightarrow 1$ transition. In a sense, this transition is a *reverse absorption.* But to undergo a reverse absorption, the atom must *emit* a photon of frequency $f = \Delta E_{atom}/h$. The end result, as seen in the right half of Figure 29.34c, is an atom in level 1 plus *two* photons! Because the first photon induced the atom to emit the second photon, this process is called **stimulated emission.**

Stimulated emission occurs only if the first photon's frequency exactly matches the $E_2 - E_1$ energy difference of the atom. This is precisely the same condition that absorption has to satisfy. More interesting, the emitted photon is *identical* to the incident photon. This means that as the two photons leave the atom they have exactly the same frequency and wavelength, are traveling in exactly the same direction, and are exactly in phase with each other. In other words, **stimulated emission produces a second photon that is an exact clone of the first.**

Stimulated emission is of no importance in most practical situations. Atoms typically spend only a few nanoseconds in an excited state before undergoing spontaneous emission, so the atom would need to be in an extremely intense light wave for stimulated emission to occur prior to spontaneous emission. Ordinary light sources are not nearly intense enough for stimulated emission to be more than a minor effect; hence it was many years before Einstein's prediction was confirmed. No one had doubted Einstein, because he had clearly demonstrated that stimulated emission was necessary to make the energy equations balance, but it seemed no more important than would pennies to a millionaire balancing her checkbook. At least, that is, until 1960, when a revolutionary invention made explicit use of stimulated emission: the laser.

A spectacular laser light show depends on three key properties of coherent laser light: It can be very intense, its color is extremely pure, and the laser beam is narrow with little divergence.

Lasers

The word **laser** is an acronym for the phrase **l**ight **a**mplification by the **s**timulated **e**mission of **r**adiation. The first laser, a ruby laser, was demonstrated in 1960, and several other kinds of lasers appeared within a few months. Today, lasers do everything from being the light source in fiber-optic communications to measuring the distance to the moon, and from playing your CD to performing delicate eye surgery.

But what is a laser? Basically it is a device that produces a beam of highly *coherent* and essentially monochromatic (single-color) light as a result of stimulated emission. **Coherent** light is light in which all the electromagnetic waves have the same phase, direction, and amplitude. It is the coherence of a laser beam that allows it to be very tightly focused or to be rapidly modulated for communications.

Let's take a brief look at how a laser works. FIGURE 29.35 represents a system of atoms that have a lower energy level E_1 and a higher energy level E_2. Suppose that there are N_1 atoms in level 1 and N_2 atoms in level 2. Left to themselves, all the atoms would soon end up in level 1 because of the spontaneous emission $2 \rightarrow 1$. To prevent this, we can imagine that some type of excitation mechanism, perhaps an electrical discharge, continuously produces new excited atoms in level 2.

Let a photon of frequency $f = (E_2 - E_1)/h$ be incident on this group of atoms. Because it has the correct frequency, it could be absorbed by one of the atoms in level 1. Another possibility is that it could cause stimulated emission from one of the level 2 atoms. Ordinarily $N_2 \ll N_1$, so absorption events far outnumber stimulated emission events. Even if a few photons were generated by stimulated emission, they would quickly be absorbed by the vastly larger group of atoms in level 1.

But what if we could somehow arrange to place *every* atom in level 2, making $N_1 = 0$? Then the incident photon, upon encountering its first atom, will cause stimulated emission. Where there was initially one photon of frequency f, now there are two. These will strike two additional excited-state atoms, again causing stimulated emission. Then there will be four photons. As FIGURE 29.36 shows, there will be a *chain reaction* of stimulated emission until all N_2 atoms emit a photon of frequency f.

FIGURE 29.35 Energy levels 1 and 2, with populations N_1 and N_2.

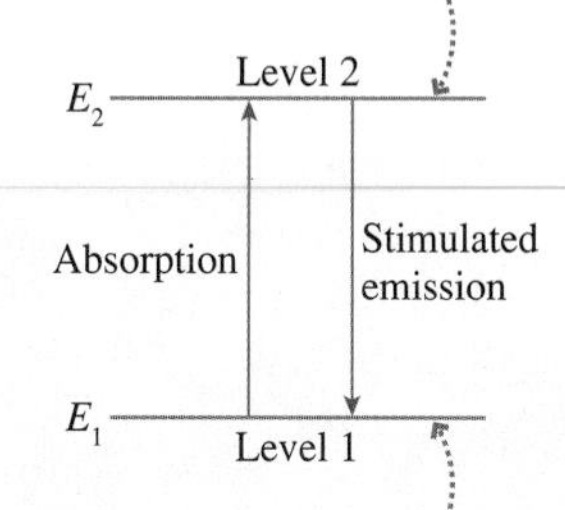

FIGURE 29.36 Stimulated emission creates a chain reaction of photon production in a population of excited atoms.

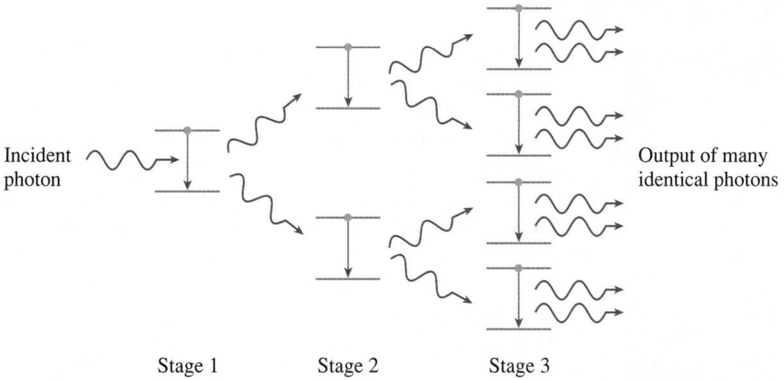

In stimulated emission, each emitted photon is *identical* to the incident photon. The chain reaction of Figure 29.36 will lead not just to N_2 photons of frequency f, but to N_2 *identical* photons, all traveling together in the same direction with the same phase. If N_2 is a large number, as would be the case in any practical device, the one initial photon will have been *amplified* into a gigantic, coherent pulse of light!

Although the chain reaction of Figure 29.36 illustrates the idea most clearly, it is not necessary for every atom to be in level 2 for amplification to occur. All that is needed is to have $N_2 > N_1$ so that stimulated emission exceeds absorption. Such a situation is called a **population inversion**. The stimulated emission is sustained by placing the *lasing medium*—the sample of atoms that emits the light—in an **optical cavity** consisting of two facing mirrors. As FIGURE 29.37 shows, the photons interact repeatedly with the atoms in the medium as they bounce back and forth. This repeated interaction is necessary for the light intensity to build up to a high level. If one of the mirrors is partially transmitting, some of the light emerges as the *laser beam*.

FIGURE 29.37 Lasing takes place in an optical cavity.

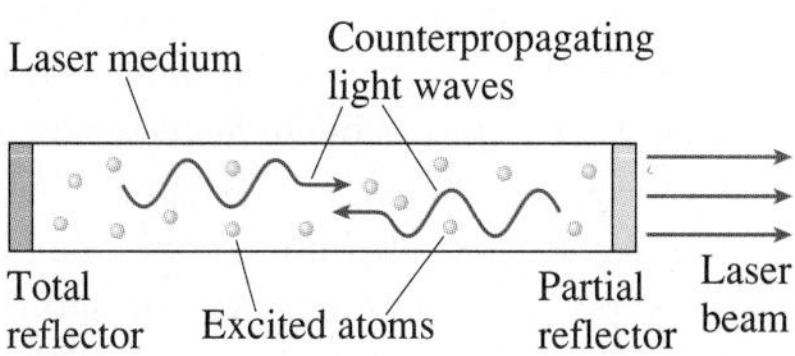

Lasers in Medicine

The invention of lasers was followed almost immediately by medical applications. Even a small-power laser beam can produce a significant amount of very localized *heating* if focused with a lens. More powerful lasers can easily *cut* through tissue by literally vaporizing it, replacing a stainless steel scalpel with a beam of light. Not only can laser surgery be very precise but it generally has less blood loss than conventional surgery because the heat of the laser seals the blood vessels and capillaries.

One common medical use of lasers is to remove plaque from artery walls, thus reducing the risk of stroke or heart attack. In this procedure, an optical fiber is threaded through arteries to reach the site. A powerful laser beam is then fired through the fiber to carefully vaporize the plaque. Laser beams traveling through optical fibers are also used to kill cancer cells in *photodynamic therapy.* In this case, light-sensitive chemicals are injected into the bloodstream and are preferentially taken up by cancer cells. The optical fiber is positioned next to the tumor and illuminates it with just the right wavelength to activate the light-sensitive chemicals and kill the cells. These procedures are minimally invasive, and they can reach areas of the body not readily accessible by conventional surgery.

◀ **Laser vision** BIO Laser-based LASIK surgery can correct for vision defects, such as near- or farsightedness, that result from an incorrect refractive power of the eye. You learned in Chapter 19 that the majority of your eye's focusing power occurs at the surface of the cornea. In LASIK, a special knife first cuts a small, thin flap in the cornea, and this flap is folded out of the way. A computer-controlled ultraviolet laser very carefully vaporizes the underlying corneal tissue to give it the desired shape; then the flap is folded back into place. The procedure takes only a few minutes and requires only a few numbing drops in the eye.

INTEGRATED EXAMPLE 29.10 Compact fluorescent lighting

You learned in Chapters 22 and 25 how an ordinary incandescent bulb works: Current passes through a filament, heating it until it glows white hot. But such bulbs are very inefficient, giving off only a few watts of visible light for every 100 W of electric power supplied to the bulb. Most of the power is, instead, converted into thermal energy.

For many years, offices and commercial buildings have used a very different type of lighting, the *fluorescent lamp.* Recently, compact fluorescent bulbs that can screw into ordinary lamp sockets have begun to make significant inroads into residential use. These bulbs are about four times more efficient than incandescent bulbs at transforming electric energy to visible light.

Inside the glass tube of a fluorescent bulb is a very small amount of mercury, which is in the form of a vapor when the bulb is on. Producing visible light occurs by a three-step process. First, a voltage of about 100 V is applied between electrodes at each end of the tube. This imparts kinetic energy to free electrons in the vapor, causing them to slam into mercury atoms and excite them by collisional excitation. Second, the excited atoms then jump to lower-energy states, emitting UV photons. Finally, the UV photons strike a *phosphor* that coats the inside of the tube, causing it to fluoresce with visible light. This is the light you see.

a. A mercury atom, after being collisionally excited by an electron, emits a photon with a wavelength of 185 nm in a quantum jump back to the ground state. If the electron starts from rest, what minimum distance must it travel to gain enough kinetic energy to cause this excitation? The 60-cm-long tube has 120 V applied between its ends.
b. After being collisionally excited, atoms sometimes emit two photons by jumping first from a high energy level to an intermediate level, giving off one photon, and then from this intermediate level to the ground state, giving off a second photon. An atom is excited to a state that is 7.79 eV above the ground state. It emits a 254-nm-wavelength photon and then a second photon. What is the wavelength of the second photon?
c. The energy-level diagram of the molecules in the phosphor is shown in **FIGURE 29.38**. After excitation by UV photons, what range of wavelengths can the phosphor emit by fluorescence?

FIGURE 29.38 Energy-level diagram of the phosphor molecules.

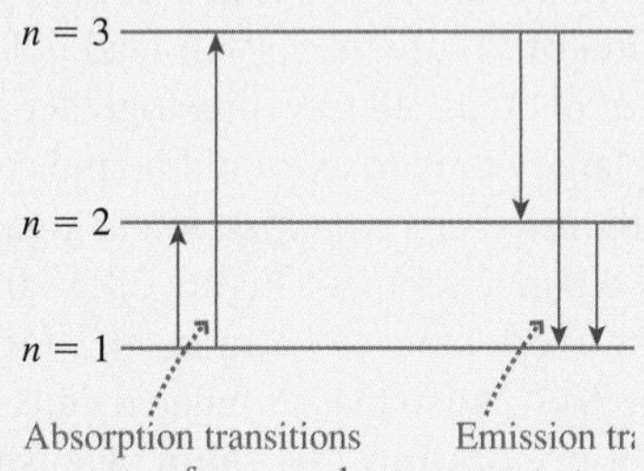

PREPARE An electron collisionally excites the mercury atom from its ground state to an excited state, increasing the atom's energy by ΔE_{atom}. Then the atom decays back to the ground state, giving off a 185-nm-wavelength photon. In order for collisional excitation to work, the kinetic energy of the incident electron must equal or exceed ΔE_{atom}; that is, it must equal or exceed the energy of a 185-nm-wavelength photon. The free electrons gain kinetic energy by accelerating through the potential difference inside the tube. We can use conservation of energy to find the distance an electron must travel to gain ΔE_{atom}, the minimum kinetic energy needed to cause an excitation.

In part b, the excited energy of the atom, 7.79 eV, is converted into photon energies. If we can find the energy of the first

254-nm-wavelength photon, the remaining energy must be that of the second photon, from which we can find its wavelength.

An inspection of Figure 29.32 shows that quantum jumps during fluorescence all begin at the bottom of the band of excited states but can end anywhere in the lower energy band. This range of energies will give us the range of wavelengths of the emitted photons.

SOLVE a. The minimum kinetic energy of the electron equals the energy of the 185-nm-wavelength photon that is subsequently emitted. This energy is

$$K_{\text{min}} = \Delta E_{\text{min}} = E_{\text{photon}} = \frac{hc}{\lambda} = \frac{1240 \text{ eV} \cdot \text{nm}}{185 \text{ nm}} = 6.7 \text{ eV}$$

where we have used the value of hc from Equation 29.23.

Recall that 1 eV is the kinetic energy gained by an electron as it accelerates through a 1 V potential difference. Here, the electron must gain a kinetic energy of 6.7 eV, so it must accelerate through a potential difference of 6.7 V. The fluorescent tube has a total potential drop of 120 V in 60 cm, or 2.0 V/cm. Thus to accelerate through a 6.7 V potential difference, the electron must travel a distance

$$\Delta x = \frac{6.7 \text{ V}}{2.0 \text{ V/cm}} = 3.4 \text{ cm}$$

b. The first emitted photon has energy

$$E_{\text{photon}} = \frac{hc}{\lambda} = \frac{1240 \text{ eV} \cdot \text{nm}}{254 \text{ nm}} = 4.88 \text{ eV}$$

The energy remaining to the second photon is then 7.79 eV − 4.88 eV = 2.91 eV; the wavelength of this photon is

$$\lambda = \frac{hc}{E_{\text{photon}}} = \frac{1240 \text{ eV} \cdot \text{nm}}{2.91 \text{ eV}} = 426 \text{ nm}$$

c. The energy of the photon emitted during a quantum jump from the bottom of the upper energy band to the top of the lower energy band is 1.8 eV, corresponding to a wavelength of

$$\lambda = \frac{hc}{E_{\text{photon}}} = \frac{1240 \text{ eV} \cdot \text{nm}}{1.8 \text{ eV}} = 690 \text{ nm}$$

The photon energy for a jump to the bottom of the lower energy band is 1.8 eV + 0.7 eV = 2.5 eV, with a wavelength of

$$\lambda = \frac{hc}{E_{\text{photon}}} = \frac{1240 \text{ eV} \cdot \text{nm}}{2.5 \text{ eV}} = 500 \text{ nm}$$

Thus this phosphor, after absorbing UV photons, emits visible light with a wavelength range of 500–690 nm.

ASSESS For part a, it seems reasonable that the electron travels only a small fraction of the tube length before gaining enough energy to collisionally excite an atom. The bulb would be very inefficient if electrons had to travel the full length before colliding with a mercury atom. The photon wavelengths in parts b and c also seem reasonable. In particular, the range of wavelengths emitted by the phosphor is a little more than half the visible spectrum, from green through red but missing blue and violet. Compact fluorescent tubes have three different phosphors, each with a somewhat different range of emission wavelengths, to give the full spectrum of white light. Slightly altering the balance between these phosphors distinguishes a "warm white" bulb from a "cool white" bulb.

SUMMARY

The goal of Chapter 29 has been to use quantum physics to understand the properties of atoms, molecules, and their spectra.

IMPORTANT CONCEPTS

The Structure of an Atom

An atom consists of a very small, positively charged nucleus, surrounded by orbiting electrons.

- The number of protons is the atom's **atomic number** Z.
- The **atomic mass number** A is the number of protons + the number of neutrons.

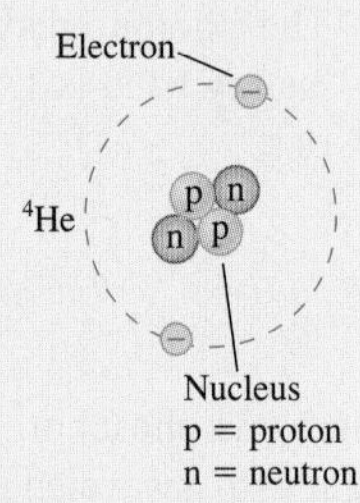

The Bohr Atom

In Bohr's model,

- The atom can exist in only certain **stationary states.** These states correspond to different electron orbits. Each state is numbered by **quantum number** $n = 1, 2, 3, \ldots$.
- Each state has a discrete, well-defined energy E_n.
- The atom can change its energy by undergoing a **quantum jump** between two states by emitting or absorbing a photon of energy $E_{\text{photon}} = \Delta E_{\text{atom}} = |E_f - E_i|$.

The Hydrogen Atom

In Bohr's model of the hydrogen atom the stationary states are found by requiring an integer number of de Broglie wavelengths to fit around the circumference of the electron's orbit: $2\pi r = n\lambda$. The integer n is the *principal quantum number.*

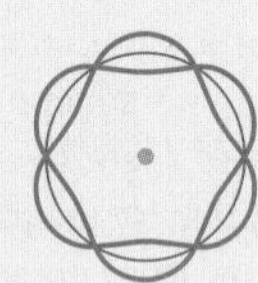

This leads to energy quantization with

$$E_n = -\frac{13.60 \text{ eV}}{n^2}$$

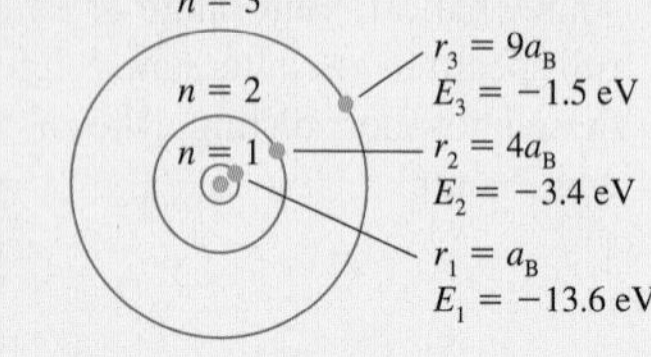

and orbit radii $r_n = n^2 a_B$, where $a_B = 0.053$ nm is the **Bohr radius.**

The wavelengths of light in the hydrogen atom spectrum are given by the **Balmer formula:**

$$\lambda_{n\to m} = \frac{91.1 \text{ nm}}{\left(\frac{1}{m^2} - \frac{1}{n^2}\right)} \qquad \begin{aligned} m &= 1, 2, 3, \ldots \\ n &= m+1, m+2, \ldots \end{aligned}$$

Beyond the Bohr model, quantum mechanics adds several quantized parameters, each with its own quantum number:

- The *orbital angular momentum,* quantum number l:

$$L = \sqrt{l(l+1)}\hbar \qquad l = 0, 1, 2, 3, \ldots, n-1$$

- The *angle of the electrons orbit,* quantum number m:

$$m = -l, -l+1, \ldots, 0, \ldots, l-1, l$$

- The *direction of the electron* **spin,** quantum number m_s:

$$m_s = -\tfrac{1}{2} \text{ or } +\tfrac{1}{2}$$

The energy of a hydrogen atom depends only on n:

0 eV
3s 3p 3d
2s 2p
All states with the same n have the same energy.
−13.60 eV 1s

Multielectron atoms

Each electron is described by the same quantum numbers (n, l, m, m_s) used for the hydrogen atom, but the energy now depends on l as well as n.

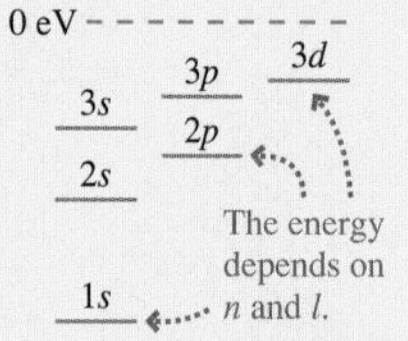

The **Pauli exclusion principle** states that no more than one electron can occupy each quantum state.

Molecules

In molecules, the states are spaced very closely into **bands** of states. Because electrons can be excited to and from many states, the spectra of molecules are broad, not discrete.

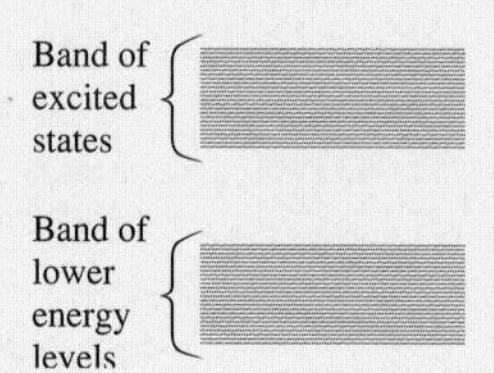

APPLICATIONS

Atomic emission spectra are generated by excitation followed by a photon-emitting quantum jump.

- **Excitation** occurs by absorption of a photon or by collision.
- A quantum jump can occur only if $\Delta l = \pm 1$.
- Quantized energies give rise to a **discrete spectrum.**

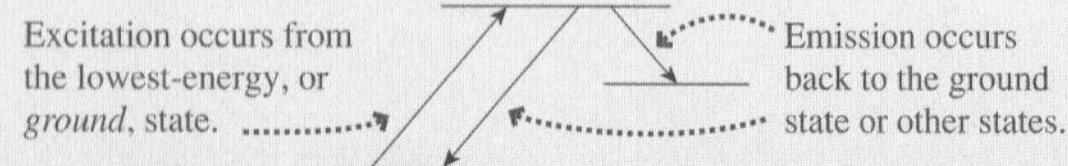

Lasers

A photon with energy $E_{\text{photon}} = E_2 - E_1$ can induce **stimulated emission** of a second photon identical to the first. These photons can then induce more atoms to emit photons. If more atoms are in state 2 than in state 1, this process can rapidly build up an intense beam of identical photons. This is the principle behind the laser.

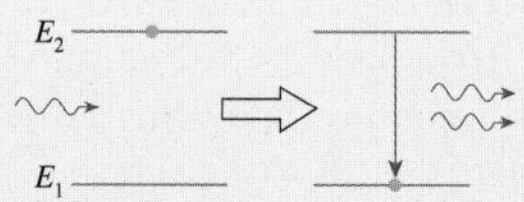

For homework assigned on MasteringPhysics, go to www.masteringphysics.com

Problems labeled INT integrate significant material from earlier chapters; BIO are of biological or medical interest.

Problem difficulty is labeled as | (straightforward) to ||||| (challenging).

QUESTIONS

Conceptual Questions

1. A neon discharge emits a bright reddish-orange spectrum. But a glass tube filled with neon is completely transparent. Why doesn't the neon in the tube absorb orange and red wavelengths?
2. The two spectra shown in Figure Q29.2 belong to the same element, a fictional Element X. Explain why they are different.

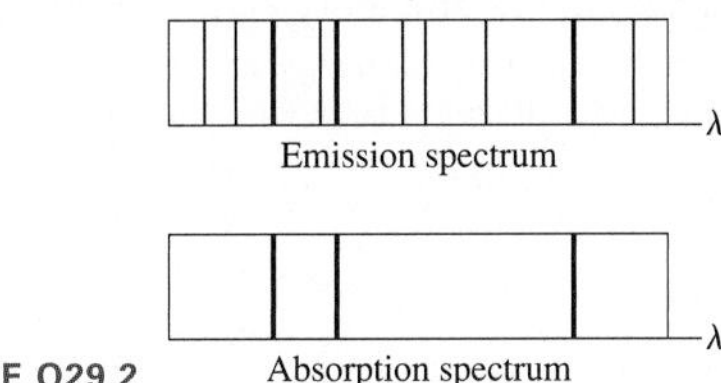

FIGURE Q29.2

3. Is a spectral line with wavelength 656.5 nm seen in the absorption spectrum of hydrogen atoms? Why or why not?
4. J. J. Thomson studied the ionization of atoms in collisions with electrons. He accelerated electrons through a potential difference, shot them into a gas of atoms, then used a mass spectrometer to detect any ions produced in the collisions. By using different gases, he found that he could produce singly ionized atoms of all the elements that he tried. When he used higher accelerating voltages, he was able to produce doubly ionized atoms of all elements *except* hydrogen.
   a. Why did Thomson have to use higher accelerating voltages to detect doubly ionized atoms than to detect singly ionized atoms?
   b. What conclusion or conclusions about hydrogen atoms can you draw from these observations? Be specific as to how your conclusions are related to the observations.
5. Bohr did not include the gravitational force in his analysis of the hydrogen atom. Is this one of the reasons that his model of the hydrogen atom had only limited success? Explain.
6. If an electron is in a *stationary state* of an atom, is the electron at rest? If not, what does the term mean?
7. The $n = 3$ state of hydrogen has $E_3 = -1.51$ eV.
   a. Why is the energy negative?
   b. What is the physical significance of the specific number 1.51 eV?
8. For a hydrogen atom, list all possible states (n, l, m, m_s) that have $E = -1.51$ eV.
9. What are the n and l values of the following states of a hydrogen atom: (a) $4d$, (b) $5f$, (c) $6s$?
10. How would you label the hydrogen-atom states with the following (n, l, m) quantum numbers: (a) (4, 3, 0), (b) (3, 2, 1), (c) (3, 2, −1)?
11. Consider the two hydrogen-atom states $5d$ and $4f$. Which has the higher energy? Explain.
12. Does each diagram in Figure Q29.12 represent a possible electron configuration of a neutral element? If so, (i) identify the element and (ii) determine if this is the ground state or an excited state. If not, why not?

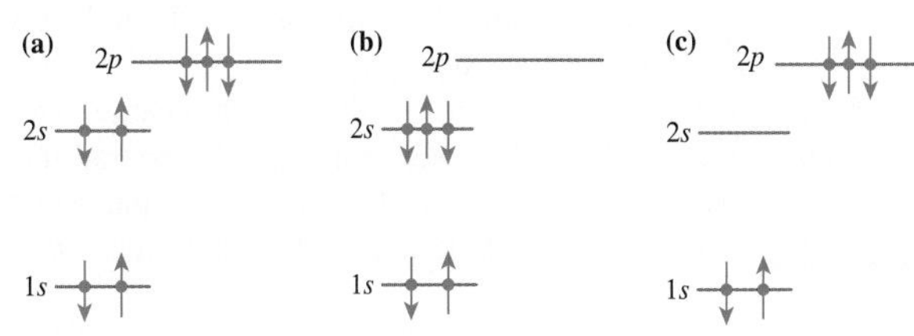

FIGURE Q29.12

13. Do the following electron configurations represent a possible state of an element? If so, (i) identify the element and (ii) determine if this is the ground state or an excited state. If not, why not?
    a. $1s^2 2s^2 2p^6 3s^2$
    b. $1s^2 2s^2 2p^7 3s$
    c. $1s^2 2s^2 2p^4 3s^2 3p^2$
14. Why is the section of the periodic table labeled as "transition elements" exactly 10 elements wide in all rows?
15. An electron is in an f state. Can it undergo a quantum jump to an s state? A p state? A d state? Explain.
16. Figure Q29.16 shows the energy-level diagram of Element X.
    a. What is the ionization energy of Element X?
    b. An atom in the ground state absorbs a photon, then emits a photon with a wavelength of 1240 nm. What conclusion can you draw about the energy of the photon that was absorbed?
    c. An atom in the ground state has a collision with an electron, then emits a photon with a wavelength of 1240 nm. What conclusion can you draw about the initial kinetic energy of the electron?

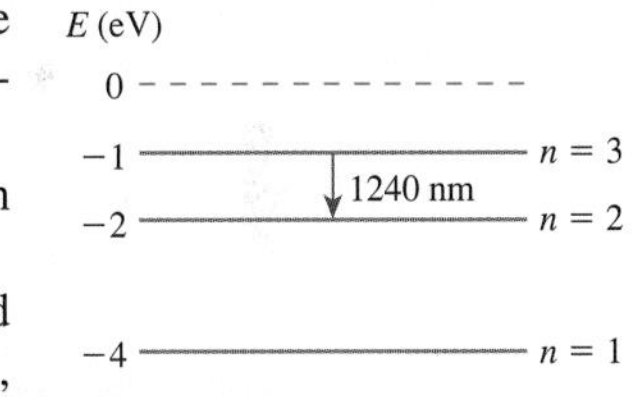

E (eV); 0; −1, $n = 3$; −2, $n = 2$; 1240 nm; −4, $n = 1$

FIGURE Q29.16

17. a. Which states of a hydrogen atom can be excited by a collision with an electron with kinetic energy $K = 12.5$ eV? Explain.
    b. After the collision the atom is not in its ground state. What happens to the electron? (i) It bounces off with $K > 12.5$ eV, (ii) It bounces off with $K = 12.5$ eV, (iii) It bounces off with $K < 12.5$ eV, (iv) It is absorbed by the atom. Explain your choice.
    c. After the collision, the atom emits a photon. List all the possible $n \to m$ transitions that might occur as a result of this collision.
18. What *is* an atom's ionization energy? In other words, if you know the ionization energy of an atom, what is it that you know about the atom?

19. Figure Q29.19 shows the energy levels of a hypothetical atom.
    a. What *minimum* kinetic energy (in eV) must an electron have to collisionally excite this atom and cause the emission of a 620 nm photon? Explain.
    b. Can an electron with $K = 6$ eV cause the emission of 620 nm light? If so, what is the final kinetic energy of the electron? If not, why not?
    c. Can a 6 eV photon cause the emission of 620 nm light from this atom? Why or why not?
    d. Can a 7 eV photon cause the emission of 620 nm light from this atom? Why or why not?

E (eV)
7 — $(n+1)p$
5 — $(n+1)s$
620 nm
3 — np
0 — ns
Ground state

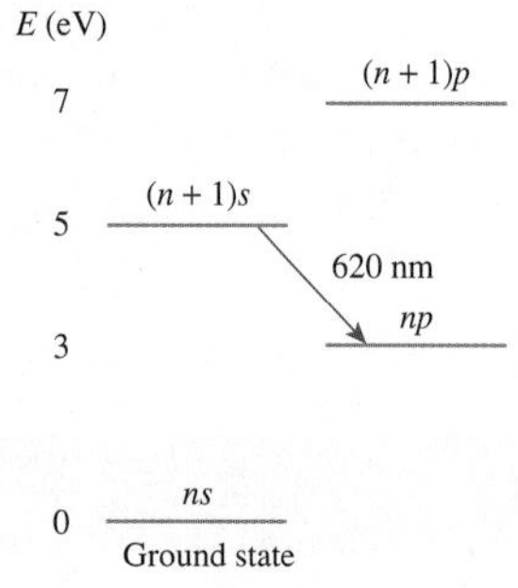

FIGURE Q29.19

20. Seven possible transitions are identified on the energy-level diagram in Figure Q29.20. For each, is this an allowed transition? If allowed, is it an emission or an absorption transition, and is the photon infrared, visible, or ultraviolet? If not allowed, why not?

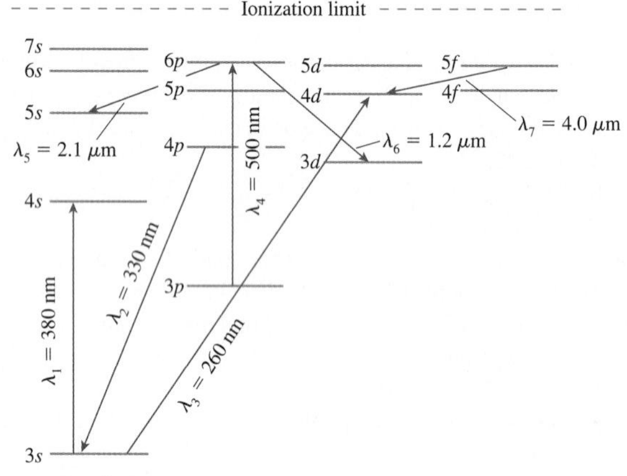

FIGURE Q29.20

21. A 2.0 eV photon is incident on an atom in the p state, as shown in the energy-level diagram in Figure Q29.21. Does the atom undergo an absorption transition, a stimulated emission transition, or neither? Explain.

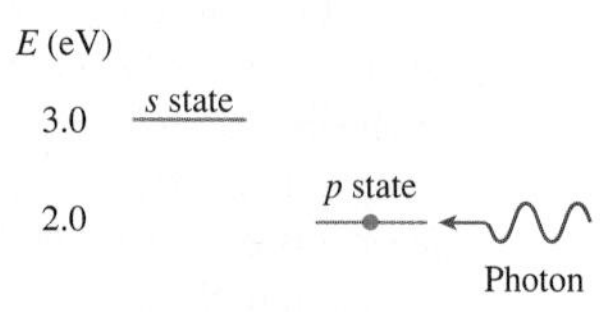

FIGURE Q29.21

22. A glass tube contains 2×10^{11} atoms, some of which are in the ground state and some of which are excited. Figure Q29.22 shows the populations for the atoms' three energy levels. Is it possible for these atoms to be a laser? If so, on which transition would laser action occur? If not, why not?

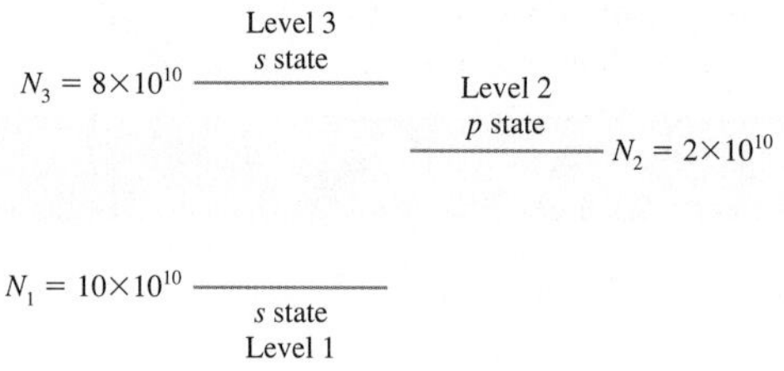

Level 3, s state — $N_3 = 8\times10^{10}$
Level 2, p state — $N_2 = 2\times10^{10}$
Level 1, s state — $N_1 = 10\times10^{10}$

FIGURE Q29.22

Multiple-Choice Questions

23. | An electron collides with an atom in its ground state. The atom then emits a photon of energy E_{photon}. In this process the *change* ΔE_{elec} in the electron's energy is
    A. Greater than E_{photon}.
    B. Greater than or equal to E_{photon}.
    C. Equal to E_{photon}.
    D. Less than or equal to E_{photon}.
    E. Less than E_{photon}.
24. || How many states are in the $l = 4$ subshell?
    A. 8 B. 9 C. 16 D. 18 E. 22
25. | What is the ground-state electron configuration of calcium ($Z = 20$)?
    A. $1s^2 2s^2 2p^6 3s^2 3p^8$ B. $1s^2 2s^2 2p^6 3s^2 3p^6 4s^1 4p^1$
    C. $1s^2 2s^2 2p^6 3s^2 3p^6 4s^2$ D. $1s^2 2s^2 2p^6 3s^2 3p^6 4p^2$
26. | An atom emits a photon with a wavelength of 275 nm. By how much does the atom's energy change?
    A. 0.72 eV B. 1.06 eV C. 2.29 eV
    D. 3.06 eV E. 4.51 eV
27. || The energy of a hydrogen atom is -3.40 eV. What is the electron's kinetic energy?
    A. 1.70 eV B. 2.62 eV C. 3.40 eV
    D. 5.73 eV E. 6.80 eV
28. | The angular momentum of an electron in a Bohr hydrogen atom is 3.18×10^{-34} kg $\cdot$ m^2/s. What is the atom's energy?
    A. -13.60 eV B. -6.73 eV C. -3.40 eV
    D. -1.51 eV E. -0.47 eV
29. | A "soft x-ray" photon with an energy of 41.8 eV is absorbed by a hydrogen atom in its ground state, knocking the atom's electron out. What is the speed of the electron as it leaves the atom?
    A. 1.84×10^5 m/s B. 3.08×10^5 m/s C. 8.16×10^5 m/s
    D. 3.15×10^6 m/s E. 3.83×10^6 m/s

PROBLEMS

Section 29.1 Spectroscopy

1. | Figure 29.2b and Table 29.1 showed the wavelengths of the first four lines in the visible spectrum of hydrogen.
    a. Determine the Balmer formula n and m values for these wavelengths.
    b. Predict the wavelength of the fifth line in the spectrum.
2. | The wavelengths in the hydrogen spectrum with $m = 1$ form a series of spectral lines called the Lyman series. Calculate the wavelengths of the first four members of the series.
3. | The Paschen series is analogous to the Balmer series, but with $m = 3$. Calculate the wavelengths of the first three members in the Paschen series. What part(s) of the electromagnetic spectrum are these in?

Section 29.2 Atoms

4. | How many electrons, protons, and neutrons are contained in the following atoms or ions: (a) ^{6}Li, (b) ^{13}C^{+}, and (c) ^{18}O^{++}?
5. | How many electrons, protons, and neutrons are contained in the following atoms or ions: (a) ^{9}Be^{+}, (b) ^{12}C, and (c) ^{15}N^{+++}?
6. | Write the symbol for an atom or ion with:
   a. four electrons, four protons, and five neutrons.
   b. six electrons, seven protons, and eight neutrons.
7. | Write the symbol for an atom or ion with:
   a. three electrons, three protons, and five neutrons.
   b. five electrons, six protons, and eight neutrons.

Section 29.3 Bohr's Model of Atomic Quantization

8. | Figure P29.8 is an energy-level diagram for a simple atom. What wavelengths appear in the atom's (a) emission spectrum and (b) absorption spectrum?

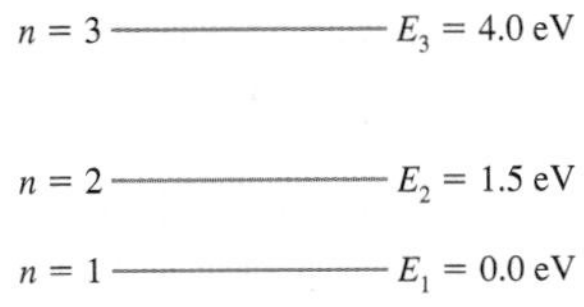

$n = 3$ ———— $E_3 = 4.0$ eV

$n = 2$ ———— $E_2 = 1.5$ eV

$n = 1$ ———— $E_1 = 0.0$ eV

FIGURE P29.8

9. | An electron with 2.0 eV of kinetic energy collides with the atom whose energy-level diagram is shown in Figure P29.8.
   a. Is the electron able to kick the atom to an excited state? Why or why not?
   b. If your answer to part a was yes, what is the electron's kinetic energy after the collision?
10. | The allowed energies of a simple atom are 0.0 eV, 4.0 eV, and 6.0 eV.
    a. Draw the atom's energy-level diagram. Label each level with the energy and the principal quantum number.
    b. What wavelengths appear in the atom's emission spectrum?
    c. What wavelengths appear in the atom's absorption spectrum?
11. ||| The allowed energies of a simple atom are 0.0 eV, 4.0 eV, and 6.0 eV. An electron traveling at a speed of 1.6×10^6 m/s collisionally excites the atom. What are the minimum and maximum speeds the electron could have after the collision?

Section 29.4 The Bohr Hydrogen Atom

12. || A researcher observes hydrogen emitting photons of energy 1.89 eV. What are the quantum numbers of the two states involved in the transition that emits these photons?
13. | A hydrogen atom is in the $n = 3$ state. In the Bohr model, how many electron wavelengths fit around this orbit?
14. || A hydrogen atom is in its $n = 1$ state. In the Bohr model, what is the ratio of its kinetic energy to its potential energy?
15. | Show, by actual calculation, that the Bohr radius is 0.0529 nm and that the ground-state energy of hydrogen is -13.60 eV.
16. | a. What quantum number of the hydrogen atom comes closest to giving a 500-nm-diameter electron orbit?
    b. What are the electron's speed and energy in this state?
17. | a. Calculate the de Broglie wavelength of the electron in the $n = 1$, 2, and 3 states of the hydrogen atom. Use the information in Table 29.2.
    b. Show numerically that the circumference of the orbit for each of these stationary states is exactly equal to n de Broglie wavelengths.
    c. Sketch the de Broglie standing wave for the $n = 3$ orbit.
18. | Show, by calculation, that the first three states of the hydrogen atom have angular momenta $\hbar$, $2\hbar$, and $3\hbar$, respectively.
19. | Determine all possible wavelengths of photons that can be emitted from the $n = 4$ state of a hydrogen atom.

Section 29.5 The Quantum-Mechanical Hydrogen Atom

20. | List the quantum numbers of (a) all possible $3p$ states and (b) all possible $3d$ states.
21. || When all quantum numbers are considered, how many different quantum states are there for a hydrogen atom with $n = 1$? With $n = 2$? With $n = 3$? List the quantum numbers of each state.
22. | What is the angular momentum of a hydrogen atom in (a) a $4p$ state and (b) a $5f$ state? Give your answers as a multiple of $\hbar$.
23. | A hydrogen atom has orbital angular momentum 3.65×10^{-34} J $\cdot$ s.
    a. What letter (s, p, d, or f) describes the electron?
    b. What is the atom's minimum possible energy? Explain.
24. || A hydrogen atom is in the $5p$ state. Determine (a) its energy, (b) its angular momentum, (c) its quantum number l, and (d) the possible values of its magnetic quantum number m.
25. ||| The angular momentum of a hydrogen atom is 4.70×10^{-34} J $\cdot$ s. What is the minimum energy, in eV, that this atom could have?

Section 29.6 Multielectron Atoms

26. | Predict the ground-state electron configurations of Mg, Sr, and Ba.
27. || Predict the ground-state electron configurations of Si, Ge, and Pb.
28. | Identify the element for each of these electron configurations. Then determine whether this configuration is the ground state or an excited state.
    a. $1s^22s^22p^5$
    b. $1s^22s^22p^63s^23p^63d^{10}4s^24p$
29. || a. With what element is the $3s$ subshell first completely filled?
    b. With what element is the $4d$ subshell first half-filled?
30. | Identify the element for each of these electron configurations. Then determine whether this configuration is the ground state or an excited state.
    a. $1s^22s^22p^63s^23p^64s^23d^9$
    b. $1s^22s^22p^63s^23p^64s^23d^{10}4p^65s^24d^{10}5p^66s^24f^{14}5d^7$
31. | Explain what is wrong with these electron configurations:
    a. $1s^22s^22p^83s^23p^4$ b. $1s^22s^32p^4$
32. || Which has higher energy: an electron in the $4f$ state or an electron in the $7s$ state? Explain.

Section 29.7 Excited States and Spectra

33. | An electron with a speed of 5.00×10^6 m/s collides with an atom. The collision excites the atom from its ground state (0 eV) to a state with an energy of 3.80 eV. What is the speed of the electron after the collision?
34. | Hydrogen gas absorbs light of wavelength 103 nm. Afterward, what wavelengths are seen in the emission spectrum?
35. || What is the minimum wavelength of light that can excite the $4s$ state of sodium?
36. || An electron with a kinetic energy of 3.90 eV collides with a sodium atom. What possible wavelengths of light are subsequently emitted?

37. | a. Is a $4p \rightarrow 4s$ transition allowed in sodium? If so, what is its wavelength? If not, why not?
b. Is a $3d \rightarrow 4s$ transition allowed in sodium? If so, what is its wavelength? If not, why not?

Section 29.8 Molecules

38. || Figure P29.38 shows a molecular energy-level diagram. What are the longest and shortest wavelengths in (a) the molecule's absorption spectrum and (b) the molecule's fluorescence spectrum?

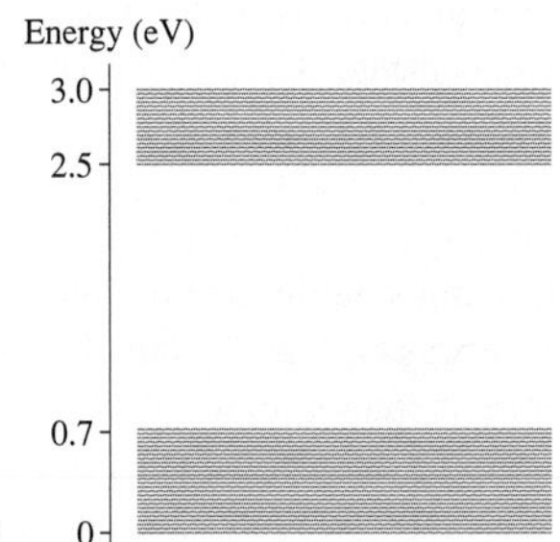

FIGURE P29.38

39. | The molecule whose energy-level diagram is shown in Figure P29.38 is illuminated by 2.7 eV photons. What is the longest wavelength of light that the molecule can emit?

Section 29.9 Stimulated Emission and Lasers

40. | A 1000 W carbon dioxide laser emits an infrared laser beam with a wavelength of 10.6 μm. How many photons are emitted per second?

41. || A 1.00 mW helium neon-laser emits a visible laser beam with a wavelength of 633 nm. How many photons are emitted per second?

42. || In LASIK surgery, a laser is used to reshape the cornea of the BIO INT eye to improve vision. The laser produces extremely short pulses of light, each containing 1.0 mJ of energy.
a. In each pulse there are 9.7×10^{14} photons. What is the wavelength of the laser?
b. Each pulse lasts only 20 ns. What is the average power delivered to the eye during a pulse?

43. || A ruby laser emits an intense pulse of light that lasts a mere 10 ns. The light has a wavelength of 690 nm, and each pulse has an energy of 500 mJ.
a. How many photons are emitted in each pulse?
b. What is the *rate* of photon emission, in photons per second, during the 10 ns that the laser is "on"?

General Problems

44. ||| A 2.55 eV photon is emitted from a hydrogen atom. What are the Balmer formula n and m values corresponding to this emission?

45. | Two of the wavelengths emitted by a hydrogen atom are 102.6 nm and 1876 nm.
a. What are the Balmer formula n and m values for each of these wavelengths?
b. For each of these wavelengths, is the light infrared, visible, or ultraviolet?

46. |||| In Example 29.2 it was assumed that the initially stationary INT gold nucleus would remain motionless during a head-on collision with an 8.3 MeV alpha particle. What is the actual recoil speed of the gold nucleus after that elastic collision? Assume that the mass of a gold nucleus is exactly 50 times the mass of an alpha particle.
Hint: Review the discussion of perfectly elastic collisions in Chapter 10.

47. | Consider the gold isotope ^{197}Au.
INT a. How many electrons, protons, and neutrons are in a neutral ^{197}Au atom?
b. The gold nucleus has a diameter of 14.0 fm. What is the density of matter in a gold nucleus?
c. The density of lead is 11,400 kg/m^3. How many times the density of lead is your answer to part b?

48. || Consider the lead isotope ^{207}Pb.
INT a. How many electrons, protons, and neutrons are in a neutral ^{207}Pb atom?
b. The lead nucleus has a diameter of 14.2 fm. What are the electric potential and the electric field strength at the surface of a lead nucleus?

49. ||| The diameter of an atom is 1.2×10^{-10} m and the diameter of its nucleus is 1.0×10^{-14} m. What percent of the atom's volume is occupied by mass and what percent is empty space?

50. | The charge-to-mass ratio of a nucleus, in units of e/u, is $q/m = Z/A$. For example, a hydrogen nucleus has $q/m = 1/1 = 1$.
a. Make a graph of charge-to-mass ratio versus proton number Z for nuclei with $Z = 5, 10, 15, 20, \ldots, 90$. For A, use the average atomic mass shown on the periodic table of elements in Appendix B. Show each of these 18 nuclei as a dot, but don't connect the dots together as a curve.
b. Describe any trend that you notice in your graph.
c. What's happening in the nuclei that is responsible for this trend?

51. | If the nucleus is a few fm in diameter, the distance between INT the centers of two protons must be $\approx$2 fm.
a. Calculate the repulsive electric force between two protons that are 2.0 fm apart.
b. Calculate the attractive gravitational force between two protons that are 2.0 fm apart. Could gravity be the force that holds the nucleus together?

52. | In a head-on collision, the closest approach of a 6.24 MeV INT alpha particle to the center of a nucleus is 6.00 fm. The nucleus is in an atom of what element? Assume that the nucleus is heavy enough to remain stationary during the collision.

53. |||| An 20 MeV alpha particle is INT fired toward a ^{238}U nucleus. It follows the path shown in Figure P29.53. What is the alpha particle's speed when it is closest to the nucleus, 20 fm from its center? Assume that the nucleus doesn't move.

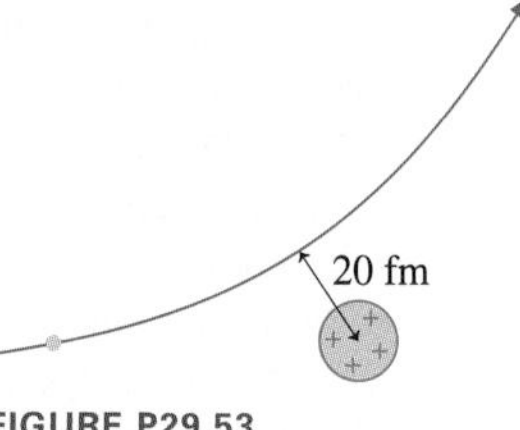

FIGURE P29.53

54. || The oxygen nucleus ^{16}O has a radius of 3.0 fm.
INT a. With what speed must a proton be fired toward an oxygen nucleus to have a turning point 1.0 fm from the surface? Assume that the nucleus is heavy enough to remain stationary during the collision.
b. What is the proton's kinetic energy in MeV?

55. || The absorption spectrum of an atom consists of the wavelengths 200 nm, 300 nm, and 500 nm.
   a. Draw the atom's energy-level diagram.
   b. What wavelengths are seen in the atom's emission spectrum?

56. || The first three energy levels of the fictitious element X are shown in Figure P29.56.
   a. What wavelengths are observed in the absorption spectrum of element X? Give your answers in nm.
   b. State whether each of your wavelengths in part a corresponds to ultraviolet, visible, or infrared light.
   c. An electron with a speed of 1.4×10^6 m/s collides with an atom of element X. Shortly afterward, the atom emits a 1240 nm photon. What was the electron's speed after the collision? Assume that, because the atom is so much more massive than the electron, the recoil of the atom is negligible. **Hint:** The energy of the photon is *not* the energy transferred to the atom in the collision.

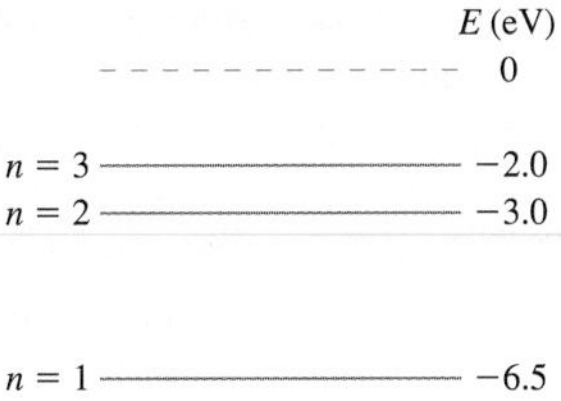

E (eV)
0
$n = 3$ −2.0
$n = 2$ −3.0
$n = 1$ −6.5

FIGURE P29.56

57. || A simple atom has four lines in its absorption spectrum. Ignoring any selection rules, how many lines will it have in its emission spectrum?

58. ||| A simple atom has only two absorption lines, at 250 nm and 600 nm. What is the wavelength of the one line in the emission spectrum that does not appear in the absorption spectrum?

59. | What is the wavelength of the series limit (i.e., the shortest possible wavelength) of the Lyman series in hydrogen?

60. || What is the energy of a Bohr hydrogen atom with a 5.18 nm diameter?

61. | A hydrogen atom in the ground state absorbs a 12.75 eV photon. Immediately after the absorption, the atom undergoes a quantum jump to the next-lowest energy level. What is the wavelength of the photon emitted in this quantum jump?

62. | a. Calculate the orbital radius and the speed of an electron in both the $n = 99$ and the $n = 100$ states of hydrogen.
INT
   b. Determine the orbital frequency of the electron in each of these states.
   c. Calculate the frequency of a photon emitted in a $100 \rightarrow 99$ transition.
   d. Compare the photon frequency of part c to the *average* of your two orbital frequencies from part b. By what percent do they differ?

63. ||| Two hydrogen atoms collide head-on. The collision brings both atoms to a halt. Immediately after the collision, both atoms emit a 121.6 nm photon. What was the speed of each atom just before the collision?
INT

64. ||| A beam of electrons is incident on a gas of hydrogen atoms.
INT
   a. What minimum speed must the electrons have to cause the emission of 656 nm light from the $3 \rightarrow 2$ transition of hydrogen?
   b. Through what potential difference must the electrons be accelerated to have this speed?

65. ||| A hydrogen atom in its fourth excited state emits a photon with a wavelength of 1282 nm. What is the atom's maximum possible orbital angular momentum after the emission? Give your answer as a multiple of $\hbar$.

66. || A particular emission line in the hydrogen spectrum has a wavelength of 656.5 nm. What are all possible transitions (e.g., $6d \rightarrow 2s$) that could give rise to this emission?

67. | a. What downward transitions are possible for a sodium atom in the $6s$ state? (See Figure 29.24.)
   b. What are the wavelengths of the photons emitted in each of these transitions?

68. || The $5d \rightarrow 3p$ transition in the emission spectrum of sodium (see Figure 29.24) has a wavelength of 499 nm. What is the energy of the $5d$ state?

69. || A sodium atom (see Figure 29.24) emits a photon with wavelength 818 nm shortly after being struck by an electron. What minimum speed did the electron have before the collision?

70. || Figure P29.70 shows a few energy levels of the mercury atom. One valence electron is always in the $6s$ state; the other electron changes states.
   a. Make a table showing all the allowed transitions in the emission spectrum. For each transition, indicate the photon wavelength, in nm.
   b. What minimum speed must an electron have to excite the 492-nm-wavelength blue emission line in the Hg spectrum?

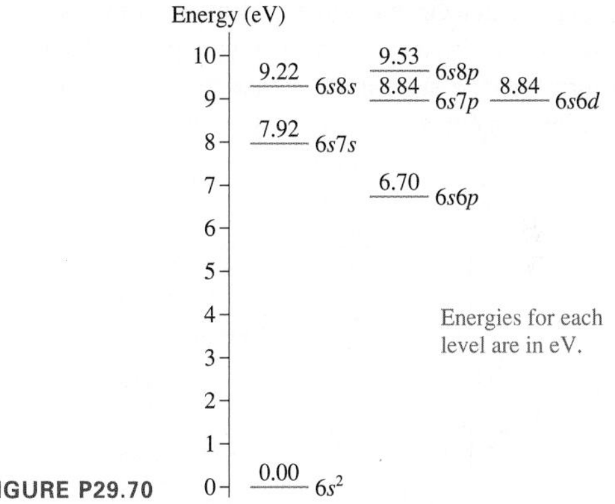

FIGURE P29.70

71. | Figure P29.71 shows the first few energy levels of the lithium atom. Make a table showing all the allowed transitions in the emission spectrum. For each transition, indicate
   a. The wavelength, in nm.
   b. Whether the transition is in the infrared, the visible, or the ultraviolet spectral region.
   c. Whether or not the transition would be observed in the lithium absorption spectrum.

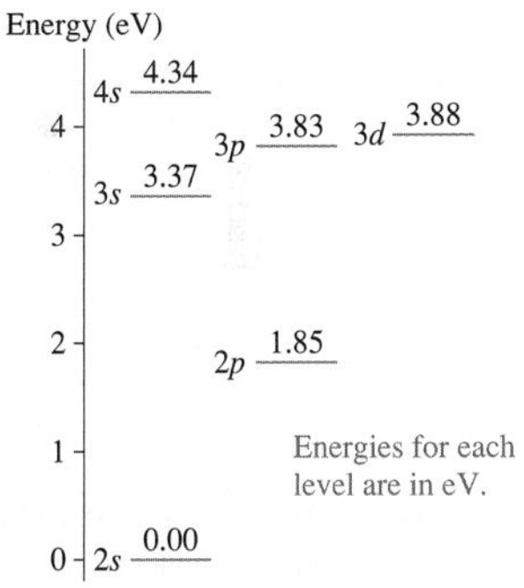

Energy (eV)
4s 4.34
3p 3.83
3d 3.88
3s 3.37
2p 1.85
2s 0.00
Energies for each level are in eV.

FIGURE P29.71

72. | A laser emits 1.00×10^{19} photons per second from an excited state with energy $E_2 = 1.17$ eV. The lower energy level is $E_1 = 0$ eV.
INT
   a. What is the wavelength of this laser?
   b. What is the power output of this laser?

73. | Fluorescence microscopy, discussed in Section 29.8, is an important tool in modern cell biology. A variation on this technique depends on a phenomenon known as two-photon excitation. If two photons are absorbed simultaneously (i.e., within about 10^{-16} s), their energies can add. A molecule that is normally excited by a 350 nm photon can be excited by two photons each having half as much energy. For this process to be useful, photons must illuminate the sample at the very high rate of at least 10^{29} photons/m$^2 \cdot$s. This is achieved by focusing a
BIO
INT

laser beam to a small spot and by concentrating the power of the laser into very short (10^{-13} s) pulses that are fired 10^8 times each second. Suppose a biologist wants to use two-photon excitation to excite a molecular species that would be excited by 500 nm light in normal one-photon fluorescence microscopy. What minimum intensity (W/m^2) must the laser beam have during each pulse?

Passage Problems

Light-Emitting Diodes

Light-emitting diodes, known by the acronym LED, produce the familiar green and red indicator lights used in a wide variety of consumer electronics. LEDs are semiconductor devices in which the electrons can exist only in certain energy levels. Much like molecules, the energy levels are packed together close enough to form what appears to be a continuous band of possible energies. Energy supplied to an LED in a circuit excites electrons from a *valence band* into a *conduction band.* An electron can emit a photon by undergoing a quantum jump from a state in the conduction band into an empty state in the valence band, as shown in Figure P29.74.

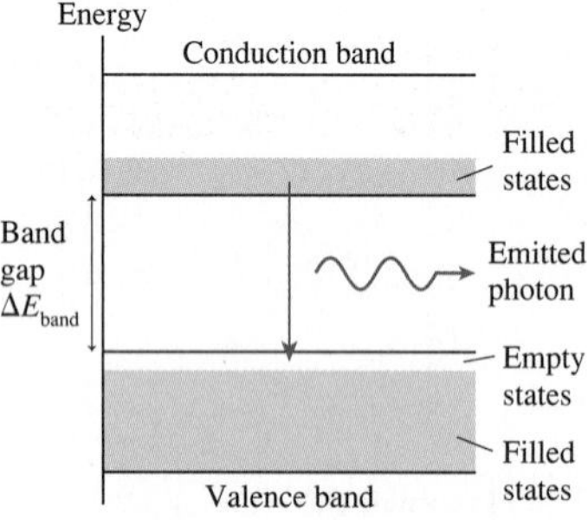

FIGURE P29.74 Energy-level diagram of an LED.

The size of the band gap ΔE_{band} determines the possible energies—and thus the wavelengths—of the emitted photons. Most LEDs emit a narrow range of wavelengths and thus have a distinct color. This makes them well-suited for traffic lights and other applications where a certain color is desired, but it makes them less desirable for general illumination. One way to make a "white" LED is to combine a blue LED with a substance that fluoresces yellow when illuminated with the blue light. The combination of the two colors makes light that appears reasonably white.

74. | An LED emits green light. Increasing the size of the band gap could change the color of the emitted light to
A. Red B. Orange
C. Yellow D. Blue

75. | Suppose the LED band gap is 2.5 eV, which corresponds to a wavelength of 500 nm. Consider the possible electron transitions in Figure P29.74. 500 nm is the
A. Maximum wavelength of the LED.
B. Average wavelength of the LED.
C. Minimum wavelength of the LED.

76. | The same kind of semiconducting material used to make an LED can also be used to convert absorbed light into electrical energy, essentially operating as an LED in reverse. In this case, the absorption of a photon causes an electron transition from a filled state in the valence band to an unfilled state in the conduction band. If $\Delta E_{band} = 1.4$ eV, what is the minimum wavelength of electromagnetic radiation that could lead to electric energy output?
A. 140 nm B. 890 nm
C. 1400 nm D. 8900 nm

77. | The efficiency of a light source is the percentage of its energy input that gets radiated as visible light. If some of the blue light in an LED is used to cause a fluorescent material to glow,
A. The overall efficiency of the LED is increased.
B. The overall efficiency of the LED does not change.
C. The overall efficiency of the LED decreases.

STOP TO THINK ANSWERS

Stop to Think 29.1: A is emission, B is absorption. All wavelengths in the absorption spectrum are seen in the emission spectrum, but not all wavelengths in the emission spectrum are seen in the absorption spectrum.

Stop to Think 29.2: 6 protons and 8 neutrons. The number of protons is the atomic number, which is 6. That leaves $14 - 6 = 8$ neutrons.

Stop to Think 29.3: In emission from the $n = 3$ to $n = 2$ transition, but not in absorption. The photon energy has to match the energy *difference* between two energy levels. Absorption is from the ground state, at $E_1 = 0$ eV. There's no energy level at 3 eV to which the atom could jump.

Stop to Think 29.4: $n = 3$. Each antinode is half a wavelength, so this standing wave has three full wavelengths in one circumference.

Stop to Think 29.5: $n = 3, l = 1$, or a $3p$ state.

Stop to Think 29.6: B. The atom would have less energy if the $3s$ electron were in a $2p$ state.

Stop to Think 29.7: C. Emission is a quantum jump to a lower-energy state. The $5p \rightarrow 4p$ transition is not allowed because $\Delta l = 0$ violates the selection rule. The lowest-energy allowed transition is $5p \rightarrow 3d$, with $E_{photon} = \Delta E_{atom} = 3.0$ eV.

30 Nuclear Physics

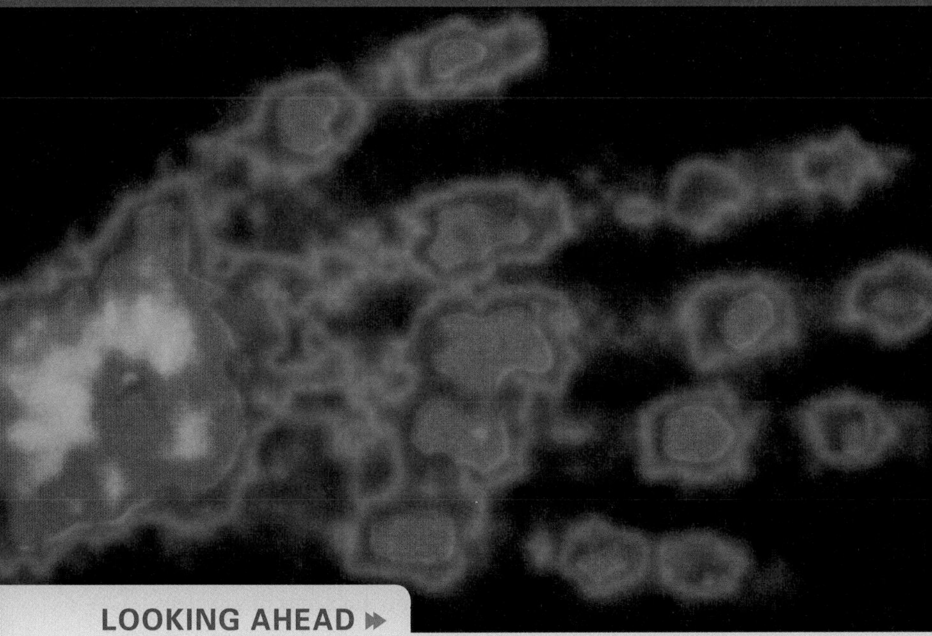

This is a *bone scan,* not an x ray. It was created using the radioactive decay of a particular type of nucleus inside the body. How can a process that occurs in the nucleus of an atom allow us to create an image of tissues of the human body?

LOOKING AHEAD ▸▸

The goals of Chapter 30 are to understand the physics of the nucleus and some of the applications of nuclear physics.

Nuclear Structure

In this chapter, we'll look at the structure of the nucleus, the small positive core of the atom. The nucleus is made of positive protons and neutral neutrons.

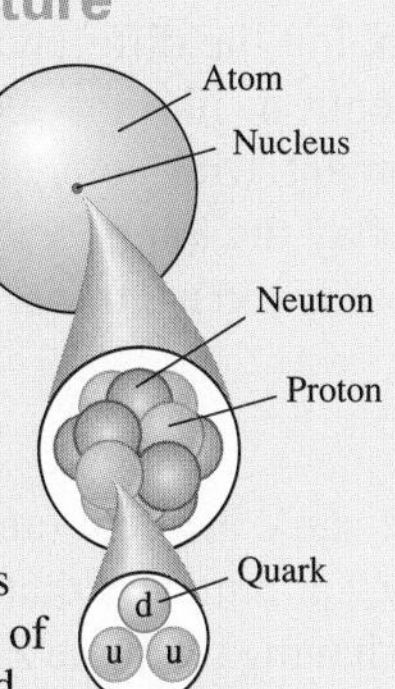

Neutrons and protons are themselves made of smaller entities called **quarks.**

Nuclear Decay

Certain nuclei **decay**, meaning they break into smaller pieces or otherwise change their structure.

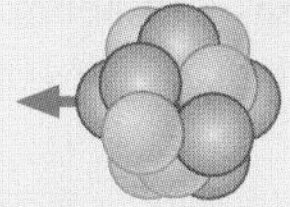

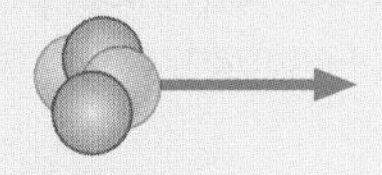

We'll develop models to explain nuclear decays. The models are similar to those that explain electron transitions in atoms.

Looking Back ◂◂
29.2 Models of the atom and the nucleus
29.6 Multielectron atoms

Nuclear transformations involve enormous energies.

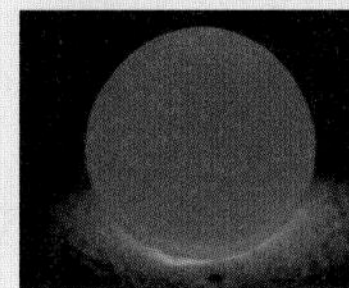

Nuclear decay will keep this plutonium sphere glowing for decades.

Looking Back ◂◂
27.10 Relativistic energy

Nuclear decay is governed by a **half-life,** the time for half of the atoms in sample to decay.

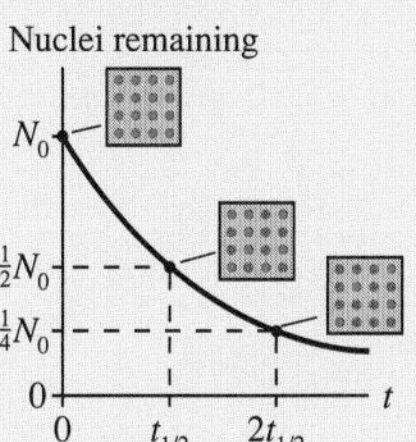

Nuclear Radiation

Nuclei decay by emitting high-energy charged particles or photons. This **nuclear radiation** comes in three main forms, corresponding to the three main types of decay.

The uranium in this mineral sample emits **alpha particles**—helium nuclei—as it decays.

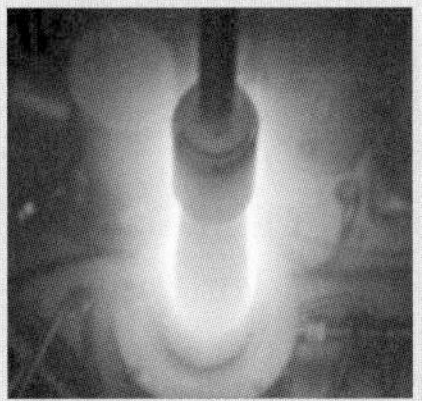

The ghostly blue glow in the water around the reactor core is produced by **beta particles**—high-energy electrons—emitted by decaying nuclei.

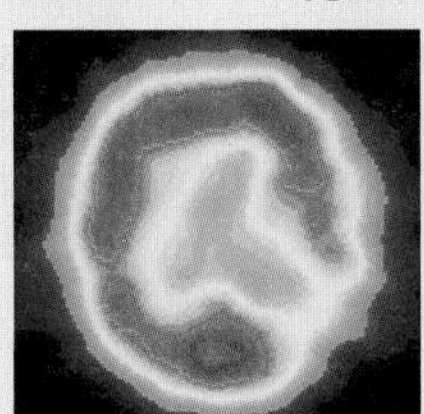

This image of the brain of a stroke patient was made with nuclei that decay by emitting **gamma rays**—high-energy photons. The area of reduced activity is clearly visible.

Applications

Nuclear physics is a practical subject with applications ranging from measuring ages to curing diseases.

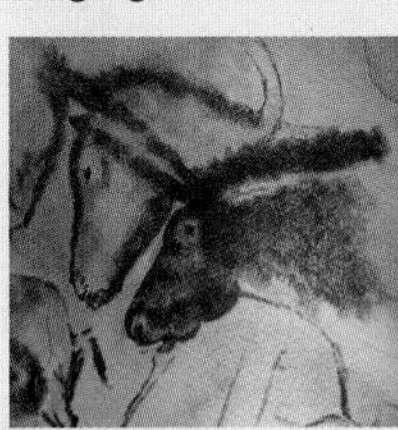

Researchers measured the fraction of different carbon isotopes in charcoal from these cave paintings to determine that they were painted 30,000 years ago.

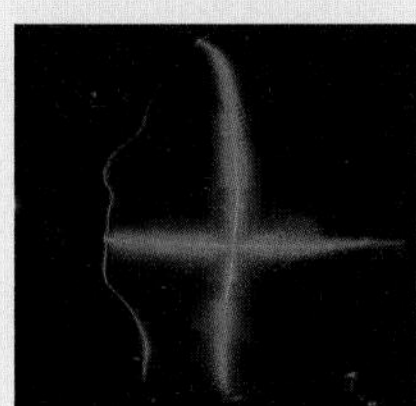

Radiation damages tissue; tumors are especially susceptible. Precisely targeted radiation can be used to treat cancer.

30.1 Nuclear Structure

FIGURE 30.1 The nucleus is a tiny speck within an atom.

This picture of an atom would need to be 10 m in diameter if it were drawn to the same scale as the dot representing the nucleus.

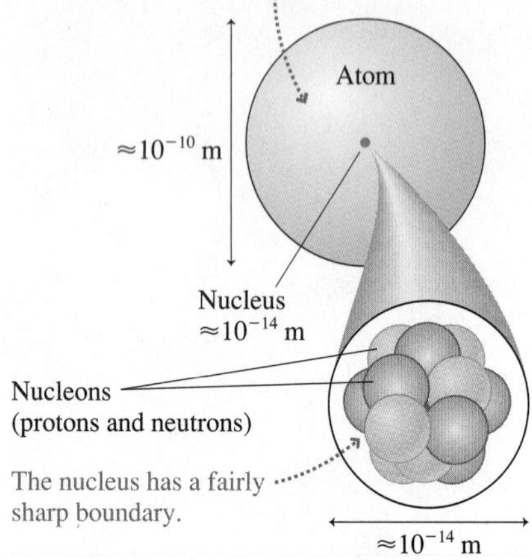

For 29 chapters, we've made frequent references to properties of atoms that are due to the negative electrons surrounding the nucleus. In this final chapter it's time to dig deeper, to talk about the nucleus itself. In particular:

- What is nuclear matter? What are its properties?
- What holds the nucleus together? Why doesn't the repulsive electrostatic force blow it apart?
- What is the connection between the nucleus and radioactivity?

These were the questions asked by the pioneers of **nuclear physics,** the study of the properties of the atomic nucleus.

Let's review some information about the atom and the nucleus. The nucleus is a tiny speck in the center of a vastly larger atom. As FIGURE 30.1 shows, the nuclear diameter of roughly 10^{-14} m is only about 1/10,000 the diameter of the atom. What we call *matter* is overwhelmingly empty space!

The nucleus is composed of two types of particles: protons and neutrons. Together, these are referred to as **nucleons.** The role of the neutrons, which have nothing to do with keeping electrons in orbit, is an important issue that we'll address in this chapter. The number of protons Z is the element's atomic number. An element is identified by the number of protons in the nucleus, not by the number of orbiting electrons. Electrons are easily added and removed, forming negative and positive ions, but doing so doesn't change the element. The mass number A is defined to be $A = Z + N$, where N is the neutron number. The mass number is the total number of nucleons in a nucleus.

NOTE ► The mass number, which is dimensionless, is *not* the same thing as the atomic mass m. We'll look at actual atomic masses later. ◄

TABLE 30.1 Protons and neutrons

	Proton	Neutron
Number	Z	N
Charge q	$+e$	0
Spin	$\frac{1}{2}$	$\frac{1}{2}$
Mass, in u	1.00728	1.00866

Protons and neutrons are virtually identical other than the fact that the proton has one unit of the fundamental charge e whereas the neutron is electrically neutral. The neutron is slightly more massive than the proton, but the difference is very small, only about 0.1%. Notice that the proton and neutron, like the electron, have an *inherent angular momentum* and magnetic moment with spin quantum number $s = \frac{1}{2}$. As a consequence, protons and neutrons obey the Pauli exclusion principle. Table 30.1 summarizes the basic properties of protons and neutrons.

Isotopes

As we learned in Chapter 29, not all atoms of the same element (and thus the same Z) have the same mass. There is a *range* of neutron numbers that happily form a nucleus with Z protons, creating a series of nuclei having the same Z-value (i.e., they are all the same chemical element) but different A-values. Each A-value in a series of nuclei with the same Z-value is called an *isotope.* Isotopes for some of the elements are given in a table in Appendix D.

FIGURE 30.2 Three isotopes of carbon.

The leading superscript gives the total number of nucleons, which is the mass number A.

$^{12}_{6}C$ $^{13}_{6}C$ $^{14}_{6}C$

The leading subscript (if included) gives the number of protons.

The three nuclei all have the same number of protons, so they are isotopes of the same element, carbon.

The notation used to label isotopes uses the mass number A as a *leading* superscript, as shown in FIGURE 30.2. Hence ordinary carbon, which has six protons and six neutrons in the nucleus (and thus has $A = 12$), is written ^{12}C and pronounced "carbon twelve." The radioactive form of carbon used in carbon dating is ^{14}C. It has six protons, making it carbon, and eight neutrons, for a total of 14 nucleons. The isotope ^{2}H is a hydrogen atom in which the nucleus is not simply a proton but a proton and a neutron. Although the isotope is a form of hydrogen, it is called **deuterium.** Sometimes, for clarity, we will find it useful to include the atomic number as a leading *subscript.* Ordinary carbon is then written as $^{12}_{6}C$; deuterium as $^{2}_{1}H$.

NOTE ► Adding the leading subscript doesn't provide any additional information. If the element is carbon, we know that it has 6 protons. But when we "balance" nuclear equations, we will find that the subscript is a useful tool and is worth including. ◄

The chemical behavior of an atom is largely determined by the orbiting electrons. Different isotopes of the same element have very similar *chemical* properties. ^{14}C will form the same chemical compounds as ^{12}C and will generally be treated the same by the body, a fact that permits the use of ^{14}C to determine the age of a sample. But the *nuclear* properties of these two isotopes are quite different, as we will see.

Most elements have multiple naturally occurring isotopes. For each element, the fraction of naturally occurring nuclei represented by one particular isotope is called the **natural abundance** of that isotope. For instance, oxygen has two primary isotopes, ^{16}O and ^{18}O. The data in Appendix D show that the natural abundance of ^{16}O is 99.76%, meaning that 9976 out of every 10,000 naturally occurring oxygen atoms are the isotope ^{16}O. Most of the remaining 0.24% of naturally occurring oxygen is the isotope ^{18}O, which has two extra neutrons.

The different masses of isotopes of an element can lead to some subtle differences in macroscopic behaviors. A water molecule made with the heavier ^{18}O isotope is slightly heavier as well, and will behave slightly differently than water made with the predominant ^{16}O isotope. Atmospheric water vapor is always deficient in ^{18}O compared to water in the ocean because the lighter molecules containing ^{16}O evaporate slightly more readily. Different atmospheric conditions lead to rain and snow (and thus lakes, rivers, and glaciers) with slightly different fractions of ^{18}O.

More than 3000 isotopes are known. The majority of these are **radioactive,** meaning that the nucleus is not stable but, after some period of time, will either fragment or emit some kind of subatomic particle in an effort to reach a more stable state. Many of these radioactive isotopes are created by nuclear reactions in the laboratory and have only a fleeting existence. Only 266 isotopes are **stable** (i.e., nonradioactive) and occur in nature. In addition, there are a handful of radioactive isotopes with such long decay times, measured in billions of years, that they also occur naturally.

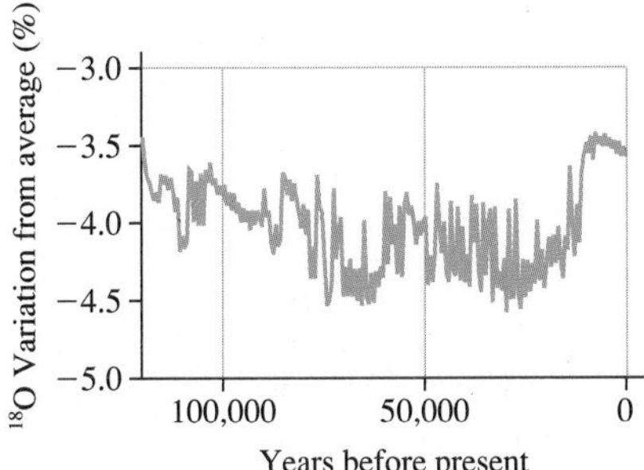

Taking the earth's temperature When water vapor condenses to make snow crystals, the fraction of molecules containing ^{18}O is greater for snow that forms at higher atmospheric temperatures. Snow accumulating over tens of thousands of years has built up a thick ice sheet in Greenland. A core sample of this ice gives a record of the isotopic composition of the snow that fell over this time period. The above graph shows the isotopic composition of a single long core taken from the ice sheet, averaged over 50-year intervals; higher numbers correspond to higher temperatures. Broad trends, such as the increase in temperature at the end of the last ice age, are clearly seen.

Atomic Mass

You learned in Chapter 12 that atomic masses are specified in terms of the *atomic mass unit* u, defined such that the atomic mass of the isotope ^{12}C is exactly 12 u. The conversion to SI units is

$$1\text{ u} = 1.6605 \times 10^{-27}\text{ kg}$$

Alternatively, as we saw in Chapter 27, we can use Einstein's $E_0 = mc^2$ to express masses in terms of their energy equivalent. The energy equivalent of 1 u of mass is

$$\begin{aligned} E_0 &= (1.6605 \times 10^{-27}\text{ kg})(2.9979 \times 10^8\text{ m/s})^2 \\ &= 1.4924 \times 10^{-10}\text{ J} = 931.49\text{ MeV} \end{aligned} \qquad (30.1)$$

By noting that Einstein's formula implies $m = E_0/c^2$, we can write 1 u in the following form:

$$1\text{ u} = \frac{E_0}{c^2} = 931.49\left(\frac{\text{MeV}}{c^2}\right) \qquad (30.2)$$

It may seem unusual, but the units MeV/c^2 are units of mass. This will be a useful unit for us when we need to compute energy equivalents. The energy equivalent of mass 1 MeV/c^2 is simply 1 MeV.

NOTE ► In the above equations, we have included more significant figures than usual. Many nuclear calculations look for the small difference between two masses that are almost the same. The two masses must be calculated or specified to four or five significant figures if their difference is to be meaningful. In calculations of nuclear energies, you should use the accurate values for nuclear masses given in Appendix D. ◄

Table 30.2 shows some important atomic mass values. Notice that the mass of a hydrogen atom is equal to the sum of the masses of a proton and an electron.

TABLE 30.2 Some atomic masses

Particle	Symbol	Mass (u)	Mass (MeV/c^2)
Electron	e	0.000549	0.51
Proton	p	1.007276	938.28
Neutron	n	1.008665	939.57
Hydrogen	^1H	1.007825	938.79
Helium	^4He	4.002602	3728.40

But a quick calculation shows that the mass of a helium atom (2 protons, 2 neutrons, and 2 electrons) is 0.03038 u *less* than the sum of the masses of its constituents. The difference is due to the *binding energy* of the nucleus, a topic we'll look at in Section 30.2.

The *chemical* atomic mass shown on the periodic table of the elements is the *weighted average* of the atomic masses of all naturally occurring isotopes. For example, chlorine has two stable isotopes: ^{35}Cl, with atomic mass $m = 34.97$ u, has an abundance of 75.8% and ^{37}Cl, at 36.97 u, has an abundance of 24.2%. The average, weighted by abundance, is $(0.758 \times 34.97 \text{ u}) + (0.242 \times 36.97 \text{ u}) = 35.45$ u. This is the value shown on the periodic table and is the correct value for most chemical calculations, but it is not the mass of any particular isotope of chlorine. Nuclear physics calculations involve the masses of specific isotopes, so you'll need to use these masses from Appendix D, not the chemical atomic masses given in the periodic table.

NOTE ► The atomic masses of the proton and the neutron are both ≈ 1 u. Consequently, the value of the mass number A is *approximately* the atomic mass in u. The approximation $m \approx A$ u is sufficient in many contexts, such as when we calculate speeds of gas molecules in Chapter 12. But in nuclear physics calculations we will need the more accurate mass values in Appendix D. ◄

STOP TO THINK 30.1 Three electrons orbit a neutral ^{6}Li atom. How many electrons orbit a neutral ^{7}Li atom?

30.2 Nuclear Stability

Because nuclei are characterized by two independent numbers, N and Z, it is useful to show the known nuclei on a plot of neutron number N versus proton number Z. **FIGURE 30.3** shows such a plot. Stable nuclei are represented by blue diamonds and unstable, radioactive nuclei by red dots.

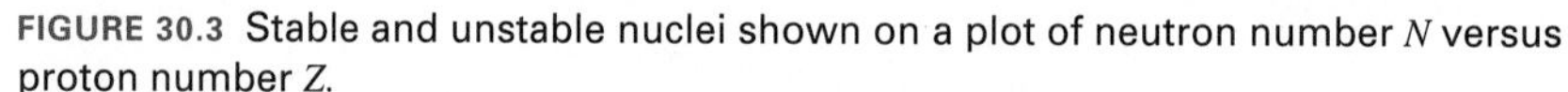

FIGURE 30.3 Stable and unstable nuclei shown on a plot of neutron number N versus proton number Z.

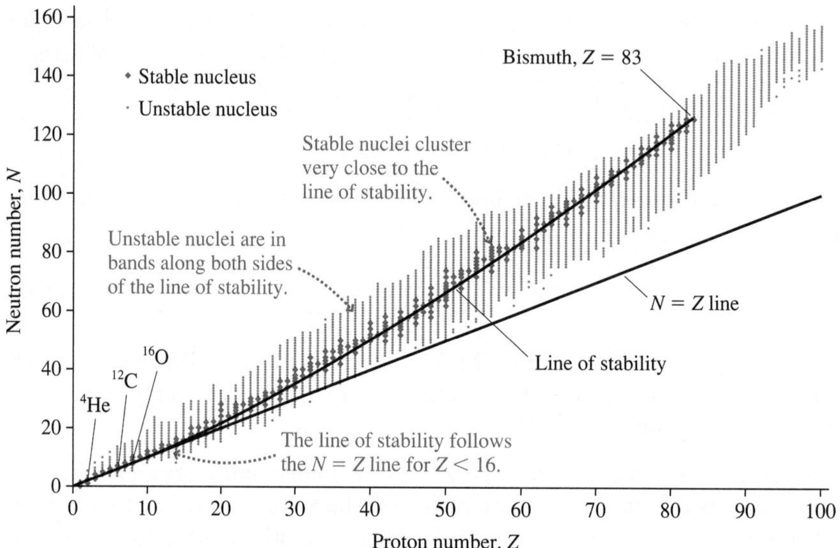

We can make several observations from this graph:

- The stable nuclei cluster very close to the curve called the **line of stability.**
- There are no stable nuclei with $Z > 83$ (bismuth). Heavier elements (up to $Z = 92$ (uranium)) are found in nature, but they are radioactive.

- Unstable nuclei are in bands along both sides of the line of stability.
- The lightest elements, with $Z < 16$, are stable when $N \approx Z$. The familiar isotopes ^{4}He, ^{12}C, and ^{16}O all have equal numbers of protons and neutrons.
- As Z increases, the number of neutrons needed for stability grows increasingly larger than the number of protons. The N/Z ratio is ≈ 1.2 at $Z = 40$ but has grown to ≈ 1.5 at $Z = 80$.

These observations—especially the fact that $N \approx Z$ for small Z but $N > Z$ for large Z—will be explained by the model of the nucleus that we'll explore in Section 30.3.

▶ **Unstable but ubiquitous uranium** All of the isotopes of uranium are unstable, but some have very long lives. Approximately half of the ^{238}U that was present at the formation of the earth 4.5 billion years ago is still around—and is all around you. Uranium is present at low concentrations in nearly all of the rocks and soil and water on the earth's surface. Much of the radiation that you are exposed to comes from this naturally occurring and widely distributed unstable element. Minerals with high concentrations of uranium are often fluorescent, as this sample of autunite ore photographed under ultraviolet light clearly shows.

Binding Energy

A nucleus is a *bound system.* That is, you would need to supply energy to disperse the nucleons by breaking the nuclear bonds between them. FIGURE 30.4 shows this idea schematically.

FIGURE 30.4 The nuclear binding energy.

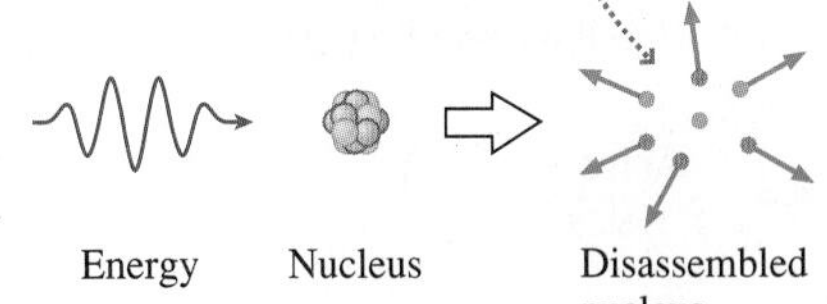

The binding energy is the energy that would be needed to disassemble a nucleus into individual nucleons.
Energy
Nucleus
Disassembled nucleus

You learned a similar idea in atomic physics. The energy levels of the hydrogen atom are negative numbers because the bound system has less energy than a free proton and electron. The energy you must supply to an atom to remove an electron is called the *ionization energy.*

In much the same way, the energy you would need to supply to a nucleus to disassemble it into individual protons and neutrons is called the **binding energy.** Whereas ionization energies of atoms are only a few eV, the binding energies of nuclei are tens or hundreds of MeV, energies large enough that their mass equivalent is not negligible.

We noted earlier that the mass of a helium atom is less than the mass of its constituents. Suppose we break a helium atom into two hydrogen atoms (taking account of the two protons and the two electrons) and two free neutrons as shown in FIGURE 30.5. The mass of the separated components is more than that of the helium atom. The difference in mass Δm arises from the energy that was put into the system to separate the tightly bound nucleons. The difference in masses is thus a measure of the binding energy:

FIGURE 30.5 The binding energy of the helium nucleus.

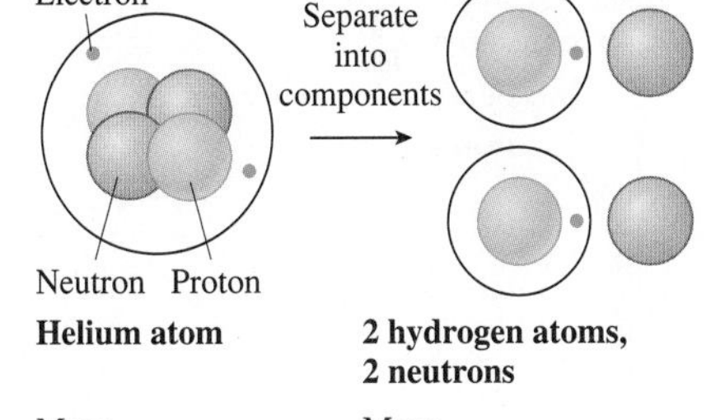

$$\text{binding energy of helium nucleus} = \Delta m \cdot c^2$$

We can evaluate this energy B by converting the mass to MeV/c^2:

$$\begin{aligned} B &= \Delta m \cdot c^2 \\ &= \left((0.03038\text{ u})\left(931.49\,\frac{\text{MeV}/c^2}{\text{u}}\right)\right)c^2 = (0.03038\text{ u})(931.49\text{ MeV/u}) \\ &= 28.30\text{ MeV} \end{aligned}$$

Generally, the nuclear binding energy is computed by considering the mass difference between the atom and its separated components, Z hydrogen atoms and N neutrons:

Activ ONLINE Physics 19.2

$$B = (Zm_{\text{H}} + Nm_{\text{n}} - m_{\text{atom}}) \times (931.49\text{ MeV/u}) \tag{30.3}$$

Nuclear binding energy for an atom of mass m_{atom} with Z protons and N neutrons

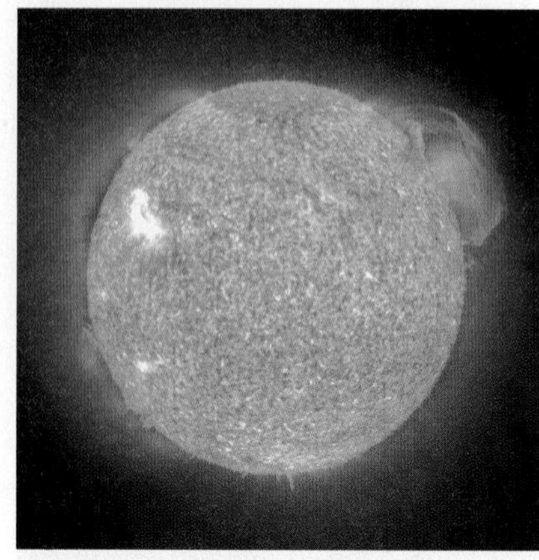

A nuclear fusion weight-loss plan The sun's energy comes from reactions that combine four hydrogen atoms to create a single atom of helium—a process called **nuclear fusion.** Because energy is released, the mass of the helium atom is less than that of the four hydrogen atoms. As the fusion reactions continue, the mass of the sun decreases—by 130 trillion tons per year! That's a lot of mass, but given the sun's enormous size, this change will amount to only a few hundredths of a percent of the sun's mass over its 10-billion-year lifetime.

EXAMPLE 30.1 Finding the binding energy of iron

What is the nuclear binding energy of ^{56}Fe to the nearest MeV?

PREPARE Appendix D gives the atomic mass of ^{56}Fe as 55.934940 u. Iron has atomic number 26, so an atom of ^{56}Fe could be separated into 26 hydrogen atoms and 30 neutrons. The mass of the separated components is more than that of the iron nucleus; the difference gives us the binding energy.

SOLVE We solve for the binding energy using Equation 30.3. The masses of the hydrogen atom and the neutron are given in Table 30.2. We find

$$B = (26(1.007825 \text{ u}) + 30(1.008665 \text{ u}) - 55.934940 \text{ u})(931.49 \text{ MeV/u})$$
$$= (0.52846 \text{ u})(931.49 \text{ MeV/u}) = 492.26 \text{ MeV} \simeq 492 \text{ MeV}$$

ASSESS The difference in mass between the nucleus and its components is a small fraction of the mass of the nucleus, so we must use several significant figures in our mass values. The mass difference is small—about half that of a proton—but the energy equivalent, the binding energy, is enormous.

How much energy is 492 MeV? To make a comparison with another energy value we have seen, the binding energy of a single iron nucleus is equivalent to the energy released in the metabolism of nearly *2 billion* molecules of ATP! The energy scale of nuclear processes is clearly quite different from that of chemical processes. In nuclear reactors, a small amount of nuclear "fuel" can produce a tremendous amount of energy.

19.3 ActivPhysics Online

As A increases, the nuclear binding energy increases, simply because there are more nuclear bonds. A more useful measure for comparing one nucleus to another is the quantity B/A, called the *binding energy per nucleon.* Iron, with $B = 492$ MeV and $A = 56$, has 8.79 MeV per nucleon. This is the amount of energy, on average, you would need to supply in order to remove *one* nucleon from the nucleus. Nuclei with larger values of B/A are more tightly held together than nuclei with smaller values of B/A.

FIGURE 30.6 The curve of binding energy for the stable nuclei.

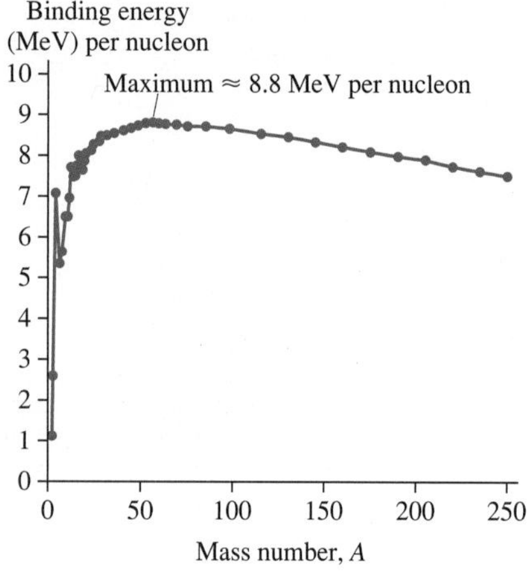

FIGURE 30.6 is a graph of the binding energy per nucleon versus mass number A. The line connecting the points is often called the **curve of binding energy.**

For small values of A, adding more nucleons increases the binding energy per nucleon. If two light nuclei can be joined together to make a single, larger nucleus, the final nucleus will have a higher binding energy per nucleon. Because the final nucleus is more tightly bound, energy will be released in this *nuclear fusion* process. Nuclear fusion of hydrogen to helium is the basic reaction that powers the sun, as we have seen.

The curve of binding energy has a broad maximum at $A \approx 60$. Nuclei with $A > 60$ become less stable as their mass increases because adding nucleons *decreases* the binding energy per nucleon. *Alpha decay*, one of the three basic types of radioactive decay that we'll examine later in this chapter, occurs when a heavy nucleus becomes more stable by ejecting a small group of nucleons in order to decrease its mass, releasing energy in the process. The decrease in binding energy per nucleon as mass increases also explains why there are no stable nuclei beyond $Z = 83$.

A few very heavy nuclei, especially some isotopes of uranium and plutonium, are so unstable that they can be induced to fragment into two lighter nuclei in the process known as *nuclear fission*. For example, the collision of a slow-moving neutron with a ^{235}U nucleus causes the reaction

$$\text{n} + {}^{235}\text{U} \rightarrow {}^{236}\text{U} \rightarrow {}^{90}\text{Sr} + {}^{144}\text{Xe} + 2\text{n} \qquad (30.4)$$

The ^{235}U nucleus absorbs the neutron to become ^{236}U, but ^{236}U is so unstable that it immediately fragments—in this case into a ^{90}Sr nucleus, a ^{144}Xe nucleus, and two neutrons. The less massive ^{90}Sr and ^{144}Xe nuclei are more tightly bound than the original ^{235}U nucleus, so a great deal of energy is released in this reaction. But the

^{90}Sr and ^{144}Xe nuclei *aren't* stable. As we've seen, nuclei with lower values of Z have relatively smaller numbers of neutrons, meaning there will be neutrons "left over" after the reaction. Equation 30.4 shows some free neutrons among the reaction products, but the two nuclear fragments have "extra" neutrons as well—they have too many neutrons and will be unstable. This is generally true for the products of a fission reaction. The fact that the waste products of nuclear fission are radioactive has important consequences for the use of nuclear fission as a source of energy.

▶ **A formidable chain reaction** For certain isotopes of uranium or plutonium, if the nucleus is struck by a neutron, it can split into two more tightly bound fragments, releasing tremendous energy. Extra neutrons are left over after the split, each of which can cause the fission of another nucleus—releasing more neutrons, which can produce further reactions. The net result is a nuclear fission **chain reaction.** A controlled reaction is the energy source of a nuclear power plant. An uncontrolled reaction is responsible for the terrible destructive power of a nuclear explosion, shown here in an above-ground test in Nevada in 1957.

30.3 Forces and Energy in the Nucleus

The nucleus of the atom is made of protons, which are positively charged, and neutrons, which have no charge. Why doesn't the repulsive force of the protons simply cause the nucleus to fly apart? This was a puzzle faced by early investigators. It soon became clear that a previously unknown force of nature operates within the nucleus to hold the nucleons together. This new force has to be stronger than the repulsive electrostatic force; hence it was named the **strong nuclear force,** or just the strong force.

The strong force has four important properties:

1. It is an *attractive* force between any two nucleons.
2. It does not act on electrons.
3. It is a *short-range* force, acting only over nuclear distances. We see no evidence for nuclear forces outside the nucleus.
4. Over the range where it acts, it is *stronger* than the electrostatic force that tries to push two protons apart.

FIGURE 30.7 summarizes the three types of interactions that take place within the nucleus. Many decades of research have shown that the strong force between two nucleons is independent of whether they are protons or neutrons. Charge is the basis for electromagnetic interactions, but it is of no relevance to the strong force. Protons and neutrons are identical as far as nuclear forces are concerned.

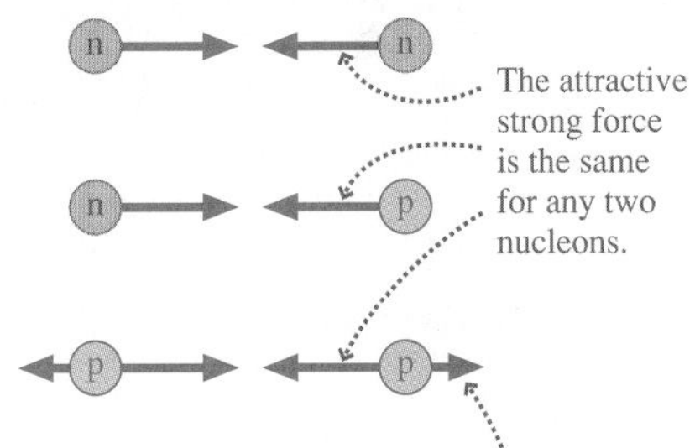

n
n
The attractive strong force is the same for any two nucleons.
n
p
p
p
Two protons also experience a smaller electrostatic repulsive force.

FIGURE 30.7 Forces between pairs of particles in the nucleus.

Protons throughout the nucleus exert repulsive electrostatic forces on each other, but, because of the short range of the strong force, a proton feels an attractive force only from the very few other protons with which it is in close contact. A nucleus with too many protons will be unstable because the repulsive electrostatic forces will overcome the attractive strong forces. Because neutrons participate in the strong force but exert no repulsive forces, **the neutrons provide the extra "glue" that holds the nucleus together.** In small nuclei, one neutron per proton is sufficient for stability. Hence small nuclei have $N \approx Z$. But as the nucleus grows, the repulsive force increases faster than the binding energy. More neutrons are needed for stability, so heavy nuclei have $N > Z$.

In Chapter 28 we learned that confining a particle to a region of space results in the quantization of energy. In Chapter 29 you saw the consequences of this quantization for electrons bound in atoms; in this chapter we will look at the quantized energy levels for protons and neutrons bound in the nucleus. Just as we did with atoms, we can "build up" the nuclear state by placing all the nucleons in the lowest energy levels consistent with the Pauli principle, which applies to nucleons just as it did to electrons. Each energy level can hold only a certain number of spin-up particles and spin-down particles, depending on the quantum numbers; additional nucleons must go into higher energy levels.

Although there is a great deal of similarity between the descriptions of nucleon energy levels in the nucleus and electron energy levels in the atom, there is one important difference: the energy scale. The electron energy levels are typically separated by a few eV, but the proton and neutron energy levels are separated by a few MeV—a million times as much.

Low-Z Nuclei

As our first example of an energy-level description of the nucleus, we'll consider the energy levels of low-Z nuclei ($Z < 8$). Because these nuclei have so few protons, we can neglect the electrostatic potential energy due to proton-proton repulsion and consider only the much larger nuclear potential energy. In that case, the energy levels of the protons and neutrons are essentially identical.

FIGURE 30.8 shows the three lowest allowed energy levels for protons and neutrons and the maximum number of protons and neutrons the Pauli principle allows in each. Energy values vary from nucleus to nucleus, but the spacing between these levels is several MeV.

FIGURE 30.8 The three lowest energy levels of a low-Z nucleus. The neutron energy levels are on the left, the proton energy levels on the right.

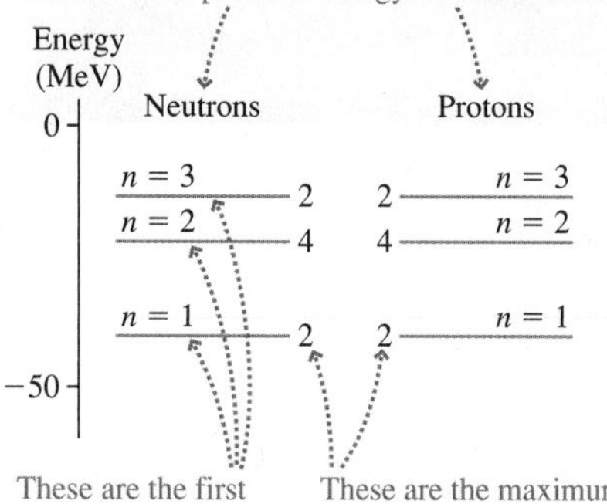
The proton potential energy is nearly identical to the neutron potential energy when Z is small.
Energy (MeV)
Neutrons
Protons
0
−50
n = 3
n = 2
n = 1
2
4
2
These are the first three allowed energy levels. They are spaced several MeV apart.
These are the maximum numbers of protons and neutrons allowed by the Pauli principle.

Suppose we look at a series of nuclei, all with $A = 12$ but with different numbers of protons and neutrons: ^{12}B, ^{12}C, ^{12}N. All have 12 nucleons, but only ^{12}C is a stable isotope; the other two are not. Why do we see this difference in stability?

FIGURE 30.9 shows the energy-level diagrams of ^{12}B, ^{12}C, and ^{12}N. Look first at ^{12}C, a nucleus with six protons and six neutrons. You can see that exactly six protons are allowed in the $n = 1$ and $n = 2$ proton energy levels. The same is true for the six neutrons. No other arrangement of the nucleons would lower the total energy, so this nucleus is stable.

^{12}B has five protons and seven neutrons. The sixth neutron fills the $n = 2$ neutron energy level, so the seventh neutron has to go into the $n = 3$ energy level. (This is just like the third electron in Li having to go into the $n = 2$ energy level because the first two electrons have filled the $n = 1$ energy level.) The $n = 2$ proton energy level has one vacancy because there are only five protons.

In atoms, electrons in higher energy levels move to lower energy levels by emitting a photon as the electron undergoes a quantum jump. That can't happen here because the higher-energy nucleon in ^{12}B is a neutron whereas the vacant lower energy level is that of a proton. But an analogous process could occur *if* a neutron could somehow turn into a proton, allowing it to move to a lower energy level. ^{12}N is just the opposite, with the seventh proton in the $n = 3$ energy level. If a proton could somehow turn into a neutron, it could move to a lower energy level, lowering the energy of the nucleus.

You can see from the diagrams that the ^{12}B and ^{12}N nuclei have significantly more energy—by several MeV—than ^{12}C. If a neutron could turn into a proton, and vice versa, these nuclei could move to a lower-energy state—that of ^{12}C. In fact, that's exactly what happens! These nuclei are not stable. We'll explore the details in Section 30.4. Both ^{12}B and ^{12}N decay into the more stable ^{12}C in the process known as *beta decay.*

FIGURE 30.9 Nuclear energy-level diagrams of ^{12}B, ^{12}C, and ^{12}N.

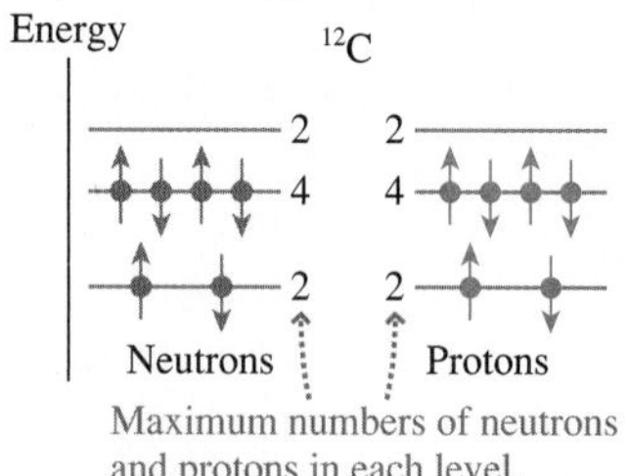
A 12C nucleus is in its lowest possible energy state.
Energy
12C
2
4
2
Neutrons
Protons
Maximum numbers of neutrons and protons in each level.

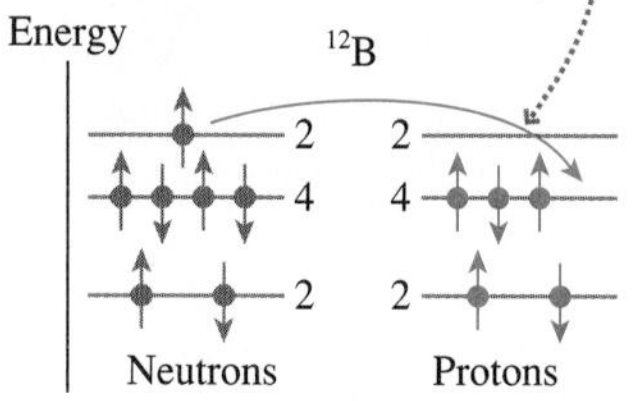
A 12B nucleus could lower its energy if a neutron could turn into a proton.
Energy
12B
2
4
2
Neutrons
Protons

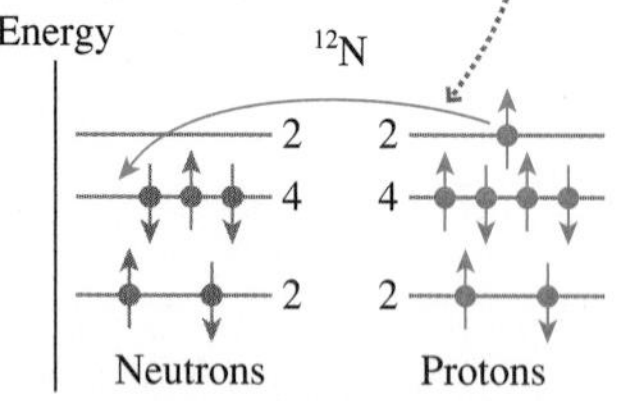
A 12N nucleus could lower its energy if a proton could turn into a neutron.
Energy
12N
2
4
2
Neutrons
Protons

High-Z Nuclei

We can use the energy levels of protons and neutrons in the nucleus to give a qualitative explanation for one more observation. **FIGURE 30.10** shows the neutron and proton energy levels of a high-Z nucleus. In a nucleus with many protons, the increasing electrostatic potential energy raises the proton energy levels but not the neutron energy levels. Protons and neutrons now have a different set of energy levels.

As a nucleus is "built," by the adding of protons and neutrons, the proton energy levels and the neutron energy levels must fill to just about the same height. If there were neutrons in energy levels above vacant proton levels, the nucleus would lower its energy by changing neutrons into protons. Similarly, a proton would turn into a neutron if there were a vacant neutron energy level beneath a filled proton level. **The**

net result is that the filled levels for protons and neutrons are at just about the same height.

Because the neutron energy levels start at a lower energy, *more neutron states* are available than proton states. Consequently, a high-Z nucleus will have more neutrons than protons. This conclusion is consistent with our observation in Figure 30.3 that $N > Z$ for heavy nuclei.

STOP TO THINK 30.2 Based on the model of nuclear energy levels and transitions you have seen, would you expect ^{13}C to be stable?

FIGURE 30.10 The proton energy levels are displaced upward in a high-Z nucleus because of the electric potential energy.

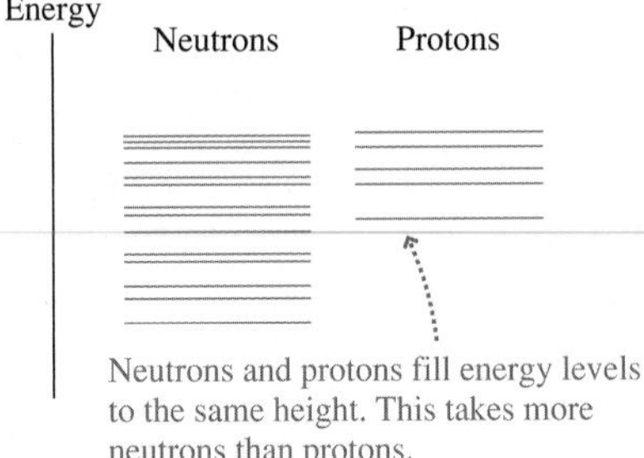

30.4 Radiation and Radioactivity

Some nuclei are unstable—they can decay—but what exactly does this mean? How do we know that they decay? What is left over when a nucleus decays?

The existence of nuclear decay was discovered in the 1890s by investigators who noticed high-energy emissions from certain atoms. Different nuclei emitted three different types of particle or ray, differentiated by their behavior in a magnetic field, as shown in **FIGURE 30.11**. Experiments to determine the nature of the emissions showed that **alpha particles** are helium nuclei and **beta particles** are electrons. **Gamma rays** were found to be similar to x rays, high-energy electromagnetic waves best described as photons. Table 30.3 summarizes the properties of alpha and beta particles and gamma rays.

TABLE 30.3 Three types of radiation

Radiation	Identification	Charge
Alpha, α	^{4}He nucleus	$+2e$
Beta, β	Electron	$-e$
Gamma, γ	High-energy photon	0

ActivPhysics ONLINE 19.4

FIGURE 30.11 Identifying radiation by its deflection in a magnetic field.

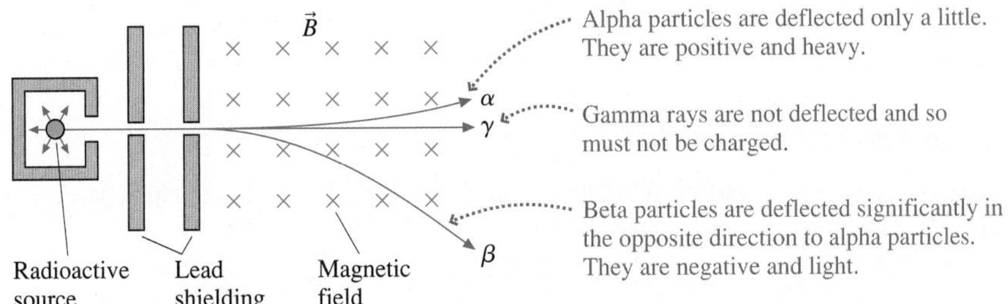

$\vec{B}$
α
γ
β
Alpha particles are deflected only a little. They are positive and heavy.
Gamma rays are not deflected and so must not be charged.
Beta particles are deflected significantly in the opposite direction to alpha particles. They are negative and light.
Radioactive source
Lead shielding
Magnetic field

We now define *radioactivity* or *radioactive decay* to be the spontaneous emission of particles or high-energy photons from unstable nuclei as they decay from higher-energy to lower-energy states. In this section, we will explore the decay mechanisms that result in the three different types of radiation, considering the changes in the nuclei that produce these decays.

Alpha Decay

We know that many large nuclei are unstable; it is energetically favorable for them to spontaneously break apart into smaller fragments. A combination of two neutrons and two protons, a ^{4}He nucleus, is an especially stable nuclear combination. When a large nucleus spontaneously decays by breaking into two smaller fragments, one of these fragments is almost always a helium nucleus—an alpha particle, symbolized by α.

An unstable nucleus that ejects an alpha particle loses two protons and two neutrons, so we can write this decay as

$$^{A}_{Z}X \rightarrow {}^{A-4}_{Z-2}Y + \alpha + \text{energy} \tag{30.5}$$

Alpha decay of a nucleus

FIGURE 30.12 Alpha decay.

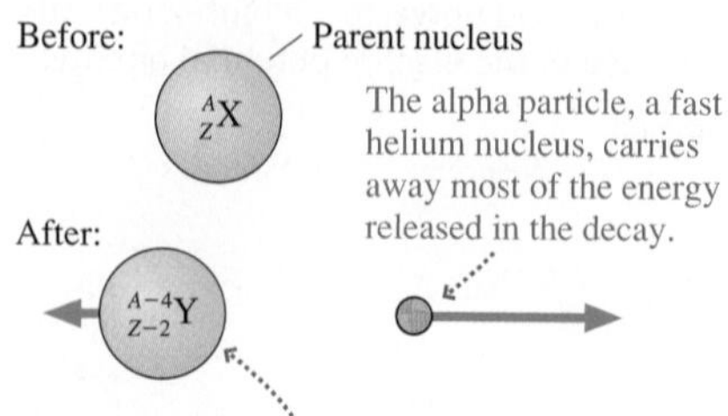
Before:
Parent nucleus
$^{A}_{Z}X$
The alpha particle, a fast helium nucleus, carries away most of the energy released in the decay.
After:
$^{A-4}_{Z-2}Y$
The daughter nucleus has two fewer protons and four fewer nucleons. It has a small recoil.

FIGURE 30.12 shows the alpha-decay process. The original nucleus X is called the **parent nucleus,** and the decay-product nucleus Y is the **daughter nucleus.**

Energy conservation tells us that an alpha decay can occur only when the mass of the parent nucleus is greater than the mass of the daughter nucleus plus the mass of the alpha particle. This requirement is often met for heavy, high-Z nuclei beyond the maximum of the curve of binding energy of Figure 30.6. It is energetically favorable for these nuclei to eject an alpha particle because the daughter nucleus is more tightly bound than the parent nucleus.

The daughter nucleus, which is much more massive than an alpha particle, undergoes only a slight recoil, as we see in Figure 30.12. Consequently, **the energy released in an alpha decay ends up mostly as the kinetic energy of the alpha particle.** We can compute this energy by looking at the mass energy difference between the initial and final states:

$$K_\alpha \approx \Delta E = (m_X - m_Y - m_{He})c^2 \tag{30.6}$$

EXAMPLE 30.2 Analyzing alpha decay in a smoke detector

Americium, atomic number 95, doesn't exist in nature; it is produced in nuclear reactors. An isotope of americium, ^{241}Am, is part of the sensing circuit in most smoke detectors. ^{241}Am decays by emitting an alpha particle. What is the daughter nucleus?

SOLVE Equation 30.5 shows that an alpha decay causes the atomic number to decrease by 2 and the atomic weight by 4. Let's write an equation for the decay showing the alpha particle as a helium nucleus, including the atomic weight superscript and the atomic number subscript for each element. There is no change in the total number of neutrons or protons, so the subscripts and superscripts must "balance" in the reaction:

$$^{241}_{95}Am \rightarrow {}^{237}_{93}? + {}^{4}_{2}He + \text{energy}$$

A quick glance at the periodic table reveals the unknown element in this equation, the daughter nucleus, to be an isotope of neptunium, $^{237}_{93}Np$.

ASSESS Balancing the two sides of the above reaction is similar to balancing the equation for a chemical reaction.

EXAMPLE 30.3 Finding the energy of an emitted alpha particle

The uranium isotope ^{238}U undergoes alpha decay to an isotope of thorium, ^{234}Th. What is the kinetic energy, in MeV, of the alpha particle?

PREPARE The decay products have less mass than the initial nucleus. This difference in mass is released as energy, most of which goes to the kinetic energy of the alpha particle. Because the energy of the alpha particle is only approximately equal to the reaction energy, we needn't use the full accuracy that the values in Appendix D provides. To 4 decimal places, the atomic mass of ^{238}U is 238.0508 u, that of ^{234}Th is 234.0436 u and that of ^{4}He—the alpha particle—is 4.0026 u.

SOLVE We can calculate the kinetic energy of the alpha particle using Equation 30.6:

$$K_\alpha = (238.0508 \text{ u} - 234.0436 \text{ u} - 4.0026 \text{ u})c^2$$
$$= (0.0046 \text{ u})c^2$$

If we convert from u to MeV/c^2, using the conversion factor 1 u = 931.49 MeV/c^2, the c^2 cancels, and we end up with

$$K_\alpha = \left(0.0046 \text{ u} \times \frac{931.49 \text{ MeV}/c^2}{1 \text{ u}}\right)c^2 = 4.3 \text{ MeV}$$

ASSESS This is a typical alpha-particle energy, corresponding to a speed of about 5% of the speed of light. Notice that with a careful use of conversion factors we never had to evaluate c^2.

Beta Decay

We identified beta decay as the emission of an electron e^-, the beta particle. A typical example of beta decay occurs in the carbon isotope ^{14}C, which undergoes the beta-decay process

$$^{14}C \rightarrow {}^{14}N + e^-$$

Carbon has $Z = 6$ and nitrogen has $Z = 7$. Because Z increases by 1 but A doesn't change, it appears that a neutron within the nucleus has changed itself into a proton by emitting an electron. That is, the basic decay process appears to be

$$n \rightarrow p + e^- \tag{30.7}$$

The electron is ejected from the nucleus but the proton is not. Thus the beta-decay process, shown in **FIGURE 30.13a**, is

$$ {}^{A}_{Z}\mathrm{X} \rightarrow {}_{Z+1}^{\;\;A}\mathrm{Y} + \mathrm{e}^- + \text{energy} \tag{30.8}$$

Beta-minus decay of a nucleus

Do neutrons *really* turn into protons? It turns out that a free neutron—one not bound in a nucleus—is *not* a stable particle. It decays into a proton and an electron, with a half-life of approximately 10 minutes. This decay conserves energy because $m_n > m_p + m_e$. Furthermore, it conserves charge.

Whether a neutron *within* a nucleus can decay depends not only on the masses of the neutron and proton but also on the masses of the parent and daughter nuclei, because energy has to be conserved for the entire nuclear system. **Beta decay occurs only if $m_X > m_Y$.** ^{14}C can undergo beta decay to ^{14}N because $m(^{14}\mathrm{C}) > m(^{14}\mathrm{N})$. But $m(^{12}\mathrm{C}) < m(^{12}\mathrm{N})$, so ^{12}C is stable and its neutrons will not decay.

A few nuclei undergo a slightly different form of beta decay by emitting a *positron.* A positron, for which we use the symbol e^+, is identical to an electron except that it has a positive charge. As we saw in Chapter 27, the positron is the *antiparticle* of the electron. To distinguish between these two forms of decay, we call the emission of an electron *beta-minus decay* and the emission of a positron *beta-plus decay.*

Inside a nucleus undergoing beta-plus decay, a proton changes into a neutron and a positron:

$$\mathrm{p}^+ \rightarrow \mathrm{n} + \mathrm{e}^+ \tag{30.9}$$

The full decay process, shown in **FIGURE 30.13b**, is

$$ {}^{A}_{Z}\mathrm{X} \rightarrow {}_{Z-1}^{\;\;A}\mathrm{Y} + \mathrm{e}^+ + \text{energy} \tag{30.10}$$

Beta-plus decay of a nucleus

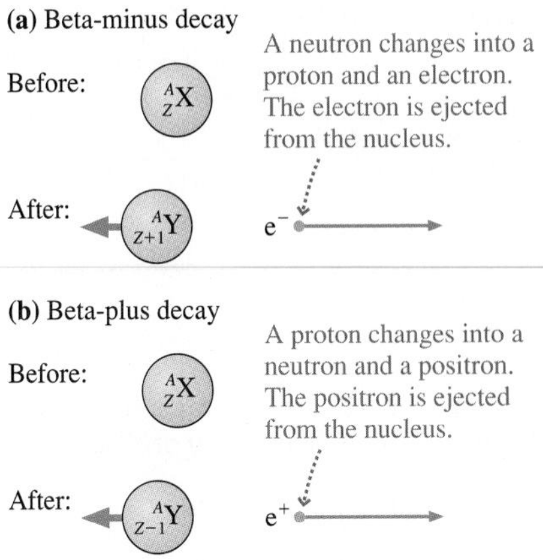

Beta-plus decay does *not* happen for a free proton because $m_p < m_n$. It *can* happen within a nucleus as long as energy is conserved for the entire nuclear system, but it is far less common than beta-minus decay.

In our earlier discussion of Section 30.3 we noted that the ^{12}B and ^{12}N nuclei could reach a lower-energy state if a proton could change into a neutron, and vice versa. Now we see that such changes can occur if the energy conditions are favorable. And, indeed, ^{12}B undergoes beta-minus decay to ^{12}C, while ^{12}N undergoes beta-plus decay to ^{12}C.

In general, beta decay is a process of nuclei with too many neutrons or too many protons that moves them closer to the line of stability in Figure 30.3.

NOTE ▶ The electron emitted in beta decay has nothing to do with the atom's valence electrons. The beta particle is created in the nucleus and ejected directly from the nucleus when a neutron is transformed into a proton and an electron. ◀

EXAMPLE 30.4 Analyzing beta decay in the human body

Your body contains several radioactive isotopes. Approximately 20% of the radiation dose you receive each year comes from the radioactive decay of these atoms. Most of this dose comes from one potassium isotope, ^{40}K, which decays by beta-minus emission. What is the daughter nucleus?

SOLVE Rewriting Equation 30.8 as

$$ {}^{40}_{19}\mathrm{K} \rightarrow {}^{40}_{20}? + \mathrm{e}^- + \text{energy}$$

we see that the daughter nucleus must be the calcium isotope ${}^{40}_{20}\mathrm{Ca}$.

FIGURE 30.13 Beta decay.

(a) Beta-minus decay

Before: ${}^{A}_{Z}\mathrm{X}$

A neutron changes into a proton and an electron. The electron is ejected from the nucleus.

After:

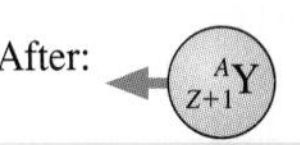

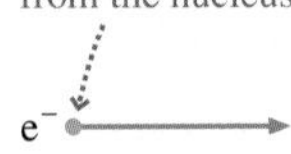

(b) Beta-plus decay

Before:

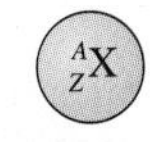

A proton changes into a neutron and a positron. The positron is ejected from the nucleus.

After:

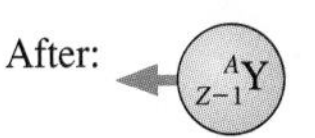

e^+

FIGURE 30.14 Gamma decay.

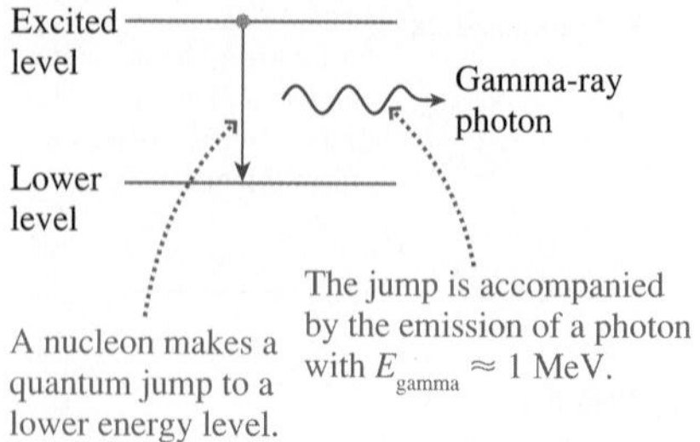
Excited level
Gamma-ray photon
Lower level
The jump is accompanied by the emission of a photon with $E_{gamma} \approx 1$ MeV.
A nucleon makes a quantum jump to a lower energy level.

FIGURE 30.15 $^{99}Tc^*$, a gamma emitter, is produced in the beta decay of ^{99}Mo.

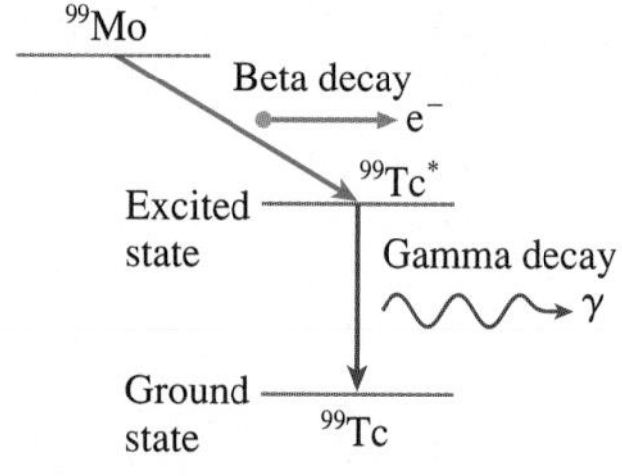
99Mo
Beta decay
e−
99Tc*
Excited state
Gamma decay
γ
Ground state
99Tc

Gamma Decay

Gamma decay is similar to quantum processes you saw in earlier chapters. In Chapter 28, you learned that an atomic system can emit a photon with $E_{photon} = \Delta E_{atom}$ when an electron undergoes a quantum jump from an excited energy level to a lower energy level. Nuclei are no different. A proton or a neutron in an excited nuclear state, such as the one shown in FIGURE 30.14, can undergo a quantum jump to a lower-energy state by emitting a high-energy photon. This is the gamma-decay process.

The spacing between atomic energy levels is only a few eV. Nuclear energy levels, by contrast, are on the order of 1 MeV apart, meaning gamma-ray photons will have energies $E_{gamma} \approx 1$ MeV. Photons with this much energy are quite penetrating and deposit an extremely large amount of energy at the point where they are finally absorbed.

Nuclei left to themselves are usually in their ground states and thus cannot emit gamma-ray photons. However, alpha and beta decay often leave the daughter nucleus in an excited nuclear state, so gamma emission is often found to accompany alpha and beta emission.

Let's look at an example. One of the most important isotopes for medical imaging is ^{99}Tc, an isotope of the element technetium. An excited state of ^{99}Tc is produced in the beta decay of the molybdenum isotope ^{99}Mo:

$$^{99}\text{Mo} \rightarrow {}^{99}\text{Tc}^* + e^- + \text{energy}$$

The asterisk signifies that the technetium nucleus is in an excited state. The excited nucleus then makes a transition to a lower-energy state via the emission of a 140 keV gamma ray:

$$^{99}\text{Tc}^* \rightarrow {}^{99}\text{Tc} + \gamma$$

The full decay process is shown in FIGURE 30.15. The final state of the technetium nucleus is much more stable than the excited state.

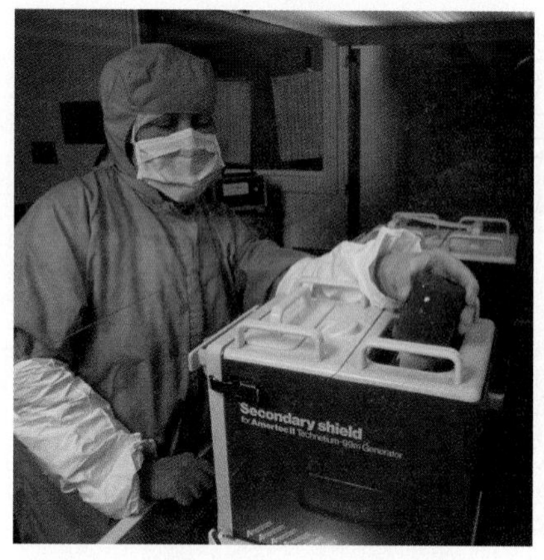
Secondary shield

◀ **Isotopes on demand** BIO Many hospitals keep a radioactive molybdenum ^{99}Mo source on hand. The ^{99}Mo undergoes beta decay, but it's not the energy or the electron released in the decay that the hospital wants—it's the technetium daughter nucleus, ^{99}Tc, which is used in medical imaging procedures. ^{99}Tc decays very rapidly, so hospitals that use it must produce it on site. This happens in a radioactive "cow," a column with chemically bound ^{99}Mo. The column is inside a shielded enclosure, which protects the technician from radiation as she extracts the ^{99}Tc produced in the decay of the ^{99}Mo source.

FIGURE 30.16 ^{235}U decay series.

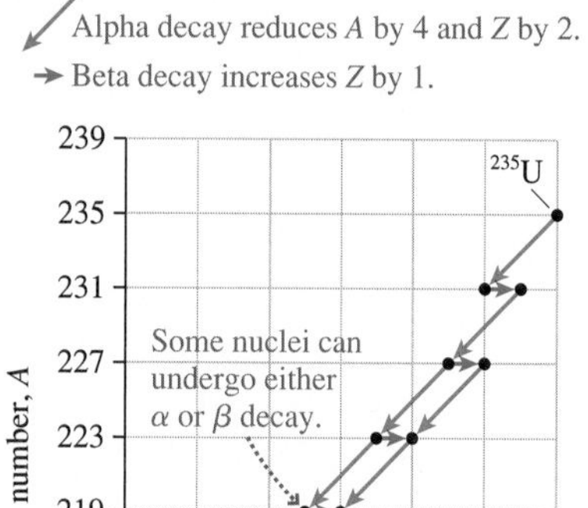
Alpha decay reduces A by 4 and Z by 2.
Beta decay increases Z by 1.
235U
Some nuclei can undergo either α or β decay.
207Pb is stable.
Mass number, A
Atomic number, Z
239
235
231
227
223
219
215
211
207
203
80
82
84
86
88
90
92

Decay Series

A radioactive nucleus decays into a daughter nucleus. In many cases, the daughter nucleus is also radioactive and decays to produce its own daughter nucleus. The process continues until reaching a daughter nucleus that is stable. The sequence of isotopes, starting with the original unstable isotope and ending with the stable isotope, is called a **decay series.**

The elements with $Z > 83$ present in the earth's crust are part of the decay series of a few long-lived isotopes of uranium and thorium. As an example, FIGURE 30.16 shows the decay series of ^{235}U, an isotope of uranium with a 700-million-year half-life. This is a very long time, but it is only about 15% the age of the earth, so most—but not all—of the ^{235}U nuclei present when the earth was formed have now decayed. There are many unstable nuclei in the decay series. Ultimately, all ^{235}U nuclei end as the ^{207}Pb isotope of lead, a stable nucleus.

Notice that some nuclei can decay by either alpha *or* beta decay. Thus there are a variety of paths that a decay can follow, but they all end at the same point.

Nuclear Radiation Is a Form of Ionizing Radiation

The energies of the alpha and beta particles and the gamma-ray photons of nuclear decay are typically in the range 0.1–10 MeV. These energies are much higher than the ionization energies of atoms and molecules, which, as we saw in Chapter 25, are

≈ 10 eV. When the sun shines on your skin, it warms it; the low-energy photons are absorbed, their energy converted to thermal energy. The much higher energies of alpha and beta particles and gamma rays cause them to interact very differently with matter, *ionizing* atoms and *breaking* molecular bonds, leaving long trails of ionized atoms and molecules behind them before finally stopping. A particle with 1 MeV of kinetic energy can ionize $\approx 100{,}000$ atoms or molecules. Alpha and beta particles and gamma rays are, like x rays, examples of ionizing radiation.

Ionization is the basis for the **Geiger counter,** the most well-known detector of nuclear radiation. FIGURE 30.17 shows how a Geiger counter detects the passage of a beta particle. A Geiger counter, like other nuclear radiation detectors, measures only *ionizing radiation.*

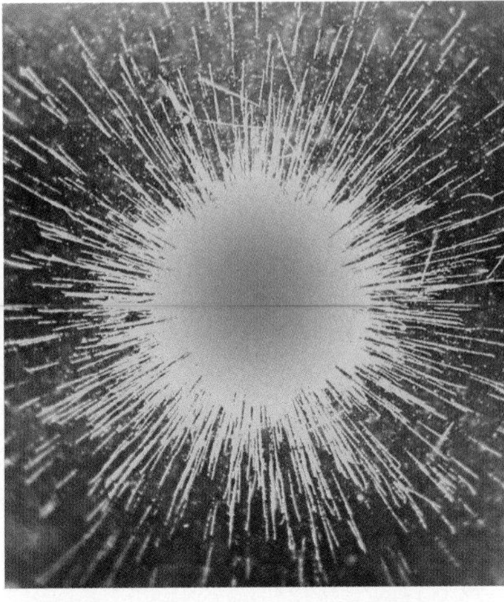

A speck of radium placed on a photographic plate emits alpha particles that leave clearly visible ionization trails.

When ionizing radiation enters the body, it causes damage in two ways. First, the ions drive chemical reactions that wouldn't otherwise occur. These reactions may damage the machinery of cells. Very large doses of ionizing radiation upset the delicate ionic balance that drives cellular transport, and can rapidly lead to cell death. For this reason, large doses of penetrating gamma rays are sometimes used to sterilize medical equipment inside and out.

Second, ionizing radiation can damage DNA molecules by ionizing them and breaking bonds. If the damage is extensive, cellular repair mechanisms will not be able to cope, and the DNA will be permanently damaged, possibly creating a mutation or a tumor. Tissues with rapidly proliferating cells, such as bone marrow, are quite sensitive to ionizing radiation. Those with less-active cell reproduction, such as the nervous system, are much less sensitive.

FIGURE 30.17 The operation of a Geiger counter.

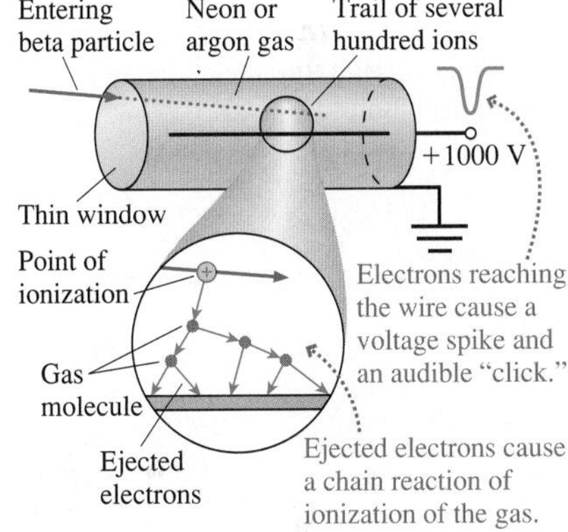

NOTE ► Ionizing radiation causes damage to materials and tissues, but **objects irradiated with alpha, beta, or gamma radiation do not become radioactive.** Ionization drives chemical processes involving the electrons. An object could become radioactive only if its nuclei were somehow changed, and that does not happen. ◄

STOP TO THINK 30.3 The cobalt isotope ^{60}Co ($Z = 27$) decays to the nickel isotope ^{60}Ni ($Z = 28$). The decay process is

A. Alpha decay.
B. Beta-minus decay.
C. Beta-plus decay.
D. Gamma decay.

30.5 Nuclear Decay and Half-Lives

The decay of nuclei is different from other types of decay you are familiar with. A tree branch that falls to the forest floor decays. It darkens, becomes soft, and crumbles. You might be able to tell, just by looking at it, about how long it had been decaying. Nuclear decay is different. The nucleus doesn't "age" in any sense. Instead, a nucleus has a certain probability that, within the next second, it will spontaneously turn into a different nucleus and, in the process, eject an alpha or beta particle or a gamma ray.

We can use an analogy here: If you toss a coin, it always has a 50% probability of showing tails, no matter what previous tosses might have been. You might "expect" heads if you've tossed 10 tails in a row, but the 11th toss still has a 50% chance of coming up tails. Likewise with nuclei. If a nucleus doesn't decay in this second, it is no more or less likely to decay in the next. The nucleus remains just as it was, without any change, until the decay finally occurs. An "old" nucleus is identical to a "young" one.

In fact, the mathematics of radioactive decay is the same as that of tossing coins. Suppose you have a large number N_0 of coins. You toss them all and then keep those that come up heads while setting aside those that come up tails. Probability dictates that about half the coins will show tails and be set aside. Now you repeat the process over and over. With each subsequent toss, about half the coins are set aside—they "decay."

After the first toss, the number of coins you have left is about $(1/2)N_0$ because you set aside about half the coins. After the second toss, when you set aside about half of that half, the number of remaining coins is about $(1/2) \times (1/2)N_0$, or $(1/2)^2N_0 = N_0/4$. Half of these coins will be set aside in the third toss, leaving you with $(1/2)^3N_0$, or 1/8 of what you started with. After m tosses—assuming you started with a very large number of coins—the number of coins left is $N = (1/2)^m N_0$.

Similarly, if you start with N_0 unstable nuclei, after an interval of time we call one *half-life*, you'll find that you have $N = (1/2)N_0$ nuclei remaining. The **half-life** $t_{1/2}$ is the average time required for one-half the nuclei to decay. This process continues, with one-half the remaining nuclei decaying in each successive half-life. The number N nuclei remaining at time t is

$$N = N_0\left(\frac{1}{2}\right)^{t/t_{1/2}} \tag{30.11}$$

Decay of nuclei in a radioactive sample in terms of half-life

Thus $N = N_0/2$ at $t = t_{1/2}$, $N = N_0/4$ at $t = 2t_{1/2}$, $N = N_0/8$ at $t = 3t_{1/2}$, and so on, with the ratio $t/t_{1/2}$ playing the role of the "number of tosses." **No matter how many nuclei there are at any point in time, the number decays by half during the next half-life.**

FIGURE 30.18 Half the nuclei decay during each half-life.

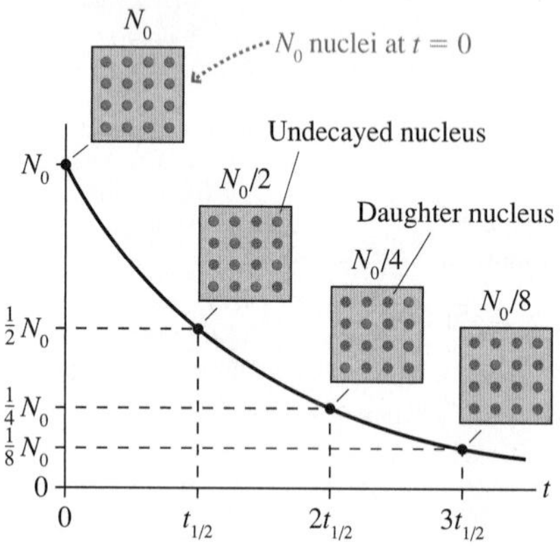

NOTE ▶ Each isotope that is unstable and decays has a characteristic half-life, which can range from a fraction of a second to billions of years. Appendix D provides nuclear data and half-lives for the isotopes referred to in this textbook. ◀

FIGURE 30.18 is a graph of the number of nuclei remaining after time t. This figure conveys two important ideas:

1. Nuclei don't vanish when they decay. The decayed nuclei have merely become some other kind of nuclei.
2. The decay process is random. We can predict that half the nuclei will decay in one half-life, but we can't predict which ones.

The graph has a form we have seen before: exponential decay. Mathematically, the number of nuclei remaining in a radioactive sample at time t is

$$N = N_0 e^{-t/\tau} \tag{30.12}$$

Exponential decay of nuclei in a radioactive sample

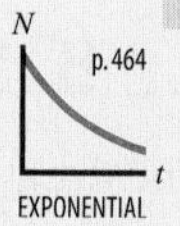

FIGURE 30.19 The number of radioactive atoms decreases exponentially with time.

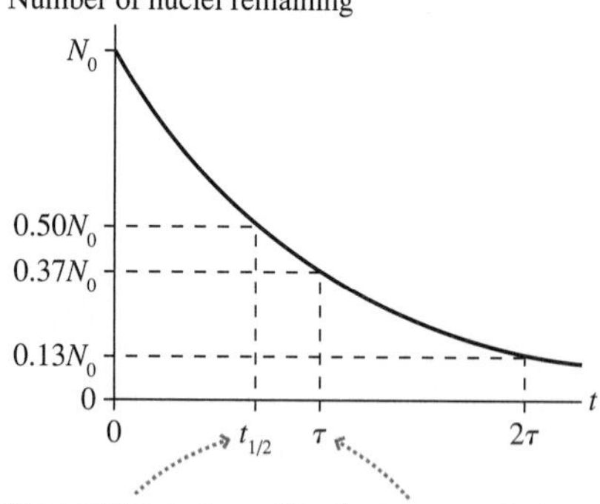
Number of nuclei remaining
N_0
$0.50N_0$
$0.37N_0$
$0.13N_0$
0
0
$t_{1/2}$
τ
2τ
t
The half-life is the time in which half the nuclei decay.
The time constant is the time at which the number of nuclei is e^{-1}, or 37% of the initial number.

where τ is the *time constant* for the decay. The exponential decay of the number of nuclei is analogous to the exponential decay of the capacitor voltage in an RC circuit or the amplitude of a damped harmonic oscillator.

FIGURE 30.19 is a graphical representation of Equation 30.12; it is the same graph as that of Figure 30.18, simply written in a different mathematical form. The number of radioactive nuclei decreases from N_0 at $t = 0$ to $N = N_0e^{-1} = 0.37N_0$ at time $t = \tau$. In practical terms, the number decreases by roughly two-thirds during one time constant.

NOTE ▶ There is no natural "starting time" for an exponential decay; you can choose any instant you wish to be $t = 0$. The number of radioactive nuclei present at that instant is N_0. If at one instant you have 10,000 radioactive nuclei whose time constant is $\tau = 10$ min, you'll have roughly 3700 nuclei 10 min later. The fact that you may have had more than 10,000 nuclei at an earlier time isn't relevant. ◀

We can relate the half-life to the time constant τ because we know, by definition, that $N = N_0/2$ at $t = t_{1/2}$. Thus, according to Equation 30.12,

$$\frac{N_0}{2} = N_0 e^{-t_{1/2}/\tau}$$

The N_0 cancels, and we can then take the natural logarithm of both sides to find

$$\ln\left(\frac{1}{2}\right) = -\ln 2 = -\frac{t_{1/2}}{\tau}$$

With one final rearrangement we have

$$t_{1/2} = \tau \ln 2 = 0.693\tau \qquad (30.13)$$

That is, the half-life of a nuclear decay is 69.3% of the time constant for the decay. Whether we use Equation 30.11 and the half-life or Equation 30.12 and the time constant is a matter of convenience; both equations describe the same decay.

EXAMPLE 30.5 Determining the decay of radioactive iodine

Patients with Graves disease have an overactive thyroid gland. A common treatment uses radioactive iodine, which is taken up by the thyroid. The radiation emitted in its decay will damage the tissues of the gland. A single pill is produced with 4.0×10^{14} atoms of the isotope ^{131}I, which has a half-life of 8.0 days.

a. How many atoms remain 24 hours after the pill's creation, when the pill is delivered to a hospital?
b. Although the iodine in the pill is constantly decaying, it is still usable as long as it contains at least 1.1×10^{14} ^{131}I atoms. What is the maximum delay before the pill is no longer usable?

PREPARE The atoms in the sample undergo exponential decay, decreasing steadily in number.

SOLVE a. The half-life is $t_{1/2} = 8.0 \text{ days} = 192 \text{ hr}$. Using Equation 30.11, we can find the number of atoms remaining after 24 hr have elapsed:

$$N = (4.0 \times 10^{14})\left(\frac{1}{2}\right)^{24/192} = 3.7 \times 10^{14} \text{ atoms}$$

b. The time after which 1.1×10^{14} atoms remain is given by

$$1.1 \times 10^{14} = (4.0 \times 10^{14})\left(\frac{1}{2}\right)^{t/192}$$

To solve for t, we write this as

$$\frac{1.1 \times 10^{14}}{4.0 \times 10^{14}} = \left(\frac{1}{2}\right)^{t/192}$$

or

$$0.275 = \left(\frac{1}{2}\right)^{t/192}$$

Now, we take the natural logarithm of both sides:

$$\ln(0.275) = \ln\left(\left(\frac{1}{2}\right)^{t/192}\right)$$

We can solve for t by using the fact that $\ln(a^x) = x\ln(a)$. This allows us to "pull out" the $t/192$ exponent to find

$$\ln(0.275) = \left(\frac{t}{192}\right)\ln\left(\frac{1}{2}\right)$$

Solving for t, we find that the pill ceases to be useful after

$$t = 192\frac{\ln(0.275)}{\ln(1/2)} = 360 \text{ hr} = 15 \text{ days}$$

ASSESS The weakest usable concentration of iodine is approximately one-fourth of the initial concentration. This means that the decay time should be approximately equal to two half-lives, which is what we found.

Activity

The **activity** R of a radioactive sample is the number of decays per second. Each decay corresponds to an alpha, beta, or gamma emission, so the activity is a measure of how much radiation is being given off. A detailed treatment of the mathematics of decay shows that the activity of a sample of N nuclei having time constant τ (and half-life $t_{1/2}$) is

$$R = \frac{N}{\tau} = \frac{0.693N}{t_{1/2}} \qquad (30.14)$$

A sample with $N = 1.0 \times 10^{10}$ nuclei decaying with time constant $\tau = 100$ s would, at that instant, have activity $R = 1.0 \times 10^8$ decays/s—in each second, 1.0×10^8 atoms would decay.

We see from Equation 30.14 that **activity is inversely proportional to the half-life.** If two samples have the same number of nuclei, the sample with the shorter half-life

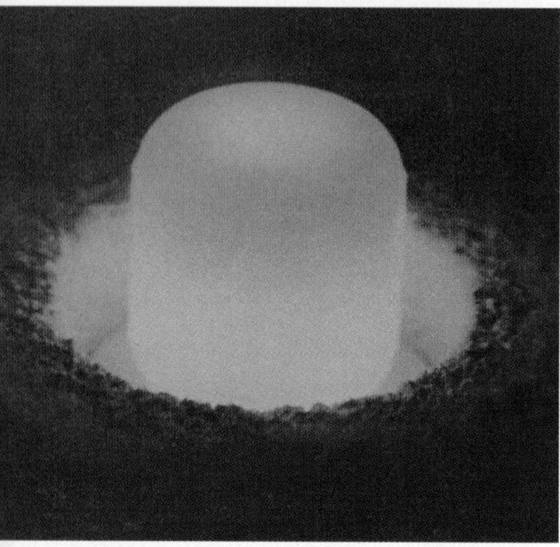

Powered by decay This pellet is made of a short-lived isotope of plutonium. The short half-life means that this isotope has a very large activity. The rate of decay is such that it warms the pellet enough to make it glow. This pellet is intended for a thermoelectric generator, which uses the heat of the pellet to produce electricity. The radioactive decay of short-lived isotopes of plutonium has been used to provide power for spacecraft on voyages far from the sun. For many years, plutonium "batteries" were used to power heart pacemakers as well.

has the larger activity. We can combine Equation 30.14 with Equations 30.11 and 30.12 to obtain an expression for the variation of activity with time:

$$R = \frac{N}{\tau} = \frac{N_0 e^{-t/\tau}}{\tau} = R_0 e^{-t/\tau} = R_0 \left(\frac{1}{2}\right)^{t/t_{1/2}} \qquad (30.15)$$

where $R_0 = N_0/\tau$ is the activity at $t = 0$. This equation has the same form as that for the decay of the sample. **The activity of a sample decreases exponentially** along with the number of remaining nuclei.

The SI unit of activity is the **becquerel,** defined as

$$1 \text{ becquerel} = 1 \text{ Bq} = 1 \text{ decay/s or } 1 \text{ s}^{-1}$$

An older unit of activity, but one that continues in widespread use, is the **curie.** The conversion factor is

$$1 \text{ curie} = 1 \text{ Ci} = 3.7 \times 10^{10} \text{ Bq}$$

1 Ci is a substantial amount of radiation. The radioactive samples used in laboratory experiments are typically $\approx 1\ \mu\text{Ci}$ or, equivalently, $\approx 40{,}000$ Bq. These samples can be handled with only minor precautions. Larger sources of activity require thick shielding and other special precautions to prevent exposure to high levels of radiation.

CONCEPTUAL EXAMPLE 30.6 Relative activities of isotopes in the body

^{40}K ($t_{1/2} = 1.3 \times 10^9$ yr) and ^{14}C ($t_{1/2} = 5.7 \times 10^3$ yr) are two radioactive isotopes found in measurable quantities in your body. Suppose you have 1 mole of each. Which is more radioactive—that is, which has a greater activity?

REASON Equation 30.14 shows that the activity of a sample is proportional to the number of atoms and inversely proportional to the half-life. Because both samples have the same number of atoms, the sample of ^{14}C, with its much shorter half-life, has a much greater activity.

EXAMPLE 30.7 Determining the decay of activity

A ^{60}Co (half-life 5.3 yr) source used to provide gamma rays to irradiate tumors has an activity of 0.43 Ci.

a. How many ^{60}Co atoms are in the source?
b. What will be the activity of the source after 10 yr?

PREPARE The activity of the source depends on the half-life and the number of atoms. If we know the activity and the half-life, we can compute the number of atoms. The number of atoms will undergo exponential decay, and so will the activity.

SOLVE a. Equation 30.14 relates the activity and the number of atoms. Rewriting the equation, we can relate the initial activity and the initial number of atoms as $N_0 = t_{1/2}R_0/0.693$. To use this equation, we need numbers in SI units. In Bq, the initial activity of the source is

$$R_0 = (0.43 \text{ Ci})\left(\frac{3.7 \times 10^{10} \text{ Bq}}{1 \text{ Ci}}\right) = 1.6 \times 10^{10} \text{ Bq}$$

The half-life $t_{1/2}$ in s is

$$t_{1/2} = (5.3 \text{ yr})\left(\frac{3.15 \times 10^7 \text{ s}}{1 \text{ yr}}\right) = 1.7 \times 10^8 \text{ s}$$

Thus the initial number of ^{60}Co atoms in the source is

$$N_0 = \frac{t_{1/2}R_0}{0.693} = \frac{(1.7 \times 10^8 \text{ s})(1.6 \times 10^{10} \text{ Bq})}{0.693}$$
$$= 3.9 \times 10^{18} \text{ atoms}$$

b. The variation of activity with time is given by Equation 30.15. After 10 yr, the activity is

$$R = R_0\left(\frac{1}{2}\right)^{t/t_{1/2}} = (1.6 \times 10^{10} \text{ Bq})\left(\frac{1}{2}\right)^{10/5.3}$$
$$= 4.3 \times 10^9 \text{ Bq} = 0.12 \text{ Ci}$$

ASSESS Although N_0 is a very large number, it is a very small fraction of a mole. The sample contains only about 400 μg of ^{60}Co. In the first part of the question we needed to convert the half-life to s. The second part of the question used a ratio of two times, so we can use any units we like as long as the units of both times are the same.

Radioactive Dating

Many geological and archeological samples can be dated by measuring the decays of naturally occurring radioactive isotopes.

The most well-known dating technique uses the radioactive carbon isotope ^{14}C and is known as carbon dating or **radiocarbon dating.** ^{14}C has a half-life of 5730 years, so any ^{14}C present when the earth formed 4.5 billion years ago has long since decayed away. Nonetheless, ^{14}C is present in atmospheric carbon dioxide because high-energy cosmic rays collide with gas molecules high in the atmosphere to create ^{14}C nuclei from nuclear reactions with nitrogen and oxygen nuclei. The creation and decay of ^{14}C have reached a steady state in which the $^{14}C/^{12}C$ ratio is relatively stable at 1.3×10^{-12}.

All living organisms constantly exchange carbon dioxide with the environment, so the $^{14}C/^{12}C$ ratio in living organisms is also 1.3×10^{-12}. As soon as an organism dies, the ^{14}C in its tissue begins to decay and no new ^{14}C is added. As time goes on, the ^{14}C decays at a well-known rate. Thus, a measurement of the activity of an ancient organic sample permits a determination of the age. "New" samples have a higher fraction of ^{14}C than "old" samples.

The first step in radiocarbon dating is to extract and purify carbon from the sample. The carbon is then placed in a shielded chamber and its activity measured. This activity is then compared to the activity of an identical modern sample. Equation 30.15 relates the activity of a sample at a time t to its initial activity. If we assume that the original activity R_0 was the same as the activity of the modern sample, we can determine the time since the decay began—and thus the age of the sample.

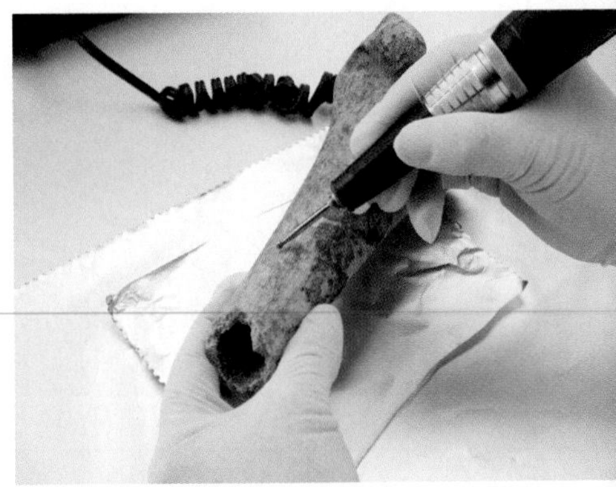

A researcher is extracting a small sample of an ancient bone. By measuring the ratio of carbon isotopes present in the sample she will determine the age of the bone.

EXAMPLE 30.8 **Carbon dating a tooth**

A rear molar from a mammoth skeleton is dated using a measurement of its ^{14}C content. Carbon from the tooth is chemically extracted and formed into benzene. The benzene sample is placed in a shielded chamber. Decays from the sample come at an average rate of 11.5 counts per minute. A modern benzene sample of the exact same size gives 54.9 counts per minute. What is the age of the skeleton?

PREPARE We can assume that, thousands of years ago, the sample had an initial activity of 54.9 counts per minute—the activity of a modern sample. The present activity is lower due to the decay of the ^{14}C since the death of the mammoth.

SOLVE Equation 30.15 gives the decrease of the activity as a function of time as $R = R_0(1/2)^{t/t_{1/2}}$. The current activity is $R = 11.5$ counts per minute, and we assume that the initial activity was $R_0 = 54.9$. t is the time since the mammoth stopped growing—the age of the skeleton. We solve for t by rearranging terms and computing a natural logarithm, as in Example 30.5:

$$\frac{R}{R_0} = \left(\frac{1}{2}\right)^{t/t_{1/2}}$$

$$\ln\left(\frac{R}{R_0}\right) = \left(\frac{t}{t_{1/2}}\right)\ln\left(\frac{1}{2}\right)$$

We then solve for the time t:

$$t = \frac{t_{1/2}}{\ln(1/2)}\ln\left(\frac{R}{R_0}\right) = \frac{5730 \text{ yr}}{\ln(1/2)}\ln\left(\frac{11.5}{54.9}\right) = 12{,}900 \text{ yr}$$

ASSESS The final time is in years, the same unit we used for the half-life. This is a realistic example of how such radiocarbon dating is done; the numbers and details used in this example come from an actual experimental measurement.

CONCEPTUAL EXAMPLE 30.9 **Source contamination**

One possible problem with carbon dating is contamination with modern carbon sources. Suppose an archaeologist has unearthed and carbon-dated a fragment of wood that has absorbed carbon of recent vintage from organic molecules in groundwater. Does he underestimate or overestimate the age of the wood?

REASON Because the wood has absorbed modern carbon, it will have more ^{14}C than it would had it decayed undisturbed. The present activity is higher than it would otherwise be. This will lead to an underestimate of the age of the wood.

▶ **Responsible dating** BIO Measuring the activity of the carbon in a sample to determine the fraction of ^{14}C can require a significant amount of organic material—perhaps 25 g. For an artifact of great historical importance, such as this parchment fragment from the Dead Sea scrolls, this would be unacceptable. Instead, dates are obtained by using a mass spectrometer to directly measure the ratio of carbon isotopes, from which the age can be determined. This can be done with excellent accuracy on as little as 0.1 g of material.

Carbon dating can be used to date skeletons, wood, paper, fur, food material, and anything else made of organic matter. It is quite accurate for ages to about 15,000 years, about three half-lives. Beyond that, the difficulty of measuring the small remaining fraction of ^{14}C and some uncertainties about the cosmic ray flux in the past combine to decrease the accuracy. Even so, items are dated to about 50,000 years with a fair degree of reliability.

Isotopes with longer half-lives are used to date geological samples. Potassium-argon dating, using ^{40}K with a half-life of 1.25 billion years, is especially useful for dating rocks of volcanic origin.

STOP TO THINK 30.4 A sample starts with 1000 radioactive atoms. How many half-lives have elapsed when 750 atoms have decayed?

A. 0.25 B. 1.5 C. 2.0 D. 2.5

30.6 Medical Applications of Nuclear Physics

Nuclear physics has brought both peril and promise to society. Radioactivity can cause tumors. At the same time, radiation can be used to diagnose and cure some cancers. This section is a brief survey of medical applications of nuclear physics.

Radiation Dose

Ionizing radiation damages cells of the body, but it also damages bacteria and other pathogens. This gamma source is used for sterilizing medical equipment. The blue glow is due to the ionization of the air around the source.

Nuclear radiation disrupts a cell's machinery by altering and damaging biological molecules, as we saw in Section 30.4. The biological effects of radiation depend on two factors. The first is the physical factor of how much energy is absorbed by the body. The second is the biological factor of how tissue reacts to different forms of radiation.

Suppose a beta particle travels through tissue, losing kinetic energy as it ionizes atoms it passes. The energy lost by the beta particle is a good measure of the number of ions produced and thus the amount of damage done. In a certain volume of tissue, more ionization means more damage. For this reason, we define the radiation **dose** as the energy from ionizing radiation absorbed by 1 kg of tissue. The SI unit for the dose is the **gray,** abbreviated Gy. The Gy is defined as

$$1 \text{ Gy} = 1.00 \text{ J/kg of absorbed energy}$$

The number of Gy depends only on the energy absorbed, not on the type of radiation or on what the absorbing material is. Another common unit for dose is the *rad;* 1 rad = 0.01 Gy.

A 1 Gy dose of gamma rays and a 1 Gy dose of alpha particles have different biological consequences. To account for such differences, the **relative biological effectiveness** (RBE) is defined as the biological effect of a given dose relative to the biological effect of an equal dose of x rays. Table 30.4 shows the relative biological effectiveness of different forms of radiation. Larger values correspond to larger biological effects.

TABLE 30.4 Relative biological effectiveness of radiation

Radiation type	RBE
X rays	1
Gamma rays	1
Beta particles	1
Protons	5
Neutrons	5–20
Alpha particles	20

The radiation **dose equivalent** is the product of the energy dose in Gy and the relative biological effectiveness. Dose equivalent is measured in **sieverts,** abbreviated Sv. To be precise,

$$\text{dose equivalent in Sv} = \text{dose in Gy} \times \text{RBE}$$

One Sv of radiation produces the same biological damage regardless of the type of radiation. Another common unit of dose equivalent (also called biologically equivalent dose) is the *rem;* 1 rem = 0.01 Sv.

NOTE ▶ In practice, the term "dose" is often used for both dose and dose equivalent. Use the units as a guide. If the unit is Sv or rem, it is a dose equivalent; if Gy or rad, a dose. ◀

EXAMPLE 30.10 Finding energy deposited in radiation exposure

A 75 kg patient is given a bone scan. A phosphorus compound containing the gamma-emitter ^{99}Tc is injected into the patient. It is taken up by the bones, and the emitted gamma rays are measured. The procedure exposes the patient to 3.6 mSv (360 mrem) of radiation. What is the total energy deposited in the patient's body, in J and in eV?

PREPARE The exposure is given in Sv, so it is a dose equivalent, a combination of deposited energy and biological effectiveness. The RBE for gamma rays is 1. Gamma rays are penetrating, and the source is distributed throughout the body, so this is a whole-body exposure. Each kg of the patient's body will receive approximately the same energy.

SOLVE The dose in Gy is the dose equivalent in Sv divided by the RBE. In this case, because RBE = 1, the dose in Gy is numerically equal to the equivalent dose in Sv. The dose is thus 3.6 mGy = 3.6×10^{-3} J/kg. The radiation energy absorbed in the patient's body is

$$\text{absorbed energy} = (3.6 \times 10^{-3}\ \text{J/kg})(75\ \text{kg}) = 0.27\ \text{J}$$

In eV, this is

$$\text{absorbed energy} = (0.27\ \text{J})(1\ \text{eV}/1.6 \times 10^{-19}\ \text{J}) = 1.7 \times 10^{18}\ \text{eV}$$

ASSESS The total energy deposited, 0.27 J, is quite small; there will be negligible heating of tissue. But radiation produces its effects in other ways, as we have seen. Because it takes only $\approx$10 eV to ionize an atom, this dose is enough energy to ionize over 10^{17} atoms, meaning it can cause significant disruption to the cells of the body.

TABLE 30.5 Radiation exposure

Radiation source	Typical exposure (mSv)
PET scan	7.0
Natural background (1 year)	3.0
Mammogram	0.70
Chest x ray	0.30
Transatlantic airplane flight	0.050
Dental x ray	0.030

The question inevitably arises: What is a safe dose? Unfortunately, there is no simple or clear definition of a safe dose. A prudent policy is to avoid unnecessary exposure, and to weigh the significance of an exposure in relation to the *natural background.* We are all exposed to continual radiation from cosmic rays and from naturally occurring radioactive atoms in the ground, the atmosphere, and even the food we eat. This background averages about 3 mSv (300 mrem) per year, although there are wide regional variations depending on the soil type and the elevation. (Higher elevations have less atmospheric shielding, and thus have a larger exposure to cosmic rays. On the moon, with no protective atmosphere, the yearly dose would be 50 mSv.)

Table 30.5 lists the expected exposure from several different sources. A dental x ray subjects a person to approximately 1% of the yearly natural background that he or she would normally receive and is likely not a cause for significant worry. Mammograms involve a much larger dose, concentrated in a small region of the body. A nuclear medicine procedure, like a PET scan (which is discussed below), may involve an exposure that is much larger than the typical yearly background dose. This significant dose must be weighed against the medical benefits of the procedure.

FIGURE 30.20 The use of gamma rays to treat a tumor in the brain.

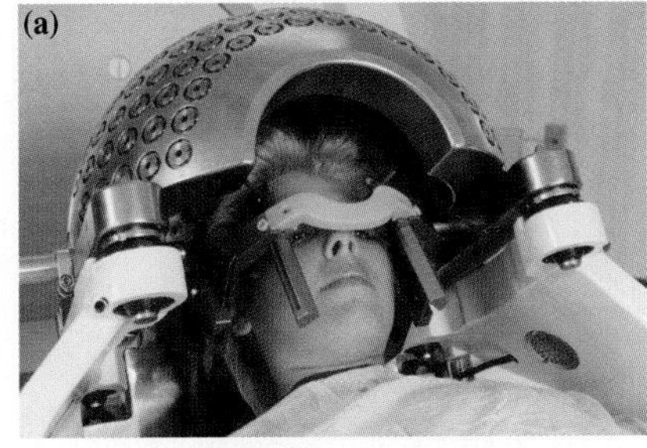
(a)

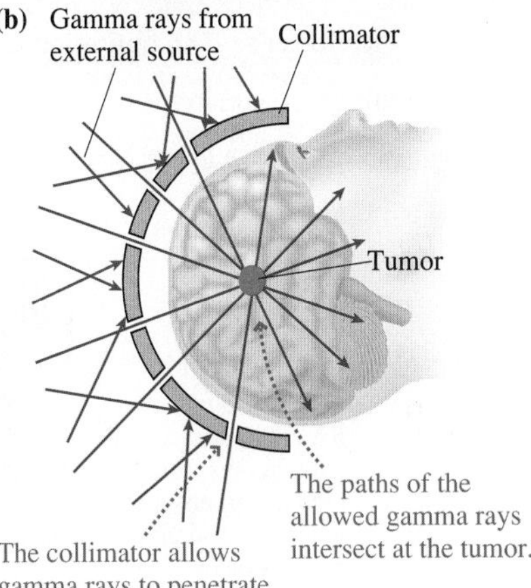
(b) Gamma rays from external source
Collimator
Tumor
The paths of the allowed gamma rays intersect at the tumor.
The collimator allows gamma rays to penetrate only along certain lines.

Nuclear Medicine

The tissues in the body most susceptible to radiation are those that are rapidly proliferating—including tumors. The goal in *radiation therapy* is to apply a large enough dose of radiation to destroy or shrink a tumor while producing minimal damage to surrounding healthy tissue.

In **FIGURE 30.20a**, a patient with a brain tumor is fitted with a metal *collimator* that absorbs gamma rays except for those traveling along desired paths. The collimator is fashioned so that gamma rays from an external source will be concentrated on the tumor, as shown in **FIGURE 30.20b**. Because the rapidly dividing cells of a tumor are much more sensitive to radiation than the tissues of the brain, and because surgical options carry risks of significant complications, radiation is a common means of treating tumors of the brain.

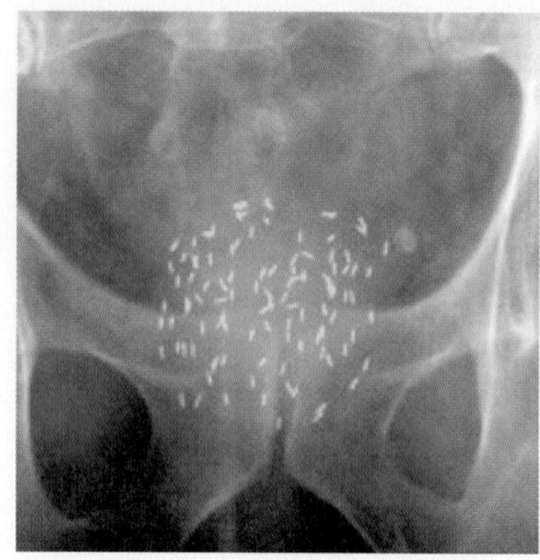

Metal "seeds" containing radioactive ^{125}I in the prostate gland can shrink a tumor.

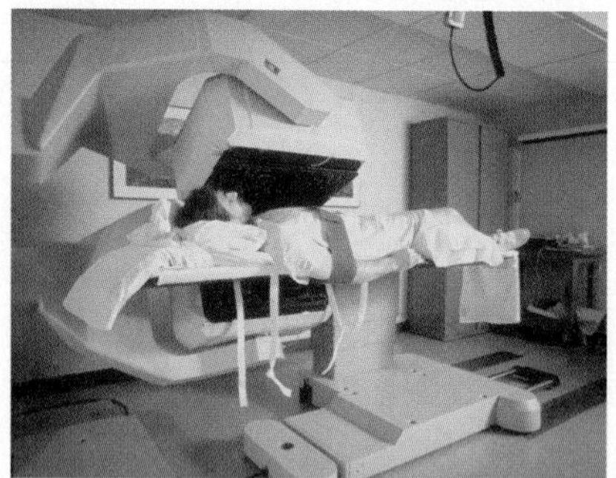

The detector is measuring gamma radiation emitted by isotopes taken up by tissues in the woman's head and neck.

Other tumors are treated by surgically implanting radioactive "seeds" within the tumor. One common type of seed contains ^{125}I, which undergoes a nuclear decay followed by emission of a 27.5 keV photon. The photon has a very short range so that it will damage only tissue close to the seed. Careful placement of the seeds permits a significant dose to a tumor with only minimal exposure to surrounding tissue.

Some tissues in the body will preferentially take up certain isotopes, allowing for treatment by isotope ingestion. A common treatment for hyperthyroidism, in which the thyroid gland is overactive, is to damage the gland with the isotope ^{131}I, a beta-emitter with a half-life of 8.0 days. A patient is given a tablet containing ^{131}I. The iodine in the blood is taken up and retained by the thyroid gland, resulting in a reduction of the gland's activity with minimal disruption of surrounding tissue.

Nuclear Imaging

X rays from an external source may be used to make an image of the body, as described in Chapter 28. *Nuclear imaging* uses an internal source—radiation from isotopes in the body—to produce an image of tissues in the body. The bone-scan image that opened the chapter is an example of nuclear imaging.

There is a key difference between x rays and nuclear imaging procedures. **An x ray is an image of anatomical structure;** it is excellent for identifying structural problems like broken bones. **Nuclear imaging creates an image of the biological activity of tissues in the body.** For example, nuclear imaging can detect reduced metabolic activity of brain tissue after a stroke.

Let's look at an example that illustrates the difference between a conventional x-ray image or CAT scan and an image made with a nuclear imaging technique. Suppose a doctor suspects a patient has cancerous tissue in the bones. An x ray does not show anything out of the ordinary; the tumors may be too small or may appear similar to normal bone. The doctor then orders a scan with a **gamma camera,** a device that can measure and produce an image from gamma rays emitted within the body.

The patient is given a dose of a phosphorus compound labeled with the gamma-emitter ^{99}Tc. This compound is taken up and retained in bone tissue where active growth is occurring. The ^{99}Tc will be concentrated in the bones where there has been recent injury or inflammation—or where a tumor is growing. The patient is then scanned with a gamma camera. FIGURE 30.21a shows how the gamma camera can pinpoint the location of the gamma-emitting isotopes in the body and produce an image that reveals their location and intensity. A typical image is shown in FIGURE 30.21b. The bright spots show high concentrations of ^{99}Tc, revealing areas of tumor growth. The tumors may be too small to show up on an x ray, but their activity is easily detected with the gamma camera. With such early detection, the patient's chance of a cure is greatly improved.

FIGURE 30.21 The operation of a gamma camera.

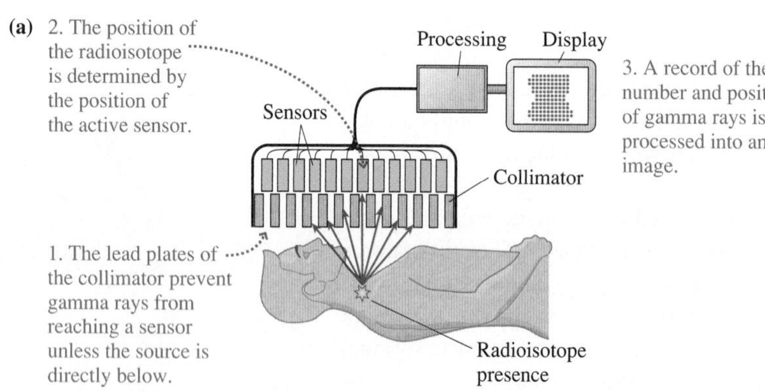

(a)
2. The position of the radioisotope is determined by the position of the active sensor.
Processing
Display
Sensors
3. A record of the number and position of gamma rays is processed into an image.
Collimator
1. The lead plates of the collimator prevent gamma rays from reaching a sensor unless the source is directly below.
Radioisotope presence

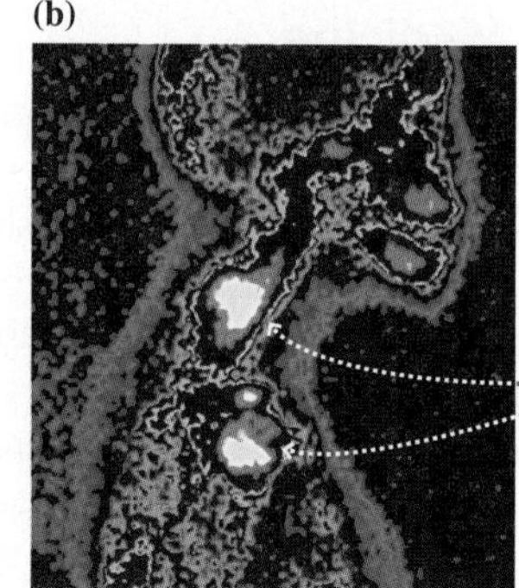

CONCEPTUAL EXAMPLE 30.11 **Using radiation to diagnose disease**

A patient suspected of having kidney disease is injected with a solution containing molecules that are taken up by healthy kidney tissue. The molecules have been "tagged" with radioactive ^{99}Tc. A gamma camera scan of the patient's abdomen gives the image in FIGURE 30.22. In this image, blue corresponds to the areas of highest activity. Which of the patient's kidneys has reduced function?

FIGURE 30.22 A gamma scan of a patient's kidneys.

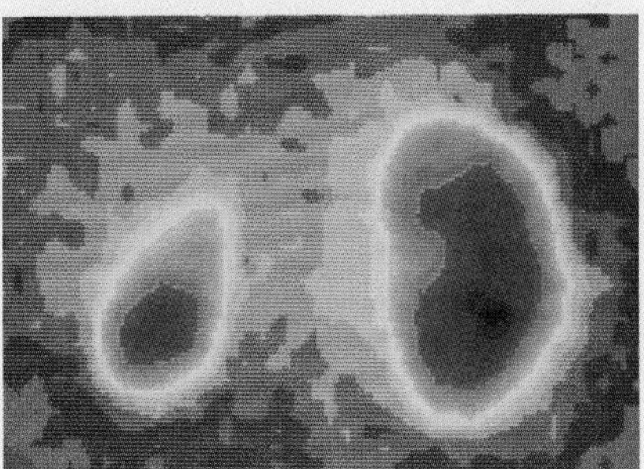

REASON Healthy tissue should show up in blue on the scan because healthy tissue will absorb molecules with the ^{99}Tc attached and will thus emit gamma rays. The kidney imaged on the right shows normal activity throughout; the kidney imaged on the left appears smaller, so it has a smaller volume of healthy tissue. The patient is ill; the problem is with the kidney imaged on the left.

ASSESS Depending on the isotope and how it is taken up by the body, either healthy tissue or damaged tissue could show up on a gamma camera scan.

FIGURE 30.23 Positron-emission tomography.

(a) When the electron and positron meet . . .

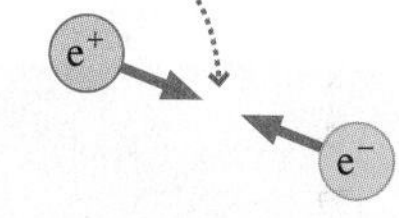

. . . the energy equivalent of their mass is converted into two gamma rays headed in opposite directions.

(b) Coincident detection of two gamma rays means that the positron source is along this line.

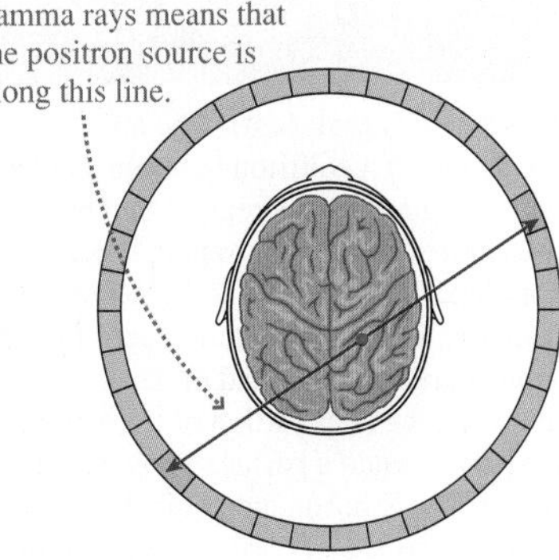

Positron-Emission Tomography

We have seen that a small number of radioactive isotopes decay by the emission of a positron. Such isotopes can be used for an imaging technique known as *positron-emission tomography,* or *PET.* PET is particularly important for imaging the brain.

The imaging process relies on the mass–energy conversion resulting from the combination of an electron and a positron. Suppose an electron and a positron are at rest, or nearly so; the combined momentum is nearly zero. The electron and positron have opposite charges and so will attract each other. When they meet, as we saw in Chapter 27, they completely annihilate—but energy and momentum are still conserved. The conservation of energy means that the annihilation will produce one or more high-energy photons—gamma rays. We learned in Chapter 27 that photons have momentum, so the annihilation can't produce a single photon because that photon would leave the scene of the annihilation with momentum. Instead the most likely result is a pair of photons directed exactly opposite each other, as shown in FIGURE 30.23a.

Most PET scans use the fluorine isotope ^{18}F, which emits a positron as it undergoes beta-plus decay to ^{18}O with a half-life of 110 minutes. ^{18}F is used to create an analog of glucose called fluorine-18 fluoro-deoxy-glucose (F-18 FDG). This compound is taken up by tissues in the brain. Areas that are more active are using more glucose, so the F-18 FDG is concentrated in active brain regions. When a fluorine atom in the F-18 FDG decays, the emitted positron immediately collides with a regular electron. The two annihilate to produce two gamma rays that travel out of the brain in opposite directions, as shown in Figure 30.23a.

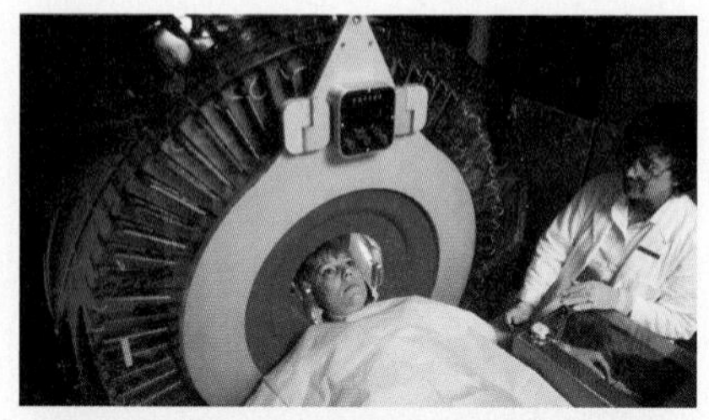

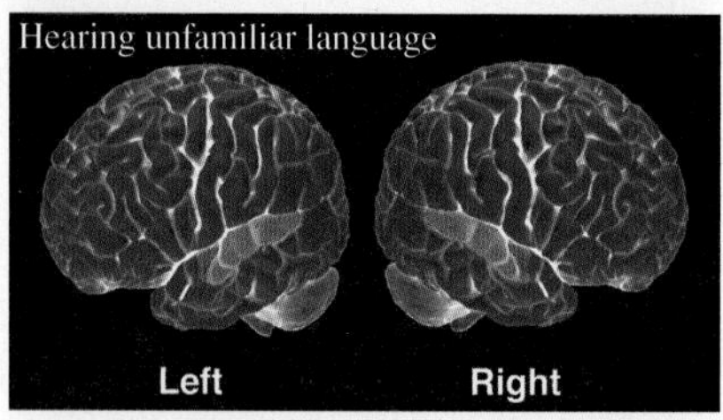

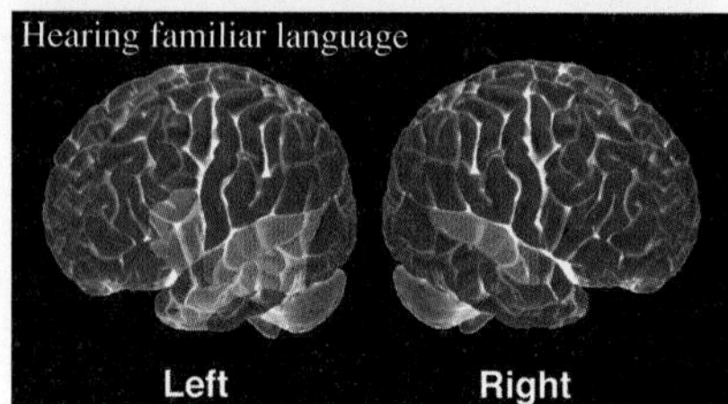

◀ **This is your brain on PET** BIO The woman in the top photo is undergoing a PET scan not to diagnose disease but to probe the workings of the brain. While undergoing a PET scan, a subject is asked to perform different mental tasks. The lower panels show functional images from a PET scan super-imposed on anatomical images from a CAT scan. The subject first listened to speech in an unfamiliar language; the active areas of the brain were those responsible for hearing. Next, she listened to speech in a familiar language, resulting in activity in the parts of the brain responsible for speech and comprehension.

FIGURE 30.23b shows a patient's head surrounded by a ring of gamma-ray detectors. Because the gamma rays from the positron's annihilation are emitted back to back, simultaneous detection of two gamma rays on opposite sides of the subject indicates that the annihilation occurred somewhere along the line between those detectors. Recording many such pairs of gamma rays shows with great accuracy where the decays are occurring. A full scan will show more activity in regions of the brain where metabolic activity is enhanced, less activity in regions where metabolic activity is depressed. An analysis of these scans can provide a conclusive diagnosis of stroke, injury, or Alzheimer's disease.

STOP TO THINK 30.5 A patient ingests a radioactive isotope to treat a tumor. The isotope provides a dose of 0.10 Gy. Which type of radiation will give the highest dose equivalent in Sv?

A. Alpha particles B. Beta particles C. Gamma rays

30.7 The Ultimate Building Blocks of Matter

19.5 

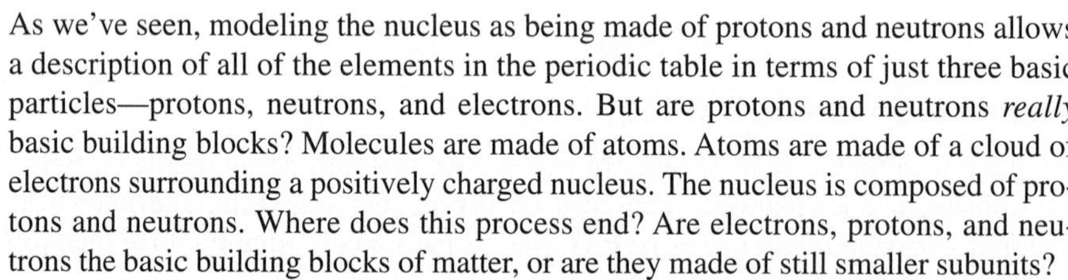

As we've seen, modeling the nucleus as being made of protons and neutrons allows a description of all of the elements in the periodic table in terms of just three basic particles—protons, neutrons, and electrons. But are protons and neutrons *really* basic building blocks? Molecules are made of atoms. Atoms are made of a cloud of electrons surrounding a positively charged nucleus. The nucleus is composed of protons and neutrons. Where does this process end? Are electrons, protons, and neutrons the basic building blocks of matter, or are they made of still smaller subunits?

This question takes us into the domain of what is known as **particle physics**—the branch of physics that deals with the basic constituents of matter and the relationships among them. Particle physics starts with the constituents of the atom, the proton, neutron, and electron, but there are many other particles below the scale of the atom. We call these particles **subatomic particles.**

Antiparticles

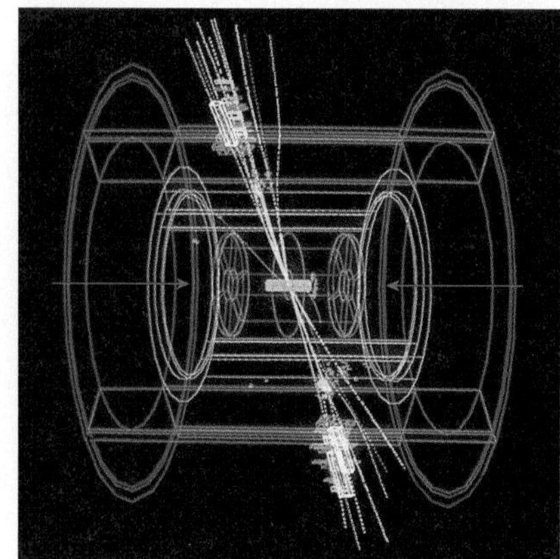

Subatomic crash tests The above picture is the record of a collision between an electron and a positron (paths represented by the green arrows) brought to high speeds in a collider. The particles annihilated in the center of a detector that measured the paths of the particles produced in the collision. In this case, the annihilation of the electron and positron created a particle known as a Z boson. The Z boson then quickly decayed into the two jets of particles seen coming out of the detector.

We've described the positron as the *antiparticle* to the electron. In what sense is a positron an *anti*electron? As we've seen, when a positron and an electron meet, they annihilate each other, turning the energy equivalent of their masses into the pure energy of two photons. Mass disappears and light appears in one of the most spectacular confirmations of Einstein's relativity.

Every subatomic particle that has been discovered has an antiparticle twin that has the same mass and the same spin but opposite charge. In addition to positrons, there are antiprotons (with $q = -e$), antineutrons (also neutral, but not the same as regular neutrons), and antimatter versions of all the various subatomic particles we will see. The notation to represent an antiparticle is a bar over the top of the symbol. A proton is represented as p, an antiproton as $\bar{\text{p}}$.

Antiparticles provide interesting opportunities for creating "exotic" subatomic particles. When a particle meets its associated antiparticle, the two annihilate, leaving nothing but their energy behind. This energy must go somewhere. Sometimes it is emitted as gamma-ray photons, but this energy can also be used to create other particles.

The major tool for creating and studying subatomic particles is the *particle collider.* These machines use electric and magnetic fields to accelerate particles and their antiparticles, such as e and $\bar{e}$, or p and $\bar{p}$, to speeds very close to the speed of light. These particles then collide head-on. As they collide and annihilate, their mass-energy and kinetic energy combine to produce exotic particles that are not part of ordinary matter. These particles come in a dizzying variety—pions, kaons, lambda particles, sigma particles, and dozens of others—each with its own antiparticle. Most live no more than a trillionth of a second.

EXAMPLE 30.12 **Determining a possible outcome of a proton-antiproton collision**

When a proton and an antiproton annihilate, the resulting energy can be used to create new particles. One possibility is the creation of electrically neutral particles called *neutral pions.* A neutral pion has a rest mass of 135 MeV/c^2. How many neutral pions could be produced in the annihilation of a proton and an antiproton? Assume the proton and antiproton are moving very slowly as they collide.

PREPARE The mass of a proton is given in Table 30.2 as 938 MeV/c^2. The mass of an antiproton is the same. Because the proton and antiproton are moving slowly, with essentially no kinetic energy, the total energy available for creating new particles is the energy equivalent of the masses of the proton and the antiproton.

SOLVE The total energy from the annihilation of a proton and an antiproton is the energy equivalent of their masses:

$$E = (m_{\text{proton}} + m_{\text{antiproton}})c^2 = (938\text{ MeV}/c^2 + 938\text{ MeV}/c^2)c^2 = 1876\text{ MeV}$$

It takes 135 MeV to create a neutral pion. The ratio

$$\frac{\text{energy available}}{\text{energy required to create a pion}} = \frac{1876\text{ MeV}}{135\text{ MeV}} = 13.9$$

tells us that we have enough energy to produce 13 neutral pions from this process, but not quite enough to produce 14.

ASSESS Because the mass of a pion is much less than that of a proton or an antiproton, the annihilation of a proton and antiproton can produce many more pions than the number of particles at the start. Though the production of 13 neutral pions is a possible outcome of a proton-antiproton interaction, it is not a likely one. In addition to the conservation of energy, there are many other physical laws that determine what types of particles, and in what quantities, are likely to be produced.

A big detector for a small particle The rubber raft in the photo is floating inside a particle detector designed to measure neutrinos. Neutrinos are so weakly interacting that a neutrino produced in a nuclear reaction in the center of the sun will likely pass through the entire mass of the sun and escape. Of course, the neutrino's weakly interacting nature also means that it is likely to pass right through a detector. The Super Kamiokande experiment in Japan monitors interactions in an enormous volume of water in order to spot a very small number of neutrino interactions.

Neutrinos

The most abundant particle in the universe is not the electron, proton, or neutron; it is a particle you may have never heard of. Vast numbers of these particles pass through your body every day, only in extremely rare cases leaving any trace of their passage. This elusive particle is the *neutrino,* a neutral, nearly massless particle that interacts only weakly with matter.

Some of the first studies of beta-minus decay, in the 1930s, found that neither energy nor momentum seemed to be conserved in the process. The physicist Wolfgang Pauli correctly suggested that, in addition to the electron, the beta-decay process emits a particle that, at that time, had not been detected. This new particle had to be electrically neutral, in order to conserve charge, and it had to be much less massive than an electron. It was later named the **neutrino,** meaning "little neutral one."

The neutrino is represented by the symbol ν, a lowercase Greek nu. There are three types of neutrinos. The neutrino involved in beta decay is the *electron neutrino* ν_e; it shows up in processes involving electrons and positrons. The electron neutrino of course has an antiparticle, the antineutrino $\overline{\nu}_e$. The full descriptions of beta-minus and beta-plus decays, including the neutrinos, are

$$n \rightarrow p^+ + e^- + \overline{\nu}_e$$

$$p^+ \rightarrow n + e^+ + \nu_e$$

FIGURE 30.24 An accurate picture of beta decay must include the antineutrino.

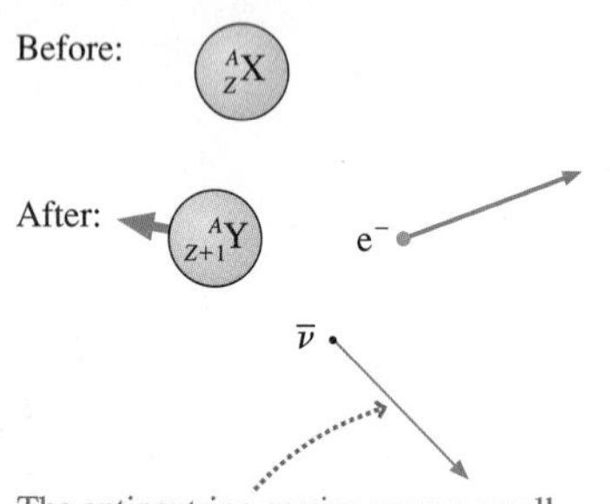

Before:
After:
The antineutrino carries away a small amount of energy and momentum.

FIGURE 30.24 shows that the electron and antineutrino (or positron and neutrino) share the energy released in the decay.

NOTE ► If we are concerned with the accurate balance of energy and momentum in a beta decay, we must include the antineutrino, but we can generally ignore the presence of this weakly interacting particle for the problems we solve in this chapter. ◄

It was initially thought that the neutrino, like the photon, was a *massless* particle. However, experiments in the last few years have shown that the neutrino mass, while tiny, is not zero. The best current evidence suggests a mass about one-millionth the mass of an electron. Experiments now under way will determine a more precise value. Because the neutrino is so abundant in the universe, this small mass is of great cosmological significance.

Quarks

The process of beta decay, in which a neutron can change into a proton, and vice versa, gives a hint that the neutron and the proton are *not* fundamental units but are made of smaller subunits.

There is another reason to imagine such subunits: the existence of dozens of subatomic particles—muons, pions, kaons, omega particles, and so on. Just as the periodic table explains the many different atomic elements in terms of three basic particles, perhaps it is possible to do something similar for this "subatomic zoo." In the 1960s, the American physicist Murray Gell-Mann and the Japanese physicist Kazuhiko Nishijima independently postulated the existence of particles that could be combined to make protons, neutrons, and other subatomic particles. Their ideas were soon confirmed in experiments showing that neutrons and protons seem to have internal structure—they appear to be made of three distinct subparticles.

We now understand protons and neutrons to be composed of smaller charged particles whimsically named **quarks.** The quarks that form protons and neutrons are called **up quarks** and **down quarks,** symbolized as u and d, respectively. The nature of these quarks and the composition of the neutron and the proton are shown in FIGURE 30.25.

FIGURE 30.25 The quark content of the proton and neutron.

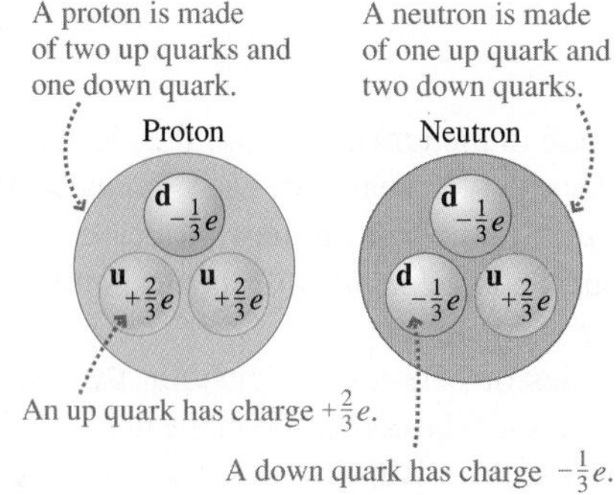

A proton is made of two up quarks and one down quark.
A neutron is made of one up quark and two down quarks.

An up quark has charge $+\frac{2}{3}e$.
A down quark has charge $-\frac{1}{3}e$.

NOTE ► It seems surprising that the charges of quarks are *fractions* of e. Don't charges have to be integer multiples of e? It's true that atoms, molecules, and all macroscopic matter must have $q = Ne$ because these entities are constructed from electrons and protons. But no law of nature prevents other types of matter from having other amounts of charge. ◄

A neutron and a proton differ by one quark. Beta decay can now be understood as a process in which a down quark changes to an up quark, or vice versa. Beta-minus decay of a neutron can be written as

$$\mathrm{d} \rightarrow \mathrm{u} + e^- + \bar{\nu}_\mathrm{e}$$

The existence of quarks thus provides an explanation of how a neutron can turn into a proton.

CONCEPTUAL EXAMPLE 30.13 Quarks and beta-plus decay

What is the quark description of beta-plus decay?

REASON In beta-plus decay, a proton turns into a neutron, with the emission of a positron and an electron neutrino. To turn a proton into a neutron requires the conversion of an up quark into a down quark; the total reaction is thus

$$\mathrm{u} \rightarrow \mathrm{d} + e^+ + \nu_\mathrm{e}$$

The quark model is very successful, but no one has ever seen a solitary quark. They don't exist alone, only inside other particles. In fact, the theory that describes quarks and their interactions specifies that we will *never* see a single quark all by itself. The force between quarks increases as they move apart, so an infinite amount of energy would be required to separate two quarks. Quarks must always be bound with other quarks, a principle known as **quark confinement.**

Fundamental Particles

Our current understanding of the truly *fundamental* particles—the ones that cannot be broken down into smaller subunits—is that they come in two basic types: **leptons** (particles like the electron and the neutrino) and quarks (which combine to form particles like the proton and the neutron). The leptons and quarks are described in Table 30.6. A few points are worthy of note:

- Each particle has an associated antiparticle.
- There are three *families* of leptons. The first is the electron and its associated neutrino, and their antiparticles. The other families are based on the muon and the tau, heavier siblings to the electron. Only the electron and positron are stable particles.
- There are also three families of quarks. The first is the up-down family that makes all "normal" matter. The other families are pairs of heavier quarks that form more exotic particles.

TABLE 30.6 Leptons and quarks

Leptons		Antileptons	
Electron	e^-	Positron	e^+
Electron neutrino	ν_e	Electron antineutrino	$\bar{\nu}_\mathrm{e}$
Muon	μ^-	Antimuon	μ^+
Muon neutrino	ν_μ	Muon antineutrino	$\bar{\nu}_\mu$
Tau	τ^-	Antitau	τ^+
Tau neutrino	ν_τ	Tau antineutrino	$\bar{\nu}_\tau$
Quarks		**Antiquarks**	
Up	u	Antiup	$\bar{\mathrm{u}}$
Down	d	Antidown	$\bar{\mathrm{d}}$
Strange	s	Antistrange	$\bar{\mathrm{s}}$
Charm	c	Anticharm	$\bar{\mathrm{c}}$
Bottom	b	Antibottom	$\bar{\mathrm{b}}$
Top	t	Antitop	$\bar{\mathrm{t}}$

As far as we know, this is where the trail ends. Matter is made of molecules; molecules of atoms; atoms of protons, neutrons, and electrons; protons and neutrons of quarks. Quarks and electrons seem to be truly fundamental. But scientists of the early 20th century thought they were at a stopping point as well—they thought that they knew all of the physics that there was to know. As we've seen over the past few chapters, this was far from true. New tools such as the next generation of particle colliders will certainly provide new discoveries and new surprises.

The early chapters of this book, in which you learned about forces and motion, had very obvious applications to things in your daily life. But in these past few chapters we see that even modern discoveries—discoveries such as antimatter that may seem like science fiction—can be put to very practical use. As we come to the close of this book, we hope that you have gained an appreciation not only for what physics tells us about the world, but also for the wide range of problems it can be used to solve.

INTEGRATED EXAMPLE 30.14 Čerenkov radiation

We know that nothing can travel faster than c, the speed of light in a vacuum. But we also know that light itself goes slower as it travels through a medium. Consequently particles moving at high speeds through a medium can be traveling faster than a light wave in the medium.

FIGURE 30.26 is a photo of the core of a nuclear reactor. The core is immersed in water, which carries away the thermal energy produced in the reactor. As high-energy electrons emerge with a speed very close to c from nuclear reactions in the core, they move through the water faster than the speed of light in water.

FIGURE 30.26 Čerenkov light illuminates the water surrounding a nuclear reactor core.

Recall that a shock wave—a sonic boom—is produced when an airplane moves faster than the speed of sound. An electron moving faster than the speed of light in water makes an electromagnetic shock wave—a light pulse analogous to a sonic boom. This particular type of electromagnetic radiation is known as **Čerenkov radiation,** or Čerenkov light, and it is responsible for the blue glow around the reactor core in Figure 30.26.

One source of high-speed electrons is the beta-minus decay of ^{133}Xe, a radioactive isotope of xenon that is produced in the fission of uranium and accumulates in the reactor core.

a. What is the daughter nucleus of this decay?
b. Assume that all of the energy released in the decay goes to kinetic energy of the emitted electron. What is the electron's kinetic energy?
c. Use the equations of special relativity to determine the electron's speed.
d. Would the emitted electron be moving at a speed high enough to cause Čerenkov light in water?
e. Based on the color of the Čerenkov light you can see in Figure 30.26, which of the following describes the spectrum of Čerenkov light?
   - The intensity of Čerenkov light is uniform at all frequencies.
   - The intensity of Čerenkov light is proportional to frequency.
   - The intensity of Čerenkov light is proportional to wavelength.

   Explain.

PREPARE ^{133}Xe undergoes beta decay, so the mass of this nucleus must be greater than that of the daughter nucleus. The "lost" mass is converted to energy, which we assume goes to the kinetic energy of the electron, so the kinetic energy of the electron will be $K = \Delta m \cdot c^2$. Once we know the electron's kinetic energy, we can determine the speed to see if it exceeds the speed of light in water. The speed of light in a medium is given by $v = c/n$, where n is the index of refraction. Water has $n = 1.33$.

SOLVE a. Equation 30.18 tells us that beta-minus decay increases Z by 1 while leaving A unchanged. Xe has $Z = 54$; $Z = 55$ is Cs (cesium), so the daughter nucleus, still with $A = 133$, is ^{133}Cs.

b. We use the data in Appendix D to find the mass difference between the parent and daughter nuclei:

$$\Delta m = m(^{133}\text{Xe}) - m(^{133}\text{Cs})$$
$$= 132.905906\ \text{u} - 132.905436\ \text{u} = 0.00047\ \text{u}$$

We assume that the energy corresponding to this mass difference is the kinetic energy of the emitted electron, so the electron's kinetic energy is

$$K = \Delta m \cdot c^2 = (0.00047\ \text{u})(931.49\ \text{MeV}/c^2)c^2 = 0.44\ \text{MeV}$$

c. The kinetic energy of the electron is large enough that we'll need to consider relativity—a classical treatment won't be sufficient. Equation 27.23 gives the relationship between an object's kinetic energy and its rest energy E_0 as $K = (\gamma - 1)E_0$. We can rearrange this to give γ in terms of a ratio of two energies that we know: the electron's rest energy, 0.51 MeV, and the electron's kinetic energy, 0.44 MeV. Because it's a ratio, we need not convert units:

$$\gamma = 1 + \frac{K}{E_0} = 1 + \frac{0.44\ \text{MeV}}{0.51\ \text{MeV}} = 1.9$$

We can now use the definition of γ to solve for the electron's speed:

$$\gamma = \frac{1}{\sqrt{1 - (v/c)^2}}$$
$$v = c\sqrt{1 - 1/\gamma^2} = 2.5 \times 10^8\ \text{m/s}$$

d. Čerenkov light will be emitted if the speed of the emitted electron is greater than the speed of a light wave in the water. The speed of light in water is

$$v = \frac{3.00 \times 10^8\ \text{m/s}}{1.33} = 2.3 \times 10^8\ \text{m/s}$$

The electron is moving faster than this, and so it will emit Čerenkov light.

e. The photo reveals that Čerenkov light appears blue, so more high-frequency blue light is emitted than low-frequency red light. The intensity is greater at higher frequencies, so the intensity is proportional to frequency—at least for visible light. The index of refraction decreases for very high frequencies, returning to $n = 1$ for x rays. Light speed at these very high frequencies is no longer slower than the particle speed, so Čerenkov light "cuts off" at very high frequencies.

ASSESS The energy of the beta particle is reasonably typical for particles emitted by nuclear decays, and we know that Čerenkov light is observed around reactor cores, so it's reasonable to expect the beta particle to be moving fast enough to emit Čerenkov light.

SUMMARY

The goals of Chapter 30 have been to understand the physics of the nucleus and some of the applications of nuclear physics.

GENERAL PRINCIPLES

The Nucleus

The nucleus is a small, dense, positive core at the center of an atom.

Z protons, charge $+e$, spin $\frac{1}{2}$
N neutrons, charge 0, spin $\frac{1}{2}$

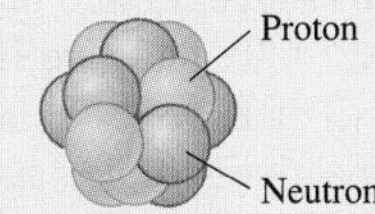

The **mass number** is

$$A = Z + N$$

Isotopes of an element have the same value of Z but different values of N.

The strong force holds nuclei together:

- It acts between any two nucleons.
- It is short range.

Adding neutrons to a nucleus allows the strong force to overcome the repulsive Coulomb force between protons.

The **binding energy** B of a nucleus depends on the mass difference between an atom and its constituents:

$$B = (Zm_{\text{H}} + Nm_{\text{n}} - m_{\text{atom}}) \times (931.49\ \text{MeV/u})$$

Nuclear Stability

Most nuclei are not **stable.** Unstable nuclei undergo **radioactive decay.** Stable nuclei cluster along the **line of stability** in a plot of the isotopes.

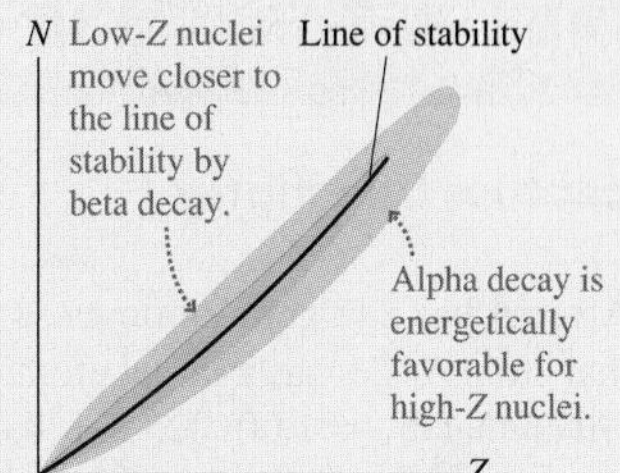

N
Low-Z nuclei move closer to the line of stability by beta decay.
Line of stability
Alpha decay is energetically favorable for high-Z nuclei.
Z

Mechanisms by which unstable nuclei decay:

Decay	Particle	Penetration
alpha	^{4}He nucleus	low
beta-minus	e^-	medium
beta-plus	e^+	medium
gamma	photon	high

Alpha and beta decays change the nucleus; the daughter nucleus is a different element.

Alpha decay:

$${}^{A}_{Z}\text{X} \rightarrow {}^{A-4}_{Z-2}\text{Y} + \alpha + \text{energy}$$

Beta-minus decay:

$${}^{A}_{Z}\text{X} \rightarrow {}_{Z+1}^{\;\;A}\text{Y} + \beta + \text{energy}$$

IMPORTANT CONCEPTS

Energy levels

Nucleons fill nuclear energy levels, similar to filling electron energy levels in atoms. Nucleons can often jump to lower energy levels by emitting beta particles or gamma photons.

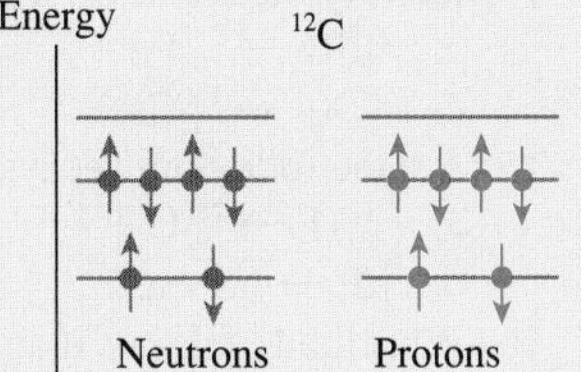

The quark model

Nucleons (and other particles) are made of quarks. Quarks and leptons are fundamental particles.

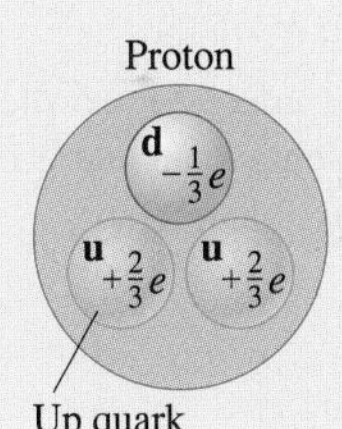

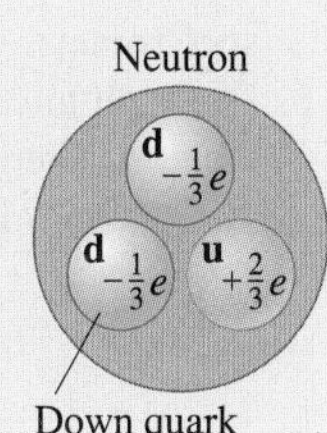

APPLICATIONS

Radioactive decay

The number of undecayed nuclei decreases exponentially with time t:

$$N = N_0 e^{-t/\tau}$$

$$N = N_0 \left(\frac{1}{2}\right)^{t/t_{1/2}}$$

The **half-life**

$$t_{1/2} = \tau \ln 2 = 0.693\tau$$

is the time in which half of any sample decays.

Measuring radiation

The **activity** of a radioactive sample is the number of decays per second. Activity is related to the half-life as

$$R = \frac{0.693N}{t_{1/2}} = \frac{N}{\tau}$$

The radiation **dose** is measured in grays, where

$$1\ \text{Gy} = 1.00\ \text{J/kg of absorbed energy}$$

The **relative biological effectiveness** (RBE) is the biological effect of a dose relative to the biological effects of x rays. The **dose equivalent** is measured in sieverts, where

$$\text{dose equivalent in Sv} = \text{dose in Gy} \times \text{RBE}$$

For homework assigned on MasteringPhysics, go to www.masteringphysics.com

Problems labeled INT integrate significant material from earlier chapters; BIO are of biological or medical interest.

Problem difficulty is labeled as | (straightforward) to ||||| (challenging).

QUESTIONS

Conceptual Questions

1. Atom A has a larger atomic mass than atom B. Does this mean that atom A also has a larger atomic number? Explain.
2. Given that $m_H = 1.007825$ u, is the mass of a hydrogen atom ^{1}H greater than, less than, or equal to 1/12 the mass of a ^{12}C atom? Explain.
3. a. Is there a stable $^{30}_{3}$Li nucleus? Explain how you made your determination.
   b. Is there a stable $^{184}_{92}$U nucleus? Explain how you made your determination.
4. Rounding slightly, the nucleus ^{3}He has a binding energy of 2.5 MeV/nucleon and the nucleus ^{6}Li has a binding energy of 5 MeV/nucleon.
   a. What is the binding energy of ^{3}He?
   b. What is the binding energy of ^{6}Li?
   c. Is it energetically possible for two ^{3}He nuclei to join or fuse together into a ^{6}Li nucleus? Explain.
   d. Is it energetically possible for a ^{6}Li nucleus to split or fission into two ^{3}He nuclei? Explain.
5. A sample contains a mix of isotopes of an element. Using a spectrometer to measure the spectrum of emitted light will not reveal the mix of isotopes; analyzing the sample with a mass spectrometer will. Explain.
6. For each nuclear energy-level diagram in Figure Q30.6, state whether it represents a nuclear ground state, an excited nuclear state, or an impossible nucleus.

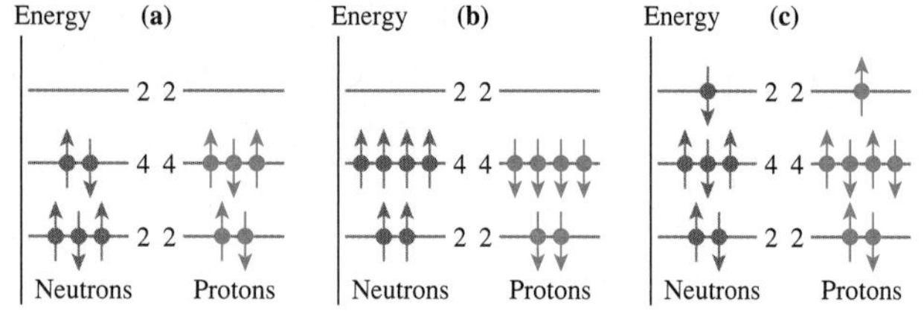

FIGURE Q30.6

7. Figure Q30.7 shows how the number of nuclei of one particular isotope varies with time. What is the half-life of the nucleus?

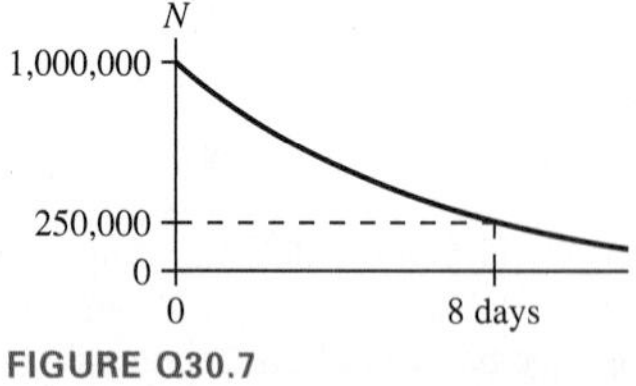

FIGURE Q30.7

8. A radioactive sample has a half-life of 10 s. 10,000 nuclei are present at $t = 20$ s.
   a. How many nuclei were there at $t = 0$ s?
   b. How many nuclei will there be at $t = 40$ s?
9. Nucleus A decays into the stable nucleus B with a half-life of 10 s. At $t = 0$ s there are 1000 A nuclei and no B nuclei. At what time will there be 750 B nuclei?
10. A radioactive sample's half-life is 1.0 min, so each nucleus in the sample has a 50% chance of undergoing a decay sometime between $t = 0$ and $t = 1$ min. One particular nucleus has not decayed at $t = 15$ min. What is the probability this nucleus will decay between $t = 15$ and $t = 16$ min?
11. Four samples each contain a single radioactive isotope. Sample A has 1 mol of matter and an activity of 100 Bq. Sample B has 10 mol and 100 Bq, sample C has 100 mol and 100 Bq, and sample D has 100 mol and 1000 Bq. Rank in order, from largest to smallest, the half-lives of these four isotopes. Explain.
12. Oil and coal generally contain no measurable ^{14}C. What does this tell us about how long they have been buried?
13. Radiocarbon dating assumes that the abundance of ^{14}C in the environment has been constant. Suppose ^{14}C was less abundant 10,000 years ago than it is today. Would this cause a lab using radiocarbon dating to overestimate or underestimate the age of a 10,000-year-old artifact? (In fact, the abundance of ^{14}C in the environment does vary slightly with time. But the issue has been well studied, and the ages of artifacts are adjusted to compensate for this variation.)
14. Identify the unknown X in the following decays:
    a. $^{222}_{86}\text{Rn} \rightarrow {}^{218}_{84}\text{Po} + X$ b. $^{228}_{88}\text{Ra} \rightarrow {}^{228}_{89}\text{Ac} + X$
    c. $^{140}_{54}\text{Xe} \rightarrow {}^{140}_{55}\text{Cs} + X$ d. $^{64}_{29}\text{Cu} \rightarrow {}^{64}_{28}\text{Ni} + X$
15. Are the following decays possible? If not, why not?
    a. $^{232}_{90}\text{Th} \rightarrow {}^{236}_{92}\text{U} + \alpha$ b. $^{238}_{94}\text{Pu} \rightarrow {}^{236}_{92}\text{U} + \alpha$
    c. $^{33}_{15}\text{P} \rightarrow {}^{32}_{16}\text{S} + e^-$
16. What kind of decay, if any, would you expect for the nuclei with the energy-level diagrams shown in Figure Q30.16?

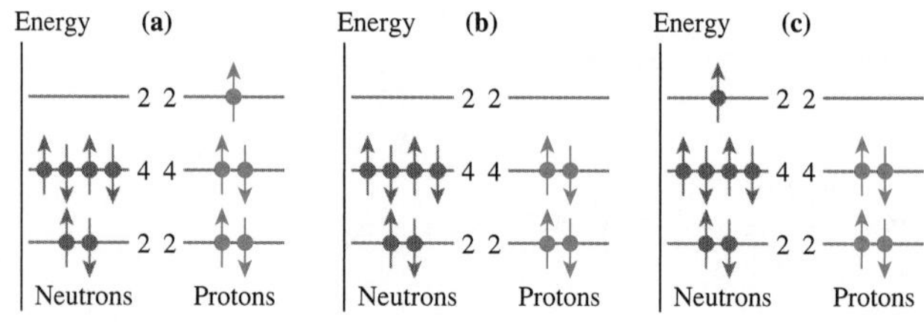

FIGURE Q30.16

17. The nuclei of ^{4}He and ^{16}O are very stable and are often referred to as "doubly magic" nuclei. Use what you know about energy levels to explain what is special about these particular nuclei.
18. BIO A and B are fresh apples. Apple A is strongly irradiated by nuclear radiation for 1 hour. Apple B is not irradiated. Afterward, in what ways are apples A and B different?
19. BIO A patient's tumor is irradiated with gamma rays from an external source. Afterward, is his body radioactive? Explain.

20. It's possible that a bone tumor will not show up on an x-ray image but will show up in a gamma scan. Explain why this is so.
BIO
21. Four radiation doses are as follows: Dose A is 10 rad with an RBE of 1, dose B is 20 rad with an RBE of 1, dose C is 10 rad with an RBE of 2, and dose D is 20 rad with an RBE of 2.
BIO
    a. Rank in order, from largest to smallest, the amount of energy delivered by these four doses.
    b. Rank in order, from largest to smallest, the biological damage caused by these four doses.
22. Two different sources of radiation give the same dose equivalent in Sv. Does this mean that the radiation from each source has the same RBE? Explain.
BIO
23. Some types of MRI can produce images of resolution and detail similar to PET. Though the images are similar, MRI is generally preferred over PET for studies of brain function involving healthy subjects. Why?
BIO
24. Sulfur colloid particles tagged with ^{99}Tc are taken up and retained by cells in the liver and spleen. A patient is suspected of having a liver tumor that would destroy these cells. Explain how a gamma camera scan could be used to confirm or rule out the existence of a tumor.
BIO
25. The first two letters in the acronym SPECT, which describes a nuclear imaging technique, stand for "single photon." Is a SPECT done with a gamma emitter or a positron emitter?
BIO

The following two questions concern an uncommon nuclear decay mode known as *electron capture*. Certain nuclei that are proton-rich but energetically prohibited from undergoing beta-plus decay can capture an electron from the 1*s* shell, which then combines with a proton to make a neutron. The basic reaction is

$$p + e^- \rightarrow n + \nu_e$$

26. Give a description of the electron capture process in terms of quarks.
27. Electron capture is usually followed by the emission of an x ray. Why?

Multiple-Choice Questions

28. | A significant fraction of the radiation dose you will receive during your life comes from radioactive materials in your body. The most important source of this radiation is the potassium isotope ^{40}K, which decays to the stable calcium isotope ^{40}Ca. What particle is emitted in the decay?
BIO
    A. A helium nucleus B. A neutron
    C. An electron D. A positron

29. | A certain watch's luminous glow is due to zinc sulfide paint that is energized by beta particles given off by *tritium*, the radioactive hydrogen isotope ^{3}H, which has a half-life of 12.3 years. This glow has about 1/10 of its initial brightness. How many years old is the watch?

    A. 20 yr B. 30 yr C. 40 yr D. 50 yr
30. | What is the unknown isotope in the following fission reaction: $n + {}^{235}U \rightarrow {}^{131}I + ? + 3n$
    A. ^{86}Rb B. ^{102}Rb C. ^{89}Y D. ^{102}Y
31. || The uranium in the earth's crust is 0.7% ^{235}U and 99.3% ^{238}U. Two billion years ago, ^{235}U comprised approximately 3% of the uranium in the earth's crust. This tells you something about the relative half-lives of the two isotopes. Suppose you have a sample of ^{235}U and a sample of ^{238}U, each with exactly the same number of atoms.
    A. The sample of ^{235}U has a higher activity.
    B. The sample of ^{238}U has a higher activity.
    C. The two samples have the same activity.
32. | Suppose you have a 1 g sample of ^{226}Ra, half-life 1600 years. How long will it be until only 0.1 g of radium is left?
    A. 1600 yr B. 3200 yr C. 5300 yr D. 16,000 yr
33. | A sample of ^{131}I, half-life 8.0 days, is registering 100 counts per second on a Geiger counter. How long will it be before the sample registers only 1 count per second?
    A. 8 days B. 53 days C. 80 days D. 800 days
34. || The complete expression for the decay of the radioactive hydrogen isotope *tritium* may be written as $^3H \rightarrow {}^3He + X + Y$. The symbols X and Y represent
    A. $X = e^+, Y = \bar{\nu}_e$ B. $X = e^-, Y = \nu_e$
    C. $X = e^+, Y = \nu_e$ D. $X = e^-, Y = \bar{\nu}_e$
35. | The quark composition of the proton and neutron are, respectively, uud and udd, where u is an up quark (charge $+\frac{2}{3}e$) and d is a down quark (charge $-\frac{1}{3}e$). There are also anti-up $\bar{u}$ (charge $-\frac{2}{3}e$) and anti-down $\bar{d}$ (charge $+\frac{1}{3}e$) quarks. The combination of a quark and an antiquark is called a *meson*. The mesons known as *pions* have the composition $\pi^+ = u\bar{d}$ and $\pi^- = \bar{u}d$. Suppose a proton collides with an antineutron. During such collisions, the various quarks and antiquarks annihilate whenever possible. When the remaining quarks combine to form a single particle, it is a
    A. Proton B. Neutron
    C. π^+ D. π^-

PROBLEMS

Section 30.1 Nuclear Structure

1. | How many protons and how many neutrons are in (a) ^{3}H, (b) ^{40}Ar, (c) ^{40}Ca, and (d) ^{239}Pu?
2. | How many protons and how many neutrons are in (a) ^{3}He, (b) ^{20}Ne, (c) ^{60}Co, and (d) ^{226}Ra?
3. || Use the data in Appendix D to calculate the chemical atomic mass of lithium, to two decimal places.
4. | Use the data in Appendix D to calculate the chemical atomic mass of neon, to two decimal places.

Section 30.2 Nuclear Stability

5. | Calculate (in MeV) the total binding energy and the binding energy per nucleon (a) for ^{3}H and (b) for ^{3}He.
6. | Calculate (in MeV) the total binding energy and the binding energy per nucleon (a) for ^{40}Ar and (b) for ^{40}K.
7. | Calculate (in MeV) the binding energy per nucleon for ^{3}He and ^{4}He. Which is more tightly bound?
8. || Calculate (in MeV) the binding energy per nucleon for ^{12}C and ^{13}C. Which is more tightly bound?

9. | Calculate (in MeV) the binding energy per nucleon for (a) ^{14}N, (b) ^{56}Fe, and (c) ^{207}Pb.
10. || When a nucleus of ^{235}U undergoes fission, it breaks into two smaller, more tightly bound fragments. Calculate the binding energy per nucleon for ^{235}U and for the fission product ^{137}Cs.
11. || When a nucleus of ^{240}Pu undergoes fission, it breaks into two smaller, more tightly bound fragments. Calculate the binding energy per nucleon for ^{240}Pu and for the fission product ^{133}Xe.

Section 30.3 Forces and Energy in the Nucleus

12. || Draw an energy-level diagram, similar to Figure 30.9 for the protons and neutrons in ^{11}Be. Do you expect this nucleus to be stable?
13. || Draw energy-level diagrams, similar to Figure 30.9, for all $A = 10$ nuclei listed in Appendix D. Show all the occupied neutron and proton levels. Which of these nuclei do you expect to be stable?
14. || Draw energy-level diagrams, similar to Figure 30.9, for all $A = 14$ nuclei listed in Appendix D. Show all the occupied neutron and proton levels. Which of these nuclei do you expect to be stable?
15. ||| You have seen that filled electron energy levels correspond to chemically stable nuclei. A similar principle holds for nuclear energy levels; nuclei with equally filled proton and neutron energy levels are especially stable. What are the three lightest isotopes whose proton and neutron energy levels are both filled, and filled equally?

Section 30.4 Radiation and Radioactivity

16. | ^{15}O and ^{131}I are isotopes used in medical imaging. ^{15}O is a beta-plus emitter, ^{131}I a beta-minus emitter. What are the daughter nuclei of the two decays? BIO
17. | Spacecraft have been powered with energy from the alpha decay of ^{238}Pu. What is the daughter nucleus?
18. | Identify the unknown isotope X in the following decays.
    a. $^{234}U \rightarrow X + \alpha$
    b. $^{32}P \rightarrow X + e^-$
    c. $X \rightarrow {}^{30}Si + e^+$
    d. $^{24}Mg \rightarrow X + \gamma$
19. | Identify the unknown isotope X in the following decays.
    a. $X \rightarrow {}^{224}Ra + \alpha$
    b. $X \rightarrow {}^{207}Pb + e^-$
    c. $^{7}Be + e^- \rightarrow X$
    d. $X \rightarrow {}^{60}Ni + \gamma$
20. | What is the energy (in MeV) released in the alpha decay of ^{239}Pu?
21. | What is the energy (in MeV) released in the alpha decay of ^{228}Th?
22. || What is the total energy (in MeV) released in the beta decay of a neutron?
23. | Medical gamma imaging is generally done with the technetium isotope $^{99}Tc^*$, which decays by emitting a gamma-ray photon with energy 140 keV. What is the mass loss of the nucleus, in u, upon emission of this gamma ray? BIO

Section 30.5 Nuclear Decay and Half-Lives

24. | The radioactive hydrogen isotope ^{3}H is called tritium. It decays by beta-minus decay with a half-life of 12.3 years.
    a. What is the daughter nucleus of tritium?
    b. A watch uses the decay of tritium to energize its glowing dial. What fraction of the tritium remains 20 years after the watch was created?
25. | The barium isotope ^{133}Ba has a half-life of 10.5 years. A sample begins with 1.0×10^{10} ^{133}Ba atoms. How many are left after (a) 2 years, (b) 20 years, and (c) 200 years?
26. | The cadmium isotope ^{109}Cd has a half-life of 462 days. A sample begins with 1.0×10^{12} ^{109}Cd atoms. How many are left after (a) 50 days, (b) 500 days, and (c) 5000 days?
27. || How many half-lives must elapse until (a) 90% and (b) 99% of a radioactive sample of atoms has decayed?
28. || The Chernobyl reactor accident in what is now Ukraine was the worst nuclear disaster of all time. Fission products from the reactor core spread over a wide area. The primary radiation exposure to people in western Europe was due to the short-lived (half-life 8.0 days) isotope ^{131}I, which fell across the landscape and was ingested by grazing cows that concentrated the isotope in their milk. Farmers couldn't sell the contaminated milk, so many opted to use the milk to make cheese, aging it until the radioactivity decayed to acceptable levels. How much time must elapse for the activity of a block of cheese containing ^{131}I to drop to 1.0% of its initial value? BIO
29. |||| What is the age in years of a bone in which the $^{14}C/^{12}C$ ratio is measured to be 1.65×10^{-13}?
30. || ^{85}Sr is a short-lived (half-life 65 days) isotope used in bone scans. A typical patient receives a dose of ^{85}Sr with an activity of 0.10 mCi. If all of the ^{85}Sr is retained by the body, what will be its activity in the patient's body after one year has passed? BIO
31. || What is the half-life in days of a radioactive sample with 5.0×10^{15} atoms and an activity of 5.0×10^{8} Bq?
32. ||| What is the activity, in Bq and Ci, of 1.0 g of ^{226}Ra? Marie Curie was the discoverer of radium; can you see where the unit of activity named after her came from? INT
33. || Many medical PET scans use the isotope ^{18}F, which has a half-life of 1.8 hr. A sample prepared at 10:00 A.M. has an activity of 20 mCi. What is the activity at 1:00 P.M., when the patient is injected? BIO
34. ||| An investigator collects a sample of a radioactive isotope with an activity of 370,000 Bq. 48 hours later, the activity is 120,000 Bq. What is the half-life of the sample?

Section 30.6 Medical Applications of Nuclear Physics

35. || A 50 kg nuclear plant worker is exposed to 20 mJ of neutron radiation with an RBE of 10. What is the dose in mSv? BIO
36. || A gamma scan showing the active volume of a patient's lungs can be created by having a patient breathe the radioactive isotope ^{133}Xe, which undergoes beta-minus decay with a subsequent gamma emission from the daughter nucleus. A typical procedure gives a dose of 0.30 rem to the lungs. How much energy is deposited in the 1.2 kg mass of a patient's lungs? BIO

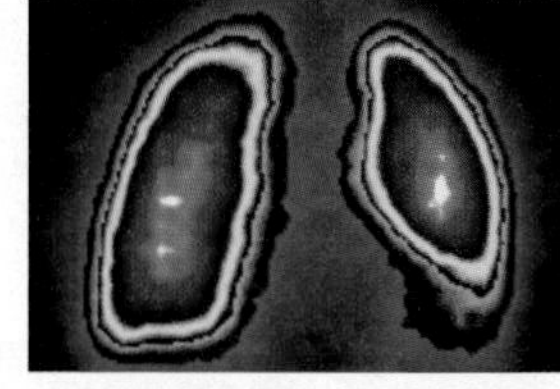

37. || How many rad of gamma-ray photons cause the same biological damage as 30 rad of alpha radiation?
38. | 150 rad of gamma radiation are directed into a 150 g tumor. How much energy does the tumor absorb? BIO
39. || During the 1950s, nuclear bombs were tested on islands in the South Pacific. In one test, personnel on a nearby island received 10 mGy per hour of beta and gamma radiation. At this rate, how long would it take to receive a potentially lethal dose equivalent of 4.5 Sv?

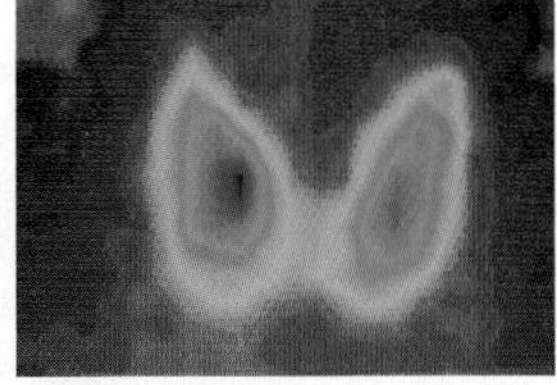

40. ||| ^{131}I undergoes beta-minus decay with a subsequent gamma emission from the daughter nucleus. Iodine in the body is almost entirely taken up by the thyroid gland, so a gamma scan using this isotope will show a bright area corresponding to the thyroid gland with the surrounding tissue appearing dark. Because the isotope is concentrated in the gland, so is the radiation dose, most of which results from the beta emission. In a typical procedure, a patient receives 0.050 mCi of ^{131}I. Assume that all of the iodine is absorbed by the 0.15 kg thyroid gland. Each ^{131}I decay produces a 0.97 MeV beta particle. Assume that half the energy of each beta particle is deposited in the gland. What dose equivalent in Sv will the gland receive in the first hour?
BIO

41. | The doctors planning a radiation therapy treatment have determined that a 100 g tumor needs to receive 0.20 J of gamma radiation. What is the dose in Gy?
BIO

42. ||| ^{90}Sr decays with the emission of a 2.8 MeV beta particle. Strontium is chemically similar to calcium and is taken up by bone. A 75 kg person exposed to waste from a nuclear accident absorbs ^{90}Sr with an activity of 370,000 Bq. Assume that all of this ^{90}Sr ends up in the skeleton. The skeleton forms 17% of the person's body mass. If 50% of the decay energy is absorbed by the skeleton, what dose equivalent in Sv will be received by the person's skeleton in the first month?
BIO

Section 30.7 The Ultimate Building Blocks of Matter

43. || What are the minimum energies of the two oppositely directed gamma rays in a PET procedure?
BIO

44. |||| Positive and negative pions, denoted π^+ and π^-, are antiparticles of each other. Each has a rest mass of 140 MeV/c^2. Suppose a collision between an electron and positron, each with kinetic energy K, produces a π^+, π^- pair. What is the smallest possible value for K?

45. || In a particular beta-minus decay of a free neutron (that is, one not part of an atomic nucleus), the emitted electron has exactly the same kinetic energy as the emitted electron antineutrino. What is the value, in MeV, of that kinetic energy? Assume that the recoiling proton has negligible kinetic energy.

46. |||| The masses of the neutrinos are still not precisely determined, but let us assume for the purpose of this problem that the mass of an electron neutrino is one millionth the mass of an electron. What is the kinetic energy, in eV, of an electron neutrino moving at 0.999c?
INT

General Problems

47. | The chemical atomic mass of hydrogen, with the two stable isotopes ^{1}H and ^{2}H (deuterium), is 1.00798 u. Use this value to determine the natural abundance of these two isotopes.

48. || You learned in Chapter 29 that the binding energy of the electron in a hydrogen atom is 13.6 eV.
   a. By how much does the mass decrease when a hydrogen atom is formed from a proton and an electron? Give your answer both in atomic mass units and as a percentage of the mass of the hydrogen atom.
   b. By how much does the mass decrease when a helium nucleus is formed from two protons and two neutrons? Give your answer both in atomic mass units and as a percentage of the mass of the helium nucleus.
   c. Compare your answers to parts a and b. Why do you hear it said that mass is "lost" in nuclear reactions but not in chemical reactions?

49. ||| Use the graph of binding energy of Figure 30.6 to estimate the total energy released if a nucleus with mass number 240 fissions into two nuclei with mass number 120.

50. ||| Could a ^{56}Fe nucleus fission into two ^{28}Al nuclei? Your answer, which should include some calculations, should be based on the curve of binding energy of Figure 30.6.

51. ||| a. What are the isotopic symbols of all isotopes in Appendix D with $A = 17$?
   b. Which of these are stable nuclei?
   c. For those that are not stable, identify both the decay mode and the daughter nucleus.

52. ||| What is the activity in Bq and in Ci of a 2.0 mg sample of ^{3}H?

53. || The activity of a sample of the cesium isotope ^{137}Cs is 2.0×10^8 Bq. Many years later, after the sample has fully decayed, how many beta particles will have been emitted?

54. ||| A 115 mCi sample of a radioactive isotope is made in a reactor. When delivered to a hospital 16 hours later, its activity is 95 mCi. The lowest usable activity level is 10 mCi.
   a. What is the isotope's half-life?
   b. For how long after delivery is the sample usable?

55. ||| You are assisting in an anthropology lab over the summer by carrying out ^{14}C dating. A graduate student found a bone he believes to be 20,000 years old. You extract the carbon from the bone and prepare an equal-mass sample of carbon from modern organic material. To determine the activity of a sample with the accuracy your supervisor demands, you need to measure the time it takes for 10,000 decays to occur.
BIO
   a. The activity of the modern sample is 1.06 Bq. How long does that measurement take?
   b. It turns out that the graduate student's estimate of the bone's age was accurate. How long does it take to measure the activity of the ancient carbon?

56. |||| A sample of wood from an archaeological excavation is dated by using a mass spectrometer to measure the fraction of ^{14}C atoms. Suppose 100 atoms of ^{14}C are found for every 1.0×10^{15} atoms of ^{12}C in the sample. What is the wood's age?
BIO

57. || A sample of 1.0×10^{10} atoms that decay by alpha emission has a half-life of 100 min. How many alpha particles are emitted between $t = 50$ min and $t = 200$ min?

58. || A sample contains radioactive atoms of two types, A and B. Initially there are five times as many A atoms as there are B atoms. Two hours later, the numbers of the two atoms are equal. The half-life of A is 0.50 hours. What is the half-life of B?

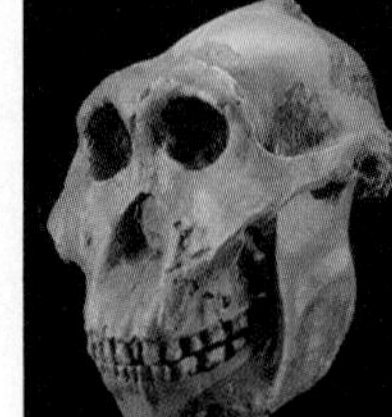

59. |||| The technique known as potassium-argon dating is used to date old lava flows and thus any fossilized skeletons found in them, like this 1.8-million-year old hominid skull. The potassium isotope ^{40}K has a 1.28-billion-year half-life and is naturally present at very low levels. ^{40}K decays by beta emission into the stable isotope ^{40}Ar. Argon is a gas, and there is no argon in flowing lava because the gas escapes. Once the lava solidifies, any argon produced in the decay of ^{40}K is trapped inside and cannot escape. What is the age of a piece of solidified lava with a ^{40}Ar/^{40}K ratio of 0.12?

60. |||| Corals take up certain elements from seawater, including uranium but not thorium. After the corals die, the uranium isotopes slowly decay into thorium isotopes. A measurement of the relative fraction of certain isotopes therefore provides a determination of the coral's age. A complicating factor is that the thorium isotopes decay as well. One scheme uses the alpha decay of ^{234}U to ^{230}Th. After a long time, the two species reach an equilibrium in which the number of ^{234}U decays per second (each producing an atom of ^{230}Th) is exactly equal to the number of ^{230}Th decays per second. What is the relative concentration of the two isotopes—the ratio of ^{234}U to ^{230}Th—when this equilibrium is reached?
BIO

61. ||| All the very heavy atoms found in the earth were created long ago by nuclear fusion reactions in a supernova, an exploding star. The debris spewed out by the supernova later coalesced to form the sun and the planets of our solar system. Nuclear physics suggests that the uranium isotopes ^{235}U ($t_{1/2} = 7.04 \times 10^8$ yr) and ^{238}U ($t_{1/2} = 4.47 \times 10^9$ yr) should have been created in roughly equal amounts. Today, 99.28% of uranium is ^{238}U and 0.72% is ^{235}U. How long ago did the supernova occur?

62. ||| ^{235}U decays to ^{207}Pb via the decay series shown in Figure 30.16. The first decay in the chain, that of ^{235}U, has a half-life of 7.0×10^8 years. The subsequent decays are much more rapid, so we can take this as the half-life for the decay of ^{235}U to ^{207}Pb. ^{238}U decays, with a half-life of 4.5×10^9 years, via a similar decay series that ends in a different lead isotope, ^{206}Pb. Again, the subsequent decays are much more rapid, so we can take this as the half-life for the decay of ^{238}U to ^{206}Pb. The two uranium decay chains can be used to precisely determine the age of certain minerals that exclude lead from their crystal structure but easily incorporate uranium. When these minerals form, they contain both ^{238}U and ^{235}U, but no lead. As time goes on, the isotopes of uranium decay, producing isotopes of lead. A measurement of the ^{238}U/^{206}Pb ratio allows a determination of the age—which can be checked using a measurement of the ^{235}U/^{207}Pb ratio. If 75% of the ^{235}U present in a particular rock has decayed to lead, what percent of the ^{238}U has decayed to lead?

63. || A 75 kg patient swallows a 30 μCi beta emitter with a half-life of 5.0 days. The beta particles are emitted with an average energy of 0.35 MeV. Ninety percent of the beta particles are absorbed within the patient's body and 10% escape. What dose equivalent does the patient receive?
BIO

64. ||| About 12% of your body mass is carbon; some of this is radioactive ^{14}C, a beta-emitter. If you absorb 100% of the 49 keV energy of each ^{14}C decay, what dose equivalent in Sv do you receive each year from the ^{14}C in your body?
BIO INT

65. ||||| Ground beef may be irradiated with high-energy electrons from a linear accelerator to kill pathogens. In a standard treatment, 1.0 kg of beef receives 4.5 kGy of radiation in 40 s.
BIO INT
   a. How much energy is deposited in the beef?
   b. What is the average rate (in W) of energy deposition?
   c. Estimate the temperature increase of the beef due to this procedure. The specific heat of beef is approximately 3/4 of that of water.

66. ||| A 70 kg human body typically contains 140 g of potassium. Potassium has a chemical atomic mass of 39.1 u and has three naturally occurring isotopes. One of those isotopes, ^{40}K, is radioactive with a half-life of 1.3 billion years and a natural abundance of 0.012%. Each ^{40}K decay deposits, on average, 1.0 MeV of energy into the body. What yearly dose in Gy does the typical person receive from the decay of ^{40}K in the body?
BIO

67. |||| What dose in rads of gamma radiation must be absorbed by a block of ice at 0°C to transform the entire block to liquid water at 0°C?
INT

68. || A chest x ray uses 10 keV photons. A 60 kg person receives a 30 mrem dose from one x ray that exposes 25% of the patient's body. How many x-ray photons are absorbed in the patient's body?
BIO

69. |||| The plutonium isotope ^{239}Pu has a half-life of 24,000 years and decays by the emission of a 5.2 MeV alpha particle. Plutonium is not especially dangerous if handled because the activity is low and the alpha radiation doesn't penetrate the skin. But the tiniest speck of plutonium can cause problems if it is inhaled and lodges deep in the lungs. Let's see why.
BIO
   a. Soot particles are roughly 1 μm in diameter, and it is known that these particles can go deep into the lungs. How many ^{239}Pu atoms are in a 1.0-μm-diameter particle of ^{239}Pu? The density of plutonium is 19,800 kg/m^3.
   b. What is the activity, in Bq, of this 1.0-μm-diameter particle?
   c. The activity of the particle is very small, but the penetrating power of alpha particles is also very small, so the damage is concentrated. The alpha particles deposit their energy in a 50-μm-diameter sphere around the plutonium particle. In one year, what is the dose equivalent in mSv to this small sphere of tissue in the lungs? Assume that the tissue density is that of water.
   d. How does the exposure to this tissue compare to the natural background exposure?

70. ||| Uranium is naturally present at low levels in many soils and rocks. The ^{238}U decay series includes the short-lived radon isotope ^{222}Rn, with $t_{1/2} = 3.82$ days. Radon is a gas, and it can seep into basements. The Environmental Protection Agency recommends that homeowners take steps to remove radon if the radon activity exceeds 4 pCi per liter of air. The daughter nuclei from radon decay are of significant concern, but the radon itself does provide some exposure.
BIO
   a. How many ^{222}Rn atoms are there in 1.0 m^3 of air if the activity is 4.0 pCi/L?
   b. The range of alpha particles in air is 3 cm. Let's model a person as a 180-cm-tall, 25-cm-diameter cylinder with a mass of 65 kg. Only decays within 3 cm of the cylinder cause exposure, and only 50% of the decays direct the alpha particle toward the person. What is the dose equivalent (in mSv) for a person who spends an entire year in a room where the activity is 4 pCi/L?

Passage Problems

Nuclear Fission

The uranium isotope ^{235}U can *fission*—break into two smaller-mass components and free neutrons—if it is struck by a free neutron. A typical reaction is

$$^{1}_{0}\text{n} + ^{235}_{92}\text{U} \rightarrow ^{141}_{56}\text{Ba} + ^{92}_{36}\text{Kr} + 3^{1}_{0}\text{n}$$

As you can see, the subscripts (the number of protons) and the superscripts (the number of nucleons) "balance" before and after the fission event; there is no change in the number of protons or neutrons. Significant energy is released in this reaction. If a fission event happens in a large chunk of ^{235}U, the neutrons released may induce the fission of other ^{235}U atoms, resulting in a chain reaction. This is how a nuclear reactor works.

The number of neutrons required to create a stable nucleus increases with atomic number. When the heavy ^{235}U nucleus fissions, the lighter reaction products are thus neutron rich and are likely unstable. Many of the short-lived radioactive nuclei used in medicine are produced in fission reactions in nuclear reactors.

71. | What statement can be made about the masses of atoms in the above reaction?
    A. $m(^{235}_{92}\text{U}) > m(^{141}_{56}\text{Ba}) + m(^{92}_{36}\text{Kr}) + 2m(^{1}_{0}\text{n})$
    B. $m(^{235}_{92}\text{U}) < m(^{141}_{56}\text{Ba}) + m(^{92}_{36}\text{Kr}) + 2m(^{1}_{0}\text{n})$
    C. $m(^{235}_{92}\text{U}) = m(^{141}_{56}\text{Ba}) + m(^{92}_{36}\text{Kr}) + 2m(^{1}_{0}\text{n})$
    D. $m(^{235}_{92}\text{U}) = m(^{141}_{56}\text{Ba}) + m(^{92}_{36}\text{Kr}) + 3m(^{1}_{0}\text{n})$
72. | Because the decay products in the above fission reaction are neutron rich, they will likely decay by what process?
    A. Alpha decay
    B. Beta decay
    C. Gamma decay
73. | ^{235}U is radioactive, with a long half-life of 704 million years. The decay products of a ^{235}U fission reaction typically have half-lives of a few minutes. This means that the decay products of a fission reaction have
    A. Much higher activity than the original uranium.
    B. Much lower activity than the original uranium.
    C. The same activity as the original uranium.
74. | If a $^{238}_{92}$U nucleus is struck by a neutron, it may absorb the neutron. The resulting nucleus then rapidly undergoes beta-minus decay. The daughter nucleus of that decay is
    A. $^{239}_{91}$Pa B. $^{239}_{92}$U C. $^{239}_{93}$Np D. $^{239}_{94}$Pu

STOP TO THINK ANSWERS

Stop to Think 30.1: Three. Different isotopes of an element have different numbers of neutrons but the same number of protons. The number of electrons in a neutral atom matches the number of protons.

Stop to Think 30.2: Yes. ^{12}C has filled levels of protons and neutrons; the neutron we add to make ^{13}C will be in a higher energy level, but there is no "hole" in a lower level for it to move to, so we expect this nucleus to be stable.

Stop to Think 30.3: B. An increase of Z with no change in A occurs when a neutron changes to a proton and an electron, ejecting the electron.

Stop to Think 30.4: C. One-quarter of the atoms are left. This is one-half of one-half, or $(1/2)^2$, so two half-lives have elapsed.

Stop to Think 30.5: A. Dose equivalent is the product of dose in Gy (the same for each) and RBE (highest for alpha particles).

PART VII SUMMARY

Modern Physics

A common theme runs through the final chapters of this book: Nature is stranger than we thought it was. From the bizarre paradoxes of special relativity to the dizzying array of subatomic particles, the physical description of the world around us has taken on an almost science-fiction air in Part VII. But this material isn't science fiction; it's real, the product of decades of careful experiments.

Relativity requires us to stretch our notions of space and time. Time really does slow down for particles moving at high speeds, as decades of experiments have shown. Classical Newtonian physics has a comforting predictability, but this is left behind in quantum theory. We simply can't know an electron's position and velocity at the same time. Many decades of clever experiments have shown conclusively that *no* underlying laws can restore the predictability of classical physics at the atomic scale. And nuclear physics has shown that the alchemist's dream is true—you *can* turn one element into another!

This new physics is surprisingly practical. Relativistic corrections allow GPS systems to give extraordinarily accurate measurements of your position anywhere on the earth. Quantum mechanics is the theory underlying the development of computer chips and other modern electronics. And your smoke detector probably contains a small amount of an element not found in nature, an element created in a nuclear reactor.

As we conclude our journey, the knowledge structure for Part VII summarizes the key ideas of relativity, quantum physics, and nuclear physics. These are the theories behind the emerging technologies of the 21st century.

KNOWLEDGE STRUCTURE VII Modern Physics

BASIC GOALS

What are the properties and characteristics of space and time?
What do we know about the nature of light and atoms?
How are atomic and nuclear phenomena explained by energy levels, wave functions, and photons?

GENERAL PRINCIPLES

Principle of relativity	All the laws of physics are the same in all inertial reference frames.
Quantization of energy	Particles of matter and photons of light have only certain allowed energies.
Uncertainty principle	$\Delta x\, \Delta p \geq h/2\pi$
Pauli exclusion principle	No more than one electron can occupy the same quantum state.

Relativity

- The speed of light c is the same in all inertial reference frames.
- No particle or causal influence can travel faster than c.
- Length contraction: The length of an object in a reference frame in which the object moves with speed v is

$$L = \sqrt{1-\beta^2}\,\ell \leq \ell$$

where ℓ is the proper length.

- Time dilation: The proper time interval $\Delta\tau$ between two events is measured in a reference frame in which the two events occur at the same position. The time interval Δt in a frame moving with relative speed v is

$$\Delta t = \Delta\tau/\sqrt{1-\beta^2} \geq \Delta\tau$$

- Particles have energy even when at rest. Mass can be transformed into energy and vice versa: $E_0 = mc^2$.

Quantum physics

- Matter has wave-like properties. A particle has a de Broglie wavelength:

$$\lambda = \frac{h}{mv}$$

- Light has particle-like properties. A photon of light of frequency f has energy:

$$E_{\text{photon}} = hf = \frac{hc}{\lambda}$$

- The wave nature of matter leads to quantized energy levels in atoms and nuclei. A transition between quantized energy levels involves the emission or absorption of a photon.

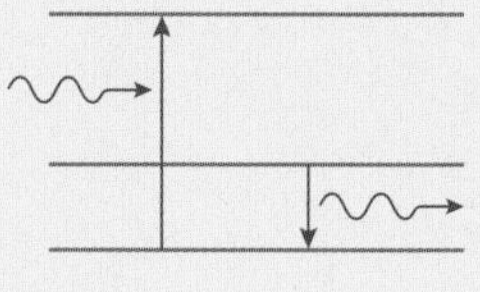

Properties of atoms

- Quantized energy levels depend on quantum numbers n and l.
- An atom can jump from one state to another by emitting or absorbing a photon of energy $E_{\text{photon}} = \Delta E_{\text{atom}}$.
- The ground-state electron configuration is the lowest-energy configuration consistent with the Pauli principle.

Properties of nuclei

- The nucleus is the small, dense, positive core at the center of an atom. The nucleus is held together by the strong force, an attractive short-range force between any two nucleons.

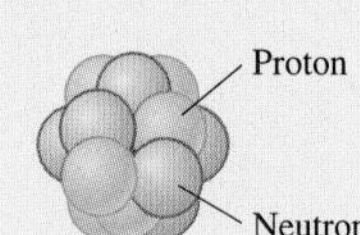

- Unstable nuclei decay by alpha, beta, or gamma decay. The number of nuclei decreases exponentially with time:

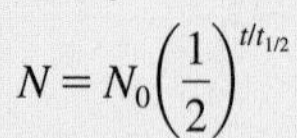

$$N = N_0\left(\frac{1}{2}\right)^{t/t_{1/2}}$$

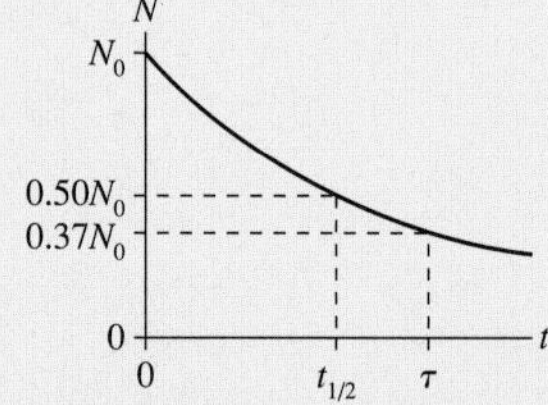

The Physics of Very Cold Atoms

Modern physics is a study of extremes. Relativity deals with the physics of objects traveling at near-light speeds. Quantum mechanics is about the physics of matter and energy at very small scales. Nuclear physics involves energies that dwarf anything dreamed of in previous centuries.

Some of the most remarkable discoveries of recent years are at another extreme—very low temperatures, mere billionths of a degree above absolute zero. Let's look at how such temperatures are achieved and some new physics that emerges.

You learned in Part III that the temperature of a gas depends on the speeds of the atoms in the gas. Suppose we start with atoms at or above room temperature. Cooling the gas means slowing the atoms down. How can we drastically reduce their speeds, bringing them nearly to a halt? The trick is to slow them, thus cooling them, using the interactions between light and atoms that we explored in Chapters 28 and 29.

Photons have momentum, and that momentum is transferred to an atom when a photon is absorbed. Part a of the figure shows an atom moving "upstream" against a laser beam tuned to an atomic transition. Photon absorptions transfer momentum, slowing the atom. Subsequent photon emissions give the atom a "kick," but in random directions, so on average the emissions won't speed up the atom in the same way that the absorptions slow it. A beam of atoms moving "upstream" against a correctly tuned laser beam is slowed down—the "hot" beam of atoms is cooled.

Once a laser cools the atoms, a different configuration of laser beams can trap them. Part b of the figure shows six overlapped laser beams, each tuned slightly *below* the frequency of an atomic absorption line. If an atom tries to leave the overlap region, it will be moving "upstream" against one of the laser beams. The atom will see that laser beam Doppler-shifted to a higher frequency, matching the transition frequency. The atom will then absorb photons from this laser beam, and the resulting kick will nudge it back into the overlap region. The atoms are trapped in what is known as *optical molasses* or, more generally, an *atom trap*. More effective traps can be made by adding magnetic fields. The final cooling of the atoms is by evaporation—letting the more energetic atoms leave the trap.

Ultimately, these techniques produce a diffuse gas of atoms moving at only $\approx$1 mm/s. This corresponds to a nearly unbelievable temperature of just a few nanokelvin—billionths of a degree above absolute zero. This is colder than outer space. The coldest spot in the universe is inside an atom trap in a physics lab.

Once the atoms are cooled, some very remarkable things happen. As we saw in Part VII, all particles, including atoms, have a wave nature. As the atoms slow, their wavelengths increase. In a correctly prepared gas at a low enough temperature, the de Broglie wavelength of an individual atom is larger than the spacing between atoms, and the wave functions of multiple atoms overlap. When this happens, some atoms undergo *Bose-Einstein condensation,* coalescing into one "super atom," with thousands of atoms occupying the same quantum state. An example of the resulting Bose-Einstein condensate is shown in part c of the figure. In the condensate, the atoms—that is, their wave functions—are all in the same place at the same time! This truly bizarre state of matter is a remarkable example of the counterintuitive nature of the quantum world.

Are there applications for Bose-Einstein condensation? Current talk of atom lasers and other futuristic concepts aside, no one really knows, just as the early architects of quantum mechanics didn't know that their theory would be used to design the chips that power personal computers.

At the start of the 20th century, there was a worry that everything in physics had been discovered. There is no such worry at the start of the 21st century, which promises to be full of wonderful discoveries and remarkable applications. What do you imagine the final chapter of a physics textbook will look like 100 years from now?

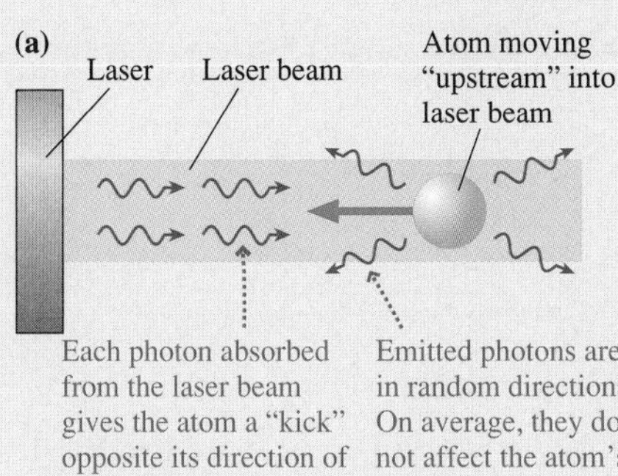

Each photon absorbed from the laser beam gives the atom a "kick" opposite its direction of travel. Thus, each absorption slightly slows the atom.

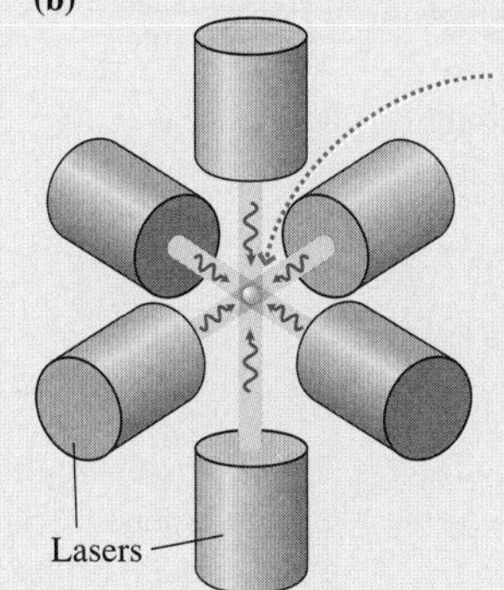

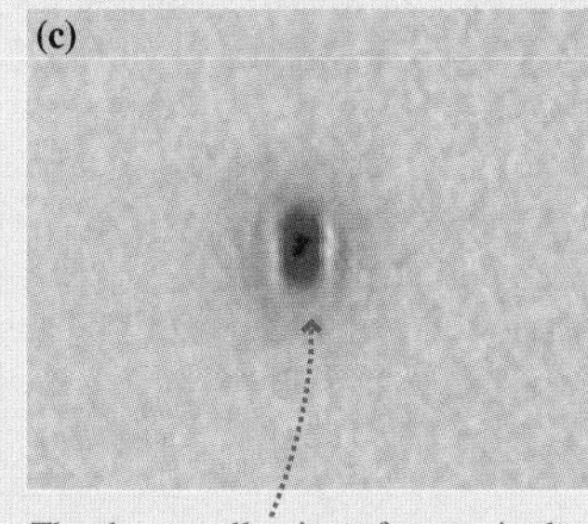

(c)
The dense collection of atoms in the center of the trap is a Bose-Einstein condensate.

Laser cooling and trapping.

PART VII PROBLEMS

The following questions are related to the passage "The Physics of Very Cold Atoms" on the preceding page.

1. Why is it useful to create a very slow-moving assembly of cold atoms?
   A. The atoms can be more easily observed at slow speeds.
   B. Lowering the temperature this way permits isotopes that normally decay in very short times to persist long enough to be studied.
   C. At low speeds the quantum nature of the atoms becomes more apparent, and new forms of matter emerge.
   D. At low speeds the quantum nature of the atoms becomes less important, and they appear more like classical particles.
2. The momentum of a photon is given by $p = h/\lambda$. Suppose an atom emits a photon. Which of the following photons will give the atom the biggest "kick"—the highest recoil speed?
   A. An infrared photon
   B. A red-light photon
   C. A blue-light photon
   D. An ultraviolet photon
3. When an atom moves "upstream" against the photons in a laser beam, the energy of the photons appears to be _______ if the atom were at rest.
   A. Greater than
   B. Less than
   C. The same as
4. A gas of cold atoms strongly absorbs light of a specific wavelength. Warming the gas causes the absorption to decrease. Which of the following is the best explanation for this reduction?
   A. Warming the gas changes the atomic energy levels.
   B. Warming the gas causes the atoms to move at higher speeds, so the atoms "see" the photons at larger Doppler shifts.
   C. Warming the gas causes more collisions between the atoms, which affects the absorption of photons.
   D. Warming the gas makes it more opaque to the photons, so fewer enter the gas.
5. A gas of cold atoms starts at a temperature of 100 nK. The average speed of the atoms is then reduced by half. What is the new temperature?
   A. 71 nK B. 50 nK
   C. 37 nK D. 25 nK
6. Rubidium is often used for the type of experiments noted in the passage. At a speed of 1.0 mm/s, what is the approximate de Broglie wavelength of an atom of ^{87}Rb?
   A. 5 nm B. 50 nm
   C. 500 nm D. 5000 nm
7. A gas of rubidium atoms and a gas of sodium atoms have been cooled to the same very low temperature. What can we say about the de Broglie wavelengths of typical atoms in the two gases?
   A. The sodium atoms have the longer wavelength.
   B. The wavelengths are the same.
   C. The rubidium atoms have the longer wavelength.

The following passages and associated questions are based on the material of Part VII.

Splitting the Atom

"Splitting" an atom in the process of nuclear fission releases a great deal of energy. If all the atoms in 1 kg of ^{235}U undergo nuclear fission, 8.0×10^{13} J will be released, equal to the energy from burning 2.3×10^{6} kg of coal. What is the source of this energy? Surprisingly, the energy from this nuclear disintegration ultimately comes from the electric potential of the positive charges that make up the nucleus.

The protons in a nucleus exert repulsive forces on each other, but this force is less than the short-range attractive nuclear force. If a nucleus breaks into two smaller nuclei, the nuclear force will hold each of the fragments together, but it won't bind the two positively charged fragments to each other. This is illustrated in Figure VII.1. The two fragments feel a strong repulsive electrostatic force. The charges are large and the distance is small (roughly equal to the sum of the radius of each of the fragments), so the force—and thus the potential energy—is quite large.

In a fission reaction, a neutron causes a nucleus of ^{235}U to split into two smaller nuclei; a typical reaction is

$$n + {}^{235}\text{U} \rightarrow {}^{87}\text{Br} + {}^{147}\text{La} + \text{neutrons} + \approx 200\ \text{MeV}$$

Right after the nucleus splits, with only the electric force now acting on the two fragments, the electrostatic potential energy of the two fragments is

$$U = \frac{kq_1q_2}{r_1 + r_2}$$

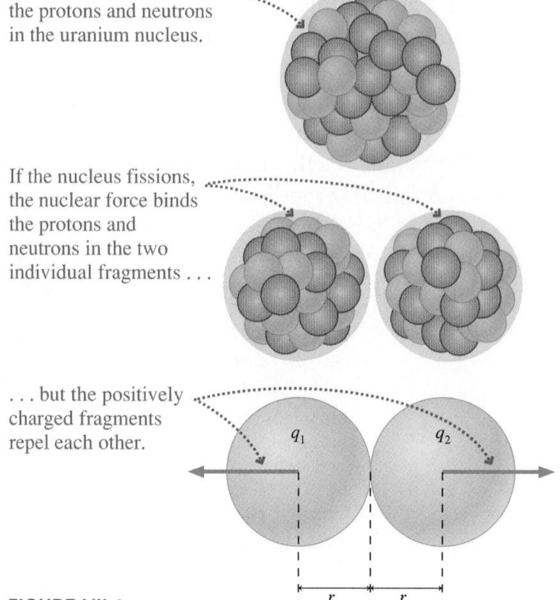

The nuclear force binds the protons and neutrons in the uranium nucleus.
If the nucleus fissions, the nuclear force binds the protons and neutrons in the two individual fragments . . .
. . . but the positively charged fragments repel each other.
q_1
q_2
r_1
r_2

FIGURE VII.1

This is the energy that will be released, transformed into kinetic energy, when the fragments fly apart. If we use reasonable estimates for the radii of the two fragments, we compute a value for the energy that is close to the experimentally observed value of 200 MeV for the energy released in the fission reaction. The energy released in this *nuclear* reaction is actually *electric* potential energy.

8. How many neutrons are "left over" in the noted fission reaction?
A. 1 B. 2 C. 3 D. 4
9. After a fission event, most of the energy released is in the form of
A. Emitted beta particles and gamma rays.
B. Kinetic energy of the emitted neutrons.
C. Nuclear energy of the two fragments.
D. Kinetic energy of the two fragments.
10. Suppose the original nucleus is at rest in the fission reaction noted above. If we neglect the momentum of the neutrons, after the two fragments fly apart,
A. The Br nucleus has more momentum.
B. The La nucleus has more momentum.
C. The momentum of the Br nucleus equals that of the La nucleus.
11. Suppose the original nucleus is at rest in the fission reaction noted above. If we neglect the kinetic energy of the neutrons, after the two fragments fly apart,
A. The Br nucleus has more kinetic energy.
B. The La nucleus has more kinetic energy.
C. The kinetic energy of the Br nucleus equals that of the La nucleus.
12. 200 MeV is a typical energy released in a fission reaction. To get a sense for the scale of the energy, if we were to use this energy to create electron-positron pairs, approximately how many pairs could we create?
A. 50 B. 100 C. 200 D. 400
13. The two fragments of a fission reaction are isotopes that are neutron-rich; each has more neutrons than the stable isotopes for their nuclear species. They will quickly decay to more stable isotopes. What is the most likely decay mode?
A. Alpha decay
B. Beta decay
C. Gamma decay

Additional Integrated Problems

14. The glow-in-the-dark dials on some watches and some keychain lights shine with energy provided by the decay of radioactive tritium, ${}^{3}_{1}\text{H}$. Tritium is a radioactive isotope of hydrogen with a half-life of 12 years. Each decay emits an electron with an energy of 19 keV. A typical new watch has tritium with a total activity of 15 MBq.

A keychain light powered by the decay of tritium.

a. What is the speed of the emitted electron? (This speed is high enough that you'll need to do a relativistic calculation.)
b. What is the power, in watts, provided by the radioactive decay process?
c. What will be the activity of the tritium in a watch after 5 years, assuming none escapes?

15. An x-ray tube is powered by a high-voltage supply that delivers 700 W to the tube. The tube converts 1% of this power into x rays of wavelength 0.030 nm.
BIO
a. Approximately how many x-ray photons are emitted per second?
b. If a 75 kg technician is accidentally exposed to the full power of the x-ray beam for 1.0 s, what dose equivalent in Sv does he receive? Assume that the x-ray energy is distributed over the body, and that 80% of the energy is absorbed.
16. Many speculative plans for spaceships capable of interstellar travel have been developed over the years. Nearly all are powered by the fusion of light nuclei, one of a very few power sources capable of providing the incredibly large energies required. A typical design for a fusion-powered craft has a 1.7×10^{6} kg ship brought up to a speed of $0.12c$ using the energy from the fusion of ${}^{2}_{1}\text{H}$ and ${}^{3}_{2}\text{He}$. Each fusion reaction produces a daughter nucleus and one free proton with a combined kinetic energy of 18 MeV; these high-speed particles are directed backward to create thrust.
a. What is the kinetic energy of the ship at the noted top speed? For the purposes of this problem you can do a nonrelativistic calculation.
b. If we assume that 50% of the energy of the fusion reactions goes into the kinetic energy of the ship (a very generous assumption), how many fusion reactions are required to get the ship up to speed?
c. How many kilograms of ${}^{2}_{1}\text{H}$ and of ${}^{3}_{2}\text{He}$ are required to produce the required number of reactions?
17. A muon is a lepton that is a higher-mass (rest mass 105 MeV/c^2) sibling to the electron. Muons are produced in the upper atmosphere when incoming cosmic rays collide with the nuclei of gas molecules. As the muons travel toward the surface of the earth, they lose energy. A muon that travels from the upper atmosphere to the surface of the earth typically begins with kinetic energy 6.0 GeV and reaches the surface of the earth with kinetic energy 4.0 GeV. The energy decreases by one-third of its initial value. By what fraction does the speed of the muon decrease?
18. A muon is a lepton that is a higher-mass (rest mass 105 MeV/c^2) sibling to the electron. Muons are produced in the upper atmosphere when incoming cosmic rays collide with the nuclei of gas molecules. The muon half-life is 1.5 μs, but atmospheric muons typically live much longer than this because of time dilation, as we saw in Chapter 27. Suppose 100,000 muons are created 120 km above the surface of the earth, each with kinetic energy 10 GeV. Assume that the muons don't lose energy but move at a constant velocity directed straight down toward the surface of the earth. How many muons survive to reach the surface?

Mathematics Review

Algebra

Using exponents: $a^{-x} = \dfrac{1}{a^x}$ $\quad a^x a^y = a^{(x+y)}$ $\quad \dfrac{a^x}{a^y} = a^{(x-y)}$ $\quad (a^x)^y = a^{xy}$

$a^0 = 1$ $\quad a^1 = a$ $\quad a^{1/n} = \sqrt[n]{a}$

Fractions: $\left(\dfrac{a}{b}\right)\left(\dfrac{c}{d}\right) = \dfrac{ac}{bd}$ $\quad \dfrac{a/b}{c/d} = \dfrac{ad}{bc}$ $\quad \dfrac{1}{1/a} = a$

Logarithms: Natural (base e) logarithms: If $a = e^x$, then $\ln(a) = x$ $\quad \ln(e^x) = x$ $\quad e^{\ln(x)} = x$

Base 10 logarithms: If $a = 10^x$, then $\log_{10}(a) = x$ $\quad \log_{10}(10^x) = x$ $\quad 10^{\log_{10}(x)} = x$

The following rules hold for both natural and base 10 algorithms:

$$\ln(ab) = \ln(a) + \ln(b) \qquad \ln\left(\frac{a}{b}\right) = \ln(a) - \ln(b) \qquad \ln(a^n) = n\ln(a)$$

The expression $\ln(a + b)$ cannot be simplified.

Linear equations: The graph of the equation $y = ax + b$ is a straight line. a is the slope of the graph. b is the y-intercept.

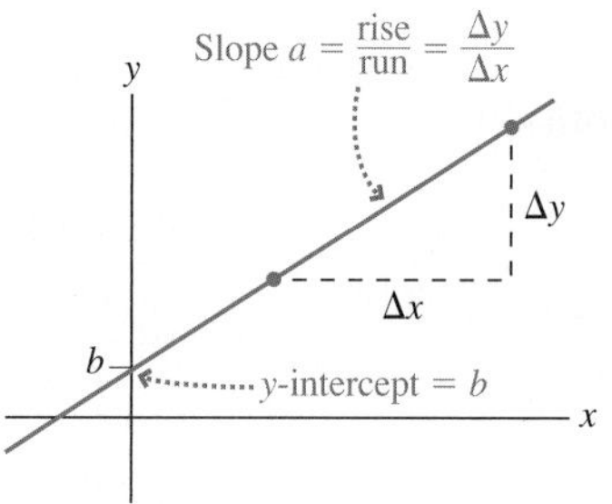

Proportionality: To say that y is proportional to x, written $y \propto x$, means that $y = ax$, where a is a constant. Proportionality is a special case of linearity. A graph of a proportional relationship is a straight line that passes through the origin. If $y \propto x$, then

$$\frac{y_1}{y_2} = \frac{x_1}{x_2}$$

Quadratic equation: The quadratic equation $ax^2 + bx + c = 0$ has the two solutions $x = \dfrac{-b \pm \sqrt{b^2 - 4ac}}{2a}$.

Geometry and Trigonometry

Area and volume:

Rectangle

$A = ab$

Rectangular box

$V = abc$

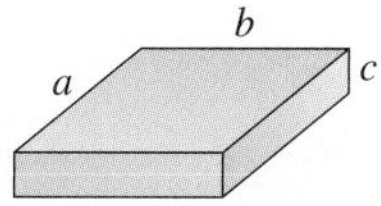

Triangle

$A = \frac{1}{2}ab$

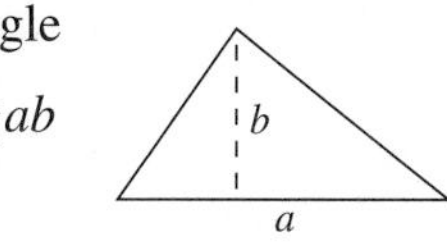

Right circular cylinder

$V = \pi r^2 l$

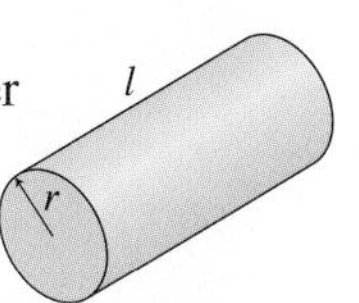

Circle

$C = 2\pi r$

$A = \pi r^2$

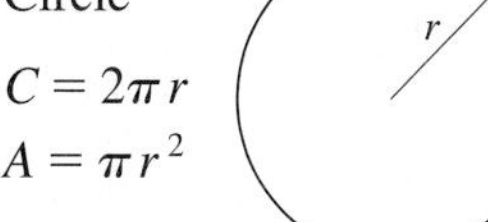

Sphere

$A = 4\pi r^2$

$V = \frac{4}{3}\pi r^3$

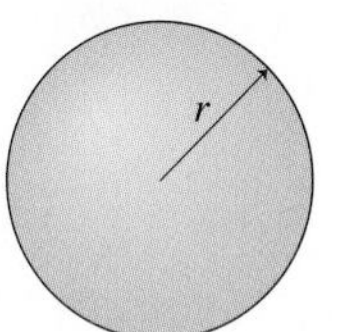

Arc length and angle: The angle θ in radians is defined as $\theta = s/r$.

The arc length that spans angle θ is $s = r\theta$.

2π rad = 360°

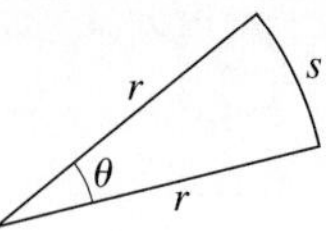

Right triangle: Pythagorean theorem $c = \sqrt{a^2 + b^2}$ or $a^2 + b^2 = c^2$

$$\sin\theta = \frac{b}{c} = \frac{\text{far side}}{\text{hypotenuse}} \qquad \theta = \sin^{-1}\left(\frac{b}{c}\right)$$

$$\cos\theta = \frac{a}{c} = \frac{\text{adjacent side}}{\text{hypotenuse}} \qquad \theta = \cos^{-1}\left(\frac{a}{c}\right)$$

$$\tan\theta = \frac{b}{a} = \frac{\text{far side}}{\text{adjacent side}} \qquad \theta = \tan^{-1}\left(\frac{b}{a}\right)$$

In general, if it is known that sine of an angle θ is x, so $x = \sin\theta$, then we can find θ by taking the *inverse sine* of x, denoted $\sin^{-1} x$. Thus $\theta = \sin^{-1} x$. Similar relations apply for cosines and tangents.

General triangle: $\alpha + \beta + \gamma = 180° = \pi$ rad

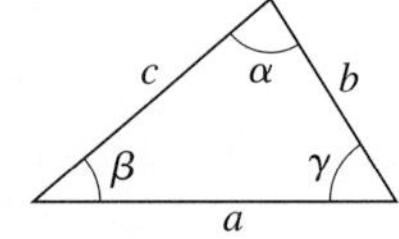

Identities:

$$\tan\alpha = \frac{\sin\alpha}{\cos\alpha} \qquad \sin^2\alpha + \cos^2\alpha = 1$$

$$\sin(-\alpha) = -\sin\alpha \qquad \cos(-\alpha) = \cos\alpha$$

$$\sin(2\alpha) = 2\sin\alpha\cos\alpha \qquad \cos(2\alpha) = \cos^2\alpha - \sin^2\alpha$$

Expansions and Approximations

Binomial approximation: $(1 + x)^n \approx 1 + nx \quad \text{if} \quad x \ll 1$

Small-angle approximation: If $\alpha \ll 1$ rad, then $\sin\alpha \approx \tan\alpha \approx \alpha$ and $\cos\alpha \approx 1$.

The small-angle approximation is excellent for $\alpha < 5°$ (≈ 0.1 rad) and generally acceptable up to $\alpha \approx 10°$.

Periodic Table of Elements

Atomic number — 27 — Co — 58.9 — Atomic mass; Co — Symbol

Period																		
1	1 H 1.0																	2 He 4.0
2	3 Li 6.9	4 Be 9.0											5 B 10.8	6 C 12.0	7 N 14.0	8 O 16.0	9 F 19.0	10 Ne 20.2
3	11 Na 23.0	12 Mg 24.3											13 Al 27.0	14 Si 28.1	15 P 31.0	16 S 32.1	17 Cl 35.5	18 Ar 39.9
4	19 K 39.1	20 Ca 40.1	21 Sc 45.0	22 Ti 47.9	23 V 50.9	24 Cr 52.0	25 Mn 54.9	26 Fe 55.8	27 Co 58.9	28 Ni 58.7	29 Cu 63.5	30 Zn 65.4	31 Ga 69.7	32 Ge 72.6	33 As 74.9	34 Se 79.0	35 Br 79.9	36 Kr 83.8
5	37 Rb 85.5	38 Sr 87.6	39 Y 88.9	40 Zr 91.2	41 Nb 92.9	42 Mo 95.9	43 Tc 96.9	44 Ru 101.1	45 Rh 102.9	46 Pd 106.4	47 Ag 107.9	48 Cd 112.4	49 In 114.8	50 Sn 118.7	51 Sb 121.8	52 Te 127.6	53 I 126.9	54 Xe 131.3
6	55 Cs 132.9	56 Ba 137.3	71 Lu 175.0	72 Hf 178.5	73 Ta 180.9	74 W 183.9	75 Re 186.2	76 Os 190.2	77 Ir 192.2	78 Pt 195.1	79 Au 197.0	80 Hg 200.6	81 Tl 204.4	82 Pb 207.2	83 Bi 209.0	84 Po 209.0	85 At 210.0	86 Rn 222.0
7	87 Fr 223.0	88 Ra 226.0	103 Lr 262.1	104 Rf 261	105 Db 262	106 Sg 263	107 Bh 264	108 Hs 269	109 Mt 268	110 Ds 271	111 Rg 272							

Transition elements: groups from Sc/Y/Lu/Lr through Zn/Cd/Hg.

Inner transition elements

Lanthanides 6	57 La 138.9	58 Ce 140.1	59 Pr 140.9	60 Nd 144.2	61 Pm 144.9	62 Sm 150.4	63 Eu 152.0	64 Gd 157.3	65 Tb 158.9	66 Dy 162.5	67 Ho 164.9	68 Er 167.3	69 Tm 168.9	70 Yb 173.0
Actinides 7	89 Ac 227.0	90 Th 232.0	91 Pa 231.0	92 U 238.0	93 Np 237.0	94 Pu 239.1	95 Am 241.1	96 Cm 244.1	97 Bk 249.1	98 Cf 252.1	99 Es 257.1	100 Fm 257.1	101 Md 258.1	102 No 259.1

APPENDIX C

ActivPhysics OnLine™ Activities

www.masteringphysics.com

1.1 Analyzing Motion Using Diagrams
1.2 Analyzing Motion Using Graphs
1.3 Predicting Motion from Graphs
1.4 Predicting Motion from Equations
1.5 Problem-Solving Strategies for Kinematics
1.6 Skier Races Downhill
1.7 Balloonist Drops Lemonade
1.8 Seat Belts Save Lives
1.9 Screeching to a Halt
1.10 Pole-Vaulter Lands
1.11 Car Starts, Then Stops
1.12 Solving Two-Vehicle Problems
1.13 Car Catches Truck
1.14 Avoiding a Rear-End Collision
2.1.1 Force Magnitudes
2.1.2 Skydiver
2.1.3 Tension Change
2.1.4 Sliding on an Incline
2.1.5 Car Race
2.2 Lifting a Crate
2.3 Lowering a Crate
2.4 Rocket Blasts Off
2.5 Truck Pulls Crate
2.6 Pushing a Crate Up a Wall
2.7 Skier Goes Down a Slope
2.8 Skier and Rope Tow
2.9 Pole-Vaulter Vaults
2.10 Truck Pulls Two Crates
2.11 Modified Atwood Machine
3.1 Solving Projectile Motion Problems
3.2 Two Balls Falling
3.3 Changing the x-Velocity
3.4 Projectile x- and y-Accelerations
3.5 Initial Velocity Components
3.6 Target Practice I
3.7 Target Practice II
4.1 Magnitude of Centripetal Acceleration
4.2 Circular Motion Problem Solving
4.3 Cart Goes Over Circular Path
4.4 Ball Swings on a String
4.5 Car Circles a Track
4.6 Satellites Orbit
5.1 Work Calculations
5.2 Upward-Moving Elevator Stops
5.3 Stopping a Downward-Moving Elevator
5.4 Inverse Bungee Jumper
5.5 Spring-Launched Bowler
5.6 Skier Speed
5.7 Modified Atwood Machine
6.1 Momentum and Energy Change
6.2 Collisions and Elasticity
6.3 Momentum Conservation and Collisions
6.4 Collision Problems
6.5 Car Collision: Two Dimensions
6.6 Saving an Astronaut
6.7 Explosion Problems
6.8 Skier and Cart
6.9 Pendulum Bashes Box
6.10 Pendulum Person-Projectile Bowling
7.1 Calculating Torques
7.2 A Tilted Beam: Torques and Equilibrium
7.3 Arm Levers
7.4 Two Painters on a Beam
7.5 Lecturing from a Beam
7.6 Rotational Inertia
7.7 Rotational Kinematics
7.8 Rotoride: Dynamics Approach
7.9 Falling Ladder
7.10 Woman and Flywheel Elevator: Dynamics Approach
7.11 Race Between a Block and a Disk
7.12 Woman and Flywheel Elevator: Energy Approach
7.13 Rotoride: Energy Approach
7.14 Ball Hits Bat
8.1 Characteristics of a Gas
8.2 Maxwell-Boltzmann Distribution: Conceptual Analysis
8.3 Maxwell-Boltzmann Distribution: Quantitative Analysis
8.4 State Variables and Ideal Gas Law
8.5 Work Done by a Gas
8.6 Heat, Internal Energy, and First Law of Thermodynamics
8.7 Heat Capacity
8.8 Isochoric Process
8.9 Isobaric Process
8.10 Isothermal Process
8.11 Adiabatic Process
8.12 Cyclic Process: Strategies
8.13 Cyclic Process: Problems
8.14 Carnot Cycle
9.1 Position Graphs and Equations
9.2 Describing Vibrational Motion
9.3 Vibrational Energy
9.4 Two Ways to Weigh Young Tarzan
9.5 Ape Drops Tarzan
9.6 Releasing a Vibrating Skier I
9.7 Releasing a Vibrating Skier II
9.8 One- and Two-Spring Vibrating Systems
9.9 Vibro-Ride
9.10 Pendulum Frequency
9.12 Risky Pendulum Walk
9.13 Physical Pendulum
10.1 Properties of Mechanical Waves
10.2 Speed of Waves on a String
10.3 Speed of Sound in a Gas
10.4 Standing Waves on Strings
10.5 Tuning a Stringed Instrument: Standing Waves
10.6 String Mass and Standing Waves
10.7 Beats and Beat Frequency
10.8 Doppler Effect: Conceptual Introduction
10.9 Doppler Effect: Problems
10.10 Complex Waves: Fourier Analysis
11.1 Electric Force: Coulomb's Law
11.2 Electric Force: Superposition Principle
11.3 Electric Force Superposition Principle (Quantitative)
11.4 Electric Field: Point Charge
11.5 Electric Field Due to a Dipole
11.6 Electric Field: Problems
11.7 Electric Flux
11.8 Gauss's Law
11.9 Motion of a Charge in an Electric Field: Introduction
11.10 Motion in an Electric Field: Problems
11.11 Electric Potential: Qualitative Introduction
11.12 Electric Potential, Field, and Force
11.13 Electrical Potential Energy and Potential
12.1 DC Series Circuits (Qualitative)
12.2 DC Parallel Circuits
12.3 DC Circuit Puzzles
12.4 Using Ammeters and Voltmeters
12.5 Using Kirchhoff's Laws
12.6 Capacitance
12.7 Series and Parallel Capacitors
12.8 *RC* Circuit Time Constants
13.1 Magnetic Field of a Wire
13.2 Magnetic Field of a Loop
13.3 Magnetic Field of a Solenoid
13.4 Magnetic Force on a Particle
13.5 Magnetic Force on a Wire
13.6 Magnetic Torque on a Loop
13.7 Mass Spectrometer
13.8 Velocity Selector
13.9 Electromagnetic Induction
13.10 Motional emf
14.1 The *RL* Circuit
14.2 The *RLC* Oscillator
14.3 The Driven Oscillator
15.1 Reflection and Refraction
15.2 Total Internal Reflection
15.3 Refraction Applications
15.4 Plane Mirrors
15.5 Spherical Mirrors: Ray Diagrams
15.6 Spherical Mirror: The Mirror Equation
15.7 Spherical Mirror: Linear Magnification
15.8 Spherical Mirror: Problems
15.9 Thin-Lens Ray Diagrams
15.10 Converging Lens Problems
15.11 Diverging Lens Problems
15.12 Two-Lens Optical Systems
16.1 Two-Source Interference: Introduction
16.2 Two-Source Interference: Qualitative Questions
16.3 Two-Source Interference: Problems
16.4 The Grating: Introduction and Qualitative Questions
16.5 The Grating: Problems
16.6 Single-Slit Diffraction
16.7 Circular Hole Diffraction
16.8 Resolving Power
16.9 Polarization
17.1 Relativity of Time
17.2 Relativity of Length
17.3 Photoelectric Effect
17.4 Compton Scattering
17.5 Electron Interference
17.6 Uncertainty Principle
17.7 Wave Packets
18.1 The Bohr Model
18.2 Spectroscopy
18.3 The Laser
19.1 Particle Scattering
19.2 Nuclear Binding Energy
19.3 Fusion
19.4 Radioactivity
19.5 Particle Physics
20.1 Potential Energy Diagrams
20.2 Particle in a Box
20.3 Potential Wells
20.4 Potential Barriers

Atomic and Nuclear Data

Atomic Number (Z)	Element	Symbol	Mass Number (A)	Atomic Mass (u)	Percent Abundance	Decay Mode	Half-Life $t_{1/2}$
0	(Neutron)	n	1	1.008 665		β^-	10.4 min
1	Hydrogen	H	1	1.007 825	99.985	stable	
	Deuterium	D	2	2.014 102	0.015	stable	
	Tritium	T	3	3.016 049		β^-	12.33 yr
2	Helium	He	3	3.016 029	0.000 1	stable	
			4	4.002 602	99.999 9	stable	
			6	6.018 886		β^-	0.81 s
3	Lithium	Li	6	6.015 121	7.50	stable	
			7	7.016 003	92.50	stable	
			8	8.022 486		β^-	0.84 s
4	Beryllium	Be	9	9.012 174	100	stable	
			10	10.013 534		β^-	1.5×10^6 yr
5	Boron	B	10	10.012 936	19.90	stable	
			11	11.009 305	80.10	stable	
			12	12.014 352		β^-	0.020 2 s
6	Carbon	C	10	10.016 854		β^+	19.3 s
			11	11.011 433		β^+	20.4 min
			12	12.000 000	98.90	stable	
			13	13.003 355	1.10	stable	
			14	14.003 242		β^-	5 730 yr
			15	15.010 599		β^-	2.45 s
7	Nitrogen	N	12	12.018 613		β^+	0.011 0 s
			13	13.005 738		β^+	9.96 min
			14	14.003 074	99.63	stable	
			15	15.000 108	0.37	stable	
			16	16.006 100		β^-	7.13 s
			17	17.008 450		β^-	4.17 s
8	Oxygen	O	15	15.003 065		β^+	122 s
			16	15.994 915	99.76	stable	
			17	16.999 132	0.04	stable	
			18	17.999 160	0.20	stable	
			19	19.003 577		β^-	26.9 s
9	Fluorine	F	18	18.000 937		β^+	109.8 min
			19	18.998 404	100	stable	
			20	19.999 982		β^-	11.0 s
10	Neon	Ne	19	19.001 880		β^+	17.2 s
			20	19.992 435	90.48	stable	
			21	20.993 841	0.27	stable	
			22	21.991 383	9.25	stable	
17	Chlorine	Cl	35	34.968 853	75.77	stable	
			36	35.968 307		β^-	3.0×10^5 yr
			37	36.965 903	24.23	stable	

Atomic Number (Z)	Element	Symbol	Mass Number (A)	Atomic Mass (u)	Percent Abundance	Decay Mode	Half-Life $t_{1/2}$
18	Argon	Ar	36	35.967 547	0.34	stable	
			38	37.962 732	0.06	stable	
			39	38.964 314		β^-	269 yr
			40	39.962 384	99.60	stable	
			42	41.963 049		β^-	33 yr
19	Potassium	K	39	38.963 708	93.26	stable	
			40	39.964 000	0.01	β^+	1.28×10^9 yr
			41	40.961 827	6.73	stable	
26	Iron	Fe	54	54.939 613	5.9	stable	
			56	55.934 940	91.72	stable	
			57	56.935 396	2.1	stable	
			58	57.933 278	0.28	stable	
			60	59.934 072		β^-	1.5×10^6 yr
27	Cobalt	Co	59	58.933 198	100	stable	
			60	59.933 820		β^-	5.27 yr
38	Strontium	Sr	84	83.913 425	0.56%	stable	
			86	85.909 262	9.86%	stable	
			87	86.908 879	7.00%	stable	
			88	87.905 614	82.58%	stable	
			89	88.907 450		β^-	50.53 days
			90	89.907 738		β^-	27.78 yr
53	Iodine	I	127	126.904 474	100	stable	
			129	128.904 984		β^-	1.6×10^7 yr
			131	130.906 124		β^-	8 days
54	Xenon	Xe	128	127.903 531	1.9	stable	
			129	128.904 779	26.4	stable	
			130	129.903 509	4.1	stable	
			131	130.905 069	21.2	stable	
			132	131.904 141	26.9	stable	
			133	132.905 906		β^-	5.4 days
			134	133.905 394	10.4	stable	
			136	135.907 215	8.9	stable	
55	Cesium	Cs	133	132.905 436	100	stable	
			137	136.907 078		β^-	30 yr
82	Lead	Pb	204	203.973 020	1.4	stable	
			206	205.974 440	24.1	stable	
			207	206.975 871	22.1	stable	
			208	207.976 627	52.4	stable	
			210	209.984 163		α, β^-	22.3 yr
			211	210.988 734		β^-	36.1 min
83	Bismuth	Bi	209	208.980 374	100	stable	
			211	210.987 254		α	2.14 min
			215	215.001 836		β^-	7.4 min
86	Radon	Rn	219	219.009 477		α	3.96 s
			220	220.011 369		α	55.6 s
			222	222.017 571		α, β^-	3.823 days

Atomic Number (Z)	Element	Symbol	Mass Number (A)	Atomic Mass (u)	Percent Abundance	Decay Mode	Half-Life $t_{1/2}$
88	Radium	Ra	223	223.018 499		α	11.43 days
			224	224.020 187		α	3.66 days
			226	226.025 402		α	1 600 yr
			228	228.031 064		β^-	5.75 yr
90	Thorium	Th	227	227.027 701		α	18.72 days
			228	228.028 716		α	1.913 yr
			229	229.031 757		α	7 300 yr
			230	230.033 127		α	75.000 yr
			231	231.036 299		α, β^-	25.52 hr
			232	232.038 051	100	α	1.40×10^{10} yr
			234	234.043 593		β^-	24.1 days
92	Uranium	U	233	233.039 630		α	1.59×10^5 yr
			234	234.040 946		α	2.45×10^5 yr
			235	235.043 924	0.72	α	7.04×10^8 yr
			236	236.045 562		α	2.34×10^7 yr
			238	238.050 784	99.28	α	4.47×10^9 yr
93	Neptunium	Np	237	237.048 168		α	2.14×10^6 yr
			238	238.050 946		β^-	2.12 days
			239	239.052 939		β^-	2.36 days
94	Plutonium	Pu	238	238.049 555		α	87.7 yr
			239	239.052 157		α	2.412×10^4 yr
			240	240.053 808		α	6 560 yr
			242	242.058 737		α	3.73×10^6 yr
			244	244.064 200		α	8.1×10^7 yr
95	Americium	Am	241	241.056 823		α	432.21 yr
			243	243.061 375		α	73.070 yr

Answers

Chapter 1

Answers to odd-numbered multiple-choice questions

21. C
23. A
25. B
27. D

Answers to odd-numbered problems

1.

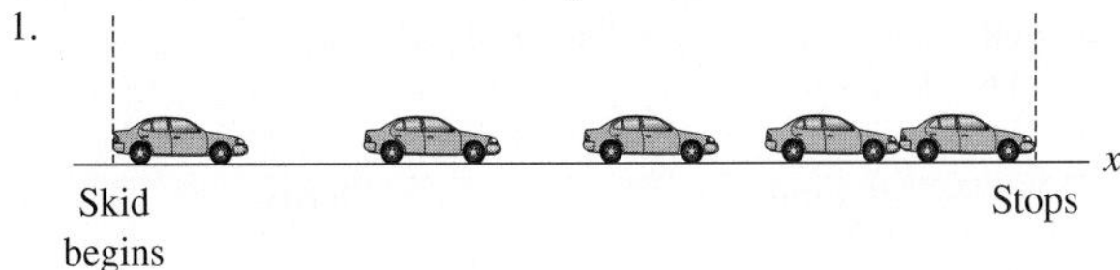

3.

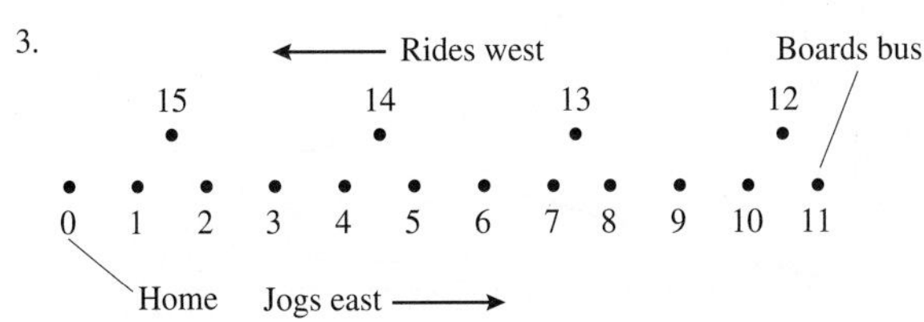

5. −22 m
7. 800 m east of starting point
9. Bike, ball, cat, toy car
11. −1.0 m/s
13. 15 s
15. a. 0.20 m b. 20 m/s c. 27 m/s
17. 1 km/ks < 1 cm/ms < 1 mm/μs
19. a. 3 b. 4 c. 5 d. 3
21. 3.81×10^2 m
23. 4.2×10^{-8} m/s
25. 1.2×10^{-9} m/s or 4.4 μm/h
27. 3.3 km
29. 10 m
31. 38 km
33. 50 yd at 32°
35. **Pictorial representation** **Motion diagram**

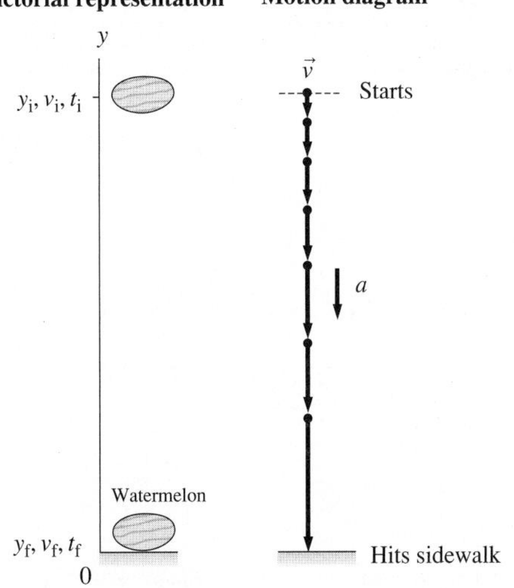

37.

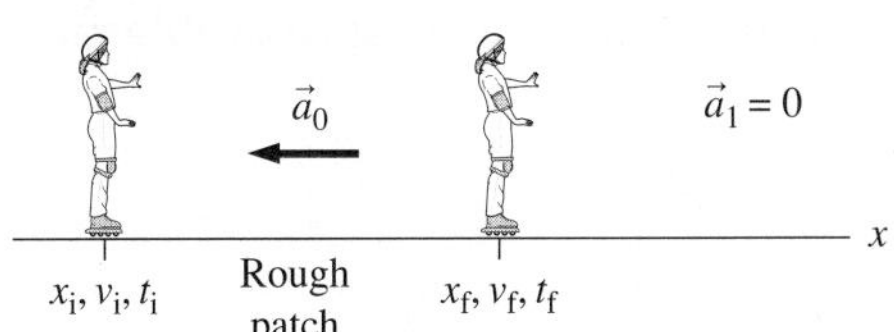

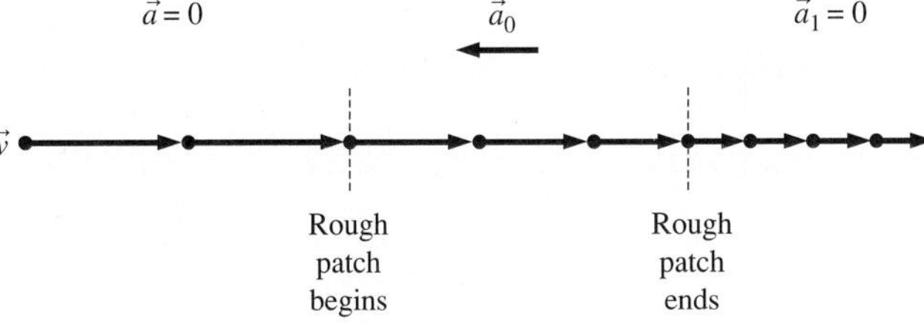

39. **Pictorial representation**

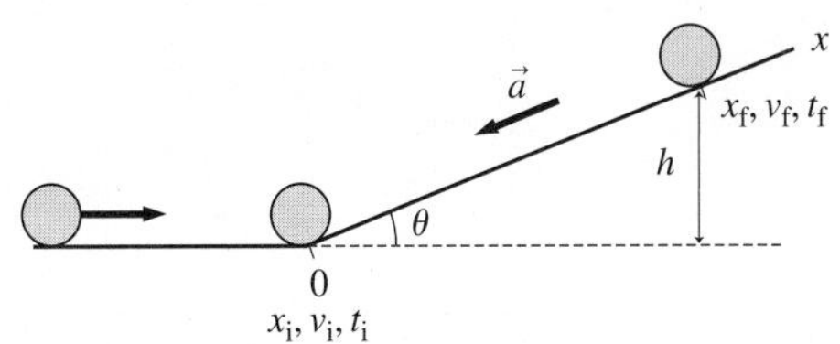

Motion diagram

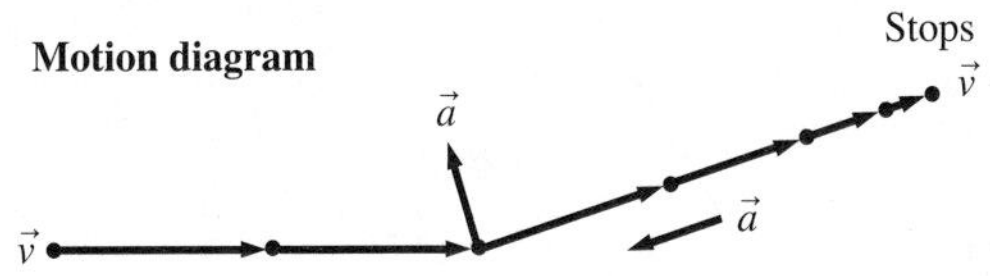

47. 12 in/ns
49. 43 mph
51. 3.0×10^4 ms
53. 40 ms
55. a.

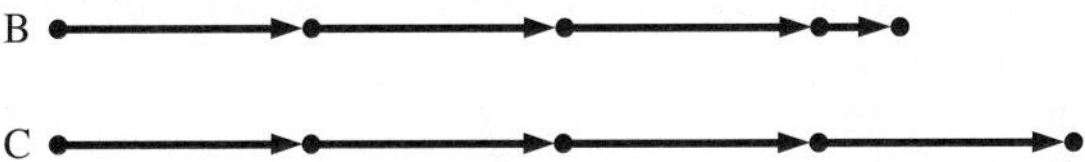

b. C
c. A

57. a. AB and CD b. All segments c. AB and CD
59. 30 m
61. 12 min
63. B
65. A

Chapter 2

Answers to odd-numbered multiple-choice questions

15. C
17. C
19. D
21. B
23. B
25. A

Answers to odd-numbered problems

1. b.

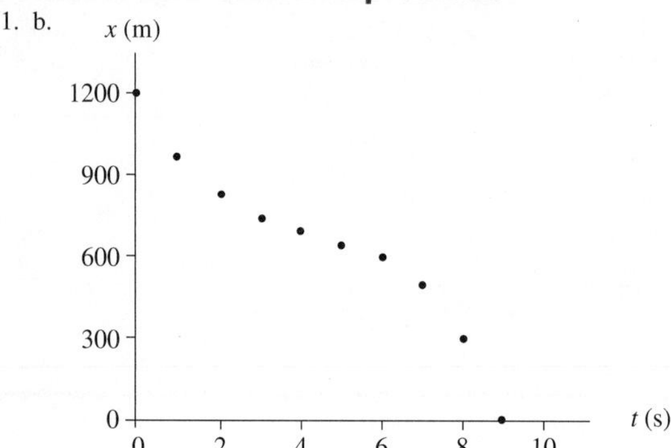

5. a. 8 s
 b.

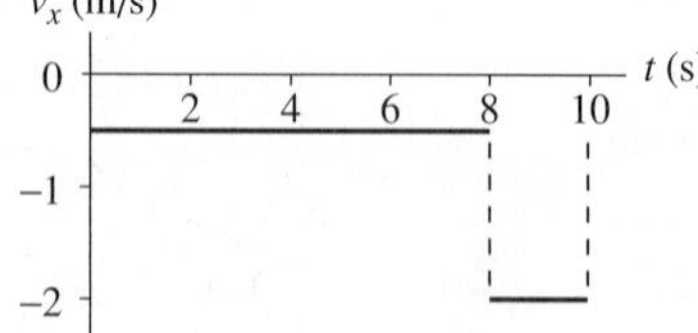

7. a.

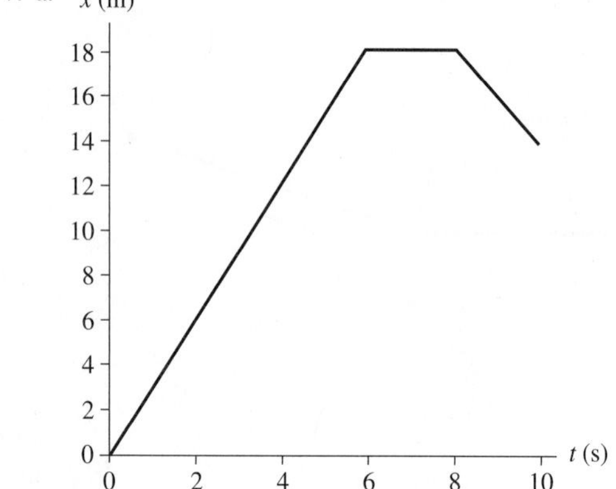

 b. 14 m
9. 2.5 m/s, 0 m/s, −10 m/s
11. a. Beth b. 20 min
13. 4.3 min
15. a. 15 m b. 90 m
17. a.

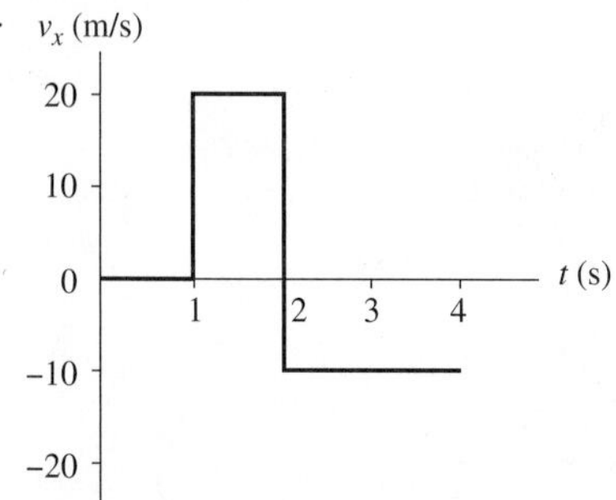

 b. Yes, at $t = 2$ s

19. a. 26 m, 28 m, 26 m b Yes, at $t = 3$ s
21.

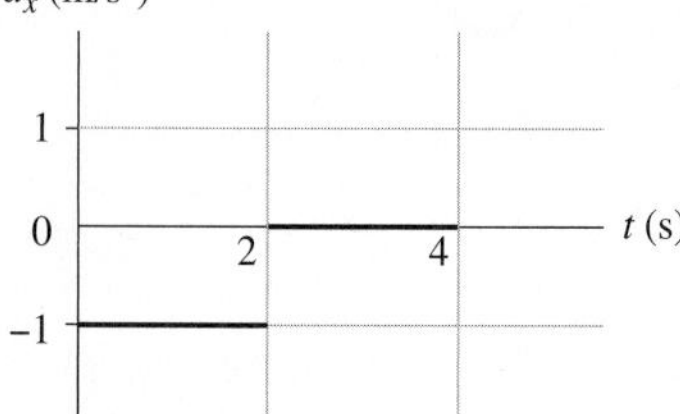

23. a. Positive b. Negative c. Negative
25. 6.1 m/s², 2.5 m/s², 1.5 m/s²
27. a. 36 m/s² b. 0.22 m
29. 26 m/s
31. 1 mi
33. Yes
35. a. 5 m b. 22 m/s
37. 10.0 s
39. 0.31 m
41. a. Both magnitudes and directions are equal b. Equal
43. a. 33 m b. 25 m/s
45. a. 3.0 s b. 15 m/s c. $(v_1)_f = 31$ m/s $(v_2)_f = 35$ m/s
47.

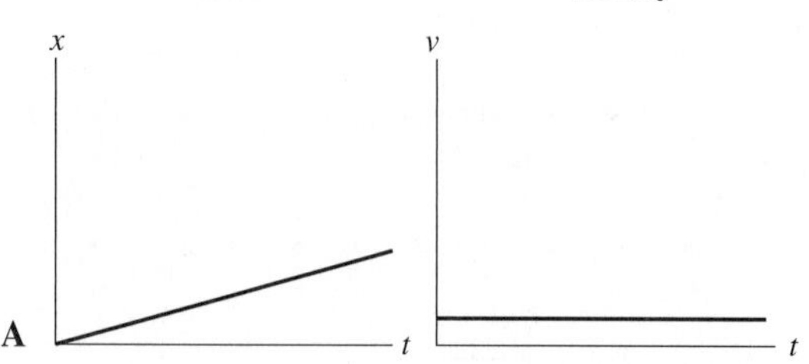

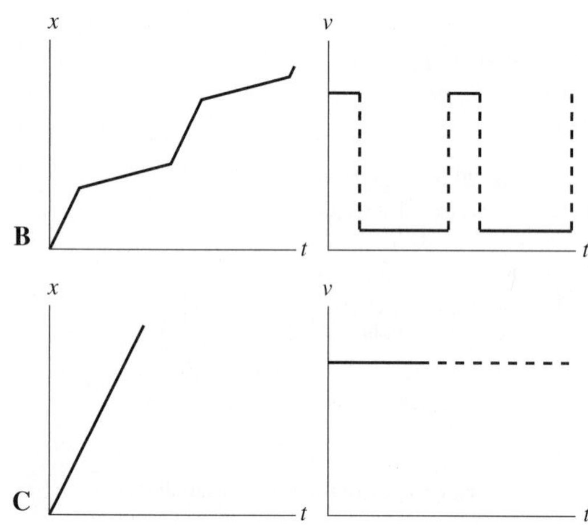

49. 57 mph
51. b.

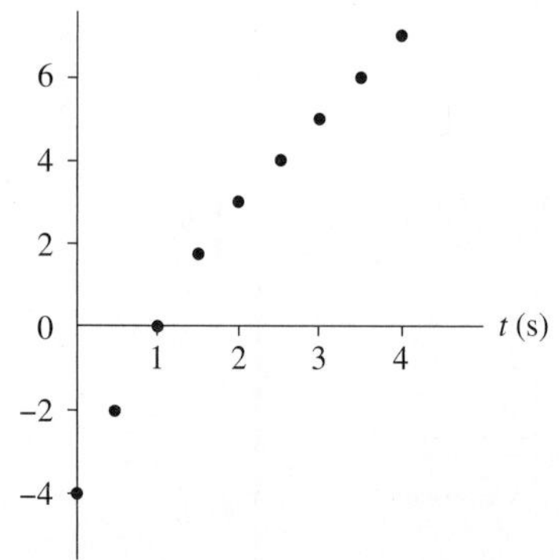

 c. 4 m d. 4 m e. 4 m/s f. 2 m/s g. −2 m/s²

53. a. 180 mph b. 2.3 m/s^2 c. 35 s d. No
55. 8.7 y
57. a. ≈1.0 cm b. 35 m/s^2 c. 0.84 m/s
59. a. 1000 m/s^2 b. 1.0 ms c. 5.1 cm
61. 33 m/s^2
63. a. 24 m/s b. 4.5 s
65. 12 m/s
67. 110 m
69. a. 4.1 s b. Same
71. a. 4.0 s b. 16 m
73. a. 100 m
b.

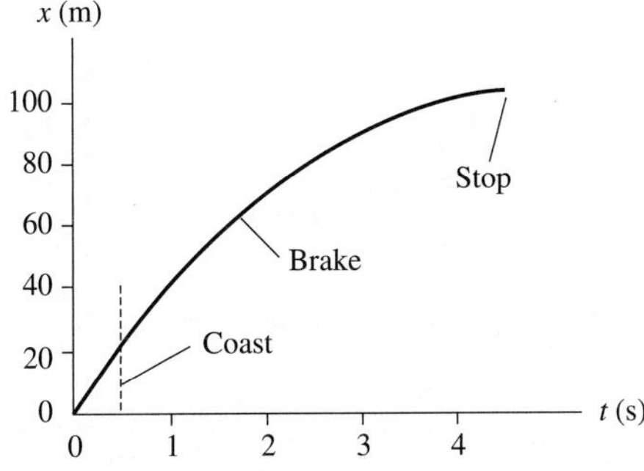

75. B
77. B

Chapter 3

Answers to odd-numbered multiple-choice questions

17. B
19. D
21. a. D b. A c. C d. A
23. a. C b. D
25. B
27. D

Answers to odd-numbered problems

7. 12 m/s
9. 87 m/s
11. a. $d_x = 71$ m, $d_y = -71$ m b. $v_x = 280$ m/s, $v_y = 100$ m/s
c. $a_x = 0.0$ m/s^2, $d_y = -5.0$ m/s^2
13. a. $v = 45$ m/s, $\theta = 63°$ b. $a = 6.3$ m/s^2, $\theta = 288°$
15. 5.0 km, 24° counterclockwise from $+x$-axis
17. 530 m
19. 1.3 s
21. Ball 1: 5 m/s; Ball 2: −15 m/s
23. 2.0 km/h
25. a. 22° b. 400 m
27. a. d. 10 m

$\vec{v}_1$ = (5, 0 m/s, horizontal)

20 m

10 m

29. 71 m
31. a. 64 ms b. 780 m/s
33. a. 15 s b. 330 m c. 280 m
35. a. 90 rev/s b. 11 ms
37. a. 170 rev/s; 6.0 ms b. 63 m/s c. 6,700 g
39. a. 4.0 m/s^2 b. 32 m/s^2
41. 27 m
43. a. $\vec{D} = (8, 7)$ c. $D = \sqrt{113}$, $\theta = \tan^{-1}(7/8) = 41°$
45. a. $\vec{B} = (-2, 2)$ b. $B = 2\sqrt{2}$, $\theta = 135°$
47. a. 4.3 m b. 5.5 m
49. 7.5 m
51. 46 min
53. a. 64 m b. 7.1 s
55. No
57. a. 39 mi b. 20 mph
59. 41° south of west
61. a. 7° south of east b. 2 h, 29 min
63. a. $(v_x)_0 = 2.0$ m/s $(v_y)_0 = 4.0$ m/s
$(v_x)_1 = 2.0$ m/s $(v_y)_0 = 2.0$ m/s
$(v_x)_2 = 2.0$ m/s $(v_y)_0 = 0.0$ m/s
$(v_x)_3 = 2.0$ m/s $(v_y)_0 = -2.0$ m/s
b. 2.0 m/s^2 c. 63°
65. a. $v = L\sqrt{g/2h}$ b. 6.0 m/s c. Yes, it is reasonable
67. 6.0 m
69. Yes, by 1.0 m
71. 3.5 s
73. a. 6.1 m/s b. 1.8 m
75. 11 mph
77. a. 2.9 m/s^2, 0.30g b. 57 mph
79. B
81. C

Chapter 4

Answers to odd-numbered multiple-choice questions

21. C
23. A
25. D
27. A
29. B

Answers to odd-numbered problems

1. First is rear-end; second is head on
5.

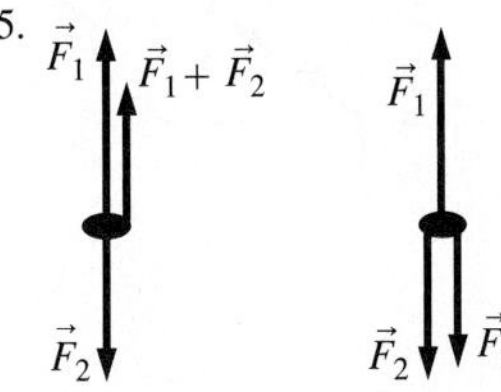

7. Weight, tension force by rope
9. Weight, normal force by ground, kinetic friction force by ground
11. Weight, normal force by ground, kinetic friction force by ground
13. $m_1 = 0.08$ kg and $m_3 = 0.50$ kg
15. a. 2.4 m/s^2 b. 0.60 m/s^2
17. a. 16 m/s^2 b. 4.0 m/s^2 c. 8.0 m/s^2 d. 32 m/s^2
19. 2.5 m/s^2
21. 0.25 kg
23. 7.5 × 10^2 N
25. 0.02 m/s^2
29.

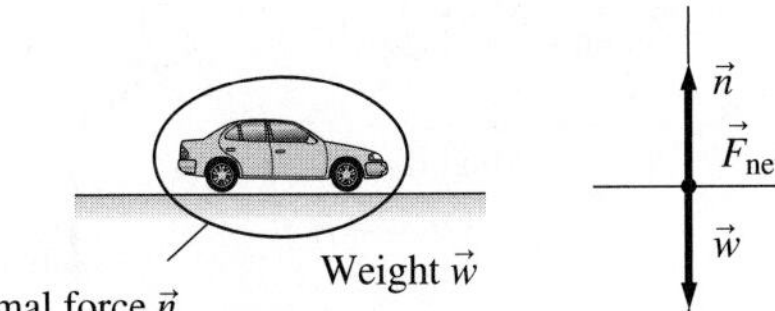

35. 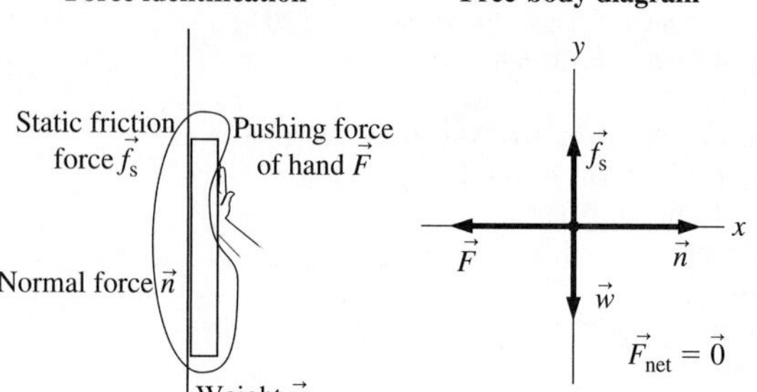

39. 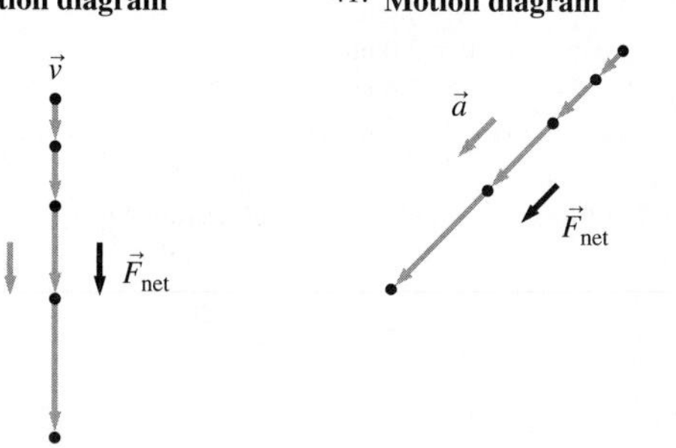

43. 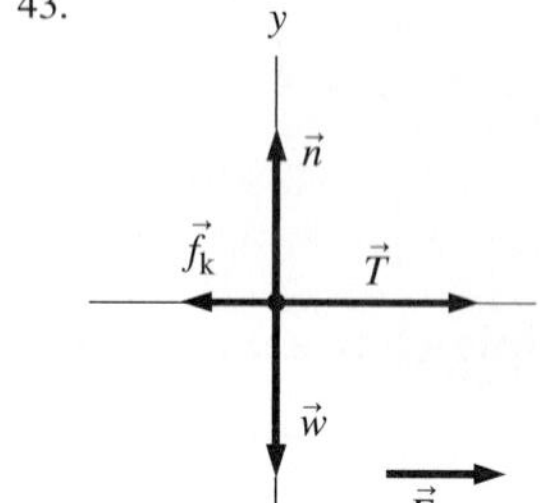

45. 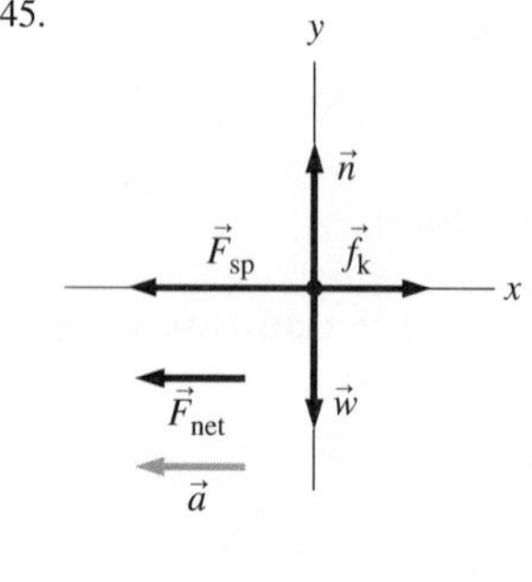

47. 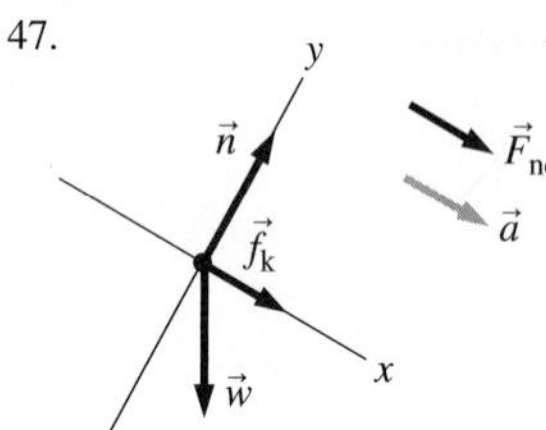

49. 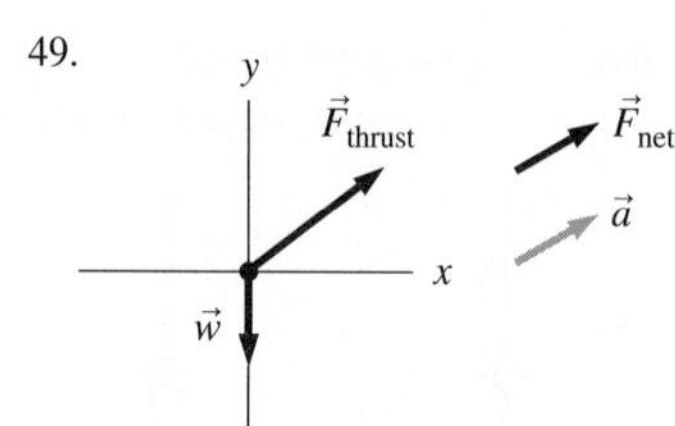

53. 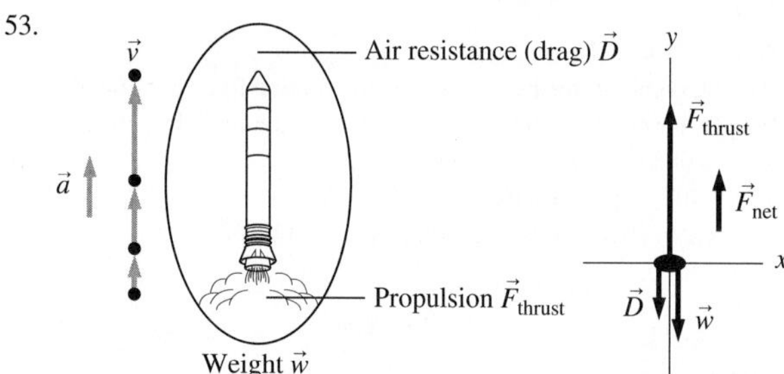

55. 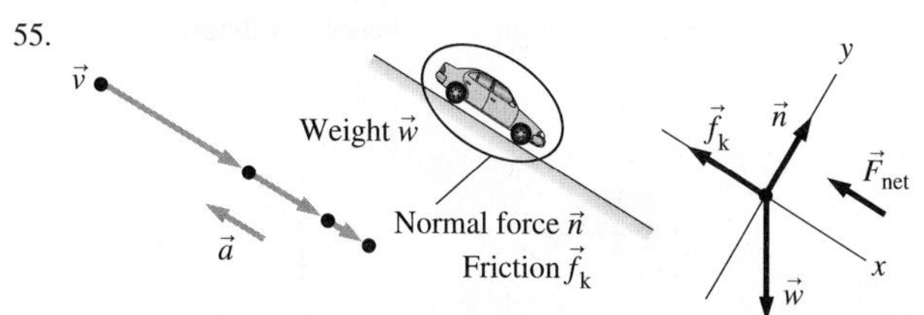

59. 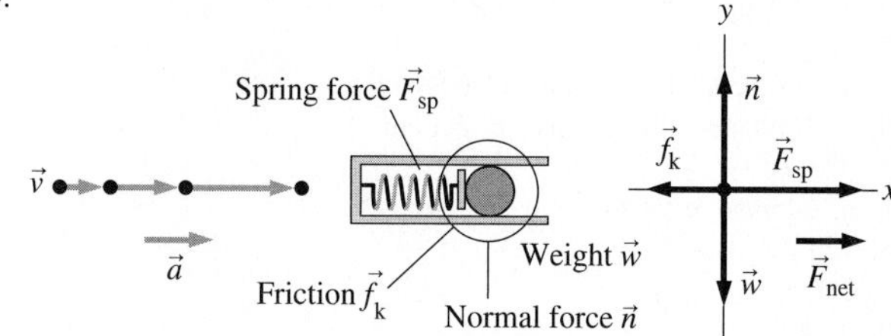

61. 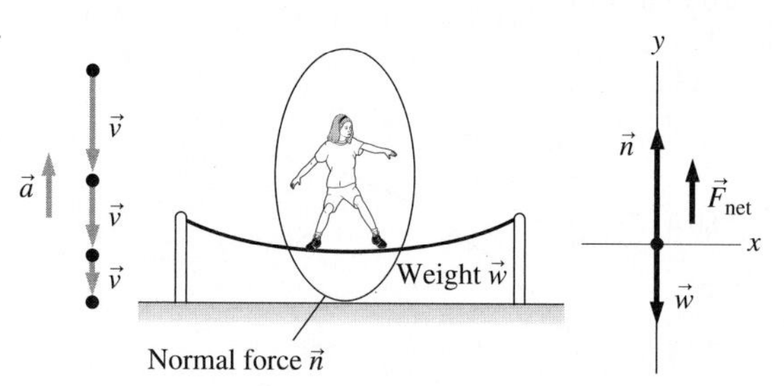

63. 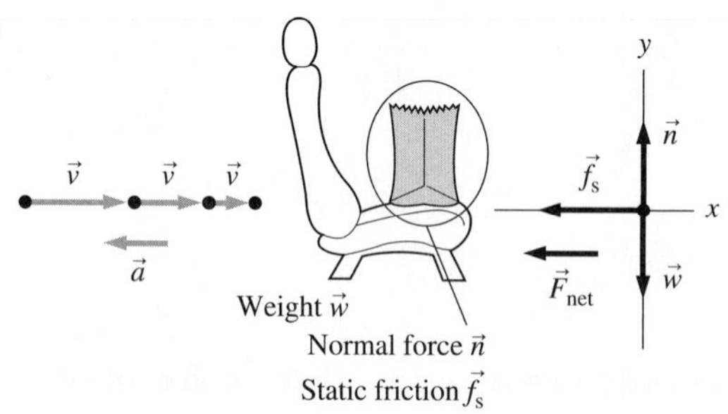

67. a. 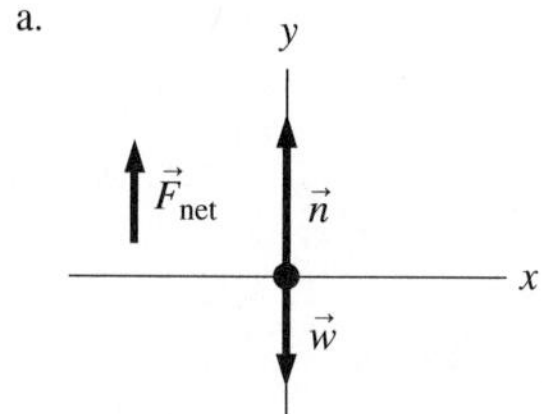

b. Greater

69. C
71. C

Chapter 5

Answers to odd-numbered multiple-choice questions

21. D
23. a. B b. D
25. a. C b. B
27. C
29. C
31. D

Answers to odd-numbered problems

1. $T_1 = 87$ N $T_2 = 50$ N
3. 110 N
5. 270 N

7. 170 kg
9.

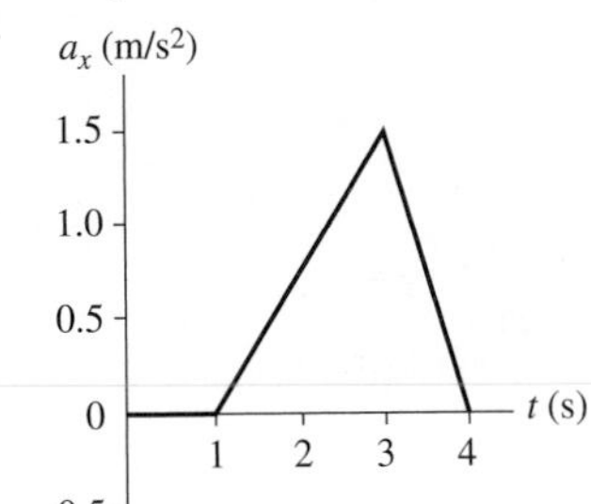

11. $a_x = 1.0\ \text{m/s}^2 \quad a_y = 0.0\ \text{m/s}^2$
13. a. 0.0 N b. 0.0 N c. 250 N
15. 9800 N, toward the rear
17. a. 539 N b. 89.1 N, 55.0 kg
19. a. 780 N b. 1600 N
21. a. 780 N b. 1100 N
23. a. 5.9 N b. 5.1 N
25. 0.25
27. 140 m
29. a. b.

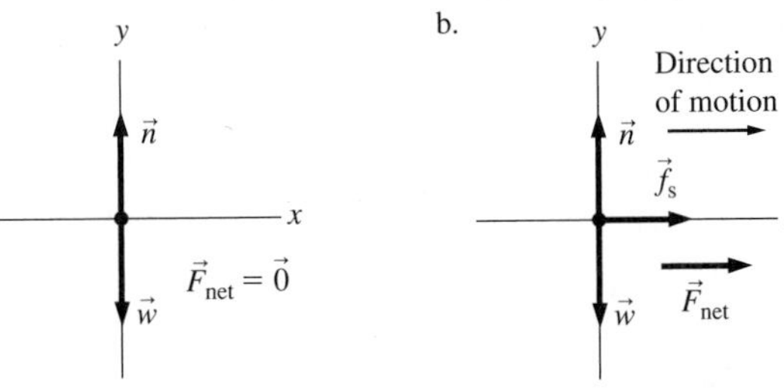

c. 4.9 m/s² d. 2.9 m/s²

31. a. 68 N b. 610 N
33. 170 m/s
35. a. 6.0 N b. 10 N
37. a. 20 N b. 21 N
39. a. 1.0 N b. 50 N
41. a. 530 N b. 5300 N
43. $F_{net}(1\ \text{s}) = 8\ \text{N} \qquad F_{net}(4\ \text{s}) = 0\ \text{N} \qquad F_{net}(7\ \text{s}) = -12\ \text{N}$
45. a. 490 N b. 740 N
47. a. $2.0 \times 10^4\ \text{m/s}^2$ b. 2800 N c. 2800 N d. Forehead no, cheek yes
49. a. 230 N b. 0.20 m/s
51. a. 3.96 N b. 2.32 N
53. a. 1.3 m/s² b. 2.0 m/s²
55. 60 m
57. a. 9.8 m/s² b. −9.8 m/s² c. 20 m/s² d. 2.5 m/s e. 31 cm
59. a. 17 m/s b. 230 m
61. Stay at rest
63. 1.9×10^5 N
65. $T_1 = 17\ \text{N} \quad T_2 = 27\ \text{N}$
67. 27°
69. a. $-5g$ b. $3g$
71. 0.93 m/s², down
73. 3.7 cm
81. A
83. D

Chapter 6

Answers to odd-numbered multiple-choice questions

21. D
23. C
25. B
27. A
29. D
31. C
33. B

Answers to odd-numbered problems

1. a. $\theta = \pi/2$ b. $\theta = 0$ c. $\theta = 4\pi/3$
3. 1.7×10^{-3} rad/s
5. a. 1.3 rad, 72° b. 4×10^{-5} rad/s
7. 3.0 rad
9. a. 25 rad b. −5 rad/s
11. 3.9 m/s
13. a. 20 s b. 2.5 m/s
15. $v = 5.7$ m/s $\quad a = 110\ \text{m/s}^2$
17. $T_3 > T_1 = T_4 > T_2$
19. 9400 N, toward center, static friction
21. a. 1700 m/s² b. 240 N
23. 12 m/s
25. 20 m/s
27. 1.6 m/s²
29. 1/2
31. 6.00×10^{-4}
33. a. 3.53×10^{22} N b. 1.99×10^{20} N c. 0.564%
35. 1600 d
37. 92 min, 7700 m/s
39. 4.2 h
43. North pole, by 2.5 N
45. 5.5 m/s
47. a. Two; upward normal force and inward b. 0.24 N
49. a. 5.0 N b. 30 rpm
51. 5.4 m/s
53. 22 m/s
55. a. 1.7 m/s², upward
b.

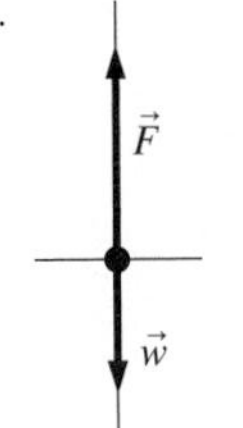

c. 12 mN, upward d. 9.8 mN, upward

57. 54 N
59. 2,400 m
61. a. 3.0×10^{24} kg b. 0.89 m/s²
63. a. 3.77 m/s² b. 1.6 m/s
65. (12 cm, 0 cm)
67. 6.5×10^{23} kg
69. 0.48 m/s
71. b. $v = 19.8$ m/s
73. b. $r = 1.00 \times 10^8$ m
75. C
77. A

Chapter 7

Answers to odd-numbered multiple-choice questions

17. C
19. A
21. A
23. C

Answers to odd-numbered problems

1. 11 m/s
3. a. 160 rad/s² b. 50 rev
5. $\tau_1 < \tau_2 = \tau_3 < \tau_4$
7. $-20\ \text{N} \cdot \text{m}$

9. 5.7 N
11. a. 0.0363 N · m b. 104 N · m
13. 5.5 N · m
15. a. 1.1 m b. 1.2 m
17. 12 N · m
19. −4.9 N · m
21. a. 34 N · m b. 24 N · m
23. a. 1.70 m b. 833 N · m
25. 7.2×10^{-7} kg · m^2
27. 120 kg · m^2
29. 1.8 kg
31. a. 6.9 kg · m^2 b. 0.75 rad/s^2
33. 1.6 kg · m^2
35. 0.047 N · m
37. 17 rad/s^2
39. 0.50 s
41. 2.2 m/s^2
43. 17 m/s
45.

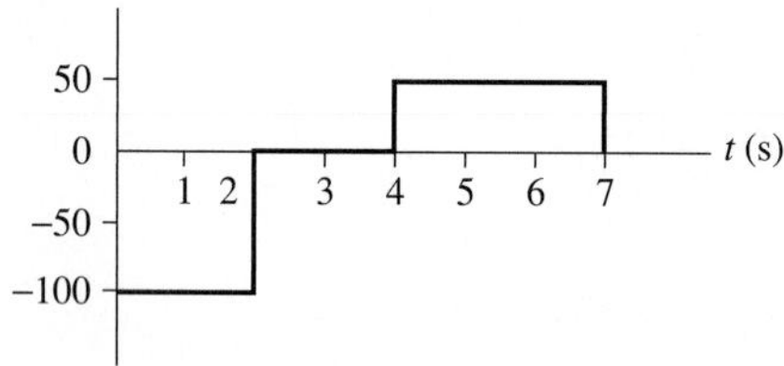

47. 55 rotations
49. −0.94 N · m
51. 7.5 cm
53. a. $x_{cg} = 6.0$ cm, $y_{cg} = 8.2$ cm c. 54°
55. 0.018 kg · m^2
57. a. 1.4 kg · m^2 b. Increase
59. 3.5 rad/s^2
61. −0.28 N · m
63. 1.1 s
65. 2.2 N
67. B
69. B

Chapter 8

Answers to odd-numbered multiple-choice questions

15. A
17. B
19. B
21. B
23. D

Answers to odd-numbered problems

1. Right 470 N; left 160 N
3. 15 cm
5. 140 N
7. No, there is a net torque
9. 590 N
11. 1:1
13. 8.9°
15. 15.0 cm
17. 830 N/m
19. a. 13 cm b. 8 cm
21. 18 cm
23. 1.5 m/s^2
25. a. 2.0 mm b. 0.25 mm
27. 7900 N
29. a. 4000 N b. 4.0 mm
31. 15 cm
33. 0.0078%
35. $F_1 = 750$ N $F_2 = 1000$ N
37. 860 N
39. a. 2000 N b. 2200 N
41. 350 N
43. 1.0 m
45. a. 49 N b. 1500 N/m c. 3.4 cm
47. 17.4 N
49. a. 1.0 cm b. 4.5 N to the left
51. 31 cm
53. a. 8.2 m/s to the right b. 30 N
55. 81 cm
57. 2.5×10^{-5}
59. a. 4.8×10^{-2} m^2 b. 70000 kg c. 14%
61. B
63. B

Part I Problems

Answers to odd-numbered problems

1. A
3. A
5. C
7. B
9. D
11. A
13. B
15. D
17. B
19. A
21. b. 2.7 N

Chapter 9

Answers to odd-numbered multiple-choice questions

19. D
21. D
23. D
25. C
27. B

Answers to odd-numbered problems

1. 75 m/s
3. 6.0 N
5. −80 N · s
7. a. 1.5 m/s to the right b. 0.5 m/s to the right
9. −110 N
11. a. 19 kN b. 300 kN
13. 0.205 m/s
15. 0.31 m/s
17. 1.43 m/s
19. 0.048 m/s
21. 4.8 m/s
23. 2.3 m/s
25. 1.0 kg
27. (−2 kg · m/s, 4 kg · m/s)
29. 14 m/s, 45° north of east
31. 510 kg · m^2/s
33. 0.025 kg · m^2/s
35. 1.3 rev/s
37. 0.2 s
39. 930 N
41. 3.9 m/s
43. 13 N · s; left, 15° above the horizontal

45. 7.5×10^{-10} N·s
47. a. 6.4 m/s b. 360 N
49. 3.6 m/s
51. 1.1 m/s
53. 2.0 m/s
55. a. 6.7×10^{-8} m/s b. 2.2×10^{-10}%
57. 13 s
59. 440 m/s
61. −20.0 m/s
63. 196 m/s
65. Forward, 1.5×10^7 m/s
67. 0.85 m/s, −72°
69. 4.5 rpm
71. a. 2.8 m/s b. 2.2 m/s, 25 cm
73. −0.90 rad/s
75. 18 rev/s, clockwise
77. B
79. C

Chapter 10

Answers to odd-numbered multiple-choice questions

23. C
25. C.
27. C.

Answers to odd-numbered problems

1. 0.0 J
3. $W_{\vec{w}} = 12.5$ kJ, $W_{\vec{T}_1} = -7.92$ kJ, $W_{\vec{T}_2} = -4.58$ kJ
5. a. 0 J b. 2.6 kJ c. −2.6 kJ
7. The bullet
9. a. 14 m/s b. Factor of 4
11. 0.0 J
13. 1.8 kg·m²
15. a. 6.8×10^5 J b. 46 m c. No
17. 63 kJ
19. 0.63 m
21. 9.7 J
23. 470 J
25. 1.0×10^2 N
27. a. 13 m/s b. 14 m/s
29. 31 m/s
31. 3.0 m/s
33. 17 m/s
35. a. The child's gravitational potential energy will be changing into kinetic energy and thermal energy. b. 550 J
37. 0.86 m/s and 2.9 m/s
39. 1/2
41. a. 1.8×10^2 J b. 59 W
43. 45 kW
45. a. 30 N b. 45 W
47. 2.0×10^4 W
49. a. 7.7 m/s b. 6.6 m/s
51. 2.3 m/s
53. a. 0.20 kJ b. 98 N c. 2.0 m d. 0.20 kJ
55. 15 m/s
57. 51 cm
59. 3.8 m, not dependent on mass
61. a. $\sqrt{\dfrac{(m-M)kd^2}{m^2}}$ b. 2.0×10^2 m/s
c. 99.8% kinetic energy transformed to thermal energy
63. 1.6 m/s
65. 8.0 m/s
67. 0.49 m
69. a. $0.048v_i$ b. 95%
71. a. 100 N b. 0.20 kW c. 1.2 kW
73. 5.5×10^4 L
75. B
77. C
79. A
81. B
83. C

Chapter 11

Answers to odd-numbered multiple-choice questions

29. C
31. B
33. D
35. C

Answers to odd-numbered problems

1. 1.7 MJ
3. 3.3%
5. 4.2 MJ
7. 230,000 J = 54,000 cal = 54 Cal
9. 1.4 km
11. 710 m or 260 flights
13. a. 200 J b. 16,000 J/day c. 0.0094 donuts
15. 313°C
17. 4.8°C
19. −700 J
21. −150 J
23. a. 15 kJ b. 27%
25. 25%
27. a. 7°C b. −4°C
29. 460°C
31. 1.5
33. a. 200 J b. 250 J
35. a. (b) only b. (a) only
37. 13%
39. 0.80 burritos
41. a. 1.4 MJ b. 1.4 MJ c. Cycling
43. a. 190 kJ b. 750 kJ c. 180 Cal d. 130 Cal e. 1200 W
45. 490 W
47. a. 300 W b. 1.1 MJ = 260 Cal
49. a. 10000 Cal b. 490 W; 390 W more
51. a. 2600 Cal b. 1.1 kg
53. 7.2×10^{22} atoms
55. a. 40% b. 220°C
57. 230 K
59. a. 60 J b. −23°C
61. 95,000
63. a. 47% b. 35%
65. a. 880 N b. 4.4 kW c. 18%
67. 8.0%
69. C
71. A
73. C

Part II Problems

Answers to odd-numbered problems

1. A
3. B
5. D
7. C
9. A
11. C
13. D

15. B
17. C
19. A
21. D
23. 0.22 m/s to the west
25. a. 3.0 m/s b. 120 N c. 60 J

Chapter 12

Answers to odd-numbered multiple-choice questions

27. D
29. D
31. A
33. E
35. B

Answers to odd-numbered problems

1. Carbon
3. 3.5×10^{24}
5. $0.024\ m^3 = 24\ L$
7. 49.7 psi
9. 1.1 N
11. a. −140°C b. 2.7×10^{-21} J
13. 1900 kPa
15. 17 L
17. a. $T_f = T_i$ b. $p_f = \frac{1}{2}p_i$
19. a. 98 cm^3
21. a. Isochoric b. 910 K, 300 K
23. a. Isobaric b. 120°C c. 0.0094 mol
25. 16°C
27. 0.30 atm
29. 4.7 m
31. 0.83%
33. 6.9 kJ
35. 2300 W
37. 250 Cal/day, 12 W
39. 260 kJ
41. 0.218 kg
43. Iron
45. 91 g
47. a. 31 J b. 60°C
49. 0.080 K
51. 41 J
53. 830 W
55. 24 W
57. 6.0 W
59. 4.7 cm^2
61. a. 50 L b. 1.3 atm
63. a. 5.4×10^{23} b. 3.6 g
65. a. 140 J b. 0.4 L
67. 35 psi
69. 174°C
71. a. 0.73 atm b. 0.52 atm
73. 60 J
75. a. 0.51 atm b. 160 K
77. 8.7 h
79. 56 J
81. 2.4×10^6 kg/min
83. 60 K
85. 0.023 K, no
87. a. 83 J/kg · K b. 200 kJ/kg
89. 970 g
91. 60 g
93. a. 3.1 atm b. 9.8 L
95. a. 150 J b. −91 J
97. 0.026 W/m · K
99. 2
101. A
103. C

Chapter 13

Answers to odd-numbered multiple-choice questions

31. A
33. D
35. B
37. B
39. A

Answers to odd-numbered problems

1. 1200 kg/m^3
3. 11 kg/m^3
5. a. 810 kg/m^3 b. 840 kg/m^3
7. 8.0%
9. a. 6.3 m^3 b. 1.2×10^5 Pa
11. 210 N
13. 3.2 km
15. a. 2.9×10^4 N b. 30 football players
17 a. 1.1×10^5 Pa b. 4400 Pa for both A–B and A–C
19. 3.68 mm
21. 89 mm
23. 0.82 m
25. 1.9 N
27. 46 kg
29. 1.0 m/s
31. 97 min
33. 110 kPa
35. 3.0 m/s
37. 260 Pa
39. 2.17 cm
41. 4.8×10^{23} atoms
43. 27 cm
45. a. 5800 N b. 6000 N
47. 1.8 cm
49. 750 kg/m^3
51. 44 N
53. a. 0.38 N b. 20 m/s
55. 2×10^{-3} m/s
57. a. p_{atmos} b. 4.6 m
59. 4.4 cm
61. −28°C
63. 13 mm Hg
65. B
67. A

Part III Problems

Answers to odd-numbered problems

1. A
3. B
5. D
7. B
9. C
11. A
13. C
15. D
17. A
19. 300 N
21. a. 0.071 mol b. $6.7 \times 10^{-4}\ m^3$ c. 0.019 mol

Chapter 14

Answers to odd-numbered multiple-choice questions

23. a. B b. C c. B d. B e. B
25. B
27. B

Answers to odd-numbered problems

1. 2.3 ms
3. 0.80 s, 1.3 Hz
5. 40 N
7. a. 13 cm b. 9.0 cm
9. a. 20 cm b. 0.25 Hz
11. a. 0.30 m b. 2.3 m/s c. 17 m/s^2
13. a. 3.0 mm b. 0.024 m/s
15. a. $U = 1/4E$ $K = 3/4E$ b. $0.707A$
17. a. 2.83 s b. 1.41 s c. 2.00 s d. 1.41 s
19. a. 2.0 cm b. 0.63 s c. 5.0 N/m d. 20 cm/s e. 1.0×10^{-3} J f. 15 cm/s
21. a. 1.00 s b. 0.628 m/s c. 0.100 J
23. a. 4.00 s b. 5.66 s c. 2.83 s d. 4.00 s
25. a. 0.10 rad b. 0.80 Hz c. 0.39 m d. −0.084 rad
27. 33.1 cm
29. 1.6 s
31. a. 0.124 rad b. 2.00 s c. 1.49 m
33. 1.3 s
35. 10.0 s
37. a. 6.2 s b. 3.4 cm
39. 2.8 s
41. 250 N/m
43. 1.1 m/s
45. a. 0.169 kg b. 0.565 m/s
47. 5.86 m/s^2
49. a. 0.25 Hz, 3.0 s b. 6.0 s, 1.5 s c. 2.25
51. a. 6.40 cm b. 28.3 cm/s
53. a. 47,000 N/m b. 1.8 Hz
55. 11 cm, 1.7 s
57. 0.64 Hz
59. 8.1 N/m
61. a. 2.00709 s b. 2.00721 s c. 4.6 h
63. a.

	f (Hz)	T (s)
24.8 cm rod	1.00	1.00
38.8 cm rod	0.800	1.25

b. 5.00 s c. 0.200 Hz

65. a. 9.5 N/m b. 0.50 m/s c. 2.9 s
67. 73 cm
69. 236 oscillations
71. 2.4 J
73. A
75. C

Chapter 15

Answers to odd-numbered multiple-choice questions

21. D
23. D
25. D

Answers to odd-numbered problems

1. 280 m/s
3. 140 m/s
5. 6000 m/s
7. History graph at $x = 6$ m

y (cm)
1 cm
0
0 1 2 3 4 5 6 t (s)

y (cm)

9. 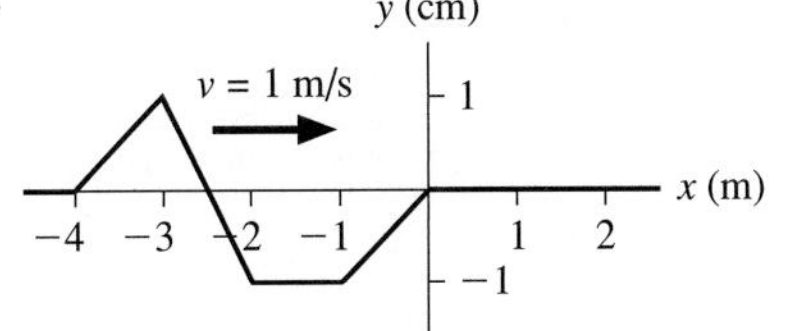

Snapshot graph at $t = 1.0$ s

11. 10 m/s
13. a. 8.6 mm b. 15 ms
15. a. 11 Hz b. 1.1 m c. 13 m/s
17. $A = 4$ cm; $\lambda = 12$ m; $f = 2.0$ Hz
19. 11 s
21. 14.8 mm
23. a. 1700 Hz b. 1.5 GHz c. 990 nm
25. a. 10 GHz b. 170 μs
27. 55 pJ
29. 600 kJ
31. 11 W
33. a. 6.7×10^4 W b. 8.5×10^{10} W/m^2
35. 25 pW/m^2, 14 dB
37. a. 3.2×10^{-3} W/m^2 b. 95 dB
39. 100 dB
41. a. 650 Hz b. 560 Hz
43. 16 m/s away
45. 4.5 m/s
47. 86 m
49. 790 m
51. 50 cm/s, 13 m/s^2
53. 890 g
55. $L_1 = 2.3$ m, $L_2 = 1.7$ m
57. a. 3.0037 m b. 11 μs c. 1.1×10^{-2}
59. 0.069°C
61. a. 3 s b. 9 m/s
63. $y(x, t) = (5.0\text{ cm})\cos\left[2\pi\left(\frac{x}{50\text{ cm}} + \frac{t}{0.125\text{ s}}\right)\right]$
65. b. 33 m/s to the right
67. 6.8 cm/s
69. 20 μW/m^2
71. a. 50 MW b. 5.0 W c. 6.4×10^{17} W/m^2 d 5.8×10^{14}
73. a. 110 dB b. 16 km
75. a. 1.9 W/m^2 b. 123 dB c. 71 m
77. 620 Hz, 580 Hz
79. 110 m/s
81. D
83. B

Chapter 16

Answers to odd-numbered multiple-choice questions

21. A
23. D
25. A
27. D

Answers to odd-numbered problems

1.

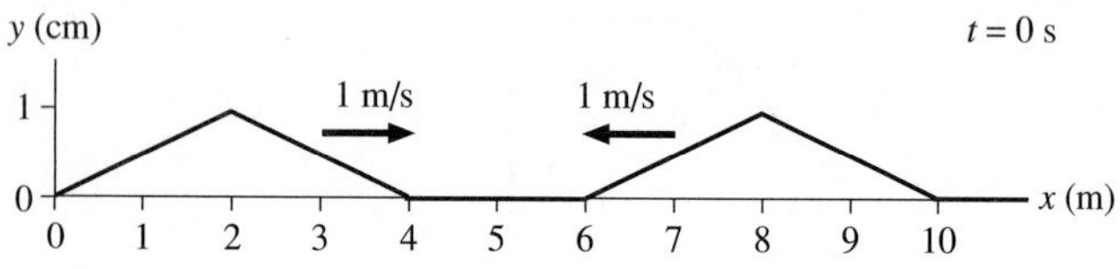

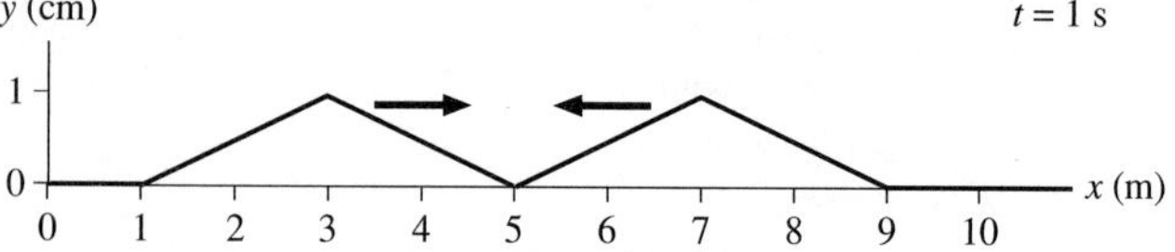

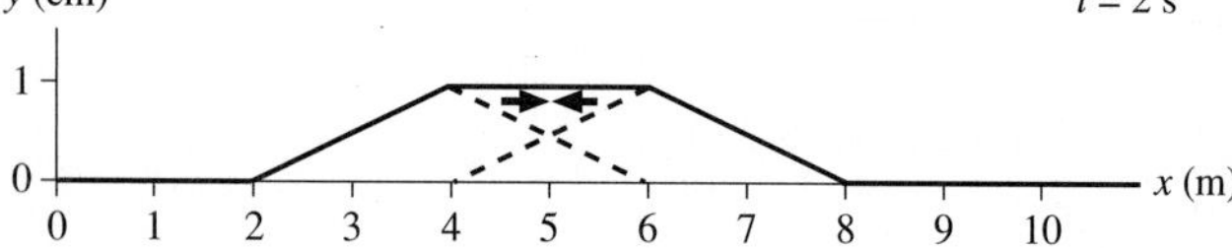

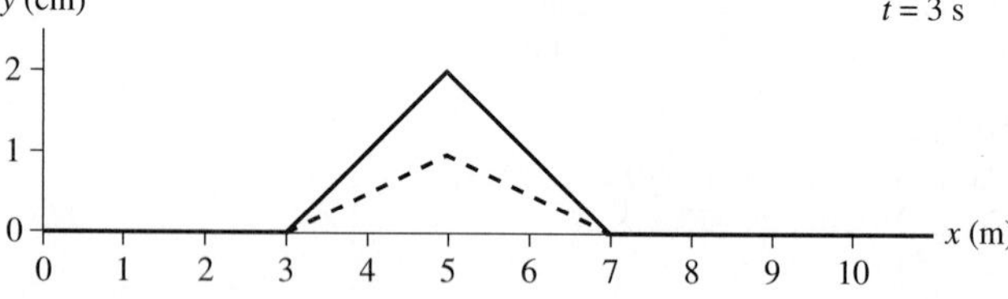

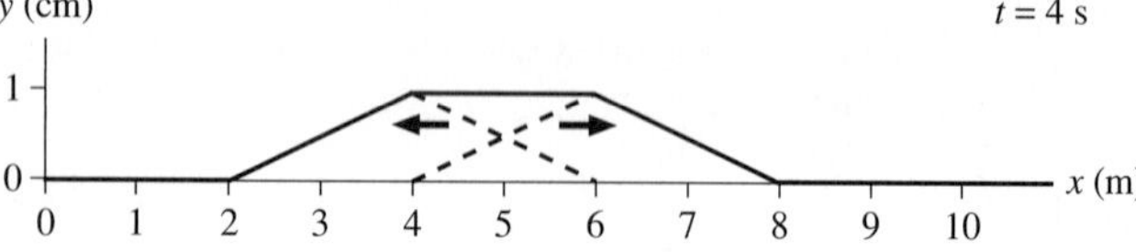

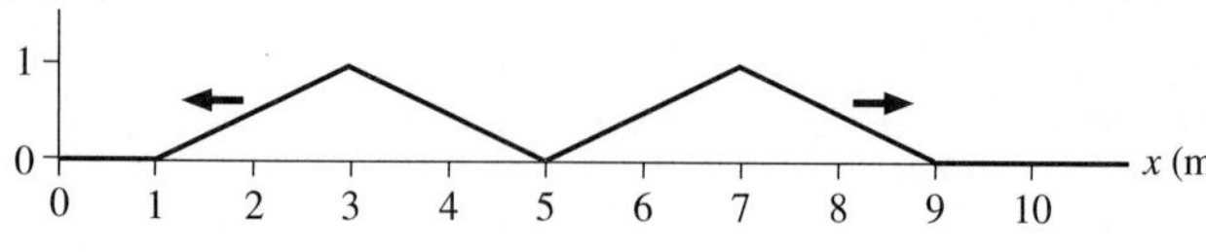

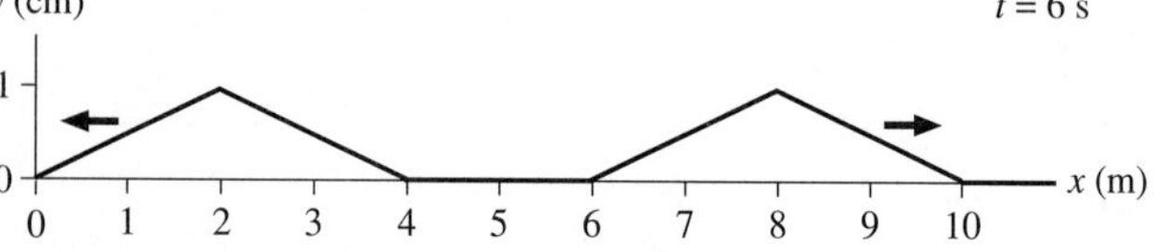

3. 4 s
5. 5.0 s
7. 60 Hz
9. 53 m/s
11. a. 4.80 m, 2.40 m, and 1.60 m b. 75.0 Hz
13. a. 700 Hz b. 56 N
15. 2180 N
17. a. 16 Hz b. 3.18 m
19. a. 2.42 m, 1.21 m, and 0.807 m b. 4.84 m, 1.61m, and 0.968 m
21. 510 Hz
23. a. 5.67 Hz b. 3.5 Hz c. The frequency will decrease.
25. 6700 Hz
27. 580 Hz and 4900 Hz
29. a. 40 cm b. 60 cm
33. 203 Hz
35. 880 m/s
37. 1.2 m
39. 65 Hz
41. 270 N
43. 12 kg
45. 8.2 m/s^2
47. 19 kN
49. 18 cm
51. b. 10.5, 13.5, 16.5, and 19.5 GHz
53. a. 0.530 m b. 2.83×10^8 Hz c. 2.83×10^8 round trips
55. 1210 Hz
57. 18 cm
59. 1400 Hz
61. a. 3.00 m b. Strong c. Weak
63. 20
65. a. 2 Hz b. 658 Hz
67. 0.81 s
69. 1750 Hz
71. 0.2 m/s
73. B
75. B

Part IV Problems

Answers to odd-numbered problems

1. $v_{light} > v_{earthquake} > v_{sound} > v_{tsunami}$
3. D
5. B
7. D
9. C
11. B
13. C
15. C
17. C
19. C
21. 500

Chapter 17

Answers to odd-numbered multiple-choice questions

19. B
21. C
23. B
25. D

Answers to odd-numbered problems

1. a. 1.5×20^{-11} s b. 3.4 mm
3. 0.40 ns
5. 670 ns
7. 0.020 rad, 1.1°
9. 0.050 rad
11. 0.40 mm
13. 2.3 μm
15. 3.2°, 6.3°
17. 221 lines/mm
19. a. 1.3 m b. 7
21. 217 nm
23. 85.9 nm
25. 121 nm
27. 97.8 nm
29. 94 μm
31. 610 nm
33. 4.9 mm
35. 78 cm
37. 6.4 cm
39. 500 nm
41. 450 nm

43.

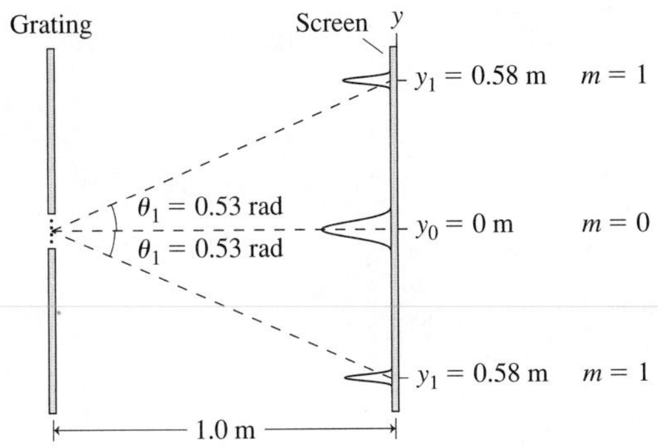

45. I_1
47. 43.2°
49. 43 cm
51. 500 nm
53. 25 cm
55. 1.8 μm
57. 500 nm
59. 667.8 nm
61. 410 nm, 690 nm, purple
63. 2.8°
65. 7.6 m
67. 0.10 mm
69. a. Double b. 0.15 mm
71. 0.1 mm
73. a. Diffraction b. 0.044° c. 4.6 mm d. 1.5 m
75. D
77. A

Chapter 18

Answers to odd-numbered multiple-choice questions

17. D
19. B
21. E
23. B
25. B
27. A

Answers to odd-numbered problems

1. 8.00 cm
3. 11 ft
5. B and C
7. 433 cm
9. a. 3 b. B = (4 m, 1 m), C = (2 m, 5 m), D = (4 m, 5 m)
   c.

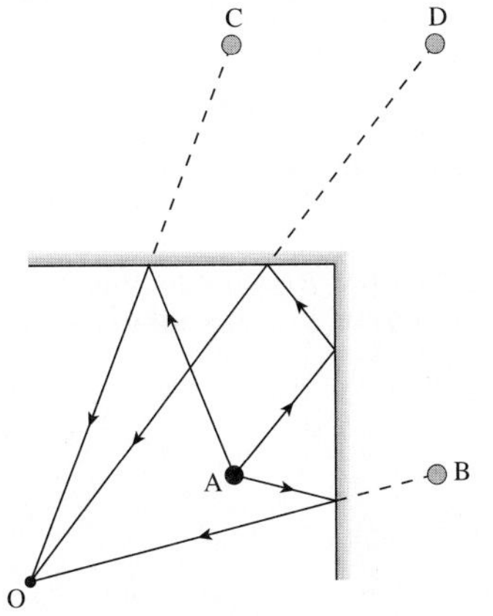

11. 1.37
13. 4.0 m
15. 76.7°
17. 23 cm
19. $s' = 15$ cm, inverted and real
21. $s' = -6.7$ cm, upright and virtual
23. $s' = -40$ cm, upright and virtual, $h' = 10$ cm
25. $s = 18$ cm
27. $s' = -1.3$ m, upright and virtual
29. $s' = 40$ cm, $h' = 2.0$ cm
31. $s' = -60$ cm, $h' = 8.0$ cm
33. $s' = -8.6$ cm, $h' = 1.1$ cm
35. $s' = -9.4$ cm, $h' = 1.9$ cm
37. $s' = -38$ cm, $h' = 7.5$ cm
39. 1.0 m
41. 42.0°
43. 60°
45. a. The angle of the beam is 32.1° from the normal. b. No
47. 82.8°
51. 35°
53. a. $\beta = \sin^{-1}\left(n \sin\frac{\alpha}{2}\right)$ b. 1.58
55. 42°
57. 5.7 m
59. 1.52
63. $2f$
65. $s' = 20$ cm, $h' = 1.0$ cm
67. $s' = -4.3$ cm, $h' = 0.57$ cm
69. $s' = -50$ cm, $h' = 40$ cm
71. $f = 44$ cm, 67 cm
73. a. 5.9 cm b. 6.0 cm
75. A
77. B

Chapter 19

Answers to odd-numbered multiple-choice questions

17. B
19. C
21. C
23. C

Answers to odd-numbered problems

1. a. -5.8 mm b. The object must be getting closer.
3. 0.30 m
5. 1/30 s
7. 22 mm, 2.8 mm
9. 2.0 D
11. 3
13. a. Farsighted b. ∞
15. 2.1 m
17. 5.0 cm
19. a. 1.20 cm b. 1.67 cm
21. 11 mm
25. $s' = -6.4$ cm to the left of the diverging lens, $h' = 1.0$ cm, same orientation as the object
27. a. 0.058 D b. 17 mm
29. a. 14° b. 6.4 mm
31. 0.28 mm, violet
33. 55 km
35. 2.7 mm
39. 0.70 cm
41. b. $s'' = -20$ cm, $h'' = 2.0$ cm

43. 5.0 mm
45. 250
47. −4
49. a. Objective: + 3.0 D; eyepiece: + 4.5 D b. $M = -1.5$ c. 0.56 m
51. 24 cm
53. 2.8×10^{-7} rad
55. 15 km
57. No
59. A
61. B

Part V Problems

Answers to odd-numbered problems

1. C
3. A
5. A
7. B
9. D
11. A
13. A
15. Aberrations are reduced.
17. 7.2 mm, 0.069 W/m^2

Chapter 20

Answers to odd-numbered multiple-choice questions

23. a. B b. C c. B d. C
25. B
27. C
29. A

Answers to odd-numbered problems

1. a. Electrons are removed. b. 3.1×10^{10}
3. 1.5×10^6 C
5. a. Electrons from sphere to rod b. 2.5×10^{10}
7. A has −160 nC; B has 0 C
9. a. 9.0 mN b. 9.0 mm/s^2
11. 2.6 cm
13. −10 nC
15. 8.0 nC
17. a. 1.7×10^{14} m/s^2, away from bead b. 3.2×10^{17} m/s^2, toward bead
19. 1.4×10^5 N/C, away from bead
21. a. 1.4×10^{-3} N/C, away from proton b. 1.4×10^{-3} N/C, toward electron
23. a. 3.6×10^4 N/C, 1.8×10^4 N/C, 1.8×10^4 N/C
b.

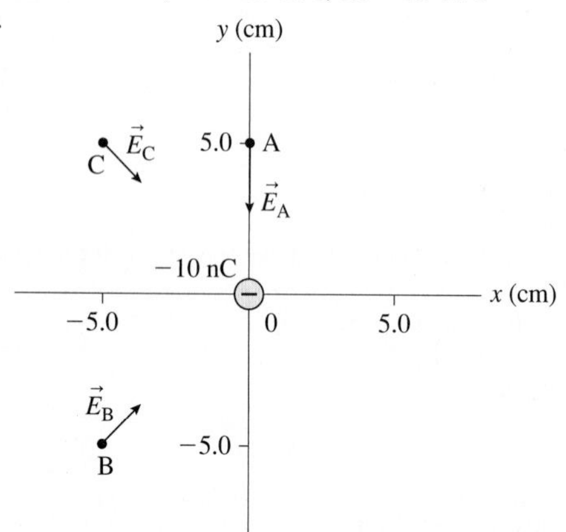

25. 2500 N/C downward
27. a. 1.0×10^{-7} N/C, upward b. 5.6×10^{-11} N/C, downward
29. a. 2 b. ¼ c. 1
31. 1.9 cm
35. 7.2×10^{-15} N
37. a. 0 b. 5.6×10^{-27} N · m
39. 1.4×10^5 C
41. a. 500 N b. 3.0×10^{29} m/s^2
43. a. 0.45 N b. $q_A = 1.0 \times 10^{-6}$ C $q_B = 0.50 \times 10^{-6}$ C
45. 4000 N/C, 9.3° above the horizontal
47. 3.1×10^{-4} N, upward
49. 0
51. 1.1×10^{-5} N, upward
53. 0.68 nC
55. b. $(2 - \sqrt{2})\frac{KQq}{L^2}$
57. 6.6×10^{15} rev/s
59. (−6.0 cm, 3.0 cm)
61. 5.9×10^5 N/C
63. 33 nC
65. 180 nC
67. 150 nC
69. 1.3 cm
71. b. 22 nC
73. D
75. A

Chapter 21

Answers to odd-numbered multiple-choice questions

21. A
23. A
25. C
27. A
29. C

Answers to odd-numbered problems

1. −3.0 μC
3. 530 V
5. a. 3000 V b. 75 μJ
7. a. 1000 eV b. 1.6×10^{-16} J c. 438 km/s
9. a. Lower b. −0.712 V c. 0.712 eV
11. 400 V
13. a. $E = 200$ kV/m, $\Delta V_C = 200$ V b. $E = 200$ kV/m, $\Delta V_C = 400$ V
15. a. 23 cm, 11 cm, 7.5 cm, 5.6 cm
17. −5.8 kV
19. a. 1.5 V b. 8.3 pC
21. −352 V,
23. 10 kV/m, left
25. a. Positive b. Negative c. Positive
27. a. 7.1 pF b. 710 pC
29. 4.8 cm
31. 24 V
33. 15 μC
35. 13 nF
37. a. $Q = 5.0 \times 10^{-10}$ C, $\Delta V_C = 9.0$ V, $E = 9.0 \times 10^4$ V/m
b. $Q = 2.5 \times 10^{-10}$ C, $\Delta V_C = 9.0$ V, $E = 9.0 \times 10^4$ V/m
39. 1.4 kV
41. 1/2
43. 1.2 kV/m
45. 32 μJ
47. a. 1.0×10^{-6} J b. −20 V
49. -2.2×10^{-19} J
51. 1.4 mN

53. ± 12 cm
55. a. 103 V b. 5.39×10^4 N/C, along the $+x$-axis
57. a. 27 V b. -4.3×10^{-18} J
59. a. 3100 V b. 5.0×10^{-16} J
61. 10 kV/m
63. 0.070 eV, increase
65. 140 km/s
67. 71 km/s
69. $\Delta V = 47$ V
71. 10 nC
73. 4.0×10^7 m/s
75. a. 2.1 nC, 3.0 kV/m, 15 V b. 1.0 nC, 1.5 kV/m, 15 V
77. a. Smaller b. Smaller c. 550 V/m, to the right
79. a. 0.80 mm b. 1.6 mm
81. 22 μF
83. b. 1.7×10^6 m/s
85. b. 0.022 mm
87. A
89. D

Chapter 22

Answers to odd-numbered multiple-choice questions

23. C
25. C
27. D
29. C

Answers to odd-numbered problems

1. 3000 C, 1.9×10^{22}
3. 1.9×10^{20} electrons
5. 9.4×10^{13} electrons
7. 13 A
9. 120 C
11. $I_B = 5$ A, $I_C = 2$ A
13. 9.4×10^{18} ions
15. 12 V
17. 32
19. a. 6.0 V, 6.0 A b. 6.0 V, 3.0 A c. 6.0 V, 2.0 A
21. 0.040 Ω
23. a. 0.087 Ω b. 0.0097 Ω
25. 4100 Ω
27. 13 mV
29. a. 50 m b. 1.0 A
31. 50 Ω
33. 3.0 mV
35. a.

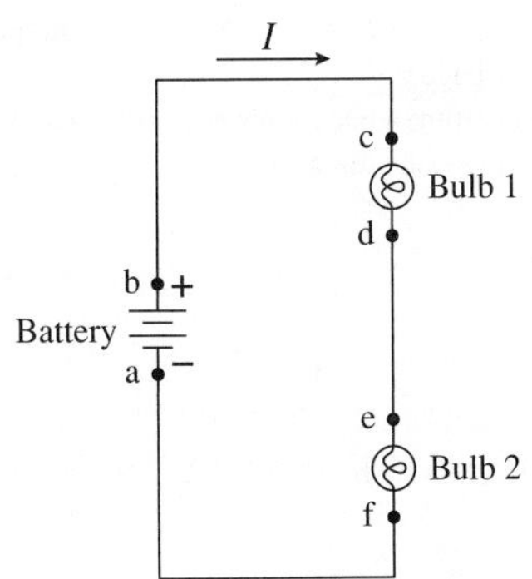

b.

V (V)
9.0 V
4.5 V
0 V
a b c d e f g s
Battery
Connecting wire
Bulb 1
Connecting wire
Bulb 2
Connecting wire

37. Yes; the hair dryer will draw 14 A.
39. a. 0.38 A b. 320 Ω
41. 15 kJ
43. 23 mA
45. 28 min
47. 20%
49. 42 GΩ
51. 33 h
53. 1.4 Ω · m
55. 1.9 V
57. a. 3.1×10^{14} b. 9.1×10^5 N/C c. 0.23 W
59. 0.62 mm
61. 28 cm
63. 8.4×10^{-8} Ω
65. a. 14 C b. 27 J
67. a. 0.24 mA, no b. 120 mA, yes
69. D
71. A

Chapter 23

Answers to odd-numbered multiple-choice questions

31. B
33. A
35. B
37. B

Answers to odd-numbered problems

1.

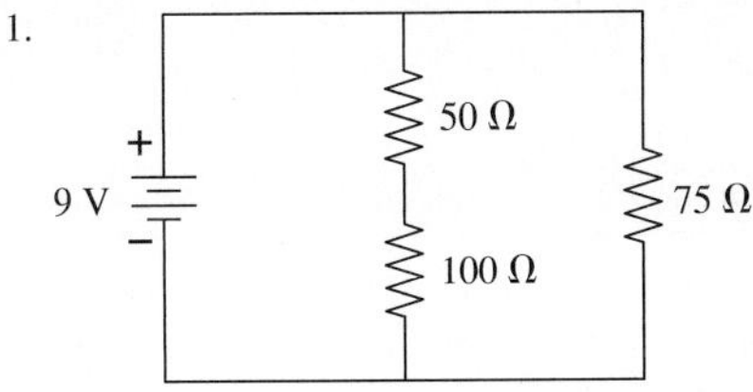

3.

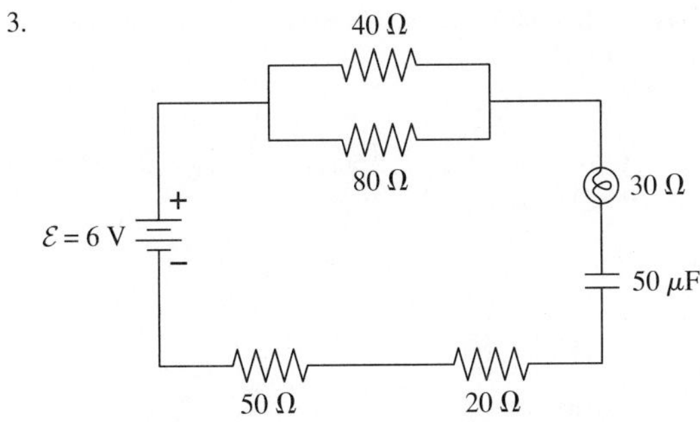

5. a. $\Delta V_{12} = -2.0$ V, $\Delta V_{23} = -1.0$ V, $\Delta V_{34} = 0$ V
   b. $\Delta V_{12} = 0$ V, $\Delta V_{23} = -3.0$ V, $\Delta V_{34} = 0$ V
7. a. 0.17 A, left
11. a. 1.0 Ω b. 1.0 Ω c. 0.5 Ω
13. Four in parallel
15. The three 6.0 Ω resistors in parallel with each other, and then that combination in series with the 3.0 Ω resistor
17. 24 Ω
19. 300 Ω
21. a. 2.0 A b. 5.0 A
23. 20 Ω, 60 V
25. Numbering the resistors from left to right:

R	***I*** **(A)**	**Δ*V*** **(V)**
R_1	1.5	7.5
R_2	0.50	2.5
R_3	0.50	2.5
R_4	0.50	2.5

27.

R	***I*** **(A)**	**Δ*V*** **(V)**
6.0 Ω (left)	2.0	12
15 Ω	0.8	12
6.0 Ω (top)	1.2	7.2
4.0 Ω	1.2	4.8

29.

R	***I*** **(A)**	**Δ*V*** **(V)**
3 Ω	2.0	6.0
4 Ω	1.5	6.0
48 Ω	0.13	6.0
16 Ω	0.38	6.0

31. a.

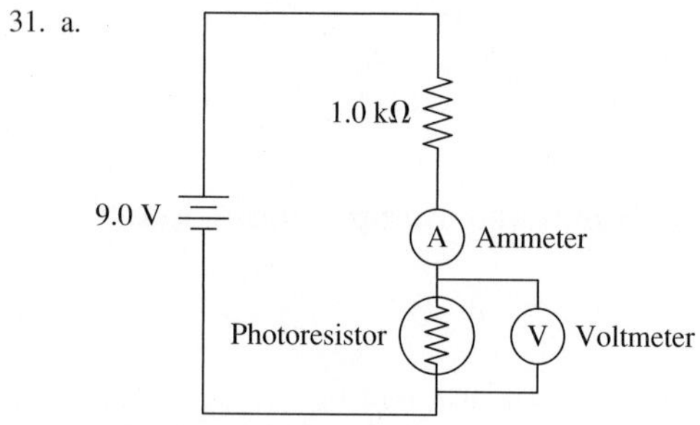

   b. 2.6 mA, 6.4 V
33. 3.0 μF
35. 150 μF in series
37. 8.9 μF
39. a. 0.57 μF b. 6.9 μC
41. 2.0 ms
43. 6.9 ms
45. 7.8 MV/m, 4.4 MV/m
47. 0.73 pF
49. a. 60 m/s b. It neglects the time between the stimulus and the peak of the action potential.
51. 1.3 W, 1.7 W
53. 93 W
55. a. $R_{0.25\,\Omega} = 0.36$ W b. $R_{0.50\,\Omega} = 0.50$ W c. $R_{1.0\,\Omega} = 0.56$ W
   d. $R_{2.0\,\Omega} = 0.50$ W e. $R_{4.0\,\Omega} = 0.36$ W
57. 1.2 A
59. 500 Ω
61. a.

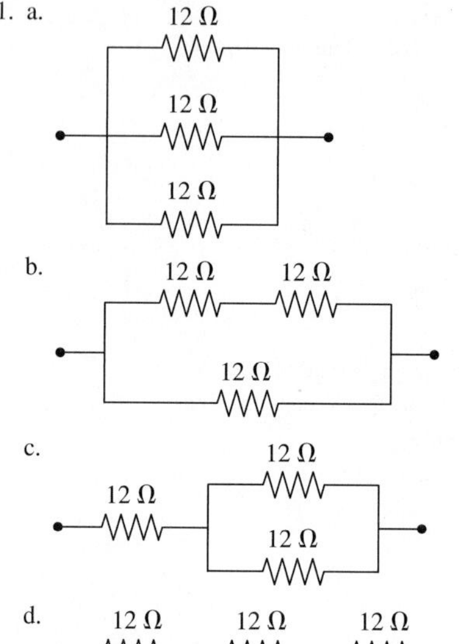

   d.

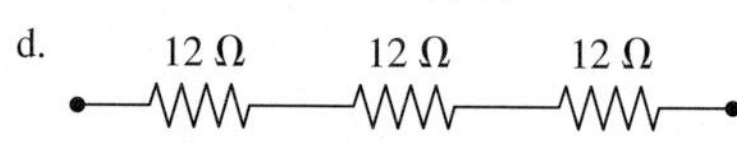

63. Put three resistors in series and then the potential difference across any one would be 3.0 V.
65. a. 0.505 Ω b. 0.500 Ω
69. $Q_1 = Q_2 = Q_3 = 750$ μC $\Delta V_1 = 50$ V $\Delta V_2 = 25$ V
   $\Delta V_3 = 25$ V
71. 870 Ω
73. 18 μF
75. a. 80 μC
   b. 0.23 ms
77. 6100
79. a. 4.8 pC b. 3.0×10^7
81. C
83. B
85. A
87. B

Chapter 24

Answers to odd-numbered multiple-choice questions

33. A
35. A
37. D
39. C

Answers to odd-numbered problems

1. $I_{\text{earthsurface}} = 2.5$ A; $I_{\text{refrigerator}} = 250$ A;
   $I_{\text{laboratory}} = 5000$ A – 50000 A; $I_{\text{magnet}} = 500{,}000$ A.
3. a. 20 A b. 1.6 mm
5. $\vec{B}_1 = (6.7 \times 10^{-5}$ T, out of page); $\vec{B}_2 = (2.0 \times 10^{-4}$ T, into page);
   $\vec{B}_3 = (6.7 \times 10^{-5}$ T, out of page)
7. a. 2 μT; 4% b. The fields from adjacent wires nearly cancel because the contributions will be in opposite directions.
9. 750 A
11. 1.6 kA
13. 3.8×10^{-7} A
15. 4.1×10^{-4} T
17. 8.6 μA
19. 13 T
21. a. $(8.0 \times 10^{-13}$ N, $-z$-direction) b. $(8.0 \times 10^{-13}$ N, 45° clockwise from $+z$-axis in yz plane)
23. a. 11 cm b. 10 m
25. 0.086 T
27. He^+
29. a. 3.2×10^7 m/s b. 3.7×10^{18} m/s² or $3.8 \times 10^{17} g$ c. 0.28 mm
   d. The force is always perpendicular to the velocity, so magnetic fields can't change the speed, only the direction.

31. $\vec{F}_{on\,1} = (2.5 \times 10^{-4}\ \text{N, up})$; $\vec{F}_{on\,2} = \vec{0}\ \text{N}$;
$\vec{F}_{on\,3} = (2.5 \times 10^{-4}\ \text{N, down})$
33. (56 N, into page)
35. a. 3.3 A b. 2.0 N c. 170 m/s^2
37. 7.5×10^{-4} N·m
39. 1.1×10^{10} A
41. 0.036 mm
43. a. 1.0×10^{-4} T; twice the earth's b. Immediately twists left, then deflection decreases as capacitor discharges.
45. 3.0 Ω
47. (2.0 mT, into page)
49. a. 1400 A b. 170 N, right c. 64 m/s
51. a. 6.0×10^{-21} N b. 38 mN/C c. 110 μV
53. 0.20 A, axis of coil north-south
55. Rapid switching doesn't allow the moments to align.
57. 1.0 T
59. 8.6 mT
61. a.

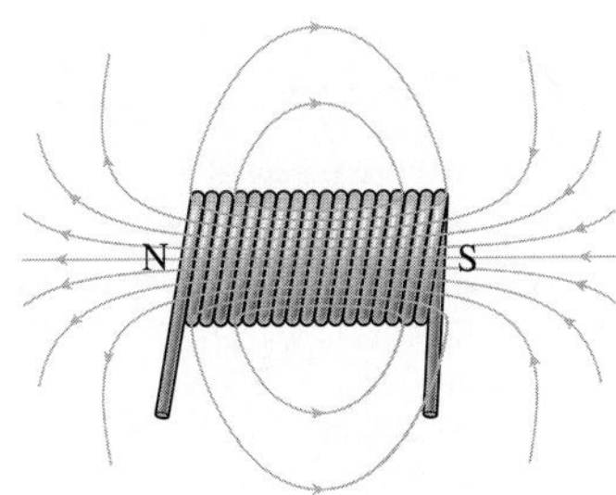

b. Attractive c. To the left, toward the solenoid
63 B
65. C
67. B
69. D

Chapter 25

Answers to odd-numbered multiple-choice questions

27. A
29. B
31. C
33. D
35. D

Answers to odd-numbered problems

1. 0.10 T, out of the page
3. a. 4.0 m/s b. 2.2 T
5. 9500 m/s
7. For $\theta = 0°$, 3.9×10^{-4} Wb; for $\theta = 30°$, 3.4×10^{-4} Wb; for $\theta = 60°$, 2.0×10^{-4} Wb; for $\theta = 90°$, 0 Wb
9. a. 6.3×10^{-5} Wb b. 6.3×10^{-5} Wb
11. a. 8.7×10^{-4} Wb b. Clockwise
13. $|\mathcal{E}_E| > |\mathcal{E}_B| > |\mathcal{E}_D| > |\mathcal{E}_A| > |\mathcal{E}_C|$.
15. 3.1 V
17. 0.44 A
19. 6.0×10^5 V/m
21. $E_0 = 1.4 \times 10^3$ V/m and $B_0 = 4.6 \times 10^{-6}$ T
23. $E_0 = 9.8 \times 10^5$ V/m and $B_0 = 3.3 \times 10^{-3}$ T
25. a. 10.0 nm b. 3.00×10^{16} Hz c. 6.67×10^{-8} T
27. a. 2.2×10^{11} V/m b. $E_0/E = 0.43$
29. a. 0.25 m b. 2500 m
31. 60°
33. 67%
35. 1200 eV
37. a. 250 keV b. 1.0×10^6 bonds
39. 1.2×10^5 J
41. 4.68×10^{21} photon/s
43. 2100 °C
45. $\lambda_{dental} = 50$ pm and $\lambda_{microtomography} = 20$ pm
47. $\Phi = Bab$
49. 1.2 V
51. 0.11 V
53. 42 mV
55. 31mA
57. 8.7 T/s
59. a. 0.20 A b. 4.0×10^{-3} N c. 11°C
61. a. 6.3×10^{-4} N b. 3.1×10^{-4} W c. 1.3×10^{-2} A, counterclockwise d. 3.1×10^{-4} W
63. $E = 550$ V/m and $B = 1.8 \times 10^{-4}$ T
65. a. 8.3×10^{-26} W/m^2 b. 7.9×10^{-12} V/m
67. $\sqrt{2}$
69. $I_0/8$
71. $f = 4.8 \times 10^{27}$ Hz and $\lambda = 6.2 \times 10^{-20}$ m
73. a. 3.4×10^{-19} J b. 1.8×10^{13} photons
75. 4.5×10^{-2} cm
77. C
79. C

Chapter 26

Answers to odd-numbered multiple-choice questions

23. D
25. A
27. A

Answers to odd-numbered problems

1. a. 50 mA b. 50 mA
3. 22.4 V
5. a. 9.0 Ω b. 19 A c. 3200 W
7. 530 A
9. 0.50 A
11. 17 mA
13. a. 47 turns b. 32 mA
15. a. 12 A b. 10 Ω
17. Yes
19. $0.15
21. 0.18 mA, no, no
23. 1000 V AC, 3000 V DC, 100000 V AC, 500000 V DC
25. 0.28 mV
27. a. 1.9 mA b. 1.9 A
29. a. 20.0 mA b. 20.0 mA c. 10.0 mA
31. 81 nF
33. 100 V
35. a. 5.00 mA b. 20.0 mA c. 40.0 mA
37. 1.4 mH
39. 7.8 min
41. 250 kHz to 360 kHz
43. 0.13 pF
45. 1/4
47. 10 mH
49. a. 200 kHz b. 200 kHz
51. 320 mA
53. 5.8 mm
55. a. −7.5 V b. 83 Ω c. 1.4 A
57. 0.63 ms
59. a. 500 Hz b. 1.0 A
61. a. 35 mH b. 4.6 A
63. A
65. B

Part VI Problems

Answers to odd-numbered problems

1. B
3. C
5. C
7. A
9. B
11. C
13. C
15. D
17. B
19. C
21. 2.0 A, clockwise
77. D
79. D

Chapter 27

Answers to odd-numbered multiple-choice questions

21. B
23. E
25. C

Answers to odd-numbered problems

1. $v_{sound} = 345$ m/s, $v_{sprinter} = 15$ m/s
3. a. 15 m/s b. 5 m/s
5. 30 s
7. 3.00×10^8 m/s
9. 3.00×10^8 m/s
11. $t_1 = 1.0\ \mu s$, $t_2 = 2.0\ \mu s$
13. She sees flash 2 40 μs after flash 1.
15. Barn 1.0 μs before tree.
17. 350 μs
19. a. 5.0 yr b. 2.2 yr c. 9.5 yr
21. 0.80c
23. 1.5 km
25. a. $(1 - 5 \times 10^{-11})c = 0.99999999995c$ b. 100,000 yr
27. 0.357c
29. 0.36c
31. 0.80c
33. $\frac{\sqrt{2}}{2}c$
35. a. 1.8×10^{16} J b. 9.0×10^9
37. $\frac{2\sqrt{2}}{3}c$
39. 0.60c
41. a. 90 m b. 0.44c
43. a. 23 μs b. 6.2 km
45. 670 ps
47. a. Less b. 14 ns
49. In the muon's frame, the atmosphere is only 1.5 km thick.
51. 1400 s
53. a. 0.99973c b. 7.6×10^{22} J c. 760 times
55. 0.71c
57. 9.00 nm/s
59. a. 0.52c b. 1 h, 40 min
61. a. $v = (1 - 4.9 \times 10^{-24})c$ b. 46 nm
63. 0.31c
65. 6.3×10^{-19} kg · m/s
67. a. 6660 MeV b. 3.63 MeV
69. 0.786c
71. a. 5.12×10^{-3} MeV b. 9.39 MeV
73. a. 1.3×10^{17} kg b. 6.7×10^{-12} % c. 1.5×10^{13} yr
75. a. 3.6×10^{-12} J b. 0.6%

Chapter 28

Answers to odd-numbered multiple-choice questions

27. B
29. A
31. B
33. D
35. A

Answers to odd-numbered problems

1. 42°
3. 69.3°
5. 4
7. 3.20 eV
9. 1.8 eV
11. a. 290 nm b. 3.7×10^5 m/s
13. a. 6.63×10^5 m/s b. 2.30 eV
15. 8.0×10^{-16} J
17. For $\lambda = 437$ nm, $E = 2.84$ eV; for $\lambda = 533$ nm, $E = 2.33$ eV; for $\lambda = 564$ nm, $E = 2.20$ eV
19. 0.57
21. 8.7×10^{31}
23. a. 2.3 eV b. 3×10^{18}
25. 9.5×10^{-7} W
27. a. 1.1×10^{-34} m b. 1.7×10^{-23} m/s
29. 1.5 eV
31. 8.2 MeV
33. 6.0×10^{-6} V
35. 0.43 nm
37. $E_1 = 1.3 \times 10^{-12}$ J; $E_2 = 5.3 \times 10^{-12}$ J; $E_3 = 1.2 \times 10^{-11}$ J
39. $\lambda = 830$ nm for the $n = 2$ to $n = 1$ transition
    $\lambda = 500$ nm for the $n = 3$ to $n = 2$ transition
    $\lambda = 310$ nm for the $n = 3$ to $n = 1$ transition
41. 4
43. 5.8 μm
45. No, somewhere in the interval -6 m/s $\leq v_x \leq 6$ m/s
47. 0.15 nm
49. a. 2.3 eV b. 240 nm
51. 4.1 eV
53. a. 250 eV, x-ray region; b. 0.060 eV
55. 1.4
57. 4.3×10^{15} photon/s
59. a. 1.7×10^{18} b. 1.7×10^{26} photons/s
61. 3.5×10^{28} photons
63. 12%
65. 0.97 mm
67. a. 3.88×10^{-12} m b. 3.71×10^{-12} m
69. a. 2.0×10^{-63} J b. 1.2×10^{-30} m/s
71. 1.4 nm
73. 1.09×10^{-9} m
75. a. $E_{min} = h^2/4mL^2$
    b. E_{min} for $n = 1$ and $\ell = 1$; $(5/2)E_{min}$ for $n = 1, \ell = 2$, and $n = 2, \ell = 1$; $4E_{min}$ for $n = 2, \ell = 2$; $5E_{min}$ for $n = 1, \ell = 3$, and $n = 3, \ell = 1$; $(13/2)E_{min}$ for $n = 2, \ell = 3$, and $n = 3, \ell = 2$
77. 18 fm
79. 19 T
81. C
83. B

Chapter 29

Answers to odd-numbered multiple-choice questions

23. B
25. C
27. C
29. D

Answers to odd-numbered problems

1. a.

Wavelength (nm)	m	n
656.6	2	3
486.3	2	4
434.2	2	5
410.3	2	6

b. 397.1 nm

3. 1876 nm, 1282 nm, 1094 nm, infrared
5. a. 3 electrons, 4 protons, 5 neutrons
b. 6 electrons, 6 protons, 6 neutrons
c. 4 electrons, 7 protons, 8 neutrons
7. a. ^{8}Li b. $^{14}C^{+}$
9. a. Yes b. 0.50 eV
11. 6.7×10^5 m/s, 1.1×10^6 m/s
13. 3
15. a. $a_B = \dfrac{4\pi\epsilon_0\hbar^2}{me^2}$ b. $E_1 = \dfrac{-e^2}{4\pi\epsilon_0(2a_B)}$
17. a. 0.332 nm, 0.665 nm, 0.997 nm b. 0.332 nm = $(2\pi(0.053\text{ nm}))/1$
0.665 nm = $(2\pi(0.212\text{ nm}))/2$ 0.997 nm = $(2\pi(0.476\text{ nm}))/3$
c.

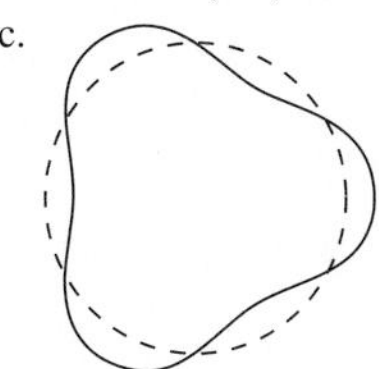

19. 97.26 nm, 486.3 nm, 1876 nm
21. $n = 1$: $\left(1, 0, 0, \pm\frac{1}{2}\right)$ $n = 2$: $\left(2, 0, 0, \pm\frac{1}{2}\right)$, $\left(2, 1, -1, \pm\frac{1}{2}\right)$, $\left(2, 1, 0, \pm\frac{1}{2}\right)$, $\left(2, 1, 1, \pm\frac{1}{2}\right)$ $n = 3$: $\left(3, 0, 0, \pm\frac{1}{2}\right)$, $\left(3, 1, -1, \pm\frac{1}{2}\right)$, $\left(3, 1, 0, \pm\frac{1}{2}\right)$, $\left(3, 1, 1, \pm\frac{1}{2}\right)$, $\left(3, 2, -2, \pm\frac{1}{2}\right)$, $\left(3, 2, -1, \pm\frac{1}{2}\right)$, $\left(3, 2, 0, \pm\frac{1}{2}\right)$, $\left(3, 2, 1, \pm\frac{1}{2}\right)$, $\left(3, 2, 2, \pm\frac{1}{2}\right)$
23. a. f b. -0.850 eV
25. -0.544 eV
27. $1s^22s^22p^63s^23p^2$, $1s^22s^22p^63s^23p^64s^23d^{10}4p^2$, $1s^22s^22p^63s^23p^64s^23d^{10}4p^65s^24d^{10}5p^66s^24f^{14}5d^{10}6p^2$
29. a. Magnesium b. Technetium
31. a. Can't have 8 electrons in a p state b. Can't have 3 electrons in an s state
33. 4.86×10^6 m/s
35. 330 nm
37. a. Yes, 2180 nm b. No, violates $\Delta l = 1$
39. 690 nm
41. 3.18×10^{15}
43. a. 1.7×10^{18} b. 1.7×10^{26} photons/s
45. a. 102.6 nm: $n = 3, m = 1$, 1876 nm: $n = 4, m = 3$
b. Ultraviolet, infrared
47. a. 79, 79, 118, b. 2.29×10^{17} kg/m^3 c. 2.01×10^{13}
49. 0.000000000058%, 99.999999999942%
51. a. 58 N b. 4.7×10^{-35} N, no
53. 1.8×10^7 m/s
55 b. $\lambda_{41} = 200$ nm $\lambda_{31} = 300$ nm $\lambda_{21} = 500$ nm $\lambda_{42} = 333$ nm $\lambda_{43} = 600$ nm $\lambda_{32} = 750$ nm
57. 10
59. 91.18 nm
61. 1876 nm
63. 44200 m/s
65. $\sqrt{6}\hbar$
67. a. $6s - 5p$, $6s - 4p$, $6s - 3p$ b. 7290 nm, 1650 nm, 515 nm
69. 1.13×10^6 m/s
71.

Transition	a. Wavelength	b. Type	c. Absorption
$2p \rightarrow 2s$	670 nm	VIS	Yes
$3s \rightarrow 2p$	816 nm	IR	No
$3p \rightarrow 2s$	324 nm	UV	Yes
$3p \rightarrow 3s$	2696 nm	IR	No
$3d \rightarrow 2p$	611 nm	VIS	No
$3d \rightarrow 3p$	25 μm	IR	No
$4s \rightarrow 2p$	498 nm	VIS	No
$4s \rightarrow 3p$	2430 nm	IR	No

73. 2×10^{23} W/m^2
75. A
77. C

Chapter 30

Answers to odd-numbered multiple-choice questions

29. C
31. A
33. B
35. C

Answers to odd-numbered problems

1. a. 1 proton, 2 neutrons b. 18 protons, 22 neutrons
c. 20 protons, 20 neutrons d. 94 protons, 145 neutrons
3. 6.94 u
5. a. ^{3}H: total is 8.48 MeV, per nucleon is 2.83 MeV
b. ^{3}He: total is 7.72 MeV, per nucleon is 2.57 MeV
7. ^{3}He: 2.57 MeV, ^{4}He: 7.07 MeV, ^{4}He
9. a. ^{14}N: 7.48 MeV b. ^{56}Fe: 8.79 MeV c. ^{207}Pb: 7.87 MeV
11. ^{240}Pu: 7.56 MeV, ^{133}Xe: 8.41 MeV
13. ^{10}B is stable
15. ^{4}He, ^{12}C, ^{16}O
17. ^{234}U
19. a. ^{228}Th b. ^{207}Tl c. ^{7}Li d. ^{60}Ni
21. 5.52 MeV
23. 1.50×10^{-4} u
25. a. 8.8×10^9 b. 2.7×10^9 c. 1.8×10^4
27. a. 3.3 b. 6.6
29. 17,000 yr
31. 80 d
33. 6.3 mCi
35. 4.0 mSv
37. 600 rad
39. 450 h
41. 2.0 Gy
43. 0.51 MeV
45. 0.391 MeV
47. ^{1}H: 99.985%, ^{2}H: 0.015%
49. 200 MeV
51. a. ^{17}N, ^{17}O b. ^{17}O c. ^{17}N decays by beta-minus decay to ^{17}O.
53. 2.7×10^{17}
55. a. 157 min b. 29.5 h
57. 4.6×10^9
59. 210 million years
61. 5.9 billion years
63. 0.47 mSv
65. a. 4.5 kJ b. 110 W c. 1.4 K
67. 3.33×10^5 Gy

69. a. 2.6×10^{10} b. 24 mBq c. 1.9×10^{8} mSv d. 64 million times greater than background level
71. A
73. C

Part VII Problems

Answers to odd-numbered problems

1. C
3. A
5. D
7. A
9. D
11. A
13. B
15. a. 1.1×10^{15} photons b. 98 mSv
17. 0.00018

Credits

COVER PHOTO: Ken Wilson, Papilio/CORBIS.

PART I Opener p. 1 Tim Davis/Corbis.

CHAPTER 1 P. 2 top Vario Images GmbH & Co.KG/Alamy; **center right** Paulo De Oliveira/Photolibrary; **bottom left** Johnny Greig/SPL/Photo Researchers, Inc.; **bottom right** Balfour Studios/Alamy; **p. 3 top left** T. Allofs/Corbis; **top right** Shutterstock; **bottom left** Richard Bartz; **bottom right** BrandX Pictures/Alamy; **p. 6** Chris Carroll/Corbis; **p. 8** Bettman/Corbis; **p. 11 top** NASA; **bottom** Dennis Kunkel/Phototake NYC; **p. 12 top** Stuart Field; **middle** Clive Rose/Getty Images; **bottom** Jed Share and Kaoru/Corbis; **p. 15** NASA; **p. 16 top right** Shutterstock; **bottom left** Shutterstock; **p. 18 top left** Royalty Free/Corbis; **top middle** BrandX/Jupiter Images; **top right** Photodisc/Getty Images; **second from top** Shutterstock; **third from top** Balfour Studios/Alamy; **bottom** Paul Chesley/Getty Images; **p. 22** Mehau Kulyk/Photo Researchers, Inc; **p. 28** Sebastian Kaulitzki/Alamy.

CHAPTER 2 P. 30 top F. Lukasseck/Corbis; **center left** A. T. Willett/Alamy; **center middle** Justin Horrocks/iStockphoto; **center right** Steve Bloom Images/Alamy; **bottom left** NASA; **bottom second from left** Lucio Villalba Parages/iStockphoto; **bottom second from right** Mark Sykes/Alamy; **bottom right** Robert Dant/iStockphoto; **p. 34** Shutterstock; **p. 36** Lucio Villalba Parages/iStockphoto; **p. 39** Shutterstock; **p. 42** Texas Transportation Institute; **p. 44** NASA; **p. 45** NASA; **p. 47** Vincent Munier/Nature Picture Library; **p. 48 top** Stephen Dalton/Photoshot/Alamy; **bottom** Paul McMullin/Construction Photography; **p. 52** NASA; **p. 53 top** Michael Brinson/Getty Images; **bottom** Richard Megna/Fundamental Photographs;
p. 55 Steve Bloom Images/Alamy; **p. 63** Krzysztof Odziomek/Shutterstock; **p. 65 top** Clem Haagner/Corbis; **bottom** Henry Hartley.

CHAPTER 3 P. 67 top Martin Harvey/Alamy; **center left** Seth Golub; **center right** Shutterstock; **bottom left** Scott Markewitz/ Getty Images; **bottom center** Kyle Krause/Photolibrary; **bottom right** Shutterstock; **p. 72 top** Gerrit Hein/iStockphoto; **bottom** Dan Schmitt/ iStockphoto; **p. 74** Tom Uhlman/Alamy; **p. 78** Don Mason/Corbis; **p. 80** Scott Markewitz/Getty Images; **p. 81** Iconica/Getty Images; **p. 82** David Madison/ Getty Images; **p. 84** Tony Freeman/PhotoEdit; **p. 85** Richard Megna/ Fundamental Photographs; **p. 86** Randy Faris/Corbis; **p. 87** Biblioteca Nazionale Centrale, Florence; **p. 88** Gustto images/Photo Researchers, Inc.; **p. 89 top** Jamie Squire/Getty Images; **bottom** Shutterstock; **p. 92 top** Shutterstock; **bottom** Jim Olive/Alamy; **p. 93** Dorling Kindersley Library; **p. 95** Andy Lyons/Getty Images **p. 97** Mike Dobel/Alamy; **p. 98** Maciej Noskowski/iStockphoto.

CHAPTER 4 P. 102 top Michael Eudenbach/Aurora; **center left to right** Edyta Linek/iStockphoto; PCN Chrome/Alamy; Ted Foxx/Alamy; Bobbieo/iStockphoto; **bottom left** David Lewis/iStockphoto; **bottom right** Michael Kappeler/AFP/Getty Images; **p. 103 top** Dimitri Vervitsiotis/AGE Fotostock; **bottom** NASA; **p. 104 top** Richard Megna/Fundamental Photographs; **bottom three** National Highway Traffic Safety Administration; **p. 105 top to bottom** Tony Freeman/PhotoEdit Inc.; Brian Drake/Photolibrary; Comstock Images/Getty Images RF; Dorling Kindersley Library; Shutterstock; Creative Digital Visions; **p. 108** Courtesy Kangoo Jumps; **p. 111** Ruben Garcia; **p. 115** Richard Megna/ Fundamental Photographs; **p. 116** Stan Honda/Getty Images; **p. 117** Martyn Hayhow/Getty Images; **p. 120** iStockphoto; **p. 122 top** Courtesy North Fork Bullets; **bottom** Raymond English; **p. 123** SCPhotos/Alamy; **p. 127** Longview/ The Image Bank/Getty Images.

CHAPTER 5 P. 131 top Alamy; **center left** Shutterstock; **center right** Özgür Donmaz/iStockphoto; **bottom left to right** Photos.com; Dattatreya/Alamy; Eric Schrader–Pearson Science; Andrew Penner/iStockphoto; NASA; **p. 132** Cesar Rangel/Getty Images; **p. 139** NASA; **p. 140** J.D. Pooley/AP Wide World; **p. 141** Taxi/Getty Images; **p. 142 top** NASA; **bottom** NASA; **p. 145** Rhoda Peacher; **p. 146** Roman Krochuk/ iStockphoto; **p. 148** Jeff Coolidge Photography; **p. 149** left Corbis Digital Stock Royalty Free; right Aurora Creative/Getty Images; **p. 161** Martin M. Rotker/Photo Researchers, Inc.; **p. 164** Richard Megna/Fundamental Photographs.

CHAPTER 6 P. 166 top John Berry/The Image Works; **center left to right** Mark Leibowitz/Corbis; David Wall/Alamy; NASA; G. Jacoby, B. Bohannan, M. Hanna/NOAO/AURA/NSF; **bottom left** Shutterstock; **bottom right** NASA; **p. 170** Dennis D. Potokar/Photo Researchers, Inc.; **p. 173** Worldwide Picture Library/Alamy; **p. 177 top** Mark Scheuern/Alamy; **bottom** Robert Laberge/ Getty Images; **p. 178** Richard Megna/Fundamental Photographs; **p. 179** David Burton; **p. 180 top** Susumu Nishinaga/Photo Researchers, Inc.; **bottom** NASA; **p. 181** Charles D. Winters/Photo Researchers, Inc.; **p. 182** Richard Megna/ Fundamental Photographs; **p. 184 top** NASA; **bottom** M.G.M./Album/Newscom; **p. 185** American Institute of Physics/Emilio Segre Visual Archives; **p. 187** NASA/JPL/University of Texas' Center for Space Research/GeoForschungsZentrum (GFZ) Potsdam; **p. 188** NASA; **p. 191** NASA/Jet Propulsion Laboratory; **p. 194** Stock Connection Distribution/Alamy; **p. 195** Atsushi Tomura/AFLO SPORT/Newscom; **p. 196** Exra Shaw/Getty Images; **p. 198** Cornell University; **p. 199** NASA.

CHAPTER 7 P. 200 top Adrian Sherratt/Alamy; **center left** iStockphoto; **center middle** Shutterstock; **center right** Steve Satushek/Photolibrary; **bottom left** Darren Baker/iStockphoto; **bottom middle** Photos.com; **bottom right** Roy Morsch/AGE Photostock/Photolibrary; **p. 206 right** STOCK IMAGE/PIXLAND/Alamy; **top left** Bob Daemmrich/The Image Works; **bottom left** Paul A. Souders/Corbis; **p. 208** Key Color/Photolibrary; **p. 209** Image 100/Alamy; **p. 211** Glidecam Industries, Inc.; **p. 212** FRANCK FIFE/AFP/Getty Images/Newscom; **p. 213** Stuart Field; **p. 214 both** Richard Megna/Fundamental Photographs; **p. 220** Richard Megna/Fundamental Photographs; **p. 221** John Mead/Photo Researchers Inc.; **p. 222** Corbis Premium RF/Alamy; **p. 225** Patricia Fenn; **p. 230 top** Dwight Whitaker and Joan Edwards/Williams College; **bottom** John Terence Turner/Alamy.

CHAPTER 8 P. 232 top Digital Vision/Getty Images; **center left** Shutterstock; **bottom left** G. Newman Lowrance/Getty Images; **bottom center** Elaine Thopson/AP Photo; **top to bottom right** Image Source Pink/Alamy; Christope Ena/AP Photo; Shutterstock; **p. 235** Stock Connection/Alamy; **p. 238** Stuart Field; **p. 239** Richard Megna/Fundamental Photographs; **p. 240** www.webshots.com; **p. 243** Javier Larrea/AGE Fotostock; **p. 244 top** Edward Parke/Alamy; **bottom** Natural History Museum, Tring UK/Dorling Kindersley; **p. 248 top** Duomo/Corbis; **bottom left** Bananastock/Alamy; **bottom right** Imageshop/Alamy; **p. 255** NASA; **p. 256 left** Owen Humphreys/PA Wire/AP Photo; **right** SPL/Photo Researchers.

PART II P. 258 Hugo Willcox/AGEfotostock.

CHAPTER 9 P. 260 top Tom & Pat Leeson/Ardea; **center left** Pete Fontaine/WireImage/Getty Images; **center right** D. Hurst/Alamy; **second from bottom left** Shutterstock; **second from bottom right** Steve Bloom Images/Alamy; **bottom left** Interfoto Pressebildagentur/Alamy; **bottom right** AP; **p. 261 left** MSC Software Corporation; **right** Russ Kinne/Comstock; **p. 263** Stephen Dalton/Photo Researchers Inc.; **p. 265 top** Richard Megna/

Fundamental Photographs; **bottom** Manfred Danegger/Peter Arnold, Inc.; **p. 273** Georgette Douwma/Nature Photo Library; **p. 275** Shutterstock; **p. 276** Digital Vision/Getty Images; **p. 278 left** Chris Trotman/Corbis; **right** Duomo/ Corbis; **p. 279** NASA; **p. 280 left** Stockfolio/Alamy; **right** Kostas Tsironis/AP; **p. 282** NASA.

CHAPTER 10 **P. 289 top** IML Image Group Ltd./Alamy; **center left to right** Steve Bloom Images/Alamy; Shutterstock; Shutterstock; David Noble Photography/Alamy; Slobo Mitic/iStockphoto; **bottom left to right** AP; AP; isifa Image Service s.r.o./Alamy; Shutterstock; **p. 290 left** Doug Martin/Photo Researchers; **middle** Image State; **right** Shutterstock; **p. 291 top left** Greg Probst/Corbis; **top middle** Brand X/Getty Images; **top right** SPL/Photo Researchers. Inc.; **bottom left, top to bottom** Stuart Franklin/Getty Images; Spencer Grant/AGE Fotostock; Scott Sroka/Getty Images; Reuters/David Gray; **p. 292 left** Reuters/Corbis; **middle** Gary Morrison/Getty Images; **right** Gary Buss/Getty Images; **p. 295** www.windpowersports.com; **p. 297** Auto Imagery, Inc. NHRA/AP; **p. 300 top** David Woodfall/Getty Images; **bottom** NASA; **p. 301** Dorling Kindersley; **p. 302** Meredith Davenport/Reuters; **p. 304 top** Corbis; **middle** Chris Salvo/Getty Images; **bottom** Chip Forelli/Getty Images **p. 305 top** Stuart Field and Brian Jones; **bottom** Richard Megna/ Fundamental Photographs; **p. 307** John Brackenbury/Photo Researchers, Inc.; **p. 310** Richard Megna/Fundamental Photographs; **p. 312** Dorling Kindersley; **p. 313 top** Lonely Planet Images/Getty Images; **bottom** David Madison/Getty Images; **p. 314** Jim Kilpatrick.

CHAPTER 11 **P. 322 top** SCPhotos/Alamy; **center left to right** Christina Richards/iStockphoto; Diane Diederich/iStockphoto; Mauro Fermariello/ Photo Researchers; **bottom left to right** Tony McConnell/Photo Researchers, Inc.; Rob Broek/iStockphoto; Christian Darkin/Photo Researchers, Inc.; Diane Diederich/iStockphoto; **p. 323 top left to right** Brian Jones; Willard Clay/Getty Images; David Woodfall/Getty Images; **bottom left to right** Brian Jones; Dorling Kindersley; BrandX/Getty Images; **p. 324** Colin Raw/Getty Images; **p. 325** Christina Richards/iStockphoto; **p. 327** Creative Digital Visions; **p. 328** Mauro Fermariello/Photo Researchers; **p. 329 top** Asgeir Helgestad/Nature Photo Library; **center** Jurgen Freund/Nature Photo Library; **bottom** BrandX/ Getty Images; **p. 331** Alamy; **p. 332** David Young-Wolff/PhotoEdit Inc.; **p. 333** SUNY; **p. 334** NASA; **p. 335** Richard Megna/Fundamental Photographs; **p. 338 top** Dorling Kindersley; **bottom** Creative Digital Visions; **p. 340 left** Chris Ladd/Getty Images; **middle** Tony McConnell/Photo Researchers, Inc.; **right** Brian Jones; **p. 342 top** Wagan Corp.; **bottom** SSPL/The Image Works; **p. 345** Ewing Galloway/Photolibrary; **p. 347** Stan Macbean; **p. 348** Mercedes Benz; **p. 349** Colin Raw/Getty Images; **p. 350** Richard Megna/Fundamental Photographs; **p. 352** Sandia National Laboratory; **p. 354** Antony Kaminju/ Reuters; **p. 355** Amrtin Harvey/Photo Researchers, Inc.; **p. 357 left** Carsten Jäger; **right** National Optical Astronomy Observatories/Photo Researchers; **p. 358 left** Flybar/SBI Industries; **right** www.consumer.org; **p. 359 both** Sea WiFS Project/NASA Goddard Space Flight Center.

PART III **P. 360** Stefano Lunardi/AGE Fotostock.

CHAPTER 12 **P. 362 top** Nutscode/T Service/Photo Researchers, Inc.; **center left** Jerry Koch/iStockphoto; **center** Corbis; **right** Alaska Railroad sun kink photo courtesy John P. Combs; **bottom left to right** Dr. Ken MacDonald/ Photo Researchers, Inc.; David Jordan/AP Photo; Clouds Hill Imaging Ltd./Photo Researchers, Inc.; **p. 363 top** Corbis; **center** Jan Tadeusz/Alamy; **bottom** Corbis; **p. 364** Richard Megna/Fundamental Photographs; **p. 366** NASA Headquarters; **p. 368** Tom McHugh/Photo Researchers, Inc.; **p. 369** Joseph Sinnot/Fundamental Photographs; **p. 370 top** Fotolia; **bottom** Roger C. Wagner; **p. 377** Department of Atmospheric Sciences at the University of Illinois at Urbana-Champaign; **p. 379** Spencer Grant/Photo Researchers, Inc.; **p. 380** Eric Schrader– Pearson Science; **p. 382** Corbis; **p. 383 top** The Open University; **bottom** Slippery Rock University; **p. 384 top** Pete Orelup/Flickr RF/Getty Images; **bottom left** Alessandro Trovati/AP Photo; **bottom middle** Stephen Oliver/Alamy; **bottom right** Photodisc/Getty Images; **p. 391 left** Cn Boon/Alamy; **middle left** Andrew Davidhazy; **middle right** Pascal Goetgheluck/Photo Researchers, Inc.; **right** Corbis; **p. 392 top** Alice Houston; **bottom** Sciencephotos/Alamy; **p. 393 top** Shutterstock; **bottom** Tony McConnell/Photo Researchers, Inc.; **p. 394 top** Johns Hopkins University Applied Physics Laboratory; **bottom** Ted Kinsman/Photo Researchers, Inc.; **p. 395** Brian Jones; **p. 398** Matthew Johnston/ Alamy; **p. 400** Brian Jones; **p. 401** Bjorn Mauck, Allgemeine Zoologie & Neurobiologie, Ruhr-Universität Bochum. The Journal of Experimental Biology 206, 1727-1738 (2003); **p. 402** Shutterstock.

CHAPTER 13 **P. 405 top** Chris A. Crumley/Alamy; **second from top left** Shutterstock; **second from top right** iStockphoto; **second from bottom left to right** JupiterImages/Brand X/Alamy; iStockphoto; AP Photo/Andres Leighton; **bottom left** AP Photo/American Society of Civil Engineers/Andy Manis; **bottom right** Dušan Zidar/Fotolia; **p. 409 both** Woods Hole Oceanographic Institution; **p. 411** NOAA; **p. 414 top** Alix/Phanie/Photo Researchers, Inc.; **bottom** Martin Harvey/Alamy; **p. 416** David Young-Wolff/Photo Edit; **p. 417** Harald Sund/Getty Images; **p. 419** Altrendo/Getty Images; **p. 420 top** Alix/Photo Researchers, Inc.; **bottom** Don Farrall/Getty Images; **p. 422** Andy Sacks/Getty Images; **p. 424** W. Perry Conway/Corbis; **p. 425** GE Medical; **p. 429** Koninklijke Philips Electronics N.V.; **p. 430** PhotoAlto/Alamy; **p. 432** Phil Dauber/Photo Researchers, Inc.; **p. 440** AP Photo/Anchorage Daily News.

PART IV **P. 442** D. Robert Franz/ImageState/Alamy.

CHAPTER 14 **P. 444 top** Tom McHugh/Photo Researchers, Inc.; **center left** Mauritius/Photolibrary; **center middle** Adam Haylock/iStockphoto; **center right top** James Stevenson/Photo Research, Inc.; **center right bottom** Alexander Sakhatovsky/iStockphoto; **bottom left** Photodisc/Alamy; **bottom right** ImageState/Alamy; **p. 446 top** Oscar Burriel/Photo Researchers; **bottom** Royalty-Free/Corbis; **p. 448** Michael Prince/Corbis; **p. 451** Corbis/AGE Fotostock; **p. 452** Shaun Cunningham/Alamy; **p. 454** Serge Kozak/AGE Fotostock; **p. 455** Barry Bland/Alamy; **p. 457** NASA; **p. 458** Craighead Research Group, Cornell University, Ithaca, NY; **p. 460 top** Tim Wright/Corbis; **bottom** SSPL/The Image Works; **p. 462** Bananastock/Alamy; **p. 464** Steve P./ Alamy; **p. 465** Nova Scotia Department of Tourism and Culture; **p. 466** Sigrid Dauth/Alamy; **p. 467** Digital Vision/Alamy; **p. 470** Martin Harvey/Alamy; **p. 473** Josep Pena Llorens/iStockphoto; **p. 476** Ted Horowitz/Corbis.

CHAPTER 15 **P. 477 top** Wildlife Pictures/Peter Arnold, Inc.; **center left to right** Richard Megna/Fundamental Photographs; Peter Arnold, Inc./Alamy; GIPhotoStock/Photo Researchers, Inc.; **bottom left** konradlew/iStockphoto; **second from bottom right** Oleksiy Maksymenko/Alamy; **bottom right** Martin Jenkinson/Alamy; **p. 478 top** Doug Wilson/Corbis; **bottom** Aflo Foto Agency/Alamy; **p. 480 top** Hugh Threlfall/Alamy; **bottom** James Gerholdt/Peter Arnold Images/Photolibrary; **p. 483** Shutterstock; **p. 488 left** Edgetech Marine; **right** B. Benoit/Photo Researchers, Inc.; **p. 490** Jeff Vanuga/Corbis; **p. 492** Mike Johnson Marine; **p. 493** David Chapman/Alamy; **p. 494** Steve Gschmeissner/Photo Researchers, Inc.; **p. 495** NASA; **p. 497 top** NASA; **bottom** NOAA/Photo Researchers, Inc.; **p. 498 left** Sound Made Visible, copyright 1997 Chris Rossi; **right** Robert Harding Picture Library Ltd/Alamy; **p. 501** David Schrader/ iStockphoto.

CHAPTER 16 **P. 507 top** Reuters/Corbis; **center left** Constance McGuire/ iStockphoto; **center right** Edward Kinsman/Photo Researchers, Inc.; **second from bottom left** Brian Jones; **bottom left to right** DSGpro/iStockphoto; Yevgeny Gultaev/iStockphoto; Shutterstock; Shutterstock; **p. 508** Brian Jones; **p. 509 top** Southern Illinois University/Photo Researchers, Inc.; **bottom** Creative Digital Visions; **p. 512** Brian Jones; **p. 513** Edward Kinsman/Photo Researchers, Inc.; **p. 515 top** Ed Elliott; **bottom** PhotoDisc/Getty Images; **p. 516** Michael Vollmer, U. Appl. Sci. Brandenburg. Germany; **p. 518** Brian Jones; **p. 519** AP Photo/Institute of Cultural Relics and Archealogy of Henan Province.; **p. 521** Royalty-Free/ Corbis; **p. 522** ImageState Royalty Free/Alamy; **p. 526** iStockphoto; **p. 527** Andrew Davidhazy/RIT/Imaging and Photo Technology; **p. 529** Anna Utekhina/ iStockphoto; **p. 531** William Manning/Corbis; **p. 533** Joe Tucciarone/Photo Researchers, Inc.; **p. 535** Leslie Garland Picture Library/Alamy; **p. 539** WL | Delft Hydraulics; **p. 541** Zoonar GmbH/Alamy.

PART V **P. 542** David Denning.

CHAPTER 17 **p. 544 top** Dave Watts/Nature Picture Library; **center left to right** Shutterstock; Jane Pang/iStockphoto; Shutterstock Dieter Zawischa - http://www.itp.uni-hannover.de/~zawischa/ITP/multibeam.html; **second from bottom right** Shutterstock; **bottom left** Ken Kay/Fundamental Photographs; **bottom right** Shutterstock; **p. 545** Steve Horrell/Photo Researchers, Inc.;

p. 546 top Todd Gipstein/Getty Images; **center and bottom** Richard Megna/Fundamental Photographs; **p. 547 top** Sciencephotos/Alamy; **bottom** David Parker/Photo Researchers, Inc.; **p. 551** Stuart Field; **p. 556 top** Andrew Syred/Photo Researchers; **bottom** Stuart Field; **p. 558 top** Andrew Bargery/Alamy; **bottom** Eric Schrader–Pearson Science; **p. 559 top** The Picture Source/Terry Oakley; **bottom** Edmund Optics; **p. 560 top** Sinclair Stammers/Photo Researchers, Inc.; **bottom** Richard Megna/Fundamental Photographs; **p. 562** Sabian Zigman; **p. 563** Ken Fay/Fundamental Photographs; **p. 566** NASA; **p. 571** Kim Taylor/Nature Picture Library; **p. 573 top** Tom Zagwodzki/Goddard Space Flight Center/NASA; **bottom left, right** Pete Vukusic, University of Exeter.

CHAPTER 18 **P. 574 top** Michael Kappeler/AFP/Getty Images; **center left to right** iStockphoto; Sciencephotos/Alamy; Leslie Garland Picture Library/Alamy; Shutterstock; Anya Levinson; **second from bottom right** David Maitland/Getty Images; **bottom left to right** Shutterstock; Piluhin/Alamy; Christopher O Driscoll/iStockphoto; **p. 576 top left** Onpoint Lasers, Inc.; top **middle left** Dynamic Graphics Group/IT Stock Free/Alamy; **top middle right** Mike Stone/Alamy; **top right** Paul Rapson/Alamy; **bottom** Stuart Field; **p. 577** Stuart Field; **p. 578 top** NASA; **bottom** Sylvester Allred/Fundamental Photographs; **p. 580** The Exploratorium; **p. 581** Creative Digital Visions; **p. 583 top** Canon; **bottom** Richard Megna/Fundamental Photographs; **p. 585 top** Pete Atkinson/Getty Images; **bottom** Custom Medical Stock Photography; **p. 586** Creative Digital Visions; **p. 587 both** Richard Megna/Fundamental Photographs; **p. 589** Creative Digital Visions; **p. 591** Photodisc Blue/Getty Images; **p. 594** Dr. Tamara Frank/Harbor Branch Oceanographic Institute at Florida Atlantic University; **p. 595** M.C. Escher's "Hand with Reflecting Sphere" ©2006 The M.C. Escher Company-Holland. All rights reserved. www.mcescher.com; **p. 596** Paul Silverman/Fundamental Photographs; **p. 601** Bajande/Photo Researchers, Inc.; **p. 603 top** AKG-Images; **bottom** Erich Lessing/Art Resource, NY; **p. 605 top** VStock/Alamy; **bottom** Nano Calvo/Alamy; **p. 607** Jason Horowitz/Corbis; **p. 608** David Hancock/Alamy.

CHAPTER 19 **P. 609 top** Juniors Bildarchiv/Alamy; **center left to right** Shutterstock; Judy Barranco/iStockphoto; Shutterstock; Richard Wainscoat/Alamy; **second from bottom middle** David Parker/SPL/Photo Researchers, Inc.; **bottom left to right** JupiterImages/Comstock Images/Alamy; Shutterstock; Springer-Verlag GmbH & Co KG; Sciencephotos/Alamy; **p. 610** Doug Perrine/Nature Photo Library; **p. 611 top** GardenWorld Images/Alamy; **bottom** NASA; **p. 612 top** Adam Black; **bottom** Richard Megna/Fundamental Photographs; **p. 614 top** Dynamic Graphics Group/IT Stock Free/Alamy; **bottom** Tetra Images/Alamy; **p. 617** Stuart Field; **p. 618** Stuart Field; **p. 619** Tetra Images/Alamy; **p. 620** Biophoto Associates/Photo Researchers, Inc.; **p. 621** NASA; **p. 625** NASA; **p. 626** Springer-Verlag GmbH & Co KG; **p. 627 left** James Cavallini/Photo Researchers, Inc.; **right** STEVE GSCHMEISSNER/Science Photo Library/Alamy; **p. 629** Shutterstock; **p. 631 top** Stuart Field; **bottom** Sam Ogden/Photo Researchers, Inc.; **p. 633** NASA; **p. 634** Alfred Pasieka/Photo Researchers, Inc.; **p. 635** NASA; **p. 637** Michael W. Davidson at Molecular Expressions; **p. 639** Jessica Riser.

PART VI **P. 640** Nonoy Tan/Alamy.

CHAPTER 20 **P. 642 top** Pacific Northwest National Laboratory; **center left** Andrew Lambert Photography/Photo Researchers, Inc.; **center middle** Eric Schrader–Pearson Science; **center right** Luminant Corporation; **bottom right** Susuma Nishinaga/Photo Researchers, Inc.; **p. 643** Creative Digital Visions; **p. 645** Creative Digital Visions; **p. 646** Allan Lee; **p. 648 top** Charles D. Winters/Photo Researchers, Inc.; **bottom** Steve Hopkin/Getty Images; **p. 652** Science Pictures Limited/Photo Researchers; **p. 655** Richard Megna/Fundamental Photographs; **p. 663 top** Brian Jones; **bottom** Reinhard H/Photolibrary; **p. 669** David Lassman/The Image Works.

CHAPTER 21 **P. 675 top** Image courtesy of the Scientific Computing and Imaging (SCI) Institute, University of Utah; **center left** MvH/iStockphoto; **center middle** Shutterstock; **center right** Photo Researchers, Inc.; **bottom left** Shutterstock; **p. 679 top** Aaron Horowitz/Corbis; **bottom** Ingram Publishing/Alamy; **p. 680 left** Stuart Field; **right** Jurg Streit/University of Bern; **p. 682** Creative Digital Visions; **p. 683** Varian Medical Systems; **p. 692** Jonathan Bird/Peter Arnold/Alamy; **p. 693** Paul Silverman/Fundamental Photographs; **p. 695 top** Lester Lefkowitz/Corbis; **bottom** Creative Digital Visions; **p. 697** Helmut Foll; **p. 699** Andrew Lambert/Alamy; **p. 700 top** Dimitri Vervits/Getty Images; **bottom** SPL/Photo Researchers, Inc.; **p. 702** NASA.

CHAPTER 22 **P. 712 top** May/Photo Researchers, Inc.; **center left** Creative Digital Visions; **center right** Creative Digital Visions; **bottom left** Fotolia; **bottom middle** iStockphoto; **bottom right** Fotolia; **p. 714** FLIR Systems, Inc. **p. 718** Brian Jones; **p. 719** Peter Scoones/Nature Picture Library; **p. 722** Dennia O'Clair/Getty Images; **p. 723** Dr. Jennifer L. Mueller; **p. 725 left** Charles Bowman/Alamy; **middle** Igor Sinicin/iStockphoto; **right** Brian Jones; **right, inset** Shutterstock; **p. 728** Brian Jones; **p. 735** Oxford Scientific/Photolibrary.

CHAPTER 23 **P. 739 top** Amazon Images/Alamy; **center left** Krzysztof Zmij/iStockphoto; **center middle** Franc Podgoršek/iStockphoto; **center right** Photos.com; **bottom left** Ace Stock Limited/Alamy; **bottom right** CNRI/Photo Researchers, Inc.; **p. 745** Dynamic Graphics Group/IT Stock Free/Alamy; **p. 747** Abelstock/Alamy; **p. 749 top** Cn Boon/Alamy; **bottom** Simon Fraser/Photo Researchers, Inc.; **p. 751** Bill Nicholson/Alamy; **p. 755** Creative Digital Visions; **p. 757** Eric Schrader–Pearson Science; **p. 758 top** SPL/Photo Researchers, Inc.; **bottom** Don W. Fawcett/Photo Researchers, Inc.; **p. 762** Stephane De Sakutin/AFP/Getty Images; **p. 763** STEVE GSCHMEISSNER/Science Photo Library/Alamy; **p. 764** SPL/Photo Researchers, Inc.

CHAPTER 24 **P. 776 top** Thierry Berrod/Mona Lisa Production/SPL/Photo Researchers, Inc.; **center left** Daniel Dempster Photography/Alamy; **center middle** Roman Krochuk/iStockphoto; **center right** Shutterstock; **bottom left** John Eisele; **bottom middle left** Urban Zone/Alamy; **bottom middle right** Pete Saloutos/Corbis RF/Alamy; **bottom right** Photos.com; **p. 779** John Eisele, Colorado State University Photography; **p. 780 top and bottom right** John Eisele, Colorado State University Photography; **bottom left** Brian Jones; **p. 781** IBM Corporation; **p. 782** John Eisele; **p. 787** Kathleen Gustafson, Ph.D., Hoglund Brain Imaging Center, University of Kansas, VSM Medtech, Ltd.; **p. 789** Charles Thatcher/Getty Images; **p. 793** Chris Madeley/Photo Researchers, Inc.; **p. 795 top** Marsh-McBirney/Hach Company; **bottom** Stuart Field; **p. 798** Prof. Brian James/Wikipedia.org; **p. 802** UHB Trust/Getty Images; **p. 804** Atsushi Arakaki et al., "Formation of magnetite by bacteria and its application," *J R Soc Interface*. 2008 September 6; 5(26): 977–999; **p. 805 top** John Eisele; **bottom** Andrew Lambert Photography/Photo Researchers, Inc.; **p. 809** 1m gradiometer results were obtained using the Bartington Grad601-2 gradiometer, courtesy of Bartington Instruments Ltd.

CHAPTER 25 **P. 816 top** Andrew Davidhazy; **center left** Richard Wheeler/Wikipedia.org; **center middle** Shutterstock; **center right** Palm, Inc.; **bottom left** Photo Researchers, Inc.; **bottom middle left** iStockphoto; **bottom middle right** Fotolia; **bottom right** iStockphoto; **p. 817** Scott McAlpine/British Columbia Institute of Technology; **p. 818** NASA; **p. 819** James Watt/Alamy; **p. 821** Eric Schrader–Pearson Science; **p. 826** James Leynse/CORBIS; **p. 828 left** Brian Jones; **right** Creative Digital Visions; **p. 830** SSPL/The Image Works; **p. 832** Tony Craddock/Photo Researchers, Inc.; **p. 833** Richard Megna/Fundamental Photographs; **p. 834 top** Innerspace Imaging/Photo Researchers, Inc.; **bottom** Brian Jones; **p. 835** Eugene Hecht, Optics 2e, p. 11, Addison Wesley; **p. 837** Brian Jones; **p. 838** Brian Jones; **p. 839 top** Ashok Captain/Alamy; **bottom** Edward Kinsman/Photo Researchers, Inc.; **p. 841 top left** NASA/JPL-Caltech/J. Keene (SSC/Caltech); **top right** NOAA; **bottom left** NASA; **bottom right** Jack O. Burns (University of Missouri) & David Clarke (St. Mary's University, Nova Scotia); **p. 845** Brian Jones; **p. 846** Dennis Nett/The Image Works.

CHAPTER 26 **P. 852 top** Duncan Usher/Alamy; **center left** Photo Researchers, Inc.; **center right** Shutterstock; **second from bottom left** Christoforos Giatzakis/Alamy; **bottom left** iStockphoto; **bottom middle** Photos.com; **bottom right** Johnny Greig RFSL/Alamy; **p. 853** Creative Digital Visions; **p. 855** iStockphoto; **p. 856 top** Eric Schrader–Pearson Science; **bottom** Creative Digital Visions; **p. 860 top** Pearson Education/PH College; **bottom** Brian Jones; **p. 861 top** Shutterstock; **bottom** Paul A. Souders/Corbis; **p. 862** Creative Digital Visions; **p. 863** Phil Degginger/Alamy; **p. 864** ViennaPhoto/Alamy; **p. 866** Eric Schrader–Pearson Science; **p. 867** Shutterstock; **p. 868** Kevin Eldredge; **p. 870** Pacific Northwest National Laboratory; **p. 879** Steve Sant/Alamy.

PART VII **P. 884** Gregory Dimijian/Photo Researchers, Inc.

CHAPTER 27 **P. 866 top** Chris Johnson/Alamy; **center left** AP Photo/Thomas Kienzle; **center right** Photos.com; **bottom left** MvH/iStockphoto; **bottom middle left** NOAA; **bottom middle right** DOE Photo; **bottom right** Fotolia; **p. 887** Topham/The Imageworks; **p. 888** Kaz Chiba/Getty Images; **p. 889** eStock Photo/Alamy; **p. 891** Jean Saint-Martin; **p. 895** Jim Corwin/Alamy; **p. 902** U.S. Department of Defense Visual Information Center; **p. 903** Eckhard Slawik/Photo Researchers, Inc.; **p. 905** Stanford Linear Accelerator Center/Photo Researchers, Inc.; **p. 908** Lawrence Berkeley/Photo Researchers, Inc.; **p. 910** Lucasfilm Ltd.; **p. 911** SSPL/The Image Works; **p. 912** SPL/Photo Researchers, Inc.; **p. 918** Dynaminc Graphics/AGE fotostock; **p. 919** NASA; **p. 920** John Chumack/Photo Researchers, Inc.

CHAPTER 28 **P. 922 top** Omikron/Photo Researchers, Inc.; **center left** Dieter Zawischa, http://www.itp.unihannover.de/~zawischa/ITP/multibeam.html; **center middle** Dr. Antoine Weis; **center right** Andrew Lambert Photography/Photo Researchers, Inc.; **bottom left** Carol & Mike Werner/Alamy; **bottom right** Ted Kinsman/Photo Researchers, Inc.; **p. 923 top** Oxford Instruments; **bottom** Neil Borden/Photo Researchers, Inc.; **p. 925** M.H.F. Wilkins; **p. 928 top** Bettman/Corbis; **bottom left** Neil McAllister/Alamy; **bottom right** Andrew Lambert/Photo Researchers, Inc.; **p. 929** William Milberry/Aluminum Studios; **p. 931 top** Dr. Antonie Weis; **bottom** Kenneth Eward/BioGrafx/Photo Researchers, Inc.; **p. 932** J. Schmitt et al., ROSAT Mission/NASA; **p. 933** Creatas/agefotostock; **p. 934** E.R. Huggins, Physics I, W.A. Benjamin, 1968, Reading, MA: Addison Wesley Longman. Reprinted with permission; **p. 935** Film Studio, Education Development Center, Newton, MA; **p. 936** Biomedical Imaging Unit, Southampton General Hospital/Photo Researchers, Inc.; **p. 937** Brian Jones; **p. 939** Bettman/Corbis; **p. 940** Ted Kinsman/Photo Researchers, Inc.; **p. 943 top** Digital Instruments; **bottom** LLNL/Photo Researchers, Inc.; **p. 944 left** Hou et al., in the 18 October 1999 issue of Physical Review Letters; **right** J. Bernholc et al., NCSU/Photo Researchers, Inc.; **p. 945** Mehau Kulyk/Photo Researchers, Inc.; **p. 946 left** Biomedical Imaging Unit, Southampton General Hospital/Photo Researchers, Inc.; **right** LLNL/Photo Researchers, Inc.; **p. 949 top** Darwin Dale/Photo Researchers, Inc.; **bottom** Bill Corwin; **p. 950 top** Frank Llosa/Frankly.com; **bottom** SPL/Photo Researchers, Inc.

CHAPTER 29 **P. 954 top** Michael W. Davidson at Molecular Expressions; **center** Ted Kinsman/Photo Researchers, Inc.; **bottom left** Shutterstock **bottom middle** NASA **bottom right** Shutterstock; **p. 955 top** Ocean Optics; **bottom** Ted Kinsman/Photo Researchers, Inc.; **p. 956** NASA; **p. 975** Courtesy National Institute of Standards and Technology; **p. 977** International Dark-Sky Association; **p. 979 top** Edward Kinsman/Photo Researchers, Inc.; **bottom** Sinclair Stammers/Photo Researchers, Inc.; **p. 980** Fotolia.com; **p. 982 top** U.S. Air Force photo/Tech. Sgt. Larry A. Simmons; **bottom** contour99/iStockphoto.

CHAPTER 30 **P. 991 top** CNRI/Photo Researchers, Inc.; **center** DOE Photo; **bottom left to right** Robert de Gugliemo/Photo Researchers, Inc.; SPL/Photo Researchers, Inc.; Photo Researchers, Inc.; "HTO"/Wikipedia; Fermilab/U.S. Department of Energy; **p. 993** British Antarctic Survey/Photo Researchers Inc.; **p. 995** Parent Gery/Wikipedia; **p. 996** NASA; **p. 997** U.S. Department of Energy; **p. 1002** David Parker/Photo Researchers, Inc.; **p. 1003** C. Powell, P. Fowler & D. Perkin/Photo Researchers, Inc.; **p. 1006** U.S. Department of Energy/Photo Researchers, Inc.; **p. 1007 top** James King-Holmes/Photo Researchers, Inc.; **bottom** Bruce E. Zuckerman/Corbis; **p. 1008** Hank Morgan/Photo Researchers, Inc.; **p. 1009** Custom Medical Stock Photo; **p. 1010 top** V. Elayne Arterbery, M.D.; **middle** SPL/Photo Researchers Inc; **bottom** CNRI/Photo Researchers, Inc.; **p. 1011** CNRI/Photo Researchers, Inc.; **p. 1012 top** Hank Morgan/Photo Researchers, Inc.; **center** Zephyr/Photo Researchers, Inc.; **bottom** CERN/Photo Researchers, Inc.; **p. 1013** ICRR Institute for Cosmic Ray Research; **p. 1016** U.S. Department of Energy; **p. 1019** Brian Jones; **p. 1020** Jean-Perrin/CNRI/Photo Researchers Inc.; **p. 1021 top** CNRI/Photo Researchers, Inc.; **bottom** Smithsonian Institute; **p. 1025** Peter Hannaford; **p. 1027** Bgran/Wikipedia.

Index

For users of the two-volume edition, pages 541 are in Volume 1 and pages 542–1027 are in Volume 2.

I

J

K

L

G

H

M

Q

R

S

T

U

V

W

X

Y

Z

Table of Problem-Solving Strategies

Note for users of the two-volume edition:
Volume 1 (pp. 1–541) includes chapters 1–16.
Volume 2 (pp. 542–1027) includes chapters 17–30.

Table of Math Relationship Boxes

Useful Data

M_e	Mass of the earth	5.98×10^{24} kg	
R_e	Radius of the earth	6.37×10^{6} m	
g	Free-fall acceleration	9.80 m/s^2	
G	Gravitational constant	6.67×10^{-11} N · m^2/kg^2	
k_B	Boltzmann's constant	1.38×10^{-23} J/K	
R	Gas constant	8.31 J/mol · K	
N_A	Avogadro's number	6.02×10^{23} particles/mol	
T_0	Absolute zero	−273°C	
p_{atm}	Standard atmosphere	101,300 Pa	
v_{sound}	Speed of sound in air at 20°C	343 m/s	
m_p	Mass of the proton (and the neutron)	1.67×10^{-27} kg	
m_e	Mass of the electron	9.11×10^{-31} kg	
K	Coulomb's law constant ($1/4\pi\epsilon_0$)	8.99×10^{9} N · m^2/C^2	
ϵ_0	Permittivity constant	8.85×10^{-12} C^2/N · m^2	
μ_0	Permeability constant	1.26×10^{-6} T · m/A	
e	Fundamental unit of charge	1.60×10^{-19} C	
c	Speed of light in vacuum	3.00×10^{8} m/s	
h	Planck's constant	6.63×10^{-34} J · s	4.14×10^{-15} eV · s
$\hbar$	Planck's constant	1.05×10^{-34} J · s	6.58×10^{-16} eV · s
a_B	Bohr radius	5.29×10^{-11} m	

Common Prefixes

Prefix	Meaning
femto-	10^{-15}
pico-	10^{-12}
nano-	10^{-9}
micro-	10^{-6}
milli-	10^{-3}
centi-	10^{-2}
kilo-	10^{3}
mega-	10^{6}
giga-	10^{9}
terra-	10^{12}

Conversion Factors

Length
1 in = 2.54 cm
1 mi = 1.609 km
1 m = 39.37 in
1 km = 0.621 mi

Velocity
1 mph = 0.447 m/s
1 m/s = 2.24 mph = 3.28 ft/s

Mass and energy
1 u = 1.661×10^{-27} kg
1 cal = 4.19 J
1 eV = 1.60×10^{-19} J

Time
1 day = 86,400 s
1 year = 3.16×10^{7} s

Force
1 lb = 4.45 N

Pressure
1 atm = 101.3 kPa = 760 mm Hg
1 atm = 14.7 lb/in^2

Rotation
1 rad = $180°/\pi = 57.3°$
1 rev = $360° = 2\pi$ rad
1 rev/s = 60 rpm

Mathematical Approximations

Binominal Approximation: $(1 + x)^n \approx 1 + nx$ if $x \ll 1$

Small-Angle Approximation: $\sin\theta \approx \tan\theta \approx \theta$ and $\cos\theta \approx 1$ if $\theta \ll 1$ radian

Greek Letters Used in Physics

Alpha		α	Nu		ν
Beta		β	Pi		π
Gamma	Γ	γ	Rho		ρ
Delta	Δ	δ	Sigma	Σ	σ
Epsilon		ϵ	Tau		τ
Eta		η	Phi	Φ	ϕ
Theta	Θ	θ	Psi		ψ
Lambda		λ	Omega	Ω	ω
Mu		μ			

Astronomical Data

Planetary body	Mean distance from sun (m)	Period (years)	Mass (kg)	Mean radius (m)
Sun	—	—	1.99×10^{30}	6.96×10^{8}
Moon	3.84×10^{8}*	27.3 days	7.36×10^{22}	1.74×10^{6}
Mercury	5.79×10^{10}	0.241	3.18×10^{23}	2.43×10^{6}
Venus	1.08×10^{11}	0.615	4.88×10^{24}	6.06×10^{6}
Earth	1.50×10^{11}	1.00	5.98×10^{24}	6.37×10^{6}
Mars	2.28×10^{11}	1.88	6.42×10^{23}	3.37×10^{6}
Jupiter	7.78×10^{11}	11.9	1.90×10^{27}	6.99×10^{7}
Saturn	1.43×10^{12}	29.5	5.68×10^{26}	5.85×10^{7}
Uranus	2.87×10^{12}	84.0	8.68×10^{25}	2.33×10^{7}
Neptune	4.50×10^{12}	165	1.03×10^{26}	2.21×10^{7}

*Distance from earth

Typical Coefficients of Friction

Material	Static μ_s	Kinetic μ_k	Rolling μ_r
Rubber on concrete	1.00	0.80	0.02
Steel on steel (dry)	0.80	0.60	0.002
Steel on steel (lubricated)	0.10	0.05	
Wood on wood	0.50	0.20	
Wood on snow	0.12	0.06	
Ice on ice	0.10	0.03	

Melting/Boiling Temperatures, Heats of Transformation

Substance	T_m (°C)	L_f (J/kg)	T_b (°C)	L_v (J/kg)
Water	0	3.33×10^{5}	100	22.6×10^{5}
Nitrogen (N_2)	−210	0.26×10^{5}	−196	1.99×10^{5}
Ethyl alcohol	−114	1.09×10^{5}	78	8.79×10^{5}
Mercury	−39	0.11×10^{5}	357	2.96×10^{5}
Lead	328	0.25×10^{5}	1750	8.58×10^{5}

Properties of Materials

Substance	ρ (kg/m^3)	c (J/kg · K)	v_{sound}(m/s)
Helium gas (1 atm, 20°C)	0.166		1010
Air (1 atm, 0°C)	1.28		331
Air (1 atm, 20°C)	1.20		343
Ethyl alcohol	790	2400	1170
Gasoline	680		
Glycerin	1260		
Mercury	13,600	140	1450
Oil (typical)	900		
Water ice	920	2090	3500
Liquid water	1000	4190	1480
Seawater	1030		1500
Blood	1060		
Muscle	1040	3600	
Fat	920	3000	
Mammalian body	1005	3400	1540
Granite	2750	790	6000
Aluminum	2700	900	5100
Copper	8920	385	
Gold	19,300	129	
Iron	7870	449	
Lead	11,300	128	1200
Diamond	3520	510	12,000
Osmium	22,610		